PASSIVE SOLAR HEATING ANALYSIS

A DESIGN MANUAL

J. DOUGLAS BALCOMB
ROBERT W. JONES
ROBERT D. McFARLAND
WILLIAM O. WRAY

CONTRIBUTORS:

Dennis Barley
Claudia Kosiewicz
Gloria Lazarus
Joseph Perry
Greg Schoenau

TECHNICAL WRITER:

Edwin Nelson, Tech. Reps., Inc.

Los Alamos National Laboratory

ISBN 0-910110-38-7

American Society of Heating, Refrigerating and Air-Conditioning Engineers
1791 Tullie Circle, NE, Atlanta, GA 30329

DISCLAIMER

This report was prepared as an account of work sponsored by an agency of the United States Government for publication by the American Society of Heating, Refrigerating and Air-Conditioning, Engineers, Inc. (ASHRAE). Neither the United States Government nor any agency thereof, nor ASHRAE, nor any of their employees, makes any warranty, express or implied, or assumes any legal liability or responsibility for the accuracy, completeness, or usefulness of any information, apparatus, product, or process disclosed, or represents that its use would not infringe privately owned rights. Reference herein to any specific commercial product, process, or service by trade name, trademark, manufacturer, or otherwise, does not necessarily constitute or imply its endorsement, recommendation, or favoring by the United States Government or any agency thereof or ASHRAE. The views and opinions of authors expressed herein do not necessarily state or reflect those of the United States Government or any agency thereof or ASHRAE.

CONTENTS

CONTENTS (continued)

CONTENTS (continued)

CONTENTS (continued)

ILLUSTRATIONS

ILLUSTRATIONS (continued)

ILLUSTRATIONS (continued)

ILLUSTRATIONS (continued)

ILLUSTRATIONS (continued)

ILLUSTRATIONS (continued)

TABLES

ACKNOWLEDGMENTS

Passive Solar Heating Analysis is the result of work at the Los Alamos National Laboratory during the period 1977-1983, performed under the auspices of the U.S. Department of Energy, Office of Solar Heat Technologies. It builds on the extensive work of Volumes II and III of the *Passive Solar Design Handbook* (subtitled *Passive Solar Design Analysis*) and the large supporting effort of the many people involved. The authors of Volume II are Douglas Balcomb; Dennis Barley; Robert McFarland; Scott Noll; Joseph Perry, Jr.; and William Wray; the authors of Volume III are Robert Jones, editor; Douglas Balcomb; Claudia Kosiewicz; Gloria Lazarus; Robert McFarland; and William Wray. Assistance in the preparation of Volumes II and III was provided by Della Baldock, Mary Beckett, Mary Blackwell, Kirk Collier, Lee Dalton, Jared Dreicer, Derrick Grimmer, James Hedstrom, Jack Hyde, Charlene McHale, Ike Meissner, Stan Moore, Lois Murray, Jan Sander, Peter Stromberg, Robert Sutton, Mark Thayer, Michael Ulibarri, and Kai Zinn. Reviewers of these volumes included Donald Abrams, Bruce Anderson, William Ashton, Bruce Baccei, Duncan Bremer, Matt Crosby, Harrison Fraker, Greg Franta, Drew Gillett, Harry Gordon, Daniel Hagan, John Hill, Mike Holtz, Bruce Hunn, Terry Ingersol, Bruce Johnson, Tim Johnson, Ove Jorgensen, Ron Judkoff, Doug Kelbaugh, William Kolar, Andrew Lau, Vivian Loftness, Fuller Moore, David Odegard, Larry Palmiter, Wayne Place, Greg Schoenau, and John Yellott.

Work on an ASHRAE book, combining and rewriting the previous work, was initiated by Michael Maybaum at the U.S. Department of Energy. Sherry Reisfeld at Los Alamos aided in obtaining a contract with Tech. Reps., Inc., of Albuquerque, New Mexico, to do the technical writing for the project.

Ed Nelson was the principal technical writer and production manager for Tech. Reps. Bruce Howden was the copy editor.

Los Alamos worked with ASHRAE from the beginning. With the assistance of Herman Barkmann and Stanley Mumma, a special ASHRAE subcommittee was formed to oversee the work, from the development of an outline through detailed review of three drafts of the manuscript. The authors are particularly indebted to the members of this subcommittee—Bruce Novell, Chairman; Steven Heibein; and Benjamin Rogers—for their diligent and thoughtful dedication to a quality product.

During the writing and editing process, Robert Jones provided liaison between Los Alamos, ASHRAE, and Tech. Reps., aided by Douglas Balcomb. Reviewers of *Passive Solar Heating Analysis* are Bruce Hunn, Harold Lorsch, Alwin Newton, and Bud Wildin. Valuable suggestions were received from William Shurcliff, based on Volumes II and III. We are especially grateful to Byron Winn who provided a detailed review of the entire manual during the last stages of its preparation.

Lastly, the authors wish to thank Donald Neeper for his help and his leadership of the Los Alamos Solar Energy Group during these years.

ASHRAE FOREWORD

This manual is a condensation of passive solar heating analysis methods developed at the Los Alamos National Laboratory. It represents a current state of the art in evaluating passive solar heating systems and is a significant addition to previously available methods.

Extensive experimental and theoretical studies were required to develop these methods, including building and testing passive "test cells," studying the dynamic behavior of these units, developing detailed hourly simulation codes to predict their behavior, and then using these to generate simpler monthly and annual analysis tools. The methods presented herein are straightforward, easy to use, and well matched to the designer's design process and required accuracy. These methods allow the building designer to choose from multiple passive system types for various types of buildings and to combine several passive systems on the same building. The designer can evaluate the effects of varying virtually all of the major passive design parameters to determine their effect on auxiliary heat savings. Design guidelines in this manual give design direction for passive solar features in buildings. Methods and recommendations are also given to optimize the balance between energy conservation and passive solar systems. Tabulated values for 223 locations in the U.S. and Canada make the methods useful virtually anywhere in the U.S. Furthermore, generalized techniques are presented that allow passive solar systems to be evaluated at any location where solar radiation and ambient temperature data are available.

The work of organizing and writing this manual was performed by Tech. Reps., Inc., of Albuquerque, New Mexico, in cooperation with the Los Alamos Solar Energy Group. The manual is written so that it is useful to building designers experienced in passive solar design and designers considering their first passive solar building. Analysis methods and data are presented in tables, graphs, and equations so that they can be used manually (with worksheets provided) or computerized.

ASHRAE members have been involved in the organization and review of this manual to help ensure its usefulness to designers and engineers. This manual has been scrutinized by these ASHRAE members and deemed to be useful and technically responsible. The manual has been formally approved by the ASHRAE technical committee on Solar Energy Utilization.

Passive Solar Heating Analysis Subcommittee
of the Solar Energy Utilization
Technical Committee, TC 6.7

Bruce Novell, Chairman
Alabama Solar Energy Center

Stephen Heibein
Innovative Design

Benjamin Rogers
Consulting Engineer

PREFACE

Passive Solar Heating Analysis represents the product of several years work at the Los Alamos National Laboratory. The object has been the development of simplified methods of analysis for passive solar heated buildings. The primary result is the solar load ratio method, which has now been applied to a variety of direct-gain, Trombe wall, water wall, and sunspace configurations. A large number of performance sensitivity curves for these systems supplement the method. Procedures for their use have been carefully tailored to integrate smoothly into the normal design process, specifically into the formative schematic design and design development phases.

The precursors to *Passive Solar Heating Analysis* are Volume Two and Volume Three of the *Passive Solar Design Handbook*, written at Los Alamos for the U.S. Department of Energy. Volume One of this series concerns passive solar design concepts. These books have served an important interim function in the evolution of passive solar practice and analysis and have been used by many designers. Volume Two, issued in January 1980, presents the solar load ratio methodology and gives results for six reference designs, specifically, direct-gain, Trombe wall, and water wall designs, both with and without night insulation. Volume Three, issued in July 1982, is a major addendum to Volume Two, expanding and modifying the original 6 reference designs to 66 and adding 28 sunspace reference designs. A procedure for balancing conservation and solar strategies is also included.

Passive Solar Heating Analysis is a complete rewrite of Volumes Two and Three by a professional technical editor. This step was taken because it was felt that the work is mature and that it deserves better presentation. The result is a smooth and consistent development with step-by-step procedures given at each stage together with convenient worksheets as appropriate. All tables and graphs required are included so that the book is self-contained, at least for users in the contiguous 48 United States and southern Canada. Acceptance within the building analysis community was assured by working closely with a special committee appointed by ASHRAE.

With minor exceptions, *Passive Solar Heating Analysis* is completely consistent with Volume Three. The solar load ratio correlation form and correlation constants are identical. The solar radiation and transmittance correlations are also identical except that azimuths north of east and west and overhang shading for vertical south-facing windows are included. A major addition is the incorporation of Chapter 2, "Passive Solar Design Guidelines," which provides specific, quantitative design guidance very early in the design process. Thirty guidelines are given that range from general rules relating to conservation, solar area, and orientation, to specific rules for the various passive solar system types. Also included are new correlations useful for estimating average inside temperature during clear days.

Designers may still wish to refer to Volume Two for some specific topics that are not covered in this book. These are chapters that pertain to design of fan-forced rock beds, diurnal heat capacity and temperature swing calculations, and economics.

Buildings cannot be engineered in the same way that a power plant, a satellite, or a ship is designed. Fine-honed precision is not justified either by the accuracy required or the design budget. This is especially true of small buildings that are normally designed using rules-of-thumb gained through experience. It is this smaller building, both residential and commercial, that is the primary target of this book, and the analysis methods given are geared principally to these applications. Little time or mathematical expertise is needed to apply the methods, especially those that are used during the critical early phases of design. Worksheets, pencil, and a handheld calculator are all that are required. It is expected that some readers will use only the simplest methods given in Part I. Others will continue to the more complex procedures in Part II.

Simplicity and speed in an analytical method necessarily involve some compromises in accuracy and generality. This is true of the methods presented here. First, the methods are based on simple correlations of complex data that are themselves estimates of building performance by computer simulation. Nevertheless, the user can expect accuracy in the 10% to 15% range, which is suitable for design estimates. Second, the methods apply to specific building configurations called reference designs. Even though there are many configurations to choose from and a large amount of sensitivity data with which to estimate the effect of modifying the principal design variables, most buildings to be analyzed will not exactly match the analytical assumptions. The designer must then use clear judgment in applying the results and candor in representing their applicability to a client.

The designer and client should understand that analytical results are not a prediction of building energy performance. And this fact is not necessarily because of shortcomings in the analytical procedure but rather because of the nature of design analysis. There are three principal reasons. First, the assumptions about the characteristics of the building cannot be expected to match the actual building. This is because of the uncertainty and variability in the properties of the building materials and in the quality of the construction. For example, the building heat loss characteristics are very sensistive to the care taken during construction to seal cracks. Second, the performance of a building depends on the actual weather rather than the long-term normals used in the analysis: the weather during any particular period will differ from the normal for that period, and the weather at a particular site will differ from that at the nearest weather station. Third, the manner in which the building is occupied and operated has a strong influence on performance and is not entirely predictable.

Often the assumptions most difficult to match are not directly related to the building design but to its operation. Important factors are the operation of vents, doors, and windows; the thermostat settings; and the rates of internal heat generation by lights, appliances, machinery, and people. The assumptions in this book are as follows. Vents in Trombe walls and sunspaces are assumed to never permit reverse flow of heat from the heated building back to the Trombe wall air space or the sunspace. The heat loss characteristics of the nonsolar parts of the building are assumed to be represented by a constant heat loss coefficient so that the building heat loss is proportional to the difference between the indoor and outdoor temperatures. (This is called a degree-day load.) The thermostat settings and the rate of internal heat generation (not including the deliberate generation of heat by space heating equipment) are assumed to be constant. A difficulty often arises with this last assumption, especially in the case of certain types of commercial buildings where a predominantly daytime occupancy leads to varying thermostat settings and internal heat generation.

Despite the importance of space heating in most U.S. locations and for most types of buildings, it is certainly not the only energy issue in a building. Space cooling is the major concern in many locations, especially in buildings with a large rate of internal heat generation. Nevertheless, this book deals only with passive solar heating. Here, there is a well-established quantitative basis, whereas natural solar cooling is less well understood. It is not that passive cooling techniques are felt to be ineffective; quite the contrary. As quantitative procedures are developed, the need for a follow-up to this book dealing with passive cooling will be evident. It should be noted, however, that the most important factor in cooling a building is not to heat it. Thus, solar gains of all kinds must be very carefully controlled during the cooling months by shades, overhangs, or vegetation.

A balanced thermal design begins from an awareness of the total energy needs of the building. But the analytical methods in this book are applicable to the passive solar heating system design even when the space heating requirement turns out to be small, provided the analytical assumptions are met. The applicability of the methods to commercial buildings is discussed further in Chapter I, "Introduction."

A difference in the analysis approach used in this book and that used in Volumes Two and Three is the comparison with a nonsolar reference building. The procedures used in the earlier volumes required defining a different degree-day base for the solar and nonsolar building. This proved awkward, confusing, and of questionable value. In this book, a single base temperature is used, defined for the solar building. The analyst can either use the simple formulation of solar savings developed, which is usually reasonably accurate, or perform two complete, separate calculations, one for the solar building and one for any arbitrarily defined nonsolar building, and then compare the auxiliary heat required by each.

Another notable change is in the terminology of the building load coefficient, termed *BLC* in Volumes Two and Three. This has caused some confusion because most analysts use *BLC* to refer to the *total* load coefficient, whereas, in Volumes Two and Three, *BLC* refers to the building *exclusive of the solar wall.* To alleviate this confusion, the term net load coefficient (*NLC*) is used instead of *BLC*, and a new term, *total* load coefficient (*TLC*), is used to refer to the entire building, including the solar wall. *BLC* is not used.

Passive solar design has become widely recognized as a very effective way to achieve large decreases in the amount of energy required for space conditioning in buildings. It has become generally accepted that these buildings work well and are cost effective. They offer ease of operation, reliability, and enhancement of livability and comfort. The quantitative basis for passive solar design has developed quickly from a rather late start. At the time of the First National Passive Solar Conference, held in Albuquerque in May 1976, there were few people in the country not present at the conference who believed that passive systems would work. This situation has changed markedly. The literature has built up rapidly, reports on the performance of successful buildings across the country have multiplied, and there is now widespread recognition that large energy savings can be obtained.

The purpose of this book is to provide a means whereby performance can be analyzed in advance. Many of the early passive solar builders designed on a rather intuitive basis, hoping for the best. Often their hopes were realized, and the buildings work well. Sometimes they did not. However, these designers represent a small number who are willing to experiment with an investment as large as a building. The vast majority of builders require reliable performance estimates, well founded in experimental observation. This book is intended to fill that need.

There is a growing body of evidence that *correctly designed* passive solar buildings have comfort characteristics that equal or exceed their contemporary counterparts. This is due to several subtle physiological factors. The most important of these is the effect of the thermal radiant environment on comfort.

Under still-air conditions, the radiant temperature is roughly equal to the actual air temperature in the individual's perception of comfort. In a passive building, the enclosing surfaces of a room are usually warmer than in a conventional building. This leads to an increase in radiant temperature. The increase in surface temperature results from two effects: well-insulated walls and multiple glazings (conservation measures) and storage of solar heat (passive solar measures). Thus, the individual can experience an equivalent comfort sensation when the air in a passive building is at 65°F as in a conventional building with a 72°F air temperature. A second effect that enhances comfort is the incredible thermal stability of a well-insulated building with large internal mass. Temperatures change very slowly, resulting in less physical stress than in a building with the same average temperature but with abrupt variations.

The apparent simplicity of passive solar buildings has misled many people into thinking that passive solar design is simple. In fact, it is a complex process, a subtle complexity consisting of consciously designing to take advantage of environmental energies and natural processes. Design decisions must factor in considerations of cost, aesthetics, use of the various spaces in the building, natural lighting, security, code restrictions, and privacy as well as thermal performance. This book provides a basis for quantifying the heating performance side of this complex situation. The procedures have been made as simple as possible without losing the fundamental relationships that influence the result. Even so, they may appear intimidating at first—as is anything new. Time spent in developing a familiarity with the procedures will lead to their comfortable and effective assimilation into the design process.

OVERVIEW

Purpose

The purpose of this book is to provide methods for quantitatively analyzing the performance of passive solar heating systems.

Design decisions must consider cost, aesthetics, function, space allocation, lighting, security, site adaptations, construction code restrictions, and privacy as well as thermal performance. This book provides a basis for quantifying the solar heating performance side of this task. The procedures have been made as simple as possible without losing the fundamental relationships that influence the result. The methods are applicable to both residential and commercial buildings. However, when being applied to commercial buildings, considerable care must be exercised to be sure the assumptions implicit in the method are appropriate to the building being analyzed.

Design Sequence

Traditionally the design process has been divided into the following sequence of steps:

a. Schematic design—review and evaluation of the building requirements and production of drawings and documents to illustrate the initial building concept.

b. Design development—preparation of more detailed drawings and documents to describe the size of the building and its architectural, structural, mechanical, and electrical systems.

c. Construction documents—preparation of finished drawings and specifications suitable for use in negotiation or bid of a construction contract.

It should be noted that this design process is a sequence of increasingly more detailed steps. Major decisions are made in the early design steps and the details developed in the later steps. Issues of energy use and energy conservation traditionally have been addressed during design development. Heating and cooling requirements have been dealt with as part of mechanical and electrical system design; insulation has been considered during detailing of wall sections. Since fuel costs were small, energy use and conservation was treated as a relatively minor constraint on overall building design and was appropriately considered during design development.

In recent years, rising fuel costs, prospective fuel shortages, and the imposition of governmental energy performance standards have combined to make energy an extremely important consideration in building design. Energy-related decisions must now be made during schematic design and their implications carried into design development. Energy considerations are no longer a minor constraint on building design that can be deferred to the detailed design stage.

To provide for these circumstances, this book is divided into two parts. Part I begins with a set of guidelines that are useful to the designer at the very beginning of schematic design. Subsequent chapters present a method for building thermal analysis that can be applied during schematic design. The method is called the annual *LCR* method. Only the general characteristics of the building need be known to apply the method. It is quick and easy to apply so that several design concepts can be analyzed for their thermal performance and the results compared before selecting a final concept for design development. Part I also treats the subject of energy conservation and presents a method for determining the optimum balance between conservation measures and solar heating measures.

Part II presents a more detailed method called the monthly *SLR* method, which is suitable for use during the design development stage. In comparison with the annual *LCR* method, the monthly *SLR* method can accommodate more design variations and evaluates building thermal performance on a month-to-month basis rather than by a single annual figure. The extra effort required is modest. Completion of the monthly *SLR* method thermal analysis is a recommended procedure after completion of design development and prior to initiating the construction documents phase.

The methods in this book lead to estimates of annual and monthly consumption of energy required to maintain comfort during the heating season. This energy will presumably be supplied by a backup or auxiliary heating system that will often be off or be operating at partial power. It is assumed that such a system is installed and that it can supply the required heat energy. Methods for sizing the capacity of this equipment are not discussed in this book. It is expected that most designers will use traditional methods, such as those described in the *ASHRAE Handbook of Fundamentals*, to determine the maximum of design-day capacity requirements.

Passive solar system energy savings estimated using the techniques in this book refer to specific passive system configurations in which all the design parameters are specified. These are called *reference designs.* There are 94 reference designs, 9 for direct gain, 42 for Trombe walls, 15 for water walls, and 28 for sunspaces. Once estimates have been made for one or a combination of these reference designs, the effect on performance of a difference in one or more of the design parameters can be estimated using sensitivity curves.

PART I
SCHEMATIC DESIGN

Scope of Part I

The methods of analysis and the supporting data tables presented in this book were developed at the Los Alamos National Laboratory as a result of a continuing research program. Much of the information has appeared previously in separate Los Alamos National Laboratory reports, conference papers, and the second and third volumes of the *Passive Solar Design Handbook*. This book integrates the previously presented material, using uniform terminology, and includes new material generated exclusively for this book. In Chapter 1, key design elements are discussed, and the physical configurations of the various passive solar systems are described and terminology is defined. Chapter 2 presents a set of design guidelines that are useful in beginning a schematic design, such as conservation levels, the size and orientation of the solar aperture, the amount and location of thermal storage mass, the orientation of the building, and other considerations. Chapter 3 discusses methods of estimating the net load coefficient (*NLC*) of a building before schematic design is completed. The computation of the degree-days (*DD*), given the building internal heat sources and thermostat setting, is discussed in Chapter 4. Chapter 5 describes the annual load collector ratio (*LCR*) method of evaluating the thermal performance of a building. This method is a simple means of estimating annual solar savings and the key parameter, auxiliary heat. Chapter 6 presents information on the optimum mix of conservation and passive solar strategies in building design. Guidance is available here on adjusting the appropriate conservation level for a given location. Chapter 7 is a collection of example problems illustrating the methods described in Part I. The appendices contain the data tables and graphs necessary to exercise the analysis methods.

CHAPTER 1
INTRODUCTION

Key Design Elements

By far the most important elements affecting the thermal performance of a passive solar building are the size of the solar aperture, the degree of energy conservation achieved, the amount and placement of thermal storage mass within the living space, and the weather. The solar aperture is the area of the glazing through which the sun's energy flows. The larger the aperture, the greater the solar gain. But bigger is not necessarily better: too large a solar aperture can cause overheating, or a waste of solar energy by ventilation to prevent overheating, and larger heat losses at night, perhaps increasing the capacity of the backup system. If the useful solar gain is not efficiently used and the heat loss potential is increased, then the large aperture is not justified.

Energy conservation means that the heat is kept within the building and prevented from being lost to the outside. The most important means for achieving this are by insulating the building "skin" (exterior surfaces) and by sealing the cracks and gaps through which air infiltrates into and escapes from the building. The more energy conserved, the less energy needed to keep the building warm. However, for very tightly sealed buildings, the air can become stale and unhealthy; some ventilation is needed. Conservation measures cost money and, above a certain conservation level, the extra cost is not justified by the potential energy savings.

Measures taken to increase the useful solar gain and to improve energy conservation both contribute to the goal of reducing the amount of auxiliary heat needed to keep the building warm, and both contribute to the cost of construction. Both of these aspects of solar design are characterized by the law of diminishing returns, e.g., the initial few square feet of solar aperture added to a building will be more effective in reducing the auxiliary heat than the last few square feet added; similarly for the insulation of a building. The most effective use of construction funds is achieved by a balance between the conservation and solar aspects.

The term "thermal storage mass" is used to describe those features such as masonry walls, brick floors, water containers, etc., that have the function of storing and releasing heat to a building. Thermal mass stabilizes the inside temperature. During the day, when there is more than enough solar energy to heat the building, the thermal mass absorbs the excess energy to prevent overheating. At night, the thermal mass releases the energy to keep the building warm. The larger the thermal mass, the less the temperature swing will be in the building through this daily cycle and the less auxiliary heat consumption there will be.

The local climate is considered in this book by the use of insolation and ambient temperature data. The appendices include insolation and temperature data for many locations in the United States and some in Canada; no additional data sources for these locations are required. Other locations can be analyzed also, but the analyst must supply the data from other sources. The climate is a major factor in determining the proper balance among the key design elements. A successful design in one location may not work well in other locations where the climate differs. The analytical tools presented here allow the balancing of the design elements to fit the climate

Physical Configurations of Passive Solar Heating Systems

A passive solar heating system is one in which the thermal energy flow is by radiation, conduction, or natural convection, as distinguished from active heating systems, which use fans or pumps for forced distribution of heat. In most cases, a passive system is intimately integrated into the architecture of the building. The elements of a building often serve a dual purpose. A window on a south-facing wall not only provides interior light and a view but also collects the sun's heat. A wall, in addition to providing structural support, can provide thermal mass for heat storage. One scheme for classifying passive solar heating systems categorizes the systems according to their physical configuration. The categories are described in the following paragraphs.

Direct Gain — The direct-gain system consists of a building with south-facing windows that admit the slanting rays of the winter sun. Thermal mass, usually in the form of masonry walls or a masonry floor plus all of the building contents, stores heat during the day and then releases this heat during the night. It is not necessary that the sun shine directly on the thermal mass, but mass should at least be visible from the directly illuminated mass so that it can receive reflected radiation and long-wave radiation.

Thermal Storage Wall — In this system, a south-facing wall is constructed of heavy masonry (Trombe wall) or of water-filled containers (water wall). The outside, south-facing surface is glazed to admit sunlight and reduce heat losses to the outside air. For Trombe walls, the heat absorbed from the sun is conducted slowly through the wall to the inner surface and is transferred to the living space by radiation and convection. Vents can be cut through the top and bottom of the Trombe wall to allow prompt transfer of heat to the living space by natural convection. For a water wall, convection currents in the water transfer heat promptly across the wall.

Sunspace — This form consists of a glassed-in sunspace (like a greenhouse, atrium, or conservatory) located on the south side of a building and separated from other building spaces by a common wall. Thermal mass, in the form of water containers or masonry walls and floors, serves to stabilize the temperature in both the sunspace and the building. The sunspace does not have its temperature regulated closely and therefore requires little, if any, auxiliary heat. The sunspace receives solar energy all year long and in warm periods should be shaded and vented to the outside to eliminate excess heat. Besides providing additional living space, the sunspace can be used as a greenhouse.

Thermal Storage Roof — This is similar to the thermal storage wall except that the thermal mass is located on the building roof. The thermal mass usually consists of water in a pond or in bags. Movable insulation is required. During winter months, the thermal mass is exposed to the sun during the day then is covered by the movable insulation at night when the stored heat is transferred through the ceiling to the living space. In summer months, the insulation is used to shield the thermal mass from the sun during the day but is removed at night to allow the thermal mass to cool. This form of passive solar heating is not treated in this book.

Convective Loop — The convective loop consists of a solar collector and a thermal storage mass, often a rockbed. The collector is at a lower level than the thermal storage. Ducts connect the top of the collector to the top of the thermal storage. The bottoms are also connected. The sun heats the air in the collector, which then expands, rises, and flows through the ducts into the top of the thermal storage. Cold air is displaced from the bottom of the thermal storage through the ducts back to the collector where the sun warms it again. The cycle continues as long as the sun shines. The heat accumulates in the thermal storage to be used in heating the building. A liquid heat transfer medium can be used instead of air, in which case the thermal storage is a water tank. Convective loops are not treated in this book.

Mixed Systems — Mixed systems combine two or more of the above categories to heat a single building.

Hybrid Systems — Combinations of active elements with passive systems are referred to as hybrid systems. An example is the use of an electric fan to force heated air from a sunspace into a building. Hybrid system performance is not treated in this book.

Terminology

The most important terms used in this volume are listed and defined here. Where symbols are used to represent the term, the symbol and its units of measurement are included. A complete glossary of terms is included in Appendix 8.

Solar Wall — The term "solar wall" refers to the glazed portions of building walls whose contributions to the solar gains of the passive solar heating system are accounted for in the analysis. Ideally, the orientation of the principal components of the solar wall is due south, or within 30 degrees of due south. (In the southern hemisphere, the solar wall should be north-facing.) Ordinary windows *may* be considered as direct-gain parts of the solar wall if the detail desired in the analysis requires it.

Net Glazing Area, A_n (ft^2) — The net glazing area is the actual area of solar wall glazing. The term "solar aperture" has the same meaning.

Tilt Angle, *TILT* (degrees) — The tilt angle is the angle between the plane of the solar aperture and the horizontal plane.

Azimuth Angle, *AZ* (degrees) — The azimuth angle is the angle between due south and the orientation of the glazing plane. The azimuth angle of a south-facing glazing is zero. Azimuth angles are negative for orientations west of south and positive for orientations east of south. Thus, the azimuth angle is −90° for a west-facing glazing and +90 degrees for an east-facing glazing. (Definitions of tilt and azimuth are also given in Chapter 9 with the help of Figure 9-4.)

Projected Area, A_p (ft^2) — The projected area, A_p, of a direct-gain window, Trombe wall, or water wall equals the net glazing area. The projected area of a sunspace is the net glazing area of the sunspace projected on a single vertical plane that has the same azimuth as the principal azimuth of the sunspace. (The concept of principal azimuth arises when the sunspace has two or more glazing planes; see Chapter 9 for details.) In a building using more than one passive system, the projected area is determined separately for each system. The total projected area of the building is the sum of the projected areas of the respective systems. It is recognized that using projected area is sometimes less convenient than using net area. The advantage occurs when one wants to analyze a sunspace configuration that does not exactly correspond to one of the five reference design types. If actual area were to be used, significant error could result, whereas using projected area gives better accuracy for a wide variety of possible configurations. It is also useful for convenience in comparing alternative systems on the same solar wall. For example, a sunspace and a Trombe wall that occupy the same area of solar wall may have different glazing areas but the same projected area. Further details on the projected area are to be found in Chapter 9.

Base Temperature, T_b (°F) — The base temperature is a fixed temperature, traditionally 65°F, against which the ambient temperature is compared. Other base temperatures may be used, however.

Degree-Days, *DD* (°F·day) — The number of degree-days is the difference between the base temperature and the mean ambient temperature for that day. If the mean ambient temperature is above the base temperature, no degree-days are counted. The number of degree-days for a month is the sum of the degree-days for each day of that month; similarly, for annual degree-days. The term "degree-days" refers to a specific base temperature.

Total Load Coefficient, *TLC* (Btu/°F·day) — The total load coefficient is the steady-state load, or heat loss, of a building per degree of inside-outside temperature difference per day. It is calculated by standard ASHRAE procedures or by the simplified procedures in Chapter 3.

Net Load Coefficient, *NLC* (Btu/°F·day) — The net load coefficient is the total load coefficient minus the load coefficient for the solar wall.

Net Reference Load, Q_{net} (Btu) — The net reference load is the building load, exclusive of losses through the solar wall, over a period of time. The net reference load is calculated from the following equation:

$$Q_{net} = NLC \cdot DD \qquad (1\text{-}1)$$

Auxiliary Heat, Q_{aux} (Btu) — The auxiliary heat is the amount of energy furnished by an auxiliary heating system (e.g., furnace) to keep a building warm.

Internal Heat, Q_{int} (Btu) — The internal heat is the amount of energy generated within a building by people, lights, appliances, etc. Internal heat does not include the heat furnished by the auxiliary heating system.

Load Collector Ratio, *LCR* (Btu/°F·day·ft²) — The load collector ratio is the quotient of the net load coefficient and the projected area:

$$LCR = NLC/A_p \quad (1\text{-}2)$$

For a given location, *LCR* is the variable that most influences the net solar heating performance of a building. It largely determines the average increase in indoor building temperature over the average ambient temperature and is the most important parameter influencing the solar savings fraction *(SSF)*. It reveals the relationship between energy conservation, which is determined by *NLC*, and the amount of solar gain, which is determined by A_p. The relationship between *LCR* and *SSF* depends on the amount of incident sunshine and the degree-days and therefore varies for each locality. Tables of *SSF* as a function of *LCR* for many localities in the United States and Canada are included in Appendix 1.

Solar Savings Fraction, *SSF* (dimensionless) — The quantity *SSF* is the ratio of the solar savings to the net reference load:

$$SSF = (Q_{net} - Q_{aux})/Q_{net} \quad (1\text{-}3)$$

The results in this volume are expressed mainly in terms of *SSF*. But *SSF* is subject to misinterpretations as an expression of the effectiveness of a solar system. For example, a possible misapplication of *SSF* is as a measure of comparison between two solar buildings. Such a comparison is valid only when the two buildings have the same *NLC*. The primary utility of *SSF* is as an intermediate step in the calculation of the auxiliary heat requirement of the solar building, which is always a valid performance measure:

$$Q_{aux} = (1 - SSF) \cdot NLC \cdot DD . \quad (1\text{-}4)$$

Solar Savings — The solar savings is defined as:

$$Q_{sav} = Q_{net} - Q_{aux} \quad (1\text{-}5)$$

Caution should be exercised in interpreting the solar savings beyond the terms of its definition. For example, the solar savings is *not* the solar energy contribution to the actual building's space heating consumption. The solar savings is, in fact, smaller than the actual solar contribution because credit is not taken for solar gains that offset losses through the solar wall nor solar gains that raise the room temperature above the thermostat setpoint. Another common misinterpretation of the solar savings is that it is the savings in the space heating requirement achieved by the solar building relative to some other "standard" building. This interpretation is correct only if the "standard" building is the same as the reference building defined above, that is, one with an energy-neutral wall in place of the solar wall. Note that solar savings refers to heat energy saved, and one must account for furnace efficiency to account for fuel savings.

Solar Load Ratio, *SLR* (dimensionless) — The solar load ratio is defined as follows:

$$SLR = Q_s/Q_{load} \quad (1\text{-}6)$$

where Q_s is a monthly solar gain and Q_{load} is a monthly heating load. The definitions of Q_s and Q_{load} are different for the various passive system types, as discussed in Part II. Using the specific definitions of Q_s and Q_{load} for the various passive system types, *SLR* can be computed on a month-by-month basis. This is useful because *SSF* has been correlated with *SLR* so that *SSF* can be computed month by month.

The correlation was established by using computer simulation to predict the thermal performance of specific reference design buildings. The computer simulation models were validated by comparing their predictions against the measured behavior of experimental buildings. An assortment of reference designs was defined for each of the passive solar heating system types. Each reference design is a detailed specification of the passive solar features of a building. The building *LCR* and the weather at the assumed building site remain variable. In the simulations, *SSF* was computed for several *LCR* values and for weather data from several different sites. An algebraic formula was then assumed that expressed *SSF* as a function of *SLR* and several adjustable coefficients (the correlation coefficients). The correlation procedure consisted of adjusting the coefficients so that the mean square deviation in the *annual SSF* was a minimum. The results of this effort are presented in this manual both as tables giving the values of the correlation coefficients for each of the specific reference designs and as graphs of *SSF* versus *SLR*.

Use of the *SLR* correlation to calculate building heating performance on a monthly basis leads to relatively high errors (standard deviations are typically 8%) when compared to hour-by-hour calculations of the computer simulation model. The errors over a 1-year period tend to cancel out so that annual performance calculations have an error of only 3%, typically. For a set of 94 reference designs, the results of the annual performance calculations have been tabulated for 223 different geographical locations. The tables are presented in Appendix 1.

Solar Gain, *S*, (Btu/ft^2 · mo) — The solar gain is the amount of solar energy absorbed by a building, per ft^2 of projected area, during a period of 1 month.

Conservation Factor, *CF*, (h·°F·ft^2/Btu)$^{1/2}$ — The conservation factor is useful in balancing conservation and solar expenditures in a building. It includes the performance and weather-dependent aspects of the desired result (wall R-value, ceiling R-value, infiltration air changes per hour, etc.).

Sensitivity Data

The monthly and annual performance calculations, the *SLR* and *LCR* methods, are based on specific reference designs. The monthly calculation allows some flexibility in the choice of certain design parameters, which are fixed in the annual method, but probably not enough to encompass all of the design variability that will be encountered. To treat off-reference designs, the sensitivity data may be used.

The sensitivity data express the effect on the annual *SSF* of varying one or two design parameters. The data are usually expressed as plots of the annual *SSF* versus one design parameter, while the other design parameters are held fixed at their reference values. Dots are plotted on the sensitivity curves to represent the reference conditions. Each set of data is presented in most cases for six different U.S. cities chosen to represent a wide geographical and climatological range. To apply the data, one of the reference cities must be chosen to represent the location of interest on the basis of climate similarity.

Two applications of sensitivity data are important. One is quantitative as a final step in a design analysis. In this application, an analysis is performed on the basis of the reference design that most nearly resembles the design of interest. Then, the annual *SSF* is corrected by use of sensitivity data.

The other application of sensitivity data is qualitative, wherein the data serve as a guide, perhaps very early in the design process, concerning the relative significance of various design parameters and their preferred values.

The sensitivity data are listed and discussed in the chapters that treat the separate passive solar system types (Chapters 12, 13, and 14). Some of the important fixed characteristics of the reference designs are

1. Glazing system configuration (number of panes, air gap, night insulation),
2. Glazing system orientation (tilt and azimuth),
3. Thermal storage mass characteristics (thickness, surface area, thermal conductivity, specific heat, surface absorptance), and
4. Common wall configuration in indirect systems (vent area and height, R-value, surface absorptance).

Units

English units are used in this volume to be consistent with customary design and engineering practice in the United States. Selected quantities are listed in Table 1-1 with their English units and the approximate conversion factor for the equivalent SI (Système International) unit.

Table 1-1

Conversion Table, English Units to SI Units

Quantity	Given English Unit	Multiply by	To Get Equivalent SI Unit
Energy	Btu	1054	J
Power	Btu/h	0.2928	W
Energy flux	Btu/h·ft^2	3.152	W/m^2
Length	ft	0.3048	m
Area	ft^2	0.09290	m^2
Volume	ft^3	0.02832	m^3
Mass	lbm	0.4536	kg
Density	lbm/ft^3	16.02	kg/m^3
Heat capacity	Btu/°F	1897	J/K
Specific heat	Btu/lbm·°F	4182	J/kg·K
Thermal conductivity (*k*)	Btu/h·°F·ft	1.729	W/K·m
Conductance (*U*)	Btu/h·°F·ft^2	5.673	W/K·m^2
Degree-days (*DD*)	°F·day	5/9	K·day
Net load coefficient (*NLC*)	Btu/°F·day	0.0220	W/K
Load collector ratio (*LCR*)	Btu/°F·day·ft^2	0.236	W/K·m^2

The SI unit of temperature, the Kelvin (K), is not as commonly used in design and engineering practice as the degree Celsius (°C). To convert a temperature in °C to K, add 273.15; no conversion is needed to convert a temperature increment (which is what occurs in combination with other units such as J/°C) from °C to K. To convert a temperature in °F to °C, subtract 32 and multiply by 5/9; to convert a temperature increment (which is what occurs in combination with other units such as Btu/°F) in °F to °C, multiply by 5/9 only.

The most customary SI unit for a building heat loss coefficient is W/°C; therefore, it is natural for *NLC* to have the unit W/°C. Similarly, a natural unit for *LCR* is W/°C·m^2. But it is more convenient in solar design analysis to adopt units that contain days as the time unit because of the frequent occurrence of combinations involving degree-days. The resulting units will not be SI units, but they will be good compromises for application to solar design analysis. The possible "SI" hybrid units are listed in Table 1-2 with the conversion factors.

The prefix k means 1,000; the abbreviation Wh means watt-hour. To convert *NLC* in W/°C to kJ/°C·day, multiply by 86.4; to Wh/°C·day, multiply by 24.

Table 1-2

Conversion Table, English Units to Hybrid SI Units

Quantity	English Unit	Multiply by	Equivalent "SI" Hybrid
NLC	Btu/°F·day	1.9	kJ/°C·day
		0.527	Wh/°C·day
		0.454	kcal/°C·day
LCR	Btu/°F·day·ft^2	20.4	kJ/°C·day·m^2
		5.67	Wh/°C·day·m^2
		4.88	kcal/°C·day·m^2

Cooling Considerations

The important issue of summer comfort and cooling loads has not been addressed in the present volume; nevertheless, *year-round* comfort and energy efficiency are the goals of a balanced passive solar design. A passive solar heating system may aggravate the summer discomfort or cooling load due to excessive solar gains through the solar aperture, but suitable summer solar controls and thoughtful use of natural convection can largely alleviate this potential problem. Indeed, some of the elements of a good passive solar design for heating also enhance summer comfort, namely, mass for thermal storage and energy conservation measures in the building envelope.

Applicability

Two methods of thermal analysis are presented in this volume: the annual *LCR* method and the monthly *SLR* method. Both are simplified methods for estimating the auxiliary heat requirements of passive solar heated buildings. The methods are specialized tools for a specialized job and certainly do not answer all of the designer's thermal questions. The user should be aware of the limits of applicability of the methods. Both use a correlation of *SSF* with *SLR* in which *SLR* is a function of the building *LCR*, the absorbed solar radiation (*S*), and the degree-days (*DD*). The correlation is established for each of 94 reference designs in which the important thermal characteristics of the building are specified at fixed values. Of the two analysis methods, the annual *LCR* method is the more restricted, and its limitations will be discussed first.

Limitations of the Annual *LCR* method — In the annual *LCR* method, the annual *SSF* is precalculated and tabulated (Appendix 1) for each of the 94 reference designs and for 223 geographical locations. The weather data used in the calculation are the long-term normals for the selected locations, and this cannot be changed. The two variables available to the analyst are *LCR* and the base temperature, T_b.

Variation of *LCR* allows the analyst to select the conservation level of the building and the sizing of the solar aperture. The fixed characteristics of the reference design cannot be altered. In addition to the reference design specifications, the orientation, tilt, ground reflectance, and solar aperture heat loss characteristics are specified, and there is no shading of the solar aperture. Some flexibility can be achieved by the use of sensitivity data, which shows how *SSF* is affected by changes in certain selected reference design characteristics.

In using the annual *LCR* method, the base temperature must be determined using the formula:

$$T_b = T_{set} - Q_{int}/TLC \qquad (1\text{-}7)$$

The *LCR* tables have been prepared for two values of base temperature, 55°F and 65°F. Interpolation may be used between these two values. The thermostat setting, T_{set}, may be placed at any value, but once selected, it cannot be varied nor can nighttime setbacks be accommodated. The internal heat source, Q_{int}, is not restricted in value, but the heat is assumed to be generated at a constant rate.

Additional Capabilities of the Monthly *SLR* Method — In the monthly *SLR* method, the analyst computes on a monthly basis the values of *LCR*, *S*, and *DD*. From these quantities, he computes monthly values of *SLR* and *SSF*. Unlike the annual method, the monthly method requires moderately extensive and repetitive computations but is free of some of the limitations of the annual method.

In computing *LCR*, the analyst can select the conservation level of the building and the size of the solar aperture just the same as can be done in the annual method. For thermal storage wall and sunspace systems, the analyst also computes the heat loss characteristics of the solar wall and is not restricted to the values assigned in the reference designs. This allows optional glazing designs for both thermal storage wall and sunspace systems and, for sunspace systems, allows choice of conservation level for the sunspace itself. In determining T_b, using Equation 1-7, the designer may vary Q_{int} and T_{set} on a month-to-month basis, but not within the month. In all other respects, T_b and Q_{int} have the same restrictions as for the annual *LCR* method.

In calculating *S*, the following characteristics of the reference designs may be varied, although when reference design values have been changed, the resulting *SSF* will not necessarily be as accurate as for the reference values:

1. Orientation of the glazing system relative to south (azimuth),
2. Tilt of the glazing (if the glazing area as well as the tilt is varied, the effect on LCR_s may need to be accounted for also),
3. Ground reflectance,
4. Shading by building overhangs or fins (fins are not treated in this book),
5. Presence of a reflector to enhance solar gain,
6. Glazing transmittance, and
7. Building absorptance.

Regarding the weather as represented by *S* and *DD*, the *SLR* correlations are expected to be valid for any location, not just the United States locations used in developing the correlations. This expectation is based on the assumption that the spectrum of weather patterns represented in the United States encompasses the weather patterns in all temperate climates where passive solar heating is of interest. Therefore, the monthly method is universally applicable, given the required weather data; whereas, the annual method is limited to just those 223 locations in the *LCR* tables. Furthermore, particular years rather than long-term normals could be used in the monthly method if some special purpose should require it and the weather data are available.

General Comments — Most *SLR* correlations are based on adequate thermal mass in the building. The correlations are not recommended for situations where there is too little heat storage. This is not to imply that the *SLR* method cannot be used for frame buildings or other lightweight construction. The mass in such buildings may be adequate for direct-gain south windows equaling 7% to 10% of the floor area. Problems tend to arise when the direct-gain percentage exceeds this figure by very much. Overheating will occur, causing discomfort; occupants will open windows in midwinter, resulting in decreased performance. It is recommended that either internal mass walls be added or that a Trombe wall or sunspace be used to extend the glazed area and thereby increase solar savings.

The analysis methods implicitly contain the assumption that the scheduling of equipment is uniform throughout the day and night. There is no thermostat setback. Equipment such as lights and appliances generates heat at a constant rate. It is unlikely that these conditions would be exactly satisfied in a given application, but the long response time in well-designed passive solar buildings reduces the effect of time-dependent scheduling. In the case of the thermostat setback, little energy is saved in a well-insulated building with abundant mass because the temperature falls very slowly at night. In the case of nonuniform internal heat generation, the mass and insulation tend to smooth out the temporal variation of the air temperature. Therefore, many cases can be accurately analyzed by ignoring the thermostat setback and by assuming that the internal heat is generated at a constant rate. In other cases, however, this is a poor assumption, e.g., a building with intermittent use and large occupancy loads.

Questions often arise about the applicability of the *SLR* techniques to commercial buildings. No general statements can be made here; in some cases, the methods will be applicable, and in others, they will not. Many commercial buildings have large internal gains that offset the need for other heat. Using a modified degree-day base allows the user to account for the effect of this internal heat. Some commercial buildings have several separate zones; however, in many commercial buildings, an air handling system is installed to continuously distribute air throughout the building. This has the effect of creating a constant temperature throughout the building, thus satisfying one of the important assumptions on which the correlations are based. If this is not the case, *SLR* methods can be applied on a zone-by-zone basis. However, in many commercial buildings, the heating, ventilating, and air-conditioning equipment introduces thermal interactions between zones and between the equipment and the passive systems. Such interactions can be either detrimental or advantageous to the overall building energy performances, but in any case, care must be exercised before applying the methods in this book.

The *SLR* method provides an estimate of the auxiliary heat only. Since heating is often not the principal thermal design issue in commercial buildings, *SLR* is only one tool. Other methods for estimating cooling needs should be used to complement *SLR* in an overall thermal design process.* ** Commercial buildings where heating and cooling systems are operated simultaneously should be analyzed by methods that account for time-of-day and building zoning. The relationship between the occupancy schedule and the timing of heat delivery also needs to be considered.

Daylighting is also an important energy design issue in commercial buildings. As windows are added to achieve daylighting, the heating requirements are increased, shifting the building toward a more skin-dominated situation and making it appropriate to use the windows for their added passive solar gain to meet the heating load. *SLR* provides the needed tool to estimate the solar savings achieved.

Summer heat avoidance is a concern to any designer, particularly in regions of summer overheating. Many design strategies are available for shading the aperture, which is essential. Good ventilation is also essential, especially for sunspace situations. The added mass in a good passive solar design aids in summer temperature stability, which may allow the building to coast comfortably through high afternoon temperatures and take advantage of cool night temperatures during seasons when they occur.

The designer may well want to use more comprehensive (and complex) calculation tools, such as computer programs on which one can perform hour-by-hour simulations of building performance. This will produce a more complete picture than *SLR* and may be warranted for certain complex designs or difficult climate situations. However, even when a designer chooses to use complex calculation tools, *SLR* provides a quick first step in the design process. *SLR* will give answers within a few percent of the hour-by-hour simulation results if the building is built and operated in accordance with the reference design assumptions and parameter choices.

**ASHRAE Handbook of Fundamentals*

** *Insulation Manual for Homes and Apartments*, National Association of Home Builders Research Foundation

CHAPTER 2
PASSIVE SOLAR DESIGN GUIDELINES

General (Rules 1-9)

Rules 1 and 2 give guidance on conservation levels and the size of the passive solar aperture. They are suitable for use during beginning schematic design and are applicable to residential buildings and commercial buildings having residential levels of internal gains, for example, commercial buildings that employ extensive daylighting.

Rule 1. Conservation Levels — Recommended levels of insulation and building air tightness can be computed-based on a *conservation factor* (*CF*). Two suggested levels of *CF*, corresponding to two different projected fuel costs, are given in Table 2-1 for 219 cities. Unless conditions indicate otherwise, use the *CF* values corresponding to high fuel cost as a starting point. A map showing the contours of *CF* as listed in Table 2-1 is given in Figure 2-1. An explanation of the development of this rule, and the Passive Solar Glazing Area Rule (Rule 2) that follows, is given in Appendix 9.

Use the following formulas to compute guidance values for insulation and air tightness levels.

$$R_{wall} = 14\,CF$$
$$R_{ceiling} = 22\,CF$$
$$R_{perimeter} = 13\,CF - 5$$
$$R_{basement} = 16\,CF - 8$$
$$N_{E,W,N} = 1.7\,CF$$
$$ACH = 0.42/CF$$

Based on guidance resulting from these formulas, select practical, buildable conservation levels as a starting point for the design, trying for the most part to stay within 20% of the guidance. For windows, choose the closest integer value.

Table 2-1

Recommended *CF, LCR* and Resulting *SSF* for Various Locations

City	low fuel cost			high fuel cost		
	CF	LCR	SSF(%)	CF	LCR	SSF(%)
ALABAMA						
BIRMINGHAM	.95	117	31	1.19	41	58
MOBILE	.75	157	36	.91	62	64
MONTGOMERY	.86	132	33	1.06	49	61
ARIZONA						
PHOENIX	.70	123	53	.83	66	74
PRESCOTT	1.07	46	63	1.26	26	80
TUCSON	.72	106	58	.85	59	77
WINSLOW	1.13	46	57	1.35	25	75
YUMA	.58	174	54	.69	96	74
ARKANSAS						
FORT SMITH	1.03	96	31	1.29	33	59
LITTLE ROCK	1.04	97	31	1.30	34	59
CALIFORNIA						
BAKERSFIELD	.83	115	42	1.01	53	64
DAGGETT	.81	95	53	.96	50	74
FRESNO	.91	108	39	1.12	47	60
LONG BEACH	.70	106	64	.81	64	80
LOS ANGELES	.70	96	70	.81	61	84

Table 2-1 (continued)

City	low fuel cost			high fuel cost		
	CF	LCR	SSF(%)	CF	LCR	SSF(%)
MOUNT SHASTA	1.28	50	44	1.60	23	63
NEEDLES	.68	138	51	.80	73	72
OAKLAND	.87	73	64	1.03	43	78
RED BLUFF	.92	116	34	1.14	47	58
SACRAMENTO	.93	102	40	1.15	46	60
SAN DIEGO	.67	106	69	.77	67	83
SAN FRANCISCO	.89	70	64	1.04	42	78
SANTA MARIA	.85	64	73	.97	42	85
COLORADO						
COLORADO SPRINGS	1.30	33	60	1.55	18	78
DENVER	1.27	36	57	1.52	19	76
EAGLE	1.52	31	49	1.88	15	68
GRAND JUNCTION	1.27	45	47	1.56	21	69
PUEBLO	1.22	41	55	1.46	21	75
CONNECTICUT						
HARTFORD	1.42	137	13	1.88	36	29
DELAWARE						
WILMINGTON	1.25	90	23	1.62	28	47
DISTRICT OF COLUMBIA						
WASHINGTON	1.26	97	21	1.64	30	45
FLORIDA						
APALACHICOLA	.69	173	39	.82	74	65
DAYTONA BEACH	.56	224	45	.66	108	70
JACKSONVILLE	.67	167	43	.80	76	68
MIAMI	.32	714	45	.37	348	69
ORLANDO	.50	272	46	.59	134	70
TALLAHASSEE	.72	152	40	.87	66	66
TAMPA	.50	278	46	.58	137	70
WEST PALM BEACH	.36	554	44	.43	264	68
GEORGIA						
ATLANTA	1.00	101	32	1.24	37	59
AUGUSTA	.91	110	36	1.12	43	62
MACON	.85	127	35	1.05	49	62
SAVANNAH	.80	137	37	.98	55	64
IDAHO						
BOISE	1.31	64	33	1.67	28	51
LEWISTON	1.30	123	18	1.69	45	32
POCATELLO	1.43	50	36	1.83	22	54
ILLINOIS						
CHICAGO	1.39	121	15	1.83	34	32
MOLINE	1.42	115	15	1.88	32	33
SPRINGFIELD	1.32	98	19	1.73	28	42
INDIANA						
EVANSVILLE	1.21	121	19	1.58	34	42
FORT WAYNE	1.40	176	10	1.85	46	24
INDIANAPOLIS	1.33	158	12	1.76	40	30
SOUTH BEND	1.42	187	9	1.88	52	21
IOWA						
BURLINGTON	1.39	91	19	1.82	27	41
DES MOINES	1.45	85	19	1.91	25	41
MASON CITY	1.56	81	18	2.07	25	36
SIOUX CITY	1.47	85	19	1.94	25	39
KANSAS						
DODGE CITY	1.21	49	45	1.49	22	69
GOODLAND	1.32	40	48	1.62	19	70
TOPEKA	1.27	71	29	1.63	24	55
WICHITA	1.19	61	38	1.48	24	64

Table 2-1 (continued)

City	low fuel cost			high fuel cost		
	CF	LCR	SSF(%)	CF	LCR	SSF(%)
KENTUCKY						
LEXINGTON	1.23	131	17	1.60	35	39
LOUISVILLE	1.22	131	17	1.59	35	40
LOUISIANA						
BATON ROUGE	.75	176	32	.92	64	61
LAKE CHARLES	.72	219	28	.89	73	57
NEW ORLEANS	.70	176	36	.85	71	64
SHREVEPORT	.85	138	32	1.04	50	61
MAINE						
CARIBOU	1.72	92	14	2.29	33	25
PORTLAND	1.52	103	15	2.02	32	30
MARYLAND						
BALTIMORE	1.23	92	24	1.58	29	48
MASSACHUSETTS						
BOSTON	1.33	113	17	1.75	32	37
MICHIGAN						
ALPENA	1.61	133	11	2.13	47	20
DETROIT	1.40	168	11	1.85	47	24
FLINT	1.48	184	9	1.96	56	19
GRAND RAPIDS	1.46	185	9	1.93	55	20
SAULT STE. MARIE	1.67	144	10	2.22	54	17
TRAVERSE CITY	1.54	191	8	2.04	63	16
MINNESOTA						
DULUTH	1.73	123	10	2.30	46	18
INTERNATIONAL FALLS	1.80	148	8	2.39	59	14
MINNEAPOLIS	1.60	126	11	2.13	39	23
ROCHESTER	1.60	124	11	2.14	38	23
MISSISSIPPI						
JACKSON	.86	133	32	1.06	48	60
MERIDIAN	.88	131	32	1.08	47	60
MISSOURI						
COLUMBIA	1.26	97	21	1.64	28	46
KANSAS CITY	1.29	80	25	1.67	25	50
SAINT LOUIS	1.22	97	23	1.58	29	48
SPRINGFIELD	1.19	82	28	1.52	27	55
MONTANA						
BILLINGS	1.47	57	29	1.91	22	48
CUT BANK	1.63	56	26	2.13	24	39
DILLON	1.55	41	37	1.98	18	55
GLASGOW	1.66	99	14	2.21	37	24
GREAT FALLS	1.51	64	25	1.97	26	41
HELENA	1.56	63	24	2.05	26	39
LEWISTOWN	1.59	55	27	2.08	23	42
MILES CITY	1.55	74	20	2.04	26	36
MISSOULA	1.55	107	15	2.04	45	24
NEBRASKA						
GRAND ISLAND	1.40	56	31	1.79	20	55
NORTH OMAHA	1.43	72	23	1.87	22	47
NORTH PLATTE	1.41	44	39	1.77	18	62
SCOTTSBLUFF	1.41	42	41	1.76	18	64
NEVADA						
ELKO	1.42	36	49	1.76	17	68
ELY	1.41	30	59	1.70	16	76
LAS VEGAS	.86	80	55	1.02	43	75
LOVELOCK	1.38	30	60	1.65	16	78
RENO	1.23	38	60	1.47	21	77
TONOPAH	1.21	37	62	1.44	21	79
WINNEMUCCA	1.33	38	52	1.63	19	70

Table 2-1 (continued)

City	low fuel cost			high fuel cost		
	CF	LCR	SSF(%)	CF	LCR	SSF(%)
NEW HAMPSHIRE						
CONCORD	1.51	123	12	2.01	36	27
NEW JERSEY						
NEWARK	1.27	99	21	1.65	29	44
NEW MEXICO						
ALBUQUERQUE	1.08	48	58	1.29	26	77
CLAYTON	1.17	40	60	1.39	22	78
FARMINGTON	1.23	38	58	1.48	21	76
LOS ALAMOS	1.29	33	61	1.53	18	79
ROSWELL	1.02	57	56	1.21	30	76
TRUTH OR CONSEQUENCES	.96	58	63	1.13	33	79
TUCUMCARI	1.05	51	58	1.25	28	77
ZUNI	1.23	37	61	1.46	20	78
NEW YORK						
ALBANY	1.47	160	10	1.95	42	24
BINGHAMTON	1.50	269	6	1.99	70	15
BUFFALO	1.47	261	7	1.94	70	15
MASSENA	1.60	205	7	2.13	59	15
NEW YORK	1.25	146	15	1.64	37	36
ROCHESTER	1.45	259	7	1.92	70	16
SYRACUSE	1.45	270	7	1.92	69	16
NORTH CAROLINA						
ASHEVILLE	1.14	67	38	1.42	27	63
CAPE HATTERAS	.94	100	37	1.15	41	63
CHARLOTTE	1.01	92	35	1.26	35	61
GREENSBORO	1.09	77	35	1.36	30	61
RALEIGH-DURHAM	1.06	88	33	1.32	33	59
NORTH DAKOTA						
BISMARCK	1.67	82	16	2.22	29	29
FARGO	1.70	128	10	2.27	45	19
MINOT	1.70	114	12	2.26	44	20
OHIO						
AKRON-CANTON	1.40	186	10	1.85	50	23
CINCINNATI	1.27	160	13	1.67	40	32
CLEVELAND	1.39	231	8	1.84	62	19
COLUMBUS	1.34	184	11	1.78	46	26
DAYTON	1.34	159	12	1.77	41	29
TOLEDO	1.42	172	10	1.87	47	23
YOUNGSTOWN	1.42	252	7	1.88	66	17
OKLAHOMA						
OKLAHOMA CITY	1.07	77	37	1.33	30	64
TULSA	1.08	92	30	1.36	31	58
OREGON						
ASTORIA	1.24	70	34	1.56	33	50
BURNS	1.45	54	33	1.86	23	50
MEDFORD	1.22	90	27	1.56	38	43
NORTH BEND	1.11	55	53	1.35	29	69
PENDLETON	1.27	120	19	1.65	46	33
PORTLAND	1.21	120	21	1.55	46	36
REDMOND	1.38	46	41	1.74	21	60
SALEM	1.21	100	25	1.55	40	42
PENNSYLVANIA						
ALLENTOWN	1.36	108	17	1.78	31	37
ERIE	1.46	216	8	1.92	64	17
HARRISBURG	1.29	120	16	1.69	32	38
PHILADELPHIA	1.25	103	20	1.62	30	44
PITTSBURGH	1.37	208	9	1.81	54	22
SCRANTON	1.40	156	11	1.86	42	26

Table 2-1 (continued)

City	low fuel cost			high fuel cost		
	CF	LCR	SSF(%)	CF	LCR	SSF(%)
RHODE ISLAND						
PROVIDENCE	1.37	92	19	1.79	28	40
SOUTH CAROLINA						
CHARLESTON	.84	129	35	1.03	51	62
COLUMBIA	.92	107	36	1.12	42	63
SPARTANBURG	1.00	92	35	1.25	35	61
SOUTH DAKOTA						
HURON	1.58	100	14	2.10	32	28
PIERRE	1.51	103	15	2.02	25	36
RAPID CITY	1.48	52	31	1.90	20	52
SIOUX FALLS	1.56	83	18	2.07	26	35
TENNESSEE						
CHATTANOOGA	1.06	121	23	1.36	37	50
KNOXVILLE	1.06	118	25	1.35	37	51
MEMPHIS	1.02	117	27	1.29	37	55
NASHVILLE	1.10	147	18	1.42	39	44
TEXAS						
ABILENE	.90	91	44	1.08	41	69
AMARILLO	1.08	52	54	1.30	26	75
AUSTIN	.76	150	36	.92	61	63
BROWNSVILLE	.48	405	33	.57	167	60
CORPUS CHRISTI	.58	260	35	.70	108	62
DALLAS	.87	118	35	1.06	46	63
DEL RIO	.72	159	38	.87	68	64
EL PASO	.88	74	58	1.04	40	77
FORT WORTH	.88	115	35	1.08	45	63
HOUSTON	.71	216	28	.87	74	58
LAREDO	.56	293	33	.68	118	60
LUBBOCK	1.00	59	56	1.19	31	76
LUFKIN	.80	145	34	.98	55	62
MIDLAND-ODESSA	.88	77	55	1.04	41	75
PORT ARTHUR	.72	187	32	.88	70	61
SAN ANGELO	.84	103	44	1.01	47	69
SAN ANTONIO	.73	159	36	.89	66	64
SHERMAN	.96	102	34	1.19	38	62
WACO	.83	131	35	1.01	52	63
WICHITA FALLS	.95	88	40	1.16	37	66
UTAH						
BRYCE CANYON	1.49	24	65	1.77	14	80
CEDAR CITY	1.27	36	58	1.53	19	76
SALT LAKE CITY	1.31	50	41	1.64	22	61
VERMONT						
BURLINGTON	1.56	229	7	2.08	64	15
VIRGINIA						
NORFOLK	1.05	85	34	1.31	33	60
RICHMOND	1.12	94	28	1.42	32	53
ROANOKE	1.16	78	31	1.47	28	57
WASHINGTON						
OLYMPIA	1.28	106	22	1.65	45	35
SEATTLE-TACOMA	1.25	112	22	1.60	48	35
SPOKANE	1.44	109	17	1.88	44	28
YAKIMA	1.35	102	20	1.75	41	33
WEST VIRGINIA						
CHARLESTON	1.21	169	14	1.58	42	34
HUNTINGTON	1.21	135	17	1.58	37	39
WISCONSIN						
EAU CLAIRE	1.62	132	11	2.16	42	21
GREEN BAY	1.58	110	13	2.10	37	25
LA CROSSE	1.53	135	11	2.03	38	25
MADISON	1.55	95	16	2.05	30	31
MILWAUKEE	1.52	99	16	2.01	33	30

Table 2-1 (continued)

City	low fuel cost			high fuel cost		
	CF	LCR	SSF(%)	CF	LCR	SSF(%)
WYOMING						
CASPER	1.44	34	50	1.78	16	70
CHEYENNE	1.39	31	56	1.68	16	75
ROCK SPRINGS	1.50	29	53	1.85	15	71
SHERIDAN	1.51	52	30	1.95	20	50
CANADA						
EDMONTON, ALBERTA	1.75	76	17	2.32	34	26
SUFFIELD, ALBERTA	1.58	44	33	2.04	19	50
NANAIMO, BC	1.27	96	25	1.63	43	38
VANCOUVER, BC	1.28	138	17	1.66	53	29
WINNIPEG, MANITOBA	1.81	60	19	2.41	25	31
DARTMOUTH, NS	1.49	66	24	1.95	23	43
MOOSONEE, ONTARIO	1.89	105	10	2.52	39	18
OTTAWA, ONTARIO	1.64	67	20	2.17	23	37
TORONTO, ONTARIO	1.46	89	18	1.92	27	37
NORMANDIN, QUEBEC	1.79	70	17	2.38	27	28

//

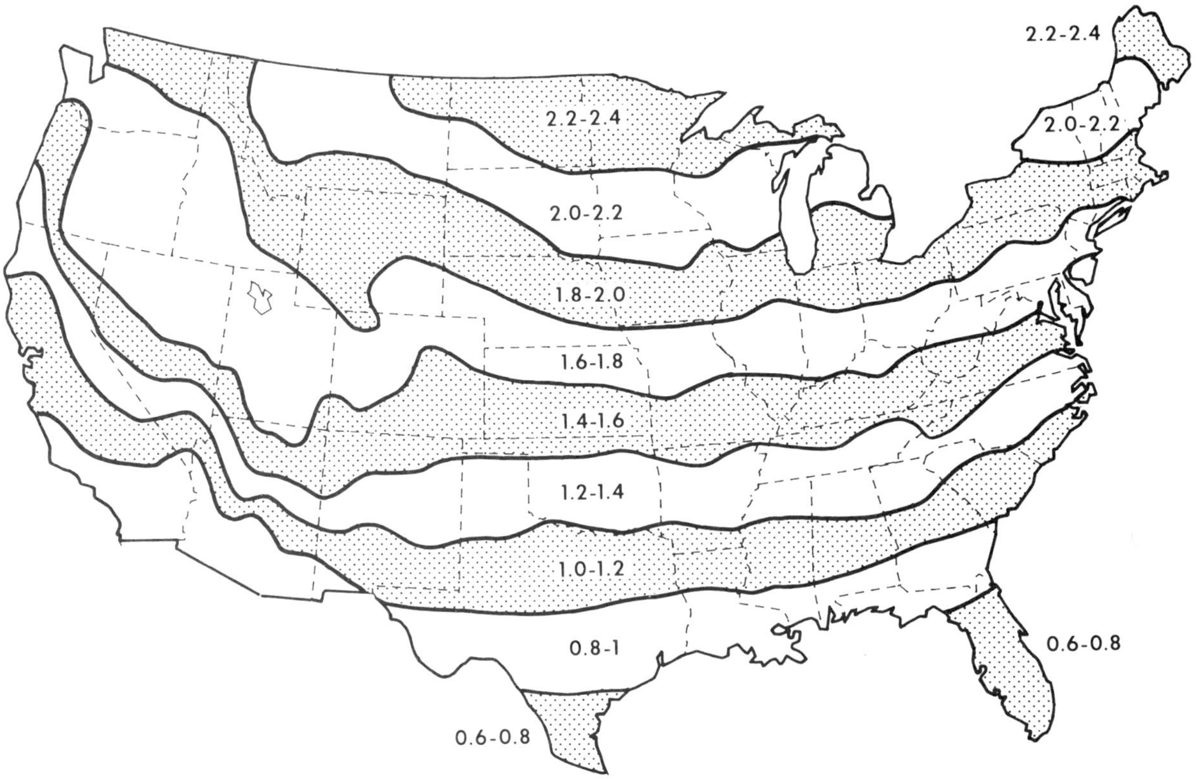

Figure 2-1. Conservation Factor (*CF*).

These guidelines pertain to heating considerations only. It may be that higher insulation levels are required to account for cooling needs.

R_{wall} and $R_{ceiling}$ are the *overall* R-values $(Btu/h \cdot °F \cdot ft^2)^{-1}$ of the opaque wall and ceiling insulation, accounting for the series effect of all air layers and materials, including the attic air space, if any.

$R_{perimeter}$ and $R_{basement}$ are the values of insulation *added to* the perimeter or basement walls. Perimeter refers to insulation that is added to the outside of foundation walls or under the slab perimeter in slab-on-grade construction. Normally this insulation is installed 2 feet down the foundation wall or 2 feet wide under the slab.

$N_{E,W,N}$ refers to the number of window glazings to be used on E,W, and N windows. However, if this number exceeds 3, consider using either double glazing combined with movable insulation or using a low conduction "superglazing" with U-value less than 0.3 $Btu/h \cdot °F \cdot ft^2$.

ACH refers to the *effective* air changes per hour caused by natural infiltration and forced ventilation. *ACH* is the sum of the natural infiltration and the non-heat-recovered portion of the forced ventilation. If a heat recovery unit is used for forced ventilation, the non-heat-recovered portion is equal to

(1 -heat recovery effectiveness factor) ACH_{forced}.

If the guidance value is less than $ACH = 0.5$ air changes per hour, a heat recovery unit is recommended for residental applications to maintain adequate indoor air quality. For commercial applications or situations with unusually high sources of indoor air pollution, higher values of ventilation may be required.

Regarding Costs

In order to obtain simple guidelines, assumed values have been used for the incremental cost of passive solar aperture, conservation improvements, and fuel; the escalation rate of fuel; and the fixed charge rate. These are outlined in Appendix 9, which also describes the methodology used to develop guidelines and explains how to modify the guidelines to account for any desired set of assumed costs. Note, however, that the guidelines depend on cost ratios and thus will not change with inflation if all costs escalate proportionally. The "high fuel cost" column in Table 2-1 is intended to represent the cost of electric resistance heat in much of the U.S. (7¢/kWh in 1983 dollars); the "low fuel cost" column is based on one-half of the high fuel cost.

Rule 2. Passive Solar Glazing Area — The recommended guideline for passive solar glazing area depends on the conservation level achieved. If the guidelines of Rule 1 have been followed, then the net load coefficient, *NLC*, will be approximately as follows:

$$NLC = \frac{(GF)(A_f)}{(CF)}$$

where *GF* is a geometry factor that accounts for the relative dimensions of the building. Suggested values of *GF* are given in Table 2-2. Alternatively, *GF* can be computed using a formula in Appendix 9. A_f is the gross floor area of the building.

Table 2-2

Geometry Factors, *GF*

Floor Area	Number of Stories 1	2	3	4
1000	7.3			
1500	6.5	6.7		
3000	5.4	5.4	5.7	
5000	4.9	4.7	4.9	5.1
10000	4.3	4.0	4.0	4.2
20000	3.9	3.5	3.5	3.5

The projected area, A_p, can now be determined, knowing the *LCR*, as determined from Table 2-1 or Figure 2-2, as follows:

$$A_p = \frac{NLC}{LCR}$$

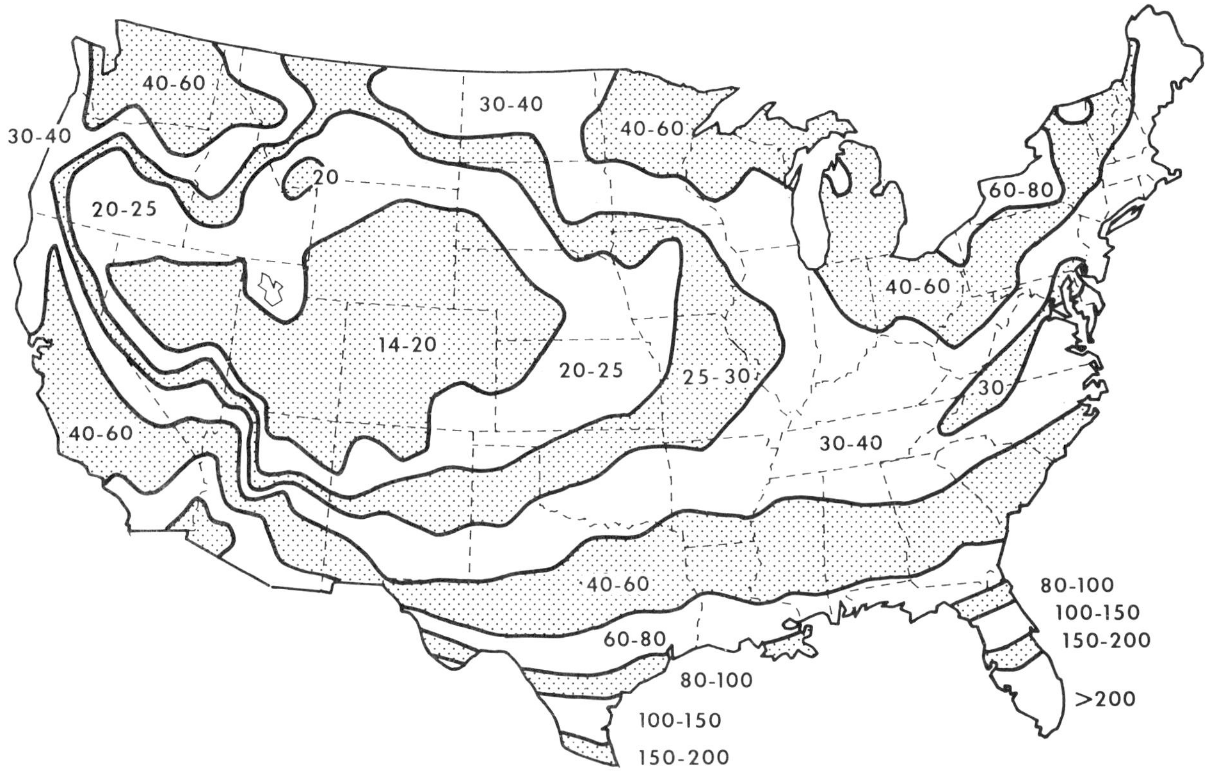

Figure 2-2. Recommended Starting-Point Load Collector Ratio (*LCR*) for Locations in the U.S. These values have been determined to yield a life-cycle optimum for selected passive solar, conservation, auxiliary heat, and financing costs. Other maps are based on these values of *LCR*.

As a starting point for design, the total projected area of all passive system types should be within 20% of this value. The recommendation is independent of the choice of passive system type.

At this point, a very rough estimate of annual auxiliary heat can be obtained from the relation

$$Q_{aux} = DD \cdot NLC \cdot (1 - SSF)$$

SSF values will depend on system type. Typical values of *SSF* for reference design SSD1 are mapped in Figure 2-3.

Figure 2-4 shows typical values of wall insulation for the U.S.

Rule 3. Distribution of Solar Glazing — Solar glazing should be distributed, space by space, in order to achieve the same ratio of load coefficient to projected area, NLC/A_p, in each zone. Where this is impractical, provide for natural convective exchange of heat from space to space and treat the combined spaces as a single zone for purposes of determining glazing area.

Rule 4. Orientation — The optimum orientation of solar glazing for all systems is within 5° of true south in most cases. If the orientation departs from true south, the penalty in *SSF* increases slowly at first and more rapidly as the departure increases. The penalty is roughly as shown in Figure 2-5.

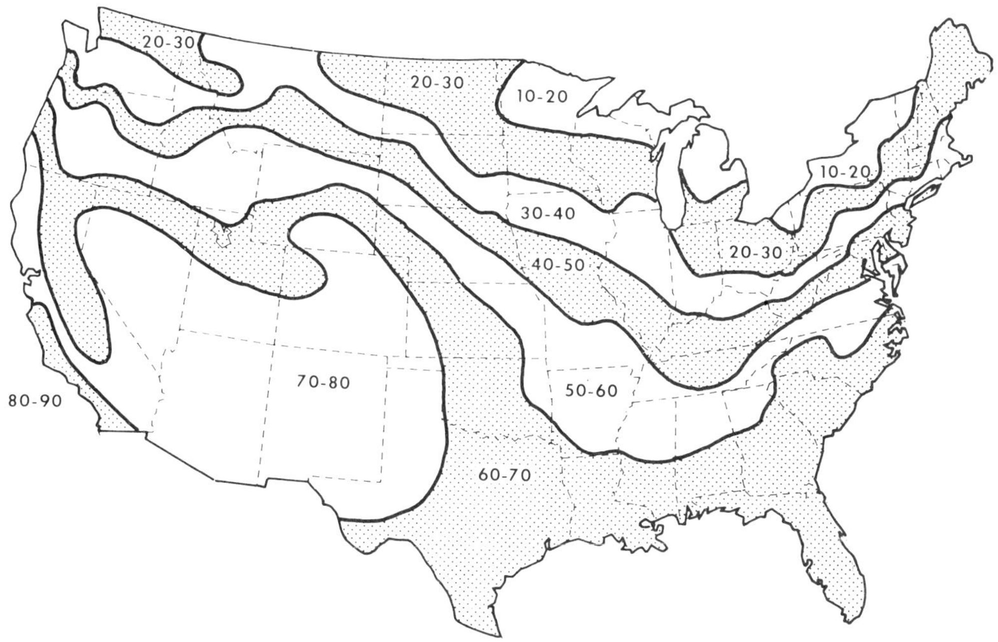

Figure 2-3. Solar Savings Fraction, in Percent, for Sunspace System SSD1 (semienclosed, 50° slope, opaque end walls, double glazing, masonry common wall, no night insulation)

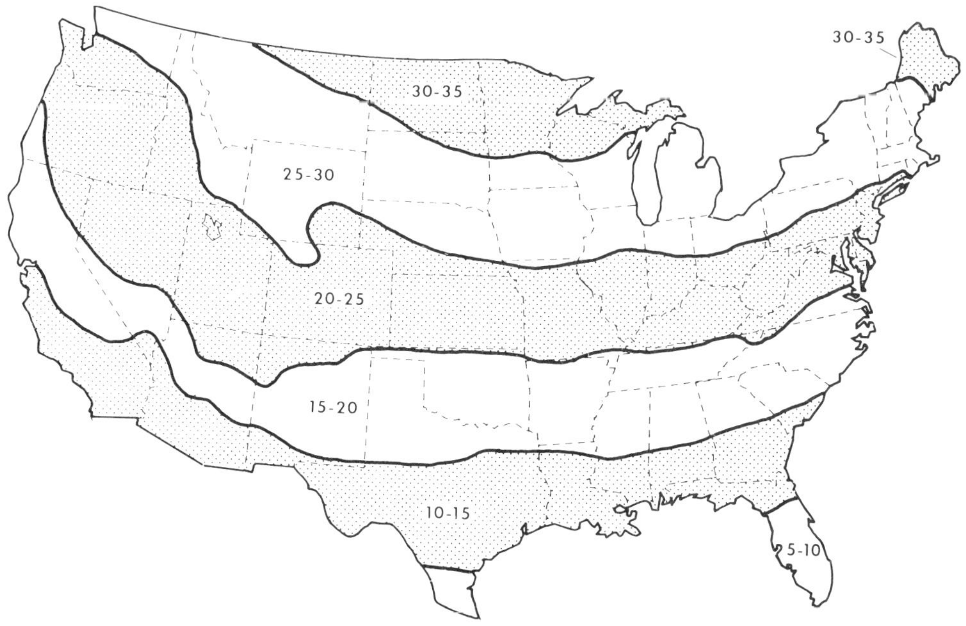

Figure 2-4. R-Value of Wall Insulation

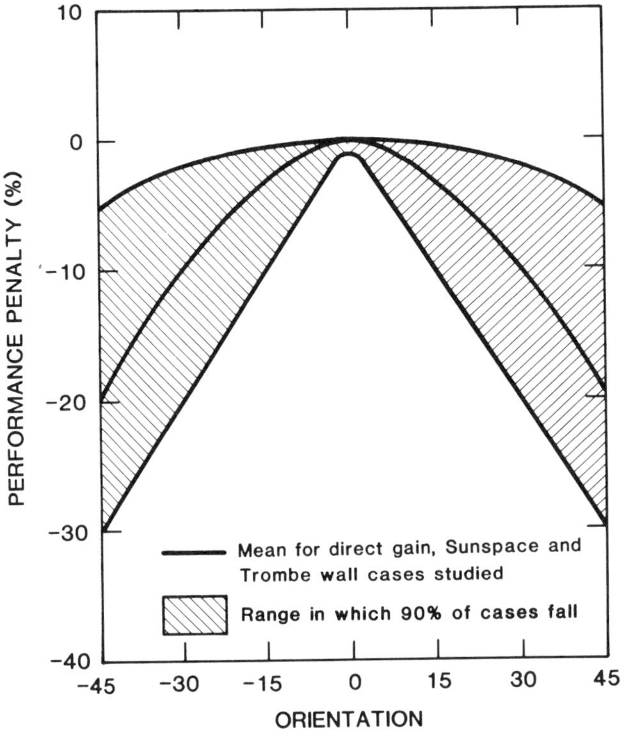

Figure 2-5. Performance Penalty for Orientations Away from Due South

Thus, one should orient within 15° of true south, if possible, and not more than 30° away from true south. Other orientation issues to be considered are as follows:

Direct Gain	The most severe performance penalties for off-south orientation are experienced by large direct-gain windows in cold, cloudy climates. Stay within 15° of south in these cases.
Length of Heating Season	Climates with a prolonged heating season and a cloudy climate (Seattle, for example) are somewhat less sensitive to orientation; relax the above rules by up to 10°.
Shading	If there is a known shading situation, orient away from the shade. For example, in a region of predominant morning fog during the heating season, shift the orientation 10° to the west; if a mountain shades in the afternoon, shift orientation 10° to the east.
Summer	Off-optimum orientation usually incurs a greater penalty in summer overheating than in winter heating performance. This is especially true at more southerly latitudes.

Rule 5. Tilt — For a given glazing area, an increase of up to 30% in performance can be achieved by tilting the solar glazing; however, this also implies

an increase in summer collection unless the aperture is shaded. The optimum tilt is in the range 50 to 65 degrees for locations in the contiguous United States.

Rule 6. Number of Glazings — The recommended number of glazings for the solar aperture is given below:

Winter Climate	Without Night Insulation or Selective Surface	With Either Night Insulation or Selective Surface
Mild (less than 2,500 *DD*)	2	1
Moderate (2,500 to 6,000 *DD*)	2	1 or 2
Severe (more than 6,000 *DD*)	3 or 4	2

Rule 7. Night Insulation — Night insulation will greatly improve the performance of any passive solar system in any climate; however, the cost effectiveness, and indeed even the necessity, for night insulation increases as the winter climate becomes colder. In addition to improving performance, night insulation improves comfort, especially in direct gain, by shielding the individual from an otherwise cold window surface at night.

The cost effectiveness of night insulation depends somewhat on passive system type. The maps in Figures 2-6 and 2-7 show the added annual energy savings for direct-gain and sunspace systems respectively, assuming an *LCR* value given in Table 2-1. These values can be weighed against the added cost of the night insulation to determine cost effectiveness. As *SSF* increases, so does the cost effectiveness of night insulation.

Although R9 night insulation has been used for the reference designs in this book, it should be noted that 73% to 82% of the benefit of R9 night insulation can be obtained with R4.

Rule 8. Mixed Systems — Mixed systems tend to improve both the comfort and the character of passive solar buildings. By choosing systems that complement one another, the designer can provide solar heat in phase with the need.

An effective procedure is to size direct-gain window area based on nonsolar or daylighting considerations, typically in the range of 5% to 10% of the floor area. Locate the direct-gain windows (subject to the restrictions of Rule 19) on the south wall or in clerestories. Double- or triple-glazed east and west windows usually have no winter thermal penalty, but the area should be located so that they will be shaded in the summer to prevent overheating. Design east, west, and north windows to achieve proper lighting, visual access to the outside, ventilation, and emergency egress. Then, if more solar glazing is indicated, use indirect systems (Trombe walls and sunspaces, for example) to achieve the desired total area.

Rule 9. Sensitivity to Design Parameter Choice — As a general rule, the performance sensitivity of passive solar buildings to variations in design parameters increases as the *SSF* increases or as the *LCR* decreases. Thus, there is wide latitude for choice at low *SSF* values, but at high *SSF* values, the designer should study the sensitivities carefully and design close to the optimum.

Direct Gain (Rules 10-19)

Rule 10. Mass Distribution — For a given quantity of thermal mass, it is normally preferable to spread the mass over as large an area as practical, consistent with locating mass within the direct-gain space. A mass surface-to-glass area ratio of at least 6:1 is recommended. Although mass located in the direct sun is perhaps 30% more effective than mass out of the sun, this should be secondary to distributing mass. As a general rule, if the point where the sun first strikes within the room can be seen from a mass surface, then this location is effective for heat storage and is called "radiatively coupled mass" (line-of-sight rule). Other mass is less effective, depending on location. If the only mechanism for heat exchange to the mass surface is by air convection, then the heat storage effectiveness of the mass is reduced by up to four times. This mass is called "convectively coupled mass." In all cases, however, increased mass increases direct-gain performance and comfort.

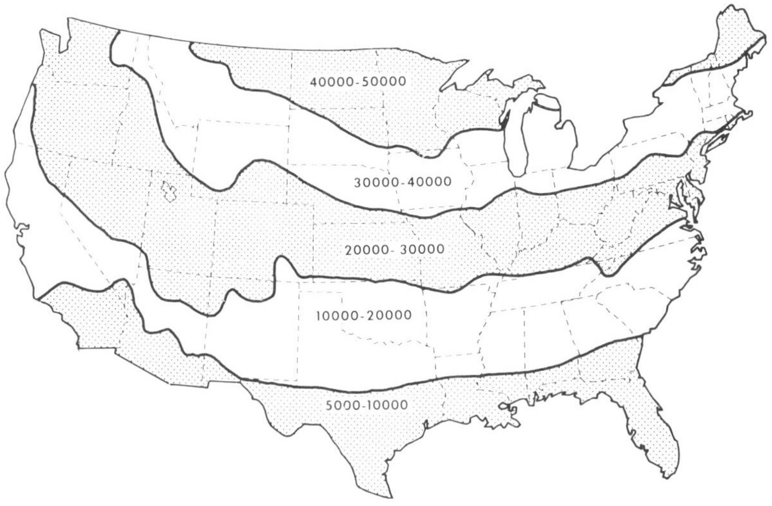

Figure 2-6. Added Annual Energy Savings, Btu per ft^2 of Projected Area, As a Result of Adding R9 Night Insulation to a Well-Designed Direct-Gain Passive Solar Heating System (System DGC3 vs. DGC1)

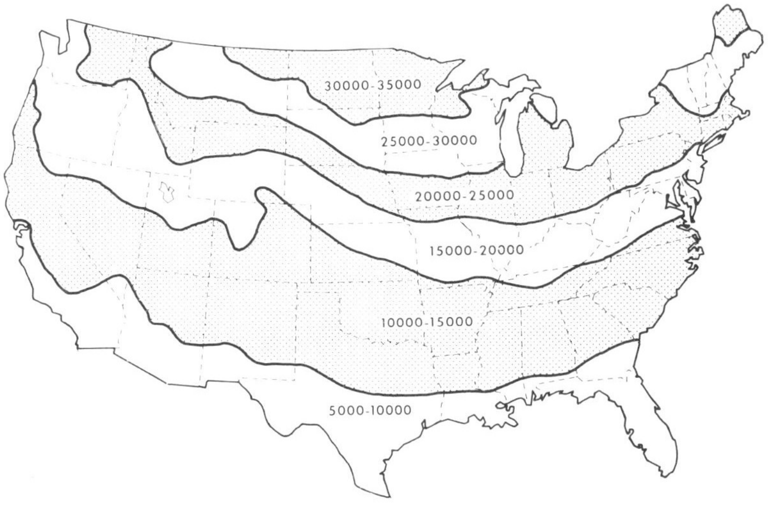

Figure 2-7. Added Annual Energy Savings, Btu per ft^2 of Projected Area, As a Result of Adding R9 Night Insulation to a Well-Designed Semienclosed Sunspace (System SSD2 vs SSD1). Note that these values are less than the added savings obtained by adding night insulation to direct gain because the performance of the non-night-insulated direct-gain case is much poorer in cold climates.

Rule 11. Mass Thickness — Per pound, mass in thin sections is more effective than mass in thick sections. For masonry materials, added mass beyond about 4 inches in thickness is less effective than mass closer to the surface. For wood, the corresponding thickness is about 1 inch. The property that influences this is the thermal diffusivity, $k/\rho c$, which measures the ease of heat penetration into the material. Values of $k/\rho c$ are given in Appendix 6, Table A6-4, for various materials. Figure 2-8 shows the heat storage effectiveness of various materials for daily cycles.

Rule 12. Internal Mass Storage Walls — Per unit of masonry, internal partition walls are more effective for heat storage than external walls, assuming both have equal access to solar heat. This is because heat can transfer to the mass through both surfaces, rather than just one surface. For greatest effectiveness, locate partition walls between two direct-gain spaces. One effective geometry is a N-S partition wall located between two direct gain rooms. Another is an E-W partition wall separating a direct-gain space on the south from a room with a clerestory window on the north.

Rule 13. Use of Color in Direct-Gain Spaces —

a. Lightweight objects should be light in color.

b. If the sun first strikes a wall or ceiling surface in a direct-gain space, then this surface should be light in color to diffuse the light and heat.

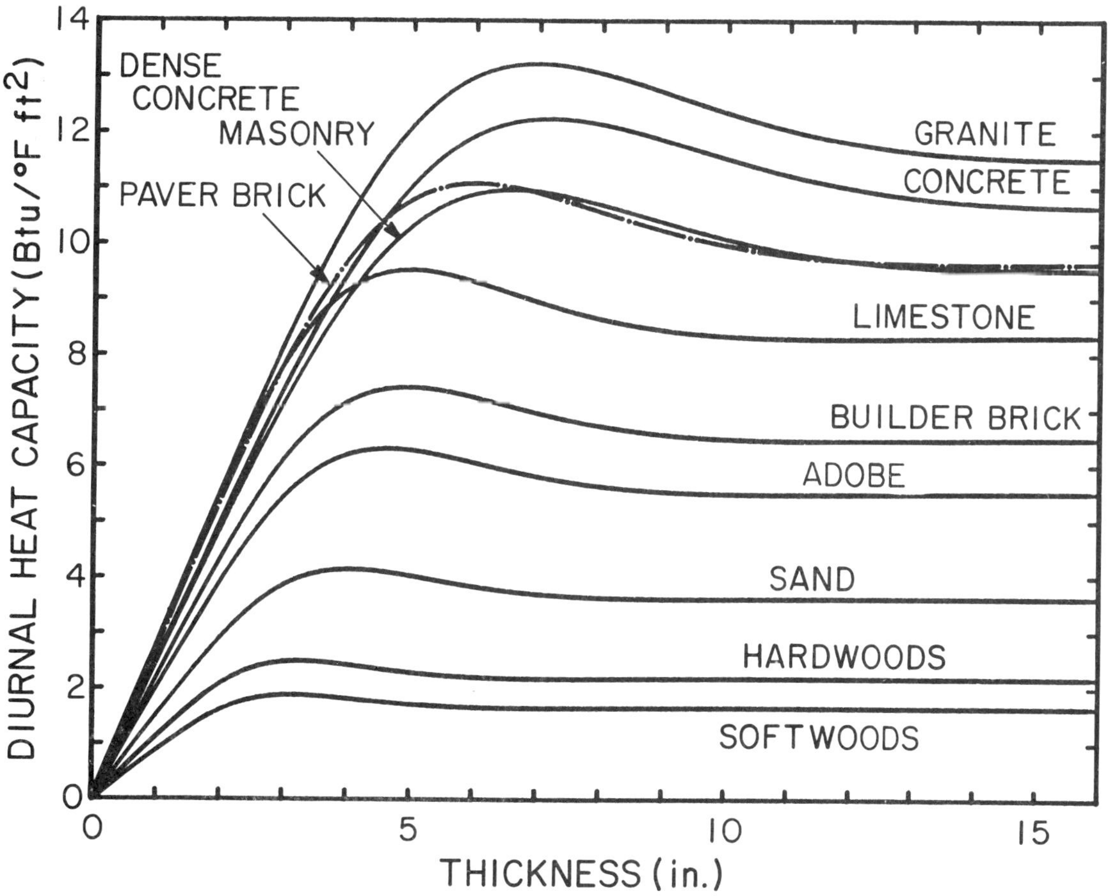

Figure 2-8. Heat Storage Effectiveness of Various Materials for Daily Cycles.

c. If more than one-half of the walls in a direct-gain space are massive, then they should be light-colored. If only one wall is massive, then it should be dark-colored (except as in 13b, above).

d. Massive floors in direct-gain spaces should be dark in color to store the heat low.

e. Clerestories should be located so that the winter sun strikes low in the space. Otherwise, the surface where the sun first strikes should be very light in color to reflect the light down into the room.

Rule 14. Surface Coverings — Any insulating cover over thermal mass decreases heat storage effectiveness and building performance. Surface coverings should have good thermal contact with the mass (no air gaps!) and an R-value less than 0.1 $ft^2 \cdot {}^\circ F \cdot h/Btu$. A surface covering R-value of 0.4 can decrease heat storage effectiveness by about 50%.

Rule 15. Concrete Masonry — When hollow-core concrete block is used for direct-gain heat storage, effectiveness can be increased by the following measures:

a. Use of a high-density block (the higher the better) with a solid density of 130 lb/ft^3 or more is preferable.

b. Use block with thick walls or grout the holes solid.

c. Use of sand as a hole-filler is effective in internal partition walls, although not as effective as grout, but not very effective in external walls.

d. If grout is used as a filler only to increase mass, then a cement-poor mix can be used to save on cost. A sand-to-cement ratio of 12:1 is recommended.

Rule 16. Brick — If brick is used for heat storage, then a dense brick is preferred. Face brick or pavers are about 36% more effective than common brick in 4-inch thicknesses or 24% more effective in 2-inch thicknesses. In the case of large mass-area-to-glazing-area ratios, Rules 15a and 16 are desensitized. Performance and comfort are not highly dependent on density.

Rule 17. Floor Materials — Rules 13d, 14, and 16 apply to floor coverings. In addition, it is desirable to effectively couple floor surface covering to a massive subfloor. Sand should not be used, or if used, it should be in as thin a layer as possible. (Sand mixed with fine dirt is less detrimental.) If insulation is used underfloor, it should be covered with a least 4 inches of mass; however, more than 6 inches of added thickness is irrelevant. Covering a mass floor with insulation, such as carpeting, essentially eliminates heat storage in the floor; furniture is almost equally detrimental (note Rule 10, above).

Rule 18. Suntempered Buildings (Large *LCR*) — In suntempered buildings, solar heat is used primarily to offset daytime heating loads. Thermal mass, although beneficial, is not as important as in buildings with small *LCR*. The above direct-gain rules may be relaxed for suntempered buildings.

Rule 19. Limits on Direct-Gain Glazing — Direct-gain glazing should be limited to prevent large temperature swings, depending on available heat storage. Approximate limits are as follows:

low-mass buildings:	7% of floor area
very high-mass buildings:	25% of floor area

To obtain a more definitive estimate, assign points for exterior walls, interior partitions, ceilings, and floors as follows:

		Points	
		Radiatively Coupled Mass	Convectively Coupled Mass
Exterior walls	Massive (6 in. or more)	8.1	3.1
	Massive (2 in.)	3.6	2.6
	Frame/sheetrock (3/8 in.)	0.3	0.3
Interior partitions	Massive (6 in.)	6.7	3.6
	Frame/sheetrock (3/8 in.)	0.4	0.4
	Brick (4 in.)	4.1	3.0
Floors (exposed floor, multiply values by 1.3 if floor is directly sunlit)	Hardwood on softwood (1 in. + 1 in.)	1.1	0
	Pavers (1-1/2 in.) on 1 in. wood	2.9	0
	Thin slab (2 in.) on 1 in. wood	3.4	0
	Thick slab (4 in. or more) on earth	6.0	0
	Carpeted	0.3	0
Ceilings	Massive (3 in.)	8.7	5.3
	Massive, extended surface	13.0	8.0
	Wood (1 in.)	1.0	1.0
	Acoustic tile	0.2	0.2
Furniture	Extensive	4.0	4.0
	Moderate	2.6	2.6
	None	0	0

The sum of the points in the "radiatively coupled" column is the maximum percentage of floor area that should be in direct gain. Thus,

$$A_p \text{ maximum} = A_{fr} \text{ (radiatively coupled points/100).}$$

where A_{fr} is the floor area of the direct-gain room.

Direct-gain A_p may be increased by no more than 5% of A_{fr} to account for convection to adjacent spaces connected by open doorways; compute points for the adjacent space using the "convectively coupled mass" column, then compute

$$A_p \text{ add} = A_{fc} \text{ (convectively coupled points/100).}$$

where A_{fc} is the floor area of convectively coupled rooms.

This rule is based on limiting clear day temperature swings to no more than 10°F. In no case should the direct-gain glazing area exceed 25% of the floor area of the space into which the direct gain enters.

If the indicated projected area from Rule 2 is greater than the maximum recommended by Rule 19, the difference should be made indirect (such as unvented Trombe wall or sunspace).

Sunspaces (Rules 20-26)

Rule 20. Use of Mass — The use of mass in a sunspace depends on the function of the space. A very lightweight sunspace is essentially an air-heating collector that can be convectively coupled to the adjacent building. The space will usually be too hot by day and too cold by night. In this case, an insulated wall separating the sunspace from the adjacent building is advised.

The alternative is to provide thermal mass to temper the sunspace. This increases the space's livability, its function as an entry, foyer, transit area, or greenhouse, and tends to reduce any tendency the space may have for summer overheating. An effective location for mass is as a thermal storage wall separating the building, water containers along the common wall, or a masonry floor.

All the direct-gain rules apply to sunspaces except that a mass-surface-to-projected-area ratio of at least 3:1 is advised rather than the 6:1 ratio recommended as a minimum for direct gain. If water is used as the primary thermal mass, then the water mass should be approximately 0.5 ft^3/ft^2 (3.8 gal/ft^2) of projected area or more. The water should be in dark-colored containers located in direct sun.

Rule 21. Do Not Glaze the End Walls — Use of glazed east and west end walls is ill-advised on the basis of both winter and summer thermal criteria. Winter heating performance is decreased in nearly every climate (although the effect is sometimes small); however, the added summer solar gain can be devastating. It is preferable to use insulated end walls, perforated with windows or doorways adequate for summer cross ventilation. Massive end walls, insulated on the exterior, are preferable to frame construction if sunspace thermal comfort is a design goal.

Rule 22. Roof — The sunspace should be designed to shade internal sunspace mass in the summer but allow sun penetration in the winter. If a portion of the roof is glazed, then provision should be made for summer shading, preferably outside the glazing. If shading is inside, provision should be made to vent the resultant hot air.

Rule 23. Common Wall — Sunspaces should be separable from the building living space, that is, the entryway or portal between the building and the sunspace should have a door that can be closed, and vents in the common wall should also be closable. The common wall is an effective and convenient location for thermal storage mass (masonry or water in containers), in which case the wall need not be insulated. For a lightweight frame common wall, the R-value should be at least R10.

Rule 24. Common-Wall Vents — The term "vents" can be confusing, so "common-wall vents," which connect the sunspace to the attached building, are distinguished from "summer vents," which connect the sunspace to the outdoors.

The primary thermal connection between sunspace and the adjacent building is normally by convection; thus, particular care must be given to this aspect of the design. The most frequently used and most practical common-wall vent is a connecting doorway or window. This method is low-cost, effective, familiar, easy to operate, and serves multiple functions—all desirable attributes. Another possibility is to provide pairs of high and low openings through the common wall. The following guidelines should be used for sizing common-wall vents:

a. Doorways (assuming 6-foot 8-inch doors) = 10% of projected area

b. Window openings (assuming 3-foot windows) = 15% of projected area

c. High and low vent pairs (assuming 8-foot separation) = 6% of projected area (combined area of both vents)

Rule 25. Summer Vents — Vents to the outside are essential to prevent sunspace overheating in summer. Both cross ventilation and stack-effect ventilation are effective, but prevailing summer wind direction should be considered in locating inlets and outlets. Inlets should be located in a region of high pressure (windward side) and outlets in a region of low pressure (leeward side or in a location of high cross-wind velocity). Vent areas should be generous. Forced ventilation (fans) can also be used effectively.

Rule 26. Wall Color — Generally, follow direct-gain rules with the following modifications:

a. If light-colored surfaces tend to reflect sunlight out of the sunspace, then they should be made darker. The color of lightweight objects does not affect performance.

b. If a sunspace is used as a greenhouse, then the surfaces in dark corners should be made light in color to improve sunlight for plants.

Trombe Walls (Rules 27-29)

Rule 27. Vents — Vents are often omitted from Trombe walls, and this is generally recommended for residential applications. Vents used in a Trombe wall result in a mixed system—thermal storage wall and convective loop. The increase in net performance due to vents is small, usually in the range of 3,000 to 5,000 Btu/yr·ft^2. Whether vents should be used depends on whether the building needs more daytime heat than is provided by direct gain. If so, then vents may be appropriate. Vents provide heated air to the building, which is useful primarily to offset immediate daytime heat loads. Vents combine with direct gain in driving building temperature swings. If used to excess, large swings will occur.

The following rules apply to vent sizing:

SSF	Area of Each Vent as a Percent of A_p
0%-25%	3
25%-50%	2
50%-90%	1
90%-100%	none

Rule 28. Wall Thickness — Trombe wall optimum thickness is given in the following table:

Material	Density (lb/ft^3)	Optimum Thickness (inches) With Vents	Optimum Thickness (inches) Without Vents
Concrete	140	16	14
Concrete Block	130	15	13
Building Brick	120	14	12
Adobe	100	12	10

Performance varies little within 20% of these optimum values.

Rule 29. Selective Surfaces — A selective surface adhered to the outside face of a thermal storage wall can be used to improve performance. Although a selective surface provides a winter heating performance enhancement nearly as great as night insulation, it has less control flexibility and may exacerbate summer overheating unless the wall is properly shaded. Normally one would use single glazing in conjunction with the selective surface since the performance of single- and double-glazed Trombe walls with a selective surface is very similar. The map in Figure 2-9 shows the performance increase for the addition of a selective surface to a single-glazed Trombe wall compared to a double-glazed, flat black Trombe wall.

If a selective surface is used on a Trombe wall or water wall, contact with the surface must be good. An R-value less than 0.03 is advised; however, this is generally the case if a continuous surface contact is made. The selective surface should have a solar absorptance greater than 0.92 and an infrared emittance less than 0.15. An increase of one percentage point in absorptance is equivalent in performance to a decrease of about 0.5 to 1.5 percentage points in emittance.

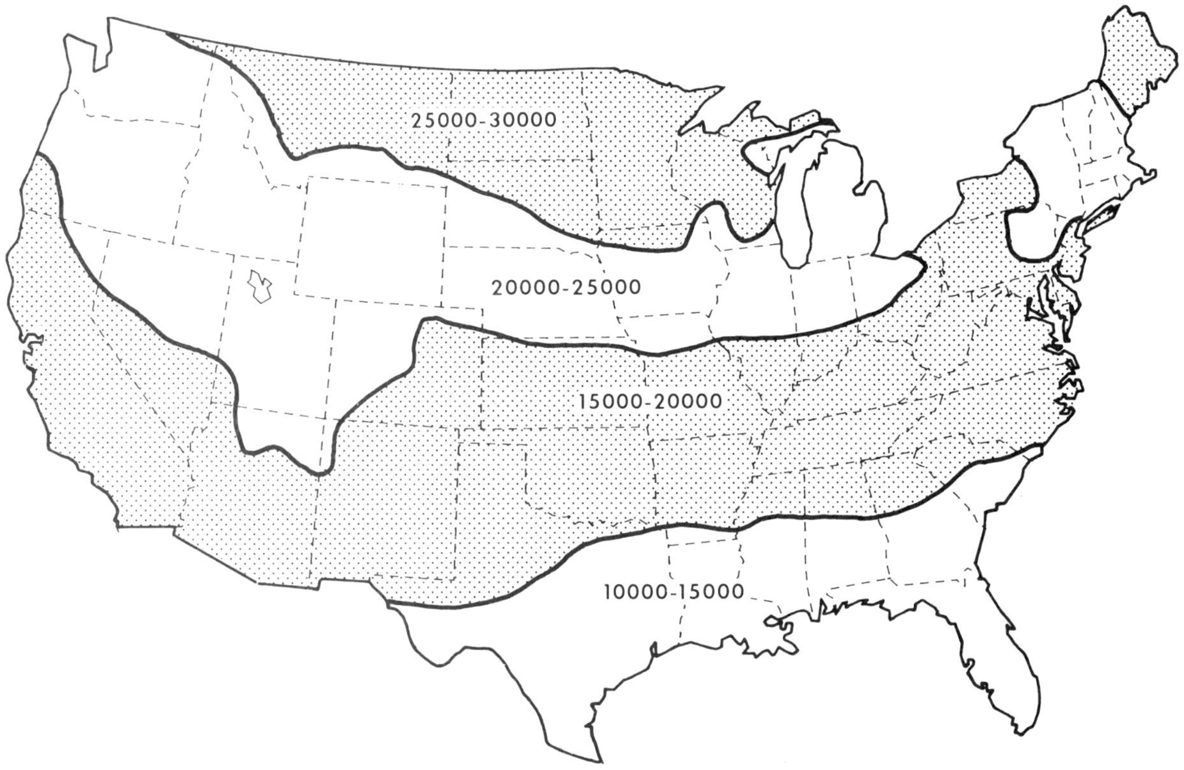

Figure 2-9. Added Annual Energy Savings, Btu per ft^2 of Projected Area, As a Result of Replacing One Glazing in a Double-Glazed Trombe Wall with a Selective Surface Adhered to the Trombe Wall Exterior Surface (System TWJ1 vs. TWF3). The contours for adding night insulation to a flat black double-glazed Trombe wall are nearly identical to these contours (TWI4 vs. TWF3).

If a selective surface is used, a vent should not be provided since dust accumulation may impair the selective surface.

Water Walls and Phase Change Material (PCM) Walls

Rule 30. The principal thermal advantage of using water or PCM in a thermal storage wall is the ability to achieve high values of thermal heat storage capacity within a smaller volume and with less weight than a masonry wall. Values of 60 to 120 Btu/°F·ft^2 of glazing are achievable and recommended. (To compute an approximate comparable storage capacity of PCM, divide the latent heat by 20°F. For example, if the PCM has a latent heat of 2,000 Btu/ft^2 of glazing surface, this is roughly equivalent to 100 Btu/°F·ft^2 of glazing in a water wall.)

Use of a selective surface on a water wall or a PCM wall is generally recommended. Follow the guidelines of Rule 29.

CHAPTER 3
THE LOAD COEFFICIENTS

Purpose of the Method

The methods commonly used to calculate the heating requirements of a building are described in the *ASHRAE Handbook of Fundamentals*. These methods picture a building as a closed but leaky box with a furnace inside that supplies heat at the exact rate necessary to keep the inside air at a constant temperature. Heat transferred to or from the building mass is not accounted for. Heat is lost from the box by conduction through the walls and by infiltration of air through the leaks. The heat loss rate for this box is the "design load" when the outside temperature is the design temperature. Solar gains are not accounted for in this calculation. In this book we are not particularly concerned with this calculation but rather with the incremental load due to a 1-degree increase in inside (or a 1-degree decrease in outside) temperature.

The Total Load Coefficient, *TLC*

The total load coefficient, *TLC*, is the additional daily total heat required to maintain the building 1°F higher than a previous level. For example, if *x* is the daily heat required to maintain a building at 69°F, then ($x + TLC \cdot 1$°F) is the daily heat required to maintain the building at 70°F. The *TLC* is related to the design load by the formula:

$$TLC = \frac{24 \times (\text{hourly design load})}{(\text{indoor temp. -design temp.})} \quad (3\text{-}1)$$

Units of *TLC* are Btu/°F·day.

The Net Load Coefficient, *NLC*

A second term must also be defined. This is the net load coefficient, *NLC*. *NLC* is the same as *TLC* except that heat losses through the solar wall are not included. The units of *NLC* are Btu/°F·day. The use of *NLC* rather than *TLC* is appropriate in *LCR* or *SLR* calculations because separate account is made of solar wall performance, including losses. Also, the use of *NLC* provides a more appropriate basis for comparison of solar buildings with nonsolar buildings. This is because the solar gains through a normal complement of windows on a south wall tend to offset losses through the rest of the wall. So a south wall on a nonsolar building tends to have a very small net energy transfer. The data in Appendix 1 have been compiled on the basis of a building load that excludes the solar wall.

Given a detailed knowledge of a building's geometry and construction, the ASHRAE methods provide comprehensive estimates of both *TLC* and *NLC*. This is of little help during the schematic design stage because the detailed specifications of the building are not yet known. Windows have not been precisely sized, wall areas and details have not been determined, and the building volume is not known. The input parameters required for a detailed calculation are unknown. Furthermore, design development is an iterative process, and few designers have the time to devote to repeated detailed calculations of the building load. A much faster procedure is needed if it is to be used.

Computation of *NLC*

The procedure presented here divides the building heat loss into seven separate elements and provides a formula for computing the load coefficient for each of the elements. The formulas used are approximations to the ASHRAE procedure. The building heat loss elements, notation, units of measurement, and formulas are displayed in Table 3-1. Depending on the house design, only one of the three elements, perimeter, basement, or floor, will normally be applicable. For slab-on-grade construction, heat loss occurs along the perimeter of the slab floor, P_f. A building with a basement will have heat loss around the perimeter of the basement, P_b. If the floor is built over a crawl-space, then the heat loss occurs over the entire area of the ground floor, A_f. In some cases a

Table 3-1

Building Elements, Notation, and Formulas for the *NLC* Computation

Building Heat Loss Element	Size of Building Element (Units)	R-Value ($h \cdot ft^2 \cdot °F/Btu$)	Load Coefficient Symbol ($Btu/°F \cdot day$)	Formula
Walls	A_w (ft^2)	R_w	$(UA)_w$	(A_w/R_w)
Roof	A_r (ft^2)	R_r	$(UA)_r$	(A_r/R_r)
E,W,N windows	A_g (ft^2)	R_g	$(UA)_g$	$1.1(A_g/N)$
Floor perimeter	P_p (ft)	R_p	$(UA)_p$	$4.17\, P_p/(R_p + 5)$
Basement perimeter	P_b (ft)	R_b	$(UA)_b$	$10.7\, P_b/(R_b + 8)$
Floor	A_f (ft^2)	R_f	$(UA)_f$	(A_f/R_f)
Infiltration	V (ft^3)	—	$(UA)_i$	$0.018\, ACH \cdot ADR \cdot V$

building may be partially over a basement and the remainder on a slab floor, in which case both P_b and P_p must be included. Infiltration heat loss is proportional to the volume, V, of the building. For single-story buildings, the volume can be taken as the floor area, A_f, times the ceiling height. For multistory buildings the volume of all rooms on the upper stories must be included. The symbol *ACH* stands for air changes per hour and represents the number of times per hour that the air in the building is replaced by infiltration from outside. The units of *ACH* are h^{-1}. The symbol *ADR* stands for air density ratio and represents the ratio between air density at the site to air density at sea level. The value of *ADR* depends on the elevation of the site and can be found from the chart in Figure 3-1. The number of glazings in the windows is N.

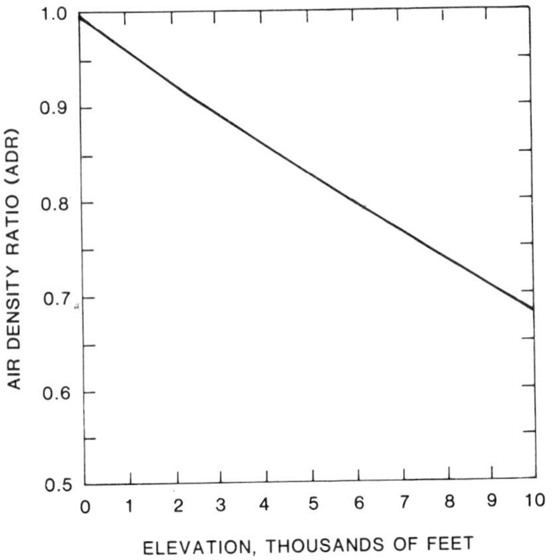

Figure 3-1. The Air Density Ratio (*ADR*) for Different Elevations. The sea level air density is 0.075 lb/ft^3.

As will be seen, the largest component of *NLC* is often $(UA)_i$, the load coefficient associated with infiltration. During design development, there is insufficient information available to estimate *ACH*. Rather, at this point, the designer should consider the determination of *ACH* to be a matter of design criteria rather than prediction. The minimum value that might be chosen will depend on several considerations:

1. The natural infiltration rate associated with normal building construction normally exceeds ½ air change per hour (½ air change/h). A typical level would be in the range of ¾ to 1¼ air change/h.

2. The minimum exchange rate of outside air recommended is ½ air change/h. Below this level the building becomes stuffy, odors build up, and high humidity due to water vapor release in the building is likely to be a problem. Additional air quality hazards include dangerous levels of formaldehydes (which emanate from resins used in building materials such as plywood, chipboard, and carpets), radon gas (which leaks up under the building through cracks in the rock and soil), oxides of nitrogen and carbon monoxide (from combustion), and tobacco smoke.

3. The natural infiltration rate can be reduced to about ½ air change/h by use of a carefully installed vapor barrier (recommended for all insulated construction anyway) and carefully sealing all cracks where air leakage might occur, especially at the sill plates.

4. Further reductions in *ACH*, to the range of 1/10 to 1/20 air change/h can be achieved by special construction techniques that include a continuous vapor barrier sealed at all overlaps, very tight-sealing windows and doors, sealing electrical outlet boxes, and special closures on vents. However, in buildings with an *ACH* of less than ½ air change/h, forced ventilation with an air-to-air heat recovery unit is recommended to maintain adequate indoor air quality. Although this approach may be routine in large commercial buildings, its use is just beginning in residential construction.

5. In special cases where pollutant source levels are known to be high, increased ventilation may be required either as a matter of common sense or building code requirement. Examples: a restaurant or lounge might require an *ACH* of 4 air change/h during periods of occupancy; buildings built in a region of high radium in the earth may require more ventilation to reduce indoor radon levels.

Therefore, it is recommended that the building designer select a minimum air infiltration rate based on a quality of sealing and construction that can economically be achieved, on the building requirements, and on whether or not an air-to-air heat exchanger is to be used. The designer must then set design details and specifications in an attempt to achieve the desired *ACH* value.

To compute *NLC,* estimate the size of each applicable building element and its R-value. Then compute the load coefficient for each building element using the formulas given in Table 3-1. The sum of the load coefficients times 24 is *NLC*. For example, in a building with a basement:

$$NLC = 24\,[(UA)_w + (UA)_r + (UA)_g + (UA)_b + (UA)_i] \qquad (3\text{-}2)$$

The Net Load Coefficient Worksheet in Appendix 7 is useful for this computation.

Note that this list of *UA* elements does not cover all possible situations. For example, earth-integrated buildings and unheated basements are not included. In these cases the designer must make a rough estimate of the effective *UA* based on ASHRAE or other procedures and good engineering judgments. Keep in mind that that ultimate use of the *UA* will be to estimate annual load by multiplying UA times the degree days times 24. Thus, if the annual load of a bermed wall were known by some other procedure, the effective *UA* of the wall could be determined by dividing this load by the product of annual degree days times 24.

Computation of *LCR*

The load collector ratio, *LCR*, is the ratio of the net

load coefficient, *NLC*, to the projected area, A_p, of the solar wall:

$$LCR = NLC/A_p \qquad (3\text{-}3)$$

This ratio reveals the extent of heat loss shown by *NLC*, in relation to the amount of solar gain, shown by A_p. For a given location, it is the most important parameter influencing the solar savings fraction, *SSF*, of a building. Appendix 1 tabulates *SSF* as a function of *LCR* for all 94 reference designs, for 223 geographic locations, and for two base temperatures

Computation of *TLC*

The total load coefficient, *TLC*, can be computed by adding the steady-state heat load coefficient of the solar wall to *NLC*.

$$TLC = NLC + 24\,U_cA_p \qquad (3\text{-}4)$$

where U_cA_p is the steady-state load coefficient of the solar wall. Values of U_c for the reference designs are shown in Tables A5-7, A5-8 and A5-9.

CHAPTER 4
DEGREE-DAYS AND BASE TEMPERATURE

Definition of Degree-Days, DD_b

The number of heating degree-days, DD_b, for one day is the difference between a base temperature, T_b, and the mean daily ambient temperature, T_m, counting only positive values.

The reported observations of mean daily ambient temperature are traditionally defined by:

$$T_m = \tfrac{1}{2}(T_{min} + T_{max}) \qquad (4\text{-}1)$$

where T_{min} and T_{max} are the minimum and maximum temperature occurrences during the day.

The number of degree-days in a month is the sum of the degree-days for each day of the month; similarly for annual degree-days.

Under steady-state conditions, the daily heat lost (Btu/day) from a building without solar gains is equal to:

$$Q_{aux} + Q_{int} = TLC(T_{set} - T_m) \qquad (4\text{-}2)$$

where Q_{int} is the daily internal heat due to appliances, lights and people; and T_{set} is the thermostat setpoint and is assumed to be the inside temperature. A typical residential value of Q_{int} is 20,000 Btu/day per occupant. For commercial applications refer to the *ASHRAE Handbook of Fundamentals*.

Rearranging this equation:

$$Q_{aux} = TLC\,(T_{set} - \frac{Q_{int}}{TLC} - T_m) \qquad (4\text{-}3)$$

If the following criterion is used to establish the base temperature,

$$T_b = T_{set} - Q_{int}TLC, \qquad (4\text{-}4)$$

then the expression for Q_{aux} simplifies to

$$Q_{aux} = TLC\,(T_b - T_m) \qquad (4\text{-}5)$$

Summing over a month or a year, counting only those days for which $T_b - T_m$ is positive,

$$Q_{aux} = TLC \cdot DD_b \qquad (4\text{-}6)$$

This expression is similar to that commonly used in the heating and ventilating industry. It should be emphasized that this equation is valid only under the condition:

$$T_b = T_{set} - Q_{int}/TLC \qquad (4\text{-}7)$$

Equation (4-6) applies to a building without solar gains. If the building has a solar wall calculate T_b and DD_b in exactly the same way. The auxiliary heat required by the building is expressed as

$$Q_{aux} = Q_{net} - Q_{sav} \qquad (4\text{-}8)$$

where

$$Q_{net} = NLC \cdot DD_b \qquad (4\text{-}9)$$

The net reference load, Q_{net}, is the heat loss from the nonsolar parts of the building and Q_{sav} is the useful heat provided by the solar wall in excess of that required to offset the solar wall losses.

By definition, the solar savings fraction, *SSF*, is the ratio of the solar savings to the net reference load:

$$SSF = Q_{sav}/Q_{net} \qquad (4\text{-}10)$$

It follows that:

$$Q_{aux} = (1 - SSF)\,NLC \cdot DD_b \qquad (4\text{-}11)$$

Determination of DD_b

Tables — Since T_b is now a variable, a method for finding DD_b for different base temperatures is required. To avoid this calculation, Appendix 3 lists values for the monthly and annual DD_b for 5 different values of T_b (50°, 55°, 60°, 65°, and 70°F) for 223 locations. For temperatures other than those tabulated, linear interpolation is adequate.

Thom Method — The tabulated values of DD_b were calculated by use of a method developed by Thom.[1,2] This method is a useful alternative to the tables and is needed for locations not listed in the tables. The normal degree-days, DD_b, below any base temperature T_b is approximated by

$$DD_b = M(T_b - T_A + I\sigma_m\sqrt{M}), \tag{4-12}$$

where

M = Number of days in month,
T_A = Monthly mean ambient temperature (°F),
σ_m = Standard deviation in monthly mean temperature (°F), and

$$I = \begin{cases} 0.34\exp(-4.7h) - 0.15\exp(-7.8h), & h \geq 0 \\ 0.34\exp(4.7h) - 0.15\exp(7.8h) - h, & h < 0 \end{cases} \tag{4-13}$$

where

$$h = (T_b - T_A)/(\sigma_m\sqrt{M}) \tag{4-14}$$

The quantities T_A and σ_m are both tabulated in Appendix 3 under the headings TA and SD, respectively. For locations where σ_m is not known, the correlation[3]

$$\sigma_m = 4.79 - 0.0337\,T_A \tag{4-15}$$

(σ_m and T_A in °F) gives accurate enough results for solar heating analysis. It is customary practice in tabulating degree-day normals to set to zero any results from Equation 4-12 that are less than 5.

An example problem of calculating DD_b is given in Chapter 7.

References

[1] H. C. S. Thom, "The Rational Relationship Between Heating Degree Days and Temperature," *Monthly Weather Review 82* (Jan 1954), pp. 1-6.

[2] H. C. S. Thom, "Normal Degree Days Below Any Base," *Monthly Weather Review 82* (May 1954), pp. 111-115.

[3] D. G. Erbs, W. A. Beckman, and S. A. Klein, "Degree-Days for Variable Base Temperatures," *Proceedings of the 6th National Passive Solar Conference*, Sept. 8-12, 1981, Portland, Ore. (AS/ISES, Newark, Delaware, 1981), pp. 387-391.

CHAPTER 5
THE ANNUAL LCR METHOD

Basis of the Method

The annual *LCR* method of estimating the performance of a passive solar heating system uses the *LCR* Tables in Appendix 1, which are calculated from *SLR* correlation equations, together with the monthly heating degree-day and solar radiation data. The tables contain the values of *SSF* associated with a particular set of 8 *LCR* values that span the range of practical interest and for each of the 94 reference designs. There is a table for each of 223 locations in the United States and Canada.

By use of the *LCR* tables, the annual *SSF* is easily found and the annual auxiliary heat requirement of the building can be calculated from

$$Q_{aux} = (1 - SSF)\, NLC \cdot DD \qquad (5\text{-}1)$$

The quantity *DD* is the same as DD_b in the previous chapter. The subscript *b* is dropped in this and following chapters to simplify the notation.

The Annual *LCR* Method

The annual *LCR* method is executed by performing the steps listed in Table 5-1. The Annual *LCR* Worksheet in Appendix 7 is useful for this computation.

Discussion of the Steps

Step 1 — Select a reference design that most nearly resembles the building to be analyzed. A summary description of all 94 designs is in Appendix 5. More detailed descriptions are in the chapters on the

Table 5-1

Procedure for the *LCR* Method

Steps		Reference
1.	Select an appropriate reference design.	Appendix 5, Tables A5-2, 3, 4, and 5
2.	Calculate the net load coefficient, *NLC*.	Chapter 3
3.	Determine the projected area of glazing, A_p.	Definition in Chapter 1
4.	Calculate the load collector ratio, *LCR*.	Chapter 3, Equation 3-3
5.	Determine the base temperature and the degree days.	Chapter 4, Equation 4-4
6.	Look up annual *SSF* in *LCR* table.	Appendix 1
7.	Compute annual auxiliary heat requirement, Q_{aux}.	Chapter 5, Equation 5-1

system types, Chapters 12 through 14. Sensitivity data can be used to correct the results for variations in certain design parameters (see the *Sensitivity Data* section below). Night insulation R-value other than 9 can be treated as described below in the *Night Insulation R-value* section. Two or more reference designs combined in the same building can be treated as described below in the *Mixed Systems* section.

Step 2 — Chapter 3 explains the calculation of *NLC*.

Step 3 — See the definition of projected area in Chapter 1.

Step 4 — $LCR = NLC/A_p$. See Chapter 1, Eq. 1-2.

Step 5 — $T_b = T_{set} - Q_{int}/TLC$. See Chapter 4, Eq. 4-4. Three levels of precision in evaluating T_b are possible, depending on the desired speed and accuracy of the estimate and on the known amount of building detail.

a. Take T_b to be a nominal 65°F. This corresponds roughly to a thermostat setting (T_{set}) of 70-72°F and an increment due to internal sources (Q_{int}/TLC) of 5-7°F. This choice simplifies the use of the *LCR* tables because no interpolation between base temperatures is required.

b. An internal heat generation rate (Q_{int}) of 20,000 Btu/day per occupant may be estimated for residential buildings. Chapter 3 explains the calculation of *TLC*.

c. The internal heat generation rate (Q_{int}) can be estimated directly from an enumeration of the heat sources—occupants, appliances, lights and machinery. The contribution of each heat source to Q_{int} is the average heat generation rate (Btu/h) of the source when it is present or operating times the number of hours per day the source is present or operating. The *ASHRAE Handbook, 1981 Fundamentals* (pp. 26.20-26.29) has information on the heat generation rate for various sources. Note that this level of detail is appropriate to the *LCR* method only when the size of Q_{int}/NLC is expected to have a major impact on the base temperature. Chapter 3 explains the calculation of *TLC*.

Step 6 — The *LCR* tables are in Appendix 1. They are arranged in alphabetical order by state then by city. Canadian locations follow the U.S. locations.

Look up *SSF* corresponding to the reference design chosen in Step 1 and *LCR* from Step 4 and the base temperature, T_b, from Step 5. If *LCR* falls between two tabulated values, interpolate linearly between them to find *SSF*. If T_b falls between the two tabulated values, use linear interpolation. When both *LCR* and T_b are intermediate to the tabulated values, calculate *SSF* for both 55°F and 65°F and then interpolate between these based on base temperature. If frequent use is to be made of the *LCR* table for a given system type and location, it may be helpful to plot a graph of *SSF* versus *LCR* for both base temperatures. Curves for intermediate base temperature can be sketched in by linear interpolation. The graph will speed up the process of estimating *SSF*.

If the location of interest has no *LCR* table, select a substitute location that is nearby, has similar solar radiation data, and similar degree-days data. If there is no such similar location, choose two locations with similar solar radiation but with degree days that bracket the desired location degree days and interpolate between the two.

Step 7 — $Q_{aux} = (1 - SSF)\, NLC \cdot DD$. See Chapter 1, Eq. 1-4. The degree-days (*DD*) are relative to the base temperature from Step 5.

Night Insulation R-Value

One of the most important design decisions is whether or not to use night insulation. The improvement in performance is dramatic, especially in cold climates where heat losses at night can be severe. In fact, under the most severe conditions, a Trombe wall or water wall without night insulation is only marginally productive; its heat losses are almost equal to its heat gains. Direct gain is even more affected, and solar savings cannot be achieved in extremely cold climates. Night insulation is particularly important under these conditions.

The reference value for night insulation is R9. This is a rather high value only occasionally achieved in practice. If the decision to use night insulation is made, then a critical design decision is the R-value to be specified.

If the R-value of the night insulation happens to be R9, then *SSF* from the tables can be used directly.

The performance of systems with night insulation R-value between 0 and 9 can be estimated as follows:

$$SSF_R = SSF_0 + Y(SSF_9 - SSF_0) \quad (5\text{-}3)$$

where

SSF_R = *SSF* for actual night insulation *R*-value

SSF_0 = *SSF* for no night insulation

SSF_9 = *SSF* for night insulation of R-value 9

and

$$Y = (1 + R_0/9)/(1 + R_0/R) \quad (5\text{-}4)$$

where

R = the actual R-value of the night insulation

and

$R_0 = 1/U_c$

where U_c is the steady-state load coefficient of the solar wall without night insulation, as listed in Appendix 5, Tables A5-7, 8, 9. For sunspaces with an insulated common wall, use U_c for the corresponding masonry-wall sunspace. For single- or triple-glazed sunspace systems, use

$R_0 = 1/U_c - 2 + R_g$

where U_c is for the corresponding double glazed sunspace system and *Rg* is the R-value of single or triple glazing (nominally 0.9 or 3.2).

A graph of *Y* versus *R* is shown in Figure 5-1. Note the diminishing returns relative to R-value shown in Figure 5-1. Half the benefit of R9 night insulation is realized for R-values 1.5 to 2.5.

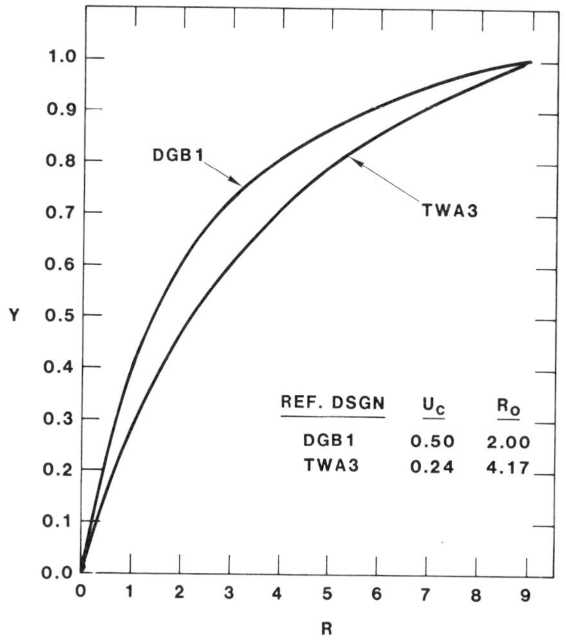

Figure 5-1. Examples of the Effectiveness of Night Insulation Compared to the Improvement Obtained with R9 Night Insulation $Y = (1 + R_0/9)/(1 + R_0/R)$

In the analysis leading to this function, the night insulation was controlled strictly by time of day, being put in place at 5:30 p.m. each day and removed at 7:30 a.m. each morning (solar time). A number of hour-by-hour water wall calculations were done to see whether performance could be materially increased using a "smart" controller. An idealized controller was employed in the analysis in which the insulation was put in place at any time when it was deemed to be thermally advantageous. The calculated increase in performance was very small. Thus, it would seem that a simple time clock could be used to determine the time of opening and closing of the night insulation.

Mixed Systems

For designs that are mixtures of two or more reference designs, calculate a single *LCR* based on the *NLC* and the total projected area for all the apertures included in the building. Then look up the *SSF* for each of the systems from the tables in Appendix 1. A weighted average of these *SSF* values will give the correct answer. The weighting factor is the proportion of the total projected area that is associated with each of the reference designs. The Annual *LCR* Worksheet in Appendix 7 is useful for performing this weighted average calculation.

Sensitivity Data

The worksheet in Appendix 7 and the tables in Appendix 1 provide a rapid and convenient method of estimating the thermal performance of a building, but the convenience is accompanied by a lack of flexibility in application. After T_{set} and Q_{int} are determined (and, therefore, T_b is determined), and for a given location and reference design, the only parameter that can be changed is *LCR*. A change in the *LCR* represents a change in the glazing area of the building and/or a change in the building's conservation level. All other building design parameters are fixed at the values specified for the reference design. To overcome this difficulty, sensitivity data are provided in Appendix 2. These data were generated by performing year-long, hour-by-hour computer simulations of building performance for different values of one of the design parameters, holding all other parameters at their reference design values. The results are generally presented in graphical form and allow the designer to see the effect of changing one particular design parameter. The sensitivity data available for each type of passive system are listed in Table 5-2.

In applying the sensitivity data, the designer executes the annual *LCR* method for the most similar reference design and then refers to the sensitivity graphs of Appendix 2 to determine the effect on performance caused by the difference between the desired building design and the reference design. For two or more design differences, the sensitivity analyses are applied one after the other as shown in Example Problem 7-6.

Average January Clear-Day Room Temperature, $T_{r,avg}$

During a succession of repetitive clear-day conditions, the 24-hour *average* temperature inside the building will stabilize. It is good design practice to limit the size of the solar collection glazing to a value that will raise the 24-hour average temperature into the comfort range on a clear January day. If more glazing than this is used, then the building will overheat on clear days. A correlation has been established that yields the average room temperature elevation due to solar gains, ΔT_{sol}, on a clear January day. The average January clear-day temperature is:

$$T_{r,avg} = T_A + \Delta T_{sol} + Q_{int}/TLC \qquad (5\text{-}5)$$

The correlations were established using computer simulations for the various reference designs.

Correlations for Room Temperature Elevation, ΔT_{sol}

A simple correlation form is as follows:

$$\Delta T_{sol} = Q_{tran}/(a + b \cdot LCR) \qquad (5\text{-}6)$$

where Q_{tran} is the solar radiation transmitted through two layers of standard 1/8-inch glass per unit projected collector area (Btu/day · ft^2), and a and b are constants for a given reference design. Q_{tran} is a function of latitude only for January clear-day conditions and is the same for all reference designs except most sunspaces. Values of Q_{tran} are given in Appendix 10, Table A10-1.

The constants a and b obtained for latitudes between 25° and 50° are listed in Appendix 10, Table A10-2.

Reference

1. J. A. Duffie and W. A. Beckman, *Solar Engineering of Thermal Processes* (New York: John Wiley & Sons, 1980), pp. 62-63.

Table 5-2

Sensitivity Studies

Design Parameter	Appendix 2 Page Number			
	Direct Gain	Suntempered	Sunspace	Thermal Storage Wall
Thermal storage mass	A2-3	A2-32	A2-38	A2-110
Thermal storage properties (ρ,*c*,*k*)	A2-14		A2-38	A2-110
Absorptance of mass surface	A2-15	A2-36	A2-74	A2-116
Mass distribution	A2-17			
Mass coverings (e.g., rugs)	A2-18			
Number of glazing layers (*N*)	A2-20	A2-34	A2-62	A2-122
Temperature bounds (max. allowable room temperature and thermostat setpoints)	A2-22		A2-102	
Nighttime thermostat setback	A2-24			
Sunspace width			A2-56	
Orientation			A2-44	A2-136
Glazing tilt			A2-46	
Sunspace vertical glazing height			A2-52	
Sunspace upper glazing tilt			A2-54	
Glazing air gap			A2-70	
Glazing thickness			A2-72	
Lightweight absorption fraction	A2-16		A2-84	
Vent area			A2-88	A2-138
Sunspace infiltration rate (*ACH*)			A2-96	
Common wall thermal resistance			A2-98	
Sunspace thermostat setpoint			A2-106	
Infrared emittance				A2-130

CHAPTER 6
BALANCING CONSERVATION AND SOLAR

Introduction

An old adage of the solar designer is "first insulate, then use solar." This is good advice, but it fails to answer the questions of how much conservation to use or how much solar to use. A simple method of determining the proper insulation levels and the degree of airtightness to use in conjunction with a passive solar design in a given location is given in this chapter.

The guidelines given as Rules 1 and 2 in Chapter 2 are based on a correct balance of conservation and solar heating with some built-in assumptions as to cost ratios and system type. They are very easy to use but not very flexible. At this late stage, toward the end of schematic design, the designer may want to factor in information as to actual costs and system selection. This chapter provides such a method.

The method considers not only the level of conservation and the amount of solar heating to be incorporated into the building, but also the costs of these features, the auxiliary heat required by the building, and the weather. The concept of the problem is illustrated in Figure 6-1, in which contours of equal energy savings are plotted against the cost of conservation and the cost of passive solar. The straight dashed line in the figure is the locus of all possible combinations of conservation and solar for a fixed budget of $6,000. The point at which the fixed

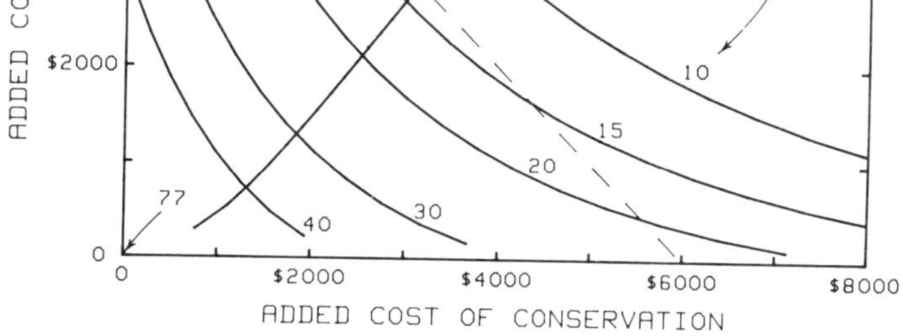

Figure 6-1. Savings in Energy, Compared with Conventional Construction, Obtainable with Different Investments in Energy Conservation (Added Insulation, Double Glazing, etc.) and Passive Solar (Trombe wall). The dashed line is for a total initial investment of $6,000 for both.

budget line is tangent to one of the energy savings contours identifies the optimum balance between conservation and solar for a fixed budget of $6,000. The method presented here finds this point without requiring the plotting of all the energy savings contours.

Costs

Energy conservation and passive solar are both strategies for saving energy and compete with one another for the builders' investment dollars. Both strategies can be characterized as behaving according to a law of diminishing returns. This is easily seen in the case of wall insulation. As an example, assume that there is a 400-ft^2 wall with a R-value of 5. For the climate involved, the dollar value of the heat lost through this wall is $200 per year. The designer decides to double the R-value from 5 to 10 for an added construction cost of $100. This halves the heat loss and reduces the heat loss cost from $200 to $100 per year, a saving of $100 per year. The payback time on the investment is 1 year. Pleased with this result, the designer increases the wall resistance factor from R10 to R20, then from R20 to R40. The corresponding costs, savings, and payback times are listed in Table 6-1.

It is easy to see that the incremental return associated with each incremental investment is decreased and that the payback time is increasing rapidly. The designer may decide to stop at R40. The overall cost of going from R5 to R40 has been $700, and the annual energy savings is $175, so the overall payback time is 4 years. If much smaller increments had been used, it would have been found that the payback time for this example increases from 1/2 year for the first small increment to 32 years for the last increment.Note that it is the *incremental* savings compared to the *incremental* cost and not the average that is important in determining when to stop.

This characteristic of decreasing annual savings associated with each increase in investment is what is meant by the law of diminishing returns. It characterizes virtually every conservation strategy, including additional wall insulation, additional ceiling insulation, increased perimeter insulation, or reduced air infiltration. It also characterizes investments in passive solar heating. The first increment of passive solar glazing area is very effective because it saves auxiliary heat during the entire heating season. Consider, however, adding glazing area to a building that already has a very high solar contribution. Since the building requires backup heat only during periods of either extremely cold weather or prolonged cloudiness, the added glazing is beneficial only during these rare periods. Although the cost of the last increment of glazing area may be the same as the first increment, its cost effectiveness is much less because the energy savings is much less. Again, we see diminishing returns and the necessity of calculating on an incremental rather than an average basis.

Table 6-1

Example Investment Costs and Payback Times for Wall Insulation

R-Value of Wall	Cost of Conservation Improvement ($)	Cost of Heat Loss ($)	Heat Loss Savings ($)	Payback Time for the Last Change (yr)
5	0	200	—	—
10	100	100	100	1
20	200	50	50	4
40	400	25	25	16

Because of these characteristics, it is usually true that a mix of conservation and passive solar strategies will produce the maximum energy savings for a given initial investment. This allows both energy conservation and passive solar to work in their most cost-effective range. It is clear from the above example that the best strategy depends on the starting point and also on the climate. For example, if the starting point is a building design that is already insulated to R40 for a location in Arizona, it is highly likely that an added passive solar element will be far more cost effective in saving additional energy than an increase in insulation. The opposite extreme might be a starting-point building design that is insulated to R5 in Wisconsin. Here it is certain that the insulation levels should be increased considerably before considering a passive solar addition.

At each point in the design, it is possible to weigh these decisions and to determine the best strategy for increasing performance. The decision always involves a trade-off between the cost of the improvement versus the increased performance. An optimal mix is achieved when the incremental cost/benefit of each conservation option is just equal to the incremental cost/benefit of the passive solar strategy being used.

The method to be described provides a technique for keeping the solar and conservation levels in proper balance with one another. The other side of the decision is determining how far to go with the combined strategies. Here there are two possibilities. The first and most likely situation is that the designer will have a budget, and this will constrain the total combined investment in conservation and passive solar. The second possibility is that life-cycle costing will determine the degree of energy savings strategies, weighing the incremental present value of future fuel savings against the cost of the additional investment. A simple procedure for determining the life-cycle optimum *SSF* will be reviewed briefly later in this chapter.

Conservation Incremental Cost, *CIC* — The construction costs assignable to improving the conservation level of a building are entered into the balancing computation in a very specific form, which, for convenience, will be called the Conservation Incremental Cost or *CIC*. This factor must be computed for each element of a building that contributes to the heat losses used in calculating the building *NLC*. Table 6-2 lists the *CIC* factors needed in the balancing computation. The heat losses from elements of the solar wall are not included in computing *NLC*, and therefore, no *CIC* factor is computed for the elements of the solar wall.

In determining *CIC*, the costs and R-values of two different building elements are determined. The ratio of the difference in cost (ΔC) to the difference in R-value (ΔR) yields the value of *CIC*:

$$CIC = \Delta C/\Delta R \tag{6-1}$$

Table 6-2

***CIC* Factors Needed for the Balancing Computation**

Building Heat Loss Element	Symbol	Units
Walls	CIC_w	$\$/R \cdot ft^2$
Roof (or ceiling)	CIC_r	$\$/R \cdot ft^2$
E, W, N windows	CIC_g	$\$/R \cdot ft^2$
Perimeter	CIC_p	$\$/R \cdot ft$
Floor	CIC_f	$\$/R \cdot ft^2$
Basement	CIC_b	$\$/R \cdot ft$
Infiltration	CIC_i	$\$/ft^3 \cdot (h/air\ change)$

For example, if the cost of an R12.7 wall is \$2.05/ft^2 and the cost of an R28 wall is \$2.69/ft^2, then $\Delta C_w = 2.69 - 2.05 = 0.64$ and $\Delta R_w = 28 - 12.7 = 14.3$. Then, C/C for walls is

$$C/C_w = \Delta C_w/\Delta R_w = 0.64/14.3 = 0.035(\$/R \cdot ft^2). \quad (6\text{-}2)$$

Incremental costs of 3.5¢/R · ft^2 in the range R12 to R20 and 5¢/R · ft^2 in the range R20 to R28 are typical for frame construction. In computing these costs, the full installed cost should be used, including materials, labor, overhead, and profit. Ceiling and floor costs are determined in the same manner as for walls.

To determine C/C for windows, C/C_g, determine the R-value ($1/U$) for different integer numbers of glazings and the corresponding costs. Then,

$$C/C_g = \Delta C_g/\Delta R_g. \quad (6\text{-}3)$$

For perimeter and basement insulation, the costs per lineal foot of outside building perimeter are used rather than per square foot.

The cost of reducing infiltration presents a special case in which it is very difficult to make any meaningful estimates. If the designer is fortunate enough to have information on actual costs and the *ACH* values corresponding to these costs, then these should be used. More commonly, rough estimates must be made instead.

Down to an infiltration *ACH* value of 1/2, reductions can be achieved simply by decreasing the natural infiltration characteristics of the house. This is done by installing a plastic vapor barrier. Further reductions can be achieved by carefully caulking all cracks, caulking overlaps in the vapor barrier, and sealing other cracks such as those that may occur around utility outlet boxes, along sill plates, and through ceiling vents. As in any conservation strategy, the first gains are the least expensive and most cost effective. As more and more care is taken in the process, costs escalate rapidly. A significant portion of the cost will be the labor cost associated with installing a very effective vapor barrier with caulked overlaps covering all surfaces. Details for installation of these vapor barriers have been developed in the Canadian plains provinces.[1] They are neither simple nor inexpensive.

In order to safely achieve lower and lower values of infiltration, a series of strategies is normally used. Little information is available on costs, but Table 6-3 shows a reasonable possibility for a 1,500-ft^2 house.

It is strongly recommended by a number of different groups that a heat recovery unit (HRU) be used in any building in which the natural infiltration rate is less than ½ air change/h.

The use of the ½ air change/h as a lower limit is somewhat arbitrary but is supported by an accumulating body of evidence from several researchers. For example, the value of ½ air change/h is recommended for adoption as a standard in Sweden.[2] By maintaining an outside air change rate at least equal to ½ air change/h, various hazards that are associated with low air makeup rates may be avoided. These hazards include formaldehydes (which emanate from the resins in plywood, chipboard, and other building materials), radon gas (which leaks up under the house through cracks in the rock and soil), oxides of nitrogen and carbon monoxide (from combustion), tobacco smoke, high humidity, and odors.[3] By installing a heat recovery unit, the outside air dilution rate can be maintained at a safe level and yet the heat losses associated with this air exchange are reduced by the effectiveness factor of the unit. The effective *ACH* is simply the sum of the natural *ACH* plus the forced *ACH* times (1 − effectiveness factor).

A simple window box heat recovery unit may be acceptable in a situation where the natural infiltration rate is reasonably close to ½ air change/h, and one assumes some mixing of the air within the building so that the room that has the heat recovery unit exchanges air with the rest of the rooms. As the building is made progressively tighter, this may not be adequate, and several heat recovery units may be required for different rooms, or an air distribution system may be required in conjunction with one main heat recovery unit. Many passive buildings do not have a central heating system, relying instead on several small unit heaters scattered throughout the building. Therefore, the cost of the distribution system must be accounted for in determining the incremental cost of the reduced *ACH*.

Table 6-3

Example of Costs Related to Reducing *ACH* for 10,000-ft³ Building

Effective *ACH*	*HAC*	Strategy	Cost of Improvement ($)	Total Cost ($)
1.5	0.667	minimal construction	0	—
1.0	1.0	plastic vapor barrier	125	125
0.5	2.	seal joints, foam cracks, more care	375	500
0.375	2.667	add window heat recovery unit (HRU)	250	750
0.300	3.333	add another HRU	250	1000
0.250	4.0	install large HRU instead, plus distribution system, 70% efficient	250	1250
0.214	4.673	larger HRU, 80% efficient, more tightening	250	1500
0.187	5.348	larger HRU, 90% efficient, more tightening	250	1750

CIC for infiltration, CIC_i, has units in terms of the reciprocal of *ACH*. Since *ACH* represents air-changes-per-hour, its reciprocal is hours-per-air-change, or *HAC*, for short. CIC_i is to be determined per cubic foot of building.

$$CIC_i = \Delta C_i/(\Delta HAC \cdot v) \qquad (6\text{-}4)$$

where v is the building volume.

In Table 6-3, the incremental cost, CIC_i, of increasing *HAC* is constant. For example, in reducing the infiltration from 0.5 to 0.375 air change/h, the computation is

$$\Delta C_i/\Delta HAC = (750 - 500)/[(2.667 - 2) \cdot 10000]$$

$$= .0375\ \$/\text{ft}^3 \cdot (\text{h/air change}) \qquad (6\text{-}5)$$

A worksheet for computing *CIC* is shown in Appendix 7.

The costs shown in Table 6-3 are for new construction. For retrofit construction, caulking and weatherstripping are practical options. These are very low cost and should bring *ACH* below 1.0 air change/h.

Solar Incremental Cost, *SIC* — The construction costs assignable to enlarging the solar system of a building are also entered into the balancing computation in a specific form. The abbreviation *SIC* will be used to identify the solar incremental cost. This factor is the cost of enlarging the solar system aperture by 1 ft² of projected area. In computing these costs, the full installed cost should be used, including materials, labor, overhead, and profit. The costs for all aspects of the solar system should be included, such as the added thermal mass in a direct-gain system, which is associated with added projected area. The best way to do this is to compute the solar costs for two systems of different solar aperture, then divide the difference in cost by the difference in the projected areas of the two apertures:

$$SIC = \Delta C/\Delta A_p \qquad (6\text{-}6)$$

The Conservation Factor, *CF*

At the start of this chapter, it was indicated that construction cost, auxiliary heat requirements, and weather data would be used in solving the problem of balancing conservation level with the amount of solar heating. All these parts of the problem are linked together by the conservation factor, *CF*, which is tabulated in Appendix 1. Appendix 1 lists the *CF* for each of the 94 reference designs at 223 locations and for 5 values of *LCR* and 2 values of base temperature. The R-values and infiltration rate that satisfy the criterion of a balanced conservation level of a building are determined by use of the following formulas:

For walls, roof, floors and E,W,N windows:

$$R_x = CF[SIC/CIC_x]^{1/2} \qquad (6\text{-}7)$$

where x denotes any of the subscripts w, r, f or g.

For perimeter:

$$R_p = 2.04\ CF[SIC/CIC_p]^{1/2} - 5 \qquad (6\text{-}8)$$

For heated basement:

$$R_b = 3.26\ CF[SIC/CIC_b]^{1/2} - 8 \qquad (6\text{-}9)$$

For infiltration:

$$1/ACH = 0.133\ CF[SIC/CIC_i]^{1/2} \qquad (6\text{-}10)$$

Note that the last formula yields *HAC*, the reciprocal of *ACH*. The formula is presented in this manner to retain the same form as for the R-value formulas.

The weather is taken into account by use of long-term weather data for each particular location in the calculation of *CF*. The ratio *SIC/CIC* incorporates cost data for both the conservation and the solar aspects of the building.

Procedure for the Balancing Calculation

The step-by-step procedure for balancing conservation and solar is shown in Table 6-4. Note that the procedure includes the *LCR* method in the final three steps.

Discussion of Procedural Steps

Step 1 — Select an initial value for *LCR*. Time can be saved by making an intelligent choice for the initial value of *LCR*. There is probably no universal best way of choosing *LCR*, but two ways of getting started are described below.

A reasonable way of selecting *LCR* is from the map shown in Figure 6-2. This is the same map as the one presented in Figure 2-2, Chapter 2, and shows the optimum value of *LCR* for a particular set of conservation and passive solar costs. It shows how climate affects the *LCR* choice if the ratio of costs is held constant. As a starting point, the designer might lean 20% to either side of the guidance given in this map if it is desired to emphasize solar or to emphasize minimum first cost. Another reason for leaning to the high or the low side may be that local energy costs are particularly high or particularly low.

Another possible way of selecting *LCR* involves the life-cycle optimum *LCR*, which will minimize the total building costs (including fuel costs during the building lifetime). This method will be discussed separately at the end of this chapter.

Step 2 — Look up *CF* in Appendix 1. Interpolate linearly for *CF* if the desired values of *LCR* or base temperature are other than the tabulated values. If the building uses more than one passive system, then calculate *CF* based on an average of the *CF*s of each reference design, weighted according to the relative projected areas, A_p, of the various types of passive systems being used.

Step 3 — Compute *CIC* for each building element and *SIC* as an area-weighted average, as described earlier in this chapter. The worksheet for *CIC* is useful.

Step 4 — Estimate the conservation add-on cost. Begin by estimating minimal R-values for each building element and the *ACH* for a roughly constructed building. Determine how much the selected

Table 6-4

Procedure for Balancing Conservation and Solar

Steps		Reference
1.	Choose a starting-point (*LCR*).	Figure 6-2
2.	Look up the value of *CF* for the selected *LCR*, location, and reference design.	Appendix 1
3.	Compute *S/C* and *C/C*.	Equations 6-1 through 6-6
4.	Calculate R-values and other conservation levels, using the formulas. Select the closest practical, buildable values.	Equations 6-7 through 6-10
5.	Estimate total add-on cost of conservation.	*C/C* Worksheet, Appendix 7
6.	Calculate the net load coefficient, *NLC*.	Chapter 3
7.	Compute the projected area of the solar aperture and the solar add-on cost.	$A_p = NLC/LCR$
8.	Estimate annual backup heating.	Appendix 1 and Equation 5-1

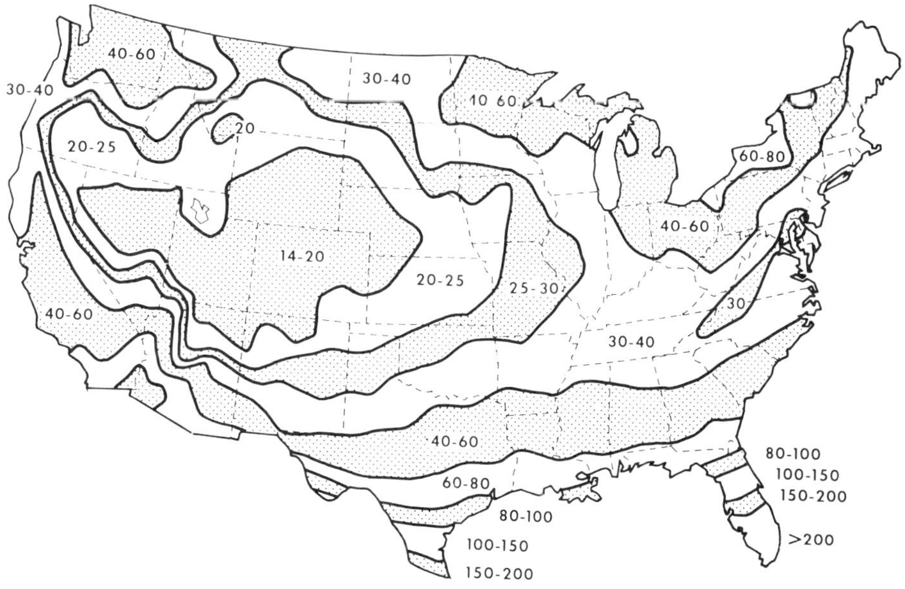

Figure 6-2. Recommended Starting-Point Load Collector Ratio (*LCR*) for Locations in the U.S. These values have been determined to yield a life-cycle optimum for selected passive solar, conservation, auxiliary heat, and financing costs.

conservation levels have been increased over the minimal case levels for each of the conservation elements of the building. This is just the difference between the selected R-values of Step 4 and the minimal case R-values and the difference between the selected $1/ACH$ and the minimal case $1/ACH$. Multiply each of these differences by its corresponding *C/C* and by the corresponding size of the element (area, length, or volume). The sum of these products is the total conservation add-on cost. The Conservation Add-on Cost Worksheet in Appendix 7 will be useful.

Step 5 — Compute the conservation levels. Use the formulas to compute the R-values for each building element and the *ACH* of the building. Typically, the formulas yield values that are in between practical choices. Select R-values and *ACH* that are the closest practical values. Note that these selected values should lie within the range of values used to compute the corresponding *C/C*. The Conservation Add-On-Cost Worksheet, Appendix 7, is useful. Note that the reciprocal of *ACH* is used in the formula for infiltration conservation level. The map in Figure 2-4 shows wall R-values obtained by this procedure, using typical cost values.

Step 6 — Compute *NLC*, using the selected R-values and the selected value for *ACH*. Use the methods explained in Chapter 3. The Net Load Coefficient worksheet in Appendix 7 is useful.

Step 7 — Compute A_p. Based on the starting point *LCR* from Step 1 and *NLC* from Step 6, calculate A_p from:

$$A_p = NLC/LCR$$

Estimate the solar add-on cost using

$$\text{Solar Add-On Cost} = A_p \cdot SIC$$

Add the solar add-on cost to the total conservation add-on cost. The sum is the total add-on cost for the balanced building with respect to a nonsolar building with minimal conservation features.

Step 8 — Estimate the auxiliary heat requirement. Return to Appendix 1, look under the proper location, reference design, and base temperatures to find the *SSF* that corresponds to the selected *LCR*. If the selected *LCR* is in between the tabulated *LCR*s, then interpolate linearly to find the *SSF* value. For a building with a mix of reference designs find the weighted average *SSF* based on the relative areas of the apertures of the reference designs, as discussed in Chapter 5. Then calculate Q_{aux} from:

$$Q_{aux} = (1 - SSF)NLC \cdot DD$$

If the costs and performance from Steps 7 and 8, respectively, are satisfactory, then one can stop. Otherwise, adjust the selected *LCR* downward to reduce auxiliary heat requirements or upward to reduce costs and repeat the computation. Each case calculated is an optimum mix. One can calculate several cases and then choose among them.

Life-Cycle Costs

In the procedures for balancing conservation and solar that have been discussed up to this point, much importance has been placed on the limitation imposed by the construction budget. Costs can be entered into the calculation in the form of life-cycle costs rather than as a construction budget limitation. The life-cycle costs include not only the initial construction costs but also the continuing costs of heating the building through its lifetime. Of particular importance are the auxiliary fuel costs. The objective of the procedures will be to identify the value of *LCR*, which minimizes the life-cycle costs. It is apparent that such a minimum exists because, for decreasing values of *LCR*, initial construction costs rise while lifetime auxiliary fuel costs drop. At some value of *LCR*, the total lifetime costs must reach a minimum.

A reasonable procedure for finding this optimum value of *LCR* starts with the evaluation of *h*, which is defined as

$$h = \frac{(\text{levelized cost of auxiliary heat})(DD)}{(FCR)} \qquad (6\text{-}11)$$

The cost of auxiliary heat is expressed in dollars per Btu and is levelized over the accounting period. For other than electric heaters located in the space, the furnace efficiency should be taken into account, that is,

Unit cost of auxiliary heat =

$$\frac{\text{unit cost of heating fuel}}{\text{furnace efficiency}} . \qquad (6\text{-}12)$$

The "fixed charge rate" (*FCR*) is an overall factor that converts an initial investment into an equivalent annual cash flow, accounting for interest, taxes, depreciation, maintenance, discount rate, and resale value. A detailed discussion of these financial factors is given in Reference 4.

The next step is to evaluate D, where D is the derivative of SSF with respect to $1/LCR$. In principle, D could be determined by differencing in the Appendix 1 tables where SSF is given for several values of LCR. This is not satisfactory in practice because roundoff results in significant error. A more satisfactory approach is to use the values of CF and SSF tabulated in Appendix 1 at the given values of LCR and calculate values of D from the formula:

$$D = (1 - SSF)/(CF^2/24 - 1/LCR). \qquad (6\text{-}13)$$

Values of D at intermediate points can be obtained by interpolation. It is helpful to plot a curve of D versus LCR.

The final step is to use the D versus LCR curve to find the LCR for which $D = S/C/h$. This is the minimum LCR that is cost effective for the particular fuel rates used. This procedure, although straightforward, does require the estimation of future fuel costs and is therefore of limited value. An advantage, however, is that no iteration is required.

It is worth noting at this point that resistance electric heat may be a reasonable choice for a passive solar building. The following factors should be considered:

1. The annual amount of auxiliary heat is small.
2. Consequently, it is doubtful that the initial cost of a heat pump would be warranted, especially since the auxiliary heat is required generally during extreme weather when a heat pump would be operating in a purely resistance mode.
3. The required capacity of the auxiliary heaters or furnace is small.
4. It is inexpensive to obtain fine zone control of resistance electric heat in small capacities. Experience shows that good zone control is essential in passive solar buildings.
5. It is well documented[5,6] that passive solar buildings intrinsically tend to demand heat during off-peak hours of the electric utility, that is, from 11:00 p.m. to 7:00 a.m. Thus, the building has a good load pattern for the utility and could qualify for off-peak rates where available. With proper control, the load could be made to be entirely off peak.[6]

It might also be reasonable to augment the auxiliary heaters with a small, efficient wood-burning stove in the main area of the building.

Cooling Issues

Only heating performance is considered in the analytical techniques developed in this chapter. It is strongly urged that sound engineering judgment be used to determine if the recommended A_p and R-values should be refined to account for cooling considerations.

As in solar heating, the adage for cooling is "conservation first." In cooling, conservation consists primarily of a series of defensive strategies that seek to minimize the cooling load by reducing heat gains from the sun and internal sources. It is also helpful to reduce heat flow inward through the walls, glazing, and roof when the outside temperature is much warmer than the inside temperature, and thus insulation and other heating season conservation strategies will help. Moreover, the effect of solar radiation absorption on outside wall and roof surfaces will generally increase their temperatures well above the building inside temperature even when the outside air temperature is low, and therefore, insulation will be beneficial.

The conclusion is that, having chosen an *LCR*, the *CF* tables should give reasonable guidance in the proper choice of conservation levels for cold climates. In the mildest heating climates, it may be that the insulation levels are too low. This question bears further investigation.

Another consideration is whether the optimum *SSF* might decrease somewhat because of cooling concerns. Solar glazing must be protected in the summer, but even if the window is shaded from direct sun, solar gains due to diffuse solar radiation can be significant. Case studies that have been made do indicate that the life-cycle optimum *SSF* values should be reduced somewhat to account for cooling considerations. Another possibility is to provide complete window sun protection such as vertical outside roll-down sunshades.

Mass that is added to a building in order to improve passive solar heating performance and comfort may also improve performance during the cooling season. The effect of mass in reducing cooling loads is most pronounced during periods when the daily outside temperature swings across the comfort range. In cases where this benefit is significant, part of the cost of mass added to improve heating performance can be justified by the improved cooling performance.

Other Advantages of a Balanced Approach

Other considerations that favor a mix of conservation and solar can be seen if we investigate problems that arise from either extreme taken alone. These problems can be largely avoided with a balanced design; the resulting building will be more comfortable, more livable, more resilient, more salable, less auxiliary heat will be needed, and the initial cost will be less.

The extreme of energy conservation attempts to isolate the inside from the outside. A very cloistered and seemingly artificial environment is created. One has to go outside to see the sky, clouds, birds, weather, and trees bending in the wind. This type of indoor environment may be exactly what some desire, but not others.

The other extreme is an overdone solar building without adequate insulation. In this case, virtually every south-facing surface is a solar collector. Storage, control, and distribution of huge solar gains become a problem, increasing cost and forcing unwelcome compromises in space allocation. Large temperature swings may be unavoidable. A traditional oversized auxiliary heating plant is needed with careful attention to distribution and zone control, and this adds to the cost. Cold wall and window surfaces require a higher inside air temperature to maintain comfort.

With a balanced approach, many of the benefits of energy conservation and passive solar remain while the disadvantages are diminished. A friendly and livable indoor environment is achieved by maintaining good visual connection with the outdoors but with comfortable shelter from its extremes. The atmosphere seems natural, working with the environment rather than against it. Since solar projected area is less than the entire south facade of the house, there is more architectural freedom in the design. Thermal storage requirements are met easily by interior mass walls serving a dual function and thus not seeming out of place. Good insulation and other conservation strategies reduce the size and cost of the auxiliary heating plant. Elaborate distribution of heat is not particularly critical. Icy drafts from cold wall and window surfaces are reduced.

Balancing energy conservation, passive solar, aesthetics, cost, and other design issues is not an easy task. However, careful consideration of design details, materials selection and thermal analysis can yield buildings that are economical, comfortable, and have low energy consumption.

Another advantage of balanced design is resilience. Suppose auxiliary heating fuel or electricity were simply not available for an extended period. Suppose a total disruption of all utility service occurred. Suppose the average outside temperature were 10°F and cloudy. The balanced design building fares better under these conditions than either of the extremes. Because of high internal heat capacity and good insulation, the building temperature drops slowly (over a period of 3 to 5 days) and eventually levels off. Because even cloudy-day solar gains are significant, this stable temperature condition is well above the outside temperature and certainly above freezing (at least in the 48 contiguous United States). The extreme energy conservation house may cool quickly or more slowly, depending on its internal heat capacity, but will eventually reach low temperatures because of the absence of a significant heat source. The extreme solar house will cool quickly due to poor insulation, and north rooms may be in danger of freezing.

Marketing is another consideration. Energy performance is of increasing concern to those who purchase buildings, both residential and commercial. Solar is visible, well publicized, and has demonstrated market appeal. Good insulation and other conservation strategies are less evident but well understood to be essential to both energy savings and comfort. Together they make a consistent and very compelling argument to an increasingly knowledgeable buying public.

References

1. "Energy Efficient Housing—A Prairie Approach," booklet prepared by Energy Research Development Group, University of Saskatchewan (Saskatchewan Mineral Resources, 1914 Hamilton St., Regina, Saskatchewan S4P 4V4 and Alberta Energy and Natural Resources, 9915 108th St., Edmonton, Alberta T5K 2C9, October 1980).

2. P. O. Fanger, "Proposed Nordic Standard for Ventilation and Thermal Comfort," *Building Energy Management, Proceedings of the International Congress,* Povoa de Varzim, Portugal, 12-16 May 1980. Pergamon Press (1981).

3. G. D. Roseme et al, "Residential Ventilation with Heat Recovery: Improving Indoor Air Quality and Saving Energy," Lawrence Berkeley Laboratory Report No. LBL-9749, Berkeley, California, October 1980.

4. J. Douglas Balcomb, et al, *Passive Solar Design Handbook, Volume Two,* DOE/CS-0127/2 (Washington, DC: U.S. Department of Energy, January 1980).

5. M. S. Baker, Mazria and Associates, "Passive Solar Technical Planning Study," Electric Power Research Institute Report EPRI EM-1591/TPS 79-750, October 1980.

6. S. E. Pyde, "Load Management and the La Vereda Passive Solar Community," *Proceedings of the Sixth National Passive Solar Conference,* 8-12 Sept. 1981, Portland, Oregon (Newark, DE: American Section of the International Solar Energy Society, 1981), pp. 20-24.

CHAPTER 7
EXAMPLE PROBLEMS, SCHEMATIC DESIGN

Introduction

This chapter presents example problems that illustrate typical computations applicable to the schematic design phase. The contents of the problems are as follows:

Example Problem 7-1 is a simple interpolation of degree-days from the weather tables. The method of linear interpolation can be applied to any other table where linear interpolation is required.

Example Problem 7-2 illustrates the computation of monthly degree-days by the Thom method.

Example Problem 7-3 develops conservation and solar guidelines for a residence.

Example Problem 7-4 applies annual *LCR* method to a house with a single reference design and an assumed 65°F base temperature.

Example Problem 7-5 extends the previous example to include calculation of *TLC*, a base temperature, and mixed design performance.

Example Problem 7-6 applies the annual *LCR* method to a mixed system in which two sensitivity analysis corrections must be made to obtain the *SSF*.

Example Problem 7-7 extends the previous example to a calculation of average January room temperature.

Example Problem 7-8 uses the Chapter 6 procedures to balance conservation and solar and to compute add-on costs.

Example Problem 7-9 uses the data from the previous example to determine the life-cycle optimum for a given fuel cost.

Example Problem 7-1

Compute the monthly degree-days for a base temperature of 63.2°F in Los Alamos, New Mexico. The nearest listed base temperatures bracketing 63.2°F are 60°F and 65°F. In Appendix 3, under Los Alamos, New Mexico, we find that $DD_{60} = 5043$ and $DD_{65} = 6437$. The formula for linear interpolation between two listed values is

$$DD_b = DD_l + (DD_u - DD_l)\,\frac{(T_b - T_l)}{(T_u - T_l)}$$

where u and l denote the upper and lower sides of the bracket, respectively. For this particular case we have

$$DD_{62} = 5043 + (6437 - 5043)\,\frac{(63.2 - 60)}{(65 - 60)} = 5935$$

Example Problem 7-2

This example may be skipped for routine applications of schematic design procedures, but if degree-day values are required for locations other than those in Appendix 3, they may be calculated by the Thom method (see Chapter 4). The method produces estimates of monthly degree-day totals from known values of the monthly mean temperature and known or estimated values of the standard deviation of the monthly mean temperature. This example illustrates the calculation for one month. Note that the mean temperature and standard deviation values are taken from Appendix 3, but if the desired location is not listed in Appendix 3 other sources of these data may be used.

Compute the monthly degree-days for the month of January for a base temperature of 62°F at Minot, North Dakota, using the Thom method.

For January $M = 31$ and, from Appendix 3, $T_A = 7.9°F$ and $\sigma_m = 7.97°F$. (Note that σ_m is listed under the column labeled SD in Appendix 3.)

From Equation 4-14,

$$h = (62 - 7.9)/(7.97\sqrt{31}) = 1.219$$

From Equation 4-13, since $h > 0$,

$$I = 0.34 \exp(-4.7 \cdot 1.219) - 0.15 \exp(-7.8 \cdot 1.219)$$

$$= 0.0011°F$$

From Equation 4-12,

$$DD_b = 31(62 - 7.9 + 0.0011 \cdot 7.97\sqrt{31})$$

$$= 1679°F \cdot \text{day}$$

Example Problem 7-3

Determine passive solar guidelines for a 1500-ft^2 residence in Los Alamos, New Mexico, using Rules 1 and 2.

Referring to Table 2-1, the following values are found under high fuel cost:

$CF =$ 1.53 (conservation factor)
$LCR =$ 18 (load collector ratio)
$SSF =$ 79 (solar savings fraction)

Conservation guidelines can be determined directly using the formulas on page 2-1.

$$R_{\text{wall}} = 14\,CF = 14 \times 1.53 = 21$$
$$R_{\text{roof}} = 22\,CF = 22 \times 1.53 = 34$$
$$R_{\text{perimeter}} = 13\,CF - 5 = 13 \times 1.53 - 5 = 15$$
$$R_{\text{basement}} = 16\,CF - 8 = 16 \times 1.53 - 8 = 16$$
$$N_{E,W,N} = 1.7\,CF = 2.6$$
$$ACH = 0.42/CF = 0.27$$

Selected values will depend on economical, buildable sections, based on these guidance values.

Solar glazing guidance follows from the following equations:

$$NLC = \frac{(GF)(A_f)}{CF}$$

$$A_p = \frac{NLC}{LCR}$$

GF (geometry factor) = 6.5 (from Table 2-2)

$$NLC = \frac{6.5 \times 1500}{1.53} = 6373 \text{ Btu/°F} \cdot \text{day}$$

$$A_p = \frac{6373}{18} = 354 \text{ ft}^2$$

An early estimate can be obtained for auxiliary heat.

$$Q_{aux} = DD \cdot NLC\,(1 - SSF)$$

$DD = 6437$ (base 65°F), from Appendix 3.

$$Q_{aux} = 6437 \times 6370 \times (1 - 0.79) = 8.6 \text{ million Btu}$$

or 0.9 Btu/°F day · ft^2

Note that the estimated NLC assumes that the guidance values have been followed and that the building geometry is reasonably compact.

Example Problem 7-4

Based on the guidelines of Example Problem 7-3, a schematic design has been developed. Rough measurements from the schematic design drawings are as follows:

Exterior opaque wall area	= 1050 ft^2
Roof area	= 1550 ft^2
Perimeter length	= 170 ft (slab on grade)
E, W, N window area	= 140 ft^2
Volume	= 12400 ft^3

Conservation values have been selected as follows:

R_{wall} = 19 (overall wall R-value)
R_{roof} = 36 (overall roof R-value)
$R_{perimeter}$ = 14 (insulation added to perimeter)
$N_{E,W,N}$ = 2 (double glazing)
ACH = 0.25 (requires heat recovery unit, see Table 6-3, for example)

Estimate *NLC, SSF*, and Q_{aux} for this design, using the Net Load Coefficient and Annual *LCR* Method Worksheet. Use T_b = 65°F.

The procedural steps in Table 5-1 are described below.

Step 1 — Select an appropriate reference design. Although the building may evolve with several passive systems, a preliminary calculation can be made assuming only a single system, usually selected to be the major system. In this case a large sunspace is contemplated and system SSD1 (semienclosed, 50° sloping glazing) is chosen.

Step 2 — The calculation of *NLC* is shown on the worksheet in Figure 7-1. The value of *ADR* = 0.77 is determined from Figure 3-1, corresponding to an elevation of 7,200 ft.

Step 3 — Determine A_p. This is most easily done by counting glazing elements and using net glazed dimensions. If 17 units each measuring 76 inches by 46 inches are to be used, then the net area is $17 \times 75 \times 45/144 = 398$ ft². One inch has been deducted from the glass dimensions for mounting the units. If the slope is 50°, then the projected area is $398 \cdot \sin 50° = 305$ ft².

Step 4 — Calculate *LCR*. See worksheet, Figure 7-1.

Step 5 — Calculate T_b and *DD*. For this calculation T_b is assumed to be 65°F and *DD* = 6437 from Appendix 3.

Step 6 — Look up *SSF*. Refer to Appendix 1 for Los Alamos, for system SSD1, for T_b = 65°F, and *LCR* = 20.

Step 7 — Compute Q_{aux}. See worksheet, Figure 7-1.

The results are as follows:

NLC = 6170 Btu/DD
SSF = 73%
Q_{aux} = 10.6 million Btu/year

These are in reasonable agreement with the guideline values. *SSF* is lower and Q_{aux} is higher because the projected area is 15% smaller than the guidance value. This is no cause for concern.

Note the level of precision carried along on the worksheet. It is misleading to list greater precision than this, and it also increases the chance of arithmetic error.

NET LOAD COEFFICIENT and ANNUAL LCR METHOD WORKSHEET

DATE 1/1/83 BY Smith LOCATION Los Alamos

NET LOAD COEFFICIENT, NLC

	BUILDING ELEMENT	FORMULA	Fill in the Blanks and Perform Arithmetic	LOAD, UA
	WALLS	: $(UA)_w = A_w/R_w$	= 1050 / 19	= 55
	ROOF	: $(UA)_r = A_r/R_r$	= 1550 / 36	= 43
	E, W, N WINDOWS	: $(UA)_g = 1.1A_x/N$	= 1.1 x 140 / 2	= 77
SELECT ONE	FLOOR	: $(UA)_f = A_f/R_f$	= ____ / ____	= ____
	PERIMETER	: $(UA)_p = 4.17\ P_p/(R_p + 5)$	= 4.17 x 170 /(14 + 5)	= 37
	BASEMENT	: $(UA)_b = 10.7\ P_b/(R_b + 8)$	= 10.7 x ____ /(____ + 8)	= ____
	INFILTRATION	: $(UA)_i = 0.018V \times ACH \times ADR$	= 0.018 x 12 400 x 0.25 x 0.77	= 43

Sum = 255 Btu/°F • h

NLC = 24 x Sum = 6120 Btu/°F • day

LOAD COLLECTOR RATIO, LCR

$LCR = NLC/A_p$ = 6120 / 305 = 20 Btu/°F • day • ft²

WEIGHTED AVERAGE, SSF

REFERENCE DESIGN	PROJECTED AREA	FRACTION OF A_p		SSF FOR EACH REFERENCE DESIGN		PRODUCT OF SSF x FRACTION	U_c	$U_c A_p$
SSD1	305	1	x	0.73	=	0.73	____	____
____	____	____	x	____	=	____	____	____
____	____	____	x	____	=	____	____	____
TOTAL A_p =	305			SSF =		0.73	TOTAL (U_cA_p) =	____

TLC = NLC + 24 x TOTAL (U_cA_p) = ____

BASE TEMPERATURE, T_b

$T_b = T_{set} - Q_{int}/TLC$ = ____ – ____ / ____ = 65 °F

AUXILIARY HEAT, Q_{aux}

$Q_{aux} = NLC \cdot DD(1 - SSF)$ = 6120 x 6437 x (1 - 0.73) x 10^{-6} = 10.6 Mega Btu/year

Figure 7-1. *NLC* and *LCR* Worksheet for Example Problem 7-4

Example Problem 7-5

This example follows the same building as in the previous problems into a later stage of schematic design. Conservation levels are the same but the passive solar design has evolved to a mixture of systems. Some of the building areas have also changed. It is now desired to estimate a base temperature assuming occupancy by three people and a thermostat setting of 70°F. Internal heat is estimated as 60,000 Btu/day.

Steps 1 and 3 — Select reference designs and calculate A_p. These are determined from the plans as follows:

	A_p (ft^2)	Reference Design
Sunspace	160	SSD1
Direct gain	60	DGC1
Trombe wall (with selective surface)	125	TWJ2
Total	345	

Step 2 — Calculate *NLC*. See worksheet, Figure 7-2.

Step 4 — Calculate *LCR*. See worksheet, Figure 7-2.

Step 5 — Determine T_{set}, Q_{int}. Calculate T_b.

First it is necessary to calculate *TLC*. Values of U_c for the reference designs are found in Tables A5-7, A5-8, and A5-9 and are entered on the worksheet. The load coefficients of the solar elements, U_cA_p, are calculated, as shown on the worksheet, summed, and added to the *UA* sum for the nonsolar elements to determine the total *UA* of the building. *TLC* is 24 times the total *UA*.

The calculation of T_b is shown on the worksheet for the assumed values of T_{set} and Q_{int}.

The value of *DD* is determined by interpolation in Appendix 3. (See Example Problem 7-1.)

Step 6 — Determine *SSF*. This is done separately for each reference design using the same value of $LCR = 17.6$. The weighted average is determined using the fraction of A_p values on the worksheet. This procedure is repeated for both $T_b = 55$°F and 65°F. The respective values are entered (parentheses are used for the 65°F values) on the worksheet. The value of *SSF* for $T_b = 63.2$°F is then determined by interpolation between the 55°F and 65°F values. The result is $SSF = 75\%$.

Step 7 — Calculate Q_{aux}. See worksheet, Figure 7-2.

NET LOAD COEFFICIENT and ANNUAL LCR METHOD WORKSHEET

DATE 1/1/83 BY Smith LOCATION Los Alamos

NET LOAD COEFFICIENT, NLC

BUILDING ELEMENT	FORMULA	Fill in the Blanks and Perform Arithmetic	LOAD, UA
WALLS	: $(UA)_w = A_w/R_w$	= 1100 / 19	= 58
ROOF	: $(UA)_r = A_r/R_r$	= 1520 / 36	= 42
E, W, N WINDOWS	: $(UA)_g = 1.1A_x/N$	= 1.1 x 130 / 2	= 71
SELECT ONE: FLOOR	: $(UA)_f = A_f/R_f$	= ____ / ____	= ____
SELECT ONE: PERIMETER	: $(UA)_p = 4.17\ P_p/(R_p + 5)$	= 4.17 x 180 /(14 + 5)	= 40
SELECT ONE: BASEMENT	: $(UA)_b = 10.7\ P_b/(R_b + 8)$	= 10.7 x ____ /(____ + 8)	= ____
INFILTRATION	: $(UA)_i = 0.018V \times ACH \times ADR$	= 0.018 x 12160 x 0.25 x 0.77	= 42

Sum = 253 Btu/°F • h

NLC = 24 x Sum = 6070 Btu/°F • day

LOAD COLLECTOR RATIO, LCR

$LCR = NLC/A_p$ = 6070 / 345 = 17.6 Btu/°F • day • ft^2

WEIGHTED AVERAGE, SSF

REFERENCE DESIGN	PROJECTED AREA	FRACTION OF A_p		SSF FOR EACH REFERENCE DESIGN 55°F (65°F)		PRODUCT OF SSF x FRACTION 55°F (65°F)	U_c	$U_c\ A_p$
SSD1	160	0.47	x	.87 (.77)	=	.41 (.36)	0.36	58
DGC1	60	0.17	x	.78 (.59)	=	.13 (.10)	0.50	30
TWJ2	125	0.36	x	.87 (.76)	=	.31 (.27)	0.20	25
TOTAL A_p =	345				SSF =	.85 (.75) (.73)	TOTAL (U_cA_p) =	113

TLC = NLC + 24 x TOTAL (U_cA_p) = 8784

BASE TEMPERATURE, T_b

$T_b = T_{set} - Q_{int}/TLC$ = 70 − 60 000 / 8784 = 63.2 °F

AUXILIARY HEAT, Q_{aux}

$Q_{aux} = NLC \cdot DD(1 - SSF)$ = 6070 x 5935 x (1 − 0.75) x 10^{-6} = 9.0 Mega Btu/year

Figure 7-2. *NLC* and *LCR* Worksheet for Example Problem 7-5

Example Problem 7-6

This problem illustrates the application of sensitivity data to the annual *LCR* method and is also a second example of mixed systems analysis. The building and the essential data are shown in the sketch in Figure 7-3. The east, west, and north windows total 80 ft^2 (A_x) and are triple-glazed. The south wall includes 90 ft^2 of triple-glazed windows (reference design DGA2) and a sunspace that corresponds to reference design SSC1 except that its east-west width is 20 feet and the floor and wall absorptance is 0.6. In the reference design specifications for SSC1, the width is 24 feet and the floor and wall absorptance is 0.8. Sensitivity data will be applied to account for these two deviations in the reference design specifications.

Step 1 — The reference designs are DGA2 and SSC1, as indicated in Figure 7-3.

Step 2 — Using the data listed in Figure 7-3, the *NLC* is calculated as shown on the worksheet in Figure 7-4.

Step 3 — The projected area is the sum of the projected area for both systems; i.e., 90 ft^2 for DGA2 and 169 ft^2 for SSC1, giving a total A_p = 259 ft^2 for the building.

Step 4 — The calculation of *LCR* is shown on the worksheet in Figure 7-4.

Step 5 — From the *LCR* tables under Nashville, Tennessee, and reference design DGA2, by interpolation, *SSF* for the DGA2 design is found to be 0.19. This is entered on the worksheet for system DGA2.

For reference design SSC1, *SSF* is 0.40 by interpolation in the *LCR* tables. This must be corrected by sensitivity analysis for the deviations from the reference design in absorptance and width. Start with the correction for absorptance. In Appendix 2, find the sensitivity curve for floor and wall absorptance for semienclosed sunspace for Nashville (Figure A2-232, which is reproduced here as Figure 7-5). Mark an "entry point" on the graph at the coordinates, absorptance = 0.8 and *SSF* = 0.40. Through this entry point, sketch a curve parallel to the solid line curves on the graph. The solid lines indicate no night insulation, which corresponds to the sunspace reference design SSC1. Where the sketched-in curve intersects the vertical through absorptance = 0.6, read the value of *SSF*. This is illustrated in Figure 7-5. The value of *SSF* is approximately 0.36.

Now correct for the deviation in east-west sunspace length. In Appendix 2, find the sensitivity curve for sunspace width for semienclosed sunspace for Nashville (Figure A2-172, which is reproduced here as Figure 7-6). The reference design width is 24 and the new *SSF* is 0.36; make an entry point at these coordinates. Sketch a curve through the entry point parallel to the solid line curves. Read the *SSF* where the sketched-in curve crosses the vertical line at width = 20. This process is illustrated in Figure 7-6. The value of *SSF* is approximately 0.37. Enter this value on the worksheet. Calculate the weighted average *SSF* for the building as shown on the worksheet.

Step 6 — The calculation of annual Q_{aux} is shown on the worksheet, Figure 7-4. This house is actually very overglazed for Nashville and would overheat badly. It was chosen to illustrate the procedure. The recommended *LCR* is 39, which is much more appropriate to this climate.

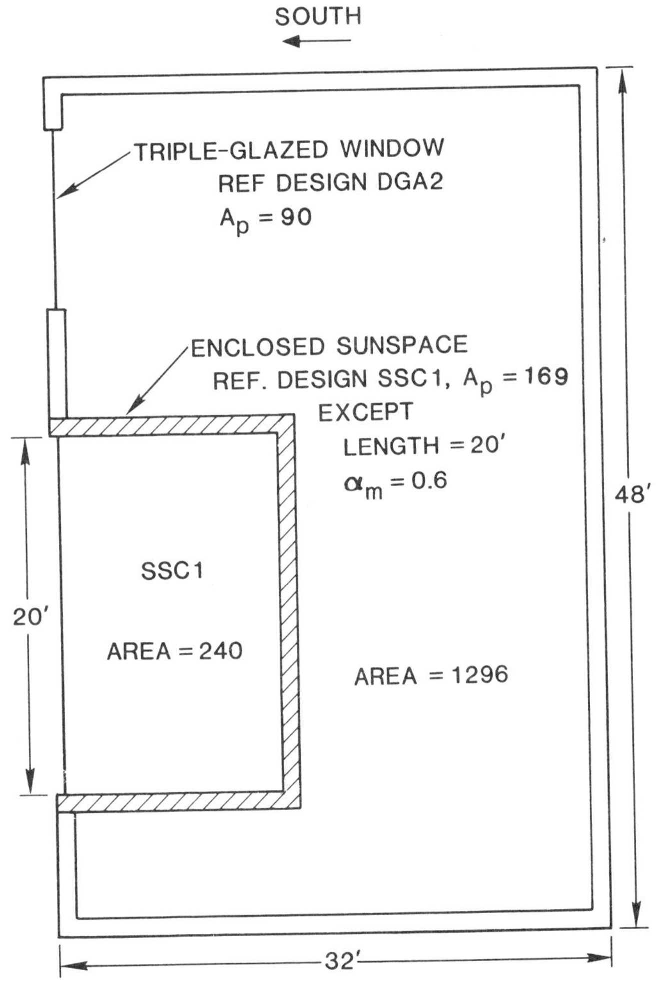

A_r = 1296
A_w = 1010
A_p = 259 (169 + 90)
A_x = 80
R_w = 24
R_r = 42
R_p = 12
P_p = 140
ACH = 0.5

LOCATION – NASHVILLE
BASE TEMPERATURE – 65°F

Figure 7-3. Sketch of Building Plan for Example Problem 7-6

NET LOAD COEFFICIENT and ANNUAL LCR METHOD WORKSHEET DATE 1/1/83 BY Smith LOCATION Nashville

NET LOAD COEFFICIENT, NLC

BUILDING ELEMENT	FORMULA	Fill in the Blanks and Perform Arithmetic	LOAD, UA
WALLS	$(UA)_w = A_w/R_w$	= 1010 / 24	= 42
ROOF	$(UA)_r = A_r/R_r$	= 1296 / 42	= 31
E, W, N WINDOWS	$(UA)_g = 1.1A_x/N$	= 1.1 x 80 / 3	= 29
SELECT ONE: FLOOR	$(UA)_f = A_f/R_f$	= ___ / ___	= ___
SELECT ONE: PERIMETER	$(UA)_p = 4.17\ F_p/(R_p + 5)$	= 4.17 x 140 /(12 + 5)	= 34
SELECT ONE: BASEMENT	$(UA)_b = 10.7\ P_b/(R_b + 8)$	= 10.7 x ___ /(___ + 8)	= ___
INFILTRATION	$(UA)_i = 0.018V \times ACH \times ADR$	= 0.018 x 10 368 x 0.5 x 0.98	= 91

Sum = 227 Btu/°F • h

NLC = 24 x Sum = 5448 Btu/°F • day

LOAD COLLECTOR RATIO, LCR

$LCR = NLC/A_p$ = 5448 / 259 = 21 Btu/°F • day • ft^2

WEIGHTED AVERAGE, SSF

REFERENCE DESIGN	PROJECTED AREA	FRACTION OF A_p		SSF FOR EACH REFERENCE DESIGN		PRODUCT OF SSF x FRACTION	U_c	$U_c A_p$
SSC1	169	0.65	x	0.37	=	0.24	0.38	64
DGA2	90	0.35	x	0.38	=	0.13	0.31	28
			x		=			
TOTAL A_p =	259			SSF =		0.37	TOTAL (U_cA_p) =	92
							TLC = NLC + 24 x TOTAL (U_cA_p) =	7656

BASE TEMPERATURE, T_b

$T_b = T_{set} - Q_{int}/TLC$ = 72 - 53 600 / 7656 = 65 °F

AUXILIARY HEAT, Q_{aux}

$Q_{aux} = NLC \bullet DD(1 - SSF)$ = 5448 x 3704 x (1 - 0.37) x 10^{-6} = 12.7 Mega Btu/year

Figure 7-4. *NLC* and *LCR* Worksheet for Example Problem 7-6

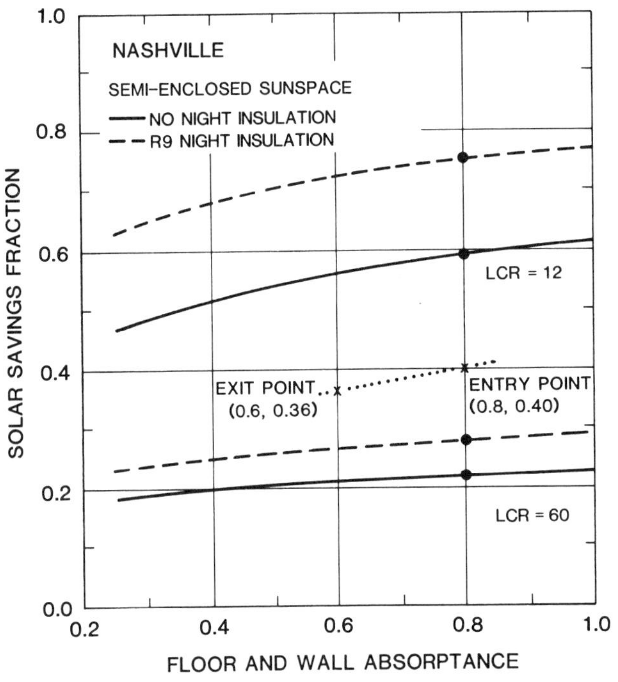

Figure 7-5. Sensitivity Analysis for Absorptance, Example Problem 7-6

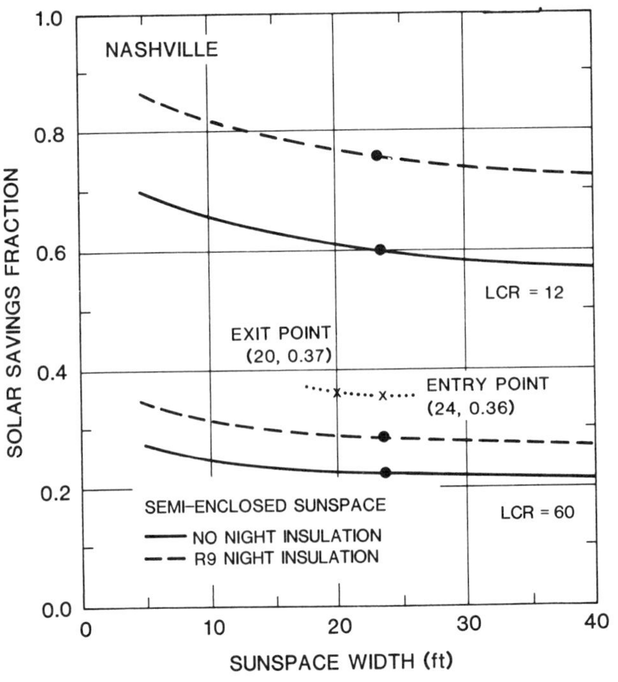

Figure 7-6. Sensitivity Analysis for Sunspace Width, Example Problem 7-6

Example Problem 7-7

The previous example in Nashville, Tennessee, has $LCR = 21$, using the following systems:

SSC1 65% of total A_p
DGA2 35% of total A_p

TLC is 7,656 Btu/°F day and $Q_{int} = 53{,}600$ Btu/day. Estimate the January average clear-day temperature.

The equations needed are 5-5 and 5-6.

$$T_{r,avg} = T_A + \Delta T_{sol} + Q_{int}/TLC \qquad (5\text{-}5)$$

$$\Delta T_{sol} = Q_{tran}/(a + b \cdot LCR) \qquad (5\text{-}6)$$

T_A for Nashville is 38.3°F in January, and the latitude is 36.1° (from Appendix 3). Values of a and b are found in Appendix A10-2 as follows:

	a	*b*
SSC1	15.496	1.11042
DGA2	7.287	.97300

A table of Q_{tran} is given in Appendix 10. For vertical glazing, $Q_{tran} = 1026$ Btu/ft$^2 \cdot$ day.

Thus,

$\Delta T_{sol} = 1026/(15.496 + 1.11042 \times 21) = 26.4°\text{F}$ SSC1
$\Delta T_{sol} = 1026/(7.287 + .97300 \times 21) = 37.0°\text{F}$ DGA2

The area-weighted average is as follows:

$\Delta T_{sol} = 26.4 \times .65 + 37.0 \times 0.35 = 30.1°\text{F}$ mixed system

Then,

$$T_{r,avg} = 38.3 + 30.1 + 53{,}600/7656 = 75.4°\text{F}$$

This is somewhat high, indicating that LCR should be increased.

Example Problem 7-8

Balance conservation and solar in a 1200-ft^2 house in Bismarck, North Dakota. Estimate add-on costs and performance. The desired passive solar system is direct gain with night insulation.

Steps from Chapter 6 are described below.

Step 1 — Choose starting point *LCR*. The map, Figure 6-2, suggests $LCR = 30$.

Step 2 — Look up *CF*. For system DGC3, $CF = 1.7$ for $LCR = 30$ (tentatively, $T_b = 65°F$; for $T_b = 55°F$, $CF = 1.6$, so it is not very sensitive).

Step 3 — Compute *S/C* and the *C/C*s. The Conservation Incremental Cost Worksheet is used to calculate *C/C* values, as shown on Figure 7-7. *S/C* is estimated at \$10/ft^2.

Step 4 — Calculate R-values and air infiltration rate. The Conservation Add-On Cost Worksheet gives the formulas and provides room for the calculation. See Figure 7-8.

Step 5 — Estimate conservation add-on cost. Continue on the worksheet, Figure 7-8. The base-case values are given as MIN. VALUE. Note that values of $HAC = 1/ACH$ are needed here.

Step 6 — Calculate *NLC*. This is done on the top of the Net Load Coefficient Worksheet. See Figure 7-9.

Step 7 — Compute A_p. The formula needed is $A_p = NLC/LCR = 4392/30 = 146$ ft^2. These values are listed on Figure 7-9. The solar add-on cost is \$10 × 146 = \$1460.

Step 8 — Estimate backup heating. Continue with the calculations to the bottom line of Figure 7-9.

Add-on costs are as follows:

Conservation add-on cost	\$3320
Solar add-on cost	1460
Total add-on cost	\$4780

This process can be repeated for another value of *LCR*; for example, $LCR = 50$. This leads to another pair of values for auxiliary heat and add-on cost. A summary of results is as follows:

LCR	Annual Q_{aux} (million Btu)	Add-on Cost (\$)	A_p (ft^2)	*SSF* (%)
30	21.2	4780	146	0.37
50	25.8	4020	91	0.26

CONSERVATION INCREMENTAL COST WORKSHEET

DATE 1/1/83 BY Smith

PROJECT —

LOCATION Bismarck, ND

	WALL		ROOF		E, W, N WINDOWS		FLOOR, PERIMETER OR BASEMENT		INFILTRATION	
	$\$/ft^2$	R VALUE	$\$/ft^2$	R VALUE	$\$/ft^2$	R VALUE	$\$/ft^2$ $/ft	R VALUE	$\$/ft^3$	1/ACH
UPPER CONSERVATION LEVEL	4.15	40	4.62	50	12.10	2.70	8.00	25	0.18	5.8
LOWER CONSERVATION LEVEL	2.70	11	4.00	19	4.20	0.91	3.00	5	0.00	1.00
DIFFERENCES: ΔC, ΔR OR $\Delta(1/ACH)$	1.45	29	0.62	31	7.90	1.79	5.00	20	0.18	4.8
CONSERVATION INCREMENTAL COST $\Delta C/\Delta R$ OR $\Delta C/\Delta(1/ACH)$	$CIC_w = 0.05$		$CIC_r = 0.02$		$CIC_g = 4.4$		$CIC_{f, p \text{ or } b} = 0.25$		$CIC_i = 0.0375$	

Figure 7-7. *CIC* Worksheet for Example Problem 7-8

CONSERVATION ADD-ON COST WORKSHEET

DATE 1/1/83 BY Smith PROJECT —

REFERENCE DESIGN(s) DGC3

LOCATION BISMARCK, ND

SIC = 10

CF = 1.7

BUILDING ELEMENT	CIC	FORMULA	CALC. VALUE	(SEL. VALUE	– MIN. VALUE)	× CIC	× SIZE	= ADD-ON COST
WALL	0.05	$R_w = CF(SIC/CIC_w)^{1/2}$	24	24	11	0.05	1000	650
ROOF	0.02	$R_r = CF(SIC/CIC_r)^{1/2}$	38	38	19	0.02	1200	460
E, W, N WINDOWS	4.4	$R_g = CF(SIC/CIC_g)^{1/2}$	2.6	2.7	0.9	4.4	100	790
FLOOR		$R_f = CF(SIC/CIC_f)^{1/2}$						
PERIMETER	0.25	$R_p = 2.04CF(SIC/CIC_p)^{1/2} - 5$	17	17	5	0.25	150	450
BASEMENT		$R_b = 3.26\ CF(SIC/CIC_b)^{1/2} - 8$						
INFILTRATION	0.0375	$1/ACH = 0.133\ CF\ (SIC/CIC_i)^{1/2}$	3.7	3.7	1	0.0375	9600	970

CONSERVATION ADD-ON COST = 3320
(SUM OF LAST COLUMN)

Figure 7-8. *CAC* Worksheet for Example Problem 7-8

NET LOAD COEFFICIENT and ANNUAL LCR METHOD WORKSHEET DATE 1/1/83 BY Smith LOCATION BISMARCK

NET LOAD COEFFICIENT, NLC

	BUILDING ELEMENT	FORMULA	Fill in the Blanks and Perform Arithmetic	LOAD, UA
	WALLS	: $(UA)_w = A_w/R_w$	= 1000 / 24	= 42
	ROOF	: $(UA)_r = A_r/R_r$	= 1200 / 38	= 32
	E, W, N WINDOWS	: $(UA)_g = 1.1A_x/N$	= 1.1 x 100 / 3	= 37
SELECT ONE	FLOOR	: $(UA)_f = A_f/R_f$	= ____ / ____	= ____
SELECT ONE	PERIMETER	: $(UA)_p = 4.17\ P_p/(R_p + 5)$	= 4.17 x 150 /(17 + 5)	= 28
SELECT ONE	BASEMENT	: $(UA)_b = 10.7\ P_b/(R_b + 8)$	= 10.7 x ____ /(____ + 8)	= ____
	INFILTRATION	: $(UA)_i = 0.018V \times ACH \times ADR$	= 0.018 x 9600 x 0.27 x 0.94	= 44

Sum = 183 Btu/°F • h

NLC = 24 x Sum = 4392 Btu/°F • day

LOAD COLLECTOR RATIO, LCR

$LCR = NLC/A_p$ = 4392 / 146 = 30 Btu/°F • day • ft^2

WEIGHTED AVERAGE, SSF

REFERENCE DESIGN	PROJECTED AREA	FRACTION OF A_p		SSF FOR EACH REFERENCE DESIGN		PRODUCT OF SSF x FRACTION	U_c	$U_c A_p$
DGC3	146	1	x	0.37	=	0.37	0.26	38
			x		=			
			x		=			
TOTAL A_p =	146			SSF =		0.37	TOTAL (U_cA_p) =	38

TLC = NLC + 24 x TOTAL (U_cA_p) = 5304

BASE TEMPERATURE, T_b

$T_b = T_{set} - Q_{int}/TLC$ = 70 − 53000 / 5304 = 60 °F

AUXILIARY HEAT, Q_{aux}

$Q_{aux} = NLC \cdot DD(1 - SSF)$ = 4392 x 7667 x (1 − 0.37) x 10^{-6} = 21.2 Mega Btu/year

Figure 7-9. *NLC* and *LCR* Worksheet for Example Problem 7-8

Example Problem 7-9

Determine the life-cycle optimum value of *LCR* for a heat cost of 7¢/kWh (levelized) and $FCR = 0.10$, based on the previous example. From Equation 6-11,

$$h = \frac{(.07)(7667)}{(3414)\,(0.10)}$$

$$= 1.57\ \$ \cdot {}^{\circ}\mathrm{F} \cdot \mathrm{day/Btu}$$

$D = SIC/h$, to achieve life-cycle optimum.

$D = 10/1.57 = 6.4$

Using Equation 6-13, and $T_b = 60°\mathrm{F}$, the following values are found:

LCR	15	20	30
CF	1.95	1.75	1.65
SSF	0.58	0.48	0.37
D	4.58	6.70	7.86

Therefore, by interpolation, $LCR = 19$.

As a practical matter, the optimum is very broad, and any *LCR* between 15 and 30 is a reasonable choice.

PART II
DESIGN DEVELOPMENT

Purpose

The purpose of Part II is to provide a method of analysis that is sensitive enough to show the relatively small changes in thermal performance that result from variations in design detail that occur during design development such as shading, orientation of glazing, etc., as discussed in Chapter 1. The method is called the monthly *SLR* method.

The 94 reference designs are retained as the starting point for the thermal analysis, but many more deviations can be accommodated by the monthly *SLR* method. Also, the thermal performance is represented on a monthly basis to show how the building responds to the annual cycle of weather changes.

The monthly *SLR* method should be executed when the designer needs performance figures on a month-by-month basis, when the design being analyzed differs from the reference designs in a way that can be accounted for only on a monthly basis, or when weather data other than that in Appendix 3 is to be used.

Scope of Part II

Chapter 8 describes the monthly solar load ratio (*SLR*) method of computing the thermal performance of a building. The solar savings fraction (*SSF*) is computed on a month-by-month basis. The flexibility of the method within certain design parameters is discussed. Chapter 9 describes the calculation of the amount of solar energy absorbed into a building through the solar aperture. Chapter 10 presents alternate methods for calculating heat loss and solar gain through windows. Chapter 11 gives methods for estimating heat flow through solar walls that differ in design detail from the reference designs. Chapters 12, 13, and 14 treat the specifics of direct-gain, sunspace, and thermal storage wall systems. The sensitivity of each system to the selection of design parameters is discussed. Chapter 15 applies the monthly *SLR* method to buildings that combine two or more passive solar systems. Some discussion is given of the suitability of the different system types and mixtures of system types for satisfying various design problems. The appendices contain the data tables, graphs, and worksheets necessary to exercise the analysis methods. Example problems are given in Chapter 16.

CHAPTER 8
THE MONTHLY SLR METHOD

Purpose of the Method

The guidelines of Chapter 2 suffice to give rough estimates of the solar heating parameters when the concept for a building is being established. During schematic design, more information is available, and the annual *LCR* method of Chapter 5 may be used. Toward the end of design development, considerable detail is available even though the construction documents have not yet been prepared. The construction document stage is expensive, and it is important that the thermal performances of the building be verified before this last stage is started. The monthly *SLR* method is suitable for this purpose. The *SLR* method provides a month-by-month picture of the thermal performance of a building that will confirm the sizing and placement of the solar features or, alternately, will reveal flaws in the solar design that should be corrected before the construction documents are generated. In the annual *LCR* method, deviations of the proposed design from the reference design are treated by sensitivity analyses; the monthly *SLR* method allows the designer to account more directly for many of these deviations and reduces or eliminates the number of sensitivity analyses. The flexibility of the monthly *SLR* method justifies the modest effort required to execute it.

Data Preparation

Building Data — Extensive preparatory work is necessary before the *SLR* method of analysis can be executed. The first step is the compilation of data and specifications concerning the building to be analyzed. Since the building is fairly well defined toward the end of the design development stage, this step involves merely assembling the data in an organized manner so that it is readily available. The major items of information are the sizes, orientations, and conductance (*U*) values of all exterior surfaces of the building (walls, roof, floor, windows, and doors), the length of the perimeter, and the *U*-value of the perimeter insulation. The data for the building features that comprise the solar wall should be listed separately.

Heat Load Parameters — The next step is to compute the building net load coefficient, *NLC*, as explained in Chapter 3, or, since the necessary data are available, the standard ASHRAE procedures may be used. In this case, *NLC* would be determined from:

$$NLC = \frac{Q_d - Q_{dsa}}{\text{design } \Delta T} \qquad (8\text{-}1)$$

where

Q_d = design load,
Q_{dsa} = design load of solar aperture.

Then, the load collector ratio, *LCR*, of the building is calculated using

$$LCR = NLC/A_p \, . \qquad (8\text{-}2)$$

For sunspace and mass wall systems, LCR_s may need to be calculated also. The procedure is analogous to the calculation of *LCR*. Details of the calculation of LCR_s for sunspace and mass wall systems are given in Chapter 11. Values of LCR_s for reference designs are listed in Chapter 11. The values of *LCR* and LCR_s, derived in this manner, will accurately represent the heat loss characteristics of the proposed building and will be directly incorporated into the *SLR* method calculations. This differs from the *LCR* method in which the correlations are based on a fixed value of LCR_s depending on the reference design.

Degree-Day Data — The next preparatory step is to compute the monthly values of *DD*, using a base temperature adjusted to account for the building's heating thermostat setpoint, T_{set}, and the internal heat rate, Q_{int}. This procedure is outlined in Chapter 4. It should be noted here that the *LCR/CF* tables were generated using two base temperatures, 55° and 65°F, so the effects of T_{set} and Q_{int} may be estimated by interpolating between these two bases, but the *SLR* method allows an arbitrary base temperature to be accounted for directly. Furthermore, a month-by-month variation in the base temperature is possible in the *SLR* method.

Solar Gain Data — The final preparatory step is the calculation of *S*, the solar radiation absorbed through the glazing of the solar wall. The procedure is described in Chapter 9. The following variables can be accounted for in the calculation of *S*: tilt and azimuth of the glazing, shading by an overhang, thickness and extinction coefficient of the glazing material, the number of layers of glazing, ground reflectance, and the absorptance of the surfaces receiving the incident radiation. This gives the *SLR* method further flexibility in adapting to variations in the design of the solar features of a building. In the *LCR* method, a separate sensitivity analysis must be applied to account for each change of these variables from the fixed values specified in the reference design.

Computing *SSF*

After completing the preparatory steps, *SSF* is computed for each month of the heating season. The equations used for computing *SSF* for direct-gain systems are different from the equations used for thermal storage wall and sunspace systems. The two different sets of equations are discussed separately below.

Equations for Direct-Gain Systems — For direct-gain systems *SSF* is computed from

$$SSF = 1 - (1 + G/LCR)\,[1 - F(SLR)] \qquad (8\text{-}3)$$

where

$$F(SLR) = \begin{cases} A \cdot SLR, \text{ for } SLR < R \\ B - C\exp(-D \cdot SLR), \text{ for } SLR \geq R \end{cases} \qquad (8\text{-}4)$$

and

$$SLR = (S/DD)/(LCR + G)\,. \qquad (8\text{-}5)$$

The values of *A, B, C, D, G,* and *R* for direct-gain systems are listed in Appendix 5, Table A5-9. Note that the proper form of Equation 8-4 is selected by comparing *SLR* with *R*. Graphs of *F*(*SLR*) versus *SLR* are shown in Appendix 5, Figures A5-19 through A5-21.

To compute *SSF*, first calculate the value of *SLR*. Next either use the graphs or use Equation (8-4) to find the value of *F*(*SLR*). Then use Equation (8-3) to calculate the value of *SSF*.

Equations for Thermal Storage Wall and Sunspace Systems — For thermal storage wall and sunspace systems *SSF* is computed from

$$SSF = 1 - C\exp(-D \cdot SLR) \qquad (8\text{-}6)$$

where

$$SLR = (S/DD - LCR_s \cdot H)/LCR \qquad (8\text{-}7)$$

The values of *C, D,* LCR_s, and *H* for thermal storage wall systems are listed in Appendix 5, Table A5-7; and for sunspace systems in Table A5-8. Graphs of *SSF* versus *SLR* for thermal storage wall systems are shown in Appendix 5, Figures A5-1 through A5-13, and for sunspace systems in Figures A5-14 through A5-18.

To compute *SSF*, first calculate *SLR*. Then either use the graphs or use Equation (8-6) to find *SSF*.

Computing Q_{aux} — The net reference load, Q_{net}, is found from

$$Q_{net} = NLC \cdot DD\,. \qquad (8\text{-}8)$$

This represents the total energy needed to heat the nonsolar portions of the building; this energy is supplied by both the auxiliary heating system and by the solar systems incorporated in the building design. The energy supplied by the auxiliary heating system is Q_{aux} and the energy supplied by the solar systems is Q_{sav}. Therefore, in equation form,

$$Q_{net} = Q_{aux} + Q_{sav} \qquad (8\text{-}9)$$

The quantity *SSF* is that fraction of Q_{net} supplied by solar energy,

$$Q_{sav} = Q_{net} \cdot SSF\,. \qquad (8\text{-}10)$$

The remainder of Q_{net} is Q_{aux}, the object of the analysis. The equation is

$$Q_{aux} = Q_{net}(1 - SSF) \qquad (8\text{-}11)$$

The annual net reference load, *annual* Q_{net}, is merely the sum of the monthly values of Q_{net}. Likewise the annual solar savings, *annual* Q_{sav}, and the annual auxiliary heat, *annual* Q_{aux}, are the sums of the monthly values of Q_{sav} and Q_{aux}, respectively. The annual solar savings fraction, *annual SSF*, is found from

$$annual\ SSF = annual\ Q_{sav}/annual\ Q_{net} \qquad (8\text{-}12)$$

Format and Step-by-Step Procedure

Format — Repetitive calculations of this sort are most efficiently and accurately done using a format that systematizes the procedure. A suggested format, the *SLR* Method Worksheet, is shown in Appendix 7.

The Monthly *SLR* Method Procedure — The step-by-step procedure for completing the *SLR* Method Worksheet is shown in Table 8-1.

Intermediate Values of Night Insulation

The R-value of the night insulation, as used herein, is the difference between the R-value of the solar wall glazing with the night insulation in place and the R-value of the solar wall glazing without night insulation. If the building design to be analyzed by the *SLR* method uses night insulation with an R-value other than 0 or 9, the calculations for the monthly values of both *SLR* and *SSF* must be done twice. For the first calculation, use the correlation coefficients, as shown in Appendix 5, Tables A5-7, 8, and 9, under the most similar reference design *without* night insulation. The monthly values of *SLR* and *SSF* that result from this calculation are identified as SLR_0 and SSF_0 to indicate no night insulation. For the second calculation, use the correlation coefficients, in Appendix 5, under the most similar reference design *with* night insulation. The resulting monthly values are identified as SLR_9 and SSF_9 to indicate R9 night insulation.

For each month, interpolate nonlinearly between SSF_0 and SSF_9 as follows:

$$SSF = SSF_0 + Y(SSF_9 - SSF_0) \qquad (8\text{-}13)$$

where

$Y = (1 + R_0/9)/(1 + R_0/R)$
R = R-value of night insulation of building being analyzed (h·°F·ft^2/Btu)
$R = 1/U_c$

where U_c is the solar wall load coefficient for the no-night-insulation case. For reference designs, U_c is tabulated in Appendix 5, Tables A5-7, 8, and 9. For sunspaces with an insulated common wall, use U_c for the corresponding masonry-wall sunspace. U_c may be calculated by ASHRAE methods for nonreference designs.

This procedure yields estimates of the monthly values of *SSF* for the building being analyzed. The *SLR* method is then completed by carrying out steps 10 through 13, as described in Table 8-1.

Table 8-1

Step-by-Step Procedure for the Monthly *SLR* Method Worksheet

Steps	Reference
1. Fill in Reference Design space.	Appendix 5, Tables A5-2 through A5-6.
2. Enter into Data block the values of *NLC*, A_p, and *LCR*.	Building design data, Chapter 3 and the *NLC* Worksheet.
3. Enter into Data block, as needed, the values of LCR_s, *A, B, C, D, G, H,* and *R*. Note that where a value is not needed for the chosen reference design the space has been crossed out.	Appendix 5, Tables A5-7, 8, and 9. For non-reference designs see Chapter 11 to calculate LCR_s.
4. Calculate $LCR_s \cdot H$, if needed, and enter into Data block.	
5. Enter monthly values of *S* into table.	Chapter 9 and Solar Gain Worksheet.
6. Enter monthly values of *DD* into table.	Chapter 4 and Appendix 3.
7. Calculate monthly values of *SLR* and enter into table.	Key Formula at top of worksheet and Chapter 8.
8. **For direct-gain systems only:**	
a. Compute *F*(*SLR*) for each month.	Key Formula at top of worksheet or graphs in Appendix 5, Figures A5-19 through A5-21.
b. Compute *SSF* for each month and enter into table.	Key Formula at top of worksheet.
9. **For thermal storage wall and sunspace systems only:**	
Compute *SSF* for each month and enter into table.	Key Formula at top of worksheet or graphs in Appendix 5, Figures A5-1 through A5-18.
10. Calculate monthly values of Q_{net} and enter into table. Note that the formula for Q_{net} printed at the head of column converts Q_{net} into millions of Btu per month (MegaBtu/mo). Sum the column to obtain *Annual* Q_{net} and enter the value into the table.	Chapter 8 and *SLR* Method Worksheet.
11. Calculate monthly values of Q_{sav} and enter into table. Sum the column to obtain *Annual* Q_{sav} and enter into the table.	Chapter 8 and *SLR* Method Worksheet.
12. Calculate monthly values of Q_{aux} and enter into table. Sum the column to obtain *Annual* Q_{aux} and enter into the table.	Chapter 8 and *SLR* Method Worksheet.
13. Calculate *Annual SSF* and enter into worksheet.	Key Formula and Chapter 8.

CHAPTER 9
SOLAR GAIN

Introduction

The monthly *SLR* method of thermal analysis requires an estimate of the monthly solar gain of the building being analyzed. The monthly solar gain, *S*, is the total monthly radiation absorbed by the building per square foot of projected area. The units of *S* are (Btu/ft^2 · month). The calculation of *S* requires the three items of solar radiation input data from Appendix 3 and involves the consideration of a large number of variables and conditions that describe the construction, materials, and orientation of the solar wall. This chapter is, therefore, necessarily complex. The following outline is provided to assist the reader in finding the particular aspect of the calculation in which he is interested:

Solar Radiation Input Data

The solar radiation input data for the computation of *S* are the daily radiation incident on a horizontal surface, *HS*; the solar noon zenith angle, (*LAT* − *DEC*); and the monthly average clearness index, K_T. These three items of data are tabulated in Appendix 3 and are defined below.

Daily Radiation Incident on a Horizontal Surface, *HS* — The values listed in Appendix 3 under HS are the monthly and annual mean values of the total daily hemispheric radiation incident on a horizontal surface. The units of *HS* are (Btu/ft^2 · day). The data for the U.S. cities are those furnished by the National Climatic Center[1] and are compiled from weather records for the period 1941 through 1970. For the Canadian cities, the source of the data was the Atmospheric Environment Service.[2]

Solar Noon Zenith Angle, *LAT* − *DEC* — The solar noon zenith angle at midmonth, *LAT* − *DEC*, is tabulated in Appendix 3 under the heading L-D. The latitude, *LAT*, in degrees, is also listed for each location. The solar declination, *DEC*, in degrees, is calculated from

$$DEC = 23.45 \sin [360(284 + n)/365] \qquad (9\text{-}1)$$

where n is the day of the year. When n is chosen as the day at the middle of the month, then *DEC* is the midmonth solar declination for that month. The value of n for the middle of each month is listed in Table 9-1.

Table 9-1

Value of *n* for Midmonth

Month	*n*
Jan	15
Feb	45
Mar	75
Apr	106
May	136
Jun	167
Jul	197
Aug	228
Sep	259
Oct	289
Nov	320
Dec	350

The Monthly Average Clearness Index, K_T — The monthly average clearness index, K_T, is the ratio of the normal monthly total hemispheric radiation on a horizontal surface, Q_h, to the monthly extraterrestrial radiation on a horizontal surface, Q_{he}:

$$K_T = Q_h/Q_{he} \,. \qquad (9\text{-}2)$$

The monthly horizontal surface radiation is determined from the normal daily value:

$$Q_h = M \cdot HS \,. \qquad (9\text{-}3)$$

where M = number of days in the month.

The monthly extraterrestrial horizontal surface radiation is determined from the daily value at midmonth:

$$Q_{he} = M \int_{\text{sunrise}}^{\text{sunset}} I \cdot \sin \theta_a = M \cdot I \cdot Y \qquad (9\text{-}4)$$

where I is the extraterrestrial solar flux at normal incidence at midmonth (Btu/h · ft^2) and Y is the integral of the sine of the solar altitude, θ_a, from sunrise to sunset. The quantity I is approximated by[3]

$$I = 428\,[1 + .033 \cos (360\, n/365)] \qquad (9\text{-}5)$$

and

$$Y + (24/\pi)[(\cos LAT \cdot \cos DEC \cdot \sin H_s) + (\pi/180)(H_s \cdot \sin LAT \cdot \sin DEC)] \qquad (9\text{-}6)$$

where H_s is the sunrise hour angle in degrees found from

$$\cos H_s = -\tan LAT \cdot \tan DEC \,. \qquad (9\text{-}7)$$

In this book, the hour angle convention is east is positive, west is negative, south is zero.

The values of n and *DEC* are for the midmonth.

Correlation Methodology

Several ratios of solar radiation are given in this chapter. Each of these ratios is given by a correlation using a solar angle, *LAT-DEC*, which is the noon solar zenith angle at midmonth, and the monthly average clearness index, K_T. The correlations have been developed by performing many hourly calculations for many locations and then determining a set of correlation coefficients which minimize the monthly mean square error between the correlation equation and the hourly values. Typical Meteorlogical Year (TMY)[4] hourly weather files were used as the starting point, usually for 26 U.S. cities representing a wide variety of climate characteristics. The hourly calculations were performed by using the total horizontal and direct normal insolation values from the weather file, assuming isotropic sky diffuse and ground reflectance. Fresnel relations were used to calculate glazing transmittance. Each glazing layer was assumed to have the optical properties of standard 1/8-inch-thick glass, with an extinction coefficient of 0.5 per inch and an index of refraction of 1.526. Transmission of diffuse radiation was calculated assuming an effective angle of incidence of 60 degrees. The reference value of ground reflectance is 0.3.

The Basic Formula for Calculating the Solar Gain, *S*

The monthly *SLR* method of analysis requires an estimate of *S*, the total monthly radiation absorbed by the building per unit of projected area (Btu/ft^2 · month). The procedure recommended to estimate *S* is based on the following equation:

$$S = M \cdot \frac{A_n}{A_p} \cdot HS \left(\frac{INC}{HOR}\right)\left(\frac{TRN}{INC}\right)\left(\frac{ABS}{TRN}\right) \tag{9-8}$$

The individual factors of this equation are defined below:

$\frac{A_n}{A_p}$ = ratio of actual net glazing area to projected area (cosecant of the tilt angle)

HS = average daily value of total radiation incident on a horizontal surface (Btu/ft^2 · day); tabulated in Appendix 3

$\left(\frac{INC}{HOR}\right)$ = ratio of monthly total solar radiation incident on a unit area of the glazing plane to that on a unit area of horizontal surface

$\left(\frac{TRN}{INC}\right)$ = ratio of monthly total solar radiation transmitted through the glazings to that incident on the glazing plane

$\left(\frac{ABS}{TRN}\right)$ = ratio of monthly total solar radiation absorbed in the building to that transmitted through the glazings

Calculation of *S* for Reference Design Systems

Simplified procedures are presented here for the calculation of *S* for systems that conform to the reference design specification.

Calculation of *S* for Direct-Gain and Thermal Storage Wall Reference Design Systems — For direct-gain and thermal storage wall reference designs, the following formula is recommended for the calculation of *S*:

$$S = M \cdot HS \left(\frac{TRN}{HOR}\right)\left(\frac{ABS}{TRN}\right) \tag{9-9}$$

where

$\left(\frac{TRN}{HOR}\right)$ = ratio of monthly total solar radiation transmitted through a unit area of glazing to that on a unit area of horizontal surface (dimensionless)

The value of (TRN/HOR) is calculated from a correlation of the form:

$$\left(\frac{TRN}{HOR}\right) = B_1 + B_2 Y + B_3 Y^2 + K_T (B_4 + B_5 Y + B_6 Y^2) \tag{9-10}$$

where

$Y = (LAT - DEC)/100.$

The parameters K_T and $LAT - DEC$ are found in Appendix 3 for the location and month under investigation. The values of the coefficients B_1 through B_6 are tabulated in Appendix 4, Table A4-3. The values of the absorption factor, (ABS/TRN), are tabulated in Appendix 4, Table A4-46.

Calculation of *S* for Sunspace Reference Design Systems — For sunspace reference designs, the following formula is recommended for the calculation of *S*:

$$S = M \cdot HS \left(\frac{ABS}{HOR}\right) \tag{9-11}$$

where

$\left(\frac{ABS}{HOR}\right)$ = ratio of monthly total solar radiation absorbed in the building per unit of projected area to that on a unit area of horizontal surface (dimensionless)

The value of (ABS/HOR) is calculated from a correlation of the same form as Equation 9-10. The values of the coefficients B_1 through B_6 are tabulated in Appendix 4, Table A4-47. Note that these coefficients vary according to the sunspace reference design under consideration.

Radiation Incident on a Vertical, South-Facing Surface, *VS*

The solar apertures for thermal storage wall and direct-gain systems are usually vertical surfaces and, in the most common case, this surface faces due south. Since this situation occurs so frequently, it is treated in more detail than other surface orientations. A diffuse ground reflectance of 0.3 is assumed. The normal value of total daily solar radiation incident on a vertical, south-facing surface, VS, is listed in the tables in Appendix 3. Both monthly and annual values are listed. The units of VS are ($Btu/ft^2 \cdot day$). The values for VS were computed using the following correlation:

$$VS/HS = 0.6866 - 0.6623\,Y + 1.3269\,Y^2 + K_T(-0.4458 + 0.3090\,Y + 4.7776\,Y^2) \quad (9\text{-}12)$$

Effect of an Overhang — Overhangs are used in most passive solar applications to reduce summer overheating. If the overhang is properly designed, there is little blockage of the midwinter sun but almost complete blockage of the direct midsummer sun. Figure 9-1 shows a simple scheme for sizing the overhang in relation to the sun angle on the summer and winter solstices. The solar zenith angle at noon on 21 June is ($LAT - 23.5°$). For noon on 21 December, it is ($LAT + 23.5°$).

A diagram for describing the geometry of an overhang is shown in Figure 9-2. The overhang and the separation are defined as fractions of the window height.

Note that the glazing should not extend to the base of the overhang because the top portion of the window would never receive direct sun but would lose as much heat as any other part of the window.

An alternative to fixed shading is movable shading. This is awkward and not much favored by designers but is quite effective. The shade can be left in place until late in the fall, thus reducing overheating substantially. The shade can then be removed and remain off until late in the spring, after the heating season is over.

Another option is to use night insulation as shading. This is a particularly attractive approach since it can be used as day insulation in the summer. It also allows a very simple and effective means of accommodating to the weather; it markedly improves performance during the winter and is especially effective at reducing summer overheating. Types of night insulation that are located outside the window are particularly effective for summer shading. If they are located inside the window the designer must be particularly careful to avoid material damage associated with buildup of heat between the glazing and the insulation by using a light-colored or reflective outer surface. Thermal stress breakage of glazing can also be a problem. Use of tempered glass will help reduce the likelihood of this occurring.

Perhaps the simplest solution to the summer overheating problem is to use one of a variety of shading devices that can be put outside the windows. Such things as roll-down bamboo or canvas shades would be particularly effective. Very low-cost solutions like greenhouse shading materials or whitewash might be used in situations where aesthetics are not of great concern.

The decrease in radiation incident on a vertical, south-facing surface due to an overhang can be obtained by multiplying the radiation incident on the surface when there is no overhang by the ratio VS_o/VS.

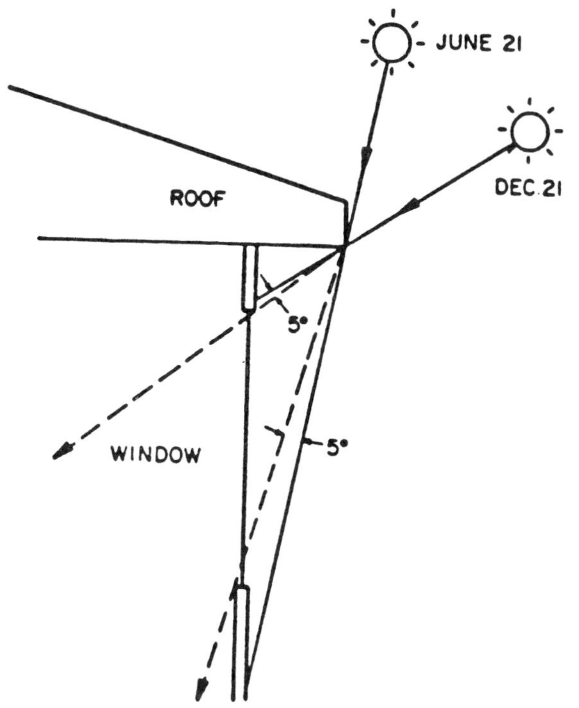

Figure 9-1. An Example Showing How Roof Overhang Geometry Can Be Modified To Provide Desired Summer Shading and Winter Solar Gain. The 5° allowances shown extend the desired effects over a 10-week season.

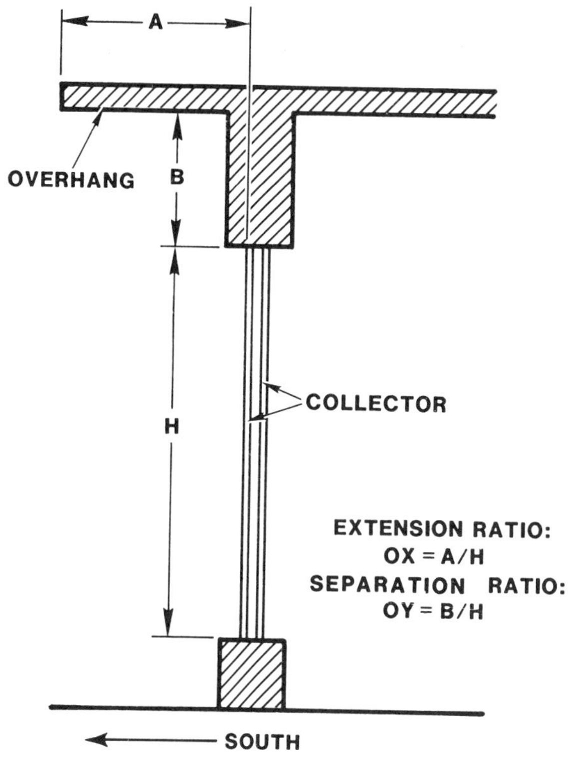

Figure 9-2. Geometry for an Overhang for a Collector on a Vertical, South-Facing Wall

For $Y < Y^*$,

$$\frac{VS_0}{VS} = B_1 + B_2Y + B_3Y^2 + B_4Y^3 + B_5Y^4 + B_6Y^5 + K_T(B_7 + B_8Y + B_9Y^2 + B_{10}Y^3 + B_{11}Y^4 + B_{12}Y^5) \qquad (9\text{-}13)$$

For $Y > Y^*$,

$$\frac{VS_0}{VS} = 0$$

where

$$Y = (LAT - DEC)/100.$$

If $VS_0/VS > 1$, set $VS_0/VS = 1$. The coefficients B_1 through B_{12} and Y^* are listed in Appendix 4, Table A4-48 and VS_0/VS is graphed in Figures A4-50 through A4-74.

Effect of a Specular Reflector — A specular reflector can increase the amount of radiation incident on a vertical, south-facing surface. The increase can be shown as the ratio VS_r/VS, where VS_r is the total radiation incident on a vertical, south-facing surface with a reflector. The geometry for a simple type of reflector is shown in

Figure 9-3. For this geometry, the correlation between VS_r/VS and $(LAT - DEC)$ has the form

$$\frac{VS_r}{VS} = B_0 + B_1 Y + B_2 Y^2 + B_3 Y^3 + B_4 Y^4 \qquad (9\text{-}14)$$

where

$$Y = (LAT - DEC)/100$$

The values of the coefficients are listed in Appendix 4, Tables A4-49 through 51, for several values of the aspect ratio, the reflector length to window height ratio, and reflector reflectance. Graphs of VS_r/VS versus $(LAT - DEC)$ are shown in Appendix 4, Figures A4-76 through 78. Inspection of the graphs shows that

a. Increase of the aspect ratio above 3 has little effect, and

b. Increase of the reflector length beyond three collector heights has little effect.

These correlations were developed in terms of the single parameter $LAT - DEC$, rather than $LAT - DEC$ and K_T, so they are not as precise as the other correlations in this chapter.

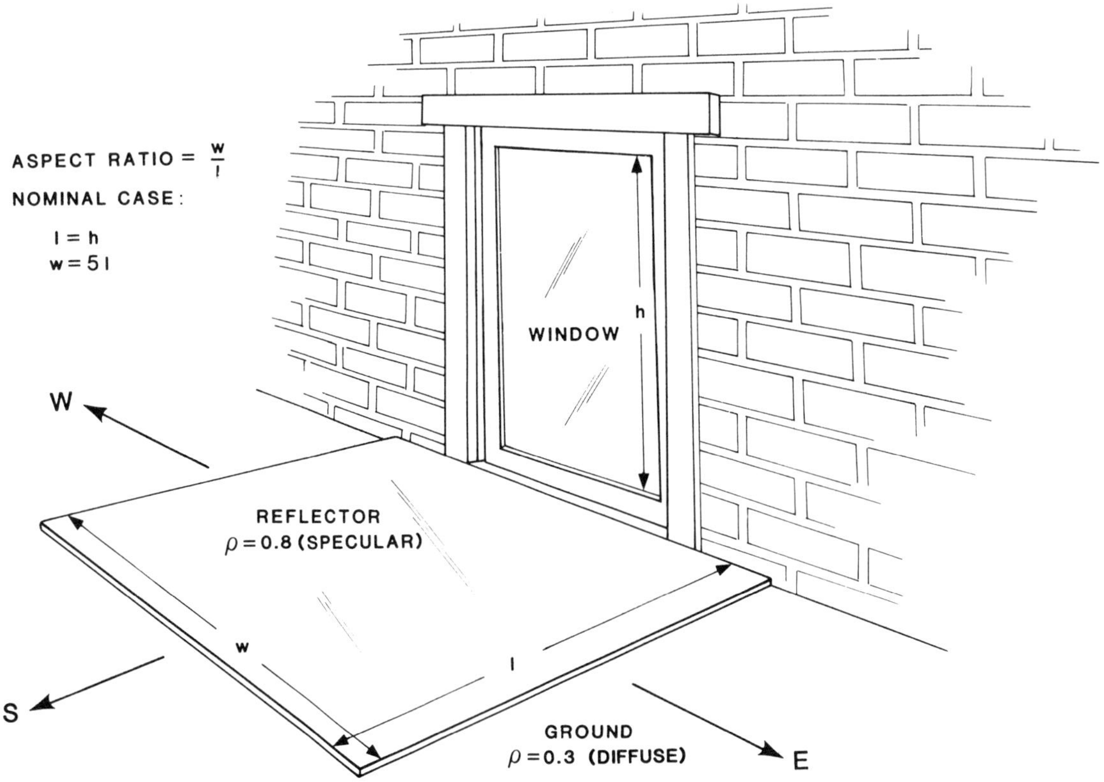

Figure 9-3. Geometry for Reflector for a Vertical, South-Facing Surface. The width of the reflector is always assumed to be the same as the width of the window.

Tilt and Azimuth of a Solar Aperture

The symbol (*INC/HOR*) represents the ratio of monthly total solar radiation incident on a unit area of the glazing plane to that on a unit area of horizontal surface.

The geometric orientation of the plane of a solar aperture can be described in terms of two angles: the tilt angle, *TILT*, and the azimuth angle, *AZ*. The angle between the plane of the solar aperture and the horizontal plane is the tilt angle, *TILT*. The azimuth angle, *AZ*, is defined as the angle between due south and the projection of the normal to the solar aperture plane onto the horizontal plane. Figure 9-4 shows these relationships. Azimuths west of south are negative, and azimuths east of south are positive.

Principal Azimuths for Sunspaces

For sunspaces with glazings that have two or more different azimuths, it is necessary to define a *principal azimuth, AZP*. This is an azimuth chosen to represent the azimuth of the whole sunspace rather than its glazing components independently. The usual case is a sunspace with a main south-facing glazing (zero azimuth) and a relatively small east-or west-facing glazing. The principal azimuth is then the azimuth angle of the main glazing, zero in this case. In the case of unusual geometries, it may be desirable to choose a principal azimuth that is not the azimuth angle of any of the glazings. For example, consider a sunspace with two glazing planes, equal in area, but one faces 30 degrees west of south (azimuth angle is −30 degrees) and one faces 30 degrees east of south (azimuth angle is +30 degrees). The principal azimuth should then be due south (0 degrees), the average of the two main glazing azimuths. The purpose of the principal azimuth is to serve as a basis for the definition of the projected area. See the definition of projected area in Chapter 1. Further discussion of projected area is at the end of the chapter.

Incident Radiation on an Arbitrarily Oriented Surface (*INC/HOR*)

A correlation was developed between (*INC/HOR*) and (*LAT* − *DEC*) and K_T. The ground reflectance

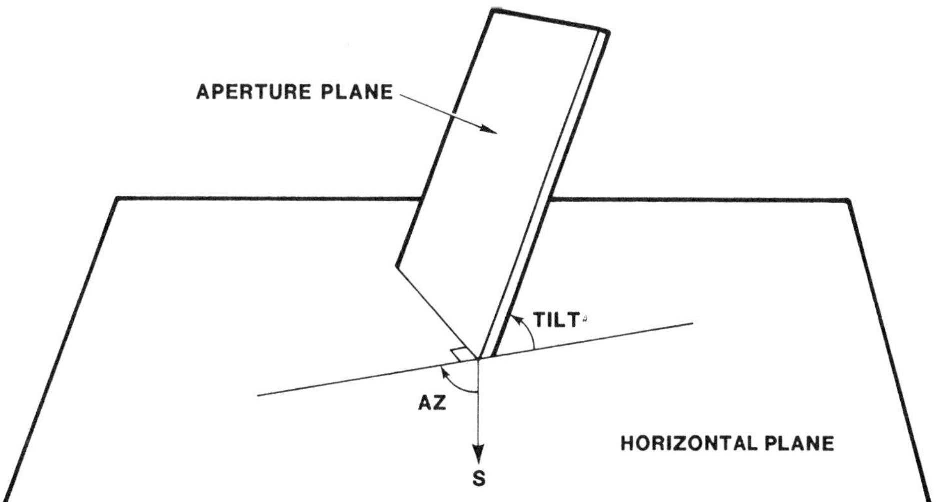

Figure 9-4. Definitions of *TILT* and *AZ*

was taken as 0.3. The correlation developed has the form:

$$\left(\frac{INC}{HOR}\right) = B_1 + B_2Y + B_3Y^2 + K_T(B_4 + B_5Y + B_6Y^2), \quad (9\text{-}15)$$

where

$$Y = (LAT - DEC)/100 .$$

The values of the coefficients were calculated for three values of *TILT* (30°, 60°, and 90°) and for seven values of *AZ* (0°, 30°, 60°, 90°, 120°, 150° and 180°) and are listed in Appendix 4, Table A4-1. Graphs of (*INC/HOR*) versus (*LAT* − *DEC*) for various values of K_T are also included in Appendix 4, Figures A4-1 through A4-21.

NOTE: Equation 9-15, or the tables or graphs in Appendix 4, are always used to obtain the values of (*INC/HOR*) for entry into the *(INC/HOR)* column of the Solar Gain Worksheet.

Radiation Correction for Ground Reflectance — If the situation under consideration involves a ground reflectance, ρ_g, other than 0.3, as used for the correlation analysis, then the corrected ratio, (*INC/HOR*), can be found as follows:

$$\left(\frac{INC}{HOR}\right) = \left(\frac{INC}{HOR}\right)_0 + F_{CG}\,(\rho_g - 0.3)\,, \quad (9\text{-}16)$$

where $(INC/HOR)_0$ is the base case value for ρ_g = 0.3 and F_{CG} is the view factor from the aperture plane to the ground. Figure 9-5 illustrates this case. In the simplest case of a uniform horizontal ground plane extending from the aperture plane to infinity,

$$F_{CG} = 1/2\,(1 - \cos TILT)\,. \quad (9\text{-}17)$$

NOTE: Equation 9-16 is *not* used to calculate values for entry into the *(INC/HOR)* column of the solar gain worksheet. Instead, Equation 9-15 is used without modification, and the effect of off-reference ground reflectance is accounted for by a modification of the ratio (*TRN/INC*). See Equation 9-19.

The radiation incident on a vertical, south-facing surface, *VS*, is listed in Appendix 3 and is calculated with the assumption that ground reflectance is 0.3. For a ground reflectance, ρ_g, other than 0.3, the corrected radiation on a vertical surface, *VS*, is found from

$$VS = VS_0 + HS \cdot F_{CG}\,(\rho_g - 0.3) \quad (9\text{-}18)$$

where VS_0 is the base case value for ρ_g = 0.3, as listed in Appendix 3. A table of representative values of ground reflectance is in Appendix 6, Table A6-3.

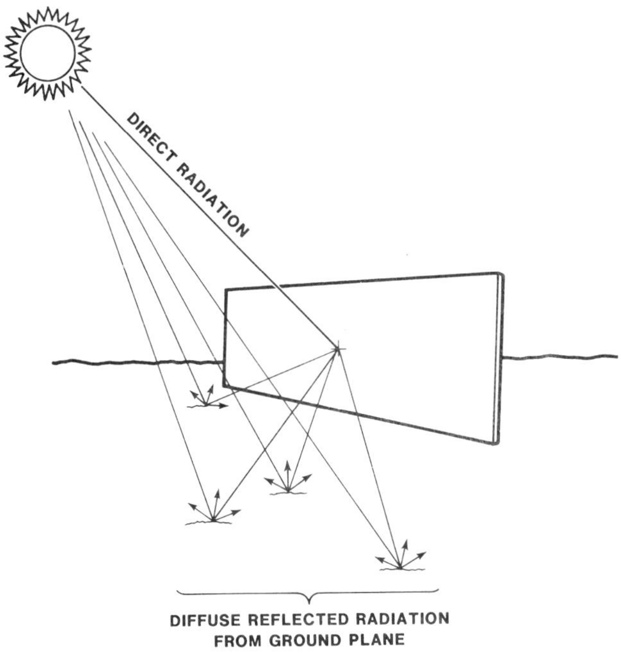

Figure 9-5. Ground Reflectance Increases the Radiation Incident on a Surface

Transmitted Radiation through Arbitrarily Oriented Glazings, (*TRN/INC*)

The symbol (*TRN/INC*) represents the ratio of monthly total solar radiation transmitted through glazings to that incident on the glazing plane. The ratio (*TRN/INC*) was correlated to the parameters *LAT* − *DEC* and K_T using a functional form similar to Equation 9-15, as was used for (*INC/HOR*), but of order Y^5 rather than Y^2. The values of the coefficients in the correlation equation for one, two, and three glazings, and for tilt angles from 0° to 90° and azimuth angles from 0° to 180° in increments of 30°, are listed in Appendix 4, Table A4-2. Graphs of (*TRN/INC*) versus *LAT* − *DEC* for various values of K_T are also included in Appendix 4, Figures A4-22 through A4-43.

Transmittance Correction for Ground Reflectance — For ground reflectance other than 0.3, a correction may be applied to the ratio (*TRN/INC*) that also accounts for the effect on (*INC/HOR*). The corrected ratio, (*TRN/INC*), may be found from:

$$\left(\frac{TRN}{INC}\right) = \left(\frac{TRN}{INC}\right)_0 + \frac{\tau_D F_{CG}(\rho_g - 0.3)}{\left(\frac{INC}{HOR}\right)_0} \quad (9\text{-}19)$$

where the subscript "*0*" refers to the base case value in which $\rho_g = 0.3$.

The diffuse transmittance, τ_D, is given in Figures 9-6 through 9-8 as a function of *E*, the extinction coefficient-thickness product, for one, two, and three glazing layers and indices of refraction of 1.3, 1.526, and 1.7. The curves were calculated by assuming vertical glazing and isotropically diffuse ground reflectance and sky radiation. They would also be good approximations for nonvertical glazing for the purpose of Equation 9-19. Note that Equation 9-19 gives the correct value of (*TRN/INC*) for use in the solar gain worksheet provided that (*INC/HOR*) is *not* corrected for off-reference ground reflectance. As such, (*TRN/INC*) no longer has the role of a transmittance; it can even have values greater than one.

Transmittance Correction for Glazing Extinction Coefficient and Thickness — For the reference designs the glazing material is assumed to have an extinction coefficient of 0.5 per inch and a thickness of 0.125 inches. The extinction coefficient-thickness product, *E*, has the value $E = 0.0625$.

If *E* is different from the reference design value, 0.0625, then the corrected value of *(TRN/INC)* can be found from

$$\left(\frac{TRN}{INC}\right) = \left(\frac{TRN}{INC}\right)_0 \exp[-1.15N(E - 0.0625)] \quad (9\text{-}20)$$

where the subscript "*0*" indicates the base case value for $E = 0.0625$ and $\rho_g = 0.3$.

Equation 9-20, with *N* as a parameter, is plotted in Figure 9-9. Equation 9-20 is valid for values of *E* in the range 0 to 0.125, and for $N = 1,2,3$.

If both *E* and ρ_g are different from the reference design values (0.0625 and 0.3 respectively) then

$$\left(\frac{TRN}{INC}\right) = \left[\left(\frac{TRN}{INC}\right)_0 + \frac{\tau_D F_{CG}(\rho_g - 0.3)}{\left(\frac{INC}{HOR}\right)_0}\right] \exp[-1.15N(E - 0.0625)] \quad (9\text{-}21)$$

NOTE: Equation 9-20 gives the correct value of *(TRN/INC)* for use in the solar gain worksheet when only the value of *E* is changed from the reference design value. Equation 9-21 gives the correct value when both *E* and ρ_g are changed.

The Absorption Factor, (*ABS/TRN*)

Once the solar radiation transmitted through the aperture has been determined, it remains to determine the absorption factor, (*ABS/TRN*), the ratio of absorbed radiation to the transmitted radiation. The discussion here, and the absorption factor tables in Appendix 4, Tables A4-5 through A4-45, are separated under three general categories: (1) absorption factors for thermal storage walls, (2) absorption factors for vertical glazing, and (3) absorption factors for tilted glazing.

Absorption Factors for Thermal Storage Walls — The simplest case is that of a thermal storage wall where the absorption factor is approximately equal to the wall absorptance. Reflections from the glazing system back to the wall slightly increase the absorption factor relative to the wall absorptance.

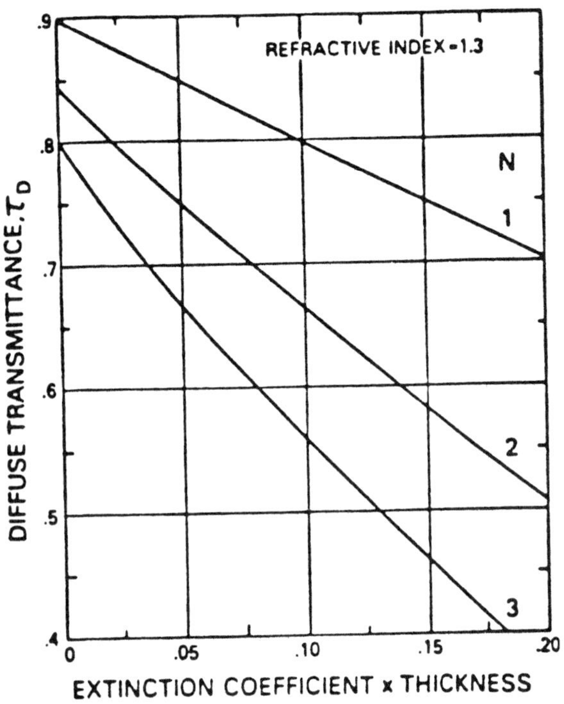

Figure 9-6

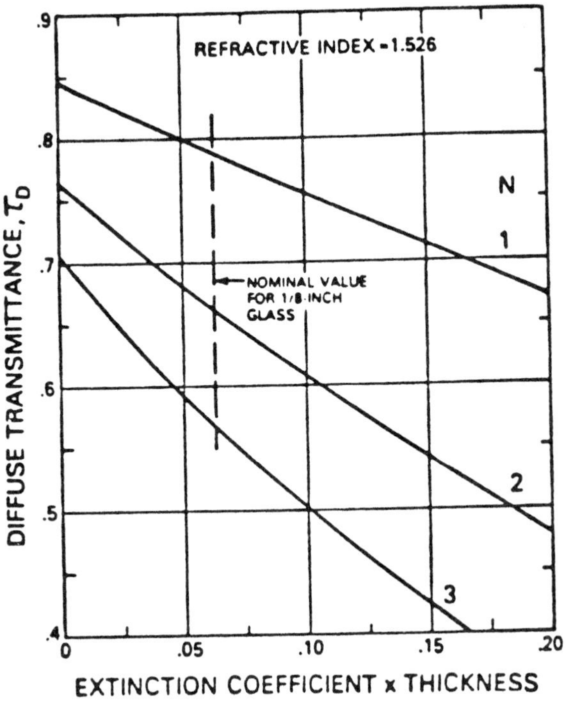

Figure 9-7

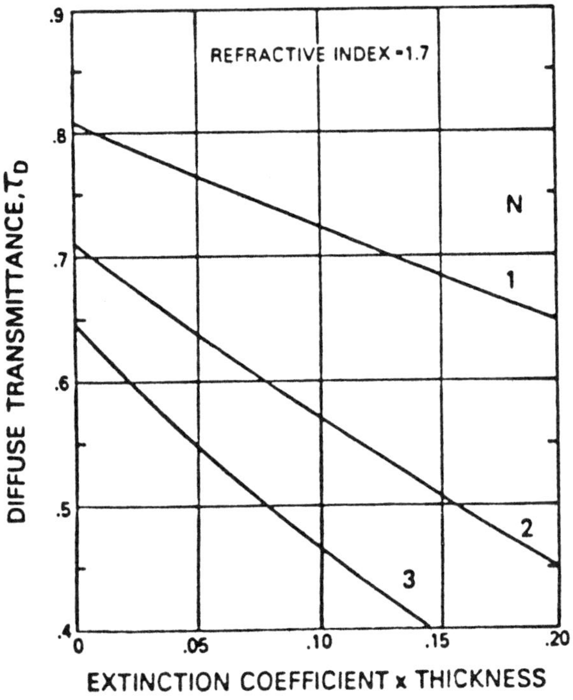

Figure 9-8

Figures 9-6 through 9-8. Diffuse Transmittance versus Extinction Coefficient-Thickness Product

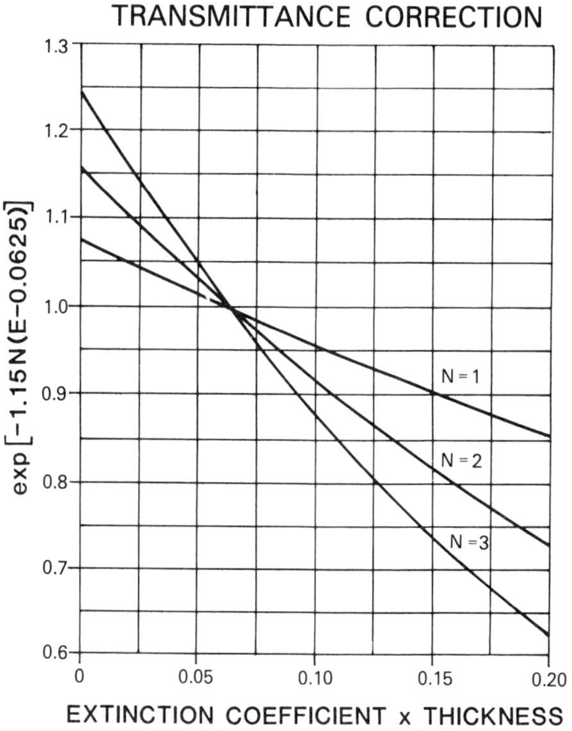

Figure 9-9. Transmittance Correction. *N* is the number of layers of glazing and *E* is the extinction coefficient-thickness product of each layer. The reference value of *E* is 0.0625.

Absorption factors for various wall absorptances, glazing extinction coefficient-thickness products, *E*, and numbers of glazing layers, *N*, are given in Appendix 4, Table A4-5.

Absorption Factors for Vertical Glazing — This category applies to direct-gain systems and to those sunspace systems that have only vertical glazing. For these systems the absorption factor depends on the interior surface absorptances and the space geometry in a complex way. The geometries for vertical glazing for which absorptance factors have been calculated are described in Figure 9-10. The end walls of sunspace designs can be either opaque or glazed. This affects the value of the overall absorption factor since glazed end walls allow more

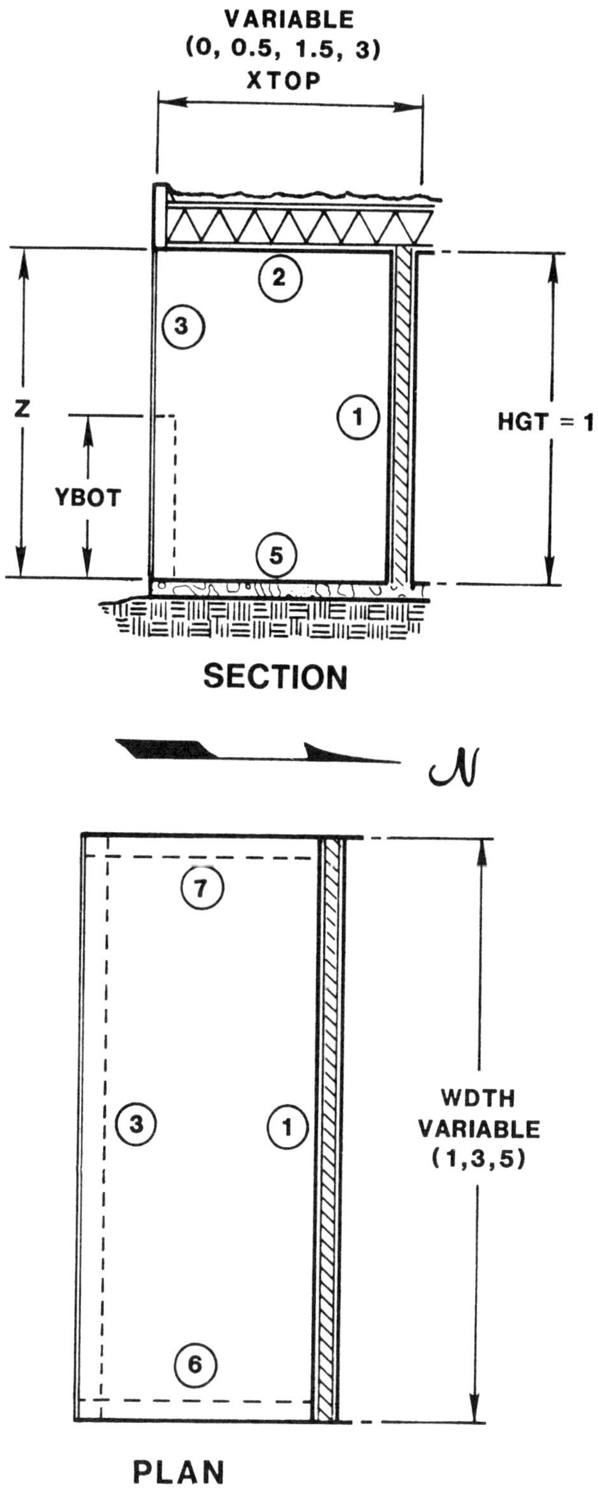

Figure 9-10. Geometries of Vertical Glazing

of the transmitted radiation, after reflection in the interior, to escape than opaque end walls do. The absorption factors are obtained from Appendix 4, Tables A4-6 through A4-17. For direct-gain reference designs, the absorption factors may be obtained from Table A4-46.

The surface absorptances of the north wall and the floor of the room are assumed to be the same and are denoted by the symbol α_m. All other surfaces are assumed to have a surface absorptance of α_o, which may be different from α_m.

Tables of absorption factors are given for two sets of conditions:

1. $\alpha_m = \alpha_o$, and α_m is varied from 0.3 to 1.0 in steps of 0.1 (as shown in Figure 9-11) and

2. $\alpha_o = 0.3$, and α_m is varied from 0.3 to 1.0 in steps of 0.1 (as shown in Figure 9-12).

Absorption factors for other combinations of surface absorptances may be estimated by performing area-weighted averages.

In calculating the absorption factors, the following assumptions were made:

1. Transmission of solar radiation through glazed surfaces is isotropically diffuse.

2. Reflection of solar radiation from interior surfaces is isotropically diffuse with a reflectance equal to 1 minus the solar absorptance (a property of the surface).

3. Reflection of radiation from the interior of glazed surfaces is for two layers of standard glass ($E = 0.0625$ and index of refraction = 1.526), about 24%. Radiation not reflected is assumed to pass through to the outside and to be lost from the system.

4. Twenty percent of the transmitted radiation is absorbed by lightweight objects before striking any interior surface. Twenty percent of the radiation in each reflection is likewise absorbed by lightweight objects.

The absorption factors in Appendix 4 are listed according to the surface through which the radiation enters the space. Surface 3 is always included. When surface 4 is glazed, the absorption factor is given for radiation that enters through surface 4. When the end walls (surfaces 6 and 7) are glazed, the absorption factor is given for surface 6, which applies to radiation that enters through either surface 6 or surface 7.

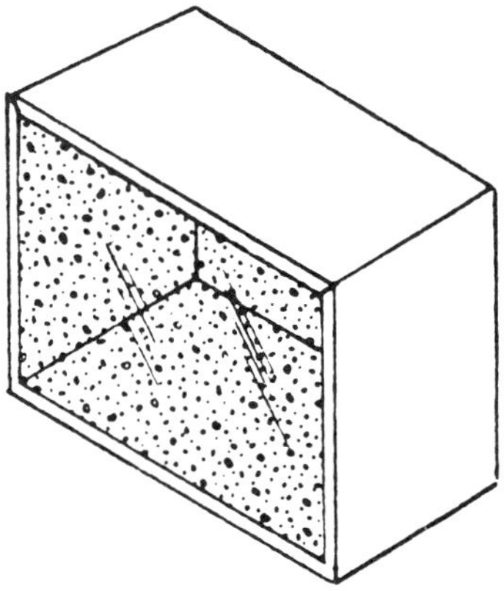

Figure 9-11. The Surface Absorptance of the Floor and North Wall, α_m, Is the Same as All Other Absorptances, α_o.

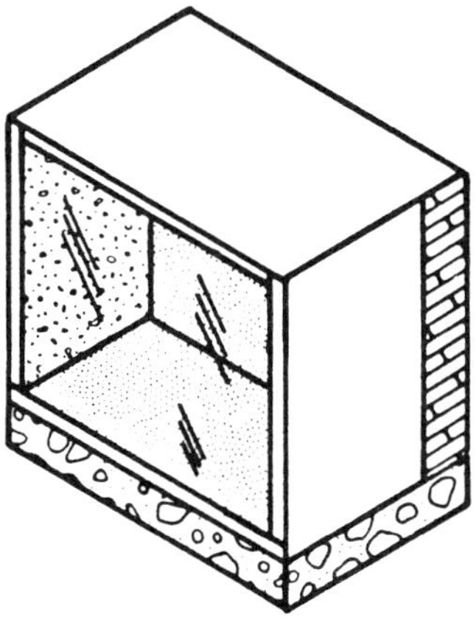

Figure 9-12. The Surface Absorptance of the Floor and North Wall, α_m, Is Different from All Other Absorptances, α_o.

The dimensions (*HGT, WDTH, XTOP,* and *Z*) shown in Figure 9-10 are all ratios to the sunspace height defined as follows:

$$HGT = \text{height/height} = 1$$
$$WDTH = \text{width/height}$$
$$XTOP = \text{depth/height}$$
$$Z = \text{glazing length/height.}$$

For vertical glazing, the glazing length is equal to the glazing height; therefore, for vertical glazing, *Z* has a maximum value of 1.

Caution should be used in applications that have substantially lower surface absorptances than in the reference designs. In such cases, a high fraction of the total absorbed radiation goes directly to the air due to the assumed lightweight absorption fraction. The resulting *S* would generally lead to an overprediction of the monthly *SSF* because solar radiation absorbed in the air is usually less effective than solar radiation absorbed on mass surfaces.

Absorption Factors for Tilted Glazing — This category applies to all sunspace systems except SSC1, SSC2, SSC3, and SSC4. This category is divided into three classes, depending on the geometry. Class I employs a single span of glazing for the south wall. Class II has a break in the glazing of the south wall, a lower vertical section, and a tilted upper section. Class III has two subdivisions, Class IIIa and Class IIIb. Both subdivisions have a tilted upper section of glazing and a lower section of opaque vertical wall. Class IIIa has a shorter glazing length, *Z*, and a different set of tilt angles from Class IIIb. The geometries are shown in Figures 9-13 through 9-16. The glazing length ratio, *Z*, pertains to the length of the tilted upper section. Otherwise the dimensions *HGT*, *WDTH*, *XTOP*, and *Z* are defined the same as for vertical glazing.

The calculations of the absorption factors for geometries with tilted glazing were done in the same manner and with the same assumptions as for vertical glazing. The same precautions apply.

The absorption factors for sunspace designs with tilted glazing are listed in Tables A4-18 through A4-45.

Correcting Double-Glazed Reference Systems to Single or Triple Glazing — It is possible to extend the *SLR* correlations for double-glazed thermal storage wall and sunspace systems to produce estimates of the performance of similar systems but with one or three glazing layers. This is most likely to be a useful procedure for sunspaces because there are only double-glazed sunspace reference systems. Readers should be warned, however, that changing the number of glazing layers is a rather radical change, and the results cannot be expected to be as precise as for the systems for which the correlations were developed.

The correction to single or triple glazing requires a modification of LCR_s, as discussed in Chapter 11, and the calculation of (*TRN/INC*) for the desired number of glazing layers, as discussed in this chapter. The third required correction is to compensate for the different amount of solar radiation, absorbed in the glazing material, which is useful in reducing heat loss through the glazings. This is done as follows:

$$\left(\frac{TRN}{INC}\right)\left(\frac{ABS}{TRN}\right) = \left(\frac{TRN}{INC}\right)\left(\frac{ABS}{TRN}\right)_o + g(N) \qquad (9\text{-}23)$$

where *N* is the number of glazing layers and *g*(*N*) is a correction term evaluated as follows:

N	*g*(*N*)
1	−0.031
2	0
3	0.025

Note that (*TRN/INC*) has already been corrected for the variation in *N*.

Computation of Monthly Solar Gain, *S*

The computation of *S* must be done for each month of the heating season and for each passive system. In this computation, it is important to distinguish three levels of structure. The first is the whole *building*. The second is the passive *system*. The third is the system *component*. A building may have only one passive system, or it may have two or more. For example, a direct-gain window

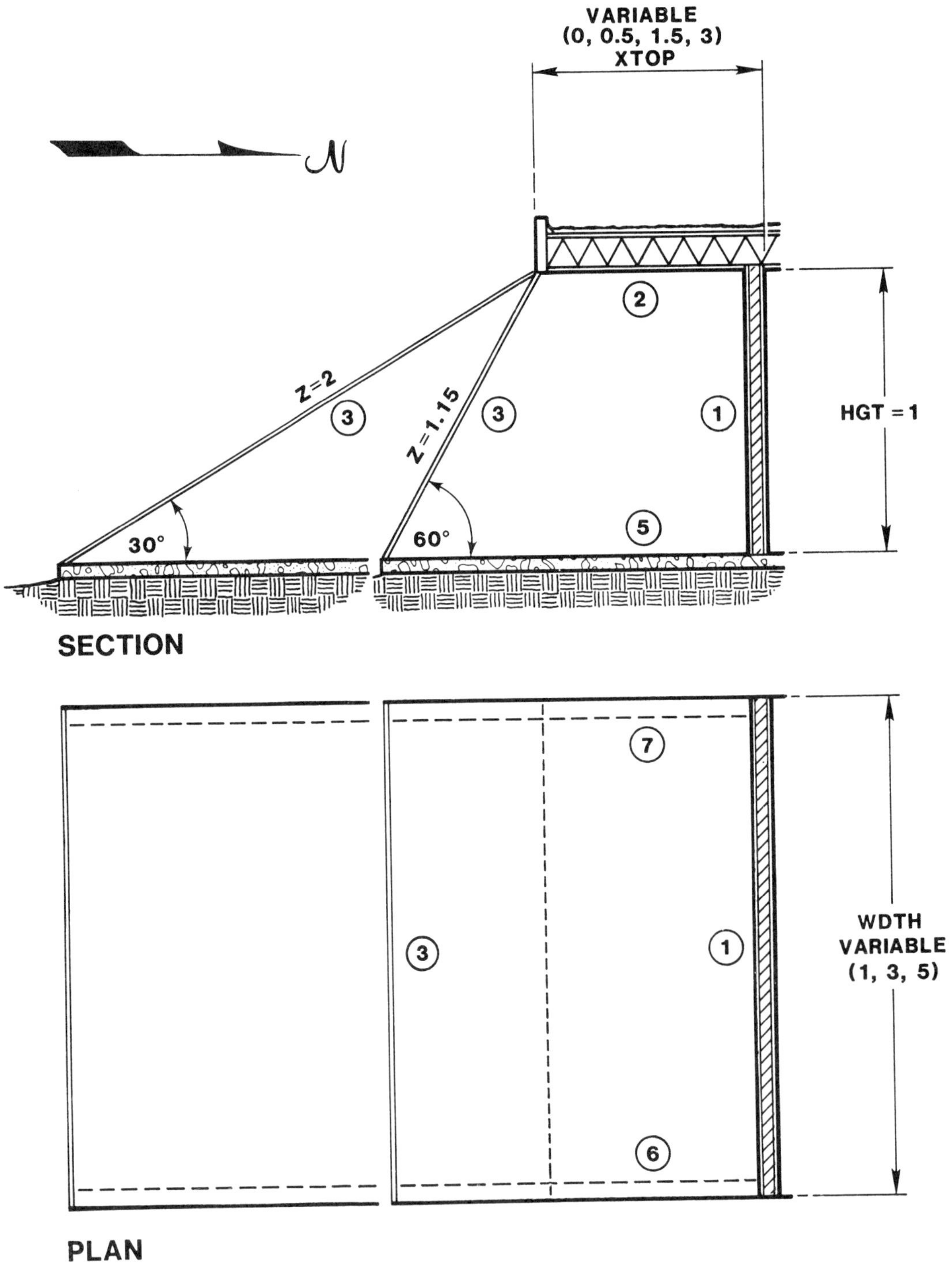

Figure 9-13. Geometries of Tilted Glazing, Class I

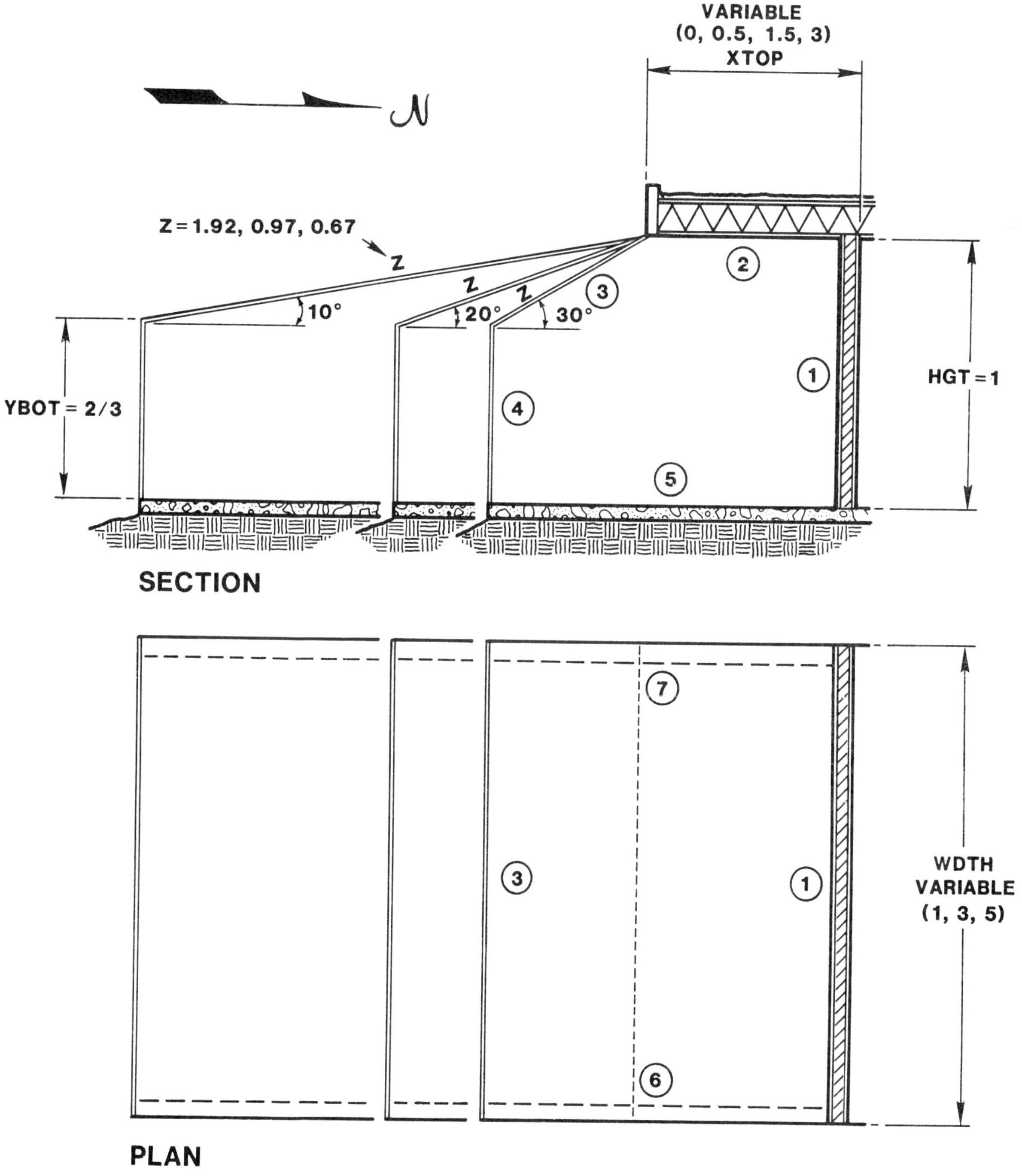

Figure 9-14. Geometries of Tilted Glazing, Class II

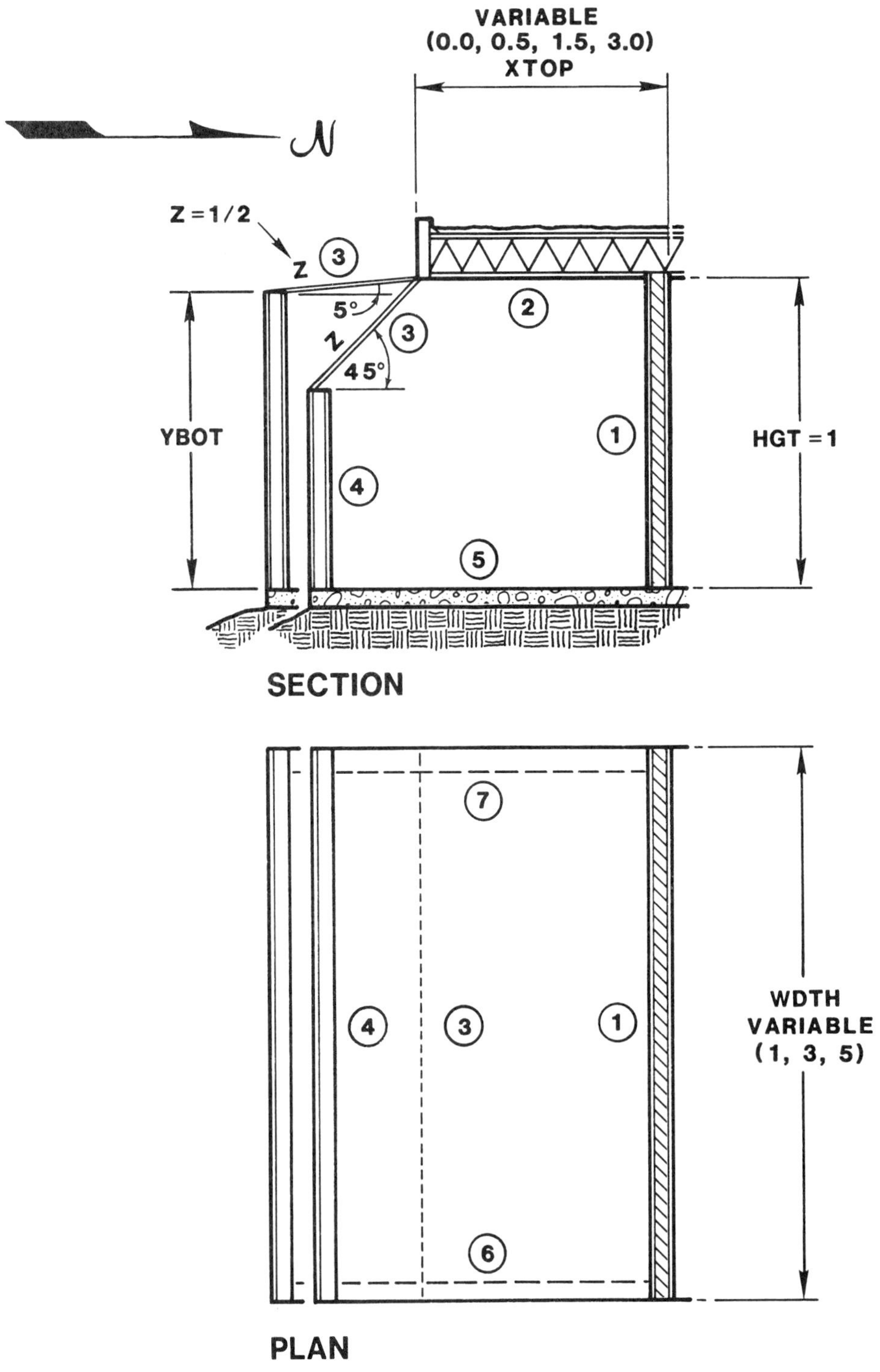

Figure 9-15. Geometries of Tilted Glazing, Class IIIa

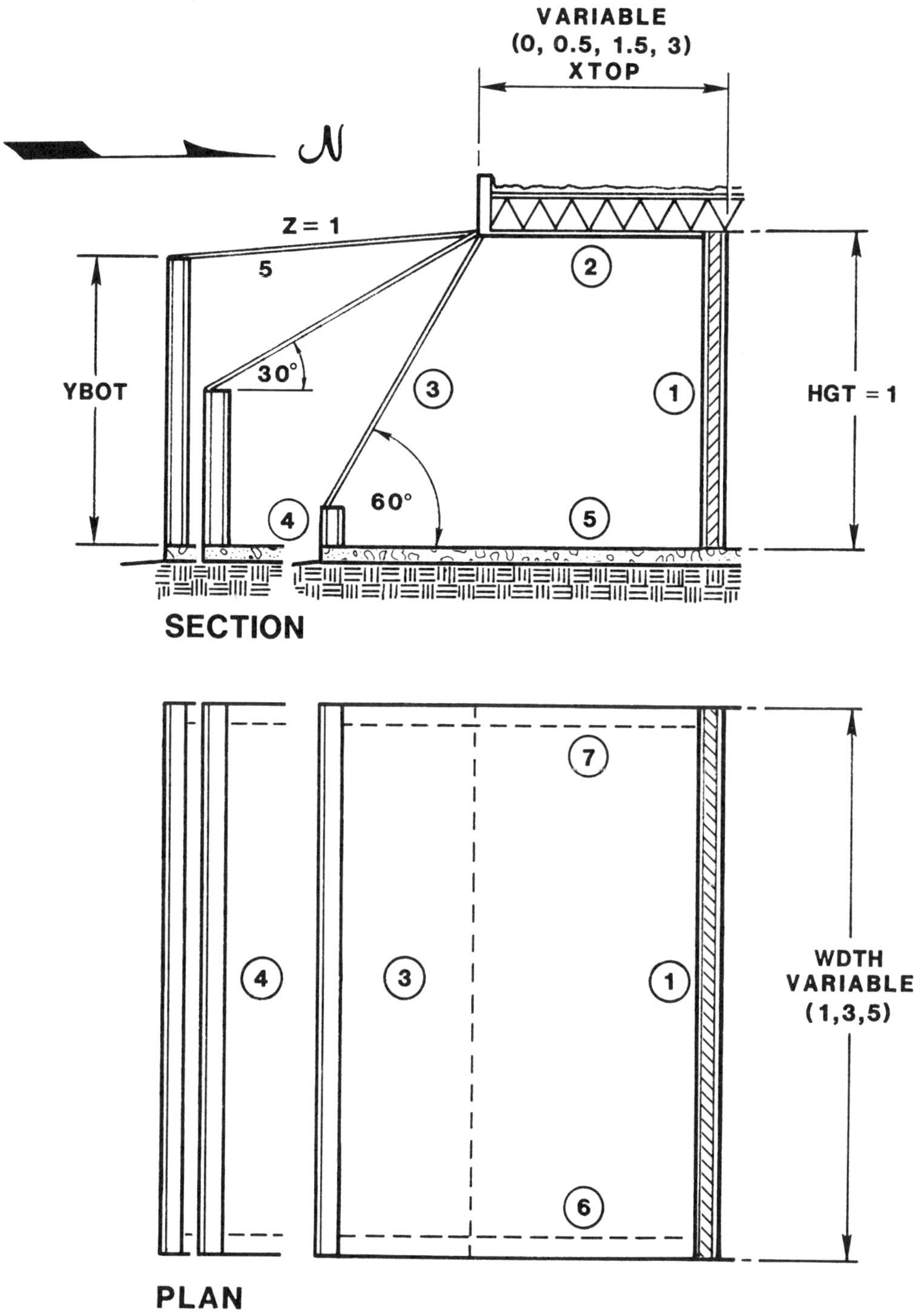

Figure 9-16. Geometries of Tilted Glazing, Class IIIb

and a sunspace are two passive systems. A given passive system may have only one component, or it may have two or more. A component is a single plane of glazing with a particular number of glazing layers and glazing material properties (thickness, extinction coefficient, and index of refraction). A direct-gain system may have just one component, for example, a south-facing, double-glazed window. Two or more similar direct-gain windows with the same orientation may be considered as one component with an area equal to the combined areas of all the windows. Similar direct-gain windows that serve the same room but have different orientations should be treated as different components of a single direct-gain system. For example, a south window and an east window in a southeast corner room are two components of a single direct-gain system. Sunspaces often have two or more components. For example, a 60-degree-tilted, south-facing principal glazing and a vertical, east-facing end-wall glazing are two components of a single sunspace system.

A separate computation must be performed for each component. The glazing system for each component must be uniform, that is, with a single orientation (for incident radiation calculation) with a single value of *N* and *E* (for transmitted radiation calculation), and must serve the same space (for absorption factor calculation). The Solar Gain Worksheet in Appendix 7 is useful for this computation. For each glazing component of a given system, the step-by-step calculation is performed as follows:

1. Determine *TILT, AZ, E, N*, ρ_g, and index of refraction for the component.
2. Calculate the net glazing area, A_n, of the component.
3. Calculate the total projected area, A_p, of the passive system.
4. List the months of the heating season and the number of days, *M*, in each month.
5. Look up in Appendix 3, for each listed month, the values of *HS*, K_T, and *LAT* − *DEC*. List these values.
6. From Appendix 4, Table A4-1, using values of *TILT* and *AZ*, determine values of correlation coefficients for (*INC*/*HOR*), and for each listed month, calculate (*INC*/*HOR*) using monthly values of *LAT* − *DEC* and K_T. For intermediate values of *TILT* and *AZ*, the values of the coefficients can be estimated by linear interpolation. Alternatively, (*INC*/*HOR*) can be calculated for both the bracketing values of *TILT* and *AZ*, and the final values of (*INC*/*HOR*) can be found by linear interpolation between the results.

   OR

   For each month, using *TILT* and *AZ* and the monthly values of *LAT* − *DEC* and K_T, read (*INC*/*HOR*) from the appropriate graph in Appendix 4, Figures A4-1 through A4-21. For intermediate values of *TILT* and *AZ*, (*INC*/*HOR*) may be estimated by linear interpolation.
7. List the monthly values of (*INC*/*HOR*).
8. From Appendix 4, Table A4-2, using values of *TILT* and *AZ*, determine values of correlation coefficients for (*TRN*/*INC*), and for each listed month, calculate (*TRN*/*INC*) using monthly values of *LAT* − *DEC* and K_T. For intermediate values of *TILT* and *AZ*, the values of the coefficients can be estimated by linear interpolation. Alternatively, the values of (*TRN*/*INC*) can be calculated for the bracketing values of *TILT* and *AZ*, and the final value of (*TRN*/*INC*) can be found by linear interpolation between the results.

   OR

   For each month, using *TILT* and *AZ* and the monthly values of *LAT* − *DEC* and K_T, read (*TRN*/*INC*) from the appropriate graph in Appendix 4, Figures A4-22 through A4-43. For intermediate values of *TILT* and *AZ*, (*INC*/*HOR*) may be estimated by linear interpolation.
9. If necessary, correct (*TRN*/*INC*) for nonreference values of ρ_g, *E* and *N*. List the values of (*TRN*/*INC*).

10. For thermal storage walls:
    Determine storage wall surface absorptance and, with values of E and N, look up absorption factor, ABS/TRN, in Appendix 4, Table A4-5.

    For direct-gain and sunspace systems:
    Determine glazing geometry, vertical or tilted, and if tilted, the class number. Determine α_m, α_o, $XTOP$, $WDTH$, and the surface number that corresponds to the glazing component. Enter appropriate table in Appendix 4, Tables A4-6 through A4-45 and look up the absorption factor, (ABS/TRN).

11. If correcting double-glazed thermal storage wall or sunspace reference designs to single or triple glazing, calculate $(TRN/INC)(ABS/TRN)$ from:

$$\left(\frac{TRN}{INC}\right)\left(\frac{ABS}{TRN}\right) = \left(\frac{TRN}{INC}\right)\left(\frac{ABS}{TRN}\right)_0 + g(N)$$

12. For each month calculate S from:

$$S = M \cdot HS \cdot \left(\frac{A_n}{A_p}\right)\left(\frac{INC}{HOR}\right)\left(\frac{TRN}{INC}\right)\left(\frac{ABS}{TRN}\right)$$

13. When the passive system has more than one glazing component, the total S for the passive system for each month is the sum of the Ss computed for each component for that month.

Remarks on the Projected Area

The projected area, A_p, is used in Step 12 of the above computation as a reference area or normalizing area, giving S units of Btu/mo · ft^2 of projected area. Refer to the definition of projected area in Chapter 1.

The projected area is a property of each passive system. For a direct-gain window, Trombe wall, or water wall, the projected area is just the net glazing area whether tilted or not tilted:

$$A_p = A_n .$$

If there are two or more similar direct-gain windows serving the same room, it is recommended that the windows be treated as two *components* of a single direct-gain system. (See the discussion of systems and components in the previous section.) In this case, the system projected area is the sum of the projected areas of the components:

$$A_p = \Sigma A_n$$

where Σ means sum over components. The same recommendation applies to two similar Trombe walls, or two similar water walls that serve the same room. ("Similar" here means that the systems are all characterized by the same reference design.) Dissimilar passive apertures serving the same room are treated as mixed systems. See Chapter 15.

The concept of projected area, as distinct from net glazing area, is really needed only in the case of sunspaces. The projected area is then the net glazing area of the sunspace projected on a single vertical plane that has the same azimuth as the principal azimuth of the sunspace. For a sunspace whose glazings have only a single azimuth angle, the principal azimuth equals the azimuth angle, and

$$A_p = \Sigma A_n \sin TILT .$$

For a sunspace whose glazing components have more than one azimuth angle, the projected area is the sum of the projected areas of the components:

$$A_p = \Sigma A_n \sin TILT \cos (AZ - AZP) .$$

The projected area was chosen as the primary glazing area parameter for sunspaces after a detailed study of sunspace sensitivity curves that showed that the projected area often characterizes the expected performance better than the actual glazing area. Another reason is that it gives a common basis to various passive solar systems that are associated with vertical wall areas. Indeed, in the simpler cases, the projected area is just the vertical wall area occupied by the system, corrected for the area of opaque mullions and other framing. Nevertheless, one may replace the projected area by any other convenient area for the purpose of the *SLR* method. The most likely substitute is the actual net glazing area, A_n. If this substitution is made, it must be done consistently everywhere the projected area appears in *SLR*. Thus, *S*, LCR_s, and *LCR* must all be redefined using A_n rather than A_p. Note, however, that the sunspace sensitivity curves have all been drawn with *LCR* defined in terms of projected area.

The best all-around approach is probably to use projected area for sunspaces and use actual net glazing area for all other apertures regardless of tilt and azimuth.

References

1. V. Cinquemani, J. R. Owenby, and R. G. Baldwin, *Input Data for Solar Systems* (Asheville, NC: National Climatic Center, November 1978).

2. D. W. Phillips and D. Aston, *Canadian Solar Radiation Data* (Downsview, Ontario: Atmospheric Environment Service, 1980).

3. J. A. Duffie and W. A. Beckman, *Solar Engineering of Thermal Processes* (New York: John Wiley & Sons, 1980).

4. I. Hall, R. Prairie, H. Anderson, and E. Boes, *Generation of Typical Meteorological Years for 26 SOLMET Stations,* SAND78-1601 (Albuquerque: Sandia Laboratories, August 1978).

CHAPTER 10
SOLAR GAIN AND HEAT LOSS THROUGH NONSOUTH WINDOWS

Introduction

In this chapter, the thermal performance of nonsouth windows is discussed and ways of estimating the net effect of their solar gains and heat loss are described.

Window Conductances

Direct-gain windows, which form a part of the solar wall of a building, are appropriately analyzed in both the annual *LCR* method and in the monthly *SLR* method by considering both the solar energy absorbed and the thermal energy lost through the glazings. In the annual *LCR* method solar gains through nonsouth windows are not accounted for, and this may cause some error in the results. For the monthly *SLR* method, as discussed in the next chapter, solar gains can be accounted for, but this requires a separate solar gain calculation for each different orientation and glazing system. This extends the computational effort to the point where it may not be practical. An alternative is to use an effective conductance defined[1] as:

$$U_e = U - Q \cdot F_u / 24\, DD \qquad (10\text{-}1)$$

where

- U = heat loss coefficient of the window ($Btu/h \cdot F \cdot ft^2$)
- Q = amount of energy absorbed by the building through the nonsouth glazing (Btu/ft^2)
- F_u = the fraction of Q used in meeting the building net space heating load ($NLC \cdot DD$)

The fraction of solar energy utilized, F_u, will depend on the portion of space heating provided by the south-wall collector (that is, *SSF*). Increasing *SSF* will result in a lower F_u because of a greater amount of heat being rejected due to a higher frequency of overheating. The basic assumption underlying the method is that the east and west glazings have the same utilization fraction as does a direct-gain south-wall system with an effective thermal storage mass per unit area of glazing of 60 $Btu/ft^2\,°F$. As a check on the sensitivity of annual F_u to south-wall system type, F_u was determined for Dodge City, Kansas, using nine different direct-gain and thermal storage wall systems. The maximum difference in F_u between system types is 0.05 for annual *SSF*s ranging from 0.2 to 0.9.

The average annual effective conductances for a number of different cities as a function of the building *SSF* are shown in Figure 10-1 for the case of a double-glazed east or west window and in Figure 10-2 for a triple-glazed east or west window. The standard heat loss coefficients, neglecting solar gain, shown in the figures are based on ordinary glass with a 1/2-inch air space between the glazings and pertain to the glazings only. Standard ASHRAE correction factors could be applied to account for frame type.

It is apparent from the graphs that for more northerly latitudes, at high *SSF* (0.8), the solar gains of the east or west glazings are small and one could use the standard conductances. In the more southerly latitudes, useful solar gains significantly reduce the effective annual conductances over nearly the complete range of *SSF*; the losses are often completely offset, as indicated by the negative values for U_e. Unfortunately, most of these cities could also experience severe cooling problems from the solar gains through the east-west glazings.

Figures 10-3 and 10-4 show the effective conductances for double-and triple-glazed north-facing windows, respectively. At high *SSF*, solar gains through north-facing windows are negligible, but over most of the range of *SSF*, the effective conductances, U_e, are significantly lower than U, the nominal heat loss coefficient.

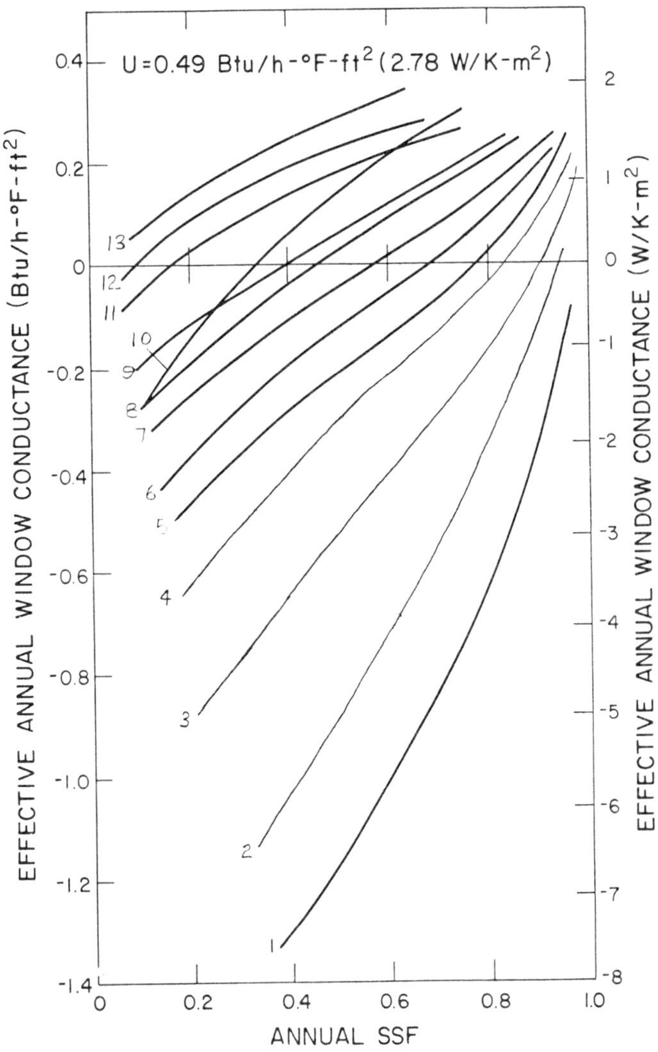

Figure 10-1. Average Annual Conductances as a Function of Building *SSF* for East- and West-Facing Double-Glazed Windows; 1-Santa Maria, CA, 2-San Francisco, CA, 3-Houston, TX, 4-Charleston, SC, 5-Albuquerque, NM, 6-Atlanta, GA, 7-Dodge City, KA, 8-Greensboro, NC, 9-St. Louis, MO, 10-Seattle, WA, 11-Boston, MA, 12-Madison, WI, 13-Edmonton, Canada

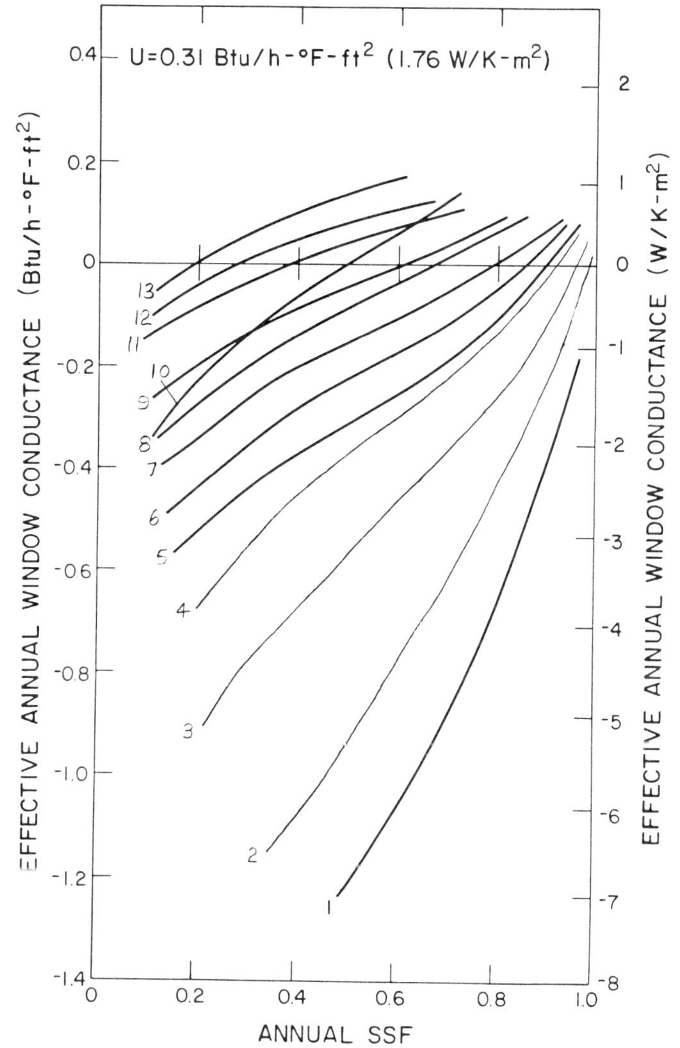

Figure 10-2. Average Annual Conductances as a Function of Building *SSF* for East- and West-Facing Triple-Glazed Windows; 1-Santa Maria, CA, 2-San Francisco, CA, 3-Houston, TX, 4-Charleston, SC, 5-Albuquerque, NM, 6-Atlanta, GA, 7-Dodge City, KA, 8-Greensboro, NC, 9-St. Louis, MO, 10-Seattle, WA, 11-Boston, MA, 12-Madison, WI, 13-Edmonton, Canada

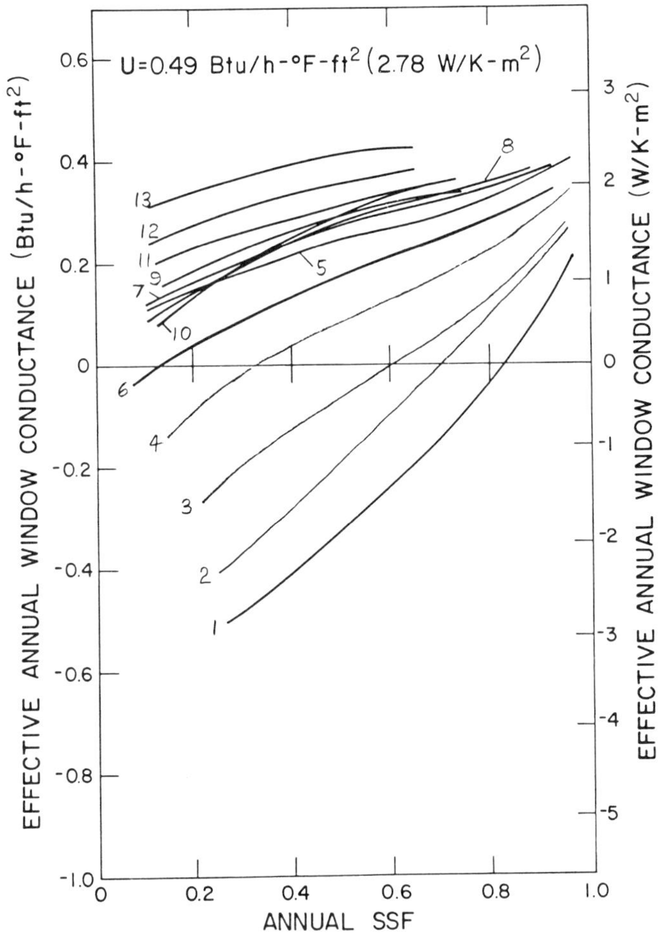

Figure 10-3. Average Annual Conductances as a Function of Building *SSF* for North-Facing Double-Glazed Windows; 1-Santa Maria, CA, 2-San Francisco, CA, 3-Houston, TX, 4-Charleston, SC, 5-Albuquerque, NM, 6-Atlanta, GA, 7-Dodge City, KA, 8-Greensboro, NC, 9-St. Louis, MO, 10-Seattle, WA, 11-Boston, MA, 12-Madison, WI, 13-Edmonton, Canada

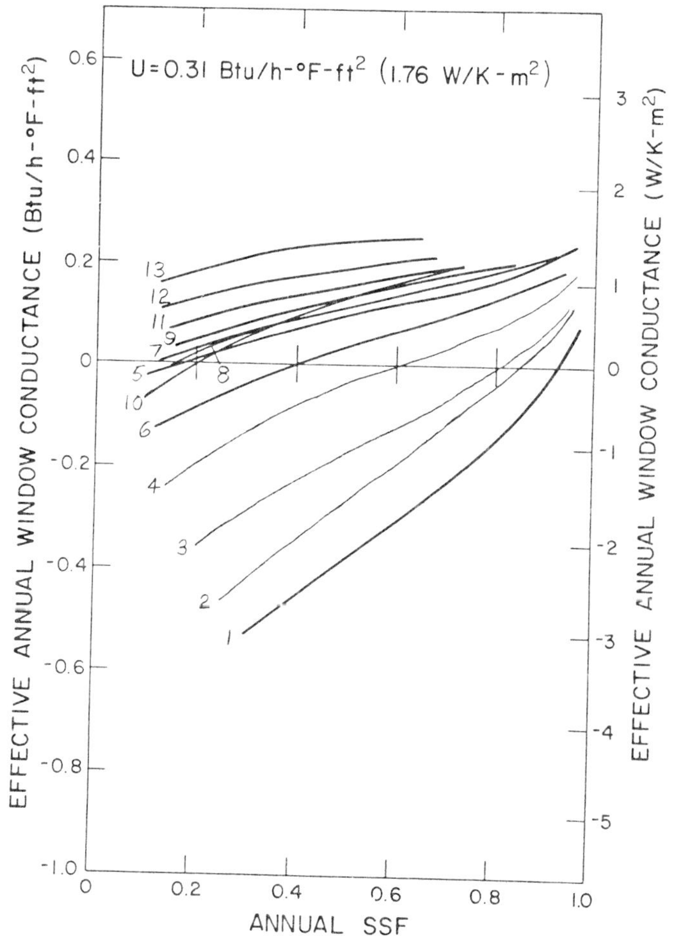

Figure 10-4. Average Annual Conductances as a Function of Building *SSF* for North-Facing Triple-Glazed Windows; 1-Santa Maria, CA, 2-San Francisco, CA, 3-Houston, TX, 4-Charleston, SC, 5-Albuquerque, NM, 6-Atlanta, GA, 7-Dodge City, KA, 8-Greensboro, NC, 9-St. Louis, MO, 10-Seattle, WA, 11-Boston, MA, 12-Madison, WI, 13-Edmonton, Canada

Recommendations — These data support the following recommendations:

1. In schematic design, double- or triple-glazed nonsouth windows may be ignored because their solar gains tend to compensate for their losses. This simplification is particularly appropriate if precise nonsouth window areas have not yet been determined. If large north window areas are anticipated, north window losses should be accounted for using one-half the standard ASHRAE heat loss coefficients.
2. In design development, east, west, and north windows should be accounted for preferably by excluding their losses from the *NLC* calculation and treating them as separate direct-gain apertures. A less laborious alternative is to use the curves in Figures 10-1 through 10-4 to account for these windows in an annual average sense through their effective conductances.

Reference

1. G. J. Schoenau, "Thermal Performance of Nonsouth Windows in Passive Solar Buildings," *Solar Energy,* Vol. 32, in press.

CHAPTER 11
SOLAR WALL LOAD COEFFICIENTS

NLC_s and LCR_s

For sunspace and thermal storage wall designs, the SLR correlations involve the quantity LCR_s. For sunspace systems, LCR_s is the load collector ratio for the sunspace itself. For thermal storage wall systems, LCR_s is the load collector ratio for the small space between the thermal storage wall and its glazing. The computation is similar to that for LCR:

$$LCR_s = NLC_s / A_{ps} \qquad (11\text{-}1)$$

where

NLC_s = load coefficient for the space

and

A_{ps} = projected area of the glazing of the space.

The calculation of NLC_s is similar to that of NLC for a building but with slightly different formulas. The main difference is that the glazing load coefficient, $(UA)_g$, is expressed in a more precise form that accounts for the thickness, t, of the air gap between glazings as well as the number, N, of glazings. This precision is desirable because the glazing load is the dominant contribution to the total load.

NLC_s and LCR_s for Sunspace Systems — Values of LCR_s for sunspace reference designs are listed in Appendix 5, Table A5-8.

For off-reference designs, the computations of the sunspace heat loss elements is analogous to those used in the computation of NLC and LCR.

Sunspace Heat Loss Element	Formula	
Glazing (without night insulation)	$(UA)_g = \dfrac{1.083\,A_n}{N + 1.517\,t - 0.4295}$	(11-2)
	where t is given in inches. If $N > 1$ then the formula is valid for t in the range 0 to 0.5 inches. If $t > 0.5$ inches set $t = 0.5$. For $N = 1$, use $t = 0.35$ inches.	
Glazing (with R9 night insulation)	$(UA)_g = 1.083\,A_n \times \left[\dfrac{1}{3} \times \dfrac{1}{N + 1.517\,t - 0.4295} + \dfrac{2}{3} \times \dfrac{1}{N + 1.517\,t + 9.318}\right]$	(11-3)
	where the factors 1/3 and 2/3 indicate the night insulation is in place for 8 hours out of 24 and removed for 16 hours out of 24. These proportions may be changed.	
Perimeter	$(UA)_p = \dfrac{4.17 P_p}{R_p + 5}$	(11-4)
Roof	$(UA)_r = A_r / R_r$	(11-5)
End Walls	$(UA)_w = A_w / R_w$	(11-6)
Infiltration	$(UA)_i = 0.018 \times V \times ACH$	(11-7)

The load coefficient for the sunspace, NLC_s, is computed from:

$$NLC_s = 24\,[(UA)_g + (UA)_p + (UA)_r + (UA)_w + (UA)_i] \qquad (11\text{-}8)$$

Note that all the UA terms have units of Btu/°F · h, while NLC_s has units of Btu/°F · day. Then use Equation 11-1 to find LCR_s.

For each sunspace reference design, the value of LCR_s has been computed and is listed in Appendix 5, Table A5-8. The method explained above for computing LCR_s is simplified relative to the one used to compute the values listed in Appendix 5. Exact agreement may not be obtained, but the difference will be less than 5%, which is sufficiently accurate for the purposes of this manual.

If the sunspace design being analyzed is the same as one of the reference designs except for the infiltration rate, ACH, then the following formula can be used to compute LCR_s:

$$LCR_s = \frac{B_0 + B_1 R_{NI}}{B_2 + R_{NI}} + B_3 \cdot ACH \qquad (11\text{-}9)$$

The values for the coefficients are listed in Table 11-1. Note that the R-value of night insulation, R_{NI}, is also a variable in this equation, but only the value of 0 or 9 should be used. Intermediate R-values are treated by interpolation between the R0 and R9 cases. See Chapter 8 for the details.

Table 11-1

Coefficients for Sunspace LCR_s Formula

Reference Design		Coefficients			
Geometry	End Walls	B_0	B_1	B_2	B_3
A	insulated	34.9	6.80	2.06	3.4
A	glazed	46.1	8.06	2.10	3.4
B	insulated	33.7	5.63	2.20	1.9
B	glazed	39.9	6.43	2.17	1.9
C	—	28.3	5.92	2.07	5.2
D	—	30.4	6.20	1.93	4.1
E	—	35.0	6.84	1.99	4.1

Some precomputed values of LCR_s for sunspace systems, computed by use of Equation 11-9, are listed in Table 11-2. Note that the values of LCR_s for ACH = 0.5 do not agree precisely with those listed for the reference designs in Table A5-8, as already explained.

Table 11-2

Precomputed Values of LCR_s for Sunspace Systems

Reference Design		*LCR*	
No Night Insulation — ACH =	0.5	1.0	2.0
SSA1, SSA5	18.6	20.3	23.7
SSA3, SSA7	23.7	25.4	28.8
SSB1, SSB5	16.5	17.2	19.1
SSB3, SSB7	19.3	20.3	22.2
SSC1, SSC3	16.3	18.9	24.1
SSD1, SSD3	17.8	19.9	24.0
SSE1, SSE3	19.6	21.7	25.8
R9 Night Insulation			
SSA2, SSA6	10.4	21.1	15.5
SSA4, SSA8	12.4	14.1	17.5
SSB2, SSB6	8.5	9.4	11.3
SSB4, SSB8	9.7	10.7	12.6
SSC2, SSC4	10.0	12.6	17.8
SSD2, SSD4	9.9	12.0	16.0
SSE2, SSE4	10.8	12.9	17.0

NLC_s and LCR_s for Thermal Storage Wall Systems — The reference design values for LCR_s are listed here in Table 11-3 for convenience. They are also listed in Appendix 5, Table A5-7.

These reference values account only for glazing losses. Off-reference glazing configurations may be treated by calculating LCR_s using the method described above for sunspaces. Additional small losses through perimeter edges, mullions, and air leakage may be accounted for. Equations 11-2 through 11-7 may be used for guidance. However LCR_s should not be used to account for major changes such as night insulation or number of glazings.

Table 11-3

LCR_s Values for Thermal Storage Wall Reference Designs

Reference Design	LCR_s
WWA1 through WWA6, TWA1 through TWC4, and TWF1 through TWH4	13.0
WWB1, TWD1, TWI1	22.0
WWB2, TWD2, TWI2	9.2
WWB3, TWD3, TWI3	8.9
WWB4, TWD4, TWI4	5.8
WWB5, TWD5, TWI5	4.5
WWC1, TWE1, TWJ1	12.0
WWC2, TWE2, TWJ2	8.7
WWC3, TWE3, TWJ3	5.5
WWC4, TWE4, TWJ4	4.3

Computation of *TLC*

Since *NLC* accounts for all heat losses from a building except those for the solar wall, *TLC* can be obtained by adding the load coefficient for the solar wall, $(UA)_{sw}$, to *NLC*:

$$TLC = NLC + 24\,(UA)_{sw} \qquad (11\text{-}10)$$

For systems that conform to the reference design specifications, $(UA)_{sw}$ is calculated from:

$$(UA)_{sw} = U_c \cdot A_p \qquad (11\text{-}11)$$

where U_c is the solar wall steady state heat transfer coefficient. Values of U_c for the reference designs are tabulated with the *SLR* correlation coefficients in Appendix 5, Tables A5-7, 8, and 9.

Computation of U_c

For systems that vary from the reference design specifications, U_c may be calculated, depending on the type of passive solar system, as described in the following paragraphs.

Direct-Gain Systems — For direct-gain systems, the solar wall consists of windows. The load coefficient of a window can be computed from:

$$U_c = (UA)_g/A_p \qquad (11\text{-}12)$$

where $(UA)_g$ is given by Eq. 11-2.

Thermal Storage Wall Systems — For thermal wall systems, heat from the building must pass first through the thermal wall itself and then through the glazing before it reaches the outside air. The solar wall load coefficient represents the series combination of the load coefficient for the thermal storage wall, $(UA)_{tw}$, and NLC_s.

For the thermal storage wall, the load coefficient is computed from:

$$(UA)_{tw} = A_{tw}/R_{tw} \qquad (11\text{-}13)$$

where

A_{tw} = the gross area of the Trombe wall

R_{tw} = the air-to-air R-value of the Trombe wall

The load coefficient for the solar wall is then computed from:

$$U_c = \frac{1}{24\,A_p}\left[\frac{(UA)_{tw} \cdot NLC_s}{(UA)_{tw} + NLC_s/24}\right] \qquad (11\text{-}14)$$

Sunspace Systems — For sunspace systems, heat from the building must pass first through the common wall into the sunspace and then by multiple parallel paths (i.e., glazing, end walls, perimeter, roof, and infiltration) to the outside air. The notation for the load coefficient for the common wall is $(UA)_{cw}$. It is computed from:

$$(UA)_{cw} = A_{cw}/R_{cw} \qquad (11\text{-}15)$$

where

A_{cw} = area of the common wall

R_{cw} = air-to-air R-value of the common wall

The series combination of $(UA)_{cw}$ and NLC_s yields U_c as follows:

$$U_c = \frac{1}{24\,A_p}\left[\frac{(UA)_{cw} \cdot NLC_s}{(UA)_{cw} + NLC_s/24}\right] \qquad (11\text{-}16)$$

CHAPTER 12
DIRECT-GAIN SYSTEMS

Introduction

In a direct-gain building, solar radiation enters the living space directly through windows, as illustrated schematically in Figure 12-1. These solar gains serve either to meet part of the current heating needs of the structure or are stored in the building mass to meet heating needs that arise later. Space heating needs not satisfied by the solar gains are met by conventional backup heating systems. Sources of internal heat displace part of the thermal load, but for many residential applications, these sources represent only a small part of the total heating requirement.

Direct-gain buildings have relatively high market appeal because they involve less departure from conventional construction than other types of passive solar heating systems. Their actual effectiveness, however, is limited to mild and moderate winter climates unless triple glazing or night insulation are employed to limit heat losses through the windows. Improperly designed direct-gain systems may overheat during the day, particularly if insufficient effective thermal storage mass has been provided. In fact, failure to provide sufficient effective thermal storage mass within the building shell is a common error that often leads to uncomfortable temperature swings and a design that yields

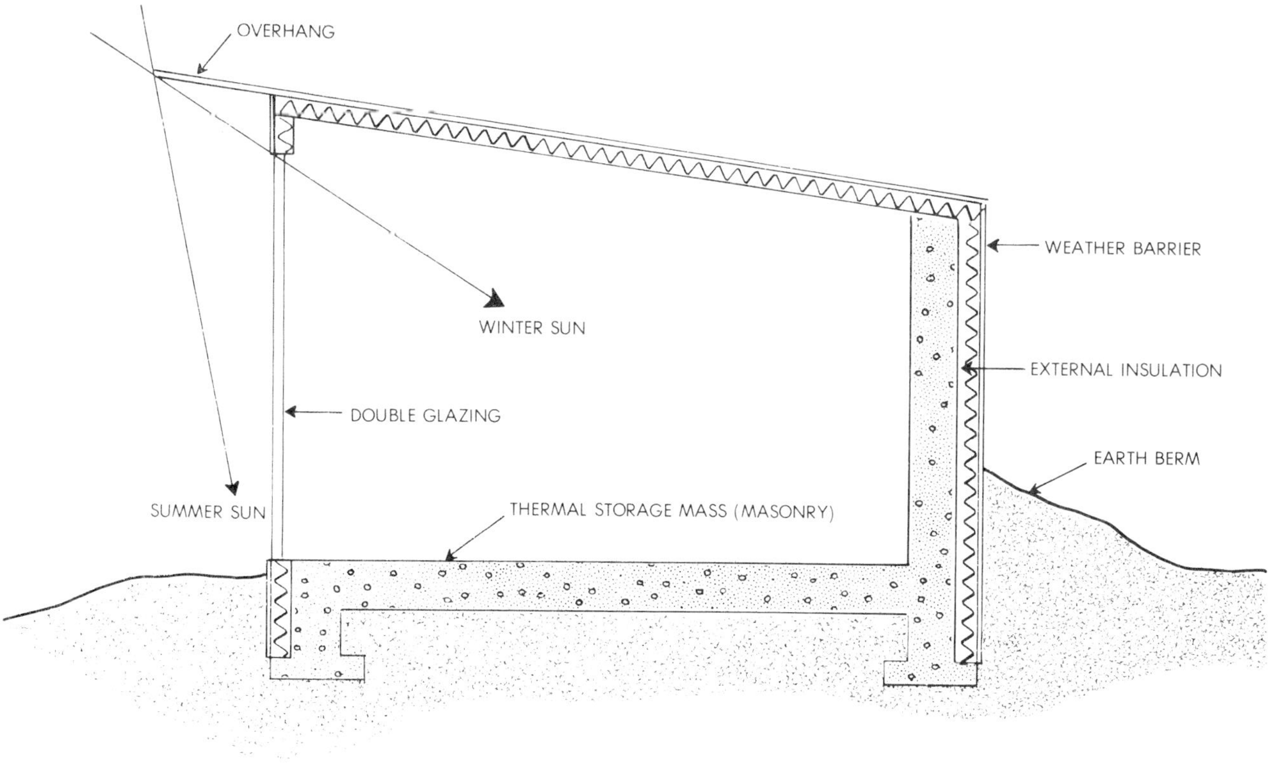

Figure 12-1. Direct-Gain Building

little energy savings. Direct-gain buildings can be susceptible to glare and fabric degradation caused by direct solar radiation within the living space. The best procedure for minimizing these problems is to distribute the transmitted solar radiation as uniformly as possible through appropriate window placement, the use of diffusive blinds or glazings, and the proper use of color in the space.

On the positive side, direct-gain buildings are attractive, provide daylight within the building, and take excellent advantage of views to the south. The incremental cost relative to an otherwise comparable nonsolar building is small, especially if the structure is inherently massive, as are the adobe architecture in the southwest and block wall on slab-on-grade buildings in the southeast and many commercial buildings of concrete construction. When designed appropriately for their locations, direct-gain buildings can provide a cost-effective means of reducing energy consumption for space heating without requiring any sacrifice of comfort or aesthetic values.

Design Elements

The basic design elements of a direct-gain system are the solar aperture and the thermal storage mass.

Solar Aperture — Windows provide the solar aperture for direct-gain systems. Usually, two glazing layers are mounted in direct-gain apertures. A single glazing layer is often undesirable because of the large heat losses. In very cold climates, triple (or even quadruple) glazing is advisable as a means of controlling heat losses. Another alternative is glazing systems with infrared reflective coatings, however the user should be aware that reduced transmittance of these coatings will tend to offset the benefit of reduced heat losses. Another method of controlling losses is the use of movable opaque insulation that covers the windows at night and is removed during the daylight hours.

An overhang can be used to shade the solar aperture from the high summer sun while permitting rays from the low winter sun to penetrate and warm the living space.

Thermal Storage Mass — Thermal storage mass plays a vital role in the effective performance of direct-gain buildings. Buildings that contain very little thermal mass are unable to store much heat for nighttime use; thus, only the daytime portion of the heat load can be met by solar gains. Overheating can be a serious problem if the solar gains are excessive. In a building whose effectiveness is impaired by insufficient thermal mass, heat losses can exceed the useful solar gain. Lightly constructed stud-wall buildings with no floor slab (or a carpeted floor slab) are the worst offenders in this category.

The effectiveness of thermal storage mass in direct-gain buildings depends on its thickness, surface area, and thermal properties (volumetric heat capacity and thermal conductivity). The best materials are those that can store large quantities of heat (high volumetric heat capacity) and then can readily transport heat (high thermal conductivity) from the mass surface to the mass interior for storage and back again to the surface to heat the space. For optimal effectiveness, thermal storage mass should be located in building zones that experience direct solar gains. The question of appropriate colors in direct-gain buildings is complex and is addressed in more detail in the section on sensitivity data. Mass located outside the zone where the direct gain enters the building will be ineffective unless care is taken to ensure free convective exchanges with directly heated zones or unless a forced-air distribution system is employed. In either case, remote thermal storage mass will be less effective than that located in direct-gain zones.

Thermal insulation forms the outer shell of a direct-gain building, enclosing the thermal storage mass, so that heat is freely transferred between the living space and the thermal mass. Floor slabs contribute to the thermal storage capacity of a structure, provided they are not isolated by carpets and cushioning pads. It is advisable to limit losses from floor slabs by insulating the perimeter of the foundation walls or underneath the floor slab, although there is ongoing debate regarding the cost effectiveness of underslab insulation. An earth berm is frequently placed on the north side of passive solar buildings to protect against cold winter winds and provide extra thermal inertia.

The Direct-Gain Reference Designs

For purposes of analysis, direct-gain buildings are divided into two broad categories; those with large amounts of thermal storage mass are called *high-mass, direct gain* buildings, and those having little mass are called *low-mass, suntempered* buildings.

All direct-gain systems are represented in this manual by eleven direct-gain reference designs that offer combinations of the values of three design parameters: (1) thermal storage capacity, (2) number of layers of glazing, and (3) whether or not night insulation is used. The eleven combinations of these parameters are listed in Table 12-1. Note that the thermal storage capacity can be expressed in terms of mass thickness and mass-area-to-glazing-area ratio. The night insulation option is included only for double glazing. Table 12-2 lists additional characteristics that apply to the direct-gain reference designs. These reference designs, in combination with the sensitivity data presented below, cover most direct-gain configurations of practical importance. The high-mass and suntempered reference designs are discussed separately.

High-Mass, Direct-Gain Reference Designs

The high-mass, direct-gain reference designs listed in Table 12-1 should be selected to meet the needs of the site and application. For example, 45 Btu/°F · ft^2 of projected area may be more thermal storage than necessary in buildings in which the solar gains are able to meet no more than daytime heating needs (i.e., low *SSF*). On the other hand, 45 Btu/°F · ft^2 may be insufficient to prevent occasional overheating in cases where the solar aperture is sized to provide a large *SSF*.

Table 12-1

Direct-Gain Reference Designs

Designation	Thermal Storage Capacity* (Btu/ft^2 · °F)	Mass Thickness** (in.)	Mass-to-Glazing-Area Ratio	No. of Glazing Layers	Night Insulation
High-Mass					
DGA1	30	2	6	2	no
DGA2	30	2	6	3	no
DGA3	30	2	6	2	yes
DGB1	45	6	3	2	no
DGB2	45	6	3	3	no
DGB3	45	6	3	2	yes
DGC1	60	4	6	2	no
DGC2	60	4	6	3	no
DGC3	60	4	6	2	yes
Suntempered					
SUNTEMP 1	10.8	0.5	20	2	no
SUNTEMP 2	10.8	0.5	20	2	yes

*Per unit of projected area
**Based on volumetric heat capacity of 30 Btu/ft^3 · °F

Table 12-2

Properties of Direct-Gain Reference Designs

	Suntempered Designs	High-Mass Designs		
Thermal Storage				
Heat capacity (Btu/°F · ft² of projected area)	10.8	30	45	60
Thickness (inches)	0.5	2	6	4
Mass-to-glazing area ratio	20.0	6	3	6
Thermal conductivity, k (Btu/h · ft · °F)	0.0923		1	
Density, ρ(lb/ft^3)	50		150	
Specific heat, c (Btu/lb · °F)	0.26		0.2	
Solar absorptance of surface	0.3		0.8*	
Infrared emittance of surface	0.9		0.9	
Number of Glazings	2		2 or 3	

	All Direct-Gain Designs
Glazing Properties	
Transmission characteristics	diffuse
Infrared emittance of surface	0.9
Index of refraction	1.526
Extinction coefficient (per inch)	0.5*
Thickness of each pane (inches)	0.125*
Air gap between layers (inches)	0.5*
Night insulation when used	R9**
Orientation	due south*
Shading	none*
Control Range of Room Air (°F)	65 to 75**
Lightweight Absorption Fraction Simulates effect of solar radiation absorption on lightweight walls or objects by transferring given fraction of transmitted and reflected solar radiation directly to room air	0.2
Ground Reflectance	0.3*
Internal Heat Generation	0**

*Indicates properties that can be varied in the monthly *SLR* method.
**Indicates properties that can be varied in both the annual *LCR* method and the monthly *SLR* method.

The design of thermal storage mass in a direct-gain building is a much more complex issue than simply the selection of a particular overall heat capacity. However, a mass-area-to-glazing-area ratio of at least 6 is recommended in any direct-gain solar building that is expected to obtain all of its heat from solar during clear winter weather.

The quantity *LCR* is the principal building variable for a given reference design. Since *LCR* is the ratio of *NLC* to A_p, it can be changed by changing the value of *NLC* (e.g., by increasing wall insulation) or A_p (e.g., by increasing window size) or both.

Many variations of the reference designs can be evaluated by use of the sensitivity studies described below. Sensitivity data may be applied in the annual *LCR* method of analysis or in the monthly *SLR* method of analysis.

High-Mass, Direct-Gain Sensitivity Data

Data presented in this section express the sensitivity of the annual *SSF* to selected design parameters. In any given set of data, one or two parameters are varied while the others are held fixed at their reference values. The reference values are indicated by dots on the curves.

Only sample sensitivity curves are presented in this chapter. The sample plots illustrate the effect of various parameter variations on performance in Albuquerque, which has a moderate winter climate. In several instances, plots for other cities are also included. The complete set of sensitivity curves is contained in Appendix 2, which includes results for six reference cities: Albuquerque, New Mexico; Boston, Massachusetts; Madison, Wisconsin; Medford, Oregon; Nashville, Tennessee; and Santa Maria, California.

The above cities were chosen to represent a wide geographical and climatological range. To apply the data, one of the reference cities must be chosen to represent the city of interest on the basis of climate similarity. Two gross measures are of use in making the selection: the 65°F-base heating degree-days and the solar radiation incident on a south-facing vertical surface. These data are available as monthly averages for numerous United States and Canadian locations in Appendix 3 where they are tabulated as D65 and VS, respectively. Figure 12-2 shows the January values of D65 and VS for each of the six reference cities.

Thermal Storage Mass — Figures 12-3 and 12-4 illustrate the sensitivity of annual heating performance to thermal storage mass thickness for a double-glazed configuration with and without night insulation. The *LCR* was held constant at 12. An additional parameter that has an important effect on the mass sensitivity is the ratio of the mass surface area to the net glazing area, A_m/A_n. Four mass-area-to-glazing-area ratios (A_m/A_n = 2, 3, 6, 10) are included on each graph, and the three reference design conditions are indicated by solid points.

It is assumed in the reference designs that the thermal storage mass is uniformly illuminated by diffusely transmitted solar radiation. In practice, the mass can be considered to be sufficiently well coupled to the solar gains if it is in the same zone as the illuminated mass, so that the various mass surfaces can exchange long-wave infrared and reflected short-wave radiation with each other. See the Mass Distribution section for details of the effect of the distribution of mass and solar radiation. Mass that is only convectively coupled to the direct-gain zone is only partially effective. A conservative analysis practice would be to disregard mass that is remote from the direct-gain zone unless special measures, such as large openings, full ceiling-height openings, or forced-air distribution, enhance the coupling.

Upon inspection of the complete set of mass sensitivity data, the following patterns emerge for high-density (130-150 lb/ft^3) masonry with a fixed A_m/A_n ratio.

- Performance sensitivity to mass thickness above 4 inches is small. The thickness may be reduced to 4 inches without incurring significant performance penalties. This generalization is independent of location, configuration, and mass surface area.

- The range of mass thicknesses between 2 and 4 inches can be considered a transition region.

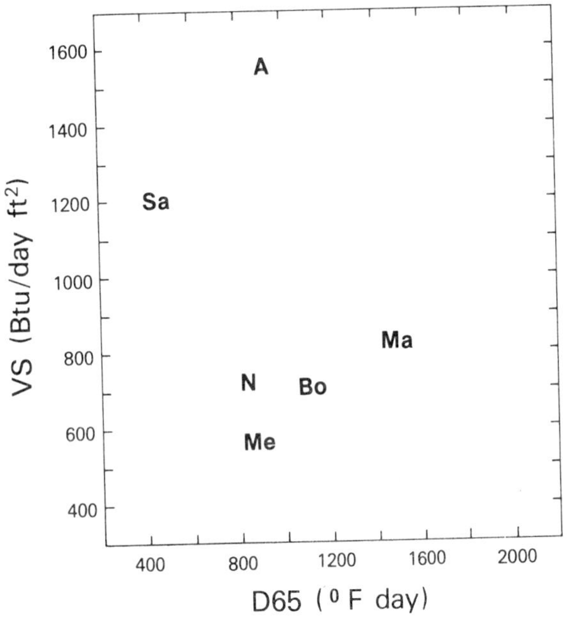

Figure 12-2. Reference Cities' Climate in Terms of the January 65°F-Base Heating Degree-Days, D65, and the Daily Average January Solar Radiation Incident on a Vertical Surface, VS. A = Albuquerque, Bo = Boston, Ma = Madison, Me = Medford, N = Nashville, and Sa = Santa Maria.

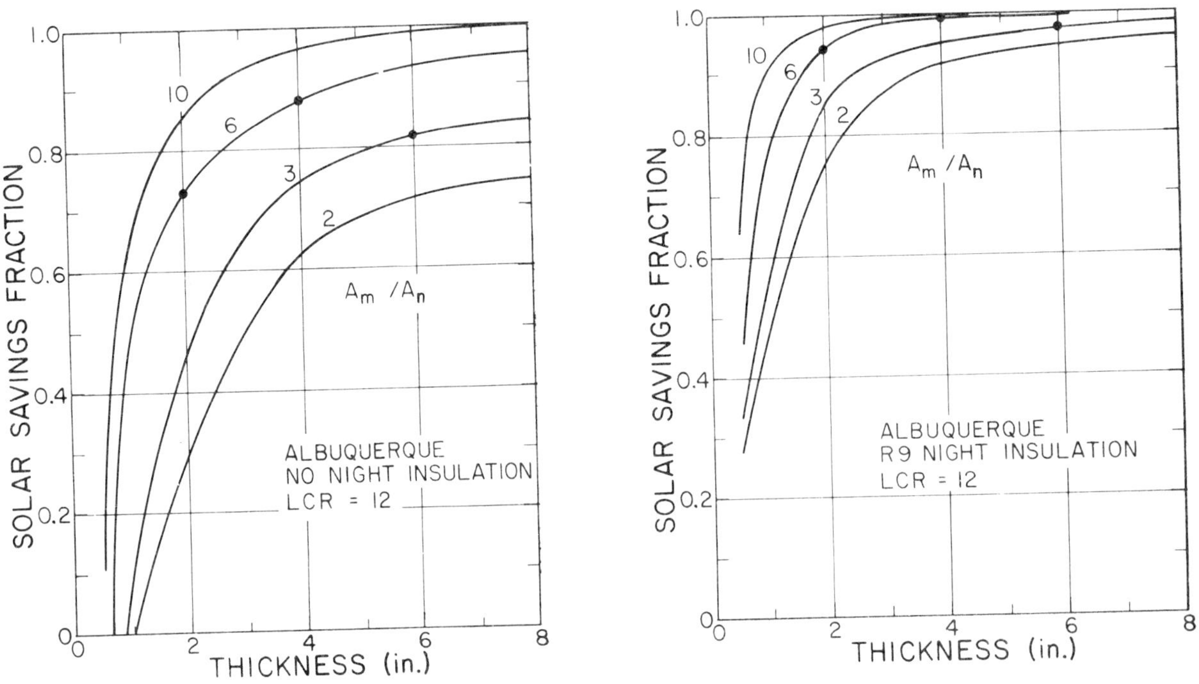

Figures 12-3 and 12-4. Annual Heating Performance Sensitivity to the Mass Thickess for a Double-Glazed Configuration for Various Values of the Mass-Area-to-Glazing-Area Ratio, A_m/A_n. The dots represent the reference conditions.

In this region, performance penalties for reduced thicknesses are becoming significant, but in some cases may be considered acceptable as design/cost trade-offs.

- For mass thicknesses below 2 inches, performance falls off much more rapidly than in the transition region. If mass thickness is less than 2 inches, then a larger surface area should be used.

- The above rules apply to high-density masonry, roughly 130 to 150 lb/ft^3. Lower density masonry has a lower thermal conductivity and therefore has a smaller effective thickness for diurnal heat storage. The same heat storage capacity must therefore be achieved with material spread over a larger area.

These same patterns are observed whether or not the system is night-insulated.

It would be a mistake to assume that all points on the sensitivity curves represent designs that can be achieved in practice. In particular, the high A_m/A_n ratios may not be obtainable for large glazing areas that cause the *LCR* to be small: there may simply not be enough surface available on which to place the mass. This problem is aggravated by the fact that only mass that is located in building zones that receive direct solar gains can be credited to the system unless special measures have been taken to enhance the convective coupling with the illuminated zones. Convective coupling can be enhanced either by a forced-air distribution system or by properly sizing the doorways and other apertures between adjacent zones.[1-3] Clerestory windows are often used to provide direct solar gains to the northern zones in a building, thereby reducing or eliminating the need for strong convective coupling to southern zones.

Properties of the Thermal Storage Medium — The effect of changes in specific heat, *c*, density, ρ, thermal conductivity, *k*, and the storage mass thickness, *t*, can be represented by two groupings of these properties. The first grouping, *ρct*, gives the thermal storage capacity of the mass layer per unit of surface area. The effect of changes in *t* was discussed in the previous section. That entire discussion can be converted to an analysis of performance sensitivity due to changes in the grouping *ρct* simply by multiplying all values of *t* by 30 Btu/ft^3 · °F, the reference value of *ρc*.

The effect of varying the magnitude of the second grouping, *ρck*, is illustrated by Figure 12-5 for Albuquerque. Three values of *LCR*, 12, 48, and 120, are included for configurations with and without night insulation. In order to facilitate interpretation of these plots, the *ρck* products for various building materials are given in Appendix 6, Table A6-4. Performance is insensitive to variations in *ρck* between 20 and 40 and only moderately sensitive between 10 and 20. However, if the value of *ρck* drops much below 10, performance begins to erode. The performance penalty is greatest for small values of *LCR* (high *SSF*) because the need for effective heat storage is most critical.

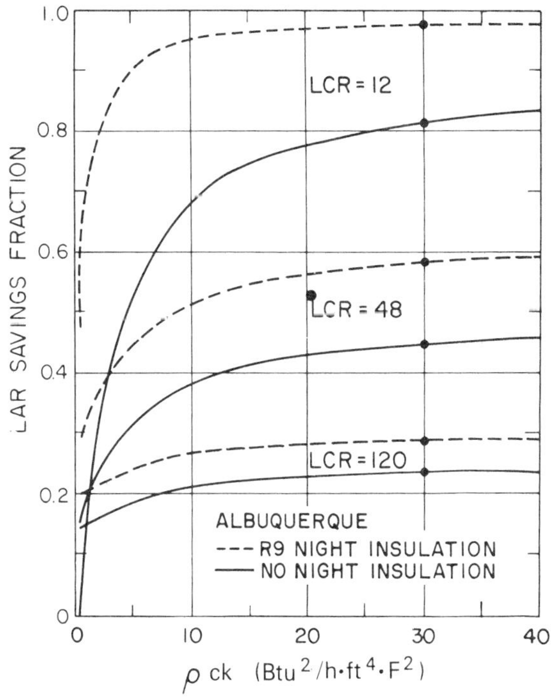

Figure 12-5. Annual Heating Performance Sensitivity to the Product of the Density, ρ(lb/ft^3), the Specific Heat, *c*(Btu/lb · °F), and the Thermal Conductivity, *k*(Btu/h ft · °F) for the Double-Glazed Configuration with a Mass Thickness of 6 inches and a Mass-Area-to-Glazing-Area Ratio of 3. The dots represent the reference conditions.

This observation does not necessarily imply that materials with ρck products less than 10 should not be used for thermal storage mass in direct-gain buildings. Adobe, for example, may have a ρck product as small as 6 but has nevertheless been successfully used as the thermal storage medium in many passive solar homes in the southwest. However, the mass-area-to-glazing-area ratios in these buildings may be larger than the reference design values. The thick adobe walls are traditional and appear frequently in the southwest, irrespective of the application of other passive solar design concepts. In a sense, therefore, the abundant thermal storage mass in adobe direct-gain buildings is free, as is also the case for any building that is inherently massive for structural reasons, provided the structural mass is not thermally isolated from the living space.

The designer is advised to adjust the amount of thermal storage mass included in direct-gain buildings on the basis of the material properties. As a general rule, materials having relatively small values of ρck will necessitate the use of larger mass surface areas.

Solar Absorptance of Mass Surfaces — The effect on performance of the average solar absorptance, α_m, of the surface of the thermal storage mass is illustrated in Figure 12-6 for Albuquerque. If the exposed mass surface is limited (i.e., if the mass-area-to-glazing-area ratio is 3 or less), it seems unwise to employ a color scheme that yields an average mass surface absorptance of less than 0.5 for direct-gain buildings. Below that value, the performance of buildings with small *LCR* and no night insulation begins to fall rapidly. This constraint is not at all severe, as indicated by the solar absorptances listed in Appendix 6, Table A6-2. A medium paint in any color and most unpainted masonry materials will exceed the suggested average absorptance of 0.5.

The results discussed above are for a direct-gain building with a mass-area-to-glazing-area ratio of 3. As the mass surface area increases relative to the glazing area, it becomes less important to employ dark colors on the mass surfaces. Indeed, for large values of A_m/A_n (about 6 or greater), it may actually be desirable to employ light colors on the mass surfaces (except floors) as a means of distributing solar radiation more uniformly about the living space. See the Mass Distribution section for more details on color in direct-gain zones.

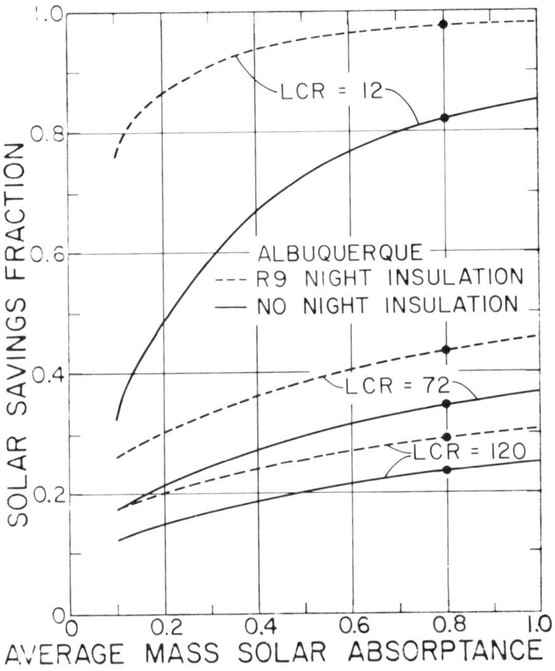

Figure 12-6. Annual Heating Performance Sensitivity to the Average Mass Solar Absorptance for the Double-Glazed Configuration with a Mass Thickness of 6 inches and a Mass-Area-to-Glazing-Area Ratio of 3. The dots represent the reference conditions.

Solar Absorption on Lightweight Surfaces — Lightweight objects, such as plants, rugs, wall hangings, furniture, or nonmassive partition walls, can impair the performance of direct-gain buildings. When solar radiation is absorbed on the surface of a lightweight object that has little heat capacity, the surface temperature rises rapidly, causing heat to be convected rapidly to the room air. This process can cause overheating, which necessitates ventilation cooling and thereby reduces the net solar gain of the structure. The extent to which performance is degraded is illustrated in Figure 12-7, where the annual *SSF* is plotted against the fraction of transmitted and reflected radiation absorbed on massless surfaces.

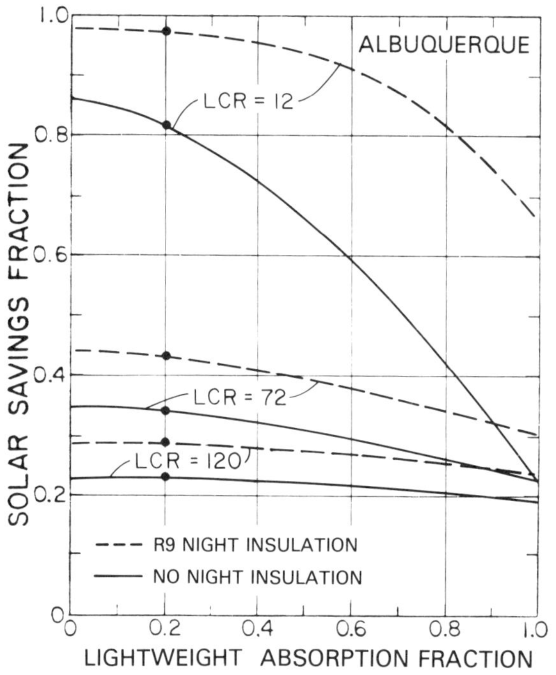

Figure 12-7. Annual Heating Performance Sensitivity to the Lightweight Absorption Fraction for a Double-Glazed Configuration with a Mass Thickness of 6 inches and a Mass-Area-to-Glazing-Area Ratio of 3. The lightweight absorption fraction is the fraction of solar radiation that directly heats the room air when it is transmitted through the glazing and when it is reflected from interior mass surfaces. It is intended to simulate the presence of lightweight surfaces and objects in the direct-gain space. The dots represent the reference conditions.

The lightweight absorption fraction simulates lightweight walls, floors, and ceiling, and also lightweight furnishings placed in the direct-gain space by the occupant. (Note that a lightweight absorption fraction is also employed in the analysis of sunspace performance, but in that case, it simulates only the presence of plants and lightweight furnishings. Lightweight walls and ceiling are accounted for explicitly in the sunspace models. See Chapter 13.)

Determination of the lightweight absorption fraction in a given direct-gain space can only be done in practice as a rough estimate. Prediction of the lightweight absorption fraction of a space under design may not be possible because it depends on the way in which the space is later furnished. But the following procedure may be useful, at least as a way of visualizing in an approximate way the meaning of the lightweight absorption fraction.

1. Stand near the south glazing, facing north. Estimate the fraction of the view that consists of lightweight floor, walls, ceiling, and other lightweight objects. If the lightweight objects have substantially different solar absorptances, estimate the fraction separately for each different absorptance.

2. Estimate the solar absorptances of the lightweight objects.

3. Multiply the fractions from step 2 by the absorptances from step 1 and add the products. The result is an estimate of the lightweight absorption fraction.

Both designer and occupant should try to minimize the occurrence of lightweight absorption to get the best possible performance from their passive solar heating system. Lightweight objects that are dark colored should be kept out of direct sunlight, and an effort should be made to avoid excessive solar-absorbing clutter in direct-gain zones. Whenever possible, use light colors on low-mass surfaces.

Mass Distribution — The effect of mass distribution in direct-gain buildings for systems with transparent or diffusing glazing is presented in Figure 12-8. The cross-hatched bars give the annual *SSF* achieved by a reference direct-gain design (mass thickness = 6 in., A_m/A_n = 3) with nondiffusing double glazing and without night insulation as the location of the thermal storage mass is progressively raised from the floor to the ceiling. The blank bars give the performance increments that result from the introduction of diffusing glazing that is otherwise assumed to have the same transmission characteristics as that of the original reference design. The dashed line, which traverses the entire graph, represents the average mass distribution assumed in the reference designs. From the results presented in Figure 12-8, it is evident that optimum performance is achieved by locating thermal storage mass on vertical surfaces and employing diffusing glazing (provided the diffusing glazing does not significantly reduce solar transmission).

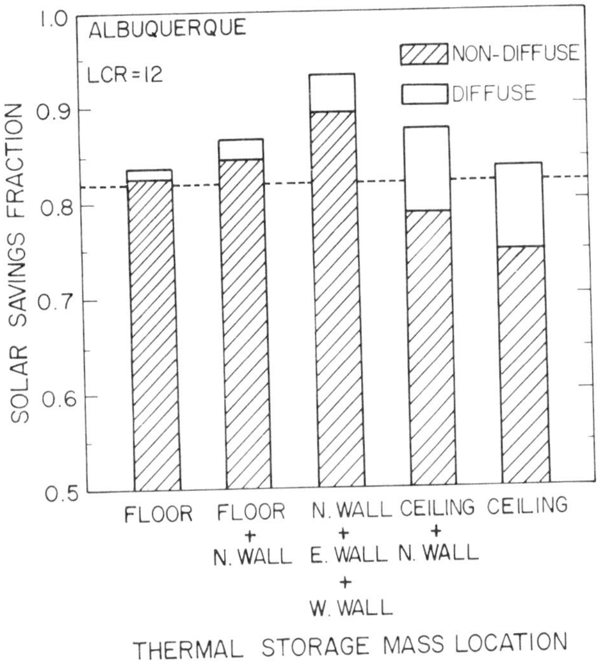

Figure 12-8. Sensitivity of Annual Heating Performance to the Distribution of Mass in the Direct-Gain Space for a Double-Glazed Configuration with No Night Insulation, a Mass Thickness of 6 inches, and a Mass-Area-to-Glazing-Area Ratio of 3.

This observation is a consequence of the low heat transfer coefficients for heat flow from the room air to the floor and ceiling, which reduce the effectiveness of horizontal surfaces. Vertical surfaces, on the other hand, easily receive and release heat by convection. The following are suggestions for placement and coloring of massive and lightweight surfaces.

- In general, surfaces in a direct-gain building should be relatively light (absorptance of 0.3 to 0.5) in color assuming there is adequate thermal storage mass. This precaution should assure adequate distribution of transmitted radiation. Massive surfaces may have an absorptance of about 0.5 while light weight surfaces should have an absorptance of 0.4 or less. An exception should be made for massive floors which should be dark in color (absorptance 0.5 to 0.7) to store the heat low.
- If the first surface struck by transmitted solar radiation is a wall or a ceiling, it should be of a light (absorptance 0.3) color. This measure will help promote uniform heating of the direct-gain zone.
- Massive surfaces should be distributed about a direct-gain zone as uniformly as possible. The use of masonry for interior partitions is especially effective because both sides of the partition can face direct-gain zones, thereby doubling the effective surface area, and because vertical surfaces are most favorably oriented for free convective heat exchange with the room air.

In connection with the above recommendations, it is suggested that the reader review the solar absorptances listed in Appendix 6, Table A6-2. Moderately light, buff bricks and concrete blocks painted a light beige color both have a solar absorptance of about 0.6, which is considerably higher than most people would expect.

Mass Coverings — Placing various types of floor or wall coverings over thermal storage mass surfaces may degrade performance. Concrete floor slabs, for example, provide the most significant storage medium in many buildings but generally require a covering of some sort for aesthetic reasons. There are many other circumstances that might lead a builder or occupant to cover mass surfaces, and the effect of this action is illustrated in Figure 12-9 and the related figures in Appendix 2. Performance sensitivity for decorative coverings with R-values from 0.0 to 2.0 is given in Figure 12-9 for Albuquerque. The covering is assumed to blanket the entire mass surface. The solar absorptance of the covering, α_c, is 0.8, the same value used for the mass surface itself in the reference design. In Appendix 2, plots for Albuquerque, Madison, and Santa Maria are presented for α_c = 0.8, and, to show the effect of employing light-colored covering materials, for α_c = 0.3.

Inspection of Figure 12-9 shows that the effect of mass coverings on performance can be quite drastic, especially for buildings with small values of *LCR* and no night insulation. The performance of buildings in this category, as we know from earlier results, is very dependent on the effective use of thermal storage mass. Penalties exceed 10% for covering R-values of 0.25 or greater on buildings with *LCR* less than 24 and with no night insulation. It is therefore suggested that 0.25 be considered a maximum acceptable R-value except in buildings that are night-insulated and/or have *LCR* of 48 or above, in which case, 0.50 might be taken as the upper limit. If, as is most likely, the entire mass surface is not to be covered, simply scale the penalty, indicated by the appropriate R-value sensitivity curve, by the ratio of the covered mass area to the total mass area to obtain an estimate of performance.

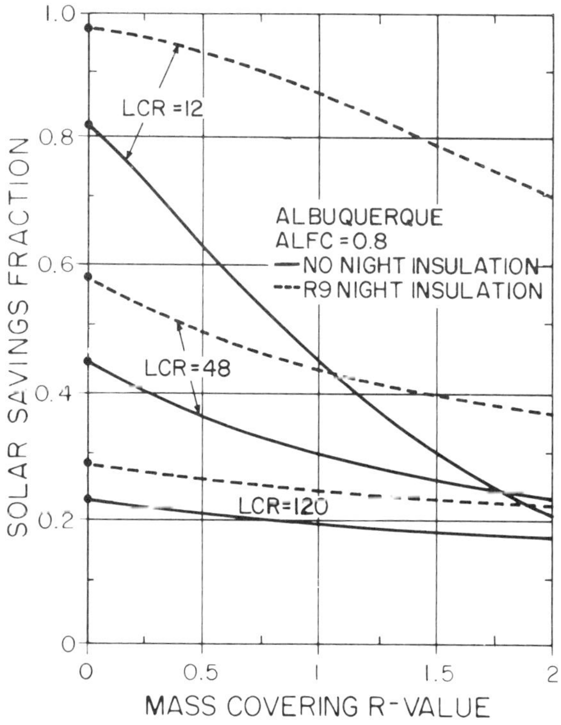

Figure 12-9. Annual Heating Performance Sensitivity to the Thermal Resistance (ft^2 · °F · h/Btu) of the Covering on the Thermal Storage Mass for a Double-Glazed Configuration with a Mass Thickness of 6 inches and a Mass-Area-to-Glazing-Area Ratio of 3. The dots represent the reference conditions. The solar absorptance of the covering is 0.8.

The R-values of various wall and floor coverings are given in Appendix 6, Table A6-1, which shows that many attractive alternatives to bare, unconditioned, concrete surfaces are available that will not significantly impair performance. Concrete surfaces that have been dyed and/or scored to resemble tile or other decorative materials can be attractive.

Number of Glazing Layers — The number of glazing layers required on direct-gain buildings depends strongly on climate and on whether or not night insulation is employed. The sensitivity of performance to the number of glazing layers is given for Albuquerque, which has a moderate climate, in Figure 12-10, and for Madison, which has a severe climate, in Figure 12-11. Note that double glazing yields no significant increase in the performance of night-insulated systems in Albuquerque but is very effective in combination with night insulation in the harsh Madison climate. For systems lacking night insulation, double glazing is a much better performer than single glazing in Albuquerque, and triple or quadruple glazing provides significant performance benefits in Madison. The number of glazings recommended on direct-gain buildings is summarized in Table 12-3 for three representative cities.

Table 12-3

Recommended Number of Glazings on Direct-Gain Buildings

Location	No Night Insulation	Night* Insulation
Santa Maria (mild climate)	1 or 2	1
Albuquerque (moderate climate)	2	1 or 2
Madison (severe climate)	3 or 4	2

*Minimum R4-R5

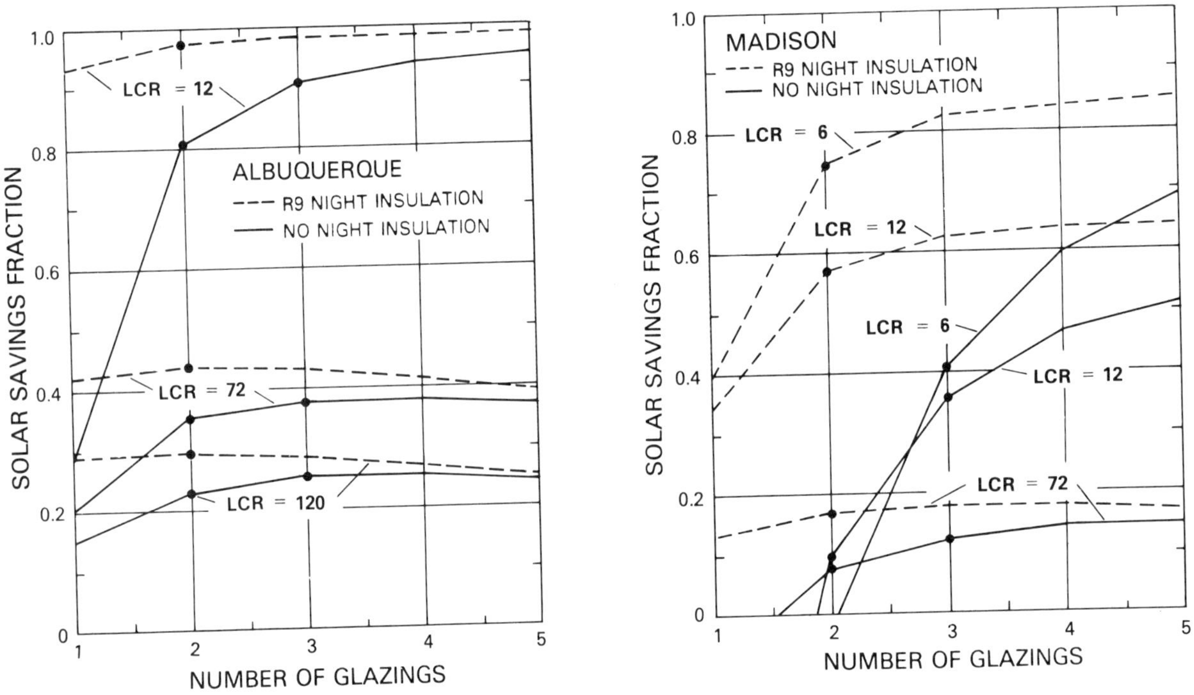

Figures 12-10 and 12-11. Annual Heating Performance Sensitivity to the Number of Panes of Glazing for a Mass Thickness of 6 inches and a Mass-Area-to-Glazing-Area Ratio of 3. The dots represent the reference conditions.

Temperature Bounds — The effect of interior temperature bounds on the performance of a direct-gain building with no night insulation located in Albuquerque is shown in Figure 12-12. The minimum room temperature, which appears on the horizontal axis, is the thermostat setpoint. These data were calculated in the absence of internal heat sources. If internal sources were included, enough heat would be introduced to the structure to increase the inside air temperature by 5° to 7°F in a typical residence without requiring any additional auxiliary heat. (The temperature rise could be considerably more in a building sealed and insulated to current standards.) The 65°F thermostat setpoint employed for the reference design would then represent a setpoint of 70° to 72°F in the presence of internal sources.

The *SSF* on the vertical axis of Figure 12-12 is calculated relative to a net reference load with a reference temperature equal to the reference thermostat setting of 65°F. The sensitivity curves, therefore, show how much energy is saved *relative to the reference setpoint* when the actual setpoint is varied. The lower boundary of the sensitivity bands gives the performance when the maximum room temperature is limited to 75°F by ventilation cooling, while the upper boundary represents performance at a maximum room temperature of 85°F. For a given site, night insulation status, and *LCR*, this set of sensitivity plots enables one to determine the effect of simultaneous variations of the minimum room air temperature from 55° to 75°F and the maximum room air temperature from 75° to 85°F. The two temperature-swing extremes that may therefore be assessed are a constant room air temperature of 75°F (zero temperature swing) and room air temperature variations from 55° to 85°F (30°F temperature swing).

Energy efficient performance is, of course, very sensitive to the minimum thermostat setting under all conditions. Note, however, that allowing a high maximum temperature inside a building (and the associated discomfort) improves performance only for small *LCR*, and then primarily when the thermostat for auxiliary heat is set relatively high. Buildings with small *LCR* often have much more transmitted solar radiation than can be utilized during the day, and a higher allowed maximum room temperature permits a larger part of the excess heat to be stored for nighttime use.

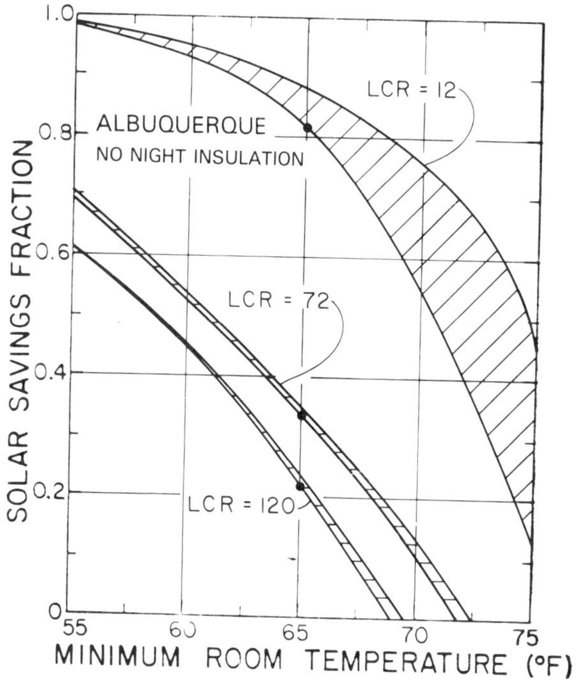

Figure 12-12. Annual Heating Performance Sensitivity to the Room Temperature Control Range for a Double-Glazed Configuration with No Night Insulation, a Mass Thickness of 6 inches, and a Mass-Area-to-Glazing-Area Ratio of 3. The minimum room temperature on the horizontal axis is the heating thermostat setpoint. The dots represent the reference conditions. The crosshatched bands correspond to a range of maximum room temperature (cooling thermostat setpoint) from 75°F to 85°F, the lower edge of the band corresponding to 75°F and the upper edge to 85°F. The reference value is 75°F.

Nighttime Thermostat Setback — The benefits obtainable from setting the thermostat back at night are not as large in a passive solar home as they would be in a home of conventional design. For much of the night, a high-solar-fraction, direct-gain building with a small *LCR* will hold the inside air temperature above the thermostat setting without the aid of auxiliary heat. Only in the early morning hours, when heat stored in the mass of the structure has been depleted, is there generally a demand for heat from the backup system. The thermostat setback, therefore, goes unnoticed much of the time. At larger values of *LCR*, less solar heat is stored for nighttime use and under these conditions, setting the thermostat back can yield significant savings.

The effect of nighttime setback on direct-gain buildings in Albuquerque is illustrated in Figure 12-13 for systems with and without night insulation. The setback period begins at 11:00 p.m. each evening and ends at 6:00 a.m. the next morning. A 10°F setback from the reference setting of 65° to 55°F can yield up to a 65% increase in solar savings or a 65% decrease in auxiliary heat. This is for the case for a building with small solar aperture and a small mass area (LCR = 120 and A_m/A_n = 3); the effect would be much less pronounced in a building with more thermal storage mass. Flat regions on these sensitivity curves indicate low-temperature ranges that are seldom, if ever, reached in practice, regardless of the thermostat setpoint.

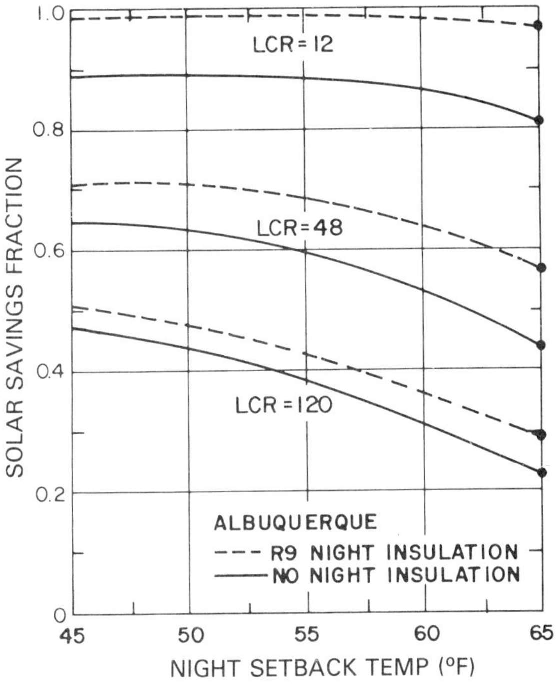

Figure 12-13. Annual Heating Performance Sensitivity to the Night Setback Temperature for a Double-Glazed Configuration with a Mass Thickness of 6 inches and a Mass-Area-to-Glazing-Area Ratio of 3. Nighttime is defined in this connection to be 11 p.m. to 6 a.m. The reference setback temperature is 65°F, that is, no reduction from the 65°F daytime setpoint. The dots represent the reference conditions.

Overheating Sensitivity Data

The inside air temperature has been assumed to be limited to 75°F in all performance data reported so far. It is assumed that the excess heat is vented to the outside air, provided that the ambient temperature is less than 75°F. If the ambient temperature is 75°F or above, an auxiliary cooling system is assumed to remove the excess heat. In practice, ventilation cooling must generally be provided by occupant intervention, that is, by opening windows, doors, or special vents whenever the temperature becomes uncomfortably warm in the living space. An alternative form of intervention would be for the occupant to reduce the solar gains by partially closing drapes or other shading devices. Since the inside air temperature was limited to 75°F in the data reported so far, the effect of occupant intervention on performance is included. However, no measure of occupant discomfort or inconvenience has been provided. In this section, we provide a relative measure of the effect of thermal storage mass and lightweight solar absorbing surfaces on thermal comfort in direct-gain buildings.

Thermal Storage Mass and Thermal Comfort — In order to quantify the tendency of a given direct-gain building to overheat, a series of data is presented for the month of January in which the inside air temperature is not limited by any form of ventilation or shading. The maximum air temperature occurring within the building under these conditions provides a relative basis for determining the effect of various parameters on the thermal comfort of the living space.

The maximum room temperature reached in Albuquerque is plotted as a function of the mass-area-to-glazing-area ratio for mass thicknesses of 2, 4, and 8 inches in Figures 12-14, 12-15, and 12-16, respectively. Each figure includes sensitivity curves

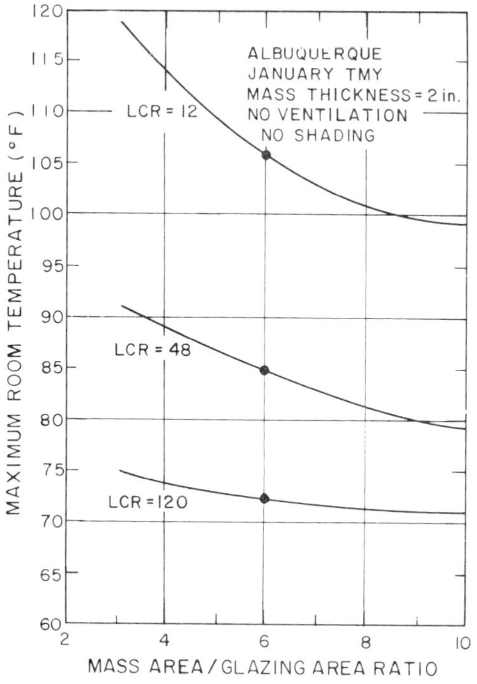

Figure 12-14

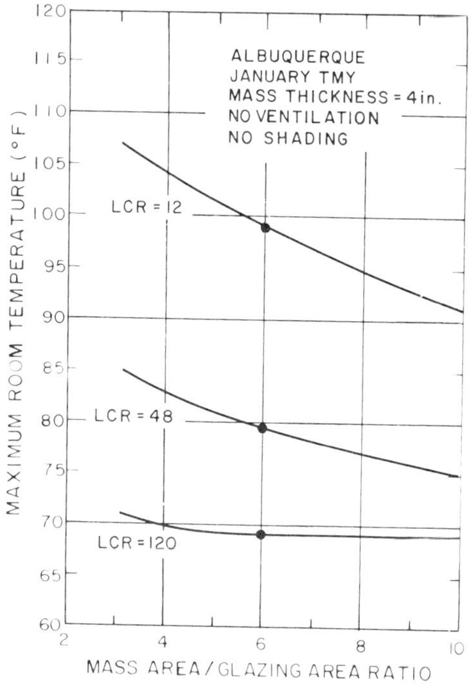

Figure 12-15

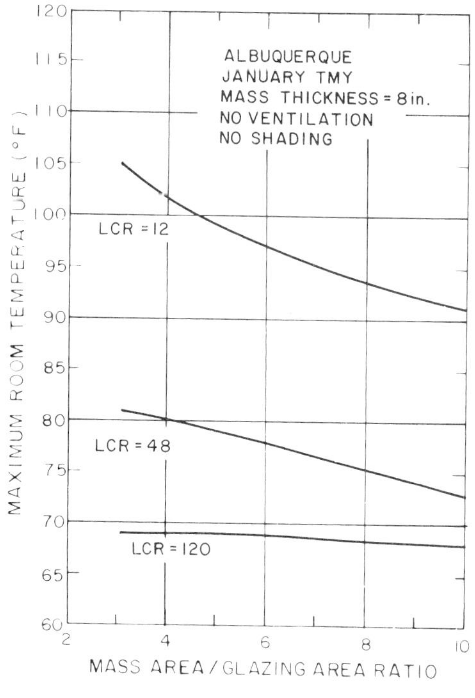

Figure 12-16

Figures 12-14 through 12-16. January Overheating Sensitivity to the Mass-Area-to-Glazing-Area Ratio for a Double-Glazed Configuration without Night Insulation. The dots represent the reference conditions.

for three values of *LCR*. The reference design mass-area-to-glazing-area ratios are indicated on each curve by a solid point, except in Figure 12-16, for which the mass thickness of 8 inches does not correspond to any of the reference designs. Remember that the temperatures indicated on these figures may be reached only once during the 1-month test period and then only if no occupant intervention is allowed. The very high temperatures indicated for some configurations are therefore not as serious a problem as one might at first suppose, but they do indicate, on a relative basis, the extent to which the occupant may be kept busy adjusting ventilation and shading in order to keep the structure comfortable. To illustrate this point, a complete temperature histogram is presented in Figure 12-17 for a building in Albuquerque during January with a thermal storage mass thickness of 4 inches, an A_m/A_n ratio of 3, and an *LCR* of 12. This is an extreme case with more than twice the recommended glazing area. This histogram corresponds to the condition represented by a maximum temperature of 108°F on the left end of the upper curve in Figure 12-15. Note that the histogram indicates that the building will have an air temperature in the range of 104° to 108°F only 1% of the time during January, and then only if no shading or ventilation is allowed.

As expected, buildings with small *LCR* (which imply large collection areas relative to the loads) are most sensitive to the amount of thermal storage mass. Note that comfort continues to improve as the mass-area-to-glazing-area ratio approaches 10 and that increasing the mass thickness from 2 to 4 inches yields a significant reduction in the maximum room temperature. However, increasing the mass thickness from 4 to 8 inches does not result in a significant improvement in the building comfort. These results indicate that one can best provide for thermal comfort in a direct-gain building by employing thermal storage mass with a thickness of about 4 inches spread over as large an area as practicable.

Lightweight Surfaces and Thermal Comfort — The discomfort caused by solar absorption on nonmassive surfaces can be indicated in the same manner used to demonstrate the beneficial effect of

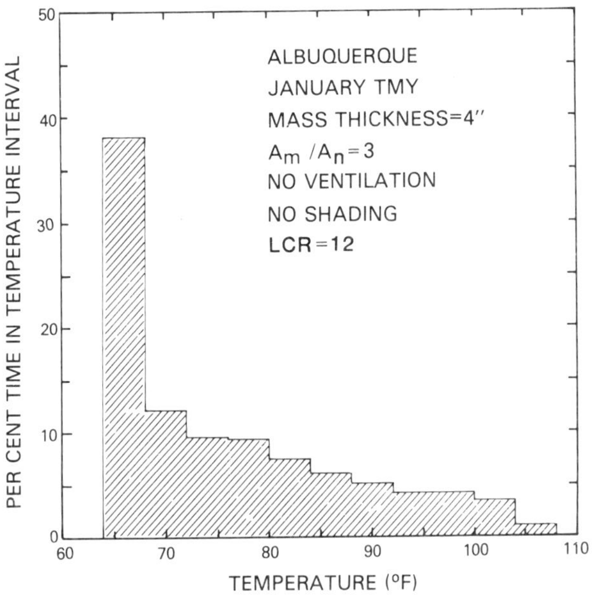

Figure 12-17. January Room Temperature Histogram for a Double-Glazed Configuration without Night Insulation, a Mass Thickness of 4 inches, and a Mass-Area-to-Glazing-Area Ratio of 3.

thermal storage mass. Figure 12-18 shows that the maximum room temperature in Albuquerque decreases rapidly as the percentage of transmitted solar radiation absorbed on lightweight surfaces decreases. This is a graphic illustration of the importance of avoiding excessive solar-absorbing clutter in direct-gain zones.

Orientation — The effect of glazing azimuth orientation on the performance of direct gain buildings depends on many factors including location, *LCR*, and whether or not night insulation is used. Buildings with small values of *LCR* and no night insulation are more sensitive to orientation than are night-insulated structures with larger *LCR* values. The effect of location on orientation sensitivity is difficult to generalize, but the performance tends to be more sensitive to orientation in colder climates. Figure 12-19 illustrates the sensitivity of performance to orientation for Albuquerque.

In most cases, the performance penalty for departing from a due-south orientation by 15 degrees is

10% or less, and by 30 degrees is 20% or less. We recommend that direct-gain buildings be oriented within 15° of due south whenever practical and certainly within 30° of due south.

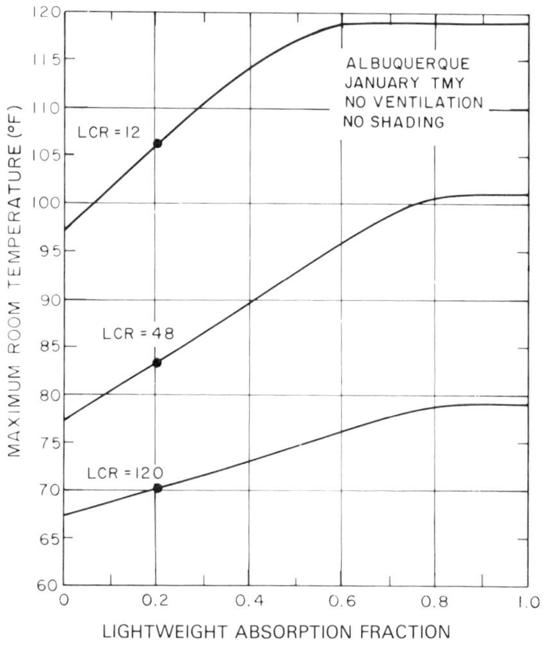

Figure 12-18. January Overheating Sensitivity to the Lightweight Absorption Fraction for a Double-Glazed Configuration without Night Insulation, a Mass Thickness of 6 inches, and a Mass-Area-to-Glazing-Area Ratio of 3. The dots represent reference design conditions.

The Low-Mass, Suntempered Concept

A suntempered building can be created by a series of minor modifications to home designs that have already been proven in the marketplace. A higher standard of insulation and sealing may also be incorporated. The second step is to move windows from the east, west, or north facade to the south facade in order to achieve substantial fraction (greater than 60%) of the window space on the south side of the building. The cost of these modifications to the builder and his customers is small, and yet, under appropriate circumstances, a significant part of the space heating load will be displaced by direct solar gains. Note that suntempering generally reduces cooling loads, and further reductions may be achieved by moving the east-facing and west-facing windows to the north side of the building or shading them during the cooling season. In order to accommodate these changes in window location, it may be convenient to orient the building with the long axis in the east-west direction, but this is not always necessary.

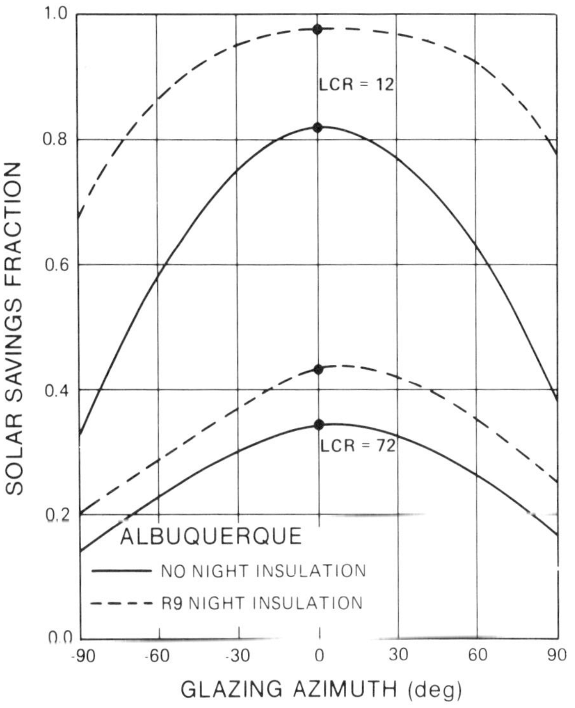

Figure 12-19. Annual Heating Performance Sensitivity to the Glazing Azimuth for Vertical Glazing, Both Without and With R9 Night Insulation in Albuquerque. The dots indicate the reference conditions.

To get an idea of the glazing areas involved in suntempering, consider a small, single-family, detached residence with a 1,500-ft^2 floor area. The total window area will be about 15% of the floor area, or 225 ft^2. Placing 60% of the window area on the south side of the building yields a projected glazing area of 135 ft^2. At this point, we can split suntempered buildings into two very different categories. Suppose, in the first case, that the building is built on a concrete floor slab but is otherwise of frame construction. The floor slab will be at least 4 inches thick, which is within the recommended mass thickness range, and will have a gross area of 1,500 ft^2. Even if only half of the gross area of the slab is available for thermal storage, we still have a mass-area-to-glazing-area ratio of

$$\frac{A_m}{A_n} = \frac{750}{135} = 5.5\,, \qquad (12\text{-}1)$$

which is slightly less than the recommended value of 6. However, furnishings and other mass in the building easily make up for this difference. Thus, although the building may be considered to be a lightweight structure, it is thermally massive relative to the small glazing area and is therefore a high-mass, direct-gain building by our definition.

If, however, the building has a wood frame floor over a crawl space, and there are no massive elements within the insulated shell of the structure, then the building represents a low-mass, suntempered building. These buildings have very little thermal storage relative to the glazing area. The most significant thermal storage medium within the structure is the gypsum board that lines the walls and the ceiling. A gypsum-board thickness of 1/2 inch is used in most wood frame houses. The properties given in Table 12-2 were obtained from the *ASHRAE Handbook, 1981 Fundamentals,* and the A_m/A_n ratio of 20 was calculated for a 1,500-ft^2 structure with a collection area of 180 ft^2, assuming only the south zones were available for thermal storage. Sensitivity data presented below allow performance estimates based on other assumptions concerning availability of gypsum board for thermal storage.

Performance of the Suntempered Reference Designs

Plots of *SSF* versus *LCR* for low-mass, suntempered buildings with and without night insulation are presented in Figures 12-20 through 12-26 for seven cities. The figures are ordered according to the increasing severity of the January climate as measured by the ratio of the solar radiation incident on a vertical, south-facing surface (Q_v) to the 65°F-base heating degree-days (D65). This ratio is tabulated in Table 12-4. The data are from Appendix 3; Q_v is obtained from VS, the average daily solar radiation incident on a vertical surface, by multiplying by the number of days in the month.

Table 12-4

January Q_v/D65 Ratio

City	Q_v/D65 (Btu/ft^2 · °F)
Santa Maria	82
Charleston	54
Albuquerque	52
Nashville	27
Medford	20
Boston	20
Madison	17

The highest solar savings fraction observed for the case with no night insulation is 18%, which occurs in Albuquerque at an *LCR* of about 72. The *SSF* decreases for values of *LCR* greater than 72 because the solar gain decreases relative to the building load. The *SSF* also decreases for values of *LCR* less than 72 because further increases in solar gain relative to the building load are not fully utilizable. It is important to note at this point that design decisions should be based on auxiliary heat requirements, not on *SSF*. The quantity *SSF* is a ratio that enables us to present performance information in a compact form; however, as seen in the following example, it is necessary to calculate the auxiliary heat to determine the effect of design decisions on energy consumption.

The *SSF* of the suntempered building with no night insulation decreases as the location is shifted to the increasingly severe climates of Charleston, Nashville, Medford, and Boston, finally dropping to a local maximum *SSF* of 3% in Madison, Wisconsin. This decrease of *SSF* in harsh winter climates was expected. The exception is Albuquerque which has the highest maximum *SSF* despite being well below Santa Maria in terms of Q_v/D65.

The addition of night insulation will, of course, improve performance. For suntempered buildings with R9 night insulation and an *LCR* of 24, *SSF* is

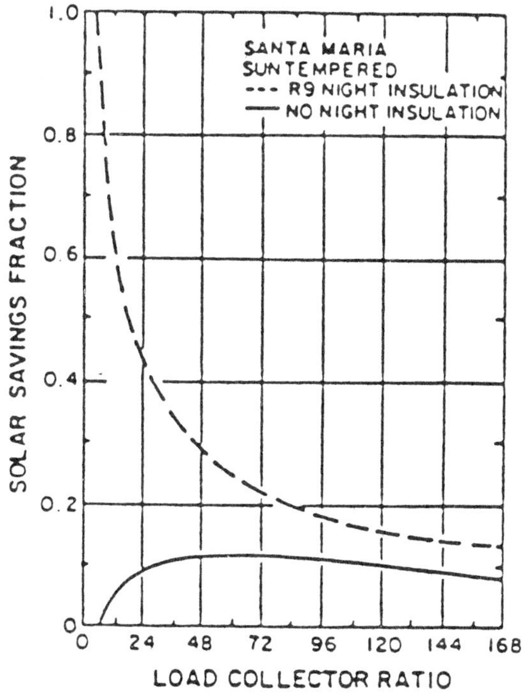

Figure 12-20

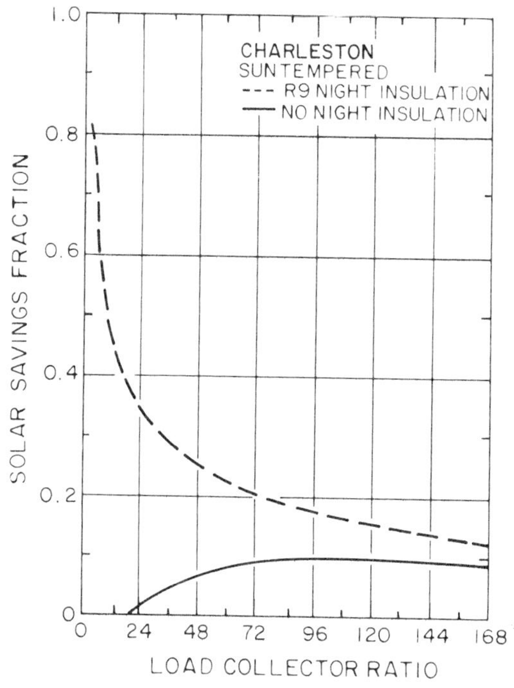

Figure 12-21

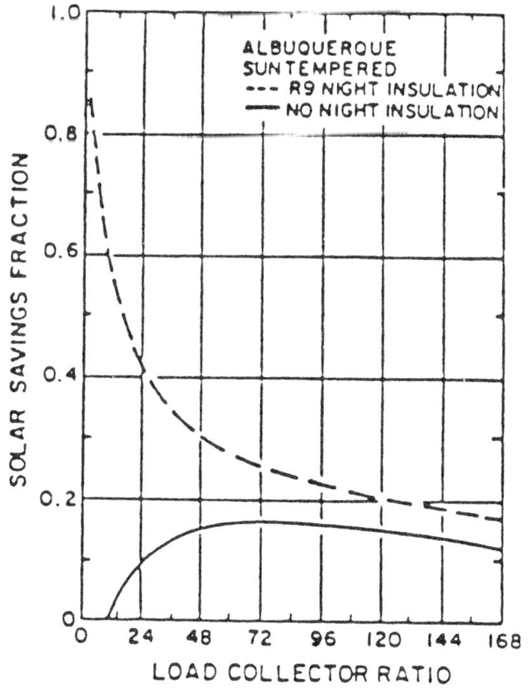

Figure 12-22

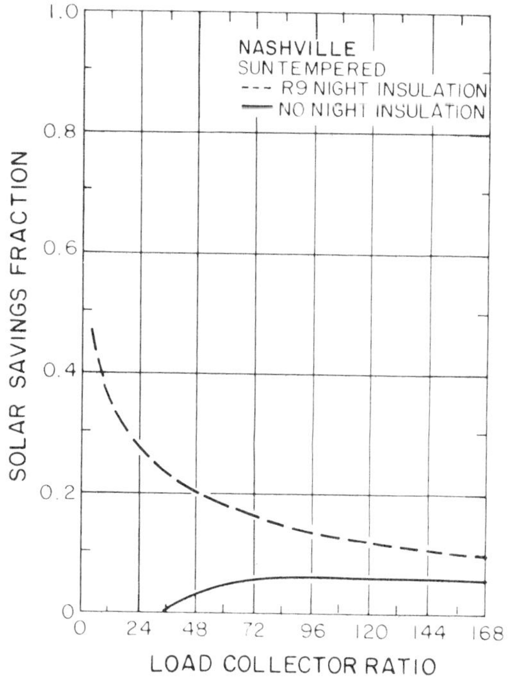

Figure 12-23

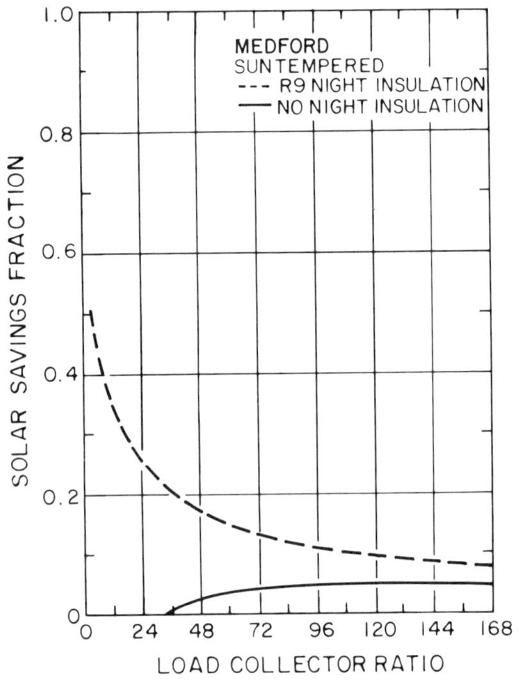

Figure 12-24

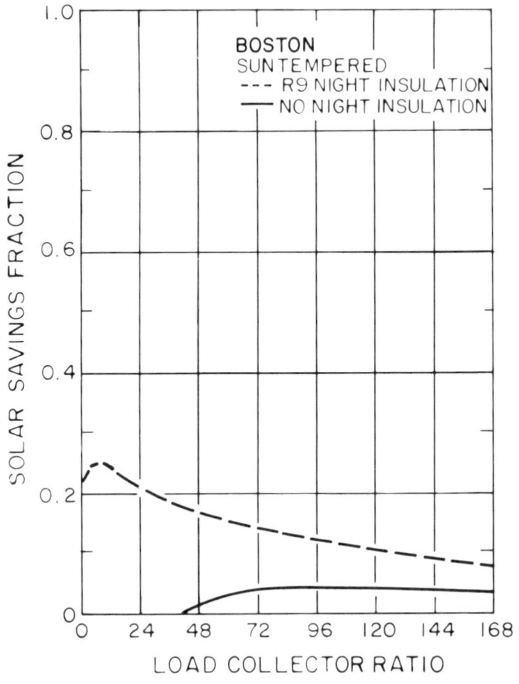

Figure 12-25

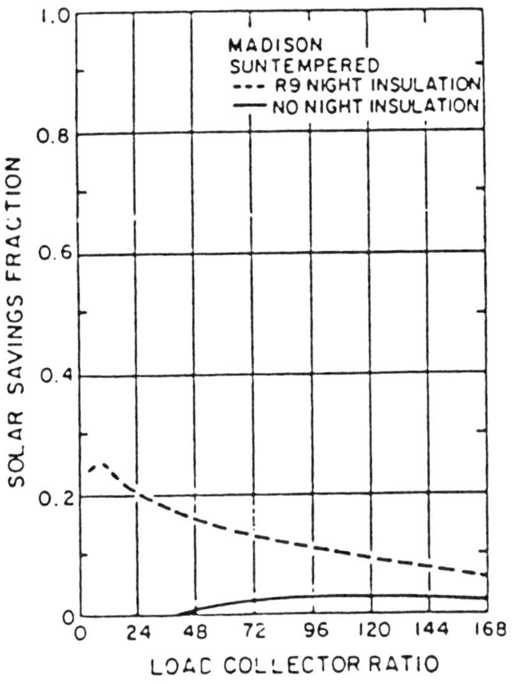

Figure 12-26

Figures 12-20 through 12-26. Annual Heating Performance as a Function of *LCR* for Low-Mass, Suntempered Buildings. The figures are ordered according to the severity of the winter climate as quantified in Table 12-4.

about 40% in Santa Maria, Charleston, and Albuquerque, about 30% in Nashville and Medford, and about 20% in Boston and Madison. Solar fractions are still far below those obtainable in higher mass structures. This is illustrated by the comparison in Table 12-5 for an *LCR* = 48 direct-gain building in Albuquerque. The high-mass *SSF*s are taken from the *LCR* tables in Appendix 1.

Table 12-5

Performance Comparisons for Albuquerque, *LCR* = 48

Direct-Gain Type	Heat Capacity (BTU/°F ft²)	*SSF* (%) No Night Insulation	*SSF* (%) R9 Night Insulation
Suntempered	10.8	15	30
High-Mass:			
DG A1, A3	30	35	50
DG B1, B3	45	38	52
DG C1, C3	60	46	58

Monthly *SLR* correlations for low-mass, suntempered buildings are not presented because the sensitivity of these structures to short-period weather patterns renders the correlations too poor to be useful.

Sensitivity Data for the Suntempered Reference Designs

The full set of suntempered sensitivity data is included in Appendix 2. In this section, the types of data are discussed.

Mass Surface Area — The effect of varying the mass-area-to-glazing-area ratio about the reference value of 20 is shown in Figure 12-27 for Albuquerque. Note that the performance at low values of *LCR* can be significantly improved by adding more 1/2-inch gypsum board. An area ratio of 40 would correspond to a 1,500-ft² frame house with 180 ft² of south-facing glazing for which the gypsum board in northern, as well as southern, zones of the building is available for thermal storage. The northern zones of the building are accessible for thermal storage if a forced-air distribution system is employed, or if wide doorways are used to enhance free convective heat exchanges with the southern zones that experience direct solar gains.[1-3] Figure 12-27 and the related graphs in Appendix 2 can also be used to account for thermal storage in furniture or other objects. Each increase of 10 in the A_m/A_n ratio is equivalent to an increase of 5.4 (Btu/°F · ft² of glazing) in the heat capacity of the structure.

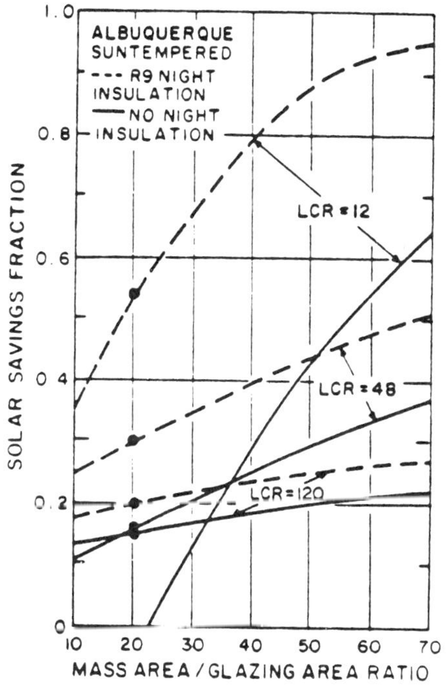

Figure 12-27. Annual Heating Performance Sensitivity to the Mass-Area-to-Glazing-Area Ratio for a Low-Mass, Suntempered Building. The dots represent the reference conditions. The *LCR* value of 12 is not characteristic of a suntempered building but is shown for comparison.

Number of Glazing Layers — The effect of varying the number of glazing layers on south windows on low-mass, suntempered buildings is illustrated in Figures 12-28 through 12-30 for Santa Maria, Albuquerque, and Madison, respectively. Note that triple glazing appears desirable at all locations for systems without night insulation and that even with night insulation, double glazing is advisable in Albuquerque and Madison.

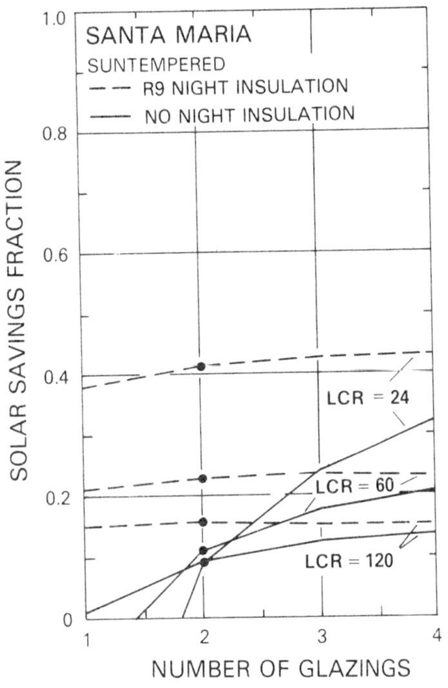

Figure 12-28

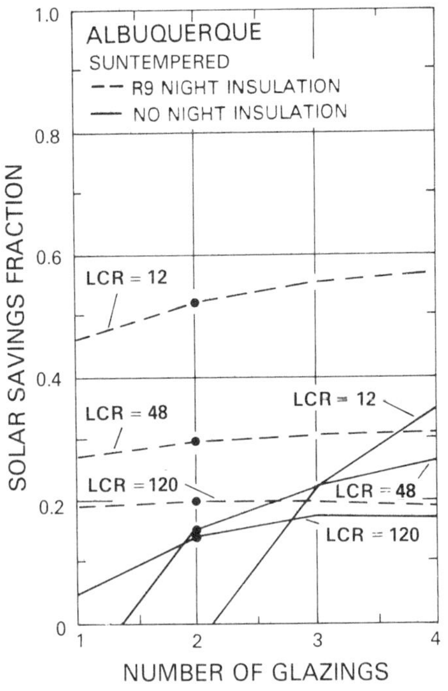

Figure 12-29

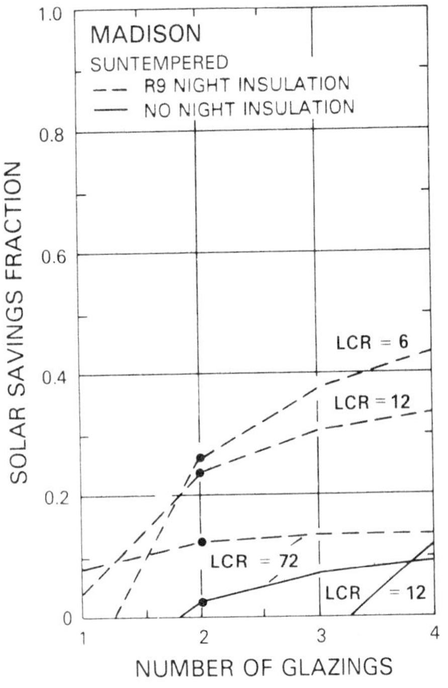

Figure 12-30

Figures 12-28 through 12-30. Annual Heating Performance Sensitivity to the Number of Glazings for a Low-Mass, Suntempered Building. The dots represent the reference conditions.

Solar Absorptance of Gypsum Board Surface — The solar absorptance of the gypsum-board surface has almost no effect on performance, as illustrated in Figure 12-31 for Albuquerque. The same lack of sensitivity is noted for the other cities in Appendix 2. For the suntempered building, the ratio of mass surface area to glazing area is so large that multiple reflections within the living space cause most of the transmitted radiation to be absorbed in the storage mass, regardless of interior color.

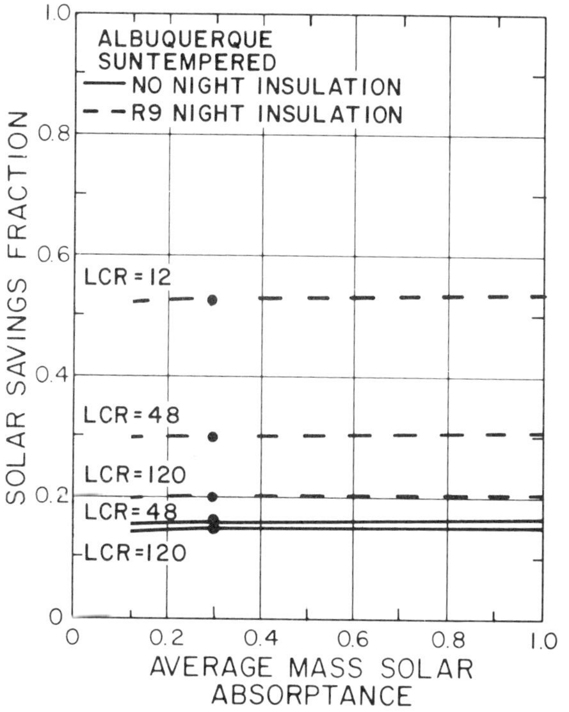

Figure 12-31. Annual Heating Performance Sensitivity to the Solar Absorptance of the Thermal Storage Mass Surface for a Low-Mass, Suntempered Building. The dots represent the reference conditions.

References

1. W. O. Wray and D. D. Weber, "LASL Similarity Studies: Part I, Hot Zone/Cold Zone: A Quantitative Study of Natural Heat Distribution Mechanisms in Passive Solar Buildings," *Proceedings of the Fourth National Passive Solar Conference,* Kansas City, Missouri, October 3-5, 1979, Vol. 4, pp 226-230.

2. D. D. Weber, W. O. Wray and R. J. Kearny, "LASL Similarity Studies: Part II, Similitude Modeling in Interzone Heat Transfer by Natural Convection," *Proceedings of the Fourth National Passive Solar Conference,* Kansas City, Missouri, October 3-5, 1979, Vol. 4, pp 231-234.

3. D. D. Weber and R. J. Kearney, "Natural Convection Heat Transfer Through an Aperture in Passive Solar Heated Buildings," *Proceedings of the Fifth National Passive Solar Conference,* Amherst, Massachusetts, October 19-26, 1980, Vol. 5.2, pp 1037-1041.

CHAPTER 13
SUNSPACE SYSTEMS

Introduction

A sunspace is a solar collector that is also a useful space capable of serving other building functions. Other terms are "solarium," "atrium," "conservatory," and "sun room." The term "attached greenhouse" is also often used but can imply a plant-growing function as well as a solar heating function. Sunspace solar heating is compatible with limited plant raising, but no account has been taken in the data presented here of the energy transfer due to the evaporation of water from plants and planters.

The sunspace performs its passive solar heating function by transmitting solar radiation through its glazing and absorbing it on its interior surfaces. The solar radiation is converted to heat upon absorption. Some of this heat is rapidly transferred by natural convection to the sunspace air, and some of it flows into massive elements in the sunspace (floor, walls, and water containers) to be returned later. The sunspace is thus a direct-gain space in which heat is used directly to maintain a temperature suitable for its intended secondary function, such as occasional living space. But the primary purpose of the sunspace as a solar heating system is to deliver heat to the adjoining rooms. This may be by conduction through a masonry common wall and by natural convection through openings (doors, windows, or special vents) in the common wall. The sunspace system then resembles an indirect-gain, thermal storage wall. Or the common wall may be insulated so that heat transfer occurs only by natural convection through openings. Then the sunspace system is in the isolated-gain category.

Some sunspaces are operated as hybrid systems in which a fan is used to transfer heated air from the sunspace to other rooms or to storage. Hybrid systems are not treated in this volume.

Sunspaces have become very popular building features, both for the attractive space they provide and for their ability to deliver solar heat to adjacent spaces. The solar heating performance can indeed be very effective, often exceeding the performance of any other passive solar system occupying the same area of south wall. The purpose of this chapter is to provide the design analysis information needed to assure that a sunspace design will be as effective as possible.

The Reference Designs

Numerous sunspace designs have been built and are of interest to designers. Any design analysis method must, then, be flexible enough to address a variety of designs. Some of the design features of interest are the glazing geometry, the common-wall configuration, the type of thermal storage, and movable night insulation. Twenty-eight reference designs have been chosen that reflect various combinations of typical choices of these features. These designs are described below and summarized in Table 13-1.

The principal sunspace glazing is assumed to face due south. Thus, wall locations are referred to by the compass directions: the principal glazing is the south wall, the principal common wall is the north wall, and the end walls are the east and west walls.

Two types of sunspace are defined according to the degree of sunspace integration with the rest of the building. One type is the *attached* sunspace, whose north wall is common with adjoining rooms and 30 feet wide in the east-west direction. The other is the *semienclosed* sunspace that has three common walls, the north, the east, and the west. The semienclosed sunspaces are all 24 feet wide (east-west) and 12 feet deep (north-south). The north common wall in each case is 9 feet high.

Table 13-1

Sunspace Reference Designs

Designation: No Night Insulation	Designation: R9 Night Insulation	Type	Tilt (deg)	Common Wall	End Walls
SSA1	SSA2	attached	50	masonry	opaque
SSA3	SSA4	attached	50	masonry	glazed
SSA5	SSA6	attached	50	insulated	opaque
SSA7	SSA8	attached	50	insulated	glazed
SSB1	SSB2	attached	90/30	masonry	opaque
SSB3	SSB4	attached	90/30	masonry	glazed
SSB5	SSB6	attached	90/30	insulated	opaque
SSB7	SSB8	attached	90/30	insulated	glazed
SSC1	SSC2	semienclosed	90	masonry	common
SSC3	SSC4	semienclosed	90	insulated	common
SSD1	SSD2	semienclosed	50	masonry	common
SSD3	SSD4	semienclosed	50	insulated	common
SSE1	SSE2	semienclosed	90/30	masonry	common
SSE3	SSE4	semienclosed	90/30	insulated	common

Two geometrical shapes of the attached sunspace and three of the semienclosed sunspace are treated. The two attached sunspace geometries are (1) a single plane of glazing on the south wall, tilted up from the horizontal by 50° and (2) two planes of south glazing, the lower one 6 feet high and vertical, the upper tilted at 30°. The three semienclosed geometries are (1) a single, vertical plane of glazing on the south wall, (2) a single, 50°-tilted plane of glazing on the south wall, and (3) two planes of south glazing, the lower one 6 feet high and vertical, the upper tilted at 30°. Altogether, these comprise five geometrical configurations, designated A through E. They are illustrated in Figure 13-1.

The reference designs include two types of common wall between the sunspace and the adjacent building. One is lightweight and insulated, corresponding to a frame wall, with a thermal resistance of R20 (20 h · °F · ft^2/Btu); and one is uninsulated, 12-inch-thick, high-density masonry with a thermal conductivity of 1.0 Btu/h · ft · °F and a volumetric heat capacity of 30 Btu/ft^3 · °F. In the lightweight-wall configuration, there is a row of water containers in the sunspace for thermal storage. The row extends the full east-west width of the sunspace. The containers are twice as high as they are deep. The water volume is 1 ft^3/ft^2 of common wall area. The containers are immediately adjacent to, but detached from, the sunspace floor and wall; they are thermally coupled to the wall and floor directly by radiation and indirectly by convection through the sunspace air.

Both wall configurations include thermocirculation vents in the common wall whose areas total 6% of the north wall area (3% for each of the top and bottom vents). The vent centers are separated by a height of 8 feet. There is no reverse thermocirculation.

For each geometry and wall configuration, movable insulation may or may not be applied at night to the sunspace glazing. When used, the night insulation has a thermal resistance of R9 and is in place from 5:30 p.m. to 7:30 a.m. solar time.

The sunspace east and west end walls may or may not be glazed. When not glazed, the end walls are insulated to a thermal resistance of R20. The

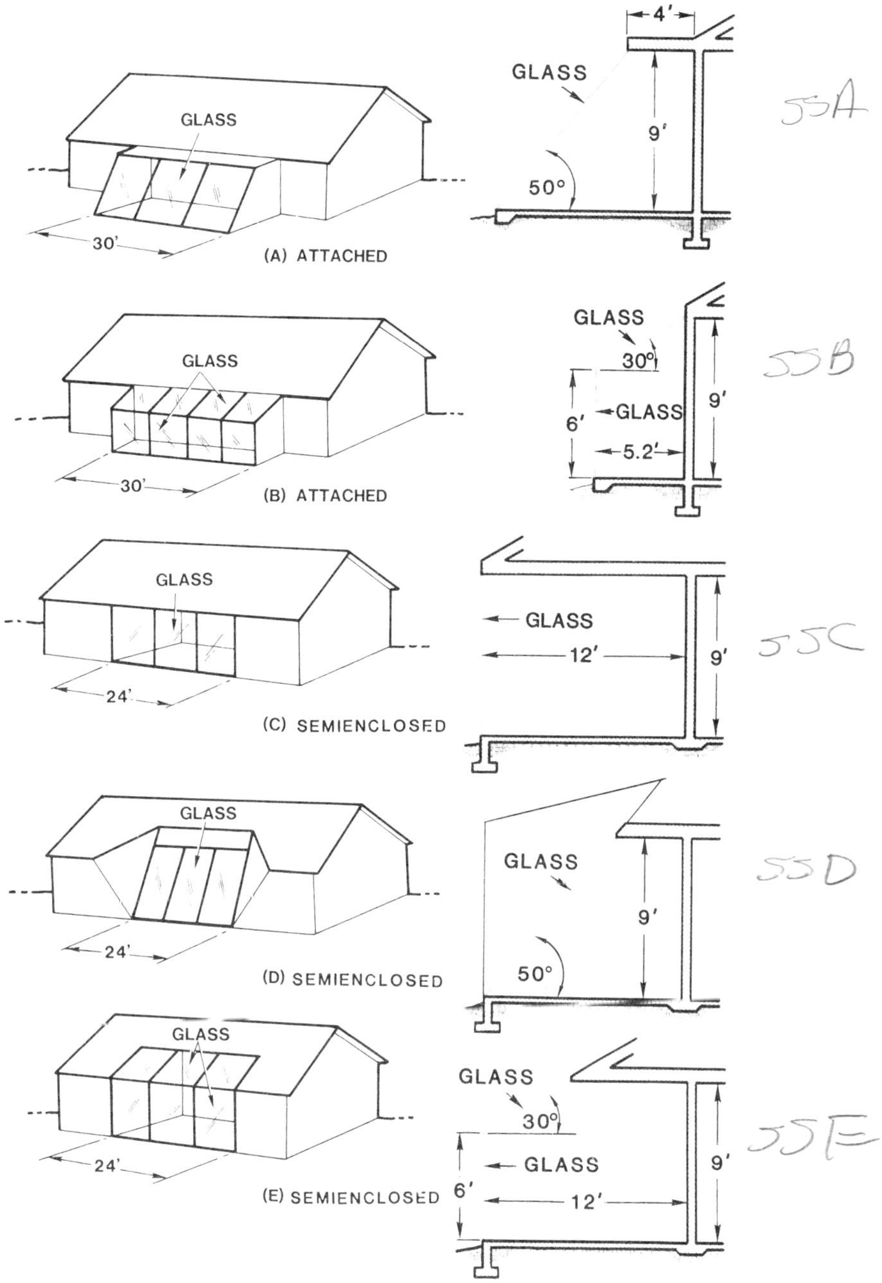

Figure 13-1. Sunspace Geometries (Not to Scale)

sunspace glazings have optical and thermal properties equivalent to two panes of ordinary, 1/8-inch window glass with a 1/2-inch air gap, with the exception that the transmitted solar radiation is assumed to be perfectly diffuse.

The sunspace floor is a 6-inch-thick slab of masonry material with a thermal conductivity of 0.5 Btu/h · ft · °F and a volumetric heat capacity of 30 Btu/ft^3 · °F. There is conduction through underlying soil to a fixed temperature deep in the earth and through perimeter insulation to the ambient air.

The floor, the common wall in the case of the masonry-wall configuration, and the water containers in the case of the lightweight-wall configuration, are capable of heat storage. The other sunspace surfaces are assumed to have no mass.

The surfaces of the common wall on the sunspace side have solar absorptances of 0.7 if they are lightweight and 0.8 if they are masonry. The water containers have a solar absorptance of 0.9. The sunspace floor has a solar absorptance of 0.8. The other surfaces (ceiling, if any, and end walls if not glazed) have solar absorptances of 0.3. The solar radiation not absorbed is assumed to be diffusely reflected. The glazed surfaces are assumed to reflect 24% of any internally reflected solar radiation that is incident on them; the other 76% of incident radiation is lost to the outside. Radiation absorbed at nonmassive surfaces is assumed to produce heat that transfers rapidly to sunspace air. It is assumed that the absorption of solar radiation on other lightweight objects, such as furniture and plants, results in 20% of transmitted and reflected radiation being transferred rapidly to the sunspace air.

Sunspace heat losses occur primarily through the glazing. There are also losses through the R20 opaque surfaces (ceiling, if any, and end walls if not glazed). A sunspace infiltration rate of 0.5 air changes per hour is assumed. No account is taken of heat transfer due to water evaporation that would occur if plants were present. (The effect of evaporation from plants can be a significant heat loss from the sunspace.)

Auxiliary heating is assumed to prevent the sunspace temperature from falling below 45°F. Ventilation cooling is assumed to limit the sunspace temperature to 95°F if possible. The heating thermostat setpoint in the adjacent building is 65°F. Ventilation or auxiliary cooling are assumed to limit the room temperature to 75°F. There are no internal heat sources in the room or sunspace.

It is important to realize that a design analysis is not necessarily restricted to the exact specifications of one of the above-described reference designs because of certain intrinsic flexibilities in the analysis method. For one thing, the sunspace dimensions that define the five geometrical configurations are specified only to establish the *relative* sunspace dimensions. It is the sunspace *shape*, not its absolute dimensions, that determines its performance per unit of projected area. For example, configuration A is nominally 30 feet wide by 9 feet high with a 4-foot ceiling and a glazing tilt of 50 degrees, but a sunspace 40 feet wide by 12 feet high with a 5-foot-4-inch ceiling and a glazing tilt of 50 degrees is also configuration A, being just one-third larger in all its dimensions but identical in shape. The effect of the absolute size of the sunspace is accounted for by the projected area, an independent parameter.

Another aspect of design flexibility concerns the use of the monthly *SLR* rather than the annual *LCR* analysis method. The essential idea is that the analyst determines the solar radiation absorbed in the sunspace. It is also possible to a limited extent to vary the sunspace heat loss coefficient. Thus, there is flexibility in the choice of such things as the sunspace glazing geometry and orientation, the sunspace glazing and opaque-wall thermal resistances, the sunspace infiltration rate, and the ground reflectance. Chapter 9 gives the details on these points.

A summary of reference design characteristics is presented in Table 13-2. The table shows those parameters that can be varied in the annual *LCR* method and in the monthly *SLR* method.

The parameters excluded from the definitions of the reference designs, the projected area (A_p) and the net load coefficient (*NLC*), are combined into $LCR = NLC/A_p$. The quantity *LCR* then becomes the principal building variable for a given reference design. Performance estimates for a given location are functions of *LCR*.

Table 13-2

Sunspace Reference Design Characteristics

Thermal storage capacity (Btu/°F · ft²) (per square foot of common wall)	
Masonry wall	30.0
Water wall	62.4
Floor	15.0
Masonry properties	
Wall thermal conductivity (Btu/h · ft · °F)	1.0
Floor thermal conductivity (Btu/h · ft · °F)	0.5
Density (lb/ft³)	150
Specific heat (Btu/lb · °F)	0.2
Infrared emittance of surface	0.9
Glazing properties	
Transmission characteristics	diffuse
Infrared emittance of surface	0.9
Orientation	due south*
Shading	none*
Number of layers	2*
Index of refraction	1.526
Extinction coefficient (per inch)	0.5*
Thickness of each pane (inches)	1/8*
Air gap between layers (inches)	0.5*
Night insulation thermal resistance, when used	R9**
Control range (°F)	
Sunspace	45 to 95
Room	65 to 75**
Thermocirculation vents	
Vent area/projected area (sum of both upper and lower vents)	0.06
Height between vents (ft)	8
Reverse flow	none
Lightweight absorption fraction Simulates effect of solar radiation absorption on lightweight objects by transferring given fraction of transmitted and reflected solar radiation directly to sunspace air.	0.20
Solar Absorptances	
Common wall, lightweight	0.7*
Common wall, masonry	0.8*
Water containers	0.9*
Floor	0.8*
Other surfaces	0.3*
Ground reflectance	0.3*
Internal heat generation	0**
Opaque wall thermal resistance (h · °F · ft²/Btu)	R20
Sunspace infiltration rate (air changes/hour)	0.5*

*Indicates properties that can be varied in the monthly *SLR* method.
**Indicates properties that can be varied in both the annual *LCR* method and the monthly *SLR* method.

Sensitivity Data

Data are presented in Appendix 2 that express the sensitivity of the annual solar savings fraction (*SSF*) to selected design parameters. In any given set of data, one or two parameters are varied while the others are held fixed at their reference values. The reference values are indicated by dots on the curves.

Each set of data is presented for six different U.S. reference cities: Albuquerque, New Mexico; Bismarck, North Dakota; Boston, Massachusetts; Ely, Nevada; Nashville, Tennessee; and Seattle, Washington. The cities were chosen to represent a wide geographical and climatological range. To apply the data, one of the reference cities must be chosen to represent the city of interest on the basis of climate similarity. Two gross measures are of use in making the choice: the 65°F-base heating degree-days and the solar radiation incident on a vertical surface. These data are tabulated as monthly averages for numerous United States and Canadian locations in Appendix 3 under the headings D65 and VS, respectively. Figure 13-2 shows the January values of D65 and VS for each of the six reference cities.

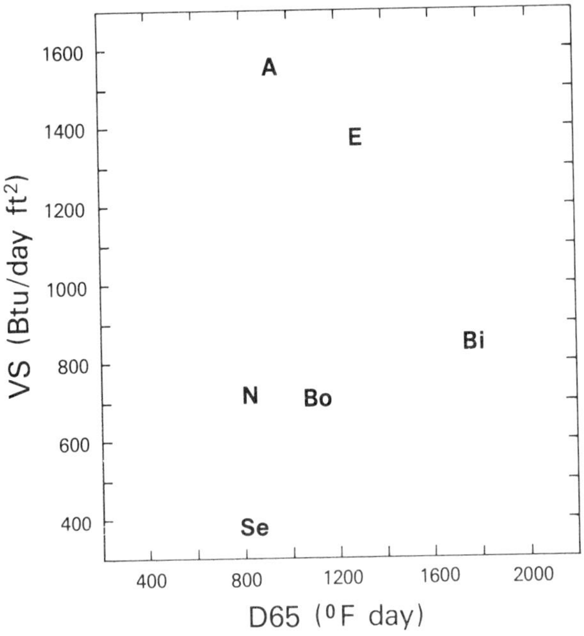

Figure 13-2. The January 65°F-Base Heating Degree-Days, D65, and the Daily Average January Solar Radiation Incident on a Vertical Surface, VS, for the Reference Cities Used for Sunspace Sensitivity Data. A = Albuquerque, Bi = Bismarck, Bo = Boston, E = Ely, N = Nashville, and Se = Seattle.

Two applications of sensitivity data are important. One is quantitative as a final step in a design analysis. In this application, an analysis is performed on the basis of the reference design that most nearly resembles the design of interest. Then the annual *SSF* is corrected by use of sensitivity data.

Another application of sensitivity data is qualitative, wherein the data serve as a guide, perhaps very early in the design process, concerning the relative significance of various design parameters and their preferred values. The discussion of the sensitivity data that follows is in this qualitative vein. Selected data illustrate the discussion, but the full set of sensitivity data are included in Appendix 2.

In this section, the various sets of sensitivity data are described using the Albuquerque data for illustration purposes.

Thermal Storage Volume — A primary element of any passive solar heating system is the thermal storage medium. In the sunspace reference designs, this is either high-density masonry or water. A design parameter of interest is the ratio of the volume of thermal storage material to the projected glazing area, or the number of cubic feet of thermal storage material per square foot of projected area, A_p. In the attached-sunspace, masonry-wall cases, the thermal storage ratio is just the wall thickness: a ratio of 1 foot implies a 1-foot-thick wall. In the lightweight, insulated-wall cases, the thermal storage ratio is the volume ratio of the water storage containers: a ratio of 1 foot implies 1 cubic foot of water per square foot of projected area. The reference design value for the attached sunspaces is 1 foot in both the masonry-wall and insulated-wall cases.

Figure 13-3 illustrates the sensitivity of the annual heating performance to the thermal storage volume ratio for geometry A. For the masonry-wall cases, there are curves for several different masonry densities. The thermal conductivity is strongly correlated with the density and is varied as a function of the density. The relationship by which this is done is $k = 0.049 \exp(0.02\ \rho)$ where k is the thermal conductivity (Btu/h · ft · °F) and ρ is the density (lb/ft³). This is a least-squares fit to data for many concretes and other masonry materials. For 150-lb/ft³ masonry, the performance generally reaches a maximum in the vicinity of a storage volume ratio of 1 foot, that is, a wall thickness of 1 foot. For lower densities, the optimum wall thickness is slightly lower. For the insulated-wall cases, there is no optimum storage volume ratio: the performance increases indefinitely with increasing water volume, although the rate of increase diminishes noticeably for water volumes above about 0.3 to 0.5 ft³/ft² of projected area. A recommended minimum is 0.3, but as much water should be used as is compatible with other design constraints.

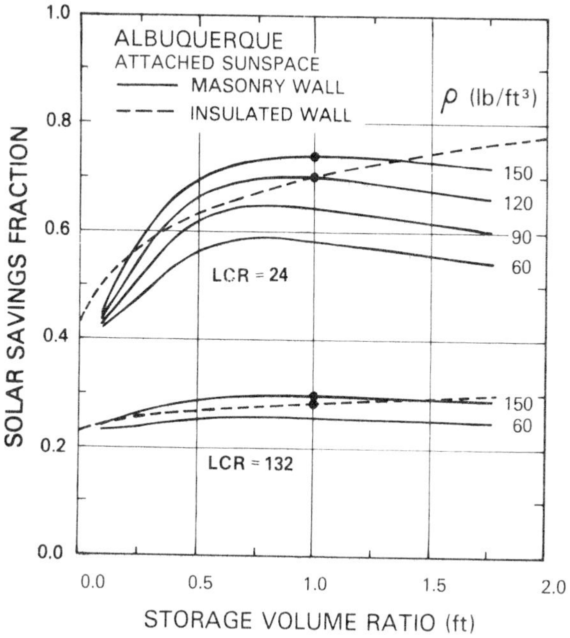

Figure 13-3. Annual Heating Performance Sensitivity to the Ratio of the Thermal Storage Volume (ft³) to the Projected Area (ft²) for Geometry A with Insulated End Walls, No Night Insulation, and Both Masonry and Insulated Walls (A1 and A5). In the masonry-wall case, the storage volume ratio is just the wall thickness. Curves for various masonry densities, ρ, (lb/ft³) are included. In the insulated-wall case, the storage volume ratio is the ratio of the volume of water containers in the sunspace to the projected area. The dots represent the reference conditions.

In the semienclosed, masonry-wall cases, all three common walls (east, north, and west) are masonry, so the volume of thermal storage mass per unit of projected area is greater than in the attached geometries. In fact, it is twice as great for geometry C, 1.69 times as great for geometry D, and 1.93 times for geometry E.

The larger masonry wall area accounts to a large extent for the performance advantage of the semienclosed over the attached geometries. The sensitivites of the performance to the masonry thickness and density are nearly the same as in the attached geometries. The reference thickness is 1 foot, and the reference density is 150 lb/ft³, as in the attached geometries. Data on the sensitivity of geometry C to the masonry storage volume are given in Appendix 2, Figures A2-108 through A2-113.

To be consistent with the semienclosed, masonry-wall cases, the reference design value of the water storage volume in semienclosed, lightweight-wall cases is also proportional to the common-wall area: 2 ft³/ft² of collection area for geometry C, 1.69 for geometry D, and 1.93 for geometry E. Data on the sensitivity to the water storage volume for the geometry C are available in Appendix 2, Figures A2-114 through A2-119.

Water Container Shape — In the insulated-wall cases, the *shape* of the water storage containers has a small influence on the average annual heating performance. One shape parameter is the overall height-to-depth ratio of the row of containers (H/D in Figure 13-4). The reference value of this parameter is 2, but slightly improved performance can be obtained with very small or very large values (always keeping the total volume constant), corresponding to tall, thin containers covering the whole wall or to low, flat containers covering the whole floor. A performance improvement of approximately 5% can be achieved by adopting one of these extreme shapes. The improved performance occurs because these extreme shapes have a higher surface-to-volume ratio, and therefore they absorb more solar radiation and release more heat by convection. Still, the volume of water has the dominant effect on performance, so the correct shape is that which allows the greatest water volume to be placed in the sunspace.

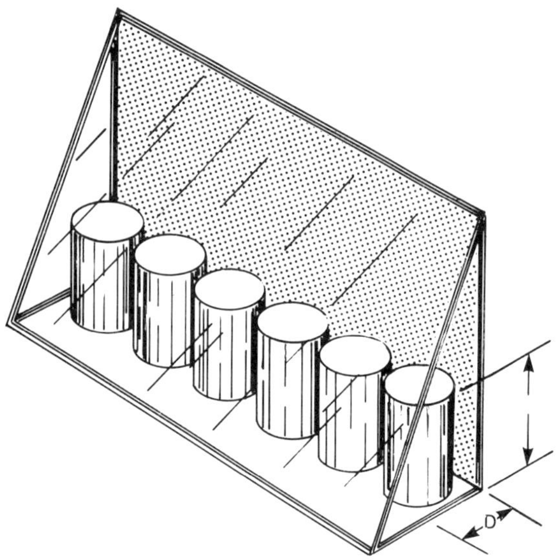

Figure 13-4. Water Container Shape. The containers have a height, H, a north-south depth, D, and a shape parameter, H/D. The reference value of H/D is 2.

Orientation — The performance of a sunspace depends on the orientation azimuth of its principal glazing relative to south. Typical performance sensitivity to orientation is illustrated by Figure 13-5, which shows plots of annual *SSF* versus orientation for geometry A in Albuquerque. The optimum orientation is always close to due south, although the performance is relatively insensitive to orientation within a few degrees of south. Taking an average over all six reference cities, the performance is reduced by about 5% when the orientation is either 30° west or east of due south. The reduction is about 10% at 40°, and 20% at 60°. This estimate pertains to both glazed-end-wall and insulated-end-wall cases. More extreme cases are represented by Seattle and Bismarck. For Seattle, the performance is reduced on the average by about 5% for orientations either 35° west or east of south, about 10% at 50°, and 20% at 70°. The greater latitude for error in the orientation in Seattle is due primarily to the cloudy climate in which a higher percentage of solar radiation is diffuse. Since the diffuse radiation is not strongly directional, optimally oriented glazing is not as important as in a sunnier climate. For Bismarck, the performance is reduced on the average by about 5% for orientations of 20° west or east of south, about 10% at 25°, and 20% at 40°. The smaller latitude for error in the orientation in Bismarck is due primarily to the extremely cold climate: glazings at off-optimum orientations suffer above-average heat losses relative to solar heat gains. As a general rule of thumb, the sunspace glazing may be oriented in a range within 25° to 35° east or west of south and with a performance reduction relative to a due south orientation of about 5% or less. The smaller range of orientation pertains to extremely cold climates, the larger range to mild, cloudy climates.

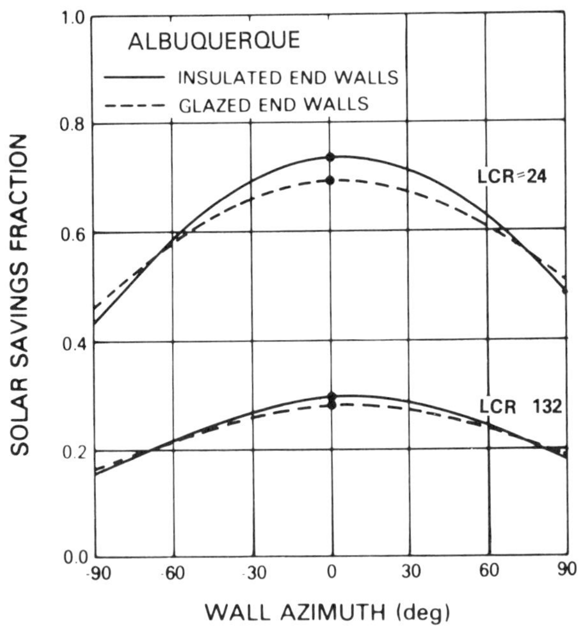

Figure 13-5. Annual Heating Performance Sensitivity to the Sunspace Orientation for Geometry A with a Masonry Common Wall, No Night Insulation, and Both Insulated and Glazed End Walls (A1 and A3). The azimuth is zero when the principal glazing faces due south, positive when east of south, and negative when west of south. The dots represent the reference conditions.

Glazing Tilt — Figure 13-6 illustrates the dependence of annual heating performance on the glazing tilt. In these sensitivity data, the glazing area is held constant while the tilt is varied. This is done by adjusting the height of an opaque knee wall at the base of the glazing. (Note that the presence of the opaque knee wall causes the projected area to vary with the tilt angle. Thus, technically, *LCR* varies with the tilt angle in this case because *NLC* is held constant. However, the *LCR* values that label Figure 13-6 and the corresponding figures in Appendix 2 refer only to the vertical-glazing configuration where the projected area and glazing area are equal.)

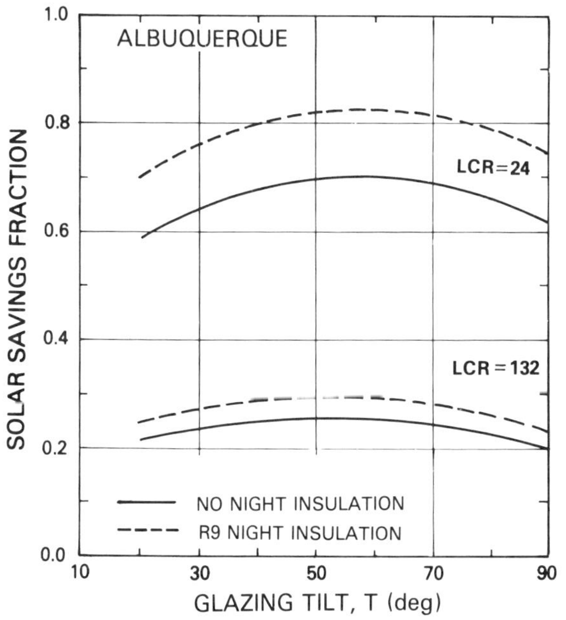

Figure 13-6. Annual Heating Performance Sensitivity to the Sunspace Glazing Tilt in the Case of a Masonry-Common Wall, Insulated-End-Wall Configuration with a Single, Tilted Glazing Plane. The glazing area is held constant as the tilt is varied by the addition of a variable-height knee wall. The drawing below illustrates the geometry.

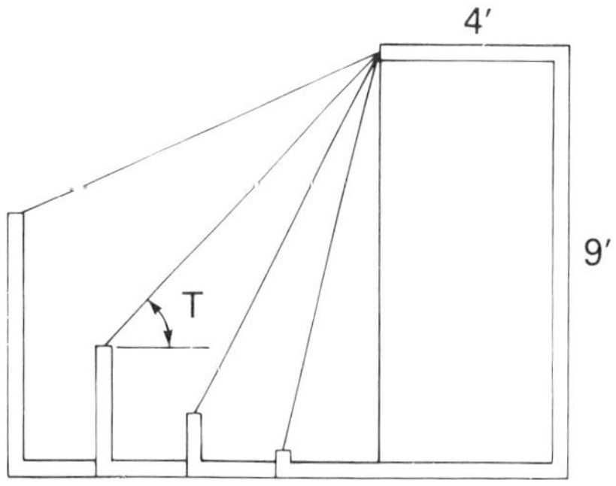

The optimum glazing tilt depends not only on the latitude but also on the climate and the building *LCR*. Therefore, there is no simple rule for tilt that depends on the latitude alone. But the performance is very insensitive to the tilt within several degrees of the optimum. As a general rule, a tilt in the range 50° to 65° is very close to the optimum. Within this range, for every case among those in the sensitivity data, the performance is at least 95% of its optimum value.

A related design issue concerns the glazing tilt without the constraint of holding the glazing area constant. The sensitivity of annual heating performance to glazing tilt in this sense is illustrated by Figures 13-7 and 13-8. Because the glazing area increases as the tilt decreases, the optimum tilt is shifted toward smaller angles. Indeed, in many cases, there is no optimum at all, but rather the heating performance continues to increase without limit as the glazing plane approaches horizontal. This does not mean, however, that very shallow glazing angles necessarily lead to good designs. On the contrary, there are important disadvantages associated with shallow glazings, including higher cost, potential snow buildup, poor water drainage, low head room, and summer overheating problems. A reasonable compromise is to stay within the 50° to 65° tilt range suggested in the previous paragraph. In fact, some designers would emphasize the advantages of steep glazing angles to the point of choosing tilts above the optimum range, perhaps even vertical.

Figures 13-9 and 13-10 illustrate sensitivity data for cases where some portion of the sunspace glazing is vertical. Figure 13-9 illustrates the effect of incorporating increasing amounts of vertical glazing; it shows essentially the transition from geometry A to geometry B. There is a heating performance penalty associated with a vertical glazing segment in a sunspace, but it may be chosen on the basis of other criteria, as discussed in the previous paragraph.

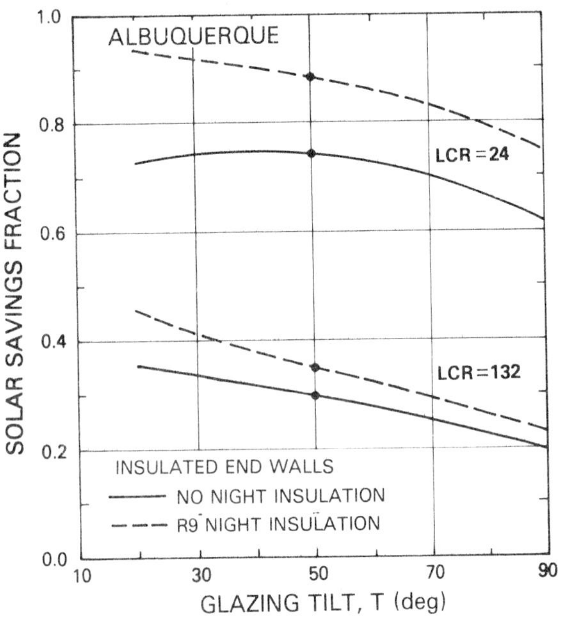

Figure 13-7

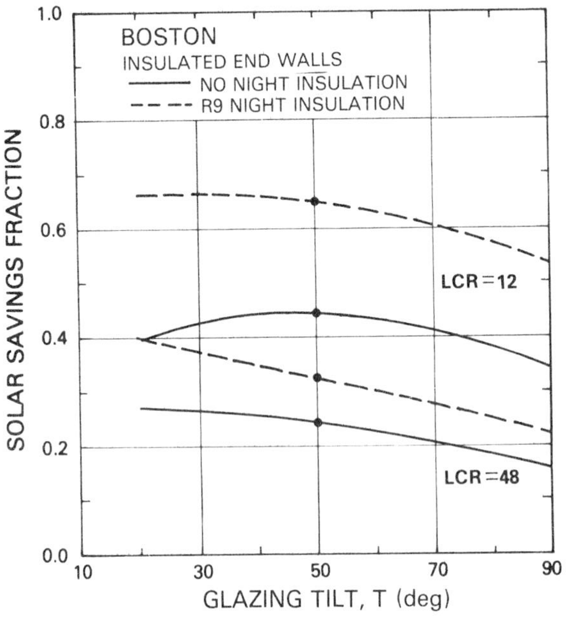

Figure 13-8

Figures 13-7 and 13-8. Annual Heating Performance Sensitivity to the Sunspace Glazing Tilt in the Case of a Masonry-Common-Wall, Insulated-End-Wall, Geometry-A-Like Configuration with a Single, Tilted Glazing Plane. The glazing area is allowed to vary as the tilt is varied. The drawing at the right illustrates the geometry. The dots represent the reference conditions.

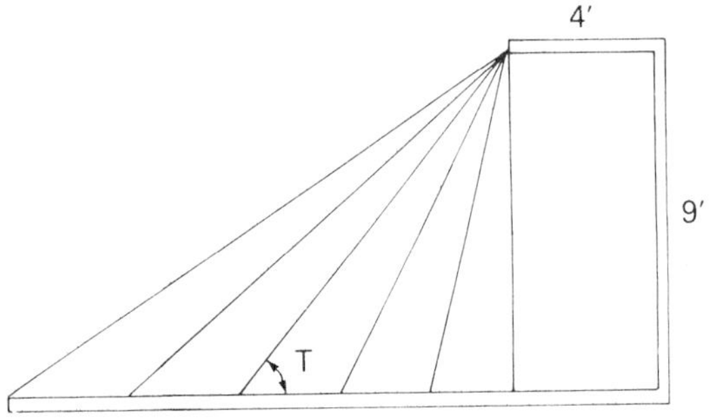

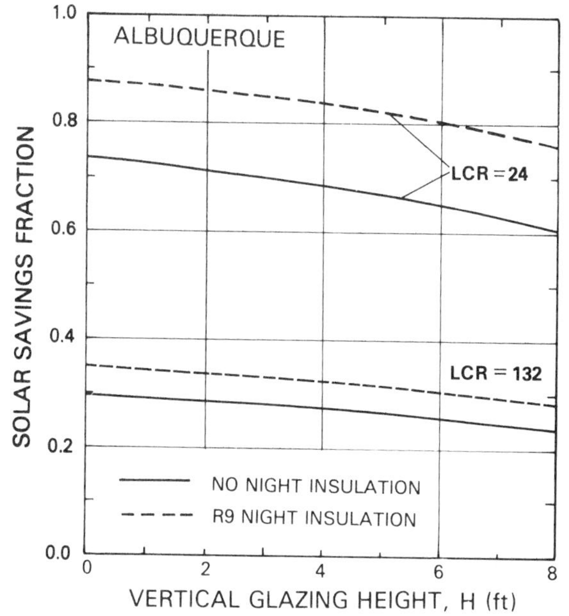

Figurc 13-9. Annual Heating Performance Sensitivity to the Sunspace Vertical Glazing Height for the Masonry-Common-Wall, Insulated-End-Wall, Attached-Sunspace Configuration. The drawing below illustrates the geometry.

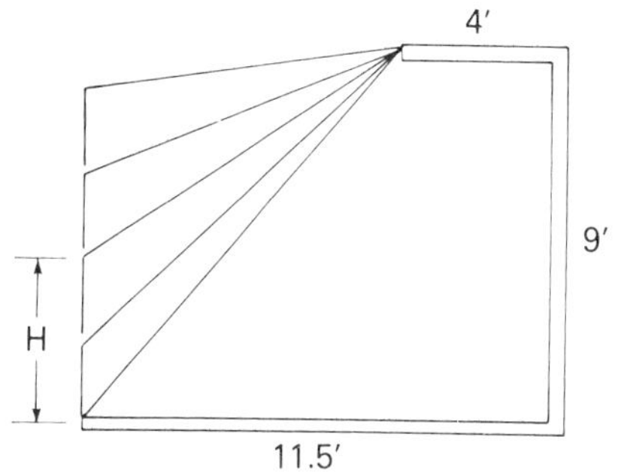

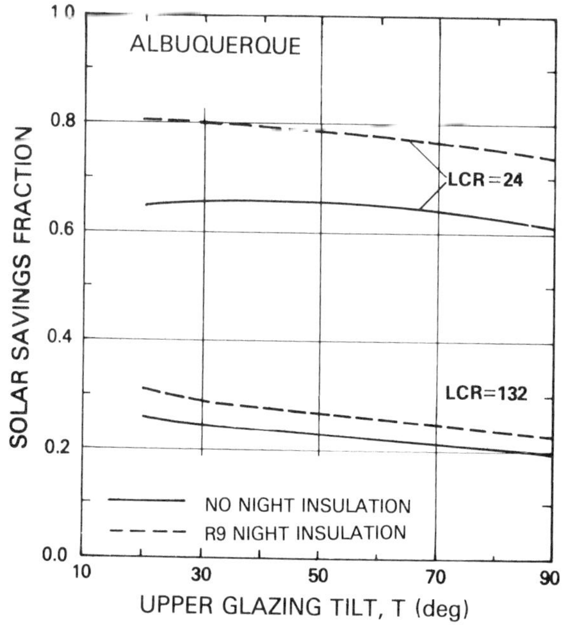

Figure 13-10. Annual Heating Performance Sensitivity to the Sunspace Upper Glazing Tilt in the Case of an Insulated-End-Wall Geometry-B-Like Configuration with a 6-foot Vertical Glazing and a Tilted Upper Glazing. The drawing below illustrates the geometry. When the tilt is 30°, the geometry resembles geometry B except that there is no 4-foot ceiling in geometry B.

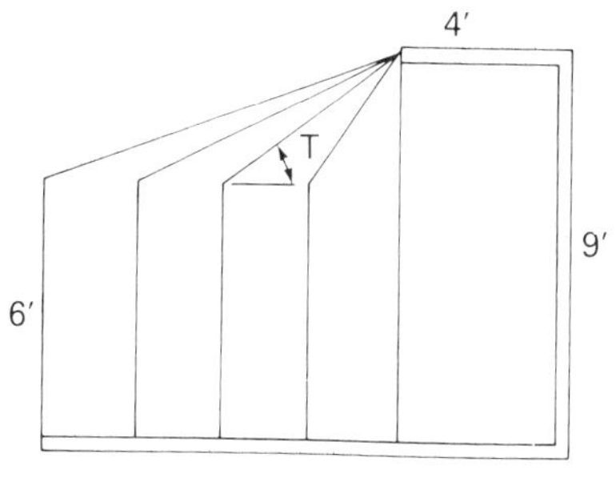

Having chosen a vertical glazing segment, another design decision concerns whether a tilted glazing segment should be placed above the vertical one, and if so, at what tilt. The consequences of this decision are illustrated in Figure 13-10, which assumes a fixed, 6-foot vertical glazing and shows the effect of varying the tilt of the upper glazing segment. The best heating performance can usually be obtained with rather low tilts. Such tilts can also be an advantage for other reasons, such as large floor area and good light distribution, but the disadvantages of low tilts mentioned above apply in this case, too. Fortunately, the sensitivity to this variation is not very great, which leaves the designer relatively free to act on criteria other than heating performance.

Sunspace Width — The east-west width of a sunspace influences its heating performance primarily because the projected area is proportional to the width. But a more subtle question concerns the effectiveness of a *unit* of projected area as a function of sunspace width. This relationship for attached sunspaces is illustrated by Figure 13-11. (Note that the curves are plotted for fixed values of *LCR*. Therefore, as the width varies, the building load is implicitly varied also so that *LCR* remains constant.) It is seen that the performance generally improves with sunspace width, but the effect is small beyond sunspace widths of 15 to 20 feet. The effect is more significant for glazed-end-wall cases than for insulated-end-wall cases because the solar radiation and heat losses through the glazed end walls are more severe. The situation is different for semienclosed sunspaces. In fact, the performance in most cases actually decreases as the sunspace width increases, as illustrated by Figure 13-12. Three reasons for this can be identified: (1) the end-wall "losses" are to the adjoining rooms and therefore not losses at all, (2) the thermal storage mass per unit of projected area decreases as the sunspace width increases because of the fixed amount of mass associated with the east and west end walls, and (3) the fraction of solar radiation transmitted through the glazing that is absorbed in the sunspace decreases as the sunspace width increases. Point 3 is due to the reduced effectiveness of the sunspace as a cavity absorber as the width increases.

Number of Glazing Layers — Figures 13-13 and 13-14 illustrate the sensitivity data on the number of layers of glazing in the sunspace. The number of glazings affects the performance in two ways. Each glazing layer reflects and absorbs a fraction of the incident solar radiation and so reduces the solar gain to the sunspace. But each glazing layer increases the thermal resistance of the glazing system and so decreases the heat loss from the sunspace. There is therefore a trade-off between these two effects, one tending to decrease heating performance and the other to increase it. Each added layer of glazing reduces the solar gain by approximately the same fraction but reduces the heat loss by a diminishing fraction. So the benefit of added glazings diminishes with number to the point where the overall performance begins to fall. The optimum number of glazings depends on the climate and the system design, especially on the presence of night insulation. If night insulation is used, double glazing is the best choice except in sunny, mild climates where single glazing may perform as well or better than double glazing. Without night insulation, a minimum of two glazings would almost always be advisable. In cold climates, three or even four glazings benefit performance.

Glazing Air Gap — The thermal resistance of double glazing depends on the width of the air gap between the layers. Therefore, the annual heating performance also depends on the air gap, as illustrated by Figure 13-15. The performance is sensitive to the air gap primarily in the range below 0.5 inch, but in very cold climates, an air gap of up to 1 inch may be worthwhile. The reference designs have a 0.5-inch air gap.

Glazing Thickness — The amount of solar radiation absorbed in the glazing material depends on the number and thickness of the layers and the transparency of the glazing material. The transparency of glazing material is expressed by the *extinction coefficient*, which is the fraction of solar radiation absorbed per unit length of travel in the material. For ordinary window glass, the extinction coefficient is about 0.5 per inch. Figure 13-16 illustrates the sensitivity of annual heating performance to glazing thickness, assuming an extinction coefficient of 0.5 per inch. The reference designs have a glazing thickness of 0.125 inch (1/8 inch).

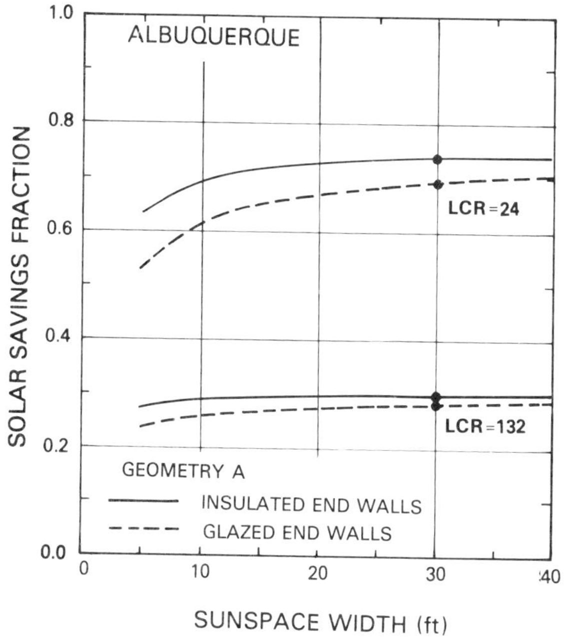

Figure 13-11. Annual Heating Performance Sensitivity to the Sunspace East-West Width for Geometry A with Masonry Common Wall, No Night Insulation, and Both Insulated and Glazed End Walls (A1 and A3). The dots represent the reference conditions. Note that *LCR* is fixed so that as the width increases the load is implicitly increased.

Figure 13-12. Annual Heating Performance Sensitivity to the Sunspace East-West Width for Geometry C with a Masonry Common Wall, Both Without and With R9 Night Insulation (C1 and C2). The dots represent the reference conditions.

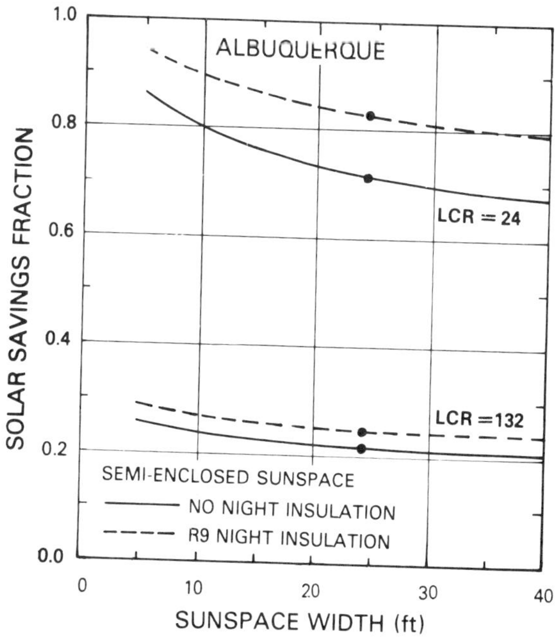

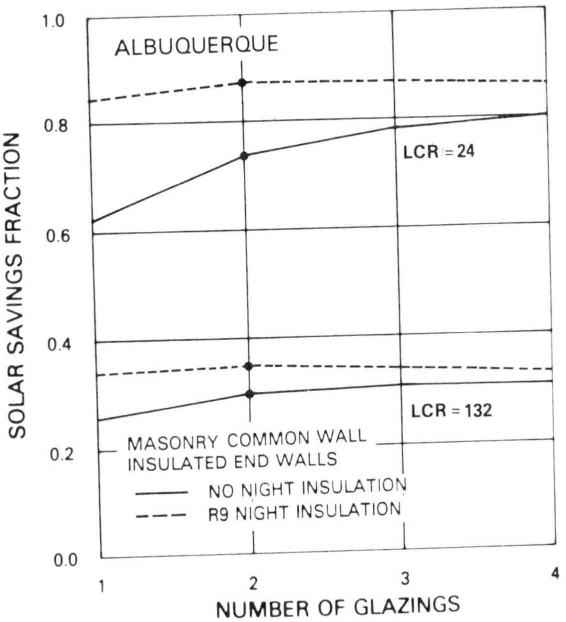

Figure 13-13

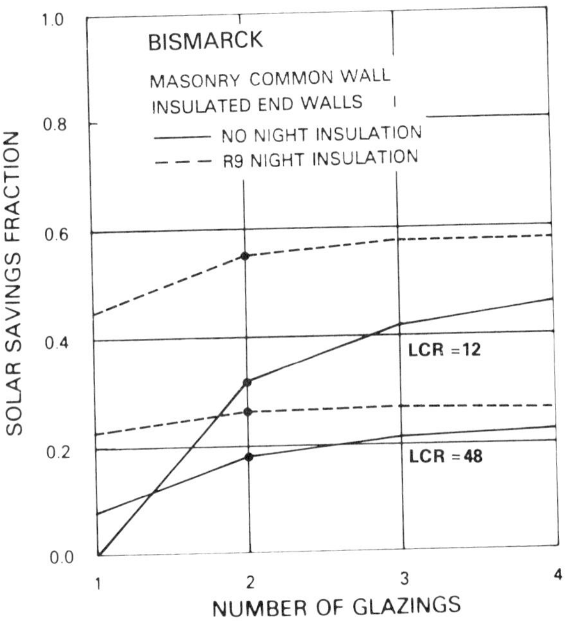

Figure 13-14

Figures 13-13 and 13-14. Annual Heating Performance Sensitivity to the Number of Layers of Sunspace Glazing for Geometry A with Masonry Common Wall, Insulated End Walls, Both Without and With R9 Night Insulation (A1 and A2). The dots represent the reference conditions.

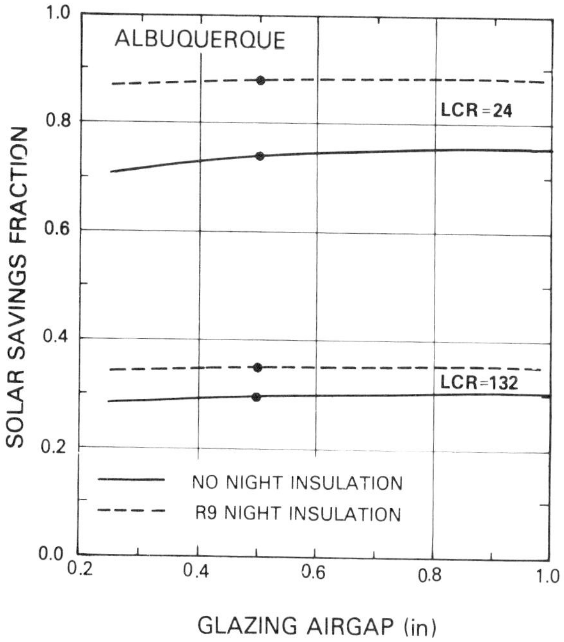

Figure 13-15. Annual Heating Performance Sensitivity to the Width of the Air Gap Between the Glazing Layers for Geometry A with Masonry Common Wall, Insulated End Walls, Both Without and With R9 Night Insulation (A1 and A2). The dots represent the reference conditions.

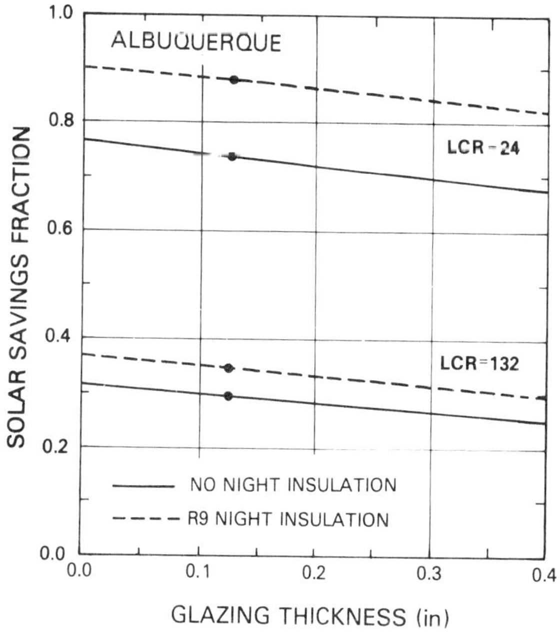

Figure 13-16. Annual Heating Performance Sensitivity to the Thickness of Each Glazing Layer for Geometry A with Masonry Common Wall, Insulated End Walls, Both Without and With R9 Night Insulation (A1 and A2). The dots represent the reference conditions.

The glazing thickness sensitivity data may be used for other extinction coefficients by using an equivalent thickness that equals the actual thickness multiplied by the ratio of the actual extinction coefficient to the reference extinction coefficient. Table 13-3 lists approximate extinction coefficients and thickness multipliers.

Table 13-3

Equivalent Thickness

Material	Extinction Coefficient (per in.)	Multiply Thickness by
Low-iron glass	0.1	0.2
Common glass	0.5	1.0
Fiberglass reinforced polyester	2.0	4.0
Acrylic	0.2	0.4
Polycarbonate	0.6	1.2

Solar Absorptances

Figures 13-17 through 13-20 illustrate the sensitivity of annual heating performance to the solar absorptances of various sunspace interior surfaces. Absorptances of various materials are listed in Appendix 6, Table A6-2. In Figure 13-17, the absorptance of the sunspace masonry common wall is varied while the absorptances of the other surfaces are held constant. The reference designs have a masonry common-wall absorptance of 0.8. In Figure 13-18, the absorptance of the sunspace floor is varied. The floor also has a reference absorptance of 0.8. In Figure 13-19, the absorptances of both the floor and wall are varied simultaneously. Figure 13-20 illustrates the sensitivity of the annual heating performance of the lightweight-wall reference designs to the absorptance of the water containers. The reference absorptance is 0.9. Dark-colored (high-absorptance) sunspace surfaces tend to perform better than light-colored ones because less solar radiation is reflected out through the glazing, but dark *mass* surfaces (wall, floor, and water containers) are especially important because solar radiation absorbed by massive elements is partially retained as stored heat to be released later when it is needed. The performance sensitivity to mass absorptance is greatest for high-*SSF* (low-*LCR*) designs where heat storage is essential in order to displace auxiliary heat during nighttime or cloudy periods. For low-*SSF* (high-*LCR*) designs, high-mass absorptance is not so important because most of the solar energy collected is needed immediately for daytime heating requirements.

The sensitivity of the annual heating performance to the absorptance of lightweight ceiling and end walls is negligible. (No data are presented.) The reason is the near cancellation of opposing effects: diffuse reflections from lightweight surfaces are partly lost from the sunspace by transmission outward through glazings and partly absorbed by massive elements in the sunspace. Therefore, criteria other than heating performance may be used to select the finishes for lightweight surfaces. It is probably good design practice to use light colors on these surfaces to improve the light distribution in the sunspace and offset the visual effect of dark colors on the mass surfaces.

Lightweight Objects — Lightweight objects such as furniture and plants in the sunspace have the effect of rapidly transferring some solar energy to the air. This is because the solar radiation absorbed by lightweight objects causes a rapid surface temperature increase and consequently a rapid transfer of heat by convection to the surrounding air. This effect has been simulated by assuming that a certain fraction, the "lightweight absorption fraction," of solar radiation directly heats the sunspace air after it is transmitted through the glazing and each time it is reflected from interior surfaces. The lightweight absorption fraction is 0.2 in all of the sunspace reference designs. This is in addition to heat transferred rapidly to the sunspace air due to absorption on the lightweight sunspace surfaces, namely, the ceiling and, when lightweight, the end walls and common wall. (Note that the lightweight absorption fraction simulates only one aspect of plants in the sunspace. The other aspect, which may have a very important effect on sunspace performance, is the transfer of latent heat due to water evaporation. The effect of evaporation is not explicitly accounted for in the analysis methods presented here.)

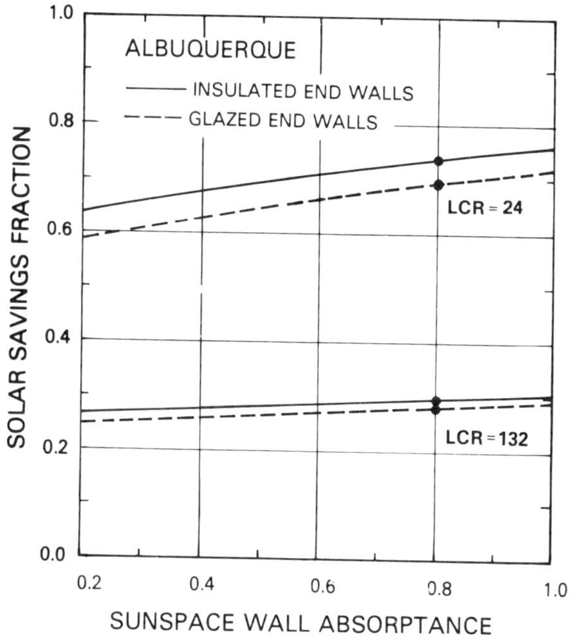

Figure 13-17. Annual Heating Performance Sensitivity to the Solar Absorptance of the Sunspace Common Wall for Geometry A, with a Masonry Common Wall, No Night Insulation, and Both Insulated and Glazed End Walls (A1 and A3). The dots represent the reference conditions.

Figure 13-18. Annual Heating Performance Sensitivity to the Solar Absorptance of the Sunspace Floor for Geometry A with a Masonry Common Wall, No Night Insulation, and Both Insulated and Glazed End Walls (A1 and A3). The dots represent the reference conditions.

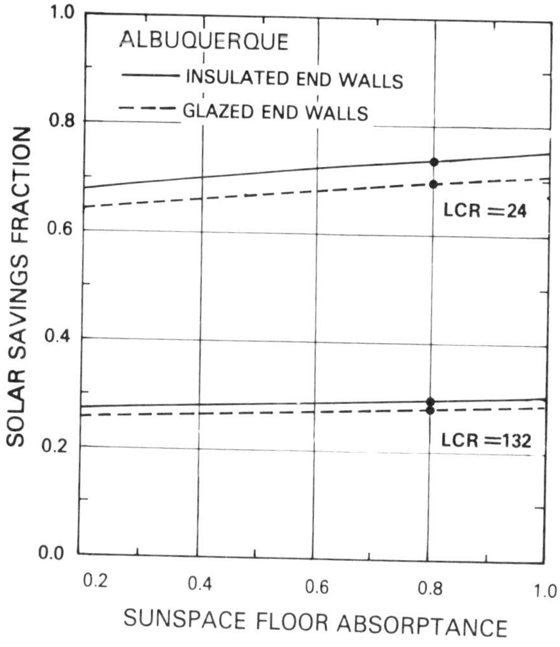

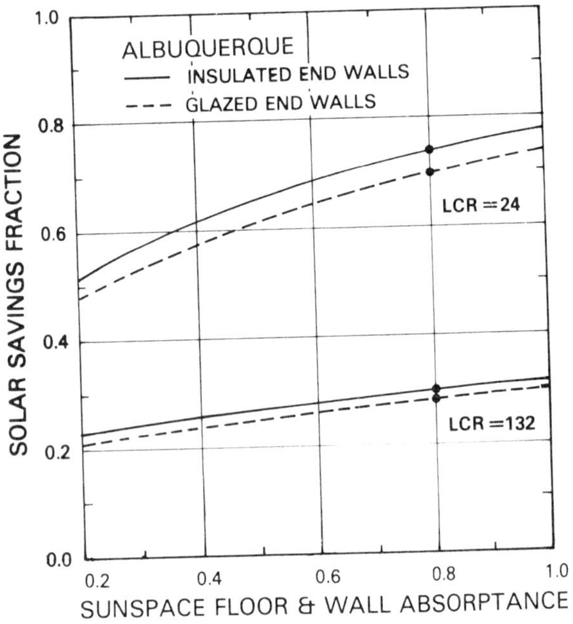

Figure 13-19. Annual Heating Performance Sensitivity to the Solar Absorptances of the Sunspace Floor and Common Wall, Varied Simultaneously, for Geometry A with Masonry Common Wall, No Night Insulation, and Both Insulated and Glazed End Walls (A1 and A3). The dots represent the reference conditions.

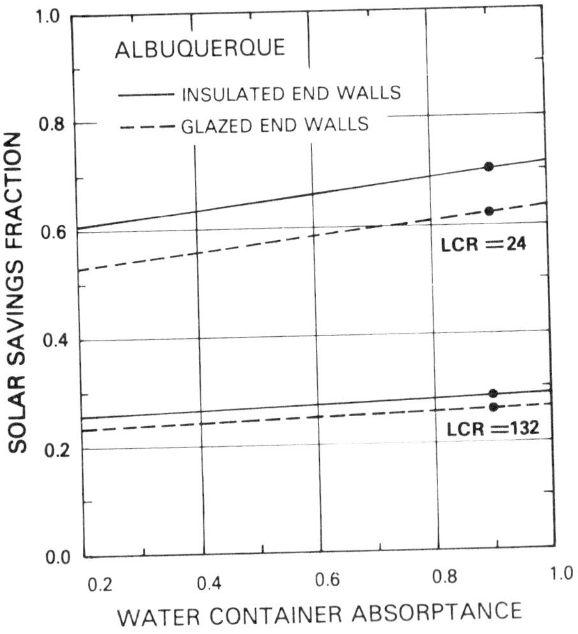

Figure 13-20. Annual Heating Performance Sensitivity to the Solar Absorptance of the Sunspace Water Containers for Geometry A with a Lightweight, Insulated Common Wall, No Night Insulation, and Both Insulated and Glazed End Walls (A5 and A7). The dots represent the reference conditions.

The sensitivity of the annual heating performance to the lightweight absorption fraction is illustrated in Figure 13-21. It is seen that the annual heating performance can be seriously degraded by the presence of lightweight objects in high-*SSF* (low-*LCR*) designs because a high *SSF* requires substantial daytime heat storage with which lightweight objects interfere. For low-*SSF* (high-*LCR*) designs, on the other hand, the performance is not very sensitive to the lightweight absorption fraction because the solar gains are mainly needed for daytime air heating. In fact, the performance in these cases can even be improved slightly by a moderate lightweight absorption fraction.

Determination of the lightweight absorption fraction in a given sunspace can only be done in practice as a rough estimate. Prediction of the lightweight absorption fraction of a sunspace under design may not be possible because it depends on the way in which the sunspace is later furnished. But the following precedure may be useful, at least as a way of visualizing in an approximate way the meaning of the lightweight absorption fraction.

1. Stand near the south glazing, facing north. Estimate the fraction of your view of the sunspace floor, north wall, and ceiling that is intercepted by lightweight objects. If the lightweight objects have substantially different solar absorptances, do step 1 separately for each different absorptance.

2. Estimate the solar absorptances of the lightweight objects.

3. Multiply the fractions from step 1 by the absorptances from step 2 and add the products. The result is an estimate of the lightweight absorption fraction.

Vent Area — It is assumed in all of the sunspace reference designs that convective heat transfer occurs from the sunspace to the adjoining rooms through openings in the sunspace common wall. This convection occurs freely due to the temperature differences between the sunspace and adjoining rooms. (Reverse flow from the rooms to the sunspace is assumed to be prevented by dampers.) The amount of convective flow through the openings depends in part on the area of the openings. The effectiveness of the area depends on the configuration of the openings, in particular on their vertical dimensions, or on the vertical distance between their centers when there are pairs of openings, one above the other. In the reference designs, it is assumed that the openings are in the form of Trombe-like thermocirculation vents, that is, pairs of vents with one near the floor and one near the ceiling. The reference height between the vent centers is 8 feet. The reference designs all have a combined vent area of 6% of the projected area, as shown in Figure 13-22. Figure 13-23 illustrates the sensitivity of the annual heating performance to the combined vent area expressed as a fraction of the projected area.

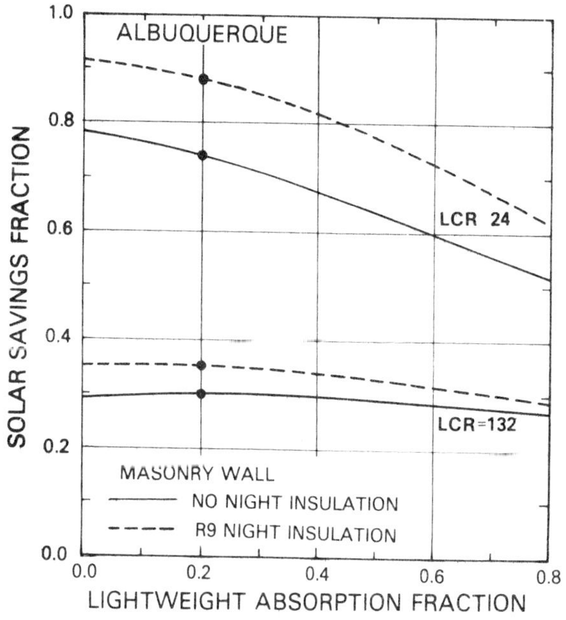

Figure 13-21. Annual Heating Performance Sensitivity to the Sunspace Lightweight Absorption Fraction for Geometry A with Masonry Common Wall, Insulated End Walls, and Both Without and With R9 Night Insulation (A1 and A2). The lightweight absorption fraction is the fraction of solar radiation that directly heats the sunspace air when it is transmitted through the glazing and when it is reflected from interior surfaces. It is intended to simulate the presence of lightweight objects in the sunspace that absorb solar radiation and rapidly convect heat to the sunspace air. The dots represent the reference conditions.

If the vertical height between the vent centers is not the 8-foot reference height, an equivalent area is obtained by multiplying the actual area by a factor from Table 13-4. (The relationship is that the parameter $A\sqrt{H}$ characterizes the vents with respect to their performance; therefore, $A' = A\sqrt{H/H'}$ where A' is the equivalent area, A is the actual area, H' is the 8-foot reference height, and H is the actual height.)

Table 13-4

Thermocirculation Vent Equivalent Area

Vertical Height Between Vent Centers (ft)	Multiply Area by
4	0.71
6	0.87
8	1.00
10	1.12
12	1.22
14	1.32
16	1.41
18	1.50
20	1.58

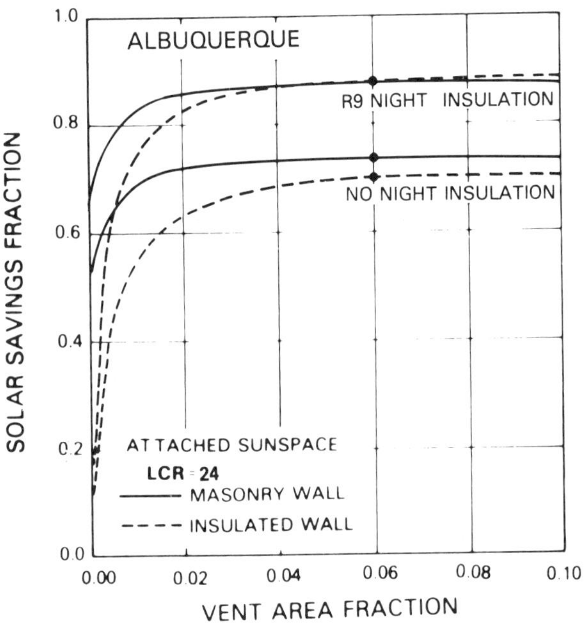

Figure 13-23. Annual Heating Performance Sensitivity to the Ratio of the Common Wall Thermocirculation Vent Area to the Total Common Wall Area for Geometry A with Insulated End Walls, with Both Masonry and Insulated Common Wall, and Both Without and With R9 Night Insulation (A1, A2, A5, and A6). The vent area is the combined area of both top and bottom vents. The dots represent the reference conditions.

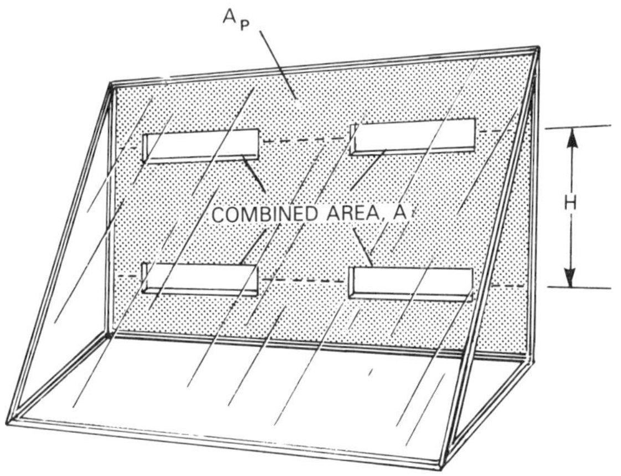

Figure 13-22. Common Wall Vents. The projected area, A_p, equals the north common-wall area in the sunspace reference designs. The combined vent area is A. The vent area fraction is A/A_p (0.06 for the reference designs). The vertical distance between the vent centers is H (8 feet for the reference designs).

If the common wall openings are single openings (windows and doors) rather than vent pairs, then the height of the opening determines the effectiveness of the area. (The "height" in this case means the vertical distance between top and bottom of the opening, not the distance from the floor to the opening.) The equivalent area multiplier is found from Table 13-5.

Table 13-5

Single Opening Equivalent Area

Height of Opening (ft)	Multiply Area by
1	0.25
2	0.35
3	0.43
4	0.50
6	0.61
8	0.71
10	0.79

Sunspace Infiltration Rate — Figure 13-24 illustrates the sensitivity of the annual heating performance to the sunspace infiltration rate. There is not as much performance sensitivity to the sunspace infiltration rate as might have been thought because the dominant sunspace heat losses are the glazing losses. Nevertheless, the infiltration component is significant and should be minimized by use of good caulking and weatherstripping practice, especially in very cold climates.

Common-Wall Thermal Resistance — The common-wall thermal resistance pertains to the lightweight, insulated-wall cases. Figures 13-25 and 13-26 illustrate the sensitivity of the annual solar savings to the common-wall thermal resistance. The R-value plotted in these figures includes only the conductive resistance; the wall-surface-to-air resistances are accounted for separately. For example, an R-value of zero refers to a wall that is an opaque film with no conductive resistance.

It is seen that the only important sensitivity of annual heating performance to the common-wall thermal resistance is below an R-value of about 5, and even then only in the relatively cold, cloudy climates. This is because daytime heat transfer from the sunspace to adjoining rooms is predominantly by convection through openings, and nighttime losses by conduction through the wall are small due to a typically small temperature difference. Caution should be used, however, in concluding on the basis of these data that insulation need not be used in a lightweight common wall. In sunny, mild climates, the heating performance will not be much affected by the amount of common-wall insulation, but insulation may be very important in these cases to minimize summer heat gain through the common wall. In cloudy, cold climates, some common-wall insulation should be used, especially if a design has less thermal storage in the sunspace or less sunspace thermal resistance to the exterior than the reference designs. There may then be significant heat loss through the common wall if it is not well insulated.

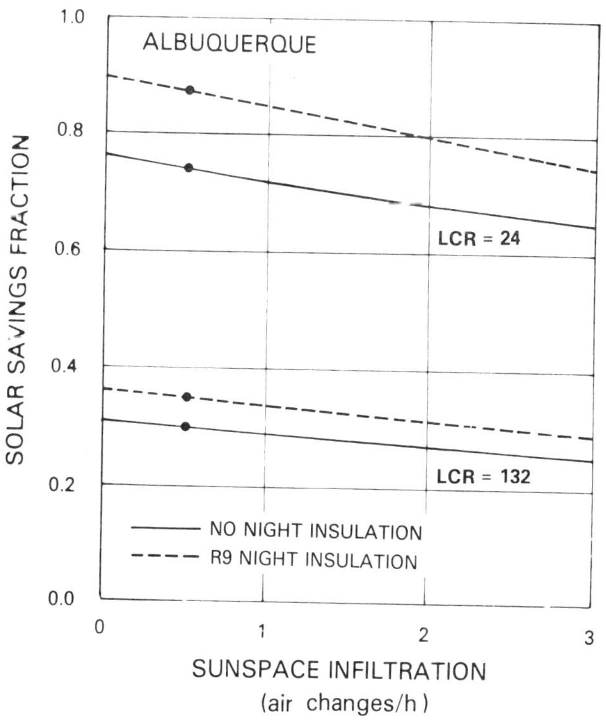

Figure 13-24. Annual Heating Performance Sensitivity to the Sunspace Infiltration Rate for Geometry A with a Masonry Common Wall, Insulated End Walls, Both Without and With R9 Night Insulation (A1 and A2). The dots represent the reference conditions.

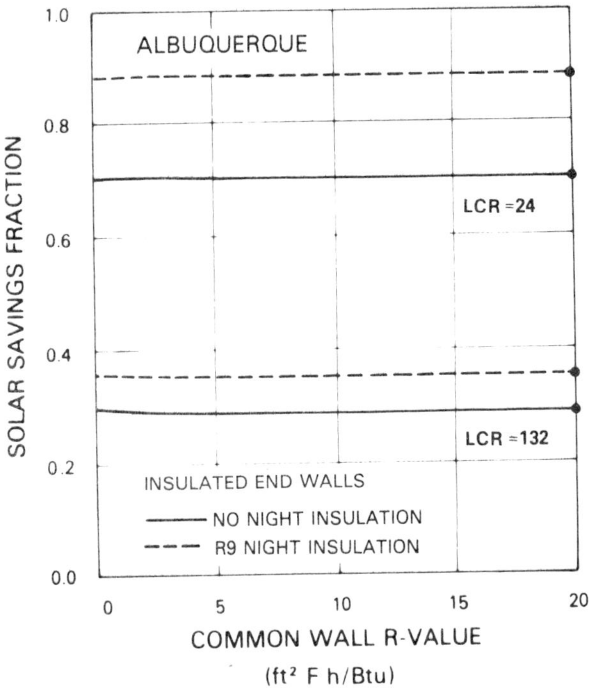

Figure 13-25

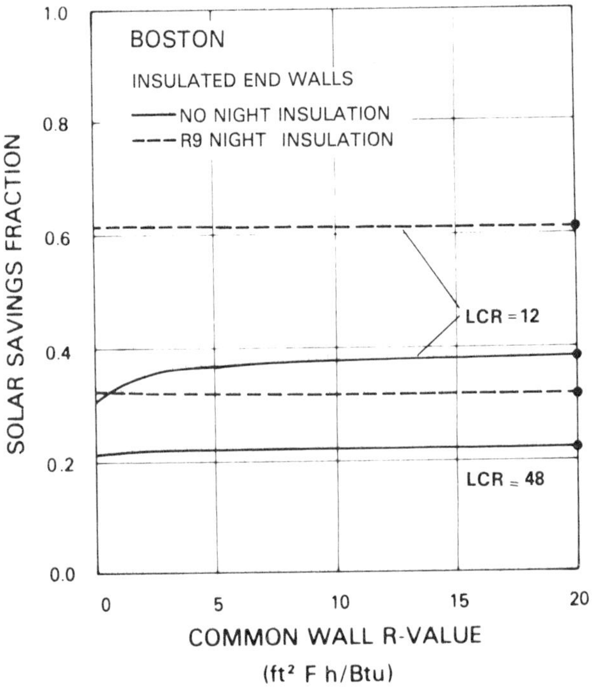

Figure 13-26

Figures 13-25 and 13-26. Annual Heating Performance Sensitivity to the Thermal Resistance of the Sunspace Common Wall for Geometry A with a Lightweight, Insulated Common Wall, Insulated End Walls, and Both Without and With R9 Night Insulation (A5 and A6). The wall-surface-to-air resistances are not included. The dots represent the reference conditions.

Thermostat Setpoints — Although the thermostat setpoints are not design parameters but rather features of occupant behavior, it is important to realize that they may have a significant effect on the building heating performance. In particular, the annual solar savings is very sensitive to the room minimum thermostat setpoint (that is, the auxiliary heat thermostat for the rooms adjoining the sunspace). This is illustrated in Figures 13-27 and 13-28. Note that *SSF* in Figure 13-27 is in relation to a reference building with the variable minimum setpoint. Therefore, the solar savings sensitivity to the setpoint does not reflect the sensitivity of the building load to the setpoint. On the other hand, *SSF* in Figure 13-28 is defined relative to a reference building with a fixed 65°F minimum setpoint. The data, therefore, reflect the sensitivity of the total energy savings to the setpoint.

No account was taken of incidental internal heat generation (appliances, people, etc.) in these data. With a normal amount of internal heat and a normal building heat loss coefficient, internal sources can be expected to raise the room temperature by 5° to 7°F. In buildings with lower than average heat loss characteristics, the effect on room temperature will be greater. This means that the room setpoint sensitivity data may be corrected to account for internal heat generation by modifying the setpoint temperatures on the horizontal axis. To make this correction, increase the temperatures by the quantity

$$\Delta T = \frac{Q_{int}}{TLC} \qquad (13\text{-}1)$$

where Q_{int} is the internal heat generation rate (Btu/day) and *TLC* is the total load coefficient (Btu/°F · day).

The total load coefficient is the net load coefficient plus the solar wall load coefficient.

The sunspace in the reference designs is heated as necessary to prevent its air temperature from falling below 45°F. The sunspace auxiliary heat is included in the definition of *SSF*:

$$SSF = 1 \times \frac{Q_{aux}}{Q_{net}} \ . \qquad (13\text{-}2)$$

Therefore, *SSF* is sensitive to the sunspace minimum thermostat setpoint to the extent that sunspace heating is needed. This sensitivity is illustrated for both Albuquerque and Bismarck in Figures 13-29 and 13-30. It is seen that there is little sensitivity to the minimum sunspace thermostat setpoint in the range 40°F to 50°F except in extremely cold climates, as represented by Bismarck. This means that an insignificant amount of auxiliary heat is required by the sunspace up to about 50°F except in as severe a climate as Bismarck. With R9 night insulation, essentially no sunspace auxiliary heat is needed up to about 55°F and very little even up to 65°F. This does not mean, however, that a sunspace temperature would never fall below the indicated ranges. There can be expected to be occasional cold, cloudy periods where some auxiliary heat is needed, for example, to protect temperature-sensitive plants. (This can normally be done by simply keeping a door or window open to the sunspace so that some auxiliary heat supplied to an adjoining room can flow to the sunspace.)

However, the total auxiliary heat required by the sunspace over a long term will be small compared to the total building auxiliary heat. All this is conditioned, of course, on the sunspace in question having a heat storage capacity and a thermal integrity comparable to the reference designs, namely, one cubic foot of water or high-density masonry per square foot of projected area, two glazing layers, R20 insulation in opaque exterior walls, and a 0.5 air changes per hour infiltration rate.

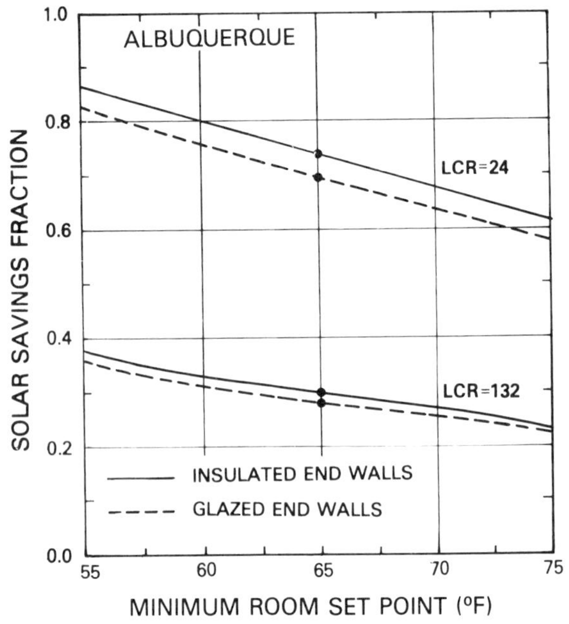

Figure 13-27. Annual Heating Performance Sensitivity to the Room Auxiliary Heat Thermostat Setpoint Temperature for Geometry A with Masonry Common Wall, No Night Insulation, and Both Insulated and Glazed End Walls (A1 and A3). The solar savings fraction in this case is defined relative to a net reference load evaluated at the variable setpoint. The dots represent the reference conditions.

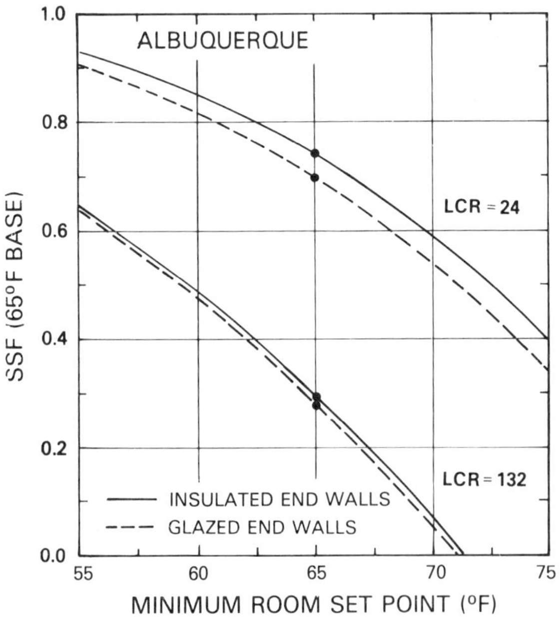

Figure 13-28. Annual Heating Performance Sensitivity to the Room Auxiliary Heat Thermostat Setpoint Temperature for Geometry A with a Masonry Common Wall, No Night Insulation, and Both Insulated and Glazed End Walls (A1 and A3). The solar savings fraction is this case is defined relative to a 65°F reference load. The dots represent the reference conditions.

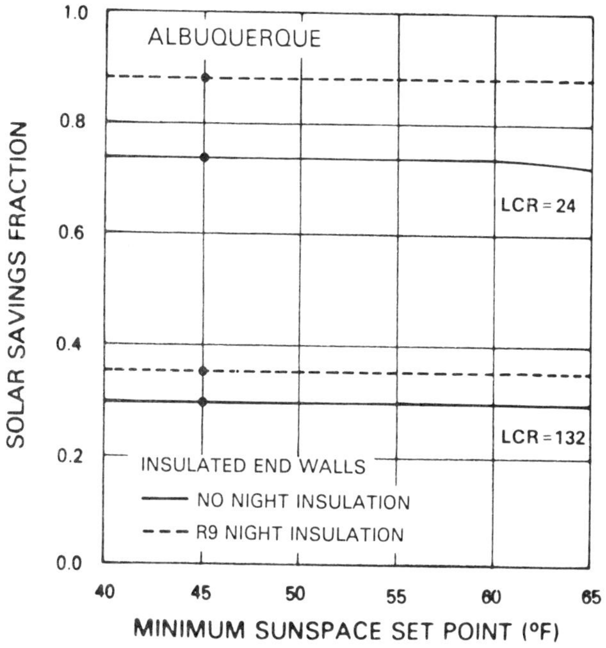

Figure 13-29

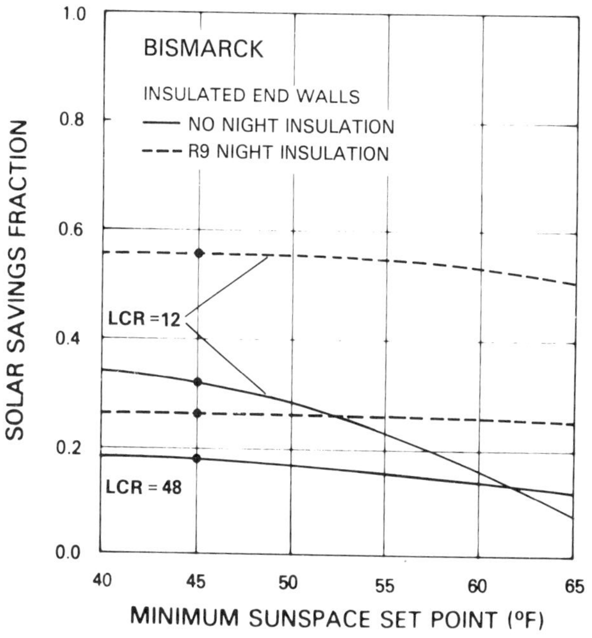

Figure 13-30

Figures 13-29 and 13-30. Annual Heating Performance Sensitivity to the Sunspace Auxiliary Heat Thermostat Setpoint Temperature for Geometry A with a Masonry Common Wall, Insulated End Walls, Both Without and With R9 Night Insulation (A1 and A2). The definition of solar savings fraction includes the sunspace auxiliary heat.

CHAPTER 14
THERMAL STORAGE WALL SYSTEMS

Introduction

A thermal storage wall consists of the wall itself, constructed of masonry (Trombe walls) or of water-filled containers (water walls), and the glazing over its outer surface. The glazing on the outer surface transmits solar radiation, which is absorbed by, and heats, the wall surface. The large mass of the wall serves to store the solar energy. Heat losses from the wall surface back to the outside are reduced by the glazing. Heat conducts slowly through a Trombe wall and then is transferred to the adjacent room by radiation and convection from the room-side surface. Water walls transfer heat rapidly, by convection currents in the water, across the container.

The glazing used for a thermal storage wall should have a high solar transmittance to allow the greatest possible fraction of the solar radiation to reach the wall surface. The solar absorptance of the wall surface should be high to assure that most of the solar radiation that reaches the wall surface is absorbed. The glazing should also have low transmittance in the infrared spectrum to trap heat reradiated from the wall surface. Movable night insulation over the glazing improves performance. A selective surface on the storage wall also improves performance.

This manual treats thermal storage wall systems with one, two, and three layers of glazing with and without night insulation; and selective-surface thermal storage walls with one and two layers of glazing with and without night insulation. A recommended minimum is two layers of glazing in most circumstances. One glazing layer may be sufficient on a selective-surface wall or in mild climates if night-insulated.

Vents may be incorporated into the top and bottom of a Trombe wall to provide convective transfer of heat to the adjacent room. The air in the space between the Trombe wall surface and the glazing will be heated and will rise, passing through the upper vent into the room. Return air will flow from the room through the lower vent. The use of vents will provide rapid daytime heating and a modest increase in the total amount of heat supplied to the building. At night, the vents must be closed; otherwise, reverse circulation will occur, warm room air will flow to the outer surface and be cooled, and cold air will be delivered to the living space through the lower vent.

Overheating during summer months can be reduced by an overhang that shades the thermal storage wall. Diffuse solar radiation will still be absorbed and this may cause overheating problems in some locations, but a cover over the glazing will block the diffuse radiation. Vents to the outside (external vents) can be used to circulate ambient air across the wall surface to cool it but these have been shown to provide little benefit. Note that summer external vents can be defeated by a southerly breeze that will flow into the upper external south vent, forcing warm air into the building. This can be prevented by using a backdraft damper or a wind turbine.

The thermal storage wall reference designs are divided into two major categories: water walls (WW) and Trombe walls (TW). The parameters that are varied among the water wall systems are the thermal storage capacity (or wall thickness), the number of layers of glazing, the wall surface optical properties (normal or selective), and night insulation (with or without). Table 14-1 lists the 15 different combinations of these parameters that make up the water wall reference designs.

The Trombe wall reference designs are split into two subcategories: vented and unvented. For both subcategories, the parameters that are varied among the Trombe wall reference designs are the thermal storage capacity (expressed also as wall thickness), the number of glazings, the wall surface (normal or selective), night insulation (with or

Table 14-1

Water Wall Reference Designs

Designation	Thermal Storage Capacity* (Btu/ft^2 · °F)	Wall Thickness (in.)	No. of Glazing Layers	Wall Surface	Night Insulation
WWA1	15.6	3	2	normal	no
WWA2	31.2	6	2	normal	no
WWA3	46.8	9	2	normal	no
WWA4	62.4	12	2	normal	no
WWA5	93.6	18	2	normal	no
WWA6	124.8	24	2	normal	no
WWB1	46.8	9	1	normal	no
WWB2	46.8	9	3	normal	no
WWB3	46.8	9	1	normal	yes
WWB4	46.8	9	2	normal	yes
WWB5	46.8	9	3	normal	yes
WWC1	46.8	9	1	selective	no
WWC2	46.8	9	2	selective	no
WWC3	46.8	9	1	selective	yes
WWC4	46.8	9	2	selective	yes

*Per unit of projected area

without), and the masonry properties (expressed as the product ρck). The 21 combinations of these parameters used for the vented and unvented Trombe wall reference designs are shown in Table 14-2.

Additional characteristics of thermal storage walls (both Trombe walls and water walls) are listed in Table 14-3.

Sensitivity Data

Data presented in this section express the sensitivity of the annual *SSF* to selected design parameters. In any given set of data, one or two parameters are varied while the others are held fixed at their reference values. The reference values are indicated by dots on the curves.

Only sample sensitivity curves are presented in this chapter. The sample plots illustrate the effect of various parameter variations on performance in Albuquerque, which has a moderate winter climate. The complete set of sensitivity curves is contained in Appendix 2, which includes results for Bismarck, North Dakota; Boston, Massachusetts; Ely, Nevada; Nashville, Tennessee; and Seattle, Washington.

The above cities were chosen to represent a wide geographical and climatological range. To apply the data, one of the reference cities must be chosen to represent the city of interest on the basis of climate similarity. Two gross measures are of use in making the selection: the 65°F-base heating degree-days and the solar radiation incident on a vertical surface. These data are available as

Table 14-2

Trombe Wall Reference Designs

Designation		Thermal Storage Capacity*	Wall** Thickness	ρck	No. of	Wall	Night
Vented	Unvented	(Btu/ft^2 · °F)	(in.)	Btu^2/h · ft^4 · °F^2	Glazings	Surface	Insulation
TWA1	TWF1	15	6	30	2	normal	no
TWA2	TWF2	22.5	9	30	2	normal	no
TWA3	TWF3	30	12	30	2	normal	no
TWA4	TWF4	45	18	30	2	normal	no
TWB1	TWG1	15	6	15	2	normal	no
TWB2	TWG2	22.5	9	15	2	normal	no
TWB3	TWG3	30	12	15	2	normal	no
TWB4	TWG4	45	18	15	2	normal	no
TWC1	TWH1	15	6	7.5	2	normal	no
TWC2	TWH2	22.5	9	7.5	2	normal	no
TWC3	TWH3	30	12	7.5	2	normal	no
TWC4	TWH4	45	18	7.5	2	normal	no
TWD1	TWI1	30	12	30	1	normal	no
TWD2	TWI2	30	12	30	3	normal	no
TWD3	TWI3	30	12	30	1	normal	yes
TWD4	TWI4	30	12	30	2	normal	yes
TWD5	TWI5	30	12	30	3	normal	yes
TWE1	TWJ1	30	12	30	1	selective	no
TWE2	TWJ2	30	12	30	2	selective	no
TWE3	TWJ3	30	12	30	1	selective	yes
TWE4	TWJ4	30	12	30	2	selective	yes

*Per unit of projected area

**For the particular case of ρc = 30 Btu/ft^3 · °F

Table 14-3

Thermal Storage Wall Reference Design Characteristics

	Trombe Wall	Water Wall
Thermal Storage		
Thermal conductivity of wall (Btu/h · ft · °F)	1	infinite
Density (lb/ft^3)	150	62.4
Specific heat (Btu/lb · °F)	0.2	1.0
Solar absorptance of wall surface	0.95*	0.95*
Solar absorptance of selective surface, when used	0.9*	0.9*
Infrared emittance of wall surface	0.9*	0.9*
Infrared emittance of selective surface, when used	0.1	0.1
Thermocirculation Vents		
Vent area, sum of upper and lower vents, when used (% of wall area)	6	not used
Vertical separation of vents, when used (ft)	8	na
Glazing		
Transmission	diffuse	
Infrared emittance of surface	0.9	
Orientation	due south*	
Shading	none*	
Number of layers	1, 2, or 3	
Index of refraction	1.526	
Extinction coefficient (per inch)	0.5*	
Thickness of each pane (inches)	1/8*	
Air gap between layers (inches)	0.5*	
Night insulation thermal resistance, when used	R9**	
Control range (°F)	65 to 75**	
Ground reflectance	0.3*	
Internal heat generation	none**	

*Indicates properties that can be varied in the monthly *SLR* method
**Indicates properties that can be varied in the annual *LCR* method and the monthly *SLR* method

monthly averages for numerous United States and Canadian locations in Appendix 3 where they are tabulated as D65 and VS, respectively.

Effects of Thermal Storage Mass — Sizing a thermal storage wall is an especially important design decision because it has major cost implications. Not only is the wall itself expensive, but the space that it takes up within the building is valuable. The general results of the sensitivity studies show that performance increases markedly with increasing thermal storage mass up to a point where the mass is sufficient to carry over excess daytime solar heat to supply to building heating load during the night. Providing for several-day heat storage is not nearly as effective in increasing annual heating performance. It is also noted that performance is more sensitive to variations in thermal storage for smaller values of *LCR*.

Figures 14-1 through 14-3 are graphs of *SSF* versus wall thickness for reference designs TWA3, TWF3, WWA3, and WWB4 for Albuquerque weather data. In the Trombe wall cases, the density of the masonry is a variable parameter, and the thermal conductivity is assumed to vary with the density. There is an optimum thickness for both the vented (TWA3) and unvented (TWF3) cases. The optimum occurs because of a trade-off between increasing performance with increasing thermal storage mass and decreasing performance because of the inability of heat to pass through the wall. For the case of a water wall, performance improves continuously with increases in thermal storage. The rate of improvement with increased thermal storage is greater for lower values of the *LCR* or higher values of *SSF*. For this reason, one should provide more thermal storage at larger values of *SSF*.

Trombe walls and water walls have comparable performance, but a water wall generally outperforms a Trombe wall of the same thickness. The difference becomes more pronounced for higher amounts of thermal storage mass. The difference is less for the night insulation case and also for low solar fractions. The water wall performance is higher for a given thickness because the volumetric heat capacity of water is greater than that of masonry and because the entire thickness of water is effective for heat storage.

Many people have supposed that Trombe walls are much less efficient than water walls because of the high daytime heat losses associated with the high surface temperatures. Although the losses are higher during the peak operating time around noon, there is a converse effect associated with the fact that the wall surface temperatures are slightly lower than those of the water wall during the majority of the time, including the entire night. Therefore, although the heat losses are higher for a short period for the Trombe wall, they are lower for a long period. These effects tend to cancel one another, and the net heat losses from both walls are very comparable.

Thermocirculation Vents — The use of thermocirculation vents is often associated with masonry thermal storage walls. This is partially due to the fact that these vents were used by Felix Trombe and Jacques Michel in the experimental house at Odeillo, and others have simply followed along. As indicated above, however, the vents may not always be desirable, and some designers prefer not to use them.

As a general rule, vents provide heat to a building during the day in the form of heated air, and if this heat is useful, then the vents may well be desired. In other circumstances, they might simply lead to building overheating.

Unvented Trombe walls provide a very attractive design alternative due to the much decreased tendency to overheat. However, performance is decreased somewhat. Figure 14-4 shows the improvement obtainable with vents.

Many designers prefer not to use vents but to use direct-gain windows instead to provide daytime building heating.

If vents are to be used, provision should be made to block them off during the summer. Vents are not recommended for thermal storage walls with selective surfaces because of dust accumulation on the selective surface.

Effect of Glazing Azimuth — Although a thermal storage wall should normally face due south, there

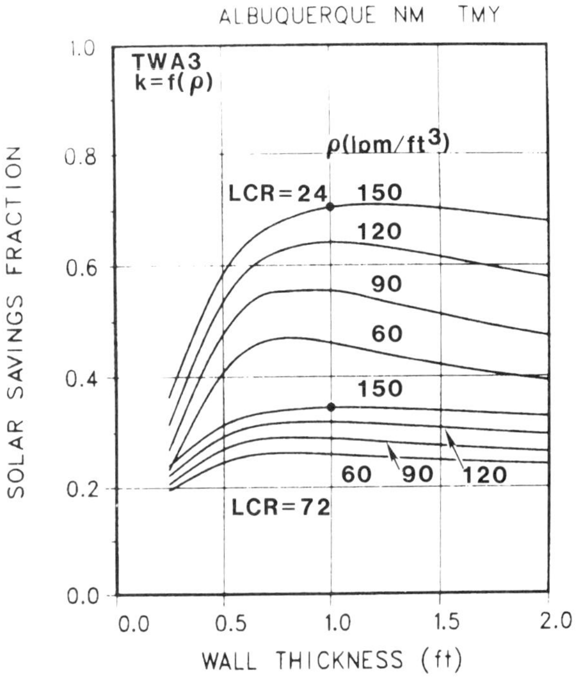

Figure 14-1. Annual Heating Performance Sensitivity to the Trombe Wall Thickness for Reference Design TWA3 (Vented)

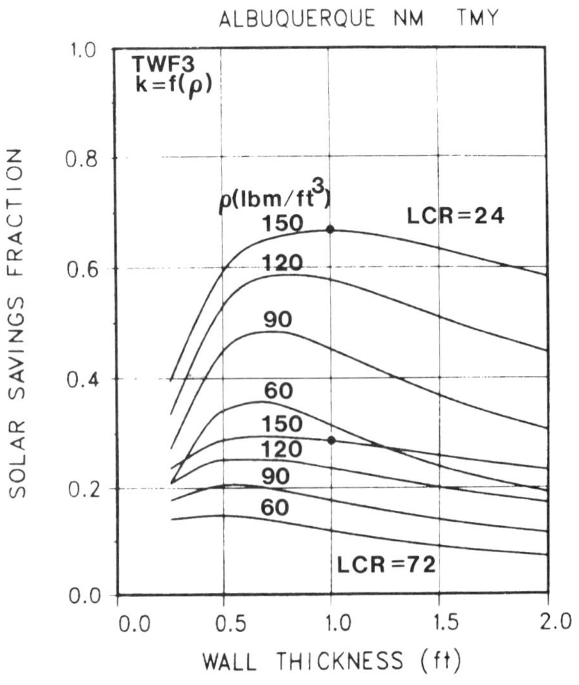

Figure 14-2. Annual Heating Performance Sensitivity to the Trombe Wall Thickness for Reference Design TWF3 (Unvented)

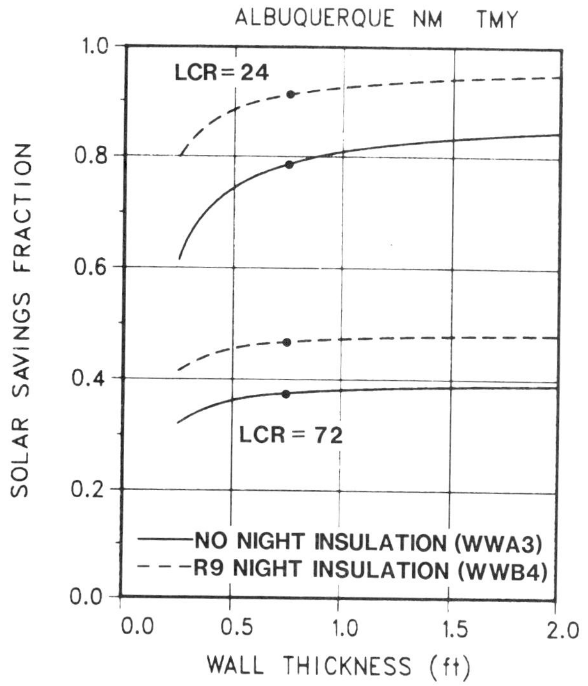

Figure 14-3. Annual Heating Performance Sensitivity to the Trombe Wall Thickness for Reference Designs WWA3 and WWB4

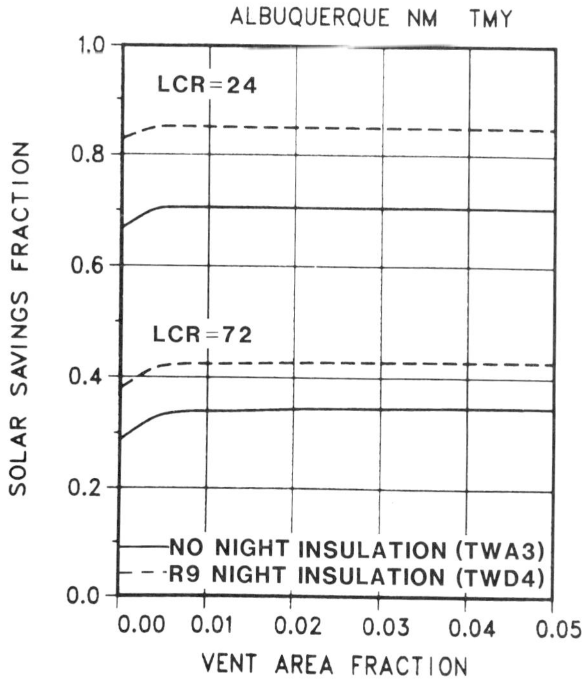

Figure 14-4. Annual Heating Performance Sensitivity to the Vent Area Fraction for Reference Designs TWA3 and TWD4

may be many instances in which the designer prefers or is constrained to use some other azimuth. The effect of varying the azimuth is shown in Figure 14-5.

The effect of wall orientation (azimuth) is relatively small for small deviations about due south. For azimuths within 15 degrees of due south, the performance penalty averages about 4% on an annual basis and is seldom more than 8%. For azimuths within 30° of due south, the performance decrease averages 10% on an annual basis and is seldom more than 20%.

Number of Glazing Layers — The performance of thermal storage walls can be improved by using additional layers of glazing to reduce heat losses from the wall to the outside. Performance increases up to four or five sheets of glass. At some point, however, the decrease in transmission associated with reflection and absorption in each added glass layer more than offsets the decrease in heat loss.

For water walls, an example of the change in *SSF* for different numbers of glazings is shown in Figure 14-6. Significant improvement in performance is achieved with the addition of the second and third layers of glazing for the case of no night insulation. With night insulation, two layers of glazing are optimum.

Figures 14-7 and 14-8 display the same type of data for vented and unvented Trombe walls, respectively. Similar conclusions can be drawn from these figures; the shapes of the curves are the same, but at a lower level of *SSF*.

Selective Surface — The use of a selective surface on a Trombe wall or a water wall eliminates the need for multiple layers of glazing, as shown in Figure 14-9. The selective surface reduces heat losses to the outside. The added layers of glazing reduce the transmittance of solar energy more than can be compensated for by reducing the already small heat losses to the outside. The selective surface often improves thermal storage wall performance as much as night insulation does. Night insulation, used with a selective surface, can be used to further increase performance and may be warranted in very cold climates.

Solar Absorptance — Thermal storage walls are very sensitive to the absorptance of the wall surface. As the absorptance decreases, performance falls off very rapidly. The data for water walls are shown in Figure 14-10. The number of glazings is used as a parameter and is varied up to three glazings. Trombe walls show similar characteristics.

Surface Emittance — The emittance of the surface of a thermal storage wall affects performance because it determines the amount of heat loss from the wall by infrared radiation. The data for water walls are shown in Figure 14-11. The number of glazings (one, two, or three) is used as a parameter in the figure. For single glazing, *SSF* drops off noticeably as the emittance increases, but this sensitivity is greatly reduced for double and triple glazings.

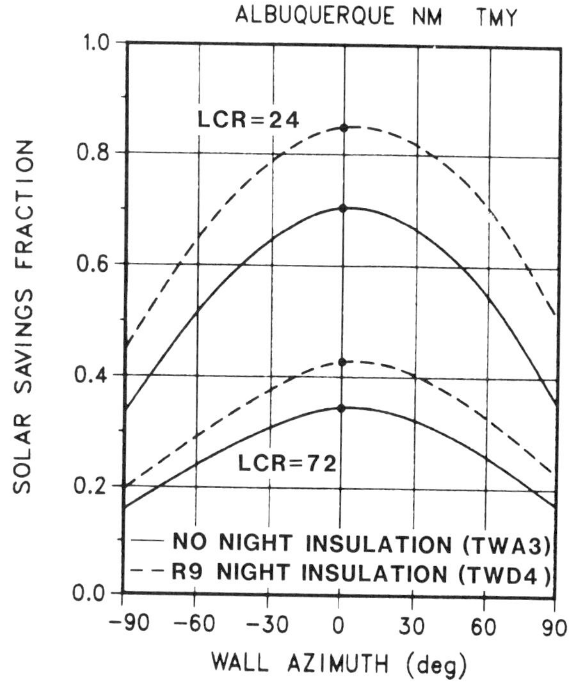

Figure 14-5. Annual Heating Performance Sensitivity to the Wall Azimuth for Reference Designs TWA3 and TWD4

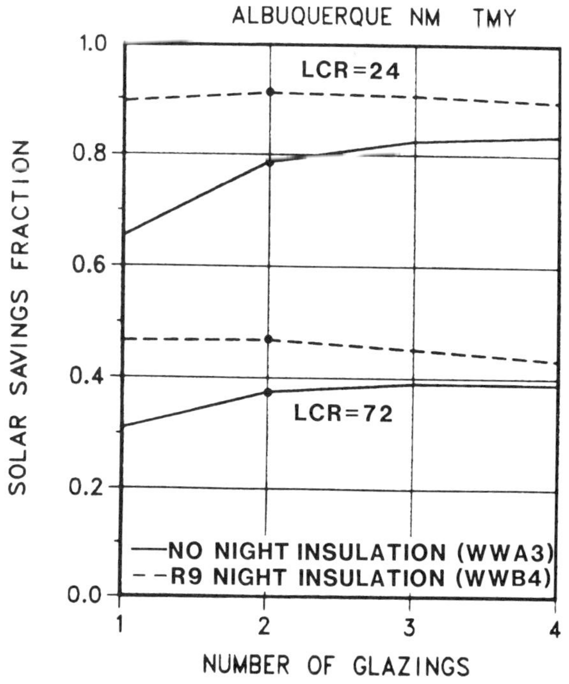

Figure 14-6. Annual Heating Performance Sensitivity to the Number of Glazings for Reference Designs WWA3 and WWB4

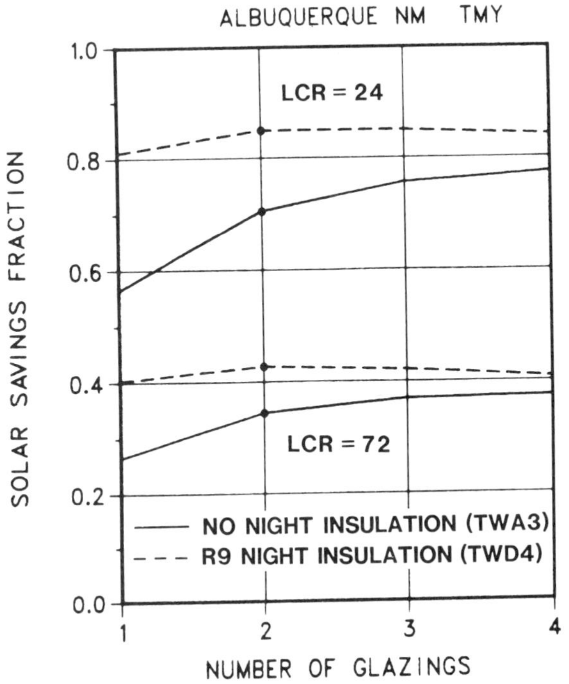

Figure 14-7. Annual Heating Performance Sensitivity to the Number of Glazings for Reference Designs TWA3 and TWD4

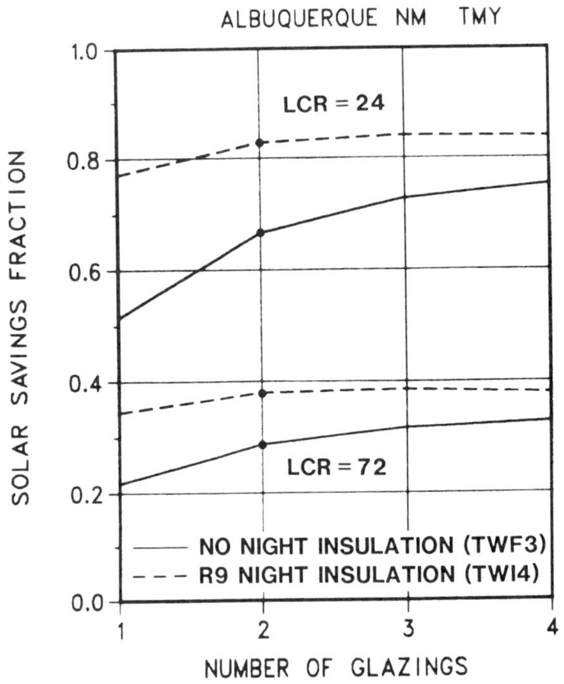

Figure 14-8. Annual Heating Performance Sensitivity to the Number of Glazings for Reference Designs TWF3 and TWI4

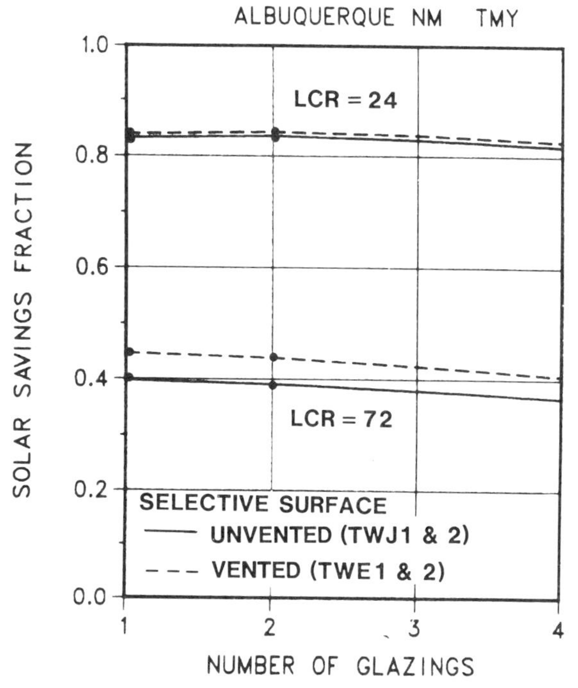

Figure 14-9. Annual Heating Performance Sensitivity to the Number of Glazings for Trombe Walls with Selective Surfaces

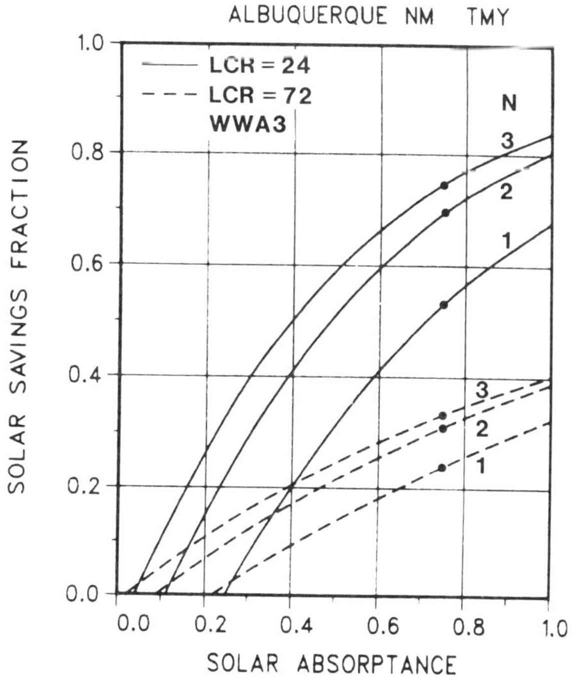

Figure 14-10. Annual Heating Performance Sensitivity to the Solar Absorptance, N is the number of glazings.

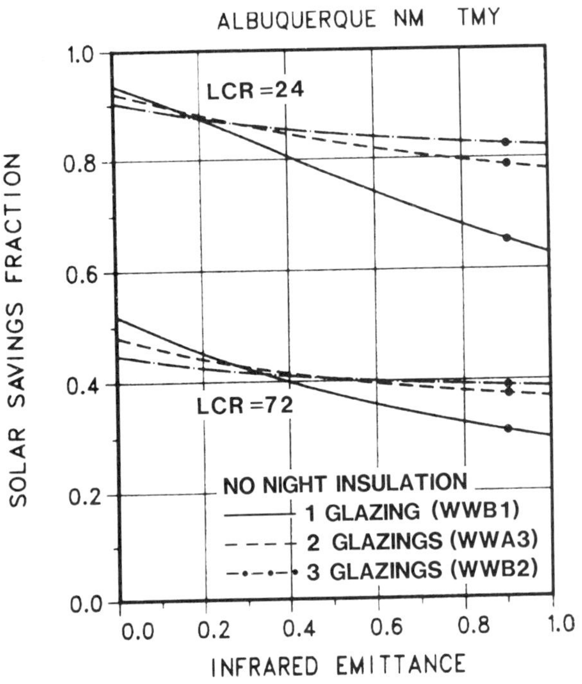

Figure 14-11. Annual Heating Performance Sensitivity to the Infrared Emittance of a Water Wall.

CHAPTER 15
MIXED SYSTEMS

Introduction

It is often the case for passive solar heated buildings that more than one passive system is used. In the first place, it is good practice to balance the heat delivery by the use of distinct passive systems that directly serve the various building zones. When this is done, it can equalize the temperatures throughout the building and reduce the dependence on convective heat transfer from one zone to another. A rough balance can be accomplished by sizing the respective apertures in proportion to the net load coefficients of each zone. Furthermore, a judicious mixture of passive system types will often lead to a more comfortable building than could be achieved by any one type alone.

A direct-gain system is usually present in each building zone because the windows are needed for view and daylight. An indirect system, such as a sunspace or a Trombe wall, is often added if the total aperture desired exceeds the recommended limit for direct gain (see Chapter 2, Rule 19). An unvented Trombe wall is a good complement to direct-gain windows because a direct-gain system delivers heat with little delay and a strong peak in midday hours, while for a Trombe wall, heat delivery peaks about 8 hours later and is much more uniform through the day and evening.

These aspects of the timing of heat delivery are important considerations in choosing passive system types. Direct-gain windows might be an appropriate choice for zones that need heat promptly in the morning and through daytime hours. A vented Trombe wall can provide prompt heat delivery in the morning by air flow through the vents plus sustained heat delivery into evening hours by conduction through the wall. An unvented Trombe wall delivers a relatively uniform supply of heat throughout the day with a weak maximum in the evening. A water wall performs similarly to an unvented Trombe wall, but the peak delivery is more pronounced and closer to midday, depending strongly on the thickness of the wall. A sunspace can combine many of the aspects of the other system types, depending on the use of doors or other openings to permit air flow, and depending on the type of common wall and thermal storage mass. Taking into account these various system characteristics, a building design may include a mixture of systems that matches the timing as well as the total amount of the heat demand.

Chapter 5 gives methods for applying the annual *LCR* method to mixed systems. This present chapter describes the procedures for applying the monthly *SLR* method to mixed systems. The basis for the procedures is the division of the building into separate zones, each of which is served by a single passive solar system. Each zone is then analyzed as if it were a separate building under the condition that the temperature is the same in each zone. The building performance is determined from the sum of the annual auxiliary heat supplied to each zone.

Notation

A supplemental notation is needed to distinguish the separate zones. The subscript *i* will be used to indicate the *i*th system, and the total number of systems will be designated as *Z*. The subscript *m* will signify the month. The subscript *B* will indicate the total building. The following samples will illustrate the use of this notation.

A_{pi} = the projected aperture area for the *i*th system

NLC_i = the net load coefficient for the zone served by the *i*th system

T_{bi} = the base temperature for the zone served by the *i*th system

$S_{i,m}$ = the absorbed solar radiation for the ith system and the mth month

$SLR_{i,m}$ = the solar load ratio for the ith system and the mth month

$SSF_{i,m}$ = the solar savings fraction for the ith system and the mth month

$Q_{auxi,m}$ = the auxiliary heat supplied to the zone served by the ith system in the mth month

The annual auxiliary heat supplied to the zone served by the ith system is given by:

$$\text{Annual } Q_{auxi} = \sum_{m=1}^{12} Q_{auxi,m} \tag{15-1}$$

The annual auxiliary heat supplied to the whole building is given by:

$$\text{Annual } Q_{auxB} = \sum_{i=1}^{Z} \text{Annual } Q_{auxi} = \sum_{i=1}^{Z} \sum_{m=1}^{12} Q_{auxi,m} \tag{15-2}$$

Separation into Zones

Distinct Zones — The simplest case occurs in buildings with internal walls that divide the building into distinct zones, each served by only one passive solar system. The sketch in Figure 15-1 illustrates this case. The internal wall is treated as a thermal barrier that prevents transfer of heat between zones 1 and 2. The two zones are then treated as separate buildings. The net load coefficient for each zone, NLC_i, is computed in the normal manner except that no heat losses are allowed through the common internal walls.

Pseudozones — In the case in which two or more passive system types serve a building that is not distinctly divided into separate zones by walls or other barriers (see Figure 15-2), there are two choices. If the passive systems are sufficiently similar to be represented by the same reference design, they should be treated as components of a single system. The usual case is two or more direct-gain windows that differ only in their orientation. Two or more components of a single system are treated by combining their solar gains, S. See Chapter 9 for the details of calculating S for multicomponent systems.

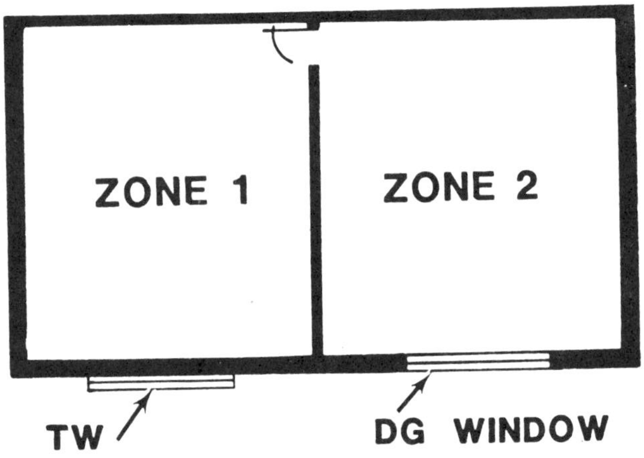

Figure 15-1. Distinct Zones Created by an Internal Wall

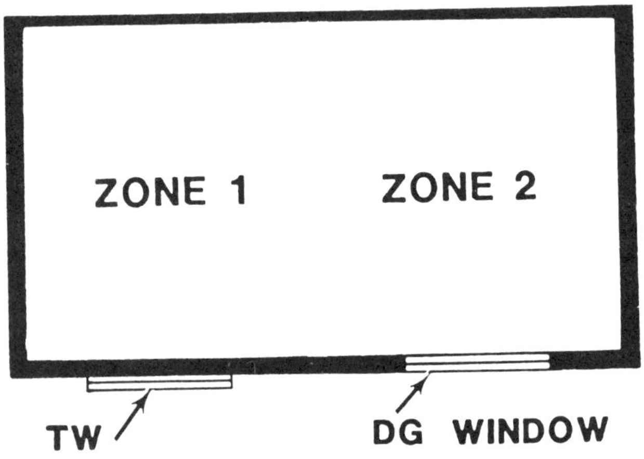

Figure 15-2. The Pseudozone Case, the Same Area Served by Two Different Passive Solar Systems

If the passive systems are not sufficiently similar to be represented by the same reference design, they must be treated as independent systems.

Such a case is a direct-gain and a Trombe wall system serving the same room as in Figure 15-2. In this case, artificial zones must be created. The physical boundaries of the zones are not important. It is only necessary to split the net load coefficient of the building (NLC_B) into components (NLC_i) corresponding to the pseudozones served by each of the passive systems. This is done by the following formula:

$$NLC_i = \frac{A_{pi}}{A_{pB}} NLC_B . \qquad (15\text{-}3)$$

where

$$A_{pB} = \sum_{i=1}^{Z} A_{pi}$$

It follows that

$$LCR_i = NLC_i/A_{pi} = NLC_B/A_{pB} = LCR_B \quad (15\text{-}4)$$

In other words, LCR has the same value for each of the pseudozones and for the whole building. Having established the value of LCR and the respective values of NLC_i, the pseudozones can be treated as separate buildings.

Total Load Coefficient for Mixed Systems

Distinct Zones — For distinct zones, the total load coefficient for the ith zone (TLC_i) is computed in the normal manner, treating each zone as a separate building:

$$TLC_i = NLC_i + 24\,(U_cA_p)_i \qquad (15\text{-}5)$$

where $(U_cA_p)_i$ is the load coefficient through the solar wall of the ith system.

The total load coefficient for the building (TLC_B) will be equal to the sum of the coefficients for the zones:

$$TLC_B = \sum_{i=1}^{Z} TLC_i \qquad (15\text{-}6)$$

Pseudozones — For pseudozones, calculate TLC_B in the ordinary way for a whole building, ignoring the pseudozones.

Internal Heat Sources for Mixed Systems

Distinct Zones — It may be possible to assign internal heat sources to specific zones rather than to the building as a whole. If so, the base temperature for the ith zone is calculated from:

$$T_{bi} = T_{seti} - Q_{inti}/TLC_i \qquad (15\text{-}7)$$

where Q_{inti} is the internal heat generation rate for zone i.

If the zones have different base temperatures, then the degree-days ($DD_{i,m}$) must be calculated for each zone and each month. If heat sources cannot be assigned to specific zones, or if the desired precision does not warrant this level of detail, then treat the internal sources as for the pseudozone case described below.

Pseudozones — For pseudozones, the base temperature is calculated from:

$$T_{bB} = T_{setB} - Q_{intB}/TLC_B \qquad (15\text{-}8)$$

and the base temperatures of all pseudozones and for the building are equal. The degree-days for all pseudozones in the building will also be equal.

Analysis Procedures for Mixed Systems with Distinct Zones

The procedures for completing the monthly SLR method for a building with mixed systems and distinct zones consists of the following steps:

Step 1. Divide the building into distinct zones. If a zone has more than one solar system, it must be divided into pseudozones as explained below. Number the zones 1 through N.

For each zone, treating each zone as a separate building:

Step 2. Compute NLC_i, TLC_i, and LCR_i.

Step 3. Determine Q_{inti}.

Step 4. Compute T_{bi}.

Step 5. Compute $DD_{i,m}$.

Step 6. Compute $S_{i,m}$.

Step 7. Complete monthly *SLR* method through the computation of Annual Q_{auxi} and Annual Q_{neti}.

For the whole building:

Step 8. Compute Annual Q_{auxB} from:

$$\text{Annual } Q_{auxB} = \sum_{i=1}^{Z} \text{Annual } Q_{auxi}\,.$$

This completes the monthly *SLR* method for a building with mixed systems and distinct zones.

If the analyst desires to determine the annual solar savings fraction for the whole building, it can be computed from:

$$\text{Annual } SSF_B = 1 - \text{Annual } Q_{auxB}/\text{Annual } Q_{netB}$$

Analysis Procedures for Mixed Systems with Pseudozones

The procedures for completing the monthly *SLR* method for a building with mixed systems serving the same space (pseudozones) consists of the following steps:

Step 1. Number the passive solar systems serving the building. These numbers, values of subscript *i*, identify the pseudozones.

Step 2. Compute NLC_B and LCR_B and TLC_B.

Step 3. Determine Q_{intB}.

Step 4. Compute T_{bB}.

Step 5. Compute DD_{Bm}.

For each pseudozone (treat each as a separate building):

Step 6. Compute $S_{i,m}$.

Step 7. Compute NLC_i.

Step 8. Complete monthly *SLR* method through the computation of Annual Q_{auxi} and Annual Q_{neti}.

For the whole building:

Step 9. Compute Annual Q_{auxB} from:

$$\text{Annual } Q_{auxB} = \sum_{i=1}^{Z} \text{Annual } Q_{auxi}\,.$$

This completes the monthly *SLR* method for a building with mixed systems serving the same space (pseudozone case).

The annual solar savings fraction for the whole building can be computed from:

$$\text{Annual } SSF_B = 1 - \text{Annual } Q_{auxB}/\text{Annual } Q_{netB}$$

CHAPTER 16
EXAMPLE PROBLEMS, DESIGN DEVELOPMENT

Introduction

This chapter presents example problems applicable to the design development phase. The general content of each problem is as follows:

Example Problem 16-1 illustrates the concept of suntempering and compares the performance of a suntempered building with its conventional counterpart.

Example Problem 16-2 demonstrates the monthly *SLR* method for a simple direct-gain reference design system and compares the results to those obtained by the annual *LCR* method.

Example Problem 16-3 uses the same direct-gain system as Example Problem 16-2 but changes ground reflectance from the standard value of 0.3 to 0.7, changes the extinction coefficient-thickness product from the standard value of 0.0625 to 0.05, and changes the refractive index from the reference value of 1.526 to 1.3. The monthly *SLR* method is used.

Example Problem 16-4 replaces the direct-gain windows of the building used in Example Problem 16-2 with a sunspace having the same projected area. The monthly *SLR* method is used.

Example Problem 16-5 changes the sunspace of Example Problem 16-4 by using a single layer of glazing, increasing the infiltration rate to 1 air change per hour, and adding R3 night insulation. The monthly *SLR* method is used.

Example Problem 16-6 uses the same building as in Example Problem 16-2 but replaces the direct-gain windows with a thermal storage wall having the same projected area. An internal heat source is assumed. The monthly *SLR* method is used.

Example Problem 16-7 uses the same building and passive heating system as Example Problem 16-6 but changes the building orientation so that the thermal storage wall aperture faces 60° east of due south. The monthly *SLR* method is used.

Example Problem 16-8 analyzes a mixed system with two distinct zones. Zone 1 is served by a Trombe wall, while Zone 2 has direct-gain windows.

Example Problem 16-9 uses the same building as in Example Problem 16-8 except that the internal wall between Zone 1 and Zone 2 is removed, creating pseudozones.

Example Problem 16-10 uses the same building as in Example Problem 16-9 but treats the windows on the west and northeast walls as part of the solar wall. The building is subdivided into two pseudozones for analysis purposes, with one pseudozone being served by three direct-gain components.

Example Problem 16-1

This example problem illustrates the concept of suntempering. Consider a typical 1,496-ft^2, single-family, detached residence in Boston, Massachusetts. Assume that the windows are double-glazed, are uniformly distributed about the perimeter of the building, and have a total area equal to 15% of the floor area, or 225 ft^2. Further, assume that the building is 34 feet deep by 44 feet wide, that the long axis is oriented in a north-south direction, that the building is of wood frame construction with a crawl space (no floor slab), and that the surface area of gypsum board available for thermal storage is 3,600 ft^2. Figure 16-1 shows a sketch of the concept. The worksheet in Figure 16-2 shows the computations of *NLC* and *LCR* for this building.

Using Figure 12-25 and extrapolating the no-night-insulation curve out to *LCR* = 201 yields an *SSF* of about 0.035. Since A_m/A_n = 73, as opposed to the reference design value of 20, the mass sensitivity curve (Figure A2-85, Appendix 2) must be used to interpret this result. The sensitivity curve (for *LCR* = 72 and no night insulation) is flat in the region near A_m/A_n = 70 and would be even flatter for larger values of *LCR*, as is the case in this example. This application of the sensitivity data indicates that thermal storage mass sensitivity can be neglected for the range of surface area ratios involved. The thermal analysis is completed by computing Q_{aux} = 53 million Btu, as shown on the worksheet, Figure 16-2. Even though this building would not ordinarily be thought of as a passive solar system, 3.5% of the heat load experienced by all elements of the shell except the south-facing windows will be met by direct solar gains.

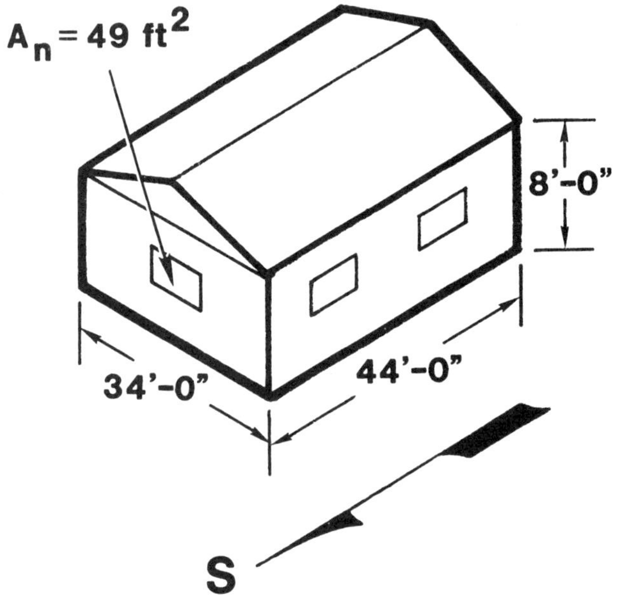

TOTAL WINDOW AREA = 225 ft^2

SOUTH WINDOW AREA(A_n) = 49 ft^2

FLOOR AREA = 1496 ft^2

MASS AREA(A_m) = 3600 ft^2

$$\frac{A_m}{A_n} = 73$$

Figure 16-1. Typical Wood Frame Building for Example Problem 16-1

NET LOAD COEFFICIENT and ANNUAL LCR METHOD WORKSHEET DATE 1/1/83 BY Smith LOCATION BOSTON

NET LOAD COEFFICIENT, NLC

BUILDING ELEMENT	FORMULA	Fill in the Blanks and Perform Arithmetic	LOAD, UA
WALLS	: $(UA)_w = A_w/R_w$	= 1023 / 15	= 68.2
ROOF	: $(UA)_r = A_r/R_r$	= 1496 / 20	= 74.8
E, W, N WINDOWS	: $(UA)_g = 1.1 A_x/N$	= 1.1 x 176 / 2	= 96.8
SELECT ONE: FLOOR	: $(UA)_f = A_f/R_f$	= 1496 / 24	= 62.3
SELECT ONE: PERIMETER	: $(UA)_p = 4.17\ P_p/(R_p + 5) = 4.17$ x	______ /(______ + 5)	= ______
SELECT ONE: BASEMENT	: $(UA)_b = 10.7\ P_b/(R_b + 8) = 10.7$ x	______ /(______ + 8)	= ______
INFILTRATION	: $(UA)_i = 0.018V \times ACH \times ADR = 0.018$ x	12000 x 1/2 x 1	= 108
		Sum =	410.1 Btu/°F • h
		NLC = 24 x Sum =	9842 Btu/°F • day

LOAD COLLECTOR RATIO, LCR

LCR = NLC/A_p = 9842 / 49 = 201 Btu/°F • day • ft^2

WEIGHTED AVERAGE, SSF

REFERENCE DESIGN	PROJECTED AREA	FRACTION OF A_p		SSF FOR EACH REFERENCE DESIGN		PRODUCT OF SSF x FRACTION	U_c	$U_c A_p$
______	______	______	x	______	=	______	______	______
______	______	______	x	______	=	______	______	______
______	______	______	x	______	=	______	______	______
TOTAL A_p =	______				SSF =	______	TOTAL (U_cA_p) =	______
							TLC = NLC + 24 x TOTAL (U_cA_p) =	______

BASE TEMPERATURE, T_b

$T_b = T_{set} - Q_{int}/TLC$ = ______ - ______ / ______ = ______ °F

AUXILIARY HEAT, Q_{aux}

Q_{aux} = NLC • DD(1 - SSF) = 9842 x 5622 x (1 - .035) x 10^{-6} = 53 Mega Btu/year

Figure 16-2. Worksheet for Typical Wood Frame Building for Example Problem 16-1

This typical wood frame building in Boston will now be suntempered by orienting the long axis in the east-west direction and placing 80% of the 225 ft^2 of window area on the south side. The solar collection area becomes 180 ft^2. A sketch and basic data are shown in Figure 16-3. The computation of *NLC* and *LCR* are shown in the worksheet in Figure 16-4. Using $LCR = 45$ and referring to Figure 12-25, we find $SSF = 0.015$. Since $A_m/A_n = 20$, the same as for the reference design, the mass sensitivity curve is not needed. The computation of Q_{aux} for the suntempered building is shown on the worksheet in Figure 16-4. Note that although the suntempered building has a lower *SSF* than the original structure, it requires less auxiliary heat. *The bottom line on performance is the auxiliary heat required.* The *SSF* should not be used to compare the performance of one building with another unless they both have the same *NLC*. The low-mass, suntempered building saves 15% of the energy that would ordinarily be required for space heating by its untempered counterpart, and this saving is achieved by simply reorienting the building and rearranging the windows.

Clearly, the suntempered building represents a step in the right direction that should not be discouraged. In fact, in mild climates, where the heating loads are small, any further steps may not be economically justifiable. However, it would be a gross oversight if we failed to point out that it is possible to greatly improve both performance and comfort by placing recommended levels of thermal storage mass within the building. The use of three or more glazings, or night insulation, can further improve performance. Multiple glazings and night insulation are, of course, desirable on all the windows of a building, not just those that face south, because they reduce heat losses through the windows. One should also consider the merits of combining direct gain with other types of passive solar systems, such as thermal storage walls, that are better suited to harsh winter climates.

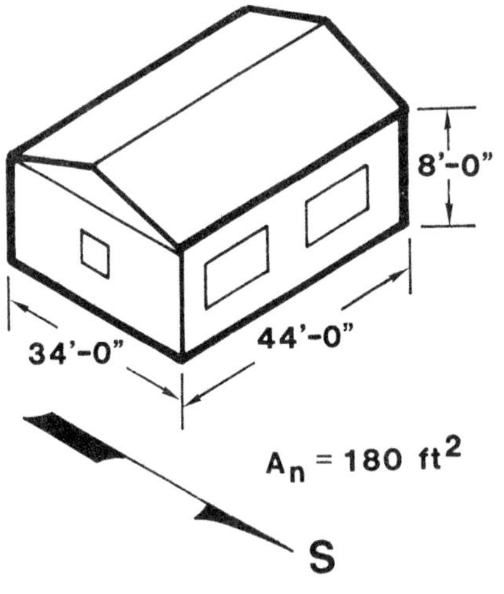

TOTAL WINDOW AREA = 225 ft^2

SOUTH WINDOW AREA(A_n) = 180 ft^2

FLOOR AREA = 1496 ft^2

MASS AREA(A_m) = 3600 ft^2

$$\frac{A_m}{A_n} = 20$$

Figure 16-3. Suntempered Wood Frame Building for Example Problem 16-1

NET LOAD COEFFICIENT and ANNUAL LCR METHOD WORKSHEET DATE 1/1/83 BY Smith LOCATION BOSTON

NET LOAD COEFFICIENT, NLC

	BUILDING ELEMENT	FORMULA	Fill in the Blanks and Perform Arithmetic	LOAD, UA
	WALLS	: $(UA)_w = A_w/R_w$	= 1023 / 15	= 68.2
	ROOF	: $(UA)_r = A_r/R_r$	= 1496 / 20	= 74.8
	E, W, N WINDOWS	: $(UA)_g = 1.1A_x/N$	= 1.1 x 45 / 2	= 24.8
SELECT ONE	FLOOR	: $(UA)_f = A_f/R_f$	= 1496 / 24	= 62.3
	PERIMETER	: $(UA)_p = 4.17\ P_p/(R_p + 5)$	= 4.17 x ____ /(____ + 5)	= ____
	BASEMENT	: $(UA)_b = 10.7\ P_b/(R_b + 8)$	= 10.7 x ____ /(____ + 8)	= ____
	INFILTRATION	: $(UA)_i$ = 0.018V x ACH x ADR	= 0.018 x 12000 x 1/2 x 1	= 108

Sum = 338.1 Btu/°F • h

NLC = 24 x Sum = 8114 Btu/°F • day

LOAD COLLECTOR RATIO, LCR

$LCR = NLC/A_p$ = 8114 / 180 = 45.1 Btu/°F • day • ft²

WEIGHTED AVERAGE, SSF

REFERENCE DESIGN	PROJECTED AREA	FRACTION OF A_p		SSF FOR EACH REFERENCE DESIGN		PRODUCT OF SSF x FRACTION	U_c	$U_c A_p$
____	____	____	x	____	=	____	____	____
____	____	____	x	____	=	____	____	____
____	____	____	x	____	=	____	____	____
TOTAL A_p =	____			SSF =		____	TOTAL (U_cA_p) =	____

TLC = NLC + 24 x TOTAL (U_cA_p) = ____

BASE TEMPERATURE, T_b

$T_b = T_{set} - Q_{int}/TLC$ = ____ - ____ / ____ = ____ °F

AUXILIARY HEAT, Q_{aux}

Q_{aux} = NLC • DD(1 - SSF) = 8114 x 5622 x (1 - .015) x 10^{-6} = 45 Mega Btu/year

Figure 16-4. Worksheet for Suntempered Wood Frame Building for Example Problem 16-1

Example Problem 16-2

This example demonstrates the monthly *SLR* method for a simple direct-gain system and compares the results to the annual *LCR* method. The house is very well insulated, a very airtight design with a direct-gain system similar to reference design DGA1 with $A_p = 270$ ft^2 on a south wall. The building has an area of 1,575 ft^2 on a slab floor. There are 30 ft^2 of triple-glazed windows on each of the east, west, and north sides. The north windows are treated at half their normal conductance values for heat loss, as recommended in Chapter 10; the east and west windows are ignored altogether. The location is St. Louis, Missouri. Assume the degree-day base temperature is is 65°F. Figure 16-5 shows the *NLC* and *LCR* worksheet for this building. Figure 16-6 shows the completed monthly *SLR* method worksheet for this example.

Step 1 — The reference design is DGA1.

Step 2 — $NLC = 5232$, $A_p = 270$, $LCR = 19.4$.

Step 3 — LCR_s and the correlation coefficients for reference design DGA1 are found in Appendix 5, Table A5-6.

Step 4 — This step not needed for direct-gain systems.

Step 5 — See Figure 16-7 for the completed solar gain worksheet. Note that the correlation for (*TRN*/*HOR*) from Appendix 4, Table A4-3, has been used, combining into one number the product (*INC*/*HOR*)(*TRN*/*INC*). The absorption factor (*ABS*/*TRN*) is the direct-gain reference design value 0.976 from Appendix 4, Table A4-46.

Step 6 — Copied from the weather data, Appendix 3, St. Louis, Missouri, base temperature of 65°F.

Steps 7 through 13 — Calculated routinely as described for the monthly *SLR* method.

Comments — Note that the annual *SSF*, as calculated by the monthly *SLR* method, agrees with that calculated by the annual *LCR* method (Figure 16-5).

NET LOAD COEFFICIENT and ANNUAL LCR METHOD WORKSHEET DATE 1/1/83 BY Smith LOCATION St. Louis

NET LOAD COEFFICIENT, NLC

BUILDING ELEMENT	FORMULA	Fill in the Blanks and Perform Arithmetic	LOAD, UA
WALLS	$(UA)_w = A_w/R_w$	= 1240 / 30	= 41
ROOF	$(UA)_r = A_r/R_r$	= 1575 / 50	= 32
E, W, N WINDOWS	$(UA)_g = 1.1A_x/N$	= 1.1 x 30 / 3 x ½	= 6
SELECT ONE: FLOOR	$(UA)_f = A_f/R_f$	= ____ / ____	=
SELECT ONE: PERIMETER	$(UA)_p = 4.17\ P_p/(R_p + 5)$	= 4.17 x 200 /(27 + 5)	= 26
SELECT ONE: BASEMENT	$(UA)_b = 10.7\ P_b/(R_b + 8)$	= 10.7 x ____ /(____ + 8)	=
INFILTRATION	$(UA)_i = 0.018V \times ACH \times ADR$	= 0.018 x 12600 x 0.5 x 1	= 113

Sum = 218 Btu/°F • h

NLC = 24 x Sum = 5232 Btu/°F • day

LOAD COLLECTOR RATIO, LCR

$LCR = NLC/A_p$ = 5232 / 270 = 19.4 Btu/°F • day • ft²

WEIGHTED AVERAGE, SSF

REFERENCE DESIGN	PROJECTED AREA	FRACTION OF A_p		SSF FOR EACH REFERENCE DESIGN		PRODUCT OF SSF x FRACTION	U_c	$U_c A_p$
DGA1	270	1	x	0.242	=	0.242		
			x		=			
			x		=			
TOTAL A_p =					SSF =	0.242	TOTAL (U_cA_p) =	

TLC = NLC + 24 x TOTAL (U_cA_p) =

BASE TEMPERATURE, T_b

$T_b = T_{set} - Q_{int}/TLC$ = ____ - ____ / ____ = 65 °F

AUXILIARY HEAT, Q_{aux}

$Q_{aux} = NLC \bullet DD(1 - SSF)$ = 5232 x 4754 x (1 - 0.242) x 10^{-6} = 18.9 Mega Btu/year

Figure 16-5. *NLC* and *LCR* Worksheet for Example Problem 16-2

SLR METHOD WORKSHEET DATE 1/1/83 BY Smith REF. DESIGN DGA1 LOCATION St. Louis

* * KEY FORMULAS		
	TW, WW, SS	SLR = (S/DD − LCR_s • H)/LCR, SSF = 1 − C exp(−D•SLR)
	DG	SLR = (S/DD)/(LCR + G), SSF = 1 − (1 + G/LCR) [1 − F(SLR)] Compute SLR then use graphs to find F(SLR) or { if SLR < R, F(SLR) = A • SLR; if SLR ≥ R, F(SLR) = B − C exp (−D • SLR)
	ALL SYSTEMS	Annual SSF = Annual Q_{sav} / Annual Q_{net}

DATA		NLC	A_p	LCR	LCR_s	A	B	C	D	G	H	R	LCR_s • H
	TW, WW, SS												
	DG	5232	270	19.4		0.565	1.009	1.044	0.7175	9.36		0.5931	

ITEM	ABSORBED SOLAR ENERGY	DEGREE-DAYS (T_b = 65 °F)	SOLAR LOAD RATIO	SOLAR SAVINGS FRACTION	NET REFERENCE LOAD	SOLAR SAVINGS	AUXILIARY ENERGY
UNITS	(Btu/ft²)	(°F • day)	Dimensionless	Dimensionless	(Mega Btu)	(Mega Btu)	(Mega Btu)
ORIGIN	SOLAR GAIN WORKSHEET	Weather Data	* *	* *	NLC • DD • 10^{-6}	Q_{net} • SSF	Q_{net} (1−SSF)
SYMBOL	S	DD	SLR	SSF	Q_{net}	Q_{sav}	Q_{aux}
JAN	19 255	1 045	0.641	0.036	5.47	0.20	5.27
FEB	20 096	838	0.834	0.163	4.38	0.71	3.67
MAR	20 964	682	1.069	0.295	3.57	1.05	2.52
APR	17 447	272	2.230	0.701	1.42	1.00	0.42
MAY	16 134	103	5.446	0.982	0.54	0.53	0.01
JUN	15 252	10	53.032	1.0	0.05	0.05	0
JUL	16 119	0		1.0	0	0	0
AUG	18 297	0		1.0	0	0	0
SEP	21 787	35	21.644	1.0	0.18	0.18	0
OCT	25 827	224	4.009	0.926	1.17	1.08	0.09
NOV	20 834	601	1.205	0.361	3.14	1.13	2.01
DEC	16 725	944	0.616	0.019	4.94	0.09	4.85
ANNUAL		Sum: 4754		* * 0.242	Sum: 24.86	Sum: 6.02	Sum: 18.84

Figure 16-6. Monthly *SLR* Method Worksheet for Example Problem 16-2

SOLAR GAIN WORKSHEET

TILT ______ AZ ______ E ______ N ______ REFR INDEX ______

TW & WW: SURFACE ABSORPTANCE = ______

DG & SS: α_m = ______ HEIGHT = ______ (ft) DEPTH = ______ (ft) WIDTH = ______ (ft) GLAZING LENGTH = ______ (ft)

α_o = ______ HGT = 1 XTOP = ______ WDTH = ______ Z = ______

DATE 1/1/83
BY Smith
REF DSGN DGA1
LOCATION St. Louis

ITEM	SOLAR NOON ZENITH ANGLE Weather Table	CLEARNESS FACTOR Weather Table	NUMBER OF DAYS IN MONTH	HORIZONTAL TOTAL INSOLATION Weather Table	GLAZED AREA TO PROJECTED AREA RATIO	(TRN/HOR) ~~INCIDENT FRACTION~~	~~TRANSMITTED FRACTION~~	ABSORBED FRACTION	PRODUCT
SYMBOL	LAT-DEC	K_T	M	HS	A_n/A_p	(INC/HOR)	(TRN/INC)	(ABS/TRN)	S
JAN	60.0	0.46	31	627	1	1.015		0.976	19 255
FEB	52.3	0.49	28	886	1	0.830		0.976	20 096
MAR	41.1	0.49	31	1 205	1	0.575		0.976	20 964
APR	28.9	0.51	30	1 564	1	0.381		0.976	17 447
MAY	19.7	0.54	31	1 871	1	0.285		0.976	16 134
JUN	15.3	0.58	30	2 092	1	0.249		0.976	15 252
JUL	17.3	0.58	31	2 049	1	0.260		0.976	16 119
AUG	25.2	0.57	31	1 816	1	0.333		0.976	18 297
SEP	36.9	0.55	30	1 459	1	0.510		0.976	21 787
OCT	48.7	0.55	31	1 100	1	0.776		0.976	25 827
NOV	58.1	0.49	30	718	1	0.991		0.976	20 834
DEC	62.1	0.43	31	531	1	1.041		0.976	16 725

Figure 16-7. Solar Gain Worksheet for Example Problem 16-2

Example Problem 16-3

For this example, the same building and location as used in Example Problem 16-2 are reanalyzed with the following change: the ground reflectance (ρ_g) is 0.7 instead of 0.3, the extinction coefficient-thickness product (E) is 0.05 instead of 0.0625, and the refractive index is 1.3 instead of 1.526. The completed monthly *SLR* method worksheet is shown in Figure 16-8.

Step-by-Step Procedures for the *SLR* Method Worksheet

Steps 1 through 4 — Same as Example Problem 16-2.

Step 5 — See Figure 16-9 for the completed solar gain worksheet. The step-by-step procedure for the solar gain worksheet is explained separately below.

Step 6 — Copied from the weather data, Appendix 3, St. Louis, Missouri; base temperature of 65°F.

Steps 7 through 13 — Calculated routinely.

Comments — In comparison to Example Problem 16-2, the improved glazing properties and ground reflectance both contribute to an increased solar gain and thus an increased performance. Improved ground reflectance can be achieved with heavy snow cover. Note that the ground reflectance can be different each month, if desired, to match expected conditions.

Step-by Step Procedures for the Solar Gain Worksheet

In Example Problem 16-3, the values of the ground reflectance, the extinction coefficient-thickness product, and the refractive index are different from the values assumed in deriving the solar radiation correlations. The effect of the changes is accounted for in the solar gain worksheet as explained in the step-by-step procedure below.

Steps 1 through 5 — Performed routinely as described in Chapter 9.

Step 6 — From Appendix 4, Table A4-1, find the values of the correlation coefficients and calculate (*INC/HOR*) for each month. Note that no correction is made for the change in ground reflectance. The correction is made in Step 8.

Step 7 — List the monthly values of (*INC/HOR*) on the solar gain worksheet.

Step 8 — Since both ground reflectance, ρ_g, and extinction coefficient-thickness product, E, have been changed, Equation 9-21 must be used. From Figure 9-6 find

$$\tau_D = 0.75 \ .$$

Note that this accounts for the change in refractive index.

Since $F_{CG} = 0.5$ and $\rho_g = 0.7$, Equation 9-21 reduces to

$$(TRN/INC) = 1.029\left[(TRN/INC)_0 + \frac{0.15}{(INC/HOR)_0}\right] .$$

The values of $(TRN/INC)_O$ are calculated using the correlation coefficients in Appendix 4, Table A4-2. The values of $(INC/HOR)_O$ are the same as those calculated in Step 6. The month-by-month values of $(TRN/INC)_O$, $(INC/HOR)_O$, and the resulting values of (TRN/INC) are listed in Table 16-1.

Table 16-1

Calculation of (*TRN/INC*) for Example Problem 16-3

Month	$(TRN/INC)_O$	$(INC/HOR)_O$	(TRN/INC)
Jan	0.708	1.438	0.836
Feb	0.694	1.204	0.842
Mar	0.660	0.878	0.855
Apr	0.611	0.628	0.874
May	0.580	0.500	0.906
Jun	0.569	0.450	0.929
Jul	0.570	0.467	0.917
Aug	0.592	0.567	0.881
Sep	0.643	0.798	0.855
Oct	0.686	1.140	0.841
Nov	0.706	1.409	0.836
Dec	0.709	1.470	0.835

SLR METHOD WORKSHEET DATE 1/1/83 BY Smith REF. DESIGN DGA1 LOCATION St. Louis

* * KEY FORMULAS	TW, WW, SS	$SLR = (S/DD - LCR_s \cdot H)/LCR$, $SSF = 1 - C \exp(-D \cdot SLR)$
	DG	$SLR = (S/DD)/(LCR + G)$, $SSF = 1 - (1 + G/LCR)\,[1 - F(SLR)]$ Compute SLR then use graphs to find F(SLR) or { if SLR < R, F(SLR) = A · SLR; if SLR ≥ R, F(SLR) = B − C exp (−D · SLR) }
	ALL SYSTEMS	Annual SSF = Annual Q_{sav} / Annual Q_{net}

DATA	NLC	Ap	LCR	LCR_s	A	B	C	D	G	H	R	$LCR_s \cdot H$
TW, WW, SS												
DG	5232	270	19.4		0.565	1.009	1.044	0.7175	9.36		0.3931	

ITEM	ABSORBED SOLAR ENERGY	DEGREE-DAYS (T_b = 65 °F)	SOLAR LOAD RATIO	SOLAR SAVINGS FRACTION	NET REFERENCE LOAD	SOLAR SAVINGS	AUXILIARY ENERGY
UNITS	(Btu/ft^2)	(°F · day)	Dimensionless	Dimensionless	(Mega Btu)	(Mega Btu)	(Mega Btu)
ORIGIN	SOLAR GAIN WORKSHEET	Weather Data	* *	* *	$NLC \cdot DD \cdot 10^{-6}$	$Q_{net} \cdot SSF$	Q_{net} (1−SSF)
SYMBOL	S	DD	SLR	SSF	Q_{net}	Q_{sav}	Q_{aux}
JAN	22 806	1 045	0.759	0.116	5.47	0.63	4.84
FEB	24 546	838	1.018	0.268	4.38	1.17	3.21
MAR	27 369	682	1.395	0.444	3.57	1.59	1.98
APR	25 135	272	3.213	0.859	1.42	1.22	0.20
MAY	25 644	103	8.657	1.0	0.54	0.54	0
JUN	25 607	10	89.037	1.0	0.05	0.05	0
JUL	26 548	0	—	1.0	0	0	0
AUG	27 446	0	—	1.0	0	0	0
SEP	29 147	35	28.956	1.0	0.18	0.18	0
OCT	31 908	224	4.953	0.969	1.17	1.13	0.04
NOV	24 764	601	1.433	0.460	3.14	1.44	1.70
DEC	19 720	944	0.726	0.094	4.94	0.46	4.48
ANNUAL		Sum: 4754		* * 0.338	Sum: 24.86	Sum: 8.41	Sum: 16.45

Figure 16-8. Monthly *SLR* Worksheet for Example Problem 16-3

SOLAR GAIN WORKSHEET

TILT 90
AZ 0
E 0.05
N 2
REFR INDEX 1.3

TW & WW: SURFACE ABSORPTANCE = ______

DG & SS:
α_m = ______ α_o = ______
HEIGHT = ______ (ft) HGT = 1
DEPTH = ______ (ft) XTOP = ______
WIDTH = ______ (ft) WDTH = ______
GLAZING LENGTH = ______ (ft) Z = ______

DATE 1/1/83
BY Smith
REF DSGN DGA1
LOCATION St. Louis

ITEM	SOLAR NOON ZENITH ANGLE Weather Table	CLEARNESS FACTOR Weather Table	NUMBER OF DAYS IN MONTH	HORIZONTAL TOTAL INSOLATION Weather Table	GLAZED AREA TO PROJECTED AREA RATIO	INCIDENT FRACTION	TRANSMITTED FRACTION	ABSORBED FRACTION	PRODUCT
SYMBOL	LAT-DEC	K_T	M	HS	A_n/A_p	(INC/HOR)	(TRN/INC)	(ABS/TRN)	S
JAN	60.0	0.46	31	627	1	1.438	0.836	0.976	22 806
FEB	52.3	0.49	28	886	1	1.204	0.842	0.976	24 546
MAR	41.1	0.49	31	1205	1	0.878	0.855	0.976	27 369
APR	28.9	0.51	30	1564	1	0.628	0.874	0.976	25 135
MAY	19.7	0.54	31	1871	1	0.500	0.906	0.976	25 644
JUN	15.3	0.58	30	2092	1	0.450	0.929	0.976	25 607
JUL	17.3	0.58	31	2049	1	0.467	0.917	0.976	26 548
AUG	25.2	0.57	31	1816	1	0.567	0.881	0.976	27 446
SEP	36.9	0.55	30	1459	1	0.798	0.855	0.976	29 147
OCT	48.7	0.55	31	1100	1	1.140	0.841	0.976	31 908
NOV	58.1	0.49	30	718	1	1.409	0.836	0.976	24 764
DEC	62.1	0.43	31	531	1	1.470	0.835	0.976	19 720

Figure 16-9. Solar Gain Worksheet for Example Problem 16-3

The (*TRN*/*INC*) column is transferred to the solar gain worksheet. This example illustrates that Equation 9-21 (and also Equation 9-19) does not yield a true transmittance value but is a special form for facilitating the calculations on the solar gain worksheet.

Steps 9 through 13 — Computed routinely.

Example Problem 16-4

This example problem considers the same building as was used in Example Problem 16-2. The sunspace reference system SSA1 replaces the direct-gain windows, and A_p = 270 ft^2, as before. The monthly *SLR* method worksheet is shown in Figure 16-10. The location is St. Louis, Missouri. The degree-days base temperature is 65°F.

Step 1 — The reference design is SSA1.

Step 2 — From Example Problem 16-2, $NLC = 5232$, $A_p = 270$, $LCR = 19.4$.

Step 3 — LCR_s and the correlation coefficients for reference design SSA1 are from Appendix 5, Table A5-8.

Step 4 — $LCR_s \cdot H = 15.438$.

Step 5 — The solar gain worksheet is shown in Figure 16-11. Since the example adheres strictly to the reference design the correlation in Table A4-47, Appendix 4 can be used. The factor (A_n/A_p) (*ABS*/*HOR*) is calculated directly.

Step 6 — The monthly degree days are copied from Appendix 3, St. Louis, Missouri, base temperature of 65°F.

Steps 7 through 13 — Calculated routinely.

Comment — Calculation of *SSF* by the annual *LCR* method (not shown) yields $SSF = 0.477$, which agrees closely with the monthly *SLR* method.

SLR METHOD WORKSHEET DATE 1/1/83 BY Smith REF. DESIGN SSA1 LOCATION St. Louis

* * KEY FORMULAS		
	TW, WW, SS	$SLR = (S/DD - LCR_s \cdot H)/LCR$, $SSF = 1 - C\ exp(-D \cdot SLR)$
	DG	$SLR = (S/DD)/(LCR + G)$, $SSF = 1 - (1 + G/LCR)\ [1 - F(SLR)]$ Compute SLR then use graphs to find F(SLR) or { if SLR < R, $F(SLR) = A \cdot SLR$; if SLR ≥ R, $F(SLR) = B - C\ exp\ (-D \cdot SLR)$
	ALL SYSTEMS	Annual SSF = Annual Q_{sav} / Annual Q_{net}

DATA	NLC	Ap	LCR	LCR_s	A	B	C	D	G	H	R	$LCR_s \cdot H$
TW, WW, SS	5232	270	19.4	18.6			0.9587	0.4770		0.83		15.438
DG												

ITEM	ABSORBED SOLAR ENERGY	DEGREE-DAYS (T_b = 65 °F)	SOLAR LOAD RATIO	SOLAR SAVINGS FRACTION	NET REFERENCE LOAD	SOLAR SAVINGS	AUXILIARY ENERGY
UNITS	(Btu/ft²)	(°F · day)	Dimensionless	Dimensionless	(Mega Btu)	(Mega Btu)	(Mega Btu)
ORIGIN	SOLAR GAIN WORKSHEET	Weather Data	* *	* *	$NLC \cdot DD \cdot 10^{-6}$	$Q_{net} \cdot SSF$	Q_{net} (1−SSF)
SYMBOL	S	DD	SLR	SSF	Q_{net}	Q_{sav}	Q_{aux}
JAN	26 862	1 045	0.529	0.255	5.47	1.39	4.08
FEB	30 464	838	1.078	0.427	4.38	1.87	2.51
MAR	37 280	682	2.022	0.635	3.57	2.27	1.30
APR	38 334	272	6.469	0.956	1.42	1.36	0.06
MAY	41 761	103	20.104	1	0.54	0.54	0
JUN	42 865	10	220.158	1	0.05	0.05	0
JUL	44 400	0	—	1	0	0	0
AUG	43 911	0	—	1	0	0	0
SEP	41 625	35	60.508	1	0.18	0.18	0
OCT	40 750	224	8.582	0.984	1.17	1.15	0.02
NOV	29 488	601	1.733	0.581	3.14	1.82	1.32
DEC	22 963	944	0.458	0.229	4.94	1.13	3.81
ANNUAL		Sum: 4754		* * 0.473	Sum: 24.86	Sum: 11.76	Sum: 13.10

Figure 16-10. Monthly *SLR* Worksheet for Example Problem 16-4

SOLAR GAIN WORKSHEET

TILT ____ AZ ____ E ____ N ____ REFR INDEX ____

TW & WW: SURFACE ABSORPTANCE = ____

DG & SS: α_m = ____ HEIGHT = ____ (ft) DEPTH = ____ (ft) WIDTH = ____ (ft) GLAZING LENGTH = ____ (ft)

α_o = ____ HGT = ____ XTOP = ____ WDTH = ____ Z = ____

DATE 1/1/83
BY Smith
REF DSGN SSA 1
LOCATION St. Louis

ITEM	SOLAR NOON ZENITH ANGLE Weather Table	CLEARNESS FACTOR Weather Table	NUMBER OF DAYS IN MONTH	HORIZONTAL TOTAL INSOLATION Weather Table	~~GLAZED AREA TO PROJECTED AREA RATIO~~	~~INCIDENT FRACTION~~	~~TRANSMITTED FRACTION~~	$(A_m/A_p)\left(\frac{ABS}{HOR}\right)$ ~~ABSORBED FRACTION~~	PRODUCT
SYMBOL	LAT-DEC	K_T	M	HS	~~A_n/A_p~~	~~(INC/HOR)~~	~~(TRN/INC)~~	~~(ABS/TRN)~~	S
JAN	60.0	0.46	31	627				1.382	26 862
FEB	52.3	0.49	28	886				1.228	30 464
MAR	41.1	0.49	31	1 205				0.998	37 280
APR	28.9	0.51	30	1 564				0.817	38 334
MAY	19.7	0.54	31	1 871				0.720	41 761
JUN	15.3	0.58	30	2 092				0.683	42 865
JUL	17.3	0.58	31	2 049				0.699	44 400
AUG	25.2	0.57	31	1 816				0.780	43 911
SEP	36.9	0.55	30	1 459				0.951	41 625
OCT	48.7	0.55	31	1 100				1.195	40 750
NOV	58.1	0.49	30	718				1.369	29 488
DEC	62.1	0.43	31	531				1.395	22 963

Figure 16-11. Solar Gain Worksheet for Example Problem 16-4

Example Problem 16-5

The same building and the same reference design (SSA1) are used in this example as were used in Example Problem 16-4. The following changes were made to the reference design: a single layer of glazing, as opposed to two layers for the reference design, and a sunspace infiltration rate of 1 air change per hour, instead of 1/2 air change per hour. Also, the sunspace uses night insulation of R3. This requires the calculation of *SSF* for each month for the case of no-night-insulation and for the case of R9 night insulation, then the correct monthly values of *SSF* are interpolated between the two. The step-by-step procedure for the *SLR* method worksheet is given immediately below, and the step-by-step procedure for the solar gain worksheet is given in the following section.

Step-by-Step Procedure for the *SLR* Method Worksheet

Step 1 — The reference design is SSA1 for the no night insulation case and SSA2 for the R9 night insulation case. Fill out a separate worksheet for each.

Step 2 — $NLC = 5232$, $A_p = 270$ and $LCR = 19.4$ for both cases.

Step 3 — The correlation coefficients for reference designs SSA1 (no night insulation) and SSA2 (R9 night insulation) are obtained from Appendix 5, Table A5-8. The values of LCR_s must be computed for the off-reference conditions.

For the no night insulation case, $LCR_s = 36.5$. The method of computation is described in Chapter 11. The computation is as follows:

$$(UA)_g = \frac{1.083\,A_n}{N + 1.517\,t - 0.4295}$$

$$= \frac{1.083 \times 352.5}{1 + 1.517 \times 0.35 - 0.4295}$$

$$= 346.6 \text{ Btu/h} \cdot {}^\circ\text{F}$$

$$(UA)_i = 0.018\,V \cdot ACH = 0.018 \times 2100 \times 1$$

$$= 37.8 \text{ Btu/h} \cdot {}^\circ\text{F}$$

$$(UA)_p + \frac{4.17\,P_p}{R_p + 5} = \frac{4.17 \times 53.1}{12 + 5}$$

$$= 13.0 \text{ Btu/h} \cdot {}^\circ\text{F}$$

$$(UA)_r = \frac{A_r}{R_s} = \frac{120}{20} = 6.0 \text{ Bth/h} \cdot {}^\circ\text{F}$$

$$(UA)_w = \frac{A_w}{R_w} = \frac{140}{20} = 7.0 \text{ Btu/h} \cdot {}^\circ\text{F}$$

Then compute NLC_s from Equation 11-8 as follows:

$$NLC_s = 24(346.6 + 37.8 + 13 + 6 + 7)$$

$$= 9850 \text{ Btu/}{}^\circ\text{F} \cdot \text{day}$$

$$LCR_s = NLC_s/A_p = 9850/270$$

$$= 36.5 \text{ Btu/}{}^\circ\text{F} \cdot \text{day} \cdot \text{ft}^2$$

For the R9 night insulation case, $LCR_s = 18.0$, calculated as follows:

$$(UA)_g = 1.083\,A_n \left[\frac{1}{3} \times \frac{1}{N + 1.517t - 0.4295} + \frac{2}{3} \times \frac{1}{N + 1.517\,t + 9.318} \right]$$

$$= 1.083 \times 352.5 \times \left[\frac{1}{3} \times \frac{1}{1 + 1.517 \times 0.35 - 0.4295} + \frac{2}{3} \times \frac{1}{1 - 1.517 \times 0.35 + 9.318} \right]$$

$$= 139.0 \text{ Btu/h} \cdot {}^\circ\text{F}$$

$(UA)_i = 37.8$ Btu/h · °F, same as for SSA1

$(UA)_p = 13.0$ Btu/h · °F, same as for SSA1

$(UA)_r$ = 6.0 Btu/h · °F, same as for SSA1

$(UA)_w$ = 7.0 Btu/h · °F, same as for SSA1

NLC_s = 4867 Btu/°F · day

LCR_s = 18.0 Btu/°F · day · ft^2

Step 4 — $LCRs \cdot H$ = 30.295 for the no-night-insulation case and 13.86 for the *R9* case.

Step 5 — Since the sunspace glazing has been changed from the reference design by using only one layer of glazing instead of two, the correlation method in Table A4-47 for calculating *S* cannot be used. The procedure for completing the solar gain worksheet is described below in the following section. The resulting values of *S* are copied onto the *SLR* method worksheet.

Step 6 — The values of *DD* were taken from Appendix 3, St. Louis, Missouri, base temperature of 65°F.

Steps 7 and 9 — These steps are executed routinely for both SSA1 and SSA2 reference designs as shown in Figures 16-12 and 16-13. The resulting values of SSF_0 and SSF_9 are listed in Table 16-2. The values of SSF_3 for R3 night insulation are calculated as shown in Table 16-2.

$$SSF_3 = SSF_0 + Y(SSF_9 - SSF_0)$$

$$Y = (1 + R_0/9)/(1 + R_0/R)$$

R_0 = 3.071 (calculated below)

R = 3 (for Example Problem 16-5)

The value of R_0 is calculated as $1/U_c$ where U_c is the solar wall load coefficient for the non-night-insulated version of the example. Values of U_c for sunspace reference designs are listed in Appendix 5, Table A5-8, but an off-reference value is required in this example because of the off-reference number of glazing layers and infiltration rate. Use Equation 11-16 to calculate the off-reference value:

Table 16-2

Calculation of SSF_3 for Example Problem 16-5

Month	SSA1 No Night Insulation SSF_0	SSA2 R9 Night Insulation SSF_9	R3 Night Insulation SSF_3
Jan	−0.013	0.384	0.250
Feb	0.240	0.587	0.470
Mar	0.539	0.793	0.707
Apr	0.956	0.992	0.980
May	1	1	1
Jun	1	1	1
Jul	1	1	1
Aug	1	1	1
Sep	1	1	1
Oct	0.985	0.998	0.994
Nov	0.458	0.742	0.646
Dec	−0.051	0.352	0.216

$$U_c = \frac{1}{24A_p}\left(\frac{(UA)_{cw} \cdot NLC_s}{(UA)_{cw} + NLC_s/24}\right)$$

where

A_p = 270 ft^2

and

$$NLC_s = LCR_s \cdot A_p = (36.5 \text{ Btu/°F} \cdot \text{day} \cdot \text{ft}^2)(270 \text{ ft}^2)$$

$$= 9855 \text{ Btu/°F} \cdot \text{day}.$$

The quantity $(UA)_{cw}$ is the value for reference design SSA1 because no modifications were made to the common wall. The value may be calculated directly using ASHRAE procedures and the properties of the common wall, but since it is a reference value, it is more convenient to base the calculation on the reference values of LCR_s and U_c. The relationship is just the general series heat transfer relationship

$$\frac{1}{U_cA_p} = \frac{1}{(UA)_{cw}} + \frac{24}{LCR_sA_p}$$

SLR METHOD WORKSHEET DATE 1/1/83 BY Smith REF. DESIGN SSA1 LOCATION St. Louis

* * KEY FORMULAS		
	TW, WW, SS	SLR = (S/DD − LCR_S • H)/LCR, SSF = 1 − C exp(−D • SLR)
	DG	SLR = (S/DD)/(LCR + G), SSF = 1 − (1 + G/LCR) [1 − F(SLR)] Compute SLR then use graphs to find F(SLR) or { if SLR < R, F(SLR) = A • SLR; if SLR ≥ R, F(SLR) = B − C exp (−D • SLR) }
	ALL SYSTEMS	Annual SSF = Annual Q_{sav} / Annual Q_{net}

DATA		NLC	Ap	LCR	LCR_S	A	B	C	D	G	H	R	LCR_S • H
	TW, WW, SS	5232	270	19.4	36.5			0.9587	0.4770		0.83		30.295
	DG												

ITEM	ABSORBED SOLAR ENERGY	DEGREE-DAYS (T_b = 65 °F)	SOLAR LOAD RATIO	SOLAR SAVINGS FRACTION	NET REFERENCE LOAD	SOLAR SAVINGS	AUXILIARY ENERGY
UNITS	(Btu/ft²)	(°F • day)	Dimensionless	Dimensionless	(Mega Btu)	(Mega Btu)	(Mega Btu)
ORIGIN	SOLAR GAIN WORKSHEET	Weather Data	* *	* *	NLC • DD • 10^{-6}	Q_{net} • SSF	Q_{net} (1−SSF)
SYMBOL	S	DD	SLR	SSF_o	Q_{net}	Q_{sav}	Q_{aux}
JAN	29 308	1 045	−0.116	−0.013			
FEB	33 294	838	0.486	0.240			
MAR	40 954	682	1.534	0.539			
APR	42 350	272	6.464	0.956			
MAY	46 337	103	21.628	1.0			
JUN	47 567	10	243.629	1.0			
JUL	49232	0	—	1.0			
AUG	48548	0	—	1.0			
SEP	45679	35	65.712	1.0			
OCT	44510	224	8.681	0.985			
NOV	32 144	601	1.195	0.458			
DEC	25059	944	−0.193	−0.051			
ANNUAL		Sum: 4754		* *	Sum:	Sum:	Sum:

Figure 16-12. Calculation of SSF_o for Example Problem 16-5

SLR METHOD WORKSHEET DATE 1/1/83 BY Smith REF. DESIGN SSA2 LOCATION ST. LOUIS

* * KEY FORMULAS	TW, WW, SS	SLR = (S/DD − LCR_s • H)/LCR, SSF = 1 − C exp(−D • SLR)
	DG	SLR = (S/DD)/(LCR + G), SSF = 1 − (1 + G/LCR) [1 − F(SLR)] Compute SLR then use graphs to find F(SLR) or { if SLR < R, F(SLR) = A • SLR; if SLR ≥ R, F(SLR) = B − C exp (−D • SLR) }
	ALL SYSTEMS	Annual SSF = Annual Q_{sav} / Annual Q_{net}

DATA		NLC	Ap	LCR	LCR_s	A	B	C	D	G	H	R	LCR_s • H
	TW, WW, SS	5232	270	19.4	18.0			0.9982	0.6614		0.77		13.86
	DG												

ITEM	ABSORBED SOLAR ENERGY	DEGREE-DAYS (T_b = 65 °F)	SOLAR LOAD RATIO	SOLAR SAVINGS FRACTION	NET REFERENCE LOAD	SOLAR SAVINGS	AUXILIARY ENERGY
UNITS	(Btu/ft²)	(°F • day)	Dimensionless	Dimensionless	(Mega Btu)	(Mega Btu)	(Mega Btu)
ORIGIN	SOLAR GAIN WORKSHEET	Weather Data	* *	* *	NLC • DD • 10^{-6}	Q_{net} • SSF	Q_{net} (1−SSF)
SYMBOL	S	DD	SLR	SSF_9	Q_{net}	Q_{sav}	Q_{aux}
JAN	29 308	1045	0.731	0.384			
FEB	33 294	838	1.334	0.587			
MAR	40 954	682	2.381	0.793			
APR	42 350	272	7.311	0.992			
MAY	46 337	103	22.475	1.0			
JUN	47 567	10	244.476	1.0			
JUL	49 232	0	—	1.0			
AUG	48 548	0	—	1.0			
SEP	45 679	35	66.559	1.0			
OCT	44 510	224	9.528	0.998			
NOV	32 144	601	2.043	0.742			
DEC	25 059	944	0.654	0.352			
ANNUAL		Sum: 4754		* *	Sum:	Sum:	Sum:

Figure 16-13. Calculation of SSF_9 for Example Problem 16-5

Solving this equation for $(UA)_{cw}$,

$$(UA)_{cw} = \frac{U_c \cdot A_p \cdot LCR_s}{LCR_s - 24U_c}$$

Substituting the SSA1 reference values of LCR_s and U_c (Appendix 5, Table A5-8),

$$(UA)_{cw} = \frac{(0.27)(270)(18.6)}{18.6 - (24)(0.27)} = 111.88 \text{ Btu/h} \cdot {}^\circ\text{F}$$

Therefore,

$$U_c = \frac{1}{(24)(270)} \left[\frac{(111.88)(9855)}{111.88 + 9855/24} \right]$$

$$= 0.3256 \text{ Btu/h} \cdot {}^\circ\text{F} \cdot \text{ft}^2 ,$$

and

$$R_o = 1/0.3256$$

$$= 3.071 \text{ h} \cdot {}^\circ\text{F} \cdot \text{ft}^2\text{/Btu} .$$

Steps 11 through 13 — Having completed the calculation of *SSF* as explained above, the values of SSF_3 as listed in Table 16-2 are copied onto the *SLR* method worksheet shown in Figure 16-14 and the computations are completed routinely.

Comments — The completed analysis shows

annual Q_{aux} = 12.49 million Btu

annual *SSF* = 0.498

Note that the use of R3 night insulation more than compensates for the change from 0.5 to 1 sunspace air change per hour and from 2 to 1 glazing layers. This is seen by comparing these results with those of Example Problem 16-4.

Step-by-Step Procedure for the Solar Gain Worksheet

Steps 1 through 5 — These steps are done routinely.

Step 6 — Since the sunspace glazing has been changed from the reference design by using only one layer of glazing instead of two, the correlation method in Table A4-47 for calculating *S* cannot be used. The *TILT* of the glazing for SSA1 and SSA2 reference designs is 50°. The closest values of *TILT* in Appendix 4, Table A4-1, are 30° and 60°; interpolation must be made to find values of (*INC/HOR*) for *TILT* = 50°.

Step 7 — The resulting values of (*INC/HOR*) are copied onto the solar gain worksheet shown in Figure 16-15.

Step 8 — To obtain values of (*TRN/INC*) for *TILT* = 50°, interpolation must be carried out in the same manner as was done for (*INC/HOR*) in Step 6. The results are listed in Table 16-3.

Table 16-3

Computation of $(TRN/INC)(ABS/TRN)_o + g(N)$ for Example Problem 16-5

Month	(*TRN/INC*)	$(ABS/TRN)_o$	$(TRN/INC)(ABS/TRN)_o + g(N)$*
Jan	0.829	0.935	0.744
Feb	0.827	0.935	0.742
Mar	0.819	0.935	0.735
Apr	0.809	0.935	0.725
May	0.799	0.935	0.716
Jun	0.793	0.935	0.710
Jul	0.796	0.935	0.713
Aug	0.806	0.935	0.723
Sep	0.817	0.935	0.733
Oct	0.827	0.935	0.742
Nov	0.829	0.935	0.744
Dec	0.828	0.935	0.743

*$N = 1, g(1) = -0.031$

Step 9 — The values of (*TRN/INC*) are not listed directly in the solar gain worksheet. Instead, a further correction must be applied to account for the

SLR METHOD WORKSHEET DATE 1/1/83 BY Smith REF. DESIGN SSA1/SSA2 with R3 night insulation LOCATION St. Louis

** KEY FORMULAS		
	TW, WW, SS	SLR = (S/DD − $LCR_s \cdot H$)/LCR, SSF = 1 − C exp(−D·SLR)
	DG	SLR = (S/DD)/(LCR + G), SSF = 1 − (1 + G/LCR) [1 − F(SLR)] Compute SLR then use graphs to find F(SLR) or { if SLR < R, F(SLR) = A · SLR; if SLR ≥ R, F(SLR) = B − C exp (−D · SLR) }
	ALL SYSTEMS	Annual SSF = Annual Q_{sav} / Annual Q_{net}

DATA		NLC	Ap	LCR	LCR_s	A	B	C	D	G	H	R	$LCR_s \cdot H$
	TW, WW, SS	5232	270	19.4									
	DG												

ITEM	ABSORBED SOLAR ENERGY	DEGREE-DAYS (T_b = ____°F)	SOLAR LOAD RATIO	SOLAR SAVINGS FRACTION	NET REFERENCE LOAD	SOLAR SAVINGS	AUXILIARY ENERGY
UNITS	(Btu/ft²)	(°F·day)	Dimensionless	Dimensionless	(Mega Btu)	(Mega Btu)	(Mega Btu)
ORIGIN	SOLAR GAIN WORKSHEET	Weather Data	**	**	NLC·DD·10^{-6}	$Q_{net} \cdot$ SSF	Q_{net} (1−SSF)
SYMBOL	S	DD	SLR	SSF_3	Q_{net}	Q_{sav}	Q_{aux}
JAN				0.250	5.47	1.37	4.10
FEB				0.470	4.38	2.06	2.32
MAR				0.707	3.57	2.52	1.05
APR				0.980	1.42	1.39	0.03
MAY				1.0	0.54	0.54	0
JUN				1.0	0.05	0.05	0
JUL				1.0	0	0	0
AUG				1.0	0	0	0
SEP				1.0	0.18	0.18	0
OCT				0.994	1.17	1.16	0.01
NOV				0.646	3.14	2.03	1.11
DEC				0.216	4.94	1.07	3.87
ANNUAL		Sum:		** 0.498	Sum: 24.86	Sum: 12.37	Sum: 12.49

Figure 16-14. Completion of Computations for Example Problem 16-5

single glazing. See the discussion of Step 11 for this correction.

Step 10 — Double linear interpolation of the values in Table A4-21 is used to determine (*ABS/TRN*) for the desired values of *TILT* and *XTOP*. The computation is shown in Table 16-4.

Table 16-4

Interpolation of (*ABS/TRN*) for Example Problem 16-5

	(*ABS/TRN*) from Table A4-21		(*ABS/TRN*) Interpolated
TILT	*XTOP* 0	*XTOP* 0.5	*XTOP* 0.44
30	0.92	0.93	0.929
60	0.92	0.94	0.938
50	—	—	0.935

The resulting value of (*ABS/TRN*) is 0.935. It is listed in Table 16-3, the same for each month. The value of (*ABS/TRN*) is not listed directly in the solar gain worksheet. Instead, a further correction must be applied to account for the single glazing.

Step 11 — The product (*TRN/INC*)(*ABS/TRN*) must be corrected for single glazing, as explained in Chapter 9. The correction is (Equation 9-23):

$$\left(\frac{TRN}{INC}\right)\left(\frac{ABS}{TRN}\right) = \left(\frac{TRN}{INC}\right)\left(\frac{ABS}{TRN}\right)_0 + g(N)$$

where $N = 1$ (single glazing) and $g(1) = -0.031$. The results are shown in Table 16-3.

The values in the far-right column are copied into the solar gain worksheet, instead of separate values for (*TRN/INC*) and (*ABS/TRN*), as shown in Figure 16-15.

Steps 12 and 13 — Calculated routinely.

SOLAR GAIN WORKSHEET

TILT 50
AZ 0
E 0.0625
N 1
REFR INDEX 1.526

TW & WW: SURFACE ABSORPTANCE = ______

DG & SS:
α_m = 0.8 — α_o = 0.3
HEIGHT = 9 (ft) — HGT = 1
DEPTH = 4 (ft) — XTOP = 0.44
WIDTH = 30 (ft) — WDTH = 3.33
GLAZING LENGTH = ______ (ft) — Z = ______

DATE 1/1/83
BY Smith
REF DSGN SSA1
LOCATION St. Louis

ITEM	SOLAR NOON ZENITH ANGLE Weather Table	CLEARNESS FACTOR Weather Table	NUMBER OF DAYS IN MONTH	HORIZONTAL TOTAL INSOLATION Weather Table	GLAZED AREA TO PROJECTED AREA RATIO	INCIDENT FRACTION	(TRN/INC)(ABS/TRN)$_o$ + g(N) ~~TRANSMITTED FRACTION~~	~~ABSORBED FRACTION~~	PRODUCT
SYMBOL	LAT-DEC	K_T	M	HS	A_n/A_p	(INC/HOR)	~~(TRN/INC)~~	~~(ABS/TRN)~~	S
JAN	60.0	0.46	31	627	1.305	1.553		0.744	29 308
FEB	52.3	0.49	28	886	1.305	1.386		0.742	33 294
MAR	41.1	0.49	31	1 205	1.305	1.143		0.735	40 954
APR	28.9	0.51	30	1 564	1.305	0.954		0.725	42 350
MAY	19.7	0.54	31	1 871	1.305	0.855		0.716	46 337
JUN	15.3	0.58	30	2 092	1.305	0.818		0.710	47 567
JUL	17.3	0.58	31	2 049	1.305	0.833		0.713	49 232
AUG	25.2	0.57	31	1 816	1.305	0.914		0.723	48 548
SEP	36.9	0.55	30	1 459	1.305	1.091		0.733	45 679
OCT	48.7	0.55	31	1 100	1.305	1.348		0.742	44 510
NOV	58.1	0.49	30	718	1.305	1.537		0.744	32 144
DEC	62.1	0.43	31	531	1.305	1.570		0.743	25 059

Figure 16-15. Solar Gain Worksheet for Example Problem 16-5

Example Problem 16-6

This example demonstrates the monthly *SLR* method for a thermal storage wall system. The same building is used in this example as was used in Example Problem 16-2 except that the direct-gain windows are replaced by a double-glazed thermal storage wall of 270 ft^2 of projected area. The wall is 12-inch-thick, high-density masonry ($\rho ck = 30$) without circulation vents. Assume the wall solar absorptance is 0.95, typical of flat-black paint. These characteristics describe the unmodified reference design TWF3. The location is St. Louis, Missouri, and the thermostat setpoint is 70°F, as opposed to 65°F for Example Problem 16-2. The following internal heat sources (in Btu/day) were assumed: people, 20,000; household appliances, 30,000; lighting 2,750. Therefore, $Q_{int} = 52{,}750$ Btu/day. $NLC = 5232$ Btu/°F · day and $LCR = 19.4$ Btu/°F · day · ft^2.

Step 1 — The reference design is TWF3.

Step 2 — $NLC = 5232, A_p = 270, LCR = 19.4$.

Step 3 — LCR_s and the correlation coefficients for reference design TWF3 are from Appendix 5, Table A5-7.

Step 4 — $LCR_s \cdot H = 11.18$

Step 5 — The solar gain is calculated using the correlation for (*TRN/HOR*) from Appendix 4, Table A4-3, which is applicable because the system being analyzed (TWF3) is an unmodified reference design with a vertical south-facing solar wall. The ratio (*ABS/TRN*) is 0.962 from Appendix 4, Table A4-46. The computations are shown on the solar gain worksheet in Figure 16-16.

Step 6 — Since an internal heat source is involved, the degree-days base temperature must be computed. The computation is performed as follows:

From Equation 11-11:

$$(UA)_{sw} = U_c \cdot A_p$$

where U_c is tabulated in Appendix 5, Table A5-7; in this case $U_c = 0.24$.

And from Equation 11-10:

$$TLC = NLC + 24(UA)_{sw}.$$

Substituting the values into these equations gives:

$$TLC = 6787 \text{ Btu/°F} \cdot \text{day}.$$

The base temperature is calculated from Equation 4-6:

$$T_b = T_{set} - Q_{int}/TLC$$

$$T_b = 70 - 52750/6787 = 62.2°F$$

The degree days may be interpolated from the weather data in Appendix 3, St. Louis, Missouri, but in this example, they are calculated using the Thom method (see Chapter 4). The results are listed on the monthly *SLR* worksheet shown in Figure 16-17.

Steps 7 through 13 — Calculated routinely.

SOLAR GAIN WORKSHEET

TILT ______ AZ ______ E ______ N ______ REFR INDEX ______

TW & WW: SURFACE ABSORPTANCE = 0.95

DG & SS: α_m = ______ HEIGHT = ______ (ft) DEPTH = ______ (ft) WIDTH = ______ (ft) GLAZING LENGTH = ______ (ft)

α_o = ______ HGT = 1 XTOP = ______ WDTH = ______ Z = ______

DATE 1/1/83 BY Smith REF DSGN TWF3 LOCATION St. Louis

ITEM	SOLAR NOON ZENITH ANGLE Weather Table	CLEARNESS FACTOR Weather Table	NUMBER OF DAYS IN MONTH	HORIZONTAL TOTAL INSOLATION Weather Table	GLAZED AREA TO PROJECTED AREA RATIO	~~INCIDENT FRACTION~~	(TRN/HOR) ~~TRANSMITTED FRACTION~~	ABSORBED FRACTION	PRODUCT
SYMBOL	LAT-DEC	K_T	M	HS	A_n/A_p	~~(INC/HOR)~~	~~(TRN/INC)~~	(ABS/TRN)	S
JAN	60.0	0.46	31	627	1		1.015	0.962	18 979
FEB	52.3	0.49	28	886	1		0.830	0.962	19 808
MAR	41.1	0.49	31	1205	1		0.575	0.962	20 663
APR	28.9	0.51	30	1564	1		0.381	0.962	17 197
MAY	19.7	0.54	31	1871	1		0.285	0.962	15 902
JUN	15.3	0.58	30	2092	1		0.249	0.962	15 033
JUL	17.3	0.58	31	2049	1		0.260	0.962	15 887
AUG	25.2	0.57	31	1816	1		0.333	0.962	18 034
SEP	36.9	0.55	30	1459	1		0.510	0.962	21 474
OCT	48.7	0.55	31	1100	1		0.776	0.962	25 456
NOV	58.1	0.49	30	718	1		0.991	0.962	20 535
DEC	62.1	0.43	31	531	1		1.041	0.962	16 485

Figure 16-16. Solar Gain Worksheet for Example Problem 16-6

SLR METHOD WORKSHEET DATE 1/1/83 BY Smith REF. DESIGN TWF3 LOCATION St. Louis

* * KEY FORMULAS		
	TW, WW, SS	SLR = (S/DD − LCR_s • H)/LCR, SSF = 1 − C exp(−D • SLR)
	DG	SLR = (S/DD)/(LCR + G), SSF = 1 − (1 + G/LCR) [1 − F(SLR)] Compute SLR then use graphs to find F(SLR) or { if SLR < R, F(SLR) = A • SLR; if SLR ≥ R, F(SLR) = B − C exp (−D • SLR) }
	ALL SYSTEMS	Annual SSF = Annual Q_{sav} / Annual Q_{net}

DATA	NLC	Ap	LCR	LCR_s	A	B	C	D	G	H	R	LCR_s • H
TW, WW, SS	5233	270	19.4	13.0			1.0189	0.6502		0.86		11.18
DG											* *	

ITEM	ABSORBED SOLAR ENERGY	DEGREE-DAYS (T_b = 62.2°F)	SOLAR LOAD RATIO	SOLAR SAVINGS FRACTION	NET REFERENCE LOAD	SOLAR SAVINGS	AUXILIARY ENERGY
UNITS	(Btu/ft²)	(°F • day)	Dimensionless	Dimensionless	(Mega Btu)	(Mega Btu)	(Mega Btu)
ORIGIN	SOLAR GAIN WORKSHEET	Weather Data	* *	* *	NLC • DD • 10^{-6}	Q_{net} • SSF	Q_{net} (1−SSF)
SYMBOL	S	DD	SLR	SSF	Q_{net}	Q_{sav}	Q_{aux}
JAN	18 979	959	0.444	0.237	5.02	1.19	3.83
FEB	19 808	760	0.767	0.381	3.98	1.52	2.46
MAR	20 663	600	1.199	0.533	3.14	1.67	1.47
APR	17 197	204	3.769	0.912	1.07	0.98	0.09
MAY	15 902	64	12.231	1	0.33	0.33	0
JUN	15 033	5	154.403	1	0.03	0.03	0
JUL	15 887	0		1	0	0	0
AUG	18 034	0		1	0	0	0
SEP	21 474	17	64.536	1	0.09	0.09	0
OCT	25 456	168	7.234	0.991	0.88	0.87	0.01
NOV	20 535	517	1.471	0.608	2.70	1.64	1.06
DEC	16 485	858	0.414	0.222	4.49	1.00	3.49
ANNUAL		Sum: 4152		* * 0.429	Sum: 21.73	Sum: 9.32	Sum: 12.41

Figure 16-17. Monthly *SLR* Worksheet for Example Problem 16-6

Example Problem 16-7

This example uses the same building and the same reference design, TWF3, as were used in Example Problem 16-6. The orientation of the building (and therefore the orientation of the solar wall) is changed to 60° to the east.

Steps 1 through 4 — These steps are the same as in Example Problem 16-6.

Step 5 — Because of the off-south azimuth, the calculation of *S* uses the separate ratios (*INC/HOR*) and (*TRN/INC*) from Appendix 4, Tables A4-1 and A4-2, rather than the overall ratio (*TRN/HOR*) as in Example Problem 16-6. The results are shown on the solar gain worksheet in Figure 16-18.

Steps 6 through 13 — With the new data for *S* and the same degree-day data as in Example Problem 16-6, these calculations are done routinely. Figure 16-19 shows the monthly *SLR* method worksheet.

Comments — The annual *SSF* of 0.262 shown on the monthly *SLR* method worksheet indicates the extent of loss in performance caused by orienting the solar wall away from true south, a 39% reduction in performance.

SOLAR GAIN WORKSHEET

TILT 90

AZ 60

E 0.0625

N 2

REFR INDEX 1.526

TW & WW: SURFACE ABSORPTANCE = 0.95

DG & SS:
- α_m = ____
- HEIGHT = ____ (ft)
- DEPTH = ____ (ft)
- WIDTH = ____ (ft)
- GLAZING LENGTH = ____ (ft)

α_o = ____

HGT = 1

XTOP = ____

WDTH = ____

Z = ____

DATE 1/1/83

BY Smith

REF DSGN TWF3

LOCATION St. Louis

ITEM	SOLAR NOON ZENITH ANGLE Weather Table	CLEARNESS FACTOR Weather Table	NUMBER OF DAYS IN MONTH	HORIZONTAL TOTAL INSOLATION Weather Table	GLAZED AREA TO PROJECTED AREA RATIO	INCIDENT FRACTION	TRANSMITTED FRACTION	ABSORBED FRACTION	PRODUCT
SYMBOL	LAT-DEC	K_T	M	HS	A_n/A_p	(INC/HOR)	(TRN/INC)	(ABS/TRN)	S
JAN	60.0	0.46	31	627	1	0.971	0.667	0.962	12 110
FEB	52.3	0.49	28	886	1	0.873	0.668	0.962	13 917
MAR	41.1	0.49	31	1 205	1	0.734	0.666	0.962	17 567
APR	28.9	0.51	30	1 564	1	0.633	0.661	0.962	18 886
MAY	19.7	0.54	31	1 871	1	0.583	0.656	0.962	21 339
JUN	15.3	0.58	30	2 092	1	0.564	0.652	0.962	22 202
JUL	17.3	0.58	31	2 049	1	0.571	0.654	0.962	22 819
AUG	25.2	0.57	31	1 816	1	0.613	0.660	0.962	21 911
SEP	36.9	0.55	30	1 459	1	0.708	0.667	0.962	19 884
OCT	48.7	0.55	31	1 100	1	0.852	0.671	0.962	18 754
NOV	58.1	0.49	30	718	1	0.960	0.668	0.962	13 288
DEC	62.1	0.43	31	531	1	0.984	0.665	0.962	10 362

Figure 16-18. Solar Gain Worksheet for Example Problem 16-7

SLR METHOD WORKSHEET DATE 1/1/83 BY Smith REF. DESIGN TWF3 LOCATION St. Louis

* * KEY FORMULAS		
	TW, WW, SS	SLR = (S/DD − LCR_s · H)/LCR, SSF = 1 − C exp(−D · SLR)
	DG	SLR = (S/DD)/(LCR + G), SSF = 1 − (1 + G/LCR) [1 − F(SLR)] Compute SLR then use graphs to find F(SLR) or { if SLR < R, F(SLR) = A · SLR; if SLR ≥ R, F(SLR) = B − C exp (−D · SLR)
	ALL SYSTEMS	Annual SSF = Annual Q_{sav} / Annual Q_{net}

DATA	NLC	Ap	LCR	LCR_s	A	B	C	D	G	H	R	LCR_s H
TW, WW, SS	5232	270	19.4	13.0			1.0189	0.6502		0.86		11.18
DG												

ITEM	ABSORBED SOLAR ENERGY	DEGREE-DAYS (T_b = 62.2°F)	SOLAR LOAD RATIO	SOLAR SAVINGS FRACTION	NET REFERENCE LOAD	SOLAR SAVINGS	AUXILIARY ENERGY
UNITS	(Btu/ft^2)	(°F · day)	Dimensionless	Dimensionless	(Mega Btu)	(Mega Btu)	(Mega Btu)
ORIGIN	SOLAR GAIN WORKSHEET	Weather Data	* *	* *	NLC · DD · 10^{-6}	Q_{net} · SSF	Q_{net} (1−SSF)
SYMBOL	S	DD	SLR	SSF	Q_{net}	Q_{sav}	Q_{aux}
JAN	12 110	959	0.075	0.030	5.02	0.15	4.87
FEB	13 917	760	0.368	0.198	3.98	0.79	3.19
MAR	17 567	600	0.933	0.445	3.14	1.40	1.74
APR	18 886	204	4.196	0.933	1.07	1.00	0.07
MAY	21 339	64	16.610	1	0.33	0.33	0
JUN	22 202	5	228.310	1	0.03	0.03	0
JUL	22 819	0		1	0	0	0
AUG	21 911	0		1	0	0	0
SEP	19 884	17	59.715	1	0.09	0.09	0
OCT	18 754	168	5.178	0.965	0.88	0.85	0.03
NOV	13 288	517	0.749	0.374	2.70	1.01	1.69
DEC	10 362	858	0.046	0.011	4.49	0.05	4.44
ANNUAL		Sum: 4152		* * 0.262	Sum: 21.73	Sum: 5.70	Sum: 16.03

Figure 16-19. Monthly *SLR* Worksheet for Example Problem 16-7

Example Problem 16-8

Consider the building shown in Figure 16-20. The internal wall creates two distinct zones. Zone 1 is served by a double-glazed, selective-surface, unvented Trombe wall. Zone 2 is served by double-glazed, direct-gain windows. The location is Colorado Springs, Colorado. The thermostat setpoint is 70°F. Zone 1 is the bedroom area. Although some body heat could be assigned to this zone, assume that there is no internal heat in Zone 1 for the purpose of this example. Zone 2 is the living-kitchen area with an internal heat generation rate of 40,000 Btu/day.

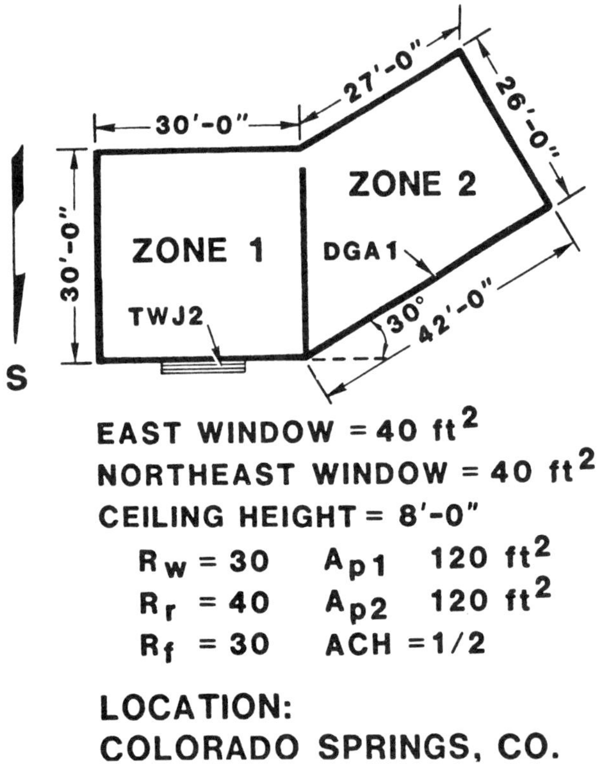

Figure 16-20. Concept for Building Analyzed in Example Problem 16-8

In addition to the two solar systems mentioned above, there are 40 ft^2 of windows on the west wall of Zone 1 and on the northeast wall of Zone 2. These windows are accounted for in an approximate way by use of information in Chapter 10. (A more precise accounting of the windows is illustrated in Example Problem 16-10.) The pertinent information for the west window is in Figure 10-1. Suppose that an *LCR* method calculation in schematic design showed an annual *SSF* of 0.40 for the building. Assume that the curves for Dodge City and Albuquerque represent bounding cases for Colorado Springs. Then the mean for these two cases is an effective annual window conductance of about -0.20 Btu/h · °F · ft^2. For the northeast window (120-degree azimuth), an interpolation is required between data for east windows (Figure 10-1) and north windows (Figure 10-3). Again using the mean of the Dodge City and Albuquerque curves, the effective conductance for north windows is about 0.22, so the interpolated estimate for a 120-degree azimuth is -0.06. The *UA* product for the two windows replaces the formula value in the *NLC* worksheets, Figure 16-21. Filled-in *SLR* worksheets are in Figures 16-22 and 16-23.

Step 1 — The reference designs are TWJ2 and DGA1.

Step 2 — Values from the *NLC* worksheets, Figure 16-21, are:

Zone 1 - $NLC_1 = 3060$ Btu/°F · day
$A_p1 = 120$ ft^2
$LCR_1 = 25.5$ Btu/°F · day · ft^2

Zone 2 - $NLC_1 = 3216$ Btu/°F · day
$A_p = 120$ ft^2
$LCR_2 = 26.8$ Btu/°F · day · ft^2

Step 3 — LCR_s and the correlation coefficients and reference designs TWJ2 and DGA1 are from Appendix 5, Tables A5-7 and A5-6, respectively.

Step 4 — For Zone 1, $LCR_s \cdot H = 5.394$
For Zone 2, this step is not applicable.

Step 5 — See Figures 16-24 and 16-25 for the completed solar gain worksheets. The (*TRN/HOR*) correlation is used for TWJ2 because it is south-facing, but the separate correlations (*INC/HOR*) and (*TRN/INC*) must be used for DGA1 because it is off-south. Table A4-5 is used for (*ABS/TRN*) in the Trombe-wall case: assume a surface absorptance of 0.95, a reference value 0.0625 of extinction coefficient-thickness product, and double

NET LOAD COEFFICIENT, NLC (ZONE 1, TWJ2)

	BUILDING ELEMENT	FORMULA	Fill in the Blanks and Perform Arithmetic	LOAD, UA
	WALLS	: $(UA)_w = A_w/R_w$	= 560 / 30	= 18.7
	ROOF	: $(UA)_r = A_r/R_r$	= 900 / 40	= 22.5
	E, W, N WINDOWS	: $(UA)_g =$ ~~$1.1A_x/N$~~	= ~~1.1 x ____ / ____~~ (-.20)(40)	= − 8.0
SELECT ONE	FLOOR	: $(UA)_f = A_f/R_f$	= 900 / 28	= 32.1
	~~PERIMETER~~	~~: $(UA)_p = 4.17\ P_p/(R_p + 5) = 4.17$ x ____ /(____ + 5)~~		= —
	~~BASEMENT~~	~~: $(UA)_b = 10.7\ P_b/(R_b + 8) = 10.7$ x ____ /(____ + 8)~~		= —
	INFILTRATION	: $(UA)_i$ = 0.018V x ACH x ADR = 0.018 x 7200 x 0.6 x 0.8		= 62.2

Sum = 127.5 Btu/°F • h

NLC = 24 x Sum = 3060 Btu/°F • day

LOAD COLLECTOR RATIO, LCR

$LCR = NLC/A_p$ = 3060 / 120 = 25.5 Btu/°F • day • ft²

NET LOAD COEFFICIENT, NLC (ZONE 2, DGA1)

	BUILDING ELEMENT	FORMULA	Fill in the Blanks and Perform Arithmetic	LOAD, UA
	WALLS	: $(UA)_w = A_w/R_w$	= 600 / 30	= 20.0
	ROOF	: $(UA)_r = A_r/R_r$	= 897 / 40	= 22.4
	E, W, N WINDOWS	: $(UA)_g = 1.1A_x/N$	= 1.1 x ____ / ____ (-.06)(40)	= − 2.4
SELECT ONE	FLOOR	: $(UA)_f = A_f/R_f$	= 897 / 28	= 32.0
	~~PERIMETER~~	~~: $(UA)_p = 4.17\ P_p/(R_p + 5) = 4.17$ x ____ /(____ + 5)~~		= —
	~~BASEMENT~~	~~: $(UA)_b = 10.7\ P_b/(R_b + 8) = 10.7$ x ____ /(____ + 8)~~		= —
	INFILTRATION	: $(UA)_i$ = 0.018V x ACH x ADR = 0.018 x 7176 x 0.6 x 0.8		= 62.0

Sum = 134.0 Btu/°F • h

NLC = 24 x Sum = 3216 Btu/°F • day

LOAD COLLECTOR RATIO, LCR

$LCR = NLC/A_p$ = 3216 / 120 = 26.8 Btu/°F • day • ft²

Figure 16-21. Computation of *NLC* and *LCR* for Zones 1 and 2 for Example Problem 16-8

SLR METHOD WORKSHEET DATE 1/1/83 BY Smith REF. DESIGN TWJ2, ZONE1 LOCATION COLORADO SPRINGS

* * KEY FORMULAS		
	TW, WW, SS	SLR = (S/DD − LCR_s • H)/LCR, SSF = 1 − C exp(−D•SLR)
	DG	SLR = (S/DD)/(LCR + G), SSF = 1 − (1 + G/LCR) [1 − F(SLR)] Compute SLR then use graphs to find F(SLR) or { if SLR < R, F(SLR) = A • SLR; if SLR ≥ R, F(SLR) = B − C exp (−D • SLR)
	ALL SYSTEMS	Annual SSF = Annual Q_{sav} / Annual Q_{net}

DATA		NLC	Ap	LCR	LCR_s	A	B	C	D	G	H	R	LCR_s • H
	TW, WW, SS	3060	120	25.5	8.7			1.0677	0.9312		0.62		5.394
	DG												

ITEM	ABSORBED SOLAR ENERGY	DEGREE-DAYS (T_b = 70 °F)	SOLAR LOAD RATIO	SOLAR SAVINGS FRACTION	NET REFERENCE LOAD	SOLAR SAVINGS	AUXILIARY ENERGY
UNITS	(Btu/ft²)	(°F • day)	Dimensionless	Dimensionless	(Mega Btu)	(Mega Btu)	(Mega Btu)
ORIGIN	SOLAR GAIN WORKSHEET	Weather Data	* *	* *	NLC • DD • 10^{-6}	Q_{net} • SSF	Q_{net} (1−SSF)
SYMBOL	S	DD	SLR	SSF	Q_{net}	Q_{sav}	Q_{aux}
JAN	32 709	1 283	0.788	0.487	3.93	1.91	2.02
FEB	30 271	1 084	0.884	0.531	3.32	1.76	1.56
MAR	28 890	1 076	0.841	0.512	3.29	1.68	1.61
APR	21 177	715	0.950	0.559	2.19	1.22	0.97
MAY	17 333	451	1.296	0.681	1.38	0.94	0.44
JUN	15 930	194	3.009	0.935	0.59	0.55	0.04
JUL	16 623	54	11.860	1	0.17	0.17	0
AUG	19 808	72	10.577	1	0.22	0.22	0
SEP	27 108	280	3.585	0.962	0.86	0.83	0.03
OCT	34 733	607	2.032	0.839	1.86	1.56	0.30
NOV	31 276	975	1.046	0.597	2.98	1.78	1.20
DEC	30 084	1 209	0.764	0.476	3.70	1.76	1.94
ANNUAL		Sum: 8000		* * 0.587	Sum: 24.49	Sum: 14.38	Sum: 10.11

Figure 16-22. *SLR* Method Worksheet for Zone 1 of Example Problem 16-8

SLR METHOD WORKSHEET DATE 1/1/83 BY Smith REF. DESIGN DGAI (ZONE 2) LOCATION COLORADO SPRINGS

* * KEY FORMULAS		
	TW, WW, SS	SLR = (S/DD − LCR_s • H)/LCR, SSF = 1 − C exp(−D • SLR)
	DG	SLR = (S/DD)/(LCR + G), SSF = 1 − (1 + G/LCR) [1 − F(SLR)] Compute SLR then use graphs to find F(SLR) or { if SLR < R, F(SLR) = A • SLR; if SLR ≥ R, F(SLR) = B − C exp (−D • SLR) }
	ALL SYSTEMS	Annual SSF = Annual Q_{sav} / Annual Q_{net}

DATA	NLC	Ap	LCR	LCR_s	A	B	C	D	G	H	R	LCR_s H
TW, WW, SS												
DG	3216	120	26.8		0.565	1.009	1.044	0.7175	9.36		0.3931	

ITEM	ABSORBED SOLAR ENERGY	DEGREE-DAYS (T_b = 61.4°F)	SOLAR LOAD RATIO	SOLAR SAVINGS FRACTION	NET REFERENCE LOAD	SOLAR SAVINGS	AUXILIARY ENERGY
UNITS	(Btu/ft^2)	(°F • day)	Dimensionless	Dimensionless	(Mega Btu)	(Mega Btu)	(Mega Btu)
ORIGIN	SOLAR GAIN WORKSHEET	Weather Data	* *	* *	NLC • DD • 10^{-6}	Q_{net} • SSF	Q_{net} (1−SSF)
SYMBOL	S	DD	SLR	SSF	Q_{net}	Q_{sav}	Q_{aux}
JAN	28 322	1 017	0.750	0.190	3.36	0.64	2.72
FEB	27 194	844	0.865	0.255	2.79	0.71	2.08
MAR	28 157	810	0.929	0.289	2.70	0.78	1.92
APR	23 340	462	1.323	0.467	1.57	0.73	0.84
MAY	21 019	202	2.572	0.790	0.73	0.58	0.15
JUN	20 392	54	8.676	1.0	0.21	0.21	0
JUL	20 706	0	—	1.0	0	0	0
AUG	22 823	0	—	1.0	0	0	0
SEP	27 549	90	7.256	1.0	0.34	0.34	0
OCT	31 883	353	2.333	0.748	1.22	0.91	0.31
NOV	27 337	718	1.015	0.332	2.40	0.80	1.60
DEC	25 933	942	0.739	0.182	3.12	0.57	2.55
ANNUAL		Sum: 5492		* * 0.340	Sum: 18.44	Sum: 6.27	Sum: 12.17

Figure 16-23. *SLR* Method Worksheet for Zone 2 of Example Problem 16-8

SOLAR GAIN WORKSHEET

TILT 90
AZ 0
E 0.0625
N 2
REFR INDEX 1.526

TW & WW: SURFACE ABSORPTANCE = 0.95

DG & SS:
α_m = ____ α_o = ____
HEIGHT = ____ (ft) HGT = 1
DEPTH = ____ (ft) XTOP = ____
WIDTH = ____ (ft) WDTH = ____
GLAZING LENGTH = ____ (ft) Z = ____

DATE 1/1/83
BY Smith
REF DSGN TWJZ (ZONE 1)
LOCATION COLORADO SPRINGS

ITEM	SOLAR NOON ZENITH ANGLE Weather Table	CLEARNESS FACTOR Weather Table	NUMBER OF DAYS IN MONTH	HORIZONTAL TOTAL INSOLATION Weather Table	GLAZED AREA TO PROJECTED AREA RATIO	~~INCIDENT FRACTION~~	(TRN/HOR) ~~TRANSMITTED FRACTION~~	ABSORBED FRACTION	PRODUCT
SYMBOL	LAT-DEC	K_T	M	HS	A_n/A_p	~~(INC/HOR)~~	~~(TRN/INC)~~	(ABS/TRN)	S
JAN	60.1	0.65	31	891	1		1.231	0.962	32 709
FEB	52.4	0.65	28	1178	1		0.954	0.962	30 271
MAR	41.2	0.64	31	1550	1		0.625	0.962	28 890
APR	29.0	0.63	30	1931	1		0.380	0.962	21 177
MAY	19.8	0.61	31	2129	1		0.273	0.962	17 333
JUN	15.4	0.65	30	2369	1		0.233	0.962	15 930
JUL	17.4	0.62	31	2212	1		0.252	0.962	16 623
AUG	25.3	0.63	31	2025	1		0.328	0.962	19 808
SEP	37.0	0.67	30	1759	1		0.534	0.962	27 108
OCT	48.8	0.68	31	1359	1		0.857	0.962	34 733
NOV	58.2	0.64	30	944	1		1.148	0.962	31 276
DEC	62.2	0.63	31	782	1		1.290	0.962	30084

Figure 16-24. Solar Gain Worksheet for Zone 1 of Example Problem 16-8

SOLAR GAIN WORKSHEET

TILT 90
AZ 30
E 0.0625
N 2
REFR INDEX 1.526

TW & WW: SURFACE ABSORPTANCE = ______

DG & SS:
- α_m = 0.7 α_o = 0.3
- HEIGHT = 9 (ft)   HGT = 1
- DEPTH = 26 (ft)   XTOP = 2.9
- WIDTH = 34.5 (ft)   WDTH = 3.8
- GLAZING LENGTH = ______ (ft)   Z = ______

DATE 1/1/83
BY Smith
REF DSGN DGA1 (ZONE 2)
LOCATION COLORADO SPRINGS

ITEM	SOLAR NOON ZENITH ANGLE Weather Table	CLEARNESS FACTOR Weather Table	NUMBER OF DAYS IN MONTH	HORIZONTAL TOTAL INSOLATION Weather Table	GLAZED AREA TO PROJECTED AREA RATIO	INCIDENT FRACTION	TRANSMITTED FRACTION	ABSORBED FRACTION	PRODUCT
SYMBOL	LAT-DEC	K_T	M	HS	A_n/A_p	(INC/HOR)	(TRN/INC)	(ABS/TRN)	S
JAN	60.1	0.65	31	891	1	1.521	0.695	0.97	28 322
FEB	52.4	0.65	28	1178	1	1.239	0.686	0.97	27 194
MAR	41.2	0.64	31	1550	1	0.903	0.669	0.97	28 157
APR	29.0	0.63	30	1931	1	0.648	0.641	0.97	23 340
MAY	19.8	0.61	31	2129	1	0.533	0.616	0.97	21 019
JUN	15.4	0.65	30	2369	1	0.493	0.600	0.97	20 392
JUL	17.4	0.62	31	2212	1	0.512	0.608	0.97	20 706
AUG	25.3	0.63	31	2025	1	0.594	0.631	0.97	22 823
SEP	37.0	0.67	30	1759	1	0.813	0.662	0.97	27 549
OCT	48.8	0.68	31	1359	1	1.144	0.682	0.97	31 883
NOV	58.2	0.64	30	944	1	1.436	0.693	0.97	27 337
DEC	62.2	0.63	31	782	1	1.580	0.698	0.97	25 933

Figure 16-25. Solar Gain Worksheet for Zone 2 of Example Problem 16-8

glazing. The resulting value of (*ABS/TRN*) is 0.962. Table A4-13 is used for the direct-gain case: assume windows that are half the wall height; a floor and north wall absorptance of 0.7; a ceiling, east wall, and west wall absorptance of 0.3; a room depth (N-S) of 26 feet; and an average width (E-W) of 34.5 feet. The parameters WDTH and XTOP are then 34.5/9 or 3.8, and 26/9 or 2.9, respectively. The value of (*ABS/TRN*) is approximately 0.97.

Step 6 — For Zone 1, no internal heat is assumed so the degree-days base temperature is the thermostat setpoint, 70°F. The corresponding degree-days are obtained from Appendix 3. For Zone 2, there is internal heat generation. The computation of the base temperature is as follows.

The first step is to calculate *TLC*, the total load coefficient. However, this can't be done in the usual way by use of Equation 11-10 because of the effective window conductance in *NLC*. Instead, the northeast window must be represented by its actual load coefficient. So we calculate *NLC* by listing the *UA* values of the various building elements including the windows. The values for walls, roof, floor, and infiltration are from the *NLC* worksheet for Zone 2 in Figure 16-21.

Zone 2 Building Element	*UA*
Walls	20.0
Roof	22.4
Floor	32.0
Infiltration	62.0
Windows (0.50)(160)	80.0
Sum =	216.4

$$TLC_2 = 24 \times 216.4 = 5194 \text{ Btu/°F} \cdot \text{day}$$

From Equation 4-7,

$$\begin{aligned} T_{b,2} &= T_{set,2} - Q_{int,2}/TLC_2 \\ &= 70 - 40{,}000/5194 \\ &= 62.3°\text{F} \end{aligned}$$

The corresponding degree-days may be obtained either by interpolating in the Appendix 3 tables or directly from the Thom method (see Chapter 4). For this example, the degree-days are obtained from the Thom method.

Steps 7 through 13 — These steps are performed routinely as if each zone were an independent building.

Comments — The total building auxiliary heat and net load are just the sums of those quantities for the two zones:

$$\begin{aligned} \text{Annual } Q_{auxB} &= 10.10 + 12.17 = 22.27 \text{ million Btu} \\ \text{Annual } Q_{netB} &= 24.49 + 18.44 = 42.93 \text{ million Btu} \\ \text{Annual } SSF_B &= 1 - 22.27/42.93 = 0.481 \end{aligned}$$

Example Problem 16-9

Consider the same building as for Example Problem 16-8 but with no internal wall. The building is shown in Figure 16-26. The two physical zones of Example Problem 16-8 are replaced with pseudozones for the purpose of the analysis (see Chapter 15). The pseudozones are defined in terms of the part of the total *NLC* assigned to each. Filled-in *SLR* method worksheets are in Figures 16-27 and 16-28.

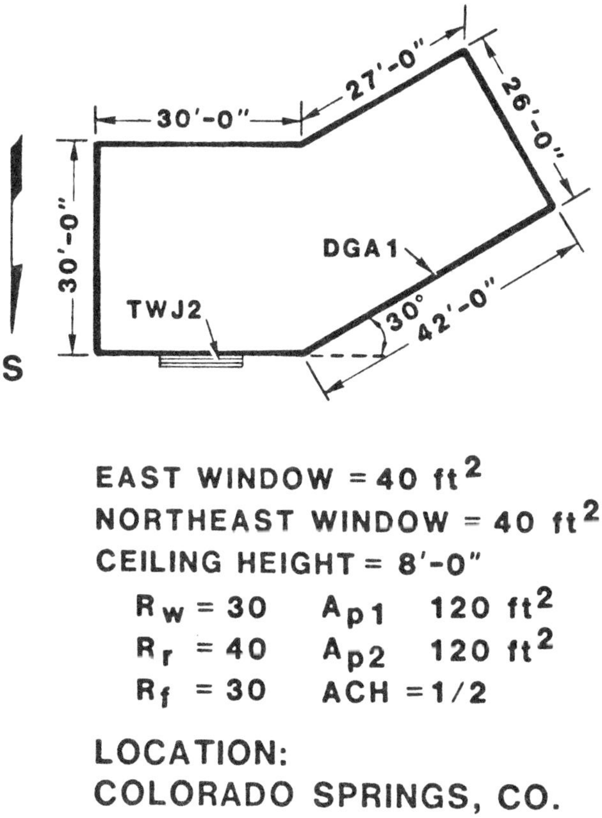

Figure 16-26. Concept for Building Analyzed in Example Problem 16-9

Step 1 — The reference designs are TWJ2 and DGA1, as in Example Problem 16-8.

Step 2 — Each zone is assigned a part of the total *NLC* in proportion to its projected area. The total *NLC* is the sum of the two zone values of Example Problem 16-8:

$$NLC_B = 3060 + 3216 = 6276 \text{ Btu/°F} \cdot \text{day}$$

Because the two projected areas are equal, the two pseudozone values of *NLC* are each just half of the total:

$$NLC_1 = NLC_2 = 6276/2 = 3138 \text{ Btu/°F} \cdot \text{day} .$$

Because of assigning the zone values of *NLC* in proportion to the projected areas, *LCR* is always the same for each zone:

$$LCR = 3138/120 = 26.2 \text{ Btu/°F} \cdot \text{day} \cdot \text{ft}^2 .$$

Steps 3 through 5 — These are the same as Steps 3 through 5 of Example Problem 16-8.

Step 6 — With no physical boundary between the zones, the internal heat may be assigned to the entire building. The resulting degree-days base temperature applies to the entire building. The calculation is as follows.

The first step is to calculate *TLC*, the total load coefficient. However, this can't be done in the usual way by use of Equation 11-10 because of the effective window conductances in *NLC*. Instead, the west and northeast windows must be represented by their actual load coefficients. So we calculate *NLC* by listing the *UA* values of the various building elements including the windows. The values for walls, roof, floor, and infiltration are the totals of Zones 1 and 2 from the *NLC* worksheets in Figure 16-21.

Building Element	***UA***
Walls	38.7
Roof	44.9
Floor	64.1
Infiltration	124.2
Windows (0.5)(200)	100.0
Trombe wall (0.2)(120)	24.0
Sum =	395.9

$$TLC_B = 24 \times 395.9 = 9502 \text{ Btu/°F} \cdot \text{day}$$

$$T_{b,B} = 70 - 40{,}000/9502 = 65.8\text{°F}$$

For this example, the degree-days are obtained by the Thom method (see Chapter 4).

SLR METHOD WORKSHEET DATE 1/1/83 BY Smith REF. DESIGN TWJ2 Pseudozone 1 LOCATION COLORADO SPRINGS

* * KEY FORMULAS	TW, WW, SS	$SLR = (S/DD - LCR_s \cdot H)/LCR$, $SSF = 1 - C \exp(-D \cdot SLR)$
	DG	$SLR = (S/DD)/(LCR + G)$, $SSF = 1 - (1 + G/LCR)[1 - F(SLR)]$ Compute SLR then use graphs to find F(SLR) or { if SLR < R, F(SLR) = A · SLR; if SLR ≥ R, F(SLR) = B − C exp (−D · SLR) }
	ALL SYSTEMS	Annual SSF = Annual Q_{sav} / Annual Q_{net}

DATA		NLC	Ap	LCR	LCR_s	A	B	C	D	G	H	R	$LCR_s \cdot H$
	TW, WW, SS	3138	120	26.2	8.7			1.0677	0.9312		0.62		5.394
	DG												

ITEM	ABSORBED SOLAR ENERGY	DEGREE-DAYS (T_b = 65.2 °F)	SOLAR LOAD RATIO	SOLAR SAVINGS FRACTION	NET REFERENCE LOAD	SOLAR SAVINGS	AUXILIARY ENERGY
UNITS	(Btu/ft²)	(°F · day)	Dimensionless	Dimensionless	(Mega Btu)	(Mega Btu)	(Mega Btu)
ORIGIN	SOLAR GAIN WORKSHEET	Weather Data	* *	* *	NLC · DD · 10^{-6}	$Q_{net} \cdot SSF$	Q_{net} (1−SSF)
SYMBOL	S	DD	SLR	SSF	Q_{net}	Q_{sav}	Q_{aux}
JAN	32 709	1 135	0.894	0.536	3.56	1.91	1.65
FEB	30 271	950	1.010	0.853	2.98	1.74	1.24
MAR	28 890	927	0.984	0.573	2.91	1.67	1.24
APR	21 177	572	1.207	0.653	1.79	1.17	0.62
MAY	17 333	307	1.949	0.826	0.96	0.79	0.17
JUN	15 930	106	5.530	0.994	0.33	0.33	0
JUL	16 623	10	63.241	1	0.03	0.03	0
AUG	19 808	14	53.796	1	0.04	0.04	0
SEP	27 108	159	6.301	0.997	0.50	0.50	0
OCT	34 733	462	2.664	0.911	1.45	1.32	0.13
NOV	31 276	831	1.231	0.661	2.61	1.73	0.88
DEC	30 084	1 060	0.877	0.528	3.33	1.76	1.57
ANNUAL		Sum: 6 533		* * 0.634	Sum: 20.49	Sum: 12.99	Sum: 7.50

Figure 16-27. *SLR* Method Worksheet for Pseudozone 1, Example Problem 16-9

SLR METHOD WORKSHEET DATE 1/1/83 BY Smith REF. DESIGN DGA1 Pseudozone 2 LOCATION COLORADO SPRINGS

* * KEY FORMULAS	TW, WW, SS	SLR = (S/DD − LCR_s • H)/LCR, SSF = 1 − C exp(−D • SLR)
	DG	SLR = (S/DD)/(LCR + G), SSF = 1 − (1 + G/LCR) [1 − F(SLR)] Compute SLR then use graphs to find F(SLR) or { if SLR < R, F(SLR) = A • SLR; if SLR ≥ R, F(SLR) = B − C exp (−D • SLR) }
	ALL SYSTEMS	Annual SSF = Annual Q_{sav} / Annual Q_{net}

DATA		NLC	Ap	LCR	LCR_s	A	B	C	D	G	H	R	LCR_s H
	TW, WW, SS												
	DG	3138	120	26.2		0.565	1.009	1.044	0.7175	9.36		0.3931	

ITEM	ABSORBED SOLAR ENERGY	DEGREE-DAYS (T_b = 65.2 °F)	SOLAR LOAD RATIO	SOLAR SAVINGS FRACTION	NET REFERENCE LOAD	SOLAR SAVINGS	AUXILIARY ENERGY
UNITS	(Btu/ft²)	(°F • day)	Dimensionless	Dimensionless	(Mega Btu)	(Mega Btu)	(Mega Btu)
ORIGIN	SOLAR GAIN WORKSHEET	Weather Data	* *	* *	NLC • DD • 10^{-6}	Q_{net} • SSF	Q_{net} (1−SSF)
SYMBOL	S	DD	SLR	SSF	Q_{net}	Q_{sav}	Q_{aux}
JAN	28 322	1 135	0.702	0.156	3.56	0.56	3.00
FEB	27 194	950	0.805	0.217	2.98	0.65	2.33
MAR	28 157	927	0.854	0.244	2.91	0.71	2.20
APR	23 340	572	1.147	0.390	1.79	0.70	1.09
MAY	21 019	307	1.925	0.656	0.96	0.63	0.33
JUN	20 392	106	5.410	0.983	0.33	0.32	0.01
JUL	20 706	10	58.228	1	0.03	0.03	0
AUG	22823	14	45.844	1	0.04	0.04	0
SEP	27 549	159	4.872	0.969	0.50	0.48	0.02
OCT	31 883	462	1.941	0.660	1.45	0.96	0.49
NOV	27 337	831	0.925	0.283	2.61	0.74	1.87
DEC	25933	1 060	0.688	0.147	3.33	0.49	2.84
ANNUAL		Sum: 6533		* * 0.308	Sum: 20.49	Sum: 6.3	Sum: 14.18

Figure 16-28. *SLR* Method Worksheet for Pseudozone 2, Example Problem 16-9

Step 7 through 13 — These steps are performed routinely as if each pseudozone were an independent building.

Comments — The total building auxiliary heat and net load are just the sums of those quantities for the two zones:

$$\textit{Annual } Q_{auxB} = 7.88 + 14.72 = 22.60 \text{ million Btu}$$
$$\textit{Annual } Q_{netB} = 21.05 + 21.05 = 42.10 \text{ million Btu}$$
$$\textit{Annual } SSF_B = 1 - 22.60/42.10 = 0.463$$

Note the auxiliary consumption is a little more than in Example Problem 16-8. This is because the internal heat is more useful when it is confined to Zone 2 whose direct-gain windows are a less effective solar system than the Trombe wall of Zone 1. However, the analysis method for mixed systems is conservative in that it does not account for heat flow across the pseudozone boundaries. Such heat flows would usually benefit the overall energy performance of the building because they allow excess heat in one zone to contribute to the heating requirement in a neighboring zone. For example, the direct-gain zone may produce excess heat by midday and require ventilation, while the Trombe wall does not reach its maximum heat delivery rate until evening. If the excess heat from the direct-gain zone is allowed to flow to the Trombe wall zone, it can contribute to the midday heating requirement of the Trombe wall zone. Therefore, the auxiliary heat estimate in this example problem is probably an overestimate.

Example Problem 16-10

Consider the same building as in Example Problem 16-9. The 40-ft^2, double-glazed windows on the west and northeast walls were treated in an approximate way by the use of effective conductances. These windows are now accounted for in more detail by considering them to be part of the solar wall. Using this approach, the load coefficient of the windows is ignored in calculating *NLC*. A filled-in *NLC* worksheet is in Figure 16-29.

Since all three of the direct-gain windows serve the same space, they should be treated as three components of a single direct-gain system. So the analysis of the building is done assuming there are two passive systems, the original 120 ft^2 of reference design TWJ2 and the three direct-gain windows as components of reference design DGA1. The method of pseudozones is used as in Example Problem 16-9 except that the two zones are now apportioned differently.

Designate Zones 1 and 2 as those served by TWJ2 and DGA1 as before. See Figures 16-30 and 16-31 for filled-in *SLR* method worksheets.

Step 1 — The two zones are TWJ2 and DGA1, respectively.

Step 2 — Each zone is assigned a part of the total *NLC* is proportion to its projected area. The two values are calculated as follows:

Total building projected area:

$$A_{pB} = 120 + 120 + 40 + 40 = 320 \text{ ft}^2$$

Zone 1:

$$A_{p1} = 120 \text{ ft}^2$$

$$NLC_1 = \frac{A_{p1}}{A_{pB}} NLC_B \qquad \text{(Equation 15-3)}$$

$$= (120/320)(6528) = 2448 \text{ Btu/°F} \cdot \text{day}$$

Zone 2:

$$A_{p2} = 120 + 40 + 40 = 200 \text{ ft}^2$$

$$NLC_2 = (200/320)(6528) = 4080 \text{ Btu/°F} \cdot \text{day}$$

Because of the manner of assigning the parts of *NLC*, *LCR* is the same for each zone:

$$LCR = 2448/120 = 4080/200$$

$$= 20.4 \text{ Btu/°F} \cdot \text{day} \cdot \text{ft}^2$$

Step 3 — LCR_s and the correlation coefficients for reference designs TWJ2 and DGA1 are the same as in Example Problem 16-9.

Step 4 — For pseudozone 1, $LCR_s \cdot H = 5.394$. For pseudozone 2, this step is not applicable.

Step 5 — Zone 1 is calculated as in Example Problem 16-9. See Figures 16-32 through 16-34 for the filled-in solar gain worksheets for Zone 2. These three worksheets show the calculation of *S* for each of the three components of the direct-gain system serving Zone 2. The total *S* for each month is the sum of the three component values. The total *S* values are entered in the *SLR* method worksheet for Zone 2, Figure 16-31. The separate correlations (*INC/HOR*) and (*TRN/INC*) are used for Zone 2 because the apertures are off-south. Assume (*ABS/TRN*) is 0.97 in both cases.

Step 6 The calculation of the degree-days base temperature is the same as in Example Problem 16-9.

For this example, the degree-days are determined by the Thom method (see Chapter 4).

Steps 7 through 13 — These steps are performed routinely, as if each pseudozone were an independent building.

Comments — The total building auxiliary heat and net load are just the sums of those quantities for the two zones:

$$\text{Annual } Q_{auxB} = 4.68 + 21.45 = 26.13 \text{ million Btu}$$

$$\text{Annual } Q_{netB} = 16.41 + 27.35 = 43.76 \text{ million Btu}$$

$$\text{Annual } SSF_B = 1 - 26.13/43.76 = 0.403$$

Note that the auxiliary heat estimate is 16% more than that of Example Problem 16-9. This difference is a result of the different analysis method: primarily the more precise treatment of the west and northeast windows. Note also that the assignment

NET LOAD COEFFICIENT and ANNUAL LCR METHOD WORKSHEET   DATE 1/1/83 BY Smith LOCATION Colorado Springs

NET LOAD COEFFICIENT, NLC

	BUILDING ELEMENT	FORMULA	Fill in the Blanks and Perform Arithmetic	LOAD, UA
	WALLS	: $(UA)_w = A_w/R_w$	= 1160 / 30	= 38.7
	ROOF	: $(UA)_r = A_r/R_r$	= 1797 / 40	= 44.9
	~~E, W, N WINDOWS~~	~~: $(UA)_g = 1.1A_x/N$~~	~~= 1.1 x ____ / ____~~	= —
SELECT ONE	FLOOR	: $(UA)_f = A_f/R_f$	= 1797 / 28	= 64.2
	~~PERIMETER~~	~~: $(UA)_p = 4.17\ P_p/(R_p + 5)$~~	~~= 4.17 x ____ /(____ + 5)~~	= —
	~~BASEMENT~~	~~: $(UA)_b = 10.7\ P_b/(R_b + 8)$~~	~~= 10.7 x ____ /(____ + 8)~~	= —
	INFILTRATION	: $(UA)_i$ = 0.018V x ACH x ADR	= 0.018 x 14376 x 0.6 x 0.8	= 124.2

Sum = 272.0 Btu/°F • h

NLC = 24 x Sum = 6528 Btu/°F • day

LOAD COLLECTOR RATIO, LCR

$LCR_B = NLC/A_p$ = 6528 / 320 = 20.4 Btu/°F • day • ft^2

WEIGHTED AVERAGE, SSF

REFERENCE DESIGN	PROJECTED AREA	FRACTION OF A_p		SSF FOR EACH REFERENCE DESIGN		PRODUCT OF SSF x FRACTION	U_c	$U_c\ A_p$
____	____	____	x	____	=	____	____	____
____	____	____	x	____	=	____	____	____
____	____	____	x	____	=	____	____	____
TOTAL A_p =	____			SSF =		____	TOTAL (U_cA_p) =	____

TLC = NLC + 24 x TOTAL (U_cA_p) = ____

BASE TEMPERATURE, T_b

$T_{bB} = T_{set} - Q_{int}/TLC_B$ = 70 − 40000 / 9504 = 65.8 °F

AUXILIARY HEAT, Q_{aux}

Q_{aux} = NLC • DD(1 − SSF) = ____ x ____ x (1 − ____) x 10^{-6} = ____ Mega Btu/year

Figure 16-29. The *NLC* Worksheet for Example Problem 16-10

SLR METHOD WORKSHEET   DATE 1/1/83   BY Smith   REF. DESIGN TWJ2 Pseudozone 1   LOCATION Colorado Springs

* * KEY FORMULAS	TW, WW, SS	SLR = (S/DD − LCR_s • H)/LCR, SSF = 1 − C exp(−D • SLR)
	DG	SLR = (S/DD)/(LCR + G), SSF = 1 − (1 + G/LCR) [1 − F(SLR)] Compute SLR then use graphs to find F(SLR) or { if SLR < R, F(SLR) = A • SLR; if SLR ≥ R, F(SLR) = B − C exp (−D • SLR) }
	ALL SYSTEMS	Annual SSF = Annual Q_{sav} / Annual Q_{net}

DATA		NLC	Ap	LCR	LCR_s	A	B	C	D	G	H	R	LCR_s • H
	TW, WW, SS	2448	120	20.4	8.7			1.0677	0.9312		0.62		5.394
	DG												

ITEM	ABSORBED SOLAR ENERGY	DEGREE-DAYS (T_b = 65.8 °F)	SOLAR LOAD RATIO	SOLAR SAVINGS FRACTION	NET REFERENCE LOAD	SOLAR SAVINGS	AUXILIARY ENERGY
UNITS	(Btu/ft²)	(°F • day)	Dimensionless	Dimensionless	(Mega Btu)	(Mega Btu)	(Mega Btu)
ORIGIN	SOLAR GAIN WORKSHEET	Weather Data	* *	* *	NLC • DD • 10^{-6}	Q_{net} • SSF	Q_{net} (1−SSF)
SYMBOL	S	DD	SLR	SSF	Q_{net}	Q_{sav}	Q_{aux}
JAN	32 709	1 153	1.126	0.626	2.82	1.77	1.05
FEB	30 271	967	1.270	0.673	2.37	1.60	0.77
MAR	28 890	946	1.233	0.661	2.32	1.53	0.79
APR	21 177	590	1.495	0.735	1.44	1.06	0.38
MAY	17 333	324	2.358	0.881	0.79	0.70	0.09
JUN	15 930	116	6.467	0.997	0.28	0.28	0
JUL	16 623	12	67.640	1	0.03	0.03	0
AUG	19 808	18	53.679	1	0.04	0.04	0
SEP	27 108	172	7.461	0.999	0.42	0.42	0
OCT	34 733	480	3.283	0.950	1.18	1.12	0.06
NOV	31 276	849	1.541	0.746	2.08	1.55	0.53
DEC	30 084	1 079	1.102	0.617	2.64	1.63	1.01
ANNUAL		Sum: 6 706		* * 0.715	Sum: 16.41	Sum: 11.73	Sum: 4.68

Figure 16-30. *SLR* Method Worksheet for Pseudozone 1, Example Problem 16-10

SLR METHOD WORKSHEET DATE 1/1/83 BY Smith REF. DESIGN DGA1 PSEUDOZONE 2 LOCATION COLORADO SPRINGS

* * KEY FORMULAS		
	TW, WW, SS	SLR = (S/DD − LCR_S • H)/LCR, SSF = 1 − C exp(−D • SLR)
	DG	SLR = (S/DD)/(LCR + G), SSF = 1 − (1 + G/LCR) [1 − F(SLR)] Compute SLR then use graphs to find F(SLR) or { if SLR < R, F(SLR) = A • SLR; if SLR ≥ R, F(SLR) = B − C exp (−D • SLR) }
	ALL SYSTEMS	Annual SSF = Annual Q_{sav} / Annual Q_{net}

DATA	NLC	Ap	LCR	LCR_S	A	B	C	D	G	H	R	LCR_S • H
TW, WW, SS												
DG	4080	200	20.4		0.565	1.009	1.044	0.7175	9.36		0.3931	

ITEM	ABSORBED SOLAR ENERGY	DEGREE-DAYS (T_b = 65.8°F)	SOLAR LOAD RATIO	SOLAR SAVINGS FRACTION	NET REFERENCE LOAD	SOLAR SAVINGS	AUXILIARY ENERGY
UNITS	(Btu/ft^2)	(°F • day)	Dimensionless	Dimensionless	(Mega Btu)	(Mega Btu)	(Mega Btu)
ORIGIN	SOLAR GAIN WORKSHEET	Weather Data	* *	* *	NLC • DD • 10^{-6}	Q_{net} • SSF	Q_{net} (1−SSF)
SYMBOL	S	DD	SLR	SSF	Q_{net}	Q_{sav}	Q_{aux}
JAN	20 409	1 153	0.595	0.019	4.70	0.09	4.61
FEB	20 497	967	0.712	0.099	3.95	0.39	3.56
MAR	23 181	946	0.823	0.169	3.86	0.65	3.21
APR	21 741	590	1.238	0.387	2.41	0.93	1.48
MAY	21 526	324	2.232	0.706	1.32	0.93	0.39
JUN	21 816	116	6.320	0.997	0.47	0.47	0
JUL	21 690	12	60.736	1.0	0.05	0.05	0
AUG	22 109	18	41.273	1.0	0.07	0.07	0
SEP	23 476	172	4.586	0.956	0.70	0.67	0.03
OCT	24 505	480	1.715	0.568	1.96	1.11	0.85
NOV	19 936	849	0.789	0.148	3.46	0.51	2.95
DEC	18 559	1 079	0.578	0.007	4.40	0.03	4.37
ANNUAL		Sum: 6706		* *0.216	Sum: 27.35	Sum: 5.90	Sum: 21.45

Figure 16-31. *SLR* Method Worksheet for Pseudozone 2, Example Problem 16-10

SOLAR GAIN WORKSHEET

TILT 90

AZ 30

E 0.0625

N 2

REFR INDEX 1.526

TW & WW: SURFACE ABSORPTANCE = ______

DG & SS: α_m = ______ ; HEIGHT = ______ (ft) ; DEPTH = ______ (ft) ; WIDTH = ______ (ft) ; GLAZING LENGTH = ______ (ft)

α_o = ______ ; HGT = 1 ; XTOP = ______ ; WDTH = ______ ; Z = ______

DATE 1/1/83

BY Smith

REF DSGN DGA1 PSEUDOZONE 2 COMPONENT 1

LOCATION COLORADO SPRINGS

ITEM	SOLAR NOON ZENITH ANGLE Weather Table	CLEARNESS FACTOR Weather Table	NUMBER OF DAYS IN MONTH	HORIZONTAL TOTAL INSOLATION Weather Table	GLAZED AREA TO PROJECTED AREA RATIO	INCIDENT FRACTION	TRANSMITTED FRACTION	ABSORBED FRACTION	PRODUCT
SYMBOL	LAT-DEC	K_T	M	HS	A_n/A_p	(INC/HOR)	(TRN/INC)	(ABS/TRN)	S
JAN	60.1	0.65	31	891	0.6	1.521	0.695	0.97	16 993
FEB	52.4	0.65	28	1178	0.6	1.239	0.686	0.97	16 316
MAR	41.2	0.64	31	1550	0.6	0.903	0.669	0.97	16 894
APR	29.0	0.63	30	1931	0.6	0.648	0.641	0.97	14 004
MAY	19.8	0.61	3-	2129	0.6	0.533	0.616	0.97	12 612
JUN	15.4	0.65	30	2367	0.6	0.493	0.600	0.97	12 235
JUL	17.4	0.62	31	2212	0.6	0.512	0.608	0.97	12 423
AUG	25.3	0.63	31	2025	0.6	0.594	0.631	0.97	13 694
SEP	37.0	0.67	30	1759	0.6	0.813	0.662	0.97	16 529
OCT	48.8	0.68	31	1357	0.6	1.144	0.682	0.97	19 130
NOV	58.2	0.64	30	944	0.6	1.436	0.693	0.97	16 402
DEC	62.2	0.63	31	782	0.6	1.580	0.698	0.97	15 560

Figure 16-32. Solar Gain Worksheet for Component 1 of Pseudozone 2, Example Problem 16-10

SOLAR GAIN WORKSHEET

TILT 90
AZ 90
E 0.0625
N 2
REFR INDEX 1.526

TW & WW: SURFACE ABSORPTANCE = ______

DG & SS:
α_m = 0.7 α_o = 0.3
HEIGHT = 9 (ft) HGT = 1
DEPTH = 26 (ft) XTOP = 2.9
WIDTH = 34.5 (ft) WDTH = 3.8
GLAZING LENGTH = ______ (ft) Z = ______

DATE 1/1/83
BY Smith
REF DSGN DGA1 PSEUDOZONE 2 COMPONENT 2
LOCATION COLORADO SPRINGS

ITEM	SOLAR NOON ZENITH ANGLE Weather Table	CLEARNESS FACTOR Weather Table	NUMBER OF DAYS IN MONTH	HORIZONTAL TOTAL INSOLATION Weather Table	GLAZED AREA TO PROJECTED AREA RATIO	INCIDENT FRACTION	TRANSMITTED FRACTION	ABSORBED FRACTION	PRODUCT
SYMBOL	LAT-DEC	K_T	M	HS	A_n/A_p	(INC/HOR)	(TRN/INC)	(ABS/TRN)	S
JAN	60.1	0.65	31	891	0.2	0.693	0.638	0.97	2 369
FEB	52.4	0.65	28	1178	0.2	0.664	0.648	0.97	2 753
MAR	41.2	0.64	31	1550	0.2	0.627	0.658	0.97	3 846
APR	29.0	0.63	30	1931	0.2	0.594	0.663	0.97	4426
MAY	19.8	0.61	31	2129	0.2	0.576	0.662	0.97	4882
JUN	15.4	0.65	30	2369	0.2	0.568	0.663	0.97	5192
JUL	17.4	0.62	31	2212	0.2	0.572	0.662	0.97	5037
AUG	25.3	0.63	31	2025	0.2	0.586	0.663	0.97	4732
SEP	37.0	0.67	30	1759	0.2	0.618	0.661	0.97	4182
OCT	48.8	0.68	31	1359	0.2	0.656	0.652	0.97	3496
NOV	58.2	0.64	30	944	0.2	0.684	0.641	0.97	2409
DEC	62.2	0.63	31	782	0.2	0.698	0.635	0.97	2084

Figure 16-33. Solar Gain Worksheet for Component 2 of Pseudozone 2, Example Problem 16-10

SOLAR GAIN WORKSHEET

TILT 90
AZ 120
E 0.0625
N 2
REFR INDEX 1.526

TW & WW: SURFACE ABSORPTANCE = ______

DG & SS:
- α_m = 0.7 — α_o = 0.3
- HEIGHT = 9 (ft) — HGT = 1
- DEPTH = 26 (ft) — XTOP = 2.9
- WIDTH = 34.5 (ft) — WDTH = 3.8
- GLAZING LENGTH = ______ (ft) — Z = ______

DATE 1/1/83
BY Smith
REF DSGN DGA1, PSEUDOZONE 2 COMPONENT 3
LOCATION COLORADO SPRINGS

ITEM	SOLAR NOON ZENITH ANGLE Weather Table	CLEARNESS FACTOR Weather Table	NUMBER OF DAYS IN MONTH	HORIZONTAL TOTAL INSOLATION Weather Table	GLAZED AREA TO PROJECTED AREA RATIO	INCIDENT FRACTION	TRANSMITTED FRACTION	ABSORBED FRACTION	PRODUCT
SYMBOL	LAT-DEC	K_T	M	HS	A_n/A_p	(INC/HOR)	(TRN/INC)	(ABS/TRN)	S
JAN	60.1	0.65	31	891	0.2	0.330	0.592	0.97	1 047
FEB	52.4	0.65	28	1178	0.2	0.367	0.608	0.97	1428
MAR	41.2	0.64	31	1550	0.2	0.415	0.631	0.97	2441
APR	29.0	0.63	30	1931	0.2	0.456	0.646	0.97	3 311
MAY	19.8	0.61	31	2 129	0.2	0.483	0.652	0.97	4 032
JUN	15.4	0.65	30	2 369	0.2	0.486	0.655	0.97	4 389
JUL	17.4	0.62	31	2 212	0.2	0.487	0.653	0.97	4 230
AUG	25.3	0.63	31	2 025	0.2	0.466	0.649	0.97	3 683
SEP	37.0	0.67	30	1 759	0.2	0.424	0.637	0.97	2765
OCT	48.8	0.68	31	1 359	0.2	0.375	0.613	0.97	1879
NOV	58.2	0.64	30	944	0.2	0.343	0.597	0.97	1125
DEC	62.2	0.63	31	782	0.2	0.328	0.593	0.97	915

Figure 16-34. Solar Gain Worksheet for Compcnent 3 of Pseudozone 2, Example Problem 16-10

of the west and northeast windows as components of Zone 2 implies a different apportionment of *NLC* between the two zones than in Example Problem 16-9. The apportionment is done according to projected area, and the projected areas of the windows are taken to be the actual glazing areas. An alternative approach is to define a principal azimuth for the direct-gain system as recommended in Chapter 9 for sunspaces. This approach might be superior for cases in which there are several windows that are far from south.

APPENDIX 1
LCR/CF TABLES

This appendix tabulates annual *SSF* for 8 selected values of *LCR* for each of the 94 reference designs and for 2 values (55°F and 65°F) of base temperature. Also, *CF* is tabulated for five selected values of *LCR* for each of the reference designs and base temperatures. There is 1 table for each of 223 U.S. and Canadian cities. The tables are arranged in alphabetical order by state followed by those for the Canadian cities.

Annual SSF is calculated from (Equation (8-12))

$$\text{annual SSF} = \frac{\text{annual } Q_{sav}}{\text{annual } Q_{net}} \qquad \text{(A1-1)}$$

which is equivalent to

$$\text{annual SSF} = \frac{\Sigma(\text{SSF} \cdot \text{DD})}{\Sigma \text{DD}} \qquad \text{(A1-2)}$$

where Σ means the sum over each of the twelve monthly values. The monthly values of *SSF* are calculated using Equation (8-3) or Equation (8-6). The monthly values of *DD* are as listed in Appendix 3 (D55 and D65 for 55°F and 65°F base temperatures, respectively). The monthly values of the solar gain, *S,* needed in the *SSF* calculations are determined from the appropriate reference design correlations (Tables A4-3 and A4-46 for thermal storage wall and direct gain designs and Table A4-47 for sunspaces). These correlations, and therefore the *LCF/CF* tables, are based on due south orientation, no shading, and a ground reflectance of 30 percent. The conservation factors, *CF,* are calculated from Equation A9-10.

BIRMINGHAM, ALABAMA

TYPE	SOLAR SAVINGS FRACTION (%) BASE TEMP: 55 F (1333 DD)								65 F (2846 DD)								CONSERVATION FACTOR 55 F					65 F				
LCR	200	100	70	50	40	30	25	20	200	100	70	50	40	30	25	20	200	100	50	30	20	200	100	50	30	20
WW A1	22	31	38	46	52	60	65	72	15	21	25	30	34	40	44	49	1.0	1.1	1.2	1.4	1.5	1.3	1.4	1.6	1.8	1.9
WW A2	23	38	47	58	65	74	79	85	14	23	30	37	43	51	56	63	0.8	0.9	1.1	1.2	1.4	1.1	1.2	1.3	1.5	1.6
WW A3	23	40	51	62	69	78	83	88	13	24	31	40	46	55	60	67	0.8	0.9	1.0	1.2	1.4	1.0	1.1	1.2	1.4	1.6
WW A4	24	41	53	64	71	80	85	90	12	24	32	41	48	57	63	70	0.7	0.8	1.0	1.2	1.3	1.0	1.1	1.2	1.4	1.5
WW A5	24	43	55	66	74	82	87	92	12	25	33	43	50	59	65	73	0.7	0.8	1.0	1.1	1.3	0.9	1.0	1.2	1.3	1.5
WW A6	24	44	56	67	75	83	88	92	12	25	34	44	51	61	67	74	0.7	0.8	1.0	1.1	1.3	0.9	1.0	1.2	1.3	1.5
WW B1	20	32	41	50	56	65	71	77	12	19	24	30	34	40	45	51	0.9	1.0	1.1	1.3	1.5	1.2	1.3	1.5	1.7	1.9
WW B2	23	42	53	65	72	81	86	91	12	25	33	43	50	59	65	72	0.7	0.8	1.0	1.2	1.3	0.9	1.0	1.2	1.3	1.5
WW B3	28	49	61	72	79	87	91	95	15	30	40	50	58	68	73	80	0.7	0.8	0.9	1.1	1.3	0.8	0.9	1.1	1.3	1.4
WW B4	27	49	61	73	80	88	92	95	13	30	41	52	60	70	76	83	0.7	0.8	0.9	1.1	1.3	0.8	0.9	1.1	1.2	1.4
WW B5	25	47	60	72	79	87	91	95	12	28	39	51	59	69	75	82	0.7	0.8	0.9	1.1	1.3	0.8	0.9	1.1	1.2	1.4
WW C1	29	50	63	74	81	88	92	96	16	32	42	53	60	70	76	82	0.7	0.8	0.9	1.1	1.3	0.8	0.9	1.1	1.2	1.4
WW C2	27	48	60	72	79	87	91	95	14	29	40	51	58	68	74	81	0.7	0.8	0.9	1.1	1.3	0.8	0.9	1.1	1.2	1.4
WW C3	30	55	68	79	86	92	95	98	16	35	47	60	68	78	83	89	0.6	0.7	0.9	1.1	1.2	0.7	0.8	1.0	1.2	1.3
WW C4	27	51	64	76	83	90	94	97	13	32	44	56	64	74	80	87	0.6	0.7	0.9	1.1	1.3	0.8	0.9	1.0	1.2	1.4
TW A1	21	30	37	45	51	59	64	71	15	21	25	30	34	39	43	48	1.0	1.1	1.2	1.4	1.6	1.3	1.4	1.6	1.8	1.9
TW A2	21	34	43	53	59	68	74	80	14	22	27	34	39	46	51	58	0.9	1.0	1.1	1.3	1.4	1.1	1.2	1.4	1.6	1.7
TW A3	21	35	45	55	63	72	77	83	12	22	28	36	41	49	55	61	0.8	0.9	1.1	1.2	1.4	1.1	1.2	1.3	1.5	1.6
TW A4	20	35	45	56	63	73	78	84	11	21	28	36	41	50	55	62	0.8	0.9	1.1	1.2	1.4	1.1	1.2	1.3	1.5	1.6
TW B1	20	30	37	46	52	60	66	72	14	20	25	30	34	40	44	50	1.0	1.1	1.2	1.4	1.5	1.3	1.4	1.6	1.7	1.9
TW B2	20	32	41	51	58	67	73	79	12	20	26	33	38	45	50	57	0.9	1.0	1.1	1.3	1.4	1.1	1.2	1.4	1.6	1.7
TW B3	19	32	41	51	58	68	73	80	11	20	26	33	38	46	51	58	0.9	1.0	1.1	1.3	1.4	1.1	1.2	1.4	1.5	1.7
TW B4	18	30	39	49	56	65	71	78	10	18	24	31	36	44	49	55	0.9	1.0	1.1	1.3	1.5	1.2	1.3	1.4	1.6	1.7
TW C1	19	29	37	45	51	60	66	72	12	19	24	29	34	40	44	50	1.0	1.1	1.2	1.4	1.5	1.3	1.4	1.5	1.7	1.8
TW C2	18	29	37	46	52	61	67	74	11	18	23	29	34	41	46	52	1.0	1.0	1.2	1.3	1.5	1.2	1.3	1.5	1.6	1.8
TW C3	17	27	35	44	50	59	65	72	11	17	22	28	33	39	44	50	1.0	1.1	1.2	1.4	1.5	1.3	1.4	1.5	1.7	1.8
TW C4	16	25	31	39	45	53	59	65	11	16	21	26	30	35	40	45	1.1	1.2	1.3	1.4	1.6	1.4	1.5	1.6	1.8	1.9
TW D1	17	27	35	43	50	58	64	71	10	16	20	25	29	35	40	45	1.0	1.1	1.2	1.4	1.5	1.3	1.4	1.6	1.8	1.9
TW D2	22	38	49	59	67	76	81	87	13	23	31	39	45	54	60	67	0.8	0.9	1.0	1.2	1.4	1.0	1.1	1.3	1.4	1.6
TW D3	24	42	53	64	72	80	85	90	13	25	34	43	50	59	65	72	0.7	0.8	1.0	1.2	1.3	0.9	1.0	1.2	1.3	1.5
TW D4	25	44	56	68	75	84	88	93	13	27	37	47	54	64	70	77	0.7	0.8	1.0	1.1	1.3	0.9	1.0	1.1	1.3	1.4
TW D5	23	43	55	67	74	83	88	92	12	26	36	47	54	64	70	77	0.7	0.8	1.0	1.1	1.3	0.9	1.0	1.1	1.3	1.4
TW E1	27	46	58	69	76	84	89	93	16	29	38	48	55	64	70	77	0.7	0.8	1.0	1.1	1.3	0.9	1.0	1.1	1.3	1.5
TW E2	26	45	57	68	76	84	88	93	14	28	37	48	55	64	70	77	0.7	0.8	1.0	1.1	1.3	0.9	1.0	1.1	1.3	1.4
TW E3	30	52	65	77	83	90	94	97	17	34	45	56	64	74	80	86	0.6	0.7	0.9	1.1	1.3	0.8	0.9	1.0	1.2	1.4
TW E4	27	49	62	74	81	88	92	96	14	31	42	53	61	71	77	84	0.7	0.8	0.9	1.1	1.3	0.8	0.9	1.1	1.2	1.4
TW F1	19	29	36	45	51	59	65	71	13	19	23	29	33	38	43	48	1.0	1.1	1.2	1.4	1.5	1.3	1.4	1.6	1.7	1.9
TW F2	19	32	40	50	57	66	72	79	11	19	25	31	36	44	49	55	0.9	1.0	1.1	1.3	1.5	1.2	1.3	1.4	1.6	1.7
TW F3	18	32	41	51	59	68	74	81	10	18	25	32	37	45	51	57	0.9	1.0	1.1	1.3	1.4	1.1	1.2	1.4	1.5	1.7
TW F4	16	30	39	50	57	67	73	79	7	16	23	30	36	44	49	56	0.9	1.0	1.1	1.3	1.4	1.1	1.2	1.4	1.5	1.7
TW G1	17	28	35	43	50	58	64	71	11	17	22	27	31	38	42	48	1.0	1.1	1.2	1.4	1.5	1.3	1.4	1.6	1.7	1.9
TW G2	16	28	37	46	53	62	68	75	8	16	22	28	33	40	45	52	0.9	1.0	1.2	1.3	1.5	1.2	1.3	1.5	1.6	1.8
TW G3	14	26	35	44	52	61	67	74	7	14	20	27	32	39	44	51	0.9	1.0	1.2	1.3	1.5	1.2	1.3	1.5	1.6	1.8
TW G4	11	22	30	39	46	55	61	69	5	11	17	23	27	34	39	46	1.0	1.1	1.2	1.4	1.5	1.3	1.4	1.5	1.7	1.8
TW H1	14	24	32	40	46	55	61	68	8	14	19	24	28	35	39	45	1.0	1.1	1.2	1.4	1.5	1.3	1.4	1.6	1.7	1.9
TW H2	11	22	29	38	44	53	59	67	5	11	16	22	26	33	38	44	1.0	1.1	1.2	1.4	1.6	1.3	1.4	1.6	1.7	1.9
TW H3	9	18	25	33	39	48	54	62	4	9	14	19	23	23	34	40	1.1	1.2	1.3	1.4	1.6	1.4	1.5	1.6	1.8	2.0
TW H4	7	14	20	26	32	39	45	52	3	7	11	15	18	24	27	33	1.2	1.3	1.4	1.6	1.7	1.6	1.7	1.8	2.0	2.1
TW I1	14	24	32	40	46	55	60	68	8	13	18	23	26	32	36	41	1.0	1.1	1.2	1.4	1.6	1.4	1.5	1.7	1.8	2.0
TW I2	19	34	45	56	63	73	79	85	10	20	27	36	42	51	56	64	0.8	0.9	1.1	1.2	1.4	1.0	1.1	1.3	1.4	1.6
TW I3	20	37	48	60	67	77	82	88	10	22	30	39	46	55	61	68	0.8	0.9	1.0	1.2	1.4	1.0	1.1	1.2	1.4	1.5
TW I4	21	40	52	64	72	81	86	91	11	24	33	43	51	60	67	74	0.7	0.8	1.0	1.2	1.3	0.9	1.0	1.2	1.3	1.5
TW I5	21	40	52	65	72	81	86	91	10	24	34	44	52	62	68	76	0.7	0.8	1.0	1.1	1.3	0.9	1.0	1.1	1.3	1.5
TW J1	24	43	55	66	74	82	87	92	13	26	35	45	52	62	68	75	0.7	0.8	1.0	1.1	1.3	0.9	1.0	1.2	1.3	1.5
TW J2	23	42	54	65	73	82	87	92	12	25	34	44	52	61	68	75	0.7	0.8	1.0	1.1	1.3	0.9	1.0	1.2	1.3	1.5
TW J3	26	49	62	74	81	88	92	96	13	30	42	53	61	71	77	84	0.7	0.8	0.9	1.1	1.3	0.8	0.9	1.1	1.2	1.4
TW J4	24	46	59	71	78	87	91	95	12	28	39	50	58	69	75	82	0.7	0.8	0.9	1.1	1.3	0.8	0.9	1.1	1.2	1.4
DG A1	18	30	39	48	53	61	65	70	9	17	22	27	31	36	40	43	0.9	1.0	1.2	1.4	1.7	1.2	1.3	1.6	1.9	2.3
DG A2	18	32	42	52	59	67	72	78	10	19	25	32	38	45	50	56	0.9	0.9	1.1	1.3	1.5	1.1	1.2	1.4	1.6	1.8
DG A3	20	37	49	61	68	77	82	86	12	23	31	40	47	56	62	69	0.8	0.9	1.0	1.2	1.4	1.0	1.1	1.2	1.4	1.6
DG B1	18	32	42	53	60	68	73	78	9	17	22	29	33	40	45	50	0.9	0.9	1.1	1.3	1.6	1.2	1.3	1.5	1.7	1.9
DG B2	18	33	44	56	65	74	79	85	10	19	26	34	40	49	55	62	0.9	0.9	1.0	1.2	1.4	1.1	1.2	1.3	1.4	1.6
DG B3	21	38	50	64	72	81	86	91	13	23	31	41	49	59	66	74	0.8	0.8	1.0	1.1	1.4	1.0	1.1	1.2	1.3	1.5
DG C1	20	37	48	61	69	77	82	86	11	20	27	34	40	49	54	61	0.8	0.9	1.0	1.2	1.5	1.1	1.2	1.3	1.5	1.7
DG C2	21	38	50	63	72	81	85	90	12	22	30	39	46	56	62	70	0.8	0.8	1.0	1.2	1.4	1.0	1.1	1.2	1.3	1.5
DG C3	23	43	56	70	79	87	91	94	14	26	35	46	55	66	73	81	0.7	0.8	0.9	1.1	1.3	0.9	1.0	1.1	1.2	1.4
SS A1	26	40	50	60	66	75	80	85	18	27	34	41	47	55	60	66	0.8	0.9	1.1	1.2	1.4	1.0	1.1	1.3	1.5	1.6
SS A2	31	49	61	71	78	85	89	93	21	34	43	52	59	68	74	80	0.7	0.8	1.0	1.1	1.3	0.9	1.0	1.1	1.3	1.4
SS A3	25	38	47	57	63	72	77	82	17	26	32	38	43	51	55	62	0.8	0.9	1.1	1.3	1.4	1.1	1.2	1.4	1.5	1.7
SS A4	31	49	60	71	77	85	89	93	20	34	42	52	58	67	73	79	0.7	0.8	1.0	1.1	1.3	0.9	1.0	1.1	1.3	1.5
SS A5	25	38	46	55	62	70	75	81	18	26	32	38	43	50	55	61	0.9	1.0	1.1	1.3	1.5	1.1	1.2	1.4	1.5	1.7
SS A6	31	49	60	71	78	85	89	93	20	34	43	52	59	68	73	80	0.7	0.8	1.0	1.1	1.3	0.9	1.0	1.1	1.3	1.4
SS A7	23	34	41	50	56	64	69	75	17	24	28	34	38	44	48	54	0.9	1.0	1.2	1.4	1.5	1.2	1.3	1.5	1.7	1.9
SS A8	30	48	59	70	76	84	88	93	20	33	42	51	57	66	71	78	0.7	0.8	1.0	1.1	1.3	0.9	1.0	1.1	1.3	1.5
SS B1	22	35	44	54	60	69	74	81	15	23	29	36	41	48	53	60	0.9	1.0	1.1	1.3	1.4	1.1	1.2	1.4	1.5	1.7
SS B2	27	44	55	65	73	81	86	90	17	30	38	47	54	62	68	75	0.7	0.8	1.0	1.2	1.3	0.9	1.0	1.2	1.3	1.5
SS B3	21	34	42	51	58	67	72	78	14	22	28	34	39	46	50	56	0.9	1.0	1.1	1.3	1.5	1.1	1.2	1.4	1.6	1.8
SS B4	26	43	54	65	72	80	85	90	17	29	37	46	53	61	67	74	0.7	0.8	1.0	1.2	1.3	0.9	1.0	1.2	1.3	1.5
SS B5	21	31	39	47	53	62	67	73	15	21	26	32	36	42	46	52	1.0	1.1	1.2	1.4	1.5	1.2	1.3	1.5	1.7	1.8
SS B6	26	43	54	65	72	80	85	90	17	29	37	46	53	61	67	74	0.7	0.8	1.0	1.2	1.3	0.9	1.0	1.2	1.3	1.5
SS B7	19	29	35	43	49	57	62	69	14	19	24	29	32	38	42	47	1.0	1.1	1.3	1.4	1.6	1.3	1.4	1.6	1.8	2.0
SS B8	25	42	53	63	70	79	84	89	16	28	36	44	51	59	65	72	0.8	0.9	1.0	1.2	1.4	0.9	1.0	1.2	1.4	1.5
SS C1	21	36	46	56	64	73	78	84	12	21	28	35	41	49	54	61	0.8	0.9	1.1	1.2	1.4	1.1	1.2	1.3	1.5	1.6
SS C2	24	42	53	64	72	81	86	91	13	25	34	43	50	59	65	72	0.7	0.8	1.0	1.2	1.3	0.9	1.0	1.2	1.3	1.5
SS C3	18	29	37	46	52	61	67	73	11	18	23	29	33	40	44	50	1.0	1.1	1.2	1.3	1.5	1.3	1.4	1.5	1.7	1.8
SS C4	21	37	47	58	66	75	80	86	12	23	30	38	45	53	59	66	0.8	0.9	1.0	1.2	1.4	1.0	1.1	1.3	1.4	1.6
SS D1	30	48	59	69	76	84	88	92	20	32	40	49	55	64	69	76	0.7	0.8	1.0	1.2	1.3	0.9	1.0	1.2	1.3	1.5
SS D2	35	57	69	79	85	92	94	97	22	39	50	60	68	77	82	87	0.6	0.7	0.9	1.1	1.3	0.8	0.9	1.0	1.2	1.4
SS D3	28	43	53	63	70	78	83	88	19	29	36	44	50	58	63	69	0.8	0.9	1.0	1.2	1.4	1.0	1.1	1.3	1.4	1.6
SS D4	34	55	67	77	84	90	93	96	22	38	48	58	65	75	80	86	0.6	0.8	0.9	1.1	1.3	0.8	0.9	1.1	1.2	1.4
SS E1	26	43	54	64	71	79	84	89	16	27	35	43	49	57	63	69	0.8	0.9	1.0	1.2	1.4	1.0	1.1	1.2	1.4	1.6
SS E2	31	51	63	74	81	88	92	95	19	34	43	54	61	70	76	82	0.7	0.8	0.9	1.1	1.3	0.8	0.9	1.1	1.2	1.4
SS E3	23	36	45	54	61	69	75	81	16	24	30	36	41	48	53	59	0.9	1.0	1.1	1.3	1.4	1.1	1.2	1.4	1.6	1.7
SS E4	28	47	58	69	76	84	88	93	18	31	40	50	56	65	71	78	0.7	0.8	1.0	1.1	1.3	0.9	1.0	1.1	1.3	1.5

MOBILE, ALABAMA

TYPE	SOLAR SAVINGS FRACTION (%)																CONSERVATION FACTOR									
	BASE TEMP: 55 F (619 DD)								65 F (1684 DD)								55 F					65 F				
LCR	200	100	70	50	40	30	25	20	200	100	70	50	40	30	25	20	200	100	50	30	20	200	100	50	30	20
WW A1	33	49	59	68	75	83	87	92	20	29	36	43	48	56	62	68	0.7	0.8	1.0	1.2	1.3	1.0	1.1	1.3	1.4	1.6
WW A2	39	60	71	81	87	92	95	98	21	35	44	54	61	70	76	82	0.6	0.7	0.9	1.1	1.3	0.9	1.0	1.1	1.2	1.4
WW A3	42	63	75	85	90	95	97	99	21	37	47	58	65	75	80	86	0.6	0.7	0.9	1.0	1.2	0.8	0.9	1.0	1.2	1.4
WW A4	43	65	77	86	91	96	98	99	21	38	49	60	67	77	82	88	0.6	0.7	0.9	1.0	1.2	0.8	0.9	1.0	1.2	1.4
WW A5	44	68	79	88	93	97	98	99	22	39	51	62	70	79	84	90	0.6	0.7	0.9	1.0	1.2	0.8	0.9	1.0	1.2	1.3
WW A6	45	69	80	89	93	97	98	99	22	40	52	63	71	80	85	91	0.6	0.7	0.8	1.0	1.2	0.7	0.8	1.0	1.2	1.3
WW B1	35	53	64	75	81	88	92	95	18	29	37	46	52	61	67	74	0.7	0.8	0.9	1.1	1.3	0.9	1.0	1.2	1.3	1.5
WW B2	42	65	77	86	91	96	98	99	21	38	49	61	68	78	83	89	0.6	0.7	0.9	1.0	1.2	0.8	0.9	1.0	1.2	1.3
WW B3	49	73	84	92	95	98	99	100	25	45	57	69	76	85	89	93	0.6	0.7	0.8	1.0	1.2	0.7	0.8	1.0	1.1	1.3
WW B4	48	73	84	92	95	98	99	100	24	45	58	70	77	86	90	94	0.5	0.7	0.8	1.0	1.2	0.7	0.8	0.9	1.1	1.3
WW B5	46	71	82	91	95	98	99	100	22	43	56	68	76	85	89	94	0.6	0.7	0.8	1.0	1.2	0.7	0.8	0.9	1.1	1.3
WW C1	51	74	85	92	96	98	99	100	27	47	59	71	78	86	90	94	0.5	0.7	0.8	1.0	1.2	0.7	0.8	0.9	1.1	1.3
WW C2	48	72	83	91	95	98	99	100	24	44	57	69	76	85	89	94	0.6	0.7	0.8	1.0	1.2	0.7	0.8	1.0	1.1	1.3
WW C3	53	78	88	95	97	99	100	100	27	51	64	77	84	91	94	97	0.5	0.6	0.8	1.0	1.2	0.6	0.7	0.9	1.1	1.3
WW C4	49	74	85	93	96	99	99	100	24	47	60	73	80	88	92	96	0.5	0.6	0.8	1.0	1.2	0.7	0.8	0.9	1.1	1.3
TW A1	32	47	57	67	74	82	86	91	20	28	35	42	47	55	61	67	0.7	0.9	1.0	1.2	1.3	1.0	1.2	1.3	1.4	1.6
TW A2	36	54	65	76	82	89	93	96	20	32	40	49	56	65	70	77	0.7	0.8	0.9	1.1	1.3	0.9	1.0	1.2	1.3	1.5
TW A3	37	57	69	79	85	91	94	97	19	33	42	52	59	68	74	81	0.6	0.8	0.9	1.1	1.3	0.9	1.0	1.1	1.3	1.4
TW A4	36	57	69	80	86	92	95	97	18	32	42	52	59	69	75	81	0.6	0.7	0.9	1.1	1.3	0.9	1.0	1.1	1.2	1.4
TW B1	32	48	58	68	75	83	87	92	19	28	35	43	48	56	62	69	0.7	0.8	1.0	1.2	1.3	1.0	1.1	1.3	1.4	1.6
TW B2	34	53	64	74	81	88	92	95	18	30	38	47	54	63	69	76	0.7	0.8	0.9	1.1	1.3	0.9	1.0	1.2	1.3	1.5
TW B3	33	53	64	75	81	89	92	96	17	30	38	48	55	64	70	77	0.7	0.8	0.9	1.1	1.3	0.9	1.0	1.1	1.3	1.5
TW B4	31	50	61	72	79	86	90	94	16	28	36	45	52	61	67	74	0.7	0.8	1.0	1.1	1.3	0.9	1.0	1.2	1.3	1.5
TW C1	30	47	57	67	74	82	87	91	17	27	34	42	48	56	62	69	0.7	0.8	1.0	1.2	1.3	1.0	1.1	1.3	1.4	1.6
TW C2	30	47	58	69	75	84	88	92	16	27	34	43	49	58	64	71	0.7	0.8	1.0	1.1	1.3	1.0	1.1	1.2	1.4	1.5
TW C3	28	45	55	66	73	81	86	91	15	25	32	40	47	55	61	68	0.7	0.8	1.0	1.2	1.3	1.0	1.1	1.3	1.4	1.6
TW C4	25	40	50	60	67	75	81	86	15	23	29	36	42	50	55	62	0.8	0.9	1.1	1.2	1.4	1.1	1.2	1.3	1.5	1.6
TW D1	29	47	57	68	75	83	87	92	15	25	32	40	46	54	60	67	0.7	0.8	1.0	1.1	1.3	1.0	1.1	1.3	1.4	1.6
TW D2	39	60	72	82	88	93	96	98	20	35	45	56	63	73	78	84	0.6	0.7	0.9	1.1	1.2	0.8	0.9	1.1	1.2	1.4
TW D3	43	65	77	86	91	96	97	99	22	38	49	61	68	77	83	88	0.6	0.7	0.9	1.0	1.2	0.8	0.9	1.0	1.2	1.3
TW D4	44	67	79	88	93	97	98	99	22	40	52	64	72	81	86	91	0.6	0.7	0.9	1.0	1.2	0.7	0.8	1.0	1.1	1.3
TW D5	42	66	78	87	92	96	98	99	21	39	51	63	71	80	85	91	0.6	0.7	0.9	1.0	1.2	0.7	0.8	1.0	1.2	1.3
TW E1	46	69	80	89	93	97	98	99	25	43	54	65	73	82	86	91	0.6	0.7	0.8	1.0	1.2	0.7	0.8	1.0	1.1	1.3
TW E2	45	68	90	89	93	97	98	99	23	42	53	65	72	81	86	91	0.6	0.7	0.9	1.0	1.2	0.7	0.8	1.0	1.1	1.3
TW E3	52	76	86	93	96	99	99	100	27	49	62	74	81	88	92	96	0.5	0.6	0.8	1.0	1.2	0.7	0.8	0.9	1.1	1.3
TW E4	48	72	84	92	95	98	99	100	25	45	58	70	78	86	91	95	0.5	0.7	0.8	1.0	1.2	0.7	0.8	0.9	1.1	1.3
TW F1	31	47	57	67	74	82	86	91	18	27	34	41	47	55	61	68	0.7	0.8	1.0	1.2	1.3	1.0	1.1	1.3	1.4	1.6
TW F2	33	52	63	74	80	88	92	95	17	29	37	46	53	62	68	75	0.7	0.8	0.9	1.1	1.3	0.9	1.0	1.2	1.3	1.5
TW F3	33	53	65	76	82	89	93	96	16	29	38	48	55	64	70	77	0.7	0.8	0.9	1.1	1.3	0.9	1.0	1.1	1.3	1.4
TW F4	31	51	63	74	81	88	92	96	14	27	36	46	53	63	69	76	0.7	0.8	0.9	1.1	1.3	0.9	1.0	1.1	1.3	1.5
TW G1	29	46	56	66	73	82	86	91	16	25	32	40	46	55	60	67	0.7	0.8	1.0	1.2	1.3	1.0	1.1	1.3	1.4	1.6
TW G2	29	48	59	70	77	85	89	94	14	26	34	43	49	59	65	72	0.7	0.8	1.0	1.1	1.3	1.0	1.1	1.2	1.3	1.5
TW G3	27	46	58	69	76	84	89	93	12	24	32	41	48	57	63	71	0.7	0.8	1.0	1.1	1.3	1.0	1.1	1.2	1.3	1.5
TW G4	23	40	52	63	70	79	84	90	10	20	27	36	42	51	57	65	0.7	0.9	1.0	1.2	1.3	1.0	1.1	1.3	1.4	1.6
TW H1	25	42	52	63	70	79	84	89	12	22	29	37	43	51	57	64	0.8	0.9	1.0	1.2	1.3	1.1	1.2	1.3	1.4	1.6
TW H2	22	39	50	61	68	77	83	88	10	19	26	34	41	49	55	63	0.8	0.9	1.0	1.2	1.3	1.1	1.2	1.3	1.4	1.6
TW H3	19	35	45	56	63	72	78	84	8	17	23	30	36	44	50	58	0.8	0.9	1.1	1.2	1.4	1.1	1.2	1.4	1.5	1.6
TW H4	14	27	36	46	53	63	69	76	6	13	18	24	29	36	41	48	0.9	1.0	1.1	1.3	1.5	1.3	1.4	1.5	1.6	1.8
TW I1	26	43	54	64	72	80	85	90	13	22	29	36	42	51	50	60	0.7	0.9	1.0	1.2	1.3	1.1	1.2	1.3	1.4	1.6
TW I2	35	57	69	79	85	92	95	97	17	31	41	52	59	69	75	82	0.6	0.7	0.9	1.1	1.3	0.8	0.9	1.1	1.2	1.4
TW I3	38	60	72	83	88	94	96	98	18	34	45	56	64	73	79	85	0.6	0.7	0.9	1.1	1.2	0.8	0.9	1.0	1.2	1.4
TW I4	40	64	76	86	91	96	97	99	19	37	48	60	68	78	83	89	0.6	0.7	0.9	1.0	1.2	0.8	0.9	1.0	1.2	1.3
TW I5	40	64	76	86	91	96	98	99	19	37	49	61	69	79	84	90	0.6	0.7	0.9	1.0	1.2	0.8	0.9	1.0	1.2	1.3
TW J1	44	67	78	87	92	96	98	99	22	40	51	63	70	80	85	90	0.6	0.7	0.9	1.0	1.2	0.8	0.8	1.0	1.2	1.3
TW J2	42	65	77	87	92	96	98	99	21	38	50	62	69	79	84	90	0.6	0.7	0.9	1.0	1.2	0.8	0.9	1.0	1.2	1.3
TW J3	48	72	84	92	95	98	99	100	24	45	58	70	78	86	91	95	0.5	0.7	0.8	1.0	1.2	0.7	0.8	0.9	1.1	1.3
TW J4	45	70	81	90	94	98	99	100	22	42	55	67	75	84	89	93	0.6	0.7	0.8	1.0	1.2	0.7	0.8	1.0	1.1	1.3
DG A1	34	53	64	73	79	85	89	92	16	28	35	44	49	56	61	66	0.7	0.8	1.0	1.2	1.4	0.9	1.0	1.3	1.5	1.8
DG A2	33	54	66	76	82	89	92	95	16	29	38	48	55	63	69	75	0.6	0.8	0.9	1.1	1.3	0.9	1.0	1.1	1.3	1.5
DG A3	37	61	73	82	87	92	94	95	19	34	45	57	65	74	79	84	0.6	0.7	0.9	1.1	1.6	0.9	0.9	1.0	1.2	1.4
DG B1	35	58	70	80	86	91	93	95	16	28	38	48	55	64	69	74	0.6	0.7	0.9	1.1	1.4	0.9	1.0	1.1	1.3	1.6
DG B2	35	59	72	82	88	93	96	98	17	30	40	52	60	70	76	82	0.6	0.7	0.9	1.1	1.3	0.9	0.9	1.1	1.2	1.4
DG B3	38	63	77	87	91	95	97	97	19	34	46	59	68	78	84	89	0.6	0.7	0.9	1.1	1.5	0.8	0.9	1.0	1.2	1.4
DG C1	40	65	78	87	91	95	96	97	18	33	44	56	64	74	79	83	0.6	0.7	0.9	1.1	1.5	0.9	0.9	1.0	1.2	1.5
DG C2	39	65	78	88	92	96	98	99	19	34	46	59	67	77	83	88	0.6	0.7	0.9	1.0	1.3	0.9	0.9	1.0	1.2	1.4
DG C3	42	70	83	91	95	97	98	98	21	39	52	66	75	85	89	93	0.6	0.6	0.8	1.1	1.9	0.8	0.8	0.9	1.1	1.3
SS A1	42	62	72	82	87	93	95	98	25	39	48	58	64	73	78	84	0.6	0.7	0.9	1.1	1.3	0.8	0.9	1.1	1.2	1.4
SS A2	50	72	82	90	94	97	99	99	30	47	58	69	76	84	88	93	0.6	0.7	0.8	1.0	1.2	0.7	0.8	1.0	1.1	1.3
SS A3	40	59	70	79	85	91	94	97	24	37	45	54	61	70	75	81	0.6	0.8	0.9	1.1	1.3	0.9	1.0	1.1	1.3	1.4
SS A4	50	71	82	90	94	97	99	99	30	47	58	69	76	84	88	92	0.6	0.7	0.8	1.0	1.2	0.7	0.8	1.0	1.1	1.3
SS A5	40	50	68	77	83	90	93	96	24	36	44	53	59	68	73	79	0.7	0.8	0.9	1.1	1.3	0.9	1.0	1.1	1.3	1.5
SS A6	50	72	82	90	94	97	99	99	30	47	58	69	76	84	88	93	0.6	0.7	0.8	1.0	1.2	0.7	0.8	1.0	1.1	1.3
SS A7	36	53	62	72	78	86	90	93	23	33	40	48	54	62	67	73	0.7	0.8	1.0	1.1	1.3	1.0	1.1	1.2	1.4	1.5
SS A8	50	71	81	89	93	97	98	99	29	46	57	68	75	83	87	92	0.6	0.7	0.9	1.0	1.2	0.7	0.8	1.0	1.2	1.3
SS B1	37	56	66	76	83	89	93	96	21	33	42	51	58	67	72	79	0.7	0.8	0.9	1.1	1.3	0.9	1.0	1.1	1.3	1.5
SS B2	44	66	77	86	91	95	97	99	25	42	52	63	70	79	84	89	0.6	0.7	0.9	1.0	1.2	0.8	0.9	1.0	1.2	1.3
SS B3	36	54	65	75	81	88	92	95	20	32	40	49	56	64	70	76	0.7	0.8	0.9	1.1	1.3	0.9	1.0	1.2	1.3	1.5
SS B4	44	65	76	85	90	95	97	99	25	41	52	62	70	78	83	89	0.6	0.7	0.9	1.0	1.2	0.8	0.9	1.0	1.2	1.4
SS B5	33	49	60	69	76	84	88	92	20	30	37	45	51	59	64	71	0.7	0.8	1.0	1.2	1.3	1.0	1.1	1.2	1.4	1.5
SS B6	44	65	76	86	91	95	97	99	25	41	52	62	70	79	84	89	0.6	0.7	0.9	1.0	1.2	0.8	0.9	1.0	1.2	1.4
SS B7	30	46	56	65	72	80	85	90	19	27	34	41	47	54	60	66	0.8	0.9	1.0	1.2	1.4	1.0	1.2	1.3	1.4	1.6
SS B8	43	64	75	85	90	95	97	99	24	40	50	61	68	77	82	88	0.6	0.7	0.9	1.1	1.2	0.8	0.9	1.0	1.2	1.4
SS C1	38	59	70	80	86	92	95	97	19	33	42	52	60	69	75	81	0.6	0.7	0.9	1.1	1.3	0.9	1.0	1.1	1.3	1.4
SS C2	42	65	77	86	91	96	98	99	22	38	49	61	68	78	83	89	0.6	0.7	0.9	1.0	1.2	0.8	0.9	1.0	1.2	1.3
SS C3	30	48	58	69	76	84	88	92	17	27	34	42	48	57	63	70	0.7	0.8	1.0	1.1	1.3	1.0	1.1	1.2	1.4	1.5
SS C4	38	59	71	81	87	93	95	98	20	34	44	54	62	71	77	84	0.6	0.7	0.9	1.1	1.2	0.8	0.9	1.1	1.2	1.4
SS D1	50	71	82	90	94	97	98	99	29	46	57	67	74	82	87	91	0.6	0.7	0.8	1.0	1.2	0.7	0.8	1.0	1.2	1.3
SS D2	57	80	89	95	97	99	100	100	34	55	67	78	84	91	94	97	0.5	0.6	0.8	1.0	1.2	0.6	0.8	0.9	1.1	1.3
SS D3	45	65	76	85	90	95	97	98	27	41	51	61	68	76	81	87	0.6	0.7	0.9	1.1	1.2	0.8	0.9	1.0	1.2	1.4
SS D4	55	78	87	94	97	99	99	100	33	53	65	76	82	89	93	96	0.5	0.6	0.8	1.0	1.2	0.7	0.8	0.9	1.1	1.3
SS E1	45	66	77	86	91	96	97	99	25	41	51	61	69	77	82	88	0.6	0.7	0.9	1.0	1.2	0.8	0.9	1.0	1.2	1.4
SS E2	52	75	85	92	96	98	99	100	29	49	61	72	79	86	90	94	0.5	0.7	0.8	1.0	1.2	0.7	0.8	0.9	1.1	1.3
SS E3	38	56	67	77	83	90	93	96	22	34	42	52	58	67	72	79	0.7	0.8	0.9	1.1	1.3	0.9	1.0	1.1	1.3	1.5
SS E4	48	70	81	89	93	97	98	99	27	45	56	67	74	83	87	92	0.6	0.7	0.8	1.0	1.2	0.7	0.8	1.0	1.1	1.3

MONTGOMERY, ALABAMA

SOLAR SAVINGS FRACTION (%) — CONSERVATION FACTOR

TYPE	BASE TEMP: 55 F (945 DD)								65 F (2272 DD)								CF 55 F					CF 65 F				
LCR	200	100	70	50	40	30	25	20	200	100	70	50	40	30	25	20	200	100	50	30	20	200	100	50	30	20
WW A1	26	39	47	57	63	71	77	82	17	24	30	36	40	47	52	58	0.9	0.9	1.1	1.3	1.4	1.2	1.3	1.4	1.6	1.7
WW A2	30	48	58	69	76	84	88	92	17	28	36	44	51	59	65	72	0.7	0.8	1.0	1.1	1.3	1.0	1.1	1.2	1.4	1.5
WW A3	31	51	62	73	80	87	91	95	17	29	38	48	55	64	70	77	0.7	0.8	0.9	1.1	1.3	0.9	1.0	1.1	1.3	1.5
WW A4	32	53	64	75	82	89	93	96	16	30	39	50	57	66	72	79	0.7	0.8	0.9	1.1	1.3	0.9	1.0	1.1	1.3	1.4
WW A5	33	55	67	78	84	91	94	97	16	31	41	52	59	69	75	81	0.6	0.7	0.9	1.1	1.3	0.8	0.9	1.1	1.2	1.4
WW A6	33	56	68	79	85	91	94	97	16	31	42	53	60	70	76	82	0.6	0.7	0.9	1.1	1.3	0.8	0.9	1.1	1.2	1.4
WW B1	26	42	52	62	69	77	82	88	14	23	30	37	42	50	55	61	0.8	0.9	1.0	1.2	1.4	1.1	1.2	1.3	1.5	1.7
WW B2	31	53	65	76	82	89	93	96	16	31	40	51	58	68	74	81	0.7	0.8	0.9	1.1	1.3	0.8	0.9	1.1	1.3	1.4
WW B3	37	60	72	83	88	94	96	98	20	37	48	59	66	76	81	87	0.6	0.7	0.9	1.1	1.2	0.8	0.9	1.0	1.2	1.4
WW B4	36	60	73	83	89	94	97	98	18	36	48	60	68	78	83	89	0.6	0.7	0.9	1.1	1.2	0.8	0.9	1.0	1.2	1.3
WW B5	34	58	71	82	88	94	96	98	16	35	47	59	67	77	82	88	0.6	0.7	0.9	1.1	1.2	0.8	0.9	1.0	1.2	1.3
WW C1	39	62	74	84	89	95	97	98	21	38	49	61	68	78	83	89	0.6	0.7	0.9	1.1	1.2	0.8	0.9	1.0	1.2	1.3
WW C2	36	59	72	82	88	94	96	98	18	36	47	59	67	76	82	88	0.6	0.7	0.9	1.1	1.2	0.8	0.9	1.0	1.2	1.4
WW C3	41	66	78	88	93	97	98	99	21	42	55	67	75	84	89	93	0.6	0.7	0.8	1.0	1.2	0.7	0.8	1.0	1.1	1.3
WW C4	37	62	75	85	90	95	97	99	18	38	51	63	72	81	86	91	0.6	0.7	0.9	1.0	1.2	0.7	0.8	1.0	1.1	1.3
TW A1	25	38	46	55	62	70	75	81	17	24	29	35	40	46	51	57	0.9	1.0	1.1	1.3	1.4	1.2	1.3	1.4	1.6	1.8
TW A2	27	43	53	64	71	79	84	89	16	26	33	41	46	54	60	67	0.8	0.9	1.0	1.2	1.4	1.0	1.1	1.3	1.4	1.6
TW A3	28	45	56	67	74	82	87	91	15	26	34	43	49	58	63	70	0.7	0.8	1.0	1.2	1.3	1.0	1.1	1.2	1.4	1.5
TW A4	27	45	57	68	75	83	87	92	14	26	34	43	49	58	64	72	0.7	0.8	1.0	1.1	1.3	0.9	1.0	1.2	1.4	1.5
TW B1	25	38	47	56	63	71	77	83	16	23	29	35	40	47	52	58	0.9	0.9	1.1	1.3	1.4	1.2	1.3	1.4	1.6	1.7
TW B2	25	41	52	62	69	78	83	88	15	24	31	39	45	53	59	66	0.8	0.9	1.0	1.2	1.4	1.0	1.1	1.3	1.4	1.6
TW B3	25	41	52	63	70	79	84	89	14	24	31	39	45	54	60	67	0.8	0.9	1.0	1.2	1.4	1.0	1.1	1.3	1.4	1.6
TW B4	23	39	49	60	67	76	81	87	13	22	29	37	43	51	57	64	0.8	0.9	1.0	1.2	1.4	1.0	1.1	1.3	1.4	1.6
TW C1	23	37	46	55	62	71	76	83	14	22	28	35	40	47	52	59	0.9	0.9	1.1	1.3	1.4	1.1	1.2	1.4	1.5	1.7
TW C2	22	37	46	57	64	73	78	84	13	22	28	35	41	48	54	61	0.8	0.9	1.1	1.2	1.4	1.1	1.2	1.3	1.5	1.7
TW C3	21	35	44	54	61	70	75	82	13	21	27	33	39	46	51	58	0.9	0.9	1.1	1.3	1.4	1.1	1.2	1.4	1.5	1.7
TW C4	19	31	39	48	55	64	69	76	12	19	24	30	35	42	46	53	0.9	1.0	1.2	1.3	1.5	1.2	1.3	1.5	1.6	1.8
TW D1	21	36	45	55	62	71	76	82	12	20	25	32	37	44	49	55	0.8	0.9	1.1	1.3	1.4	1.2	1.3	1.4	1.6	1.7
TW D2	29	48	60	71	78	85	90	94	16	28	37	46	53	63	68	75	0.7	0.8	1.0	1.1	1.3	0.9	1.0	1.2	1.3	1.5
TW D3	32	53	64	75	82	89	93	96	17	31	41	51	58	68	73	80	0.7	0.8	0.9	1.1	1.3	0.9	1.0	1.1	1.3	1.4
TW D4	33	55	67	78	84	91	94	97	17	33	43	54	62	72	78	84	0.6	0.7	0.9	1.1	1.3	0.8	0.9	1.1	1.2	1.4
TW D5	31	53	66	77	84	91	94	97	16	32	43	54	62	72	77	84	0.6	0.7	0.9	1.1	1.3	0.8	0.9	1.1	1.2	1.4
TW E1	35	57	69	79	85	92	95	97	19	35	45	56	63	72	78	84	0.6	0.7	0.9	1.1	1.3	0.8	0.9	1.1	1.2	1.4
TW E2	34	56	68	79	85	91	94	97	18	34	44	55	63	72	78	84	0.6	0.7	0.9	1.1	1.3	0.8	0.9	1.1	1.2	1.4
TW E3	40	64	76	86	91	96	97	99	21	40	52	64	72	81	86	91	0.6	0.7	0.9	1.0	1.2	0.7	0.8	1.0	1.1	1.3
TW E4	36	60	73	83	89	94	97	98	19	37	49	61	69	79	84	89	0.6	0.7	0.9	1.1	1.2	0.7	0.8	1.0	1.2	1.3
TW F1	24	37	46	55	62	70	76	82	15	22	28	34	39	46	51	57	0.9	0.9	1.1	1.3	1.4	1.2	1.3	1.4	1.6	1.7
TW F2	25	41	51	62	69	77	82	88	14	23	30	38	44	52	58	64	0.8	0.9	1.0	1.2	1.4	1.0	1.1	1.3	1.4	1.6
TW F3	24	41	52	63	70	79	84	89	12	23	30	39	45	54	60	67	0.8	0.9	1.0	1.2	1.3	1.0	1.1	1.3	1.4	1.6
TW F4	22	39	50	61	69	78	83	89	10	21	28	37	43	52	58	66	0.8	0.9	1.0	1.2	1.3	1.0	1.1	1.3	1.4	1.6
TW G1	22	35	44	54	61	70	75	82	13	21	26	33	38	45	50	57	0.9	0.9	1.1	1.3	1.4	1.2	1.3	1.4	1.6	1.7
TW G2	21	37	47	58	65	74	79	85	11	20	27	34	40	48	54	61	0.8	0.9	1.1	1.2	1.4	1.1	1.2	1.3	1.5	1.6
TW G3	19	35	45	56	63	73	78	85	9	18	25	33	39	47	53	60	0.8	0.9	1.1	1.2	1.4	1.1	1.2	1.3	1.5	1.6
TW G4	15	30	39	50	57	67	73	80	7	15	21	28	34	42	47	54	0.9	1.0	1.1	1.3	1.4	1.2	1.2	1.4	1.5	1.7
TW H1	18	32	41	51	58	67	73	79	10	17	23	30	35	42	47	54	0.9	1.0	1.1	1.3	1.4	1.2	1.3	1.4	1.6	1.7
TW H2	15	29	38	48	56	65	71	78	7	15	21	27	33	40	45	52	0.9	1.0	1.1	1.3	1.4	1.2	1.3	1.4	1.6	1.7
TW H3	13	25	34	43	50	60	66	73	5	12	18	24	29	36	41	48	0.9	1.0	1.2	1.3	1.5	1.3	1.4	1.5	1.7	1.8
TW H4	10	19	27	35	41	50	56	64	4	9	14	19	23	29	33	39	1.1	1.1	1.3	1.4	1.6	1.4	1.5	1.7	1.8	2.0
TW I1	19	32	42	51	58	67	73	80	10	17	22	29	33	40	45	51	0.9	1.0	1.1	1.3	1.4	1.2	1.3	1.5	1.6	1.8
TW I2	26	44	56	67	75	83	88	92	13	25	33	43	50	59	65	72	0.7	0.8	1.0	1.1	1.3	0.9	1.0	1.2	1.3	1.5
TW I3	28	48	60	71	78	86	90	94	14	27	36	46	54	63	69	77	0.7	0.8	1.0	1.1	1.3	0.9	1.0	1.1	1.3	1.5
TW I4	29	51	64	75	82	89	93	96	14	30	40	51	59	69	75	81	0.7	0.8	0.9	1.1	1.3	0.8	0.9	1.1	1.2	1.4
TW I5	29	51	64	75	82	89	93	96	14	30	40	51	59	69	76	82	0.7	0.8	0.9	1.1	1.3	0.8	0.9	1.1	1.2	1.4
TW J1	32	54	66	77	84	90	94	97	17	32	42	53	61	70	76	83	0.6	0.8	0.9	1.1	1.3	0.8	0.9	1.1	1.2	1.4
TW J2	31	53	65	76	83	90	93	96	15	31	41	52	60	70	76	82	0.6	0.8	0.9	1.1	1.3	0.8	0.9	1.1	1.2	1.4
TW J3	36	60	73	83	89	94	97	98	18	37	49	61	69	79	84	90	0.6	0.7	0.9	1.0	1.2	0.7	0.8	1.0	1.2	1.3
TW J4	33	57	70	81	87	93	96	98	16	34	46	58	66	76	82	88	0.6	0.7	0.9	1.1	1.2	0.8	0.9	1.0	1.2	1.3
DG A1	24	41	50	60	66	73	77	82	12	21	28	34	39	45	49	53	0.8	0.9	1.1	1.3	1.5	1.1	1.2	1.4	1.7	2.0
DG A2	24	42	53	64	70	78	83	87	13	23	31	39	45	53	58	64	0.7	0.8	1.0	1.2	1.4	1.0	1.1	1.3	1.5	1.7
DG A3	27	48	60	72	78	85	89	92	15	27	37	47	55	64	70	77	0.7	0.8	1.0	1.2	1.4	0.9	1.0	1.1	1.3	1.5
DG B1	25	43	55	67	73	81	85	88	12	22	29	37	43	51	56	62	0.7	0.8	1.0	1.2	1.5	1.0	1.2	1.3	1.5	1.8
DG B2	25	44	57	70	77	85	89	92	13	24	32	41	49	58	65	71	0.7	0.8	1.0	1.1	1.3	1.0	1.1	1.2	1.3	1.5
DG B3	28	49	63	76	83	89	93	95	15	28	38	49	57	68	75	82	0.7	0.8	0.9	1.1	1.4	0.9	1.0	1.1	1.2	1.4
DG C1	29	50	63	75	81	88	91	93	14	25	34	43	50	60	66	72	0.7	0.8	0.9	1.2	1.4	1.0	1.1	1.2	1.3	1.6
DG C2	28	50	64	76	83	90	93	96	15	27	36	47	55	66	72	79	0.7	0.7	0.9	1.1	1.3	0.9	1.0	1.1	1.2	1.5
DG C3	31	55	69	82	88	93	96	97	18	32	42	55	64	75	81	87	0.6	0.7	0.9	1.1	1.4	0.9	0.9	1.0	1.2	1.4
SS A1	33	50	60	71	77	84	88	93	21	32	40	48	55	63	68	75	0.7	0.8	1.0	1.1	1.3	0.9	1.0	1.2	1.3	1.5
SS A2	39	60	71	81	86	92	95	97	24	40	50	60	67	76	81	86	0.6	0.7	0.9	1.1	1.3	0.8	0.9	1.0	1.2	1.4
SS A3	31	48	58	68	74	82	86	91	20	30	37	45	51	59	64	71	0.7	0.8	1.0	1.2	1.3	1.0	1.1	1.2	1.4	1.6
SS A4	39	60	71	80	86	92	95	97	24	39	49	59	66	75	80	86	0.6	0.7	0.9	1.1	1.3	0.8	0.9	1.1	1.2	1.4
SS A5	31	47	56	66	72	80	85	89	21	31	37	45	50	58	63	69	0.8	0.9	1.0	1.2	1.4	1.0	1.1	1.3	1.4	1.6
SS A6	39	60	71	81	86	92	95	97	24	40	49	60	67	75	80	86	0.6	0.7	0.9	1.1	1.3	0.8	0.9	1.1	1.2	1.4
SS A7	28	42	51	60	67	75	79	85	19	28	33	40	45	52	56	62	0.8	0.9	1.1	1.2	1.4	1.1	1.2	1.4	1.5	1.7
SS A8	39	59	70	80	85	92	94	97	24	39	48	58	65	74	79	85	0.6	0.7	0.9	1.1	1.3	0.8	0.9	1.1	1.2	1.4
SS B1	28	44	54	65	71	80	84	89	17	28	35	42	48	56	62	68	0.8	0.9	1.0	1.2	1.4	1.0	1.1	1.3	1.4	1.6
SS B2	34	54	65	76	82	89	92	96	21	35	44	54	61	70	76	82	0.7	0.8	0.9	1.1	1.3	0.8	0.9	1.1	1.3	1.4
SS B3	27	43	53	63	69	78	83	88	17	26	33	40	46	54	59	66	0.8	0.9	1.0	1.2	1.4	1.0	1.1	1.3	1.4	1.6
SS B4	34	53	65	75	81	88	92	95	20	34	43	53	60	69	75	81	0.7	0.8	0.9	1.1	1.3	0.8	1.0	1.1	1.3	1.4
SS B5	25	39	48	58	64	73	78	83	17	25	31	37	42	49	54	61	0.8	0.9	1.1	1.3	1.4	1.1	1.2	1.4	1.5	1.7
SS B6	34	53	65	75	82	89	92	95	20	34	43	53	60	69	75	81	0.7	0.8	0.9	1.1	1.3	0.8	0.9	1.1	1.3	1.4
SS B7	23	36	44	53	60	68	73	79	16	23	28	34	39	45	50	55	0.9	1.0	1.1	1.3	1.5	1.2	1.3	1.5	1.6	1.8
SS B8	33	52	63	74	80	88	91	95	20	33	42	52	58	68	73	80	0.7	0.8	0.9	1.1	1.3	0.9	1.0	1.1	1.3	1.4
SS C1	28	46	57	68	75	83	88	92	15	26	34	43	49	58	64	71	0.7	0.8	1.0	1.1	1.3	1.0	1.1	1.2	1.4	1.5
SS C2	32	53	65	76	82	89	93	96	17	31	41	51	58	68	74	80	0.7	0.8	0.9	1.1	1.3	0.8	1.0	1.1	1.3	1.4
SS C3	23	37	47	57	63	72	78	84	14	22	28	35	40	47	52	59	0.8	0.9	1.1	1.2	1.4	1.1	1.2	1.4	1.5	1.7
SS C4	28	47	58	69	76	84	89	93	15	28	36	45	52	61	67	74	0.7	0.8	1.0	1.1	1.3	0.9	1.0	1.2	1.3	1.5
SS D1	39	59	70	80	86	92	94	97	24	38	47	57	64	73	78	84	0.6	0.7	0.9	1.1	1.3	0.8	0.9	1.1	1.2	1.4
SS D2	45	68	79	88	92	96	98	99	27	46	57	68	75	83	88	92	0.6	0.7	0.9	1.0	1.2	0.7	0.8	1.0	1.1	1.3
SS D3	35	53	64	74	80	87	91	94	23	35	43	51	58	66	72	78	0.7	0.8	1.0	1.1	1.3	0.9	1.0	1.2	1.3	1.5
SS D4	44	66	77	86	91	96	97	99	26	44	55	66	73	82	86	91	0.6	0.7	0.9	1.0	1.2	0.7	0.8	1.0	1.2	1.3
SS E1	34	54	65	75	81	88	92	95	20	33	42	51	58	67	72	78	0.7	0.8	0.9	1.1	1.3	0.9	1.0	1.1	1.3	1.5
SS E2	40	62	74	84	89	94	96	98	23	40	51	62	69	78	83	88	0.6	0.7	0.9	1.1	1.2	0.8	0.9	1.0	1.2	1.4
SS E3	29	45	55	65	72	80	84	89	18	28	35	43	48	56	62	68	0.8	0.9	1.0	1.2	1.4	1.0	1.1	1.3	1.4	1.6
SS E4	37	58	69	80	85	92	95	97	22	37	47	57	64	73	79	85	0.6	0.7	0.9	1.1	1.3	0.8	0.9	1.1	1.2	1.4

PHOENIX, ARIZONA

TYPE	SOLAR SAVINGS FRACTION (%) BASE TEMP: 55 F (436 DD)								65 F (1556 DD)								CONSERVATION FACTOR 55 F					65 F				
LCR	200	100	70	50	40	30	25	20	200	100	70	50	40	30	25	20	200	100	50	30	20	200	100	50	30	20
WW A1	53	72	82	90	94	97	98	99	27	41	50	60	66	75	80	85	0.6	0.7	0.9	1.0	1.2	0.8	0.9	1.1	1.2	1.4
WW A2	63	84	92	97	98	100	100	100	32	50	62	72	79	87	90	94	0.5	0.6	0.8	1.0	1.2	0.7	0.8	1.0	1.1	1.3
WW A3	67	87	94	98	99	100	100	100	33	54	66	77	83	90	93	96	0.5	0.6	0.8	1.0	1.2	0.7	0.8	0.9	1.1	1.3
WW A4	69	89	95	98	99	100	100	100	34	56	68	79	85	91	94	97	0.5	0.6	0.8	1.0	1.2	0.6	0.7	0.9	1.1	1.3
WW A5	71	90	96	99	100	100	100	100	35	58	70	81	87	93	95	98	0.5	0.6	0.8	1.0	-	0.6	0.7	0.9	1.1	1.2
WW A6	72	91	97	99	100	100	100	100	36	59	71	82	88	93	96	98	0.5	0.6	0.8	1.0	-	0.6	0.7	0.9	1.1	1.2
WW B1	58	79	88	94	97	99	99	100	28	44	55	65	72	81	85	90	0.5	0.6	0.8	1.0	1.2	0.7	0.8	1.0	1.2	1.3
WW B2	69	89	95	98	99	100	100	100	34	56	68	79	85	92	95	97	0.5	0.6	0.8	1.0	1.2	0.6	0.7	0.9	1.1	1.3
WW B3	75	93	98	99	100	100	100	100	40	64	76	85	90	95	97	99	0.5	0.6	0.8	1.0	-	0.6	0.7	0.9	1.0	1.2
WW B4	75	93	98	99	100	100	100	100	39	64	76	86	91	96	98	99	0.5	0.6	0.8	1.0	-	0.6	0.7	0.9	1.0	1.2
WW B5	73	92	97	99	100	100	100	100	37	61	74	85	90	95	97	99	0.5	0.6	0.8	1.0	-	0.6	0.7	0.9	1.0	1.2
WW C1	77	94	98	99	100	100	100	100	41	65	77	87	91	96	98	99	0.5	0.6	0.8	1.0	-	0.6	0.7	0.9	1.0	1.2
WW C2	74	93	97	99	100	100	100	100	39	63	75	85	90	95	97	99	0.5	0.6	0.8	1.0	-	0.6	0.7	0.9	1.0	1.2
WW C3	80	95	99	100	100	100	100	100	43	69	81	90	94	98	99	100	0.4	0.6	0.8	0.9	-	0.6	0.7	0.8	1.0	1.2
WW C4	76	94	98	100	100	100	100	100	40	65	78	88	92	97	98	99	0.5	0.6	0.8	0.9	-	0.6	0.7	0.9	1.0	1.2
TW A1	51	71	81	89	93	97	98	99	26	40	49	58	65	73	79	84	0.6	0.7	0.9	1.0	1.2	0.8	0.9	1.1	1.2	1.4
TW A2	58	79	88	94	97	99	100	100	29	46	56	67	74	82	87	91	0.5	0.6	0.8	1.0	1.2	0.7	0.8	1.0	1.2	1.3
TW A3	61	82	90	96	98	99	100	100	29	48	59	70	77	85	89	93	0.5	0.6	0.8	1.0	1.2	0.7	0.8	1.0	1.1	1.3
TW A4	61	82	91	96	98	99	100	100	29	48	60	71	78	86	90	94	0.5	0.6	0.8	1.0	1.2	0.7	0.8	1.0	1.1	1.3
TW B1	51	72	82	90	94	97	98	99	26	40	49	59	66	75	80	85	0.6	0.7	0.9	1.0	1.2	0.8	0.9	1.1	1.2	1.4
TW B2	56	77	87	94	97	99	99	100	27	44	55	65	72	81	86	91	0.5	0.7	0.8	1.0	1.2	0.7	0.8	1.0	1.2	1.3
TW B3	56	78	87	94	97	99	99	100	26	44	55	66	73	82	86	91	0.5	0.6	0.8	1.0	1.2	0.7	0.8	1.0	1.2	1.3
TW B4	54	75	85	92	96	98	99	100	25	41	52	63	70	79	84	89	0.6	0.7	0.8	1.0	1.2	0.8	0.9	1.0	1.2	1.3
TW C1	50	71	81	89	93	97	98	99	24	39	48	58	65	74	80	85	0.6	0.7	0.9	1.0	1.2	0.8	0.9	1.1	1.2	1.4
TW C2	51	72	82	90	94	97	99	99	24	39	49	60	67	76	81	87	0.6	0.7	0.8	1.0	1.2	0.8	0.9	1.0	1.2	1.4
TW C3	48	69	80	88	93	97	98	99	22	37	47	57	64	73	79	85	0.6	0.7	0.9	1.0	1.2	0.8	0.9	1.1	1.2	1.4
TW C4	43	63	74	83	89	94	96	98	20	33	41	51	58	67	73	79	0.6	0.7	0.9	1.1	1.2	0.9	1.0	1.1	1.3	1.4
TW D1	51	72	82	90	94	98	99	100	23	38	48	58	65	74	80	85	0.6	0.7	0.8	1.0	1.2	0.8	0.9	1.1	1.2	1.4
TW D2	64	85	92	97	99	100	100	100	31	51	63	74	81	88	92	95	0.5	0.6	0.8	1.0	1.2	0.7	0.8	0.9	1.1	1.3
TW D3	68	88	95	98	99	100	100	100	34	56	68	78	85	91	94	97	0.5	0.6	0.8	1.0	1.2	0.6	0.7	0.9	1.1	1.3
TW D4	70	90	96	99	99	100	100	100	35	58	70	81	87	93	96	98	0.5	0.6	0.8	1.0	-	0.6	0.7	0.9	1.1	1.2
TW D5	68	89	95	98	99	100	100	100	34	57	69	80	86	92	95	98	0.5	0.6	0.8	1.0	1.2	0.6	0.7	0.9	1.1	1.2
TW E1	72	91	96	99	100	100	100	100	38	60	72	82	88	93	96	98	0.5	0.6	0.8	1.0	-	0.6	0.7	0.9	1.1	1.2
TW E2	71	90	96	99	100	100	100	100	36	59	71	82	88	93	96	98	0.5	0.6	0.8	1.0	-	0.6	0.7	0.9	1.1	1.2
TW E3	78	94	98	100	100	100	100	100	42	67	79	88	93	97	98	99	0.5	0.6	0.8	0.9	-	0.6	0.7	0.9	1.0	1.2
TW E4	74	93	97	99	100	100	100	100	39	63	76	86	91	96	98	99	0.5	0.6	0.8	1.0	-	0.6	0.7	0.9	1.0	1.2
TW F1	51	71	81	89	93	97	98	99	25	39	48	58	65	74	79	85	0.6	0.7	0.9	1.0	1.2	0.8	0.9	1.1	1.2	1.4
TW F2	56	77	87	94	96	99	99	100	26	43	54	65	72	81	85	90	0.5	0.7	0.8	1.0	1.2	0.8	0.8	1.0	1.2	1.3
TW F3	57	79	88	95	97	99	100	100	26	44	55	67	74	82	87	92	0.5	0.6	0.8	1.0	1.2	0.7	0.8	1.0	1.1	1.3
TW F4	55	77	87	94	97	99	99	100	23	42	53	65	72	81	86	91	0.5	0.6	0.8	1.0	1.2	0.7	0.8	1.0	1.2	1.3
TW G1	49	70	81	89	93	97	98	99	23	37	47	57	64	73	79	85	0.6	0.7	0.9	1.0	1.2	0.8	0.9	1.1	1.2	1.4
TW G2	52	74	84	92	95	98	99	100	23	39	50	61	68	77	83	88	0.6	0.7	0.8	1.0	1.2	0.8	0.9	1.0	1.2	1.4
TW G3	50	72	83	91	95	98	99	100	21	37	48	59	67	76	82	87	0.6	0.7	0.8	1.0	1.2	0.8	0.9	1.0	1.2	1.4
TW G4	44	66	78	87	92	96	98	99	17	32	42	53	61	70	76	83	0.6	0.7	0.9	1.0	1.2	0.8	0.9	1.1	1.2	1.4
TW H1	46	67	78	87	92	96	98	99	19	34	43	54	61	70	76	82	0.6	0.7	0.9	1.0	1.2	0.8	0.9	1.1	1.2	1.4
TW H2	43	65	76	86	91	95	97	99	17	31	41	51	59	69	74	81	0.6	0.7	0.9	1.0	1.2	0.9	0.9	1.1	1.2	1.4
TW H3	38	59	71	81	87	93	96	98	14	27	36	46	53	63	69	76	0.6	0.7	0.9	1.1	1.2	0.9	1.0	1.1	1.3	1.5
TW H4	31	50	61	72	79	87	91	95	10	21	29	37	44	53	59	67	0.7	0.8	0.9	1.1	1.3	1.0	1.1	1.2	1.4	1.5
TW I1	48	69	80	88	93	97	98	99	20	35	44	55	62	71	77	83	0.6	0.7	0.9	1.0	1.2	0.8	0.9	1.1	1.2	1.4
TW I2	60	82	91	96	98	99	100	100	27	47	59	71	78	86	90	94	0.5	0.6	0.8	1.0	1.2	0.7	0.8	1.0	1.1	1.3
TW I3	63	85	93	97	99	100	100	100	30	51	63	74	81	89	92	96	0.5	0.6	0.8	1.0	1.2	0.7	0.8	0.9	1.1	1.3
TW I4	66	87	94	98	99	100	100	100	32	54	67	78	85	91	94	97	0.5	0.6	0.8	1.0	1.2	0.6	0.7	0.9	1.1	1.3
TW I5	66	88	95	98	99	100	100	100	31	54	67	78	85	92	95	97	0.5	0.6	0.8	1.0	1.2	0.6	0.7	0.9	1.1	1.3
TW J1	69	89	95	99	99	100	100	100	35	57	70	80	86	92	95	98	0.5	0.6	0.8	1.0	1.2	0.6	0.7	0.9	1.1	1.2
TW J2	68	88	95	98	99	100	100	100	33	56	68	79	86	92	95	97	0.5	0.6	0.8	1.0	1.2	0.6	0.7	0.9	1.1	1.3
TW J3	74	93	97	99	100	100	100	100	38	63	76	86	91	96	98	99	0.5	0.6	0.8	1.0	-	0.6	0.7	0.9	1.0	1.2
TW J4	72	91	97	99	100	100	100	100	35	60	73	84	80	95	97	99	0.5	0.6	0.8	1.0	-	0.6	0.7	0.9	1.0	1.2
DG A1	59	80	88	94	97	99	99	100	26	43	54	64	70	77	81	85	0.5	0.6	0.8	1.0	1.1	0.7	0.8	1.0	1.3	1.5
DC A2	58	80	89	95	97	99	100	100	26	45	56	67	74	82	86	90	0.5	0.6	0.8	1.0	-	0.7	0.8	1.0	1.2	1.4
DG A3	63	85	92	95	96	97	97	97	29	51	64	75	81	87	91	93	0.5	0.6	0.9	1.9	-	0.7	0.8	0.9	1.2	1.4
DG B1	64	86	93	97	98	99	99	99	27	47	60	71	77	84	87	91	0.5	0.6	0.8	1.2	2.8	0.7	0.8	1.0	1.2	1.4
DG B2	63	86	94	98	99	100	100	100	27	48	61	73	80	88	91	94	0.5	0.6	0.8	1.0	1.2	0.7	0.8	0.9	1.1	1.3
DG B3	66	89	95	97	98	98	98	98	29	53	67	79	85	91	94	96	0.5	0.6	0.9	2.2	-	0.7	0.7	0.9	1.1	1.4
DG C1	71	91	96	98	98	98	98	98	31	54	68	79	85	90	93	95	0.5	0.6	0.9	-	-	0.6	0.7	0.9	1.1	1.4
DG C2	69	91	96	99	100	100	100	100	30	54	68	80	86	92	95	97	0.5	0.6	0.8	1.0	-	0.7	0.7	0.9	1.1	1.3
DG C3	72	93	97	98	98	98	98	98	33	59	74	85	90	95	96	98	0.4	0.6	1.0	-	-	0.6	0.7	0.9	1.1	1.4
SS A1	62	83	91	96	98	99	100	100	34	52	62	73	79	86	90	94	0.5	0.6	0.8	1.0	1.2	0.7	0.8	1.0	1.1	1.3
SS A2	72	90	96	99	99	100	100	100	41	62	73	83	88	93	96	98	0.5	0.6	0.8	1.0	-	0.6	0.7	0.9	1.1	1.2
SS A3	60	80	89	95	97	99	100	100	32	49	59	69	76	83	87	92	0.5	0.6	0.8	1.0	1.2	0.7	0.8	1.0	1.2	1.3
SS A4	71	90	96	99	99	100	100	100	41	61	72	82	87	93	95	98	0.5	0.6	0.8	1.0	1.2	0.6	0.7	0.9	1.1	1.3
SS A5	59	79	87	94	97	99	99	100	33	48	58	68	74	82	86	91	0.5	0.7	0.8	1.0	1.2	0.7	0.8	1.0	1.2	1.3
SS A6	72	90	96	99	99	100	100	100	41	62	73	83	88	93	96	98	0.5	0.6	0.8	1.0	-	0.6	0.7	0.9	1.1	1.2
SS A7	54	73	83	90	94	97	99	99	29	43	53	62	68	76	81	86	0.6	0.7	0.8	1.0	1.2	0.8	0.9	1.1	1.2	1.4
SS A8	71	89	95	98	99	100	100	100	40	61	72	81	87	92	95	97	0.5	0.6	0.8	1.0	1.2	0.6	0.7	0.9	1.1	1.3
SS B1	57	78	87	94	96	99	99	100	29	46	56	67	73	81	86	90	0.5	0.7	0.8	1.0	1.2	0.7	0.8	1.0	1.2	1.3
SS B2	66	86	93	97	99	100	100	100	36	56	67	78	84	90	93	96	0.5	0.6	0.8	1.0	1.2	0.7	0.8	0.9	1.1	1.3
SS B3	55	76	86	93	96	98	99	100	28	44	54	64	71	79	84	89	0.6	0.7	0.8	1.0	1.2	0.8	0.9	1.0	1.2	1.4
SS B4	65	86	93	97	99	100	100	100	35	55	66	77	83	90	93	96	0.5	0.6	0.8	1.0	1.2	0.7	0.8	0.9	1.1	1.3
SS B5	51	71	81	89	93	97	98	99	27	41	50	60	66	75	80	85	0.6	0.7	0.9	1.0	1.2	0.8	0.9	1.1	1.2	1.4
SS B6	66	86	93	97	99	100	100	100	35	55	67	77	83	90	93	96	0.5	0.6	0.8	1.0	1.2	0.7	0.8	0.9	1.1	1.3
SS B7	47	67	77	86	91	95	97	99	25	37	46	55	62	70	75	81	0.6	0.7	0.9	1.1	1.2	0.9	1.0	1.1	1.3	1.4
SS B8	64	85	93	97	99	100	100	100	34	54	65	76	82	89	92	95	0.5	0.6	0.8	1.0	1.2	0.7	0.8	0.9	1.1	1.3
SS C1	62	83	92	97	98	100	100	100	30	49	61	72	78	86	90	94	0.5	0.6	0.8	1.0	1.2	0.7	0.8	1.0	1.1	1.3
SS C2	68	88	95	98	99	100	100	100	34	56	68	79	85	91	94	97	0.5	0.6	0.8	1.0	1.2	0.6	0.7	0.9	1.1	1.3
SS C3	52	72	83	91	94	98	99	100	24	39	49	60	67	76	81	86	0.6	0.7	0.8	1.0	1.2	0.8	0.9	1.0	1.2	1.4
SS C4	62	83	92	96	98	100	100	100	30	50	62	73	79	87	91	95	0.5	0.6	0.8	1.0	1.2	0.7	0.8	0.9	1.1	1.3
SS D1	72	90	96	99	99	100	100	100	41	61	72	82	87	93	95	97	0.5	0.6	0.8	1.0	-	0.6	0.7	0.9	1.1	1.3
SS D2	80	95	98	100	100	100	100	100	47	70	81	89	93	97	98	99	0.5	0.6	0.8	0.9	-	0.6	0.7	0.9	1.0	1.2
SS D3	66	86	93	97	99	100	100	100	37	55	66	76	82	89	92	95	0.5	0.6	0.8	1.0	1.2	0.7	0.8	0.9	1.1	1.3
SS D4	78	94	98	99	100	100	100	100	46	68	79	88	92	96	98	99	0.5	0.6	0.8	1.0	-	0.6	0.7	0.9	1.0	1.2
SS E1	68	87	94	98	99	100	100	100	36	56	67	77	83	90	93	96	0.5	0.6	0.8	1.0	1.2	0.7	0.8	0.9	1.1	1.3
SS E2	75	93	97	99	100	100	100	100	42	64	76	85	90	95	97	99	0.5	0.6	0.8	1.0	-	0.6	0.7	0.9	1.1	1.2
SS E3	58	78	87	94	97	99	99	100	30	47	57	67	74	82	86	91	0.5	0.7	0.8	1.0	1.2	0.7	0.8	1.0	1.2	1.3
SS E4	71	90	96	99	99	100	100	100	39	60	71	81	87	93	95	98	0.5	0.6	0.8	1.0	1.2	0.6	0.7	0.9	1.1	1.3

PRESCOTT, ARIZONA

TYPE	SOLAR SAVINGS FRACTION (%) — BASE TEMP: 55 F (2325 DD)								65 F (4458 DD)								CONSERVATION FACTOR — 55 F					65 F				
LCR	100	70	50	40	30	25	20	15	100	70	50	40	30	25	20	15	100	50	30	20	15	100	50	30	20	15
WW A1	34	42	50	57	66	71	78	85	23	29	35	40	47	52	58	67	1.0	1.2	1.3	1.5	1.6	1.3	1.4	1.6	1.7	1.8
WW A2	42	52	63	70	79	84	90	95	27	35	44	51	60	66	73	82	0.9	1.0	1.2	1.3	1.5	1.1	1.2	1.3	1.5	1.6
WW A3	44	56	67	75	83	88	93	97	28	38	48	55	65	71	78	86	0.8	1.0	1.1	1.3	1.5	1.0	1.1	1.3	1.4	1.6
WW A4	46	58	70	77	85	90	94	97	29	39	50	57	67	73	80	88	0.8	1.0	1.1	1.3	1.4	1.0	1.1	1.3	1.4	1.6
WW A5	48	60	72	79	87	91	95	98	30	41	52	60	70	76	83	90	0.8	0.9	1.1	1.3	1.4	0.9	1.1	1.2	1.4	1.5
WW A6	49	62	73	81	88	92	96	98	31	41	53	61	71	77	84	91	0.8	0.9	1.1	1.3	1.4	0.9	1.1	1.2	1.4	1.5
WW B1	36	45	55	62	72	77	83	90	22	29	36	42	50	56	63	72	0.9	1.1	1.2	1.4	1.5	1.2	1.3	1.5	1.6	1.7
WW B2	46	58	70	78	86	90	94	98	30	40	51	59	69	75	82	89	0.8	0.9	1.1	1.3	1.4	0.9	1.1	1.2	1.4	1.5
WW B3	54	67	78	85	91	94	97	99	36	48	59	67	77	83	89	94	0.7	0.9	1.1	1.3	1.4	0.9	1.0	1.2	1.3	1.5
WW B4	54	67	79	85	92	95	98	99	36	48	60	69	79	84	90	95	0.7	0.9	1.1	1.2	1.4	0.8	1.0	1.2	1.3	1.5
WW B5	52	65	77	84	91	94	97	99	34	46	59	67	78	83	89	95	0.7	0.9	1.1	1.2	1.4	0.9	1.0	1.2	1.3	1.5
WW C1	56	68	80	86	92	95	98	99	38	50	61	69	79	84	90	95	0.7	0.9	1.1	1.2	1.4	0.9	1.0	1.2	1.3	1.5
WW C2	53	66	78	84	91	94	97	99	35	47	59	67	77	83	89	94	0.7	0.9	1.1	1.2	1.4	0.9	1.0	1.2	1.3	1.5
WW C3	60	73	84	90	95	97	99	100	42	55	68	76	85	90	94	98	0.7	0.9	1.0	1.2	1.4	0.8	0.9	1.1	1.3	1.4
WW C4	56	69	81	87	94	96	98	100	38	51	64	72	82	87	92	97	0.7	0.9	1.1	1.2	1.4	0.8	1.0	1.1	1.3	1.4
TW A1	33	41	49	56	64	70	77	84	23	28	34	39	46	51	57	66	1.0	1.2	1.3	1.5	1.6	1.3	1.4	1.6	1.7	1.8
TW A2	38	47	58	65	74	80	86	92	25	32	40	46	55	60	68	77	0.9	1.1	1.2	1.4	1.5	1.1	1.3	1.4	1.6	1.7
TW A3	39	50	61	68	78	83	88	94	25	33	42	49	58	64	72	80	0.9	1.0	1.2	1.3	1.5	1.1	1.2	1.3	1.5	1.6
TW A4	39	50	61	69	78	84	89	94	25	33	42	49	59	65	73	82	0.9	1.0	1.2	1.3	1.5	1.0	1.2	1.3	1.5	1.6
TW B1	33	41	50	57	66	71	78	86	22	28	34	40	47	52	59	68	1.0	1.1	1.3	1.5	1.6	1.3	1.4	1.5	1.7	1.8
TW B2	36	46	56	63	73	78	85	91	23	30	38	45	53	59	67	76	0.9	1.1	1.2	1.4	1.5	1.1	1.3	1.4	1.6	1.7
TW B3	36	46	56	64	73	79	85	92	23	30	39	45	54	60	68	77	0.9	1.1	1.2	1.4	1.5	1.1	1.2	1.4	1.5	1.7
TW B4	33	43	54	61	71	76	83	90	21	28	36	43	51	57	65	74	0.9	1.1	1.2	1.4	1.5	1.2	1.3	1.4	1.6	1.7
TW C1	32	40	49	56	65	71	78	86	21	27	34	39	47	52	59	68	1.0	1.1	1.3	1.4	1.6	1.3	1.4	1.5	1.7	1.8
TW C2	32	41	50	57	67	73	80	87	21	27	34	40	48	54	61	70	1.0	1.1	1.3	1.4	1.6	1.2	1.3	1.5	1.6	1.8
TW C3	30	38	48	55	64	70	77	85	20	25	32	38	46	51	59	68	1.0	1.1	1.3	1.5	1.6	1.3	1.4	1.5	1.7	1.8
TW C4	27	34	43	49	58	64	71	80	18	23	29	34	41	46	53	62	1.1	1.2	1.4	1.5	1.7	1.4	1.5	1.6	1.7	1.9
TW D1	30	39	48	55	65	71	77	85	19	24	31	36	44	49	56	65	1.0	1.1	1.3	1.5	1.6	1.3	1.4	1.5	1.7	1.8
TW D2	42	53	65	72	81	86	91	96	27	36	46	53	63	69	77	85	0.8	1.0	1.1	1.3	1.5	1.0	1.1	1.3	1.4	1.6
TW D3	46	58	70	77	85	90	94	97	30	40	51	59	69	75	82	89	0.8	1.0	1.1	1.3	1.4	0.9	1.1	1.2	1.4	1.5
TW D4	48	61	73	80	88	92	96	98	32	43	54	62	73	79	85	92	0.8	0.9	1.1	1.3	1.4	0.9	1.0	1.2	1.4	1.5
TW D5	47	60	72	79	88	92	95	98	31	42	54	62	72	78	85	92	0.8	0.9	1.1	1.3	1.4	0.9	1.0	1.2	1.4	1.5
TW E1	51	63	74	81	89	92	96	98	34	45	56	64	74	79	86	92	0.8	0.9	1.1	1.3	1.4	0.9	1.0	1.2	1.4	1.5
TW E2	50	62	74	81	89	92	96	98	33	44	55	63	73	79	86	92	0.8	0.9	1.1	1.3	1.4	0.9	1.0	1.2	1.4	1.5
TW E3	57	71	82	88	94	96	98	99	40	52	65	73	82	87	92	97	0.7	0.9	1.1	1.2	1.4	0.8	1.0	1.1	1.3	1.4
TW E4	54	67	79	85	92	95	98	99	37	49	61	70	79	85	90	96	0.7	0.9	1.1	1.2	1.4	0.8	1.0	1.1	1.3	1.5
TW F1	32	40	49	56	65	70	77	85	21	27	33	38	46	51	57	66	1.0	1.2	1.3	1.5	1.6	1.3	1.4	1.6	1.7	1.8
TW F2	35	45	55	63	72	78	84	91	22	29	37	43	52	58	65	75	0.9	1.1	1.2	1.4	1.5	1.1	1.3	1.4	1.6	1.7
TW F3	35	46	57	64	74	80	86	92	22	29	38	45	54	60	68	77	0.9	1.0	1.2	1.4	1.5	1.1	1.2	1.4	1.5	1.7
TW F4	33	44	55	63	73	79	85	92	20	27	36	43	52	59	67	76	0.9	1.1	1.2	1.4	1.5	1.1	1.2	1.4	1.5	1.7
TW G1	30	39	48	55	64	70	77	85	20	25	32	37	45	50	57	66	1.0	1.2	1.3	1.5	1.6	1.3	1.4	1.5	1.7	1.8
TW G2	31	41	51	59	68	74	81	89	19	26	34	40	48	54	62	71	1.0	1.1	1.2	1.4	1.6	1.2	1.3	1.4	1.6	1.7
TW G3	29	39	49	57	67	73	80	88	17	24	32	38	47	53	61	70	1.0	1.1	1.3	1.4	1.6	1.2	1.3	1.5	1.6	1.7
TW G4	25	34	44	51	61	67	75	83	14	20	27	33	41	47	55	65	1.0	1.2	1.3	1.5	1.6	1.3	1.4	1.5	1.7	1.8
TW H1	27	35	44	51	61	67	74	83	16	22	29	34	42	47	54	64	1.0	1.2	1.3	1.5	1.6	1.3	1.4	1.6	1.7	1.8
TW H2	24	33	42	49	59	65	73	82	14	19	26	32	40	45	53	62	1.1	1.2	1.3	1.5	1.6	1.3	1.4	1.6	1.7	1.8
TW H3	21	28	37	44	53	60	67	77	11	17	23	28	35	41	48	57	1.1	1.2	1.4	1.5	1.7	1.4	1.5	1.6	1.8	1.9
TW H4	16	22	30	35	44	50	57	67	8	13	18	22	28	33	39	48	1.2	1.4	1.5	1.6	1.8	1.6	1.7	1.8	1.9	2.0
TW I1	27	36	45	52	61	67	74	83	16	21	28	33	40	45	52	61	1.0	1.2	1.3	1.5	1.6	1.3	1.5	1.6	1.7	1.9
TW I2	38	50	61	69	78	84	89	95	24	33	42	50	59	66	73	82	0.9	1.0	1.2	1.3	1.5	1.0	1.2	1.3	1.5	1.6
TW I3	42	53	65	73	82	87	92	96	26	36	46	54	64	70	78	86	0.8	1.0	1.1	1.3	1.5	1.0	1.1	1.3	1.4	1.6
TW I4	45	57	70	77	86	90	94	98	29	39	51	59	69	76	83	90	0.8	0.9	1.1	1.3	1.4	0.9	1.1	1.2	1.4	1.5
TW I5	45	58	70	78	86	91	95	98	29	40	51	60	70	76	83	91	0.8	0.9	1.1	1.3	1.4	0.9	1.1	1.2	1.4	1.5
TW J1	48	60	72	79	87	91	95	98	31	42	53	61	71	77	84	91	0.8	0.9	1.1	1.3	1.4	0.9	1.1	1.2	1.4	1.5
TW J2	46	59	71	78	87	91	95	98	30	41	52	60	70	77	83	91	0.8	0.9	1.1	1.3	1.4	0.9	1.1	1.2	1.4	1.5
TW J3	54	67	79	86	92	95	98	99	36	49	61	70	80	85	91	96	0.7	0.9	1.1	1.2	1.4	0.8	1.0	1.1	1.3	1.5
TW J4	51	64	76	83	91	94	97	99	33	46	58	67	77	83	89	94	0.7	0.9	1.1	1.3	1.4	0.9	1.0	1.2	1.3	1.5
DG A1	34	44	53	59	67	72	77	82	20	27	34	39	45	49	54	60	0.9	1.1	1.4	1.6	1.8	1.2	1.4	1.7	2.0	2.3
DG A2	36	46	57	64	73	78	84	89	22	30	39	45	53	59	65	73	0.9	1.1	1.2	1.4	1.6	1.1	1.2	1.4	1.6	1.8
DG A3	41	54	66	74	82	86	90	93	26	36	47	55	65	71	78	85	0.8	1.0	1.2	1.4	1.7	1.0	1.1	1.3	1.5	1.7
DG B1	36	48	60	67	75	80	84	89	20	27	36	43	52	57	63	70	0.9	1.0	1.2	1.5	1.7	1.1	1.2	1.4	1.7	2.0
DG B2	37	50	63	71	80	85	90	94	22	31	41	49	59	66	73	80	0.8	1.0	1.2	1.4	1.5	1.1	1.2	1.3	1.5	1.7
DG B3	42	56	70	78	86	90	94	96	27	37	49	58	69	76	83	90	0.8	0.9	1.1	1.3	1.6	1.0	1.1	1.2	1.4	1.6
DG C1	42	55	69	76	84	88	91	94	24	33	43	51	62	68	74	80	0.8	0.9	1.2	1.4	1.7	1.0	1.1	1.3	1.6	1.8
DG C2	42	56	70	78	86	90	94	97	26	35	47	56	67	73	80	87	0.8	0.9	1.1	1.3	1.5	1.0	1.1	1.2	1.4	1.6
DG C3	47	63	77	84	91	94	96	98	30	41	55	65	77	83	89	94	0.7	0.9	1.1	1.3	1.7	0.9	1.0	1.1	1.3	1.5
SS A1	44	54	64	71	80	84	89	94	32	40	49	55	64	70	76	84	0.9	1.0	1.2	1.4	1.5	1.0	1.2	1.3	1.5	1.6
SS A2	54	65	76	82	89	93	96	98	40	50	61	68	77	82	87	93	0.8	0.9	1.1	1.3	1.4	0.9	1.0	1.2	1.4	1.5
SS A3	42	51	61	68	76	81	87	92	30	38	46	52	60	65	72	80	0.9	1.1	1.2	1.4	1.5	1.1	1.2	1.4	1.5	1.7
SS A4	53	64	75	81	88	92	95	98	40	50	60	67	76	81	86	92	0.8	0.9	1.1	1.3	1.4	0.9	1.0	1.2	1.4	1.5
SS A5	41	50	60	66	75	80	85	91	30	37	45	51	59	64	71	79	0.9	1.1	1.2	1.4	1.5	1.1	1.2	1.4	1.6	1.7
SS A6	54	65	76	82	89	92	96	98	40	50	61	68	77	82	87	93	0.8	0.9	1.1	1.3	1.4	0.9	1.0	1.2	1.4	1.5
SS A7	37	45	54	60	68	73	80	86	27	33	40	45	52	57	63	71	1.0	1.1	1.3	1.5	1.6	1.2	1.3	1.5	1.7	1.8
SS A8	52	64	74	81	88	91	95	98	39	49	59	66	75	80	86	92	0.8	0.9	1.1	1.3	1.4	0.9	1.1	1.2	1.4	1.5
SS B1	38	48	58	65	74	79	85	91	27	34	43	49	57	63	69	78	0.9	1.1	1.2	1.4	1.5	1.1	1.2	1.4	1.6	1.7
SS B2	48	59	70	77	85	89	93	97	34	44	54	62	71	77	83	90	0.8	1.0	1.1	1.3	1.5	0.9	1.1	1.2	1.4	1.5
SS B3	37	46	56	63	71	77	83	89	26	33	40	46	54	60	67	75	0.9	1.1	1.2	1.4	1.6	1.1	1.3	1.4	1.6	1.7
SS B4	47	58	69	76	84	88	93	97	34	43	54	61	70	75	82	89	0.8	1.0	1.1	1.3	1.5	0.9	1.1	1.3	1.4	1.6
SS B5	34	42	51	58	66	72	78	86	24	30	37	42	50	55	61	70	1.0	1.1	1.3	1.5	1.6	1.2	1.4	1.5	1.7	1.8
SS B6	47	58	69	76	84	89	93	97	34	43	54	61	70	76	82	89	0.8	1.0	1.1	1.3	1.5	0.9	1.1	1.2	1.4	1.6
SS B7	31	39	47	53	61	67	73	81	22	28	34	38	45	50	56	64	1.1	1.2	1.4	1.5	1.7	1.3	1.4	1.6	1.8	1.9
SS B8	46	57	68	75	83	87	92	96	33	42	52	59	68	74	80	88	0.8	1.0	1.1	1.3	1.5	1.0	1.1	1.3	1.4	1.6
SS C1	40	51	62	70	79	84	89	94	25	33	42	49	59	65	72	81	0.9	1.0	1.2	1.3	1.5	1.1	1.2	1.3	1.5	1.6
SS C2	46	58	70	77	86	90	94	97	30	40	51	59	69	75	82	89	0.8	0.9	1.1	1.3	1.4	0.9	1.1	1.2	1.4	1.5
SS C3	32	41	50	57	67	72	79	87	21	27	34	39	47	53	60	69	1.0	1.1	1.3	1.4	1.6	1.2	1.4	1.5	1.6	1.8
SS C4	41	52	63	71	80	85	90	95	27	35	45	52	62	68	76	84	0.9	1.0	1.2	1.3	1.5	1.0	1.2	1.3	1.5	1.6
SS D1	53	64	74	81	88	92	95	98	38	48	58	65	74	79	85	91	0.8	0.9	1.1	1.3	1.4	0.9	1.1	1.2	1.4	1.5
SS D2	62	74	84	89	94	96	98	99	46	58	69	76	85	89	93	97	0.7	0.9	1.1	1.2	1.4	0.8	1.0	1.1	1.3	1.5
SS D3	47	58	68	75	83	87	91	96	35	43	52	59	67	73	79	86	0.8	1.0	1.2	1.3	1.5	1.0	1.1	1.3	1.5	1.6
SS D4	60	72	82	87	93	96	98	99	45	56	67	74	83	87	92	96	0.7	0.9	1.1	1.2	1.4	0.8	1.0	1.1	1.3	1.5
SS E1	47	58	69	76	84	88	92	96	33	42	52	59	68	73	80	87	0.8	1.0	1.1	1.3	1.5	1.0	1.1	1.3	1.4	1.6
SS E2	56	68	79	85	91	94	97	99	40	51	63	70	79	84	89	95	0.7	0.9	1.1	1.3	1.4	0.9	1.0	1.2	1.3	1.5
SS E3	39	49	58	65	74	79	85	91	28	35	43	49	57	63	69	78	0.9	1.1	1.2	1.4	1.5	1.1	1.2	1.4	1.6	1.7
SS E4	51	63	74	81	88	92	95	98	37	47	58	65	74	80	86	92	0.8	0.9	1.1	1.3	1.4	0.9	1.0	1.2	1.4	1.5

TUCSON, ARIZONA

SOLAR SAVINGS FRACTION (%) / CONSERVATION FACTOR

TYPE	BASE TEMP: 55 F (541 DD)								65 F (1758 DD)								CONSERVATION FACTOR 55 F					65 F				
LCR	200	100	70	50	40	30	25	20	200	100	70	50	40	30	25	20	200	100	50	30	20	200	100	50	30	20
WW A1	50	70	81	89	93	97	98	99	26	39	48	58	65	73	79	85	0.6	0.7	0.9	1.0	1.2	0.9	0.9	1.1	1.2	1.4
WW A2	61	82	91	96	98	99	100	100	30	48	60	71	78	86	90	94	0.5	0.6	0.8	1.0	1.2	0.7	0.8	1.0	1.1	1.3
WW A3	65	86	93	97	99	100	100	100	31	52	64	75	82	89	93	96	0.5	0.6	0.8	1.0	1.2	0.7	0.8	0.9	1.1	1.3
WW A4	67	88	94	98	99	100	100	100	32	54	66	78	84	91	94	97	0.5	0.6	0.8	1.0	1.2	0.7	0.7	0.9	1.1	1.3
WW A5	69	89	95	99	99	100	100	100	33	56	69	80	86	92	95	98	0.5	0.6	0.8	1.0	-	0.6	0.7	0.9	1.1	1.2
WW A6	70	90	96	99	100	100	100	100	34	57	70	81	87	93	96	98	0.5	0.6	0.8	1.0	-	0.6	0.7	0.9	1.1	1.2
WW B1	56	77	87	93	96	99	99	100	26	42	53	64	71	80	85	90	0.5	0.7	0.8	1.0	1.2	0.8	0.9	1.0	1.2	1.3
WW B2	66	87	94	98	99	100	100	100	32	54	67	78	84	91	94	97	0.5	0.6	0.8	1.0	1.2	0.6	0.7	0.9	1.1	1.2
WW B3	74	92	97	99	100	100	100	100	38	62	75	85	90	95	97	99	0.5	0.6	0.8	1.0	-	0.6	0.7	0.9	1.0	1.2
WW B4	73	92	97	99	100	100	100	100	37	62	75	85	91	96	98	99	0.5	0.6	0.8	1.0	-	0.6	0.7	0.9	1.0	1.2
WW B5	71	91	97	99	100	100	100	100	35	60	73	84	90	95	97	99	0.5	0.6	0.8	1.0	-	0.6	0.7	0.9	1.0	1.2
WW C1	75	93	97	99	100	100	100	100	39	64	76	86	91	96	98	99	0.5	0.6	0.8	1.0	-	0.6	0.7	0.9	1.0	1.2
WW C2	73	92	97	99	100	100	100	100	37	61	74	84	90	95	97	99	0.5	0.6	0.8	1.0	-	0.6	0.7	0.9	1.0	1.2
WW C3	78	95	98	100	100	100	100	100	41	68	80	90	94	98	99	100	0.4	0.6	0.8	0.9	-	0.6	0.7	0.8	1.0	1.2
WW C4	74	93	98	99	100	100	100	100	37	64	77	87	92	97	98	99	0.5	0.6	0.8	1.0	-	0.6	0.7	0.8	1.0	1.2
TW A1	49	69	79	88	92	96	98	99	25	38	47	56	63	72	78	84	0.6	0.7	0.9	1.0	1.2	0.9	0.9	1.1	1.2	1.4
TW A2	56	77	87	94	96	99	99	100	27	44	55	65	73	81	86	91	0.5	0.7	0.8	1.0	1.2	0.8	0.8	1.0	1.2	1.3
TW A3	58	80	89	95	98	99	100	100	28	46	58	69	76	84	89	93	0.5	0.6	0.8	1.0	1.2	0.7	0.8	1.0	1.1	1.3
TW A4	59	81	90	95	98	99	100	100	27	46	58	69	77	85	89	94	0.5	0.6	0.8	1.0	1.2	0.7	0.8	1.0	1.1	1.3
TW B1	49	70	80	88	93	97	98	99	25	38	47	57	64	73	79	85	0.6	0.7	0.9	1.0	1.2	0.8	0.9	1.1	1.2	1.4
TW B2	54	76	86	93	96	98	99	100	26	42	53	64	71	80	85	90	0.5	0.7	0.8	1.0	1.2	0.8	0.9	1.0	1.2	1.3
TW B3	54	76	86	93	96	99	99	100	25	42	53	64	72	81	86	91	0.5	0.7	0.8	1.0	1.2	0.8	0.8	1.0	1.2	1.3
TW B4	51	73	84	91	95	98	99	100	23	39	50	61	69	78	83	89	0.6	0.7	0.8	1.0	1.2	0.8	0.9	1.0	1.2	1.3
TW C1	48	69	80	88	92	96	98	99	23	37	46	57	64	73	79	85	0.6	0.7	0.9	1.0	1.2	0.8	0.9	1.1	1.2	1.4
TW C2	48	70	81	89	93	97	98	99	23	37	47	58	65	75	80	86	0.6	0.7	0.9	1.0	1.2	0.8	0.9	1.0	1.2	1.4
TW C3	45	67	78	87	92	96	98	99	21	35	45	55	62	72	77	84	0.6	0.7	0.9	1.0	1.2	0.9	0.9	1.1	1.2	1.4
TW C4	40	61	72	82	87	93	96	98	20	31	40	49	56	65	71	78	0.6	0.7	0.9	1.1	1.2	0.9	1.0	1.1	1.3	1.4
TW D1	48	70	81	89	93	97	98	99	22	36	46	56	64	73	79	85	0.6	0.7	0.9	1.0	1.2	0.8	0.9	1.1	1.2	1.4
TW D2	61	83	91	96	98	100	100	100	29	49	61	73	80	87	91	95	0.5	0.6	0.8	1.0	1.2	0.7	0.8	0.9	1.1	1.3
TW D3	66	87	94	98	99	100	100	100	32	54	66	77	84	91	94	97	0.5	0.6	0.8	1.0	1.2	0.7	0.8	0.9	1.1	1.3
TW D4	68	88	95	98	99	100	100	100	33	56	69	80	86	93	96	98	0.5	0.6	0.8	1.0	1.2	0.6	0.7	0.9	1.1	1.2
TW D5	66	87	94	98	99	100	100	100	32	55	68	79	86	92	95	98	0.5	0.6	0.8	1.0	1.2	0.6	0.7	0.9	1.1	1.2
TW E1	70	90	96	99	99	100	100	100	36	58	71	81	87	93	96	98	0.5	0.6	0.8	1.0	-	0.6	0.7	0.9	1.1	1.2
TW E2	69	89	95	98	99	100	100	100	35	57	70	81	87	93	96	98	0.5	0.6	0.8	1.0	1.2	0.6	0.7	0.9	1.1	1.2
TW E3	76	93	98	99	100	100	100	100	40	65	78	88	92	97	98	99	0.5	0.6	0.8	1.0	-	0.6	0.7	0.8	1.0	1.2
TW E4	73	92	97	99	100	100	100	100	37	62	75	85	91	96	98	99	0.5	0.6	0.8	1.0	-	0.6	0.7	0.9	1.0	1.2
TW F1	48	69	80	88	92	96	98	99	24	37	46	56	63	72	78	84	0.6	0.7	0.9	1.0	1.2	0.9	0.9	1.1	1.2	1.4
TW F2	53	76	86	93	96	98	99	100	25	41	52	63	71	80	85	90	0.5	0.7	0.8	1.0	1.2	0.8	0.9	1.0	1.2	1.3
TW F3	54	77	87	94	97	99	99	100	24	42	53	65	73	81	86	91	0.5	0.6	0.8	1.0	1.2	0.7	0.8	1.0	1.1	1.3
TW F4	52	76	86	93	96	99	99	100	22	40	51	63	71	80	85	91	0.5	0.7	0.8	1.0	1.2	0.8	0.8	1.0	1.2	1.3
TW G1	47	68	79	88	92	96	98	99	22	36	45	55	63	72	78	84	0.6	0.7	0.9	1.0	1.2	0.9	0.9	1.1	1.2	1.4
TW G2	49	72	83	91	94	98	99	100	21	37	48	59	67	76	82	88	0.6	0.7	0.8	1.0	1.2	0.8	0.9	1.0	1.2	1.4
TW G3	47	70	81	90	94	97	99	99	19	35	46	57	65	75	81	87	0.6	0.7	0.8	1.0	1.2	0.8	0.9	1.0	1.2	1.4
TW G4	41	64	76	86	91	95	97	99	16	30	40	51	59	69	75	82	0.6	0.7	0.9	1.0	1.2	0.9	0.9	1.1	1.2	1.4
TW H1	43	65	76	86	91	95	97	99	18	32	41	52	59	69	75	82	0.6	0.7	0.9	1.0	1.2	0.9	1.0	1.1	1.2	1.4
TW H2	40	62	74	84	89	95	97	99	16	29	39	49	57	67	73	80	0.6	0.7	0.9	1.0	1.2	0.9	1.0	1.1	1.3	1.4
TW H3	35	57	69	79	86	92	95	97	13	25	34	44	51	61	68	75	0.6	0.7	0.9	1.1	1.2	0.9	1.0	1.2	1.3	1.5
TW H4	28	47	59	70	77	85	90	94	10	20	27	35	42	51	58	65	0.7	0.8	1.0	1.1	1.3	1.1	1.1	1.3	1.4	1.5
TW I1	45	67	78	87	92	96	98	99	19	33	42	53	60	70	75	82	0.6	0.7	0.9	1.0	1.2	0.9	0.9	1.1	1.2	1.4
TW I2	58	80	90	95	98	[illegible]	100	100	26	45	57	69	77	85	90	94	0.5	0.6	0.8	1.0	1.2	0.7	0.8	1.0	1.1	1.3
TW I3	61	83	92	97	98	100	100	100	28	49	61	73	80	88	92	96	0.5	0.6	0.8	1.0	1.2	0.7	0.8	0.9	1.1	1.3
TW I4	64	86	94	98	99	100	100	100	30	52	65	77	84	91	94	97	0.5	0.6	0.8	1.0	1.2	0.6	0.7	0.9	1.1	1.3
TW I5	64	86	94	98	99	100	100	100	29	52	65	77	84	91	94	97	0.5	0.6	0.8	1.0	1.2	0.6	0.7	0.9	1.1	1.2
TW J1	67	88	95	98	99	100	100	100	33	55	68	79	86	92	95	98	0.5	0.6	0.8	1.0	1.2	0.6	0.7	0.9	1.1	1.2
TW J2	66	87	94	98	99	100	100	100	31	54	67	78	85	92	95	97	0.5	0.6	0.8	1.0	1.2	0.6	0.7	0.9	1.1	1.2
TW J3	73	92	97	99	100	100	100	100	36	62	75	85	91	96	98	99	0.5	0.6	0.8	1.0	-	0.6	0.7	0.9	1.0	1.2
TW J4	70	90	96	99	100	100	100	100	33	58	72	83	[illegible]	94	97	99	0.5	0.6	0.8	1.0	-	0.6	0.7	0.9	1.0	1.2
DG A1	56	78	87	93	96	[illegible]	99	100	24	41	52	62	68	76	80	84	0.5	0.7	0.8	1.0	1.1	0.8	0.9	1.1	1.3	1.5
DG A2	56	79	88	94	97	99	100	100	24	43	54	65	72	81	85	89	0.5	0.6	0.8	1.0	1.1	0.7	0.8	1.0	1.2	1.4
DG A3	61	83	91	95	96	97	97	97	27	49	62	74	80	87	90	93	0.5	0.6	0.9	1.8	-	0.7	0.8	0.9	1.1	1.4
DG B1	61	84	92	96	98	99	99	99	24	44	58	69	76	83	87	91	0.5	0.6	0.8	1.1	2.3	0.7	0.8	1.0	1.2	1.4
DG B2	61	85	93	97	99	100	100	100	25	45	59	72	79	87	91	94	0.5	0.6	0.8	1.0	1.2	0.7	0.8	0.9	1.1	1.3
DG B3	64	87	94	97	98	98	98	98	27	50	65	78	85	91	94	96	0.5	0.6	0.9	1.9	-	0.7	0.7	0.9	1.1	1.4
DG C1	69	90	96	98	98	98	98	98	28	51	66	78	84	90	93	95	0.5	0.6	0.9	3.1	-	0.7	0.7	0.9	1.1	1.4
DG C2	67	90	96	99	99	100	100	100	28	51	66	79	85	92	95	97	0.5	0.6	0.8	1.0	1.8	0.7	0.7	0.9	1.1	1.3
DG C3	70	92	96	98	98	98	98	98	31	56	72	84	90	95	96	98	0.5	0.6	0.9	-	-	0.6	0.7	0.9	1.1	1.4
SS A1	61	82	90	96	98	99	100	100	33	51	62	72	79	86	90	94	0.5	0.6	0.8	1.0	1.2	0.7	0.8	1.0	1.1	1.3
SS A2	71	90	96	99	99	100	100	100	40	61	73	83	88	94	96	98	0.5	0.6	0.8	1.0	1.2	0.6	0.7	0.9	1.1	1.2
SS A3	59	79	88	94	97	99	99	100	32	49	59	69	76	84	88	92	0.5	0.6	0.8	1.0	1.2	0.7	0.8	1.0	1.1	1.3
SS A4	71	89	95	98	99	100	100	100	40	61	72	82	88	93	96	98	0.5	0.6	0.8	1.0	1.2	0.6	0.7	0.9	1.1	1.2
SS A5	58	78	87	93	96	99	99	100	32	48	58	68	74	82	87	91	0.5	0.7	0.8	1.0	1.2	0.7	0.8	1.0	1.2	1.3
SS A6	71	90	96	99	99	100	100	100	40	61	73	83	88	94	96	98	0.5	0.6	0.8	1.0	1.2	0.6	0.7	0.9	1.1	1.2
SS A7	52	72	82	90	94	97	98	99	29	43	52	62	68	76	81	87	0.6	0.7	0.9	1.0	1.2	0.8	0.9	1.1	1.2	1.4
SS A8	70	89	95	98	99	100	100	100	40	60	72	82	87	93	95	98	0.5	0.6	0.8	1.0	1.2	0.6	0.7	0.9	1.1	1.3
SS B1	55	77	86	93	96	98	99	100	29	45	55	66	73	81	86	91	0.5	0.7	0.8	1.0	1.2	0.7	0.8	1.0	1.2	1.3
SS B2	65	85	93	97	99	100	100	100	35	55	67	77	84	90	94	97	0.5	0.6	0.8	1.0	1.2	0.7	0.8	0.9	1.1	1.3
SS B3	54	75	85	92	95	98	99	100	27	43	53	64	71	79	84	89	0.6	0.7	0.8	1.0	1.2	0.8	0.9	1.0	1.2	1.4
SS B4	64	85	93	97	99	100	100	100	34	54	66	77	83	90	93	96	0.5	0.6	0.8	1.0	1.2	0.7	0.8	0.9	1.1	1.3
SS B5	49	70	80	88	93	96	98	99	26	40	49	59	66	74	79	85	0.6	0.7	0.9	1.0	1.2	0.8	0.9	1.1	1.2	1.4
SS B6	65	85	93	97	99	100	100	100	34	54	66	77	83	90	93	96	0.5	0.6	0.8	1.0	1.2	0.7	0.8	0.9	1.1	1.3
SS B7	46	66	76	85	90	95	97	99	24	36	45	54	61	70	75	81	0.6	0.7	0.9	1.1	1.2	0.9	1.0	1.1	1.3	1.4
SS B8	63	84	92	97	98	100	100	100	33	53	65	75	82	89	93	96	0.5	0.6	0.8	1.0	1.2	0.7	0.8	0.9	1.1	1.3
SS C1	60	82	91	96	98	99	100	100	28	47	59	70	77	86	90	94	0.5	0.6	0.8	1.0	1.2	0.7	0.8	1.0	1.1	1.3
SS C2	66	87	94	98	99	100	100	100	32	54	66	78	84	91	94	97	0.5	0.6	0.8	1.0	1.2	0.7	0.7	0.9	1.1	1.3
SS C3	49	71	81	89	93	97	98	99	23	37	47	58	65	74	80	86	0.6	0.7	0.9	1.0	1.2	0.8	0.9	1.1	1.2	1.4
SS C4	60	82	90	96	98	99	100	100	28	48	60	71	78	86	91	95	0.5	0.6	0.8	1.0	1.2	0.7	0.8	0.9	1.1	1.3
SS D1	71	90	96	99	99	100	100	100	40	61	72	82	87	93	96	98	0.5	0.6	0.8	1.0	1.2	0.6	0.7	0.9	1.1	1.2
SS D2	79	95	98	100	100	100	100	100	46	70	81	90	94	97	99	99	0.5	0.6	0.8	0.9	-	0.6	0.7	0.8	1.0	1.2
SS D3	65	85	92	97	99	100	100	100	36	55	66	76	82	89	92	96	0.5	0.6	0.8	1.0	1.2	0.7	0.8	0.9	1.1	1.3
SS D4	77	94	98	99	100	100	100	100	45	68	79	88	92	96	98	99	0.5	0.6	0.8	1.0	-	0.6	0.7	0.9	1.0	1.2
SS E1	67	87	94	98	99	100	100	100	35	55	67	77	83	90	93	96	0.5	0.6	0.8	1.0	1.2	0.7	0.8	0.9	1.1	1.3
SS E2	74	92	97	99	100	100	100	100	41	64	76	85	91	95	97	99	0.5	0.6	0.8	1.0	-	0.6	0.7	0.9	1.0	1.2
SS E3	56	77	87	93	96	99	99	100	30	46	56	67	73	82	86	91	0.5	0.7	0.8	1.0	1.2	0.7	0.8	1.0	1.2	1.3
SS E4	70	89	95	98	99	100	100	100	38	59	71	81	87	93	96	98	0.5	0.6	0.8	1.0	1.2	0.6	0.7	0.9	1.1	1.2

WINSLOW, ARIZONA

TYPE	SOLAR SAVINGS FRACTION (%) BASE TEMP: 55 F (2693 DD)								65 F (4738 DD)								CONSERVATION FACTOR 55 F					65 F				
LCR	100	70	50	40	30	25	20	15	100	70	50	40	30	25	20	15	100	50	30	20	15	100	50	30	20	15
WW A1	30	36	44	50	57	63	69	77	22	26	32	36	43	47	53	61	1.1	1.3	1.4	1.6	1.7	1.4	1.5	1.7	1.8	1.9
WW A2	35	45	55	62	71	76	83	89	24	32	40	46	55	60	68	76	0.9	1.1	1.2	1.4	1.6	1.1	1.3	1.4	1.6	1.7
WW A3	38	48	59	66	75	81	86	92	25	34	43	50	59	65	72	81	0.9	1.0	1.2	1.4	1.5	1.0	1.2	1.3	1.5	1.6
WW A4	39	50	61	69	78	83	88	94	26	35	45	52	62	68	75	83	0.9	1.0	1.2	1.4	1.5	1.0	1.2	1.3	1.5	1.6
WW A5	40	52	63	71	80	85	90	95	27	36	47	54	64	70	78	86	0.8	1.0	1.2	1.3	1.5	1.0	1.1	1.3	1.4	1.6
WW A6	41	53	65	72	81	86	91	95	27	37	48	55	65	72	79	87	0.8	1.0	1.2	1.3	1.5	1.0	1.1	1.3	1.4	1.6
WW B1	30	38	47	53	62	68	74	82	20	26	32	37	45	50	56	65	1.0	1.2	1.3	1.5	1.7	1.3	1.4	1.6	1.7	1.9
WW B2	39	51	62	70	79	84	89	94	26	36	46	54	64	70	77	85	0.9	1.0	1.2	1.3	1.5	1.0	1.1	1.3	1.4	1.6
WW B3	46	58	70	77	85	89	94	97	32	43	54	62	72	78	84	91	0.8	1.0	1.1	1.3	1.5	0.9	1.1	1.2	1.4	1.5
WW B4	46	59	71	78	86	90	94	98	32	44	56	64	74	80	86	93	0.8	0.9	1.1	1.3	1.4	0.9	1.0	1.2	1.4	1.5
WW B5	44	57	69	77	85	90	94	97	30	42	54	63	73	79	86	92	0.8	0.9	1.1	1.3	1.4	0.9	1.0	1.2	1.4	1.5
WW C1	48	60	72	79	87	91	94	98	34	45	57	64	74	80	86	92	0.8	0.9	1.1	1.3	1.4	0.9	1.0	1.2	1.4	1.5
WW C2	45	58	70	77	85	90	94	97	32	43	54	63	73	79	85	92	0.8	0.9	1.1	1.3	1.4	0.9	1.0	1.2	1.4	1.5
WW C3	52	65	77	84	91	94	97	99	38	51	63	72	81	86	92	96	0.7	0.9	1.1	1.3	1.4	0.8	1.0	1.1	1.3	1.5
WW C4	48	61	74	81	89	92	96	98	34	47	59	68	78	84	89	95	0.8	0.9	1.1	1.3	1.4	0.8	1.0	1.2	1.3	1.5
TW A1	29	35	43	48	56	62	68	77	21	26	31	36	42	46	52	60	1.1	1.3	1.4	1.6	1.7	1.4	1.5	1.7	1.8	1.9
TW A2	32	41	50	57	66	71	78	85	22	29	36	42	50	55	62	71	1.0	1.1	1.3	1.5	1.6	1.2	1.3	1.5	1.6	1.8
TW A3	33	43	53	60	69	75	81	88	23	30	38	44	53	59	66	75	1.0	1.1	1.3	1.4	1.6	1.1	1.3	1.4	1.6	1.7
TW A4	33	43	53	61	70	76	82	89	22	30	38	45	54	60	67	76	0.9	1.1	1.2	1.4	1.6	1.1	1.2	1.4	1.5	1.7
TW B1	29	36	43	49	58	63	70	78	21	26	31	36	43	48	54	62	1.1	1.2	1.4	1.6	1.7	1.3	1.5	1.6	1.8	1.9
TW B2	31	39	48	55	64	70	77	85	21	27	35	41	49	54	61	70	1.0	1.1	1.3	1.5	1.6	1.2	1.3	1.5	1.6	1.8
TW B3	30	39	49	56	65	71	78	85	20	27	35	41	49	55	62	72	1.0	1.1	1.3	1.5	1.6	1.2	1.3	1.5	1.6	1.7
TW B4	28	37	46	53	62	68	75	83	19	26	33	39	47	53	60	69	1.0	1.2	1.3	1.5	1.6	1.2	1.3	1.5	1.6	1.8
TW C1	27	35	43	49	57	63	70	78	19	25	31	36	43	48	54	63	1.1	1.2	1.4	1.5	1.7	1.3	1.5	1.6	1.7	1.9
TW C2	27	35	44	50	59	65	72	80	19	24	31	36	44	49	56	65	1.1	1.2	1.4	1.5	1.7	1.3	1.4	1.5	1.7	1.8
TW C3	26	33	41	48	56	62	69	78	18	23	30	35	42	47	54	63	1.1	1.2	1.4	1.5	1.7	1.3	1.4	1.6	1.7	1.9
TW C4	23	30	37	43	51	56	63	72	17	21	27	31	38	42	48	57	1.2	1.3	1.5	1.6	1.8	1.4	1.6	1.7	1.8	2.0
TW D1	25	33	41	47	55	61	68	77	16	21	27	32	39	44	50	58	1.1	1.2	1.4	1.6	1.7	1.4	1.5	1.7	1.8	1.9
TW D2	36	46	57	64	74	79	85	91	25	33	42	49	58	64	72	80	0.9	1.1	1.2	1.4	1.5	1.1	1.2	1.3	1.5	1.6
TW D3	39	50	62	69	78	83	89	94	27	36	46	54	63	69	77	85	0.9	1.0	1.2	1.4	1.5	1.0	1.1	1.3	1.5	1.6
TW D4	41	53	65	73	82	86	91	96	29	39	50	58	68	74	81	89	0.8	1.0	1.1	1.3	1.5	0.9	1.1	1.2	1.4	1.5
TW D5	40	52	64	72	81	86	91	96	28	38	50	58	68	74	81	89	0.8	1.0	1.1	1.3	1.5	0.9	1.1	1.2	1.4	1.5
TW E1	43	55	66	74	82	87	91	96	31	41	51	59	69	74	81	89	0.8	1.0	1.2	1.3	1.5	0.9	1.1	1.2	1.4	1.6
TW E2	43	54	66	73	82	87	92	96	30	40	51	59	68	75	81	89	0.8	1.0	1.1	1.3	1.5	0.9	1.1	1.2	1.4	1.5
TW E3	50	63	74	81	89	92	96	98	36	48	60	68	78	83	89	94	0.8	0.9	1.1	1.3	1.4	0.9	1.0	1.2	1.3	1.5
TW E4	46	59	71	79	87	91	95	98	33	45	57	65	75	81	87	93	0.8	0.9	1.1	1.3	1.4	0.9	1.0	1.2	1.3	1.5
TW F1	28	34	42	48	56	62	69	77	20	24	30	35	41	46	52	61	1.1	1.3	1.4	1.6	1.7	1.4	1.5	1.6	1.8	1.9
TW F2	30	38	48	54	64	69	76	84	20	26	34	39	47	53	60	69	1.0	1.2	1.3	1.5	1.6	1.2	1.3	1.5	1.6	1.8
TW F3	30	39	49	56	65	71	78	86	19	26	34	40	49	55	62	72	1.0	1.1	1.3	1.4	1.6	1.2	1.3	1.4	1.6	1.7
TW F4	27	37	47	54	64	70	77	85	17	24	32	39	48	54	61	71	1.0	1.1	1.3	1.5	1.6	1.2	1.3	1.5	1.6	1.7
TW G1	26	33	41	47	56	61	68	77	18	23	29	34	41	45	52	61	1.1	1.2	1.4	1.6	1.7	1.4	1.5	1.6	1.8	1.9
TW G2	26	34	44	50	60	66	73	81	17	23	30	36	44	49	56	66	1.0	1.2	1.3	1.5	1.6	1.3	1.4	1.5	1.7	1.8
TW G3	24	33	42	49	58	64	72	80	15	21	29	34	42	48	55	65	1.1	1.2	1.3	1.5	1.6	1.3	1.4	1.5	1.7	1.8
TW G4	20	28	37	43	52	59	66	75	12	18	24	30	37	43	50	59	1.1	1.2	1.4	1.6	1.7	1.3	1.5	1.6	1.7	1.9
TW H1	22	30	38	44	53	58	66	75	14	20	26	30	38	43	49	58	1.1	1.3	1.4	1.6	1.7	1.4	1.5	1.6	1.8	1.9
TW H2	20	27	35	42	51	57	64	73	12	17	23	28	36	41	48	57	1.1	1.3	1.4	1.6	1.7	1.4	1.5	1.6	1.8	1.9
TW H3	17	23	31	37	46	51	59	68	10	15	20	25	32	36	43	52	1.2	1.3	1.5	1.6	1.8	1.5	1.6	1.7	1.9	2.0
TW H4	13	18	25	30	37	42	49	59	7	11	16	19	25	29	35	43	1.4	1.5	1.6	1.8	1.9	1.7	1.8	1.9	2.0	2.1
TW I1	22	29	37	43	52	57	64	73	14	19	24	29	35	40	46	55	1.1	1.3	1.4	1.6	1.7	1.4	1.6	1.7	1.9	2.0
TW I2	32	42	53	61	70	76	83	90	21	29	38	45	55	61	68	78	0.9	1.1	1.2	1.4	1.6	1.1	1.2	1.4	1.5	1.7
TW I3	35	46	57	65	74	80	86	92	23	32	42	49	59	65	73	82	0.9	1.0	1.2	1.4	1.5	1.0	1.2	1.3	1.5	1.6
TW I4	38	50	61	69	79	84	89	95	25	36	46	54	65	71	78	86	0.8	1.0	1.2	1.3	1.5	1.0	1.1	1.3	1.4	1.6
TW I5	38	50	62	70	79	85	90	95	26	36	47	55	66	72	79	87	0.8	1.0	1.2	1.3	1.5	1.0	1.1	1.3	1.4	1.6
TW J1	41	52	64	71	80	85	90	95	28	38	49	56	66	72	79	87	0.8	1.0	1.2	1.3	1.5	1.0	1.1	1.3	1.4	1.6
TW J2	39	51	63	71	80	85	90	95	27	37	48	55	66	72	79	87	0.8	1.0	1.2	1.3	1.5	1.0	1.1	1.3	1.4	1.6
TW J3	46	59	71	79	87	91	95	98	33	45	57	65	75	81	87	93	0.8	0.9	1.1	1.3	1.4	0.9	1.0	1.2	1.3	1.5
TW J4	43	56	69	76	85	89	94	97	30	42	54	62	73	79	85	92	0.8	0.9	1.1	1.3	1.4	0.9	1.0	1.2	1.4	1.5
DG A1	28	36	45	50	58	62	67	72	18	23	30	34	40	44	48	53	1.0	1.2	1.5	1.8	2.0	1.3	1.5	1.8	2.1	2.5
DG A2	30	39	49	56	65	70	76	82	19	27	35	41	49	54	60	67	1.0	1.1	1.3	1.5	1.7	1.2	1.3	1.5	1.7	1.9
DG A3	35	46	58	66	75	80	85	90	24	32	43	50	60	66	73	81	0.9	1.0	1.2	1.4	1.7	1.1	1.1	1.3	1.5	1.7
DG B1	29	39	49	57	65	70	75	81	18	24	32	37	45	51	56	63	1.0	1.1	1.3	1.6	1.9	1.2	1.3	1.5	1.8	2.1
DG B2	31	42	53	62	71	77	83	88	20	27	36	44	53	60	67	75	0.9	1.0	1.2	1.4	1.6	1.1	1.2	1.4	1.6	1.8
DG B3	35	48	61	69	79	84	89	94	24	33	44	53	64	71	78	86	0.9	1.0	1.2	1.4	1.6	1.0	1.1	1.2	1.4	1.6
DG C1	34	46	57	65	75	79	84	88	21	29	38	45	55	61	67	74	0.9	1.0	1.2	1.5	1.8	1.1	1.2	1.4	1.6	1.9
DG C2	35	47	60	69	78	83	88	93	23	32	42	50	61	68	75	83	0.9	1.0	1.2	1.4	1.6	1.1	1.1	1.3	1.5	1.7
DG C3	40	54	67	76	85	89	93	96	28	38	50	59	71	78	85	91	0.8	0.9	1.1	1.3	1.6	1.0	1.0	1.2	1.4	1.6
SS A1	38	47	56	63	71	76	82	88	29	36	44	50	58	63	70	78	0.9	1.1	1.3	1.4	1.6	1.1	1.2	1.4	1.6	1.7
SS A2	46	57	68	74	82	87	91	95	36	46	56	62	71	77	83	89	0.8	1.0	1.2	1.3	1.5	0.9	1.1	1.2	1.4	1.6
SS A3	35	44	52	59	67	72	78	85	27	34	41	46	54	59	65	73	1.0	1.1	1.3	1.5	1.6	1.1	1.3	1.5	1.6	1.8
SS A4	46	56	66	73	81	86	90	95	36	45	55	61	70	75	81	88	0.8	1.0	1.2	1.3	1.5	0.9	1.1	1.3	1.4	1.6
SS A5	35	43	52	58	66	71	77	84	28	34	41	46	53	58	64	72	1.0	1.2	1.3	1.5	1.6	1.2	1.3	1.5	1.7	1.8
SS A6	46	57	67	74	82	87	91	95	36	45	55	62	71	76	82	89	0.8	1.0	1.2	1.3	1.5	0.9	1.1	1.3	1.4	1.6
SS A7	32	38	46	52	59	64	70	78	25	30	36	40	47	51	57	64	1.1	1.2	1.4	1.6	1.7	1.3	1.4	1.6	1.8	1.9
SS A8	45	55	65	72	80	85	90	94	35	44	53	60	69	74	80	87	0.8	1.0	1.2	1.4	1.5	1.0	1.1	1.3	1.5	1.6
SS B1	33	41	50	56	65	70	77	84	24	31	38	44	51	57	63	72	1.0	1.2	1.3	1.5	1.6	1.2	1.3	1.5	1.6	1.8
SS B2	41	51	62	69	77	82	88	93	31	40	50	56	66	71	78	85	0.9	1.0	1.2	1.4	1.5	1.0	1.1	1.3	1.5	1.6
SS B3	31	39	48	54	62	68	74	82	23	29	36	41	49	54	60	68	1.0	1.2	1.3	1.5	1.7	1.2	1.4	1.5	1.7	1.8
SS B4	40	50	61	68	76	81	87	92	30	39	49	55	64	70	77	84	0.9	1.0	1.2	1.4	1.5	1.0	1.1	1.3	1.5	1.6
SS B5	29	36	44	50	58	63	69	77	22	27	33	38	45	49	55	63	1.1	1.3	1.4	1.6	1.7	1.3	1.4	1.6	1.8	1.9
SS B6	40	50	61	68	77	82	87	93	30	39	49	56	65	70	77	85	0.9	1.0	1.2	1.4	1.5	1.0	1.1	1.3	1.5	1.6
SS B7	27	33	40	45	53	58	64	72	20	25	30	34	40	45	50	58	1.2	1.3	1.5	1.6	1.8	1.4	1.5	1.7	1.9	2.0
SS B8	39	49	59	66	75	80	86	92	29	38	47	54	63	68	75	83	0.9	1.1	1.2	1.4	1.5	1.0	1.2	1.3	1.5	1.6
SS C1	34	43	54	61	70	76	82	89	22	30	38	44	53	59	66	75	0.9	1.1	1.2	1.4	1.6	1.1	1.3	1.4	1.6	1.7
SS C2	39	50	62	69	78	84	89	94	27	36	46	54	64	70	77	85	0.9	1.0	1.2	1.3	1.5	1.0	1.1	1.3	1.4	1.6
SS C3	27	35	43	49	58	64	71	79	19	24	31	35	43	48	54	63	1.1	1.2	1.4	1.5	1.7	1.3	1.4	1.6	1.7	1.9
SS C4	35	45	56	63	72	78	84	91	24	32	41	48	57	63	71	79	0.9	1.1	1.2	1.4	1.5	1.1	1.2	1.4	1.5	1.7
SS D1	45	55	65	72	80	85	90	94	34	43	53	59	68	73	79	86	0.9	1.0	1.2	1.4	1.5	1.0	1.1	1.3	1.5	1.6
SS D2	53	65	76	82	89	92	96	98	42	53	64	71	80	84	89	94	0.8	0.9	1.1	1.3	1.4	0.8	1.0	1.2	1.3	1.5
SS D3	41	50	59	66	74	79	85	91	31	39	47	53	61	67	73	81	0.9	1.1	1.2	1.4	1.6	1.0	1.2	1.4	1.5	1.7
SS D4	52	63	74	80	87	91	95	98	40	51	62	69	78	83	88	93	0.8	0.9	1.1	1.3	1.4	0.9	1.0	1.2	1.4	1.5
SS E1	40	50	60	67	75	80	86	92	29	37	46	52	61	67	73	81	0.9	1.1	1.2	1.4	1.5	1.0	1.2	1.4	1.5	1.7
SS E2	47	59	70	77	85	89	93	97	36	46	57	64	73	79	85	91	0.8	1.0	1.1	1.3	1.5	0.9	1.1	1.2	1.4	1.5
SS E3	33	41	50	57	65	70	77	84	25	31	38	44	51	57	63	71	1.0	1.2	1.3	1.5	1.6	1.2	1.3	1.5	1.7	1.8
SS E4	44	55	65	72	81	85	90	95	33	42	53	60	69	74	81	88	0.8	1.0	1.2	1.3	1.5	0.9	1.1	1.3	1.4	1.6

YUMA, ARIZONA

TYPE	SOLAR SAVINGS FRACTION (%) BASE TEMP: 55 F (211 DD)								65 F (1010 DD)								CONSERVATION FACTOR 55 F					65 F				
LCR	200	100	70	50	40	30	25	20	200	100	70	50	40	30	25	20	200	100	50	30	20	200	100	50	30	20
WW A1	71	90	95	98	99	100	100	100	36	53	64	73	80	87	90	94	0.5	0.6	0.8	1.0	1.2	0.7	0.8	1.0	1.1	1.3
WW A2	83	96	99	100	100	100	100	100	44	65	76	85	90	95	97	99	0.4	0.6	0.8	0.9	-	0.6	0.7	0.9	1.1	1.2
WW A3	87	98	99	100	100	100	100	100	46	69	80	88	93	97	98	99	0.4	0.6	0.7	-	-	0.6	0.7	0.9	1.0	1.2
WW A4	88	98	100	100	100	100	100	100	48	71	82	90	94	97	99	99	0.4	0.6	0.7	-	-	0.6	0.7	0.8	1.0	1.2
WW A5	90	99	100	100	100	100	100	100	50	73	84	91	95	98	99	100	0.4	0.5	0.7	-	-	0.5	0.7	0.8	1.0	1.2
WW A6	91	99	100	100	100	100	100	100	50	74	85	92	96	98	99	100	0.4	0.5	0.7	-	-	0.5	0.7	0.8	1.0	1.2
WW B1	78	94	98	99	100	100	100	100	39	59	70	80	85	92	94	97	0.5	0.6	0.8	1.0	-	0.6	0.7	0.9	1.1	1.3
WW B2	88	98	100	100	100	100	100	100	48	71	82	90	94	97	99	100	0.4	0.6	0.7	-	-	0.6	0.7	0.8	1.0	1.2
WW B3	93	99	100	100	100	100	100	100	55	78	88	94	97	99	100	100	0.4	0.5	0.7	-	-	0.5	0.6	0.8	1.0	1.2
WW B4	92	99	100	100	100	100	100	100	54	78	88	94	97	99	100	100	0.4	0.5	0.7	-	-	0.5	0.6	0.8	1.0	1.2
WW B5	91	99	100	100	100	100	100	100	51	76	86	93	97	99	99	100	0.4	0.5	0.7	-	-	0.5	0.6	0.8	1.0	1.2
WW C1	93	99	100	100	100	100	100	100	56	79	89	95	97	99	100	100	0.4	0.5	0.7	-	-	0.5	0.6	0.8	1.0	1.2
WW C2	92	99	100	100	100	100	100	100	53	77	87	94	97	99	99	100	0.4	0.5	0.7	-	-	0.5	0.6	0.8	1.0	1.2
WW C3	95	100	100	100	100	100	100	100	59	83	91	96	98	100	100	100	0.4	0.5	0.7	-	-	0.5	0.6	0.8	1.0	1.2
WW C4	93	99	100	100	100	100	100	100	55	79	89	95	98	99	100	100	0.4	0.5	0.7	-	-	0.5	0.6	0.8	1.0	1.2
TW A1	70	89	95	98	99	100	100	100	35	52	62	72	78	86	90	94	0.5	0.6	0.8	1.0	1.2	0.7	0.8	1.0	1.1	1.3
TW A2	78	94	98	99	100	100	100	100	39	60	71	81	86	92	95	97	0.5	0.6	0.8	1.0	-	0.6	0.7	0.9	1.1	1.3
TW A3	81	96	99	100	100	100	100	100	41	63	74	83	89	94	96	98	0.4	0.6	0.8	0.9	-	0.6	0.7	0.9	1.1	1.2
TW A4	82	96	99	100	100	100	100	100	41	63	74	84	89	94	97	98	0.4	0.6	0.8	0.9	-	0.6	0.7	0.9	1.1	1.2
TW B1	71	89	95	98	99	100	100	100	35	52	63	73	79	87	90	94	0.5	0.6	0.8	1.0	1.2	0.7	0.8	1.0	1.1	1.3
TW B2	77	93	98	99	100	100	100	100	38	58	69	79	85	91	94	97	0.5	0.6	0.8	1.0	-	0.6	0.7	0.9	1.1	1.3
TW B3	77	94	98	99	100	100	100	100	37	58	70	80	86	92	95	97	0.5	0.6	0.8	1.0	-	0.6	0.7	0.9	1.1	1.3
TW B4	74	92	97	99	100	100	100	100	35	55	67	77	83	90	93	96	0.5	0.6	0.8	1.0	-	0.7	0.8	0.9	1.1	1.3
TW C1	70	89	95	98	99	100	100	100	33	51	62	72	79	86	90	94	0.5	0.6	0.8	1.0	1.2	0.7	0.8	1.0	1.1	1.3
TW C2	71	90	96	99	99	100	100	100	33	52	63	74	80	87	91	95	0.5	0.6	0.8	1.0	1.2	0.7	0.8	0.9	1.1	1.3
TW C3	68	88	94	98	99	100	100	100	31	49	61	71	78	85	90	94	0.5	0.6	0.8	1.0	1.2	0.7	0.8	1.0	1.1	1.3
TW C4	61	82	91	96	98	99	100	100	28	44	54	65	72	80	85	90	0.5	0.6	0.8	1.0	1.2	0.8	0.9	1.0	1.2	1.3
TW D1	71	90	96	99	99	100	100	100	33	52	63	73	80	87	91	94	0.5	0.6	0.8	1.0	1.2	0.7	0.8	1.0	1.1	1.3
TW D2	84	97	99	100	100	100	100	100	43	66	77	86	91	96	97	99	0.4	0.6	0.8	0.9	-	0.6	0.7	0.9	1.0	1.2
TW D3	87	98	100	100	100	100	100	100	47	70	81	89	94	97	98	99	0.4	0.6	0.7	-	-	0.6	0.7	0.8	1.0	1.2
TW D4	89	98	100	100	100	100	100	100	49	73	83	91	95	98	99	100	0.4	0.6	0.7	-	-	0.6	0.7	0.8	1.0	1.2
TW D5	88	98	100	100	100	100	100	100	47	71	82	90	94	98	99	100	0.4	0.6	0.7	-	-	0.6	0.7	0.8	1.0	1.2
TW E1	90	99	100	100	100	100	100	100	51	74	85	92	95	98	99	100	0.4	0.5	0.7	-	-	0.5	0.7	0.8	1.0	1.2
TW E2	90	99	100	100	100	100	100	100	50	74	84	92	95	98	99	100	0.4	0.6	0.7	-	-	0.5	0.7	0.8	1.0	1.2
TW E3	94	99	100	100	100	100	100	100	57	80	90	95	98	99	100	100	0.4	0.5	0.7	-	-	0.5	0.6	0.8	1.0	1.2
TW E4	92	99	100	100	100	100	100	100	53	77	87	94	97	99	100	100	0.4	0.5	0.7	-	-	0.5	0.6	0.8	1.0	1.2
TW F1	70	89	95	98	99	100	100	100	34	52	62	72	79	86	90	94	0.5	0.6	0.8	1.0	1.2	0.7	0.8	1.0	1.1	1.3
TW F2	77	93	98	99	100	100	100	100	37	57	69	79	85	91	94	97	0.5	0.6	0.8	1.0	-	0.6	0.7	0.9	1.1	1.3
TW F3	78	94	98	100	100	100	100	100	37	59	70	81	86	92	95	98	0.5	0.6	0.8	1.0	-	0.6	0.7	0.9	1.1	1.2
TW F4	76	94	98	99	100	100	100	100	35	57	69	79	85	92	95	97	0.5	0.6	0.8	1.0	-	0.6	0.7	0.9	1.1	1.3
TW G1	69	89	95	98	99	100	100	100	32	50	61	72	78	86	90	94	0.5	0.6	0.8	1.0	1.2	0.7	0.8	1.0	1.1	1.3
TW G2	73	91	97	99	100	100	100	100	33	53	65	75	82	89	92	96	0.5	0.6	0.8	1.0	-	0.7	0.8	0.9	1.1	1.3
TW G3	71	90	96	99	100	100	100	100	31	51	63	74	81	88	92	95	0.5	0.6	0.8	1.0	-	0.7	0.8	0.9	1.1	1.3
TW G4	65	86	94	98	99	100	100	100	26	45	57	68	75	84	88	93	0.5	0.6	0.8	1.0	1.2	0.7	0.8	1.0	1.1	1.3
TW H1	66	86	94	98	99	100	100	100	28	47	58	69	76	84	88	92	0.5	0.6	0.8	1.0	1.2	0.7	0.8	1.0	1.1	1.3
TW H2	63	85	93	97	99	100	100	100	26	44	55	67	74	82	87	92	0.5	0.6	0.8	1.0	1.2	0.7	0.8	1.0	1.2	1.3
TW H3	57	80	90	95	98	99	100	100	22	39	50	61	68	78	83	88	0.5	0.6	0.8	1.0	1.2	0.8	0.9	1.0	1.2	1.4
TW H4	48	71	82	90	94	98	99	100	17	31	41	51	59	68	74	81	0.6	0.7	0.8	1.0	1.2	0.9	[illegible]	1.1	1.3	1.4
TW I1	68	88	95	98	99	100	100	100	30	48	59	70	77	85	89	93	0.5	0.6	0.8	1.0	1.2	0.7	0.8	1.0	1.1	1.3
TW I2	81	96	99	100	100	100	100	100	40	62	74	84	89	94	97	98	0.4	0.6	0.8	0.9	-	0.6	0.7	0.9	1.1	1.2
TW I3	84	97	99	100	100	100	100	100	43	66	77	87	91	96	98	99	0.4	0.6	0.8	-	-	0.6	0.7	0.9	1.0	1.2
TW I4	87	98	99	100	100	100	100	100	45	69	81	89	93	97	99	99	0.4	0.6	0.7	-	-	0.6	0.7	0.8	1.0	1.2
TW I5	87	98	99	100	100	100	100	100	45	69	81	89	94	97	99	99	0.4	0.6	0.7	-	-	0.6	0.7	0.8	1.0	1.2
TW J1	89	98	100	100	100	100	100	100	49	72	83	91	94	98	99	100	0.4	0.6	0.7	-	-	0.6	0.7	0.8	1.0	1.2
TW J2	88	98	100	100	100	100	100	100	47	71	82	90	94	97	99	100	0.4	0.6	0.7	-	-	0.6	0.7	0.8	1.0	1.2
TW J3	92	99	100	100	100	100	100	100	53	77	87	94	97	99	100	100	0.4	0.5	0.7	-	-	0.5	0.6	0.8	1.0	1.2
TW J4	90	99	100	100	100	100	100	100	50	75	85	93	96	99	99	100	0.4	0.5	0.7	-	-	0.5	0.7	0.8	1.0	1.2
DG A1	80	95	99	100	100	100	100	100	38	60	69	79	84	89	92	95	0.4	0.6	0.7	-	-	0.6	0.7	0.9	1.1	1.3
DG A2	80	95	99	100	100	100	100	100	38	60	71	81	86	92	95	97	0.4	0.6	0.7	-	-	0.6	0.7	0.9	1.1	1.3
DG A3	84	95	97	97	97	97	97	97	42	66	78	86	90	94	95	96	0.4	0.6	2.1	-	-	0.6	0.7	0.9	1.2	1.7
DG B1	87	97	99	99	99	99	99	99	40	65	76	85	90	94	95	97	0.4	0.6	1.1	-	-	0.6	0.7	0.9	1.1	1.4
DG B2	86	98	99	100	100	100	100	100	40	65	78	87	92	96	97	99	0.4	0.6	0.8	-	-	0.6	0.7	0.9	1.1	1.3
DG B3	88	97	98	98	98	98	98	98	43	69	82	90	94	96	97	98	0.4	0.6	2.8	-	-	0.6	0.7	0.9	1.1	1.7
DG C1	92	98	99	99	99	99	98	98	46	72	84	91	94	96	97	98	0.4	0.6	-	-	-	0.6	0.6	0.9	1.1	1.8
DG C2	91	99	100	100	100	100	100	100	44	71	83	91	95	98	99	99	0.4	0.5	0.9	-	-	0.6	0.6	0.8	1.0	1.2
DG C3	92	98	98	99	99	98	98	98	48	76	87	94	96	98	98	98	0.4	0.7	-	-	-	0.5	0.6	0.8	1.2	2.7
SS A1	82	96	99	100	100	100	100	100	45	65	75	84	89	94	96	98	0.4	0.6	0.8	0.9	-	0.6	0.7	0.9	1.1	1.2
SS A2	90	98	100	100	100	100	100	100	53	75	84	92	95	98	99	100	0.4	0.6	0.7	-	-	0.6	0.7	0.8	1.0	1.2
SS A3	79	94	98	99	100	100	100	100	43	62	72	82	87	92	95	97	0.5	0.6	0.8	1.0	-	0.6	0.7	0.9	1.1	1.3
SS A4	89	98	100	100	100	100	100	100	53	74	84	91	95	98	99	100	0.4	0.6	0.7	-	-	0.6	0.7	0.8	1.0	1.2
SS A5	78	93	98	99	100	100	100	100	42	61	71	80	86	91	94	97	0.5	0.6	0.8	1.0	-	0.6	0.8	0.9	1.1	1.3
SS A6	90	98	100	100	100	100	100	100	53	75	85	92	95	98	99	100	0.4	0.6	0.7	-	-	0.6	0.7	0.8	1.0	1.2
SS A7	72	90	96	98	99	100	100	100	38	55	65	75	81	87	91	94	0.5	0.6	0.8	1.0	1.2	0.7	0.8	1.0	1.1	1.3
SS A8	89	98	100	100	100	100	100	100	52	73	83	91	94	98	99	100	0.4	0.6	0.7	-	-	0.6	0.7	0.8	1.0	1.2
SS B1	77	93	97	99	100	100	100	100	39	59	70	79	85	91	94	97	0.5	0.6	0.8	1.0	-	0.6	0.8	0.9	1.1	1.3
SS B2	85	97	99	100	100	100	100	100	47	69	80	88	92	96	98	99	0.4	0.6	0.7	-	-	0.6	0.7	0.9	1.0	1.2
SS B3	75	92	97	99	100	100	100	100	38	57	68	78	83	90	93	96	0.5	0.6	0.8	1.0	-	0.7	0.8	0.9	1.1	1.3
SS B4	85	97	99	100	100	100	100	100	47	68	79	87	92	96	98	99	0.4	0.6	0.8	-	-	0.6	0.7	0.9	1.0	1.2
SS B5	70	89	95	98	99	100	100	100	35	53	63	73	79	86	90	94	0.5	0.6	0.8	1.0	1.2	0.7	0.8	1.0	1.1	1.3
SS B6	85	97	99	100	100	100	100	100	47	69	79	88	92	96	98	99	0.4	0.6	0.7	-	-	0.6	0.7	0.9	1.0	1.2
SS B7	66	86	93	97	99	100	100	100	32	49	59	68	75	82	87	91	0.5	0.6	0.8	1.0	1.2	0.7	0.8	1.0	1.2	1.3
SS B8	84	97	99	100	100	100	100	100	46	67	78	87	91	96	97	99	0.4	0.6	0.8	0.9	-	0.6	0.7	0.9	1.0	1.2
SS C1	83	96	99	100	100	100	100	100	42	64	75	85	90	95	97	99	0.4	0.6	0.8	0.9	-	0.6	0.7	0.9	1.1	1.2
SS C2	88	98	100	100	100	100	100	100	47	70	81	90	94	97	99	99	0.4	0.6	0.7	-	-	0.6	0.7	0.8	1.0	1.2
SS C3	72	90	96	99	99	100	100	100	34	53	64	74	80	88	91	95	0.5	0.6	0.8	1.0	1.2	0.7	0.8	0.9	1.1	1.3
SS C4	83	96	99	100	100	100	100	100	42	64	76	85	90	95	97	99	0.4	0.6	0.8	0.9	-	0.6	0.7	0.9	1.0	1.2
SS D1	90	99	100	100	100	100	100	100	53	74	84	91	95	98	99	100	0.4	0.6	0.7	-	-	0.6	0.7	0.8	1.0	1.2
SS D2	94	100	100	100	100	100	100	100	61	82	91	96	98	99	100	100	0.4	0.5	0.7	-	-	0.5	0.6	0.8	1.0	1.2
SS D3	85	97	99	100	100	100	100	100	48	68	79	87	91	96	97	99	0.4	0.6	0.8	0.9	-	0.6	0.7	0.9	1.0	1.2
SS D4	93	99	100	100	100	100	100	100	59	80	89	95	97	99	100	100	0.4	0.5	0.7	-	-	0.5	0.6	0.8	1.0	1.2
SS E1	87	98	99	100	100	100	100	100	48	70	80	88	93	97	98	99	0.4	0.6	0.7	-	-	0.6	0.7	0.9	1.0	1.2
SS E2	92	99	100	100	100	100	100	100	55	78	87	94	97	99	99	100	0.4	0.5	0.7	-	-	0.5	0.6	0.8	1.0	1.2
SS E3	78	94	98	99	100	100	100	100	40	60	70	80	85	91	94	97	0.5	0.6	0.8	1.0	-	0.6	0.8	0.9	1.1	1.3
SS E4	89	98	100	100	100	100	100	100	51	73	83	91	95	98	99	100	0.4	0.6	0.7	-	-	0.6	0.7	0.8	1.0	1.2

FORT SMITH, ARKANSAS

TYPE	SOLAR SAVINGS FRACTION (%)																CONSERVATION FACTOR									
BASE TEMP:	55 F (1690 DD)								65 F (3338 DD)								55 F					65 F				
LCR	200	100	70	50	40	30	25	20	200	100	70	50	40	30	25	20	200	100	50	30	20	200	100	50	30	20
WW A1	20	28	34	41	47	55	60	66	14	20	23	28	31	37	40	45	1.1	1.2	1.3	1.4	1.6	1.4	1.5	1.7	1.8	2.0
WW A2	20	33	42	52	59	68	74	80	13	21	27	34	39	47	52	59	0.9	1.0	1.1	1.3	1.4	1.1	1.2	1.4	1.5	1.7
WW A3	20	35	45	56	63	73	78	84	11	21	29	37	43	51	57	64	0.8	0.9	1.1	1.2	1.4	1.0	1.1	1.3	1.4	1.6
WW A4	20	36	47	58	66	75	81	86	11	22	29	38	44	53	59	66	0.8	0.9	1.0	1.2	1.4	1.0	1.1	1.3	1.4	1.6
WW A5	20	38	49	60	68	77	83	88	10	22	30	39	46	56	62	69	0.8	0.9	1.0	1.2	1.4	1.0	1.1	1.2	1.4	1.5
WW A6	20	38	50	62	69	79	84	89	10	22	31	40	47	57	63	71	0.8	0.8	1.0	1.2	1.3	0.9	1.0	1.2	1.3	1.5
WW B1	17	28	35	44	50	59	65	71	10	17	21	26	31	36	41	46	1.0	1.1	1.2	1.4	1.5	1.3	1.4	1.6	1.8	1.9
WW B2	20	37	48	59	67	76	82	87	11	22	30	40	46	56	62	69	0.8	0.9	1.0	1.2	1.4	1.0	1.1	1.2	1.4	1.5
WW B3	24	43	55	67	75	83	88	92	13	27	37	47	54	64	70	78	0.7	0.8	1.0	1.1	1.3	0.9	1.0	1.1	1.3	1.4
WW B4	22	43	56	68	76	84	89	93	12	27	37	49	57	67	73	80	0.7	0.8	1.0	1.1	1.3	0.8	0.9	1.1	1.2	1.4
WW B5	21	41	54	67	75	84	88	93	10	26	36	48	56	66	73	80	0.7	0.8	1.0	1.1	1.3	0.9	0.9	1.1	1.2	1.4
WW C1	25	45	57	69	76	85	89	93	14	29	39	49	57	67	73	80	0.7	0.8	1.0	1.1	1.3	0.9	1.0	1.1	1.3	1.4
WW C2	23	43	55	67	75	83	88	93	12	27	37	48	55	65	72	79	0.7	0.8	1.0	1.1	1.3	0.9	1.0	1.1	1.3	1.4
WW C3	26	49	63	75	82	90	93	96	13	32	44	56	65	75	81	87	0.7	0.8	0.9	1.1	1.3	0.8	0.9	1.0	1.2	1.3
WW C4	23	45	59	71	79	87	91	95	11	29	40	53	61	72	78	85	0.7	0.8	0.9	1.1	1.3	0.8	0.9	1.0	1.2	1.4
TW A1	19	27	33	41	46	54	59	66	14	19	23	27	31	36	40	45	1.1	1.2	1.3	1.5	1.6	1.4	1.5	1.7	1.8	2.0
TW A2	19	30	38	47	54	63	69	75	12	20	25	31	36	43	48	54	0.9	1.0	1.2	1.3	1.5	1.2	1.3	1.5	1.6	1.8
TW A3	18	31	40	50	57	66	72	79	11	19	26	33	38	46	51	58	0.9	1.0	1.1	1.3	1.4	1.1	1.2	1.4	1.5	1.7
TW A4	17	31	40	50	58	67	73	80	10	19	25	33	38	46	52	59	0.9	1.0	1.1	1.3	1.4	1.1	1.2	1.3	1.5	1.7
TW B1	18	27	33	41	47	55	60	67	13	19	23	28	31	37	41	47	1.1	1.2	1.3	1.4	1.6	1.4	1.5	1.6	1.8	1.9
TW B2	17	29	37	46	53	62	67	74	11	18	24	30	35	42	47	53	1.0	1.0	1.2	1.3	1.5	1.2	1.3	1.5	1.6	1.8
TW B3	16	28	37	46	53	62	68	75	10	18	23	30	35	43	48	54	0.9	1.0	1.2	1.3	1.5	1.2	1.3	1.4	1.6	1.7
TW B4	15	26	35	44	50	60	66	73	9	17	22	28	33	40	45	52	1.0	1.1	1.2	1.3	1.5	1.2	1.3	1.5	1.6	1.8
TW C1	17	26	33	40	46	55	60	67	11	17	22	27	31	37	41	47	1.1	1.1	1.3	1.4	1.6	1.3	1.4	1.6	1.7	1.9
TW C2	15	25	33	41	47	56	62	69	10	16	21	27	31	38	43	49	1.0	1.1	1.2	1.4	1.5	1.3	1.4	1.5	1.7	1.8
TW C3	15	24	31	39	45	54	59	66	10	16	20	26	30	36	41	47	1.1	1.1	1.3	1.4	1.6	1.3	1.4	1.6	1.7	1.9
TW C4	14	22	28	35	40	48	53	60	10	15	19	24	27	33	37	42	1.2	1.2	1.4	1.5	1.6	1.5	1.5	1.7	1.8	2.0
TW D1	14	23	30	38	44	52	58	65	8	14	18	23	26	32	36	41	1.1	1.2	1.3	1.4	1.6	1.4	1.5	1.7	1.9	2.0
TW D2	19	34	44	54	62	71	77	83	11	21	28	36	42	51	57	64	0.8	0.9	1.1	1.2	1.4	1.0	1.1	1.3	1.4	1.6
TW D3	20	37	48	59	66	76	81	87	12	23	31	40	46	55	61	69	0.8	0.9	1.0	1.2	1.4	1.0	1.1	1.2	1.4	1.5
TW D4	21	39	51	62	70	79	85	90	12	25	34	44	51	61	67	75	0.8	0.8	1.0	1.2	1.3	0.9	1.0	1.2	1.3	1.5
TW D5	20	38	50	62	70	79	84	90	11	24	33	44	51	61	67	75	0.8	0.8	1.0	1.2	1.3	0.9	1.0	1.1	1.3	1.5
TW E1	23	41	52	64	71	80	85	90	14	26	35	45	51	61	67	74	0.8	0.8	1.0	1.2	1.3	0.9	1.0	1.2	1.3	1.5
TW E2	22	40	52	63	71	80	85	90	13	26	35	44	52	61	67	75	0.8	0.8	1.0	1.2	1.3	0.9	1.0	1.2	1.3	1.5
TW E3	26	47	60	72	79	87	91	95	14	31	42	53	61	71	77	84	0.7	0.8	0.9	1.1	1.3	0.8	0.9	1.1	1.2	1.4
TW E4	23	44	57	69	76	85	89	94	13	28	39	50	58	69	75	82	0.7	0.8	1.0	1.1	1.3	0.8	0.9	1.1	1.2	1.4
TW F1	17	26	32	40	46	54	59	66	12	17	21	26	30	35	39	45	1.1	1.2	1.3	1.4	1.6	1.4	1.5	1.7	1.8	2.0
TW F2	16	28	36	45	52	61	66	73	10	17	22	29	33	40	45	52	1.0	1.0	1.2	1.3	1.5	1.2	1.3	1.5	1.6	1.8
TW F3	15	27	36	46	53	63	69	76	8	16	22	29	34	42	47	54	0.9	1.0	1.1	1.3	1.5	1.2	1.3	1.4	1.6	1.7
TW F4	13	25	34	44	51	61	67	75	6	14	20	27	33	40	46	53	0.9	1.0	1.2	1.3	1.5	1.2	1.3	1.4	1.6	1.7
TW G1	15	24	31	39	45	53	59	66	10	16	20	25	29	35	39	44	1.1	1.2	1.3	1.4	1.6	1.4	1.5	1.6	1.8	1.9
TW G2	13	24	32	41	48	57	63	70	7	14	19	25	30	37	42	48	1.0	1.1	1.2	1.4	1.5	1.3	1.4	1.5	1.7	1.8
TW G3	11	22	30	39	46	55	62	69	5	12	18	24	29	36	41	47	1.0	1.1	1.2	1.4	1.5	1.3	1.4	1.5	1.7	1.8
TW G4	9	19	26	34	41	50	56	63	4	10	15	20	25	31	36	42	1.1	1.2	1.3	1.4	1.6	1.4	1.4	1.6	1.7	1.9
TW H1	12	21	27	35	41	50	55	63	7	12	17	22	26	32	36	42	1.1	1.2	1.3	1.4	1.6	1.4	1.5	1.6	1.8	2.0
TW H2	9	18	25	33	39	48	54	61	4	10	14	20	24	30	34	40	1.1	1.2	1.3	1.5	1.6	1.4	1.5	1.6	1.8	1.9
TW H3	7	15	22	29	35	43	49	56	3	8	12	17	21	27	31	36	1.2	1.3	1.4	1.5	1.7	1.5	1.6	1.7	1.9	2.0
TW H4	5	12	17	23	28	35	40	47	2	6	9	13	16	21	25	30	1.3	1.4	1.5	1.7	1.8	1.7	1.8	1.9	2.0	2.2
TW I1	12	21	27	35	40	49	54	61	7	12	15	20	23	29	32	37	1.1	1.2	1.3	1.5	1.6	1.5	1.6	1.8	1.9	2.1
TW I2	16	30	40	50	58	68	74	80	8	18	25	33	39	47	53	60	0.9	1.0	1.1	1.3	1.4	1.1	1.2	1.3	1.5	1.6
TW I3	17	33	43	54	62	72	77	84	9	20	27	36	42	51	57	65	0.8	0.9	1.1	1.2	1.4	1.0	1.1	1.3	1.4	1.6
TW I4	18	35	47	59	67	76	82	88	9	21	30	40	47	57	64	71	0.8	0.9	1.0	1.2	1.3	0.9	1.0	1.2	1.3	1.5
TW I5	17	35	47	59	67	77	83	88	8	22	31	41	48	59	65	73	0.8	0.9	1.0	1.2	1.3	0.9	1.0	1.2	1.3	1.5
TW J1	21	38	49	61	69	78	83	89	11	24	32	42	49	59	65	72	0.8	0.9	1.0	1.2	1.3	0.9	1.0	1.2	1.3	1.5
TW J2	19	37	48	60	68	77	83	88	10	23	32	41	48	58	65	72	0.8	0.9	1.0	1.2	1.3	0.9	1.0	1.2	1.3	1.5
TW J3	22	43	56	69	76	85	89	94	11	28	38	50	58	69	75	82	0.7	0.8	0.9	1.1	1.3	0.8	0.9	1.1	1.2	1.4
TW J4	20	41	53	66	74	83	88	92	10	25	36	47	55	66	72	80	0.7	0.8	1.0	1.1	1.3	0.9	0.9	1.1	1.2	1.4
DG A1	15	26	34	42	47	55	59	64	8	14	19	24	28	33	36	39	1.0	1.1	1.3	1.5	1.8	1.3	1.4	1.7	2.0	2.5
DG A2	15	28	37	46	53	62	67	73	9	17	23	29	35	42	46	52	0.9	1.0	1.2	1.4	1.6	1.2	1.3	1.4	1.6	1.9
DG A3	18	32	43	55	63	72	78	83	11	21	28	37	44	53	59	66	0.9	0.9	1.0	1.2	1.5	1.1	1.1	1.2	1.4	1.6
DG B1	15	27	36	46	53	62	67	72	8	15	19	25	30	36	40	46	1.0	1.0	1.1	1.4	1.6	1.3	1.4	1.5	1.7	2.0
DG B2	16	29	39	50	58	68	74	80	9	17	23	31	36	45	51	58	0.9	1.0	1.1	1.3	1.5	1.2	1.3	1.4	1.5	1.7
DG B3	18	33	45	58	67	77	82	88	11	21	29	38	45	56	63	71	0.9	0.9	1.0	1.2	1.4	1.1	1.1	1.2	1.3	1.5
DG C1	17	32	42	54	62	71	77	82	10	18	24	30	36	44	50	56	0.9	0.9	1.0	1.2	1.5	1.1	1.3	1.4	1.5	1.7
DG C2	18	33	44	57	65	76	81	86	11	20	27	35	42	52	59	66	0.9	0.9	1.0	1.2	1.4	1.1	1.2	1.3	1.4	1.6
DG C3	20	38	50	64	73	83	88	92	13	24	32	43	51	63	70	78	0.8	0.8	0.9	1.1	1.3	1.0	1.1	1.1	1.2	1.4
SS A1	22	35	44	54	60	69	74	80	16	25	31	38	43	50	55	62	0.9	1.0	1.1	1.3	1.4	1.1	1.2	1.3	1.5	1.7
SS A2	27	44	55	66	73	81	86	90	18	31	39	49	55	64	70	77	0.7	0.8	1.0	1.2	1.3	0.9	1.0	1.2	1.3	1.5
SS A3	21	33	41	50	57	65	71	77	15	23	28	35	39	46	51	57	0.9	1.0	1.2	1.3	1.5	1.1	1.3	1.4	1.6	1.8
SS A4	26	43	54	65	72	80	85	90	18	30	39	48	54	63	69	75	0.7	0.8	1.0	1.2	1.4	0.9	1.0	1.2	1.3	1.5
SS A5	22	33	41	49	56	64	69	76	17	24	29	35	39	46	50	56	0.9	1.0	1.2	1.3	1.5	1.2	1.3	1.5	1.6	1.8
SS A6	27	44	55	65	72	81	85	90	18	31	39	48	55	64	69	76	0.7	0.8	1.0	1.2	1.3	0.9	1.0	1.2	1.3	1.5
SS A7	20	30	36	44	49	57	62	69	16	21	26	31	34	40	44	49	1.0	1.1	1.3	1.4	1.6	1.3	1.4	1.6	1.8	1.9
SS A8	26	43	53	64	71	79	84	89	18	30	38	47	53	62	67	74	0.8	0.9	1.0	1.2	1.4	0.9	1.0	1.2	1.3	1.5
SS B1	19	31	39	48	54	63	69	75	13	21	26	33	37	44	49	55	0.9	1.0	1.2	1.3	1.5	1.2	1.3	1.4	1.6	1.8
SS B2	23	39	49	60	67	76	81	87	15	27	35	43	50	59	64	71	0.8	0.9	1.0	1.2	1.4	1.0	1.1	1.2	1.4	1.5
SS B3	18	29	37	45	52	60	66	72	12	19	25	30	35	41	46	52	1.0	1.1	1.2	1.4	1.5	1.2	1.3	1.5	1.7	1.8
SS B4	22	38	48	59	66	75	80	86	15	26	34	42	49	57	63	70	0.8	0.9	1.0	1.2	1.4	1.0	1.1	1.2	1.4	1.5
SS B5	18	27	34	42	47	55	61	67	13	19	24	29	33	38	42	48	1.1	1.1	1.3	1.4	1.6	1.3	1.4	1.6	1.8	1.9
SS B6	22	38	48	59	66	75	81	86	15	26	34	42	49	58	63	70	0.8	0.9	1.0	1.2	1.4	1.0	1.1	1.2	1.4	1.5
SS B7	17	25	31	38	43	51	56	62	12	18	21	26	29	34	38	42	1.1	1.2	1.3	1.5	1.7	1.4	1.5	1.7	1.9	2.1
SS B8	22	37	47	57	64	73	79	85	14	25	32	41	47	55	61	68	0.8	0.9	1.1	1.2	1.4	1.0	1.1	1.2	1.4	1.6
SS C1	18	31	41	51	58	67	73	80	10	19	25	32	37	45	51	57	0.9	1.0	1.1	1.3	1.4	1.1	1.2	1.4	1.5	1.7
SS C2	20	37	48	59	67	76	81	87	11	23	31	40	47	56	62	69	0.8	0.9	1.0	1.2	1.4	1.0	1.1	1.2	1.4	1.5
SS C3	16	25	33	41	47	55	61	68	10	16	21	26	30	36	41	47	1.0	1.1	1.3	1.4	1.6	1.3	1.4	1.6	1.7	1.9
SS C4	18	33	42	53	60	70	76	82	11	21	27	35	41	50	56	63	0.9	0.9	1.1	1.2	1.4	1.1	1.2	1.3	1.4	1.6
SS D1	26	42	53	63	70	79	83	88	17	29	36	45	51	60	65	72	0.8	0.9	1.0	1.2	1.4	0.9	1.1	1.2	1.4	1.5
SS D2	30	51	63	74	81	88	92	95	20	35	46	56	64	73	79	85	0.7	0.8	0.9	1.1	1.3	0.8	0.9	1.1	1.2	1.4
SS D3	24	38	47	57	64	72	78	83	17	27	33	40	46	53	59	65	0.8	0.9	1.1	1.3	1.4	1.0	1.1	1.3	1.5	1.6
SS D4	29	49	61	72	79	86	90	94	19	34	44	54	62	71	77	83	0.7	0.8	1.0	1.1	1.3	0.8	0.9	1.1	1.2	1.4
SS E1	22	37	47	58	65	73	79	84	14	24	31	39	45	53	58	65	0.8	0.9	1.1	1.2	1.4	1.0	1.1	1.3	1.4	1.6
SS E2	26	45	57	68	75	84	88	92	16	30	39	50	57	66	72	79	0.7	0.8	1.0	1.1	1.3	0.9	1.0	1.1	1.3	1.4
SS E3	20	31	39	48	54	63	69	75	14	21	27	33	37	44	49	55	0.9	1.0	1.2	1.3	1.5	1.2	1.3	1.5	1.6	1.8
SS E4	24	41	52	64	71	79	84	89	15	28	36	46	52	61	67	74	0.8	0.8	1.0	1.2	1.3	0.9	1.0	1.2	1.3	1.5

LITTLE ROCK, ARKANSAS

TYPE	SOLAR SAVINGS FRACTION (%)																CONSERVATION FACTOR									
BASE TEMP:	55 F (1687 DD)								65 F (3357 DD)								55 F					65 F				
LCR	200	100	70	50	40	30	25	20	200	100	70	50	40	30	25	20	200	100	50	30	20	200	100	50	30	20
WW A1	19	27	33	40	46	53	59	65	14.	19	23	27	31	36	39	44	1.1	1.2	1.3	1.5	1.6	1.4	1.5	1.7	1.9	2.0
WW A2	20	32	41	51	58	67	72	79	12	21	26	33	38	46	51	58	0.9	1.0	1.1	1.3	1.4	1.1	1.2	1.4	1.5	1.7
WW A3	20	34	44	55	62	71	77	83	11	21	28	36	41	50	55	62	0.8	0.9	1.1	1.2	1.4	1.1	1.2	1.3	1.5	1.6
WW A4	19	35	46	57	64	74	79	85	10	21	28	37	43	52	58	65	0.8	0.9	1.0	1.2	1.4	1.0	1.1	1.3	1.4	1.6
WW A5	19	36	48	59	67	76	81	87	10	21	29	38	45	54	60	68	0.8	0.9	1.0	1.2	1.4	1.0	1.1	1.2	1.4	1.5
WW A6	19	37	49	60	68	77	83	88	9	21	30	39	46	55	62	69	0.8	0.9	1.0	1.2	1.3	1.0	1.1	1.2	1.4	1.5
WW B1	17	27	34	43	49	58	63	70	10	16	21	26	30	35	39	45	1.0	1.1	1.2	1.4	1.6	1.3	1.4	1.6	1.8	2.0
WW B2	19	36	47	58	65	75	80	86	10	22	30	39	45	54	61	68	0.8	0.9	1.0	1.2	1.4	1.0	1.1	1.2	1.4	1.5
WW B3	23	42	54	66	73	82	87	92	13	26	36	46	53	63	69	76	0.7	0.8	1.0	1.1	1.3	0.9	1.0	1.1	1.3	1.5
WW B4	22	42	55	67	75	83	88	93	11	26	37	48	55	66	72	79	0.7	0.8	1.0	1.1	1.3	0.9	1.0	1.1	1.3	1.4
WW B5	20	40	53	65	73	82	87	92	10	25	35	47	55	65	71	79	0.7	0.8	1.0	1.1	1.3	0.9	1.0	1.1	1.3	1.4
WW C1	24	44	56	68	75	84	88	93	14	28	38	48	56	65	72	79	0.7	0.8	1.0	1.1	1.3	0.9	1.0	1.1	1.3	1.4
WW C2	22	42	54	66	73	82	87	92	12	26	36	46	54	64	70	78	0.7	0.8	1.0	1.1	1.3	0.9	1.0	1.1	1.3	1.4
WW C3	25	48	61	74	81	89	92	96	13	31	43	55	64	74	80	86	0.7	0.8	0.9	1.1	1.3	0.8	0.9	1.0	1.2	1.4
WW C4	22	44	57	70	78	86	91	95	11	28	40	52	60	71	77	84	0.7	0.8	0.9	1.1	1.3	0.8	0.9	1.1	1.2	1.4
TW A1	19	27	33	40	45	52	58	64	14	19	23	27	30	35	39	44	1.1	1.2	1.3	1.5	1.6	1.4	1.6	1.7	1.9	2.0
TW A2	18	29	37	46	53	62	67	74	12	19	24	30	35	42	47	53	1.0	1.0	1.2	1.3	1.5	1.2	1.3	1.5	1.6	1.8
TW A3	18	30	39	49	56	65	71	78	11	19	25	32	37	45	50	56	0.9	1.0	1.1	1.3	1.5	1.1	1.2	1.4	1.6	1.7
TW A4	17	30	39	49	56	66	72	79	9	18	24	32	37	45	51	58	0.9	1.0	1.1	1.3	1.4	1.1	1.2	1.4	1.5	1.7
TW B1	18	26	33	40	46	54	59	66	13	18	22	27	31	36	40	45	1.1	1.2	1.3	1.4	1.6	1.4	1.5	1.6	1.8	2.0
TW B2	17	28	36	45	51	60	66	73	11	18	23	29	34	41	45	52	1.0	1.1	1.2	1.3	1.5	1.2	1.3	1.5	1.6	1.8
TW B3	16	27	36	45	52	61	67	74	10	17	23	29	34	42	47	53	1.0	1.0	1.2	1.3	1.5	1.2	1.3	1.4	1.6	1.8
TW B4	15	26	34	43	49	58	64	71	9	16	22	28	33	40	44	51	1.0	1.1	1.2	1.4	1.5	1.2	1.3	1.5	1.6	1.8
TW C1	16	25	32	39	45	53	59	66	11	17	21	26	30	36	40	46	1.1	1.2	1.3	1.4	1.6	1.4	1.5	1.6	1.8	1.9
TW C2	15	25	32	40	46	55	61	68	10	16	21	26	31	37	42	48	1.0	1.1	1.3	1.4	1.6	1.3	1.4	1.6	1.7	1.9
TW C3	14	23	30	38	44	52	58	65	10	16	20	25	29	36	40	46	1.1	1.2	1.3	1.4	1.6	1.4	1.4	1.6	1.7	1.9
TW C4	14	22	27	34	39	47	52	59	10	15	19	23	27	32	36	41	1.2	1.3	1.4	1.5	1.7	1.5	1.6	1.7	1.9	2.0
TW D1	14	23	29	37	43	51	56	63	8	14	17	22	26	31	35	40	1.1	1.2	1.3	1.5	1.6	1.4	1.5	1.7	1.9	2.1
TW D2	19	33	42	53	60	70	75	82	11	21	28	35	41	50	55	63	0.9	0.9	1.1	1.2	1.4	1.1	1.2	1.3	1.5	1.6
TW D3	20	36	46	57	65	74	80	86	11	22	30	39	45	54	60	68	0.8	0.9	1.0	1.2	1.4	1.0	1.1	1.2	1.4	1.5
TW D4	20	38	49	61	69	78	83	89	11	24	33	43	50	60	66	73	0.8	0.9	1.0	1.2	1.3	0.9	1.0	1.2	1.3	1.5
TW D5	19	37	43	60	68	78	83	89	10	23	33	43	50	60	66	74	0.8	0.9	1.0	1.2	1.3	0.9	1.0	1.2	1.3	1.5
TW E1	23	40	51	62	70	79	84	89	13	26	34	44	50	60	66	73	0.8	0.9	1.0	1.2	1.3	0.9	1.0	1.2	1.3	1.5
TW E2	21	39	50	62	70	79	84	89	12	25	34	43	50	60	66	74	0.8	0.9	1.0	1.2	1.3	0.9	1.0	1.2	1.3	1.5
TW E3	25	46	59	71	78	86	90	94	14	30	41	52	60	70	76	83	0.7	0.8	0.9	1.1	1.3	0.8	0.9	1.1	1.2	1.4
TW E4	22	43	55	68	75	84	89	93	12	28	38	49	57	68	74	81	0.7	0.8	1.0	1.1	1.3	0.8	0.9	1.1	1.2	1.4
TW F1	17	25	32	39	44	52	58	64	12	17	21	26	29	35	38	44	1.1	1.2	1.3	1.5	1.6	1.4	1.5	1.7	1.8	2.0
TW F2	16	27	35	44	50	59	65	72	10	17	22	28	33	39	44	50	1.0	1.1	1.2	1.4	1.5	1.2	1.3	1.5	1.7	1.8
TW F3	15	27	35	45	52	61	67	74	8	16	22	28	33	41	46	52	0.9	1.0	1.2	1.3	1.5	1.2	1.3	1.4	1.6	1.7
TW F4	12	25	33	43	50	60	66	73	6	14	20	26	32	39	45	51	1.0	1.0	1.2	1.3	1.5	1.2	1.3	1.4	1.6	1.7
TW G1	15	24	30	38	43	52	57	64	10	15	19	24	28	34	38	43	1.1	1.2	1.3	1.4	1.6	1.4	1.5	1.7	1.8	2.0
TW G2	13	23	31	40	46	55	61	69	7	14	19	25	29	36	41	47	1.0	1.1	1.2	1.4	1.5	1.3	1.4	1.5	1.7	1.8
TW G3	11	22	29	38	45	54	60	68	5	12	17	23	28	35	40	46	1.0	1.1	1.2	1.4	1.5	1.3	1.4	1.5	1.7	1.8
TW G4	9	18	25	33	39	48	54	62	3	10	14	20	24	31	35	41	1.1	1.2	1.3	1.4	1.6	1.4	1.5	1.6	1.8	1.9
TW H1	12	20	27	34	40	48	54	61	7	12	16	21	25	31	35	41	1.1	1.2	1.3	1.5	1.6	1.4	1.5	1.7	1.8	2.0
TW H2	9	18	24	32	38	46	52	60	4	10	14	19	23	29	34	39	1.1	1.2	1.3	1.5	1.6	1.4	1.5	1.7	1.8	2.0
TW H3	7	15	21	28	34	42	47	55	3	8	12	16	20	26	30	36	1.2	1.3	1.4	1.5	1.7	1.5	1.6	1.7	1.9	2.0
TW H4	5	11	16	22	27	34	39	45	2	6	9	13	16	21	24	29	1.3	1.4	1.6	1.7	1.8	1.7	1.8	1.9	2.1	2.2
TW I1	12	20	26	34	39	47	53	60	6	11	15	19	23	28	31	36	1.1	1.2	1.3	1.5	1.7	1.5	1.6	1.8	2.0	2.1
TW I2	15	29	39	49	57	66	72	79	8	17	24	32	38	46	52	59	0.9	1.0	1.1	1.3	1.4	1.1	1.2	1.3	1.5	1.6
TW I3	17	32	42	53	61	70	76	83	9	19	27	35	41	50	56	64	0.8	0.9	1.1	1.2	1.4	1.0	1.1	1.3	1.4	1.6
TW I4	17	34	46	57	65	75	81	87	9	21	30	39	46	56	63	70	0.8	0.9	1.0	1.2	1.4	1.0	1.1	1.2	1.3	1.5
TW I5	17	34	46	58	66	76	82	88	8	21	30	40	47	57	64	72	0.8	0.9	1.0	1.2	1.3	0.9	1.0	1.2	1.3	1.5
TW J1	20	37	48	60	67	77	82	88	11	23	32	41	48	57	64	71	0.8	0.9	1.0	1.2	1.4	0.9	1.0	1.2	1.3	1.5
TW J2	19	36	47	59	67	76	82	88	10	22	31	40	47	57	63	71	0.8	0.9	1.0	1.2	1.4	0.9	1.0	1.2	1.3	1.5
TW J3	21	42	55	67	75	84	89	93	11	27	38	49	57	67	74	81	0.7	0.8	1.0	1.1	1.3	0.8	0.9	1.1	1.2	1.4
TW J4	19	39	52	65	73	82	87	92	10	25	35	46	54	65	71	70	0.7	0.8	1.0	1.1	1.3	0.9	1.0	1.1	1.3	1.4
DG A1	14	25	33	41	40	53	57	62	8	14	19	24	27	32	35	38	1.0	1.1	1.3	1.5	1.8	1.3	1.4	1.7	2.1	2.5
DG A2	15	27	36	45	52	60	66	72	9	16	22	29	34	41	45	51	1.0	1.0	1.2	1.4	1.6	1.2	1.3	1.4	1.6	1.9
DG A3	17	32	42	54	62	71	76	82	11	20	27	36	43	52	58	65	0.9	0.9	1.1	1.2	1.5	1.1	1.2	1.3	1.4	1.6
DG B1	14	26	35	44	52	60	65	71	8	14	19	24	29	33	39	44	1.0	1.0	1.2	1.4	1.7	1.3	1.4	1.6	1.8	2.0
DG B2	15	28	38	49	57	67	73	79	9	17	23	30	35	44	49	56	0.9	1.0	1.1	1.3	1.5	1.2	1.3	1.4	1.5	1.7
DG B3	17	32	43	56	65	76	81	87	11	21	28	37	44	55	62	70	0.9	0.9	1.0	1.2	1.4	1.1	1.1	1.2	1.3	1.5
DG C1	17	30	41	52	60	70	75	80	10	17	23	30	35	43	48	54	0.9	0.9	1.1	1.3	1.5	1.2	1.3	1.4	1.6	1.8
DG C2	17	32	43	55	64	74	80	85	11	20	26	35	41	51	57	65	0.9	0.9	1.0	1.2	1.4	1.1	1.2	1.3	1.4	1.6
DG C3	20	36	49	63	72	82	87	92	13	24	32	42	50	61	69	77	0.8	0.8	1.0	1.1	1.3	1.0	1.1	1.1	1.2	1.4
SS A1	22	35	43	53	59	68	73	80	16	25	31	37	43	50	55	61	0.9	1.0	1.1	1.3	1.5	1.1	1.2	1.4	1.5	1.7
SS A2	26	43	54	65	72	80	85	90	18	31	39	48	55	64	69	76	0.8	0.8	1.0	1.2	1.3	0.9	1.0	1.2	1.3	1.5
SS A3	21	33	41	49	56	64	70	76	15	23	28	34	39	46	50	56	0.9	1.0	1.2	1.3	1.5	1.1	1.3	1.4	1.6	1.8
SS A4	26	43	53	64	71	79	84	89	18	30	38	47	54	62	68	75	0.8	0.9	1.0	1.2	1.4	0.9	1.0	1.2	1.3	1.5
SS A5	22	33	40	49	55	63	68	75	17	24	29	35	39	45	50	56	1.0	1.0	1.2	1.3	1.5	1.2	1.3	1.5	1.6	1.8
SS A6	26	43	54	64	71	80	85	90	18	30	39	48	54	63	69	75	0.8	0.8	1.0	1.2	1.3	0.9	1.0	1.2	1.3	1.5
SS A7	20	25	36	43	49	56	62	68	16	21	26	30	34	39	43	48	1.0	1.1	1.3	1.4	1.6	1.3	1.4	1.6	1.8	2.0
SS A8	26	42	52	63	70	78	83	88	18	30	37	46	52	61	67	73	0.8	0.9	1.0	1.2	1.4	0.9	1.0	1.2	1.4	1.5
SS B1	19	30	38	47	53	62	67	74	13	21	26	32	37	44	48	54	1.0	1.0	1.2	1.3	1.5	1.2	1.3	1.5	1.6	1.8
SS B2	22	38	48	59	66	75	80	86	15	26	34	43	49	58	64	71	0.8	0.9	1.0	1.2	1.4	1.0	1.1	1.2	1.4	1.5
SS B3	18	28	36	45	51	59	65	71	12	19	24	30	34	41	45	51	1.0	1.1	1.2	1.4	1.5	1.2	1.3	1.5	1.7	1.8
SS B4	22	37	47	58	65	74	79	85	15	26	33	42	48	57	62	69	0.8	0.9	1.1	1.2	1.4	1.0	1.1	1.2	1.4	1.5
SS B5	18	27	33	41	47	55	60	66	13	19	23	28	32	38	42	47	1.1	1.2	1.3	1.4	1.6	1.3	1.4	1.6	1.8	1.9
SS B6	22	37	47	58	65	74	80	85	15	26	33	42	48	57	63	70	0.8	0.9	1.0	1.2	1.4	1.0	1.1	1.2	1.4	1.5
SS B7	17	24	30	37	42	50	55	61	13	17	21	25	29	34	37	42	1.1	1.2	1.4	1.5	1.7	1.4	1.6	1.7	1.9	2.1
SS B8	21	36	46	56	64	73	78	84	14	25	32	40	46	55	60	67	0.8	0.9	1.1	1.2	1.4	1.0	1.1	1.3	1.4	1.6
SS C1	17	30	39	49	56	66	72	78	10	18	24	31	36	44	49	56	0.9	1.0	1.1	1.3	1.5	1.1	1.2	1.4	1.6	1.7
SS C2	20	36	46	58	65	75	80	86	11	22	30	39	46	55	61	68	0.8	0.9	1.0	1.2	1.4	1.0	1.1	1.2	1.4	1.5
SS C3	16	25	32	40	46	54	60	67	10	16	20	26	30	36	40	45	1.1	1.1	1.3	1.4	1.6	1.4	1.5	1.6	1.8	1.9
SS C4	18	32	41	52	59	68	74	81	11	20	27	35	41	49	55	62	0.9	1.0	1.1	1.3	1.4	1.1	1.2	1.3	1.5	1.6
SS D1	26	42	52	62	69	78	83	88	17	29	36	44	50	59	64	71	0.8	0.9	1.0	1.2	1.4	0.9	1.1	1.2	1.4	1.5
SS D2	30	50	62	73	80	87	91	95	19	35	45	56	63	72	78	84	0.7	0.8	0.9	1.1	1.3	0.8	0.9	1.1	1.2	1.4
SS D3	24	37	47	56	63	71	77	82	17	26	33	40	45	53	58	64	0.8	0.9	1.1	1.3	1.4	1.0	1.2	1.3	1.5	1.6
SS D4	29	48	60	71	78	86	90	94	19	34	44	54	61	70	76	82	0.7	0.8	1.0	1.1	1.3	0.8	0.9	1.1	1.2	1.4
SS E1	22	37	46	57	64	72	78	84	14	24	31	38	44	52	57	64	0.8	0.9	1.1	1.2	1.4	1.0	1.1	1.3	1.5	1.6
SS E2	25	44	56	67	74	83	87	92	16	30	39	49	56	65	71	78	0.7	0.8	1.0	1.1	1.3	0.9	1.0	1.1	1.3	1.5
SS E3	20	31	39	47	54	62	67	74	14	21	26	32	37	43	48	54	1.0	1.0	1.2	1.3	1.5	1.2	1.3	1.5	1.6	1.8
SS E4	24	41	52	63	70	79	83	89	15	28	36	45	52	61	66	73	0.8	0.9	1.0	1.2	1.4	0.9	1.0	1.2	1.3	1.5

BAKERSFIELD, CALIFORNIA

TYPE	SOLAR SAVINGS FRACTION (%) BASE TEMP: 55 F (760 DD) LCR 200	100	70	50	40	30	25	20	65 F (2196 DD) 200	100	70	50	40	30	25	20	CONSERVATION FACTOR 55 F 200	100	50	30	20	65 F 200	100	50	30	20
WW A1	34	49	57	66	72	79	84	88	21	30	36	43	48	56	61	67	0.7	0.9	1.1	1.2	1.4	1.0	1.1	1.3	1.5	1.6
WW A2	40	58	68	77	83	89	93	96	22	35	44	53	60	68	74	80	0.6	0.8	0.9	1.1	1.3	0.9	1.0	1.1	1.3	1.5
WW A3	42	61	72	81	86	92	95	97	22	37	47	57	64	73	78	83	0.6	0.7	0.9	1.1	1.3	0.8	0.9	1.1	1.2	1.4
WW A4	43	63	73	83	88	93	96	98	22	38	48	59	66	75	80	85	0.6	0.7	0.9	1.1	1.2	0.8	0.9	1.0	1.2	1.4
WW A5	44	65	75	85	90	95	97	98	22	39	50	61	68	77	82	87	0.6	0.7	0.9	1.1	1.2	0.8	0.9	1.0	1.2	1.4
WW A6	45	66	76	85	90	95	97	99	22	40	51	62	69	78	83	88	0.6	0.7	0.9	1.0	1.2	0.7	0.8	1.0	1.2	1.4
WW B1	36	52	62	71	77	84	88	92	19	30	37	46	52	60	65	71	0.7	0.8	1.0	1.2	1.3	0.9	1.1	1.2	1.4	1.6
WW B2	43	63	74	83	88	94	96	98	22	38	49	60	67	76	81	86	0.6	0.7	0.9	1.1	1.2	0.8	0.9	1.0	1.2	1.4
WW B3	48	70	80	88	93	97	98	99	26	45	57	68	74	82	87	91	0.6	0.7	0.9	1.0	1.2	0.7	0.8	1.0	1.2	1.3
WW B4	48	70	80	89	93	97	98	99	24	45	57	69	76	84	88	92	0.6	0.7	0.9	1.0	1.2	0.7	0.8	1.0	1.1	1.3
WW B5	46	68	79	88	92	96	98	99	23	43	55	67	75	83	87	92	0.6	0.7	0.9	1.0	1.2	0.7	0.8	1.0	1.1	1.3
WW C1	50	71	81	89	93	97	98	99	27	47	58	69	76	84	88	92	0.6	0.7	0.9	1.0	1.2	0.7	0.8	1.0	1.2	1.3
WW C2	47	69	80	88	92	96	98	99	25	44	56	68	75	83	87	92	0.6	0.7	0.9	1.0	1.2	0.7	0.8	1.0	1.2	1.3
WW C3	52	75	85	92	96	98	99	100	28	51	63	75	82	89	92	96	0.5	0.7	0.8	1.0	1.2	0.6	0.8	0.9	1.1	1.3
WW C4	48	71	82	90	94	98	99	100	25	47	60	72	79	86	90	94	0.6	0.7	0.8	1.0	1.2	0.7	0.8	0.9	1.1	1.3
TW A1	33	48	56	65	71	78	83	87	20	29	35	42	48	55	60	66	0.8	0.9	1.1	1.2	1.4	1.0	1.2	1.3	1.5	1.6
TW A2	37	54	63	73	79	86	89	93	20	32	40	49	55	64	69	75	0.7	0.8	1.0	1.1	1.3	0.9	1.0	1.2	1.3	1.5
TW A3	38	56	66	76	81	88	92	95	20	33	42	51	58	67	72	78	0.7	0.8	1.0	1.1	1.3	0.9	1.0	1.1	1.3	1.5
TW A4	37	56	66	76	82	89	92	95	19	33	42	52	59	68	73	79	0.7	0.8	0.9	1.1	1.3	0.9	1.0	1.1	1.3	1.5
TW B1	33	48	57	66	72	79	84	88	19	29	35	43	48	56	61	67	0.7	0.9	1.0	1.2	1.4	1.0	1.1	1.3	1.4	1.6
TW B2	35	52	62	71	77	85	89	93	19	30	38	47	54	62	68	74	0.7	0.8	1.0	1.2	1.3	0.9	1.0	1.2	1.3	1.5
TW B3	34	52	62	72	78	85	89	93	18	30	39	48	54	63	69	75	0.7	0.8	1.0	1.1	1.3	0.9	1.0	1.2	1.3	1.5
TW B4	32	50	60	69	75	83	87	92	17	28	36	45	52	60	66	73	0.7	0.8	1.0	1.2	1.3	0.9	1.0	1.2	1.4	1.5
TW C1	31	47	56	65	71	79	83	88	18	28	34	42	48	56	61	68	0.7	0.9	1.0	1.2	1.4	1.0	1.1	1.3	1.4	1.6
TW C2	31	47	57	66	72	80	85	89	17	27	35	43	49	57	63	69	0.7	0.9	1.0	1.2	1.4	1.0	1.1	1.2	1.4	1.6
TW C3	29	45	54	64	70	78	83	88	16	26	33	41	47	55	60	67	0.7	0.9	1.0	1.2	1.4	1.0	1.1	1.3	1.4	1.6
TW C4	27	41	49	58	65	72	77	83	15	24	30	37	42	50	55	61	0.8	0.9	1.1	1.3	1.4	1.1	1.2	1.4	1.5	1.7
TW D1	30	46	56	65	71	79	83	88	16	25	32	40	46	54	59	65	0.7	0.9	1.0	1.2	1.4	1.0	1.1	1.3	1.5	1.6
TW D2	40	59	69	79	84	91	94	97	21	35	45	55	62	71	76	82	0.6	0.8	0.9	1.1	1.3	0.8	0.9	1.1	1.3	1.4
TW D3	43	63	73	82	88	93	96	98	22	39	49	60	67	75	80	86	0.6	0.7	0.9	1.1	1.2	0.8	0.9	1.0	1.2	1.4
TW D4	44	65	76	85	90	95	97	99	23	41	52	63	70	79	84	89	0.6	0.7	0.9	1.1	1.2	0.7	0.8	1.0	1.2	1.4
TW D5	42	64	75	84	89	94	97	98	22	40	51	62	70	79	84	89	0.6	0.7	0.9	1.1	1.2	0.7	0.8	1.0	1.2	1.4
TW E1	46	66	77	86	90	95	97	99	25	43	54	64	71	80	84	89	0.6	0.7	0.9	1.1	1.2	0.7	0.8	1.0	1.2	1.4
TW E2	45	66	76	85	90	95	97	99	24	42	53	64	71	80	84	89	0.6	0.7	0.9	1.1	1.2	0.7	0.8	1.0	1.2	1.4
TW E3	51	72	83	90	94	98	99	100	28	49	61	72	79	86	90	94	0.6	0.7	0.8	1.0	1.2	0.7	0.8	0.9	1.1	1.3
TW E4	48	70	80	89	93	97	98	99	25	46	58	69	76	84	89	93	0.6	0.7	0.9	1.0	1.2	0.7	0.8	1.0	1.1	1.3
TW F1	32	47	56	65	71	78	83	88	19	28	34	42	47	55	60	66	0.7	0.9	1.1	1.2	1.4	1.0	1.1	1.3	1.5	1.6
TW F2	34	51	61	71	77	84	88	92	18	30	38	46	53	61	67	73	0.7	0.8	1.0	1.2	1.3	0.9	1.0	1.2	1.4	1.5
TW F3	34	52	63	72	79	86	90	94	17	29	38	48	54	63	69	75	0.7	0.8	1.0	1.1	1.3	0.9	1.0	1.2	1.3	1.5
TW F4	32	50	61	71	77	85	89	93	15	27	36	46	53	62	68	74	0.7	0.8	1.0	1.1	1.3	0.9	1.0	1.2	1.3	1.5
TW G1	30	46	55	64	70	78	83	88	16	26	33	40	46	54	59	66	0.7	0.9	1.0	1.2	1.4	1.0	1.1	1.3	1.4	1.6
TW G2	30	48	58	67	74	82	86	91	15	26	34	43	49	58	63	70	0.7	0.8	1.0	1.2	1.3	1.0	1.1	1.2	1.4	1.6
TW G3	28	46	56	66	73	81	85	90	13	24	32	41	48	57	62	69	0.7	0.8	1.0	1.2	1.3	1.0	1.1	1.2	1.4	1.6
TW G4	24	41	51	61	67	76	81	86	10	20	28	36	42	51	57	64	0.7	0.9	1.1	1.2	1.4	1.0	1.1	1.3	1.4	1.6
TW H1	27	42	52	61	67	76	80	86	13	23	29	37	43	51	57	63	0.8	0.9	1.1	1.2	1.4	1.0	1.2	1.3	1.5	1.6
TW H2	24	40	49	59	66	74	79	85	10	20	27	35	41	49	55	62	0.8	0.9	1.1	1.2	1.4	1.1	1.2	1.3	1.5	1.6
TW H3	20	35	45	54	61	69	75	81	8	17	23	31	36	45	50	57	0.8	0.9	1.1	1.3	1.4	1.1	1.2	1.4	1.5	1.7
TW H4	16	29	37	46	52	61	66	73	6	13	18	25	29	37	42	48	0.9	1.0	1.2	1.4	1.5	1.3	1.4	1.5	1.7	1.8
TW I1	28	43	52	62	68	76	81	86	13	23	29	37	42	50	55	62	0.8	0.9	1.1	1.2	1.4	1.1	1.2	1.3	1.5	1.7
TW I2	36	55	66	76	82	89	92	96	18	32	42	52	59	68	74	80	0.6	0.8	0.9	1.1	1.3	0.8	0.9	1.1	1.3	1.5
TW I3	38	58	69	79	85	91	94	97	19	35	45	55	63	72	77	83	0.6	0.8	0.9	1.1	1.3	0.8	0.9	1.1	1.2	1.4
TW I4	40	62	73	82	88	93	96	98	20	37	48	60	67	76	81	87	0.6	0.7	0.9	1.1	1.2	0.8	0.9	1.0	1.2	1.4
TW I5	40	62	73	82	88	93	96	98	19	37	49	60	68	77	82	88	0.6	0.7	0.9	1.1	1.2	0.8	0.9	1.0	1.2	1.4
TW J1	43	64	75	84	89	94	96	98	23	40	51	62	69	78	83	88	0.6	0.7	0.9	1.1	1.2	0.8	0.9	1.0	1.2	1.4
TW J2	42	63	74	83	88	94	96	98	21	39	50	61	68	77	82	88	0.6	0.7	0.9	1.1	1.2	0.8	0.9	1.0	1.2	1.4
TW J3	47	69	80	89	93	97	98	99	24	45	57	69	76	84	89	93	0.6	0.7	0.9	1.0	1.2	0.7	0.8	1.0	1.1	1.3
TW J4	44	67	78	87	92	96	98	99	22	42	55	67	74	82	87	92	0.6	0.7	0.9	1.0	1.2	0.7	0.8	1.0	1.1	1.3
DG A1	35	52	61	70	75	81	85	88	17	28	36	44	49	56	60	64	0.7	0.8	1.0	1.2	1.5	0.9	1.1	1.3	1.5	1.9
DG A2	34	53	63	73	79	85	89	92	17	30	39	48	54	63	67	73	0.7	0.8	1.0	1.2	1.4	0.9	1.0	1.2	1.4	1.6
DG A3	38	58	69	79	84	90	92	94	19	35	45	56	63	72	77	82	0.6	0.7	0.9	1.2	1.5	0.8	0.9	1.1	1.3	1.5
DG B1	36	55	66	76	81	87	90	93	17	29	38	48	54	62	67	72	0.7	0.8	1.0	1.2	1.4	0.9	1.0	1.2	1.4	1.7
DG B2	36	56	68	78	84	90	93	96	17	31	41	52	59	68	74	79	0.7	0.8	0.9	1.1	1.3	0.9	0.9	1.1	1.3	1.5
DG B3	39	60	72	83	88	93	95	97	20	35	47	59	67	76	81	87	0.6	0.7	0.9	1.1	1.4	0.8	0.9	1.0	1.2	1.4
DG C1	41	61	73	83	88	92	94	96	19	34	45	55	62	71	75	80	0.6	0.7	0.9	1.1	1.5	0.9	0.9	1.1	1.3	1.6
DG C2	40	61	74	84	89	94	96	98	20	35	46	58	66	75	80	85	0.6	0.7	0.9	1.1	1.3	0.8	0.9	1.0	1.2	1.4
DG C3	43	65	78	88	92	96	97	98	22	40	52	65	73	82	87	91	0.6	0.7	0.9	1.1	1.5	0.8	0.8	1.0	1.1	1.4
SS A1	41	58	67	76	82	88	92	95	25	38	47	55	61	69	74	80	0.7	0.8	1.0	1.1	1.3	0.8	1.0	1.1	1.3	1.5
SS A2	47	67	76	85	90	95	97	98	30	46	56	66	73	80	85	89	0.6	0.7	0.9	1.1	1.2	0.7	0.8	1.0	1.2	1.4
SS A3	39	55	64	73	79	86	89	93	24	36	44	52	58	66	70	76	0.7	0.8	1.0	1.1	1.3	0.9	1.0	1.2	1.4	1.5
SS A4	47	66	76	84	89	94	96	98	29	46	56	65	72	79	84	88	0.6	0.7	0.9	1.1	1.2	0.7	0.9	1.0	1.2	1.4
SS A5	39	54	63	72	78	84	88	92	25	36	43	51	57	65	69	75	0.7	0.8	1.0	1.2	1.3	0.9	1.0	1.2	1.4	1.6
SS A6	47	67	76	85	90	95	97	98	30	46	56	66	72	80	84	89	0.6	0.7	0.9	1.1	1.2	0.7	0.8	1.0	1.2	1.4
SS A7	35	50	58	66	72	79	84	88	23	32	39	46	51	58	63	69	0.7	0.9	1.1	1.2	1.4	1.0	1.1	1.3	1.5	1.6
SS A8	47	65	75	84	89	94	96	98	29	45	55	64	71	78	83	88	0.6	0.7	0.9	1.1	1.2	0.7	0.9	1.0	1.2	1.4
SS B1	36	53	62	71	77	84	88	92	21	33	41	50	56	64	69	75	0.7	0.8	1.0	1.2	1.3	0.9	1.0	1.2	1.4	1.5
SS B2	43	62	72	81	86	92	95	97	25	41	51	61	68	76	81	86	0.6	0.8	0.9	1.1	1.3	0.8	0.9	1.1	1.2	1.4
SS B3	35	51	60	69	75	82	87	91	21	32	39	47	53	61	66	72	0.7	0.8	1.0	1.2	1.3	0.9	1.0	1.2	1.4	1.6
SS B4	42	61	71	80	86	91	94	97	25	40	50	60	67	75	80	85	0.6	0.8	0.9	1.1	1.3	0.8	0.9	1.1	1.2	1.4
SS B5	33	47	56	65	71	78	83	88	20	30	37	44	49	57	62	68	0.7	0.9	1.1	1.2	1.4	1.0	1.1	1.3	1.5	1.6
SS B6	42	61	71	80	86	92	95	97	25	40	50	60	67	75	80	85	0.6	0.8	0.9	1.1	1.3	0.8	0.9	1.1	1.2	1.4
SS B7	30	44	52	61	67	74	79	84	19	27	33	40	45	52	57	63	0.8	0.9	1.1	1.3	1.4	1.1	1.2	1.3	1.5	1.7
SS B8	41	60	70	79	85	91	94	97	24	39	49	59	65	73	78	84	0.6	0.8	0.9	1.1	1.3	0.8	0.9	1.1	1.3	1.4
SS C1	38	57	67	77	83	89	92	96	20	33	42	52	59	67	73	79	0.6	0.8	0.9	1.1	1.3	0.9	1.0	1.1	1.3	1.5
SS C2	43	63	73	83	88	93	96	98	22	39	49	60	67	76	81	86	0.6	0.7	0.9	1.1	1.2	0.8	0.9	1.0	1.2	1.4
SS C3	32	48	57	66	72	80	84	89	17	27	34	42	48	56	62	68	0.7	0.9	1.0	1.2	1.4	1.0	1.1	1.3	1.4	1.6
SS C4	38	57	68	77	83	90	93	96	20	34	44	54	61	70	75	81	0.6	0.8	0.9	1.1	1.3	0.8	0.9	1.1	1.3	1.4
SS D1	47	66	76	84	89	94	96	98	29	45	55	64	70	78	82	87	0.6	0.7	0.9	1.1	1.2	0.7	0.9	1.0	1.2	1.4
SS D2	53	74	83	91	95	98	99	100	33	53	64	74	80	87	90	94	0.6	0.7	0.8	1.0	1.2	0.7	0.8	1.0	1.1	1.3
SS D3	43	61	70	79	84	90	93	96	27	41	50	58	65	72	77	82	0.6	0.8	0.9	1.1	1.3	0.8	0.9	1.1	1.3	1.5
SS D4	52	72	82	90	93	97	98	99	32	51	62	72	78	85	89	93	0.6	0.7	0.9	1.0	1.2	0.7	0.8	1.0	1.2	1.3
SS E1	43	62	72	81	86	92	95	97	25	40	49	59	65	73	78	83	0.6	0.8	0.9	1.1	1.3	0.8	0.9	1.1	1.3	1.4
SS E2	49	69	79	88	92	96	98	99	29	47	58	68	75	83	87	91	0.6	0.7	0.9	1.0	1.2	0.7	0.8	1.0	1.2	1.3
SS E3	37	53	63	72	77	84	88	92	22	34	42	50	56	64	69	74	0.7	0.8	1.0	1.2	1.3	0.9	1.0	1.2	1.4	1.6
SS E4	46	65	75	84	89	94	96	98	27	44	54	64	71	79	83	88	0.6	0.7	0.9	1.1	1.2	0.7	0.9	1.0	1.2	1.4

DAGGETT, CALIFORNIA

TYPE	SOLAR SAVINGS FRACTION (%)																CONSERVATION FACTOR									
BASE TEMP:	55 F (829 DD)								65 F (2206 DD)								55 F					65 F				
LCR	200	100	70	50	40	30	25	20	200	100	70	50	40	30	25	20	200	100	50	30	20	200	100	50	30	20
WW A1	39	56	67	76	82	88	92	95	23	34	42	50	56	65	70	77	0.7	0.8	1.0	1.1	1.3	0.9	1.0	1.2	1.3	1.5
WW A2	47	68	78	87	91	96	98	99	25	41	52	62	69	78	83	88	0.6	0.7	0.9	1.0	1.2	0.8	0.9	1.0	1.2	1.4
WW A3	50	72	82	90	94	97	99	99	26	44	55	67	74	82	87	91	0.6	0.7	0.8	1.0	1.2	0.7	0.8	1.0	1.2	1.3
WW A4	51	74	84	91	95	98	99	100	27	46	57	69	76	84	89	93	0.5	0.7	0.8	1.0	1.2	0.7	0.8	1.0	1.1	1.3
WW A5	53	76	86	93	96	98	99	100	27	47	60	71	78	86	90	94	0.5	0.7	0.8	1.0	1.2	0.7	0.8	0.9	1.1	1.3
WW A6	54	77	87	93	96	99	99	100	27	48	61	72	79	87	91	95	0.5	0.7	0.8	1.0	1.2	0.7	0.8	0.9	1.1	1.3
WW B1	42	62	73	82	87	93	95	98	22	36	45	55	61	70	76	82	0.6	0.7	0.9	1.1	1.3	0.8	0.9	1.1	1.3	1.4
WW B2	51	74	84	91	95	98	99	100	26	46	58	69	77	85	89	93	0.5	0.7	0.8	1.0	1.2	0.7	0.8	1.0	1.1	1.3
WW B3	58	80	89	95	97	99	100	100	31	53	66	77	83	90	94	96	0.5	0.6	0.8	1.0	1.2	0.6	0.8	0.9	1.1	1.3
WW B4	57	80	89	95	98	99	100	100	30	53	66	78	84	91	94	97	0.5	0.6	0.8	1.0	1.2	0.6	0.7	0.9	1.1	1.3
WW B5	55	79	88	94	97	99	100	100	28	51	64	76	83	90	94	97	0.5	0.6	0.8	1.0	1.2	0.6	0.7	0.9	1.1	1.3
WW C1	59	81	90	96	98	99	100	100	33	55	67	78	85	91	94	97	0.5	0.6	0.8	1.0	1.2	0.6	0.7	0.9	1.1	1.3
WW C2	57	80	89	95	97	99	100	100	30	52	65	77	83	90	94	97	0.5	0.6	0.8	1.0	1.2	0.6	0.7	0.9	1.1	1.3
WW C3	62	85	92	97	99	100	100	100	34	59	72	83	89	94	97	98	0.5	0.6	0.8	1.0	1.2	0.6	0.7	0.9	1.1	1.2
WW C4	58	82	91	96	98	99	100	100	31	55	69	80	86	93	95	98	0.5	0.6	0.8	1.0	1.2	0.6	0.7	0.9	1.1	1.2
TW A1	38	55	65	75	81	87	91	95	22	33	41	49	55	64	69	76	0.7	0.8	1.0	1.1	1.3	0.9	1.0	1.2	1.3	1.5
TW A2	43	63	73	83	88	93	96	98	23	37	47	57	64	73	78	84	0.6	0.7	0.9	1.1	1.2	0.8	0.9	1.1	1.2	1.4
TW A3	44	66	76	85	90	95	97	99	24	39	49	60	67	76	82	87	0.6	0.7	0.9	1.1	1.2	0.8	0.9	1.0	1.2	1.4
TW A4	44	66	77	86	91	95	97	99	23	39	50	61	68	77	82	88	0.6	0.7	0.9	1.0	1.2	0.8	0.9	1.0	1.2	1.4
TW B1	38	56	66	76	82	88	92	95	22	33	41	50	56	65	70	77	0.7	0.8	0.9	1.1	1.3	0.9	1.0	1.2	1.3	1.5
TW B2	41	61	72	81	87	93	95	98	22	36	45	55	63	72	77	83	0.6	0.7	0.9	1.1	1.3	0.8	0.9	1.1	1.2	1.4
TW B3	41	61	72	82	87	93	96	98	21	36	45	56	63	72	78	84	0.6	0.7	0.9	1.1	1.2	0.8	0.9	1.1	1.2	1.4
TW B4	38	58	70	79	85	91	94	97	20	33	43	53	60	70	75	82	0.6	0.7	0.9	1.1	1.3	0.9	0.9	1.1	1.2	1.4
TW C1	36	55	65	75	81	88	92	95	20	32	40	49	56	65	70	77	0.7	0.8	0.9	1.1	1.3	0.9	1.0	1.2	1.3	1.5
TW C2	36	55	66	76	82	89	93	96	19	32	40	50	57	66	72	79	0.7	0.8	0.9	1.1	1.3	0.9	1.0	1.1	1.3	1.4
TW C3	34	53	64	74	80	87	91	95	18	30	38	48	54	63	69	76	0.7	0.8	1.0	1.1	1.3	0.9	1.0	1.2	1.3	1.5
TW C4	31	47	58	68	74	82	87	91	17	27	34	42	49	57	63	70	0.7	0.8	1.0	1.2	1.3	1.0	1.1	1.2	1.4	1.5
TW D1	36	55	66	76	82	89	92	95	18	30	39	48	55	64	69	76	0.7	0.8	0.9	1.1	1.3	0.9	1.0	1.2	1.3	1.5
TW D2	47	69	79	88	92	96	98	99	25	42	53	64	71	80	85	90	0.6	0.7	0.9	1.0	1.2	0.8	0.8	1.0	1.2	1.3
TW D3	51	73	83	91	95	98	99	100	27	46	58	69	76	84	89	93	0.6	0.7	0.8	1.0	1.2	0.7	0.8	1.0	1.1	1.3
TW D4	52	75	85	92	96	98	99	100	28	48	60	72	79	87	91	95	0.5	0.7	0.8	1.0	1.2	0.7	0.8	0.9	1.1	1.3
TW D5	51	74	84	92	95	98	99	100	26	47	59	71	78	87	91	94	0.5	0.7	0.8	1.0	1.2	0.7	0.8	0.9	1.1	1.3
TW E1	55	77	86	93	96	99	99	100	30	50	62	73	80	88	91	95	0.5	0.7	0.8	1.0	1.2	0.7	0.8	0.9	1.1	1.3
TW E2	54	76	86	93	96	98	99	100	29	49	61	73	80	87	91	95	0.5	0.7	0.8	1.0	1.2	0.7	0.8	0.9	1.1	1.3
TW E3	60	82	91	96	98	99	100	100	34	57	70	81	87	93	95	98	0.5	0.6	0.8	1.0	1.2	0.6	0.7	0.9	1.1	1.2
TW E4	57	80	89	95	97	99	100	100	30	53	66	78	84	91	94	97	0.5	0.6	0.8	1.0	1.2	0.6	0.7	0.9	1.1	1.3
TW F1	37	55	65	75	81	88	91	95	21	32	40	49	55	64	69	76	0.7	0.8	1.0	1.1	1.3	0.9	1.0	1.2	1.3	1.5
TW F2	40	61	72	81	87	92	95	97	21	35	44	55	62	71	77	83	0.6	0.7	0.9	1.1	1.3	0.8	0.9	1.1	1.2	1.4
TW F3	41	62	73	83	88	93	96	98	20	35	45	56	64	73	79	85	0.6	0.7	0.9	1.1	1.2	0.8	0.9	1.1	1.2	1.4
TW F4	38	60	72	81	87	93	95	98	18	33	43	54	62	72	77	84	0.6	0.7	0.9	1.1	1.2	0.8	0.9	1.1	1.2	1.4
TW G1	35	54	64	74	81	88	91	95	19	30	38	48	54	63	69	76	0.7	0.8	1.0	1.1	1.3	0.9	1.0	1.2	1.3	1.5
TW G2	36	56	68	78	84	90	94	96	18	31	40	51	58	67	73	80	0.6	0.8	0.9	1.1	1.3	0.9	1.0	1.1	1.3	1.4
TW G3	34	55	66	77	83	90	93	96	16	29	39	49	56	66	72	79	0.6	0.8	0.9	1.1	1.3	0.9	1.0	1.1	1.3	1.4
TW G4	29	49	60	71	78	86	90	94	13	25	33	43	50	60	66	74	0.7	0.8	1.0	1.1	1.3	0.9	1.0	1.2	1.3	1.5
TW H1	31	50	61	71	78	86	90	94	15	27	35	44	51	60	66	73	0.7	0.8	1.0	1.1	1.3	1.0	1.0	1.2	1.3	1.5
TW H2	28	47	59	70	76	84	89	93	13	24	32	42	49	58	64	72	0.7	0.8	1.0	1.1	1.3	1.0	1.1	1.2	1.3	1.5
TW H3	25	42	53	64	71	80	85	90	10	21	28	37	44	53	59	66	0.7	0.8	1.0	1.2	1.3	1.0	1.1	1.2	1.4	1.6
TW H4	19	34	44	55	62	71	77	83	8	16	22	30	35	44	49	57	0.8	0.9	1.1	1.2	1.4	1.1	1.2	1.4	1.5	1.7
TW I1	33	51	62	73	79	86	90	94	16	27	35	44	51	60	66	70	0.7	0.8	1.0	1.1	1.3	1.0	1.0	1.2	1.3	1.5
TW I2	43	65	77	86	91	95	97	99	21	38	49	61	68	77	83	88	0.6	0.7	0.9	1.0	1.2	0.8	0.9	1.0	1.2	1.4
TW I3	46	69	80	88	93	97	98	99	23	41	53	64	72	81	86	91	0.6	0.7	0.9	1.0	1.2	0.7	0.8	1.0	1.2	1.3
TW I4	49	72	83	91	94	98	99	100	24	44	57	69	76	85	89	93	0.6	0.7	0.8	1.0	1.2	0.7	0.8	1.0	1.1	1.3
TW I5	48	72	83	91	95	98	99	100	24	44	57	69	77	85	90	94	0.6	0.7	0.8	1.0	1.2	0.7	0.8	1.0	1.1	1.3
TW J1	52	75	85	92	95	98	99	100	27	47	59	71	78	86	90	94	0.5	0.7	0.8	1.0	1.2	0.7	0.8	0.9	1.1	1.3
TW J2	50	73	84	91	95	98	99	100	26	46	58	70	77	86	90	94	0.5	0.7	0.8	1.0	1.2	0.7	0.8	1.0	1.1	1.3
TW J3	56	80	89	95	97	99	100	100	30	53	66	78	84	91	94	97	0.5	0.6	0.8	1.0	1.2	0.6	0.7	0.9	1.1	1.3
TW J4	54	77	87	94	97	99	99	100	27	50	63	75	82	90	93	96	0.5	0.6	0.8	1.0	1.2	0.7	0.8	0.9	1.1	1.3
DG A1	42	62	72	81	86	91	93	96	20	34	43	53	59	66	70	75	0.6	0.7	0.9	1.1	1.3	0.8	0.9	1.1	1.4	1.6
DG A2	41	63	74	83	88	93	95	98	20	36	46	57	64	72	77	82	0.6	0.7	0.9	1.1	1.3	0.8	0.9	1.1	1.3	1.5
DG A3	45	69	80	88	91	94	95	96	23	41	53	65	73	81	85	89	0.6	0.7	0.9	1.2	1.8	0.8	0.8	1.0	1.2	1.4
DG B1	44	68	79	87	91	95	96	97	20	36	47	59	66	74	78	83	0.6	0.7	0.9	1.1	1.4	0.8	0.9	1.0	1.3	1.5
DG B2	44	69	80	89	93	96	98	99	21	37	49	62	70	79	84	88	0.6	0.7	0.9	1.0	1.2	0.8	0.9	1.0	1.2	1.4
DG B3	47	72	84	91	94	97	98	98	23	42	55	69	77	85	89	93	0.6	0.7	0.8	1.1	1.9	0.8	0.8	0.9	1.1	1.4
DG C1	50	75	86	92	95	97	98	98	23	42	54	67	75	82	86	90	0.5	0.6	0.9	1.2	1.9	0.8	0.8	1.0	1.2	1.5
DG C2	48	74	85	93	96	98	99	100	23	42	56	69	77	85	89	93	0.5	0.6	0.8	1.0	1.2	0.8	0.8	0.9	1.1	1.3
DG C3	52	78	89	95	97	98	98	98	26	47	62	75	83	90	93	96	0.5	0.6	0.8	1.2	3.4	0.7	0.8	0.9	1.1	1.3
SS A1	47	67	77	85	90	95	97	98	28	44	53	63	70	78	82	88	0.6	0.7	0.9	1.1	1.2	0.8	0.9	1.0	1.2	1.4
SS A2	55	76	86	92	96	98	99	100	34	53	64	74	80	87	91	95	0.5	0.7	0.8	1.0	1.2	0.7	0.8	0.9	1.1	1.3
SS A3	45	64	74	83	88	93	96	98	27	41	50	60	66	74	79	85	0.6	0.7	0.9	1.1	1.3	0.8	0.9	1.1	1.2	1.4
SS A4	55	75	85	92	95	98	99	100	34	52	63	73	80	87	90	94	0.5	0.7	0.8	1.0	1.2	0.7	0.8	1.0	1.1	1.3
SS A5	44	63	73	82	87	92	95	97	27	41	49	59	65	73	78	84	0.6	0.7	0.9	1.1	1.3	0.8	0.9	1.1	1.3	1.4
SS A6	55	76	86	92	96	98	99	100	34	53	64	74	80	87	91	95	0.5	0.7	0.8	1.0	1.2	0.7	0.8	0.9	1.1	1.3
SS A7	40	57	67	76	82	88	92	95	25	37	44	53	59	67	72	78	0.7	0.8	1.0	1.1	1.3	0.9	1.0	1.2	1.3	1.5
SS A8	54	75	84	91	95	98	99	100	33	51	62	72	79	86	90	94	0.6	0.7	0.8	1.0	1.2	0.7	0.8	1.0	1.1	1.3
SS B1	42	61	72	81	86	92	95	97	24	38	47	57	64	72	77	83	0.6	0.7	0.9	1.1	1.3	0.8	0.9	1.1	1.2	1.4
SS B2	50	71	81	89	93	97	98	99	29	47	58	69	75	83	88	92	0.6	0.7	0.9	1.0	1.2	0.7	0.8	1.0	1.2	1.3
SS B3	40	59	70	79	85	91	94	97	23	36	45	55	61	70	75	81	0.6	0.8	0.9	1.1	1.3	0.8	1.0	1.1	1.3	1.4
SS B4	49	70	80	89	93	97	98	99	29	46	57	68	74	82	87	91	0.6	0.7	0.9	1.0	1.2	0.7	0.8	1.0	1.2	1.3
SS B5	37	55	65	75	80	87	91	94	22	34	42	50	57	65	70	76	0.7	0.8	1.0	1.1	1.3	0.9	1.0	1.2	1.3	1.5
SS B6	49	71	81	89	93	97	98	99	29	46	57	68	75	83	87	92	0.6	0.7	0.9	1.0	1.2	0.7	0.8	1.0	1.2	1.3
SS B7	34	51	61	70	76	84	88	92	21	31	38	46	52	60	65	72	0.7	0.8	1.0	1.2	1.3	1.0	1.1	1.2	1.4	1.6
SS B8	48	69	79	88	92	96	98	99	28	45	56	66	73	81	86	91	0.6	0.7	0.9	1.0	1.2	0.7	0.8	1.0	1.2	1.3
SS C1	46	67	78	87	91	96	97	99	24	40	50	61	69	77	83	88	0.6	0.7	0.9	1.0	1.2	0.8	0.9	1.0	1.2	1.4
SS C2	51	73	83	91	95	98	99	100	27	46	58	69	76	85	89	93	0.6	0.7	0.8	1.0	1.2	0.7	0.8	1.0	1.1	1.3
SS C3	37	56	67	77	83	89	93	96	20	32	40	50	57	66	71	78	0.7	0.8	0.9	1.1	1.3	0.9	1.0	1.1	1.3	1.5
SS C4	45	67	78	87	92	96	98	99	24	41	52	63	70	79	84	89	0.6	0.7	0.9	1.0	1.2	0.8	0.9	1.0	1.2	1.3
SS D1	55	76	85	92	95	98	99	100	33	52	62	73	79	86	90	94	0.5	0.7	0.8	1.0	1.2	0.7	0.8	1.0	1.1	1.3
SS D2	63	83	91	96	98	99	100	100	39	60	72	82	87	93	96	98	0.5	0.6	0.8	1.0	1.2	0.6	0.7	0.9	1.1	1.3
SS D3	50	70	80	88	92	96	98	99	31	47	57	66	73	81	85	90	0.6	0.7	0.9	1.0	1.2	0.7	0.9	1.0	1.2	1.4
SS D4	61	82	90	95	98	99	100	100	37	58	70	80	86	92	95	97	0.5	0.6	0.8	1.0	1.2	0.6	0.7	0.9	1.1	1.3
SS E1	51	72	82	89	93	97	98	99	29	46	57	67	74	82	86	91	0.6	0.7	0.9	1.0	1.2	0.7	0.8	1.0	1.2	1.3
SS E2	58	79	88	94	97	99	100	100	34	55	66	77	83	90	93	96	0.5	0.6	0.8	1.0	1.2	0.6	0.8	0.9	1.1	1.3
SS E3	43	62	72	81	87	92	95	97	25	39	48	57	64	72	78	83	0.6	0.7	0.9	1.1	1.3	0.8	0.9	1.1	1.3	1.4
SS E4	54	75	85	92	95	98	99	100	31	50	62	72	79	86	90	94	0.5	0.7	0.8	1.0	1.2	0.7	0.8	1.0	1.1	1.3

FRESNO, CALIFORNIA

TYPE	SOLAR SAVINGS FRACTION (%) — BASE TEMP: 55 F (1000 DD)								65 F (2657 DD)								CONSERVATION FACTOR — 55 F					65 F				
LCR	200	100	70	50	40	30	25	20	200	100	70	50	40	30	25	20	200	100	50	30	20	200	100	50	30	20
WW A1	29	41	49	57	63	70	74	80	19	26	32	38	42	48	53	58	0.8	1.0	1.2	1.3	1.5	1.1	1.2	1.4	1.6	1.8
WW A2	33	49	53	68	74	81	85	90	19	30	38	46	52	60	65	71	0.7	0.8	1.0	1.2	1.4	0.9	1.0	1.2	1.4	1.6
WW A3	34	52	62	71	77	84	88	92	18	31	40	49	56	64	69	75	0.7	0.8	1.0	1.2	1.3	0.9	1.0	1.2	1.3	1.5
WW A4	35	53	64	73	79	86	90	94	18	32	41	51	57	66	71	77	0.7	0.8	1.0	1.1	1.3	0.8	1.0	1.1	1.3	1.5
WW A5	36	55	65	75	81	88	91	95	18	33	43	53	59	68	73	79	0.6	0.8	1.0	1.1	1.3	0.8	0.9	1.1	1.3	1.5
WW A6	36	56	66	76	82	88	92	95	18	33	43	54	61	69	74	80	0.6	0.8	0.9	1.1	1.3	0.8	0.9	1.1	1.3	1.5
WW B1	29	44	52	61	67	74	79	84	16	25	32	38	44	51	55	61	0.8	0.9	1.1	1.3	1.4	1.0	1.2	1.4	1.5	1.8
WW B2	34	54	64	74	80	87	90	94	18	33	42	52	59	68	73	79	0.7	0.8	1.0	1.1	1.3	0.8	0.9	1.1	1.3	1.5
WW B3	40	60	70	80	85	91	94	97	21	38	49	59	66	75	80	85	0.6	0.8	0.9	1.1	1.3	0.8	0.9	1.0	1.2	1.4
WW B4	39	60	71	80	86	92	95	97	20	38	49	61	68	77	82	87	0.6	0.7	0.9	1.1	1.3	0.7	0.9	1.0	1.2	1.4
WW B5	37	58	69	79	85	91	94	97	18	37	48	59	67	76	81	87	0.6	0.7	0.9	1.1	1.3	0.8	0.9	1.0	1.2	1.4
WW C1	41	61	72	81	86	92	95	97	23	40	51	61	68	77	81	87	0.6	0.7	0.9	1.1	1.3	0.7	0.9	1.0	1.2	1.4
WW C2	39	59	70	80	85	91	94	97	20	38	49	60	67	76	81	86	0.6	0.7	0.9	1.1	1.3	0.8	0.9	1.0	1.2	1.4
WW C3	42	65	76	85	90	95	97	99	23	44	56	67	75	83	87	92	0.6	0.7	0.9	1.1	1.2	0.7	0.8	1.0	1.2	1.3
WW C4	39	62	73	83	88	93	96	98	20	40	52	64	71	80	85	90	0.6	0.7	0.9	1.1	1.2	0.7	0.8	1.0	1.2	1.3
TW A1	28	40	48	56	62	69	73	79	18	26	31	37	41	48	52	58	0.8	1.0	1.2	1.3	1.5	1.1	1.3	1.4	1.6	1.8
TW A2	30	45	54	63	69	77	81	86	18	28	35	42	48	55	60	67	0.7	0.9	1.1	1.2	1.4	1.0	1.1	1.3	1.5	1.6
TW A3	31	47	57	66	72	79	84	89	17	28	36	44	50	58	64	70	0.7	0.9	1.0	1.2	1.4	0.9	1.1	1.2	1.4	1.6
TW A4	30	47	57	66	72	80	84	89	16	28	36	44	51	59	64	71	0.7	0.8	1.0	1.2	1.4	0.9	1.0	1.2	1.4	1.6
TW B1	27	40	49	57	62	70	74	80	17	25	31	37	42	49	53	59	0.8	0.9	1.1	1.3	1.5	1.1	1.2	1.4	1.6	1.8
TW B2	28	44	53	62	68	75	80	85	16	26	33	41	47	54	59	66	0.8	0.9	1.1	1.3	1.4	1.0	1.1	1.3	1.5	1.6
TW B3	28	44	53	62	68	76	81	86	15	26	33	41	47	55	60	67	0.7	0.9	1.1	1.2	1.4	1.0	1.1	1.3	1.4	1.6
TW B4	26	41	51	60	66	74	79	84	14	24	31	39	45	53	58	64	0.8	0.9	1.1	1.3	1.4	1.0	1.1	1.3	1.5	1.6
TW C1	26	39	48	56	62	70	74	80	16	24	30	37	42	49	53	59	0.8	0.9	1.1	1.3	1.5	1.1	1.2	1.4	1.6	1.7
TW C2	25	39	48	57	63	71	76	81	14	23	30	37	42	50	55	61	0.8	0.9	1.1	1.3	1.5	1.1	1.2	1.3	1.5	1.7
TW C3	24	38	46	55	61	69	73	79	14	22	28	35	40	48	53	59	0.8	0.9	1.1	1.3	1.5	1.1	1.2	1.4	1.5	1.7
TW C4	22	34	42	50	56	63	68	74	13	21	26	32	37	43	48	54	0.9	1.0	1.2	1.4	1.6	1.2	1.3	1.5	1.6	1.8
TW D1	24	38	47	55	61	69	73	79	13	21	27	34	38	45	50	55	0.8	0.9	1.1	1.3	1.5	1.1	1.2	1.4	1.6	1.8
TW D2	32	50	60	69	75	83	87	91	18	30	39	48	54	63	68	75	0.7	0.8	1.0	1.2	1.3	0.9	1.0	1.2	1.3	1.5
TW D3	35	53	64	73	79	86	90	94	19	33	42	52	59	67	73	79	0.7	0.8	1.0	1.1	1.3	0.8	1.0	1.1	1.3	1.5
TW D4	36	55	66	76	82	88	92	95	19	35	45	56	63	72	77	83	0.6	0.8	0.9	1.1	1.3	0.8	0.9	1.1	1.2	1.4
TW D5	34	54	65	75	81	88	92	95	18	34	44	55	62	71	77	83	0.6	0.8	0.9	1.1	1.3	0.8	0.9	1.1	1.2	1.4
TW E1	38	57	67	77	82	89	92	95	21	37	46	57	63	72	77	83	0.6	0.8	0.9	1.1	1.3	0.8	0.9	1.1	1.3	1.4
TW E2	37	56	67	77	82	89	92	95	20	36	46	56	63	72	77	83	0.6	0.8	0.9	1.1	1.3	0.8	0.9	1.1	1.3	1.4
TW E3	42	63	74	83	88	93	96	98	23	42	53	65	72	80	85	89	0.6	0.7	0.9	1.1	1.2	0.7	0.8	1.0	1.2	1.4
TW E4	39	60	71	81	86	92	95	97	21	39	51	62	69	78	83	88	0.6	0.7	0.9	1.1	1.3	0.7	0.8	1.0	1.2	1.4
TW F1	26	40	48	56	62	69	74	79	16	24	30	36	41	47	52	58	0.8	1.0	1.1	1.3	1.5	1.1	1.2	1.4	1.6	1.8
TW F2	28	43	52	61	67	75	80	85	15	25	32	40	45	53	58	64	0.8	0.9	1.1	1.3	1.4	1.0	1.1	1.3	1.5	1.7
TW F3	27	43	53	63	69	77	81	86	14	25	32	40	47	55	60	66	0.7	0.9	1.1	1.2	1.4	1.0	1.1	1.3	1.4	1.6
TW F4	25	41	51	61	67	75	80	86	12	23	30	39	45	53	59	65	0.7	0.9	1.1	1.2	1.4	1.0	1.1	1.3	1.4	1.6
TW G1	24	38	46	55	61	68	73	79	14	22	28	35	40	47	51	57	0.8	1.0	1.1	1.3	1.5	1.1	1.2	1.4	1.6	1.8
TW G2	24	39	49	58	64	72	77	82	12	22	29	36	42	50	55	61	0.8	0.9	1.1	1.3	1.4	1.0	1.2	1.3	1.5	1.7
TW G3	22	37	47	56	63	71	76	81	10	20	27	35	40	48	54	60	0.8	0.9	1.1	1.3	1.4	1.0	1.2	1.3	1.5	1.7
TW G4	18	33	42	51	57	66	71	77	8	17	23	30	36	43	49	55	0.8	1.0	1.1	1.3	1.5	1.1	1.2	1.4	1.6	1.7
TW H1	21	34	43	52	58	66	71	77	11	19	25	32	37	44	49	55	0.8	1.0	1.2	1.3	1.5	1.1	1.2	1.4	1.6	1.8
TW H2	18	32	41	50	56	64	69	75	8	17	23	29	35	42	47	53	0.8	1.0	1.2	1.3	1.5	1.1	1.3	1.4	1.6	1.8
TW H3	15	28	36	45	51	60	65	71	6	14	20	26	31	38	43	49	0.9	1.0	1.2	1.4	1.6	1.2	1.3	1.5	1.7	1.8
TW H4	12	22	30	38	43	51	57	63	5	11	15	21	25	31	35	41	1.0	1.1	1.3	1.5	1.7	1.4	1.5	1.6	1.8	2.0
TW I1	22	35	43	52	58	65	70	76	11	19	24	31	35	42	46	52	0.8	1.0	1.2	1.4	1.5	1.1	1.3	1.5	1.7	1.8
TW I2	29	46	57	66	73	80	85	89	14	27	35	44	51	60	65	72	0.7	0.8	1.0	1.2	1.4	0.9	1.0	1.2	1.4	1.5
TW I3	31	49	59	69	76	83	87	92	16	29	38	48	55	63	69	75	0.7	0.8	1.0	1.2	1.3	0.9	1.0	1.1	1.3	1.5
TW I4	32	52	63	73	79	86	90	94	16	32	42	52	59	69	74	80	0.7	0.8	1.0	1.1	1.3	0.8	0.9	1.1	1.3	1.4
TW I5	32	52	63	73	80	87	91	94	16	32	42	53	60	70	75	81	0.7	0.8	1.0	1.1	1.3	0.8	0.9	1.1	1.3	1.4
TW J1	35	55	65	75	81	88	91	95	19	34	44	54	61	70	75	81	0.6	0.8	1.0	1.1	1.3	0.8	0.9	1.1	1.3	1.5
TW J2	34	54	64	74	80	87	91	94	17	33	43	53	60	69	75	81	0.7	0.8	1.0	1.1	1.3	0.8	0.9	1.1	1.3	1.4
TW J3	38	60	71	81	86	92	95	97	20	39	50	62	69	78	83	88	0.6	0.7	0.9	1.1	1.3	0.7	0.8	1.0	1.2	1.4
TW J4	36	57	69	79	84	91	94	97	18	36	47	59	66	76	81	86	0.6	0.7	0.9	1.1	1.3	0.8	0.9	1.0	1.2	1.4
DG A1	27	43	51	60	65	71	74	78	14	23	30	36	41	47	50	54	0.8	0.9	1.1	1.4	1.7	1.0	1.2	1.4	1.7	2.1
DG A2	28	44	54	63	69	76	80	84	14	25	33	41	47	54	59	65	0.7	0.9	1.1	1.3	1.5	1.0	1.1	1.3	1.5	1.7
DG A3	31	49	60	70	76	83	86	90	17	30	39	49	56	65	70	76	0.7	0.8	1.0	1.2	1.4	0.9	1.0	1.1	1.3	1.5
DG B1	29	45	55	64	70	77	81	85	14	24	31	39	45	52	56	61	0.7	0.9	1.0	1.3	1.5	1.0	1.1	1.3	1.5	1.9
DG B2	29	47	57	67	74	81	85	90	15	26	35	44	50	59	64	70	0.7	0.8	1.0	1.2	1.4	1.0	1.0	1.2	1.4	1.6
DG B3	32	50	61	73	79	87	90	93	17	30	40	51	58	68	74	80	0.7	0.8	0.9	1.1	1.4	0.9	1.0	1.1	1.3	1.5
DG C1	33	50	60	71	77	84	87	91	16	28	37	46	52	60	64	70	0.7	0.8	1.0	1.2	1.5	0.9	1.0	1.2	1.4	1.7
DG C2	33	51	62	73	80	87	90	94	17	30	39	49	56	65	71	77	0.7	0.8	0.9	1.1	1.3	0.9	1.0	1.1	1.3	1.5
DG C3	35	55	66	78	85	91	94	96	19	34	45	56	64	74	79	85	0.7	0.8	0.9	1.1	1.3	0.9	0.9	1.0	1.2	1.4
SS A1	34	49	58	67	73	80	84	88	22	34	41	49	54	61	66	72	0.7	0.9	1.0	1.2	1.4	0.9	1.0	1.2	1.4	1.6
SS A2	40	58	68	77	82	88	92	95	26	41	50	59	65	73	78	83	0.7	0.8	1.0	1.1	1.3	0.8	0.9	1.1	1.3	1.5
SS A3	32	47	55	64	70	77	81	86	21	32	38	46	51	58	62	68	0.7	0.9	1.1	1.3	1.4	0.9	1.1	1.3	1.5	1.7
SS A4	40	57	67	76	81	88	91	95	26	40	49	58	64	72	77	82	0.7	0.8	1.0	1.1	1.3	0.8	0.9	1.1	1.3	1.5
SS A5	33	47	55	63	69	76	80	85	22	32	38	45	50	57	62	67	0.8	0.9	1.1	1.3	1.4	1.0	1.1	1.3	1.5	1.7
SS A6	40	58	68	76	82	88	92	95	26	41	50	59	65	73	78	83	0.7	0.8	1.0	1.1	1.3	0.8	0.9	1.1	1.3	1.5
SS A7	30	42	50	58	63	70	74	80	21	29	35	41	45	51	55	60	0.8	1.0	1.2	1.3	1.5	1.0	1.2	1.4	1.6	1.8
SS A8	39	57	66	75	81	87	91	94	26	40	48	57	63	71	76	81	0.7	0.8	1.0	1.1	1.3	0.8	0.9	1.1	1.3	1.5
SS B1	30	45	53	62	68	75	80	85	19	29	36	43	49	56	61	66	0.8	0.9	1.1	1.3	1.4	1.0	1.1	1.3	1.5	1.7
SS B2	36	53	63	72	78	85	89	93	22	36	45	54	60	69	74	79	0.7	0.8	1.0	1.2	1.3	0.8	0.9	1.1	1.3	1.5
SS B3	29	43	51	60	66	73	78	83	18	28	34	41	46	53	58	64	0.8	0.9	1.1	1.3	1.5	1.0	1.1	1.3	1.5	1.7
SS B4	35	52	62	71	77	84	88	92	22	35	44	53	59	67	72	78	0.7	0.8	1.0	1.2	1.3	0.8	1.0	1.1	1.3	1.5
SS B5	28	40	48	56	62	69	73	79	18	27	32	39	43	50	54	60	0.8	1.0	1.2	1.3	1.5	1.1	1.2	1.4	1.6	1.8
SS B6	35	53	62	72	78	85	88	92	22	35	44	53	60	68	73	79	0.7	0.8	1.0	1.2	1.3	0.8	1.0	1.1	1.3	1.5
SS B7	25	37	45	52	58	65	69	75	17	24	29	35	39	45	49	55	0.9	1.0	1.2	1.4	1.6	1.1	1.3	1.5	1.7	1.9
SS B8	34	51	61	70	76	83	87	91	21	34	43	52	58	66	71	77	0.7	0.8	1.0	1.2	1.3	0.8	1.0	1.2	1.3	1.5
SS C1	31	48	57	67	73	80	85	89	16	28	36	44	50	59	64	70	0.7	0.8	1.0	1.2	1.4	0.9	1.0	1.2	1.4	1.6
SS C2	35	53	64	73	79	86	90	94	18	33	42	52	59	68	73	79	0.7	0.8	1.0	1.1	1.3	0.8	0.9	1.1	1.3	1.5
SS C3	26	40	48	57	63	70	75	81	15	24	30	36	42	49	54	60	0.8	0.9	1.1	1.3	1.5	1.1	1.2	1.4	1.5	1.7
SS C4	31	49	59	68	74	82	86	90	17	30	38	47	53	62	67	74	0.7	0.8	1.0	1.2	1.4	0.9	1.0	1.2	1.3	1.5
SS D1	40	57	66	75	81	87	91	94	25	39	48	56	62	70	74	80	0.7	0.8	1.0	1.1	1.3	0.8	0.9	1.1	1.3	1.5
SS D2	45	65	75	83	88	93	96	98	29	46	57	67	73	80	85	89	0.6	0.7	0.9	1.1	1.2	0.7	0.8	1.0	1.2	1.4
SS D3	37	52	61	70	75	82	86	91	24	36	43	51	57	64	69	74	0.7	0.8	1.0	1.2	1.4	0.9	1.0	1.2	1.4	1.6
SS D4	44	63	73	82	87	92	95	97	28	45	55	65	71	79	83	88	0.6	0.7	0.9	1.1	1.3	0.7	0.9	1.0	1.2	1.4
SS E1	36	53	62	71	77	84	88	92	22	34	42	51	57	64	69	75	0.7	0.8	1.0	1.2	1.3	0.9	1.0	1.2	1.4	1.6
SS E2	41	60	70	79	85	91	94	97	25	41	51	61	67	75	80	85	0.6	0.8	0.9	1.1	1.3	0.8	0.9	1.1	1.2	1.4
SS E3	31	45	54	62	68	75	80	85	20	30	36	44	49	56	60	66	0.8	0.9	1.1	1.3	1.4	1.0	1.1	1.3	1.5	1.7
SS E4	38	56	66	75	81	88	91	95	23	38	47	57	63	71	76	82	0.7	0.8	1.0	1.1	1.3	0.8	0.9	1.1	1.3	1.5

LONG BEACH, CALIFORNIA

TYPE	SOLAR SAVINGS FRACTION (%) BASE TEMP: 55 F (292 DD)								65 F (1607 DD)								CONSERVATION FACTOR 55 F					65 F				
LCR	200	100	70	50	40	30	25	20	200	100	70	50	40	30	25	20	200	100	50	30	20	200	100	50	30	20
WW A1	64	84	91	96	98	99	100	100	29	43	52	62	69	77	82	88	0.5	0.6	0.8	1.0	1.2	0.8	0.9	1.0	1.2	1.4
WW A2	76	92	97	99	100	100	100	100	33	53	64	75	82	89	93	96	0.5	0.6	0.8	1.0	-	0.7	0.8	0.9	1.1	1.3
WW A3	80	95	98	100	100	100	100	100	35	56	68	79	86	92	95	97	0.5	0.6	0.8	0.9	-	0.6	0.7	0.9	1.1	1.2
WW A4	81	96	99	100	100	100	100	100	36	58	71	81	87	93	96	98	0.4	0.6	0.8	0.9	-	0.6	0.7	0.9	1.1	1.2
WW A5	83	96	99	100	100	100	100	100	37	61	73	84	89	95	97	99	0.4	0.6	0.8	0.9	-	0.6	0.7	0.9	1.0	1.2
WW A6	84	97	99	100	100	100	100	100	38	62	74	85	90	95	97	99	0.4	0.6	0.7	-	-	0.6	0.7	0.9	1.0	1.2
WW B1	71	89	95	98	99	100	100	100	30	47	58	69	75	84	88	93	0.5	0.6	0.8	1.0	1.2	0.7	0.8	1.0	1.1	1.3
WW B2	81	95	99	100	100	100	100	100	36	58	71	82	88	93	96	98	0.4	0.6	0.8	0.9	-	0.6	0.7	0.9	1.1	1.2
WW B3	87	98	99	100	100	100	100	100	43	67	79	88	93	97	98	99	0.4	0.6	0.7	-	-	0.6	0.7	0.8	1.0	1.2
WW B4	87	98	99	100	100	100	100	100	41	66	79	88	93	97	98	99	0.4	0.6	0.7	-	-	0.6	0.7	0.8	1.0	1.2
WW B5	85	97	99	100	100	100	100	100	39	64	77	87	92	96	98	99	0.4	0.6	0.7	-	-	0.6	0.7	0.9	1.0	1.2
WW C1	88	98	100	100	100	100	100	100	44	68	80	89	93	97	99	99	0.4	0.6	0.7	-	-	0.6	0.7	0.8	1.0	1.2
WW C2	86	97	99	100	100	100	100	100	41	65	78	87	92	97	98	99	0.4	0.6	0.7	-	-	0.6	0.7	0.8	1.0	1.2
WW C3	90	99	100	100	100	100	100	100	46	72	84	92	96	98	99	100	0.4	0.5	0.7	-	-	0.5	0.7	0.8	1.0	1.2
WW C4	88	98	100	100	100	100	100	100	42	68	81	90	94	98	99	100	0.4	0.6	0.7	-	-	0.6	0.7	0.8	1.0	1.2
TW A1	63	82	90	95	98	99	100	100	28	42	51	61	67	76	81	87	0.5	0.6	0.8	1.0	1.2	0.8	0.9	1.1	1.2	1.4
TW A2	71	89	95	98	99	100	100	100	30	48	59	70	77	85	89	93	0.5	0.6	0.8	1.0	1.2	0.7	0.8	1.0	1.1	1.3
TW A3	74	91	96	99	100	100	100	100	31	50	62	73	80	88	91	95	0.5	0.6	0.8	1.0	-	0.7	0.8	0.9	1.1	1.3
TW A4	74	92	97	99	100	100	100	100	31	50	62	74	81	88	92	96	0.5	0.6	0.8	1.0	-	0.7	0.8	0.9	1.1	1.3
TW B1	63	83	91	96	98	99	100	100	27	42	51	61	68	77	82	88	0.5	0.6	0.8	1.0	1.2	0.8	0.9	1.0	1.2	1.4
TW B2	69	88	94	98	99	100	100	100	29	46	57	68	75	84	88	93	0.5	0.6	0.8	1.0	1.2	0.7	0.8	1.0	1.1	1.3
TW B3	69	88	95	98	99	100	100	100	28	46	57	69	76	84	89	93	0.5	0.6	0.8	1.0	1.2	0.7	0.8	1.0	1.1	1.3
TW B4	66	86	93	97	99	100	100	100	26	43	54	66	73	82	87	91	0.5	0.6	0.8	1.0	1.2	0.7	0.8	1.0	1.1	1.3
TW C1	62	82	91	96	98	99	100	100	26	41	50	61	68	77	82	88	0.5	0.6	0.8	1.0	1.2	0.8	0.9	1.0	1.2	1.4
TW C2	63	83	91	96	98	99	100	100	25	41	51	62	69	78	84	89	0.5	0.6	0.8	1.0	1.2	0.8	0.9	1.0	1.2	1.3
TW C3	60	81	90	95	97	99	100	100	24	39	49	59	67	76	81	87	0.5	0.6	0.8	1.0	1.2	0.8	0.9	1.0	1.2	1.4
TW C4	54	75	85	92	95	98	99	100	22	35	43	53	60	69	75	82	0.6	0.7	0.8	1.0	1.2	0.9	1.0	1.1	1.3	1.4
TW D1	64	84	92	96	98	99	100	100	25	40	51	61	68	77	83	88	0.5	0.6	0.8	1.0	1.2	0.8	0.9	1.0	1.2	1.4
TW D2	76	93	97	99	100	100	100	100	33	53	65	76	83	90	94	97	0.5	0.6	0.8	1.0	-	0.7	0.8	0.9	1.1	1.3
TW D3	81	95	98	100	100	100	100	100	36	58	71	81	87	93	96	98	0.4	0.6	0.8	0.9	-	0.6	0.7	0.9	1.1	1.2
TW D4	82	96	99	100	100	100	100	100	37	60	73	84	89	95	97	99	0.4	0.6	0.8	0.9	-	0.6	0.7	0.9	1.0	1.2
TW D5	81	95	99	100	100	100	100	100	36	59	72	83	88	94	97	98	0.4	0.6	0.8	0.9	-	0.6	0.7	0.9	1.0	1.2
TW E1	84	97	99	100	100	100	100	100	40	63	75	85	90	95	97	99	0.4	0.6	0.8	0.9	-	0.6	0.7	0.9	1.0	1.2
TW E2	83	96	99	100	100	100	100	100	39	62	74	84	90	95	97	99	0.4	0.6	0.8	0.9	-	0.6	0.7	0.9	1.0	1.2
TW E3	88	98	100	100	100	100	100	100	45	70	82	90	94	98	99	100	0.4	0.6	0.7	-	-	0.6	0.7	0.8	1.0	1.2
TW E4	86	98	99	100	100	100	100	100	41	66	78	88	93	97	98	99	0.4	0.6	0.7	-	-	0.6	0.7	0.8	1.0	1.2
TW F1	63	83	91	96	98	99	100	100	26	41	50	61	68	76	82	87	0.5	0.6	0.8	1.0	1.2	0.8	0.9	1.0	1.2	1.4
TW F2	69	88	94	98	99	100	100	100	28	45	56	67	75	83	88	92	0.5	0.6	0.8	1.0	1.2	0.7	0.8	1.0	1.1	1.3
TW F3	70	89	95	98	99	100	100	100	28	46	58	69	77	85	89	94	0.5	0.6	0.8	1.0	1.2	0.7	0.8	1.0	1.1	1.3
TW F4	68	88	95	98	99	100	100	100	25	44	56	68	75	84	88	93	0.5	0.6	0.8	1.0	1.2	0.7	0.8	1.0	1.1	1.3
TW G1	61	82	90	96	98	99	100	100	24	39	49	60	67	76	81	87	0.5	0.6	0.8	1.0	1.2	0.8	0.9	1.0	1.2	1.4
TW G2	65	85	93	97	99	100	100	100	24	41	52	64	71	80	85	90	0.5	0.6	0.8	1.0	1.2	0.8	0.9	1.0	1.2	1.3
TW G3	63	84	92	97	98	100	100	100	22	39	50	62	70	79	84	90	0.5	0.6	0.8	1.0	1.2	0.8	0.9	1.0	1.2	1.3
TW G4	57	79	88	94	97	99	100	100	18	34	45	56	63	73	79	85	0.5	0.6	0.8	1.0	1.2	0.8	0.9	1.0	1.2	1.4
TW H1	58	79	88	94	97	99	100	100	21	36	46	56	64	73	79	85	0.5	0.6	0.8	1.0	1.2	0.8	0.9	1.1	1.2	1.4
TW H2	55	77	87	94	97	99	99	100	18	33	43	54	61	71	77	84	0.5	0.6	0.8	1.0	1.2	0.8	0.9	1.1	1.2	1.4
TW H3	50	72	83	91	95	98	99	100	15	29	38	48	56	66	72	79	0.6	0.7	0.8	1.0	1.2	0.9	1.0	1.1	1.3	1.4
TW H4	41	63	74	84	89	94	97	98	11	22	30	40	46	56	62	70	0.6	0.7	0.9	1.1	1.2	1.0	1.1	1.2	1.4	1.5
TW I1	60	81	90	95	98	99	100	100	22	37	47	58	65	74	80	86	0.5	0.6	0.8	1.0	1.2	0.8	0.9	1.1	1.2	1.4
TW I2	73	91	97	99	100	100	100	100	29	50	62	73	80	88	92	96	0.5	0.6	0.8	1.0	-	0.7	0.8	0.9	1.1	1.3
TW I3	77	93	98	99	100	100	100	100	32	54	66	77	84	91	94	97	0.5	0.6	0.8	1.0	-	0.7	0.8	0.9	1.1	1.3
TW I4	79	95	98	100	100	100	100	100	34	57	70	81	87	93	96	98	0.5	0.6	0.8	0.9	-	0.6	0.7	0.9	1.1	1.2
TW I5	79	95	98	100	100	100	100	100	33	57	70	81	87	93	96	98	0.5	0.6	0.8	0.9	-	0.6	0.7	0.9	1.1	1.2
TW J1	82	96	99	100	100	100	100	100	37	60	72	83	89	94	97	98	0.4	0.6	0.8	0.9	-	0.6	0.7	0.9	1.0	1.2
TW J2	81	95	99	100	100	100	100	100	35	58	71	82	88	94	96	98	0.4	0.6	0.8	0.9	-	0.6	0.7	0.9	1.1	1.2
TW J3	86	98	99	100	100	100	100	100	41	66	79	88	93	97	98	99	0.4	0.6	0.7		-	0.6	0.7	0.8	1.0	1.2
TW J4	84	97	99	100	100	100	100	100	38	63	76	86	91	96	98	99	0.4	0.6	0.7	-	-	0.6	0.7	0.9	1.0	1.2
DG A1	72	90	96	99	100	100	100	100	28	46	56	66	73	80	84	88	0.5	0.6	0.8	-	-	0.7	0.8	1.0	1.2	1.4
DG A2	72	90	96	99	100	100	100	100	28	47	59	70	76	84	88	92	0.5	0.6	0.8	-	-	0.7	0.8	1.0	1.2	1.3
DG A3	77	92	95	97	97	97	97	97	31	53	66	77	83	89	92	94	0.5	0.6	1.2	-	-	0.7	0.7	0.9	1.1	1.5
DG B1	79	94	97	99	99	99	99	99	28	50	63	74	80	87	90	93	0.5	0.6	0.9	2.5	-	0.7	0.8	0.9	1.2	1.4
DG B2	78	95	98	100	100	100	100	100	28	50	64	76	83	90	93	96	0.4	0.6	0.8	1.0	-	0.7	0.7	0.9	1.1	1.3
DG B3	80	95	97	98	98	98	98	98	31	55	70	82	88	93	95	97	0.4	0.6	1.2	-	-	0.7	0.7	0.9	1.1	1.4
DG C1	85	97	98	99	99	99	98	98	32	57	71	82	88	93	95	96	0.4	0.6	1.6	-	-	0.6	0.7	0.9	1.1	1.4
DG C2	83	97	99	100	100	100	100	100	32	56	71	83	89	94	96	98	0.4	0.6	0.8	-	-	0.6	0.7	0.9	1.1	1.3
DG C3	86	97	98	98	98	98	98	98	35	62	77	88	92	96	97	98	0.4	0.6	2.1	-	-	0.6	0.7	0.8	1.1	1.6
SS A1	76	92	97	99	100	100	100	100	38	57	68	77	83	90	93	96	0.5	0.6	0.8	1.0	-	0.7	0.8	0.9	1.1	1.3
SS A2	84	96	99	100	100	100	100	100	46	67	78	87	91	96	97	99	0.4	0.6	0.8	0.9	-	0.6	0.7	0.9	1.0	1.2
SS A3	73	90	96	99	99	100	100	100	37	54	65	75	81	88	91	95	0.5	0.6	0.8	1.0	1.2	0.7	0.8	1.0	1.1	1.3
SS A4	84	96	99	100	100	100	100	100	46	67	77	86	91	95	97	99	0.4	0.6	0.8	0.9	-	0.6	0.7	0.9	1.0	1.2
SS A5	72	89	95	98	99	100	100	100	36	53	63	73	79	86	90	94	0.5	0.6	0.8	1.0	1.2	0.7	0.8	1.0	1.1	1.3
SS A6	84	96	99	100	100	100	100	100	46	67	78	87	91	96	97	99	0.4	0.6	0.8	0.9	-	0.6	0.7	0.9	1.0	1.2
SS A7	66	85	92	97	98	99	100	100	33	48	58	67	74	81	86	90	0.5	0.6	0.8	1.0	1.2	0.7	0.9	1.0	1.2	1.4
SS A8	83	96	99	100	100	100	100	100	45	66	77	86	91	95	97	99	0.4	0.6	0.8	0.9	-	0.6	0.7	0.9	1.1	1.2
SS B1	70	89	95	98	99	100	100	100	33	50	61	72	78	86	90	94	0.5	0.6	0.8	1.0	1.2	0.7	0.8	1.0	1.1	1.3
SS B2	79	94	98	99	100	100	100	100	40	61	72	82	87	93	96	98	0.5	0.6	0.8	1.0	-	0.6	0.7	0.9	1.1	1.2
SS B3	69	87	94	98	99	100	100	100	32	49	59	70	76	84	88	92	0.5	0.6	0.8	1.0	1.2	0.7	0.8	1.0	1.1	1.3
SS B4	79	94	98	99	100	100	100	100	39	60	71	81	87	93	95	98	0.5	0.6	0.8	1.0	-	0.6	0.7	0.9	1.1	1.3
SS B5	64	83	91	96	98	99	100	100	30	45	54	64	71	79	84	89	0.5	0.6	0.8	1.0	1.2	0.8	0.9	1.0	1.2	1.4
SS B6	79	94	98	99	100	100	100	100	39	60	72	82	87	93	96	98	0.5	0.6	0.8	1.0	-	0.6	0.7	0.9	1.1	1.2
SS B7	60	80	88	94	97	99	99	100	27	41	51	60	67	75	80	86	0.5	0.7	0.8	1.0	1.2	0.8	0.9	1.1	1.2	1.4
SS B8	78	93	98	99	100	100	100	100	38	59	70	80	86	92	95	97	0.5	0.6	0.8	1.0	-	0.6	0.7	0.9	1.1	1.3
SS C1	75	92	97	99	100	100	100	100	32	52	63	74	81	89	92	96	0.5	0.6	0.8	1.0	-	0.7	0.8	0.9	1.1	1.3
SS C2	81	95	98	100	100	100	100	100	36	58	71	81	87	93	96	98	0.4	0.6	0.8	0.9	-	0.6	0.7	0.9	1.1	1.2
SS C3	64	84	92	96	98	99	100	100	26	41	52	62	69	78	83	89	0.5	0.6	0.8	1.0	1.2	0.8	0.9	1.0	1.2	1.3
SS C4	75	92	97	99	100	100	100	100	32	52	64	75	82	89	93	96	0.5	0.6	0.8	1.0	-	0.7	0.8	0.9	1.1	1.3
SS D1	84	96	99	100	100	100	100	100	45	66	77	86	91	95	97	99	0.4	0.6	0.8	0.9	-	0.6	0.7	0.9	1.1	1.2
SS D2	90	99	100	100	100	100	100	100	52	75	85	92	96	98	99	100	0.4	0.6	0.7	-	-	0.5	0.7	0.8	1.0	1.2
SS D3	79	94	98	99	100	100	100	100	41	60	71	81	86	92	95	97	0.5	0.6	0.8	1.0	-	0.6	0.7	0.9	1.1	1.3
SS D4	89	98	100	100	100	100	100	100	51	73	84	91	95	98	99	100	0.4	0.6	0.7	-	-	0.6	0.7	0.8	1.0	1.2
SS E1	81	95	98	100	100	100	100	100	40	61	72	82	87	93	96	98	0.5	0.6	0.8	1.0	-	0.6	0.7	0.9	1.1	1.2
SS E2	87	98	99	100	100	100	100	100	47	69	81	89	93	97	98	99	0.4	0.6	0.7	-	-	0.6	0.7	0.9	1.0	1.2
SS E3	71	89	95	98	99	100	100	100	34	51	62	72	78	86	90	94	0.5	0.6	0.8	1.0	1.2	0.7	0.8	1.0	1.1	1.3
SS E4	83	96	99	100	100	100	100	100	43	65	76	86	90	95	97	99	0.4	0.6	0.8	0.9	-	0.6	0.7	0.9	1.0	1.2

LOS ANGELES, CALIFORNIA

TYPE	SOLAR SAVINGS FRACTION (%)																CONSERVATION FACTOR									
BASE TEMP:	55 F (283 DD)								65 F (1793 DD)								55 F					65 F				
LCR	200	100	70	50	40	30	25	20	200	100	70	50	40	30	25	20	200	100	50	30	20	200	100	50	30	20
WW A1	67	86	93	97	98	99	100	100	30	44	54	63	70	79	83	89	0.5	0.6	0.8	1.0	1.2	0.8	0.9	1.0	1.2	1.4
WW A2	78	94	98	99	100	100	100	100	35	54	66	76	83	90	93	96	0.5	0.6	0.8	1.0	-	0.7	0.8	0.9	1.1	1.3
WW A3	82	96	99	100	100	100	100	100	37	58	70	81	87	93	95	98	0.4	0.6	0.8	0.9	-	0.6	0.7	0.9	1.1	1.2
WW A4	84	96	99	100	100	100	100	100	38	60	72	83	88	94	96	98	0.4	0.6	0.8	0.9	-	0.6	0.7	0.9	1.1	1.2
WW A5	86	97	99	100	100	100	100	100	39	62	74	85	90	95	97	99	0.4	0.6	0.7	-	-	0.6	0.7	0.9	1.0	1.2
WW A6	86	97	99	100	100	100	100	100	40	63	76	86	91	96	98	99	0.4	0.6	0.7	-	-	0.6	0.7	0.9	1.0	1.2
WW B1	74	91	96	99	99	100	100	100	31	49	60	70	77	85	89	93	0.5	0.6	0.8	1.0	1.2	0.7	0.8	1.0	1.1	1.3
WW B2	83	96	99	100	100	100	100	100	37	60	72	83	88	94	96	98	0.4	0.6	0.8	0.9	-	0.6	0.7	0.9	1.1	1.2
WW B3	89	98	100	100	100	100	100	100	44	68	80	89	93	97	99	99	0.4	0.6	0.7	-	-	0.6	0.7	0.8	1.0	1.2
WW B4	88	98	100	100	100	100	100	100	43	68	80	89	93	97	99	99	0.4	0.6	0.7	-	-	0.6	0.7	0.8	1.0	1.2
WW B5	87	98	99	100	100	100	100	100	40	66	78	88	93	97	98	99	0.4	0.6	0.7	-	-	0.6	0.7	0.8	1.0	1.2
WW C1	90	98	100	100	100	100	100	100	46	70	81	90	94	98	99	100	0.4	0.6	0.7	-	-	0.6	0.7	0.8	1.0	1.2
WW C2	88	98	100	100	100	100	100	100	42	67	79	88	93	97	98	99	0.4	0.6	0.7	-	-	0.6	0.7	0.8	1.0	1.2
WW C3	92	99	100	100	100	100	100	100	48	74	85	93	96	99	99	100	0.4	0.5	0.7	-	-	0.5	0.6	0.8	1.0	1.2
WW C4	89	98	100	100	100	100	100	100	44	70	82	91	95	98	99	100	0.4	0.6	0.7	-	-	0.6	0.7	0.8	1.0	1.2
TW A1	65	84	92	96	98	99	100	100	29	43	52	62	69	77	82	88	0.5	0.6	0.8	1.0	1.2	0.8	0.9	1.0	1.2	1.4
TW A2	73	91	96	99	99	100	100	100	32	49	60	71	78	86	90	94	0.5	0.6	0.8	1.0	1.2	0.7	0.8	1.0	1.1	1.3
TW A3	76	93	97	99	100	100	100	100	33	52	63	74	81	88	92	96	0.5	0.6	0.8	1.0	-	0.7	0.8	0.9	1.1	1.3
TW A4	77	93	97	99	100	100	100	100	32	52	64	75	82	89	93	96	0.5	0.6	0.8	1.0	-	0.7	0.8	0.9	1.1	1.3
TW B1	66	85	92	97	98	99	100	100	28	43	53	63	70	78	83	89	0.5	0.6	0.8	1.0	1.2	0.8	0.9	1.0	1.2	1.4
TW B2	72	90	95	98	99	100	100	100	30	47	59	69	76	85	89	93	0.5	0.6	0.8	1.0	1.2	0.7	0.8	1.0	1.1	1.3
TW B3	72	90	96	98	99	100	100	100	29	48	59	70	77	85	90	94	0.5	0.6	0.8	1.0	1.2	0.7	0.8	1.0	1.1	1.3
TW B4	69	88	94	98	99	100	100	100	27	45	56	67	74	83	87	92	0.5	0.6	0.8	1.0	1.2	0.7	0.8	1.0	1.1	1.3
TW C1	65	85	92	96	98	99	100	100	27	42	52	62	69	78	83	89	0.5	0.6	0.8	1.0	1.2	0.8	0.9	1.0	1.2	1.3
TW C2	66	86	93	97	99	100	100	100	26	42	53	64	71	80	85	90	0.5	0.6	0.8	1.0	1.2	0.8	0.9	1.0	1.2	1.3
TW C3	63	83	91	96	98	99	100	100	25	40	50	61	68	77	82	88	0.5	0.6	0.8	1.0	1.2	0.8	0.9	1.0	1.2	1.4
TW C4	57	78	87	93	96	98	99	100	23	36	45	55	62	71	76	83	0.5	0.7	0.8	1.0	1.2	0.9	1.0	1.1	1.2	1.4
TW D1	67	86	93	97	99	100	100	100	26	42	52	63	70	79	84	89	0.5	0.6	0.8	1.0	1.2	0.8	0.9	1.0	1.2	1.3
TW D2	79	94	98	99	100	100	100	100	34	55	67	78	84	91	94	97	0.5	0.6	0.8	1.0	-	0.7	0.8	0.9	1.1	1.3
TW D3	83	96	99	100	100	100	100	100	38	60	72	83	88	94	96	98	0.4	0.6	0.8	0.9	-	0.6	0.7	0.9	1.1	1.2
TW D4	84	97	99	100	100	100	100	100	39	62	74	85	90	95	97	99	0.4	0.6	0.8	0.9	-	0.6	0.7	0.9	1.0	1.2
TW D5	83	96	99	100	100	100	100	100	37	61	73	84	89	95	97	99	0.4	0.6	0.8	0.9	-	0.6	0.7	0.9	1.0	1.2
TW E1	86	97	99	100	100	100	100	100	42	65	76	86	91	96	98	99	0.4	0.6	0.7	-	-	0.6	0.7	0.9	1.0	1.2
TW E2	85	97	99	100	100	100	100	100	40	63	75	85	91	95	97	99	0.4	0.6	0.7	-	-	0.6	0.7	0.9	1.0	1.2
TW E3	90	99	100	100	100	100	100	100	47	71	83	91	95	98	99	100	0.4	0.6	0.7	-	-	0.6	0.7	0.8	1.0	1.2
TW E4	88	98	100	100	100	100	100	100	43	67	80	89	93	97	99	99	0.4	0.6	0.7	-	-	0.6	0.7	0.8	1.0	1.2
TW F1	65	85	92	96	98	99	100	100	28	42	52	62	69	78	83	88	0.5	0.6	0.8	1.0	1.2	0.8	0.9	1.0	1.2	1.4
TW F2	72	90	95	98	99	100	100	100	29	47	58	69	76	84	89	93	0.5	0.6	0.8	1.0	1.2	0.7	0.8	1.0	1.1	1.3
TW F3	73	91	96	99	99	100	100	100	29	48	59	71	78	86	90	94	0.5	0.6	0.8	1.0	-	0.7	0.8	1.0	1.1	1.3
TW F4	71	90	96	98	99	100	100	100	27	46	58	69	76	85	89	94	0.5	0.6	0.8	1.0	1.2	0.7	0.8	1.0	1.1	1.3
TW G1	64	84	92	96	98	99	100	100	26	41	51	61	68	77	82	88	0.5	0.6	0.8	1.0	1.2	0.8	0.9	1.0	1.2	1.4
TW G2	68	87	94	98	99	100	100	100	25	43	54	65	72	81	86	91	0.5	0.6	0.8	1.0	1.2	0.7	0.8	1.0	1.2	1.3
TW G3	66	86	93	97	99	100	100	100	23	41	52	63	71	80	85	90	0.5	0.6	0.8	1.0	1.2	0.8	0.8	1.0	1.2	1.3
TW G4	60	82	90	95	98	99	100	100	19	35	46	57	65	75	80	86	0.5	0.6	0.8	1.0	1.2	0.8	0.9	1.0	1.2	1.4
TW H1	61	82	90	95	98	99	100	100	22	37	47	58	65	74	80	86	0.5	0.6	0.8	1.0	1.2	0.8	0.9	1.0	1.2	1.4
TW H2	58	80	89	95	97	99	100	100	19	34	45	55	63	73	78	85	0.5	0.6	0.8	1.0	1.2	0.8	0.9	1.1	1.2	1.4
TW H3	53	75	85	92	95	98	99	100	16	30	40	50	58	67	73	80	0.5	0.7	0.8	1.0	1.2	0.9	1.0	1.1	1.3	1.4
TW H4	43	66	77	86	91	95	97	99	12	24	32	41	48	57	64	71	0.6	0.7	0.9	1.0	1.2	1.0	1.1	1.2	1.3	1.5
TW I1	63	84	91	96	98	99	100	100	23	39	49	59	67	76	81	87	0.5	0.6	0.8	1.0	1.2	0.8	0.9	1.0	1.2	1.4
TW I2	76	93	97	99	100	100	100	100	31	51	63	75	82	89	93	96	0.5	0.6	0.8	1.0	-	0.7	0.8	0.9	1.1	1.3
TW I3	79	94	98	99	100	100	100	100	34	55	68	79	85	92	95	97	0.5	0.6	0.8	1.0	-	0.6	0.7	0.9	1.1	1.2
TW I4	82	96	99	100	100	100	100	100	35	58	71	82	88	94	96	98	0.4	0.6	0.8	0.9	-	0.6	0.7	0.9	1.1	1.2
TW I5	82	96	99	100	100	100	100	100	35	58	71	82	88	94	96	98	0.4	0.6	0.8	0.9	-	0.6	0.7	0.9	1.0	1.2
TW J1	84	97	99	100	100	100	100	100	39	62	74	84	90	95	97	99	0.4	0.6	0.8	0.9	-	0.6	0.7	0.9	1.0	1.2
TW J2	83	96	99	100	100	100	100	100	37	60	72	83	89	94	97	98	0.4	0.6	0.8	0.9	-	0.6	0.7	0.9	1.0	1.2
TW J3	88	98	100	100	100	100	100	100	43	68	80	89	94	97	99	100	0.4	0.6	0.7	-	-	0.6	0.7	0.8	1.0	1.2
TW J4	86	97	99	100	100	100	100	100	39	64	77	87	92	96	98	99	0.4	0.6	0.7	-	-	0.6	0.7	0.9	1.0	1.2
DG A1	75	92	97	99	100	100	100	100	29	47	58	68	74	81	85	89	0.5	0.6	0.8	-	-	0.7	0.8	1.0	1.2	1.4
DG A2	75	92	97	99	100	100	100	100	29	49	60	71	78	85	89	93	0.5	0.6	0.8	-	-	0.7	0.8	1.0	1.1	1.3
DG A3	79	93	96	97	97	97	97	97	32	55	68	79	84	90	93	95	0.5	0.6	1.3	-	-	0.7	0.7	0.9	1.1	1.5
DG B1	81	95	98	99	99	99	99	99	30	52	65	76	82	88	91	94	0.4	0.6	0.9	-	-	0.7	0.7	0.9	1.2	1.4
DG B2	81	95	99	100	100	100	100	100	30	52	66	78	84	91	94	96	0.4	0.6	0.8	1.0	-	0.7	0.7	0.9	1.1	1.3
DG B3	83	96	98	98	98	98	98	98	33	57	72	83	89	94	96	97	0.4	0.6	1.4	-	-	0.6	0.7	0.9	1.1	1.4
DG C1	87	97	98	99	99	99	98	98	34	59	73	84	89	93	95	96	0.4	0.6	2.1	-	-	0.6	0.7	0.9	1.1	1.5
DG C2	86	98	99	100	100	100	100	100	34	58	73	84	90	95	97	98	0.4	0.6	0.8	-	-	0.6	0.7	0.9	1.1	1.3
DG C3	88	97	98	98	99	98	98	98	37	63	78	89	93	96	97	98	0.4	0.6	2.7	-	-	0.6	0.7	0.8	1.1	1.6
SS A1	79	93	97	99	100	100	100	100	41	60	71	80	86	92	95	97	0.5	0.6	0.8	1.0	-	0.6	0.7	0.9	1.1	1.3
SS A2	87	97	99	100	100	100	100	100	49	70	81	89	93	97	98	99	0.4	0.6	0.8	0.9	-	0.6	0.7	0.9	1.0	1.2
SS A3	77	92	97	99	100	100	100	100	39	58	68	78	84	90	93	96	0.5	0.6	0.8	1.0	-	0.7	0.8	0.9	1.1	1.3
SS A4	86	97	99	100	100	100	100	100	49	70	81	89	93	97	98	99	0.4	0.6	0.8	0.9	-	0.6	0.7	0.9	1.0	1.2
SS A5	75	91	96	98	99	100	100	100	38	56	66	76	82	89	92	95	0.5	0.6	0.8	1.0	1.2	0.7	0.8	0.9	1.1	1.3
SS A6	87	97	99	100	100	100	100	100	49	70	81	89	93	97	98	99	0.4	0.6	0.8	0.9	-	0.6	0.7	0.9	1.0	1.2
SS A7	70	87	94	97	99	100	100	100	35	51	61	71	77	84	88	92	0.5	0.6	0.8	1.0	1.2	0.7	0.8	1.0	1.2	1.3
SS A8	86	97	99	100	100	100	100	100	48	69	80	88	92	96	98	99	0.4	0.6	0.8	0.9	-	0.6	0.7	0.9	1.0	1.2
SS B1	74	90	96	98	99	100	100	100	35	53	64	74	81	88	91	95	0.5	0.6	0.8	1.0	1.2	0.7	0.8	0.9	1.1	1.3
SS B2	82	95	98	100	100	100	100	100	42	64	75	84	89	94	97	98	0.5	0.6	0.8	1.0	-	0.6	0.7	0.9	1.1	1.2
SS B3	73	89	95	98	99	100	100	100	34	52	63	73	79	87	90	94	0.5	0.6	0.8	1.0	1.2	0.7	0.8	1.0	1.1	1.3
SS B4	82	95	98	100	100	100	100	100	42	63	74	84	89	94	96	98	0.5	0.6	0.8	1.0	-	0.6	0.7	0.9	1.1	1.2
SS B5	68	86	92	97	98	99	100	100	32	47	57	67	74	82	86	91	0.5	0.6	0.8	1.0	1.2	0.7	0.8	1.0	1.2	1.3
SS B6	82	95	98	100	100	100	100	100	42	63	74	84	89	94	97	98	0.5	0.6	0.8	1.0	-	0.6	0.7	0.9	1.1	1.2
SS B7	64	83	90	95	97	99	100	100	29	44	54	63	70	78	83	88	0.5	0.6	0.8	1.0	1.2	0.8	0.9	1.0	1.2	1.4
SS B8	81	95	98	99	100	100	100	100	41	62	73	83	88	94	96	98	0.5	0.6	0.8	1.0	-	0.6	0.7	0.9	1.1	1.2
SS C1	78	94	98	99	100	100	100	100	33	53	65	76	83	90	93	96	0.5	0.6	0.8	1.0	-	0.7	0.8	0.9	1.1	1.3
SS C2	83	96	99	100	100	100	100	100	38	60	72	83	88	94	96	98	0.4	0.6	0.8	0.9	-	0.6	0.7	0.9	1.1	1.2
SS C3	67	86	93	97	99	100	100	100	27	43	53	64	71	80	85	90	0.5	0.6	0.8	1.0	1.2	0.8	0.9	1.0	1.2	1.3
SS C4	78	93	98	99	100	100	100	100	33	54	66	77	83	90	94	97	0.5	0.6	0.8	1.0	-	0.7	0.8	0.9	1.1	1.3
SS D1	87	97	99	100	100	100	100	100	48	70	80	88	93	96	98	99	0.4	0.6	0.8	0.9	-	0.6	0.7	0.9	1.0	1.2
SS D2	92	99	100	100	100	100	100	100	56	78	88	94	97	99	99	100	0.4	0.5	0.7	-	-	0.5	0.6	0.8	1.0	1.2
SS D3	82	95	98	100	100	100	100	100	44	64	74	83	88	94	96	98	0.5	0.6	0.8	1.0	-	0.6	0.7	0.9	1.1	1.3
SS D4	91	99	100	100	100	100	100	100	54	76	86	93	96	98	99	100	0.4	0.6	0.7	-	-	0.5	0.7	0.8	1.0	1.2
SS E1	84	96	99	100	100	100	100	100	43	64	75	84	90	94	97	98	0.4	0.6	0.8	0.9	-	0.6	0.7	0.9	1.1	1.2
SS E2	89	98	100	100	100	100	100	100	50	73	83	91	95	98	99	100	0.4	0.6	0.7	-	-	0.6	0.7	0.8	1.0	1.2
SS E3	75	91	96	99	99	100	100	100	36	54	65	75	81	88	92	95	0.5	0.6	0.8	1.0	1.2	0.7	0.8	0.9	1.1	1.3
SS E4	86	97	99	100	100	100	100	100	46	68	79	88	92	96	98	99	0.4	0.6	0.8	0.9	-	0.6	0.7	0.9	1.0	1.2

MOUNT SHASTA, CALIFORNIA

TYPE	SOLAR SAVINGS FRACTION (%) BASE TEMP: 55 F (3220 DD) LCR 100	70	50	40	30	25	20	15	65 F (5896 DD) 100	70	50	40	30	25	20	15	CONSERVATION FACTOR 55 F 100	50	30	20	15	65 F 100	50	30	20	15
WW A1	24	29	34	38	45	49	54	62	18	21	25	28	32	35	39	45	1.3	1.5	1.7	1.8	2.0	1.7	1.9	2.1	2.3	2.4
WW A2	27	34	42	48	56	61	68	76	19	24	30	34	41	45	51	59	1.1	1.3	1.4	1.6	1.7	1.3	1.5	1.7	1.8	2.0
WW A3	28	36	45	52	60	66	72	80	19	25	32	37	44	49	55	64	1.0	1.2	1.4	1.5	1.7	1.2	1.4	1.6	1.8	1.9
WW A4	28	37	47	54	63	68	75	83	19	25	32	38	46	51	58	66	1.0	1.2	1.3	1.5	1.7	1.2	1.4	1.5	1.7	1.9
WW A5	29	39	49	56	65	71	77	85	19	26	34	40	48	53	60	69	1.0	1.1	1.3	1.5	1.6	1.2	1.3	1.5	1.7	1.8
WW A6	30	39	50	57	66	72	78	86	19	26	34	40	49	55	61	70	1.0	1.1	1.3	1.5	1.6	1.1	1.3	1.5	1.6	1.8
WW B1	22	28	35	39	46	51	57	65	15	18	23	26	30	34	38	43	1.2	1.4	1.6	1.8	1.9	1.6	1.8	2.0	2.3	2.5
WW B2	29	38	48	55	64	70	77	85	19	26	34	40	49	54	61	70	1.0	1.1	1.3	1.5	1.6	1.1	1.3	1.5	1.6	1.8
WW B3	35	45	56	63	72	78	84	90	24	32	41	48	57	63	70	78	0.9	1.1	1.2	1.4	1.6	1.0	1.2	1.4	1.5	1.7
WW B4	34	46	57	65	74	80	86	92	23	33	43	50	60	66	74	82	0.9	1.0	1.2	1.4	1.5	1.0	1.2	1.3	1.5	1.6
WW B5	33	44	56	64	74	79	85	92	22	32	42	50	60	66	73	82	0.9	1.0	1.2	1.4	1.5	1.0	1.2	1.3	1.5	1.6
WW C1	36	47	58	65	74	79	85	91	25	34	43	50	60	65	72	81	0.9	1.0	1.2	1.4	1.5	1.0	1.2	1.3	1.5	1.7
WW C2	34	45	56	64	73	79	85	91	23	32	42	49	58	65	72	80	0.9	1.1	1.2	1.4	1.5	1.0	1.2	1.3	1.5	1.7
WW C3	40	52	64	72	81	86	91	95	28	39	51	59	69	75	82	89	0.8	1.0	1.2	1.3	1.5	0.9	1.1	1.2	1.4	1.6
WW C4	36	49	61	69	78	83	89	94	25	36	47	55	65	72	79	87	0.8	1.0	1.2	1.3	1.5	0.9	1.1	1.3	1.4	1.6
TW A1	23	28	34	38	44	48	54	61	18	21	25	28	32	35	39	45	1.3	1.5	1.7	1.8	2.0	1.7	1.9	2.0	2.2	2.4
TW A2	25	31	39	44	52	57	63	71	18	22	27	32	37	42	47	54	1.2	1.3	1.5	1.7	1.8	1.4	1.6	1.8	1.9	2.1
TW A3	25	32	41	46	55	60	67	75	17	22	28	33	40	44	50	58	1.1	1.3	1.4	1.6	1.7	1.3	1.5	1.7	1.8	2.0
TW A4	24	32	41	47	55	61	68	76	16	22	28	33	40	45	51	60	1.1	1.2	1.4	1.6	1.7	1.3	1.5	1.6	1.8	1.9
TW B1	23	28	34	38	45	50	55	63	17	21	25	28	33	36	41	47	1.3	1.5	1.6	1.8	1.9	1.6	1.8	2.0	2.1	2.3
TW B2	23	30	37	43	51	56	62	71	16	21	26	31	37	41	46	54	1.2	1.3	1.5	1.7	1.8	1.4	1.6	1.7	1.9	2.1
TW B3	23	30	37	43	51	57	63	72	16	21	26	31	37	42	47	55	1.1	1.3	1.5	1.6	1.8	1.4	1.5	1.7	1.9	2.0
TW B4	21	28	35	41	49	54	61	70	15	20	25	29	35	40	46	54	1.2	1.3	1.5	1.7	1.8	1.4	1.6	1.7	1.9	2.1
TW C1	22	27	33	38	45	50	56	64	16	20	24	28	33	36	41	48	1.3	1.4	1.6	1.8	1.9	1.6	1.7	1.9	2.1	2.2
TW C2	21	27	34	39	46	51	58	66	15	19	24	28	34	38	43	50	1.2	1.4	1.6	1.7	1.9	1.5	1.7	1.8	2.0	2.1
TW C3	20	26	32	37	44	49	55	64	14	18	23	27	32	36	41	49	1.3	1.4	1.6	1.7	1.9	1.5	1.7	1.9	2.0	2.2
TW C4	19	23	29	33	40	44	50	58	14	17	21	25	29	33	38	44	1.4	1.5	1.7	1.8	2.0	1.6	1.8	2.0	2.1	2.3
TW D1	19	24	30	34	41	45	51	59	12	16	19	22	27	30	34	40	1.3	1.5	1.7	1.8	2.0	1.7	1.9	2.1	2.3	2.5
TW D2	27	35	44	51	60	65	72	80	19	25	32	37	45	50	57	65	1.0	1.2	1.4	1.5	1.7	1.2	1.4	1.5	1.7	1.9
TW D3	30	39	48	55	64	70	76	84	20	27	35	41	49	54	61	70	1.0	1.1	1.3	1.5	1.6	1.2	1.3	1.5	1.7	1.8
TW D4	31	41	52	59	69	74	81	88	22	30	39	45	54	60	68	77	0.9	1.1	1.3	1.4	1.6	1.1	1.2	1.4	1.6	1.7
TW D5	31	41	52	59	69	75	81	88	21	29	39	46	55	61	69	78	0.9	1.1	1.2	1.4	1.6	1.1	1.2	1.4	1.5	1.7
TW E1	33	43	53	60	69	74	81	88	23	31	39	46	54	60	67	75	0.9	1.1	1.3	1.4	1.6	1.1	1.2	1.4	1.6	1.7
TW E2	32	42	53	60	69	75	81	88	23	30	39	46	55	61	68	76	0.9	1.1	1.3	1.4	1.6	1.1	1.2	1.4	1.6	1.7
TW E3	38	50	61	69	78	83	88	94	27	37	48	55	65	71	78	86	0.9	1.0	1.2	1.4	1.5	1.0	1.1	1.3	1.4	1.6
TW E4	36	47	58	66	76	81	87	93	25	35	45	53	63	69	76	84	0.9	1.0	1.2	1.4	1.5	1.0	1.1	1.3	1.5	1.6
TW F1	22	27	33	37	44	48	54	62	16	19	23	26	31	34	38	45	1.3	1.5	1.6	1.8	2.0	1.6	1.8	2.0	2.2	2.4
TW F2	22	29	36	42	49	54	61	69	15	20	25	29	35	39	44	52	1.2	1.3	1.5	1.7	1.8	1.4	1.6	1.8	2.0	2.1
TW F3	22	29	37	43	51	56	63	72	14	19	25	29	36	40	46	54	1.1	1.3	1.5	1.6	1.8	1.4	1.5	1.7	1.9	2.0
TW F4	20	27	35	41	49	55	62	71	12	17	23	28	35	39	45	54	1.1	1.3	1.5	1.6	1.8	1.4	1.5	1.7	1.9	2.0
TW G1	20	25	31	36	43	47	53	62	14	18	22	25	30	34	38	45	1.3	1.5	1.6	1.8	1.9	1.6	1.8	2.0	2.1	2.3
TW G2	19	25	33	38	46	51	58	66	12	17	22	26	32	36	42	49	1.2	1.4	1.5	1.7	1.8	1.5	1.6	1.8	2.0	2.1
TW G3	17	24	31	37	45	50	57	66	11	15	21	25	31	35	41	49	1.2	1.4	1.5	1.7	1.8	1.5	1.6	1.8	2.0	2.1
TW G4	14	20	27	32	40	45	51	60	8	13	17	21	27	31	37	44	1.3	1.4	1.6	1.8	1.9	1.5	1.7	1.9	2.0	2.2
TW H1	17	22	28	33	40	45	51	59	11	15	19	22	27	31	36	43	1.3	1.5	1.6	1.8	1.9	1.6	1.8	2.0	2.1	2.3
TW H2	14	20	26	31	38	43	49	58	9	12	17	21	26	30	35	42	1.3	1.5	1.6	1.8	1.9	1.6	1.8	1.9	2.1	2.3
TW H3	12	17	23	27	34	39	45	54	7	10	15	18	23	27	31	38	1.4	1.5	1.7	1.9	2.0	1.7	1.9	2.0	2.2	2.3
TW H4	9	13	18	22	28	32	37	45	5	8	11	14	18	21	26	32	1.6	1.7	1.9	2.0	2.2	1.9	2.0	2.2	2.4	2.5
TW I1	16	21	27	31	38	42	48	55	10	13	17	20	24	27	31	36	1.4	1.5	1.7	1.9	2.1	1.7	2.0	2.2	2.4	2.6
TW I2	24	32	41	47	56	62	69	78	16	22	28	34	41	46	53	62	1.1	1.2	1.4	1.5	1.7	1.3	1.4	1.6	1.7	1.9
TW I3	26	35	44	51	60	66	73	81	17	24	31	37	45	51	58	67	1.0	1.2	1.3	1.5	1.7	1.2	1.4	1.5	1.7	1.8
TW I4	28	38	48	56	65	71	78	86	19	27	35	42	51	57	64	74	1.0	1.1	1.3	1.4	1.6	1.1	1.3	1.4	1.6	1.7
TW I5	28	38	49	57	67	73	79	87	19	27	36	43	53	59	66	76	0.9	1.1	1.3	1.4	1.6	1.1	1.2	1.4	1.6	1.7
TW J1	31	40	50	57	67	72	79	86	21	29	37	43	52	58	65	74	1.0	1.1	1.3	1.5	1.6	1.1	1.3	1.4	1.6	1.7
TW J2	29	39	50	57	66	72	79	86	20	28	36	43	52	58	65	74	1.0	1.1	1.3	1.4	1.6	1.1	1.3	1.4	1.6	1.7
TW J3	35	46	58	66	75	81	87	93	24	34	45	52	62	69	76	84	0.9	1.0	1.2	1.4	1.5	1.0	1.1	1.3	1.5	1.6
TW J4	33	44	55	63	73	79	85	91	22	32	42	50	60	66	74	82	0.9	1.0	1.2	1.4	1.5	1.0	1.2	1.3	1.5	1.6
DG A1	20	26	32	37	42	46	50	54	13	16	20	23	27	29	32	35	1.2	1.5	1.8	2.2	2.6	1.6	1.9	2.3	3.0	3.8
DG A2	22	29	37	43	51	55	61	68	15	20	26	30	36	40	45	52	1.1	1.3	1.5	1.8	2.0	1.4	1.5	1.8	2.0	2.3
DG A3	26	35	45	52	61	67	73	80	19	25	33	39	47	53	60	68	1.0	1.2	1.3	1.5	1.7	1.2	1.3	1.5	1.7	1.9
DG B1	21	27	35	40	47	52	57	63	13	17	21	24	29	33	37	41	1.2	1.3	1.6	1.9	2.2	1.5	1.8	2.0	2.4	2.8
DG B2	23	30	39	46	55	61	67	75	15	21	27	31	39	44	50	57	1.1	1.2	1.4	1.6	1.8	1.4	1.5	1.6	1.8	2.0
DG B3	27	36	47	54	65	71	78	85	19	26	34	40	50	56	64	73	1.0	1.1	1.3	1.5	1.6	1.2	1.3	1.4	1.6	1.7
DG C1	24	32	41	47	56	61	67	73	15	20	26	30	36	41	46	52	1.1	1.2	1.4	1.7	2.0	1.4	1.6	1.7	2.0	2.3
DG C2	26	35	45	52	62	68	75	82	18	24	31	37	45	51	57	66	1.0	1.1	1.3	1.5	1.7	1.3	1.4	1.5	1.7	1.9
DG C3	30	40	52	61	71	78	84	90	22	29	38	45	55	62	70	80	1.0	1.0	1.2	1.4	1.6	1.1	1.2	1.3	1.5	1.7
SS A1	30	37	45	50	57	62	68	75	24	29	35	40	46	50	55	62	1.1	1.3	1.5	1.6	1.8	1.3	1.5	1.7	1.9	2.0
SS A2	37	46	55	62	70	75	81	87	29	37	45	51	59	64	70	77	0.9	1.1	1.3	1.5	1.6	1.1	1.2	1.4	1.6	1.7
SS A3	28	35	41	46	53	58	63	71	22	27	32	36	42	45	50	56	1.1	1.3	1.5	1.7	1.9	1.3	1.6	1.8	2.0	2.2
SS A4	36	45	54	60	69	74	79	86	29	36	44	50	57	62	68	76	1.0	1.1	1.3	1.5	1.6	1.1	1.3	1.4	1.6	1.8
SS A5	29	35	41	46	53	57	63	70	23	28	33	37	42	46	50	57	1.2	1.4	1.5	1.7	1.9	1.4	1.6	1.8	2.0	2.2
SS A6	37	46	55	61	69	75	80	87	29	36	45	50	58	63	69	77	0.9	1.1	1.3	1.5	1.6	1.1	1.2	1.4	1.6	1.8
SS A7	26	31	37	41	47	51	56	63	21	25	29	32	37	40	44	49	1.3	1.5	1.7	1.9	2.0	1.5	1.7	2.0	2.2	2.4
SS A8	36	44	53	59	67	72	78	85	28	35	43	48	56	61	67	74	1.0	1.1	1.3	1.5	1.6	1.1	1.3	1.5	1.7	1.8
SS B1	26	32	39	45	52	57	63	70	20	25	30	34	40	44	49	56	1.2	1.3	1.5	1.7	1.8	1.3	1.5	1.7	1.9	2.1
SS B2	33	41	50	57	65	70	77	84	25	32	40	46	54	59	65	73	1.0	1.2	1.3	1.5	1.6	1.1	1.3	1.5	1.6	1.8
SS B3	25	31	37	42	49	54	59	67	19	23	28	32	37	41	46	52	1.2	1.4	1.6	1.8	1.9	1.4	1.6	1.8	2.1	2.2
SS B4	32	40	49	55	64	69	75	83	25	32	39	45	52	57	64	72	1.0	1.2	1.3	1.5	1.7	1.1	1.3	1.5	1.7	1.8
SS B5	24	29	35	39	46	50	55	63	19	23	27	30	35	38	43	49	1.3	1.5	1.6	1.8	2.0	1.5	1.7	1.9	2.1	2.3
SS B6	32	40	49	56	64	70	76	83	25	31	39	45	53	58	64	72	1.0	1.2	1.3	1.5	1.7	1.1	1.3	1.5	1.7	1.8
SS B7	22	26	32	36	41	45	50	57	17	21	24	27	31	34	38	43	1.4	1.6	1.8	1.9	2.1	1.6	1.9	2.1	2.3	2.5
SS B8	31	39	48	54	62	67	74	81	24	30	38	43	50	55	62	69	1.0	1.2	1.4	1.5	1.7	1.1	1.3	1.5	1.7	1.9
SS C1	25	32	40	46	55	60	67	75	16	22	28	32	39	43	49	57	1.1	1.3	1.4	1.6	1.7	1.3	1.5	1.7	1.9	2.0
SS C2	29	39	48	55	64	70	77	84	20	27	35	41	49	55	62	71	1.0	1.1	1.3	1.5	1.6	1.1	1.3	1.5	1.6	1.8
SS C3	21	27	33	38	45	50	56	64	15	19	23	27	32	35	40	47	1.3	1.4	1.6	1.8	1.9	1.5	1.7	1.9	2.1	2.2
SS C4	26	34	43	50	58	64	71	79	18	24	31	37	44	49	56	64	1.0	1.2	1.4	1.5	1.7	1.2	1.4	1.6	1.7	1.9
SS D1	35	43	52	58	66	71	77	83	27	34	41	46	53	58	63	70	1.0	1.2	1.3	1.5	1.7	1.1	1.3	1.5	1.7	1.9
SS D2	42	52	63	69	78	82	87	93	33	42	51	58	66	72	78	85	0.9	1.0	1.2	1.4	1.5	1.0	1.1	1.3	1.5	1.6
SS D3	32	40	47	53	60	65	71	78	25	31	37	42	48	52	58	65	1.1	1.2	1.4	1.6	1.8	1.2	1.4	1.6	1.8	2.0
SS D4	41	51	61	67	76	81	86	91	32	41	50	56	65	70	76	83	0.9	1.1	1.2	1.4	1.6	1.0	1.2	1.3	1.5	1.7
SS E1	31	38	47	53	60	65	71	79	23	29	35	40	46	51	56	63	1.0	1.2	1.4	1.6	1.7	1.2	1.4	1.6	1.8	2.0
SS E2	37	47	57	64	72	77	83	89	28	36	45	51	60	65	71	79	0.9	1.1	1.3	1.4	1.6	1.0	1.2	1.4	1.6	1.7
SS E3	27	33	40	45	52	56	62	70	21	25	30	34	40	44	48	55	1.2	1.3	1.5	1.7	1.9	1.4	1.6	1.8	2.0	2.2
SS E4	34	43	53	59	68	73	79	86	26	34	42	48	56	61	67	75	0.9	1.1	1.3	1.5	1.6	1.1	1.3	1.4	1.6	1.8

NEEDLES, CALIFORNIA

TYPE	SOLAR SAVINGS FRACTION (%)																CONSERVATION FACTOR									
	BASE TEMP: 55 F (416 DD)								65 F (1431 DD)								55 F					65 F				
LCR	200	100	70	50	40	30	25	20	200	100	70	50	40	30	25	20	200	100	50	30	20	200	100	50	30	20
WW A1	52	72	82	90	94	97	98	99	29	43	52	62	69	77	81	87	0.6	0.7	0.9	1.0	1.2	0.8	0.9	1.1	1.2	1.4
WW A2	63	84	92	97	98	99	100	100	34	53	64	75	81	88	91	95	0.5	0.6	0.8	1.0	1.2	0.7	0.8	0.9	1.1	1.3
WW A3	67	87	94	98	99	100	100	100	36	57	68	79	85	91	94	97	0.5	0.6	0.8	1.0	1.2	0.6	0.7	0.9	1.1	1.3
WW A4	69	89	95	98	99	100	100	100	37	59	70	81	86	92	95	97	0.5	0.6	0.8	1.0	1.2	0.6	0.7	0.9	1.1	1.3
WW A5	71	90	96	99	100	100	100	100	38	61	73	83	88	94	96	98	0.5	0.6	0.8	1.0	-	0.6	0.7	0.9	1.1	1.2
WW A6	72	91	96	99	100	100	100	100	39	62	74	84	89	94	96	98	0.5	0.6	0.8	1.0	-	0.6	0.7	0.9	1.1	1.2
WW B1	57	79	88	94	97	99	99	100	30	47	57	68	74	82	87	91	0.5	0.6	0.8	1.0	1.2	0.7	0.8	1.0	1.2	1.3
WW B2	68	89	95	98	99	100	100	100	36	59	71	81	87	93	95	98	0.5	0.6	0.8	1.0	1.2	0.6	0.7	0.9	1.1	1.2
WW B3	75	93	97	99	100	100	100	100	43	66	78	87	91	96	98	99	0.5	0.6	0.8	1.0	-	0.6	0.7	0.9	1.0	1.2
WW B4	75	93	98	99	100	100	100	100	42	66	78	87	92	96	98	99	0.5	0.6	0.8	1.0	-	0.6	0.7	0.9	1.0	1.2
WW B5	73	92	97	99	100	100	100	100	39	64	76	86	91	96	98	99	0.5	0.6	0.8	1.0	-	0.6	0.7	0.9	1.0	1.2
WW C1	76	94	98	99	100	100	100	100	44	68	79	88	92	96	98	99	0.5	0.6	0.8	1.0	-	0.6	0.7	0.9	1.0	1.2
WW C2	74	93	97	99	100	100	100	100	41	65	77	86	91	96	98	99	0.5	0.6	0.8	1.0	-	0.6	0.7	0.9	1.0	1.2
WW C3	79	95	99	100	100	100	100	100	46	72	83	91	95	98	99	100	0.4	0.6	0.8	0.9	-	0.5	0.7	0.8	1.0	1.2
WW C4	76	94	98	100	100	100	100	100	42	68	80	89	93	97	98	99	0.5	0.6	0.8	0.9	-	0.6	0.7	0.8	1.0	1.2
TW A1	51	71	81	89	93	97	98	99	28	42	51	61	67	76	80	86	0.6	0.7	0.9	1.0	1.2	0.8	0.9	1.1	1.2	1.4
TW A2	58	79	88	94	97	99	99	100	31	48	59	69	76	84	88	92	0.5	0.6	0.8	1.0	1.2	0.7	0.8	1.0	1.2	1.3
TW A3	60	82	90	96	98	99	100	100	32	51	62	73	79	87	90	94	0.5	0.6	0.8	1.0	1.2	0.7	0.8	1.0	1.1	1.3
TW A4	61	82	91	96	98	99	100	100	31	51	62	73	80	87	91	95	0.5	0.6	0.8	1.0	1.2	0.7	0.8	0.9	1.1	1.3
TW B1	51	71	82	90	94	97	98	99	28	42	52	62	68	77	81	87	0.6	0.7	0.9	1.0	1.2	0.8	0.9	1.0	1.2	1.4
TW B2	56	77	87	94	96	99	99	100	29	46	57	68	75	83	87	92	0.5	0.7	0.8	1.0	1.2	0.7	0.8	1.0	1.2	1.3
TW B3	56	78	87	94	97	99	99	100	29	47	58	68	75	83	88	92	0.5	0.6	0.8	1.0	1.2	0.7	0.8	1.0	1.1	1.3
TW B4	53	75	85	92	96	98	99	100	27	44	55	66	73	81	86	90	0.6	0.7	0.8	1.0	1.2	0.7	0.8	1.0	1.2	1.3
TW C1	50	71	81	89	93	97	98	99	26	41	51	61	68	76	81	87	0.6	0.7	0.9	1.0	1.2	0.8	0.9	1.0	1.2	1.4
TW C2	50	72	82	90	94	97	99	99	25	41	52	62	69	78	83	88	0.6	0.7	0.8	1.0	1.2	0.8	0.9	1.0	1.2	1.4
TW C3	48	69	80	88	93	97	98	99	24	39	49	59	66	75	80	86	0.6	0.7	0.9	1.0	1.2	0.8	0.9	1.0	1.2	1.4
TW C4	43	63	74	83	88	94	96	98	22	35	44	54	60	69	75	81	0.6	0.7	0.9	1.1	1.2	0.9	1.0	1.1	1.3	1.4
TW D1	50	72	82	90	94	97	99	99	25	40	51	61	68	76	81	87	0.6	0.7	0.8	1.0	1.2	0.8	0.9	1.0	1.2	1.4
TW D2	63	84	92	97	99	100	100	100	33	54	65	76	82	89	93	96	0.5	0.6	0.8	1.0	1.2	0.7	0.8	0.9	1.1	1.3
TW D3	68	88	95	98	99	100	100	100	37	58	70	80	86	92	95	97	0.5	0.6	0.8	1.0	1.2	0.6	0.7	0.9	1.1	1.3
TW D4	70	90	96	99	99	100	100	100	38	61	73	83	88	94	96	98	0.5	0.6	0.8	1.0	-	0.6	0.7	0.9	1.1	1.2
TW D5	68	89	95	98	99	100	100	100	36	59	72	82	88	93	96	98	0.5	0.6	0.8	1.0	1.2	0.6	0.7	0.9	1.1	1.2
TW E1	72	91	96	99	100	100	100	100	40	62	74	84	89	94	96	98	0.5	0.6	0.8	1.0	-	0.6	0.7	0.9	1.1	1.2
TW E2	71	90	96	99	100	100	100	100	39	62	74	83	89	94	96	98	0.5	0.6	0.8	1.0	-	0.6	0.7	0.9	1.1	1.2
TW E3	77	94	98	100	100	100	100	100	45	69	81	89	93	97	98	99	0.5	0.6	0.8	0.9	-	0.6	0.7	0.8	1.0	1.2
TW E4	74	93	97	99	100	100	100	100	42	66	78	87	92	96	98	99	0.5	0.6	0.8	1.0	-	0.6	0.7	0.9	1.0	1.2
TW F1	50	71	81	89	93	97	98	99	27	41	51	61	67	76	81	86	0.6	0.7	0.9	1.0	1.2	0.8	0.9	1.1	1.2	1.4
TW F2	55	77	87	93	96	99	99	100	28	46	57	67	74	82	87	91	0.5	0.7	0.8	1.0	1.2	0.7	0.8	1.0	1.2	1.3
TW F3	57	79	88	94	97	99	100	100	28	47	58	69	76	84	88	93	0.5	0.6	0.8	1.0	1.2	0.7	0.8	1.0	1.1	1.3
TW F4	54	77	87	94	97	99	99	100	26	45	56	67	75	83	87	92	0.5	0.6	0.8	1.0	1.2	0.7	0.8	1.0	1.1	1.3
TW G1	49	70	81	89	93	97	98	99	25	40	50	60	67	75	80	86	0.6	0.7	0.9	1.0	1.2	0.8	0.9	1.1	1.2	1.4
TW G2	51	74	84	92	95	98	99	100	25	42	53	64	71	79	84	89	0.6	0.7	0.8	1.0	1.2	0.8	0.9	1.0	1.2	1.4
TW G3	49	72	83	91	95	98	99	100	23	40	51	62	69	78	83	89	0.6	0.7	0.8	1.0	1.2	0.8	0.9	1.0	1.2	1.4
TW G4	44	66	77	87	92	96	98	99	19	34	45	56	63	73	78	84	0.6	0.7	0.9	1.0	1.2	0.8	0.9	1.1	1.2	1.4
TW H1	45	67	78	87	92	96	98	99	21	36	46	56	64	73	78	84	0.6	0.7	0.9	1.0	1.2	0.8	0.9	1.1	1.2	1.4
TW H2	43	64	76	85	91	95	97	99	18	33	44	54	62	71	77	83	0.6	0.7	0.9	1.0	1.2	0.8	0.9	1.1	1.2	1.4
TW H3	38	59	71	81	87	93	96	98	15	29	39	49	56	66	72	78	0.6	0.7	0.9	1.1	1.2	0.9	1.0	1.1	1.3	1.4
TW H4	30	49	61	72	79	87	91	95	12	23	31	40	47	56	62	69	0.7	0.8	0.9	1.1	1.3	1.0	1.1	1.2	1.4	1.5
TW I1	47	69	79	88	93	97	98	99	22	37	47	57	64	73	78	84	0.6	0.7	0.9	1.0	1.2	0.8	0.9	1.1	1.2	1.4
TW I2	60	82	91	96	98	99	100	100	30	50	62	73	80	87	91	95	0.5	0.6	0.8	1.0	1.2	0.7	0.8	0.9	1.1	1.3
TW I3	63	85	93	97	99	100	100	100	32	53	65	76	83	90	93	96	0.5	0.6	0.8	1.0	1.2	0.6	0.8	0.9	1.1	1.3
TW I4	66	87	94	98	99	100	100	100	34	57	69	80	86	92	95	97	0.5	0.6	0.8	1.0	1.2	0.6	0.7	0.9	1.1	1.3
TW I5	66	87	94	98	99	100	100	100	34	57	70	80	86	92	95	98	0.5	0.6	0.8	1.0	1.2	0.6	0.7	0.9	1.1	1.2
TW J1	69	89	95	98	99	100	100	100	37	60	72	82	87	93	96	98	0.5	0.6	0.8	1.0	1.2	0.6	0.7	0.9	1.1	1.2
TW J2	68	88	95	98	99	100	100	100	36	58	71	81	87	93	95	98	0.5	0.6	0.8	1.0	1.2	0.6	0.7	0.9	1.1	1.2
TW J3	74	93	97	99	100	100	100	100	41	66	78	87	92	96	98	99	0.5	0.6	0.8	1.0	-	0.6	0.7	0.9	1.0	1.2
TW J4	71	91	97	99	100	100	100	100	38	63	75	85	90	95	97	99	0.5	0.6	0.8	1.0	-	0.6	0.7	0.9	1.0	1.2
DG A1	58	80	88	94	97	99	99	100	28	46	57	66	72	79	83	87	0.5	0.6	0.8	1.0	1.1	0.7	0.8	1.0	1.3	1.5
DG A2	58	80	89	95	97	99	100	100	28	48	59	70	76	83	87	91	0.5	0.6	0.8	1.0	-	0.7	0.8	1.0	1.2	1.4
DG A3	63	85	91	95	96	97	97	97	32	54	66	77	83	88	91	94	0.5	0.6	0.9	1.9	-	0.7	0.7	0.9	1.2	1.5
DG B1	64	86	93	97	98	99	99	99	29	50	62	73	79	86	89	92	0.5	0.6	0.8	1.2	2.7	0.7	0.8	1.0	1.2	1.4
DG B2	63	86	94	98	99	100	100	100	29	51	64	76	82	89	92	95	0.5	0.6	0.8	1.0	1.2	0.7	0.8	0.9	1.1	1.3
DG B3	66	89	95	97	98	98	98	98	32	55	69	81	87	92	95	96	0.5	0.6	0.9	2.1	-	0.6	0.7	0.9	1.1	1.4
DG C1	71	91	96	98	98	98	98	98	34	57	70	81	86	91	93	95	0.5	0.6	0.9	5.0	-	0.6	0.7	0.9	1.1	1.4
DG C2	69	91	96	99	100	100	100	100	33	57	71	82	88	93	95	97	0.5	0.6	0.8	1.0	-	0.6	0.7	0.9	1.1	1.3
DG C3	72	93	97	98	98	98	98	98	36	61	76	86	91	95	97	98	0.4	0.6	1.0	-	-	0.6	0.7	0.9	1.1	1.5
SS A1	61	82	90	96	98	99	100	100	35	53	64	74	80	87	90	94	0.5	0.6	0.8	1.0	1.2	0.7	0.8	1.0	1.1	1.3
SS A2	71	90	96	98	99	100	100	100	42	63	74	83	88	94	96	98	0.5	0.6	0.8	1.0	1.2	0.6	0.7	0.9	1.1	1.2
SS A3	59	79	88	94	97	99	99	100	33	50	60	70	76	84	88	92	0.5	0.6	0.8	1.0	1.2	0.7	0.8	1.0	1.2	1.3
SS A4	70	89	95	98	99	100	100	100	42	62	73	82	88	93	96	98	0.5	0.6	0.8	1.0	1.2	0.6	0.7	0.9	1.1	1.3
SS A5	58	78	87	93	96	99	99	100	34	50	59	69	75	83	87	91	0.5	0.7	0.8	1.0	1.2	0.7	0.8	1.0	1.2	1.3
SS A6	71	90	96	99	99	100	100	100	42	63	74	83	88	94	96	98	0.5	0.6	0.8	1.0	1.2	0.6	0.7	0.9	1.1	1.2
SS A7	52	72	82	90	94	97	98	99	30	45	54	63	69	77	82	87	0.6	0.7	0.9	1.0	1.2	0.8	0.9	1.1	1.2	1.4
SS A8	70	89	95	98	99	100	100	100	41	62	72	82	87	93	95	97	0.5	0.6	0.8	1.0	1.2	0.6	0.7	0.9	1.1	1.3
SS B1	56	77	86	93	96	98	99	100	31	48	58	68	74	82	87	91	0.5	0.7	0.8	1.0	1.2	0.7	0.8	1.0	1.2	1.3
SS B2	65	86	93	97	99	100	100	100	37	57	69	79	84	91	94	97	0.5	0.6	0.8	1.0	1.2	0.6	0.8	0.9	1.1	1.3
SS B3	54	75	85	92	95	98	99	100	29	46	56	66	72	80	85	89	0.6	0.7	0.8	1.0	1.2	0.7	0.9	1.0	1.2	1.4
SS B4	64	85	93	97	99	100	100	100	36	56	68	78	84	90	93	96	0.5	0.6	0.8	1.0	1.2	0.6	0.8	0.9	1.1	1.3
SS B5	50	70	80	89	93	97	98	99	28	42	51	61	68	76	80	86	0.6	0.7	0.9	1.0	1.2	0.8	0.9	1.1	1.2	1.4
SS B6	65	85	93	97	99	100	100	100	37	57	68	78	84	90	94	96	0.5	0.6	0.8	1.0	1.2	0.6	0.8	0.9	1.1	1.3
SS B7	46	66	76	85	90	95	97	99	26	39	47	56	63	71	76	82	0.6	0.7	0.9	1.1	1.2	0.8	0.9	1.1	1.3	1.4
SS B8	63	84	92	97	98	100	100	100	35	55	66	77	83	89	93	96	0.5	0.6	0.8	1.0	1.2	0.7	0.8	0.9	1.1	1.3
SS C1	62	83	92	96	98	99	100	100	32	52	63	74	80	88	91	95	0.5	0.6	0.8	1.0	1.2	0.7	0.8	0.9	1.1	1.3
SS C2	68	88	95	98	99	100	100	100	36	58	70	80	86	92	95	97	0.5	0.6	0.8	1.0	1.2	0.6	0.7	0.9	1.1	1.3
SS C3	51	72	83	90	94	98	99	99	26	42	52	62	69	78	83	88	0.6	0.7	0.8	1.0	1.2	0.8	0.9	1.0	1.2	1.4
SS C4	62	83	91	96	98	99	100	100	32	52	64	75	81	88	92	95	0.5	0.6	0.8	1.0	1.2	0.7	0.8	0.9	1.1	1.3
SS D1	71	90	96	98	99	100	100	100	42	62	73	82	88	93	95	98	0.5	0.6	0.8	1.0	1.2	0.6	0.7	0.9	1.1	1.3
SS D2	79	95	98	100	100	100	100	100	49	71	82	90	94	97	98	99	0.5	0.6	0.8	0.9	-	0.6	0.7	0.9	1.0	1.2
SS D3	65	85	92	97	98	100	100	100	38	57	67	77	83	89	92	95	0.5	0.6	0.8	1.0	1.2	0.7	0.8	0.9	1.1	1.3
SS D4	77	93	98	99	100	100	100	100	47	69	80	88	92	96	98	99	0.5	0.6	0.8	1.0	-	0.6	0.7	0.9	1.0	1.2
SS E1	67	87	94	98	99	100	100	100	38	57	68	78	84	90	93	96	0.5	0.6	0.8	1.0	1.2	0.6	0.8	0.9	1.1	1.3
SS E2	74	92	97	99	100	100	100	100	44	66	77	86	91	95	97	99	0.5	0.6	0.8	1.0	-	0.6	0.7	0.9	1.0	1.2
SS E3	57	78	87	93	96	99	99	100	32	48	59	68	75	82	87	91	0.5	0.7	0.8	1.0	1.2	0.7	0.8	1.0	1.2	1.3
SS E4	70	89	95	98	99	100	100	100	40	61	73	82	88	93	96	98	0.5	0.6	0.8	1.0	1.2	0.6	0.7	0.9	1.1	1.2

OAKLAND, CALIFORNIA

SOLAR SAVINGS FRACTION (%)

TYPE	55 F (719 DD) LCR 200	100	70	50	40	30	25	20	65 F (2915 DD) 200	100	70	50	40	30	25	20
WW A1	40	57	67	76	82	88	91	95	22	33	40	48	54	62	67	73
WW A2	48	69	79	87	91	95	97	99	24	39	49	59	66	75	80	85
WW A3	51	72	82	90	93	97	98	99	25	42	53	63	70	79	83	89
WW A4	52	74	84	91	94	97	99	99	25	43	54	65	72	81	85	90
WW A5	54	76	86	92	95	98	99	100	26	45	56	67	75	83	87	92
WW A6	55	77	86	93	96	98	99	100	26	46	57	69	76	84	88	92
WW B1	43	63	73	82	87	92	95	97	21	34	43	52	59	67	72	78
WW B2	52	74	84	91	95	98	99	99	25	44	55	66	73	82	86	91
WW B3	59	80	89	95	97	99	99	100	30	51	63	74	80	88	91	95
WW B4	58	80	89	95	97	99	100	100	28	50	63	74	81	88	92	95
WW B5	56	79	88	94	97	99	99	100	27	49	61	73	80	88	91	95
WW C1	60	81	90	95	97	99	100	100	32	53	65	75	82	89	92	95
WW C2	58	80	88	94	97	99	99	100	29	50	62	73	80	88	91	95
WW C3	63	84	92	97	98	100	100	100	33	57	69	80	86	92	95	97
WW C4	59	82	90	96	98	99	100	100	29	52	65	77	84	90	94	97
TW A1	39	56	66	75	81	87	91	94	21	32	39	47	53	61	66	72
TW A2	44	64	74	83	88	93	95	97	22	36	45	54	61	70	75	81
TW A3	46	66	77	85	90	95	97	98	22	37	47	57	64	73	78	84
TW A4	45	67	77	86	90	95	97	98	21	37	47	58	65	74	79	85
TW B1	38	57	67	76	82	88	91	95	21	32	39	48	54	62	67	73
TW B2	42	62	72	82	87	92	95	97	21	34	43	53	59	68	74	80
TW B3	42	62	73	82	87	93	95	97	20	34	43	53	60	69	74	81
TW B4	39	59	70	80	85	91	94	97	19	32	41	51	57	66	72	78
TW C1	37	56	66	76	81	88	91	95	19	30	38	47	53	61	67	73
TW C2	37	56	67	77	83	89	92	95	18	30	39	48	54	63	69	75
TW C3	35	54	64	74	80	87	91	94	17	29	37	45	52	60	66	73
TW C4	31	48	59	68	75	82	86	91	16	26	33	41	47	55	60	67
TW D1	37	56	67	76	82	88	92	95	17	29	37	46	52	60	66	72
TW D2	48	69	80	88	92	96	98	99	23	40	50	61	68	77	82	87
TW D3	52	74	83	91	94	97	99	99	26	44	55	66	73	81	86	91
TW D4	53	76	85	92	95	98	99	100	26	46	58	69	76	84	88	93
TW D5	52	74	84	92	95	98	99	100	25	45	57	68	75	84	88	92
TW E1	56	77	86	93	96	98	99	100	29	48	60	70	77	85	89	93
TW E2	55	77	86	92	96	98	99	100	27	47	59	70	77	85	89	93
TW E3	61	83	91	96	98	99	100	100	32	55	67	78	84	91	94	96
TW E4	58	80	89	95	97	99	100	100	29	51	63	75	81	89	92	96
TW F1	38	56	66	75	81	88	91	94	20	31	38	46	52	61	66	72
TW F2	41	61	72	81	87	92	95	97	20	33	42	52	59	67	73	79
TW F3	42	63	74	83	88	93	96	98	19	34	43	53	60	69	75	81
TW F4	39	61	72	82	87	92	95	97	17	31	41	51	59	68	74	80
TW G1	36	55	65	75	81	87	91	94	18	29	37	45	52	60	65	72
TW G2	37	58	69	78	84	90	93	96	16	30	39	48	55	64	70	76
TW G3	35	56	67	77	83	89	93	96	15	28	37	46	53	63	68	75
TW G4	30	50	61	72	78	86	89	93	11	23	32	41	48	57	63	70
TW H1	32	51	62	72	78	85	89	93	14	25	33	42	48	57	63	69
TW H2	29	48	60	70	77	84	88	92	12	23	31	40	46	55	61	68
TW H3	25	43	54	65	72	80	85	90	9	20	27	35	41	50	56	63
TW H4	20	35	45	56	63	72	77	83	7	15	21	28	34	41	47	54
TW I1	34	53	63	73	79	86	90	94	15	26	34	42	49	57	62	69
TW I2	44	66	77	86	90	95	97	98	20	36	47	57	65	74	79	85
TW I3	47	69	80	88	92	96	98	99	22	40	51	62	69	78	83	88
TW I4	50	73	83	90	94	97	99	99	23	42	54	65	73	81	86	91
TW I5	50	73	83	91	94	97	99	99	22	42	54	66	73	82	87	92
TW J1	53	75	85	92	95	98	99	100	26	45	57	68	75	83	88	92
TW J2	51	74	84	91	95	98	99	100	24	44	55	67	74	82	87	92
TW J3	58	80	89	95	97	99	100	100	29	51	64	75	82	89	92	96
TW J4	55	78	87	94	96	99	99	100	26	48	60	72	79	87	91	95
DG A1	40	63	73	81	86	90	93	95	19	32	41	50	55	62	66	71
DG A2	42	64	75	83	88	93	95	97	19	34	44	54	60	69	73	79
DG A3	47	70	80	87	91	94	95	96	22	39	50	62	69	78	82	87
DG B1	46	68	79	87	90	94	96	97	19	34	44	55	61	70	74	79
DG B2	45	69	80	88	92	96	98	99	20	36	47	58	66	75	80	85
DG B3	48	73	83	91	94	96	97	98	22	40	52	65	73	82	86	91
DG C1	51	75	85	92	94	96	97	98	22	39	51	62	70	78	82	86
DG C2	50	75	85	92	95	98	99	99	23	40	52	65	73	81	86	90
DG C3	53	78	88	94	96	98	98	98	25	45	58	71	79	87	91	94
SS A1	49	68	77	85	90	94	96	98	30	45	54	63	69	77	81	86
SS A2	57	77	85	92	95	98	99	100	36	54	64	74	79	86	90	93
SS A3	47	65	75	83	87	92	95	97	29	43	52	61	66	74	78	83
SS A4	57	76	85	91	95	98	99	99	36	54	64	73	79	85	89	93
SS A5	46	64	73	82	86	92	94	97	29	43	51	59	65	72	77	82
SS A6	57	77	85	92	95	98	99	100	36	54	64	73	79	86	89	93
SS A7	42	59	68	77	82	88	91	94	27	39	47	54	60	67	71	77
SS A8	56	75	84	91	94	97	99	99	36	53	63	72	78	85	88	92
SS B1	44	63	73	81	86	92	94	97	26	40	49	58	64	71	76	82
SS B2	52	72	81	89	93	96	98	99	31	48	59	68	75	82	86	90
SS B3	42	61	71	80	85	90	93	96	25	39	47	56	62	69	74	79
SS B4	51	71	81	88	92	96	98	99	31	48	58	68	74	81	85	90
SS B5	39	57	66	75	81	87	90	94	24	36	43	51	57	65	70	75
SS B6	51	72	81	89	93	96	98	99	30	48	58	68	74	81	86	90
SS B7	36	53	62	71	77	84	88	92	22	33	40	48	53	61	65	71
SS B8	50	70	80	88	92	96	97	99	30	47	57	66	73	80	84	89
SS C1	47	68	78	86	91	95	97	99	22	38	48	58	65	74	79	85
SS G2	52	74	83	91	94	97	99	99	25	44	55	66	73	81	86	91
SS C3	38	57	67	77	83	89	92	95	19	31	39	48	54	62	68	74
SS C4	47	68	78	87	91	95	97	99	23	39	49	60	67	76	81	86
SS D1	57	76	85	91	95	98	99	99	36	53	63	72	78	84	88	92
SS D2	64	83	91	96	98	99	100	100	41	61	71	81	86	91	94	97
SS D3	52	71	80	87	92	96	97	99	33	48	58	66	72	79	83	88
SS D4	62	82	90	95	97	99	100	100	39	59	70	79	84	90	93	96
SS E1	53	73	82	89	93	96	98	99	31	48	57	67	73	80	84	89
SS E2	59	80	88	94	97	99	99	100	36	55	66	76	82	88	91	95
SS E3	45	63	73	82	87	92	94	97	27	41	49	58	64	72	76	81
SS E4	55	76	85	91	95	98	99	100	33	52	62	72	78	85	89	92

CONSERVATION FACTOR

TYPE	55 F 200	100	50	30	20	65 F 200	100	50	30	20
WW A1	0.7	0.8	1.0	1.1	1.3	1.0	1.1	1.2	1.4	1.5
WW A2	0.6	0.7	0.9	1.1	1.2	0.8	0.9	1.1	1.2	1.4
WW A3	0.6	0.7	0.9	1.0	1.2	0.7	0.9	1.0	1.2	1.4
WW A4	0.5	0.7	0.8	1.0	1.2	0.7	0.8	1.0	1.2	1.3
WW A5	0.5	0.7	0.8	1.0	1.2	0.7	0.8	1.0	1.2	1.3
WW A6	0.5	0.7	0.8	1.0	1.2	0.7	0.8	1.0	1.1	1.3
WW B1	0.6	0.7	0.9	1.1	1.3	0.9	1.0	1.1	1.3	1.5
WW B2	0.5	0.7	0.8	1.0	1.2	0.7	0.8	1.0	1.2	1.3
WW B3	0.5	0.6	0.8	1.0	1.2	0.7	0.8	0.9	1.1	1.3
WW B4	0.5	0.6	0.8	1.0	1.2	0.7	0.8	0.9	1.1	1.3
WW B5	0.5	0.6	0.8	1.0	1.2	0.7	0.8	0.9	1.1	1.3
WW C1	0.5	0.6	0.8	1.0	1.2	0.7	0.8	0.9	1.1	1.3
WW C2	0.5	0.6	0.8	1.0	1.2	0.7	0.8	0.9	1.1	1.3
WW C3	0.5	0.6	0.8	1.0	1.2	0.6	0.7	0.9	1.1	1.3
WW C4	0.5	0.6	0.8	1.0	1.2	0.6	0.7	0.9	1.1	1.3
TW A1	0.7	0.8	1.0	1.1	1.3	1.0	1.1	1.2	1.4	1.6
TW A2	0.6	0.7	0.9	1.1	1.3	0.8	0.9	1.1	1.3	1.4
TW A3	0.6	0.7	0.9	1.1	1.2	0.8	0.9	1.1	1.2	1.4
TW A4	0.6	0.7	0.9	1.1	1.2	0.8	0.9	1.1	1.2	1.4
TW B1	0.7	0.8	1.0	1.1	1.3	0.9	1.0	1.2	1.4	1.5
TW B2	0.6	0.7	0.9	1.1	1.3	0.9	1.0	1.1	1.3	1.5
TW B3	0.6	0.7	0.9	1.1	1.3	0.8	0.9	1.1	1.3	1.4
TW B4	0.6	0.7	0.9	1.1	1.3	0.9	1.0	1.1	1.3	1.5
TW C1	0.7	0.8	1.0	1.1	1.3	0.9	1.0	1.2	1.4	1.5
TW C2	0.7	0.8	0.9	1.1	1.3	0.9	1.0	1.2	1.3	1.5
TW C3	0.7	0.8	1.0	1.1	1.3	0.9	1.0	1.2	1.4	1.5
TW C4	0.7	0.8	1.0	1.2	1.4	1.0	1.1	1.3	1.4	1.6
TW D1	0.7	0.8	1.0	1.1	1.3	0.9	1.0	1.2	1.4	1.5
TW D2	0.6	0.7	0.9	1.0	1.2	0.8	0.9	1.0	1.2	1.4
TW D3	0.6	0.7	0.9	1.0	1.2	0.7	0.8	1.0	1.2	1.3
TW D4	0.5	0.7	0.8	1.0	1.2	0.7	0.8	1.0	1.1	1.3
TW D5	0.5	0.7	0.8	1.0	1.2	0.7	0.8	1.0	1.1	1.3
TW E1	0.5	0.7	0.8	1.0	1.2	0.7	0.8	1.0	1.1	1.3
TW E2	0.5	0.7	0.8	1.0	1.2	0.7	0.8	1.0	1.1	1.3
TW E3	0.5	0.6	0.8	1.0	1.2	0.6	0.7	0.9	1.1	1.3
TW E4	0.5	0.6	0.8	1.0	1.2	0.7	0.8	0.9	1.1	1.3
TW F1	0.7	0.8	1.0	1.1	1.3	1.0	1.1	1.2	1.4	1.5
TW F2	0.6	0.7	0.9	1.1	1.3	0.9	1.0	1.1	1.3	1.5
TW F3	0.6	0.7	0.9	1.1	1.3	0.8	0.9	1.1	1.3	1.4
TW F4	0.6	0.7	0.9	1.1	1.3	0.8	0.9	1.1	1.3	1.4
TW G1	0.7	0.8	1.0	1.1	1.3	0.9	1.0	1.2	1.4	1.5
TW G2	0.6	0.8	0.9	1.1	1.3	0.9	1.0	1.1	1.3	1.5
TW G3	0.6	0.8	0.9	1.1	1.3	0.9	1.0	1.2	1.3	1.5
TW G4	0.7	0.8	1.0	1.1	1.3	1.0	1.0	1.2	1.4	1.5
TW H1	0.7	0.8	1.0	1.1	1.3	1.0	1.1	1.2	1.4	1.6
TW H2	0.7	0.8	1.0	1.2	1.3	1.0	1.1	1.2	1.4	1.6
TW H3	0.7	0.8	1.0	1.2	1.4	1.0	1.1	1.3	1.4	1.6
TW H4	0.8	0.9	1.1	1.2	1.4	1.2	1.3	1.4	1.6	1.7
TW I1	0.7	0.8	1.0	1.1	1.3	1.0	1.1	1.2	1.4	1.6
TW I2	0.6	0.7	0.9	1.1	1.2	0.8	0.9	1.1	1.2	1.4
TW I3	0.6	0.7	0.9	1.0	1.2	0.7	0.9	1.0	1.2	1.4
TW I4	0.5	0.7	0.8	1.0	1.2	0.7	0.8	1.0	1.2	1.3
TW I5	0.5	0.7	0.8	1.0	1.2	0.7	0.8	1.0	1.1	1.3
TW J1	0.5	0.7	0.8	1.0	1.2	0.7	0.8	1.0	1.1	1.3
TW J2	0.5	0.7	0.8	1.0	1.2	0.7	0.8	1.0	1.1	1.0
TW J3	0.5	0.6	0.8	1.0	1.2	0.6	0.8	0.9	1.1	1.3
TW J4	0.5	0.6	0.8	1.0	1.2	0.7	0.8	0.9	1.1	1.3
DG A1	0.6	0.7	0.9	1.2	1.4	0.8	1.0	1.2	1.4	1.7
DG A2	0.6	0.7	0.9	1.1	1.3	0.8	0.9	1.1	1.3	1.5
DG A3	0.6	0.7	0.9	1.2	1.7	0.8	0.8	1.0	1.2	1.4
DG B1	0.6	0.7	0.9	1.1	1.4	0.8	0.9	1.1	1.3	1.6
DG B2	0.6	0.7	0.9	1.1	1.3	0.8	0.9	1.0	1.2	1.4
DG B3	0.6	0.7	0.9	1.1	1.7	0.8	0.8	1.0	1.2	1.4
DG C1	0.5	0.7	0.9	1.2	1.8	0.8	0.9	1.0	1.2	1.5
DG C2	0.5	0.6	0.8	1.0	1.2	0.8	0.8	1.0	1.2	1.4
DG C3	0.5	0.6	0.8	1.2	2.6	0.7	0.8	0.9	1.1	1.3
SS A1	0.6	0.7	0.9	1.1	1.3	0.7	0.9	1.1	1.3	1.4
SS A2	0.5	0.7	0.8	1.0	1.2	0.7	0.8	1.0	1.2	1.3
SS A3	0.6	0.7	0.9	1.1	1.3	0.8	0.9	1.1	1.3	1.5
SS A4	0.5	0.7	0.9	1.0	1.2	0.7	0.8	1.0	1.2	1.3
SS A5	0.6	0.8	0.9	1.1	1.3	0.8	0.9	1.1	1.3	1.5
SS A6	0.5	0.7	0.8	1.0	1.2	0.7	0.8	1.0	1.2	1.3
SS A7	0.7	0.8	1.0	1.1	1.3	0.8	1.0	1.2	1.4	1.6
SS A8	0.6	0.7	0.9	1.0	1.2	0.7	0.8	1.0	1.2	1.4
SS B1	0.6	0.7	0.9	1.1	1.3	0.8	0.9	1.1	1.3	1.5
SS B2	0.6	0.7	0.9	1.0	1.2	0.7	0.8	1.0	1.2	1.4
SS B3	0.6	0.8	0.9	1.1	1.3	0.8	0.9	1.1	1.3	1.5
SS B4	0.6	0.7	0.9	1.0	1.2	0.7	0.8	1.0	1.2	1.4
SS B5	0.7	0.8	1.0	1.1	1.3	0.9	1.0	1.2	1.4	1.6
SS B6	0.6	0.7	0.9	1.0	1.2	0.7	0.8	1.0	1.2	1.4
SS B7	0.7	0.8	1.0	1.2	1.4	0.9	1.0	1.2	1.4	1.6
SS B8	0.6	0.7	0.9	1.1	1.2	0.7	0.8	1.0	1.2	1.4
SS C1	0.6	0.7	0.9	1.1	1.2	0.8	0.9	1.1	1.2	1.4
SS G2	0.5	0.7	0.9	1.0	1.2	0.7	0.8	1.0	1.2	1.3
SS C3	0.7	0.8	0.9	1.1	1.3	0.9	1.0	1.2	1.3	1.5
SS C4	0.6	0.7	0.9	1.1	1.2	0.8	0.9	1.0	1.2	1.4
SS D1	0.5	0.7	0.9	1.0	1.2	0.7	0.8	1.0	1.2	1.4
SS D2	0.5	0.6	0.8	1.0	1.2	0.6	0.7	0.9	1.1	1.3
SS D3	0.6	0.7	0.9	1.1	1.2	0.7	0.9	1.0	1.2	1.4
SS D4	0.5	0.6	0.8	1.0	1.2	0.6	0.8	0.9	1.1	1.3
SS E1	0.6	0.7	0.9	1.0	1.2	0.7	0.8	1.0	1.2	1.4
SS E2	0.5	0.7	0.8	1.0	1.2	0.6	0.8	1.0	1.1	1.3
SS E3	0.6	0.7	0.9	1.1	1.3	0.8	0.9	1.1	1.3	1.5
SS E4	0.5	0.7	0.8	1.0	1.2	0.7	0.8	1.0	1.2	1.3

RED BLUFF, CALIFORNIA

TYPE	SOLAR SAVINGS FRACTION (%) BASE TEMP: 55 F (1028 DD) LCR 200	100	70	50	40	30	25	20	65 F (2693 DD) 200	100	70	50	40	30	25	20	CONSERVATION FACTOR 55 F 200	100	50	30	20	65 F 200	100	50	30	20
WW A1	28	41	50	58	63	70	75	80	18	25	30	36	40	47	51	57	0.8	0.9	1.1	1.3	1.5	1.2	1.3	1.5	1.6	1.8
WW A2	33	50	59	68	74	81	85	90	18	29	36	44	50	58	64	70	0.7	0.8	1.0	1.2	1.4	0.9	1.1	1.2	1.4	1.6
WW A3	34	52	63	72	78	85	88	92	17	30	38	47	54	62	68	74	0.7	0.8	1.0	1.2	1.3	0.9	1.0	1.2	1.3	1.5
WW A4	35	54	64	74	80	86	90	94	17	30	39	49	56	65	70	76	0.7	0.8	1.0	1.1	1.3	0.9	1.0	1.1	1.3	1.5
WW A5	36	56	66	76	81	88	91	95	17	31	41	51	58	67	72	78	0.6	0.8	1.0	1.1	1.3	0.8	1.0	1.1	1.3	1.5
WW A6	36	56	67	77	82	89	92	95	17	32	42	52	59	68	73	80	0.6	0.8	0.9	1.1	1.3	0.8	0.9	1.1	1.3	1.5
WW B1	29	44	53	62	68	75	79	84	15	24	30	37	42	49	53	59	0.8	0.9	1.1	1.3	1.4	1.1	1.2	1.4	1.6	1.8
WW B2	35	54	65	74	80	87	91	94	17	31	40	50	57	66	72	78	0.6	0.8	1.0	1.1	1.3	0.8	1.0	1.1	1.3	1.5
WW B3	40	61	71	80	86	91	94	97	20	37	47	58	65	74	79	84	0.6	0.7	0.9	1.1	1.3	0.8	0.9	1.1	1.2	1.4
WW B4	39	61	72	81	86	92	95	97	19	36	48	59	67	76	81	86	0.6	0.7	0.9	1.1	1.3	0.8	0.9	1.0	1.2	1.4
WW B5	37	59	70	80	85	91	94	97	17	35	46	58	66	75	80	86	0.6	0.7	0.9	1.1	1.3	0.8	0.9	1.0	1.2	1.4
WW C1	41	62	72	81	87	92	95	97	21	38	49	60	67	76	81	86	0.6	0.7	0.9	1.1	1.3	0.8	0.9	1.0	1.2	1.4
WW C2	39	60	71	80	86	91	94	97	19	36	47	58	65	74	80	85	0.6	0.7	0.9	1.1	1.3	0.8	0.9	1.0	1.2	1.4
WW C3	43	66	76	85	90	95	97	99	21	42	54	66	73	82	87	91	0.6	0.7	0.9	1.1	1.2	0.7	0.8	1.0	1.2	1.3
WW C4	40	62	74	83	88	94	96	98	19	38	51	63	70	79	84	89	0.6	0.7	0.9	1.1	1.2	0.7	0.8	1.0	1.2	1.3
TW A1	28	40	48	57	62	70	74	79	18	25	30	35	40	46	50	56	0.8	1.0	1.1	1.3	1.5	1.2	1.3	1.5	1.6	1.8
TW A2	30	46	55	64	70	77	82	86	17	26	33	41	46	54	59	65	0.7	0.9	1.1	1.2	1.4	1.0	1.1	1.3	1.5	1.6
TW A3	30	48	57	67	73	80	84	89	16	27	34	43	49	57	62	69	0.7	0.8	1.0	1.2	1.4	1.0	1.1	1.2	1.4	1.6
TW A4	30	47	58	67	73	81	85	89	15	26	34	43	49	58	63	70	0.7	0.8	1.0	1.2	1.4	0.9	1.1	1.2	1.4	1.6
TW B1	27	40	49	57	63	71	75	80	17	24	30	36	40	47	52	58	0.8	0.9	1.1	1.3	1.5	1.1	1.3	1.4	1.6	1.8
TW B2	28	44	53	63	69	76	81	86	15	25	32	39	45	53	58	64	0.7	0.9	1.1	1.2	1.4	1.0	1.1	1.3	1.5	1.6
TW B3	28	44	54	63	69	77	81	86	14	25	32	39	45	53	59	65	0.7	0.9	1.1	1.2	1.4	1.0	1.1	1.3	1.4	1.6
TW B4	26	42	51	61	67	74	79	84	13	23	30	37	43	51	56	63	0.8	0.9	1.1	1.3	1.4	1.0	1.1	1.3	1.5	1.6
TW C1	26	39	48	57	63	70	75	80	15	23	29	35	40	47	52	58	0.8	0.9	1.1	1.3	1.5	1.1	1.2	1.4	1.6	1.7
TW C2	25	39	49	58	64	72	76	82	14	22	28	35	41	48	53	60	0.8	0.9	1.1	1.3	1.5	1.1	1.2	1.4	1.5	1.7
TW C3	24	38	46	55	62	69	74	80	13	21	27	34	39	46	51	57	0.8	0.9	1.1	1.3	1.5	1.1	1.2	1.4	1.6	1.7
TW C4	22	34	42	50	56	64	69	75	13	20	25	31	35	42	46	52	0.9	1.0	1.2	1.4	1.5	1.2	1.3	1.5	1.7	1.8
TW D1	24	38	47	56	62	69	74	79	12	20	26	32	37	43	48	54	0.8	0.9	1.1	1.3	1.5	1.1	1.3	1.5	1.6	1.8
TW D2	32	50	60	70	76	83	87	91	17	29	37	46	53	62	67	74	0.7	0.8	1.0	1.2	1.3	0.9	1.0	1.2	1.3	1.5
TW D3	35	54	64	74	80	86	90	94	18	31	40	50	57	66	71	78	0.7	0.8	1.0	1.1	1.3	0.9	1.0	1.1	1.3	1.5
TW D4	36	56	67	76	82	89	92	95	18	33	43	54	61	70	76	82	0.6	0.8	0.9	1.1	1.3	0.8	0.9	1.1	1.3	1.4
TW D5	34	55	66	76	82	88	92	95	17	32	43	54	61	70	76	82	0.6	0.8	0.9	1.1	1.3	0.8	0.9	1.1	1.2	1.4
TW E1	38	58	68	77	83	89	92	95	20	35	45	55	62	71	76	82	0.6	0.8	0.9	1.1	1.3	0.8	0.9	1.1	1.3	1.4
TW E2	37	57	68	77	83	89	92	96	19	34	44	55	62	71	76	82	0.6	0.8	0.9	1.1	1.3	0.8	0.9	1.1	1.3	1.4
TW E3	42	64	74	83	88	94	96	98	22	40	52	63	70	79	84	89	0.6	0.7	0.9	1.1	1.2	0.7	0.8	1.0	1.2	1.4
TW E4	39	61	72	81	87	92	95	97	19	37	49	60	68	77	82	88	0.6	0.7	0.9	1.1	1.3	0.8	0.9	1.0	1.2	1.4
TW F1	26	40	48	56	62	70	74	80	16	23	28	34	39	46	50	56	0.8	0.9	1.1	1.3	1.5	1.2	1.3	1.4	1.6	1.8
TW F2	27	43	53	62	68	76	80	85	14	24	31	38	44	51	57	63	0.8	0.9	1.1	1.3	1.4	1.0	1.1	1.3	1.5	1.7
TW F3	27	44	54	63	69	77	82	87	13	23	31	39	45	53	58	65	0.7	0.9	1.1	1.2	1.4	1.0	1.1	1.3	1.4	1.6
TW F4	25	42	52	62	68	76	81	86	11	21	29	37	43	52	57	64	0.7	0.9	1.1	1.2	1.4	1.0	1.1	1.3	1.4	1.6
TW G1	24	38	47	56	62	69	74	79	14	21	27	33	38	45	50	56	0.8	0.9	1.1	1.3	1.5	1.1	1.3	1.4	1.6	1.8
TW G2	24	39	49	58	65	73	77	83	11	21	27	35	40	48	53	60	0.8	0.9	1.1	1.3	1.4	1.1	1.2	1.3	1.5	1.7
TW G3	22	38	47	57	63	71	76	82	10	19	26	33	39	47	52	59	0.8	0.9	1.1	1.3	1.5	1.1	1.2	1.3	1.5	1.7
TW G4	18	33	42	52	58	67	72	78	7	16	22	29	34	42	47	54	0.8	0.9	1.1	1.3	1.5	1.1	1.2	1.4	1.6	1.7
TW H1	21	34	43	52	59	67	71	77	10	18	24	30	35	42	47	53	0.8	1.0	1.1	1.3	1.5	1.2	1.3	1.4	1.6	1.8
TW H2	18	32	41	50	57	65	70	76	8	16	21	28	33	40	45	52	0.8	1.0	1.1	1.3	1.5	1.2	1.3	1.4	1.6	1.8
TW H3	15	28	37	46	52	60	66	72	6	13	18	24	29	36	41	47	0.9	1.0	1.2	1.4	1.6	1.2	1.4	1.5	1.7	1.8
TW H4	11	22	30	38	44	52	57	64	4	10	14	19	23	29	34	39	1.0	1.1	1.3	1.5	1.7	1.4	1.5	1.7	1.8	2.0
TW I1	21	35	44	53	59	66	71	77	10	18	23	29	33	40	44	50	0.8	1.0	1.2	1.4	1.5	1.2	1.3	1.5	1.7	1.9
TW I2	29	47	57	67	73	81	85	90	13	26	34	43	49	58	64	71	0.7	0.8	1.0	1.2	1.4	0.9	1.0	1.2	1.4	1.5
TW I3	31	50	60	70	76	83	87	92	14	28	36	46	53	62	68	74	0.7	0.8	1.0	1.2	1.3	0.9	1.0	1.2	1.3	1.5
TW I4	32	53	64	74	80	87	90	94	15	30	40	50	58	67	73	79	0.6	0.8	1.0	1.1	1.3	0.8	0.9	1.1	1.3	1.4
TW I5	32	53	64	74	80	87	91	94	14	30	40	51	59	68	74	81	0.6	0.8	1.0	1.1	1.3	0.8	0.9	1.1	1.3	1.4
TW J1	35	55	66	75	81	88	91	95	18	32	42	52	59	68	74	80	0.6	0.8	1.0	1.1	1.3	0.8	0.9	1.1	1.3	1.5
TW J2	34	54	65	75	81	87	91	95	16	31	41	52	59	68	74	80	0.6	0.8	1.0	1.1	1.3	0.8	0.9	1.1	1.3	1.4
TW J3	38	60	71	81	86	92	95	97	18	37	48	60	68	77	82	87	0.6	0.7	0.9	1.1	1.3	0.7	0.9	1.0	1.2	1.4
TW J4	36	58	69	79	85	91	94	97	17	34	46	57	65	75	80	86	0.6	0.7	0.9	1.1	1.3	0.8	0.9	1.0	1.2	1.4
DG A1	27	43	52	60	65	72	75	79	13	22	28	35	39	45	48	52	0.7	0.9	1.1	1.4	1.7	1.1	1.2	1.4	1.8	2.1
DG A2	27	44	54	64	70	77	81	85	13	24	31	39	45	53	57	63	0.7	0.8	1.1	1.3	1.5	1.0	1.1	1.3	1.5	1.7
DG A3	31	49	60	70	76	83	86	90	16	28	37	47	54	63	69	75	0.7	0.8	1.0	1.2	1.4	0.9	1.0	1.1	1.3	1.5
DG B1	29	46	56	65	71	78	81	85	13	22	29	37	43	50	54	59	0.7	0.9	1.0	1.3	1.5	1.1	1.1	1.3	1.6	1.9
DG B2	29	47	58	68	75	82	86	90	14	25	33	42	48	57	63	69	0.7	0.8	1.0	1.2	1.4	1.0	1.1	1.2	1.4	1.6
DG B3	32	51	62	73	80	87	90	93	16	29	38	49	57	67	73	79	0.7	0.8	0.9	1.1	1.4	0.9	1.0	1.1	1.3	1.5
DG C1	33	51	61	72	78	84	88	91	15	26	35	44	50	58	63	69	0.7	0.8	1.0	1.2	1.5	1.0	1.0	1.2	1.4	1.7
DG C2	33	52	63	74	80	87	90	94	16	28	37	48	55	64	70	76	0.7	0.8	0.9	1.1	1.3	0.9	1.0	1.1	1.3	1.5
DG C3	36	56	67	79	85	91	94	96	18	32	43	55	63	73	79	85	0.6	0.8	0.9	1.1	1.4	0.9	0.9	1.0	1.2	1.4
SS A1	34	49	58	67	72	79	83	88	21	32	39	46	52	59	64	70	0.7	0.9	1.1	1.2	1.4	0.9	1.1	1.3	1.4	1.6
SS A2	40	58	67	76	82	88	91	95	24	38	47	57	63	71	76	82	0.6	0.8	1.0	1.1	1.3	0.8	0.9	1.1	1.3	1.5
SS A3	32	47	55	64	69	76	80	85	20	29	36	43	48	55	59	65	0.7	0.9	1.1	1.3	1.4	1.0	1.1	1.3	1.5	1.7
SS A4	39	57	67	75	81	87	91	94	24	38	47	56	62	70	75	80	0.7	0.8	1.0	1.1	1.3	0.8	0.9	1.1	1.3	1.5
SS A5	32	47	55	63	68	75	79	84	21	30	36	43	48	55	59	65	0.8	0.9	1.1	1.3	1.5	1.0	1.1	1.3	1.5	1.7
SS A6	40	58	67	76	82	88	91	95	24	38	47	56	63	71	76	81	0.6	0.8	1.0	1.1	1.3	0.8	0.9	1.1	1.3	1.5
SS A7	30	42	50	58	63	70	74	79	19	27	32	38	42	48	53	58	0.8	1.0	1.2	1.4	1.5	1.1	1.2	1.4	1.6	1.8
SS A8	39	56	66	75	80	87	90	94	24	37	46	55	61	69	74	79	0.7	0.8	1.0	1.1	1.3	0.8	1.0	1.1	1.3	1.5
SS B1	30	45	54	62	68	75	80	85	18	27	34	41	46	54	58	64	0.8	0.9	1.1	1.3	1.4	1.0	1.1	1.3	1.5	1.7
SS B2	35	53	63	72	78	85	89	92	21	34	43	52	58	67	72	78	0.7	0.8	1.0	1.2	1.3	0.9	1.0	1.1	1.3	1.5
SS B3	29	43	52	60	66	73	77	82	17	26	32	39	44	51	55	61	0.8	0.9	1.1	1.3	1.5	1.0	1.2	1.4	1.5	1.7
SS B4	35	52	62	71	77	84	88	92	20	33	42	51	57	65	70	76	0.7	0.8	1.0	1.2	1.3	0.9	1.0	1.2	1.3	1.5
SS B5	27	40	48	56	62	69	73	79	17	25	30	36	41	47	52	57	0.8	1.0	1.2	1.3	1.5	1.1	1.2	1.4	1.6	1.8
SS B6	35	53	63	72	77	84	88	92	20	33	42	51	57	66	71	77	0.7	0.8	1.0	1.2	1.3	0.9	1.0	1.1	1.3	1.5
SS B7	25	37	45	52	58	65	69	74	16	23	28	33	37	43	47	52	0.9	1.0	1.2	1.4	1.6	1.2	1.3	1.5	1.7	1.9
SS B8	34	51	61	70	76	83	87	91	20	32	40	49	55	64	69	75	0.7	0.8	1.0	1.2	1.4	0.9	1.0	1.2	1.3	1.5
SS C1	31	48	58	67	73	81	85	89	15	27	34	43	49	57	62	69	0.7	0.8	1.0	1.2	1.4	1.0	1.1	1.2	1.4	1.6
SS C2	35	54	64	74	80	87	90	94	17	31	41	50	57	66	72	78	0.7	0.8	1.0	1.1	1.3	0.9	1.0	1.1	1.3	1.5
SS C3	25	40	49	58	64	71	76	81	14	22	28	35	40	47	52	58	0.8	0.9	1.1	1.3	1.5	1.1	1.2	1.4	1.6	1.7
SS C4	31	49	59	69	75	82	86	91	16	28	36	45	52	60	66	73	0.7	0.8	1.0	1.2	1.4	0.9	1.0	1.2	1.4	1.5
SS D1	40	57	66	75	80	87	90	94	24	37	45	54	60	68	72	78	0.7	0.8	1.0	1.2	1.3	0.8	1.0	1.1	1.3	1.5
SS D2	45	64	74	83	88	93	96	98	27	44	54	64	71	79	83	88	0.6	0.7	0.9	1.1	1.3	0.7	0.9	1.0	1.2	1.4
SS D3	36	52	61	70	75	82	86	90	23	34	41	49	54	62	67	72	0.7	0.8	1.0	1.2	1.4	0.9	1.0	1.2	1.4	1.6
SS D4	44	63	73	81	86	92	95	97	26	42	52	62	69	77	82	87	0.6	0.7	0.9	1.1	1.3	0.7	0.9	1.0	1.2	1.4
SS E1	36	53	62	71	77	83	87	91	20	32	40	48	54	62	67	73	0.7	0.8	1.0	1.2	1.3	0.9	1.0	1.2	1.4	1.6
SS E2	41	60	70	79	85	90	93	96	23	39	49	59	65	74	78	84	0.6	0.8	0.9	1.1	1.3	0.8	0.9	1.1	1.2	1.4
SS E3	31	45	54	63	68	75	80	85	19	28	34	41	46	53	58	64	0.8	0.9	1.1	1.3	1.4	1.0	1.1	1.3	1.5	1.7
SS E4	38	56	66	75	81	87	91	94	22	36	45	55	61	69	74	80	0.6	0.8	1.0	1.1	1.3	0.8	0.9	1.1	1.3	1.5

SACRAMENTO, CALIFORNIA

TYPE	SOLAR SAVINGS FRACTION (%)																CONSERVATION FACTOR									
BASE TEMP:	55 F (1048 DD)								65 F (2845 DD)								55 F					65 F				
LCR	200	100	70	50	40	30	25	20	200	100	70	50	40	30	25	20	200	100	50	30	20	200	100	50	30	20
WW A1	29	42	49	57	63	70	74	79	18	26	31	37	41	48	52	58	0.8	1.0	1.2	1.3	1.5	1.1	1.3	1.4	1.6	1.8
WW A2	33	49	59	68	73	80	85	89	18	29	37	45	51	59	64	70	0.7	0.8	1.0	1.2	1.4	0.9	1.1	1.2	1.4	1.6
WW A3	34	52	62	71	77	84	88	92	18	31	39	48	55	63	68	74	0.7	0.8	1.0	1.2	1.3	0.9	1.0	1.2	1.3	1.5
WW A4	35	54	64	73	79	85	89	93	17	31	40	50	57	65	70	77	0.7	0.8	1.0	1.2	1.3	0.9	1.0	1.1	1.3	1.5
WW A5	36	55	65	75	80	87	91	94	17	32	42	52	59	67	73	79	0.6	0.8	1.0	1.1	1.3	0.8	0.9	1.1	1.3	1.5
WW A6	36	56	66	76	81	88	91	95	17	33	43	53	60	69	74	80	0.6	0.8	1.0	1.1	1.3	0.8	0.9	1.1	1.3	1.5
WW B1	29	44	53	61	67	74	78	83	16	25	31	38	43	50	54	60	0.8	0.9	1.1	1.3	1.5	1.0	1.2	1.4	1.6	1.8
WW B2	35	54	64	74	79	86	90	94	17	32	41	51	58	67	72	78	0.7	0.8	1.0	1.1	1.3	0.8	0.9	1.1	1.3	1.5
WW B3	40	60	70	79	85	91	94	96	21	38	48	59	66	74	79	85	0.6	0.8	0.9	1.1	1.3	0.8	0.9	1.1	1.2	1.4
WW B4	39	60	71	80	86	91	94	97	19	37	49	60	67	76	81	87	0.6	0.7	0.9	1.1	1.3	0.7	0.9	1.0	1.2	1.4
WW B5	37	58	69	79	85	91	94	97	18	36	47	59	66	75	81	86	0.6	0.7	0.9	1.1	1.3	0.8	0.9	1.0	1.2	1.4
WW C1	41	61	72	81	86	92	94	97	22	39	50	61	68	76	81	86	0.6	0.7	0.9	1.1	1.3	0.8	0.9	1.0	1.2	1.4
WW C2	39	59	70	79	85	91	94	97	20	37	48	59	66	75	80	85	0.6	0.8	0.9	1.1	1.3	0.8	0.9	1.0	1.2	1.4
WW C3	43	65	76	85	90	95	97	98	22	43	55	67	74	82	87	91	0.6	0.7	0.9	1.1	1.2	0.7	0.8	1.0	1.2	1.3
WW C4	40	62	73	82	88	93	96	98	19	39	52	63	71	80	85	90	0.6	0.7	0.9	1.1	1.2	0.7	0.8	1.0	1.2	1.4
TW A1	28	41	48	56	62	69	73	78	18	25	30	36	41	47	51	57	0.8	1.0	1.2	1.4	1.5	1.2	1.3	1.4	1.6	1.8
TW A2	30	46	55	63	69	76	81	86	17	27	34	42	47	55	60	66	0.7	0.9	1.1	1.3	1.4	1.0	1.1	1.3	1.5	1.6
TW A3	31	47	57	66	72	79	83	88	16	28	35	44	50	58	63	69	0.7	0.9	1.0	1.2	1.4	0.9	1.1	1.2	1.4	1.6
TW A4	30	47	57	66	72	80	84	89	15	27	35	44	50	58	64	70	0.7	0.8	1.0	1.2	1.4	0.9	1.0	1.2	1.4	1.6
TW B1	28	41	49	57	63	70	74	79	17	25	30	37	41	48	53	58	0.8	0.9	1.1	1.3	1.5	1.1	1.2	1.4	1.6	1.8
TW B2	29	44	53	62	68	75	80	85	16	26	32	40	46	54	59	65	0.8	0.9	1.1	1.3	1.4	1.0	1.1	1.3	1.5	1.6
TW B3	28	44	53	62	68	76	80	85	15	25	32	40	46	54	60	66	0.7	0.9	1.1	1.3	1.4	1.0	1.1	1.3	1.4	1.6
TW B4	26	42	51	60	66	74	78	84	14	24	31	38	44	52	57	64	0.8	0.9	1.1	1.3	1.4	1.0	1.1	1.3	1.5	1.6
TW C1	26	40	48	56	62	70	74	79	15	23	29	36	41	48	53	59	0.8	0.9	1.1	1.3	1.5	1.1	1.2	1.4	1.6	1.7
TW C2	25	40	49	57	63	71	75	81	14	23	29	36	42	49	54	60	0.8	0.9	1.1	1.3	1.5	1.1	1.2	1.3	1.5	1.7
TW C3	24	38	46	55	61	69	73	79	13	22	28	35	40	47	52	58	0.8	0.9	1.1	1.3	1.5	1.1	1.2	1.4	1.5	1.7
TW C4	22	34	42	50	56	64	68	74	13	20	25	31	36	43	47	53	0.9	1.0	1.2	1.4	1.6	1.2	1.3	1.5	1.6	1.8
TW D1	25	38	47	55	61	68	73	78	13	21	27	33	38	44	49	55	0.8	0.9	1.2	1.3	1.5	1.1	1.2	1.4	1.6	1.8
TW D2	32	50	60	69	75	82	86	91	17	30	38	47	54	62	68	74	0.7	0.8	1.0	1.2	1.4	0.9	1.0	1.2	1.3	1.5
TW D3	35	54	64	73	79	86	89	93	18	32	42	51	58	67	72	78	0.7	0.8	1.0	1.2	1.3	0.8	1.0	1.1	1.3	1.5
TW D4	36	56	66	76	81	88	91	95	18	34	44	55	62	71	76	82	0.6	0.8	1.0	1.1	1.3	0.8	0.9	1.1	1.3	1.4
TW D5	35	55	65	75	81	88	91	95	17	33	44	54	62	71	76	82	0.6	0.8	1.0	1.1	1.3	0.8	0.9	1.1	1.2	1.4
TW E1	38	57	67	77	82	88	92	95	21	36	46	56	63	71	76	82	0.6	0.8	1.0	1.1	1.3	0.8	0.9	1.1	1.3	1.4
TW E2	37	57	67	76	82	88	92	95	20	35	45	56	63	71	77	82	0.6	0.8	1.0	1.1	1.3	0.8	0.9	1.1	1.3	1.4
TW E3	42	63	74	83	88	93	96	98	23	41	53	64	71	79	84	89	0.6	0.7	0.9	1.1	1.2	0.7	0.8	1.0	1.2	1.4
TW E4	39	60	71	80	86	92	95	97	20	38	50	61	69	77	82	88	0.6	0.7	0.9	1.1	1.3	0.7	0.9	1.0	1.2	1.4
TW F1	27	40	48	56	62	69	73	79	16	24	29	35	40	47	51	57	0.8	1.0	1.2	1.3	1.5	1.1	1.3	1.4	1.6	1.8
TW F2	28	43	52	61	67	75	79	84	15	25	31	39	45	52	57	64	0.8	0.9	1.1	1.3	1.4	1.0	1.1	1.3	1.5	1.7
TW F3	27	44	53	63	69	76	81	86	13	24	32	40	46	54	59	66	0.7	0.9	1.1	1.2	1.4	1.0	1.1	1.3	1.4	1.6
TW F4	25	42	52	61	67	75	80	85	11	22	30	38	44	53	58	65	0.7	0.9	1.1	1.2	1.4	1.0	1.1	1.3	1.4	1.6
TW G1	25	38	47	55	61	68	73	78	14	22	27	34	39	46	51	57	0.8	0.9	1.1	1.3	1.5	1.1	1.2	1.4	1.6	1.8
TW G2	24	40	49	58	64	72	76	82	12	21	28	36	41	49	54	61	0.8	0.9	1.1	1.3	1.5	1.0	1.2	1.3	1.5	1.7
TW G3	22	38	47	57	63	71	75	81	10	19	26	34	40	48	53	60	0.8	0.9	1.1	1.3	1.5	1.1	1.2	1.3	1.5	1.7
TW G4	19	33	42	51	58	66	71	77	7	16	22	30	35	43	48	55	0.8	1.0	1.1	1.3	1.5	1.1	1.2	1.4	1.6	1.7
TW H1	21	35	43	52	58	66	71	76	10	18	24	31	36	43	48	54	0.8	1.0	1.2	1.4	1.5	1.2	1.3	1.4	1.6	1.8
TW H2	19	32	41	50	56	64	69	75	8	16	22	29	34	41	46	53	0.8	1.0	1.2	1.4	1.5	1.2	1.3	1.4	1.6	1.8
TW H3	16	28	37	46	52	60	65	71	6	13	19	25	30	37	42	48	0.9	1.0	1.2	1.4	1.6	1.2	1.3	1.5	1.7	1.8
TW H4	12	23	30	38	44	52	57	63	4	10	15	20	24	30	35	40	1.0	1.1	1.3	1.5	1.7	1.4	1.5	1.6	1.8	2.0
TW I1	22	36	44	52	58	65	70	76	11	18	24	30	35	41	46	51	0.8	1.0	1.2	1.4	1.5	1.2	1.3	1.5	1.7	1.9
TW I2	29	47	57	66	72	80	84	89	14	26	35	44	50	59	65	71	0.7	0.8	1.0	1.2	1.4	0.9	1.0	1.2	1.4	1.6
TW I3	31	49	60	69	75	83	87	91	15	29	38	47	54	63	68	75	0.7	0.8	1.0	1.2	1.3	0.9	1.0	1.2	1.3	1.5
TW I4	33	52	63	73	79	86	90	94	15	31	41	51	59	68	73	80	0.7	0.8	1.0	1.1	1.3	0.8	0.9	1.1	1.3	1.5
TW I5	32	53	63	73	79	86	90	94	15	31	41	52	60	69	75	81	0.7	0.8	1.0	1.1	1.3	0.8	0.9	1.1	1.3	1.4
TW J1	36	55	65	75	80	87	91	94	18	33	43	53	60	69	74	80	0.6	0.8	1.0	1.1	1.3	0.8	0.9	1.1	1.3	1.5
TW J2	34	54	64	74	80	87	90	94	17	32	42	53	60	69	74	80	0.7	0.8	1.0	1.1	1.3	0.8	0.9	1.1	1.3	1.5
TW J3	38	60	71	80	86	92	95	97	19	38	50	61	68	77	82	88	0.6	0.7	0.9	1.1	1.3	0.7	0.8	1.0	1.2	1.4
TW J4	36	57	69	78	84	90	93	96	17	35	47	58	66	75	80	86	0.6	0.8	0.9	1.1	1.3	0.8	0.9	1.0	1.2	1.4
DG A1	28	43	52	60	65	71	74	78	13	23	29	36	40	46	49	53	0.7	0.9	1.2	1.4	1.7	1.0	1.2	1.4	1.7	2.1
DG A2	28	44	54	63	69	76	80	84	14	25	32	40	46	54	58	64	0.7	0.9	1.1	1.3	1.5	1.0	1.1	1.3	1.5	1.7
DG A3	31	49	60	69	75	82	86	89	16	29	38	48	55	64	69	75	0.7	0.8	1.0	1.2	1.4	0.9	1.0	1.1	1.3	1.6
DG B1	29	46	55	64	70	77	80	84	13	23	31	39	44	51	55	60	0.7	0.9	1.1	1.3	1.6	1.0	1.1	1.3	1.6	1.9
DG B2	30	47	57	67	74	81	85	89	14	25	34	43	50	58	64	70	0.7	0.8	1.0	1.2	1.4	1.0	1.0	1.2	1.4	1.6
DG B3	32	51	61	72	79	86	90	93	16	30	39	50	58	67	73	79	0.7	0.8	1.0	1.1	1.4	0.9	1.0	1.1	1.3	1.5
DG C1	33	51	60	71	77	83	87	90	16	27	36	45	51	59	63	69	0.7	0.8	1.0	1.2	1.5	0.9	1.0	1.2	1.4	1.7
DG C2	33	52	62	73	79	86	90	93	16	29	39	49	56	65	70	76	0.7	0.8	1.0	1.1	1.3	0.9	1.0	1.1	1.3	1.5
DG C3	36	55	66	78	84	90	93	96	19	33	44	56	63	73	79	85	0.6	0.8	0.9	1.1	1.3	0.9	0.9	1.0	1.2	1.4
SS A1	35	50	58	67	72	79	83	88	22	33	41	48	54	61	65	71	0.7	0.9	1.1	1.2	1.4	0.9	1.0	1.2	1.4	1.6
SS A2	40	58	67	76	81	88	91	94	26	41	50	59	65	72	77	82	0.7	0.8	1.0	1.1	1.3	0.8	0.9	1.1	1.3	1.5
SS A3	33	47	55	64	69	76	80	85	21	31	38	45	50	57	61	67	0.7	0.9	1.1	1.3	1.4	0.9	1.1	1.3	1.5	1.7
SS A4	40	57	67	75	81	87	90	94	26	40	49	58	64	71	76	81	0.7	0.8	1.0	1.1	1.3	0.8	0.9	1.1	1.3	1.5
SS A5	33	47	55	63	68	75	79	84	22	32	38	45	50	56	61	66	0.8	0.9	1.1	1.3	1.5	1.0	1.1	1.3	1.5	1.7
SS A6	40	58	67	76	81	88	91	94	26	40	49	58	64	72	77	82	0.7	0.8	1.0	1.1	1.3	0.8	0.9	1.1	1.3	1.5
SS A7	30	43	50	58	63	69	74	79	20	29	34	40	45	51	55	60	0.8	1.0	1.2	1.4	1.5	1.0	1.2	1.4	1.6	1.8
SS A8	40	57	66	74	80	86	90	93	25	39	48	57	63	70	75	80	0.7	0.8	1.0	1.2	1.3	0.8	0.9	1.1	1.3	1.5
SS B1	31	45	54	62	68	75	79	84	19	29	36	43	48	55	60	66	0.8	0.9	1.1	1.3	1.4	1.0	1.1	1.3	1.5	1.7
SS B2	36	53	63	72	78	84	88	92	22	36	45	54	60	68	73	79	0.7	0.8	1.0	1.2	1.3	0.8	1.0	1.1	1.3	1.5
SS B3	29	43	52	60	65	73	77	82	18	28	34	41	46	53	57	63	0.8	0.9	1.1	1.3	1.5	1.0	1.1	1.3	1.5	1.7
SS B4	36	53	62	71	77	84	87	92	22	35	44	53	59	67	72	77	0.7	0.8	1.0	1.2	1.3	0.8	1.0	1.1	1.3	1.5
SS B5	28	41	48	56	62	69	73	78	18	26	32	38	43	49	53	59	0.8	1.0	1.2	1.4	1.5	1.1	1.2	1.4	1.6	1.8
SS B6	36	53	62	71	77	84	88	92	21	35	44	53	59	67	72	78	0.7	0.8	1.0	1.2	1.3	0.8	1.0	1.1	1.3	1.5
SS B7	26	38	45	52	58	64	69	74	17	24	29	35	39	45	49	54	0.9	1.0	1.2	1.4	1.6	1.1	1.3	1.5	1.7	1.9
SS B8	35	51	61	70	76	82	86	91	21	34	42	51	57	65	70	76	0.7	0.8	1.0	1.2	1.4	0.8	1.0	1.2	1.3	1.5
SS C1	31	48	58	67	73	80	84	89	16	27	35	44	50	58	63	69	0.7	0.8	1.0	1.2	1.4	0.9	1.1	1.2	1.4	1.6
SS C2	35	54	64	73	79	86	90	93	18	32	42	51	58	67	72	78	0.7	0.8	1.0	1.1	1.3	0.8	0.9	1.1	1.3	1.5
SS C3	26	40	49	57	63	70	75	80	14	23	29	36	41	48	53	59	0.8	0.9	1.1	1.3	1.5	1.1	1.2	1.4	1.6	1.7
SS C4	32	49	59	68	74	81	85	90	16	29	37	46	53	61	67	73	0.7	0.8	1.0	1.2	1.4	0.9	1.0	1.2	1.4	1.5
SS D1	40	57	66	75	80	86	90	93	25	39	47	56	61	69	73	79	0.7	0.8	1.0	1.2	1.3	0.8	0.9	1.1	1.3	1.5
SS D2	45	64	74	83	88	93	95	98	28	46	56	66	72	80	84	88	0.6	0.7	0.9	1.1	1.3	0.7	0.8	1.0	1.2	1.4
SS D3	37	52	61	69	75	82	85	90	24	36	43	51	56	64	68	73	0.7	0.8	1.0	1.2	1.4	0.9	1.0	1.2	1.4	1.6
SS D4	44	63	72	81	86	92	94	97	28	44	54	64	70	78	82	87	0.6	0.8	0.9	1.1	1.3	0.7	0.9	1.0	1.2	1.4
SS E1	36	53	62	71	76	83	87	91	21	34	42	50	56	64	68	74	0.7	0.8	1.0	1.2	1.4	0.9	1.0	1.2	1.4	1.6
SS E2	41	60	70	79	84	90	93	96	24	41	50	60	67	75	79	84	0.6	0.8	0.9	1.1	1.3	0.8	0.9	1.1	1.3	1.4
SS E3	31	46	54	62	68	75	79	84	20	30	36	43	48	55	60	65	0.8	0.9	1.1	1.3	1.4	1.0	1.1	1.3	1.5	1.7
SS E4	38	56	66	75	81	87	91	94	23	38	47	56	63	71	75	81	0.7	0.8	1.0	1.1	1.3	0.8	0.9	1.1	1.3	1.5

SAN DIEGO, CALIFORNIA

TYPE	SOLAR SAVINGS FRACTION (%) BASE TEMP: 55 F (214 DD)								65 F (1512 DD)								CONSERVATION FACTOR 55 F					65 F				
LCR	200	100	70	50	40	30	25	20	200	100	70	50	40	30	25	20	200	100	50	30	20	200	100	50	30	20
WW A1	71	87	93	97	98	99	100	100	31	47	56	67	73	82	86	91	0.5	0.6	0.8	1.0	1.2	0.8	0.9	1.0	1.2	1.3
WW A2	81	94	98	99	100	100	100	100	37	57	69	80	86	92	95	97	0.5	0.6	0.8	1.0	-	0.6	0.7	0.9	1.1	1.2
WW A3	84	96	99	100	100	100	100	100	39	61	73	83	89	94	97	98	0.4	0.6	0.8	0.9	-	0.6	0.7	0.9	1.0	1.2
WW A4	86	97	99	100	100	100	100	100	40	63	75	85	91	95	97	99	0.4	0.6	0.8	0.9	-	0.6	0.7	0.9	1.0	1.2
WW A5	87	97	99	100	100	100	100	100	42	66	78	87	92	96	98	99	0.4	0.6	0.8	-	-	0.6	0.7	0.9	1.0	1.2
WW A6	88	98	99	100	100	100	100	100	43	67	79	88	93	97	98	99	0.4	0.6	0.7	-	-	0.6	0.7	0.8	1.0	1.2
WW B1	77	92	96	99	99	100	100	100	33	52	63	74	80	88	91	95	0.5	0.6	0.8	1.0	1.2	0.7	0.8	0.9	1.1	1.3
WW B2	85	97	99	100	100	100	100	100	40	63	76	86	91	96	97	99	0.4	0.6	0.8	0.9	-	0.6	0.7	0.9	1.0	1.2
WW B3	90	98	100	100	100	100	100	100	47	72	83	91	95	98	99	100	0.4	0.6	0.7	-	-	0.6	0.7	0.8	1.0	1.2
WW B4	90	98	100	100	100	100	100	100	46	71	83	91	95	98	99	100	0.4	0.6	0.7	-	-	0.6	0.7	0.8	1.0	1.2
WW B5	89	98	99	100	100	100	100	100	43	69	81	90	94	98	99	100	0.4	0.6	0.7	-	-	0.6	0.7	0.8	1.0	1.2
WW C1	91	98	100	100	100	100	100	100	49	73	84	92	96	98	99	100	0.4	0.6	0.7	-	-	0.5	0.7	0.8	1.0	1.2
WW C2	89	98	100	100	100	100	100	100	45	70	82	91	95	98	99	100	0.4	0.6	0.7	-	-	0.6	0.7	0.8	1.0	1.2
WW C3	92	99	100	100	100	100	100	100	51	77	88	94	97	99	100	100	0.4	0.5	0.7	-	-	0.5	0.6	0.8	1.0	1.2
WW C4	91	98	100	100	100	100	100	100	47	73	84	92	96	99	99	100	0.4	0.6	0.7	-	-	0.5	0.6	0.8	1.0	1.2
TW A1	69	86	93	97	98	99	100	100	30	45	55	65	72	80	85	90	0.5	0.6	0.8	1.0	1.2	0.8	0.9	1.0	1.2	1.3
TW A2	77	92	96	99	99	100	100	100	34	52	64	74	81	88	92	95	0.5	0.6	0.8	1.0	1.2	0.7	0.8	0.9	1.1	1.3
TW A3	79	93	97	99	100	100	100	100	35	55	67	78	84	91	94	97	0.5	0.6	0.8	1.0	-	0.7	0.8	0.9	1.1	1.3
TW A4	80	94	97	99	100	100	100	100	34	55	67	78	85	91	94	97	0.5	0.6	0.8	1.0	-	0.6	0.7	0.9	1.1	1.3
TW B1	70	87	93	97	98	99	100	100	30	46	56	66	73	81	86	91	0.5	0.6	0.8	1.0	1.2	0.8	0.9	1.0	1.2	1.3
TW B2	75	91	96	98	99	100	100	100	32	50	62	73	80	87	91	95	0.5	0.6	0.8	1.0	1.2	0.7	0.8	0.9	1.1	1.3
TW B3	76	91	96	99	99	100	100	100	31	51	62	73	80	88	92	95	0.5	0.6	0.8	1.0	1.2	0.7	0.8	0.9	1.1	1.3
TW B4	73	90	95	98	99	100	100	100	29	48	59	70	77	86	90	94	0.5	0.6	0.8	1.0	1.2	0.7	0.8	1.0	1.1	1.3
TW C1	69	86	93	97	98	99	100	100	28	44	55	66	73	81	86	91	0.5	0.6	0.8	1.0	1.2	0.8	0.9	1.0	1.2	1.3
TW C2	70	87	93	97	99	100	100	100	28	45	56	67	74	83	87	92	0.5	0.6	0.8	1.0	1.2	0.7	0.8	1.0	1.1	1.3
TW C3	67	85	92	96	98	99	100	100	26	43	53	64	71	80	85	90	0.5	0.6	0.8	1.0	1.2	0.8	0.9	1.0	1.2	1.3
TW C4	61	81	88	94	96	98	99	100	24	38	47	58	65	74	79	85	0.5	0.7	0.8	1.0	1.2	0.8	0.9	1.1	1.2	1.4
TW D1	71	88	94	97	99	100	100	100	28	45	56	66	74	82	87	92	0.5	0.6	0.8	1.0	1.2	0.7	0.8	1.0	1.1	1.3
TW D2	82	95	98	99	100	100	100	100	37	58	70	81	87	93	96	98	0.5	0.6	0.8	1.0	-	0.6	0.7	0.9	1.1	1.2
TW D3	85	96	99	100	100	100	100	100	41	63	75	85	91	95	97	99	0.4	0.6	0.8	0.9	-	0.6	0.7	0.9	1.0	1.2
TW D4	86	97	99	100	100	100	100	100	41	65	77	87	92	96	98	99	0.4	0.6	0.8	0.9	-	0.6	0.7	0.9	1.0	1.2
TW D5	85	97	99	100	100	100	100	100	40	64	76	86	91	96	98	99	0.4	0.6	0.8	0.9	-	0.6	0.7	0.9	1.0	1.2
TW E1	88	97	99	100	100	100	100	100	44	68	79	88	93	97	98	99	0.4	0.6	0.8	-	-	0.6	0.7	0.8	1.0	1.2
TW E2	87	97	99	100	100	100	100	100	43	66	78	88	92	97	98	99	0.4	0.6	0.8	0.9	-	0.6	0.7	0.9	1.0	1.2
TW E3	91	99	100	100	100	100	100	100	50	75	85	93	96	99	99	100	0.4	0.6	0.7	-	-	0.5	0.6	0.8	1.0	1.2
TW E4	89	98	100	100	100	100	100	100	46	71	83	91	95	98	99	100	0.4	0.6	0.7	-	-	0.6	0.7	0.8	1.0	1.2
TW F1	69	87	93	97	98	99	100	100	29	45	55	65	72	81	86	91	0.5	0.6	0.8	1.0	1.2	0.8	0.9	1.0	1.2	1.3
TW F2	75	91	96	98	99	100	100	100	31	50	61	72	79	87	91	95	0.5	0.6	0.8	1.0	1.2	0.7	0.8	0.9	1.1	1.3
TW F3	76	92	96	99	99	100	100	100	31	51	63	74	81	89	92	96	0.5	0.6	0.8	1.0	1.2	0.7	0.8	0.9	1.1	1.3
TW F4	75	91	96	99	99	100	100	100	29	49	61	73	80	88	92	95	0.5	0.6	0.8	1.0	1.2	0.7	0.8	0.9	1.1	1.3
TW G1	68	86	93	97	98	99	100	100	27	43	54	65	72	80	85	90	0.5	0.6	0.8	1.0	1.2	0.8	0.9	1.0	1.2	1.3
TW G2	72	89	94	98	99	100	100	100	27	46	57	68	76	84	89	93	0.5	0.6	0.8	1.0	1.2	0.7	0.8	1.0	1.1	1.3
TW G3	70	88	94	97	99	100	100	100	25	44	55	67	74	83	88	93	0.5	0.6	0.8	1.0	1.2	0.7	0.8	1.0	1.1	1.3
TW G4	64	84	91	96	98	99	100	100	21	38	49	61	68	78	83	89	0.5	0.6	0.8	1.0	1.2	0.8	0.9	1.0	1.2	1.3
TW H1	65	84	91	96	98	99	100	100	24	40	50	61	69	78	83	89	0.5	0.6	0.8	1.0	1.2	0.8	0.9	1.0	1.2	1.3
TW H2	63	83	90	95	97	99	100	100	21	37	48	59	66	76	82	87	0.5	0.6	0.8	1.0	1.2	0.8	0.9	1.0	1.2	1.4
TW H3	57	78	87	93	96	98	99	100	18	32	42	53	61	71	77	83	0.5	0.7	0.8	1.0	1.2	0.8	0.9	1.1	1.2	1.4
TW H4	48	70	80	88	92	96	97	99	13	25	34	44	51	61	67	74	0.6	0.7	0.9	1.1	1.2	0.9	1.0	1.2	1.3	1.5
TW I1	67	86	92	96	98	99	100	100	25	41	52	63	70	79	84	90	0.5	0.6	0.8	1.0	1.2	0.8	0.9	1.0	1.2	1.3
TW I2	79	93	97	99	100	100	100	100	33	54	67	78	84	91	94	97	0.5	0.6	0.8	1.0	-	0.6	0.7	0.9	1.1	1.3
TW I3	82	95	98	99	100	100	100	100	36	59	71	82	88	94	96	98	0.5	0.6	0.8	1.0	-	0.6	0.7	0.9	1.1	1.2
TW I4	84	96	99	100	100	100	100	100	38	62	74	85	90	95	97	99	0.4	0.6	0.8	0.9	-	0.6	0.7	0.9	1.0	1.2
TW I5	84	96	99	100	100	100	100	100	37	62	74	85	90	95	97	99	0.4	0.6	0.8	0.9	-	0.6	0.7	0.9	1.0	1.2
TW J1	86	97	99	100	100	100	100	100	42	65	77	87	92	96	98	99	0.4	0.6	0.8	0.9	-	0.6	0.7	0.9	1.0	1.2
TW J2	85	96	99	100	100	100	100	100	39	63	76	86	91	96	98	99	0.4	0.6	0.8	0.9	-	0.6	0.7	0.9	1.0	1.2
TW J3	89	98	100	100	100	100	100	100	46	71	83	91	95	98	99	100	0.4	0.6	0.7	-	-	0.6	0.7	0.8	1.0	1.2
TW J4	88	98	99	100	100	100	100	100	42	68	80	89	94	97	99	100	0.4	0.6	0.7	-	-	0.6	0.7	0.8	1.0	1.2
DG A1	78	93	97	99	100	100	100	100	31	51	62	72	78	84	88	91	0.5	0.6	0.8	-	-	0.7	0.8	1.0	1.2	1.4
DG A2	78	93	97	99	100	100	100	100	31	52	64	74	81	88	91	95	0.5	0.6	0.8	-	-	0.7	0.8	0.9	1.1	1.3
DG A3	81	93	96	97	97	97	97	97	34	58	71	81	87	92	94	95	0.5	0.6	1.3	-	-	0.6	0.7	0.9	1.1	1.5
DG B1	83	95	98	99	99	99	99	99	32	56	69	79	85	90	93	95	0.4	0.6	0.9	3.8	-	0.6	0.7	0.9	1.1	1.4
DG B2	83	96	99	100	100	100	100	100	32	56	70	81	87	93	95	97	0.4	0.6	0.8	1.0	-	0.6	0.7	0.9	1.1	1.3
DG B3	85	96	98	98	98	98	98	98	35	61	75	86	91	95	96	97	0.4	0.6	1.3	-	-	0.6	0.7	0.9	1.1	1.5
DG C1	88	97	98	99	99	99	98	98	37	63	77	87	91	95	96	97	0.4	0.6	2.0	-	-	0.6	0.7	0.9	1.1	1.5
DG C2	87	98	99	100	100	100	100	100	36	63	77	87	92	96	98	99	0.4	0.6	0.8	-	-	0.6	0.7	0.9	1.0	1.3
DG C3	89	97	98	98	99	98	98	98	39	68	82	91	94	97	98	98	0.4	0.6	2.5	-	-	0.6	0.6	0.8	1.1	1.8
SS A1	81	94	97	99	100	100	100	100	43	62	73	82	88	93	95	98	0.5	0.6	0.8	1.0	-	0.6	0.7	0.9	1.1	1.3
SS A2	88	97	99	100	100	100	100	100	51	72	83	90	94	97	99	99	0.4	0.6	0.8	0.9	-	0.6	0.7	0.8	1.0	1.2
SS A3	79	92	97	99	100	100	100	100	41	60	71	80	86	91	94	97	0.5	0.6	0.8	1.0	-	0.6	0.8	0.9	1.1	1.3
SS A4	87	97	99	100	100	100	100	100	51	72	82	90	94	97	98	99	0.4	0.6	0.8	0.9	-	0.6	0.7	0.9	1.0	1.2
SS A5	77	91	96	98	99	100	100	100	40	58	69	78	84	90	93	96	0.5	0.6	0.8	1.0	1.2	0.7	0.8	0.9	1.1	1.3
SS A6	88	97	99	100	100	100	100	100	51	73	83	90	94	97	99	99	0.4	0.6	0.8	0.9	-	0.6	0.7	0.8	1.0	1.2
SS A7	73	88	94	97	99	100	100	100	37	53	63	73	79	86	90	93	0.5	0.6	0.8	1.0	1.2	0.7	0.8	1.0	1.1	1.3
SS A8	87	97	99	100	100	100	100	100	50	72	82	90	94	97	98	99	0.4	0.6	0.8	0.9	-	0.6	0.7	0.9	1.0	1.2
SS B1	76	91	96	98	99	100	100	100	37	56	67	77	83	89	93	96	0.5	0.6	0.8	1.0	1.2	0.7	0.8	0.9	1.1	1.3
SS B2	84	95	98	100	100	100	100	100	45	66	77	86	91	95	97	99	0.5	0.6	0.8	1.0	-	0.6	0.7	0.9	1.0	1.2
SS B3	75	90	95	98	99	100	100	100	36	54	65	75	81	88	92	95	0.5	0.6	0.8	1.0	1.2	0.7	0.8	0.9	1.1	1.3
SS B4	83	95	98	100	100	100	100	100	44	66	77	86	91	95	97	99	0.5	0.6	0.8	1.0	-	0.6	0.7	0.9	1.1	1.2
SS B5	70	87	93	97	98	99	100	100	33	50	60	70	76	84	88	92	0.5	0.6	0.8	1.0	1.2	0.7	0.8	1.0	1.2	1.3
SS B6	83	95	98	100	100	100	100	100	44	66	77	86	91	95	97	99	0.5	0.6	0.8	1.0	-	0.6	0.7	0.9	1.0	1.2
SS B7	67	84	91	95	97	99	100	100	31	46	56	66	72	80	85	90	0.5	0.6	0.8	1.0	1.2	0.8	0.9	1.0	1.2	1.4
SS B8	82	95	98	99	100	100	100	100	43	65	76	85	90	95	97	98	0.5	0.6	0.8	1.0	-	0.6	0.7	0.9	1.1	1.2
SS C1	81	94	98	99	100	100	100	100	36	56	68	79	85	92	95	97	0.5	0.6	0.8	1.0	-	0.6	0.7	0.9	1.1	1.2
SS C2	85	96	99	100	100	100	100	100	40	63	75	85	91	95	97	99	0.4	0.6	0.8	0.9	-	0.6	0.7	0.9	1.0	1.2
SS C3	71	88	94	97	99	100	100	100	29	46	56	67	74	83	87	92	0.5	0.6	0.8	1.0	1.2	0.7	0.8	1.0	1.1	1.3
SS C4	80	94	98	99	100	100	100	100	36	57	69	80	86	92	95	98	0.5	0.6	0.8	1.0	-	0.6	0.7	0.9	1.1	1.2
SS D1	88	97	99	100	100	100	100	100	51	72	82	90	94	97	98	99	0.4	0.6	0.8	0.9	-	0.6	0.7	0.9	1.0	1.2
SS D2	92	99	100	100	100	100	100	100	58	80	89	95	97	99	100	100	0.4	0.6	0.7	-	-	0.5	0.6	0.8	1.0	1.2
SS D3	83	95	98	99	100	100	100	100	46	66	76	85	90	95	97	98	0.5	0.6	0.8	1.0	-	0.6	0.7	0.9	1.1	1.2
SS D4	91	99	100	100	100	100	100	100	56	78	88	94	97	99	99	100	0.4	0.6	0.7	-	-	0.5	0.6	0.8	1.0	1.2
SS E1	85	96	99	100	100	100	100	100	45	67	78	86	91	95	97	99	0.5	0.6	0.8	0.9	-	0.6	0.7	0.9	1.0	1.2
SS E2	90	98	100	100	100	100	100	100	52	75	85	92	96	98	99	100	0.4	0.6	0.7	-	-	0.5	0.7	0.8	1.0	1.2
SS E3	77	91	96	98	99	100	100	100	38	57	67	77	83	90	93	96	0.5	0.6	0.8	1.0	1.2	0.7	0.8	0.9	1.1	1.3
SS E4	87	97	99	100	100	100	100	100	48	70	81	89	93	97	98	99	0.4	0.6	0.8	0.9	-	0.6	0.7	0.9	1.0	1.2

SAN FRANCISCO, CALIFORNIA

TYPE	SSF (%) 55 F (779 DD) LCR 200	100	70	50	40	30	25	20	SSF (%) 65 F (3050 DD) LCR 200	100	70	50	40	30	25	20	CF 55 F 200	100	50	30	20	CF 65 F 200	100	50	30	20
WW A1	38	56	66	75	81	87	91	94	22	32	39	47	52	60	65	72	0.7	0.8	1.0	1.1	1.3	1.0	1.1	1.2	1.4	1.6
WW A2	46	67	77	86	90	95	97	98	23	38	48	58	64	73	78	84	0.6	0.7	0.9	1.1	1.2	0.8	0.9	1.1	1.2	1.4
WW A3	49	71	81	89	93	96	98	99	24	41	51	62	69	77	82	88	0.6	0.7	0.9	1.0	1.2	0.8	0.9	1.0	1.2	1.4
WW A4	51	73	83	90	94	97	98	99	24	42	53	64	71	79	84	89	0.6	0.7	0.9	1.0	1.2	0.7	0.8	1.0	1.2	1.4
WW A5	52	75	84	92	95	98	99	100	25	43	55	66	73	82	86	91	0.5	0.7	0.8	1.0	1.2	0.7	0.8	1.0	1.2	1.3
WW A6	53	76	85	92	95	98	99	100	25	44	56	67	74	83	87	92	0.5	0.7	0.8	1.0	1.2	0.7	0.8	1.0	1.2	1.3
WW B1	41	61	72	81	86	92	94	97	20	33	42	51	57	65	71	77	0.6	0.7	0.9	1.1	1.3	0.9	1.0	1.2	1.3	1.5
WW B2	50	73	83	90	94	97	99	99	24	42	53	65	72	80	85	90	0.6	0.7	0.9	1.0	1.2	0.7	0.8	1.0	1.2	1.3
WW B3	57	79	88	94	97	99	99	100	29	50	61	72	79	87	91	94	0.5	0.6	0.8	1.0	1.2	0.7	0.8	0.9	1.1	1.3
WW B4	56	79	88	94	97	99	99	100	27	49	61	73	80	87	91	95	0.5	0.6	0.8	1.0	1.2	0.7	0.8	0.9	1.1	1.3
WW B5	54	77	87	94	96	99	99	100	25	47	60	71	79	87	91	94	0.5	0.6	0.8	1.0	1.2	0.7	0.8	0.9	1.1	1.3
WW C1	59	80	89	95	97	99	100	100	30	51	63	74	81	88	92	95	0.5	0.6	0.8	1.0	1.2	0.7	0.8	0.9	1.1	1.3
WW C2	56	78	88	94	97	99	99	100	27	48	61	72	79	87	91	94	0.5	0.6	0.8	1.0	1.2	0.7	0.8	0.9	1.1	1.3
WW C3	61	83	91	96	98	99	100	100	31	55	68	79	85	92	95	97	0.5	0.6	0.8	1.0	1.2	0.6	0.7	0.9	1.1	1.3
WW C4	57	81	89	95	97	99	100	100	28	51	64	76	82	90	93	96	0.5	0.6	0.8	1.0	1.2	0.6	0.8	0.9	1.1	1.3
TW A1	37	54	64	74	80	86	90	94	21	31	38	46	51	59	64	71	0.7	0.8	1.0	1.1	1.3	1.0	1.1	1.2	1.4	1.6
TW A2	42	62	73	82	87	92	95	97	22	35	44	53	59	68	73	80	0.6	0.7	0.9	1.1	1.3	0.9	1.0	1.1	1.3	1.5
TW A3	44	65	75	84	89	94	96	98	21	36	46	56	63	71	77	83	0.6	0.7	0.9	1.1	1.2	0.8	0.9	1.1	1.2	1.4
TW A4	44	65	76	85	90	94	96	98	20	36	46	56	63	72	78	84	0.6	0.7	0.9	1.1	1.2	0.8	0.9	1.1	1.2	1.4
TW B1	37	55	65	75	81	87	91	94	20	31	38	46	52	60	66	72	0.7	0.8	1.0	1.1	1.3	1.0	1.1	1.2	1.4	1.5
TW B2	40	60	71	80	86	91	94	97	20	33	42	51	58	67	72	79	0.6	0.7	0.9	1.1	1.3	0.9	1.0	1.1	1.3	1.5
TW B3	40	61	71	81	86	92	95	97	19	33	42	52	59	68	73	79	0.6	0.7	0.9	1.1	1.3	0.9	1.0	1.1	1.3	1.4
TW B4	38	58	69	78	84	90	93	96	18	31	40	49	56	65	71	77	0.6	0.8	0.9	1.1	1.3	0.9	1.0	1.1	1.3	1.5
TW C1	36	54	64	74	80	87	91	94	19	30	37	46	52	60	65	72	0.7	0.8	1.0	1.1	1.3	1.0	1.1	1.2	1.4	1.5
TW C2	36	55	65	75	81	88	91	95	18	29	37	46	53	62	67	74	0.7	0.8	0.9	1.1	1.3	0.9	1.0	1.2	1.3	1.5
TW C3	34	52	63	73	79	86	90	94	17	28	36	44	50	59	65	71	0.7	0.8	1.0	1.1	1.3	1.0	1.1	1.2	1.4	1.5
TW C4	30	47	57	67	73	81	86	90	16	25	32	40	45	53	59	65	0.7	0.8	1.0	1.2	1.4	1.0	1.1	1.3	1.4	1.6
TW D1	35	54	65	75	81	87	91	94	17	28	36	44	51	59	64	71	0.7	0.8	1.0	1.1	1.3	0.9	1.0	1.2	1.4	1.6
TW D2	46	68	78	87	91	95	97	99	22	39	49	60	67	75	81	86	0.6	0.7	0.9	1.1	1.2	0.8	0.9	1.0	1.2	1.4
TW D3	50	72	82	90	94	97	98	99	25	43	54	65	72	80	85	90	0.6	0.7	0.9	1.0	1.2	0.7	0.8	1.0	1.2	1.4
TW D4	52	74	84	91	95	98	99	100	25	44	56	67	75	83	87	92	0.5	0.7	0.8	1.0	1.2	0.7	0.8	1.0	1.1	1.3
TW D5	50	73	83	91	94	98	99	99	24	43	55	67	74	83	87	92	0.5	0.7	0.8	1.0	1.2	0.7	0.8	1.0	1.1	1.3
TW E1	54	76	85	92	95	98	99	100	28	47	58	69	76	84	88	92	0.5	0.7	0.8	1.0	1.2	0.7	0.8	1.0	1.1	1.3
TW E2	53	75	85	92	95	98	99	100	26	46	57	68	75	83	88	92	0.5	0.7	0.8	1.0	1.2	0.7	0.8	1.0	1.1	1.3
TW E3	59	81	90	95	98	99	100	100	31	53	66	77	83	90	93	96	0.5	0.6	0.8	1.0	1.2	0.6	0.8	0.9	1.1	1.3
TW E4	56	79	88	94	97	99	99	100	28	49	62	73	80	88	92	95	0.5	0.6	0.8	1.0	1.2	0.7	0.8	0.9	1.1	1.3
TW F1	36	54	64	74	80	87	90	94	19	30	37	45	51	59	64	71	0.7	0.8	1.0	1.1	1.3	1.0	1.1	1.2	1.4	1.6
TW F2	40	60	71	80	86	91	94	97	19	32	41	50	57	66	72	78	0.6	0.7	0.9	1.1	1.3	0.9	1.0	1.1	1.3	1.5
TW F3	40	61	72	82	87	92	95	97	18	32	42	52	59	68	73	80	0.6	0.7	0.9	1.1	1.3	0.8	0.9	1.1	1.3	1.4
TW F4	38	59	71	80	86	92	95	97	16	30	40	50	57	66	72	79	0.6	0.7	0.9	1.1	1.3	0.9	1.0	1.1	1.3	1.4
TW G1	35	53	64	73	80	86	90	94	17	28	36	44	50	59	64	71	0.7	0.8	1.0	1.1	1.3	1.0	1.1	1.2	1.4	1.6
TW G2	36	56	67	77	83	89	93	96	16	29	37	47	53	62	68	75	0.6	0.8	0.9	1.1	1.3	0.9	1.0	1.2	1.3	1.5
TW G3	34	54	65	76	82	89	92	95	14	27	35	45	52	61	67	74	0.7	0.8	0.9	1.1	1.3	0.9	1.0	1.2	1.3	1.5
TW G4	29	48	59	70	77	85	89	93	11	22	31	40	46	55	61	69	0.7	0.8	1.0	1.1	1.3	1.0	1.1	1.2	1.4	1.5
TW H1	31	49	60	71	77	84	88	93	14	25	32	41	47	55	61	68	0.7	0.8	1.0	1.2	1.3	1.0	1.1	1.2	1.4	1.6
TW H2	28	47	58	69	75	83	87	92	11	22	30	38	45	54	59	67	0.7	0.8	1.0	1.2	1.3	1.0	1.1	1.2	1.4	1.6
TW H3	24	42	53	63	70	79	84	89	9	19	26	34	40	49	54	62	0.7	0.8	1.0	1.2	1.4	1.1	1.2	1.3	1.5	1.6
TW H4	19	34	44	54	61	70	76	82	6	14	20	27	32	40	45	52	0.8	0.9	1.1	1.3	1.4	1.2	1.3	1.4	1.6	1.7
TW I1	32	51	62	72	78	85	89	93	14	25	33	41	47	55	61	67	0.7	0.8	1.0	1.1	1.3	1.0	1.1	1.2	1.4	1.6
TW I2	42	64	76	85	90	94	97	98	19	35	45	56	60	73	78	84	0.6	0.7	0.9	1.1	1.2	0.8	0.9	1.1	1.2	1.4
TW I3	45	60	79	87	92	96	97	99	21	38	49	60	68	76	82	87	0.6	0.7	0.9	1.0	1.2	0.8	0.9	1.0	1.2	1.4
TW I4	48	71	82	90	94	97	98	99	22	41	53	64	71	80	85	90	0.6	0.7	0.9	1.0	1.2	0.7	0.8	1.0	1.2	1.3
TW I5	48	71	82	90	94	97	98	99	21	41	53	65	72	81	86	91	0.6	0.7	0.9	1.0	1.2	0.7	0.8	1.0	1.2	1.3
TW J1	51	74	84	91	94	98	99	99	25	44	56	67	74	82	87	91	0.5	0.7	0.8	1.0	1.2	0.7	0.8	1.0	1.2	1.3
TW J2	50	72	83	90	94	97	99	99	23	42	54	65	73	81	86	91	0.6	0.7	0.8	1.0	1.2	0.7	0.8	1.0	1.2	1.3
TW J3	56	79	88	94	97	99	99	100	27	50	62	74	81	88	92	95	0.5	0.6	0.8	1.0	1.2	0.7	0.8	0.9	1.1	1.3
TW J4	53	76	86	93	96	98	99	100	25	46	59	71	78	86	90	94	0.5	0.7	0.8	1.0	1.2	0.7	0.8	0.9	1.1	1.3
DG A1	41	61	71	80	85	90	92	95	18	31	40	48	54	61	65	69	0.6	0.7	1.0	1.2	1.4	0.9	1.0	1.2	1.5	1.8
DG A2	41	62	73	82	87	92	95	97	18	33	42	52	59	67	72	78	0.6	0.7	0.9	1.1	1.3	0.9	0.9	1.1	1.3	1.5
DG A3	45	68	79	86	90	94	95	96	21	38	49	60	68	76	81	86	0.6	0.7	0.9	1.2	1.7	0.8	0.9	1.0	1.2	1.4
DG B1	43	67	78	86	90	94	95	97	18	33	43	53	60	68	72	77	0.6	0.7	0.9	1.1	1.4	0.8	0.9	1.1	1.3	1.6
DG B2	43	67	79	87	92	96	97	98	19	34	45	56	64	73	79	84	0.6	0.7	0.9	1.1	1.3	0.8	0.9	1.0	1.2	1.4
DG B3	46	71	82	90	94	96	97	98	21	39	51	63	72	81	85	90	0.6	0.7	0.9	1.1	1.7	0.8	0.8	1.0	1.2	1.4
DG C1	49	74	84	91	94	96	97	98	21	38	49	61	68	76	81	85	0.5	0.7	0.9	1.1	1.7	0.8	0.9	1.0	1.2	1.5
DG C2	48	73	84	92	95	98	99	99	22	39	51	63	71	80	85	89	0.5	0.7	0.8	1.0	1.3	0.8	0.9	1.0	1.2	1.4
DG C3	51	77	87	94	96	98	98	98	24	43	57	70	78	86	90	94	0.5	0.6	0.8	1.2	2.3	0.7	0.8	0.9	1.1	1.3
SS A1	48	67	76	84	89	94	96	98	30	45	54	63	68	76	80	85	0.6	0.7	0.9	1.1	1.3	0.8	0.9	1.1	1.3	1.4
SS A2	56	76	85	91	95	98	99	99	35	53	63	73	79	85	89	93	0.5	0.7	0.9	1.0	1.2	0.7	0.8	1.0	1.2	1.3
SS A3	46	64	74	82	87	92	95	97	29	43	51	60	66	73	77	82	0.6	0.7	0.9	1.1	1.3	0.8	0.9	1.1	1.3	1.5
SS A4	55	75	84	91	94	97	99	99	35	53	63	72	78	85	88	92	0.6	0.7	0.9	1.0	1.2	0.7	0.8	1.0	1.2	1.4
SS A5	45	63	72	81	86	91	94	96	29	42	50	59	64	72	76	81	0.6	0.8	0.9	1.1	1.3	0.8	0.9	1.1	1.3	1.5
SS A6	56	76	85	91	95	98	99	99	35	53	63	73	79	85	89	93	0.5	0.7	0.9	1.0	1.2	0.7	0.8	1.0	1.2	1.3
SS A7	41	58	67	76	81	87	90	94	26	38	46	54	59	66	71	76	0.7	0.8	1.0	1.2	1.3	0.9	1.0	1.2	1.4	1.6
SS A8	55	75	84	90	94	97	98	99	35	52	62	71	77	84	88	92	0.6	0.7	0.9	1.0	1.2	0.7	0.8	1.0	1.2	1.4
SS B1	42	61	72	80	85	91	94	96	25	39	48	57	63	70	75	81	0.6	0.8	0.9	1.1	1.3	0.8	0.9	1.1	1.3	1.5
SS B2	50	71	81	88	92	96	98	99	30	47	58	67	74	81	86	90	0.6	0.7	0.9	1.0	1.2	0.7	0.8	1.0	1.2	1.4
SS B3	41	60	70	79	84	90	93	96	24	38	46	55	61	68	73	79	0.6	0.8	0.9	1.1	1.3	0.8	1.0	1.1	1.3	1.5
SS B4	50	70	80	88	92	96	97	99	30	47	57	67	73	81	85	89	0.6	0.7	0.9	1.1	1.2	0.7	0.8	1.0	1.2	1.4
SS B5	38	55	65	74	80	86	90	93	23	35	43	51	56	64	69	74	0.7	0.8	1.0	1.2	1.3	0.9	1.0	1.2	1.4	1.6
SS B6	50	71	80	88	92	96	98	99	30	47	57	67	73	81	85	90	0.6	0.7	0.9	1.0	1.2	0.7	0.8	1.0	1.2	1.4
SS B7	35	52	61	70	76	83	87	91	21	32	39	47	52	60	64	70	0.7	0.8	1.0	1.2	1.4	0.9	1.1	1.2	1.4	1.6
SS B8	49	69	79	87	91	95	97	99	29	46	56	66	72	79	84	88	0.6	0.7	0.9	1.1	1.2	0.7	0.8	1.0	1.2	1.4
SS C1	45	66	77	85	90	95	97	98	21	37	47	57	64	72	78	84	0.6	0.7	0.9	1.1	1.2	0.8	0.9	1.1	1.2	1.4
SS C2	50	72	82	90	94	97	98	99	24	42	54	64	72	80	85	90	0.6	0.7	0.9	1.0	1.2	0.7	0.8	1.0	1.2	1.3
SS C3	36	55	66	76	82	88	91	95	18	30	38	46	53	61	66	73	0.7	0.8	0.9	1.1	1.3	0.9	1.0	1.2	1.4	1.5
SS C4	45	66	77	86	90	95	97	99	22	38	48	58	66	74	80	85	0.6	0.7	0.9	1.1	1.2	0.8	0.9	1.1	1.2	1.4
SS D1	56	75	84	91	94	97	99	99	35	52	62	71	77	84	87	91	0.6	0.7	0.9	1.0	1.2	0.7	0.8	1.0	1.2	1.4
SS D2	63	83	90	95	98	99	100	100	40	60	71	80	85	91	94	96	0.5	0.6	0.8	1.0	1.2	0.6	0.7	0.9	1.1	1.3
SS D3	51	70	79	87	91	95	97	99	32	48	57	66	71	79	83	87	0.6	0.7	0.9	1.1	1.2	0.7	0.9	1.0	1.2	1.4
SS D4	61	81	89	95	97	99	99	100	39	58	69	78	84	90	93	96	0.5	0.6	0.8	1.0	1.2	0.6	0.8	0.9	1.1	1.3
SS E1	51	72	81	88	92	96	98	99	30	47	57	66	72	79	84	88	0.6	0.7	0.9	1.0	1.2	0.7	0.9	1.0	1.2	1.4
SS E2	58	79	87	93	96	98	99	100	35	54	65	75	81	87	91	94	0.5	0.7	0.8	1.0	1.2	0.7	0.8	1.0	1.1	1.3
SS E3	43	62	72	81	86	91	94	97	26	40	48	57	63	71	75	81	0.6	0.7	0.9	1.1	1.3	0.8	0.9	1.1	1.3	1.5
SS E4	54	75	84	91	94	97	99	99	32	51	61	71	77	84	88	92	0.5	0.7	0.9	1.0	1.2	0.7	0.8	1.0	1.2	1.3

SANTA MARIA, CALIFORNIA

SOLAR SAVINGS FRACTION (%) — CONSERVATION FACTOR

TYPE	55 F (,690 DD)								65 F (3061 DD)								CF 55 F					CF 65 F				
BASE TEMP: LCR	200	100	70	50	40	30	25	20	200	100	70	50	40	30	25	20	200	100	50	30	20	200	100	50	30	20
WW A1	46	66	77	86	90	95	97	99	22	33	41	50	56	65	70	77	0.6	0.7	0.9	1.1	1.2	0.9	1.0	1.2	1.3	1.5
WW A2	56	78	88	94	97	99	99	100	24	41	51	62	69	78	83	89	0.5	0.6	0.8	1.0	1.2	0.8	0.9	1.0	1.2	1.3
WW A3	60	82	91	96	98	99	100	100	25	44	55	67	74	82	87	92	0.5	0.6	0.8	1.0	1.2	0.7	0.8	1.0	1.1	1.3
WW A4	62	84	92	97	99	100	100	100	26	45	57	69	76	85	89	93	0.5	0.6	0.8	1.0	1.2	0.7	0.8	1.0	1.1	1.3
WW A5	64	86	93	98	99	100	100	100	26	47	60	71	78	87	91	95	0.5	0.6	0.8	1.0	1.2	0.7	0.8	0.9	1.1	1.3
WW A6	65	87	94	98	99	100	100	100	27	48	61	72	80	88	91	95	0.5	0.6	0.8	1.0	1.2	0.7	0.8	0.9	1.1	1.3
WW B1	51	73	83	91	95	98	99	100	21	36	45	55	62	71	77	83	0.6	0.7	0.8	1.0	1.2	0.8	0.9	1.1	1.2	1.4
WW B2	62	84	92	97	99	100	100	100	26	46	58	69	77	85	90	94	0.5	0.6	0.8	1.0	1.2	0.7	0.8	1.0	1.1	1.3
WW B3	69	90	96	99	99	100	100	100	31	54	67	78	84	91	94	97	0.5	0.6	0.8	1.0	1.2	0.6	0.7	0.9	1.1	1.3
WW B4	68	89	96	99	99	100	100	100	29	53	66	78	84	91	95	97	0.5	0.6	0.8	1.0	-	0.6	0.7	0.9	1.1	1.2
WW B5	66	88	95	98	99	100	100	100	27	51	64	76	83	90	94	97	0.5	0.6	0.8	1.0	1.2	0.6	0.7	0.9	1.1	1.3
WW C1	71	90	96	99	100	100	100	100	33	56	68	79	86	92	95	98	0.5	0.6	0.8	1.0	-	0.6	0.7	0.9	1.1	1.2
WW C2	68	89	95	98	99	100	100	100	29	52	65	77	84	91	94	97	0.5	0.6	0.8	1.0	1.2	0.6	0.7	0.9	1.1	1.3
WW C3	74	93	97	99	100	100	100	100	34	60	73	84	90	95	97	99	0.5	0.6	0.8	1.0	-	0.6	0.7	0.9	1.0	1.2
WW C4	70	90	96	99	100	100	100	100	30	55	69	80	87	93	96	98	0.5	0.6	0.8	1.0	-	0.6	0.7	0.9	1.1	1.2
TW A1	45	65	75	84	89	94	97	98	21	32	40	49	55	64	69	76	0.6	0.7	0.9	1.1	1.2	1.0	1.0	1.2	1.3	1.5
TW A2	51	73	83	91	95	98	99	100	22	37	47	57	64	73	79	85	0.6	0.7	0.8	1.0	1.2	0.8	0.9	1.1	1.2	1.4
TW A3	54	76	86	93	96	99	99	100	22	39	49	60	67	77	82	88	0.5	0.7	0.8	1.0	1.2	0.8	0.9	1.0	1.2	1.4
TW A4	54	77	87	94	96	99	99	100	22	38	49	61	68	77	83	88	0.5	0.6	0.8	1.0	1.2	0.8	0.9	1.0	1.2	1.3
TW B1	45	65	76	85	90	95	97	99	21	32	40	49	56	65	70	77	0.6	0.7	0.9	1.1	1.2	0.9	1.0	1.2	1.3	1.5
TW B2	49	71	82	90	94	97	99	99	21	35	45	55	62	72	77	84	0.6	0.7	0.8	1.0	1.2	0.8	0.9	1.1	1.2	1.4
TW B3	50	72	83	91	94	98	99	100	20	35	45	56	63	73	78	84	0.6	0.7	0.8	1.0	1.2	0.8	0.9	1.1	1.2	1.4
TW B4	47	69	80	89	93	97	98	99	19	33	42	53	60	70	75	82	0.6	0.7	0.9	1.0	1.2	0.9	0.9	1.1	1.2	1.4
TW C1	44	64	75	85	90	95	97	99	19	31	39	49	55	64	70	77	0.6	0.7	0.9	1.1	1.2	0.9	1.0	1.1	1.3	1.5
TW C2	44	65	77	86	91	95	97	99	18	31	40	50	57	66	72	79	0.6	0.7	0.9	1.0	1.2	0.9	1.0	1.1	1.3	1.4
TW C3	42	63	74	84	89	94	97	98	17	29	38	47	54	63	69	76	0.6	0.7	0.9	1.1	1.2	0.9	1.0	1.2	1.3	1.5
TW C4	37	56	67	78	84	91	94	97	16	26	33	42	48	57	63	70	0.7	0.8	0.9	1.1	1.3	1.0	1.1	1.2	1.4	1.5
TW D1	44	66	77	86	91	96	97	99	17	30	39	48	55	65	70	77	0.6	0.7	0.9	1.0	1.2	0.9	1.0	1.1	1.3	1.5
TW D2	57	79	89	95	97	99	100	100	24	41	53	64	71	80	85	90	0.5	0.6	0.8	1.0	1.2	0.7	0.8	1.0	1.2	1.3
TW D3	61	83	92	97	98	100	100	100	27	46	58	70	77	85	90	94	0.5	0.6	0.8	1.0	1.2	0.7	0.8	1.0	1.1	1.3
TW D4	63	85	93	97	99	100	100	100	27	48	60	72	79	87	91	95	0.5	0.6	0.8	1.0	1.2	0.7	0.8	0.9	1.1	1.3
TW D5	61	84	92	97	99	100	100	100	25	47	59	71	79	87	91	95	0.5	0.6	0.8	1.0	1.2	0.7	0.8	0.9	1.1	1.3
TW E1	66	87	94	98	99	100	100	100	30	51	63	74	81	89	92	96	0.5	0.6	0.8	1.0	1.2	0.7	0.8	0.9	1.1	1.3
TW E2	64	86	93	98	99	100	100	100	28	49	61	73	80	88	92	95	0.5	0.6	0.8	1.0	1.2	0.7	0.8	0.9	1.1	1.3
TW E3	72	91	97	99	100	100	100	100	34	58	70	81	88	93	96	98	0.5	0.6	0.8	1.0	-	0.6	0.7	0.9	1.1	1.2
TW E4	68	89	95	99	99	100	100	100	30	53	66	78	85	91	95	97	0.5	0.6	0.8	1.0	1.2	0.6	0.7	0.9	1.1	1.2
TW F1	44	65	76	85	90	95	97	99	20	31	39	48	55	64	69	76	0.6	0.7	0.9	1.1	1.2	0.9	1.0	1.2	1.3	1.5
TW F2	49	71	82	90	94	97	99	99	20	34	44	54	62	71	77	83	0.6	0.7	0.8	1.0	1.2	0.8	0.9	1.1	1.2	1.4
TW F3	50	73	83	91	95	98	99	100	19	35	45	56	64	73	79	85	0.6	0.7	0.8	1.0	1.2	0.8	0.9	1.1	1.2	1.4
TW F4	48	71	82	90	94	98	99	100	17	32	43	54	62	72	78	84	0.6	0.7	0.8	1.0	1.2	0.8	0.9	1.1	1.2	1.4
TW G1	43	64	75	84	90	95	97	98	18	30	38	47	54	63	69	76	0.6	0.7	0.9	1.1	1.2	0.9	1.0	1.2	1.3	1.5
TW G2	45	67	79	88	92	96	98	99	16	30	40	50	58	67	73	80	0.6	0.7	0.9	1.0	1.2	0.9	1.0	1.1	1.3	1.4
TW G3	43	65	77	87	91	96	98	99	15	28	38	49	56	66	72	79	0.6	0.7	0.9	1.0	1.2	0.9	1.0	1.1	1.3	1.4
TW G4	37	59	71	82	88	93	96	98	11	24	33	43	50	60	66	74	0.6	0.7	0.9	1.1	1.2	0.9	1.0	1.2	1.3	1.5
TW H1	39	60	72	82	88	93	96	98	14	26	34	44	51	60	66	73	0.6	0.7	0.9	1.1	1.2	1.0	1.0	1.2	1.3	1.5
TW H2	36	58	70	80	86	92	95	98	11	23	32	41	48	58	64	72	0.6	0.7	0.9	1.1	1.2	1.0	1.0	1.2	1.3	1.5
TW H3	32	52	64	75	82	89	93	96	9	20	27	36	43	53	59	66	0.7	0.8	0.9	1.1	1.3	1.0	1.1	1.2	1.4	1.5
TW H4	25	43	54	65	73	82	86	91	7	15	21	29	35	43	49	56	0.7	0.8	1.0	1.1	1.3	1.2	1.2	1.4	1.5	1.7
TW I1	41	62	74	84	89	94	97	98	15	27	35	45	52	61	67	74	0.6	0.7	0.9	1.1	1.2	0.9	1.0	1.2	1.3	1.5
TW I2	53	76	86	93	96	99	99	100	20	38	49	60	68	78	83	89	0.5	0.6	0.8	1.0	1.2	0.8	0.9	1.0	1.2	1.3
TW I3	57	80	89	95	97	99	100	100	23	42	54	65	73	82	87	92	0.5	0.6	0.8	1.0	1.2	0.7	0.8	1.0	1.1	1.3
TW I4	59	82	91	96	98	100	100	100	23	44	57	69	76	85	89	94	0.5	0.6	0.8	1.0	1.2	0.7	0.8	1.0	1.1	1.3
TW I5	59	82	91	96	98	100	100	100	23	44	57	69	77	85	90	94	0.5	0.6	0.8	1.0	1.2	0.7	0.8	0.9	1.1	1.3
TW J1	63	85	93	97	99	100	100	100	27	48	60	72	79	87	91	95	0.5	0.6	0.8	1.0	1.2	0.7	0.8	0.9	1.1	1.3
TW J2	61	84	92	97	99	100	100	100	25	46	58	70	77	86	90	94	0.5	0.6	0.8	1.0	1.2	0.7	0.8	0.9	1.1	1.3
TW J3	68	89	95	99	99	100	100	100	30	54	67	79	85	92	95	98	0.5	0.6	0.8	1.0	1.2	0.6	0.7	0.9	1.1	1.2
TW J4	65	87	94	98	99	100	100	100	26	50	63	75	82	90	93	97	0.5	0.6	0.8	1.0	1.2	0.6	0.8	0.9	1.1	1.3
DG A1	51	73	83	90	94	97	98	99	19	34	43	52	59	66	71	75	0.6	0.7	0.9	1.0	1.2	0.8	0.9	1.1	1.4	1.6
DG A2	51	74	84	92	95	98	99	100	19	35	46	56	64	72	77	83	0.5	0.7	0.8	1.0	1.2	0.8	0.9	1.1	1.3	1.5
DG A3	56	80	88	93	95	96	97	97	22	40	53	65	73	81	85	89	0.5	0.6	0.9	1.4	-	0.8	0.8	1.0	1.2	1.4
DG B1	56	80	89	94	97	98	99	99	19	35	47	59	66	74	79	83	0.5	0.6	0.8	1.1	1.7	0.8	0.9	1.0	1.3	1.5
DG B2	55	81	90	96	98	99	100	100	20	37	49	62	70	79	84	89	0.5	0.6	0.8	1.0	1.2	0.8	0.9	1.0	1.2	1.4
DG B3	58	84	92	96	97	98	98	98	22	41	55	69	77	86	90	93	0.5	0.6	0.8	1.5	-	0.7	0.8	0.9	1.1	1.3
DG C1	63	87	94	97	98	98	98	98	22	41	54	67	75	83	87	90	0.5	0.6	0.9	1.6	-	0.7	0.8	1.0	1.2	1.4
DG C2	61	86	94	98	99	100	100	100	23	42	55	69	77	85	90	93	0.5	0.6	0.8	1.0	1.4	0.7	0.8	0.9	1.1	1.3
DG C3	64	89	95	98	98	98	98	98	25	47	62	76	83	91	94	96	0.5	0.6	0.9	2.4	-	0.7	0.8	0.9	1.1	1.3
SS A1	59	79	88	94	97	99	99	100	32	49	60	70	76	83	87	92	0.5	0.7	0.8	1.0	1.2	0.7	0.8	1.0	1.2	1.3
SS A2	68	87	94	98	99	100	100	100	39	59	70	80	86	91	94	97	0.5	0.6	0.8	1.0	1.2	0.6	0.7	0.9	1.1	1.3
SS A3	57	77	86	93	96	98	99	100	31	47	57	67	73	81	85	90	0.5	0.7	0.8	1.0	1.2	0.7	0.8	1.0	1.2	1.4
SS A4	68	87	94	97	99	100	100	100	39	59	70	80	85	91	94	97	0.5	0.6	0.8	1.0	1.2	0.6	0.7	0.9	1.1	1.3
SS A5	55	75	85	91	95	98	99	99	30	46	55	65	71	79	84	88	0.6	0.7	0.8	1.0	1.2	0.8	0.9	1.0	1.2	1.4
SS A6	69	88	94	98	99	100	100	100	39	59	70	80	85	91	94	97	0.5	0.6	0.8	1.0	1.2	0.6	0.7	0.9	1.1	1.3
SS A7	51	70	80	88	92	96	97	99	28	42	51	60	66	74	79	84	0.6	0.7	0.9	1.1	1.2	0.8	0.9	1.1	1.3	1.4
SS A8	68	87	93	97	99	100	100	100	39	59	69	79	85	91	94	96	0.5	0.6	0.8	1.0	1.2	0.6	0.8	0.9	1.1	1.3
SS B1	53	74	84	91	95	98	99	99	27	43	53	63	70	78	83	88	0.6	0.7	0.8	1.0	1.2	0.8	0.9	1.0	1.2	1.4
SS B2	62	83	91	96	98	99	100	100	33	52	64	74	80	88	91	95	0.5	0.6	0.8	1.0	1.2	0.7	0.8	0.9	1.1	1.3
SS B3	52	72	82	90	94	97	98	99	26	41	51	61	68	76	81	86	0.6	0.7	0.9	1.0	1.2	0.8	0.9	1.0	1.2	1.4
SS B4	62	83	91	96	98	99	100	100	33	52	63	74	80	87	91	94	0.5	0.6	0.8	1.0	1.2	0.7	0.8	1.0	1.1	1.3
SS B5	47	67	78	86	91	95	97	99	24	38	47	56	62	71	76	82	0.6	0.7	0.9	1.1	1.2	0.8	0.9	1.1	1.3	1.4
SS B6	62	83	91	96	98	99	100	100	32	52	63	74	80	87	91	95	0.5	0.6	0.8	1.0	1.2	0.7	0.8	0.9	1.1	1.3
SS B7	44	63	74	83	88	93	96	98	22	35	43	52	58	67	72	78	0.6	0.7	0.9	1.1	1.3	0.9	1.0	1.1	1.3	1.5
SS B8	61	82	90	95	98	99	100	100	32	51	62	73	79	86	90	94	0.5	0.6	0.8	1.0	1.2	0.7	0.8	1.0	1.1	1.3
SS C1	55	78	88	94	97	99	99	100	23	40	50	61	69	78	83	89	0.5	0.6	0.8	1.0	1.2	0.8	0.9	1.0	1.2	1.3
SS C2	61	83	92	97	98	100	100	100	26	46	58	69	77	85	89	94	0.5	0.6	0.8	1.0	1.2	0.7	0.8	1.0	1.1	1.3
SS C3	45	66	77	86	91	96	97	99	19	31	40	50	57	66	72	78	0.6	0.7	0.9	1.0	1.2	0.9	1.0	1.1	1.3	1.4
SS C4	55	78	88	94	97	99	99	100	23	40	52	63	70	79	85	90	0.5	0.6	0.8	1.0	1.2	0.8	0.8	1.0	1.2	1.3
SS D1	69	87	94	98	99	100	100	100	38	58	69	79	84	91	94	96	0.5	0.6	0.8	1.0	1.2	0.6	0.8	0.9	1.1	1.3
SS D2	76	93	97	99	100	100	100	100	44	67	78	87	91	96	97	99	0.5	0.6	0.8	1.0	-	0.6	0.7	0.9	1.0	1.2
SS D3	63	82	90	95	98	99	100	100	35	53	63	73	79	86	90	93	0.5	0.6	0.8	1.0	1.2	0.7	0.8	1.0	1.1	1.3
SS D4	74	92	97	99	100	100	100	100	43	65	76	85	90	95	97	98	0.5	0.6	0.8	1.0	-	0.6	0.7	0.9	1.1	1.2
SS E1	64	84	92	96	98	99	100	100	33	52	63	73	80	87	91	94	0.5	0.6	0.8	1.0	1.2	0.7	0.8	1.0	1.1	1.3
SS E2	71	90	96	98	99	100	100	100	38	61	72	82	88	93	96	98	0.5	0.6	0.8	1.0	1.2	0.6	0.7	0.9	1.1	1.3
SS E3	54	75	84	91	95	98	99	99	28	43	53	63	70	78	83	88	0.6	0.7	0.8	1.0	1.2	0.8	0.9	1.0	1.2	1.4
SS E4	67	87	94	97	99	100	100	100	35	56	68	78	84	90	93	96	0.5	0.6	0.8	1.0	1.2	0.6	0.8	0.9	1.1	1.3

COLORADO SPRINGS, COLORADO

TYPE	SOLAR SAVINGS FRACTION (%) BASE TEMP: 55 F (3967 DD) LCR 100	70	50	40	30	25	20	15	65 F (6477 DD) 100	70	50	40	30	25	20	15	CONSERVATION FACTOR 55 F 100	50	30	20	15	65 F 100	50	30	20	15
WW A1	25	30	37	42	49	54	60	69	19	23	27	31	36	40	45	53	1.3	1.4	1.5	1.7	1.8	1.5	1.7	1.8	2.0	2.1
WW A2	29	37	46	53	62	68	75	84	20	27	34	39	47	53	60	69	1.0	1.2	1.3	1.5	1.6	1.2	1.4	1.5	1.6	1.8
WW A3	30	40	50	57	67	73	80	88	21	28	36	43	52	57	65	74	1.0	1.1	1.3	1.4	1.6	1.1	1.3	1.4	1.6	1.7
WW A4	31	41	52	59	69	75	82	90	21	29	38	44	54	60	68	77	0.9	1.1	1.2	1.4	1.5	1.1	1.2	1.4	1.5	1.7
WW A5	32	43	54	62	72	78	85	91	21	30	39	47	56	63	70	80	0.9	1.1	1.2	1.4	1.5	1.1	1.2	1.3	1.5	1.6
WW A6	33	44	55	63	73	79	86	92	22	30	40	48	58	64	72	81	0.9	1.0	1.2	1.4	1.5	1.0	1.2	1.3	1.5	1.6
WW B1	24	30	38	44	52	58	65	74	16	21	26	30	37	41	47	55	1.2	1.3	1.4	1.6	1.7	1.4	1.6	1.7	1.9	2.0
WW B2	32	42	53	61	71	77	84	91	22	30	40	47	56	63	70	80	0.9	1.1	1.2	1.4	1.5	1.1	1.2	1.3	1.5	1.6
WW B3	38	49	61	69	79	84	90	95	27	37	47	55	65	71	79	87	0.9	1.0	1.2	1.3	1.5	1.0	1.1	1.3	1.4	1.6
WW B4	38	50	62	71	80	86	91	96	27	37	49	57	68	74	81	89	0.8	1.0	1.1	1.3	1.5	0.9	1.1	1.2	1.4	1.5
WW B5	36	48	61	69	79	85	90	96	25	36	48	56	67	73	81	89	0.8	1.0	1.1	1.3	1.5	0.9	1.1	1.2	1.4	1.5
WW C1	39	51	63	71	80	86	91	96	28	39	50	57	67	74	81	89	0.8	1.0	1.1	1.3	1.5	0.9	1.1	1.2	1.4	1.5
WW C2	37	49	61	69	79	85	90	95	26	37	48	56	66	72	80	88	0.8	1.0	1.2	1.3	1.5	1.0	1.1	1.2	1.4	1.5
WW C3	43	57	70	78	86	91	95	98	32	44	57	65	76	82	88	94	0.8	0.9	1.1	1.3	1.4	0.9	1.0	1.2	1.3	1.5
WW C4	40	53	66	74	84	89	93	97	28	40	53	61	72	79	85	92	0.8	1.0	1.1	1.3	1.4	0.9	1.0	1.2	1.4	1.5
TW A1	24	30	36	41	48	53	59	68	19	22	27	30	36	40	45	53	1.3	1.4	1.5	1.7	1.8	1.5	1.7	1.8	2.0	2.1
TW A2	26	34	42	48	57	63	70	79	19	24	31	36	43	48	55	64	1.1	1.2	1.4	1.5	1.7	1.3	1.4	1.6	1.7	1.9
TW A3	27	35	44	51	60	66	74	82	19	25	32	38	46	51	59	68	1.0	1.2	1.3	1.5	1.6	1.2	1.4	1.5	1.7	1.8
TW A4	26	35	45	52	61	67	75	83	18	25	32	38	47	52	60	69	1.0	1.2	1.3	1.5	1.6	1.2	1.3	1.5	1.6	1.8
TW B1	24	30	36	42	49	54	61	70	18	22	27	31	37	41	47	55	1.2	1.4	1.5	1.7	1.8	1.5	1.6	1.8	1.9	2.0
TW B2	25	32	40	47	55	61	69	78	18	23	30	35	42	47	54	63	1.1	1.2	1.4	1.5	1.7	1.3	1.4	1.6	1.7	1.9
TW B3	24	32	41	47	56	62	70	79	17	23	30	35	43	48	55	64	1.1	1.2	1.4	1.5	1.7	1.3	1.4	1.6	1.7	1.8
TW B4	23	30	38	45	54	60	67	76	16	21	28	33	40	46	52	62	1.1	1.3	1.4	1.5	1.7	1.3	1.5	1.6	1.7	1.9
TW C1	23	29	36	41	49	54	61	70	17	21	26	31	37	41	47	56	1.2	1.4	1.5	1.6	1.8	1.5	1.6	1.7	1.9	2.0
TW C2	22	29	36	42	50	56	63	73	16	21	26	31	38	43	49	58	1.2	1.3	1.5	1.6	1.7	1.4	1.5	1.7	1.8	1.9
TW C3	21	27	34	40	48	54	61	70	15	20	25	30	36	41	47	56	1.2	1.3	1.5	1.6	1.8	1.4	1.6	1.7	1.8	2.0
TW C4	19	25	31	36	43	48	55	64	15	18	23	27	32	37	42	50	1.3	1.4	1.6	1.7	1.8	1.6	1.7	1.8	2.0	2.1
TW D1	20	26	33	38	46	51	58	67	13	17	22	26	32	36	41	49	1.2	1.4	1.5	1.7	1.8	1.5	1.7	1.8	2.0	2.1
TW D2	29	38	48	56	65	71	79	87	21	28	36	42	51	57	65	74	1.0	1.1	1.3	1.4	1.6	1.1	1.3	1.4	1.6	1.7
TW D3	32	42	53	60	70	76	83	90	22	31	40	47	56	62	70	79	0.9	1.1	1.2	1.4	1.5	1.1	1.2	1.4	1.5	1.6
TW D4	34	45	57	65	75	80	87	93	24	33	44	51	61	68	75	84	0.9	1.0	1.2	1.4	1.5	1.0	1.1	1.3	1.4	1.6
TW D5	33	44	56	64	74	80	87	93	23	33	43	51	61	68	76	85	0.9	1.0	1.2	1.3	1.5	1.0	1.1	1.3	1.4	1.6
TW E1	36	46	58	65	75	81	87	93	26	35	45	52	62	68	75	84	0.9	1.0	1.2	1.4	1.5	1.0	1.2	1.3	1.5	1.6
TW E2	35	46	57	65	75	81	87	93	25	34	44	52	62	68	76	84	0.9	1.0	1.2	1.3	1.5	1.0	1.1	1.3	1.4	1.6
TW E3	42	54	66	74	83	88	93	97	31	42	54	62	72	78	85	92	0.8	1.0	1.1	1.3	1.4	0.9	1.0	1.2	1.4	1.5
TW E4	39	51	63	71	81	86	92	96	28	39	51	59	69	76	83	90	0.8	1.0	1.1	1.3	1.5	0.9	1.1	1.2	1.4	1.5
TW F1	23	28	35	40	48	53	60	68	17	21	26	29	35	39	45	53	1.3	1.4	1.5	1.7	1.8	1.5	1.6	1.8	1.9	2.1
TW F2	24	31	39	46	54	60	68	77	17	22	28	33	40	45	52	61	1.1	1.3	1.4	1.5	1.7	1.3	1.5	1.6	1.8	1.9
TW F3	23	31	40	47	56	62	70	79	16	22	28	34	42	47	54	64	1.1	1.2	1.4	1.5	1.6	1.3	1.4	1.5	1.7	1.8
TW F4	21	29	38	45	55	61	69	78	14	20	27	32	40	46	53	63	1.1	1.2	1.4	1.5	1.6	1.3	1.4	1.5	1.7	1.8
TW G1	21	27	34	39	47	52	59	68	15	19	24	28	34	39	44	53	1.2	1.4	1.5	1.7	1.8	1.5	1.6	1.8	1.9	2.0
TW G2	20	27	36	42	51	57	64	74	14	19	25	30	37	42	49	58	1.2	1.3	1.4	1.6	1.7	1.4	1.5	1.6	1.8	1.9
TW G3	19	26	34	40	49	55	63	73	12	17	23	28	36	41	48	57	1.2	1.3	1.4	1.6	1.7	1.4	1.5	1.6	1.8	1.9
TW G4	15	22	29	35	44	49	57	67	9	14	20	24	31	36	43	52	1.2	1.4	1.5	1.6	1.8	1.5	1.6	1.7	1.9	2.0
TW H1	18	24	31	36	44	49	56	66	12	16	21	25	31	36	42	50	1.3	1.4	1.5	1.7	1.8	1.5	1.6	1.8	1.9	2.0
TW H2	15	21	28	34	42	47	55	65	9	14	19	23	30	34	40	49	1.3	1.4	1.5	1.7	1.8	1.5	1.6	1.8	1.9	2.0
TW H3	13	18	25	30	37	43	50	59	8	11	16	20	26	30	36	45	1.3	1.5	1.6	1.7	1.9	1.6	1.7	1.9	2.0	2.1
TW H4	9	14	19	23	30	35	41	50	5	9	12	16	21	24	29	37	1.5	1.6	1.8	1.9	2.0	1.8	1.9	2.0	2.2	2.3
TW I1	17	23	30	35	42	47	54	63	11	15	19	23	29	32	38	45	1.3	1.4	1.6	1.7	1.8	1.6	1.7	1.9	2.0	2.2
TW I2	26	35	45	52	62	68	76	84	17	24	32	39	48	54	61	71	1.0	1.2	1.3	1.5	1.6	1.2	1.3	1.4	1.6	1.7
TW I3	28	38	48	56	66	72	79	87	19	27	36	42	52	58	66	76	1.0	1.1	1.3	1.4	1.6	1.1	1.2	1.4	1.5	1.7
TW I4	30	41	53	61	71	77	84	91	21	30	40	48	58	64	72	82	0.9	1.1	1.2	1.4	1.5	1.0	1.2	1.3	1.5	1.6
TW I5	31	42	54	62	72	78	85	92	21	30	41	49	59	66	74	83	0.9	1.0	1.2	1.4	1.5	1.0	1.2	1.3	1.5	1.6
TW J1	33	44	55	63	73	79	85	92	23	32	42	49	59	66	73	82	0.9	1.0	1.2	1.4	1.5	1.0	1.2	1.3	1.5	1.6
TW J2	32	43	54	62	72	78	85	92	22	31	41	49	59	65	73	82	0.9	1.0	1.2	1.4	1.5	1.0	1.2	1.3	1.5	1.6
TW J3	38	51	63	71	81	86	92	96	27	38	50	59	69	76	83	90	0.8	1.0	1.1	1.3	1.5	0.9	1.1	1.2	1.4	1.5
TW J4	35	48	60	69	79	84	90	95	25	36	47	56	67	73	81	89	0.8	1.0	1.1	1.3	1.5	0.9	1.1	1.2	1.4	1.5
DG A1	22	29	36	41	48	52	57	62	14	19	24	28	33	36	40	44	1.1	1.4	1.6	1.9	2.2	1.4	1.6	2.0	2.4	2.9
DG A2	24	32	41	47	56	61	67	75	16	22	29	34	42	47	52	60	1.1	1.2	1.4	1.6	1.8	1.3	1.4	1.6	1.8	2.0
DG A3	28	38	49	57	67	73	79	86	20	27	36	44	54	60	67	76	1.0	1.1	1.3	1.5	1.7	1.1	1.2	1.4	1.6	1.7
DG B1	22	30	39	46	55	60	66	72	14	19	25	29	36	41	47	53	1.1	1.2	1.4	1.7	1.9	1.4	1.5	1.7	2.0	2.3
DG B2	24	33	43	51	62	68	75	82	16	23	30	36	45	51	59	68	1.1	1.1	1.3	1.5	1.7	1.3	1.4	1.4	1.7	1.9
DG B3	28	39	51	60	72	78	84	91	20	28	38	45	56	64	72	81	1.0	1.0	1.2	1.4	1.6	1.1	1.2	1.3	1.5	1.6
DG C1	26	35	46	54	64	70	76	83	17	23	30	36	45	51	58	66	1.0	1.1	1.3	1.5	1.8	1.3	1.4	1.5	1.7	2.0
DG C2	28	38	49	58	70	76	82	89	19	26	35	42	53	60	67	76	1.0	1.1	1.2	1.4	1.6	1.2	1.2	1.3	1.5	1.7
DG C3	32	43	57	67	79	84	90	94	23	32	43	51	64	71	79	87	0.9	1.0	1.1	1.3	1.5	1.1	1.1	1.2	1.4	1.6
SS A1	31	39	48	54	63	68	75	83	25	31	38	44	51	56	63	71	1.0	1.2	1.3	1.5	1.6	1.2	1.3	1.5	1.6	1.8
SS A2	39	49	60	67	76	81	87	93	31	40	49	56	65	71	77	85	0.9	1.0	1.2	1.4	1.5	1.0	1.1	1.3	1.5	1.6
SS A3	29	36	44	50	58	64	70	78	23	28	35	40	47	51	58	66	1.1	1.2	1.4	1.6	1.7	1.2	1.4	1.6	1.7	1.9
SS A4	38	48	58	65	74	80	86	92	31	39	48	55	64	69	76	84	0.9	1.1	1.2	1.4	1.5	1.0	1.2	1.3	1.5	1.6
SS A5	30	36	44	50	58	63	69	78	24	29	35	40	47	51	57	65	1.1	1.3	1.4	1.6	1.7	1.3	1.4	1.6	1.8	1.9
SS A6	39	49	59	66	75	81	87	93	31	39	49	56	65	70	77	85	0.9	1.0	1.2	1.4	1.5	1.0	1.1	1.3	1.5	1.6
SS A7	26	32	39	44	51	55	62	70	21	26	31	35	40	44	49	57	1.2	1.4	1.5	1.7	1.8	1.4	1.6	1.8	1.9	2.1
SS A8	38	47	57	64	73	79	85	91	30	38	47	53	62	68	74	82	0.9	1.1	1.2	1.4	1.5	1.0	1.2	1.3	1.5	1.6
SS B1	27	34	42	48	56	62	69	77	21	26	33	38	45	50	56	65	1.1	1.2	1.4	1.6	1.7	1.3	1.4	1.6	1.7	1.9
SS B2	34	44	54	61	70	76	83	90	27	35	44	50	59	65	72	81	0.9	1.1	1.2	1.4	1.5	1.0	1.2	1.3	1.5	1.6
SS B3	25	32	40	45	53	59	66	74	19	24	30	35	42	46	52	61	1.1	1.3	1.4	1.6	1.7	1.3	1.5	1.6	1.8	1.9
SS B4	33	43	53	60	69	75	81	88	26	34	42	49	58	64	71	79	1.0	1.1	1.3	1.4	1.6	1.1	1.2	1.4	1.5	1.7
SS B5	24	30	37	42	49	54	61	69	19	23	29	33	39	43	48	56	1.2	1.4	1.5	1.7	1.8	1.4	1.6	1.7	1.9	2.0
SS B6	33	43	53	60	69	75	82	89	26	34	43	49	58	64	71	80	0.9	1.1	1.2	1.4	1.6	1.1	1.2	1.4	1.5	1.7
SS B7	22	27	33	38	44	49	55	63	17	21	26	29	34	38	43	50	1.3	1.5	1.6	1.8	1.9	1.5	1.7	1.9	2.0	2.2
SS B8	32	41	51	58	67	73	80	87	25	32	41	47	56	62	69	77	1.0	1.1	1.3	1.4	1.6	1.1	1.2	1.4	1.5	1.7
SS C1	27	35	45	51	61	67	74	83	18	24	32	37	46	51	58	68	1.0	1.2	1.3	1.5	1.6	1.2	1.4	1.5	1.7	1.8
SS C2	32	42	53	61	71	77	83	90	22	30	40	47	56	63	70	80	0.9	1.1	1.2	1.4	1.5	1.1	1.2	1.3	1.5	1.6
SS C3	22	28	36	41	49	55	62	71	16	20	26	30	36	41	47	55	1.2	1.3	1.5	1.6	1.8	1.4	1.6	1.7	1.9	2.0
SS C4	28	37	47	54	64	70	77	86	20	27	35	41	50	56	64	73	1.0	1.1	1.3	1.4	1.6	1.2	1.3	1.4	1.6	1.7
SS D1	37	47	57	64	73	78	84	91	29	37	46	52	61	66	73	81	0.9	1.1	1.2	1.4	1.5	1.0	1.2	1.4	1.5	1.7
SS D2	45	57	68	75	84	88	93	97	36	46	57	65	74	79	85	92	0.8	1.0	1.1	1.3	1.5	0.9	1.0	1.2	1.4	1.5
SS D3	34	42	51	57	66	72	78	86	27	33	41	46	54	60	66	74	1.0	1.1	1.3	1.5	1.6	1.1	1.3	1.4	1.6	1.7
SS D4	44	55	66	73	82	87	92	96	34	45	55	62	72	77	84	90	0.8	1.0	1.1	1.3	1.5	0.9	1.1	1.2	1.4	1.5
SS E1	32	41	51	58	67	73	79	87	24	31	39	45	53	59	66	74	1.0	1.1	1.3	1.4	1.6	1.1	1.3	1.4	1.6	1.7
SS E2	39	51	62	69	79	84	89	94	30	40	50	57	67	73	80	87	0.9	1.0	1.2	1.3	1.5	1.0	1.1	1.3	1.4	1.6
SS E3	27	34	42	48	56	62	69	77	21	27	33	38	44	49	55	64	1.1	1.3	1.4	1.6	1.7	1.3	1.4	1.6	1.8	1.9
SS E4	36	46	57	65	74	79	86	92	[illegible]	37	46	53	62	68	75	83	0.9	1.1	1.2	1.4	1.5	1.0	1.2	1.3	1.5	1.6

DENVER, COLORADO

TYPE	SOLAR SAVINGS FRACTION (%) BASE TEMP: 55 F (3609 DD) LCR 100	70	50	40	30	25	20	15	65 F (6022 DD) 100	70	50	40	30	25	20	15	CONSERVATION FACTOR 55 F 100	50	30	20	15	65 F 100	50	30	20	15
WW A1	25	31	37	42	50	55	61	70	19	23	28	32	37	41	46	54	1.2	1.4	1.5	1.7	1.8	1.5	1.7	1.8	1.9	2.1
WW A2	30	38	47	54	63	69	76	85	21	27	35	40	48	54	61	70	1.0	1.2	1.3	1.5	1.6	1.2	1.3	1.5	1.6	1.8
WW A3	31	41	51	58	68	74	81	88	21	29	37	43	52	58	66	75	1.0	1.1	1.2	1.4	1.6	1.1	1.3	1.4	1.6	1.7
WW A4	32	42	53	61	70	76	83	90	22	30	39	45	55	61	68	78	0.9	1.1	1.2	1.4	1.5	1.1	1.2	1.4	1.5	1.7
WW A5	33	44	55	63	73	79	85	92	22	31	40	47	57	63	71	80	0.9	1.0	1.2	1.4	1.5	1.0	1.2	1.3	1.5	1.6
WW A6	34	45	56	64	74	80	87	93	22	31	41	48	58	65	73	82	0.9	1.0	1.2	1.4	1.5	1.0	1.2	1.3	1.5	1.6
WW B1	24	31	39	45	53	59	66	75	17	21	27	31	37	42	48	56	1.1	1.3	1.4	1.6	1.7	1.4	1.6	1.7	1.9	2.0
WW B2	33	43	54	62	72	78	85	91	22	31	40	47	57	63	71	80	0.9	1.1	1.2	1.4	1.5	1.1	1.2	1.3	1.5	1.6
WW B3	39	50	62	70	80	85	90	95	27	37	48	56	66	72	79	87	0.8	1.0	1.2	1.3	1.5	1.0	1.1	1.3	1.4	1.6
WW B4	39	51	63	72	81	86	92	96	27	38	50	58	68	75	82	89	0.8	1.0	1.1	1.3	1.5	0.9	1.1	1.2	1.4	1.5
WW B5	37	49	62	70	80	86	91	96	26	37	48	57	68	74	81	89	0.8	1.0	1.1	1.3	1.5	0.9	1.1	1.2	1.4	1.5
WW C1	40	52	64	72	81	86	92	96	29	39	50	58	68	74	81	89	0.8	1.0	1.1	1.3	1.5	0.9	1.1	1.2	1.4	1.5
WW C2	38	50	62	70	80	85	91	96	27	37	48	56	67	73	80	88	0.8	1.0	1.1	1.3	1.5	1.0	1.1	1.2	1.4	1.5
WW C3	44	58	71	78	87	91	95	98	32	45	57	66	76	82	88	94	0.8	0.9	1.1	1.3	1.4	0.9	1.0	1.2	1.3	1.5
WW C4	41	54	67	75	84	89	94	97	29	41	54	62	73	79	86	93	0.8	0.9	1.1	1.3	1.4	0.9	1.0	1.2	1.3	1.5
TW A1	25	30	37	42	49	54	60	69	19	23	27	31	37	41	46	53	1.3	1.4	1.5	1.7	1.8	1.5	1.7	1.8	1.9	2.1
TW A2	27	34	43	49	58	64	71	80	19	25	31	36	44	49	55	64	1.1	1.2	1.4	1.5	1.7	1.3	1.4	1.6	1.7	1.8
TW A3	28	36	45	52	61	67	75	83	19	26	33	39	47	52	59	69	1.0	1.2	1.3	1.5	1.6	1.2	1.4	1.5	1.6	1.8
TW A4	27	36	45	53	62	68	76	84	18	25	33	39	47	53	61	70	1.0	1.2	1.3	1.5	1.6	1.2	1.3	1.5	1.6	1.7
TW B1	24	30	37	42	50	55	62	71	18	22	28	32	38	42	47	56	1.2	1.4	1.5	1.6	1.8	1.5	1.6	1.7	1.9	2.0
TW B2	25	33	41	48	57	62	70	79	18	24	30	35	43	48	55	64	1.1	1.2	1.4	1.5	1.7	1.3	1.4	1.6	1.7	1.8
TW B3	25	33	42	48	57	63	71	80	17	23	30	36	43	49	56	65	1.1	1.2	1.4	1.5	1.6	1.3	1.4	1.5	1.7	1.8
TW B4	23	31	39	46	55	61	68	77	16	22	28	34	41	46	53	63	1.1	1.2	1.4	1.5	1.7	1.3	1.4	1.6	1.7	1.9
TW C1	23	29	36	42	50	55	62	71	17	22	27	31	37	42	48	56	1.2	1.3	1.5	1.6	1.7	1.4	1.6	1.7	1.9	2.0
TW C2	23	29	37	43	51	57	64	74	16	21	27	32	39	43	50	59	1.2	1.3	1.4	1.6	1.7	1.4	1.5	1.6	1.8	1.9
TW C3	22	28	35	41	49	54	62	71	15	20	26	30	37	41	48	56	1.2	1.3	1.5	1.6	1.7	1.4	1.6	1.7	1.8	2.0
TW C4	20	25	32	36	44	49	56	65	15	19	23	27	33	37	43	51	1.3	1.4	1.6	1.7	1.8	1.5	1.7	1.8	1.9	2.1
TW D1	20	26	34	39	47	52	59	68	14	18	23	27	33	37	42	50	1.2	1.4	1.5	1.7	1.8	1.5	1.7	1.8	2.0	2.1
TW D2	30	39	49	57	66	73	80	87	21	28	37	43	52	58	65	75	1.0	1.1	1.3	1.4	1.6	1.1	1.3	1.4	1.6	1.7
TW D3	33	43	54	61	71	77	84	91	23	31	40	47	57	63	71	80	0.9	1.1	1.2	1.4	1.5	1.1	1.2	1.4	1.5	1.6
TW D4	35	46	58	66	75	81	87	93	25	34	44	52	62	68	76	85	0.9	1.0	1.2	1.3	1.5	1.0	1.1	1.3	1.4	1.6
TW D5	34	45	57	65	75	81	87	93	24	34	44	52	62	69	76	85	0.9	1.0	1.2	1.3	1.5	1.0	1.1	1.3	1.4	1.6
TW E1	37	47	59	66	76	82	88	93	26	35	45	53	62	69	76	84	0.9	1.0	1.2	1.3	1.5	1.0	1.1	1.3	1.5	1.6
TW E2	36	47	58	66	76	82	88	94	26	35	45	52	62	69	76	85	0.9	1.0	1.2	1.3	1.5	1.0	1.1	1.3	1.4	1.6
TW E3	42	55	67	75	84	89	93	97	31	42	54	62	73	79	85	92	0.8	1.0	1.1	1.3	1.4	0.9	1.0	1.2	1.4	1.5
TW E4	39	52	64	72	82	87	92	96	29	39	51	59	70	76	83	91	0.8	1.0	1.1	1.3	1.4	0.9	1.1	1.2	1.4	1.5
TW F1	23	29	36	41	49	54	61	70	17	21	26	30	36	40	46	54	1.2	1.4	1.5	1.7	1.8	1.5	1.6	1.8	1.9	2.0
TW F2	24	32	40	47	55	61	69	78	17	22	29	34	41	46	53	62	1.1	1.2	1.4	1.5	1.7	1.3	1.5	1.6	1.7	1.9
TW F3	24	32	41	48	57	64	71	80	16	22	29	35	43	48	55	65	1.1	1.2	1.3	1.5	1.6	1.3	1.4	1.5	1.7	1.8
TW F4	22	30	39	46	56	62	70	79	14	20	27	33	41	47	54	64	1.1	1.2	1.4	1.5	1.6	1.3	1.4	1.5	1.7	1.8
TW G1	22	28	35	40	48	53	60	70	15	20	25	29	35	39	45	54	1.2	1.4	1.5	1.6	1.8	1.5	1.6	1.7	1.9	2.0
TW G2	21	28	36	43	52	58	65	75	14	19	26	30	38	43	50	59	1.1	1.3	1.4	1.6	1.7	1.3	1.5	1.6	1.8	1.9
TW G3	19	27	35	41	50	56	64	74	12	18	24	29	36	42	49	58	1.1	1.3	1.4	1.6	1.7	1.4	1.5	1.6	1.8	1.9
TW G4	16	22	30	36	45	50	58	68	10	14	20	25	32	37	43	53	1.2	1.3	1.5	1.6	1.8	1.4	1.6	1.7	1.8	2.0
TW H1	18	24	31	37	45	50	57	67	12	16	22	26	32	37	43	51	1.2	1.4	1.5	1.7	1.8	1.5	1.6	1.8	1.9	2.0
TW H2	16	22	29	35	43	49	56	66	10	14	20	24	30	35	41	50	1.3	1.4	1.5	1.7	1.8	1.5	1.6	1.8	1.9	2.0
TW H3	13	19	25	31	38	44	51	61	8	12	17	21	27	31	37	46	1.3	1.5	1.6	1.7	1.9	1.6	1.7	1.8	2.0	2.1
TW H4	10	14	20	24	31	35	42	51	6	9	13	16	21	25	30	37	1.5	1.6	1.7	1.9	2.0	1.8	1.9	2.0	2.2	2.3
TW I1	18	24	30	36	43	48	55	65	11	15	20	24	29	33	39	46	1.3	1.4	1.6	1.7	1.8	1.6	1.7	1.9	2.0	2.2
TW I2	26	35	46	53	63	69	77	85	18	25	33	39	48	54	62	72	1.0	1.1	1.3	1.4	1.6	1.2	1.3	1.4	1.6	1.7
TW I3	29	38	49	57	67	73	80	88	20	27	36	43	52	59	66	76	1.0	1.1	1.2	1.4	1.5	1.1	1.2	1.4	1.5	1.7
TW I4	31	42	54	62	72	78	85	92	21	31	41	48	58	65	73	82	0.9	1.0	1.2	1.4	1.5	1.0	1.2	1.3	1.5	1.6
TW I5	31	43	55	63	73	79	86	92	22	31	42	49	60	66	74	83	0.9	1.0	1.2	1.4	1.5	1.0	1.1	1.3	1.5	1.6
TW J1	34	45	56	64	74	80	86	92	24	33	43	50	60	66	74	83	0.9	1.0	1.2	1.4	1.5	1.0	1.2	1.3	1.5	1.6
TW J2	33	44	55	63	73	79	86	92	23	32	42	49	59	66	74	83	0.9	1.0	1.2	1.4	1.5	1.0	1.2	1.3	1.5	1.6
TW J3	39	51	64	72	82	87	92	96	28	39	51	59	70	76	83	91	0.8	1.0	1.1	1.3	1.4	0.9	1.1	1.2	1.4	1.5
TW J4	36	49	61	70	80	85	91	96	26	36	48	56	67	74	81	89	0.8	1.0	1.1	1.3	1.5	0.9	1.1	1.2	1.4	1.5
DG A1	23	30	37	42	49	53	58	64	14	19	25	28	33	37	40	45	1.1	1.3	1.6	1.9	2.2	1.4	1.6	1.9	2.4	2.8
DG A2	24	33	42	48	57	62	68	76	16	22	30	35	43	47	53	61	1.1	1.2	1.4	1.6	1.8	1.3	1.4	1.6	1.8	2.0
DG A3	29	39	50	58	68	74	80	87	20	28	37	44	54	61	68	76	1.0	1.1	1.3	1.5	1.7	1.1	1.2	1.4	1.6	1.7
DG B1	23	31	40	47	56	61	67	74	14	19	25	30	37	42	48	54	1.1	1.2	1.4	1.7	1.9	1.4	1.5	1.7	1.9	2.3
DG B2	25	34	44	53	63	69	76	83	17	23	31	37	46	52	60	68	1.0	1.1	1.3	1.5	1.7	1.2	1.3	1.4	1.6	1.8
DG B3	29	39	52	61	73	79	85	91	21	29	38	46	57	65	73	82	1.0	1.0	1.2	1.4	1.6	1.1	1.2	1.3	1.5	1.6
DG C1	27	36	47	55	66	72	78	84	17	24	31	37	46	52	59	66	1.0	1.1	1.3	1.5	1.8	1.2	1.4	1.5	1.7	2.0
DG C2	29	39	51	60	71	77	83	90	20	27	36	43	53	60	68	77	1.0	1.0	1.2	1.4	1.6	1.2	1.2	1.3	1.5	1.7
DG C3	33	44	58	68	80	85	91	95	24	33	43	52	64	72	80	88	0.9	1.0	1.1	1.3	1.5	1.0	1.1	1.2	1.4	1.6
SS A1	32	39	48	54	63	69	75	83	25	31	38	44	51	56	63	71	1.0	1.2	1.3	1.5	1.6	1.2	1.3	1.5	1.7	1.8
SS A2	39	49	60	67	76	81	87	93	31	40	49	56	65	71	77	85	0.9	1.0	1.2	1.4	1.5	1.0	1.1	1.3	1.5	1.6
SS A3	29	36	44	50	59	64	71	79	23	28	35	40	47	51	57	65	1.1	1.2	1.4	1.5	1.7	1.2	1.4	1.6	1.7	1.9
SS A4	38	48	59	66	75	80	86	92	30	39	48	55	63	69	76	83	0.9	1.1	1.2	1.4	1.5	1.0	1.2	1.3	1.5	1.6
SS A5	30	37	44	50	58	63	70	78	24	29	35	40	47	51	57	65	1.1	1.3	1.4	1.6	1.7	1.3	1.4	1.6	1.8	1.9
SS A6	39	49	60	67	76	81	87	93	31	39	49	56	64	70	77	85	0.9	1.0	1.2	1.4	1.5	1.0	1.1	1.3	1.5	1.6
SS A7	26	32	39	44	51	56	62	70	21	26	31	35	40	44	49	56	1.2	1.4	1.5	1.7	1.8	1.4	1.6	1.8	1.9	2.1
SS A8	38	47	58	65	74	79	85	91	30	38	47	53	62	68	74	82	0.9	1.1	1.2	1.4	1.5	1.0	1.2	1.3	1.5	1.6
SS B1	27	34	42	48	57	62	69	78	21	26	33	38	45	50	56	65	1.1	1.2	1.4	1.5	1.7	1.3	1.4	1.6	1.7	1.9
SS B2	34	44	54	61	71	77	83	90	27	35	44	50	59	65	72	81	0.9	1.1	1.2	1.4	1.5	1.0	1.2	1.3	1.5	1.6
SS B3	25	32	40	46	54	59	66	75	19	25	31	35	42	46	52	61	1.1	1.3	1.4	1.6	1.7	1.3	1.5	1.6	1.8	1.9
SS B4	33	43	53	60	69	75	82	89	26	34	43	49	58	64	71	79	0.9	1.1	1.3	1.4	1.6	1.1	1.2	1.4	1.5	1.7
SS B5	24	30	37	42	50	55	61	70	19	24	29	33	39	43	48	56	1.2	1.4	1.5	1.7	1.8	1.4	1.6	1.7	1.9	2.0
SS B6	34	43	53	61	70	76	82	89	26	34	43	49	58	64	71	80	0.9	1.1	1.2	1.4	1.5	1.1	1.2	1.4	1.5	1.7
SS B7	22	27	33	38	45	49	55	64	17	21	26	29	34	38	43	50	1.3	1.5	1.6	1.8	1.9	1.5	1.7	1.9	2.0	2.2
SS B8	32	41	51	58	68	74	80	88	25	32	41	47	56	62	69	77	1.0	1.1	1.3	1.4	1.6	1.1	1.2	1.4	1.6	1.7
SS C1	28	36	46	53	62	68	75	84	19	25	33	38	46	52	59	68	1.0	1.2	1.3	1.5	1.6	1.2	1.4	1.5	1.6	1.8
SS C2	33	43	54	62	72	78	84	91	23	31	41	48	57	63	71	80	0.9	1.1	1.2	1.4	1.5	1.1	1.2	1.3	1.5	1.6
SS C3	23	29	36	42	50	56	63	72	16	21	26	31	37	42	48	56	1.2	1.3	1.5	1.6	1.7	1.4	1.6	1.7	1.8	2.0
SS C4	29	38	48	55	65	71	78	86	20	28	36	42	51	57	64	74	1.0	1.1	1.3	1.4	1.6	1.1	1.3	1.4	1.6	1.7
SS D1	38	47	57	64	73	79	85	91	29	37	46	52	60	66	73	81	0.9	1.1	1.2	1.4	1.5	1.0	1.2	1.4	1.5	1.7
SS D2	45	57	69	76	84	89	93	97	36	46	57	64	74	79	85	92	0.8	1.0	1.1	1.3	1.5	0.9	1.1	1.2	1.4	1.5
SS D3	34	42	51	58	67	72	79	86	27	33	41	46	54	59	66	74	1.0	1.1	1.3	1.5	1.6	1.1	1.3	1.5	1.6	1.8
SS D4	44	55	66	74	82	87	92	96	34	45	55	62	72	77	83	90	0.8	1.0	1.1	1.3	1.5	0.9	1.1	1.2	1.4	1.5
SS E1	33	42	51	58	67	73	80	87	24	31	39	45	54	59	66	74	1.0	1.1	1.3	1.4	1.6	1.1	1.3	1.4	1.6	1.7
SS E2	40	51	62	70	79	84	90	95	30	40	50	57	67	73	80	87	0.9	1.0	1.2	1.3	1.5	1.0	1.1	1.3	1.4	1.6
SS E3	28	35	43	49	57	62	69	78	21	27	33	38	45	49	56	64	1.1	1.2	1.4	1.6	1.7	1.3	1.4	1.6	1.8	1.9
SS E4	37	47	58	65	74	80	86	92	28	37	46	53	62	68	75	83	0.9	1.0	1.2	1.4	1.5	1.0	1.2	1.3	1.5	1.6

EAGLE, COLORADO

TYPE	SOLAR SAVINGS FRACTION (%) BASE TEMP: 55 F (5511 DD)								65 F (8426 DD)								CONSERVATION FACTOR 55 F					65 F				
LCR	100	70	50	40	30	25	20	15	100	70	50	40	30	25	20	15	100	50	30	20	15	100	50	30	20	15
WW A1	19	22	26	30	35	38	43	49	16	18	21	24	27	30	33	39	1.6	1.7	1.9	2.1	2.2	1.9	2.1	2.2	2.4	2.6
WW A2	20	25	32	37	44	49	56	64	15	20	25	29	35	39	45	53	1.3	1.4	1.6	1.7	1.9	1.4	1.6	1.8	1.9	2.1
WW A3	20	27	34	40	48	54	61	69	15	20	26	31	38	43	49	58	1.2	1.3	1.5	1.6	1.8	1.3	1.5	1.7	1.8	2.0
WW A4	20	27	36	42	50	56	63	72	14	20	27	32	40	45	52	60	1.1	1.3	1.4	1.6	1.7	1.3	1.4	1.6	1.8	1.9
WW A5	20	28	37	43	52	58	66	75	14	21	28	34	42	47	54	63	1.1	1.2	1.4	1.6	1.7	1.2	1.4	1.5	1.7	1.9
WW A6	20	29	38	44	54	60	67	76	14	21	29	34	43	48	55	65	1.1	1.2	1.4	1.6	1.7	1.2	1.4	1.5	1.7	1.8
WW B1	16	20	24	28	34	38	43	50	11	14	18	20	24	27	31	36	1.5	1.7	1.9	2.0	2.2	1.8	2.0	2.2	2.5	2.7
WW B2	21	29	37	44	53	59	66	75	15	22	29	35	43	49	56	65	1.1	1.2	1.4	1.6	1.7	1.2	1.4	1.5	1.7	1.8
WW B3	25	34	44	52	61	67	74	83	19	27	36	42	51	57	65	74	1.0	1.2	1.3	1.5	1.6	1.1	1.3	1.4	1.6	1.7
WW B4	25	35	46	54	64	71	78	86	19	28	38	45	55	61	69	79	1.0	1.1	1.3	1.4	1.6	1.1	1.2	1.3	1.5	1.7
WW B5	24	34	45	53	64	70	78	86	18	27	37	45	55	61	69	79	1.0	1.1	1.3	1.4	1.6	1.0	1.2	1.3	1.5	1.6
WW C1	27	36	47	54	64	70	77	85	21	29	38	45	54	60	68	77	1.0	1.1	1.3	1.5	1.6	1.1	1.2	1.4	1.5	1.7
WW C2	25	35	45	53	63	69	76	84	19	27	37	44	53	60	67	77	1.0	1.1	1.3	1.5	1.6	1.1	1.2	1.4	1.5	1.7
WW C3	30	42	54	62	73	79	85	92	23	34	46	54	65	71	78	87	0.9	1.0	1.2	1.4	1.5	1.0	1.1	1.3	1.4	1.6
WW C4	27	38	50	59	69	75	82	90	21	31	42	50	61	68	75	84	0.9	1.1	1.2	1.4	1.5	1.0	1.1	1.3	1.4	1.6
TW A1	18	22	26	29	34	38	42	49	15	18	21	23	27	30	34	39	1.6	1.7	1.9	2.1	2.2	1.9	2.0	2.2	2.4	2.6
TW A2	19	24	29	34	41	45	51	59	15	18	23	27	32	36	41	48	1.3	1.5	1.7	1.8	2.0	1.5	1.7	1.9	2.0	2.2
TW A3	18	24	31	36	43	48	55	63	14	18	24	28	34	38	44	52	1.3	1.4	1.6	1.7	1.9	1.4	1.6	1.8	1.9	2.1
TW A4	17	23	31	36	44	49	56	65	13	18	23	28	35	39	45	54	1.2	1.4	1.5	1.7	1.8	1.4	1.6	1.7	1.9	2.0
TW B1	18	22	26	30	35	39	44	51	15	17	21	24	28	31	35	41	1.5	1.7	1.8	2.0	2.1	1.8	1.9	2.1	2.3	2.4
TW B2	17	22	28	33	40	44	51	59	13	17	22	26	31	35	41	48	1.4	1.5	1.7	1.8	1.9	1.5	1.7	1.9	2.0	2.2
TW B3	17	22	28	33	40	45	52	60	13	17	22	26	32	36	42	50	1.3	1.5	1.6	1.8	1.9	1.5	1.7	1.8	2.0	2.1
TW B4	16	21	27	31	38	43	49	58	12	16	21	25	30	34	40	48	1.4	1.5	1.7	1.8	1.9	1.5	1.7	1.8	2.0	2.1
TW C1	17	21	25	29	35	39	45	52	13	16	20	23	28	31	36	42	1.5	1.6	1.8	2.0	2.1	1.7	1.9	2.0	2.2	2.3
TW C2	16	20	26	30	36	41	46	55	12	16	20	24	29	32	38	45	1.4	1.6	1.7	1.9	2.0	1.6	1.8	1.9	2.1	2.2
TW C3	15	19	24	28	34	39	45	53	12	15	19	23	28	31	36	43	1.5	1.6	1.8	1.9	2.1	1.7	1.8	2.0	2.1	2.3
TW C4	14	18	22	26	31	35	40	48	12	15	18	21	25	28	33	39	1.6	1.7	1.9	2.0	2.2	1.8	2.0	2.1	2.3	2.4
TW D1	13	17	21	24	29	33	38	45	9	12	15	18	21	24	27	32	1.6	1.8	1.9	2.1	2.3	1.9	2.1	2.3	2.5	2.7
TW D2	20	27	34	40	48	54	61	70	15	21	27	32	39	44	51	60	1.2	1.3	1.5	1.6	1.8	1.3	1.5	1.6	1.8	1.9
TW D3	21	29	38	44	53	59	66	75	16	23	30	35	43	49	56	65	1.1	1.3	1.4	1.6	1.7	1.2	1.4	1.5	1.7	1.8
TW D4	23	32	42	49	58	64	72	81	18	25	34	40	49	55	63	72	1.0	1.2	1.3	1.5	1.6	1.1	1.3	1.4	1.6	1.7
TW D5	23	32	42	49	59	65	73	81	17	25	34	41	50	56	64	74	1.0	1.2	1.3	1.5	1.6	1.1	1.3	1.4	1.6	1.7
TW E1	25	33	42	49	58	64	71	80	19	26	34	40	49	55	62	71	1.0	1.2	1.4	1.5	1.7	1.2	1.3	1.5	1.6	1.8
TW E2	24	33	42	49	59	65	72	81	18	26	34	41	49	55	63	72	1.0	1.2	1.3	1.5	1.6	1.1	1.3	1.4	1.6	1.7
TW E3	29	40	51	59	69	75	82	89	23	32	43	50	60	67	74	83	0.9	1.1	1.2	1.4	1.5	1.0	1.1	1.3	1.5	1.6
TW E4	27	37	48	56	66	73	80	88	21	30	40	48	58	64	72	81	0.9	1.1	1.3	1.4	1.6	1.0	1.2	1.3	1.5	1.6
TW F1	17	20	25	28	33	37	42	49	13	16	19	22	26	29	33	38	1.6	1.7	1.9	2.0	2.2	1.8	2.0	2.2	2.4	2.5
TW F2	16	21	27	31	38	43	49	57	12	16	21	24	29	33	38	46	1.4	1.5	1.7	1.8	2.0	1.6	1.7	1.9	2.1	2.2
TW F3	15	21	27	32	39	44	51	60	11	15	20	24	30	35	40	48	1.3	1.5	1.6	1.8	1.9	1.5	1.7	1.8	2.0	2.1
TW F4	13	19	25	30	38	43	50	59	9	13	19	23	29	33	39	48	1.3	1.5	1.6	1.8	1.9	1.5	1.6	1.8	2.0	2.1
TW G1	15	19	23	27	33	36	42	49	11	14	18	21	25	28	33	39	1.5	1.7	1.8	2.0	2.1	1.8	1.9	2.1	2.3	2.4
TW G2	13	18	24	28	35	39	46	54	9	13	18	21	27	31	36	43	1.4	1.6	1.7	1.9	2.0	1.6	1.8	1.9	2.1	2.2
TW G3	12	16	22	27	34	38	45	53	8	12	16	20	26	30	35	43	1.4	1.6	1.7	1.9	2.0	1.6	1.8	1.9	2.1	2.2
TW G4	9	14	19	23	29	34	40	48	6	9	14	17	22	26	31	38	1.5	1.6	1.8	1.9	2.1	1.7	1.8	2.0	2.1	2.3
TW H1	12	16	20	24	30	34	39	47	8	12	15	18	23	26	30	37	1.6	1.7	1.9	2.0	2.2	1.8	1.9	2.1	2.3	2.4
TW H2	9	13	18	22	28	32	38	46	6	9	13	16	21	25	29	36	1.5	1.7	1.8	2.0	2.1	1.8	1.9	2.1	2.2	2.4
TW H3	7	11	16	19	25	29	34	42	5	8	11	14	19	22	26	33	1.6	1.8	1.9	2.1	2.2	1.9	2.0	2.2	2.3	2.4
TW H4	5	8	12	15	20	23	28	35	3	6	8	11	15	17	21	27	1.8	2.0	2.1	2.3	2.4	2.1	2.2	2.4	2.5	2.6
TW I1	11	14	18	22	26	30	34	41	7	10	13	15	19	21	24	29	1.7	1.8	2.0	2.2	2.4	2.0	2.2	2.4	2.6	2.8
TW I2	17	23	31	37	45	50	58	67	12	17	24	29	36	41	47	57	1.2	1.4	1.5	1.7	1.8	1.4	1.5	1.6	1.8	1.9
TW I3	18	25	34	40	49	55	62	71	13	19	26	32	40	45	52	61	1.1	1.3	1.4	1.6	1.7	1.3	1.4	1.6	1.7	1.9
TW I4	20	29	38	45	55	61	69	78	15	22	30	37	46	52	60	69	1.1	1.2	1.4	1.5	1.7	1.2	1.3	1.5	1.6	1.8
TW I5	20	29	39	46	56	63	70	80	15	23	31	38	48	54	62	72	1.0	1.2	1.3	1.5	1.6	1.1	1.3	1.4	1.6	1.7
TW J1	22	30	40	46	56	62	69	78	17	24	32	38	47	52	60	69	1.1	1.2	1.4	1.5	1.7	1.2	1.3	1.5	1.6	1.8
TW J2	21	30	39	46	56	62	69	79	16	23	31	38	47	53	60	70	1.1	1.2	1.4	1.5	1.7	1.2	1.3	1.5	1.6	1.8
TW J3	26	36	48	56	66	72	80	87	20	29	40	47	58	64	72	81	0.9	1.1	1.3	1.4	1.6	1.0	1.2	1.3	1.5	1.6
TW J4	24	34	45	53	64	70	77	86	18	27	37	45	55	62	70	79	1.0	1.1	1.3	1.4	1.6	1.0	1.2	1.3	1.5	1.6
DG A1	13	18	22	26	30	[illegible]	38	40	9	12	16	18	21	23	25	28	1.5	1.7	2.1	2.6	3.2	1.8	2.1	2.5	3.3	4.5
DG A2	16	21	28	33	39	44	49	56	12	16	21	25	31	35	40	46	1.3	1.5	1.7	1.9	2.2	1.5	1.6	1.8	2.1	2.4
DG A3	20	27	35	42	51	57	64	72	15	21	28	34	42	48	55	64	1.2	1.3	1.4	1.6	1.8	1.3	1.4	1.5	1.7	1.9
DG B1	13	18	23	27	33	37	42	48	9	12	16	19	23	26	29	34	1.4	1.6	1.8	2.1	2.5	1.7	2.0	2.2	2.6	3.0
DG B2	16	22	29	34	42	48	55	63	12	16	22	26	33	37	43	51	1.3	1.4	1.5	1.7	1.9	1.5	1.6	1.7	1.9	2.1
DG B3	20	27	36	43	53	60	68	78	16	22	29	35	44	50	58	69	1.1	1.2	1.3	1.5	1.7	1.3	1.4	1.5	1.6	1.8
DG C1	16	22	28	34	41	46	52	59	11	16	20	24	30	33	38	45	1.3	1.4	1.6	1.8	2.1	1.5	1.7	1.9	2.1	2.4
DG C2	19	26	34	40	49	55	63	72	14	20	26	31	39	44	51	60	1.2	1.3	1.4	1.6	1.8	1.4	1.5	1.6	1.7	1.9
DG C3	23	31	41	49	60	67	75	84	18	25	33	40	50	57	65	76	1.1	1.2	1.3	1.4	1.6	1.2	1.3	1.4	1.5	1.7
SS A1	23	29	35	40	47	51	57	65	20	25	31	35	41	45	50	57	1.2	1.4	1.6	1.8	1.9	1.3	1.5	1.7	1.9	2.1
SS A2	29	37	46	52	60	66	72	80	25	32	40	46	54	59	66	74	1.0	1.2	1.4	1.5	1.7	1.1	1.3	1.4	1.6	1.8
SS A3	21	26	32	36	42	46	52	59	19	23	28	31	36	40	44	50	1.3	1.5	1.7	1.9	2.0	1.4	1.7	1.9	2.1	2.3
SS A4	28	36	44	50	59	64	70	78	25	32	39	45	52	57	64	72	1.1	1.2	1.4	1.6	1.7	1.1	1.3	1.5	1.7	1.8
SS A5	23	27	33	37	43	47	52	59	20	24	29	32	37	40	45	51	1.3	1.5	1.7	1.9	2.0	1.5	1.7	1.9	2.1	2.3
SS A6	29	37	45	51	60	65	72	80	25	32	40	45	53	59	65	73	1.0	1.2	1.4	1.5	1.7	1.1	1.3	1.5	1.6	1.8
SS A7	20	24	28	32	37	40	44	51	18	21	25	28	32	34	38	43	1.5	1.7	1.9	2.1	2.3	1.7	1.9	2.1	2.4	2.6
SS A8	28	35	43	49	57	62	69	77	24	31	38	43	51	56	62	70	1.1	1.2	1.4	1.6	1.7	1.1	1.3	1.5	1.7	1.8
SS B1	19	24	30	35	41	45	51	59	17	21	26	29	35	39	44	50	1.3	1.5	1.7	1.9	2.0	1.5	1.7	1.8	2.0	2.2
SS B2	25	32	41	47	55	61	67	76	21	28	35	41	49	54	61	69	1.1	1.3	1.4	1.6	1.7	1.2	1.3	1.5	1.7	1.8
SS B3	18	23	28	32	38	42	47	54	15	19	24	27	32	35	40	46	1.4	1.6	1.8	1.9	2.1	1.5	1.7	2.0	2.2	2.4
SS B4	24	31	39	45	54	59	66	74	21	27	34	40	47	52	59	67	1.1	1.3	1.4	1.6	1.7	1.2	1.4	1.5	1.7	1.9
SS B5	18	22	27	30	36	39	44	51	16	19	23	26	30	33	37	43	1.5	1.7	1.9	2.0	2.2	1.7	1.9	2.1	2.3	2.5
SS B6	24	31	40	46	54	60	66	75	21	27	34	40	48	53	59	68	1.1	1.3	1.4	1.6	1.7	1.2	1.4	1.5	1.7	1.8
SS B7	17	20	24	27	31	35	39	45	14	17	20	23	26	29	32	37	1.6	1.8	2.0	2.2	2.4	1.8	2.0	2.3	2.5	2.7
SS B8	23	30	38	44	52	57	64	72	20	26	33	38	45	50	56	65	1.1	1.3	1.5	1.6	1.8	1.2	1.4	1.6	1.7	1.9
SS C1	18	23	30	35	42	48	54	63	13	17	23	27	33	37	43	50	1.3	1.4	1.6	1.7	1.9	1.5	1.6	1.8	2.0	2.1
SS C2	21	29	38	44	53	59	66	75	16	22	30	36	44	49	56	66	1.1	1.2	1.4	1.6	1.7	1.2	1.4	1.5	1.7	1.8
SS C3	16	20	25	29	34	38	44	52	12	15	19	22	27	30	34	41	1.5	1.6	1.8	2.0	2.1	1.7	1.9	2.1	2.2	2.4
SS C4	19	26	34	39	47	53	60	69	15	20	27	32	39	44	50	59	1.2	1.3	1.5	1.6	1.8	1.3	1.5	1.6	1.8	1.9
SS D1	27	34	42	47	55	60	66	74	23	29	36	41	48	52	58	66	1.1	1.3	1.4	1.6	1.8	1.2	1.4	1.6	1.8	1.9
SS D2	33	43	53	60	69	74	81	88	28	37	47	53	62	68	74	82	0.9	1.1	1.3	1.4	1.6	1.0	1.2	1.3	1.5	1.7
SS D3	25	31	38	43	50	54	60	68	22	27	33	37	43	47	52	60	1.2	1.4	1.5	1.7	1.9	1.3	1.5	1.7	1.9	2.1
SS D4	32	41	51	58	67	72	79	86	28	36	45	51	60	66	72	80	1.0	1.1	1.3	1.5	1.6	1.0	1.2	1.4	1.5	1.7
SS E1	22	29	36	41	48	53	59	68	18	24	30	34	41	45	50	58	1.2	1.4	1.5	1.7	1.9	1.3	1.5	1.7	1.9	2.1
SS E2	28	37	46	53	62	68	74	82	23	31	40	46	54	60	67	75	1.0	1.2	1.3	1.5	1.6	1.1	1.2	1.4	1.6	1.7
SS E3	20	25	30	35	41	45	50	58	17	21	26	29	34	38	43	49	1.4	1.5	1.7	1.9	2.0	1.5	1.7	1.9	2.1	2.3
SS E4	26	34	42	49	57	63	70	78	22	29	37	42	50	56	62	71	1.1	1.2	1.4	1.6	1.7	1.1	1.3	1.5	1.7	1.8

GRAND JUNCTION, COLORADO

TYPE	SOLAR SAVINGS FRACTION (%) BASE TEMP: 55 F (3440 DD)								65 F (5607 DD)								CONSERVATION FACTOR 55 F					65 F				
LCR	100	70	50	40	30	25	20	15	100	70	50	40	30	25	20	15	100	50	30	20	15	100	50	30	20	15
WW A1	24	29	35	39	46	51	56	64	19	23	27	30	35	39	44	50	1.3	1.5	1.6	1.8	1.9	1.6	1.7	1.9	2.1	2.2
WW A2	27	35	43	50	58	64	71	79	20	26	33	38	45	50	57	65	1.1	1.2	1.4	1.5	1.7	1.2	1.4	1.6	1.7	1.9
WW A3	29	37	47	54	63	68	75	83	20	27	35	41	49	55	62	71	1.0	1.2	1.3	1.5	1.6	1.2	1.3	1.5	1.6	1.8
WW A4	29	38	48	56	65	71	78	85	20	28	36	43	51	57	64	73	1.0	1.1	1.3	1.5	1.6	1.1	1.3	1.4	1.6	1.7
WW A5	30	40	50	58	67	73	80	87	21	29	38	44	54	60	67	76	0.9	1.1	1.3	1.4	1.6	1.1	1.2	1.4	1.6	1.7
WW A6	30	41	51	59	69	75	81	88	21	29	39	45	55	61	68	77	0.9	1.1	1.2	1.4	1.6	1.1	1.2	1.4	1.5	1.7
WW B1	22	28	35	41	48	53	60	68	16	20	25	29	35	39	44	51	1.2	1.4	1.5	1.7	1.8	1.5	1.6	1.8	2.0	2.2
WW B2	30	39	50	57	67	73	80	87	21	29	38	45	54	60	67	76	1.0	1.1	1.3	1.4	1.6	1.1	1.2	1.4	1.5	1.7
WW B3	35	46	57	65	74	80	86	92	26	35	45	53	62	68	75	84	0.9	1.0	1.2	1.4	1.5	1.0	1.1	1.3	1.5	1.6
WW B4	35	47	59	67	77	82	88	93	26	36	47	55	65	71	79	87	0.9	1.0	1.2	1.4	1.5	1.0	1.1	1.3	1.4	1.6
WW B5	34	46	58	66	76	81	87	93	25	35	46	54	65	71	78	86	0.9	1.0	1.2	1.4	1.5	1.0	1.1	1.3	1.4	1.6
WW C1	37	48	59	67	76	82	87	93	27	37	47	55	65	71	78	86	0.9	1.0	1.2	1.4	1.5	1.0	1.1	1.3	1.4	1.6
WW C2	35	46	58	66	75	81	87	93	26	35	46	54	64	70	77	85	0.9	1.0	1.2	1.4	1.5	1.0	1.1	1.3	1.4	1.6
WW C3	41	54	66	74	83	88	92	96	31	42	55	63	73	79	86	92	0.8	1.0	1.1	1.3	1.5	0.9	1.0	1.2	1.4	1.5
WW C4	37	50	62	71	80	85	91	95	28	39	51	59	70	76	83	90	0.8	1.0	1.1	1.3	1.5	0.9	1.1	1.2	1.4	1.5
TW A1	23	28	34	39	45	50	56	64	19	22	26	30	35	38	43	50	1.3	1.5	1.6	1.8	1.9	1.6	1.7	1.9	2.0	2.2
TW A2	25	32	40	45	53	59	66	74	19	24	30	35	41	46	52	61	1.1	1.3	1.4	1.6	1.7	1.3	1.5	1.6	1.8	1.9
TW A3	26	33	42	48	57	62	69	78	19	25	31	37	44	49	56	65	1.1	1.2	1.4	1.6	1.7	1.3	1.4	1.6	1.7	1.9
TW A4	25	33	42	48	57	63	70	79	18	24	31	37	45	50	57	66	1.1	1.2	1.4	1.5	1.7	1.2	1.4	1.5	1.7	1.8
TW B1	23	28	35	39	46	51	57	66	18	22	27	30	36	40	45	52	1.3	1.4	1.6	1.7	1.9	1.5	1.7	1.8	2.0	2.1
TW B2	24	30	38	44	52	58	65	73	18	23	29	34	40	45	51	60	1.1	1.3	1.4	1.6	1.7	1.3	1.5	1.6	1.8	1.9
TW B3	23	30	38	44	53	59	66	74	17	22	29	34	41	46	52	61	1.1	1.3	1.4	1.6	1.7	1.3	1.5	1.6	1.8	1.9
TW B4	22	28	36	42	50	56	63	72	16	21	27	32	39	44	50	59	1.2	1.3	1.5	1.6	1.8	1.3	1.5	1.6	1.8	1.9
TW C1	22	27	34	39	46	51	58	66	17	21	26	30	36	40	45	53	1.3	1.4	1.6	1.7	1.8	1.5	1.6	1.8	1.9	2.1
TW C2	21	27	34	40	47	53	60	68	16	20	26	30	37	41	47	56	1.2	1.4	1.5	1.7	1.8	1.4	1.6	1.7	1.9	2.0
TW C3	20	26	33	38	45	50	57	66	15	20	25	29	35	39	45	53	1.3	1.4	1.5	1.7	1.8	1.5	1.6	1.7	1.9	2.0
TW C4	19	24	29	34	41	45	52	60	15	18	23	26	32	36	41	48	1.4	1.5	1.6	1.8	1.9	1.6	1.7	1.9	2.0	2.1
TW D1	19	24	30	35	42	47	53	62	13	17	21	25	30	34	39	46	1.3	1.5	1.6	1.8	1.9	1.6	1.7	1.9	2.1	2.2
TW D2	28	36	46	52	62	68	75	83	20	27	35	41	49	55	62	71	1.0	1.2	1.3	1.5	1.6	1.2	1.3	1.5	1.6	1.8
TW D3	30	39	49	57	66	72	79	86	22	30	38	45	54	59	67	76	1.0	1.1	1.3	1.4	1.6	1.1	1.2	1.4	1.6	1.7
TW D4	32	42	53	61	71	77	83	90	24	32	42	49	59	65	73	81	0.9	1.1	1.2	1.4	1.5	1.0	1.2	1.3	1.5	1.6
TW D5	31	42	53	61	71	77	83	90	23	32	42	50	59	66	73	82	0.9	1.1	1.2	1.4	1.5	1.0	1.2	1.3	1.5	1.6
TW E1	34	44	54	62	71	77	83	90	25	34	43	50	59	65	72	81	0.9	1.1	1.2	1.4	1.6	1.0	1.2	1.3	1.5	1.6
TW E2	33	43	54	62	71	77	83	90	25	33	43	50	59	66	73	82	0.9	1.1	1.2	1.4	1.5	1.0	1.2	1.3	1.5	1.6
TW E3	39	51	63	71	80	85	90	95	30	40	52	59	70	76	82	90	0.8	1.0	1.2	1.3	1.5	0.9	1.1	1.2	1.4	1.5
TW E4	36	48	60	68	78	83	89	94	27	38	49	57	67	73	80	88	0.9	1.0	1.2	1.3	1.5	0.9	1.1	1.2	1.4	1.6
TW F1	22	27	33	38	45	50	56	64	17	21	25	29	34	38	43	50	1.3	1.4	1.6	1.8	1.9	1.5	1.7	1.9	2.0	2.2
TW F2	23	29	37	43	51	56	63	72	16	21	27	32	39	44	50	58	1.2	1.3	1.5	1.6	1.8	1.4	1.5	1.7	1.8	2.0
TW F3	22	29	38	44	53	58	66	74	15	21	28	33	40	45	52	61	1.1	1.3	1.4	1.6	1.7	1.3	1.4	1.6	1.8	1.9
TW F4	20	27	36	42	51	57	64	74	13	19	26	31	39	44	51	60	1.1	1.3	1.4	1.6	1.7	1.3	1.4	1.6	1.8	1.9
TW G1	20	26	32	37	44	49	55	64	15	19	24	28	33	37	43	50	1.3	1.4	1.6	1.7	1.9	1.5	1.7	1.8	2.0	2.1
TW G2	19	26	33	39	47	53	60	69	14	18	24	29	36	40	46	55	1.2	1.3	1.5	1.6	1.8	1.4	1.5	1.7	1.9	2.0
TW G3	18	24	32	38	46	52	59	68	12	17	23	27	34	39	46	54	1.2	1.3	1.5	1.6	1.8	1.4	1.5	1.7	1.8	2.0
TW G4	14	20	27	33	41	46	53	63	9	14	19	24	30	35	41	49	1.3	1.4	1.6	1.7	1.8	1.5	1.6	1.8	1.9	2.1
TW H1	17	22	29	34	41	46	53	62	12	16	21	25	30	35	40	48	1.3	1.4	1.6	1.8	1.9	1.5	1.7	1.8	2.0	2.1
TW H2	14	20	27	32	39	44	51	60	9	14	19	23	29	33	39	47	1.3	1.5	1.6	1.8	1.9	1.5	1.7	1.8	2.0	2.1
TW H3	12	17	23	28	35	40	47	56	8	11	16	20	25	29	35	43	1.4	1.5	1.7	1.8	1.9	1.6	1.8	1.9	2.1	2.2
TW H4	9	13	18	22	28	33	38	47	6	9	12	15	20	24	28	35	1.5	1.7	1.8	2.0	2.1	1.8	2.0	2.1	2.2	2.4
TW I1	16	21	27	32	39	44	50	58	11	15	19	22	27	31	35	42	1.3	1.5	1.7	1.8	2.0	1.6	1.8	2.0	2.2	2.3
TW I2	24	33	42	49	58	64	71	80	17	24	31	37	46	51	59	68	1.0	1.2	1.3	1.5	1.7	1.2	1.3	1.5	1.6	1.8
TW I3	26	35	45	52	62	68	75	83	19	26	34	41	50	55	63	72	1.0	1.2	1.3	1.5	1.6	1.1	1.3	1.4	1.6	1.7
TW I4	29	39	50	58	68	74	80	88	20	29	39	46	56	62	70	79	0.9	1.1	1.2	1.4	1.6	1.1	1.2	1.4	1.5	1.7
TW I5	29	39	51	58	69	75	82	89	21	30	40	47	57	63	71	80	0.9	1.1	1.2	1.4	1.6	1.0	1.2	1.3	1.5	1.6
TW J1	31	41	52	59	69	75	81	88	23	31	40	47	57	63	70	79	0.9	1.1	1.2	1.4	1.6	1.1	1.2	1.4	1.5	1.7
TW J2	30	40	51	59	69	75	81	88	22	30	40	47	57	63	70	79	0.9	1.1	1.2	1.4	1.6	1.1	1.2	1.3	1.5	1.6
TW J3	36	48	60	68	77	83	89	94	26	37	48	56	67	73	80	88	0.9	1.0	1.2	1.3	1.5	0.9	1.1	1.2	1.4	1.6
TW J4	33	45	57	65	75	81	87	93	24	35	46	54	64	71	78	86	0.9	1.0	1.2	1.4	1.5	1.0	1.1	1.3	1.4	1.6
DG A1	21	27	33	38	44	48	52	57	14	18	23	26	31	34	37	41	1.2	1.4	1.7	2.1	2.4	1.5	1.7	2.1	2.6	3.1
DG A2	22	30	38	44	52	57	63	70	16	22	28	33	40	45	50	57	1.1	1.3	1.5	1.7	1.9	1.3	1.4	1.7	1.9	2.1
DG A3	27	36	46	54	63	69	76	83	20	27	36	42	52	58	65	73	1.0	1.1	1.3	1.5	1.7	1.2	1.3	1.4	1.6	1.8
DG B1	21	28	36	42	50	55	60	66	14	18	24	28	34	38	43	49	1.2	1.3	1.5	1.8	2.1	1.4	1.6	1.8	2.1	2.4
DG B2	23	31	41	48	57	63	70	78	16	22	29	35	43	49	56	64	1.1	1.2	1.3	1.6	1.8	1.3	1.4	1.5	1.7	1.9
DG B3	27	37	48	56	67	74	80	87	20	28	37	44	54	61	69	78	1.0	1.1	1.2	1.4	1.6	1.1	1.2	1.3	1.5	1.7
DG C1	25	33	42	49	59	64	70	77	17	22	29	34	42	47	53	61	1.1	1.2	1.4	1.6	1.9	1.3	1.4	1.6	1.8	2.1
DG C2	27	36	46	54	65	71	78	85	19	26	34	41	50	56	64	73	1.0	1.1	1.3	1.5	1.7	1.2	1.3	1.4	1.6	1.8
DG C3	31	41	54	63	74	80	86	92	23	31	41	49	61	68	76	85	0.9	1.0	1.2	1.4	1.6	1.1	1.2	1.3	1.4	1.6
SS A1	29	36	44	50	58	63	69	77	24	29	36	41	48	52	58	66	1.1	1.3	1.4	1.6	1.7	1.2	1.4	1.6	1.7	1.9
SS A2	36	45	55	62	71	76	82	89	29	38	46	53	61	67	73	81	0.9	1.1	1.3	1.4	1.6	1.0	1.2	1.4	1.5	1.7
SS A3	27	33	40	46	53	58	64	72	22	27	32	37	43	47	53	60	1.1	1.3	1.5	1.7	1.8	1.3	1.5	1.7	1.9	2.0
SS A4	35	44	54	61	69	75	81	87	29	37	45	51	59	65	71	79	1.0	1.1	1.3	1.4	1.6	1.0	1.2	1.4	1.6	1.7
SS A5	28	34	41	46	53	58	64	72	23	28	33	37	43	48	53	60	1.2	1.3	1.5	1.7	1.8	1.3	1.5	1.7	1.9	2.0
SS A6	36	45	55	62	70	76	82	89	29	37	46	52	61	66	73	81	0.9	1.1	1.3	1.4	1.6	1.0	1.2	1.4	1.5	1.7
SS A7	25	30	36	40	46	50	56	64	21	24	29	32	37	41	45	52	1.3	1.5	1.6	1.8	2.0	1.5	1.7	1.9	2.1	2.2
SS A8	35	43	53	59	68	73	80	87	28	36	44	50	58	63	70	78	1.0	1.1	1.3	1.5	1.6	1.1	1.2	1.4	1.6	1.7
SS B1	25	31	39	44	52	57	64	72	20	25	31	35	42	46	52	60	1.2	1.3	1.5	1.6	1.8	1.3	1.5	1.7	1.8	2.0
SS B2	32	41	50	57	66	71	78	85	25	33	41	47	56	62	68	77	1.0	1.1	1.3	1.5	1.6	1.1	1.2	1.4	1.6	1.7
SS B3	23	29	36	41	49	54	60	68	18	23	29	33	39	43	48	56	1.2	1.4	1.5	1.7	1.8	1.4	1.6	1.7	1.9	2.1
SS B4	31	39	49	55	64	70	76	84	25	32	40	46	54	60	67	75	1.0	1.2	1.3	1.5	1.6	1.1	1.3	1.4	1.6	1.7
SS B5	23	28	34	39	45	50	56	64	19	23	27	31	36	40	45	52	1.3	1.4	1.6	1.8	1.9	1.5	1.7	1.8	2.0	2.2
SS B6	31	40	49	56	65	71	77	85	25	32	40	46	55	60	67	76	1.0	1.1	1.3	1.5	1.6	1.1	1.3	1.4	1.6	1.7
SS B7	21	25	31	35	41	45	50	58	17	20	24	27	32	35	40	46	1.4	1.6	1.7	1.9	2.0	1.6	1.8	2.0	2.2	2.3
SS B8	30	38	47	54	62	68	75	83	24	31	38	44	52	58	65	73	1.0	1.2	1.3	1.5	1.6	1.1	1.3	1.5	1.6	1.8
SS C1	25	33	42	48	57	63	69	78	18	24	31	36	43	49	55	64	1.1	1.2	1.4	1.5	1.7	1.3	1.4	1.6	1.7	1.9
SS C2	30	40	50	57	67	73	79	87	22	30	38	45	54	60	67	76	1.0	1.1	1.3	1.4	1.6	1.1	1.2	1.4	1.6	1.7
SS C3	21	27	34	39	46	51	58	67	16	20	25	29	35	39	45	53	1.2	1.4	1.5	1.7	1.8	1.5	1.6	1.8	1.9	2.1
SS C4	27	35	44	51	60	66	73	82	20	26	34	40	48	54	61	70	1.0	1.2	1.3	1.5	1.6	1.2	1.3	1.5	1.6	1.8
SS D1	34	43	52	59	67	72	79	86	27	35	42	48	56	61	68	75	1.0	1.1	1.3	1.5	1.6	1.1	1.3	1.4	1.6	1.8
SS D2	42	52	63	70	79	84	89	94	33	43	54	61	70	75	81	88	0.9	1.0	1.2	1.4	1.5	0.9	1.1	1.3	1.4	1.6
SS D3	31	39	47	53	61	66	72	80	25	31	38	43	50	55	61	69	1.0	1.2	1.4	1.5	1.7	1.2	1.4	1.5	1.7	1.9
SS D4	40	51	61	68	77	82	88	93	32	42	52	59	68	73	80	87	0.9	1.0	1.2	1.4	1.5	1.0	1.1	1.3	1.5	1.6
SS E1	30	38	47	53	61	67	73	81	23	29	37	42	49	54	61	69	1.0	1.2	1.4	1.5	1.7	1.2	1.3	1.5	1.7	1.8
SS E2	36	47	57	64	74	79	85	91	28	37	47	54	63	69	75	83	0.9	1.1	1.2	1.4	1.5	1.0	1.2	1.3	1.5	1.6
SS E3	26	32	39	44	52	57	63	71	20	25	31	35	41	46	51	59	1.2	1.3	1.5	1.7	1.8	1.3	1.5	1.7	1.9	2.0
SS E4	34	43	53	60	69	74	81	88	26	34	43	50	58	64	71	79	0.9	1.1	1.3	1.4	1.6	1.0	1.2	1.4	1.5	1.7

PUEBLO, COLORADO

TYPE	SSF (%) 55 F (3223 DD) LCR 100	70	50	40	30	25	20	15	SSF (%) 65 F (5398 DD) LCR 100	70	50	40	30	25	20	15	CF 55 F 100	50	30	20	15	CF 65 F 100	50	30	20	15
WW A1	26	32	39	44	52	57	63	72	20	24	29	33	39	43	48	56	1.2	1.3	1.5	1.6	1.8	1.5	1.6	1.8	1.9	2.0
WW A2	31	40	49	56	65	71	78	86	22	28	36	42	50	56	63	72	1.0	1.1	1.3	1.4	1.6	1.2	1.3	1.5	1.6	1.7
WW A3	33	42	53	60	70	76	83	90	23	30	39	45	54	60	68	77	0.9	1.1	1.2	1.4	1.5	1.1	1.2	1.4	1.5	1.7
WW A4	34	44	55	63	72	78	85	91	23	31	40	47	57	63	70	79	0.9	1.1	1.2	1.4	1.5	1.1	1.2	1.4	1.5	1.6
WW A5	35	46	57	65	75	81	87	93	23	32	42	49	59	65	73	82	0.9	1.0	1.2	1.4	1.5	1.0	1.2	1.3	1.5	1.6
WW A6	35	47	58	66	76	82	88	94	24	33	43	50	60	67	74	83	0.9	1.0	1.2	1.3	1.5	1.0	1.2	1.3	1.5	1.6
WW B1	26	33	41	47	56	61	68	77	17	22	28	33	39	44	50	58	1.1	1.3	1.4	1.6	1.7	1.4	1.5	1.7	1.8	2.0
WW B2	34	45	56	64	74	80	86	92	24	32	42	49	59	65	73	82	0.9	1.0	1.2	1.4	1.5	1.0	1.2	1.3	1.5	1.6
WW B3	40	52	64	72	81	86	91	96	29	39	50	57	67	74	81	88	0.8	1.0	1.1	1.3	1.5	0.9	1.1	1.2	1.4	1.5
WW B4	40	53	65	73	83	88	92	97	28	39	51	59	70	76	83	90	0.8	1.0	1.1	1.3	1.4	0.9	1.1	1.2	1.4	1.5
WW B5	39	51	64	72	82	87	92	96	27	38	50	58	69	76	83	90	0.8	1.0	1.1	1.3	1.5	0.9	1.1	1.2	1.4	1.5
WW C1	42	54	66	74	83	88	92	97	30	41	52	60	70	76	83	90	0.8	1.0	1.1	1.3	1.5	0.9	1.1	1.2	1.4	1.5
WW C2	40	52	64	72	82	87	92	96	28	39	50	58	68	75	82	89	0.8	1.0	1.1	1.3	1.5	0.9	1.1	1.2	1.4	1.5
WW C3	46	60	72	80	88	92	96	98	34	46	59	67	78	83	89	95	0.8	0.9	1.1	1.3	1.4	0.8	1.0	1.2	1.3	1.5
WW C4	42	56	69	77	86	90	94	98	30	43	55	64	74	80	87	93	0.8	0.9	1.1	1.3	1.4	0.9	1.0	1.2	1.3	1.5
TW A1	26	31	38	43	51	56	63	71	20	24	28	32	38	42	48	55	1.2	1.4	1.5	1.6	1.8	1.5	1.6	1.8	1.9	2.0
TW A2	28	36	45	51	60	66	73	81	20	26	33	38	45	51	57	66	1.1	1.2	1.3	1.5	1.6	1.3	1.4	1.5	1.7	1.8
TW A3	29	38	47	54	64	70	77	85	20	27	34	40	48	54	61	71	1.0	1.1	1.3	1.5	1.6	1.2	1.3	1.5	1.6	1.8
TW A4	28	37	47	55	64	70	78	86	19	26	34	40	49	55	62	72	1.0	1.1	1.3	1.4	1.6	1.2	1.3	1.4	1.6	1.7
TW B1	25	31	39	44	52	57	64	73	19	23	29	33	39	43	49	57	1.2	1.3	1.5	1.6	1.7	1.4	1.6	1.7	1.9	2.0
TW B2	27	34	43	50	59	64	72	80	19	25	31	37	44	50	56	66	1.1	1.2	1.4	1.5	1.6	1.3	1.4	1.5	1.7	1.8
TW B3	26	34	43	50	59	65	73	81	18	24	32	37	45	50	58	67	1.1	1.2	1.3	1.5	1.6	1.2	1.4	1.5	1.7	1.8
TW B4	24	32	41	47	57	63	70	79	17	23	30	35	43	48	55	64	1.1	1.2	1.4	1.5	1.7	1.3	1.4	1.6	1.7	1.8
TW C1	24	30	38	44	52	57	64	73	18	22	28	32	39	44	50	58	1.2	1.3	1.4	1.6	1.7	1.4	1.5	1.7	1.8	1.9
TW C2	24	31	39	45	53	59	66	75	17	22	28	33	40	45	52	61	1.1	1.3	1.4	1.6	1.7	1.4	1.5	1.6	1.8	1.9
TW C3	22	29	37	42	51	56	64	73	16	21	27	31	38	43	49	58	1.2	1.3	1.4	1.6	1.7	1.4	1.5	1.7	1.8	1.9
TW C4	21	26	33	38	45	51	57	67	16	20	24	28	34	39	44	53	1.3	1.4	1.5	1.7	1.8	1.5	1.6	1.8	1.9	2.0
TW D1	21	28	35	41	49	54	62	71	14	19	24	28	34	38	44	52	1.2	1.3	1.5	1.6	1.8	1.5	1.6	1.8	1.9	2.0
TW D2	31	41	51	59	68	74	81	89	22	30	38	45	54	60	67	77	1.0	1.1	1.2	1.4	1.5	1.1	1.2	1.4	1.5	1.7
TW D3	34	45	56	63	73	79	85	92	24	33	42	49	59	65	72	81	0.9	1.1	1.2	1.4	1.5	1.0	1.2	1.3	1.5	1.6
TW D4	36	48	59	67	77	83	89	94	26	35	46	54	64	70	77	86	0.9	1.0	1.2	1.3	1.5	1.0	1.1	1.3	1.4	1.6
TW D5	35	47	59	67	77	82	88	94	25	35	46	53	64	70	78	86	0.9	1.0	1.2	1.3	1.5	1.0	1.1	1.3	1.4	1.6
TW E1	38	49	61	68	78	83	89	94	28	37	47	54	64	70	77	86	0.9	1.0	1.2	1.3	1.5	1.0	1.1	1.3	1.4	1.6
TW E2	37	49	60	68	78	83	89	94	27	36	47	54	64	70	78	86	0.9	1.0	1.2	1.3	1.5	1.0	1.1	1.3	1.4	1.6
TW E3	44	57	69	77	86	90	94	98	32	44	56	64	74	80	86	93	0.8	0.9	1.1	1.3	1.4	0.9	1.0	1.2	1.4	1.5
TW E4	41	54	66	74	83	88	93	97	30	41	53	61	71	78	84	91	0.8	1.0	1.1	1.3	1.4	0.9	1.0	1.2	1.4	1.5
TW F1	24	30	37	43	51	56	63	72	18	22	27	31	37	42	47	56	1.2	1.3	1.5	1.6	1.8	1.5	1.6	1.7	1.9	2.0
TW F2	26	33	42	48	58	63	71	80	18	23	30	35	43	48	55	64	1.1	1.2	1.4	1.5	1.7	1.3	1.4	1.6	1.7	1.8
TW F3	25	34	43	50	59	66	73	82	17	23	31	36	44	50	57	67	1.0	1.2	1.3	1.5	1.6	1.2	1.4	1.5	1.7	1.8
TW F4	23	32	41	48	58	64	72	81	15	21	29	35	43	49	56	66	1.1	1.2	1.3	1.5	1.6	1.2	1.4	1.5	1.7	1.8
TW G1	22	29	36	42	50	55	62	72	16	21	26	30	37	41	47	56	1.2	1.3	1.5	1.6	1.7	1.4	1.6	1.7	1.9	2.0
TW G2	22	30	38	45	54	60	67	77	15	20	27	32	39	45	51	61	1.1	1.2	1.4	1.5	1.7	1.3	1.5	1.6	1.7	1.9
TW G3	20	28	37	43	52	58	66	76	13	19	25	30	38	43	50	60	1.1	1.3	1.4	1.5	1.7	1.3	1.5	1.6	1.7	1.9
TW G4	17	24	32	38	47	53	60	70	10	15	21	26	33	38	45	54	1.2	1.3	1.5	1.6	1.7	1.4	1.5	1.7	1.8	1.9
TW H1	19	25	33	38	47	52	60	69	13	17	23	27	34	38	44	53	1.2	1.3	1.5	1.6	1.8	1.5	1.6	1.7	1.9	2.0
TW H2	16	23	30	36	45	51	58	68	10	15	21	25	32	37	43	52	1.2	1.4	1.5	1.6	1.8	1.5	1.6	1.7	1.9	2.0
TW H3	14	20	27	32	40	46	53	63	9	13	18	22	28	33	39	47	1.3	1.4	1.6	1.7	1.8	1.5	1.7	1.8	1.0	2.1
TW H4	10	15	21	25	32	37	44	53	6	10	14	17	22	26	31	39	1.5	1.6	1.7	1.8	2.0	1.7	1.9	2.0	2.1	2.2
TW I1	19	25	32	37	45	51	58	67	12	16	21	25	31	35	41	49	1.2	1.4	1.5	1.7	1.8	1.5	1.7	1.8	2.0	2.1
TW I2	28	37	47	55	05	71	78	86	19	26	35	41	50	56	64	73	1.0	1.1	1.3	1.4	1.6	1.1	1.3	1.4	1.6	1.7
TW I3	30	40	51	59	69	75	82	89	21	29	38	45	54	60	68	78	0.9	1.1	1.2	1.4	1.5	1.1	1.2	1.4	1.5	1.7
TW I4	33	44	56	64	74	80	86	93	23	32	42	50	60	67	74	83	0.9	1.0	1.2	1.4	1.5	1.0	1.1	1.3	1.5	1.6
TW I5	33	44	56	65	75	81	87	93	23	32	43	51	61	68	76	85	0.9	1.0	1.2	1.3	1.5	1.0	1.1	1.3	1.4	1.6
TW J1	35	46	58	66	76	81	87	93	25	34	44	52	62	68	75	84	0.9	1.0	1.2	1.4	1.5	1.0	1.1	1.3	1.5	1.6
TW J2	34	45	57	65	75	81	87	93	24	33	43	51	61	68	75	84	0.9	1.0	1.2	1.3	1.5	1.0	1.1	1.0	1.5	1.6
TW J3	40	53	66	74	83	88	93	97	29	40	52	61	71	78	84	91	0.8	1.0	1.1	1.3	1.4	0.9	1.0	1.2	1.4	1.5
TW J4	38	50	63	71	81	86	92	96	27	38	50	58	69	75	82	90	0.8	1.0	1.1	1.3	1.5	0.9	1.1	1.2	1.4	1.5
DG A1	24	31	39	44	51	55	60	66	15	20	26	30	35	39	43	47	1.1	1.3	1.6	1.9	2.1	1.4	1.6	1.9	2.3	2.7
DG A2	26	34	44	50	59	64	70	78	17	24	31	37	44	49	55	63	1.0	1.2	1.4	1.6	1.8	1.2	1.4	1.6	1.8	2.0
DG A3	30	40	52	60	70	76	81	88	21	29	39	46	56	62	69	78	0.9	1.1	1.2	1.5	1.7	1.1	1.2	1.3	1.5	1.7
DG B1	24	33	42	49	58	64	69	76	15	21	27	32	40	44	50	57	1.1	1.2	1.4	1.6	1.9	1.3	1.5	1.6	1.9	2.2
DG B2	26	36	47	55	65	71	78	85	18	24	32	39	48	54	62	70	1.0	1.1	1.3	1.5	1.7	1.2	1.3	1.4	1.6	1.8
DG B3	30	41	54	63	74	80	86	92	22	30	40	48	59	66	74	83	0.9	1.0	1.2	1.4	1.6	1.1	1.2	1.3	1.5	1.6
DG C1	28	38	50	58	68	74	80	85	19	25	33	39	48	54	61	69	1.0	1.1	1.3	1.5	1.8	1.2	1.3	1.4	1.7	1.9
DG C2	30	41	53	62	73	79	85	91	21	28	38	45	56	62	70	79	0.9	1.0	1.2	1.4	1.6	1.1	1.2	1.3	1.5	1.7
DG C3	34	47	61	71	81	87	91	95	25	34	45	54	66	74	81	89	0.9	1.0	1.1	1.3	1.6	1.0	1.1	1.2	1.4	1.6
SS A1	33	41	50	56	65	70	77	85	25	32	39	45	52	58	64	73	1.0	1.2	1.3	1.5	1.6	1.2	1.3	1.5	1.6	1.8
SS A2	41	51	62	69	78	83	88	94	32	41	50	57	66	72	79	86	0.9	1.0	1.2	1.4	1.5	1.0	1.1	1.3	1.5	1.6
SS A3	30	38	46	52	60	66	72	80	23	29	36	41	48	53	59	67	1.1	1.2	1.4	1.5	1.7	1.2	1.4	1.6	1.7	1.9
SS A4	40	50	60	67	76	81	87	93	31	40	49	56	65	70	77	85	0.9	1.0	1.2	1.4	1.5	1.0	1.1	1.3	1.5	1.6
SS A5	31	38	46	52	60	65	72	80	25	30	36	41	48	52	59	67	1.1	1.2	1.4	1.5	1.7	1.3	1.4	1.6	1.7	1.9
SS A6	40	51	61	68	77	83	88	93	32	40	50	57	66	71	78	86	0.9	1.0	1.2	1.4	1.5	1.0	1.1	1.3	1.5	1.6
SS A7	27	33	40	45	53	58	64	72	22	26	31	35	41	45	51	58	1.2	1.3	1.5	1.6	1.8	1.4	1.6	1.7	1.9	2.0
SS A8	39	49	59	66	75	80	86	92	31	39	48	55	63	69	76	83	0.9	1.0	1.2	1.4	1.5	1.0	1.2	1.3	1.5	1.6
SS B1	28	36	44	50	59	64	71	80	21	27	34	39	46	51	58	66	1.1	1.2	1.4	1.5	1.7	1.3	1.4	1.6	1.7	1.8
SS B2	36	45	56	63	72	78	84	91	28	36	45	51	61	66	73	82	0.9	1.1	1.2	1.4	1.5	1.0	1.2	1.3	1.5	1.6
SS B3	26	33	41	47	56	61	68	77	20	25	31	36	43	48	54	62	1.1	1.3	1.4	1.6	1.7	1.3	1.5	1.6	1.8	1.9
SS B4	35	44	55	62	71	77	83	90	27	35	44	50	59	65	72	80	0.9	1.1	1.2	1.4	1.5	1.1	1.2	1.4	1.5	1.7
SS B5	25	31	38	44	51	56	63	72	20	24	30	34	40	44	50	58	1.2	1.3	1.5	1.6	1.8	1.4	1.6	1.7	1.9	2.0
SS B6	35	45	55	62	72	77	84	90	27	35	44	50	59	65	72	81	0.9	1.1	1.2	1.4	1.5	1.0	1.2	1.3	1.5	1.6
SS B7	23	28	34	39	46	51	57	66	18	22	26	30	35	39	44	51	1.3	1.4	1.6	1.7	1.9	1.5	1.7	1.8	2.0	2.1
SS B8	33	43	53	60	69	75	82	89	26	33	42	48	57	63	70	79	0.9	1.1	1.3	1.4	1.6	1.1	1.2	1.4	1.5	1.7
SS C1	29	38	48	55	64	70	77	85	20	26	34	40	48	54	61	70	1.0	1.1	1.3	1.4	1.6	1.2	1.3	1.5	1.6	1.8
SS C2	34	45	56	64	74	79	86	92	24	32	42	49	59	65	73	82	0.9	1.0	1.2	1.4	1.5	1.0	1.2	1.3	1.5	1.6
SS C3	24	30	38	44	52	58	65	74	17	22	27	32	39	43	50	58	1.2	1.3	1.4	1.6	1.7	1.4	1.5	1.7	1.8	1.9
SS C4	30	40	50	57	67	73	80	88	21	29	37	44	53	59	66	75	1.0	1.1	1.3	1.4	1.6	1.1	1.3	1.4	1.6	1.7
SS D1	39	49	59	66	75	80	86	92	30	38	47	53	62	67	74	82	0.9	1.1	1.2	1.4	1.5	1.0	1.2	1.3	1.5	1.6
SS D2	47	59	70	77	85	90	94	97	37	47	58	66	75	80	86	92	0.8	1.0	1.1	1.3	1.4	0.9	1.0	1.2	1.4	1.5
SS D3	35	44	53	60	68	74	80	87	27	34	42	48	56	61	67	76	1.0	1.1	1.3	1.4	1.6	1.1	1.3	1.4	1.6	1.7
SS D4	45	57	68	75	84	88	93	97	35	46	56	64	73	78	85	91	0.8	1.0	1.1	1.3	1.5	0.9	1.1	1.2	1.4	1.5
SS E1	34	43	53	60	69	75	81	88	25	32	41	47	55	61	67	76	1.0	1.1	1.3	1.4	1.6	1.1	1.3	1.4	1.6	1.7
SS E2	41	53	64	72	81	86	91	95	31	41	51	59	68	74	81	88	0.8	1.0	1.2	1.3	1.5	0.9	1.1	1.3	1.4	1.6
SS E3	29	36	44	50	59	64	71	79	22	28	34	39	46	51	57	65	1.1	1.2	1.4	1.5	1.7	1.3	1.4	1.6	1.7	1.9
SS E4	38	48	59	67	76	81	87	93	29	38	47	54	64	69	76	84	0.9	1.0	1.2	1.4	1.5	1.0	1.1	1.3	1.5	1.6

HARTFORD, CONNECTICUT

TYPE	SOLAR SAVINGS FRACTION (%)																CONSERVATION FACTOR									
	BASE TEMP: 55 F (3976 DD)								65 F (6354 DD)								55 F					65 F				
LCR	100	70	50	40	30	25	20	15	100	70	50	40	30	25	20	15	100	50	30	20	15	100	50	30	20	15
WW A1	14	16	18	20	22	23	26	29	12	13	14	14	16	16	17	18	2.2	2.5	2.8	3.1	3.4	3.0	3.5	4.2	-	-
WW A2	13	16	20	23	27	30	34	40	9	11	14	16	18	20	23	26	1.7	1.9	2.1	2.3	2.5	2.1	2.3	2.7	3.0	3.3
WW A3	12	16	20	24	29	33	38	45	7	10	13	16	19	22	25	30	1.5	1.7	1.9	2.1	2.3	1.9	2.1	2.4	2.7	2.9
WW A4	11	15	21	25	30	34	40	47	6	10	13	16	20	23	27	32	1.5	1.7	1.8	2.0	2.2	1.8	2.0	2.2	2.5	2.7
WW A5	11	15	21	25	32	36	42	50	6	9	13	16	21	24	28	34	1.4	1.6	1.8	1.9	2.1	1.7	1.9	2.1	2.4	2.6
WW A6	10	15	21	26	33	37	43	51	5	9	13	16	21	25	29	35	1.4	1.6	1.7	1.9	2.1	1.6	1.9	2.1	2.3	2.5
WW B1	9	11	13	14	16	18	19	21	6	6	7	7	8	8	7	6	2.2	2.7	3.2	3.7	4.3	3.5	-	-	-	-
WW B2	12	17	23	28	34	39	45	54	7	11	16	20	25	28	33	40	1.4	1.5	1.7	1.9	2.0	1.6	1.8	2.0	2.2	2.3
WW B3	15	21	28	34	42	47	54	63	9	15	20	25	31	36	41	49	1.2	1.4	1.6	1.7	1.9	1.4	1.6	1.8	2.0	2.1
WW B4	15	22	31	37	46	52	60	70	9	16	23	28	36	42	49	58	1.2	1.3	1.5	1.6	1.8	1.3	1.4	1.6	1.8	1.9
WW B5	14	22	31	37	46	53	61	70	9	16	23	29	37	43	50	60	1.1	1.3	1.4	1.6	1.7	1.3	1.4	1.6	1.7	1.9
WW C1	16	23	31	36	45	50	57	66	11	16	23	27	34	39	45	54	1.2	1.4	1.5	1.7	1.8	1.3	1.5	1.7	1.9	2.0
WW C2	15	22	30	36	44	50	58	67	10	15	22	27	35	40	46	55	1.2	1.3	1.5	1.7	1.8	1.3	1.5	1.7	1.8	2.0
WW C3	18	28	38	46	56	63	71	80	13	21	30	37	47	53	61	71	1.0	1.2	1.3	1.5	1.6	1.1	1.3	1.4	1.6	1.7
WW C4	16	25	35	43	53	59	68	77	11	19	28	34	44	50	58	68	1.1	1.2	1.4	1.5	1.7	1.2	1.3	1.5	1.6	1.8
TW A1	14	16	18	20	22	24	26	30	12	13	14	15	16	17	18	20	2.2	2.5	2.7	3.0	3.2	2.9	3.3	3.8	4.4	-
TW A2	13	15	19	21	25	28	32	37	9	11	13	15	18	20	22	25	1.8	2.0	2.2	2.4	2.6	2.2	2.5	2.8	3.1	3.4
TW A3	11	15	19	22	27	30	34	41	8	10	13	15	18	21	24	28	1.6	1.9	2.0	2.2	2.4	2.0	2.2	2.5	2.8	3.0
TW A4	10	14	18	22	27	31	35	42	6	9	12	15	18	21	24	29	1.6	1.8	2.0	2.1	2.3	1.9	2.1	2.4	2.6	2.8
TW B1	13	15	18	20	23	25	28	32	11	12	14	15	17	18	19	22	2.0	2.3	2.5	2.8	2.9	2.6	2.9	3.3	3.7	4.1
TW B2	11	14	18	21	25	28	32	38	8	10	13	15	18	20	22	26	1.7	2.0	2.2	2.3	2.5	2.1	2.4	2.6	2.9	3.1
TW B3	11	14	18	21	25	28	33	39	7	10	12	15	18	20	23	28	1.7	1.9	2.1	2.3	2.4	2.0	2.3	2.5	2.7	3.0
TW B4	10	13	17	20	24	28	32	38	7	9	12	14	17	20	23	27	1.7	1.9	2.1	2.3	2.4	2.1	2.3	2.5	2.7	2.9
TW C1	12	14	17	19	23	25	29	34	9	11	13	14	17	18	21	24	1.9	2.2	2.4	2.6	2.7	2.4	2.6	2.9	3.2	3.5
TW C2	11	13	17	19	23	26	30	36	8	10	12	14	17	19	22	26	1.8	2.0	2.2	2.4	2.5	2.2	2.4	2.6	2.9	3.1
TW C3	10	13	16	19	23	25	29	35	8	10	12	14	17	19	22	26	1.9	2.1	2.3	2.4	2.6	2.2	2.5	2.7	2.9	3.1
TW C4	11	13	16	18	21	24	27	32	9	10	12	14	16	18	21	25	2.0	2.2	2.4	2.6	2.7	2.4	2.6	2.8	3.0	3.2
TW D1	8	9	11	13	15	16	18	21	5	5	6	7	8	8	8	8	2.3	2.7	3.1	3.6	4.0	3.4	4.3	-	-	-
TW D2	12	17	22	26	32	36	42	49	9	12	16	19	23	27	31	37	1.5	1.6	1.8	2.0	2.1	1.7	1.9	2.1	2.3	2.5
TW D3	13	18	24	28	35	39	45	53	9	12	17	20	25	29	34	41	1.4	1.6	1.7	1.9	2.0	1.6	1.8	2.0	2.2	2.4
TW D4	14	20	28	33	41	47	54	63	10	15	21	25	32	37	43	52	1.2	1.4	1.6	1.7	1.8	1.4	1.6	1.7	1.9	2.0
TW D5	14	21	28	34	43	48	56	65	10	15	22	27	34	39	46	55	1.2	1.4	1.5	1.7	1.8	1.3	1.5	1.7	1.8	1.9
TW E1	15	21	28	33	40	45	52	60	11	15	21	25	31	35	40	48	1.3	1.5	1.6	1.8	1.9	1.5	1.6	1.8	2.0	2.2
TW E2	15	21	28	33	41	46	53	62	10	15	21	26	32	37	43	51	1.3	1.4	1.6	1.7	1.9	1.4	1.6	1.7	1.9	2.1
TW E3	18	26	36	42	52	58	66	75	13	20	28	34	43	48	56	65	1.1	1.2	1.4	1.6	1.7	1.2	1.4	1.5	1.7	1.8
TW E4	17	25	34	40	50	56	64	74	12	19	26	32	41	47	54	64	1.1	1.3	1.4	1.6	1.7	1.2	1.4	1.5	1.7	1.8
TW F1	12	14	16	18	21	23	25	29	10	11	12	13	14	16	17	19	2.1	2.4	2.7	2.9	3.1	2.8	3.2	3.7	4.3	4.9
TW F2	10	13	16	19	23	26	29	35	7	9	11	13	15	17	19	23	1.8	2.0	2.2	2.4	2.6	2.2	2.5	2.8	3.2	3.5
TW F3	9	12	16	19	23	26	31	37	5	7	10	12	15	17	20	24	1.7	1.9	2.1	2.3	2.5	2.1	2.3	2.6	2.9	3.1
TW F4	7	10	14	17	22	25	30	37	3	6	9	11	14	17	20	25	1.7	1.9	2.1	2.3	2.4	2.0	2.3	2.5	2.7	3.0
TW G1	10	12	15	17	20	22	25	29	7	9	10	12	13	15	17	19	2.0	2.3	2.5	2.8	3.0	2.6	2.9	3.3	3.7	4.1
TW G2	8	11	14	17	20	23	27	33	5	7	9	11	13	15	18	22	1.8	2.1	2.3	2.4	2.6	2.2	2.5	2.8	3.1	3.3
TW G3	6	9	13	15	20	23	27	33	3	5	8	10	13	15	18	22	1.8	2.0	2.2	2.4	2.5	2.2	2.4	2.7	2.9	3.1
TW G4	4	7	10	13	17	20	24	30	2	4	6	8	11	13	16	20	1.9	2.1	2.3	2.5	2.6	2.3	2.5	2.7	2.9	3.1
TW H1	7	9	12	14	18	20	23	28	5	6	8	9	12	13	15	18	2.0	2.3	2.5	2.7	2.9	2.5	2.8	3.1	3.4	3.7
TW H2	5	7	10	13	16	19	22	28	2	4	6	8	10	12	15	18	2.0	2.2	2.4	2.6	2.7	2.4	2.6	2.9	3.1	3.4
TW H3	4	6	9	11	14	17	20	25	1	3	5	7	9	11	13	17	2.1	2.3	2.5	2.7	2.8	2.5	2.7	2.9	3.2	3.4
TW H4	3	4	7	9	12	14	17	21	1	2	4	5	7	9	11	14	2.3	2.5	2.7	2.9	3.0	2.8	3.0	3.2	3.4	3.6
TW I1	6	7	9	10	12	14	15	18	3	4	4	5	6	6	6	6	2.4	2.8	3.3	3.8	4.2	3.6	4.6	-	-	-
TW I2	10	14	19	23	28	33	38	46	6	9	13	16	20	23	28	34	1.5	1.7	1.9	2.0	2.2	1.8	2.0	2.1	2.3	2.5
TW I3	10	15	21	25	32	36	42	50	6	10	14	18	23	27	31	38	1.4	1.6	1.8	1.9	2.1	1.6	1.8	2.0	2.2	2.4
TW I4	12	18	25	30	38	43	50	60	7	12	18	23	29	34	40	49	1.3	1.4	1.6	1.7	1.9	1.4	1.6	1.8	1.9	2.1
TW I5	12	18	26	32	40	46	53	63	7	13	19	24	31	37	43	53	1.2	1.4	1.5	1.7	1.8	1.4	1.5	1.7	1.8	2.0
TW J1	13	19	26	31	38	43	50	59	9	13	19	23	29	33	39	46	1.3	1.5	1.6	1.8	1.9	1.5	1.7	1.8	2.0	2.2
TW J2	13	19	25	31	39	44	51	60	8	13	19	23	30	34	40	49	1.3	1.4	1.6	1.7	1.9	1.4	1.6	1.8	1.9	2.1
TW J3	16	24	33	40	49	56	64	73	11	18	26	32	40	46	54	64	1.1	1.3	1.4	1.6	1.7	1.2	1.4	1.5	1.7	1.8
TW J4	14	22	31	38	47	54	61	71	9	16	24	30	39	45	52	62	1.1	1.3	1.4	1.6	1.7	1.2	1.4	1.5	1.7	1.8
DG A1	7	9	11	13	14	15	16	17	4	5	6	6	7	7	6	5	2.2	2.7	3.5	-	-	3.1	4.2	-	-	-
DG A2	10	13	17	20	24	27	31	36	7	9	12	14	17	19	22	26	1.7	1.9	2.1	2.4	2.8	2.0	2.3	2.5	2.9	3.4
DG A3	13	18	24	28	35	40	46	55	10	14	18	22	28	32	37	45	1.4	1.6	1.7	1.8	2.0	1.6	1.8	1.9	2.0	2.2
DG B1	7	9	11	13	15	17	18	20	4	5	6	6	7	7	7	7	2.2	2.6	3.0	3.7	4.8	3.1	3.9	-	-	-
DG B2	10	13	17	21	25	29	34	40	7	9	12	15	18	20	23	28	1.7	1.9	2.0	2.2	2.4	2.0	2.2	2.5	2.7	3.0
DG B3	14	19	24	29	37	42	49	59	10	14	19	23	29	33	39	48	1.4	1.5	1.6	1.7	1.9	1.6	1.7	1.8	1.9	2.1
DG C1	9	12	15	17	21	23	26	30	5	7	9	10	12	13	14	15	1.9	2.2	2.4	2.8	3.2	2.4	2.8	3.7	-	-
DG C2	12	16	21	25	31	35	41	48	9	12	15	18	23	26	30	36	1.6	1.7	1.8	2.0	2.2	1.8	1.9	2.2	2.4	2.5
DG C3	16	21	28	33	42	47	55	66	12	17	22	27	33	38	45	54	1.3	1.4	1.5	1.6	1.8	1.5	1.6	1.7	1.8	1.9
SS A1	16	20	24	27	31	34	38	44	13	16	19	21	25	27	30	34	1.6	1.8	2.0	2.2	2.4	1.8	2.1	2.4	2.7	3.0
SS A2	20	26	32	37	44	49	55	64	16	21	27	31	37	41	47	54	1.3	1.4	1.6	1.8	1.9	1.4	1.6	1.8	1.9	2.1
SS A3	15	17	21	23	27	29	32	37	12	14	16	18	20	22	24	26	1.8	2.0	2.3	2.6	2.8	2.1	2.4	2.9	3.4	4.0
SS A4	19	25	31	36	42	47	53	61	16	20	26	30	35	39	44	51	1.3	1.5	1.6	1.8	2.0	1.4	1.6	1.8	2.0	2.2
SS A5	16	19	22	25	28	31	34	39	14	16	18	20	22	24	26	29	1.8	2.0	2.3	2.5	2.7	2.1	2.4	2.8	3.2	3.6
SS A6	20	25	31	36	43	48	54	62	16	21	26	30	36	40	45	52	1.3	1.4	1.6	1.8	1.9	1.4	1.6	1.8	2.0	2.1
SS A7	15	17	19	21	23	25	27	31	13	14	16	17	18	19	20	22	2.0	2.4	2.7	3.1	3.4	2.5	3.0	3.6	4.6	-
SS A8	19	24	30	34	41	45	51	58	15	20	24	28	33	37	42	48	1.3	1.5	1.7	1.9	2.0	1.5	1.7	1.9	2.1	2.3
SS B1	13	16	20	23	26	29	33	38	11	13	15	17	20	22	25	28	1.7	2.0	2.2	2.4	2.6	2.0	2.3	2.6	3.0	3.3
SS B2	17	23	29	33	40	45	51	59	14	18	24	27	33	37	43	50	1.3	1.5	1.7	1.8	2.0	1.4	1.6	1.8	2.0	2.1
SS B3	12	15	18	20	23	26	29	33	10	11	14	15	17	19	21	23	1.8	2.1	2.4	2.7	2.9	2.2	2.6	3.0	3.5	4.0
SS B4	17	22	27	32	38	43	49	57	13	17	22	26	32	35	40	47	1.3	1.5	1.7	1.9	2.0	1.5	1.7	1.9	2.1	2.2
SS B5	13	16	18	20	23	25	28	32	11	13	15	16	18	19	21	23	2.0	2.3	2.5	2.8	3.0	2.4	2.7	3.1	3.6	4.1
SS B6	16	22	27	32	39	43	49	57	13	17	22	26	32	36	41	48	1.3	1.5	1.7	1.8	2.0	1.5	1.7	1.9	2.0	2.2
SS B7	12	14	16	18	20	21	24	26	10	11	13	14	15	16	17	18	2.2	2.6	2.9	3.3	3.6	2.7	3.2	3.8	4.7	-
SS B8	15	20	26	30	36	40	46	54	12	16	21	24	29	33	38	44	1.4	1.6	1.7	1.9	2.1	1.5	1.7	1.9	2.2	2.3
SS C1	10	13	17	20	24	27	32	37	6	8	11	13	15	17	20	23	1.7	1.9	2.1	2.3	2.5	2.1	2.4	2.8	3.2	3.6
SS C2	13	18	24	28	35	40	46	55	8	12	17	21	26	30	35	42	1.4	1.5	1.7	1.9	2.0	1.6	1.8	1.9	2.1	2.3
SS C3	10	13	16	18	21	23	26	31	8	9	11	12	14	16	18	20	2.0	2.2	2.5	2.7	2.9	2.5	2.8	3.2	3.6	4.0
SS C4	12	17	22	26	31	36	41	49	8	12	16	19	23	27	31	37	1.5	1.7	1.8	2.0	2.1	1.7	1.9	2.1	2.3	2.5
SS D1	17	22	27	31	36	40	45	52	13	17	21	24	28	31	34	39	1.4	1.6	1.8	2.0	2.2	1.6	1.9	2.2	2.5	2.7
SS D2	22	29	37	43	51	57	64	73	17	23	30	36	43	48	54	63	1.1	1.3	1.4	1.6	1.8	1.2	1.4	1.6	1.8	1.9
SS D3	17	21	25	28	33	36	40	46	14	17	20	22	26	28	31	35	1.5	1.8	2.0	2.2	2.4	1.8	2.1	2.4	2.7	3.0
SS D4	21	28	36	41	50	55	62	70	17	23	29	34	41	46	52	60	1.1	1.3	1.5	1.6	1.8	1.3	1.4	1.6	1.8	2.0
SS E1	14	17	22	25	30	33	37	43	10	13	16	18	21	23	26	29	1.6	1.8	2.0	2.2	2.4	1.8	2.2	2.5	3.0	3.4
SS E2	17	24	31	36	44	49	56	64	13	18	24	29	35	40	45	53	1.2	1.4	1.6	1.7	1.9	1.3	1.5	1.7	1.9	2.1
SS E3	14	17	20	22	26	28	32	36	11	13	15	17	19	21	23	26	1.8	2.0	2.3	2.5	2.7	2.1	2.5	2.9	3.3	3.7
SS E4	17	22	29	34	40	45	51	60	13	18	23	27	33	37	42	49	1.3	1.5	1.6	1.8	2.0	1.4	1.6	1.8	2.0	2.2

WILMINGTON, DELAWARE

TYPE	SOLAR SAVINGS FRACTION (%) BASE TEMP: 55 F (2847 DD) LCR 100	70	50	40	30	25	20	15	65 F (4943 DD) 100	70	50	40	30	25	20	15	CONSERVATION FACTOR 55 F 100	50	30	20	15	65 F 100	50	30	20	15
WW A1	19	23	27	30	35	38	43	49	14	17	19	21	24	26	29	33	1.6	1.8	1.9	2.1	2.2	2.1	2.3	2.5	2.7	3.0
WW A2	20	26	32	37	44	49	56	64	13	17	22	25	31	34	39	46	1.3	1.4	1.6	1.7	1.9	1.6	1.7	1.9	2.1	2.3
WW A3	20	27	34	40	48	53	60	69	13	17	23	27	33	37	43	51	1.2	1.3	1.5	1.7	1.8	1.4	1.6	1.8	2.0	2.1
WW A4	20	28	36	42	50	56	63	72	12	17	23	28	35	39	45	53	1.1	1.3	1.5	1.6	1.8	1.4	1.5	1.7	1.9	2.0
WW A5	21	28	37	43	52	58	65	74	12	18	24	29	36	41	47	56	1.1	1.3	1.4	1.6	1.7	1.3	1.5	1.7	1.8	2.0
WW A6	21	29	38	44	53	59	67	76	12	18	24	30	37	42	49	58	1.1	1.2	1.4	1.6	1.7	1.3	1.5	1.6	1.8	1.9
WW B1	16	20	25	28	34	37	42	49	10	12	15	17	20	22	24	28	1.5	1.7	1.9	2.1	2.2	2.0	2.3	2.6	3.0	3.3
WW B2	21	29	38	44	53	59	66	75	13	19	26	31	38	43	50	59	1.1	1.3	1.4	1.6	1.7	1.3	1.5	1.6	1.8	1.9
WW B3	26	35	44	51	61	67	74	82	16	23	32	38	46	52	59	68	1.0	1.2	1.3	1.5	1.6	1.2	1.3	1.5	1.7	1.8
WW B4	26	35	46	54	64	70	77	86	16	25	34	41	50	56	64	74	1.0	1.1	1.3	1.4	1.6	1.1	1.2	1.4	1.6	1.7
WW B5	24	34	45	53	63	70	77	86	15	24	33	40	50	57	65	74	1.0	1.1	1.3	1.4	1.6	1.1	1.2	1.4	1.6	1.7
WW C1	27	36	47	54	63	69	76	84	18	25	34	40	49	55	62	71	1.0	1.1	1.3	1.5	1.6	1.1	1.3	1.4	1.6	1.8
WW C2	25	35	45	53	62	69	76	84	16	24	33	39	48	54	62	72	1.0	1.1	1.3	1.5	1.6	1.1	1.3	1.4	1.6	1.7
WW C3	30	42	54	62	72	78	85	91	20	30	41	49	60	66	74	83	0.9	1.0	1.2	1.4	1.5	1.0	1.1	1.3	1.5	1.6
WW C4	27	38	50	59	69	75	82	90	18	28	38	46	56	63	71	81	0.9	1.1	1.2	1.4	1.5	1.0	1.2	1.3	1.5	1.6
TW A1	19	22	26	30	34	38	43	49	14	16	19	21	24	26	29	34	1.6	1.8	1.9	2.1	2.2	2.0	2.2	2.4	2.7	2.9
TW A2	19	24	30	34	41	45	51	59	13	16	20	24	28	32	36	42	1.4	1.5	1.7	1.8	2.0	1.7	1.8	2.0	2.2	2.4
TW A3	19	24	31	36	43	48	55	63	12	16	21	24	30	34	39	46	1.3	1.4	1.6	1.7	1.9	1.6	1.7	1.9	2.1	2.2
TW A4	18	24	31	36	44	49	56	65	11	15	20	24	30	34	40	47	1.2	1.4	1.6	1.7	1.9	1.5	1.7	1.8	2.0	2.2
TW B1	18	22	26	30	35	39	44	51	13	16	19	21	25	27	31	36	1.5	1.7	1.9	2.0	2.2	1.9	2.1	2.3	2.5	2.7
TW B2	18	23	29	33	40	44	50	59	12	15	20	23	28	31	36	42	1.4	1.5	1.7	1.8	2.0	1.7	1.8	2.0	2.2	2.3
TW B3	17	22	29	33	40	45	51	60	11	15	19	23	28	32	37	44	1.3	1.5	1.6	1.8	1.9	1.6	1.8	1.9	2.1	2.3
TW B4	16	21	27	32	38	43	49	58	11	14	18	22	27	31	35	42	1.4	1.5	1.7	1.8	2.0	1.6	1.8	2.0	2.1	2.3
TW C1	17	21	26	30	35	39	45	52	12	15	18	21	25	28	32	37	1.5	1.7	1.8	2.0	2.1	1.8	2.0	2.2	2.4	2.5
TW C2	16	21	26	30	36	41	46	54	11	14	18	21	26	29	33	40	1.4	1.6	1.7	1.9	2.0	1.7	1.9	2.1	2.2	2.4
TW C3	15	20	25	29	35	39	45	52	11	14	17	20	25	28	32	38	1.5	1.6	1.8	1.9	2.1	1.8	1.9	2.1	2.3	2.4
TW C4	15	19	23	26	32	35	40	48	11	14	17	19	23	26	30	35	1.6	1.7	1.9	2.0	2.2	1.9	2.1	2.2	2.4	2.5
TW D1	13	17	21	25	30	33	38	44	8	10	13	15	17	19	22	26	1.6	1.8	2.0	2.1	2.3	2.1	2.4	2.6	3.0	3.2
TW D2	20	27	34	40	48	54	61	70	13	18	24	29	35	40	46	54	1.2	1.3	1.5	1.6	1.8	1.4	1.5	1.7	1.9	2.0
TW D3	22	29	38	44	53	58	65	74	14	20	26	31	39	44	50	59	1.1	1.3	1.4	1.6	1.7	1.3	1.5	1.6	1.8	1.9
TW D4	23	32	42	49	58	64	72	80	15	22	30	36	45	51	58	67	1.0	1.2	1.3	1.5	1.6	1.2	1.3	1.5	1.7	1.8
TW D5	23	32	42	49	59	65	72	81	15	22	31	37	46	52	59	69	1.0	1.2	1.3	1.5	1.6	1.2	1.3	1.5	1.6	1.8
TW E1	25	33	42	49	58	64	71	79	17	23	31	36	44	49	56	65	1.1	1.2	1.4	1.5	1.7	1.2	1.4	1.5	1.7	1.8
TW E2	24	33	42	49	58	64	72	80	16	23	31	37	45	50	58	67	1.0	1.2	1.3	1.5	1.6	1.2	1.3	1.5	1.7	1.8
TW E3	29	40	51	59	68	75	81	89	20	29	39	46	56	62	70	79	0.9	1.1	1.2	1.4	1.6	1.0	1.2	1.4	1.5	1.7
TW E4	27	37	48	56	66	72	79	87	18	27	37	44	53	60	68	77	1.0	1.1	1.3	1.4	1.6	1.1	1.2	1.4	1.5	1.7
TW F1	17	21	25	29	34	37	42	49	12	15	17	20	23	25	29	33	1.6	1.7	1.9	2.1	2.2	2.0	2.2	2.4	2.6	2.8
TW F2	17	21	27	32	38	43	49	57	11	14	18	21	26	29	33	40	1.4	1.5	1.7	1.9	2.0	1.7	1.9	2.1	2.2	2.4
TW F3	16	21	27	32	39	44	51	59	9	13	18	21	26	30	35	42	1.3	1.5	1.6	1.8	1.9	1.6	1.8	2.0	2.1	2.3
TW F4	14	19	26	31	38	43	50	59	7	11	16	20	25	29	34	42	1.3	1.5	1.6	1.8	1.9	1.6	1.8	1.9	2.1	2.3
TW G1	15	19	24	27	33	37	42	49	10	13	16	18	22	25	28	33	1.5	1.7	1.9	2.0	2.2	1.9	2.1	2.3	2.5	2.7
TW G2	14	18	24	28	35	39	46	54	8	11	15	18	23	26	31	37	1.4	1.6	1.7	1.9	2.0	1.7	1.9	2.1	2.3	2.4
TW G3	12	17	23	27	34	38	45	53	6	10	14	17	22	26	30	37	1.4	1.6	1.7	1.9	2.0	1.7	1.9	2.1	2.2	2.4
TW G4	10	14	19	23	30	34	40	48	5	8	11	14	19	22	27	34	1.5	1.6	1.8	1.9	2.1	1.8	2.0	2.1	2.3	2.4
TW H1	12	16	21	24	30	34	39	47	7	10	13	16	20	22	26	32	1.6	1.7	1.9	2.0	2.2	1.9	2.1	2.3	2.5	2.6
TW H2	10	14	19	23	28	33	38	46	5	8	11	14	18	21	25	31	1.5	1.7	1.9	2.0	2.1	1.9	2.1	2.2	2.4	2.5
TW H3	8	12	16	20	25	29	34	42	4	6	9	12	16	19	23	29	1.6	1.8	1.9	2.1	2.2	2.0	2.1	2.3	2.5	2.6
TW H4	6	9	13	16	20	23	28	35	3	5	7	9	13	15	18	23	1.8	2.0	2.1	2.3	2.4	2.2	2.4	2.5	2.7	2.8
TW I1	11	15	19	22	26	30	34	41	6	8	10	12	15	17	19	23	1.7	1.9	2.0	2.2	2.4	2.2	2.5	2.8	3.1	3.4
TW I2	17	21	31	37	45	50	57	67	10	15	21	25	32	36	42	51	1.2	1.4	1.5	1.7	1.8	1.4	1.6	1.8	1.9	2.1
TW I3	19	26	34	40	49	54	62	71	11	17	23	28	35	40	47	55	1.2	1.3	1.5	1.6	1.8	1.4	1.5	1.7	1.8	2.0
TW I4	20	29	38	45	55	61	68	78	13	19	27	33	41	47	55	64	1.1	1.2	1.4	1.5	1.7	1.2	1.4	1.5	1.7	1.8
TW I5	21	29	39	46	56	62	70	79	13	20	28	34	43	49	57	67	1.0	1.2	1.3	1.5	1.6	1.2	1.3	1.5	1.6	1.8
TW J1	23	31	40	46	56	62	69	78	15	21	28	34	42	47	54	63	1.1	1.2	1.4	1.5	1.7	1.2	1.4	1.5	1.7	1.9
TW J2	22	30	39	46	56	62	69	78	14	20	28	34	42	48	55	65	1.1	1.2	1.4	1.5	1.7	1.2	1.4	1.5	1.7	1.8
TW J3	26	36	48	56	66	72	79	87	17	26	36	43	53	60	67	77	0.9	1.1	1.2	1.4	1.6	1.1	1.2	1.4	1.5	1.7
TW J4	24	34	45	53	63	70	77	86	16	24	34	41	51	57	65	75	1.0	1.1	1.3	1.4	1.6	1.1	1.2	1.4	1.5	1.7
DG A1	14	18	23	26	30	33	36	40	8	10	13	15	17	19	20	22	1.5	1.7	2.1	2.6	3.3	2.0	2.3	2.9	4.1	-
DG A2	16	22	28	33	39	44	49	56	10	14	19	22	27	31	35	41	1.3	1.5	1.7	1.9	2.2	1.6	1.8	2.0	2.3	2.6
DG A3	20	27	35	42	51	57	64	72	14	19	25	31	38	43	50	59	1.2	1.3	1.4	1.6	1.8	1.4	1.5	1.6	1.8	2.0
DG B1	14	18	24	28	33	37	42	47	8	10	13	15	18	21	23	26	1.5	1.6	1.8	2.1	2.5	1.9	2.2	2.5	3.1	3.7
DG B2	17	22	29	34	42	48	54	63	11	14	19	23	28	33	38	45	1.3	1.4	1.6	1.7	2.0	1.6	1.7	1.9	2.1	2.3
DG B3	20	28	36	43	53	60	68	77	14	20	26	32	40	46	53	63	1.2	1.2	1.4	1.5	1.7	1.3	1.4	1.5	1.7	1.8
DG C1	17	22	29	34	41	45	51	59	10	13	17	20	24	27	31	36	1.3	1.4	1.6	1.8	2.1	1.6	1.9	2.1	2.4	2.8
DG C2	19	26	34	40	49	55	63	71	13	17	23	27	34	39	45	54	1.2	1.3	1.4	1.6	1.8	1.4	1.6	1.7	1.9	2.0
DG C3	23	31	41	49	59	66	75	84	17	23	30	36	45	51	60	70	1.1	1.2	1.3	1.4	1.6	1.2	1.4	1.4	1.6	1.7
SS A1	23	29	35	40	46	51	57	65	17	22	26	30	35	39	44	50	1.3	1.4	1.6	1.8	1.9	1.5	1.7	1.9	2.1	2.2
SS A2	29	37	45	52	60	66	72	80	22	28	36	41	49	54	60	69	1.0	1.2	1.4	1.5	1.7	1.2	1.3	1.5	1.7	1.8
SS A3	21	26	32	36	42	46	52	59	16	20	24	27	31	34	38	44	1.3	1.5	1.7	1.9	2.0	1.6	1.8	2.0	2.3	2.5
SS A4	28	36	44	50	59	64	71	79	21	28	35	40	47	52	58	66	1.1	1.2	1.4	1.6	1.7	1.2	1.4	1.5	1.7	1.9
SS A5	22	27	32	36	42	46	52	59	18	21	25	28	32	35	39	45	1.4	1.5	1.7	1.9	2.0	1.6	1.8	2.0	2.3	2.4
SS A6	28	36	45	51	59	65	72	80	22	28	35	40	48	53	59	68	1.1	1.2	1.4	1.5	1.7	1.2	1.4	1.5	1.7	1.8
SS A7	20	24	28	31	36	40	44	51	16	19	21	24	27	29	32	37	1.5	1.7	1.9	2.1	2.3	1.8	2.1	2.4	2.6	2.9
SS A8	27	35	43	49	57	63	69	77	21	27	33	38	46	50	56	64	1.1	1.2	1.4	1.6	1.7	1.2	1.4	1.6	1.8	1.9
SS B1	20	24	30	35	41	45	51	59	14	18	22	25	30	34	38	44	1.3	1.5	1.7	1.8	2.0	1.6	1.8	2.0	2.2	2.4
SS B2	25	32	41	47	55	61	68	76	19	25	32	37	44	49	56	64	1.1	1.3	1.4	1.6	1.7	1.2	1.4	1.6	1.7	1.9
SS B3	18	23	28	32	38	42	47	55	13	16	20	23	27	30	34	39	1.4	1.6	1.8	1.9	2.1	1.7	1.9	2.1	2.4	2.6
SS B4	24	31	39	45	54	59	66	74	18	24	30	35	43	47	54	62	1.1	1.3	1.4	1.6	1.7	1.3	1.4	1.6	1.8	1.9
SS B5	18	22	27	30	35	39	44	51	14	17	20	22	26	29	32	37	1.5	1.7	1.9	2.0	2.2	1.8	2.0	2.3	2.5	2.7
SS B6	24	31	40	46	54	59	66	75	18	24	30	35	43	48	54	62	1.1	1.3	1.4	1.6	1.7	1.3	1.4	1.6	1.8	1.9
SS B7	17	20	24	27	31	35	39	45	13	15	18	20	23	25	28	32	1.6	1.8	2.0	2.2	2.4	2.0	2.3	2.5	2.8	3.0
SS B8	23	30	38	43	52	57	64	72	17	23	29	34	40	45	51	59	1.1	1.3	1.5	1.6	1.8	1.3	1.5	1.6	1.8	2.0
SS C1	18	24	30	35	42	47	54	62	11	15	19	23	28	32	36	43	1.3	1.4	1.6	1.8	1.9	1.6	1.8	2.0	2.1	2.3
SS C2	22	29	38	44	53	59	66	75	14	20	26	31	39	44	51	60	1.1	1.3	1.4	1.6	1.7	1.3	1.4	1.6	1.8	1.9
SS C3	16	20	25	29	34	39	44	51	11	14	17	19	23	26	30	35	1.5	1.7	1.8	2.0	2.1	1.9	2.0	2.2	2.4	2.6
SS C4	20	26	34	39	47	53	60	69	13	18	24	28	35	39	45	54	1.2	1.3	1.5	1.7	1.8	1.4	1.6	1.7	1.9	2.0
SS D1	26	33	41	47	55	60	66	74	19	25	31	35	42	46	52	59	1.1	1.3	1.4	1.6	1.8	1.3	1.5	1.7	1.9	2.0
SS D2	32	42	52	59	69	74	81	88	24	32	41	48	57	62	69	78	0.9	1.1	1.3	1.4	1.6	1.1	1.2	1.4	1.6	1.7
SS D3	25	31	37	42	49	54	60	68	19	23	28	32	37	41	46	53	1.2	1.4	1.5	1.7	1.9	1.4	1.6	1.8	2.0	2.2
SS D4	31	41	50	57	67	72	79	86	24	31	40	46	55	60	67	75	1.0	1.1	1.3	1.5	1.6	1.1	1.2	1.4	1.6	1.7
SS E1	22	29	36	41	48	53	60	68	16	20	25	29	35	39	43	50	1.2	1.4	1.5	1.7	1.9	1.4	1.6	1.8	2.0	2.2
SS E2	28	36	46	53	62	68	75	83	20	27	35	41	49	54	61	70	1.0	1.2	1.3	1.5	1.6	1.1	1.3	1.5	1.7	1.8
SS E3	20	25	30	35	41	45	50	58	15	18	22	25	30	33	37	42	1.4	1.5	1.7	1.9	2.0	1.6	1.8	2.1	2.3	2.5
SS E4	26	34	42	49	57	63	70	78	19	25	32	38	45	50	57	65	1.1	1.2	1.4	1.5	1.7	1.2	1.4	1.5	1.7	1.9

WASHINGTON, DC

SOLAR SAVINGS FRACTION (%)

TYPE	BASE TEMP: 55 F (2897 DD)								65 F (5010 DD)							
LCR	100	70	50	40	30	25	20	15	100	70	50	40	30	25	20	15
WW A1	19	22	26	29	33	36	40	47	14	16	18	20	23	25	27	31
WW A2	19	25	31	35	42	47	53	61	13	16	21	24	29	32	37	43
WW A3	19	26	33	38	46	51	57	66	12	16	22	25	31	35	40	48
WW A4	19	26	34	39	48	53	60	69	11	16	22	26	33	37	43	50
WW A5	20	27	35	41	50	55	63	72	11	16	23	27	34	39	45	53
WW A6	20	27	36	42	51	57	64	73	11	16	23	28	35	40	46	55
WW B1	15	19	23	27	31	35	39	46	9	11	14	15	18	20	22	25
WW B2	20	27	36	42	50	56	63	73	12	18	24	29	36	41	48	56
WW B3	24	33	42	49	59	65	72	80	15	22	30	36	44	50	56	66
WW B4	24	34	44	52	62	68	75	84	15	23	32	39	48	54	62	72
WW B5	23	33	44	51	61	68	75	84	15	23	32	39	48	55	63	72
WW C1	26	35	45	52	61	67	74	83	17	24	32	38	47	53	60	69
WW C2	24	33	43	50	60	66	74	82	15	23	31	38	46	52	60	69
WW C3	29	40	52	60	70	76	83	90	19	29	40	48	58	65	72	82
WW C4	26	37	48	56	67	73	81	88	17	26	37	44	55	61	69	79
TW A1	18	22	25	28	33	36	40	47	14	16	18	20	23	25	28	32
TW A2	18	23	28	33	39	43	49	56	13	16	20	22	27	30	34	40
TW A3	18	23	29	34	41	46	52	60	12	15	20	23	28	32	37	43
TW A4	17	23	29	34	42	47	53	62	10	14	19	23	29	33	38	45
TW B1	18	21	25	29	34	37	42	49	13	15	18	20	24	26	29	34
TW B2	17	22	27	32	38	42	48	56	11	15	19	22	26	30	34	40
TW B3	16	21	27	32	38	43	49	57	11	14	18	22	27	30	35	42
TW B4	15	20	26	30	37	41	47	55	10	13	18	21	26	29	34	40
TW C1	16	20	25	28	34	37	43	50	12	14	18	20	24	27	30	36
TW C2	15	20	25	29	35	39	44	52	11	14	17	20	24	28	32	38
TW C3	15	19	24	28	33	37	43	50	10	13	17	19	24	27	31	37
TW C4	14	18	22	25	30	34	39	46	11	13	16	18	22	25	28	34
TW D1	13	16	20	23	28	31	35	41	8	10	12	14	16	18	20	23
TW D2	19	26	33	38	46	52	58	67	13	17	23	27	33	38	44	52
TW D3	21	28	36	42	50	56	63	72	13	19	25	30	37	42	48	56
TW D4	22	31	40	47	56	62	69	78	15	21	29	35	43	49	56	65
TW D5	22	30	40	47	56	63	70	79	14	21	29	36	44	50	58	67
TW E1	24	32	40	47	56	62	69	77	16	22	29	35	42	47	54	63
TW E2	23	31	40	47	56	62	69	78	15	22	29	35	43	49	56	65
TW E3	28	38	49	56	66	73	80	87	19	28	37	44	54	60	68	77
TW E4	26	36	46	54	64	70	78	86	17	26	35	42	52	58	66	75
TW F1	17	20	24	27	32	35	40	46	12	14	17	19	22	24	27	31
TW F2	16	21	26	30	36	40	46	54	10	13	17	20	24	27	31	37
TW F3	15	20	26	31	37	42	48	57	9	12	16	20	25	28	33	40
TW F4	13	18	24	29	36	41	47	56	7	10	15	18	24	27	32	39
TW G1	15	18	23	26	31	35	40	47	10	12	15	17	21	23	27	31
TW G2	13	18	23	27	33	37	43	51	8	11	14	17	22	25	29	35
TW G3	11	16	21	26	32	36	42	51	6	9	13	16	21	24	29	35
TW G4	9	13	18	22	28	32	38	46	4	7	11	14	18	21	25	32
TW H1	12	15	20	23	29	32	37	44	7	10	13	15	19	21	25	30
TW H2	9	13	18	21	27	31	36	44	5	7	11	13	17	20	24	30
TW H3	7	11	15	19	24	28	33	40	3	6	9	11	15	18	21	27
TW H4	6	8	12	15	19	22	27	33	2	4	7	9	12	14	17	22
TW I1	11	14	18	20	25	28	32	38	6	8	10	11	14	15	17	20
TW I2	16	22	30	35	43	48	55	64	10	14	20	24	30	34	40	48
TW I3	18	24	32	38	46	52	59	68	11	16	22	27	33	38	45	53
TW I4	19	27	36	43	53	59	66	76	12	18	26	32	40	45	53	62
TW I5	20	28	37	44	54	60	68	77	12	19	27	33	42	47	55	65
TW J1	21	29	38	44	53	59	67	76	14	20	27	32	40	45	52	61
TW J2	21	29	38	44	53	59	67	76	13	19	27	32	40	46	53	62
TW J3	25	35	46	54	64	70	77	86	16	25	34	42	51	58	66	75
TW J4	23	33	43	51	61	68	75	84	15	23	32	39	49	55	63	73
DG A1	13	17	21	24	28	31	34	37	7	9	12	14	16	17	18	19
DG A2	16	21	27	31	38	42	47	54	10	13	18	21	26	29	33	39
DG A3	19	26	34	40	49	54	61	70	13	18	25	29	37	42	48	57
DG B1	13	17	22	26	31	34	39	44	7	9	12	14	17	19	21	23
DG B2	16	21	28	33	40	45	52	60	10	14	18	22	27	31	36	42
DG B3	20	26	35	41	51	57	66	75	14	19	25	31	38	44	51	61
DG C1	16	21	27	31	38	42	48	55	9	12	16	19	23	25	29	33
DG C2	19	25	32	38	46	52	60	69	12	17	22	26	33	37	43	51
DG C3	22	30	39	46	57	64	72	82	16	22	29	35	43	50	58	68
SS A1	22	28	34	38	45	49	55	63	17	21	26	29	34	38	42	49
SS A2	28	35	44	50	58	64	71	79	21	28	35	40	48	53	59	67
SS A3	21	25	31	35	40	44	50	57	16	19	23	26	30	33	37	42
SS A4	27	35	43	49	57	62	69	77	21	27	34	39	46	51	57	65
SS A5	22	26	31	35	41	45	50	57	17	20	24	27	31	34	38	43
SS A6	28	35	43	49	58	63	70	78	21	27	34	39	47	52	58	66
SS A7	20	23	27	30	35	38	43	49	16	18	21	23	26	28	31	35
SS A8	27	34	42	47	56	61	67	76	20	26	33	37	44	49	55	63
SS B1	19	24	29	33	39	44	49	57	14	17	21	25	29	32	36	42
SS B2	24	31	39	45	53	59	66	74	18	24	31	36	43	48	54	63
SS B3	18	22	27	31	36	40	45	53	13	16	20	22	26	29	33	38
SS B4	24	30	38	44	52	57	64	73	18	23	30	34	41	46	52	60
SS B5	18	22	26	29	34	38	42	49	14	16	19	22	25	28	31	36
SS B6	24	30	38	44	52	58	65	73	17	23	30	34	41	46	52	61
SS B7	16	20	23	26	30	33	37	43	13	15	17	19	22	24	26	30
SS B8	22	29	36	42	50	55	62	71	17	22	28	33	39	44	50	58
SS C1	17	22	29	33	40	45	51	59	10	14	18	21	26	30	34	40
SS C2	21	28	36	42	51	56	64	73	13	19	25	30	37	42	48	57
SS C3	15	19	24	27	33	37	42	49	10	13	16	18	22	25	28	33
SS C4	19	25	32	38	45	51	58	66	12	17	23	27	33	37	43	51
SS D1	26	32	40	45	53	58	64	72	19	24	30	34	40	44	50	57
SS D2	31	41	51	58	67	73	79	86	24	32	40	47	55	61	68	76
SS D3	24	30	36	41	47	52	58	66	18	23	27	31	36	40	44	51
SS D4	30	39	49	56	65	71	77	85	23	30	39	45	53	59	65	74
SS E1	22	27	34	39	46	51	57	65	15	19	24	28	33	37	41	48
SS E2	27	35	44	51	60	66	73	81	19	26	34	39	47	53	59	68
SS E3	20	24	29	33	39	43	48	56	15	18	22	24	29	31	35	41
SS E4	25	33	41	47	56	61	68	77	18	24	31	36	44	49	55	63

CONSERVATION FACTOR

TYPE	55 F					65 F				
	100	50	30	20	15	100	50	30	20	15
WW A1	1.6	1.8	2.0	2.2	2.3	2.1	2.4	2.6	2.9	3.1
WW A2	1.3	1.5	1.6	1.8	1.9	1.6	1.8	2.0	2.2	2.4
WW A3	1.2	1.4	1.5	1.7	1.8	1.5	1.7	1.8	2.0	2.2
WW A4	1.2	1.3	1.5	1.7	1.8	1.4	1.6	1.8	2.0	2.1
WW A5	1.1	1.3	1.4	1.6	1.8	1.4	1.5	1.7	1.9	2.0
WW A6	1.1	1.3	1.4	1.6	1.7	1.3	1.5	1.7	1.9	2.0
WW B1	1.5	1.8	2.0	2.2	2.3	2.1	2.4	2.8	3.3	3.7
WW B2	1.1	1.3	1.4	1.6	1.7	1.3	1.5	1.7	1.8	2.0
WW B3	1.0	1.2	1.3	1.5	1.7	1.2	1.4	1.5	1.7	1.8
WW B4	1.0	1.1	1.3	1.5	1.6	1.1	1.3	1.4	1.6	1.7
WW B5	1.0	1.1	1.3	1.5	1.6	1.1	1.3	1.4	1.6	1.7
WW C1	1.0	1.2	1.3	1.5	1.6	1.2	1.3	1.5	1.6	1.8
WW C2	1.0	1.2	1.3	1.5	1.6	1.2	1.3	1.5	1.6	1.8
WW C3	0.9	1.1	1.2	1.4	1.5	1.0	1.2	1.3	1.5	1.6
WW C4	0.9	1.1	1.2	1.4	1.6	1.0	1.2	1.3	1.5	1.6
TW A1	1.6	1.8	2.0	2.2	2.3	2.1	2.3	2.6	2.8	3.0
TW A2	1.4	1.6	1.7	1.9	2.0	1.7	1.9	2.1	2.3	2.5
TW A3	1.3	1.5	1.6	1.8	1.9	1.6	1.8	2.0	2.1	2.3
TW A4	1.3	1.4	1.6	1.8	1.9	1.5	1.7	1.9	2.1	2.2
TW B1	1.6	1.7	1.9	2.1	2.2	2.0	2.2	2.4	2.6	2.8
TW B2	1.4	1.6	1.7	1.9	2.0	1.7	1.9	2.1	2.2	2.4
TW B3	1.4	1.5	1.7	1.8	2.0	1.6	1.8	2.0	2.2	2.3
TW B4	1.4	1.6	1.7	1.9	2.0	1.7	1.8	2.0	2.2	2.3
TW C1	1.5	1.7	1.9	2.0	2.2	1.9	2.1	2.3	2.4	2.6
TW C2	1.5	1.6	1.8	1.9	2.1	1.8	2.0	2.1	2.3	2.5
TW C3	1.5	1.7	1.8	2.0	2.1	1.8	2.0	2.2	2.3	2.5
TW C4	1.6	1.8	1.9	2.1	2.2	2.0	2.1	2.3	2.5	2.6
TW D1	1.6	1.9	2.1	2.2	2.4	2.2	2.5	2.8	3.2	3.5
TW D2	1.2	1.4	1.5	1.7	1.8	1.4	1.6	1.8	1.9	2.1
TW D3	1.1	1.3	1.5	1.6	1.8	1.3	1.5	1.7	1.8	2.0
TW D4	1.1	1.2	1.4	1.5	1.7	1.2	1.4	1.5	1.7	1.8
TW D5	1.1	1.2	1.4	1.5	1.7	1.2	1.3	1.5	1.6	1.8
TW E1	1.1	1.2	1.4	1.5	1.7	1.2	1.4	1.6	1.7	1.9
TW E2	1.1	1.2	1.4	1.5	1.7	1.2	1.4	1.5	1.7	1.8
TW E3	1.0	1.1	1.3	1.4	1.6	1.1	1.2	1.4	1.5	1.7
TW E4	1.0	1.1	1.3	1.4	1.6	1.1	1.2	1.4	1.5	1.7
TW F1	1.6	1.8	2.0	2.1	2.3	2.1	2.3	2.5	2.7	3.0
TW F2	1.4	1.6	1.7	1.9	2.0	1.8	1.9	2.1	2.3	2.5
TW F3	1.4	1.5	1.7	1.8	2.0	1.7	1.8	2.0	2.2	2.4
TW F4	1.4	1.5	1.7	1.8	2.0	1.6	1.8	2.0	2.2	2.3
TW G1	1.6	1.8	1.9	2.1	2.2	2.0	2.2	2.4	2.6	2.8
TW G2	1.4	1.6	1.8	1.9	2.1	1.8	2.0	2.1	2.3	2.5
TW G3	1.4	1.6	1.8	1.9	2.1	1.8	1.9	2.1	2.3	2.5
TW G4	1.5	1.7	1.8	2.0	2.1	1.9	2.0	2.2	2.4	2.5
TW H1	1.6	1.8	1.9	2.1	2.2	2.0	2.2	2.4	2.6	2.7
TW H2	1.6	1.7	1.9	2.1	2.2	1.9	2.1	2.3	2.5	2.6
TW H3	1.7	1.8	2.0	2.1	2.3	2.0	2.2	2.4	2.5	2.7
TW H4	1.8	2.0	2.2	2.3	2.5	2.3	2.4	2.6	2.8	2.9
TW I1	1.7	1.9	2.1	2.3	2.5	2.3	2.6	3.0	3.3	3.7
TW I2	1.2	1.4	1.6	1.7	1.8	1.5	1.6	1.8	2.0	2.1
TW I3	1.2	1.3	1.5	1.6	1.8	1.4	1.5	1.7	1.9	2.0
TW I4	1.1	1.2	1.4	1.6	1.7	1.2	1.4	1.6	1.7	1.8
TW I5	1.1	1.2	1.4	1.5	1.7	1.2	1.4	1.5	1.7	1.8
TW J1	1.1	1.2	1.4	1.6	1.7	1.3	1.4	1.6	1.7	1.9
TW J2	1.1	1.2	1.4	1.5	1.7	1.2	1.4	1.6	1.7	1.9
TW J3	1.0	1.1	1.3	1.4	1.6	1.1	1.2	1.4	1.5	1.7
TW J4	1.0	1.1	1.3	1.5	1.6	1.1	1.3	1.4	1.6	1.7
DG A1	1.5	1.8	2.2	2.8	3.5	2.0	2.4	3.1	4.6	-
DG A2	1.4	1.5	1.7	2.0	2.2	1.7	1.8	2.0	2.3	2.7
DG A3	1.2	1.3	1.4	1.6	1.8	1.4	1.5	1.6	1.8	2.0
DG B1	1.5	1.7	1.9	2.2	2.6	2.0	2.3	2.7	3.3	4.2
DG B2	1.3	1.5	1.6	1.8	2.0	1.6	1.8	1.9	2.1	2.3
DG B3	1.2	1.3	1.4	1.5	1.7	1.4	1.5	1.6	1.7	1.8
DG C1	1.4	1.5	1.7	1.9	2.2	1.7	2.0	2.3	2.6	3.0
DG C2	1.3	1.3	1.5	1.6	1.8	1.5	1.6	1.8	1.9	2.1
DG C3	1.1	1.2	1.3	1.5	1.6	1.3	1.4	1.5	1.6	1.7
SS A1	1.3	1.5	1.6	1.8	1.9	1.5	1.7	1 9	2.1	2.3
SS A2	1.1	1.2	1.4	1.5	1.7	1.2	1.4	1.5	1.7	1.9
SS A3	1.4	1.6	1.7	1.9	2.1	1.6	1.9	2.1	2.4	2.6
SS A4	1.1	1.2	1.4	1.6	1.7	1.2	1.4	1.6	1.8	1.9
SS A5	1.4	1.6	1.8	1.9	2.1	1.7	1.9	2.1	2.3	2.5
SS A6	1.1	1.2	1.4	1.6	1.7	1.2	1.4	1.6	1.7	1.9
SS A7	1.5	1.8	2.0	2.2	2.3	1.9	2.2	2.4	2.8	3.0
SS A8	1.1	1.3	1.4	1.6	1.7	1.2	1.4	1.6	1.8	1.9
SS B1	1.4	1.6	1.7	1.9	2.0	1.6	1.8	2.0	2.3	2.4
SS B2	1.1	1.3	1.4	1.6	1.7	1.3	1.4	1.6	1.8	1.9
SS B3	1.4	1.6	1.8	2.0	2.1	1.7	2.0	2.2	2.4	2.7
SS B4	1.1	1.3	1.5	1.6	1.8	1.3	1.5	1.6	1.8	1.9
SS B5	1.5	1.7	1.9	2.1	2.2	1.9	2.1	2.3	2.6	2.8
SS B6	1.1	1.3	1.4	1.6	1.8	1.3	1.4	1.6	1.8	1.9
SS B7	1.7	1.9	2.1	2.3	2.4	2.1	2.3	2.6	2.9	3.2
SS B8	1.2	1.3	1.5	1.6	1.8	1.3	1.5	1.7	1.9	2.0
SS C1	1.3	1.5	1.6	1.8	2.0	1.6	1.8	2.0	2.2	2.4
SS C2	1.1	1.3	1.4	1.6	1.7	1.3	1.5	1.6	1.8	2.0
SS C3	1.5	1.7	1.9	2.0	2.2	1.9	2.1	2.3	2.5	2.7
SS C4	1.2	1.4	1.5	1.7	1.8	1.4	1.6	1.8	1.9	2.1
SS D1	1.1	1.3	1.5	1.6	1.8	1.3	1.5	1.7	1.9	2.1
SS D2	1.0	1.1	1.3	1.4	1.6	1.1	1.2	1.4	1.6	1.7
SS D3	1.2	1.4	1.6	1.7	1.9	1.4	1.7	1.9	2.1	2.2
SS D4	1.0	1.1	1.3	1.5	1.6	1.1	1.3	1.4	1.6	1.8
SS E1	1.2	1.4	1.6	1.7	1.9	1.5	1.7	1.9	2.1	2.3
SS E2	1.0	1.2	1.3	1.5	1.7	1.2	1.3	1.5	1.7	1.8
SS E3	1.4	1.6	1.8	1.9	2.1	1.7	1.9	2.1	2.4	2.6
SS E4	1.1	1.2	1.4	1.6	1.7	1.2	1.4	1.6	1.8	1.9

APALACHICOLA, FLORIDA

TYPE	SSF (%) 55 F (426 DD) LCR 200	100	70	50	40	30	25	20	SSF (%) 65 F (1361 DD) LCR 200	100	70	50	40	30	25	20	CF 55 F 200	100	50	30	20	CF 65 F 200	100	50	30	20
WW A1	41	59	70	79	85	91	94	97	23	34	41	50	56	64	70	76	0.7	0.8	0.9	1.1	1.3	0.9	1.0	1.2	1.3	1.5
WW A2	50	71	82	90	94	97	99	100	25	41	51	62	69	78	83	88	0.6	0.7	0.8	1.0	1.2	0.8	0.9	1.0	1.2	1.4
WW A3	53	75	86	93	96	99	99	100	26	44	55	66	73	82	87	91	0.5	0.7	0.8	1.0	1.2	0.7	0.8	1.0	1.2	1.3
WW A4	54	77	87	94	97	99	100	100	26	45	57	68	76	84	88	93	0.5	0.6	0.8	1.0	1.2	0.7	0.8	1.0	1.1	1.3
WW A5	56	79	89	95	98	99	100	100	27	47	59	71	78	86	90	94	0.5	0.6	0.8	1.0	1.2	0.7	0.8	0.9	1.1	1.3
WW A6	57	80	90	96	98	99	100	100	27	48	60	72	79	87	91	95	0.5	0.6	0.8	1.0	1.2	0.7	0.8	0.9	1.1	1.3
WW B1	45	65	76	86	91	95	97	99	22	35	45	54	61	70	76	82	0.6	0.7	0.9	1.0	1.2	0.9	1.0	1.1	1.3	1.4
WW B2	54	77	87	94	97	99	100	100	26	45	57	69	76	85	89	93	0.5	0.6	0.8	1.0	1.2	0.7	0.8	1.0	1.1	1.3
WW B3	61	84	92	97	99	100	100	100	31	53	65	77	83	90	94	97	0.5	0.6	0.8	1.0	1.2	0.7	0.8	0.9	1.1	1.3
WW B4	61	84	92	97	99	100	100	100	30	53	66	77	84	91	94	97	0.5	0.6	0.8	1.0	1.2	0.6	0.7	0.9	1.1	1.3
WW B5	58	82	91	96	98	100	100	100	28	51	64	76	83	90	93	97	0.5	0.6	0.8	1.0	1.2	0.7	0.8	0.9	1.1	1.3
WW C1	63	85	93	97	99	100	100	100	33	55	67	78	85	91	94	97	0.5	0.6	0.8	1.0	1.2	0.6	0.7	0.9	1.1	1.3
WW C2	60	83	92	97	99	100	100	100	30	52	65	76	83	90	94	97	0.5	0.6	0.8	1.0	1.2	0.7	0.8	0.9	1.1	1.3
WW C3	66	88	95	98	99	100	100	100	34	59	72	83	89	94	97	99	0.5	0.6	0.8	1.0	-	0.6	0.7	0.9	1.0	1.2
WW C4	62	85	93	98	99	100	100	100	30	55	68	80	86	93	95	98	0.5	0.6	0.8	1.0	1.2	0.6	0.7	0.9	1.1	1.2
TW A1	40	58	68	78	84	91	94	97	23	33	40	49	55	63	69	75	0.7	0.8	0.9	1.1	1.3	1.0	1.0	1.2	1.3	1.5
TW A2	45	66	77	86	91	96	97	99	23	37	47	57	64	73	78	84	0.6	0.7	0.9	1.0	1.2	0.8	0.9	1.1	1.2	1.4
TW A3	47	69	80	89	93	97	98	99	23	39	49	60	67	76	81	87	0.6	0.7	0.8	1.0	1.2	0.8	0.9	1.0	1.2	1.4
TW A4	47	69	80	89	93	97	99	99	22	39	49	60	68	77	82	88	0.6	0.7	0.8	1.0	1.2	0.8	0.9	1.0	1.2	1.4
TW B1	40	58	69	79	85	91	94	97	22	33	41	49	56	64	70	76	0.7	0.8	0.9	1.1	1.3	0.9	1.0	1.2	1.3	1.5
TW B2	43	64	75	85	90	95	97	99	22	35	45	55	62	71	77	83	0.6	0.7	0.9	1.0	1.2	0.8	0.9	1.1	1.2	1.4
TW B3	43	64	76	85	91	95	97	99	21	35	45	55	63	72	78	84	0.6	0.7	0.9	1.0	1.2	0.8	0.9	1.1	1.2	1.4
TW B4	41	61	73	83	88	94	96	98	20	33	43	53	60	69	75	81	0.6	0.7	0.9	1.1	1.2	0.9	1.0	1.1	1.3	1.4
TW C1	39	57	68	78	85	91	94	97	20	32	40	49	55	64	70	76	0.7	0.8	0.9	1.1	1.3	0.9	1.0	1.2	1.3	1.5
TW C2	39	58	69	80	86	92	95	97	19	32	40	50	56	66	71	78	0.6	0.8	0.9	1.1	1.2	0.9	1.0	1.1	1.3	1.5
TW C3	37	56	67	77	83	90	94	97	18	30	38	47	54	63	69	76	0.7	0.8	0.9	1.1	1.3	0.9	1.0	1.2	1.3	1.5
TW C4	33	50	60	71	78	85	90	94	17	27	34	42	48	57	63	70	0.7	0.8	1.0	1.1	1.3	1.0	1.1	1.2	1.4	1.5
TW D1	39	58	69	80	86	92	95	97	18	30	38	48	54	63	69	76	0.6	0.8	0.9	1.1	1.2	0.9	1.0	1.2	1.3	1.5
TW D2	50	72	83	91	95	98	99	100	25	41	52	64	71	80	85	90	0.6	0.7	0.8	1.0	1.2	0.8	0.9	1.0	1.2	1.3
TW D3	54	77	87	94	97	99	99	100	27	46	57	69	76	84	89	93	0.5	0.6	0.8	1.0	1.2	0.7	0.8	1.0	1.1	1.3
TW D4	55	79	88	95	97	99	100	100	27	48	60	72	79	87	91	95	0.5	0.6	0.8	1.0	1.2	0.7	0.8	0.9	1.1	1.3
TW D5	54	77	87	94	97	99	100	100	26	46	59	71	78	86	90	94	0.5	0.6	0.8	1.0	1.2	0.7	0.8	0.9	1.1	1.3
TW E1	58	80	90	96	98	99	100	100	30	50	62	73	80	88	91	95	0.5	0.6	0.8	1.0	1.2	0.7	0.8	0.9	1.1	1.3
TW E2	57	80	89	95	98	99	100	100	29	49	61	72	79	87	91	95	0.5	0.6	0.8	1.0	1.2	0.7	0.8	0.9	1.1	1.3
TW E3	64	86	94	98	99	100	100	100	33	57	69	80	87	93	96	98	0.5	0.6	0.8	1.0	1.2	0.6	0.7	0.9	1.1	1.2
TW E4	60	83	92	97	99	100	100	100	30	53	66	77	84	91	94	97	0.5	0.6	0.8	1.0	1.2	0.6	0.7	0.9	1.1	1.3
TW F1	39	58	68	78	84	91	94	97	21	32	40	48	55	63	69	76	0.7	0.8	0.9	1.1	1.3	0.9	1.0	1.2	1.3	1.5
TW F2	43	64	75	85	90	95	97	99	21	35	44	54	61	71	76	82	0.6	0.7	0.9	1.0	1.2	0.8	0.9	1.1	1.2	1.4
TW F3	43	65	77	86	91	96	98	99	20	35	45	56	63	73	78	84	0.6	0.7	0.9	1.0	1.2	0.8	0.9	1.1	1.2	1.4
TW F4	41	63	75	85	90	95	97	99	18	33	43	54	61	71	77	83	0.6	0.7	0.9	1.0	1.2	0.8	0.9	1.1	1.2	1.4
TW G1	38	56	68	78	84	91	94	97	19	30	38	47	54	63	68	75	0.7	0.8	0.9	1.1	1.3	0.9	1.0	1.2	1.3	1.5
TW G2	39	60	71	82	87	93	96	98	18	31	40	50	57	67	73	80	0.6	0.7	0.9	1.1	1.2	0.9	1.0	1.1	1.3	1.4
TW G3	37	58	70	80	86	93	95	98	16	29	38	49	56	66	72	79	0.6	0.7	0.9	1.1	1.2	0.9	1.0	1.1	1.3	1.4
TW G4	32	52	64	75	82	89	93	96	13	25	33	43	50	60	66	73	0.7	0.8	0.9	1.1	1.3	0.9	1.0	1.2	1.3	1.5
TW H1	34	53	64	75	82	89	93	96	15	27	35	44	50	60	65	73	0.7	0.8	0.9	1.1	1.3	1.0	1.1	1.2	1.3	1.5
TW H2	31	50	62	73	80	88	92	95	13	24	32	41	48	58	64	71	0.7	0.8	0.9	1.1	1.3	1.0	1.1	1.2	1.3	1.5
TW H3	27	45	56	68	75	83	88	93	10	21	28	37	43	52	58	66	0.7	0.8	1.0	1.1	1.3	1.0	1.1	1.3	1.4	1.6
TW H4	21	37	47	58	65	75	80	86	8	16	22	29	35	43	49	56	0.8	0.9	1.0	1.2	1.4	1.2	1.2	1.4	1.5	1.7
TW I1	35	55	66	77	83	90	93	96	16	27	35	44	51	60	65	73	0.7	0.8	0.9	1.1	1.3	1.0	1.1	1.2	1.3	1.5
TW I2	46	69	80	80	00	97	99	99	21	38	49	60	68	77	82	88	0.6	0.7	0.8	1.0	1.2	0.8	0.9	1.0	1.2	1.4
TW I3	49	72	83	91	95	98	99	100	23	41	53	64	72	81	86	91	0.6	0.7	0.8	1.0	1.2	0.7	0.8	1.0	1.2	1.3
TW I4	52	76	86	93	97	99	99	100	24	44	56	68	76	84	89	93	0.5	0.6	0.8	1.0	1.2	0.7	0.8	1.0	1.1	1.3
TW I5	51	76	86	93	97	99	99	100	24	44	57	69	76	85	89	94	0.5	0.6	0.8	1.0	1.2	0.7	0.8	1.0	1.1	1.3
TW J1	55	78	88	95	97	99	100	100	27	47	59	71	78	86	90	94	0.5	0.6	0.8	1.0	1.2	0.7	0.8	0.9	1.1	1.3
TW J2	53	77	87	94	97	99	100	100	26	46	58	70	77	85	90	94	0.5	0.6	0.8	1.0	1.2	0.7	0.8	1.0	1.1	1.3
TW J3	60	83	92	97	99	100	100	100	30	53	66	78	84	91	94	97	0.5	0.6	0.8	1.0	1.2	0.0	0.7	0.9	1.1	1.2
TW J4	57	81	90	96	98	99	100	100	27	50	63	75	82	00	93	96	0.5	0.6	0.8	1.0	1.2	0.7	0.8	0.9	1.1	1.3
DG A1	45	66	70	85	89	94	96	98	20	34	43	52	58	65	70	75	0.6	0.7	0.9	1.1	1.3	0.8	1.0	1.2	1.4	1.6
DG A2	44	66	78	87	91	96	97	99	20	35	46	56	63	72	76	82	0.6	0.7	0.9	1.0	1.2	0.8	0.9	1.1	1.3	1.5
DG A3	48	73	83	90	93	95	96	97	23	40	53	65	72	80	85	89	0.5	0.7	0.9	1.2	2.5	0.8	0.8	1.0	1.2	1.4
DG B1	47	72	83	91	94	97	98	98	20	36	47	58	65	73	78	83	0.5	0.7	0.9	1.1	1.4	0.8	0.9	1.0	1.3	1.5
DG B2	46	73	84	92	95	98	99	100	21	37	49	61	69	78	83	88	0.6	0.7	0.8	1.0	1.2	0.8	0.9	1.0	1.2	1.4
DG B3	49	77	87	94	96	98	98	98	23	42	55	68	76	85	89	93	0.5	0.6	0.8	1.2	2.9	0.8	0.8	0.9	1.1	1.4
DG C1	53	80	90	95	97	98	98	98	23	41	54	66	74	82	86	90	0.5	0.6	0.8	1.2	-	0.8	0.8	1.0	1.2	1.4
DG C2	51	79	89	95	98	99	100	100	23	42	55	68	76	85	89	93	0.5	0.6	0.8	1.0	1.3	0.8	0.8	0.9	1.1	1.3
DG C3	55	83	92	96	98	98	98	98	26	47	61	75	83	90	93	96	0.5	0.6	0.8	1.5	-	0.7	0.8	0.9	1.1	1.3
SS A1	52	73	83	91	95	98	99	100	29	45	55	65	72	80	85	90	0.6	0.7	0.8	1.0	1.2	0.8	0.9	1.0	1.2	1.4
SS A2	61	82	91	96	98	99	100	100	35	55	66	77	83	90	93	96	0.5	0.6	0.8	1.0	1.2	0.7	0.8	0.9	1.1	1.3
SS A3	50	70	81	89	93	97	98	99	28	43	53	63	69	78	82	88	0.6	0.7	0.9	1.0	1.2	0.8	0.9	1.0	1.2	1.4
SS A4	61	82	91	96	98	99	100	100	35	55	66	76	82	89	93	96	0.5	0.6	0.8	1.0	1.2	0.7	0.8	0.9	1.1	1.3
SS A5	49	68	79	87	92	96	98	99	28	42	51	61	67	76	81	86	0.6	0.7	0.9	1.0	1.2	0.8	0.9	1.1	1.2	1.4
SS A6	61	82	91	96	98	99	100	100	35	55	66	76	83	89	93	96	0.5	0.6	0.8	1.0	1.2	0.7	0.8	0.9	1.1	1.3
SS A7	45	63	74	83	88	94	96	98	26	38	46	55	61	70	75	81	0.6	0.7	0.9	1.1	1.2	0.9	1.0	1.1	1.3	1.5
SS A8	60	82	90	96	98	99	100	100	35	54	65	75	82	89	92	95	0.5	0.6	0.8	1.0	1.2	0.7	0.8	0.9	1.1	1.3
SS B1	46	67	78	87	91	96	98	99	25	39	49	59	66	74	80	85	0.6	0.7	0.9	1.0	1.2	0.8	0.9	1.1	1.2	1.4
SS B2	55	77	87	94	97	99	99	100	30	48	60	71	77	85	89	94	0.5	0.7	0.8	1.0	1.2	0.7	0.8	1.0	1.1	1.3
SS B3	45	65	76	85	91	95	97	99	24	38	47	57	64	73	78	84	0.6	0.7	0.9	1.0	1.2	0.8	0.9	1.1	1.2	1.4
SS B4	55	76	86	93	96	99	99	100	30	48	59	70	77	85	89	93	0.5	0.7	0.8	1.0	1.2	0.7	0.8	1.0	1.1	1.3
SS B5	41	60	71	80	86	92	95	97	23	35	43	52	58	67	72	79	0.6	0.8	0.9	1.1	1.3	0.9	1.0	1.1	1.3	1.5
SS B6	54	77	86	93	96	99	99	100	30	48	59	70	77	85	89	93	0.5	0.7	0.8	1.0	1.2	0.7	0.8	1.0	1.1	1.3
SS B7	39	56	67	77	83	90	93	96	21	32	40	48	54	63	68	74	0.7	0.8	0.9	1.1	1.3	0.9	1.0	1.2	1.3	1.5
SS B8	53	75	86	93	96	99	99	100	29	47	58	69	76	84	88	92	0.5	0.7	0.8	1.0	1.2	0.7	0.8	1.0	1.1	1.3
SS C1	48	71	81	90	94	97	99	100	23	39	50	61	68	77	82	88	0.6	0.7	0.8	1.0	1.2	0.8	0.9	1.0	1.2	1.4
SS C2	54	77	87	94	97	99	100	100	26	45	57	69	76	84	89	93	0.5	0.6	0.8	1.0	1.2	0.7	0.8	1.0	1.1	1.3
SS C3	39	59	70	80	86	92	95	97	20	32	40	49	56	65	71	78	0.6	0.7	0.9	1.1	1.2	0.9	1.0	1.1	1.3	1.5
SS C4	48	71	82	90	94	98	99	100	24	40	51	62	70	79	84	89	0.6	0.7	0.8	1.0	1.2	0.8	0.9	1.0	1.2	1.3
SS D1	61	82	91	96	98	99	100	100	35	54	65	75	81	88	92	95	0.5	0.6	0.8	1.0	1.2	0.7	0.8	0.9	1.1	1.3
SS D2	69	89	96	99	99	100	100	100	40	63	74	84	89	95	97	98	0.5	0.6	0.8	1.0	-	0.6	0.7	0.9	1.1	1.2
SS D3	55	76	86	93	96	99	99	100	32	48	59	69	76	83	88	92	0.6	0.7	0.8	1.0	1.2	0.7	0.8	1.0	1.2	1.3
SS D4	67	88	95	98	99	100	100	100	39	61	72	82	88	93	96	98	0.5	0.6	0.8	1.0	1.2	0.5	0.7	0.9	1.1	1.2
SS E1	56	78	88	94	97	99	100	100	30	48	59	70	77	84	89	93	0.5	0.6	0.8	1.0	1.2	0.7	0.8	1.0	1.1	1.3
SS E2	64	85	93	97	99	100	100	100	35	57	69	79	85	92	95	97	0.5	0.6	0.8	1.0	1.2	0.6	0.7	0.9	1.1	1.3
SS E3	47	68	78	87	92	96	98	99	26	40	49	59	66	75	80	85	0.6	0.7	0.9	1.0	1.2	0.8	0.9	1.1	1.2	1.4
SS E4	59	81	90	96	98	99	100	100	32	52	64	75	81	88	92	95	0.5	0.6	0.8	1.0	1.2	0.7	0.8	0.9	1.1	1.3

DAYTONA BEACH, FLORIDA

TYPE	SOLAR SAVINGS FRACTION (%) BASE TEMP: 55 F (215 DD)								65 F (902 DD)								CONSERVATION FACTOR 55 F					65 F				
LCR	200	100	70	50	40	30	25	20	200	100	70	50	40	30	25	20	200	100	50	30	20	200	100	50	30	20
WW A1	63	83	91	96	98	100	100	100	30	45	55	66	73	81	86	91	0.5	0.6	0.8	1.0	1.2	0.8	0.9	1.0	1.2	1.3
WW A2	75	93	97	99	100	100	100	100	36	56	68	79	85	92	95	97	0.5	0.6	0.8	1.0	-	0.7	0.7	0.9	1.1	1.3
WW A3	79	95	99	100	100	100	100	100	38	60	72	83	89	94	96	98	0.5	0.6	0.8	0.9	-	0.6	0.7	0.9	1.1	1.2
WW A4	81	96	99	100	100	100	100	100	39	62	75	85	90	95	97	99	0.4	0.6	0.7	0.9	-	0.6	0.7	0.9	1.0	1.2
WW A5	83	97	99	100	100	100	100	100	41	65	77	87	92	96	98	99	0.4	0.6	0.7	-	-	0.6	0.7	0.9	1.0	1.2
WW A6	84	97	99	100	100	100	100	100	41	66	78	88	92	97	98	99	0.4	0.6	0.7	-	-	0.6	0.7	0.8	1.0	1.2
WW B1	70	89	96	99	99	100	100	100	32	50	62	73	79	87	91	95	0.5	0.6	0.8	1.0	-	0.7	0.8	0.9	1.1	1.3
WW B2	81	96	99	100	100	100	100	100	39	62	75	85	90	95	97	99	0.4	0.6	0.7	0.9	-	0.6	0.7	0.9	1.0	1.2
WW B3	87	98	100	100	100	100	100	100	46	70	82	91	95	98	99	100	0.4	0.6	0.7	-	-	0.6	0.7	0.8	1.0	1.2
WW B4	87	98	100	100	100	100	100	100	45	70	82	91	95	98	99	100	0.4	0.6	0.7	-	-	0.6	0.7	0.8	1.0	1.2
WW B5	85	98	99	100	100	100	100	100	42	68	80	90	94	98	99	100	0.4	0.6	0.7	-	-	0.6	0.7	0.8	1.0	1.2
WW C1	88	98	100	100	100	100	100	100	47	72	83	91	95	98	99	100	0.4	0.5	0.7	-	-	0.6	0.7	0.8	1.0	1.2
WW C2	86	98	100	100	100	100	100	100	44	69	81	90	94	98	99	100	0.4	0.6	0.7	-	-	0.6	0.7	0.8	1.0	1.2
WW C3	90	99	100	100	100	100	100	100	50	76	87	94	97	99	100	100	0.4	0.5	0.7	-	-	0.5	0.6	0.8	1.0	1.2
WW C4	88	98	100	100	100	100	100	100	46	72	84	92	96	98	99	100	0.4	0.5	0.7	-	-	0.5	0.7	0.8	1.0	1.2
TW A1	62	82	90	96	98	99	100	100	29	44	54	64	71	80	85	90	0.5	0.6	0.8	1.0	1.2	0.8	0.9	1.0	1.2	1.3
TW A2	70	89	95	99	99	100	100	100	33	51	63	73	80	88	92	95	0.5	0.6	0.8	1.0	-	0.7	0.8	0.9	1.1	1.3
TW A3	73	91	97	99	100	100	100	100	34	54	66	77	83	90	94	97	0.5	0.6	0.8	1.0	-	0.7	0.8	0.9	1.1	1.3
TW A4	73	92	97	99	100	100	100	100	33	54	66	77	84	91	94	97	0.5	0.6	0.8	1.0	-	0.7	0.8	0.9	1.1	1.3
TW B1	62	83	91	96	98	99	100	100	29	44	55	65	72	81	86	91	0.5	0.6	0.8	1.0	1.2	0.8	0.9	1.0	1.2	1.3
TW B2	68	88	95	98	99	100	100	100	31	49	61	72	79	87	91	95	0.5	0.6	0.8	1.0	1.2	0.7	0.8	0.9	1.1	1.3
TW B3	68	89	95	98	99	100	100	100	30	49	61	73	80	87	91	95	0.5	0.6	0.8	1.0	-	0.7	0.8	0.9	1.1	1.3
TW B4	65	86	94	98	99	100	100	100	28	47	58	69	77	85	89	94	0.5	0.6	0.8	1.0	1.2	0.7	0.8	1.0	1.1	1.3
TW C1	61	82	91	96	98	99	100	100	27	43	54	65	72	80	85	90	0.5	0.6	0.8	1.0	1.2	0.8	0.9	1.0	1.2	1.3
TW C2	62	83	92	97	98	100	100	100	27	44	55	66	73	82	87	92	0.5	0.6	0.8	1.0	1.2	0.8	0.8	1.0	1.1	1.3
TW C3	59	81	90	96	98	99	100	100	25	41	52	63	70	79	84	90	0.5	0.6	0.8	1.0	1.2	0.8	0.9	1.0	1.2	1.3
TW C4	53	74	85	92	96	98	99	100	23	37	46	57	64	73	79	85	0.6	0.7	0.8	1.0	1.2	0.8	0.9	1.1	1.2	1.4
TW D1	63	84	92	97	99	100	100	100	26	43	54	65	72	81	86	91	0.5	0.6	0.8	1.0	1.2	0.8	0.8	1.0	1.2	1.3
TW D2	76	93	98	99	100	100	100	100	35	57	69	80	86	92	95	98	0.5	0.6	0.8	0.9	-	0.6	0.7	0.9	1.1	1.2
TW D3	80	96	99	100	100	100	100	100	39	62	74	85	90	95	97	99	0.4	0.6	0.8	0.9	-	0.6	0.7	0.9	1.0	1.2
TW D4	82	96	99	100	100	100	100	100	40	64	77	87	92	96	98	99	0.4	0.6	0.7	-	-	0.6	0.7	0.9	1.0	1.2
TW D5	80	96	99	100	100	100	100	100	39	63	75	86	91	96	98	99	0.4	0.6	0.7	0.9	-	0.6	0.7	0.9	1.0	1.2
TW E1	84	97	99	100	100	100	100	100	43	67	78	88	93	97	98	99	0.4	0.6	0.7	-	-	0.6	0.7	0.9	1.0	1.2
TW E2	83	97	99	100	100	100	100	100	42	66	78	87	92	96	98	99	0.4	0.6	0.7	-	-	0.6	0.7	0.9	1.0	1.2
TW E3	89	99	100	100	100	100	100	100	48	74	85	93	96	99	99	100	0.4	0.5	0.7	-	-	0.5	0.7	0.8	1.0	1.2
TW E4	86	98	100	100	100	100	100	100	44	70	82	91	95	98	99	100	0.4	0.6	0.7	-	-	0.6	0.7	0.8	1.0	1.2
TW F1	62	82	91	96	98	99	100	100	28	44	54	64	71	80	85	90	0.5	0.6	0.8	1.0	1.2	0.8	0.9	1.0	1.2	1.3
TW F2	68	88	95	98	99	100	100	100	30	49	60	71	78	86	91	94	0.5	0.6	0.8	1.0	1.2	0.7	0.8	0.9	1.1	1.3
TW F3	69	89	96	99	100	100	100	100	30	50	62	73	80	88	92	95	0.5	0.6	0.8	1.0	-	0.7	0.8	0.9	1.1	1.3
TW F4	67	88	95	98	99	100	100	100	27	48	60	72	79	87	91	95	0.5	0.6	0.8	1.0	-	0.7	0.8	0.9	1.1	1.3
TW G1	60	82	90	96	98	99	100	100	26	42	53	64	71	80	85	90	0.5	0.6	0.8	1.0	1.2	0.8	0.9	1.0	1.2	1.3
TW G2	64	85	93	97	99	100	100	100	26	44	56	68	75	84	88	93	0.5	0.6	0.8	1.0	1.2	0.7	0.8	1.0	1.1	1.3
TW G3	62	84	92	97	99	100	100	100	24	42	54	66	74	82	87	92	0.5	0.6	0.8	1.0	1.2	0.7	0.8	1.0	1.1	1.3
TW G4	56	79	88	95	97	99	100	100	20	37	48	60	67	77	83	88	0.5	0.6	0.8	1.0	1.2	0.8	0.9	1.0	1.2	1.3
TW H1	57	79	89	95	97	99	100	100	23	38	49	60	68	77	82	88	0.5	0.6	0.8	1.0	1.2	0.8	0.9	1.0	1.2	1.3
TW H2	54	77	87	94	97	99	100	100	20	36	46	58	66	75	81	87	0.5	0.6	0.8	1.0	1.2	0.8	0.9	1.0	1.2	1.4
TW H3	49	72	83	91	95	98	99	100	17	31	41	52	60	70	76	83	0.6	0.7	0.8	1.0	1.2	0.9	0.9	1.1	1.2	1.4
TW H4	40	62	74	84	89	95	97	99	12	24	33	43	50	60	66	74	0.6	0.7	0.9	1.0	1.2	1.0	1.0	1.2	1.3	1.5
TW I1	59	81	90	96	98	99	100	100	24	40	51	62	69	78	83	89	0.5	0.6	0.8	1.0	1.2	0.8	0.9	1.0	1.2	1.3
TW I2	72	92	97	99	100	100	100	100	32	53	66	77	84	91	94	97	0.5	0.6	0.8	1.0	-	0.7	0.8	0.9	1.1	1.3
TW I3	76	94	98	100	100	100	100	100	35	57	70	81	87	93	96	98	0.5	0.6	0.8	0.9	-	0.6	0.7	0.9	1.1	1.2
TW I4	79	95	99	100	100	100	100	100	37	61	73	84	90	95	97	99	0.4	0.6	0.8	0.9	-	0.6	0.7	0.9	1.0	1.2
TW I5	79	95	99	100	100	100	100	100	36	61	74	84	90	95	97	99	0.4	0.6	0.8	0.9	-	0.6	0.7	0.9	1.0	1.2
TW J1	82	96	99	100	100	100	100	100	40	64	76	86	91	96	98	99	0.4	0.6	0.7	-	-	0.6	0.7	0.9	1.0	1.2
TW J2	80	96	99	100	100	100	100	100	38	62	75	85	90	95	97	99	0.4	0.6	0.7	0.9	-	0.6	0.7	0.9	1.0	1.2
TW J3	86	98	100	100	100	100	100	100	44	70	82	91	95	98	99	100	0.4	0.6	0.7	-	-	0.6	0.7	0.8	1.0	1.2
TW J4	84	97	99	100	100	100	100	100	41	67	79	89	93	97	99	99	0.4	0.6	0.7	-	-	0.6	0.7	0.8	1.0	1.2
DG A1	71	90	96	99	100	100	100	100	30	50	61	71	77	84	87	91	0.5	0.6	0.8	-	-	0.7	0.8	1.0	1.2	1.4
DG A2	71	91	96	99	100	100	100	100	30	51	63	74	80	87	91	94	0.5	0.6	0.8	-	-	0.7	0.8	0.9	1.1	1.3
DG A3	76	92	96	97	97	97	97	97	33	57	70	81	86	91	94	95	0.5	0.6	1.3	-	-	0.6	0.7	0.9	1.1	1.5
DG B1	78	95	98	99	99	99	99	99	31	54	68	78	84	90	93	95	0.4	0.6	0.9	-	-	0.7	0.7	0.9	1.1	1.4
DG B2	78	95	98	100	100	100	100	100	31	55	69	81	87	92	95	97	0.4	0.6	0.8	1.0	-	0.7	0.7	0.9	1.1	1.3
DG B3	80	95	98	98	98	98	98	98	34	60	74	85	90	95	96	97	0.4	0.6	1.4	-	-	0.6	0.7	0.9	1.1	1.5
DG C1	85	97	98	99	99	99	98	98	36	62	76	86	91	94	96	97	0.4	0.6	2.5	-	-	0.6	0.7	0.9	1.1	1.5
DG C2	83	97	99	100	100	100	100	100	35	61	76	86	91	96	98	99	0.4	0.6	0.8	-	-	0.6	0.7	0.9	1.0	1.2
DG C3	86	97	98	99	99	98	98	98	38	67	81	90	94	97	98	98	0.4	0.6	3.3	-	-	0.6	0.6	0.8	1.1	1.8
SS A1	77	93	98	99	100	100	100	100	40	60	71	81	87	93	95	98	0.5	0.6	0.8	0.9	-	0.6	0.7	0.9	1.1	1.2
SS A2	86	98	99	100	100	100	100	100	48	70	81	90	94	97	99	99	0.4	0.6	0.7	-	-	0.6	0.7	0.8	1.0	1.2
SS A3	74	92	97	99	100	100	100	100	38	57	69	79	85	91	94	97	0.5	0.6	0.8	1.0	-	0.7	0.8	0.9	1.1	1.3
SS A4	85	98	99	100	100	100	100	100	48	70	81	89	94	97	99	99	0.4	0.6	0.7	-	-	0.6	0.7	0.8	1.0	1.2
SS A5	72	91	96	99	100	100	100	100	37	56	66	77	83	89	93	96	0.5	0.6	0.8	1.0	-	0.7	0.8	0.9	1.1	1.3
SS A6	86	98	99	100	100	100	100	100	48	71	81	90	94	97	99	99	0.4	0.6	0.7	-	-	0.6	0.7	0.8	1.0	1.2
SS A7	67	87	94	98	99	100	100	100	34	51	61	71	78	85	89	93	0.5	0.6	0.8	1.0	1.2	0.7	0.8	1.0	1.1	1.3
SS A8	85	97	99	100	100	100	100	100	47	70	81	89	93	97	98	99	0.4	0.6	0.7	-	-	0.6	0.7	0.9	1.0	1.2
SS B1	71	90	96	99	100	100	100	100	34	53	65	75	82	89	92	96	0.5	0.6	0.8	1.0	-	0.7	0.8	0.9	1.1	1.3
SS B2	80	95	99	100	100	100	100	100	42	64	76	85	90	95	97	99	0.4	0.6	0.8	0.9	-	0.6	0.7	0.9	1.0	1.2
SS B3	69	89	95	99	99	100	100	100	33	52	63	74	80	88	91	95	0.5	0.6	0.8	1.0	-	0.7	0.8	0.9	1.1	1.3
SS B4	80	95	99	100	100	100	100	100	41	63	75	85	90	95	97	99	0.4	0.6	0.8	0.9	-	0.6	0.7	0.9	1.0	1.2
SS B5	64	84	92	97	99	100	100	100	31	47	57	68	75	83	87	92	0.5	0.6	0.8	1.0	1.2	0.7	0.8	1.0	1.2	1.3
SS B6	80	95	99	100	100	100	100	100	41	63	75	85	90	95	97	99	0.4	0.6	0.8	0.9	-	0.6	0.7	0.9	1.0	1.2
SS B7	60	81	90	95	98	99	100	100	28	43	53	64	70	79	84	89	0.5	0.6	0.8	1.0	1.2	0.8	0.9	1.0	1.2	1.4
SS B8	79	95	98	100	100	100	100	100	40	62	74	84	89	94	97	98	0.5	0.6	0.8	0.9	-	0.6	0.7	0.9	1.1	1.2
SS C1	75	93	97	99	100	100	100	100	34	55	67	78	85	91	94	97	0.5	0.6	0.8	1.0	-	0.7	0.7	0.9	1.1	1.3
SS C2	80	96	99	100	100	100	100	100	39	62	74	85	90	95	97	99	0.4	0.6	0.8	0.9	-	0.6	0.7	0.9	1.0	1.2
SS C3	63	84	92	97	99	100	100	100	28	44	55	66	73	82	87	91	0.5	0.6	0.8	1.0	1.2	0.8	0.8	1.0	1.1	1.3
SS C4	74	92	97	99	100	100	100	100	34	56	68	79	85	92	95	97	0.5	0.6	0.8	1.0	-	0.6	0.7	0.9	1.1	1.2
SS D1	86	98	99	100	100	100	100	100	48	70	81	89	93	97	98	99	0.4	0.6	0.7	-	-	0.6	0.7	0.8	1.0	1.2
SS D2	92	99	100	100	100	100	100	100	55	79	88	95	97	99	100	100	0.4	0.5	0.7	-	-	0.5	0.6	0.8	1.0	1.2
SS D3	80	95	99	100	100	100	100	100	43	64	75	84	89	94	97	98	0.5	0.6	0.8	0.9	-	0.6	0.7	0.9	1.1	1.2
SS D4	90	99	100	100	100	100	100	100	53	77	87	94	97	99	99	100	0.4	0.5	0.7	-	-	0.5	0.6	0.8	1.0	1.2
SS E1	82	96	99	100	100	100	100	100	43	65	76	85	91	95	97	99	0.4	0.6	0.7	-	-	0.6	0.7	0.9	1.0	1.2
SS E2	88	98	100	100	100	100	100	100	49	73	84	92	95	98	99	100	0.4	0.5	0.7	-	-	0.5	0.7	0.8	1.0	1.2
SS E3	72	90	96	99	100	100	100	100	35	54	65	76	82	89	93	96	0.5	0.6	0.8	1.0	-	0.7	0.8	0.9	1.1	1.3
SS E4	84	97	99	100	100	100	100	100	45	68	80	89	93	97	98	99	0.4	0.6	0.7	-	-	0.6	0.7	0.9	1.0	1.2

JACKSONVILLE, FLORIDA

TYPE	SOLAR SAVINGS FRACTION (%) BASE TEMP: 55 F (429 DD)								65 F (1327 DD)								CONSERVATION FACTOR 55 F					65 F				
LCR	200	100	70	50	40	30	25	20	200	100	70	50	40	30	25	20	200	100	50	30	20	200	100	50	30	20
WW A1	44	63	73	83	88	94	96	98	24	35	43	52	59	68	73	80	0.6	0.7	0.9	1.1	1.2	0.9	1.0	1.1	1.3	1.4
WW A2	53	75	85	93	96	98	99	100	27	43	54	65	73	81	86	91	0.6	0.7	0.8	1.0	1.2	0.8	0.8	1.0	1.2	1.3
WW A3	56	79	89	95	97	99	100	100	28	47	58	70	77	85	90	94	0.5	0.6	0.8	1.0	1.2	0.7	0.8	1.0	1.1	1.3
WW A4	58	81	90	96	98	99	100	100	28	48	60	72	79	87	91	95	0.5	0.6	0.8	1.0	1.2	0.7	0.8	0.9	1.1	1.3
WW A5	60	83	92	97	99	100	100	100	29	50	63	74	81	89	93	96	0.5	0.6	0.8	1.0	1.2	0.7	0.8	0.9	1.1	1.3
WW A6	61	84	92	97	99	100	100	100	29	51	64	76	82	90	93	96	0.5	0.6	0.8	1.0	1.2	0.7	0.8	0.9	1.1	1.3
WW B1	48	69	80	89	93	97	98	99	23	38	47	58	65	74	80	85	0.6	0.7	0.8	1.0	1.2	0.8	0.9	1.1	1.2	1.4
WW B2	58	81	90	96	98	99	100	100	28	48	61	72	80	88	92	95	0.5	0.6	0.8	1.0	1.2	0.7	0.8	0.9	1.1	1.3
WW B3	65	87	94	98	99	100	100	100	33	56	69	80	86	93	95	98	0.5	0.6	0.8	1.0	1.2	0.6	0.7	0.9	1.1	1.2
WW B4	65	87	94	98	99	100	100	100	32	56	69	81	87	93	96	98	0.5	0.6	0.8	1.0	1.2	0.6	0.7	0.9	1.1	1.2
WW B5	62	85	93	98	99	100	100	100	30	54	67	79	86	92	95	98	0.5	0.6	0.8	1.0	1.2	0.6	0.7	0.9	1.1	1.2
WW C1	67	88	95	98	99	100	100	100	35	58	71	82	88	93	96	98	0.5	0.6	0.8	1.0	-	0.6	0.7	0.9	1.1	1.2
WW C2	64	86	94	98	99	100	100	100	32	55	68	80	86	92	95	98	0.5	0.6	0.8	1.0	1.2	0.6	0.7	0.9	1.1	1.2
WW C3	70	91	97	99	100	100	100	100	36	62	75	86	91	96	98	99	0.5	0.6	0.8	1.0	-	0.6	0.7	0.9	1.0	1.2
WW C4	66	88	95	99	99	100	100	100	33	58	72	83	89	94	97	99	0.5	0.6	0.8	1.0	-	0.6	0.7	0.9	1.0	1.2
TW A1	42	61	72	81	87	93	96	98	23	34	42	51	58	67	72	79	0.6	0.7	0.9	1.1	1.2	0.9	1.0	1.1	1.3	1.5
TW A2	48	70	80	89	93	97	98	99	25	39	49	60	67	76	81	87	0.6	0.7	0.8	1.0	1.2	0.8	0.9	1.0	1.2	1.4
TW A3	51	73	83	91	95	98	99	100	25	41	52	63	70	80	85	90	0.6	0.7	0.8	1.0	1.2	0.8	0.9	1.0	1.2	1.3
TW A4	50	73	84	92	95	98	99	100	24	41	52	64	71	80	85	91	0.6	0.7	0.8	1.0	1.2	0.8	0.8	1.0	1.2	1.3
TW B1	42	62	73	82	88	94	96	98	23	34	43	52	59	68	73	80	0.6	0.7	0.9	1.1	1.2	0.9	1.0	1.1	1.3	1.4
TW B2	46	68	79	88	92	97	98	99	23	38	47	58	65	75	80	86	0.6	0.7	0.9	1.0	1.2	0.8	0.9	1.0	1.2	1.4
TW B3	46	68	79	88	93	97	98	99	22	37	48	59	66	76	81	87	0.6	0.7	0.8	1.0	1.2	0.8	0.9	1.0	1.2	1.4
TW B4	44	65	77	86	91	96	98	99	21	35	45	56	63	73	78	85	0.6	0.7	0.9	1.0	1.2	0.8	0.9	1.1	1.2	1.4
TW C1	41	61	72	82	88	93	96	98	21	33	42	51	58	67	73	80	0.6	0.7	0.9	1.1	1.2	0.9	1.0	1.1	1.3	1.4
TW C2	41	62	73	83	89	94	97	98	20	33	42	52	60	69	75	81	0.6	0.7	0.9	1.1	1.2	0.9	1.0	1.1	1.2	1.4
TW C3	39	59	70	81	87	93	95	98	19	32	40	50	57	66	72	79	0.6	0.7	0.9	1.1	1.2	0.9	1.0	1.1	1.3	1.4
TW C4	35	53	64	74	81	88	92	96	18	28	36	44	51	60	66	73	0.7	0.8	0.9	1.1	1.3	1.0	1.1	1.2	1.3	1.5
TW D1	41	62	73	83	89	94	97	98	19	32	41	51	58	67	73	80	0.6	0.7	0.9	1.0	1.2	0.9	1.0	1.1	1.3	1.4
TW D2	53	76	86	93	96	99	99	100	26	44	55	67	74	83	88	92	0.5	0.7	0.8	1.0	1.2	0.7	0.8	1.0	1.1	1.3
TW D3	58	80	90	96	98	99	100	100	29	48	61	72	79	87	91	95	0.5	0.6	0.8	1.0	1.2	0.7	0.8	0.9	1.1	1.3
TW D4	59	82	91	96	98	100	100	100	29	51	63	75	82	89	93	96	0.5	0.6	0.8	1.0	1.2	0.7	0.8	0.9	1.1	1.3
TW D5	58	81	90	96	98	99	100	100	28	49	62	74	81	89	93	96	0.5	0.6	0.8	1.0	1.2	0.7	0.8	0.9	1.1	1.3
TW E1	62	84	92	97	99	100	100	100	32	53	65	76	83	90	94	97	0.5	0.6	0.8	1.0	1.2	0.7	0.8	0.9	1.1	1.3
TW E2	61	83	92	97	99	100	100	100	31	52	64	76	83	90	93	96	0.5	0.6	0.8	1.0	1.2	0.7	0.8	0.9	1.1	1.3
TW E3	68	89	96	99	100	100	100	100	36	60	73	83	89	95	97	99	0.5	0.6	0.8	1.0	-	0.6	0.7	0.9	1.0	1.2
TW E4	64	86	94	98	99	100	100	100	32	56	69	81	87	93	96	98	0.5	0.6	0.8	1.0	1.2	0.6	0.7	0.9	1.1	1.2
TW F1	42	61	72	82	88	93	96	98	22	34	42	51	58	67	72	79	0.6	0.7	0.9	1.1	1.2	0.9	1.0	1.1	1.3	1.4
TW F2	46	68	79	88	92	97	98	99	22	37	47	57	65	74	80	86	0.6	0.7	0.9	1.0	1.2	0.8	0.9	1.1	1.2	1.4
TW F3	47	69	80	89	94	97	99	99	21	37	48	59	67	76	82	88	0.6	0.7	0.8	1.0	1.2	0.8	0.9	1.0	1.2	1.4
TW F4	44	67	79	88	93	97	98	99	19	35	46	57	65	75	80	87	0.6	0.7	0.8	1.0	1.2	0.8	0.9	1.0	1.2	1.4
TW G1	40	60	71	81	87	93	96	98	20	32	40	50	57	66	72	79	0.6	0.7	0.9	1.1	1.2	0.9	1.0	1.1	1.3	1.4
TW G2	42	63	75	85	90	95	97	99	19	33	43	53	61	71	76	83	0.6	0.7	0.9	1.0	1.2	0.9	0.9	1.1	1.2	1.4
TW G3	40	62	74	84	89	95	97	99	17	31	41	52	59	69	75	82	0.6	0.7	0.9	1.0	1.2	0.9	0.9	1.1	1.2	1.4
TW G4	34	56	68	79	85	92	95	97	14	26	35	46	53	63	69	77	0.6	0.7	0.9	1.1	1.2	0.9	1.0	1.1	1.3	1.4
TW H1	36	57	68	79	85	92	95	97	16	28	37	46	53	63	69	76	0.6	0.8	0.9	1.1	1.2	0.9	1.0	1.2	1.3	1.5
TW H2	33	54	66	77	83	90	94	97	14	26	34	44	51	61	67	75	0.7	0.8	0.9	1.1	1.3	0.9	1.0	1.2	1.3	1.5
TW H3	29	49	60	72	79	87	91	95	11	22	30	39	46	56	62	70	0.7	0.8	0.9	1.1	1.3	1.0	1.1	1.2	1.4	1.5
TW H4	23	40	51	62	69	78	84	89	8	17	24	31	37	46	52	59	0.8	0.9	1.0	1.2	1.3	1.1	1.2	1.3	1.5	1.6
TW I1	38	59	70	80	86	93	95	98	17	29	37	47	54	63	69	76	0.6	0.7	0.9	1.1	1.2	0.9	1.0	1.1	1.3	1.5
TW I2	49	73	84	92	95	98	99	100	23	40	52	63	71	80	85	91	0.6	0.7	0.8	1.0	1.2	0.8	0.8	1.0	1.2	1.3
TW I3	53	76	87	94	97	99	100	100	25	44	56	68	75	84	88	93	0.5	0.6	0.8	1.0	1.2	0.7	0.8	1.0	1.1	1.3
TW I4	56	79	89	95	98	99	100	100	26	47	60	72	79	87	91	95	0.5	0.6	0.8	1.0	1.2	0.7	0.8	0.9	1.1	1.3
TW I5	55	79	89	95	98	99	100	100	25	47	60	72	79	88	92	95	0.5	0.6	0.8	1.0	1.2	0.7	0.8	0.9	1.1	1.3
TW J1	59	82	91	96	98	100	100	100	29	50	62	74	81	89	92	96	0.5	0.6	0.8	1.0	1.2	0.7	0.8	0.9	1.1	1.3
TW J2	57	81	90	96	98	99	100	100	27	48	61	73	80	88	92	96	0.5	0.6	0.8	1.0	1.2	0.7	0.8	0.9	1.1	1.3
TW J3	64	87	94	98	99	100	100	100	32	56	69	81	87	93	96	98	0.5	0.6	0.8	1.0	1.2	0.6	0.7	0.9	1.1	1.2
TW J4	61	84	93	97	99	100	100	100	29	53	66	78	85	92	95	97	0.5	0.6	0.8	1.0	1.2	0.6	0.7	0.9	1.1	1.2
DG A1	48	70	80	88	92	96	98	99	21	36	46	55	62	69	74	79	0.6	0.7	0.9	1.0	1.2	0.8	0.9	1.1	1.3	1.6
DG A2	48	70	81	90	94	97	99	100	21	38	49	59	67	75	80	85	0.6	0.7	0.8	1.0	1.2	0.8	0.9	1.0	1.2	1.4
DG A3	52	76	86	92	95	96	97	97	24	43	56	68	75	83	87	91	0.5	0.6	0.9	1.3	4.5	0.7	0.8	1.0	1.2	1.4
DG B1	51	77	87	93	96	98	98	99	21	38	50	62	70	78	82	86	0.5	0.6	0.8	1.1	1.5	0.8	0.8	1.0	1.2	1.5
DG B2	50	77	88	94	97	99	99	100	22	40	52	66	73	82	87	91	0.5	0.6	0.8	1.0	1.2	0.8	0.8	1.0	1.2	1.4
DG B3	53	81	90	95	97	98	98	98	24	44	59	72	80	88	91	94	0.5	0.6	0.8	1.3	-	0.7	0.8	0.9	1.1	1.3
DG C1	57	84	92	96	98	98	98	98	24	44	58	71	79	86	89	92	0.5	0.6	0.8	1.4	-	0.7	0.8	0.9	1.2	1.4
DG C2	56	83	92	97	99	100	100	100	24	45	59	73	80	88	92	95	0.5	0.6	0.8	1.0	1.3	0.7	0.8	0.9	1.1	1.3
DG C3	59	87	94	97	98	98	98	98	27	50	65	79	86	92	95	97	0.5	0.6	0.8	2.0	-	0.7	0.7	0.9	1.1	1.4
SS A1	55	76	86	93	96	99	99	100	30	47	58	68	75	83	87	92	0.6	0.7	0.8	1.0	1.2	0.7	0.8	1.0	1.2	1.3
SS A2	64	85	93	97	99	100	100	100	36	57	69	79	85	92	94	97	0.5	0.6	0.8	1.0	1.2	0.6	0.7	0.9	1.1	1.3
SS A3	53	74	84	91	95	98	99	100	29	45	55	65	72	80	85	90	0.6	0.7	0.8	1.0	1.2	0.8	0.9	1.0	1.2	1.3
SS A4	64	85	93	97	99	100	100	100	36	57	68	79	85	91	94	97	0.5	0.6	0.8	1.0	1.2	0.6	0.7	0.9	1.1	1.3
SS A5	51	72	82	90	94	97	99	99	29	44	53	63	70	78	83	88	0.6	0.7	0.8	1.0	1.2	0.8	0.9	1.0	1.2	1.4
SS A6	64	85	93	97	99	100	100	100	36	57	69	79	85	91	94	97	0.5	0.6	0.8	1.0	1.2	0.6	0.7	0.9	1.1	1.3
SS A7	47	66	77	86	90	95	97	99	27	39	48	57	64	73	78	84	0.6	0.7	0.9	1.0	1.2	0.9	0.9	1.1	1.2	1.4
SS A8	64	84	92	97	99	100	100	100	36	56	68	78	84	91	94	97	0.5	0.6	0.8	1.0	1.2	0.7	0.8	0.9	1.1	1.3
SS B1	49	70	81	89	93	97	99	99	26	41	51	62	69	77	82	88	0.6	0.7	0.8	1.0	1.2	0.8	0.9	1.0	1.2	1.4
SS B2	58	80	89	95	98	99	100	100	31	51	62	73	80	88	91	95	0.5	0.6	0.8	1.0	1.2	0.7	0.8	0.9	1.1	1.3
SS B3	48	69	79	88	93	97	98	99	25	39	49	60	67	75	81	86	0.6	0.7	0.9	1.0	1.2	0.8	0.9	1.0	1.2	1.4
SS B4	58	79	89	95	98	99	100	100	31	50	62	73	80	87	91	95	0.5	0.6	0.8	1.0	1.2	0.7	0.8	0.9	1.1	1.3
SS B5	44	63	74	83	89	94	96	98	24	36	45	54	61	70	75	82	0.6	0.7	0.9	1.1	1.2	0.9	1.0	1.1	1.3	1.4
SS B6	58	80	89	95	98	99	100	100	31	50	62	73	80	87	91	95	0.5	0.6	0.8	1.0	1.2	0.7	0.8	0.9	1.1	1.3
SS B7	41	59	70	80	86	92	95	97	22	33	41	50	57	65	71	77	0.6	0.8	0.9	1.1	1.2	0.9	1.0	1.2	1.3	1.5
SS B8	57	79	88	95	97	99	100	100	30	49	60	71	78	86	90	94	0.5	0.6	0.8	1.0	1.2	0.7	0.8	1.0	1.1	1.3
SS C1	52	74	85	92	96	98	99	100	25	42	53	64	72	81	86	91	0.6	0.7	0.8	1.0	1.2	0.8	0.8	1.0	1.2	1.3
SS C2	58	80	90	96	98	99	100	100	28	48	60	72	79	87	91	95	0.5	0.6	0.8	1.0	1.2	0.7	0.8	0.9	1.1	1.3
SS C3	42	63	74	83	89	94	97	98	21	34	43	52	59	69	75	81	0.6	0.7	0.9	1.0	1.2	0.9	1.0	1.1	1.3	1.4
SS C4	52	74	85	92	96	98	99	100	25	43	54	66	73	82	87	92	0.5	0.7	0.8	1.0	1.2	0.7	0.8	1.0	1.1	1.3
SS D1	64	85	93	97	99	100	100	100	36	56	68	78	84	91	94	97	0.5	0.6	0.8	1.0	1.2	0.7	0.8	0.9	1.1	1.3
SS D2	72	91	97	99	100	100	100	100	42	65	77	87	91	96	98	99	0.5	0.6	0.8	1.0	-	0.6	0.7	0.9	1.0	1.2
SS D3	58	79	88	95	97	99	100	100	33	50	61	72	78	86	90	94	0.5	0.6	0.8	1.0	1.2	0.7	0.8	1.0	1.1	1.3
SS D4	70	90	96	99	100	100	100	100	41	63	75	85	90	95	97	99	0.5	0.6	0.8	1.0	-	0.6	0.7	0.9	1.0	1.2
SS E1	59	81	90	96	98	99	100	100	32	50	62	73	80	87	91	95	0.5	0.6	0.8	1.0	1.2	0.7	0.8	0.9	1.1	1.3
SS E2	67	88	95	98	99	100	100	100	37	59	71	82	88	93	96	98	0.5	0.6	0.8	1.0	1.2	0.6	0.7	0.9	1.1	1.2
SS E3	50	71	81	90	94	97	99	99	27	42	52	62	69	78	83	88	0.6	0.7	0.8	1.0	1.2	0.8	0.9	1.0	1.2	1.4
SS E4	63	84	92	97	99	100	100	100	34	55	67	77	84	91	94	97	0.5	0.6	0.8	1.0	1.2	0.7	0.8	0.9	1.1	1.3

MIAMI, FLORIDA

TYPE	SOLAR SAVINGS FRACTION (%) BASE TEMP: 55 F (13 DD)								65 F (206 DD)								CONSERVATION FACTOR 55 F					65 F				
LCR	200	100	70	50	40	30	25	20	200	100	70	50	40	30	25	20	200	100	50	30	20	200	100	50	30	20
WW A1	100	100	100	100	100	100	100	100	66	85	93	97	99	100	100	100	0.4	-	-	-	-	0.5	0.6	0.8	1.0	1.2
WW A2	100	100	100	100	100	100	100	100	78	94	98	99	100	100	100	100	0.4	-	-	-	-	0.5	0.6	0.8	1.0	-
WW A3	100	100	100	100	100	100	100	100	82	96	99	100	100	100	100	100	0.4	-	-	-	-	0.4	0.6	0.8	0.9	-
WW A4	100	100	100	100	100	100	100	100	84	97	99	100	100	100	100	100	0.4	-	-	-	-	0.4	0.6	0.8	0.9	-
WW A5	100	100	100	100	100	100	100	100	86	97	99	100	100	100	100	100	-	-	-	-	-	0.4	0.6	0.7	-	-
WW A6	100	100	100	100	100	100	100	100	86	98	99	100	100	100	100	100	-	-	-	-	-	0.4	0.6	0.7	-	-
WW B1	100	100	100	100	100	100	100	100	73	91	96	99	100	100	100	100	0.4	-	-	-	-	0.5	0.6	0.8	1.0	-
WW B2	100	100	100	100	100	100	100	100	83	97	99	100	100	100	100	100	0.4	-	-	-	-	0.4	0.6	0.8	0.9	-
WW B3	100	100	100	100	100	100	100	100	89	99	100	100	100	100	100	100	-	-	-	-	-	0.4	0.5	0.7	-	-
WW B4	100	100	100	100	100	100	100	100	89	98	100	100	100	100	100	100	-	-	-	-	-	0.4	0.6	0.7	-	-
WW B5	100	100	100	100	100	100	100	100	87	98	100	100	100	100	100	100	-	-	-	-	-	0.4	0.6	0.7	-	-
WW C1	100	100	100	100	100	100	100	100	90	99	100	100	100	100	100	100	-	-	-	-	-	0.4	0.5	0.7	-	-
WW C2	100	100	100	100	100	100	100	100	88	98	100	100	100	100	100	100	-	-	-	-	-	0.4	0.6	0.7	-	-
WW C3	100	100	100	100	100	100	100	100	92	99	100	100	100	100	100	100	-	-	-	-	-	0.4	0.5	0.7	-	-
WW C4	100	100	100	100	100	100	100	100	90	99	100	100	100	100	100	100	-	-	-	-	-	0.4	0.5	0.7	-	-
TW A1	100	100	100	100	100	100	100	100	64	84	92	97	98	100	100	100	0.4	-	-	-	-	0.5	0.6	0.8	1.0	1.2
TW A2	100	100	100	100	100	100	100	100	73	91	96	99	100	100	100	100	0.4	-	-	-	-	0.5	0.6	0.8	1.0	-
TW A3	100	100	100	100	100	100	100	100	76	93	97	99	100	100	100	100	0.4	-	-	-	-	0.5	0.6	0.8	1.0	-
TW A4	100	100	100	100	100	100	100	100	76	93	98	99	100	100	100	100	0.4	-	-	-	-	0.5	0.6	0.8	1.0	-
TW B1	100	100	100	100	100	100	100	100	65	85	93	97	99	100	100	100	0.4	-	-	-	-	0.5	0.6	0.8	1.0	1.2
TW B2	100	100	100	100	100	100	100	100	71	90	96	99	99	100	100	100	0.4	-	-	-	-	0.5	0.6	0.8	1.0	1.2
TW B3	100	100	100	100	100	100	100	100	71	90	96	99	100	100	100	100	0.4	-	-	-	-	0.5	0.6	0.8	1.0	-
TW B4	100	100	100	100	100	100	100	100	68	88	95	98	99	100	100	100	0.4	-	-	-	-	0.5	0.6	0.8	1.0	1.2
TW C1	100	100	100	100	100	100	100	100	64	84	92	97	98	100	100	100	0.4	-	-	-	-	0.5	0.6	0.8	1.0	1.2
TW C2	100	100	100	100	100	100	100	100	65	86	93	97	99	100	100	100	0.4	-	-	-	-	0.5	0.6	0.8	1.0	1.2
TW C3	100	100	100	100	100	100	100	100	62	83	91	96	98	99	100	100	0.4	0.5	-	-	-	0.5	0.6	0.8	1.0	1.2
TW C4	99	100	100	100	100	100	100	100	56	77	87	93	96	99	99	100	0.4	0.5	-	-	-	0.5	0.7	0.8	1.0	1.2
TW D1	100	100	100	100	100	100	100	100	66	86	94	98	99	100	100	100	0.4	-	-	-	-	0.5	0.6	0.8	1.0	1.2
TW D2	100	100	100	100	100	100	100	100	78	94	98	100	100	100	100	100	0.4	-	-	-	-	0.5	0.6	0.8	1.0	-
TW D3	100	100	100	100	100	100	100	100	83	96	99	100	100	100	100	100	0.4	-	-	-	-	0.4	0.6	0.8	0.9	-
TW D4	100	100	100	100	100	100	100	100	84	97	99	100	100	100	100	100	-	-	-	-	-	0.4	0.6	0.7	-	-
TW D5	100	100	100	100	100	100	100	100	83	97	99	100	100	100	100	100	0.4	-	-	-	-	0.4	0.6	0.8	0.9	-
TW E1	100	100	100	100	100	100	100	100	86	98	99	100	100	100	100	100	-	-	-	-	-	0.4	0.6	0.7	-	-
TW E2	100	100	100	100	100	100	100	100	85	97	99	100	100	100	100	100	-	-	-	-	-	0.4	0.6	0.7	-	-
TW E3	100	100	100	100	100	100	100	100	90	99	100	100	100	100	100	100	-	-	-	-	-	0.4	0.5	0.7	-	-
TW E4	100	100	100	100	100	100	100	100	88	98	100	100	100	100	100	100	-	-	-	-	-	0.4	0.6	0.7	-	-
TW F1	100	100	100	100	100	100	100	100	65	85	92	97	99	100	100	100	0.4	-	-	-	-	0.5	0.6	0.8	1.0	1.2
TW F2	100	100	100	100	100	100	100	100	71	90	96	99	99	100	100	100	0.4	-	-	-	-	0.5	0.6	0.8	1.0	1.2
TW F3	100	100	100	100	100	100	100	100	72	91	96	99	100	100	100	100	0.4	-	-	-	-	0.5	0.6	0.8	1.0	-
TW F4	100	100	100	100	100	100	100	100	71	90	96	99	100	100	100	100	0.4	-	-	-	-	0.5	0.6	0.8	1.0	-
TW G1	100	100	100	100	100	100	100	100	63	84	92	97	98	100	100	100	0.4	-	-	-	-	0.5	0.6	0.8	1.0	1.2
TW G2	100	100	100	100	100	100	100	100	67	87	94	98	99	100	100	100	0.4	-	-	-	-	0.5	0.6	0.8	1.0	1.2
TW G3	100	100	100	100	100	100	100	100	65	86	94	98	99	100	100	100	0.4	-	-	-	-	0.5	0.6	0.8	1.0	1.2
TW G4	100	100	100	100	100	100	100	100	59	81	90	96	98	99	100	100	0.4	0.5	-	-	-	0.5	0.6	0.8	1.0	1.2
TW H1	100	100	100	100	100	100	100	100	60	82	90	96	98	99	100	100	0.4	0.5	-	-	-	0.5	0.6	0.8	1.0	1.2
TW H2	100	100	100	100	100	100	100	100	57	80	89	95	97	99	100	100	0.4	0.5	-	-	-	0.5	0.6	0.8	1.0	1.2
TW H3	99	100	100	100	100	100	100	100	52	75	85	92	96	98	99	100	0.4	0.5	-	-	-	0.5	0.7	0.8	1.0	1.2
TW H4	97	100	100	100	100	100	100	100	42	65	76	86	91	96	98	99	0.4	0.5	-	-	-	0.6	0.7	0.9	1.0	1.2
TW I1	100	100	100	100	100	100	100	100	63	84	92	97	98	100	100	100	0.4	-	-	-	-	0.5	0.6	0.8	1.0	1.2
TW I2	100	100	100	100	100	100	100	100	75	93	97	99	100	100	100	100	0.4	-	-	-	-	0.5	0.6	0.8	1.0	-
TW I3	100	100	100	100	100	100	100	100	79	95	98	100	100	100	100	100	0.4	-	-	-	-	0.5	0.6	0.8	0.9	-
TW I4	100	100	100	100	100	100	100	100	82	96	99	100	100	100	100	100	0.4	-	-	-	-	0.4	0.6	0.8	0.9	-
TW I5	100	100	100	100	100	100	100	100	81	96	99	100	100	100	100	100	0.4	-	-	-	-	0.4	0.6	0.8	0.9	-
TW J1	100	100	100	100	100	100	100	100	84	97	99	100	100	100	100	100	-	-	-	-	-	0.4	0.6	0.7	-	-
TW J2	100	100	100	100	100	100	100	100	83	96	99	100	100	100	100	100	0.4	-	-	-	-	0.4	0.6	0.8	0.9	-
TW J3	100	100	100	100	100	100	100	100	88	98	100	100	100	100	100	100	-	-	-	-	-	0.4	0.6	0.7	-	-
TW J4	100	100	100	100	100	100	100	100	86	98	99	100	100	100	100	100	-	-	-	-	-	0.4	0.6	0.7	-	-
DG A1	100	100	100	100	100	100	100	100	75	92	97	99	100	100	100	100	-	-	-	-	-	0.5	0.6	0.8	-	-
DG A2	100	100	100	100	100	100	100	100	74	92	97	99	100	100	100	100	-	-	-	-	-	0.5	0.6	0.8	-	-
DG A3	97	97	97	97	97	97	97	97	79	93	96	97	97	97	97	97	-	-	-	-	-	0.5	0.6	1.4	-	-
DG B1	99	99	99	99	99	99	99	99	81	95	98	99	99	99	99	99	-	-	-	-	-	0.4	0.6	0.9	-	-
DG B2	100	100	100	100	100	100	100	100	80	96	99	100	100	100	100	100	0.4	-	-	-	-	0.4	0.6	0.8	1.0	-
DG B3	98	98	98	98	98	98	98	98	83	96	98	98	98	98	98	98	-	-	-	-	-	0.4	0.6	1.5	-	-
DG C1	99	99	99	99	99	99	98	98	88	97	98	99	99	99	98	98	-	-	-	-	-	0.4	0.6	3.2	-	-
DG C2	100	100	100	100	100	100	100	100	86	98	99	100	100	100	100	100	-	-	-	-	-	0.4	0.6	0.8	-	-
DG C3	99	99	99	99	99	98	98	98	88	98	98	99	99	98	98	98	-	-	-	-	-	0.4	0.6	3.7	-	-
SS A1	100	100	100	100	100	100	100	100	81	95	99	100	100	100	100	100	0.4	-	-	-	-	0.5	0.6	0.8	0.9	-
SS A2	100	100	100	100	100	100	100	100	89	98	100	100	100	100	100	100	-	-	-	-	-	0.4	0.5	0.7	-	-
SS A3	100	100	100	100	100	100	100	100	79	95	98	100	100	100	100	100	0.4	-	-	-	-	0.5	0.6	0.8	0.9	-
SS A4	100	100	100	100	100	100	100	100	89	99	100	100	100	100	100	100	-	-	-	-	-	0.4	0.5	0.7	-	-
SS A5	100	100	100	100	100	100	100	100	77	93	98	99	100	100	100	100	0.4	-	-	-	-	0.5	0.6	0.8	1.0	-
SS A6	100	100	100	100	100	100	100	100	89	99	100	100	100	100	100	100	-	-	-	-	-	0.4	0.5	0.7	-	-
SS A7	100	100	100	100	100	100	100	100	72	90	96	99	100	100	100	100	0.4	-	-	-	-	0.5	0.6	0.8	1.0	-
SS A8	100	100	100	100	100	100	100	100	89	98	100	100	100	100	100	100	-	-	-	-	-	0.4	0.5	0.7	-	-
SS B1	100	100	100	100	100	100	100	100	75	92	97	99	100	100	100	100	0.4	-	-	-	-	0.5	0.6	0.8	1.0	-
SS B2	100	100	100	100	100	100	100	100	84	97	99	100	100	100	100	100	-	-	-	-	-	0.4	0.6	0.7	-	-
SS B3	100	100	100	100	100	100	100	100	74	92	97	99	100	100	100	100	0.4	-	-	-	-	0.5	0.6	0.8	1.0	-
SS B4	100	100	100	100	100	100	100	100	84	97	99	100	100	100	100	100	-	-	-	-	-	0.4	0.6	0.7	-	-
SS B5	100	100	100	100	100	100	100	100	68	87	94	98	99	100	100	100	0.4	-	-	-	-	0.5	0.6	0.8	1.0	1.2
SS B6	100	100	100	100	100	100	100	100	84	97	99	100	100	100	100	100	-	-	-	-	-	0.4	0.6	0.7	-	-
SS B7	100	100	100	100	100	100	100	100	64	85	92	97	99	100	100	100	0.4	-	-	-	-	0.5	0.6	0.8	1.0	1.2
SS B8	100	100	100	100	100	100	100	100	83	96	99	100	100	100	100	100	-	-	-	-	-	0.4	0.6	0.7	-	-
SS C1	100	100	100	100	100	100	100	100	78	94	98	99	100	100	100	100	0.4	-	-	-	-	0.5	0.6	0.8	1.0	-
SS C2	100	100	100	100	100	100	100	100	83	96	99	100	100	100	100	100	0.4	-	-	-	-	0.4	0.6	0.8	0.9	-
SS C3	100	100	100	100	100	100	100	100	66	86	93	97	99	100	100	100	0.4	-	-	-	-	0.5	0.6	0.8	1.0	1.2
SS C4	100	100	100	100	100	100	100	100	77	94	98	99	100	100	100	100	0.4	-	-	-	-	0.5	0.6	0.8	1.0	-
SS D1	100	100	100	100	100	100	100	100	89	99	100	100	100	100	100	100	-	-	-	-	-	0.4	0.5	0.7	-	-
SS D2	100	100	100	100	100	100	100	100	94	100	100	100	100	100	100	100	-	-	-	-	-	0.4	0.5	0.7	-	-
SS D3	100	100	100	100	100	100	100	100	84	97	99	100	100	100	100	100	-	-	-	-	-	0.4	0.6	0.7	-	-
SS D4	100	100	100	100	100	100	100	100	93	99	100	100	100	100	100	100	-	-	-	-	-	0.4	0.5	0.7	-	-
SS E1	100	100	100	100	100	100	100	100	85	97	99	100	100	100	100	100	-	-	-	-	-	0.4	0.6	0.7	-	-
SS E2	100	100	100	100	100	100	100	100	91	99	100	100	100	100	100	100	-	-	-	-	-	0.4	0.5	0.7	-	-
SS E3	100	100	100	100	100	100	100	100	76	93	97	99	100	100	100	100	0.4	-	-	-	-	0.5	0.6	0.8	1.0	-
SS E4	100	100	100	100	100	100	100	100	88	98	100	100	100	100	100	100	-	-	-	-	-	0.4	0.6	0.7	-	-

ORLANDO, FLORIDA

TYPE	SOLAR SAVINGS FRACTION (%)																CONSERVATION FACTOR									
	BASE TEMP: 55 F (151 DD)								65 F (733 DD)								55 F					65 F				
LCR	200	100	70	50	40	30	25	20	200	100	70	50	40	30	25	20	200	100	50	30	20	200	100	50	30	20
WW A1	74	91	96	99	100	100	100	100	35	52	63	73	79	87	91	95	0.5	0.6	0.8	1.0	-	0.7	0.8	1.0	1.1	1.3
WW A2	85	97	99	100	100	100	100	100	42	64	76	85	90	95	97	99	0.4	0.6	0.7	-	-	0.6	0.7	0.9	1.0	1.2
WW A3	88	98	100	100	100	100	100	100	45	68	80	89	93	97	98	99	0.4	0.6	0.7	-	-	0.6	0.7	0.8	1.0	1.2
WW A4	90	99	100	100	100	100	100	100	46	70	82	90	94	98	99	100	0.4	0.5	0.7	-	-	0.6	0.7	0.8	1.0	1.2
WW A5	91	99	100	100	100	100	100	100	48	73	84	92	95	98	99	100	0.4	0.5	0.7	-	-	0.5	0.7	0.8	1.0	1.2
WW A6	92	99	100	100	100	100	100	100	49	74	85	92	96	98	99	100	0.4	0.5	0.7	-	-	0.5	0.7	0.8	1.0	1.2
WW B1	81	95	98	100	100	100	100	100	37	58	70	80	86	92	95	97	0.5	0.6	0.8	0.9	-	0.6	0.7	0.9	1.1	1.2
WW B2	90	99	100	100	100	100	100	100	46	70	82	90	94	98	99	100	0.4	0.6	0.7	-	-	0.6	0.7	0.8	1.0	1.2
WW B3	94	99	100	100	100	100	100	100	54	78	88	95	97	99	100	100	0.4	0.5	0.7	-	-	0.5	0.6	0.8	1.0	1.2
WW B4	94	99	100	100	100	100	100	100	52	78	88	95	97	99	100	100	0.4	0.5	0.7	-	-	0.5	0.6	0.8	1.0	1.2
WW B5	93	99	100	100	100	100	100	100	50	76	86	94	97	99	100	100	0.4	0.5	0.7	-	-	0.5	0.6	0.8	1.0	1.2
WW C1	94	100	100	100	100	100	100	100	55	79	89	95	98	99	100	100	0.4	0.5	0.7	-	-	0.5	0.6	0.8	1.0	1.2
WW C2	93	99	100	100	100	100	100	100	52	77	87	94	97	99	100	100	0.4	0.5	0.7	-	-	0.5	0.6	0.8	1.0	1.2
WW C3	96	100	100	100	100	100	100	100	58	83	92	97	99	100	100	100	0.4	0.5	-	-	-	0.5	0.6	0.8	1.0	1.2
WW C4	94	100	100	100	100	100	100	100	54	79	89	95	98	99	100	100	0.4	0.5	0.7	-	-	0.5	0.6	0.8	1.0	1.2
TW A1	73	90	96	99	99	100	100	100	34	51	61	72	78	86	90	94	0.5	0.6	0.8	1.0	1.2	0.7	0.8	1.0	1.1	1.3
TW A2	81	95	98	100	100	100	100	100	38	59	70	80	86	92	95	98	0.5	0.6	0.8	0.9	-	0.6	0.7	0.9	1.1	1.2
TW A3	83	96	99	100	100	100	100	100	40	62	73	83	89	94	97	98	0.4	0.6	0.8	0.9	-	0.6	0.7	0.9	1.1	1.2
TW A4	84	97	99	100	100	100	100	100	39	62	74	84	90	95	97	99	0.4	0.6	0.8	0.9	-	0.6	0.7	0.9	1.0	1.2
TW B1	73	91	96	99	100	100	100	100	33	51	62	73	79	87	91	94	0.5	0.6	0.8	1.0	-	0.7	0.8	1.0	1.1	1.3
TW B2	79	94	98	100	100	100	100	100	36	57	68	79	85	92	95	97	0.5	0.6	0.8	1.0	-	0.6	0.7	0.9	1.1	1.3
TW B3	80	95	98	100	100	100	100	100	36	57	69	80	86	92	95	97	0.5	0.6	0.8	1.0	-	0.6	0.7	0.9	1.1	1.2
TW B4	77	93	98	99	100	100	100	100	33	54	66	77	83	90	94	97	0.5	0.6	0.8	1.0	-	0.7	0.8	0.9	1.1	1.3
TW C1	72	90	96	99	99	100	100	100	32	50	61	72	79	86	90	94	0.5	0.6	0.8	1.0	1.2	0.7	0.8	1.0	1.1	1.3
TW C2	73	91	96	99	100	100	100	100	32	51	62	73	80	88	92	95	0.5	0.6	0.8	1.0	-	0.7	0.8	0.9	1.1	1.3
TW C3	71	89	95	98	99	100	100	100	30	48	59	70	77	86	90	94	0.5	0.6	0.8	1.0	1.2	0.7	0.8	1.0	1.1	1.3
TW C4	64	84	92	97	98	100	100	100	27	43	53	64	71	80	85	90	0.5	0.6	0.8	1.0	1.2	0.8	0.9	1.0	1.2	1.3
TW D1	74	92	97	99	100	100	100	100	31	51	62	73	80	87	91	95	0.5	0.6	0.8	1.0	-	0.7	0.8	0.9	1.1	1.3
TW D2	86	97	99	100	100	100	100	100	42	65	77	86	91	96	98	99	0.4	0.6	0.7	-	-	0.6	0.7	0.9	1.0	1.2
TW D3	89	98	100	100	100	100	100	100	46	70	81	90	94	98	99	100	0.4	0.6	0.7	-	-	0.6	0.7	0.8	1.0	1.2
TW D4	90	99	100	100	100	100	100	100	47	72	83	91	95	98	99	100	0.4	0.5	0.7	-	-	0.6	0.7	0.8	1.0	1.2
TW D5	89	99	100	100	100	100	100	100	46	71	82	91	95	98	99	100	0.4	0.6	0.7	-	-	0.6	0.7	0.8	1.0	1.2
TW E1	92	99	100	100	100	100	100	100	50	74	85	92	96	98	99	100	0.4	0.5	0.7	-	-	0.5	0.7	0.8	1.0	1.2
TW E2	91	99	100	100	100	100	100	100	49	73	84	92	96	98	99	100	0.4	0.5	0.7	-	-	0.5	0.7	0.8	1.0	1.2
TW E3	95	100	100	100	100	100	100	100	56	81	90	96	98	99	100	100	0.4	0.5	0.7	-	-	0.5	0.6	0.8	1.0	1.2
TW E4	93	99	100	100	100	100	100	100	52	77	88	94	97	99	100	100	0.4	0.5	0.7	-	-	0.5	0.6	0.8	1.0	1.2
TW F1	73	91	96	99	99	100	100	100	33	50	61	72	79	86	90	94	0.5	0.6	0.8	1.0	1.2	0.7	0.8	1.0	1.1	1.3
TW F2	79	94	98	100	100	100	100	100	35	56	68	79	85	91	95	97	0.5	0.6	0.8	1.0	-	0.6	0.7	0.9	1.1	1.3
TW F3	81	95	98	100	100	100	100	100	36	58	70	80	87	93	95	98	0.4	0.6	0.8	0.9	-	0.6	0.7	0.9	1.1	1.2
TW F4	79	95	98	100	100	100	100	100	33	55	68	79	85	92	95	97	0.5	0.6	0.8	0.9	-	0.6	0.7	0.9	1.1	1.2
TW G1	72	90	96	99	99	100	100	100	31	49	60	71	78	86	90	94	0.5	0.6	0.8	1.0	1.2	0.7	0.8	1.0	1.1	1.3
TW G2	75	93	97	99	100	100	100	100	31	52	64	75	82	89	93	96	0.5	0.6	0.8	1.0	-	0.7	0.8	0.9	1.1	1.3
TW G3	74	92	97	99	100	100	100	100	29	50	62	74	81	88	92	96	0.5	0.6	0.8	1.0	-	0.7	0.8	0.9	1.1	1.3
TW G4	68	88	95	98	99	100	100	100	25	44	56	68	75	84	88	93	0.5	0.6	0.8	1.0	1.2	0.7	0.8	1.0	1.1	1.3
TW H1	69	88	95	98	99	100	100	100	27	45	57	68	75	84	88	93	0.5	0.6	0.8	1.0	1.2	0.7	0.8	1.0	1.1	1.3
TW H2	66	87	94	98	99	100	100	100	24	42	54	66	73	82	87	92	0.5	0.6	0.8	1.0	1.2	0.7	0.8	1.0	1.1	1.3
TW H3	61	83	91	96	98	99	100	100	20	37	48	60	68	77	83	89	0.5	0.6	0.8	1.0	1.2	0.8	0.9	1.0	1.2	1.3
TW H4	51	74	84	92	95	98	99	100	15	29	39	50	58	68	74	81	0.6	0.7	0.8	1.0	1.2	0.9	1.0	1.1	1.2	1.4
TW I1	71	90	96	99	99	100	100	100	28	47	58	70	77	85	89	94	0.5	0.6	0.8	1.0	1.2	0.7	0.8	1.0	1.1	1.3
TW I2	83	96	99	100	100	100	100	100	38	61	73	84	89	95	97	99	0.4	0.6	0.8	0.9	-	0.6	0.7	0.9	1.0	1.2
TW I3	86	98	99	100	100	100	100	100	41	65	77	87	92	96	98	99	0.4	0.6	0.7	-	-	0.6	0.7	0.9	1.0	1.2
TW I4	88	98	100	100	100	100	100	100	44	69	80	89	94	97	99	100	0.4	0.6	0.7	-	-	0.6	0.7	0.8	1.0	1.2
TW I5	88	98	100	100	100	100	100	100	43	68	81	90	94	97	99	100	0.4	0.6	0.7	-	-	0.6	0.7	0.8	1.0	1.2
TW J1	90	99	100	100	100	100	100	100	47	72	83	91	95	98	99	100	0.4	0.5	0.7	-	-	0.6	0.7	0.8	1.0	1.2
TW J2	89	99	100	100	100	100	100	100	45	70	82	90	94	98	99	100	0.4	0.6	0.7	-	-	0.6	0.7	0.8	1.0	1.2
TW J3	93	99	100	100	100	100	100	100	52	77	88	95	97	99	100	100	0.4	0.5	0.7	-	-	0.5	0.6	0.8	1.0	1.2
TW J4	92	99	100	100	100	100	100	100	49	74	86	93	96	99	99	100	0.4	0.5	0.7	-	-	0.5	0.6	0.8	1.0	1.2
DG A1	83	96	99	100	100	100	100	100	37	58	69	78	84	90	93	95	0.4	0.6	0.7	-	-	0.6	0.7	0.9	1.1	1.3
DG A2	82	96	99	100	100	100	100	100	36	59	71	81	86	92	95	97	0.4	0.6	0.7	-	-	0.6	0.7	0.9	1.1	1.3
DG A3	86	95	97	97	97	97	97	97	40	66	77	86	90	94	95	96	0.4	0.7	2.9	-	-	0.6	0.7	0.9	1.2	1.8
DG B1	88	98	99	99	99	99	99	99	38	64	76	85	90	94	96	97	0.4	0.6	1.2	-	-	0.6	0.7	0.9	1.1	1.4
DG B2	88	98	100	100	100	100	100	100	38	64	77	87	92	96	98	99	0.4	0.6	0.7	-	-	0.6	0.7	0.9	1.0	1.2
DG B3	90	97	98	98	98	98	98	98	41	69	82	90	94	97	97	98	0.4	0.6	4.8	-	-	0.6	0.7	0.8	1.1	1.8
DG C1	93	98	99	99	99	99	98	98	44	72	84	91	94	97	97	98	0.4	0.7	-	-	-	0.6	0.6	0.9	1.1	1.9
DG C2	92	99	100	100	100	100	100	100	42	71	83	92	95	98	99	100	0.4	0.5	0.9	-	-	0.6	0.6	0.8	1.0	1.2
DG C3	93	98	99	99	99	98	98	98	46	76	88	94	96	98	98	98	0.4	0.7	-	-	-	0.5	0.6	0.8	1.2	3.2
SS A1	87	98	99	100	100	100	100	100	46	67	78	87	92	96	98	99	0.4	0.6	0.7	-	-	0.6	0.7	0.9	1.0	1.2
SS A2	93	99	100	100	100	100	100	100	55	78	87	94	97	99	99	100	0.4	0.5	0.7	-	-	0.5	0.6	0.8	1.0	1.2
SS A3	85	97	99	100	100	100	100	100	44	65	76	85	90	95	97	99	0.4	0.6	0.7	-	-	0.6	0.7	0.9	1.0	1.2
SS A4	93	99	100	100	100	100	100	100	55	77	87	94	97	99	99	100	0.4	0.5	0.7	-	-	0.5	0.6	0.8	1.0	1.2
SS A5	83	96	99	100	100	100	100	100	43	63	74	83	88	94	96	98	0.4	0.6	0.8	0.9	-	0.6	0.7	0.9	1.1	1.2
SS A6	93	99	100	100	100	100	100	100	55	78	87	94	97	99	100	100	0.4	0.5	0.7	-	-	0.5	0.6	0.8	1.0	1.2
SS A7	78	94	93	99	100	100	100	100	39	58	68	78	84	90	94	97	0.5	0.6	0.8	1.0	-	0.7	0.8	0.9	1.1	1.3
SS A8	93	99	100	100	100	100	100	100	55	77	87	93	96	99	99	100	0.4	0.5	0.7	-	-	0.5	0.7	0.8	1.0	1.2
SS B1	82	96	99	100	100	100	100	100	40	61	72	82	88	93	96	98	0.4	0.6	0.8	0.9	-	0.6	0.7	0.9	1.1	1.2
SS B2	89	99	100	100	100	100	100	100	48	71	82	90	94	98	99	100	0.4	0.5	0.7	-	-	0.6	0.7	0.8	1.0	1.2
SS B3	81	95	99	100	100	100	100	100	39	59	71	81	86	92	95	98	0.5	0.6	0.8	0.9	-	0.6	0.7	0.9	1.1	1.2
SS B4	89	99	100	100	100	100	100	100	48	71	82	90	94	97	99	100	0.4	0.5	0.7	-	-	0.6	0.7	0.8	1.0	1.2
SS B5	75	92	97	99	100	100	100	100	35	54	65	75	81	89	92	95	0.5	0.6	0.8	1.0	-	0.7	0.8	0.9	1.1	1.3
SS B6	89	99	100	100	100	100	100	100	48	71	82	90	94	98	99	100	0.4	0.5	0.7	-	-	0.6	0.7	0.8	1.0	1.2
SS B7	72	90	96	99	99	100	100	100	33	50	61	71	78	85	89	94	0.5	0.6	0.8	1.0	-	0.7	0.8	1.0	1.1	1.3
SS B8	89	98	100	100	100	100	100	100	47	70	81	89	94	97	99	99	0.4	0.6	0.7	-	-	0.6	0.7	0.8	1.0	1.2
SS C1	85	97	99	100	100	100	100	100	41	63	75	85	90	95	97	99	0.4	0.6	0.7	-	-	0.6	0.7	0.9	1.0	1.2
SS C2	89	98	100	100	100	100	100	100	46	70	81	90	94	98	99	100	0.4	0.6	0.7	-	-	0.6	0.7	0.8	1.0	1.2
SS C3	74	92	97	99	100	100	100	100	32	51	63	74	80	88	92	95	0.5	0.6	0.8	1.0	-	0.7	0.8	0.9	1.1	1.3
SS C4	85	97	99	100	100	100	100	100	41	63	75	85	90	95	97	99	0.4	0.6	0.8	-	-	0.6	0.7	0.9	1.0	1.2
SS D1	94	99	100	100	100	100	100	100	55	77	87	94	97	99	99	100	0.4	0.5	0.7	-	-	0.5	0.6	0.8	1.0	1.2
SS D2	97	100	100	100	100	100	100	100	63	85	93	97	99	100	100	100	0.4	0.5	-	-	-	0.5	0.6	0.8	1.0	1.2
SS D3	89	99	100	100	100	100	100	100	49	71	82	90	94	97	99	99	0.4	0.6	0.7	-	-	0.6	0.7	0.8	1.0	1.2
SS D4	96	100	100	100	100	100	100	100	61	83	92	97	98	100	100	100	0.4	0.5	-	-	-	0.5	0.6	0.8	1.0	1.2
SS E1	91	99	100	100	100	100	100	100	50	72	83	91	95	98	99	100	0.4	0.5	0.7	-	-	0.6	0.7	0.8	1.0	1.2
SS E2	95	100	100	100	100	100	100	100	57	80	90	95	98	99	100	100	0.4	0.5	0.7	-	-	0.5	0.6	0.8	1.0	1.2
SS E3	82	96	99	100	100	100	100	100	41	62	73	82	88	93	96	98	0.4	0.6	0.8	0.9	-	0.6	0.7	0.9	1.1	1.2
SS E4	93	99	100	100	100	100	100	100	53	76	86	93	96	99	99	100	0.4	0.5	0.7	-	-	0.5	0.7	0.8	1.0	1.2

TALLAHASSEE, FLORIDA

TYPE	SOLAR SAVINGS FRACTION (%) BASE TEMP: 55 F (550 DD) LCR 200	100	70	50	40	30	25	20	65 F (1563 DD) 200	100	70	50	40	30	25	20	CONSERVATION FACTOR 55 F 200	100	50	30	20	65 F 200	100	50	30	20
WW A1	37	53	64	74	80	87	91	95	22	32	39	47	53	61	67	73	0.7	0.8	1.0	1.1	1.3	1.0	1.1	1.2	1.4	1.5
WW A2	44	65	76	85	91	95	97	99	23	38	48	59	66	75	80	86	0.6	0.7	0.9	1.0	1.2	0.8	0.9	1.0	1.2	1.4
WW A3	47	69	80	89	93	97	98	99	24	41	52	63	70	79	84	90	0.6	0.7	0.8	1.0	1.2	0.8	0.9	1.0	1.2	1.3
WW A4	48	71	82	90	94	98	99	100	24	42	53	65	72	81	86	91	0.6	0.7	0.8	1.0	1.2	0.7	0.8	1.0	1.1	1.3
WW A5	50	73	84	92	96	98	99	100	25	44	56	67	75	84	88	93	0.6	0.7	0.8	1.0	1.2	0.7	0.8	1.0	1.1	1.3
WW A6	51	74	85	93	96	99	99	100	25	44	57	69	76	85	89	93	0.5	0.7	0.8	1.0	1.2	0.7	0.8	1.0	1.1	1.3
WW B1	39	59	70	80	86	92	95	98	20	33	41	51	58	67	72	79	0.6	0.7	0.9	1.1	1.2	0.9	1.0	1.1	1.3	1.4
WW B2	48	71	82	91	95	98	99	100	24	42	54	66	73	82	87	92	0.6	0.7	0.8	1.0	1.2	0.7	0.8	1.0	1.1	1.3
WW B3	55	78	88	95	97	99	100	100	29	50	62	74	81	88	92	96	0.5	0.6	0.8	1.0	1.2	0.7	0.8	0.9	1.1	1.3
WW B4	54	78	88	95	97	99	100	100	27	49	62	74	82	89	93	96	0.5	0.6	0.8	1.0	1.2	0.7	0.8	0.9	1.1	1.3
WW B5	52	76	87	94	97	99	100	100	25	47	61	73	80	88	92	96	0.5	0.6	0.8	1.0	1.2	0.7	0.8	0.9	1.1	1.3
WW C1	56	79	89	95	98	99	100	100	30	51	64	75	82	90	93	96	0.5	0.6	0.8	1.0	1.2	0.7	0.8	0.9	1.1	1.3
WW C2	53	77	87	94	97	99	100	100	27	49	61	73	81	88	92	96	0.5	0.6	0.8	1.0	1.2	0.6	0.7	0.9	1.1	1.2
WW C3	59	83	92	97	99	100	100	100	31	56	69	81	87	93	96	98	0.5	0.6	0.8	1.0	1.2	0.6	0.7	0.9	1.1	1.2
WW C4	55	80	90	96	98	99	100	100	28	51	65	77	84	91	94	97	0.5	0.6	0.8	1.0	1.2	0.6	0.7	0.9	1.1	1.2
TW A1	36	52	62	72	79	86	90	94	21	31	38	46	52	60	65	72	0.7	0.8	1.0	1.1	1.3	1.0	1.1	1.2	1.4	1.5
TW A2	40	60	71	81	87	93	95	98	22	35	44	53	60	70	75	82	0.6	0.7	0.9	1.1	1.2	0.9	1.0	1.1	1.3	1.4
TW A3	42	63	74	84	89	95	97	99	22	36	46	56	64	73	79	85	0.6	0.7	0.9	1.0	1.2	0.8	0.9	1.1	1.2	1.4
TW A4	41	63	74	84	90	95	97	99	21	36	46	57	64	74	79	86	0.6	0.7	0.9	1.0	1.2	0.8	0.9	1.0	1.2	1.4
TW B1	35	53	63	73	80	87	91	95	20	31	38	46	53	61	67	74	0.7	0.8	1.0	1.1	1.3	1.0	1.1	1.2	1.3	1.5
TW B2	38	58	69	79	85	92	95	97	20	33	42	52	59	68	74	80	0.6	0.8	0.9	1.1	1.2	0.9	1.0	1.1	1.3	1.4
TW B3	38	58	70	80	86	92	95	98	19	33	42	52	59	69	75	81	0.6	0.7	0.9	1.1	1.2	0.9	1.0	1.1	1.3	1.4
TW B4	35	55	67	77	84	90	94	97	18	31	40	49	57	66	72	79	0.7	0.8	0.9	1.1	1.3	0.9	1.0	1.1	1.3	1.4
TW C1	34	52	62	73	79	87	91	95	19	30	37	46	52	61	67	74	0.7	0.8	1.0	1.1	1.3	1.0	1.1	1.2	1.3	1.5
TW C2	34	52	63	74	80	88	92	95	18	29	37	47	53	62	68	75	0.7	0.8	0.9	1.1	1.3	0.9	1.0	1.2	1.3	1.5
TW C3	32	50	61	71	78	86	90	94	17	28	36	44	51	60	66	73	0.7	0.8	1.0	1.1	1.3	1.0	1.1	1.2	1.3	1.5
TW C4	29	44	55	65	72	80	85	90	16	25	32	40	45	54	59	66	0.8	0.9	1.0	1.2	1.3	1.1	1.1	1.3	1.4	1.6
TW D1	33	52	63	73	80	88	92	95	17	28	35	44	51	60	66	73	0.7	0.8	0.9	1.1	1.3	1.0	1.1	1.2	1.3	1.5
TW D2	44	66	77	87	91	96	98	99	23	39	49	60	68	77	82	88	0.6	0.7	0.9	1.0	1.2	0.8	0.9	1.0	1.2	1.4
TW D3	48	70	82	90	94	98	99	100	25	43	54	65	73	82	87	91	0.6	0.7	0.8	1.0	1.2	0.7	0.8	1.0	1.1	1.3
TW D4	49	73	84	92	95	98	99	100	25	45	57	69	76	85	89	93	0.6	0.7	0.8	1.0	1.2	0.7	0.8	1.0	1.1	1.3
TW D5	47	71	82	91	95	98	99	100	24	43	56	68	75	84	89	93	0.6	0.7	0.8	1.0	1.2	0.7	0.8	1.0	1.1	1.3
TW E1	52	74	85	93	96	99	99	100	28	47	59	70	77	85	90	94	0.5	0.7	0.8	1.0	1.2	0.7	0.8	1.0	1.1	1.3
TW E2	50	74	84	92	96	98	99	100	26	46	58	69	77	85	90	94	0.6	0.7	0.8	1.0	1.2	0.7	0.8	1.0	1.1	1.3
TW E3	57	81	90	96	98	99	100	100	31	53	66	78	84	91	94	97	0.5	0.6	0.8	1.0	1.2	0.6	0.7	0.9	1.1	1.2
TW E4	54	78	88	95	97	99	100	100	28	50	63	75	82	89	93	96	0.5	0.6	0.8	1.0	1.2	0.7	0.8	0.9	1.1	1.3
TW F1	35	52	62	72	79	87	91	94	20	30	37	45	51	60	66	73	0.7	0.8	1.0	1.1	1.3	1.0	1.1	1.2	1.4	1.5
TW F2	37	57	69	79	85	92	95	97	19	32	41	51	58	67	73	80	0.6	0.8	0.9	1.1	1.2	0.9	1.0	1.1	1.3	1.4
TW F3	38	59	70	81	87	93	96	98	18	32	42	52	60	69	75	82	0.6	0.7	0.9	1.1	1.2	0.9	0.9	1.1	1.2	1.4
TW F4	35	57	69	79	86	92	95	98	16	30	40	50	58	68	74	81	0.6	0.7	0.9	1.1	1.2	0.9	1.0	1.1	1.2	1.4
TW G1	33	51	61	72	78	86	90	94	17	28	36	44	51	59	65	72	0.7	0.8	1.0	1.1	1.3	1.0	1.1	1.2	1.4	1.5
TW G2	34	53	65	76	82	89	93	96	16	28	37	47	54	64	70	77	0.7	0.8	0.9	1.1	1.3	0.9	1.0	1.1	1.3	1.5
TW G3	31	51	63	74	81	89	92	96	14	27	35	45	52	62	68	76	0.7	0.8	0.9	1.1	1.3	0.9	1.0	1.2	1.3	1.5
TW G4	27	46	57	68	76	84	89	93	11	22	31	40	47	56	62	70	0.7	0.8	1.0	1.1	1.3	1.0	1.1	1.2	1.4	1.5
TW H1	29	47	58	69	76	84	89	93	14	25	32	41	47	56	62	70	0.7	0.8	1.0	1.1	1.3	1.0	1.1	1.2	1.4	1.5
TW H2	26	44	55	67	74	82	87	92	11	22	30	38	45	54	60	68	0.7	0.8	1.0	1.1	1.3	1.0	1.1	1.2	1.4	1.5
TW H3	22	39	50	61	68	78	83	89	9	19	26	34	40	49	55	63	0.8	0.9	1.0	1.2	1.3	1.1	1.2	1.3	1.4	1.6
TW H4	17	32	41	51	59	68	74	81	7	14	20	27	32	40	46	53	0.8	0.9	1.1	1.3	1.4	1.2	1.3	1.4	1.6	1.7
TW I1	30	48	59	70	77	85	90	94	14	25	32	41	47	56	62	69	0.7	0.8	1.0	1.1	1.3	1.0	1.1	1.2	1.4	1.5
TW I2	40	62	74	84	90	95	97	99	19	35	46	57	64	74	80	86	0.6	0.7	0.9	1.0	1.2	0.8	0.9	1.0	1.2	1.4
TW I3	43	66	78	87	92	96	98	99	21	38	49	61	69	78	83	89	0.6	0.7	0.9	1.0	1.2	0.8	0.9	1.0	1.2	1.3
TW I4	45	69	81	90	94	98	99	100	22	41	53	65	73	82	87	92	0.6	0.7	0.8	1.0	1.2	0.7	0.8	1.0	1.1	1.3
TW I5	45	69	81	90	94	98	99	100	21	41	53	66	73	83	87	92	0.6	0.7	0.8	1.0	1.2	0.7	0.8	1.0	1.1	1.3
TW J1	49	72	83	91	95	98	99	100	25	44	56	68	75	84	88	93	0.6	0.7	0.8	1.0	1.2	0.7	0.8	1.0	1.1	1.3
TW J2	47	71	82	91	95	98	99	100	23	42	54	66	74	83	88	92	0.6	0.7	0.8	1.0	1.2	0.7	0.8	1.0	1.1	1.3
TW J3	53	78	88	95	97	99	100	100	27	50	63	75	82	90	93	96	0.5	0.6	0.8	1.0	1.2	0.7	0.8	0.9	1.1	1.3
TW J4	50	75	86	93	97	99	99	100	25	46	60	72	79	88	92	95	0.5	0.6	0.8	1.0	1.2	0.7	0.8	0.9	1.1	1.3
DG A1	39	59	69	79	84	90	93	95	18	31	40	48	54	62	66	71	0.6	0.8	0.9	1.1	1.3	0.9	1.0	1.2	1.4	1.7
DG A2	38	60	71	81	87	92	95	97	18	33	42	53	60	68	74	79	0.6	0.7	0.9	1.1	1.3	0.9	0.9	1.1	1.3	1.5
DG A3	42	66	78	87	91	94	95	96	21	37	49	62	69	78	83	87	0.6	0.7	0.9	1.2	1.8	0.8	0.9	1.0	1.2	1.4
DG B1	40	65	77	86	90	94	96	97	18	32	43	54	61	70	75	80	0.6	0.7	0.9	1.1	1.4	0.9	0.9	1.1	1.3	1.5
DG B2	40	65	78	87	92	96	98	99	19	34	45	58	66	75	81	86	0.6	0.7	0.9	1.0	1.2	0.9	0.9	1.0	1.2	1.4
DG B3	42	70	82	91	94	97	98	98	21	38	51	65	73	83	87	92	0.6	0.7	0.8	1.1	1.9	0.8	0.8	0.9	1.1	1.4
DG C1	45	73	84	92	95	97	98	98	21	38	49	62	71	79	83	88	0.6	0.6	0.8	1.1	1.9	0.8	0.9	1.0	1.2	1.5
DG C2	44	71	84	92	96	98	99	100	21	38	51	65	73	82	87	91	0.6	0.6	0.8	1.0	1.2	0.8	0.8	1.0	1.1	1.3
DG C3	47	76	88	94	97	98	98	98	24	43	57	72	80	88	92	95	0.5	0.6	0.8	1.2	3.6	0.8	0.8	0.9	1.1	1.3
SS A1	46	66	77	86	91	96	98	99	27	42	52	62	69	77	82	88	0.6	0.7	0.9	1.0	1.2	0.8	0.9	1.0	1.2	1.4
SS A2	55	76	86	93	96	99	99	100	32	51	63	73	80	87	91	95	0.5	0.7	0.8	1.0	1.2	0.7	0.8	0.9	1.1	1.3
SS A3	44	64	75	84	89	95	97	98	26	40	49	59	66	74	79	85	0.6	0.7	0.9	1.1	1.2	0.8	0.9	1.1	1.2	1.4
SS A4	54	76	86	93	96	99	99	100	32	51	62	73	80	87	91	95	0.5	0.7	0.8	1.0	1.2	0.7	0.8	0.9	1.1	1.3
SS A5	43	62	73	82	88	93	96	98	26	39	48	57	64	72	77	83	0.6	0.7	0.9	1.1	1.2	0.8	0.9	1.1	1.3	1.4
SS A6	55	77	86	93	96	99	99	100	32	51	63	73	80	87	91	95	0.5	0.7	0.8	1.0	1.2	0.7	0.8	0.9	1.1	1.3
SS A7	39	57	67	77	83	90	93	96	24	35	43	52	58	66	71	78	0.7	0.8	0.9	1.1	1.3	0.9	1.0	1.2	1.3	1.5
SS A8	54	76	86	93	96	99	99	100	32	50	61	72	79	86	90	94	0.6	0.7	0.8	1.0	1.2	0.7	0.8	1.0	1.1	1.3
SS B1	41	60	71	81	87	93	96	98	23	36	46	55	62	71	77	83	0.6	0.7	0.9	1.1	1.2	0.8	0.9	1.1	1.2	1.4
SS B2	49	71	81	90	94	97	99	100	28	45	56	67	74	83	87	92	0.6	0.7	0.8	1.0	1.2	0.7	0.8	1.0	1.1	1.3
SS B3	39	59	70	80	86	92	95	97	22	35	44	53	60	69	74	81	0.6	0.8	0.9	1.1	1.2	0.9	1.0	1.1	1.3	1.4
SS B4	48	70	81	89	94	97	99	99	27	45	56	67	74	82	87	91	0.6	0.7	0.8	1.0	1.2	0.7	0.8	1.0	1.2	1.3
SS B5	36	54	64	74	81	88	92	95	22	32	40	49	55	63	69	76	0.7	0.8	0.9	1.1	1.3	0.9	1.0	1.2	1.3	1.5
SS B6	48	70	81	90	94	97	99	100	27	45	56	67	74	82	87	92	0.6	0.7	0.8	1.0	1.2	0.7	0.8	1.0	1.2	1.3
SS B7	34	50	60	70	77	85	89	93	20	30	37	45	51	59	64	71	0.7	0.8	1.0	1.1	1.3	1.0	1.1	1.2	1.4	1.5
SS B8	47	69	80	89	93	97	98	99	27	43	54	65	72	81	86	91	0.6	0.7	0.8	1.0	1.2	0.7	0.8	1.0	1.2	1.3
SS C1	43	64	75	85	90	95	97	99	21	37	47	57	65	74	80	86	0.6	0.7	0.9	1.0	1.2	0.8	0.9	1.0	1.2	1.4
SS C2	48	70	82	90	94	98	99	100	24	42	54	66	73	82	87	92	0.6	0.7	0.8	1.0	1.2	0.7	0.8	1.0	1.1	1.3
SS C3	34	53	64	74	81	88	92	95	18	30	37	46	53	62	68	75	0.7	0.8	0.9	1.1	1.3	0.9	1.0	1.2	1.3	1.5
SS C4	43	64	76	85	91	96	97	99	22	38	48	59	67	76	81	87	0.6	0.7	0.9	1.0	1.2	0.8	0.9	1.0	1.2	1.4
SS D1	55	76	86	93	96	99	99	100	32	50	61	72	78	86	90	94	0.5	0.7	0.8	1.0	1.2	0.7	0.8	1.0	1.1	1.3
SS D2	62	84	92	97	99	100	100	100	37	59	71	82	87	93	96	98	0.5	0.6	0.8	1.0	1.2	0.6	0.7	0.9	1.1	1.2
SS D3	49	70	80	89	93	97	98	99	29	45	55	65	72	80	85	90	0.6	0.7	0.9	1.0	1.2	0.8	0.9	1.0	1.2	1.3
SS D4	60	82	91	96	98	100	100	100	36	57	69	80	86	92	95	97	0.5	0.6	0.8	1.0	1.2	0.6	0.7	0.9	1.1	1.3
SS E1	50	71	82	90	94	98	99	100	28	45	55	66	73	81	86	91	0.6	0.7	0.8	1.0	1.2	0.7	0.8	1.0	1.2	1.3
SS E2	57	80	89	95	98	99	100	100	32	53	65	76	83	90	93	96	0.5	0.6	0.8	1.0	1.2	0.7	0.8	0.9	1.1	1.3
SS E3	42	61	72	82	87	93	96	98	24	37	46	56	63	71	77	83	0.6	0.7	0.9	1.1	1.2	0.8	0.9	1.1	1.3	1.4
SS E4	53	75	85	93	96	99	99	100	30	49	60	71	78	86	90	94	0.5	0.7	0.8	1.0	1.2	0.7	0.8	1.0	1.1	1.3

TAMPA, FLORIDA

TYPE	SOLAR SAVINGS FRACTION (%) BASE TEMP: 55 F (151 DD) LCR 200	100	70	50	40	30	25	20	65 F (718 DD) 200	100	70	50	40	30	25	20	CONSERVATION FACTOR 55 F 200	100	50	30	20	65 F 200	100	50	30	20
WW A1	73	91	96	99	100	100	100	100	35	52	62	73	79	87	91	94	0.5	0.6	0.8	1.0	-	0.7	0.8	1.0	1.1	1.3
WW A2	85	97	99	100	100	100	100	100	42	64	75	85	90	95	97	99	0.4	0.6	0.7	-	-	0.6	0.7	0.9	1.0	1.2
WW A3	88	98	100	100	100	100	100	100	45	68	80	89	93	97	98	99	0.4	0.6	0.7	-	-	0.6	0.7	0.8	1.0	1.2
WW A4	90	99	100	100	100	100	100	100	46	70	82	90	94	98	99	100	0.4	0.5	0.7	-	-	0.6	0.7	0.8	1.0	1.2
WW A5	91	99	100	100	100	100	100	100	48	73	84	92	95	98	99	100	0.4	0.5	0.7	-	-	0.5	0.7	0.8	1.0	1.2
WW A6	92	99	100	100	100	100	100	100	49	74	85	92	96	98	99	100	0.4	0.5	0.7	-	-	0.5	0.7	0.8	1.0	1.2
WW B1	80	95	99	100	100	100	100	100	37	58	69	80	86	92	95	97	0.5	0.6	0.8	0.9	-	0.6	0.7	0.9	1.1	1.2
WW B2	89	99	100	100	100	100	100	100	46	70	82	90	94	98	99	100	0.4	0.5	0.7	-	-	0.6	0.7	0.8	1.0	1.2
WW B3	94	100	100	100	100	100	100	100	53	78	88	94	97	99	100	100	0.4	0.5	0.7	-	-	0.5	0.6	0.8	1.0	1.2
WW B4	93	99	100	100	100	100	100	100	52	77	88	95	97	99	100	100	0.4	0.5	0.7	-	-	0.5	0.6	0.8	1.0	1.2
WW B5	92	99	100	100	100	100	100	100	50	75	86	94	97	99	100	100	0.4	0.5	0.7	-	-	0.5	0.6	0.8	1.0	1.2
WW C1	94	100	100	100	100	100	100	100	55	79	89	95	98	99	100	100	0.4	0.5	0.7	-	-	0.5	0.6	0.8	1.0	1.2
WW C2	93	99	100	100	100	100	100	100	52	77	87	94	97	99	100	100	0.4	0.5	0.7	-	-	0.5	0.6	0.8	1.0	1.2
WW C3	96	100	100	100	100	100	100	100	58	83	92	97	99	100	100	100	0.4	0.5	-	-	-	0.5	0.6	0.8	1.0	1.2
WW C4	94	100	100	100	100	100	100	100	53	79	89	95	98	99	100	100	0.4	0.5	0.7	-	-	0.5	0.6	0.8	1.0	1.2
TW A1	72	90	96	99	99	100	100	100	33	50	61	71	78	86	90	94	0.5	0.6	0.8	1.0	-	0.7	0.8	1.0	1.1	1.3
TW A2	80	95	98	100	100	100	100	100	38	58	70	80	86	92	95	98	0.5	0.6	0.8	0.9	-	0.6	0.7	0.9	1.1	1.2
TW A3	83	96	99	100	100	100	100	100	39	61	73	83	89	94	97	98	0.4	0.6	0.7	0.9	-	0.6	0.7	0.9	1.1	1.2
TW A4	83	97	99	100	100	100	100	100	39	62	74	84	89	95	97	99	0.4	0.6	0.7	-	-	0.6	0.7	0.9	1.0	1.2
TW B1	72	91	96	99	100	100	100	100	33	51	62	72	79	87	91	94	0.5	0.6	0.8	1.0	-	0.7	0.8	1.0	1.1	1.3
TW B2	78	94	98	100	100	100	100	100	36	56	68	79	85	92	95	97	0.5	0.6	0.8	0.9	-	0.6	0.7	0.9	1.1	1.3
TW B3	79	95	98	100	100	100	100	100	36	57	69	79	86	92	95	97	0.5	0.6	0.8	0.9	-	0.6	0.7	0.9	1.1	1.2
TW B4	76	93	98	99	100	100	100	100	33	54	66	77	83	90	94	97	0.5	0.6	0.8	1.0	-	0.7	0.8	0.9	1.1	1.3
TW C1	72	90	96	99	100	100	100	100	32	50	61	72	79	86	90	94	0.5	0.6	0.8	1.0	-	0.7	0.8	1.0	1.1	1.3
TW C2	73	91	96	99	100	100	100	100	32	51	62	73	80	88	91	95	0.5	0.6	0.8	1.0	-	0.7	0.8	0.9	1.1	1.3
TW C3	70	89	95	98	99	100	100	100	30	48	59	70	77	85	90	94	0.5	0.6	0.8	1.0	1.2	0.7	0.8	1.0	1.1	1.3
TW C4	63	84	92	97	98	100	100	100	27	42	53	64	71	80	85	90	0.5	0.6	0.8	1.0	1.2	0.8	0.9	1.0	1.2	1.3
TW D1	74	92	97	99	100	100	100	100	31	50	62	73	80	87	91	95	0.5	0.6	0.8	1.0	-	0.7	0.8	0.9	1.1	1.3
TW D2	85	97	99	100	100	100	100	100	42	65	76	86	91	96	98	99	0.4	0.6	0.7	-	-	0.6	0.7	0.9	1.0	1.2
TW D3	89	99	100	100	100	100	100	100	46	70	81	90	94	97	99	100	0.4	0.5	0.7	-	-	0.6	0.7	0.8	1.0	1.2
TW D4	90	99	100	100	100	100	100	100	47	72	83	91	95	98	99	100	0.4	0.5	0.7	-	-	0.6	0.7	0.8	1.0	1.2
TW D5	89	99	100	100	100	100	100	100	45	70	82	91	95	98	99	100	0.4	0.5	0.7	-	-	0.6	0.7	0.8	1.0	1.2
TW E1	91	99	100	100	100	100	100	100	50	74	85	92	96	98	99	100	0.4	0.5	0.7	-	-	0.5	0.7	0.8	1.0	1.2
TW E2	91	99	100	100	100	100	100	100	49	73	84	92	95	98	99	100	0.4	0.5	0.7	-	-	0.5	0.7	0.8	1.0	1.2
TW E3	95	100	100	100	100	100	100	100	56	80	90	96	98	99	100	100	0.4	0.5	-	-	-	0.5	0.6	0.8	1.0	1.2
TW E4	93	99	100	100	100	100	100	100	52	77	88	94	97	99	100	100	0.4	0.5	0.7	-	-	0.5	0.6	0.8	1.0	1.2
TW F1	72	90	96	99	100	100	100	100	32	50	61	72	78	86	90	94	0.5	0.6	0.8	1.0	-	0.7	0.8	1.0	1.1	1.3
TW F2	78	94	98	100	100	100	100	100	35	56	68	79	85	91	94	97	0.5	0.6	0.8	0.9	-	0.6	0.7	0.9	1.1	1.3
TW F3	80	95	99	100	100	100	100	100	36	57	70	80	86	93	95	98	0.4	0.6	0.8	0.9	-	0.6	0.7	0.9	1.1	1.2
TW F4	78	95	98	100	100	100	100	100	33	55	68	79	85	92	95	97	0.5	0.6	0.8	0.9	-	0.6	0.7	0.9	1.1	1.2
TW G1	71	90	96	99	99	100	100	100	31	49	60	71	78	86	90	94	0.5	0.6	0.8	1.0	-	0.7	0.8	1.0	1.1	1.3
TW G2	75	92	97	99	100	100	100	100	31	52	64	75	82	89	93	96	0.5	0.6	0.8	1.0	-	0.7	0.8	0.9	1.1	1.3
TW G3	73	92	97	99	100	100	100	100	29	50	62	74	81	88	92	96	0.5	0.6	0.8	1.0	-	0.7	0.8	0.9	1.1	1.3
TW G4	67	88	95	98	99	100	100	100	24	44	56	67	75	84	88	93	0.5	0.6	0.8	1.0	1.2	0.7	0.8	1.0	1.1	1.3
TW H1	68	88	95	98	99	100	100	100	27	45	56	68	75	84	88	93	0.5	0.6	0.8	1.0	1.2	0.7	0.8	1.0	1.1	1.3
TW H2	65	86	94	98	99	100	100	100	24	42	54	66	73	82	87	92	0.5	0.6	0.8	1.0	1.2	0.7	0.8	1.0	1.1	1.3
TW H3	60	82	91	96	98	99	100	100	20	37	48	60	68	77	83	88	0.5	0.6	0.8	1.0	1.2	0.8	0.9	1.0	1.2	1.3
TW H4	50	73	84	91	95	98	99	100	15	29	39	50	57	67	74	81	0.6	0.7	0.8	1.0	1.2	0.9	1.0	1.1	1.2	1.4
TW I1	70	90	96	99	99	100	100	100	28	47	58	70	77	85	89	93	0.5	0.6	0.8	1.0	-	0.7	0.8	1.0	1.1	1.3
TW I2	83	96	99	100	100	100	100	100	38	61	73	84	89	95	97	99	0.4	0.6	0.7	-	-	0.6	0.7	0.9	1.0	1.2
TW I3	86	98	99	100	100	100	100	100	41	65	77	87	92	96	98	99	0.4	0.6	0.7	-	-	0.6	0.7	0.9	1.0	1.2
TW I4	88	98	100	100	100	100	100	100	43	68	80	89	94	97	99	100	0.4	0.6	0.7	-	-	0.6	0.7	0.8	1.0	1.2
TW I5	88	98	100	100	100	100	100	100	43	68	80	89	94	97	99	100	0.4	0.6	0.7	-	-	0.6	0.7	0.8	1.0	1.2
TW J1	90	99	100	100	100	100	100	100	47	72	83	91	95	98	99	100	0.4	0.5	0.7	-	-	0.6	0.7	0.8	1.0	1.2
TW J2	89	99	100	100	100	100	100	100	45	70	82	90	94	98	99	100	0.4	0.5	0.7	-	-	0.6	0.7	0.8	1.0	1.2
TW J3	93	99	100	100	100	100	100	100	52	77	88	95	97	99	100	100	0.4	0.5	0.7	-	-	0.5	0.6	0.8	1.0	1.2
TW J4	92	99	100	100	100	100	100	100	48	74	85	93	96	99	99	100	0.4	0.5	0.7	-	-	0.5	0.6	0.8	1.0	1.2
DG A1	82	96	99	100	100	100	100	100	36	57	69	78	84	90	92	95	0.4	0.6	-	-	-	0.6	0.7	0.9	1.1	1.3
DG A2	82	96	99	100	100	100	100	100	36	58	70	81	86	92	95	97	0.4	0.6	-	-	-	0.6	0.7	0.9	1.1	1.3
DG A3	86	95	97	97	97	97	97	97	40	65	77	86	90	94	95	96	0.4	0.6	3.4	-	-	0.6	0.7	0.9	1.2	1.8
DG B1	88	98	99	99	99	99	99	99	38	64	76	85	90	94	96	97	0.4	0.6	1.3	-	-	0.6	0.7	0.9	1.1	1.4
DG B2	88	98	100	100	100	100	100	100	38	64	77	87	92	96	98	99	0.4	0.6	0.7	-	-	0.6	0.7	0.9	1.0	1.2
DG B3	89	98	98	98	98	98	98	98	41	69	82	90	94	97	97	98	0.4	0.6	-	-	-	0.6	0.7	0.8	1.1	1.8
DG C1	93	98	99	99	99	99	98	98	44	72	84	91	94	97	97	98	0.4	0.7	-	-	-	0.6	0.6	0.9	1.1	1.9
DG C2	92	99	100	100	100	100	100	100	42	71	83	92	95	98	99	100	0.4	0.5	1.0	-	-	0.6	0.6	0.8	1.0	1.2
DG C3	93	98	99	99	99	98	98	98	46	76	87	94	96	98	98	98	0.4	0.7	-	-	-	0.5	0.6	0.8	1.2	3.3
SS A1	86	98	99	100	100	100	100	100	46	67	78	87	92	96	98	99	0.4	0.6	0.7	-	-	0.6	0.7	0.9	1.0	1.2
SS A2	93	99	100	100	100	100	100	100	55	78	87	94	97	99	99	100	0.4	0.5	0.7	-	-	0.5	0.6	0.8	1.0	1.2
SS A3	85	97	99	100	100	100	100	100	44	65	76	85	90	95	97	99	0.4	0.6	0.7	-	-	0.6	0.7	0.9	1.0	1.2
SS A4	93	99	100	100	100	100	100	100	55	78	87	94	97	99	99	100	0.4	0.5	0.7	-	-	0.5	0.6	0.8	1.0	1.2
SS A5	83	96	99	100	100	100	100	100	43	63	74	83	88	94	96	98	0.4	0.6	0.8	0.9	-	0.6	0.7	0.9	1.1	1.2
SS A6	93	99	100	100	100	100	100	100	55	78	87	94	97	99	99	100	0.4	0.5	0.7	-	-	0.5	0.6	0.8	1.0	1.2
SS A7	78	94	98	99	100	100	100	100	39	58	68	78	84	90	94	96	0.5	0.6	0.8	0.9	-	0.7	0.8	0.9	1.1	1.3
SS A8	93	99	100	100	100	100	100	100	55	77	87	93	96	99	99	100	0.4	0.5	0.7	-	-	0.5	0.7	0.8	1.0	1.2
SS B1	81	96	99	100	100	100	100	100	40	61	72	82	88	93	96	98	0.4	0.6	0.8	0.9	-	0.6	0.7	0.9	1.1	1.2
SS B2	89	99	100	100	100	100	100	100	48	71	82	90	94	98	99	100	0.4	0.5	0.7	-	-	0.6	0.7	0.8	1.0	1.2
SS B3	80	95	99	100	100	100	100	100	39	59	71	81	86	92	95	98	0.5	0.6	0.8	0.9	-	0.6	0.7	0.9	1.1	1.2
SS B4	89	98	100	100	100	100	100	100	48	71	82	90	94	97	99	100	0.4	0.5	0.7	-	-	0.6	0.7	0.8	1.0	1.2
SS B5	75	92	97	99	100	100	100	100	35	54	65	75	81	88	92	95	0.5	0.6	0.8	1.0	-	0.7	0.8	0.9	1.1	1.3
SS B6	89	98	100	100	100	100	100	100	48	71	82	90	94	97	99	100	0.4	0.5	0.7	-	-	0.6	0.7	0.8	1.0	1.2
SS B7	71	90	96	99	99	100	100	100	33	50	61	71	78	85	89	94	0.5	0.6	0.8	1.0	-	0.7	0.8	1.0	1.1	1.3
SS B8	88	98	100	100	100	100	100	100	47	70	81	89	94	97	99	99	0.4	0.6	0.7	-	-	0.6	0.7	0.8	1.0	1.2
SS C1	85	97	99	100	100	100	100	100	41	63	75	85	90	95	97	99	0.4	0.6	0.7	-	-	0.6	0.7	0.9	1.0	1.2
SS C2	89	99	100	100	100	100	100	100	46	70	81	90	94	98	99	100	0.4	0.5	0.7	-	-	0.6	0.7	0.8	1.0	1.2
SS C3	73	91	97	99	100	100	100	100	32	51	63	73	80	88	92	95	0.5	0.6	0.8	1.0	-	0.7	0.8	0.9	1.1	1.3
SS C4	84	97	99	100	100	100	100	100	40	63	75	85	90	95	97	99	0.4	0.6	0.7	-	-	0.6	0.7	0.9	1.0	1.2
SS D1	93	99	100	100	100	100	100	100	55	77	87	94	97	99	99	100	0.4	0.5	0.7	-	-	0.5	0.7	0.8	1.0	1.2
SS D2	97	100	100	100	100	100	100	100	63	85	93	97	99	100	100	100	0.4	0.5	-	-	-	0.5	0.6	0.8	1.0	1.2
SS D3	89	98	100	100	100	100	100	100	50	71	82	90	94	97	99	99	0.4	0.5	0.7	-	-	0.6	0.7	0.8	1.0	1.2
SS D4	96	100	100	100	100	100	100	100	61	83	92	97	98	100	100	100	0.4	0.5	-	-	-	0.5	0.6	0.8	1.0	1.2
SS E1	90	99	100	100	100	100	100	100	50	72	83	91	95	98	99	100	0.4	0.5	0.7	-	-	0.6	0.7	0.8	1.0	1.2
SS E2	95	100	100	100	100	100	100	100	57	80	90	95	98	99	100	100	0.4	0.5	-	-	-	0.5	0.6	0.8	1.0	1.2
SS E3	82	96	99	100	100	100	100	100	41	62	73	82	88	93	96	98	0.4	0.6	0.8	0.9	-	0.6	0.7	0.9	1.1	1.2
SS E4	92	99	100	100	100	100	100	100	53	76	86	93	96	99	99	100	0.4	0.5	0.7	-	-	0.5	0.7	0.8	1.0	1.2

WEST PALM BEACH, FLORIDA

TYPE	SOLAR SAVINGS FRACTION (%)																CONSERVATION FACTOR									
BASE TEMP:	55 F (27 DD)								65 F (299 DD)								55 F					65 F				
LCR	200	100	70	50	40	30	25	20	200	100	70	50	40	30	25	20	200	100	50	30	20	200	100	50	30	20
WW A1	99	100	100	100	100	100	100	100	54	75	84	92	95	98	99	100	0.4	0.5	-	-	-	0.6	0.7	0.8	1.0	1.2
WW A2	100	100	100	100	100	100	100	100	65	86	93	97	99	100	100	100	0.4	-	-	-	-	0.5	0.6	0.8	1.0	1.2
WW A3	100	100	100	100	100	100	100	100	69	89	95	99	99	100	100	100	0.4	-	-	-	-	0.5	0.6	0.8	1.0	1.2
WW A4	100	100	100	100	100	100	100	100	71	91	96	99	100	100	100	100	0.4	-	-	-	-	0.5	0.6	0.8	1.0	-
WW A5	100	100	100	100	100	100	100	100	74	92	97	99	100	100	100	100	0.4	-	-	-	-	0.5	0.6	0.8	1.0	-
WW A6	100	100	100	100	100	100	100	100	75	93	97	99	100	100	100	100	0.4	-	-	-	-	0.5	0.6	0.8	1.0	-
WW B1	100	100	100	100	100	100	100	100	60	82	90	96	98	99	100	100	0.4	0.5	-	-	-	0.5	0.6	0.8	1.0	1.2
WW B2	100	100	100	100	100	100	100	100	71	90	96	99	100	100	100	100	0.4	-	-	-	-	0.5	0.6	0.8	1.0	-
WW B3	100	100	100	100	100	100	100	100	78	95	98	100	100	100	100	100	0.4	-	-	-	-	0.5	0.6	0.8	0.9	-
WW B4	100	100	100	100	100	100	100	100	78	94	98	100	100	100	100	100	0.4	-	-	-	-	0.5	0.6	0.8	0.9	-
WW B5	100	100	100	100	100	100	100	100	75	93	98	99	100	100	100	100	0.4	-	-	-	-	0.5	0.6	0.8	0.9	-
WW C1	100	100	100	100	100	100	100	100	80	95	99	100	100	100	100	100	0.4	-	-	-	-	0.4	0.6	0.8	0.9	-
WW C2	100	100	100	100	100	100	100	100	77	94	98	100	100	100	100	100	0.4	-	-	-	-	0.5	0.6	0.8	0.9	-
WW C3	100	100	100	100	100	100	100	100	82	97	99	100	100	100	100	100	0.4	-	-	-	-	0.4	0.6	0.7	-	-
WW C4	100	100	100	100	100	100	100	100	79	95	99	100	100	100	100	100	0.4	-	-	-	-	0.4	0.6	0.8	0.9	-
TW A1	98	100	100	100	100	100	100	100	52	73	83	91	94	98	99	100	0.4	0.5	-	-	-	0.6	0.7	0.8	1.0	1.2
TW A2	100	100	100	100	100	100	100	100	60	81	90	96	98	99	100	100	0.4	0.5	-	-	-	0.5	0.6	0.8	1.0	1.2
TW A3	100	100	100	100	100	100	100	100	63	84	92	97	99	100	100	100	0.4	0.5	-	-	-	0.5	0.6	0.8	1.0	1.2
TW A4	100	100	100	100	100	100	100	100	63	85	93	97	99	100	100	100	0.4	0.5	-	-	-	0.5	0.6	0.8	1.0	1.2
TW B1	99	100	100	100	100	100	100	100	53	74	84	91	95	98	99	100	0.4	0.5	-	-	-	0.6	0.7	0.8	1.0	1.2
TW B2	99	100	100	100	100	100	100	100	58	80	89	95	97	99	100	100	0.4	0.5	-	-	-	0.5	0.6	0.8	1.0	1.2
TW B3	99	100	100	100	100	100	100	100	58	80	89	95	98	99	100	100	0.4	0.5	-	-	-	0.5	0.6	0.8	1.0	1.2
TW B4	99	100	100	100	100	100	100	100	55	78	87	94	97	99	99	100	0.4	0.5	-	-	-	0.5	0.6	0.8	1.0	1.2
TW C1	98	100	100	100	100	100	100	100	52	73	83	91	95	98	99	100	0.4	0.5	-	-	-	0.6	0.7	0.8	1.0	1.2
TW C2	99	100	100	100	100	100	100	100	52	74	84	92	95	98	99	100	0.4	0.5	-	-	-	0.6	0.7	0.8	1.0	1.2
TW C3	98	100	100	100	100	100	100	100	50	71	82	90	94	97	99	100	0.4	0.5	-	-	-	0.6	0.7	0.8	1.0	1.2
TW C4	96	100	100	100	100	100	100	100	44	65	76	85	90	95	97	99	0.4	0.5	-	-	-	0.6	0.7	0.9	1.0	1.2
TW D1	99	100	100	100	100	100	100	100	53	75	85	92	96	98	99	100	0.4	0.5	-	-	-	0.5	0.7	0.8	1.0	1.2
TW D2	100	100	100	100	100	100	100	100	66	87	94	98	99	100	100	100	0.4	-	-	-	-	0.5	0.6	0.8	1.0	1.2
TW D3	100	100	100	100	100	100	100	100	71	90	96	99	100	100	100	100	0.4	-	-	-	-	0.5	0.6	0.8	1.0	-
TW D4	100	100	100	100	100	100	100	100	72	91	97	99	100	100	100	100	0.4	-	-	-	-	0.5	0.6	0.8	1.0	-
TW D5	100	100	100	100	100	100	100	100	71	90	96	99	100	100	100	100	0.4	-	-	-	-	0.5	0.6	0.8	1.0	-
TW E1	100	100	100	100	100	100	100	100	75	93	97	99	100	100	100	100	0.4	-	-	-	-	0.5	0.6	0.8	1.0	-
TW E2	100	100	100	100	100	100	100	100	73	92	97	99	100	100	100	100	0.4	-	-	-	-	0.5	0.6	0.8	1.0	-
TW E3	100	100	100	100	100	100	100	100	80	96	99	100	100	100	100	100	0.4	-	-	-	-	0.4	0.6	0.8	0.9	-
TW E4	100	100	100	100	100	100	100	100	77	94	98	100	100	100	100	100	0.4	-	-	-	-	0.5	0.6	0.8	0.9	-
TW F1	98	100	100	100	100	100	100	100	52	73	83	91	95	98	99	100	0.4	0.5	-	-	-	0.6	0.7	0.8	1.0	1.2
TW F2	99	100	100	100	100	100	100	100	58	80	89	95	97	99	100	100	0.4	0.5	-	-	-	0.5	0.6	0.8	1.0	1.2
TW F3	100	100	100	100	100	100	100	100	59	81	90	96	98	99	100	100	0.4	0.5	-	-	-	0.5	0.6	0.8	1.0	1.2
TW F4	99	100	100	100	100	100	100	100	57	80	89	95	98	99	100	100	0.4	0.5	-	-	-	0.5	0.6	0.8	1.0	1.2
TW G1	98	100	100	100	100	100	100	100	51	73	83	91	95	98	99	100	0.4	0.5	-	-	-	0.6	0.7	0.8	1.0	1.2
TW G2	99	100	100	100	100	100	100	100	53	76	86	93	96	99	99	100	0.4	0.5	-	-	-	0.5	0.7	0.8	1.0	1.2
TW G3	99	100	100	100	100	100	100	100	52	75	85	93	96	98	99	100	0.4	0.5	-	-	-	0.5	0.7	0.8	1.0	1.2
TW G4	98	100	100	100	100	100	100	100	45	69	80	89	93	97	98	99	0.4	0.5	-	-	-	0.6	0.7	0.8	1.0	1.2
TW H1	98	100	100	100	100	100	100	100	47	69	80	89	93	97	98	99	0.4	0.5	-	-	-	0.6	0.7	0.8	1.0	1.2
TW H2	97	100	100	100	100	100	100	100	44	67	78	88	92	96	98	99	0.4	0.5	-	-	-	0.6	0.7	0.9	1.0	1.2
TW H3	96	100	100	100	100	100	100	100	39	61	73	83	89	94	97	98	0.4	0.5	-	-	-	0.6	0.7	0.9	1.0	1.2
TW H4	91	99	100	100	100	100	100	100	31	51	63	75	81	89	92	96	0.4	0.6	0.7	-	-	0.7	0.8	0.9	1.1	1.3
TW I1	98	100	100	100	100	100	100	100	49	72	82	90	94	98	99	100	0.4	0.5	-	-	-	0.6	0.7	0.8	1.0	1.2
TW I2	100	100	100	100	100	100	100	100	62	84	92	97	99	100	100	100	0.4	0.5	-	-	-	0.5	0.6	0.8	1.0	1.2
TW I3	100	100	100	100	100	100	100	100	66	87	94	98	99	100	100	100	0.4	-	-	-	-	0.5	0.6	0.8	1.0	1.2
TW I4	100	100	100	100	100	100	100	100	69	89	96	99	100	100	100	100	0.4	-	-	-	-	0.5	0.6	0.8	1.0	-
TW I5	100	100	100	100	100	100	100	100	69	89	96	99	100	100	100	100	0.4	-	-	-	-	0.5	0.6	0.8	1.0	-
TW J1	100	100	100	100	100	100	100	100	72	91	97	99	100	100	100	100	0.4	-	-	-	-	0.5	0.6	0.8	1.0	-
TW J2	100	100	100	100	100	100	100	100	70	90	96	99	100	100	100	100	0.4	-	-	-	-	0.5	0.6	0.8	1.0	-
TW J3	100	100	100	100	100	100	100	100	77	94	98	100	100	100	100	100	0.4	-	-	-	-	0.5	0.6	0.8	0.9	-
TW J4	100	100	100	100	100	100	100	100	74	93	98	99	100	100	100	100	0.4	-	-	-	-	0.5	0.6	0.8	1.0	-
DG A1	100	100	100	100	100	100	100	100	61	82	90	96	98	99	100	100	-	-	-	-	-	0.5	0.6	0.8	1.0	-
DG A2	100	100	100	100	100	100	100	100	61	83	91	96	98	100	100	100	0.4	-	-	-	-	0.5	0.6	0.8	1.0	-
DG A3	97	97	97	97	97	97	97	97	66	87	93	96	96	97	97	97	1.0	-	-	-	-	0.5	0.6	0.9	2.6	-
DG B1	99	99	99	99	99	99	99	99	67	88	94	98	98	99	99	99	0.6	-	-	-	-	0.5	0.6	0.8	1.3	-
DG B2	100	100	100	100	100	100	100	100	66	88	95	98	99	100	100	100	0.4	-	-	-	-	0.5	0.6	0.8	1.0	1.2
DG B3	98	98	98	98	98	98	98	98	69	91	96	98	98	98	98	98	1.3	-	-	-	-	0.5	0.6	0.9	3.4	-
DG C1	99	99	99	99	99	99	98	98	74	93	97	98	98	98	98	98	-	-	-	-	-	0.4	0.6	1.0	-	-
DG C2	100	100	100	100	100	100	100	100	72	93	97	99	100	100	100	100	0.5	-	-	-	-	0.5	0.6	0.8	1.1	-
DG C3	99	99	99	99	99	98	98	98	76	94	97	98	98	98	98	98	-	-	-	-	-	0.4	0.6	1.1	-	-
SS A1	100	100	100	100	100	100	100	100	69	88	95	98	99	100	100	100	0.4	-	-	-	-	0.5	0.6	0.8	1.0	1.2
SS A2	100	100	100	100	100	100	100	100	78	94	98	100	100	100	100	100	0.4	-	-	-	-	0.5	0.6	0.8	0.9	-
SS A3	100	100	100	100	100	100	100	100	67	86	94	98	99	100	100	100	0.4	-	-	-	-	0.5	0.6	0.8	1.0	1.2
SS A4	100	100	100	100	100	100	100	100	78	94	98	100	100	100	100	100	0.4	-	-	-	-	0.5	0.6	0.8	0.9	-
SS A5	100	100	100	100	100	100	100	100	64	84	92	97	99	100	100	100	0.4	0.5	-	-	-	0.5	0.6	0.8	1.0	1.2
SS A6	100	100	100	100	100	100	100	100	78	94	98	100	100	100	100	100	0.4	-	-	-	-	0.5	0.6	0.8	0.9	-
SS A7	100	100	100	100	100	100	100	100	60	80	89	95	97	99	100	100	0.4	0.5	-	-	-	0.5	0.6	0.8	1.0	1.2
SS A8	100	100	100	100	100	100	100	100	78	94	98	100	100	100	100	100	0.4	-	-	-	-	0.5	0.6	0.8	0.9	-
SS B1	100	100	100	100	100	100	100	100	62	83	91	96	98	100	100	100	0.4	0.5	-	-	-	0.5	0.6	0.8	1.0	1.2
SS B2	100	100	100	100	100	100	100	100	72	91	96	99	100	100	100	100	0.4	-	-	-	-	0.5	0.6	0.8	1.0	-
SS B3	100	100	100	100	100	100	100	100	61	82	91	96	98	99	100	100	0.4	0.5	-	-	-	0.5	0.6	0.8	1.0	1.2
SS B4	100	100	100	100	100	100	100	100	72	91	96	99	100	100	100	100	0.4	-	-	-	-	0.5	0.6	0.8	1.0	-
SS B5	99	100	100	100	100	100	100	100	56	77	86	93	96	99	99	100	0.4	0.5	-	-	-	0.5	0.7	0.8	1.0	1.2
SS B6	100	100	100	100	100	100	100	100	72	91	96	99	100	100	100	100	0.4	-	-	-	-	0.5	0.6	0.8	1.0	-
SS B7	99	100	100	100	100	100	100	100	52	73	83	91	95	98	99	100	0.4	0.5	-	-	-	0.6	0.7	0.8	1.0	1.2
SS B8	100	100	100	100	100	100	100	100	71	90	96	99	100	100	100	100	0.4	-	-	-	-	0.5	0.6	0.8	1.0	-
SS C1	100	100	100	100	100	100	100	100	65	86	93	97	99	100	100	100	0.4	-	-	-	-	0.5	0.6	0.8	1.0	1.2
SS C2	100	100	100	100	100	100	100	100	71	90	96	99	100	100	100	100	0.4	-	-	-	-	0.5	0.6	0.8	1.0	-
SS C3	99	100	100	100	100	100	100	100	53	75	85	92	96	98	99	100	0.4	0.5	-	-	-	0.6	0.7	0.8	1.0	1.2
SS C4	100	100	100	100	100	100	100	100	64	86	93	97	99	100	100	100	0.4	0.5	-	-	-	0.5	0.6	0.8	1.0	1.2
SS D1	100	100	100	100	100	100	100	100	78	94	98	100	100	100	100	100	0.4	-	-	-	-	0.5	0.6	0.8	0.9	-
SS D2	100	100	100	100	100	100	100	100	86	98	99	100	100	100	100	100	-	-	-	-	-	0.4	0.6	0.7	-	-
SS D3	100	100	100	100	100	100	100	100	72	91	96	99	100	100	100	100	0.4	-	-	-	-	0.5	0.6	0.8	1.0	-
SS D4	100	100	100	100	100	100	100	100	84	97	99	100	100	100	100	100	0.4	-	-	-	-	0.4	0.6	0.7	-	-
SS E1	100	100	100	100	100	100	100	100	74	92	97	99	100	100	100	100	0.4	-	-	-	-	0.5	0.6	0.8	1.0	-
SS E2	100	100	100	100	100	100	100	100	81	96	99	100	100	100	100	100	0.4	-	-	-	-	0.4	0.6	0.7	0.9	-
SS E3	100	100	100	100	100	100	100	100	63	84	92	97	98	100	100	100	0.4	0.5	-	-	-	0.5	0.6	0.8	1.0	1.2
SS E4	100	100	100	100	100	100	100	100	77	94	98	99	100	100	100	100	0.4	-	-	-	-	0.5	0.6	0.8	0.9	-

ATLANTA, GEORGIA

SOLAR SAVINGS FRACTION (%)

TYPE	LCR	BASE TEMP: 55 F (1461 DD)								65 F (3105 DD)							
		200	100	70	50	40	30	25	20	200	100	70	50	40	30	25	20
WW A1		20	29	36	43	49	57	62	68	15	20	24	29	32	37	41	46
WW A2		21	35	44	54	61	70	76	82	13	22	28	35	40	48	53	60
WW A3		21	37	47	58	65	75	80	86	12	22	29	37	43	52	57	65
WW A4		21	38	49	60	68	77	82	88	11	22	30	39	45	54	60	67
WW A5		22	40	51	63	70	79	84	90	11	23	31	40	47	56	62	70
WW A6		22	40	52	64	71	80	85	91	10	23	32	41	48	58	64	71
WW B1		18	29	37	46	53	61	67	74	11	17	22	27	32	38	42	47
WW B2		21	39	50	61	69	78	83	89	11	23	31	40	47	56	62	70
WW B3		25	45	57	69	76	85	89	93	14	28	38	48	55	65	71	78
WW B4		24	45	58	70	77	86	90	94	12	28	38	49	57	67	74	81
WW B5		22	43	56	69	76	85	89	94	11	27	37	48	56	67	73	80
WW C1		27	47	59	71	78	86	90	94	15	30	40	50	58	67	73	80
WW C2		24	45	57	69	76	85	89	94	13	28	38	48	56	66	72	79
WW C3		27	51	65	77	83	91	94	97	14	33	45	57	65	76	81	87
WW C4		24	47	61	73	80	88	92	96	12	30	41	53	62	72	78	85
TW A1		20	28	35	42	48	56	61	67	15	20	24	28	32	37	41	46
TW A2		20	32	40	49	56	65	71	77	13	20	26	32	37	44	49	55
TW A3		19	33	42	52	59	68	74	81	12	20	26	33	39	47	52	59
TW A4		18	32	42	52	60	69	75	82	10	19	26	33	39	47	53	60
TW B1		19	28	35	43	49	57	62	69	13	19	23	28	32	38	42	47
TW B2		18	30	38	48	54	64	69	76	11	19	24	31	36	43	48	54
TW B3		17	30	39	48	55	64	70	77	10	18	24	31	36	43	49	55
TW B4		16	28	36	46	52	62	67	74	10	17	23	29	34	41	46	53
TW C1		17	27	34	42	48	57	62	69	12	18	22	28	32	38	42	48
TW C2		16	27	34	43	49	58	64	71	10	17	22	28	32	39	44	50
TW C3		16	25	32	41	47	56	61	68	10	16	21	27	31	37	42	48
TW C4		15	23	29	36	42	50	55	62	10	16	20	24	28	34	38	43
TW D1		15	25	32	40	46	55	60	67	9	15	19	24	27	33	37	42
TW D2		20	35	45	56	63	73	78	85	12	22	29	37	43	52	57	65
TW D3		22	39	50	61	68	77	83	88	12	24	32	41	47	56	62	69
TW D4		22	41	53	64	72	81	86	91	12	25	35	45	52	61	68	75
TW D5		21	40	52	63	71	80	85	91	11	25	34	44	52	62	68	75
TW E1		25	43	54	66	73	82	86	91	14	27	36	45	52	62	68	75
TW E2		24	42	54	65	73	81	86	91	13	26	35	45	52	62	68	75
TW E3		27	49	62	74	81	88	92	96	15	32	43	54	62	72	78	84
TW E4		25	46	59	71	78	86	91	94	13	29	40	51	59	69	75	82
TW F1		18	27	34	42	47	56	61	68	13	18	22	27	31	36	40	46
TW F2		17	29	37	47	54	63	68	75	10	18	23	29	34	41	46	52
TW F3		16	29	38	48	55	65	71	77	9	17	23	30	35	43	48	55
TW F4		14	27	36	46	53	63	69	76	7	15	21	28	33	41	47	53
TW G1		16	25	32	40	46	55	61	67	10	16	20	26	30	35	40	45
TW G2		14	26	34	43	50	59	65	72	8	15	20	26	31	38	43	49
TW G3		12	24	32	41	48	57	64	71	6	13	18	25	30	37	42	48
TW G4		10	20	27	36	43	52	58	65	4	10	15	21	26	32	37	43
TW H1		13	22	29	37	43	52	58	65	7	13	17	23	27	33	37	43
TW H2		10	20	27	35	41	50	56	63	5	11	15	20	25	31	35	41
TW H3		8	17	23	31	36	45	51	58	3	9	13	18	22	27	32	37
TW H4		6	13	18	24	29	36	42	49	2	6	10	14	17	22	26	31
TW I1		13	22	29	37	43	51	57	64	7	12	16	21	24	30	34	39
TW I2		17	32	42	52	60	70	75	82	9	19	26	34	40	48	54	61
TW I3		18	34	45	56	64	73	79	85	10	20	28	37	43	52	58	66
TW I4		19	37	49	61	69	78	83	89	10	22	31	41	48	58	64	72
TW I5		19	37	49	61	69	79	84	90	9	22	32	42	49	59	66	73
TW J1		22	40	51	63	71	80	85	90	12	25	33	43	50	59	65	73
TW J2		21	39	50	62	70	79	84	90	11	23	32	42	49	59	65	73
TW J3		24	45	58	71	78	86	91	95	12	28	39	51	59	69	75	82
TW J4		22	43	55	68	75	84	89	93	11	26	37	48	56	67	73	80
DG A1		10	28	36	44	50	57	61	66	9	15	20	25	29	34	37	40
DG A2		16	29	39	48	55	64	69	75	10	17	23	30	35	43	47	53
DG A3		19	34	45	57	65	74	79	84	12	21	29	38	44	54	60	67
DG B1		16	29	38	48	56	64	69	75	9	15	20	26	31	37	42	47
DG B2		17	31	41	52	61	70	76	82	10	18	24	32	37	46	52	59
DG B3		19	35	47	60	69	79	84	89	12	22	30	39	46	57	64	72
DG C1		18	34	44	56	64	74	79	83	10	19	25	32	37	45	50	57
DG C2		19	35	46	59	68	77	83	88	11	21	28	36	43	53	59	67
DG C3		21	39	52	66	75	85	89	93	13	25	33	44	52	63	71	79
SS A1		24	38	47	57	63	72	77	83	17	26	32	40	45	52	57	64
SS A2		29	47	58	68	75	83	87	92	20	33	41	50	57	66	72	78
SS A3		23	35	44	53	60	68	74	80	16	24	30	37	42	48	53	59
SS A4		29	46	57	68	74	82	87	91	19	32	41	50	56	65	70	77
SS A5		24	35	43	52	58	67	72	78	18	25	31	37	41	48	52	58
SS A6		29	46	57	68	75	83	87	92	19	32	41	50	57	66	71	78
SS A7		22	32	39	47	52	60	66	72	16	23	27	32	36	42	46	51
SS A8		28	45	56	67	73	82	86	91	19	31	40	49	55	64	69	76
SS B1		20	33	41	50	57	66	71	78	14	22	28	34	39	46	51	57
SS B2		25	41	52	62	70	78	83	89	16	28	36	45	52	60	66	73
SS B3		20	31	39	48	55	63	69	75	13	21	26	32	37	43	48	54
SS B4		24	40	51	62	69	77	82	88	16	28	35	44	50	59	65	72
SS B5		19	29	36	44	50	58	64	70	14	20	25	30	34	40	44	50
SS B6		24	40	51	62	69	78	83	88	16	27	35	44	51	59	65	72
SS B7		18	27	33	40	46	54	59	65	13	19	23	27	31	36	40	45
SS B8		23	39	49	60	67	76	81	87	15	26	34	43	49	57	63	70
SS C1		19	33	43	53	60	69	75	81	11	20	26	33	38	46	51	58
SS C2		22	39	50	61	69	78	83	89	12	24	32	41	47	57	63	70
SS C3		17	27	34	43	49	57	63	70	11	17	22	27	31	37	42	47
SS C4		20	34	44	55	62	72	77	84	11	21	28	36	42	51	56	64
SS D1		28	45	56	66	73	81	86	90	19	31	38	47	53	62	67	73
SS D2		33	54	66	77	83	90	93	96	21	37	48	58	65	75	80	86
SS D3		26	41	50	60	67	75	80	86	18	28	35	42	48	55	61	67
SS D4		32	52	64	74	81	88	92	95	21	36	46	56	63	72	78	84
SS E1		24	40	50	61	68	76	81	87	15	26	33	41	47	55	60	67
SS E2		28	48	60	71	78	86	90	94	17	32	41	51	58	68	74	80
SS E3		21	33	42	51	57	66	71	78	15	23	28	34	39	46	51	57
SS E4		26	44	55	66	73	82	86	91	17	29	38	47	54	63	69	76

CONSERVATION FACTOR

TYPE	55 F					65 F				
	200	100	50	30	20	200	100	50	30	20
WW A1	1.0	1.1	1.3	1.4	1.6	1.4	1.5	1.7	1.8	2.0
WW A2	0.9	0.9	1.1	1.3	1.4	1.1	1.2	1.4	1.5	1.7
WW A3	0.8	0.9	1.0	1.2	1.4	1.0	1.1	1.3	1.4	1.6
WW A4	0.8	0.9	1.0	1.2	1.4	1.0	1.1	1.2	1.4	1.6
WW A5	0.8	0.8	1.0	1.2	1.3	0.9	1.1	1.2	1.4	1.5
WW A6	0.7	0.8	1.0	1.2	1.3	0.9	1.0	1.2	1.3	1.5
WW B1	1.0	1.0	1.2	1.3	1.5	1.3	1.4	1.6	1.8	1.9
WW B2	0.8	0.9	1.0	1.2	1.3	1.0	1.1	1.2	1.4	1.5
WW B3	0.7	0.8	1.0	1.1	1.3	0.9	1.0	1.1	1.3	1.4
WW B4	0.7	0.8	0.9	1.1	1.3	0.8	0.9	1.1	1.2	1.4
WW B5	0.7	0.8	1.0	1.1	1.3	0.8	0.9	1.1	1.2	1.4
WW C1	0.7	0.8	0.9	1.1	1.3	0.8	0.9	1.1	1.3	1.4
WW C2	0.7	0.8	1.0	1.1	1.3	0.9	1.0	1.1	1.3	1.4
WW C3	0.6	0.7	0.9	1.1	1.3	0.8	0.9	1.0	1.2	1.3
WW C4	0.7	0.8	0.9	1.1	1.3	0.8	0.9	1.0	1.2	1.4
TW A1	1.1	1.2	1.3	1.4	1.6	1.4	1.5	1.7	1.8	2.0
TW A2	0.9	1.0	1.1	1.3	1.5	1.2	1.3	1.4	1.6	1.8
TW A3	0.9	1.0	1.1	1.3	1.4	1.1	1.2	1.4	1.5	1.7
TW A4	0.9	0.9	1.1	1.2	1.4	1.1	1.1	1.3	1.5	1.7
TW B1	1.0	1.1	1.3	1.4	1.6	1.3	1.5	1.6	1.8	1.9
TW B2	0.9	1.0	1.2	1.3	1.5	1.2	1.3	1.4	1.6	1.8
TW B3	0.9	1.0	1.1	1.3	1.5	1.2	1.3	1.4	1.6	1.7
TW B4	0.9	1.0	1.2	1.3	1.5	1.2	1.3	1.4	1.6	1.8
TW C1	1.0	1.1	1.2	1.4	1.6	1.3	1.4	1.6	1.7	1.9
TW C2	1.0	1.1	1.2	1.4	1.5	1.3	1.4	1.5	1.7	1.8
TW C3	1.0	1.1	1.2	1.4	1.6	1.3	1.4	1.6	1.7	1.9
TW C4	1.1	1.2	1.3	1.5	1.6	1.4	1.5	1.7	1.8	2.0
TW D1	1.0	1.1	1.3	1.4	1.6	1.4	1.5	1.7	1.8	2.0
TW D2	0.8	0.9	1.1	1.2	1.4	1.0	1.1	1.3	1.4	1.6
TW D3	0.8	0.9	1.0	1.2	1.4	1.0	1.1	1.2	1.4	1.5
TW D4	0.7	0.8	1.0	1.2	1.3	0.9	1.0	1.2	1.3	1.5
TW D5	0.7	0.8	1.0	1.2	1.3	0.9	1.0	1.1	1.3	1.5
TW E1	0.7	0.8	1.0	1.2	1.3	0.9	1.0	1.2	1.3	1.5
TW E2	0.7	0.8	1.0	1.1	1.3	0.9	1.0	1.2	1.3	1.5
TW E3	0.7	0.8	0.9	1.1	1.3	0.8	0.9	1.1	1.2	1.4
TW E4	0.7	0.8	0.9	1.1	1.3	0.8	0.9	1.1	1.2	1.4
TW F1	1.0	1.1	1.3	1.4	1.6	1.4	1.5	1.6	1.8	2.0
TW F2	0.9	1.0	1.2	1.3	1.5	1.2	1.3	1.5	1.6	1.8
TW F3	0.9	1.0	1.1	1.3	1.5	1.2	1.3	1.4	1.6	1.7
TW F4	0.9	1.0	1.1	1.3	1.5	1.2	1.3	1.4	1.6	1.7
TW G1	1.0	1.1	1.3	1.4	1.6	1.4	1.5	1.6	1.8	1.9
TW G2	1.0	1.1	1.2	1.3	1.5	1.2	1.3	1.5	1.7	1.8
TW G3	1.0	1.1	1.2	1.3	1.5	1.2	1.3	1.5	1.6	1.8
TW G4	1.0	1.1	1.3	1.4	1.6	1.3	1.4	1.6	1.7	1.9
TW H1	1.1	1.2	1.3	1.4	1.6	1.4	1.5	1.6	1.8	1.9
TW H2	1.1	1.2	1.3	1.4	1.6	1.4	1.5	1.6	1.8	1.9
TW H3	1.1	1.2	1.4	1.5	1.6	1.5	1.6	1.7	1.9	2.0
TW H4	1.3	1.4	1.5	1.6	1.8	1.7	1.7	1.9	2.0	2.2
TW I1	1.1	1.2	1.3	1.4	1.6	1.4	1.6	1.7	1.9	2.1
TW I2	0.8	0.9	1.1	1.2	1.4	1.1	1.2	1.3	1.5	1.6
TW I3	0.8	0.9	1.0	1.2	1.4	1.0	1.1	1.3	1.4	1.6
TW I4	0.8	0.9	1.0	1.2	1.3	0.9	1.0	1.2	1.3	1.5
TW I5	0.8	0.8	1.0	1.2	1.3	0.9	1.0	1.2	1.3	1.5
TW J1	0.8	0.8	1.0	1.2	1.3	0.9	1.0	1.2	1.3	1.5
TW J2	0.8	0.8	1.0	1.2	1.3	0.9	1.0	1.2	1.3	1.6
TW J3	0.7	0.8	0.9	1.1	1.3	0.8	0.9	1.1	1.2	1.4
TW J4	0.7	0.8	1.0	1.1	1.3	0.8	0.9	1.1	1.2	1.4
DG A1	0.9	1.0	1.2	1.5	1.8	1.3	1.4	1.6	2.0	2.4
DG A2	0.9	1.0	1.1	1.3	1.6	1.2	1.3	1.4	1.6	1.9
DG A3	0.8	0.9	1.0	1.2	1.4	1.1	1.1	1.2	1.4	1.6
DG B1	0.9	1.0	1.1	1.3	1.6	1.2	1.4	1.5	1.7	2.0
DG B2	0.9	0.9	1.1	1.2	1.4	1.2	1.2	1.4	1.5	1.7
DG B3	0.8	0.9	1.0	1.2	1.4	1.0	1.1	1.2	1.3	1.5
DG C1	0.8	0.9	1.0	1.2	1.5	1.1	1.2	1.4	1.5	1.7
DG C2	0.8	0.9	1.0	1.2	1.4	1.1	1.2	1.3	1.4	1.6
DG C3	0.8	0.8	0.9	1.1	1.3	1.0	1.0	1.1	1.2	1.4
SS A1	0.8	0.9	1.1	1.3	1.4	1.0	1.2	1.3	1.5	1.6
SS A2	0.7	0.8	1.0	1.2	1.3	0.9	1.0	1.1	1.3	1.5
SS A3	0.9	1.0	1.1	1.3	1.5	1.1	1.2	1.4	1.6	1.7
SS A4	0.7	0.8	1.0	1.2	1.3	0.9	1.0	1.2	1.3	1.5
SS A5	0.9	1.0	1.1	1.3	1.5	1.1	1.2	1.4	1.6	1.8
SS A6	0.7	0.8	1.0	1.2	1.3	0.9	1.0	1.1	1.3	1.5
SS A7	1.0	1.1	1.2	1.4	1.6	1.2	1.4	1.6	1.7	1.9
SS A8	0.7	0.8	1.0	1.2	1.3	0.9	1.0	1.2	1.3	1.5
SS B1	0.9	1.0	1.1	1.3	1.5	1.1	1.2	1.4	1.6	1.7
SS B2	0.8	0.9	1.0	1.2	1.4	0.9	1.0	1.2	1.3	1.5
SS B3	0.9	1.0	1.2	1.3	1.5	1.2	1.3	1.5	1.6	1.8
SS B4	0.8	0.9	1.0	1.2	1.4	0.9	1.0	1.2	1.4	1.5
SS B5	1.0	1.1	1.2	1.4	1.6	1.3	1.4	1.6	1.7	1.9
SS B6	0.8	0.9	1.0	1.2	1.4	0.9	1.0	1.2	1.4	1.5
SS B7	1.1	1.2	1.3	1.5	1.6	1.4	1.5	1.7	1.9	2.0
SS B8	0.8	0.9	1.0	1.2	1.4	1.0	1.1	1.2	1.4	1.5
SS C1	0.9	0.9	1.1	1.3	1.4	1.1	1.2	1.4	1.5	1.7
SS C2	0.8	0.9	1.0	1.2	1.3	1.0	1.1	1.2	1.4	1.5
SS C3	1.0	1.1	1.2	1.4	1.5	1.3	1.4	1.6	1.7	1.9
SS C4	0.8	0.9	1.1	1.2	1.4	1.0	1.1	1.3	1.4	1.6
SS D1	0.7	0.8	1.0	1.2	1.3	0.9	1.0	1.2	1.4	1.5
SS D2	0.6	0.8	0.9	1.1	1.3	0.8	0.9	1.0	1.2	1.4
SS D3	0.8	0.9	1.1	1.2	1.4	1.0	1.1	1.3	1.4	1.6
SS D4	0.7	0.8	0.9	1.1	1.3	0.8	0.9	1.1	1.2	1.4
SS E1	0.8	0.9	1.0	1.2	1.4	1.0	1.1	1.3	1.4	1.6
SS E2	0.7	0.8	1.0	1.1	1.3	0.8	0.9	1.1	1.3	1.4
SS E3	0.9	1.0	1.1	1.3	1.5	1.1	1.3	1.4	1.6	1.8
SS E4	0.7	0.8	1.0	1.2	1.3	0.9	1.0	1.2	1.3	1.5

AUGUSTA, GEORGIA

TYPE	SOLAR SAVINGS FRACTION (%) BASE TEMP: 55 F (1106 DD)								65 F (2549 DD)								CONSERVATION FACTOR 55 F					65 F				
LCR	200	100	70	50	40	30	25	20	200	100	70	50	40	30	25	20	200	100	50	30	20	200	100	50	30	20
WW A1	24	36	44	53	60	68	74	80	17	23	28	34	38	45	49	55	0.9	1.0	1.1	1.3	1.5	1.2	1.3	1.5	1.6	1.8
WW A2	27	44	55	66	73	81	86	91	16	26	34	42	48	57	62	69	0.7	0.8	1.0	1.2	1.3	1.0	1.1	1.2	1.4	1.5
WW A3	28	47	59	70	77	85	89	94	15	28	36	45	52	61	67	74	0.7	0.8	1.0	1.1	1.3	0.9	1.0	1.2	1.3	1.5
WW A4	29	49	61	72	79	87	91	95	15	28	37	47	54	64	70	77	0.7	0.8	0.9	1.1	1.3	0.9	1.0	1.1	1.3	1.5
WW A5	29	51	63	75	81	89	92	96	15	29	39	49	56	66	72	79	0.7	0.8	0.9	1.1	1.3	0.9	1.0	1.1	1.3	1.4
WW A6	30	52	64	76	83	90	93	96	14	29	39	50	58	67	73	80	0.7	0.8	0.9	1.1	1.3	0.8	0.9	1.1	1.3	1.4
WW B1	23	38	48	58	65	74	79	85	13	22	28	34	40	47	52	58	0.8	0.9	1.1	1.2	1.4	1.1	1.2	1.4	1.5	1.7
WW B2	28	49	61	73	80	88	91	95	15	29	38	48	56	65	71	78	0.7	0.8	0.9	1.1	1.3	0.9	1.0	1.1	1.3	1.4
WW B3	34	57	69	80	86	92	95	98	18	34	45	56	64	74	79	86	0.6	0.7	0.9	1.1	1.2	0.8	0.9	1.0	1.2	1.4
WW B4	33	57	70	81	87	93	96	98	16	34	46	58	66	76	81	87	0.6	0.7	0.9	1.1	1.2	0.8	0.9	1.0	1.2	1.4
WW B5	31	55	68	79	86	92	95	98	15	33	44	56	64	75	80	87	0.6	0.7	0.9	1.1	1.2	0.8	0.9	1.0	1.2	1.4
WW C1	35	58	71	81	87	93	96	98	19	36	47	58	66	76	81	87	0.6	0.7	0.9	1.1	1.2	0.8	0.9	1.0	1.2	1.4
WW C2	33	56	69	80	86	92	95	98	17	34	45	56	64	74	80	86	0.6	0.7	0.9	1.1	1.2	0.8	0.9	1.0	1.2	1.4
WW C3	37	63	76	86	91	96	98	99	19	40	52	65	73	82	87	92	0.6	0.7	0.9	1.0	1.2	0.7	0.8	1.0	1.1	1.3
WW C4	33	59	72	83	89	94	97	99	16	36	49	61	69	79	85	90	0.6	0.7	0.9	1.0	1.2	0.7	0.8	1.0	1.1	1.3
TW A1	23	35	43	52	58	67	72	79	16	23	28	33	38	44	49	54	0.9	1.0	1.1	1.3	1.5	1.2	1.3	1.5	1.6	1.8
TW A2	25	40	50	60	68	76	82	87	15	24	31	38	44	52	57	64	0.8	0.9	1.0	1.2	1.4	1.1	1.2	1.3	1.5	1.6
TW A3	25	42	53	64	71	80	85	90	14	25	32	40	47	55	61	68	0.8	0.9	1.0	1.2	1.3	1.0	1.1	1.2	1.4	1.6
TW A4	24	42	53	64	72	80	85	90	13	24	32	41	47	56	62	69	0.7	0.8	1.0	1.2	1.3	1.0	1.1	1.2	1.4	1.5
TW B1	23	35	43	53	59	68	74	80	15	22	28	34	38	45	50	56	0.9	1.0	1.1	1.3	1.4	1.2	1.3	1.4	1.6	1.7
TW B2	23	38	48	59	66	75	80	86	14	23	29	37	43	51	56	63	0.8	0.9	1.0	1.2	1.4	1.1	1.2	1.3	1.5	1.6
TW B3	22	38	49	59	67	76	81	87	13	22	29	37	43	52	57	64	0.8	0.9	1.0	1.2	1.4	1.0	1.1	1.3	1.4	1.6
TW B4	21	36	46	56	64	73	79	85	12	21	28	35	41	49	55	62	0.8	0.9	1.1	1.2	1.4	1.1	1.2	1.3	1.5	1.6
TW C1	21	34	43	52	59	68	74	80	14	21	27	33	38	45	50	56	0.9	1.0	1.1	1.3	1.4	1.2	1.3	1.4	1.6	1.7
TW C2	20	34	43	53	60	70	75	82	12	20	26	33	39	46	51	58	0.9	0.9	1.1	1.3	1.4	1.1	1.2	1.4	1.5	1.7
TW C3	19	32	41	51	58	67	73	79	12	19	25	32	37	44	49	56	0.9	1.0	1.1	1.3	1.4	1.2	1.3	1.4	1.6	1.7
TW C4	18	28	36	45	52	61	66	73	12	18	23	29	33	40	44	50	1.0	1.1	1.2	1.3	1.5	1.3	1.4	1.5	1.7	1.8
TW D1	19	32	41	51	58	67	73	80	11	18	23	30	34	41	46	52	0.9	1.0	1.1	1.3	1.4	1.2	1.3	1.5	1.6	1.8
TW D2	26	45	56	68	75	83	88	92	15	27	35	44	51	60	66	73	0.7	0.8	1.0	1.1	1.3	0.9	1.0	1.2	1.3	1.5
TW D3	29	49	61	72	79	87	91	95	16	29	38	48	56	65	71	78	0.7	0.8	0.9	1.1	1.3	0.9	1.0	1.1	1.3	1.4
TW D4	30	51	64	75	82	89	93	96	16	31	41	52	60	70	76	82	0.7	0.8	0.9	1.1	1.3	0.8	0.9	1.1	1.2	1.4
TW D5	28	50	63	74	81	89	93	96	15	30	41	52	59	69	76	82	0.7	0.8	0.9	1.1	1.3	0.8	0.9	1.1	1.2	1.4
TW E1	32	53	66	77	83	90	93	96	18	33	43	53	61	70	76	82	0.7	0.8	0.9	1.1	1.3	0.8	0.9	1.1	1.2	1.4
TW E2	31	53	65	76	83	90	93	96	17	32	42	53	60	70	76	83	0.7	0.8	0.9	1.1	1.3	0.8	0.9	1.1	1.2	1.4
TW E3	36	60	73	83	89	94	97	99	20	38	50	62	70	79	85	90	0.6	0.7	0.9	1.0	1.2	0.7	0.8	1.0	1.2	1.3
TW E4	33	57	70	81	87	93	96	98	17	35	47	59	67	77	82	88	0.6	0.7	0.9	1.1	1.2	0.8	0.9	1.0	1.2	1.3
TW F1	22	34	42	52	58	67	73	79	14	21	26	32	37	44	48	54	0.9	1.0	1.1	1.3	1.5	1.2	1.3	1.5	1.6	1.8
TW F2	22	37	48	58	65	74	80	86	13	22	28	36	42	50	55	62	0.8	0.9	1.0	1.2	1.4	1.1	1.2	1.3	1.5	1.6
TW F3	21	38	49	60	67	76	82	87	11	21	28	37	43	51	57	64	0.8	0.9	1.0	1.2	1.4	1.0	1.1	1.3	1.4	1.6
TW F4	19	36	47	58	66	75	81	87	9	19	26	35	41	50	56	63	0.8	0.9	1.0	1.2	1.4	1.0	1.1	1.3	1.4	1.6
TW G1	20	32	41	51	58	67	72	79	12	19	25	31	36	43	48	54	0.9	1.0	1.1	1.3	1.4	1.2	1.3	1.5	1.6	1.8
TW G2	19	33	43	54	61	71	77	83	10	19	25	32	38	46	51	58	0.8	0.9	1.1	1.2	1.4	1.1	1.2	1.4	1.5	1.7
TW G3	17	32	42	52	60	70	75	82	8	17	23	31	36	45	50	57	0.8	0.9	1.1	1.2	1.4	1.1	1.2	1.4	1.5	1.7
TW G4	13	27	36	46	54	64	70	77	6	14	20	27	32	40	45	52	0.9	1.0	1.1	1.3	1.4	1.2	1.3	1.4	1.6	1.7
TW H1	16	29	37	47	54	64	70	77	9	16	22	28	33	40	45	51	0.9	1.0	1.1	1.3	1.5	1.2	1.3	1.5	1.6	1.8
TW H2	13	26	35	45	52	62	68	75	6	14	19	26	31	38	43	50	0.9	1.0	1.2	1.3	1.5	1.2	1.3	1.5	1.6	1.8
TW H3	11	22	31	40	47	56	63	70	5	11	16	22	27	34	39	45	1.0	1.1	1.2	1.4	1.5	1.3	1.4	1.5	1.7	1.8
TW H4	8	17	24	32	38	47	53	60	3	8	12	17	21	27	31	37	1.1	1.2	1.3	1.5	1.6	1.5	1.6	1.7	1.8	2.0
TW I1	17	29	38	48	55	64	70	77	9	16	21	27	31	38	42	48	0.9	1.0	1.1	1.3	1.5	1.3	1.4	1.5	1.7	1.8
TW I2	23	41	53	64	72	81	86	91	12	23	31	41	47	57	63	70	0.7	0.8	1.0	1.2	1.3	1.0	1.1	1.2	1.4	1.5
TW I3	25	44	56	68	75	84	88	93	13	25	34	44	51	61	67	74	0.7	0.8	1.0	1.1	1.3	0.9	1.0	1.2	1.3	1.5
TW I4	26	48	60	72	79	87	91	95	13	28	38	49	56	66	73	80	0.7	0.8	0.9	1.1	1.3	0.9	1.0	1.1	1.3	1.4
TW I5	26	48	61	72	80	88	92	95	12	28	38	49	57	67	74	81	0.7	0.8	0.9	1.1	1.3	0.8	0.9	1.1	1.2	1.4
TW J1	29	51	63	74	81	89	92	96	16	30	40	51	58	68	74	81	0.7	0.8	0.9	1.1	1.3	0.8	0.9	1.1	1.3	1.4
TW J2	28	49	62	73	80	88	92	95	14	29	39	50	57	67	73	80	0.7	0.8	0.9	1.1	1.3	0.8	0.9	1.1	1.3	1.4
TW J3	32	57	70	81	87	93	96	98	16	35	47	59	67	77	82	88	0.6	0.7	0.9	1.1	1.2	0.8	0.9	1.0	1.2	1.3
TW J4	30	54	67	78	85	92	95	97	14	32	44	56	64	74	80	86	0.6	0.7	0.9	1.1	1.2	0.8	0.9	1.0	1.2	1.4
DG A1	21	37	47	56	62	70	74	79	11	20	26	32	37	43	46	50	0.8	0.9	1.1	1.3	1.6	1.1	1.2	1.5	1.7	2.1
DG A2	21	39	49	60	67	76	80	85	12	22	29	37	43	51	56	62	0.8	0.9	1.0	1.2	1.4	1.1	1.1	1.3	1.5	1.7
DG A3	24	44	57	69	76	83	87	91	14	26	35	45	52	62	68	75	0.7	0.8	1.0	1.2	1.4	1.0	1.0	1.1	1.3	1.5
DG B1	22	39	51	63	70	78	82	86	11	20	26	34	40	48	53	59	0.8	0.8	1.0	1.2	1.5	1.1	1.2	1.3	1.5	1.8
DG B2	22	41	53	66	74	82	87	91	12	22	30	39	46	56	62	69	0.8	0.8	1.0	1.2	1.4	1.0	1.1	1.2	1.4	1.6
DG B3	25	45	59	73	80	88	91	94	14	26	35	46	55	66	73	80	0.7	0.8	0.9	1.1	1.3	0.9	1.0	1.1	1.2	1.4
DG C1	25	45	59	72	79	86	89	92	13	24	31	40	47	57	63	69	0.7	0.8	0.9	1.2	1.4	1.0	1.1	1.2	1.4	1.6
DG C2	25	46	60	73	80	88	92	95	14	26	34	45	52	63	70	77	0.7	0.8	0.9	1.1	1.3	1.0	1.0	1.1	1.3	1.5
DG C3	28	51	66	79	86	92	95	97	16	30	40	52	61	73	79	86	0.7	0.7	0.9	1.1	1.4	0.9	1.0	1.0	1.2	1.4
SS A1	30	46	57	67	74	82	86	91	20	30	38	46	52	60	65	72	0.7	0.8	1.0	1.2	1.3	0.9	1.1	1.2	1.4	1.5
SS A2	36	56	68	78	84	91	94	97	23	38	47	57	64	73	79	85	0.6	0.8	0.9	1.1	1.3	0.8	0.9	1.1	1.2	1.4
SS A3	28	44	54	64	71	79	84	89	18	28	35	43	48	56	61	68	0.8	0.9	1.0	1.2	1.4	1.0	1.1	1.3	1.4	1.6
SS A4	36	56	67	77	84	90	93	96	23	37	47	57	63	72	78	84	0.6	0.8	0.9	1.1	1.3	0.8	0.9	1.1	1.2	1.4
SS A5	28	43	53	62	69	77	82	87	20	29	35	42	48	55	60	67	0.8	0.9	1.0	1.2	1.4	1.0	1.1	1.3	1.5	1.6
SS A6	36	56	68	78	84	90	94	97	23	37	47	57	64	73	78	84	0.6	0.8	0.9	1.1	1.3	0.8	0.9	1.1	1.2	1.4
SS A7	26	39	47	57	63	71	76	82	18	26	31	38	42	49	53	59	0.9	0.9	1.1	1.3	1.4	1.1	1.2	1.4	1.6	1.7
SS A8	35	55	66	77	83	90	93	96	22	36	46	55	62	71	77	83	0.7	0.8	0.9	1.1	1.3	0.8	0.9	1.1	1.2	1.4
SS B1	25	41	51	61	68	76	81	87	16	26	33	40	46	54	59	66	0.8	0.9	1.0	1.2	1.4	1.0	1.1	1.3	1.4	1.6
SS B2	31	50	62	73	79	87	91	94	19	33	42	51	58	68	73	80	0.7	0.8	1.0	1.1	1.3	0.9	1.0	1.1	1.3	1.4
SS B3	24	39	49	59	66	74	80	85	15	24	31	38	43	51	56	63	0.8	0.9	1.1	1.2	1.4	1.1	1.2	1.3	1.5	1.6
SS B4	30	50	61	72	79	86	90	94	19	32	41	51	57	67	72	79	0.7	0.8	1.0	1.1	1.3	0.9	1.0	1.1	1.3	1.4
SS B5	23	36	45	54	61	69	74	81	16	23	29	35	40	47	52	58	0.9	1.0	1.1	1.3	1.4	1.1	1.3	1.4	1.6	1.7
SS B6	30	50	61	72	79	86	90	94	19	32	41	51	58	67	72	79	0.7	0.8	1.0	1.1	1.3	0.9	1.0	1.1	1.3	1.4
SS B7	21	33	41	50	56	65	70	76	15	21	26	32	36	42	47	52	0.9	1.0	1.2	1.3	1.5	1.2	1.3	1.5	1.7	1.8
SS B8	29	48	60	71	77	85	89	93	18	31	40	49	56	65	71	77	0.7	0.8	1.0	1.1	1.3	0.9	1.0	1.1	1.3	1.5
SS C1	25	43	54	65	72	81	86	91	14	24	32	40	47	55	61	68	0.7	0.8	1.0	1.2	1.3	1.0	1.1	1.2	1.4	1.6
SS C2	29	49	61	73	80	87	91	95	15	29	38	49	56	65	71	78	0.7	0.8	0.9	1.1	1.3	0.9	1.0	1.1	1.3	1.4
SS C3	21	34	43	53	60	69	75	81	13	20	26	33	38	45	50	56	0.9	1.0	1.1	1.3	1.4	1.2	1.3	1.4	1.6	1.7
SS C4	25	44	55	66	74	82	87	92	14	26	34	43	50	59	65	72	0.7	0.8	1.0	1.1	1.3	0.9	1.0	1.2	1.3	1.5
SS D1	35	55	67	77	83	90	93	96	22	36	45	54	61	70	75	81	0.7	0.8	0.9	1.1	1.3	0.8	0.9	1.1	1.3	1.4
SS D2	41	64	76	85	90	95	97	99	25	43	54	66	73	81	86	91	0.6	0.7	0.9	1.0	1.2	0.7	0.8	1.0	1.2	1.3
SS D3	32	50	60	71	77	85	89	93	21	33	40	49	55	63	69	75	0.7	0.8	1.0	1.1	1.3	0.9	1.0	1.2	1.3	1.5
SS D4	40	62	74	84	89	94	97	98	25	42	53	63	71	79	84	90	0.6	0.7	0.9	1.1	1.2	0.7	0.9	1.0	1.2	1.3
SS E1	31	50	61	72	78	86	90	94	19	31	39	48	55	64	69	76	0.7	0.8	1.0	1.1	1.3	0.9	1.0	1.2	1.3	1.5
SS E2	36	59	71	81	87	93	95	98	21	38	48	59	66	76	81	87	0.6	0.7	0.9	1.1	1.2	0.8	0.9	1.0	1.2	1.4
SS E3	26	41	51	62	68	77	82	87	17	27	33	40	46	54	59	65	0.8	0.9	1.0	1.2	1.4	1.0	1.1	1.3	1.5	1.6
SS E4	33	54	66	77	83	90	93	96	20	35	44	55	62	71	77	83	0.7	0.8	0.9	1.1	1.3	0.8	0.9	1.1	1.2	1.4

MACON, GEORGIA

TYPE	SOLAR SAVINGS FRACTION (%) BASE TEMP: 55 F (934 DD) LCR 200	100	70	50	40	30	25	20	65 F (2243 DD) 200	100	70	50	40	30	25	20	CONSERVATION FACTOR 55 F 200	100	50	30	20	65 F 200	100	50	30	20
WW A1	27	40	49	58	65	74	79	84	18	25	30	37	41	48	53	59	0.8	0.9	1.1	1.2	1.4	1.2	1.3	1.4	1.6	1.7
WW A2	31	49	61	71	78	86	90	94	18	29	37	46	52	61	67	73	0.7	0.8	1.0	1.1	1.3	0.9	1.0	1.2	1.3	1.5
WW A3	32	53	65	75	82	89	93	96	17	30	39	49	56	65	71	78	0.7	0.8	0.9	1.1	1.3	0.9	1.0	1.1	1.3	1.4
WW A4	33	55	67	78	84	91	94	97	17	31	41	51	58	68	74	80	0.6	0.7	0.9	1.1	1.3	0.8	1.0	1.1	1.3	1.4
WW A5	34	57	69	80	86	92	95	98	17	32	42	53	61	70	76	83	0.6	0.7	0.9	1.1	1.2	0.8	0.9	1.1	1.2	1.4
WW A6	35	58	70	81	87	93	96	98	17	33	43	54	62	72	77	84	0.6	0.7	0.9	1.1	1.2	0.8	0.9	1.1	1.2	1.4
WW B1	27	43	54	64	71	80	84	90	15	24	31	38	43	51	56	63	0.8	0.9	1.0	1.2	1.3	1.1	1.2	1.3	1.5	1.6
WW B2	33	55	67	78	84	91	94	97	17	32	42	52	60	69	75	82	0.6	0.7	0.9	1.1	1.3	0.8	0.9	1.1	1.2	1.4
WW B3	39	63	75	85	90	95	97	99	20	38	49	60	68	77	83	88	0.6	0.7	0.9	1.0	1.2	0.8	0.9	1.0	1.2	1.3
WW B4	38	62	75	85	90	95	97	99	19	37	49	61	69	79	84	90	0.6	0.7	0.9	1.0	1.2	0.7	0.8	1.0	1.2	1.3
WW B5	36	60	73	84	89	95	97	99	17	36	48	60	68	78	83	89	0.6	0.7	0.9	1.0	1.2	0.8	0.9	1.0	1.2	1.3
WW C1	40	64	76	86	91	96	98	99	21	39	51	62	70	79	84	90	0.6	0.7	0.9	1.0	1.2	0.7	0.9	1.0	1.2	1.3
WW C2	38	62	74	84	90	95	97	99	19	37	48	60	68	78	83	89	0.6	0.7	0.9	1.0	1.2	0.8	0.9	1.0	1.2	1.3
WW C3	42	68	80	89	94	97	99	100	21	43	56	69	76	85	90	94	0.6	0.7	0.8	1.0	1.2	0.7	0.8	0.9	1.1	1.3
WW C4	39	64	77	87	92	96	98	99	19	39	52	65	73	82	87	92	0.6	0.7	0.9	1.0	1.2	0.7	0.8	1.0	1.1	1.3
TW A1	26	39	48	57	64	72	78	83	17	25	30	36	41	47	52	58	0.8	0.9	1.1	1.2	1.4	1.2	1.3	1.4	1.6	1.7
TW A2	28	45	55	66	73	81	86	91	17	27	34	42	48	56	61	68	0.7	0.8	1.0	1.2	1.3	1.0	1.1	1.3	1.4	1.6
TW A3	29	47	58	69	76	84	89	93	16	27	35	44	50	59	65	72	0.7	0.8	1.0	1.1	1.3	1.0	1.1	1.2	1.4	1.5
TW A4	28	47	59	70	77	85	89	93	15	27	35	44	51	60	66	73	0.7	0.8	1.0	1.1	1.3	0.9	1.0	1.2	1.3	1.5
TW B1	25	39	48	58	65	74	79	85	16	24	30	36	41	48	53	60	0.8	0.9	1.1	1.2	1.4	1.1	1.2	1.4	1.5	1.7
TW B2	26	43	54	64	71	80	85	90	15	25	32	40	46	55	60	67	0.8	0.9	1.0	1.2	1.3	1.0	1.1	1.3	1.4	1.6
TW B3	26	43	54	65	72	81	86	91	14	25	32	40	47	55	61	68	0.7	0.8	1.0	1.2	1.3	1.0	1.1	1.2	1.4	1.5
TW B4	24	40	51	62	69	78	83	89	13	23	30	38	44	53	58	66	0.8	0.9	1.0	1.2	1.3	1.0	1.1	1.3	1.4	1.6
TW C1	24	38	47	57	64	73	79	84	15	23	29	36	41	48	53	60	0.8	0.9	1.1	1.2	1.4	1.1	1.2	1.4	1.5	1.7
TW C2	23	38	48	59	66	75	80	86	13	22	29	36	42	50	55	62	0.8	0.9	1.1	1.2	1.4	1.1	1.2	1.3	1.5	1.6
TW C3	22	36	46	56	63	72	78	84	13	21	27	34	40	47	53	60	0.8	0.9	1.1	1.2	1.4	1.1	1.2	1.4	1.5	1.7
TW C4	20	32	41	50	57	66	71	78	13	20	25	31	36	43	48	54	0.9	1.0	1.1	1.3	1.5	1.2	1.3	1.5	1.6	1.8
TW D1	22	37	47	57	64	73	79	85	12	20	26	33	38	45	50	57	0.8	0.9	1.1	1.2	1.4	1.1	1.2	1.4	1.6	1.7
TW D2	30	50	62	73	80	87	91	95	16	29	38	48	55	64	70	77	0.7	0.8	0.9	1.1	1.3	0.9	1.0	1.1	1.3	1.5
TW D3	33	55	67	77	84	91	94	97	18	32	42	52	59	69	75	81	0.6	0.8	0.9	1.1	1.3	0.8	0.9	1.1	1.2	1.4
TW D4	34	57	69	80	86	92	95	98	18	34	45	56	63	73	79	85	0.6	0.7	0.9	1.1	1.2	0.8	0.9	1.0	1.2	1.4
TW D5	33	55	68	79	85	92	95	97	17	33	44	55	63	73	79	85	0.6	0.7	0.9	1.1	1.2	0.8	0.9	1.0	1.2	1.4
TW E1	37	59	71	81	87	93	96	98	20	36	46	57	64	74	79	85	0.6	0.7	0.9	1.1	1.2	0.8	0.9	1.1	1.2	1.4
TW E2	36	58	70	81	87	93	96	98	19	35	45	56	64	74	79	86	0.6	0.7	0.9	1.1	1.2	0.8	0.9	1.0	1.2	1.4
TW E3	41	66	78	87	92	96	98	99	22	41	53	65	73	82	87	92	0.6	0.7	0.9	1.0	1.2	0.7	0.8	1.0	1.1	1.3
TW E4	38	62	75	85	90	95	97	99	19	38	50	62	70	80	85	90	0.6	0.7	0.9	1.0	1.2	0.7	0.8	1.0	1.2	1.3
TW F1	24	38	47	57	64	73	78	84	15	23	29	35	40	47	52	58	0.8	0.9	1.1	1.2	1.4	1.2	1.3	1.4	1.6	1.7
TW F2	25	42	53	64	71	80	84	90	14	24	31	39	45	53	59	66	0.8	0.9	1.0	1.2	1.3	1.0	1.1	1.3	1.4	1.6
TW F3	25	43	54	65	73	81	86	91	13	24	31	40	46	55	61	68	0.7	0.8	1.0	1.2	1.3	1.0	1.1	1.2	1.4	1.5
TW F4	23	41	52	64	71	80	85	90	11	22	29	38	45	54	60	67	0.7	0.8	1.0	1.2	1.3	1.0	1.1	1.2	1.4	1.5
TW G1	22	36	46	56	63	72	78	84	13	21	27	34	39	46	51	58	0.8	0.9	1.1	1.2	1.4	1.1	1.2	1.4	1.5	1.7
TW G2	22	38	49	60	67	76	82	87	11	21	28	35	41	50	55	63	0.8	0.9	1.0	1.2	1.4	1.1	1.2	1.3	1.5	1.6
TW G3	20	36	47	58	66	75	81	86	9	19	26	34	40	48	54	62	0.8	0.9	1.0	1.2	1.4	1.1	1.2	1.3	1.5	1.6
TW G4	16	31	41	52	60	69	75	82	7	16	22	29	35	43	49	56	0.8	0.9	1.1	1.2	1.4	1.1	1.2	1.4	1.5	1.7
TW H1	19	33	42	53	60	69	75	81	10	18	24	31	36	43	48	55	0.9	0.9	1.1	1.3	1.4	1.2	1.3	1.4	1.6	1.7
TW H2	16	30	40	50	58	67	73	80	7	15	21	28	34	41	47	54	0.9	1.0	1.1	1.3	1.4	1.2	1.3	1.4	1.6	1.7
TW H3	13	26	35	45	52	62	68	75	6	13	18	25	30	37	42	49	0.9	1.0	1.1	1.3	1.5	1.2	1.3	1.5	1.6	1.8
TW H4	10	20	28	36	43	52	58	66	4	10	14	19	24	30	34	41	1.0	1.1	1.2	1.4	1.0	1.4	1.5	1.6	1.8	1.9
TW I1	20	34	43	54	61	70	75	82	10	18	23	30	34	42	47	53	0.8	0.9	1.1	1.3	1.4	1.2	1.3	1.5	1.6	1.8
TW I2	27	46	58	70	77	85	89	94	13	26	34	44	51	60	67	74	0.7	0.8	1.0	1.1	1.3	0.9	1.0	1.2	1.3	1.5
TW I3	29	50	62	73	80	88	92	95	14	28	37	48	55	65	71	78	0.7	0.8	0.9	1.1	1.3	0.9	1.0	1.1	1.3	1.4
TW I4	31	53	66	77	84	91	94	97	15	30	41	52	60	70	76	83	0.6	0.7	0.9	1.1	1.3	0.8	0.9	1.1	1.2	1.4
TW I5	30	53	66	77	84	91	94	97	14	31	41	53	61	71	77	84	0.6	0.7	0.9	1.1	1.3	0.8	0.9	1.1	1.2	1.4
TW J1	34	56	68	79	85	92	95	97	18	33	43	54	62	72	77	84	0.6	0.7	0.9	1.1	1.2	0.8	0.9	1.1	1.2	1.4
TW J2	32	55	67	78	85	91	94	97	16	32	42	53	61	71	77	84	0.6	0.7	0.9	1.1	1.3	0.8	0.9	1.1	1.2	1.4
TW J3	37	62	75	85	90	95	97	99	19	38	50	62	70	80	85	90	0.6	0.7	0.9	1.0	1.2	0.7	0.8	1.0	1.1	1.3
TW J4	34	60	72	00	89	94	97	98	17	35	47	59	67	77	83	89	0.6	0.7	0.9	1.0	1.2	0.8	0.9	1.0	1.2	1.3
DG A1	25	42	53	62	68	76	80	84	12	22	29	36	40	47	51	55	0.7	0.9	1.1	1.3	1.5	1.0	1.2	1.4	1.7	2.0
DG A2	25	44	55	66	73	81	85	89	13	24	32	40	46	55	60	66	0.7	0.8	1.0	1.2	1.4	1.0	1.1	1.2	1.4	1.7
DG A3	28	50	63	74	80	87	90	93	16	28	38	48	56	66	72	78	0.7	0.8	0.9	1.2	1.4	0.9	1.0	1.1	1.3	1.5
DG B1	26	45	58	69	76	83	87	90	13	22	30	38	44	53	58	64	0.7	0.8	1.0	1.2	1.4	1.0	1.1	1.3	1.5	1.7
DG B2	26	46	60	72	79	87	90	94	14	25	33	43	50	60	66	73	0.7	0.8	0.9	1.1	1.3	1.0	1.1	1.2	1.3	1.5
DG B3	29	51	66	78	85	91	94	96	16	29	39	50	59	70	76	83	0.7	0.7	0.9	1.1	1.4	0.9	1.0	1.1	1.2	1.4
DG C1	30	52	66	78	84	90	92	95	15	26	35	45	52	62	68	74	0.7	0.7	0.9	1.1	1.4	1.0	1.0	1.1	1.3	1.6
DG C2	30	52	67	79	85	92	94	97	16	28	38	49	57	68	74	80	0.7	0.7	0.9	1.1	1.3	0.9	1.0	1.1	1.2	1.4
DG C3	33	57	72	84	90	95	96	98	18	32	43	56	65	77	83	88	0.6	0.7	0.9	1.1	1.4	0.9	0.9	1.0	1.2	1.4
SS A1	33	51	62	72	79	86	90	94	21	33	41	49	55	64	69	76	0.7	0.8	1.0	1.1	1.3	0.9	1.0	1.2	1.3	1.5
SS A2	40	62	73	83	88	93	96	98	25	40	50	61	68	77	82	87	0.6	0.7	0.9	1.1	1.2	0.8	0.9	1.0	1.2	1.4
SS A3	32	49	59	69	76	84	88	92	20	31	38	46	52	60	65	72	0.7	0.8	1.0	1.2	1.3	0.9	1.1	1.2	1.4	1.5
SS A4	40	61	72	82	88	93	96	98	25	40	50	60	67	76	81	87	0.6	0.7	0.9	1.1	1.2	0.8	0.9	1.0	1.2	1.4
SS A5	32	48	58	68	74	82	86	91	21	31	38	45	51	59	64	70	0.7	0.8	1.0	1.2	1.3	1.0	1.1	1.2	1.4	1.6
SS A6	40	62	73	82	88	93	96	98	25	40	50	61	68	76	81	87	0.6	0.7	0.9	1.1	1.2	0.8	0.9	1.0	1.2	1.4
SS A7	29	43	52	62	68	76	81	87	20	28	34	40	45	53	57	64	0.8	0.9	1.1	1.2	1.4	1.1	1.2	1.3	1.5	1.7
SS A8	40	61	72	81	87	93	95	98	24	39	49	59	66	75	80	86	0.6	0.7	0.9	1.1	1.2	0.8	0.9	1.1	1.2	1.4
SS B1	29	46	56	66	73	81	86	91	18	28	35	43	49	57	63	70	0.7	0.8	1.0	1.2	1.3	1.0	1.1	1.2	1.4	1.6
SS B2	35	56	67	78	84	90	94	96	21	35	45	55	62	71	77	83	0.6	0.8	0.9	1.1	1.3	0.8	0.9	1.1	1.2	1.4
SS B3	28	44	54	65	71	80	84	89	17	27	33	41	47	55	60	67	0.8	0.9	1.0	1.2	1.4	1.0	1.1	1.3	1.4	1.6
SS B4	35	55	66	77	83	90	93	96	21	35	44	54	61	70	76	82	0.7	0.8	0.9	1.1	1.3	0.8	0.9	1.1	1.3	1.4
SS B5	26	40	50	59	66	74	79	85	17	25	31	38	43	50	55	62	0.8	0.9	1.1	1.2	1.4	1.1	1.2	1.4	1.5	1.7
SS B6	35	55	67	77	83	90	93	96	21	35	44	54	61	70	76	82	0.7	0.8	0.9	1.1	1.3	0.8	0.9	1.1	1.2	1.4
SS B7	24	37	46	55	62	70	75	81	16	23	28	35	39	46	50	57	0.9	1.0	1.1	1.3	1.4	1.2	1.3	1.4	1.6	1.8
SS B8	34	54	65	76	82	89	92	96	20	34	43	52	59	69	74	81	0.7	0.8	0.9	1.1	1.3	0.8	1.0	1.1	1.3	1.4
SS C1	29	48	60	71	77	85	90	94	15	27	35	44	51	60	65	72	0.7	0.8	1.0	1.1	1.3	0.9	1.0	1.2	1.3	1.5
SS C2	33	55	67	78	84	91	94	97	17	32	42	52	60	69	75	82	0.6	0.7	0.9	1.1	1.3	0.8	0.9	1.1	1.2	1.4
SS C3	24	38	48	59	66	75	80	86	14	22	29	36	41	49	54	61	0.8	0.9	1.1	1.2	1.4	1.1	1.2	1.4	1.5	1.7
SS C4	29	49	60	72	78	86	90	94	16	28	37	47	54	63	69	76	0.7	0.8	1.0	1.1	1.3	0.9	1.0	1.2	1.3	1.5
SS D1	40	61	72	82	87	93	95	98	24	39	48	58	65	74	79	85	0.6	0.7	0.9	1.1	1.2	0.8	0.9	1.1	1.2	1.4
SS D2	47	70	81	89	93	97	98	99	28	47	58	69	76	84	89	93	0.6	0.7	0.8	1.0	1.2	0.7	0.8	1.0	1.1	1.3
SS D3	36	55	66	76	82	89	92	95	23	35	43	52	59	67	73	79	0.7	0.8	0.9	1.1	1.3	0.9	1.0	1.1	1.3	1.5
SS D4	45	68	79	88	92	96	98	99	27	45	56	67	74	83	87	92	0.6	0.7	0.9	1.0	1.2	0.7	0.8	1.0	1.1	1.3
SS E1	36	56	67	77	83	90	93	96	21	34	43	52	59	68	73	80	0.6	0.8	0.9	1.1	1.3	0.9	1.0	1.1	1.3	1.4
SS E2	41	64	76	85	90	95	97	99	23	41	52	63	70	79	84	89	0.6	0.7	0.9	1.0	1.2	0.7	0.9	1.0	1.2	1.3
SS E3	30	46	57	67	74	82	86	91	19	29	36	44	49	57	63	69	0.7	0.8	1.0	1.2	1.3	1.0	1.1	1.3	1.4	1.6
SS E4	38	60	71	81	87	93	96	98	22	38	48	58	65	75	80	86	0.6	0.7	0.9	1.1	1.2	0.8	0.9	1.0	1.2	1.4

SAVANNAH, GEORGIA

TYPE	SOLAR SAVINGS FRACTION (%)																CONSERVATION FACTOR									
	BASE TEMP: 55 F (751 DD)								65 F (1952 DD)								55 F					65 F				
LCR	200	100	70	50	40	30	25	20	200	100	70	50	40	30	25	20	200	100	50	30	20	200	100	50	30	20
WW A1	30	45	55	65	72	80	84	89	19	27	33	40	46	53	58	65	0.8	0.9	1.0	1.2	1.3	1.1	1.2	1.3	1.5	1.6
WW A2	36	56	67	78	84	90	94	97	20	32	41	51	57	66	72	79	0.7	0.8	0.9	1.1	1.3	0.9	1.0	1.1	1.3	1.4
WW A3	38	60	71	82	87	93	96	98	20	34	44	54	62	71	77	83	0.6	0.7	0.9	1.1	1.2	0.8	0.9	1.1	1.2	1.4
WW A4	39	62	74	84	89	94	97	98	19	35	45	56	64	73	79	85	0.6	0.7	0.9	1.0	1.2	0.8	0.9	1.1	1.2	1.4
WW A5	40	64	76	86	91	95	97	99	19	36	47	59	66	76	81	87	0.6	0.7	0.9	1.0	1.2	0.8	0.9	1.0	1.2	1.4
WW A6	41	65	77	86	91	96	98	99	19	37	48	60	67	77	82	88	0.6	0.7	0.9	1.0	1.2	0.8	0.9	1.0	1.2	1.3
WW B1	31	50	61	71	78	86	90	94	17	27	35	43	49	57	63	70	0.7	0.8	1.0	1.1	1.3	1.0	1.1	1.2	1.4	1.5
WW B2	39	62	74	84	89	95	97	99	19	36	46	57	65	75	80	86	0.6	0.7	0.9	1.0	1.2	0.8	0.9	1.0	1.2	1.4
WW B3	45	69	81	89	94	97	99	99	23	42	54	66	73	82	87	92	0.6	0.7	0.8	1.0	1.2	0.7	0.8	1.0	1.1	1.3
WW B4	44	69	81	90	94	97	99	100	22	42	54	66	74	83	88	93	0.6	0.7	0.8	1.0	1.2	0.7	0.8	1.0	1.1	1.3
WW B5	42	67	79	89	93	97	98	99	20	40	53	65	73	82	87	92	0.6	0.7	0.8	1.0	1.2	0.7	0.8	1.0	1.1	1.3
WW C1	47	71	82	90	94	98	99	100	25	44	56	67	75	84	88	93	0.6	0.7	0.8	1.0	1.2	0.7	0.8	1.0	1.1	1.3
WW C2	44	68	80	89	93	97	99	99	22	41	53	65	73	82	87	92	0.6	0.7	0.8	1.0	1.2	0.7	0.8	1.0	1.1	1.3
WW C3	49	75	86	93	96	99	99	100	25	48	61	73	81	89	92	96	0.5	0.6	0.8	1.0	1.2	0.7	0.8	0.9	1.1	1.3
WW C4	45	71	83	91	95	98	99	100	22	44	57	70	77	86	90	94	0.5	0.7	0.8	1.0	1.2	0.7	0.8	0.9	1.1	1.3
TW A1	29	44	54	64	70	79	83	89	18	27	33	40	45	52	57	64	0.8	0.9	1.0	1.2	1.4	1.1	1.2	1.3	1.5	1.6
TW A2	32	51	62	72	79	87	91	94	18	29	37	46	52	61	67	74	0.7	0.8	1.0	1.1	1.3	0.9	1.0	1.2	1.3	1.5
TW A3	34	53	65	76	82	89	93	96	18	30	39	49	55	65	70	77	0.7	0.8	0.9	1.1	1.3	0.9	1.0	1.1	1.3	1.5
TW A4	33	54	65	76	83	90	93	96	17	30	39	49	56	65	71	78	0.7	0.8	0.9	1.1	1.3	0.9	1.0	1.1	1.3	1.4
TW B1	29	44	54	65	71	80	84	89	18	26	33	40	46	53	59	65	0.8	0.9	1.0	1.2	1.3	1.1	1.2	1.3	1.5	1.6
TW B2	30	49	60	71	78	85	90	94	17	28	36	44	51	60	66	73	0.7	0.8	1.0	1.1	1.3	1.0	1.1	1.2	1.3	1.5
TW B3	30	49	61	72	78	86	90	94	16	28	36	45	51	61	66	74	0.7	0.8	1.0	1.1	1.3	0.9	1.0	1.2	1.3	1.5
TW B4	28	46	58	69	76	84	88	93	15	26	34	42	49	58	64	71	0.7	0.8	1.0	1.1	1.3	1.0	1.1	1.2	1.4	1.5
TW C1	27	43	53	64	71	79	84	89	16	25	32	39	45	53	59	65	0.8	0.9	1.0	1.2	1.3	1.1	1.2	1.3	1.4	1.6
TW C2	27	44	54	65	72	81	86	91	15	25	32	40	46	55	60	67	0.7	0.8	1.0	1.2	1.3	1.0	1.1	1.3	1.4	1.6
TW C3	25	41	52	62	69	78	83	89	14	24	30	38	44	52	58	65	0.8	0.9	1.0	1.2	1.3	1.1	1.2	1.3	1.4	1.6
TW C4	23	37	46	56	63	72	78	84	14	22	27	34	39	47	52	59	0.8	0.9	1.1	1.2	1.4	1.2	1.2	1.4	1.5	1.7
TW D1	26	43	54	64	71	80	85	90	14	23	29	37	43	51	56	63	0.8	0.9	1.0	1.2	1.3	1.1	1.2	1.3	1.5	1.6
TW D2	35	57	68	79	85	92	95	97	18	33	42	52	60	69	75	82	0.6	0.7	0.9	1.1	1.3	0.8	0.9	1.1	1.2	1.4
TW D3	39	61	73	83	89	94	97	98	20	36	46	57	65	74	80	86	0.6	0.7	0.9	1.1	1.2	0.8	0.9	1.0	1.2	1.4
TW D4	40	64	76	85	91	95	97	99	20	38	49	61	68	78	83	89	0.6	0.7	0.9	1.0	1.2	0.8	0.9	1.0	1.2	1.3
TW D5	38	62	74	85	90	95	97	99	19	37	48	60	68	77	83	89	0.6	0.7	0.9	1.0	1.2	0.8	0.9	1.0	1.2	1.3
TW E1	43	66	77	87	91	96	98	99	23	40	51	62	70	79	84	89	0.6	0.7	0.9	1.0	1.2	0.8	0.9	1.0	1.2	1.3
TW E2	41	65	77	86	91	96	98	99	22	39	50	62	69	78	84	89	0.6	0.7	0.9	1.0	1.2	0.8	0.9	1.0	1.2	1.3
TW E3	48	72	83	92	95	98	99	100	25	46	58	70	78	86	90	94	0.5	0.7	0.8	1.0	1.2	0.7	0.8	0.9	1.1	1.3
TW E4	44	69	81	90	94	97	99	100	22	43	55	67	75	84	89	93	0.6	0.7	0.8	1.0	1.2	0.7	0.8	1.0	1.1	1.3
TW F1	28	44	54	64	71	79	84	89	17	25	32	39	44	52	57	64	0.8	0.9	1.0	1.2	1.4	1.1	1.2	1.3	1.5	1.6
TW F2	30	48	60	70	77	85	89	94	16	27	35	43	50	59	65	72	0.7	0.8	1.0	1.1	1.3	1.0	1.1	1.2	1.4	1.5
TW F3	30	49	61	72	79	87	91	95	15	27	35	45	51	61	67	74	0.7	0.8	0.9	1.1	1.3	0.9	1.0	1.2	1.3	1.5
TW F4	27	47	59	71	78	86	90	94	12	25	33	43	50	59	65	73	0.7	0.8	1.0	1.1	1.3	0.9	1.0	1.2	1.3	1.5
TW G1	26	42	52	63	70	79	83	89	15	24	30	38	43	51	57	64	0.8	0.9	1.0	1.2	1.3	1.1	1.2	1.3	1.5	1.6
TW G2	26	44	56	67	74	82	87	92	13	24	31	40	46	55	61	68	0.7	0.8	1.0	1.1	1.3	1.0	1.1	1.2	1.4	1.5
TW G3	24	42	54	65	72	81	86	91	11	22	29	38	45	54	60	67	0.7	0.8	1.0	1.2	1.3	1.0	1.1	1.2	1.4	1.5
TW G4	20	37	48	59	67	76	81	87	8	18	25	33	39	48	54	61	0.8	0.9	1.0	1.2	1.4	1.1	1.2	1.3	1.5	1.6
TW H1	22	38	49	59	67	76	81	87	11	20	27	34	40	48	54	61	0.8	0.9	1.0	1.2	1.4	1.1	1.2	1.3	1.5	1.6
TW H2	19	36	46	57	65	74	80	86	9	18	24	32	38	46	52	59	0.8	0.9	1.0	1.2	1.4	1.1	1.2	1.3	1.5	1.6
TW H3	16	31	41	52	59	69	75	81	7	15	21	28	34	42	47	54	0.8	0.9	1.1	1.2	1.4	1.2	1.3	1.4	1.5	1.7
TW H4	12	24	33	42	49	59	65	73	5	11	16	22	27	34	39	45	1.0	1.0	1.2	1.3	1.5	1.3	1.4	1.5	1.7	1.8
TW I1	23	40	50	61	68	77	82	88	11	20	27	34	39	47	52	59	0.8	0.9	1.0	1.2	1.4	1.1	1.2	1.4	1.5	1.7
TW I2	32	53	65	76	83	90	93	96	15	29	39	49	56	66	72	79	0.7	0.8	0.9	1.1	1.3	0.9	1.0	1.1	1.3	1.4
TW I3	34	57	69	80	86	92	95	98	17	32	42	53	60	70	76	83	0.6	0.7	0.9	1.1	1.2	0.8	0.9	1.1	1.2	1.4
TW I4	36	60	72	83	89	94	97	98	17	34	45	57	65	75	81	87	0.6	0.7	0.9	1.0	1.2	0.8	0.9	1.0	1.2	1.4
TW I5	36	60	73	83	89	94	97	99	17	34	46	58	66	76	81	87	0.6	0.7	0.9	1.0	1.2	0.8	0.9	1.0	1.2	1.3
TW J1	40	63	75	85	90	95	97	99	20	37	48	59	67	77	82	88	0.6	0.7	0.9	1.0	1.2	0.8	0.9	1.0	1.2	1.4
TW J2	38	62	74	84	89	95	97	99	19	36	47	58	66	76	81	87	0.6	0.7	0.9	1.0	1.2	0.8	0.9	1.0	1.2	1.4
TW J3	44	69	81	90	94	97	99	100	22	42	55	67	75	84	89	93	0.6	0.7	0.8	1.0	1.2	0.7	0.8	1.0	1.1	1.3
TW J4	41	66	78	88	93	97	98	99	19	39	52	64	72	82	87	92	0.6	0.7	0.8	1.0	1.2	0.7	0.8	1.0	1.1	1.3
DG A1	30	49	60	70	76	82	86	90	14	25	33	40	46	53	57	62	0.7	0.8	1.0	1.2	1.4	1.0	1.1	1.3	1.6	1.9
DG A2	30	50	62	73	79	86	90	93	15	27	36	45	51	60	65	71	0.7	0.8	1.0	1.1	1.3	0.9	1.0	1.2	1.4	1.6
DG A3	33	57	69	80	85	91	93	95	17	32	42	53	61	71	76	82	0.6	0.7	0.9	1.1	1.5	0.9	0.9	1.1	1.2	1.5
DG B1	31	54	66	77	83	89	91	94	14	26	34	44	51	60	65	70	0.7	0.7	0.9	1.1	1.4	1.0	1.0	1.2	1.4	1.7
DG B2	31	54	68	79	85	91	94	97	15	28	37	48	56	66	72	78	0.7	0.7	0.9	1.1	1.3	0.9	1.0	1.1	1.3	1.5
DG B3	34	59	73	84	89	94	96	97	18	32	43	56	65	75	81	87	0.6	0.7	0.9	1.1	1.4	0.9	0.9	1.0	1.2	1.4
DG C1	36	61	74	85	89	94	95	97	17	30	40	51	59	69	75	80	0.6	0.7	0.9	1.1	1.5	0.9	1.0	1.1	1.3	1.5
DG C2	35	60	74	85	90	95	97	98	18	32	42	55	63	74	79	85	0.6	0.7	0.9	1.1	1.3	0.9	0.9	1.0	1.2	1.4
DG C3	39	65	80	89	93	97	98	98	20	36	48	62	71	82	87	92	0.6	0.7	0.8	1.1	1.7	0.8	0.9	1.0	1.1	1.3
SS A1	38	58	69	78	84	91	94	97	23	36	45	54	61	69	75	81	0.6	0.8	0.9	1.1	1.3	0.9	1.0	1.1	1.3	1.4
SS A2	46	68	79	87	92	96	98	99	28	44	55	66	73	81	86	91	0.6	0.7	0.9	1.0	1.2	0.7	0.8	1.0	1.2	1.3
SS A3	37	55	66	76	82	89	92	95	22	34	42	51	57	66	71	77	0.7	0.8	0.9	1.1	1.3	0.9	1.0	1.2	1.3	1.5
SS A4	46	68	78	87	92	96	98	99	28	44	55	65	72	80	85	90	0.6	0.7	0.9	1.0	1.2	0.7	0.9	1.0	1.2	1.3
SS A5	36	54	64	74	80	87	91	94	23	34	42	50	56	64	69	76	0.7	0.8	1.0	1.1	1.3	0.9	1.0	1.2	1.3	1.5
SS A6	46	68	79	87	92	96	98	99	28	44	55	65	72	81	86	90	0.6	0.7	0.9	1.0	1.2	0.7	0.8	1.0	1.2	1.3
SS A7	33	49	59	68	75	82	87	91	21	31	37	45	50	58	63	69	0.7	0.8	1.0	1.2	1.3	1.0	1.1	1.3	1.4	1.6
SS A8	46	67	78	87	91	96	98	99	27	43	54	64	71	80	84	89	0.6	0.7	0.9	1.0	1.2	0.8	0.9	1.0	1.2	1.4
SS B1	33	52	63	73	79	87	91	94	20	31	39	48	54	63	68	75	0.7	0.8	1.0	1.1	1.3	0.9	1.0	1.2	1.3	1.5
SS B2	40	62	73	83	88	94	96	98	24	39	49	60	67	76	81	87	0.6	0.7	0.9	1.1	1.2	0.8	0.9	1.0	1.2	1.4
SS B3	32	50	61	71	78	85	89	93	19	30	37	46	52	60	66	72	0.7	0.8	1.0	1.1	1.3	0.9	1.1	1.2	1.4	1.5
SS B4	40	61	73	82	88	94	96	98	23	38	48	59	66	75	80	86	0.6	0.7	0.9	1.1	1.2	0.8	0.9	1.0	1.2	1.4
SS B5	30	46	56	66	72	80	85	90	19	28	35	42	48	55	61	67	0.7	0.9	1.0	1.2	1.3	1.0	1.1	1.3	1.4	1.6
SS B6	40	61	73	83	88	94	96	98	23	38	48	59	66	75	80	86	0.6	0.7	0.9	1.1	1.2	0.8	0.9	1.0	1.2	1.4
SS B7	27	42	52	61	68	76	81	87	17	26	32	38	43	51	56	62	0.8	0.9	1.1	1.2	1.4	1.1	1.2	1.4	1.5	1.7
SS B8	39	60	72	81	87	93	96	98	22	37	47	57	64	74	79	85	0.6	0.7	0.9	1.1	1.2	0.8	0.9	1.1	1.2	1.4
SS C1	34	55	66	77	83	90	94	97	17	30	39	49	56	65	71	78	0.7	0.8	0.9	1.1	1.3	0.9	1.0	1.1	1.3	1.4
SS C2	39	61	73	83	89	94	97	98	20	36	46	57	65	74	80	86	0.6	0.7	0.9	1.1	1.2	0.8	0.9	1.0	1.2	1.4
SS C3	27	44	55	65	72	81	85	90	15	25	32	40	45	54	59	66	0.7	0.8	1.0	1.2	1.3	1.0	1.1	1.3	1.4	1.6
SS C4	34	55	67	78	84	91	94	97	18	32	41	51	59	68	74	81	0.6	0.8	0.9	1.1	1.3	0.9	1.0	1.1	1.3	1.4
SS D1	46	67	78	87	92	96	98	99	27	43	53	63	70	79	84	89	0.6	0.7	0.9	1.0	1.2	0.8	0.9	1.0	1.2	1.4
SS D2	53	76	86	93	96	99	99	100	31	51	63	74	81	88	92	95	0.5	0.7	0.8	1.0	1.2	0.7	0.8	0.9	1.1	1.3
SS D3	41	61	72	82	87	93	95	98	25	39	48	57	64	73	78	84	0.6	0.7	0.9	1.1	1.2	0.8	0.9	1.1	1.2	1.4
SS D4	51	74	84	92	95	98	99	100	30	49	61	72	79	87	91	94	0.5	0.7	0.8	1.0	1.2	0.7	0.8	0.9	1.1	1.3
SS E1	41	62	74	83	88	94	96	98	23	38	47	57	64	73	79	84	0.6	0.7	0.9	1.1	1.2	0.8	0.9	1.1	1.2	1.4
SS E2	48	71	82	90	94	97	99	100	27	45	57	68	75	84	88	92	0.6	0.7	0.8	1.0	1.2	0.7	0.8	1.0	1.1	1.3
SS E3	34	52	63	73	80	87	91	94	21	32	40	48	54	63	68	75	0.7	0.8	1.0	1.1	1.3	0.9	1.0	1.2	1.3	1.5
SS E4	44	66	77	87	91	96	98	99	25	42	52	63	71	79	84	89	0.6	0.7	0.9	1.0	1.2	0.8	0.9	1.0	1.2	1.3

BOISE, IDAHO

TYPE	SOLAR SAVINGS FRACTION (%)																CONSERVATION FACTOR									
	BASE TEMP: 55 F (3408 DD)								65 F (5837 DD)								55 F					65 F				
LCR	100	70	50	40	30	25	20	15	100	70	50	40	30	25	20	15	100	50	30	20	15	100	50	30	20	15
WW A1	22	26	31	34	39	43	47	53	18	20	24	26	30	33	36	41	1.4	1.7	1.9	2.1	2.3	1.7	2.0	2.2	2.5	2.7
WW A2	24	30	37	42	48	53	58	66	18	23	28	32	38	41	46	53	1.2	1.4	1.6	1.8	2.0	1.4	1.6	1.8	2.0	2.2
WW A3	25	32	39	44	52	57	62	70	18	23	29	34	40	45	50	57	1.1	1.3	1.5	1.7	1.9	1.3	1.5	1.7	1.9	2.1
WW A4	25	32	40	46	54	59	65	72	17	23	30	35	42	46	52	59	1.1	1.3	1.5	1.7	1.8	1.2	1.4	1.6	1.9	2.1
WW A5	25	33	42	48	56	61	67	74	17	24	31	36	43	48	54	61	1.1	1.2	1.4	1.6	1.8	1.2	1.4	1.6	1.8	2.0
WW A6	25	34	42	48	57	62	68	75	17	24	32	37	44	49	55	63	1.0	1.2	1.4	1.6	1.8	1.2	1.4	1.6	1.8	2.0
WW B1	20	24	29	33	38	42	46	52	14	17	21	23	27	30	33	36	1.4	1.6	1.9	2.1	2.3	1.6	1.9	2.3	2.7	3.1
WW B2	26	34	42	48	56	61	68	75	18	25	32	38	45	50	56	64	1.1	1.2	1.4	1.6	1.8	1.2	1.4	1.6	1.8	1.9
WW B3	30	39	48	55	63	68	75	82	22	30	38	44	52	57	64	72	1.0	1.2	1.3	1.5	1.7	1.1	1.3	1.5	1.7	1.8
WW B4	30	40	50	57	66	72	78	85	22	31	40	47	56	62	69	77	0.9	1.1	1.3	1.5	1.6	1.0	1.2	1.4	1.6	1.7
WW B5	29	39	49	57	66	72	78	85	21	30	40	47	56	62	69	77	0.9	1.1	1.3	1.5	1.6	1.0	1.2	1.4	1.6	1.7
WW C1	32	41	50	57	65	71	77	84	24	32	40	47	55	60	67	74	1.0	1.1	1.3	1.5	1.7	1.1	1.2	1.4	1.6	1.8
WW C2	30	39	49	56	65	70	77	84	22	30	39	46	54	60	67	75	1.0	1.1	1.3	1.5	1.7	1.1	1.2	1.4	1.6	1.8
WW C3	35	46	57	64	74	79	85	91	26	37	48	55	65	70	77	85	0.9	1.0	1.2	1.4	1.6	0.9	1.1	1.3	1.5	1.6
WW C4	32	43	54	61	71	76	83	89	24	34	44	52	62	68	75	83	0.9	1.1	1.2	1.4	1.6	1.0	1.1	1.3	1.5	1.6
TW A1	22	26	30	34	39	42	47	53	17	20	24	26	30	33	36	41	1.4	1.7	1.9	2.1	2.3	1.7	1.9	2.2	2.4	2.7
TW A2	23	28	34	39	45	49	55	62	17	21	26	30	35	38	43	49	1.3	1.5	1.7	1.9	2.0	1.4	1.7	1.9	2.1	2.3
TW A3	23	29	36	40	47	52	58	65	16	21	27	31	37	41	46	53	1.2	1.4	1.6	1.8	1.9	1.4	1.6	1.8	2.0	2.2
TW A4	22	28	35	41	48	53	59	66	15	21	26	31	37	41	47	54	1.2	1.4	1.6	1.7	1.9	1.3	1.5	1.7	2.0	2.1
TW B1	21	26	31	34	40	43	48	55	17	20	24	27	31	34	38	43	1.4	1.6	1.8	2.0	2.2	1.6	1.9	2.1	2.3	2.5
TW B2	21	27	33	38	44	48	54	61	16	20	25	29	34	38	43	49	1.3	1.5	1.6	1.8	2.0	1.4	1.6	1.9	2.1	2.3
TW B3	21	27	33	38	45	49	55	62	15	20	25	29	35	39	44	50	1.2	1.4	1.6	1.8	2.0	1.4	1.6	1.8	2.0	2.2
TW B4	20	25	31	36	43	47	53	61	14	19	24	28	33	37	42	49	1.3	1.5	1.6	1.8	2.0	1.4	1.6	1.8	2.0	2.2
TW C1	20	25	30	34	40	44	49	55	15	19	23	26	31	34	38	44	1.4	1.6	1.8	2.0	2.1	1.6	1.8	2.0	2.2	2.4
TW C2	19	24	30	34	41	45	50	58	14	18	23	26	32	35	40	46	1.3	1.5	1.7	1.9	2.1	1.5	1.7	1.9	2.1	2.3
TW C3	18	23	29	33	39	43	49	56	14	18	22	25	31	34	39	45	1.4	1.5	1.7	1.9	2.1	1.6	1.7	1.9	2.1	2.3
TW C4	18	22	27	30	36	40	45	51	14	17	21	24	28	31	36	42	1.5	1.6	1.8	2.0	2.2	1.7	1.9	2.1	2.3	2.4
TW D1	17	21	26	29	34	38	42	48	12	15	18	21	24	27	30	34	1.4	1.7	1.9	2.2	2.4	1.7	2.0	2.3	2.7	3.0
TW D2	24	31	39	45	52	57	63	71	18	24	30	35	42	46	52	60	1.1	1.3	1.5	1.7	1.8	1.3	1.4	1.6	1.8	2.0
TW D3	26	34	42	48	56	61	67	75	19	25	33	38	45	50	56	64	1.1	1.3	1.4	1.6	1.8	1.2	1.4	1.6	1.8	2.0
TW D4	28	37	46	52	61	66	73	81	20	28	36	43	51	56	63	71	1.0	1.2	1.4	1.5	1.7	1.1	1.3	1.5	1.6	1.8
TW D5	27	36	46	53	62	67	74	82	20	28	37	43	52	58	64	73	1.0	1.2	1.3	1.5	1.7	1.1	1.3	1.4	1.6	1.8
TW E1	29	38	46	52	61	66	72	79	22	29	37	43	50	55	62	69	1.0	1.2	1.4	1.6	1.7	1.1	1.3	1.5	1.7	1.9
TW E2	29	37	46	53	61	67	73	80	21	29	37	43	51	56	63	71	1.0	1.2	1.4	1.5	1.7	1.1	1.3	1.5	1.7	1.8
TW E3	34	44	54	61	70	76	82	88	26	35	45	52	61	67	73	81	0.9	1.1	1.3	1.4	1.6	1.0	1.2	1.3	1.5	1.7
TW E4	31	42	52	59	68	74	80	87	24	33	43	50	59	65	72	80	0.9	1.1	1.3	1.5	1.6	1.0	1.2	1.3	1.5	1.7
TW F1	20	24	29	33	38	42	46	53	15	19	22	25	29	32	36	41	1.4	1.6	1.8	2.1	2.2	1.7	1.9	2.1	2.4	2.6
TW F2	20	26	32	36	43	47	52	59	15	19	24	27	32	36	41	47	1.3	1.5	1.7	1.9	2.0	1.5	1.7	1.9	2.1	2.3
TW F3	19	25	32	37	44	48	54	62	13	18	23	27	33	37	42	49	1.2	1.4	1.6	1.8	2.0	1.4	1.6	1.8	2.0	2.2
TW F4	17	24	30	35	42	47	53	61	11	16	22	26	32	36	41	49	1.2	1.4	1.6	1.8	2.0	1.4	1.6	1.8	2.0	2.2
TW G1	18	23	28	32	37	41	46	53	13	17	21	24	28	31	35	41	1.4	1.6	1.8	2.0	2.2	1.6	1.9	2.1	2.3	2.5
TW G2	17	23	29	33	40	44	49	57	12	16	21	24	30	33	38	45	1.3	1.5	1.7	1.9	2.1	1.5	1.7	1.9	2.1	2.3
TW G3	15	21	27	32	38	43	49	56	10	14	19	23	29	32	37	44	1.3	1.5	1.7	1.9	2.0	1.5	1.7	1.9	2.1	2.3
TW G4	13	18	24	28	34	39	44	52	8	12	16	20	25	29	34	41	1.4	1.6	1.8	1.9	2.1	1.6	1.8	2.0	2.2	2.3
TW H1	15	20	25	29	35	39	44	51	10	14	18	21	26	29	33	39	1.4	1.6	1.8	2.0	2.2	1.7	1.9	2.1	2.3	2.5
TW H2	13	17	23	27	33	37	43	50	8	12	16	19	24	28	32	39	1.4	1.6	1.8	2.0	2.2	1.6	1.8	2.0	2.2	2.4
TW H3	11	15	20	24	30	34	39	46	6	10	14	17	22	25	29	36	1.5	1.7	1.9	2.1	2.2	1.7	1.9	2.1	2.3	2.5
TW H4	8	12	16	19	24	28	33	39	5	7	11	13	17	20	24	30	1.6	1.8	2.0	2.2	2.4	1.9	2.1	2.3	2.5	2.6
TW I1	14	19	23	27	31	35	39	45	9	12	16	18	22	24	27	31	1.5	1.7	2.0	2.2	2.4	1.8	2.1	2.4	2.7	3.1
TW I2	21	28	36	41	49	54	60	68	15	20	27	32	38	43	49	57	1.1	1.3	1.5	1.7	1.9	1.3	1.5	1.7	1.9	2.0
TW I3	23	00	38	44	52	57	64	72	16	22	29	35	42	47	53	61	1.1	1.3	1.5	1.6	1.8	1.2	1.4	1.6	1.8	2.0
TW I4	25	33	43	49	58	63	70	78	18	25	33	39	48	53	60	69	1.0	1.2	1.4	1.6	1.7	1.1	1.3	1.5	1.7	1.8
TW I5	25	34	43	50	59	65	72	80	18	26	34	41	49	55	62	71	1.0	1.2	1.4	1.5	1.7	1.1	1.3	1.4	1.6	1.8
TW J1	27	35	44	50	59	64	70	78	20	27	35	40	48	53	60	68	1.0	1.2	1.4	1.6	1.7	1.1	1.3	1.5	1.7	1.9
TW J2	26	34	44	50	59	64	71	79	19	26	34	40	48	54	61	69	1.0	1.2	1.4	1.6	1.7	1.1	1.3	1.5	1.7	1.8
TW J3	31	41	51	59	68	73	80	87	23	32	42	49	58	64	71	80	0.9	1.1	1.3	1.5	1.6	1.0	1.2	1.3	1.5	1.7
TW J4	29	39	49	56	66	72	78	85	21	30	40	47	56	62	69	78	0.9	1.1	1.0	1.5	1.6	1.0	1.2	1.4	1.5	1.7
DG A1	18	23	28	31	35	38	41	44	12	15	19	21	25	26	28	30	1.3	1.7	2.1	2.7	3.3	1.6	2.0	2.5	3.5	-
DG A2	20	26	33	37	44	48	53	59	14	19	24	28	34	38	42	48	1.2	1.4	1.7	1.9	2.2	1.4	1.6	1.9	2.2	2.5
DG A3	24	32	40	46	54	59	66	73	18	24	32	37	45	50	56	64	1.1	1.3	1.4	1.7	1.9	1.2	1.4	1.6	1.8	2.0
DG B1	18	24	29	33	39	42	46	50	12	16	20	23	27	29	32	35	1.3	1.5	1.9	2.3	2.6	1.6	1.9	2.2	2.8	3.7
DG B2	21	27	34	40	47	52	57	65	15	20	26	30	36	40	46	52	1.2	1.4	1.6	1.8	2.0	1.4	1.5	1.7	2.0	2.3
DG B3	25	33	41	48	57	62	69	77	19	25	33	39	47	52	59	68	1.1	1.2	1.4	1.5	1.7	1.2	1.3	1.5	1.7	1.8
DG C1	22	28	35	40	46	50	54	60	15	19	25	28	33	37	41	45	1.2	1.4	1.7	2.0	2.3	1.4	1.6	1.9	2.4	2.9
DG C2	24	31	39	45	53	58	64	72	17	23	30	35	42	47	53	60	1.1	1.3	1.5	1.7	1.9	1.3	1.4	1.6	1.9	2.1
DG C3	28	36	46	53	62	68	75	83	21	28	37	43	52	58	65	74	1.0	1.1	1.3	1.5	1.7	1.1	1.3	1.4	1.6	1.7
SS A1	26	32	38	42	48	52	58	64	22	26	32	35	41	44	48	54	1.2	1.4	1.6	1.8	2.0	1.3	1.6	1.8	2.1	2.3
SS A2	32	40	48	53	61	66	72	79	27	33	41	46	53	58	63	70	1.0	1.2	1.4	1.6	1.7	1.1	1.3	1.5	1.7	1.9
SS A3	24	29	35	38	44	48	52	58	20	24	29	32	36	39	43	48	1.3	1.5	1.8	2.0	2.2	1.4	1.7	2.0	2.3	2.6
SS A4	31	38	46	52	59	64	70	77	26	32	39	44	51	56	61	68	1.0	1.2	1.4	1.6	1.8	1.1	1.3	1.6	1.8	1.9
SS A5	25	30	35	39	45	48	53	59	21	25	30	33	37	40	44	49	1.3	1.5	1.8	2.0	2.2	1.4	1.7	2.0	2.3	2.5
SS A6	32	39	47	53	60	65	71	78	26	33	40	45	52	57	62	69	1.0	1.2	1.4	1.6	1.8	1.1	1.3	1.5	1.7	1.9
SS A7	23	27	31	34	39	42	46	51	19	23	26	29	32	35	38	41	1.4	1.7	2.0	2.2	2.4	1.6	1.9	2.2	2.6	3.0
SS A8	30	38	45	50	58	62	68	75	25	32	38	43	50	54	59	66	1.1	1.3	1.5	1.6	1.8	1.2	1.4	1.6	1.8	2.0
SS B1	23	28	34	38	44	48	53	59	18	22	27	31	36	39	43	49	1.3	1.5	1.7	1.9	2.1	1.4	1.7	1.9	2.2	2.4
SS B2	28	36	43	49	57	62	68	75	23	30	37	42	49	54	59	67	1.1	1.3	1.4	1.6	1.8	1.2	1.4	1.6	1.8	1.9
SS B3	21	26	31	35	41	44	49	55	17	21	25	28	33	36	39	44	1.3	1.6	1.8	2.0	2.2	1.5	1.8	2.0	2.3	2.6
SS B4	27	35	42	48	55	60	66	73	22	29	35	40	47	52	57	65	1.1	1.3	1.5	1.7	1.8	1.2	1.4	1.6	1.8	2.0
SS B5	21	25	30	34	39	42	46	52	17	21	25	27	32	34	38	43	1.4	1.6	1.9	2.1	2.3	1.6	1.9	2.1	2.4	2.7
SS B6	28	35	43	48	56	61	67	74	22	29	36	41	48	52	58	65	1.1	1.3	1.5	1.6	1.8	1.2	1.4	1.6	1.8	1.9
SS B7	19	23	27	30	35	38	41	47	16	19	22	24	28	30	33	37	1.5	1.8	2.0	2.3	2.5	1.7	2.0	2.3	2.7	3.0
SS B8	26	33	41	46	53	58	64	71	21	27	34	39	45	50	55	62	1.1	1.3	1.5	1.7	1.8	1.2	1.4	1.6	1.8	2.0
SS C1	22	28	35	40	47	51	57	64	15	20	26	30	35	39	44	50	1.2	1.4	1.6	1.8	2.0	1.4	1.6	1.8	2.1	2.3
SS C2	26	34	42	48	56	62	68	76	19	25	33	38	46	51	57	65	1.1	1.2	1.4	1.6	1.8	1.2	1.4	1.6	1.8	1.9
SS C3	19	24	29	33	39	43	48	55	14	18	22	25	30	33	37	43	1.4	1.6	1.8	2.0	2.2	1.6	1.8	2.0	2.3	2.5
SS C4	24	31	38	44	51	56	62	70	17	23	29	34	41	46	52	59	1.1	1.3	1.5	1.7	1.8	1.3	1.5	1.6	1.8	2.0
SS D1	30	37	44	49	56	60	65	72	24	30	36	41	46	50	55	61	1.1	1.3	1.5	1.7	1.9	1.2	1.4	1.7	1.9	2.1
SS D2	36	45	54	60	68	73	79	86	29	38	46	52	60	65	71	78	0.9	1.1	1.3	1.5	1.6	1.0	1.2	1.4	1.6	1.8
SS D3	28	34	40	45	51	55	60	67	23	28	33	37	43	46	50	56	1.2	1.4	1.6	1.8	2.0	1.3	1.5	1.8	2.1	2.3
SS D4	35	43	52	58	66	71	77	84	29	37	45	51	58	63	69	76	1.0	1.1	1.3	1.5	1.7	1.0	1.2	1.4	1.6	1.8
SS E1	26	32	39	44	50	54	60	66	20	25	31	35	40	44	48	54	1.2	1.4	1.6	1.8	2.0	1.3	1.5	1.8	2.1	2.3
SS E2	31	40	48	54	62	68	73	81	25	32	40	46	53	58	64	71	1.0	1.2	1.4	1.6	1.7	1.1	1.3	1.5	1.7	1.9
SS E3	23	28	34	38	43	47	52	58	19	23	27	31	35	38	42	47	1.3	1.5	1.7	2.0	2.1	1.5	1.7	2.0	2.3	2.5
SS E4	29	37	45	51	59	64	70	77	24	30	38	43	50	54	60	67	1.0	1.2	1.4	1.6	1.8	1.1	1.3	1.5	1.8	1.9

LEWISTON, IDAHO

TYPE	SOLAR SAVINGS FRACTION (%)																CONSERVATION FACTOR									
BASE TEMP:	55 F (3057 DD)								65 F (5472 DD)								55 F					65 F				
LCR	100	70	50	40	30	25	20	15	100	70	50	40	30	25	20	15	100	50	30	20	15	100	50	30	20	15
WW A1	19	22	25	27	31	33	36	39	15	17	19	21	23	24	26	28	1.7	2.0	2.3	2.7	3.0	2.1	2.5	3.0	3.6	4.3
WW A2	19	24	29	32	37	40	45	50	14	17	21	24	27	30	33	37	1.4	1.6	1.9	2.2	2.4	1.6	1.9	2.3	2.7	3.1
WW A3	19	24	30	34	39	43	48	54	13	17	21	25	29	32	35	40	1.3	1.5	1.8	2.0	2.2	1.5	1.8	2.1	2.5	2.8
WW A4	19	24	30	35	41	45	50	56	12	17	22	25	30	33	37	42	1.3	1.5	1.7	2.0	2.2	1.4	1.7	2.0	2.4	2.7
WW A5	19	25	31	36	42	46	51	58	12	17	22	26	31	34	38	43	1.2	1.5	1.7	1.9	2.1	1.4	1.7	2.0	2.3	2.6
WW A6	18	25	31	36	43	47	53	60	12	17	22	26	31	35	39	44	1.2	1.4	1.7	1.9	2.1	1.4	1.6	1.9	2.2	2.5
WW B1	15	18	21	24	27	29	31	34	10	12	14	15	17	18	18	18	1.7	2.0	2.5	3.1	3.7	2.1	2.7	3.8	-	-
WW B2	19	26	32	37	44	49	54	62	13	18	24	28	34	38	43	49	1.2	1.4	1.6	1.8	2.0	1.4	1.6	1.8	2.1	2.3
WW B3	23	30	38	43	51	55	61	69	16	22	29	34	40	44	50	56	1.1	1.3	1.5	1.7	1.9	1.2	1.5	1.7	1.9	2.1
WW B4	23	31	40	46	55	60	67	75	16	23	31	37	45	50	56	64	1.0	1.2	1.4	1.6	1.8	1.2	1.3	1.5	1.7	1.9
WW B5	22	31	40	46	55	60	67	75	15	23	31	37	45	51	58	66	1.0	1.2	1.4	1.6	1.7	1.1	1.3	1.5	1.7	1.9
WW C1	24	32	40	46	53	58	64	72	18	24	31	36	43	47	53	60	1.1	1.3	1.5	1.7	1.8	1.2	1.4	1.6	1.9	2.1
WW C2	23	31	39	45	53	58	65	73	16	23	30	36	43	48	54	62	1.1	1.3	1.5	1.6	1.8	1.2	1.4	1.6	1.8	2.0
WW C3	27	37	47	54	63	69	75	83	20	29	38	45	54	60	67	75	1.0	1.1	1.3	1.5	1.6	1.0	1.2	1.4	1.6	1.7
WW C4	25	34	44	51	60	66	73	81	18	26	36	42	51	57	64	73	1.0	1.2	1.3	1.5	1.7	1.1	1.2	1.4	1.6	1.8
TW A1	19	22	25	27	31	33	36	40	15	17	19	21	23	25	27	29	1.7	2.0	2.3	2.6	2.9	2.0	2.4	2.8	3.3	3.9
TW A2	18	22	27	30	35	38	42	47	14	17	20	23	26	28	31	35	1.4	1.7	2.0	2.2	2.5	1.7	2.0	2.3	2.7	3.1
TW A3	18	22	28	31	36	40	44	50	13	16	20	23	27	30	33	38	1.4	1.6	1.9	2.1	2.3	1.6	1.9	2.2	2.5	2.8
TW A4	17	22	27	31	37	41	45	52	11	15	20	23	27	30	34	39	1.3	1.6	1.8	2.1	2.2	1.5	1.8	2.1	2.4	2.7
TW B1	18	21	25	28	31	34	37	42	14	16	19	21	24	26	28	31	1.6	1.9	2.2	2.5	2.7	1.9	2.2	2.6	3.0	3.4
TW B2	17	21	26	29	34	38	42	48	12	16	19	22	26	28	32	36	1.4	1.7	1.9	2.2	2.4	1.7	2.0	2.3	2.6	2.9
TW B3	16	21	26	29	35	38	43	49	12	15	19	22	26	29	32	37	1.4	1.7	1.9	2.1	2.3	1.6	1.9	2.2	2.5	2.8
TW B4	15	20	25	28	33	37	42	48	11	14	18	21	25	28	32	37	1.4	1.7	1.9	2.1	2.3	1.7	1.9	2.2	2.5	2.7
TW C1	17	20	24	27	31	34	38	43	13	15	18	21	24	26	29	33	1.6	1.8	2.1	2.3	2.6	1.9	2.1	2.4	2.7	3.0
TW C2	16	20	24	27	32	35	40	45	12	15	18	21	24	27	30	35	1.5	1.7	2.0	2.2	2.4	1.8	2.0	2.3	2.6	2.8
TW C3	15	19	23	26	31	34	38	44	11	14	18	20	24	26	30	34	1.5	1.8	2.0	2.2	2.4	1.8	2.0	2.3	2.6	2.8
TW C4	15	18	22	25	29	32	36	41	12	14	17	19	23	25	28	33	1.7	1.9	2.1	2.3	2.5	1.9	2.2	2.4	2.6	2.9
TW D1	13	16	19	21	25	27	29	32	9	11	13	14	16	17	18	19	1.7	2.1	2.5	3.0	3.4	2.2	2.7	3.4	4.5	-
TW D2	19	25	31	35	41	45	51	58	14	18	23	27	32	36	40	46	1.3	1.5	1.7	1.9	2.1	1.4	1.7	1.9	2.2	2.4
TW D3	20	26	33	37	44	48	54	61	14	19	25	29	34	38	43	49	1.2	1.4	1.6	1.8	2.0	1.4	1.6	1.8	2.1	2.3
TW D4	22	29	37	42	50	55	61	69	16	22	29	34	41	45	51	59	1.1	1.3	1.5	1.7	1.8	1.2	1.4	1.6	1.8	2.0
TW D5	21	29	37	43	51	56	63	71	15	22	29	35	42	47	54	62	1.1	1.3	1.5	1.6	1.8	1.2	1.4	1.6	1.8	1.9
TW E1	23	30	37	42	49	54	59	67	17	22	29	33	39	43	49	55	1.1	1.4	1.6	1.8	1.9	1.3	1.5	1.7	2.0	2.2
TW E2	22	30	37	43	50	55	61	69	16	22	29	34	40	45	51	58	1.1	1.3	1.5	1.7	1.9	1.2	1.5	1.7	1.9	2.1
TW E3	26	35	44	51	59	65	71	79	20	27	36	42	50	55	62	70	1.0	1.2	1.4	1.6	1.7	1.1	1.3	1.5	1.7	1.8
TW E4	25	33	42	49	58	63	70	78	18	26	34	40	49	54	61	69	1.0	1.2	1.4	1.6	1.7	1.1	1.3	1.5	1.7	1.8
TW F1	17	20	23	26	30	32	35	39	13	15	18	19	22	24	26	28	1.7	1.9	2.3	2.6	2.9	2.0	2.4	2.8	3.2	3.8
TW F2	16	20	25	28	32	36	40	45	11	14	18	20	24	26	29	33	1.5	1.7	2.0	2.3	2.5	1.7	2.0	2.4	2.7	3.1
TW F3	15	19	24	28	33	37	41	47	10	13	17	20	24	27	30	35	1.4	1.7	1.9	2.2	2.4	1.7	1.9	2.2	2.6	2.9
TW F4	13	17	23	27	32	36	40	47	8	11	15	19	23	26	30	35	1.4	1.6	1.9	2.1	2.3	1.6	1.9	2.2	2.5	2.7
TW G1	15	18	22	25	29	31	35	40	11	13	16	18	21	23	25	29	1.6	1.9	2.2	2.5	2.7	1.9	2.3	2.6	3.0	3.4
TW G2	13	17	22	25	30	33	37	43	9	12	15	18	21	24	27	32	1.5	1.7	2.0	2.3	2.5	1.8	2.0	2.3	2.7	3.0
TW G3	11	16	20	24	29	32	37	43	7	10	14	17	21	23	27	32	1.5	1.7	2.0	2.2	2.4	1.7	2.0	2.3	2.6	2.8
TW G4	9	13	18	21	26	29	34	40	5	8	12	14	18	21	25	29	1.6	1.8	2.0	2.3	2.5	1.8	2.1	2.3	2.6	2.8
TW H1	12	15	19	22	26	29	33	38	8	11	13	16	19	21	24	28	1.6	1.9	2.1	2.4	2.6	1.9	2.2	2.5	2.9	3.2
TW H2	9	13	17	20	25	28	32	38	6	8	12	14	17	20	23	28	1.6	1.8	2.1	2.3	2.5	1.9	2.1	2.4	2.7	3.0
TW H3	8	11	15	18	23	26	30	35	4	7	10	12	16	18	21	26	1.7	1.9	2.1	2.4	2.6	2.0	2.2	2.5	2.7	3.0
TW H4	6	9	12	15	19	21	25	30	3	5	8	10	13	15	18	22	1.9	2.1	2.3	2.5	2.7	2.2	2.4	2.6	2.9	3.1
TW I1	11	14	17	19	22	24	26	30	7	9	10	12	14	15	16	17	1.8	2.1	2.6	3.0	3.5	2.2	2.8	3.5	4.7	-
TW I2	16	22	28	32	38	42	48	55	11	15	20	24	29	32	37	43	1.3	1.5	1.7	1.9	2.1	1.5	1.7	1.9	2.2	2.4
TW I3	17	23	30	34	41	45	51	58	12	16	22	26	32	36	40	47	1.2	1.5	1.7	1.9	2.0	1.4	1.6	1.9	2.1	2.3
TW I4	19	26	34	39	47	52	58	67	13	19	26	31	38	42	48	56	1.1	1.3	1.5	1.7	1.9	1.3	1.5	1.7	1.9	2.0
TW I5	19	26	35	41	49	54	61	69	13	19	27	32	40	45	51	59	1.1	1.3	1.5	1.7	1.8	1.2	1.4	1.6	1.8	2.0
TW J1	21	27	35	40	47	52	58	65	15	20	26	31	37	42	47	54	1.2	1.4	1.6	1.8	1.9	1.3	1.5	1.7	2.0	2.2
TW J2	20	27	34	40	48	53	59	67	14	20	26	31	38	43	49	56	1.1	1.3	1.5	1.7	1.9	1.3	1.5	1.7	1.9	2.1
TW J3	24	32	42	48	57	63	69	78	17	25	33	40	48	54	60	69	1.0	1.2	1.4	1.6	1.7	1.1	1.3	1.5	1.7	1.8
TW J4	22	31	40	46	55	61	68	76	16	23	32	38	46	52	59	68	1.0	1.2	1.4	1.6	1.7	1.1	1.3	1.5	1.7	1.8
DG A1	13	17	20	22	25	26	28	29	8	11	13	14	16	16	17	17	1.6	2.1	2.7	4.0	-	2.0	2.7	4.0	-	-
DG A2	16	20	25	29	34	37	41	46	11	15	19	22	26	28	31	35	1.4	1.7	2.0	2.3	2.6	1.6	1.9	2.2	2.7	3.1
DG A3	20	26	32	37	44	49	54	62	15	20	25	30	36	40	45	52	1.2	1.4	1.6	1.8	2.0	1.4	1.6	1.8	2.0	2.2
DG B1	14	17	21	23	27	29	31	33	9	11	13	15	17	18	19	20	1.6	2.0	2.4	3.3	4.4	2.0	2.4	3.3	-	-
DG B2	17	21	27	31	36	40	44	50	11	15	20	23	27	30	34	38	1.4	1.6	1.8	2.1	2.4	1.6	1.8	2.1	2.5	2.8
DG B3	20	26	33	39	46	51	57	65	15	20	26	31	38	42	48	55	1.2	1.4	1.5	1.7	1.9	1.4	1.5	1.7	1.9	2.1
DG C1	17	21	25	28	33	35	38	42	11	14	17	20	23	24	26	28	1.4	1.8	2.1	2.6	3.3	1.7	2.1	2.6	3.7	-
DG C2	19	25	31	35	41	46	51	57	14	18	23	27	32	36	40	46	1.3	1.5	1.7	2.0	2.2	1.5	1.6	1.9	2.2	2.5
DG C3	23	30	37	43	51	56	63	71	17	23	30	35	42	47	53	61	1.2	1.3	1.5	1.7	1.8	1.3	1.4	1.6	1.8	2.0
SS A1	21	26	30	34	38	41	45	50	18	21	25	28	31	34	37	40	1.4	1.7	2.0	2.2	2.4	1.6	1.9	2.2	2.6	3.0
SS A2	26	32	39	44	50	55	60	67	21	27	33	37	43	47	52	58	1.2	1.4	1.6	1.8	1.9	1.3	1.5	1.7	2.0	2.2
SS A3	20	23	27	30	34	36	39	43	16	19	22	24	27	29	31	33	1.5	1.8	2.2	2.5	2.8	1.7	2.1	2.6	3.2	4.0
SS A4	25	31	37	42	48	52	58	65	21	26	32	36	41	45	49	55	1.2	1.4	1.6	1.8	2.0	1.3	1.5	1.8	2.1	2.3
SS A5	21	25	29	31	35	38	41	45	18	21	24	26	29	31	33	36	1.5	1.8	2.1	2.5	2.7	1.7	2.1	2.5	3.0	3.5
SS A6	26	32	38	43	49	54	59	66	21	26	32	36	42	46	51	57	1.2	1.4	1.6	1.8	2.0	1.3	1.5	1.8	2.0	2.2
SS A7	19	22	25	27	30	32	35	38	16	19	21	22	25	26	27	29	1.7	2.0	2.5	2.9	3.3	1.9	2.4	3.0	3.9	-
SS A8	25	30	36	41	47	51	56	62	20	25	30	34	39	43	47	53	1.2	1.4	1.7	1.9	2.0	1.3	1.6	1.9	2.1	2.4
SS B1	18	22	27	30	34	37	41	46	15	18	21	24	27	29	32	36	1.5	1.7	2.0	2.3	2.5	1.7	2.0	2.3	2.8	3.2
SS B2	23	29	36	40	47	51	57	64	19	24	30	34	40	44	49	55	1.2	1.4	1.6	1.8	2.0	1.3	1.5	1.8	2.0	2.2
SS B3	17	21	24	27	31	34	37	41	14	16	19	21	24	26	28	31	1.5	1.9	2.2	2.5	2.8	1.8	2.2	2.6	3.2	3.8
SS B4	22	28	34	39	45	49	55	62	18	23	28	32	38	42	46	53	1.2	1.4	1.7	1.9	2.0	1.3	1.6	1.8	2.0	2.3
SS B5	18	21	24	27	30	33	36	40	15	17	20	22	24	26	28	31	1.6	1.9	2.3	2.6	2.9	1.9	2.2	2.7	3.2	3.7
SS B6	22	28	34	39	45	50	55	62	18	23	28	33	38	42	47	53	1.2	1.4	1.6	1.8	2.0	1.3	1.6	1.8	2.0	2.2
SS B7	16	19	22	24	27	29	31	35	13	15	17	19	21	23	24	26	1.8	2.1	2.5	2.9	3.3	2.1	2.5	3.0	3.7	4.6
SS B8	21	27	33	37	43	47	52	59	17	22	27	31	36	39	44	50	1.2	1.5	1.7	1.9	2.1	1.4	1.6	1.9	2.1	2.4
SS C1	17	21	26	30	35	38	42	48	11	15	18	21	25	27	30	33	1.4	1.7	1.9	2.2	2.4	1.6	2.0	2.3	2.8	3.3
SS C2	20	26	33	38	45	49	55	62	14	19	25	29	35	39	44	50	1.2	1.4	1.6	1.8	2.0	1.3	1.6	1.8	2.1	2.3
SS C3	16	19	23	26	30	33	36	41	11	14	17	19	22	24	27	30	1.6	1.8	2.1	2.4	2.7	1.9	2.2	2.6	3.0	3.4
SS C4	19	24	30	35	41	45	50	57	13	18	23	26	32	35	40	46	1.3	1.5	1.7	1.9	2.1	1.5	1.7	1.9	2.2	2.4
SS D1	23	29	34	38	43	47	51	57	19	23	28	31	35	38	41	45	1.3	1.5	1.8	2.0	2.2	1.4	1.7	2.1	2.5	2.9
SS D2	28	36	44	49	57	62	68	75	23	30	37	42	49	53	59	65	1.1	1.3	1.5	1.6	1.8	1.1	1.4	1.6	1.8	2.0
SS D3	23	27	32	35	40	43	47	52	19	22	26	29	33	35	38	41	1.4	1.6	1.9	2.2	2.4	1.5	1.9	2.2	2.7	3.1
SS D4	28	35	42	48	55	60	66	73	23	29	36	41	47	52	57	64	1.1	1.3	1.5	1.7	1.8	1.2	1.4	1.6	1.8	2.0
SS E1	20	25	30	33	38	41	45	50	15	19	23	25	29	31	34	36	1.3	1.6	1.9	2.2	2.5	1.5	1.9	2.3	2.9	3.6
SS E2	24	31	38	44	51	55	61	68	19	25	31	36	42	46	51	57	1.1	1.3	1.5	1.7	1.9	1.2	1.5	1.7	2.0	2.2
SS E3	19	23	27	30	34	36	40	44	15	18	21	23	26	28	31	34	1.5	1.8	2.1	2.4	2.7	1.7	2.1	2.5	3.0	3.6
SS E4	23	29	36	41	47	52	57	64	18	24	29	34	39	43	48	54	1.2	1.4	1.6	1.8	2.0	1.3	1.5	1.8	2.0	2.3

POCATELLO, IDAHO

TYPE	SOLAR SAVINGS FRACTION (%)																CONSERVATION FACTOR									
BASE TEMP:	55 F (4468 DD)								65 F (7068 DD)								55 F					65 F				
LCR	100	70	50	40	30	25	20	15	100	70	50	40	30	25	20	15	100	50	30	20	15	100	50	30	20	15
WW A1	20	23	27	30	35	38	42	48	16	18	21	24	27	30	33	37	1.6	1.8	2.0	2.2	2.4	1.9	2.1	2.3	2.6	2.9
WW A2	21	26	33	37	44	48	54	61	16	20	25	29	34	38	43	49	1.3	1.5	1.6	1.8	2.0	1.4	1.6	1.9	2.1	2.3
WW A3	21	27	35	40	47	52	58	65	15	20	26	31	37	41	46	53	1.2	1.4	1.6	1.8	1.9	1.3	1.5	1.7	2.0	2.2
WW A4	21	28	36	41	49	54	60	68	15	20	27	32	38	43	48	56	1.1	1.3	1.5	1.7	1.9	1.3	1.5	1.7	1.9	2.1
WW A5	21	29	37	43	51	56	62	70	15	21	28	33	40	44	50	58	1.1	1.3	1.5	1.7	1.8	1.2	1.4	1.6	1.8	2.0
WW A6	21	29	37	43	52	57	63	72	14	21	28	33	41	46	52	59	1.1	1.3	1.5	1.6	1.8	1.2	1.4	1.6	1.8	2.0
WW B1	17	21	25	28	33	36	40	46	12	15	18	20	24	26	28	32	1.5	1.7	2.0	2.2	2.5	1.8	2.1	2.4	2.9	3.3
WW B2	22	29	37	43	52	57	64	72	16	22	29	34	42	47	53	61	1.1	1.3	1.5	1.6	1.8	1.2	1.4	1.6	1.8	1.9
WW B3	26	35	44	50	59	64	71	79	19	27	35	41	49	54	61	69	1.0	1.2	1.4	1.6	1.7	1.1	1.3	1.5	1.7	1.8
WW B4	26	36	46	53	62	68	75	83	19	28	37	44	53	59	66	75	1.0	1.1	1.3	1.5	1.6	1.1	1.2	1.4	1.6	1.7
WW B5	25	35	45	52	62	68	75	83	18	27	37	44	53	59	67	75	1.0	1.1	1.3	1.5	1.6	1.1	1.2	1.4	1.6	1.7
WW C1	28	36	46	53	61	67	73	81	21	29	37	43	52	57	64	72	1.0	1.2	1.3	1.5	1.7	1.1	1.3	1.4	1.6	1.8
WW C2	26	35	45	52	61	66	73	81	19	27	36	43	51	57	64	73	1.0	1.2	1.3	1.5	1.7	1.1	1.3	1.4	1.6	1.8
WW C3	31	42	53	61	70	76	82	89	23	34	45	52	62	68	75	83	0.9	1.1	1.2	1.4	1.6	1.0	1.1	1.3	1.5	1.6
WW C4	28	39	50	57	67	73	80	87	21	31	41	49	59	65	73	81	0.9	1.1	1.3	1.4	1.6	1.0	1.2	1.3	1.5	1.6
TW A1	19	23	27	30	35	38	42	48	16	18	21	24	27	30	33	37	1.6	1.8	2.0	2.2	2.4	1.8	2.1	2.3	2.6	2.8
TW A2	20	25	30	34	40	44	50	57	15	19	23	27	32	35	40	46	1.3	1.5	1.7	1.9	2.1	1.5	1.7	2.0	2.2	2.4
TW A3	19	25	31	36	43	47	53	60	14	19	24	28	33	37	42	49	1.3	1.5	1.6	1.8	2.0	1.4	1.6	1.8	2.1	2.2
TW A4	18	24	31	36	43	48	54	62	13	18	24	28	34	38	43	50	1.2	1.4	1.6	1.8	2.0	1.4	1.6	1.8	2.0	2.2
TW B1	19	23	27	30	35	39	43	50	15	18	21	24	28	31	34	39	1.5	1.7	1.9	2.1	2.3	1.8	2.0	2.2	2.4	2.6
TW B2	18	23	29	33	40	44	49	57	14	18	22	26	31	35	39	46	1.3	1.5	1.7	1.9	2.1	1.5	1.7	1.9	2.1	2.3
TW B3	18	23	29	34	40	44	50	58	13	17	22	26	31	35	40	47	1.3	1.5	1.7	1.9	2.0	1.5	1.7	1.9	2.1	2.3
TW B4	17	22	28	32	38	43	48	56	12	16	21	25	30	34	39	46	1.3	1.5	1.7	1.9	2.0	1.5	1.7	1.9	2.1	2.3
TW C1	17	22	26	30	35	39	44	51	14	17	21	24	28	31	35	41	1.5	1.7	1.8	2.0	2.2	1.7	1.9	2.1	2.3	2.5
TW C2	17	21	26	30	36	40	46	53	13	16	20	24	29	32	37	43	1.4	1.6	1.8	2.0	2.1	1.6	1.8	2.0	2.2	2.4
TW C3	16	20	25	29	35	39	44	51	12	16	20	23	28	31	35	42	1.5	1.6	1.8	2.0	2.1	1.7	1.8	2.0	2.2	2.4
TW C4	15	19	23	27	32	36	40	47	12	15	19	21	26	29	33	39	1.6	1.7	1.9	2.1	2.3	1.8	2.0	2.1	2.3	2.5
TW D1	14	18	22	25	30	33	37	42	10	12	15	18	21	23	26	30	1.6	1.8	2.0	2.3	2.5	1.9	2.1	2.5	2.8	3.1
TW D2	21	27	35	40	48	53	59	67	15	21	27	32	39	43	49	57	1.2	1.4	1.5	1.7	1.9	1.3	1.5	1.7	1.9	2.0
TW D3	22	30	38	43	51	56	63	71	17	23	30	35	42	47	53	61	1.1	1.3	1.5	1.7	1.8	1.2	1.4	1.6	1.8	2.0
TW D4	24	32	41	48	57	63	69	78	18	25	33	39	48	53	60	69	1.0	1.2	1.4	1.6	1.7	1.1	1.3	1.5	1.7	1.8
TW D5	24	32	42	48	57	63	70	79	18	25	34	40	49	55	62	71	1.0	1.2	1.4	1.5	1.7	1.1	1.3	1.5	1.6	1.8
TW E1	26	33	42	48	56	62	68	76	19	26	34	39	47	52	59	67	1.1	1.2	1.4	1.6	1.7	1.2	1.3	1.5	1.7	1.9
TW E2	25	33	42	48	57	63	69	77	19	26	34	40	48	53	60	69	1.0	1.2	1.4	1.6	1.7	1.1	1.3	1.5	1.7	1.8
TW E3	30	40	50	57	67	72	79	86	23	32	42	49	58	64	71	79	0.9	1.1	1.3	1.5	1.6	1.0	1.2	1.4	1.5	1.7
TW E4	28	37	48	55	65	70	77	85	21	30	40	47	56	62	69	78	1.0	1.1	1.3	1.5	1.6	1.0	1.2	1.4	1.5	1.7
TW F1	18	21	26	29	34	37	41	48	14	17	20	22	26	29	32	37	1.5	1.7	1.9	2.2	2.3	1.8	2.0	2.3	2.5	2.7
TW F2	17	22	28	32	38	42	47	55	13	16	21	24	29	33	37	43	1.4	1.6	1.7	1.9	2.1	1.6	1.8	2.0	2.2	2.4
TW F3	16	22	28	32	39	43	49	57	11	16	20	24	30	34	39	45	1.3	1.5	1.7	1.9	2.0	1.5	1.7	1.9	2.1	2.3
TW F4	14	20	26	31	38	42	48	56	9	14	19	23	29	33	38	45	1.3	1.5	1.7	1.9	2.0	1.5	1.7	1.9	2.1	2.2
TW G1	16	20	24	28	33	36	41	48	12	15	18	21	25	28	32	37	1.5	1.7	1.9	2.1	2.3	1.8	2.0	2.2	2.4	2.6
TW G2	14	19	25	29	35	39	45	52	10	14	18	21	26	30	35	41	1.4	1.6	1.8	2.0	2.1	1.6	1.8	2.0	2.2	2.4
TW G3	13	17	23	28	34	38	44	51	8	12	17	20	25	29	34	41	1.4	1.6	1.8	1.9	2.1	1.6	1.8	2.0	2.2	2.3
TW G4	10	15	20	24	30	34	40	47	6	10	14	17	22	26	30	37	1.5	1.6	1.8	2.0	2.2	1.7	1.9	2.0	2.2	2.4
TW H1	13	17	21	25	30	34	39	46	9	12	16	18	23	26	30	36	1.5	1.7	1.9	2.1	2.3	1.8	2.0	2.2	2.4	2.6
TW H2	10	14	19	23	29	33	38	45	6	10	14	17	21	24	29	35	1.5	1.7	1.9	2.1	2.2	1.8	1.9	2.1	2.3	2.5
TW H3	8	12	17	20	26	29	34	42	5	8	12	14	19	22	26	32	1.6	1.8	2.0	2.1	2.3	1.8	2.0	2.2	2.4	2.5
TW H4	6	9	13	16	21	24	28	35	4	6	9	11	15	18	21	27	1.8	2.0	2.1	2.3	2.5	2.1	2.2	2.4	2.6	2.7
TW I1	12	15	19	22	27	30	34	39	8	10	13	15	18	21	23	27	1.6	1.9	2.1	2.4	2.6	2.0	2.2	2.6	2.9	3.3
TW I2	18	24	31	37	44	49	56	64	12	18	24	28	35	40	46	54	1.2	1.4	1.6	1.7	1.9	1.4	1.5	1.7	1.9	2.1
TW I3	19	26	34	40	48	53	60	68	14	20	26	31	39	43	50	58	1.2	1.3	1.5	1.7	1.8	1.3	1.5	1.6	1.8	2.0
TW I4	21	29	38	45	54	59	66	75	15	22	30	36	45	50	57	66	1.1	1.2	1.4	1.6	1.7	1.2	1.3	1.5	1.7	1.8
TW I5	21	30	39	46	55	61	68	77	15	23	31	38	46	52	59	69	1.0	1.2	1.4	1.6	1.7	1.1	1.3	1.5	1.6	1.8
TW J1	23	31	40	46	54	60	66	75	17	24	32	37	45	50	57	65	1.1	1.2	1.4	1.6	1.8	1.2	1.4	1.5	1.7	1.9
TW J2	22	30	39	46	54	60	67	75	16	23	31	37	45	51	58	66	1.1	1.2	1.4	1.6	1.7	1.2	1.3	1.5	1.7	1.8
TW J3	27	37	47	55	64	70	77	85	20	29	39	46	56	62	69	78	1.0	1.1	1.3	1.5	1.6	1.0	1.2	1.4	1.6	1.7
TW J4	25	34	45	52	62	68	75	83	18	27	37	44	54	60	67	76	1.0	1.1	1.3	1.5	1.6	1.1	1.2	1.4	1.6	1.7
DG A1	15	10	20	26	30	33	35	38	10	13	16	18	21	23	24	26	1.5	1.8	2.2	2.9	3.7	1.7	2.1	2.7	3.8	-
DG A2	17	22	29	33	39	43	48	54	12	17	22	25	31	34	39	44	1.3	1.5	1.7	2.0	2.3	1.5	1.7	1.9	2.2	2.5
DG A3	21	28	36	42	50	55	62	70	16	22	29	34	42	47	53	61	1.2	1.3	1.5	1.7	1.9	1.3	1.4	1.6	1.8	2.0
DG B1	15	19	24	28	33	36	40	44	10	13	17	19	23	25	28	31	1.4	1.6	2.0	2.4	2.8	1.7	2.0	2.4	2.9	3.8
DG B2	18	23	30	35	42	47	53	60	13	17	23	27	33	37	42	48	1.3	1.4	1.6	1.8	2.0	1.5	1.6	1.8	2.1	2.3
DG B3	21	29	37	43	52	58	65	74	16	23	30	35	44	49	56	65	1.1	1.3	1.4	1.6	1.8	1.3	1.4	1.5	1.7	1.9
DG C1	18	23	29	34	40	44	49	54	12	16	21	24	29	32	36	41	1.3	1.5	1.7	2.1	2.4	1.5	1.7	2.0	2.4	2.9
DG C2	20	27	34	40	48	54	60	68	15	20	27	31	38	43	49	57	1.2	1.3	1.5	1.7	1.9	1.3	1.5	1.6	1.9	2.1
DG C3	24	32	41	49	58	64	72	80	19	26	34	40	49	55	62	71	1.1	1.2	1.3	1.5	1.7	1.2	1.3	1.4	1.6	1.8
SS A1	24	29	35	39	45	49	54	61	20	24	29	33	38	42	46	52	1.3	1.5	1.7	1.9	2.1	1.4	1.6	1.9	2.1	2.3
SS A2	29	36	44	50	58	63	69	76	25	31	39	44	51	56	61	69	1.1	1.2	1.4	1.6	1.8	1.1	1.3	1.5	1.7	1.9
SS A3	22	26	31	35	40	44	48	54	18	22	26	30	34	37	40	45	1.4	1.6	1.8	2.1	2.3	1.5	1.8	2.1	2.4	2.6
SS A4	28	35	43	48	56	61	66	74	24	30	37	42	49	54	59	66	1.1	1.3	1.5	1.7	1.8	1.2	1.4	1.6	1.8	2.0
SS A5	23	27	32	36	41	45	49	55	20	23	28	31	35	38	42	47	1.4	1.6	1.8	2.0	2.2	1.5	1.8	2.0	2.3	2.6
SS A6	29	36	44	49	57	62	68	75	24	31	38	43	50	55	60	68	1.1	1.3	1.4	1.6	1.8	1.1	1.3	1.5	1.7	1.9
SS A7	21	24	28	31	36	38	42	47	18	21	24	27	30	32	35	39	1.5	1.8	2.0	2.3	2.5	1.7	2.0	2.3	2.7	3.0
SS A8	28	34	42	47	54	59	65	72	23	30	36	41	47	52	57	64	1.1	1.3	1.5	1.7	1.8	1.2	1.4	1.6	1.8	2.0
SS B1	20	25	30	34	40	44	49	55	16	20	25	28	33	36	41	46	1.3	1.6	1.8	2.0	2.2	1.5	1.7	2.0	2.2	2.4
SS B2	25	32	40	46	53	58	65	72	21	27	34	39	47	51	57	65	1.1	1.3	1.5	1.7	1.8	1.2	1.4	1.6	1.8	1.9
SS B3	19	23	28	32	37	40	45	51	15	19	23	26	30	33	37	42	1.4	1.6	1.9	2.1	2.3	1.6	1.8	2.1	2.4	2.7
SS B4	25	31	39	44	52	56	63	70	20	26	33	38	45	49	55	63	1.1	1.3	1.5	1.7	1.8	1.2	1.4	1.6	1.8	2.0
SS B5	19	23	27	30	35	38	42	48	16	19	23	25	29	32	35	40	1.5	1.7	2.0	2.2	2.4	1.7	1.9	2.2	2.5	2.7
SS B6	25	31	39	45	52	57	63	71	20	26	33	38	45	50	56	63	1.1	1.3	1.5	1.7	1.8	1.2	1.4	1.6	1.8	2.0
SS B7	17	20	24	27	31	34	37	42	14	17	20	22	25	28	31	34	1.6	1.9	2.1	2.4	2.6	1.8	2.1	2.4	2.8	3.1
SS B8	23	30	37	42	50	54	60	68	19	25	32	36	43	47	53	60	1.2	1.3	1.5	1.7	1.9	1.2	1.4	1.7	1.9	2.0
SS C1	19	24	30	35	42	46	52	59	13	18	23	26	32	35	40	46	1.3	1.5	1.7	1.9	2.0	1.5	1.7	1.9	2.1	2.3
SS C2	22	30	38	44	52	57	64	72	16	23	30	35	42	47	54	62	1.1	1.3	1.5	1.6	1.8	1.2	1.4	1.6	1.8	1.9
SS C3	17	21	26	29	35	38	43	50	12	16	19	22	27	30	33	39	1.5	1.7	1.9	2.1	2.2	1.7	1.9	2.1	2.4	2.6
SS C4	20	27	34	39	47	52	58	66	15	20	27	31	38	42	48	56	1.2	1.4	1.5	1.7	1.9	1.3	1.5	1.7	1.9	2.0
SS D1	27	33	40	45	52	56	62	69	22	28	34	38	44	48	53	59	1.1	1.3	1.6	1.8	1.9	1.2	1.5	1.7	2.0	2.2
SS D2	32	41	50	57	65	70	76	83	27	35	44	50	58	63	69	77	1.0	1.1	1.3	1.5	1.7	1.0	1.2	1.4	1.6	1.8
SS D3	25	31	37	41	47	51	56	63	21	26	31	35	40	44	48	54	1.2	1.4	1.7	1.9	2.0	1.3	1.6	1.8	2.1	2.3
SS D4	31	40	49	55	63	68	75	82	27	34	43	48	56	61	67	75	1.0	1.2	1.4	1.5	1.7	1.1	1.2	1.4	1.6	1.8
SS E1	23	29	35	40	46	50	55	62	18	23	28	32	37	41	45	51	1.2	1.4	1.6	1.9	2.0	1.3	1.6	1.9	2.1	2.4
SS E2	28	36	45	51	59	64	70	78	23	30	38	43	51	56	62	69	1.0	1.2	1.4	1.6	1.7	1.1	1.3	1.5	1.7	1.9
SS E3	21	25	30	34	39	43	48	54	17	21	25	28	33	36	39	45	1.4	1.6	1.8	2.0	2.2	1.5	1.8	2.0	2.3	2.6
SS E4	26	34	41	47	55	60	66	74	21	28	35	40	47	52	58	65	1.1	1.3	1.5	1.6	1.8	1.2	1.4	1.6	1.8	1.9

CHICAGO, ILLINOIS

TYPE	SOLAR SAVINGS FRACTION (%) BASE TEMP: 55 F (3913 DD)								65 F (6125 DD)								CONSERVATION FACTOR 55 F					65 F				
LCR	100	70	50	40	30	25	20	15	100	70	50	40	30	25	20	15	100	50	30	20	15	100	50	30	20	15
WW A1	15	18	20	22	25	27	30	33	13	14	16	17	18	19	21	23	2.0	2.3	2.6	2.8	3.0	2.6	3.0	3.5	4.1	4.7
WW A2	15	18	23	26	31	34	39	45	11	13	16	19	22	24	27	32	1.5	1.8	2.0	2.1	2.3	1.9	2.1	2.4	2.7	3.0
WW A3	14	18	24	28	33	37	43	50	9	13	16	19	23	26	30	35	1.4	1.6	1.8	2.0	2.2	1.7	1.9	2.2	2.4	2.7
WW A4	13	18	24	28	35	39	45	53	8	12	16	20	24	28	32	38	1.4	1.6	1.7	1.9	2.1	1.6	1.8	2.1	2.3	2.5
WW A5	13	19	25	29	36	41	47	55	8	12	17	20	25	29	33	40	1.3	1.5	1.7	1.9	2.0	1.5	1.8	2.0	2.2	2.4
WW A6	13	19	25	30	37	42	48	57	7	12	17	20	26	30	34	41	1.3	1.5	1.7	1.8	2.0	1.5	1.7	1.9	2.2	2.4
WW B1	11	13	16	18	20	22	24	27	7	8	10	10	11	12	12	12	2.0	2.4	2.7	3.1	3.5	2.8	3.5	5.0	-	-
WW B2	14	20	26	31	39	43	50	58	9	14	19	23	29	33	38	45	1.3	1.5	1.6	1.8	2.0	1.5	1.7	1.9	2.1	2.2
WW B3	17	24	32	38	46	51	58	67	12	17	24	28	35	40	46	54	1.2	1.3	1.5	1.7	1.8	1.3	1.5	1.7	1.9	2.0
WW B4	17	25	34	41	50	56	64	73	12	19	26	32	40	46	53	62	1.1	1.3	1.4	1.6	1.7	1.2	1.4	1.5	1.7	1.9
WW B5	17	25	34	41	50	57	64	74	11	18	26	32	41	47	54	64	1.1	1.2	1.4	1.6	1.7	1.2	1.4	1.5	1.7	1.8
WW C1	19	26	34	40	49	54	61	70	13	19	26	31	38	43	50	58	1.1	1.3	1.5	1.6	1.8	1.3	1.5	1.6	1.8	2.0
WW C2	17	25	33	40	48	54	62	71	12	18	25	31	38	44	51	60	1.1	1.3	1.4	1.6	1.8	1.3	1.4	1.6	1.8	1.9
WW C3	21	31	42	49	60	66	74	82	15	24	34	41	50	57	65	74	1.0	1.1	1.3	1.5	1.6	1.1	1.2	1.4	1.6	1.7
WW C4	19	28	39	46	56	63	71	80	13	22	31	38	47	54	62	72	1.0	1.2	1.3	1.5	1.6	1.1	1.3	1.4	1.6	1.7
TW A1	15	18	20	22	25	27	30	34	13	14	16	17	19	20	22	24	2.0	2.2	2.5	2.7	2.9	2.5	2.9	3.2	3.7	4.2
TW A2	14	18	21	24	29	32	36	42	11	13	16	18	21	23	26	30	1.6	1.9	2.1	2.3	2.4	2.0	2.2	2.5	2.8	3.1
TW A3	13	17	22	25	30	34	39	46	9	12	16	18	22	25	28	33	1.5	1.7	1.9	2.1	2.3	1.8	2.1	2.3	2.6	2.8
TW A4	12	16	21	25	31	35	40	47	8	11	15	18	22	25	29	34	1.5	1.7	1.9	2.0	2.2	1.8	2.0	2.2	2.4	2.6
TW B1	14	17	20	22	26	28	31	36	12	13	15	17	19	21	23	26	1.9	2.1	2.3	2.6	2.7	2.3	2.6	2.9	3.2	3.6
TW B2	13	17	21	24	28	32	36	42	10	12	15	17	21	23	26	31	1.6	1.8	2.0	2.2	2.4	1.9	2.2	2.4	2.7	2.9
TW B3	12	16	20	24	29	32	37	44	9	11	15	17	21	24	27	32	1.6	1.8	2.0	2.1	2.3	1.9	2.1	2.3	2.5	2.8
TW B4	12	15	19	23	28	31	36	42	8	11	14	17	20	23	27	32	1.6	1.8	2.0	2.2	2.3	1.9	2.1	2.3	2.5	2.7
TW C1	13	16	19	22	26	29	32	38	11	12	15	17	19	21	24	28	1.8	2.0	2.2	2.4	2.6	2.2	2.4	2.7	2.9	3.2
TW C2	12	15	19	22	26	30	34	40	9	12	14	16	20	22	25	30	1.7	1.9	2.1	2.3	2.4	2.0	2.2	2.4	2.7	2.9
TW C3	12	15	18	21	26	29	33	39	9	11	14	16	19	22	25	29	1.7	1.9	2.1	2.3	2.5	2.1	2.3	2.5	2.7	2.9
TW C4	12	15	18	20	24	27	30	36	10	12	14	16	19	21	24	28	1.9	2.1	2.3	2.4	2.6	2.2	2.4	2.6	2.8	3.0
TW D1	9	11	14	16	18	20	22	26	6	7	8	9	11	11	12	13	2.0	2.4	2.7	3.1	3.4	2.8	3.4	4.2	-	-
TW D2	15	19	25	29	36	40	46	54	10	14	18	22	27	30	35	42	1.4	1.6	1.7	1.9	2.0	1.6	1.8	2.0	2.2	2.3
TW D3	15	21	27	32	39	43	50	58	11	15	20	24	29	33	38	45	1.3	1.5	1.7	1.8	2.0	1.5	1.7	1.9	2.1	2.2
TW D4	17	23	31	37	45	51	58	67	12	17	24	29	36	41	47	56	1.2	1.4	1.5	1.7	1.8	1.3	1.5	1.7	1.8	2.0
TW D5	16	23	31	37	46	52	59	69	12	18	25	30	38	43	50	59	1.2	1.3	1.5	1.6	1.8	1.3	1.4	1.6	1.8	1.9
TW E1	18	24	31	36	44	49	56	65	13	18	24	28	35	39	45	52	1.2	1.4	1.6	1.7	1.9	1.4	1.6	1.7	1.9	2.1
TW E2	17	24	31	37	45	50	57	66	12	18	24	29	36	40	47	55	1.2	1.4	1.5	1.7	1.8	1.3	1.5	1.7	1.9	2.0
TW E3	21	30	39	46	55	62	69	78	15	23	31	37	46	52	59	69	1.1	1.2	1.4	1.5	1.7	1.2	1.3	1.5	1.6	1.8
TW E4	19	28	37	44	53	60	67	77	14	21	30	36	45	51	58	68	1.1	1.2	1.4	1.5	1.7	1.2	1.3	1.5	1.6	1.8
TW F1	13	16	19	21	24	26	29	33	11	12	14	15	17	19	20	23	1.9	2.2	2.5	2.7	2.9	2.5	2.8	3.1	3.6	4.0
TW F2	12	15	19	22	26	29	34	40	8	11	13	15	19	21	24	27	1.7	1.9	2.1	2.3	2.5	2.0	2.3	2.6	2.9	3.1
TW F3	10	14	19	22	27	30	35	42	7	9	13	15	19	21	25	29	1.6	1.8	2.0	2.2	2.3	1.9	2.1	2.4	2.6	2.9
TW F4	9	12	17	21	26	29	34	42	5	8	11	14	18	20	24	29	1.6	1.8	2.0	2.1	2.3	1.9	2.1	2.3	2.5	2.7
TW G1	11	14	17	19	23	25	29	34	9	10	12	14	16	18	20	23	1.9	2.1	2.4	2.6	2.7	2.3	2.6	2.9	3.3	3.6
TW G2	9	13	16	19	24	27	31	37	6	8	11	13	16	19	22	26	1.7	1.9	2.1	2.3	2.5	2.0	2.3	2.5	2.8	3.0
TW G3	8	11	15	18	23	26	31	37	4	7	10	12	16	18	21	26	1.7	1.9	2.1	2.3	2.4	2.0	2.2	2.4	2.7	2.9
TW G4	6	9	13	16	20	23	28	34	3	5	8	10	14	16	19	24	1.8	2.0	2.2	2.3	2.5	2.1	2.3	2.5	2.7	2.9
TW H1	8	11	14	17	21	23	27	32	6	8	10	12	14	16	19	22	1.9	2.1	2.3	2.5	2.7	2.3	2.5	2.8	3.1	3.3
TW H2	6	9	12	15	19	22	26	32	3	6	8	10	13	15	18	22	1.8	2.0	2.3	2.4	2.6	2.2	2.4	2.6	2.9	3.1
TW H3	5	7	11	13	17	20	24	29	2	4	7	9	11	13	16	21	1.9	2.1	2.3	2.5	2.7	2.3	2.5	2.7	2.9	3.1
TW H4	3	6	8	11	14	16	19	24	2	3	5	7	9	11	13	17	2.1	2.3	2.5	2.7	2.9	2.5	2.7	2.9	3.1	3.3
TW I1	7	9	12	13	16	17	20	23	4	5	7	7	9	9	10	11	2.1	2.5	2.9	3.2	3.5	2.9	3.6	4.4	-	-
TW I2	12	16	22	26	32	37	42	50	8	11	15	19	24	27	32	38	1.4	1.6	1.8	1.9	2.1	1.6	1.8	2.0	2.2	2.4
TW I3	12	18	24	29	35	40	46	55	8	12	17	21	27	31	36	43	1.3	1.5	1.7	1.9	2.0	1.5	1.7	1.9	2.1	2.3
TW I4	14	20	28	34	42	47	54	64	9	15	21	26	33	38	44	53	1.2	1.4	1.5	1.7	1.8	1.4	1.5	1.7	1.9	2.0
TW I5	14	21	29	35	44	49	57	66	9	15	22	27	35	40	47	56	1.2	1.3	1.5	1.7	1.8	1.3	1.5	1.6	1.8	1.9
TW J1	16	22	29	34	42	47	54	63	11	16	22	26	33	37	43	51	1.2	1.4	1.6	1.7	1.9	1.4	1.6	1.8	1.9	2.1
TW J2	15	21	29	34	42	48	55	64	10	15	22	26	33	38	44	53	1.2	1.4	1.5	1.7	1.8	1.4	1.5	1.7	1.9	2.0
TW J3	18	27	36	43	53	59	67	76	13	20	29	35	44	50	58	67	1.1	1.2	1.4	1.5	1.7	1.2	1.3	1.5	1.6	1.8
TW J4	17	25	34	41	51	57	65	75	12	19	27	33	42	48	56	66	1.1	1.2	1.4	1.6	1.7	1.2	1.3	1.5	1.7	1.8
DG A1	9	11	14	16	18	20	21	22	5	7	8	9	10	10	11	10	1.9	2.4	3.0	4.4	-	2.6	3.3	-	-	-
DG A2	12	15	20	23	28	31	35	41	8	11	14	17	20	23	26	30	1.6	1.8	2.0	2.3	2.6	1.9	2.1	2.4	2.7	3.1
DG A3	15	20	26	31	39	44	50	59	12	16	21	25	31	35	41	48	1.4	1.5	1.6	1.8	2.0	1.5	1.7	1.8	2.0	2.2
DG B1	9	11	14	16	19	21	24	26	5	7	8	9	11	11	12	12	1.9	2.3	2.6	3.2	4.0	2.5	3.1	4.3	-	-
DG B2	12	16	20	24	29	33	38	45	8	11	15	17	21	24	28	33	1.6	1.8	1.9	2.1	2.3	1.8	2.0	2.3	2.5	2.8
DG B3	16	21	27	32	40	46	53	63	12	16	22	26	32	37	43	52	1.3	1.5	1.6	1.7	1.8	1.5	1.6	1.8	1.9	2.0
DG C1	11	14	18	21	25	28	32	36	7	9	12	13	16	17	19	21	1.7	2.0	2.1	2.5	2.9	2.1	2.5	3.1	3.7	4.7
DG C2	14	19	24	28	35	39	45	53	10	14	18	21	26	30	34	41	1.4	1.6	1.7	1.9	2.1	1.6	1.8	2.0	2.2	2.4
DG C3	18	24	31	37	46	52	59	70	14	19	25	30	37	42	49	58	1.3	1.4	1.5	1.6	1.7	1.4	1.5	1.6	1.8	1.9
SS A1	18	22	26	29	34	37	42	48	15	18	21	24	27	30	33	38	1.5	1.7	1.9	2.1	2.3	1.7	2.0	2.3	2.6	2.8
SS A2	22	28	35	40	47	52	58	67	18	23	29	34	40	44	50	57	1.2	1.4	1.5	1.7	1.9	1.3	1.5	1.7	1.9	2.0
SS A3	16	19	23	26	30	32	36	41	13	16	18	20	23	25	27	30	1.6	1.9	2.2	2.4	2.6	1.9	2.2	2.6	3.1	3.5
SS A4	21	27	33	38	45	50	56	64	18	23	28	32	38	42	47	54	1.2	1.4	1.6	1.8	1.9	1.4	1.6	1.8	2.0	2.1
SS A5	18	21	25	27	31	34	37	42	15	18	20	22	25	27	30	33	1.7	1.9	2.2	2.4	2.6	1.9	2.2	2.6	2.9	3.3
SS A6	22	27	34	39	46	51	57	66	18	23	29	33	39	43	48	56	1.2	1.4	1.6	1.7	1.9	1.3	1.5	1.7	1.9	2.1
SS A7	16	19	21	23	26	28	31	34	14	16	18	19	21	22	23	25	1.9	2.2	2.6	2.9	3.1	2.3	2.7	3.2	3.9	4.7
SS A8	21	26	32	37	44	48	54	62	17	22	27	31	36	40	45	52	1.3	1.5	1.6	1.8	2.0	1.4	1.6	1.8	2.0	2.2
SS B1	15	18	22	25	29	32	37	42	12	15	18	20	23	25	28	32	1.6	1.9	2.1	2.3	2.4	1.9	2.1	2.4	2.8	3.0
SS B2	19	25	31	36	43	48	54	62	16	20	26	30	36	40	46	53	1.3	1.4	1.6	1.8	1.9	1.4	1.6	1.8	1.9	2.1
SS B3	14	17	20	23	26	29	32	37	11	13	16	17	20	22	24	27	1.7	2.0	2.3	2.5	2.7	2.0	2.3	2.7	3.2	3.6
SS B4	18	24	30	35	41	46	52	60	15	19	25	29	34	38	44	51	1.3	1.5	1.6	1.8	2.0	1.4	1.6	1.8	2.0	2.2
SS B5	15	17	20	22	26	28	31	36	12	14	16	18	20	22	24	27	1.8	2.1	2.4	2.6	2.8	2.2	2.5	2.8	3.3	3.6
SS B6	18	24	30	35	42	46	52	61	15	19	25	29	35	39	44	51	1.3	1.5	1.6	1.8	1.9	1.4	1.6	1.8	2.0	2.1
SS B7	13	15	18	20	22	24	27	30	11	13	14	16	17	18	20	22	2.0	2.4	2.7	3.0	3.2	2.5	2.9	3.4	4.0	4.7
SS B8	17	22	28	33	39	44	49	57	14	18	23	27	32	36	41	48	1.3	1.5	1.7	1.9	2.0	1.5	1.7	1.9	2.1	2.3
SS C1	12	16	20	24	28	32	36	43	8	11	14	16	19	21	24	28	1.6	1.8	2.0	2.2	2.4	1.9	2.2	2.5	2.8	3.1
SS C2	15	21	27	32	39	44	50	59	10	15	20	24	30	34	39	46	1.3	1.5	1.6	1.8	1.9	1.5	1.7	1.9	2.0	2.2
SS C3	12	15	18	20	24	27	30	35	9	11	13	15	17	19	21	24	1.8	2.1	2.3	2.5	2.7	2.2	2.5	2.8	3.2	3.5
SS C4	14	19	24	29	35	39	45	53	10	14	18	22	27	30	35	42	1.4	1.6	1.7	1.9	2.1	1.6	1.8	2.0	2.2	2.3
SS D1	19	24	30	34	40	44	49	56	15	19	24	27	31	34	38	43	1.3	1.5	1.7	1.9	2.1	1.5	1.8	2.0	2.3	2.6
SS D2	24	32	40	46	55	60	67	75	19	26	33	39	46	51	58	66	1.1	1.2	1.4	1.6	1.7	1.2	1.4	1.5	1.7	1.9
SS D3	19	23	27	31	36	39	44	50	16	19	22	25	29	31	35	39	1.5	1.7	1.9	2.1	2.3	1.7	1.9	2.2	2.5	2.8
SS D4	23	31	39	45	53	58	65	73	19	25	32	37	45	49	56	64	1.1	1.3	1.4	1.6	1.8	1.2	1.4	1.6	1.8	1.9
SS E1	16	20	25	28	33	37	41	48	12	15	18	21	25	27	30	34	1.5	1.7	1.9	2.1	2.3	1.7	2.0	2.4	2.7	3.1
SS E2	20	26	34	39	47	53	59	68	15	21	27	32	39	43	49	57	1.2	1.3	1.5	1.7	1.8	1.3	1.5	1.7	1.9	2.0
SS E3	15	19	22	25	29	32	35	41	12	15	17	19	22	24	27	30	1.7	1.9	2.2	2.4	2.6	2.0	2.3	2.6	3.0	3.4
SS E4	19	25	32	37	44	49	55	63	15	20	25	30	36	40	45	53	1.2	1.4	1.6	1.8	1.9	1.4	1.6	1.8	2.0	2.1

MOLINE, ILLINOIS

TYPE	SOLAR SAVINGS FRACTION (%) BASE TEMP: 55 F (4177 DD)								65 F (6399 DD)								CONSERVATION FACTOR 55 F					65 F				
LCR	100	70	50	40	30	25	20	15	100	70	50	40	30	25	20	15	100	50	30	20	15	100	50	30	20	15
WW A1	15	17	20	22	24	26	29	33	13	14	16	17	18	19	21	23	2.0	2.3	2.6	2.9	3.1	2.6	3.0	3.5	4.0	4.7
WW A2	14	18	22	25	30	34	38	45	11	13	16	19	22	24	28	32	1.6	1.8	2.0	2.2	2.3	1.9	2.1	2.4	2.7	2.9
WW A3	13	18	23	27	33	37	42	49	9	13	16	19	24	26	30	36	1.4	1.6	1.8	2.0	2.2	1.7	1.9	2.2	2.4	2.6
WW A4	13	18	23	28	34	38	44	52	8	12	16	20	24	28	32	38	1.4	1.6	1.8	1.9	2.1	1.6	1.8	2.1	2.3	2.5
WW A5	12	18	24	29	35	40	46	55	8	12	17	20	25	29	34	40	1.3	1.5	1.7	1.9	2.0	1.5	1.8	2.0	2.2	2.4
WW A6	12	18	24	29	36	41	48	56	7	12	17	20	26	30	35	41	1.3	1.5	1.7	1.8	2.0	1.5	1.7	1.9	2.2	2.3
WW B1	10	13	15	17	19	21	23	26	7	8	10	10	11	12	12	12	2.0	2.4	2.8	3.2	3.5	2.7	3.5	4.9	-	-
WW B2	14	19	26	31	38	43	49	58	9	14	19	23	29	33	38	46	1.3	1.5	1.6	1.8	1.9	1.5	1.7	1.9	2.1	2.2
WW B3	17	24	31	37	45	51	58	67	12	17	24	29	35	40	46	54	1.2	1.4	1.5	1.7	1.8	1.3	1.5	1.7	1.9	2.0
WW B4	17	25	34	40	49	56	63	73	12	19	26	32	40	46	53	63	1.1	1.3	1.4	1.6	1.7	1.2	1.4	1.5	1.7	1.9
WW B5	16	24	33	40	50	56	64	73	11	18	26	32	41	47	54	64	1.1	1.3	1.4	1.6	1.7	1.2	1.4	1.5	1.7	1.8
WW C1	18	25	33	40	48	54	61	70	13	19	26	31	39	44	50	59	1.2	1.3	1.5	1.6	1.8	1.3	1.4	1.6	1.8	1.9
WW C2	17	24	33	39	48	54	61	70	12	18	25	31	39	44	51	60	1.1	1.3	1.5	1.6	1.8	1.3	1.4	1.6	1.8	1.9
WW C3	21	30	41	49	59	66	73	82	15	24	34	41	51	57	65	74	1.0	1.2	1.3	1.5	1.6	1.1	1.2	1.4	1.6	1.7
WW C4	18	28	38	46	56	62	70	80	13	22	31	38	47	54	62	72	1.0	1.2	1.3	1.5	1.6	1.1	1.3	1.4	1.6	1.7
TW A1	15	17	20	22	24	27	29	33	13	14	16	17	19	20	22	24	2.0	2.3	2.5	2.8	3.0	2.5	2.8	3.2	3.7	4.1
TW A2	14	17	21	24	28	31	35	41	11	13	16	18	21	23	26	30	1.7	1.9	2.1	2.3	2.4	2.0	2.2	2.5	2.8	3.0
TW A3	13	17	21	25	30	33	38	45	9	12	16	18	22	25	28	33	1.6	1.8	1.9	2.1	2.3	1.8	2.1	2.3	2.6	2.8
TW A4	12	16	21	24	30	34	39	46	8	11	15	18	22	25	29	34	1.5	1.7	1.9	2.0	2.2	1.7	2.0	2.2	2.4	2.6
TW B1	14	17	19	22	25	27	31	35	12	13	15	17	19	21	23	26	1.9	2.1	2.4	2.6	2.8	2.3	2.6	2.9	3.2	3.5
TW B2	13	16	20	23	28	31	35	42	10	12	15	17	21	23	26	31	1.6	1.9	2.0	2.2	2.4	1.9	2.2	2.4	2.7	2.9
TW B3	12	15	20	23	28	31	36	43	9	11	15	17	21	24	27	32	1.6	1.8	2.0	2.2	2.3	1.9	2.1	2.3	2.5	2.7
TW B4	11	15	19	22	27	30	35	42	8	11	14	17	20	23	27	32	1.6	1.8	2.0	2.2	2.3	1.9	2.1	2.3	2.5	2.7
TW C1	13	16	19	21	25	28	32	37	11	12	15	17	19	21	24	28	1.8	2.0	2.2	2.4	2.6	2.2	2.4	2.7	2.9	3.1
TW C2	12	15	19	21	26	29	33	39	9	12	14	17	20	22	25	30	1.7	1.9	2.1	2.3	2.4	2.0	2.2	2.4	2.7	2.9
TW C3	11	14	18	21	25	28	32	38	9	11	14	16	19	22	25	29	1.8	2.0	2.2	2.3	2.5	2.1	2.3	2.5	2.7	2.9
TW C4	12	14	17	20	23	26	30	35	10	12	14	16	19	21	24	28	1.9	2.1	2.3	2.5	2.6	2.2	2.4	2.6	2.8	3.0
TW D1	9	11	13	15	17	19	22	25	6	7	8	9	11	11	12	13	2.1	2.5	2.8	3.1	3.4	2.8	3.4	4.2	-	-
TW D2	14	19	24	29	35	39	45	53	10	14	19	22	27	31	35	42	1.4	1.6	1.7	1.9	2.0	1.6	1.8	2.0	2.2	2.3
TW D3	15	20	26	31	38	43	49	57	11	15	20	24	29	33	38	46	1.3	1.5	1.7	1.8	2.0	1.5	1.7	1.9	2.1	2.2
TW D4	16	23	30	36	44	50	57	66	12	17	24	29	36	41	47	56	1.2	1.4	1.5	1.7	1.8	1.3	1.5	1.7	1.8	2.0
TW D5	16	23	31	37	45	51	59	68	12	18	25	30	38	43	50	59	1.2	1.3	1.5	1.6	1.8	1.3	1.4	1.6	1.8	1.9
TW E1	17	23	30	36	43	48	55	64	13	18	24	28	35	39	45	53	1.2	1.4	1.6	1.7	1.9	1.4	1.6	1.7	1.9	2.1
TW E2	17	23	31	36	44	50	57	66	12	18	24	29	36	41	47	55	1.2	1.4	1.5	1.7	1.8	1.3	1.5	1.7	1.9	2.0
TW E3	20	29	38	45	55	61	69	78	15	23	31	38	46	52	60	69	1.1	1.2	1.4	1.5	1.7	1.2	1.3	1.5	1.6	1.8
TW E4	19	27	36	43	53	59	67	76	14	21	30	36	45	51	58	68	1.1	1.2	1.4	1.5	1.7	1.2	1.3	1.5	1.6	1.8
TW F1	13	15	18	20	23	25	28	33	11	12	14	15	17	19	20	23	2.0	2.2	2.5	2.7	2.9	2.4	2.8	3.1	3.6	4.0
TW F2	11	15	18	21	26	29	33	39	8	11	13	15	19	21	24	28	1.7	1.9	2.1	2.3	2.5	2.0	2.3	2.6	2.8	3.1
TW F3	10	14	18	21	26	30	34	41	7	9	13	15	19	21	25	29	1.6	1.8	2.0	2.2	2.3	1.9	2.1	2.4	2.6	2.8
TW F4	8	12	16	20	25	29	34	41	5	8	11	14	18	20	24	29	1.6	1.8	2.0	2.2	2.3	1.9	2.1	2.3	2.5	2.7
TW G1	11	14	17	19	22	25	28	33	9	10	12	14	16	18	20	23	1.9	2.1	2.4	2.6	2.8	2.3	2.6	2.9	3.3	3.6
TW G2	9	12	16	19	23	26	31	37	6	8	11	13	16	19	22	26	1.7	1.9	2.1	2.3	2.5	2.0	2.3	2.5	2.8	3.0
TW G3	7	11	15	18	22	25	30	36	4	7	10	12	16	18	21	26	1.7	1.9	2.1	2.3	2.4	2.0	2.2	2.4	2.7	2.9
TW G4	5	8	12	15	19	23	27	33	3	5	8	10	14	16	19	24	1.8	2.0	2.2	2.4	2.5	2.1	2.3	2.5	2.7	2.9
TW H1	8	11	14	16	20	23	26	31	6	8	10	12	14	16	19	22	1.9	2.1	2.3	2.5	2.7	2.3	2.5	2.8	3.1	3.3
TW H2	6	9	12	15	19	21	25	31	3	6	8	10	13	15	18	22	1.9	2.1	2.3	2.5	2.6	2.2	2.4	2.6	2.9	3.1
TW H3	4	7	10	13	16	19	23	28	2	4	7	9	11	14	16	21	1.9	2.1	2.3	2.5	2.7	2.3	2.5	2.7	2.9	3.1
TW H4	3	5	8	10	13	16	19	24	2	3	5	7	9	11	14	17	2.2	2.4	2.6	2.7	2.9	2.5	2.7	2.9	3.1	3.3
TW I1	7	9	11	13	15	17	19	22	4	5	7	7	9	9	10	11	2.2	2.6	2.9	3.3	3.6	2.9	3.5	4.4	-	-
TW I2	11	16	21	25	31	36	42	50	8	11	15	19	24	27	32	39	1.4	1.6	1.8	1.9	2.1	1.6	1.8	2.0	2.2	2.4
TW I3	12	17	23	28	35	39	46	54	8	12	17	21	27	31	36	43	1.4	1.5	1.7	1.9	2.0	1.5	1.7	1.9	2.1	2.2
TW I4	13	20	27	33	41	47	54	63	9	15	21	26	33	38	44	53	1.2	1.4	1.5	1.7	1.8	1.4	1.5	1.7	1.9	2.0
TW I5	13	20	28	34	43	49	56	66	9	15	22	27	35	40	47	57	1.2	1.4	1.5	1.7	1.8	1.3	1.5	1.6	1.8	1.9
TW J1	15	21	28	33	41	46	53	62	11	16	22	26	33	37	43	51	1.3	1.4	1.6	1.7	1.9	1.4	1.6	1.7	1.9	2.1
TW J2	14	21	28	34	42	47	54	64	10	15	22	26	33	38	44	53	1.2	1.4	1.5	1.7	1.8	1.4	1.5	1.7	1.9	2.0
TW J3	18	26	36	43	52	59	66	76	13	20	29	35	44	50	58	67	1.1	1.2	1.4	1.5	1.7	1.2	1.3	1.5	1.6	1.8
TW J4	16	24	34	41	50	57	64	74	12	19	27	33	42	48	56	66	1.1	1.2	1.4	1.6	1.7	1.2	1.3	1.5	1.7	1.8
DG A1	8	11	13	15	17	19	20	21	5	7	8	9	10	11	11	10	2.0	2.4	3.1	4.6	-	2.5	3.3	-	-	-
DG A2	11	15	19	22	27	30	35	40	8	11	14	17	20	23	26	30	1.6	1.8	2.0	2.3	2.6	1.9	2.1	2.4	2.7	3.1
DG A3	15	20	26	31	38	43	49	58	12	16	21	25	31	35	41	49	1.4	1.5	1.6	1.8	2.0	1.5	1.7	1.8	2.0	2.1
DG B1	9	11	13	15	18	20	23	25	5	7	8	9	11	11	12	12	2.0	2.3	2.6	3.2	4.1	2.5	3.2	4.3	-	-
DG B2	12	15	20	23	28	32	37	44	8	11	15	17	21	24	28	33	1.6	1.8	1.9	2.1	2.3	1.8	2.0	2.3	2.5	2.8
DG B3	15	20	26	32	39	45	52	62	12	16	22	26	32	37	43	52	1.4	1.5	1.6	1.7	1.8	1.5	1.6	1.8	1.9	2.0
DG C1	11	14	17	20	24	27	31	35	7	9	12	13	16	17	19	21	1.7	2.0	2.1	2.5	2.9	2.1	2.5	3.1	3.7	4.5
DG C2	14	18	23	27	34	39	45	53	10	14	18	21	26	29	34	41	1.5	1.6	1.7	1.9	2.1	1.6	1.8	2.0	2.2	2.4
DG C3	17	23	30	36	45	51	59	69	14	19	25	30	37	42	49	59	1.3	1.4	1.5	1.6	1.7	1.4	1.5	1.6	1.8	1.9
SS A1	17	21	25	29	33	37	41	47	15	18	21	24	27	30	33	38	1.5	1.8	2.0	2.1	2.3	1.7	2.0	2.3	2.5	2.8
SS A2	21	27	34	39	47	52	58	66	18	23	29	34	40	44	50	57	1.2	1.4	1.5	1.7	1.9	1.3	1.5	1.7	1.9	2.0
SS A3	16	19	22	25	29	31	35	40	13	16	18	20	23	25	27	30	1.7	1.9	2.2	2.4	2.6	1.9	2.3	2.6	3.1	3.5
SS A4	21	26	33	38	45	49	55	64	18	22	28	32	38	42	47	54	1.2	1.4	1.6	1.8	1.9	1.4	1.6	1.8	2.0	2.1
SS A5	17	20	24	26	30	33	37	42	15	17	20	22	25	27	29	33	1.7	1.9	2.2	2.4	2.6	1.9	2.2	2.6	2.9	3.3
SS A6	21	27	33	38	46	50	57	65	18	23	28	33	39	43	48	56	1.2	1.4	1.6	1.7	1.9	1.3	1.5	1.7	1.9	2.1
SS A7	16	18	21	23	25	27	30	33	14	16	17	19	21	22	23	25	1.9	2.3	2.6	2.9	3.1	2.3	2.7	3.2	3.9	4.7
SS A8	20	25	32	36	43	47	53	61	17	22	27	31	36	40	45	52	1.3	1.5	1.6	1.8	2.0	1.4	1.6	1.8	2.0	2.2
SS B1	14	18	21	24	29	32	36	42	12	14	17	20	23	25	28	32	1.6	1.9	2.1	2.3	2.5	1.9	2.1	2.4	2.7	3.0
SS B2	19	24	30	35	42	47	53	62	15	20	26	30	36	40	46	53	1.3	1.5	1.6	1.8	1.9	1.4	1.6	1.8	1.9	2.1
SS B3	13	16	19	22	26	28	32	37	11	13	15	17	20	22	24	27	1.8	2.0	2.3	2.5	2.7	2.0	2.4	2.7	3.2	3.5
SS B4	18	23	29	34	40	45	51	60	15	19	25	29	34	38	43	51	1.3	1.5	1.7	1.8	2.0	1.4	1.6	1.8	2.0	2.2
SS B5	14	17	20	22	25	27	30	35	12	14	16	18	20	22	24	27	1.9	2.1	2.4	2.6	2.8	2.2	2.5	2.9	3.3	3.6
SS B6	18	23	29	34	41	46	52	60	15	19	25	29	34	39	44	51	1.3	1.5	1.6	1.8	1.9	1.4	1.6	1.8	2.0	2.1
SS B7	13	15	17	19	22	23	26	29	11	13	14	15	17	18	20	22	2.1	2.4	2.7	3.0	3.3	2.5	2.9	3.4	4.0	4.7
SS B8	17	22	28	32	38	43	49	57	14	18	23	27	32	36	41	48	1.3	1.5	1.7	1.9	2.0	1.5	1.7	1.9	2.1	2.2
SS C1	12	15	20	23	28	31	36	42	8	11	14	16	19	21	24	29	1.6	1.8	2.0	2.2	2.4	1.9	2.2	2.5	2.8	3.1
SS C2	14	20	26	31	38	43	50	58	10	15	20	24	30	34	39	47	1.3	1.5	1.6	1.8	1.9	1.5	1.7	1.8	2.0	2.2
SS C3	12	14	17	20	23	26	30	35	9	11	13	15	17	19	21	25	1.8	2.1	2.3	2.5	2.7	2.2	2.5	2.8	3.2	3.4
SS C4	14	18	24	28	34	39	44	53	10	14	18	22	27	30	35	42	1.4	1.6	1.8	1.9	2.1	1.6	1.8	2.0	2.2	2.3
SS D1	19	24	29	33	39	43	48	56	15	19	24	27	31	34	38	44	1.4	1.6	1.7	1.9	2.1	1.5	1.8	2.0	2.3	2.5
SS D2	23	31	39	45	54	60	67	75	19	26	33	39	46	51	58	66	1.1	1.3	1.4	1.6	1.7	1.2	1.4	1.5	1.7	1.9
SS D3	18	22	27	30	35	39	43	50	16	19	22	25	29	31	35	39	1.5	1.7	1.9	2.1	2.3	1.7	1.9	2.2	2.5	2.8
SS D4	23	30	38	44	52	58	64	73	19	25	32	37	45	49	56	64	1.1	1.3	1.4	1.6	1.8	1.2	1.4	1.6	1.8	1.9
SS E1	15	19	24	27	32	36	41	47	12	15	18	21	24	27	30	34	1.5	1.7	1.9	2.1	2.3	1.7	2.0	2.3	2.7	3.0
SS E2	19	26	33	39	47	52	59	67	15	21	27	32	38	43	49	57	1.2	1.4	1.5	1.7	1.8	1.3	1.5	1.7	1.9	2.0
SS E3	15	18	21	24	28	31	34	40	12	15	17	19	22	24	27	30	1.7	2.0	2.2	2.4	2.6	2.0	2.3	2.6	3.0	3.3
SS E4	18	24	31	36	43	48	54	63	15	20	25	30	36	40	45	53	1.3	1.4	1.6	1.8	1.9	1.4	1.6	1.8	2.0	2.1

SPRINGFIELD, ILLINOIS

TYPE	SOLAR SAVINGS FRACTION (%) BASE TEMP: 55 F (3474 DD) LCR 100	70	50	40	30	25	20	15	65 F (5563 DD) 100	70	50	40	30	25	20	15	CONSERVATION FACTOR 55 F 100	50	30	20	15	65 F 100	50	30	20	15
WW A1	17	20	23	26	30	33	36	42	14	16	18	20	22	24	26	29	1.8	2.0	2.2	2.3	2.5	2.2	2.5	2.8	3.1	3.4
WW A2	17	22	28	32	38	42	48	56	13	16	20	23	28	31	35	41	1.4	1.6	1.7	1.9	2.0	1.7	1.9	2.1	2.3	2.5
WW A3	17	23	29	34	41	46	53	61	12	16	21	24	30	33	38	45	1.3	1.5	1.6	1.8	1.9	1.5	1.7	1.9	2.1	2.3
WW A4	17	23	30	35	43	48	55	64	11	16	21	25	31	35	40	48	1.2	1.4	1.6	1.7	1.9	1.5	1.6	1.8	2.0	2.2
WW A5	17	23	31	37	45	51	58	67	11	16	21	26	32	37	42	50	1.2	1.4	1.5	1.7	1.8	1.4	1.6	1.8	2.0	2.1
WW A6	17	24	32	38	46	52	59	68	10	16	22	26	33	38	43	52	1.2	1.3	1.5	1.7	1.8	1.4	1.5	1.7	1.9	2.1
WW B1	13	17	20	23	27	30	34	40	9	11	13	15	17	18	20	22	1.7	1.9	2.1	2.3	2.5	2.2	2.6	3.1	3.7	4.3
WW B2	18	24	32	38	46	52	59	68	12	17	23	28	35	40	46	54	1.2	1.3	1.5	1.7	1.8	1.4	1.5	1.7	1.9	2.0
WW B3	21	29	38	45	54	60	67	76	15	21	29	34	42	47	54	63	1.1	1.2	1.4	1.6	1.7	1.2	1.4	1.6	1.7	1.9
WW B4	21	30	40	48	58	64	72	81	15	23	31	38	46	53	60	70	1.0	1.2	1.3	1.5	1.6	1.1	1.3	1.5	1.6	1.8
WW B5	21	30	40	47	57	64	72	81	14	22	31	38	47	53	61	71	1.0	1.2	1.3	1.5	1.6	1.1	1.3	1.4	1.6	1.7
WW C1	23	31	41	47	57	63	70	79	16	23	31	37	45	51	58	67	1.1	1.2	1.4	1.5	1.7	1.2	1.3	1.5	1.7	1.8
WW C2	21	30	39	46	56	62	70	79	15	22	30	36	45	50	58	67	1.1	1.2	1.4	1.5	1.7	1.2	1.3	1.5	1.7	1.8
WW C3	26	36	48	56	67	73	80	88	19	28	39	46	56	63	71	80	0.9	1.1	1.2	1.4	1.6	1.0	1.2	1.3	1.5	1.6
WW C4	23	33	45	53	63	70	77	86	17	26	36	43	53	60	68	77	1.0	1.1	1.3	1.4	1.6	1.1	1.2	1.4	1.5	1.7
TW A1	17	20	23	26	30	33	36	42	14	16	18	20	22	24	27	30	1.7	2.0	2.1	2.3	2.5	2.2	2.4	2.7	3.0	3.2
TW A2	17	21	26	29	35	39	44	52	13	16	19	22	26	29	32	38	1.5	1.7	1.8	2.0	2.1	1.8	2.0	2.2	2.4	2.6
TW A3	16	21	26	31	37	42	47	56	12	15	19	22	27	30	35	41	1.4	1.6	1.7	1.9	2.0	1.6	1.8	2.0	2.2	2.4
TW A4	15	20	26	31	38	42	49	57	10	14	19	22	27	31	36	43	1.4	1.5	1.7	1.8	2.0	1.6	1.8	2.0	2.1	2.3
TW B1	16	19	23	26	31	34	38	44	13	15	18	20	23	25	28	32	1.7	1.9	2.0	2.2	2.3	2.0	2.3	2.5	2.7	2.9
TW B2	15	20	25	28	34	38	44	51	11	15	18	21	25	28	32	38	1.5	1.6	1.8	2.0	2.1	1.7	1.9	2.1	2.3	2.5
TW B3	15	19	24	29	35	39	45	53	11	14	18	21	26	29	33	39	1.4	1.6	1.8	1.9	2.1	1.7	1.9	2.1	2.3	2.4
TW B4	14	18	23	27	33	37	43	51	10	13	17	20	25	28	32	38	1.5	1.6	1.8	1.9	2.1	1.7	1.9	2.1	2.3	2.4
TW C1	15	19	23	26	31	34	39	45	12	14	17	20	23	26	29	34	1.6	1.8	2.0	2.1	2.3	1.9	2.1	2.3	2.6	2.7
TW C2	14	18	22	26	31	35	40	48	11	14	17	20	24	27	30	36	1.6	1.7	1.9	2.0	2.2	1.8	2.0	2.2	2.4	2.5
TW C3	14	17	22	25	30	34	39	46	11	13	16	19	23	26	30	35	1.6	1.8	1.9	2.1	2.2	1.9	2.0	2.2	2.4	2.6
TW C4	13	17	20	23	28	31	35	42	11	13	16	18	22	24	28	33	1.7	1.9	2.1	2.2	2.3	2.0	2.2	2.4	2.6	2.7
TW D1	11	14	17	20	24	27	30	36	8	9	11	13	15	17	19	21	1.8	2.0	2.2	2.4	2.6	2.3	2.6	3.0	3.5	3.9
TW D2	17	23	30	35	42	47	54	63	13	17	22	26	32	36	42	50	1.3	1.4	1.6	1.7	1.9	1.5	1.6	1.8	2.0	2.1
TW D3	18	25	32	38	46	52	59	68	13	18	24	29	35	40	46	54	1.2	1.4	1.5	1.7	1.8	1.4	1.5	1.7	1.9	2.0
TW D4	20	28	36	43	52	58	66	75	14	21	28	34	42	47	54	63	1.1	1.3	1.4	1.6	1.7	1.2	1.4	1.6	1.7	1.8
TW D5	20	28	37	43	53	59	67	76	14	21	29	34	43	49	56	65	1.1	1.2	1.4	1.5	1.7	1.2	1.4	1.5	1.7	1.8
TW E1	21	29	37	43	52	57	64	73	16	22	28	33	41	46	52	61	1.1	1.3	1.4	1.6	1.7	1.3	1.4	1.6	1.8	1.9
TW E2	21	28	37	43	52	58	66	75	15	21	28	34	42	47	54	63	1.1	1.3	1.4	1.6	1.7	1.2	1.4	1.6	1.7	1.9
TW E3	25	35	45	53	63	69	76	85	19	27	36	43	52	58	66	75	1.0	1.1	1.3	1.5	1.6	1.1	1.2	1.4	1.6	1.7
TW E4	23	33	43	50	60	67	74	83	17	25	34	41	50	56	64	74	1.0	1.2	1.3	1.5	1.6	1.1	1.3	1.4	1.6	1.7
TW F1	15	18	22	25	29	32	36	42	12	14	16	18	21	23	26	29	1.7	1.9	2.1	2.3	2.4	2.1	2.4	2.6	2.9	3.1
TW F2	14	18	23	27	32	36	42	49	10	13	17	19	23	26	30	35	1.5	1.7	1.8	2.0	2.1	1.8	2.0	2.2	2.4	2.6
TW F3	13	18	23	27	33	38	43	52	9	12	16	19	24	27	31	37	1.4	1.6	1.8	1.9	2.1	1.7	1.9	2.1	2.3	2.5
TW F4	11	16	21	26	32	37	43	51	7	10	14	18	22	26	31	37	1.4	1.6	1.8	1.9	2.0	1.7	1.9	2.1	2.3	2.4
TW G1	13	17	20	23	28	31	36	42	10	12	15	17	20	22	25	30	1.7	1.9	2.1	2.2	2.4	2.0	2.3	2.5	2.7	2.9
TW G2	11	15	20	24	29	33	39	46	8	11	14	17	21	24	28	33	1.5	1.7	1.9	2.0	2.2	1.8	2.0	2.2	2.4	2.6
TW G3	10	14	19	23	28	32	38	46	6	9	13	16	20	23	27	33	1.5	1.7	1.9	2.0	2.2	1.8	2.0	2.2	2.4	2.5
TW G4	8	11	16	20	25	29	34	42	4	7	11	13	17	20	24	30	1.6	1.8	1.9	2.1	2.2	1.9	2.1	2.3	2.4	2.6
TW H1	10	14	18	21	25	29	33	40	7	10	12	15	18	20	24	28	1.7	1.9	2.0	2.2	2.3	2.0	2.2	2.4	2.7	2.9
TW H2	8	11	16	19	24	27	32	39	5	7	10	13	17	19	23	28	1.7	1.9	2.0	2.2	2.3	2.0	2.2	2.4	2.6	2.7
TW H3	6	10	13	16	21	24	29	36	4	6	9	11	15	17	21	26	1.8	1.9	2.1	2.3	2.4	2.1	2.3	2.4	2.6	2.8
TW H4	5	7	10	13	17	20	24	30	3	5	7	9	12	14	17	21	1.9	2.1	2.3	2.5	2.6	2.3	2.5	2.7	2.8	3.0
TW I1	9	12	15	17	21	24	27	32	6	7	9	11	13	14	16	18	1.9	2.1	2.3	2.5	2.7	2.4	2.8	3.2	3.7	4.2
TW I2	14	20	26	31	39	44	51	60	10	14	19	23	29	33	38	46	1.3	1.5	1.6	1.8	1.9	1.5	1.7	1.8	2.0	2.2
TW I3	15	22	29	34	42	48	55	64	11	15	21	26	32	37	43	51	1.2	1.4	1.5	1.7	1.8	1.4	1.6	1.8	1.9	2.1
TW I4	17	25	33	40	49	55	62	72	12	18	25	30	38	44	51	60	1.1	1.3	1.4	1.6	1.7	1.3	1.4	1.6	1.7	1.9
TW I5	17	25	34	41	50	57	64	74	12	18	26	32	40	46	53	63	1.1	1.3	1.4	1.6	1.7	1.2	1.4	1.5	1.7	1.8
TW J1	19	26	34	41	49	55	62	72	13	19	26	31	38	43	50	59	1.1	1.3	1.4	1.6	1.7	1.3	1.5	1.6	1.8	1.9
TW J2	18	26	34	40	49	55	63	72	13	19	26	31	39	44	51	60	1.1	1.3	1.4	1.6	1.7	1.3	1.4	1.6	1.7	1.9
TW J3	22	32	42	50	60	66	74	83	16	24	33	40	50	56	64	74	1.0	1.1	1.3	1.5	1.6	1.1	1.3	1.4	1.6	1.7
TW J4	20	30	40	47	58	64	72	81	14	22	31	38	48	54	62	72	1.0	1.2	1.3	1.5	1.6	1.1	1.3	1.4	1.6	1.7
DG A1	11	15	18	21	24	26	29	31	7	9	11	13	15	16	17	17	1.7	2.0	2.4	3.1	4.1	2.1	2.6	3.4	-	-
DG A2	14	18	24	28	34	38	43	49	10	13	17	20	25	28	32	37	1.4	1.6	1.8	2.1	2.3	1.7	1.9	2.1	2.4	2.8
DG A3	17	23	31	36	45	51	58	66	14	18	24	28	35	40	47	55	1.3	1.4	1.5	1.7	1.9	1.4	1.6	1.7	1.8	2.0
DG B1	11	15	19	22	26	29	33	38	7	9	12	13	16	17	19	21	1.7	1.8	2.1	2.5	2.8	2.1	2.5	3.0	3.7	4.8
DG B2	14	19	24	29	36	41	47	55	10	14	18	21	26	29	34	40	1.4	1.5	1.7	1.9	2.1	1.6	1.8	2.0	2.2	2.4
DG B3	18	24	32	38	47	53	61	71	14	19	25	30	37	42	49	59	1.3	1.3	1.4	1.6	1.7	1.4	1.5	1.6	1.7	1.9
DG C1	14	18	23	27	33	37	42	49	9	12	15	18	21	23	26	30	1.5	1.6	1.8	2.1	2.3	1.8	2.1	2.4	2.7	3.2
DG C2	17	22	29	34	42	48	55	64	12	16	21	25	31	35	41	49	1.3	1.4	1.5	1.7	1.9	1.5	1.7	1.8	2.0	2.2
DG C3	20	27	36	43	53	60	68	78	16	22	28	34	42	48	56	66	1.2	1.2	1.3	1.5	1.7	1.3	1.4	1.5	1.6	1.8
SS A1	20	25	30	34	40	45	50	58	17	20	25	28	32	36	40	46	1.4	1.5	1.7	1.9	2.0	1.6	1.8	2.0	2.2	2.4
SS A2	25	32	40	46	54	60	66	75	21	27	33	38	45	50	56	65	1.1	1.3	1.4	1.6	1.7	1.2	1.4	1.6	1.7	1.9
SS A3	19	23	27	31	36	40	44	51	15	18	22	24	28	31	34	38	1.5	1.7	1.9	2.0	2.2	1.7	2.0	2.2	2.5	2.7
SS A4	25	31	39	45	52	58	64	73	20	26	32	37	44	48	54	62	1.1	1.3	1.5	1.6	1.8	1.3	1.4	1.6	1.8	1.9
SS A5	20	24	28	32	37	40	45	52	17	20	23	26	29	32	35	40	1.5	1.7	1.9	2.0	2.2	1.7	2.0	2.2	2.5	2.7
SS A6	25	32	40	45	53	59	66	74	20	26	33	38	44	49	55	63	1.1	1.3	1.4	1.6	1.8	1.2	1.4	1.6	1.8	1.9
SS A7	18	21	25	27	31	34	38	43	16	18	20	22	25	27	29	32	1.7	1.9	2.1	2.3	2.5	2.0	2.3	2.6	3.0	3.3
SS A8	24	30	38	43	51	56	63	71	20	25	31	35	42	46	52	60	1.2	1.3	1.5	1.6	1.8	1.3	1.5	1.7	1.8	2.0
SS B1	17	21	26	30	35	39	44	52	14	17	21	23	28	30	34	40	1.5	1.7	1.8	2.0	2.1	1.7	1.9	2.1	2.4	2.6
SS B2	22	28	36	41	49	55	62	70	18	23	29	34	41	46	52	60	1.2	1.3	1.5	1.6	1.8	1.3	1.5	1.6	1.8	1.9
SS B3	16	20	24	27	32	36	40	47	13	15	19	21	25	27	30	35	1.5	1.8	1.9	2.1	2.3	1.8	2.1	2.3	2.6	2.8
SS B4	21	27	35	40	48	53	60	68	17	22	28	33	39	44	50	58	1.2	1.4	1.5	1.7	1.8	1.3	1.5	1.7	1.9	2.0
SS B5	16	20	23	26	31	34	38	44	14	16	19	21	24	26	29	33	1.7	1.9	2.1	2.2	2.4	1.9	2.2	2.5	2.7	3.0
SS B6	21	27	35	40	48	53	60	69	17	22	28	33	40	44	50	58	1.2	1.3	1.5	1.7	1.8	1.3	1.5	1.7	1.8	2.0
SS B7	15	18	21	23	27	29	33	38	13	14	17	18	21	22	25	28	1.8	2.1	2.3	2.5	2.6	2.2	2.5	2.8	3.2	3.5
SS B8	20	26	33	38	46	51	57	66	16	21	27	31	37	41	47	55	1.2	1.4	1.5	1.7	1.8	1.4	1.6	1.7	1.9	2.1
SS C1	15	20	25	30	36	40	46	54	10	14	17	20	25	28	32	38	1.4	1.6	1.7	1.9	2.0	1.7	1.9	2.1	2.3	2.5
SS C2	18	25	33	38	47	52	59	68	13	18	24	29	36	40	46	55	1.2	1.3	1.5	1.7	1.8	1.4	1.5	1.7	1.9	2.0
SS C3	14	17	21	25	29	33	37	44	11	13	16	18	21	24	27	31	1.6	1.8	2.0	2.2	2.3	2.0	2.2	2.4	2.7	2.9
SS C4	17	23	29	34	42	47	53	62	12	17	22	26	32	36	41	49	1.3	1.4	1.6	1.7	1.9	1.5	1.6	1.8	2.0	2.1
SS D1	23	29	36	41	48	53	59	67	18	23	28	32	38	42	47	53	1.2	1.4	1.5	1.7	1.9	1.4	1.6	1.8	2.0	2.2
SS D2	28	37	46	53	63	68	75	83	23	30	38	44	53	58	65	73	1.0	1.2	1.3	1.5	1.6	1.1	1.3	1.4	1.6	1.8
SS D3	22	27	32	37	43	47	53	61	18	22	26	29	34	37	42	48	1.3	1.5	1.7	1.8	2.0	1.5	1.7	2.0	2.2	2.3
SS D4	27	36	45	51	60	66	73	81	22	29	37	43	51	56	63	71	1.0	1.2	1.3	1.5	1.7	1.1	1.3	1.5	1.6	1.8
SS E1	19	24	30	35	41	46	52	60	14	18	23	26	31	34	38	44	1.3	1.5	1.7	1.8	2.0	1.5	1.8	2.0	2.2	2.4
SS E2	24	31	40	47	55	61	68	77	18	25	32	37	45	50	57	65	1.1	1.2	1.4	1.6	1.7	1.2	1.4	1.6	1.7	1.9
SS E3	18	22	26	30	35	38	43	50	14	17	21	23	27	29	33	38	1.5	1.7	1.9	2.0	2.2	1.7	2.0	2.3	2.5	2.7
SS E4	22	29	37	43	51	57	64	72	18	23	30	35	41	46	52	61	1.1	1.3	1.5	1.6	1.8	1.3	1.5	1.6	1.8	1.9

EVANSVILLE, INDIANA

TYPE	SOLAR SAVINGS FRACTION (%) BASE TEMP: 55 F (2689 DD)								65 F (4633 DD)								CONSERVATION FACTOR 55 F					65 F				
LCR	100	70	50	40	30	25	20	15	100	70	50	40	30	25	20	15	100	50	30	20	15	100	50	30	20	15
WW A1	19	22	26	29	34	37	41	48	15	17	19	21	24	26	29	32	1.6	1.8	2.0	2.1	2.3	2.0	2.3	2.6	2.9	3.1
WW A2	20	25	31	36	43	48	54	62	14	18	22	25	30	33	38	44	1.3	1.5	1.6	1.8	1.9	1.6	1.8	2.0	2.2	2.3
WW A3	20	26	33	39	47	52	59	68	13	18	23	27	32	36	42	49	1.2	1.4	1.5	1.7	1.8	1.4	1.6	1.8	2.0	2.2
WW A4	20	27	34	40	49	54	61	70	13	18	23	28	34	38	44	51	1.2	1.3	1.5	1.6	1.8	1.4	1.6	1.8	2.0	2.1
WW A5	20	27	36	42	51	57	64	73	12	18	24	29	35	40	46	54	1.1	1.3	1.4	1.6	1.7	1.3	1.5	1.7	1.9	2.0
WW A6	20	28	36	43	52	58	65	74	12	18	24	29	36	41	47	55	1.1	1.3	1.4	1.6	1.7	1.3	1.5	1.7	1.9	2.0
WW B1	16	19	24	27	32	36	41	47	10	12	15	17	19	21	23	26	1.5	1.7	1.9	2.1	2.3	2.0	2.4	2.8	3.2	3.7
WW B2	21	28	36	43	51	57	65	74	14	19	26	30	38	42	49	57	1.1	1.3	1.4	1.6	1.7	1.3	1.5	1.7	1.8	2.0
WW B3	25	33	43	50	60	66	73	81	17	24	31	37	45	50	57	66	1.0	1.2	1.3	1.5	1.6	1.2	1.4	1.5	1.7	1.8
WW B4	25	34	45	53	63	69	76	85	17	25	33	40	49	55	63	72	1.0	1.1	1.3	1.5	1.6	1.1	1.3	1.4	1.6	1.7
WW B5	24	33	44	52	62	69	76	85	16	24	33	40	49	56	63	73	1.0	1.1	1.3	1.4	1.6	1.1	1.3	1.4	1.6	1.7
WW C1	26	35	45	53	62	68	75	83	18	25	33	39	48	53	61	70	1.0	1.1	1.3	1.5	1.6	1.1	1.3	1.5	1.6	1.8
WW C2	25	34	44	51	61	67	75	83	17	24	32	39	47	53	61	70	1.0	1.1	1.3	1.5	1.6	1.1	1.3	1.5	1.6	1.8
WW C3	29	41	53	61	71	77	84	91	21	30	41	49	59	65	73	82	0.9	1.0	1.2	1.4	1.5	1.0	1.2	1.3	1.5	1.6
WW C4	26	37	49	57	68	74	81	89	18	28	38	45	55	62	70	79	0.9	1.1	1.2	1.4	1.5	1.0	1.2	1.3	1.5	1.6
TW A1	19	22	26	29	33	37	41	48	15	17	19	21	24	26	29	33	1.6	1.8	2.0	2.1	2.3	2.0	2.3	2.5	2.8	3.0
TW A2	19	23	29	33	39	44	50	58	14	17	21	24	28	31	35	41	1.4	1.5	1.7	1.9	2.0	1.7	1.9	2.1	2.3	2.5
TW A3	18	24	30	35	42	47	53	62	13	17	21	24	30	33	38	44	1.3	1.5	1.6	1.8	1.9	1.6	1.8	2.0	2.1	2.3
TW A4	17	23	30	35	42	48	54	63	11	16	21	24	30	34	39	46	1.3	1.4	1.6	1.7	1.9	1.5	1.7	1.9	2.1	2.2
TW B1	18	22	26	29	34	38	43	50	14	16	19	22	25	27	30	35	1.6	1.7	1.9	2.0	2.2	1.9	2.1	2.4	2.6	2.8
TW B2	17	22	28	32	39	43	49	57	12	16	20	23	28	31	35	41	1.4	1.5	1.7	1.8	2.0	1.7	1.8	2.0	2.2	2.4
TW B3	17	22	28	32	39	44	50	59	12	15	20	23	28	31	36	43	1.4	1.5	1.7	1.8	1.9	1.6	1.8	2.0	2.2	2.3
TW B4	16	21	26	31	37	42	48	56	11	14	19	22	27	30	35	41	1.4	1.5	1.7	1.8	2.0	1.6	1.8	2.0	2.2	2.3
TW C1	17	21	25	29	34	38	43	51	13	15	19	21	25	28	31	37	1.5	1.7	1.8	2.0	2.1	1.8	2.0	2.2	2.4	2.6
TW C2	16	20	25	29	35	39	45	53	12	15	18	21	26	29	33	39	1.5	1.6	1.8	1.9	2.1	1.7	1.9	2.1	2.3	2.5
TW C3	15	19	24	28	34	38	43	51	11	14	18	21	25	28	32	38	1.5	1.7	1.8	2.0	2.1	1.8	2.0	2.1	2.3	2.5
TW C4	15	18	23	26	31	34	39	46	11	14	17	19	23	26	30	35	1.6	1.8	1.9	2.1	2.2	1.9	2.1	2.3	2.5	2.6
TW D1	13	17	21	24	28	32	36	43	8	11	13	15	17	19	22	25	1.6	1.8	2.0	2.2	2.3	2.1	2.4	2.8	3.2	3.5
TW D2	20	26	33	39	47	53	60	69	14	19	24	28	35	39	45	53	1.2	1.4	1.5	1.7	1.8	1.4	1.6	1.8	1.9	2.1
TW D3	21	28	37	43	51	57	64	73	14	20	26	31	38	43	49	57	1.1	1.3	1.4	1.6	1.7	1.3	1.5	1.7	1.8	2.0
TW D4	23	31	41	47	57	63	70	79	16	23	30	36	44	50	57	66	1.1	1.2	1.4	1.5	1.7	1.2	1.4	1.5	1.7	1.8
TW D5	22	31	41	48	57	64	71	80	16	23	31	37	45	51	58	68	1.0	1.2	1.3	1.5	1.6	1.2	1.3	1.5	1.6	1.8
TW E1	24	32	41	48	57	63	70	78	17	23	30	36	43	48	55	64	1.1	1.2	1.4	1.5	1.7	1.2	1.4	1.6	1.7	1.9
TW E2	24	32	41	48	57	63	71	79	17	23	31	36	44	49	56	65	1.1	1.2	1.4	1.5	1.7	1.2	1.4	1.5	1.7	1.8
TW E3	28	39	50	57	67	74	80	88	20	29	38	45	55	61	68	77	0.9	1.1	1.2	1.4	1.6	1.1	1.2	1.4	1.5	1.7
TW E4	26	36	47	55	65	71	78	86	19	27	36	43	53	59	66	76	1.0	1.1	1.3	1.4	1.6	1.1	1.2	1.4	1.5	1.7
TW F1	17	20	25	28	33	36	41	47	13	15	18	20	23	25	28	32	1.6	1.8	1.9	2.1	2.2	2.0	2.2	2.5	2.7	2.9
TW F2	16	21	26	31	37	41	47	55	11	14	18	21	26	29	33	38	1.4	1.6	1.7	1.9	2.0	1.7	1.9	2.1	2.3	2.5
TW F3	15	20	26	31	38	43	49	58	10	13	18	21	26	29	34	41	1.3	1.5	1.7	1.8	2.0	1.6	1.8	2.0	2.2	2.4
TW F4	13	19	25	30	37	42	48	57	8	12	16	20	25	29	33	40	1.3	1.5	1.7	1.8	1.9	1.6	1.8	2.0	2.2	2.3
TW G1	15	19	23	27	32	35	40	48	11	13	16	19	22	25	28	33	1.6	1.7	1.9	2.1	2.2	1.9	2.1	2.4	2.6	2.8
TW G2	14	18	23	28	34	38	44	52	9	12	16	19	23	26	30	36	1.4	1.6	1.8	1.9	2.0	1.7	1.9	2.1	2.3	2.5
TW G3	12	16	22	26	33	37	43	52	7	10	14	17	22	25	30	36	1.4	1.6	1.7	1.9	2.0	1.7	1.9	2.1	2.3	2.5
TW G4	9	14	19	23	29	33	39	47	5	8	12	15	19	22	27	33	1.5	1.7	1.8	2.0	2.1	1.8	2.0	2.2	2.4	2.5
TW H1	12	16	20	24	29	33	38	45	8	10	14	16	20	22	26	31	1.6	1.8	1.9	2.1	2.2	1.9	2.1	2.3	2.5	2.7
TW H2	10	14	18	22	27	31	37	45	5	8	12	14	18	21	25	31	1.6	1.7	1.9	2.0	2.2	1.9	2.1	2.3	2.5	2.6
TW H3	8	12	16	19	24	28	33	41	4	7	10	12	16	19	23	28	1.6	1.8	2.0	2.1	2.2	2.0	2.2	2.3	2.5	2.7
TW H4	6	9	12	15	20	23	27	34	3	5	8	10	13	15	19	23	1.8	2.0	2.2	2.3	2.4	2.2	2.4	2.5	2.7	2.9
TW I1	11	14	18	21	25	29	33	39	7	9	11	13	15	17	19	22	1.7	1.9	2.1	2.3	2.4	2.2	2.5	2.9	3.3	3.7
TW I2	17	23	30	36	44	49	56	65	11	15	21	25	31	36	41	49	1.2	1.4	1.5	1.7	1.8	1.4	1.6	1.8	2.0	2.1
TW I3	18	25	33	39	47	53	60	70	12	17	23	28	35	39	45	54	1.2	1.3	1.5	1.6	1.8	1.4	1.5	1.7	1.9	2.0
TW I4	20	28	37	44	53	60	67	77	13	20	27	33	41	46	53	63	1.1	1.2	1.4	1.5	1.7	1.2	1.4	1.5	1.7	1.8
TW I5	20	28	38	45	55	61	69	78	13	20	28	34	43	48	56	66	1.1	1.2	1.4	1.5	1.7	1.2	1.4	1.5	1.7	1.8
TW J1	22	30	39	45	54	60	68	77	15	21	28	33	41	46	53	62	1.1	1.2	1.4	1.5	1.7	1.2	1.4	1.6	1.7	1.9
TW J2	21	29	38	45	54	60	68	77	14	21	28	33	41	47	54	63	1.1	1.2	1.4	1.5	1.7	1.2	1.4	1.6	1.7	1.9
TW J3	25	35	47	54	65	71	78	86	18	26	36	43	52	58	66	76	1.0	1.1	1.3	1.4	1.6	1.1	1.2	1.4	1.5	1.7
TW J4	23	33	44	52	62	69	76	85	16	24	33	40	50	56	64	74	1.0	1.1	1.3	1.4	1.6	1.1	1.2	1.4	1.6	1.7
DG A1	14	18	22	25	29	32	35	38	8	11	13	15	17	19	20	21	1.5	1.8	2.2	2.7	3.4	2.0	2.4	3.1	4.6	-
DG A2	16	21	27	32	38	43	48	55	11	15	19	22	27	30	34	40	1.4	1.5	1.7	2.0	2.2	1.6	1.8	2.0	2.3	2.7
DG A3	20	26	34	40	49	55	62	71	15	20	26	30	38	43	49	58	1.2	1.3	1.4	1.6	1.8	1.4	1.5	1.6	1.8	2.0
DG B1	14	18	23	26	32	35	40	46	8	11	13	15	18	20	22	25	1.5	1.7	1.9	2.2	2.6	1.9	2.3	2.7	3.3	4.2
DG B2	16	21	28	33	41	46	53	61	11	15	20	23	28	32	37	43	1.4	1.4	1.6	1.8	2.0	1.6	1.8	1.9	2.1	2.3
DG B3	20	27	35	42	52	58	67	76	15	20	27	32	39	45	52	62	1.2	1.3	1.4	1.5	1.7	1.3	1.5	1.6	1.7	1.8
DG C1	16	21	28	32	39	44	50	57	11	14	17	20	24	27	30	34	1.4	1.5	1.7	1.9	2.1	1.7	2.0	2.3	2.6	3.0
DG C2	19	25	33	39	48	53	61	70	13	18	23	27	34	38	44	52	1.3	1.3	1.5	1.6	1.8	1.4	1.6	1.7	1.9	2.1
DG C3	23	30	40	47	58	65	74	83	17	23	30	36	44	50	58	69	1.1	1.2	1.3	1.4	1.6	1.3	1.4	1.5	1.6	1.7
SS A1	23	28	34	39	46	50	56	64	18	22	27	30	35	39	43	49	1.3	1.4	1.6	1.8	1.9	1.5	1.7	1.9	2.1	2.3
SS A2	28	36	45	51	60	65	72	80	22	29	36	41	48	53	60	68	1.1	1.2	1.4	1.5	1.7	1.2	1.4	1.5	1.7	1.8
SS A3	21	26	31	35	41	46	51	59	17	20	24	27	31	34	38	43	1.4	1.5	1.7	1.9	2.0	1.6	1.9	2.1	2.4	2.5
SS A4	28	35	44	50	58	64	70	78	22	28	35	40	47	51	58	66	1.1	1.2	1.4	1.6	1.7	1.2	1.4	1.6	1.7	1.9
SS A5	22	27	32	36	42	46	51	58	18	21	25	28	32	35	39	44	1.4	1.6	1.7	1.9	2.0	1.6	1.9	2.1	2.3	2.5
SS A6	28	36	44	50	59	65	71	79	22	28	35	40	47	52	58	67	1.1	1.2	1.4	1.5	1.7	1.2	1.4	1.6	1.7	1.9
SS A7	20	24	28	31	36	39	44	50	17	19	22	24	27	29	32	36	1.6	1.8	1.9	2.1	2.3	1.8	2.1	2.4	2.8	3.0
SS A8	27	34	42	48	57	62	69	77	21	27	34	38	45	50	56	63	1.1	1.2	1.4	1.6	1.7	1.2	1.4	1.6	1.8	1.9
SS B1	19	24	30	34	40	45	50	58	15	18	23	26	30	33	37	43	1.4	1.5	1.7	1.9	2.0	1.6	1.8	2.0	2.3	2.4
SS B2	25	32	40	46	54	60	67	76	19	25	32	37	44	49	55	63	1.1	1.3	1.4	1.6	1.7	1.3	1.4	1.6	1.8	1.9
SS B3	18	22	27	31	37	41	47	54	14	17	21	23	27	30	34	39	1.4	1.6	1.8	1.9	2.1	1.7	1.9	2.2	2.4	2.6
SS B4	24	31	39	45	53	59	65	74	19	24	31	35	42	47	53	61	1.1	1.3	1.4	1.6	1.7	1.3	1.5	1.6	1.8	1.9
SS B5	18	22	26	30	35	38	43	50	15	17	20	23	26	29	32	37	1.5	1.7	1.9	2.0	2.2	1.8	2.1	2.3	2.6	2.8
SS B6	24	31	39	45	53	59	66	74	18	24	31	35	42	47	53	61	1.1	1.3	1.4	1.6	1.7	1.3	1.4	1.6	1.8	1.9
SS B7	17	20	24	26	31	34	38	44	13	16	18	20	23	25	28	31	1.7	1.9	2.1	2.2	2.4	2.0	2.3	2.6	2.9	3.2
SS B8	23	29	37	43	51	56	63	72	18	23	29	34	40	45	50	58	1.2	1.3	1.5	1.6	1.8	1.3	1.5	1.7	1.8	2.0
SS C1	18	23	29	34	41	46	52	61	11	15	19	23	28	31	35	41	1.3	1.5	1.6	1.8	1.9	1.6	1.8	2.0	2.2	2.4
SS C2	21	28	37	43	52	58	65	74	14	20	26	31	38	43	49	58	1.1	1.3	1.4	1.6	1.7	1.3	1.5	1.6	1.8	2.0
SS C3	16	20	24	28	33	37	43	50	11	14	17	20	23	26	29	34	1.5	1.7	1.8	2.0	2.1	1.9	2.1	2.3	2.5	2.7
SS C4	19	26	33	38	46	52	59	68	13	18	24	28	34	38	44	52	1.2	1.4	1.5	1.7	1.8	1.4	1.6	1.8	1.9	2.1
SS D1	26	33	40	46	54	59	66	74	20	25	31	35	41	45	50	58	1.1	1.3	1.4	1.6	1.8	1.3	1.5	1.7	1.9	2.1
SS D2	32	41	52	59	68	74	80	87	25	33	41	47	56	61	68	77	0.9	1.1	1.3	1.4	1.6	1.1	1.2	1.4	1.6	1.7
SS D3	24	30	37	41	49	53	59	67	19	24	28	32	37	41	45	52	1.2	1.4	1.5	1.7	1.9	1.4	1.6	1.9	2.1	2.2
SS D4	31	40	50	57	66	72	78	86	24	31	40	46	54	59	66	74	1.0	1.1	1.3	1.5	1.6	1.1	1.3	1.4	1.6	1.7
SS E1	22	28	35	40	48	53	59	67	16	20	25	29	34	38	42	49	1.2	1.4	1.5	1.7	1.9	1.4	1.7	1.9	2.1	2.3
SS E2	27	36	45	52	61	67	74	82	20	27	35	40	48	53	60	69	1.0	1.2	1.3	1.5	1.6	1.2	1.3	1.5	1.7	1.8
SS E3	20	25	30	34	40	44	50	57	16	19	23	25	30	32	36	41	1.4	1.6	1.7	1.9	2.0	1.6	1.9	2.1	2.4	2.5
SS E4	25	33	42	48	57	62	69	78	19	25	32	37	45	49	56	64	1.1	1.2	1.4	1.6	1.7	1.2	1.4	1.6	1.7	1.9

FORT WAYNE, INDIANA

TYPE	SOLAR SAVINGS FRACTION (%) BASE TEMP: 55 F (3937 DD)								65 F (6213 DD)								CONSERVATION FACTOR 55 F					65 F				
LCR	100	70	50	40	30	25	20	15	100	70	50	40	30	25	20	15	100	50	30	20	15	100	50	30	20	15
WW A1	14	16	18	19	21	22	24	27	12	12	13	14	15	16	16	17	2.2	2.6	3.1	3.5	3.9	3.1	3.8	4.8	-	-
WW A2	12	16	19	22	26	28	32	37	9	11	13	15	17	19	21	24	1.7	2.0	2.2	2.5	2.6	2.1	2.5	2.9	3.4	3.9
WW A3	11	15	19	23	27	31	35	41	7	10	13	15	18	20	23	26	1.6	1.8	2.0	2.2	2.4	1.9	2.2	2.6	3.0	3.3
WW A4	11	15	20	23	28	32	37	44	6	9	12	15	19	21	24	28	1.5	1.7	1.9	2.1	2.3	1.8	2.1	2.4	2.8	3.1
WW A5	10	15	20	24	30	33	39	46	5	9	12	15	19	22	25	30	1.4	1.7	1.9	2.1	2.2	1.7	2.0	2.3	2.6	2.9
WW A6	10	15	20	24	30	34	40	47	5	8	12	15	19	22	26	31	1.4	1.6	1.8	2.0	2.2	1.7	1.9	2.2	2.5	2.8
WW B1	9	11	12	13	15	16	17	18	6	6	7	7	7	6	5	3	2.3	3.0	3.8	4.9	-	3.8	-	-	-	-
WW B2	12	17	22	26	33	37	43	50	7	11	15	19	23	27	31	37	1.4	1.6	1.8	2.0	2.1	1.6	1.8	2.1	2.3	2.5
WW B3	14	20	27	32	39	44	51	59	9	14	19	23	29	33	39	46	1.3	1.4	1.6	1.8	1.9	1.4	1.6	1.8	2.1	2.2
WW B4	14	21	29	35	44	50	57	67	9	15	22	27	35	40	46	55	1.2	1.3	1.5	1.7	1.8	1.3	1.5	1.6	1.8	2.0
WW B5	14	21	29	36	45	51	58	68	9	15	22	28	36	41	48	58	1.2	1.3	1.5	1.6	1.8	1.3	1.4	1.6	1.8	1.9
WW C1	16	22	29	35	42	48	54	63	11	16	22	26	33	37	43	50	1.2	1.4	1.6	1.7	1.9	1.4	1.6	1.8	2.0	2.1
WW C2	14	21	29	34	42	48	55	64	9	15	21	26	33	38	44	52	1.2	1.4	1.5	1.7	1.8	1.4	1.5	1.7	1.9	2.0
WW C3	18	27	37	44	54	60	68	78	12	20	29	36	45	51	59	69	1.0	1.2	1.4	1.5	1.7	1.1	1.3	1.5	1.6	1.8
WW C4	16	24	34	41	51	57	65	75	11	18	27	33	42	48	56	66	1.1	1.2	1.4	1.5	1.7	1.2	1.3	1.5	1.6	1.8
TW A1	14	16	18	19	21	23	25	28	12	13	14	15	16	16	17	18	2.2	2.6	2.9	3.3	3.6	2.9	3.5	4.2	-	-
TW A2	12	15	18	21	24	27	30	35	9	11	13	15	17	19	20	23	1.8	2.1	2.3	2.6	2.8	2.2	2.6	3.0	3.4	3.9
TW A3	11	15	18	21	25	28	32	38	8	10	13	15	17	19	22	25	1.7	1.9	2.2	2.4	2.5	2.1	2.4	2.7	3.0	3.4
TW A4	10	14	18	21	26	29	33	39	6	9	12	14	17	20	23	27	1.6	1.9	2.1	2.3	2.4	1.9	2.2	2.5	2.8	3.1
TW B1	13	15	18	19	22	24	26	30	11	12	13	14	16	17	18	20	2.1	2.4	2.7	3.0	3.2	2.6	3.1	3.5	4.1	4.7
TW B2	11	14	18	20	24	27	30	35	8	10	13	14	17	19	21	24	1.8	2.0	2.3	2.5	2.7	2.2	2.5	2.8	3.1	3.5
TW B3	10	14	17	20	24	27	31	37	7	9	12	14	17	19	22	25	1.7	2.0	2.2	2.4	2.5	2.1	2.3	2.6	2.9	3.2
TW B4	10	13	17	19	23	26	30	36	7	9	12	14	17	19	22	26	1.7	2.0	2.2	2.4	2.6	2.1	2.3	2.6	2.9	3.1
TW C1	12	14	17	19	22	24	27	32	9	11	13	14	16	18	20	22	2.0	2.2	2.5	2.7	2.9	2.4	2.7	3.1	3.4	3.8
TW C2	11	13	16	19	23	25	29	34	8	10	12	14	16	18	21	24	1.8	2.1	2.3	2.5	2.7	2.2	2.5	2.8	3.1	3.3
TW C3	10	13	16	18	22	25	28	33	8	10	12	14	16	18	21	24	1.9	2.1	2.3	2.5	2.7	2.3	2.5	2.8	3.0	3.3
TW C4	11	13	16	18	21	23	26	31	9	10	12	14	16	18	20	24	2.0	2.2	2.5	2.7	2.8	2.4	2.7	2.9	3.2	3.4
TW D1	7	9	11	12	14	15	16	18	5	5	6	6	7	7	7	6	2.3	2.9	3.5	4.2	4.9	3.5	4.9	-	-	-
TW D2	12	16	21	25	30	34	39	46	8	12	15	18	22	25	29	35	1.5	1.7	1.9	2.1	2.2	1.8	2.0	2.2	2.4	2.6
TW D3	13	17	23	27	33	37	43	50	8	12	16	19	24	27	32	38	1.4	1.6	1.8	2.0	2.1	1.6	1.8	2.1	2.3	2.5
TW D4	14	20	27	32	39	45	51	60	9	14	20	25	31	35	41	49	1.3	1.4	1.6	1.8	1.9	1.4	1.6	1.8	1.9	2.1
TW D5	14	20	27	33	41	46	53	63	9	15	21	26	33	38	44	53	1.2	1.4	1.6	1.7	1.8	1.4	1.5	1.7	1.9	2.0
TW E1	15	21	27	31	38	43	49	57	11	15	20	24	29	33	38	45	1.3	1.5	1.7	1.8	2.0	1.5	1.7	1.9	2.1	2.3
TW E2	15	20	27	32	39	44	51	60	10	15	20	25	31	35	40	48	1.3	1.5	1.6	1.8	1.9	1.4	1.6	1.8	2.0	2.1
TW E3	18	26	34	41	50	56	63	73	13	19	27	33	41	46	53	63	1.1	1.3	1.4	1.6	1.7	1.2	1.4	1.5	1.7	1.9
TW E4	16	24	33	39	48	54	62	71	11	18	26	31	40	45	52	62	1.1	1.3	1.4	1.6	1.7	1.2	1.4	1.6	1.7	1.9
TW F1	12	14	16	18	20	22	24	27	10	11	12	13	14	15	16	17	2.1	2.5	2.9	3.2	3.5	2.9	3.4	4.0	4.9	-
TW F2	10	13	16	18	22	24	27	32	7	9	11	12	14	16	18	21	1.8	2.1	2.4	2.6	2.8	2.3	2.6	3.0	3.5	4.0
TW F3	8	12	15	18	22	25	29	34	5	7	10	12	14	16	19	22	1.7	2.0	2.2	2.4	2.6	2.1	2.4	2.8	3.1	3.5
TW F4	7	10	14	17	21	24	28	34	3	6	8	10	13	16	18	22	1.7	2.0	2.2	2.4	2.5	2.1	2.3	2.6	2.9	3.2
TW G1	10	12	15	16	19	21	24	27	7	9	10	11	13	14	15	17	2.1	2.4	2.7	3.0	3.2	2.7	3.1	3.5	4.1	4.7
TW G2	8	10	14	16	20	22	26	30	5	6	9	10	13	14	17	20	1.9	2.1	2.4	2.6	2.8	2.3	2.6	2.9	3.3	3.7
TW G3	6	9	12	15	19	21	25	30	3	5	7	9	12	14	16	20	1.8	2.1	2.3	2.5	2.7	2.2	2.5	2.8	3.1	3.4
TW G4	4	7	10	13	16	19	23	28	2	4	6	8	10	12	15	19	1.9	2.1	2.4	2.6	2.7	2.3	2.6	2.8	3.1	3.3
TW H1	7	9	12	14	17	19	22	26	5	6	8	9	11	12	14	17	2.0	2.3	2.6	2.9	3.1	2.6	2.9	3.3	3.7	4.1
TW H2	5	7	10	12	16	18	21	26	2	4	6	8	10	12	14	17	2.0	2.2	2.5	2.7	2.9	2.4	2.7	3.0	3.3	3.6
TW H3	4	6	9	11	14	16	19	24	1	3	5	6	9	10	13	16	2.1	2.3	2.5	2.8	2.9	2.5	2.8	3.0	3.3	3.5
TW H4	2	4	7	9	11	13	16	20	1	2	4	5	7	9	11	14	2.3	2.5	2.7	3.0	3.1	2.8	3.0	3.3	3.5	3.7
TW I1	6	7	9	10	12	13	14	15	3	4	4	5	5	5	5	4	2.4	3.0	3.7	4.5	-	3.8	-	-	-	-
TW I2	9	13	18	22	27	31	36	43	6	9	12	15	19	22	26	31	1.5	1.7	1.9	2.1	2.3	1.8	2.0	2.2	2.5	2.7
TW I3	10	15	20	24	30	34	40	48	6	10	14	17	22	25	30	36	1.4	1.6	1.8	2.0	2.1	1.7	1.9	2.1	2.3	2.5
TW I4	11	17	24	29	36	42	48	57	7	12	17	22	28	33	38	47	1.3	1.5	1.6	1.8	1.9	1.5	1.6	1.8	2.0	2.1
TW I5	11	18	25	30	38	44	51	60	7	12	19	23	30	35	42	50	1.3	1.4	1.6	1.7	1.9	1.4	1.6	1.7	1.9	2.0
TW J1	13	18	25	29	36	41	47	56	9	13	18	22	27	31	36	44	1.3	1.5	1.7	1.8	2.0	1.5	1.7	1.9	2.1	2.3
TW J2	12	18	25	30	37	42	48	57	8	13	18	22	28	33	38	46	1.3	1.5	1.6	1.8	1.9	1.5	1.6	1.8	2.0	2.2
TW J3	15	23	32	38	47	54	61	71	10	17	25	31	39	45	52	61	1.1	1.3	1.4	1.6	1.7	1.2	1.4	1.6	1.7	1.9
TW J4	14	21	30	36	45	52	59	69	9	16	23	29	37	43	50	60	1.2	1.3	1.5	1.6	1.8	1.3	1.4	1.6	1.7	1.9
DG A1	7	9	11	12	13	14	15	15	4	5	5	6	6	6	5	3	2.2	2.8	4.0	-	-	3.2	4.8	-	-	-
DG A2	10	13	17	19	23	26	30	34	7	9	12	14	17	19	21	24	1.7	2.0	2.2	2.6	2.9	2.1	2.4	2.7	3.1	3.6
DG A3	13	18	23	27	34	38	44	52	10	14	18	22	27	31	36	43	1.5	1.6	1.7	1.9	2.1	1.6	1.8	1.9	2.1	2.3
DG B1	7	9	11	12	14	15	16	18	4	5	5	6	6	6	6	5	2.2	2.9	3.3	4.6	-	3.2	4.4	-	-	-
DG B2	10	13	17	20	24	27	32	37	7	9	12	14	17	19	22	26	1.7	2.0	2.1	2.4	2.6	2.1	2.3	2.6	2.9	3.2
DG B3	14	19	24	28	35	40	47	56	11	14	19	23	28	32	38	46	1.4	1.6	1.7	1.8	1.9	1.6	1.7	1.9	2.0	2.2
DG C1	9	12	14	16	19	21	24	27	5	7	9	10	11	12	12	12	1.9	2.4	2.6	3.1	3.7	2.5	3.0	4.2	-	-
DG C2	12	16	20	24	29	33	38	45	9	12	15	18	22	24	28	33	1.6	1.8	1.9	2.1	2.3	1.8	2.0	2.3	2.5	2.7
DG C3	16	21	27	32	40	46	53	63	12	16	22	26	32	37	43	52	1.3	1.5	1.6	1.7	1.8	1.5	1.6	1.8	1.9	2.0
SS A1	16	19	23	26	30	33	37	42	13	16	19	21	24	26	29	32	1.6	1.9	2.1	2.3	2.5	1.8	2.2	2.5	2.9	3.3
SS A2	20	25	31	36	43	47	53	61	16	21	26	30	36	40	45	52	1.3	1.5	1.6	1.8	1.9	1.4	1.6	1.8	2.0	2.2
SS A3	15	17	20	23	26	28	31	35	12	14	16	18	20	21	23	25	1.8	2.1	2.4	2.7	3.0	2.1	2.5	3.1	3.8	4.7
SS A4	19	24	30	35	41	45	51	59	16	20	25	29	34	38	42	49	1.3	1.5	1.7	1.9	2.0	1.4	1.7	1.9	2.1	2.3
SS A5	16	19	22	24	27	30	33	37	14	16	18	20	22	23	25	28	1.8	2.1	2.4	2.7	2.9	2.1	2.5	2.9	3.5	4.1
SS A6	19	25	31	35	42	46	52	60	16	20	26	29	35	39	43	50	1.3	1.5	1.7	1.8	2.0	1.4	1.6	1.8	2.1	2.2
SS A7	15	17	19	21	23	25	26	29	13	14	16	17	18	19	20	20	2.1	2.5	2.9	3.4	3.8	2.5	3.1	3.9	-	-
SS A8	19	24	29	33	39	43	49	56	15	19	24	27	32	36	40	46	1.3	1.5	1.7	1.9	2.1	1.5	1.7	2.0	2.2	2.4
SS B1	13	16	20	22	26	28	32	36	11	13	15	17	20	22	24	27	1.7	2.0	2.3	2.5	2.7	2.0	2.4	2.8	3.2	3.6
SS B2	17	22	28	33	39	43	49	57	14	18	23	27	32	36	41	48	1.3	1.5	1.7	1.9	2.0	1.5	1.7	1.9	2.1	2.2
SS B3	12	15	18	20	23	25	28	31	10	11	13	15	17	18	20	22	1.9	2.2	2.5	2.8	3.1	2.2	2.7	3.2	3.9	4.7
SS B4	17	21	27	31	37	41	47	55	13	17	22	26	31	34	39	46	1.4	1.6	1.7	1.9	2.1	1.5	1.7	1.9	2.1	2.3
SS B5	13	16	18	20	23	25	27	30	11	13	15	16	18	19	20	22	2.0	2.3	2.7	3.0	3.2	2.4	2.8	3.3	3.9	4.6
SS B6	16	21	27	31	37	42	47	55	13	17	22	26	31	34	39	46	1.4	1.6	1.7	1.9	2.0	1.5	1.7	1.9	2.1	2.3
SS B7	12	14	16	17	20	21	23	25	10	11	13	14	15	16	17	17	2.2	2.6	3.1	3.5	3.9	2.7	3.3	4.1	-	-
SS B8	16	20	25	29	35	39	44	52	12	16	20	24	29	32	36	42	1.4	1.6	1.8	2.0	2.1	1.6	1.8	2.0	2.2	2.4
SS C1	10	13	16	19	23	25	29	34	6	8	10	12	14	16	17	20	1.7	2.0	2.3	2.5	2.7	2.2	2.6	3.0	3.6	4.3
SS C2	12	17	23	27	33	38	43	51	8	12	16	20	25	28	32	39	1.4	1.6	1.8	1.9	2.1	1.6	1.8	2.0	2.3	2.4
SS C3	10	13	15	17	20	22	25	29	8	9	11	12	14	15	16	18	2.0	2.3	2.6	2.9	3.1	2.6	3.0	3.4	4.0	4.6
SS C4	12	16	21	25	30	34	39	46	8	12	15	18	22	25	29	35	1.5	1.7	1.9	2.1	2.2	1.8	2.0	2.2	2.4	2.6
SS D1	17	21	26	30	35	38	43	49	13	17	20	23	27	29	32	36	1.4	1.7	1.9	2.1	2.3	1.6	2.0	2.3	2.7	3.0
SS D2	21	28	36	42	50	55	62	70	17	23	30	35	42	46	52	60	1.1	1.3	1.5	1.6	1.8	1.2	1.4	1.6	1.8	2.0
SS D3	17	21	24	27	32	35	38	44	14	17	20	22	25	27	29	33	1.6	1.8	2.1	2.3	2.5	1.8	2.1	2.5	2.9	3.3
SS D4	21	27	35	40	48	53	60	68	17	22	29	33	40	45	50	58	1.2	1.3	1.5	1.7	1.8	1.3	1.5	1.7	1.9	2.0
SS E1	14	17	21	24	28	31	35	40	10	13	15	17	20	22	24	26	1.6	1.9	2.1	2.4	2.6	1.9	2.3	2.8	3.4	4.1
SS E2	17	23	30	35	42	47	53	62	13	18	24	28	34	38	43	50	1.2	1.4	1.6	1.8	1.9	1.4	1.6	1.8	2.0	2.2
SS E3	14	17	20	22	25	27	30	34	11	13	15	17	19	20	22	24	1.8	2.1	2.4	2.7	2.9	2.1	2.6	3.0	3.7	4.3
SS E4	17	22	28	32	39	43	49	57	13	17	22	26	31	35	40	46	1.3	1.5	1.7	1.9	2.0	1.5	1.7	1.9	2.1	2.3

INDIANAPOLIS, INDIANA

TYPE	SOLAR SAVINGS FRACTION (%) BASE TEMP: 55 F (3440 DD) LCR 100	70	50	40	30	25	20	15	65 F (5585 DD) 100	70	50	40	30	25	20	15	CONSERVATION FACTOR 55 F 100	50	30	20	15	65 F 100	50	30	20	15
WW A1	15	18	20	22	25	27	29	33	13	14	15	16	18	19	20	21	2.0	2.3	2.6	2.8	3.0	2.6	3.1	3.6	4.3	-
WW A2	14	18	23	26	31	34	39	45	10	13	16	18	21	23	26	30	1.6	1.8	2.0	2.1	2.3	1.9	2.2	2.5	2.8	3.1
WW A3	14	18	23	27	33	37	43	50	9	12	16	19	22	25	29	34	1.4	1.6	1.8	2.0	2.1	1.7	2.0	2.2	2.5	2.7
WW A4	13	18	24	28	35	39	45	53	8	12	16	19	23	26	30	36	1.4	1.6	1.7	1.9	2.1	1.6	1.9	2.1	2.4	2.6
WW A5	13	18	25	29	36	41	47	55	7	11	16	19	24	28	32	38	1.3	1.5	1.7	1.9	2.0	1.6	1.8	2.0	2.3	2.5
WW A6	13	18	25	30	37	42	48	57	7	11	16	19	25	28	33	39	1.3	1.5	1.6	1.8	2.0	1.5	1.8	2.0	2.2	2.4
WW B1	11	13	15	17	20	22	24	27	7	8	9	10	11	11	11	11	2.0	2.4	2.7	3.1	3.4	2.8	3.8	-	-	-
WW B2	14	20	26	31	38	43	50	59	9	13	18	22	28	32	37	44	1.3	1.5	1.6	1.8	1.9	1.5	1.7	1.9	2.1	2.3
WW B3	17	24	32	38	46	51	58	67	11	17	23	28	34	39	45	53	1.2	1.3	1.5	1.7	1.8	1.3	1.5	1.7	1.9	2.1
WW B4	17	25	34	41	50	56	64	73	11	18	25	31	39	45	52	61	1.1	1.3	1.4	1.6	1.7	1.2	1.4	1.6	1.7	1.9
WW B5	16	25	34	41	50	56	64	74	11	18	26	32	40	46	53	63	1.1	1.3	1.4	1.6	1.7	1.2	1.4	1.5	1.7	1.8
WW C1	19	26	34	40	49	54	62	71	13	19	25	30	37	42	49	57	1.1	1.3	1.5	1.6	1.8	1.3	1.5	1.7	1.8	2.0
WW C2	17	25	33	39	48	54	62	71	11	18	25	30	37	43	49	58	1.1	1.3	1.4	1.6	1.8	1.3	1.4	1.6	1.8	1.9
WW C3	21	31	42	49	60	66	74	83	15	23	33	40	50	56	64	73	1.0	1.1	1.3	1.5	1.6	1.1	1.2	1.4	1.6	1.7
WW C4	19	28	38	46	56	63	71	80	13	21	30	37	46	53	61	71	1.0	1.2	1.3	1.5	1.6	1.1	1.3	1.4	1.6	1.7
TW A1	15	17	20	22	25	27	30	34	12	14	15	17	18	19	21	23	2.0	2.3	2.5	2.7	2.9	2.5	2.9	3.4	3.9	4.4
TW A2	14	17	21	24	29	32	36	42	10	13	15	17	20	22	25	29	1.7	1.9	2.1	2.2	2.4	2.0	2.3	2.6	2.9	3.2
TW A3	13	17	22	25	30	34	39	46	9	12	15	18	21	24	27	31	1.5	1.7	1.9	2.1	2.3	1.9	2.1	2.4	2.6	2.9
TW A4	12	16	21	25	31	34	40	47	8	11	14	17	21	24	28	33	1.5	1.7	1.9	2.0	2.2	1.8	2.0	2.2	2.5	2.7
TW B1	14	17	20	22	25	28	31	36	12	13	15	17	19	20	22	25	1.9	2.1	2.4	2.6	2.7	2.3	2.7	3.0	3.4	3.7
TW B2	13	16	20	24	28	31	36	42	9	12	15	17	20	22	25	30	1.6	1.8	2.0	2.2	2.4	2.0	2.2	2.5	2.7	3.0
TW B3	12	16	20	23	28	32	37	44	8	11	14	17	20	23	26	31	1.6	1.8	2.0	2.1	2.3	1.9	2.1	2.4	2.6	2.8
TW B4	12	15	19	23	27	31	36	42	8	11	14	16	20	22	26	31	1.6	1.8	2.0	2.2	2.3	1.9	2.1	2.4	2.6	2.8
TW C1	13	16	19	22	26	28	32	37	10	12	14	16	19	21	23	27	1.8	2.0	2.2	2.4	2.6	2.2	2.5	2.7	3.0	3.2
TW C2	12	15	19	22	26	29	34	40	9	11	14	16	19	21	25	29	1.7	1.9	2.1	2.3	2.4	2.0	2.3	2.5	2.7	2.9
TW C3	12	15	18	21	25	28	33	39	9	11	14	16	19	21	24	28	1.7	2.0	2.1	2.3	2.4	2.1	2.3	2.5	2.8	2.9
TW C4	12	15	18	20	24	26	30	36	10	11	14	15	18	20	23	27	1.9	2.1	2.3	2.4	2.6	2.2	2.4	2.7	2.9	3.1
TW D1	9	11	14	15	18	20	22	26	6	7	8	9	10	11	11	12	2.1	2.4	2.8	3.1	3.3	2.8	3.5	4.5	-	-
TW D2	14	19	25	29	35	40	46	54	10	14	18	21	26	30	34	40	1.4	1.6	1.7	1.9	2.0	1.6	1.8	2.0	2.2	2.4
TW D3	15	20	27	32	38	43	50	58	10	14	19	23	28	32	37	44	1.3	1.5	1.6	1.8	2.0	1.5	1.7	1.9	2.1	2.3
TW D4	16	23	31	37	45	50	58	67	11	17	23	28	35	40	46	55	1.2	1.4	1.5	1.7	1.8	1.3	1.5	1.7	1.8	2.0
TW D5	16	23	31	37	46	52	59	69	11	17	24	29	37	42	49	58	1.2	1.3	1.5	1.6	1.8	1.3	1.5	1.6	1.8	1.9
TW E1	18	24	31	36	44	49	56	65	13	17	23	27	34	38	44	51	1.2	1.4	1.5	1.7	1.9	1.4	1.6	1.8	2.0	2.1
TW E2	17	24	31	37	45	50	57	66	12	17	23	28	35	39	46	54	1.2	1.4	1.5	1.7	1.8	1.4	1.5	1.7	1.9	2.0
TW E3	21	29	39	46	55	62	69	78	15	22	31	37	45	51	58	68	1.1	1.2	1.4	1.5	1.7	1.2	1.3	1.5	1.6	1.8
TW E4	19	27	37	44	53	60	67	77	14	21	29	35	44	50	57	67	1.1	1.2	1.4	1.5	1.7	1.2	1.3	1.5	1.7	1.8
TW F1	13	16	18	20	24	26	29	33	10	12	14	15	17	18	20	22	1.9	2.2	2.5	2.7	2.9	2.5	2.9	3.3	3.8	4.3
TW F2	12	15	19	22	26	29	33	40	8	10	13	15	18	20	23	26	1.7	1.9	2.1	2.3	2.4	2.1	2.3	2.6	3.0	3.2
TW F3	10	14	18	22	27	30	35	42	6	9	12	14	18	20	23	28	1.6	1.8	2.0	2.2	2.3	1.9	2.2	2.4	2.7	3.0
TW F4	8	12	17	20	26	29	34	41	5	7	11	13	17	20	23	28	1.6	1.8	2.0	2.1	2.3	1.9	2.1	2.4	2.6	2.8
TW G1	11	14	17	19	23	25	29	33	8	10	12	13	16	17	19	22	1.9	2.1	2.4	2.6	2.7	2.4	2.7	3.0	3.4	3.7
TW G2	9	12	16	19	24	27	31	37	6	8	11	13	16	18	21	25	1.7	1.9	2.1	2.3	2.4	2.1	2.3	2.6	2.9	3.1
TW G3	8	11	15	18	23	26	30	37	4	7	9	12	15	17	20	25	1.7	1.9	2.1	2.3	2.4	2.0	2.3	2.5	2.8	3.0
TW G4	6	9	13	15	20	23	27	34	3	5	8	10	13	15	18	23	1.8	2.0	2.2	2.3	2.5	2.1	2.3	2.6	2.8	3.0
TW H1	8	11	14	17	20	23	27	32	6	7	9	11	14	15	18	21	1.9	2.1	2.3	2.5	2.7	2.3	2.6	2.9	3.2	3.4
TW H2	6	9	12	15	19	22	26	32	3	5	8	10	12	14	17	21	1.8	2.1	2.3	2.4	2.6	2.2	2.5	2.7	3.0	3.2
TW H3	5	7	11	13	17	20	23	29	2	4	6	8	11	13	16	20	1.9	2.1	2.3	2.5	2.7	2.3	2.5	2.8	3.0	3.2
TW H4	3	6	8	10	14	16	19	24	2	3	5	6	9	11	13	17	2.1	2.3	2.5	2.7	2.9	2.6	2.8	3.0	3.2	3.4
TW I1	7	9	11	13	15	17	19	23	4	5	6	7	8	9	9	10	2.2	2.5	2.9	3.2	3.5	3.0	3.7	4.8	-	-
TW I2	11	16	22	26	32	36	42	50	7	11	15	18	23	26	31	37	1.4	1.6	1.8	1.9	2.1	1.7	1.9	2.1	2.3	2.4
TW I3	12	18	24	28	35	40	46	55	8	12	17	20	26	30	35	42	1.4	1.5	1.7	1.8	2.0	1.6	1.7	1.9	2.1	2.3
TW I4	14	20	28	33	42	47	54	64	9	14	20	25	32	37	43	52	1.2	1.4	1.5	1.7	1.8	1.4	1.5	1.7	1.9	2.0
TW I5	14	21	29	35	43	49	57	66	9	15	21	27	34	39	46	55	1.2	1.3	1.5	1.7	1.8	1.3	1.5	1.6	1.8	1.9
TW J1	15	22	29	34	42	47	54	63	10	15	21	25	32	36	42	50	1.2	1.4	1.6	1.7	1.9	1.4	1.6	1.8	2.0	2.1
TW J2	15	21	28	34	42	48	55	64	10	15	21	26	32	37	43	52	1.2	1.4	1.5	1.7	1.8	1.4	1.5	1.7	1.9	2.0
TW J3	18	27	36	43	53	59	67	77	12	20	28	34	43	49	57	66	1.1	1.2	1.4	1.5	1.7	1.2	1.3	1.5	1.7	1.8
TW J4	17	25	34	41	51	57	65	75	11	18	26	32	41	47	55	65	1.1	1.2	1.4	1.6	1.7	1.2	1.4	1.5	1.7	1.8
DG A1	9	11	14	16	18	19	21	22	5	6	8	8	9	10	10	9	1.9	2.4	3.0	4.4	-	2.6	3.4	-	-	-
DG A2	11	15	19	23	28	31	35	41	8	11	14	16	20	22	25	29	1.6	1.8	2.0	2.3	2.6	1.9	2.1	2.4	2.8	3.2
DG A3	15	20	26	31	38	43	50	59	11	15	20	24	30	34	40	47	1.4	1.5	1.6	1.8	2.0	1.5	1.7	1.8	2.0	2.2
DG B1	9	11	14	16	19	21	23	26	5	6	8	9	10	10	11	11	2.0	2.3	2.6	3.2	3.9	2.5	3.3	4.8	-	-
DG B2	12	16	20	23	29	33	38	45	8	11	14	17	20	23	26	31	1.6	1.8	1.9	2.1	2.3	1.8	2.1	2.3	2.6	2.9
DG B3	16	21	27	32	40	46	53	63	12	16	21	25	31	36	42	51	1.4	1.5	1.6	1.7	1.8	1.5	1.6	1.8	1.9	2.0
DG C1	11	14	18	20	25	28	31	36	7	9	11	13	15	16	17	19	1.7	2.0	2.1	2.5	2.9	2.1	2.6	3.3	4.1	-
DG C2	14	18	24	28	35	39	45	53	10	13	17	21	25	28	33	39	1.5	1.6	1.7	1.9	2.1	1.7	1.9	2.1	2.2	2.5
DG C3	18	23	30	36	45	51	59	70	13	18	24	29	36	41	48	57	1.3	1.4	1.5	1.6	1.7	1.4	1.5	1.7	1.8	1.9
SS A1	18	22	26	30	35	38	43	49	15	18	21	24	27	30	33	38	1.5	1.7	1.9	2.1	2.2	1.7	2.0	2.3	2.6	2.8
SS A2	22	28	35	40	48	53	60	68	18	23	29	34	40	44	50	57	1.2	1.4	1.5	1.7	1.8	1.3	1.5	1.7	1.9	2.0
SS A3	16	20	23	26	30	33	37	43	13	16	18	20	23	25	27	30	1.6	1.9	2.1	2.3	2.5	1.9	2.2	2.6	3.1	3.5
SS A4	22	27	34	39	46	51	57	66	18	23	28	32	38	42	47	54	1.2	1.4	1.6	1.7	1.9	1.4	1.6	1.8	2.0	2.1
SS A5	18	21	25	28	32	34	38	44	15	18	20	22	25	27	29	33	1.6	1.9	2.1	2.3	2.5	1.9	2.2	2.8	3.0	3.3
SS A6	22	28	35	40	47	52	58	67	18	23	28	33	39	43	48	56	1.2	1.4	1.5	1.7	1.9	1.3	1.5	1.7	1.9	2.1
SS A7	16	19	22	24	27	29	31	35	14	16	18	19	21	22	24	25	1.9	2.2	2.5	2.7	3.0	2.2	2.7	3.2	3.9	4.7
SS A8	21	27	33	38	45	49	55	63	17	22	27	31	36	40	45	52	1.3	1.4	1.6	1.8	1.9	1.4	1.6	1.8	2.0	2.2
SS B1	15	18	22	25	30	33	37	43	12	15	17	20	23	25	28	32	1.6	1.8	2.0	2.2	2.4	1.9	2.2	2.5	2.8	3.0
SS B2	19	25	31	36	44	48	55	63	15	20	26	30	36	40	46	53	1.3	1.4	1.6	1.7	1.9	1.4	1.6	1.8	1.9	2.1
SS B3	14	17	20	23	27	30	33	39	11	13	16	17	20	22	24	27	1.7	2.0	2.2	2.4	2.6	2.0	2.4	2.7	3.2	3.6
SS B4	19	24	30	35	42	47	53	61	15	19	25	29	34	38	43	51	1.3	1.5	1.6	1.8	1.9	1.4	1.6	1.8	2.0	2.2
SS B5	15	17	20	23	26	29	32	36	12	14	16	18	20	22	24	27	1.8	2.1	2.3	2.6	2.7	2.2	2.5	2.9	3.3	3.7
SS B6	18	24	30	35	42	47	53	62	15	19	25	29	34	38	44	51	1.3	1.5	1.6	1.8	1.9	1.4	1.6	1.8	2.0	2.1
SS B7	13	16	18	20	23	25	27	31	11	13	14	16	17	18	20	22	2.0	2.3	2.6	2.9	3.1	2.4	2.9	3.4	4.0	4.7
SS B8	18	23	29	33	40	44	50	59	14	18	23	27	32	36	41	48	1.3	1.5	1.7	1.8	2.0	1.5	1.7	1.9	2.1	2.2
SS C1	12	16	20	23	28	32	36	43	8	10	13	15	18	20	23	27	1.6	1.8	2.0	2.2	2.3	1.9	2.2	2.6	2.9	3.2
SS C2	15	20	27	32	39	44	50	59	10	14	19	23	29	33	38	45	1.3	1.5	1.6	1.8	1.9	1.5	1.7	1.9	2.1	2.2
SS C3	12	15	18	20	24	27	30	35	9	11	13	14	17	18	20	23	1.8	2.1	2.3	2.5	2.7	2.3	2.6	2.9	3.3	3.6
SS C4	14	19	24	29	35	39	45	53	10	13	18	21	26	29	34	40	1.4	1.6	1.7	1.9	2.0	1.6	1.8	2.0	2.2	2.4
SS D1	20	25	30	35	41	45	50	58	15	19	24	27	31	34	38	43	1.3	1.5	1.7	1.9	2.1	1.5	1.8	2.0	2.3	2.5
SS D2	24	32	41	47	56	61	68	77	19	26	33	39	46	51	58	66	1.1	1.2	1.4	1.6	1.7	1.2	1.4	1.5	1.7	1.9
SS D3	19	23	28	31	37	40	45	52	16	19	22	25	29	31	34	39	1.5	1.7	1.9	2.0	2.2	1.7	1.9	2.2	2.5	2.8
SS D4	24	31	39	45	54	59	66	75	19	25	32	37	45	49	56	64	1.1	1.3	1.4	1.6	1.7	1.2	1.4	1.6	1.8	1.9
SS E1	16	20	25	29	34	38	43	49	12	15	18	21	24	27	30	34	1.5	1.7	1.9	2.1	2.2	1.7	2.0	2.4	2.7	3.0
SS E2	20	27	34	40	48	54	60	69	15	21	27	32	38	43	49	56	1.2	1.3	1.5	1.7	1.8	1.3	1.5	1.7	1.9	2.0
SS E3	16	19	22	25	29	32	36	42	12	15	17	19	22	24	26	30	1.7	1.9	2.1	2.3	2.5	1.9	2.3	2.6	3.0	3.4
SS E4	19	25	32	37	44	49	56	64	15	20	25	30	36	40	45	52	1.2	1.4	1.6	1.7	1.9	1.4	1.6	1.8	2.0	2.1

SOUTH BEND, INDIANA

TYPE	SOLAR SAVINGS FRACTION (%) BASE TEMP: 55 F (4124 DD) LCR								65 F (6465 DD)								CONSERVATION FACTOR 55 F					65 F				
	100	70	50	40	30	25	20	15	100	70	50	40	30	25	20	15	100	50	30	20	15	100	50	30	20	15
WW A1	13	15	17	18	19	21	22	24	12	12	13	14	14	15	15	15	2.4	2.9	3.5	4.1	4.8	3.3	4.3	-	-	-
WW A2	12	14	17	20	23	25	28	33	9	10	12	14	16	17	19	21	1.8	2.1	2.4	2.7	3.0	2.2	2.7	3.2	3.9	4.9
WW A3	10	14	18	20	24	27	31	36	7	9	12	14	16	18	20	23	1.6	1.9	2.2	2.5	2.7	2.0	2.4	2.8	3.4	3.9
WW A4	10	13	18	21	25	28	33	38	6	9	11	14	17	19	21	25	1.6	1.9	2.1	2.3	2.5	1.9	2.2	2.6	3.1	3.6
WW A5	9	13	18	21	26	30	34	40	5	8	11	14	17	19	22	26	1.5	1.8	2.0	2.2	2.4	1.8	2.1	2.5	2.9	3.3
WW A6	9	13	18	21	27	30	35	42	4	8	11	14	17	20	23	27	1.5	1.7	2.0	2.2	2.4	1.8	2.1	2.4	2.8	3.2
WW B1	8	10	11	12	12	13	13	13	6	6	6	6	5	5	3	0	2.5	3.5	-	-	-	4.3	-	-	-	-
WW B2	11	15	20	24	30	34	39	46	7	10	14	17	22	25	29	34	1.5	1.7	1.9	2.1	2.2	1.7	1.9	2.2	2.4	2.7
WW B3	13	19	25	29	36	41	47	55	9	13	18	22	27	31	36	42	1.3	1.5	1.7	1.9	2.0	1.5	1.7	1.9	2.2	2.4
WW B4	13	20	27	33	41	47	54	63	9	14	21	26	33	38	44	52	1.2	1.4	1.6	1.7	1.9	1.3	1.5	1.7	1.9	2.0
WW B5	12	19	27	33	42	47	55	64	8	14	21	27	34	39	46	55	1.2	1.4	1.5	1.7	1.8	1.3	1.5	1.6	1.8	2.0
WW C1	14	20	27	32	39	44	50	59	10	15	21	25	31	35	40	47	1.3	1.5	1.6	1.8	2.0	1.4	1.6	1.8	2.0	2.2
WW C2	13	19	26	32	39	44	51	60	9	14	20	25	31	36	41	49	1.2	1.4	1.6	1.8	1.9	1.4	1.6	1.8	2.0	2.1
WW C3	16	25	34	41	51	57	65	74	12	19	28	34	43	49	57	66	1.1	1.2	1.4	1.6	1.7	1.2	1.3	1.5	1.7	1.8
WW C4	14	23	32	39	48	54	62	72	10	17	26	32	41	46	54	64	1.1	1.3	1.4	1.6	1.7	1.2	1.4	1.5	1.7	1.8
TW A1	13	15	17	18	20	21	23	25	12	12	13	14	15	16	16	17	2.3	2.7	3.2	3.8	4.2	3.1	3.8	4.8	-	-
TW A2	12	14	17	19	22	24	27	31	9	11	13	14	16	17	19	21	1.9	2.2	2.5	2.8	3.1	2.3	2.8	3.2	3.9	4.6
TW A3	10	13	17	19	23	25	29	34	8	10	12	14	16	18	20	23	1.8	2.1	2.3	2.6	2.8	2.1	2.5	2.9	3.4	3.9
TW A4	9	13	16	19	23	26	30	35	6	9	11	13	16	18	21	24	1.7	2.0	2.2	2.4	2.6	2.0	2.3	2.7	3.1	3.4
TW B1	12	14	17	18	20	22	24	27	11	12	13	14	15	16	17	19	2.2	2.5	2.9	3.3	3.6	2.7	3.3	3.9	4.7	-
TW B2	11	13	16	19	22	24	27	32	8	10	12	14	16	17	20	22	1.8	2.1	2.4	2.7	2.9	2.2	2.6	3.0	3.4	3.9
TW B3	10	13	16	18	22	25	28	33	7	9	12	13	16	18	20	23	1.8	2.1	2.3	2.5	2.8	2.1	2.5	2.8	3.2	3.5
TW B4	9	12	15	18	22	24	28	33	7	9	11	13	16	18	20	24	1.8	2.1	2.3	2.6	2.7	2.2	2.5	2.8	3.1	3.4
TW C1	11	13	16	18	21	23	25	29	9	11	12	14	16	17	19	21	2.0	2.3	2.7	2.9	3.2	2.5	2.9	3.3	3.7	4.2
TW C2	10	13	15	18	21	23	26	31	8	10	12	13	16	17	20	23	1.9	2.2	2.5	2.7	2.9	2.3	2.6	2.9	3.3	3.6
TW C3	10	12	15	17	21	23	26	30	8	10	12	13	16	17	20	23	2.0	2.2	2.5	2.7	2.9	2.3	2.6	2.9	3.2	3.5
TW C4	10	12	15	17	20	22	25	29	9	10	12	14	16	17	19	22	2.1	2.3	2.6	2.8	3.0	2.5	2.8	3.0	3.3	3.6
TW D1	7	8	10	11	12	13	13	14	5	5	6	6	6	6	5	4	2.5	3.2	4.2	-	-	3.8	-	-	-	-
TW D2	11	15	20	23	28	31	36	42	8	11	15	17	21	24	27	32	1.6	1.8	2.0	2.2	2.4	1.8	2.1	2.3	2.6	2.8
TW D3	12	16	21	25	30	34	39	46	8	12	15	18	23	26	29	35	1.5	1.7	1.9	2.1	2.2	1.7	1.9	2.2	2.5	2.7
TW D4	13	18	25	30	37	42	48	57	9	14	19	23	29	34	39	47	1.3	1.5	1.7	1.8	2.0	1.5	1.6	1.8	2.0	2.2
TW D5	13	19	26	31	38	44	50	59	9	14	20	25	32	36	42	51	1.3	1.4	1.6	1.8	1.9	1.4	1.6	1.7	1.9	2.1
TW E1	14	19	25	29	35	40	45	53	10	14	19	23	28	31	36	42	1.4	1.6	1.7	1.9	2.1	1.5	1.8	2.0	2.2	2.4
TW E2	14	19	25	30	37	41	47	56	10	14	19	23	29	33	38	45	1.3	1.5	1.7	1.9	2.0	1.5	1.7	1.9	2.1	2.2
TW E3	17	24	32	38	47	53	60	69	12	19	26	31	39	44	51	60	1.1	1.3	1.5	1.6	1.8	1.2	1.4	1.6	1.8	1.9
TW E4	15	22	31	37	45	51	58	68	11	17	25	30	38	43	50	59	1.2	1.3	1.5	1.6	1.8	1.3	1.4	1.6	1.8	1.9
TW F1	11	13	15	16	18	20	21	24	10	10	12	12	13	14	15	15	2.3	2.7	3.2	3.6	4.1	3.0	3.7	4.6	-	-
TW F2	9	12	15	17	20	22	25	28	7	8	10	12	14	15	16	18	1.9	2.3	2.6	2.9	3.1	2.4	2.8	3.3	4.0	4.7
TW F3	8	11	14	16	20	22	25	30	5	7	9	11	13	15	17	20	1.8	2.1	2.4	2.7	2.9	2.2	2.6	3.0	3.5	4.0
TW F4	6	9	12	15	19	21	25	30	3	5	8	10	12	14	17	20	1.8	2.1	2.3	2.6	2.8	2.1	2.5	2.8	3.2	3.5
TW G1	9	11	13	15	17	19	21	24	7	9	10	11	12	13	14	16	2.2	2.5	2.9	3.3	3.6	2.8	3.3	3.9	4.7	-
TW G2	7	9	12	14	18	20	23	27	5	6	8	10	12	13	15	18	1.9	2.2	2.5	2.8	3.0	2.4	2.7	3.1	3.6	4.1
TW G3	5	8	11	13	17	19	22	27	3	5	7	9	11	13	15	18	1.9	2.2	2.5	2.7	2.9	2.3	2.6	3.0	3.3	3.7
TW G4	4	6	9	11	15	17	20	25	2	4	6	7	10	11	14	17	2.0	2.2	2.5	2.7	2.9	2.4	2.7	3.0	3.3	3.6
TW H1	6	9	11	13	15	17	20	23	5	6	8	9	10	12	13	15	2.1	2.4	2.8	3.1	3.4	2.6	3.1	3.5	4.0	4.6
TW H2	4	6	9	11	14	16	19	23	2	4	6	7	9	11	13	16	2.1	2.3	2.6	2.9	3.1	2.5	2.8	3.2	3.5	3.9
TW H3	3	5	8	10	13	15	18	22	2	3	5	6	8	10	12	15	2.2	2.4	2.7	2.9	3.1	2.6	2.9	3.2	3.5	3.8
TW H4	2	4	6	8	10	12	15	18	1	2	4	5	7	8	10	13	2.4	2.6	2.9	3.1	3.3	2.8	3.1	3.4	3.7	3.9
TW I1	5	6	8	9	10	10	11	12	3	4	4	4	4	4	3	2	2.6	3.4	4.5	-	-	4.0	-	-	-	-
TW I2	9	12	17	20	25	28	33	39	6	8	12	14	18	21	24	29	1.6	1.8	2.0	2.2	2.4	1.9	2.1	2.4	2.6	2.9
TW I3	9	14	18	22	28	31	36	44	6	9	13	16	21	24	28	33	1.5	1.7	1.9	2.1	2.3	1.7	1.9	2.2	2.4	2.6
TW I4	10	16	22	27	34	39	45	54	7	11	17	21	27	31	36	44	1.3	1.5	1.7	1.9	2.0	1.5	1.7	1.9	2.0	2.2
TW I5	10	16	23	28	36	41	48	57	7	12	18	22	29	34	40	48	1.3	1.5	1.6	1.8	1.9	1.4	1.6	1.8	1.9	2.1
TW J1	12	17	23	27	33	38	43	51	8	12	17	21	26	29	34	41	1.4	1.6	1.8	1.9	2.1	1.6	1.8	2.0	2.2	2.4
TW J2	11	17	23	27	34	39	45	53	8	12	17	21	27	31	36	43	1.3	1.5	1.7	1.9	2.0	1.5	1.7	1.9	2.1	2.2
TW J3	14	21	30	36	45	51	58	68	10	16	24	29	37	43	50	59	1.2	1.3	1.5	1.6	1.8	1.3	1.4	1.6	1.8	1.9
TW J4	13	20	28	34	43	49	56	66	9	15	22	28	36	41	48	57	1.2	1.3	1.5	1.7	1.8	1.3	1.4	1.6	1.8	1.9
DG A1	6	8	9	10	11	12	12	11	4	4	5	5	5	5	4	1	2.4	3.2	-	-	-	3.4	-	-	-	-
DG A2	9	12	15	18	22	24	27	31	7	9	11	13	16	17	20	22	1.8	2.1	2.4	2.7	3.1	2.1	2.5	2.8	3.4	4.0
DG A3	13	17	22	26	32	36	41	49	10	14	18	21	26	29	34	41	1.5	1.7	1.8	2.0	2.1	1.7	1.8	2.0	2.2	2.4
DG B1	6	8	10	11	12	13	13	14	4	4	5	5	5	5	5	3	2.3	3.2	4.1	-	-	3.5	-	-	-	-
DG B2	10	13	16	18	22	25	29	33	7	9	12	14	16	18	21	24	1.8	2.1	2.2	2.5	2.8	2.1	2.4	2.7	3.2	3.5
DG B3	13	18	23	27	33	38	44	52	10	14	19	22	27	31	36	44	1.5	1.6	1.7	1.9	2.0	1.6	1.8	1.9	2.1	2.2
DG C1	8	11	13	14	17	18	20	22	5	7	8	9	10	10	10	10	2.0	2.6	3.0	3.7	5.0	2.7	3.3	-	-	-
DG C2	12	15	19	22	27	31	35	41	9	11	15	17	21	23	26	31	1.6	1.9	2.0	2.2	2.4	1.9	2.1	2.4	2.7	2.9
DG C3	15	20	26	30	38	43	50	59	12	16	21	25	31	35	41	49	1.4	1.5	1.6	1.8	1.9	1.5	1.7	1.8	1.9	2.1
SS A1	15	18	22	24	28	30	34	38	13	15	18	20	23	25	27	29	1.7	2.0	2.3	2.5	2.8	1.9	2.3	2.7	3.2	3.7
SS A2	19	24	30	34	40	45	50	58	16	20	25	29	35	38	43	49	1.3	1.5	1.7	1.9	2.0	1.4	1.6	1.9	2.1	2.3
SS A3	14	16	19	21	24	25	28	31	12	14	16	17	19	20	21	22	1.9	2.3	2.7	3.1	3.5	2.2	2.7	3.4	4.6	-
SS A4	18	23	29	33	38	42	48	55	16	20	24	28	33	36	40	46	1.4	1.6	1.8	2.0	2.1	1.5	1.7	2.0	2.2	2.4
SS A5	16	18	21	23	26	27	30	33	14	16	18	19	21	22	24	25	1.9	2.2	2.6	3.0	3.3	2.2	2.6	3.2	4.0	5.0
SS A6	18	23	29	33	39	43	49	56	16	20	25	28	33	37	41	47	1.3	1.5	1.7	1.9	2.1	1.5	1.7	1.9	2.2	2.3
SS A7	15	16	18	20	21	23	24	26	13	14	15	16	17	18	18	19	2.2	2.7	3.3	4.0	4.7	2.6	3.3	4.5	-	-
SS A8	18	22	27	31	37	40	45	52	15	19	23	26	31	34	38	43	1.4	1.6	1.8	2.0	2.2	1.5	1.8	2.1	2.3	2.6
SS B1	13	15	18	21	24	26	29	33	11	13	15	17	19	20	22	25	1.8	2.1	2.5	2.8	3.0	2.1	2.5	2.9	3.5	4.1
SS B2	16	21	27	31	37	41	46	54	14	18	23	26	31	35	39	46	1.4	1.6	1.8	1.9	2.1	1.5	1.7	1.9	2.1	2.3
SS B3	12	14	16	18	21	23	25	28	10	11	13	14	16	17	18	19	1.9	2.4	2.8	3.2	3.5	2.3	2.8	3.5	4.6	-
SS B4	16	20	25	29	35	39	44	51	13	17	21	25	30	33	37	43	1.4	1.6	1.8	2.0	2.2	1.5	1.8	2.0	2.2	2.4
SS B5	13	15	17	19	21	23	25	27	11	13	14	15	17	18	19	21	2.1	2.5	2.9	3.3	3.7	2.5	3.0	3.6	4.5	-
SS B6	16	20	25	29	35	39	44	52	13	17	21	25	30	33	37	43	1.4	1.6	1.8	2.0	2.1	1.5	1.8	2.0	2.2	2.4
SS B7	12	13	15	16	18	19	21	22	10	11	13	13	14	15	15	16	2.3	2.8	3.4	4.1	4.8	2.9	3.5	4.6	-	-
SS B8	15	19	24	27	33	36	41	48	12	16	20	23	27	30	34	40	1.5	1.7	1.9	2.1	2.2	1.6	1.8	2.1	2.3	2.6
SS C1	9	12	15	17	20	22	25	29	6	8	10	11	13	14	15	17	1.8	2.2	2.5	2.8	3.1	2.3	2.8	3.4	4.4	-
SS C2	11	16	21	25	31	35	40	47	8	11	15	18	23	26	30	36	1.4	1.7	1.9	2.0	2.2	1.7	1.9	2.1	2.4	2.6
SS C3	10	12	14	16	18	20	22	25	8	9	10	11	13	14	15	16	2.1	2.5	2.8	3.2	3.5	2.7	3.2	3.8	4.6	-
SS C4	11	15	19	23	28	31	36	42	8	11	15	17	21	24	27	32	1.6	1.8	2.0	2.2	2.3	1.8	2.0	2.3	2.5	2.8
SS D1	16	20	24	27	32	35	39	44	13	16	20	22	25	27	30	33	1.5	1.8	2.0	2.3	2.5	1.7	2.1	2.5	3.0	3.5
SS D2	20	26	34	39	47	52	58	66	16	22	28	33	40	44	50	57	1.2	1.4	1.5	1.7	1.9	1.3	1.5	1.7	1.9	2.1
SS D3	16	19	23	26	29	32	35	39	14	16	19	21	24	25	27	30	1.6	2.0	2.3	2.5	2.8	1.9	2.3	2.7	3.3	3.9
SS D4	20	26	33	38	45	50	56	64	16	22	28	32	38	42	48	55	1.2	1.4	1.6	1.8	1.9	1.3	1.5	1.7	1.9	2.1
SS E1	13	16	19	22	25	28	31	35	10	12	15	16	18	20	21	23	1.7	2.0	2.4	2.7	3.0	2.0	2.4	3.1	4.2	-
SS E2	16	22	28	32	39	44	49	57	13	17	23	26	32	36	40	47	1.3	1.5	1.7	1.9	2.0	1.4	1.6	1.9	2.1	2.3
SS E3	13	16	18	20	23	25	27	30	11	13	15	16	18	19	20	22	1.9	2.3	2.7	3.0	3.3	2.2	2.7	3.3	4.3	-
SS E4	16	21	26	30	36	40	46	53	13	17	21	25	30	33	38	44	1.4	1.6	1.8	2.0	2.1	1.5	1.7	2.0	2.2	2.4

BURLINGTON, IOWA

TYPE	SSF (%) Base temp 55 F (3978 DD) LCR 100	70	50	40	30	25	20	15	SSF (%) 65 F (6154 DD) 100	70	50	40	30	25	20	15	Conservation factor 55 F 100	50	30	20	15	Conservation factor 65 F 100	50	30	20	15
WW A1	17	19	22	24	28	30	34	38	14	15	17	19	21	23	25	27	1.8	2.1	2.3	2.5	2.7	2.3	2.6	2.9	3.3	3.6
WW A2	16	21	26	29	35	39	45	52	12	15	19	22	26	29	33	38	1.4	1.6	1.8	2.0	2.1	1.7	1.9	2.1	2.4	2.6
WW A3	16	21	27	31	38	43	49	57	11	15	19	23	28	31	36	42	1.3	1.5	1.7	1.8	2.0	1.6	1.8	2.0	2.2	2.4
WW A4	15	21	28	33	40	45	51	60	10	15	20	24	29	33	38	45	1.3	1.5	1.6	1.8	1.9	1.5	1.7	1.9	2.1	2.3
WW A5	15	21	28	34	42	47	53	62	10	15	20	24	30	34	40	47	1.2	1.4	1.6	1.7	1.9	1.4	1.6	1.8	2.0	2.2
WW A6	15	22	29	35	43	48	55	64	9	15	20	25	31	35	41	49	1.2	1.4	1.5	1.7	1.9	1.4	1.6	1.8	2.0	2.1
WW B1	12	15	18	21	24	27	30	35	9	10	12	13	15	16	18	19	1.8	2.1	2.3	2.6	2.8	2.3	2.8	3.4	4.3	-
WW B2	16	23	30	35	43	49	55	65	11	16	22	27	33	38	44	52	1.2	1.4	1.5	1.7	1.8	1.4	1.6	1.7	1.9	2.1
WW B3	20	27	36	42	51	57	64	73	14	20	27	33	40	45	52	61	1.1	1.3	1.4	1.6	1.7	1.2	1.4	1.6	1.8	1.9
WW B4	20	28	38	45	55	61	69	78	14	21	30	36	45	51	58	68	1.1	1.2	1.4	1.5	1.7	1.2	1.3	1.5	1.6	1.8
WW B5	19	28	38	45	55	61	69	78	13	21	30	36	45	51	59	69	1.1	1.2	1.4	1.5	1.7	1.2	1.3	1.5	1.6	1.8
WW C1	21	29	38	45	54	60	67	76	16	22	30	35	43	49	55	64	1.1	1.2	1.4	1.6	1.7	1.2	1.4	1.5	1.7	1.9
WW C2	20	28	37	44	53	59	67	76	14	21	29	35	43	49	56	65	1.1	1.2	1.4	1.6	1.7	1.2	1.4	1.5	1.7	1.8
WW C3	24	34	45	53	64	70	78	86	18	27	37	45	55	61	69	79	1.0	1.1	1.3	1.4	1.6	1.0	1.2	1.3	1.5	1.7
WW C4	21	31	42	50	61	67	75	84	16	24	34	42	52	58	66	76	1.0	1.1	1.3	1.5	1.6	1.1	1.2	1.4	1.5	1.7
TW A1	16	19	22	24	28	30	34	39	14	15	17	19	21	23	25	28	1.8	2.1	2.3	2.5	2.6	2.2	2.5	2.8	3.1	3.4
TW A2	16	19	24	27	32	36	41	48	12	15	18	21	24	27	31	36	1.5	1.7	1.9	2.1	2.2	1.8	2.0	2.3	2.5	2.7
TW A3	15	19	25	28	34	39	44	52	11	14	18	21	26	29	33	39	1.4	1.6	1.8	2.0	2.1	1.7	1.9	2.1	2.3	2.5
TW A4	14	19	24	28	35	39	45	53	10	13	18	21	26	29	34	40	1.4	1.6	1.7	1.9	2.0	1.6	1.8	2.0	2.2	2.4
TW B1	16	18	22	25	29	31	35	41	13	15	17	19	22	24	27	31	1.7	2.0	2.1	2.3	2.5	2.1	2.3	2.6	2.8	3.1
TW B2	14	18	23	27	32	36	41	48	11	14	17	20	24	27	31	36	1.5	1.7	1.9	2.0	2.2	1.8	2.0	2.2	2.4	2.6
TW B3	14	18	23	27	32	36	42	49	10	13	17	20	24	27	32	37	1.5	1.7	1.8	2.0	2.1	1.7	1.9	2.1	2.3	2.5
TW B4	13	17	22	25	31	35	40	48	10	13	16	19	23	27	31	37	1.5	1.7	1.9	2.0	2.2	1.8	2.0	2.1	2.3	2.5
TW C1	14	17	21	24	29	32	36	42	11	14	17	19	22	24	28	32	1.7	1.9	2.1	2.2	2.4	2.0	2.2	2.4	2.6	2.8
TW C2	13	17	21	24	29	33	38	45	10	13	16	19	23	25	29	34	1.6	1.8	2.0	2.1	2.3	1.9	2.1	2.3	2.5	2.6
TW C3	13	16	20	23	28	32	36	43	10	13	16	18	22	25	28	34	1.6	1.8	2.0	2.2	2.3	1.9	2.1	2.3	2.5	2.6
TW C4	13	16	19	22	26	29	33	39	11	13	15	18	21	23	26	31	1.8	2.0	2.1	2.3	2.4	2.1	2.2	2.4	2.6	2.8
TW D1	10	13	16	18	22	24	27	32	7	9	11	12	14	15	17	19	1.9	2.2	2.4	2.6	2.8	2.4	2.8	3.3	3.8	4.4
TW D2	16	22	28	33	40	44	51	59	12	16	21	25	31	35	40	48	1.3	1.5	1.6	1.8	1.9	1.5	1.7	1.8	2.0	2.2
TW D3	17	23	30	35	43	48	55	64	12	17	23	27	34	38	44	52	1.2	1.4	1.6	1.7	1.9	1.4	1.6	1.8	1.9	2.1
TW D4	19	26	34	40	49	55	62	72	14	20	27	32	40	45	52	61	1.1	1.3	1.4	1.6	1.7	1.3	1.4	1.6	1.7	1.9
TW D5	18	26	34	41	50	56	64	73	13	20	27	33	41	47	54	64	1.1	1.3	1.4	1.6	1.7	1.2	1.4	1.5	1.7	1.8
TW E1	20	27	34	40	49	54	61	70	15	20	27	32	39	44	50	58	1.2	1.3	1.5	1.6	1.8	1.3	1.5	1.6	1.8	2.0
TW E2	19	27	35	41	49	55	62	72	14	20	27	32	40	45	52	61	1.1	1.3	1.5	1.6	1.8	1.3	1.4	1.6	1.8	1.9
TW E3	23	33	43	50	60	66	73	82	18	26	35	41	51	57	64	74	1.0	1.2	1.3	1.5	1.6	1.1	1.3	1.4	1.6	1.7
TW E4	22	31	40	48	58	64	72	81	16	24	33	39	49	55	62	72	1.0	1.2	1.3	1.5	1.6	1.1	1.3	1.4	1.6	1.7
TW F1	14	17	20	23	27	29	33	38	12	14	16	17	20	22	24	28	1.8	2.0	2.2	2.4	2.6	2.2	2.5	2.7	3.0	3.3
TW F2	13	17	21	25	30	34	38	45	10	12	16	18	22	25	28	33	1.6	1.8	1.9	2.1	2.3	1.8	2.1	2.3	2.5	2.7
TW F3	12	16	21	25	31	35	40	48	8	11	15	18	22	25	29	35	1.5	1.7	1.8	2.0	2.2	1.7	2.0	2.2	2.4	2.6
TW F4	10	14	20	24	29	34	39	47	6	10	13	17	21	24	29	35	1.5	1.7	1.8	2.0	2.1	1.7	1.9	2.1	2.3	2.5
TW G1	12	15	19	22	26	29	33	39	10	12	14	16	19	21	24	28	1.7	2.0	2.2	2.3	2.5	2.1	2.3	2.6	2.9	3.1
TW G2	11	14	19	22	27	31	36	43	7	10	13	16	20	22	26	31	1.6	1.8	2.0	2.1	2.3	1.9	2.1	2.3	2.5	2.7
TW G3	9	13	17	21	26	30	35	42	6	8	12	15	19	22	26	31	1.6	1.8	1.9	2.1	2.2	1.9	2.0	2.2	2.5	2.6
TW G4	7	10	15	18	23	26	31	38	4	7	10	12	16	19	23	28	1.7	1.9	2.0	2.2	2.3	1.9	2.1	2.3	2.5	2.7
TW H1	9	13	16	19	23	26	31	37	7	9	12	14	17	19	22	27	1.7	2.0	2.1	2.3	2.5	2.1	2.3	2.5	2.8	3.0
TW H2	7	10	14	17	22	25	30	36	4	7	10	12	15	18	21	27	1.7	1.9	2.1	2.3	2.4	2.0	2.2	2.4	2.6	2.8
TW H3	6	9	12	15	19	23	27	33	3	5	8	10	14	16	19	24	1.8	2.0	2.2	2.3	2.5	2.1	2.3	2.5	2.7	2.9
TW H4	4	7	10	12	16	18	22	27	2	4	6	8	11	13	16	20	2.0	2.2	2.4	2.6	2.7	2.4	2.5	2.7	2.9	3.1
TW I1	8	11	14	16	19	21	24	28	5	7	9	10	12	13	14	16	2.0	2.2	2.5	2.8	3.0	2.5	2.9	3.4	4.0	4.7
TW I2	13	18	25	29	36	41	47	56	9	13	18	22	27	31	37	44	1.4	1.5	1.7	1.8	2.0	1.5	1.7	1.9	2.1	2.2
TW I3	14	20	27	32	39	45	51	60	10	15	20	24	31	35	41	49	1.3	1.4	1.6	1.8	1.9	1.4	1.6	1.8	2.0	2.1
TW I4	16	23	31	37	46	52	59	69	11	17	24	29	37	42	49	58	1.2	1.3	1.5	1.6	1.8	1.3	1.4	1.6	1.8	1.9
TW I5	16	23	32	38	48	54	61	71	11	18	25	31	39	44	52	61	1.1	1.3	1.4	1.6	1.7	1.3	1.4	1.6	1.7	1.9
TW J1	18	24	32	38	46	52	59	68	13	18	25	30	37	42	48	57	1.2	1.3	1.5	1.7	1.8	1.3	1.5	1.7	1.8	2.0
TW J2	17	24	32	38	47	52	60	69	12	18	25	30	37	43	49	58	1.2	1.3	1.5	1.6	1.8	1.3	1.5	1.6	1.8	1.9
TW J3	20	30	40	47	57	64	71	80	15	23	32	39	48	54	62	72	1.0	1.2	1.3	1.5	1.6	1.1	1.3	1.4	1.6	1.7
TW J4	19	28	38	46	55	61	69	79	14	21	30	37	46	52	60	70	1.1	1.2	1.3	1.5	1.6	1.1	1.3	1.4	1.6	1.7
DG A1	10	13	17	19	22	24	26	28	7	8	10	12	14	14	15	15	1.7	2.1	2.6	3.5	4.8	2.2	2.7	3.7	-	-
DG A2	13	17	22	26	31	35	40	46	9	13	16	19	24	26	30	35	1.5	1.7	1.9	2.2	2.4	1.7	1.9	2.2	2.5	2.8
DG A3	17	22	29	34	42	48	55	63	13	18	23	27	34	39	45	53	1.3	1.4	1.5	1.7	1.9	1.4	1.6	1.7	1.9	2.1
DG B1	10	13	17	19	23	26	29	33	7	9	11	12	14	15	17	19	1.8	2.0	2.3	2.7	3.1	2.1	2.6	3.3	4.1	-
DG B2	13	18	23	27	33	38	43	51	10	13	17	20	25	28	32	38	1.5	1.6	1.8	2.0	2.1	1.7	1.9	2.1	2.3	2.5
DG B3	17	23	30	35	44	50	58	68	13	18	24	29	36	41	48	57	1.3	1.4	1.5	1.6	1.8	1.4	1.5	1.7	1.8	1.9
DG C1	13	17	21	24	30	33	38	44	9	11	14	17	20	22	24	28	1.6	1.7	1.9	2.2	2.5	1.8	2.2	2.6	2.9	3.5
DG C2	16	21	27	32	39	44	51	60	12	16	20	24	30	34	39	46	1.4	1.5	1.6	1.8	1.9	1.5	1.7	1.9	2.0	2.2
DG C3	19	26	34	40	50	56	65	75	15	21	27	32	40	46	54	64	1.2	1.3	1.4	1.5	1.7	1.3	1.4	1.5	1.7	1.8
SS A1	19	23	28	32	38	41	47	54	16	19	23	26	31	34	38	43	1.4	1.6	1.8	2.0	2.1	1.6	1.8	2.1	2.3	2.5
SS A2	24	30	38	43	51	56	63	72	20	26	32	37	44	48	54	62	1.2	1.3	1.5	1.6	1.8	1.3	1.4	1.6	1.8	1.9
SS A3	17	21	25	28	33	36	41	47	15	17	21	23	26	29	32	36	1.5	1.8	2.0	2.2	2.3	1.8	2.1	2.4	2.7	2.9
SS A4	23	29	36	42	49	54	61	69	19	25	31	35	42	46	52	59	1.2	1.4	1.5	1.7	1.8	1.3	1.5	1.7	1.9	2.0
SS A5	19	23	27	30	34	37	42	48	16	19	22	25	28	30	34	38	1.6	1.8	2.0	2.2	2.3	1.8	2.0	2.3	2.6	2.8
SS A6	23	30	37	43	50	56	62	71	19	25	31	36	43	47	53	61	1.2	1.3	1.5	1.7	1.8	1.3	1.5	1.6	1.8	2.0
SS A7	17	20	23	25	29	31	34	39	15	17	19	21	23	25	27	30	1.8	2.0	2.3	2.5	2.7	2.0	2.4	2.8	3.2	3.7
SS A8	22	28	35	40	48	53	59	67	19	24	30	34	40	44	50	57	1.2	1.4	1.5	1.7	1.9	1.3	1.5	1.7	1.9	2.1
SS B1	16	20	24	28	33	36	41	48	13	16	20	22	26	29	32	37	1.5	1.7	1.9	2.1	2.2	1.7	2.0	2.2	2.5	2.7
SS B2	21	27	34	39	47	52	58	67	17	22	28	33	39	44	50	58	1.2	1.4	1.5	1.7	1.8	1.3	1.5	1.7	1.8	2.0
SS B3	15	18	22	25	30	33	37	43	12	15	18	20	23	25	28	32	1.6	1.9	2.1	2.3	2.4	1.9	2.1	2.4	2.8	3.0
SS B4	20	26	32	38	45	50	56	65	16	21	27	31	38	42	48	55	1.2	1.4	1.6	1.7	1.9	1.4	1.5	1.7	1.9	2.0
SS B5	16	19	22	25	28	31	35	40	13	15	18	20	23	25	28	31	1.7	2.0	2.2	2.4	2.5	2.0	2.3	2.6	2.9	3.1
SS B6	20	26	33	38	45	50	57	66	16	21	27	32	38	42	48	56	1.2	1.4	1.6	1.7	1.9	1.3	1.5	1.7	1.9	2.0
SS B7	14	17	19	22	25	27	30	34	12	14	16	17	20	21	23	26	1.9	2.2	2.4	2.7	2.9	2.2	2.6	3.0	3.4	3.8
SS B8	19	24	31	36	43	48	54	62	15	20	25	30	36	40	45	52	1.3	1.4	1.6	1.8	1.9	1.4	1.6	1.8	2.0	2.1
SS C1	14	18	23	27	33	37	42	50	10	13	16	19	23	26	30	35	1.5	1.6	1.8	2.0	2.1	1.7	2.0	2.2	2.4	2.6
SS C2	17	23	30	36	44	49	56	65	12	17	23	27	34	38	44	53	1.2	1.4	1.5	1.7	1.8	1.4	1.6	1.7	1.9	2.1
SS C3	13	16	20	23	27	30	35	41	10	12	15	17	20	22	25	29	1.7	1.9	2.1	2.3	2.4	2.0	2.3	2.5	2.8	3.0
SS C4	16	21	27	32	39	44	50	59	12	16	21	25	30	34	40	47	1.3	1.5	1.6	1.8	1.9	1.5	1.7	1.9	2.0	2.2
SS D1	21	27	33	38	44	49	55	63	17	22	27	30	36	39	44	50	1.3	1.4	1.6	1.8	2.0	1.4	1.6	1.9	2.1	2.3
SS D2	26	35	44	50	59	65	72	80	21	29	37	43	51	56	63	71	1.0	1.2	1.4	1.5	1.7	1.1	1.3	1.5	1.6	1.8
SS D3	20	25	30	34	40	44	49	56	17	21	25	28	32	35	39	45	1.4	1.6	1.8	1.9	2.1	1.5	1.8	2.0	2.3	2.5
SS D4	26	33	42	48	57	63	70	78	21	28	35	41	49	54	60	69	1.1	1.2	1.4	1.5	1.7	1.2	1.3	1.5	1.7	1.8
SS E1	17	22	28	32	38	42	47	55	14	17	21	24	29	32	36	41	1.4	1.6	1.8	1.9	2.1	1.6	1.8	2.1	2.4	2.6
SS E2	22	29	37	43	52	58	64	73	17	23	30	36	43	48	54	62	1.1	1.3	1.4	1.6	1.8	1.2	1.4	1.6	1.8	1.9
SS E3	17	20	24	28	32	35	40	46	14	16	20	22	25	28	31	35	1.6	1.8	2.0	2.2	2.3	1.8	2.1	2.4	2.6	2.9
SS E4	21	27	35	40	48	53	60	69	17	22	28	33	40	44	50	58	1.2	1.3	1.5	1.7	1.8	1.3	1.5	1.7	1.8	2.0

DES MOINES, IOWA

TYPE	SOLAR SAVINGS FRACTION (%) — BASE TEMP: 55 F (4475 DD)								65 F (6718 DD)								CONSERVATION FACTOR — 55 F					65 F				
LCR	100	70	50	40	30	25	20	15	100	70	50	40	30	25	20	15	100	50	30	20	15	100	50	30	20	15
WW A1	16	18	21	23	26	28	31	36	13	15	17	18	20	22	23	26	1.9	2.2	2.4	2.6	2.8	2.4	2.7	3.1	3.5	3.9
WW A2	15	19	24	27	33	37	42	49	12	15	18	21	25	28	31	36	1.5	1.7	1.9	2.0	2.2	1.7	2.0	2.2	2.5	2.6
WW A3	15	19	25	29	36	40	46	54	11	14	19	22	27	30	34	41	1.4	1.6	1.7	1.9	2.0	1.6	1.8	2.0	2.2	2.4
WW A4	14	19	26	30	37	42	48	57	10	14	19	22	28	31	36	43	1.3	1.5	1.7	1.8	2.0	1.5	1.7	1.9	2.1	2.3
WW A5	14	20	26	31	39	44	50	59	9	14	19	23	29	33	38	45	1.3	1.5	1.6	1.8	1.9	1.5	1.7	1.9	2.1	2.2
WW A6	14	20	27	32	40	45	52	61	9	14	19	23	29	34	39	47	1.3	1.4	1.6	1.8	1.9	1.4	1.6	1.8	2.0	2.2
WW B1	11	14	17	19	22	24	27	31	8	10	11	12	14	15	16	17	1.9	2.2	2.5	2.8	3.0	2.4	3.0	3.7	4.9	-
WW B2	15	21	28	33	41	46	53	62	11	16	21	26	32	36	42	50	1.3	1.4	1.6	1.7	1.9	1.4	1.6	1.8	2.0	2.1
WW B3	18	25	34	40	48	54	61	71	13	19	26	31	39	44	50	59	1.1	1.3	1.5	1.6	1.8	1.3	1.4	1.6	1.8	1.9
WW B4	18	27	36	43	52	59	66	76	13	21	29	35	43	49	57	66	1.1	1.2	1.4	1.5	1.7	1.2	1.3	1.5	1.7	1.8
WW B5	17	26	36	43	52	59	67	76	13	20	29	35	44	50	58	68	1.1	1.2	1.4	1.5	1.7	1.2	1.3	1.5	1.6	1.8
WW C1	20	27	36	42	51	57	64	74	15	21	29	34	42	47	54	63	1.1	1.3	1.4	1.6	1.7	1.2	1.4	1.6	1.7	1.9
WW C2	18	26	35	41	51	57	64	74	14	20	28	34	42	47	54	64	1.1	1.3	1.4	1.6	1.7	1.2	1.4	1.5	1.7	1.8
WW C3	22	32	43	51	62	68	76	85	17	26	36	43	53	60	68	77	1.0	1.1	1.3	1.4	1.6	1.1	1.2	1.4	1.5	1.7
WW C4	20	30	40	48	58	65	73	82	15	24	33	40	50	57	65	75	1.0	1.2	1.3	1.5	1.6	1.1	1.2	1.4	1.5	1.7
TW A1	16	18	21	23	26	29	32	36	13	15	17	18	21	22	24	27	1.9	2.2	2.4	2.6	2.7	2.3	2.6	2.9	3.3	3.6
TW A2	15	18	22	26	30	34	38	45	12	14	17	20	23	26	29	34	1.6	1.8	2.0	2.1	2.3	1.9	2.1	2.3	2.6	2.8
TW A3	14	18	23	27	32	36	41	49	11	14	17	20	24	27	31	37	1.5	1.7	1.8	2.0	2.2	1.7	1.9	2.2	2.4	2.5
TW A4	13	17	22	27	33	37	42	50	9	13	17	20	25	28	32	39	1.4	1.6	1.8	2.0	2.1	1.7	1.9	2.1	2.3	2.4
TW B1	15	17	21	23	27	29	33	38	12	14	17	18	21	23	26	29	1.8	2.0	2.2	2.4	2.6	2.1	2.4	2.7	3.0	3.2
TW B2	13	17	21	25	30	33	38	45	11	13	17	19	23	26	29	35	1.6	1.8	1.9	2.1	2.3	1.8	2.0	2.3	2.5	2.7
TW B3	13	17	21	25	30	34	39	46	10	13	16	19	23	26	30	36	1.5	1.7	1.9	2.1	2.2	1.8	2.0	2.2	2.4	2.6
TW B4	12	16	20	24	29	33	38	45	9	12	16	18	23	25	29	35	1.6	1.8	1.9	2.1	2.2	1.8	2.0	2.2	2.4	2.6
TW C1	14	17	20	23	27	30	34	40	11	13	16	18	21	24	27	31	1.7	2.0	2.1	2.3	2.4	2.0	2.3	2.5	2.7	2.9
TW C2	12	16	20	23	28	31	36	42	10	13	16	18	22	24	28	33	1.7	1.9	2.0	2.2	2.3	1.9	2.1	2.3	2.5	2.7
TW C3	12	15	19	22	27	30	34	41	10	12	15	18	21	24	27	32	1.7	1.9	2.1	2.2	2.4	2.0	2.1	2.3	2.5	2.7
TW C4	12	15	18	21	25	28	32	37	10	12	15	17	20	22	26	30	1.8	2.0	2.2	2.4	2.5	2.1	2.3	2.5	2.7	2.8
TW D1	9	12	14	16	20	22	25	29	7	8	10	11	13	14	15	17	2.0	2.3	2.6	2.8	3.0	2.5	2.9	3.5	4.2	5.0
TW D2	15	20	26	31	37	42	48	57	11	16	20	24	30	33	39	46	1.4	1.5	1.7	1.8	2.0	1.5	1.7	1.9	2.1	2.2
TW D3	16	22	28	33	41	46	52	61	12	17	22	26	32	36	42	50	1.3	1.4	1.6	1.8	1.9	1.4	1.6	1.8	2.0	2.1
TW D4	17	24	32	38	47	53	60	70	13	19	26	31	39	44	51	60	1.2	1.3	1.5	1.6	1.8	1.3	1.4	1.6	1.8	1.9
TW D5	17	24	33	39	48	54	62	71	13	19	27	32	40	46	53	62	1.1	1.3	1.4	1.6	1.7	1.2	1.4	1.6	1.7	1.8
TW E1	19	25	32	38	46	52	59	68	14	20	26	31	38	42	48	57	1.2	1.4	1.5	1.7	1.8	1.3	1.5	1.7	1.8	2.0
TW E2	18	25	33	39	47	53	60	69	14	20	26	31	39	44	50	59	1.2	1.3	1.5	1.6	1.8	1.3	1.5	1.6	1.8	1.9
TW E3	22	31	41	48	58	64	71	80	17	25	34	40	49	55	63	72	1.0	1.2	1.3	1.5	1.6	1.1	1.3	1.4	1.6	1.7
TW E4	20	29	39	46	56	62	70	79	15	23	32	38	47	54	61	71	1.1	1.2	1.3	1.5	1.6	1.1	1.3	1.4	1.6	1.7
TW F1	14	16	19	22	25	27	31	36	11	13	15	17	19	21	23	26	1.9	2.1	2.3	2.5	2.7	2.3	2.6	2.9	3.2	3.5
TW F2	12	16	20	23	28	31	36	42	9	12	15	17	21	23	27	31	1.6	1.8	2.0	2.2	2.3	1.9	2.1	2.4	2.6	2.8
TW F3	11	15	20	23	29	32	38	45	8	11	14	17	21	24	28	33	1.5	1.7	1.9	2.1	2.2	1.8	2.0	2.2	2.4	2.6
TW F4	9	13	18	22	27	31	37	44	6	9	13	16	20	23	27	33	1.5	1.7	1.9	2.1	2.2	1.8	2.0	2.2	2.4	2.5
TW G1	12	14	18	20	24	27	31	36	9	11	14	15	18	20	23	26	1.8	2.0	2.2	2.4	2.6	2.2	2.4	2.7	3.0	3.2
TW G2	10	13	17	20	25	29	33	40	7	10	13	15	19	21	25	30	1.7	1.9	2.0	2.2	2.3	1.9	2.1	2.4	2.6	2.8
TW G3	8	12	16	19	24	28	33	40	5	8	11	14	18	21	24	30	1.6	1.8	2.0	2.2	2.3	1.9	2.1	2.3	2.5	2.7
TW G4	6	9	13	16	21	25	29	36	4	6	9	12	15	18	22	27	1.7	1.9	2.1	2.3	2.4	2.0	2.2	2.4	2.6	2.7
TW H1	9	12	15	18	22	25	29	34	6	9	11	13	16	18	21	25	1.8	2.0	2.2	2.4	2.5	2.1	2.4	2.6	2.8	3.1
TW H2	6	9	13	16	20	23	28	34	4	6	9	11	15	17	20	25	1.8	2.0	2.2	2.3	2.5	2.1	2.3	2.5	2.7	2.9
TW H3	5	8	11	14	18	21	25	31	3	5	8	10	13	15	19	23	1.9	2.1	2.3	2.4	2.6	2.2	2.4	2.6	2.8	2.9
TW H4	4	6	9	11	14	17	20	26	2	4	6	8	10	12	15	19	2.1	2.3	2.5	2.6	2.8	2.4	2.6	2.8	3.0	3.1
TW I1	8	10	12	14	17	19	22	25	5	6	8	9	11	12	13	14	2.1	2.4	2.7	2.9	3.2	2.6	3.1	3.7	4.5	-
TW I2	12	17	23	27	34	39	45	53	9	13	17	21	26	30	35	43	1.4	1.6	1.7	1.9	2.0	1.6	1.7	1.9	2.1	2.3
TW I3	13	19	25	30	37	42	49	58	9	14	19	23	29	34	39	47	1.3	1.5	1.6	1.8	1.9	1.5	1.6	1.8	2.0	2.1
TW I4	15	21	29	35	44	49	57	67	11	16	23	28	36	41	48	57	1.2	1.4	1.5	1.7	1.8	1.3	1.5	1.6	1.8	1.9
TW I5	15	22	30	36	45	51	59	69	11	17	24	30	38	43	50	60	1.2	1.3	1.5	1.6	1.8	1.3	1.4	1.6	1.7	1.9
TW J1	16	23	30	36	44	49	56	66	12	18	24	29	35	40	47	55	1.2	1.4	1.5	1.7	1.8	1.3	1.5	1.7	1.8	2.0
TW J2	16	22	30	36	44	50	57	67	11	17	24	29	36	41	48	57	1.2	1.3	1.5	1.7	1.8	1.3	1.5	1.6	1.8	1.9
TW J3	19	28	38	45	55	61	69	79	14	22	31	38	47	53	61	71	1.1	1.2	1.3	1.5	1.6	1.1	1.3	1.4	1.6	1.7
TW J4	17	26	36	43	53	59	67	77	13	21	29	36	45	51	59	69	1.1	1.2	1.4	1.5	1.7	1.2	1.3	1.5	1.6	1.8
DG A1	9	12	15	17	20	21	23	25	6	8	10	11	12	13	14	14	1.8	2.2	2.8	3.8	-	2.3	2.9	4.0	-	-
DG A2	12	16	21	24	29	33	37	43	9	12	16	18	23	25	29	34	1.6	1.7	1.9	2.2	2.5	1.8	2.0	2.2	2.5	2.9
DG A3	16	21	27	32	40	45	52	61	13	17	22	27	33	38	44	52	1.3	1.4	1.6	1.7	1.9	1.5	1.6	1.7	1.9	2.1
DG B1	9	12	15	17	21	23	26	30	6	8	10	11	13	14	15	17	1.8	2.2	2.4	2.9	3.4	2.2	2.8	3.5	4.5	-
DG B2	12	16	21	25	31	35	40	48	9	13	16	19	23	27	31	36	1.5	1.7	1.8	2.0	2.2	1.7	1.9	2.1	2.3	2.6
DG B3	16	21	28	34	42	48	55	66	13	18	23	28	34	39	46	56	1.3	1.4	1.5	1.6	1.8	1.4	1.6	1.7	1.8	1.9
DG C1	12	15	19	22	27	30	35	40	8	11	14	15	18	20	22	26	1.6	1.9	2.0	2.3	2.6	1.9	2.3	2.7	3.1	3.7
DG C2	15	19	25	30	37	42	48	57	11	15	20	23	29	32	38	45	1.4	1.5	1.6	1.8	2.0	1.6	1.8	1.9	2.1	2.3
DG C3	18	24	32	38	47	54	62	73	15	20	26	31	39	45	52	62	1.2	1.3	1.4	1.6	1.7	1.3	1.5	1.6	1.7	1.8
SS A1	18	22	27	30	35	39	44	51	16	19	23	25	29	32	36	41	1.5	1.7	1.9	2.0	2.2	1.6	1.9	2.1	2.4	2.6
SS A2	22	29	36	41	49	54	61	69	19	25	31	36	42	47	53	61	1.2	1.4	1.5	1.7	1.8	1.3	1.5	1.6	1.8	2.0
SS A3	16	20	24	26	31	34	38	43	14	17	20	22	25	27	30	34	1.6	1.9	2.1	2.3	2.4	1.8	2.1	2.5	2.8	3.1
SS A4	22	28	34	40	47	52	58	67	19	24	30	34	40	45	50	58	1.2	1.4	1.5	1.7	1.9	1.3	1.5	1.7	1.9	2.0
SS A5	18	21	25	28	32	35	39	45	16	19	21	24	27	29	32	36	1.6	1.9	2.1	2.3	2.4	1.8	2.1	2.4	2.7	3.0
SS A6	22	28	35	40	48	53	60	68	19	24	30	35	41	46	51	59	1.2	1.4	1.5	1.7	1.8	1.3	1.5	1.7	1.8	2.0
SS A7	16	19	22	24	27	29	32	36	15	16	19	20	22	24	26	28	1.9	2.2	2.4	2.7	2.9	2.1	2.5	3.0	3.5	3.9
SS A8	21	27	33	38	45	50	56	64	18	23	28	33	38	43	48	55	1.2	1.4	1.6	1.7	1.9	1.4	1.6	1.8	1.9	2.1
SS B1	15	19	23	26	31	34	38	45	13	16	19	21	25	27	31	35	1.6	1.8	2.0	2.2	2.3	1.8	2.0	2.3	2.6	2.8
SS B2	19	25	32	37	44	49	56	65	16	22	27	32	38	43	48	56	1.2	1.4	1.6	1.7	1.9	1.3	1.5	1.7	1.9	2.0
SS B3	14	17	21	23	27	30	34	40	12	14	17	19	22	24	27	30	1.7	1.9	2.2	2.4	2.5	1.9	2.2	2.6	2.9	3.2
SS B4	19	24	31	36	43	47	54	62	16	21	26	30	36	41	46	54	1.3	1.4	1.6	1.8	1.9	1.4	1.6	1.8	1.9	2.1
SS B5	15	17	21	23	27	29	33	38	13	15	17	19	22	24	26	30	1.8	2.1	2.3	2.5	2.7	2.1	2.4	2.7	3.0	3.3
SS B6	19	24	31	36	43	48	54	63	16	21	26	30	37	41	47	54	1.3	1.4	1.6	1.7	1.9	1.4	1.6	1.7	1.9	2.1
SS B7	13	16	18	20	23	25	28	32	12	13	15	17	19	20	22	24	2.0	2.3	2.6	2.8	3.0	2.3	2.7	3.1	3.6	4.0
SS B8	18	23	29	34	40	45	51	60	15	19	25	28	34	38	43	51	1.3	1.5	1.6	1.8	1.9	1.4	1.6	1.8	2.0	2.2
SS C1	13	17	21	25	30	34	39	46	9	12	16	18	22	25	28	33	1.5	1.7	1.9	2.1	2.2	1.8	2.0	2.3	2.5	2.7
SS C2	16	22	28	34	41	46	53	62	12	16	22	26	33	37	43	51	1.3	1.4	1.6	1.7	1.9	1.4	1.6	1.8	1.9	2.1
SS C3	12	15	19	21	25	28	32	38	10	12	14	16	19	21	24	28	1.8	2.0	2.2	2.4	2.5	2.1	2.3	2.6	2.9	3.1
SS C4	15	20	26	30	37	41	48	56	11	15	20	24	29	33	38	46	1.4	1.5	1.7	1.8	2.0	1.5	1.7	1.9	2.1	2.2
SS D1	20	25	31	35	42	46	52	59	17	21	26	29	34	37	42	48	1.3	1.5	1.7	1.8	2.0	1.4	1.7	1.9	2.1	2.3
SS D2	25	33	41	48	57	62	69	78	21	28	35	41	49	54	61	70	1.1	1.2	1.4	1.5	1.7	1.1	1.3	1.5	1.7	1.8
SS D3	19	24	28	32	37	41	46	53	17	20	24	27	31	34	38	43	1.4	1.6	1.8	2.0	2.2	1.6	1.8	2.1	2.3	2.5
SS D4	24	32	40	46	55	60	67	76	20	27	34	40	47	52	59	67	1.1	1.2	1.4	1.6	1.7	1.2	1.4	1.5	1.7	1.8
SS E1	16	21	26	29	35	39	44	51	13	16	20	23	27	30	34	39	1.4	1.6	1.8	2.0	2.2	1.6	1.9	2.2	2.4	2.7
SS E2	20	27	35	41	49	55	62	71	16	22	29	34	41	46	52	61	1.1	1.3	1.5	1.6	1.8	1.2	1.4	1.6	1.8	1.9
SS E3	16	19	23	26	30	33	37	43	13	16	19	21	24	26	29	33	1.6	1.9	2.1	2.3	2.4	1.8	2.1	2.4	2.7	3.0
SS E4	19	26	33	38	45	51	57	66	16	21	27	32	38	43	48	56	1.2	1.4	1.5	1.7	1.8	1.3	1.5	1.7	1.9	2.0

MASON CITY, IOWA

TYPE	SOLAR SAVINGS FRACTION (%) BASE TEMP: 55 F (5438 DD) LCR 100	70	50	40	30	25	20	15	65 F (7906 DD) 100	70	50	40	30	25	20	15	CONSERVATION FACTOR 55 F 100	50	30	20	15	65 F 100	50	30	20	15
WW A1	14	16	18	20	22	24	26	30	13	14	15	17	18	19	20	22	2.2	2.5	2.8	3.1	3.3	2.6	3.0	3.6	4.2	5.0
WW A2	13	16	20	23	28	31	35	41	10	13	16	18	22	24	27	31	1.6	1.9	2.1	2.3	2.4	1.9	2.2	2.5	2.8	3.0
WW A3	12	16	21	25	30	34	39	46	9	12	16	19	23	26	29	35	1.5	1.7	1.9	2.1	2.2	1.7	2.0	2.2	2.5	2.7
WW A4	11	16	21	25	31	35	41	48	8	12	16	19	24	27	31	37	1.4	1.6	1.8	2.0	2.2	1.6	1.9	2.1	2.4	2.5
WW A5	11	16	22	26	32	37	43	51	8	12	16	20	25	28	33	39	1.4	1.6	1.8	1.9	2.1	1.6	1.8	2.0	2.2	2.4
WW A6	11	16	22	27	33	38	44	52	7	11	16	20	25	29	34	40	1.4	1.5	1.7	1.9	2.0	1.5	1.7	2.0	2.2	2.4
WW B1	9	11	13	15	17	18	20	22	7	8	9	10	11	11	11	11	2.2	2.6	3.1	3.7	4.2	2.8	3.7	-	-	-
WW B2	12	18	24	28	35	40	46	55	9	14	19	23	28	32	37	45	1.4	1.5	1.7	1.9	2.0	1.5	1.7	1.9	2.1	2.2
WW B3	15	22	29	34	42	48	54	63	12	17	23	28	35	39	45	53	1.2	1.4	1.6	1.7	1.9	1.3	1.5	1.7	1.9	2.0
WW B4	15	23	31	38	47	53	60	70	12	18	26	32	40	45	52	62	1.1	1.3	1.5	1.6	1.7	1.2	1.4	1.6	1.7	1.9
WW B5	15	22	31	38	47	53	61	71	11	18	26	32	41	46	54	64	1.1	1.3	1.4	1.6	1.7	1.2	1.4	1.5	1.7	1.8
WW C1	17	23	31	37	45	51	58	67	13	19	26	31	38	43	49	58	1.2	1.3	1.5	1.7	1.8	1.3	1.5	1.6	1.8	2.0
WW C2	15	22	30	36	45	51	58	68	12	18	25	30	38	43	50	59	1.2	1.3	1.5	1.6	1.8	1.3	1.4	1.6	1.8	1.9
WW C3	19	28	39	46	56	63	71	80	15	24	33	40	50	56	64	74	1.0	1.2	1.3	1.5	1.6	1.1	1.2	1.4	1.6	1.7
WW C4	17	26	36	43	53	60	68	78	13	21	30	37	47	53	61	71	1.1	1.2	1.4	1.5	1.7	1.1	1.3	1.4	1.6	1.7
TW A1	14	16	18	20	23	24	27	30	13	14	16	17	19	20	21	23	2.1	2.4	2.7	3.0	3.2	2.5	2.9	3.3	3.8	4.3
TW A2	13	16	19	22	26	29	33	38	11	13	16	18	21	23	26	29	1.7	2.0	2.2	2.4	2.5	2.0	2.3	2.6	2.9	3.1
TW A3	12	15	19	23	27	31	35	41	9	12	15	18	21	24	27	32	1.6	1.8	2.0	2.2	2.4	1.8	2.1	2.3	2.6	2.8
TW A4	11	14	19	22	28	31	36	43	8	11	15	18	22	24	28	34	1.6	1.8	2.0	2.1	2.3	1.8	2.0	2.2	2.5	2.7
TW B1	13	16	18	20	23	25	28	32	12	13	15	17	19	21	23	25	2.0	2.2	2.5	2.7	2.9	2.3	2.6	3.0	3.3	3.7
TW B2	12	15	18	21	26	29	33	39	10	12	15	17	20	23	26	30	1.7	1.9	2.1	2.3	2.5	1.9	2.2	2.5	2.7	2.9
TW B3	11	14	18	21	26	29	33	40	9	11	15	17	21	23	27	31	1.7	1.9	2.1	2.2	2.4	1.9	2.1	2.3	2.6	2.8
TW B4	10	14	17	20	25	28	32	39	8	11	14	16	20	23	26	31	1.7	1.9	2.1	2.3	2.4	1.9	2.1	2.4	2.6	2.8
TW C1	12	15	18	20	23	26	29	34	11	12	15	17	19	21	24	27	1.9	2.1	2.3	2.5	2.7	2.2	2.4	2.7	3.0	3.2
TW C2	11	14	17	20	24	27	31	36	9	11	14	16	20	22	25	29	1.8	2.0	2.2	2.4	2.5	2.0	2.3	2.5	2.7	2.9
TW C3	11	13	17	19	23	26	30	35	9	11	14	16	19	21	24	29	1.8	2.0	2.2	2.4	2.6	2.1	2.3	2.5	2.7	2.9
TW C4	11	13	16	18	22	24	28	33	10	12	14	16	19	21	23	27	2.0	2.2	2.4	2.6	2.7	2.2	2.4	2.6	2.9	3.0
TW D1	8	10	12	13	15	17	19	21	6	7	8	9	10	11	12	12	2.3	2.6	3.1	3.5	3.9	2.8	3.5	4.4	-	-
TW D2	13	17	22	27	32	37	42	50	10	14	18	22	27	30	35	41	1.5	1.6	1.8	2.0	2.1	1.6	1.8	2.0	2.2	2.4
TW D3	13	18	24	29	35	40	46	54	10	15	20	23	29	33	38	45	1.4	1.5	1.7	1.9	2.0	1.5	1.7	1.9	2.1	2.3
TW D4	15	21	28	34	42	47	54	64	12	17	23	28	35	40	47	55	1.2	1.4	1.5	1.7	1.8	1.3	1.5	1.7	1.8	2.0
TW D5	14	21	29	35	43	49	56	66	12	17	24	30	37	42	49	59	1.2	1.4	1.5	1.7	1.8	1.3	1.5	1.6	1.8	1.9
TW E1	16	22	28	33	41	46	52	61	13	18	23	28	34	38	44	52	1.3	1.4	1.6	1.8	1.9	1.4	1.6	1.8	1.9	2.1
TW E2	15	22	29	34	42	47	54	63	12	18	24	28	35	40	46	54	1.2	1.4	1.6	1.7	1.9	1.4	1.5	1.7	1.9	2.0
TW E3	19	27	36	43	52	58	66	75	15	23	31	37	46	51	59	68	1.1	1.2	1.4	1.6	1.7	1.2	1.3	1.5	1.6	1.8
TW E4	17	25	34	41	50	57	64	74	14	21	29	35	44	50	57	67	1.1	1.3	1.4	1.6	1.7	1.2	1.3	1.5	1.6	1.8
TW F1	12	14	17	19	21	23	26	30	11	12	14	15	17	18	20	22	2.1	2.4	2.6	2.9	3.1	2.4	2.8	3.2	3.7	4.2
TW F2	10	13	17	19	23	26	30	36	8	11	13	15	18	20	23	27	1.8	2.0	2.2	2.4	2.6	2.0	2.3	2.6	2.9	3.2
TW F3	9	12	16	19	24	27	31	38	7	9	12	15	18	21	24	29	1.7	1.9	2.1	2.3	2.4	1.9	2.2	2.4	2.7	2.9
TW F4	7	10	15	18	23	26	31	37	5	8	11	13	17	20	24	29	1.7	1.9	2.1	2.2	2.4	1.9	2.1	2.3	2.6	2.8
TW G1	10	12	15	17	20	23	26	30	9	10	12	14	16	18	20	23	2.0	2.3	2.5	2.7	2.9	2.3	2.6	3.0	3.3	3.7
TW G2	8	11	14	17	21	24	28	34	6	8	11	13	16	18	21	25	1.8	2.0	2.2	2.4	2.6	2.0	2.3	2.6	2.8	3.1
TW G3	6	9	13	16	20	23	27	33	4	7	10	12	15	18	21	26	1.8	2.0	2.2	2.4	2.5	2.0	2.2	2.5	2.7	2.9
TW G4	5	7	11	13	18	20	24	30	3	5	8	10	13	16	19	23	1.9	2.1	2.3	2.4	2.6	2.1	2.3	2.5	2.8	3.0
TW H1	7	10	13	15	18	21	24	29	6	8	10	11	14	16	18	22	2.0	2.2	2.5	2.7	2.8	2.3	2.6	2.8	3.1	3.4
TW H2	5	8	11	13	17	19	23	28	3	5	8	10	13	15	18	22	2.0	2.2	2.4	2.6	2.7	2.2	2.4	2.7	2.9	3.1
TW H3	4	6	9	11	15	17	21	26	2	4	7	9	11	13	16	20	2.0	2.2	2.4	2.6	2.8	2.3	2.5	2.7	3.0	3.2
TW H4	3	5	7	9	12	14	17	22	2	3	5	7	9	11	13	17	2.3	2.5	2.6	2.8	3.0	2.5	2.7	3.0	3.2	3.4
TW I1	6	8	9	11	13	14	16	18	4	5	6	7	8	9	9	10	2.4	2.8	3.2	3.7	4.1	2.9	3.7	4.7	-	-
TW I2	10	14	19	23	29	33	39	47	7	11	15	19	23	27	31	38	1.5	1.7	1.8	2.0	2.1	1.7	1.9	2.0	2.2	2.4
TW I3	11	16	21	26	32	37	43	51	8	12	17	21	26	30	35	42	1.4	1.6	1.8	1.9	2.1	1.5	1.7	1.9	2.1	2.3
TW I4	12	18	25	31	39	44	51	61	9	14	21	25	32	37	44	53	1.3	1.4	1.6	1.7	1.9	1.4	1.5	1.7	1.9	2.0
TW I5	12	19	26	32	41	46	54	63	9	15	22	27	35	40	47	56	1.2	1.4	1.5	1.7	1.8	1.3	1.5	1.6	1.8	1.9
TW J1	14	19	26	31	38	44	50	59	11	16	21	26	32	36	42	50	1.3	1.5	1.6	1.8	1.9	1.4	1.6	1.8	1.9	2.1
TW J2	13	19	26	31	39	44	51	61	10	15	21	26	33	38	44	52	1.3	1.4	1.6	1.7	1.9	1.4	1.5	1.7	1.9	2.0
TW J3	16	24	33	40	50	56	64	74	13	20	28	35	43	49	57	67	1.1	1.3	1.4	1.6	1.7	1.2	1.3	1.5	1.6	1.8
TW J4	15	23	32	38	48	54	62	72	11	19	27	33	42	48	55	65	1.1	1.3	1.4	1.6	1.7	1.2	1.3	1.5	1.7	1.8
DG A1	7	9	12	13	15	10	17	18	5	7	8	9	10	10	10	9	2.1	2.6	3.4	-	-	2.6	3.4	-	-	-
DG A2	10	13	17	21	25	28	32	37	8	11	14	16	20	22	26	30	1.7	1.9	2.1	2.4	2.7	1.9	2.1	2.4	2.8	3.2
DG A3	14	18	24	29	36	40	47	55	12	16	21	25	30	35	40	48	1.4	1.6	1.7	1.8	2.0	1.5	1.7	1.8	2.0	2.2
DG B1	7	9	12	13	16	17	19	21	5	7	8	9	10	11	11	11	2.1	2.6	2.9	3.6	4.7	2.6	3.3	4.6	-	-
DG B2	10	14	18	21	26	30	34	41	8	11	15	17	21	23	27	32	1.7	1.9	2.0	2.2	2.4	1.9	2.1	2.3	2.6	2.8
DG B3	14	19	25	30	37	42	50	59	12	16	22	26	32	36	42	51	1.4	1.5	1.6	1.7	1.9	1.5	1.6	1.8	1.9	2.0
DG C1	10	12	15	18	21	24	27	31	7	9	11	13	15	16	18	20	1.8	2.1	2.4	2.7	3.1	2.1	2.6	3.2	4.0	4.7
DG C2	13	17	22	25	31	36	41	49	10	14	18	21	26	29	33	40	1.5	1.7	1.8	2.0	2.1	1.7	1.9	2.1	2.2	2.4
DG C3	16	22	28	34	42	48	56	66	14	19	25	29	36	41	48	58	1.3	1.4	1.5	1.6	1.8	1.4	1.5	1.7	1.8	1.9
SS A1	16	19	23	26	31	34	38	43	14	17	21	23	26	29	32	36	1.6	1.8	2.1	2.3	2.4	1.7	2.0	2.3	2.6	2.9
SS A2	20	25	32	37	44	48	55	63	18	23	28	33	39	43	49	56	1.3	1.4	1.6	1.8	1.9	1.3	1.5	1.7	1.9	2.1
SS A3	14	17	20	23	26	28	31	36	13	15	18	20	22	24	26	28	1.8	2.1	2.4	2.6	2.9	2.0	2.3	2.8	3.3	3.8
SS A4	19	24	30	35	41	46	52	60	17	22	27	31	37	41	46	53	1.3	1.5	1.7	1.8	2.0	1.4	1.6	1.8	2.0	2.2
SS A5	16	19	22	25	28	30	33	38	15	17	20	22	24	26	28	31	1.8	2.1	2.3	2.6	2.8	2.0	2.3	2.7	3.1	3.4
SS A6	19	25	31	36	43	47	53	61	17	22	28	32	38	42	47	54	1.3	1.5	1.6	1.8	1.9	1.4	1.6	1.8	2.0	2.1
SS A7	15	17	19	21	23	25	27	30	14	15	17	18	20	21	22	24	2.1	2.4	2.8	3.2	3.6	2.3	2.8	3.4	4.3	-
SS A8	19	24	29	34	40	44	50	57	17	21	26	30	35	39	43	50	1.3	1.5	1.7	1.9	2.0	1.4	1.6	1.9	2.1	2.3
SS B1	13	16	20	22	26	29	33	38	12	14	17	19	22	24	27	31	1.7	2.0	2.2	2.4	2.6	1.9	2.2	2.5	2.8	3.1
SS B2	17	22	28	33	40	44	50	59	15	20	25	29	35	39	45	52	1.3	1.5	1.7	1.8	2.0	1.4	1.6	1.8	2.0	2.1
SS B3	12	15	18	20	23	25	28	32	11	13	15	17	19	21	23	26	1.9	2.2	2.4	2.7	2.9	2.1	2.4	2.9	3.3	3.8
SS B4	16	21	27	31	38	42	48	56	15	19	24	28	33	37	42	49	1.4	1.5	1.7	1.9	2.0	1.4	1.7	1.8	2.0	2.2
SS B5	13	16	18	20	23	25	28	32	12	14	16	18	20	21	23	26	2.0	2.3	2.6	2.8	3.1	2.2	2.6	3.0	3.4	3.8
SS B6	16	21	27	32	38	43	49	57	14	19	24	28	34	38	43	50	1.3	1.5	1.7	1.9	2.0	1.4	1.6	1.8	2.0	2.2
SS B7	12	14	16	17	20	21	23	26	11	12	14	15	17	18	19	20	2.2	2.6	3.0	3.3	3.7	2.5	3.0	3.6	4.3	-
SS B8	15	20	25	30	36	40	45	53	14	18	22	26	31	35	39	46	1.4	1.6	1.8	1.9	2.1	1.5	1.7	1.9	2.1	2.3
SS C1	10	14	18	21	25	28	32	38	8	10	13	16	19	21	24	28	1.7	1.9	2.1	2.3	2.5	1.9	2.2	2.5	2.9	3.2
SS C2	13	18	24	29	36	40	47	55	10	15	20	24	29	33	38	46	1.4	1.5	1.7	1.9	2.0	1.5	1.7	1.9	2.1	2.2
SS C3	11	13	16	18	21	24	27	32	9	11	13	15	17	19	21	24	1.9	2.2	2.4	2.7	2.8	2.2	2.6	2.9	3.2	3.6
SS C4	12	17	22	26	32	36	42	49	10	14	18	21	26	30	34	41	1.5	1.6	1.8	2.0	2.1	1.6	1.8	2.0	2.2	2.4
SS D1	17	22	27	30	36	39	44	51	15	19	23	26	30	33	37	42	1.4	1.6	1.8	2.0	2.2	1.5	1.8	2.1	2.4	2.6
SS D2	21	29	37	43	51	56	63	72	19	25	32	38	45	50	56	65	1.1	1.3	1.5	1.6	1.8	1.2	1.4	1.6	1.7	1.9
SS D3	17	21	25	28	32	35	40	45	15	18	22	24	28	30	33	37	1.6	1.8	2.0	2.2	2.4	1.7	2.0	2.3	2.6	2.9
SS D4	21	28	35	41	49	54	61	70	18	25	31	36	43	48	54	62	1.2	1.3	1.5	1.7	1.8	1.2	1.4	1.6	1.8	1.9
SS E1	14	17	21	25	29	32	36	42	12	14	18	20	23	26	28	32	1.6	1.8	2.1	2.3	2.5	1.7	2.1	2.4	2.8	3.2
SS E2	17	24	31	36	43	48	55	64	15	20	26	31	37	42	47	55	1.2	1.4	1.6	1.7	1.9	1.3	1.5	1.7	1.9	2.1
SS E3	14	17	20	22	26	28	31	36	12	15	17	19	21	23	25	28	1.8	2.1	2.3	2.6	2.8	2.0	2.3	2.7	3.1	3.5
SS E4	17	22	28	33	40	45	51	59	15	19	25	29	35	39	44	51	1.3	1.5	1.7	1.8	2.0	1.4	1.6	1.8	2.0	2.2

SIOUX CITY, IOWA

TYPE	SOLAR SAVINGS FRACTION (%) BASE TEMP: 55 F (4676 DD)								65 F (6957 DD)								CONSERVATION FACTOR 55 F					65 F				
LCR	100	70	50	40	30	25	20	15	100	70	50	40	30	25	20	15	100	50	30	20	15	100	50	30	20	15
WW A1	16	18	20	22	25	27	30	34	13	15	17	18	20	21	23	25	2.0	2.2	2.5	2.7	2.9	2.4	2.7	3.1	3.6	4.0
WW A2	15	19	23	27	32	35	40	47	11	14	18	21	24	27	31	36	1.5	1.7	1.9	2.1	2.2	1.8	2.0	2.3	2.5	2.7
WW A3	14	19	24	28	34	39	44	52	10	14	18	21	26	29	34	40	1.4	1.6	1.8	1.9	2.1	1.6	1.8	2.1	2.3	2.5
WW A4	14	19	25	29	36	40	46	55	10	14	18	22	27	31	35	42	1.4	1.5	1.7	1.9	2.0	1.5	1.8	2.0	2.2	2.4
WW A5	13	19	25	30	37	42	49	57	9	14	19	23	28	32	37	44	1.3	1.5	1.6	1.8	2.0	1.5	1.7	1.9	2.1	2.3
WW A6	13	19	26	31	38	43	50	59	9	13	19	23	29	33	38	46	1.3	1.5	1.6	1.8	1.9	1.4	1.6	1.9	2.0	2.2
WW B1	11	13	16	18	21	23	25	29	8	10	11	12	14	14	15	16	2.0	2.3	2.6	2.9	3.2	2.5	3.1	3.9	-	-
WW B2	14	20	27	32	40	45	51	60	11	15	21	25	31	36	41	49	1.3	1.4	1.6	1.8	1.9	1.4	1.6	1.8	2.0	2.1
WW B3	18	25	32	38	47	53	60	69	13	19	26	31	38	43	49	58	1.2	1.3	1.5	1.6	1.8	1.3	1.5	1.6	1.8	2.0
WW B4	18	26	35	42	51	57	65	75	13	20	28	34	43	49	56	66	1.1	1.2	1.4	1.6	1.7	1.2	1.3	1.5	1.7	1.8
WW B5	17	25	35	42	51	58	66	75	13	20	28	35	44	50	57	67	1.1	1.2	1.4	1.5	1.7	1.2	1.3	1.5	1.6	1.8
WW C1	19	26	35	41	50	56	63	72	15	21	28	34	41	46	53	62	1.1	1.3	1.4	1.6	1.7	1.2	1.4	1.6	1.7	1.9
WW C2	18	25	34	40	49	55	63	72	13	20	27	33	41	47	54	63	1.1	1.3	1.4	1.6	1.7	1.2	1.4	1.6	1.7	1.9
WW C3	21	31	42	50	61	67	75	84	17	26	36	43	53	59	67	77	1.0	1.1	1.3	1.5	1.6	1.1	1.2	1.4	1.5	1.7
WW C4	19	29	39	47	57	64	72	81	15	23	33	40	50	56	64	74	1.0	1.2	1.3	1.5	1.6	1.1	1.2	1.4	1.6	1.7
TW A1	15	18	20	22	25	28	31	35	13	15	17	18	20	22	24	27	2.0	2.2	2.4	2.7	2.8	2.3	2.6	3.0	3.4	3.7
TW A2	14	18	22	25	29	33	37	44	12	14	17	20	23	26	29	33	1.6	1.8	2.0	2.2	2.3	1.9	2.1	2.4	2.6	2.8
TW A3	13	17	22	26	31	35	40	47	10	14	17	20	24	27	31	36	1.5	1.7	1.9	2.1	2.2	1.7	2.0	2.2	2.4	2.6
TW A4	12	17	22	26	31	36	41	49	9	13	17	20	24	28	32	38	1.5	1.7	1.8	2.0	2.1	1.7	1.9	2.1	2.3	2.5
TW B1	15	17	20	22	26	29	32	37	12	14	17	18	21	23	25	29	1.9	2.1	2.3	2.5	2.6	2.2	2.4	2.7	3.0	3.3
TW B2	13	17	21	24	29	32	37	44	10	13	16	19	23	25	29	34	1.6	1.8	2.0	2.2	2.3	1.8	2.1	2.3	2.5	2.7
TW B3	12	16	21	24	29	33	38	45	10	13	16	19	23	26	30	35	1.6	1.8	1.9	2.1	2.2	1.8	2.0	2.2	2.4	2.6
TW B4	12	15	20	23	28	32	37	44	9	12	16	18	22	25	29	35	1.6	1.8	2.0	2.1	2.3	1.8	2.0	2.2	2.4	2.6
TW C1	13	16	20	22	26	29	33	39	11	13	16	18	21	23	26	30	1.8	2.0	2.2	2.4	2.5	2.0	2.3	2.5	2.8	3.0
TW C2	12	15	19	22	27	30	34	41	10	12	15	18	21	24	28	33	1.7	1.9	2.1	2.2	2.4	1.9	2.1	2.3	2.5	2.7
TW C3	12	15	19	22	26	29	33	40	10	12	15	17	21	23	27	32	1.7	1.9	2.1	2.3	2.4	2.0	2.2	2.4	2.6	2.7
TW C4	12	15	18	20	24	27	31	36	10	12	15	17	20	22	25	30	1.9	2.1	2.2	2.4	2.5	2.1	2.3	2.5	2.7	2.9
TW D1	9	11	14	16	19	21	23	27	7	8	10	11	13	14	15	16	2.0	2.4	2.7	2.9	3.2	2.5	3.0	3.6	4.4	-
TW D2	15	20	25	30	36	41	47	55	11	15	20	24	29	33	38	45	1.4	1.5	1.7	1.9	2.0	1.5	1.7	1.9	2.1	2.2
TW D3	15	21	27	32	39	44	51	60	12	16	22	26	32	36	41	49	1.3	1.5	1.6	1.8	1.9	1.4	1.6	1.8	2.0	2.1
TW D4	17	24	31	37	46	52	59	68	13	19	26	31	38	43	50	59	1.2	1.3	1.5	1.6	1.8	1.3	1.4	1.6	1.8	1.9
TW D5	16	24	32	38	47	53	60	70	13	19	26	32	40	45	52	62	1.2	1.3	1.5	1.6	1.7	1.3	1.4	1.6	1.7	1.9
TW E1	18	24	32	37	45	50	57	66	14	19	26	30	37	42	48	56	1.2	1.4	1.5	1.7	1.8	1.3	1.5	1.7	1.9	2.0
TW E2	18	24	32	38	46	51	59	68	14	19	26	31	38	43	50	58	1.2	1.3	1.5	1.7	1.8	1.3	1.5	1.6	1.8	1.9
TW E3	21	30	40	47	56	63	70	79	17	25	33	40	49	55	62	72	1.0	1.2	1.3	1.5	1.6	1.1	1.3	1.4	1.6	1.7
TW E4	20	28	38	45	54	61	69	78	15	23	32	38	47	53	61	70	1.1	1.2	1.4	1.5	1.7	1.1	1.3	1.4	1.6	1.7
TW F1	13	16	19	21	24	27	30	34	11	13	15	17	19	21	23	26	1.9	2.2	2.4	2.6	2.8	2.3	2.6	2.9	3.3	3.6
TW F2	12	15	19	22	27	30	35	41	9	12	15	17	21	23	26	31	1.7	1.9	2.1	2.2	2.4	1.9	2.2	2.4	2.7	2.9
TW F3	11	14	19	22	27	31	36	43	8	11	14	17	21	24	27	33	1.6	1.8	1.9	2.1	2.3	1.8	2.0	2.3	2.5	2.7
TW F4	9	13	17	21	26	30	35	43	6	9	13	15	20	23	27	33	1.6	1.8	1.9	2.1	2.2	1.8	2.0	2.2	2.4	2.6
TW G1	11	14	17	20	23	26	29	35	9	11	14	15	18	20	22	26	1.9	2.1	2.3	2.5	2.7	2.2	2.4	2.7	3.0	3.3
TW G2	9	13	17	20	24	28	32	39	7	9	12	15	18	21	24	29	1.7	1.9	2.1	2.2	2.4	1.9	2.2	2.4	2.6	2.8
TW G3	8	11	15	18	23	27	31	38	5	8	11	14	17	20	24	29	1.7	1.9	2.0	2.2	2.4	1.9	2.1	2.3	2.6	2.7
TW G4	6	9	13	16	20	24	28	35	4	6	9	12	15	18	21	27	1.8	1.9	2.1	2.3	2.4	2.0	2.2	2.4	2.6	2.8
TW H1	8	11	15	17	21	24	27	33	6	8	11	13	16	18	21	25	1.9	2.1	2.3	2.5	2.6	2.1	2.4	2.6	2.9	3.1
TW H2	6	9	13	15	19	22	27	33	4	6	9	11	15	17	20	25	1.8	2.0	2.2	2.4	2.5	2.1	2.3	2.5	2.7	2.9
TW H3	5	7	11	13	17	20	24	30	3	5	8	10	13	15	18	23	1.9	2.1	2.3	2.5	2.6	2.2	2.4	2.6	2.8	3.0
TW H4	3	6	8	11	14	16	20	25	2	4	6	8	10	12	15	19	2.1	2.3	2.5	2.7	2.8	2.4	2.6	2.8	3.0	3.2
TW I1	7	9	12	13	16	18	20	24	5	6	8	9	10	11	12	14	2.1	2.5	2.8	3.1	3.3	2.6	3.1	3.8	4.7	-
TW I2	12	16	22	26	33	37	44	52	8	12	17	21	26	30	35	42	1.4	1.6	1.7	1.9	2.0	1.6	1.8	1.9	2.1	2.3
TW I3	13	18	24	29	36	41	47	56	9	14	19	23	29	33	39	46	1.3	1.5	1.7	1.8	2.0	1.5	1.7	1.8	2.0	2.2
TW I4	14	21	28	34	43	48	56	65	10	16	23	28	35	40	47	56	1.2	1.4	1.5	1.7	1.8	1.3	1.5	1.6	1.8	1.9
TW I5	14	21	29	36	44	50	58	68	10	17	24	29	37	43	50	59	1.2	1.3	1.5	1.6	1.8	1.3	1.4	1.6	1.7	1.9
TW J1	16	22	29	35	43	48	55	64	12	17	23	28	35	40	46	54	1.2	1.4	1.5	1.7	1.8	1.3	1.5	1.7	1.9	2.0
TW J2	15	22	29	35	43	49	56	66	11	17	23	28	36	41	47	56	1.2	1.4	1.5	1.7	1.8	1.3	1.5	1.7	1.8	2.0
TW J3	18	27	37	44	54	60	68	78	14	22	31	37	46	52	60	70	1.1	1.2	1.4	1.5	1.7	1.1	1.3	1.4	1.6	1.7
TW J4	17	25	35	42	52	58	66	76	13	20	29	35	44	51	58	68	1.1	1.2	1.4	1.5	1.7	1.2	1.3	1.5	1.6	1.8
DG A1	9	11	14	16	19	20	22	23	6	8	10	11	12	13	13	13	1.9	2.3	2.9	4.1	-	2.3	2.9	4.1	-	-
DG A2	12	15	20	23	28	32	36	42	9	12	16	18	22	25	28	33	1.6	1.8	2.0	2.3	2.5	1.8	2.0	2.2	2.6	2.9
DG A3	15	20	27	32	39	44	51	60	13	17	22	26	33	37	43	51	1.4	1.5	1.6	1.8	1.9	1.5	1.6	1.7	1.9	2.1
DG B1	9	11	14	16	20	22	25	28	6	8	10	11	13	14	15	16	1.9	2.2	2.5	3.0	3.6	2.3	2.8	3.7	4.7	-
DG B2	12	16	20	24	30	34	39	46	9	12	16	19	23	26	30	36	1.6	1.7	1.9	2.0	2.2	1.7	2.0	2.2	2.4	2.6
DG B3	16	21	27	33	41	47	54	64	13	18	23	28	34	39	46	55	1.3	1.4	1.5	1.7	1.8	1.4	1.6	1.7	1.8	2.0
DG C1	11	15	18	21	26	29	33	38	8	11	13	15	18	20	22	25	1.7	1.9	2.1	2.4	2.8	1.9	2.3	2.8	3.2	3.9
DG C2	14	19	24	29	35	40	47	55	11	15	20	23	28	32	37	44	1.4	1.6	1.7	1.9	2.0	1.6	1.8	1.9	2.1	2.3
DG C3	18	24	31	37	46	53	61	71	15	20	26	31	39	44	52	61	1.3	1.4	1.4	1.6	1.7	1.3	1.5	1.6	1.7	1.8
SS A1	17	21	26	29	34	37	42	49	15	18	22	25	29	31	35	40	1.5	1.7	1.9	2.1	2.3	1.7	1.9	2.2	2.4	2.7
SS A2	22	28	35	40	47	52	59	67	19	24	30	35	41	46	52	59	1.2	1.4	1.5	1.7	1.8	1.3	1.5	1.7	1.8	2.0
SS A3	16	19	23	25	29	32	36	41	14	16	19	21	24	26	29	32	1.7	1.9	2.2	2.4	2.5	1.8	2.2	2.5	2.9	3.3
SS A4	21	27	33	38	45	50	56	65	18	23	29	33	39	43	49	56	1.2	1.4	1.6	1.7	1.9	1.3	1.5	1.7	1.9	2.1
SS A5	18	21	24	27	31	34	37	43	16	18	21	23	26	28	31	35	1.7	1.9	2.1	2.3	2.5	1.8	2.2	2.5	2.8	3.1
SS A6	21	27	34	39	46	51	58	66	18	24	30	34	40	45	50	58	1.2	1.4	1.6	1.7	1.9	1.3	1.5	1.7	1.9	2.0
SS A7	16	18	21	23	26	28	30	34	14	16	18	20	22	23	25	27	1.9	2.2	2.5	2.8	3.0	2.2	2.6	3.1	3.7	4.2
SS A8	20	26	32	37	44	48	54	62	18	22	28	32	38	41	47	54	1.3	1.5	1.6	1.8	1.9	1.4	1.6	1.8	2.0	2.1
SS B1	15	18	22	25	29	32	37	43	13	15	18	21	24	27	30	34	1.6	1.9	2.1	2.2	2.4	1.8	2.1	2.4	2.6	2.9
SS B2	19	24	31	36	43	48	54	63	16	21	27	31	37	42	47	55	1.3	1.4	1.6	1.7	1.9	1.4	1.5	1.7	1.9	2.0
SS B3	13	16	20	22	26	29	32	38	11	14	16	18	21	23	26	29	1.7	2.0	2.2	2.4	2.6	1.9	2.3	2.6	3.0	3.3
SS B4	18	23	30	34	41	46	52	61	16	20	26	30	36	40	45	53	1.3	1.5	1.6	1.8	1.9	1.4	1.6	1.8	2.0	2.1
SS B5	14	17	20	22	26	28	31	36	13	15	17	19	21	23	25	29	1.9	2.1	2.4	2.6	2.8	2.1	2.4	2.7	3.1	3.4
SS B6	18	23	30	35	42	46	53	61	15	20	26	30	36	40	46	53	1.3	1.5	1.6	1.8	1.9	1.4	1.6	1.8	1.9	2.1
SS B7	13	15	18	19	22	24	26	30	12	13	15	16	18	20	21	23	2.1	2.4	2.7	3.0	3.2	2.3	2.7	3.2	3.8	4.3
SS B8	17	22	28	32	39	44	50	58	14	19	24	28	33	37	42	49	1.3	1.5	1.7	1.8	2.0	1.4	1.6	1.8	2.0	2.2
SS C1	12	16	21	24	29	33	38	45	9	12	15	18	22	24	28	32	1.5	1.8	1.9	2.1	2.3	1.8	2.0	2.3	2.6	2.8
SS C2	15	21	28	33	40	45	52	61	11	16	22	26	32	36	42	50	1.3	1.4	1.6	1.8	1.9	1.4	1.6	1.8	2.0	2.1
SS C3	12	15	18	21	25	27	31	36	10	12	14	16	19	21	24	27	1.8	2.0	2.2	2.4	2.6	2.1	2.4	2.6	2.9	3.2
SS C4	14	19	25	29	36	40	46	55	11	15	20	24	29	33	38	45	1.4	1.6	1.7	1.9	2.0	1.5	1.7	1.9	2.1	2.2
SS D1	19	24	30	34	40	44	50	57	16	20	25	28	33	36	41	46	1.3	1.5	1.7	1.9	2.1	1.5	1.7	2.0	2.2	2.4
SS D2	24	31	40	46	55	61	68	76	20	27	35	40	48	53	60	68	1.1	1.2	1.4	1.6	1.7	1.2	1.3	1.5	1.7	1.8
SS D3	19	23	27	31	36	40	44	51	16	20	23	26	30	33	36	41	1.5	1.7	1.9	2.1	2.2	1.6	1.9	2.1	2.4	2.6
SS D4	23	30	39	45	53	59	66	74	20	26	33	39	46	51	58	66	1.1	1.3	1.4	1.6	1.7	1.2	1.4	1.5	1.7	1.9
SS E1	15	20	24	28	33	37	42	49	13	16	20	22	26	29	32	37	1.5	1.7	1.9	2.1	2.2	1.6	1.9	2.2	2.5	2.8
SS E2	20	26	34	40	48	53	60	69	16	22	28	33	40	45	51	59	1.2	1.3	1.5	1.7	1.8	1.3	1.5	1.6	1.8	2.0
SS E3	15	18	22	25	29	32	35	41	13	16	18	21	24	26	28	32	1.7	1.9	2.1	2.3	2.5	1.9	2.2	2.5	2.8	3.1
SS E4	19	25	31	37	44	49	55	64	16	21	27	31	37	42	47	55	1.2	1.4	1.6	1.7	1.9	1.3	1.5	1.7	1.9	2.1

DODGE CITY, KANSAS

TYPE	SOLAR SAVINGS FRACTION (%) — BASE TEMP: 55 F (3017 DD) LCR 100	70	50	40	30	25	20	15	65 F (5053 DD) 100	70	50	40	30	25	20	15	CONSERVATION FACTOR — 55 F 100	50	30	20	15	65 F 100	50	30	20	15
WW A1	25	30	36	41	48	53	59	68	19	23	27	30	35	39	44	51	1.3	1.4	1.6	1.7	1.8	1.6	1.7	1.9	2.0	2.2
WW A2	28	36	45	52	61	67	74	82	20	26	33	38	46	51	58	67	1.0	1.2	1.3	1.5	1.6	1.2	1.4	1.5	1.7	1.8
WW A3	30	39	49	56	66	72	79	86	21	27	35	41	50	55	63	72	1.0	1.1	1.3	1.4	1.6	1.2	1.3	1.5	1.6	1.7
WW A4	30	40	51	58	68	74	81	88	21	28	37	43	52	58	65	75	1.0	1.1	1.2	1.4	1.6	1.1	1.3	1.4	1.6	1.7
WW A5	31	42	53	61	71	77	83	90	21	29	38	45	54	60	68	77	0.9	1.1	1.2	1.4	1.5	1.1	1.2	1.4	1.5	1.7
WW A6	32	43	54	62	72	78	84	91	21	30	39	46	56	62	69	79	0.9	1.1	1.2	1.4	1.5	1.1	1.2	1.4	1.5	1.7
WW B1	23	30	37	43	51	56	63	72	16	20	25	29	35	39	45	52	1.2	1.3	1.5	1.6	1.8	1.5	1.6	1.8	2.0	2.1
WW B2	31	41	52	60	70	76	82	90	21	29	38	45	54	61	68	78	0.9	1.1	1.2	1.4	1.5	1.1	1.2	1.4	1.5	1.7
WW B3	37	48	60	68	77	83	89	94	26	35	46	53	63	69	77	85	0.9	1.0	1.2	1.3	1.5	1.0	1.1	1.3	1.4	1.6
WW B4	37	49	61	70	79	85	90	95	26	36	47	55	66	72	80	88	0.8	1.0	1.2	1.3	1.5	1.0	1.1	1.3	1.4	1.6
WW B5	35	48	60	68	78	84	90	95	25	35	46	55	65	72	79	87	0.9	1.0	1.2	1.3	1.5	1.0	1.1	1.3	1.4	1.6
WW C1	39	50	62	70	79	85	90	95	28	37	48	55	65	72	79	87	0.8	1.0	1.2	1.3	1.5	1.0	1.1	1.3	1.4	1.6
WW C2	36	48	60	68	78	83	89	95	26	36	46	54	64	71	78	86	0.9	1.0	1.2	1.3	1.5	1.0	1.1	1.3	1.4	1.6
WW C3	42	56	69	77	85	90	94	98	31	43	55	64	74	80	87	93	0.8	0.9	1.1	1.3	1.4	0.9	1.0	1.2	1.3	1.5
WW C4	39	52	65	73	83	88	93	97	28	39	52	60	71	77	84	91	0.8	1.0	1.1	1.3	1.4	0.9	1.0	1.2	1.4	1.5
TW A1	24	29	35	40	47	52	58	67	19	22	27	30	35	39	44	51	1.3	1.4	1.6	1.7	1.8	1.6	1.7	1.9	2.0	2.2
TW A2	26	33	41	47	56	62	69	77	19	24	30	35	42	47	53	61	1.1	1.3	1.4	1.5	1.7	1.3	1.5	1.6	1.8	1.9
TW A3	27	34	43	50	59	65	72	81	19	25	32	37	44	50	57	66	1.1	1.2	1.3	1.5	1.6	1.3	1.4	1.5	1.7	1.8
TW A4	26	34	44	51	60	66	73	82	18	24	31	37	45	51	58	67	1.0	1.2	1.3	1.5	1.6	1.2	1.4	1.5	1.7	1.8
TW B1	24	29	36	41	48	53	60	69	18	22	27	30	36	40	45	53	1.3	1.4	1.5	1.7	1.8	1.5	1.7	1.8	2.0	2.1
TW B2	24	32	40	46	54	60	67	76	18	23	29	34	41	46	52	61	1.1	1.3	1.4	1.6	1.7	1.3	1.5	1.6	1.8	1.9
TW B3	24	31	40	46	55	61	69	78	17	23	29	34	41	46	53	62	1.1	1.2	1.4	1.5	1.7	1.3	1.4	1.6	1.7	1.9
TW B4	23	30	38	44	53	58	66	75	16	21	27	32	39	44	51	60	1.1	1.3	1.4	1.6	1.7	1.3	1.5	1.6	1.8	1.9
TW C1	22	28	35	40	48	53	60	69	17	21	26	30	36	40	46	54	1.2	1.4	1.5	1.7	1.8	1.5	1.6	1.8	1.9	2.0
TW C2	22	28	36	41	49	55	62	71	16	21	26	31	37	42	48	56	1.2	1.3	1.5	1.6	1.7	1.4	1.6	1.7	1.8	2.0
TW C3	21	27	34	39	47	53	60	69	15	20	25	29	35	40	46	54	1.2	1.4	1.5	1.6	1.8	1.5	1.6	1.7	1.9	2.0
TW C4	19	24	30	35	42	47	54	63	15	18	23	27	32	36	41	49	1.3	1.5	1.6	1.7	1.9	1.6	1.7	1.9	2.0	2.1
TW D1	20	25	32	37	45	50	57	66	13	17	22	25	31	34	40	47	1.3	1.4	1.5	1.7	1.8	1.6	1.7	1.9	2.1	2.2
TW D2	29	38	47	55	64	70	77	86	20	27	35	41	50	55	63	72	1.0	1.1	1.3	1.4	1.6	1.2	1.3	1.4	1.6	1.7
TW D3	31	41	52	59	69	75	82	89	22	30	39	45	54	60	68	77	0.9	1.1	1.2	1.4	1.5	1.1	1.2	1.4	1.5	1.7
TW D4	33	44	56	63	73	79	86	92	24	33	43	50	60	66	74	82	0.9	1.0	1.2	1.4	1.5	1.0	1.2	1.3	1.5	1.6
TW D5	32	43	55	63	73	79	86	92	23	32	42	50	60	66	74	83	0.9	1.0	1.2	1.4	1.5	1.0	1.2	1.3	1.5	1.6
TW E1	35	46	57	64	74	80	86	92	25	34	43	50	60	66	73	82	0.9	1.0	1.2	1.4	1.5	1.0	1.2	1.3	1.5	1.6
TW E2	34	45	56	64	74	80	86	92	25	33	43	50	60	66	74	83	0.9	1.0	1.2	1.4	1.5	1.0	1.2	1.3	1.5	1.6
TW E3	41	53	65	73	82	87	92	96	30	41	52	60	70	76	83	90	0.8	1.0	1.1	1.3	1.5	0.9	1.1	1.2	1.4	1.5
TW E4	38	50	62	70	80	85	91	96	27	38	49	57	68	74	81	89	0.8	1.0	1.1	1.3	1.5	0.9	1.1	1.2	1.4	1.5
TW F1	23	28	34	39	47	52	58	67	17	21	25	29	34	38	43	51	1.3	1.4	1.5	1.7	1.8	1.5	1.7	1.8	2.0	2.1
TW F2	23	30	39	45	53	59	66	75	17	22	28	32	39	44	50	59	1.1	1.3	1.4	1.6	1.7	1.4	1.5	1.6	1.8	1.9
TW F3	23	31	39	46	55	61	69	78	16	21	28	33	40	46	52	62	1.1	1.2	1.4	1.5	1.7	1.3	1.4	1.6	1.7	1.9
TW F4	21	29	38	44	54	60	67	77	14	19	26	31	39	44	51	61	1.1	1.2	1.4	1.5	1.7	1.3	1.4	1.6	1.7	1.9
TW G1	21	26	33	38	46	51	58	67	15	19	24	28	33	38	43	51	1.3	1.4	1.5	1.7	1.8	1.5	1.7	1.8	2.0	2.1
TW G2	20	27	35	41	50	55	63	72	14	19	24	29	36	41	47	56	1.2	1.3	1.4	1.6	1.7	1.4	1.5	1.7	1.8	2.0
TW G3	18	25	33	39	48	54	62	71	12	17	23	28	35	40	46	55	1.2	1.3	1.4	1.6	1.7	1.4	1.5	1.7	1.8	2.0
TW G4	15	21	29	34	43	48	56	66	9	14	19	24	30	35	41	50	1.2	1.4	1.5	1.7	1.8	1.5	1.6	1.8	1.9	2.0
TW H1	17	23	30	35	43	48	55	65	12	16	21	25	31	35	41	49	1.3	1.4	1.5	1.7	1.8	1.5	1.7	1.8	2.0	2.1
TW H2	15	21	28	33	41	47	54	63	10	14	19	23	29	33	39	48	1.3	1.4	1.6	1.7	1.8	1.5	1.7	1.8	2.0	2.1
TW H3	13	18	24	29	37	42	49	58	8	12	16	20	26	30	35	43	1.4	1.5	1.6	1.8	1.9	1.6	1.8	1.9	2.0	2.2
TW H4	9	14	19	23	29	34	40	49	6	9	13	16	20	24	29	36	1.5	1.7	1.8	1.9	2.0	1.8	1.9	2.1	2.2	2.3
TW I1	17	22	29	34	41	46	53	62	11	15	19	22	27	31	36	43	1.3	1.5	1.6	1.7	1.9	1.6	1.8	2.0	2.1	2.3
TW I2	25	34	44	51	61	67	74	83	17	24	32	38	46	52	59	69	1.0	1.2	1.3	1.5	1.6	1.2	1.3	1.5	1.6	1.8
TW I3	27	37	47	55	65	71	78	86	19	26	35	41	50	56	64	73	1.0	1.1	1.3	1.4	1.6	1.1	1.3	1.4	1.6	1.7
TW I4	30	41	52	60	70	76	83	90	21	29	39	46	56	63	70	80	0.9	1.1	1.2	1.4	1.5	1.1	1.2	1.3	1.5	1.6
TW I5	30	41	53	61	71	77	84	91	21	30	40	47	57	64	72	81	0.9	1.1	1.2	1.4	1.5	1.0	1.2	1.3	1.5	1.6
TW J1	32	43	54	62	72	78	84	91	23	31	41	48	57	64	71	80	0.9	1.1	1.2	1.4	1.5	1.1	1.2	1.3	1.5	1.6
TW J2	31	42	53	61	71	77	84	91	22	30	40	47	57	63	71	80	0.9	1.1	1.2	1.4	1.5	1.0	1.2	1.3	1.5	1.6
TW J3	37	50	62	70	80	85	91	96	27	37	49	57	68	74	81	89	0.8	1.0	1.1	1.3	1.5	0.9	1.1	1.2	1.4	1.5
TW J4	35	47	59	68	78	83	89	95	24	35	46	54	65	71	79	87	0.9	1.0	1.2	1.3	1.5	1.0	1.1	1.3	1.4	1.6
DG A1	21	20	35	40	47	51	55	61	14	18	23	27	31	34	38	42	1.2	1.4	1.6	2.0	2.3	1.5	1.7	2.0	2.5	3.0
DG A2	23	31	40	46	55	60	66	73	16	22	28	33	41	45	51	58	1.1	1.2	1.4	1.6	1.8	1.3	1.4	1.6	1.9	2.1
DG A3	27	37	48	56	66	72	78	85	20	27	36	42	52	58	65	74	1.0	1.1	1.3	1.5	1.7	1.2	1.2	1.4	1.6	1.8
DG B1	22	29	38	44	53	58	64	71	14	18	24	28	35	39	44	50	1.1	1.2	1.4	1.7	2.0	1.4	1.6	1.7	2.0	2.4
DG B2	24	32	42	50	61	67	74	81	17	22	29	35	44	49	57	65	1.1	1.1	1.3	1.5	1.7	1.3	1.4	1.5	1.7	1.9
DG B3	28	38	50	59	70	77	83	90	20	28	37	44	55	62	70	79	1.0	1.0	1.2	1.4	1.6	1.1	1.2	1.3	1.5	1.7
DG C1	25	34	45	53	63	69	75	81	17	22	29	34	42	48	55	62	1.0	1.1	1.3	1.6	1.8	1.3	1.4	1.5	1.7	2.0
DG C2	27	37	49	57	68	74	81	88	19	26	34	41	51	57	65	74	1.0	1.1	1.2	1.4	1.6	1.2	1.3	1.4	1.6	1.8
DG C3	32	43	56	66	77	83	89	94	23	31	41	50	61	69	77	86	0.9	1.0	1.1	1.3	1.5	1.1	1.1	1.2	1.4	1.6
SS A1	31	38	47	53	61	67	73	81	24	30	37	42	49	54	60	68	1.1	1.2	1.4	1.5	1.7	1.2	1.4	1.5	1.7	1.8
SS A2	38	48	58	65	74	80	86	92	30	38	47	54	63	68	75	83	0.9	1.1	1.2	1.4	1.5	1.0	1.2	1.3	1.5	1.6
SS A3	28	35	43	49	57	62	69	77	22	27	33	38	44	49	54	62	1.1	1.3	1.4	1.6	1.7	1.3	1.5	1.6	1.8	1.9
SS A4	37	47	57	64	73	79	85	91	29	37	46	52	61	67	73	81	0.9	1.1	1.2	1.4	1.5	1.0	1.2	1.4	1.5	1.7
SS A5	29	35	43	48	56	61	68	76	23	28	34	38	44	49	54	62	1.1	1.3	1.4	1.6	1.7	1.3	1.5	1.7	1.8	2.0
SS A6	38	48	58	65	74	80	85	92	30	38	47	53	62	68	74	82	0.9	1.1	1.2	1.4	1.5	1.0	1.2	1.3	1.5	1.6
SS A7	26	31	38	42	49	54	60	68	21	25	30	33	38	42	47	53	1.2	1.4	1.5	1.7	1.8	1.5	1.7	1.8	2.0	2.2
SS A8	36	46	56	63	72	77	84	90	29	36	45	51	60	65	72	80	0.9	1.1	1.2	1.4	1.6	1.1	1.2	1.4	1.5	1.7
SS B1	26	33	41	47	55	61	67	76	20	25	31	36	43	47	53	62	1.1	1.3	1.4	1.6	1.7	1.3	1.5	1.6	1.8	1.9
SS B2	33	43	53	60	69	75	81	89	26	33	42	48	57	63	70	78	1.0	1.1	1.3	1.4	1.6	1.1	1.2	1.4	1.5	1.7
SS B3	25	31	38	44	52	57	64	73	19	24	29	33	40	44	50	57	1.2	1.3	1.5	1.6	1.8	1.4	1.5	1.7	1.9	2.0
SS B4	32	41	51	58	68	73	80	87	25	33	41	47	55	61	68	77	1.0	1.1	1.3	1.4	1.6	1.1	1.2	1.4	1.6	1.7
SS B5	24	29	36	41	48	53	59	68	19	23	28	31	37	41	46	53	1.2	1.4	1.5	1.7	1.8	1.5	1.6	1.8	2.0	2.1
SS B6	33	42	52	59	68	74	81	88	25	33	41	47	56	62	69	77	1.0	1.1	1.3	1.4	1.6	1.1	1.2	1.4	1.6	1.7
SS B7	22	26	32	37	43	48	54	62	17	21	25	28	33	36	41	47	1.3	1.5	1.6	1.8	1.9	1.6	1.8	2.0	2.1	2.3
SS B8	31	40	50	57	66	72	79	86	24	31	39	45	54	59	66	75	1.0	1.1	1.3	1.4	1.6	1.1	1.3	1.4	1.6	1.7
SS C1	26	35	44	50	60	66	73	82	18	24	31	36	44	49	56	65	1.0	1.2	1.3	1.5	1.6	1.3	1.4	1.5	1.7	1.8
SS C2	31	41	52	60	69	75	82	89	22	30	39	45	55	61	68	77	0.9	1.1	1.2	1.4	1.5	1.1	1.2	1.4	1.5	1.7
SS C3	22	28	35	40	48	54	61	70	16	20	25	29	35	40	45	53	1.2	1.4	1.5	1.6	1.8	1.5	1.6	1.8	1.9	2.0
SS C4	28	37	46	53	63	69	76	85	20	27	34	40	49	54	62	71	1.0	1.1	1.3	1.5	1.6	1.2	1.3	1.5	1.6	1.7
SS D1	36	45	55	62	71	77	83	89	28	35	43	49	58	63	70	78	0.9	1.1	1.2	1.4	1.6	1.1	1.2	1.4	1.6	1.7
SS D2	44	55	67	74	83	87	92	96	34	44	55	62	71	77	83	90	0.8	1.0	1.1	1.3	1.5	0.9	1.1	1.2	1.4	1.5
SS D3	33	41	50	56	65	70	77	84	26	32	39	44	52	57	63	71	1.0	1.2	1.3	1.5	1.6	1.2	1.3	1.5	1.7	1.8
SS D4	42	53	65	72	81	86	91	95	33	43	53	60	69	75	81	88	0.8	1.0	1.2	1.3	1.5	0.9	1.1	1.3	1.4	1.6
SS E1	31	40	50	56	65	71	78	85	23	30	37	43	51	56	63	71	1.0	1.1	1.3	1.5	1.6	1.2	1.3	1.5	1.6	1.8
SS E2	38	49	60	68	77	83	88	94	29	38	48	55	64	70	77	85	0.9	1.0	1.2	1.4	1.5	1.0	1.1	1.3	1.5	1.6
SS E3	27	34	41	47	55	60	67	76	21	26	32	36	42	47	53	61	1.1	1.3	1.4	1.6	1.7	1.3	1.5	1.7	1.8	2.0
SS E4	35	45	56	63	73	78	84	91	27	35	44	51	60	65	72	81	0.9	1.1	1.2	1.4	1.5	1.0	1.2	1.3	1.5	1.6

GOODLAND, KANSAS

TYPE	SOLAR SAVINGS FRACTION (%) BASE TEMP: 55 F (3811 DD)								65 F (6127 DD)								CONSERVATION FACTOR 55 F					65 F				
LCR	100	70	50	40	30	25	20	15	100	70	50	40	30	25	20	15	100	50	30	20	15	100	50	30	20	15
WW A1	23	27	33	37	43	48	54	62	18	21	25	28	32	36	40	47	1.4	1.5	1.7	1.8	1.9	1.7	1.8	2.0	2.1	2.3
WW A2	25	33	41	47	55	61	68	77	18	24	30	35	42	47	53	62	1.1	1.3	1.4	1.5	1.7	1.3	1.5	1.6	1.8	1.9
WW A3	26	35	44	51	60	66	73	82	18	25	32	38	46	51	58	67	1.0	1.2	1.3	1.5	1.6	1.2	1.4	1.5	1:7	1.8
WW A4	27	36	46	53	62	68	76	84	18	25	33	39	48	53	61	70	1.0	1.2	1.3	1.5	1.6	1.2	1.3	1.5	1.6	1.8
WW A5	28	37	47	55	65	71	78	86	18	26	34	41	50	56	63	73	1.0	1.1	1.3	1.4	1.6	1.1	1.3	1.4	1.6	1.7
WW A6	28	38	48	56	66	72	80	87	18	26	35	42	51	57	65	74	1.0	1.1	1.3	1.4	1.6	1.1	1.3	1.4	1.6	1.7
WW B1	21	26	33	38	45	50	57	65	14	18	22	26	31	35	40	47	1.3	1.4	1.6	1.7	1.8	1.6	1.7	1.9	2.1	2.2
WW B2	28	37	47	54	64	71	78	86	19	26	35	41	50	57	64	74	1.0	1.1	1.3	1.4	1.6	1.1	1.3	1.4	1.6	1.7
WW B3	33	43	55	63	72	78	85	91	23	32	42	49	59	65	73	82	0.9	1.1	1.2	1.4	1.5	1.0	1.2	1.3	1.5	1.6
WW B4	33	44	56	64	75	81	87	93	23	33	44	52	62	69	76	85	0.9	1.0	1.2	1.3	1.5	1.0	1.1	1.3	1.4	1.6
WW B5	31	43	55	63	74	80	86	93	22	32	43	51	62	68	76	85	0.9	1.0	1.2	1.4	1.5	1.0	1.1	1.3	1.4	1.6
WW C1	35	45	57	65	75	80	86	93	25	34	44	52	62	68	75	84	0.9	1.0	1.2	1.4	1.5	1.0	1.1	1.3	1.5	1.6
WW C2	33	44	55	63	73	79	86	92	23	32	43	50	60	67	74	83	0.9	1.0	1.2	1.4	1.5	1.0	1.1	1.3	1.5	1.6
WW C3	38	51	64	72	82	87	92	96	28	40	52	60	71	77	84	91	0.8	1.0	1.1	1.3	1.5	0.9	1.0	1.2	1.4	1.5
WW C4	35	47	60	68	79	84	90	95	25	36	48	56	67	74	81	89	0.8	1.0	1.2	1.3	1.5	0.9	1.1	1.2	1.4	1.5
TW A1	22	27	32	36	43	47	53	61	17	20	24	27	32	35	40	47	1.4	1.5	1.7	1.8	1.9	1.7	1.8	2.0	2.1	2.3
TW A2	23	30	37	43	51	56	63	72	17	22	27	32	38	43	49	57	1.2	1.3	1.5	1.6	1.7	1.4	1.5	1.7	1.8	2.0
TW A3	24	31	39	45	54	60	67	76	17	22	29	33	41	46	52	61	1.1	1.3	1.4	1.6	1.7	1.3	1.5	1.6	1.8	1.9
TW A4	23	31	39	46	55	61	68	77	16	22	28	34	41	46	53	63	1.1	1.2	1.4	1.5	1.7	1.3	1.4	1.6	1.7	1.9
TW B1	22	27	32	37	44	48	55	63	17	20	24	28	33	37	42	49	1.3	1.5	1.6	1.8	1.9	1.6	1.7	1.9	2.0	2.2
TW B2	22	28	36	41	49	55	62	71	16	21	26	31	37	42	48	56	1.2	1.3	1.5	1.6	1.8	1.4	1.5	1.7	1.8	2.0
TW B3	22	28	36	42	50	56	63	72	15	20	26	31	38	43	49	58	1.2	1.3	1.5	1.6	1.7	1.4	1.5	1.7	1.8	1.9
TW B4	20	27	34	40	48	53	60	70	14	19	25	29	36	41	47	56	1.2	1.3	1.5	1.6	1.8	1.4	1.5	1.7	1.8	2.0
TW C1	20	26	32	37	44	48	55	64	15	19	24	27	33	37	42	50	1.3	1.5	1.6	1.7	1.9	1.5	1.7	1.8	2.0	2.1
TW C2	20	25	32	37	45	50	57	66	14	19	24	28	34	38	44	52	1.3	1.4	1.5	1.7	1.8	1.5	1.6	1.8	1.9	2.0
TW C3	19	24	31	35	43	48	54	64	14	18	23	27	32	37	42	50	1.3	1.4	1.6	1.7	1.9	1.5	1.7	1.8	2.0	2.1
TW C4	18	22	28	32	38	43	49	58	13	17	21	24	29	33	38	45	1.4	1.5	1.7	1.8	2.0	1.7	1.8	1.9	2.1	2.2
TW D1	17	22	28	33	39	44	50	59	12	15	19	22	27	30	35	42	1.4	1.5	1.7	1.8	1.9	1.7	1.8	2.0	2.2	2.3
TW D2	26	34	43	50	59	65	72	81	18	25	32	38	46	52	59	68	1.1	1.2	1.3	1.5	1.6	1.2	1.4	1.5	1.6	1.8
TW D3	28	37	47	54	64	70	77	85	20	27	35	42	50	56	64	73	1.0	1.1	1.3	1.4	1.6	1.1	1.3	1.4	1.6	1.7
TW D4	30	40	51	58	69	75	82	89	21	30	39	46	56	62	70	79	0.9	1.1	1.2	1.4	1.5	1.1	1.2	1.3	1.5	1.6
TW D5	29	39	50	58	68	75	82	89	21	29	39	46	56	63	71	80	0.9	1.1	1.2	1.4	1.5	1.1	1.2	1.3	1.5	1.6
TW E1	31	41	52	59	69	75	82	89	23	31	40	47	56	62	69	78	0.9	1.1	1.2	1.4	1.5	1.1	1.2	1.4	1.5	1.7
TW E2	31	41	52	59	69	75	82	89	22	31	40	47	56	63	70	79	0.9	1.1	1.2	1.4	1.5	1.1	1.2	1.4	1.5	1.6
TW E3	37	48	60	69	78	84	89	95	27	38	49	57	67	73	80	88	0.9	1.0	1.2	1.3	1.5	0.9	1.1	1.2	1.4	1.5
TW E4	34	46	58	66	76	82	88	94	25	35	46	54	64	71	78	87	0.9	1.0	1.2	1.3	1.5	1.0	1.1	1.3	1.4	1.6
TW F1	21	25	31	36	42	47	53	61	16	19	23	26	31	35	39	46	1.4	1.5	1.6	1.8	1.9	1.6	1.8	1.9	2.1	2.2
TW F2	21	27	35	40	48	54	61	70	15	19	25	29	36	40	46	54	1.2	1.3	1.5	1.6	1.8	1.4	1.6	1.7	1.9	2.0
TW F3	20	27	35	41	50	56	63	72	14	19	25	30	37	42	48	57	1.2	1.3	1.4	1.6	1.7	1.4	1.5	1.7	1.8	1.9
TW F4	18	25	33	39	48	54	62	71	12	17	23	28	35	40	47	56	1.2	1.3	1.4	1.6	1.7	1.4	1.5	1.7	1.8	1.9
TW G1	19	24	30	34	41	46	53	61	14	17	22	25	30	34	39	46	1.3	1.5	1.6	1.8	1.9	1.6	1.8	1.9	2.1	2.2
TW G2	18	24	31	37	44	50	57	67	12	16	22	26	32	37	43	51	1.2	1.4	1.5	1.7	1.8	1.5	1.6	1.8	1.9	2.0
TW G3	16	22	29	35	43	49	56	66	10	15	20	25	31	36	42	51	1.2	1.4	1.5	1.7	1.8	1.5	1.6	1.8	1.9	2.0
TW G4	13	19	25	30	38	43	50	60	8	12	17	21	27	32	37	46	1.3	1.5	1.6	1.7	1.9	1.5	1.7	1.8	2.0	2.1
TW H1	15	21	27	31	38	43	50	59	10	14	19	22	28	31	37	44	1.4	1.5	1.6	1.8	1.9	1.6	1.8	1.9	2.1	2.2
TW H2	13	18	24	29	37	42	48	58	8	12	17	20	26	30	36	43	1.4	1.5	1.6	1.8	1.9	1.6	1.7	1.9	2.0	2.2
TW H3	11	16	21	26	32	37	44	53	6	10	14	18	23	27	32	39	1.4	1.6	1.7	1.9	2.0	1.7	1.8	2.0	2.1	2.2
TW H4	8	12	17	20	26	30	36	44	5	7	11	14	18	21	26	32	1.6	1.7	1.9	2.0	2.1	1.9	2.0	2.2	2.3	2.4
TW I1	15	20	25	30	36	41	47	55	10	13	16	19	24	27	32	38	1.4	1.6	1.7	1.9	2.0	1.7	1.9	2.1	2.3	2.4
TW I2	22	30	39	46	55	62	69	78	15	21	29	34	42	48	55	65	1.1	1.2	1.4	1.5	1.7	1.2	1.4	1.5	1.7	1.8
TW I3	24	33	43	50	59	66	73	82	17	24	32	38	46	52	60	69	1.0	1.2	1.3	1.5	1.6	1.2	1.3	1.5	1.6	1.8
TW I4	27	36	47	55	65	72	79	87	18	26	36	43	52	59	67	76	1.0	1.1	1.3	1.4	1.6	1.1	1.2	1.4	1.5	1.7
TW I5	27	37	48	56	66	73	80	88	18	27	37	44	54	60	68	78	1.0	1.1	1.2	1.4	1.5	1.1	1.2	1.4	1.5	1.6
TW J1	29	39	49	57	67	73	80	88	20	28	37	44	53	60	67	77	1.0	1.1	1.3	1.4	1.6	1.1	1.2	1.4	1.5	1.7
TW J2	28	38	48	56	66	72	80	88	19	28	37	44	53	60	67	77	1.0	1.1	1.3	1.4	1.6	1.1	1.2	1.4	1.5	1.7
TW J3	33	45	57	65	76	81	88	94	24	34	45	53	64	71	78	87	0.9	1.0	1.2	1.3	1.5	1.0	1.1	1.3	1.4	1.6
TW J4	31	42	54	63	73	79	86	92	22	32	43	51	61	68	76	85	0.9	1.0	1.2	1.4	1.5	1.0	1.1	1.3	1.4	1.6
DG A1	19	24	31	35	41	45	49	54	12	16	20	23	28	30	33	37	1.2	1.5	1.8	2.1	2.5	1.5	1.8	2.2	2.7	3.4
DG A2	21	28	36	42	49	55	61	68	14	19	25	30	37	41	47	54	1.2	1.3	1.5	1.7	1.9	1.4	1.5	1.7	2.0	2.2
DG A3	25	33	43	51	61	67	74	81	18	25	33	39	48	55	62	71	1.1	1.1	1.3	1.5	1.7	1.2	1.3	1.4	1.6	1.8
DG B1	19	25	32	38	46	51	57	64	12	16	21	24	30	34	39	45	1.2	1.3	1.5	1.8	2.1	1.5	1.7	1.9	2.1	2.5
DG B2	21	28	37	44	54	61	68	76	15	20	26	31	39	45	52	61	1.1	1.2	1.4	1.6	1.8	1.3	1.5	1.6	1.7	2.0
DG B3	25	34	45	53	65	72	79	87	18	25	34	40	51	58	66	76	1.0	1.1	1.2	1.4	1.6	1.2	1.3	1.4	1.5	1.7
DG C1	22	30	38	46	55	61	68	75	15	20	26	30	37	42	49	57	1.1	1.2	1.4	1.6	1.9	1.4	1.5	1.6	1.8	2.1
DG C2	24	33	43	51	62	69	76	83	17	23	31	37	46	52	60	70	1.1	1.1	1.3	1.5	1.7	1.2	1.3	1.4	1.6	1.8
DG C3	28	38	50	60	72	79	85	92	21	29	38	46	57	65	74	83	1.0	1.0	1.2	1.4	1.6	1.1	1.2	1.3	1.4	1.6
SS A1	28	35	42	48	56	61	68	76	22	28	34	39	45	50	56	64	1.1	1.3	1.4	1.6	1.7	1.3	1.4	1.6	1.8	1.9
SS A2	35	44	54	61	70	75	82	89	28	36	44	51	59	65	72	80	0.9	1.1	1.3	1.4	1.6	1.1	1.2	1.4	1.5	1.7
SS A3	26	32	39	44	51	56	63	71	20	25	31	35	41	45	50	58	1.2	1.3	1.5	1.7	1.8	1.4	1.5	1.7	1.9	2.0
SS A4	34	43	52	59	68	74	80	87	27	35	43	49	58	63	70	78	1.0	1.1	1.3	1.4	1.6	1.1	1.2	1.4	1.6	1.7
SS A5	27	32	39	44	51	56	62	71	22	26	31	35	41	45	51	58	1.2	1.4	1.5	1.7	1.8	1.4	1.6	1.7	1.9	2.0
SS A6	34	43	53	60	69	75	81	88	27	35	44	50	59	64	71	79	1.0	1.1	1.3	1.4	1.6	1.1	1.2	1.4	1.5	1.7
SS A7	24	29	34	38	45	49	54	62	19	23	27	31	35	39	43	49	1.3	1.5	1.7	1.8	2.0	1.5	1.7	1.9	2.1	2.3
SS A8	33	42	51	58	67	72	79	86	27	34	42	48	56	61	68	76	1.0	1.1	1.3	1.4	1.6	1.1	1.3	1.4	1.6	1.7
SS B1	24	30	37	42	50	55	62	71	18	23	29	33	39	44	50	57	1.2	1.3	1.5	1.6	1.8	1.4	1.5	1.7	1.9	2.0
SS B2	30	39	48	55	64	70	77	85	24	31	39	45	54	59	66	75	1.0	1.1	1.3	1.5	1.6	1.1	1.3	1.4	1.6	1.7
SS B3	22	28	35	40	47	52	58	67	17	22	27	31	36	40	46	53	1.2	1.4	1.6	1.7	1.8	1.4	1.6	1.8	2.0	2.1
SS B4	29	38	47	54	63	69	75	84	23	30	38	44	52	58	65	73	1.0	1.2	1.3	1.5	1.6	1.1	1.3	1.4	1.6	1.7
SS B5	22	27	32	37	43	48	54	62	17	21	26	29	34	38	42	49	1.3	1.5	1.6	1.8	1.9	1.5	1.7	1.9	2.1	2.2
SS B6	29	38	47	54	63	69	76	84	23	30	38	44	53	58	65	74	1.0	1.2	1.3	1.5	1.6	1.1	1.3	1.4	1.6	1.7
SS B7	20	24	29	33	39	43	48	56	16	19	23	26	30	33	37	43	1.4	1.6	1.8	1.9	2.0	1.7	1.9	2.1	2.2	2.4
SS B8	28	36	45	52	61	67	74	82	22	29	36	42	50	56	62	71	1.0	1.2	1.3	1.5	1.6	1.2	1.3	1.5	1.6	1.8
SS C1	23	31	39	45	54	60	67	76	16	21	28	33	40	45	51	60	1.1	1.3	1.4	1.6	1.7	1.3	1.5	1.6	1.8	1.9
SS C2	28	37	47	54	64	70	78	86	20	27	35	42	51	57	64	74	1.0	1.1	1.3	1.4	1.6	1.1	1.3	1.4	1.6	1.7
SS C3	20	25	31	36	43	48	55	64	14	18	23	27	32	36	41	49	1.3	1.4	1.6	1.7	1.9	1.5	1.7	1.9	2.0	2.1
SS C4	25	33	42	49	58	64	71	80	18	24	31	37	45	51	58	67	1.1	1.2	1.4	1.5	1.6	1.2	1.4	1.5	1.7	1.8
SS D1	33	41	50	57	66	71	78	85	26	32	40	46	54	59	65	74	1.0	1.1	1.3	1.5	1.6	1.1	1.3	1.5	1.6	1.8
SS D2	40	51	62	69	78	83	89	94	32	41	51	59	68	74	80	88	0.9	1.0	1.2	1.3	1.5	0.9	1.1	1.3	1.4	1.6
SS D3	30	37	45	51	59	65	71	79	24	30	36	41	48	53	59	67	1.1	1.2	1.4	1.5	1.7	1.2	1.4	1.6	1.7	1.9
SS D4	38	49	60	67	76	82	87	93	30	40	50	56	66	71	78	86	0.9	1.0	1.2	1.4	1.5	1.0	1.1	1.3	1.5	1.6
SS E1	28	36	45	51	60	65	72	80	21	27	34	39	47	52	58	66	1.1	1.2	1.4	1.5	1.7	1.2	1.4	1.5	1.7	1.9
SS E2	34	45	55	63	72	78	84	91	26	35	45	51	61	66	74	82	0.9	1.1	1.2	1.4	1.5	1.0	1.2	1.3	1.5	1.6
SS E3	24	30	37	42	50	55	61	70	19	24	29	33	39	43	49	56	1.2	1.4	1.5	1.7	1.8	1.4	1.6	1.7	1.9	2.0
SS E4	32	41	51	58	67	73	80	87	25	32	41	47	56	62	69	77	1.0	1.1	1.3	1.4	1.6	1.1	1.2	1.4	1.5	1.7

TOPEKA, KANSAS

TYPE	SOLAR SAVINGS FRACTION (%) BASE TEMP: 55 F (3208 DD)								65 F (5243 DD)								CONSERVATION FACTOR 55 F					65 F				
LCR	100	70	50	40	30	25	20	15	100	70	50	40	30	25	20	15	100	50	30	20	15	100	50	30	20	15
WW A1	20	24	29	32	37	41	46	53	16	19	22	24	28	30	34	39	1.5	1.7	1.8	2.0	2.1	1.8	2.0	2.3	2.5	2.6
WW A2	22	28	35	40	48	53	60	68	16	20	26	30	35	40	45	52	1.2	1.4	1.5	1.7	1.8	1.4	1.6	1.8	2.0	2.1
WW A3	22	29	37	43	52	58	65	73	16	21	27	32	38	43	49	57	1.1	1.3	1.4	1.6	1.7	1.3	1.5	1.7	1.8	2.0
WW A4	22	30	39	45	54	60	67	76	15	21	28	33	40	45	52	60	1.1	1.2	1.4	1.6	1.7	1.3	1.4	1.6	1.8	1.9
WW A5	23	31	40	47	56	62	70	79	15	21	29	34	42	47	54	63	1.1	1.2	1.4	1.5	1.7	1.2	1.4	1.6	1.7	1.9
WW A6	23	32	41	48	58	64	71	80	15	22	29	35	43	48	55	65	1.0	1.2	1.3	1.5	1.7	1.2	1.4	1.5	1.7	1.8
WW B1	17	22	27	31	37	41	47	55	12	15	18	21	25	27	31	35	1.4	1.6	1.8	1.9	2.1	1.8	2.0	2.3	2.5	2.7
WW B2	23	31	41	47	57	63	70	79	16	23	30	35	43	49	56	65	1.1	1.2	1.4	1.5	1.7	1.2	1.4	1.5	1.7	1.8
WW B3	28	37	48	55	65	71	78	86	20	27	36	42	51	57	65	74	1.0	1.1	1.3	1.4	1.6	1.1	1.3	1.4	1.6	1.7
WW B4	28	38	49	57	68	74	81	88	20	28	38	45	55	61	69	78	0.9	1.1	1.2	1.4	1.6	1.1	1.2	1.4	1.5	1.7
WW B5	27	37	49	57	67	73	80	88	19	28	38	45	55	61	69	79	0.9	1.1	1.2	1.4	1.6	1.0	1.2	1.3	1.5	1.6
WW C1	29	39	50	57	67	73	80	87	21	29	38	45	54	60	67	76	1.0	1.1	1.3	1.4	1.6	1.1	1.2	1.4	1.6	1.7
WW C2	28	38	48	56	66	72	79	87	20	28	37	44	53	60	67	76	1.0	1.1	1.3	1.4	1.6	1.1	1.2	1.4	1.5	1.7
WW C3	33	45	57	65	75	81	87	93	24	34	46	54	64	71	78	86	0.9	1.0	1.2	1.3	1.5	1.0	1.1	1.3	1.4	1.6
WW C4	30	41	53	62	72	78	85	92	21	31	42	50	61	68	75	84	0.9	1.0	1.2	1.4	1.5	1.0	1.1	1.3	1.4	1.6
TW A1	20	24	28	32	37	41	46	53	16	19	22	24	28	31	34	39	1.5	1.7	1.8	2.0	2.1	1.8	2.0	2.2	2.4	2.6
TW A2	20	26	32	37	44	49	55	63	15	19	24	27	33	36	41	48	1.3	1.4	1.6	1.8	1.9	1.5	1.7	1.9	2.1	2.2
TW A3	20	26	33	39	47	52	59	67	15	19	24	29	35	39	44	52	1.2	1.4	1.5	1.7	1.8	1.4	1.6	1.8	1.9	2.1
TW A4	19	26	33	39	47	53	60	69	14	18	24	29	35	40	45	54	1.2	1.3	1.5	1.6	1.8	1.4	1.6	1.7	1.9	2.0
TW B1	19	23	28	32	38	42	47	55	15	18	22	24	29	32	36	41	1.5	1.6	1.8	1.9	2.1	1.7	1.9	2.1	2.3	2.5
TW B2	19	25	31	36	43	48	54	63	14	18	23	27	32	36	41	48	1.3	1.4	1.6	1.7	1.9	1.5	1.7	1.9	2.0	2.2
TW B3	19	24	31	36	43	49	55	64	13	18	23	27	32	36	42	50	1.3	1.4	1.6	1.7	1.9	1.5	1.7	1.8	2.0	2.1
TW B4	17	23	29	34	41	46	53	62	13	17	22	25	31	35	40	48	1.3	1.5	1.6	1.8	1.9	1.5	1.7	1.8	2.0	2.2
TW C1	18	23	28	32	38	42	48	56	14	17	21	24	29	32	36	43	1.4	1.6	1.7	1.9	2.0	1.7	1.9	2.0	2.2	2.4
TW C2	17	22	28	32	39	44	50	58	13	17	21	24	29	33	38	45	1.4	1.5	1.7	1.8	2.0	1.6	1.8	1.9	2.1	2.2
TW C3	17	21	27	31	37	42	48	56	13	16	20	23	28	32	37	43	1.4	1.6	1.7	1.9	2.0	1.7	1.8	2.0	2.1	2.3
TW C4	16	20	24	28	34	38	43	51	12	15	19	22	26	29	33	40	1.5	1.7	1.8	2.0	2.1	1.8	1.9	2.1	2.3	2.4
TW D1	15	19	23	27	33	37	42	49	10	13	16	18	22	24	28	32	1.5	1.7	1.9	2.0	2.2	1.9	2.1	2.4	2.6	2.8
TW D2	22	29	37	43	52	58	65	74	16	21	28	33	40	45	51	60	1.1	1.3	1.4	1.6	1.7	1.3	1.5	1.6	1.8	1.9
TW D3	24	32	40	47	56	62	69	78	17	23	30	36	43	49	56	65	1.1	1.2	1.4	1.5	1.7	1.2	1.4	1.5	1.7	1.8
TW D4	26	35	45	52	62	68	75	83	18	26	34	41	49	55	63	72	1.0	1.2	1.3	1.5	1.6	1.1	1.3	1.4	1.6	1.7
TW D5	25	34	44	52	62	68	75	84	18	26	35	41	50	56	64	74	1.0	1.1	1.3	1.5	1.6	1.1	1.3	1.4	1.6	1.7
TW E1	27	36	45	52	62	67	75	83	20	27	35	41	49	55	62	71	1.0	1.2	1.3	1.5	1.6	1.2	1.3	1.5	1.6	1.8
TW E2	26	35	45	52	62	68	75	84	19	27	35	41	50	55	63	72	1.0	1.2	1.3	1.5	1.6	1.1	1.3	1.4	1.6	1.7
TW E3	32	42	54	62	72	78	84	91	23	33	43	50	60	67	74	83	0.9	1.1	1.2	1.4	1.5	1.0	1.2	1.3	1.5	1.6
TW E4	29	40	51	59	69	75	82	90	21	31	41	48	58	64	72	81	0.9	1.1	1.2	1.4	1.5	1.0	1.2	1.3	1.5	1.6
TW F1	18	22	27	31	36	40	46	53	14	17	20	23	27	30	33	39	1.5	1.7	1.8	2.0	2.1	1.8	2.0	2.2	2.4	2.6
TW F2	18	23	30	34	41	46	52	61	13	17	21	25	30	34	39	46	1.3	1.5	1.6	1.8	1.9	1.6	1.7	1.9	2.1	2.2
TW F3	17	23	30	35	43	48	55	64	12	16	21	25	31	35	40	48	1.3	1.4	1.6	1.7	1.9	1.5	1.7	1.8	2.0	2.1
TW F4	15	21	28	33	41	47	54	63	10	14	19	24	30	34	40	48	1.3	1.4	1.6	1.7	1.9	1.5	1.6	1.8	2.0	2.1
TW G1	16	21	26	30	35	40	45	53	12	15	19	22	26	29	33	39	1.5	1.6	1.8	1.9	2.1	1.8	1.9	2.1	2.3	2.5
TW G2	15	20	26	31	38	43	49	58	10	14	19	22	27	31	36	43	1.3	1.5	1.7	1.8	1.9	1.6	1.8	1.9	2.1	2.3
TW G3	13	19	25	30	37	42	48	57	9	12	17	21	26	30	35	43	1.4	1.5	1.7	1.8	1.9	1.6	1.8	1.9	2.1	2.2
TW G4	11	16	21	26	32	37	43	52	6	10	14	18	23	27	32	39	1.4	1.6	1.7	1.9	2.0	1.7	1.8	2.0	2.2	2.3
TW H1	13	18	23	27	33	37	43	51	9	12	16	19	23	27	31	37	1.5	1.6	1.8	1.9	2.1	1.8	1.9	2.1	2.3	2.4
TW H2	11	15	21	25	31	35	41	50	7	10	14	17	22	25	30	37	1.5	1.6	1.8	1.9	2.1	1.7	1.9	2.1	2.2	2.4
TW H3	9	13	18	22	28	32	37	45	5	8	12	15	19	23	27	33	1.6	1.7	1.9	2.0	2.1	1.8	2.0	2.2	2.3	2.5
TW H4	7	10	14	17	22	26	31	38	4	6	9	12	15	18	22	28	1.7	1.9	2.0	2.2	2.3	2.0	2.2	2.4	2.5	2.7
TW I1	12	16	21	24	29	33	38	45	8	11	13	16	19	21	25	29	1.6	1.8	1.9	2.1	2.2	2.0	2.2	2.5	2.7	2.9
TW I2	19	26	34	40	48	54	61	71	13	18	24	29	36	41	48	56	1.2	1.3	1.5	1.6	1.8	1.3	1.5	1.7	1.8	2.0
TW I3	20	28	37	43	52	58	65	75	14	20	27	32	40	45	52	61	1.1	1.3	1.4	1.6	1.7	1.3	1.4	1.6	1.7	1.9
TW I4	22	31	41	48	58	64	72	81	16	23	31	37	46	52	60	69	1.0	1.2	1.3	1.5	1.6	1.2	1.3	1.5	1.6	1.8
TW I5	23	32	42	49	59	66	73	82	16	23	32	39	48	54	62	71	1.0	1.2	1.3	1.5	1.6	1.1	1.3	1.4	1.6	1.7
TW J1	25	33	43	50	59	65	73	81	17	24	32	38	47	52	59	69	1.0	1.2	1.3	1.5	1.6	1.2	1.3	1.5	1.6	1.8
TW J2	24	32	42	49	59	65	73	81	17	24	32	38	47	53	60	70	1.0	1.2	1.3	1.5	1.6	1.2	1.3	1.5	1.6	1.8
TW J3	28	39	51	59	69	75	82	89	20	30	40	47	58	64	72	81	0.9	1.1	1.2	1.4	1.5	1.0	1.2	1.3	1.5	1.6
TW J4	26	37	48	56	67	73	80	88	19	28	38	45	55	62	70	79	0.9	1.1	1.2	1.4	1.6	1.0	1.2	1.3	1.5	1.6
DG A1	15	20	25	29	34	37	40	44	10	13	16	19	22	24	26	28	1.4	1.6	2.0	2.4	2.9	1.7	2.1	2.6	3.4	4.7
DG A2	18	24	30	30	43	47	53	60	13	17	22	26	32	35	40	46	1.3	1.4	1.6	1.9	2.1	1.5	1.7	1.9	2.1	2.4
DG A3	21	29	38	44	54	60	67	75	16	22	29	34	42	48	55	64	1.1	1.2	1.4	1.6	1.8	1.3	1.4	1.5	1.7	1.9
DG B1	15	20	26	31	37	42	47	53	10	13	17	19	23	26	29	34	1.4	1.5	1.7	2.0	2.4	1.7	2.0	2.2	2.6	3.1
DG B2	18	24	32	37	46	52	59	67	13	17	23	27	33	38	43	51	1.3	1.3	1.5	1.7	1.9	1.5	1.6	1.7	1.9	2.1
DG B3	22	30	39	46	57	64	72	80	17	23	30	36	44	50	58	69	1.1	1.2	1.3	1.5	1.7	1.3	1.4	1.5	1.6	1.8
DG C1	19	24	32	37	45	51	57	64	13	17	21	25	30	34	38	44	1.3	1.4	1.5	1.7	2.0	1.5	1.7	1.9	2.2	2.5
DG C2	21	28	37	43	53	59	67	75	15	21	27	32	39	44	51	60	1.2	1.2	1.4	1.6	1.8	1.3	1.5	1.6	1.8	1.9
DG C3	25	33	44	52	63	71	79	86	19	26	34	40	50	57	65	76	1.1	1.1	1.2	1.4	1.6	1.2	1.3	1.4	1.5	1.7
SS A1	24	30	37	42	49	54	60	69	20	24	30	33	39	43	48	55	1.2	1.4	1.5	1.7	1.8	1.4	1.6	1.8	1.9	2.1
SS A2	30	39	48	54	63	69	75	83	25	31	39	45	53	58	65	73	1.0	1.2	1.3	1.5	1.6	1.1	1.3	1.5	1.6	1.8
SS A3	22	28	34	38	45	49	55	63	18	22	26	30	35	38	43	49	1.3	1.5	1.6	1.8	1.9	1.5	1.7	1.9	2.1	2.3
SS A4	30	38	46	53	61	67	74	81	24	31	38	43	51	56	62	71	1.0	1.2	1.3	1.5	1.7	1.2	1.3	1.5	1.7	1.8
SS A5	24	29	34	38	45	49	55	63	20	23	28	31	36	39	43	50	1.3	1.5	1.6	1.8	2.0	1.5	1.7	1.9	2.1	2.3
SS A6	30	38	47	54	62	68	75	83	24	31	39	44	52	57	64	72	1.0	1.2	1.3	1.5	1.6	1.1	1.3	1.5	1.6	1.8
SS A7	21	25	30	33	39	42	47	54	18	21	24	27	30	33	36	41	1.5	1.7	1.8	2.0	2.1	1.7	2.0	2.2	2.4	2.6
SS A8	29	37	45	51	60	66	72	80	23	30	37	42	49	54	61	69	1.1	1.2	1.4	1.5	1.7	1.2	1.4	1.5	1.7	1.8
SS B1	21	26	32	37	43	48	54	62	16	20	25	29	34	38	42	49	1.3	1.5	1.6	1.8	1.9	1.5	1.7	1.9	2.1	2.2
SS B2	27	34	43	49	58	64	70	79	21	28	35	40	48	53	60	68	1.1	1.2	1.4	1.5	1.7	1.2	1.4	1.5	1.7	1.8
SS B3	19	24	30	34	40	45	51	58	15	19	23	26	31	34	38	44	1.4	1.5	1.7	1.9	2.0	1.6	1.8	2.0	2.2	2.4
SS B4	26	33	41	48	56	62	69	77	20	27	34	39	46	51	58	66	1.1	1.2	1.4	1.6	1.7	1.2	1.4	1.5	1.7	1.9
SS B5	19	24	28	32	38	42	47	54	16	19	23	25	29	32	36	42	1.5	1.6	1.8	2.0	2.1	1.7	1.9	2.1	2.3	2.5
SS B6	26	33	42	48	57	62	69	78	20	27	34	39	46	52	58	67	1.1	1.2	1.4	1.5	1.7	1.2	1.4	1.5	1.7	1.8
SS B7	18	21	25	29	33	37	41	48	14	17	20	22	26	28	31	36	1.6	1.8	1.9	2.1	2.2	1.8	2.1	2.3	2.6	2.8
SS B8	25	32	40	46	54	60	67	75	19	25	32	37	44	49	55	64	1.1	1.3	1.4	1.6	1.7	1.2	1.4	1.6	1.7	1.9
SS C1	20	26	33	38	46	51	58	67	14	18	23	27	33	37	43	50	1.2	1.4	1.5	1.7	1.8	1.4	1.6	1.8	2.0	2.1
SS C2	24	32	41	47	57	63	70	79	17	23	30	36	44	49	56	65	1.1	1.2	1.4	1.5	1.7	1.2	1.4	1.5	1.7	1.8
SS C3	17	22	27	31	37	42	47	55	13	16	20	23	27	30	35	41	1.4	1.6	1.7	1.9	2.0	1.7	1.9	2.1	2.2	2.4
SS C4	21	28	36	42	51	56	64	73	16	21	27	32	39	44	50	59	1.1	1.3	1.4	1.6	1.7	1.3	1.5	1.6	1.8	1.9
SS D1	28	36	44	50	58	64	70	78	22	28	34	39	46	51	57	65	1.1	1.2	1.4	1.6	1.7	1.2	1.4	1.6	1.8	1.9
SS D2	34	45	55	62	72	77	83	90	27	36	45	52	61	66	73	81	0.9	1.1	1.2	1.4	1.6	1.0	1.2	1.3	1.5	1.7
SS D3	26	32	39	45	52	57	64	72	21	26	31	36	41	46	51	58	1.2	1.3	1.5	1.7	1.8	1.3	1.5	1.7	1.9	2.1
SS D4	33	43	53	60	70	75	81	88	27	35	44	50	59	64	71	79	0.9	1.1	1.3	1.4	1.6	1.0	1.2	1.4	1.5	1.7
SS E1	24	31	38	44	52	57	64	72	18	23	29	33	39	43	49	56	1.2	1.3	1.5	1.6	1.8	1.3	1.5	1.7	1.9	2.1
SS E2	30	39	49	56	65	71	78	85	23	30	39	45	53	59	66	74	1.0	1.1	1.3	1.5	1.6	1.1	1.3	1.4	1.6	1.7
SS E3	21	26	32	37	43	48	54	62	17	21	25	29	33	37	41	48	1.3	1.5	1.6	1.8	1.9	1.5	1.7	1.9	2.1	2.3
SS E4	28	36	45	51	60	66	73	81	21	28	36	41	49	55	61	70	1.0	1.2	1.3	1.5	1.7	1.2	1.3	1.5	1.7	1.8

WICHITA, KANSAS

TYPE	SOLAR SAVINGS FRACTION (%) BASE TEMP: 55 F (2756 DD)								65 F (4688 DD)								CONSERVATION FACTOR 55 F					65 F				
LCR	100	70	50	40	30	25	20	15	100	70	50	40	30	25	20	15	100	50	30	20	15	100	50	30	20	15
WW A1	24	29	35	40	46	51	57	65	19	22	26	29	34	37	42	49	1.3	1.4	1.6	1.7	1.9	1.6	1.8	1.9	2.1	2.2
WW A2	28	35	44	50	59	65	72	80	20	25	32	37	44	49	55	64	1.1	1.2	1.4	1.5	1.7	1.3	1.4	1.6	1.7	1.9
WW A3	29	38	47	54	64	69	76	84	20	26	34	39	47	53	60	69	1.0	1.1	1.3	1.5	1.6	1.2	1.3	1.5	1.7	1.8
WW A4	29	39	49	56	66	72	79	86	20	27	35	41	50	55	63	72	1.0	1.1	1.3	1.4	1.6	1.1	1.3	1.5	1.6	1.7
WW A5	30	40	51	59	68	74	81	88	20	28	36	43	52	58	65	74	0.9	1.1	1.2	1.4	1.6	1.1	1.3	1.4	1.6	1.7
WW A6	31	41	52	60	70	76	82	89	20	28	37	44	53	59	67	76	0.9	1.1	1.2	1.4	1.5	1.1	1.2	1.4	1.5	1.7
WW B1	23	29	36	41	49	54	61	69	15	19	24	28	33	37	42	49	1.2	1.4	1.5	1.7	1.8	1.5	1.7	1.9	2.0	2.2
WW B2	30	40	50	58	68	74	81	88	20	28	37	43	52	58	66	75	1.0	1.1	1.3	1.4	1.6	1.1	1.3	1.4	1.6	1.7
WW B3	36	47	58	66	75	81	87	93	25	34	44	51	61	67	74	83	0.9	1.0	1.2	1.4	1.5	1.0	1.2	1.3	1.5	1.6
WW B4	36	48	60	68	77	83	89	94	25	35	46	53	64	70	77	86	0.9	1.0	1.2	1.3	1.5	1.0	1.1	1.3	1.4	1.6
WW B5	34	46	58	67	77	82	88	94	24	34	45	53	63	70	77	86	0.9	1.0	1.2	1.3	1.5	1.0	1.1	1.3	1.4	1.6
WW C1	37	49	60	68	77	83	88	94	27	36	46	53	63	69	77	85	0.9	1.0	1.2	1.3	1.5	1.0	1.1	1.3	1.5	1.6
WW C2	35	47	58	66	76	82	88	93	25	34	45	52	62	68	76	84	0.9	1.0	1.2	1.4	1.5	1.0	1.1	1.3	1.5	1.6
WW C3	41	54	67	75	84	89	93	97	30	41	53	62	72	78	85	92	0.8	1.0	1.1	1.3	1.4	0.9	1.0	1.2	1.4	1.5
WW C4	38	50	63	71	81	86	91	96	27	38	50	58	69	75	82	90	0.8	1.0	1.1	1.3	1.5	0.9	1.1	1.2	1.4	1.5
TW A1	24	29	34	39	46	50	56	65	18	22	26	29	34	37	42	48	1.3	1.5	1.6	1.7	1.9	1.6	1.8	1.9	2.1	2.2
TW A2	25	32	40	46	54	60	66	75	18	23	29	33	40	45	51	59	1.1	1.3	1.4	1.6	1.7	1.4	1.5	1.7	1.8	2.0
TW A3	26	33	42	48	57	63	70	79	18	24	30	35	43	48	54	63	1.1	1.2	1.4	1.5	1.7	1.3	1.4	1.6	1.7	1.9
TW A4	25	33	42	49	58	64	71	80	17	23	30	35	43	48	55	64	1.1	1.2	1.4	1.5	1.7	1.3	1.4	1.6	1.7	1.8
TW B1	23	29	35	40	47	52	58	67	18	21	26	29	35	38	43	51	1.3	1.4	1.6	1.7	1.8	1.5	1.7	1.9	2.0	2.2
TW B2	24	31	38	44	53	58	65	74	17	22	28	32	39	44	50	58	1.1	1.3	1.4	1.6	1.7	1.4	1.5	1.7	1.8	2.0
TW B3	23	31	39	45	54	59	66	75	16	22	28	33	40	44	51	60	1.1	1.3	1.4	1.6	1.7	1.3	1.5	1.6	1.8	1.9
TW B4	22	29	37	43	51	57	64	73	15	20	26	31	38	42	49	57	1.2	1.3	1.4	1.6	1.7	1.4	1.5	1.7	1.8	1.9
TW C1	22	28	34	39	47	52	58	67	16	20	25	29	35	39	44	52	1.3	1.4	1.5	1.7	1.8	1.5	1.7	1.8	2.0	2.1
TW C2	21	27	35	40	48	53	60	69	15	20	25	29	36	40	46	54	1.2	1.4	1.5	1.6	1.8	1.4	1.6	1.7	1.9	2.0
TW C3	20	26	33	38	46	51	58	67	15	19	24	28	34	38	44	52	1.3	1.4	1.5	1.7	1.8	1.5	1.6	1.8	1.9	2.1
TW C4	19	24	30	34	41	46	52	61	14	18	22	26	31	35	40	47	1.4	1.5	1.6	1.8	1.9	1.6	1.8	1.9	2.1	2.2
TW D1	19	24	31	36	43	48	54	63	13	16	21	24	29	32	37	44	1.3	1.4	1.6	1.7	1.9	1.6	1.8	2.0	2.1	2.3
TW D2	28	36	46	53	62	68	75	84	20	26	34	40	48	53	60	70	1.0	1.2	1.3	1.5	1.6	1.2	1.3	1.5	1.6	1.8
TW D3	30	40	50	57	67	73	80	87	21	29	37	43	52	58	65	74	1.0	1.1	1.3	1.4	1.6	1.1	1.3	1.4	1.6	1.7
TW D4	32	43	54	62	72	77	84	91	23	31	41	48	58	64	71	80	0.9	1.1	1.2	1.4	1.5	1.0	1.2	1.3	1.5	1.6
TW D5	32	42	54	61	71	77	84	91	22	31	41	48	58	64	72	81	0.9	1.1	1.2	1.4	1.5	1.0	1.2	1.3	1.5	1.6
TW E1	34	44	55	62	72	78	84	91	24	33	42	48	58	64	71	80	0.9	1.1	1.2	1.4	1.5	1.1	1.2	1.4	1.5	1.7
TW E2	33	44	55	62	72	78	84	91	24	32	42	49	58	64	72	80	0.9	1.1	1.2	1.4	1.5	1.0	1.2	1.3	1.5	1.6
TW E3	40	52	64	72	81	86	91	96	29	39	50	58	68	74	81	89	0.8	1.0	1.1	1.3	1.5	0.9	1.1	1.2	1.4	1.5
TW E4	37	49	61	69	78	84	89	95	26	37	48	55	66	72	79	87	0.9	1.0	1.2	1.3	1.5	1.0	1.1	1.3	1.4	1.6
TW F1	22	27	34	38	45	50	57	65	17	20	24	28	33	37	41	48	1.3	1.4	1.6	1.7	1.9	1.6	1.7	1.9	2.1	2.2
TW F2	23	30	37	43	52	57	64	73	16	21	26	31	37	42	48	56	1.2	1.3	1.4	1.6	1.7	1.4	1.5	1.7	1.9	2.0
TW F3	22	30	38	44	53	59	67	76	15	20	27	31	39	44	50	59	1.1	1.3	1.4	1.6	1.7	1.3	1.5	1.6	1.8	1.9
TW F4	20	28	36	43	52	58	65	75	13	18	25	30	37	42	49	58	1.1	1.3	1.4	1.6	1.7	1.3	1.5	1.6	1.8	1.9
TW G1	20	26	32	37	45	50	56	65	15	18	23	27	32	36	41	48	1.3	1.4	1.6	1.7	1.8	1.6	1.7	1.9	2.0	2.2
TW G2	20	26	34	40	48	54	61	70	13	18	23	28	34	39	45	53	1.2	1.3	1.5	1.6	1.8	1.4	1.6	1.7	1.9	2.0
TW G3	18	25	32	38	47	52	60	69	11	16	22	26	33	38	44	53	1.2	1.3	1.5	1.6	1.8	1.4	1.6	1.7	1.9	2.0
TW G4	15	21	28	33	41	47	54	64	9	13	19	23	29	33	39	48	1.3	1.4	1.5	1.7	1.8	1.5	1.7	1.8	2.0	2.1
TW H1	17	23	29	34	41	47	53	63	11	15	20	24	29	33	39	46	1.3	1.4	1.6	1.7	1.9	1.6	1.7	1.9	2.0	2.2
TW H2	15	20	27	32	40	45	52	61	9	13	18	22	28	32	37	45	1.3	1.4	1.6	1.7	1.9	1.6	1.7	1.9	2.0	2.1
TW H3	12	17	23	28	35	40	47	56	7	11	15	19	24	28	34	41	1.4	1.5	1.7	1.8	1.9	1.6	1.8	1.9	2.1	2.2
TW H4	9	13	18	22	28	33	39	47	5	8	12	15	19	23	27	34	1.5	1.7	1.8	2.0	2.1	1.8	2.0	2.1	2.3	2.4
TW I1	17	22	28	33	39	44	51	59	11	14	18	21	26	29	34	40	1.3	1.5	1.6	1.8	1.9	1.7	1.9	2.0	2.2	2.4
TW I2	24	33	42	49	59	65	72	81	16	23	30	36	44	50	57	66	1.0	1.2	1.3	1.5	1.6	1.2	1.4	1.5	1.7	1.8
TW I3	27	36	46	53	63	69	76	84	18	25	33	39	48	54	61	71	1.0	1.1	1.3	1.5	1.6	1.2	1.3	1.5	1.6	1.7
TW I4	29	39	50	58	68	74	81	89	20	28	38	45	54	60	68	78	0.9	1.1	1.2	1.4	1.5	1.1	1.2	1.4	1.5	1.7
TW I5	29	40	51	59	69	76	82	90	20	29	38	46	56	62	70	79	0.9	1.1	1.2	1.4	1.5	1.1	1.2	1.3	1.5	1.6
TW J1	31	41	52	60	70	76	82	89	22	30	39	46	55	61	69	78	0.9	1.1	1.2	1.4	1.5	1.1	1.2	1.4	1.5	1.7
TW J2	30	41	52	59	69	75	82	89	21	29	39	46	55	61	69	78	0.9	1.1	1.2	1.4	1.5	1.1	1.2	1.4	1.5	1.7
TW J3	36	48	60	68	78	84	89	95	25	36	47	55	66	72	79	87	0.8	1.0	1.2	1.3	1.5	1.0	1.1	1.3	1.4	1.6
TW J4	34	45	58	66	76	82	88	94	23	34	45	53	63	70	77	86	0.9	1.0	1.2	1.3	1.5	1.0	1.1	1.3	1.4	1.6
DG A1	21	27	34	39	45	49	53	58	13	17	22	25	30	32	35	39	1.2	1.4	1.7	2.0	2.4	1.5	1.7	2.1	2.6	3.2
DG A2	23	30	39	45	53	58	64	71	15	21	27	32	39	43	49	56	1.1	1.3	1.5	1.7	1.9	1.3	1.5	1.7	1.9	2.2
DG A3	27	36	47	54	64	70	76	83	19	26	34	41	50	56	63	72	1.0	1.1	1.3	1.5	1.7	1.2	1.3	1.4	1.6	1.8
DG B1	21	28	36	42	51	56	62	68	13	18	23	27	32	36	41	47	1.2	1.3	1.5	1.8	2.0	1.5	1.6	1.8	2.1	2.5
DG B2	23	31	41	48	58	64	71	79	16	21	28	33	41	47	54	62	1.1	1.2	1.3	1.5	1.7	1.3	1.4	1.5	1.7	1.9
DG B3	27	37	48	57	68	75	81	88	20	27	35	42	52	59	68	77	1.0	1.1	1.2	1.4	1.6	1.2	1.3	1.3	1.5	1.7
DG C1	25	33	43	50	60	66	72	78	16	21	28	33	40	45	51	59	1.1	1.2	1.3	1.6	1.8	1.3	1.5	1.6	1.8	2.1
DG C2	27	36	47	55	66	72	79	86	19	25	33	39	48	54	62	71	1.0	1.1	1.2	1.5	1.6	1.2	1.3	1.4	1.6	1.8
DG C3	31	42	54	63	75	81	87	93	23	30	40	48	59	66	75	84	0.9	1.0	1.2	1.4	1.6	1.1	1.2	1.3	1.4	1.6
SS A1	30	37	45	51	59	65	71	79	23	29	35	40	47	51	58	66	1.1	1.2	1.4	1.5	1.7	1.2	1.4	1.6	1.7	1.9
SS A2	37	46	57	64	73	78	84	90	29	37	46	52	61	66	73	81	0.9	1.1	1.2	1.4	1.5	1.0	1.2	1.4	1.5	1.7
SS A3	28	34	42	47	55	60	67	75	21	26	32	36	42	47	52	60	1.1	1.3	1.4	1.6	1.8	1.3	1.5	1.7	1.9	2.0
SS A4	36	45	55	62	71	77	83	89	28	36	44	51	59	64	71	79	0.9	1.1	1.2	1.4	1.6	1.1	1.2	1.4	1.5	1.7
SS A5	28	34	42	47	55	60	66	74	23	27	33	37	43	47	52	60	1.2	1.3	1.5	1.6	1.8	1.3	1.5	1.7	1.9	2.0
SS A6	37	46	56	63	72	78	84	90	29	36	45	51	60	65	72	80	0.9	1.1	1.2	1.4	1.6	1.0	1.2	1.4	1.5	1.7
SS A7	25	30	36	41	48	52	58	66	20	24	28	32	37	40	45	51	1.3	1.4	1.6	1.7	1.9	1.5	1.7	1.9	2.1	2.2
SS A8	35	45	54	61	70	76	82	89	28	35	43	49	57	63	70	78	0.9	1.1	1.3	1.4	1.6	1.1	1.2	1.4	1.6	1.7
SS B1	26	32	40	45	53	59	65	74	19	24	30	35	41	45	51	59	1.1	1.3	1.4	1.6	1.7	1.3	1.5	1.7	1.8	2.0
SS B2	32	41	51	58	67	73	80	87	25	32	41	47	55	61	68	76	1.0	1.1	1.3	1.4	1.6	1.1	1.3	1.4	1.6	1.7
SS B3	24	30	37	43	50	56	62	71	18	23	28	32	38	42	47	55	1.2	1.3	1.5	1.7	1.8	1.4	1.6	1.8	1.9	2.1
SS B4	32	40	50	57	66	72	78	86	24	31	39	45	54	59	66	75	1.0	1.1	1.3	1.5	1.6	1.1	1.3	1.4	1.6	1.7
SS B5	23	29	35	40	46	51	57	66	18	22	27	30	35	39	44	51	1.3	1.4	1.6	1.7	1.9	1.5	1.7	1.9	2.0	2.2
SS B6	32	41	50	57	67	72	79	86	24	31	40	46	54	60	67	75	1.0	1.1	1.3	1.4	1.6	1.1	1.3	1.4	1.6	1.7
SS B7	21	26	31	35	42	46	52	60	17	20	24	27	31	35	39	45	1.4	1.5	1.7	1.8	2.0	1.6	1.8	2.0	2.2	2.4
SS B8	30	39	48	55	64	70	77	85	23	30	38	44	52	57	64	73	1.0	1.1	1.3	1.5	1.6	1.1	1.3	1.5	1.6	1.8
SS C1	26	33	42	49	58	63	71	79	17	23	29	35	42	47	53	62	1.1	1.2	1.4	1.5	1.7	1.3	1.4	1.6	1.8	1.9
SS C2	30	40	50	58	67	73	80	88	21	29	37	44	53	58	66	75	1.0	1.1	1.3	1.4	1.6	1.1	1.3	1.4	1.6	1.7
SS C3	21	27	34	39	47	52	59	68	15	19	24	28	34	38	43	51	1.2	1.4	1.5	1.7	1.8	1.5	1.7	1.8	2.0	2.1
SS C4	27	35	45	52	61	67	74	83	19	26	33	39	47	52	59	69	1.0	1.2	1.3	1.5	1.6	1.2	1.3	1.5	1.6	1.8
SS D1	35	44	54	61	69	75	81	88	27	34	42	47	55	61	67	75	1.0	1.1	1.3	1.4	1.6	1.1	1.3	1.4	1.6	1.7
SS D2	43	54	65	72	81	86	91	95	33	42	53	60	69	75	81	88	0.8	1.0	1.2	1.3	1.5	0.9	1.1	1.3	1.4	1.6
SS D3	32	40	48	54	63	68	75	82	25	31	38	43	50	54	61	69	1.0	1.2	1.3	1.5	1.7	1.2	1.4	1.5	1.7	1.8
SS D4	41	52	63	70	79	84	89	94	32	41	51	58	67	73	79	87	0.9	1.0	1.2	1.4	1.5	1.0	1.1	1.3	1.4	1.6
SS E1	30	39	48	55	63	69	76	83	22	29	36	41	48	53	60	68	1.0	1.2	1.3	1.5	1.6	1.2	1.4	1.5	1.7	1.8
SS E2	37	48	59	66	75	81	87	92	28	36	46	53	62	68	75	83	0.9	1.0	1.2	1.4	1.5	1.0	1.2	1.3	1.5	1.6
SS E3	26	33	40	45	53	59	65	74	20	25	30	35	41	45	50	58	1.2	1.3	1.5	1.6	1.8	1.4	1.5	1.7	1.9	2.0
SS E4	34	44	54	62	71	76	83	89	26	34	42	49	58	63	70	79	0.9	1.1	1.2	1.4	1.6	1.1	1.2	1.4	1.5	1.7

LEXINGTON, KENTUCKY

SOLAR SAVINGS FRACTION (%) / CONSERVATION FACTOR

TYPE	BASE TEMP: 55 F (2743 DD)								65 F (4732 DD)								CF 55 F					CF 65 F				
LCR	100	70	50	40	30	25	20	15	100	70	50	40	30	25	20	15	100	50	30	20	15	100	50	30	20	15
WW A1	18	21	25	28	32	35	39	45	14	16	18	20	22	24	26	30	1.7	1.9	2.1	2.2	2.4	2.2	2.4	2.8	3.1	3.4
WW A2	19	24	30	34	41	45	51	60	13	16	20	23	28	31	35	41	1.3	1.5	1.7	1.8	2.0	1.6	1.8	2.1	2.3	2.5
WW A3	19	25	31	37	44	49	56	65	12	16	21	25	30	34	39	45	1.2	1.4	1.6	1.7	1.9	1.5	1.7	1.9	2.1	2.3
WW A4	18	25	32	38	46	51	58	67	11	16	21	25	31	35	41	48	1.2	1.4	1.5	1.7	1.8	1.4	1.6	1.8	2.0	2.2
WW A5	19	26	34	40	48	54	61	70	11	16	22	26	33	37	43	50	1.2	1.3	1.5	1.6	1.8	1.4	1.6	1.8	2.0	2.1
WW A6	19	26	34	41	49	55	62	72	11	16	22	27	33	38	44	52	1.1	1.3	1.4	1.6	1.8	1.4	1.5	1.7	1.9	2.1
WW B1	15	18	22	25	30	33	38	44	9	11	13	15	17	19	20	23	1.6	1.8	2.0	2.2	2.4	2.2	2.6	3.1	3.7	4.3
WW B2	19	26	34	40	49	55	62	71	12	18	24	28	35	40	46	54	1.2	1.3	1.5	1.6	1.8	1.3	1.5	1.7	1.9	2.0
WW B3	23	32	41	48	57	63	71	79	15	22	29	35	42	48	55	64	1.1	1.2	1.4	1.5	1.7	1.2	1.4	1.6	1.7	1.9
WW B4	23	33	43	50	60	67	74	83	15	23	31	38	47	53	60	70	1.0	1.1	1.3	1.5	1.6	1.1	1.3	1.5	1.6	1.8
WW B5	22	32	42	50	60	67	74	83	15	22	31	38	47	53	61	71	1.0	1.1	1.3	1.5	1.6	1.1	1.3	1.4	1.6	1.7
WW C1	25	34	43	50	60	66	73	82	17	24	31	37	45	51	58	67	1.0	1.2	1.3	1.5	1.6	1.2	1.3	1.5	1.7	1.8
WW C2	23	32	42	49	59	65	73	81	15	22	30	37	45	51	58	67	1.0	1.2	1.3	1.5	1.6	1.2	1.3	1.5	1.7	1.8
WW C3	28	39	51	59	69	75	82	90	19	29	39	47	57	63	71	80	0.9	1.1	1.2	1.4	1.5	1.0	1.2	1.3	1.5	1.6
WW C4	25	36	47	55	66	72	80	88	17	26	36	43	53	60	68	78	0.9	1.1	1.2	1.4	1.6	1.1	1.2	1.4	1.5	1.7
TW A1	18	21	25	28	32	35	39	45	14	16	18	20	23	25	27	31	1.7	1.9	2.1	2.2	2.4	2.1	2.4	2.7	3.0	3.2
TW A2	18	22	27	31	37	42	47	55	13	16	19	22	26	29	33	38	1.4	1.6	1.7	1.9	2.0	1.7	2.0	2.2	2.4	2.6
TW A3	17	22	28	33	40	44	51	59	12	15	19	23	27	31	35	41	1.3	1.5	1.7	1.8	2.0	1.6	1.8	2.0	2.2	2.4
TW A4	16	22	28	33	40	45	52	60	11	14	19	23	28	31	36	43	1.3	1.5	1.6	1.8	1.9	1.6	1.8	2.0	2.1	2.3
TW B1	17	21	25	28	33	36	41	47	13	16	18	20	23	25	28	33	1.6	1.8	2.0	2.1	2.3	2.0	2.2	2.5	2.7	2.9
TW B2	17	21	26	30	37	41	47	55	12	15	18	21	26	29	33	39	1.4	1.6	1.7	1.9	2.0	1.7	1.9	2.1	2.3	2.5
TW B3	16	21	26	31	37	42	48	56	11	14	18	21	26	29	34	40	1.4	1.6	1.7	1.9	2.0	1.7	1.9	2.1	2.3	2.4
TW B4	15	20	25	29	35	40	46	54	10	14	17	20	25	28	33	39	1.4	1.6	1.7	1.9	2.0	1.7	1.9	2.1	2.3	2.4
TW C1	16	20	24	27	33	36	41	48	12	15	18	20	23	26	29	34	1.6	1.7	1.9	2.1	2.2	1.9	2.1	2.3	2.5	2.7
TW C2	15	19	24	28	33	37	43	51	11	14	17	20	24	27	31	36	1.5	1.7	1.8	2.0	2.1	1.8	2.0	2.2	2.4	2.5
TW C3	15	18	23	27	32	36	41	49	11	13	17	19	23	26	30	36	1.5	1.7	1.9	2.0	2.1	1.8	2.0	2.2	2.4	2.6
TW C4	14	18	22	25	29	33	38	44	11	13	16	18	22	24	28	33	1.7	1.8	2.0	2.1	2.3	2.0	2.2	2.4	2.5	2.7
TW D1	12	15	19	22	26	29	34	40	8	10	12	13	16	17	19	22	1.7	1.9	2.1	2.3	2.4	2.2	2.6	3.0	3.5	3.9
TW D2	19	25	32	37	45	50	57	66	13	17	23	27	33	37	42	50	1.2	1.4	1.5	1.7	1.8	1.5	1.6	1.8	2.0	2.1
TW D3	20	27	35	41	49	55	62	71	13	19	24	29	36	40	46	54	1.2	1.3	1.5	1.6	1.8	1.4	1.5	1.7	1.9	2.0
TW D4	22	30	39	45	55	61	68	77	15	21	28	34	42	47	54	64	1.1	1.2	1.4	1.5	1.7	1.2	1.4	1.6	1.7	1.8
TW D5	21	29	39	46	55	61	69	78	14	21	29	35	43	49	56	66	1.1	1.2	1.4	1.5	1.7	1.2	1.4	1.5	1.7	1.8
TW E1	23	31	39	46	54	60	67	76	16	22	29	34	41	46	52	61	1.1	1.2	1.4	1.6	1.7	1.3	1.4	1.6	1.8	1.9
TW E2	22	30	39	46	55	61	68	77	15	22	29	34	42	47	54	63	1.1	1.2	1.4	1.5	1.7	1.2	1.4	1.6	1.7	1.9
TW E3	27	37	48	55	65	71	79	87	19	27	36	43	52	59	66	75	1.0	1.1	1.3	1.4	1.6	1.1	1.2	1.4	1.6	1.7
TW E4	25	35	45	53	63	69	77	85	17	25	34	41	50	57	64	74	1.0	1.1	1.3	1.4	1.6	1.1	1.3	1.4	1.6	1.7
TW F1	16	19	23	26	31	34	38	45	12	14	17	19	21	23	26	30	1.6	1.8	2.0	2.2	2.3	2.1	2.3	2.6	2.9	3.1
TW F2	15	20	25	29	35	39	45	52	10	13	17	20	24	26	30	36	1.4	1.6	1.8	1.9	2.1	1.8	2.0	2.2	2.4	2.6
TW F3	14	19	25	29	36	40	47	55	9	12	16	19	24	27	32	38	1.4	1.6	1.7	1.9	2.0	1.7	1.9	2.1	2.3	2.5
TW F4	12	17	23	28	35	39	46	54	7	11	15	18	23	26	31	37	1.4	1.5	1.7	1.9	2.0	1.7	1.9	2.1	2.2	2.4
TW G1	14	18	22	25	30	33	38	45	10	12	15	17	20	23	26	30	1.6	1.8	2.0	2.1	2.3	2.0	2.2	2.5	2.7	2.9
TW G2	13	17	22	26	32	36	42	50	8	11	14	17	21	24	28	34	1.5	1.7	1.8	2.0	2.1	1.8	2.0	2.2	2.4	2.6
TW G3	11	15	21	25	31	35	41	49	6	9	13	16	20	23	27	33	1.5	1.6	1.8	2.0	2.1	1.8	2.0	2.2	2.4	2.5
TW G4	9	13	17	21	27	31	37	44	4	7	11	14	18	21	25	30	1.6	1.7	1.9	2.0	2.2	1.9	2.1	2.2	2.4	2.6
TW H1	11	15	19	22	27	31	36	43	7	10	13	15	18	21	24	29	1.6	1.8	2.0	2.1	2.3	2.0	2.2	2.4	2.7	2.8
TW H2	9	13	17	21	26	30	35	42	5	7	11	13	17	20	23	29	1.6	1.8	1.9	2.1	2.2	2.0	2.2	2.4	2.6	2.7
TW H3	7	11	15	18	23	26	31	39	4	6	9	11	15	17	21	26	1.7	1.9	2.0	2.2	2.3	2.0	2.2	2.4	2.6	2.8
TW H4	5	8	12	14	18	21	26	32	3	5	7	9	12	14	17	22	1.9	2.1	2.2	2.4	2.5	2.3	2.5	2.6	2.8	3.0
TW I1	10	13	17	19	23	26	30	36	6	8	10	11	13	15	16	19	1.8	2.0	2.2	2.4	2.5	2.3	2.7	3.1	3.7	4.1
TW I2	16	22	28	34	41	47	54	63	10	14	19	23	29	33	39	46	1.3	1.4	1.6	1.7	1.9	1.5	1.7	1.8	2.0	2.2
TW I3	17	24	31	37	45	51	58	67	11	16	22	26	32	37	43	51	1.2	1.4	1.5	1.7	1.8	1.4	1.6	1.7	1.9	2.1
TW I4	19	26	35	42	51	57	65	74	12	18	25	31	39	44	51	60	1.1	1.3	1.4	1.6	1.7	1.3	1.4	1.6	1.7	1.9
TW I5	19	27	36	43	53	59	67	76	12	19	26	32	41	46	54	63	1.1	1.2	1.4	1.5	1.7	1.2	1.4	1.5	1.7	1.8
TW J1	21	28	37	43	52	58	65	74	14	20	26	31	39	44	50	59	1.1	1.3	1.4	1.6	1.7	1.3	1.5	1.6	1.8	1.9
TW J2	20	28	36	43	52	58	66	75	13	19	26	32	39	45	51	61	1.1	1.3	1.4	1.6	1.7	1.3	1.4	1.6	1.7	1.9
TW J3	24	34	45	52	63	69	76	85	16	24	34	41	50	56	64	74	1.0	1.1	1.3	1.4	1.6	1.1	1.2	1.4	1.6	1.7
TW J4	22	32	42	50	60	67	74	83	15	23	32	38	48	54	62	72	1.0	1.1	1.3	1.5	1.6	1.1	1.3	1.4	1.6	1.7
DG A1	10	16	20	23	27	29	32	35	7	9	12	13	15	16	17	18	1.6	1.9	2.3	2.9	3.7	2.1	2.6	3.3	-	-
DG A2	15	20	26	30	36	40	46	52	10	14	17	21	25	28	32	37	1.4	1.6	1.8	2.0	2.3	1.7	1.9	2.1	2.4	2.8
DG A3	19	25	33	39	47	53	60	69	14	19	24	29	36	41	47	55	1.2	1.3	1.5	1.7	1.8	1.4	1.5	1.7	1.8	2.0
DG B1	13	16	21	24	29	33	37	42	7	10	12	14	16	17	19	21	1.6	1.7	2.0	2.3	2.7	2.0	2.5	3.0	3.7	4.8
DG B2	15	20	26	31	38	43	50	58	10	14	18	21	26	30	34	40	1.4	1.5	1.6	1.8	2.0	1.6	1.8	2.0	2.2	2.4
DG B3	19	25	34	40	49	56	64	74	14	19	25	30	37	43	50	59	1.2	1.3	1.4	1.5	1.7	1.4	1.5	1.6	1.7	1.9
DG C1	15	20	25	30	36	41	46	53	10	13	16	18	21	24	27	31	1.4	1.5	1.7	2.0	2.2	1.7	2.1	2.4	2.8	3.2
DG C2	18	24	31	37	45	51	58	67	12	17	22	26	31	36	41	49	1.3	1.4	1.5	1.6	1.8	1.5	1.7	1.8	2.0	2.2
DG C3	22	29	38	45	55	63	71	81	16	22	29	34	42	48	56	66	1.2	1.2	1.3	1.5	1.6	1.3	1.4	1.5	1.6	1.8
SS A1	22	27	33	37	44	48	54	62	17	21	25	29	34	37	41	47	1.3	1.5	1.6	1.8	1.9	1.5	1.7	2.0	2.2	2.3
SS A2	27	35	43	49	58	63	70	78	21	28	34	39	47	51	58	66	1.1	1.2	1.4	1.6	1.7	1.2	1.4	1.6	1.7	1.9
SS A3	20	25	30	34	40	44	49	56	16	19	23	25	29	32	36	40	1.4	1.6	1.7	1.9	2.1	1.6	1.9	2.2	2.4	2.7
SS A4	27	34	42	48	56	62	68	77	21	27	33	38	45	50	56	64	1.1	1.2	1.4	1.6	1.7	1.2	1.4	1.6	1.8	1.9
SS A5	22	26	31	34	40	44	49	56	17	21	24	27	31	33	37	42	1.4	1.6	1.8	1.9	2.1	1.7	1.9	2.2	2.4	2.6
SS A6	27	34	42	49	57	63	69	78	21	27	34	39	46	50	57	65	1.1	1.2	1.4	1.6	1.7	1.2	1.4	1.6	1.8	1.9
SS A7	20	23	27	30	34	37	42	48	16	18	21	23	26	28	30	34	1.6	1.8	2.0	2.2	2.3	1.9	2.2	2.5	2.9	3.2
SS A8	26	33	41	47	55	60	67	75	20	26	32	37	43	48	54	61	1.1	1.3	1.4	1.6	1.7	1.3	1.5	1.6	1.8	2.0
SS B1	19	23	28	32	38	43	48	56	14	18	21	24	29	32	35	41	1.4	1.6	1.7	1.9	2.0	1.6	1.9	2.1	2.3	2.5
SS B2	24	31	38	44	53	58	65	74	18	24	30	35	42	47	53	61	1.1	1.3	1.4	1.6	1.7	1.3	1.5	1.6	1.8	1.9
SS B3	17	22	26	30	35	39	44	52	13	16	20	22	26	28	32	36	1.5	1.7	1.8	2.0	2.1	1.7	2.0	2.3	2.5	2.8
SS B4	23	30	37	43	51	57	63	72	18	23	29	34	41	45	51	59	1.2	1.3	1.5	1.6	1.8	1.3	1.5	1.7	1.8	2.0
SS B5	18	21	25	29	33	37	41	48	14	17	20	22	25	27	30	35	1.6	1.8	1.9	2.1	2.3	1.9	2.1	2.4	2.7	2.9
SS B6	23	30	37	43	51	57	64	73	18	23	29	34	41	45	51	59	1.2	1.3	1.5	1.6	1.8	1.3	1.5	1.7	1.8	2.0
SS B7	16	19	23	25	29	32	36	42	13	15	17	19	22	24	26	29	1.7	1.9	2.1	2.3	2.5	2.1	2.4	2.7	3.1	3.3
SS B8	22	28	36	41	49	55	61	70	17	22	28	32	38	43	48	56	1.2	1.3	1.5	1.7	1.8	1.3	1.5	1.7	1.9	2.0
SS C1	16	21	27	32	39	43	49	58	10	14	18	21	25	28	32	38	1.3	1.5	1.7	1.8	2.0	1.7	1.9	2.1	2.3	2.5
SS C2	20	27	35	41	49	55	62	71	13	18	25	29	36	41	47	55	1.2	1.3	1.5	1.6	1.8	1.3	1.5	1.7	1.9	2.0
SS C3	15	19	23	26	32	35	40	47	11	13	16	18	22	24	27	32	1.6	1.7	1.9	2.1	2.2	1.9	2.2	2.4	2.7	2.9
SS C4	18	24	31	36	44	49	56	65	13	17	22	26	32	36	42	49	1.2	1.4	1.5	1.7	1.8	1.5	1.6	1.8	2.0	2.1
SS D1	25	31	39	44	52	57	64	72	19	24	29	33	39	43	48	55	1.2	1.3	1.5	1.7	1.8	1.3	1.6	1.8	2.0	2.1
SS D2	31	40	50	57	66	72	79	86	23	31	39	46	54	59	66	75	1.0	1.1	1.3	1.5	1.6	1.1	1.3	1.4	1.6	1.7
SS D3	24	29	35	40	47	51	57	65	18	23	27	30	35	39	43	49	1.3	1.4	1.6	1.8	1.9	1.5	1.7	1.9	2.1	2.3
SS D4	30	38	48	55	64	70	77	84	23	30	38	44	52	57	64	73	1.0	1.1	1.3	1.5	1.6	1.1	1.3	1.5	1.6	1.8
SS E1	21	27	33	38	45	50	56	64	15	19	24	27	32	35	40	46	1.2	1.4	1.6	1.8	1.9	1.5	1.7	2.0	2.2	2.4
SS E2	26	34	43	50	59	65	72	80	19	26	33	38	46	51	58	66	1.0	1.2	1.3	1.5	1.7	1.2	1.4	1.5	1.7	1.8
SS E3	19	24	29	32	38	42	47	55	15	18	21	24	28	31	34	39	1.4	1.6	1.8	1.9	2.1	1.7	1.9	2.2	2.5	2.7
SS E4	24	32	40	46	55	60	67	76	18	24	31	36	43	47	54	62	1.1	1.3	1.4	1.6	1.7	1.2	1.4	1.6	1.8	1.9

LOUISVILLE, KENTUCKY

TYPE	SOLAR SAVINGS FRACTION (%) BASE TEMP: 55 F (2682 DD) LCR 100	70	50	40	30	25	20	15	65 F (4649 DD) 100	70	50	40	30	25	20	15	CONSERVATION FACTOR 55 F 100	50	30	20	15	65 F 100	50	30	20	15
WW A1	19	22	26	29	33	36	40	46	14	16	19	20	23	25	27	31	1.6	1.8	2.0	2.2	2.3	2.1	2.4	2.7	3.0	3.3
WW A2	19	25	31	35	42	47	53	61	13	17	21	24	29	32	36	42	1.3	1.5	1.6	1.8	1.9	1.6	1.8	2.0	2.2	2.4
WW A3	19	25	32	38	45	51	57	66	12	17	22	25	31	35	40	47	1.2	1.4	1.5	1.7	1.8	1.5	1.7	1.9	2.1	2.2
WW A4	19	26	34	39	47	53	60	69	12	17	22	26	32	36	42	49	1.2	1.3	1.5	1.7	1.8	1.4	1.6	1.8	2.0	2.1
WW A5	19	27	35	41	50	55	62	71	11	17	23	27	34	38	44	52	1.1	1.3	1.4	1.6	1.8	1.4	1.6	1.7	1.9	2.1
WW A6	19	27	35	42	51	57	64	73	11	17	23	28	34	39	45	53	1.1	1.3	1.4	1.6	1.7	1.3	1.5	1.7	1.9	2.0
WW B1	15	19	23	26	31	35	39	46	10	12	14	16	18	20	21	24	1.6	1.8	2.0	2.1	2.3	2.1	2.5	2.9	3.5	4.0
WW B2	20	27	35	42	50	56	63	72	13	18	24	29	36	41	47	55	1.1	1.3	1.4	1.6	1.7	1.3	1.5	1.7	1.9	2.0
WW B3	24	33	42	49	58	65	72	80	16	22	30	36	43	49	56	65	1.0	1.2	1.3	1.5	1.7	1.2	1.4	1.5	1.7	1.8
WW B4	24	34	44	52	62	68	75	84	16	23	32	39	48	54	61	71	1.0	1.1	1.3	1.5	1.6	1.1	1.3	1.4	1.6	1.7
WW B5	23	33	43	51	61	68	75	84	15	23	32	39	48	54	62	72	1.0	1.1	1.3	1.5	1.6	1.1	1.3	1.4	1.6	1.7
WW C1	26	35	44	51	61	67	74	83	17	24	32	38	46	52	59	68	1.0	1.2	1.3	1.5	1.6	1.2	1.3	1.5	1.7	1.8
WW C2	24	33	43	50	60	66	74	82	16	23	31	37	46	52	59	68	1.0	1.2	1.3	1.5	1.6	1.2	1.3	1.5	1.6	1.8
WW C3	29	40	52	60	70	76	83	90	20	29	40	47	58	64	72	81	0.9	1.1	1.2	1.4	1.5	1.0	1.2	1.3	1.5	1.6
WW C4	26	37	48	56	67	73	80	88	17	26	37	44	54	61	69	78	0.9	1.1	1.2	1.4	1.6	1.0	1.2	1.3	1.5	1.6
TW A1	18	22	25	28	33	36	40	46	14	16	19	21	23	25	28	32	1.6	1.8	2.0	2.2	2.3	2.1	2.3	2.6	2.9	3.1
TW A2	18	23	28	32	38	43	48	56	13	16	20	23	27	30	34	39	1.4	1.6	1.7	1.9	2.0	1.7	1.9	2.1	2.4	2.5
TW A3	18	23	29	34	41	46	52	60	12	16	20	23	28	32	36	43	1.3	1.5	1.6	1.8	1.9	1.6	1.8	2.0	2.2	2.4
TW A4	17	23	29	34	41	46	53	62	11	15	20	23	28	32	37	44	1.3	1.4	1.6	1.8	1.9	1.5	1.7	1.9	2.1	2.3
TW B1	18	21	25	29	34	37	42	48	13	16	18	21	24	26	29	34	1.6	1.8	1.9	2.1	2.2	2.0	2.2	2.4	2.7	2.9
TW B2	17	22	27	31	38	42	48	56	12	15	19	22	26	29	34	40	1.4	1.6	1.7	1.9	2.0	1.7	1.9	2.1	2.3	2.5
TW B3	16	21	27	32	38	43	49	57	11	14	19	22	27	30	34	41	1.4	1.5	1.7	1.8	2.0	1.6	1.8	2.0	2.2	2.4
TW B4	15	20	26	30	36	41	47	55	10	14	18	21	26	29	33	40	1.4	1.6	1.7	1.9	2.0	1.7	1.9	2.0	2.2	2.4
TW C1	17	20	25	28	34	37	42	50	12	15	18	20	24	27	30	35	1.5	1.7	1.9	2.0	2.2	1.9	2.1	2.3	2.5	2.7
TW C2	16	20	25	29	34	39	44	52	11	14	18	20	25	28	32	37	1.5	1.6	1.8	1.9	2.1	1.8	2.0	2.2	2.3	2.5
TW C3	15	19	24	27	33	37	42	50	11	14	17	20	24	27	31	36	1.5	1.7	1.8	2.0	2.1	1.8	2.0	2.2	2.4	2.5
TW C4	15	18	22	25	30	34	39	46	11	13	16	19	22	25	28	34	1.6	1.8	2.0	2.1	2.2	2.0	2.1	2.3	2.5	2.7
TW D1	13	16	20	23	27	31	35	41	8	10	12	14	16	18	20	23	1.7	1.9	2.1	2.2	2.4	2.2	2.5	2.9	3.3	3.7
TW D2	19	26	33	38	46	51	58	67	13	18	23	27	33	38	43	51	1.2	1.4	1.5	1.7	1.8	1.4	1.6	1.8	2.0	2.1
TW D3	21	28	36	42	50	56	63	72	14	19	25	30	36	41	47	55	1.1	1.3	1.4	1.6	1.8	1.3	1.5	1.7	1.9	2.0
TW D4	22	30	40	46	56	62	69	78	15	22	29	35	43	48	55	64	1.1	1.2	1.4	1.5	1.7	1.2	1.4	1.5	1.7	1.8
TW D5	22	30	40	47	56	63	70	79	15	22	30	35	44	50	57	67	1.1	1.2	1.4	1.5	1.7	1.2	1.3	1.5	1.7	1.8
TW E1	24	32	40	47	56	61	69	77	16	22	29	34	42	47	53	62	1.1	1.2	1.4	1.5	1.7	1.3	1.4	1.6	1.8	1.9
TW E2	23	31	40	47	56	62	69	78	16	22	29	35	43	48	55	64	1.1	1.2	1.4	1.5	1.7	1.2	1.4	1.6	1.7	1.9
TW E3	28	38	49	56	66	73	80	87	19	28	37	44	53	59	67	76	1.0	1.1	1.3	1.4	1.6	1.1	1.2	1.4	1.5	1.7
TW E4	26	35	46	54	64	70	78	86	18	26	35	42	51	58	65	75	1.0	1.1	1.3	1.4	1.6	1.1	1.2	1.4	1.6	1.7
TW F1	17	20	24	27	32	35	40	46	12	14	17	19	22	24	27	31	1.6	1.8	2.0	2.1	2.3	2.0	2.3	2.6	2.8	3.1
TW F2	16	20	26	30	36	40	46	54	11	14	17	20	24	27	31	37	1.4	1.6	1.8	1.9	2.0	1.7	2.0	2.2	2.4	2.6
TW F3	15	20	26	30	37	42	48	56	9	13	17	20	25	28	32	39	1.4	1.5	1.7	1.8	2.0	1.7	1.9	2.1	2.3	2.4
TW F4	13	18	24	29	36	41	47	56	7	11	15	19	24	27	32	39	1.4	1.5	1.7	1.8	2.0	1.6	1.8	2.0	2.2	2.4
TW G1	15	18	23	26	31	35	39	46	10	13	16	18	21	23	27	31	1.6	1.8	1.9	2.1	2.2	2.0	2.2	2.4	2.7	2.9
TW G2	13	18	23	27	33	37	43	51	8	11	15	18	22	25	29	35	1.5	1.6	1.8	1.9	2.1	1.8	2.0	2.2	2.4	2.6
TW G3	11	16	21	26	32	36	42	50	6	10	13	16	21	24	28	34	1.5	1.6	1.8	1.9	2.1	1.8	1.9	2.1	2.3	2.5
TW G4	9	13	18	22	28	32	38	46	5	8	11	14	18	21	25	31	1.5	1.7	1.9	2.0	2.1	1.8	2.0	2.2	2.4	2.6
TW H1	12	15	20	23	28	32	37	44	7	10	13	15	19	21	25	30	1.6	1.8	1.9	2.1	2.2	2.0	2.2	2.4	2.6	2.8
TW H2	9	13	18	21	27	31	36	43	5	8	11	13	17	20	24	29	1.6	1.8	1.9	2.1	2.2	1.9	2.1	2.3	2.5	2.7
TW H3	8	11	15	19	24	27	32	40	4	6	9	12	15	18	22	27	1.7	1.8	2.0	2.2	2.3	2.0	2.2	2.4	2.6	2.7
TW H4	6	9	12	15	19	22	27	33	3	5	7	9	12	14	18	22	1.8	2.0	2.2	2.3	2.5	2.3	2.4	2.6	2.8	2.9
TW I1	11	14	18	20	25	28	32	38	6	8	10	12	14	15	17	20	1.7	2.0	2.1	2.3	2.5	2.3	2.6	3.0	3.5	3.9
TW I2	16	22	29	35	43	48	55	64	10	15	20	24	30	34	40	48	1.3	1.4	1.6	1.7	1.8	1.5	1.6	1.8	2.0	2.1
TW I3	18	24	32	38	46	52	59	68	11	16	22	27	33	38	44	52	1.2	1.3	1.5	1.6	1.8	1.4	1.6	1.7	1.9	2.0
TW I4	19	27	36	43	52	59	66	76	12	19	26	32	39	45	52	61	1.1	1.2	1.4	1.5	1.7	1.3	1.4	1.6	1.7	1.9
TW I5	20	28	37	44	54	60	68	77	12	19	27	33	41	47	54	64	1.1	1.2	1.4	1.5	1.7	1.2	1.4	1.5	1.7	1.8
TW J1	21	29	38	44	53	59	67	76	14	20	27	32	40	45	51	60	1.1	1.2	1.4	1.6	1.7	1.3	1.4	1.6	1.8	1.9
TW J2	21	28	37	44	53	59	67	76	13	20	27	32	40	45	52	62	1.1	1.2	1.4	1.5	1.7	1.3	1.4	1.6	1.7	1.9
TW J3	25	35	46	53	64	70	77	86	17	25	34	41	51	57	65	75	1.0	1.1	1.3	1.4	1.6	1.1	1.2	1.4	1.6	1.7
TW J4	23	33	43	51	61	68	75	84	15	23	32	39	49	55	63	73	1.0	1.1	1.3	1.5	1.6	1.1	1.3	1.4	1.6	1.7
DG A1	13	17	21	24	28	31	33	36	8	10	12	14	16	17	18	19	1.5	1.8	2.2	2.8	3.5	2.0	2.5	3.2	-	-
DG A2	16	21	26	31	37	42	47	53	10	14	18	21	26	29	33	38	1.4	1.5	1.7	2.0	2.2	1.7	1.8	2.1	2.4	2.7
DG A3	19	26	33	40	48	54	61	70	14	19	25	29	36	41	48	56	1.2	1.3	1.4	1.6	1.8	1.4	1.5	1.7	1.8	2.0
DG B1	13	17	22	25	31	34	38	44	8	10	12	14	17	18	20	23	1.6	1.7	1.9	2.2	2.6	2.0	2.4	2.9	3.6	4.5
DG B2	16	21	27	32	40	45	51	60	11	14	19	22	27	30	35	42	1.4	1.5	1.6	1.8	2.0	1.6	1.8	2.0	2.2	2.4
DG B3	20	26	35	41	51	57	65	75	14	20	26	31	38	43	50	60	1.2	1.3	1.4	1.5	1.7	1.4	1.5	1.6	1.7	1.9
DG C1	16	21	27	31	38	42	48	55	10	13	16	19	22	25	28	32	1.4	1.5	1.7	1.9	2.2	1.7	2.0	2.4	2.7	3.1
DG C2	18	24	32	38	46	52	60	68	13	17	22	26	32	37	42	50	1.3	1.3	1.5	1.6	1.8	1.5	1.6	1.8	2.0	2.1
DG C3	22	30	39	46	57	64	72	82	16	22	29	35	43	49	57	67	1.1	1.2	1.3	1.5	1.6	1.3	1.4	1.5	1.6	1.7
SS A1	22	28	34	38	45	49	55	63	17	21	26	29	34	37	42	48	1.3	1.5	1.6	1.8	1.9	1.5	1.7	1.9	2.1	2.3
SS A2	28	35	44	50	59	64	71	79	22	28	35	40	47	52	58	67	1.1	1.2	1.4	1.5	1.7	1.2	1.4	1.6	1.7	1.9
SS A3	21	25	31	35	41	45	50	57	16	19	23	26	30	33	36	41	1.4	1.6	1.7	1.9	2.1	1.6	1.9	2.2	2.4	2.6
SS A4	27	35	43	49	57	63	69	78	21	27	34	39	46	50	56	64	1.1	1.2	1.4	1.6	1.7	1.2	1.4	1.6	1.8	1.9
SS A5	22	26	31	35	41	45	50	57	18	21	24	27	31	34	37	42	1.4	1.6	1.7	1.9	2.1	1.7	1.9	2.1	2.4	2.6
SS A6	28	35	43	50	58	64	70	78	21	27	34	39	46	51	57	65	1.1	1.2	1.4	1.6	1.7	1.2	1.4	1.6	1.7	1.9
SS A7	20	23	27	30	35	38	43	49	16	19	21	23	26	28	31	35	1.6	1.8	2.0	2.1	2.3	1.9	2.2	2.5	2.8	3.1
SS A8	27	34	42	48	56	61	68	76	21	26	33	37	44	48	54	62	1.1	1.3	1.4	1.6	1.7	1.3	1.4	1.6	1.8	2.0
SS B1	19	24	29	33	39	44	49	57	14	18	22	25	29	32	36	42	1.4	1.6	1.7	1.9	2.0	1.6	1.9	2.1	2.3	2.5
SS B2	24	31	39	45	54	59	66	75	19	24	31	36	43	47	54	62	1.1	1.3	1.4	1.6	1.7	1.3	1.4	1.6	1.8	1.9
SS B3	18	22	27	31	36	40	46	53	13	16	20	22	26	29	32	37	1.4	1.6	1.8	2.0	2.1	1.7	2.0	2.2	2.5	2.7
SS B4	24	30	38	44	52	58	64	73	18	24	30	34	41	46	52	60	1.1	1.3	1.4	1.6	1.8	1.3	1.5	1.6	1.8	2.0
SS B5	18	22	26	29	34	38	42	49	14	17	20	22	25	28	31	35	1.5	1.8	1.9	2.1	2.2	1.9	2.1	2.4	2.6	2.8
SS B6	24	30	38	44	52	58	65	73	18	23	30	34	41	46	52	60	1.1	1.3	1.4	1.6	1.7	1.3	1.5	1.6	1.8	1.9
SS B7	17	20	23	26	30	33	37	43	13	15	18	19	22	24	26	30	1.7	1.9	2.1	2.3	2.4	2.0	2.3	2.7	3.0	3.3
SS B8	23	29	36	42	50	56	62	71	17	22	28	33	39	43	49	57	1.2	1.3	1.5	1.6	1.8	1.3	1.5	1.7	1.9	2.0
SS C1	17	22	28	33	40	45	51	59	11	14	18	22	26	29	34	39	1.3	1.5	1.6	1.8	2.0	1.6	1.8	2.1	2.3	2.5
SS C2	21	28	36	42	51	56	64	73	13	19	25	30	37	42	48	56	1.1	1.3	1.4	1.6	1.7	1.3	1.5	1.7	1.8	2.0
SS C3	15	19	24	27	33	36	41	49	11	13	16	19	22	25	28	33	1.5	1.7	1.9	2.0	2.2	1.9	2.1	2.4	2.6	2.8
SS C4	19	25	32	37	45	51	57	66	13	17	23	27	33	37	43	50	1.2	1.4	1.5	1.7	1.8	1.4	1.6	1.8	2.0	2.1
SS D1	26	32	40	45	53	58	65	73	19	24	30	34	40	44	49	56	1.1	1.3	1.5	1.6	1.8	1.3	1.5	1.7	1.9	2.1
SS D2	31	41	51	58	67	73	79	87	24	32	40	46	55	60	67	75	1.0	1.1	1.3	1.4	1.6	1.1	1.2	1.4	1.6	1.7
SS D3	24	30	36	41	48	52	58	66	19	23	27	31	36	39	44	50	1.2	1.4	1.6	1.7	1.9	1.5	1.7	1.9	2.1	2.3
SS D4	30	39	49	56	65	71	77	85	23	31	39	45	53	58	65	73	1.0	1.1	1.3	1.5	1.6	1.1	1.3	1.4	1.6	1.8
SS E1	22	27	34	39	46	51	58	66	15	20	24	28	33	36	41	47	1.2	1.4	1.6	1.7	1.9	1.5	1.7	1.9	2.2	2.3
SS E2	27	35	44	51	60	66	73	81	19	26	34	39	47	52	59	67	1.0	1.2	1.3	1.5	1.7	1.2	1.3	1.5	1.7	1.8
SS E3	20	24	29	33	39	43	48	56	15	18	22	25	29	31	35	40	1.4	1.6	1.7	1.9	2.1	1.7	1.9	2.2	2.4	2.6
SS E4	25	32	41	47	56	61	68	77	19	25	31	36	43	48	54	63	1.1	1.2	1.4	1.6	1.7	1.2	1.4	1.6	1.8	1.9

BATON ROUGE, LOUISIANA

TYPE	SOLAR SAVINGS FRACTION (%) BASE TEMP: 55 F (587 DD)								65 F (1670 DD)								CONSERVATION FACTOR 55 F					65 F				
LCR	200	100	70	50	40	30	25	20	200	100	70	50	40	30	25	20	200	100	50	30	20	200	100	50	30	20
WW A1	33	49	58	68	74	82	86	91	20	28	35	42	47	55	60	66	0.7	0.9	1.0	1.2	1.3	1.1	1.2	1.3	1.5	1.6
WW A2	39	59	70	80	86	92	95	97	21	34	42	52	59	68	73	80	0.6	0.7	0.9	1.1	1.3	0.9	1.0	1.1	1.3	1.4
WW A3	42	63	74	84	89	94	96	98	21	35	45	56	63	72	78	84	0.6	0.7	0.9	1.1	1.2	0.8	0.9	1.1	1.2	1.4
WW A4	43	65	76	86	91	95	97	99	21	36	47	58	65	75	80	86	0.6	0.7	0.9	1.0	1.2	0.8	0.9	1.0	1.2	1.4
WW A5	44	67	78	87	92	96	98	99	21	38	49	60	68	77	82	88	0.6	0.7	0.9	1.0	1.2	0.8	0.9	1.0	1.2	1.4
WW A6	45	68	79	88	93	97	98	99	21	38	50	61	69	78	83	89	0.6	0.7	0.9	1.0	1.2	0.8	0.9	1.0	1.2	1.3
WW B1	35	53	64	74	80	87	91	95	18	28	36	44	50	59	65	71	0.7	0.8	1.0	1.1	1.3	1.0	1.1	1.2	1.4	1.5
WW B2	43	65	76	86	91	95	97	99	20	37	48	59	66	76	81	87	0.6	0.7	0.9	1.0	1.2	0.8	0.9	1.0	1.2	1.4
WW B3	49	72	83	91	95	98	99	100	25	44	55	67	74	83	88	92	0.6	0.7	0.8	1.0	1.2	0.7	0.8	1.0	1.1	1.3
WW B4	48	72	83	91	95	98	99	100	23	43	56	68	75	84	89	93	0.6	0.7	0.8	1.0	1.2	0.7	0.8	1.0	1.1	1.3
WW B5	46	70	81	90	94	97	99	100	21	42	54	66	74	83	88	93	0.6	0.7	0.8	1.0	1.2	0.7	0.8	1.0	1.1	1.3
WW C1	51	74	84	92	95	98	99	100	26	45	57	69	76	85	89	93	0.5	0.7	0.8	1.0	1.2	0.7	0.8	1.0	1.1	1.3
WW C2	48	71	82	90	94	98	99	100	23	43	55	67	74	83	88	92	0.6	0.7	0.8	1.0	1.2	0.7	0.8	1.0	1.1	1.3
WW C3	53	77	87	94	97	99	100	100	26	49	63	75	82	89	93	96	0.5	0.6	0.8	1.0	1.2	0.7	0.8	0.9	1.1	1.3
WW C4	49	74	85	92	96	98	99	100	23	45	59	71	79	87	91	95	0.5	0.7	0.8	1.0	1.2	0.7	0.8	0.9	1.1	1.3
TW A1	32	48	57	67	73	81	85	90	19	28	34	41	46	54	59	65	0.8	0.9	1.0	1.2	1.3	1.1	1.2	1.3	1.5	1.6
TW A2	36	54	65	75	81	88	92	95	19	31	39	47	54	63	68	75	0.7	0.8	0.9	1.1	1.3	0.9	1.0	1.2	1.3	1.5
TW A3	37	57	68	78	84	91	94	97	19	32	40	50	57	66	72	79	0.6	0.8	0.9	1.1	1.3	0.9	1.0	1.1	1.3	1.4
TW A4	37	57	68	79	85	91	94	97	18	31	40	50	57	67	73	79	0.6	0.8	0.9	1.1	1.3	0.9	1.0	1.1	1.3	1.4
TW B1	32	48	58	67	74	82	86	91	18	28	34	41	47	55	60	67	0.7	0.8	1.0	1.2	1.3	1.0	1.1	1.3	1.4	1.6
TW B2	34	52	63	74	80	87	91	95	18	29	37	46	52	61	67	74	0.7	0.8	1.0	1.1	1.3	0.9	1.0	1.2	1.3	1.5
TW B3	34	53	64	74	81	88	92	95	17	29	37	46	53	62	68	75	0.7	0.8	0.9	1.1	1.3	0.9	1.0	1.2	1.3	1.5
TW B4	32	50	61	71	78	86	90	94	16	27	35	44	50	59	65	72	0.7	0.8	1.0	1.1	1.3	1.0	1.1	1.2	1.3	1.5
TW C1	31	47	57	67	73	81	86	91	17	26	33	41	46	55	60	67	0.7	0.8	1.0	1.2	1.3	1.0	1.1	1.3	1.4	1.6
TW C2	30	47	58	68	75	83	87	92	16	26	33	41	47	56	62	69	0.7	0.8	1.0	1.2	1.3	1.0	1.1	1.2	1.4	1.5
TW C3	29	45	55	65	72	80	85	90	15	25	31	39	45	54	59	66	0.7	0.9	1.0	1.2	1.3	1.0	1.1	1.3	1.4	1.6
TW C4	26	40	50	59	66	75	80	86	14	23	28	35	41	48	53	60	0.8	0.9	1.1	1.2	1.4	1.1	1.2	1.4	1.5	1.7
TW D1	30	47	57	67	74	82	87	91	15	24	31	38	44	52	58	65	0.7	0.8	1.0	1.2	1.3	1.0	1.1	1.3	1.4	1.6
TW D2	39	60	71	81	87	93	95	98	20	34	44	54	61	71	76	83	0.6	0.7	0.9	1.1	1.3	0.8	0.9	1.1	1.2	1.4
TW D3	43	64	76	85	90	95	97	99	21	37	48	59	66	75	81	87	0.6	0.7	0.9	1.0	1.2	0.8	0.9	1.0	1.2	1.4
TW D4	44	67	78	87	92	95	98	99	22	39	51	62	70	79	84	90	0.6	0.7	0.9	1.0	1.2	0.8	0.9	1.0	1.2	1.3
TW D5	42	65	77	86	91	96	98	99	20	38	50	61	69	79	84	89	0.6	0.7	0.9	1.0	1.2	0.8	0.9	1.0	1.2	1.3
TW E1	46	69	80	88	93	97	98	99	24	41	52	63	71	80	85	90	0.6	0.7	0.9	1.0	1.2	0.7	0.9	1.0	1.2	1.3
TW E2	45	68	79	88	92	96	98	99	23	40	52	63	70	80	85	90	0.6	0.7	0.9	1.0	1.2	0.8	0.9	1.0	1.2	1.3
TW E3	51	75	85	93	96	98	99	100	26	47	60	72	79	87	91	95	0.5	0.7	0.8	1.0	1.2	0.7	0.8	0.9	1.1	1.3
TW E4	48	72	83	91	95	98	99	100	24	44	56	69	76	85	89	94	0.6	0.7	0.8	1.0	1.2	0.7	0.8	1.0	1.1	1.3
TW F1	31	47	57	67	73	81	86	90	18	27	33	40	46	53	59	65	0.7	0.9	1.0	1.2	1.3	1.0	1.2	1.3	1.5	1.6
TW F2	33	52	63	73	80	87	91	95	17	28	36	45	51	60	66	73	0.7	0.8	1.0	1.1	1.3	0.9	1.0	1.2	1.3	1.5
TW F3	33	53	64	75	81	89	92	95	16	28	37	46	53	62	68	75	0.7	0.8	0.9	1.1	1.3	0.9	1.0	1.2	1.3	1.5
TW F4	31	51	63	73	80	88	91	95	13	26	35	44	51	61	67	74	0.7	0.8	0.9	1.1	1.3	0.9	1.0	1.2	1.3	1.5
TW G1	29	46	56	66	73	81	85	90	15	25	31	39	45	53	58	65	0.7	0.9	1.0	1.2	1.3	1.0	1.2	1.3	1.4	1.6
TW G2	30	48	59	70	76	84	89	93	14	25	32	41	48	57	62	70	0.7	0.8	1.0	1.1	1.3	1.0	1.1	1.2	1.4	1.5
TW G3	28	46	57	68	75	83	88	92	12	23	31	39	46	55	61	69	0.7	0.8	1.0	1.1	1.3	1.0	1.1	1.2	1.4	1.5
TW G4	23	41	51	62	69	78	83	89	9	19	26	34	41	49	55	63	0.8	0.9	1.0	1.2	1.3	1.0	1.1	1.3	1.4	1.6
TW H1	26	42	52	63	70	78	83	89	12	21	28	36	41	50	55	62	0.8	0.9	1.0	1.2	1.4	1.1	1.2	1.3	1.5	1.6
TW H2	23	39	50	60	68	77	82	88	9	19	26	33	39	48	53	61	0.8	0.9	1.0	1.2	1.4	1.1	1.2	1.3	1.5	1.6
TW H3	20	35	45	55	62	72	77	84	7	16	22	29	35	43	48	56	0.8	0.9	1.1	1.2	1.4	1.1	1.2	1.4	1.5	1.7
TW H4	15	28	37	46	53	62	68	75	5	12	17	23	28	35	40	47	0.9	1.0	1.2	1.3	1.5	1.3	1.4	1.5	1.7	1.8
TW I1	27	43	54	64	71	79	84	89	12	21	28	35	41	49	54	61	0.7	0.9	1.0	1.2	1.3	1.1	1.2	1.3	1.5	1.6
TW I2	36	56	68	78	85	91	94	97	18	30	40	50	58	67	73	80	0.6	0.8	0.9	1.1	1.3	0.9	1.0	1.1	1.3	1.4
TW I3	38	60	72	82	88	93	96	98	18	33	43	54	62	71	77	84	0.6	0.7	0.9	1.1	1.2	0.8	0.9	1.1	1.2	1.4
TW I4	40	63	75	85	90	95	97	99	18	36	47	59	66	76	82	87	0.6	0.7	0.9	1.0	1.2	0.8	0.9	1.0	1.2	1.4
TW I5	40	63	75	85	90	95	97	99	18	36	47	59	67	77	82	88	0.6	0.7	0.9	1.0	1.2	0.8	0.9	1.0	1.2	1.3
TW J1	44	66	77	87	91	96	98	99	21	38	50	61	69	78	83	89	0.6	0.7	0.9	1.0	1.2	0.8	0.9	1.0	1.2	1.3
TW J2	42	65	76	86	91	96	97	99	20	37	48	60	68	77	82	88	0.6	0.7	0.9	1.0	1.2	0.8	0.9	1.0	1.2	1.3
TW J3	48	72	83	91	95	98	99	100	23	44	56	69	76	85	89	94	0.6	0.7	0.8	1.0	1.2	0.7	0.8	1.0	1.1	1.3
TW J4	45	69	80	89	94	97	99	99	21	41	53	66	73	83	87	92	0.6	0.7	0.8	1.0	1.2	0.7	0.8	1.0	1.1	1.3
DG A1	34	53	63	72	78	84	88	91	15	27	34	42	47	54	58	63	0.7	0.8	1.0	1.2	1.4	0.9	1.1	1.3	1.5	1.8
DG A2	34	54	65	75	81	88	91	94	16	28	37	46	53	61	67	73	0.7	0.8	0.9	1.1	1.3	0.9	1.0	1.2	1.4	1.6
DG A3	37	60	72	82	87	91	93	95	18	33	43	55	63	72	77	83	0.6	0.7	0.9	1.1	1.5	0.9	0.9	1.0	1.2	1.5
DG B1	35	57	69	79	85	90	93	95	15	27	36	46	53	62	67	72	0.6	0.7	0.9	1.1	1.4	1.0	1.0	1.1	1.4	1.6
DG B2	25	58	71	82	87	93	95	97	16	29	39	50	58	68	74	80	0.6	0.7	0.9	1.1	1.3	0.9	1.0	1.1	1.3	1.5
DG B3	38	62	76	86	91	95	96	97	19	33	45	57	66	76	82	87	0.6	0.7	0.9	1.1	1.5	0.9	0.9	1.0	1.2	1.4
DG C1	40	64	77	86	91	94	96	97	18	32	42	53	61	71	76	81	0.6	0.7	0.9	1.1	1.5	0.9	0.9	1.1	1.3	1.5
DG C2	39	64	77	87	92	96	97	99	19	33	44	56	65	75	80	86	0.6	0.7	0.9	1.1	1.3	0.9	0.9	1.0	1.2	1.4
DG C3	42	69	82	91	94	97	98	98	21	38	50	64	73	83	88	92	0.6	0.7	0.8	1.1	1.7	0.8	0.8	0.9	1.1	1.3
SS A1	42	61	72	81	87	92	95	97	24	38	47	56	63	71	76	82	0.6	0.7	0.9	1.1	1.3	0.8	0.9	1.1	1.3	1.4
SS A2	50	71	81	90	94	97	98	99	29	46	57	68	75	83	87	92	0.6	0.7	0.9	1.0	1.2	0.7	0.8	1.0	1.2	1.3
SS A3	41	59	69	79	85	91	94	97	23	36	44	53	59	68	73	79	0.6	0.8	0.9	1.1	1.3	0.9	1.0	1.1	1.3	1.5
SS A4	50	71	81	89	93	97	98	99	29	46	57	67	74	82	87	91	0.6	0.7	0.9	1.0	1.2	0.7	0.8	1.0	1.2	1.3
SS A5	40	57	67	77	83	89	93	96	24	35	43	52	58	66	71	78	0.7	0.8	0.9	1.1	1.3	0.9	1.0	1.2	1.3	1.5
SS A6	50	71	82	90	94	97	98	99	29	46	57	67	74	82	87	91	0.6	0.7	0.9	1.0	1.2	0.7	0.8	1.0	1.2	1.3
SS A7	36	53	62	72	78	85	89	93	22	32	39	47	52	60	65	71	0.7	0.8	1.0	1.1	1.3	1.0	1.1	1.2	1.4	1.6
SS A8	50	70	81	89	93	97	98	99	29	45	56	66	73	81	86	91	0.6	0.7	0.9	1.0	1.2	0.7	0.8	1.0	1.2	1.3
SS B1	37	55	66	76	82	89	92	95	21	33	41	50	56	65	70	77	0.7	0.8	0.9	1.1	1.3	0.9	1.0	1.2	1.3	1.5
SS B2	44	65	76	85	90	95	97	99	25	41	51	62	69	77	83	88	0.6	0.7	0.9	1.1	1.2	0.8	0.9	1.0	1.2	1.4
SS B3	36	54	64	74	81	88	91	95	20	31	39	48	54	63	68	75	0.7	0.8	1.0	1.1	1.3	0.9	1.0	1.2	1.3	1.5
SS B4	44	65	76	85	90	95	97	99	25	40	50	61	68	77	82	87	0.6	0.7	0.9	1.1	1.2	0.8	0.9	1.0	1.2	1.4
SS B5	33	50	59	69	75	83	87	92	20	29	36	44	49	57	63	69	0.7	0.8	1.0	1.2	1.3	1.0	1.1	1.3	1.4	1.6
SS B6	44	65	76	85	90	95	97	99	24	40	50	61	68	77	82	87	0.6	0.7	0.9	1.1	1.2	0.8	0.9	1.0	1.2	1.4
SS B7	31	46	55	65	71	79	84	89	18	27	33	40	45	53	58	64	0.8	0.9	1.0	1.2	1.4	1.1	1.2	1.3	1.5	1.6
SS B8	43	64	75	84	89	94	97	98	24	39	49	59	67	75	81	86	0.6	0.7	0.9	1.1	1.2	0.8	0.9	1.0	1.2	1.4
SS C1	38	58	69	79	85	92	95	97	19	32	41	51	58	67	73	79	0.6	0.7	0.9	1.1	1.3	0.9	1.0	1.1	1.3	1.4
SS C2	42	65	76	85	90	95	97	99	21	37	48	59	66	76	81	87	0.6	0.7	0.9	1.0	1.2	0.8	0.9	1.0	1.2	1.4
SS C3	31	48	58	68	75	83	87	92	16	26	33	41	47	55	61	68	0.7	0.8	1.0	1.2	1.3	1.0	1.1	1.3	1.4	1.6
SS C4	38	59	70	80	86	92	95	97	19	33	43	53	60	69	75	82	0.6	0.7	0.9	1.1	1.3	0.8	0.9	1.1	1.2	1.4
SS D1	50	71	81	89	93	97	98	99	28	45	55	66	72	81	85	90	0.6	0.7	0.9	1.0	1.2	0.7	0.8	1.0	1.2	1.4
SS D2	57	79	88	94	97	99	100	100	33	53	65	76	82	89	93	96	0.5	0.6	0.8	1.0	1.2	0.7	0.8	0.9	1.1	1.3
SS D3	45	65	75	84	89	94	96	98	26	40	50	59	66	74	80	85	0.6	0.7	0.9	1.1	1.2	0.8	0.9	1.1	1.2	1.4
SS D4	55	77	87	93	96	99	99	100	32	51	63	74	81	88	92	95	0.5	0.7	0.8	1.0	1.2	0.7	0.8	0.9	1.1	1.3
SS E1	45	66	77	86	90	95	97	99	24	40	49	60	67	75	80	86	0.6	0.7	0.9	1.1	1.2	0.8	0.9	1.0	1.2	1.4
SS E2	52	74	84	92	95	98	99	100	28	47	59	70	77	85	89	93	0.6	0.7	0.8	1.0	1.2	0.7	0.8	1.0	1.1	1.3
SS E3	38	56	67	76	82	89	92	96	22	33	41	50	56	65	70	77	0.7	0.8	0.9	1.1	1.3	0.9	1.0	1.2	1.3	1.5
SS E4	48	70	80	89	93	97	98	99	26	44	54	65	72	81	86	90	0.6	0.7	0.9	1.0	1.2	0.7	0.8	1.0	1.2	1.3

LAKE CHARLES, LOUISIANA

TYPE	SOLAR SAVINGS FRACTION (%)																CONSERVATION FACTOR									
	BASE TEMP: 55 F (500 DD)								65 F (1498 DD)								55 F					65 F				
LCR	200	100	70	50	40	30	25	20	200	100	70	50	40	30	25	20	200	100	50	30	20	200	100	50	30	20
WW A1	34	50	60	69	76	83	87	92	20	29	35	42	47	55	60	66	0.7	0.8	1.0	1.2	1.3	1.1	1.2	1.3	1.5	1.6
WW A2	41	61	72	81	87	92	95	97	21	34	42	52	59	68	73	80	0.6	0.7	0.9	1.1	1.3	0.9	1.0	1.1	1.3	1.4
WW A3	43	65	76	85	90	95	97	98	21	35	45	56	63	72	78	84	0.6	0.7	0.9	1.1	1.2	0.8	0.9	1.1	1.2	1.4
WW A4	44	67	78	87	91	96	97	99	21	36	47	58	65	75	80	86	0.6	0.7	0.9	1.0	1.2	0.8	0.9	1.0	1.2	1.4
WW A5	46	69	80	88	93	96	98	99	21	38	49	60	68	77	82	88	0.6	0.7	0.9	1.0	1.2	0.8	0.9	1.0	1.2	1.4
WW A6	47	70	81	89	93	97	98	99	21	38	50	61	69	78	83	89	0.6	0.7	0.9	1.0	1.2	0.8	0.9	1.0	1.2	1.3
WW B1	36	55	66	76	82	88	92	95	18	28	36	44	50	59	65	71	0.7	0.8	0.9	1.1	1.3	1.0	1.1	1.2	1.4	1.5
WW B2	44	67	78	87	91	96	97	99	21	37	48	59	66	76	81	87	0.6	0.7	0.9	1.0	1.2	0.8	0.9	1.0	1.2	1.4
WW B3	51	74	84	92	95	98	99	100	25	44	55	67	74	83	88	92	0.5	0.7	0.8	1.0	1.2	0.7	0.8	1.0	1.1	1.3
WW B4	50	74	84	92	95	98	99	100	23	43	56	68	75	84	89	93	0.5	0.7	0.8	1.0	1.2	0.7	0.8	1.0	1.1	1.3
WW B5	48	72	83	91	95	98	99	100	21	42	54	66	74	83	88	92	0.6	0.7	0.8	1.0	1.2	0.7	0.8	1.0	1.1	1.3
WW C1	52	75	85	92	96	98	99	100	26	45	57	69	76	84	89	93	0.5	0.7	0.8	1.0	1.2	0.7	0.8	1.0	1.1	1.3
WW C2	49	73	83	91	95	98	99	100	23	43	55	67	74	83	88	92	0.5	0.7	0.8	1.0	1.2	0.7	0.8	1.0	1.1	1.3
WW C3	55	79	88	95	97	99	100	100	26	49	63	75	82	89	93	96	0.5	0.6	0.8	1.0	1.2	0.7	0.8	0.9	1.1	1.3
WW C4	51	75	86	93	96	98	99	100	23	45	59	71	78	87	91	95	0.5	0.6	0.8	1.0	1.2	0.7	0.8	0.9	1.1	1.3
TW A1	33	49	59	68	74	82	86	91	20	28	34	41	46	54	59	65	0.7	0.8	1.0	1.2	1.3	1.1	1.2	1.3	1.5	1.6
TW A2	37	56	67	77	83	89	92	96	19	31	39	47	54	63	68	75	0.7	0.8	0.9	1.1	1.3	0.9	1.0	1.2	1.3	1.5
TW A3	39	59	70	80	85	91	94	97	19	32	40	50	57	66	72	78	0.6	0.7	0.9	1.1	1.3	0.9	1.0	1.1	1.3	1.5
TW A4	38	59	70	80	86	92	95	97	18	31	40	50	57	67	73	79	0.6	0.7	0.9	1.1	1.3	0.9	1.0	1.1	1.3	1.4
TW B1	33	49	59	69	75	83	87	92	19	28	34	41	47	55	60	67	0.7	0.8	1.0	1.2	1.3	1.0	1.2	1.3	1.4	1.6
TW B2	35	54	65	75	81	88	92	95	18	29	37	46	52	61	67	74	0.7	0.8	0.9	1.1	1.3	0.9	1.0	1.2	1.3	1.5
TW B3	35	54	65	76	82	89	92	95	17	29	37	46	53	62	68	75	0.7	0.8	0.9	1.1	1.3	0.9	1.0	1.2	1.3	1.5
TW B4	33	52	63	73	79	87	91	94	16	27	35	44	50	59	65	72	0.7	0.8	1.0	1.1	1.3	1.0	1.1	1.2	1.3	1.5
TW C1	32	48	58	68	75	83	87	91	17	26	33	41	46	55	60	67	0.7	0.8	1.0	1.2	1.3	1.0	1.1	1.3	1.4	1.6
TW C2	31	49	59	70	76	84	88	92	16	26	33	41	47	56	62	69	0.7	0.8	1.0	1.1	1.3	1.0	1.1	1.2	1.4	1.6
TW C3	29	46	57	67	74	82	86	91	15	25	31	39	45	54	59	66	0.7	0.8	1.0	1.2	1.3	1.0	1.1	1.3	1.4	1.6
TW C4	27	42	51	61	68	76	81	87	15	23	28	35	41	48	53	60	0.8	0.9	1.1	1.2	1.4	1.1	1.2	1.4	1.5	1.7
TW D1	31	48	59	69	76	83	88	92	15	24	31	38	44	52	58	65	0.7	0.8	1.0	1.2	1.3	1.0	1.2	1.3	1.4	1.6
TW D2	40	62	73	82	88	93	96	98	20	34	44	54	61	71	76	83	0.6	0.7	0.9	1.1	1.3	0.8	0.9	1.1	1.2	1.4
TW D3	44	66	77	86	91	95	97	99	21	37	48	59	66	75	81	87	0.6	0.7	0.9	1.1	1.2	0.8	0.9	1.0	1.2	1.4
TW D4	45	68	79	88	92	96	98	99	22	39	51	62	70	79	84	89	0.6	0.7	0.9	1.0	1.2	0.8	0.9	1.0	1.2	1.3
TW D5	44	67	78	87	92	96	98	99	20	38	50	61	69	78	84	89	0.6	0.7	0.9	1.0	1.2	0.8	0.9	1.0	1.2	1.3
TW E1	48	70	81	89	93	97	98	99	24	41	52	64	71	80	85	90	0.6	0.7	0.9	1.0	1.2	0.8	0.8	1.0	1.2	1.3
TW E2	47	69	80	89	93	97	98	99	23	40	52	63	70	79	84	90	0.6	0.7	0.9	1.0	1.2	0.8	0.9	1.0	1.2	1.3
TW E3	53	77	86	93	96	99	99	100	26	47	60	72	79	87	91	95	0.5	0.6	0.8	1.0	1.2	0.7	0.8	0.9	1.1	1.3
TW E4	50	73	84	92	95	98	99	100	24	44	56	69	76	85	89	93	0.5	0.7	0.8	1.0	1.2	0.7	0.8	1.0	1.1	1.3
TW F1	32	48	58	68	75	82	87	91	18	27	33	40	46	53	59	65	0.7	0.8	1.0	1.2	1.3	1.1	1.2	1.3	1.5	1.6
TW F2	35	54	65	75	81	88	92	95	17	28	36	45	51	60	66	73	0.7	0.8	0.9	1.1	1.3	0.9	1.1	1.2	1.3	1.5
TW F3	35	55	66	76	83	89	93	96	16	28	37	46	53	62	68	75	0.7	0.8	0.9	1.1	1.3	0.9	1.0	1.2	1.3	1.5
TW F4	32	53	64	75	81	88	92	95	14	26	35	44	51	61	67	74	0.7	0.8	0.9	1.1	1.3	0.9	1.0	1.2	1.3	1.5
TW G1	30	47	57	68	74	82	86	91	16	25	31	39	45	53	58	65	0.7	0.8	1.0	1.2	1.3	1.0	1.2	1.3	1.4	1.6
TW G2	31	50	61	71	78	85	89	93	14	25	32	41	48	57	62	70	0.7	0.8	1.0	1.1	1.3	1.0	1.1	1.2	1.4	1.5
TW G3	29	48	59	70	77	85	89	93	12	23	31	39	46	55	61	69	0.7	0.8	1.0	1.1	1.3	1.0	1.1	1.2	1.4	1.5
TW G4	24	42	53	64	71	80	85	90	9	19	26	34	41	50	55	63	0.7	0.8	1.0	1.2	1.3	1.0	1.2	1.3	1.4	1.6
TW H1	27	43	54	64	71	80	84	89	12	21	28	36	41	50	55	62	0.7	0.9	1.0	1.2	1.4	1.1	1.2	1.3	1.5	1.6
TW H2	24	41	52	62	69	78	83	88	10	19	26	33	39	48	53	61	0.8	0.9	1.0	1.2	1.4	1.1	1.2	1.3	1.5	1.6
TW H3	20	36	46	57	64	73	79	85	8	16	22	29	35	43	49	56	0.8	0.9	1.1	1.2	1.4	1.1	1.3	1.4	1.5	1.7
TW H4	15	29	38	48	55	64	70	77	6	12	17	23	28	35	40	47	0.9	1.0	1.1	1.3	1.5	1.3	1.4	1.5	1.7	1.8
TW I1	28	45	55	66	72	81	85	90	13	21	28	35	41	49	54	61	0.7	0.8	1.0	1.2	1.4	1.1	1.2	1.3	1.5	1.6
TW I2	37	58	70	80	86	92	95	97	17	30	40	50	58	67	73	80	0.6	0.7	0.9	1.1	1.3	0.9	1.0	1.1	1.3	1.4
TW I3	40	62	73	83	88	94	96	98	18	33	43	54	62	71	77	84	0.6	0.7	0.9	1.1	1.2	0.8	0.9	1.1	1.2	1.4
TW I4	42	65	77	86	91	95	97	99	19	36	47	59	66	76	81	87	0.6	0.7	0.9	1.0	1.2	0.8	0.9	1.0	1.2	1.4
TW I5	42	65	77	86	91	96	97	99	18	36	47	59	67	77	82	88	0.6	0.7	0.9	1.0	1.2	0.8	0.9	1.0	1.2	1.3
TW J1	45	68	79	88	92	96	98	99	21	38	50	61	69	78	83	88	0.6	0.7	0.9	1.0	1.2	0.8	0.9	1.0	1.2	1.4
TW J2	43	66	78	87	92	96	98	99	20	37	48	60	68	77	82	88	0.6	0.7	0.9	1.0	1.2	0.8	0.9	1.0	1.2	1.4
TW J3	49	73	84	92	95	98	99	100	23	44	56	69	76	85	89	94	0.5	0.7	0.8	1.0	1.2	0.7	0.8	1.0	1.1	1.3
TW J4	46	71	82	90	94	97	99	99	21	41	53	66	73	83	87	92	0.6	0.7	0.8	1.0	1.2	0.7	0.8	1.0	1.1	1.3
DG A1	35	55	65	74	80	86	89	92	16	27	34	42	47	54	58	63	0.7	0.8	1.0	1.2	1.4	1.0	1.1	1.3	1.5	1.8
DG A2	35	56	67	77	83	89	92	95	16	28	37	46	53	61	67	73	0.6	0.8	0.9	1.1	1.3	0.9	1.0	1.2	1.4	1.6
DG A3	39	62	73	83	87	92	94	95	18	33	43	55	63	72	77	83	0.6	0.7	0.9	1.2	1.6	0.9	0.9	1.0	1.2	1.5
DG B1	37	59	71	81	86	91	93	95	16	27	36	46	53	61	66	72	0.6	0.7	0.9	1.1	1.4	1.0	1.0	1.1	1.4	1.6
DG B2	37	60	73	83	88	93	95	97	16	29	39	50	58	68	73	80	0.6	0.7	0.9	1.1	1.3	0.9	1.0	1.1	1.3	1.5
DG B3	40	64	77	87	91	95	96	97	19	33	45	57	66	76	82	87	0.6	0.7	0.9	1.1	1.5	0.9	0.9	1.0	1.2	1.4
DG C1	42	66	79	87	91	95	96	97	18	32	42	53	61	71	76	81	0.6	0.7	0.9	1.1	1.5	0.9	0.9	1.1	1.3	1.5
DG C2	41	66	79	88	92	96	98	99	19	33	44	56	65	75	80	86	0.6	0.7	0.9	1.1	1.3	0.9	0.9	1.0	1.2	1.4
DG C3	44	70	83	91	94	97	98	98	21	38	50	64	73	83	87	92	0.6	0.7	0.8	1.1	1.7	0.8	0.8	0.9	1.1	1.3
SS A1	44	63	74	83	88	93	96	98	25	38	47	56	63	72	77	83	0.6	0.7	0.9	1.1	1.3	0.8	0.9	1.1	1.3	1.4
SS A2	52	73	83	91	94	97	99	99	29	47	57	68	75	83	87	92	0.6	0.7	0.8	1.0	1.2	0.7	0.8	1.0	1.2	1.3
SS A3	42	61	71	81	86	92	94	97	24	36	44	54	60	68	74	80	0.6	0.8	0.9	1.1	1.3	0.9	1.0	1.1	1.3	1.5
SS A4	52	73	83	90	94	97	98	99	29	46	57	68	75	82	87	91	0.6	0.7	0.9	1.0	1.2	0.7	0.8	1.0	1.2	1.3
SS A5	41	59	69	79	84	90	93	96	24	36	43	52	58	67	72	78	0.7	0.8	0.9	1.1	1.3	0.9	1.0	1.2	1.3	1.5
SS A6	52	73	83	91	94	97	99	99	29	46	57	68	75	83	87	92	0.6	0.7	0.8	1.0	1.2	0.7	0.8	1.0	1.2	1.3
SS A7	38	55	64	74	80	86	90	94	22	32	39	47	53	61	66	72	0.7	0.8	1.0	1.1	1.3	1.0	1.1	1.2	1.4	1.6
SS A8	52	72	82	90	94	97	98	99	29	46	56	67	74	82	86	91	0.6	0.7	0.9	1.0	1.2	0.7	0.8	1.0	1.2	1.3
SS B1	39	57	68	78	84	90	93	96	21	33	41	50	57	65	71	77	0.7	0.8	0.9	1.1	1.3	0.9	1.0	1.1	1.3	1.5
SS B2	46	67	78	87	91	96	97	99	25	41	51	62	69	78	83	88	0.6	0.7	0.9	1.1	1.2	0.8	0.9	1.0	1.2	1.4
SS B3	38	56	66	76	82	89	92	95	20	31	39	48	55	63	69	75	0.7	0.8	0.9	1.1	1.3	0.9	1.0	1.2	1.3	1.5
SS B4	46	67	78	86	91	95	97	99	25	40	51	61	68	77	82	88	0.6	0.7	0.9	1.1	1.2	0.8	0.9	1.0	1.2	1.4
SS B5	35	51	61	71	77	85	89	93	20	29	36	44	50	58	63	70	0.7	0.8	1.0	1.2	1.3	1.0	1.1	1.2	1.4	1.6
SS B6	46	67	78	86	91	95	97	99	24	40	51	61	68	77	82	88	0.6	0.7	0.9	1.1	1.2	0.8	0.9	1.0	1.2	1.4
SS B7	32	48	58	67	73	81	85	90	18	27	33	40	46	53	59	65	0.7	0.9	1.0	1.2	1.4	1.1	1.2	1.3	1.5	1.6
SS B8	45	66	77	85	90	95	97	98	24	39	49	60	67	76	81	86	0.6	0.7	0.9	1.1	1.2	0.8	0.9	1.0	1.2	1.4
SS C1	39	60	71	81	86	92	95	97	19	32	41	51	58	67	73	79	0.6	0.7	0.9	1.1	1.3	0.9	1.0	1.1	1.3	1.4
SS C2	44	66	77	86	91	96	97	99	21	37	48	59	66	76	81	87	0.6	0.7	0.9	1.0	1.2	0.8	0.9	1.0	1.2	1.4
SS C3	32	49	60	70	76	84	88	92	17	26	33	41	47	55	61	68	0.7	0.8	1.0	1.2	1.3	1.0	1.1	1.3	1.4	1.6
SS C4	39	60	72	81	87	93	95	97	19	33	43	53	60	69	75	82	0.6	0.7	0.9	1.1	1.3	0.9	0.9	1.1	1.2	1.4
SS D1	52	73	83	90	94	97	98	99	29	45	56	66	73	81	85	90	0.6	0.7	0.9	1.0	1.2	0.7	0.8	1.0	1.2	1.4
SS D2	59	81	89	95	97	99	100	100	33	54	66	76	83	89	93	96	0.5	0.6	0.8	1.0	1.2	0.7	0.8	0.9	1.1	1.3
SS D3	47	67	77	86	90	95	97	98	27	41	50	60	67	75	80	85	0.6	0.7	0.9	1.1	1.2	0.8	0.9	1.1	1.2	1.4
SS D4	57	79	88	94	97	99	99	100	32	52	64	74	81	88	92	95	0.5	0.6	0.8	1.0	1.2	0.7	0.8	0.9	1.1	1.3
SS E1	47	68	78	87	91	96	97	99	25	40	50	60	67	76	81	86	0.6	0.7	0.9	1.1	1.2	0.8	0.9	1.0	1.2	1.4
SS E2	54	76	86	93	96	98	99	100	28	48	59	70	77	85	89	93	0.5	0.7	0.8	1.0	1.2	0.7	0.8	1.0	1.1	1.3
SS E3	40	58	69	78	84	90	93	96	22	33	42	50	57	65	71	77	0.6	0.8	0.9	1.1	1.3	0.9	1.0	1.2	1.3	1.5
SS E4	50	72	82	90	94	97	98	99	26	44	55	66	73	81	86	91	0.6	0.7	0.9	1.0	1.2	0.7	0.8	1.0	1.2	1.3

NEW ORLEANS, LOUISIANA

TYPE	SOLAR SAVINGS FRACTION (%) BASE TEMP: 55 F (492 DD) LCR 200	100	70	50	40	30	25	20	65 F (1465 DD) 200	100	70	50	40	30	25	20	CONSERVATION FACTOR 55 F 200	100	50	30	20	65 F 200	100	50	30	20
WW A1	38	55	65	74	80	87	91	95	22	32	39	47	53	61	67	73	0.7	0.8	1.0	1.1	1.3	1.0	1.1	1.2	1.4	1.5
WW A2	45	66	77	86	91	95	97	99	24	39	48	59	66	75	80	86	0.6	0.7	0.9	1.1	1.2	0.8	0.9	1.1	1.2	1.4
WW A3	48	70	81	89	93	97	98	99	24	41	52	63	70	79	84	89	0.6	0.7	0.9	1.0	1.2	0.8	0.9	1.0	1.2	1.3
WW A4	49	72	83	90	94	97	99	99	24	42	54	65	72	81	86	91	0.6	0.7	0.8	1.0	1.2	0.7	0.8	1.0	1.2	1.3
WW A5	51	74	84	92	95	98	99	100	25	44	56	67	75	83	88	92	0.5	0.7	0.8	1.0	1.2	0.7	0.8	1.0	1.1	1.3
WW A6	52	75	85	93	96	98	99	100	25	45	57	69	76	84	89	93	0.5	0.7	0.8	1.0	1.2	0.7	0.8	1.0	1.1	1.3
WW B1	41	60	71	81	86	92	95	97	21	33	42	51	58	67	72	79	0.6	0.7	0.9	1.1	1.3	0.9	1.0	1.1	1.3	1.5
WW B2	49	72	83	91	94	98	99	99	24	43	54	66	73	82	87	92	0.6	0.7	0.8	1.0	1.2	0.7	0.8	1.0	1.1	1.3
WW B3	56	79	88	94	97	99	100	100	29	50	62	74	81	88	92	95	0.5	0.6	0.8	1.0	1.2	0.7	0.8	0.9	1.1	1.3
WW B4	55	79	88	95	97	99	100	100	28	50	63	74	81	89	93	96	0.5	0.6	0.8	1.0	1.2	0.7	0.8	0.9	1.1	1.3
WW B5	53	77	87	94	97	99	99	100	26	48	61	73	80	88	92	96	0.5	0.6	0.8	1.0	1.2	0.7	0.8	0.9	1.1	1.3
WW C1	58	80	89	95	97	99	100	100	30	52	64	75	82	89	93	96	0.5	0.6	0.8	1.0	1.2	0.7	0.8	0.9	1.1	1.3
WW C2	55	78	88	94	97	99	99	100	28	49	62	73	80	88	92	95	0.5	0.6	0.8	1.0	1.2	0.7	0.8	0.9	1.1	1.3
WW C3	60	83	92	97	98	100	100	100	31	56	69	81	87	93	96	98	0.5	0.6	0.8	1.0	1.2	0.6	0.7	0.9	1.1	1.2
WW C4	56	80	90	95	98	99	100	100	28	52	65	77	84	91	94	97	0.5	0.6	0.8	1.0	1.2	0.6	0.7	0.9	1.1	1.3
TW A1	36	53	63	73	79	86	90	94	21	31	38	46	52	60	65	72	0.7	0.8	1.0	1.1	1.3	1.0	1.1	1.2	1.4	1.5
TW A2	41	61	72	81	87	93	95	97	22	35	44	54	60	70	75	81	0.6	0.7	0.9	1.1	1.3	0.9	1.0	1.1	1.3	1.4
TW A3	43	64	75	84	89	94	97	98	22	36	46	57	64	73	79	85	0.6	0.7	0.9	1.1	1.2	0.8	0.9	1.1	1.2	1.4
TW A4	43	64	75	85	90	95	97	98	21	36	46	57	64	74	79	85	0.6	0.7	0.9	1.1	1.2	0.8	0.9	1.1	1.2	1.4
TW B1	36	54	64	74	80	87	91	94	21	31	38	47	53	61	67	74	0.7	0.8	1.0	1.1	1.3	1.0	1.1	1.2	1.4	1.5
TW B2	39	59	70	80	86	92	95	97	20	33	42	52	59	68	74	80	0.6	0.7	0.9	1.1	1.3	0.9	1.0	1.1	1.3	1.4
TW B3	39	59	71	81	86	92	95	97	20	33	42	52	60	69	75	81	0.6	0.7	0.9	1.1	1.3	0.9	1.0	1.1	1.3	1.4
TW B4	37	57	68	78	84	91	94	96	18	31	40	50	57	66	72	79	0.6	0.8	0.9	1.1	1.3	0.9	1.0	1.1	1.3	1.4
TW C1	35	53	63	73	80	87	91	94	19	30	37	46	52	61	67	74	0.7	0.8	1.0	1.1	1.3	1.0	1.1	1.2	1.3	1.5
TW C2	35	54	64	75	81	88	92	95	18	30	38	47	53	63	68	75	0.7	0.8	0.9	1.1	1.3	0.9	1.0	1.2	1.3	1.5
TW C3	33	51	62	72	79	86	90	94	17	28	36	45	51	60	66	73	0.7	0.8	1.0	1.1	1.3	1.0	1.1	1.2	1.3	1.5
TW C4	29	46	56	66	73	81	85	90	16	26	32	40	46	54	60	67	0.7	0.8	1.0	1.2	1.3	1.0	1.1	1.3	1.4	1.6
TW D1	35	53	64	74	81	88	91	95	17	28	36	45	51	60	66	73	0.7	0.8	0.9	1.1	1.3	1.0	1.1	1.2	1.3	1.5
TW D2	45	67	78	87	91	96	98	99	23	39	49	60	68	77	82	88	0.6	0.7	0.9	1.0	1.2	0.8	0.9	1.0	1.2	1.4
TW D3	49	72	82	90	94	97	99	99	25	43	54	66	73	82	86	91	0.6	0.7	0.8	1.0	1.2	0.7	0.8	1.0	1.2	1.3
TW D4	50	74	84	92	95	98	99	100	25	45	57	69	76	85	89	93	0.5	0.7	0.8	1.0	1.2	0.7	0.8	1.0	1.1	1.3
TW D5	49	72	83	91	95	98	99	100	24	44	56	68	75	84	89	93	0.6	0.7	0.8	1.0	1.2	0.7	0.8	1.0	1.1	1.3
TW E1	53	75	85	92	96	98	99	100	28	47	59	70	77	85	90	94	0.5	0.7	0.8	1.0	1.2	0.7	0.8	1.0	1.1	1.3
TW E2	52	75	85	92	95	98	99	100	27	46	58	69	77	85	89	94	0.5	0.7	0.8	1.0	1.2	0.7	0.8	1.0	1.1	1.3
TW E3	59	81	90	96	98	99	100	100	31	54	66	78	84	91	94	97	0.5	0.6	0.8	1.0	1.2	0.6	0.7	0.9	1.1	1.3
TW E4	55	78	88	94	97	99	100	100	28	50	63	75	82	89	93	96	0.5	0.6	0.8	1.0	1.2	0.7	0.8	0.9	1.1	1.3
TW F1	36	53	63	73	80	87	91	94	20	30	37	46	52	60	66	72	0.7	0.8	1.0	1.1	1.3	1.0	1.1	1.2	1.4	1.5
TW F2	39	59	70	80	86	92	95	97	20	33	41	51	58	67	73	80	0.6	0.7	0.9	1.1	1.3	0.9	1.0	1.1	1.3	1.4
TW F3	39	60	71	81	87	93	95	98	19	33	42	53	60	69	75	82	0.6	0.7	0.9	1.1	1.3	0.8	0.9	1.1	1.2	1.4
TW F4	37	58	70	80	86	92	95	97	16	31	40	51	58	68	74	81	0.6	0.7	0.9	1.1	1.3	0.9	1.0	1.1	1.3	1.4
TW G1	34	52	63	73	79	86	90	94	18	28	36	44	51	60	65	72	0.7	0.8	1.0	1.1	1.3	1.0	1.1	1.2	1.4	1.5
TW G2	35	55	66	76	83	90	93	96	16	29	38	47	54	64	70	77	0.7	0.8	0.9	1.1	1.3	0.9	1.0	1.1	1.3	1.5
TW G3	33	53	64	75	82	89	92	96	14	27	36	46	53	62	68	76	0.7	0.8	0.9	1.1	1.3	0.9	1.0	1.2	1.3	1.5
TW G4	28	47	58	69	76	84	89	93	11	23	31	40	47	56	62	70	0.7	0.8	1.0	1.1	1.3	1.0	1.1	1.2	1.4	1.5
TW H1	30	48	59	70	76	84	89	93	14	25	32	41	47	56	62	69	0.7	0.8	1.0	1.1	1.3	1.0	1.1	1.2	1.4	1.5
TW H2	27	46	57	68	75	83	87	92	12	22	30	39	45	54	60	68	0.7	0.8	1.0	1.1	1.3	1.0	1.1	1.2	1.4	1.5
TW H3	23	41	51	62	70	78	84	89	9	19	26	34	40	49	55	63	0.7	0.9	1.0	1.2	1.3	1.1	1.2	1.3	1.4	1.6
TW H4	18	33	43	53	60	69	75	82	7	15	21	27	33	40	46	53	0.8	0.9	1.1	1.3	1.4	1.2	1.3	1.4	1.6	1.7
TW I1	32	50	61	71	78	85	90	93	15	25	33	41	47	56	62	69	0.7	0.8	1.0	1.1	1.3	1.0	1.1	1.2	1.4	1.5
TW I2	41	63	75	85	90	95	97	98	20	35	46	57	64	74	80	86	0.6	0.7	0.9	1.1	1.2	0.8	0.9	1.0	1.2	1.4
TW I3	44	67	78	87	92	96	98	99	21	38	50	61	69	78	83	89	0.6	0.7	0.9	1.0	1.2	0.8	0.9	1.0	1.2	1.3
TW I4	47	70	81	90	94	97	99	99	22	41	53	65	73	82	87	92	0.6	0.7	0.8	1.0	1.2	0.7	0.8	1.0	1.1	1.3
TW I5	47	70	81	90	94	97	99	99	22	41	53	66	73	82	87	92	0.6	0.7	0.8	1.0	1.2	0.7	0.8	1.0	1.1	1.3
TW J1	50	73	84	91	95	98	99	100	25	44	56	68	75	84	88	93	0.6	0.7	0.8	1.0	1.2	0.7	0.8	1.0	1.1	1.3
TW J2	48	72	83	91	94	98	99	100	24	43	55	66	74	83	88	92	0.6	0.7	0.8	1.0	1.2	0.7	0.8	1.0	1.1	1.3
TW J3	55	78	88	95	97	99	100	100	27	50	63	75	82	89	93	96	0.5	0.6	0.8	1.0	1.2	0.7	0.8	0.9	1.1	1.3
TW J4	52	76	86	93	96	99	99	100	25	47	60	72	79	87	92	95	0.5	0.6	0.8	1.0	1.2	0.7	0.8	0.9	1.1	1.3
DG A1	40	60	71	79	85	90	93	95	18	32	40	49	55	62	66	71	0.6	0.7	0.9	1.1	1.3	0.9	1.0	1.2	1.4	1.7
DG A2	40	61	72	82	87	92	95	97	19	33	43	53	60	68	73	79	0.6	0.7	0.9	1.1	1.3	0.9	0.9	1.1	1.3	1.5
DG A3	43	67	78	87	91	94	95	96	21	38	50	62	69	78	83	87	0.6	0.7	0.9	1.2	1.7	0.8	0.9	1.0	1.2	1.4
DG B1	42	66	77	86	90	94	96	97	18	33	43	54	61	70	75	80	0.6	0.7	0.9	1.1	1.4	0.9	0.9	1.1	1.3	1.6
DG B2	41	66	78	88	92	96	97	99	19	34	45	58	66	75	80	86	0.6	0.7	0.9	1.1	1.3	0.9	0.9	1.0	1.2	1.4
DG B3	44	70	82	90	94	96	97	98	22	39	51	65	73	83	87	91	0.6	0.7	0.9	1.1	1.7	0.8	0.8	0.9	1.1	1.4
DG C1	47	73	84	91	94	96	97	98	21	38	50	62	70	79	83	87	0.5	0.6	0.9	1.2	1.8	0.8	0.9	1.0	1.2	1.5
DG C2	46	72	84	92	95	98	99	99	22	39	51	65	73	82	87	91	0.6	0.6	0.8	1.0	1.2	0.8	0.8	1.0	1.1	1.4
DG C3	49	77	88	94	96	98	98	98	24	44	57	72	80	88	92	95	0.5	0.6	0.8	1.2	2.6	0.8	0.8	0.9	1.1	1.3
SS A1	48	68	78	87	91	96	97	99	28	43	52	62	69	78	82	88	0.6	0.7	0.9	1.1	1.2	0.8	0.9	1.0	1.2	1.4
SS A2	56	78	87	93	96	99	99	100	33	52	63	74	80	88	91	95	0.5	0.7	0.8	1.0	1.2	0.7	0.8	0.9	1.1	1.3
SS A3	46	66	76	85	90	94	97	98	26	40	50	59	66	75	80	85	0.6	0.7	0.9	1.1	1.2	0.8	0.9	1.1	1.2	1.4
SS A4	56	77	87	93	96	98	99	100	33	52	63	73	80	87	91	95	0.5	0.7	0.8	1.0	1.2	0.7	0.8	0.9	1.1	1.3
SS A5	45	64	74	83	88	93	96	98	27	40	48	58	64	73	78	84	0.6	0.7	0.9	1.1	1.3	0.8	0.9	1.1	1.2	1.4
SS A6	56	78	87	93	96	99	99	100	33	62	60	74	80	87	91	95	0.5	0.7	0.8	1.0	1.2	0.7	0.8	0.9	1.1	1.3
SS A7	41	59	69	78	84	90	93	96	25	36	44	52	58	67	72	78	0.7	0.8	0.9	1.1	1.3	0.9	1.0	1.2	1.3	1.5
SS A8	56	77	86	93	96	98	99	100	33	51	62	73	79	87	90	94	0.5	0.7	0.8	1.0	1.2	0.7	0.8	1.0	1.1	1.3
SS B1	42	62	73	82	87	93	95	98	24	37	46	56	63	72	77	83	0.6	0.7	0.9	1.1	1.3	0.8	0.9	1.1	1.2	1.4
SS B2	51	72	82	90	94	97	99	99	28	46	57	68	75	83	87	92	0.6	0.7	0.8	1.0	1.2	0.7	0.8	1.0	1.1	1.3
SS B3	41	60	71	81	86	92	95	97	23	36	44	54	61	69	75	81	0.6	0.7	0.9	1.1	1.3	0.9	1.0	1.1	1.3	1.4
SS B4	50	72	82	90	94	97	98	99	28	45	56	67	74	82	87	92	0.6	0.7	0.9	1.0	1.2	0.7	0.8	1.0	1.2	1.3
SS B5	38	56	66	76	82	88	92	95	22	33	41	49	56	64	69	76	0.7	0.8	0.9	1.1	1.3	0.9	1.0	1.2	1.3	1.5
SS B6	50	72	82	90	94	97	98	99	28	45	56	67	74	83	87	92	0.6	0.7	0.8	1.0	1.2	0.7	0.8	1.0	1.2	1.3
SS B7	35	52	62	72	78	85	89	93	20	30	37	45	51	59	65	71	0.7	0.8	1.0	1.1	1.3	1.0	1.1	1.2	1.4	1.5
SS B8	49	70	81	89	93	97	98	99	27	44	55	66	73	81	86	91	0.6	0.7	0.9	1.0	1.2	0.7	0.8	1.0	1.2	1.3
SS C1	44	65	76	85	90	95	97	99	22	37	47	58	65	74	79	85	0.6	0.7	0.9	1.1	1.2	0.8	0.9	1.1	1.2	1.4
SS C2	49	72	82	90	94	97	99	99	25	43	54	66	73	82	87	91	0.6	0.7	0.8	1.0	1.2	0.7	0.8	1.0	1.1	1.3
SS C3	36	54	65	75	81	88	92	95	19	30	38	47	53	62	68	75	0.7	0.8	0.9	1.1	1.3	0.9	1.0	1.2	1.3	1.5
SS C4	44	66	77	86	91	95	97	99	22	38	48	59	67	76	81	87	0.6	0.7	0.9	1.0	1.2	0.8	0.9	1.0	1.2	1.4
SS D1	56	77	87	93	96	98	99	100	33	51	62	72	79	86	90	94	0.5	0.7	0.8	1.0	1.2	0.7	0.8	1.0	1.1	1.3
SS D2	64	85	92	97	99	100	100	100	38	60	72	82	87	93	96	98	0.5	0.6	0.8	1.0	1.2	0.6	0.7	0.9	1.1	1.2
SS D3	51	71	81	89	93	97	98	99	30	46	56	66	73	81	85	90	0.6	0.7	0.9	1.0	1.2	0.8	0.9	1.0	1.2	1.4
SS D4	62	83	91	96	98	99	100	100	36	58	70	80	86	92	95	97	0.5	0.6	0.8	1.0	1.2	0.6	0.7	0.9	1.1	1.3
SS E1	52	73	83	91	94	97	99	99	28	45	56	67	74	82	86	91	0.6	0.7	0.8	1.0	1.2	0.7	0.8	1.0	1.2	1.3
SS E2	59	80	89	95	97	99	100	100	33	53	66	76	83	90	93	96	0.5	0.6	0.8	1.0	1.2	0.7	0.8	0.9	1.1	1.3
SS E3	43	63	73	83	88	93	96	98	25	38	47	56	63	72	77	83	0.6	0.7	0.9	1.1	1.3	0.8	0.9	1.1	1.2	1.4
SS E4	54	76	86	93	96	98	99	100	30	49	61	72	79	86	90	94	0.5	0.7	0.8	1.0	1.2	0.7	0.8	1.0	1.1	1.3

SHREVEPORT, LOUISIANA

TYPE	SOLAR SAVINGS FRACTION (%) BASE TEMP: 55 F (883 DD)								65 F (2175 DD)								CONSERVATION FACTOR 55 F					65 F				
LCR	200	100	70	50	40	30	25	20	200	100	70	50	40	30	25	20	200	100	50	30	20	200	100	50	30	20
WW A1	27	41	50	60	66	75	80	85	18	25	31	37	41	48	53	59	0.8	0.9	1.1	1.2	1.4	1.2	1.3	1.4	1.6	1.7
WW A2	32	51	62	73	79	87	91	94	18	29	37	46	52	61	67	73	0.7	0.8	1.0	1.1	1.3	0.9	1.0	1.2	1.3	1.5
WW A3	34	54	66	77	83	90	93	96	17	30	39	49	56	65	71	78	0.7	0.8	0.9	1.1	1.3	0.9	1.0	1.1	1.3	1.5
WW A4	34	56	68	79	85	91	94	97	17	31	41	51	58	68	74	80	0.6	0.7	0.9	1.1	1.3	0.8	1.0	1.1	1.3	1.4
WW A5	36	58	70	81	87	93	95	98	17	32	42	53	61	70	76	83	0.6	0.7	0.9	1.1	1.2	0.8	0.9	1.1	1.2	1.4
WW A6	36	59	72	82	88	93	96	98	17	33	43	54	62	72	77	84	0.6	0.7	0.9	1.1	1.2	0.8	0.9	1.1	1.2	1.4
WW B1	28	45	55	66	73	81	85	90	15	24	31	38	43	51	57	63	0.7	0.8	1.0	1.2	1.3	1.0	1.2	1.3	1.5	1.6
WW B2	34	56	68	79	85	92	95	97	17	32	42	52	60	69	75	82	0.6	0.7	0.9	1.1	1.3	0.8	0.9	1.1	1.2	1.4
WW B3	40	64	76	85	91	95	97	99	20	38	49	60	68	77	83	88	0.6	0.7	0.9	1.0	1.2	0.8	0.9	1.0	1.2	1.3
WW B4	39	64	76	86	91	96	98	99	19	38	49	61	69	79	84	90	0.6	0.7	0.9	1.0	1.2	0.7	0.8	1.0	1.2	1.3
WW B5	37	62	74	85	90	95	97	99	17	36	48	60	68	78	83	89	0.6	0.7	0.9	1.0	1.2	0.8	0.9	1.0	1.2	1.3
WW C1	42	65	77	87	92	96	98	99	22	39	51	62	70	79	84	90	0.6	0.7	0.9	1.0	1.2	0.7	0.9	1.0	1.2	1.3
WW C2	39	63	75	85	90	95	97	99	19	37	49	60	68	78	83	89	0.6	0.7	0.9	1.0	1.2	0.8	0.9	1.0	1.2	1.3
WW C3	44	70	81	90	94	98	99	100	22	43	56	69	76	85	90	94	0.6	0.7	0.8	1.0	1.2	0.7	0.8	0.9	1.1	1.3
WW C4	40	66	78	88	92	97	98	99	19	39	52	65	73	82	87	92	0.6	0.7	0.8	1.0	1.2	0.7	0.8	1.0	1.1	1.3
TW A1	27	40	49	59	65	74	79	84	17	25	30	36	41	48	52	58	0.8	0.9	1.1	1.2	1.4	1.2	1.3	1.4	1.6	1.7
TW A2	29	46	57	67	74	82	87	91	17	27	34	42	48	56	61	68	0.7	0.8	1.0	1.2	1.3	1.0	1.1	1.3	1.4	1.6
TW A3	30	48	60	71	77	85	89	93	16	27	35	44	50	59	65	72	0.7	0.8	1.0	1.1	1.3	1.0	1.1	1.2	1.4	1.5
TW A4	29	48	60	71	78	86	90	94	15	27	35	44	51	60	66	73	0.7	0.8	1.0	1.1	1.3	0.9	1.0	1.2	1.3	1.5
TW B1	26	40	50	59	66	75	80	86	16	24	30	36	41	49	54	60	0.8	0.9	1.1	1.2	1.4	1.1	1.2	1.4	1.5	1.7
TW B2	27	44	55	66	73	81	86	91	15	25	32	40	46	55	60	67	0.7	0.8	1.0	1.2	1.3	1.0	1.1	1.3	1.4	1.6
TW B3	27	44	55	66	73	82	86	91	14	25	32	40	47	55	61	68	0.7	0.8	1.0	1.2	1.3	1.0	1.1	1.2	1.4	1.6
TW B4	25	42	52	63	70	79	84	89	13	23	30	38	44	53	59	66	0.8	0.9	1.0	1.2	1.3	1.0	1.1	1.3	1.4	1.6
TW C1	25	39	49	59	66	74	80	85	15	23	29	36	41	48	54	60	0.8	0.9	1.1	1.2	1.4	1.1	1.2	1.4	1.5	1.7
TW C2	24	39	49	60	67	76	81	87	14	22	29	36	42	50	55	62	0.8	0.9	1.0	1.2	1.4	1.1	1.2	1.3	1.5	1.6
TW C3	23	37	47	57	64	73	79	85	13	21	27	34	40	48	53	60	0.8	0.9	1.1	1.2	1.4	1.1	1.2	1.4	1.5	1.7
TW C4	21	33	42	51	58	67	73	79	13	20	25	31	36	43	48	54	0.9	1.0	1.1	1.3	1.4	1.2	1.3	1.5	1.6	1.8
TW D1	23	38	48	59	66	74	80	86	12	20	26	33	38	45	50	57	0.8	0.9	1.1	1.2	1.4	1.1	1.2	1.4	1.6	1.7
TW D2	31	51	63	74	81	88	92	95	17	29	38	48	55	64	70	77	0.7	0.8	0.9	1.1	1.3	0.9	1.0	1.1	1.3	1.5
TW D3	34	56	68	79	85	91	94	97	18	32	42	52	60	69	75	81	0.6	0.7	0.9	1.1	1.3	0.8	0.9	1.1	1.2	1.4
TW D4	35	58	71	81	87	93	96	98	18	34	45	56	63	73	79	85	0.6	0.7	0.9	1.1	1.2	0.8	0.9	1.0	1.2	1.4
TW D5	34	57	69	80	86	93	95	98	17	33	44	55	63	73	79	85	0.6	0.7	0.9	1.1	1.2	0.8	0.9	1.0	1.2	1.4
TW E1	38	60	72	82	88	94	96	98	20	36	46	57	64	74	79	85	0.6	0.7	0.9	1.1	1.2	0.8	0.9	1.0	1.2	1.4
TW E2	37	59	72	82	88	93	96	98	19	35	46	57	64	74	79	86	0.6	0.7	0.9	1.1	1.2	0.8	0.9	1.0	1.2	1.4
TW E3	43	67	79	88	93	97	98	99	22	41	54	66	73	82	87	92	0.6	0.7	0.9	1.0	1.2	0.7	0.8	1.0	1.1	1.3
TW E4	39	64	76	86	91	96	98	99	20	38	50	62	70	80	85	90	0.6	0.7	0.9	1.0	1.2	0.7	0.8	1.0	1.2	1.3
TW F1	25	39	49	58	65	74	79	85	16	23	29	35	40	47	52	58	0.8	0.9	1.1	1.2	1.4	1.1	1.3	1.4	1.6	1.7
TW F2	26	44	54	65	72	81	85	90	14	24	31	39	45	53	59	66	0.7	0.8	1.0	1.2	1.3	1.0	1.1	1.3	1.4	1.6
TW F3	26	44	56	67	74	82	87	92	13	24	31	40	46	55	61	68	0.7	0.8	1.0	1.1	1.3	1.0	1.1	1.2	1.4	1.5
TW F4	24	42	54	65	73	81	86	91	11	22	29	38	45	54	60	67	0.7	0.8	1.0	1.2	1.3	1.0	1.1	1.2	1.4	1.5
TW G1	23	38	47	58	65	73	79	85	13	21	27	34	39	46	52	58	0.8	0.9	1.1	1.2	1.4	1.1	1.2	1.4	1.5	1.7
TW G2	23	39	50	61	68	77	83	88	11	21	28	36	41	50	56	63	0.8	0.9	1.0	1.2	1.4	1.1	1.2	1.3	1.5	1.6
TW G3	21	38	48	60	67	76	82	87	9	19	26	34	40	48	54	62	0.8	0.9	1.0	1.2	1.4	1.1	1.2	1.3	1.5	1.6
TW G4	17	32	43	53	61	71	76	83	7	16	22	29	35	43	49	56	0.8	0.9	1.1	1.2	1.4	1.1	1.2	1.4	1.5	1.7
TW H1	20	34	44	54	61	71	76	83	10	18	24	31	36	43	49	55	0.8	0.9	1.1	1.2	1.4	1.2	1.3	1.4	1.6	1.7
TW H2	17	31	41	52	59	69	75	81	7	16	21	28	34	41	47	54	0.9	0.9	1.1	1.3	1.4	1.2	1.3	1.4	1.6	1.7
TW H3	14	27	36	46	54	63	69	77	6	13	18	25	30	37	42	49	0.9	1.0	1.1	1.3	1.5	1.2	1.3	1.5	1.6	1.8
TW H4	10	21	29	38	44	54	60	67	4	10	14	20	24	30	35	41	1.0	1.1	1.2	1.4	1.5	1.4	1.5	1.6	1.8	1.9
TW I1	20	35	45	55	62	71	77	83	10	18	23	30	35	42	47	53	0.8	0.9	1.1	1.2	1.4	1.2	1.3	1.5	1.6	1.8
TW I2	28	48	60	71	78	86	90	94	13	26	35	44	51	61	67	74	0.7	0.8	1.0	1.1	1.3	0.9	1.0	1.2	1.3	1.5
TW I3	30	51	63	75	81	89	92	96	14	28	38	48	55	65	71	78	0.7	0.8	0.9	1.1	1.3	0.9	1.0	1.1	1.3	1.4
TW I4	32	55	67	78	85	91	94	97	15	31	41	52	60	70	76	83	0.6	0.7	0.9	1.1	1.3	0.8	0.9	1.1	1.2	1.4
TW I5	32	55	67	79	85	92	95	97	15	31	41	53	61	71	77	84	0.6	0.7	0.9	1.1	1.3	0.8	0.9	1.1	1.2	1.4
TW J1	35	58	70	80	86	92	95	98	18	33	44	54	62	72	77	84	0.6	0.7	0.9	1.1	1.2	0.8	0.9	1.1	1.2	1.4
TW J2	33	56	69	79	86	92	95	97	16	32	42	53	61	71	77	84	0.6	0.7	0.9	1.1	1.3	0.8	0.9	1.1	1.2	1.4
TW J3	39	64	76	86	91	96	98	99	19	38	50	62	70	80	85	90	0.6	0.7	0.9	1.0	1.2	0.7	0.8	1.0	1.2	1.3
TW J4	36	61	73	84	89	95	97	99	17	35	47	59	68	77	83	89	0.6	0.7	0.9	1.0	1.2	0.8	0.9	1.0	1.2	1.3
DG A1	26	44	54	64	70	77	81	85	13	22	29	36	41	47	51	55	0.7	0.8	1.0	1.3	1.5	1.0	1.2	1.4	1.7	2.0
DG A2	26	45	57	67	74	82	86	90	13	24	32	40	47	55	60	66	0.7	0.8	1.0	1.2	1.4	1.0	1.1	1.2	1.4	1.7
DG A3	29	51	64	75	81	88	91	93	16	28	38	48	56	66	72	78	0.7	0.8	0.9	1.2	1.4	0.9	1.0	1.1	1.3	1.5
DG B1	27	47	60	71	77	84	87	91	13	22	30	38	45	53	58	64	0.7	0.8	1.0	1.2	1.4	1.0	1.1	1.3	1.5	1.7
DG B2	27	48	62	74	80	88	91	94	14	25	33	43	50	60	66	73	0.7	0.8	0.9	1.1	1.3	1.0	1.1	1.2	1.3	1.5
DG B3	30	53	67	79	85	91	94	96	16	29	39	50	59	70	76	83	0.7	0.7	0.9	1.1	1.4	0.9	1.0	1.1	1.2	1.4
DG C1	31	54	68	79	85	90	93	95	15	26	35	45	52	62	68	74	0.7	0.7	0.9	1.1	1.4	1.0	1.0	1.1	1.3	1.6
DG C2	31	54	68	80	86	92	95	97	16	28	38	49	57	68	74	80	0.7	0.7	0.9	1.1	1.3	0.9	1.0	1.1	1.2	1.4
DG C3	34	59	74	85	90	95	96	98	18	32	44	56	66	77	83	88	0.6	0.7	0.9	1.1	1.4	0.9	0.9	1.0	1.2	1.4
SS A1	35	53	64	74	80	87	91	94	21	33	41	49	56	64	70	76	0.7	0.8	1.0	1.1	1.3	0.9	1.0	1.2	1.3	1.5
SS A2	42	63	74	84	89	94	96	98	25	41	51	61	68	77	82	87	0.6	0.7	0.9	1.1	1.2	0.8	0.9	1.0	1.2	1.4
SS A3	33	50	61	71	77	85	89	93	20	31	38	46	52	60	66	72	0.7	0.8	1.0	1.1	1.3	0.9	1.1	1.2	1.4	1.5
SS A4	42	63	74	83	88	94	96	98	25	40	50	60	67	76	81	87	0.6	0.7	0.9	1.1	1.2	0.8	0.9	1.0	1.2	1.4
SS A5	33	49	59	69	75	83	87	92	21	31	38	46	51	59	64	71	0.7	0.8	1.0	1.2	1.3	1.0	1.1	1.2	1.4	1.6
SS A6	42	63	74	84	89	94	96	98	25	40	50	61	68	77	82	87	0.6	0.7	0.9	1.1	1.2	0.8	0.9	1.0	1.2	1.4
SS A7	30	44	54	63	70	78	82	87	20	28	34	41	46	53	58	64	0.8	0.9	1.0	1.2	1.4	1.1	1.2	1.3	1.5	1.7
SS A8	41	62	73	83	88	93	96	98	25	39	49	59	66	75	80	86	0.6	0.7	0.9	1.1	1.2	0.8	0.9	1.1	1.2	1.4
SS B1	30	47	58	68	75	82	87	91	18	28	35	43	49	58	63	70	0.7	0.8	1.0	1.2	1.3	1.0	1.1	1.2	1.4	1.6
SS B2	36	57	68	79	85	91	94	97	21	36	45	55	62	71	77	83	0.6	0.7	0.9	1.1	1.3	0.8	0.9	1.1	1.2	1.4
SS B3	29	45	56	66	73	81	85	90	17	27	34	41	47	55	60	67	0.7	0.8	1.0	1.2	1.3	1.0	1.1	1.3	1.4	1.6
SS B4	36	56	68	78	84	91	94	97	21	35	44	54	61	70	76	82	0.6	0.8	0.9	1.1	1.3	0.8	0.9	1.1	1.3	1.4
SS B5	27	41	51	61	67	76	81	86	17	26	31	38	43	51	56	62	0.8	0.9	1.1	1.2	1.4	1.1	1.2	1.4	1.5	1.7
SS B6	36	56	68	78	84	91	94	97	21	35	44	54	61	70	76	82	0.6	0.8	0.9	1.1	1.3	0.8	0.9	1.1	1.2	1.4
SS B7	25	38	47	56	63	71	77	82	16	23	29	35	39	46	51	57	0.8	0.9	1.1	1.3	1.4	1.2	1.3	1.4	1.6	1.7
SS B8	35	55	67	77	83	90	93	96	20	34	43	53	60	69	74	81	0.7	0.8	0.9	1.1	1.3	0.8	1.0	1.1	1.3	1.4
SS C1	30	49	61	72	79	86	90	94	15	27	35	44	51	60	66	73	0.7	0.8	1.0	1.1	1.3	0.9	1.0	1.2	1.3	1.5
SS C2	34	56	68	79	85	91	94	97	17	32	42	52	60	69	75	82	0.6	0.7	0.9	1.1	1.3	0.8	0.9	1.1	1.2	1.4
SS C3	25	40	50	60	67	76	81	87	14	23	29	36	41	49	54	61	0.8	0.9	1.0	1.2	1.4	1.1	1.2	1.4	1.5	1.7
SS C4	30	50	62	73	80	87	91	95	16	29	37	47	54	63	69	76	0.7	0.8	0.9	1.1	1.3	0.9	1.0	1.2	1.3	1.5
SS D1	42	62	73	83	88	93	96	98	24	39	48	58	65	74	79	85	0.6	0.7	0.9	1.1	1.2	0.8	0.9	1.1	1.2	1.4
SS D2	48	71	82	90	94	97	99	99	28	47	58	69	76	85	89	93	0.6	0.7	0.8	1.0	1.2	0.7	0.8	1.0	1.1	1.3
SS D3	37	56	67	77	83	89	93	96	23	35	44	53	59	68	73	79	0.7	0.8	0.9	1.1	1.3	0.9	1.0	1.1	1.3	1.5
SS D4	47	69	80	89	93	97	98	99	27	45	56	67	74	83	87	92	0.6	0.7	0.9	1.0	1.2	0.7	0.8	1.0	1.1	1.3
SS E1	37	57	68	78	84	91	94	97	21	34	43	52	59	68	73	80	0.6	0.8	0.9	1.1	1.3	0.9	1.0	1.1	1.3	1.5
SS E2	43	66	77	86	91	96	97	99	24	41	52	63	70	79	84	89	0.6	0.7	0.9	1.0	1.2	0.7	0.9	1.0	1.2	1.3
SS E3	31	48	58	68	75	83	87	91	19	29	36	44	50	58	63	70	0.7	0.8	1.0	1.2	1.3	1.0	1.1	1.2	1.4	1.6
SS E4	39	61	73	82	88	93	96	98	22	38	48	58	66	75	80	86	0.6	0.7	0.9	1.1	1.2	0.8	0.9	1.0	1.2	1.4

CARIBOU, MAINE

TYPE	SOLAR SAVINGS FRACTION (%) BASE TEMP: 55 F (6644 DD) LCR 100	70	50	40	30	25	20	15	65 F (9637 DD) 100	70	50	40	30	25	20	15	CONSERVATION FACTOR 55 F 100	50	30	20	15	65 F 100	50	30	20	15
WW A1	12	13	14	15	16	17	18	20	11	11	12	13	13	14	14	13	2.8	3.3	3.9	4.8	-	3.4	4.4	-	-	-
WW A2	9	12	15	17	20	22	24	28	8	9	11	13	15	16	18	19	2.0	2.3	2.6	3.0	3.4	2.3	2.7	3.3	4.0	4.9
WW A3	8	11	14	17	21	23	27	31	6	8	11	13	15	17	19	22	1.8	2.0	2.3	2.7	3.0	2.0	2.4	2.8	3.4	4.0
WW A4	7	10	14	17	21	24	28	33	5	7	10	13	16	18	20	23	1.7	1.9	2.2	2.5	2.8	1.9	2.3	2.6	3.1	3.5
WW A5	6	10	14	18	22	25	29	35	4	7	10	13	16	18	21	25	1.6	1.9	2.1	2.4	2.7	1.8	2.2	2.5	2.9	3.4
WW A6	6	10	14	18	23	26	30	36	3	6	10	13	16	19	22	26	1.6	1.8	2.1	2.3	2.6	1.8	2.1	2.4	2.8	3.2
WW B1	6	7	8	8	9	9	9	7	5	5	5	5	4	3	2	-1	3.2	4.5	-	-	-	4.7	-	-	-	-
WW B2	8	12	17	21	26	30	35	41	6	9	13	16	21	24	28	33	1.5	1.7	1.9	2.2	2.4	1.7	1.9	2.2	2.4	2.7
WW B3	10	15	21	26	32	37	42	50	8	12	17	21	26	30	35	41	1.4	1.6	1.8	2.0	2.2	1.5	1.7	1.9	2.2	2.4
WW B4	10	17	24	30	38	43	50	59	8	13	20	25	32	37	43	52	1.3	1.4	1.6	1.8	1.9	1.3	1.5	1.7	1.9	2.0
WW B5	10	17	24	30	38	44	51	61	7	13	20	26	33	38	45	54	1.2	1.4	1.6	1.7	1.9	1.3	1.5	1.6	1.8	2.0
WW C1	12	17	24	28	35	40	46	54	9	14	19	23	29	33	39	46	1.3	1.5	1.7	1.9	2.1	1.4	1.6	1.8	2.1	2.3
WW C2	10	16	23	28	36	41	47	56	8	13	19	24	30	35	40	48	1.3	1.5	1.6	1.8	2.0	1.4	1.6	1.8	2.0	2.1
WW C3	13	22	31	38	48	54	62	71	11	18	27	33	42	48	56	65	1.1	1.3	1.4	1.6	1.7	1.2	1.3	1.5	1.7	1.8
WW C4	12	20	29	35	45	51	59	69	9	16	25	31	40	46	53	63	1.1	1.3	1.5	1.6	1.8	1.2	1.4	1.5	1.7	1.8
TW A1	12	13	15	16	17	18	19	21	11	12	13	13	14	14	15	15	2.7	3.1	3.6	4.2	4.9	3.2	3.9	-	-	-
TW A2	10	12	14	16	19	21	23	27	8	10	12	13	15	16	18	20	2.1	2.4	2.7	3.0	3.4	2.4	2.8	3.3	3.9	4.7
TW A3	8	11	14	16	20	22	25	29	7	9	11	13	15	17	19	21	1.9	2.2	2.5	2.8	3.1	2.2	2.5	2.9	3.4	3.9
TW A4	7	10	13	16	20	22	26	31	5	8	10	12	15	17	19	23	1.8	2.1	2.3	2.6	2.8	2.0	2.4	2.7	3.1	3.5
TW B1	11	12	14	16	17	19	21	23	10	11	12	13	14	15	16	17	2.5	2.8	3.2	3.6	4.0	2.8	3.3	4.0	4.9	-
TW B2	9	11	14	16	19	21	24	28	7	9	11	13	15	16	18	21	2.1	2.3	2.6	2.8	3.1	2.3	2.6	3.0	3.5	3.9
TW B3	8	10	13	15	19	21	24	29	6	8	11	12	15	17	19	22	2.0	2.2	2.4	2.7	3.0	2.2	2.5	2.8	3.2	3.6
TW B4	8	10	13	15	18	21	24	29	6	8	10	12	15	17	19	23	2.0	2.2	2.4	2.7	2.9	2.2	2.5	2.8	3.1	3.4
TW C1	10	12	14	15	18	19	22	25	9	10	12	13	15	16	17	20	2.3	2.6	2.8	3.1	3.4	2.6	2.9	3.3	3.8	4.3
TW C2	8	11	13	15	18	20	23	27	7	9	11	12	15	16	19	22	2.1	2.3	2.6	2.8	3.1	2.3	2.6	3.0	3.3	3.6
TW C3	8	10	13	15	18	20	23	27	7	9	11	12	15	16	19	22	2.2	2.4	2.6	2.8	3.0	2.4	2.7	3.0	3.3	3.5
TW C4	9	11	13	15	17	19	22	26	8	10	11	13	15	16	18	21	2.3	2.5	2.7	3.0	3.2	2.5	2.8	3.1	3.4	3.6
TW D1	5	6	7	8	8	9	9	9	4	4	5	5	5	5	4	2	3.1	3.9	-	-	-	4.1	-	-	-	-
TW D2	9	13	17	20	25	28	32	38	7	10	14	16	20	23	26	31	1.7	1.9	2.1	2.3	2.5	1.8	2.1	2.3	2.6	2.8
TW D3	9	13	18	21	27	30	35	41	7	10	14	17	21	24	28	33	1.6	1.8	2.0	2.2	2.4	1.7	2.0	2.2	2.5	2.7
TW D4	10	16	22	27	33	38	44	53	8	13	18	22	28	33	38	46	1.4	1.5	1.7	1.9	2.0	1.5	1.7	1.8	2.0	2.2
TW D5	10	16	23	28	35	40	47	56	8	13	19	24	31	35	41	50	1.3	1.5	1.6	1.8	2.0	1.4	1.6	1.7	1.9	2.1
TW E1	11	16	22	26	32	36	41	48	9	13	18	21	26	30	35	41	1.4	1.6	1.8	2.0	2.2	1.6	1.8	2.0	2.2	2.4
TW E2	11	16	22	27	33	38	44	51	9	13	18	22	28	32	37	44	1.4	1.6	1.7	1.9	2.1	1.5	1.7	1.9	2.1	2.3
TW E3	14	21	29	35	43	49	56	65	11	18	25	30	38	43	50	59	1.2	1.3	1.5	1.7	1.8	1.3	1.4	1.6	1.8	1.9
TW E4	13	20	28	34	42	48	55	65	10	16	24	29	37	42	49	59	1.2	1.4	1.5	1.7	1.8	1.3	1.4	1.6	1.8	1.9
TW F1	10	11	13	14	15	17	18	20	9	10	11	11	12	13	13	14	2.6	3.0	3.5	4.0	4.7	3.1	3.8	4.8	-	-
TW F2	7	10	12	14	16	18	21	24	6	8	9	11	12	14	15	17	2.2	2.4	2.7	3.1	3.5	2.4	2.9	3.4	4.0	4.8
TW F3	6	8	11	13	16	19	22	26	4	6	8	10	12	14	16	18	2.0	2.3	2.5	2.8	3.1	2.2	2.6	3.0	3.5	4.0
TW F4	4	6	9	12	15	18	21	26	2	4	7	9	11	13	15	19	2.0	2.2	2.4	2.7	2.9	2.2	2.5	2.8	3.2	3.6
TW G1	8	9	11	12	14	16	18	20	7	8	9	10	11	12	13	14	2.5	2.8	3.2	3.6	4.0	2.8	3.3	4.0	4.9	-
TW G2	5	7	10	12	14	16	19	23	4	5	7	9	11	12	14	16	2.2	2.4	2.7	3.0	3.3	2.4	2.8	3.2	3.7	4.2
TW G3	4	6	8	10	14	16	19	23	2	4	6	8	10	12	14	17	2.1	2.3	2.6	2.9	3.1	2.3	2.7	3.0	3.4	3.7
TW G4	2	4	7	9	12	14	17	21	1	3	5	6	9	10	13	16	2.2	2.4	2.6	2.9	3.1	2.4	2.7	3.0	3.3	3.6
TW H1	5	7	9	10	12	14	16	19	4	5	7	8	9	11	12	14	2.4	2.7	3.0	3.3	3.6	2.7	3.1	3.6	4.1	4.7
TW H2	3	5	7	9	11	13	16	20	2	3	5	6	8	10	12	14	2.3	2.6	2.8	3.1	3.3	2.6	2.9	3.2	3.6	3.9
TW H3	2	4	6	7	10	12	14	18	1	2	4	5	7	9	11	14	2.4	2.6	2.9	3.1	3.3	2.6	2.9	3.2	3.5	3.8
TW H4	1	3	4	6	8	10	12	15	0	1	3	4	6	7	9	12	2.7	2.9	3.1	3.3	3.5	2.9	3.2	3.4	3.7	4.0
TW I1	3	4	5	6	6	7	7	7	2	3	3	3	3	3	2	0	3.3	4.2	-	-	-	4.4	-	-	-	-
TW I2	6	10	14	17	21	25	29	35	5	7	11	13	17	20	23	28	1.7	1.9	2.1	2.3	2.5	1.9	2.1	2.4	2.6	2.9
TW I3	7	11	16	10	24	28	32	39	5	8	12	15	19	22	26	32	1.6	1.8	2.0	2.2	2.4	1.7	2.0	2.2	2.4	2.7
TW I4	8	13	19	24	30	35	41	50	6	10	16	20	26	30	35	43	1.4	1.6	1.7	1.9	2.1	1.5	1.7	1.9	2.1	2.2
TW I5	8	14	20	25	33	38	44	53	6	11	17	21	28	33	39	47	1.4	1.5	1.7	1.8	2.0	1.4	1.6	1.8	1.9	2.1
TW J1	9	14	20	24	30	34	39	47	7	11	16	19	25	28	33	39	1.5	1.6	1.8	2.0	2.2	1.6	1.8	2.0	2.2	2.4
TW J2	9	14	20	24	31	35	41	49	7	11	16	20	26	30	35	43	1.4	1.6	1.8	1.9	2.1	1.5	1.7	1.9	2.1	2.3
TW J3	11	18	26	33	41	47	55	64	9	15	23	28	36	42	49	58	1.2	1.4	1.5	1.7	1.8	1.3	1.4	1.6	1.8	1.9
TW J4	10	17	25	31	40	46	53	63	8	14	21	27	35	40	47	57	1.2	1.4	1.5	1.7	1.0	1.3	1.4	1.6	1.8	1.9
DG A1	4	5	7	7	8	8	8	7	3	3	4	4	4	4	3	0	2.8	3.8	-	-	-	3.6	-	-	-	-
DG A2	7	10	13	15	18	21	24	27	6	8	10	12	15	16	19	21	2.0	2.2	2.5	2.9	3.3	2.2	2.5	2.8	3.4	4.0
DG A3	11	15	19	23	29	33	38	46	9	13	17	20	25	29	33	40	1.6	1.7	1.9	2.0	2.2	1.7	1.8	2.0	2.2	2.4
DG B1	4	5	7	7	8	9	9	9	3	3	4	4	4	4	3	2	2.8	3.5	4.9	-	-	3.6	-	-	-	-
DG B2	7	10	13	16	19	22	25	30	6	8	11	12	15	17	20	23	1.9	2.1	2.3	2.7	3.0	2.1	2.4	2.7	3.1	3.5
DG B3	11	15	20	24	30	35	41	49	10	13	17	21	26	30	36	43	1.6	1.7	1.8	1.9	2.1	1.7	1.8	1.9	2.1	2.2
DG C1	6	8	10	11	13	15	15	17	4	6	7	8	8	9	9	9	2.3	2.6	3.2	4.4	-	2.7	3.7	-	-	-
DG C2	9	12	16	19	24	27	32	37	8	10	13	16	20	22	26	30	1.7	1.9	2.1	2.3	2.6	1.9	2.1	2.3	2.6	2.9
DG C3	13	17	23	28	35	40	46	55	11	15	20	24	30	35	41	49	1.4	1.5	1.7	1.8	2.0	1.5	1.7	1.8	1.9	2.1
SS A1	13	15	18	21	24	26	28	32	12	14	17	18	21	22	24	27	1.9	2.2	2.5	2.9	3.3	2.0	2.4	2.8	3.4	4.1
SS A2	16	21	26	30	36	.40	45	52	15	19	24	27	32	36	41	47	1.4	1.6	1.8	2.0	2.2	1.5	1.7	1.9	2.1	2.4
SS A3	11	13	16	17	19	20	22	24	11	12	14	15	16	17	18	19	2.1	2.6	3.1	3.9	4.9	2.3	3.0	3.9	-	-
SS A4	15	20	25	28	34	37	42	48	14	18	22	26	30	33	37	43	1.4	1.7	1.9	2.1	2.3	1.5	1.8	2.0	2.3	2.5
SS A5	14	16	18	19	22	23	25	27	13	15	16	17	19	20	21	23	2.1	2.5	2.9	3.4	4.0	2.3	2.8	3.5	4.4	-
SS A6	16	20	25	29	35	38	43	50	14	18	23	26	31	35	39	45	1.4	1.6	1.8	2.1	2.3	1.5	1.7	2.0	2.2	2.4
SS A7	13	14	15	16	17	18	19	20	12	13	14	15	15	16	16	16	2.6	3.2	4.0	-	-	2.8	3.8	-	-	-
SS A8	15	19	23	27	32	35	39	45	14	17	21	24	28	31	35	40	1.5	1.7	2.0	2.2	2.4	1.6	1.9	2.1	2.4	2.7
SS B1	10	13	15	17	20	22	24	27	10	11	13	15	17	18	20	22	2.0	2.4	2.7	3.1	3.5	2.2	2.6	3.1	3.8	4.5
SS B2	14	18	23	27	33	36	42	49	12	16	21	24	29	33	37	44	1.5	1.7	1.8	2.0	2.2	1.5	1.8	2.0	2.2	2.4
SS B3	9	11	13	15	17	18	20	22	9	10	12	13	14	15	16	17	2.2	2.7	3.2	3.8	4.6	2.4	3.1	3.9	-	-
SS B4	13	17	22	25	31	34	39	46	12	16	20	23	27	31	35	41	1.5	1.7	1.9	2.1	2.3	1.6	1.8	2.1	2.3	2.5
SS B5	11	13	14	16	17	19	20	22	10	12	13	14	15	16	17	18	2.4	2.8	3.2	3.8	4.4	2.6	3.2	3.9	5.0	-
SS B6	13	17	22	26	31	35	40	46	12	15	20	23	28	31	35	41	1.5	1.7	1.9	2.1	2.3	1.6	1.8	2.0	2.3	2.5
SS B7	10	11	12	13	14	15	16	17	9	10	11	12	13	13	13	13	2.8	3.3	4.1	-	-	3.0	3.9	-	-	-
SS B8	12	16	20	23	28	32	36	42	11	14	18	21	25	28	32	37	1.6	1.8	2.0	2.2	2.4	1.7	1.9	2.2	2.4	2.7
SS C1	7	9	12	14	17	18	21	24	5	7	9	10	12	13	14	15	2.0	2.3	2.7	3.1	3.6	2.3	2.9	3.5	4.5	-
SS C2	9	13	18	22	27	31	36	42	7	10	14	17	22	25	29	35	1.5	1.7	1.9	2.1	2.3	1.7	1.9	2.1	2.4	2.6
SS C3	8	10	12	13	15	17	19	21	7	8	9	10	12	13	14	15	2.4	2.7	3.1	3.5	3.9	2.7	3.2	3.9	4.7	-
SS C4	9	12	17	20	24	28	32	38	7	10	14	16	20	23	26	31	1.7	1.9	2.1	2.3	2.5	1.8	2.1	2.3	2.6	2.8
SS D1	13	16	20	23	27	29	32	36	12	15	18	20	23	24	27	29	1.6	1.9	2.3	2.7	3.1	1.8	2.2	2.6	3.2	4.0
SS D2	17	23	29	35	42	46	52	60	15	20	26	31	37	42	47	54	1.2	1.4	1.6	1.8	2.0	1.3	1.5	1.7	2.0	2.1
SS D3	14	16	19	22	25	27	29	33	13	15	17	19	21	23	25	27	1.8	2.1	2.5	2.9	3.3	2.0	2.4	2.9	3.6	4.4
SS D4	16	22	29	33	40	45	50	58	15	20	26	30	36	40	45	52	1.3	1.5	1.7	1.9	2.0	1.3	1.6	1.8	2.0	2.2
SS E1	10	12	15	17	20	22	24	27	8	11	13	14	16	17	18	19	1.9	2.2	2.7	3.3	4.0	2.1	2.6	3.4	4.8	-
SS E2	13	18	24	28	34	38	43	50	11	16	21	24	30	33	38	44	1.4	1.6	1.8	2.0	2.2	1.5	1.7	1.9	2.2	2.4
SS E3	11	13	15	17	19	20	22	25	10	12	13	14	16	17	18	19	2.2	2.5	3.0	3.5	4.2	2.4	2.9	3.6	4.8	-
SS E4	13	17	22	26	32	35	40	47	11	15	20	23	28	31	35	41	1.5	1.7	1.9	2.1	2.3	1.5	1.8	2.0	2.3	2.5

PORTLAND, MAINE

TYPE	SOLAR SAVINGS FRACTION (%) BASE TEMP: 55 F (4767 DD)								65 F (7499 DD)								CONSERVATION FACTOR 55 F					65 F				
LCR	100	70	50	40	30	25	20	15	100	70	50	40	30	25	20	15	100	50	30	20	15	100	50	30	20	15
WW A1	13	15	17	18	20	22	23	26	12	13	14	14	15	16	16	17	2.4	2.7	3.1	3.5	3.9	3.0	3.7	4.7	-	-
WW A2	11	15	18	21	25	27	31	36	9	11	13	15	18	19	22	25	1.7	2.0	2.2	2.5	2.6	2.1	2.4	2.8	3.2	3.6
WW A3	10	14	18	22	26	30	34	40	7	10	13	15	18	21	24	28	1.6	1.8	2.0	2.2	2.4	1.9	2.2	2.5	2.8	3.1
WW A4	10	14	19	22	27	31	36	43	6	9	13	15	19	22	25	30	1.5	1.7	1.9	2.1	2.3	1.8	2.1	2.4	2.6	2.9
WW A5	9	14	19	23	29	33	38	45	5	9	13	15	20	22	26	31	1.5	1.7	1.9	2.1	2.2	1.7	2.0	2.2	2.5	2.8
WW A6	9	14	19	23	29	34	39	47	5	8	12	16	20	23	27	33	1.4	1.6	1.8	2.0	2.2	1.7	1.9	2.2	2.4	2.7
WW B1	8	10	11	12	14	15	16	17	6	6	7	7	7	7	6	4	2.4	3.0	3.8	4.9	-	3.6	-	-	-	-
WW B2	11	16	21	25	32	36	42	50	7	11	15	19	24	27	32	38	1.4	1.6	1.8	1.9	2.1	1.6	1.8	2.0	2.2	2.4
WW B3	13	19	26	31	39	44	50	59	9	14	20	24	30	34	40	47	1.3	1.4	1.6	1.8	1.9	1.4	1.6	1.8	2.0	2.2
WW B4	13	20	29	35	43	49	57	66	9	15	22	28	35	41	47	57	1.2	1.3	1.5	1.7	1.8	1.3	1.5	1.6	1.8	1.9
WW B5	13	20	29	35	44	50	58	68	9	15	22	28	36	42	49	59	1.2	1.3	1.5	1.6	1.8	1.3	1.4	1.6	1.7	1.9
WW C1	15	21	28	34	42	47	54	63	11	16	22	27	33	38	44	52	1.2	1.4	1.6	1.7	1.9	1.4	1.6	1.7	1.9	2.1
WW C2	13	20	28	33	42	47	54	64	9	15	21	26	33	38	45	53	1.2	1.4	1.5	1.7	1.8	1.3	1.5	1.7	1.9	2.0
WW C3	17	26	36	43	53	60	68	77	12	20	29	36	46	52	60	70	1.1	1.2	1.4	1.5	1.7	1.1	1.3	1.4	1.6	1.7
WW C4	15	23	33	40	50	57	65	75	11	18	27	33	43	49	57	67	1.1	1.2	1.4	1.5	1.7	1.2	1.3	1.5	1.6	1.8
TW A1	13	15	17	18	21	22	24	27	12	13	14	15	16	17	18	19	2.3	2.6	2.9	3.3	3.6	2.9	3.4	4.1	5.0	-
TW A2	12	14	17	20	23	26	29	34	9	11	13	15	17	19	21	24	1.9	2.1	2.3	2.6	2.8	2.2	2.6	2.9	3.3	3.7
TW A3	10	14	17	20	24	27	31	37	8	10	13	15	18	20	22	26	1.7	1.9	2.2	2.4	2.5	2.0	2.3	2.6	2.9	3.2
TW A4	9	13	17	20	25	28	32	39	6	9	12	14	18	20	23	28	1.7	1.9	2.1	2.3	2.4	1.9	2.2	2.5	2.7	3.0
TW B1	12	14	17	18	21	23	25	29	11	12	13	15	16	17	19	21	2.2	2.4	2.7	3.0	3.2	2.6	3.0	3.5	4.0	4.5
TW B2	10	13	17	19	23	26	29	35	8	10	13	14	17	19	21	25	1.8	2.0	2.3	2.5	2.7	2.1	2.4	2.7	3.0	3.3
TW B3	10	13	16	19	23	26	30	36	7	10	12	14	17	19	22	26	1.8	2.0	2.2	2.4	2.6	2.1	2.3	2.6	2.9	3.1
TW B4	9	12	16	18	22	25	29	35	7	9	12	14	17	19	22	26	1.8	2.0	2.2	2.4	2.6	2.1	2.3	2.6	2.8	3.0
TW C1	11	13	16	18	21	23	26	31	9	11	13	14	16	18	20	23	2.0	2.3	2.5	2.7	2.9	2.4	2.7	3.0	3.4	3.7
TW C2	10	12	16	18	22	24	28	33	8	10	12	14	17	19	21	25	1.9	2.1	2.3	2.5	2.7	2.2	2.5	2.7	3.0	3.2
TW C3	10	12	15	17	21	24	27	32	8	10	12	14	16	18	21	25	2.0	2.1	2.3	2.5	2.7	2.2	2.5	2.7	3.0	3.2
TW C4	10	12	15	17	20	22	25	30	9	10	12	14	16	18	20	24	2.1	2.3	2.5	2.7	2.8	2.4	2.6	2.9	3.1	3.3
TW D1	7	8	10	11	13	14	15	17	5	5	6	7	7	7	7	7	2.5	2.9	3.5	4.2	5.0	3.4	4.7	-	-	-
TW D2	11	15	20	24	30	33	39	46	9	12	15	18	23	26	30	35	1.5	1.7	1.9	2.1	2.2	1.7	2.0	2.2	2.4	2.6
TW D3	12	16	22	26	32	36	42	50	8	12	16	20	25	28	32	39	1.4	1.6	1.8	2.0	2.1	1.6	1.8	2.0	2.3	2.4
TW D4	13	19	26	31	39	44	51	60	10	15	20	25	31	36	42	50	1.3	1.4	1.6	1.8	1.9	1.4	1.6	1.8	1.9	2.1
TW D5	13	19	27	32	40	46	53	62	10	15	21	26	33	38	45	54	1.2	1.4	1.6	1.7	1.8	1.4	1.5	1.7	1.8	2.0
TW E1	14	20	26	31	37	42	48	57	11	15	20	24	30	34	39	46	1.3	1.5	1.7	1.8	2.0	1.5	1.7	1.9	2.1	2.2
TW E2	14	20	26	31	39	44	50	59	10	15	21	25	31	35	41	49	1.3	1.5	1.6	1.8	1.9	1.4	1.6	1.8	2.0	2.1
TW E3	17	25	33	40	49	55	63	72	13	20	27	33	41	47	54	64	1.1	1.3	1.4	1.6	1.7	1.2	1.4	1.5	1.7	1.8
TW E4	15	23	32	38	47	54	61	71	12	18	26	32	40	46	53	63	1.1	1.3	1.4	1.6	1.7	1.2	1.4	1.5	1.7	1.8
TW F1	11	13	15	17	19	21	23	26	10	11	12	13	14	15	16	17	2.3	2.5	2.9	3.2	3.5	2.8	3.3	3.9	4.7	-
TW F2	9	12	15	17	21	23	27	31	7	9	11	12	15	16	18	21	1.9	2.1	2.4	2.6	2.8	2.3	2.6	3.0	3.4	3.8
TW F3	8	11	14	17	21	24	28	33	5	7	10	12	15	17	19	23	1.8	2.0	2.2	2.4	2.6	2.1	2.4	2.7	3.0	3.3
TW F4	6	9	13	16	20	23	27	33	3	6	8	10	14	16	19	23	1.8	2.0	2.2	2.4	2.5	2.1	2.3	2.6	2.9	3.1
TW G1	9	11	14	15	18	20	23	26	7	9	10	11	13	14	16	18	2.2	2.4	2.7	3.0	3.2	2.6	3.0	3.5	4.0	4.5
TW G2	7	9	12	15	19	21	25	30	5	7	9	10	13	15	17	20	1.9	2.1	2.4	2.6	2.8	2.3	2.6	2.9	3.2	3.5
TW G3	5	8	11	14	18	20	24	30	3	5	7	9	12	14	17	20	1.9	2.1	2.3	2.5	2.7	2.2	2.5	2.8	3.0	3.3
TW G4	3	6	9	12	15	18	22	27	2	4	6	8	11	12	15	19	2.0	2.2	2.4	2.6	2.7	2.3	2.5	2.8	3.0	3.3
TW H1	6	8	11	13	16	18	21	25	5	6	8	9	11	13	14	17	2.1	2.4	2.6	2.9	3.1	2.5	2.9	3.2	3.6	3.9
TW H2	4	6	9	11	15	17	20	25	2	4	6	8	10	12	14	17	2.1	2.3	2.5	2.7	2.9	2.4	2.7	3.0	3.3	3.5
TW H3	3	5	8	10	13	15	18	23	1	3	5	7	9	11	13	16	2.2	2.4	2.6	2.8	2.9	2.5	2.8	3.0	3.3	3.5
TW H4	2	4	6	8	10	12	15	19	1	2	4	5	7	9	11	14	2.4	2.6	2.8	3.0	3.1	2.7	3.0	3.2	3.5	3.7
TW I1	5	6	8	9	10	11	13	14	3	4	4	5	5	5	5	4	2.6	3.1	3.7	4.5	-	3.6	-	-	-	-
TW I2	8	12	17	21	26	30	35	42	6	9	12	15	20	23	27	32	1.6	1.7	1.9	2.1	2.3	1.8	2.0	2.2	2.4	2.6
TW I3	9	14	19	23	29	33	39	47	6	10	14	17	22	26	30	37	1.5	1.6	1.8	2.0	2.1	1.7	1.9	2.1	2.3	2.4
TW I4	10	16	23	28	36	41	48	57	7	12	18	22	28	33	39	47	1.3	1.5	1.6	1.8	1.9	1.5	1.6	1.8	2.0	2.1
TW I5	10	17	24	30	38	43	50	60	7	13	19	24	31	36	42	51	1.3	1.4	1.6	1.7	1.9	1.4	1.5	1.7	1.9	2.0
TW J1	12	17	24	28	35	40	46	55	9	13	18	22	28	32	37	45	1.3	1.5	1.7	1.8	2.0	1.5	1.7	1.9	2.1	2.2
TW J2	11	17	24	29	36	41	48	57	8	13	18	22	29	33	39	47	1.3	1.5	1.6	1.8	1.9	1.5	1.6	1.8	2.0	2.1
TW J3	14	22	31	37	47	53	61	71	10	17	25	31	39	45	53	62	1.1	1.3	1.4	1.6	1.7	1.2	1.4	1.5	1.7	1.8
TW J4	13	20	29	36	45	51	59	69	9	16	23	29	38	43	51	61	1.2	1.3	1.5	1.6	1.8	1.3	1.4	1.6	1.7	1.9
DG A1	6	8	10	11	12	13	14	14	4	5	5	6	6	6	6	4	2.3	2.9	4.0	-	-	3.1	4.6	-	-	-
DG A2	9	12	16	18	22	25	29	33	7	9	12	14	17	19	21	25	1.8	2.0	2.2	2.5	2.9	2.1	2.3	2.6	3.1	3.5
DG A3	12	17	22	26	33	38	44	52	10	14	18	22	27	31	36	43	1.5	1.6	1.7	1.9	2.1	1.6	1.8	1.9	2.1	2.3
DG B1	6	8	10	11	13	14	15	16	4	5	5	6	6	7	6	5	2.2	2.8	3.5	4.5	-	3.1	4.4	-	-	-
DG B2	9	12	16	19	23	26	31	36	7	9	12	14	17	20	22	26	1.7	1.9	2.1	2.3	2.6	2.0	2.3	2.5	2.9	3.1
DG B3	13	18	23	28	34	39	46	55	11	14	19	23	28	33	38	46	1.4	1.6	1.7	1.8	1.9	1.6	1.7	1.9	2.0	2.1
DG C1	8	11	13	15	18	20	23	26	5	7	8	10	11	12	13	13	1.9	2.3	2.7	3.1	3.7	2.5	3.1	4.0	-	-
DG C2	11	15	19	23	28	32	38	45	9	12	15	18	22	25	29	34	1.6	1.8	1.9	2.1	2.3	1.8	2.0	2.2	2.4	2.6
DG C3	15	20	26	31	39	45	52	62	12	16	22	26	32	37	43	53	1.3	1.5	1.6	1.7	1.8	1.5	1.6	1.7	1.9	2.0
SS A1	15	18	22	25	29	32	35	41	13	16	19	21	24	26	29	32	1.7	1.9	2.1	2.4	2.6	1.9	2.2	2.5	2.8	3.2
SS A2	19	24	30	35	42	46	52	60	16	21	26	30	36	40	45	52	1.3	1.5	1.6	1.8	2.0	1.4	1.6	1.8	2.0	2.2
SS A3	14	16	19	21	25	27	29	33	12	14	16	18	20	21	23	25	1.8	2.1	2.5	2.8	3.1	2.1	2.5	3.0	3.7	4.5
SS A4	18	23	29	34	40	44	50	57	16	20	25	29	34	38	43	49	1.3	1.5	1.7	1.9	2.0	1.4	1.6	1.9	2.1	2.3
SS A5	16	18	21	23	26	29	31	35	14	16	18	20	22	24	25	28	1.8	2.1	2.4	2.7	3.0	2.1	2.5	2.9	3.4	3.9
SS A6	18	24	30	34	41	45	51	59	16	20	26	29	35	39	44	51	1.3	1.5	1.7	1.9	2.0	1.4	1.6	1.8	2.0	2.2
SS A7	14	16	18	20	22	23	25	27	13	14	16	17	18	19	20	21	2.2	2.5	3.0	3.5	4.0	2.5	3.1	3.9	-	-
SS A8	18	22	28	32	38	42	47	55	15	19	24	28	33	36	40	46	1.4	1.6	1.8	1.9	2.1	1.5	1.7	1.9	2.2	2.4
SS B1	12	15	18	21	25	27	30	35	11	13	15	17	20	22	24	27	1.8	2.0	2.3	2.6	2.8	2.0	2.4	2.7	3.1	3.5
SS B2	16	21	27	31	38	42	48	56	14	18	23	27	33	37	42	49	1.4	1.5	1.7	1.9	2.0	1.5	1.7	1.8	2.0	2.2
SS B3	11	14	16	19	22	24	26	30	10	11	13	15	17	18	20	22	1.9	2.2	2.6	2.9	3.2	2.2	2.7	3.1	3.8	4.5
SS B4	15	20	26	30	36	40	46	54	13	17	22	26	31	35	39	46	1.4	1.6	1.8	1.9	2.1	1.5	1.7	1.9	2.1	2.3
SS B5	13	15	17	19	22	23	26	29	11	13	15	16	18	19	20	22	2.1	2.4	2.7	3.0	3.3	2.4	2.8	3.3	3.8	4.4
SS B6	15	20	26	30	36	41	46	54	13	17	22	26	31	35	40	46	1.4	1.6	1.7	1.9	2.1	1.5	1.7	1.9	2.1	2.3
SS B7	11	13	15	16	18	20	21	24	10	11	13	14	15	16	17	18	2.3	2.7	3.1	3.6	4.0	2.7	3.3	4.1	-	-
SS B8	14	19	24	28	34	38	43	50	12	16	20	24	29	32	37	43	1.4	1.6	1.8	2.0	2.2	1.6	1.8	2.0	2.2	2.4
SS C1	9	12	15	18	22	25	28	33	6	8	10	12	14	16	18	21	1.8	2.0	2.3	2.5	2.7	2.2	2.5	3.0	3.4	3.9
SS C2	11	16	22	26	33	37	43	51	8	12	16	20	25	28	33	40	1.4	1.6	1.8	1.9	2.1	1.6	1.8	2.0	2.2	2.4
SS C3	10	12	14	16	19	21	24	28	8	9	11	12	14	15	17	19	2.1	2.3	2.6	2.9	3.1	2.5	2.9	3.4	3.8	4.3
SS C4	11	15	20	24	29	33	38	45	8	12	15	18	23	26	30	36	1.5	1.7	1.9	2.1	2.2	1.7	2.0	2.2	2.4	2.5
SS D1	16	20	25	28	33	37	41	47	13	17	21	23	27	30	33	37	1.5	1.7	1.9	2.2	2.3	1.6	1.9	2.3	2.6	2.9
SS D2	20	27	35	41	49	54	60	69	17	23	30	35	42	47	53	61	1.1	1.3	1.5	1.7	1.8	1.2	1.4	1.6	1.8	2.0
SS D3	16	19	23	26	30	33	37	42	14	17	20	22	25	27	30	33	1.6	1.9	2.1	2.4	2.6	1.8	2.1	2.5	2.9	3.2
SS D4	20	26	34	39	47	52	58	67	17	22	29	34	40	45	51	59	1.2	1.4	1.5	1.7	1.8	1.3	1.5	1.7	1.8	2.0
SS E1	13	16	20	23	27	30	33	38	10	13	15	17	20	22	24	27	1.6	1.9	2.2	2.5	2.7	1.9	2.3	2.7	3.2	3.9
SS E2	16	22	29	34	41	46	52	60	13	18	24	28	34	38	44	51	1.3	1.4	1.6	1.8	1.9	1.4	1.6	1.8	2.0	2.2
SS E3	13	16	18	21	24	26	29	33	11	13	15	17	19	20	22	25	1.9	2.2	2.5	2.8	3.0	2.2	2.6	3.0	3.5	4.1
SS E4	16	21	27	31	38	42	48	56	13	17	22	26	32	35	40	47	1.3	1.5	1.7	1.9	2.0	1.5	1.7	1.9	2.1	2.3

BALTIMORE, MARYLAND

SOLAR SAVINGS FRACTION (%)

TYPE	55 F (2688 DD) LCR 100	70	50	40	30	25	20	15	65 F (4731 DD) 100	70	50	40	30	25	20	15
WW A1	20	24	28	31	37	40	45	52	15	17	20	22	25	27	31	35
WW A2	21	27	34	39	47	52	58	67	14	18	23	27	32	36	41	48
WW A3	22	29	36	42	50	56	63	72	13	18	24	28	35	39	45	53
WW A4	22	29	38	44	53	58	65	74	13	18	25	29	36	41	47	56
WW A5	22	30	39	46	55	61	68	77	13	19	25	31	38	43	50	58
WW A6	22	31	40	47	56	62	69	78	13	19	26	31	39	44	51	60
WW B1	17	21	26	30	36	40	45	52	10	13	16	18	21	23	26	30
WW B2	23	31	40	46	55	61	68	77	14	20	27	32	40	45	52	61
WW B3	27	36	47	54	63	69	76	84	17	25	33	39	48	54	61	70
WW B4	27	37	48	56	66	72	79	87	17	26	35	42	52	58	66	75
WW B5	26	36	47	55	66	72	79	87	16	25	35	42	52	58	66	76
WW C1	29	38	49	56	66	72	79	86	19	27	35	42	51	57	64	73
WW C2	27	37	47	55	65	71	78	86	17	25	34	41	50	56	64	73
WW C3	32	44	56	64	74	80	86	93	21	32	43	51	61	68	76	84
WW C4	29	40	52	61	71	77	84	91	19	29	40	47	58	65	72	82
TW A1	20	23	28	31	36	40	45	52	15	17	20	22	25	28	31	36
TW A2	20	25	31	36	43	47	54	62	14	17	21	25	30	33	38	44
TW A3	20	26	33	38	45	50	57	66	13	17	22	26	31	35	41	48
TW A4	19	25	33	38	46	51	58	67	12	16	22	26	32	36	42	50
TW B1	19	23	28	32	37	41	46	54	14	16	20	22	26	29	32	38
TW B2	19	24	30	35	42	47	53	61	12	16	20	24	29	33	37	44
TW B3	18	24	30	35	42	47	54	62	12	16	20	24	29	33	38	46
TW B4	17	22	29	33	40	45	52	60	11	15	19	23	28	32	37	44
TW C1	18	22	27	31	37	41	47	54	13	16	19	22	26	29	33	39
TW C2	17	22	27	32	38	43	49	57	12	15	19	22	27	30	35	41
TW C3	16	21	26	30	36	41	47	55	11	14	18	21	26	29	34	40
TW C4	16	19	24	28	33	37	42	50	11	14	17	20	24	27	31	37
TW D1	14	18	23	26	32	35	40	47	9	11	13	16	19	21	24	28
TW D2	22	28	36	42	51	56	63	72	14	19	25	30	37	41	48	56
TW D3	23	31	40	46	55	61	68	77	15	21	28	33	40	45	52	61
TW D4	25	34	44	51	60	66	74	82	16	23	31	38	46	52	59	69
TW D5	24	33	43	51	61	67	74	83	16	23	32	38	47	53	61	71
TW E1	26	35	44	51	60	66	73	81	18	24	32	38	46	51	58	67
TW E2	26	35	44	51	61	67	74	82	17	24	32	38	46	52	59	69
TW E3	31	42	53	61	70	76	83	90	21	30	40	47	57	63	71	80
TW E4	29	39	50	58	68	74	81	89	19	28	38	45	55	61	69	78
TW F1	18	22	26	30	35	39	44	51	13	15	18	21	24	27	30	35
TW F2	18	23	29	33	40	45	51	59	11	15	19	22	27	30	35	42
TW F3	17	22	29	34	41	47	53	62	10	14	19	22	28	32	37	44
TW F4	15	20	27	32	40	45	52	61	8	12	17	21	26	31	36	44
TW G1	16	20	25	29	35	39	44	52	11	13	17	19	23	26	30	35
TW G2	15	20	26	30	37	42	48	56	9	12	16	19	24	28	33	39
TW G3	13	18	24	29	36	41	47	56	7	11	15	18	23	27	32	39
TW G4	10	15	21	25	31	36	42	51	5	8	12	15	20	24	28	35
TW H1	13	17	22	26	32	36	42	49	8	11	14	17	21	24	28	34
TW H2	10	15	20	24	30	34	40	48	5	8	12	15	19	22	27	33
TW H3	9	13	17	21	27	31	36	44	4	7	10	13	17	20	24	30
TW H4	6	10	14	17	21	25	30	37	3	5	8	10	13	16	19	25
TW I1	12	16	20	23	28	32	37	44	7	9	11	13	16	18	21	25
TW I2	18	25	30	39	47	53	60	69	11	16	22	26	33	38	44	53
TW I3	20	27	36	42	51	57	64	73	12	18	24	29	37	42	48	57
TW I4	22	30	40	47	57	63	71	80	13	20	28	34	43	49	56	66
TW I5	22	31	41	48	58	65	72	81	14	21	29	36	45	51	58	68
TW J1	24	32	42	48	58	64	71	80	15	22	29	35	43	49	56	65
TW J2	23	32	41	48	58	64	71	80	14	21	29	35	44	49	57	66
TW J3	28	38	50	58	68	74	81	89	18	27	37	44	55	61	69	78
TW J4	26	36	47	55	65	72	79	87	16	25	35	42	52	60	67	70
DG A1	15	10	24	28	33	35	39	43	8	11	14	16	19	20	22	24
DG A2	17	23	30	35	42	46	52	58	11	15	19	23	28	32	37	42
DG A3	21	28	37	43	53	59	66	74	15	20	26	32	39	45	52	60
DG B1	15	20	26	30	36	40	45	51	8	11	14	17	20	22	25	29
DG B2	18	24	31	36	45	50	57	65	11	15	20	24	30	34	40	47
DG B3	22	29	38	45	55	62	70	79	15	21	27	33	41	47	55	65
DG C1	18	24	31	36	44	49	55	62	11	14	18	22	26	29	34	39
DG C2	20	27	36	42	51	58	65	74	13	18	24	29	36	41	47	56
DG C3	24	33	43	51	62	69	77	85	17	24	31	37	47	53	62	72
SS A1	24	30	36	41	48	53	59	67	18	22	27	31	37	41	46	52
SS A2	30	38	47	53	62	68	74	82	23	29	37	42	50	55	62	70
SS A3	22	27	33	38	44	48	54	62	17	20	25	28	32	36	40	46
SS A4	29	37	46	52	61	66	73	81	22	29	36	41	49	54	60	68
SS A5	23	28	34	38	44	48	54	62	18	22	26	29	33	37	41	47
SS A6	30	38	47	53	62	67	74	82	22	29	36	42	49	55	61	69
SS A7	21	25	30	33	38	42	47	53	16	19	22	25	28	31	34	38
SS A8	29	36	45	51	59	65	71	79	22	28	35	40	47	52	58	66
SS B1	21	26	32	36	43	47	53	61	15	19	23	27	31	35	40	46
SS B2	26	34	42	48	57	63	70	78	19	26	33	38	45	51	57	66
SS B3	19	24	29	34	40	44	50	57	14	17	21	24	29	32	36	41
SS B4	26	33	41	47	56	61	68	76	19	25	32	37	44	49	55	64
SS B5	19	23	28	32	37	41	46	53	15	17	21	23	27	30	34	39
SS B6	25	33	41	47	56	62	68	77	19	25	31	37	44	49	56	64
SS B7	18	21	25	28	33	36	41	47	13	16	18	21	24	26	29	33
SS B8	24	31	39	45	54	59	66	75	18	23	30	35	42	47	53	61
SS C1	19	25	32	37	45	50	56	65	12	16	20	24	30	33	38	46
SS C2	23	31	40	46	55	61	68	77	15	21	28	33	41	46	52	61
SS C3	17	21	26	30	36	41	46	54	11	14	18	20	25	27	31	37
SS C4	21	28	35	41	50	55	62	71	14	19	25	29	36	41	47	55
SS D1	28	35	43	49	57	62	69	77	20	26	32	37	43	48	54	61
SS D2	34	44	54	62	71	76	82	89	25	34	43	49	58	64	71	79
SS D3	26	32	39	44	51	56	62	70	19	24	29	33	39	43	48	55
SS D4	33	42	52	60	69	74	81	88	25	33	41	48	56	62	69	77
SS E1	24	30	38	43	51	56	62	71	16	21	26	31	36	40	46	53
SS E2	29	38	48	55	64	70	77	84	21	28	36	42	51	56	63	72
SS E3	21	26	32	36	43	47	53	61	16	19	23	26	31	34	38	44
SS E4	27	35	44	51	60	65	72	80	20	26	33	39	47	52	58	67

CONSERVATION FACTOR

TYPE	55 F 100	50	30	20	15	65 F 100	50	30	20	15
WW A1	1.5	1.7	1.9	2.0	2.2	2.0	2.2	2.4	2.6	2.8
WW A2	1.2	1.4	1.5	1.7	1.8	1.5	1.7	1.9	2.1	2.2
WW A3	1.1	1.3	1.5	1.6	1.8	1.4	1.6	1.7	1.9	2.1
WW A4	1.1	1.3	1.4	1.6	1.7	1.3	1.5	1.7	1.9	2.0
WW A5	1.1	1.2	1.4	1.6	1.7	1.3	1.5	1.6	1.8	1.9
WW A6	1.1	1.2	1.4	1.5	1.7	1.3	1.4	1.6	1.8	1.9
WW B1	1.4	1.6	1.8	2.0	2.1	1.9	2.2	2.5	2.8	3.1
WW B2	1.1	1.2	1.4	1.5	1.7	1.3	1.4	1.6	1.8	1.9
WW B3	1.0	1.1	1.3	1.5	1.6	1.1	1.3	1.5	1.6	1.8
WW B4	1.0	1.1	1.3	1.4	1.6	1.1	1.2	1.4	1.5	1.7
WW B5	1.0	1.1	1.3	1.4	1.6	1.1	1.2	1.4	1.5	1.7
WW C1	1.0	1.1	1.3	1.4	1.6	1.1	1.3	1.4	1.6	1.7
WW C2	1.0	1.1	1.3	1.4	1.6	1.1	1.3	1.4	1.6	1.7
WW C3	0.9	1.0	1.2	1.4	1.5	1.0	1.1	1.3	1.4	1.6
WW C4	0.9	1.1	1.2	1.4	1.5	1.0	1.2	1.3	1.5	1.6
TW A1	1.5	1.7	1.9	2.0	2.2	2.0	2.2	2.4	2.6	2.8
TW A2	1.3	1.5	1.6	1.8	1.9	1.6	1.8	2.0	2.2	2.3
TW A3	1.2	1.4	1.5	1.7	1.8	1.5	1.7	1.9	2.0	2.2
TW A4	1.2	1.4	1.5	1.7	1.8	1.5	1.6	1.8	2.0	2.1
TW B1	1.5	1.6	1.8	2.0	2.1	1.9	2.0	2.2	2.4	2.6
TW B2	1.3	1.5	1.6	1.8	1.9	1.6	1.8	1.9	2.1	2.3
TW B3	1.3	1.4	1.6	1.8	1.9	1.6	1.7	1.9	2.1	2.2
TW B4	1.3	1.5	1.6	1.8	1.9	1.6	1.8	1.9	2.1	2.2
TW C1	1.4	1.6	1.8	1.9	2.1	1.8	2.0	2.1	2.3	2.5
TW C2	1.4	1.5	1.7	1.9	2.0	1.7	1.9	2.0	2.2	2.3
TW C3	1.4	1.6	1.7	1.9	2.0	1.7	1.9	2.1	2.2	2.4
TW C4	1.5	1.7	1.9	2.0	2.1	1.9	2.0	2.2	2.4	2.5
TW D1	1.5	1.7	1.9	2.1	2.2	2.0	2.3	2.5	2.8	3.1
TW D2	1.2	1.3	1.5	1.6	1.8	1.4	1.5	1.7	1.8	2.0
TW D3	1.1	1.2	1.4	1.6	1.7	1.3	1.4	1.6	1.8	1.9
TW D4	1.0	1.2	1.3	1.5	1.6	1.2	1.3	1.5	1.6	1.8
TW D5	1.0	1.2	1.3	1.5	1.6	1.2	1.3	1.4	1.6	1.7
TW E1	1.0	1.2	1.3	1.5	1.6	1.2	1.4	1.5	1.7	1.8
TW E2	1.0	1.2	1.3	1.5	1.6	1.2	1.3	1.5	1.6	1.8
TW E3	0.9	1.1	1.2	1.4	1.5	1.0	1.2	1.3	1.5	1.6
TW E4	0.9	1.1	1.2	1.4	1.6	1.1	1.2	1.4	1.5	1.7
TW F1	1.5	1.7	1.8	2.0	2.1	1.9	2.1	2.3	2.5	2.7
TW F2	1.3	1.5	1.6	1.8	1.9	1.7	1.8	2.0	2.2	2.3
TW F3	1.3	1.4	1.6	1.8	1.9	1.6	1.7	1.9	2.1	2.2
TW F4	1.3	1.4	1.6	1.7	1.9	1.6	1.7	1.9	2.1	2.2
TW G1	1.5	1.6	1.8	2.0	2.1	1.9	2.1	2.2	2.4	2.6
TW G2	1.4	1.5	1.7	1.8	2.0	1.7	1.9	2.0	2.2	2.4
TW G3	1.4	1.5	1.7	1.8	2.0	1.7	1.8	2.0	2.2	2.3
TW G4	1.4	1.6	1.8	1.9	2.0	1.8	1.9	2.1	2.3	2.4
TW H1	1.5	1.7	1.8	2.0	2.1	1.9	2.0	2.2	2.4	2.6
TW H2	1.5	1.7	1.8	2.0	2.1	1.9	2.0	2.2	2.3	2.5
TW H3	1.6	1.7	1.9	2.0	2.2	1.9	2.1	2.2	2.4	2.6
TW H4	1.8	1.9	2.1	2.2	2.3	2.2	2.3	2.5	2.6	2.8
TW I1	1.0	1.8	2.0	2.1	2.3	2.1	2.4	2.6	2.9	3.2
TW I2	1.2	1.3	1.5	1.6	1.8	1.4	1.6	1.7	1.9	2.0
TW I3	1.1	1.3	1.4	1.6	1.7	1.3	1.5	1.6	1.8	1.9
TW I4	1.0	1.2	1.3	1.5	1.6	1.2	1.3	1.5	1.7	1.8
TW I5	1.0	1.2	1.3	1.5	1.6	1.2	1.3	1.5	1.6	1.8
TW J1	1.0	1.2	1.4	1.5	1.7	1.2	1.4	1.5	1.7	1.8
TW J2	1.0	1.2	1.3	1.5	1.6	1.2	1.3	1.5	1.7	1.8
TW J3	0.9	1.1	1.2	1.4	1.6	1.0	1.2	1.4	1.5	1.7
TW J4	1.0	1.1	1.3	1.4	1.6	1.1	1.2	1.4	1.5	1.7
DG A1	1.4	1.7	2.0	2.5	3.1	1.9	2.2	2.8	3.8	-
DG A2	1.3	1.4	1.6	1.9	2.1	1.6	1.7	1.9	2.2	2.5
DG A3	1.1	1.3	1.4	1.6	1.8	1.3	1.5	1.6	1.8	1.9
DG B1	1.4	1.6	1.8	2.1	2.4	1.8	2.1	2.4	2.9	3.5
DG B2	1.3	1.4	1.5	1.7	1.9	1.5	1.7	1.8	2.0	2.2
DG B3	1.1	1.2	1.3	1.5	1.7	1.3	1.4	1.5	1.7	1.8
DG C1	1.3	1.4	1.6	1.8	2.1	1.6	1.8	2.0	2.3	2.7
DG C2	1.2	1.3	1.4	1.6	1.8	1.4	1.5	1.7	1.8	2.0
DG C3	1.1	1.1	1.3	1.4	1.6	1.2	1.3	1.4	1.6	1.7
SS A1	1.2	1.4	1.5	1.7	1.9	1.4	1.6	1.8	2.0	2.2
SS A2	1.0	1.2	1.3	1.5	1.7	1.2	1.3	1.5	1.7	1.8
SS A3	1.3	1.5	1.6	1.8	2.0	1.5	1.8	2.0	2.2	2.4
SS A4	1.0	1.2	1.4	1.5	1.7	1.2	1.4	1.5	1.7	1.8
SS A5	1.3	1.5	1.7	1.8	2.0	1.6	1.8	2.0	2.2	2.4
SS A6	1.0	1.2	1.3	1.5	1.7	1.2	1.3	1.5	1.7	1.8
SS A7	1.5	1.7	1.8	2.0	2.2	1.8	2.0	2.3	2.6	2.8
SS A8	1.1	1.2	1.4	1.5	1.7	1.2	1.4	1.6	1.7	1.9
SS B1	1.3	1.5	1.6	1.8	1.9	1.6	1.7	1.9	2.1	2.3
SS B2	1.1	1.2	1.4	1.5	1.7	1.2	1.4	1.5	1.7	1.9
SS B3	1.4	1.5	1.7	1.9	2.0	1.6	1.9	2.1	2.3	2.5
SS B4	1.1	1.2	1.4	1.6	1.7	1.2	1.4	1.6	1.7	1.9
SS B5	1.5	1.6	1.8	2.0	2.1	1.8	2.0	2.2	2.4	2.6
SS B6	1.1	1.2	1.4	1.6	1.7	1.2	1.4	1.6	1.7	1.9
SS B7	1.6	1.8	2.0	2.1	2.3	2.0	2.2	2.4	2.7	2.9
SS B8	1.1	1.3	1.4	1.6	1.7	1.3	1.4	1.6	1.8	1.9
SS C1	1.2	1.4	1.6	1.7	1.9	1.5	1.7	1.9	2.1	2.2
SS C2	1.1	1.2	1.4	1.5	1.7	1.3	1.4	1.6	1.7	1.9
SS C3	1.4	1.6	1.8	1.9	2.1	1.8	2.0	2.2	2.4	2.5
SS C4	1.2	1.3	1.5	1.6	1.8	1.4	1.5	1.7	1.9	2.0
SS D1	1.1	1.2	1.4	1.6	1.7	1.3	1.5	1.6	1.8	2.0
SS D2	0.9	1.1	1.2	1.4	1.6	1.0	1.2	1.4	1.5	1.7
SS D3	1.2	1.3	1.5	1.7	1.8	1.4	1.6	1.8	2.0	2.1
SS D4	0.9	1.1	1.3	1.4	1.6	1.1	1.2	1.4	1.6	1.7
SS E1	1.2	1.3	1.5	1.7	1.8	1.4	1.6	1.8	2.0	2.2
SS E2	1.0	1.1	1.3	1.5	1.6	1.1	1.3	1.5	1.6	1.8
SS E3	1.3	1.5	1.7	1.8	2.0	1.6	1.8	2.0	2.2	2.4
SS E4	1.0	1.2	1.4	1.5	1.7	1.2	1.4	1.5	1.7	1.8

BOSTON, MASSACHUSETTS

TYPE	SOLAR SAVINGS FRACTION (%) — BASE TEMP: 55 F (3320 DD)								65 F (5622 DD)								CONSERVATION FACTOR — 55 F					65 F				
LCR	100	70	50	40	30	25	20	15	100	70	50	40	30	25	20	15	100	50	30	20	15	100	50	30	20	15
WW A1	16	19	22	24	28	30	33	38	13	15	16	18	20	21	23	25	1.9	2.1	2.3	2.5	2.7	2.4	2.8	3.1	3.5	3.9
WW A2	16	20	25	29	35	39	44	52	11	14	18	20	24	27	30	35	1.5	1.6	1.8	2.0	2.1	1.8	2.0	2.2	2.5	2.7
WW A3	16	21	27	31	38	43	49	57	10	14	18	21	26	29	33	40	1.3	1.5	1.7	1.8	2.0	1.6	1.8	2.1	2.3	2.4
WW A4	15	21	27	32	40	45	51	60	9	13	18	22	27	30	35	42	1.3	1.5	1.6	1.8	1.9	1.6	1.8	2.0	2.2	2.3
WW A5	15	21	28	34	41	47	53	62	9	13	18	22	28	32	37	44	1.2	1.4	1.6	1.7	1.9	1.5	1.7	1.9	2.1	2.2
WW A6	15	21	29	34	42	48	55	64	8	13	18	23	29	33	38	46	1.2	1.4	1.5	1.7	1.9	1.5	1.6	1.8	2.0	2.2
WW B1	12	15	18	20	24	27	30	35	8	9	11	12	13	14	15	16	1.8	2.1	2.3	2.6	2.8	2.5	3.1	3.8	-	-
WW B2	16	22	30	35	43	48	55	64	10	15	21	25	31	35	41	49	1.2	1.4	1.5	1.7	1.8	1.4	1.6	1.8	2.0	2.1
WW B3	19	27	36	42	51	57	64	73	13	19	26	31	38	43	50	58	1.1	1.3	1.4	1.6	1.7	1.3	1.5	1.6	1.8	1.9
WW B4	20	28	38	45	55	61	69	78	13	20	28	34	43	49	56	66	1.1	1.2	1.4	1.5	1.7	1.2	1.3	1.5	1.7	1.8
WW B5	19	27	37	45	55	61	69	78	12	20	28	34	43	49	57	67	1.1	1.2	1.4	1.5	1.7	1.2	1.3	1.5	1.6	1.8
WW C1	21	29	38	45	54	60	67	76	14	21	28	33	41	46	53	62	1.1	1.2	1.4	1.6	1.7	1.2	1.4	1.6	1.7	1.9
WW C2	19	28	37	44	53	59	67	76	13	20	27	33	41	47	54	63	1.1	1.2	1.4	1.6	1.7	1.2	1.4	1.5	1.7	1.8
WW C3	24	34	45	53	64	70	78	86	16	25	35	43	53	59	67	77	1.0	1.1	1.3	1.4	1.6	1.1	1.2	1.4	1.5	1.7
WW C4	21	31	42	50	61	67	75	84	14	23	33	40	50	56	64	74	1.0	1.1	1.3	1.5	1.6	1.1	1.2	1.4	1.5	1.7
TW A1	16	19	22	24	28	30	34	39	13	15	16	18	20	21	23	26	1.8	2.1	2.3	2.4	2.6	2.4	2.7	3.0	3.3	3.6
TW A2	15	19	24	27	32	36	41	48	11	14	17	19	23	25	28	33	1.5	1.7	1.9	2.1	2.2	1.9	2.1	2.4	2.6	2.8
TW A3	15	19	24	28	34	38	44	52	10	13	17	20	24	27	31	36	1.5	1.6	1.8	2.0	2.1	1.8	2.0	2.2	2.4	2.6
TW A4	14	18	24	28	35	39	45	53	9	12	16	19	24	27	32	38	1.4	1.6	1.7	1.9	2.0	1.7	1.9	2.1	2.3	2.5
TW B1	15	18	22	24	28	31	35	41	12	14	16	18	20	22	25	28	1.8	2.0	2.1	2.3	2.5	2.2	2.5	2.7	3.0	3.2
TW B2	14	18	23	26	32	35	41	48	10	13	16	19	22	25	29	34	1.5	1.7	1.9	2.0	2.2	1.9	2.1	2.3	2.5	2.7
TW B3	13	18	22	26	32	36	42	49	9	12	16	18	23	26	29	35	1.5	1.7	1.8	2.0	2.1	1.8	2.0	2.2	2.4	2.6
TW B4	13	17	21	25	31	35	40	47	9	12	15	18	22	25	29	34	1.5	1.7	1.9	2.0	2.2	1.8	2.0	2.2	2.4	2.6
TW C1	14	17	21	24	28	32	36	42	11	13	16	18	21	23	26	30	1.7	1.9	2.1	2.2	2.4	2.1	2.3	2.5	2.7	2.9
TW C2	13	17	21	24	29	33	38	45	10	12	15	18	21	24	27	32	1.6	1.8	2.0	2.1	2.3	1.9	2.1	2.3	2.5	2.7
TW C3	13	16	20	23	28	31	36	43	10	12	15	17	21	23	27	32	1.7	1.8	2.0	2.2	2.3	2.0	2.2	2.4	2.6	2.7
TW C4	13	16	19	22	26	29	33	39	10	12	15	17	20	22	25	30	1.8	2.0	2.1	2.3	2.4	2.1	2.3	2.5	2.7	2.9
TW D1	10	13	16	18	21	24	27	32	7	8	9	11	12	13	15	16	1.9	2.2	2.4	2.6	2.8	2.6	3.0	3.6	4.3	-
TW D2	16	21	28	32	39	44	51	59	11	15	20	23	29	33	38	45	1.3	1.5	1.6	1.8	1.9	1.5	1.7	1.9	2.1	2.2
TW D3	17	23	30	35	43	48	55	64	11	16	21	25	32	36	41	49	1.2	1.4	1.6	1.7	1.9	1.4	1.6	1.8	2.0	2.1
TW D4	18	26	34	40	49	55	62	72	13	18	25	31	38	43	50	59	1.1	1.3	1.4	1.6	1.7	1.3	1.4	1.6	1.8	1.9
TW D5	18	26	34	41	50	56	64	73	12	19	26	32	40	45	52	62	1.1	1.3	1.4	1.6	1.7	1.3	1.4	1.6	1.7	1.9
TW E1	20	26	34	40	48	54	61	70	14	19	25	30	37	42	48	56	1.2	1.3	1.5	1.6	1.8	1.3	1.5	1.7	1.8	2.0
TW E2	19	26	34	41	49	55	62	72	13	19	26	31	38	43	50	58	1.2	1.3	1.4	1.6	1.8	1.3	1.5	1.6	1.8	1.9
TW E3	23	32	43	50	60	66	73	82	17	24	33	40	49	55	62	72	1.0	1.2	1.3	1.5	1.6	1.1	1.3	1.4	1.6	1.7
TW E4	21	30	40	48	58	64	72	81	15	23	31	38	47	53	61	70	1.0	1.2	1.3	1.5	1.6	1.1	1.3	1.4	1.6	1.7
TW F1	14	17	20	23	26	29	33	38	11	13	15	16	19	20	22	25	1.8	2.0	2.2	2.4	2.6	2.3	2.6	2.9	3.2	3.5
TW F2	13	17	21	25	30	33	38	45	9	11	14	17	20	23	26	31	1.6	1.8	1.9	2.1	2.2	1.9	2.2	2.4	2.6	2.9
TW F3	12	16	21	25	30	35	40	48	7	10	14	16	20	23	27	33	1.5	1.7	1.8	2.0	2.2	1.8	2.0	2.2	2.5	2.7
TW F4	10	14	19	23	29	34	39	47	5	9	12	15	19	22	27	32	1.5	1.7	1.8	2.0	2.1	1.8	2.0	2.2	2.4	2.6
TW G1	12	15	19	21	26	29	33	38	9	11	13	15	18	19	22	26	1.8	2.0	2.2	2.3	2.5	2.2	2.5	2.7	3.0	3.2
TW G2	10	14	18	22	27	31	36	43	7	9	12	14	18	21	24	29	1.6	1.8	2.0	2.1	2.3	2.0	2.2	2.4	2.6	2.8
TW G3	9	13	17	21	26	30	35	42	5	8	11	13	17	20	23	29	1.6	1.8	1.9	2.1	2.2	1.9	2.1	2.3	2.5	2.7
TW G4	7	10	14	18	23	26	31	38	3	6	9	11	15	17	21	26	1.7	1.9	2.0	2.2	2.3	2.0	2.2	2.4	2.6	2.8
TW H1	9	12	16	19	23	26	31	37	6	8	11	13	15	18	20	25	1.8	2.0	2.1	2.3	2.4	2.2	2.4	2.6	2.9	3.1
TW H2	7	10	14	17	22	25	30	36	4	6	9	11	14	16	20	24	1.7	1.9	2.1	2.3	2.4	2.1	2.3	2.5	2.7	2.9
TW H3	6	9	12	15	19	22	27	33	3	5	7	9	12	15	18	22	1.8	2.0	2.2	2.3	2.5	2.2	2.4	2.6	2.8	3.0
TW H4	4	6	9	12	15	18	22	27	2	4	6	7	10	12	15	19	2.0	2.2	2.4	2.6	2.7	2.4	2.6	2.8	3.0	3.2
TW I1	8	11	13	15	19	21	24	28	5	6	7	9	10	11	12	14	2.0	2.3	2.5	2.7	2.9	2.7	3.2	3.8	4.6	-
TW I2	13	18	24	29	36	41	47	56	8	12	17	20	26	29	34	42	1.4	1.5	1.7	1.8	2.0	1.6	1.8	1.9	2.1	2.3
TW I3	14	20	27	32	39	45	51	60	9	13	19	23	29	33	38	46	1.3	1.4	1.6	1.8	1.9	1.5	1.7	1.8	2.0	2.2
TW I4	16	23	31	37	46	52	59	69	10	16	22	27	35	40	47	56	1.2	1.3	1.5	1.6	1.8	1.3	1.5	1.6	1.8	1.9
TW I5	16	23	32	38	47	54	61	71	10	16	23	29	37	42	50	59	1.1	1.3	1.4	1.6	1.7	1.3	1.4	1.6	1.7	1.9
TW J1	17	24	32	38	46	52	59	68	12	17	23	28	35	40	46	54	1.2	1.3	1.5	1.7	1.8	1.4	1.5	1.7	1.9	2.0
TW J2	17	24	32	38	46	52	60	69	11	17	23	28	35	41	47	56	1.2	1.3	1.5	1.6	1.8	1.3	1.5	1.6	1.8	1.9
TW J3	20	29	40	47	57	64	71	80	14	22	30	37	46	52	60	70	1.0	1.2	1.3	1.5	1.6	1.1	1.3	1.4	1.6	1.7
TW J4	19	27	37	45	55	61	69	79	12	20	29	35	44	50	58	68	1.1	1.2	1.3	1.5	1.6	1.2	1.3	1.5	1.6	1.8
DG A1	10	13	16	19	22	23	25	27	6	7	9	10	12	12	13	13	1.8	2.1	2.6	3.5	4.8	2.4	3.0	4.1	-	-
DG A2	13	17	22	26	31	35	40	46	9	12	15	18	22	25	28	33	1.5	1.7	1.9	2.1	2.4	1.8	2.0	2.3	2.6	2.9
DG A3	16	22	29	34	42	48	55	63	12	17	22	26	32	37	43	51	1.3	1.4	1.5	1.7	1.9	1.5	1.6	1.7	1.9	2.1
DG B1	10	13	17	19	23	26	29	33	6	7	9	11	12	13	14	16	1.8	2.0	2.3	2.7	3.1	2.4	2.9	3.6	4.6	-
DG B2	13	17	22	27	33	37	43	51	9	12	16	18	23	26	30	36	1.5	1.6	1.8	1.9	2.1	1.8	2.0	2.2	2.4	2.6
DG B3	17	22	30	35	44	50	58	68	13	17	23	27	34	39	46	55	1.3	1.4	1.5	1.6	1.8	1.5	1.6	1.7	1.8	1.9
DG C1	12	16	21	24	30	33	38	43	8	10	13	15	17	19	22	25	1.6	1.7	1.9	2.2	2.5	2.0	2.3	2.7	3.2	3.8
DG C2	15	20	27	32	39	44	51	60	11	15	19	23	28	32	37	44	1.4	1.5	1.6	1.8	1.9	1.6	1.8	1.9	2.1	2.3
DG C3	19	25	34	40	50	56	65	75	14	20	26	31	38	44	51	62	1.2	1.3	1.4	1.5	1.7	1.4	1.5	1.6	1.7	1.8
SS A1	19	23	28	32	38	42	47	54	15	19	22	25	29	32	36	41	1.4	1.6	1.8	2.0	2.1	1.6	1.9	2.1	2.4	2.6
SS A2	24	30	38	43	51	57	63	72	19	25	31	36	42	47	53	61	1.2	1.3	1.5	1.6	1.8	1.3	1.5	1.6	1.8	2.0
SS A3	17	21	25	29	33	37	41	47	14	17	20	22	25	27	30	34	1.5	1.8	2.0	2.2	2.3	1.8	2.1	2.4	2.8	3.1
SS A4	23	29	37	42	50	55	61	69	19	24	30	34	40	45	50	58	1.2	1.3	1.5	1.7	1.8	1.3	1.5	1.7	1.9	2.0
SS A5	19	23	27	30	34	38	42	48	16	18	21	24	27	29	32	36	1.6	1.8	2.0	2.2	2.3	1.8	2.1	2.4	2.7	2.9
SS A6	23	30	37	43	50	56	62	71	19	24	30	35	41	46	51	59	1.2	1.3	1.5	1.7	1.8	1.3	1.5	1.7	1.8	2.0
SS A7	17	20	23	25	29	31	35	39	14	16	18	20	22	24	26	28	1.8	2.0	2.3	2.5	2.7	2.1	2.5	2.9	3.4	3.9
SS A8	22	29	35	41	48	53	59	67	18	23	29	33	39	43	48	55	1.2	1.4	1.5	1.7	1.9	1.3	1.5	1.7	1.9	2.1
SS B1	16	20	24	28	33	36	41	48	13	15	19	21	25	27	31	36	1.5	1.7	1.9	2.1	2.2	1.8	2.0	2.3	2.5	2.8
SS B2	21	27	34	39	47	52	59	67	16	21	27	32	38	43	48	56	1.2	1.4	1.5	1.7	1.8	1.3	1.5	1.7	1.9	2.0
SS B3	15	18	22	25	30	33	37	43	11	14	17	19	22	24	27	31	1.6	1.8	2.0	2.2	2.4	1.9	2.2	2.5	2.9	3.1
SS B4	20	26	33	38	45	50	57	65	16	21	26	30	37	41	46	54	1.2	1.4	1.6	1.7	1.9	1.4	1.6	1.7	1.9	2.1
SS B5	16	19	22	25	28	31	35	41	13	15	17	19	22	24	26	30	1.7	2.0	2.2	2.4	2.5	2.1	2.4	2.7	3.0	3.3
SS B6	20	26	33	38	45	51	57	66	15	20	26	30	37	41	47	54	1.2	1.4	1.5	1.7	1.9	1.4	1.6	1.7	1.9	2.1
SS B7	14	17	19	22	25	27	30	35	12	13	15	17	19	20	22	24	1.9	2.2	2.4	2.6	2.8	2.3	2.7	3.1	3.5	4.0
SS B8	19	24	31	36	43	48	54	63	15	19	24	29	34	38	44	51	1.3	1.4	1.6	1.8	1.9	1.4	1.6	1.8	2.0	2.1
SS C1	14	18	23	27	33	37	42	50	9	12	15	17	21	24	27	32	1.5	1.6	1.8	2.0	2.1	1.8	2.1	2.3	2.6	2.8
SS C2	17	23	30	36	43	49	56	65	11	16	21	26	32	36	42	50	1.2	1.4	1.5	1.7	1.8	1.4	1.6	1.8	1.9	2.1
SS C3	13	16	20	23	27	30	34	41	9	12	14	16	19	21	23	27	1.7	1.9	2.1	2.3	2.4	2.1	2.4	2.6	2.9	3.1
SS C4	16	21	27	32	39	44	50	59	11	15	19	23	29	32	38	45	1.3	1.5	1.6	1.8	1.9	1.6	1.7	1.9	2.1	2.2
SS D1	21	27	33	38	45	49	55	63	16	21	25	29	34	38	42	48	1.3	1.4	1.6	1.8	2.0	1.4	1.7	1.9	2.1	2.3
SS D2	26	35	44	50	59	65	72	80	20	28	35	41	49	54	61	69	1.0	1.2	1.4	1.5	1.7	1.1	1.3	1.5	1.7	1.8
SS D3	20	25	30	34	40	44	49	56	16	20	24	27	31	34	38	43	1.4	1.6	1.7	1.9	2.1	1.6	1.8	2.1	2.3	2.5
SS D4	25	33	42	49	57	63	70	78	20	27	34	40	47	52	59	67	1.1	1.2	1.4	1.6	1.7	1.2	1.3	1.5	1.7	1.8
SS E1	17	22	28	32	38	42	48	55	13	16	20	23	27	30	34	39	1.4	1.6	1.7	1.9	2.1	1.6	1.9	2.2	2.4	2.7
SS E2	22	29	38	44	52	58	65	73	16	22	29	34	41	46	52	61	1.1	1.3	1.4	1.6	1.8	1.2	1.4	1.6	1.8	1.9
SS E3	17	20	24	28	32	36	40	46	13	16	19	21	24	26	29	33	1.6	1.8	2.0	2.2	2.3	1.9	2.1	2.4	2.7	3.0
SS E4	21	27	35	40	48	53	60	69	16	21	27	32	38	43	48	56	1.2	1.3	1.5	1.7	1.8	1.3	1.5	1.7	1.9	2.0

ALPENA, MICHIGAN

TYPE	SOLAR SAVINGS FRACTION (%) BASE TEMP: 55 F (5637 DD)								65 F (8519 DD)								CONSERVATION FACTOR 55 F					65 F				
LCR	100	70	50	40	30	25	20	15	100	70	50	40	30	25	20	15	100	50	30	20	15	100	50	30	20	15
WW A1	12	13	14	15	16	16	17	17	11	12	12	13	13	13	13	12	3.0	3.6	4.7	-	-	3.5	-	-	-	-
WW A2	9	11	14	16	18	20	21	24	8	9	11	12	14	14	15	16	2.1	2.4	2.9	3.5	4.3	2.3	3.0	3.9	-	-
WW A3	8	11	14	16	19	21	23	26	6	8	10	12	14	15	16	17	1.9	2.2	2.6	3.1	3.6	2.1	2.6	3.3	4.5	-
WW A4	7	10	13	16	19	21	24	28	5	7	10	11	14	15	17	19	1.8	2.1	2.5	2.9	3.3	2.0	2.5	3.1	4.0	-
WW A5	6	9	13	16	20	22	25	29	4	6	9	11	14	15	17	19	1.7	2.0	2.3	2.7	3.1	1.9	2.3	2.9	3.7	4.8
WW A6	5	9	13	16	20	23	26	30	3	6	9	11	14	16	18	20	1.7	1.9	2.3	2.7	3.0	1.8	2.3	2.8	3.5	4.4
WW B1	6	7	7	8	7	7	6	3	5	5	5	4	3	2	0	-4	3.5	-	-	-	-	-	-	-	-	-
WW B2	8	12	16	19	24	27	31	37	6	9	13	15	19	21	25	29	1.6	1.8	2.1	2.4	2.6	1.7	2.1	2.4	2.7	3.1
WW B3	10	15	20	24	30	33	38	45	8	12	16	19	24	27	31	36	1.4	1.6	1.9	2.1	2.3	1.5	1.8	2.1	2.4	2.7
WW B4	10	16	23	28	35	40	46	55	7	13	19	23	30	34	40	48	1.3	1.5	1.7	1.9	2.0	1.4	1.6	1.8	2.0	2.2
WW B5	9	16	23	28	36	42	48	57	7	13	19	24	31	36	42	51	1.3	1.4	1.6	1.8	1.9	1.3	1.5	1.7	1.9	2.1
WW C1	11	16	22	27	33	37	42	49	9	13	18	22	27	31	35	41	1.4	1.6	1.8	2.0	2.2	1.5	1.7	2.0	2.2	2.5
WW C2	10	16	22	27	33	38	44	52	8	13	18	22	28	32	37	44	1.3	1.5	1.7	1.9	2.1	1.4	1.7	1.9	2.1	2.3
WW C3	13	21	30	36	45	51	59	68	10	18	26	32	40	46	53	62	1.1	1.3	1.5	1.6	1.8	1.2	1.4	1.5	1.7	1.9
WW C4	11	19	27	34	43	49	56	66	9	16	23	29	38	43	50	60	1.2	1.3	1.5	1.7	1.8	1.2	1.4	1.6	1.7	1.9
TW A1	12	13	14	15	16	17	18	19	11	12	12	13	14	14	14	14	2.8	3.4	4.1	-	-	3.2	4.3	-	-	-
TW A2	10	12	14	15	18	19	21	24	8	10	11	13	14	15	16	17	2.2	2.5	3.0	3.5	4.1	2.4	3.0	3.8	-	-
TW A3	8	11	13	15	18	20	22	26	7	9	11	12	14	15	17	18	2.0	2.3	2.7	3.1	3.6	2.2	2.7	3.3	4.2	-
TW A4	7	10	13	15	18	20	23	27	5	7	10	12	14	15	17	20	1.9	2.2	2.5	2.9	3.2	2.1	2.5	3.0	3.6	4.3
TW B1	11	12	14	15	17	18	19	21	10	11	12	13	14	14	15	16	2.6	3.0	3.5	4.1	5.0	2.9	3.6	4.6	-	-
TW B2	9	11	13	15	18	19	22	25	7	9	11	12	14	15	17	18	2.1	2.4	2.8	3.2	3.6	2.3	2.8	3.4	4.1	5.0
TW B3	8	10	13	15	18	20	22	26	6	8	10	12	14	15	17	20	2.0	2.3	2.6	3.0	3.3	2.2	2.6	3.1	3.7	4.3
TW B4	8	10	12	14	17	20	22	26	6	8	10	12	14	16	18	20	2.1	2.3	2.6	2.9	3.2	2.2	2.6	3.0	3.5	3.9
TW C1	10	11	13	15	17	19	20	23	9	10	11	13	14	15	16	18	2.4	2.7	3.0	3.5	3.9	2.6	3.1	3.7	4.4	-
TW C2	8	10	13	15	17	19	22	25	7	9	11	12	14	16	17	20	2.2	2.4	2.7	3.1	3.4	2.4	2.7	3.2	3.7	4.2
TW C3	8	10	13	14	17	19	21	25	7	9	11	12	14	16	18	20	2.2	2.5	2.7	3.0	3.3	2.4	2.7	3.1	3.6	4.0
TW C4	9	11	13	14	17	19	21	24	8	10	11	13	15	16	18	20	2.4	2.6	2.9	3.1	3.4	2.5	2.9	3.2	3.6	3.9
TW D1	5	6	7	7	8	8	8	7	4	4	5	5	4	4	3	0	3.3	4.5	-	-	-	4.2	-	-	-	-
TW D2	9	12	16	19	23	26	30	35	7	10	13	15	19	21	24	28	1.7	1.9	2.2	2.5	2.7	1.9	2.2	2.5	2.9	3.2
TW D3	9	13	17	20	25	28	32	37	7	10	14	16	20	22	26	30	1.6	1.8	2.1	2.4	2.6	1.7	2.1	2.4	2.7	3.1
TW D4	10	15	21	25	32	36	41	49	8	13	18	21	27	31	36	42	1.4	1.6	1.8	2.0	2.1	1.5	1.7	1.9	2.1	2.3
TW D5	10	16	22	27	34	38	44	53	8	13	19	23	29	33	39	47	1.4	1.5	1.7	1.9	2.0	1.4	1.6	1.8	2.0	2.2
TW E1	11	16	21	24	30	33	38	44	9	13	17	20	25	28	32	37	1.5	1.7	1.9	2.2	2.3	1.6	1.9	2.1	2.4	2.7
TW E2	11	16	21	25	31	35	41	48	9	13	18	21	26	30	34	41	1.4	1.6	1.8	2.0	2.2	1.5	1.8	2.0	2.2	2.4
TW E3	13	20	28	33	41	46	53	62	11	17	24	29	36	41	47	56	1.2	1.4	1.6	1.7	1.9	1.3	1.5	1.7	1.8	2.0
TW E4	12	19	26	32	40	45	52	61	10	16	23	28	35	40	47	55	1.2	1.4	1.6	1.7	1.9	1.3	1.5	1.7	1.8	2.0
TW F1	10	11	12	13	15	16	17	18	9	10	11	11	12	12	12	12	2.7	3.2	3.9	5.0	-	3.1	4.1	-	-	-
TW F2	7	9	11	13	15	17	19	21	6	8	9	10	12	12	13	14	2.2	2.6	3.0	3.6	4.2	2.5	3.1	3.9	-	-
TW F3	6	8	10	12	15	17	19	22	4	6	8	9	11	12	14	15	2.1	2.4	2.8	3.2	3.7	2.3	2.8	3.4	4.2	-
TW F4	4	6	9	11	14	16	19	23	2	4	6	8	10	12	14	16	2.0	2.3	2.6	3.0	3.3	2.2	2.6	3.1	3.7	4.3
TW G1	8	9	11	12	14	15	16	18	7	8	9	10	11	11	12	13	2.6	3.0	3.5	4.1	4.9	2.9	3.6	4.6	-	-
TW G2	5	7	9	11	14	15	17	20	4	5	7	8	10	11	13	14	2.2	2.5	2.9	3.4	3.8	2.5	2.9	3.6	4.4	-
TW G3	4	6	8	10	13	15	17	21	2	4	6	7	9	11	12	15	2.2	2.5	2.8	3.1	3.5	2.4	2.8	3.3	3.8	4.5
TW G4	2	4	7	8	11	13	16	19	1	3	5	6	8	10	12	14	2.2	2.5	2.8	3.1	3.4	2.4	2.8	3.2	3.6	4.1
TW H1	5	7	8	10	12	13	15	17	4	5	7	8	9	10	11	12	2.5	2.8	3.2	3.7	4.2	2.7	3.2	3.9	4.8	-
TW H2	3	5	7	8	11	12	15	18	2	3	5	6	8	9	11	13	2.4	2.7	3.0	3.3	3.6	2.6	3.0	3.4	4.0	4.5
TW H3	2	4	6	7	10	11	14	17	1	2	4	5	7	8	10	13	2.5	2.7	3.0	3.3	3.6	2.6	3.0	3.4	3.8	4.2
TW H4	1	3	4	6	8	9	11	14	0	2	3	4	6	7	9	11	2.7	2.9	3.2	3.4	3.7	2.9	3.2	3.5	3.9	4.2
TW I1	4	4	5	5	6	6	6	5	2	3	3	3	3	2	1	-1	3.5	4.9	-	-	-	4.5	-	-	-	-
TW I2	6	9	10	10	20	23	27	32	5	7	10	13	16	18	21	25	1.8	2.0	2.2	2.5	2.7	1.9	2.2	2.5	2.9	3.2
TW I3	7	10	15	18	23	26	30	36	5	8	12	14	18	21	24	29	1.6	1.9	2.1	2.4	2.6	1.8	2.1	2.3	2.7	2.9
TW I4	8	13	18	23	29	33	39	46	6	10	15	19	24	28	33	40	1.4	1.6	1.8	2.0	2.2	1.5	1.7	2.0	2.2	2.3
TW I5	8	13	19	24	31	36	42	50	6	11	16	20	26	31	36	44	1.4	1.6	1.7	1.9	2.1	1.5	1.7	1.8	2.0	2.2
TW J1	9	14	19	22	28	32	36	43	7	11	15	18	23	26	30	36	1.5	1.7	1.9	2.1	2.3	1.6	1.9	2.1	2.4	2.6
TW J2	9	13	19	23	29	33	38	46	7	11	15	19	24	28	33	39	1.4	1.6	1.8	2.0	2.2	1.5	1.8	2.0	2.2	2.4
TW J3	11	18	25	31	39	45	52	61	9	15	22	27	34	40	46	55	1.2	1.4	1.6	1.7	1.9	1.3	1.5	1.7	1.8	2.0
TW J4	10	16	24	30	38	43	50	59	8	14	20	26	33	38	45	54	1.2	1.4	1.6	1.8	1.9	1.3	1.5	1.7	1.8	2.0
DG A1	4	5	6	7	7	7	6	4	3	4	4	4	3	3	1	-1	3.0	4.4	-	-	-	3.7	-	-	-	-
DG A2	7	10	12	15	17	19	22	25	6	8	10	12	14	16	17	19	2.0	2.3	2.7	3.2	3.8	2.2	2.6	3.1	3.8	4.7
DG A3	11	15	19	23	28	31	36	43	9	13	17	20	24	27	32	37	1.6	1.8	1.9	2.1	2.4	1.7	1.9	2.1	2.3	2.5
DG B1	4	5	6	7	8	8	7	6	3	4	4	4	4	3	2	0	2.8	4.0	-	-	-	3.6	-	-	-	-
DG B2	8	10	13	15	18	20	23	27	6	8	11	12	15	16	18	21	1.9	2.2	2.6	2.9	3.3	2.1	2.5	2.9	3.5	4.1
DG B3	11	15	20	24	29	33	38	46	10	13	17	21	26	29	34	40	1.6	1.7	1.9	2.0	2.2	1.7	1.8	2.0	2.2	2.4
DG C1	6	8	10	11	12	13	13	14	5	6	7	7	8	8	8	7	2.2	2.9	4.3	-	-	2.8	4.2	-	-	-
DG C2	9	12	16	19	23	26	29	34	8	10	13	16	19	21	24	28	1.7	2.0	2.3	2.5	2.8	1.9	2.2	2.5	2.9	3.3
DG C3	13	17	23	27	33	38	43	52	11	15	20	24	29	33	38	46	1.4	1.6	1.8	1.9	2.1	1.6	1.7	1.9	2.1	2.2
SS A1	13	16	19	21	23	25	27	30	12	15	17	19	21	22	24	25	1.9	2.2	2.7	3.2	3.7	2.0	2.5	3.1	3.9	-
SS A2	16	21	26	30	35	39	43	50	15	19	24	27	32	35	39	45	1.4	1.6	1.9	2.1	2.3	1.5	1.7	2.0	2.3	2.5
SS A3	12	14	16	17	19	20	22	23	11	13	15	16	17	17	18	18	2.1	2.6	3.4	4.6	-	2.3	3.0	4.3	-	-
SS A4	16	20	25	28	33	36	41	46	15	19	23	26	30	33	36	41	1.5	1.7	2.0	2.2	2.4	1.5	1.8	2.1	2.4	2.7
SS A5	14	16	18	20	22	23	24	26	13	15	17	18	19	20	21	22	2.1	2.5	3.1	3.9	4.9	2.3	2.9	3.8	-	-
SS A6	16	20	25	29	34	37	42	48	15	19	23	26	31	34	38	43	1.4	1.7	1.9	2.1	2.3	1.5	1.8	2.1	2.4	2.6
SS A7	13	14	16	17	18	18	19	19	12	14	15	15	16	16	16	15	2.5	3.2	4.3	-	-	2.7	3.8	-	-	-
SS A8	15	19	24	27	31	34	38	43	14	18	22	24	28	31	34	38	1.5	1.8	2.0	2.3	2.6	1.6	1.9	2.2	2.6	3.0
SS B1	11	13	15	17	20	21	23	26	10	12	14	15	17	18	19	21	2.0	2.4	2.9	3.4	4.0	2.2	2.7	3.4	4.3	-
SS B2	14	18	23	27	32	36	40	47	13	17	21	24	29	32	36	42	1.5	1.7	1.9	2.1	2.3	1.5	1.8	2.0	2.3	2.5
SS B3	10	12	14	15	17	18	19	20	9	11	12	13	14	15	15	16	2.2	2.7	3.4	4.4	-	2.4	3.2	4.3	-	-
SS B4	13	17	22	25	30	34	38	44	12	16	20	23	27	30	34	39	1.5	1.7	2.0	2.2	2.4	1.6	1.9	2.1	2.4	2.7
SS B5	11	13	15	16	18	19	20	22	11	12	13	14	16	16	17	18	2.4	2.8	3.4	4.3	-	2.5	3.2	4.2	-	-
SS B6	13	17	22	25	30	34	38	44	12	16	20	23	27	30	34	39	1.5	1.7	2.0	2.2	2.4	1.6	1.8	2.1	2.4	2.6
SS B7	10	12	13	14	15	15	16	17	10	11	12	12	13	13	13	13	2.7	3.4	4.3	-	-	2.9	4.0	-	-	-
SS B8	12	16	20	24	28	31	35	41	11	15	18	21	25	28	31	36	1.6	1.8	2.1	2.3	2.5	1.7	1.9	2.2	2.6	2.9
SS C1	7	9	11	13	15	16	18	20	5	7	8	9	10	11	11	11	2.1	2.5	3.1	3.9	4.9	2.4	3.2	4.4	-	-
SS C2	9	13	17	20	25	28	33	38	7	10	14	16	20	23	26	31	1.6	1.8	2.1	2.3	2.5	1.7	2.0	2.3	2.6	2.9
SS C3	8	10	11	13	14	16	17	19	7	8	9	10	11	12	13	13	2.5	2.9	3.4	4.0	4.8	2.8	3.4	4.5	-	-
SS C4	9	12	16	19	23	26	30	35	7	10	13	16	19	21	24	28	1.7	1.9	2.2	2.5	2.7	1.9	2.2	2.5	2.8	3.1
SS D1	14	17	20	23	26	28	30	33	12	15	18	20	22	23	25	26	1.7	2.0	2.5	3.0	3.5	1.8	2.3	2.9	3.9	-
SS D2	17	23	29	34	40	45	50	57	15	21	26	31	36	40	45	51	1.3	1.5	1.7	1.9	2.1	1.3	1.6	1.8	2.0	2.3
SS D3	14	17	20	22	24	26	28	31	13	15	18	19	21	22	24	25	1.8	2.2	2.7	3.2	3.9	2.0	2.5	3.2	4.3	-
SS D4	17	22	28	33	39	43	48	56	15	20	26	30	35	39	44	50	1.3	1.5	1.7	1.9	2.1	1.4	1.6	1.8	2.1	2.3
SS E1	10	13	15	17	19	21	22	24	9	11	13	14	15	16	17	16	1.9	2.4	3.0	4.0	-	2.1	2.8	4.0	-	-
SS E2	13	18	23	27	33	37	41	48	12	16	20	24	29	32	36	41	1.4	1.6	1.9	2.1	2.3	1.5	1.7	2.0	2.3	2.6
SS E3	11	13	15	17	19	20	21	23	11	12	14	15	16	17	17	18	2.2	2.6	3.2	4.1	-	2.3	3.0	4.1	-	-
SS E4	13	17	22	26	31	34	39	44	12	16	20	23	27	30	34	39	1.5	1.7	2.0	2.2	2.4	1.6	1.8	2.1	2.4	2.7

DETROIT, MICHIGAN

SOLAR SAVINGS FRACTION (%) — CONSERVATION FACTOR

TYPE	BASE TEMP: 55 F (3940 DD)								65 F (6234 DD)								CF 55 F					CF 65 F				
LCR	100	70	50	40	30	25	20	15	100	70	50	40	30	25	20	15	100	50	30	20	15	100	50	30	20	15
WW A1	14	16	18	19	21	23	24	27	12	13	14	14	15	16	16	17	2.2	2.6	3.1	3.5	3.9	3.1	3.8	4.8	-	-
WW A2	12	16	19	22	26	28	32	37	9	11	13	15	17	19	21	24	1.7	2.0	2.2	2.5	2.7	2.1	2.5	2.9	3.4	3.9
WW A3	11	15	19	23	27	31	35	41	7	10	13	15	18	20	23	26	1.6	1.8	2.0	2.3	2.4	1.9	2.2	2.6	3.0	3.4
WW A4	11	15	20	23	28	32	37	43	6	9	13	15	19	21	24	28	1.5	1.7	1.9	2.2	2.3	1.8	2.1	2.4	2.8	3.1
WW A5	10	15	20	24	30	34	39	46	5	9	12	15	19	22	25	30	1.4	1.7	1.9	2.1	2.3	1.7	2.0	2.3	2.6	2.9
WW A6	10	15	20	24	30	34	40	47	5	8	12	15	20	22	26	31	1.4	1.6	1.8	2.0	2.2	1.7	1.9	2.2	2.6	2.8
WW B1	9	11	12	13	15	16	17	18	6	6	7	7	7	6	6	3	2.3	2.9	3.8	4.9	-	3.7	-	-	-	-
WW B2	12	17	22	26	33	37	43	50	7	11	15	19	23	27	31	37	1.4	1.6	1.8	2.0	2.1	1.6	1.8	2.1	2.3	2.5
WW B3	14	20	27	32	39	44	51	59	9	14	19	24	29	33	39	46	1.3	1.4	1.6	1.8	2.0	1.4	1.6	1.9	2.1	2.2
WW B4	14	21	29	36	44	50	57	66	9	15	22	27	35	40	46	55	1.2	1.3	1.5	1.7	1.8	1.3	1.5	1.6	1.8	2.0
WW B5	14	21	29	36	45	51	58	68	9	15	22	28	36	41	48	58	1.2	1.3	1.5	1.6	1.8	1.3	1.4	1.6	1.8	1.9
WW C1	16	22	29	35	42	48	54	63	11	16	22	26	33	37	43	50	1.2	1.4	1.6	1.7	1.9	1.4	1.6	1.8	2.0	2.1
WW C2	14	21	29	34	42	48	55	64	9	15	21	26	33	38	44	52	1.2	1.4	1.5	1.7	1.9	1.4	1.5	1.7	1.9	2.0
WW C3	18	27	37	44	54	60	68	77	12	20	29	36	45	51	59	69	1.0	1.2	1.4	1.5	1.7	1.1	1.3	1.5	1.6	1.8
WW C4	16	24	34	41	51	57	65	75	11	18	27	33	42	48	56	66	1.1	1.2	1.4	1.5	1.7	1.2	1.3	1.5	1.7	1.8
TW A1	14	16	18	19	22	23	25	28	12	13	14	15	16	16	17	18	2.2	2.5	2.9	3.3	3.6	2.9	3.5	4.2	-	-
TW A2	12	15	18	21	24	27	30	35	9	11	13	15	17	19	21	23	1.8	2.1	2.3	2.6	2.8	2.2	2.6	3.0	3.4	3.9
TW A3	11	15	18	21	25	28	32	38	8	10	13	15	17	19	22	25	1.7	1.9	2.2	2.4	2.6	2.0	2.3	2.7	3.1	3.4
TW A4	10	14	18	21	26	29	33	39	7	9	12	14	17	20	23	27	1.6	1.9	2.1	2.3	2.4	1.9	2.2	2.5	2.8	3.1
TW B1	13	15	18	19	22	24	26	30	11	12	13	15	16	17	19	20	2.0	2.4	2.7	3.0	3.2	2.6	3.1	3.5	4.1	4.8
TW B2	11	14	18	20	24	27	30	35	8	10	13	14	17	19	21	24	1.8	2.0	2.3	2.5	2.7	2.2	2.5	2.8	3.1	3.5
TW B3	10	14	17	20	24	27	31	37	7	10	12	14	17	19	22	26	1.7	2.0	2.2	2.4	2.6	2.1	2.3	2.6	2.9	3.2
TW B4	10	13	17	19	23	26	30	36	7	9	12	14	17	19	22	26	1.7	2.0	2.2	2.4	2.6	2.1	2.3	2.6	2.9	3.1
TW C1	12	14	17	19	22	24	27	32	9	11	13	14	16	18	20	22	1.9	2.2	2.5	2.7	2.9	2.4	2.7	3.1	3.4	3.8
TW C2	11	13	17	19	23	25	29	34	8	10	12	14	17	18	21	24	1.8	2.1	2.3	2.5	2.7	2.2	2.5	2.8	3.1	3.3
TW C3	10	13	16	18	22	25	28	33	8	10	12	14	16	18	21	24	1.9	2.1	2.3	2.5	2.7	2.3	2.5	2.8	3.0	3.3
TW C4	11	13	16	18	21	23	26	31	9	11	12	14	16	18	20	24	2.0	2.2	2.5	2.7	2.8	2.4	2.7	2.9	3.2	3.4
TW D1	7	9	11	12	14	15	16	18	5	6	6	7	7	7	7	6	2.3	2.9	3.5	4.2	5.0	3.5	4.9	-	-	-
TW D2	12	16	21	25	30	34	39	46	9	12	15	18	22	25	29	35	1.5	1.7	1.9	2.1	2.2	1.8	2.0	2.2	2.4	2.6
TW D3	13	17	23	27	33	37	43	50	9	12	16	20	24	27	32	38	1.4	1.6	1.8	2.0	2.1	1.6	1.8	2.1	2.3	2.5
TW D4	14	20	27	32	39	45	51	60	10	15	20	25	31	35	41	49	1.3	1.4	1.6	1.8	1.9	1.4	1.6	1.8	2.0	2.1
TW D5	14	20	27	33	41	46	53	63	10	15	21	26	33	38	44	53	1.2	1.4	1.6	1.7	1.9	1.4	1.5	1.7	1.9	2.0
TW E1	15	21	27	31	38	43	49	57	11	15	20	24	29	33	38	45	1.3	1.5	1.7	1.8	2.0	1.5	1.7	1.9	2.1	2.3
TW E2	15	21	27	32	39	44	51	59	10	15	20	25	31	35	40	48	1.3	1.5	1.6	1.8	1.9	1.4	1.6	1.8	2.0	2.2
TW E3	18	26	34	41	50	56	63	72	13	20	27	33	41	46	53	63	1.1	1.3	1.4	1.6	1.7	1.2	1.4	1.6	1.7	1.9
TW E4	16	24	33	39	48	54	62	71	12	18	26	31	40	45	52	62	1.1	1.3	1.4	1.6	1.7	1.2	1.4	1.6	1.7	1.9
TW F1	12	14	16	18	20	22	24	27	10	11	12	13	14	15	16	17	2.1	2.5	2.9	3.2	3.5	2.8	3.4	4.0	4.9	-
TW F2	10	13	16	18	22	24	28	32	7	9	11	12	15	16	18	21	1.8	2.1	2.4	2.6	2.8	2.3	2.6	3.0	3.5	4.0
TW F3	8	12	15	18	22	25	29	34	5	7	10	12	14	16	19	22	1.7	2.0	2.2	2.4	2.6	2.1	2.4	2.8	3.1	3.5
TW F4	7	10	14	17	21	24	28	34	3	6	8	10	14	16	18	22	1.7	2.0	2.2	2.4	2.6	2.1	2.3	2.6	2.9	3.2
TW G1	10	12	15	16	19	21	24	27	8	9	10	11	13	14	16	17	2.1	2.4	2.7	3.0	3.2	2.6	3.1	3.5	4.1	4.8
TW G2	8	10	14	16	20	22	26	30	5	7	9	10	13	14	17	20	1.8	2.1	2.4	2.6	2.8	2.3	2.6	2.9	3.3	3.7
TW G3	6	9	12	15	19	21	25	30	3	5	7	9	12	14	16	20	1.8	2.1	2.3	2.5	2.7	2.2	2.5	2.8	3.1	3.4
TW G4	4	7	10	13	16	19	23	28	2	4	6	8	10	12	15	19	1.9	2.1	2.4	2.6	2.7	2.3	2.6	2.8	3.1	3.3
TW H1	7	9	12	14	17	19	22	26	5	6	8	9	11	13	14	17	2.0	2.3	2.6	2.9	3.1	2.6	2.9	3.3	3.7	4.1
TW H2	5	7	10	12	16	18	21	26	2	4	6	8	10	12	14	17	2.0	2.2	2.5	2.7	2.9	2.4	2.7	3.0	3.3	3.6
TW H3	3	6	9	11	14	16	19	24	2	3	5	7	9	11	13	16	2.1	2.3	2.5	2.8	2.9	2.5	2.8	3.0	3.3	3.5
TW H4	2	4	7	9	11	13	16	20	1	2	4	5	7	9	11	14	2.3	2.5	2.7	3.0	3.1	2.7	3.0	3.3	3.5	3.7
TW I1	6	7	9	10	12	13	14	15	3	4	4	5	5	5	5	4	2.4	3.0	3.7	4.5	-	3.7	-	-	-	-
TW I2	9	13	18	22	27	31	36	43	6	9	12	15	19	22	26	31	1.5	1.7	1.9	2.1	2.3	1.8	2.0	2.2	2.5	2.7
TW I3	10	15	20	24	30	34	40	47	6	10	14	17	22	25	30	36	1.4	1.6	1.8	2.0	2.2	1.7	1.9	2.1	2.3	2.5
TW I4	11	17	24	29	36	42	48	57	7	12	18	22	28	33	38	46	1.3	1.5	1.6	1.8	1.9	1.5	1.6	1.8	2.0	2.1
TW I5	11	18	25	30	38	44	51	60	7	13	19	23	30	35	42	50	1.3	1.4	1.6	1.7	1.9	1.4	1.6	1.7	1.9	2.0
TW J1	13	18	25	29	36	41	47	55	9	13	18	22	27	31	36	43	1.3	1.5	1.7	1.9	2.0	1.5	1.7	1.9	2.1	2.3
TW J2	12	18	25	30	37	42	48	57	8	13	18	22	28	33	38	46	1.3	1.5	1.6	1.8	1.9	1.5	1.6	1.8	2.0	2.2
TW J3	15	23	32	38	47	54	61	71	10	17	25	31	39	45	52	61	1.1	1.3	1.4	1.6	1.7	1.2	1.4	1.6	1.7	1.9
TW J4	14	21	30	36	45	52	59	69	9	16	23	29	37	43	50	60	1.1	1.3	1.5	1.6	1.8	1.3	1.4	1.6	1.7	1.9
DG A1	7	9	11	12	13	14	15	15	4	5	6	6	6	6	5	4	2.2	2.8	4.0	-	-	3.2	4.8	-	-	-
DG A2	10	13	17	19	23	26	30	34	7	9	12	14	17	19	21	24	1.7	2.0	2.2	2.6	2.9	2.1	2.3	2.7	3.1	3.7
DG A3	13	18	23	27	34	38	44	52	10	14	18	22	27	31	36	43	1.5	1.6	1.7	1.9	2.1	1.6	1.8	1.9	2.1	2.3
DG B1	7	9	11	12	14	15	17	18	4	5	6	6	7	7	6	5	2.2	2.8	3.3	4.6	-	3.2	4.3	-	-	-
DG B2	10	13	17	20	24	28	32	37	7	9	12	14	17	20	22	26	1.7	1.9	2.1	2.4	2.6	2.1	2.3	2.6	2.9	3.2
DG B3	14	19	24	28	35	40	47	56	11	14	19	23	28	32	38	46	1.4	1.6	1.7	1.8	2.0	1.6	1.7	1.9	2.0	2.2
DG C1	9	12	14	16	19	21	24	27	5	7	9	10	11	12	12	13	1.9	2.3	2.6	3.1	3.8	2.5	3.0	4.1	-	-
DG C2	12	16	21	24	30	33	39	45	9	12	15	18	22	25	28	33	1.6	1.8	1.9	2.1	2.3	1.8	2.0	2.3	2.5	2.7
DG C3	16	21	27	32	40	46	53	63	12	17	22	26	32	37	43	52	1.3	1.5	1.6	1.7	1.8	1.5	1.6	1.8	1.9	2.0
SS A1	16	19	23	26	30	33	36	41	13	16	19	21	24	26	28	32	1.6	1.9	2.1	2.4	2.6	1.9	2.2	2.5	3.0	3.3
SS A2	20	25	31	36	43	47	53	61	16	21	26	30	36	40	45	51	1.3	1.5	1.6	1.8	2.0	1.4	1.6	1.8	2.0	2.2
SS A3	15	17	20	22	26	28	30	34	12	14	16	18	20	21	22	24	1.8	2.1	2.5	2.8	3.1	2.1	2.6	3.1	4.0	-
SS A4	19	24	30	34	41	45	50	58	16	20	25	29	34	37	42	48	1.3	1.5	1.7	1.9	2.0	1.4	1.7	1.9	2.1	2.3
SS A5	16	19	22	24	27	29	32	36	14	16	18	20	22	23	25	27	1.8	2.1	2.4	2.7	3.0	2.1	2.5	3.0	3.6	4.2
SS A6	19	25	31	35	42	46	52	59	16	20	25	29	35	38	43	49	1.3	1.5	1.7	1.8	2.0	1.4	1.6	1.9	2.1	2.3
SS A7	15	17	19	21	23	24	26	28	13	14	16	17	18	19	19	20	2.1	2.5	3.0	3.5	3.9	2.5	3.1	4.0	-	-
SS A8	19	23	29	33	39	43	48	55	15	19	24	27	32	35	40	45	1.3	1.6	1.7	1.9	2.1	1.5	1.7	2.0	2.2	2.4
SS B1	13	16	19	22	26	28	31	36	11	13	15	17	20	21	24	26	1.7	2.0	2.3	2.6	2.8	2.0	2.4	2.8	3.2	3.7
SS B2	17	22	28	32	39	43	49	57	14	18	23	27	32	36	41	48	1.3	1.5	1.7	1.9	2.0	1.5	1.7	1.9	2.1	2.2
SS B3	12	15	18	20	23	25	27	31	10	11	13	15	17	18	20	21	1.9	2.2	2.5	2.9	3.1	2.2	2.7	3.2	4.0	5.0
SS B4	16	21	27	31	37	41	47	54	13	17	22	26	31	34	39	45	1.4	1.6	1.8	1.9	2.1	1.5	1.7	1.9	2.2	2.3
SS B5	13	16	18	20	23	24	27	30	11	13	15	16	18	19	20	22	2.0	2.3	2.7	3.0	3.3	2.4	2.8	3.3	4.0	4.7
SS B6	16	21	27	31	37	41	47	55	13	17	22	25	31	34	39	45	1.4	1.6	1.7	1.9	2.1	1.5	1.7	1.9	2.1	2.3
SS B7	12	14	16	17	19	21	23	25	10	11	13	14	15	16	16	17	2.2	2.6	3.1	3.6	4.0	2.8	3.3	4.2	-	-
SS B8	15	20	25	29	35	39	44	51	12	16	20	24	28	32	36	42	1.4	1.6	1.8	2.0	2.1	1.6	1.8	2.0	2.3	2.5
SS C1	10	13	16	19	23	26	29	34	6	8	10	12	14	16	18	20	1.7	2.0	2.3	2.5	2.8	2.2	2.6	3.0	3.7	4.4
SS C2	12	17	23	27	33	38	43	51	8	12	16	20	25	28	32	39	1.4	1.6	1.8	1.9	2.1	1.6	1.8	2.0	2.3	2.5
SS C3	10	13	15	17	20	22	25	29	8	9	11	12	14	15	17	18	2.0	2.3	2.6	2.9	3.1	2.5	3.0	3.4	4.0	4.6
SS C4	12	16	21	25	30	34	39	46	8	12	15	18	22	25	29	35	1.5	1.7	1.9	2.1	2.2	1.8	2.0	2.2	2.4	2.6
SS D1	17	21	26	29	34	38	42	48	13	17	20	23	26	29	32	35	1.4	1.7	1.9	2.1	2.3	1.6	2.0	2.3	2.7	3.1
SS D2	21	28	36	41	49	55	61	70	17	23	30	34	41	46	52	59	1.1	1.3	1.5	1.7	1.8	1.2	1.4	1.6	1.8	2.0
SS D3	17	20	24	27	31	34	38	43	14	17	20	22	25	27	29	32	1.6	1.9	2.1	2.3	2.5	1.8	2.2	2.5	3.0	3.4
SS D4	21	27	35	40	47	53	59	67	17	22	29	33	40	44	50	57	1.2	1.4	1.5	1.7	1.9	1.3	1.5	1.7	1.9	2.0
SS E1	14	17	21	24	28	31	34	39	10	13	15	17	20	21	23	25	1.6	1.9	2.2	2.4	2.7	1.9	2.3	2.8	3.5	4.4
SS E2	17	23	30	35	42	47	53	61	13	18	24	28	33	37	43	49	1.2	1.4	1.6	1.8	1.9	1.4	1.6	1.8	2.0	2.2
SS E3	14	17	19	22	25	27	30	34	11	13	15	17	19	20	22	24	1.8	2.1	2.4	2.7	3.0	2.2	2.6	3.1	3.8	4.5
SS E4	17	22	28	32	39	43	49	57	13	17	22	26	31	35	40	46	1.3	1.5	1.7	1.9	2.0	1.5	1.7	1.9	2.1	2.3

FLINT, MICHIGAN

SOLAR SAVINGS FRACTION (%)

TYPE	55 F (4549 DD) 100	70	50	40	30	25	20	15	65 F (7043 DD) 100	70	50	40	30	25	20	15
BASE TEMP: / LCR																
WW A1	12	14	15	16	17	18	19	20	11	12	12	13	13	13	13	12
WW A2	10	13	15	17	20	22	24	28	8	9	11	12	14	15	16	17
WW A3	9	12	15	18	21	23	26	31	6	8	10	12	14	15	17	19
WW A4	8	11	15	18	22	24	28	33	5	7	10	12	14	16	18	20
WW A5	7	11	15	18	22	25	29	34	4	7	9	12	14	16	19	21
WW A6	7	11	15	18	23	26	30	36	3	6	9	11	15	17	19	22
WW B1	7	8	9	9	10	9	9	7	5	5	5	4	3	2	0	-3
WW B2	9	13	18	21	26	30	34	41	6	9	13	15	19	22	25	30
WW B3	11	16	22	26	32	37	42	50	8	12	16	20	25	28	32	38
WW B4	11	17	25	30	38	43	50	59	8	13	19	24	30	35	41	49
WW B5	11	17	25	30	38	44	51	61	7	13	19	25	32	37	43	52
WW C1	13	18	24	29	35	40	46	54	9	13	19	22	28	32	36	43
WW C2	11	17	24	29	36	41	47	56	8	13	18	22	29	33	38	46
WW C3	14	23	32	38	48	54	61	71	11	18	26	32	41	47	54	63
WW C4	13	20	29	36	45	51	59	69	9	16	24	30	38	44	51	61
TW A1	12	14	15	16	18	19	20	21	11	12	13	13	14	14	14	14
TW A2	10	13	15	17	20	21	24	27	9	10	12	13	14	15	16	18
TW A3	9	12	15	17	20	22	25	29	7	9	11	12	14	16	17	19
TW A4	8	11	14	17	20	23	26	30	5	8	10	12	14	16	18	21
TW B1	11	13	15	16	18	20	21	23	10	11	12	13	14	15	15	16
TW B2	9	12	14	16	19	21	24	28	8	9	11	12	14	16	17	19
TW B3	8	11	14	16	20	22	25	29	7	8	10	12	14	16	18	20
TW B4	8	11	14	16	19	21	24	29	6	8	10	12	14	16	18	21
TW C1	10	12	14	16	18	20	22	25	9	10	12	13	14	15	17	18
TW C2	9	11	14	16	19	21	24	27	7	9	11	12	14	16	18	20
TW C3	9	11	14	16	18	21	23	27	8	9	11	12	14	16	18	21
TW C4	10	11	14	15	18	20	22	26	8	10	11	13	15	16	18	21
TW D1	6	7	8	9	9	10	10	10	4	4	5	5	4	4	3	1
TW D2	10	13	18	21	25	28	32	38	8	10	13	16	19	21	24	29
TW D3	10	14	19	22	27	30	35	41	7	10	14	17	20	23	26	31
TW D4	11	17	23	27	34	38	44	52	8	13	18	22	27	31	36	44
TW D5	11	17	23	28	36	40	47	56	8	13	19	23	29	34	40	48
TW E1	12	17	22	26	32	36	41	48	9	13	17	21	25	28	33	38
TW E2	12	17	23	27	33	38	43	51	9	13	18	21	27	30	35	42
TW E3	15	22	29	35	44	49	56	65	11	17	24	29	37	42	48	57
TW E4	13	20	28	34	42	48	55	64	10	16	23	28	36	41	48	57
TW F1	10	12	13	15	16	17	19	20	9	10	11	11	12	12	13	13
TW F2	8	10	13	15	17	19	21	24	6	8	9	10	12	13	14	15
TW F3	6	9	12	14	17	19	22	26	4	6	8	9	11	13	14	16
TW F4	4	7	10	13	16	18	22	26	2	4	7	8	11	12	14	17
TW G1	8	10	12	13	15	17	18	21	7	8	9	10	11	12	12	13
TW G2	6	8	10	12	15	17	20	23	4	5	7	8	10	11	13	15
TW G3	4	6	9	11	14	17	19	23	2	4	6	7	10	11	13	15
TW G4	3	5	8	10	13	15	18	22	1	3	5	6	8	10	12	15
TW H1	6	7	9	11	13	15	17	20	4	5	7	8	9	10	11	13
TW H2	3	5	8	9	12	14	16	20	2	3	5	6	8	9	11	13
TW H3	2	4	6	8	11	13	15	19	1	2	4	5	7	9	10	13
TW H4	1	3	5	6	9	10	13	16	1	2	3	4	6	7	9	11
TW I1	4	6	6	7	7	8	8	8	3	3	3	3	3	2	1	0
TW I2	7	11	15	18	22	25	29	35	5	7	10	13	16	18	22	26
TW I3	8	12	16	20	25	28	33	39	5	8	12	15	18	21	25	30
TW I4	9	14	20	24	31	35	41	50	6	10	15	19	25	29	34	41
TW I5	9	14	21	26	33	38	44	53	6	11	16	21	27	31	37	45
TW J1	10	15	20	24	30	34	39	47	7	11	15	19	24	27	31	37
TW J2	10	15	20	25	31	35	41	49	7	11	16	19	25	28	33	40
TW J3	12	19	27	33	41	47	54	64	9	15	22	27	35	40	47	56
TW J4	11	10	26	31	40	45	53	62	8	14	21	26	33	39	46	55
DG A1	5	6	8	8	9	9	9	7	3	4	4	4	3	3	2	0
DG A2	8	11	14	16	19	21	24	27	6	8	10	12	14	16	18	20
DG A3	12	16	20	24	29	33	38	45	10	13	17	20	24	28	32	38
DG B1	5	6	8	8	9	9	9	9	3	3	4	4	4	3	2	0
DG B2	8	11	14	16	20	22	25	29	6	8	11	12	15	16	18	21
DG B3	12	16	21	25	31	35	41	49	10	13	17	21	26	29	34	41
DG C1	7	9	11	12	14	15	16	17	5	6	7	7	8	8	8	7
DG C2	10	14	17	20	24	27	31	37	8	10	13	16	19	21	24	28
DG C3	14	18	24	28	35	39	46	55	11	15	20	24	29	33	39	46
SS A1	14	17	20	22	25	27	30	34	12	15	17	19	21	22	24	26
SS A2	17	22	28	32	37	41	46	53	15	19	24	27	32	36	40	46
SS A3	13	15	17	19	21	22	24	26	11	13	14	16	17	18	18	19
SS A4	17	21	26	30	35	39	44	50	15	19	23	26	30	34	37	43
SS A5	15	17	19	21	23	25	27	29	13	15	17	18	19	20	21	22
SS A6	17	22	27	31	36	40	45	52	15	19	23	26	31	34	38	44
SS A7	14	15	17	18	19	20	21	22	12	13	14	15	16	16	16	16
SS A8	16	20	25	29	34	37	41	47	14	18	22	25	29	31	35	40
SS B1	12	14	17	19	21	23	26	29	10	12	14	15	17	18	20	22
SS B2	15	19	25	28	34	38	43	50	13	17	21	24	29	33	37	43
SS B3	10	12	15	16	18	20	21	24	9	10	12	13	14	15	16	16
SS B4	14	19	23	27	32	36	41	47	12	16	20	23	28	31	35	40
SS B5	12	14	16	17	19	20	22	24	11	12	13	14	16	16	17	18
SS B6	14	18	23	27	32	36	41	48	12	16	20	23	28	31	35	41
SS B7	11	12	14	15	16	17	18	19	10	11	12	12	13	13	14	14
SS B8	13	17	22	25	30	33	38	44	11	15	18	21	25	28	32	37
SS C1	8	10	13	14	17	19	21	24	5	7	8	9	11	11	12	13
SS C2	10	14	19	22	27	31	36	42	7	10	14	17	21	23	27	32
SS C3	9	10	12	14	16	17	19	22	7	8	9	10	11	12	13	14
SS C4	10	13	17	20	25	28	32	38	7	10	13	16	19	22	25	29
SS D1	15	18	22	24	28	31	34	38	12	15	18	20	23	24	26	28
SS D2	18	24	31	36	43	48	54	62	15	21	27	31	37	41	46	53
SS D3	15	18	21	23	26	28	31	35	13	15	18	19	22	23	25	26
SS D4	18	24	30	35	41	46	52	60	15	20	26	30	36	40	45	52
SS E1	11	14	17	19	22	24	26	29	9	11	13	14	16	17	18	18
SS E2	14	19	25	29	35	40	45	52	11	16	21	24	29	33	37	43
SS E3	12	14	17	18	20	22	24	26	10	12	14	15	16	17	18	19
SS E4	14	19	24	28	33	37	42	49	12	16	20	23	28	31	35	40

CONSERVATION FACTOR

TYPE	55 F 100	50	30	20	15	65 F 100	50	30	20	15
WW A1	2.6	3.2	4.0	-	-	3.6	-	-	-	-
WW A2	1.9	2.3	2.7	3.1	3.5	2.4	2.9	3.7	4.9	-
WW A3	1.7	2.1	2.4	2.7	3.0	2.1	2.6	3.2	4.0	-
WW A4	1.7	2.0	2.3	2.6	2.8	2.0	2.4	2.9	3.6	4.4
WW A5	1.6	1.9	2.2	2.5	2.7	1.9	2.3	2.7	3.3	4.0
WW A6	1.6	1.9	2.1	2.4	2.6	1.9	2.2	2.6	3.2	3.7
WW B1	2.9	4.4	-	-	-	-	-	-	-	-
WW B2	1.5	1.8	2.0	2.2	2.4	1.8	2.0	2.3	2.6	2.9
WW B3	1.4	1.6	1.8	2.0	2.2	1.5	1.8	2.0	2.3	2.5
WW B4	1.3	1.4	1.6	1.8	1.9	1.4	1.6	1.8	1.9	2.1
WW B5	1.2	1.4	1.6	1.7	1.9	1.3	1.5	1.7	1.9	2.0
WW C1	1.3	1.5	1.7	1.9	2.1	1.5	1.7	1.9	2.2	2.4
WW C2	1.3	1.5	1.7	1.8	2.0	1.4	1.6	1.8	2.0	2.2
WW C3	1.1	1.3	1.4	1.6	1.7	1.2	1.4	1.5	1.7	1.8
WW C4	1.1	1.3	1.5	1.6	1.8	1.2	1.4	1.5	1.7	1.9
TW A1	2.6	3.0	3.6	4.5	-	3.3	4.4	-	-	-
TW A2	2.0	2.4	2.8	3.2	3.5	2.5	3.0	3.7	4.6	-
TW A3	1.9	2.2	2.5	2.9	3.1	2.3	2.7	3.2	3.9	4.7
TW A4	1.8	2.1	2.4	2.7	2.9	2.1	2.5	2.9	3.4	3.9
TW B1	2.3	2.7	3.2	3.7	4.2	2.9	3.6	4.5	-	-
TW B2	2.0	2.3	2.6	2.9	3.2	2.4	2.8	3.3	3.9	4.5
TW B3	1.9	2.2	2.5	2.8	3.0	2.3	2.6	3.0	3.5	4.0
TW B4	1.9	2.2	2.5	2.8	3.0	2.3	2.6	3.0	3.3	3.7
TW C1	2.2	2.5	2.8	3.2	3.5	2.6	3.1	3.6	4.2	4.9
TW C2	2.0	2.3	2.6	2.9	3.1	2.4	2.8	3.1	3.6	4.0
TW C3	2.1	2.3	2.6	2.9	3.1	2.4	2.8	3.1	3.5	3.8
TW C4	2.2	2.5	2.7	3.0	3.2	2.6	2.9	3.2	3.5	3.8
TW D1	2.9	3.7	-	-	-	4.5	-	-	-	-
TW D2	1.6	1.9	2.1	2.3	2.5	1.9	2.2	2.4	2.8	3.0
TW D3	1.5	1.8	2.0	2.2	2.4	1.8	2.0	2.3	2.6	2.9
TW D4	1.4	1.5	1.7	1.9	2.0	1.5	1.7	1.9	2.1	2.3
TW D5	1.3	1.5	1.7	1.8	2.0	1.4	1.6	1.8	2.0	2.1
TW E1	1.4	1.6	1.8	2.0	2.2	1.6	1.8	2.1	2.3	2.5
TW E2	1.4	1.6	1.8	1.9	2.1	1.5	1.7	2.0	2.2	2.3
TW E3	1.2	1.4	1.5	1.7	1.8	1.3	1.5	1.6	1.8	2.0
TW E4	1.2	1.4	1.5	1.7	1.8	1.3	1.5	1.6	1.8	2.0
TW F1	2.5	2.9	3.5	4.3	-	3.2	4.2	-	-	-
TW F2	2.1	2.4	2.8	3.2	3.6	2.5	3.1	3.8	4.7	-
TW F3	1.9	2.2	2.6	2.9	3.2	2.3	2.8	3.3	3.9	4.7
TW F4	1.9	2.2	2.5	2.8	3.0	2.2	2.6	3.0	3.5	4.0
TW G1	2.4	2.7	3.2	3.7	4.2	2.9	3.6	4.5	-	-
TW G2	2.1	2.4	2.7	3.1	3.4	2.5	3.0	3.5	4.1	4.8
TW G3	2.0	2.3	2.6	2.9	3.2	2.4	2.8	3.2	3.7	4.2
TW G4	2.1	2.4	2.6	2.9	3.2	2.5	2.8	3.2	3.5	3.9
TW H1	2.3	2.6	3.0	3.4	3.8	2.8	3.3	3.9	4.6	-
TW H2	2.2	2.5	2.8	3.1	3.4	2.6	3.0	3.4	3.9	4.3
TW H3	2.3	2.6	2.8	3.1	3.4	2.7	3.0	3.4	3.8	4.1
TW H4	2.6	2.8	3.0	3.3	3.5	2.9	3.2	3.6	3.9	4.2
TW I1	3.0	3.9	-	-	-	4.8	-	-	-	-
TW I2	1.7	1.9	2.2	2.4	2.6	1.9	2.2	2.5	2.8	3.1
TW I3	1.6	1.8	2.0	2.2	2.4	1.8	2.0	2.3	2.6	2.8
TW I4	1.4	1.6	1.8	1.9	2.1	1.5	1.7	1.9	2.1	2.3
TW I5	1.3	1.5	1.7	1.9	2.0	1.5	1.6	1.8	2.0	2.2
TW J1	1.4	1.6	1.8	2.0	2.2	1.6	1.8	2.1	2.3	2.5
TW J2	1.4	1.6	1.8	2.0	2.1	1.6	1.8	2.0	2.2	2.3
TW J3	1.2	1.4	1.5	1.7	1.8	1.3	1.5	1.6	1.8	2.0
TW J4	1.2	1.4	1.5	1.7	1.8	1.3	1.5	1.6	1.8	2.0
DG A1	2.6	3.7	-	-	-	3.9	-	-	-	-
DG A2	1.9	2.2	2.5	3.0	3.4	2.2	2.6	3.0	3.7	4.5
DG A3	1.5	1.7	1.9	2.1	2.2	1.7	1.9	2.1	2.3	2.5
DG B1	2.5	3.7	-	-	-	4.1	-	-	-	-
DG B2	1.8	2.2	2.5	2.7	3.1	2.2	2.5	2.9	3.5	4.0
DG B3	1.5	1.7	1.8	2.0	2.1	1.7	1.8	2.0	2.2	2.3
DG C1	2.1	2.9	3.8	4.6	-	3.0	3.7	-	-	-
DG C2	1.7	1.9	2.2	2.4	2.6	2.0	2.2	2.5	2.9	3.2
DG C3	1.4	1.6	1.7	1.8	2.0	1.6	1.7	1.9	2.0	2.2
SS A1	1.8	2.1	2.5	2.8	3.1	2.0	2.4	2.9	3.6	4.4
SS A2	1.4	1.6	1.8	2.0	2.1	1.5	1.7	1.9	2.2	2.4
SS A3	2.0	2.5	3.0	3.7	4.3	2.3	3.0	4.0	-	-
SS A4	1.4	1.6	1.9	2.1	2.2	1.5	1.8	2.1	2.3	2.6
SS A5	2.0	2.4	2.9	3.4	3.9	2.3	2.8	3.6	4.8	-
SS A6	1.4	1.6	1.8	2.0	2.2	1.5	1.8	2.0	2.3	2.5
SS A7	2.3	2.9	3.8	-	-	2.8	3.8	-	-	-
SS A8	1.4	1.7	1.9	2.2	2.3	1.6	1.9	2.2	2.5	2.7
SS B1	1.9	2.3	2.7	3.1	3.4	2.2	2.7	3.2	4.0	4.9
SS B2	1.4	1.6	1.8	2.0	2.2	1.6	1.8	2.0	2.2	2.4
SS B3	2.1	2.5	3.1	3.7	4.3	2.5	3.1	4.0	-	-
SS B4	1.5	1.7	1.9	2.1	2.3	1.6	1.8	2.1	2.3	2.5
SS B5	2.2	2.7	3.2	3.8	4.3	2.6	3.2	4.0	-	-
SS B6	1.5	1.7	1.9	2.1	2.2	1.6	1.8	2.1	2.3	2.5
SS B7	2.5	3.1	3.8	5.0	-	3.1	4.0	-	-	-
SS B8	1.5	1.8	2.0	2.2	2.4	1.7	1.9	2.2	2.5	2.7
SS C1	2.0	2.4	2.8	3.3	3.7	2.5	3.1	4.1	-	-
SS C2	1.5	1.7	2.0	2.2	2.3	1.7	2.0	2.3	2.5	2.8
SS C3	2.3	2.6	3.1	3.6	4.1	2.8	3.5	4.3	-	-
SS C4	1.6	1.9	2.1	2.3	2.5	1.9	2.1	2.4	2.7	3.0
SS D1	1.6	1.9	2.2	2.6	2.8	1.8	2.2	2.7	3.4	4.3
SS D2	1.2	1.4	1.6	1.8	1.9	1.3	1.5	1.8	2.0	2.1
SS D3	1.7	2.1	2.5	2.8	3.1	2.0	2.4	3.0	3.8	4.9
SS D4	1.3	1.5	1.6	1.8	2.0	1.3	1.6	1.8	2.0	2.2
SS E1	1.8	2.2	2.7	3.2	3.6	2.1	2.7	3.6	-	-
SS E2	1.3	1.6	1.8	2.0	2.1	1.5	1.7	2.0	2.2	2.5
SS E3	2.0	2.4	2.9	3.5	4.0	2.4	3.0	3.8	-	-
SS E4	1.4	1.7	1.9	2.1	2.2	1.6	1.8	2.1	2.3	2.6

GRAND RAPIDS, MICHIGAN

TYPE	SOLAR SAVINGS FRACTION (%) BASE TEMP: 55 F (4389 DD) LCR 100	70	50	40	30	25	20	15	65 F (6803 DD) 100	70	50	40	30	25	20	15	CONSERVATION FACTOR 55 F 100	50	30	20	15	65 F 100	50	30	20	15
WW A1	13	14	16	17	18	19	20	21	11	12	13	13	13	14	14	13	2.5	3.1	3.8	4.9	-	3.5	4.8	-	-	-
WW A2	11	13	16	18	21	23	26	29	8	10	12	13	15	16	17	18	1.9	2.2	2.6	3.0	3.4	2.3	2.8	3.5	4.6	-
WW A3	9	13	16	19	22	25	28	32	6	9	11	13	15	16	18	20	1.7	2.0	2.4	2.7	3.0	2.1	2.5	3.0	3.8	4.9
WW A4	9	12	16	19	23	26	29	34	5	8	10	12	15	17	19	21	1.6	1.9	2.2	2.5	2.8	2.0	2.4	2.8	3.5	4.2
WW A5	8	12	16	19	24	27	31	36	4	7	10	12	15	17	20	22	1.6	1.9	2.1	2.4	2.7	1.9	2.2	2.7	3.2	3.9
WW A6	7	11	16	19	24	27	31	37	4	7	10	12	16	18	20	23	1.5	1.8	2.1	2.3	2.6	1.8	2.2	2.6	3.1	3.7
WW B1	7	8	10	10	11	11	10	9	5	5	5	5	4	3	1	-2	2.7	4.0	-	-	-	-	-	-	-	-
WW B2	10	14	19	22	27	31	36	42	6	10	13	16	20	23	26	31	1.5	1.7	2.0	2.2	2.4	1.7	2.0	2.3	2.6	2.9
WW B3	12	17	23	27	33	38	43	51	8	12	17	20	26	29	33	39	1.4	1.6	1.8	2.0	2.1	1.5	1.8	2.0	2.3	2.5
WW B4	12	18	25	31	39	44	51	60	8	13	20	24	31	36	42	50	1.2	1.4	1.6	1.8	1.9	1.4	1.5	1.7	1.9	2.1
WW B5	11	18	26	31	39	45	52	61	8	13	20	25	33	38	44	53	1.2	1.4	1.6	1.7	1.9	1.3	1.5	1.7	1.9	2.0
WW C1	13	19	25	30	37	41	47	55	9	14	19	23	29	33	37	44	1.3	1.5	1.7	1.9	2.0	1.5	1.7	1.9	2.1	2.3
WW C2	12	18	25	30	37	42	48	57	8	13	19	23	29	34	39	47	1.3	1.5	1.7	1.8	2.0	1.4	1.6	1.8	2.0	2.2
WW C3	15	23	33	39	49	55	62	72	11	18	27	33	42	47	55	64	1.1	1.3	1.4	1.6	1.7	1.2	1.3	1.5	1.7	1.8
WW C4	13	21	30	37	46	52	60	69	9	16	24	30	39	45	52	62	1.1	1.3	1.5	1.6	1.8	1.2	1.4	1.5	1.7	1.9
TW A1	13	14	16	17	19	20	21	23	11	12	13	13	14	15	15	15	2.4	2.9	3.5	4.2	-	3.2	4.2	-	-	-
TW A2	11	13	16	18	20	22	25	28	9	10	12	13	15	16	17	19	2.0	2.3	2.7	3.1	3.4	2.4	2.9	3.5	4.4	-
TW A3	10	12	16	18	21	23	26	30	7	9	11	13	15	16	18	20	1.8	2.1	2.5	2.8	3.1	2.2	2.6	3.1	3.7	4.5
TW A4	8	11	15	17	21	24	27	32	6	8	10	12	15	17	19	22	1.7	2.0	2.3	2.6	2.8	2.1	2.4	2.8	3.3	3.8
TW B1	12	14	16	17	19	20	22	24	10	11	13	13	14	15	16	17	2.3	2.6	3.1	3.6	4.0	2.9	3.5	4.3	-	-
TW B2	10	12	15	17	20	22	25	29	8	9	11	13	15	16	18	20	1.9	2.2	2.6	2.9	3.1	2.3	2.7	3.2	3.7	4.3
TW B3	9	12	15	17	20	23	26	30	7	9	11	12	15	16	19	21	1.9	2.1	2.4	2.7	3.0	2.2	2.6	3.0	3.4	3.9
TW B4	9	11	14	17	20	22	26	30	7	9	11	12	15	17	19	22	1.9	2.2	2.4	2.7	2.9	2.2	2.5	2.9	3.3	3.6
TW C1	11	13	15	17	19	21	23	26	9	10	12	13	15	16	17	19	2.1	2.4	2.8	3.1	3.4	2.6	3.0	3.5	4.1	4.7
TW C2	9	12	14	17	20	22	24	28	8	9	11	13	15	16	18	21	2.0	2.3	2.6	2.8	3.1	2.4	2.7	3.1	3.5	3.8
TW C3	9	12	14	16	19	21	24	28	8	9	11	13	15	16	18	21	2.0	2.3	2.6	2.8	3.1	2.4	2.7	3.0	3.4	3.7
TW C4	10	12	14	16	19	21	23	27	9	10	12	13	15	17	18	21	2.2	2.4	2.7	2.9	3.2	2.5	2.8	3.1	3.5	3.7
TW D1	6	7	9	9	10	11	11	11	4	5	5	5	5	5	4	2	2.7	3.5	-	-	-	4.2	-	-	-	-
TW D2	10	14	18	21	26	29	33	39	8	11	14	16	20	22	25	30	1.6	1.8	2.1	2.3	2.5	1.9	2.1	2.4	2.7	3.0
TW D3	11	15	19	23	28	31	36	42	8	11	14	17	21	24	27	32	1.5	1.8	2.0	2.2	2.4	1.7	2.0	2.3	2.6	2.9
TW D4	12	17	23	28	35	39	45	54	9	13	18	22	28	32	37	44	1.3	1.5	1.7	1.9	2.0	1.5	1.7	1.9	2.1	2.3
TW D5	12	18	24	29	36	41	48	57	9	14	19	24	30	35	41	49	1.3	1.5	1.6	1.8	2.0	1.4	1.6	1.8	2.0	2.1
TW E1	13	18	23	27	33	37	42	49	10	14	18	21	26	29	33	39	1.4	1.6	1.8	2.0	2.2	1.6	1.8	2.1	2.3	2.5
TW E2	13	18	24	28	34	39	45	52	9	14	18	22	27	31	36	43	1.4	1.6	1.7	1.9	2.1	1.5	1.7	1.9	2.1	2.3
TW E3	15	22	30	36	45	50	57	66	12	18	25	30	37	43	49	58	1.2	1.3	1.5	1.7	1.8	1.3	1.4	1.6	1.8	2.0
TW E4	14	21	29	35	43	49	56	65	10	17	24	29	36	42	48	57	1.2	1.4	1.5	1.7	1.8	1.3	1.4	1.6	1.8	1.9
TW F1	11	12	14	15	17	18	20	21	9	10	11	12	12	13	13	13	2.4	2.8	3.4	4.1	4.8	3.1	4.0	-	-	-
TW F2	9	11	13	15	18	20	22	25	6	8	10	11	12	13	15	16	2.0	2.4	2.7	3.1	3.5	2.5	3.0	3.6	4.5	-
TW F3	7	10	13	15	18	20	23	27	5	6	8	10	12	13	15	17	1.9	2.2	2.5	2.9	3.1	2.3	2.7	3.2	3.8	4.5
TW F4	5	8	11	14	17	19	23	27	3	5	7	9	11	13	15	18	1.8	2.1	2.4	2.7	3.0	2.2	2.6	2.9	3.4	3.9
TW G1	9	10	12	14	16	17	19	22	7	8	9	10	11	12	13	14	2.3	2.7	3.1	3.6	4.1	2.9	3.5	4.3	-	-
TW G2	6	9	11	13	16	18	21	24	4	6	8	9	11	12	14	16	2.0	2.3	2.7	3.0	3.3	2.5	2.9	3.4	4.0	4.6
TW G3	5	7	10	12	15	17	20	24	3	4	6	8	10	12	14	16	2.0	2.3	2.6	2.9	3.1	2.4	2.7	3.1	3.6	4.0
TW G4	3	5	8	10	13	16	19	23	2	3	5	7	9	10	13	16	2.1	2.3	2.6	2.9	3.1	2.4	2.8	3.1	3.5	3.8
TW H1	6	8	10	12	14	16	18	21	4	6	7	8	10	11	12	14	2.2	2.6	2.9	3.3	3.6	2.7	3.2	3.7	4.4	-
TW H2	4	6	8	10	13	15	17	21	2	4	5	7	9	10	12	14	2.2	2.4	2.7	3.1	3.3	2.6	2.9	3.3	3.8	4.2
TW H3	3	5	7	9	11	13	16	20	1	3	4	6	8	9	11	14	2.2	2.5	2.8	3.1	3.3	2.6	3.0	3.3	3.7	4.0
TW H4	2	3	5	7	9	11	13	17	1	2	4	5	6	8	9	12	2.5	2.7	3.0	3.2	3.5	2.9	3.2	3.5	3.8	4.1
TW I1	4	6	7	7	8	9	9	9	3	3	3	3	3	3	2	0	2.8	3.7	-	-	-	4.5	-	-	-	-
TW I2	8	11	15	18	23	26	30	36	5	8	11	13	17	19	22	27	1.6	1.9	2.1	2.3	2.5	1.9	2.2	2.4	2.7	3.0
TW I3	8	12	17	20	26	29	34	40	6	9	12	15	19	22	26	31	1.5	1.8	2.0	2.2	2.4	1.8	2.0	2.3	2.5	2.8
TW I4	9	15	21	25	32	36	42	51	6	11	16	20	25	29	35	42	1.4	1.6	1.7	1.9	2.1	1.5	1.7	1.9	2.1	2.3
TW I5	9	15	22	27	34	39	45	54	6	11	17	21	28	32	38	46	1.3	1.5	1.7	1.8	2.0	1.5	1.6	1.8	2.0	2.1
TW J1	11	16	21	25	31	35	41	48	8	12	16	19	24	28	32	38	1.4	1.6	1.8	2.0	2.2	1.6	1.8	2.0	2.3	2.5
TW J2	10	15	21	26	32	37	42	50	7	11	16	20	25	29	34	41	1.4	1.6	1.8	1.9	2.1	1.5	1.7	1.9	2.1	2.3
TW J3	13	20	28	34	42	48	55	65	9	15	23	28	36	41	48	57	1.2	1.4	1.5	1.7	1.8	1.3	1.4	1.6	1.8	1.9
TW J4	12	19	26	32	41	46	54	63	8	14	21	27	34	40	46	55	1.2	1.4	1.5	1.7	1.8	1.3	1.5	1.6	1.8	2.0
DG A1	6	7	8	9	10	10	10	8	3	4	4	4	4	4	2	0	2.5	3.5	-	-	-	3.7	-	-	-	-
DG A2	9	11	14	17	20	22	25	28	6	8	11	12	15	16	18	21	1.9	2.1	2.5	2.9	3.3	2.2	2.5	3.0	3.6	4.3
DG A3	12	16	21	25	30	34	39	46	10	13	17	20	25	28	33	39	1.5	1.7	1.9	2.1	2.2	1.7	1.9	2.1	2.3	2.5
DG B1	6	7	8	9	10	11	11	11	3	4	4	5	4	4	3	1	2.5	3.5	-	-	-	3.9	-	-	-	-
DG B2	9	12	15	17	21	23	26	31	7	9	11	13	15	17	19	22	1.8	2.1	2.4	2.7	3.0	2.2	2.4	2.8	3.3	3.8
DG B3	13	17	22	26	32	36	42	50	10	14	18	21	26	30	35	42	1.5	1.7	1.8	1.9	2.1	1.7	1.8	2.0	2.1	2.3
DG C1	8	10	12	13	15	16	17	19	5	6	7	8	9	9	9	8	2.1	2.8	3.4	4.2	-	2.9	3.5	-	-	-
DG C2	11	14	18	21	25	29	33	38	8	11	14	16	20	22	25	29	1.7	1.9	2.1	2.3	2.6	1.9	2.1	2.4	2.8	3.1
DG C3	14	19	25	29	36	41	47	56	12	16	21	24	30	34	40	47	1.4	1.6	1.7	1.8	2.0	1.6	1.7	1.9	2.0	2.2
SS A1	15	17	20	23	26	28	31	35	13	15	17	19	21	23	25	27	1.7	2.1	2.4	2.8	3.1	2.0	2.4	2.9	3.5	4.2
SS A2	18	23	28	32	38	42	47	54	15	20	24	28	33	36	41	47	1.3	1.6	1.8	2.0	2.1	1.5	1.7	1.9	2.2	2.4
SS A3	13	15	18	20	22	23	25	27	11	13	15	16	17	18	19	20	1.9	2.4	2.9	3.5	4.1	2.3	2.9	3.8	-	-
SS A4	17	22	27	31	36	40	45	51	15	19	23	26	31	34	38	43	1.4	1.6	1.8	2.0	2.2	1.5	1.8	2.0	2.3	2.5
SS A5	15	17	20	22	24	26	28	30	14	15	17	18	20	21	22	23	1.9	2.4	2.8	3.3	3.7	2.3	2.8	3.5	4.6	-
SS A6	17	22	27	31	37	41	46	53	15	19	24	27	32	35	39	45	1.4	1.6	1.8	2.0	2.2	1.5	1.7	2.0	2.2	2.5
SS A7	14	16	17	18	20	21	22	23	13	14	15	15	16	17	17	17	2.3	2.9	3.6	4.8	-	2.8	3.6	-	-	-
SS A8	17	21	26	29	34	38	42	49	14	18	22	25	29	32	36	40	1.4	1.7	1.9	2.1	2.3	1.6	1.8	2.1	2.5	2.7
SS B1	12	14	17	19	22	24	26	30	10	12	14	16	18	19	21	22	1.9	2.2	2.6	3.0	3.3	2.2	2.6	3.1	3.9	4.8
SS B2	15	20	25	29	35	39	44	51	13	17	22	25	30	33	38	44	1.4	1.6	1.8	2.0	2.2	1.5	1.8	2.0	2.2	2.4
SS B3	11	13	15	17	19	21	22	25	9	11	12	13	15	16	17	17	2.0	2.5	3.0	3.6	4.1	2.4	3.0	3.9	-	-
SS B4	15	19	24	28	33	37	41	48	13	16	20	24	28	31	35	41	1.4	1.7	1.9	2.1	2.2	1.6	1.8	2.1	2.3	2.5
SS B5	12	14	16	18	20	21	23	25	11	12	14	15	16	17	18	19	2.2	2.6	3.1	3.6	4.2	2.6	3.1	3.9	-	-
SS B6	15	19	24	28	33	37	42	49	12	16	20	24	28	31	36	41	1.4	1.7	1.9	2.1	2.2	1.6	1.8	2.0	2.3	2.5
SS B7	11	13	14	15	17	18	19	20	10	11	12	13	13	14	14	14	2.4	3.0	3.7	4.7	-	3.0	3.9	-	-	-
SS B8	14	18	22	26	31	34	39	45	12	15	19	22	26	29	32	37	1.5	1.7	2.0	2.2	2.3	1.6	1.9	2.2	2.5	2.7
SS C1	8	11	13	15	18	20	22	25	5	7	9	10	11	12	13	14	1.9	2.3	2.7	3.2	3.6	2.4	3.0	3.8	-	-
SS C2	10	15	19	23	28	32	37	43	7	11	14	17	21	24	28	33	1.5	1.7	1.9	2.1	2.3	1.7	2.0	2.2	2.5	2.8
SS C3	9	11	13	15	17	18	20	23	7	9	10	11	12	13	14	15	2.2	2.6	3.0	3.5	3.9	2.8	3.4	4.1	-	-
SS C4	10	14	18	21	26	29	33	39	8	11	14	16	20	22	26	30	1.6	1.8	2.1	2.3	2.5	1.9	2.1	2.4	2.7	2.9
SS D1	15	19	23	25	29	32	35	40	13	15	18	20	23	25	27	29	1.6	1.9	2.2	2.5	2.8	1.8	2.2	2.7	3.4	4.2
SS D2	19	25	32	37	44	49	55	63	16	21	27	32	38	42	47	54	1.2	1.4	1.6	1.8	1.9	1.3	1.5	1.7	2.0	2.1
SS D3	15	18	22	24	27	29	32	36	13	16	18	20	22	24	25	27	1.7	2.1	2.4	2.8	3.1	1.9	2.4	2.9	3.7	4.7
SS D4	18	24	31	36	42	47	53	61	15	21	26	31	36	40	45	52	1.2	1.4	1.6	1.8	2.0	1.3	1.6	1.8	2.0	2.2
SS E1	12	15	18	20	23	25	27	30	9	11	13	15	17	18	19	19	1.7	2.2	2.6	3.1	3.5	2.1	2.6	3.5	-	-
SS E2	15	20	26	30	36	41	46	53	12	16	21	25	30	33	38	44	1.3	1.5	1.8	2.0	2.1	1.4	1.7	2.0	2.2	2.5
SS E3	12	15	17	19	21	23	25	27	11	12	14	15	17	18	19	20	2.0	2.4	2.9	3.4	3.9	2.3	2.9	3.7	-	-
SS E4	15	19	25	28	34	38	43	50	12	16	20	24	28	31	35	41	1.4	1.6	1.8	2.0	2.2	1.5	1.8	2.1	2.3	2.6

SAULT STE. MARIE, MICHIGAN

TYPE	SOLAR SAVINGS FRACTION (%) BASE TEMP: 55 F (6224 DD) LCR 100	70	50	40	30	25	20	15	65 F (9201 DD) 100	70	50	40	30	25	20	15	CONSERVATION FACTOR 55 F 100	50	30	20	15	65 F 100	50	30	20	15
WW A1	11	12	13	13	14	14	14	13	10	11	11	11	11	11	10	9	3.4	4.6	-	-	-	4.1	-	-	-	-
WW A2	8	10	12	13	15	16	17	18	7	8	9	10	11	12	12	12	2.3	2.8	3.4	4.7	-	2.5	3.3	4.8	-	-
WW A3	7	9	11	13	15	17	19	20	5	7	8	10	11	12	13	13	2.0	2.5	3.0	3.9	-	2.3	2.9	3.9	-	-
WW A4	5	8	11	13	16	17	19	22	4	6	8	9	11	12	13	14	1.9	2.3	2.8	3.5	4.5	2.1	2.7	3.5	-	-
WW A5	5	7	10	13	16	18	20	23	3	5	7	9	11	12	13	14	1.8	2.2	2.7	3.3	4.1	2.0	2.6	3.3	4.8	-
WW A6	4	7	10	13	16	18	21	23	2	4	7	9	11	12	14	15	1.8	2.1	2.6	3.2	3.9	2.0	2.5	3.2	4.4	-
WW B1	5	5	6	5	5	4	2	-1	4	4	3	3	1	0	-2	-8	4.9	-	-	-	-	-	-	-	-	-
WW B2	7	10	14	17	21	23	27	31	5	8	11	13	16	19	21	25	1.7	2.0	2.3	2.6	2.9	1.8	2.2	2.5	3.0	3.5
WW B3	8	12	17	21	26	29	33	39	6	10	14	17	21	24	27	32	1.5	1.8	2.0	2.3	2.6	1.6	1.9	2.2	2.6	3.0
WW B4	8	14	20	25	32	36	42	50	6	11	17	21	27	31	37	44	1.4	1.5	1.7	1.9	2.1	1.4	1.6	1.9	2.1	2.3
WW B5	8	14	21	26	33	38	44	53	6	11	17	22	29	33	39	47	1.3	1.5	1.7	1.9	2.0	1.4	1.6	1.8	2.0	2.1
WW C1	10	14	20	23	29	33	37	44	8	12	16	20	24	28	32	37	1.4	1.7	1.9	2.2	2.4	1.5	1.8	2.1	2.4	2.7
WW C2	9	14	19	24	30	34	39	47	7	11	16	20	25	29	34	40	1.4	1.6	1.8	2.0	2.2	1.5	1.7	1.9	2.2	2.4
WW C3	11	19	27	33	42	47	55	64	9	16	24	29	38	43	50	59	1.2	1.3	1.5	1.7	1.9	1.2	1.4	1.6	1.8	1.9
WW C4	10	17	25	31	39	45	52	62	8	14	22	27	35	41	48	57	1.2	1.4	1.5	1.7	1.9	1.3	1.4	1.6	1.8	1.9
TW A1	11	12	13	14	14	15	15	15	10	11	11	12	12	12	12	11	3.2	4.0	-	-	-	3.7	-	-	-	-
TW A2	9	10	12	13	15	16	18	19	8	9	10	11	12	13	13	14	2.4	2.9	3.4	4.4	-	2.7	3.3	4.5	-	-
TW A3	7	9	11	13	15	17	19	21	6	7	9	10	12	13	14	15	2.2	2.6	3.0	3.7	4.6	2.4	3.0	3.8	-	-
TW A4	6	8	11	13	15	17	19	22	4	6	8	10	12	13	14	16	2.1	2.4	2.8	3.3	3.9	2.2	2.7	3.3	4.2	-
TW B1	10	11	13	14	15	16	16	17	9	10	11	12	12	13	13	13	2.8	3.4	4.1	-	-	3.2	4.0	-	-	-
TW B2	8	10	12	13	15	17	18	21	7	8	10	11	12	13	14	15	2.3	2.7	3.1	3.7	4.4	2.5	3.0	3.7	4.9	-
TW B3	7	9	11	13	15	17	19	22	6	7	9	10	12	13	15	16	2.2	2.5	2.9	3.4	3.9	2.4	2.8	3.4	4.2	-
TW B4	7	9	11	13	15	17	19	22	5	7	9	10	12	14	15	18	2.2	2.5	2.8	3.2	3.6	2.4	2.8	3.2	3.8	4.5
TW C1	9	11	12	13	15	16	18	20	8	9	10	11	13	13	14	15	2.6	3.0	3.4	4.0	4.6	2.8	3.3	4.1	-	-
TW C2	8	9	11	13	15	17	19	22	7	8	10	11	13	14	15	17	2.3	2.7	3.0	3.4	3.8	2.5	2.9	3.4	4.1	4.8
TW C3	8	9	11	13	15	17	19	22	7	8	10	11	13	14	16	18	2.4	2.7	3.0	3.3	3.7	2.5	2.9	3.4	3.9	4.4
TW C4	9	10	12	13	15	17	19	22	8	9	10	12	13	15	16	18	2.5	2.8	3.1	3.4	3.7	2.7	3.0	3.4	3.8	4.3
TW D1	4	5	5	5	5	5	4	2	3	3	3	3	2	2	0	-2	4.1	-	-	-	-	-	-	-	-	-
TW D2	8	11	14	17	20	23	26	30	6	9	12	14	17	19	21	24	1.8	2.1	2.4	2.7	3.0	2.0	2.3	2.7	3.1	3.6
TW D3	8	11	15	17	21	24	28	32	6	9	12	14	18	20	22	26	1.7	2.0	2.3	2.6	2.9	1.8	2.2	2.6	3.0	3.5
TW D4	9	13	19	23	28	32	38	45	7	11	16	19	25	28	33	39	1.5	1.7	1.9	2.1	2.3	1.5	1.8	2.0	2.2	2.4
TW D5	9	14	20	24	31	35	41	49	7	12	17	21	27	31	36	44	1.4	1.6	1.8	2.0	2.1	1.5	1.7	1.9	2.1	2.2
TW E1	10	14	18	22	26	30	34	39	8	12	15	18	22	25	28	33	1.6	1.8	2.1	2.3	2.6	1.7	1.9	2.3	2.6	2.9
TW E2	10	14	19	23	28	32	36	43	8	12	16	19	24	27	31	37	1.5	1.7	1.9	2.2	2.4	1.6	1.8	2.1	2.3	2.6
TW E3	12	18	25	30	38	43	49	57	10	16	22	27	34	38	44	52	1.3	1.4	1.6	1.8	2.0	1.3	1.5	1.7	1.9	2.1
TW E4	11	17	24	29	37	42	49	57	9	14	21	26	33	38	44	52	1.3	1.4	1.6	1.8	2.0	1.3	1.5	1.7	1.9	2.1
TW F1	9	10	11	12	13	13	14	14	8	9	9	10	10	10	10	9	3.1	3.8	5.0	-	-	3.5	4.8	-	-	-
TW F2	7	8	10	11	13	14	15	17	5	6	8	9	10	10	11	11	2.5	2.9	3.5	4.4	-	2.7	3.4	4.6	-	-
TW F3	5	7	9	10	12	14	16	18	3	5	6	8	9	10	11	12	2.3	2.6	3.1	3.8	4.6	2.5	3.0	3.8	-	-
TW F4	3	5	7	9	12	13	16	18	1	3	5	7	8	10	11	13	2.2	2.5	2.9	3.4	3.9	2.4	2.8	3.4	4.2	-
TW G1	7	8	10	10	12	13	13	14	6	7	8	8	9	10	10	10	2.8	3.4	4.1	-	-	3.2	4.0	-	-	-
TW G2	4	6	8	9	11	12	14	16	3	4	6	7	8	9	10	11	2.4	2.8	3.3	3.9	4.6	2.6	3.2	4.0	-	-
TW G3	3	5	7	8	10	12	14	17	1	3	5	6	8	9	10	12	2.3	2.7	3.1	3.5	4.0	2.5	3.0	3.6	4.3	-
TW G4	2	3	5	7	9	11	13	16	0	2	4	5	7	8	10	12	2.4	2.7	3.0	3.4	3.8	2.6	3.0	3.4	4.0	4.5
TW H1	4	6	7	8	10	11	12	14	3	4	6	6	8	8	9	10	2.7	3.1	3.6	4.3	-	3.0	3.5	4.4	-	-
TW H2	2	4	5	7	9	10	12	15	1	2	4	5	6	8	9	11	2.6	2.9	3.3	3.7	4.1	2.8	3.2	3.7	4.4	-
TW H3	1	3	5	6	8	9	11	14	0	2	3	4	6	7	9	11	2.6	2.9	3.2	3.6	3.9	2.8	3.2	3.6	4.1	4.6
TW H4	1	2	4	5	7	8	10	12	0	1	2	3	5	6	7	10	2.9	3.2	3.4	3.7	4.0	3.1	3.4	3.7	4.1	4.5
TW I1	3	3	4	4	3	3	2	0	2	2	2	2	1	0	0	-4	4.4	-	-	-	-	-	-	-	-	-
TW I2	5	8	11	14	17	20	23	27	4	6	9	11	14	16	18	22	1.9	2.1	2.4	2.7	3.0	2.0	2.3	2.7	3.1	3.6
TW I3	6	9	13	15	20	22	26	31	4	7	10	13	16	18	21	25	1.7	2.0	2.3	2.6	2.8	1.9	2.2	2.5	2.9	3.2
TW I4	6	11	16	20	26	30	35	42	5	9	13	17	22	26	30	37	1.5	1.7	1.9	2.1	2.3	1.6	1.8	2.0	2.3	2.5
TW I5	7	11	17	22	28	32	38	46	5	9	15	19	24	28	34	41	1.4	1.6	1.8	2.0	2.2	1.5	1.7	1.9	2.1	2.3
TW J1	8	12	16	20	25	28	32	38	6	10	14	17	21	24	27	32	1.6	1.8	2.0	2.3	2.5	1.7	1.9	2.2	2.5	2.8
TW J2	7	12	17	20	26	30	35	41	6	9	14	17	22	25	30	36	1.5	1.7	1.9	2.2	2.3	1.6	1.8	2.1	2.3	2.5
TW J3	9	16	23	28	36	41	48	57	7	13	20	25	32	37	43	52	1.3	1.4	1.6	1.8	2.0	1.3	1.5	1.7	1.9	2.1
TW J4	8	15	22	27	35	40	47	56	6	12	19	24	31	36	42	51	1.3	1.5	1.6	1.8	2.0	1.3	1.5	1.7	1.9	2.1
DG A1	4	4	5	5	5	4	3	0	2	3	3	2	2	1	0	-4	3.6	-	-	-	-	4.5	-	-	-	-
DG A2	7	9	11	13	15	17	19	21	5	7	9	11	13	14	15	17	2.2	2.5	2.9	3.5	4.4	2.3	2.7	3.3	4.3	-
DG A3	10	13	17	21	25	29	33	39	9	12	15	18	23	26	30	35	1.7	1.8	2.0	2.3	2.5	1.8	2.0	2.2	2.4	2.7
DG B1	3	4	5	5	5	5	4	2	2	2	3	2	2	1	0	-2	3.5	-	-	-	-	4.4	-	-	-	-
DG B2	7	9	11	13	16	18	20	23	5	7	9	11	13	14	16	18	2.1	2.4	2.8	3.3	3.9	2.3	2.7	3.1	3.8	4.7
DG B3	10	14	18	22	27	31	35	42	9	12	16	19	24	27	32	38	1.6	1.8	2.0	2.2	2.3	1.7	1.9	2.1	2.3	2.5
DG C1	5	6	8	9	9	10	10	9	4	5	5	6	6	6	5	3	2.7	3.4	-	-	-	3.1	-	-	-	-
DG C2	8	11	14	17	20	23	26	30	7	9	12	14	17	19	22	25	1.9	2.1	2.4	2.8	3.2	2.0	2.3	2.6	3.1	3.6
DG C3	12	16	21	25	31	35	40	48	10	14	19	22	28	31	36	43	1.5	1.7	1.8	2.0	2.2	1.6	1.8	1.9	2.1	2.3
SS A1	12	14	17	18	21	22	24	25	12	13	15	17	19	20	21	22	2.0	2.4	3.0	3.8	4.8	2.1	2.7	3.4	4.8	-
SS A2	15	19	24	27	32	35	39	45	14	18	22	25	30	33	36	41	1.5	1.7	2.0	2.2	2.5	1.5	1.8	2.1	2.4	2.7
SS A3	11	13	14	15	16	17	18	18	10	12	13	14	15	15	15	14	2.4	3.0	4.2	-	-	2.4	3.4	-	-	-
SS A4	14	18	22	25	30	33	36	41	14	17	21	24	28	30	33	37	1.5	1.8	2.1	2.4	2.7	1.6	1.9	2.2	2.6	3.0
SS A5	13	15	16	18	19	20	21	22	13	14	15	16	17	18	18	18	2.3	2.9	3.7	-	-	2.4	3.2	4.5	-	-
SS A6	14	18	23	26	31	34	38	43	14	17	21	24	28	31	34	39	1.5	1.8	2.0	2.3	2.6	1.6	1.9	2.2	2.5	2.8
SS A7	12	13	14	15	15	16	16	15	12	13	13	14	14	14	14	12	2.9	3.9	-	-	-	3.0	4.5	-	-	-
SS A8	14	17	21	24	28	31	34	38	13	16	20	22	26	28	31	34	1.6	1.9	2.2	2.6	2.9	1.7	2.0	2.4	2.9	3.4
SS B1	10	12	14	15	17	18	20	21	9	11	12	14	15	16	17	18	2.2	2.7	3.3	4.1	-	2.3	2.9	3.8	-	-
SS B2	13	17	21	24	29	32	37	42	12	16	20	23	27	30	34	39	1.6	1.8	2.0	2.3	2.5	1.6	1.9	2.1	2.4	2.6
SS B3	9	10	12	13	14	15	15	16	8	9	11	11	12	13	13	12	2.5	3.1	4.1	-	-	2.6	3.5	-	-	-
SS B4	12	16	20	23	27	30	34	39	11	15	18	21	25	28	31	36	1.6	1.8	2.1	2.4	2.6	1.6	1.9	2.2	2.6	2.9
SS B5	11	12	13	14	15	16	17	18	10	11	12	13	14	14	15	15	2.7	3.2	4.1	-	-	2.7	3.6	5.0	-	-
SS B6	12	16	20	23	27	31	35	40	11	14	18	21	25	28	31	36	1.6	1.8	2.1	2.3	2.6	1.6	1.9	2.2	2.5	2.8
SS B7	10	11	12	12	13	13	13	13	9	10	11	11	12	12	11	11	3.1	4.0	-	-	-	3.2	4.5	-	-	-
SS B8	11	14	18	21	25	28	31	36	10	13	17	19	23	25	28	32	1.7	1.9	2.2	2.5	2.8	1.7	2.0	2.4	2.8	3.1
SS C1	6	7	9	10	12	13	14	14	4	5	7	7	8	8	8	7	2.4	2.9	3.8	-	-	2.7	3.6	-	-	-
SS C2	7	11	15	18	22	25	28	33	6	9	12	14	18	20	23	27	1.7	1.9	2.2	2.5	2.8	1.8	2.1	2.5	2.9	3.3
SS C3	8	9	10	11	12	13	14	15	6	7	8	9	10	10	10	10	2.7	3.3	4.0	-	-	3.0	3.9	-	-	-
SS C4	8	11	14	17	20	23	26	30	6	9	12	14	17	19	21	25	1.8	2.1	2.4	2.7	3.0	2.0	2.3	2.6	3.0	3.4
SS D1	12	15	18	19	22	24	25	27	11	13	16	17	19	20	21	22	1.8	2.2	2.8	3.8	-	1.9	2.5	3.4	-	-
SS D2	15	20	26	30	36	40	45	52	14	19	24	28	33	37	41	47	1.3	1.5	1.8	2.0	2.2	1.4	1.6	1.9	2.2	2.4
SS D3	13	15	17	19	21	22	24	25	12	14	16	17	19	20	21	21	2.0	2.4	3.1	4.1	-	2.1	2.7	3.7	-	-
SS D4	15	20	25	30	35	39	44	50	14	19	24	27	32	36	40	46	1.4	1.6	1.8	2.1	2.3	1.4	1.7	1.9	2.2	2.5
SS E1	9	11	13	14	16	17	17	17	8	9	11	12	13	13	13	11	2.1	2.7	3.8	-	-	2.3	3.2	-	-	-
SS E2	11	16	21	24	29	32	36	42	10	14	18	22	26	29	32	36	1.5	1.7	2.0	2.3	2.5	1.5	1.8	2.2	2.5	2.9
SS E3	10	12	14	15	16	17	18	18	10	11	12	13	14	15	15	14	2.4	3.0	3.9	-	-	2.5	3.4	-	-	-
SS E4	12	15	20	23	27	30	34	39	11	14	18	21	25	27	30	35	1.6	1.8	2.1	2.4	2.7	1.6	1.9	2.3	2.6	3.0

TRAVERSE CITY, MICHIGAN

TYPE	SOLAR SAVINGS FRACTION (%) BASE TEMP: 55 F (5039 DD) LCR 100	70	50	40	30	25	20	15	65 F (7703 DD) 100	70	50	40	30	25	20	15	CONSERVATION FACTOR 55 F 100	50	30	20	15	65 F 100	50	30	20	15
WW A1	12	13	14	14	15	15	15	15	11	11	12	12	12	12	11	9	3.1	3.9	-	-	-	3.9	-	-	-	-
WW A2	9	11	13	15	17	18	19	21	7	9	10	11	12	12	13	12	2.1	2.6	3.3	4.3	-	2.5	3.4	4.9	-	-
WW A3	7	10	13	14	17	19	20	23	6	7	9	10	12	13	13	13	1.9	2.3	2.9	3.6	4.5	2.3	2.9	4.0	-	-
WW A4	6	9	12	14	17	19	21	24	4	6	8	10	12	13	14	14	1.8	2.2	2.7	3.3	4.0	2.1	2.7	3.6	-	-
WW A5	5	9	12	14	18	20	22	25	3	5	8	9	12	13	14	15	1.7	2.1	2.6	3.1	3.7	2.0	2.6	3.3	4.8	-
WW A6	5	8	12	14	18	20	23	26	3	5	8	9	12	13	14	15	1.7	2.1	2.5	3.0	3.5	2.0	2.5	3.2	4.4	-
WW B1	6	6	7	7	6	5	4	0	4	4	4	3	2	0	-2	-8	3.7	-	-	-	-	-	-	-	-	-
WW B2	7	11	15	18	22	25	29	33	5	8	11	14	17	19	22	25	1.6	1.9	2.2	2.5	2.8	1.8	2.2	2.6	3.0	3.5
WW B3	9	14	19	22	27	31	35	41	7	11	15	18	22	24	28	32	1.5	1.7	2.0	2.2	2.5	1.6	1.9	2.2	2.6	3.0
WW B4	9	15	21	26	33	38	44	52	7	12	17	22	28	32	37	44	1.3	1.5	1.7	1.9	2.1	1.4	1.6	1.9	2.1	2.3
WW B5	9	15	22	27	34	39	46	54	7	12	18	23	29	34	40	48	1.3	1.5	1.7	1.8	2.0	1.4	1.6	1.8	2.0	2.1
WW C1	11	16	21	25	31	34	39	46	8	12	17	20	25	28	32	37	1.4	1.6	1.9	2.1	2.3	1.5	1.8	2.1	2.4	2.7
WW C2	9	15	21	25	31	36	41	49	7	12	17	20	26	30	34	41	1.4	1.6	1.8	2.0	2.2	1.5	1.7	1.9	2.2	2.4
WW C3	12	20	28	35	43	49	56	65	10	16	24	30	38	43	50	59	1.1	1.3	1.5	1.7	1.8	1.2	1.4	1.6	1.8	1.9
WW C4	11	18	26	32	41	46	54	63	8	15	22	28	36	41	48	57	1.2	1.4	1.5	1.7	1.8	1.3	1.4	1.6	1.8	1.9
TW A1	12	13	14	15	16	16	17	17	11	11	12	12	13	13	13	12	2.9	3.5	4.6	-	-	3.5	-	-	-	-
TW A2	9	11	13	15	17	18	19	21	8	9	11	12	13	13	14	14	2.2	2.7	3.3	4.1	-	2.6	3.4	4.6	-	-
TW A3	8	10	13	14	17	18	20	23	7	8	10	11	12	13	14	15	2.0	2.4	2.9	3.5	4.2	2.4	3.0	3.8	-	-
TW A4	7	9	12	14	17	19	21	24	5	7	9	10	12	14	15	17	1.9	2.3	2.7	3.2	3.7	2.2	2.7	3.3	4.2	-
TW B1	11	12	13	15	16	17	18	19	10	11	12	12	13	13	14	13	2.6	3.1	3.8	4.9	-	3.1	4.0	-	-	-
TW B2	8	10	13	14	17	18	20	22	7	9	10	11	13	14	15	16	2.2	2.5	3.0	3.5	4.1	2.5	3.0	3.8	4.9	-
TW B3	7	10	12	14	17	18	21	23	6	8	10	11	13	14	15	17	2.1	2.4	2.8	3.3	3.7	2.4	2.8	3.4	4.2	-
TW B4	7	9	12	14	16	18	21	24	6	8	10	11	13	14	16	18	2.1	2.4	2.7	3.1	3.5	2.3	2.8	3.3	3.9	4.5
TW C1	9	11	13	14	16	18	19	21	9	10	11	12	13	14	15	16	2.4	2.8	3.2	3.8	4.4	2.7	3.3	4.1	-	-
TW C2	8	10	12	14	16	18	20	23	7	9	10	12	13	14	16	18	2.2	2.5	2.9	3.3	3.7	2.5	2.9	3.5	4.1	4.8
TW C3	8	10	12	14	16	18	20	23	7	9	10	12	13	15	16	18	2.3	2.5	2.9	3.2	3.6	2.5	2.9	3.4	3.9	4.4
TW C4	9	11	12	14	16	18	20	23	8	10	11	12	14	15	17	19	2.4	2.7	3.0	3.3	3.6	2.6	3.0	3.4	3.9	4.3
TW D1	5	6	6	6	7	6	6	4	4	4	4	4	3	2	1	-2	3.5	-	-	-	-	-	-	-	-	-
TW D2	9	12	15	18	22	24	27	32	7	9	12	14	17	19	21	25	1.8	2.0	2.3	2.6	2.9	2.0	2.3	2.7	3.1	3.6
TW D3	9	12	16	19	23	26	29	34	7	10	13	15	18	20	23	26	1.6	1.9	2.2	2.5	2.8	1.8	2.2	2.6	3.0	3.5
TW D4	10	14	20	24	30	34	39	46	8	12	16	20	25	28	33	39	1.4	1.6	1.9	2.1	2.2	1.6	1.8	2.0	2.2	2.4
TW D5	10	15	21	25	32	36	42	50	8	12	18	22	27	31	37	44	1.4	1.6	1.8	1.9	2.1	1.5	1.7	1.9	2.1	2.2
TW E1	11	15	20	23	28	31	35	41	9	12	16	19	23	25	29	33	1.5	1.8	2.0	2.3	2.5	1.7	2.0	2.3	2.6	2.9
TW E2	10	15	20	24	29	33	38	45	8	12	16	20	24	28	32	37	1.5	1.7	1.9	2.1	2.3	1.6	1.8	2.1	2.3	2.6
TW E3	13	19	26	32	39	44	51	59	10	16	22	27	34	39	45	52	1.2	1.4	1.6	1.8	1.9	1.3	1.5	1.7	1.9	2.1
TW E4	12	18	25	31	38	43	50	59	9	15	21	26	33	38	44	53	1.2	1.4	1.6	1.8	1.9	1.3	1.5	1.7	1.9	2.0
TW F1	10	11	12	13	14	15	15	16	9	9	10	10	11	11	11	10	2.8	3.4	4.4	-	-	3.4	4.9	-	-	-
TW F2	7	9	11	12	14	15	17	18	6	7	8	9	10	11	11	12	2.3	2.7	3.3	4.2	-	2.7	3.4	4.7	-	-
TW F3	5	7	10	11	14	15	17	20	4	5	7	8	10	11	12	13	2.1	2.5	3.0	3.6	4.3	2.5	3.1	3.9	-	-
TW F4	3	6	8	10	13	15	17	20	2	4	6	7	9	10	12	13	2.1	2.4	2.8	3.3	3.7	2.3	2.8	3.4	4.2	-
TW G1	8	9	10	11	13	14	15	16	7	8	8	9	10	10	11	10	2.6	3.1	3.8	4.8	-	3.1	4.0	-	-	-
TW G2	5	7	9	10	12	14	16	18	4	5	6	8	9	10	11	12	2.3	2.6	3.1	3.7	4.4	2.6	3.2	4.0	-	-
TW G3	3	5	8	9	12	13	16	18	2	4	5	7	8	9	11	12	2.2	2.5	2.9	3.4	3.9	2.5	3.0	3.6	4.4	-
TW G4	2	4	6	8	10	12	14	18	1	3	4	6	7	9	10	12	2.3	2.6	2.9	3.3	3.7	2.5	3.0	3.4	4.0	4.5
TW H1	5	6	8	9	11	12	14	16	4	5	6	7	8	9	10	10	2.6	2.9	3.4	4.1	4.8	2.9	3.6	4.4	-	-
TW H2	2	4	6	8	10	11	13	16	2	3	4	6	7	8	9	11	2.4	2.7	3.1	3.5	4.0	2.7	3.2	3.8	4.4	-
TW H3	2	3	5	7	9	10	13	15	1	2	4	5	6	8	9	11	2.5	2.8	3.1	3.5	3.8	2.7	3.2	3.6	4.1	4.7
TW H4	1	2	4	5	7	9	11	14	0	2	3	4	6	7	8	10	2.8	3.0	3.3	3.6	3.9	3.0	3.3	3.7	4.2	4.6
TW I1	3	4	4	5	5	5	4	2	2	2	2	2	1	1	0	-3	3.7	-	-	-	-	-	-	-	-	-
TW I2	6	9	12	15	19	21	24	29	4	7	9	11	14	16	19	22	1.8	2.1	2.4	2.7	3.0	2.0	2.4	2.7	3.1	3.6
TW I3	6	10	14	17	21	24	28	33	5	8	11	13	16	19	22	26	1.7	1.9	2.2	2.5	2.7	1.9	2.2	2.5	2.9	3.2
TW I4	7	12	17	21	27	31	36	44	5	9	14	17	22	26	31	37	1.5	1.7	1.9	2.1	2.3	1.6	1.8	2.0	2.3	2.5
TW I5	7	12	18	23	29	34	40	48	6	10	15	19	25	29	34	41	1.4	1.6	1.8	2.0	2.1	1.5	1.7	1.9	2.1	2.3
TW J1	9	13	18	21	26	30	34	40	7	10	14	17	21	24	28	32	1.5	1.8	2.0	2.3	2.5	1.7	1.9	2.2	2.5	2.8
TW J2	8	13	18	22	27	31	36	43	6	10	14	18	22	26	30	36	1.5	1.7	1.9	2.1	2.3	1.6	1.8	2.1	2.3	2.5
TW J3	10	17	24	30	37	43	49	58	8	14	20	25	33	37	44	52	1.2	1.4	1.6	1.8	1.9	1.3	1.5	1.7	1.9	2.1
TW J4	9	16	23	28	36	41	48	57	7	13	19	24	31	36	42	51	1.3	1.4	1.6	1.8	1.9	1.3	1.5	1.7	1.9	2.1
DG A1	4	5	6	6	6	6	5	2	3	3	3	3	2	1	0	-3	3.1	-	-	-	-	4.3	-	-	-	-
DG A2	7	9	12	14	17	18	20	23	6	8	10	11	13	14	16	17	2.0	2.4	2.8	3.4	4.2	2.3	2.7	3.3	4.3	-
DG A3	11	14	18	22	27	30	35	41	9	12	16	19	23	26	30	35	1.6	1.8	2.0	2.2	2.4	1.8	2.0	2.2	2.4	2.7
DG B1	4	5	6	6	6	6	5	4	3	3	3	3	2	2	0	-2	2.9	4.6	-	-	-	4.6	-	-	-	-
DG B2	7	10	12	14	17	19	21	24	6	8	10	11	14	15	17	19	1.9	2.3	2.7	3.2	3.7	2.3	2.7	3.1	3.9	4.8
DG B3	11	15	19	23	28	32	37	44	10	13	17	20	24	28	32	38	1.6	1.8	1.9	2.1	2.3	1.7	1.9	2.1	2.3	2.5
DG C1	6	7	9	10	11	11	11	11	4	5	6	6	6	6	6	4	2.3	3.2	-	-	-	3.2	4.9	-	-	-
DG C2	9	12	15	18	21	24	27	31	8	10	13	15	18	20	22	25	1.7	2.0	2.4	2.7	3.0	2.0	2.3	2.6	3.1	3.6
DG C3	13	17	22	26	32	36	41	49	11	15	19	23	28	32	36	43	1.5	1.7	1.8	2.0	2.1	1.6	1.8	1.9	2.1	2.3
SS A1	13	15	18	20	22	24	26	28	12	14	16	17	19	20	22	23	1.9	2.3	2.9	3.5	4.2	2.1	2.6	3.4	4.6	-
SS A2	16	20	25	29	34	37	41	47	15	18	23	26	30	33	37	42	1.4	1.7	1.9	2.2	2.4	1.5	1.8	2.1	2.4	2.6
SS A3	12	14	15	17	18	19	20	20	11	12	14	14	15	16	16	15	2.2	2.8	3.8	-	-	2.4	3.4	-	-	-
SS A4	15	19	24	27	32	35	39	44	14	18	22	24	28	31	34	38	1.5	1.8	2.0	2.3	2.5	1.6	1.9	2.2	2.6	2.9
SS A5	14	16	18	19	21	22	23	24	13	15	16	17	18	19	19	19	2.2	2.7	3.4	4.6	-	2.4	3.1	4.3	-	-
SS A6	15	20	24	28	32	36	40	45	14	18	22	25	29	32	35	40	1.5	1.7	2.0	2.2	2.4	1.6	1.8	2.2	2.5	2.8
SS A7	13	14	15	16	17	17	18	17	12	13	14	14	15	15	14	13	2.6	3.4	-	-	-	3.0	4.4	-	-	-
SS A8	15	19	23	26	30	33	36	41	14	17	20	23	26	29	32	35	1.5	1.8	2.1	2.4	2.7	1.6	2.0	2.4	2.8	3.2
SS B1	11	13	15	16	19	20	22	23	10	11	13	14	16	17	18	18	2.1	2.5	3.1	3.8	4.7	2.3	2.9	3.7	-	-
SS B2	14	18	22	26	31	34	38	45	12	16	20	23	28	30	34	39	1.5	1.7	2.0	2.2	2.4	1.6	1.9	2.1	2.4	2.6
SS B3	9	11	13	14	16	17	17	18	9	10	11	12	13	13	14	13	2.3	2.9	3.8	-	-	2.6	3.5	-	-	-
SS B4	13	17	21	24	29	32	36	42	12	15	19	22	26	29	32	37	1.5	1.8	2.1	2.3	2.5	1.6	1.9	2.2	2.5	2.8
SS B5	11	13	14	15	17	18	19	20	11	12	13	14	15	15	15	16	2.4	3.0	3.7	5.0	-	2.7	3.6	4.9	-	-
SS B6	13	17	21	24	29	32	36	42	12	15	19	22	26	29	32	37	1.5	1.8	2.0	2.3	2.5	1.6	1.9	2.2	2.5	2.8
SS B7	10	11	12	13	14	15	15	15	10	11	11	12	12	12	12	11	2.8	3.5	4.9	-	-	3.2	4.5	-	-	-
SS B8	12	16	20	23	27	30	33	38	11	14	17	20	24	26	29	33	1.6	1.9	2.2	2.4	2.7	1.7	2.0	2.4	2.7	3.1
SS C1	6	8	10	12	13	14	15	16	5	6	7	8	9	9	9	8	2.2	2.7	3.5	5.0	-	2.7	3.7	-	-	-
SS C2	8	12	16	19	23	26	30	35	6	9	13	15	18	21	24	27	1.6	1.9	2.2	2.5	2.7	1.8	2.1	2.5	2.9	3.3
SS C3	8	9	11	12	14	15	16	17	7	8	9	10	10	11	11	11	2.5	3.0	3.7	4.8	-	3.0	3.9	-	-	-
SS C4	9	12	15	18	22	24	28	32	7	10	12	14	17	19	22	25	1.8	2.0	2.3	2.6	2.9	2.0	2.3	2.6	3.0	3.4
SS D1	13	16	19	21	24	26	28	30	12	14	16	18	20	21	22	23	1.7	2.1	2.7	3.3	4.1	1.9	2.4	3.3	-	-
SS D2	16	22	28	32	38	42	48	55	14	19	25	29	34	38	42	48	1.3	1.5	1.7	2.0	2.1	1.4	1.6	1.9	2.1	2.4
SS D3	14	16	19	21	23	24	26	28	13	15	17	18	20	21	21	22	1.9	2.3	2.9	3.7	4.6	2.1	2.7	3.6	-	-
SS D4	16	21	27	31	37	41	46	53	14	19	24	28	33	37	41	46	1.3	1.6	1.8	2.0	2.2	1.4	1.7	1.9	2.2	2.4
SS E1	10	12	14	16	18	19	20	20	8	10	12	13	14	14	14	13	1.9	2.5	3.5	-	-	2.3	3.1	-	-	-
SS E2	13	17	22	26	31	34	39	45	11	15	19	22	27	29	33	37	1.4	1.7	2.0	2.2	2.4	1.5	1.8	2.1	2.5	2.8
SS E3	11	13	15	16	18	19	20	21	10	12	13	14	15	15	16	15	2.2	2.7	3.6	4.9	-	2.5	3.3	4.8	-	-
SS E4	13	17	21	25	29	32	36	42	11	15	19	21	25	28	31	35	1.5	1.8	2.0	2.3	2.5	1.6	1.9	2.2	2.6	3.0

DULUTH, MINNESOTA

TYPE	SOLAR SAVINGS FRACTION (%)																CONSERVATION FACTOR									
BASE TEMP:	55 F (6800 DD)								65 F (9761 DD)								55 F					65 F				
LCR	100	70	50	40	30	25	20	15	100	70	50	40	30	25	20	15	100	50	30	20	15	100	50	30	20	15
WW A1	11	12	13	13	14	14	15	14	10	11	11	11	11	11	11	10	3.3	4.3	-	-	-	4.1	-	-	-	-
WW A2	8	10	12	14	16	17	18	20	7	8	10	11	12	13	14	14	2.2	2.7	3.2	4.0	-	2.5	3.2	4.1	-	-
WW A3	7	9	12	14	16	18	20	22	5	7	9	10	12	13	14	16	2.0	2.4	2.8	3.4	4.1	2.2	2.7	3.4	4.7	-
WW A4	6	8	11	13	16	18	21	24	4	6	8	10	12	13	15	17	1.9	2.2	2.6	3.1	3.7	2.1	2.6	3.1	4.1	-
WW A5	5	8	11	13	17	19	22	25	3	5	8	10	12	14	16	18	1.8	2.1	2.5	3.0	3.4	2.0	2.4	3.0	3.8	5.0
WW A6	4	7	11	13	17	19	22	26	2	5	7	9	12	14	16	18	1.8	2.1	2.4	2.8	3.2	1.9	2.3	2.8	3.6	4.6
WW B1	5	6	6	6	5	4	3	0	4	4	3	3	2	0	-1	-6	4.5	-	-	-	-	-	-	-	-	-
WW B2	7	10	14	17	21	24	28	34	5	8	11	14	17	20	23	27	1.7	1.9	2.2	2.5	2.7	1.8	2.1	2.4	2.8	3.1
WW B3	8	13	18	21	27	30	35	41	6	10	15	18	22	25	29	35	1.5	1.7	2.0	2.2	2.4	1.6	1.9	2.1	2.4	2.7
WW B4	8	14	21	26	33	37	44	52	6	11	17	22	28	33	39	46	1.3	1.5	1.7	1.9	2.0	1.4	1.6	1.8	2.0	2.2
WW B5	8	14	21	26	34	39	46	55	6	12	18	23	30	35	41	50	1.3	1.5	1.6	1.8	2.0	1.4	1.5	1.7	1.9	2.1
WW C1	10	15	20	24	30	34	39	46	8	12	17	20	26	29	34	40	1.4	1.6	1.9	2.1	2.3	1.5	1.7	2.0	2.3	2.5
WW C2	9	14	20	24	31	35	41	49	7	11	17	21	27	31	36	43	1.4	1.6	1.8	2.0	2.1	1.5	1.7	1.9	2.1	2.3
WW C3	11	19	27	34	43	49	56	66	9	16	24	30	39	44	52	61	1.2	1.3	1.5	1.7	1.8	1.2	1.4	1.5	1.7	1.9
WW C4	10	17	25	32	40	46	54	63	8	14	22	28	36	42	49	59	1.2	1.4	1.5	1.7	1.8	1.2	1.4	1.6	1.7	1.9
TW A1	11	12	13	14	15	15	16	16	10	11	12	12	12	12	12	12	3.1	3.8	4.9	-	-	3.7	4.9	-	-	-
TW A2	9	11	12	14	16	17	18	20	8	9	10	11	13	13	14	15	2.3	2.8	3.3	3.9	4.7	2.6	3.2	4.0	-	-
TW A3	7	9	12	13	16	18	20	22	6	8	9	11	12	14	15	16	2.1	2.5	2.9	3.4	4.0	2.4	2.8	3.4	4.3	-
TW A4	6	8	11	13	16	18	20	24	4	6	9	10	12	14	16	18	2.0	2.3	2.7	3.1	3.5	2.2	2.6	3.1	3.7	4.4
TW B1	10	12	13	14	15	16	17	18	9	10	11	12	13	13	14	14	2.8	3.3	3.9	4.8	-	3.2	3.9	-	-	-
TW B2	8	10	12	13	16	17	19	22	7	8	10	11	13	14	15	17	2.3	2.6	3.0	3.5	3.9	2.5	2.9	3.5	4.2	-
TW B3	7	9	11	13	16	17	20	23	6	7	9	11	13	14	16	18	2.2	2.5	2.8	3.2	3.6	2.4	2.8	3.2	3.8	4.4
TW B4	7	9	11	13	16	17	20	23	5	7	9	11	13	14	16	19	2.2	2.5	2.8	3.1	3.4	2.4	2.7	3.1	3.5	4.0
TW C1	9	11	12	14	15	17	18	20	8	9	11	11	13	14	15	16	2.5	2.9	3.3	3.8	4.3	2.8	3.3	3.9	4.6	-
TW C2	8	10	12	13	16	17	19	22	7	8	10	11	13	14	16	18	2.3	2.6	2.9	3.3	3.6	2.5	2.9	3.3	3.8	4.3
TW C3	8	9	12	13	15	17	19	22	7	8	10	11	13	14	16	19	2.3	2.6	2.9	3.2	3.5	2.5	2.9	3.3	3.7	4.1
TW C4	9	10	12	13	16	17	19	22	8	9	11	12	14	15	17	19	2.5	2.8	3.0	3.3	3.6	2.7	3.0	3.3	3.7	4.0
TW D1	4	5	5	6	6	5	5	3	3	3	3	3	3	2	1	-1	3.9	-	-	-	-	-	-	-	-	-
TW D2	8	11	14	17	21	23	27	32	6	9	12	14	17	19	22	26	1.8	2.0	2.3	2.6	2.8	2.0	2.2	2.5	2.9	3.2
TW D3	8	11	15	18	22	25	29	34	6	9	12	15	18	21	24	28	1.7	1.9	2.2	2.5	2.7	1.8	2.1	2.4	2.8	3.1
TW D4	9	14	19	23	29	33	39	46	7	11	16	20	25	29	34	41	1.5	1.6	1.8	2.0	2.2	1.5	1.7	1.9	2.2	2.3
TW D5	9	14	20	25	31	36	42	50	7	12	17	22	28	32	38	46	1.4	1.6	1.7	1.9	2.1	1.5	1.6	1.8	2.0	2.2
TW E1	10	14	19	22	27	31	35	41	8	12	16	19	23	26	30	35	1.5	1.8	2.0	2.2	2.4	1.6	1.9	2.2	2.4	2.7
TW E2	10	14	19	23	29	33	38	45	8	12	16	20	25	28	33	39	1.5	1.7	1.9	2.1	2.3	1.6	1.8	2.0	2.2	2.4
TW E3	12	18	25	31	39	44	51	59	10	16	22	28	35	40	46	54	1.3	1.4	1.6	1.8	1.9	1.3	1.5	1.7	1.9	2.0
TW E4	11	17	24	30	38	43	50	59	9	15	21	27	34	39	46	54	1.3	1.4	1.6	1.8	1.9	1.3	1.5	1.7	1.8	2.0
TW F1	9	10	11	12	13	14	14	15	8	9	10	10	10	11	11	10	3.0	3.7	4.6	-	-	3.5	4.7	-	-	-
TW F2	7	8	10	11	13	14	16	18	5	7	8	9	10	11	12	13	2.4	2.8	3.3	4.0	4.8	2.7	3.3	4.1	-	-
TW F3	5	7	9	11	13	15	17	19	3	5	7	8	10	11	12	14	2.2	2.6	3.0	3.5	4.0	2.5	2.9	3.5	4.4	-
TW F4	3	5	7	9	12	14	16	20	1	3	5	7	9	10	12	14	2.1	2.5	2.8	3.2	3.6	2.4	2.8	3.2	3.8	4.4
TW G1	7	8	10	11	12	13	14	15	6	7	8	9	9	10	11	11	2.8	3.3	3.9	4.8	-	3.2	3.9	-	-	-
TW G2	4	6	8	9	12	13	15	17	3	4	6	7	9	10	11	12	2.4	2.7	3.2	3.7	4.2	2.6	3.1	3.7	4.5	-
TW G3	3	5	7	8	11	12	15	18	2	3	5	6	8	9	11	13	2.3	2.6	3.0	3.4	3.8	2.5	2.9	3.4	4.0	4.6
TW G4	2	3	5	7	9	11	14	17	0	2	4	5	7	8	10	13	2.4	2.7	3.0	3.3	3.6	2.6	2.9	3.3	3.7	4.1
TW H1	4	6	7	8	10	11	13	15	3	4	6	6	8	9	10	11	2.7	3.1	3.5	4.1	4.6	3.0	3.5	4.2	-	-
TW H2	2	4	6	7	9	11	12	15	1	2	4	5	7	8	9	11	2.5	2.8	3.2	3.6	3.9	2.8	3.2	3.6	4.1	4.6
TW H3	1	3	5	6	8	10	12	15	0	2	3	4	6	7	9	11	2.6	2.9	3.2	3.5	3.8	2.8	3.2	3.5	3.9	4.3
TW H4	1	2	4	5	7	8	10	13	0	1	2	3	5	6	8	10	2.8	3.1	3.4	3.7	3.9	3.1	3.4	3.7	4.0	4.4
TW I1	3	3	4	4	4	4	3	1	2	2	2	2	1	0	0	-3	4.2	-	-	-	-	-	-	-	-	-
TW I2	5	8	12	14	18	20	24	29	4	6	9	11	14	17	19	23	1.9	2.1	2.4	2.6	2.9	2.0	2.3	2.6	2.9	3.2
TW I3	6	9	13	16	20	23	27	32	4	7	10	13	17	19	23	27	1.7	2.0	2.2	2.5	2.7	1.8	2.1	2.4	2.7	3.0
TW I4	7	11	16	20	26	31	36	44	5	9	14	17	23	27	32	39	1.5	1.7	1.9	2.1	2.2	1.6	1.8	2.0	2.2	2.4
TW I5	7	12	18	22	29	33	39	48	5	10	15	19	25	29	35	43	1.4	1.6	1.8	2.0	2.1	1.5	1.7	1.9	2.0	2.2
TW J1	8	12	17	20	25	29	33	40	6	10	14	17	22	25	29	34	1.6	1.8	2.0	2.2	2.4	1.7	1.9	2.1	2.4	2.6
TW J2	7	12	17	21	27	31	36	43	6	10	14	18	23	26	31	38	1.5	1.7	1.9	2.1	2.3	1.6	1.8	2.0	2.2	2.4
TW J3	9	16	23	29	37	42	49	58	7	13	20	26	33	38	45	54	1.3	1.4	1.6	1.8	1.9	1.3	1.5	1.7	1.8	2.0
TW J4	9	15	22	28	35	41	48	57	7	12	19	24	32	37	44	53	1.3	1.4	1.6	1.8	1.9	1.3	1.5	1.7	1.8	2.0
DG A1	4	4	5	5	5	4	4	1	2	3	3	2	2	1	0	-3	3.4	-	-	-	-	4.6	-	-	-	-
DG A2	7	9	11	13	16	17	19	22	5	7	9	11	13	14	16	18	2.1	2.4	2.8	3.4	4.0	2.3	2.7	3.2	3.9	4.8
DG A3	10	13	18	21	26	29	34	40	9	12	15	18	23	26	30	36	1.7	1.8	2.0	2.2	2.4	1.8	1.9	2.1	2.3	2.6
DG B1	3	4	5	5	5	5	4	3	2	2	3	2	2	1	0	-1	3.4	-	-	-	-	4.7	-	-	-	-
DG B2	7	9	11	13	16	18	21	24	5	7	9	11	13	15	17	19	2.1	2.4	2.7	3.1	3.6	2.3	2.6	3.0	3.6	4.2
DG B3	10	14	18	22	27	31	36	43	9	12	16	19	24	28	33	39	1.6	1.8	1.9	2.1	2.2	1.7	1.9	2.0	2.2	2.4
DG C1	5	6	8	9	10	10	10	10	4	4	5	6	6	6	6	5	2.6	3.4	-	-	-	3.2	4.5	-	-	-
DG C2	8	11	14	17	21	23	26	31	7	9	12	14	17	20	22	26	1.8	2.1	2.4	2.7	3.0	2.0	2.3	2.5	2.9	3.3
DG C3	12	16	21	25	31	35	41	49	10	14	19	22	28	32	37	44	1.5	1.7	1.8	2.0	2.1	1.6	1.7	1.9	2.1	2.2
SS A1	12	14	16	18	20	22	24	26	11	13	15	17	18	20	21	22	2.0	2.5	2.9	3.6	4.3	2.1	2.6	3.3	4.3	-
SS A2	15	19	23	27	32	35	40	46	14	18	22	25	30	33	37	42	1.5	1.7	1.9	2.2	2.4	1.5	1.8	2.0	2.3	2.6
SS A3	11	12	14	15	16	17	17	18	10	11	13	13	14	15	15	14	2.4	3.1	4.2	-	-	2.5	3.4	-	-	-
SS A4	14	18	22	25	30	33	36	42	13	17	21	23	28	30	34	38	1.5	1.8	2.1	2.3	2.6	1.6	1.9	2.2	2.5	2.8
SS A5	13	14	16	17	19	20	21	22	12	14	15	16	17	18	19	19	2.3	2.9	3.6	4.8	-	2.5	3.2	4.2	-	-
SS A6	14	18	23	26	31	34	38	44	13	17	21	24	28	31	35	40	1.5	1.8	2.0	2.3	2.5	1.6	1.8	2.1	2.4	2.7
SS A7	12	13	14	14	15	15	15	15	11	12	13	13	14	14	13	12	2.9	4.0	-	-	-	3.1	4.6	-	-	-
SS A8	13	17	21	24	28	30	34	39	13	16	19	22	26	28	31	35	1.6	1.9	2.2	2.5	2.7	1.7	2.0	2.3	2.7	3.1
SS B1	10	11	13	15	17	18	20	21	9	10	12	13	15	16	17	18	2.2	2.7	3.2	3.9	4.8	2.4	2.9	3.6	4.8	-
SS B2	13	16	21	24	29	32	37	43	12	15	19	22	27	30	34	40	1.6	1.8	2.0	2.2	2.4	1.6	1.8	2.1	2.3	2.5
SS B3	9	10	11	12	14	15	15	16	8	9	10	11	12	12	13	13	2.5	3.1	4.1	-	-	2.7	3.5	-	-	-
SS B4	12	15	20	23	27	30	34	40	11	14	18	21	25	28	32	37	1.6	1.8	2.1	2.3	2.5	1.7	1.9	2.2	2.5	2.7
SS B5	10	12	13	14	15	16	17	18	10	11	12	13	14	14	15	15	2.7	3.2	4.0	-	-	2.8	3.6	4.8	-	-
SS B6	12	15	20	23	27	31	35	41	11	14	18	21	25	28	32	37	1.6	1.8	2.1	2.3	2.5	1.7	1.9	2.2	2.4	2.7
SS B7	10	10	11	12	12	13	13	13	9	10	10	11	11	11	11	11	3.2	4.1	-	-	-	3.4	4.7	-	-	-
SS B8	11	14	18	21	25	28	31	36	10	13	17	19	23	25	29	33	1.7	1.9	2.2	2.5	2.7	1.7	2.0	2.3	2.6	2.9
SS C1	6	7	9	11	12	14	15	16	4	5	7	8	9	9	10	10	2.3	2.8	3.5	4.6	-	2.6	3.4	4.7	-	-
SS C2	8	11	15	18	23	26	30	35	6	9	12	15	19	21	25	29	1.7	1.9	2.1	2.4	2.6	1.8	2.1	2.3	2.7	3.0
SS C3	7	9	10	11	13	14	15	16	6	7	8	9	10	11	11	12	2.7	3.2	3.8	4.7	-	3.0	3.8	4.9	-	-
SS C4	8	11	14	17	21	23	27	32	6	9	12	14	17	20	23	27	1.8	2.0	2.3	2.6	2.8	1.9	2.2	2.5	2.8	3.1
SS D1	12	14	17	19	22	24	26	28	11	13	16	17	19	21	22	23	1.8	2.2	2.7	3.4	4.2	1.9	2.4	3.1	4.4	-
SS D2	15	20	26	30	37	41	46	53	13	19	24	28	34	38	43	49	1.3	1.5	1.7	2.0	2.1	1.4	1.6	1.8	2.1	2.3
SS D3	13	15	17	19	21	22	24	26	12	14	16	17	19	20	21	22	2.0	2.4	3.0	3.8	4.7	2.1	2.6	3.4	4.8	-
SS D4	15	20	25	29	35	39	44	51	13	18	23	27	33	36	41	47	1.4	1.6	1.8	2.0	2.2	1.4	1.6	1.9	2.1	2.4
SS E1	9	11	13	14	16	17	18	18	7	9	11	12	13	13	14	13	2.1	2.7	3.6	-	-	2.3	3.1	4.6	-	-
SS E2	11	16	20	24	29	33	37	43	10	14	18	22	26	29	33	38	1.5	1.7	2.0	2.2	2.4	1.5	1.8	2.1	2.4	2.7
SS E3	10	12	13	14	16	17	18	19	9	11	12	13	14	15	15	15	2.4	3.0	3.8	-	-	2.6	3.3	4.6	-	-
SS E4	11	15	20	23	27	31	35	40	10	14	18	21	25	28	31	36	1.6	1.8	2.1	2.3	2.5	1.6	1.9	2.2	2.5	2.8

INTERNATIONAL FALLS, MINNESOTA

SOLAR SAVINGS FRACTION (%) / CONSERVATION FACTOR

TYPE	SSF BASE TEMP: 55 F (7629 DD)								SSF 65 F (10553 DD)								CF 55 F					CF 65 F				
LCR	100	70	50	40	30	25	20	15	100	70	50	40	30	25	20	15	100	50	30	20	15	100	50	30	20	15
WW A1	11	11	12	12	13	12	12	11	10	10	11	11	11	10	10	8	3.7	-	-	-	-	4.5	-	-	-	-
WW A2	8	9	11	12	13	14	15	15	6	8	9	10	11	11	11	11	2.4	3.0	4.1	-	-	2.7	3.6	-	-	-
WW A3	6	8	10	11	13	14	15	16	5	6	8	9	10	11	12	11	2.1	2.7	3.5	4.8	-	2.4	3.1	4.2	-	-
WW A4	5	7	9	11	13	15	16	18	3	5	7	8	10	11	12	12	2.0	2.5	3.2	4.2	-	2.2	2.8	3.7	-	-
WW A5	4	6	9	11	13	15	17	18	2	4	6	8	10	11	12	13	1.9	2.4	3.0	3.9	-	2.1	2.7	3.5	-	-
WW A6	3	6	9	11	13	15	17	19	1	4	6	8	10	11	12	13	1.9	2.3	2.9	3.6	4.7	2.0	2.6	3.3	4.9	-
WW B1	4	5	4	4	3	1	0	-5	4	3	3	2	0	-1	-4	-10	-	-	-	-	-	-	-	-	-	-
WW B2	6	9	12	15	19	21	24	28	4	7	10	12	16	18	20	23	1.8	2.1	2.4	2.8	3.1	1.9	2.2	2.6	3.1	3.6
WW B3	7	11	15	19	23	26	30	35	6	9	13	16	20	23	26	30	1.6	1.8	2.1	2.5	2.7	1.7	2.0	2.3	2.7	3.1
WW B4	7	13	18	23	29	34	39	47	6	11	16	20	26	30	36	43	1.4	1.6	1.8	2.0	2.2	1.4	1.7	1.9	2.1	2.3
WW B5	7	13	19	24	31	36	42	50	5	11	17	21	28	33	39	46	1.3	1.5	1.7	1.9	2.1	1.4	1.6	1.8	2.0	2.2
WW C1	9	13	18	21	26	30	34	40	7	11	15	19	23	26	30	35	1.5	1.7	2.0	2.3	2.5	1.6	1.8	2.1	2.4	2.8
WW C2	8	12	18	22	27	31	37	43	6	10	15	19	24	28	33	39	1.4	1.7	1.9	2.1	2.3	1.5	1.7	2.0	2.2	2.5
WW C3	10	17	25	31	39	45	52	61	8	15	23	28	37	42	49	58	1.2	1.4	1.6	1.7	1.9	1.2	1.4	1.6	1.8	1.9
WW C4	9	15	23	29	37	43	50	59	7	13	21	26	34	40	47	56	1.2	1.4	1.6	1.8	1.9	1.3	1.4	1.6	1.8	2.0
TW A1	11	12	12	13	13	14	14	13	10	11	11	11	12	11	11	10	3.4	4.4	-	-	-	4.0	-	-	-	-
TW A2	8	10	11	12	14	14	15	16	7	8	10	10	11	12	12	12	2.5	3.1	3.9	-	-	2.8	3.6	4.9	-	-
TW A3	7	8	10	12	14	15	16	17	6	7	9	10	11	12	13	13	2.3	2.8	3.4	4.4	-	2.5	3.1	4.0	-	-
TW A4	5	7	10	11	13	15	17	19	4	6	8	9	11	12	13	15	2.1	2.6	3.1	3.7	4.5	2.3	2.8	3.5	4.5	-
TW B1	10	11	12	13	14	14	15	15	9	10	11	11	12	12	12	12	3.0	3.7	4.8	-	-	3.3	4.4	-	-	-
TW B2	7	9	11	12	14	15	16	18	6	8	9	10	12	12	13	14	2.4	2.8	3.4	4.2	-	2.6	3.2	3.9	-	-
TW B3	6	8	10	12	14	15	17	19	5	7	9	10	11	13	14	15	2.3	2.7	3.2	3.8	4.5	2.5	2.9	3.6	4.4	-
TW B4	6	8	10	12	14	15	17	20	5	7	9	10	12	13	15	17	2.3	2.6	3.0	3.5	4.0	2.4	2.9	3.4	4.0	4.7
TW C1	9	10	11	12	14	15	16	17	8	9	10	11	12	13	14	15	2.7	3.1	3.7	4.5	-	2.9	3.5	4.3	-	-
TW C2	7	9	11	12	14	15	17	19	6	8	9	10	12	13	15	16	2.4	2.8	3.2	3.8	4.3	2.6	3.1	3.6	4.2	-
TW C3	7	9	11	12	14	15	17	20	7	8	9	11	12	13	15	17	2.5	2.8	3.2	3.6	4.1	2.6	3.0	3.5	4.0	4.6
TW C4	8	10	11	13	14	16	18	20	8	9	10	11	13	14	16	18	2.6	2.9	3.2	3.6	4.0	2.8	3.1	3.5	4.0	4.4
TW D1	4	4	4	4	4	3	2	0	3	3	3	3	2	1	0	-4	4.8	-	-	-	-	-	-	-	-	-
TW D2	7	10	13	15	18	21	23	27	6	8	11	13	16	18	20	23	1.9	2.2	2.5	2.9	3.3	2.0	2.4	2.7	3.2	3.7
TW D3	7	10	13	16	19	22	25	28	6	8	11	13	17	19	21	24	1.8	2.1	2.4	2.8	3.2	1.9	2.2	2.6	3.1	3.6
TW D4	8	12	17	21	26	30	35	42	7	11	15	19	24	27	32	38	1.5	1.7	1.9	2.2	2.3	1.6	1.8	2.0	2.3	2.5
TW D5	8	13	18	23	29	33	39	46	7	11	16	20	26	30	36	43	1.4	1.6	1.8	2.0	2.2	1.5	1.7	1.9	2.1	2.3
TW E1	9	13	17	20	24	27	31	36	8	11	15	17	21	24	27	31	1.6	1.9	2.2	2.5	2.7	1.7	2.0	2.3	2.7	3.0
TW E2	9	13	17	21	26	29	34	40	7	11	15	18	23	26	30	36	1.5	1.8	2.0	2.3	2.5	1.6	1.9	2.1	2.4	2.6
TW E3	11	17	23	28	35	40	46	55	9	15	21	26	33	37	43	51	1.3	1.5	1.7	1.9	2.0	1.3	1.5	1.7	1.9	2.1
TW E4	10	16	22	27	35	40	46	55	8	14	20	25	32	37	43	51	1.3	1.5	1.7	1.8	2.0	1.3	1.5	1.7	1.9	2.1
TW F1	9	10	10	11	12	12	12	12	8	9	9	9	10	10	9	8	3.3	4.3	-	-	-	3.8	-	-	-	-
TW F2	6	7	9	10	11	12	13	14	5	6	7	8	9	10	10	10	2.6	3.1	4.0	-	-	2.9	3.6	5.0	-	-
TW F3	4	6	8	9	11	12	13	15	3	4	6	7	8	9	10	11	2.4	2.8	3.5	4.4	-	2.6	3.2	4.1	-	-
TW F4	2	4	6	8	10	11	13	15	1	3	5	6	8	9	10	12	2.3	2.7	3.2	3.8	4.5	2.5	2.9	3.5	4.4	-
TW G1	7	8	9	10	11	11	12	12	6	7	7	8	9	9	9	9	3.0	3.7	4.7	-	-	3.4	4.4	-	-	-
TW G2	4	5	7	8	10	11	12	14	3	4	5	6	8	8	9	10	2.5	3.0	3.6	4.5	-	2.8	3.4	4.2	-	-
TW G3	2	4	6	7	9	10	12	14	1	3	4	5	7	8	9	11	2.4	2.8	3.3	3.9	4.6	2.6	3.1	3.7	4.6	-
TW G4	1	3	5	6	8	9	11	14	0	2	3	4	6	7	9	11	2.5	2.8	3.2	3.7	4.1	2.7	3.1	3.5	4.1	4.7
TW H1	4	5	6	7	9	10	11	12	3	4	5	6	7	8	8	9	2.8	3.3	4.0	5.0	-	3.1	3.8	4.7	-	-
TW H2	2	3	5	6	8	9	10	12	1	2	3	4	6	7	8	10	2.7	3.0	3.5	4.1	4.6	2.9	3.3	3.9	4.6	-
TW H3	1	2	4	5	7	8	10	12	0	1	3	4	5	6	8	10	2.7	3.0	3.4	3.9	4.3	2.9	3.3	3.8	4.3	4.8
TW H4	0	2	3	4	6	7	9	11	0	1	2	3	5	6	7	9	3.0	3.3	3.6	3.9	4.3	3.2	3.5	3.9	4.3	4.7
TW I1	2	2	3	3	2	1	0	-2	1	1	1	1	0	0	-2	-5	-	-	-	-	-	-	-	-	-	-
TW I2	5	7	10	12	16	18	21	24	4	6	8	10	13	15	17	20	1.9	2.2	2.6	2.9	3.3	2.1	2.4	2.8	3.2	3.7
TW I3	5	8	11	14	17	20	23	28	4	6	9	12	15	17	20	24	1.8	2.1	2.4	2.7	3.0	1.9	2.2	2.5	2.9	3.3
TW I4	6	10	15	18	24	28	32	39	4	8	13	16	21	25	29	36	1.5	1.8	2.0	2.2	2.4	1.6	1.8	2.1	2.3	2.5
TW I5	6	11	16	20	26	30	36	44	5	9	14	18	24	28	33	40	1.5	1.7	1.9	2.0	2.2	1.5	1.7	1.9	2.1	2.3
TW J1	7	11	15	18	22	25	29	35	6	9	13	16	20	22	26	31	1.6	1.9	2.1	2.4	2.7	1.7	2.0	2.3	2.6	2.9
TW J2	6	11	15	19	24	27	32	38	5	9	13	16	21	24	29	34	1.6	1.8	2.0	2.2	2.4	1.6	1.9	2.1	2.4	2.6
TW J3	8	14	21	26	34	39	45	54	7	13	19	24	31	36	42	51	1.3	1.5	1.7	1.8	2.0	1.3	1.5	1.7	1.9	2.1
TW J4	7	13	20	25	33	38	44	53	6	12	18	23	30	35	41	50	1.3	1.5	1.7	1.9	2.0	1.4	1.5	1.7	1.9	2.1
DG A1	3	3	4	4	3	2	1	-1	2	2	2	2	1	0	-1	-5	3.9	-	-	-	-	-	-	-	-	-
DG A2	6	8	10	12	14	15	17	19	5	7	9	10	12	13	14	16	2.2	2.6	3.1	3.9	4.9	2.4	2.8	3.4	4.4	-
DG A3	9	13	17	20	24	27	31	37	8	11	15	18	22	25	29	34	1.7	1.9	2.1	2.4	2.6	1.8	2.0	2.2	2.5	2.7
DG B1	3	3	4	4	3	3	2	0	2	2	2	2	1	0	0	-4	3.9	-	-	-	-	-	-	-	-	-
DG B2	6	8	10	12	15	16	18	20	5	7	9	10	12	14	15	17	2.2	2.5	3.0	3.6	4.4	2.4	2.7	3.2	4.0	5.0
DG B3	10	13	17	21	25	29	33	40	9	12	16	19	23	27	31	37	1.7	1.8	2.0	2.2	2.4	1.8	1.9	2.1	2.3	2.5
DG C1	4	6	7	7	8	8	7	6	3	4	5	5	5	5	4	2	2.8	3.9	-	-	-	3.5	-	-	-	-
DG C2	8	10	13	16	19	21	23	27	7	9	11	14	16	18	21	24	1.9	2.2	2.6	3.0	3.4	2.1	2.3	2.6	3.2	3.7
DG C3	11	15	20	24	29	33	38	45	10	14	18	22	27	31	35	42	1.6	1.7	1.9	2.1	2.2	1.7	1.8	1.9	2.2	2.4
SS A1	11	13	15	16	18	18	20	20	11	12	14	15	16	17	18	18	2.2	2.8	3.6	4.9	-	2.3	2.9	3.9	-	-
SS A2	13	17	21	24	29	32	35	40	13	16	20	23	27	30	33	38	1.6	1.8	2.1	2.4	2.6	1.6	1.9	2.2	2.5	2.8
SS A3	10	11	12	13	13	13	13	12	9	10	11	12	12	12	12	10	2.7	3.8	-	-	-	2.8	4.2	-	-	-
SS A4	13	16	20	22	26	29	32	36	12	15	19	21	25	27	30	34	1.6	2.0	2.3	2.6	2.9	1.7	2.0	2.4	2.8	3.2
SS A5	12	13	15	15	16	17	17	17	12	13	14	15	15	16	16	15	2.6	3.4	4.8	-	-	2.7	3.6	-	-	-
SS A6	13	16	20	23	27	30	33	38	12	16	19	22	26	28	32	36	1.6	1.9	2.2	2.5	2.8	1.6	1.9	2.3	2.7	3.0
SS A7	11	12	12	13	13	13	12	10	11	12	12	12	12	12	11	9	3.4	-	-	-	-	3.6	-	-	-	-
SS A8	12	15	18	21	24	26	29	33	12	15	18	20	23	25	27	30	1.7	2.1	2.4	2.9	3.2	1.8	2.1	2.6	3.1	3.7
SS B1	9	10	12	13	14	15	16	17	8	10	11	12	13	14	14	15	2.4	3.0	3.9	-	-	2.5	3.3	4.4	-	-
SS B2	12	15	19	22	26	29	33	38	11	14	18	21	25	28	31	36	1.6	1.9	2.1	2.4	2.6	1.7	1.9	2.2	2.5	2.7
SS B3	8	9	10	11	11	12	12	11	7	8	9	10	10	10	10	9	2.8	3.8	-	-	-	2.9	4.2	-	-	-
SS B4	11	14	18	20	24	27	30	35	10	13	17	19	23	25	29	33	1.7	2.0	2.3	2.5	2.8	1.7	2.0	2.3	2.7	3.0
SS B5	10	11	12	13	13	14	14	14	9	10	11	12	12	13	13	12	2.9	3.7	-	-	-	3.1	4.2	-	-	-
SS B6	11	14	18	20	24	27	31	36	10	13	17	19	23	26	29	33	1.7	1.9	2.2	2.5	2.7	1.7	2.0	2.3	2.6	2.9
SS B7	9	10	10	10	11	11	10	9	9	9	10	10	10	10	9	8	3.6	-	-	-	-	3.8	-	-	-	-
SS B8	10	13	16	18	22	24	27	31	9	12	15	17	21	23	26	29	1.8	2.1	2.4	2.7	3.1	1.8	2.1	2.5	2.9	3.3
SS C1	5	6	8	9	10	10	11	11	4	5	6	7	7	7	7	6	2.5	3.3	4.7	-	-	2.8	3.9	-	-	-
SS C2	7	10	13	16	20	22	26	30	5	8	11	14	17	19	22	26	1.7	2.0	2.3	2.7	3.0	1.9	2.2	2.5	2.9	3.4
SS C3	7	8	9	10	11	12	12	13	6	7	8	8	9	9	10	9	2.9	3.5	4.6	-	-	3.2	4.2	-	-	-
SS C4	7	10	13	15	18	21	24	27	6	8	11	13	16	18	20	24	1.9	2.2	2.5	2.9	3.2	2.0	2.3	2.7	3.1	3.5
SS D1	10	13	15	16	18	19	20	21	10	12	14	15	17	17	18	17	2.0	2.6	3.4	-	-	2.1	2.7	4.0	-	-
SS D2	13	18	23	27	32	36	41	47	12	17	22	26	31	34	38	44	1.4	1.6	1.9	2.1	2.3	1.4	1.7	2.0	2.3	2.5
SS D3	12	13	15	16	18	19	19	20	11	13	14	15	17	17	18	17	2.2	2.8	3.8	-	-	2.3	3.0	4.3	-	-
SS D4	13	18	23	26	31	35	39	45	12	17	21	25	30	33	37	42	1.4	1.7	1.9	2.2	2.4	1.5	1.7	2.0	2.3	2.6
SS E1	7	9	10	11	12	12	12	11	7	8	9	10	10	10	10	7	2.3	3.3	-	-	-	2.5	3.7	-	-	-
SS E2	10	14	18	21	25	28	32	37	9	13	16	19	23	26	29	33	1.6	1.9	2.2	2.5	2.8	1.6	1.9	2.3	2.7	3.1
SS E3	9	11	12	13	13	14	14	14	9	10	11	12	12	12	12	11	2.6	3.5	-	-	-	2.8	3.9	-	-	-
SS E4	10	14	17	20	24	27	30	34	9	13	16	19	22	25	28	31	1.6	1.9	2.3	2.6	2.9	1.7	2.0	2.4	2.8	3.2

MINNEAPOLIS, MINNESOTA

TYPE	SOLAR SAVINGS FRACTION (%) BASE TEMP: 55 F (5693 DD) LCR								65 F (8165 DD)								CONSERVATION FACTOR 55 F					65 F				
	100	70	50	40	30	25	20	15	100	70	50	40	30	25	20	15	100	50	30	20	15	100	50	30	20	15
WW A1	13	14	15	16	18	19	20	22	11	12	13	14	14	15	15	15	2.6	3.1	3.7	4.4	-	3.2	4.0	-	-	-
WW A2	10	13	16	18	21	24	26	30	8	10	12	14	16	18	19	22	1.9	2.2	2.5	2.8	3.1	2.2	2.6	3.1	3.7	4.4
WW A3	9	12	16	19	23	25	29	34	7	9	12	14	17	19	21	24	1.7	2.0	2.3	2.5	2.7	2.0	2.3	2.7	3.2	3.6
WW A4	8	12	16	19	23	26	30	36	6	9	12	14	17	19	22	26	1.6	1.9	2.1	2.4	2.6	1.9	2.2	2.5	2.9	3.3
WW A5	8	12	16	19	24	28	32	38	5	8	11	14	18	20	23	27	1.6	1.8	2.0	2.3	2.5	1.8	2.1	2.4	2.8	3.1
WW A6	7	11	16	20	25	28	33	39	4	8	11	14	18	21	24	28	1.5	1.8	2.0	2.2	2.4	1.7	2.0	2.3	2.7	3.0
WW B1	7	8	9	10	10	11	11	10	5	6	6	6	6	5	4	1	2.8	3.9	-	-	-	4.1	-	-	-	-
WW B2	9	14	19	22	28	32	37	44	7	10	15	18	22	25	29	35	1.5	1.7	1.9	2.1	2.3	1.7	1.9	2.1	2.4	2.6
WW B3	11	17	23	28	34	39	45	53	9	13	18	22	28	32	37	43	1.3	1.5	1.7	1.9	2.1	1.5	1.7	1.9	2.1	2.3
WW B4	12	18	26	31	39	45	52	61	9	14	21	26	33	38	45	54	1.2	1.4	1.6	1.7	1.9	1.3	1.5	1.7	1.9	2.0
WW B5	11	18	26	32	40	46	53	63	8	14	21	27	35	40	47	56	1.2	1.4	1.5	1.7	1.8	1.3	1.5	1.6	1.8	1.9
WW C1	13	19	25	30	37	42	48	57	10	15	21	25	31	35	41	48	1.3	1.5	1.7	1.8	2.0	1.4	1.6	1.8	2.0	2.2
WW C2	12	18	25	30	38	43	49	58	9	14	20	25	32	36	42	50	1.3	1.4	1.6	1.8	1.9	1.4	1.6	1.7	1.9	2.1
WW C3	15	23	33	40	49	56	64	73	12	19	28	35	44	50	57	67	1.1	1.2	1.4	1.6	1.7	1.2	1.3	1.5	1.6	1.8
WW C4	13	21	30	37	47	53	61	71	10	17	26	32	41	47	55	65	1.1	1.3	1.4	1.6	1.7	1.2	1.3	1.5	1.7	1.8
TW A1	13	14	16	17	18	20	21	23	11	12	13	14	15	16	16	17	2.5	2.9	3.4	3.9	4.5	3.0	3.7	4.5	-	-
TW A2	11	13	16	18	21	23	25	29	9	11	13	14	16	18	19	21	2.0	2.3	2.6	2.9	3.2	2.3	2.7	3.1	3.7	4.3
TW A3	9	12	15	18	21	24	27	32	7	10	12	14	16	18	20	23	1.8	2.1	2.4	2.6	2.9	2.1	2.4	2.8	3.2	3.6
TW A4	8	11	15	17	21	24	28	33	6	8	11	13	16	19	21	25	1.8	2.0	2.3	2.5	2.7	2.0	2.3	2.6	3.0	3.3
TW B1	12	13	15	17	19	20	22	25	10	12	13	14	15	16	18	19	2.3	2.6	3.0	3.4	3.7	2.7	3.2	3.7	4.5	-
TW B2	10	12	15	17	20	23	26	30	8	10	12	14	16	18	20	23	2.0	2.2	2.5	2.7	3.0	2.2	2.5	2.9	3.3	3.7
TW B3	9	11	14	17	20	23	26	31	7	9	12	13	16	18	21	24	1.9	2.1	2.4	2.6	2.8	2.1	2.4	2.7	3.1	3.4
TW B4	8	11	14	16	20	22	26	31	7	9	11	13	16	18	21	24	1.9	2.1	2.4	2.6	2.8	2.1	2.4	2.7	3.0	3.3
TW C1	10	12	15	16	19	21	23	27	9	11	12	14	16	17	19	21	2.2	2.4	2.7	3.0	3.3	2.5	2.8	3.2	3.6	4.0
TW C2	9	11	14	16	19	22	25	29	8	10	12	13	16	18	20	23	2.0	2.3	2.5	2.8	2.9	2.3	2.6	2.9	3.2	3.5
TW C3	9	11	14	16	19	21	24	29	8	10	12	13	16	17	20	23	2.1	2.3	2.5	2.8	3.0	2.3	2.6	2.9	3.2	3.4
TW C4	10	12	14	16	18	20	23	27	9	10	12	13	16	17	19	23	2.2	2.4	2.7	2.9	3.1	2.5	2.7	3.0	3.2	3.5
TW D1	6	7	8	9	10	11	11	12	4	5	6	6	6	6	5	4	2.8	3.5	4.7	-	-	3.8	-	-	-	-
TW D2	10	14	18	21	26	30	34	41	8	11	15	17	21	24	28	33	1.6	1.8	2.0	2.2	2.4	1.8	2.0	2.3	2.5	2.7
TW D3	10	14	19	23	28	32	37	44	8	11	15	18	23	26	30	35	1.5	1.7	1.9	2.1	2.3	1.7	1.9	2.1	2.4	2.6
TW D4	12	17	23	28	35	40	46	55	9	14	19	24	30	34	40	48	1.3	1.5	1.7	1.8	2.0	1.4	1.6	1.8	2.0	2.1
TW D5	11	17	24	29	37	42	49	58	9	14	20	25	32	37	43	52	1.3	1.5	1.6	1.8	1.9	1.4	1.6	1.7	1.9	2.0
TW E1	13	18	23	27	33	38	43	51	10	14	19	23	28	32	36	43	1.4	1.6	1.8	2.0	2.1	1.5	1.7	1.9	2.2	2.3
TW E2	12	18	24	28	35	40	46	54	10	14	20	24	29	34	39	46	1.4	1.5	1.7	1.9	2.0	1.5	1.7	1.8	2.0	2.2
TW E3	15	22	30	37	45	51	58	68	12	19	26	32	40	45	52	61	1.2	1.3	1.5	1.7	1.8	1.2	1.4	1.6	1.7	1.9
TW E4	14	21	29	35	44	50	57	67	11	17	25	30	38	44	51	60	1.2	1.3	1.5	1.7	1.8	1.3	1.4	1.6	1.7	1.9
TW F1	11	12	14	15	17	18	20	22	9	10	11	12	13	14	15	16	2.5	2.8	3.3	3.8	4.3	3.0	3.5	4.3	-	-
TW F2	8	11	13	15	18	20	23	26	7	8	10	12	14	15	17	19	2.0	2.3	2.6	3.0	3.2	2.4	2.7	3.2	3.8	4.4
TW F3	7	9	12	15	18	20	24	28	5	7	9	11	13	15	17	20	1.9	2.2	2.5	2.7	2.9	2.2	2.5	2.9	3.3	3.7
TW F4	5	8	11	13	17	20	23	28	3	5	8	10	13	15	17	21	1.9	2.1	2.4	2.6	2.8	2.1	2.4	2.7	3.1	3.4
TW G1	9	10	12	14	16	17	19	22	7	8	10	11	12	13	15	16	2.3	2.7	3.0	3.4	3.8	2.7	3.2	3.7	4.5	-
TW G2	6	8	11	13	16	18	21	25	4	6	8	10	12	13	15	18	2.1	2.3	2.6	2.9	3.1	2.3	2.7	3.1	3.5	3.9
TW G3	4	7	10	12	15	18	21	25	3	5	7	9	11	13	15	19	2.0	2.3	2.5	2.8	3.0	2.3	2.6	2.9	3.2	3.6
TW G4	3	5	8	10	13	16	19	23	2	3	6	7	10	12	14	17	2.1	2.3	2.6	2.8	3.0	2.4	2.6	2.9	3.2	3.5
TW H1	6	8	10	11	14	16	18	21	4	6	7	9	10	12	13	16	2.3	2.6	2.9	3.2	3.5	2.6	3.0	3.4	3.9	4.4
TW H2	3	5	8	10	13	15	17	21	2	4	6	7	9	11	13	16	2.2	2.4	2.7	3.0	3.2	2.5	2.8	3.1	3.4	3.8
TW H3	2	4	7	8	11	13	16	20	1	3	5	6	8	10	12	15	2.3	2.5	2.8	3.0	3.2	2.6	2.8	3.1	3.4	3.7
TW H4	2	3	5	7	9	11	13	17	1	2	4	5	7	8	10	13	2.5	2.8	3.0	3.2	3.4	2.8	3.1	3.3	3.6	3.8
TW I1	4	5	6	7	8	9	9	9	3	3	4	4	4	4	4	2	3.0	3.7	5.0	-	-	4.0	-	-	-	-
TW I2	7	11	15	18	23	27	31	37	5	8	12	14	18	21	25	30	1.7	1.9	2.1	2.3	2.4	1.8	2.1	2.3	2.6	2.8
TW I3	8	12	17	20	26	29	34	42	6	9	13	16	21	24	28	34	1.6	1.8	1.9	2.1	2.3	1.7	1.9	2.1	2.4	2.6
TW I4	9	14	21	25	32	37	43	52	7	11	17	21	27	31	37	45	1.4	1.5	1.7	1.9	2.0	1.5	1.7	1.8	2.0	2.2
TW I5	9	15	22	27	34	39	46	55	7	12	18	23	29	34	40	49	1.3	1.5	1.6	1.8	1.9	1.4	1.6	1.7	1.9	2.1
TW J1	11	15	21	25	31	36	42	49	8	12	17	21	26	30	35	41	1.4	1.6	1.8	2.0	2.1	1.5	1.7	1.9	2.2	2.3
TW J2	10	15	21	26	33	37	43	52	7	12	17	21	27	31	37	44	1.4	1.6	1.7	1.9	2.0	1.5	1.7	1.9	2.0	2.2
TW J3	12	20	28	34	43	49	56	66	10	16	24	29	38	43	50	60	1.2	1.3	1.5	1.7	1.8	1.2	1.4	1.6	1.7	1.9
TW J4	11	18	26	33	41	47	55	65	9	15	22	28	36	42	49	58	1.2	1.4	1.5	1.7	1.8	1.3	1.4	1.6	1.8	1.9
DG A1	5	7	8	9	10	10	10	9	4	4	5	5	5	5	4	2	2.6	3.4	-	-	-	3.3	-	-	-	-
DG A2	8	11	14	16	20	22	25	29	7	9	11	13	16	18	20	23	1.9	2.1	2.4	2.8	3.2	2.1	2.4	2.8	3.3	3.8
DG A3	12	16	21	24	30	34	40	47	10	13	18	21	26	30	35	41	1.5	1.7	1.8	2.0	2.2	1.7	1.8	2.0	2.2	2.4
DG B1	5	7	8	9	10	10	11	11	3	4	5	5	6	5	5	3	2.5	3.4	4.7	-	-	3.3	4.8	-	-	-
DG B2	8	11	14	17	21	23	27	32	7	9	12	14	17	19	21	25	1.8	2.1	2.3	2.5	2.8	2.1	2.3	2.7	3.0	3.4
DG B3	12	16	21	25	32	36	42	51	10	14	19	22	28	31	37	44	1.5	1.7	1.8	1.9	2.0	1.6	1.8	1.9	2.1	2.2
DG C1	7	9	11	13	15	16	18	20	5	6	8	9	10	11	11	11	2.1	2.7	3.3	3.9	4.9	2.6	3.2	4.6	-	-
DG C2	10	14	18	21	25	29	33	40	8	11	15	17	21	24	27	32	1.7	1.9	2.0	2.2	2.5	1.9	2.0	2.3	2.6	2.8
DG C3	14	19	24	29	36	41	48	57	12	16	21	25	31	36	42	50	1.4	1.6	1.6	1.8	1.9	1.5	1.6	1.8	1.9	2.0
SS A1	14	16	19	22	25	27	30	34	12	15	17	19	21	23	25	28	1.8	2.1	2.4	2.7	3.0	2.0	2.3	2.8	3.3	3.8
SS A2	17	22	27	31	37	41	47	54	15	19	24	28	33	37	42	48	1.4	1.6	1.8	1.9	2.1	1.5	1.7	1.9	2.1	2.3
SS A3	12	14	16	18	20	22	24	26	11	13	14	16	17	18	19	20	2.1	2.5	3.0	3.6	4.1	2.3	2.9	3.7	-	-
SS A4	16	21	26	29	35	39	44	50	15	18	23	26	31	34	38	44	1.4	1.6	1.8	2.0	2.2	1.5	1.8	2.0	2.2	2.5
SS A5	14	16	19	20	23	24	26	29	13	15	17	18	20	21	22	24	2.1	2.4	2.8	3.3	3.7	2.3	2.7	3.3	4.2	-
SS A6	16	21	26	30	36	40	45	52	15	19	24	27	32	35	40	46	1.4	1.6	1.8	2.0	2.1	1.5	1.7	1.9	2.2	2.4
SS A7	13	14	16	17	18	19	20	21	12	13	14	15	16	16	17	17	2.5	3.0	3.8	4.9	-	2.8	3.6	-	-	-
SS A8	16	20	24	28	33	37	41	47	14	18	22	25	29	32	36	41	1.5	1.7	1.9	2.1	2.3	1.6	1.8	2.1	2.4	2.6
SS B1	11	13	16	18	21	23	25	29	10	12	14	16	18	19	21	23	2.0	2.3	2.6	3.0	3.3	2.2	2.6	3.0	3.6	4.2
SS B2	15	19	24	28	34	38	43	50	13	17	22	25	30	34	38	45	1.4	1.6	1.8	2.0	2.1	1.5	1.7	1.9	2.1	2.3
SS B3	10	12	14	16	18	19	21	23	9	10	12	13	15	16	17	18	2.2	2.6	3.0	3.6	4.1	2.4	3.0	3.7	4.9	-
SS B4	14	18	23	26	32	36	41	47	12	16	20	24	28	32	36	42	1.5	1.7	1.9	2.1	2.2	1.6	1.8	2.0	2.3	2.4
SS B5	12	13	15	17	19	20	22	24	11	12	13	14	16	17	18	19	2.3	2.7	3.1	3.6	4.1	2.6	3.1	3.7	4.7	-
SS B6	14	18	23	27	32	36	41	48	12	16	20	24	28	32	36	42	1.5	1.7	1.9	2.0	2.2	1.6	1.8	2.0	2.2	2.4
SS B7	11	12	13	14	16	16	17	19	10	11	12	12	13	14	14	14	2.7	3.2	3.9	4.8	-	3.0	3.8	-	-	-
SS B8	13	17	21	25	30	33	38	44	11	15	19	22	26	29	33	38	1.5	1.8	2.0	2.2	2.3	1.6	1.9	2.1	2.4	2.6
SS C1	8	10	13	15	18	20	23	27	6	8	10	11	13	14	16	18	1.9	2.2	2.6	2.9	3.2	2.2	2.7	3.3	4.1	-
SS C2	10	14	19	23	29	33	38	45	8	11	15	19	23	27	31	37	1.5	1.7	1.9	2.1	2.2	1.6	1.9	2.1	2.3	2.5
SS C3	9	11	13	14	17	18	20	23	7	9	10	11	13	14	15	17	2.3	2.6	2.9	3.3	3.6	2.6	3.1	3.6	4.3	-
SS C4	10	14	18	21	26	29	34	40	8	11	15	17	21	24	28	33	1.6	1.8	2.0	2.2	2.4	1.8	2.0	2.2	2.5	2.7
SS D1	14	17	21	24	28	31	34	39	12	15	18	21	24	26	28	31	1.6	1.9	2.2	2.5	2.7	1.7	2.1	2.5	3.0	3.6
SS D2	17	24	31	36	43	48	54	63	15	21	27	32	38	43	48	56	1.2	1.4	1.6	1.8	1.9	1.3	1.5	1.7	1.9	2.1
SS D3	14	17	20	23	26	28	31	35	13	15	18	20	22	24	26	28	1.8	2.1	2.4	2.7	3.0	1.9	2.3	2.8	3.4	4.0
SS D4	17	23	30	35	42	46	52	60	15	21	26	31	37	41	47	54	1.3	1.4	1.6	1.8	2.0	1.3	1.5	1.7	2.0	2.1
SS E1	11	13	16	19	22	24	26	30	9	11	13	15	17	18	20	21	1.8	2.2	2.6	3.0	3.4	2.0	2.5	3.2	4.3	-
SS E2	14	19	25	29	36	40	46	53	12	16	21	25	31	34	39	46	1.3	1.5	1.7	1.9	2.1	1.4	1.7	1.9	2.1	2.3
SS E3	12	14	16	18	20	22	24	26	10	12	14	15	17	18	19	20	2.1	2.4	2.9	3.3	3.7	2.3	2.8	3.5	4.4	-
SS E4	14	18	23	27	33	37	42	49	12	16	20	24	29	32	36	42	1.4	1.6	1.8	2.0	2.2	1.5	1.8	2.0	2.2	2.4

ROCHESTER, MINNESOTA

SOLAR SAVINGS FRACTION (%)

TYPE	BASE TEMP: 55 F (5692 DD) LCR 100	70	50	40	30	25	20	15	65 F (8231 DD) 100	70	50	40	30	25	20	15
WW A1	12	14	15	16	17	18	19	21	11	12	13	13	14	14	15	15
WW A2	10	13	15	18	21	23	26	30	8	10	12	14	16	17	19	21
WW A3	9	12	15	18	22	25	28	33	7	9	12	14	16	18	20	23
WW A4	8	11	15	18	23	26	30	35	5	8	11	13	17	19	21	25
WW A5	7	11	15	19	24	27	31	37	5	8	11	13	17	19	22	27
WW A6	7	11	15	19	24	28	32	39	4	7	11	13	17	20	23	28
WW B1	7	8	9	9	10	10	10	9	5	6	6	6	5	4	3	0
WW B2	9	13	18	22	27	31	36	43	7	10	14	17	22	25	29	34
WW B3	11	16	22	27	34	38	44	52	8	13	18	22	27	31	36	43
WW B4	11	18	25	31	39	44	51	61	8	14	21	26	33	38	44	53
WW B5	11	17	25	31	40	45	53	63	8	14	21	26	34	39	46	56
WW C1	12	18	25	30	37	42	48	56	10	15	20	24	30	35	40	48
WW C2	11	17	24	29	37	42	49	58	9	14	20	24	31	36	42	50
WW C3	14	23	32	39	49	55	63	73	11	19	28	34	43	49	57	67
WW C4	13	21	30	36	46	52	60	70	10	17	25	32	41	47	54	64
TW A1	12	14	15	16	18	19	20	22	11	12	13	14	15	15	16	16
TW A2	10	13	15	17	20	22	25	28	9	10	12	14	16	17	19	21
TW A3	9	12	15	17	21	23	26	31	7	9	12	13	16	18	20	23
TW A4	8	11	14	17	21	23	27	32	6	8	11	13	16	18	21	24
TW B1	11	13	15	16	18	20	22	24	10	11	13	14	15	16	17	18
TW B2	9	12	14	17	20	22	25	29	8	10	12	13	16	17	19	22
TW B3	8	11	14	16	20	22	26	30	7	9	11	13	16	17	20	23
TW B4	8	11	14	16	19	22	25	30	7	9	11	13	16	17	20	24
TW C1	10	12	14	16	19	20	23	26	9	10	12	13	15	17	18	21
TW C2	9	11	14	16	19	21	24	28	8	9	11	13	15	17	19	23
TW C3	9	11	14	16	19	21	24	28	8	9	11	13	15	17	19	23
TW C4	9	11	14	15	18	20	23	27	9	10	12	13	15	17	19	22
TW D1	6	7	8	9	9	10	11	11	4	5	5	6	6	5	5	4
TW D2	10	13	18	21	26	29	34	40	8	11	14	17	21	24	27	32
TW D3	10	14	19	22	28	31	36	43	8	11	15	18	22	25	29	35
TW D4	11	17	23	28	35	39	46	54	9	14	19	23	29	34	39	47
TW D5	11	17	24	29	36	42	48	58	9	14	20	25	31	36	42	51
TW E1	12	17	23	27	33	37	43	51	10	14	19	22	27	31	36	42
TW E2	12	17	23	28	34	39	45	53	10	14	19	23	29	33	38	46
TW E3	15	22	30	36	45	50	58	67	12	18	26	31	39	44	51	60
TW E4	13	20	28	35	43	49	57	66	11	17	24	30	38	43	50	60
TW F1	10	12	13	15	16	18	19	21	9	10	11	12	13	14	14	15
TW F2	8	10	13	15	17	19	22	26	7	8	10	11	13	14	16	18
TW F3	6	9	12	14	17	20	23	27	5	7	9	11	13	15	17	20
TW F4	4	7	10	13	16	19	23	27	3	5	7	9	12	14	17	20
TW G1	8	10	12	13	15	17	19	22	7	8	10	11	12	13	14	15
TW G2	6	8	10	12	15	18	20	24	4	6	8	9	11	13	15	18
TW G3	4	6	9	11	15	17	20	24	3	5	7	8	11	12	15	18
TW G4	3	5	7	10	13	15	18	22	1	3	5	7	9	11	14	17
TW H1	5	7	9	11	13	15	17	21	4	6	7	8	10	11	13	15
TW H2	3	5	7	9	12	14	17	21	2	4	5	7	9	10	12	15
TW H3	2	4	6	8	11	13	15	19	1	3	4	6	8	10	12	15
TW H4	1	3	5	6	9	10	13	16	1	2	4	5	7	8	10	13
TW I1	4	5	6	7	7	8	8	9	3	3	4	4	4	4	3	2
TW I2	7	11	15	18	23	26	30	37	5	8	11	14	18	21	24	29
TW I3	8	12	16	20	25	29	34	41	6	9	13	16	20	23	27	33
TW I4	9	14	20	25	32	36	43	52	6	11	16	20	27	31	36	44
TW I5	9	15	21	26	34	39	46	55	7	12	18	22	29	33	40	48
TW J1	10	15	20	25	31	35	41	49	8	12	17	20	26	29	34	41
TW J2	10	15	21	25	32	37	43	51	7	12	17	21	27	31	36	44
TW J3	12	19	27	34	42	48	56	66	9	16	23	29	37	43	50	59
TW J4	11	18	26	32	41	47	54	64	8	15	22	28	36	41	48	58
DG A1	5	6	7	8	9	9	9	8	3	4	5	5	5	4	4	1
DG A2	8	10	14	16	19	22	25	29	6	9	11	13	15	17	19	22
DG A3	11	15	20	24	30	34	39	47	10	13	17	21	26	29	34	41
DG B1	5	6	7	8	9	10	10	10	3	4	5	5	5	5	4	2
DG B2	8	11	14	16	20	22	26	31	7	9	11	13	16	18	20	24
DG B3	12	16	21	25	31	35	42	50	10	14	18	22	27	31	36	44
DG C1	7	9	11	12	14	15	17	19	5	6	8	8	9	10	10	10
DG C2	10	13	17	20	25	28	32	39	8	11	14	17	20	23	26	31
DG C3	13	18	24	28	35	40	47	57	12	16	21	25	31	35	41	49
SS A1	13	16	19	21	25	27	30	34	12	15	17	19	21	23	25	28
SS A2	17	21	27	31	37	41	47	54	15	19	24	28	33	37	42	48
SS A3	12	14	16	18	20	22	24	26	11	13	15	16	17	18	19	20
SS A4	16	20	26	29	35	39	44	51	15	19	23	26	31	34	39	44
SS A5	14	16	18	20	23	24	26	29	13	15	17	18	20	21	22	24
SS A6	16	21	26	30	36	40	45	52	15	19	24	27	32	36	40	46
SS A7	13	14	16	17	18	19	20	21	12	13	14	15	16	16	17	17
SS A8	15	20	24	28	33	37	41	48	14	18	22	25	29	32	36	41
SS B1	11	13	16	18	21	23	25	29	10	12	14	16	18	19	21	23
SS B2	14	19	24	28	34	38	43	50	13	17	22	25	30	34	38	45
SS B3	10	12	14	15	18	19	21	23	9	10	12	13	15	16	17	18
SS B4	14	18	23	26	32	36	41	48	12	16	20	24	28	32	36	42
SS B5	12	13	15	16	18	20	22	24	11	12	13	15	16	17	18	19
SS B6	13	18	23	27	32	36	41	48	12	16	20	24	28	32	36	42
SS B7	10	12	13	14	15	16	17	19	10	11	12	12	13	14	14	14
SS B8	13	17	21	24	30	33	38	44	11	15	19	22	26	29	33	38
SS C1	7	10	13	15	18	20	22	26	6	7	9	11	13	14	15	17
SS C2	10	14	19	23	28	32	37	45	7	11	15	18	23	26	30	36
SS C3	9	10	12	14	16	18	20	23	7	9	10	11	13	14	15	16
SS C4	10	13	17	21	25	29	33	40	8	11	14	17	21	24	27	32
SS D1	14	17	21	24	28	31	34	39	12	15	18	21	24	26	28	31
SS D2	17	24	31	36	43	48	55	63	15	21	27	32	38	43	49	56
SS D3	14	17	20	23	26	28	31	35	13	15	18	20	22	24	26	28
SS D4	17	23	30	35	42	46	52	61	15	21	26	31	37	41	47	54
SS E1	11	13	16	18	22	24	26	30	9	11	13	15	17	18	20	21
SS E2	14	19	25	29	36	40	46	53	12	16	21	25	31	34	39	46
SS E3	12	14	16	18	20	22	24	26	11	12	14	15	17	18	19	20
SS E4	13	18	23	27	33	37	42	49	12	16	20	24	29	32	36	42

CONSERVATION FACTOR

TYPE	55 F 100	50	30	20	15	65 F 100	50	30	20	15
WW A1	2.7	3.1	3.8	4.5	-	3.3	4.3	-	-	-
WW A2	1.9	2.2	2.5	2.8	3.1	2.2	2.6	3.1	3.8	4.5
WW A3	1.7	2.0	2.3	2.5	2.7	2.0	2.4	2.8	3.2	3.7
WW A4	1.7	1.9	2.2	2.4	2.6	1.9	2.2	2.6	3.0	3.3
WW A5	1.6	1.8	2.1	2.3	2.5	1.8	2.1	2.4	2.8	3.1
WW A6	1.5	1.8	2.0	2.2	2.4	1.8	2.0	2.4	2.7	3.0
WW B1	2.9	4.0	-	-	-	4.5	-	-	-	-
WW B2	1.5	1.7	1.9	2.1	2.3	1.7	1.9	2.1	2.4	2.6
WW B3	1.4	1.5	1.7	1.9	2.1	1.5	1.7	1.9	2.1	2.3
WW B4	1.2	1.4	1.6	1.7	1.9	1.3	1.5	1.7	1.9	2.0
WW B5	1.2	1.4	1.5	1.7	1.8	1.3	1.5	1.6	1.8	1.9
WW C1	1.3	1.5	1.7	1.8	2.0	1.4	1.6	1.8	2.0	2.2
WW C2	1.3	1.5	1.6	1.8	1.9	1.4	1.6	1.7	1.9	2.1
WW C3	1.1	1.3	1.4	1.6	1.7	1.2	1.3	1.5	1.6	1.8
WW C4	1.1	1.3	1.4	1.6	1.7	1.2	1.3	1.5	1.7	1.8
TW A1	2.6	3.0	3.5	4.0	4.6	3.1	3.8	4.8	-	-
TW A2	2.0	2.3	2.6	2.9	3.2	2.4	2.8	3.2	3.8	4.4
TW A3	1.9	2.1	2.4	2.7	2.9	2.1	2.5	2.9	3.3	3.7
TW A4	1.8	2.0	2.3	2.5	2.7	2.0	2.3	2.6	3.0	3.3
TW B1	2.4	2.7	3.1	3.5	3.8	2.8	3.3	3.9	4.6	-
TW B2	2.0	2.2	2.5	2.8	3.0	2.3	2.6	3.0	3.4	3.7
TW B3	1.9	2.2	2.4	2.6	2.8	2.2	2.5	2.8	3.1	3.4
TW B4	1.9	2.2	2.4	2.6	2.8	2.2	2.5	2.7	3.0	3.3
TW C1	2.2	2.5	2.8	3.1	3.3	2.5	2.9	3.3	3.7	4.1
TW C2	2.1	2.3	2.5	2.8	3.0	2.3	2.6	2.9	3.2	3.5
TW C3	2.1	2.3	2.6	2.8	3.0	2.3	2.6	2.9	3.2	3.5
TW C4	2.3	2.5	2.7	2.9	3.1	2.5	2.8	3.0	3.3	3.5
TW D1	2.9	3.7	4.9	-	-	3.9	-	-	-	-
TW D2	1.6	1.8	2.0	2.2	2.4	1.8	2.0	2.3	2.5	2.7
TW D3	1.5	1.7	1.9	2.1	2.3	1.7	1.9	2.2	2.4	2.6
TW D4	1.3	1.5	1.7	1.8	2.0	1.5	1.6	1.8	2.0	2.1
TW D5	1.3	1.5	1.6	1.8	1.9	1.4	1.6	1.7	1.9	2.0
TW E1	1.4	1.6	1.8	2.0	2.1	1.5	1.8	2.0	2.2	2.3
TW E2	1.4	1.5	1.7	1.9	2.0	1.5	1.7	1.9	2.0	2.2
TW E3	1.2	1.3	1.5	1.6	1.8	1.2	1.4	1.6	1.7	1.9
TW E4	1.2	1.3	1.5	1.7	1.8	1.3	1.4	1.6	1.7	1.9
TW F1	2.5	2.9	3.4	3.9	4.4	3.0	3.7	4.6	-	-
TW F2	2.1	2.4	2.7	3.0	3.3	2.4	2.8	3.3	3.9	4.5
TW F3	1.9	2.2	2.5	2.7	3.0	2.2	2.6	2.9	3.4	3.8
TW F4	1.9	2.1	2.4	2.6	2.8	2.2	2.5	2.8	3.1	3.4
TW G1	2.4	2.7	3.1	3.5	3.8	2.8	3.3	3.9	4.6	-
TW G2	2.1	2.4	2.6	2.9	3.1	2.4	2.7	3.1	3.6	4.0
TW G3	2.0	2.3	2.5	2.8	3.0	2.3	2.6	3.0	3.3	3.6
TW G4	2.1	2.4	2.6	2.8	3.0	2.4	2.7	3.0	3.3	3.5
TW H1	2.3	2.6	2.9	3.2	3.5	2.7	3.1	3.5	4.0	4.5
TW H2	2.2	2.5	2.7	3.0	3.2	2.5	2.8	3.2	3.5	3.8
TW H3	2.3	2.6	2.8	3.0	3.2	2.6	2.9	3.2	3.5	3.7
TW H4	2.6	2.8	3.0	3.2	3.4	2.8	3.1	3.4	3.7	3.9
TW I1	3.1	3.9	-	-	-	4.2	-	-	-	-
TW I2	1.7	1.9	2.1	2.3	2.4	1.9	2.1	2.3	2.6	2.8
TW I3	1.6	1.8	2.0	2.1	2.3	1.7	1.9	2.2	2.4	2.6
TW I4	1.4	1.6	1.7	1.9	2.0	1.5	1.7	1.8	2.0	2.2
TW I5	1.3	1.5	1.7	1.8	1.9	1.4	1.6	1.8	1.9	2.1
TW J1	1.4	1.6	1.8	2.0	2.1	1.6	1.8	2.0	2.2	2.3
TW J2	1.4	1.6	1.7	1.9	2.0	1.5	1.7	1.9	2.1	2.2
TW J3	1.2	1.3	1.5	1.7	1.8	1.3	1.4	1.6	1.7	1.9
TW J4	1.2	1.4	1.5	1.7	1.8	1.3	1.4	1.6	1.8	1.9
DG A1	2.7	3.5	-	-	-	3.5	-	-	-	-
DG A2	1.9	2.2	2.4	2.8	3.2	2.1	2.4	2.8	3.3	3.9
DG A3	1.6	1.7	1.8	2.0	2.2	1.7	1.8	2.0	2.2	2.4
DG B1	2.6	3.5	-	-	-	3.5	-	-	-	-
DG B2	1.9	2.1	2.4	2.6	2.8	2.1	2.4	2.7	3.1	3.5
DG B3	1.5	1.7	1.8	1.9	2.0	1.6	1.8	1.9	2.1	2.2
DG C1	2.1	2.7	3.4	3.9	-	2.7	3.3	-	-	-
DG C2	1.7	1.9	2.1	2.2	2.5	1.9	2.1	2.3	2.6	2.9
DG C3	1.4	1.6	1.7	1.8	1.9	1.5	1.6	1.8	1.9	2.0
SS A1	1.8	2.1	2.4	2.7	3.0	2.0	2.3	2.8	3.3	3.7
SS A2	1.4	1.6	1.8	1.9	2.1	1.5	1.7	1.9	2.1	2.3
SS A3	2.1	2.5	2.9	3.5	4.0	2.3	2.9	3.7	-	-
SS A4	1.4	1.6	1.8	2.0	2.2	1.5	1.7	2.0	2.2	2.4
SS A5	2.1	2.4	2.8	3.2	3.6	2.3	2.7	3.3	4.1	-
SS A6	1.4	1.6	1.8	2.0	2.1	1.5	1.7	1.9	2.2	2.3
SS A7	2.5	3.0	3.7	4.8	-	2.8	3.6	-	-	-
SS A8	1.5	1.7	1.9	2.1	2.3	1.6	1.8	2.1	2.4	2.6
SS B1	2.0	2.3	2.6	2.9	3.2	2.2	2.6	3.0	3.6	4.2
SS B2	1.4	1.6	1.8	2.0	2.1	1.5	1.7	1.9	2.1	2.3
SS B3	2.2	2.6	3.0	3.5	4.0	2.4	3.0	3.7	4.9	-
SS B4	1.5	1.7	1.9	2.1	2.2	1.6	1.8	2.0	2.2	2.4
SS B5	2.3	2.7	3.1	3.6	4.0	2.6	3.1	3.8	4.7	-
SS B6	1.5	1.7	1.9	2.0	2.2	1.6	1.8	2.0	2.2	2.4
SS B7	2.7	3.2	3.9	4.7	-	3.0	3.8	-	-	-
SS B8	1.5	1.8	2.0	2.2	2.3	1.6	1.9	2.1	2.4	2.6
SS C1	2.0	2.3	2.6	3.0	3.3	2.3	2.8	3.4	4.2	-
SS C2	1.5	1.7	1.9	2.1	2.2	1.7	1.9	2.1	2.3	2.5
SS C3	2.3	2.6	3.0	3.4	3.7	2.7	3.2	3.7	4.5	-
SS C4	1.6	1.8	2.0	2.2	2.4	1.8	2.0	2.3	2.5	2.7
SS D1	1.6	1.9	2.2	2.4	2.7	1.8	2.1	2.5	3.0	3.5
SS D2	1.2	1.4	1.6	1.8	1.9	1.3	1.5	1.7	1.9	2.1
SS D3	1.8	2.1	2.4	2.7	3.0	1.9	2.3	2.8	3.3	3.9
SS D4	1.3	1.4	1.6	1.8	1.9	1.3	1.5	1.7	1.9	2.1
SS E1	1.8	2.2	2.5	2.9	3.3	2.0	2.5	3.2	4.2	-
SS E2	1.3	1.5	1.7	1.9	2.1	1.4	1.7	1.9	2.1	2.3
SS E3	2.1	2.5	2.9	3.3	3.7	2.3	2.8	3.5	4.4	-
SS E4	1.4	1.6	1.8	2.0	2.2	1.5	1.8	2.0	2.2	2.4

JACKSON, MISSISSIPPI

TYPE	SOLAR SAVINGS FRACTION (%) BASE TEMP: 55 F (993 DD) LCR 200	100	70	50	40	30	25	20	65 F (2300 DD) 200	100	70	50	40	30	25	20	CONSERVATION FACTOR 55 F 200	100	50	30	20	65 F 200	100	50	30	20
WW A1	25	38	46	55	62	70	75	81	17	24	29	35	40	46	51	57	0.9	1.0	1.1	1.3	1.4	1.2	1.3	1.5	1.6	1.8
WW A2	29	46	57	68	75	83	87	92	17	28	35	44	50	58	64	71	0.7	0.8	1.0	1.2	1.3	1.0	1.1	1.2	1.4	1.5
WW A3	30	50	61	72	79	86	90	94	16	29	37	47	54	63	69	76	0.7	0.8	1.0	1.1	1.3	0.9	1.0	1.2	1.3	1.5
WW A4	31	51	63	74	81	88	92	95	16	29	39	49	56	65	71	78	0.7	0.8	0.9	1.1	1.3	0.9	1.0	1.1	1.3	1.4
WW A5	32	53	65	76	83	90	93	96	16	30	40	51	58	68	74	80	0.6	0.8	0.9	1.1	1.3	0.8	0.9	1.1	1.3	1.4
WW A6	32	54	66	78	84	91	94	97	15	31	41	52	59	69	75	82	0.6	0.7	0.9	1.1	1.3	0.8	0.9	1.1	1.2	1.4
WW B1	25	40	50	61	67	76	81	87	14	23	29	36	41	49	54	60	0.8	0.9	1.0	1.2	1.4	1.1	1.2	1.4	1.5	1.7
WW B2	30	51	63	75	81	89	92	96	16	30	40	50	57	67	73	80	0.7	0.8	0.9	1.1	1.3	0.9	1.0	1.1	1.3	1.4
WW B3	36	59	71	82	87	93	96	98	19	36	47	58	66	75	81	87	0.6	0.7	0.9	1.1	1.2	0.8	0.9	1.0	1.2	1.4
WW B4	35	59	71	82	88	94	96	98	17	36	47	59	67	77	82	88	0.6	0.7	0.9	1.1	1.2	0.8	0.9	1.0	1.2	1.3
WW B5	33	57	70	81	87	93	96	98	16	34	46	58	66	76	82	88	0.6	0.7	0.9	1.1	1.2	0.8	0.9	1.0	1.2	1.3
WW C1	38	61	73	83	89	94	96	98	20	38	49	60	68	77	82	88	0.6	0.7	0.9	1.1	1.2	0.8	0.9	1.0	1.2	1.4
WW C2	35	58	71	81	87	93	96	98	18	35	46	58	66	76	81	87	0.6	0.7	0.9	1.1	1.2	0.8	0.9	1.0	1.2	1.4
WW C3	39	65	77	87	92	96	98	99	20	41	54	67	74	84	88	93	0.6	0.7	0.9	1.0	1.2	0.7	0.8	1.0	1.1	1.3
WW C4	36	61	74	84	90	95	97	99	17	38	50	63	71	80	86	91	0.6	0.7	0.9	1.0	1.2	0.7	0.8	1.0	1.1	1.3
TW A1	25	37	45	54	61	69	74	80	17	24	29	34	39	45	50	56	0.9	1.0	1.1	1.3	1.4	1.2	1.3	1.5	1.6	1.8
TW A2	26	42	52	63	69	78	83	88	16	25	32	40	46	54	59	66	0.8	0.9	1.0	1.2	1.4	1.0	1.1	1.3	1.4	1.6
TW A3	27	44	55	66	73	81	86	91	15	26	33	42	48	57	63	70	0.7	0.8	1.0	1.2	1.3	1.0	1.1	1.2	1.4	1.5
TW A4	26	44	55	66	73	82	87	91	14	25	33	42	49	58	64	71	0.7	0.8	1.0	1.2	1.3	1.0	1.1	1.2	1.4	1.5
TW B1	24	37	45	55	61	70	75	82	16	23	29	35	40	47	51	58	0.9	1.0	1.1	1.3	1.4	1.2	1.3	1.4	1.6	1.7
TW B2	25	40	50	61	68	77	82	87	14	24	31	38	44	52	58	65	0.8	0.9	1.0	1.2	1.4	1.0	1.1	1.3	1.4	1.6
TW B3	24	40	51	61	69	77	83	88	13	23	31	39	45	53	59	66	0.8	0.9	1.0	1.2	1.4	1.0	1.1	1.3	1.4	1.6
TW B4	22	38	48	59	66	75	80	86	12	22	29	36	42	51	56	63	0.8	0.9	1.0	1.2	1.4	1.1	1.2	1.3	1.5	1.6
TW C1	23	36	45	54	61	70	75	81	14	22	28	34	39	46	51	58	0.9	1.0	1.1	1.3	1.4	1.2	1.3	1.4	1.6	1.7
TW C2	22	36	45	55	62	71	77	83	13	21	27	35	40	48	53	60	0.8	0.9	1.1	1.2	1.4	1.1	1.2	1.4	1.5	1.7
TW C3	21	34	43	53	60	69	74	81	12	20	26	33	38	46	51	57	0.9	1.0	1.1	1.3	1.4	1.2	1.2	1.4	1.5	1.7
TW C4	19	30	38	47	54	63	68	75	12	19	24	30	34	41	46	52	1.0	1.0	1.2	1.3	1.5	1.3	1.3	1.5	1.6	1.8
TW D1	21	34	44	54	61	69	75	81	11	19	25	31	36	43	48	54	0.9	0.9	1.1	1.3	1.4	1.2	1.3	1.4	1.6	1.8
TW D2	28	47	58	69	76	84	89	93	15	28	36	46	52	62	68	75	0.7	0.8	1.0	1.1	1.3	0.9	1.0	1.2	1.3	1.5
TW D3	31	51	63	74	81	88	92	95	17	30	40	50	57	67	73	79	0.7	0.8	0.9	1.1	1.3	0.9	1.0	1.1	1.3	1.4
TW D4	32	54	66	77	83	90	94	97	17	32	43	54	61	71	77	83	0.6	0.8	0.9	1.1	1.3	0.8	0.9	1.1	1.2	1.4
TW D5	30	52	65	76	83	90	93	96	16	31	42	53	61	71	77	83	0.7	0.8	0.9	1.1	1.3	0.8	0.9	1.1	1.2	1.4
TW E1	34	56	68	78	84	91	94	97	19	34	44	55	62	72	77	84	0.6	0.7	0.9	1.1	1.3	0.8	0.9	1.1	1.2	1.4
TW E2	33	55	67	78	84	91	94	97	18	33	44	54	62	72	77	84	0.6	0.7	0.9	1.1	1.3	0.8	0.9	1.1	1.2	1.4
TW E3	38	62	75	85	90	95	97	99	21	40	52	63	71	80	86	91	0.6	0.7	0.9	1.0	1.2	0.7	0.8	1.0	1.2	1.3
TW E4	35	59	72	82	88	94	96	98	18	37	48	60	68	78	83	89	0.6	0.7	0.9	1.1	1.2	0.8	0.9	1.0	1.2	1.3
TW F1	23	36	45	54	61	69	75	81	15	22	27	34	38	45	50	56	0.9	1.0	1.1	1.3	1.4	1.2	1.3	1.4	1.6	1.7
TW F2	24	39	50	60	67	76	81	87	13	23	30	37	43	51	57	64	0.8	0.9	1.0	1.2	1.4	1.0	1.2	1.3	1.5	1.6
TW F3	23	40	51	62	69	78	83	89	12	22	30	38	44	53	59	66	0.8	0.9	1.0	1.2	1.4	1.0	1.1	1.3	1.4	1.6
TW F4	21	38	49	60	68	77	82	88	10	20	28	36	43	51	57	65	0.8	0.9	1.0	1.2	1.4	1.0	1.1	1.3	1.4	1.6
TW G1	21	34	43	53	60	69	74	81	13	20	26	32	37	44	49	56	0.9	1.0	1.1	1.3	1.4	1.2	1.3	1.4	1.6	1.7
TW G2	20	36	46	56	64	73	78	84	10	20	26	34	39	48	53	60	0.8	0.9	1.1	1.2	1.4	1.1	1.2	1.3	1.5	1.6
TW G3	18	34	44	55	62	72	77	84	9	18	24	32	38	46	52	59	0.8	0.9	1.1	1.2	1.4	1.1	1.2	1.3	1.5	1.6
TW G4	15	29	38	49	56	66	72	79	6	15	21	28	33	41	47	54	0.9	1.0	1.1	1.3	1.4	1.2	1.3	1.4	1.6	1.7
TW H1	18	31	40	49	56	66	71	78	9	17	23	29	34	41	46	53	0.9	1.0	1.1	1.3	1.5	1.2	1.3	1.4	1.6	1.7
TW H2	15	28	37	47	54	64	70	77	7	14	20	27	32	39	45	51	0.9	1.0	1.1	1.3	1.5	1.2	1.3	1.4	1.6	1.7
TW H3	12	24	33	42	49	58	65	72	5	12	17	23	28	35	40	47	1.0	1.0	1.2	1.3	1.5	1.3	1.4	1.5	1.7	1.8
TW H4	9	19	26	34	40	49	55	62	4	9	13	18	22	28	33	39	1.1	1.2	1.3	1.4	1.6	1.5	1.5	1.7	1.8	2.0
TW I1	18	31	40	50	57	66	72	78	9	17	22	28	33	39	44	50	0.9	1.0	1.1	1.3	1.5	1.2	1.3	1.5	1.7	1.8
TW I2	25	43	55	66	73	82	87	91	12	24	33	42	49	58	64	72	0.7	0.8	1.0	1.1	1.3	0.9	1.0	1.2	1.3	1.5
TW I3	27	47	59	70	77	85	89	94	13	27	36	46	53	62	69	76	0.7	0.8	1.0	1.1	1.3	0.9	1.0	1.1	1.3	1.5
TW I4	28	50	62	74	81	88	92	96	14	29	39	50	58	68	74	81	0.7	0.8	0.9	1.1	1.3	0.8	0.9	1.1	1.2	1.4
TW I5	28	50	63	74	81	89	92	96	13	29	40	51	59	69	75	82	0.7	0.8	0.9	1.1	1.3	0.8	0.9	1.1	1.2	1.4
TW J1	31	53	65	76	83	90	93	96	17	32	42	52	60	69	75	82	0.7	0.8	0.9	1.1	1.3	0.8	0.9	1.1	1.2	1.4
TW J2	30	51	64	75	82	89	93	96	15	30	40	51	59	69	75	82	0.7	0.8	0.9	1.1	1.3	0.8	0.9	1.1	1.2	1.4
TW J3	34	59	72	82	88	94	96	98	17	36	48	60	68	78	83	89	0.6	0.7	0.9	1.1	1.2	0.7	0.9	1.0	1.2	1.3
TW J4	32	56	69	80	86	92	95	98	15	34	45	57	65	75	81	87	0.6	0.7	0.9	1.1	1.2	0.8	0.9	1.0	1.2	1.3
DG A1	23	39	49	59	65	72	76	81	12	21	27	34	38	44	48	52	0.8	0.9	1.1	1.3	1.6	1.1	1.2	1.4	1.7	2.1
DG A2	23	41	52	62	69	77	82	87	12	23	30	38	44	52	57	63	0.8	0.8	1.0	1.2	1.4	1.0	1.1	1.3	1.5	1.7
DG A3	26	46	59	70	77	84	88	91	15	27	36	46	54	64	69	76	0.7	0.8	1.0	1.2	1.4	0.9	1.0	1.1	1.3	1.5
DG B1	24	42	54	65	72	79	83	87	12	21	28	36	42	50	55	61	0.7	0.8	1.0	1.2	1.5	1.1	1.2	1.3	1.5	1.8
DG B2	24	43	56	68	76	84	88	92	13	23	31	41	48	57	64	70	0.7	0.8	1.0	1.2	1.4	1.0	1.1	1.2	1.3	1.6
DG B3	27	48	62	74	82	89	92	95	15	27	37	48	56	68	74	81	0.7	0.8	0.9	1.1	1.4	0.9	1.0	1.1	1.2	1.4
DG C1	27	48	61	74	80	87	90	93	14	25	33	42	49	59	65	71	0.7	0.8	0.9	1.2	1.4	1.0	1.1	1.2	1.3	1.6
DG C2	27	48	62	75	82	89	92	95	15	27	36	46	54	65	71	78	0.7	0.8	0.9	1.1	1.3	0.9	1.0	1.1	1.3	1.5
DG C3	30	53	68	81	87	93	95	97	17	31	41	54	63	74	81	87	0.7	0.7	0.9	1.1	1.4	0.9	0.9	1.0	1.2	1.4
SS A1	32	49	59	70	76	84	88	92	20	32	39	48	54	62	68	74	0.7	0.8	1.0	1.2	1.3	0.9	1.0	1.2	1.4	1.5
SS A2	38	59	70	80	86	92	95	97	24	39	49	59	66	75	80	86	0.6	0.7	0.9	1.1	1.3	0.8	0.9	1.1	1.2	1.4
SS A3	30	47	57	67	73	81	85	90	19	30	37	45	50	58	64	70	0.7	0.8	1.0	1.2	1.3	1.0	1.1	1.2	1.4	1.6
SS A4	38	59	70	80	85	91	94	97	24	39	49	59	66	74	80	85	0.6	0.7	0.9	1.1	1.3	0.8	0.9	1.1	1.2	1.4
SS A5	30	46	55	65	71	79	84	89	20	30	37	44	50	57	62	69	0.8	0.9	1.0	1.2	1.4	1.0	1.1	1.3	1.4	1.6
SS A6	38	59	70	80	86	92	95	97	24	39	49	59	66	75	80	86	0.6	0.7	0.9	1.1	1.3	0.8	0.9	1.1	1.2	1.4
SS A7	28	41	50	59	66	74	79	84	19	27	33	39	44	51	56	62	0.8	0.9	1.1	1.3	1.4	1.1	1.2	1.4	1.5	1.7
SS A8	38	58	69	79	85	91	94	97	24	38	48	58	64	73	78	84	0.6	0.8	0.9	1.1	1.3	0.8	0.9	1.1	1.2	1.4
SS B1	27	43	53	64	70	79	83	88	17	27	34	42	48	56	61	68	0.8	0.9	1.0	1.2	1.4	1.0	1.1	1.3	1.4	1.6
SS B2	33	53	64	75	81	88	92	95	20	34	43	53	60	69	75	81	0.7	0.8	0.9	1.1	1.3	0.8	0.9	1.1	1.3	1.4
SS B3	26	42	52	62	68	77	82	87	16	26	32	40	45	53	58	65	0.8	0.9	1.0	1.2	1.4	1.0	1.1	1.3	1.5	1.6
SS B4	33	52	64	74	81	88	91	95	20	34	43	52	59	68	74	81	0.7	0.8	0.9	1.1	1.3	0.8	1.0	1.1	1.3	1.4
SS B5	25	38	47	57	63	72	77	83	17	25	30	37	42	49	54	60	0.8	0.9	1.1	1.3	1.4	1.1	1.2	1.4	1.5	1.7
SS B6	33	52	64	74	81	88	92	95	20	34	43	52	59	69	74	81	0.7	0.8	0.9	1.1	1.3	0.8	1.0	1.1	1.3	1.4
SS B7	23	35	43	52	59	67	72	78	15	22	28	33	38	44	49	55	0.9	1.0	1.1	1.3	1.5	1.2	1.3	1.5	1.6	1.8
SS B8	32	51	62	73	79	87	91	94	19	32	41	51	58	67	73	79	0.7	0.8	1.0	1.1	1.3	0.9	1.0	1.1	1.3	1.4
SS C1	27	45	56	67	74	82	87	92	14	26	33	42	48	57	63	70	0.7	0.8	1.0	1.2	1.3	1.0	1.1	1.2	1.4	1.5
SS C2	31	51	63	74	81	88	92	95	16	30	40	50	57	67	73	80	0.7	0.8	0.9	1.1	1.3	0.9	1.0	1.1	1.3	1.4
SS C3	22	36	45	55	62	71	77	83	13	21	27	34	39	47	52	58	0.8	0.9	1.1	1.2	1.4	1.1	1.2	1.4	1.5	1.7
SS C4	27	46	57	68	75	83	88	92	15	27	35	45	51	61	67	74	0.7	0.8	1.0	1.1	1.3	0.9	1.0	1.2	1.3	1.5
SS D1	38	58	69	79	85	91	94	97	23	37	47	56	63	72	77	83	0.6	0.8	0.9	1.1	1.3	0.8	0.9	1.1	1.2	1.4
SS D2	44	67	78	87	92	96	98	99	27	45	56	68	75	83	88	92	0.6	0.7	0.9	1.0	1.2	0.7	0.8	1.0	1.1	1.3
SS D3	34	52	63	73	79	86	90	94	22	34	42	51	57	66	71	77	0.7	0.8	1.0	1.1	1.3	0.9	1.0	1.2	1.3	1.5
SS D4	43	65	76	85	90	95	97	99	26	44	55	65	73	81	86	91	0.6	0.7	0.9	1.0	1.2	0.7	0.8	1.0	1.2	1.3
SS E1	34	53	64	74	81	88	91	95	20	33	41	50	57	66	71	78	0.7	0.8	0.9	1.1	1.3	0.9	1.0	1.1	1.3	1.5
SS E2	39	61	73	83	88	94	96	98	23	39	50	61	68	77	83	88	0.6	0.7	0.9	1.1	1.2	0.8	0.9	1.0	1.2	1.4
SS E3	28	44	54	64	71	79	84	89	18	28	35	42	48	56	61	68	0.8	0.9	1.0	1.2	1.4	1.0	1.1	1.3	1.4	1.6
SS E4	36	57	68	79	85	91	94	97	21	36	46	57	64	73	78	84	0.6	0.7	0.9	1.1	1.3	0.8	0.9	1.1	1.2	1.4

MERIDIAN, MISSISSIPPI

TYPE	SOLAR SAVINGS FRACTION (%) BASE TEMP: 55 F (1047 DD) LCR 200	100	70	50	40	30	25	20	65 F (2393 DD) 200	100	70	50	40	30	25	20	CONSERVATION FACTOR 55 F 200	100	50	30	20	65 F 200	100	50	30	20
WW A1	25	36	45	53	60	68	73	80	17	23	28	34	38	45	49	55	0.9	1.0	1.1	1.3	1.5	1.2	1.3	1.5	1.6	1.8
WW A2	28	45	55	66	73	81	86	91	16	27	34	42	48	57	62	69	0.7	0.8	1.0	1.2	1.3	1.0	1.1	1.2	1.4	1.6
WW A3	29	48	59	70	77	85	89	93	15	28	36	45	52	61	67	74	0.7	0.8	1.0	1.1	1.3	0.9	1.0	1.2	1.3	1.5
WW A4	29	49	61	72	79	87	91	94	15	28	37	47	54	64	69	76	0.7	0.8	0.9	1.1	1.3	0.9	1.0	1.1	1.3	1.5
WW A5	30	51	63	74	81	89	92	96	15	29	39	49	56	66	72	79	0.7	0.8	0.9	1.1	1.3	0.9	1.0	1.1	1.3	1.4
WW A6	30	52	64	76	82	89	93	96	14	29	39	50	58	67	73	80	0.7	0.8	0.9	1.1	1.3	0.8	0.9	1.1	1.3	1.4
WW B1	24	39	48	58	65	74	79	85	13	22	28	35	40	47	52	58	0.8	0.9	1.1	1.2	1.4	1.1	1.2	1.4	1.5	1.7
WW B2	29	49	61	73	80	87	91	95	15	29	38	48	56	65	71	78	0.7	0.8	0.9	1.1	1.3	0.9	1.0	1.1	1.3	1.4
WW B3	35	57	69	80	86	92	95	97	18	35	45	56	64	74	79	85	0.6	0.7	0.9	1.1	1.2	0.8	0.9	1.0	1.2	1.4
WW B4	33	57	70	80	87	93	95	98	16	34	46	58	66	75	81	87	0.6	0.7	0.9	1.1	1.2	0.8	0.9	1.0	1.2	1.4
WW B5	31	55	68	79	85	92	95	97	15	33	44	56	64	74	80	87	0.6	0.7	0.9	1.1	1.2	0.8	0.9	1.0	1.2	1.4
WW C1	36	59	71	81	87	93	96	98	19	36	47	58	66	76	81	87	0.6	0.7	0.9	1.1	1.2	0.8	0.9	1.0	1.2	1.4
WW C2	33	56	69	80	86	92	95	97	17	34	45	56	64	74	80	86	0.6	0.7	0.9	1.1	1.2	0.8	0.9	1.0	1.2	1.4
WW C3	37	63	76	86	91	96	98	99	19	40	53	65	73	82	87	92	0.6	0.7	0.9	1.0	1.2	0.7	0.8	1.0	1.1	1.3
WW C4	34	59	72	83	89	94	97	98	16	36	49	61	69	79	85	90	0.5	0.7	0.9	1.0	1.2	0.7	0.8	1.0	1.2	1.3
TW A1	24	35	43	52	59	67	72	79	16	23	28	33	38	44	49	54	0.9	1.0	1.1	1.3	1.5	1.2	1.3	1.5	1.6	1.8
TW A2	25	40	50	61	68	76	81	87	15	24	31	38	44	52	57	64	0.8	0.9	1.0	1.2	1.4	1.1	1.2	1.3	1.5	1.6
TW A3	26	42	53	64	71	79	84	89	14	25	32	41	47	55	61	68	0.8	0.9	1.0	1.2	1.3	1.0	1.1	1.2	1.4	1.6
TW A4	25	42	53	64	72	80	85	90	13	24	32	41	47	56	62	69	0.7	0.8	1.0	1.2	1.3	1.0	1.1	1.2	1.4	1.5
TW B1	23	35	44	53	60	68	74	80	15	22	28	34	38	45	50	56	0.9	1.0	1.1	1.3	1.5	1.2	1.3	1.4	1.6	1.8
TW B2	24	39	49	59	66	75	80	86	14	23	30	37	43	51	56	63	0.8	0.9	1.1	1.2	1.4	1.1	1.2	1.3	1.5	1.6
TW B3	23	39	49	60	67	76	81	87	13	22	29	37	43	52	57	64	0.8	0.9	1.0	1.2	1.4	1.0	1.1	1.3	1.4	1.6
TW B4	21	36	46	57	64	73	78	84	12	21	28	35	41	49	55	62	0.8	0.9	1.1	1.2	1.4	1.1	1.2	1.3	1.5	1.6
TW C1	22	34	43	52	59	68	73	80	14	21	27	33	38	45	50	56	0.9	1.0	1.1	1.3	1.4	1.2	1.3	1.4	1.6	1.7
TW C2	21	34	44	53	60	69	75	81	12	20	26	33	39	46	51	58	0.9	1.0	1.1	1.3	1.4	1.1	1.2	1.4	1.5	1.7
TW C3	20	32	41	51	58	67	72	79	12	19	25	32	37	44	49	56	0.9	1.0	1.1	1.3	1.4	1.2	1.3	1.4	1.6	1.7
TW C4	19	29	37	46	52	61	66	73	12	18	23	29	33	40	44	50	1.0	1.1	1.2	1.3	1.5	1.3	1.4	1.5	1.7	1.8
TW D1	20	33	42	52	58	67	73	79	11	18	24	30	34	41	46	52	0.9	1.0	1.1	1.3	1.4	1.2	1.3	1.5	1.6	1.8
TW D2	27	45	56	68	75	83	87	92	15	27	35	44	51	60	66	73	0.7	0.8	1.0	1.1	1.3	0.9	1.0	1.2	1.3	1.5
TW D3	29	49	61	72	79	87	91	95	16	29	39	49	56	65	71	78	0.7	0.8	0.9	1.1	1.3	0.9	1.0	1.1	1.3	1.4
TW D4	30	52	64	75	82	89	93	96	16	31	41	52	60	70	76	82	0.7	0.8	0.9	1.1	1.3	0.8	0.9	1.1	1.2	1.4
TW D5	29	50	63	74	81	89	92	96	15	30	41	52	59	69	75	82	0.7	0.8	0.9	1.1	1.3	0.8	0.9	1.1	1.2	1.4
TW E1	33	54	66	76	83	90	93	96	18	33	43	53	61	70	76	82	0.7	0.8	0.9	1.1	1.3	0.8	0.9	1.1	1.2	1.4
TW E2	32	53	65	76	83	90	93	96	17	32	42	53	60	70	76	83	0.7	0.8	0.9	1.1	1.3	0.8	0.9	1.1	1.2	1.4
TW E3	37	60	73	83	89	94	97	98	20	38	50	62	70	79	84	90	0.6	0.7	0.9	1.1	1.2	0.7	0.8	1.0	1.2	1.3
TW E4	33	57	70	81	87	93	96	98	17	35	47	59	67	77	82	88	0.6	0.7	0.9	1.1	1.2	0.8	0.9	1.0	1.2	1.3
TW F1	22	35	43	52	59	67	73	79	14	21	26	32	37	44	48	54	0.9	1.0	1.1	1.3	1.5	1.2	1.3	1.5	1.6	1.8
TW F2	23	38	48	58	65	74	80	85	13	22	28	36	42	50	55	62	0.8	0.9	1.1	1.2	1.4	1.1	1.2	1.3	1.5	1.6
TW F3	22	38	49	60	67	76	81	87	11	21	29	37	43	51	57	64	0.8	0.9	1.0	1.2	1.4	1.0	1.1	1.3	1.4	1.6
TW F4	20	36	47	58	66	75	80	86	9	19	27	35	41	50	56	63	0.8	0.9	1.0	1.2	1.4	1.0	1.1	1.3	1.4	1.6
TW G1	20	33	42	51	58	67	72	79	12	19	25	31	36	43	48	54	0.9	1.0	1.1	1.3	1.5	1.2	1.3	1.4	1.6	1.8
TW G2	19	34	44	54	62	71	76	83	10	19	25	32	38	46	51	58	0.8	0.9	1.1	1.2	1.4	1.1	1.2	1.4	1.5	1.7
TW G3	18	32	42	53	60	70	75	82	8	17	23	31	37	45	50	57	0.8	0.9	1.1	1.2	1.4	1.1	1.2	1.4	1.5	1.7
TW G4	14	27	37	47	54	64	70	77	6	14	20	27	32	40	45	52	0.9	1.0	1.1	1.3	1.5	1.2	1.3	1.4	1.6	1.7
TW H1	17	29	38	48	54	64	70	76	9	16	22	28	33	40	45	51	0.9	1.0	1.1	1.3	1.5	1.2	1.3	1.5	1.6	1.8
TW H2	14	27	35	45	52	62	68	75	6	14	19	26	31	38	43	50	0.9	1.0	1.2	1.3	1.5	1.2	1.3	1.5	1.6	1.8
TW H3	12	23	31	40	47	56	63	70	5	11	16	22	27	34	39	45	1.0	1.1	1.2	1.4	1.5	1.3	1.4	1.5	1.7	1.8
TW H4	9	18	25	32	38	47	53	60	3	8	12	17	21	27	32	37	1.1	1.2	1.3	1.5	1.6	1.5	1.6	1.7	1.8	2.0
TW I1	18	30	39	48	55	64	70	76	9	16	21	27	31	38	42	49	0.9	1.0	1.1	1.3	1.5	1.3	1.4	1.5	1.7	1.8
TW I2	24	41	53	64	71	80	85	90	12	23	32	41	47	57	63	70	0.7	0.8	1.0	1.2	1.3	1.0	1.1	1.2	1.4	1.5
TW I3	25	45	57	68	75	84	88	93	13	26	34	44	51	61	67	74	0.7	0.8	1.0	1.1	1.3	0.9	1.0	1.2	1.3	1.5
TW I4	27	48	60	72	79	87	91	95	13	28	38	49	56	66	72	80	0.7	0.8	0.9	1.1	1.3	0.9	1.0	1.1	1.3	1.4
TW I5	26	48	61	72	80	87	91	95	12	28	38	49	57	67	73	81	0.7	0.8	0.9	1.1	1.3	0.8	0.9	1.1	1.2	1.4
TW J1	30	51	63	74	81	88	92	95	16	30	40	51	58	68	74	81	0.7	0.8	0.9	1.1	1.3	0.8	0.9	1.1	1.3	1.4
TW J2	28	49	62	73	80	88	92	95	14	29	39	50	57	67	73	80	0.7	0.8	0.9	1.1	1.3	0.8	0.9	1.1	1.3	1.4
TW J3	33	57	70	81	87	93	96	98	16	35	47	59	67	77	82	88	0.6	0.7	0.9	1.1	1.2	0.8	0.9	1.0	1.2	1.3
TW J4	30	54	67	78	85	91	95	97	15	32	44	56	64	74	80	86	0.6	0.7	0.9	1.1	1.3	0.8	0.9	1.0	1.2	1.4
DG A1	22	38	47	56	62	70	74	79	11	20	26	32	37	43	46	50	0.8	0.9	1.1	1.3	1.6	1.1	1.2	1.5	1.8	2.1
DG A2	22	39	50	60	67	75	80	85	12	22	29	37	43	51	56	62	0.8	0.9	1.0	1.2	1.4	1.1	1.1	1.3	1.5	1.7
DG A3	25	44	57	69	75	83	87	91	14	26	35	45	52	62	68	75	0.7	0.8	1.0	1.2	1.4	1.0	1.0	1.1	1.3	1.5
DG B1	22	40	51	63	70	77	82	86	11	20	27	34	40	48	53	59	0.8	0.8	1.0	1.2	1.5	1.1	1.2	1.3	1.5	1.8
DG B2	23	41	54	66	74	82	86	91	12	22	30	39	46	56	62	69	0.8	0.8	1.0	1.2	1.4	1.0	1.1	1.2	1.4	1.6
DG B3	25	46	59	72	80	87	91	94	14	26	36	46	55	66	72	80	0.7	0.8	0.9	1.1	1.4	0.9	1.0	1.1	1.2	1.4
DG C1	26	46	59	71	78	85	89	92	13	24	31	41	47	57	63	69	0.7	0.8	0.9	1.2	1.4	1.0	1.1	1.2	1.4	1.6
DG C2	26	46	60	73	80	88	91	95	14	26	34	45	52	63	69	76	0.7	0.8	0.9	1.1	1.3	0.9	1.0	1.1	1.3	1.5
DG C3	29	51	66	79	86	92	95	97	16	30	40	52	61	73	79	86	0.7	0.7	0.9	1.1	1.4	0.9	1.0	1.0	1.2	1.4
SS A1	31	47	58	68	74	82	86	91	20	31	38	46	52	61	66	72	0.7	0.8	1.0	1.2	1.3	0.9	1.1	1.2	1.4	1.5
SS A2	37	57	68	78	84	91	94	97	23	38	48	58	65	74	79	85	0.6	0.8	0.9	1.1	1.3	0.8	0.9	1.1	1.2	1.4
SS A3	29	45	55	65	71	79	84	89	19	29	36	43	49	57	62	68	0.8	0.9	1.0	1.2	1.4	1.0	1.1	1.3	1.4	1.6
SS A4	37	57	68	78	84	90	93	96	23	38	47	57	64	73	78	84	0.6	0.8	0.9	1.1	1.3	0.8	0.9	1.1	1.2	1.4
SS A5	29	44	53	63	70	78	82	87	20	29	36	43	48	56	61	67	0.8	0.9	1.0	1.2	1.4	1.0	1.1	1.3	1.5	1.6
SS A6	37	57	68	78	84	91	94	97	23	38	47	58	64	73	79	84	0.6	0.8	0.9	1.1	1.3	0.8	0.9	1.1	1.2	1.4
SS A7	27	40	48	57	64	72	77	82	18	26	32	38	43	50	54	60	0.8	0.9	1.1	1.3	1.4	1.1	1.2	1.4	1.6	1.7
SS A8	36	56	67	77	83	90	93	96	23	37	46	56	63	72	77	83	0.7	0.8	0.9	1.1	1.3	0.8	0.9	1.1	1.2	1.4
SS B1	26	42	52	62	68	77	82	87	16	26	33	41	46	54	59	66	0.8	0.9	1.0	1.2	1.4	1.0	1.1	1.3	1.4	1.6
SS B2	32	51	62	73	80	87	91	94	20	33	42	52	59	68	74	80	0.7	0.8	1.0	1.1	1.3	0.9	1.0	1.1	1.3	1.4
SS B3	25	40	50	60	66	75	80	85	16	25	31	39	44	52	57	63	0.8	0.9	1.1	1.2	1.4	1.0	1.2	1.3	1.5	1.6
SS B4	32	51	62	72	79	86	90	94	19	33	42	51	58	67	73	79	0.7	0.8	1.0	1.1	1.3	0.9	1.0	1.1	1.3	1.4
SS B5	24	37	45	55	61	70	75	81	16	24	29	36	40	47	52	58	0.9	1.0	1.1	1.3	1.4	1.1	1.2	1.4	1.6	1.7
SS B6	31	51	62	72	79	87	90	94	19	32	41	51	58	67	73	79	0.7	0.8	1.0	1.1	1.3	0.9	1.0	1.1	1.3	1.4
SS B7	22	34	42	51	57	65	70	77	15	22	27	32	37	43	47	53	0.9	1.0	1.2	1.3	1.5	1.2	1.3	1.5	1.7	1.8
SS B8	31	49	60	71	78	85	89	93	18	31	40	50	56	65	71	78	0.7	0.8	1.0	1.1	1.3	0.9	1.0	1.1	1.3	1.5
SS C1	26	43	54	65	72	81	85	90	14	24	32	41	47	55	61	68	0.7	0.8	1.0	1.2	1.3	1.0	1.1	1.2	1.4	1.6
SS C2	29	49	61	72	79	87	91	95	15	29	38	49	56	65	71	78	0.7	0.8	0.9	1.1	1.3	0.9	1.0	1.1	1.3	1.4
SS C3	22	35	44	53	60	69	75	81	13	20	26	33	38	45	50	56	0.9	1.0	1.1	1.3	1.4	1.2	1.3	1.4	1.6	1.7
SS C4	26	44	55	66	73	82	87	91	14	26	34	43	50	59	65	72	0.7	0.8	1.0	1.2	1.3	0.9	1.0	1.2	1.3	1.5
SS D1	36	56	67	77	83	90	93	96	23	36	45	55	62	70	76	82	0.7	0.8	0.9	1.1	1.3	0.8	0.9	1.1	1.3	1.4
SS D2	42	65	76	86	91	95	97	99	26	44	55	66	73	82	86	91	0.6	0.7	0.9	1.0	1.2	0.7	0.8	1.0	1.2	1.3
SS D3	33	51	61	71	77	85	89	93	21	33	41	49	56	64	69	76	0.7	0.8	1.0	1.1	1.3	0.9	1.0	1.2	1.3	1.5
SS D4	41	63	74	84	89	94	97	98	25	42	53	64	71	80	85	90	0.6	0.7	0.9	1.1	1.2	0.7	0.8	1.0	1.2	1.3
SS E1	32	51	62	72	79	86	90	94	19	31	40	49	55	64	70	76	0.7	0.8	1.0	1.1	1.3	0.9	1.0	1.2	1.3	1.5
SS E2	37	59	71	81	87	93	95	98	22	38	49	60	67	76	81	87	0.6	0.7	0.9	1.1	1.2	0.8	0.9	1.0	1.2	1.4
SS E3	27	42	52	62	69	77	82	87	17	27	33	41	46	54	59	66	0.8	0.9	1.0	1.2	1.4	1.0	1.1	1.3	1.5	1.6
SS E4	34	55	67	77	83	90	93	96	20	35	45	55	62	71	77	83	0.7	0.8	0.9	1.1	1.3	0.8	0.9	1.1	1.2	1.4

COLUMBIA, MISSOURI

TYPE	SSF (%) 55 F (3072 DD) LCR 100	70	50	40	30	25	20	15	SSF (%) 65 F (5085 DD) 100	70	50	40	30	25	20	15	CF 55 F 100	50	30	20	15	CF 65 F 100	50	30	20	15
WW A1	19	22	26	29	34	37	41	47	15	17	20	22	25	27	29	33	1.6	1.8	2.0	2.1	2.3	2.0	2.3	2.5	2.8	3.0
WW A2	20	25	31	36	43	48	54	62	14	18	22	26	31	34	39	45	1.3	1.5	1.6	1.8	1.9	1.5	1.7	1.9	2.1	2.3
WW A3	20	26	33	39	46	52	59	67	13	18	23	27	33	37	43	50	1.2	1.4	1.5	1.7	1.8	1.4	1.6	1.8	2.0	2.1
WW A4	20	26	34	40	48	54	61	70	13	18	24	28	35	39	45	53	1.2	1.3	1.5	1.6	1.8	1.4	1.6	1.7	1.9	2.1
WW A5	20	27	36	42	51	56	64	73	13	18	24	29	36	41	47	56	1.1	1.3	1.4	1.6	1.7	1.3	1.5	1.7	1.9	2.0
WW A6	20	28	36	43	52	58	65	74	12	18	25	30	37	42	48	57	1.1	1.3	1.4	1.6	1.7	1.3	1.5	1.6	1.8	2.0
WW B1	15	19	24	27	32	36	40	47	10	13	15	17	20	22	24	28	1.5	1.7	1.9	2.1	2.3	2.0	2.3	2.7	3.1	3.4
WW B2	20	28	36	43	51	57	64	74	14	20	26	31	38	43	50	59	1.1	1.3	1.4	1.6	1.7	1.3	1.5	1.6	1.8	1.9
WW B3	25	33	43	50	59	66	73	81	17	24	32	38	46	51	58	68	1.0	1.2	1.3	1.5	1.6	1.2	1.3	1.5	1.7	1.8
WW B4	25	34	45	52	63	69	76	85	17	25	34	41	50	56	64	73	1.0	1.1	1.3	1.5	1.6	1.1	1.3	1.4	1.6	1.7
WW B5	24	33	44	52	62	69	76	85	16	24	34	41	50	56	64	74	1.0	1.1	1.3	1.4	1.6	1.1	1.2	1.4	1.6	1.7
WW C1	26	35	45	52	62	68	75	83	19	26	34	40	49	54	62	71	1.0	1.1	1.3	1.5	1.6	1.1	1.3	1.5	1.6	1.8
WW C2	24	34	44	51	61	67	75	83	17	25	33	39	48	54	62	71	1.0	1.2	1.3	1.5	1.6	1.1	1.3	1.4	1.6	1.7
WW C3	29	40	52	61	71	77	84	91	21	31	42	49	60	66	74	83	0.9	1.0	1.2	1.4	1.5	1.0	1.1	1.3	1.5	1.6
WW C4	26	37	49	57	68	74	81	89	19	28	38	46	56	63	71	80	0.9	1.1	1.2	1.4	1.5	1.0	1.2	1.3	1.5	1.6
TW A1	19	22	26	29	33	37	41	47	15	17	20	22	25	27	30	34	1.6	1.8	2.0	2.1	2.3	2.0	2.2	2.5	2.7	2.9
TW A2	19	23	29	33	39	44	50	58	14	17	21	24	29	32	36	42	1.4	1.5	1.7	1.9	2.0	1.7	1.9	2.1	2.2	2.4
TW A3	18	24	30	35	42	47	53	62	13	17	21	25	30	34	39	46	1.3	1.5	1.6	1.8	1.9	1.5	1.7	1.9	2.1	2.3
TW A4	17	23	30	35	42	47	54	63	12	16	21	25	31	34	40	47	1.3	1.4	1.6	1.7	1.9	1.5	1.7	1.9	2.0	2.2
TW B1	18	21	26	29	34	38	43	50	14	17	20	22	25	28	31	36	1.6	1.7	1.9	2.1	2.2	1.9	2.1	2.3	2.5	2.7
TW B2	17	22	28	32	38	43	49	57	13	16	20	23	28	31	36	42	1.4	1.5	1.7	1.8	2.0	1.6	1.8	2.0	2.2	2.4
TW B3	17	22	28	32	39	44	50	58	12	16	20	23	28	32	37	44	1.4	1.5	1.7	1.8	2.0	1.6	1.8	2.0	2.1	2.3
TW B4	16	20	26	31	37	42	48	56	11	15	19	22	27	31	36	42	1.4	1.5	1.7	1.8	2.0	1.6	1.8	2.0	2.2	2.3
TW C1	17	21	25	29	34	38	43	51	13	16	19	22	25	28	32	38	1.5	1.7	1.8	2.0	2.1	1.8	2.0	2.2	2.4	2.6
TW C2	16	20	25	29	35	39	45	53	12	15	19	22	26	29	34	40	1.5	1.6	1.8	1.9	2.1	1.7	1.9	2.1	2.3	2.4
TW C3	15	19	24	28	34	38	43	51	11	14	18	21	25	28	32	39	1.5	1.7	1.8	2.0	2.1	1.8	1.9	2.1	2.3	2.4
TW C4	15	18	22	26	31	34	39	46	12	14	17	20	24	26	30	36	1.6	1.8	1.9	2.1	2.2	1.9	2.1	2.3	2.4	2.6
TW D1	13	16	20	24	28	32	36	42	9	11	13	15	18	20	22	26	1.6	1.8	2.0	2.2	2.3	2.1	2.4	2.7	3.0	3.3
TW D2	20	26	33	39	47	52	59	68	14	19	25	29	35	40	46	54	1.2	1.4	1.5	1.7	1.8	1.4	1.6	1.7	1.9	2.0
TW D3	21	28	36	42	51	57	64	73	15	20	27	32	39	43	50	58	1.1	1.3	1.4	1.6	1.7	1.3	1.5	1.6	1.8	2.0
TW D4	23	31	40	47	57	63	70	79	16	23	31	36	45	50	58	67	1.1	1.2	1.4	1.5	1.7	1.2	1.3	1.5	1.7	1.8
TW D5	22	31	40	47	57	63	71	80	16	23	31	37	46	52	59	69	1.0	1.2	1.3	1.5	1.6	1.2	1.3	1.5	1.6	1.8
TW E1	24	32	41	47	57	62	70	78	17	24	31	36	44	49	56	65	1.1	1.2	1.4	1.5	1.7	1.2	1.4	1.5	1.7	1.8
TW E2	24	32	41	48	57	63	70	79	17	24	31	37	45	50	57	67	1.1	1.2	1.4	1.5	1.7	1.2	1.4	1.5	1.7	1.8
TW E3	28	39	49	57	67	73	80	88	21	29	39	46	55	62	69	78	0.9	1.1	1.2	1.4	1.6	1.1	1.2	1.4	1.5	1.7
TW E4	26	36	47	55	65	71	78	86	19	27	37	44	53	60	67	77	1.0	1.1	1.3	1.4	1.6	1.1	1.2	1.4	1.5	1.7
TW F1	17	20	24	28	32	36	41	47	13	15	18	20	23	26	29	33	1.6	1.8	1.9	2.1	2.2	2.0	2.2	2.4	2.7	2.9
TW F2	16	21	26	30	37	41	47	55	11	15	19	22	26	29	33	40	1.4	1.6	1.7	1.9	2.0	1.7	1.9	2.1	2.3	2.4
TW F3	15	20	26	31	38	43	49	58	10	14	18	22	27	30	35	42	1.3	1.5	1.7	1.8	2.0	1.6	1.8	2.0	2.2	2.3
TW F4	13	18	25	29	37	41	48	57	8	12	17	20	25	29	34	41	1.3	1.5	1.7	1.8	1.9	1.6	1.8	2.0	2.1	2.3
TW G1	15	19	23	26	32	35	40	47	11	14	17	19	23	25	29	33	1.6	1.7	1.9	2.1	2.2	1.9	2.1	2.3	2.5	2.7
TW G2	13	18	23	27	34	38	44	52	9	12	16	19	24	27	31	37	1.4	1.6	1.8	1.9	2.0	1.7	1.9	2.1	2.3	2.4
TW G3	12	16	22	26	32	37	43	52	7	11	15	18	23	26	30	37	1.4	1.6	1.7	1.9	2.0	1.7	1.9	2.1	2.3	2.4
TW G4	9	13	19	23	28	33	39	47	5	8	12	15	20	23	27	34	1.5	1.7	1.8	2.0	2.1	1.8	2.0	2.1	2.3	2.5
TW H1	12	16	20	24	29	33	38	45	8	11	14	16	20	23	27	32	1.6	1.8	1.9	2.1	2.2	1.9	2.1	2.3	2.5	2.7
TW H2	9	13	18	22	27	31	37	44	6	9	12	15	19	22	26	32	1.6	1.7	1.9	2.0	2.2	1.9	2.1	2.2	2.4	2.6
TW H3	8	11	16	19	24	28	33	41	4	7	10	13	17	19	23	29	1.7	1.8	2.0	2.1	2.2	2.0	2.1	2.3	2.5	2.6
TW H4	6	9	12	15	19	23	27	34	3	5	8	10	13	16	19	24	1.8	2.0	2.2	2.3	2.4	2.2	2.4	2.5	2.7	2.8
TW I1	11	14	18	21	25	28	33	39	7	9	11	13	15	17	20	23	1.7	1.9	2.1	2.3	2.4	2.2	2.5	2.8	3.2	3.5
TW I2	17	23	30	35	43	49	56	65	11	16	21	26	32	36	42	50	1.2	1.4	1.5	1.7	1.8	1.4	1.6	1.8	1.9	2.1
TW I3	18	25	33	39	47	53	60	69	12	17	24	28	35	40	46	55	1.2	1.3	1.5	1.6	1.8	1.3	1.5	1.7	1.8	2.0
TW I4	20	28	37	44	53	60	67	76	13	20	28	33	42	47	54	64	1.1	1.2	1.4	1.5	1.7	1.2	1.4	1.5	1.7	1.8
TW I5	20	28	38	45	55	61	69	78	13	21	29	35	43	49	57	67	1.1	1.2	1.4	1.5	1.7	1.2	1.3	1.5	1.7	1.8
TW J1	22	30	38	45	54	60	67	76	15	21	29	34	42	47	54	63	1.1	1.2	1.4	1.5	1.7	1.2	1.4	1.6	1.7	1.9
TW J2	21	29	38	45	54	60	68	77	14	21	28	34	42	48	55	64	1.1	1.2	1.4	1.5	1.7	1.2	1.4	1.5	1.7	1.8
TW J3	25	35	46	54	65	71	78	86	18	26	36	43	53	59	67	77	1.0	1.1	1.3	1.4	1.6	1.1	1.2	1.4	1.5	1.7
TW J4	23	33	44	52	62	69	76	85	16	25	34	41	51	57	65	75	1.0	1.1	1.3	1.4	1.6	1.1	1.2	1.4	1.5	1.7
DG A1	13	17	22	25	29	31	34	38	8	11	14	15	18	19	21	22	1.5	1.8	2.2	2.7	3.4	1.9	2.3	3.0	4.3	-
DG A2	16	21	27	32	38	42	48	54	11	15	19	23	28	31	35	41	1.4	1.5	1.7	2.0	2.2	1.6	1.8	2.0	2.3	2.6
DG A3	19	26	34	40	49	55	62	71	15	20	26	31	38	43	50	59	1.2	1.3	1.4	1.6	1.8	1.4	1.5	1.6	1.8	2.0
DG B1	13	17	22	26	32	35	40	45	8	11	14	16	19	21	23	26	1.5	1.7	1.9	2.2	2.6	1.9	2.3	2.6	3.2	3.9
DG B2	16	21	28	33	41	46	53	61	11	15	20	23	29	33	38	45	1.3	1.4	1.6	1.8	2.0	1.6	1.7	1.9	2.1	2.3
DG B3	20	27	35	42	52	58	67	76	15	21	27	32	40	45	53	63	1.2	1.3	1.4	1.5	1.7	1.3	1.5	1.6	1.7	1.8
DG C1	16	21	27	32	39	44	49	57	11	14	18	21	25	27	31	36	1.4	1.5	1.7	1.9	2.1	1.7	1.9	2.2	2.5	2.8
DG C2	19	25	33	39	47	53	61	70	14	18	24	28	34	39	45	53	1.2	1.3	1.5	1.6	1.8	1.4	1.6	1.7	1.9	2.1
DG C3	22	30	40	47	58	65	73	82	17	23	31	36	45	51	59	70	1.1	1.2	1.3	1.4	1.6	1.3	1.4	1.5	1.6	1.7
SS A1	22	28	34	38	45	50	56	64	18	22	27	30	35	39	44	50	1.3	1.4	1.6	1.8	1.9	1.5	1.7	1.9	2.1	2.2
SS A2	28	36	44	50	59	65	71	79	23	29	36	41	49	54	60	68	1.1	1.2	1.4	1.5	1.7	1.2	1.4	1.5	1.7	1.8
SS A3	21	25	31	35	41	45	50	58	17	20	24	27	31	34	38	43	1.4	1.6	1.7	1.9	2.0	1.6	1.8	2.1	2.3	2.5
SS A4	27	35	43	49	57	63	70	78	22	28	35	40	47	52	58	66	1.1	1.2	1.4	1.6	1.7	1.2	1.4	1.6	1.7	1.9
SS A5	22	26	31	35	41	45	50	58	18	22	25	28	32	35	39	44	1.4	1.6	1.7	1.9	2.1	1.6	1.9	2.1	2.3	2.5
SS A6	28	35	44	50	58	64	71	79	22	28	35	41	48	53	59	67	1.1	1.2	1.4	1.5	1.7	1.2	1.4	1.5	1.7	1.9
SS A7	20	23	27	31	35	38	43	49	17	19	22	24	27	30	32	36	1.6	1.8	1.9	2.1	2.3	1.8	2.1	2.4	2.7	3.0
SS A8	27	34	42	48	56	61	68	76	21	27	34	39	45	50	56	64	1.1	1.3	1.4	1.6	1.7	1.2	1.4	1.6	1.8	1.9
SS B1	19	24	29	33	40	44	50	57	15	19	23	26	30	34	38	44	1.4	1.5	1.7	1.9	2.0	1.6	1.8	2.0	2.2	2.4
SS B2	24	31	39	45	54	59	66	75	19	25	32	37	44	49	55	64	1.1	1.3	1.4	1.6	1.7	1.2	1.4	1.6	1.7	1.9
SS B3	18	22	27	31	37	41	46	53	14	17	21	23	27	30	34	39	1.4	1.6	1.8	2.0	2.1	1.7	1.9	2.2	2.4	2.6
SS B4	24	30	38	44	52	58	65	73	19	24	31	36	42	47	53	62	1.1	1.3	1.4	1.6	1.8	1.3	1.4	1.6	1.8	1.9
SS B5	18	22	26	29	34	38	43	49	15	17	21	23	27	29	32	37	1.5	1.7	1.9	2.1	2.2	1.8	2.1	2.3	2.5	2.7
SS B6	24	31	38	44	53	58	65	74	19	24	31	36	43	47	54	62	1.1	1.3	1.4	1.6	1.7	1.3	1.4	1.6	1.8	1.9
SS B7	16	20	23	26	30	33	37	43	13	16	18	20	23	25	28	31	1.7	1.9	2.1	2.3	2.4	2.0	2.3	2.6	2.9	3.1
SS B8	23	29	37	42	50	56	63	71	18	23	29	34	40	45	51	59	1.2	1.3	1.5	1.6	1.8	1.3	1.5	1.7	1.8	2.0
SS C1	17	23	29	34	41	46	52	60	12	15	20	23	28	32	36	43	1.3	1.5	1.6	1.8	1.9	1.6	1.8	2.0	2.2	2.3
SS C2	21	28	37	43	52	57	65	74	14	20	27	32	39	44	50	59	1.1	1.3	1.4	1.6	1.7	1.3	1.5	1.6	1.8	1.9
SS C3	16	20	24	28	33	37	42	50	12	14	17	20	24	26	30	35	1.5	1.7	1.9	2.0	2.1	1.8	2.0	2.3	2.5	2.7
SS C4	19	25	33	38	46	51	58	67	14	19	24	28	35	39	45	53	1.2	1.4	1.5	1.7	1.8	1.4	1.6	1.7	1.9	2.0
SS D1	26	32	40	46	53	59	65	73	20	25	31	35	42	46	51	58	1.1	1.3	1.5	1.6	1.8	1.3	1.5	1.7	1.9	2.0
SS D2	32	41	51	58	68	73	80	87	25	33	41	48	56	62	69	77	1.0	1.1	1.3	1.4	1.6	1.1	1.2	1.4	1.6	1.7
SS D3	24	30	36	41	48	53	59	67	19	24	29	32	37	41	46	52	1.2	1.4	1.6	1.7	1.9	1.4	1.6	1.8	2.0	2.2
SS D4	31	40	49	56	65	71	78	85	24	32	40	46	54	60	67	75	1.0	1.1	1.3	1.5	1.6	1.1	1.3	1.4	1.6	1.7
SS E1	22	28	34	40	47	52	58	66	16	21	26	29	35	38	43	50	1.2	1.4	1.5	1.7	1.9	1.4	1.7	1.9	2.1	2.2
SS E2	27	35	45	51	61	66	73	81	20	27	35	41	49	54	61	69	1.0	1.2	1.3	1.5	1.7	1.1	1.3	1.5	1.7	1.8
SS E3	20	24	29	33	39	43	49	56	16	19	23	26	30	33	37	42	1.4	1.6	1.7	1.9	2.1	1.6	1.9	2.1	2.3	2.5
SS E4	25	33	41	47	56	62	69	77	19	26	32	38	45	50	56	65	1.1	1.2	1.4	1.6	1.7	1.2	1.4	1.6	1.7	1.9

KANSAS CITY, MISSOURI

SOLAR SAVINGS FRACTION (%) / CONSERVATION FACTOR

TYPE	BASE TEMP: 55 F (3319 DD)								65 F (5369 DD)								CF 55 F					CF 65 F				
LCR	100	70	50	40	30	25	20	15	100	70	50	40	30	25	20	15	100	50	30	20	15	100	50	30	20	15
WW A1	19	22	26	30	34	38	42	49	15	18	20	22	25	28	31	35	1.6	1.8	1.9	2.1	2.2	2.0	2.2	2.4	2.7	2.9
WW A2	20	25	32	37	44	49	55	64	15	19	23	27	32	36	41	48	1.3	1.4	1.6	1.7	1.9	1.5	1.7	1.9	2.1	2.2
WW A3	20	27	34	40	48	53	60	69	14	19	24	29	35	39	45	53	1.2	1.3	1.5	1.7	1.8	1.4	1.6	1.8	1.9	2.1
WW A4	20	27	35	41	50	55	63	72	13	19	25	30	36	41	47	56	1.1	1.3	1.5	1.6	1.8	1.3	1.5	1.7	1.9	2.0
WW A5	20	28	37	43	52	58	65	74	13	19	26	31	38	43	49	58	1.1	1.3	1.4	1.6	1.7	1.3	1.5	1.6	1.8	1.9
WW A6	20	28	37	44	53	59	67	76	13	19	26	31	39	44	51	60	1.1	1.2	1.4	1.6	1.7	1.3	1.4	1.6	1.8	1.9
WW B1	16	20	24	28	33	37	42	49	11	13	16	18	21	24	26	30	1.5	1.7	1.9	2.0	2.2	1.9	2.2	2.5	2.8	3.1
WW B2	21	28	37	44	53	58	66	75	14	20	27	32	40	45	52	61	1.1	1.3	1.4	1.6	1.7	1.3	1.4	1.6	1.8	1.9
WW B3	25	34	44	51	61	67	74	83	18	25	33	39	48	53	61	70	1.0	1.2	1.3	1.5	1.6	1.1	1.3	1.5	1.6	1.8
WW B4	25	35	46	54	64	70	77	86	18	26	35	42	52	58	66	75	1.0	1.1	1.3	1.4	1.6	1.1	1.2	1.4	1.5	1.7
WW B5	24	34	45	53	63	70	77	86	17	25	35	42	52	58	66	76	1.0	1.1	1.3	1.4	1.6	1.1	1.2	1.4	1.5	1.7
WW C1	27	36	46	53	63	69	76	85	19	27	35	42	51	56	64	73	1.0	1.1	1.3	1.5	1.6	1.1	1.3	1.4	1.6	1.7
WW C2	25	34	45	52	62	68	76	84	18	25	34	41	50	56	64	73	1.0	1.1	1.3	1.5	1.6	1.1	1.3	1.4	1.6	1.7
WW C3	30	41	53	62	72	78	85	92	22	32	43	51	61	68	75	84	0.9	1.0	1.2	1.4	1.5	1.0	1.1	1.3	1.4	1.6
WW C4	27	38	50	58	69	75	82	90	19	29	40	47	58	65	72	82	0.9	1.1	1.2	1.4	1.5	1.0	1.2	1.3	1.5	1.6
TW A1	19	22	26	29	34	37	42	49	15	17	20	22	26	28	31	36	1.6	1.8	1.9	2.1	2.2	2.0	2.2	2.4	2.6	2.8
TW A2	19	24	29	34	40	45	51	59	14	18	22	25	30	33	38	44	1.4	1.5	1.7	1.8	2.0	1.6	1.8	2.0	2.2	2.3
TW A3	19	24	31	35	43	48	54	63	13	17	22	26	32	35	41	48	1.3	1.4	1.6	1.7	1.9	1.5	1.7	1.9	2.0	2.2
TW A4	18	23	30	36	43	49	55	64	12	17	22	26	32	36	42	49	1.3	1.4	1.6	1.7	1.8	1.5	1.6	1.8	2.0	2.1
TW B1	18	22	26	30	35	39	44	51	14	17	20	23	26	29	32	38	1.5	1.7	1.9	2.0	2.2	1.9	2.1	2.2	2.4	2.6
TW B2	18	22	28	33	39	44	50	58	13	17	21	24	29	33	38	44	1.4	1.5	1.7	1.8	2.0	1.6	1.8	2.0	2.1	2.3
TW B3	17	22	28	33	40	45	51	60	12	16	21	24	30	33	38	46	1.3	1.5	1.6	1.8	1.9	1.6	1.7	1.9	2.1	2.2
TW B4	16	21	27	31	38	43	49	58	12	15	20	23	28	32	37	44	1.4	1.5	1.7	1.8	2.0	1.6	1.8	1.9	2.1	2.2
TW C1	17	21	26	29	35	39	44	52	13	16	19	22	26	29	33	39	1.5	1.7	1.8	2.0	2.1	1.8	2.0	2.1	2.3	2.5
TW C2	16	20	26	30	36	40	46	54	12	15	19	22	27	30	35	41	1.4	1.6	1.7	1.9	2.0	1.7	1.9	2.0	2.2	2.3
TW C3	15	19	24	28	34	38	44	52	12	15	19	22	26	29	34	40	1.5	1.6	1.8	1.9	2.1	1.7	1.9	2.1	2.2	2.4
TW C4	15	18	23	26	31	35	40	47	12	14	18	20	24	27	31	37	1.6	1.8	1.9	2.1	2.2	1.9	2.0	2.2	2.4	2.5
TW D1	13	17	21	24	29	32	37	44	9	11	14	16	19	21	24	28	1.6	1.8	2.0	2.1	2.3	2.0	2.3	2.6	2.9	3.1
TW D2	20	27	34	40	48	54	61	70	14	20	25	30	37	41	48	56	1.2	1.3	1.5	1.6	1.8	1.4	1.5	1.7	1.8	2.0
TW D3	21	29	37	43	52	58	65	74	15	21	28	33	40	45	52	60	1.1	1.3	1.4	1.6	1.7	1.3	1.4	1.6	1.8	1.9
TW D4	23	32	41	48	58	64	71	80	17	24	32	38	46	52	59	69	1.0	1.2	1.3	1.5	1.6	1.2	1.3	1.5	1.6	1.8
TW D5	23	31	41	48	58	65	72	81	16	24	32	38	47	53	61	71	1.0	1.2	1.3	1.5	1.6	1.2	1.3	1.5	1.6	1.7
TW E1	25	33	42	48	58	64	71	80	18	25	32	38	46	51	58	67	1.1	1.2	1.4	1.5	1.7	1.2	1.4	1.5	1.7	1.8
TW E2	24	32	42	49	58	64	72	80	18	24	32	38	46	52	59	69	1.0	1.2	1.3	1.5	1.6	1.2	1.3	1.5	1.6	1.8
TW E3	29	39	50	58	68	74	81	89	21	30	40	47	57	63	71	80	0.9	1.1	1.2	1.4	1.6	1.0	1.2	1.3	1.5	1.6
TW E4	27	37	48	56	66	72	79	87	20	28	38	45	55	61	69	78	1.0	1.1	1.3	1.4	1.6	1.1	1.2	1.4	1.5	1.7
TW F1	17	21	25	28	33	37	42	49	13	16	19	21	24	27	30	35	1.6	1.7	1.9	2.1	2.2	1.9	2.1	2.3	2.6	2.7
TW F2	16	21	27	31	38	42	48	56	12	15	19	23	27	31	35	42	1.4	1.5	1.7	1.9	2.0	1.6	1.8	2.0	2.2	2.4
TW F3	15	21	27	32	39	44	50	59	10	14	19	23	28	32	37	44	1.3	1.5	1.6	1.8	1.9	1.6	1.7	1.9	2.1	2.2
TW F4	13	19	25	30	37	43	49	58	8	12	17	21	27	31	36	44	1.3	1.5	1.6	1.8	1.9	1.6	1.7	1.9	2.1	2.2
TW G1	15	19	23	27	32	36	41	49	11	14	17	20	23	26	30	35	1.5	1.7	1.9	2.0	2.2	1.9	2.1	2.3	2.5	2.6
TW G2	14	18	24	28	34	39	45	53	9	13	17	20	25	28	33	39	1.4	1.6	1.7	1.9	2.0	1.7	1.9	2.0	2.2	2.4
TW G3	12	17	22	27	33	38	44	53	8	11	15	19	24	27	32	39	1.4	1.6	1.7	1.9	2.0	1.7	1.8	2.0	2.2	2.3
TW G4	9	14	19	23	29	34	40	48	6	9	13	16	21	24	29	35	1.5	1.7	1.8	2.0	2.1	1.8	1.9	2.1	2.3	2.4
TW H1	12	16	20	24	30	34	39	46	8	11	14	17	21	24	28	34	1.6	1.7	1.9	2.0	2.2	1.9	2.0	2.2	2.4	2.6
TW H2	10	14	18	22	28	32	38	46	6	9	12	15	20	23	27	33	1.6	1.7	1.9	2.0	2.1	1.8	2.0	2.2	2.4	2.5
TW H3	8	11	16	19	25	29	34	42	5	7	11	13	17	20	24	30	1.6	1.8	1.9	2.1	2.2	1.9	2.1	2.3	2.4	2.6
TW H4	6	9	12	15	20	23	28	34	3	6	8	10	14	16	20	25	1.8	2.0	2.1	2.3	2.4	2.1	2.3	2.5	2.6	2.8
TW I1	11	14	18	21	26	29	34	40	7	9	12	14	16	18	21	25	1.7	1.9	2.1	2.2	2.4	2.1	2.4	2.7	3.0	3.2
TW I2	17	23	31	36	44	50	57	67	12	16	22	27	33	38	44	53	1.2	1.4	1.5	1.7	1.8	1.4	1.6	1.7	1.9	2.0
TW I3	18	25	33	39	48	54	61	71	13	18	25	30	37	42	48	57	1.2	1.3	1.5	1.6	1.7	1.3	1.5	1.6	1.8	1.9
TW I4	20	28	38	45	54	61	68	78	14	21	29	35	43	49	56	66	1.1	1.2	1.4	1.5	1.7	1.2	1.4	1.5	1.7	1.8
TW I5	20	29	39	46	56	62	70	79	14	21	30	36	45	51	58	68	1.1	1.2	1.3	1.5	1.6	1.2	1.3	1.5	1.6	1.8
TW J1	22	30	39	46	55	61	69	78	16	22	30	35	43	49	56	65	1.1	1.2	1.4	1.5	1.7	1.2	1.4	1.5	1.7	1.8
TW J2	21	30	39	46	55	61	69	78	15	22	29	35	44	49	57	66	1.1	1.2	1.4	1.5	1.7	1.2	1.4	1.5	1.7	1.8
TW J3	26	36	47	55	66	72	79	87	19	27	37	45	54	61	69	78	1.0	1.1	1.3	1.4	1.6	1.1	1.2	1.4	1.5	1.7
TW J4	24	34	45	53	63	70	77	86	17	26	35	42	52	59	67	76	1.0	1.1	1.3	1.4	1.6	1.1	1.2	1.4	1.5	1.7
DG A1	14	18	22	26	30	33	36	39	9	11	14	16	19	21	22	24	1.5	1.8	2.1	2.6	3.2	1.9	2.2	2.8	3.9	-
DG A2	16	21	28	32	39	43	49	56	11	15	20	23	29	32	37	42	1.3	1.5	1.7	1.9	2.2	1.6	1.7	1.9	2.2	2.5
DG A3	20	26	35	41	50	56	63	72	15	20	27	32	39	45	52	60	1.2	1.3	1.4	1.6	1.8	1.3	1.5	1.6	1.8	1.9
DG B1	14	18	23	27	33	37	41	47	9	11	15	17	20	22	25	28	1.5	1.6	1.8	2.1	2.5	1.8	2.2	2.5	2.9	3.5
DG B2	16	22	28	34	42	47	54	63	12	16	21	24	30	34	40	47	1.3	1.4	1.5	1.7	1.9	1.5	1.7	1.8	2.0	2.2
DG B3	20	27	36	43	53	60	68	77	16	21	28	33	41	47	55	65	1.2	1.2	1.4	1.5	1.7	1.3	1.4	1.5	1.7	1.8
DG C1	16	22	28	33	40	45	51	59	11	15	19	22	26	29	33	39	1.4	1.4	1.6	1.8	2.1	1.6	1.9	2.1	2.4	2.7
DG C2	19	25	33	39	49	55	62	71	14	19	24	29	36	41	47	56	1.2	1.3	1.4	1.6	1.8	1.4	1.6	1.7	1.8	2.0
DG C3	23	31	40	48	59	66	75	84	18	24	31	37	46	53	61	72	1.1	1.2	1.3	1.4	1.6	1.2	1.3	1.4	1.6	1.7
SS A1	23	28	34	39	46	50	56	65	18	23	27	31	36	40	45	52	1.3	1.4	1.6	1.8	1.9	1.4	1.7	1.8	2.0	2.2
SS A2	28	36	44	51	60	65	72	80	23	29	37	42	50	55	61	70	1.1	1.2	1.4	1.5	1.7	1.2	1.3	1.5	1.7	1.8
SS A3	21	25	31	35	41	45	51	58	17	20	24	28	32	35	39	45	1.4	1.5	1.7	1.9	2.0	1.6	1.8	2.0	2.3	2.4
SS A4	27	35	43	49	58	63	70	78	22	29	35	41	48	53	59	67	1.1	1.2	1.4	1.5	1.7	1.2	1.4	1.5	1.7	1.8
SS A5	22	27	32	36	41	46	51	58	18	22	26	29	33	36	40	46	1.4	1.6	1.7	1.9	2.0	1.6	1.8	2.0	2.2	2.4
SS A6	28	35	44	50	59	65	71	80	23	29	36	41	49	54	60	69	1.1	1.2	1.4	1.5	1.7	1.2	1.4	1.5	1.7	1.8
SS A7	20	23	28	31	35	39	43	50	17	19	22	25	28	30	33	37	1.5	1.8	1.9	2.1	2.3	1.8	2.1	2.4	2.6	2.9
SS A8	27	34	42	48	56	62	69	77	22	28	34	39	46	51	57	65.	1.1	1.2	1.4	1.6	1.7	1.2	1.4	1.6	1.7	1.9
SS B1	19	24	29	34	40	44	50	58	15	19	23	26	31	35	39	45	1.4	1.5	1.7	1.8	2.0	1.6	1.8	2.0	2.2	2.3
SS B2	25	32	40	46	54	60	67	76	20	26	33	38	45	50	57	65	1.1	1.3	1.4	1.6	1.7	1.2	1.4	1.6	1.7	1.9
SS B3	18	22	27	31	37	41	46	54	14	17	21	24	28	31	35	41	1.4	1.6	1.8	1.9	2.1	1.7	1.9	2.1	2.3	2.5
SS B4	24	31	39	44	53	58	65	74	19	25	31	36	43	48	55	63	1.1	1.3	1.4	1.6	1.7	1.3	1.4	1.6	1.8	1.9
SS B5	18	22	26	30	35	38	43	50	15	18	21	24	27	30	33	38	1.5	1.7	1.9	2.0	2.2	1.8	2.0	2.2	2.5	2.6
SS B6	24	31	39	45	53	59	66	75	19	25	31	36	44	49	55	63	1.1	1.3	1.4	1.6	1.7	1.3	1.4	1.6	1.7	1.9
SS B7	16	20	23	26	31	34	38	44	14	16	19	21	24	26	29	33	1.7	1.9	2.1	2.2	2.4	2.0	2.2	2.5	2.8	3.0
SS B8	23	29	37	43	51	56	63	72	18	23	30	34	41	46	52	60	1.2	1.3	1.5	1.6	1.8	1.3	1.5	1.6	1.8	1.9
SS C1	18	23	30	35	42	47	53	62	12	16	21	24	30	33	38	45	1.3	1.4	1.6	1.8	1.9	1.5	1.7	1.9	2.1	2.2
SS C2	21	29	37	44	53	59	66	75	15	21	28	33	41	46	52	61	1.1	1.3	1.4	1.6	1.7	1.3	1.4	1.6	1.8	1.9
SS C3	16	20	25	28	34	38	43	51	12	15	18	21	25	28	32	37	1.5	1.7	1.8	2.0	2.1	1.8	2.0	2.2	2.4	2.5
SS C4	19	26	33	39	47	53	60	69	14	19	25	30	36	41	47	55	1.2	1.3	1.5	1.6	1.8	1.4	1.5	1.7	1.9	2.0
SS D1	26	33	40	46	54	59	66	74	20	26	32	36	43	47	53	60	1.1	1.3	1.4	1.6	1.8	1.3	1.5	1.7	1.8	2.0
SS D2	32	41	52	59	68	74	80	88	25	33	42	49	58	63	70	79	1.0	1.1	1.3	1.4	1.6	1.0	1.2	1.4	1.5	1.7
SS D3	24	30	36	41	48	53	59	68	20	24	29	33	38	42	47	54	1.2	1.4	1.5	1.7	1.9	1.4	1.6	1.8	2.0	2.1
SS D4	31	40	50	57	66	72	78	86	25	32	41	47	56	61	68	76	1.0	1.1	1.3	1.5	1.6	1.1	1.2	1.4	1.6	1.7
SS E1	22	28	35	40	48	53	59	67	17	21	26	30	36	40	45	52	1.2	1.4	1.5	1.7	1.8	1.4	1.6	1.8	2.0	2.2
SS E2	27	36	45	52	61	67	74	82	21	28	36	42	50	55	62	71	1.0	1.2	1.3	1.5	1.6	1.1	1.3	1.5	1.6	1.8
SS E3	20	24	30	34	40	44	50	57	16	19	23	26	31	34	38	44	1.4	1.6	1.7	1.9	2.0	1.6	1.8	2.0	2.3	2.4
SS E4	25	33	42	48	57	63	70	78	20	26	33	39	46	51	58	66	1.1	1.2	1.4	1.5	1.7	1.2	1.4	1.5	1.7	1.8

SAINT LOUIS, MISSOURI

TYPE	SOLAR SAVINGS FRACTION (%) BASE TEMP: 55 F (2804 DD)								65 F (4754 DD)								CONSERVATION FACTOR 55 F					65 F				
LCR	100	70	50	40	30	25	20	15	100	70	50	40	30	25	20	15	100	50	30	20	15	100	50	30	20	15
WW A1	20	24	28	32	37	40	45	52	16	18	21	23	27	29	32	37	1.5	1.7	1.9	2.0	2.2	1.9	2.1	2.4	2.6	2.8
WW A2	22	28	34	39	47	52	59	67	15	20	24	28	34	38	43	50	1.2	1.4	1.5	1.7	1.8	1.5	1.7	1.9	2.0	2.2
WW A3	22	29	37	43	51	56	63	72	15	20	26	30	36	41	47	55	1.2	1.3	1.5	1.6	1.8	1.4	1.5	1.7	1.9	2.0
WW A4	22	30	38	44	53	59	66	75	14	20	26	31	38	43	49	57	1.1	1.3	1.4	1.6	1.7	1.3	1.5	1.7	1.8	2.0
WW A5	22	30	39	46	55	61	69	77	14	20	27	32	40	45	51	60	1.1	1.2	1.4	1.5	1.7	1.3	1.4	1.6	1.8	1.9
WW A6	23	31	40	47	57	63	70	79	14	20	28	33	41	46	53	61	1.1	1.2	1.4	1.5	1.7	1.2	1.4	1.6	1.7	1.9
WW B1	17	22	27	31	36	40	46	53	12	14	17	20	23	25	28	32	1.4	1.6	1.8	2.0	2.1	1.8	2.1	2.4	2.7	3.0
WW B2	23	31	40	46	56	62	69	78	15	22	29	34	41	47	53	62	1.1	1.2	1.4	1.5	1.7	1.2	1.4	1.6	1.7	1.9
WW B3	27	37	47	54	64	70	77	85	19	26	34	41	49	55	62	71	1.0	1.1	1.3	1.5	1.6	1.1	1.3	1.5	1.6	1.8
WW B4	27	38	49	56	67	73	80	87	19	27	37	44	53	59	67	76	1.0	1.1	1.3	1.4	1.6	1.1	1.2	1.4	1.5	1.7
WW B5	26	37	48	56	66	72	79	87	18	27	36	43	53	59	67	77	1.0	1.1	1.3	1.4	1.6	1.1	1.2	1.4	1.5	1.7
WW C1	29	39	49	56	66	72	79	87	20	28	37	43	52	58	65	74	1.0	1.1	1.3	1.4	1.6	1.1	1.3	1.4	1.6	1.7
WW C2	27	37	47	55	65	71	78	86	19	27	36	42	51	57	65	74	1.0	1.1	1.3	1.4	1.6	1.1	1.3	1.4	1.6	1.7
WW C3	32	44	56	64	75	80	87	93	23	33	44	52	62	69	76	85	0.9	1.0	1.2	1.4	1.5	1.0	1.1	1.3	1.4	1.6
WW C4	29	41	53	61	71	77	84	91	20	30	41	49	59	66	73	82	0.9	1.0	1.2	1.4	1.5	1.0	1.1	1.3	1.5	1.6
TW A1	20	24	28	31	36	40	45	52	16	18	21	23	27	29	32	37	1.5	1.7	1.9	2.0	2.2	1.9	2.1	2.3	2.5	2.7
TW A2	20	25	31	36	43	48	54	62	15	19	23	26	31	35	39	46	1.3	1.5	1.6	1.8	1.9	1.6	1.8	1.9	2.1	2.3
TW A3	20	26	33	38	46	51	58	66	14	18	23	27	33	37	42	50	1.2	1.4	1.5	1.7	1.8	1.5	1.7	1.8	2.0	2.2
TW A4	19	26	33	38	46	52	59	68	13	18	23	27	33	38	43	51	1.2	1.4	1.5	1.7	1.8	1.4	1.6	1.8	1.9	2.1
TW B1	19	23	28	32	37	41	46	54	15	18	21	24	27	30	34	39	1.5	1.6	1.8	1.9	2.1	1.8	2.0	2.2	2.4	2.6
TW B2	19	24	30	35	42	47	53	62	14	17	22	25	31	34	39	46	1.3	1.5	1.6	1.8	1.9	1.6	1.7	1.9	2.1	2.2
TW B3	18	24	30	35	43	48	54	63	13	17	22	25	31	35	40	47	1.3	1.4	1.6	1.7	1.9	1.5	1.7	1.9	2.0	2.2
TW B4	17	23	29	34	41	45	52	61	12	16	21	24	30	33	39	46	1.3	1.5	1.6	1.8	1.9	1.6	1.7	1.9	2.1	2.2
TW C1	18	22	27	31	37	41	47	55	14	17	20	23	28	31	35	41	1.4	1.6	1.8	1.9	2.0	1.7	1.9	2.1	2.3	2.4
TW C2	17	22	27	32	38	43	49	57	13	16	20	23	28	32	36	43	1.4	1.5	1.7	1.8	2.0	1.6	1.8	2.0	2.2	2.3
TW C3	17	21	26	30	37	41	47	55	12	15	19	23	27	31	35	42	1.4	1.6	1.7	1.9	2.0	1.7	1.9	2.0	2.2	2.3
TW C4	16	20	24	28	33	37	42	50	12	15	18	21	25	28	32	38	1.5	1.7	1.8	2.0	2.1	1.8	2.0	2.2	2.3	2.5
TW D1	15	18	23	27	32	36	41	48	10	12	15	17	20	23	26	30	1.5	1.7	1.9	2.0	2.2	1.9	2.2	2.5	2.8	3.0
TW D2	22	29	36	42	51	57	64	73	15	21	27	31	38	43	49	57	1.2	1.3	1.4	1.6	1.7	1.3	1.5	1.7	1.8	2.0
TW D3	23	31	40	46	55	61	68	77	16	22	29	34	42	47	53	62	1.1	1.2	1.4	1.5	1.7	1.3	1.4	1.6	1.8	1.9
TW D4	25	34	44	51	61	67	74	83	18	25	33	39	48	53	61	70	1.0	1.2	1.3	1.5	1.6	1.2	1.3	1.5	1.6	1.8
TW D5	25	34	44	51	61	67	75	83	17	25	33	40	49	55	62	72	1.0	1.2	1.3	1.5	1.6	1.1	1.3	1.4	1.6	1.7
TW E1	27	35	44	51	61	66	74	82	19	26	33	39	47	52	59	68	1.0	1.2	1.3	1.5	1.6	1.2	1.3	1.5	1.7	1.8
TW E2	26	35	44	51	61	67	74	83	18	26	33	39	48	53	61	70	1.0	1.2	1.3	1.5	1.6	1.2	1.3	1.5	1.6	1.8
TW E3	31	42	53	61	71	77	83	90	23	32	41	49	58	65	72	81	0.9	1.1	1.2	1.4	1.5	1.0	1.2	1.3	1.5	1.6
TW E4	29	39	50	58	68	75	81	89	21	29	39	46	56	62	70	79	0.9	1.1	1.2	1.4	1.5	1.0	1.2	1.3	1.5	1.6
TW F1	18	22	27	30	36	39	45	52	14	16	20	22	26	28	32	37	1.5	1.7	1.8	2.0	2.1	1.9	2.1	2.3	2.5	2.7
TW F2	18	23	29	34	40	45	51	60	12	16	20	24	29	32	37	43	1.3	1.5	1.6	1.8	1.9	1.6	1.8	2.0	2.2	2.3
TW F3	17	23	29	34	42	47	54	62	11	15	20	24	29	33	38	46	1.3	1.4	1.6	1.7	1.9	1.5	1.7	1.9	2.1	2.2
TW F4	15	21	27	33	40	46	52	62	9	13	18	22	28	32	38	45	1.3	1.4	1.6	1.7	1.9	1.5	1.7	1.9	2.0	2.2
TW G1	16	20	25	29	35	39	44	52	12	15	18	21	25	28	31	37	1.5	1.7	1.8	2.0	2.1	1.8	2.0	2.2	2.4	2.6
TW G2	15	20	26	30	37	42	48	57	10	13	18	21	26	29	34	41	1.4	1.5	1.7	1.8	2.0	1.6	1.8	2.0	2.2	2.3
TW G3	13	18	24	29	36	41	47	56	8	12	16	20	25	29	34	41	1.4	1.5	1.7	1.8	2.0	1.6	1.8	2.0	2.2	2.3
TW G4	11	15	21	25	32	36	42	51	6	10	14	17	22	25	30	37	1.4	1.6	1.7	1.9	2.0	1.7	1.9	2.1	2.2	2.4
TW H1	13	17	22	26	32	36	42	50	9	12	15	18	22	25	29	35	1.5	1.7	1.8	2.0	2.1	1.8	2.0	2.2	2.4	2.5
TW H2	11	15	20	24	30	35	41	49	6	10	13	16	21	24	28	35	1.5	1.7	1.8	2.0	2.1	1.8	2.0	2.1	2.3	2.5
TW H3	9	13	18	21	27	31	37	45	5	8	11	14	18	21	26	32	1.6	1.7	1.9	2.0	2.2	1.9	2.0	2.2	2.4	2.5
TW H4	7	10	14	17	22	25	30	37	4	6	9	11	15	17	21	26	1.7	1.9	2.1	2.2	2.3	2.1	2.3	2.4	2.6	2.7
TW I1	12	16	20	24	29	32	37	44	8	10	13	15	18	20	23	27	1.6	1.8	2.0	2.1	2.3	2.0	2.3	2.6	2.9	3.1
TW I2	19	25	33	39	47	53	60	70	12	17	23	28	35	39	46	54	1.2	1.3	1.5	1.6	1.8	1.4	1.5	1.7	1.9	2.0
TW I3	20	27	36	42	51	57	64	74	13	19	26	31	38	43	50	59	1.1	1.3	1.4	1.6	1.7	1.3	1.5	1.6	1.8	1.9
TW I4	22	31	40	47	57	63	71	80	15	22	30	36	44	50	58	67	1.0	1.2	1.3	1.5	1.6	1.2	1.3	1.5	1.7	1.8
TW I5	22	31	41	49	58	65	73	81	15	22	31	37	46	52	60	69	1.0	1.2	1.3	1.5	1.6	1.2	1.3	1.5	1.6	1.8
TW J1	24	33	42	49	58	64	72	80	17	23	31	37	45	50	57	66	1.0	1.2	1.3	1.5	1.7	1.2	1.4	1.5	1.7	1.8
TW J2	23	32	41	48	58	64	72	81	16	23	31	37	45	51	58	67	1.0	1.2	1.3	1.5	1.6	1.2	1.3	1.5	1.7	1.8
TW J3	28	38	50	58	68	74	81	89	20	29	39	46	56	62	70	79	0.9	1.1	1.2	1.4	1.5	1.0	1.2	1.3	1.5	1.6
TW J4	26	36	47	55	66	72	79	87	18	27	36	44	53	60	68	77	1.0	1.1	1.3	1.4	1.6	1.1	1.2	1.4	1.5	1.7
DG A1	15	20	25	28	33	36	39	43	9	12	15	18	20	22	24	26	1.4	1.7	2.0	2.5	3.0	1.8	2.2	2.7	3.7	-
DG A2	17	23	30	35	42	46	52	59	12	16	21	25	30	34	38	44	1.3	1.4	1.6	1.9	2.1	1.5	1.7	1.9	2.2	2.5
DG A3	21	28	37	44	53	59	66	74	16	21	28	33	41	46	53	62	1.1	1.2	1.4	1.6	1.8	1.3	1.4	1.6	1.7	1.9
DG B1	15	20	26	30	36	40	45	51	10	13	16	18	22	24	27	31	1.4	1.5	1.8	2.0	2.4	1.8	2.1	2.4	2.8	3.4
DG B2	18	24	31	37	45	51	58	66	12	17	22	26	32	36	41	49	1.3	1.4	1.5	1.7	1.9	1.5	1.7	1.8	2.0	2.2
DG B3	22	29	38	45	56	63	71	80	16	22	29	34	43	48	56	66	1.1	1.2	1.3	1.5	1.7	1.3	1.4	1.5	1.6	1.8
DG C1	18	24	31	36	44	49	55	63	12	16	20	23	28	31	35	41	1.3	1.4	1.6	1.8	2.1	1.6	1.8	2.0	2.3	2.6
DG C2	21	27	36	42	52	58	66	74	15	20	26	30	37	42	49	57	1.2	1.3	1.4	1.6	1.8	1.4	1.5	1.6	1.8	2.0
DG C3	24	33	43	51	62	69	77	86	19	25	33	39	48	54	63	73	1.1	1.1	1.3	1.4	1.6	1.2	1.3	1.4	1.5	1.7
SS A1	24	30	36	42	49	54	60	68	19	24	29	32	38	42	47	53	1.2	1.4	1.5	1.7	1.9	1.4	1.6	1.8	2.0	2.2
SS A2	30	38	47	54	63	68	75	83	24	31	38	44	51	56	63	71	1.0	1.2	1.3	1.5	1.6	1.1	1.3	1.5	1.6	1.8
SS A3	22	27	33	38	44	49	55	62	18	21	26	29	34	37	41	47	1.3	1.5	1.6	1.8	2.0	1.5	1.8	2.0	2.2	2.4
SS A4	29	37	46	52	61	67	73	81	23	30	37	42	50	54	61	69	1.0	1.2	1.4	1.5	1.7	1.2	1.3	1.5	1.7	1.8
SS A5	24	28	34	38	44	49	54	62	19	23	27	30	35	38	42	48	1.3	1.5	1.7	1.8	2.0	1.5	1.8	2.0	2.2	2.4
SS A6	30	38	47	53	62	68	74	82	24	30	37	43	50	55	62	70	1.0	1.2	1.3	1.5	1.7	1.2	1.3	1.5	1.7	1.8
SS A7	21	25	30	33	38	42	47	54	17	20	23	26	29	32	35	39	1.5	1.7	1.8	2.0	2.2	1.7	2.0	2.3	2.5	2.8
SS A8	29	36	45	51	60	65	72	80	23	29	36	41	48	53	59	67	1.1	1.2	1.4	1.5	1.7	1.2	1.4	1.6	1.7	1.9
SS B1	21	26	32	36	43	48	54	62	16	20	24	28	33	36	41	47	1.3	1.5	1.6	1.8	1.9	1.5	1.7	1.9	2.1	2.3
SS B2	26	34	42	49	57	63	70	78	21	27	34	39	46	52	58	67	1.1	1.2	1.4	1.5	1.7	1.2	1.4	1.5	1.7	1.8
SS B3	19	24	29	34	40	44	50	58	15	18	22	25	30	33	37	43	1.4	1.5	1.7	1.9	2.0	1.6	1.8	2.1	2.3	2.5
SS B4	26	33	41	47	56	61	68	77	20	26	33	38	45	50	56	64	1.1	1.2	1.4	1.6	1.7	1.2	1.4	1.6	1.7	1.9
SS B5	19	23	28	32	37	41	46	54	16	19	22	25	29	31	35	40	1.5	1.6	1.8	2.0	2.1	1.7	2.0	2.2	2.4	2.6
SS B6	26	33	41	48	56	62	69	77	20	26	33	38	45	50	56	65	1.1	1.2	1.4	1.6	1.7	1.2	1.4	1.6	1.7	1.9
SS B7	18	21	25	28	33	36	41	48	14	17	20	22	25	27	30	34	1.6	1.8	2.0	2.1	2.3	1.9	2.2	2.4	2.7	2.9
SS B8	24	31	39	45	54	59	66	75	19	25	31	36	43	48	54	62	1.1	1.3	1.4	1.6	1.7	1.3	1.4	1.6	1.8	1.9
SS C1	19	25	32	38	45	50	57	66	13	17	22	26	31	35	40	47	1.2	1.4	1.5	1.7	1.8	1.5	1.7	1.9	2.1	2.2
SS C2	23	31	40	47	56	62	69	78	16	22	29	34	42	47	54	63	1.1	1.2	1.4	1.5	1.7	1.2	1.4	1.6	1.7	1.9
SS C3	17	22	27	31	37	41	46	54	12	15	19	22	26	29	33	39	1.4	1.6	1.8	1.9	2.0	1.7	1.9	2.1	2.3	2.5
SS C4	21	28	36	42	50	55	63	72	15	20	26	31	38	42	48	57	1.2	1.3	1.5	1.6	1.8	1.3	1.5	1.7	1.8	2.0
SS D1	28	35	43	49	58	63	69	77	22	27	33	38	44	49	55	62	1.1	1.2	1.4	1.6	1.7	1.2	1.4	1.6	1.8	2.0
SS D2	34	44	55	62	71	77	83	89	27	35	44	50	59	65	72	80	0.9	1.1	1.2	1.4	1.6	1.0	1.2	1.4	1.5	1.7
SS D3	26	32	39	44	52	57	63	71	21	25	30	34	40	44	49	56	1.2	1.3	1.5	1.7	1.8	1.4	1.6	1.8	2.0	2.1
SS D4	33	43	53	60	69	75	81	88	26	34	42	49	57	63	69	78	0.9	1.1	1.3	1.4	1.6	1.1	1.2	1.4	1.6	1.7
SS E1	24	30	38	43	51	56	63	71	18	22	28	32	38	42	47	54	1.2	1.3	1.5	1.7	1.8	1.4	1.6	1.8	2.0	2.1
SS E2	29	38	48	55	65	70	77	85	22	29	37	43	52	57	64	73	1.0	1.1	1.3	1.5	1.6	1.1	1.3	1.5	1.6	1.8
SS E3	21	26	32	36	43	47	53	61	17	20	24	28	32	35	40	46	1.3	1.5	1.7	1.8	2.0	1.6	1.8	2.0	2.2	2.4
SS E4	27	35	44	51	60	66	73	81	21	27	35	40	48	53	59	68	1.0	1.2	1.3	1.5	1.7	1.2	1.4	1.5	1.7	1.8

SPRINGFIELD, MISSOURI

TYPE	SSF 55 F (2618 DD) LCR 100	70	50	40	30	25	20	15	SSF 65 F (4571 DD) LCR 100	70	50	40	30	25	20	15	CF 55 F 100	50	30	20	15	CF 65 F 100	50	30	20	15
WW A1	22	26	31	35	41	45	51	58	17	20	23	25	29	32	36	41	1.4	1.6	1.7	1.9	2.0	1.8	2.0	2.2	2.4	2.5
WW A2	24	31	39	44	53	58	65	74	17	21	27	31	37	41	47	55	1.1	1.3	1.4	1.6	1.7	1.4	1.6	1.7	1.9	2.0
WW A3	25	33	41	48	57	63	70	78	16	22	28	33	40	45	52	60	1.1	1.2	1.4	1.5	1.7	1.3	1.5	1.6	1.8	1.9
WW A4	25	34	43	50	59	65	72	81	16	22	29	35	42	47	54	63	1.0	1.2	1.3	1.5	1.6	1.2	1.4	1.6	1.7	1.9
WW A5	26	35	45	52	62	68	75	83	16	23	30	36	44	50	57	66	1.0	1.2	1.3	1.5	1.6	1.2	1.4	1.5	1.7	1.8
WW A6	26	36	46	53	63	69	76	84	16	23	31	37	45	51	58	67	1.0	1.1	1.3	1.5	1.6	1.2	1.3	1.5	1.7	1.8
WW B1	20	25	31	35	42	47	53	61	13	16	20	22	26	29	33	39	1.3	1.5	1.6	1.8	1.9	1.7	1.9	2.2	2.4	2.6
WW B2	26	35	45	52	61	67	75	83	17	24	31	37	45	51	58	67	1.0	1.2	1.3	1.5	1.6	1.2	1.3	1.5	1.7	1.8
WW B3	31	41	52	60	69	75	82	89	21	29	38	44	53	59	67	76	0.9	1.1	1.2	1.4	1.5	1.1	1.2	1.4	1.6	1.7
WW B4	31	42	54	62	72	78	84	91	21	30	40	47	57	63	71	80	0.9	1.1	1.2	1.4	1.5	1.0	1.2	1.3	1.5	1.6
WW B5	30	41	53	61	71	77	84	91	20	29	39	47	57	63	71	80	0.9	1.1	1.2	1.4	1.5	1.0	1.2	1.3	1.5	1.6
WW C1	33	43	54	62	72	77	84	91	22	31	40	47	56	62	70	79	0.9	1.1	1.2	1.4	1.5	1.1	1.2	1.4	1.5	1.7
WW C2	31	41	53	60	70	76	83	90	21	29	39	46	55	62	69	78	0.9	1.1	1.2	1.4	1.5	1.1	1.2	1.4	1.5	1.7
WW C3	36	49	61	69	79	85	90	95	25	36	47	56	66	73	80	88	0.8	1.0	1.2	1.3	1.5	0.9	1.1	1.2	1.4	1.6
WW C4	33	45	57	66	76	82	88	94	22	33	44	52	63	69	77	85	0.9	1.0	1.2	1.3	1.5	1.0	1.1	1.3	1.4	1.6
TW A1	21	26	31	35	41	45	50	58	17	19	23	25	29	32	36	41	1.4	1.6	1.7	1.9	2.0	1.8	2.0	2.2	2.3	2.5
TW A2	23	28	35	40	48	53	60	69	16	20	25	29	34	38	43	51	1.2	1.4	1.5	1.7	1.8	1.5	1.7	1.8	2.0	2.1
TW A3	23	29	37	43	51	57	64	73	15	20	26	30	36	41	46	55	1.2	1.3	1.5	1.6	1.7	1.4	1.6	1.7	1.9	2.0
TW A4	22	29	37	43	52	58	65	74	14	19	25	30	37	41	48	56	1.1	1.3	1.4	1.6	1.7	1.4	1.5	1.7	1.8	2.0
TW B1	21	26	31	35	42	46	52	60	16	19	23	25	30	33	37	43	1.4	1.5	1.7	1.8	2.0	1.7	1.9	2.1	2.2	2.4
TW B2	21	27	34	39	47	52	59	68	15	19	24	28	33	38	43	50	1.2	1.4	1.5	1.7	1.8	1.5	1.7	1.8	2.0	2.1
TW B3	21	27	34	40	48	53	60	69	14	18	24	28	34	38	44	52	1.2	1.4	1.5	1.7	1.8	1.5	1.6	1.8	1.9	2.1
TW B4	19	25	32	38	45	51	57	67	13	17	23	27	32	37	42	50	1.2	1.4	1.5	1.7	1.8	1.5	1.6	1.8	2.0	2.1
TW C1	20	25	30	35	41	46	52	61	15	18	22	25	30	33	38	45	1.4	1.5	1.6	1.8	1.9	1.7	1.8	2.0	2.2	2.3
TW C2	19	24	31	35	43	48	54	63	14	17	22	25	31	34	40	47	1.3	1.5	1.6	1.7	1.9	1.6	1.7	1.9	2.1	2.2
TW C3	18	23	29	34	41	46	52	61	13	17	21	24	29	33	38	45	1.3	1.5	1.6	1.8	1.9	1.6	1.8	1.9	2.1	2.2
TW C4	17	21	27	31	37	41	47	55	13	16	20	23	27	30	35	41	1.4	1.6	1.7	1.9	2.0	1.8	1.9	2.1	2.2	2.4
TW D1	16	21	27	31	37	41	47	55	11	14	17	19	23	26	30	35	1.4	1.6	1.7	1.9	2.0	1.8	2.0	2.3	2.5	2.6
TW D2	24	32	41	47	56	62	69	78	17	22	29	34	41	47	53	62	1.1	1.2	1.4	1.5	1.7	1.3	1.4	1.6	1.7	1.9
TW D3	26	35	44	51	61	67	74	82	18	24	32	37	45	51	58	67	1.0	1.2	1.3	1.5	1.6	1.2	1.4	1.5	1.7	1.8
TW D4	28	38	48	56	66	72	79	87	19	27	36	42	51	57	65	74	1.0	1.1	1.3	1.4	1.6	1.1	1.3	1.4	1.6	1.7
TW D5	28	37	48	56	66	72	79	87	19	27	36	43	52	58	66	75	1.0	1.1	1.3	1.4	1.6	1.1	1.2	1.4	1.5	1.7
TW E1	30	39	49	56	66	72	79	87	21	28	36	42	51	57	64	73	1.0	1.1	1.3	1.4	1.6	1.1	1.3	1.4	1.6	1.7
TW E2	29	39	49	57	66	72	79	87	20	28	36	43	51	57	65	74	1.0	1.1	1.3	1.4	1.6	1.1	1.3	1.4	1.6	1.7
TW E3	35	46	58	66	76	81	87	93	25	34	45	52	62	68	76	84	0.9	1.0	1.2	1.4	1.5	1.0	1.1	1.3	1.5	1.6
TW E4	32	43	55	63	73	79	86	92	22	32	42	50	60	66	74	83	0.9	1.0	1.2	1.4	1.5	1.0	1.2	1.3	1.5	1.6
TW F1	20	24	30	34	40	44	50	58	15	18	21	24	28	31	35	41	1.4	1.6	1.7	1.9	2.0	1.8	1.9	2.1	2.3	2.5
TW F2	20	26	33	38	45	51	57	66	14	18	22	26	32	36	41	48	1.2	1.4	1.5	1.7	1.8	1.5	1.7	1.9	2.0	2.2
TW F3	19	26	33	39	47	53	60	69	12	17	22	26	33	37	43	51	1.2	1.3	1.5	1.6	1.8	1.5	1.6	1.8	1.9	2.1
TW F4	17	24	31	37	46	51	59	68	10	15	21	25	31	36	42	50	1.2	1.3	1.5	1.6	1.8	1.4	1.6	1.8	1.9	2.1
TW G1	18	23	28	33	39	44	50	58	13	16	20	23	27	30	35	41	1.4	1.5	1.7	1.8	2.0	1.7	1.9	2.1	2.2	2.4
TW G2	17	23	29	35	42	47	54	63	11	15	20	23	29	33	38	45	1.3	1.4	1.6	1.7	1.9	1.6	1.7	1.9	2.1	2.2
TW G3	15	21	28	33	41	46	53	62	9	13	18	22	28	32	37	45	1.3	1.4	1.6	1.7	1.9	1.6	1.7	1.9	2.0	2.2
TW G4	12	18	24	29	36	41	48	57	7	11	15	19	24	28	33	41	1.4	1.5	1.6	1.8	1.9	1.6	1.8	2.0	2.1	2.2
TW H1	15	20	25	30	36	41	47	56	10	13	17	20	25	28	33	39	1.4	1.6	1.7	1.8	2.0	1.7	1.9	2.1	2.2	2.4
TW H2	12	17	23	28	34	39	46	55	7	11	15	18	23	27	31	38	1.4	1.6	1.7	1.8	2.0	1.7	1.9	2.0	2.2	2.3
TW H3	10	15	20	24	31	35	41	50	6	9	13	16	20	24	28	35	1.5	1.6	1.8	1.9	2.0	1.8	1.9	2.1	2.3	2.4
TW H4	8	11	16	19	25	29	34	42	4	7	10	12	16	19	23	29	1.6	1.8	1.9	2.1	2.2	2.0	2.2	2.3	2.5	2.6
TW I1	14	19	24	28	34	38	43	51	9	11	14	17	21	23	27	32	1.5	1.6	1.8	1.9	2.1	1.9	2.1	2.3	2.6	2.8
TW I2	21	29	37	44	53	59	66	75	14	19	26	31	38	43	50	59	1.1	1.3	1.4	1.6	1.7	1.3	1.5	1.6	1.8	1.9
TW I3	23	31	40	47	57	63	70	79	15	21	28	34	42	47	54	63	1.1	1.2	1.4	1.5	1.7	1.2	1.4	1.6	1.7	1.8
TW I4	25	35	45	52	62	69	76	85	16	24	32	39	48	54	62	71	1.0	1.1	1.3	1.4	1.6	1.1	1.3	1.4	1.6	1.7
TW I5	25	35	46	53	64	70	77	86	17	24	33	40	49	56	63	73	1.0	1.1	1.3	1.4	1.6	1.1	1.3	1.4	1.6	1.7
TW J1	27	37	47	54	64	70	77	85	18	26	34	40	49	54	62	71	1.0	1.1	1.3	1.4	1.6	1.1	1.3	1.5	1.6	1.7
TW J2	26	36	46	53	63	70	77	85	18	25	33	40	49	55	62	72	1.0	1.1	1.3	1.4	1.6	1.1	1.3	1.4	1.6	1.7
TW J3	32	43	55	63	73	79	86	92	22	31	42	49	59	66	74	83	0.9	1.0	1.2	1.4	1.5	1.0	1.2	1.3	1.5	1.6
TW J4	29	40	52	60	71	77	84	91	20	29	39	47	57	63	71	81	0.9	1.1	1.2	1.4	1.5	1.0	1.2	1.3	1.5	1.6
DG A1	18	23	29	33	38	42	46	50	11	14	18	20	24	26	28	30	1.3	1.5	1.9	2.2	2.6	1.7	2.0	2.5	3.2	4.2
DG A2	20	26	34	39	47	52	58	65	13	18	23	27	33	37	42	48	1.2	1.3	1.5	1.8	2.0	1.5	1.6	1.8	2.1	2.4
DG A3	24	32	41	49	58	64	71	79	17	23	30	36	44	50	57	66	1.1	1.2	1.3	1.5	1.7	1.3	1.4	1.5	1.7	1.9
DG B1	18	23	30	35	43	48	53	59	11	14	18	21	25	28	32	37	1.3	1.4	1.6	1.9	2.2	1.7	1.9	2.1	2.5	2.9
DG B2	20	27	35	42	51	57	64	72	14	18	24	28	35	39	46	54	1.2	1.3	1.4	1.6	1.8	1.4	1.6	1.7	1.9	2.1
DG B3	24	32	43	50	62	68	76	84	17	24	31	37	46	52	60	71	1.1	1.1	1.3	1.5	1.6	1.2	1.4	1.4	1.6	1.7
DG C1	21	28	36	42	51	57	64	71	13	18	22	26	32	36	41	47	1.2	1.3	1.4	1.7	1.9	1.5	1.7	1.8	2.1	2.3
DG C2	23	31	41	48	58	65	72	80	16	21	28	33	41	46	54	63	1.1	1.2	1.3	1.5	1.7	1.3	1.4	1.6	1.7	1.9
DG C3	27	37	48	57	68	75	83	90	20	27	35	42	52	59	67	78	1.0	1.1	1.2	1.4	1.6	1.2	1.3	1.4	1.5	1.7
SS A1	27	34	41	46	54	60	66	74	21	26	31	36	42	46	51	59	1.1	1.3	1.4	1.6	1.8	1.3	1.5	1.7	1.9	2.0
SS A2	34	42	52	59	68	74	80	87	26	33	41	47	55	61	67	76	1.0	1.1	1.3	1.4	1.6	1.1	1.3	1.4	1.6	1.7
SS A3	25	31	38	43	50	55	61	69	19	23	28	32	37	41	46	53	1.2	1.4	1.5	1.7	1.8	1.4	1.6	1.8	2.0	2.2
SS A4	33	42	51	58	67	72	79	86	25	32	40	46	54	59	66	74	1.0	1.1	1.3	1.5	1.6	1.1	1.3	1.4	1.6	1.8
SS A5	26	31	38	43	50	54	61	69	21	25	29	33	38	41	46	53	1.2	1.4	1.5	1.7	1.8	1.5	1.7	1.8	2.0	2.2
SS A6	33	42	52	59	68	73	80	87	26	33	41	46	55	60	67	75	1.0	1.1	1.3	1.4	1.6	1.1	1.3	1.4	1.6	1.7
SS A7	23	28	33	37	43	47	53	61	19	22	26	28	32	35	39	45	1.4	1.5	1.7	1.8	2.0	1.6	1.9	2.1	2.3	2.5
SS A8	32	41	50	57	65	71	78	85	25	32	39	44	52	57	64	72	1.0	1.1	1.3	1.5	1.6	1.1	1.3	1.5	1.6	1.8
SS B1	23	29	36	41	48	53	60	68	17	22	27	30	36	40	45	52	1.2	1.4	1.5	1.7	1.8	1.4	1.6	1.8	2.0	2.1
SS B2	29	38	47	54	63	68	75	83	22	29	37	42	50	55	62	71	1.0	1.2	1.3	1.5	1.6	1.2	1.3	1.5	1.6	1.8
SS B3	22	27	33	38	45	50	56	65	16	20	25	28	33	37	41	48	1.3	1.4	1.6	1.7	1.9	1.5	1.7	1.9	2.1	2.3
SS B4	29	37	46	52	61	67	74	82	22	28	35	41	49	54	60	69	1.0	1.2	1.3	1.5	1.6	1.2	1.3	1.5	1.7	1.8
SS B5	21	26	31	36	42	46	52	60	17	20	24	27	31	34	39	45	1.4	1.5	1.7	1.8	2.0	1.6	1.8	2.0	2.2	2.4
SS B6	29	37	46	53	62	67	74	82	22	28	35	41	49	54	61	70	1.0	1.2	1.3	1.5	1.6	1.2	1.3	1.5	1.7	1.8
SS B7	19	24	28	32	38	41	47	54	15	18	21	24	28	30	34	39	1.5	1.6	1.8	2.0	2.1	1.8	2.0	2.2	2.4	2.6
SS B8	27	35	44	51	59	65	72	80	21	27	34	39	47	52	58	67	1.1	1.2	1.4	1.5	1.7	1.2	1.4	1.5	1.7	1.8
SS C1	22	29	37	43	51	57	64	73	14	19	25	29	35	39	45	53	1.2	1.3	1.4	1.6	1.7	1.4	1.6	1.8	1.9	2.1
SS C2	26	35	45	52	61	67	75	83	18	24	32	38	46	51	58	68	1.0	1.2	1.3	1.5	1.6	1.2	1.3	1.5	1.7	1.8
SS C3	19	24	30	34	41	46	52	61	13	17	21	24	29	32	37	43	1.3	1.5	1.6	1.8	1.9	1.7	1.8	2.0	2.2	2.3
SS C4	24	31	40	46	55	61	68	77	16	22	28	34	41	46	52	61	1.1	1.2	1.4	1.5	1.7	1.3	1.4	1.6	1.8	1.9
SS D1	32	40	49	55	64	69	76	83	24	30	37	42	49	54	60	68	1.0	1.2	1.3	1.5	1.6	1.2	1.4	1.5	1.7	1.8
SS D2	38	49	60	68	77	82	87	93	29	38	48	54	64	69	76	84	0.9	1.0	1.2	1.4	1.5	1.0	1.1	1.3	1.5	1.6
SS D3	29	36	44	50	58	63	69	77	22	28	33	38	44	48	54	62	1.1	1.3	1.4	1.6	1.7	1.3	1.5	1.7	1.8	2.0
SS D4	37	47	58	65	75	80	86	92	28	37	46	53	61	67	74	82	0.9	1.0	1.2	1.4	1.5	1.0	1.2	1.3	1.5	1.6
SS E1	27	35	43	49	58	63	70	78	19	25	31	36	42	47	52	60	1.1	1.2	1.4	1.6	1.7	1.3	1.5	1.7	1.8	2.0
SS E2	33	43	54	61	70	76	82	89	24	32	41	47	56	62	69	77	0.9	1.1	1.2	1.4	1.6	1.1	1.2	1.4	1.6	1.7
SS E3	24	29	36	41	48	53	59	68	18	22	27	30	36	39	44	51	1.2	1.4	1.5	1.7	1.8	1.5	1.7	1.9	2.0	2.2
SS E4	31	40	49	56	66	71	78	86	23	30	38	44	52	57	64	73	1.0	1.1	1.3	1.5	1.6	1.1	1.3	1.5	1.6	1.8

BILLINGS, MONTANA

SOLAR SAVINGS FRACTION (%)

TYPE	55 F (4704 DD) 100	70	50	40	30	25	20	15	65 F (7279 DD) 100	70	50	40	30	25	20	15
WW A1	18	21	25	27	31	34	37	43	15	17	19	21	24	26	29	32
WW A2	18	23	29	33	39	43	49	56	14	17	22	25	30	33	38	44
WW A3	18	24	30	35	42	47	53	61	13	17	23	27	32	36	41	48
WW A4	18	24	31	36	44	49	55	63	12	17	23	27	34	38	43	50
WW A5	18	25	32	38	46	51	57	66	12	18	24	28	35	39	45	53
WW A6	18	25	33	39	47	52	59	67	12	18	24	29	36	40	46	54
WW B1	14	18	21	24	28	31	35	40	10	12	15	17	19	21	23	25
WW B2	19	26	33	39	47	52	59	68	13	19	26	30	37	42	48	57
WW B3	23	31	39	46	54	60	67	76	16	23	31	37	44	50	56	65
WW B4	23	32	41	49	58	64	71	80	16	24	33	40	49	55	62	71
WW B5	22	31	41	48	58	64	72	80	16	24	33	40	49	55	63	72
WW C1	24	32	41	48	57	63	70	78	18	25	33	39	47	53	60	68
WW C2	23	31	40	47	56	62	70	78	16	24	32	39	47	53	60	69
WW C3	27	37	49	57	67	73	80	87	20	30	41	48	58	65	72	81
WW C4	24	35	46	53	64	70	77	85	18	27	38	45	55	62	69	79
TW A1	18	21	24	27	31	34	38	43	14	17	19	21	24	26	29	33
TW A2	18	22	27	31	36	40	45	52	13	17	21	24	28	31	35	41
TW A3	17	22	28	32	38	42	48	56	12	16	21	24	29	33	38	44
TW A4	16	21	27	32	39	43	49	57	11	15	20	24	30	34	39	45
TW B1	17	20	24	27	32	35	39	45	14	16	19	21	25	27	30	35
TW B2	16	21	26	30	35	39	44	52	12	16	20	23	27	31	35	41
TW B3	16	20	26	30	36	40	45	53	11	15	19	23	28	31	36	42
TW B4	15	19	24	28	34	38	44	51	11	14	19	22	27	30	35	41
TW C1	16	19	24	27	32	35	40	46	12	15	18	21	25	28	31	37
TW C2	15	19	24	27	33	36	41	48	11	14	18	21	25	29	33	39
TW C3	14	18	23	26	31	35	40	47	11	14	18	20	25	28	32	38
TW C4	14	17	21	24	29	32	37	43	11	14	17	19	23	26	29	35
TW D1	12	15	19	21	25	28	31	36	8	10	13	15	17	19	21	24
TW D2	18	24	31	36	43	48	55	63	14	18	24	28	35	39	45	52
TW D3	20	26	33	39	47	52	58	67	14	20	26	31	38	42	48	56
TW D4	21	29	37	44	53	58	66	74	16	22	30	36	44	49	56	65
TW D5	21	29	38	44	53	59	67	76	15	22	31	37	45	51	58	67
TW E1	22	30	38	44	52	57	64	73	17	23	30	35	43	48	54	63
TW E2	22	30	38	44	53	58	65	74	16	23	30	36	44	49	56	65
TW E3	26	36	46	53	63	69	76	84	20	29	38	45	54	60	68	76
TW E4	24	34	44	51	61	67	74	83	18	27	36	43	52	58	66	75
TW F1	16	19	23	26	30	33	37	42	13	15	18	20	23	25	28	32
TW F2	15	19	24	28	34	37	42	49	11	14	18	21	25	28	32	38
TW F3	14	19	24	28	34	39	44	52	9	13	18	21	26	29	34	40
TW F4	12	17	23	27	33	38	43	51	8	11	16	20	25	28	33	40
TW G1	14	17	21	25	29	32	37	43	11	13	16	18	22	24	28	32
TW G2	12	17	21	25	31	35	40	47	8	12	15	18	23	26	30	36
TW G3	11	15	20	24	30	34	39	46	7	10	14	17	22	25	30	36
TW G4	8	12	17	21	26	30	35	42	5	8	12	15	19	22	27	33
TW H1	11	15	19	22	27	30	34	41	8	10	13	16	20	22	26	31
TW H2	9	12	17	20	25	29	33	40	5	8	11	14	18	21	25	31
TW H3	7	10	14	18	22	26	30	37	4	7	10	12	16	19	23	28
TW H4	5	8	11	14	18	21	25	31	3	5	7	10	13	15	18	23
TW I1	10	13	16	19	22	25	28	33	6	8	11	12	15	16	19	21
TW I2	15	21	28	33	40	45	51	60	11	15	21	25	31	36	41	49
TW I3	17	23	30	35	43	48	55	64	11	17	23	28	34	39	45	53
TW I4	18	26	34	41	49	55	62	72	13	19	27	33	41	46	53	62
TW I5	18	26	35	42	51	57	64	74	13	20	28	34	43	48	56	65
TW J1	20	27	35	41	50	55	62	71	15	21	28	33	41	46	52	61
TW J2	19	27	35	41	50	56	63	72	14	20	28	33	41	47	53	62
TW J3	23	33	43	50	60	66	74	82	17	26	35	42	52	58	66	75
TW J4	22	31	41	48	58	64	72	81	16	24	33	40	50	56	64	73
DG A1	12	16	20	22	26	28	30	32	8	10	13	15	17	18	20	21
DG A2	15	20	25	29	35	39	44	50	11	14	19	22	27	30	34	39
DG A3	19	25	32	38	46	51	58	66	14	20	26	31	38	43	49	57
DG B1	13	16	20	23	28	31	34	38	8	11	14	16	18	20	22	25
DG B2	15	20	26	30	37	42	47	55	11	15	20	23	28	32	37	43
DG B3	19	25	33	39	48	54	61	71	15	20	27	32	39	45	52	61
DG C1	15	20	25	29	34	38	43	48	10	14	18	20	24	27	30	34
DG C2	18	24	30	35	43	49	55	63	13	18	23	28	34	38	44	52
DG C3	22	29	37	44	53	60	68	77	17	23	30	36	44	50	58	68
SS A1	21	25	30	34	40	44	48	55	17	21	25	29	33	36	40	46
SS A2	25	32	40	45	53	58	64	72	21	28	34	39	46	51	57	64
SS A3	19	23	27	30	35	38	42	48	16	19	22	25	29	31	34	38
SS A4	25	31	38	43	51	56	62	69	21	27	33	38	44	48	54	61
SS A5	21	24	29	32	36	40	44	49	17	21	24	27	30	33	36	41
SS A6	25	32	39	45	52	57	63	71	21	27	34	38	45	50	55	63
SS A7	19	21	25	27	31	33	36	41	16	18	21	23	26	27	30	33
SS A8	24	30	37	42	49	54	60	67	20	26	32	36	42	47	52	59
SS B1	18	22	26	30	35	39	43	50	14	18	22	24	29	32	35	41
SS B2	22	29	36	41	49	54	60	68	18	24	31	35	42	47	53	60
SS B3	16	20	24	27	32	35	39	45	13	16	19	22	26	28	31	35
SS B4	22	28	35	40	47	52	58	66	18	23	29	34	40	45	50	58
SS B5	17	20	24	27	31	34	37	43	14	17	20	22	25	27	30	34
SS B6	22	28	35	40	47	52	59	67	18	23	29	34	41	45	51	59
SS B7	15	18	21	23	27	29	32	37	13	15	17	19	22	23	26	29
SS B8	21	26	33	38	45	49	56	64	17	22	28	32	38	42	48	55
SS C1	16	21	27	31	37	41	46	54	11	15	19	23	27	31	35	41
SS C2	19	26	34	39	47	53	59	68	14	20	26	31	38	43	49	57
SS C3	15	18	23	26	31	34	38	45	11	14	17	20	23	26	29	34
SS C4	18	24	30	35	42	47	54	62	13	18	24	28	34	38	44	52
SS D1	23	29	35	40	46	51	56	63	19	24	29	33	39	42	47	53
SS D2	28	37	46	52	61	66	72	80	23	31	39	45	53	58	65	73
SS D3	22	27	32	36	42	46	51	57	18	23	27	30	35	38	42	48
SS D4	28	35	44	50	59	64	70	78	23	30	38	44	51	56	63	70
SS E1	19	25	30	34	40	44	49	56	15	19	24	27	32	35	39	44
SS E2	24	32	40	46	54	59	66	74	19	26	33	38	46	51	57	65
SS E3	18	22	26	30	35	38	42	48	15	18	22	24	28	31	34	39
SS E4	23	29	37	42	50	55	61	70	18	24	31	36	43	47	53	61

CONSERVATION FACTOR

TYPE	55 F 100	50	30	20	15	65 F 100	50	30	20	15
WW A1	1.7	1.9	2.2	2.4	2.5	2.0	2.3	2.6	2.9	3.2
WW A2	1.4	1.5	1.7	1.9	2.1	1.6	1.8	2.0	2.2	2.4
WW A3	1.3	1.5	1.6	1.8	2.0	1.4	1.6	1.9	2.1	2.2
WW A4	1.2	1.4	1.6	1.8	1.9	1.4	1.6	1.8	2.0	2.2
WW A5	1.2	1.4	1.5	1.7	1.9	1.3	1.5	1.7	1.9	2.1
WW A6	1.2	1.3	1.5	1.7	1.8	1.3	1.5	1.7	1.9	2.1
WW B1	1.6	1.9	2.2	2.4	2.6	2.0	2.4	2.8	3.4	4.0
WW B2	1.2	1.3	1.5	1.7	1.8	1.3	1.5	1.7	1.8	2.0
WW B3	1.1	1.2	1.4	1.6	1.7	1.2	1.4	1.5	1.7	1.9
WW B4	1.0	1.2	1.3	1.5	1.7	1.1	1.3	1.4	1.6	1.8
WW B5	1.0	1.2	1.3	1.5	1.6	1.1	1.3	1.4	1.6	1.7
WW C1	1.0	1.2	1.4	1.5	1.7	1.1	1.3	1.5	1.7	1.8
WW C2	1.0	1.2	1.4	1.5	1.7	1.1	1.3	1.5	1.6	1.8
WW C3	0.9	1.1	1.3	1.4	1.6	1.0	1.2	1.3	1.5	1.6
WW C4	1.0	1.1	1.3	1.4	1.6	1.0	1.2	1.3	1.5	1.7
TW A1	1.7	1.9	2.1	2.3	2.5	2.0	2.2	2.5	2.8	3.1
TW A2	1.4	1.6	1.8	2.0	2.2	1.7	1.9	2.1	2.3	2.5
TW A3	1.4	1.5	1.7	1.9	2.1	1.5	1.7	2.0	2.2	2.4
TW A4	1.3	1.5	1.7	1.9	2.0	1.5	1.7	1.9	2.1	2.3
TW B1	1.6	1.8	2.0	2.2	2.4	1.9	2.1	2.4	2.6	2.8
TW B2	1.4	1.6	1.8	2.0	2.1	1.6	1.8	2.0	2.3	2.4
TW B3	1.4	1.6	1.8	1.9	2.1	1.6	1.8	2.0	2.2	2.4
TW B4	1.4	1.6	1.8	2.0	2.1	1.6	1.8	2.0	2.2	2.4
TW C1	1.6	1.8	2.0	2.1	2.3	1.8	2.0	2.2	2.5	2.6
TW C2	1.5	1.7	1.9	2.1	2.2	1.7	1.9	2.1	2.3	2.5
TW C3	1.6	1.7	1.9	2.1	2.2	1.8	1.9	2.1	2.3	2.5
TW C4	1.7	1.9	2.0	2.2	2.3	1.9	2.1	2.3	2.5	2.6
TW D1	1.7	2.0	2.2	2.5	2.7	2.1	2.4	2.8	3.2	3.7
TW D2	1.2	1.4	1.6	1.8	1.9	1.4	1.6	1.8	1.9	2.1
TW D3	1.2	1.4	1.5	1.7	1.8	1.3	1.5	1.7	1.9	2.0
TW D4	1.1	1.3	1.4	1.6	1.7	1.2	1.4	1.5	1.7	1.8
TW D5	1.1	1.2	1.4	1.6	1.7	1.2	1.3	1.5	1.7	1.8
TW E1	1.1	1.3	1.4	1.6	1.8	1.2	1.4	1.6	1.8	1.9
TW E2	1.1	1.3	1.4	1.6	1.7	1.2	1.4	1.5	1.7	1.9
TW E3	1.0	1.1	1.3	1.5	1.6	1.1	1.2	1.4	1.6	1.7
TW E4	1.0	1.2	1.3	1.5	1.6	1.1	1.2	1.4	1.6	1.7
TW F1	1.7	1.9	2.1	2.3	2.5	2.0	2.2	2.5	2.7	3.0
TW F2	1.5	1.7	1.9	2.0	2.2	1.7	1.9	2.1	2.4	2.5
TW F3	1.4	1.6	1.8	2.0	2.1	1.6	1.8	2.0	2.2	2.4
TW F4	1.4	1.6	1.8	1.9	2.1	1.6	1.8	2.0	2.2	2.4
TW G1	1.6	1.8	2.0	2.2	2.4	1.9	2.1	2.4	2.6	2.8
TW G2	1.5	1.7	1.9	2.1	2.2	1.7	1.9	2.1	2.3	2.5
TW G3	1.5	1.7	1.9	2.0	2.2	1.7	1.9	2.1	2.3	2.5
TW G4	1.6	1.8	1.9	2.1	2.3	1.8	2.0	2.2	2.4	2.5
TW H1	1.6	1.8	2.0	2.2	2.4	1.9	2.1	2.3	2.5	2.7
TW H2	1.6	1.8	2.0	2.2	2.3	1.9	2.1	2.3	2.5	2.6
TW H3	1.7	1.9	2.1	2.3	2.4	2.0	2.1	2.3	2.5	2.7
TW H4	1.9	2.1	2.3	2.4	2.6	2.2	2.4	2.6	2.7	2.9
TW I1	1.8	2.1	2.3	2.6	2.8	2.2	2.5	2.9	3.4	3.8
TW I2	1.3	1.5	1.6	1.8	1.9	1.4	1.6	1.8	2.0	2.1
TW I3	1.2	1.4	1.6	1.7	1.9	1.4	1.5	1.7	1.9	2.0
TW I4	1.1	1.3	1.4	1.6	1.7	1.2	1.4	1.6	1.7	1.9
TW I5	1.1	1.3	1.4	1.6	1.7	1.2	1.3	1.5	1.7	1.8
TW J1	1.1	1.3	1.5	1.6	1.8	1.2	1.4	1.6	1.8	1.9
TW J2	1.1	1.3	1.4	1.6	1.8	1.2	1.4	1.6	1.7	1.9
TW J3	1.0	1.2	1.3	1.5	1.6	1.1	1.2	1.4	1.6	1.7
TW J4	1.0	1.2	1.3	1.5	1.6	1.1	1.2	1.4	1.6	1.7
DG A1	1.6	1.9	2.4	3.2	4.3	1.9	2.4	3.1	4.7	-
DG A2	1.4	1.6	1.8	2.1	2.4	1.6	1.8	2.0	2.3	2.7
DG A3	1.2	1.4	1.5	1.7	1.9	1.4	1.5	1.6	1.8	2.0
DG B1	1.6	1.9	2.1	2.5	3.0	1.9	2.3	2.7	3.4	4.4
DG B2	1.4	1.5	1.7	1.9	2.1	1.6	1.7	1.9	2.2	2.4
DG B3	1.2	1.3	1.4	1.6	1.8	1.3	1.5	1.6	1.7	1.9
DG C1	1.4	1.6	1.8	2.1	2.5	1.7	1.9	2.3	2.6	3.1
DG C2	1.3	1.4	1.5	1.7	1.9	1.4	1.6	1.8	1.9	2.2
DG C3	1.2	1.3	1.4	1.5	1.7	1.3	1.4	1.5	1.6	1.8
SS A1	1.4	1.6	1.8	2.0	2.2	1.5	1.7	2.0	2.3	2.5
SS A2	1.1	1.3	1.5	1.7	1.8	1.2	1.4	1.6	1.8	1.9
SS A3	1.5	1.7	2.0	2.2	2.4	1.6	1.9	2.3	2.6	2.9
SS A4	1.2	1.3	1.5	1.7	1.8	1.2	1.4	1.6	1.8	2.0
SS A5	1.5	1.7	2.0	2.2	2.3	1.7	1.9	2.2	2.5	2.8
SS A6	1.1	1.3	1.5	1.7	1.8	1.2	1.4	1.6	1.8	2.0
SS A7	1.7	2.0	2.2	2.5	2.8	1.9	2.2	2.6	3.1	3.5
SS A8	1.2	1.4	1.6	1.7	1.9	1.3	1.5	1.7	1.9	2.1
SS B1	1.5	1.7	1.9	2.1	2.3	1.6	1.9	2.1	2.4	2.6
SS B2	1.2	1.4	1.5	1.7	1.8	1.3	1.5	1.6	1.8	2.0
SS B3	1.5	1.8	2.0	2.2	2.4	1.7	2.0	2.3	2.6	2.9
SS B4	1.2	1.4	1.6	1.7	1.9	1.3	1.5	1.7	1.9	2.0
SS B5	1.6	1.9	2.1	2.3	2.5	1.9	2.1	2.4	2.7	3.0
SS B6	1.2	1.4	1.5	1.7	1.9	1.3	1.5	1.7	1.9	2.0
SS B7	1.8	2.1	2.3	2.6	2.8	2.1	2.4	2.7	3.1	3.5
SS B8	1.2	1.4	1.6	1.8	1.9	1.3	1.5	1.7	1.9	2.1
SS C1	1.4	1.6	1.8	1.9	2.1	1.6	1.8	2.0	2.3	2.5
SS C2	1.2	1.3	1.5	1.7	1.8	1.3	1.5	1.7	1.8	2.0
SS C3	1.6	1.8	2.0	2.2	2.3	1.9	2.1	2.3	2.5	2.8
SS C4	1.3	1.4	1.6	1.8	1.9	1.4	1.6	1.8	2.0	2.1
SS D1	1.2	1.4	1.6	1.8	2.0	1.3	1.6	1.8	2.1	2.3
SS D2	1.0	1.2	1.4	1.5	1.7	1.1	1.3	1.5	1.6	1.8
SS D3	1.3	1.5	1.7	1.9	2.1	1.5	1.7	2.0	2.2	2.4
SS D4	1.0	1.2	1.4	1.6	1.7	1.1	1.3	1.5	1.7	1.8
SS E1	1.3	1.5	1.7	1.9	2.1	1.5	1.7	2.0	2.3	2.5
SS E2	1.1	1.3	1.4	1.6	1.8	1.2	1.4	1.6	1.8	1.9
SS E3	1.5	1.7	1.9	2.2	2.3	1.7	1.9	2.2	2.5	2.8
SS E4	1.1	1.3	1.5	1.7	1.8	1.2	1.4	1.6	1.8	2.0

CUT BANK, MONTANA

SOLAR SAVINGS FRACTION (%) / CONSERVATION FACTOR

TYPE	BASE TEMP: 55 F (6112 DD)								65 F (9061 DD)								CONSERVATION FACTOR 55 F					CONSERVATION FACTOR 65 F				
LCR	100	70	50	40	30	25	20	15	100	70	50	40	30	25	20	15	100	50	30	20	15	100	50	30	20	15
WW A1	16	18	21	23	26	28	30	34	14	15	17	19	21	22	24	26	1.9	2.2	2.6	2.9	3.3	2.3	2.6	3.1	3.6	4.2
WW A2	15	19	24	27	32	35	39	44	12	15	19	21	25	28	31	35	1.5	1.7	2.0	2.3	2.5	1.7	2.0	2.3	2.6	2.9
WW A3	15	19	25	28	34	38	42	48	11	15	19	22	27	30	34	39	1.4	1.6	1.9	2.1	2.3	1.6	1.8	2.1	2.4	2.7
WW A4	14	19	25	29	35	39	44	51	10	14	19	23	28	31	35	4[illegible]	1.3	1.6	1.8	2.0	2.2	1.5	1.7	2.0	2.3	2.6
WW A5	14	20	26	30	36	41	46	53	10	14	19	23	29	32	37	43	1.3	1.5	1.7	2.0	2.2	1.4	1.7	1.9	2.2	2.5
WW A6	14	20	26	31	37	42	47	54	9	14	20	24	29	33	38	44	1.3	1.5	1.7	1.9	2.1	1.4	1.6	1.9	2.2	2.4
WW B1	11	14	17	18	21	23	24	27	8	10	12	13	14	15	16	16	1.9	2.3	2.8	3.5	4.4	2.3	2.9	3.9	-	-
WW B2	15	21	27	32	39	44	49	57	11	16	22	26	32	36	41	48	1.3	1.5	1.7	1.9	2.0	1.4	1.6	1.8	2.0	2.2
WW B3	18	25	33	38	45	50	57	65	14	20	26	31	38	43	49	56	1.2	1.4	1.6	1.7	1.9	1.3	1.5	1.7	1.9	2.1
WW B4	18	26	35	41	50	56	63	71	14	21	29	35	43	49	55	64	1.1	1.3	1.4	1.6	1.8	1.2	1.3	1.5	1.7	1.9
WW B5	18	26	35	41	50	56	63	72	13	21	29	35	44	50	57	66	1.1	1.3	1.4	1.6	1.8	1.2	1.3	1.5	1.7	1.8
WW C1	20	27	35	40	48	53	60	68	15	22	29	34	41	46	52	60	1.1	1.3	1.5	1.7	1.9	1.2	1.4	1.6	1.8	2.0
WW C2	18	26	34	40	48	54	60	69	14	21	28	34	41	47	53	62	1.1	1.3	1.5	1.7	1.8	1.2	1.4	1.6	1.8	1.9
WW C3	22	32	42	49	59	65	72	81	17	26	36	43	53	59	66	75	1.0	1.2	1.3	1.5	1.7	1.0	1.2	1.4	1.6	1.7
WW C4	20	29	39	46	56	62	70	78	15	24	33	40	50	56	64	73	1.0	1.2	1.4	1.5	1.7	1.1	1.2	1.4	1.6	1.7
TW A1	16	18	21	23	26	28	31	34	13	15	17	19	21	23	24	27	1.9	2.2	2.5	2.8	3.1	2.2	2.6	2.9	3.3	3.8
TW A2	15	18	22	25	30	33	37	42	12	15	18	20	24	26	29	34	1.6	1.8	2.1	2.3	2.6	1.8	2.1	2.4	2.7	3.0
TW A3	14	18	23	26	31	35	39	45	11	14	18	21	25	28	31	36	1.5	1.7	2.0	2.2	2.4	1.7	1.9	2.2	2.5	2.7
TW A4	13	17	22	26	31	35	40	46	9	13	17	20	25	28	32	38	1.4	1.7	1.9	2.1	2.3	1.6	1.9	2.1	2.4	2.6
TW B1	15	18	21	23	27	29	32	36	13	15	17	19	21	23	26	29	1.8	2.1	2.3	2.6	2.9	2.1	2.4	2.7	3.0	3.3
TW B2	14	17	21	25	29	32	37	42	11	14	17	20	23	26	30	34	1.6	1.8	2.0	2.3	2.5	1.8	2.0	2.3	2.5	2.8
TW B3	13	17	21	25	30	33	37	43	10	13	17	19	24	26	30	35	1.5	1.8	2.0	2.2	2.4	1.7	2.0	2.2	2.4	2.7
TW B4	12	16	20	24	28	32	36	42	9	12	16	19	23	26	30	35	1.6	1.8	2.0	2.2	2.4	1.8	2.0	2.2	2.4	2.7
TW C1	14	17	20	23	27	29	33	38	11	14	16	18	22	24	27	31	1.7	2.0	2.2	2.5	2.7	2.0	2.2	2.5	2.8	3.0
TW C2	13	16	20	23	27	30	34	40	10	13	16	18	22	25	28	33	1.6	1.9	2.1	2.3	2.5	1.9	2.1	2.3	2.6	2.8
TW C3	12	15	19	22	26	29	33	39	10	12	16	18	22	24	28	32	1.7	1.9	2.1	2.3	2.5	1.9	2.1	2.3	2.6	2.8
TW C4	12	15	18	21	25	28	31	36	10	13	15	17	20	23	26	30	1.8	2.0	2.2	2.4	2.6	2.1	2.3	2.5	2.7	2.9
TW D1	10	12	15	16	19	21	23	26	7	9	10	12	13	14	16	17	2.0	2.3	2.8	3.3	3.8	2.4	2.9	3.5	4.5	-
TW D2	15	20	26	30	36	40	46	53	12	16	21	24	30	34	38	45	1.4	1.6	1.8	2.0	2.1	1.5	1.7	1.9	2.1	2.3
TW D3	16	21	28	32	39	43	49	56	12	17	22	26	32	36	41	48	1.3	1.5	1.7	1.9	2.1	1.4	1.6	1.8	2.1	2.3
TW D4	17	24	32	37	45	50	57	65	13	19	26	31	39	44	50	58	1.2	1.4	1.5	1.7	1.9	1.3	1.4	1.6	1.8	2.0
TW D5	17	24	32	38	46	52	59	68	13	20	27	32	40	46	52	61	1.2	1.3	1.5	1.7	1.8	1.2	1.4	1.6	1.8	1.9
TW E1	19	25	32	37	44	49	55	63	15	20	26	31	37	42	47	55	1.2	1.4	1.6	1.8	1.9	1.3	1.5	1.7	1.9	2.1
TW E2	18	25	32	38	45	50	57	65	14	20	27	32	39	43	49	57	1.2	1.4	1.6	1.7	1.9	1.3	1.5	1.6	1.8	2.0
TW E3	22	30	39	46	55	61	68	76	17	25	34	40	49	54	61	70	1.0	1.2	1.4	1.6	1.7	1.1	1.3	1.5	1.6	1.8
TW E4	20	29	38	44	53	59	66	75	16	23	32	38	47	53	60	69	1.1	1.2	1.4	1.6	1.7	1.1	1.3	1.5	1.6	1.8
TW F1	14	16	19	22	25	27	30	34	11	13	15	17	20	21	23	26	1.9	2.1	2.4	2.8	3.0	2.2	2.5	2.9	3.3	3.7
TW F2	12	16	20	23	27	30	34	39	9	12	15	18	21	24	27	31	1.6	1.9	2.1	2.4	2.6	1.9	2.1	2.4	2.7	3.0
TW F3	11	15	19	23	28	31	35	41	8	11	15	17	21	24	28	33	1.5	1.8	2.0	2.3	2.5	1.7	2.0	2.2	2.5	2.8
TW F4	9	13	18	21	27	30	35	41	6	9	13	16	20	23	27	33	1.5	1.8	2.0	2.2	2.4	1.7	2.0	2.2	2.4	2.7
TW G1	12	15	18	20	24	26	30	34	9	11	14	16	19	20	23	26	1.8	2.1	2.3	2.6	2.9	2.1	2.4	2.7	3.0	3.3
TW G2	10	13	17	20	25	28	32	37	7	10	13	15	19	22	25	30	1.6	1.9	2.1	2.4	2.6	1.9	2.1	2.4	2.6	2.9
TW G3	8	12	16	19	24	27	31	37	5	8	12	14	18	21	24	30	1.6	1.9	2.1	2.3	2.5	1.8	2.1	2.3	2.6	2.8
TW G4	6	10	13	17	21	24	28	34	4	6	10	12	16	18	22	27	1.7	1.9	2.1	2.4	2.5	1.9	2.1	2.4	2.6	2.8
TW H1	9	12	15	18	22	24	28	33	7	9	11	13	16	19	21	25	1.8	2.0	2.3	2.5	2.8	2.1	2.3	2.6	2.9	3.1
TW H2	6	10	13	16	20	23	27	32	4	7	9	12	15	18	21	25	1.8	2.0	2.2	2.5	2.7	2.0	2.3	2.5	2.7	2.9
TW H3	5	8	11	14	18	21	25	30	3	5	8	10	13	16	19	23	1.9	2.1	2.3	2.5	2.7	2.1	2.3	2.5	2.8	3.0
TW H4	4	6	9	11	15	17	20	25	2	4	6	8	11	13	16	20	2.1	2.3	2.5	2.7	2.9	2.4	2.5	2.8	3.0	3.1
TW I1	8	10	12	14	17	18	20	23	5	7	8	9	11	12	13	14	2.0	2.4	2.9	3.4	3.9	2.5	3.0	3.7	4.7	-
TW I2	12	17	23	27	33	37	43	50	9	13	18	21	27	30	35	42	1.4	1.6	1.8	2.0	2.2	1.5	1.7	1.9	2.2	2.4
TW I3	13	19	25	29	36	40	46	54	10	14	20	24	29	34	39	46	1.3	1.5	1.7	1.9	2.1	1.5	1.6	1.9	2.1	2.3
TW I4	15	21	29	34	42	47	54	63	11	17	23	28	36	41	47	56	1.2	1.4	1.6	1.7	1.9	1.3	1.5	1.7	1.8	2.0
TW I5	15	22	30	36	44	50	56	65	11	17	24	30	38	43	50	59	1.2	1.3	1.5	1.7	1.8	1.3	1.4	1.6	1.8	1.9
TW J1	16	23	30	35	42	47	53	61	12	18	24	29	35	40	46	53	1.2	1.4	1.6	1.8	2.0	1.3	1.5	1.7	1.9	2.1
TW J2	16	22	30	35	43	48	54	63	12	17	24	29	36	41	47	55	1.2	1.4	1.6	1.8	1.9	1.3	1.5	1.7	1.9	2.0
TW J3	19	28	37	44	53	59	66	75	15	22	31	38	47	52	60	69	1.1	1.2	1.4	1.6	1.7	1.1	1.3	1.5	1.6	1.8
TW J4	18	26	35	42	51	57	64	73	13	21	29	36	45	51	58	67	1.1	1.2	1.4	1.6	1.7	1.2	1.3	1.5	1.7	1.8
DG A1	10	12	15	17	19	20	22	22	6	8	10	12	13	14	14	14	1.8	2.3	3.1	4.8	-	2.2	2.8	4.1	-	-
DG A2	12	16	21	24	29	32	36	41	9	13	16	19	23	26	29	33	1.5	1.8	2.0	2.4	2.7	1.7	2.0	2.2	2.6	3.0
DG A3	16	21	28	32	39	44	50	58	13	17	23	27	34	38	44	51	1.3	1.5	1.6	1.9	2.0	1.4	1.6	1.8	2.0	2.2
DG B1	10	13	16	18	21	22	24	27	6	8	11	12	14	15	16	17	1.8	2.2	2.7	3.4	4.9	2.2	2.6	3.4	5.0	-
DG B2	13	17	22	25	31	34	39	45	9	13	17	20	24	28	31	37	1.5	1.7	1.9	2.2	2.5	1.7	1.9	2.1	2.4	2.7
DG B3	16	22	29	34	41	46	53	61	13	18	24	29	35	40	46	55	1.3	1.4	1.6	1.8	1.9	1.4	1.5	1.7	1.9	2.0
DG C1	12	16	20	23	26	29	32	36	8	11	14	17	20	21	24	26	1.6	1.9	2.2	2.7	3.3	1.9	2.2	2.7	3.4	4.6
DG C2	15	20	26	30	36	40	46	52	11	16	20	24	30	33	38	44	1.4	1.6	1.7	2.0	2.2	1.5	1.7	1.9	2.2	2.4
DG C3	19	25	32	38	46	52	59	68	15	21	27	33	40	45	52	61	1.2	1.4	1.5	1.7	1.8	1.3	1.4	1.6	1.8	1.9
SS A1	18	22	26	29	33	36	40	45	16	19	23	25	29	32	35	39	1.5	1.8	2.1	2.3	2.6	1.6	1.9	2.2	2.6	3.0
SS A2	22	28	34	39	46	50	56	63	19	25	31	35	41	45	51	57	1.2	1.4	1.6	1.8	2.0	1.3	1.5	1.7	1.9	2.1
SS A3	16	19	23	25	29	31	34	37	14	17	20	22	25	26	29	31	1.7	2.0	2.4	2.8	3.1	1.8	2.2	2.6	3.2	4.0
SS A4	21	27	33	37	43	47	53	59	19	24	29	33	39	43	48	54	1.3	1.5	1.7	1.9	2.1	1.3	1.6	1.8	2.0	2.2
SS A5	18	21	25	27	31	33	36	40	16	19	22	24	27	29	31	34	1.7	2.0	2.3	2.6	2.9	1.8	2.1	2.5	3.0	3.5
SS A6	22	27	34	38	45	49	54	61	19	24	30	34	40	44	49	56	1.2	1.4	1.6	1.9	2.0	1.3	1.5	1.7	2.0	2.2
SS A7	16	19	21	23	26	27	29	32	15	17	19	20	22	23	25	26	1.9	2.3	2.7	3.3	3.8	2.1	2.5	3.1	4.0	-
SS A8	21	26	32	36	42	46	51	57	18	23	28	32	37	41	45	51	1.3	1.5	1.7	2.0	2.1	1.4	1.6	1.9	2.1	2.4
SS B1	15	18	22	25	29	32	35	40	13	16	19	21	25	27	30	34	1.6	1.9	2.2	2.5	2.7	1.8	2.1	2.4	2.7	3.1
SS B2	19	25	31	36	42	47	52	60	17	22	27	32	38	42	47	54	1.3	1.5	1.7	1.9	2.0	1.3	1.5	1.8	2.0	2.1
SS B3	14	17	20	23	26	28	31	35	12	14	17	19	22	24	26	29	1.7	2.0	2.4	2.7	3.1	1.9	2.2	2.7	3.2	3.8
SS B4	19	24	30	34	40	44	50	57	16	21	26	30	36	40	45	51	1.3	1.5	1.7	1.9	2.1	1.4	1.6	1.8	2.0	2.2
SS B5	15	17	20	23	26	28	31	34	13	15	18	19	22	24	26	29	1.8	2.1	2.5	2.8	3.1	2.0	2.4	2.7	3.2	3.7
SS B6	18	24	30	34	41	45	50	58	16	21	26	30	36	40	45	52	1.3	1.5	1.7	1.9	2.1	1.4	1.6	1.8	2.0	2.2
SS B7	13	16	18	20	22	24	26	29	12	14	15	17	19	20	22	24	2.0	2.4	2.8	3.3	3.7	2.3	2.7	3.2	3.9	4.7
SS B8	17	22	28	32	38	42	47	54	15	19	25	28	34	37	42	48	1.3	1.6	1.8	2.0	2.2	1.4	1.6	1.9	2.1	2.3
SS C1	13	17	21	25	29	32	36	42	9	12	16	19	22	25	28	32	1.5	1.8	2.0	2.3	2.5	1.7	2.0	2.3	2.7	3.1
SS C2	16	22	28	33	39	44	50	58	12	17	22	27	33	37	42	49	1.3	1.5	1.7	1.9	2.0	1.4	1.6	1.8	2.0	2.2
SS C3	12	15	19	21	25	28	31	36	10	12	15	17	20	22	24	28	1.8	2.0	2.3	2.6	2.8	2.0	2.3	2.6	2.9	3.3
SS C4	15	20	25	30	36	40	45	52	11	16	20	24	29	33	38	45	1.4	1.6	1.8	2.0	2.1	1.5	1.7	1.9	2.1	2.3
SS D1	20	24	29	33	38	41	45	51	17	21	25	29	33	36	39	43	1.4	1.6	1.9	2.2	2.4	1.5	1.7	2.1	2.5	2.8
SS D2	24	31	39	45	52	57	63	71	21	28	35	40	47	52	58	65	1.1	1.3	1.5	1.7	1.8	1.2	1.4	1.6	1.8	2.0
SS D3	19	23	27	30	35	38	41	46	17	20	24	27	30	33	36	40	1.5	1.7	2.0	2.3	2.6	1.6	1.9	2.2	2.6	3.0
SS D4	23	30	38	43	50	55	61	69	20	27	34	39	46	50	56	63	1.1	1.3	1.5	1.7	1.9	1.2	1.4	1.6	1.8	2.0
SS E1	16	20	24	28	32	35	38	43	13	17	20	23	27	29	32	35	1.5	1.8	2.1	2.4	2.7	1.6	1.9	2.3	2.9	3.4
SS E2	20	26	33	38	45	50	56	63	17	23	29	34	40	44	50	56	1.2	1.4	1.6	1.8	2.0	1.3	1.5	1.7	1.9	2.1
SS E3	16	19	22	25	28	31	34	38	13	16	19	21	24	26	29	32	1.7	2.0	2.3	2.6	2.9	1.8	2.2	2.5	3.0	3.5
SS E4	19	25	31	36	42	47	52	59	16	21	27	31	37	41	46	53	1.2	1.5	1.7	1.9	2.0	1.3	1.5	1.8	2.0	2.2

DILLON, MONTANA

TYPE	SOLAR SAVINGS FRACTION (%) BASE TEMP: 55 F (5468 DD) LCR 100	70	50	40	30	25	20	15	65 F (8362 DD) 100	70	50	40	30	25	20	15	CONSERVATION FACTOR 55 F 100	50	30	20	15	65 F 100	50	30	20	15
WW A1	18	21	25	28	32	35	39	45	15	17	20	22	25	27	30	34	1.6	1.8	2.1	2.3	2.5	2.0	2.2	2.5	2.7	3.0
WW A2	19	24	30	34	41	45	51	58	14	18	23	26	31	35	40	46	1.3	1.5	1.7	1.9	2.0	1.5	1.7	1.9	2.1	2.3
WW A3	19	25	32	37	44	49	55	63	13	18	24	28	34	38	43	51	1.2	1.4	1.6	1.8	1.9	1.4	1.6	1.8	2.0	2.2
WW A4	19	25	33	38	46	51	57	66	13	18	24	29	35	40	46	53	1.2	1.4	1.5	1.7	1.9	1.4	1.5	1.7	1.9	2.1
WW A5	19	26	34	40	48	53	60	68	13	18	25	30	37	42	48	56	1.1	1.3	1.5	1.7	1.8	1.3	1.5	1.7	1.9	2.0
WW A6	19	26	35	41	49	54	61	70	12	18	25	30	38	43	49	57	1.1	1.3	1.5	1.7	1.8	1.3	1.4	1.6	1.8	2.0
WW B1	15	18	23	26	30	33	37	43	10	13	15	18	21	23	25	28	1.6	1.8	2.1	2.3	2.5	2.0	2.2	2.6	3.0	3.4
WW B2	20	27	35	41	49	54	61	70	14	20	26	32	39	44	51	59	1.1	1.3	1.5	1.6	1.8	1.3	1.4	1.6	1.8	2.0
WW B3	24	32	41	48	56	62	69	77	17	24	32	38	47	52	59	68	1.0	1.2	1.4	1.6	1.7	1.2	1.3	1.5	1.7	1.8
WW B4	24	33	43	50	60	66	73	82	17	25	35	41	51	57	64	73	1.0	1.2	1.3	1.5	1.6	1.1	1.2	1.4	1.6	1.7
WW B5	23	32	43	50	60	66	73	82	16	25	34	41	51	57	65	74	1.0	1.2	1.3	1.5	1.6	1.1	1.2	1.4	1.6	1.7
WW C1	25	34	43	50	59	65	72	80	19	26	35	41	49	55	62	71	1.0	1.2	1.4	1.5	1.7	1.1	1.3	1.5	1.6	1.8
WW C2	23	32	42	49	58	64	71	80	17	25	34	40	49	55	62	71	1.0	1.2	1.3	1.5	1.7	1.1	1.3	1.4	1.6	1.8
WW C3	28	39	50	58	68	74	81	88	21	31	42	50	60	67	74	83	0.9	1.1	1.2	1.4	1.6	1.0	1.1	1.3	1.5	1.6
WW C4	25	36	47	55	65	72	79	86	19	28	39	47	57	63	71	80	0.9	1.1	1.3	1.4	1.6	1.0	1.2	1.3	1.5	1.6
TW A1	18	21	25	28	32	35	39	45	15	17	20	22	25	27	30	35	1.6	1.8	2.0	2.3	2.4	2.0	2.2	2.4	2.7	2.9
TW A2	18	22	28	32	38	42	47	54	14	17	21	24	29	32	37	43	1.4	1.6	1.8	2.0	2.1	1.6	1.8	2.0	2.2	2.4
TW A3	17	23	29	33	40	44	50	58	13	17	22	25	31	34	39	46	1.3	1.5	1.7	1.9	2.0	1.5	1.7	1.9	2.1	2.3
TW A4	16	22	29	33	40	45	51	59	12	16	21	25	31	35	41	48	1.3	1.5	1.6	1.8	2.0	1.5	1.6	1.8	2.0	2.2
TW B1	17	21	25	28	33	36	41	47	14	16	20	22	26	28	32	37	1.6	1.8	2.0	2.2	2.3	1.9	2.1	2.3	2.5	2.7
TW B2	17	21	27	31	37	41	46	54	12	16	20	24	28	32	37	43	1.4	1.6	1.8	1.9	2.1	1.6	1.8	2.0	2.2	2.3
TW B3	16	21	27	31	37	42	47	55	12	15	20	24	29	33	37	44	1.4	1.5	1.7	1.9	2.0	1.6	1.7	1.9	2.1	2.3
TW B4	15	20	25	30	36	40	46	53	11	15	19	22	28	31	36	43	1.4	1.6	1.7	1.9	2.1	1.6	1.8	2.0	2.1	2.3
TW C1	16	20	24	28	33	37	41	48	13	15	19	22	26	29	33	38	1.5	1.7	1.9	2.1	2.2	1.8	2.0	2.2	2.4	2.5
TW C2	15	19	24	28	34	38	43	50	12	15	19	22	26	30	34	40	1.5	1.6	1.8	2.0	2.1	1.7	1.9	2.0	2.2	2.4
TW C3	14	19	23	27	32	36	41	49	11	14	18	21	25	29	33	39	1.5	1.7	1.9	2.0	2.2	1.8	1.9	2.1	2.3	2.4
TW C4	14	18	22	25	30	33	38	45	11	14	17	20	24	26	30	36	1.6	1.8	2.0	2.1	2.3	1.9	2.1	2.2	2.4	2.5
TW D1	12	16	19	22	27	30	34	39	9	11	13	15	18	20	23	27	1.7	1.9	2.1	2.4	2.6	2.1	2.3	2.6	2.9	3.3
TW D2	19	25	32	38	45	50	57	65	14	19	25	29	36	40	46	54	1.2	1.4	1.6	1.7	1.9	1.4	1.5	1.7	1.9	2.0
TW D3	20	27	35	41	49	54	61	69	15	20	27	32	39	44	50	59	1.2	1.3	1.5	1.7	1.8	1.3	1.5	1.6	1.8	2.0
TW D4	22	30	39	45	54	60	67	76	16	23	31	37	45	51	58	67	1.1	1.2	1.4	1.6	1.7	1.2	1.3	1.5	1.7	1.8
TW D5	21	30	39	46	55	61	68	77	16	23	31	38	47	52	60	69	1.1	1.2	1.4	1.5	1.7	1.2	1.3	1.5	1.6	1.8
TW E1	23	31	39	45	54	59	66	75	17	24	31	37	45	50	56	65	1.1	1.3	1.4	1.6	1.7	1.2	1.4	1.5	1.7	1.9
TW E2	23	31	39	46	55	60	67	76	17	24	31	37	45	51	58	67	1.1	1.2	1.4	1.6	1.7	1.2	1.3	1.5	1.7	1.8
TW E3	27	37	48	55	65	70	77	85	21	30	39	47	56	62	69	78	1.0	1.1	1.3	1.5	1.6	1.0	1.2	1.4	1.5	1.7
TW E4	25	35	45	53	62	69	76	84	19	28	37	44	54	60	68	77	1.0	1.1	1.3	1.5	1.6	1.1	1.2	1.4	1.5	1.7
TW F1	16	20	24	27	31	34	39	45	13	15	18	20	24	26	29	34	1.6	1.8	2.0	2.2	2.4	1.9	2.2	2.4	2.6	2.8
TW F2	15	20	25	29	35	39	44	52	11	15	19	22	26	30	34	40	1.4	1.6	1.8	2.0	2.1	1.7	1.9	2.0	2.2	2.4
TW F3	14	19	25	30	36	41	46	54	10	14	18	22	27	31	36	43	1.4	1.5	1.7	1.9	2.1	1.6	1.8	1.9	2.1	2.3
TW F4	12	18	24	28	35	39	45	54	8	12	17	20	26	30	35	42	1.4	1.5	1.7	1.9	2.0	1.6	1.7	1.9	2.1	2.3
TW G1	14	18	22	25	30	34	38	45	11	13	17	19	23	25	29	34	1.6	1.8	2.0	2.2	2.3	1.9	2.1	2.3	2.5	2.7
TW G2	13	17	22	26	32	36	42	49	9	12	16	19	24	27	32	38	1.5	1.6	1.8	2.0	2.2	1.7	1.9	2.1	2.3	2.4
TW G3	11	15	21	25	31	35	41	49	7	10	15	18	23	26	31	38	1.4	1.6	1.8	2.0	2.1	1.7	1.9	2.0	2.2	2.4
TW G4	9	13	18	22	27	31	37	44	5	8	12	15	20	23	28	34	1.5	1.7	1.9	2.1	2.2	1.8	1.9	2.1	2.3	2.4
TW H1	11	15	19	23	28	31	36	43	8	11	14	16	20	23	27	33	1.6	1.8	2.0	2.2	2.3	1.9	2.1	2.3	2.4	2.6
TW H2	9	13	17	21	26	30	35	42	5	8	12	15	19	22	26	32	1.6	1.8	1.9	2.1	2.3	1.9	2.0	2.2	2.4	2.5
TW H3	7	11	15	18	23	27	32	39	4	7	10	13	17	20	24	29	1.7	1.8	2.0	2.2	2.3	1.9	2.1	2.3	2.5	2.6
TW H4	5	8	12	14	19	22	26	32	3	5	8	10	13	16	19	24	1.9	2.0	2.2	2.4	2.5	2.2	2.3	2.5	2.7	2.8
TW I1	10	13	17	20	24	27	30	36	7	9	11	13	16	10	20	24	1.7	2.0	2.2	2.4	2.7	2.2	2.4	2.7	3.1	3.4
TW I2	16	22	29	34	42	47	53	62	11	16	22	26	33	37	43	51	1.2	1.4	1.6	1.8	1.9	1.4	1.6	1.7	1.9	2.1
TW I3	17	24	31	37	45	50	57	66	12	17	24	29	36	41	47	56	1.2	1.4	1.5	1.7	1.8	1.3	1.5	1.7	1.8	2.0
TW I4	19	27	36	42	51	57	64	73	13	20	28	34	42	48	55	64	1.1	1.3	1.4	1.6	1.7	1.2	1.4	1.5	1.7	1.8
TW I5	19	27	37	43	53	59	66	75	13	21	29	35	44	50	57	67	1.1	1.2	1.4	1.6	1.7	1.2	1.3	1.5	1.6	1.8
TW J1	21	28	37	43	52	57	64	73	15	22	29	35	42	48	54	63	1.1	1.3	1.4	1.6	1.8	1.2	1.4	1.6	1.7	1.9
TW J2	20	28	37	43	52	58	65	74	14	21	29	35	43	48	55	65	1.1	1.3	1.4	1.6	1.7	1.2	1.4	1.5	1.7	1.8
TW J3	24	34	45	52	62	68	75	84	18	27	37	44	54	60	67	77	1.0	1.1	1.3	1.5	1.6	1.1	1.2	1.4	1.5	1.7
TW J4	22	32	42	50	60	66	73	82	16	25	31	42	51	58	65	75	1.0	1.2	1.3	1.5	1.6	1.1	1.2	1.4	1.6	1.7
DG A1	13	17	21	24	27	30	32	35	8	11	14	16	18	20	21	23	1.5	1.9	2.3	3.0	3.9	1.9	2.3	2.9	4.1	-
DG A2	15	20	26	30	37	41	46	52	11	15	19	23	28	31	36	41	1.4	1.5	1.8	2.0	2.3	1.6	1.7	2.0	2.3	2.6
DG A3	19	26	33	39	47	53	60	68	15	20	26	31	39	44	51	59	1.2	1.3	1.5	1.7	1.9	1.4	1.5	1.6	1.8	2.0
DG B1	13	17	22	25	30	33	37	41	8	11	14	16	20	22	24	27	1.5	1.8	2.0	2.4	2.9	1.9	2.1	2.5	3.0	3.9
DG B2	16	21	27	32	39	44	50	57	11	15	20	24	30	34	39	46	1.3	1.5	1.6	1.9	2.1	1.6	1.7	1.9	2.1	2.3
DG B3	20	26	34	40	50	56	63	73	15	20	27	33	41	47	54	63	1.2	1.3	1.4	1.6	1.7	1.3	1.4	1.5	1.7	1.9
DG C1	16	21	26	31	37	41	46	51	10	14	18	21	26	29	33	37	1.3	1.6	1.8	2.1	2.4	1.6	1.8	2.1	2.5	2.9
DG C2	18	24	32	37	45	51	57	66	13	18	24	29	35	40	46	54	1.2	1.4	1.5	1.7	1.9	1.4	1.5	1.7	1.9	2.1
DG C3	22	30	39	45	56	62	70	79	17	23	31	37	46	52	60	70	1.1	1.2	1.3	1.5	1.7	1.2	1.3	1.4	1.6	1.8
SS A1	22	27	32	36	42	46	51	58	18	22	27	31	36	39	43	49	1.3	1.5	1.7	1.9	2.1	1.5	1.7	1.9	2.1	2.4
SS A2	27	34	42	47	55	60	66	74	23	29	36	41	49	53	59	67	1.1	1.3	1.5	1.6	1.8	1.2	1.4	1.5	1.7	1.9
SS A3	20	24	29	32	37	41	45	51	17	20	24	27	31	34	37	42	1.4	1.7	1.9	2.1	2.3	1.6	1.8	2.1	2.4	2.7
SS A4	26	33	40	46	53	58	64	72	22	28	35	40	47	51	57	64	1.1	1.3	1.5	1.7	1.8	1.2	1.4	1.6	1.8	2.0
SS A5	21	25	30	34	38	42	46	52	18	22	25	28	32	35	39	44	1.4	1.7	1.9	2.1	2.3	1.6	1.8	2.1	2.4	2.6
SS A6	26	33	41	47	54	59	66	73	22	28	35	41	48	52	58	66	1.1	1.3	1.5	1.6	1.8	1.2	1.4	1.6	1.8	1.9
SS A7	19	22	26	29	33	35	39	44	16	19	22	24	27	29	32	36	1.6	1.9	2.1	2.4	2.6	1.8	2.1	2.4	2.8	3.2
SS A8	25	32	39	44	52	56	62	70	21	27	34	38	45	49	55	62	1.1	1.3	1.5	1.7	1.9	1.2	1.4	1.6	1.8	2.0
SS B1	18	23	28	32	37	41	46	52	15	18	23	26	31	34	38	43	1.4	1.6	1.8	2.0	2.2	1.6	1.8	2.0	2.3	2.5
SS B2	23	30	38	43	51	56	62	70	19	25	32	37	44	49	55	63	1.1	1.3	1.5	1.7	1.8	1.2	1.4	1.6	1.8	1.9
SS B3	17	21	26	29	34	37	42	48	14	17	21	23	27	30	34	39	1.5	1.7	1.9	2.2	2.4	1.7	1.9	2.2	2.5	2.7
SS B4	23	29	36	42	49	54	60	68	19	24	31	36	42	47	53	60	1.2	1.3	1.5	1.7	1.9	1.3	1.4	1.6	1.8	2.0
SS B5	17	21	25	28	32	36	40	45	15	17	21	23	27	29	32	37	1.6	1.8	2.0	2.2	2.4	1.8	2.0	2.3	2.5	2.8
SS B6	23	29	37	42	49	55	61	69	18	24	31	36	43	47	53	61	1.2	1.3	1.5	1.7	1.8	1.3	1.4	1.6	1.8	2.0
SS B7	16	19	22	25	28	31	34	39	13	16	18	20	23	25	28	31	1.7	2.0	2.2	2.5	2.7	2.0	2.3	2.5	2.9	3.2
SS B8	21	28	35	40	47	52	58	66	18	23	29	34	40	45	50	58	1.2	1.4	1.6	1.7	1.9	1.3	1.5	1.7	1.9	2.0
SS C1	17	22	28	32	39	43	49	56	12	15	20	24	29	32	37	44	1.3	1.5	1.7	1.9	2.1	1.5	1.7	1.9	2.2	2.4
SS C2	20	27	35	41	49	55	61	70	14	20	27	32	40	45	51	60	1.1	1.3	1.5	1.7	1.8	1.3	1.4	1.6	1.8	1.9
SS C3	15	19	23	27	32	36	40	47	11	14	17	20	24	27	31	36	1.5	1.7	1.9	2.1	2.3	1.8	2.0	2.2	2.4	2.6
SS C4	18	25	31	37	44	49	56	64	14	19	24	29	35	40	46	54	1.2	1.4	1.6	1.7	1.9	1.4	1.5	1.7	1.9	2.0
SS D1	24	31	37	42	49	53	59	66	20	25	31	35	41	45	50	56	1.2	1.4	1.6	1.8	1.9	1.3	1.5	1.7	2.0	2.2
SS D2	30	39	48	54	63	68	74	82	25	33	41	48	56	61	67	75	1.0	1.2	1.3	1.5	1.7	1.1	1.2	1.4	1.6	1.8
SS D3	23	28	34	38	44	48	53	60	19	24	29	32	37	41	45	51	1.3	1.5	1.7	1.9	2.1	1.4	1.6	1.9	2.1	2.3
SS D4	29	37	46	52	61	66	72	80	24	32	40	46	54	59	65	73	1.0	1.2	1.4	1.5	1.7	1.1	1.3	1.5	1.6	1.8
SS E1	21	26	32	37	43	47	52	59	16	21	26	29	35	38	42	48	1.3	1.5	1.7	1.9	2.1	1.4	1.6	1.9	2.2	2.4
SS E2	25	33	42	48	56	62	68	76	20	27	35	41	48	53	60	68	1.1	1.2	1.4	1.6	1.7	1.1	1.3	1.5	1.7	1.9
SS E3	19	23	28	32	37	40	45	51	15	19	23	26	30	33	37	42	1.4	1.7	1.9	2.1	2.3	1.6	1.8	2.1	2.4	2.6
SS E4	24	31	39	44	52	57	64	72	19	26	33	38	45	50	56	63	1.1	1.3	1.5	1.7	1.8	1.2	1.4	1.6	1.8	1.9

GLASGOW, MONTANA

TYPE	SOLAR SAVINGS FRACTION (%)																CONSERVATION FACTOR									
	BASE TEMP: 55 F (6352 DD)								65 F (8986 DD)								55 F					65 F				
LCR	100	70	50	40	30	25	20	15	100	70	50	40	30	25	20	15	100	50	30	20	15	100	50	30	20	15
WW A1	14	15	17	18	19	20	22	23	12	13	14	15	16	17	17	18	2.4	2.9	3.6	4.4	-	2.8	3.4	4.4	-	-
WW A2	12	14	17	20	23	25	28	31	9	12	14	16	19	20	22	24	1.8	2.1	2.5	2.9	3.2	2.0	2.4	2.9	3.5	4.3
WW A3	11	14	18	20	24	27	30	34	8	11	14	16	19	21	24	27	1.7	2.0	2.3	2.6	2.9	1.8	2.2	2.6	3.1	3.7
WW A4	10	13	17	20	25	28	31	36	7	10	14	16	20	22	25	28	1.6	1.9	2.2	2.5	2.7	1.7	2.1	2.4	2.9	3.4
WW A5	9	13	18	21	26	29	33	38	6	10	14	16	20	23	26	30	1.5	1.8	2.1	2.3	2.6	1.7	2.0	2.3	2.8	3.2
WW A6	9	13	18	21	26	29	34	40	6	9	14	17	21	23	27	31	1.5	1.8	2.0	2.3	2.5	1.6	1.9	2.3	2.7	3.0
WW B1	8	10	11	11	12	12	12	11	6	7	8	8	8	8	7	4	2.6	3.7	-	-	-	3.3	-	-	-	-
WW B2	11	15	20	24	29	33	38	45	8	12	17	20	25	28	32	38	1.5	1.7	1.9	2.1	2.3	1.6	1.8	2.1	2.3	2.6
WW B3	13	18	24	29	35	39	45	52	10	15	20	24	30	34	39	45	1.3	1.5	1.7	1.9	2.1	1.4	1.6	1.9	2.1	2.3
WW B4	13	20	27	33	40	46	52	61	10	16	23	28	36	40	47	55	1.2	1.4	1.6	1.8	1.9	1.3	1.5	1.7	1.9	2.0
WW B5	13	19	27	33	41	47	54	63	10	16	23	29	37	42	49	58	1.2	1.4	1.5	1.7	1.9	1.2	1.4	1.6	1.8	1.9
WW C1	14	20	27	31	38	43	49	57	11	17	23	27	33	37	43	50	1.3	1.5	1.7	1.9	2.0	1.3	1.6	1.8	2.0	2.2
WW C2	13	19	26	31	39	44	50	58	10	16	22	27	34	38	44	52	1.3	1.4	1.6	1.8	2.0	1.3	1.5	1.7	1.9	2.1
WW C3	16	25	34	41	50	56	64	73	13	21	30	37	45	51	59	68	1.1	1.3	1.4	1.6	1.7	1.1	1.3	1.5	1.6	1.8
WW C4	15	23	32	38	47	53	61	71	12	19	28	34	43	49	56	66	1.1	1.3	1.4	1.6	1.8	1.2	1.3	1.5	1.7	1.8
TW A1	14	15	17	18	20	21	23	24	12	13	14	15	17	18	19	20	2.3	2.8	3.3	3.9	4.6	2.7	3.2	3.9	-	-
TW A2	12	14	17	19	22	24	27	30	10	12	14	16	18	20	22	24	1.9	2.2	2.6	3.0	3.3	2.1	2.5	2.9	3.5	4.1
TW A3	11	14	17	19	23	25	28	32	8	11	14	16	19	21	23	26	1.8	2.1	2.4	2.7	3.0	1.9	2.3	2.7	3.1	3.6
TW A4	9	13	16	19	23	26	29	34	7	10	13	15	19	21	24	28	1.7	2.0	2.3	2.5	2.8	1.9	2.1	2.5	2.9	3.2
TW B1	13	15	17	18	20	22	24	26	11	12	14	15	17	18	20	21	2.2	2.6	3.0	3.4	3.8	2.5	2.9	3.4	4.0	4.9
TW B2	11	13	16	19	22	24	27	31	9	11	13	15	18	20	22	25	1.9	2.2	2.5	2.8	3.0	2.1	2.4	2.7	3.2	3.6
TW B3	10	13	16	18	22	24	28	32	8	10	13	15	18	20	23	26	1.8	2.1	2.4	2.6	2.9	2.0	2.3	2.6	3.0	3.3
TW B4	9	12	15	18	21	24	27	32	7	10	13	15	18	20	23	27	1.8	2.1	2.4	2.6	2.8	2.0	2.3	2.6	2.9	3.2
TW C1	11	14	16	18	21	22	25	28	10	11	13	15	17	19	21	24	2.1	2.4	2.7	3.0	3.3	2.3	2.6	3.0	3.4	3.8
TW C2	10	13	16	18	21	23	26	30	8	10	13	15	18	20	22	25	1.9	2.2	2.5	2.8	3.0	2.1	2.4	2.7	3.0	3.4
TW C3	10	12	15	17	21	23	26	30	8	10	13	15	17	19	22	25	2.0	2.2	2.5	2.8	3.0	2.2	2.4	2.7	3.0	3.3
TW C4	11	13	15	17	20	22	25	29	9	11	13	15	17	19	21	25	2.1	2.4	2.6	2.9	3.1	2.3	2.6	2.8	3.1	3.3
TW D1	7	8	10	11	12	12	13	13	5	6	7	7	8	8	8	7	2.6	3.3	4.6	-	-	3.2	4.2	-	-	-
TW D2	12	15	20	23	28	31	35	41	9	13	16	19	24	26	30	35	1.6	1.8	2.0	2.2	2.4	1.7	1.9	2.2	2.5	2.7
TW D3	12	16	21	24	29	33	38	44	9	13	17	21	25	28	32	38	1.5	1.7	1.9	2.2	2.3	1.6	1.8	2.1	2.4	2.6
TW D4	13	18	25	29	36	41	47	55	10	15	21	26	32	36	42	49	1.3	1.5	1.7	1.9	2.0	1.4	1.6	1.8	2.0	2.1
TW D5	13	19	26	31	38	43	50	58	10	16	22	27	34	39	45	53	1.3	1.5	1.6	1.8	1.9	1.3	1.5	1.7	1.9	2.0
TW E1	14	19	24	29	35	39	44	51	11	16	21	25	30	34	38	45	1.4	1.6	1.8	2.0	2.2	1.5	1.7	1.9	2.2	2.4
TW E2	14	19	25	30	36	41	46	54	11	16	21	26	32	36	41	48	1.3	1.5	1.7	1.9	2.1	1.4	1.6	1.8	2.0	2.2
TW E3	17	24	32	38	46	51	58	67	14	20	28	34	41	47	53	62	1.2	1.3	1.5	1.7	1.8	1.2	1.4	1.6	1.8	1.9
TW E4	15	22	30	36	45	50	57	67	12	19	27	32	40	46	53	62	1.2	1.3	1.5	1.7	1.8	1.2	1.4	1.6	1.7	1.9
TW F1	12	13	15	17	18	20	21	23	10	11	13	14	15	16	17	18	2.3	2.7	3.2	3.8	4.4	2.7	3.1	3.8	4.8	-
TW F2	10	12	15	17	20	22	24	27	7	9	12	13	16	17	19	22	1.9	2.3	2.6	3.0	3.3	2.2	2.5	3.0	3.5	4.2
TW F3	8	11	14	16	20	22	25	29	6	8	11	13	16	17	20	23	1.8	2.1	2.4	2.8	3.0	2.0	2.3	2.7	3.2	3.6
TW F4	6	9	12	15	19	21	25	29	4	6	9	11	15	17	20	23	1.8	2.1	2.4	2.6	2.9	2.0	2.2	2.6	3.0	3.3
TW G1	10	11	14	15	17	19	21	24	8	9	11	12	14	15	17	19	2.2	2.6	3.0	3.4	3.8	2.5	2.9	3.4	4.0	4.9
TW G2	7	10	12	15	18	20	22	26	5	7	9	11	14	16	18	21	1.9	2.3	2.6	2.9	3.2	2.2	2.5	2.9	3.3	3.8
TW G3	6	8	11	13	17	19	22	26	3	6	8	10	13	15	18	21	1.9	2.2	2.5	2.8	3.0	2.1	2.4	2.7	3.1	3.5
TW G4	4	6	9	12	15	17	20	25	2	4	7	9	11	13	16	20	2.0	2.3	2.5	2.8	3.0	2.2	2.4	2.7	3.1	3.3
TW H1	7	9	11	13	15	17	20	23	5	7	9	10	12	14	16	18	2.1	2.5	2.8	3.2	3.5	2.4	2.7	3.1	3.6	4.1
TW H2	4	7	9	11	14	16	19	23	3	5	7	8	11	13	15	18	2.1	2.4	2.7	3.0	3.2	2.3	2.6	2.9	3.3	3.6
TW H3	3	5	8	10	13	15	18	21	2	3	6	7	10	12	14	17	2.2	2.4	2.7	3.0	3.2	2.4	2.6	2.9	3.2	3.5
TW H4	2	4	6	8	10	12	15	18	1	3	4	6	8	9	12	15	2.4	2.6	2.9	3.2	3.4	2.7	2.9	3.1	3.4	3.6
TW I1	5	6	8	9	10	10	11	11	3	4	5	6	6	6	6	5	2.7	3.5	4.8	-	-	3.4	4.4	-	-	-
TW I2	9	12	17	20	25	28	32	38	6	10	13	16	21	23	27	32	1.6	1.8	2.1	2.3	2.5	1.7	2.0	2.2	2.5	2.7
TW I3	9	13	18	22	27	31	35	42	7	11	15	18	23	26	30	36	1.5	1.7	2.0	2.2	2.3	1.6	1.8	2.1	2.4	2.6
TW I4	10	16	22	27	33	38	44	52	8	13	19	23	29	34	39	47	1.4	1.5	1.7	1.9	2.0	1.4	1.6	1.8	2.0	2.2
TW I5	11	16	23	28	36	41	47	56	8	13	20	25	31	36	42	51	1.3	1.5	1.7	1.8	2.0	1.4	1.5	1.7	1.9	2.1
TW J1	12	17	22	27	33	37	42	50	9	14	19	23	28	32	37	43	1.4	1.6	1.8	2.0	2.2	1.5	1.7	1.9	2.2	2.3
TW J2	11	17	23	27	34	38	44	52	9	14	19	23	29	34	39	46	1.4	1.5	1.7	1.9	2.1	1.4	1.6	1.8	2.0	2.2
TW J3	14	21	29	35	44	50	57	66	11	18	26	31	40	45	52	61	1.2	1.3	1.5	1.7	1.8	1.2	1.4	1.6	1.7	1.9
TW J4	13	20	28	34	42	48	55	65	10	17	24	30	38	44	51	60	1.2	1.4	1.5	1.7	1.8	1.2	1.4	1.6	1.8	1.9
DG A1	7	8	10	10	11	12	12	11	4	6	7	7	7	7	7	5	2.4	3.3	-	-	-	2.8	4.1	-	-	-
DG A2	10	12	16	18	22	24	27	30	7	10	13	15	18	20	22	25	1.8	2.1	2.4	2.8	3.2	2.0	2.2	2.6	3.1	3.7
DG A3	13	17	22	26	32	36	41	48	11	15	19	23	28	32	37	43	1.5	1.7	1.8	2.0	2.2	1.6	1.7	1.9	2.1	2.4
DG B1	7	8	10	11	12	12	13	13	4	6	7	8	8	8	8	7	2.3	3.2	4.6	-	-	2.7	3.6	-	-	-
DG B2	10	13	16	19	22	25	28	33	8	10	13	16	19	21	24	27	1.8	2.0	2.3	2.6	2.9	1.9	2.1	2.5	2.9	3.4
DG B3	13	18	23	27	33	37	43	51	11	15	20	24	30	34	39	46	1.5	1.6	1.8	1.9	2.1	1.5	1.7	1.9	2.0	2.2
DG C1	9	11	13	15	17	18	19	21	6	8	10	11	13	14	14	14	2.0	2.6	3.2	4.1	-	2.2	2.7	3.9	-	-
DG C2	12	15	19	23	27	30	35	40	9	13	17	19	24	26	30	35	1.6	1.8	2.0	2.3	2.5	1.7	1.9	2.2	2.5	2.8
DG C3	15	20	26	30	37	42	49	58	13	18	23	28	34	38	44	52	1.4	1.5	1.6	1.8	2.0	1.4	1.6	1.8	1.9	2.1
SS A1	14	17	20	22	25	27	30	33	13	15	18	20	23	24	26	29	1.8	2.1	2.5	2.9	3.2	1.9	2.3	2.8	3.4	4.0
SS A2	18	22	28	31	37	41	46	53	16	20	25	29	34	38	42	48	1.4	1.6	1.8	2.0	2.2	1.4	1.7	1.9	2.1	2.3
SS A3	13	15	17	19	21	22	23	25	12	13	15	17	18	19	20	21	2.0	2.6	3.2	4.0	-	2.2	2.7	3.7	-	-
SS A4	17	21	26	29	35	38	43	49	15	19	24	27	32	35	39	44	1.4	1.7	1.9	2.1	2.3	1.5	1.7	2.0	2.3	2.5
SS A5	15	17	19	21	23	25	27	29	14	16	18	19	21	22	23	25	2.0	2.4	2.9	3.5	4.1	2.1	2.6	3.3	4.2	-
SS A6	17	22	27	31	36	40	44	51	16	20	25	28	33	36	41	46	1.4	1.6	1.8	2.1	2.2	1.4	1.7	2.0	2.2	2.4
SS A7	14	15	17	18	19	20	20	21	13	14	15	16	17	17	18	18	2.4	3.1	4.0	-	-	2.6	3.3	4.8	-	-
SS A8	16	20	25	28	33	36	40	46	15	19	23	26	30	33	36	41	1.5	1.7	2.0	2.2	2.4	1.5	1.8	2.1	2.4	2.7
SS B1	12	14	17	19	22	23	26	29	11	13	15	17	19	21	22	25	1.9	2.3	2.7	3.1	3.5	2.0	2.4	2.9	3.6	4.3
SS B2	15	20	25	29	34	38	43	50	14	18	23	26	31	35	39	46	1.4	1.6	1.8	2.0	2.2	1.5	1.7	1.9	2.2	2.4
SS B3	11	13	15	16	18	20	21	23	9	11	13	14	16	17	18	19	2.1	2.6	3.1	3.8	4.6	2.3	2.8	3.6	4.9	-
SS B4	15	19	23	27	32	36	40	47	13	17	21	25	29	33	37	42	1.5	1.7	1.9	2.1	2.3	1.5	1.8	2.0	2.3	2.5
SS B5	12	14	16	17	19	21	22	24	11	13	14	16	17	18	19	21	2.2	2.7	3.2	3.8	4.4	2.4	2.9	3.5	4.5	-
SS B6	15	19	24	27	33	36	41	48	13	17	21	25	30	33	37	43	1.5	1.7	1.9	2.1	2.3	1.5	1.7	2.0	2.2	2.4
SS B7	11	13	14	15	16	17	18	19	10	11	12	13	14	15	15	16	2.5	3.2	4.0	-	-	2.8	3.4	4.6	-	-
SS B8	14	17	22	25	30	33	37	43	12	16	20	23	27	30	34	39	1.5	1.8	2.0	2.2	2.4	1.6	1.8	2.1	2.4	2.6
SS C1	9	12	15	17	20	22	24	28	7	9	11	13	15	17	19	20	1.8	2.2	2.6	3.0	3.4	2.1	2.5	3.0	3.8	5.0
SS C2	11	16	21	25	30	34	39	46	9	13	17	21	26	29	33	39	1.5	1.7	1.9	2.1	2.3	1.6	1.8	2.0	2.3	2.5
SS C3	10	12	14	16	18	20	22	25	8	10	12	13	15	16	18	20	2.1	2.5	2.9	3.3	3.7	2.4	2.8	3.3	4.0	4.8
SS C4	11	15	19	23	27	31	35	41	9	12	16	19	23	26	30	35	1.6	1.8	2.0	2.2	2.4	1.7	1.9	2.2	2.4	2.7
SS D1	15	18	22	24	28	30	33	37	13	16	19	22	25	27	29	31	1.6	1.9	2.3	2.7	3.0	1.7	2.1	2.6	3.2	4.0
SS D2	18	24	31	36	43	47	53	61	16	22	28	33	39	43	49	56	1.2	1.4	1.6	1.8	2.0	1.3	1.5	1.7	1.9	2.1
SS D3	15	18	21	23	26	28	31	34	14	16	19	21	23	25	27	29	1.8	2.1	2.5	2.9	3.3	1.8	2.3	2.8	3.5	4.4
SS D4	18	24	30	35	41	46	51	59	16	22	28	32	38	42	47	54	1.3	1.5	1.7	1.9	2.0	1.3	1.5	1.8	2.0	2.2
SS E1	12	14	17	19	22	24	26	29	10	12	15	16	19	20	21	22	1.8	2.2	2.7	3.3	3.9	1.9	2.4	3.2	4.6	-
SS E2	15	20	25	30	36	40	45	52	13	18	23	27	32	35	40	46	1.3	1.6	1.8	2.0	2.2	1.4	1.7	1.9	2.2	2.4
SS E3	13	15	17	19	21	22	24	26	11	13	15	16	18	19	21	22	2.0	2.5	3.0	3.5	4.2	2.2	2.7	3.3	4.4	-
SS E4	15	19	24	28	33	37	42	48	13	17	22	25	30	33	37	43	1.4	1.7	1.9	2.1	2.3	1.5	1.7	2.0	2.3	2.5

GREAT FALLS, MONTANA

TYPE	SOLAR SAVINGS FRACTION (%)																CONSERVATION FACTOR									
BASE TEMP:	55 F (5028 DD)								65 F (7674 DD)								55 F					65 F				
LCR	100	70	50	40	30	25	20	15	100	70	50	40	30	25	20	15	100	50	30	20	15	100	50	30	20	15
WW A1	17	20	23	25	28	31	33	38	14	16	18	20	22	24	26	29	1.8	2.1	2.4	2.7	2.9	2.2	2.5	2.8	3.3	3.7
WW A2	17	21	26	30	35	39	43	49	13	16	20	23	27	30	34	39	1.4	1.7	1.9	2.1	2.3	1.6	1.9	2.1	2.4	2.7
WW A3	17	22	27	32	37	42	47	54	12	16	21	24	29	33	37	42	1.3	1.6	1.8	2.0	2.2	1.5	1.7	2.0	2.3	2.5
WW A4	16	22	28	32	39	43	49	56	11	16	21	25	30	34	38	45	1.3	1.5	1.7	1.9	2.1	1.4	1.7	1.9	2.2	2.4
WW A5	16	22	29	34	40	45	51	58	11	16	22	26	31	35	40	47	1.2	1.5	1.7	1.9	2.0	1.4	1.6	1.9	2.1	2.3
WW A6	16	22	29	34	41	46	52	60	11	16	22	26	32	36	41	48	1.2	1.4	1.6	1.8	2.0	1.4	1.6	1.8	2.1	2.3
WW B1	13	16	19	21	24	26	29	32	9	11	13	15	17	18	19	20	1.8	2.1	2.5	2.9	3.4	2.2	2.6	3.3	4.6	-
WW B2	17	23	30	35	42	47	53	61	12	18	24	28	34	39	44	51	1.2	1.4	1.6	1.8	2.0	1.4	1.5	1.8	2.0	2.2
WW B3	20	28	36	41	49	55	61	69	15	22	29	34	41	46	52	59	1.1	1.3	1.5	1.7	1.8	1.2	1.4	1.6	1.8	2.0
WW B4	21	29	38	44	53	59	66	75	15	23	31	37	46	51	58	67	1.1	1.2	1.4	1.6	1.7	1.1	1.3	1.5	1.7	1.8
WW B5	20	28	38	44	54	60	67	76	14	22	31	37	46	52	59	68	1.1	1.2	1.4	1.6	1.7	1.1	1.3	1.5	1.6	1.8
WW C1	22	30	38	44	52	57	64	72	17	23	31	36	44	49	55	63	1.1	1.3	1.5	1.6	1.8	1.2	1.4	1.6	1.8	1.9
WW C2	20	28	37	43	52	57	64	73	15	22	30	36	44	49	56	64	1.1	1.3	1.4	1.6	1.8	1.2	1.4	1.5	1.7	1.9
WW C3	24	34	45	52	62	68	75	83	19	28	38	45	55	61	69	77	1.0	1.1	1.3	1.5	1.6	1.0	1.2	1.4	1.5	1.7
WW C4	22	32	42	50	59	65	73	81	17	26	35	43	52	58	66	75	1.0	1.2	1.3	1.5	1.6	1.1	1.2	1.4	1.6	1.7
TW A1	17	20	23	25	28	31	34	38	14	16	18	20	23	24	27	30	1.8	2.1	2.3	2.6	2.8	2.1	2.4	2.7	3.1	3.5
TW A2	16	20	25	28	33	36	40	46	13	16	19	22	26	28	32	37	1.5	1.7	2.0	2.2	2.4	1.7	2.0	2.2	2.5	2.8
TW A3	16	20	25	29	34	38	43	49	12	15	19	22	27	30	34	39	1.4	1.6	1.9	2.1	2.3	1.6	1.8	2.1	2.4	2.6
TW A4	15	19	25	29	35	39	44	51	10	14	19	22	27	31	35	41	1.4	1.6	1.8	2.0	2.2	1.6	1.8	2.0	2.3	2.5
TW B1	16	19	23	25	29	32	35	40	13	15	18	20	23	25	28	32	1.7	2.0	2.2	2.4	2.6	2.0	2.2	2.5	2.8	3.1
TW B2	15	19	24	27	32	36	40	46	12	15	18	21	25	28	32	37	1.5	1.7	1.9	2.2	2.3	1.7	1.9	2.2	2.4	2.7
TW B3	14	19	23	27	32	36	41	47	11	14	18	21	26	29	33	38	1.5	1.7	1.9	2.1	2.3	1.7	1.9	2.1	2.3	2.6
TW B4	14	18	22	26	31	35	40	46	10	13	17	20	25	28	32	38	1.5	1.7	1.9	2.1	2.3	1.7	1.9	2.1	2.3	2.5
TW C1	15	18	22	25	29	32	36	42	12	14	17	20	23	26	29	33	1.7	1.9	2.1	2.3	2.5	1.9	2.1	2.4	2.6	2.9
TW C2	14	18	22	25	30	33	38	44	11	14	17	20	24	27	30	36	1.6	1.8	2.0	2.2	2.4	1.8	2.0	2.2	2.4	2.7
TW C3	13	17	21	24	29	32	36	42	11	13	17	19	23	26	30	35	1.6	1.8	2.0	2.2	2.4	1.9	2.0	2.2	2.5	2.7
TW C4	13	16	20	23	27	30	34	39	11	13	16	18	22	24	28	33	1.7	1.9	2.1	2.3	2.5	2.0	2.2	2.4	2.6	2.8
TW D1	11	14	17	19	22	24	27	30	8	9	12	13	15	17	18	20	1.8	2.2	2.5	2.9	3.2	2.3	2.6	3.1	3.9	4.8
TW D2	17	22	28	33	39	44	50	57	13	17	22	26	32	36	41	48	1.3	1.5	1.7	1.9	2.0	1.4	1.6	1.8	2.1	2.2
TW D3	18	24	30	35	42	47	53	61	13	18	24	28	35	39	44	51	1.2	1.4	1.6	1.8	2.0	1.4	1.6	1.8	2.0	2.2
TW D4	19	26	34	40	48	54	61	69	15	21	28	34	41	46	53	61	1.1	1.3	1.5	1.7	1.8	1.2	1.4	1.6	1.8	1.9
TW D5	19	27	35	41	50	55	62	71	14	21	29	34	43	48	55	64	1.1	1.3	1.5	1.6	1.8	1.2	1.4	1.5	1.7	1.9
TW E1	21	27	35	40	47	52	59	67	16	22	28	33	40	44	50	58	1.2	1.4	1.5	1.7	1.9	1.3	1.5	1.7	1.9	2.0
TW E2	20	27	35	41	48	54	60	69	15	22	28	34	41	46	52	60	1.1	1.3	1.5	1.7	1.8	1.2	1.4	1.6	1.8	2.0
TW E3	24	33	42	49	58	64	71	79	19	27	36	42	51	57	64	72	1.0	1.2	1.4	1.5	1.7	1.1	1.3	1.4	1.6	1.8
TW E4	22	31	40	47	57	62	70	78	17	25	34	41	49	55	62	71	1.0	1.2	1.4	1.5	1.7	1.1	1.3	1.4	1.6	1.8
TW F1	15	18	21	24	27	30	33	38	12	14	17	18	21	23	26	29	1.8	2.0	2.3	2.5	2.8	2.1	2.3	2.7	3.0	3.4
TW F2	14	18	22	25	30	33	38	44	10	13	17	19	23	26	29	34	1.5	1.8	2.0	2.2	2.4	1.8	2.0	2.3	2.6	2.8
TW F3	13	17	22	25	31	34	39	46	9	12	16	19	24	27	31	36	1.5	1.7	1.9	2.1	2.3	1.7	1.9	2.1	2.4	2.6
TW F4	11	15	20	24	30	33	39	45	7	10	15	18	23	26	30	36	1.5	1.7	1.9	2.1	2.3	1.7	1.9	2.1	2.3	2.6
TW G1	13	16	20	22	26	29	33	38	10	12	15	17	20	22	25	29	1.7	2.0	2.2	2.4	2.7	2.0	2.2	2.5	2.8	3.1
TW G2	11	15	19	23	28	31	35	42	8	11	14	17	21	24	27	32	1.6	1.8	2.0	2.2	2.4	1.8	2.0	2.3	2.5	2.8
TW G3	10	13	18	22	27	30	35	41	6	9	13	16	20	23	27	32	1.6	1.8	2.0	2.2	2.4	1.8	2.0	2.2	2.5	2.7
TW G4	7	11	15	19	23	27	31	38	4	7	11	13	17	20	24	30	1.6	1.8	2.1	2.3	2.4	1.9	2.1	2.3	2.5	2.7
TW H1	10	13	17	20	24	27	31	36	7	10	12	15	18	20	24	28	1.7	1.9	2.2	2.4	2.6	2.0	2.2	2.5	2.7	3.0
TW H2	8	11	15	18	23	26	30	36	5	7	11	13	17	19	23	28	1.7	1.9	2.1	2.3	2.5	2.0	2.2	2.4	2.6	2.8
TW H3	6	9	13	16	20	23	27	33	4	6	9	11	15	17	21	26	1.8	2.0	2.2	2.4	2.6	2.1	2.2	2.4	2.7	2.8
TW H4	5	7	10	13	16	19	23	28	3	5	7	9	12	14	17	22	2.0	2.2	2.4	2.6	2.7	2.2	2.5	2.6	2.8	3.0
TW I1	9	11	14	16	19	21	24	27	6	8	10	11	13	14	16	18	1.9	2.2	2.6	3.0	3.3	2.4	2.7	3.3	4.0	-
TW I2	14	19	25	30	36	41	46	54	10	14	19	23	29	33	38	45	1.3	1.5	1.7	1.9	2.1	1.5	1.7	1.9	2.1	2.3
TW I3	15	21	27	32	39	44	50	58	11	16	21	26	32	36	41	49	1.3	1.5	1.7	1.8	2.0	1.4	1.6	1.8	2.0	2.2
TW I4	17	24	31	37	45	51	58	66	12	18	25	30	38	43	50	58	1.2	1.3	1.5	1.7	1.8	1.3	1.4	1.6	1.8	2.0
TW I5	17	24	32	39	47	53	60	69	12	19	26	32	40	45	52	61	1.1	1.3	1.5	1.6	1.8	1.2	1.4	1.6	1.7	1.9
TW J1	18	25	32	38	45	50	57	65	14	19	26	31	38	42	48	56	1.2	1.4	1.5	1.7	1.9	1.3	1.5	1.7	1.9	2.0
TW J2	18	24	32	38	46	51	58	67	13	19	26	31	38	43	50	58	1.2	1.3	1.5	1.7	1.8	1.3	1.4	1.6	1.8	2.0
TW J3	21	30	40	47	56	62	69	78	16	24	33	40	49	55	62	71	1.0	1.2	1.4	1.5	1.7	1.1	1.3	1.4	1.6	1.8
TW J4	20	28	38	45	54	60	67	76	15	20	31	38	47	53	60	69	1.1	1.2	1.4	1.6	1.7	1.1	1.3	1.5	1.6	1.8
DG A1	11	14	17	20	22	24	25	27	7	9	12	13	15	16	17	17	1.7	2.1	2.8	3.9	-	2.1	2.6	3.6	-	-
DG A2	14	18	23	27	32	35	39	45	10	14	18	21	25	28	32	36	1.5	1.7	1.9	2.2	2.6	1.7	1.9	2.1	2.5	2.9
DG A3	17	23	30	35	42	47	53	61	14	19	24	29	36	40	46	53	1.3	1.4	1.6	1.8	2.0	1.4	1.5	1.7	1.9	2.1
DG B1	11	15	18	20	24	26	29	32	7	10	12	14	16	18	19	21	1.7	2.0	2.4	3.0	4.0	2.0	2.4	3.1	4.2	-
DG B2	14	19	24	28	34	38	43	49	10	14	18	22	27	30	34	40	1.4	1.6	1.8	2.1	2.3	1.6	1.8	2.0	2.3	2.6
DG B3	18	24	31	36	44	50	57	65	14	19	26	30	37	42	49	57	1.3	1.4	1.5	1.7	1.9	1.4	1.5	1.6	1.8	2.0
DG C1	14	18	22	25	30	33	37	41	10	13	16	19	22	24	27	29	1.5	1.8	2.0	2.4	3.0	1.7	2.0	2.5	3.1	3.8
DG C2	17	22	28	32	39	44	50	57	12	17	22	26	32	36	41	47	1.3	1.5	1.7	1.9	2.1	1.5	1.6	1.8	2.1	2.3
DG C3	20	27	35	41	49	55	63	71	16	22	29	34	42	48	55	63	1.2	1.3	1.4	1.6	1.8	1.3	1.4	1.6	1.7	1.9
SS A1	19	23	28	31	36	39	43	49	16	20	24	27	31	34	37	42	1.4	1.7	1.9	2.2	2.4	1.6	1.8	2.1	2.5	2.8
SS A2	24	30	37	42	49	53	59	67	20	26	32	37	43	48	53	60	1.2	1.4	1.6	1.8	1.9	1.2	1.5	1.7	1.9	2.1
SS A3	18	21	25	27	31	34	37	42	15	18	21	23	26	28	31	34	1.6	1.9	2.2	2.5	2.8	1.7	2.1	2.5	3.0	3.5
SS A4	23	29	35	40	46	51	56	64	20	25	31	35	41	45	50	56	1.2	1.4	1.6	1.8	2.0	1.3	1.5	1.7	2.0	2.2
SS A5	19	23	26	29	33	36	39	44	17	20	23	25	28	31	33	37	1.6	1.9	2.1	2.4	2.7	1.7	2.0	2.4	2.8	3.2
SS A6	23	29	36	41	48	52	58	65	20	26	32	36	42	46	52	58	1.2	1.4	1.6	1.8	1.9	1.3	1.5	1.7	1.9	2.1
SS A7	17	20	23	25	28	30	32	36	15	17	20	21	24	25	27	29	1.8	2.1	2.5	2.9	3.3	2.0	2.4	2.9	3.5	4.4
SS A8	22	28	34	39	45	49	54	61	19	24	30	34	39	43	48	54	1.2	1.4	1.7	1.9	2.0	1.3	1.5	1.8	2.0	2.2
SS B1	16	20	24	27	32	35	39	44	14	17	20	23	27	29	32	37	1.5	1.8	2.0	2.3	2.5	1.7	2.0	2.3	2.6	2.9
SS B2	21	27	33	38	45	50	55	63	18	23	29	33	40	44	49	56	1.2	1.4	1.6	1.8	1.9	1.3	1.5	1.7	1.9	2.1
SS B3	15	18	22	25	29	31	35	39	12	15	18	20	23	26	28	31	1.6	1.9	2.2	2.5	2.8	1.8	2.1	2.5	2.9	3.4
SS B4	20	26	32	36	43	47	53	61	17	22	28	32	38	42	47	54	1.3	1.5	1.6	1.8	2.0	1.3	1.5	1.8	2.0	2.2
SS B5	16	19	22	24	28	30	34	38	14	16	19	21	23	25	28	31	1.7	2.0	2.3	2.6	2.8	2.0	2.2	2.6	3.0	3.4
SS B6	20	26	32	37	44	48	54	62	17	22	28	32	38	42	47	55	1.2	1.4	1.6	1.8	2.0	1.3	1.5	1.7	2.0	2.1
SS B7	14	17	20	22	24	26	29	32	12	14	16	18	20	22	23	26	1.9	2.2	2.6	2.9	3.3	2.2	2.5	3.0	3.5	4.1
SS B8	19	24	30	35	41	45	51	58	16	21	26	30	36	39	44	51	1.3	1.5	1.7	1.9	2.1	1.4	1.6	1.8	2.1	2.2
SS C1	15	19	24	27	33	36	41	47	10	14	18	21	25	27	31	36	1.4	1.7	1.9	2.1	2.4	1.7	1.9	2.2	2.5	2.8
SS C2	18	24	31	36	43	48	54	62	13	18	24	29	35	39	45	52	1.2	1.4	1.6	1.8	1.9	1.3	1.5	1.7	2.0	2.1
SS C3	14	17	21	24	28	31	34	40	11	13	16	18	21	24	27	31	1.7	1.9	2.1	2.4	2.6	1.9	2.2	2.5	2.8	3.1
SS C4	16	22	28	32	39	43	49	56	12	17	22	26	32	36	41	47	1.3	1.5	1.7	1.9	2.0	1.5	1.6	1.9	2.1	2.2
SS D1	21	26	32	36	41	45	50	56	18	22	27	30	35	38	42	47	1.3	1.5	1.8	2.0	2.2	1.4	1.7	2.0	2.3	2.6
SS D2	26	34	42	48	56	61	67	75	22	29	37	42	49	54	60	68	1.1	1.3	1.4	1.6	1.8	1.1	1.3	1.5	1.7	1.9
SS D3	21	25	30	33	38	41	45	51	18	21	25	28	32	35	38	43	1.4	1.7	1.9	2.2	2.4	1.5	1.8	2.1	2.5	2.8
SS D4	25	33	40	46	54	59	65	73	22	28	36	41	48	53	58	66	1.1	1.3	1.5	1.7	1.8	1.1	1.4	1.6	1.8	1.9
SS E1	18	22	27	31	36	39	43	49	14	18	22	25	29	31	35	38	1.4	1.7	1.9	2.2	2.4	1.5	1.8	2.2	2.6	3.0
SS E2	22	29	36	42	49	54	60	68	18	24	31	36	42	47	52	59	1.1	1.3	1.5	1.7	1.9	1.2	1.4	1.7	1.9	2.1
SS E3	17	20	24	27	31	34	38	42	14	17	20	23	26	28	31	34	1.6	1.9	2.1	2.4	2.6	1.8	2.0	2.4	2.8	3.2
SS E4	21	27	34	39	46	50	56	64	17	23	29	33	40	44	49	56	1.2	1.4	1.6	1.8	2.0	1.3	1.5	1.7	1.9	2.1

HELENA, MONTANA

SOLAR SAVINGS FRACTION (%) / CONSERVATION FACTOR

TYPE	SSF BASE TEMP: 55 F (5408 DD)								SSF 65 F (8197 DD)								CF 55 F					CF 65 F				
LCR	100	70	50	40	30	25	20	15	100	70	50	40	30	25	20	15	100	50	30	20	15	100	50	30	20	15
WW A1	16	19	21	23	26	28	31	35	13	15	17	18	20	22	24	26	1.9	2.2	2.5	2.8	3.2	2.3	2.7	3.0	3.6	4.2
WW A2	16	20	24	28	32	36	40	46	12	15	18	21	25	28	31	35	1.5	1.7	2.0	2.2	2.4	1.7	2.0	2.3	2.6	2.9
WW A3	15	20	25	29	35	38	43	50	11	14	19	22	27	30	34	39	1.4	1.6	1.8	2.1	2.3	1.6	1.8	2.1	2.4	2.7
WW A4	15	20	26	30	36	40	45	52	10	14	19	22	28	31	35	41	1.3	1.6	1.8	2.0	2.2	1.5	1.7	2.0	2.3	2.5
WW A5	14	20	26	31	37	42	47	54	9	14	19	23	29	32	37	43	1.3	1.5	1.7	1.9	2.1	1.5	1.7	1.9	2.2	2.4
WW A6	14	20	27	31	38	43	48	56	9	14	19	23	29	33	38	44	1.3	1.5	1.7	1.9	2.1	1.4	1.6	1.9	2.2	2.4
WW B1	12	14	17	19	22	23	25	28	8	10	12	13	14	15	16	16	1.9	2.3	2.7	3.3	4.1	2.4	2.9	3.8	-	-
WW B2	16	21	28	33	40	44	50	58	11	16	21	26	32	36	41	48	1.3	1.5	1.7	1.9	2.0	1.4	1.6	1.8	2.0	2.2
WW B3	19	26	33	39	46	51	58	66	14	20	26	31	38	43	49	56	1.2	1.3	1.5	1.7	1.9	1.3	1.5	1.7	1.9	2.1
WW B4	19	27	36	42	51	56	63	72	14	21	29	35	43	49	55	64	1.1	1.3	1.4	1.6	1.8	1.2	1.3	1.5	1.7	1.9
WW B5	18	26	35	42	51	57	64	73	13	20	29	35	44	50	57	66	1.1	1.2	1.4	1.6	1.7	1.2	1.3	1.5	1.7	1.8
WW C1	20	27	35	41	49	54	61	69	15	21	29	34	41	46	52	60	1.1	1.3	1.5	1.7	1.8	1.2	1.4	1.6	1.8	2.0
WW C2	19	26	35	41	49	55	61	70	14	20	28	34	41	47	53	62	1.1	1.3	1.5	1.7	1.8	1.2	1.4	1.6	1.8	1.9
WW C3	23	32	43	50	60	66	73	81	17	26	36	43	53	59	66	75	1.0	1.2	1.3	1.5	1.7	1.0	1.2	1.4	1.6	1.7
WW C4	20	30	40	47	57	63	70	79	15	24	33	40	50	56	64	73	1.0	1.2	1.3	1.5	1.7	1.1	1.2	1.4	1.6	1.7
TW A1	16	18	21	23	26	29	31	35	13	15	17	19	21	22	24	27	1.9	2.1	2.4	2.7	3.0	2.3	2.6	2.9	3.3	3.8
TW A2	15	19	23	26	30	33	37	43	12	14	18	20	24	26	29	33	1.6	1.8	2.1	2.3	2.5	1.8	2.1	2.3	2.7	2.9
TW A3	14	19	23	27	32	35	40	46	11	14	18	20	25	28	31	36	1.5	1.7	1.9	2.2	2.4	1.7	1.9	2.2	2.5	2.7
TW A4	13	18	23	27	32	36	41	47	9	13	17	20	25	28	32	38	1.4	1.7	1.9	2.1	2.3	1.6	1.8	2.1	2.3	2.6
TW B1	15	18	21	24	27	30	33	37	12	14	17	19	21	23	26	29	1.8	2.0	2.3	2.6	2.8	2.1	2.4	2.7	3.0	3.3
TW B2	14	18	22	25	30	33	37	43	11	13	17	19	23	26	29	34	1.6	1.8	2.0	2.2	2.4	1.8	2.0	2.3	2.5	2.8
TW B3	13	17	22	25	30	34	38	44	10	13	16	19	23	26	30	35	1.5	1.7	2.0	2.2	2.4	1.8	2.0	2.2	2.4	2.7
TW B4	12	16	21	24	29	33	37	43	9	12	16	19	23	26	30	35	1.6	1.8	2.0	2.2	2.4	1.8	2.0	2.2	2.4	2.6
TW C1	14	17	21	23	27	30	34	39	11	14	16	18	21	24	27	31	1.7	1.9	2.2	2.4	2.6	2.0	2.2	2.5	2.7	3.0
TW C2	13	16	20	23	28	31	35	41	10	13	16	18	22	25	28	33	1.6	1.9	2.1	2.3	2.5	1.9	2.1	2.3	2.5	2.8
TW C3	12	16	20	23	27	30	34	40	10	12	15	18	21	24	27	32	1.7	1.9	2.1	2.3	2.5	1.9	2.1	2.3	2.6	2.8
TW C4	12	15	19	21	25	28	32	37	10	12	15	17	20	23	26	30	1.8	2.0	2.2	2.4	2.6	2.1	2.3	2.5	2.7	2.9
TW D1	10	12	15	17	20	22	24	27	7	8	10	11	13	14	16	17	1.9	2.3	2.7	3.2	3.6	2.5	2.9	3.5	4.4	-
TW D2	16	21	26	31	37	41	47	54	12	16	21	24	30	33	38	45	1.3	1.5	1.7	1.9	2.1	1.5	1.7	1.9	2.1	2.3
TW D3	16	22	28	33	40	44	50	57	12	17	22	26	32	36	41	48	1.3	1.5	1.7	1.9	2.0	1.4	1.6	1.8	2.1	2.2
TW D4	18	25	32	38	46	51	58	66	13	19	26	31	39	44	50	58	1.2	1.3	1.5	1.7	1.9	1.3	1.4	1.6	1.8	2.0
TW D5	18	25	33	39	47	53	60	68	13	19	27	32	40	46	52	61	1.1	1.3	1.5	1.7	1.8	1.2	1.4	1.6	1.7	1.9
TW E1	19	25	32	38	45	50	56	64	14	20	26	31	37	42	47	55	1.2	1.4	1.6	1.8	1.9	1.3	1.5	1.7	1.9	2.1
TW E2	19	25	33	38	46	51	57	66	14	20	26	31	39	43	49	57	1.2	1.4	1.5	1.7	1.9	1.3	1.5	1.6	1.8	2.0
TW E3	22	31	40	47	56	62	68	77	17	25	34	40	49	54	61	70	1.0	1.2	1.4	1.6	1.7	1.1	1.3	1.5	1.6	1.8
TW E4	21	29	38	45	54	60	67	76	16	23	32	38	47	53	60	69	1.1	1.2	1.4	1.6	1.7	1.1	1.3	1.5	1.6	1.8
TW F1	14	17	20	22	25	28	31	35	11	13	15	17	19	21	23	26	1.8	2.1	2.4	2.7	3.0	2.2	2.5	2.8	3.2	3.7
TW F2	13	16	20	23	28	31	35	40	9	12	15	18	21	24	27	31	1.6	1.8	2.1	2.3	2.6	1.9	2.1	2.4	2.7	3.0
TW F3	11	15	20	23	28	32	36	42	8	11	14	17	21	24	28	33	1.5	1.8	2.0	2.2	2.4	1.8	2.0	2.2	2.5	2.8
TW F4	9	14	18	22	27	31	36	42	6	9	13	16	20	23	27	33	1.5	1.7	2.0	2.2	2.4	1.7	2.0	2.2	2.4	2.7
TW G1	12	15	18	21	24	27	30	35	9	11	14	16	18	20	23	26	1.8	2.0	2.3	2.6	2.8	2.1	2.4	2.7	3.0	3.3
TW G2	10	14	18	21	25	29	33	38	7	10	13	15	19	21	25	29	1.6	1.9	2.1	2.3	2.5	1.9	2.1	2.4	2.6	2.9
TW G3	8	12	16	20	24	28	32	38	5	8	11	14	18	21	24	29	1.6	1.8	2.1	2.3	2.5	1.9	2.1	2.3	2.6	2.8
TW G4	6	10	14	17	22	25	29	35	4	6	9	12	16	18	22	27	1.7	1.9	2.1	2.3	2.5	2.0	2.2	2.4	2.6	2.8
TW H1	9	12	16	18	22	25	28	33	6	9	11	13	16	18	21	25	1.8	2.0	2.3	2.5	2.7	2.1	2.3	2.6	2.9	3.1
TW H2	7	10	14	16	21	24	28	33	4	6	9	12	15	17	21	25	1.8	2.0	2.2	2.4	2.6	2.1	2.3	2.5	2.7	2.9
TW H3	5	8	12	14	18	21	25	31	3	5	8	10	13	16	19	23	1.8	2.1	2.3	2.5	2.7	2.2	2.3	2.5	2.8	3.0
TW H4	4	6	9	11	15	17	21	26	2	4	6	8	11	13	15	20	2.0	2.2	2.5	2.7	2.8	2.4	2.6	2.8	3.0	3.1
TW I1	8	10	13	15	17	19	21	24	5	6	8	9	11	12	13	14	2.0	2.4	2.8	3.3	3.8	2.6	3.0	3.6	4.6	-
TW I2	13	18	23	28	34	38	43	51	9	13	17	21	27	30	35	42	1.4	1.6	1.8	2.0	2.1	1.6	1.7	1.9	2.2	2.4
TW I3	14	19	25	30	37	41	47	55	9	14	19	24	29	33	39	46	1.3	1.5	1.7	1.9	2.1	1.5	1.6	1.8	2.1	2.2
TW I4	15	22	29	35	43	48	55	64	11	16	23	28	36	41	47	56	1.2	1.4	1.6	1.7	1.9	1.3	1.5	1.7	1.8	2.0
TW I5	15	22	30	36	45	50	57	66	11	17	24	30	38	43	50	59	1.2	1.3	1.5	1.7	1.8	1.3	1.4	1.6	1.8	1.9
TW J1	17	23	30	35	43	48	54	62	12	18	24	29	35	40	46	53	1.2	1.4	1.6	1.8	1.9	1.3	1.5	1.7	1.9	2.1
TW J2	16	23	30	36	43	49	55	64	12	17	24	29	36	41	47	55	1.2	1.4	1.6	1.7	1.9	1.3	1.5	1.7	1.9	2.0
TW J3	19	28	37	44	53	59	67	75	15	22	31	38	47	52	60	69	1.1	1.2	1.4	1.6	1.7	1.1	1.3	1.5	1.6	1.8
TW J4	18	26	36	42	52	58	65	74	13	21	29	36	45	51	58	67	1.1	1.2	1.4	1.6	1.7	1.2	1.3	1.5	1.7	1.8
DG A1	10	13	16	17	20	21	23	23	6	8	10	11	13	13	14	14	1.8	2.3	3.0	4.5	-	2.2	2.8	4.1	-	-
DG A2	13	17	21	25	30	33	37	42	9	12	16	19	23	26	29	33	1.5	1.7	2.0	2.3	2.7	1.7	2.0	2.2	2.6	3.0
DG A3	16	22	28	33	40	45	51	58	13	17	23	27	34	38	44	51	1.3	1.5	1.6	1.8	2.0	1.5	1.6	1.8	2.0	2.2
DG B1	10	13	16	18	21	23	25	28	6	8	10	12	14	15	16	17	1.7	2.2	2.6	3.3	4.7	2.2	2.6	3.4	5.0	-
DG B2	13	17	22	26	31	35	40	46	9	13	17	20	24	27	31	37	1.5	1.7	1.9	2.2	2.4	1.7	1.9	2.1	2.4	2.7
DG B3	17	22	29	34	42	47	54	62	13	18	24	28	35	40	46	55	1.3	1.4	1.6	1.8	1.9	1.4	1.5	1.7	1.9	2.0
DG C1	13	16	20	23	27	30	33	37	8	11	14	16	19	21	23	26	1.6	1.9	2.2	2.6	3.3	1.8	2.1	2.6	3.4	4.5
DG C2	15	20	26	30	37	41	47	53	11	15	20	24	30	33	38	44	1.4	1.6	1.7	2.0	2.2	1.5	1.7	1.9	2.2	2.4
DG C3	19	25	33	38	47	53	60	68	15	20	27	32	40	45	52	61	1.2	1.4	1.5	1.7	1.8	1.3	1.4	1.6	1.7	1.9
SS A1	18	22	27	30	34	37	41	46	16	19	23	25	29	32	35	39	1.5	1.7	2.0	2.3	2.5	1.6	1.9	2.2	2.6	2.9
SS A2	23	29	35	40	47	51	57	64	19	25	31	35	42	46	51	58	1.2	1.4	1.6	1.8	2.0	1.3	1.5	1.7	1.9	2.1
SS A3	17	20	23	26	30	32	35	39	14	17	20	22	25	27	29	32	1.6	1.9	2.3	2.6	3.0	1.8	2.1	2.6	3.1	3.8
SS A4	22	27	34	38	45	49	54	61	19	24	30	34	39	43	48	54	1.2	1.4	1.7	1.9	2.0	1.3	1.5	1.8	2.0	2.2
SS A5	18	22	25	28	31	34	37	41	16	19	22	24	27	29	31	35	1.6	1.9	2.2	2.5	2.8	1.8	2.1	2.5	2.9	3.4
SS A6	22	28	35	39	46	50	56	63	19	24	30	35	41	45	50	56	1.2	1.4	1.6	1.8	2.0	1.3	1.5	1.7	2.0	2.2
SS A7	17	19	22	24	26	28	30	33	15	17	19	20	22	24	25	27	1.9	2.2	2.6	3.1	3.6	2.1	2.5	3.0	3.8	4.9
SS A8	21	27	33	37	43	47	52	59	18	23	28	32	38	41	46	52	1.3	1.5	1.7	1.9	2.1	1.3	1.6	1.8	2.1	2.3
SS B1	15	19	23	26	30	33	36	41	13	16	19	21	25	27	30	34	1.6	1.9	2.1	2.4	2.6	1.8	2.0	2.3	2.7	3.1
SS B2	20	25	32	36	43	48	53	61	17	22	27	32	38	42	47	54	1.3	1.5	1.6	1.8	2.0	1.3	1.5	1.7	2.0	2.1
SS B3	14	17	21	23	27	29	32	36	12	14	17	19	22	24	26	29	1.7	2.0	2.3	2.6	3.0	1.9	2.2	2.6	3.1	3.6
SS B4	19	24	30	35	41	45	51	58	16	21	26	30	36	40	45	52	1.3	1.5	1.7	1.9	2.1	1.4	1.6	1.8	2.0	2.2
SS B5	15	18	21	23	26	29	32	36	13	15	18	19	22	24	26	29	1.8	2.1	2.4	2.7	3.0	2.0	2.3	2.7	3.1	3.6
SS B6	19	24	31	35	42	46	52	59	16	21	26	30	36	40	46	52	1.3	1.5	1.7	1.9	2.0	1.4	1.6	1.8	2.0	2.2
SS B7	14	16	18	20	23	25	27	30	12	13	15	17	19	20	22	24	2.0	2.3	2.7	3.1	3.5	2.3	2.7	3.1	3.8	4.5
SS B8	18	23	29	33	39	43	48	55	15	19	25	29	34	38	42	49	1.3	1.5	1.7	1.9	2.1	1.4	1.6	1.9	2.1	2.3
SS C1	13	17	22	25	30	33	37	43	9	12	16	18	22	25	28	32	1.5	1.8	2.0	2.3	2.5	1.8	2.0	2.3	2.7	3.0
SS C2	16	22	29	33	40	45	51	58	12	17	22	27	33	37	42	49	1.3	1.5	1.7	1.8	2.0	1.4	1.6	1.8	2.0	2.2
SS C3	13	16	19	22	26	28	32	36	10	12	15	17	19	21	24	28	1.7	2.0	2.2	2.5	2.7	2.1	2.3	2.6	2.9	3.3
SS C4	15	20	26	30	36	41	46	53	11	15	20	24	29	33	38	45	1.4	1.6	1.8	1.9	2.1	1.5	1.7	1.9	2.1	2.3
SS D1	20	25	30	34	39	43	47	53	17	21	26	29	33	36	40	44	1.3	1.6	1.8	2.1	2.3	1.4	1.7	2.0	2.4	2.7
SS D2	25	32	40	46	53	58	65	72	21	28	35	40	48	52	58	66	1.1	1.3	1.5	1.7	1.8	1.1	1.4	1.6	1.8	1.9
SS D3	20	24	28	31	36	39	43	48	17	20	24	27	31	33	36	40	1.4	1.7	2.0	2.3	2.5	1.6	1.9	2.2	2.6	2.9
SS D4	24	31	39	44	52	57	63	70	20	27	34	39	46	51	56	64	1.1	1.3	1.5	1.7	1.9	1.2	1.4	1.6	1.8	2.0
SS E1	17	21	25	29	33	36	40	45	13	17	20	23	27	29	32	35	1.4	1.7	2.0	2.3	2.6	1.6	1.9	2.3	2.8	3.3
SS E2	21	27	34	39	47	51	57	65	17	23	29	34	40	45	50	57	1.2	1.4	1.6	1.8	1.9	1.2	1.5	1.7	1.9	2.1
SS E3	16	19	23	26	29	32	35	40	13	16	19	21	24	26	29	32	1.6	1.9	2.2	2.5	2.8	1.8	2.1	2.5	3.0	3.4
SS E4	20	26	32	37	43	48	54	61	16	21	27	32	38	42	47	53	1.2	1.4	1.6	1.8	2.0	1.3	1.5	1.8	2.0	2.2

LEWISTOWN, MONTANA

SOLAR SAVINGS FRACTION (%)

TYPE	BASE TEMP: 55 F (5691 DD)								65 F (8601 DD)							
LCR	100	70	50	40	30	25	20	15	100	70	50	40	30	25	20	15
WW A1	16	19	21	24	27	29	32	35	13	15	17	18	20	22	24	26
WW A2	16	20	24	28	33	36	41	47	12	15	18	21	25	28	31	36
WW A3	15	20	26	30	35	39	44	51	11	14	19	22	27	30	34	40
WW A4	15	20	26	30	37	41	46	54	10	14	19	23	28	31	36	42
WW A5	15	20	27	32	38	43	48	56	9	14	19	23	29	33	37	44
WW A6	14	20	27	32	39	44	50	57	9	14	19	24	29	33	38	45
WW B1	12	14	17	19	22	24	26	29	8	10	12	13	14	15	16	17
WW B2	16	22	28	33	40	45	51	59	11	16	21	26	32	36	42	49
WW B3	19	26	34	39	47	53	59	67	14	20	26	32	39	43	50	57
WW B4	19	27	36	43	52	57	65	73	14	21	29	35	43	49	56	65
WW B5	18	27	36	43	52	58	65	74	13	20	29	35	44	50	57	67
WW C1	20	28	36	42	50	55	62	71	15	22	29	34	42	47	53	61
WW C2	19	27	35	41	50	55	62	71	14	20	28	34	42	47	54	62
WW C3	23	33	43	51	61	67	74	82	17	26	36	43	53	59	67	76
WW C4	21	30	40	48	58	64	71	80	15	24	33	41	50	57	64	74
TW A1	16	18	21	24	27	29	32	36	13	15	17	19	21	22	24	27
TW A2	15	19	23	26	31	34	38	44	12	14	18	20	24	26	29	34
TW A3	14	19	24	27	32	36	41	47	11	14	18	20	25	28	31	37
TW A4	13	18	23	27	33	37	42	49	9	13	17	20	25	28	32	38
TW B1	15	18	21	24	27	30	33	38	12	14	17	19	21	23	26	29
TW B2	14	18	22	25	30	34	38	44	11	13	17	19	23	26	30	35
TW B3	13	17	22	25	31	34	39	45	10	13	16	19	23	26	30	36
TW B4	12	16	21	24	29	33	38	44	9	12	16	19	23	26	30	35
TW C1	14	17	21	23	28	30	34	40	11	13	16	18	21	24	27	31
TW C2	13	16	20	24	28	31	36	42	10	13	16	18	22	25	28	33
TW C3	12	16	20	23	27	30	35	41	10	12	15	18	21	24	27	32
TW C4	12	15	19	21	25	28	32	38	10	12	15	17	20	23	26	30
TW D1	10	12	15	17	20	22	25	28	7	8	10	11	13	14	16	17
TW D2	16	21	27	31	37	42	47	55	12	16	21	24	30	34	39	45
TW D3	16	22	29	34	40	45	51	59	12	17	22	26	32	36	42	49
TW D4	18	25	33	39	47	52	59	68	13	19	26	31	39	44	50	59
TW D5	18	25	33	39	48	53	60	70	13	19	27	32	40	46	53	62
TW E1	19	26	33	38	46	51	57	65	14	20	26	31	38	42	48	56
TW E2	19	26	33	39	47	52	58	67	14	20	26	32	39	44	50	58
TW E3	23	31	41	47	57	62	70	78	17	25	34	40	49	55	62	71
TW E4	21	29	39	46	55	61	68	77	16	23	32	38	47	53	61	70
TW F1	14	17	20	22	26	28	31	35	11	13	15	17	19	21	23	26
TW F2	13	16	21	24	28	31	36	41	9	12	15	18	21	24	27	31
TW F3	11	16	20	24	29	32	37	43	8	11	14	17	21	24	28	33
TW F4	9	14	19	22	28	31	36	43	6	9	13	16	20	23	27	33
TW G1	12	15	18	21	25	27	31	36	9	11	14	16	18	20	23	27
TW G2	10	14	18	21	26	29	33	39	7	10	13	15	19	21	25	30
TW G3	8	12	17	20	25	28	33	39	5	8	11	14	18	21	24	30
TW G4	6	10	14	17	22	25	30	36	4	6	9	12	16	18	22	27
TW H1	9	12	16	18	22	25	29	34	6	9	11	13	16	18	21	25
TW H2	7	10	14	17	21	24	28	34	4	6	9	12	15	17	21	25
TW H3	5	8	12	15	19	22	26	31	3	5	8	10	13	16	19	23
TW H4	4	6	9	11	15	18	21	26	2	4	6	8	11	13	15	20
TW I1	8	10	13	15	18	19	22	25	5	6	8	9	11	12	13	15
TW I2	13	18	24	28	34	39	44	52	9	13	17	21	27	30	35	42
TW I3	14	19	26	30	37	42	48	56	9	14	19	24	30	34	39	46
TW I4	15	22	30	35	43	49	56	65	11	16	23	28	36	41	47	56
TW I5	15	23	31	37	45	51	58	67	11	17	24	30	38	43	50	59
TW J1	17	23	31	36	43	49	55	63	12	18	24	29	36	40	46	54
TW J2	16	23	30	36	44	49	56	65	11	17	24	29	36	41	48	56
TW J3	20	28	38	45	54	60	68	77	15	22	31	38	47	53	60	70
TW J4	18	27	36	43	52	58	66	75	13	21	29	36	45	51	58	68
DG A1	10	13	16	18	20	22	23	24	6	8	10	11	13	13	14	14
DG A2	13	17	21	25	30	33	37	43	9	12	16	19	23	26	29	34
DG A3	16	22	28	33	40	45	51	60	13	17	23	27	34	38	44	51
DG B1	10	13	16	19	22	24	26	29	6	8	10	12	14	15	16	17
DG B2	13	17	22	26	31	35	40	47	9	13	17	20	24	27	31	37
DG B3	17	23	29	34	42	48	55	63	13	18	24	28	35	40	46	55
DG C1	13	16	20	23	28	30	34	38	8	11	14	16	19	21	23	26
DG C2	15	20	26	31	37	42	47	55	11	15	20	24	29	33	38	45
DG C3	19	26	33	39	47	53	61	70	15	20	27	32	40	45	52	61
SS A1	19	23	27	30	35	38	42	48	16	19	23	26	30	32	36	40
SS A2	23	29	36	41	48	52	58	66	20	25	31	36	42	46	52	59
SS A3	17	20	24	27	31	33	36	40	14	17	20	22	25	27	30	33
SS A4	22	28	34	39	46	50	55	63	19	24	30	34	40	44	49	56
SS A5	19	22	26	28	32	35	38	43	16	19	22	24	27	29	32	36
SS A6	23	29	35	40	47	51	57	64	19	25	31	35	41	45	51	57
SS A7	17	20	22	24	27	29	31	35	15	17	19	21	23	24	26	28
SS A8	22	27	33	38	44	48	53	60	18	23	29	33	38	42	47	53
SS B1	16	19	23	26	31	34	37	43	13	16	19	22	25	28	31	35
SS B2	20	26	32	37	44	48	54	62	17	22	28	32	38	43	48	55
SS B3	14	18	21	24	28	30	33	38	12	14	17	19	22	24	27	30
SS B4	19	25	31	36	42	46	52	60	16	21	27	31	37	41	46	53
SS B5	15	18	21	24	27	29	32	37	13	15	18	20	22	24	27	30
SS B6	19	25	31	36	42	47	53	60	16	21	27	31	37	41	46	53
SS B7	14	16	19	21	24	25	28	31	12	14	16	17	19	20	22	24
SS B8	18	24	29	34	40	44	50	57	15	20	25	29	34	38	43	50
SS C1	13	17	22	26	31	34	38	44	9	12	16	18	22	25	28	33
SS C2	16	22	29	34	41	46	52	60	12	17	22	27	33	37	43	50
SS C3	13	16	19	22	26	29	32	37	10	12	14	16	19	22	24	28
SS C4	15	20	26	31	37	41	47	54	11	15	20	24	29	33	38	45
SS D1	21	26	31	35	40	44	49	55	17	21	26	29	34	37	41	46
SS D2	25	33	41	47	55	60	66	74	21	28	36	41	48	53	59	67
SS D3	20	24	29	32	37	40	44	50	17	20	24	27	31	34	37	42
SS D4	25	32	40	45	53	58	64	72	21	27	34	40	47	52	57	65
SS E1	17	21	26	29	34	37	42	47	13	17	21	24	27	30	33	37
SS E2	21	28	35	40	48	53	59	67	17	23	29	34	41	46	51	58
SS E3	16	20	23	26	30	33	36	41	13	16	19	21	25	27	29	33
SS E4	20	26	33	38	44	49	55	62	16	22	28	32	38	42	48	55

CONSERVATION FACTOR

TYPE	55 F					65 F				
	100	50	30	20	15	100	50	30	20	15
WW A1	1.9	2.1	2.5	2.8	3.0	2.3	2.7	3.0	3.5	4.0
WW A2	1.5	1.7	1.9	2.2	2.4	1.7	2.0	2.2	2.5	2.8
WW A3	1.4	1.6	1.8	2.0	2.2	1.6	1.8	2.1	2.3	2.6
WW A4	1.3	1.5	1.7	1.9	2.1	1.5	1.7	2.0	2.2	2.5
WW A5	1.3	1.5	1.7	1.9	2.1	1.5	1.7	1.9	2.1	2.4
WW A6	1.2	1.5	1.7	1.9	2.0	1.4	1.6	1.9	2.1	2.3
WW B1	1.8	2.2	2.6	3.1	3.6	2.4	2.9	3.7	-	-
WW B2	1.3	1.4	1.6	1.8	2.0	1.4	1.6	1.8	2.0	2.2
WW B3	1.1	1.3	1.5	1.7	1.9	1.3	1.4	1.6	1.8	2.0
WW B4	1.1	1.3	1.4	1.6	1.7	1.2	1.3	1.5	1.7	1.8
WW B5	1.1	1.2	1.4	1.6	1.7	1.2	1.3	1.5	1.7	1.8
WW C1	1.1	1.3	1.5	1.7	1.8	1.2	1.4	1.6	1.8	1.9
WW C2	1.1	1.3	1.5	1.6	1.8	1.2	1.4	1.6	1.7	1.9
WW C3	1.0	1.1	1.3	1.5	1.6	1.0	1.2	1.4	1.5	1.7
WW C4	1.0	1.2	1.3	1.5	1.6	1.1	1.2	1.4	1.6	1.7
TW A1	1.8	2.1	2.4	2.7	2.9	2.3	2.6	2.9	3.3	3.7
TW A2	1.6	1.8	2.0	2.3	2.4	1.8	2.1	2.3	2.6	2.9
TW A3	1.5	1.7	1.9	2.1	2.3	1.7	1.9	2.2	2.4	2.7
TW A4	1.4	1.6	1.8	2.0	2.2	1.6	1.8	2.1	2.3	2.5
TW B1	1.8	2.0	2.3	2.5	2.7	2.1	2.4	2.7	3.0	3.3
TW B2	1.5	1.8	2.0	2.2	2.4	1.8	2.0	2.3	2.5	2.7
TW B3	1.5	1.7	1.9	2.1	2.3	1.7	2.0	2.2	2.4	2.6
TW B4	1.5	1.7	2.0	2.2	2.3	1.8	2.0	2.2	2.4	2.6
TW C1	1.7	1.9	2.2	2.4	2.6	2.0	2.2	2.5	2.7	2.9
TW C2	1.6	1.8	2.0	2.3	2.4	1.9	2.1	2.3	2.5	2.7
TW C3	1.7	1.9	2.1	2.3	2.4	1.9	2.1	2.3	2.5	2.7
TW C4	1.8	2.0	2.2	2.4	2.6	2.1	2.3	2.5	2.7	2.8
TW D1	1.9	2.2	2.6	3.0	3.4	2.4	2.9	3.4	4.3	-
TW D2	1.3	1.5	1.7	1.9	2.1	1.5	1.7	1.9	2.1	2.3
TW D3	1.3	1.5	1.7	1.8	2.0	1.4	1.6	1.8	2.0	2.2
TW D4	1.2	1.3	1.5	1.7	1.8	1.3	1.4	1.6	1.8	1.9
TW D5	1.1	1.3	1.5	1.6	1.8	1.2	1.4	1.6	1.7	1.9
TW E1	1.2	1.4	1.6	1.7	1.9	1.3	1.5	1.7	1.9	2.1
TW E2	1.2	1.3	1.5	1.7	1.8	1.3	1.5	1.6	1.8	2.0
TW E3	1.0	1.2	1.4	1.5	1.7	1.1	1.3	1.4	1.6	1.8
TW E4	1.0	1.2	1.4	1.5	1.7	1.1	1.3	1.5	1.6	1.8
TW F1	1.8	2.1	2.4	2.6	2.9	2.2	2.5	2.8	3.2	3.6
TW F2	1.6	1.8	2.1	2.3	2.5	1.9	2.1	2.4	2.6	2.9
TW F3	1.5	1.7	2.0	2.2	2.4	1.8	2.0	2.2	2.5	2.7
TW F4	1.5	1.7	1.9	2.1	2.3	1.7	2.0	2.2	2.4	2.6
TW G1	1.8	2.0	2.3	2.5	2.7	2.1	2.4	2.7	3.0	3.3
TW G2	1.6	1.8	2.1	2.3	2.5	1.9	2.1	2.4	2.6	2.8
TW G3	1.6	1.8	2.0	2.2	2.4	1.9	2.1	2.3	2.5	2.7
TW G4	1.7	1.9	2.1	2.3	2.5	2.0	2.2	2.4	2.6	2.8
TW H1	1.8	2.0	2.2	2.5	2.7	2.1	2.3	2.6	2.8	3.1
TW H2	1.7	2.0	2.2	2.4	2.6	2.0	2.3	2.5	2.7	2.9
TW H3	1.8	2.0	2.2	2.5	2.6	2.1	2.3	2.5	2.8	2.9
TW H4	2.0	2.2	2.4	2.6	2.8	2.4	2.6	2.8	3.0	3.1
TW I1	2.0	2.3	2.7	3.2	3.6	2.5	3.0	3.6	4.5	
TW I2	1.4	1.6	1.8	1.9	2.1	1.6	1.7	1.9	2.1	2.3
TW I3	1.3	1.5	1.7	1.9	2.0	1.5	1.6	1.8	2.0	2.2
TW I4	1.2	1.4	1.5	1.7	1.9	1.3	1.5	1.6	1.8	2.0
TW I5	1.2	1.3	1.5	1.7	1.8	1.3	1.4	1.6	1.8	1.9
TW J1	1.2	1.4	1.6	1.7	1.9	1.3	1.5	1.7	1.9	2.1
TW J2	1.2	1.4	1.5	1.7	1.9	1.3	1.5	1.6	1.8	2.0
TW J3	1.0	1.2	1.4	1.5	1.7	1.1	1.3	1.5	1.6	1.8
TW J4	1.1	1.2	1.4	1.6	1.7	1.2	1.2	1.5	1.6	1.8
DG A1	1.8	2.2	2.9	4.3	-	2.2	2.8	4.0	-	-
DG A2	1.5	1.7	2.0	2.3	2.6	1.7	2.0	2.2	2.6	3.0
DG A3	1.3	1.5	1.6	1.8	2.0	1.5	1.6	1.7	1.9	2.1
DG B1	1.7	2.1	2.5	3.2	4.2	2.2	2.6	3.4	4.7	-
DG B2	1.5	1.7	1.9	2.1	2.4	1.7	1.9	2.1	2.4	2.7
DG B3	1.3	1.4	1.6	1.7	1.9	1.4	1.5	1.7	1.8	2.0
DG C1	1.5	1.9	2.2	2.5	3.0	1.9	2.2	2.6	3.3	4.2
DG C2	1.3	1.5	1.7	1.9	2.1	1.6	1.7	1.9	2.1	2.3
DG C3	1.2	1.3	1.5	1.6	1.8	1.3	1.4	1.6	1.7	1.9
SS A1	1.5	1.7	2.0	2.2	2.4	1.6	1.9	2.2	2.5	2.8
SS A2	1.2	1.4	1.6	1.8	1.9	1.3	1.5	1.7	1.9	2.1
SS A3	1.6	1.9	2.2	2.5	2.8	1.8	2.1	2.5	3.0	3.6
SS A4	1.2	1.4	1.6	1.8	2.0	1.3	1.5	1.7	2.0	2.2
SS A5	1.6	1.9	2.2	2.5	2.7	1.8	2.1	2.4	2.8	3.2
SS A6	1.2	1.4	1.6	1.8	1.9	1.3	1.5	1.7	1.9	2.1
SS A7	1.8	2.2	2.6	3.0	3.4	2.1	2.5	3.0	3.6	4.5
SS A8	1.2	1.5	1.7	1.9	2.0	1.3	1.6	1.8	2.0	2.3
SS B1	1.6	1.8	2.1	2.3	2.5	1.8	2.0	2.3	2.6	2.9
SS B2	1.2	1.4	1.6	1.8	2.0	1.3	1.5	1.7	1.9	2.1
SS B3	1.7	2.0	2.3	2.6	2.8	1.9	2.2	2.6	3.0	3.5
SS B4	1.3	1.5	1.7	1.9	2.0	1.4	1.6	1.8	2.0	2.2
SS B5	1.8	2.0	2.3	2.6	2.9	2.0	2.3	2.7	3.0	3.4
SS B6	1.3	1.5	1.6	1.8	2.0	1.4	1.6	1.8	2.0	2.1
SS B7	1.9	2.3	2.6	3.0	3.4	2.3	2.6	3.1	3.6	4.3
SS B8	1.3	1.5	1.7	1.9	2.1	1.4	1.6	1.8	2.1	2.3
SS C1	1.5	1.7	2.0	2.2	2.4	1.8	2.0	2.3	2.6	2.9
SS C2	1.3	1.4	1.6	1.8	2.0	1.4	1.6	1.8	2.0	2.2
SS C3	1.7	2.0	2.2	2.5	2.7	2.1	2.3	2.6	2.9	3.2
SS C4	1.3	1.5	1.7	1.9	2.1	1.5	1.7	1.9	2.1	2.3
SS D1	1.3	1.6	1.8	2.0	2.2	1.4	1.7	2.0	2.3	2.6
SS D2	1.1	1.3	1.4	1.6	1.8	1.1	1.3	1.5	1.7	1.9
SS D3	1.4	1.7	1.9	2.2	2.4	1.6	1.8	2.1	2.5	2.8
SS D4	1.1	1.3	1.5	1.7	1.8	1.2	1.4	1.6	1.8	1.9
SS E1	1.4	1.7	2.0	2.2	2.5	1.6	1.9	2.2	2.7	3.1
SS E2	1.2	1.4	1.5	1.7	1.9	1.2	1.4	1.7	1.9	2.1
SS E3	1.6	1.9	2.2	2.5	2.7	1.8	2.1	2.5	2.9	3.3
SS E4	1.2	1.4	1.6	1.8	2.0	1.3	1.5	1.7	2.0	2.1

MILES CITY, MONTANA

TYPE	SOLAR SAVINGS FRACTION (%)																CONSERVATION FACTOR									
BASE TEMP:	55 F (5406 DD)								65 F (7904 DD)								55 F					65 F				
LCR	100	70	50	40	30	25	20	15	100	70	50	40	30	25	20	15	100	50	30	20	15	100	50	30	20	15
WW A1	16	18	21	23	26	27	30	33	13	15	17	19	21	22	24	26	1.9	2.3	2.6	2.9	3.2	2.3	2.6	3.1	3.6	4.2
WW A2	15	19	23	27	31	35	39	45	12	15	18	21	25	27	31	35	1.5	1.8	2.0	2.2	2.4	1.7	2.0	2.3	2.6	2.9
WW A3	15	19	24	28	34	37	42	49	11	15	19	22	26	30	33	39	1.4	1.6	1.8	2.1	2.2	1.6	1.8	2.1	2.4	2.6
WW A4	14	19	25	29	35	39	44	52	10	14	19	22	27	31	35	41	1.4	1.6	1.8	2.0	2.1	1.5	1.8	2.0	2.3	2.5
WW A5	14	19	25	30	36	41	46	54	10	14	19	23	28	32	37	43	1.3	1.5	1.7	1.9	2.1	1.4	1.7	1.9	2.2	2.4
WW A6	14	19	26	30	37	42	48	55	9	14	19	23	29	33	38	44	1.3	1.5	1.7	1.9	2.0	1.4	1.7	1.9	2.1	2.3
WW B1	11	14	16	18	21	22	24	27	8	10	12	13	14	15	16	16	1.9	2.4	2.8	3.4	3.9	2.3	3.0	4.0	-	-
WW B2	15	21	27	32	39	44	50	58	11	16	22	26	32	36	41	48	1.3	1.5	1.7	1.8	2.0	1.4	1.6	1.8	2:0	2.2
WW B3	18	25	32	38	46	51	57	66	14	20	26	31	38	43	49	57	1.2	1.4	1.5	1.7	1.9	1.3	1.5	1.7	1.9	2.0.
WW B4	18	26	35	41	50	56	63	72	14	21	29	35	43	49	55	64	1.1	1.3	1.4	1.6	1.8	1.2	1.4	1.5	1.7	1.9
WW B5	17	26	35	41	50	56	64	73	13	21	29	35	44	49	57	66	1.1	1.3	1.4	1.6	1.7	1.2	1.3	1.5	1.7	1.8
WW C1	19	27	34	40	48	54	60	69	15	22	29	34	41	46	52	60	1.1	1.3	1.5	1.7	1.8	1.2	1.4	1.6	1.8	2.0
WW C2	18	26	34	40	48	54	61	70	14	20	28	33	41	46	53	62	1.1	1.3	1.5	1.6	1.8	1.2	1.4	1.6	1.8	1.9
WW C3	22	31	42	49	59	65	73	81	17	26	36	43	53	59	66	76	1.0	1.2	1.3	1.5	1.6	1.1	1.2	1.4	1.6	1.7
WW C4	20	29	39	46	56	63	70	79	15	24	33	40	50	56	64	73	1.0	1.2	1.3	1.5	1.7	1.1	1.2	1.4	1.6	1.7
TW A1	16	18	21	23	26	28	30	34	13	15	17	19	21	22	24	27	1.9	2.2	2.5	2.8	3.0	2.2	2.6	2.9	3.4	3.8
TW A2	15	18	22	25	29	32	36	42	12	15	18	20	24	26	29	33	1.6	1.9	2.1	2.3	2.5	1.8	2.1	2.4	2.7	2.9
TW A3	14	18	22	26	31	34	39	45	11	14	18	21	25	27	31	36	1.5	1.7	2.0	2.2	2.3	1.7	1.9	2.2	2.5	2.7
TW A4	13	17	22	26	31	35	40	47	9	13	17	20	25	28	32	38	1.5	1.7	1.9	2.1	2.3	1.6	1.9	2.1	2.4	2.6
TW B1	15	18	21	23	26	29	32	36	13	15	17	19	21	23	26	29	1.8	2.1	2.4	2.6	2.8	2.1	2.4	2.7	3.0	3.4
TW B2	14	17	21	24	29	32	36	42	11	14	17	20	23	26	29	34	1.6	1.8	2.0	2.3	2.4	1.8	2.0	2.3	2.6	2.8
TW B3	13	17	21	24	29	33	37	43	10	13	17	19	24	26	30	35	1.6	1.8	2.0	2.2	2.4	1.7	2.0	2.2	2.5	2.7
TW B4	12	16	20	23	28	32	36	42	9	12	16	19	23	26	29	35	1.6	1.8	2.0	2.2	2.4	1.8	2.0	2.2	2.5	2.7
TW C1	14	17	20	23	26	29	33	38	11	14	16	18	22	24	27	31	1.8	2.0	2.2	2.4	2.6	2.0	2.2	2.5	2.8	3.0
TW C2	13	16	20	23	27	30	34	40	10	13	16	18	22	25	28	33	1.7	1.9	2.1	2.3	2.5	1.9	2.1	2.3	2.6	2.8
TW C3	12	16	19	22	26	29	33	39	10	12	15	18	21	24	27	32	1.7	1.9	2.1	2.3	2.5	1.9	2.1	2.3	2.6	2.8
TW C4	12	15	18	21	25	27	31	36	10	12	15	17	20	23	26	30	1.8	2.1	2.3	2.5	2.6	2.1	2.3	2.5	2.7	2.9
TW D1	10	12	14	16	19	20	23	26	7	8	10	11	13	14	15	17	2.0	2.4	2.8	3.2	3.6	2.4	2.9	3.6	4.6	-
TW D2	15	20	26	30	36	40	46	53	12	16	21	24	30	33	38	45	1.4	1.6	1.8	1.9	2.1	1.5	1.7	1.9	2.1	2.3
TW D3	16	21	27	32	39	43	49	57	12	17	22	26	32	36	41	48	1.3	1.5	1.7	1.9	2.0	1.4	1.6	1.8	2.0	2.2
TW D4	17	24	31	37	45	50	57	66	13	19	26	31	39	44	50	58	1.2	1.4	1.5	1.7	1.8	1.3	1.4	1.6	1.8	2.0
TW D5	17	24	32	38	46	52	59	68	13	20	27	32	40	46	52	61	1.2	1.3	1.5	1.7	1.8	1.2	1.4	1.6	1.7	1.9
TW E1	18	25	31	37	44	49	55	63	15	20	26	31	37	42	47	55	1.2	1.4	1.6	1.8	1.9	1.3	1.5	1.7	1.9	2.1
TW E2	18	25	32	37	45	50	57	66	14	20	27	31	38	43	49	58	1.2	1.4	1.5	1.7	1.9	1.3	1.5	1.7	1.8	2.0
TW E3	22	30	39	46	55	61	68	77	17	25	34	40	49	54	61	70	1.1	1.2	1.4	1.5	1.7	1.1	1.3	1.5	1.6	1.8
TW E4	20	28	37	44	53	59	67	76	16	23	32	38	47	53	60	69	1.1	1.2	1.4	1.6	1.7	1.1	1.3	1.5	1.6	1.8
TW F1	14	16	19	21	24	27	30	34	11	13	15	17	19	21	23	26	1.9	2.2	2.5	2.7	3.0	2.2	2.5	2.9	3.3	3.7
TW F2	12	16	20	23	27	30	34	39	9	12	15	18	21	24	27	31	1.6	1.9	2.1	2.3	2.5	1.9	2.1	2.4	2.7	3.0
TW F3	11	15	19	23	27	31	35	42	8	11	15	17	21	24	28	33	1.6	1.8	2.0	2.2	2.4	1.8	2.0	2.3	2.5	2.8
TW F4	9	13	18	21	26	30	35	41	6	9	13	16	20	23	27	33	1.6	1.8	2.0	2.2	2.3	1.7	2.0	2.2	2.5	2.7
TW G1	12	15	18	20	24	26	29	34	9	11	14	16	19	20	23	26	1.8	2.1	2.4	2.6	2.8	2.1	2.4	2.7	3.0	3.4
TW G2	10	13	17	20	24	28	32	37	7	10	13	15	19	21	25	29	1.7	1.9	2.1	2.3	2.5	1.9	2.1	2.4	2.7	2.9
TW G3	8	12	16	19	24	27	31	37	5	8	12	14	18	21	24	29	1.7	1.9	2.1	2.3	2.5	1.9	2.1	2.3	2.6	2.8
TW G4	6	10	13	16	21	24	28	34	4	6	10	12	16	18	22	27	1.7	2.0	2.2	2.4	2.5	1.9	2.1	2.4	2.6	2.8
TW H1	9	12	15	18	21	24	27	32	6	9	11	13	16	19	21	25	1.8	2.1	2.3	2.5	2.7	2.1	2.3	2.6	2.9	3.2
TW H2	7	10	13	16	20	23	27	32	4	7	9	12	15	17	21	25	1.8	2.0	2.3	2.5	2.6	2.0	2.2	2.5	2.7	3.0
TW H3	5	8	11	14	18	21	24	30	3	5	8	10	13	16	19	23	1.9	2.1	2.3	2.5	2.7	2.1	2.3	2.5	2.8	3.0
TW H4	4	6	9	11	14	17	20	25	2	4	6	8	11	13	16	20	2.1	2.3	2.5	2.7	2.9	2.4	2.6	2.7	3.0	3.2
TW I1	8	10	12	14	16	18	20	23	5	7	8	9	11	12	13	14	2.1	2.5	2.9	3.4	3.8	2.5	3.0	3.7	4.9	-
TW I2	12	17	22	27	33	37	43	50	9	13	18	21	27	30	35	42	1.4	1.6	1.8	2.0	2.1	1.5	1.7	2.0	2.2	2.3
TW I3	13	18	24	29	36	40	46	54	9	14	19	23	29	33	38	46	1.3	1.5	1.7	1.9	2.0	1.5	1.7	1.9	2.1	2.2
TW I4	15	21	28	34	42	47	54	63	11	17	23	28	36	41	47	56	1.2	1.4	1.6	1.7	1.9	1.3	1.5	1.7	1.8	2.0
TW I5	15	22	30	35	44	49	57	66	11	17	24	30	38	43	50	59	1.2	1.3	1.5	1.7	1.8	1.3	1.4	1.6	1.8	1.9
TW J1	16	22	29	34	42	47	53	62	12	18	24	29	35	40	45	53	1.2	1.4	1.6	1.8	1.9	1.3	1.5	1.7	1.9	2.1
TW J2	16	22	29	35	42	48	55	63	12	17	24	29	36	41	47	55	1.2	1.4	1.6	1.7	1.9	1.3	1.5	1.7	1.9	2.0
TW J3	19	27	37	43	53	59	66	75	15	22	31	37	46	52	60	69	1.1	1.2	1.4	1.6	1.7	1.1	1.3	1.5	1.6	1.8
TW J4	17	26	35	42	51	57	64	74	13	21	29	36	45	51	58	67	1.1	1.2	1.4	1.6	1.7	1.2	1.3	1.5	1.6	1.8
DG A1	10	12	15	17	19	20	21	22	6	8	10	11	13	14	14	14	1.9	2.3	3.1	4.6	-	2.2	2.8	4.2	-	-
DG A2	12	16	21	24	29	32	36	41	9	12	16	19	23	26	29	33	1.6	1.8	2.0	2.3	2.7	1.7	2.0	2.3	2.6	3.0
DG A3	16	21	27	32	39	44	50	58	13	17	23	27	33	38	43	51	1.3	1.5	1.6	1.8	2.0	1.4	1.6	1.8	2.0	2.1
DG B1	10	12	15	17	20	22	24	26	6	8	11	12	14	15	16	16	1.9	2.3	2.7	3.4	4.5	2.2	2.7	3.6	-	-
DG B2	13	17	21	25	30	34	38	45	10	13	17	20	24	27	31	36	1.5	1.7	1.9	2.1	2.4	1.7	1.9	2.2	2.4	2.7
DG B3	16	22	28	33	41	46	53	62	13	18	24	28	35	40	46	54	1.3	1.5	1.6	1.7	1.9	1.4	1.6	1.7	1.9	2.0
DG C1	12	15	19	22	26	28	32	36	9	11	14	16	19	21	23	25	1.6	2.0	2.2	2.6	3.1	1.9	2.2	2.8	3.4	4.4
DG C2	15	20	25	29	35	40	45	53	12	16	20	24	29	33	38	44	1.4	1.6	1.7	1.9	2.1	1.5	1.7	1.9	2.2	2.4
DG C3	19	25	32	37	46	52	59	69	15	21	27	32	40	45	52	61	1.2	1.4	1.5	1.6	1.8	1.3	1.5	1.6	1.7	1.9
SS A1	18	21	25	28	33	36	40	45	15	19	22	25	28	31	34	38	1.5	1.8	2.0	2.3	2.5	1.6	1.9	2.3	2.6	2.9
SS A2	22	27	34	39	45	50	56	64	19	24	30	35	41	45	50	57	1.2	1.4	1.6	1.8	1.9	1.3	1.5	1.7	1.9	2.1
SS A3	16	19	22	25	28	30	33	37	14	·16	19	21	24	26	28	30	1.7	2.0	2.3	2.7	2.9	1.8	2.2	2.7	3.2	3.8
SS A4	21	26	32	37	43	47	53	60	18	23	29	33	38	42	47	54	1.3	1.5	1.7	1.9	2.0	1.3	1.6	1.8	2.0	2.2
SS A5	18	21	24	27	30	32	36	40	16	18	21	23	26	28	30	33	1.7	2.0	2.3	2.6	2.8	1.8	2.2	2.6	3.0	3.4
SS A6	21	27	33	38	44	49	55	62	19	24	29	34	40	44	49	56	1.2	1.4	1.6	1.8	2.0	1.3	1.5	1.7	2.0	2.1
SS A7	16	18	21	23	25	27	29	31	14	16	18	20	21	23	24	26	2.0	2.3	2.7	3.2	3.6	2.1	2.6	3.2	4.1	-
SS A8	20	25	31	35	41	46	51	58	18	22	28	31	37	40	45	51	1.3	1.5	1.7	1.9	2.1	1.4	1.6	1.9	2.1	2.3
SS B1	15	18	22	25	29	31	35	40	13	15	19	21	24	26	29	33	1.6	1.9	2.2	2.4	2.6	1.8	2.1	2.4	2.8	3.1
SS B2	19	24	31	35	42	46	52	60	16	21	27	31	37	41	47	54	1.3	1.5	1.7	1.8	2.0	1.4	1.6	1.8	2.0	2.1
SS B3	14	16	20	22	25	28	31	35	11	14	16	18	21	23	25	28	1.8	2.1	2.4	2.7	3.0	1.9	2.3	2.7	3.2	3.7
SS B4	18	23	29	34	40	44	50	57	16	20	26	30	35	39	44	51	1.3	1.5	1.7	1.9	2.0	1.4	1.6	1.8	2.0	2.2
SS B5	15	17	20	22	25	27	30	34	13	15	17	19	21	23	25	28	1.9	2.2	2.5	2.8	3.0	2.1	2.4	2.8	3.2	3.7
SS B6	18	23	29	34	40	45	50	58	15	20	26	30	36	40	45	52	1.3	1.5	1.7	1.9	2.0	1.4	1.6	1.8	2.0	2.2
SS B7	13	15	18	19	22	23	25	28	12	13	15	16	18	19	21	23	2.1	2.4	2.8	3.2	3.6	2.3	2.7	3.3	4.0	4.8
SS B8	17	22	28	32	38	42	47	54	15	19	24	28	33	37	41	48	1.4	1.6	1.8	1.9	2.1	1.4	1.7	1.9	2.1	2.3
SS C1	13	17	21	24	29	32	36	42	9	12	16	18	22	25	28	32	1.5	1.8	2.0	2.3	2.5	1.7	2.0	2.3	2.7	3.0
SS C2	16	21	28	32	39	44	50	58	12	17	22	27	33	37	42	49	1.3	1.5	1.7	1.8	2.0	1.4	1.6	1.8	2.0	2.2
SS C3	13	15	19	21	25	27	31	35	10	12	15	17	20	22	24	28	1.8	2.0	2.3	2.5	2.7	2.0	2.3	2.6	3.0	3.3
SS C4	15	20	25	29	35	40	45	53	11	16	20	24	29	33	38	44	1.4	1.6	1.8	1.9	2.1	1.5	1.7	1.9	2.1	2.3
SS D1	19	24	29	32	38	41	46	52	16	20	25	28	32	35	39	43	1.4	1.6	1.8	2.1	2.3	1.5	1.8	2.1	2.4	2.7
SS D2	23	31	39	44	52	58	64	72	20	27	34	39	47	51	57	65	1.1	1.3	1.5	1.6	1.8	1.2	1.4	1.6	1.8	1.9
SS D3	19	23	27	30	34	38	41	47	16	20	23	26	30	32	35	39	1.5	1.8	2.0	2.2	2.5	1.6	1.9	2.3	2.6	2.9
SS D4	23	30	37	43	51	56	62	70	20	26	33	38	45	50	56	63	1.1	1.3	1.5	1.7	1.8	1.2	1.4	1.6	1.8	2.0
SS E1	16	20	24	27	32	35	39	44	13	16	20	22	26	28	31	34	1.5	1.8	2.0	2.3	2.5	1.6	2.0	2.4	2.8	3.2
SS E2	19	26	33	38	45	50	56	65	16	22	28	33	39	44	49	57	1.2	1.4	1.6	1.8	1.9	1.3	1.5	1.7	1.9	2.1
SS E3	15	18	22	24	28	30	34	38	13	16	19	21	23	25	28	31	1.7	2.0	2.3	2.6	2.8	1.9	2.2	2.6	3.0	3.5
SS E4	19	24	31	35	42	47	53	60	16	21	27	31	37	41	46	53	1.3	1.5	1.6	1.8	2.0	1.3	1.6	1.8	2.0	2.2

MISSOULA, MONTANA

TYPE	SSF 55 F (5078 DD) LCR 100	70	50	40	30	25	20	15	SSF 65 F (7936 DD) 100	70	50	40	30	25	20	15	CF 55 F 100	50	30	20	15	CF 65 F 100	50	30	20	15
WW A1	14	16	18	19	20	21	23	24	12	13	14	15	16	16	17	17	2.3	2.8	3.5	4.6	-	2.9	3.6	4.8	-	-
WW A2	12	15	18	21	24	26	28	31	9	12	14	16	18	19	21	22	1.8	2.1	2.6	3.1	3.7	2.0	2.5	3.0	4.0	-
WW A3	11	15	19	21	25	27	30	33	8	11	14	16	19	20	22	24	1.6	2.0	2.3	2.8	3.3	1.9	2.2	2.7	3.5	4.8
WW A4	11	14	18	21	25	28	31	35	7	10	13	16	19	21	23	26	1.6	1.9	2.2	2.7	3.1	1.8	2.1	2.6	3.3	4.3
WW A5	10	14	19	22	26	29	32	37	6	10	13	16	19	21	24	27	1.5	1.8	2.2	2.6	3.0	1.7	2.0	2.5	3.1	4.0
WW A6	10	14	19	22	27	29	33	37	6	9	13	16	20	22	24	27	1.5	1.8	2.1	2.5	2.9	1.6	2.0	2.4	3.0	3.8
WW B1	9	10	12	13	13	13	13	12	6	7	8	8	8	7	6	2	2.4	3.4	-	-	-	3.3	-	-	-	-
WW B2	12	16	21	25	30	33	38	43	8	12	16	19	24	27	30	35	1.4	1.7	2.0	2.2	2.5	1.6	1.9	2.2	2.5	2.9
WW B3	14	19	25	30	35	39	44	51	10	15	20	24	29	33	37	42	1.3	1.6	1.8	2.1	2.3	1.4	1.7	2.0	2.3	2.6
WW B4	14	21	28	33	41	45	52	59	10	16	23	28	35	39	45	52	1.2	1.4	1.6	1.8	2.0	1.3	1.5	1.7	1.9	2.1
WW B5	13	20	28	34	42	47	53	62	10	16	23	28	36	41	47	55	1.2	1.4	1.6	1.8	1.9	1.3	1.4	1.6	1.9	2.0
WW C1	15	21	28	32	38	43	48	55	12	17	22	27	32	36	41	47	1.3	1.5	1.7	2.0	2.2	1.4	1.6	1.9	2.1	2.4
WW C2	14	20	27	32	39	43	49	57	10	16	22	26	33	37	42	49	1.2	1.5	1.7	1.9	2.1	1.3	1.5	1.8	2.0	2.2
WW C3	17	26	35	41	50	56	63	71	13	21	30	36	44	50	57	66	1.1	1.3	1.4	1.6	1.8	1.1	1.3	1.5	1.7	1.9
WW C4	15	24	32	39	48	53	60	69	11	19	27	34	42	48	55	64	1.1	1.3	1.5	1.6	1.8	1.2	1.3	1.5	1.7	1.9
TW A1	14	16	18	19	21	22	23	25	12	13	14	15	17	17	18	19	2.2	2.7	3.2	4.0	-	2.7	3.3	4.1	-	-
TW A2	13	15	18	20	23	25	27	30	10	12	14	16	18	19	21	23	1.8	2.2	2.6	3.1	3.6	2.2	2.5	3.1	3.8	4.9
TW A3	11	14	18	20	24	26	29	32	9	11	14	15	18	20	22	24	1.7	2.0	2.4	2.8	3.3	2.0	2.3	2.8	3.4	4.2
TW A4	10	13	17	20	24	26	30	34	7	10	13	15	18	20	23	26	1.7	2.0	2.3	2.7	3.0	1.9	2.2	2.6	3.1	3.7
TW B1	13	15	17	19	21	23	25	27	11	13	14	15	17	18	19	21	2.1	2.5	2.9	3.5	4.1	2.5	2.9	3.5	4.4	-
TW B2	11	14	17	20	23	25	28	31	9	11	13	15	18	19	22	24	1.8	2.1	2.5	2.9	3.3	2.1	2.4	2.8	3.4	4.0
TW B3	11	14	17	19	23	25	28	32	8	10	13	15	18	20	22	25	1.8	2.1	2.4	2.7	3.1	2.0	2.3	2.7	3.1	3.6
TW B4	10	13	16	19	22	25	28	32	8	10	13	15	18	20	22	26	1.8	2.1	2.4	2.7	3.0	2.0	2.3	2.6	3.0	3.4
TW C1	12	14	17	19	22	23	26	29	10	12	14	15	17	19	20	23	2.0	2.3	2.7	3.1	3.4	2.3	2.7	3.1	3.6	4.2
TW C2	11	13	16	19	22	24	27	31	9	11	13	15	17	19	22	25	1.9	2.2	2.5	2.8	3.1	2.1	2.4	2.8	3.2	3.6
TW C3	11	13	16	18	21	24	27	31	9	11	13	15	17	19	21	25	1.9	2.2	2.5	2.8	3.1	2.2	2.5	2.8	3.1	3.5
TW C4	11	13	16	18	21	23	26	29	9	11	13	15	17	19	21	24	2.0	2.3	2.6	2.9	3.1	2.3	2.6	2.9	3.2	3.5
TW D1	8	9	11	12	13	13	14	14	5	6	7	7	8	8	8	6	2.4	3.2	4.4	-	-	3.2	4.5	-	-	-
TW D2	12	16	21	24	28	32	36	41	9	12	16	19	23	26	29	33	1.5	1.8	2.1	2.3	2.6	1.7	2.0	2.3	2.6	2.9
TW D3	13	17	22	25	30	34	38	43	9	13	17	20	25	27	31	35	1.5	1.7	2.0	2.3	2.5	1.6	1.9	2.2	2.5	2.9
TW D4	14	19	26	30	37	41	47	54	10	15	21	25	31	35	40	47	1.3	1.5	1.7	1.9	2.1	1.4	1.6	1.8	2.1	2.3
TW D5	14	20	26	32	39	43	49	58	10	16	22	27	33	38	44	51	1.3	1.5	1.6	1.8	2.0	1.4	1.5	1.7	1.9	2.1
TW E1	15	20	25	30	35	39	44	50	12	16	21	24	29	33	37	42	1.4	1.6	1.8	2.1	2.3	1.5	1.7	2.0	2.3	2.6
TW E2	15	20	26	30	37	41	46	53	11	16	21	25	31	35	39	46	1.3	1.5	1.8	2.0	2.2	1.4	1.6	1.9	2.1	2.3
TW E3	18	25	33	38	46	51	58	66	14	20	28	33	41	46	52	60	1.1	1.3	1.5	1.7	1.9	1.2	1.4	1.6	1.8	2.0
TW E4	16	23	31	37	45	50	57	65	12	19	26	32	40	45	51	60	1.2	1.3	1.5	1.7	1.9	1.2	1.4	1.6	1.8	2.0
TW F1	12	14	16	17	19	21	22	24	10	11	13	14	15	16	17	17	2.2	2.6	3.2	3.9	4.8	2.7	3.2	4.0	-	-
TW F2	10	13	16	18	21	22	25	28	8	10	12	13	15	17	18	20	1.9	2.2	2.6	3.1	3.7	2.2	2.6	3.1	3.9	5.0
TW F3	9	12	15	17	21	23	26	29	6	8	11	13	15	17	19	22	1.8	2.1	2.5	2.9	3.3	2.0	2.4	2.8	3.4	4.2
TW F4	7	10	13	16	20	22	25	29	4	6	9	11	14	16	19	22	1.7	2.0	2.4	2.7	3.1	2.0	2.3	2.7	3.1	3.7
TW G1	10	12	14	16	18	20	22	24	8	9	11	12	14	15	16	18	2.1	2.5	2.9	3.5	4.1	2.5	3.0	3.5	4.4	-
TW G2	8	10	13	15	19	21	23	27	5	7	10	11	14	15	17	20	1.9	2.2	2.6	3.0	3.4	2.2	2.5	3.0	3.5	4.2
TW G3	6	9	12	14	18	20	23	27	4	6	8	10	13	15	17	20	1.9	2.2	2.5	2.8	3.2	2.1	2.4	2.8	3.2	3.7
TW G4	5	7	10	12	16	18	21	25	2	4	7	9	11	13	16	19	1.9	2.2	2.5	2.8	3.1	2.2	2.5	2.8	3.2	3.5
TW H1	7	9	12	14	16	18	21	24	5	7	9	10	12	13	15	17	2.1	2.4	2.8	3.2	3.6	2.4	2.8	3.2	3.8	4.4
TW H2	5	7	10	12	15	17	20	24	3	5	7	8	11	13	15	18	2.0	2.3	2.6	3.0	3.3	2.3	2.6	3.0	3.4	3.8
TW H3	4	6	9	11	14	16	19	22	2	4	6	7	10	11	14	17	2.1	2.4	2.7	3.0	3.2	2.4	2.7	3.0	3.3	3.6
TW H4	3	5	7	9	11	13	16	19	1	3	4	6	8	10	12	15	2.3	2.5	2.8	3.1	3.3	2.6	2.9	3.2	3.5	3.7
TW I1	6	7	9	10	11	11	12	12	4	4	5	6	6	6	6	4	2.5	3.3	4.6	-	-	3.3	4.7	-	-	-
TW I2	10	13	18	21	25	29	33	38	9	10	13	16	20	23	26	31	1.6	1.8	2.1	2.4	2.6	1.8	2.0	2.3	2.6	2.9
TW I3	10	14	19	23	28	31	36	42	7	11	15	18	22	25	29	34	1.5	1.7	2.0	2.2	2.5	1.6	1.9	2.2	2.5	2.8
TW I4	11	17	23	28	34	38	44	52	8	13	18	23	29	33	38	45	1.3	1.5	1.7	2.0	2.1	1.4	1.6	1.8	2.1	2.3
TW I5	11	17	24	29	36	41	47	55	8	13	19	24	31	35	41	49	1.3	1.5	1.7	1.9	2.0	1.4	1.6	1.8	2.0	2.1
TW J1	13	18	23	28	33	37	42	49	9	14	19	22	28	31	36	41	1.4	1.6	1.8	2.1	2.3	1.5	1.7	2.0	2.3	2.5
TW J2	12	18	24	28	34	39	44	51	9	13	19	23	29	33	38	44	1.3	1.5	1.8	2.0	2.2	1.4	1.6	1.9	2.1	2.3
TW J3	15	22	30	36	44	50	56	65	11	18	25	31	39	44	50	59	1.2	1.3	1.5	1.7	1.9	1.2	1.4	1.6	1.8	2.0
TW J4	14	21	29	35	43	48	55	64	10	17	24	30	37	43	49	58	1.2	1.4	1.6	1.7	1.9	1.2	1.4	1.6	1.8	2.0
DG A1	7	9	11	12	13	13	10	12	5	6	7	7	7	7	6	4	2.2	3.1	-	-	-	2.8	4.3	-	-	-
DG A2	10	13	17	19	23	25	28	31	8	10	13	15	18	20	22	25	1.7	2.0	2.4	2.9	3.4	2.0	2.3	2.7	3.3	4.0
DG A3	14	18	23	27	33	37	41	48	11	15	19	23	28	32	36	42	1.4	1.6	1.9	2.1	2.3	1.6	1.7	2.0	2.2	2.5
DG B1	7	9	11	12	14	14	15	14	5	6	7	8	8	8	8	6	2.2	2.9	4.3	-	-	2.7	3.7	-	-	-
DG B2	10	14	17	20	24	26	29	33	8	10	13	16	19	21	24	27	1.7	2.0	2.3	2.7	3.1	1.9	2.2	2.5	3.0	3.7
DG B3	14	19	24	28	34	39	44	51	11	15	20	24	30	34	39	45	1.4	1.6	1.8	2.0	2.2	1.5	1.7	1.9	2.1	2.3
DG C1	9	12	15	17	19	20	21	22	6	8	10	11	13	14	14	14	1.9	2.4	3.2	-	-	2.2	2.8	3.9	-	-
DG C2	12	16	21	24	29	32	35	40	9	13	16	19	23	26	30	34	1.6	1.8	2.1	2.4	2.7	1.7	1.9	2.2	2.6	3.1
DG C3	16	21	28	32	39	43	49	57	13	18	23	28	34	38	44	51	1.3	1.5	1.7	1.9	2.1	1.4	1.6	1.8	2.0	2.2
SS A1	16	19	22	24	27	29	32	35	14	16	19	21	24	25	27	29	1.7	2.1	2.5	3.0	3.5	1.8	2.2	2.7	3.5	4.5
SS A2	19	24	29	33	39	43	47	53	17	22	27	30	35	39	43	48	1.3	1.6	1.8	2.1	2.3	1.4	1.6	1.9	2.2	2.5
SS A3	14	17	19	21	23	24	26	27	13	15	17	18	20	21	22	22	1.9	2.4	3.0	4.0	-	2.0	2.6	3.6	-	-
SS A4	19	23	28	32	37	40	44	50	16	21	25	29	33	36	40	44	1.4	1.6	1.9	2.2	2.4	1.4	1.7	2.0	2.4	2.7
SS A5	16	19	21	23	25	27	29	31	15	17	19	20	22	23	25	26	1.9	2.3	2.8	3.5	4.4	2.0	2.5	3.2	4.4	-
SS A6	19	24	29	32	38	41	46	51	17	21	26	29	34	37	41	46	1.4	1.6	1.9	2.1	2.3	1.4	1.7	2.0	2.3	2.6
SS A7	15	17	18	20	21	22	23	24	13	15	16	17	18	19	19	19	2.2	2.8	3.7	-	-	2.4	3.1	4.4	-	-
SS A8	18	22	27	30	35	38	42	47	16	20	24	27	31	34	37	41	1.4	1.7	2.0	2.3	2.5	1.5	1.8	2.1	2.5	2.9
SS B1	13	16	19	21	24	25	28	31	11	13	16	18	20	22	23	25	1.8	2.2	2.6	3.1	3.7	2.0	2.4	2.9	3.7	4.8
SS B2	17	21	27	30	36	40	44	51	15	19	24	27	32	36	40	46	1.4	1.6	1.8	2.1	2.3	1.5	1.7	1.9	2.2	2.4
SS B3	12	14	17	18	21	22	24	25	10	12	14	15	17	18	19	20	2.0	2.4	3.0	3.8	5.0	2.2	2.7	3.5	-	-
SS B4	16	21	25	29	34	38	42	48	14	18	23	26	30	34	38	43	1.4	1.7	1.9	2.2	2.4	1.5	1.8	2.0	2.3	2.6
SS B5	13	15	18	19	21	23	24	26	12	13	15	16	18	19	20	22	2.1	2.5	3.0	3.7	4.5	2.3	2.8	3.4	4.5	-
SS B6	16	20	25	29	34	38	42	48	14	18	22	26	31	34	38	43	1.4	1.7	1.9	2.1	2.3	1.5	1.7	2.0	2.3	2.5
SS B7	12	14	16	17	18	19	20	21	11	12	13	14	15	16	17	17	2.3	2.9	3.6	4.8	-	2.6	3.3	4.3	-	-
SS B8	15	19	24	27	32	35	39	45	13	17	21	24	28	31	35	39	1.5	1.7	2.0	2.3	2.5	1.6	1.8	2.1	2.5	2.8
SS C1	10	13	16	18	21	23	25	27	7	9	11	13	15	16	17	18	1.8	2.2	2.7	3.3	4.1	2.1	2.6	3.3	4.6	-
SS C2	12	17	22	26	31	34	39	44	9	13	17	20	25	28	32	37	1.4	1.7	1.9	2.2	2.4	1.6	1.8	2.1	2.4	2.8
SS C3	11	13	15	17	19	21	23	25	8	10	12	13	15	16	17	19	2.0	2.4	2.9	3.4	4.0	2.4	2.9	3.4	4.3	-
SS C4	12	16	20	24	28	31	35	41	9	12	16	19	23	26	29	34	1.5	1.8	2.1	2.3	2.6	1.7	2.0	2.2	2.6	2.9
SS D1	16	20	24	27	30	32	35	38	14	17	21	23	26	27	29	31	1.5	1.9	2.3	2.8	3.4	1.6	2.1	2.7	3.6	-
SS D2	20	26	33	38	44	48	54	60	18	23	29	34	40	44	48	55	1.2	1.4	1.7	1.9	2.1	1.2	1.5	1.8	2.0	2.2
SS D3	17	20	23	25	28	30	32	35	15	17	20	22	25	26	28	29	1.7	2.0	2.5	3.0	3.7	1.8	2.2	2.8	3.8	-
SS D4	20	26	32	36	43	47	52	59	17	23	29	33	39	42	47	53	1.2	1.5	1.7	1.9	2.1	1.3	1.5	1.8	2.1	2.3
SS E1	13	16	19	21	24	26	28	29	11	13	16	18	19	21	21	22	1.7	2.1	2.7	3.6	5.0	1.9	2.4	3.3	-	-
SS E2	16	22	27	32	37	41	46	52	14	19	24	27	33	36	40	45	1.3	1.6	1.8	2.1	2.3	1.4	1.7	2.0	2.3	2.6
SS E3	14	16	19	20	23	24	26	28	12	14	16	17	19	20	21	22	1.9	2.3	2.9	3.6	4.6	2.1	2.6	3.3	4.7	-
SS E4	16	21	26	30	35	38	43	49	14	18	23	26	31	34	38	43	1.4	1.6	1.9	2.2	2.4	1.5	1.7	2.0	2.4	2.7

GRAND ISLAND, NEBRASKA

SOLAR SAVINGS FRACTION (%) — CONSERVATION FACTOR

TYPE	BASE TEMP: 55 F (4171 DD)								65 F (6430 DD)								CF 55 F					CF 65 F				
LCR	100	70	50	40	30	25	20	15	100	70	50	40	30	25	20	15	100	50	30	20	15	100	50	30	20	15
WW A1	18	22	25	28	33	36	41	47	15	18	20	22	26	28	31	35	1.6	1.8	2.0	2.2	2.3	2.0	2.2	2.4	2.7	2.9
WW A2	19	24	31	35	42	47	53	62	15	19	23	27	32	36	41	48	1.3	1.5	1.6	1.8	1.9	1.5	1.7	1.9	2.1	2.2
WW A3	19	25	33	38	46	51	58	67	14	19	24	29	35	39	45	53	1.2	1.4	1.5	1.7	1.8	1.4	1.6	1.7	1.9	2.1
WW A4	19	26	34	40	48	53	61	70	14	19	25	30	36	41	47	56	1.2	1.3	1.5	1.6	1.8	1.3	1.5	1.7	1.9	2.0
WW A5	19	27	35	41	50	56	63	72	13	19	26	31	38	43	50	58	1.1	1.3	1.4	1.6	1.7	1.3	1.5	1.6	1.8	1.9
WW A6	19	27	36	42	51	57	65	74	13	19	26	31	39	44	51	60	1.1	1.3	1.4	1.6	1.7	1.3	1.4	1.6	1.8	1.9
WW B1	15	19	23	26	31	35	40	46	11	13	16	18	21	24	26	30	1.6	1.8	1.9	2.1	2.3	1.9	2.2	2.5	2.8	3.1
WW B2	20	27	36	42	51	57	64	73	14	20	27	33	40	45	52	61	1.1	1.3	1.4	1.6	1.7	1.3	1.4	1.6	1.8	1.9
WW B3	24	33	42	49	59	65	72	81	18	25	33	39	48	53	61	70	1.0	1.2	1.3	1.5	1.6	1.1	1.3	1.5	1.6	1.8
WW B4	24	34	44	52	62	69	76	85	18	26	35	42	52	58	66	75	1.0	1.1	1.3	1.4	1.6	1.1	1.2	1.4	1.5	1.7
WW B5	23	33	44	51	62	68	76	85	17	25	35	42	52	58	66	76	1.0	1.1	1.3	1.4	1.6	1.1	1.2	1.4	1.5	1.7
WW C1	26	35	45	52	61	68	75	83	19	27	35	42	51	56	64	73	1.0	1.2	1.3	1.5	1.6	1.1	1.3	1.4	1.6	1.7
WW C2	24	33	43	51	61	67	74	83	18	26	34	41	50	56	64	73	1.0	1.2	1.3	1.5	1.6	1.1	1.3	1.4	1.6	1.7
WW C3	29	40	52	60	71	77	84	91	22	32	43	51	61	68	75	84	0.9	1.1	1.2	1.4	1.5	1.0	1.1	1.3	1.4	1.6
WW C4	26	37	48	57	67	74	81	89	19	29	40	48	58	65	72	82	0.9	1.1	1.2	1.4	1.5	1.0	1.2	1.3	1.5	1.6
TW A1	18	21	25	28	33	36	40	47	15	17	20	22	26	28	31	36	1.6	1.8	2.0	2.2	2.3	1.9	2.2	2.4	2.6	2.8
TW A2	18	23	28	32	39	43	49	57	14	18	22	25	30	33	38	44	1.4	1.6	1.7	1.9	2.0	1.6	1.8	2.0	2.2	2.3
TW A3	18	23	29	34	41	46	52	61	13	17	22	26	32	35	41	48	1.3	1.5	1.6	1.8	1.9	1.5	1.7	1.9	2.0	2.2
TW A4	17	22	29	34	42	47	54	63	12	17	22	26	32	36	42	50	1.3	1.4	1.6	1.7	1.9	1.5	1.6	1.8	2.0	2.1
TW B1	17	21	25	29	34	37	42	49	14	17	20	23	26	29	33	38	1.6	1.7	1.9	2.1	2.2	1.8	2.0	2.2	2.4	2.6
TW B2	17	22	27	31	38	42	48	57	13	17	21	24	29	33	38	44	1.4	1.5	1.7	1.9	2.0	1.6	1.8	2.0	2.1	2.3
TW B3	16	21	27	32	38	43	49	58	12	16	21	24	30	33	39	46	1.4	1.5	1.7	1.8	2.0	1.6	1.7	1.9	2.1	2.2
TW B4	15	20	26	30	37	41	47	56	12	15	20	23	28	32	37	44	1.4	1.6	1.7	1.9	2.0	1.6	1.8	1.9	2.1	2.2
TW C1	16	20	25	28	34	37	43	50	13	16	19	22	26	29	33	39	1.5	1.7	1.9	2.0	2.1	1.8	2.0	2.1	2.3	2.5
TW C2	15	20	25	29	35	39	44	52	12	15	19	22	27	30	35	42	1.5	1.6	1.8	1.9	2.1	1.7	1.9	2.0	2.2	2.3
TW C3	15	19	24	27	33	37	43	50	12	15	19	22	26	29	34	40	1.5	1.7	1.8	2.0	2.1	1.7	1.9	2.1	2.2	2.4
TW C4	14	18	22	25	30	34	39	46	12	14	18	20	24	27	31	37	1.6	1.8	1.9	2.1	2.2	1.9	2.0	2.2	2.4	2.5
TW D1	12	16	20	23	28	31	35	42	9	11	14	16	19	21	24	28	1.7	1.9	2.0	2.2	2.4	2.0	2.3	2.6	2.8	3.1
TW D2	19	26	33	38	46	52	59	68	14	20	26	30	37	42	48	56	1.2	1.4	1.5	1.7	1.8	1.4	1.5	1.7	1.8	2.0
TW D3	21	28	36	42	50	56	63	73	15	21	28	33	40	45	52	61	1.1	1.3	1.4	1.6	1.7	1.3	1.4	1.6	1.8	1.9
TW D4	22	31	40	47	56	62	70	79	17	24	32	38	46	52	60	69	1.1	1.2	1.4	1.5	1.7	1.2	1.3	1.5	1.6	1.8
TW D5	22	30	40	47	57	63	71	80	16	24	32	39	47	53	61	71	1.1	1.2	1.3	1.5	1.6	1.2	1.3	1.4	1.6	1.7
TW E1	24	31	40	47	56	62	69	78	18	25	32	38	46	51	58	67	1.1	1.2	1.4	1.5	1.7	1.2	1.4	1.5	1.7	1.8
TW E2	23	31	40	47	57	63	70	79	18	24	32	38	47	52	59	69	1.1	1.2	1.4	1.5	1.7	1.2	1.3	1.5	1.6	1.8
TW E3	28	38	49	57	67	73	80	88	21	30	40	47	57	63	71	80	1.0	1.1	1.3	1.4	1.6	1.0	1.2	1.3	1.5	1.6
TW E4	26	36	46	54	64	71	78	86	20	28	38	45	55	61	69	78	1.0	1.1	1.3	1.4	1.6	1.1	1.2	1.4	1.5	1.7
TW F1	16	20	24	27	32	35	40	47	13	16	19	21	24	27	30	35	1.6	1.8	2.0	2.1	2.3	1.9	2.1	2.3	2.5	2.7
TW F2	16	20	26	30	36	41	46	55	12	15	19	23	27	31	35	42	1.4	1.6	1.7	1.9	2.0	1.6	1.8	2.0	2.2	2.4
TW F3	15	20	26	30	37	42	48	57	10	14	19	23	28	32	37	44	1.4	1.5	1.7	1.8	2.0	1.6	1.7	1.9	2.1	2.2
TW F4	13	18	24	29	36	41	47	56	8	13	17	21	27	31	36	44	1.4	1.5	1.7	1.8	1.9	1.5	1.7	1.9	2.1	2.2
TW G1	14	18	22	26	31	35	40	47	11	14	17	20	24	26	30	35	1.6	1.8	1.9	2.1	2.2	1.9	2.1	2.3	2.4	2.6
TW G2	13	17	23	27	33	37	43	52	9	13	17	20	25	28	33	39	1.4	1.6	1.8	1.9	2.0	1.7	1.9	2.0	2.2	2.4
TW G3	11	16	21	26	32	36	42	51	7	11	15	19	24	27	32	39	1.4	1.6	1.8	1.9	2.0	1.7	1.8	2.0	2.2	2.3
TW G4	9	13	18	22	28	32	38	46	6	9	13	16	21	24	29	35	1.5	1.7	1.8	2.0	2.1	1.8	1.9	2.1	2.3	2.4
TW H1	11	15	20	23	28	32	37	45	8	11	14	17	21	24	28	34	1.6	1.8	1.9	2.1	2.2	1.9	2.0	2.2	2.4	2.6
TW H2	9	13	18	21	27	31	36	44	6	9	12	15	20	23	27	33	1.6	1.7	1.9	2.1	2.2	1.8	2.0	2.2	2.4	2.5
TW H3	7	11	15	19	24	27	33	40	5	7	11	13	17	20	24	30	1.7	1.8	2.0	2.1	2.3	1.9	2.1	2.3	2.4	2.6
TW H4	5	8	12	15	19	22	27	33	3	5	8	10	14	16	20	25	1.9	2.0	2.2	2.3	2.5	2.1	2.3	2.5	2.6	2.8
TW I1	10	14	17	20	25	28	32	38	7	9	12	14	16	18	21	25	1.7	1.9	2.1	2.3	2.4	2.1	2.4	2.7	3.0	3.2
TW I2	16	22	29	35	43	48	55	65	12	16	22	27	33	38	44	53	1.2	1.4	1.5	1.7	1.8	1.4	1.6	1.7	1.9	2.0
TW I3	17	24	32	38	46	52	60	69	13	18	25	30	37	42	48	57	1.2	1.3	1.5	1.6	1.8	1.3	1.5	1.6	1.8	1.9
TW I4	19	27	36	43	53	59	67	76	14	21	29	35	43	49	56	66	1.1	1.2	1.4	1.5	1.7	1.2	1.4	1.5	1.7	1.8
TW I5	19	28	37	44	54	61	69	78	14	21	30	36	45	51	59	68	1.1	1.2	1.4	1.5	1.7	1.2	1.3	1.5	1.6	1.8
TW J1	21	29	38	44	54	60	67	76	16	22	30	35	43	49	56	65	1.1	1.2	1.4	1.5	1.7	1.2	1.4	1.5	1.7	1.8
TW J2	20	28	37	44	54	60	67	77	15	22	29	35	44	49	57	66	1.1	1.2	1.4	1.5	1.7	1.2	1.4	1.5	1.7	1.8
TW J3	25	35	46	54	64	70	78	86	19	27	37	45	55	61	69	78	1.0	1.1	1.3	1.4	1.6	1.1	1.2	1.4	1.5	1.7
TW J4	23	33	43	51	62	68	76	85	17	26	35	42	52	59	67	76	1.0	1.1	1.3	1.4	1.6	1.1	1.2	1.4	1.5	1.7
DG A1	13	17	21	24	28	31	34	37	9	11	14	16	19	21	22	24	1.5	1.8	2.2	2.8	3.4	1.9	2.2	2.8	3.9	-
DG A2	15	20	26	31	38	42	47	54	11	15	20	24	29	32	37	43	1.4	1.5	1.7	2.0	2.2	1.6	1.7	1.9	2.2	2.5
DG A3	19	26	33	40	49	55	62	70	15	20	27	32	40	45	52	60	1.2	1.3	1.4	1.6	1.8	1.3	1.5	1.6	1.8	1.9
DG B1	13	17	22	25	31	34	39	45	9	12	15	17	20	22	25	29	1.5	1.7	1.9	2.2	2.6	1.8	2.2	2.5	2.9	3.5
DG B2	16	21	27	32	40	45	52	61	12	16	21	24	30	34	40	47	1.3	1.5	1.6	1.8	2.0	1.5	1.7	1.8	2.0	2.2
DG B3	19	26	34	41	51	58	66	76	15	21	28	33	41	47	55	65	1.2	1.3	1.4	1.5	1.7	1.3	1.4	1.5	1.7	1.8
DG C1	16	21	26	31	38	43	49	56	11	15	19	22	26	29	33	39	1.4	1.5	1.7	1.9	2.1	1.6	1.9	2.1	2.3	2.7
DG C2	18	24	32	38	47	53	60	69	14	19	25	29	36	41	47	56	1.2	1.3	1.5	1.6	1.8	1.4	1.5	1.7	1.8	2.0
DG C3	22	30	39	46	57	65	73	82	18	24	31	37	47	53	61	72	1.1	1.2	1.3	1.4	1.6	1.2	1.3	1.4	1.6	1.7
SS A1	22	27	33	37	44	48	54	62	18	22	27	31	36	40	44	51	1.3	1.5	1.6	1.8	1.9	1.5	1.7	1.9	2.1	2.2
SS A2	27	34	43	49	57	63	70	78	23	29	36	42	49	54	61	69	1.1	1.2	1.4	1.5	1.7	1.2	1.3	1.5	1.7	1.8
SS A3	20	24	29	33	39	43	48	55	17	20	24	27	31	34	38	44	1.4	1.6	1.8	1.9	2.1	1.6	1.8	2.1	2.3	2.5
SS A4	26	33	41	47	55	61	68	76	22	28	35	40	47	52	58	67	1.1	1.3	1.4	1.6	1.7	1.2	1.4	1.6	1.7	1.9
SS A5	21	26	30	34	40	43	49	56	18	22	26	29	33	36	40	45	1.4	1.6	1.8	1.9	2.1	1.6	1.8	2.0	2.3	2.4
SS A6	27	34	42	48	57	62	69	77	22	29	36	41	48	53	60	68	1.1	1.2	1.4	1.6	1.7	1.2	1.4	1.5	1.7	1.8
SS A7	19	22	26	29	34	37	41	47	17	19	22	24	28	30	33	37	1.6	1.8	2.0	2.2	2.4	1.8	2.1	2.4	2.7	2.9
SS A8	26	33	40	46	54	59	66	74	22	27	34	39	46	50	56	64	1.1	1.3	1.4	1.6	1.7	1.2	1.4	1.6	1.8	1.9
SS B1	18	23	28	32	38	42	48	56	15	19	23	26	31	34	39	45	1.4	1.6	1.7	1.9	2.0	1.6	1.8	2.0	2.2	2.4
SS B2	24	30	38	44	52	58	65	74	20	26	32	37	45	50	56	65	1.1	1.3	1.4	1.6	1.7	1.2	1.4	1.6	1.7	1.9
SS B3	17	21	26	30	35	39	44	51	14	17	21	24	28	31	35	40	1.5	1.7	1.8	2.0	2.2	1.7	1.9	2.1	2.4	2.6
SS B4	23	29	37	43	51	56	63	72	19	25	31	36	43	48	54	62	1.2	1.3	1.5	1.6	1.8	1.3	1.4	1.6	1.8	1.9
SS B5	17	21	25	28	33	37	41	48	15	18	21	23	27	30	33	38	1.6	1.8	1.9	2.1	2.3	1.8	2.0	2.3	2.5	2.7
SS B6	23	30	37	43	51	57	64	73	19	25	31	36	43	48	55	63	1.1	1.3	1.5	1.6	1.8	1.3	1.4	1.6	1.8	1.9
SS B7	16	19	22	25	29	32	36	41	13	16	18	20	23	25	28	32	1.7	1.9	2.1	2.3	2.5	2.0	2.2	2.5	2.8	3.0
SS B8	22	28	35	41	49	54	61	70	18	23	29	34	41	46	52	60	1.2	1.3	1.5	1.7	1.8	1.3	1.5	1.6	1.8	2.0
SS C1	17	22	28	33	40	45	51	60	12	16	21	25	30	34	39	46	1.3	1.5	1.6	1.8	1.9	1.5	1.7	1.9	2.1	2.2
SS C2	20	28	36	42	51	57	64	73	15	21	28	33	41	46	53	61	1.1	1.3	1.4	1.6	1.7	1.3	1.4	1.6	1.7	1.9
SS C3	15	19	24	27	33	37	42	49	12	15	18	21	25	28	32	37	1.5	1.7	1.9	2.0	2.2	1.8	2.0	2.2	2.4	2.5
SS C4	19	25	32	38	45	51	58	67	14	19	25	30	36	41	47	55	1.2	1.4	1.5	1.7	1.8	1.4	1.5	1.7	1.9	2.0
SS D1	25	31	39	44	52	57	63	72	20	26	32	36	42	47	52	60	1.2	1.3	1.5	1.6	1.8	1.3	1.5	1.7	1.9	2.0
SS D2	30	40	50	57	66	72	78	86	25	33	42	48	57	63	70	78	1.0	1.1	1.3	1.4	1.6	1.1	1.2	1.4	1.6	1.7
SS D3	23	29	35	40	46	51	57	65	20	24	29	33	38	42	47	53	1.3	1.4	1.6	1.8	1.9	1.4	1.6	1.8	2.0	2.2
SS D4	29	38	48	55	64	70	76	84	24	32	41	47	55	61	67	76	1.0	1.1	1.3	1.5	1.6	1.1	1.2	1.4	1.6	1.7
SS E1	21	27	33	38	45	50	56	64	16	21	26	30	35	39	44	51	1.2	1.4	1.6	1.7	1.9	1.4	1.6	1.8	2.0	2.2
SS E2	26	34	43	50	59	65	72	80	21	28	36	41	50	55	62	70	1.0	1.2	1.3	1.5	1.7	1.1	1.3	1.5	1.6	1.8
SS E3	19	23	28	32	38	42	47	55	16	19	23	26	30	33	37	43	1.4	1.6	1.8	1.9	2.1	1.6	1.8	2.1	2.3	2.5
SS E4	24	32	40	46	55	60	67	76	20	26	33	38	46	51	57	66	1.1	1.3	1.4	1.6	1.7	1.2	1.4	1.5	1.7	1.9

NORTH PLATTE, NEBRASKA

TYPE	SOLAR SAVINGS FRACTION (%) BASE TEMP: 55 F (4351 DD) LCR								65 F (6746 DD)								CONSERVATION FACTOR 55 F					65 F				
	100	70	50	40	30	25	20	15	100	70	50	40	30	25	20	15	100	50	30	20	15	100	50	30	20	15
WW A1	20	23	28	31	36	40	45	52	16	18	22	24	28	30	34	39	1.5	1.7	1.9	2.0	2.2	1.8	2.0	2.2	2.4	2.6
WW A2	21	27	34	39	46	51	58	67	16	20	25	29	35	40	45	53	1.2	1.4	1.5	1.7	1.8	1.4	1.6	1.8	1.9	2.1
WW A3	21	28	36	42	50	56	63	72	15	21	27	32	38	43	49	58	1.2	1.3	1.5	1.6	1.7	1.3	1.5	1.7	1.8	2.0
WW A4	21	29	37	44	52	58	66	75	15	21	28	33	40	45	52	61	1.1	1.3	1.4	1.6	1.7	1.3	1.4	1.6	1.8	1.9
WW A5	22	30	39	46	55	61	68	77	15	21	29	34	42	47	54	64	1.1	1.2	1.4	1.5	1.7	1.2	1.4	1.5	1.7	1.8
WW A6	22	30	40	47	56	62	70	79	15	21	29	35	43	49	56	65	1.1	1.2	1.4	1.5	1.7	1.2	1.4	1.5	1.7	1.8
WW B1	17	21	26	30	36	40	45	52	12	15	18	21	24	27	31	36	1.4	1.6	1.8	2.0	2.1	1.8	2.0	2.2	2.5	2.7
WW B2	22	30	39	46	55	61	69	78	16	22	30	35	44	49	56	66	1.1	1.2	1.4	1.5	1.7	1.2	1.4	1.5	1.7	1.8
WW B3	27	36	46	53	63	69	77	85	19	27	36	43	52	58	65	74	1.0	1.1	1.3	1.5	1.6	1.1	1.3	1.4	1.6	1.7
WW B4	27	37	48	56	66	72	80	88	19	28	38	45	55	62	69	79	1.0	1.1	1.3	1.4	1.6	1.0	1.2	1.3	1.5	1.6
WW B5	26	36	47	55	66	72	79	87	18	27	38	45	55	62	70	79	1.0	1.1	1.3	1.4	1.6	1.0	1.2	1.3	1.5	1.6
WW C1	28	38	48	56	66	72	79	87	21	29	38	45	54	60	68	77	1.0	1.1	1.3	1.4	1.6	1.1	1.2	1.4	1.5	1.7
WW C2	27	36	47	55	65	71	78	86	19	28	37	44	54	60	67	77	1.0	1.1	1.3	1.4	1.6	1.1	1.2	1.4	1.5	1.7
WW C3	31	43	56	64	74	80	87	93	24	34	46	54	65	71	79	87	0.9	1.0	1.2	1.4	1.5	1.0	1.1	1.3	1.4	1.6
WW C4	29	40	52	60	71	77	84	91	21	31	42	50	61	68	76	84	0.9	1.0	1.2	1.4	1.5	1.0	1.1	1.3	1.4	1.6
TW A1	19	23	27	31	36	39	44	51	16	18	21	24	28	30	34	39	1.5	1.7	1.9	2.0	2.2	1.8	2.0	2.2	2.4	2.6
TW A2	20	25	31	36	42	47	53	62	15	19	24	27	33	36	41	49	1.3	1.5	1.6	1.8	1.9	1.5	1.7	1.9	2.0	2.2
TW A3	20	25	32	37	45	50	57	66	14	19	24	28	35	39	45	53	1.2	1.4	1.5	1.7	1.8	1.4	1.6	1.8	1.9	2.1
TW A4	19	25	32	38	46	51	58	67	13	18	24	28	35	40	46	54	1.2	1.4	1.5	1.7	1.8	1.4	1.6	1.7	1.9	2.0
TW B1	19	23	27	31	37	41	46	53	15	18	21	24	28	31	35	41	1.5	1.6	1.8	2.0	2.1	1.8	1.9	2.1	2.3	2.4
TW B2	18	24	30	34	41	46	53	61	14	18	23	26	32	36	41	49	1.3	1.5	1.6	1.8	1.9	1.5	1.7	1.9	2.0	2.2
TW B3	18	23	30	35	42	47	54	63	13	17	22	26	32	36	42	50	1.3	1.4	1.6	1.7	1.9	1.5	1.6	1.8	2.0	2.1
TW B4	17	22	28	33	40	45	51	60	12	16	21	25	31	35	40	48	1.3	1.5	1.6	1.8	1.9	1.5	1.7	1.8	2.0	2.1
TW C1	17	22	27	31	37	41	46	54	14	17	21	24	28	32	36	43	1.5	1.6	1.8	1.9	2.0	1.7	1.9	2.0	2.2	2.3
TW C2	17	21	27	31	38	42	48	57	13	16	21	24	29	33	38	45	1.4	1.5	1.7	1.8	2.0	1.6	1.8	1.9	2.1	2.2
TW C3	16	20	26	30	36	40	46	55	12	16	20	23	28	32	36	43	1.4	1.6	1.7	1.9	2.0	1.7	1.8	2.0	2.1	2.3
TW C4	15	19	24	27	33	37	42	49	12	15	19	21	26	29	33	40	1.6	1.7	1.9	2.0	2.1	1.8	1.9	2.1	2.3	2.4
TW D1	14	18	22	26	31	35	40	47	10	12	15	18	22	24	28	33	1.5	1.7	1.9	2.1	2.2	1.9	2.1	2.3	2.6	2.7
TW D2	21	28	36	42	50	56	63	72	16	21	28	33	40	45	51	60	1.2	1.3	1.4	1.6	1.7	1.3	1.5	1.6	1.8	1.9
TW D3	23	30	39	46	55	61	68	77	17	23	30	36	44	49	56	65	1.1	1.2	1.4	1.5	1.7	1.2	1.4	1.5	1.7	1.8
TW D4	24	33	43	50	60	66	74	83	18	26	34	41	50	56	63	73	1.0	1.2	1.3	1.5	1.6	1.1	1.3	1.4	1.6	1.7
TW D5	24	33	43	50	60	67	74	83	18	26	34	41	50	57	64	74	1.0	1.2	1.3	1.5	1.6	1.1	1.3	1.4	1.6	1.7
TW E1	26	34	44	51	60	66	73	82	20	27	35	41	49	55	62	71	1.0	1.2	1.3	1.5	1.6	1.2	1.3	1.5	1.6	1.8
TW E2	25	34	44	51	60	67	74	83	19	26	35	41	50	56	63	72	1.0	1.2	1.3	1.5	1.6	1.1	1.3	1.4	1.6	1.7
TW E3	30	41	52	60	70	77	83	90	23	33	43	50	60	67	74	83	0.9	1.1	1.2	1.4	1.5	1.0	1.1	1.3	1.5	1.6
TW E4	28	39	50	58	68	74	81	89	21	30	41	48	58	65	72	81	0.9	1.1	1.2	1.4	1.5	1.0	1.2	1.3	1.5	1.6
TW F1	18	21	26	30	35	39	44	51	14	17	20	23	27	29	33	39	1.5	1.7	1.8	2.0	2.1	1.8	2.0	2.2	2.4	2.5
TW F2	17	22	28	33	40	45	51	59	13	16	21	25	30	34	39	46	1.3	1.5	1.6	1.8	1.9	1.6	1.7	1.9	2.1	2.2
TW F3	16	22	29	34	41	46	53	62	11	16	21	25	31	35	41	48	1.3	1.4	1.6	1.7	1.9	1.5	1.7	1.8	2.0	2.1
TW F4	14	20	27	32	40	45	52	61	9	14	19	23	29	34	40	48	1.3	1.4	1.6	1.7	1.9	1.5	1.6	1.8	2.0	2.1
TW G1	16	20	25	28	34	38	44	51	12	15	18	21	26	29	33	39	1.5	1.7	1.8	2.0	2.1	1.8	1.9	2.1	2.3	2.4
TW G2	14	19	25	30	37	41	48	56	10	14	18	22	27	31	36	43	1.4	1.5	1.7	1.8	2.0	1.6	1.8	1.9	2.1	2.2
TW G3	13	18	24	28	35	40	47	56	8	12	17	20	26	30	35	43	1.4	1.5	1.7	1.8	2.0	1.6	1.7	1.9	2.1	2.2
TW G4	10	15	20	25	31	36	42	51	6	10	14	17	23	26	31	39	1.5	1.6	1.8	1.9	2.0	1.7	1.8	2.0	2.2	2.3
TW H1	13	17	22	26	31	36	41	49	9	12	16	19	23	26	31	37	1.5	1.7	1.8	2.0	2.1	1.8	1.9	2.1	2.3	2.4
TW H2	10	14	20	24	30	34	40	48	6	10	14	17	22	25	30	37	1.5	1.7	1.8	2.0	2.1	1.7	1.9	2.1	2.2	2.4
TW H3	8	12	17	21	26	30	36	44	5	8	12	15	19	22	27	33	1.6	1.7	1.9	2.0	2.2	1.8	2.0	2.2	2.3	2.4
TW H4	6	9	13	16	21	25	29	36	4	6	9	11	15	18	22	27	1.8	1.9	2.1	2.2	2.3	2.1	2.2	2.4	2.5	2.0
TW I1	12	15	20	23	28	32	36	43	8	10	13	16	19	21	24	29	1.6	1.8	2.0	2.1	2.3	2.0	2.2	2.4	2.7	2.9
TW I2	18	25	32	38	47	52	60	69	13	18	24	29	36	41	48	57	1.2	1.3	1.5	1.6	1.8	1.3	1.5	1.6	1.8	1.9
TW I3	19	27	35	42	50	56	64	73	14	20	27	32	40	45	52	61	1.1	1.3	1.4	1.6	1.7	1.3	1.4	1.6	1.7	1.9
TW I4	21	30	40	47	57	63	71	80	15	23	31	37	46	52	60	70	1.0	1.2	1.3	1.5	1.6	1.2	1.3	1.5	1.6	1.7
TW I5	22	31	41	48	58	65	72	81	15	23	32	39	48	54	62	72	1.0	1.2	1.3	1.5	1.6	1.1	1.3	1.4	1.6	1.7
TW J1	23	32	41	48	58	64	71	80	17	24	32	38	47	53	60	69	1.0	1.2	1.3	1.5	1.6	1.2	1.3	1.5	1.6	1.8
TW J2	23	31	41	48	58	64	71	81	16	24	32	38	47	53	60	70	1.0	1.2	1.3	1.5	1.6	1.2	1.3	1.5	1.6	1.7
TW J3	27	38	49	57	68	74	81	89	20	30	40	48	58	64	72	81	0.9	1.1	1.2	1.4	1.5	1.0	1.2	1.3	1.5	1.6
TW J4	25	36	47	55	65	72	79	87	18	28	38	45	55	62	70	79	1.0	1.1	1.3	1.4	1.6	1.0	1.2	1.3	1.5	1.6
DG A1	15	19	24	28	32	35	38	42	10	13	16	19	22	24	26	28	1.4	1.7	2.0	2.5	3.0	1.7	2.0	2.5	3.3	4.5
DG A2	17	22	29	34	41	46	51	58	12	16	22	26	31	35	40	46	1.3	1.4	1.6	1.9	2.1	1.5	1.6	1.8	2.1	2.4
DG A3	21	28	36	43	53	59	66	74	16	22	29	34	42	48	55	64	1.2	1.2	1.4	1.6	1.8	1.3	1.4	1.5	1.7	1.9
DG B1	15	19	25	29	35	40	45	51	10	13	16	19	23	26	30	34	1.4	1.6	1.8	2.0	2.4	1.7	2.0	2.2	2.6	3.0
DG B2	17	23	30	36	44	50	57	66	12	17	22	27	33	38	44	52	1.3	1.4	1.5	1.7	1.9	1.5	1.6	1.7	1.9	2.1
DG B3	21	28	37	45	55	62	70	80	16	22	30	35	44	50	59	69	1.1	1.2	1.3	1.5	1.7	1.3	1.4	1.5	1.6	1.8
DG C1	18	23	30	35	43	49	55	63	12	16	21	24	30	33	38	45	1.3	1.4	1.5	1.7	2.0	1.5	1.7	1.9	2.1	2.4
DG C2	20	27	35	42	51	58	66	74	15	20	26	32	39	45	52	61	1.2	1.3	1.4	1.6	1.8	1.4	1.5	1.6	1.7	1.9
DG C3	24	32	42	50	62	69	77	86	19	25	34	40	50	57	66	76	1.1	1.1	1.2	1.4	1.6	1.2	1.3	1.4	1.5	1.7
SS A1	23	29	36	40	47	52	58	66	19	24	29	33	39	43	48	55	1.2	1.4	1.6	1.7	1.9	1.4	1.6	1.8	1.9	2.1
SS A2	29	37	46	52	61	67	74	82	24	31	39	45	53	58	64	73	1.0	1.2	1.3	1.5	1.6	1.1	1.3	1.5	1.6	1.8
SS A3	22	26	32	36	43	47	52	60	18	22	26	30	34	38	42	48	1.3	1.5	1.7	1.8	2.0	1.5	1.7	1.9	2.1	2.3
SS A4	28	36	45	51	59	65	71	80	24	30	38	43	51	56	62	70	1.1	1.2	1.4	1.5	1.7	1.2	1.3	1.5	1.7	1.8
SS A5	23	28	33	37	43	47	53	60	19	23	27	31	35	39	43	49	1.3	1.5	1.7	1.9	2.0	1.5	1.7	1.9	2.1	2.3
SS A6	29	37	46	52	61	66	73	81	24	31	38	44	52	57	64	72	1.0	1.2	1.4	1.5	1.7	1.1	1.3	1.5	1.6	1.8
SS A7	20	24	29	32	37	40	45	51	17	20	24	26	30	33	36	41	1.5	1.7	1.9	2.1	2.2	1.7	2.0	2.2	2.5	2.7
SS A8	28	35	43	50	58	63	70	78	23	29	36	42	49	54	60	68	1.1	1.2	1.4	1.6	1.7	1.2	1.4	1.5	1.7	1.8
SS B1	20	25	31	35	42	46	52	60	16	20	25	28	34	37	42	49	1.3	1.5	1.7	1.8	2.0	1.5	1.7	1.9	2.1	2.2
SS B2	25	33	41	47	56	62	69	77	21	27	34	40	48	53	60	68	1.1	1.2	1.4	1.6	1.7	1.2	1.4	1.5	1.7	1.8
SS B3	19	23	28	33	39	43	48	56	15	18	23	26	31	34	38	44	1.4	1.6	1.7	1.9	2.0	1.6	1.8	2.0	2.2	2.4
SS B4	25	32	40	46	54	60	67	75	20	26	33	38	46	51	57	66	1.1	1.3	1.4	1.6	1.7	1.2	1.4	1.5	1.7	1.9
SS B5	19	23	27	31	36	40	45	52	16	19	22	25	29	32	36	42	1.5	1.7	1.8	2.0	2.1	1.7	1.9	2.1	2.3	2.5
SS B6	25	32	40	46	55	61	68	76	20	26	33	39	46	51	58	67	1.1	1.3	1.4	1.6	1.7	1.2	1.4	1.5	1.7	1.8
SS B7	17	20	24	27	32	35	40	46	14	17	20	22	25	28	31	36	1.6	1.8	2.0	2.2	2.3	1.9	2.1	2.3	2.6	2.8
SS B8	24	30	38	44	52	58	65	73	19	25	32	37	44	49	55	64	1.1	1.3	1.4	1.6	1.7	1.2	1.4	1.6	1.8	1.9
SS C1	19	25	32	37	44	50	56	65	13	18	23	27	33	37	43	51	1.2	1.4	1.5	1.7	1.8	1.4	1.6	1.8	2.0	2.1
SS C2	23	31	39	46	55	61	69	78	16	23	30	36	44	49	57	66	1.1	1.2	1.4	1.5	1.7	1.2	1.4	1.5	1.7	1.8
SS C3	16	21	26	30	36	40	46	54	12	16	20	23	27	30	35	41	1.4	1.6	1.8	1.9	2.0	1.7	1.9	2.1	2.2	2.4
SS C4	21	27	35	41	49	55	62	71	15	21	27	32	39	44	51	59	1.2	1.3	1.5	1.6	1.7	1.3	1.5	1.6	1.8	1.9
SS D1	27	34	42	48	56	61	68	76	22	28	34	39	46	51	57	64	1.1	1.3	1.4	1.6	1.7	1.2	1.4	1.6	1.8	1.9
SS D2	33	43	53	60	70	75	82	89	27	36	45	52	61	66	73	81	0.9	1.1	1.2	1.4	1.6	1.0	1.2	1.3	1.5	1.7
SS D3	25	31	38	43	50	55	61	69	21	26	31	35	41	45	51	58	1.2	1.4	1.5	1.7	1.8	1.3	1.5	1.7	1.9	2.1
SS D4	32	41	51	58	68	73	80	87	26	34	43	50	59	64	71	79	1.0	1.1	1.3	1.4	1.6	1.0	1.2	1.4	1.5	1.7
SS E1	23	29	36	42	49	55	61	70	18	23	29	33	39	43	49	56	1.2	1.3	1.5	1.7	1.8	1.3	1.5	1.7	1.9	2.1
SS E2	28	37	47	54	63	69	76	84	22	30	38	45	53	59	66	74	1.0	1.2	1.3	1.5	1.6	1.1	1.3	1.4	1.6	1.7
SS E3	21	25	31	35	41	46	51	59	17	20	25	28	33	37	41	48	1.3	1.5	1.7	1.8	2.0	1.5	1.7	1.9	2.1	2.3
SS E4	26	34	43	50	59	64	71	80	21	28	35	41	49	54	61	70	1.1	1.2	1.4	1.5	1.7	1.2	1.3	1.5	1.7	1.8

OMAHA, NEBRASKA

TYPE	SOLAR SAVINGS FRACTION (%) BASE TEMP: 55 F (4353 DD)								65 F (6606 DD)								CONSERVATION FACTOR 55 F					65 F				
LCR	100	70	50	40	30	25	20	15	100	70	50	40	30	25	20	15	100	50	30	20	15	100	50	30	20	15
WW A1	17	19	22	25	28	31	35	40	14	16	18	20	22	24	26	29	1.8	2.0	2.2	2.4	2.6	2.2	2.5	2.8	3.1	3.3
WW A2	16	21	26	30	36	41	46	54	13	16	20	23	27	31	35	41	1.4	1.6	1.8	1.9	2.0	1.7	1.9	2.1	2.3	2.4
WW A3	16	21	28	32	39	44	51	59	12	16	21	24	30	33	38	45	1.3	1.5	1.6	1.8	1.9	1.5	1.7	1.9	2.1	2.2
WW A4	16	22	28	34	41	46	53	62	11	16	21	25	31	35	40	48	1.3	1.4	1.6	1.7	1.9	1.5	1.6	1.8	2.0	2.2
WW A5	16	22	29	35	43	49	56	65	11	16	21	26	32	37	43	51	1.2	1.4	1.5	1.7	1.8	1.4	1.6	1.8	1.9	2.1
WW A6	15	22	30	36	44	50	57	67	10	16	22	26	33	38	44	52	1.2	1.4	1.5	1.7	1.8	1.4	1.5	1.7	1.9	2.0
WW B1	12	15	19	21	25	28	32	37	9	11	13	14	17	18	20	22	1.8	2.0	2.2	2.4	2.6	2.2	2.6	3.1	3.6	4.2
WW B2	17	23	31	36	45	50	57	67	12	17	23	28	35	40	46	54	1.2	1.4	1.5	1.7	1.8	1.4	1.5	1.7	1.9	2.0
WW B3	20	28	37	43	52	58	66	75	15	21	29	34	42	47	54	64	1.1	1.2	1.4	1.6	1.7	1.2	1.4	1.6	1.7	1.9
WW B4	20	29	39	46	56	63	70	80	15	22	31	37	47	53	60	70	1.0	1.2	1.3	1.5	1.6	1.1	1.3	1.5	1.6	1.7
WW B5	19	28	38	46	56	63	71	80	14	22	31	38	47	53	61	71	1.0	1.2	1.3	1.5	1.6	1.1	1.3	1.4	1.6	1.7
WW C1	22	30	39	46	55	61	69	78	16	23	31	37	45	51	58	67	1.1	1.2	1.4	1.5	1.7	1.2	1.3	1.5	1.7	1.8
WW C2	20	28	38	45	54	61	68	78	15	22	30	36	45	51	58	68	1.1	1.2	1.4	1.5	1.7	1.2	1.3	1.5	1.6	1.8
WW C3	24	35	46	55	65	72	79	87	19	28	39	46	56	63	71	80	1.0	1.1	1.2	1.4	1.6	1.0	1.2	1.3	1.5	1.6
WW C4	22	32	43	51	62	69	76	85	16	25	36	43	53	60	68	78	1.0	1.1	1.3	1.4	1.6	1.1	1.2	1.4	1.5	1.6
TW A1	16	19	22	25	28	31	35	40	14	16	18	20	22	24	27	30	1.8	2.0	2.2	2.4	2.5	2.2	2.4	2.7	3.0	3.2
TW A2	16	20	24	28	33	37	42	50	13	15	19	22	26	29	32	38	1.5	1.7	1.9	2.0	2.1	1.8	2.0	2.2	2.4	2.5
TW A3	15	20	25	29	35	40	46	54	11	15	19	22	27	30	35	41	1.4	1.6	1.7	1.9	2.0	1.6	1.8	2.0	2.2	2.4
TW A4	14	19	25	29	36	40	47	55	10	14	19	22	27	31	36	43	1.4	1.5	1.7	1.9	2.0	1.6	1.8	2.0	2.1	2.3
TW B1	16	19	22	25	29	32	36	42	13	15	18	20	23	25	28	32	1.7	1.9	2.1	2.3	2.4	2.0	2.3	2.5	2.7	2.9
TW B2	15	19	23	27	33	37	42	50	11	14	18	21	25	28	32	38	1.5	1.7	1.8	2.0	2.1	1.7	1.9	2.1	2.3	2.5
TW B3	14	18	23	27	33	37	43	51	11	14	18	21	25	29	33	40	1.5	1.6	1.8	1.9	2.1	1.7	1.9	2.1	2.2	2.4
TW B4	13	17	22	26	32	36	41	49	10	13	17	20	25	28	32	39	1.5	1.7	1.8	2.0	2.1	1.7	1.9	2.1	2.3	2.4
TW C1	14	18	21	25	29	32	37	44	12	14	17	20	23	26	29	34	1.7	1.9	2.0	2.2	2.3	1.9	2.1	2.3	2.5	2.7
TW C2	13	17	21	25	30	34	39	46	11	13	17	20	24	26	30	36	1.6	1.8	1.9	2.1	2.2	1.8	2.0	2.2	2.4	2.5
TW C3	13	16	21	24	29	32	37	44	10	13	16	19	23	26	30	35	1.6	1.8	2.0	2.1	2.2	1.9	2.1	2.2	2.4	2.6
TW C4	13	16	19	22	27	30	34	40	11	13	16	18	22	24	27	33	1.8	1.9	2.1	2.2	2.4	2.0	2.2	2.4	2.6	2.7
TW D1	10	13	16	19	22	25	28	34	8	9	11	13	15	17	18	21	1.9	2.1	2.3	2.5	2.7	2.3	2.7	3.1	3.5	3.9
TW D2	16	22	28	33	41	46	52	61	12	17	22	26	32	36	42	50	1.3	1.5	1.6	1.7	1.9	1.5	1.6	1.8	2.0	2.1
TW D3	17	24	31	36	44	50	57	66	13	18	24	28	35	40	46	54	1.2	1.4	1.5	1.7	1.8	1.4	1.5	1.7	1.9	2.0
TW D4	19	26	35	41	50	57	64	74	14	21	28	34	42	47	54	64	1.1	1.3	1.4	1.6	1.7	1.2	1.4	1.5	1.7	1.8
TW D5	19	26	35	42	51	58	65	75	14	21	29	34	43	49	56	66	1.1	1.3	1.4	1.6	1.7	1.2	1.4	1.5	1.7	1.8
TW E1	20	27	35	41	50	56	63	72	16	21	28	33	41	46	52	61	1.2	1.3	1.4	1.6	1.7	1.3	1.4	1.6	1.8	1.9
TW E2	20	27	35	42	51	57	64	73	15	21	28	34	42	47	54	63	1.1	1.3	1.4	1.6	1.7	1.3	1.4	1.6	1.7	1.9
TW E3	24	33	44	51	61	68	75	84	19	27	36	43	52	58	66	76	1.0	1.1	1.3	1.5	1.6	1.1	1.2	1.4	1.5	1.7
TW E4	22	31	41	49	59	65	73	82	17	25	34	41	50	57	64	74	1.0	1.2	1.3	1.5	1.6	1.1	1.3	1.4	1.6	1.7
TW F1	15	17	21	23	27	30	34	40	12	14	16	18	21	23	26	29	1.8	2.0	2.2	2.3	2.5	2.1	2.4	2.6	2.9	3.1
TW F2	13	17	22	25	31	35	40	47	10	13	16	19	23	26	30	35	1.5	1.7	1.9	2.0	2.2	1.8	2.0	2.2	2.4	2.6
TW F3	12	16	22	26	32	36	42	50	9	12	16	19	24	27	31	37	1.5	1.6	1.8	2.0	2.1	1.7	1.9	2.1	2.3	2.4
TW F4	10	15	20	24	30	35	41	49	7	10	14	18	22	26	31	37	1.5	1.6	1.8	1.9	2.1	1.7	1.9	2.1	2.2	2.4
TW G1	13	16	19	22	26	30	34	40	10	12	15	17	20	22	25	30	1.7	1.9	2.1	2.3	2.4	2.0	2.3	2.5	2.7	2.9
TW G2	11	14	19	23	28	32	37	44	8	11	14	17	21	24	28	33	1.6	1.8	1.9	2.1	2.2	1.8	2.0	2.2	2.4	2.6
TW G3	9	13	18	21	27	31	36	44	6	9	13	16	20	23	27	33	1.6	1.7	1.9	2.1	2.2	1.8	2.0	2.2	2.4	2.5
TW G4	7	11	15	18	23	27	32	40	4	7	10	13	17	20	24	30	1.7	1.8	2.0	2.1	2.3	1.9	2.1	2.3	2.4	2.6
TW H1	10	13	16	19	24	27	32	38	7	9	12	14	18	20	24	28	1.7	1.9	2.1	2.3	2.4	2.0	2.2	2.5	2.7	2.8
TW H2	7	10	14	18	22	26	31	38	5	7	10	13	16	19	23	28	1.7	1.9	2.1	2.2	2.3	2.0	2.2	2.4	2.6	2.7
TW H3	6	9	12	15	20	23	28	34	4	6	9	11	14	17	21	26	1.8	2.0	2.1	2.3	2.4	2.1	2.3	2.4	2.6	2.8
TW H4	4	7	10	12	16	19	22	28	3	4	7	9	12	14	17	21	2.0	2.2	2.4	2.5	2.6	2.3	2.5	2.7	2.8	3.0
TW I1	8	11	14	16	19	22	25	30	6	7	9	11	13	14	16	18	1.9	2.2	2.4	2.6	2.8	2.4	2.8	3.2	3.7	4.1
TW I2	13	19	25	30	37	42	49	58	10	14	19	23	29	33	39	46	1.3	1.5	1.6	1.8	1.9	1.5	1.7	1.8	2.0	2.1
TW I3	14	20	27	33	41	46	53	62	10	15	21	25	32	37	43	51	1.3	1.4	1.6	1.7	1.9	1.4	1.6	1.7	1.9	2.0
TW I4	16	23	32	38	47	53	61	71	12	18	25	30	38	44	51	60	1.2	1.3	1.4	1.6	1.7	1.3	1.4	1.6	1.7	1.9
TW I5	16	24	33	39	49	55	63	73	12	18	26	32	40	46	53	63	1.1	1.3	1.4	1.6	1.7	1.2	1.4	1.5	1.7	1.8
TW J1	18	25	33	39	47	53	61	70	13	19	26	31	38	43	50	59	1.2	1.3	1.5	1.6	1.8	1.3	1.5	1.6	1.8	1.9
TW J2	17	24	33	39	48	54	61	71	13	19	26	31	39	44	51	61	1.2	1.3	1.4	1.6	1.7	1.3	1.4	1.6	1.7	1.9
TW J3	21	30	41	48	58	65	73	82	16	24	33	40	50	56	64	74	1.0	1.2	1.3	1.5	1.6	1.1	1.3	1.4	1.6	1.7
TW J4	19	28	38	46	56	63	71	80	14	22	31	38	48	54	62	72	1.0	1.2	1.3	1.5	1.6	1.1	1.3	1.4	1.6	1.7
DG A1	10	13	17	19	23	25	27	29	7	9	11	13	15	16	17	17	1.7	2.0	2.5	3.3	4.3	2.1	2.6	3.4	-	-
DG A2	13	17	22	26	32	36	41	47	10	13	17	20	25	28	32	37	1.5	1.6	1.8	2.1	2.4	1.7	1.9	2.1	2.4	2.7
DG A3	17	22	29	35	43	49	56	65	13	18	24	28	35	40	47	55	1.3	1.4	1.5	1.7	1.9	1.4	1.6	1.7	1.8	2.0
DG B1	10	13	17	20	24	27	31	35	7	9	11	13	15	17	19	21	1.7	2.0	2.2	2.5	2.9	2.1	2.5	3.1	3.6	4.6
DG B2	13	18	23	27	34	39	45	53	10	14	18	21	26	29	34	40	1.5	1.6	1.7	1.9	2.1	1.7	1.9	2.0	2.2	2.4
DG B3	17	23	30	36	45	51	60	70	14	19	25	29	37	42	49	59	1.3	1.4	1.4	1.6	1.7	1.4	1.5	1.6	1.7	1.9
DG C1	13	17	21	25	31	35	40	46	9	12	15	18	21	23	26	30	1.5	1.7	1.8	2.1	2.3	1.8	2.1	2.5	2.7	3.1
DG C2	16	21	27	32	40	46	53	62	12	16	21	25	31	35	41	49	1.4	1.5	1.6	1.7	1.9	1.5	1.7	1.8	2.0	2.1
DG C3	19	26	34	41	51	58	67	77	16	21	28	33	42	48	56	66	1.2	1.3	1.4	1.5	1.7	1.3	1.4	1.5	1.6	1.7
SS A1	19	24	29	32	38	42	48	55	16	20	24	27	32	35	39	45	1.4	1.6	1.8	1.9	2.1	1.6	1.8	2.0	2.2	2.4
SS A2	24	31	38	44	52	57	64	73	20	26	33	38	45	50	56	64	1.1	1.3	1.5	1.6	1.8	1.2	1.4	1.6	1.8	1.9
SS A3	17	21	25	29	33	37	41	48	15	18	21	24	27	30	33	37	1.5	1.8	1.9	2.1	2.3	1.7	2.0	2.3	2.6	2.8
SS A4	23	29	37	42	50	55	62	70	20	25	31	36	43	47	53	61	1.2	1.3	1.5	1.6	1.8	1.3	1.5	1.6	1.8	2.0
SS A5	19	23	27	30	35	38	42	49	17	20	23	25	29	31	35	39	1.6	1.8	1.9	2.1	2.3	1.7	2.0	2.3	2.5	2.7
SS A6	23	30	38	43	51	57	63	72	20	26	32	37	44	48	55	63	1.2	1.3	1.5	1.6	1.8	1.3	1.4	1.6	1.8	1.9
SS A7	17	20	23	26	29	32	35	40	15	17	20	22	24	26	28	31	1.8	2.0	2.3	2.5	2.6	2.0	2.4	2.7	3.1	3.4
SS A8	22	29	36	41	48	53	60	68	19	24	30	35	41	45	51	59	1.2	1.4	1.5	1.7	1.8	1.3	1.5	1.7	1.9	2.0
SS B1	16	20	24	28	33	37	42	49	14	17	20	23	27	30	34	39	1.5	1.7	1.9	2.1	2.2	1.7	1.9	2.2	2.4	2.6
SS B2	21	27	34	39	47	53	59	68	18	23	29	34	40	45	51	59	1.2	1.4	1.5	1.7	1.8	1.3	1.5	1.6	1.8	2.0
SS B3	15	18	22	25	30	33	38	44	12	15	18	21	24	26	29	34	1.6	1.8	2.0	2.2	2.3	1.8	2.1	2.4	2.7	2.9
SS B4	20	26	33	38	45	51	57	66	17	22	28	32	39	43	49	57	1.2	1.4	1.5	1.7	1.8	1.3	1.5	1.7	1.9	2.0
SS B5	16	19	22	25	29	32	36	41	14	16	19	21	24	26	29	33	1.7	1.9	2.1	2.3	2.5	2.0	2.2	2.5	2.8	3.0
SS B6	20	26	33	38	46	51	58	67	17	22	28	32	39	44	49	58	1.2	1.4	1.5	1.7	1.8	1.3	1.5	1.7	1.8	2.0
SS B7	14	17	19	22	25	27	30	35	12	14	16	18	20	22	24	27	1.9	2.2	2.4	2.6	2.8	2.2	2.5	2.9	3.2	3.6
SS B8	19	25	31	36	43	48	55	64	16	21	26	30	37	41	46	54	1.3	1.4	1.6	1.7	1.9	1.4	1.6	1.7	1.9	2.1
SS C1	14	18	24	28	34	38	44	52	10	13	17	20	25	28	32	38	1.4	1.6	1.8	1.9	2.1	1.7	1.9	2.1	2.3	2.5
SS C2	17	24	31	37	45	50	58	67	13	18	24	29	36	40	47	55	1.2	1.4	1.5	1.7	1.8	1.4	1.5	1.7	1.9	2.0
SS C3	13	16	20	23	28	31	36	42	10	13	16	18	21	24	27	31	1.7	1.9	2.0	2.2	2.3	2.0	2.2	2.4	2.7	2.8
SS C4	16	21	28	33	40	45	52	61	12	17	22	26	32	36	41	49	1.3	1.5	1.6	1.8	1.9	1.5	1.6	1.8	2.0	2.1
SS D1	21	27	33	38	45	50	56	64	18	22	28	31	37	41	46	52	1.2	1.4	1.6	1.7	1.9	1.4	1.6	1.8	2.0	2.2
SS D2	26	35	44	51	60	66	73	82	22	30	38	44	52	57	64	73	1.0	1.2	1.3	1.5	1.6	1.1	1.3	1.4	1.6	1.8
SS D3	20	25	30	35	41	45	50	58	18	21	26	29	33	37	41	47	1.4	1.6	1.7	1.9	2.0	1.5	1.8	2.0	2.2	2.4
SS D4	26	34	43	49	58	64	71	79	22	29	36	42	50	55	62	71	1.1	1.2	1.4	1.5	1.7	1.1	1.3	1.5	1.6	1.8
SS E1	18	22	28	32	39	43	49	57	14	18	22	25	30	33	37	43	1.4	1.5	1.7	1.9	2.0	1.5	1.8	2.0	2.2	2.4
SS E2	22	30	38	44	53	59	66	75	18	24	31	37	44	49	56	65	1.1	1.3	1.4	1.6	1.7	1.2	1.4	1.6	1.7	1.9
SS E3	17	20	25	28	33	36	41	47	14	17	20	23	26	29	32	37	1.6	1.8	1.9	2.1	2.3	1.8	2.0	2.3	2.5	2.7
SS E4	21	27	35	41	49	54	61	70	17	23	29	34	41	46	52	60	1.2	1.3	1.5	1.6	1.8	1.3	1.5	1.6	1.8	1.9

SCOTTSBLUFF, NEBRASKA

TYPE	SOLAR SAVINGS FRACTION (%)																CONSERVATION FACTOR									
BASE TEMP:	55 F (4314 DD)								65 F (6782 DD)								55 F					65 F				
LCR	100	70	50	40	30	25	20	15	100	70	50	40	30	25	20	15	100	50	30	20	15	100	50	30	20	15
WW A1	20	24	29	32	38	41	46	54	16	19	22	24	28	31	35	40	1.5	1.7	1.8	2.0	2.1	1.8	2.0	2.2	2.4	2.5
WW A2	22	28	35	40	48	53	60	69	16	20	26	30	36	41	46	54	1.2	1.4	1.5	1.7	1.8	1.4	1.6	1.7	1.9	2.0
WW A3	22	29	38	44	52	58	65	74	15	21	27	32	39	44	51	59	1.1	1.3	1.4	1.6	1.7	1.3	1.5	1.6	1.8	1.9
WW A4	22	30	39	45	54	60	68	77	15	21	28	34	41	46	53	62	1.1	1.2	1.4	1.5	1.7	1.3	1.4	1.6	1.7	1.9
WW A5	23	31	41	47	57	63	70	79	15	22	29	35	43	49	56	65	1.1	1.2	1.4	1.5	1.7	1.2	1.4	1.5	1.7	1.8
WW A6	23	32	41	48	58	64	72	81	15	22	30	36	44	50	57	67	1.0	1.2	1.3	1.5	1.6	1.2	1.3	1.5	1.7	1.8
WW B1	17	22	27	31	37	42	47	55	12	15	19	21	25	28	32	37	1.4	1.6	1.7	1.9	2.0	1.7	2.0	2.2	2.4	2.6
WW B2	23	32	41	48	57	63	70	79	16	23	30	36	44	50	57	67	1.1	1.2	1.4	1.5	1.7	1.2	1.4	1.5	1.7	1.8
WW B3	28	37	48	55	65	71	78	86	20	28	37	43	53	59	66	75	1.0	1.1	1.3	1.4	1.6	1.1	1.2	1.4	1.6	1.7
WW B4	28	38	50	58	68	74	81	89	20	29	39	46	56	63	70	80	0.9	1.1	1.2	1.4	1.5	1.0	1.2	1.3	1.5	1.6
WW B5	27	37	49	57	67	74	81	89	19	28	38	46	56	63	71	80	0.9	1.1	1.2	1.4	1.5	1.0	1.2	1.3	1.5	1.6
WW C1	30	39	50	58	67	73	80	88	21	30	39	46	55	61	69	78	0.9	1.1	1.3	1.4	1.6	1.1	1.2	1.4	1.5	1.7
WW C2	28	38	49	56	66	73	80	87	20	28	38	45	55	61	69	78	1.0	1.1	1.3	1.4	1.6	1.1	1.2	1.4	1.5	1.7
WW C3	33	45	57	65	76	82	88	94	24	35	47	55	65	72	79	87	0.9	1.0	1.2	1.3	1.5	0.9	1.1	1.2	1.4	1.6
WW C4	30	41	54	62	72	79	85	92	21	32	43	51	62	69	76	85	0.9	1.0	1.2	1.4	1.5	1.0	1.1	1.3	1.4	1.6
TW A1	20	24	28	32	37	41	46	53	16	18	22	24	28	31	35	40	1.5	1.7	1.8	2.0	2.1	1.8	2.0	2.2	2.4	2.5
TW A2	20	26	32	37	44	49	55	64	15	19	24	28	33	37	42	50	1.3	1.4	1.6	1.7	1.9	1.5	1.7	1.8	2.0	2.1
TW A3	20	26	34	39	47	52	59	68	14	19	25	29	35	40	46	54	1.2	1.4	1.5	1.7	1.8	1.4	1.6	1.7	1.9	2.0
TW A4	19	26	34	39	47	53	60	69	13	18	24	29	36	40	47	55	1.2	1.3	1.5	1.6	1.8	1.4	1.5	1.7	1.8	2.0
TW B1	19	23	28	32	38	42	48	55	15	18	22	25	29	32	36	42	1.5	1.6	1.8	1.9	2.0	1.7	1.9	2.1	2.2	2.4
TW B2	19	24	31	36	43	48	54	63	14	18	23	27	32	37	42	50	1.3	1.4	1.6	1.7	1.9	1.5	1.7	1.8	2.0	2.1
TW B3	18	24	31	36	44	49	55	64	13	18	23	27	33	37	43	51	1.3	1.4	1.6	1.7	1.8	1.5	1.6	1.8	1.9	2.1
TW B4	17	23	29	34	41	47	53	62	12	17	22	26	31	36	41	49	1.3	1.4	1.6	1.7	1.9	1.5	1.7	1.8	2.0	2.1
TW C1	18	22	28	32	38	42	48	56	14	17	21	24	29	32	37	44	1.4	1.6	1.7	1.9	2.0	1.7	1.8	2.0	2.2	2.3
TW C2	17	22	28	32	39	44	50	59	13	16	21	24	30	34	39	46	1.4	1.5	1.7	1.8	1.9	1.6	1.7	1.9	2.1	2.2
TW C3	16	21	27	31	37	42	48	56	12	16	20	23	29	32	37	44	1.4	1.6	1.7	1.9	2.0	1.6	1.8	1.9	2.1	2.2
TW C4	16	20	24	28	34	38	43	51	12	15	19	22	26	29	34	40	1.5	1.7	1.8	2.0	2.1	1.8	1.9	2.1	2.2	2.4
TW D1	14	19	23	27	33	37	42	49	10	13	16	18	22	25	29	34	1.5	1.7	1.8	2.0	2.1	1.8	2.1	2.3	2.5	2.7
TW D2	22	29	37	43	52	58	65	74	16	21	28	33	41	46	52	61	1.1	1.3	1.4	1.6	1.7	1.3	1.4	1.6	1.7	1.9
TW D3	24	32	41	47	56	62	70	79	17	23	31	36	44	50	57	66	1.1	1.2	1.4	1.5	1.7	1.2	1.4	1.5	1.7	1.8
TW D4	26	35	45	52	62	68	75	84	18	26	35	41	50	57	64	74	1.0	1.1	1.3	1.5	1.6	1.1	1.3	1.4	1.6	1.7
TW D5	25	34	45	52	62	68	76	84	18	26	35	42	51	58	65	75	1.0	1.1	1.3	1.4	1.6	1.1	1.2	1.4	1.6	1.7
TW E1	27	36	45	52	62	68	75	83	20	27	35	41	50	56	63	72	1.0	1.2	1.3	1.5	1.6	1.1	1.3	1.4	1.6	1.7
TW E2	26	35	45	53	62	68	76	84	19	27	35	42	51	57	64	73	1.0	1.1	1.3	1.5	1.6	1.1	1.3	1.4	1.6	1.7
TW E3	32	43	54	62	72	78	85	91	24	33	44	51	61	68	75	84	0.9	1.1	1.2	1.4	1.5	1.0	1.1	1.3	1.5	1.6
TW E4	29	40	51	59	70	76	83	90	21	31	41	49	59	65	73	82	0.9	1.1	1.2	1.4	1.5	1.0	1.2	1.3	1.5	1.6
TW F1	18	22	27	31	36	40	46	53	14	17	20	23	27	30	34	40	1.5	1.6	1.8	2.0	2.1	1.8	2.0	2.1	2.3	2.5
TW F2	18	23	30	34	41	46	53	61	13	17	21	25	31	35	40	47	1.3	1.5	1.6	1.8	1.9	1.5	1.7	1.9	2.0	2.2
TW F3	17	23	30	35	43	48	55	64	11	16	21	25	31	36	42	50	1.3	1.4	1.6	1.7	1.8	1.5	1.6	1.8	1.9	2.1
TW F4	15	21	28	34	41	47	54	63	9	14	20	24	30	35	41	49	1.3	1.4	1.6	1.7	1.8	1.5	1.6	1.8	1.9	2.1
TW G1	16	21	26	30	36	40	45	53	12	15	19	22	26	29	34	40	1.5	1.6	1.8	1.9	2.0	1.7	1.9	2.1	2.3	2.4
TW G2	15	20	26	31	38	43	49	58	10	14	19	22	28	32	37	45	1.3	1.5	1.6	1.8	1.9	1.6	1.7	1.9	2.1	2.2
TW G3	13	19	25	30	37	42	48	58	8	12	17	21	27	31	36	44	1.3	1.5	1.6	1.8	1.9	1.6	1.7	1.9	2.0	2.2
TW G4	11	15	21	26	32	37	44	52	6	10	14	18	23	27	32	40	1.4	1.6	1.7	1.9	2.0	1.7	1.8	2.0	2.1	2.3
TW H1	13	17	23	27	33	37	43	51	9	12	16	19	24	27	32	38	1.5	1.6	1.8	1.9	2.1	1.7	1.9	2.1	2.2	2.4
TW H2	11	15	20	25	31	36	42	50	6	10	14	17	22	26	30	38	1.5	1.6	1.8	1.9	2.0	1.7	1.9	2.0	2.2	2.3
TW H3	9	13	18	22	28	32	37	46	5	8	12	15	19	23	27	34	1.6	1.7	1.8	2.0	2.1	1.8	2.0	2.1	2.3	2.4
TW H4	7	10	14	17	22	26	31	38	4	6	9	11	15	18	22	28	1.7	1.9	2.0	2.2	2.3	2.1	2.2	2.3	2.6	[illegible]
TW I1	12	16	21	24	30	33	38	46	8	10	13	16	19	22	25	30	1.6	1.7	1.9	2.1	2.2	1.9	2.1	2.4	2.6	2.8
TW I2	19	26	34	40	48	54	62	71	13	18	25	30	37	42	49	58	1.2	1.3	1.5	1.6	1.7	1.3	1.5	1.6	1.8	1.9
TW I3	20	28	37	43	52	58	66	75	14	20	27	33	41	46	53	63	1.1	1.3	1.4	1.6	1.7	1.3	1.4	1.6	1.7	1.9
TW I4	22	31	41	48	58	65	72	81	15	23	31	38	47	53	61	71	1.0	1.2	1.3	1.5	1.6	1.2	1.3	1.4	1.6	1.7
TW I5	23	32	42	50	60	66	74	83	16	23	32	39	49	55	63	73	1.0	1.2	1.3	1.5	1.6	1.1	1.3	1.4	1.6	1.7
TW J1	25	33	43	50	59	66	73	82	17	25	33	39	48	54	61	70	1.0	1.2	1.3	1.5	1.6	1.2	1.3	1.5	1.6	1.8
TW J2	24	32	42	49	59	66	73	82	17	24	32	39	48	54	61	71	1.0	1.2	1.3	1.5	1.6	1.1	1.3	1.4	1.6	1.7
TW J3	28	39	51	59	69	76	82	90	21	30	41	48	59	65	73	82	0.9	1.1	1.2	1.4	1.5	1.0	1.2	1.3	1.5	1.6
TW J4	26	37	48	56	67	73	81	88	19	28	38	46	56	63	71	80	0.9	1.1	1.2	1.4	1.5	1.0	1.2	1.3	1.5	1.6
DG A1	15	20	25	29	34	37	40	45	10	13	16	19	22	25	27	29	1.4	1.6	2.0	2.4	2.9	1.7	2.0	2.5	3.2	4.3
DG A2	17	23	30	36	43	48	53	60	12	17	22	26	32	36	41	47	1.3	1.4	1.6	1.8	2.1	1.5	1.6	1.8	2.1	2.4
DG A3	21	29	38	45	54	60	67	76	16	22	29	35	43	49	56	65	1.1	1.2	1.4	1.6	1.8	1.3	1.4	1.5	1.7	1.9
DG B1	15	20	26	31	38	42	47	53	10	13	17	20	24	27	31	35	1.4	1.5	1.7	2.0	2.3	1.7	1.9	2.2	2.5	2.9
DG B2	18	24	32	38	46	52	59	68	13	17	23	27	34	39	45	53	1.2	1.3	1.5	1.7	1.9	1.5	1.6	1.7	1.9	2.1
DG B3	22	30	39	46	57	64	72	81	16	23	30	36	45	52	60	70	1.1	1.2	1.3	1.5	1.7	1.3	1.4	1.4	1.6	1.7
DG C1	19	24	32	37	46	51	58	65	12	16	21	25	31	35	40	47	1.3	1.4	1.5	1.7	2.0	1.5	1.7	1.9	2.1	2.3
DG C2	21	28	37	43	53	60	68	76	15	20	27	32	40	46	53	62	1.2	1.2	1.4	1.5	1.7	1.3	1.5	1.6	1.7	1.9
DG C3	25	33	44	52	64	71	79	87	19	26	34	41	51	58	67	77	1.1	1.1	1.2	1.4	1.6	1.2	1.3	1.4	1.5	1.7
SS A1	24	30	37	42	49	54	60	68	20	24	30	34	40	44	49	56	1.2	1.4	1.5	1.7	1.8	1.4	1.6	1.7	1.9	2.1
SS A2	30	38	47	54	63	68	75	83	25	32	40	45	53	59	65	74	1.0	1.2	1.3	1.5	1.6	1.1	1.3	1.4	1.6	1.8
SS A3	22	27	33	38	44	48	54	62	18	22	27	30	35	39	43	50	1.3	1.5	1.6	1.8	2.0	1.5	1.7	1.9	2.1	2.3
SS A4	29	37	46	52	61	66	73	81	24	31	38	44	51	57	63	71	1.0	1.2	1.4	1.5	1.7	1.1	1.3	1.5	1.7	1.8
SS A5	23	28	34	38	45	49	54	62	19	23	28	31	36	40	44	51	1.3	1.5	1.7	1.8	2.0	1.5	1.7	1.9	2.1	2.2
SS A6	30	38	47	53	62	68	74	82	24	31	39	45	53	58	64	73	1.0	1.2	1.3	1.5	1.6	1.1	1.3	1.5	1.6	1.8
SS A7	21	25	30	33	38	42	47	53	18	21	24	27	31	33	37	42	1.5	1.7	1.8	2.0	2.2	1.7	1.9	2.2	2.4	2.6
SS A8	29	36	45	51	59	65	71	79	23	30	37	42	50	55	61	69	1.1	1.2	1.4	1.5	1.7	1.2	1.3	1.5	1.7	1.8
SS B1	21	26	32	37	43	48	54	62	16	20	25	29	34	38	43	50	1.3	1.5	1.6	1.8	1.9	1.5	1.7	1.8	2.0	2.2
SS B2	26	34	43	49	57	63	70	79	21	28	35	41	48	54	60	69	1.1	1.2	1.4	1.5	1.7	1.2	1.3	1.5	1.7	1.8
SS B3	19	24	30	34	40	44	50	58	15	19	23	26	31	35	39	45	1.4	1.5	1.7	1.9	2.0	1.6	1.8	2.0	2.2	2.3
SS B4	26	33	41	47	56	61	68	77	20	27	34	39	47	52	58	67	1.1	1.2	1.4	1.6	1.7	1.2	1.4	1.5	1.7	1.8
SS B5	19	23	28	32	38	41	47	54	16	19	23	25	30	33	37	43	1.5	1.6	1.8	2.0	2.1	1.7	1.9	2.1	2.3	2.4
SS B6	26	33	42	48	56	62	69	78	20	27	34	39	47	52	59	68	1.1	1.2	1.4	1.6	1.7	1.2	1.4	1.5	1.7	1.8
SS B7	17	21	25	28	33	37	41	48	14	17	20	22	26	28	32	37	1.6	1.8	1.9	2.1	2.3	1.8	2.1	2.3	2.5	2.7
SS B8	24	32	40	46	54	59	66	75	19	25	32	37	45	50	56	65	1.1	1.3	1.4	1.6	1.7	1.2	1.4	1.6	1.7	1.9
SS C1	20	26	33	39	46	52	59	68	13	18	24	28	34	38	44	52	1.2	1.4	1.5	1.7	1.8	1.4	1.6	1.8	1.9	2.1
SS C2	24	32	41	48	57	63	70	79	17	23	31	37	45	50	58	67	1.1	1.2	1.4	1.5	1.7	1.2	1.4	1.5	1.7	1.8
SS C3	17	22	27	31	37	42	48	56	13	16	20	23	28	31	36	42	1.4	1.6	1.7	1.9	2.0	1.7	1.8	2.0	2.2	2.3
SS C4	21	28	36	42	51	57	64	73	15	21	27	33	40	45	52	61	1.1	1.3	1.4	1.6	1.7	1.3	1.5	1.6	1.8	1.9
SS D1	28	35	44	49	58	63	69	78	22	28	35	40	47	52	58	66	1.1	1.2	1.4	1.6	1.7	1.2	1.4	1.6	1.8	1.9
SS D2	34	44	55	62	71	77	83	90	28	36	46	53	62	67	74	82	0.9	1.1	1.2	1.4	1.6	1.0	1.2	1.3	1.5	1.6
SS D3	26	32	39	44	52	57	63	71	21	26	32	36	42	46	52	59	1.2	1.3	1.5	1.7	1.8	1.3	1.5	1.7	1.9	2.0
SS D4	33	43	53	60	69	75	81	88	27	35	44	51	59	65	72	80	0.9	1.1	1.3	1.4	1.6	1.0	1.2	1.4	1.5	1.7
SS E1	24	31	38	43	51	56	63	71	18	23	29	34	40	44	50	58	1.1	1.3	1.5	1.6	1.8	1.3	1.5	1.7	1.9	2.0
SS E2	29	39	48	55	65	71	77	85	23	31	39	45	54	60	67	75	1.0	1.1	1.3	1.5	1.6	1.1	1.2	1.4	1.6	1.7
SS E3	21	26	32	37	43	47	53	61	17	21	25	29	34	37	42	49	1.3	1.5	1.7	1.8	2.0	1.5	1.7	1.9	2.1	2.3
SS E4	27	36	45	51	60	66	73	81	21	28	36	42	50	55	62	71	1.0	1.2	1.3	1.5	1.7	1.1	1.3	1.5	1.6	1.8

ELKO, NEVADA

SOLAR SAVINGS FRACTION (%) — CONSERVATION FACTOR

TYPE	BASE TEMP: 55 F (4725 DD)								65 F (7490 DD)								CF 55 F					CF 65 F				
LCR	100	70	50	40	30	25	20	15	100	70	50	40	30	25	20	15	100	50	30	20	15	100	50	30	20	15
WW A1	22	26	31	34	40	44	49	56	17	20	24	26	31	34	38	43	1.4	1.6	1.8	1.9	2.1	1.7	1.9	2.1	2.3	2.4
WW A2	24	30	38	43	51	56	62	71	17	22	28	33	39	44	50	57	1.2	1.3	1.5	1.7	1.8	1.4	1.5	1.7	1.9	2.0
WW A3	24	32	40	46	55	60	67	75	17	23	30	35	43	48	54	62	1.1	1.3	1.4	1.6	1.7	1.3	1.4	1.6	1.8	1.9
WW A4	25	33	42	48	57	62	69	78	17	24	31	36	44	50	56	65	1.1	1.2	1.4	1.6	1.7	1.2	1.4	1.5	1.7	1.9
WW A5	25	34	43	50	59	65	72	80	17	24	32	38	46	52	59	68	1.0	1.2	1.4	1.5	1.7	1.2	1.3	1.5	1.7	1.8
WW A6	25	34	44	51	60	66	73	81	17	24	33	39	47	53	60	69	1.0	1.2	1.3	1.5	1.7	1.1	1.3	1.5	1.6	1.8
WW B1	19	24	30	34	40	45	50	57	13	17	21	24	29	32	36	42	1.3	1.5	1.7	1.9	2.1	1.6	1.8	2.0	2.3	2.5
WW B2	25	34	43	50	59	65	72	80	18	25	33	39	47	53	60	69	1.0	1.2	1.3	1.5	1.7	1.2	1.3	1.5	1.6	1.8
WW B3	30	40	50	58	67	73	79	87	22	30	40	47	56	62	69	77	0.9	1.1	1.3	1.4	1.6	1.1	1.2	1.4	1.5	1.7
WW B4	30	41	52	60	70	75	82	89	22	31	42	49	59	65	73	81	0.9	1.1	1.2	1.4	1.6	1.0	1.2	1.3	1.5	1.6
WW B5	29	40	51	59	69	75	82	89	21	30	41	48	59	65	73	82	0.9	1.1	1.2	1.4	1.6	1.0	1.2	1.3	1.5	1.6
WW C1	32	42	53	60	69	75	81	88	24	32	42	49	58	64	71	80	0.9	1.1	1.3	1.4	1.6	1.0	1.2	1.3	1.5	1.7
WW C2	30	40	51	58	68	74	81	88	22	31	40	48	57	63	71	80	0.9	1.1	1.3	1.4	1.6	1.0	1.2	1.3	1.5	1.7
WW C3	35	47	59	67	77	82	88	94	27	38	49	58	68	74	81	89	0.9	1.0	1.2	1.4	1.5	0.9	1.1	1.2	1.4	1.6
WW C4	32	44	56	64	74	80	86	92	24	34	46	54	64	71	78	86	0.9	1.0	1.2	1.4	1.5	1.0	1.1	1.3	1.4	1.6
TW A1	21	25	30	34	40	43	49	56	17	20	23	26	30	33	38	43	1.4	1.6	1.8	1.9	2.1	1.7	1.9	2.1	2.3	2.4
TW A2	22	28	34	39	47	51	58	66	16	21	26	30	36	40	46	53	1.2	1.4	1.6	1.7	1.9	1.4	1.6	1.8	1.9	2.1
TW A3	22	29	36	42	49	55	61	70	16	21	27	32	38	43	49	57	1.2	1.3	1.5	1.7	1.8	1.4	1.5	1.7	1.8	2.0
TW A4	21	28	36	42	50	55	62	71	15	20	27	32	39	44	50	58	1.1	1.3	1.5	1.6	1.8	1.3	1.5	1.6	1.8	2.0
TW B1	21	25	30	34	40	45	50	58	16	19	23	26	31	35	39	46	1.4	1.6	1.7	1.9	2.0	1.6	1.8	2.0	2.2	2.3
TW B2	21	26	33	38	46	50	57	65	15	20	25	29	35	39	45	53	1.2	1.4	1.6	1.7	1.9	1.4	1.6	1.8	1.9	2.1
TW B3	20	26	33	39	46	51	58	66	14	19	25	29	36	40	46	54	1.2	1.4	1.5	1.7	1.8	1.4	1.6	1.7	1.9	2.0
TW B4	19	25	31	37	44	49	56	64	14	18	24	28	34	38	44	52	1.3	1.4	1.6	1.7	1.9	1.4	1.6	1.8	1.9	2.1
TW C1	19	24	30	34	40	45	51	58	15	18	23	26	31	35	40	47	1.4	1.5	1.7	1.9	2.0	1.6	1.8	1.9	2.1	2.2
TW C2	19	24	30	35	42	46	52	61	14	18	23	26	32	36	41	49	1.3	1.5	1.6	1.8	1.9	1.5	1.7	1.8	2.0	2.1
TW C3	18	23	29	33	40	44	50	59	13	17	22	25	31	35	40	47	1.4	1.5	1.7	1.8	2.0	1.6	1.7	1.9	2.0	2.2
TW C4	17	21	26	30	36	40	46	53	13	16	20	23	28	32	36	43	1.5	1.6	1.8	1.9	2.1	1.7	1.9	2.0	2.2	2.3
TW D1	16	21	26	30	36	40	45	52	11	14	18	21	25	28	32	38	1.4	1.6	1.8	2.0	2.1	1.7	1.9	2.1	2.3	2.5
TW D2	24	31	40	46	54	60	67	75	17	23	30	36	43	49	55	64	1.1	1.3	1.4	1.6	1.7	1.2	1.4	1.6	1.7	1.9
TW D3	26	34	43	50	59	64	71	80	19	26	33	39	48	53	60	69	1.0	1.2	1.4	1.5	1.7	1.2	1.3	1.5	1.7	1.8
TW D4	28	37	47	54	64	70	77	85	20	28	37	44	53	59	67	76	1.0	1.1	1.3	1.5	1.6	1.1	1.2	1.4	1.6	1.7
TW D5	27	37	47	54	64	70	77	85	20	28	37	44	54	60	68	77	1.0	1.1	1.3	1.5	1.6	1.1	1.2	1.4	1.5	1.7
TW E1	29	38	48	55	64	70	76	84	22	29	38	44	53	59	66	75	1.0	1.1	1.3	1.5	1.6	1.1	1.3	1.4	1.6	1.7
TW E2	29	38	48	55	64	70	77	85	21	29	38	44	53	59	67	76	1.0	1.1	1.3	1.5	1.6	1.1	1.2	1.4	1.6	1.7
TW E3	34	45	56	64	74	79	85	91	26	36	46	54	64	70	77	85	0.9	1.0	1.2	1.4	1.5	1.0	1.1	1.3	1.4	1.6
TW E4	32	42	54	61	71	77	83	90	23	33	44	51	61	68	75	84	0.9	1.1	1.2	1.4	1.5	1.0	1.1	1.3	1.5	1.6
TW F1	20	24	29	33	39	43	48	56	15	18	22	25	29	33	37	43	1.4	1.6	1.8	1.9	2.1	1.7	1.9	2.0	2.2	2.4
TW F2	20	25	32	37	44	49	55	64	14	18	23	27	33	38	43	51	1.3	1.4	1.6	1.8	1.9	1.5	1.6	1.8	2.0	2.1
TW F3	19	25	32	38	45	51	57	66	13	18	23	28	34	39	45	53	1.2	1.4	1.5	1.7	1.8	1.4	1.6	1.7	1.9	2.0
TW F4	17	23	31	36	44	49	56	65	11	16	22	26	33	38	44	52	1.2	1.4	1.5	1.7	1.8	1.4	1.6	1.7	1.9	2.0
TW G1	18	22	28	32	38	42	48	56	13	16	20	24	29	32	37	43	1.4	1.6	1.7	1.9	2.0	1.7	1.8	2.0	2.2	2.3
TW G2	17	22	29	34	41	46	52	61	11	15	21	24	30	35	40	48	1.3	1.5	1.6	1.8	1.9	1.5	1.7	1.8	2.0	2.1
TW G3	15	20	27	32	39	45	51	60	10	14	19	23	29	34	39	47	1.3	1.5	1.6	1.8	1.9	1.5	1.7	1.8	2.0	2.1
TW G4	12	17	23	28	35	40	46	55	7	11	16	20	25	30	35	43	1.4	1.5	1.7	1.8	2.0	1.6	1.7	1.9	2.1	2.2
TW H1	14	19	25	29	35	40	46	54	10	13	18	21	26	30	34	41	1.4	1.6	1.7	1.9	2.0	1.7	1.8	2.0	2.2	2.3
TW H2	12	17	23	27	34	38	44	53	8	11	16	19	24	28	33	41	1.4	1.6	1.7	1.9	2.0	1.7	1.8	2.0	2.1	2.3
TW H3	10	14	20	24	30	34	40	48	6	9	13	17	21	25	30	37	1.5	1.6	1.8	2.0	2.1	1.7	1.9	2.0	2.2	2.3
TW H4	7	11	15	19	24	28	33	40	4	7	10	13	17	20	24	30	1.7	1.8	2.0	2.1	2.3	2.0	2.1	2.2	2.4	2.5
TW I1	14	18	23	27	32	36	41	49	9	12	15	18	22	25	29	34	1.5	1.7	1.8	2.0	2.2	1.8	2.0	2.2	2.4	2.6
TW I2	21	28	36	42	51	57	64	73	14	20	27	32	40	45	52	61	1.1	1.3	1.4	1.6	1.7	1.3	1.4	1.6	1.8	1.9
TW I3	22	30	39	46	55	61	68	76	16	22	30	36	44	49	56	65	1.1	1.2	1.4	1.6	1.7	1.2	1.4	1.5	1.7	1.8
TW I4	24	34	44	51	60	67	74	82	17	25	34	41	50	56	63	73	1.0	1.2	1.3	1.5	1.6	1.1	1.3	1.4	1.6	1.7
TW I5	25	34	44	52	62	68	75	84	17	26	35	42	51	58	65	75	1.0	1.1	1.3	1.5	1.6	1.1	1.2	1.4	1.6	1.7
TW J1	27	36	45	52	62	67	74	82	19	27	36	42	51	57	64	73	1.0	1.2	1.3	1.5	1.6	1.1	1.3	1.4	1.6	1.7
TW J2	26	35	45	52	61	67	74	83	18	26	35	41	51	57	64	73	1.0	1.2	1.3	1.5	1.6	1.1	1.3	1.4	1.6	1.7
TW J3	31	42	53	61	71	77	83	90	23	33	43	51	61	68	75	84	0.9	1.1	1.2	1.4	1.5	1.0	1.1	1.3	1.5	1.6
TW J4	29	39	51	59	69	75	81	89	21	30	41	48	59	65	73	82	0.9	1.1	1.2	1.4	1.6	1.0	1.2	1.3	1.5	1.6
DG A1	17	22	28	32	37	40	43	48	11	15	19	22	25	28	30	33	1.3	1.6	1.9	2.4	2.9	1.6	1.9	2.3	3.0	3.9
DG A2	19	26	33	38	45	50	56	62	14	18	24	29	35	39	44	50	1.2	1.4	1.6	1.8	2.1	1.4	1.6	1.8	2.0	2.3
DG A3	23	31	40	47	56	62	69	77	17	24	31	37	46	52	59	67	1.1	1.2	1.4	1.6	1.8	1.2	1.3	1.5	1.7	1.9
DG B1	17	23	29	34	41	45	50	56	11	15	19	23	28	31	35	40	1.3	1.4	1.7	2.0	2.3	1.6	1.8	2.0	2.4	2.8
DG B2	20	27	34	41	49	55	61	69	14	19	25	30	37	42	48	56	1.2	1.3	1.5	1.7	1.9	1.4	1.5	1.6	1.9	2.0
DG B3	24	32	42	49	59	66	73	82	18	24	32	39	48	55	63	72	1.1	1.2	1.3	1.5	1.7	1.2	1.3	1.4	1.6	1.7
DG C1	21	27	35	41	49	54	60	67	14	19	24	28	35	39	44	51	1.2	1.3	1.5	1.7	2.0	1.4	1.6	1.8	2.0	2.3
DG C2	23	31	40	47	56	62	69	77	16	22	30	35	43	49	56	65	1.1	1.2	1.4	1.6	1.8	1.3	1.4	1.5	1.7	1.9
DG C3	27	36	47	55	66	72	80	87	20	28	37	44	54	61	69	79	1.0	1.1	1.2	1.4	1.6	1.1	1.2	1.3	1.5	1.7
SS A1	27	33	40	45	52	57	63	70	22	27	33	38	44	48	54	61	1.2	1.3	1.5	1.7	1.9	1.3	1.5	1.7	1.9	2.0
SS A2	33	42	50	57	65	70	76	84	28	35	43	49	57	62	69	76	1.0	1.2	1.3	1.5	1.7	1.1	1.2	1.4	1.6	1.8
SS A3	25	30	37	41	48	52	57	64	20	25	30	34	40	43	48	54	1.2	1.4	1.6	1.8	2.0	1.4	1.6	1.8	2.0	2.2
SS A4	32	41	49	55	63	68	75	82	27	34	42	48	56	61	67	74	1.0	1.2	1.4	1.5	1.7	1.1	1.3	1.5	1.6	1.8
SS A5	26	31	37	42	48	52	57	64	22	26	31	35	40	44	49	55	1.2	1.4	1.6	1.8	2.0	1.4	1.6	1.8	2.0	2.2
SS A6	33	41	50	56	65	70	76	83	27	35	43	49	57	62	68	76	1.0	1.2	1.3	1.5	1.7	1.1	1.2	1.4	1.6	1.8
SS A7	23	28	33	36	42	45	50	56	20	23	27	30	35	38	42	47	1.4	1.6	1.8	2.0	2.2	1.5	1.8	2.0	2.3	2.5
SS A8	32	40	48	54	62	67	73	81	26	33	41	47	54	59	65	73	1.0	1.2	1.4	1.6	1.7	1.1	1.3	1.5	1.7	1.8
SS B1	23	29	35	40	46	51	57	64	18	23	28	32	38	42	47	54	1.2	1.4	1.6	1.8	1.9	1.4	1.6	1.8	2.0	2.1
SS B2	29	37	46	52	60	65	72	80	24	31	38	44	52	57	64	72	1.0	1.2	1.4	1.5	1.7	1.1	1.3	1.5	1.6	1.8
SS B3	21	27	33	37	43	48	53	60	17	21	26	30	35	39	44	50	1.3	1.5	1.7	1.9	2.0	1.4	1.6	1.9	2.1	2.3
SS B4	28	36	44	50	59	64	70	78	23	30	37	43	51	56	62	70	1.1	1.2	1.4	1.6	1.7	1.1	1.3	1.5	1.7	1.8
SS B5	21	26	31	35	41	45	50	57	17	21	25	28	33	37	41	47	1.4	1.6	1.8	1.9	2.1	1.6	1.8	2.0	2.2	2.3
SS B6	28	36	45	51	59	64	71	79	23	30	37	43	51	56	63	71	1.0	1.2	1.4	1.6	1.7	1.1	1.3	1.5	1.7	1.8
SS B7	19	23	28	31	36	40	44	51	16	19	23	25	29	32	36	41	1.5	1.7	1.9	2.1	2.3	1.7	1.9	2.1	2.4	2.6
SS B8	27	35	43	49	57	62	68	76	22	28	36	41	49	54	60	68	1.1	1.2	1.4	1.6	1.7	1.2	1.3	1.5	1.7	1.9
SS C1	22	28	36	41	49	54	61	69	15	20	26	31	37	42	48	56	1.2	1.3	1.5	1.7	1.8	1.4	1.5	1.7	1.9	2.0
SS C2	26	34	43	50	59	65	72	80	18	25	33	39	48	54	61	70	1.0	1.2	1.4	1.5	1.7	1.2	1.3	1.5	1.6	1.8
SS C3	19	23	29	34	40	44	50	58	14	17	22	25	30	34	39	46	1.4	1.5	1.7	1.9	2.0	1.6	1.8	1.9	2.1	2.3
SS C4	23	31	39	45	53	59	66	74	17	23	30	35	43	48	55	63	1.1	1.3	1.4	1.6	1.7	1.3	1.4	1.6	1.7	1.9
SS D1	31	39	47	53	60	65	71	79	25	32	39	44	51	56	62	69	1.0	1.2	1.4	1.6	1.7	1.1	1.3	1.5	1.7	1.9
SS D2	38	48	58	64	73	78	84	90	31	40	50	56	65	70	77	84	0.9	1.1	1.2	1.4	1.6	1.0	1.1	1.3	1.5	1.6
SS D3	29	35	42	48	55	60	65	73	24	29	36	40	46	51	56	63	1.1	1.3	1.5	1.7	1.8	1.2	1.4	1.6	1.8	2.0
SS D4	36	46	56	63	71	76	82	89	30	39	48	55	63	68	75	82	0.9	1.1	1.3	1.4	1.6	1.0	1.2	1.3	1.5	1.7
SS E1	27	34	41	47	54	59	65	73	21	27	33	38	44	49	54	62	1.1	1.3	1.5	1.7	1.8	1.2	1.4	1.6	1.8	2.0
SS E2	32	42	51	58	67	72	79	86	26	34	43	49	58	63	70	78	1.0	1.1	1.3	1.5	1.6	1.0	1.2	1.4	1.6	1.7
SS E3	23	29	35	40	46	51	56	64	19	23	28	32	38	42	47	53	1.2	1.4	1.6	1.8	2.0	1.4	1.6	1.8	2.0	2.2
SS E4	30	39	48	54	63	68	74	82	24	32	40	46	54	59	66	74	1.0	1.2	1.3	1.5	1.7	1.1	1.3	1.4	1.6	1.8

ELY, NEVADA

TYPE	SOLAR SAVINGS FRACTION (%) BASE TEMP: 55 F (5011 DD)								65 F (7818 DD)								CONSERVATION FACTOR 55 F					65 F				
LCR	100	70	50	40	30	25	20	15	100	70	50	40	30	25	20	15	100	50	30	20	15	100	50	30	20	15
WW A1	22	27	32	37	43	47	53	61	17	20	24	27	32	35	40	47	1.4	1.5	1.7	1.8	2.0	1.7	1.8	2.0	2.1	2.3
WW A2	25	32	40	47	55	61	68	76	18	23	30	35	42	47	53	62	1.1	1.3	1.4	1.6	1.7	1.3	1.4	1.6	1.8	1.9
WW A3	26	34	44	50	59	65	72	81	18	24	32	37	46	51	58	67	1.0	1.2	1.3	1.5	1.6	1.2	1.4	1.5	1.7	1.8
WW A4	27	35	45	52	62	68	75	83	18	25	33	39	48	53	61	70	1.0	1.2	1.3	1.5	1.6	1.2	1.3	1.5	1.6	1.8
WW A5	27	37	47	54	64	70	77	85	18	26	34	41	50	56	64	73	1.0	1.1	1.3	1.4	1.6	1.1	1.3	1.4	1.6	1.7
WW A6	28	38	48	56	66	72	79	86	18	26	35	42	51	57	65	75	1.0	1.1	1.3	1.4	1.6	1.1	1.2	1.4	1.6	1.7
WW B1	21	26	33	38	45	50	56	65	14	18	22	26	31	35	40	47	1.3	1.4	1.6	1.7	1.9	1.6	1.7	1.9	2.1	2.2
WW B2	27	37	47	54	64	70	77	85	19	26	35	41	50	57	64	74	1.0	1.1	1.3	1.4	1.6	1.1	1.3	1.4	1.6	1.7
WW B3	33	43	55	62	72	78	84	91	23	32	42	50	59	66	73	82	0.9	1.1	1.2	1.4	1.5	1.0	1.2	1.3	1.5	1.6
WW B4	33	44	56	64	74	80	86	92	23	33	44	52	62	69	76	85	0.9	1.0	1.2	1.4	1.5	1.0	1.1	1.3	1.4	1.6
WW B5	31	43	55	63	73	79	86	92	22	32	43	51	62	68	76	85	0.9	1.0	1.2	1.4	1.5	1.0	1.1	1.3	1.4	1.6
WW C1	34	45	57	64	74	80	86	92	25	34	45	52	62	68	76	84	0.9	1.0	1.2	1.4	1.5	1.0	1.1	1.3	1.5	1.6
WW C2	32	43	55	63	73	79	85	91	23	32	43	50	60	67	75	83	0.9	1.0	1.2	1.4	1.5	1.0	1.1	1.3	1.5	1.6
WW C3	38	51	63	72	81	86	91	96	28	40	52	60	71	77	84	91	0.8	1.0	1.1	1.3	1.5	0.9	1.0	1.2	1.4	1.5
WW C4	35	47	60	68	78	84	89	95	25	36	48	56	67	74	81	89	0.8	1.0	1.2	1.3	1.5	0.9	1.1	1.2	1.4	1.5
TW A1	22	26	32	36	42	47	53	61	17	20	24	27	32	35	40	47	1.4	1.5	1.7	1.8	2.0	1.7	1.8	2.0	2.1	2.3
TW A2	23	30	37	42	50	56	62	71	17	21	27	31	38	43	49	57	1.2	1.3	1.5	1.6	1.8	1.4	1.5	1.7	1.8	2.0
TW A3	23	31	39	45	53	59	66	75	16	22	28	33	41	46	52	61	1.1	1.3	1.4	1.6	1.7	1.3	1.5	1.6	1.8	1.9
TW A4	23	30	39	45	54	60	67	76	15	21	28	33	41	46	53	63	1.1	1.2	1.4	1.6	1.7	1.3	1.4	1.6	1.7	1.9
TW B1	21	26	32	37	43	48	54	63	16	20	24	27	33	36	41	49	1.3	1.5	1.6	1.8	1.9	1.6	1.7	1.9	2.0	2.2
TW B2	22	28	35	41	49	55	61	70	16	20	26	30	37	42	48	57	1.2	1.3	1.5	1.6	1.8	1.4	1.5	1.7	1.8	2.0
TW B3	21	28	36	41	50	55	62	72	15	20	26	31	38	43	49	58	1.2	1.3	1.5	1.6	1.8	1.4	1.5	1.6	1.8	1.9
TW B4	20	26	34	39	47	53	60	69	14	19	24	29	36	41	47	56	1.2	1.3	1.5	1.6	1.8	1.4	1.5	1.7	1.8	2.0
TW C1	20	25	32	36	43	48	55	63	15	19	23	27	33	37	42	50	1.3	1.5	1.6	1.7	1.9	1.6	1.7	1.8	2.0	2.1
TW C2	19	25	32	37	45	50	56	65	14	18	23	27	34	38	44	52	1.3	1.4	1.6	1.7	1.8	1.5	1.6	1.8	1.9	2.0
TW C3	19	24	30	35	42	48	54	63	14	17	22	26	32	36	42	50	1.3	1.4	1.6	1.7	1.9	1.5	1.7	1.8	1.9	2.1
TW C4	17	22	27	32	38	43	49	57	13	17	21	24	29	33	38	45	1.4	1.5	1.7	1.8	2.0	1.7	1.8	1.9	2.1	2.2
TW D1	17	22	28	33	39	44	50	58	11	15	19	22	27	31	35	42	1.4	1.5	1.7	1.8	2.0	1.7	1.8	2.0	2.2	2.3
TW D2	25	33	43	49	59	64	72	80	18	24	32	38	46	52	59	68	1.1	1.2	1.3	1.5	1.6	1.2	1.3	1.5	1.6	1.8
TW D3	28	37	47	54	63	69	76	84	20	27	35	42	51	57	64	73	1.0	1.1	1.3	1.5	1.6	1.1	1.3	1.4	1.6	1.7
TW D4	30	40	50	58	68	74	81	88	21	29	39	46	56	62	70	79	0.9	1.1	1.2	1.4	1.6	1.1	1.2	1.3	1.5	1.6
TW D5	29	39	50	58	68	74	81	89	20	29	39	46	56	63	71	80	0.9	1.1	1.2	1.4	1.5	1.1	1.2	1.3	1.5	1.6
TW E1	31	41	52	59	68	74	81	88	23	31	40	47	56	62	70	79	0.9	1.1	1.3	1.4	1.6	1.1	1.2	1.4	1.5	1.7
TW E2	31	41	51	59	69	75	81	89	22	30	40	47	56	63	70	79	0.9	1.1	1.2	1.4	1.6	1.1	1.2	1.3	1.5	1.6
TW E3	37	48	60	68	78	83	89	94	27	38	49	57	67	74	81	88	0.9	1.0	1.2	1.3	1.5	0.9	1.1	1.2	1.4	1.5
TW E4	34	45	57	65	75	81	87	93	24	35	46	54	64	71	78	87	0.9	1.0	1.2	1.4	1.5	1.0	1.1	1.3	1.4	1.6
TW F1	20	25	31	35	42	46	53	61	15	19	23	26	31	34	39	46	1.4	1.5	1.6	1.8	1.9	1.7	1.8	1.9	2.1	2.2
TW F2	21	27	34	40	48	53	60	69	14	19	25	29	35	40	46	55	1.2	1.3	1.5	1.7	1.8	1.4	1.6	1.7	1.9	2.0
TW F3	20	27	35	41	49	55	62	72	13	18	25	29	37	42	48	57	1.2	1.3	1.5	1.6	1.7	1.4	1.5	1.6	1.8	1.9
TW F4	18	25	33	39	48	54	61	71	11	17	23	28	35	40	47	56	1.2	1.3	1.5	1.6	1.7	1.4	1.5	1.6	1.8	1.9
TW G1	18	24	30	34	41	46	52	61	13	17	21	25	30	34	39	47	1.3	1.5	1.6	1.8	1.9	1.6	1.8	1.9	2.0	2.2
TW G2	18	24	31	36	44	50	57	66	12	16	21	26	32	37	43	51	1.2	1.4	1.5	1.7	1.8	1.5	1.6	1.7	1.9	2.0
TW G3	16	22	29	35	43	48	56	65	10	14	20	24	31	36	42	51	1.2	1.4	1.5	1.7	1.8	1.5	1.6	1.7	1.9	2.0
TW G4	13	18	25	30	38	43	50	60	8	12	17	21	27	31	37	46	1.3	1.5	1.6	1.8	1.9	1.6	1.7	1.8	2.0	2.1
TW H1	15	20	26	31	38	43	49	58	10	14	18	22	27	31	37	44	1.4	1.5	1.6	1.8	1.9	1.6	1.8	1.9	2.1	2.2
TW H2	13	18	24	29	36	41	48	57	8	12	16	20	26	30	35	43	1.4	1.5	1.6	1.8	1.9	1.6	1.8	1.9	2.0	2.2
TW H3	10	15	21	26	32	37	43	52	6	10	14	17	22	26	32	39	1.4	1.6	1.7	1.9	2.0	1.7	1.8	2.0	2.1	2.2
TW H4	8	12	16	20	26	30	36	44	4	7	10	13	18	21	25	32	1.6	1.7	1.9	2.0	2.1	1.9	2.1	2.2	2.3	2.4
TW I1	15	19	25	29	36	40	46	55	9	12	16	19	24	27	32	38	1.4	1.6	1.7	1.9	2.0	1.8	1.9	2.1	2.2	2.4
TW I2	22	30	39	46	55	61	68	78	15	21	28	34	42	48	55	65	1.1	1.2	1.4	1.5	1.7	1.3	1.4	1.5	1.7	1.8
TW I3	24	33	42	49	59	65	73	81	16	23	31	38	47	53	60	70	1.0	1.2	1.3	1.5	1.6	1.2	1.3	1.5	1.6	1.8
TW I4	26	36	47	55	65	71	78	86	18	26	36	43	52	59	67	77	1.0	1.1	1.3	1.4	1.6	1.1	1.2	1.4	1.5	1.7
TW I5	26	37	48	55	66	72	79	87	18	27	36	44	54	60	69	78	1.0	1.1	1.3	1.4	1.6	1.1	1.2	1.4	1.5	1.6
TW J1	29	38	49	56	66	72	79	87	20	28	37	44	54	60	68	77	1.0	1.1	1.3	1.4	1.6	1.1	1.2	1.4	1.5	1.7
TW J2	28	37	48	56	66	72	79	87	19	27	37	44	53	60	68	77	1.0	1.1	1.3	1.4	1.6	1.1	1.2	1.4	1.5	1.7
TW J3	33	45	57	65	75	81	87	93	24	34	46	54	64	71	78	87	0.9	1.0	1.2	1.4	1.5	1.0	1.1	1.3	1.4	1.6
TW J4	31	42	54	62	73	79	85	92	22	32	43	51	62	68	76	85	0.9	1.0	1.2	1.4	1.5	1.0	1.1	1.3	1.4	1.6
DG A1	18	24	30	35	41	44	48	53	11	15	20	23	27	30	33	37	1.2	1.5	1.8	2.1	2.5	1.6	1.8	2.2	2.7	3.3
DG A2	20	27	35	41	49	54	60	67	14	19	25	30	37	41	47	54	1.2	1.3	1.5	1.7	1.9	1.4	1.5	1.7	1.9	2.2
DG A3	24	33	43	51	61	67	73	81	18	24	32	39	48	55	62	71	1.1	1.1	1.3	1.5	1.7	1.2	1.3	1.4	1.6	1.8
DG B1	19	25	32	38	46	51	56	63	11	15	20	24	30	34	39	45	1.2	1.3	1.5	1.8	2.1	1.5	1.7	1.8	2.1	2.5
DG B2	21	28	37	44	54	60	67	75	14	19	26	31	39	45	52	61	1.1	1.2	1.4	1.6	1.8	1.4	1.5	1.6	1.7	2.0
DG B3	25	34	45	53	64	71	78	86	18	25	33	40	51	58	66	76	1.0	1.1	1.2	1.4	1.6	1.2	1.3	1.4	1.5	1.7
DG C1	22	30	38	45	55	60	67	74	14	19	25	30	38	43	49	57	1.1	1.2	1.4	1.6	1.9	1.4	1.5	1.6	1.8	2.1
DG C2	24	33	43	51	61	68	75	82	17	23	31	37	46	53	60	70	1.1	1.1	1.3	1.5	1.7	1.2	1.3	1.4	1.6	1.8
DG C3	28	38	50	59	71	78	85	91	21	29	38	46	57	65	74	83	1.0	1.0	1.2	1.4	1.6	1.1	1.2	1.3	1.4	1.6
SS A1	29	36	44	49	57	62	68	76	23	29	36	40	48	52	58	66	1.1	1.3	1.4	1.6	1.7	1.2	1.4	1.6	1.7	1.9
SS A2	36	45	55	61	70	75	82	88	29	37	46	53	61	67	73	81	0.9	1.1	1.3	1.4	1.6	1.0	1.2	1.4	1.5	1.7
SS A3	27	33	40	45	53	57	63	71	21	26	32	37	43	47	53	60	1.2	1.3	1.5	1.7	1.8	1.3	1.5	1.7	1.9	2.0
SS A4	35	44	54	60	69	74	80	87	29	36	45	51	60	65	71	79	1.0	1.1	1.3	1.5	1.6	1.0	1.2	1.4	1.6	1.7
SS A5	28	34	40	45	52	57	63	71	22	27	33	37	43	47	53	60	1.2	1.3	1.5	1.7	1.8	1.3	1.5	1.7	1.9	2.0
SS A6	36	45	54	61	70	75	81	88	29	37	46	52	61	66	73	80	0.9	1.1	1.3	1.4	1.6	1.0	1.2	1.4	1.5	1.7
SS A7	25	30	35	40	46	50	56	63	20	24	29	32	37	41	46	52	1.3	1.5	1.6	1.8	2.0	1.5	1.7	1.9	2.1	2.2
SS A8	34	43	52	59	67	73	79	86	28	36	44	50	58	64	70	78	1.0	1.1	1.3	1.5	1.6	1.1	1.2	1.4	1.6	1.7
SS B1	24	31	38	43	51	56	62	71	19	24	30	34	41	46	51	59	1.2	1.3	1.5	1.7	1.8	1.3	1.5	1.7	1.8	2.0
SS B2	31	40	49	56	65	70	77	85	25	32	41	47	55	61	68	76	1.0	1.1	1.3	1.5	1.6	1.1	1.2	1.4	1.6	1.7
SS B3	23	29	36	41	48	53	59	67	18	22	28	32	38	42	48	55	1.2	1.4	1.6	1.7	1.9	1.4	1.6	1.7	1.9	2.1
SS B4	30	39	48	55	63	69	75	83	24	31	40	46	54	60	66	75	1.0	1.2	1.3	1.5	1.6	1.1	1.3	1.4	1.6	1.7
SS B5	22	28	33	38	44	49	55	63	18	22	26	30	35	39	44	51	1.3	1.5	1.6	1.8	1.9	1.5	1.7	1.8	2.0	2.2
SS B6	30	39	48	55	64	69	76	84	24	31	40	46	54	60	67	75	1.0	1.2	1.3	1.5	1.6	1.1	1.3	1.4	1.6	1.7
SS B7	20	25	30	34	40	44	49	57	16	20	24	27	31	35	39	45	1.4	1.6	1.7	1.9	2.1	1.6	1.8	2.0	2.2	2.3
SS B8	29	37	46	53	62	67	74	82	23	30	38	44	52	57	64	73	1.0	1.2	1.3	1.5	1.7	1.1	1.3	1.5	1.6	1.8
SS C1	23	30	39	45	54	59	66	75	16	21	27	32	40	45	51	60	1.1	1.3	1.4	1.6	1.7	1.3	1.5	1.6	1.8	1.9
SS C2	28	37	47	54	64	70	77	85	19	27	35	42	51	57	64	74	1.0	1.1	1.3	1.5	1.6	1.1	1.3	1.4	1.6	1.7
SS C3	19	25	31	36	43	48	55	63	14	18	22	26	32	36	41	49	1.3	1.4	1.6	1.7	1.9	1.6	1.7	1.8	2.0	2.1
SS C4	25	33	42	48	57	63	71	79	17	24	31	37	45	51	58	67	1.1	1.2	1.4	1.5	1.7	1.2	1.4	1.5	1.7	1.8
SS D1	34	43	52	58	66	72	78	85	27	34	42	48	56	61	67	75	1.0	1.1	1.3	1.5	1.6	1.1	1.3	1.4	1.6	1.8
SS D2	41	52	63	70	78	83	88	94	33	43	53	61	70	75	81	88	0.9	1.0	1.2	1.4	1.5	0.9	1.1	1.3	1.4	1.6
SS D3	31	38	46	52	60	65	72	79	25	31	38	43	50	55	61	69	1.1	1.2	1.4	1.6	1.7	1.2	1.3	1.5	1.7	1.9
SS D4	40	50	61	68	76	81	87	92	32	42	52	59	68	73	79	87	0.9	1.0	1.2	1.4	1.5	0.9	1.1	1.3	1.5	1.6
SS E1	29	37	46	52	60	65	72	80	22	28	36	41	49	54	60	68	1.0	1.2	1.4	1.5	1.7	1.2	1.3	1.5	1.7	1.9
SS E2	36	46	56	63	72	78	84	90	27	37	46	53	62	68	75	83	0.9	1.1	1.2	1.4	1.6	1.0	1.2	1.3	1.5	1.6
SS E3	25	31	38	43	51	56	62	70	20	24	30	34	41	45	51	58	1.2	1.3	1.5	1.7	1.8	1.4	1.5	1.7	1.9	2.0
SS E4	33	42	52	59	68	73	80	87	25	34	43	49	58	63	70	79	1.0	1.1	1.3	1.4	1.6	1.1	1.2	1.4	1.5	1.7

LAS VEGAS, NEVADA

SOLAR SAVINGS FRACTION (%)

TYPE	BASE TEMP: 55 F (1120 DD)								65 F (2602 DD)							
LCR	200	100	70	50	40	30	25	20	200	100	70	50	40	30	25	20
WW A1	34	50	60	70	76	83	88	92	22	32	39	47	53	62	67	74
WW A2	40	61	72	82	87	93	95	98	24	39	49	59	66	75	81	86
WW A3	43	65	76	85	90	95	97	99	24	41	52	63	71	80	85	90
WW A4	44	67	78	87	92	96	98	99	24	43	54	66	73	82	86	91
WW A5	46	69	80	89	93	97	98	99	25	44	56	68	75	84	88	93
WW A6	46	70	81	89	94	97	98	99	25	45	57	69	77	85	89	93
WW B1	36	55	66	76	82	88	92	95	20	33	42	51	58	67	73	79
WW B2	44	67	78	87	92	96	98	99	24	43	55	66	74	83	87	92
WW B3	50	74	84	92	95	98	99	100	29	50	63	74	81	88	92	95
WW B4	50	74	84	92	95	98	99	100	28	50	63	75	82	89	93	96
WW B5	48	72	83	91	95	98	99	100	26	48	61	73	81	88	92	96
WW C1	52	75	85	92	96	98	99	100	30	52	64	76	82	89	93	96
WW C2	49	73	84	91	95	98	99	100	28	49	62	74	81	88	92	96
WW C3	54	79	88	95	97	99	100	10C	31	56	69	81	87	93	96	98
WW C4	51	76	86	93	96	99	99	100	28	52	66	78	84	91	94	97
TW A1	33	49	59	68	75	82	87	91	21	31	38	46	52	61	66	73
TW A2	37	56	67	77	83	89	93	96	22	35	44	54	61	70	76	82
TW A3	38	59	70	80	86	92	95	97	22	37	47	57	64	74	79	85
TW A4	38	59	70	80	86	92	95	97	21	36	47	58	65	74	80	86
TW B1	33	49	59	69	76	83	87	92	20	31	39	47	53	62	67	74
TW B2	35	54	65	75	82	88	92	95	20	33	43	52	59	69	74	81
TW B3	35	54	66	76	82	89	92	96	19	33	43	53	60	69	75	82
TW B4	32	51	63	73	80	87	91	95	18	31	40	50	57	67	72	79
TW C1	31	48	58	69	75	83	87	92	19	30	38	46	53	62	67	74
TW C2	31	49	59	70	77	84	88	93	18	30	38	47	54	63	69	76
TW C3	29	46	57	67	74	82	87	91	17	28	36	45	51	60	66	73
TW C4	26	41	51	61	68	76	81	87	16	25	32	40	46	54	60	67
TW D1	30	48	59	69	76	83	88	92	17	28	36	45	51	60	66	73
TW D2	40	62	73	83	88	94	96	98	23	39	50	61	69	78	83	88
TW D3	44	66	77	86	91	96	97	99	25	43	54	66	73	82	87	91
TW D4	45	68	80	88	93	97	98	99	25	45	57	69	77	85	89	93
TW D5	44	67	79	88	92	96	98	99	24	44	56	68	76	84	89	93
TW E1	48	70	81	89	93	97	98	99	28	47	59	70	77	85	90	94
TW E2	47	70	81	89	93	97	98	99	27	46	58	70	77	85	90	94
TW E3	53	76	86	93	96	99	99	100	31	54	67	78	85	91	94	97
TW E4	49	74	84	92	95	98	99	100	28	50	63	75	82	90	93	96
TW F1	32	48	58	68	75	83	87	91	20	30	37	46	52	61	66	73
TW F2	34	53	65	75	81	88	92	95	19	33	42	52	59	68	74	80
TW F3	34	55	66	77	83	90	93	96	18	33	43	53	60	70	76	82
TW F4	32	53	64	75	82	89	92	96	16	31	41	51	59	68	74	81
TW G1	30	47	57	68	74	82	87	91	17	28	36	45	51	60	66	73
TW G2	30	49	61	71	78	86	90	94	16	29	38	48	55	64	70	77
TW G3	28	47	59	70	77	85	89	93	14	27	36	46	53	63	69	76
TW G4	24	42	53	64	71	80	85	90	11	23	31	40	47	57	63	71
TW H1	26	43	54	65	72	80	85	90	14	25	33	41	48	57	63	70
TW H2	23	40	51	62	70	78	83	89	11	22	30	39	46	55	61	68
TW H3	20	36	46	57	64	74	79	85	9	19	26	34	41	50	56	63
TW H4	15	28	38	48	55	64	70	77	7	14	20	27	33	41	46	54
TW I1	27	44	55	66	72	81	85	90	14	25	33	41	48	56	62	69
TW I2	37	58	70	80	86	92	95	97	19	36	46	57	65	75	80	86
TW I3	39	61	73	83	89	94	96	98	21	38	50	61	69	78	83	89
TW I4	42	65	77	86	91	96	98	99	22	42	54	66	73	82	87	92
TW I5	41	65	77	86	91	96	98	99	22	42	54	66	74	83	88	92
TW J1	45	68	79	88	92	96	98	99	25	44	56	68	75	84	88	93
TW J2	43	67	78	87	92	96	98	99	24	43	55	67	75	83	88	92
TW J3	49	73	84	92	95	98	99	100	27	50	63	75	82	90	93	96
TW J4	46	71	82	90	94	98	99	100	25	47	60	72	80	88	92	95
DG A1	35	55	65	74	80	86	89	92	18	32	40	49	55	63	67	72
DG A2	35	56	67	77	83	89	92	95	18	33	43	53	60	69	74	80
DG A3	38	62	74	83	88	92	94	95	21	38	50	62	70	78	83	88
DG B1	37	60	72	81	86	91	93	95	18	33	44	55	62	71	75	80
DG B2	36	60	73	83	89	94	96	98	19	35	46	58	67	76	81	86
DG B3	39	65	78	87	91	95	97	98	21	39	52	65	74	83	87	92
DG C1	42	67	79	88	92	95	96	97	21	38	51	63	71	79	84	88
DG C2	41	66	79	88	93	96	98	99	21	39	52	65	74	83	87	91
DG C3	44	71	83	91	95	97	98	98	24	44	58	72	80	88	92	95
SS A1	40	60	70	79	85	91	94	97	26	40	50	59	66	74	79	85
SS A2	48	69	80	88	92	96	98	99	31	49	60	71	77	85	89	93
SS A3	38	56	67	76	82	89	92	95	25	38	47	56	62	71	76	82
SS A4	48	69	79	87	92	96	98	99	31	49	59	70	76	84	88	92
SS A5	38	56	66	75	81	88	91	95	25	38	46	55	61	70	75	81
SS A6	48	69	80	88	92	96	98	99	31	49	60	71	77	85	89	93
SS A7	34	50	60	69	75	83	87	91	23	34	41	49	55	63	68	74
SS A8	47	68	78	87	91	96	97	99	30	48	58	69	75	83	88	92
SS B1	36	54	65	75	81	88	91	95	22	35	44	53	60	69	74	80
SS B2	43	64	75	84	89	94	96	98	27	44	54	65	72	81	85	90
SS B3	34	52	62	72	79	86	90	94	21	34	42	51	57	66	71	78
SS B4	42	63	74	83	88	94	96	98	26	43	53	64	71	80	84	89
SS B5	32	48	58	68	74	82	86	91	21	31	39	47	53	61	67	73
SS B6	42	64	75	84	89	94	96	98	26	43	54	65	72	80	85	90
SS B7	29	44	53	63	69	77	82	87	19	29	35	43	48	56	61	68
SS B8	41	62	73	82	88	93	96	98	25	42	52	63	70	78	83	88
SS C1	39	60	71	81	87	93	95	98	22	37	47	58	65	75	80	86
SS C2	44	66	78	87	91	96	98	99	24	43	55	66	73	82	87	92
SS C3	31	49	60	70	77	84	88	93	18	30	38	47	54	63	68	75
SS C4	39	60	72	82	87	93	96	98	22	38	49	60	67	76	82	87
SS D1	48	69	79	88	92	96	98	99	31	48	59	69	76	83	88	92
SS D2	55	77	87	93	96	99	99	100	36	57	69	79	85	91	94	97
SS D3	44	63	73	82	88	93	96	98	28	43	53	63	69	78	82	88
SS D4	53	75	85	92	96	98	99	100	34	55	66	77	83	90	93	96
SS E1	44	64	75	84	89	94	96	98	27	43	53	64	71	79	84	89
SS E2	50	73	83	91	94	98	99	100	31	51	63	74	80	88	91	95
SS E3	37	55	65	75	81	88	91	95	23	36	45	54	60	69	74	80
SS E4	46	68	79	87	92	96	98	99	29	47	58	69	76	84	88	93

CONSERVATION FACTOR

TYPE	55 F					65 F				
	200	100	50	30	20	200	100	50	30	20
WW A1	0.7	0.8	1.0	1.2	1.3	1.0	1.1	1.2	1.4	1.5
WW A2	0.6	0.7	0.9	1.1	1.3	0.8	0.9	1.0	1.2	1.4
WW A3	0.6	0.7	0.9	1.1	1.2	0.8	0.9	1.0	1.2	1.3
WW A4	0.6	0.7	0.9	1.0	1.2	0.7	0.8	1.0	1.2	1.3
WW A5	0.6	0.7	0.9	1.0	1.2	0.7	0.8	1.0	1.1	1.3
WW A6	0.6	0.7	0.8	1.0	1.2	0.7	0.8	1.0	1.1	1.3
WW B1	0.7	0.8	0.9	1.1	1.3	0.9	1.0	1.1	1.3	1.5
WW B2	0.6	0.7	0.9	1.0	1.2	0.7	0.8	1.0	1.1	1.3
WW B3	0.5	0.7	0.8	1.0	1.2	0.7	0.8	0.9	1.1	1.3
WW B4	0.5	0.7	0.8	1.0	1.2	0.7	0.8	0.9	1.1	1.3
WW B5	0.5	0.7	0.8	1.0	1.2	0.7	0.8	0.9	1.1	1.3
WW C1	0.5	0.7	0.8	1.0	1.2	0.7	0.8	0.9	1.1	1.3
WW C2	0.5	0.7	0.8	1.0	1.2	0.7	0.8	0.9	1.1	1.3
WW C3	0.5	0.6	0.8	1.0	1.2	0.6	0.7	0.9	1.1	1.2
WW C4	0.5	0.6	0.8	1.0	1.2	0.6	0.7	0.9	1.1	1.3
TW A1	0.7	0.8	1.0	1.2	1.3	1.0	1.1	1.2	1.4	1.5
TW A2	0.7	0.8	0.9	1.1	1.3	0.9	1.0	1.1	1.3	1.4
TW A3	0.6	0.7	0.9	1.1	1.3	0.8	0.9	1.1	1.2	1.4
TW A4	0.6	0.7	0.9	1.1	1.3	0.8	0.9	1.0	1.2	1.4
TW B1	0.7	0.8	1.0	1.2	1.3	1.0	1.1	1.2	1.3	1.5
TW B2	0.7	0.8	0.9	1.1	1.3	0.9	1.0	1.1	1.3	1.4
TW B3	0.7	0.8	0.9	1.1	1.3	0.9	0.9	1.1	1.2	1.4
TW B4	0.7	0.8	1.0	1.1	1.3	0.9	1.0	1.1	1.3	1.4
TW C1	0.7	0.8	1.0	1.2	1.3	1.0	1.0	1.2	1.3	1.5
TW C2	0.7	0.8	1.0	1.1	1.3	0.9	1.0	1.2	1.3	1.5
TW C3	0.7	0.8	1.0	1.2	1.3	1.0	1.1	1.2	1.3	1.5
TW C4	0.8	0.9	1.0	1.2	1.4	1.0	1.1	1.3	1.4	1.6
TW D1	0.7	0.8	1.0	1.2	1.3	1.0	1.1	1.2	1.3	1.5
TW D2	0.6	0.7	0.9	1.1	1.2	0.8	0.9	1.0	1.2	1.4
TW D3	0.6	0.7	0.9	1.0	1.2	0.7	0.8	1.0	1.2	1.3
TW D4	0.6	0.7	0.9	1.0	1.2	0.7	0.8	1.0	1.1	1.3
TW D5	0.6	0.7	0.9	1.0	1.2	0.7	0.8	1.0	1.1	1.3
TW E1	0.6	0.7	0.9	1.0	1.2	0.7	0.8	1.0	1.1	1.3
TW E2	0.6	0.7	0.9	1.0	1.2	0.7	0.8	1.0	1.1	1.3
TW E3	0.5	0.6	0.8	1.0	1.2	0.6	0.7	0.9	1.1	1.3
TW E4	0.5	0.7	0.8	1.0	1.2	0.7	0.8	0.9	1.1	1.3
TW F1	0.7	0.8	1.0	1.2	1.3	1.0	1.1	1.2	1.4	1.5
TW F2	0.7	0.8	0.9	1.1	1.3	0.9	1.0	1.1	1.3	1.4
TW F3	0.7	0.8	0.9	1.1	1.3	0.8	0.9	1.1	1.2	1.4
TW F4	0.7	0.8	0.9	1.1	1.3	0.9	0.9	1.1	1.2	1.4
TW G1	0.7	0.8	1.0	1.2	1.3	1.0	1.1	1.2	1.3	1.5
TW G2	0.7	0.8	1.0	1.1	1.3	0.9	1.0	1.1	1.3	1.5
TW G3	0.7	0.8	1.0	1.1	1.3	0.9	1.0	1.1	1.3	1.5
TW G4	0.7	0.8	1.0	1.2	1.3	1.0	1.1	1.2	1.3	1.5
TW H1	0.7	0.8	1.0	1.2	1.3	1.0	1.1	1.2	1.4	1.5
TW H2	0.8	0.9	1.0	1.2	1.4	1.0	1.1	1.2	1.4	1.5
TW H3	0.8	0.9	1.1	1.2	1.4	1.1	1.2	1.3	1.4	1.6
TW H4	0.9	1.0	1.1	1.3	1.5	1.2	1.3	1.4	1.6	1.7
TW I1	0.7	0.8	1.0	1.2	1.3	1.0	1.1	1.2	1.4	1.5
TW I2	0.6	0.7	0.9	1.1	1.3	0.8	0.9	1.0	1.2	1.4
TW I3	0.6	0.7	0.9	1.1	1.2	0.8	0.9	1.0	1.2	1.3
TW I4	0.6	0.7	0.9	1.0	1.2	0.7	0.8	1.0	1.1	1.3
TW I5	0.6	0.7	0.9	1.0	1.2	0.7	0.8	1.0	1.1	1.3
TW J1	0.6	0.7	0.9	1.0	1.2	0.7	0.8	1.0	1.1	1.3
TW J2	0.6	0.7	0.9	1.0	1.2	0.7	0.8	1.0	1.1	1.3
TW J3	0.5	0.7	0.8	1.0	1.2	0.7	0.8	0.9	1.1	1.3
TW J4	0.6	0.7	0.8	1.0	1.2	0.7	0.8	0.9	1.1	1.3
DG A1	0.7	0.8	1.0	1.2	1.4	0.9	1.0	1.2	1.4	1.7
DG A2	0.6	0.8	0.9	1.1	1.3	0.8	0.9	1.1	1.3	1.5
DG A3	0.6	0.7	0.9	1.2	1.6	0.8	0.8	1.0	1.2	1.4
DG B1	0.6	0.7	0.9	1.1	1.4	0.9	0.9	1.1	1.3	1.6
DG B2	0.6	0.7	0.9	1.1	1.3	0.8	0.9	1.0	1.2	1.4
DG B3	0.6	0.7	0.9	1.1	1.5	0.8	0.8	0.9	1.1	1.4
DG C1	0.6	0.7	0.9	1.1	1.5	0.8	0.8	1.0	1.2	1.5
DG C2	0.6	0.7	0.9	1.0	1.3	0.8	0.8	1.0	1.1	1.4
DG C3	0.6	0.6	0.8	1.1	1.9	0.7	0.8	0.9	1.1	1.3
SS A1	0.6	0.8	0.9	1.1	1.3	0.8	0.9	1.1	1.2	1.4
SS A2	0.6	0.7	0.9	1.0	1.2	0.7	0.8	1.0	1.1	1.3
SS A3	0.7	0.8	0.9	1.1	1.3	0.8	0.9	1.1	1.3	1.4
SS A4	0.6	0.7	0.9	1.0	1.2	0.7	0.8	1.0	1.1	1.3
SS A5	0.7	0.8	1.0	1.1	1.3	0.9	1.0	1.1	1.3	1.5
SS A6	0.6	0.7	0.9	1.0	1.2	0.7	0.8	1.0	1.1	1.3
SS A7	0.7	0.8	1.0	1.2	1.3	0.9	1.0	1.2	1.4	1.5
SS A8	0.6	0.7	0.9	1.0	1.2	0.7	0.8	1.0	1.2	1.3
SS B1	0.7	0.8	1.0	1.1	1.3	0.9	1.0	1.1	1.3	1.4
SS B2	0.6	0.7	0.9	1.1	1.2	0.7	0.8	1.0	1.2	1.3
SS B3	0.7	0.8	1.0	1.1	1.3	0.9	1.0	1.1	1.3	1.5
SS B4	0.6	0.7	0.9	1.1	1.2	0.7	0.9	1.0	1.2	1.4
SS B5	0.7	0.8	1.0	1.2	1.3	0.9	1.1	1.2	1.4	1.5
SS B6	0.6	0.7	0.9	1.1	1.2	0.7	0.8	1.0	1.2	1.3
SS B7	0.8	0.9	1.0	1.2	1.4	1.0	1.1	1.3	1.4	1.6
SS B8	0.6	0.7	0.9	1.1	1.2	0.8	0.9	1.0	1.2	1.4
SS C1	0.6	0.7	0.9	1.1	1.3	0.8	0.9	1.0	1.2	1.4
SS C2	0.6	0.7	0.9	1.0	1.2	0.7	0.8	1.0	1.1	1.3
SS C3	0.7	0.8	1.0	1.1	1.3	0.9	1.0	1.2	1.3	1.5
SS C4	0.6	0.7	0.9	1.1	1.2	0.8	0.9	1.0	1.2	1.4
SS D1	0.6	0.7	0.9	1.0	1.2	0.7	0.8	1.0	1.2	1.3
SS D2	0.5	0.7	0.8	1.0	1.2	0.6	0.7	0.9	1.1	1.3
SS D3	0.6	0.7	0.9	1.1	1.3	0.8	0.9	1.0	1.2	1.4
SS D4	0.5	0.7	0.8	1.0	1.2	0.7	0.8	0.9	1.1	1.3
SS E1	0.6	0.7	0.9	1.1	1.2	0.8	0.9	1.0	1.2	1.4
SS E2	0.6	0.7	0.8	1.0	1.2	0.7	0.8	0.9	1.1	1.3
SS E3	0.7	0.8	1.0	1.1	1.3	0.9	1.0	1.1	1.3	1.5
SS E4	0.6	0.7	0.9	1.0	1.2	0.7	0.8	1.0	1.1	1.3

LOVELOCK, NEVADA

TYPE	SOLAR SAVINGS FRACTION (%)																CONSERVATION FACTOR									
	BASE TEMP: 55 F (3564 DD)								65 F (5998 DD)								55 F					65 F				
LCR	100	70	50	40	30	25	20	15	100	70	50	40	30	25	20	15	100	50	30	20	15	100	50	30	20	15
WW A1	27	33	40	45	53	57	64	72	21	25	30	34	40	44	50	57	1.2	1.3	1.5	1.7	1.8	1.4	1.6	1.7	1.9	2.0
WW A2	32	41	50	57	65	71	77	85	23	30	38	43	52	57	64	72	1.0	1.1	1.3	1.5	1.6	1.2	1.3	1.5	1.6	1.8
WW A3	34	44	54	61	70	75	81	88	24	32	41	47	56	61	68	77	0.9	1.1	1.3	1.4	1.6	1.1	1.2	1.4	1.6	1.7
WW A4	35	45	56	63	72	77	84	90	24	33	42	49	58	64	71	79	0.9	1.1	1.2	1.4	1.6	1.0	1.2	1.4	1.5	1.7
WW A5	36	47	58	65	74	80	86	92	25	34	44	51	60	66	73	82	0.9	1.0	1.2	1.4	1.5	1.0	1.2	1.3	1.5	1.6
WW A6	37	48	59	66	75	81	87	92	25	35	45	52	62	68	75	83	0.9	1.0	1.2	1.4	1.5	1.0	1.1	1.3	1.5	1.6
WW B1	27	34	42	48	56	61	68	76	19	24	30	35	41	46	52	60	1.1	1.3	1.4	1.6	1.7	1.3	1.5	1.7	1.8	2.0
WW B2	35	46	57	64	73	79	85	91	25	34	44	51	60	66	73	82	0.9	1.1	1.2	1.4	1.5	1.0	1.2	1.3	1.5	1.6
WW B3	42	53	64	72	80	85	90	95	30	41	51	59	68	74	81	88	0.8	1.0	1.2	1.3	1.5	0.9	1.1	1.3	1.4	1.6
WW B4	42	54	66	73	82	87	91	96	30	41	53	61	71	77	83	90	0.8	1.0	1.1	1.3	1.5	0.9	1.1	1.2	1.4	1.5
WW B5	40	52	64	72	81	86	91	96	29	40	52	60	70	76	83	90	0.8	1.0	1.1	1.3	1.5	0.9	1.1	1.2	1.4	1.5
WW C1	43	55	66	73	82	87	91	96	32	42	53	61	71	76	83	90	0.8	1.0	1.2	1.3	1.5	0.9	1.1	1.2	1.4	1.6
WW C2	41	53	64	72	81	86	91	95	30	40	52	59	69	75	82	89	0.8	1.0	1.2	1.3	1.5	0.9	1.1	1.2	1.4	1.6
WW C3	47	60	72	79	87	91	95	98	35	48	60	68	78	84	89	94	0.8	0.9	1.1	1.3	1.4	0.8	1.0	1.2	1.3	1.5
WW C4	44	56	69	76	85	89	94	97	32	44	57	65	75	81	87	93	0.8	1.0	1.1	1.3	1.5	0.9	1.0	1.2	1.4	1.5
TW A1	27	33	39	44	52	57	63	71	21	25	30	34	40	44	49	57	1.2	1.4	1.5	1.7	1.8	1.4	1.6	1.7	1.9	2.0
TW A2	29	37	46	52	60	66	72	80	22	28	34	40	47	52	59	67	1.1	1.2	1.4	1.5	1.7	1.2	1.4	1.5	1.7	1.8
TW A3	30	39	48	55	64	69	76	84	22	28	36	42	50	56	62	71	1.0	1.2	1.3	1.5	1.6	1.2	1.3	1.5	1.6	1.8
TW A4	30	39	48	55	64	70	77	85	21	28	36	42	51	56	64	73	1.0	1.1	1.3	1.5	1.6	1.1	1.3	1.4	1.6	1.7
TW B1	26	33	40	45	53	58	64	73	20	25	30	34	41	45	51	59	1.2	1.3	1.5	1.6	1.8	1.4	1.5	1.7	1.8	2.0
TW B2	28	36	44	50	59	65	71	80	20	26	33	38	46	51	58	67	1.1	1.2	1.4	1.5	1.7	1.2	1.4	1.5	1.7	1.8
TW B3	28	36	44	51	60	65	72	80	20	26	33	39	47	52	59	68	1.0	1.2	1.4	1.5	1.7	1.2	1.4	1.5	1.7	1.8
TW B4	26	34	42	48	57	63	70	78	18	24	31	37	44	50	56	65	1.1	1.2	1.4	1.5	1.7	1.2	1.4	1.5	1.7	1.8
TW C1	25	32	39	45	53	58	64	73	19	24	30	34	41	45	51	60	1.2	1.3	1.5	1.6	1.8	1.4	1.5	1.7	1.8	1.9
TW C2	25	32	40	46	54	59	66	75	18	23	30	35	42	47	53	62	1.1	1.3	1.4	1.6	1.7	1.3	1.5	1.6	1.8	1.9
TW C3	24	30	38	44	52	57	64	73	17	22	28	33	40	45	51	60	1.2	1.3	1.5	1.6	1.8	1.4	1.5	1.6	1.8	1.9
TW C4	22	27	34	39	47	52	58	67	16	21	26	30	36	40	46	54	1.2	1.4	1.5	1.7	1.8	1.5	1.6	1.7	1.9	2.0
TW D1	23	29	37	42	50	55	61	70	16	20	26	30	36	40	46	54	1.2	1.3	1.5	1.7	1.8	1.4	1.6	1.7	1.9	2.1
TW D2	33	42	52	59	68	74	80	88	24	31	40	46	55	61	68	77	1.0	1.1	1.3	1.4	1.6	1.1	1.2	1.4	1.6	1.7
TW D3	36	46	56	63	73	78	84	90	26	34	44	51	60	66	73	81	0.9	1.1	1.2	1.4	1.6	1.0	1.2	1.3	1.5	1.6
TW D4	38	49	60	67	77	82	88	93	27	37	48	55	65	71	78	86	0.9	1.0	1.2	1.4	1.5	1.0	1.1	1.3	1.4	1.6
TW D5	37	48	59	67	76	82	88	93	27	37	47	55	65	71	78	86	0.9	1.0	1.2	1.4	1.5	1.0	1.1	1.3	1.4	1.6
TW E1	39	50	61	68	77	82	88	93	29	38	49	56	65	71	78	85	0.9	1.0	1.2	1.4	1.5	1.0	1.1	1.3	1.5	1.6
TW E2	39	50	61	68	77	82	88	93	28	38	48	56	65	71	78	86	0.9	1.0	1.2	1.4	1.5	1.0	1.1	1.3	1.4	1.6
TW E3	45	57	69	76	85	89	93	97	34	46	57	65	75	80	86	92	0.8	1.0	1.1	1.3	1.5	0.9	1.0	1.2	1.4	1.5
TW E4	42	55	66	74	83	87	92	96	31	43	54	62	72	78	85	91	0.8	1.0	1.1	1.3	1.5	0.9	1.0	1.2	1.4	1.5
TW F1	25	32	39	44	51	57	63	71	19	24	29	33	39	43	49	57	1.2	1.3	1.5	1.7	1.8	1.4	1.6	1.7	1.9	2.0
TW F2	27	35	43	49	58	64	70	79	19	25	32	37	45	50	56	65	1.1	1.2	1.4	1.5	1.7	1.3	1.4	1.6	1.7	1.9
TW F3	27	35	44	51	60	66	73	81	18	25	32	38	46	52	59	68	1.0	1.2	1.3	1.5	1.7	1.2	1.4	1.5	1.7	1.8
TW F4	25	33	42	49	58	64	71	80	16	23	31	36	45	50	57	67	1.0	1.2	1.4	1.5	1.7	1.2	1.4	1.5	1.7	1.8
TW G1	24	30	37	43	51	56	63	71	17	22	27	32	38	43	49	57	1.2	1.3	1.5	1.6	1.8	1.4	1.5	1.7	1.9	2.0
TW G2	24	31	39	46	54	60	67	76	16	22	28	34	41	46	53	62	1.1	1.3	1.4	1.6	1.7	1.3	1.4	1.6	1.7	1.9
TW G3	22	29	38	44	53	59	66	75	14	20	27	32	40	45	52	61	1.1	1.3	1.4	1.6	1.7	1.3	1.4	1.6	1.7	1.9
TW G4	18	25	33	39	48	53	60	70	11	17	23	28	35	40	47	56	1.2	1.3	1.5	1.6	1.8	1.4	1.5	1.7	1.8	1.9
TW H1	20	27	34	40	48	53	60	69	14	19	24	29	35	40	46	55	1.2	1.3	1.5	1.7	1.8	1.4	1.6	1.7	1.9	2.0
TW H2	18	24	32	38	46	51	58	68	11	16	22	27	34	38	45	54	1.2	1.3	1.5	1.7	1.8	1.4	1.6	1.7	1.9	2.0
TW H3	15	21	28	33	41	47	53	63	9	14	19	23	30	34	40	49	1.3	1.4	1.6	1.7	1.9	1.5	1.6	1.8	1.9	2.1
TW H4	11	16	22	27	34	38	45	54	7	10	15	18	24	28	33	41	1.4	1.6	1.7	1.9	2.0	1.7	1.8	1.9	2.1	2.2
TW I1	20	26	33	39	46	51	58	66	13	18	23	27	33	37	42	50	1.2	1.4	1.6	1.7	1.8	1.5	1.6	1.8	2.0	2.1
TW I2	29	38	48	56	65	71	78	85	20	28	36	43	52	58	65	74	1.0	1.1	1.3	1.5	1.6	1.1	1.3	1.4	1.6	1.7
TW I3	31	41	52	59	69	74	81	88	22	30	40	46	56	62	69	78	0.9	1.1	1.3	1.4	1.6	1.1	1.2	1.4	1.5	1.7
TW I4	34	45	56	64	74	79	85	92	24	34	44	51	61	68	75	84	0.9	1.0	1.2	1.4	1.5	1.0	1.1	1.3	1.5	1.6
TW I5	34	46	57	65	75	80	86	92	24	34	45	52	63	69	76	85	0.9	1.0	1.2	1.4	1.5	1.0	1.1	1.3	1.4	1.6
TW J1	37	47	58	66	75	80	86	92	27	36	46	53	63	69	76	84	0.9	1.0	1.2	1.4	1.5	1.0	1.1	1.3	1.5	1.6
TW J2	36	46	58	65	75	80	86	92	25	35	45	53	62	69	76	84	0.9	1.0	1.2	1.4	1.5	1.0	1.1	1.3	1.5	1.6
TW J3	42	54	66	74	82	87	92	96	31	42	54	62	72	78	85	91	0.8	1.0	1.1	1.3	1.5	0.9	1.0	1.2	1.4	1.5
TW J4	39	51	63	71	80	86	91	95	28	39	51	59	70	76	83	90	0.8	1.0	1.2	1.3	1.5	0.9	1.1	1.2	1.4	1.5
DG A1	25	33	40	45	52	56	60	66	17	22	28	32	37	41	44	49	1.1	1.3	1.6	1.9	2.2	1.3	1.6	1.9	2.3	2.7
DG A2	27	36	45	51	59	64	70	77	19	25	33	38	46	51	57	64	1.0	1.2	1.4	1.6	1.8	1.2	1.4	1.6	1.8	2.0
DG A3	32	42	53	60	70	75	81	87	23	31	41	48	57	63	70	78	0.9	1.1	1.3	1.5	1.7	1.1	1.2	1.4	1.6	1.7
DG B1	26	34	44	50	58	63	69	74	17	23	29	35	42	46	52	58	1.0	1.2	1.4	1.7	2.0	1.3	1.4	1.6	1.9	2.3
DG B2	28	37	48	55	65	71	77	84	19	26	34	41	50	56	63	71	1.0	1.1	1.3	1.5	1.7	1.2	1.3	1.4	1.6	1.8
DG B3	32	43	55	63	74	80	85	91	23	32	42	50	60	67	75	83	0.9	1.0	1.2	1.4	1.6	1.1	1.1	1.3	1.5	1.6
DG C1	30	40	51	58	67	73	78	83	20	27	35	41	50	56	62	69	1.0	1.1	1.3	1.6	1.8	1.2	1.3	1.5	1.7	2.0
DG C2	32	42	54	62	72	78	84	89	22	30	40	47	57	64	71	79	0.9	1.0	1.2	1.4	1.6	1.1	1.2	1.3	1.5	1.7
DG C3	36	48	61	70	80	85	90	94	27	36	47	56	67	74	81	89	0.9	1.0	1.1	1.4	1.6	1.0	1.1	1.2	1.4	1.6
SS A1	34	42	50	56	64	69	75	82	27	34	41	46	54	59	65	72	1.0	1.2	1.4	1.5	1.7	1.1	1.3	1.5	1.7	1.8
SS A2	42	51	61	68	76	81	86	92	34	43	52	58	67	72	78	85	0.9	1.0	1.2	1.4	1.5	1.0	1.1	1.3	1.5	1.6
SS A3	32	39	46	52	60	65	71	78	25	31	38	42	49	54	59	67	1.1	1.2	1.4	1.6	1.7	1.2	1.4	1.6	1.8	1.9
SS A4	41	50	60	66	75	80	85	91	33	42	51	57	65	70	76	84	0.9	1.1	1.2	1.4	1.6	1.0	1.2	1.3	1.5	1.7
SS A5	32	39	46	52	59	64	70	78	26	32	38	43	49	54	59	67	1.1	1.3	1.4	1.6	1.7	1.2	1.4	1.6	1.8	1.9
SS A6	41	51	61	68	76	81	86	92	34	42	51	58	66	72	78	85	0.9	1.1	1.2	1.4	1.5	1.0	1.1	1.3	1.5	1.6
SS A7	29	35	41	46	53	57	63	70	23	28	33	37	43	47	52	59	1.2	1.4	1.5	1.7	1.9	1.3	1.5	1.7	1.9	2.1
SS A8	40	49	59	65	74	79	84	90	32	41	49	56	64	69	75	82	0.9	1.1	1.2	1.4	1.6	1.0	1.2	1.3	1.5	1.7
SS B1	29	37	45	51	58	64	70	78	23	29	36	41	48	52	59	66	1.1	1.2	1.4	1.6	1.7	1.2	1.4	1.6	1.7	1.9
SS B2	37	46	56	63	72	77	83	89	29	37	46	53	62	67	74	81	0.9	1.1	1.3	1.4	1.6	1.0	1.2	1.3	1.5	1.7
SS B3	28	35	42	48	55	60	67	75	22	27	33	38	45	49	55	63	1.1	1.3	1.4	1.6	1.8	1.3	1.4	1.6	1.8	2.0
SS B4	36	45	55	61	70	75	81	88	29	37	45	52	60	65	72	80	0.9	1.1	1.3	1.4	1.6	1.0	1.2	1.4	1.5	1.7
SS B5	27	33	39	45	52	56	63	70	21	26	31	36	42	46	51	59	1.2	1.3	1.5	1.7	1.8	1.4	1.5	1.7	1.9	2.0
SS B6	36	45	55	62	71	76	82	89	28	37	46	52	61	66	73	81	0.9	1.1	1.3	1.4	1.6	1.0	1.2	1.4	1.5	1.7
SS B7	24	30	36	40	47	51	57	65	19	23	28	32	37	41	46	53	1.3	1.4	1.6	1.8	1.9	1.5	1.6	1.8	2.0	2.2
SS B8	35	44	53	60	69	74	80	87	27	35	44	50	58	64	70	78	1.0	1.1	1.3	1.5	1.6	1.1	1.2	1.4	1.6	1.7
SS C1	30	39	48	55	64	70	76	84	21	28	36	42	50	55	62	71	1.0	1.1	1.3	1.5	1.6	1.2	1.3	1.5	1.6	1.8
SS C2	36	46	57	64	73	79	85	91	25	34	44	51	60	66	73	82	0.9	1.1	1.2	1.4	1.5	1.0	1.2	1.3	1.5	1.6
SS C3	25	32	39	45	53	58	65	73	18	23	29	34	40	45	51	59	1.1	1.3	1.5	1.6	1.8	1.3	1.5	1.7	1.8	1.9
SS C4	32	41	51	58	67	73	79	87	23	30	39	45	54	60	67	76	1.0	1.1	1.3	1.4	1.6	1.1	1.2	1.4	1.6	1.7
SS D1	40	49	59	65	73	78	84	90	32	40	48	54	62	68	74	81	0.9	1.1	1.3	1.4	1.6	1.0	1.2	1.4	1.5	1.7
SS D2	48	59	69	76	84	88	92	96	39	49	59	66	75	80	86	91	0.8	1.0	1.2	1.3	1.5	0.9	1.1	1.2	1.4	1.6
SS D3	36	45	53	59	67	72	78	85	29	36	44	49	57	62	68	75	1.0	1.1	1.3	1.5	1.6	1.1	1.3	1.4	1.6	1.8
SS D4	46	57	67	74	82	86	91	95	37	47	57	64	73	78	84	90	0.8	1.0	1.2	1.3	1.5	0.9	1.1	1.2	1.4	1.6
SS E1	35	44	53	60	68	73	79	86	27	34	42	48	56	61	68	75	1.0	1.1	1.3	1.5	1.6	1.1	1.3	1.4	1.6	1.8
SS E2	42	53	64	71	79	84	89	94	33	43	53	60	69	74	80	87	0.9	1.0	1.2	1.4	1.5	0.9	1.1	1.3	1.4	1.6
SS E3	30	37	45	51	59	64	70	78	24	29	36	41	48	52	58	66	1.1	1.2	1.4	1.6	1.7	1.2	1.4	1.6	1.8	1.9
SS E4	39	49	59	66	75	80	85	91	31	39	49	56	64	70	76	84	0.9	1.1	1.2	1.4	1.5	1.0	1.1	1.3	1.5	1.6

RENO, NEVADA

TYPE	SSF (%) 55 F (3394 DD) LCR 100	70	50	40	30	25	20	15	SSF (%) 65 F (6027 DD) LCR 100	70	50	40	30	25	20	15	CONSERVATION FACTOR 55 F 100	50	30	20	15	CONSERVATION FACTOR 65 F 100	50	30	20	15
WW A1	29	35	42	47	55	60	66	74	21	26	31	35	41	45	51	59	1.1	1.3	1.5	1.6	1.8	1.4	1.6	1.7	1.9	2.0
WW A2	34	43	52	59	68	74	80	87	24	31	39	44	53	58	65	74	1.0	1.1	1.3	1.4	1.6	1.1	1.3	1.4	1.6	1.7
WW A3	36	46	56	63	72	78	84	90	25	33	41	48	57	63	70	78	0.9	1.1	1.2	1.4	1.6	1.1	1.2	1.4	1.5	1.7
WW A4	37	47	58	66	75	80	86	92	25	34	43	50	59	65	72	81	0.9	1.0	1.2	1.4	1.5	1.0	1.2	1.3	1.5	1.6
WW A5	38	49	61	68	77	82	88	93	26	35	45	52	62	68	75	83	0.9	1.0	1.2	1.4	1.5	1.0	1.1	1.3	1.5	1.6
WW A6	39	50	62	69	78	83	89	94	26	36	46	53	63	69	76	84	0.9	1.0	1.2	1.4	1.5	1.0	1.1	1.3	1.5	1.6
WW B1	29	36	45	51	59	64	71	79	19	25	31	36	43	47	53	62	1.1	1.2	1.4	1.6	1.7	1.3	1.5	1.6	1.8	1.9
WW B2	37	48	59	67	76	81	87	93	26	35	45	52	61	67	75	83	0.9	1.0	1.2	1.4	1.5	1.0	1.2	1.3	1.5	1.6
WW B3	44	56	67	74	83	87	92	96	31	42	53	60	70	76	82	89	0.8	1.0	1.1	1.3	1.5	0.9	1.1	1.2	1.4	1.6
WW B4	44	56	68	76	84	89	93	97	31	42	54	62	72	78	84	91	0.8	1.0	1.1	1.3	1.5	0.9	1.0	1.2	1.4	1.5
WW B5	42	55	67	74	83	88	92	96	29	41	53	61	71	77	84	91	0.8	1.0	1.1	1.3	1.5	0.9	1.1	1.2	1.4	1.5
WW C1	46	57	69	76	84	88	93	97	33	44	55	62	72	78	84	91	0.8	1.0	1.1	1.3	1.5	0.9	1.1	1.2	1.4	1.5
WW C2	43	55	67	74	83	88	92	96	31	41	53	60	70	76	83	90	0.8	1.0	1.1	1.3	1.5	0.9	1.1	1.2	1.4	1.5
WW C3	50	63	75	82	89	93	96	98	36	49	62	70	79	85	90	95	0.8	0.9	1.1	1.3	1.4	0.8	1.0	1.2	1.3	1.5
WW C4	46	59	71	79	87	91	95	98	33	45	58	66	76	82	88	94	0.8	0.9	1.1	1.3	1.4	0.9	1.0	1.2	1.3	1.5
TW A1	28	34	41	47	54	59	65	74	21	25	30	35	41	45	50	58	1.2	1.3	1.5	1.6	1.8	1.4	1.6	1.7	1.9	2.0
TW A2	31	39	48	54	63	69	75	83	22	28	35	40	48	53	60	69	1.0	1.2	1.3	1.5	1.6	1.2	1.4	1.5	1.7	1.8
TW A3	32	41	51	57	66	72	78	86	22	29	37	43	51	57	64	73	1.0	1.1	1.3	1.5	1.6	1.2	1.3	1.4	1.6	1.7
TW A4	32	41	51	58	67	73	79	87	21	29	37	43	52	58	65	74	1.0	1.1	1.3	1.4	1.6	1.1	1.3	1.4	1.6	1.7
TW B1	28	34	42	47	55	60	67	75	20	25	31	35	42	46	52	60	1.1	1.3	1.4	1.6	1.7	1.4	1.5	1.7	1.8	2.0
TW B2	29	37	46	53	62	67	74	82	21	27	34	39	47	52	59	68	1.0	1.2	1.3	1.5	1.6	1.2	1.4	1.5	1.7	1.8
TW B3	29	37	47	53	62	68	75	83	20	26	34	40	48	53	60	69	1.0	1.2	1.3	1.5	1.6	1.2	1.3	1.5	1.6	1.8
TW B4	27	35	44	51	60	65	72	81	19	25	32	37	45	51	58	67	1.0	1.2	1.3	1.5	1.7	1.2	1.4	1.5	1.7	1.8
TW C1	26	33	41	47	55	60	67	76	19	24	30	35	41	46	52	61	1.1	1.3	1.4	1.6	1.7	1.3	1.5	1.6	1.8	1.9
TW C2	26	33	42	48	56	62	69	77	18	24	30	35	43	48	54	63	1.1	1.2	1.4	1.6	1.7	1.3	1.4	1.6	1.7	1.9
TW C3	25	32	40	46	54	60	66	75	18	23	29	34	41	46	52	61	1.1	1.3	1.4	1.6	1.7	1.3	1.5	1.6	1.8	1.9
TW C4	23	29	36	41	49	54	61	69	17	21	26	30	37	41	47	55	1.2	1.4	1.5	1.7	1.8	1.5	1.6	1.7	1.9	2.0
TW D1	24	31	39	45	53	58	65	73	16	21	27	31	37	42	48	56	1.1	1.3	1.5	1.6	1.8	1.4	1.5	1.7	1.9	2.0
TW D2	34	44	54	62	71	76	83	89	24	32	41	47	56	62	69	78	0.9	1.1	1.2	1.4	1.6	1.1	1.2	1.4	1.5	1.7
TW D3	38	48	59	66	75	81	86	92	26	35	45	52	61	67	74	83	0.9	1.0	1.2	1.4	1.5	1.0	1.2	1.3	1.5	1.6
TW D4	40	51	62	70	79	84	89	94	28	38	48	56	66	72	79	87	0.8	1.0	1.2	1.3	1.5	1.0	1.1	1.3	1.4	1.6
TW D5	39	50	62	69	79	84	89	94	27	37	48	56	66	72	79	87	0.8	1.0	1.2	1.3	1.5	1.0	1.1	1.3	1.4	1.6
TW E1	42	53	64	71	80	84	89	94	30	39	50	57	66	72	79	87	0.8	1.0	1.2	1.3	1.5	1.0	1.1	1.3	1.4	1.6
TW E2	41	52	63	71	80	85	90	95	29	39	49	57	66	72	79	87	0.8	1.0	1.2	1.3	1.5	1.0	1.1	1.3	1.4	1.6
TW E3	48	60	72	79	87	91	94	98	35	47	58	66	76	82	87	93	0.8	0.9	1.1	1.3	1.4	0.9	1.0	1.2	1.4	1.5
TW E4	45	57	69	76	85	89	93	97	32	43	55	63	73	79	86	92	0.8	1.0	1.1	1.3	1.5	0.9	1.0	1.2	1.4	1.5
TW F1	27	33	41	46	54	59	66	74	19	24	29	34	40	44	50	58	1.1	1.3	1.5	1.6	1.8	1.4	1.5	1.7	1.8	2.0
TW F2	28	36	45	52	61	66	73	81	20	26	33	38	46	51	58	66	1.0	1.2	1.3	1.5	1.7	1.2	1.4	1.5	1.7	1.8
TW F3	28	37	47	53	63	68	75	83	19	25	33	39	47	53	60	69	1.0	1.2	1.3	1.5	1.6	1.2	1.3	1.5	1.6	1.8
TW F4	26	35	45	52	61	67	74	82	17	23	31	37	46	51	59	68	1.0	1.2	1.3	1.5	1.6	1.2	1.3	1.5	1.6	1.8
TW G1	25	32	39	45	53	59	65	74	17	22	28	33	39	44	50	58	1.1	1.3	1.4	1.6	1.7	1.4	1.5	1.7	1.8	2.0
TW G2	25	33	42	48	57	63	70	78	16	22	29	34	42	47	54	63	1.1	1.2	1.4	1.5	1.7	1.3	1.4	1.6	1.7	1.9
TW G3	23	31	40	47	56	62	69	78	15	21	28	33	41	46	53	62	1.1	1.2	1.4	1.5	1.7	1.3	1.4	1.6	1.7	1.9
TW G4	19	27	35	41	50	56	63	73	12	17	24	29	36	41	48	57	1.1	1.3	1.4	1.6	1.7	1.4	1.5	1.6	1.8	1.9
TW H1	21	28	36	42	50	56	63	72	14	19	25	29	36	41	47	56	1.2	1.3	1.5	1.6	1.8	1.4	1.5	1.7	1.8	2.0
TW H2	19	26	34	40	48	54	61	70	12	17	23	27	34	39	46	55	1.2	1.3	1.5	1.6	1.8	1.4	1.5	1.7	1.8	2.0
TW H3	16	22	30	35	43	49	56	66	10	14	20	24	31	35	41	50	1.2	1.4	1.5	1.7	1.8	1.5	1.6	1.8	1.9	2.0
TW H4	12	17	23	28	35	41	47	56	7	11	15	19	24	28	34	42	1.4	1.5	1.7	1.8	1.9	1.7	1.8	1.9	2.1	2.2
TW I1	21	28	36	41	49	54	61	70	14	18	24	28	34	38	44	52	1.2	1.3	1.5	1.7	1.8	1.4	1.6	1.8	1.9	2.1
TW I2	31	41	51	58	68	73	80	87	21	28	37	44	53	59	66	75	1.0	1.1	1.3	1.4	1.6	1.1	1.2	1.4	1.6	1.7
TW I3	33	44	54	62	71	77	83	90	23	31	41	48	57	63	70	79	0.9	1.1	1.2	1.4	1.6	1.1	1.2	1.4	1.5	1.7
TW I4	36	47	59	67	76	82	87	93	25	34	45	52	63	69	76	85	0.9	1.0	1.2	1.4	1.5	1.0	1.1	1.3	1.5	1.6
TW I5	36	48	60	67	77	82	88	94	25	35	46	53	64	70	77	86	0.9	1.0	1.2	1.4	1.5	1.0	1.1	1.3	1.4	1.6
TW J1	39	50	61	69	78	83	88	93	27	37	47	54	64	70	77	85	0.9	1.0	1.2	1.4	1.5	1.0	1.1	1.3	1.5	1.6
TW J2	38	49	60	68	77	82	88	93	26	36	46	54	64	70	77	85	0.9	1.0	1.2	1.4	1.5	1.0	1.1	1.3	1.4	1.6
TW J3	44	57	69	76	85	89	93	97	32	43	55	63	73	79	86	92	0.8	1.0	1.1	1.3	1.5	0.9	1.0	1.2	1.4	1.5
TW J4	41	54	66	74	83	87	92	96	29	40	52	60	71	77	84	91	0.8	1.0	1.1	1.3	1.5	0.9	1.1	1.2	1.4	1.5
DG A1	27	35	43	48	55	59	64	69	17	23	29	33	38	42	46	50	1.1	1.3	1.5	1.8	2.1	1.3	1.5	1.8	2.2	2.6
DG A2	29	38	47	54	62	67	73	79	19	26	34	39	47	52	58	65	1.0	1.2	1.4	1.6	1.8	1.2	1.3	1.5	1.8	2.0
DG A3	33	44	55	63	72	77	83	88	23	32	41	49	58	64	71	79	0.9	1.0	1.2	1.5	1.7	1.1	1.2	1.3	1.5	1.7
DG B1	28	37	47	53	62	67	72	78	17	23	30	36	43	48	53	60	1.0	1.2	1.4	1.7	1.9	1.3	1.4	1.6	1.9	2.2
DG B2	30	40	51	58	68	74	80	86	20	27	35	42	51	57	64	72	1.0	1.1	1.3	1.5	1.7	1.2	1.3	1.4	1.6	1.8
DG B3	34	45	58	66	76	82	87	92	24	32	43	51	62	69	76	84	0.9	1.0	1.2	1.4	1.6	1.1	1.1	1.3	1.5	1.6
DG C1	32	43	54	62	71	76	81	86	21	28	36	43	52	57	64	71	0.9	1.1	1.3	1.5	1.8	1.1	1.3	1.4	1.7	1.9
DG C2	34	45	57	65	75	80	86	91	23	31	41	48	59	65	72	80	0.9	1.0	1.2	1.4	1.6	1.1	1.2	1.3	1.5	1.7
DG C3	38	51	64	73	83	87	92	95	27	37	48	57	69	75	83	89	0.8	1.0	1.1	1.3	1.6	1.0	1.1	1.2	1.4	1.6
SS A1	36	45	53	60	68	73	78	85	29	36	43	49	56	61	67	75	1.0	1.1	1.3	1.5	1.6	1.1	1.3	1.4	1.6	1.8
SS A2	44	55	65	71	79	84	89	94	36	45	54	61	69	74	80	87	0.9	1.0	1.2	1.4	1.5	0.9	1.1	1.3	1.5	1.6
SS A3	34	42	50	56	63	68	74	81	27	33	40	45	52	57	62	70	1.0	1.2	1.4	1.5	1.7	1.2	1.3	1.5	1.7	1.9
SS A4	44	53	63	70	78	83	88	93	35	44	53	59	68	73	79	86	0.9	1.0	1.2	1.4	1.5	1.0	1.1	1.3	1.5	1.6
SS A5	34	42	50	55	63	68	74	81	27	33	40	45	52	56	62	69	1.0	1.2	1.4	1.6	1.7	1.2	1.4	1.5	1.7	1.9
SS A6	44	54	64	71	79	84	88	93	35	44	54	60	69	74	80	87	0.9	1.0	1.2	1.4	1.5	0.9	1.1	1.3	1.5	1.6
SS A7	31	37	44	49	56	61	67	74	25	30	35	40	46	50	55	62	1.1	1.3	1.5	1.7	1.8	1.3	1.5	1.7	1.9	2.0
SS A8	43	53	62	69	77	82	87	92	34	43	52	58	67	72	78	84	0.9	1.0	1.2	1.4	1.5	1.0	1.1	1.3	1.5	1.7
SS B1	31	39	48	54	62	67	73	81	24	30	37	43	50	55	61	69	1.0	1.2	1.4	1.5	1.7	1.2	1.3	1.5	1.7	1.8
SS B2	39	49	59	66	75	80	85	91	31	39	48	55	64	69	76	83	0.9	1.1	1.2	1.4	1.6	1.0	1.2	1.3	1.5	1.6
SS B3	30	37	45	51	59	64	70	78	23	29	35	40	47	52	58	65	1.1	1.2	1.4	1.6	1.7	1.2	1.4	1.6	1.8	1.9
SS B4	38	48	58	65	73	78	84	90	30	38	47	54	62	68	74	82	0.9	1.1	1.2	1.4	1.6	1.0	1.2	1.3	1.5	1.7
SS B5	28	35	42	47	55	60	66	74	22	27	33	37	44	48	54	61	1.1	1.3	1.5	1.6	1.8	1.3	1.5	1.7	1.8	2.0
SS B6	39	48	58	65	74	79	85	91	30	38	48	54	63	68	75	82	0.9	1.1	1.2	1.4	1.6	1.0	1.2	1.3	1.5	1.7
SS B7	26	32	38	43	50	55	61	68	20	25	30	34	39	43	48	55	1.2	1.4	1.5	1.7	1.9	1.4	1.6	1.8	2.0	2.1
SS B8	37	47	57	63	72	77	83	89	29	37	46	52	61	66	73	80	0.9	1.1	1.3	1.4	1.6	1.0	1.2	1.4	1.5	1.7
SS C1	32	41	51	58	67	73	79	86	22	29	37	43	51	57	64	73	1.0	1.1	1.3	1.5	1.6	1.1	1.3	1.4	1.6	1.8
SS C2	37	48	59	67	76	81	87	92	26	35	45	52	61	67	75	83	0.9	1.0	1.2	1.4	1.5	1.0	1.2	1.3	1.5	1.6
SS C3	26	33	41	47	56	61	68	76	18	24	30	34	41	46	52	61	1.1	1.3	1.4	1.6	1.7	1.3	1.5	1.6	1.8	1.9
SS C4	33	43	53	60	70	75	82	89	23	31	40	46	55	61	68	77	0.9	1.1	1.3	1.4	1.6	1.1	1.2	1.4	1.5	1.7
SS D1	43	53	62	69	77	81	86	92	34	42	51	57	65	70	76	83	0.9	1.1	1.2	1.4	1.6	1.0	1.2	1.3	1.5	1.7
SS D2	51	62	72	79	86	90	94	97	41	51	62	69	77	82	87	93	0.8	1.0	1.1	1.3	1.5	0.9	1.0	1.2	1.4	1.5
SS D3	39	48	57	63	71	76	81	88	31	38	46	52	59	64	70	78	0.9	1.1	1.3	1.5	1.6	1.1	1.2	1.4	1.6	1.8
SS D4	49	60	71	77	85	89	93	96	39	50	60	67	75	80	86	91	0.8	1.0	1.1	1.3	1.5	0.9	1.1	1.2	1.4	1.6
SS E1	38	47	57	63	72	77	82	89	29	36	45	51	59	64	70	78	0.9	1.1	1.3	1.4	1.6	1.0	1.2	1.4	1.6	1.7
SS E2	45	56	67	74	82	86	91	95	35	45	55	62	71	76	82	89	0.8	1.0	1.2	1.3	1.5	0.9	1.1	1.3	1.4	1.6
SS E3	32	40	48	54	62	67	73	81	25	31	38	43	50	55	61	68	1.0	1.2	1.4	1.5	1.7	1.2	1.4	1.5	1.7	1.9
SS E4	42	52	63	69	78	83	88	93	32	41	51	58	67	72	78	85	0.9	1.0	1.2	1.4	1.5	1.0	1.1	1.3	1.5	1.6

TONOPAH, NEVADA

TYPE	SOLAR SAVINGS FRACTION (%)																CONSERVATION FACTOR									
	BASE TEMP: 55 F (3502 DD)								65 F (5902 DD)								55 F					65 F				
LCR	100	70	50	40	30	25	20	15	100	70	50	40	30	25	20	15	100	50	30	20	15	100	50	30	20	15
WW A1	29	35	43	48	56	61	68	76	22	26	32	36	42	46	52	60	1.1	1.3	1.4	1.6	1.7	1.4	1.5	1.7	1.8	1.9
WW A2	34	44	53	60	70	75	82	89	24	31	39	46	54	60	67	76	1.0	1.1	1.3	1.4	1.6	1.1	1.3	1.4	1.6	1.7
WW A3	36	47	57	65	74	80	86	92	25	33	43	49	59	65	72	81	0.9	1.1	1.2	1.4	1.5	1.1	1.2	1.3	1.5	1.6
WW A4	38	48	59	67	76	82	87	93	26	34	44	51	61	67	75	83	0.9	1.0	1.2	1.4	1.5	1.0	1.2	1.3	1.5	1.6
WW A5	39	50	62	69	79	84	89	94	26	36	46	54	64	70	77	85	0.9	1.0	1.2	1.3	1.5	1.0	1.1	1.3	1.4	1.6
WW A6	40	51	63	71	80	85	90	95	27	36	47	55	65	71	78	86	0.8	1.0	1.2	1.3	1.5	1.0	1.1	1.3	1.4	1.6
WW B1	29	37	45	52	61	66	73	81	20	25	32	37	44	49	55	64	1.0	1.2	1.4	1.5	1.7	1.3	1.4	1.6	1.7	1.9
WW B2	38	49	60	68	78	83	88	94	26	36	46	53	63	69	77	85	0.9	1.0	1.2	1.3	1.5	1.0	1.1	1.3	1.4	1.6
WW B3	45	57	68	76	84	89	93	97	32	43	54	62	72	77	84	91	0.8	1.0	1.1	1.3	1.5	0.9	1.1	1.2	1.4	1.5
WW B4	45	57	69	77	85	90	94	97	32	43	55	63	74	80	86	92	0.8	0.9	1.1	1.3	1.4	0.9	1.0	1.2	1.4	1.5
WW B5	43	56	68	76	85	89	94	97	30	42	54	62	73	79	85	92	0.8	1.0	1.1	1.3	1.4	0.9	1.0	1.2	1.4	1.5
WW C1	46	59	70	77	86	90	94	97	34	45	56	64	74	79	86	92	0.8	1.0	1.1	1.3	1.4	0.9	1.0	1.2	1.4	1.5
WW C2	44	56	68	76	84	89	93	97	31	42	54	62	72	78	85	91	0.8	1.0	1.1	1.3	1.5	0.9	1.0	1.2	1.4	1.5
WW C3	50	64	76	83	90	94	97	99	37	50	63	71	81	86	91	96	0.7	0.9	1.1	1.3	1.4	0.8	1.0	1.1	1.3	1.5
WW C4	47	60	72	80	88	92	96	98	34	46	59	67	78	83	89	95	0.8	0.9	1.1	1.3	1.4	0.8	1.0	1.2	1.3	1.5
TW A1	28	34	42	47	55	60	67	75	21	26	31	35	41	46	52	60	1.1	1.3	1.4	1.6	1.7	1.4	1.5	1.7	1.8	2.0
TW A2	31	40	49	55	64	70	77	84	22	29	36	41	49	55	62	71	1.0	1.2	1.3	1.5	1.6	1.2	1.3	1.5	1.6	1.8
TW A3	32	41	51	58	68	74	80	88	23	30	38	44	53	58	66	75	1.0	1.1	1.3	1.4	1.6	1.1	1.3	1.4	1.6	1.7
TW A4	32	41	52	59	69	74	81	88	22	29	38	44	53	59	67	76	0.9	1.1	1.3	1.4	1.6	1.1	1.2	1.4	1.6	1.7
TW B1	28	35	42	48	56	62	68	77	21	25	31	36	42	47	53	62	1.1	1.3	1.4	1.6	1.7	1.4	1.5	1.6	1.8	1.9
TW B2	30	38	47	54	63	69	76	84	21	27	34	40	48	54	61	70	1.0	1.2	1.3	1.5	1.6	1.2	1.3	1.5	1.6	1.8
TW B3	29	38	47	54	64	70	76	85	20	27	35	41	49	55	62	71	1.0	1.1	1.3	1.5	1.6	1.2	1.3	1.5	1.6	1.7
TW B4	27	36	45	52	61	67	74	82	19	25	33	38	46	52	59	69	1.0	1.2	1.3	1.5	1.6	1.2	1.4	1.5	1.6	1.8
TW C1	27	34	42	48	56	62	68	77	19	24	31	35	42	47	54	62	1.1	1.3	1.4	1.6	1.7	1.3	1.5	1.6	1.8	1.9
TW C2	26	34	42	49	57	63	70	79	19	24	31	36	44	49	56	65	1.1	1.2	1.4	1.5	1.7	1.3	1.4	1.6	1.7	1.8
TW C3	25	32	40	46	55	61	68	77	18	23	29	34	42	47	53	62	1.1	1.3	1.4	1.6	1.7	1.3	1.5	1.6	1.7	1.9
TW C4	23	29	36	41	49	55	62	71	17	21	27	31	37	42	48	56	1.2	1.3	1.5	1.6	1.8	1.4	1.6	1.7	1.8	2.0
TW D1	24	31	39	45	54	59	66	75	16	21	27	32	38	43	49	58	1.1	1.3	1.4	1.6	1.7	1.4	1.5	1.7	1.8	2.0
TW D2	35	45	55	63	72	78	84	91	24	32	42	48	58	64	71	80	0.9	1.1	1.2	1.4	1.5	1.1	1.2	1.3	1.5	1.6
TW D3	38	49	60	68	77	82	88	93	27	36	46	53	63	69	76	84	0.9	1.0	1.2	1.4	1.5	1.0	1.1	1.3	1.5	1.6
TW D4	40	52	64	71	80	85	91	95	29	39	50	57	67	74	81	88	0.8	1.0	1.2	1.3	1.5	0.9	1.1	1.2	1.4	1.6
TW D5	39	51	63	71	80	85	90	95	28	38	49	57	67	74	81	89	0.8	1.0	1.2	1.3	1.5	0.9	1.1	1.2	1.4	1.5
TW E1	42	53	65	72	81	86	91	95	31	40	51	58	68	74	81	88	0.8	1.0	1.2	1.3	1.5	1.0	1.1	1.3	1.4	1.6
TW E2	41	53	64	72	81	86	91	96	30	40	50	58	68	74	81	89	0.8	1.0	1.2	1.3	1.5	0.9	1.1	1.2	1.4	1.6
TW E3	48	61	73	80	88	92	95	98	36	48	60	68	78	83	89	94	0.8	0.9	1.1	1.3	1.4	0.9	1.0	1.2	1.3	1.5
TW E4	45	58	70	78	86	90	94	98	33	44	56	65	75	81	87	93	0.8	0.9	1.1	1.3	1.4	0.9	1.0	1.2	1.3	1.5
TW F1	27	33	41	47	55	60	67	76	20	24	30	34	41	46	52	60	1.1	1.3	1.4	1.6	1.7	1.4	1.5	1.7	1.8	1.9
TW F2	29	37	46	53	62	68	75	83	20	26	33	39	47	52	59	69	1.0	1.2	1.3	1.5	1.6	1.2	1.4	1.5	1.7	1.8
TW F3	29	38	47	54	64	70	77	85	19	26	34	40	48	54	62	71	1.0	1.1	1.3	1.5	1.6	1.2	1.3	1.5	1.6	1.7
TW F4	26	36	46	53	62	69	76	84	17	24	32	38	47	53	60	70	1.0	1.1	1.3	1.5	1.6	1.2	1.3	1.5	1.6	1.7
TW G1	25	32	40	46	54	60	67	76	18	23	29	33	40	45	51	60	1.1	1.3	1.4	1.6	1.7	1.4	1.5	1.6	1.8	1.9
TW G2	25	33	42	49	58	64	71	80	17	23	30	35	43	49	56	65	1.1	1.2	1.4	1.5	1.6	1.3	1.4	1.5	1.7	1.8
TW G3	23	31	41	47	57	63	70	79	15	21	28	34	42	47	55	64	1.1	1.2	1.4	1.5	1.7	1.3	1.4	1.5	1.7	1.8
TW G4	19	27	35	42	51	57	65	74	12	17	24	29	37	42	49	59	1.1	1.3	1.4	1.6	1.7	1.3	1.5	1.6	1.8	1.9
TW H1	22	29	36	43	51	57	64	73	14	19	25	30	37	42	49	58	1.1	1.3	1.4	1.6	1.7	1.4	1.5	1.7	1.8	1.9
TW H2	19	26	34	40	49	55	62	72	12	17	23	28	35	40	47	56	1.2	1.3	1.4	1.6	1.7	1.4	1.5	1.7	1.8	1.9
TW H3	16	22	30	36	44	50	57	67	10	14	20	25	31	36	42	52	1.2	1.4	1.5	1.7	1.8	1.5	1.6	1.7	1.9	2.0
TW H4	12	17	24	29	36	41	48	57	7	11	15	19	25	29	35	43	1.4	1.5	1.6	1.8	1.9	1.7	1.8	1.9	2.0	2.2
TW I1	22	28	36	42	50	56	63	72	14	19	24	29	35	39	46	54	1.2	1.3	1.5	1.6	1.8	1.4	1.6	1.7	1.9	2.0
TW I2	31	41	52	59	69	75	82	89	21	29	38	45	54	60	68	77	0.9	1.1	1.2	1.4	1.6	1.1	1.2	1.4	1.5	1.7
TW I3	34	44	55	63	73	79	85	91	23	32	42	49	58	65	72	81	0.9	1.1	1.2	1.4	1.5	1.0	1.2	1.3	1.5	1.6
TW I4	37	48	60	68	78	83	89	94	25	35	46	54	64	70	78	86	0.9	1.0	1.2	1.3	1.5	1.0	1.1	1.3	1.4	1.6
TW I5	37	49	61	69	78	84	89	95	25	36	47	55	65	72	79	87	0.9	1.0	1.2	1.3	1.5	1.0	1.1	1.3	1.4	1.6
TW J1	39	51	62	70	79	84	90	95	28	38	48	56	66	72	79	87	0.9	1.0	1.2	1.3	1.5	1.0	1.1	1.3	1.4	1.6
TW J2	38	50	61	69	79	84	89	95	27	36	47	55	65	71	79	87	0.9	1.0	1.2	1.3	1.5	1.0	1.1	1.3	1.4	1.6
TW J3	45	58	70	77	86	90	94	98	32	44	56	65	75	81	87	93	0.8	0.9	1.1	1.3	1.4	0.9	1.0	1.2	1.3	1.5
TW J4	42	55	67	75	84	89	93	97	30	41	53	62	72	78	85	92	0.8	1.0	1.1	1.3	1.4	0.9	1.0	1.2	1.4	1.5
DG A1	27	35	43	49	56	60	65	71	17	23	29	34	40	43	47	52	1.0	1.3	1.5	1.8	2.1	1.3	1.5	1.8	2.1	2.5
DG A2	29	38	48	55	63	68	74	81	19	26	34	40	48	53	59	67	1.0	1.1	1.3	1.6	1.7	1.2	1.3	1.5	1.7	1.9
DG A3	34	45	56	64	73	79	84	89	24	32	42	50	60	66	73	81	0.9	1.0	1.2	1.4	1.7	1.1	1.1	1.3	1.5	1.7
DG B1	28	37	48	55	64	69	74	80	18	24	31	37	45	50	56	62	1.0	1.1	1.3	1.6	1.9	1.2	1.4	1.5	1.8	2.1
DG B2	30	40	52	60	70	75	81	88	20	27	36	43	53	59	66	75	1.0	1.1	1.2	1.5	1.6	1.1	1.2	1.4	1.6	1.8
DG B3	34	46	59	68	78	83	89	93	24	33	44	52	64	70	78	86	0.9	1.0	1.2	1.4	1.6	1.0	1.1	1.2	1.4	1.6
DG C1	33	44	55	64	73	78	83	88	21	28	37	44	54	60	67	74	0.9	1.0	1.2	1.5	1.8	1.1	1.2	1.4	1.6	1.9
DG C2	34	46	58	67	77	82	88	93	23	31	42	50	60	67	74	82	0.9	1.0	1.2	1.4	1.6	1.1	1.1	1.3	1.5	1.7
DG C3	39	52	66	75	84	89	93	96	27	37	49	59	71	78	84	91	0.8	0.9	1.1	1.3	1.6	1.0	1.0	1.2	1.4	1.6
SS A1	37	45	54	61	69	74	80	87	29	36	44	49	57	63	69	77	1.0	1.1	1.3	1.5	1.6	1.1	1.2	1.4	1.6	1.7
SS A2	45	55	66	73	81	85	90	95	36	45	55	62	71	76	82	89	0.8	1.0	1.2	1.3	1.5	0.9	1.1	1.3	1.4	1.6
SS A3	34	42	51	57	65	70	76	84	27	33	40	46	53	58	64	72	1.0	1.2	1.3	1.5	1.7	1.1	1.3	1.5	1.7	1.8
SS A4	44	54	65	71	80	84	89	94	35	44	54	61	69	74	81	87	0.9	1.0	1.2	1.4	1.5	0.9	1.1	1.3	1.4	1.6
SS A5	34	42	50	56	64	69	76	83	27	34	40	45	53	58	64	71	1.0	1.2	1.3	1.5	1.7	1.2	1.3	1.5	1.7	1.8
SS A6	45	55	66	72	81	85	90	95	36	45	55	61	70	76	82	88	0.8	1.0	1.2	1.3	1.5	0.9	1.1	1.3	1.4	1.6
SS A7	31	37	45	50	57	62	68	76	25	30	35	40	46	50	56	63	1.1	1.3	1.4	1.6	1.8	1.3	1.5	1.6	1.8	2.0
SS A8	43	54	64	70	79	83	88	94	35	43	53	59	68	73	79	86	0.9	1.0	1.2	1.4	1.5	1.0	1.1	1.3	1.5	1.6
SS B1	32	40	48	55	63	69	75	83	24	31	38	43	51	56	63	71	1.0	1.2	1.3	1.5	1.6	1.2	1.3	1.5	1.7	1.8
SS B2	40	50	60	67	76	81	87	92	31	40	49	56	65	71	77	85	0.9	1.0	1.2	1.4	1.5	1.0	1.1	1.3	1.5	1.6
SS B3	30	38	46	52	60	66	72	80	23	29	36	41	48	53	59	67	1.0	1.2	1.4	1.5	1.7	1.2	1.4	1.5	1.7	1.9
SS B4	39	49	59	66	75	80	86	92	30	39	48	55	64	69	76	84	0.9	1.1	1.2	1.4	1.5	1.0	1.2	1.3	1.5	1.6
SS B5	28	35	43	48	56	61	68	76	22	27	33	38	44	49	55	63	1.1	1.3	1.4	1.6	1.7	1.3	1.5	1.6	1.8	1.9
SS B6	39	49	59	66	75	81	86	92	30	39	48	55	64	70	76	84	0.9	1.0	1.2	1.4	1.5	1.0	1.1	1.3	1.5	1.6
SS B7	26	32	39	44	51	56	62	70	20	25	30	34	40	44	49	57	1.2	1.4	1.5	1.7	1.8	1.4	1.6	1.7	1.9	2.0
SS B8	38	47	57	64	73	79	84	91	29	37	46	53	62	67	74	82	0.9	1.1	1.2	1.4	1.5	1.0	1.2	1.3	1.5	1.6
SS C1	33	42	52	59	69	74	81	88	22	29	38	44	53	59	66	75	1.0	1.1	1.3	1.4	1.6	1.1	1.3	1.4	1.6	1.7
SS C2	38	49	60	68	77	83	88	94	27	36	46	53	63	69	76	85	0.9	1.0	1.2	1.4	1.5	1.0	1.1	1.3	1.5	1.6
SS C3	26	34	42	48	57	62	69	78	19	24	30	35	42	47	54	63	1.1	1.2	1.4	1.5	1.7	1.3	1.5	1.6	1.7	1.9
SS C4	34	44	54	61	71	77	83	90	24	32	41	47	57	63	70	79	0.9	1.1	1.2	1.4	1.5	1.1	1.2	1.4	1.5	1.7
SS D1	44	54	64	70	79	83	88	93	34	43	52	58	67	72	78	85	0.9	1.0	1.2	1.4	1.5	1.0	1.1	1.3	1.5	1.6
SS D2	52	63	74	81	88	91	95	98	41	52	63	70	79	84	89	94	0.8	0.9	1.1	1.3	1.4	0.9	1.0	1.2	1.4	1,5
SS D3	39	48	58	64	72	78	83	89	31	38	47	53	61	66	72	80	0.9	1.1	1.3	1.4	1.6	1.1	1.2	1.4	1.6	1.7
SS D4	50	61	72	79	86	90	94	97	40	50	61	68	77	82	87	93	0.8	1.0	1.1	1.3	1.5	0.9	1.0	1.2	1.4	1.5
SS E1	38	48	58	65	73	79	84	91	29	37	45	52	60	66	72	80	0.9	1.1	1.2	1.4	1.6	1.0	1.2	1.4	1.5	1.7
SS E2	46	57	68	75	84	88	92	96	35	46	56	63	73	78	84	91	0.8	1.0	1.1	1.3	1.5	0.9	1.1	1.2	1.4	1.5
SS E3	32	40	49	55	63	69	75	83	25	31	38	43	51	56	62	70	1.0	1.2	1.3	1.5	1.7	1.2	1.3	1.5	1.7	1.8
SS E4	42	53	64	71	79	84	89	94	33	42	52	59	68	74	80	87	0.9	1.0	1.2	1.4	1.5	1.0	1.1	1.3	1.4	1.6

WINNEMUCCA, NEVADA

TYPE	SOLAR SAVINGS FRACTION (%) BASE TEMP: 55 F (4006 DD)								65 F (6636 DD)								CONSERVATION FACTOR 55 F					65 F				
LCR	100	70	50	40	30	25	20	15	100	70	50	40	30	25	20	15	100	50	30	20	15	100	50	30	20	15
WW A1	24	29	35	39	46	50	56	64	18	22	26	29	34	38	42	49	1.3	1.5	1.6	1.8	1.9	1.6	1.8	1.9	2.1	2.2
WW A2	27	35	43	50	58	63	70	78	19	25	32	37	44	49	55	64	1.1	1.2	1.4	1.6	1.7	1.3	1.4	1.6	1.7	1.9
WW A3	29	37	47	53	62	68	74	82	20	26	34	40	48	53	60	69	1.0	1.2	1.3	1.5	1.7	1.2	1.3	1.5	1.7	1.8
WW A4	29	38	48	55	64	70	77	84	20	27	35	41	50	55	62	71	1.0	1.1	1.3	1.5	1.6	1.1	1.3	1.5	1.6	1.8
WW A5	30	40	50	57	67	73	79	86	20	28	37	43	52	58	65	74	0.9	1.1	1.3	1.5	1.6	1.1	1.3	1.4	1.6	1.7
WW A6	31	41	51	59	68	74	80	87	20	28	37	44	53	59	66	75	0.9	1.1	1.3	1.4	1.6	1.1	1.2	1.4	1.6	1.7
WW B1	23	29	36	41	48	53	59	67	15	20	24	28	34	37	42	49	1.2	1.4	1.5	1.7	1.9	1.5	1.7	1.9	2.1	2.2
WW B2	30	40	50	57	66	72	79	86	20	28	37	44	53	58	66	75	1.0	1.1	1.3	1.4	1.6	1.1	1.2	1.4	1.6	1.7
WW B3	36	46	57	65	74	79	85	91	25	34	44	51	61	67	74	82	0.9	1.0	1.2	1.4	1.5	1.0	1.2	1.3	1.5	1.6
WW B4	35	47	59	66	76	81	87	93	25	35	46	54	64	70	77	85	0.9	1.0	1.2	1.4	1.5	1.0	1.1	1.3	1.4	1.6
WW B5	34	46	58	65	75	81	87	93	24	34	45	53	63	70	77	85	0.9	1.0	1.2	1.4	1.5	1.0	1.1	1.3	1.4	1.6
WW C1	37	48	59	67	76	81	87	92	27	36	47	54	63	69	76	84	0.9	1.0	1.2	1.4	1.5	1.0	1.1	1.3	1.5	1.6
WW C2	35	46	58	65	75	80	86	92	25	34	45	52	62	68	76	84	0.9	1.0	1.2	1.4	1.5	1.0	1.1	1.3	1.5	1.6
WW C3	41	54	66	73	82	87	92	96	30	42	54	62	72	78	85	91	0.8	1.0	1.1	1.3	1.5	0.9	1.0	1.2	1.4	1.5
WW C4	37	50	62	70	80	85	90	95	27	38	50	58	69	75	82	89	0.8	1.0	1.2	1.3	1.5	0.9	1.1	1.2	1.4	1.5
TW A1	24	29	34	39	45	50	56	63	18	22	26	29	34	37	42	49	1.3	1.5	1.6	1.8	1.9	1.6	1.8	1.9	2.1	2.2
TW A2	25	32	40	45	53	58	65	73	18	23	29	34	40	45	51	59	1.1	1.3	1.5	1.6	1.8	1.4	1.5	1.7	1.8	2.0
TW A3	26	33	42	48	56	62	69	77	18	24	30	35	43	48	54	63	1.1	1.2	1.4	1.6	1.7	1.3	1.4	1.6	1.7	1.9
TW A4	25	33	42	48	57	63	70	78	17	23	30	36	43	49	55	64	1.1	1.2	1.4	1.5	1.7	1.2	1.4	1.6	1.7	1.9
TW B1	23	28	35	39	46	51	57	65	17	21	26	29	35	39	44	51	1.3	1.4	1.6	1.8	1.9	1.5	1.7	1.8	2.0	2.2
TW B2	24	31	38	44	52	57	64	73	17	22	28	33	39	44	50	58	1.1	1.3	1.5	1.6	1.8	1.4	1.5	1.7	1.8	2.0
TW B3	23	30	38	44	53	58	65	74	16	22	28	33	40	45	51	60	1.1	1.3	1.4	1.6	1.7	1.3	1.5	1.6	1.8	1.9
TW B4	22	29	36	42	50	56	63	71	15	20	26	31	38	43	49	58	1.2	1.3	1.5	1.6	1.8	1.4	1.5	1.7	1.8	2.0
TW C1	22	27	34	39	46	51	57	66	16	20	25	29	35	39	44	52	1.3	1.4	1.6	1.7	1.9	1.5	1.6	1.8	2.0	2.1
TW C2	21	27	34	40	47	53	59	68	15	20	25	29	36	40	46	54	1.2	1.4	1.5	1.7	1.8	1.4	1.6	1.7	1.9	2.0
TW C3	20	26	33	38	45	50	57	66	15	19	24	28	34	38	44	52	1.3	1.4	1.6	1.7	1.9	1.5	1.6	1.8	1.9	2.1
TW C4	19	24	30	34	41	46	52	60	14	18	22	26	31	35	40	47	1.4	1.5	1.6	1.8	1.9	1.6	1.7	1.9	2.0	2.2
TW D1	19	24	31	36	42	47	53	61	13	16	21	24	29	33	38	44	1.3	1.5	1.6	1.8	1.9	1.6	1.8	1.9	2.1	2.3
TW D2	28	36	46	52	61	67	74	82	20	26	34	40	48	54	61	70	1.0	1.2	1.3	1.5	1.6	1.2	1.3	1.5	1.6	1.8
TW D3	30	40	50	57	66	71	78	85	21	29	37	44	52	58	65	74	1.0	1.1	1.3	1.5	1.6	1.1	1.3	1.4	1.6	1.7
TW D4	32	42	53	61	70	76	82	89	23	32	41	48	58	64	71	80	0.9	1.1	1.2	1.4	1.6	1.0	1.2	1.3	1.5	1.6
TW D5	31	42	53	61	70	76	83	89	22	31	41	48	58	65	72	81	0.9	1.1	1.2	1.4	1.6	1.0	1.2	1.3	1.5	1.6
TW E1	34	44	54	61	71	76	82	89	25	33	42	49	58	64	71	79	0.9	1.1	1.2	1.4	1.6	1.0	1.2	1.4	1.5	1.7
TW E2	33	43	54	61	71	76	83	89	24	32	42	49	58	64	71	80	0.9	1.1	1.2	1.4	1.6	1.0	1.2	1.3	1.5	1.7
TW E3	39	51	63	70	79	84	90	95	29	40	51	58	68	74	81	89	0.8	1.0	1.2	1.3	1.5	0.9	1.1	1.2	1.4	1.6
TW E4	36	48	60	68	77	82	88	94	26	37	48	56	66	72	79	87	0.9	1.0	1.2	1.4	1.5	1.0	1.1	1.3	1.4	1.6
TW F1	22	27	33	38	45	50	55	64	16	20	24	28	33	37	42	49	1.3	1.4	1.6	1.8	1.9	1.6	1.7	1.9	2.1	2.2
TW F2	23	29	37	43	51	56	63	71	16	21	27	31	38	42	48	56	1.2	1.3	1.5	1.6	1.8	1.4	1.5	1.7	1.9	2.0
TW F3	22	30	38	44	52	58	65	74	15	20	27	32	39	44	50	59	1.1	1.3	1.4	1.6	1.7	1.3	1.5	1.6	1.8	1.9
TW F4	20	28	36	42	51	57	64	73	13	18	25	30	37	43	49	58	1.1	1.3	1.4	1.6	1.7	1.3	1.5	1.6	1.8	1.9
TW G1	20	26	32	37	44	49	55	64	14	18	23	27	32	36	41	49	1.3	1.4	1.6	1.8	1.9	1.5	1.7	1.9	2.0	2.2
TW G2	20	26	34	39	47	53	60	68	13	18	23	28	34	39	45	54	1.2	1.3	1.5	1.7	1.8	1.4	1.6	1.7	1.9	2.0
TW G3	18	24	32	38	46	51	58	68	11	16	22	27	33	38	44	53	1.2	1.3	1.5	1.7	1.8	1.4	1.6	1.7	1.9	2.0
TW G4	15	21	28	33	41	46	53	62	9	13	19	23	29	34	40	48	1.3	1.4	1.6	1.7	1.9	1.5	1.6	1.8	1.9	2.1
TW H1	17	22	29	34	41	46	53	61	11	15	20	24	29	34	39	47	1.3	1.4	1.6	1.8	1.9	1.6	1.7	1.9	2.0	2.2
TW H2	14	20	27	32	39	44	51	60	9	13	18	22	28	32	38	46	1.3	1.5	1.6	1.8	1.9	1.6	1.7	1.8	2.0	2.1
TW H3	12	17	23	28	35	40	47	55	7	11	15	19	25	29	34	42	1.4	1.5	1.7	1.8	2.0	1.6	1.8	1.9	2.1	2.2
TW H4	9	13	18	22	28	33	38	47	5	8	12	15	19	23	28	34	1.5	1.7	1.8	2.0	2.1	1.8	2.0	2.1	2.3	2.4
TW I1	17	22	28	32	39	44	50	58	11	14	18	21	26	30	34	41	1.3	1.5	1.7	1.8	2.0	1.7	1.8	2.0	2.2	2.4
TW I2	24	33	42	49	58	64	71	79	16	23	30	36	44	50	57	66	1.0	1.2	1.4	1.5	1.7	1.2	1.4	1.5	1.7	1.8
TW I3	26	35	45	52	62	68	74	83	18	25	34	40	48	54	62	71	1.0	1.2	1.3	1.5	1.6	1.1	1.3	1.5	1.6	1.8
TW I4	29	39	50	57	67	73	80	87	20	28	38	45	54	61	68	77	0.9	1.1	1.3	1.4	1.6	1.1	1.2	1.4	1.5	1.7
TW I5	29	39	51	58	68	74	81	88	20	29	39	46	56	62	70	79	0.9	1.1	1.2	1.4	1.6	1.0	1.2	1.3	1.5	1.6
TW J1	31	41	52	59	68	74	81	88	22	30	40	46	56	62	69	78	0.9	1.1	1.3	1.4	1.6	1.1	1.2	1.4	1.5	1.7
TW J2	30	40	51	58	68	74	81	88	21	29	39	46	55	61	69	78	0.9	1.1	1.3	1.4	1.6	1.1	1.2	1.4	1.5	1.7
TW J3	36	48	60	67	77	82	88	93	26	36	48	55	66	72	79	87	0.9	1.0	1.2	1.4	1.5	0.9	1.1	1.3	1.4	1.6
TW J4	33	45	57	65	75	80	86	92	24	34	45	53	63	70	77	85	0.9	1.0	1.2	1.4	1.5	1.0	1.1	1.3	1.4	1.6
DG A1	21	27	33	38	44	48	52	57	13	17	22	25	30	33	36	39	1.2	1.4	1.7	2.1	2.5	1.5	1.7	2.1	2.6	3.3
DG A2	23	30	38	44	52	57	63	70	15	21	27	32	39	43	49	56	1.1	1.3	1.5	1.7	1.9	1.3	1.5	1.7	1.9	2.2
DG A3	27	36	46	54	63	69	75	82	19	26	35	41	50	56	63	72	1.0	1.1	1.3	1.5	1.7	1.2	1.3	1.4	1.6	1.8
DG B1	21	28	36	42	49	54	59	65	13	18	23	27	33	37	41	47	1.2	1.3	1.5	1.8	2.1	1.5	1.6	1.8	2.2	2.5
DG B2	23	31	41	48	57	63	69	77	16	22	28	34	42	47	54	62	1.1	1.2	1.4	1.6	1.8	1.3	1.4	1.5	1.7	2.0
DG B3	28	37	48	56	67	73	80	87	20	27	36	43	53	60	68	77	1.0	1.1	1.2	1.4	1.6	1.2	1.2	1.4	1.5	1.7
DG C1	25	33	42	49	58	63	69	76	16	22	28	33	41	46	51	58	1.1	1.2	1.4	1.6	1.9	1.3	1.5	1.6	1.9	2.1
DG C2	27	36	46	54	64	70	77	84	19	25	33	40	49	55	62	71	1.0	1.1	1.3	1.5	1.7	1.2	1.3	1.4	1.6	1.8
DG C3	31	42	54	62	73	79	86	91	23	31	41	48	59	66	75	83	0.9	1.0	1.2	1.4	1.6	1.1	1.2	1.3	1.4	1.6
SS A1	30	37	45	51	58	63	69	77	24	30	36	41	48	53	58	66	1.1	1.3	1.4	1.6	1.8	1.2	1.4	1.6	1.8	1.9
SS A2	37	47	56	63	71	76	82	88	30	38	47	53	61	67	73	80	0.9	1.1	1.3	1.4	1.6	1.0	1.2	1.4	1.5	1.7
SS A3	28	35	42	47	54	59	64	72	22	28	33	38	44	48	53	60	1.1	1.3	1.5	1.7	1.8	1.3	1.5	1.7	1.9	2.1
SS A4	37	45	55	61	69	74	80	87	29	37	46	52	60	65	71	78	0.9	1.1	1.3	1.5	1.6	1.0	1.2	1.4	1.6	1.7
SS A5	29	35	42	47	54	58	64	71	23	28	34	38	44	48	53	60	1.2	1.3	1.5	1.7	1.8	1.3	1.5	1.7	1.9	2.1
SS A6	37	46	56	62	71	76	81	88	30	38	46	52	61	66	72	80	0.9	1.1	1.3	1.5	1.6	1.0	1.2	1.4	1.6	1.7
SS A7	26	31	37	41	47	51	57	64	21	25	30	33	38	41	46	52	1.3	1.5	1.6	1.8	2.0	1.5	1.7	1.9	2.1	2.3
SS A8	36	45	54	60	68	73	79	86	29	36	45	50	58	63	69	77	1.0	1.1	1.3	1.5	1.6	1.1	1.2	1.4	1.6	1.8
SS B1	26	33	40	45	53	58	64	72	20	25	31	36	42	46	52	60	1.1	1.3	1.5	1.7	1.8	1.3	1.5	1.7	1.9	2.0
SS B2	33	42	51	57	66	71	78	85	26	33	42	48	56	61	68	76	1.0	1.1	1.3	1.5	1.6	1.1	1.2	1.4	1.6	1.7
SS B3	24	31	37	43	50	54	60	68	19	24	29	33	39	43	48	55	1.2	1.4	1.5	1.7	1.9	1.4	1.6	1.7	1.9	2.1
SS B4	32	41	50	56	65	70	76	84	25	32	41	46	55	60	66	74	1.0	1.2	1.3	1.5	1.6	1.1	1.3	1.4	1.6	1.8
SS B5	24	29	35	40	46	51	56	64	19	23	28	31	37	40	45	52	1.3	1.4	1.6	1.8	1.9	1.5	1.6	1.8	2.0	2.2
SS B6	32	41	50	57	65	71	77	84	25	32	41	47	55	60	67	75	1.0	1.1	1.3	1.5	1.6	1.1	1.3	1.4	1.6	1.7
SS B7	22	26	32	36	42	46	51	58	17	21	25	28	33	36	40	46	1.4	1.5	1.7	1.9	2.1	1.6	1.8	2.0	2.2	2.4
SS B8	31	39	48	54	63	68	75	82	24	31	39	45	53	58	64	72	1.0	1.2	1.3	1.5	1.7	1.1	1.3	1.5	1.6	1.8
SS C1	25	33	42	48	56	62	69	77	17	23	30	35	42	47	53	62	1.1	1.2	1.4	1.6	1.7	1.3	1.4	1.6	1.8	1.9
SS C2	30	40	50	57	66	72	79	86	21	29	37	44	53	59	66	75	1.0	1.1	1.3	1.5	1.6	1.1	1.3	1.4	1.6	1.7
SS C3	21	27	34	39	46	51	58	66	15	19	24	28	34	38	44	51	1.2	1.4	1.6	1.7	1.9	1.5	1.6	1.8	2.0	2.1
SS C4	27	35	44	51	60	66	73	81	19	26	33	39	47	53	60	69	1.0	1.2	1.3	1.5	1.7	1.2	1.3	1.5	1.7	1.8
SS D1	36	44	53	59	67	72	78	85	28	35	43	48	56	61	67	74	1.0	1.1	1.3	1.5	1.7	1.1	1.3	1.5	1.6	1.8
SS D2	43	53	64	70	79	83	88	93	34	44	54	61	69	75	81	87	0.9	1.0	1.2	1.4	1.5	0.9	1.1	1.3	1.5	1.6
SS D3	32	40	48	54	61	66	72	79	26	32	39	44	51	55	61	68	1.0	1.2	1.4	1.6	1.7	1.2	1.4	1.5	1.7	1.9
SS D4	41	52	62	69	77	82	87	92	33	42	52	59	67	73	79	86	0.9	1.0	1.2	1.4	1.5	0.9	1.1	1.3	1.5	1.6
SS E1	31	39	47	53	62	67	73	80	23	30	37	42	49	54	60	68	1.0	1.2	1.4	1.6	1.7	1.2	1.3	1.5	1.7	1.9
SS E2	37	47	58	65	73	78	84	90	29	38	47	54	63	68	75	82	0.9	1.1	1.2	1.4	1.6	1.0	1.2	1.3	1.5	1.7
SS E3	27	33	40	45	53	57	63	71	21	26	31	36	42	46	51	58	1.2	1.3	1.5	1.7	1.8	1.3	1.5	1.7	1.9	2.1
SS E4	35	44	54	60	69	74	80	87	27	35	44	50	58	64	70	78	0.9	1.1	1.3	1.5	1.6	1.0	1.2	1.4	1.6	1.7

CONCORD, NEW HAMPSHIRE

SOLAR SAVINGS FRACTION (%)

TYPF	BASE TEMP: 55 F (4761 DD)								65 F (7360 DD)							
LCR	100	70	50	40	30	25	20	15	100	70	50	40	30	25	20	15
WW A1	13	14	16	17	19	20	22	24	11	12	13	13	14	15	15	15
WW A2	11	14	17	20	23	26	29	34	8	10	12	14	16	18	20	23
WW A3	10	13	17	21	25	28	32	38	7	9	12	14	17	19	22	25
WW A4	9	13	18	21	26	29	34	40	6	8	12	14	17	20	23	27
WW A5	8	13	18	22	27	31	36	43	5	8	11	14	18	21	24	29
WW A6	8	13	18	22	28	32	37	44	4	7	11	14	18	21	25	30
WW B1	8	9	10	11	13	13	14	15	5	6	6	6	6	5	4	2
WW B2	10	15	20	24	30	35	40	48	7	10	14	18	22	26	30	36
WW B3	13	18	25	30	37	42	48	57	9	13	18	22	28	32	38	45
WW B4	13	20	27	33	42	48	55	65	8	14	21	26	34	39	46	55
WW B5	12	19	27	34	43	49	56	66	8	14	21	27	35	40	48	57
WW C1	14	20	27	32	40	45	52	61	10	15	21	25	32	36	42	50
WW C2	13	19	27	32	40	46	53	62	9	14	20	25	32	37	43	52
WW C3	16	25	35	42	52	58	66	76	12	19	28	35	44	51	58	68
WW C4	14	23	32	39	49	55	63	73	10	17	26	32	41	48	55	66
TW A1	13	14	16	18	20	21	23	25	11	12	13	14	15	16	16	17
TW A2	11	14	17	19	22	25	28	32	9	11	12	14	16	18	19	22
TW A3	10	13	16	19	23	26	30	35	7	9	12	14	16	18	21	24
TW A4	9	12	16	19	23	26	31	37	6	8	11	13	16	19	22	26
TW B1	12	14	16	18	20	22	24	27	10	11	13	14	15	16	18	19
TW B2	10	13	16	18	22	24	28	33	8	10	12	14	16	18	20	23
TW B3	9	12	15	18	22	25	29	34	7	9	11	13	16	18	21	24
TW B4	9	12	15	18	21	24	28	33	7	9	11	13	16	18	21	25
TW C1	11	13	15	17	20	22	25	29	9	11	12	14	16	17	19	21
TW C2	10	12	15	17	21	23	27	31	8	9	12	13	16	17	20	23
TW C3	9	12	15	17	20	23	26	31	8	9	12	13	16	17	20	23
TW C4	10	12	14	16	19	22	25	29	9	10	12	13	16	17	19	23
TW D1	6	8	9	10	12	13	14	15	4	5	5	6	6	6	6	5
TW D2	11	15	19	23	28	32	37	44	8	11	15	17	21	24	28	34
TW D3	11	16	21	25	31	35	40	48	8	11	15	19	23	26	31	37
TW D4	12	18	25	30	37	42	49	58	9	14	19	24	30	35	40	49
TW D5	12	18	26	31	39	44	51	61	9	14	20	25	32	37	43	52
TW E1	14	19	25	29	36	41	47	55	10	14	19	23	28	32	37	44
TW E2	13	19	25	30	37	42	49	57	10	14	19	24	30	34	40	47
TW E3	16	24	32	39	48	54	61	71	12	19	26	32	40	46	53	62
TW E4	15	22	31	37	46	52	60	69	11	17	25	31	39	44	52	61
TW F1	11	13	15	16	18	20	22	24	9	10	11	12	13	14	15	16
TW F2	9	11	14	16	20	22	25	29	7	8	10	12	14	15	17	19
TW F3	7	10	13	16	20	23	26	31	5	7	9	11	13	15	18	21
TW F4	5	8	12	15	19	22	26	31	3	5	8	10	12	15	17	21
TW G1	9	11	13	15	17	19	21	25	7	8	10	11	12	13	15	16
TW G2	6	9	12	14	18	20	23	28	4	6	8	10	12	13	16	19
TW G3	5	7	11	13	17	19	23	28	3	5	7	9	11	13	15	19
TW G4	3	6	9	11	15	17	20	26	1	3	5	7	10	11	14	18
TW H1	6	8	10	12	15	17	20	24	4	6	7	9	10	12	13	16
TW H2	4	6	9	11	14	16	19	24	2	4	5	7	9	11	13	16
TW H3	3	5	7	9	12	14	17	22	1	3	5	6	8	10	12	15
TW H4	2	3	6	7	10	12	14	18	1	2	4	5	7	8	10	13
TW I1	5	6	7	8	10	10	11	12	3	3	4	4	4	4	4	3
TW I2	8	12	16	20	25	29	34	41	5	8	12	14	18	21	25	31
TW I3	9	13	10	22	28	32	37	45	6	9	13	16	21	24	29	35
TW I4	10	15	22	27	34	39	46	55	7	11	17	21	27	32	38	46
TW I5	10	16	23	28	36	42	49	58	7	12	18	22	29	34	41	50
TW J1	11	17	23	27	34	39	45	53	8	12	17	21	26	30	36	43
TW J2	11	16	23	28	35	40	46	55	7	12	17	21	27	32	37	45
TW J3	14	21	30	36	45	52	59	69	10	16	24	30	38	44	51	61
TW J4	12	20	28	34	43	50	57	67	9	15	22	28	36	42	49	59
DG A1	6	7	9	10	11	12	12	12	3	4	5	5	5	5	4	2
DG A2	9	12	15	18	21	24	27	32	6	9	11	13	16	18	20	23
DG A3	12	16	22	26	32	36	42	50	10	13	17	21	26	30	35	42
DG B1	6	7	9	10	12	13	14	14	3	4	5	5	5	5	5	4
DG B2	9	12	15	18	22	25	29	35	7	9	11	13	16	18	21	25
DG B3	13	17	22	27	33	38	45	54	10	14	18	22	27	31	37	45
DG C1	8	10	12	14	17	18	21	23	5	6	8	9	10	10	11	11
DG C2	11	14	19	22	27	31	36	43	8	11	14	17	21	24	27	32
DG C3	14	19	26	30	38	43	51	61	12	16	21	25	31	36	42	51
SS A1	15	18	21	24	28	30	34	39	13	15	18	20	23	25	27	31
SS A2	18	23	29	34	40	45	51	58	16	20	25	29	35	39	44	51
SS A3	13	16	18	20	23	25	28	31	11	13	15	17	19	20	21	23
SS A4	18	23	28	32	38	43	48	55	15	19	24	28	33	36	41	47
SS A5	15	18	20	22	25	27	30	34	13	15	17	19	21	22	24	26
SS A6	18	23	29	33	39	44	49	57	15	19	24	28	34	37	42	49
SS A7	14	16	17	19	21	22	24	26	12	14	15	16	17	18	18	19
SS A8	17	22	27	31	37	41	46	53	15	18	23	26	31	34	39	44
SS B1	12	15	18	20	23	26	29	33	10	12	14	16	19	20	22	25
SS B2	16	21	26	30	37	41	47	54	13	17	22	26	31	35	40	47
SS B3	11	13	16	18	20	22	25	28	9	11	13	14	16	17	18	20
SS B4	15	20	25	29	35	39	44	52	13	17	21	25	30	33	38	44
SS B5	12	14	16	18	21	22	25	28	11	12	14	15	17	18	19	21
SS B6	15	19	25	29	35	39	45	52	12	16	21	24	30	33	38	45
SS B7	11	13	14	16	17	19	20	22	10	11	12	13	14	15	15	16
SS B8	14	18	23	27	33	37	42	49	12	15	19	23	27	31	35	41
SS C1	8	11	15	17	21	23	26	31	6	7	9	11	13	15	16	19
SS C2	11	16	21	25	31	35	41	49	8	11	15	19	24	27	31	38
SS C3	9	11	14	16	18	20	23	26	7	9	10	11	13	14	15	17
SS C4	11	15	19	23	28	32	37	44	8	11	14	17	21	24	28	34
SS D1	15	19	24	27	32	35	39	45	13	16	19	22	25	28	31	35
SS D2	19	26	34	39	47	52	59	67	16	22	29	33	40	45	51	59
SS D3	16	19	22	25	29	32	35	40	13	16	19	21	24	26	28	31
SS D4	19	25	32	38	45	50	57	65	16	21	28	32	39	43	49	57
SS E1	12	15	19	22	25	28	31	36	9	12	14	16	19	20	22	24
SS E2	15	21	28	32	39	44	50	58	12	17	22	27	32	37	42	49
SS E3	13	15	18	20	23	25	27	31	11	12	14	16	18	19	21	23
SS E4	15	20	26	30	36	41	46	54	12	16	21	25	30	34	39	45

CONSERVATION FACTOR

TYPF	55 F					65 F				
	100	50	30	20	15	100	50	30	20	15
WW A1	2.4	2.8	3.3	3.7	4.2	3.3	4.1	-	-	-
WW A2	1.8	2.0	2.3	2.6	2.8	2.2	2.6	3.0	3.4	3.9
WW A3	1.6	1.9	2.1	2.3	2.5	2.0	2.3	2.6	3.0	3.3
WW A4	1.6	1.8	2.0	2.2	2.4	1.9	2.1	2.4	2.8	3.1
WW A5	1.5	1.7	1.9	2.1	2.3	1.8	2.0	2.3	2.6	2.9
WW A6	1.5	1.7	1.9	2.1	2.2	1.7	2.0	2.2	2.5	2.8
WW B1	2.6	3.2	4.3	-	-	4.3	-	-	-	-
WW B2	1.4	1.6	1.8	2.0	2.1	1.7	1.9	2.1	2.3	2.5
WW B3	1.3	1.5	1.7	1.8	2.0	1.5	1.7	1.9	2.1	2.2
WW B4	1.2	1.4	1.5	1.7	1.8	1.3	1.5	1.6	1.8	2.0
WW B5	1.2	1.3	1.5	1.7	1.8	1.3	1.4	1.6	1.8	1.9
WW C1	1.3	1.4	1.6	1.8	1.9	1.4	1.6	1.8	2.0	2.1
WW C2	1.2	1.4	1.6	1.7	1.9	1.4	1.5	1.7	1.9	2.0
WW C3	1.1	1.2	1.4	1.5	1.7	1.1	1.3	1.5	1.6	1.8
WW C4	1.1	1.2	1.4	1.6	1.7	1.2	1.3	1.5	1.6	1.8
TW A1	2.4	2.7	3.1	3.5	3.8	3.1	3.7	4.5	-	-
TW A2	1.9	2.2	2.4	2.7	2.9	2.3	2.7	3.1	3.5	3.9
TW A3	1.8	2.0	2.2	2.5	2.6	2.1	2.4	2.7	3.1	3.4
TW A4	1.7	1.9	2.1	2.3	2.5	2.0	2.3	2.5	2.8	3.1
TW B1	2.2	2.5	2.8	3.1	3.4	2.8	3.2	3.7	4.3	4.9
TW B2	1.9	2.1	2.3	2.6	2.7	2.2	2.5	2.8	3.2	3.5
TW B3	1.8	2.0	2.2	2.5	2.6	2.1	2.4	2.7	3.0	3.2
TW B4	1.8	2.0	2.3	2.5	2.6	2.2	2.4	2.7	2.9	3.2
TW C1	2.1	2.3	2.6	2.8	3.0	2.5	2.8	3.2	3.5	3.8
TW C2	2.0	2.2	2.4	2.6	2.8	2.3	2.6	2.8	3.1	3.3
TW C3	2.0	2.2	2.4	2.6	2.8	2.3	2.6	2.8	3.1	3.3
TW C4	2.1	2.3	2.5	2.8	2.9	2.5	2.7	3.0	3.2	3.4
TW D1	2.6	3.1	3.8	4.8	-	3.8	-	-	-	-
TW D2	1.6	1.7	1.9	2.1	2.3	1.8	2.0	2.2	2.4	2.6
TW D3	1.5	1.7	1.8	2.0	2.2	1.7	1.9	2.1	2.3	2.5
TW D4	1.3	1.5	1.6	1.8	1.9	1.4	1.6	1.8	1.9	2.1
TW D5	1.3	1.4	1.6	1.7	1.9	1.4	1.5	1.7	1.9	2.0
TW E1	1.4	1.5	1.7	1.9	2.0	1.5	1.7	1.9	2.1	2.3
TW E2	1.3	1.5	1.6	1.8	2.0	1.5	1.6	1.8	2.0	2.1
TW E3	1.1	1.3	1.4	1.6	1.7	1.2	1.4	1.6	1.7	1.9
TW E4	1.2	1.3	1.5	1.6	1.8	1.3	1.4	1.6	1.7	1.9
TW F1	2.3	2.6	3.0	3.4	3.7	3.0	3.6	4.3	-	-
TW F2	1.9	2.2	2.5	2.7	2.9	2.4	2.7	3.1	3.6	4.0
TW F3	1.8	2.1	2.3	2.5	2.7	2.2	2.5	2.8	3.2	3.5
TW F4	1.8	2.0	2.2	2.5	2.6	2.1	2.4	2.7	3.0	3.2
TW G1	2.2	2.5	2.8	3.1	3.4	2.8	3.2	3.7	4.3	4.9
TW G2	2.0	2.2	2.4	2.7	2.9	2.4	2.7	3.0	3.4	3.7
TW G3	1.9	2.2	2.4	2.6	2.8	2.3	2.6	2.9	3.2	3.4
TW G4	2.0	2.2	2.4	2.7	2.8	2.4	2.6	2.9	3.1	3.4
TW H1	2.2	2.4	2.7	3.0	3.2	2.6	3.0	3.4	3.8	4.1
TW H2	2.1	2.3	2.6	2.8	3.0	2.5	2.8	3.1	3.4	3.6
TW H3	2.2	2.4	2.6	2.8	3.0	2.6	2.9	3.1	3.4	3.6
TW H4	2.5	2.7	2.9	3.1	3.2	2.8	3.1	3.3	3.6	3.8
TW I1	2.7	3.3	4.0	-	-	1.1		-	-	-
TW I2	1.6	1.8	2.0	2.2	2.3	1.8	2.1	2.3	2.5	2.7
TW I3	1.5	1.7	1.9	2.0	2.2	1.7	1.9	2.1	2.3	2.5
TW I4	1.3	1.5	1.7	1.8	2.0	1.5	1.6	1.8	2.0	2.1
TW I5	1.3	1.4	1.6	1.8	1.9	1.4	1.6	1.7	1.9	2.0
TW J1	1.4	1.5	1.7	1.9	2.0	1.5	1.7	1.9	2.1	2.3
TW J2	1.3	1.5	1.7	1.8	2.0	1.5	1.7	1.8	2.0	2.2
TW J3	1.1	1.3	1.5	1.6	1.8	1.2	1.4	1.6	1.7	1.9
TW J4	1.2	1.3	1.5	1.6	1.8	1.3	1.4	1.6	1.7	1.9
DG A1	2.4	3.0	4.4	-	-	3.4	-	-	-	-
DG A2	1.8	2.0	2.3	2.6	3.0	2.1	2.4	2.7	3.2	3.7
DG A3	1.5	1.6	1.8	1.9	2.1	1.7	1.8	2.0	2.1	2.3
DG B1	2.4	3.0	3.9	-	-	3.6	4.9	-	-	-
DG B2	1.8	2.0	2.2	2.4	2.7	2.1	2.3	2.6	3.0	3.3
DG B3	1.5	1.6	1.7	1.8	2.0	1.7	1.8	1.9	2.0	2.2
DG C1	2.0	2.4	2.9	3.3	4.1	2.7	3.2	4.4	-	-
DG C2	1.6	1.8	2.0	2.1	2.3	1.9	2.0	2.3	2.5	2.7
DG C3	1.4	1.5	1.6	1.7	1.8	1.5	1.6	1.8	1.9	2.0
SS A1	1.7	2.0	2.2	2.5	2.7	1.9	2.2	2.6	3.0	3.3
SS A2	1.3	1.5	1.7	1.9	2.0	1.4	1.6	1.8	2.0	2.2
SS A3	1.9	2.2	2.6	3.0	3.3	2.2	2.6	3.2	4.0	-
SS A4	1.3	1.5	1.7	1.9	2.1	1.5	1.7	1.9	2.1	2.3
SS A5	1.9	2.2	2.5	2.8	3.1	2.2	2.6	3.0	3.6	4.2
SS A6	1.3	1.5	1.7	1.9	2.0	1.4	1.7	1.9	2.1	2.3
SS A7	2.2	2.7	3.1	3.7	4.3	2.7	3.3	4.2	-	-
SS A8	1.4	1.6	1.8	2.0	2.1	1.5	1.7	2.0	2.2	2.4
SS B1	1.8	2.1	2.4	2.7	2.9	2.1	2.5	2.8	3.3	3.7
SS B2	1.4	1.6	1.7	1.9	2.1	1.5	1.7	1.9	2.1	2.2
SS B3	2.0	2.3	2.7	3.0	3.4	2.4	2.8	3.3	4.1	5.0
SS B4	1.4	1.6	1.8	2.0	2.1	1.5	1.7	2.0	2.2	2.3
SS B5	2.1	2.5	2.8	3.2	3.5	2.5	3.0	3.5	4.1	4.8
SS B6	1.4	1.6	1.8	1.9	2.1	1.5	1.7	1.9	2.1	2.3
SS B7	2.4	2.8	3.3	3.8	4.4	2.9	3.6	4.4	-	-
SS B8	1.5	1.7	1.9	2.0	2.2	1.6	1.8	2.0	2.3	2.5
SS C1	1.8	2.1	2.4	2.6	2.9	2.3	2.7	3.1	3.7	4.3
SS C2	1.4	1.6	1.8	2.0	2.1	1.6	1.8	2.0	2.3	2.4
SS C3	2.1	2.4	2.7	3.0	3.3	2.7	3.1	3.6	4.1	4.7
SS C4	1.6	1.8	1.9	2.1	2.3	1.8	2.0	2.2	2.4	2.6
SS D1	1.5	1.7	2.0	2.2	2.4	1.7	2.0	2.3	2.7	3.1
SS D2	1.2	1.3	1.5	1.7	1.8	1.3	1.4	1.6	1.8	2.0
SS D3	1.6	1.9	2.2	2.4	2.7	1.9	2.2	2.6	3.0	3.4
SS D4	1.2	1.4	1.6	1.7	1.9	1.3	1.5	1.7	1.9	2.0
SS E1	1.7	2.0	2.3	2.6	2.8	2.0	2.4	2.9	3.5	4.3
SS E2	1.3	1.5	1.6	1.8	2.0	1.4	1.6	1.8	2.0	2.2
SS E3	1.9	2.2	2.6	2.9	3.2	2.3	2.7	3.2	3.8	4.5
SS E4	1.4	1.5	1.7	1.9	2.1	1.5	1.7	1.9	2.1	2.3

NEWARK, NEW JERSEY

TYPE	SOLAR SAVINGS FRACTION (%) BASE TEMP: 55 F (2924 DD)								65 F (5042 DD)								CONSERVATION FACTOR 55 F					65 F				
LCR	100	70	50	40	30	25	20	15	100	70	50	40	30	25	20	15	100	50	30	20	15	100	50	30	20	15
WW A1	18	22	25	28	33	36	40	47	14	16	18	20	23	25	27	31	1.7	1.8	2.0	2.2	2.3	2.1	2.4	2.6	2.9	3.1
WW A2	19	24	30	35	42	47	53	61	13	16	21	24	29	32	37	43	1.3	1.5	1.6	1.8	1.9	1.6	1.8	2.0	2.2	2.3
WW A3	19	25	32	38	46	51	58	66	12	16	22	26	31	35	41	48	1.2	1.4	1.5	1.7	1.8	1.5	1.7	1.8	2.0	2.2
WW A4	19	26	34	39	48	53	60	69	11	16	22	26	33	37	43	51	1.2	1.3	1.5	1.7	1.8	1.4	1.6	1.8	1.9	2.1
WW A5	19	27	35	41	50	55	63	72	11	16	23	27	34	39	45	53	1.1	1.3	1.4	1.6	1.8	1.4	1.5	1.7	1.9	2.0
WW A6	19	27	36	42	51	57	64	73	11	16	23	28	35	40	46	55	1.1	1.3	1.4	1.6	1.7	1.3	1.5	1.7	1.8	2.0
WW B1	15	19	23	26	31	35	39	46	9	11	14	16	18	20	22	25	1.6	1.8	1.9	2.1	2.3	2.1	2.4	2.8	3.2	3.6
WW B2	20	27	36	42	51	56	64	73	12	18	24	29	37	41	48	57	1.1	1.3	1.4	1.6	1.7	1.3	1.5	1.7	1.8	2.0
WW B3	24	33	42	49	59	65	72	80	16	22	30	36	44	50	57	66	1.0	1.2	1.3	1.5	1.6	1.2	1.4	1.5	1.7	1.8
WW B4	24	34	44	52	62	68	76	84	15	23	32	39	48	55	62	72	1.0	1.1	1.3	1.5	1.6	1.1	1.3	1.4	1.6	1.7
WW B5	23	33	43	51	61	68	75	84	15	23	32	39	49	55	63	73	1.0	1.1	1.3	1.5	1.6	1.1	1.3	1.4	1.6	1.7
WW C1	26	35	44	52	61	67	74	83	17	24	32	38	47	53	60	69	1.0	1.2	1.3	1.5	1.6	1.2	1.3	1.5	1.6	1.8
WW C2	24	33	43	50	60	66	74	82	15	23	31	38	47	53	60	70	1.0	1.2	1.3	1.5	1.6	1.2	1.3	1.5	1.6	1.8
WW C3	28	40	52	60	70	77	83	90	19	29	40	48	58	65	73	82	0.9	1.1	1.2	1.4	1.5	1.0	1.2	1.3	1.5	1.6
WW C4	26	37	48	56	67	73	81	88	17	26	37	44	55	61	70	79	0.9	1.1	1.2	1.4	1.5	1.0	1.2	1.3	1.5	1.6
TW A1	18	21	25	28	33	36	40	47	14	16	18	20	23	25	28	32	1.7	1.8	2.0	2.2	2.3	2.1	2.3	2.6	2.8	3.0
TW A2	18	23	28	32	39	43	49	57	13	16	20	23	27	30	34	40	1.4	1.6	1.7	1.9	2.0	1.7	1.9	2.1	2.3	2.5
TW A3	18	23	29	34	41	46	52	61	12	15	20	23	28	32	37	44	1.3	1.5	1.6	1.8	1.9	1.6	1.8	2.0	2.1	2.3
TW A4	17	22	29	34	42	47	53	62	10	15	19	23	29	33	38	45	1.3	1.4	1.6	1.8	1.9	1.5	1.7	1.9	2.1	2.2
TW B1	17	21	25	28	33	37	42	49	13	15	18	20	24	26	29	34	1.6	1.8	1.9	2.1	2.2	2.0	2.2	2.4	2.6	2.8
TW B2	17	21	27	31	38	42	48	56	12	15	19	22	26	30	34	40	1.4	1.6	1.7	1.9	2.0	1.7	1.9	2.1	2.2	2.4
TW B3	16	21	27	31	38	43	49	57	11	14	18	22	27	30	35	42	1.4	1.5	1.7	1.8	2.0	1.6	1.8	2.0	2.2	2.3
TW B4	15	20	26	30	36	41	47	55	10	13	18	21	26	29	34	40	1.4	1.6	1.7	1.9	2.0	1.7	1.8	2.0	2.2	2.3
TW C1	16	20	25	28	33	37	42	50	12	14	18	20	24	27	30	36	1.5	1.7	1.9	2.0	2.2	1.9	2.1	2.3	2.4	2.6
TW C2	15	19	25	28	34	39	44	52	11	14	17	20	24	28	32	38	1.5	1.6	1.8	1.9	2.1	1.8	2.0	2.1	2.3	2.5
TW C3	15	19	23	27	33	37	42	50	10	13	17	19	24	27	31	37	1.5	1.7	1.8	2.0	2.1	1.8	2.0	2.2	2.3	2.5
TW C4	14	18	22	25	30	34	39	46	11	13	16	18	22	25	28	34	1.6	1.8	1.9	2.1	2.2	2.0	2.1	2.3	2.5	2.6
TW D1	12	16	20	23	27	31	35	41	8	10	12	14	16	18	20	23	1.7	1.9	2.0	2.2	2.4	2.2	2.5	2.8	3.2	3.5
TW D2	19	25	33	38	46	52	59	67	13	17	23	27	34	38	44	52	1.2	1.4	1.5	1.7	1.8	1.4	1.6	1.8	1.9	2.1
TW D3	20	28	36	42	50	56	63	72	13	19	25	30	37	42	48	56	1.1	1.3	1.4	1.6	1.8	1.3	1.5	1.7	1.8	2.0
TW D4	22	30	40	47	56	62	70	79	15	21	29	35	43	49	56	65	1.1	1.2	1.4	1.5	1.7	1.2	1.4	1.5	1.7	1.8
TW D5	22	30	40	47	56	63	70	79	14	21	30	36	44	50	58	67	1.1	1.2	1.3	1.5	1.6	1.2	1.3	1.5	1.6	1.8
TW E1	24	31	40	47	56	62	69	77	16	22	29	35	42	48	54	63	1.1	1.2	1.4	1.5	1.7	1.2	1.4	1.6	1.7	1.9
TW E2	23	31	40	47	56	62	70	78	16	22	29	35	43	49	56	65	1.1	1.2	1.4	1.5	1.7	1.2	1.4	1.5	1.7	1.8
TW E3	28	38	49	56	66	73	80	87	19	28	37	44	54	60	68	77	1.0	1.1	1.3	1.4	1.6	1.1	1.2	1.4	1.5	1.7
TW E4	25	35	46	54	64	70	78	86	17	26	35	42	52	58	66	76	1.0	1.1	1.3	1.4	1.6	1.1	1.2	1.4	1.5	1.7
TW F1	16	20	24	27	32	35	40	46	12	14	17	19	22	24	27	31	1.6	1.8	2.0	2.1	2.3	2.1	2.3	2.5	2.7	2.9
TW F2	16	20	26	30	36	40	46	54	10	13	17	20	24	27	32	37	1.4	1.6	1.7	1.9	2.0	1.8	1.9	2.1	2.3	2.5
TW F3	15	20	26	30	37	42	48	57	9	12	17	20	25	28	33	40	1.4	1.5	1.7	1.8	2.0	1.7	1.8	2.0	2.2	2.4
TW F4	13	18	24	29	36	41	47	56	7	11	15	18	24	27	32	39	1.4	1.5	1.7	1.8	2.0	1.6	1.8	2.0	2.2	2.3
TW G1	14	18	22	26	31	35	39	47	10	12	15	17	21	23	27	31	1.6	1.8	1.9	2.1	2.2	2.0	2.2	2.4	2.6	2.8
TW G2	13	17	23	27	33	37	43	51	8	11	14	17	22	25	29	35	1.5	1.6	1.8	1.9	2.1	1.8	2.0	2.1	2.3	2.5
TW G3	11	16	21	25	32	36	42	51	6	9	13	16	21	24	29	35	1.5	1.6	1.8	1.9	2.1	1.8	1.9	2.1	2.3	2.4
TW G4	9	13	18	22	28	32	38	46	4	7	11	14	18	21	25	32	1.5	1.7	1.8	2.0	2.1	1.9	2.0	2.2	2.4	2.5
TW H1	11	15	19	23	28	32	37	44	7	10	13	15	19	21	25	30	1.6	1.8	1.9	2.1	2.2	2.0	2.2	2.4	2.6	2.7
TW H2	9	13	17	21	27	31	36	44	5	7	11	13	17	20	24	30	1.6	1.8	1.9	2.1	2.2	1.9	2.1	2.3	2.5	2.6
TW H3	7	11	15	18	24	27	32	40	4	6	9	11	15	18	22	27	1.7	1.8	2.0	2.1	2.3	2.0	2.2	2.4	2.5	2.7
TW H4	5	8	12	15	19	22	26	33	2	4	7	9	12	14	17	22	1.9	2.0	2.2	2.3	2.5	2.3	2.4	2.6	2.8	2.9
TW I1	10	13	17	20	24	28	32	38	6	8	10	11	14	15	17	20	1.7	1.9	2.1	2.3	2.5	2.3	2.6	2.9	3.3	3.7
TW I2	16	22	29	35	43	48	55	64	10	14	20	24	30	35	40	49	1.2	1.4	1.5	1.7	1.8	1.5	1.6	1.8	2.0	2.1
TW I3	17	24	32	38	46	52	59	69	11	16	22	27	34	38	45	53	1.2	1.3	1.5	1.6	1.8	1.4	1.5	1.7	1.9	2.0
TW I4	19	27	36	43	53	59	66	76	12	18	26	32	40	46	53	62	1.1	1.2	1.4	1.5	1.7	1.2	1.4	1.5	1.7	1.8
TW I5	19	28	37	44	54	60	68	78	12	19	27	33	42	48	55	65	1.1	1.2	1.4	1.5	1.7	1.2	1.4	1.5	1.7	1.8
TW J1	21	29	38	44	53	59	67	76	14	20	27	32	40	45	52	61	1.1	1.2	1.4	1.6	1.7	1.3	1.4	1.6	1.7	1.9
TW J2	20	28	37	44	53	60	67	76	13	19	27	32	40	46	53	63	1.1	1.2	1.4	1.5	1.7	1.2	1.4	1.6	1.7	1.8
TW J3	25	35	46	53	64	70	77	86	16	25	35	42	51	58	66	75	1.0	1.1	1.3	1.4	1.6	1.1	1.2	1.4	1.5	1.7
TW J4	23	33	43	51	61	68	75	84	15	23	32	39	49	56	64	73	1.0	1.1	1.3	1.5	1.6	1.1	1.2	1.4	1.6	1.7
DG A1	13	17	21	24	28	31	34	37	7	9	12	14	16	17	19	19	1.5	1.8	2.2	2.8	3.5	2.0	2.4	3.1	4.6	-
DG A2	15	20	26	31	37	42	47	54	10	13	18	21	26	29	33	39	1.4	1.5	1.7	2.0	2.2	1.7	1.8	2.0	2.3	2.6
DG A3	19	26	33	40	49	54	62	70	14	18	25	29	37	42	48	57	1.2	1.3	1.4	1.6	1.8	1.4	1.5	1.6	1.8	2.0
DG B1	13	17	22	25	31	34	39	44	7	9	12	14	17	19	21	24	1.5	1.7	1.9	2.2	2.6	2.0	2.4	2.7	3.3	4.1
DG B2	16	21	27	32	40	45	52	60	10	14	18	22	27	31	36	43	1.3	1.5	1.6	1.8	2.0	1.6	1.8	1.9	2.1	2.3
DG B3	19	26	35	41	51	57	66	75	14	19	25	30	38	44	51	61	1.2	1.3	1.4	1.5	1.7	1.4	1.5	1.6	1.7	1.8
DG C1	16	21	27	31	38	42	48	55	9	13	16	19	23	25	29	33	1.4	1.5	1.7	1.9	2.2	1.7	2.0	2.3	2.6	2.9
DG C2	18	24	32	38	46	52	60	69	12	17	22	26	33	37	43	51	1.2	1.3	1.5	1.6	1.8	1.5	1.6	1.7	1.9	2.1
DG C3	22	30	39	46	57	64	73	82	16	22	29	35	43	50	58	68	1.1	1.2	1.3	1.5	1.6	1.3	1.4	1.5	1.6	1.7
SS A1	22	27	33	38	44	49	54	62	17	21	25	29	34	37	42	48	1.3	1.5	1.6	1.8	2.0	1.5	1.7	1.9	2.1	2.3
SS A2	27	35	43	49	58	63	70	78	21	27	34	40	47	52	58	67	1.1	1.2	1.4	1.6	1.7	1.2	1.4	1.5	1.7	1.9
SS A3	20	25	30	34	40	44	49	56	15	19	22	25	29	32	36	41	1.4	1.6	1.8	1.9	2.1	1.6	1.9	2.1	2.4	2.6
SS A4	27	34	42	48	56	62	68	76	21	27	33	38	45	50	56	64	1.1	1.2	1.4	1.6	1.7	1.2	1.4	1.6	1.8	1.9
SS A5	21	26	31	34	40	44	49	56	17	20	24	27	31	33	37	42	1.4	1.6	1.8	1.9	2.1	1.7	1.9	2.1	2.4	2.6
SS A6	27	34	43	49	57	63	69	78	21	27	34	39	46	51	57	66	1.1	1.2	1.4	1.6	1.7	1.2	1.4	1.6	1.7	1.9
SS A7	19	23	27	30	34	37	42	48	16	18	21	23	26	28	30	34	1.6	1.8	2.0	2.2	2.3	1.9	2.2	2.5	2.8	3.1
SS A8	26	33	41	47	55	60	67	75	20	26	32	37	44	48	54	62	1.1	1.3	1.4	1.6	1.7	1.3	1.4	1.6	1.8	2.0
SS B1	19	23	28	33	39	43	48	56	14	17	21	24	29	32	36	42	1.4	1.6	1.7	1.9	2.0	1.6	1.8	2.1	2.3	2.4
SS B2	24	31	39	45	53	58	65	74	18	24	30	35	42	47	54	62	1.1	1.3	1.4	1.6	1.7	1.3	1.4	1.6	1.8	1.9
SS B3	17	21	26	30	36	40	45	52	13	16	19	22	26	28	32	37	1.5	1.6	1.8	2.0	2.1	1.7	2.0	2.2	2.5	2.7
SS B4	23	30	37	43	51	57	64	72	17	23	29	34	41	46	52	60	1.1	1.3	1.5	1.6	1.8	1.3	1.5	1.6	1.8	2.0
SS B5	18	21	25	29	34	37	42	48	14	16	19	22	25	27	31	35	1.6	1.8	1.9	2.1	2.3	1.9	2.1	2.4	2.6	2.8
SS B6	23	30	38	43	52	57	64	73	17	23	29	34	41	46	52	60	1.1	1.3	1.5	1.6	1.8	1.3	1.5	1.6	1.8	1.9
SS B7	16	19	23	25	29	32	36	42	12	15	17	19	22	23	26	30	1.7	1.9	2.1	2.3	2.5	2.1	2.4	2.6	3.0	3.2
SS B8	22	28	36	41	49	55	61	70	16	22	28	32	39	43	49	57	1.2	1.3	1.5	1.7	1.8	1.3	1.5	1.7	1.9	2.0
SS C1	17	22	28	33	40	45	51	59	10	14	18	22	26	30	34	41	1.3	1.5	1.6	1.8	1.9	1.6	1.8	2.0	2.2	2.4
SS C2	20	28	36	42	51	57	64	73	13	19	25	30	37	42	49	57	1.1	1.3	1.4	1.6	1.7	1.3	1.5	1.6	1.8	2.0
SS C3	15	19	24	27	33	36	42	49	11	13	16	19	22	25	28	33	1.5	1.7	1.9	2.0	2.2	1.9	2.1	2.3	2.5	2.7
SS C4	19	25	32	37	45	51	58	67	12	17	23	27	33	38	43	51	1.2	1.4	1.5	1.7	1.8	1.4	1.6	1.8	1.9	2.1
SS D1	25	32	39	44	52	57	64	72	19	24	29	34	40	44	49	56	1.1	1.3	1.5	1.7	1.8	1.3	1.5	1.7	1.9	2.1
SS D2	31	40	50	57	66	72	79	86	23	31	40	46	55	60	67	76	1.0	1.1	1.3	1.5	1.6	1.1	1.2	1.4	1.6	1.7
SS D3	23	29	35	40	47	51	57	65	18	22	27	31	36	39	44	50	1.3	1.4	1.6	1.8	1.9	1.5	1.7	1.9	2.1	2.3
SS D4	30	39	48	55	64	70	77	84	23	30	38	44	53	58	65	73	1.0	1.1	1.3	1.5	1.6	1.1	1.3	1.4	1.6	1.8
SS E1	21	27	33	39	46	50	57	65	15	19	24	27	33	36	41	47	1.2	1.4	1.6	1.8	1.9	1.5	1.7	1.9	2.1	2.3
SS E2	26	34	44	50	59	65	72	80	19	26	33	39	47	52	59	68	1.0	1.2	1.3	1.5	1.7	1.2	1.3	1.5	1.7	1.8
SS E3	19	24	29	33	38	42	48	55	14	18	21	24	28	31	35	40	1.4	1.6	1.8	1.9	2.1	1.7	1.9	2.1	2.4	2.6
SS E4	24	32	40	46	55	61	68	76	18	24	31	36	43	48	55	63	1.1	1.2	1.4	1.6	1.7	1.2	1.4	1.6	1.8	1.9

ALBUQUERQUE, NEW MEXICO

TYPE	SOLAR SAVINGS FRACTION (%) BASE TEMP: 55 F (2332 DD)								65 F (4293 DD)								CONSERVATION FACTOR 55 F					65 F				
LCR	100	70	50	40	30	25	20	15	100	70	50	40	30	25	20	15	100	50	30	20	15	100	50	30	20	15
WW A1	32	39	48	54	62	68	74	82	23	28	34	39	46	50	57	65	1.1	1.2	1.3	1.5	1.6	1.3	1.4	1.6	1.7	1.9
WW A2	39	49	60	67	76	81	87	93	26	34	43	49	58	64	72	80	0.9	1.0	1.2	1.4	1.5	1.1	1.2	1.4	1.5	1.6
WW A3	41	53	64	71	80	85	90	95	28	37	46	54	63	69	76	85	0.8	1.0	1.2	1.3	1.5	1.0	1.1	1.3	1.5	1.6
WW A4	43	55	66	74	83	87	92	96	28	38	48	56	66	72	79	87	0.8	1.0	1.1	1.3	1.5	1.0	1.1	1.3	1.4	1.6
WW A5	45	57	69	76	85	89	93	97	29	39	50	58	68	74	81	89	0.8	1.0	1.1	1.3	1.4	0.9	1.1	1.2	1.4	1.5
WW A6	45	58	70	77	86	90	94	97	30	40	51	59	70	76	83	90	0.8	0.9	1.1	1.3	1.4	0.9	1.1	1.2	1.4	1.5
WW B1	33	42	52	59	68	74	80	87	22	28	35	40	48	54	61	70	1.0	1.1	1.3	1.4	1.6	1.2	1.3	1.5	1.6	1.8
WW B2	43	55	67	75	83	88	93	97	29	39	50	57	67	73	81	88	0.8	1.0	1.1	1.3	1.5	1.0	1.1	1.2	1.4	1.5
WW B3	50	63	75	82	89	93	96	98	35	46	58	66	76	81	87	93	0.8	0.9	1.1	1.3	1.4	0.9	1.0	1.2	1.3	1.5
WW B4	50	64	76	83	90	94	97	99	35	47	59	67	77	83	89	94	0.8	0.9	1.1	1.3	1.4	0.9	1.0	1.2	1.3	1.5
WW B5	48	62	74	81	89	93	96	99	33	45	58	66	76	82	88	94	0.8	0.9	1.1	1.3	1.4	0.9	1.0	1.2	1.3	1.5
WW C1	52	65	76	83	90	94	97	99	37	48	60	68	78	83	89	94	0.8	0.9	1.1	1.3	1.4	0.9	1.0	1.2	1.3	1.5
WW C2	50	63	75	82	89	93	96	99	34	46	58	66	76	82	88	94	0.8	0.9	1.1	1.3	1.4	0.9	1.0	1.2	1.3	1.5
WW C3	56	70	81	88	94	96	98	99	40	54	67	75	84	89	93	97	0.7	0.9	1.1	1.2	1.4	0.8	1.0	1.1	1.3	1.4
WW C4	52	66	78	85	92	95	97	99	37	50	63	71	81	86	92	96	0.7	0.9	1.1	1.2	1.4	0.8	1.0	1.1	1.3	1.5
TW A1	31	38	46	53	61	67	73	82	22	27	33	38	45	50	56	64	1.1	1.2	1.4	1.5	1.6	1.3	1.5	1.6	1.7	1.9
TW A2	35	44	54	61	71	76	83	90	24	31	39	45	53	59	66	75	0.9	1.1	1.2	1.4	1.6	1.1	1.3	1.4	1.6	1.7
TW A3	37	47	58	65	74	80	86	92	25	32	41	48	57	63	70	79	0.9	1.0	1.2	1.4	1.5	1.1	1.2	1.4	1.5	1.7
TW A4	36	47	58	66	75	81	87	93	24	32	41	48	58	64	71	80	0.9	1.0	1.2	1.4	1.5	1.1	1.2	1.3	1.5	1.6
TW B1	31	38	47	54	62	68	75	83	22	27	34	39	46	51	57	66	1.1	1.2	1.3	1.5	1.6	1.3	1.4	1.6	1.7	1.8
TW B2	33	43	53	60	69	75	82	89	23	30	37	43	52	58	65	74	1.0	1.1	1.2	1.4	1.6	1.2	1.3	1.4	1.6	1.7
TW B3	33	43	53	61	70	76	83	90	22	29	38	44	53	59	66	75	0.9	1.1	1.2	1.4	1.5	1.1	1.3	1.4	1.6	1.7
TW B4	31	40	50	58	67	73	80	88	21	28	35	42	50	56	63	73	1.0	1.1	1.3	1.4	1.6	1.2	1.3	1.4	1.6	1.7
TW C1	30	38	46	53	62	68	75	83	21	26	33	38	46	51	58	67	1.0	1.2	1.3	1.5	1.6	1.3	1.4	1.5	1.7	1.8
TW C2	29	38	47	54	64	70	77	85	20	26	33	39	47	53	60	69	1.0	1.2	1.3	1.5	1.6	1.2	1.4	1.5	1.6	1.8
TW C3	28	36	45	52	61	67	74	82	19	25	32	37	45	50	57	67	1.0	1.2	1.3	1.5	1.6	1.3	1.4	1.5	1.7	1.8
TW C4	25	32	40	46	55	60	68	77	18	23	29	33	40	45	51	60	1.1	1.3	1.4	1.6	1.7	1.4	1.5	1.6	1.8	1.9
TW D1	28	36	45	52	61	67	74	82	18	23	30	35	42	47	54	63	1.0	1.2	1.3	1.5	1.6	1.3	1.4	1.6	1.7	1.9
TW D2	39	50	62	69	78	84	89	94	27	35	45	52	62	68	75	84	0.9	1.0	1.2	1.3	1.5	1.0	1.2	1.3	1.5	1.6
TW D3	43	55	66	74	83	87	92	96	29	39	49	57	67	73	80	88	0.8	1.0	1.1	1.3	1.5	1.0	1.1	1.3	1.4	1.6
TW D4	45	58	70	77	86	90	94	97	31	42	53	61	71	77	84	91	0.8	0.9	1.1	1.3	1.4	0.9	1.1	1.2	1.4	1.5
TW D5	44	57	69	77	85	90	94	97	30	41	53	61	71	77	84	91	0.8	0.9	1.1	1.3	1.4	0.9	1.1	1.2	1.4	1.5
TW E1	47	59	71	78	86	90	94	98	33	43	54	62	72	78	84	91	0.8	0.9	1.1	1.3	1.4	0.9	1.1	1.2	1.4	1.5
TW E2	47	59	71	78	86	90	94	98	32	43	54	62	72	78	84	91	0.8	0.9	1.1	1.3	1.4	0.9	1.1	1.2	1.4	1.5
TW E3	54	67	79	85	92	95	97	99	39	51	63	71	81	86	91	96	0.7	0.9	1.1	1.2	1.4	0.8	1.0	1.1	1.3	1.5
TW E4	51	64	76	83	90	94	97	99	36	48	60	68	78	84	90	95	0.8	0.9	1.1	1.3	1.4	0.9	1.0	1.2	1.3	1.5
TW F1	30	37	46	52	61	67	74	82	21	26	32	37	44	49	56	65	1.1	1.2	1.3	1.5	1.6	1.3	1.4	1.6	1.7	1.9
TW F2	32	42	52	59	69	74	81	88	22	28	36	42	51	57	64	73	1.0	1.1	1.3	1.4	1.6	1.2	1.3	1.4	1.6	1.7
TW F3	33	43	53	61	71	77	83	90	21	29	37	44	53	59	66	76	0.9	1.1	1.2	1.4	1.5	1.1	1.3	1.4	1.6	1.7
TW F4	31	41	52	59	69	75	82	89	19	27	35	42	51	57	65	75	0.9	1.1	1.2	1.4	1.5	1.1	1.3	1.4	1.6	1.7
TW G1	28	36	45	51	61	66	74	82	19	25	31	36	44	49	56	65	1.1	1.2	1.3	1.5	1.6	1.3	1.4	1.6	1.7	1.8
TW G2	29	38	48	55	65	71	78	86	18	25	33	39	47	53	60	70	1.0	1.1	1.3	1.4	1.6	1.2	1.3	1.5	1.6	1.7
TW G3	27	36	46	54	63	70	77	85	17	23	31	37	46	52	59	69	1.0	1.1	1.3	1.4	1.6	1.2	1.3	1.5	1.6	1.8
TW G4	23	31	40	48	57	64	71	80	13	19	27	32	40	46	53	63	1.1	1.2	1.3	1.5	1.6	1.3	1.4	1.5	1.7	1.8
TW H1	25	32	41	48	57	63	71	80	16	21	28	33	41	46	53	62	1.1	1.2	1.4	1.5	1.6	1.3	1.4	1.6	1.7	1.9
TW H2	22	30	39	46	55	62	69	78	13	19	26	31	39	44	51	61	1.1	1.2	1.4	1.5	1.7	1.3	1.4	1.6	1.7	1.9
TW H3	19	26	34	41	50	56	64	74	11	16	22	27	34	39	46	56	1.2	1.3	1.4	1.6	1.7	1.4	1.5	1.7	1.8	1.9
TW H4	14	20	27	33	41	47	54	64	8	12	17	21	27	32	38	46	1.3	1.4	1.5	1.7	1.8	1.6	1.7	1.8	2.0	2.1
TW I1	25	33	41	48	57	63	70	79	15	21	27	32	39	44	50	59	1.1	1.2	1.4	1.5	1.7	1.4	1.5	1.6	1.8	1.9
TW I2	36	46	58	66	75	81	87	00	23	32	41	48	58	64	72	81	0.9	1.0	1.2	1.4	1.5	1.0	1.2	1.3	1.5	1.6
TW I3	38	50	62	70	79	84	90	95	25	35	45	52	62	69	76	85	0.9	1.0	1.2	1.3	1.5	1.0	1.1	1.3	1.4	1.6
TW I4	42	54	66	74	83	88	93	97	28	38	50	58	68	74	81	89	0.8	1.0	1.1	1.3	1.5	0.9	1.1	1.2	1.4	1.5
TW I5	42	54	67	75	84	88	93	97	28	39	50	58	69	75	82	90	0.8	1.0	1.1	1.3	1.4	0.9	1.1	1.2	1.4	1.5
TW J1	44	57	69	76	85	89	93	97	30	41	52	60	70	76	83	90	0.8	1.0	1.1	1.3	1.4	0.9	1.1	1.2	1.4	1.5
TW J2	43	56	68	75	84	89	93	97	29	40	51	59	69	75	82	90	0.8	1.0	1.1	1.3	1.4	0.9	1.1	1.2	1.4	1.5
TW J3	50	64	76	83	90	94	97	99	35	47	60	68	78	84	90	95	0.7	0.9	1.1	1.3	1.4	0.8	1.0	1.2	1.3	1.5
TW J4	47	61	73	81	89	92	96	98	32	45	57	65	76	82	88	04	0.8	0.9	1.1	1.3	1.4	0.9	1.0	1.2	1.3	1.5
DG A1	32	40	50	56	63	68	73	78	19	26	33	37	44	48	52	58	1.0	1.2	1.4	1.7	1.9	1.2	1.4	1.7	2.0	2.3
DG A2	33	43	54	61	70	75	81	87	21	29	38	44	52	57	64	71	0.9	1.1	1.3	1.5	1.7	1.1	1.3	1.5	1.7	1.9
DG A3	38	51	63	70	79	84	88	92	25	35	46	54	64	70	76	84	0.8	1.0	1.2	1.4	1.7	1.0	1.1	1.3	1.5	1.7
DG B1	33	44	56	63	72	76	81	86	20	27	35	41	50	55	61	68	0.9	1.0	1.3	1.5	1.8	1.2	1.3	1.5	1.7	2.0
DG B2	34	46	59	67	77	82	87	92	22	30	40	47	58	64	71	79	0.9	1.0	1.2	1.4	1.6	1.1	1.2	1.3	1.5	1.7
DG B3	39	52	66	75	84	88	92	95	26	36	47	56	68	75	82	89	0.8	0.9	1.1	1.4	1.6	1.0	1.1	1.2	1.4	1.6
DG C1	38	51	64	72	81	85	89	92	23	32	42	49	60	66	72	79	0.8	1.0	1.2	1.5	1.7	1.1	1.2	1.3	1.6	1.8
DG C2	39	52	66	75	83	88	92	96	25	34	46	54	65	72	79	86	0.8	0.9	1.1	1.3	1.5	1.0	1.1	1.2	1.4	1.6
DG C3	44	59	73	81	89	93	95	97	30	40	53	63	75	82	88	93	0.8	0.9	1.1	1.3	1.6	0.9	1.0	1.2	1.4	1.5
SS A1	40	50	60	67	76	81	86	92	30	38	46	52	61	66	73	81	0.9	1.1	1.2	1.4	1.5	1.0	1.2	1.4	1.5	1.7
SS A2	49	61	72	78	86	90	94	97	38	48	58	65	74	79	85	92	0.8	1.0	1.1	1.3	1.5	0.9	1.1	1.2	1.4	1.5
SS A3	38	47	56	63	72	77	83	89	28	35	43	49	57	62	68	77	0.9	1.1	1.3	1.4	1.6	1.1	1.3	1.4	1.6	1.7
SS A4	49	60	71	77	85	89	93	97	37	47	57	64	73	78	84	91	0.8	1.0	1.1	1.3	1.5	0.9	1.1	1.2	1.4	1.5
SS A5	38	46	55	62	71	76	82	88	29	35	43	48	56	61	68	76	1.0	1.1	1.3	1.4	1.6	1.1	1.3	1.4	1.6	1.7
SS A6	49	61	72	78	86	90	94	97	37	47	58	65	74	79	85	91	0.8	1.0	1.1	1.3	1.5	0.9	1.1	1.2	1.4	1.5
SS A7	33	41	49	55	64	69	75	83	25	31	37	42	49	54	60	68	1.0	1.2	1.3	1.5	1.7	1.2	1.4	1.6	1.7	1.9
SS A8	48	59	70	77	84	89	93	97	36	46	56	63	72	77	83	90	0.8	1.0	1.1	1.3	1.5	0.9	1.1	1.2	1.4	1.6
SS B1	35	44	54	61	70	75	81	88	25	32	40	46	54	60	67	75	1.0	1.1	1.3	1.4	1.6	1.1	1.3	1.4	1.6	1.7
SS B2	44	55	66	73	82	86	91	96	32	42	52	59	68	74	81	88	0.8	1.0	1.2	1.3	1.5	1.0	1.1	1.3	1.4	1.6
SS B3	33	42	51	58	67	72	79	86	24	30	38	43	51	57	63	72	1.0	1.1	1.3	1.5	1.6	1.2	1.3	1.5	1.6	1.8
SS B4	43	54	65	72	81	85	90	95	32	41	51	58	67	73	79	87	0.8	1.0	1.2	1.3	1.5	1.0	1.1	1.3	1.4	1.6
SS B5	31	39	47	53	62	67	74	82	23	29	35	40	47	52	58	67	1.1	1.2	1.3	1.5	1.6	1.3	1.4	1.6	1.7	1.8
SS B6	43	54	65	72	81	86	91	95	32	41	51	58	67	73	80	87	0.8	1.0	1.2	1.3	1.5	1.0	1.1	1.3	1.4	1.6
SS B7	28	35	43	48	57	62	69	77	21	26	31	36	42	47	53	61	1.1	1.3	1.4	1.6	1.7	1.3	1.5	1.6	1.8	1.9
SS B8	42	52	63	71	79	84	89	94	31	39	49	56	65	71	78	86	0.9	1.0	1.2	1.3	1.5	1.0	1.1	1.3	1.5	1.6
SS C1	37	48	58	66	75	81	87	93	24	32	41	48	57	63	71	79	0.9	1.0	1.2	1.4	1.5	1.1	1.2	1.4	1.5	1.6
SS C2	43	55	67	74	83	88	92	96	29	39	50	57	67	73	80	88	0.8	1.0	1.1	1.3	1.5	1.0	1.1	1.2	1.4	1.6
SS C3	30	38	47	54	63	69	76	84	20	26	33	38	46	51	58	67	1.0	1.2	1.3	1.5	1.6	1.3	1.4	1.5	1.7	1.8
SS C4	38	49	60	68	77	83	88	94	26	34	44	51	61	67	74	83	0.9	1.0	1.2	1.3	1.5	1.0	1.2	1.3	1.5	1.6
SS D1	48	59	70	77	85	89	93	97	36	45	55	62	71	76	82	89	0.8	1.0	1.1	1.3	1.5	0.9	1.1	1.3	1.4	1.6
SS D2	57	69	80	86	92	95	97	99	44	55	66	74	82	87	92	96	0.7	0.9	1.1	1.3	1.4	0.8	1.0	1.1	1.3	1.5
SS D3	43	53	64	70	79	84	89	94	33	41	49	56	64	70	76	84	0.9	1.0	1.2	1.4	1.5	1.0	1.2	1.3	1.5	1.6
SS D4	55	67	78	84	91	94	97	99	42	53	64	72	80	85	90	95	0.7	0.9	1.1	1.3	1.4	0.8	1.0	1.2	1.3	1.5
SS E1	43	54	64	72	80	85	90	95	31	39	49	55	64	70	77	85	0.9	1.0	1.2	1.3	1.5	1.0	1.1	1.3	1.5	1.6
SS E2	51	63	75	81	89	92	96	98	38	49	60	67	76	82	88	93	0.8	0.9	1.1	1.3	1.4	0.9	1.0	1.2	1.4	1.5
SS E3	36	44	54	61	70	75	81	88	26	33	40	46	54	60	66	75	1.0	1.1	1.3	1.4	1.6	1.1	1.3	1.4	1.6	1.7
SS E4	47	59	70	77	85	89	93	97	35	45	55	62	72	77	84	90	0.8	1.0	1.1	1.3	1.5	0.9	1.1	1.2	1.4	1.5

CLAYTON, NEW MEXICO

SOLAR SAVINGS FRACTION (%) — CONSERVATION FACTOR

TYPE	BASE TEMP: 55 F (2972 DD)								65 F (5213 DD)								CF 55 F					CF 65 F				
LCR	100	70	50	40	30	25	20	15	100	70	50	40	30	25	20	15	100	50	30	20	15	100	50	30	20	15
WW A1	28	35	42	48	56	61	68	77	21	26	31	35	41	45	51	59	1.1	1.3	1.4	1.6	1.7	1.4	1.5	1.7	1.8	2.0
WW A2	34	43	53	61	70	76	82	90	24	30	38	45	53	59	66	75	1.0	1.1	1.2	1.4	1.5	1.1	1.3	1.4	1.6	1.7
WW A3	36	46	57	65	75	80	87	93	24	32	42	48	58	64	71	80	0.9	1.0	1.2	1.4	1.5	1.1	1.2	1.3	1.5	1.6
WW A4	37	48	60	67	77	83	88	94	25	33	43	50	60	66	74	83	0.9	1.0	1.2	1.3	1.5	1.0	1.2	1.3	1.5	1.6
WW A5	38	50	62	70	80	85	90	95	25	35	45	53	63	69	76	85	0.8	1.0	1.2	1.3	1.5	1.0	1.1	1.3	1.4	1.6
WW A6	39	51	63	71	81	86	91	96	26	35	46	54	64	70	78	86	0.8	1.0	1.1	1.3	1.5	1.0	1.1	1.3	1.4	1.6
WW B1	28	36	45	52	61	67	74	82	19	24	31	36	43	48	54	63	1.0	1.2	1.3	1.5	1.6	1.3	1.4	1.6	1.7	1.9
WW B2	37	49	61	69	78	84	89	95	25	35	45	52	62	68	76	85	0.9	1.0	1.2	1.3	1.5	1.0	1.1	1.3	1.4	1.6
WW B3	44	57	69	77	85	90	94	97	31	42	53	61	71	77	84	91	0.8	0.9	1.1	1.3	1.4	0.9	1.1	1.2	1.4	1.5
WW B4	44	57	70	78	86	91	95	98	31	42	54	62	73	79	86	92	0.8	0.9	1.1	1.3	1.4	0.9	1.0	1.2	1.4	1.5
WW B5	42	56	68	76	85	90	94	98	29	41	53	61	72	78	85	92	0.8	0.9	1.1	1.3	1.4	0.9	1.0	1.2	1.4	1.5
WW C1	46	59	71	78	87	91	95	98	33	43	55	63	73	79	85	92	0.8	0.9	1.1	1.3	1.4	0.9	1.0	1.2	1.4	1.5
WW C2	44	56	69	76	85	90	94	98	30	41	53	61	71	78	84	91	0.8	0.9	1.1	1.3	1.4	0.9	1.1	1.2	1.4	1.5
WW C3	50	64	77	84	91	94	97	99	36	49	62	70	80	86	91	96	0.7	0.9	1.1	1.2	1.4	0.8	1.0	1.1	1.3	1.5
WW C4	46	60	73	81	89	93	96	99	33	45	58	67	77	83	89	95	0.8	0.9	1.1	1.3	1.4	0.9	1.0	1.2	1.3	1.5
TW A1	28	34	41	47	55	60	67	76	21	25	30	34	40	45	51	59	1.2	1.3	1.4	1.6	1.7	1.4	1.6	1.7	1.8	2.0
TW A2	31	39	48	55	64	70	77	85	22	28	35	40	48	54	61	70	1.0	1.1	1.3	1.4	1.6	1.2	1.4	1.5	1.6	1.8
TW A3	32	41	51	59	68	74	81	89	22	29	37	43	52	57	65	74	1.0	1.1	1.2	1.4	1.6	1.2	1.3	1.4	1.6	1.7
TW A4	31	41	52	59	69	75	82	89	21	28	37	43	52	58	66	75	1.0	1.1	1.2	1.4	1.5	1.1	1.3	1.4	1.6	1.7
TW B1	27	34	42	48	56	62	69	78	20	25	30	35	41	46	52	61	1.1	1.3	1.4	1.5	1.7	1.4	1.5	1.6	1.8	1.9
TW B2	29	37	47	54	63	69	76	85	20	26	34	39	47	53	60	69	1.0	1.2	1.3	1.5	1.6	1.2	1.4	1.5	1.6	1.8
TW B3	29	37	47	54	64	70	77	85	20	26	34	39	48	54	61	70	1.0	1.1	1.3	1.4	1.6	1.2	1.3	1.5	1.6	1.8
TW B4	27	35	45	51	61	67	75	83	19	25	32	37	45	51	58	68	1.0	1.2	1.3	1.5	1.6	1.2	1.4	1.5	1.7	1.8
TW C1	26	33	41	47	56	62	69	78	19	24	30	34	41	46	53	62	1.1	1.3	1.4	1.5	1.7	1.4	1.5	1.6	1.8	1.9
TW C2	26	33	42	48	57	63	71	80	18	24	30	35	43	48	55	64	1.1	1.2	1.4	1.5	1.6	1.3	1.4	1.6	1.7	1.8
TW C3	24	31	40	46	55	61	68	77	17	22	29	33	41	46	52	61	1.1	1.3	1.4	1.5	1.7	1.3	1.5	1.6	1.7	1.9
TW C4	22	28	35	41	49	55	62	71	16	21	26	30	36	41	47	55	1.2	1.3	1.5	1.6	1.7	1.5	1.6	1.7	1.9	2.0
TW D1	24	31	39	45	54	60	67	76	16	21	26	31	37	42	48	57	1.1	1.3	1.4	1.6	1.7	1.4	1.5	1.7	1.8	2.0
TW D2	34	44	55	63	73	79	85	92	24	32	41	47	57	63	70	79	0.9	1.1	1.2	1.4	1.5	1.1	1.2	1.4	1.5	1.6
TW D3	38	49	60	68	78	83	89	94	26	35	45	52	62	68	75	84	0.9	1.0	1.2	1.3	1.5	1.0	1.2	1.3	1.5	1.6
TW D4	40	52	64	72	81	86	91	96	28	38	49	56	67	73	80	88	0.8	1.0	1.1	1.3	1.5	1.0	1.1	1.2	1.4	1.5
TW D5	39	51	63	71	81	86	91	96	27	37	48	56	67	73	80	88	0.8	1.0	1.1	1.3	1.5	1.0	1.1	1.2	1.4	1.5
TW E1	42	53	65	73	82	87	92	96	30	39	50	57	67	73	80	88	0.8	1.0	1.1	1.3	1.5	1.0	1.1	1.3	1.4	1.6
TW E2	41	53	65	72	82	87	92	96	29	39	49	57	67	73	80	88	0.8	1.0	1.1	1.3	1.5	1.0	1.1	1.2	1.4	1.5
TW E3	48	61	73	81	89	93	96	99	35	47	59	67	77	83	89	94	0.8	0.9	1.1	1.3	1.4	0.9	1.0	1.2	1.3	1.5
TW E4	45	58	70	78	87	91	95	98	32	43	56	64	74	80	87	93	0.8	0.9	1.1	1.3	1.4	0.9	1.0	1.2	1.3	1.5
TW F1	26	33	41	46	55	60	68	76	19	24	29	33	40	44	51	59	1.1	1.3	1.4	1.6	1.7	1.4	1.5	1.7	1.8	1.9
TW F2	28	36	46	53	62	68	75	84	19	25	32	38	46	51	58	68	1.0	1.2	1.3	1.5	1.6	1.2	1.4	1.5	1.7	1.8
TW F3	28	37	47	54	64	70	78	86	19	25	33	39	47	53	61	70	1.0	1.1	1.3	1.4	1.6	1.2	1.3	1.5	1.6	1.7
TW F4	26	35	45	53	63	69	77	85	16	23	31	37	46	52	59	69	1.0	1.1	1.3	1.4	1.6	1.2	1.3	1.5	1.6	1.7
TW G1	24	31	39	45	54	60	67	76	17	22	28	32	39	44	50	59	1.1	1.3	1.4	1.6	1.7	1.4	1.5	1.7	1.8	1.9
TW G2	25	33	42	49	58	64	72	81	16	22	29	34	42	48	55	64	1.1	1.2	1.3	1.5	1.6	1.3	1.4	1.5	1.7	1.8
TW G3	23	31	40	47	57	63	71	80	14	20	27	33	41	46	54	63	1.1	1.2	1.3	1.5	1.6	1.3	1.4	1.5	1.7	1.8
TW G4	19	26	35	41	51	57	65	75	12	17	23	28	36	41	48	58	1.1	1.3	1.4	1.5	1.7	1.4	1.5	1.6	1.8	1.9
TW H1	21	28	36	42	51	57	64	74	14	19	25	29	36	41	47	57	1.2	1.3	1.4	1.6	1.7	1.4	1.5	1.7	1.8	1.9
TW H2	18	25	34	40	49	55	63	73	12	16	22	27	34	39	46	55	1.2	1.3	1.4	1.6	1.7	1.4	1.5	1.7	1.8	1.9
TW H3	16	22	29	35	44	50	57	67	9	14	19	24	30	35	41	50	1.2	1.4	1.5	1.6	1.8	1.5	1.6	1.7	1.9	2.0
TW H4	12	17	23	28	35	41	48	57	7	11	15	19	24	28	34	42	1.4	1.5	1.6	1.8	1.9	1.7	1.8	1.9	2.1	2.2
TW I1	21	28	36	42	50	56	63	73	13	18	23	27	34	38	44	53	1.2	1.3	1.4	1.6	1.7	1.5	1.6	1.7	1.9	2.0
TW I2	31	41	52	59	69	76	82	90	20	28	37	44	53	59	67	77	0.9	1.1	1.2	1.4	1.5	1.1	1.2	1.4	1.5	1.7
TW I3	33	44	55	63	73	79	86	92	22	31	41	48	57	64	72	81	0.9	1.0	1.2	1.4	1.5	1.1	1.2	1.3	1.5	1.6
TW I4	36	48	60	68	78	84	89	95	24	34	45	53	63	70	77	86	0.9	1.0	1.2	1.3	1.5	1.0	1.1	1.3	1.4	1.6
TW I5	36	48	61	69	79	84	90	95	25	35	46	54	64	71	78	87	0.9	1.0	1.2	1.3	1.5	1.0	1.1	1.3	1.4	1.6
TW J1	39	51	62	70	80	85	90	95	27	37	47	55	65	71	78	87	0.8	1.0	1.2	1.3	1.5	1.0	1.1	1.3	1.4	1.6
TW J2	38	49	61	69	79	85	90	95	26	35	46	54	64	71	78	86	0.9	1.0	1.2	1.3	1.5	1.0	1.1	1.3	1.4	1.6
TW J3	44	58	70	78	87	91	95	98	31	43	55	64	74	80	87	93	0.8	0.9	1.1	1.3	1.4	0.9	1.0	1.2	1.3	1.5
TW J4	41	55	67	75	85	89	94	97	29	40	52	61	71	78	85	92	0.8	0.9	1.1	1.3	1.4	0.9	1.0	1.2	1.4	1.5
DG A1	27	35	43	49	56	61	66	72	17	22	28	33	38	42	46	51	1.0	1.2	1.5	1.7	2.0	1.3	1.5	1.8	2.2	2.5
DG A2	28	38	48	54	63	69	75	82	19	25	33	39	47	52	58	66	1.0	1.1	1.3	1.5	1.7	1.2	1.3	1.5	1.7	1.9
DG A3	33	44	56	64	74	79	85	90	23	31	41	49	59	65	72	80	0.9	1.0	1.2	1.4	1.6	1.1	1.2	1.3	1.5	1.7
DG B1	27	37	48	55	64	69	75	81	17	23	30	35	43	49	54	61	1.0	1.1	1.3	1.6	1.8	1.3	1.4	1.5	1.8	2.1
DG B2	29	39	52	60	70	76	82	89	19	26	35	42	52	58	65	74	1.0	1.0	1.2	1.4	1.6	1.2	1.3	1.4	1.6	1.8
DG B3	33	45	59	68	79	84	89	94	23	32	42	51	63	70	77	85	0.9	1.0	1.1	1.4	1.6	1.1	1.1	1.2	1.4	1.6
DG C1	32	43	56	65	74	79	84	89	20	27	36	43	52	59	66	73	0.9	1.0	1.2	1.5	1.7	1.2	1.3	1.4	1.6	1.9
DG C2	33	45	59	68	78	83	89	93	22	30	40	48	59	66	74	82	0.9	1.0	1.2	1.4	1.5	1.1	1.2	1.3	1.5	1.7
DG C3	38	51	66	76	85	90	94	96	27	36	48	57	70	77	84	91	0.8	0.9	1.1	1.3	1.6	1.0	1.1	1.2	1.4	1.6
SS A1	36	45	55	62	70	76	82	89	28	35	43	49	57	62	69	77	1.0	1.1	1.3	1.4	1.6	1.1	1.3	1.4	1.6	1.7
SS A2	45	56	67	74	82	87	92	96	35	44	54	61	70	76	82	89	0.8	1.0	1.1	1.3	1.5	0.9	1.1	1.2	1.4	1.6
SS A3	34	42	51	58	66	72	78	86	26	32	39	45	52	58	64	72	1.0	1.1	1.3	1.5	1.6	1.2	1.3	1.5	1.6	1.8
SS A4	44	55	66	73	81	86	91	95	34	43	53	60	69	75	81	88	0.8	1.0	1.2	1.3	1.5	1.0	1.1	1.3	1.4	1.6
SS A5	34	42	50	57	65	71	77	85	27	33	39	45	52	57	63	71	1.0	1.2	1.3	1.5	1.6	1.2	1.3	1.5	1.7	1.8
SS A6	45	56	67	74	82	87	92	96	35	44	54	61	70	76	82	89	0.8	1.0	1.1	1.3	1.5	0.9	1.1	1.3	1.4	1.6
SS A7	30	37	45	50	58	63	70	78	24	29	35	39	45	50	56	63	1.1	1.3	1.4	1.6	1.7	1.3	1.5	1.6	1.8	1.9
SS A8	43	54	65	72	80	85	90	95	34	43	52	59	68	73	80	87	0.9	1.0	1.2	1.3	1.5	1.0	1.1	1.3	1.4	1.6
SS B1	31	39	48	55	64	70	77	85	23	30	37	42	50	55	62	71	1.0	1.2	1.3	1.5	1.6	1.2	1.3	1.5	1.6	1.8
SS B2	39	50	61	68	77	83	88	94	30	39	48	55	64	70	77	85	0.9	1.0	1.2	1.4	1.5	1.0	1.1	1.3	1.5	1.6
SS B3	30	37	46	52	61	67	74	82	22	28	35	40	47	52	59	67	1.0	1.2	1.3	1.5	1.6	1.2	1.4	1.5	1.7	1.8
SS B4	39	49	60	67	76	81	87	93	29	38	47	54	63	69	76	84	0.9	1.0	1.2	1.4	1.5	1.0	1.2	1.3	1.5	1.6
SS B5	28	35	42	48	56	62	69	77	21	26	32	37	43	48	54	62	1.1	1.3	1.4	1.6	1.7	1.3	1.5	1.6	1.8	1.9
SS B6	39	49	60	67	77	82	88	93	29	38	47	54	63	69	76	84	0.9	1.0	1.2	1.4	1.5	1.0	1.2	1.3	1.5	1.6
SS B7	25	31	38	44	51	57	63	72	20	24	29	33	39	43	49	56	1.2	1.3	1.5	1.6	1.8	1.4	1.6	1.7	1.9	2.0
SS B8	37	47	58	65	75	80	86	92	28	36	46	52	61	67	74	82	0.9	1.0	1.2	1.4	1.5	1.0	1.2	1.3	1.5	1.6
SS C1	32	42	52	59	69	75	82	89	21	28	37	43	52	57	65	74	1.0	1.1	1.2	1.4	1.5	1.1	1.3	1.4	1.6	1.7
SS C2	37	49	60	68	78	83	89	95	26	35	45	52	62	68	76	84	0.9	1.0	1.2	1.3	1.5	1.0	1.1	1.3	1.4	1.6
SS C3	26	33	41	48	57	63	70	79	18	23	29	34	41	46	53	62	1.1	1.2	1.4	1.5	1.7	1.3	1.5	1.6	1.8	1.9
SS C4	33	43	54	62	71	77	84	91	23	31	40	46	56	62	69	79	0.9	1.1	1.2	1.4	1.5	1.1	1.2	1.4	1.5	1.7
SS D1	43	54	65	72	80	85	90	95	33	42	51	58	67	72	79	86	0.9	1.0	1.2	1.3	1.5	1.0	1.1	1.3	1.5	1.6
SS D2	52	64	75	82	89	93	96	99	40	51	63	70	79	84	89	95	0.8	0.9	1.1	1.3	1.4	0.9	1.0	1.2	1.3	1.5
SS D3	39	48	58	65	74	79	85	91	30	37	46	52	60	66	72	80	0.9	1.1	1.2	1.4	1.5	1.1	1.2	1.4	1.5	1.7
SS D4	50	62	73	80	88	92	95	98	39	49	60	68	77	82	88	93	0.8	0.9	1.1	1.3	1.4	0.9	1.0	1.2	1.4	1.5
SS E1	38	48	59	66	75	80	86	92	28	36	45	51	60	65	72	81	0.9	1.0	1.2	1.4	1.5	1.0	1.2	1.4	1.5	1.7
SS E2	46	58	69	77	85	89	94	97	34	45	55	63	72	78	84	91	0.8	1.0	1.1	1.3	1.4	0.9	1.1	1.2	1.4	1.5
SS E3	32	40	49	55	64	70	77	84	24	30	37	42	50	55	62	70	1.0	1.2	1.3	1.5	1.6	1.2	1.3	1.5	1.7	1.8
SS E4	42	53	65	72	81	86	91	96	32	41	51	58	68	74	80	88	0.8	1.0	1.2	1.3	1.5	1.0	1.1	1.3	1.4	1.6

FARMINGTON, NEW MEXICO

TYPE	SOLAR SAVINGS FRACTION (%) BASE TEMP: 55 F (3424 DD) LCR 100	70	50	40	30	25	20	15	65 F (5717 DD) 100	70	50	40	30	25	20	15	CONSERVATION FACTOR 55 F 100	50	30	20	15	65 F 100	50	30	20	15
WW A1	26	32	38	43	51	56	63	71	20	24	29	33	38	42	48	56	1.2	1.4	1.5	1.6	1.8	1.5	1.6	1.8	1.9	2.0
WW A2	30	39	48	55	64	70	77	85	22	28	36	42	50	55	62	71	1.0	1.1	1.3	1.5	1.6	1.2	1.3	1.5	1.6	1.8
WW A3	32	42	52	59	69	75	81	89	22	30	39	45	54	60	67	76	0.9	1.1	1.2	1.4	1.6	1.1	1.2	1.4	1.6	1.7
WW A4	33	43	54	62	71	77	84	90	23	31	40	47	56	62	70	79	0.9	1.1	1.2	1.4	1.5	1.1	1.2	1.4	1.5	1.7
WW A5	34	45	56	64	74	80	86	92	23	32	42	49	59	65	73	81	0.9	1.0	1.2	1.4	1.5	1.0	1.2	1.3	1.5	1.6
WW A6	35	46	58	65	75	81	87	93	23	33	43	50	60	66	74	83	0.9	1.0	1.2	1.4	1.5	1.0	1.2	1.3	1.5	1.6
WW B1	25	32	40	46	55	60	67	76	17	22	28	33	39	44	50	58	1.1	1.3	1.4	1.6	1.7	1.4	1.5	1.7	1.8	2.0
WW B2	33	44	55	63	73	79	85	91	23	32	42	49	59	65	72	81	0.9	1.1	1.2	1.4	1.5	1.0	1.2	1.3	1.5	1.6
WW B3	40	52	63	71	80	85	90	95	29	39	50	57	67	73	80	88	0.8	1.0	1.2	1.3	1.5	0.9	1.1	1.2	1.4	1.6
WW B4	40	52	65	72	82	87	92	96	28	39	51	59	70	76	83	90	0.8	1.0	1.1	1.3	1.5	0.9	1.1	1.2	1.4	1.5
WW B5	38	51	63	71	81	86	91	96	27	38	50	58	69	75	82	90	0.8	1.0	1.1	1.3	1.5	0.9	1.1	1.2	1.4	1.5
WW C1	41	53	65	73	82	87	92	96	30	41	52	60	70	76	82	90	0.8	1.0	1.1	1.3	1.5	0.9	1.1	1.2	1.4	1.5
WW C2	39	51	63	71	81	86	91	96	28	39	50	58	68	74	81	89	0.8	1.0	1.1	1.3	1.5	0.9	1.1	1.2	1.4	1.5
WW C3	45	59	72	79	87	91	95	98	34	46	59	67	78	83	89	95	0.8	0.9	1.1	1.3	1.4	0.8	1.0	1.2	1.3	1.5
WW C4	42	55	68	76	85	89	94	97	30	42	55	64	74	80	87	93	0.8	0.9	1.1	1.3	1.4	0.9	1.0	1.2	1.3	1.5
TW A1	25	31	38	43	50	55	62	70	20	24	28	32	38	42	47	55	1.2	1.4	1.5	1.7	1.8	1.5	1.6	1.8	1.9	2.0
TW A2	28	35	44	50	59	65	72	80	20	26	33	38	45	50	57	66	1.1	1.2	1.4	1.5	1.7	1.3	1.4	1.5	1.7	1.8
TW A3	28	37	46	53	63	69	76	84	20	27	34	40	48	54	61	70	1.0	1.2	1.3	1.5	1.6	1.2	1.3	1.5	1.6	1.8
TW A4	28	37	47	54	63	69	77	85	19	26	34	40	49	55	62	72	1.0	1.1	1.3	1.5	1.6	1.2	1.3	1.4	1.6	1.7
TW B1	25	31	38	43	51	56	63	72	19	23	29	33	39	43	49	57	1.2	1.3	1.5	1.6	1.8	1.4	1.6	1.7	1.9	2.0
TW B2	26	34	42	49	58	64	71	79	19	25	31	36	44	49	56	65	1.1	1.2	1.4	1.5	1.7	1.3	1.4	1.5	1.7	1.8
TW B3	26	34	43	49	58	64	72	80	18	24	31	37	45	50	57	67	1.1	1.2	1.3	1.5	1.6	1.3	1.4	1.5	1.7	1.8
TW B4	24	32	40	47	56	62	69	78	17	23	30	35	43	48	55	64	1.1	1.2	1.4	1.5	1.7	1.3	1.4	1.6	1.7	1.8
TW C1	24	30	37	43	51	56	63	72	18	22	28	32	39	43	49	58	1.2	1.3	1.5	1.6	1.7	1.4	1.5	1.7	1.8	2.0
TW C2	23	30	38	44	52	58	65	74	17	22	28	33	40	45	51	60	1.2	1.3	1.4	1.6	1.7	1.4	1.5	1.6	1.8	1.9
TW C3	22	28	36	42	50	56	63	72	16	21	27	31	38	43	49	58	1.2	1.3	1.5	1.6	1.7	1.4	1.5	1.7	1.8	1.9
TW C4	20	26	32	37	45	50	57	66	15	19	24	28	34	38	44	52	1.3	1.4	1.6	1.7	1.8	1.5	1.6	1.8	1.9	2.0
TW D1	21	27	35	40	48	54	61	69	14	19	24	28	34	38	44	52	1.2	1.3	1.5	1.6	1.8	1.5	1.6	1.8	1.9	2.1
TW D2	31	40	50	58	67	73	80	88	22	29	38	44	53	59	67	76	1.0	1.1	1.3	1.4	1.6	1.1	1.2	1.4	1.5	1.7
TW D3	34	44	55	62	72	78	84	91	24	32	42	49	58	65	72	81	0.9	1.1	1.2	1.4	1.5	1.0	1.2	1.3	1.5	1.6
TW D4	36	47	59	67	76	82	88	93	26	35	46	53	63	70	77	86	0.9	1.0	1.2	1.4	1.5	1.0	1.1	1.3	1.4	1.6
TW D5	35	46	58	66	76	82	88	93	25	35	45	53	63	70	77	86	0.9	1.0	1.2	1.3	1.5	1.0	1.1	1.3	1.4	1.6
TW E1	38	49	60	67	77	82	88	93	27	37	47	54	64	70	77	85	0.9	1.0	1.2	1.4	1.5	1.0	1.1	1.3	1.4	1.6
TW E2	37	48	59	67	77	82	88	94	27	36	46	54	64	70	77	86	0.9	1.0	1.2	1.4	1.5	1.0	1.1	1.3	1.4	1.6
TW E3	44	56	68	76	85	89	93	97	32	44	56	64	74	80	86	93	0.8	1.0	1.1	1.3	1.4	0.9	1.0	1.2	1.4	1.5
TW E4	40	53	65	73	82	87	92	96	30	41	53	61	71	77	84	91	0.8	1.0	1.1	1.3	1.5	0.9	1.1	1.2	1.4	1.5
TW F1	24	30	37	42	50	55	62	71	18	22	27	31	37	41	47	55	1.2	1.3	1.5	1.6	1.8	1.5	1.6	1.7	1.9	2.0
TW F2	25	33	41	48	57	63	70	78	18	23	30	35	43	48	55	64	1.1	1.2	1.4	1.5	1.7	1.3	1.4	1.6	1.7	1.8
TW F3	25	33	42	49	59	65	72	81	17	23	30	36	44	50	57	66	1.1	1.2	1.3	1.5	1.6	1.2	1.4	1.5	1.7	1.8
TW F4	23	31	40	47	57	63	71	80	15	21	29	34	43	48	56	65	1.1	1.2	1.3	1.5	1.6	1.2	1.4	1.5	1.7	1.8
TW G1	22	28	35	41	49	55	62	71	16	20	26	30	36	41	47	55	1.2	1.3	1.5	1.6	1.8	1.4	1.6	1.7	1.9	2.0
TW G2	22	29	38	44	53	59	66	75	15	20	27	32	39	44	51	60	1.1	1.3	1.4	1.6	1.7	1.3	1.5	1.6	1.7	1.9
TW G3	20	27	36	42	51	58	65	75	13	19	25	30	38	43	50	60	1.1	1.3	1.4	1.6	1.7	1.3	1.5	1.6	1.7	1.9
TW G4	16	23	31	37	46	52	59	69	10	15	21	26	33	38	45	54	1.2	1.3	1.5	1.6	1.8	1.4	1.5	1.7	1.8	1.9
TW H1	19	25	32	38	46	51	59	68	13	17	23	27	33	38	44	53	1.2	1.4	1.5	1.6	1.8	1.5	1.6	1.7	1.9	2.0
TW H2	16	22	30	36	44	50	57	67	10	15	21	25	32	36	43	52	1.2	1.4	1.5	1.7	1.8	1.5	1.6	1.7	1.9	2.0
TW H3	14	19	26	31	39	45	52	62	8	13	18	22	28	32	38	47	1.3	1.4	1.6	1.7	1.8	1.6	1.7	1.8	1.9	2.1
TW H4	10	15	20	25	32	36	43	52	6	10	14	17	22	26	31	39	1.5	1.6	1.7	1.9	2.0	1.7	1.9	2.0	2.1	2.2
TW I1	18	24	31	37	44	50	57	66	12	16	21	25	31	35	40	48	1.3	1.4	1.5	1.7	1.8	1.5	1.7	1.8	2.0	2.1
TW I2	27	36	47	54	64	70	77	85	19	26	34	41	50	56	63	73	1.0	1.1	1.3	1.4	1.6	1.1	1.3	1.4	1.6	1.7
TW I3	30	40	50	58	68	74	81	88	21	29	38	45	54	60	68	77	0.9	1.1	1.2	1.4	1.6	1.1	1.2	1.4	1.5	1.7
TW I4	32	43	55	63	73	79	85	92	22	32	42	50	60	66	74	83	0.9	1.0	1.2	1.4	1.5	1.0	1.2	1.3	1.5	1.6
TW I5	32	44	56	64	74	80	86	92	23	32	43	51	61	68	75	84	0.9	1.0	1.2	1.4	1.5	1.0	1.1	1.3	1.4	1.6
TW J1	35	46	57	65	75	80	86	92	25	34	44	52	61	68	75	84	0.9	1.0	1.2	1.4	1.5	1.0	1.1	1.3	1.5	1.6
TW J2	34	45	56	64	74	80	86	92	24	33	43	51	61	67	75	84	0.9	1.0	1.2	1.4	1.5	1.0	1.1	1.3	1.5	1.6
TW J3	40	53	65	73	82	87	92	96	29	40	52	61	71	77	84	91	0.8	1.0	1.1	1.3	1.5	0.9	1.0	1.2	1.4	1.5
TW J4	37	50	62	70	80	85	91	96	27	38	49	58	68	75	82	90	0.8	1.0	1.1	1.3	1.5	0.9	1.1	1.2	1.4	1.5
DG A1	23	30	38	43	50	54	59	65	15	20	26	30	35	38	42	47	1.1	1.3	1.6	1.9	2.2	1.4	1.6	1.9	2.3	2.7
DG A2	25	33	43	49	58	63	69	77	17	23	31	36	44	49	55	62	1.0	1.2	1.4	1.6	1.8	1.3	1.4	1.6	1.8	2.0
DG A3	29	40	51	59	69	75	81	87	21	29	39	46	56	62	69	77	0.9	1.1	1.3	1.5	1.7	1.1	1.2	1.4	1.6	1.7
DG B1	24	32	42	48	57	62	68	74	15	20	27	32	39	44	50	56	1.1	1.2	1.4	1.7	1.9	1.3	1.5	1.6	1.9	2.2
DG B2	26	35	46	54	64	70	77	84	18	24	32	39	48	54	61	70	1.0	1.1	1.3	1.5	1.7	1.2	1.3	1.4	1.6	1.8
DG B3	30	41	53	62	73	79	85	91	22	30	40	48	59	66	74	83	0.9	1.0	1.2	1.4	1.6	1.1	1.2	1.3	1.5	1.6
DG C1	28	38	49	57	67	72	78	84	18	25	33	39	48	54	61	68	1.0	1.1	1.3	1.5	1.8	1.2	1.3	1.5	1.7	2.0
DG C2	29	40	52	61	72	78	84	90	21	28	37	45	55	62	70	78	0.9	1.0	1.2	1.4	1.6	1.1	1.2	1.3	1.5	1.7
DG C3	34	46	60	69	80	85	90	95	25	34	45	54	66	73	81	88	0.9	1.0	1.1	1.3	1.6	1.0	1.1	1.2	1.4	1.6
SS A1	33	41	50	56	65	70	76	84	26	33	40	46	53	59	65	73	1.0	1.2	1.3	1.5	1.6	1.1	1.3	1.5	1.6	1.8
SS A2	41	51	62	69	77	82	88	93	33	42	51	58	67	73	79	86	0.9	1.0	1.2	1.4	1.5	1.0	1.1	1.3	1.5	1.6
SS A3	31	38	46	52	60	66	72	80	24	30	37	42	49	54	60	68	1.1	1.2	1.4	1.5	1.7	1.2	1.4	1.5	1.7	1.9
SS A4	40	50	60	67	76	81	86	92	32	41	50	57	66	71	77	85	0.9	1.0	1.2	1.4	1.5	1.0	1.1	1.3	1.5	1.6
SS A5	31	38	46	52	60	65	71	79	25	31	37	42	49	53	59	67	1.1	1.2	1.4	1.6	1.7	1.2	1.4	1.6	1.7	1.9
SS A6	41	51	61	68	77	82	87	93	32	41	51	58	67	72	79	86	0.9	1.0	1.2	1.4	1.5	1.0	1.1	1.3	1.5	1.6
SS A7	28	34	40	46	53	58	64	72	23	27	32	36	42	46	52	59	1.2	1.3	1.5	1.7	1.8	1.4	1.5	1.7	1.9	2.0
SS A8	39	49	59	66	75	80	86	91	31	40	49	56	64	70	76	84	0.9	1.1	1.2	1.4	1.5	1.0	1.2	1.3	1.5	1.6
SS B1	28	36	44	50	58	64	71	79	22	28	34	39	47	52	58	67	1.1	1.2	1.4	1.5	1.7	1.2	1.4	1.5	1.7	1.9
SS B2	36	45	56	63	72	77	84	90	28	36	45	52	61	67	74	82	0.9	1.1	1.2	1.4	1.6	1.0	1.2	1.3	1.5	1.6
SS B3	27	34	41	47	55	61	67	76	21	26	32	37	44	49	55	63	1.1	1.3	1.4	1.6	1.7	1.3	1.4	1.6	1.8	1.9
SS B4	35	44	55	62	71	76	82	89	27	35	44	51	60	65	72	81	0.9	1.1	1.2	1.4	1.6	1.0	1.2	1.4	1.5	1.7
SS B5	25	31	38	44	51	56	63	71	20	25	30	34	41	45	51	58	1.2	1.3	1.5	1.7	1.8	1.4	1.5	1.7	1.9	2.0
SS B6	35	45	55	62	71	77	83	90	27	35	45	51	60	66	73	81	0.9	1.1	1.2	1.4	1.6	1.0	1.2	1.3	1.5	1.7
SS B7	23	28	35	39	46	51	57	65	18	22	27	31	36	40	45	52	1.3	1.4	1.6	1.7	1.9	1.5	1.7	1.8	2.0	2.1
SS B8	34	43	53	60	69	75	81	88	26	34	43	49	58	64	71	79	0.9	1.1	1.3	1.4	1.6	1.1	1.2	1.4	1.5	1.7
SS C1	28	37	47	54	63	69	76	84	20	26	34	40	48	54	61	70	1.0	1.1	1.3	1.5	1.6	1.2	1.3	1.5	1.6	1.8
SS C2	34	44	55	63	73	78	85	91	24	32	42	49	59	65	72	81	0.9	1.1	1.2	1.4	1.5	1.0	1.2	1.3	1.5	1.6
SS C3	23	30	37	43	51	57	64	73	17	22	27	32	38	43	49	58	1.2	1.3	1.4	1.6	1.7	1.4	1.5	1.7	1.8	1.9
SS C4	30	39	49	56	66	72	79	87	21	29	37	43	52	58	66	75	1.0	1.1	1.3	1.4	1.6	1.1	1.3	1.4	1.6	1.7
SS D1	39	49	59	66	74	79	85	91	31	39	48	54	63	68	75	82	0.9	1.1	1.2	1.4	1.6	1.0	1.2	1.3	1.5	1.7
SS D2	47	59	70	77	85	89	93	97	38	48	59	67	76	81	86	92	0.8	1.0	1.1	1.3	1.5	0.9	1.0	1.2	1.4	1.5
SS D3	35	44	53	60	68	73	79	86	28	35	43	49	57	62	68	76	1.0	1.1	1.3	1.5	1.6	1.1	1.3	1.4	1.6	1.7
SS D4	46	57	68	75	83	87	92	96	36	47	57	64	73	79	85	91	0.8	1.0	1.1	1.3	1.5	0.9	1.1	1.2	1.4	1.5
SS E1	34	43	53	60	69	74	80	87	26	33	41	47	56	61	68	76	1.0	1.1	1.3	1.4	1.6	1.1	1.2	1.4	1.6	1.7
SS E2	41	53	64	71	80	85	90	95	32	42	52	59	69	75	81	88	0.8	1.0	1.2	1.3	1.5	0.9	1.1	1.3	1.4	1.6
SS E3	29	36	44	50	58	64	70	79	23	28	35	40	47	52	58	66	1.1	1.2	1.4	1.6	1.7	1.3	1.4	1.6	1.7	1.9
SS E4	38	48	59	66	75	81	86	92	29	38	48	55	64	70	77	84	0.9	1.0	1.2	1.4	1.5	1.0	1.1	1.3	1.5	1.6

LOS ALAMOS, NEW MEXICO

SOLAR SAVINGS FRACTION (%) — CONSERVATION FACTOR

TYPE	BASE TEMP: 55 F (3851 DD) LCR 100	70	50	40	30	25	20	15	65 F (6437 DD) 100	70	50	40	30	25	20	15	CF 55 F 100	50	30	20	15	CF 65 F 100	50	30	20	15
WW A1	24	29	36	40	47	52	59	67	18	22	26	30	35	38	43	51	1.3	1.4	1.6	1.7	1.8	1.6	1.7	1.9	2.0	2.1
WW A2	28	36	45	51	61	67	74	82	20	25	32	38	45	51	57	67	1.0	1.2	1.3	1.5	1.6	1.3	1.4	1.5	1.7	1.8
WW A3	29	38	48	56	65	71	78	86	20	27	35	41	49	55	62	72	1.0	1.1	1.3	1.4	1.6	1.2	1.3	1.4	1.6	1.7
WW A4	30	40	50	58	68	74	81	88	20	27	36	42	52	58	65	75	1.0	1.1	1.2	1.4	1.5	1.1	1.3	1.4	1.6	1.7
WW A5	31	41	52	60	70	76	83	90	20	28	38	44	54	60	68	77	0.9	1.1	1.2	1.4	1.5	1.1	1.2	1.4	1.5	1.7
WW A6	31	42	54	62	72	78	84	91	20	29	38	45	55	62	69	79	0.9	1.1	1.2	1.4	1.5	1.1	1.2	1.4	1.5	1.6
WW B1	23	29	37	43	51	56	63	72	15	20	25	29	35	39	45	53	1.2	1.3	1.5	1.6	1.7	1.5	1.6	1.8	1.9	2.1
WW B2	30	41	52	59	69	75	82	90	21	29	38	45	54	60	68	78	0.9	1.1	1.2	1.4	1.5	1.1	1.2	1.4	1.5	1.7
WW B3	37	48	60	68	77	83	89	94	26	35	46	53	63	70	77	85	0.9	1.0	1.2	1.3	1.5	1.0	1.1	1.3	1.4	1.6
WW B4	36	49	61	69	79	85	90	95	25	36	47	55	66	72	80	88	0.8	1.0	1.1	1.3	1.5	1.0	1.1	1.2	1.4	1.5
WW B5	35	47	60	68	78	84	90	95	24	34	46	54	65	72	79	87	0.9	1.0	1.2	1.3	1.5	1.0	1.1	1.2	1.4	1.5
WW C1	38	50	62	70	79	85	90	95	27	37	48	55	66	72	79	87	0.8	1.0	1.2	1.3	1.5	1.0	1.1	1.3	1.4	1.6
WW C2	36	48	60	68	78	83	89	95	25	35	46	54	64	71	78	86	0.9	1.0	1.2	1.3	1.5	1.0	1.1	1.3	1.4	1.6
WW C3	42	56	68	76	85	90	94	98	30	43	55	64	74	80	87	93	0.8	0.9	1.1	1.3	1.4	0.9	1.0	1.2	1.3	1.5
WW C4	38	52	64	73	82	88	93	97	27	39	51	60	71	77	84	91	0.8	1.0	1.1	1.3	1.4	0.9	1.0	1.2	1.4	1.5
TW A1	23	29	35	40	47	52	58	67	18	22	26	29	34	38	43	50	1.3	1.4	1.6	1.7	1.8	1.6	1.7	1.9	2.0	2.1
TW A2	25	33	41	47	55	61	68	77	18	23	29	34	41	46	52	61	1.1	1.3	1.4	1.5	1.7	1.3	1.5	1.6	1.8	1.9
TW A3	26	34	43	50	59	65	72	81	18	24	31	36	44	49	56	66	1.1	1.2	1.3	1.5	1.6	1.3	1.4	1.5	1.7	1.8
TW A4	25	34	43	50	60	66	73	82	17	23	31	36	45	50	57	67	1.0	1.2	1.3	1.5	1.6	1.2	1.4	1.5	1.7	1.8
TW B1	23	29	35	40	48	53	60	68	17	21	26	30	35	39	45	52	1.3	1.4	1.5	1.7	1.8	1.5	1.7	1.8	1.9	2.1
TW B2	24	31	39	45	54	60	67	76	17	22	28	33	40	45	52	61	1.1	1.3	1.4	1.6	1.7	1.3	1.5	1.6	1.8	1.9
TW B3	23	31	39	46	55	61	68	77	16	22	28	33	41	46	53	62	1.1	1.2	1.4	1.5	1.7	1.3	1.5	1.6	1.7	1.9
TW B4	22	29	37	43	52	58	65	75	15	20	27	32	39	44	50	60	1.1	1.3	1.4	1.6	1.7	1.4	1.5	1.6	1.8	1.9
TW C1	22	28	35	40	48	53	60	69	16	20	25	29	35	40	45	53	1.2	1.4	1.5	1.7	1.8	1.5	1.6	1.8	1.9	2.0
TW C2	21	28	35	41	49	55	62	71	15	20	25	30	36	41	47	56	1.2	1.3	1.5	1.6	1.7	1.4	1.6	1.7	1.8	2.0
TW C3	20	26	33	39	47	52	59	68	15	19	24	28	35	39	45	54	1.2	1.4	1.5	1.7	1.8	1.5	1.6	1.7	1.9	2.0
TW C4	19	24	30	35	42	47	53	62	14	18	22	26	31	35	41	48	1.3	1.5	1.6	1.7	1.9	1.6	1.7	1.9	2.0	2.1
TW D1	19	25	32	37	44	50	56	66	13	17	21	25	30	34	39	47	1.3	1.4	1.5	1.7	1.8	1.6	1.7	1.9	2.0	2.2
TW D2	28	37	47	54	64	70	77	85	20	26	34	41	49	55	62	72	1.0	1.1	1.3	1.4	1.6	1.2	1.3	1.4	1.6	1.7
TW D3	31	41	51	59	69	75	82	89	22	29	38	45	54	60	68	77	0.9	1.1	1.2	1.4	1.5	1.1	1.2	1.4	1.5	1.7
TW D4	33	44	55	63	73	79	86	92	23	32	42	49	59	66	74	83	0.9	1.0	1.2	1.4	1.5	1.0	1.2	1.3	1.5	1.6
TW D5	32	43	55	63	73	79	85	92	22	32	42	49	60	66	74	83	0.9	1.0	1.2	1.4	1.5	1.0	1.2	1.3	1.5	1.6
TW E1	35	45	56	64	74	80	86	92	25	33	43	50	60	66	74	82	0.9	1.0	1.2	1.4	1.5	1.0	1.2	1.3	1.5	1.6
TW E2	34	45	56	64	74	80	86	92	24	33	43	50	60	66	74	83	0.9	1.0	1.2	1.4	1.5	1.0	1.2	1.3	1.5	1.6
TW E3	40	53	65	73	82	87	92	97	29	40	52	60	70	77	84	91	0.8	1.0	1.1	1.3	1.5	0.9	1.1	1.2	1.4	1.5
TW E4	37	50	62	70	80	85	91	96	27	37	49	57	68	74	81	89	0.8	1.0	1.1	1.3	1.5	0.9	1.1	1.2	1.4	1.5
TW F1	22	28	34	39	46	51	58	67	16	20	25	28	34	38	43	50	1.3	1.4	1.5	1.7	1.8	1.6	1.7	1.8	2.0	2.1
TW F2	23	30	38	44	53	59	66	75	16	21	27	32	39	43	50	59	1.1	1.3	1.4	1.6	1.7	1.4	1.5	1.6	1.8	1.9
TW F3	22	30	39	46	55	61	68	78	15	20	27	32	40	45	52	62	1.1	1.2	1.4	1.5	1.7	1.3	1.4	1.6	1.7	1.9
TW F4	20	28	37	44	53	59	67	77	13	19	25	31	38	44	51	61	1.1	1.2	1.4	1.5	1.7	1.3	1.4	1.6	1.7	1.9
TW G1	20	26	33	38	46	51	58	67	14	18	23	27	33	37	42	50	1.3	1.4	1.5	1.7	1.8	1.5	1.7	1.8	2.0	2.1
TW G2	20	26	34	40	49	55	62	72	13	18	24	28	35	40	47	56	1.2	1.3	1.4	1.6	1.7	1.4	1.5	1.7	1.8	1.9
TW G3	18	25	33	39	48	54	61	71	11	16	22	27	34	39	46	55	1.2	1.3	1.5	1.6	1.7	1.4	1.5	1.7	1.8	1.9
TW G4	14	21	28	34	42	48	55	65	9	13	19	23	30	34	41	50	1.3	1.4	1.5	1.7	1.8	1.5	1.6	1.8	1.9	2.0
TW H1	17	23	29	35	42	48	55	64	11	15	20	24	30	34	40	48	1.3	1.4	1.6	1.7	1.8	1.6	1.7	1.8	2.0	2.1
TW H2	14	20	27	33	41	46	53	63	9	13	18	22	28	33	39	47	1.3	1.4	1.6	1.7	1.8	1.5	1.7	1.8	2.0	2.1
TW H3	12	17	24	29	36	41	48	58	7	11	15	19	25	29	35	43	1.4	1.5	1.6	1.8	1.9	1.6	1.8	1.9	2.0	2.2
TW H4	9	13	18	22	29	33	40	48	5	8	12	15	20	23	28	35	1.5	1.7	1.8	1.9	2.0	1.8	2.0	2.1	2.2	2.3
TW I1	17	22	28	33	41	46	53	62	11	14	18	22	27	31	36	43	1.3	1.5	1.6	1.7	1.9	1.6	1.8	2.0	2.1	2.2
TW I2	25	33	43	50	60	67	74	83	16	23	31	37	45	51	59	69	1.0	1.2	1.3	1.5	1.6	1.2	1.3	1.5	1.6	1.8
TW I3	27	37	47	54	65	71	78	86	18	26	34	41	50	56	64	74	1.0	1.1	1.3	1.4	1.6	1.1	1.3	1.4	1.6	1.7
TW I4	29	40	51	59	70	76	83	90	20	29	38	46	56	62	70	80	0.9	1.1	1.2	1.4	1.5	1.1	1.2	1.3	1.5	1.6
TW I5	29	40	52	60	71	77	84	91	20	29	39	47	57	64	72	81	0.9	1.1	1.2	1.4	1.5	1.0	1.2	1.3	1.5	1.6
TW J1	32	43	54	62	72	78	84	91	22	31	40	48	57	64	71	81	0.9	1.1	1.2	1.4	1.5	1.1	1.2	1.3	1.5	1.6
TW J2	31	41	53	61	71	77	84	91	21	30	40	47	57	63	71	80	0.9	1.1	1.2	1.4	1.5	1.1	1.2	1.3	1.5	1.6
TW J3	37	49	62	70	80	85	91	96	26	37	49	57	68	74	81	89	0.8	1.0	1.1	1.3	1.5	0.9	1.1	1.2	1.4	1.5
TW J4	34	46	59	67	77	83	89	95	24	34	46	54	65	71	79	87	0.9	1.0	1.2	1.3	1.5	1.0	1.1	1.2	1.4	1.5
DG A1	21	27	35	40	46	50	55	60	13	17	22	26	31	34	37	41	1.2	1.4	1.6	2.0	2.3	1.5	1.7	2.0	2.5	3.0
DG A2	23	31	39	46	54	59	66	73	15	21	28	33	40	45	50	58	1.1	1.2	1.4	1.6	1.8	1.3	1.4	1.6	1.9	2.1
DG A3	27	37	48	56	66	72	78	85	19	26	35	42	52	58	65	74	1.0	1.1	1.3	1.5	1.7	1.2	1.2	1.4	1.6	1.8
DG B1	21	28	37	44	53	58	64	70	13	18	23	27	34	38	44	50	1.1	1.2	1.4	1.7	2.0	1.5	1.6	1.7	2.0	2.4
DG B2	23	32	42	50	60	66	73	81	16	21	28	34	43	49	56	65	1.1	1.1	1.3	1.5	1.7	1.3	1.4	1.5	1.7	1.9
DG B3	27	37	49	59	70	77	83	90	20	27	36	43	54	62	70	80	1.0	1.0	1.2	1.4	1.6	1.2	1.2	1.3	1.5	1.7
DG C1	25	33	44	52	62	68	75	81	16	22	28	34	42	48	55	63	1.1	1.1	1.3	1.5	1.8	1.3	1.4	1.5	1.7	2.0
DG C2	27	36	48	57	68	74	81	88	18	25	33	40	50	57	65	74	1.0	1.1	1.2	1.4	1.6	1.2	1.3	1.4	1.6	1.7
DG C3	31	42	56	66	77	83	89	94	22	31	41	49	61	69	77	86	0.9	1.0	1.1	1.3	1.5	1.1	1.1	1.2	1.4	1.6
SS A1	32	39	48	54	63	68	75	83	25	31	38	44	51	56	63	71	1.0	1.2	1.3	1.5	1.6	1.2	1.3	1.5	1.7	1.8
SS A2	39	49	60	67	76	81	87	93	31	40	49	56	65	71	77	85	0.9	1.0	1.2	1.4	1.5	1.0	1.1	1.3	1.5	1.6
SS A3	29	37	45	51	59	64	71	79	23	29	35	40	47	52	58	66	1.1	1.2	1.4	1.6	1.7	1.2	1.4	1.6	1.7	1.9
SS A4	39	49	59	66	75	80	86	92	31	39	48	55	64	69	76	84	0.9	1.1	1.2	1.4	1.5	1.0	1.2	1.3	1.5	1.6
SS A5	30	37	44	50	58	63	70	78	24	29	35	40	47	51	57	65	1.1	1.3	1.4	1.6	1.7	1.3	1.4	1.6	1.8	1.9
SS A6	39	49	60	67	76	81	87	92	31	39	49	56	65	70	77	85	0.9	1.0	1.2	1.4	1.5	1.0	1.1	1.3	1.5	1.6
SS A7	27	32	39	44	51	56	62	70	22	26	31	35	41	45	50	57	1.2	1.4	1.5	1.7	1.8	1.4	1.6	1.7	1.9	2.1
SS A8	38	48	58	65	74	79	85	91	30	38	47	54	63	68	75	83	0.9	1.1	1.2	1.4	1.5	1.0	1.2	1.3	1.5	1.6
SS B1	27	34	42	48	56	62	69	77	21	26	32	37	44	49	56	64	1.1	1.2	1.4	1.6	1.7	1.3	1.4	1.6	1.7	1.9
SS B2	34	44	54	61	70	76	82	89	27	34	43	50	59	65	72	80	0.9	1.1	1.2	1.4	1.6	1.1	1.2	1.4	1.5	1.7
SS B3	25	32	40	45	53	59	66	74	19	24	30	35	42	46	52	60	1.1	1.3	1.4	1.6	1.7	1.3	1.5	1.6	1.8	1.9
SS B4	33	43	53	60	69	75	81	88	26	34	42	49	58	63	70	79	1.0	1.1	1.3	1.4	1.6	1.1	1.2	1.4	1.5	1.7
SS B5	24	30	37	42	49	54	61	69	19	23	29	32	38	43	48	56	1.2	1.4	1.5	1.7	1.8	1.4	1.6	1.7	1.9	2.0
SS B6	33	43	53	60	69	75	82	89	26	34	42	49	58	64	71	79	0.9	1.1	1.3	1.4	1.6	1.1	1.2	1.4	1.5	1.7
SS B7	22	27	33	38	44	49	55	63	17	21	26	29	34	38	43	50	1.3	1.5	1.6	1.8	1.9	1.5	1.7	1.9	2.0	2.2
SS B8	32	41	51	58	67	73	80	87	25	32	41	47	56	62	69	77	1.0	1.1	1.3	1.4	1.6	1.1	1.2	1.4	1.6	1.7
SS C1	26	34	43	50	59	65	73	81	17	23	30	36	43	49	56	65	1.0	1.2	1.3	1.5	1.6	1.3	1.4	1.5	1.7	1.8
SS C2	31	41	52	59	69	75	82	89	21	29	38	45	54	61	68	78	0.9	1.1	1.2	1.4	1.5	1.1	1.2	1.4	1.5	1.7
SS C3	21	27	34	40	48	53	60	69	15	20	25	29	35	39	45	53	1.2	1.4	1.5	1.6	1.8	1.5	1.6	1.8	1.9	2.0
SS C4	27	36	46	53	63	69	76	84	19	26	34	40	48	54	62	71	1.0	1.1	1.3	1.5	1.6	1.2	1.3	1.5	1.6	1.7
SS D1	38	47	57	64	73	78	84	91	29	37	46	52	61	66	73	81	0.9	1.1	1.2	1.4	1.5	1.0	1.2	1.4	1.5	1.7
SS D2	45	57	68	76	84	88	93	97	36	46	57	65	74	79	85	92	0.8	1.0	1.1	1.3	1.5	0.9	1.0	1.2	1.4	1.5
SS D3	34	42	51	58	66	72	78	86	27	33	41	47	54	60	66	74	1.0	1.1	1.3	1.5	1.6	1.1	1.3	1.4	1.6	1.8
SS D4	44	55	66	73	82	87	92	96	35	45	55	62	72	77	84	90	0.8	1.0	1.1	1.3	1.5	0.9	1.1	1.2	1.4	1.5
SS E1	32	41	51	58	67	72	79	86	24	31	39	45	53	59	65	74	1.0	1.1	1.3	1.4	1.6	1.1	1.3	1.4	1.6	1.7
SS E2	39	51	62	69	78	84	89	94	30	39	50	57	67	73	79	87	0.9	1.0	1.2	1.3	1.5	1.0	1.1	1.3	1.4	1.6
SS E3	27	34	42	48	56	62	69	77	21	27	33	37	44	49	55	63	1.1	1.3	1.4	1.6	1.7	1.3	1.4	1.6	1.8	1.9
SS E4	36	46	57	64	74	79	85	92	28	36	46	53	62	68	75	83	0.9	1.1	1.2	1.4	1.5	1.0	1.2	1.3	1.5	1.6

ROSWELL, NEW MEXICO

TYPE	SOLAR SAVINGS FRACTION (%) BASE TEMP: 55 F (1899 DD) LCR 200	100	70	50	40	30	25	20	65 F (3701 DD) 200	100	70	50	40	30	25	20	CONSERVATION FACTOR 55 F 200	100	50	30	20	65 F 200	100	50	30	20
WW A1	24	35	43	52	59	67	73	79	17	24	30	36	41	48	53	60	0.9	1.0	1.1	1.3	1.5	1.2	1.3	1.4	1.5	1.7
WW A2	27	43	54	65	72	81	85	90	17	28	37	46	52	62	68	75	0.8	0.9	1.0	1.2	1.3	0.9	1.0	1.2	1.3	1.5
WW A3	28	46	58	69	76	85	89	93	16	30	39	49	57	66	72	79	0.7	0.8	1.0	1.1	1.3	0.9	1.0	1.1	1.3	1.4
WW A4	28	48	60	71	79	86	91	94	16	31	41	51	59	69	75	82	0.7	0.8	0.9	1.1	1.3	0.9	0.9	1.1	1.2	1.4
WW A5	29	50	62	74	81	88	92	96	16	32	42	54	61	71	77	84	0.7	0.8	0.9	1.1	1.3	0.8	0.9	1.1	1.2	1.4
WW A6	29	51	63	75	82	89	93	96	16	32	43	55	63	73	79	85	0.7	0.8	0.9	1.1	1.3	0.8	0.9	1.0	1.2	1.4
WW B1	23	37	47	57	64	73	79	85	14	23	30	38	44	52	57	64	0.8	0.9	1.1	1.2	1.4	1.1	1.2	1.3	1.4	1.6
WW B2	28	48	60	72	79	87	91	95	16	31	41	53	60	70	76	83	0.7	0.8	0.9	1.1	1.3	0.8	0.9	1.1	1.2	1.4
WW B3	33	56	68	79	86	92	95	97	20	37	49	61	69	78	84	89	0.6	0.7	0.9	1.1	1.2	0.8	0.9	1.0	1.2	1.3
WW B4	32	55	69	80	86	93	95	98	18	37	50	62	70	80	85	91	0.6	0.7	0.9	1.1	1.2	0.7	0.8	1.0	1.1	1.3
WW B5	30	53	67	79	85	92	95	97	16	36	48	61	69	79	84	90	0.6	0.7	0.9	1.1	1.2	0.8	0.8	1.0	1.2	1.3
WW C1	34	57	70	81	87	93	96	98	21	39	51	63	71	80	85	91	0.6	0.7	0.9	1.1	1.2	0.7	0.8	1.0	1.2	1.3
WW C2	32	55	68	79	85	92	95	97	18	37	49	61	69	79	84	90	0.6	0.7	0.9	1.1	1.2	0.8	0.9	1.0	1.2	1.3
WW C3	36	62	75	85	91	96	97	99	21	43	57	69	77	86	90	95	0.6	0.7	0.9	1.0	1.2	0.7	0.8	0.9	1.1	1.3
WW C4	32	58	71	82	88	94	96	98	18	39	52	65	74	83	88	93	0.6	0.7	0.9	1.0	1.2	0.7	0.8	1.0	1.1	1.3
TW A1	23	34	42	51	57	66	72	78	17	24	29	35	40	48	53	59	0.9	1.0	1.2	1.3	1.5	1.2	1.3	1.4	1.5	1.7
TW A2	24	39	49	59	67	76	81	87	16	26	33	41	48	56	62	69	0.8	0.9	1.0	1.2	1.4	1.0	1.1	1.2	1.4	1.5
TW A3	25	41	52	63	70	79	84	89	15	27	35	44	51	60	66	73	0.8	0.9	1.0	1.2	1.3	1.0	1.1	1.2	1.4	1.5
TW A4	24	41	52	63	71	80	85	90	14	26	34	44	51	61	67	74	0.8	0.9	1.0	1.2	1.3	1.0	1.0	1.2	1.3	1.5
TW B1	23	34	43	52	58	67	73	79	16	23	29	36	41	49	54	61	0.8	0.9	1.0	1.2	1.3	0.9	1.0	1.2	1.3	1.5
TW B2	23	37	47	58	65	74	80	86	14	24	32	40	46	55	61	68	0.9	1.0	1.1	1.3	1.4	1.2	1.2	1.4	1.5	1.7
TW B3	22	37	47	58	66	75	80	86	14	24	32	40	47	56	62	69	0.8	0.9	1.1	1.2	1.4	1.0	1.1	1.2	1.4	1.5
TW B4	21	35	45	55	63	72	78	84	13	22	30	38	44	53	59	66	0.8	0.9	1.0	1.2	1.4	1.0	1.1	1.2	1.4	1.5
TW C1	21	33	42	51	58	67	73	79	14	22	28	35	41	48	54	61	0.8	0.9	1.1	1.2	1.4	1.0	1.1	1.3	1.4	1.6
TW C2	20	33	42	52	59	69	74	81	13	22	28	36	42	50	56	63	0.9	1.0	1.1	1.3	1.4	1.1	1.2	1.4	1.5	1.6
TW C3	19	31	40	50	56	66	72	79	12	21	27	34	39	48	53	60	0.9	1.0	1.1	1.3	1.4	1.1	1.2	1.3	1.5	1.6
TW C4	18	28	36	44	51	59	65	72	12	19	24	30	35	42	48	54	0.9	1.0	1.1	1.3	1.4	1.1	1.2	1.4	1.5	1.6
TW D1	19	32	40	50	57	66	72	79	12	19	25	32	38	45	51	58	1.0	1.1	1.2	1.3	1.5	1.3	1.3	1.5	1.6	1.7
TW D2	26	44	55	67	74	83	87	92	16	29	38	48	55	65	71	78	0.9	1.0	1.1	1.3	1.4	1.2	1.2	1.4	1.5	1.7
TW D3	28	48	60	71	78	86	91	94	17	32	42	52	60	70	76	83	0.7	0.8	1.0	1.1	1.3	0.9	1.0	1.1	1.3	1.4
TW D4	29	50	63	74	81	89	93	96	17	33	45	56	64	74	80	86	0.7	0.8	0.9	1.1	1.3	0.8	0.9	1.1	1.2	1.4
TW D5	28	49	62	74	81	88	92	96	16	33	44	55	64	74	80	86	0.7	0.8	0.9	1.1	1.3	0.8	0.9	1.0	1.2	1.4
TW E1	31	52	64	76	82	90	93	96	19	35	46	57	65	75	80	86	0.7	0.8	0.9	1.1	1.3	0.8	0.9	1.0	1.2	1.4
TW E2	30	51	64	75	82	89	93	96	18	35	46	57	65	75	80	86	0.7	0.8	0.9	1.1	1.3	0.8	0.9	1.0	1.2	1.4
TW E3	35	59	72	83	89	94	96	98	21	41	54	66	74	83	88	93	0.6	0.7	0.9	1.1	1.2	0.7	0.8	1.0	1.1	1.3
TW E4	32	56	69	80	86	93	95	98	19	38	51	63	71	81	86	91	0.6	0.7	0.9	1.1	1.2	0.7	0.8	1.0	1.1	1.3
TW F1	22	33	42	51	57	66	72	79	15	22	28	35	40	47	52	59	0.9	1.0	1.1	1.3	1.5	1.2	1.3	1.4	1.5	1.7
TW F2	22	37	46	57	64	74	79	85	14	23	31	39	45	54	60	67	0.8	0.9	1.1	1.2	1.4	1.0	1.1	1.3	1.4	1.6
TW F3	21	37	48	59	66	76	81	87	12	23	31	40	46	56	62	69	0.8	0.9	1.0	1.2	1.4	1.0	1.1	1.2	1.4	1.5
TW F4	19	35	46	57	65	74	80	86	10	21	29	38	45	54	61	68	0.8	0.9	1.0	1.2	1.4	1.0	1.1	1.2	1.4	1.5
TW G1	20	32	40	50	57	66	72	78	13	21	26	33	39	46	52	59	0.8	0.9	1.0	1.2	1.4	1.0	1.1	1.2	1.4	1.5
TW G2	19	33	42	53	60	70	76	82	11	20	27	35	41	50	56	63	0.9	1.0	1.1	1.3	1.4	1.2	1.2	1.4	1.5	1.7
TW G3	17	31	41	51	59	69	75	81	9	18	25	33	40	49	55	62	0.9	0.9	1.1	1.2	1.4	1.1	1.2	1.3	1.4	1.6
TW G4	13	26	35	45	53	63	69	76	6	15	21	29	35	43	49	57	0.9	1.0	1.1	1.2	1.4	1.1	1.2	1.3	1.4	1.6
TW H1	16	28	37	46	53	63	69	76	10	17	23	30	35	43	49	56	0.9	1.0	1.1	1.3	1.5	1.2	1.2	1.4	1.5	1.7
TW H2	13	26	34	44	51	61	67	74	7	15	21	28	33	41	47	54	0.9	1.0	1.2	1.3	1.5	1.2	1.3	1.4	1.5	1.7
TW H3	11	22	30	39	46	55	61	69	5	12	18	24	29	37	42	49	0.9	1.0	1.2	1.3	1.5	1.2	1.3	1.4	1.5	1.7
TW H4	8	17	23	31	37	46	52	59	4	9	13	19	23	29	34	40	1.0	1.1	1.2	1.4	1.5	1.3	1.4	1.5	1.6	1.8
TW I1	17	29	37	46	53	63	69	76	10	17	23	29	34	42	47	54	1.1	1.2	1.3	1.5	1.6	1.4	1.5	1.6	1.8	1.9
TW I2	23	40	51	63	71	80	85	90	13	25	34	44	51	61	67	75	0.9	1.0	1.2	1.3	1.5	1.2	1.3	1.4	1.6	1.7
TW I3	24	43	55	67	74	83	88	92	14	28	37	48	55	66	72	79	0.8	0.8	1.0	1.2	1.3	0.9	1.0	1.2	1.3	1.5
TW I4	26	46	59	71	79	87	91	95	14	30	41	52	60	71	77	84	0.7	0.8	1.0	1.1	1.3	0.9	1.0	1.1	1.3	1.4
TW I5	25	47	59	72	79	87	91	95	14	30	41	53	61	72	78	85	0.7	0.8	0.9	1.1	1.3	0.8	0.9	1.1	1.2	1.4
TW J1	29	49	62	73	80	88	92	95	17	33	43	55	63	72	78	85	0.7	0.8	0.9	1.1	1.3	0.8	0.9	1.0	1.2	1.4
TW J2	27	48	61	72	80	88	91	95	15	31	42	54	62	72	78	84	0.7	0.8	0.9	1.1	1.3	0.8	0.9	1.1	1.2	1.4
TW J3	31	55	69	80	86	93	96	98	18	38	50	63	71	81	86	91	0.7	0.8	0.9	1.1	1.3	0.8	0.9	1.1	1.2	1.4
TW J4	29	52	66	78	84	91	94	97	16	35	47	60	68	78	84	90	0.6	0.7	0.9	1.1	1.2	0.7	0.8	1.0	1.1	1.3
DG A1	21	36	46	55	61	69	73	78	12	21	28	35	41	47	51	56	0.6	0.7	0.9	1.1	1.3	0.8	0.8	1.0	1.2	1.3
DG A2	21	38	48	59	66	75	80	85	12	23	31	40	47	55	60	67	0.8	0.9	1.1	1.3	1.6	1.1	1.1	1.4	1.6	1.9
DG A3	24	43	56	68	75	83	87	90	15	27	37	49	57	67	72	79	0.8	0.9	1.0	1.2	1.4	1.0	1.1	1.2	1.4	1.6
DG B1	21	38	50	62	69	77	81	86	12	22	29	38	45	54	59	65	0.7	0.8	1.0	1.2	1.4	0.9	1.0	1.1	1.3	1.5
DG B2	22	39	52	65	73	82	86	90	13	24	32	43	51	61	67	74	0.8	0.9	1.0	1.2	1.5	1.1	1.1	1.2	1.4	1.7
DG B3	24	44	58	72	79	87	91	94	15	28	38	50	60	71	77	84	0.8	0.8	1.0	1.2	1.4	1.0	1.1	1.1	1.3	1.5
DG C1	24	44	58	71	78	85	88	92	14	26	35	45	53	64	70	76	0.7	0.8	0.9	1.1	1.4	0.9	1.0	1.0	1.2	1.4
DG C2	24	45	59	72	80	87	91	94	15	27	37	49	58	69	75	82	0.7	0.8	0.9	1.2	1.4	1.0	1.0	1.1	1.3	1.5
DG C3	27	50	65	78	85	92	94	97	17	32	43	57	67	78	84	89	0.7	0.8	0.9	1.1	1.3	0.9	1.0	1.1	1.2	1.4
SS A1	29	45	55	65	72	80	85	90	21	32	40	49	56	64	70	76	0.7	0.7	0.9	1.1	1.4	0.9	0.9	1.0	1.1	1.3
SS A2	35	54	66	76	83	90	93	96	24	40	50	61	68	77	82	88	0.8	0.9	1.0	1.2	1.4	0.9	1.0	1.2	1.3	1.5
SS A3	27	42	52	62	68	77	82	87	19	30	38	46	52	60	65	72	0.7	0.8	0.9	1.1	1.3	0.8	0.9	1.0	1.2	1.4
SS A4	34	54	65	76	82	89	92	96	24	40	50	60	67	76	81	87	0.8	0.9	1.0	1.2	1.4	1.0	1.1	1.2	1.4	1.5
SS A5	28	42	51	61	67	76	80	86	20	30	37	45	51	59	64	71	0.7	0.8	0.9	1.1	1.3	0.8	0.9	1.0	1.2	1.4
SS A6	35	54	66	76	83	89	93	96	24	40	50	61	68	77	82	87	0.8	0.9	1.1	1.2	1.4	1.0	1.1	1.2	1.4	1.6
SS A7	25	37	45	54	61	69	74	80	19	27	33	40	45	52	57	63	0.7	0.8	0.9	1.1	1.3	0.8	0.9	1.0	1.2	1.4
SS A8	34	53	64	75	81	88	92	95	24	39	49	59	66	75	80	86	0.9	1.0	1.1	1.3	1.5	1.1	1.2	1.3	1.5	1.7
SS B1	25	39	49	59	66	75	80	86	17	27	35	43	49	57	63	70	0.7	0.8	0.9	1.1	1.3	0.8	0.9	1.1	1.2	1.4
SS B2	30	48	60	71	78	85	90	94	20	35	44	55	62	71	77	83	0.8	0.9	1.1	1.2	1.4	1.0	1.1	1.2	1.4	1.6
SS B3	24	37	47	57	63	72	78	84	16	26	33	41	46	55	60	67	0.7	0.8	1.0	1.1	1.3	0.8	0.9	1.1	1.2	1.4
SS B4	29	48	59	70	77	85	89	93	20	34	44	54	61	70	76	82	0.8	0.9	1.1	1.2	1.4	1.0	1.1	1.3	1.4	1.6
SS B5	23	35	43	52	58	67	73	79	16	25	31	37	43	50	55	62	0.7	0.8	1.0	1.1	1.3	0.8	0.9	1.1	1.3	1.4
SS B6	29	48	59	70	77	85	89	93	20	34	44	54	61	70	76	82	0.9	1.0	1.1	1.3	1.5	1.1	1.2	1.4	1.5	1.7
SS B7	21	32	39	47	54	62	68	74	15	22	28	34	39	45	50	56	0.7	0.8	1.0	1.1	1.3	0.8	0.9	1.1	1.2	1.4
SS B8	28	46	57	68	75	84	88	92	19	33	42	52	59	68	74	81	1.0	1.1	1.2	1.3	1.5	1.2	1.3	1.4	1.6	1.8
SS C1	25	42	53	64	71	80	85	90	15	26	35	44	51	60	66	74	0.7	0.8	1.0	1.1	1.3	0.9	1.0	1.1	1.3	1.4
SS C2	28	48	60	72	79	87	91	95	16	31	42	53	60	70	76	83	0.8	0.9	1.0	1.2	1.3	1.0	1.0	1.2	1.3	1.5
SS C3	21	33	42	52	59	68	74	80	13	22	28	35	41	49	54	61	0.7	0.8	0.9	1.1	1.3	0.8	0.9	1.1	1.2	1.4
SS C4	25	43	54	65	73	81	86	91	15	28	37	47	54	64	70	77	0.9	1.0	1.1	1.3	1.4	1.1	1.2	1.3	1.5	1.6
SS D1	34	53	65	75	81	88	92	95	24	39	48	58	65	74	79	85	0.7	0.8	1.0	1.2	1.3	0.9	1.0	1.1	1.3	1.4
SS D2	40	63	74	84	89	94	97	98	27	47	58	69	77	85	89	93	0.7	0.8	0.9	1.1	1.3	0.8	0.9	1.1	1.2	1.4
SS D3	31	48	58	69	75	83	87	92	22	35	43	52	59	67	73	79	0.6	0.7	0.9	1.1	1.2	0.7	0.8	1.0	1.1	1.3
SS D4	38	60	72	82	88	93	96	98	26	45	56	67	74	83	87	92	0.7	0.8	1.0	1.2	1.3	0.9	1.0	1.1	1.3	1.5
SS E1	30	48	59	70	77	85	89	93	20	33	42	52	59	68	73	80	0.6	0.7	0.9	1.1	1.2	0.7	0.8	1.0	1.1	1.3
SS E2	35	57	69	79	85	92	95	97	23	40	52	63	70	79	84	90	0.7	0.8	1.0	1.1	1.3	0.9	1.0	1.1	1.3	1.4
SS E3	26	40	49	59	66	75	80	86	18	28	35	43	49	57	63	69	0.6	0.7	0.9	1.1	1.3	0.7	0.8	1.0	1.2	1.3
SS E4	32	52	64	75	81	89	92	95	21	37	47	58	65	75	80	86	0.8	0.9	1.1	1.2	1.4	1.0	1.1	1.2	1.4	1.6

TRUTH OR CONSEQ., NEW MEXICO

TYPE	SOLAR SAVINGS FRACTION (%) BASE TEMP: 55 F (1639 DD)								65 F (3396 DD)								CONSERVATION FACTOR 55 F					65 F				
LCR	200	100	70	50	40	30	25	20	200	100	70	50	40	30	25	20	200	100	50	30	20	200	100	50	30	20
WW A1	27	40	49	59	66	74	80	85	19	27	33	40	46	54	59	66	0.8	0.9	1.1	1.2	1.4	1.1	1.2	1.3	1.4	1.6
WW A2	31	50	61	72	79	86	90	94	19	32	41	51	58	68	74	80	0.7	0.8	1.0	1.1	1.3	0.9	1.0	1.1	1.3	1.4
WW A3	33	53	65	76	83	90	93	96	19	34	44	55	63	72	78	85	0.7	0.8	0.9	1.1	1.3	0.8	0.9	1.1	1.2	1.4
WW A4	34	55	68	78	85	91	94	97	19	35	46	57	65	75	81	87	0.6	0.7	0.9	1.1	1.3	0.8	0.9	1.0	1.2	1.4
WW A5	35	57	70	81	87	93	95	98	19	36	48	60	68	77	83	89	0.6	0.7	0.9	1.1	1.2	0.8	0.9	1.0	1.2	1.3
WW A6	35	59	71	82	88	93	96	98	19	37	49	61	69	79	84	90	0.6	0.7	0.9	1.1	1.2	0.8	0.9	1.0	1.2	1.3
WW B1	27	44	54	65	72	80	85	90	16	27	35	43	50	59	64	71	0.8	0.8	1.0	1.2	1.3	1.0	1.1	1.2	1.4	1.5
WW B2	33	55	68	79	85	92	95	97	19	36	47	58	66	76	82	88	0.6	0.7	0.9	1.1	1.3	0.8	0.9	1.0	1.2	1.3
WW B3	39	63	75	85	90	95	97	99	23	42	55	67	74	83	88	93	0.6	0.7	0.9	1.0	1.2	0.7	0.8	1.0	1.1	1.3
WW B4	38	63	76	86	91	96	98	99	21	42	55	68	76	85	89	94	0.6	0.7	0.9	1.0	1.2	0.7	0.8	1.0	1.1	1.3
WW B5	36	61	74	84	90	95	97	99	20	40	53	66	74	83	88	93	0.6	0.7	0.9	1.0	1.2	0.7	0.8	1.0	1.1	1.3
WW C1	41	65	77	86	91	96	98	99	24	44	57	69	76	85	89	94	0.6	0.7	0.9	1.0	1.2	0.7	0.8	1.0	1.1	1.3
WW C2	38	62	75	85	90	95	97	99	22	42	54	67	74	83	88	93	0.6	0.7	0.9	1.0	1.2	0.7	0.8	1.0	1.1	1.3
WW C3	43	69	81	90	94	98	99	100	25	48	62	75	82	90	93	97	0.6	0.7	0.8	1.0	1.2	0.7	0.8	0.9	1.1	1.3
WW C4	39	65	78	88	92	97	98	99	22	44	58	71	79	87	91	95	0.6	0.7	0.8	1.0	1.2	0.7	0.8	0.9	1.1	1.3
TW A1	26	39	48	58	65	73	78	84	18	26	32	40	45	53	58	65	0.8	0.9	1.1	1.2	1.4	1.1	1.2	1.3	1.5	1.6
TW A2	28	45	56	67	74	82	87	91	18	29	37	46	53	62	68	75	0.7	0.8	1.0	1.2	1.3	1.0	1.0	1.2	1.3	1.5
TW A3	29	47	59	70	77	85	89	93	17	30	39	49	56	66	72	79	0.7	0.8	1.0	1.1	1.3	0.9	1.0	1.1	1.3	1.4
TW A4	28	48	59	71	78	86	90	94	16	30	39	49	57	67	73	80	0.7	0.8	1.0	1.1	1.3	0.9	1.0	1.1	1.3	1.4
TW B1	26	39	49	59	66	74	80	85	17	26	33	40	46	54	60	67	0.8	0.9	1.1	1.2	1.4	1.1	1.2	1.3	1.4	1.6
TW B2	27	43	54	65	72	81	86	91	16	28	36	45	52	61	67	74	0.8	0.8	1.0	1.2	1.3	1.0	1.0	1.2	1.3	1.5
TW B3	26	43	55	66	73	82	86	91	15	27	36	45	52	62	68	75	0.7	0.8	1.0	1.2	1.3	0.9	1.0	1.2	1.3	1.5
TW B4	24	41	52	63	70	79	84	89	14	25	34	43	49	59	65	72	0.8	0.9	1.0	1.2	1.3	1.0	1.1	1.2	1.3	1.5
TW C1	24	38	48	58	65	74	79	85	16	25	32	39	45	54	59	67	0.8	0.9	1.1	1.2	1.4	1.1	1.1	1.3	1.4	1.6
TW C2	23	38	49	59	66	76	81	87	14	24	32	40	46	55	61	69	0.8	0.9	1.0	1.2	1.4	1.0	1.1	1.2	1.4	1.5
TW C3	22	36	46	56	64	73	78	85	14	23	30	38	44	53	59	66	0.8	0.9	1.1	1.2	1.4	1.1	1.1	1.3	1.4	1.6
TW C4	20	32	41	50	57	67	72	79	13	21	27	34	39	47	53	60	0.9	1.0	1.1	1.3	1.4	1.2	1.2	1.4	1.5	1.6
TW D1	23	37	47	58	65	74	79	85	13	22	29	37	43	52	57	65	0.8	0.9	1.1	1.2	1.4	1.1	1.2	1.3	1.4	1.6
TW D2	31	51	62	74	81	88	92	95	18	32	42	53	61	70	76	83	0.7	0.8	0.9	1.1	1.3	0.8	0.9	1.1	1.2	1.4
TW D3	33	55	67	78	85	91	94	97	20	36	47	58	66	76	81	87	0.6	0.7	0.9	1.1	1.3	0.8	0.9	1.0	1.2	1.4
TW D4	35	57	70	81	87	93	96	98	20	38	50	62	70	79	85	90	0.6	0.7	0.9	1.1	1.2	0.8	0.9	1.0	1.2	1.3
TW D5	33	56	69	80	86	92	95	98	19	37	49	61	69	79	84	90	0.6	0.7	0.9	1.1	1.2	0.8	0.9	1.0	1.2	1.3
TW E1	37	60	72	82	88	93	96	98	22	40	51	63	71	80	85	90	0.6	0.7	0.9	1.1	1.2	0.8	0.8	1.0	1.2	1.3
TW E2	36	59	71	82	88	93	96	98	21	39	51	63	70	80	85	90	0.6	0.7	0.9	1.1	1.2	0.8	0.8	1.0	1.2	1.3
TW E3	42	66	79	88	93	97	98	99	25	46	59	72	79	87	91	95	0.6	0.7	0.9	1.0	1.2	0.7	0.8	0.9	1.1	1.3
TW E4	38	63	76	86	91	96	98	99	22	43	56	68	76	85	90	94	0.6	0.7	0.9	1.0	1.2	0.7	0.8	0.9	1.1	1.3
TW F1	25	38	48	58	65	74	79	85	16	25	31	39	45	53	58	65	0.8	0.9	1.1	1.2	1.4	1.1	1.2	1.3	1.4	1.6
TW F2	26	43	53	64	72	80	85	90	15	27	35	44	51	60	66	73	0.8	0.8	1.0	1.2	1.3	1.0	1.1	1.2	1.3	1.5
TW F3	25	43	55	66	74	82	87	92	14	26	35	45	52	62	68	75	0.7	0.8	1.0	1.1	1.3	0.9	1.0	1.2	1.3	1.5
TW F4	23	41	53	65	72	81	86	91	12	24	33	43	50	60	67	74	0.7	0.8	1.0	1.2	1.3	0.9	1.0	1.2	1.3	1.5
TW G1	23	37	47	57	64	73	78	85	14	23	30	38	44	52	58	65	0.8	0.9	1.1	1.2	1.4	1.1	1.2	1.3	1.4	1.6
TW G2	22	39	49	60	68	77	82	88	12	23	31	40	47	56	62	70	0.8	0.9	1.0	1.2	1.4	1.0	1.1	1.2	1.4	1.5
TW G3	20	37	48	59	66	76	81	87	10	21	29	38	45	55	61	69	0.8	0.9	1.0	1.2	1.4	1.0	1.1	1.2	1.4	1.5
TW G4	16	31	42	53	60	70	76	83	8	18	25	33	39	49	55	63	0.8	0.9	1.1	1.2	1.4	1.1	1.2	1.3	1.4	1.6
TW H1	19	33	43	53	61	70	76	82	11	20	26	34	40	49	55	62	0.9	0.9	1.1	1.2	1.4	1.1	1.2	1.3	1.5	1.6
TW H2	16	30	40	51	58	68	74	81	8	17	24	32	38	47	53	60	0.9	1.0	1.1	1.2	1.4	1.1	1.2	1.3	1.5	1.6
TW H3	14	26	35	46	53	63	69	76	6	14	21	28	34	42	48	55	0.9	1.0	1.1	1.3	1.5	1.2	1.3	1.4	1.5	1.7
TW H4	10	20	28	37	43	53	59	67	4	11	16	22	26	34	39	46	1.0	1.1	1.2	1.4	1.5	1.4	1.4	1.5	1.7	1.8
TW I1	20	34	44	54	61	71	76	83	11	20	26	34	40	48	54	61	0.8	0.9	1.1	1.2	1.4	1.1	1.2	1.3	1.5	1.6
TW I2	27	47	59	70	78	86	90	94	15	29	39	49	57	67	73	80	0.7	0.8	1.0	1.1	1.3	0.9	1.0	1.1	1.2	1.4
TW I3	29	50	63	74	81	88	92	96	16	32	42	53	61	71	77	84	0.7	0.8	0.9	1.1	1.3	0.8	0.9	1.1	1.2	1.4
TW I4	31	54	67	78	85	91	94	97	17	34	46	58	66	76	82	88	0.6	0.7	0.9	1.1	1.3	0.8	0.9	1.0	1.2	1.3
TW I5	31	54	67	78	85	92	95	97	16	34	46	58	67	77	83	89	0.6	0.7	0.9	1.1	1.3	0.8	0.9	1.0	1.2	1.3
TW J1	34	57	69	80	86	92	95	98	20	37	49	60	68	78	83	89	0.6	0.7	0.9	1.1	1.2	0.8	0.9	1.0	1.2	1.3
TW J2	33	55	68	79	85	92	95	97	18	36	47	59	67	77	83	89	0.6	0.7	0.9	1.1	1.2	0.8	0.9	1.0	1.2	1.3
TW J3	38	63	76	86	91	96	98	99	21	42	56	68	76	85	90	94	0.6	0.7	0.9	1.0	1.2	0.7	0.8	0.9	1.1	1.3
TW J4	35	60	73	84	89	95	97	99	19	40	53	65	73	83	88	93	0.6	0.7	0.9	1.0	1.2	0.7	0.8	1.0	1.1	1.3
DG A1	25	43	53	63	69	77	81	85	14	25	33	41	46	54	58	63	0.7	0.9	1.0	1.3	1.5	1.0	1.1	1.3	1.5	1.8
DG A2	25	44	56	67	74	81	86	90	14	27	36	45	52	61	66	73	0.7	0.8	1.0	1.2	1.4	1.0	1.0	1.2	1.4	1.6
DG A3	28	50	63	75	81	87	91	93	17	31	42	54	62	72	77	83	0.7	0.8	0.9	1.2	1.4	0.9	0.9	1.0	1.2	1.4
DG B1	26	46	59	71	77	84	87	91	14	25	35	45	52	61	67	72	0.7	0.8	1.0	1.2	1.4	1.0	1.0	1.1	1.4	1.6
DG B2	26	47	61	73	80	88	91	94	15	27	37	49	57	68	74	80	0.7	0.8	0.9	1.1	1.3	0.9	1.0	1.1	1.2	1.5
DG B3	29	52	67	79	85	91	94	96	17	32	43	57	66	77	82	88	0.7	0.7	0.9	1.1	1.4	0.9	0.9	1.0	1.2	1.4
DG C1	30	53	67	79	85	90	93	95	16	30	41	53	61	71	77	82	0.7	0.7	0.9	1.1	1.4	0.9	0.9	1.0	1.2	1.5
DG C2	30	53	68	80	86	92	95	97	17	31	43	56	65	75	81	87	0.7	0.7	0.9	1.1	1.3	0.9	0.9	1.0	1.2	1.4
DG C3	33	58	73	85	90	95	96	98	19	36	49	64	73	83	88	93	0.6	0.7	0.9	1.1	1.4	0.8	0.9	0.9	1.1	1.3
SS A1	33	51	62	72	78	86	90	93	23	36	45	54	61	70	75	81	0.7	0.8	1.0	1.1	1.3	0.9	1.0	1.1	1.3	1.4
SS A2	40	61	72	82	88	93	96	98	27	44	55	66	73	82	86	91	0.6	0.7	0.9	1.1	1.2	0.7	0.8	1.0	1.2	1.3
SS A3	31	48	58	69	75	83	87	91	21	34	42	51	57	66	71	77	0.7	0.8	1.0	1.2	1.3	0.9	1.0	1.2	1.3	1.5
SS A4	39	60	72	81	87	93	95	98	27	44	54	65	72	81	85	90	0.6	0.7	0.9	1.1	1.3	0.7	0.8	1.0	1.2	1.3
SS A5	31	47	57	67	74	82	86	91	22	34	41	50	56	65	70	76	0.7	0.8	1.0	1.2	1.3	0.9	1.0	1.2	1.3	1.5
SS A6	40	61	72	82	88	93	96	98	27	44	55	66	73	81	86	91	0.6	0.7	0.9	1.1	1.2	0.7	0.8	1.0	1.2	1.3
SS A7	28	42	51	61	67	76	80	86	21	30	37	44	50	58	63	69	0.8	0.9	1.1	1.2	1.4	1.0	1.1	1.3	1.4	1.6
SS A8	39	60	71	81	86	92	95	97	27	43	54	64	71	80	84	90	0.6	0.7	0.9	1.1	1.3	0.8	0.9	1.0	1.2	1.3
SS B1	28	45	55	66	73	81	85	90	19	31	39	48	54	63	69	75	0.8	0.8	1.0	1.2	1.3	0.9	1.0	1.2	1.3	1.5
SS B2	35	55	66	77	83	90	93	96	23	39	49	60	67	76	81	87	0.7	0.8	0.9	1.1	1.3	0.8	0.9	1.0	1.2	1.4
SS B3	27	43	53	64	70	79	84	89	18	29	37	45	52	60	66	73	0.8	0.9	1.0	1.2	1.4	1.0	1.1	1.2	1.4	1.5
SS B4	34	54	66	76	82	89	93	96	22	38	48	59	66	75	80	86	0.7	0.8	0.9	1.1	1.3	0.8	0.9	1.0	1.2	1.4
SS B5	26	40	49	59	65	74	79	85	18	27	34	42	47	55	61	67	0.8	0.9	1.1	1.2	1.4	1.0	1.1	1.3	1.4	1.6
SS B6	34	54	66	77	83	90	93	96	22	38	48	59	66	75	81	87	0.7	0.8	0.9	1.1	1.3	0.8	0.9	1.0	1.2	1.4
SS B7	24	36	45	54	61	69	74	81	17	25	31	38	43	51	56	62	0.9	1.0	1.1	1.3	1.4	1.1	1.2	1.3	1.5	1.7
SS B8	33	53	64	75	81	88	92	95	22	37	47	57	64	74	79	85	0.7	0.8	0.9	1.1	1.3	0.8	0.9	1.1	1.2	1.4
SS C1	29	49	60	71	78	86	90	94	17	30	40	50	57	67	73	80	0.7	0.8	1.0	1.1	1.3	0.9	1.0	1.1	1.3	1.4
SS C2	33	55	67	78	85	91	94	97	19	36	47	58	66	76	81	87	0.6	0.7	0.9	1.1	1.3	0.8	0.9	1.0	1.2	1.3
SS C3	24	39	49	59	66	75	81	86	15	25	32	40	46	54	60	67	0.8	0.9	1.0	1.2	1.4	1.0	1.1	1.3	1.4	1.6
SS C4	30	49	61	72	79	87	91	95	17	32	41	52	59	69	75	82	0.7	0.8	0.9	1.1	1.3	0.9	0.9	1.1	1.2	1.4
SS D1	40	60	72	81	87	92	95	97	27	43	53	64	71	79	84	89	0.6	0.7	0.9	1.1	1.3	0.8	0.9	1.0	1.2	1.4
SS D2	46	69	80	89	93	97	98	99	31	52	64	75	81	89	92	96	0.6	0.7	0.9	1.0	1.2	0.7	0.8	0.9	1.1	1.3
SS D3	36	54	65	75	81	88	92	95	25	39	48	58	64	73	78	84	0.7	0.8	0.9	1.1	1.3	0.8	0.9	1.1	1.2	1.4
SS D4	44	67	79	87	92	96	98	99	30	50	61	73	79	87	91	95	0.6	0.7	0.9	1.0	1.2	0.7	0.8	0.9	1.1	1.3
SS E1	35	55	66	77	83	90	93	96	22	37	47	58	65	74	79	85	0.7	0.8	0.9	1.1	1.3	0.8	0.9	1.1	1.2	1.4
SS E2	41	64	75	85	90	95	97	99	26	45	57	68	76	84	88	93	0.6	0.7	0.9	1.1	1.2	0.7	0.8	1.0	1.1	1.3
SS E3	29	46	56	66	73	81	86	90	20	31	39	48	54	63	69	75	0.7	0.8	1.0	1.2	1.3	0.9	1.0	1.2	1.3	1.5
SS E4	37	59	71	81	87	92	95	98	24	41	52	63	71	80	85	90	0.6	0.7	0.9	1.1	1.3	0.7	0.9	1.0	1.2	1.3

TUCUMCARI, NEW MEXICO

TYPE	SOLAR SAVINGS FRACTION (%)																CONSERVATION FACTOR									
	BASE TEMP: 55 F (2142 DD)								65 F (4052 DD)								55 F					65 F				
LCR	100	70	50	40	30	25	20	15	100	70	50	40	30	25	20	15	100	50	30	20	15	100	50	30	20	15
WW A1	34	42	50	57	66	71	78	85	24	29	35	40	47	52	59	67	1.0	1.2	1.3	1.5	1.6	1.3	1.4	1.6	1.7	1.8
WW A2	41	52	63	70	79	84	90	95	28	36	45	51	60	66	73	82	0.9	1.0	1.2	1.3	1.5	1.0	1.2	1.3	1.5	1.6
WW A3	44	56	67	75	83	88	93	97	29	38	48	55	65	71	78	86	0.8	1.0	1.1	1.3	1.5	1.0	1.1	1.3	1.4	1.6
WW A4	46	58	70	77	85	90	94	97	30	39	50	58	68	74	81	88	0.8	1.0	1.1	1.3	1.4	1.0	1.1	1.2	1.4	1.5
WW A5	48	60	72	79	87	91	95	98	31	41	52	60	70	76	83	90	0.8	0.9	1.1	1.3	1.4	0.9	1.1	1.2	1.4	1.5
WW A6	49	61	73	81	88	92	96	98	31	42	53	61	71	78	84	91	0.8	0.9	1.1	1.3	1.4	0.9	1.1	1.2	1.4	1.5
WW B1	36	45	55	62	71	77	83	90	23	29	37	42	50	56	63	72	0.9	1.1	1.2	1.4	1.5	1.2	1.3	1.5	1.6	1.7
WW B2	46	58	70	78	86	90	94	98	30	40	51	59	69	75	82	90	0.8	0.9	1.1	1.3	1.4	0.9	1.1	1.2	1.4	1.5
WW B3	54	66	78	84	91	94	97	99	36	48	60	68	77	83	89	94	0.7	0.9	1.1	1.3	1.4	0.9	1.0	1.2	1.3	1.5
WW B4	54	67	79	85	92	95	98	99	36	48	61	69	79	84	90	95	0.7	0.9	1.1	1.2	1.4	0.8	1.0	1.2	1.3	1.5
WW B5	52	65	77	84	91	94	97	99	35	47	59	68	78	84	89	95	0.7	0.9	1.1	1.2	1.4	0.9	1.0	1.2	1.3	1.5
WW C1	55	68	79	86	92	95	98	99	38	50	62	70	79	85	90	95	0.7	0.9	1.1	1.2	1.4	0.9	1.0	1.2	1.3	1.5
WW C2	53	56	78	84	91	94	97	99	36	47	60	68	78	83	89	95	0.7	0.9	1.1	1.2	1.4	0.9	1.0	1.2	1.3	1.5
WW C3	60	73	84	90	95	97	99	100	42	55	68	76	85	90	94	98	0.7	0.9	1.0	1.2	1.4	0.8	0.9	1.1	1.3	1.4
WW C4	56	69	81	87	94	96	98	99	38	51	64	73	82	87	92	97	0.7	0.9	1.1	1.2	1.4	0.8	1.0	1.1	1.3	1.4
TW A1	33	41	49	56	64	70	77	84	23	28	35	39	46	51	58	66	1.0	1.2	1.3	1.5	1.6	1.3	1.4	1.6	1.7	1.8
TW A2	38	47	58	65	74	79	85	92	25	32	40	46	55	61	68	77	0.9	1.1	1.2	1.4	1.5	1.1	1.3	1.4	1.5	1.7
TW A3	39	50	61	68	77	83	88	94	26	34	43	49	59	65	72	81	0.9	1.0	1.2	1.3	1.5	1.1	1.2	1.3	1.5	1.6
TW A4	39	50	61	69	78	84	89	94	25	33	43	50	59	66	73	82	0.9	1.0	1.2	1.3	1.5	1.0	1.2	1.3	1.5	1.6
TW B1	33	41	50	57	66	71	78	86	23	28	35	40	47	53	59	68	1.0	1.1	1.3	1.5	1.6	1.3	1.4	1.5	1.7	1.8
TW B2	36	45	56	63	73	78	84	91	24	31	39	45	54	60	67	76	0.9	1.1	1.2	1.4	1.5	1.1	1.3	1.4	1.6	1.7
TW B3	36	46	56	64	73	79	85	92	23	31	39	46	55	61	68	77	0.9	1.1	1.2	1.4	1.5	1.1	1.2	1.4	1.5	1.7
TW B4	33	43	53	61	70	76	83	90	22	29	37	43	52	58	65	75	0.9	1.1	1.2	1.4	1.5	1.1	1.3	1.4	1.6	1.7
TW C1	32	40	49	56	65	71	78	86	22	27	34	40	47	53	60	69	1.0	1.1	1.3	1.4	1.6	1.2	1.4	1.5	1.7	1.8
TW C2	32	41	50	57	67	73	80	87	21	27	35	40	49	54	62	71	1.0	1.1	1.3	1.4	1.6	1.2	1.3	1.5	1.6	1.7
TW C3	30	38	48	55	64	70	77	85	20	26	33	38	46	52	59	68	1.0	1.1	1.3	1.5	1.6	1.2	1.4	1.5	1.7	1.8
TW C4	27	34	43	49	58	64	71	80	19	24	30	34	41	46	53	62	1.1	1.2	1.4	1.5	1.7	1.4	1.5	1.6	1.7	1.9
TW D1	30	39	48	55	64	70	77	85	19	25	31	36	44	49	56	65	1.0	1.1	1.3	1.5	1.6	1.3	1.4	1.5	1.7	1.8
TW D2	42	53	65	72	81	86	91	96	28	37	47	54	64	70	77	85	0.8	1.0	1.1	1.3	1.5	1.0	1.1	1.3	1.4	1.6
TW D3	46	58	70	77	85	90	94	97	31	41	51	59	69	75	82	89	0.8	1.0	1.1	1.3	1.4	0.9	1.1	1.2	1.4	1.5
TW D4	48	61	73	80	88	92	95	98	33	43	55	63	73	79	85	92	0.8	0.9	1.1	1.3	1.4	0.9	1.0	1.2	1.4	1.5
TW D5	47	60	72	79	88	92	95	98	32	43	54	62	73	79	85	92	0.8	0.9	1.1	1.3	1.4	0.9	1.0	1.2	1.4	1.5
TW E1	50	63	74	81	89	92	96	98	34	45	56	64	74	79	86	92	0.8	0.9	1.1	1.3	1.4	0.9	1.0	1.2	1.4	1.5
TW E2	50	62	74	81	89	92	96	98	34	44	55	64	74	79	86	92	0.8	0.9	1.1	1.3	1.4	0.9	1.0	1.2	1.4	1.5
TW E3	57	70	81	88	94	96	98	99	40	53	65	73	82	87	92	97	0.7	0.9	1.1	1.2	1.4	0.8	1.0	1.1	1.3	1.4
TW E4	54	67	79	85	92	95	98	99	37	49	62	70	80	85	91	96	0.7	0.9	1.1	1.2	1.4	0.8	1.0	1.1	1.3	1.5
TW F1	32	40	49	56	64	70	77	85	22	27	34	39	46	51	58	67	1.0	1.2	1.3	1.5	1.6	1.3	1.4	1.5	1.7	1.8
TW F2	35	45	55	62	72	78	84	91	23	30	38	44	53	58	66	75	0.9	1.1	1.2	1.4	1.5	1.1	1.3	1.4	1.6	1.7
TW F3	35	46	57	64	74	80	86	92	22	30	39	45	54	61	68	78	0.9	1.0	1.2	1.4	1.5	1.1	1.2	1.4	1.5	1.7
TW F4	33	44	55	63	73	78	85	91	20	28	37	44	53	59	67	77	0.9	1.1	1.2	1.4	1.5	1.1	1.2	1.4	1.5	1.7
TW G1	30	39	48	55	64	70	77	85	20	26	32	38	45	51	57	67	1.0	1.2	1.3	1.5	1.6	1.3	1.4	1.5	1.7	1.8
TW G2	31	41	51	58	68	74	81	89	19	26	34	40	49	55	62	72	1.0	1.1	1.2	1.4	1.6	1.2	1.3	1.4	1.6	1.7
TW G3	29	39	49	57	67	73	80	88	18	24	32	39	47	53	61	71	1.0	1.1	1.3	1.4	1.6	1.2	1.3	1.5	1.6	1.7
TW G4	25	34	43	51	61	67	75	83	14	21	28	34	42	48	55	65	1.0	1.2	1.3	1.5	1.6	1.3	1.4	1.5	1.7	1.8
TW H1	27	35	44	51	61	67	74	83	17	22	29	34	42	48	55	64	1.0	1.2	1.3	1.5	1.6	1.3	1.4	1.6	1.7	1.8
TW H2	24	32	42	49	59	65	73	82	14	20	27	32	40	46	53	63	1.1	1.2	1.3	1.5	1.6	1.3	1.4	1.6	1.7	1.8
TW H3	21	28	37	44	53	60	67	77	12	17	23	28	36	41	48	58	1.1	1.2	1.4	1.5	1.7	1.4	1.5	1.6	1.8	1.9
TW H4	16	22	30	35	44	50	57	67	9	13	18	22	29	33	39	48	1.2	1.4	1.5	1.6	1.8	1.6	1.7	1.8	1.9	2.0
TW I1	27	35	45	51	61	67	74	82	16	22	28	33	40	46	52	62	1.0	1.2	1.3	1.5	1.6	1.3	1.5	1.6	1.7	1.9
TW I2	38	50	61	69	78	84	89	95	24	33	43	50	60	66	74	83	0.9	1.0	1.2	1.3	1.5	1.0	1.2	1.3	1.5	1.6
TW I3	41	50	65	73	82	87	92	96	27	36	47	54	64	71	78	86	0.8	1.0	1.1	1.3	1.5	1.0	1.1	1.3	1.4	1.6
TW I4	45	57	69	77	86	90	94	98	29	40	51	59	70	76	83	90	0.8	0.9	1.1	1.3	1.4	0.9	1.1	1.2	1.4	1.5
TW I5	45	58	70	78	86	90	95	98	29	40	52	60	71	77	84	91	0.8	0.9	1.1	1.3	1.4	0.9	1.1	1.2	1.4	1.5
TW J1	48	60	72	79	87	91	95	98	32	42	53	61	71	77	84	91	0.8	0.9	1.1	1.3	1.4	0.9	1.1	1.2	1.4	1.5
TW J2	46	59	71	78	87	91	95	98	30	41	53	61	71	77	84	91	0.8	0.9	1.1	1.3	1.4	0.9	1.1	1.2	1.4	1.5
TW J3	53	67	79	85	92	95	98	99	37	49	62	70	80	85	91	96	0.7	0.9	1.1	1.2	1.4	0.8	1.0	1.1	1.3	1.6
TW J4	50	64	76	83	91	94	97	99	34	46	59	67	77	83	89	95	0.7	0.9	1.1	1.3	1.4	0.9	1.0	1.2	1.3	1.5
DG A1	34	44	53	59	67	71	76	82	21	27	34	39	46	50	55	60	0.9	1.1	1.4	1.6	1.8	1.2	1.4	1.7	2.0	2.3
DG A2	30	40	57	64	73	78	83	89	22	30	39	45	54	59	65	73	0.9	1.1	1.2	1.4	1.6	1.1	1.2	1.4	1.6	1.8
DG A3	41	54	66	73	82	86	90	93	27	36	47	55	66	71	78	85	0.8	1.0	1.2	1.4	1.7	1.0	1.1	1.3	1.5	1.7
DG B1	36	48	60	67	75	80	84	89	21	28	37	44	52	58	64	70	0.9	1.0	1.2	1.5	1.7	1.1	1.2	1.4	1.7	2.0
DG B2	37	50	63	71	80	85	90	94	23	31	41	49	60	66	73	81	0.8	1.0	1.2	1.4	1.5	1.1	1.1	1.3	1.5	1.7
DG B3	42	56	70	78	86	90	94	96	27	37	49	58	70	76	83	90	0.8	0.9	1.1	1.3	1.6	1.0	1.0	1.2	1.4	1.6
DG C1	42	55	69	76	84	88	91	94	25	33	44	52	62	68	74	81	0.8	0.9	1.2	1.4	1.7	1.1	1.1	1.3	1.5	1.8
DG C2	42	56	70	78	86	90	94	97	26	36	48	56	68	74	81	88	0.8	0.9	1.1	1.3	1.5	1.0	1.1	1.2	1.4	1.6
DG C3	47	63	77	84	91	94	96	98	31	42	55	65	77	83	89	94	0.7	0.9	1.1	1.3	1.7	0.9	1.0	1.1	1.3	1.5
SS A1	43	53	63	70	79	83	89	94	31	39	48	54	63	68	75	83	0.9	1.0	1.2	1.4	1.5	1.0	1.2	1.3	1.5	1.6
SS A2	52	64	75	81	88	92	95	98	39	49	59	67	76	81	87	93	0.8	0.9	1.1	1.3	1.4	0.9	1.0	1.2	1.4	1.5
SS A3	40	50	59	66	75	80	86	92	29	36	44	50	58	64	70	78	0.9	1.1	1.2	1.4	1.5	1.1	1.2	1.4	1.6	1.7
SS A4	52	63	74	80	88	91	95	98	38	48	58	65	74	80	86	92	0.8	0.9	1.1	1.3	1.4	0.9	1.1	1.2	1.4	1.5
SS A5	40	49	58	65	74	79	85	91	29	36	44	50	58	63	69	77	0.9	1.1	1.2	1.4	1.5	1.1	1.3	1.4	1.6	1.7
SS A6	52	64	75	81	88	92	95	98	39	49	59	66	75	81	86	92	0.8	0.9	1.1	1.3	1.4	0.9	1.0	1.2	1.4	1.5
SS A7	36	43	52	58	67	72	78	86	26	32	39	44	51	56	62	70	1.0	1.1	1.3	1.5	1.6	1.2	1.4	1.5	1.7	1.8
SS A8	51	62	73	79	87	91	94	98	38	47	57	64	73	79	85	91	0.8	0.9	1.1	1.3	1.4	0.9	1.1	1.2	1.4	1.5
SS B1	37	47	57	61	73	78	84	91	26	33	42	48	56	61	68	77	0.9	1.1	1.2	1.4	1.5	1.1	1.3	1.4	1.6	1.7
SS B2	47	58	69	76	84	89	93	97	34	43	53	61	70	76	82	89	0.8	1.0	1.1	1.3	1.5	0.9	1.1	1.2	1.4	1.6
SS B3	36	45	54	61	70	76	82	89	25	32	39	45	53	58	65	74	1.0	1.1	1.3	1.4	1.6	1.1	1.3	1.4	1.6	1.7
SS B4	46	57	68	75	83	88	92	96	33	42	52	59	69	74	81	88	0.8	1.0	1.1	1.3	1.5	1.0	1.1	1.3	1.4	1.6
SS B5	33	41	50	56	65	71	77	85	24	30	36	41	49	54	60	69	1.0	1.2	1.3	1.5	1.6	1.2	1.4	1.5	1.7	1.8
SS B6	46	57	68	75	84	88	93	97	33	42	53	60	69	75	81	89	0.8	1.0	1.1	1.3	1.5	0.9	1.1	1.3	1.4	1.6
SS B7	30	37	46	52	60	66	72	80	22	27	33	37	44	49	55	63	1.1	1.2	1.4	1.5	1.7	1.3	1.5	1.6	1.8	1.9
SS B8	44	55	66	74	82	87	91	96	32	41	51	58	67	73	79	87	0.8	1.0	1.2	1.3	1.5	1.0	1.1	1.3	1.4	1.6
SS C1	40	51	62	69	79	84	89	94	26	34	43	50	59	65	73	81	0.9	1.0	1.2	1.3	1.5	1.1	1.2	1.3	1.5	1.6
SS C2	46	58	70	77	86	90	94	97	30	40	51	59	69	75	82	89	0.8	0.9	1.1	1.3	1.4	0.9	1.1	1.2	1.4	1.5
SS C3	32	41	50	57	66	72	79	87	21	27	34	40	48	53	60	69	1.0	1.1	1.3	1.4	1.6	1.2	1.4	1.5	1.6	1.8
SS C4	41	52	63	71	80	85	90	95	27	36	46	53	62	69	76	84	0.9	1.0	1.2	1.3	1.5	1.0	1.2	1.3	1.5	1.6
SS D1	51	63	73	80	87	91	95	98	37	47	57	64	73	78	84	90	0.8	0.9	1.1	1.3	1.4	0.9	1.1	1.2	1.4	1.5
SS D2	60	72	83	88	94	96	98	99	45	57	68	75	84	88	93	97	0.7	0.9	1.1	1.2	1.4	0.8	1.0	1.1	1.3	1.5
SS D3	46	56	67	74	82	86	91	95	34	42	51	57	66	72	78	85	0.8	1.0	1.2	1.3	1.5	1.0	1.1	1.3	1.5	1.6
SS D4	58	70	81	87	93	95	98	99	43	55	66	73	82	87	91	96	0.7	0.9	1.1	1.2	1.4	0.8	1.0	1.2	1.3	1.5
SS E1	46	57	68	75	83	88	92	96	32	41	50	57	66	72	79	86	0.8	1.0	1.1	1.3	1.5	1.0	1.1	1.3	1.4	1.6
SS E2	54	67	78	84	91	94	97	99	39	50	61	69	78	83	89	94	0.7	0.9	1.1	1.3	1.4	0.9	1.0	1.2	1.3	1.5
SS E3	38	47	57	64	73	78	84	91	27	34	42	48	56	61	68	77	0.9	1.1	1.2	1.4	1.5	1.1	1.3	1.4	1.6	1.7
SS E4	50	62	73	80	87	91	95	98	36	46	57	64	73	79	85	91	0.8	0.9	1.1	1.3	1.4	0.9	1.1	1.2	1.4	1.5

ZUNI, NEW MEXICO

SOLAR SAVINGS FRACTION (%) — CONSERVATION FACTOR

TYPE	55 F (3389 DD)								65 F (5826 DD)								CF 55 F					CF 65 F				
LCR	100	70	50	40	30	25	20	15	100	70	50	40	30	25	20	15	100	50	30	20	15	100	50	30	20	15
WW A1	26	32	39	44	52	57	63	72	20	24	29	32	38	42	48	55	1.2	1.3	1.5	1.6	1.8	1.5	1.6	1.8	1.9	2.0
WW A2	31	39	49	56	65	71	78	86	22	28	35	41	49	55	62	71	1.0	1.1	1.3	1.4	1.6	1.2	1.3	1.5	1.6	1.7
WW A3	33	42	53	60	70	76	83	90	22	30	38	45	54	60	67	77	0.9	1.1	1.2	1.4	1.5	1.1	1.2	1.4	1.5	1.7
WW A4	33	44	55	63	72	78	85	91	22	30	40	47	56	62	70	79	0.9	1.1	1.2	1.4	1.5	1.1	1.2	1.4	1.5	1.6
WW A5	35	46	57	65	75	81	87	93	23	32	41	49	59	65	73	82	0.9	1.0	1.2	1.4	1.5	1.0	1.2	1.3	1.5	1.6
WW A6	35	47	58	66	76	82	88	94	23	32	42	50	60	66	74	83	0.9	1.0	1.2	1.3	1.5	1.0	1.2	1.3	1.5	1.6
WW B1	26	33	41	47	56	62	69	77	17	22	28	32	39	44	50	58	1.1	1.2	1.4	1.5	1.7	1.4	1.5	1.7	1.8	1.9
WW B2	34	45	56	64	74	80	86	92	23	32	41	48	58	65	72	81	0.9	1.0	1.2	1.4	1.5	1.0	1.2	1.3	1.5	1.6
WW B3	41	52	64	72	81	86	91	96	28	38	49	57	67	74	81	88	0.8	1.0	1.1	1.3	1.5	1.0	1.1	1.2	1.4	1.5
WW B4	40	53	65	73	83	88	93	97	28	39	51	59	70	76	83	90	0.8	1.0	1.1	1.3	1.4	0.9	1.1	1.2	1.4	1.5
WW B5	39	51	64	72	82	87	92	96	27	38	50	58	69	75	82	90	0.8	1.0	1.1	1.3	1.5	0.9	1.1	1.2	1.4	1.5
WW C1	42	54	66	74	83	88	93	97	30	40	52	60	70	76	83	90	0.8	1.0	1.1	1.3	1.5	0.9	1.1	1.2	1.4	1.5
WW C2	40	52	64	72	82	87	92	96	28	38	50	58	68	74	82	89	0.8	1.0	1.1	1.3	1.5	0.9	1.1	1.2	1.4	1.5
WW C3	46	60	73	80	88	92	96	98	33	46	59	67	78	83	89	95	0.8	0.9	1.1	1.3	1.4	0.9	1.0	1.2	1.3	1.5
WW C4	42	56	69	77	86	90	94	98	30	42	55	63	74	80	87	93	0.8	0.9	1.1	1.3	1.4	0.9	1.0	1.2	1.3	1.5
TW A1	26	31	38	43	51	56	63	71	19	23	28	32	38	42	47	55	1.2	1.4	1.5	1.6	1.8	1.5	1.6	1.8	1.9	2.0
TW A2	28	36	44	51	60	66	73	81	20	26	32	37	45	50	57	66	1.1	1.2	1.3	1.5	1.6	1.3	1.4	1.5	1.7	1.8
TW A3	29	38	47	54	64	70	77	85	20	26	34	40	48	54	61	70	1.0	1.1	1.3	1.5	1.6	1.2	1.3	1.5	1.6	1.8
TW A4	28	37	47	55	64	70	78	86	19	26	34	40	49	55	62	72	1.0	1.1	1.3	1.4	1.6	1.2	1.3	1.4	1.6	1.7
TW B1	25	31	39	44	52	57	64	73	19	23	28	32	38	43	49	57	1.2	1.3	1.5	1.6	1.7	1.4	1.6	1.7	1.9	2.0
TW B2	26	34	43	49	59	64	72	80	19	24	31	36	44	49	56	65	1.1	1.2	1.4	1.5	1.6	1.3	1.4	1.5	1.7	1.8
TW B3	26	34	43	50	59	65	73	81	18	24	31	36	44	50	57	67	1.1	1.2	1.3	1.5	1.6	1.3	1.4	1.5	1.7	1.8
TW B4	24	32	41	47	57	63	70	79	17	23	29	34	42	48	55	64	1.1	1.2	1.4	1.5	1.7	1.3	1.4	1.6	1.7	1.8
TW C1	24	30	38	44	52	57	64	73	18	22	28	32	38	43	49	58	1.2	1.3	1.4	1.6	1.7	1.4	1.5	1.7	1.8	1.9
TW C2	23	30	38	45	53	59	66	75	17	22	28	33	40	45	51	60	1.1	1.3	1.4	1.6	1.7	1.4	1.5	1.6	1.8	1.9
TW C3	22	29	37	42	51	56	64	73	16	21	26	31	38	42	49	58	1.2	1.3	1.4	1.6	1.7	1.4	1.5	1.7	1.8	1.9
TW C4	20	26	33	38	45	51	57	67	15	19	24	28	34	38	44	52	1.3	1.4	1.5	1.7	1.8	1.5	1.7	1.8	1.9	2.0
TW D1	21	28	35	41	49	55	62	71	14	19	24	28	34	38	44	52	1.2	1.3	1.5	1.6	1.8	1.5	1.6	1.8	1.9	2.0
TW D2	31	41	51	58	68	74	81	89	22	29	38	44	53	59	67	76	1.0	1.1	1.2	1.4	1.6	1.1	1.3	1.4	1.5	1.7
TW D3	34	45	56	64	73	79	85	92	24	32	42	49	58	65	72	81	0.9	1.0	1.2	1.4	1.5	1.0	1.2	1.3	1.5	1.6
TW D4	36	48	59	67	77	83	89	94	25	35	45	53	63	70	77	86	0.9	1.0	1.2	1.3	1.5	1.0	1.1	1.3	1.4	1.6
TW D5	35	47	59	67	77	82	88	94	25	34	45	53	63	70	77	86	0.9	1.0	1.2	1.3	1.5	1.0	1.1	1.3	1.4	1.6
TW E1	38	49	61	68	78	83	89	94	27	37	47	54	64	70	77	86	0.9	1.0	1.2	1.3	1.5	1.0	1.1	1.3	1.4	1.6
TW E2	37	49	60	68	78	83	89	94	26	36	46	54	64	70	77	86	0.9	1.0	1.2	1.3	1.5	1.0	1.1	1.3	1.4	1.6
TW E3	44	57	69	77	86	90	94	98	32	44	56	64	74	80	86	93	0.8	0.9	1.1	1.3	1.4	0.9	1.0	1.2	1.4	1.5
TW E4	41	54	66	74	83	88	93	97	29	40	52	61	71	77	84	91	0.8	1.0	1.1	1.3	1.4	0.9	1.1	1.2	1.4	1.5
TW F1	24	30	37	43	51	56	63	72	18	22	27	31	37	41	47	55	1.2	1.3	1.5	1.6	1.8	1.5	1.6	1.7	1.9	2.0
TW F2	26	33	42	48	58	63	71	80	18	23	30	35	42	48	54	64	1.1	1.2	1.4	1.5	1.6	1.3	1.4	1.6	1.7	1.8
TW F3	25	34	43	50	59	66	73	82	17	23	30	36	44	49	57	66	1.0	1.2	1.3	1.5	1.6	1.2	1.4	1.5	1.7	1.8
TW F4	23	32	41	48	58	64	72	81	15	21	28	34	42	48	56	65	1.0	1.2	1.3	1.5	1.6	1.2	1.4	1.5	1.7	1.8
TW G1	22	29	36	42	50	55	62	72	16	20	25	30	36	41	47	55	1.2	1.3	1.5	1.6	1.7	1.5	1.6	1.7	1.9	2.0
TW G2	22	30	38	45	54	60	67	77	15	20	26	31	39	44	51	60	1.1	1.2	1.4	1.5	1.7	1.3	1.5	1.6	1.7	1.9
TW G3	20	28	36	43	52	58	66	76	13	18	25	30	37	43	50	59	1.1	1.3	1.4	1.5	1.7	1.3	1.5	1.6	1.7	1.9
TW G4	17	24	32	38	47	53	60	70	10	15	21	26	33	38	45	54	1.2	1.3	1.5	1.6	1.7	1.4	1.6	1.7	1.8	1.9
TW H1	19	25	33	38	47	52	60	69	13	17	22	27	33	38	44	53	1.2	1.3	1.5	1.6	1.8	1.5	1.6	1.7	1.9	2.0
TW H2	16	23	30	36	45	51	58	68	10	15	20	25	31	36	42	52	1.2	1.4	1.5	1.6	1.8	1.5	1.6	1.7	1.9	2.0
TW H3	14	20	27	32	40	46	53	63	8	12	17	21	28	32	38	47	1.3	1.4	1.6	1.7	1.8	1.6	1.7	1.8	1.9	2.1
TW H4	10	15	21	25	32	37	44	53	6	9	13	17	22	26	31	38	1.4	1.6	1.7	1.8	2.0	1.8	1.9	2.0	2.1	2.2
TW I1	19	25	32	38	45	51	58	67	12	16	21	25	31	35	40	48	1.2	1.4	1.5	1.7	1.8	1.5	1.7	1.8	2.0	2.1
TW I2	28	37	47	55	65	71	78	86	18	26	34	40	49	56	63	73	1.0	1.1	1.3	1.4	1.6	1.2	1.3	1.4	1.6	1.7
TW I3	30	40	51	59	69	75	82	89	20	28	38	44	54	60	68	78	0.9	1.1	1.2	1.4	1.5	1.1	1.2	1.4	1.5	1.7
TW I4	33	44	56	64	74	80	86	93	22	31	42	49	60	66	74	83	0.9	1.0	1.2	1.4	1.5	1.0	1.2	1.3	1.5	1.6
TW I5	33	44	56	65	75	81	87	93	22	32	43	50	61	68	75	84	0.9	1.0	1.2	1.3	1.5	1.0	1.1	1.3	1.4	1.6
TW J1	36	47	58	66	76	81	87	93	25	34	44	51	61	68	75	84	0.9	1.0	1.2	1.3	1.5	1.0	1.1	1.3	1.5	1.6
TW J2	34	45	57	65	75	81	87	93	23	33	43	51	61	67	75	84	0.9	1.0	1.2	1.3	1.5	1.0	1.1	1.3	1.5	1.6
TW J3	41	53	66	74	83	88	93	97	29	40	52	61	71	78	84	91	0.8	1.0	1.1	1.3	1.4	0.9	1.0	1.2	1.4	1.5
TW J4	38	50	63	71	81	86	92	96	26	37	49	58	68	75	82	90	0.8	1.0	1.1	1.3	1.5	0.9	1.1	1.2	1.4	1.5
DG A1	24	31	39	44	51	55	60	66	15	20	25	29	35	38	42	47	1.1	1.3	1.6	1.9	2.1	1.4	1.6	1.9	2.3	2.7
DG A2	25	34	43	50	59	64	70	78	17	23	30	36	44	48	55	62	1.0	1.2	1.4	1.6	1.8	1.3	1.4	1.6	1.8	2.0
DG A3	30	40	52	60	70	76	82	88	21	29	38	46	56	62	69	77	0.9	1.1	1.2	1.5	1.7	1.1	1.2	1.3	1.5	1.7
DG B1	24	32	42	49	58	64	70	76	15	20	26	31	39	44	50	56	1.1	1.2	1.4	1.6	1.9	1.4	1.5	1.6	1.9	2.2
DG B2	26	35	47	55	65	71	78	85	17	24	32	38	47	54	61	70	1.0	1.1	1.3	1.5	1.7	1.2	1.3	1.4	1.6	1.8
DG B3	30	41	54	64	75	80	86	92	21	29	39	47	59	66	74	83	0.9	1.0	1.2	1.4	1.6	1.1	1.2	1.3	1.5	1.6
DG C1	28	38	50	58	68	74	80	85	18	24	32	38	48	54	61	68	1.0	1.1	1.3	1.5	1.8	1.2	1.3	1.4	1.7	1.9
DG C2	30	41	53	62	73	79	85	91	20	28	37	44	55	62	70	78	0.9	1.0	1.2	1.4	1.6	1.1	1.2	1.3	1.5	1.7
DG C3	34	47	61	71	81	87	91	95	24	33	44	53	66	73	81	89	0.9	1.0	1.1	1.3	1.6	1.0	1.1	1.2	1.4	1.6
SS A1	34	43	52	58	67	72	79	86	27	34	41	47	55	60	67	75	1.0	1.1	1.3	1.5	1.6	1.1	1.3	1.4	1.6	1.7
SS A2	42	53	64	71	79	84	89	94	34	43	53	60	69	74	81	88	0.9	1.0	1.2	1.3	1.5	1.0	1.1	1.3	1.4	1.6
SS A3	32	40	48	54	63	68	75	82	25	31	38	43	51	56	62	70	1.0	1.2	1.3	1.5	1.7	1.2	1.3	1.5	1.7	1.8
SS A4	42	52	63	70	78	83	88	94	33	42	52	59	67	73	79	87	0.9	1.0	1.2	1.4	1.5	1.0	1.1	1.3	1.5	1.6
SS A5	32	40	48	54	62	67	74	81	26	32	38	43	50	55	61	69	1.1	1.2	1.4	1.5	1.7	1.2	1.4	1.5	1.7	1.8
SS A6	42	53	63	70	79	84	89	94	33	43	52	59	68	74	80	87	0.9	1.0	1.2	1.3	1.5	1.0	1.1	1.3	1.4	1.6
SS A7	29	35	42	48	55	60	66	74	23	28	34	38	44	48	54	61	1.1	1.3	1.5	1.6	1.8	1.3	1.5	1.7	1.8	2.0
SS A8	41	51	62	69	77	82	88	93	33	41	51	57	66	72	78	85	0.9	1.0	1.2	1.4	1.5	1.0	1.1	1.3	1.5	1.6
SS B1	29	37	45	52	60	66	73	81	22	28	35	41	48	53	60	68	1.1	1.2	1.4	1.5	1.7	1.2	1.4	1.5	1.7	1.8
SS B2	37	47	57	65	74	79	85	92	29	37	47	53	62	68	75	83	0.9	1.1	1.2	1.4	1.5	1.0	1.2	1.3	1.5	1.6
SS B3	28	35	43	49	58	63	70	78	21	27	33	38	45	50	56	65	1.1	1.2	1.4	1.5	1.7	1.3	1.4	1.6	1.7	1.9
SS B4	36	46	56	64	73	78	84	91	28	36	46	52	61	67	74	82	0.9	1.1	1.2	1.4	1.5	1.0	1.2	1.3	1.5	1.6
SS B5	26	33	40	45	53	58	65	73	21	25	31	35	42	46	52	60	1.2	1.3	1.5	1.6	1.8	1.4	1.5	1.7	1.8	2.0
SS B6	36	46	57	64	73	79	85	91	28	36	46	52	61	67	74	82	0.9	1.1	1.2	1.4	1.5	1.0	1.2	1.3	1.5	1.6
SS B7	24	29	36	41	48	53	59	68	19	23	28	32	37	41	47	54	1.2	1.4	1.5	1.7	1.8	1.5	1.6	1.8	1.9	2.1
SS B8	35	45	55	62	71	77	83	90	27	35	44	51	59	65	72	80	0.9	1.1	1.2	1.4	1.6	1.0	1.2	1.4	1.5	1.7
SS C1	29	38	48	55	64	70	77	85	19	26	34	39	48	53	61	70	1.0	1.1	1.3	1.4	1.6	1.2	1.3	1.5	1.6	1.8
SS C2	34	45	56	64	74	79	86	92	23	32	42	49	59	65	72	81	0.9	1.0	1.2	1.4	1.5	1.0	1.2	1.3	1.5	1.6
SS C3	24	30	38	44	52	58	65	74	17	21	27	31	38	43	49	58	1.2	1.3	1.4	1.6	1.7	1.4	1.5	1.7	1.8	1.9
SS C4	30	40	50	57	67	73	80	88	21	28	37	43	52	58	66	75	1.0	1.1	1.3	1.4	1.6	1.1	1.3	1.4	1.6	1.7
SS D1	41	51	61	68	77	82	87	93	32	40	49	56	65	70	77	84	0.9	1.0	1.2	1.4	1.5	1.0	1.2	1.3	1.5	1.6
SS D2	49	61	72	79	87	91	95	98	39	50	61	68	77	82	88	93	0.8	0.9	1.1	1.3	1.4	0.9	1.0	1.2	1.4	1.5
SS D3	37	46	55	62	70	76	82	89	29	36	44	50	58	64	70	78	1.0	1.1	1.3	1.4	1.6	1.1	1.2	1.4	1.6	1.7
SS D4	47	59	70	77	85	89	93	97	37	48	59	66	75	80	86	92	0.8	1.0	1.1	1.3	1.5	0.9	1.0	1.2	1.4	1.5
SS E1	36	45	55	62	71	77	83	89	27	34	43	49	57	63	70	78	0.9	1.1	1.2	1.4	1.6	1.1	1.2	1.4	1.6	1.7
SS E2	43	54	66	73	82	87	91	96	33	43	53	61	70	76	82	89	0.8	1.0	1.1	1.3	1.5	0.9	1.1	1.2	1.4	1.6
SS E3	30	37	46	52	61	66	73	81	23	29	36	41	48	53	59	68	1.1	1.2	1.4	1.5	1.7	1.2	1.4	1.5	1.7	1.8
SS E4	40	50	61	68	77	83	88	94	30	39	49	56	66	71	78	86	0.9	1.0	1.2	1.4	1.5	1.0	1.1	1.3	1.5	1.6

ALBANY, NEW YORK

SOLAR SAVINGS FRACTION (%)

TYPE	BASE TEMP: 55 F (4456 DD) LCR 100	70	50	40	30	25	20	15	65 F (6891 DD) 100	70	50	40	30	25	20	15
WW A1	13	15	16	17	19	20	22	24	11	12	13	13	14	14	15	15
WW A2	11	14	17	20	23	26	29	33	8	10	12	14	16	17	19	22
WW A3	10	13	17	20	25	28	32	37	7	9	12	14	17	18	21	24
WW A4	9	13	17	21	25	29	33	40	5	8	11	14	17	19	22	26
WW A5	9	13	18	21	27	30	35	42	4	8	11	14	17	20	23	28
WW A6	8	13	18	22	27	31	36	43	4	7	11	14	18	20	24	29
WW B1	8	9	10	11	12	13	13	14	5	5	6	6	5	5	4	1
WW B2	10	15	20	24	30	34	39	47	7	10	14	17	22	25	29	35
WW B3	13	18	25	29	36	41	48	56	8	13	18	22	28	32	37	44
WW B4	13	19	27	33	41	47	54	64	8	14	21	26	33	38	45	54
WW B5	12	19	27	33	42	48	56	65	8	14	21	27	34	40	47	56
WW C1	14	20	27	32	40	45	51	60	10	15	20	25	31	35	41	49
WW C2	13	19	26	32	40	45	52	61	9	14	20	25	31	36	42	51
WW C3	16	25	34	42	51	58	66	76	11	19	28	34	44	50	58	68
WW C4	14	22	32	39	48	55	63	73	10	17	25	32	41	47	55	65
TW A1	13	15	16	18	20	21	23	25	11	12	13	14	15	15	16	17
TW A2	11	14	17	19	22	24	27	32	9	10	12	14	16	17	19	21
TW A3	10	13	16	19	23	26	29	34	7	9	12	13	16	18	20	23
TW A4	9	12	16	19	23	26	30	36	6	8	11	13	16	18	21	25
TW B1	12	14	16	18	20	22	24	27	10	11	13	14	15	16	17	19
TW B2	10	13	16	18	22	24	28	32	8	10	12	13	16	17	20	23
TW B3	9	12	16	18	22	25	28	33	7	9	11	13	16	18	20	24
TW B4	9	12	15	18	21	24	28	33	7	9	11	13	16	18	20	24
TW C1	11	13	16	17	20	22	25	29	9	10	12	13	15	17	18	21
TW C2	10	12	15	17	21	23	26	31	8	9	11	13	15	17	20	23
TW C3	10	12	15	17	20	23	26	30	8	9	11	13	15	17	19	23
TW C4	10	12	15	16	19	22	24	29	9	10	12	13	15	17	19	22
TW D1	6	8	9	10	12	12	13	14	4	5	5	6	6	6	5	4
TW D2	11	15	19	23	28	32	36	43	8	11	14	17	21	24	28	33
TW D3	11	16	21	25	30	34	40	47	8	11	15	18	23	26	30	36
TW D4	12	18	25	30	37	42	49	58	9	13	19	23	30	34	40	48
TW D5	12	18	25	31	39	44	51	60	9	14	20	25	32	36	43	52
TW E1	14	19	25	29	36	40	46	54	10	14	19	22	28	31	36	43
TW E2	13	19	25	30	37	42	48	57	9	14	19	23	29	33	39	46
TW E3	16	24	32	38	47	53	61	70	12	18	26	31	40	45	52	61
TW E4	15	22	30	37	46	52	59	69	11	17	24	30	38	44	51	61
TW F1	11	13	15	16	18	20	21	24	9	10	11	12	13	14	14	15
TW F2	9	11	14	16	20	22	25	29	6	8	10	11	13	15	16	19
TW F3	7	10	13	16	20	22	26	31	5	7	9	11	13	15	17	20
TW F4	5	8	12	15	19	21	25	31	3	5	7	9	12	14	17	21
TW G1	9	11	13	15	17	19	21	24	7	8	10	11	12	13	14	16
TW G2	6	9	12	14	17	20	23	27	4	6	8	9	12	13	15	18
TW G3	5	8	11	13	17	19	22	27	3	4	7	8	11	13	15	18
TW G4	3	6	9	11	14	17	20	25	1	3	5	7	9	11	14	17
TW H1	6	8	11	12	15	17	20	23	4	6	7	8	10	11	13	15
TW H2	4	6	9	11	14	16	19	23	2	4	5	7	9	10	13	16
TW H3	3	5	7	9	12	14	17	22	1	3	4	6	8	9	12	15
TW H4	2	4	6	7	10	12	14	18	1	2	3	5	6	8	10	13
TW I1	6	6	7	8	9	10	11	12	3	3	4	4	4	4	3	2
TW I2	8	12	16	20	25	28	33	40	5	8	11	14	18	21	24	30
TW I3	9	13	18	22	28	32	37	44	6	9	13	16	21	24	28	34
TW I4	10	15	22	27	34	39	46	55	6	11	16	21	27	31	37	45
TW I5	10	16	23	28	36	41	48	58	7	12	18	22	29	34	40	49
TW J1	11	17	22	27	34	38	44	52	8	12	17	20	26	30	35	42
TW J2	11	16	22	27	34	39	46	54	7	12	17	21	27	31	37	45
TW J3	14	21	30	36	45	51	59	69	9	16	24	29	38	43	50	60
TW J4	12	20	28	34	43	49	57	67	8	15	22	28	36	42	49	59
DG A1	6	7	9	10	11	12	12	11	3	4	5	5	5	4	4	2
DG A2	9	12	15	18	21	24	27	31	6	8	11	13	15	17	20	23
DG A3	12	16	21	25	32	36	42	50	10	13	17	21	26	29	34	41
DG B1	6	7	9	10	11	12	13	14	3	4	4	5	5	5	4	3
DG B2	9	12	15	18	22	25	29	34	6	9	11	13	16	18	21	24
DG B3	13	17	22	26	33	38	44	53	10	14	18	22	27	31	36	44
DG C1	8	10	12	14	16	18	20	23	5	6	7	8	10	10	10	10
DG C2	11	15	19	22	27	31	35	42	8	11	14	17	20	23	27	32
DG C3	15	19	25	30	37	43	50	60	12	16	21	25	31	35	41	50
SS A1	15	18	21	24	27	30	33	38	12	15	17	19	22	24	27	30
SS A2	18	23	29	34	40	44	50	58	15	20	25	29	34	38	43	50
SS A3	13	16	18	20	23	25	27	31	11	13	15	16	18	19	21	22
SS A4	18	22	28	32	38	42	48	55	15	19	24	27	32	36	40	46
SS A5	15	18	20	22	25	27	30	33	13	15	17	18	20	22	23	25
SS A6	18	23	28	33	39	43	49	56	15	19	24	28	33	37	41	48
SS A7	14	16	18	19	21	22	23	25	12	14	15	16	17	17	18	18
SS A8	17	22	27	31	36	40	45	52	14	18	23	26	30	34	38	43
SS B1	12	15	18	20	23	26	29	33	10	12	14	16	18	20	22	25
SS B2	16	21	26	30	36	41	46	54	13	17	22	26	31	35	39	46
SS B3	11	13	16	18	20	22	24	28	9	11	12	14	15	16	18	19
SS B4	15	20	25	29	35	39	44	51	12	16	21	24	29	33	37	43
SS B5	12	14	17	18	21	22	24	27	11	12	14	15	16	17	19	20
SS B6	15	19	25	29	35	39	44	52	12	16	21	24	29	33	37	44
SS B7	11	13	14	16	17	19	20	22	10	11	12	13	14	14	15	15
SS B8	14	18	23	27	32	36	41	48	11	15	19	22	27	30	34	40
SS C1	9	11	14	17	20	23	26	30	6	7	9	11	13	14	16	18
SS C2	11	16	21	25	31	35	40	48	7	11	15	18	23	26	31	37
SS C3	9	11	14	16	18	20	22	26	7	9	10	11	13	14	15	17
SS C4	11	15	19	23	28	31	36	43	8	11	14	17	21	24	28	33
SS D1	15	19	24	27	31	35	39	45	12	16	19	21	25	27	30	33
SS D2	19	26	33	39	47	52	58	67	16	22	28	33	40	44	50	58
SS D3	16	19	22	25	29	31	35	40	13	16	18	20	23	25	27	30
SS D4	19	25	32	37	45	50	56	65	16	21	27	32	38	43	48	56
SS E1	12	15	19	21	25	27	31	35	9	11	14	16	18	20	21	23
SS E2	15	21	27	32	39	44	50	58	12	17	22	26	32	36	41	48
SS E3	13	15	18	20	22	24	27	30	11	12	14	16	17	19	20	22
SS E4	15	20	26	30	36	40	46	54	12	16	21	25	30	33	38	44

CONSERVATION FACTOR

TYPE	55 F 100	50	30	20	15	65 F 100	50	30	20	15
WW A1	2.4	2.9	3.4	3.9	4.4	3.4	4.2	-	-	-
WW A2	1.8	2.1	2.4	2.6	2.8	2.2	2.6	3.0	3.6	4.1
WW A3	1.6	1.9	2.1	2.4	2.5	2.0	2.3	2.7	3.1	3.5
WW A4	1.6	1.8	2.0	2.2	2.4	1.9	2.2	2.5	2.9	3.2
WW A5	1.5	1.7	1.9	2.1	2.3	1.8	2.1	2.4	2.7	3.0
WW A6	1.5	1.7	1.9	2.1	2.3	1.7	2.0	2.3	2.6	2.9
WW B1	2.6	3.4	4.6	-	-	4.5	-	-	-	-
WW B2	1.5	1.7	1.8	2.0	2.2	1.7	1.9	2.1	2.3	2.5
WW B3	1.3	1.5	1.7	1.8	2.0	1.5	1.7	1.9	2.1	2.3
WW B4	1.2	1.4	1.5	1.7	1.8	1.3	1.5	1.7	1.8	2.0
WW B5	1.2	1.3	1.5	1.7	1.8	1.3	1.5	1.6	1.8	1.9
WW C1	1.3	1.4	1.6	1.8	1.9	1.4	1.6	1.8	2.0	2.1
WW C2	1.2	1.4	1.6	1.7	1.9	1.4	1.6	1.7	1.9	2.1
WW C3	1.1	1.2	1.4	1.5	1.7	1.2	1.3	1.5	1.6	1.8
WW C4	1.1	1.3	1.4	1.6	1.7	1.2	1.3	1.5	1.7	1.8
TW A1	2.4	2.7	3.2	3.6	4.0	3.2	3.8	4.6	-	-
TW A2	1.9	2.2	2.5	2.7	2.9	2.4	2.7	3.1	3.6	4.1
TW A3	1.8	2.0	2.3	2.5	2.7	2.1	2.5	2.8	3.2	3.5
TW A4	1.7	1.9	2.2	2.4	2.5	2.0	2.3	2.6	2.9	3.2
TW B1	2.2	2.5	2.8	3.2	3.4	2.8	3.3	3.8	4.4	-
TW B2	1.9	2.1	2.4	2.6	2.8	2.3	2.6	2.9	3.3	3.6
TW B3	1.8	2.0	2.3	2.5	2.7	2.2	2.4	2.7	3.0	3.3
TW B4	1.8	2.1	2.3	2.5	2.7	2.2	2.4	2.7	3.0	3.2
TW C1	2.1	2.3	2.6	2.9	3.1	2.5	2.9	3.2	3.6	4.0
TW C2	1.9	2.2	2.4	2.6	2.8	2.3	2.6	2.9	3.2	3.4
TW C3	2.0	2.2	2.4	2.7	2.8	2.4	2.6	2.9	3.1	3.4
TW C4	2.1	2.3	2.6	2.8	3.0	2.5	2.8	3.0	3.3	3.5
TW D1	2.6	3.2	4.0	-	-	4.0	-	-	-	-
TW D2	1.6	1.8	2.0	2.1	2.3	1.8	2.0	2.2	2.5	2.7
TW D3	1.5	1.7	1.9	2.0	2.2	1.7	1.9	2.1	2.4	2.6
TW D4	1.3	1.5	1.6	1.8	1.9	1.5	1.6	1.8	2.0	2.1
TW D5	1.3	1.4	1.6	1.7	1.9	1.4	1.6	1.7	1.9	2.0
TW E1	1.4	1.5	1.7	1.9	2.0	1.5	1.7	1.9	2.1	2.3
TW E2	1.3	1.5	1.7	1.8	2.0	1.5	1.7	1.8	2.0	2.2
TW E3	1.1	1.3	1.5	1.6	1.8	1.2	1.4	1.6	1.7	1.9
TW E4	1.2	1.3	1.5	1.6	1.8	1.3	1.4	1.6	1.7	1.9
TW F1	2.3	2.7	3.1	3.5	3.8	3.1	3.7	4.4	-	-
TW F2	1.9	2.2	2.5	2.8	3.0	2.4	2.8	3.2	3.7	4.2
TW F3	1.8	2.1	2.3	2.6.	2.8	2.2	2.5	2.9	3.3	3.6
TW F4	1.8	2.0	2.3	2.5	2.7	2.2	2.4	2.7	3.0	3.3
TW G1	2.2	2.5	2.9	3.2	3.4	2.8	3.3	3.8	4.4	-
TW G2	2.0	2.2	2.5	2.7	2.9	2.4	2.7	3.1	3.5	3.8
TW G3	1.9	2.2	2.4	2.6	2.8	2.3	2.6	2.9	3.2	3.5
TW G4	2.0	2.2	2.5	2.7	2.9	2.4	2.7	2.9	3.2	3.4
TW H1	2.2	2.5	2.7	3.0	3.2	2.7	3.1	3.4	3.9	4.3
TW H2	2.1	2.4	2.6	2.8	3.0	2.5	2.8	3.1	3.4	3.7
TW H3	2.2	2.4	2.7	2.9	3.1	2.6	2.9	3.2	3.4	3.7
TW H4	2.4	2.6	2.9	3.1	3.3	2.9	3.1	3.4	3.6	[illegible]
TW I1	2.7	3.4	4.2	-	-	4.3	-	-	-	-
TW I2	1.6	1.8	2.0	2.2	2.3	1.9	2.1	2.3	2.5	2.7
TW I3	1.5	1.7	1.9	2.1	2.2	1.7	1.9	2.1	2.4	2.5
TW I4	1.3	1.5	1.7	1.8	2.0	1.5	1.7	1.8	2.0	2.2
TW I5	1.3	1.5	1.6	1.8	1.9	1.4	1.6	1.7	1.9	2.1
TW J1	1.4	1.6	1.7	1.9	2.0	1.5	1.7	1.9	2.1	2.3
TW J2	1.3	1.5	1.7	1.8	2.0	1.5	1.7	1.9	2.0	2.2
TW J3	1.2	1.3	1.5	1.6	1.8	1.3	1.4	1.6	1.7	1.9
TW J4	1.2	1.3	1.5	1.6	1.8	1.3	1.4	1.6	1.7	1.9
DG A1	2.4	3.1	4.7	-	-	3.5	-	-	-	-
DG A2	1.8	2.1	2.3	2.7	3.0	2.2	2.4	2.8	3.2	3.8
DG A3	1.5	1.7	1.8	2.0	2.1	1.7	1.8	2.0	2.1	2.3
DG B1	2.4	3.1	4.0	-	-	3.6	-	-	-	-
DG B2	1.8	2.0	2.2	2.5	2.7	2.1	2.3	2.7	3.0	3.4
DG B3	1.5	1.6	1.7	1.9	2.0	1.7	1.8	1.9	2.1	2.2
DG C1	2.0	2.5	2.9	3.4	4.3	2.7	3.3	4.7	-	-
DG C2	1.6	1.8	2.0	2.2	2.3	1.9	2.1	2.3	2.6	2.8
DG C3	1.4	1.5	1.6	1.7	1.9	1.5	1.6	1.8	1.9	2.0
SS A1	1.7	2.0	2.2	2.5	2.7	1.9	2.3	2.6	3.1	3.4
SS A2	1.3	1.5	1.7	1.9	2.0	1.4	1.6	1.8	2.0	2.2
SS A3	1.9	2.3	2.6	3.0	3.3	2.2	2.7	3.3	4.3	-
SS A4	1.4	1.6	1.8	1.9	2.1	1.5	1.7	1.9	2.2	2.3
SS A5	1.9	2.2	2.6	2.9	3.2	2.2	2.6	3.1	3.8	4.5
SS A6	1.3	1.5	1.7	1.9	2.0	1.5	1.7	1.9	2.1	2.3
SS A7	2.2	2.7	3.2	3.9	4.5	2.7	3.4	4.5	-	-
SS A8	1.4	1.6	1.8	2.0	2.2	1.5	1.8	2.0	2.3	2.5
SS B1	1.8	2.1	2.4	2.7	2.9	2.1	2.5	2.9	3.4	3.8
SS B2	1.4	1.6	1.7	1.9	2.1	1.5	1.7	1.9	2.1	2.3
SS B3	2.0	2.4	2.7	3.1	3.4	2.4	2.8	3.4	4.3	-
SS B4	1.4	1.6	1.8	2.0	2.1	1.6	1.8	2.0	2.2	2.4
SS B5	2.1	2.5	2.9	3.2	3.5	2.6	3.0	3.5	4.3	-
SS B6	1.4	1.6	1.8	2.0	2.1	1.5	1.7	2.0	2.2	2.3
SS B7	2.4	2.9	3.4	4.0	4.5	3.0	3.6	4.6	-	-
SS B8	1.5	1.7	1.9	2.1	2.2	1.6	1.8	2.1	2.3	2.5
SS C1	1.8	2.1	2.4	2.7	2.9	2.3	2.7	3.2	3.9	4.7
SS C2	1.4	1.6	1.8	2.0	2.1	1.7	1.9	2.1	2.3	2.5
SS C3	2.1	2.4	2.8	3.1	3.3	2.7	3.1	3.6	4.3	5.0
SS C4	1.6	1.8	2.0	2.1	2.3	1.8	2.0	2.2	2.5	2.7
SS D1	1.5	1.8	2.0	2.2	2.4	1.7	2.0	2.4	2.8	3.2
SS D2	1.2	1.4	1.5	1.7	1.8	1.3	1.5	1.7	1.9	2.0
SS D3	1.7	1.9	2.2	2.5	2.7	1.9	2.2	2.6	3.1	3.6
SS D4	1.2	1.4	1.6	1.7	1.9	1.3	1.5	1.7	1.9	2.1
SS E1	1.7	2.0	2.3	2.6	2.9	2.0	2.4	3.0	3.7	4.6
SS E2	1.3	1.5	1.7	1.8	2.0	1.4	1.6	1.8	2.1	2.2
SS E3	1.9	2.3	2.6	2.9	3.2	2.3	2.7	3.3	4.0	4.8
SS E4	1.4	1.6	1.7	1.9	2.1	1.5	1.7	1.9	2.2	2.3

BINGHAMTON, NEW YORK

TYPE	SOLAR SAVINGS FRACTION (%) BASE TEMP: 55 F (4719 DD)								65 F (7288 DD)								CONSERVATION FACTOR 55 F					65 F				
LCR	100	70	50	40	30	25	20	15	100	70	50	40	30	25	20	15	100	50	30	20	15	100	50	30	20	15
WW A1	11	12	13	13	14	14	14	14	10	10	11	11	10	10	9	7	3.3	4.3	-	-	-	-	-	-	-	-
WW A2	8	10	12	14	15	17	18	20	6	7	8	9	10	10	11	10	2.2	2.7	3.3	4.0	4.7	2.8	3.8	-	-	-
WW A3	7	9	12	13	16	18	20	23	4	6	7	8	10	10	11	12	2.0	2.4	2.9	3.3	3.7	2.5	3.2	4.2	-	-
WW A4	6	8	11	13	16	18	21	24	3	5	7	8	10	11	12	13	1.9	2.3	2.7	3.0	3.4	2.3	2.9	3.7	-	-
WW A5	5	8	11	13	17	19	22	26	2	4	6	7	9	11	12	13	1.8	2.2	2.5	2.8	3.1	2.2	2.7	3.4	4.5	-
WW A6	4	7	11	13	17	19	23	27	1	3	6	7	9	11	12	14	1.8	2.1	2.4	2.7	3.0	2.1	2.6	3.2	4.1	-
WW B1	5	6	6	6	5	4	3	0	4	3	2	2	0	-1	-4	-10	4.3	-	-	-	-	-	-	-	-	-
WW B2	7	10	14	17	21	24	28	34	4	7	10	12	15	17	20	24	1.7	1.9	2.2	2.4	2.6	1.9	2.3	2.6	3.0	3.3
WW B3	8	13	18	21	27	31	35	42	5	9	13	16	20	23	26	31	1.5	1.7	1.9	2.1	2.3	1.7	2.0	2.2	2.5	2.8
WW B4	8	14	20	25	32	37	44	53	5	10	16	20	26	30	36	43	1.3	1.5	1.7	1.9	2.0	1.5	1.7	1.9	2.0	2.2
WW B5	8	14	21	26	34	39	46	55	5	10	16	21	28	32	38	47	1.3	1.5	1.6	1.8	1.9	1.4	1.6	1.8	1.9	2.1
WW C1	10	15	20	24	30	34	40	47	7	11	15	18	23	26	31	37	1.4	1.6	1.8	2.0	2.2	1.6	1.8	2.1	2.3	2.5
WW C2	9	14	20	24	31	35	41	49	6	10	15	19	24	28	33	40	1.4	1.6	1.8	1.9	2.1	1.5	1.7	2.0	2.2	2.3
WW C3	11	19	27	34	43	49	57	66	8	15	22	28	37	42	49	59	1.2	1.3	1.5	1.6	1.8	1.2	1.4	1.6	1.7	1.9
WW C4	10	17	25	31	40	46	54	64	7	13	20	26	34	40	47	56	1.2	1.4	1.5	1.7	1.8	1.3	1.4	1.6	1.8	1.9
TW A1	11	12	13	14	15	15	16	16	10	11	11	11	11	11	11	10	3.1	3.8	5.0	-	-	4.3	-	-	-	-
TW A2	9	11	12	14	16	17	18	20	7	8	9	10	11	11	12	12	2.3	2.8	3.3	4.0	4.6	2.9	3.8	-	-	-
TW A3	7	9	12	13	16	17	19	22	6	7	8	9	11	11	12	13	2.1	2.5	2.9	3.4	3.8	2.6	3.2	4.1	-	-
TW A4	6	8	11	13	16	18	20	24	4	6	7	9	10	12	13	15	2.0	2.4	2.7	3.1	3.4	2.4	2.9	3.5	4.3	-
TW B1	10	11	13	14	15	16	17	18	9	10	10	11	11	12	12	12	2.8	3.3	4.0	4.9	-	3.5	4.7	-	-	-
TW B2	8	10	12	13	16	17	19	22	6	8	9	10	11	12	13	14	2.2	2.6	3.0	3.5	3.9	2.7	3.3	4.1	-	-
TW B3	7	9	11	13	16	17	20	23	5	7	8	9	11	12	13	15	2.1	2.5	2.8	3.2	3.5	2.6	3.1	3.6	4.4	-
TW B4	7	9	11	13	15	17	20	23	5	7	8	10	11	13	14	16	2.2	2.5	2.8	3.1	3.4	2.5	3.0	3.4	4.0	4.6
TW C1	9	11	12	13	15	17	18	20	8	9	10	11	12	12	13	14	2.5	2.9	3.3	3.8	4.3	3.0	3.7	4.5	-	-
TW C2	8	9	12	13	15	17	19	22	6	8	9	10	12	13	14	16	2.3	2.6	3.0	3.3	3.6	2.7	3.2	3.7	4.3	4.9
TW C3	8	9	11	13	15	17	19	22	7	8	9	10	12	13	14	16	2.3	2.6	2.9	3.3	3.6	2.7	3.1	3.6	4.1	4.6
TW C4	9	10	12	13	15	17	19	22	8	9	10	11	13	14	15	17	2.5	2.8	3.0	3.4	3.6	2.8	3.2	3.6	4.1	4.4
TW D1	4	5	5	6	6	5	5	3	3	3	3	2	1	0	0	-4	3.8	-	-	-	-	-	-	-	-	-
TW D2	8	11	14	17	21	23	27	32	6	8	11	13	15	17	20	23	1.8	2.1	2.3	2.6	2.8	2.1	2.4	2.8	3.1	3.5
TW D3	8	11	15	18	22	25	29	34	6	8	11	13	16	18	21	25	1.7	1.9	2.2	2.4	2.6	1.9	2.3	2.6	3.0	3.3
TW D4	9	14	19	23	29	33	39	47	6	10	15	18	23	27	31	38	1.5	1.6	1.8	2.0	2.2	1.6	1.8	2.0	2.2	2.4
TW D5	9	14	20	24	31	36	42	50	7	11	16	20	26	30	35	43	1.4	1.6	1.7	1.9	2.0	1.5	1.7	1.9	2.1	2.2
TW E1	10	14	19	22	27	31	35	42	7	11	14	17	21	24	27	32	1.5	1.8	2.0	2.2	2.3	1.7	2.0	2.3	2.5	2.8
TW E2	10	14	19	23	29	32	38	45	7	11	15	18	23	26	30	36	1.5	1.7	1.9	2.1	2.2	1.7	1.9	2.1	2.3	2.5
TW E3	12	18	25	31	39	44	51	60	9	14	21	26	33	37	44	52	1.3	1.4	1.6	1.8	1.9	1.4	1.5	1.7	1.9	2.0
TW E4	11	17	24	30	38	43	50	59	8	13	20	25	32	37	43	52	1.3	1.4	1.6	1.8	1.9	1.4	1.5	1.7	1.9	2.0
TW F1	9	10	11	12	13	14	14	15	8	8	9	9	9	9	9	8	3.0	3.7	4.7	-	-	4.1	-	-	-	-
TW F2	7	8	10	11	13	14	16	18	5	6	7	8	9	9	9	10	2.4	2.8	3.4	4.1	4.7	3.0	3.9	-	-	-
TW F3	5	7	9	10	13	14	16	19	3	4	6	7	8	9	10	10	2.2	2.6	3.0	3.5	3.9	2.7	3.3	4.2	-	-
TW F4	3	5	7	9	12	14	16	19	1	3	4	5	7	8	10	11	2.1	2.5	2.8	3.2	3.5	2.6	3.0	3.6	4.4	-
TW G1	7	8	10	11	12	13	14	15	6	7	7	8	8	8	9	8	2.8	3.3	4.0	4.9	-	3.6	4.7	-	-	-
TW G2	4	6	8	9	11	13	15	17	3	4	5	6	7	8	9	10	2.4	2.7	3.2	3.7	4.2	2.9	3.5	4.4	-	-
TW G3	3	5	7	8	11	12	14	17	1	3	4	5	7	8	9	10	2.3	2.6	3.0	3.4	3.7	2.7	3.2	3.8	4.6	-
TW G4	1	3	5	7	9	11	13	17	0	2	3	4	6	7	8	10	2.4	2.7	3.0	3.3	3.6	2.8	3.2	3.6	4.2	4.7
TW H1	4	6	7	8	10	11	13	15	3	4	5	6	7	7	8	8	2.6	3.1	3.5	4.1	4.7	3.2	4.0	5.0	-	-
TW H2	2	4	5	7	9	10	12	15	1	2	3	4	6	6	8	9	2.5	2.8	3.2	3.6	4.0	3.0	3.5	4.0	4.7	-
TW H3	1	3	5	6	8	10	12	14	0	1	3	4	5	6	7	9	2.6	2.9	3.2	3.5	3.8	3.0	3.4	3.9	4.4	4.9
TW H4	1	2	4	5	7	8	10	13	0	1	2	3	4	5	7	8	2.8	3.1	3.4	3.7	3.9	3.2	3.6	4.0	4.4	4.8
TW I1	3	3	4	4	4	4	3	1	1	1	1	1	0	0	-2	-5	4.1	-	-	-	-	-	-	-	-	-
TW I2	5	8	11	14	18	20	24	29	3	6	8	10	13	14	17	20	1.9	2.1	2.4	2.6	2.8	2.1	2.5	2.8	3.2	3.5
TW I3	6	9	13	16	20	23	27	33	4	6	9	11	15	17	20	24	1.7	2.0	2.2	2.4	2.6	2.0	2.2	2.5	2.9	3.1
TW I4	7	11	16	20	26	30	36	44	4	8	12	16	21	24	29	36	1.5	1.7	1.9	2.0	2.2	1.6	1.8	2.0	2.2	2.4
TW I5	7	12	17	22	28	33	39	48	4	9	13	17	23	27	33	40	1.4	1.6	1.8	1.9	2.1	1.6	1.7	1.9	2.1	2.3
TW J1	8	12	17	20	25	29	34	40	6	9	12	15	19	22	26	31	1.6	1.8	2.0	2.2	2.3	1.7	2.0	2.2	2.5	2.7
TW J2	7	12	17	21	26	30	36	43	5	9	13	16	21	24	28	35	1.5	1.7	1.9	2.1	2.2	1.7	1.9	2.1	2.3	2.5
TW J3	10	16	23	29	37	42	49	59	7	12	19	24	31	36	42	51	1.3	1.4	1.6	1.7	1.9	1.4	1.5	1.7	1.9	2.0
TW J4	8	15	22	27	35	41	48	57	6	11	18	22	30	35	41	50	1.3	1.4	1.6	1.8	1.9	1.4	1.5	1.7	1.9	2.0
DG A1	3	4	5	5	5	4	3	1	2	2	2	1	0	0	-2	-6	3.5	-	-	-	-	-	-	-	-	-
DG A2	7	9	11	13	15	17	19	22	5	7	8	9	11	12	14	15	2.1	2.5	2.9	3.4	4.0	2.5	3.0	3.6	4.4	-
DG A3	10	13	17	21	25	29	34	40	8	11	15	17	21	24	28	33	1.7	1.9	2.0	2.2	2.4	1.9	2.0	2.2	2.4	2.6
DG B1	3	4	5	5	5	4	4	2	2	2	2	1	0	0	-1	-5	3.3	-	-	-	-	-	-	-	-	-
DG B2	7	9	11	13	16	17	20	23	5	7	8	10	12	13	14	16	2.1	2.4	2.8	3.2	3.6	2.5	2.9	3.4	4.2	-
DG B3	10	14	18	22	27	30	36	43	9	12	15	18	23	26	30	36	1.6	1.8	2.0	2.1	2.2	1.8	2.0	2.1	2.3	2.5
DG C1	5	6	8	8	9	9	9	9	3	4	4	4	4	4	3	1	2.5	3.7	-	-	-	3.8	-	-	-	-
DG C2	8	11	14	17	20	22	26	30	7	9	11	13	16	17	20	22	1.8	2.1	2.4	2.6	2.9	2.2	2.4	2.8	3.3	3.7
DG C3	12	16	21	25	30	34	40	49	10	13	18	21	26	30	34	41	1.5	1.7	1.8	1.9	2.1	1.7	1.8	2.0	2.2	2.3
SS A1	12	15	17	19	22	23	25	28	11	13	15	16	18	19	20	22	2.0	2.3	2.8	3.2	3.6	2.2	2.7	3.4	4.3	-
SS A2	15	20	24	28	33	37	42	49	13	17	21	25	29	32	36	42	1.4	1.7	1.9	2.1	2.2	1.6	1.8	2.1	2.3	2.5
SS A3	11	13	15	16	18	19	20	21	10	11	12	13	14	14	15	14	2.2	2.8	3.6	4.7	-	2.6	3.5	-	-	-
SS A4	15	19	23	27	32	35	39	45	13	16	20	23	27	30	34	38	1.5	1.7	2.0	2.2	2.4	1.6	1.9	2.2	2.5	2.7
SS A5	13	15	17	18	20	21	22	24	12	13	15	16	17	17	18	18	2.2	2.7	3.3	4.1	4.9	2.6	3.3	4.4	-	-
SS A6	15	19	24	27	32	36	40	47	13	17	21	24	28	31	34	40	1.5	1.7	1.9	2.1	2.3	1.6	1.9	2.1	2.4	2.6
SS A7	12	14	15	15	16	17	17	17	11	12	13	13	13	13	13	12	2.7	3.5	5.0	-	-	3.3	4.9	-	-	-
SS A8	14	18	22	25	30	33	37	42	13	16	19	22	25	28	31	35	1.5	1.8	2.1	2.3	2.5	1.7	2.0	2.3	2.7	3.0
SS B1	10	12	14	16	18	19	21	24	9	10	12	13	14	15	16	17	2.1	2.6	3.0	3.6	4.1	2.5	3.1	3.8	-	-
SS B2	13	17	22	25	30	34	39	45	11	15	19	22	26	29	33	39	1.5	1.7	1.9	2.1	2.3	1.6	1.9	2.1	2.3	2.5
SS B3	9	11	12	14	15	16	17	18	8	9	10	11	12	12	12	12	2.4	2.9	3.7	4.7	-	2.8	3.7	-	-	-
SS B4	13	16	21	24	29	32	36	43	11	14	18	21	25	27	31	36	1.6	1.8	2.0	2.2	2.4	1.7	1.9	2.2	2.5	2.7
SS B5	11	12	14	15	16	17	18	19	10	11	12	12	13	14	14	14	2.5	3.1	3.8	4.7	-	3.0	3.9	-	-	-
SS B6	12	16	20	24	29	32	36	43	11	14	18	20	25	27	31	36	1.6	1.8	2.0	2.2	2.4	1.7	1.9	2.2	2.4	2.7
SS B7	10	11	12	13	14	14	14	15	9	10	10	11	11	11	11	10	2.9	3.7	5.0	-	-	3.6	-	-	-	-
SS B8	12	15	19	22	26	29	33	39	10	13	16	19	22	25	28	32	1.6	1.9	2.1	2.3	2.5	1.8	2.1	2.3	2.6	2.9
SS C1	6	7	9	11	12	13	15	16	4	5	6	6	7	7	7	6	2.3	2.8	3.5	4.5	-	3.0	4.2	-	-	-
SS C2	8	11	15	18	22	26	30	35	5	8	11	13	17	19	22	26	1.7	1.9	2.1	2.4	2.5	1.9	2.2	2.5	2.8	3.1
SS C3	7	9	10	11	13	13	15	16	6	7	8	8	9	9	9	9	2.7	3.2	3.8	4.8	-	3.4	4.5	-	-	-
SS C4	8	11	14	17	21	23	27	32	6	8	11	13	16	18	20	24	1.8	2.1	2.3	2.6	2.7	2.1	2.4	2.7	3.0	3.4
SS D1	12	15	18	21	24	26	28	32	10	13	15	17	19	20	21	23	1.7	2.1	2.5	2.9	3.2	2.0	2.5	3.2	4.3	-
SS D2	16	21	27	32	39	43	49	57	13	18	24	28	33	37	42	49	1.3	1.5	1.7	1.9	2.0	1.4	1.6	1.8	2.1	2.2
SS D3	13	16	18	20	22	24	26	29	12	13	15	17	18	19	20	21	1.9	2.3	2.8	3.3	3.7	2.2	2.7	3.5	4.9	-
SS D4	15	21	27	31	37	42	47	55	13	18	23	27	32	36	41	47	1.3	1.5	1.7	1.9	2.1	1.4	1.7	1.9	2.1	2.3
SS E1	9	11	14	15	17	19	20	22	7	9	10	11	12	13	13	12	2.0	2.5	3.2	4.0	5.0	2.4	3.3	-	-	-
SS E2	12	16	22	25	31	35	40	47	10	13	18	21	25	28	32	38	1.4	1.7	1.9	2.1	2.2	1.6	1.8	2.1	2.4	2.6
SS E3	11	12	14	15	17	18	19	21	9	10	12	12	13	14	14	14	2.3	2.8	3.5	4.4	-	2.7	3.6	-	-	-
SS E4	12	16	21	24	29	32	37	43	10	13	17	20	24	27	30	35	1.5	1.8	2.0	2.2	2.4	1.7	1.9	2.2	2.5	2.8

BUFFALO, NEW YORK

TYPE	LCR 55 F (4434 DD) SSF 100	70	50	40	30	25	20	15	LCR 65 F (6931 DD) SSF 100	70	50	40	30	25	20	15	CF 55 F 100	50	30	20	15	CF 65 F 100	50	30	20	15
WW A1	12	13	14	14	15	16	16	16	10	11	11	11	11	11	10	8	3.0	3.8	-	-	-	4.6	-	-	-	-
WW A2	9	11	13	15	17	18	20	22	7	8	9	10	11	11	12	12	2.1	2.6	3.1	3.8	4.5	2.7	3.5	5.0	-	-
WW A3	7	10	13	15	18	19	22	25	5	6	8	9	11	11	12	13	1.9	2.3	2.8	3.2	3.7	2.4	3.0	3.9	-	-
WW A4	6	9	12	15	18	20	23	26	4	5	7	9	11	12	13	14	1.8	2.2	2.6	3.0	3.3	2.2	2.8	3.5	4.9	-
WW A5	6	9	12	15	18	21	24	28	2	5	7	8	10	12	13	15	1.7	2.1	2.4	2.8	3.1	2.1	2.6	3.3	4.4	-
WW A6	5	8	12	15	19	21	24	29	2	4	6	8	11	12	14	15	1.7	2.0	2.4	2.7	3.0	2.1	2.5	3.1	4.1	-
WW B1	6	6	7	7	6	6	5	2	4	4	3	2	1	0	-3	-9	3.6	-	-	-	-	-	-	-	-	-
WW B2	8	11	15	18	23	26	30	35	5	7	10	13	16	18	21	25	1.6	1.9	2.2	2.4	2.6	1.9	2.2	2.5	2.9	3.3
WW B3	9	14	19	23	28	32	37	44	6	10	14	17	21	24	27	32	1.4	1.7	1.9	2.1	2.3	1.7	1.9	2.2	2.5	2.8
WW B4	9	15	22	27	34	39	45	54	6	11	16	21	27	31	37	44	1.3	1.5	1.7	1.9	2.0	1.4	1.6	1.8	2.0	2.2
WW B5	9	15	22	27	35	40	47	56	6	11	17	22	28	33	39	48	1.3	1.5	1.6	1.8	1.9	1.4	1.6	1.8	1.9	2.1
WW C1	11	16	21	26	32	36	41	48	7	11	16	19	24	27	32	37	1.4	1.6	1.8	2.0	2.2	1.6	1.8	2.1	2.3	2.6
WW C2	10	15	21	26	32	37	42	51	6	11	16	19	25	29	34	41	1.4	1.6	1.8	1.9	2.1	1.5	1.7	1.9	2.2	2.3
WW C3	12	20	29	35	44	50	58	67	9	16	23	29	37	43	50	59	1.1	1.3	1.5	1.6	1.8	1.2	1.4	1.6	1.7	1.9
WW C4	11	18	26	33	41	47	55	65	7	14	21	27	35	40	48	57	1.2	1.3	1.5	1.7	1.8	1.3	1.4	1.6	1.8	1.9
TW A1	12	13	14	15	16	16	17	18	10	11	11	12	12	12	11	11	2.8	3.5	4.5	-	-	4.0	-	-	-	-
TW A2	9	11	13	15	17	18	20	22	8	9	10	11	12	12	13	13	2.2	2.6	3.2	3.8	4.4	2.8	3.6	4.7	-	-
TW A3	8	10	13	15	17	19	21	24	6	7	9	10	11	12	13	14	2.0	2.4	2.8	3.3	3.7	2.5	3.1	3.9	-	-
TW A4	7	9	12	14	17	19	22	25	4	6	8	9	11	12	14	16	1.9	2.3	2.6	3.0	3.3	2.3	2.8	3.4	4.1	-
TW B1	11	12	14	15	16	17	18	20	9	10	11	11	12	12	13	12	2.6	3.0	3.7	4.6	-	3.4	4.4	-	-	-
TW B2	8	10	13	14	17	18	21	23	7	8	9	10	12	13	14	15	2.1	2.5	2.9	3.4	3.7	2.6	3.2	3.9	4.9	-
TW B3	7	10	12	14	17	19	21	24	6	7	9	10	12	13	14	16	2.0	2.4	2.7	3.1	3.4	2.5	3.0	3.5	4.2	-
TW B4	7	9	12	14	17	19	21	25	6	7	9	10	12	13	15	17	2.1	2.4	2.7	3.0	3.3	2.5	2.9	3.3	3.9	4.4
TW C1	10	11	13	14	16	18	19	22	8	9	10	11	12	13	14	15	2.4	2.7	3.2	3.7	4.1	2.9	3.5	4.3	-	-
TW C2	8	10	12	14	17	18	21	24	7	8	10	11	12	13	15	17	2.2	2.5	2.8	3.2	3.5	2.6	3.1	3.6	4.1	4.7
TW C3	8	10	12	14	16	18	21	24	7	8	10	11	13	14	15	17	2.2	2.5	2.8	3.2	3.5	2.6	3.0	3.5	4.0	4.4
TW C4	9	11	13	14	16	18	20	23	8	9	11	12	13	14	16	18	2.4	2.6	2.9	3.2	3.5	2.8	3.2	3.5	3.9	4.3
TW D1	5	6	6	7	7	7	6	5	3	3	3	3	2	1	0	-3	3.3	5.0	-	-	-	-	-	-	-	-
TW D2	9	12	15	18	22	25	28	33	6	9	11	13	16	18	21	24	1.7	2.0	2.3	2.5	2.7	2.0	2.4	2.7	3.1	3.4
TW D3	9	12	16	19	24	27	30	36	6	9	12	14	17	19	22	26	1.6	1.9	2.2	2.4	2.6	1.9	2.2	2.5	2.9	3.3
TW D4	10	15	20	24	30	35	40	48	7	11	15	19	24	28	32	39	1.4	1.6	1.8	2.0	2.1	1.6	1.8	2.0	2.2	2.4
TW D5	10	15	21	26	32	37	43	52	7	11	17	21	27	31	36	44	1.4	1.6	1.7	1.9	2.0	1.5	1.7	1.9	2.1	2.2
TW E1	11	15	20	23	28	32	37	43	8	11	15	18	22	25	28	33	1.5	1.7	2.0	2.2	2.3	1.7	2.0	2.2	2.5	2.8
TW E2	10	15	20	24	30	34	39	46	8	11	15	19	23	27	31	37	1.4	1.7	1.9	2.0	2.2	1.6	1.8	2.1	2.3	2.5
TW E3	13	19	27	32	40	45	52	61	10	15	22	26	33	38	44	53	1.2	1.4	1.6	1.7	1.9	1.3	1.5	1.7	1.9	2.0
TW E4	12	18	25	31	39	44	51	60	8	14	20	25	33	37	44	53	1.2	1.4	1.6	1.7	1.9	1.3	1.5	1.7	1.9	2.0
TW F1	10	11	12	13	14	15	16	16	8	9	9	10	10	10	10	9	2.8	3.4	4.3	-	-	3.8	-	-	-	-
TW F2	7	9	11	12	14	16	17	19	5	6	8	8	9	10	10	11	2.2	2.7	3.2	3.9	4.5	2.9	3.7	4.8	-	-
TW F3	5	8	10	12	14	16	18	21	3	5	6	7	9	10	11	12	2.1	2.5	2.9	3.4	3.8	2.6	3.2	4.0	-	-
TW F4	3	6	9	11	13	15	18	21	2	3	5	6	8	9	11	12	2.0	2.4	2.7	3.1	3.4	2.5	2.9	3.5	4.2	-
TW G1	8	9	10	12	13	14	15	17	6	7	8	8	9	9	9	9	2.6	3.1	3.7	4.6	-	3.4	4.4	-	-	-
TW G2	5	7	9	10	13	14	16	19	3	4	6	7	8	9	10	11	2.2	2.6	3.1	3.6	4.0	2.8	3.4	4.2	-	-
TW G3	3	5	8	9	12	14	16	19	2	3	5	6	7	8	10	11	2.2	2.5	2.9	3.3	3.6	2.7	3.1	3.7	4.4	-
TW G4	2	4	6	8	11	12	15	18	1	2	4	5	6	8	9	11	2.3	2.5	2.9	3.2	3.5	2.7	3.1	3.5	4.0	4.5
TW H1	5	6	8	9	11	13	14	16	3	4	5	6	7	8	9	9	2.5	2.9	3.4	3.9	4.5	3.1	3.8	4.7	-	-
TW H2	2	4	6	8	10	12	14	17	1	2	4	5	6	7	8	10	2.4	2.7	3.1	3.5	3.8	2.9	3.4	3.9	4.5	-
TW H3	2	3	5	7	9	11	13	16	1	2	3	4	6	7	8	10	2.5	2.7	3.1	3.4	3.7	2.9	3.3	3.8	4.2	4.7
TW H4	1	2	4	5	7	9	11	14	0	1	2	3	5	6	7	9	2.7	3.0	3.2	3.5	3.8	3.1	3.5	3.9	4.3	4.6
TW I1	3	4	5	5	5	5	5	3	2	2	2	1	0	0	-1	-4	3.5	-	-	-	-	-	-	-	-	-
TW I2	6	9	13	15	19	22	25	30	4	6	9	11	13	16	10	21	1.8	2.1	2.3	2.6	2.8	2.1	2.4	2.7	3.1	3.5
TW I3	6	10	14	17	21	25	29	34	4	7	10	12	16	18	21	25	1.7	1.9	2.2	2.4	2.6	1.9	2.2	2.5	2.8	3.1
TW I4	7	12	18	22	28	32	37	45	5	9	13	16	22	25	30	37	1.5	1.7	1.8	2.0	2.2	1.6	1.8	2.0	2.2	2.4
TW I5	7	13	19	23	30	34	41	49	5	9	14	18	24	28	33	41	1.4	1.6	1.8	1.9	2.1	1.5	1.7	1.9	2.1	2.3
TW J1	9	13	18	21	27	30	35	42	6	9	13	16	20	23	27	32	1.5	1.7	2.0	2.2	2.3	1.7	2.0	2.2	2.5	2.7
TW J2	8	13	18	22	28	32	37	45	5	9	13	17	22	25	29	36	1.5	1.7	1.9	2.1	2.2	1.6	1.9	2.1	2.3	2.5
TW J3	11	17	25	30	38	44	51	60	7	13	19	25	32	37	43	52	1.2	1.4	1.6	1.7	1.9	1.3	1.5	1.7	1.9	2.0
TW J4	9	16	23	29	37	42	49	59	6	12	18	23	30	35	42	51	1.3	1.4	1.6	1.8	1.9	1.4	1.5	1.7	1.9	2.0
DG A1	4	5	6	6	6	6	5	3	2	2	2	2	1	0	-1	-4	3.0	4.9	-	-	-	-	-	-	-	-
DG A2	7	9	12	14	17	18	21	23	5	7	9	10	12	13	14	16	2.0	2.4	2.8	3.3	3.9	2.4	2.9	3.5	4.3	-
DG A3	11	14	18	22	27	30	35	41	9	12	15	18	22	25	29	34	1.6	1.8	2.0	2.2	2.4	1.8	2.0	2.2	2.4	2.6
DG B1	4	5	6	6	6	6	6	4	2	2	2	2	1	0	0	-4	2.9	4.9	-	-	-	-	-	-	-	-
DG B2	7	10	12	14	17	19	22	25	5	7	9	10	12	14	15	17	2.0	2.3	2.7	3.0	3.5	2.4	2.7	3.3	4.0	4.9
DG B3	11	15	19	23	28	32	37	44	9	12	16	19	23	27	31	37	1.6	1.8	1.9	2.1	2.2	1.8	1.9	2.1	2.3	2.5
DG C1	6	8	9	10	11	11	11	11	4	4	5	5	5	5	4	2	2.4	3.3	-	-	-	3.7	-	-	-	-
DG C2	9	12	15	18	21	24	27	32	7	9	12	14	16	18	21	24	1.8	2.1	2.4	2.6	2.9	2.1	2.3	2.7	3.2	3.6
DG C3	13	17	22	26	32	36	42	50	10	14	18	22	27	31	35	42	1.5	1.7	1.8	1.9	2.1	1.7	1.8	2.0	2.2	2.3
SS A1	13	16	18	20	23	25	27	30	11	13	15	17	19	20	21	22	1.9	2.3	2.7	3.2	3.6	2.2	2.6	3.3	4.3	-
SS A2	16	21	26	29	35	38	43	50	14	18	22	25	30	33	37	42	1.4	1.7	1.9	2.1	2.2	1.5	1.8	2.0	2.3	2.5
SS A3	12	14	16	17	19	20	21	22	10	12	13	14	15	15	15	15	2.1	2.7	3.5	4.5	-	2.5	3.4	-	-	-
SS A4	16	20	24	28	33	36	40	47	13	17	21	24	28	31	34	39	1.5	1.7	2.0	2.2	2.4	1.6	1.9	2.2	2.5	2.8
SS A5	14	16	18	19	21	22	24	25	13	14	15	16	17	18	19	19	2.1	2.6	3.2	4.0	4.8	2.5	3.2	4.2	-	-
SS A6	16	20	25	28	33	37	41	48	13	17	21	24	28	31	35	40	1.4	1.7	1.9	2.1	2.3	1.6	1.8	2.1	2.4	2.7
SS A7	13	14	16	16	17	18	18	19	12	13	13	14	14	14	14	13	2.5	3.3	4.6	-	-	3.2	4.5	-	-	-
SS A8	15	19	23	26	31	34	38	44	13	16	20	22	26	29	32	36	1.5	1.8	2.0	2.3	2.5	1.7	2.0	2.3	2.7	3.0
SS B1	11	13	15	17	19	21	23	25	9	11	12	13	15	16	17	18	2.0	2.5	2.9	3.5	4.0	2.4	3.0	3.7	4.9	-
SS B2	14	18	23	26	31	35	40	46	12	15	19	23	27	30	34	40	1.5	1.7	1.9	2.1	2.3	1.6	1.9	2.1	2.3	2.5
SS B3	10	11	13	15	16	17	19	20	8	9	11	11	12	13	13	13	2.2	2.8	3.5	4.5	-	2.7	3.5	-	-	-
SS B4	13	17	22	25	30	33	38	44	11	15	18	21	25	28	32	37	1.5	1.8	2.0	2.2	2.4	1.7	1.9	2.2	2.5	2.7
SS B5	11	13	14	16	17	18	19	21	10	11	12	13	14	14	15	15	2.4	2.9	3.6	4.5	-	2.9	3.7	4.9	-	-
SS B6	13	17	22	25	30	33	38	44	11	14	18	21	25	28	32	37	1.5	1.8	2.0	2.2	2.4	1.7	1.9	2.2	2.4	2.7
SS B7	10	11	13	14	15	15	16	16	9	10	11	11	11	12	11	11	2.7	3.4	4.6	-	-	3.5	4.8	-	-	-
SS B8	12	16	20	23	28	31	35	40	10	13	17	19	23	26	29	33	1.6	1.8	2.1	2.3	2.5	1.7	2.0	2.3	2.6	2.9
SS C1	6	8	10	12	14	15	16	18	4	5	6	7	8	8	8	7	2.1	2.7	3.3	4.2	-	2.9	3.9	-	-	-
SS C2	8	12	16	19	24	27	31	37	6	8	12	14	17	20	23	27	1.6	1.9	2.1	2.3	2.5	1.9	2.2	2.5	2.8	3.1
SS C3	8	9	11	12	14	15	16	18	6	7	8	9	9	10	10	10	2.5	3.0	3.6	4.4	-	3.3	4.2	-	-	-
SS C4	9	12	15	18	22	25	28	33	6	9	11	13	16	18	21	25	1.7	2.0	2.3	2.5	2.7	2.0	2.3	2.6	3.0	3.3
SS D1	13	16	20	22	25	27	30	33	11	13	16	17	20	21	22	23	1.7	2.1	2.5	2.9	3.2	1.9	2.4	3.1	4.3	-
SS D2	17	22	29	33	40	44	50	58	14	19	24	28	34	38	43	49	1.3	1.5	1.7	1.9	2.0	1.4	1.6	1.8	2.1	2.3
SS D3	14	16	19	21	24	25	27	30	12	14	16	17	19	20	21	22	1.8	2.3	2.7	3.2	3.7	2.1	2.7	3.4	4.8	-
SS D4	16	22	28	32	38	43	48	56	14	18	24	27	33	37	41	47	1.3	1.5	1.7	1.9	2.1	1.4	1.6	1.9	2.1	2.3
SS E1	10	12	15	17	19	20	22	23	8	9	11	12	13	13	14	13	1.9	2.4	3.1	3.9	4.9	2.3	3.1	4.8	-	-
SS E2	13	18	23	27	32	36	41	48	10	14	18	22	26	29	33	38	1.4	1.6	1.9	2.1	2.3	1.5	1.8	2.1	2.4	2.7
SS E3	11	13	15	16	18	19	21	22	10	11	12	13	14	15	15	15	2.2	2.7	3.3	4.2	-	2.6	3.4	4.7	-	-
SS E4	13	17	22	25	30	34	38	44	10	14	18	21	25	28	31	36	1.5	1.7	2.0	2.2	2.4	1.6	1.9	2.2	2.5	2.8

MASSENA, NEW YORK

TYPE	SSF 55 F (5602 DD) LCR 100	70	50	40	30	25	20	15	SSF 65 F (8299 DD) 100	70	50	40	30	25	20	15	CF 55 F 100	50	30	20	15	CF 65 F 100	50	30	20	15
WW A1	11	12	13	13	14	14	14	14	11	11	11	11	11	11	10	9	3.3	4.3	-	-	-	4.1	-	-	-	-
WW A2	8	10	12	13	15	17	18	20	7	8	10	10	11	12	12	12	2.2	2.7	3.3	4.0	4.9	2.6	3.5	4.8	-	-
WW A3	6	9	11	13	16	17	20	22	5	7	9	10	11	12	13	14	2.0	2.4	2.9	3.4	3.9	2.3	3.0	3.8	-	-
WW A4	5	8	11	13	16	18	21	24	4	6	8	9	11	12	14	15	1.9	2.2	2.7	3.1	3.5	2.2	2.7	3.4	4.5	-
WW A5	4	7	11	13	16	19	21	25	3	5	7	9	11	12	14	16	1.8	2.1	2.5	2.9	3.3	2.1	2.6	3.2	4.1	-
WW A6	4	7	10	13	17	19	22	26	2	4	7	9	11	13	14	16	1.8	2.1	2.4	2.8	3.1	2.0	2.5	3.0	3.8	5.0
WW B1	5	5	5	5	5	4	2	0	4	4	3	3	1	0	-2	-8	4.5	-	-	-	-	-	-	-	-	-
WW B2	6	10	14	17	21	24	28	33	5	8	11	13	17	19	22	26	1.7	1.9	2.2	2.4	2.7	1.9	2.2	2.5	2.9	3.2
WW B3	8	12	17	21	26	30	35	42	6	10	14	17	21	24	28	33	1.5	1.7	2.0	2.2	2.4	1.6	1.9	2.2	2.5	2.8
WW B4	8	14	20	25	32	37	43	52	6	11	17	21	27	32	37	45	1.3	1.5	1.7	1.9	2.0	1.4	1.6	1.8	2.0	2.2
WW B5	8	14	21	26	34	39	46	55	6	11	17	22	29	34	40	48	1.3	1.5	1.6	1.8	2.0	1.4	1.6	1.7	1.9	2.1
WW C1	10	14	20	24	30	34	39	46	8	12	16	20	25	28	32	38	1.4	1.6	1.8	2.1	2.2	1.6	1.8	2.0	2.3	2.5
WW C2	8	14	19	24	30	35	41	49	7	11	16	20	26	29	35	42	1.4	1.6	1.8	2.0	2.1	1.5	1.7	1.9	2.1	2.3
WW C3	11	19	27	34	43	49	56	66	9	16	24	30	38	43	51	60	1.2	1.3	1.5	1.7	1.8	1.2	1.4	1.6	1.7	1.9
WW C4	10	17	25	31	40	46	53	63	8	14	22	27	35	41	48	58	1.2	1.4	1.5	1.7	1.8	1.3	1.4	1.6	1.8	1.9
TW A1	11	12	13	14	14	15	16	16	11	11	12	12	12	12	12	11	3.1	3.8	4.9	-	-	3.7	-	-	-	-
TW A2	9	10	12	13	15	17	18	20	8	9	10	11	12	13	13	14	2.4	2.8	3.3	4.0	4.7	2.7	3.5	4.6	-	-
TW A3	7	9	11	13	16	17	19	22	6	8	9	10	12	13	14	15	2.1	2.5	2.9	3.4	3.9	2.4	3.0	3.8	4.9	-
TW A4	6	8	11	13	16	17	20	23	5	7	8	10	12	13	15	16	2.0	2.3	2.7	3.1	3.4	2.3	2.8	3.3	4.0	4.8
TW B1	10	11	13	13	15	16	17	18	10	10	11	12	12	13	13	13	2.8	3.3	3.9	4.9	-	3.2	4.2	-	-	-
TW B2	8	9	12	13	15	17	19	21	7	8	10	11	12	13	14	16	2.3	2.6	3.0	3.5	3.9	2.5	3.1	3.8	4.7	-
TW B3	7	9	11	13	15	17	19	23	6	7	9	10	12	13	15	17	2.2	2.5	2.8	3.2	3.6	2.4	2.9	3.4	4.1	4.8
TW B4	6	8	11	13	15	17	20	23	6	7	9	11	12	14	15	18	2.2	2.5	2.8	3.1	3.4	2.4	2.8	3.3	3.8	4.3
TW C1	9	10	12	13	15	16	18	20	8	9	11	12	13	14	14	16	2.5	2.9	3.3	3.8	4.3	2.8	3.4	4.2	-	-
TW C2	7	9	11	13	15	17	19	22	7	8	10	11	13	14	15	17	2.3	2.6	2.9	3.3	3.6	2.5	3.0	3.5	4.1	4.6
TW C3	8	9	11	13	15	17	19	22	7	8	10	11	13	14	16	18	2.4	2.6	2.9	3.3	3.6	2.5	3.0	3.4	3.9	4.3
TW C4	8	10	12	13	15	17	19	22	8	9	11	12	14	15	16	18	2.5	2.8	3.0	3.3	3.6	2.7	3.1	3.5	3.9	4.3
TW D1	4	5	5	5	5	5	4	3	3	4	4	3	2	2	0	-2	4.0	-	-	-	-	-	-	-	-	-
TW D2	8	11	14	17	20	23	26	31	7	9	12	14	17	19	21	25	1.8	2.1	2.3	2.6	2.8	2.0	2.3	2.7	3.0	3.4
TW D3	8	11	15	18	22	25	28	34	6	9	12	14	18	20	23	27	1.7	1.9	2.2	2.5	2.7	1.9	2.2	2.5	2.9	3.2
TW D4	9	13	19	23	29	33	39	46	7	11	16	19	25	28	33	40	1.5	1.6	1.8	2.0	2.2	1.6	1.8	2.0	2.2	2.4
TW D5	9	14	20	24	31	36	42	50	7	12	17	21	27	31	37	45	1.4	1.6	1.7	1.9	2.1	1.5	1.7	1.9	2.0	2.2
TW E1	10	14	18	22	27	30	35	41	8	12	15	18	22	25	29	34	1.5	1.8	2.0	2.2	2.4	1.7	2.0	2.2	2.5	2.7
TW E2	9	14	19	23	28	32	37	45	8	12	16	19	24	27	32	38	1.5	1.7	1.9	2.1	2.2	1.6	1.8	2.1	2.3	2.5
TW E3	12	18	25	31	38	44	50	60	10	16	22	27	34	39	45	53	1.3	1.4	1.6	1.8	1.9	1.3	1.5	1.7	1.9	2.0
TW E4	11	17	24	30	37	43	50	59	9	14	21	26	33	38	44	53	1.3	1.4	1.6	1.8	1.9	1.3	1.5	1.7	1.9	2.0
TW F1	9	10	11	12	13	13	14	14	8	9	10	10	10	10	10	10	3.0	3.7	4.7	-	-	3.5	-	-	-	-
TW F2	6	8	10	11	13	14	16	17	6	7	8	9	10	10	11	11	2.4	2.8	3.4	4.1	4.8	2.7	3.5	4.7	-	-
TW F3	4	6	9	10	13	14	16	19	4	5	7	8	9	10	11	12	2.2	2.6	3.0	3.5	4.0	2.5	3.1	3.9	5.0	-
TW F4	3	5	7	9	12	14	16	19	2	3	5	7	8	10	11	13	2.2	2.5	2.8	3.2	3.6	2.4	2.9	3.4	4.1	4.8
TW G1	7	8	9	10	12	13	14	15	6	7	8	9	9	10	10	10	2.8	3.3	3.9	4.9	-	3.2	4.2	-	-	-
TW G2	4	6	8	9	11	13	14	17	3	5	6	7	8	9	10	11	2.4	2.7	3.2	3.7	4.2	2.7	3.3	4.1	-	-
TW G3	2	4	6	8	11	12	14	17	2	3	5	6	8	9	10	12	2.3	2.6	3.0	3.4	3.8	2.5	3.1	3.6	4.3	-
TW G4	1	3	5	7	9	11	13	16	1	2	4	5	7	8	10	12	2.4	2.7	3.0	3.3	3.6	2.6	3.0	3.5	4.0	4.4
TW H1	4	5	7	8	10	11	13	14	4	5	6	7	8	8	9	10	2.7	3.1	3.5	4.1	4.7	3.0	3.7	4.6	-	-
TW H2	2	3	5	7	9	10	12	15	1	3	4	5	7	8	9	11	2.5	2.8	3.2	3.6	3.9	2.7	3.2	3.8	4.5	-
TW H3	1	3	4	6	8	9	11	14	1	2	3	4	6	7	9	11	2.6	2.9	3.2	3.5	3.8	2.8	3.2	3.7	4.2	4.6
TW H4	1	2	3	5	6	8	10	12	0	1	3	4	5	6	8	10	2.9	3.1	3.4	3.7	3.9	3.0	3.4	3.8	4.2	4.6
TW I1	2	3	3	3	3	3	3	1	2	2	2	2	1	0	0	-4	4.3	-	-	-	-	-	-	-	-	-
TW I2	5	8	11	14	17	20	24	28	4	6	9	11	14	16	18	22	1.9	2.1	2.4	2.6	2.9	2.0	2.4	2.7	3.1	3.4
TW I3	5	9	13	16	20	23	27	32	4	7	10	13	16	18	21	26	1.7	2.0	2.2	2.4	2.6	1.9	2.2	2.5	2.8	3.1
TW I4	6	11	16	20	26	30	36	43	5	9	13	17	22	26	31	37	1.5	1.7	1.9	2.1	2.2	1.6	1.8	2.0	2.2	2.4
TW I5	6	11	17	22	28	33	39	47	5	10	15	19	24	29	34	42	1.4	1.6	1.8	1.9	2.1	1.5	1.7	1.9	2.1	2.2
TW J1	8	12	16	20	25	29	33	40	6	10	14	16	21	24	28	33	1.6	1.8	2.0	2.2	2.4	1.7	2.0	2.2	2.5	2.7
TW J2	7	11	17	21	26	30	35	43	6	10	14	17	22	25	30	36	1.5	1.7	1.9	2.1	2.2	1.6	1.8	2.1	2.3	2.5
TW J3	9	16	23	29	37	42	49	58	8	13	20	25	32	37	44	53	1.3	1.4	1.6	1.8	1.9	1.3	1.5	1.7	1.9	2.0
TW J4	8	14	22	27	35	41	48	57	7	12	19	24	31	36	43	51	1.3	1.4	1.6	1.8	1.9	1.4	1.5	1.7	1.9	2.0
DG A1	3	4	4	5	5	4	3	1	2	3	3	2	2	1	0	-4	3.5	-	-	-	-	4.8	-	-	-	-
DG A2	6	8	11	13	15	17	19	22	6	7	9	11	12	14	15	17	2.1	2.5	2.8	3.4	4.0	2.4	2.8	3.4	4.2	-
DG A3	10	13	17	20	25	29	33	40	9	12	15	18	22	25	29	35	1.7	1.8	2.0	2.2	2.4	1.8	2.0	2.2	2.4	2.6
DG B1	3	4	5	5	5	4	4	2	2	3	3	2	2	1	0	-3	3.3	-	-	-	-	-	-	-	-	-
DG B2	6	9	11	13	16	18	20	23	6	7	9	11	13	14	16	18	2.1	2.4	2.8	3.1	3.6	2.3	2.8	3.2	3.9	4.6
DG B3	10	14	18	22	27	30	36	43	9	12	16	19	24	27	31	38	1.6	1.8	2.0	2.1	2.2	1.8	1.9	2.1	2.3	2.4
DG C1	5	6	7	8	9	9	9	9	4	5	5	5	6	5	5	3	2.5	3.5	-	-	-	3.4	-	-	-	-
DG C2	8	11	14	17	20	22	26	30	7	9	12	14	17	19	21	24	1.8	2.1	2.4	2.6	2.9	2.1	2.4	2.7	3.1	3.5
DG C3	12	16	21	25	30	35	40	49	11	14	18	22	27	31	36	43	1.5	1.7	1.8	2.0	2.1	1.6	1.8	1.9	2.1	2.3
SS A1	12	14	16	18	20	22	24	26	12	13	15	16	18	19	20	22	2.0	2.4	2.9	3.4	4.0	2.2	2.7	3.4	4.5	-
SS A2	15	19	23	27	32	36	40	47	14	18	22	25	29	32	36	42	1.5	1.7	1.9	2.1	2.3	1.6	1.8	2.1	2.3	2.5
SS A3	11	12	14	15	16	17	18	18	10	12	13	13	14	14	15	14	2.3	3.0	3.9	-	-	2.6	3.6	-	-	-
SS A4	14	18	22	25	30	33	37	43	13	17	21	23	27	30	33	38	1.5	1.8	2.0	2.3	2.5	1.6	1.9	2.2	2.5	2.8
SS A5	13	14	16	17	19	20	21	22	13	14	15	16	17	18	18	18	2.3	2.8	3.5	4.5	-	2.5	3.3	4.5	-	-
SS A6	14	18	23	26	31	34	39	45	13	17	21	24	28	31	34	39	1.5	1.7	2.0	2.2	2.4	1.6	1.9	2.1	2.4	2.7
SS A7	12	13	14	14	15	15	16	15	12	13	13	13	14	14	13	12	2.9	3.8	-	-	-	3.2	-	-	-	-
SS A8	13	17	21	24	28	31	35	40	13	16	19	22	25	28	31	35	1.6	1.9	2.1	2.4	2.6	1.7	2.0	2.4	2.7	3.1
SS B1	10	11	13	15	17	18	20	22	9	11	12	13	15	16	17	17	2.2	2.6	3.2	3.8	4.5	2.4	3.1	3.9	-	-
SS B2	12	16	21	24	29	33	37	44	12	15	19	22	27	30	34	39	1.5	1.8	2.0	2.2	2.3	1.6	1.9	2.1	2.4	2.6
SS B3	8	10	11	13	14	15	16	16	8	9	10	11	12	12	12	12	2.5	3.1	3.9	-	-	2.7	3.8	-	-	-
SS B4	12	15	20	23	27	31	35	41	11	14	18	21	25	28	31	36	1.6	1.8	2.1	2.3	2.5	1.7	2.0	2.2	2.5	2.7
SS B5	10	12	13	14	15	16	17	18	10	11	12	13	14	14	14	15	2.6	3.2	3.9	-	-	2.9	3.8	-	-	-
SS B6	12	15	20	23	28	31	35	41	11	14	18	21	25	28	31	36	1.6	1.8	2.0	2.3	2.4	1.7	1.9	2.2	2.5	2.7
SS B7	9	10	11	12	13	13	13	13	9	10	11	11	11	11	11	10	3.1	4.0	-	-	-	3.4	-	-	-	-
SS B8	11	14	18	21	25	28	32	37	10	13	17	19	23	25	28	32	1.7	1.9	2.2	2.4	2.6	1.8	2.1	2.4	2.7	3.0
SS C1	5	7	9	10	12	13	14	16	4	5	7	7	8	8	8	8	2.3	2.8	3.5	4.6	-	2.7	3.8	-	-	-
SS C2	7	11	15	18	22	25	29	35	6	9	12	14	18	20	23	28	1.7	1.9	2.2	2.4	2.6	1.8	2.1	2.4	2.8	3.1
SS C3	7	8	10	11	12	13	14	16	7	8	8	9	10	10	10	11	2.7	3.2	3.8	4.7	-	3.1	4.0	-	-	-
SS C4	8	11	14	17	20	23	27	32	7	9	12	14	17	19	22	25	1.8	2.1	2.3	2.6	2.8	2.0	2.3	2.6	2.9	3.3
SS D1	12	14	17	19	22	24	26	29	11	13	15	17	19	20	21	22	1.8	2.2	2.6	3.2	3.7	2.0	2.5	3.2	4.5	-
SS D2	15	20	26	31	37	41	47	55	14	18	24	28	33	37	42	49	1.3	1.5	1.7	1.9	2.1	1.4	1.6	1.8	2.1	2.3
SS D3	13	15	17	19	21	23	24	27	12	14	16	17	19	19	20	21	2.0	2.4	2.9	3.6	4.2	2.2	2.8	3.6	-	-
SS D4	15	20	25	30	36	40	45	52	14	18	23	27	32	36	41	47	1.3	1.6	1.8	2.0	2.1	1.4	1.7	1.9	2.1	2.3
SS E1	8	11	13	14	16	17	18	19	8	9	11	12	12	13	13	12	2.1	2.6	3.4	4.7	-	2.4	3.3	-	-	-
SS E2	11	16	21	24	30	33	38	44	10	14	18	21	26	29	33	38	1.5	1.7	1.9	2.2	2.3	1.6	1.8	2.1	2.4	2.7
SS E3	10	12	13	14	16	17	18	19	10	11	12	13	14	14	15	14	2.4	2.9	3.7	4.9	-	2.6	3.6	-	-	-
SS E4	11	15	20	23	28	31	35	41	10	14	18	20	24	27	31	35	1.5	1.8	2.0	2.3	2.5	1.7	1.9	2.2	2.5	2.8

NEW YORK, NEW YORK

SOLAR SAVINGS FRACTION (%)

TYPE	BASE TEMP: 55 F (2775 DD) LCR 100	70	50	40	30	25	20	15	65 F (4851 DD) 100	70	50	40	30	25	20	15
WW A1	17	20	23	26	30	32	36	42	13	15	17	18	20	22	24	26
WW A2	17	22	27	32	38	42	48	56	11	14	18	21	25	28	32	37
WW A3	17	22	29	34	41	46	52	61	10	14	19	22	27	30	35	41
WW A4	16	23	30	35	43	48	55	63	10	14	19	22	28	32	37	44
WW A5	16	23	31	37	45	50	57	66	9	14	19	23	29	33	39	46
WW A6	16	23	31	37	46	51	58	67	9	14	19	24	30	34	40	48
WW B1	13	16	20	23	27	30	34	39	8	10	11	13	14	15	16	18
WW B2	17	24	32	38	46	52	59	68	11	16	21	26	32	37	42	51
WW B3	21	29	38	45	54	60	67	76	13	19	26	32	39	44	51	60
WW B4	21	30	40	48	57	64	71	80	13	21	29	35	44	50	57	67
WW B5	20	29	40	47	57	64	71	80	13	20	29	35	44	50	58	68
WW C1	23	31	40	47	57	63	70	79	15	21	29	34	42	48	55	64
WW C2	21	30	39	46	56	62	69	78	13	20	28	34	42	48	55	64
WW C3	25	36	48	56	66	73	80	88	17	26	36	44	54	61	68	78
WW C4	23	33	45	53	63	70	77	86	15	24	33	41	51	57	65	75
TW A1	17	20	23	26	30	32	36	42	13	15	17	18	21	22	24	27
TW A2	16	20	25	29	35	39	44	51	11	14	17	20	23	26	30	35
TW A3	16	20	26	30	37	41	47	55	10	14	17	20	25	28	32	38
TW A4	15	20	26	30	37	42	48	57	9	13	17	20	25	28	33	39
TW B1	16	19	23	26	30	33	38	44	12	14	16	18	21	23	26	29
TW B2	15	19	24	28	34	38	43	51	10	13	17	19	23	26	30	35
TW B3	14	19	24	28	34	39	44	52	9	13	16	19	23	26	30	36
TW B4	14	18	23	27	33	37	43	51	9	12	16	18	23	26	30	36
TW C1	15	18	22	25	30	34	38	45	11	13	16	18	21	24	27	31
TW C2	14	18	22	26	31	35	40	47	10	12	15	18	22	24	28	33
TW C3	13	17	21	25	30	34	39	46	10	12	15	17	21	24	27	33
TW C4	13	16	20	23	27	31	35	42	10	12	15	17	20	22	26	30
TW D1	11	14	17	20	24	26	30	36	7	8	10	11	13	14	16	18
TW D2	17	23	29	35	42	47	54	63	11	15	20	24	30	34	39	46
TW D3	18	25	32	38	46	51	58	67	12	16	22	26	33	37	43	51
TW D4	20	27	36	43	52	58	65	74	13	19	26	31	39	44	51	60
TW D5	19	27	36	43	53	59	66	76	13	19	27	32	40	46	53	63
TW E1	21	28	37	43	51	57	64	73	14	20	26	31	38	43	49	58
TW E2	21	28	37	43	52	58	65	74	14	20	26	32	39	44	51	60
TW E3	25	34	45	52	62	69	76	84	17	25	34	40	50	56	63	73
TW E4	23	32	43	50	60	66	74	83	15	23	32	39	48	54	62	71
TW F1	15	18	21	24	28	32	36	41	11	13	15	17	19	21	23	26
TW F2	14	18	23	27	32	36	41	49	9	12	15	17	21	24	27	32
TW F3	13	17	23	27	33	37	43	51	7	11	14	17	21	24	28	34
TW F4	11	15	21	25	32	36	42	51	6	9	13	16	20	23	28	34
TW G1	13	16	20	23	28	31	35	42	9	11	13	15	18	20	23	27
TW G2	11	15	20	24	29	33	39	46	7	9	12	15	19	21	25	30
TW G3	9	14	18	22	28	32	38	46	5	8	11	14	18	21	24	30
TW G4	7	11	16	19	25	28	34	41	3	6	9	12	15	18	22	27
TW H1	10	13	17	20	25	28	33	40	6	8	11	13	16	18	21	26
TW H2	8	11	15	18	23	27	32	39	4	6	9	11	15	17	20	25
TW H3	6	9	13	16	21	24	29	36	3	5	8	10	13	15	19	23
TW H4	5	7	10	13	17	19	23	29	2	4	6	8	10	12	15	19
TW I1	9	12	15	17	21	23	27	32	5	6	8	9	11	12	13	15
TW I2	14	20	26	31	39	44	50	59	8	12	17	21	26	30	36	43
TW I3	15	21	29	34	42	48	54	64	9	14	19	23	30	34	40	48
TW I4	17	24	33	39	48	54	62	72	10	16	23	28	36	41	48	57
TW I5	17	25	34	41	50	56	64	74	10	17	24	30	38	43	51	60
TW J1	19	26	34	40	49	55	62	71	12	17	24	29	36	41	47	56
TW J2	18	25	34	40	49	55	63	72	11	17	24	29	36	42	48	57
TW J3	22	31	42	50	60	66	74	83	14	22	31	38	47	54	61	71
TW J4	20	29	40	47	57	64	72	81	13	21	29	36	45	51	59	69
DG A1	11	14	18	21	24	26	29	31	6	8	10	11	12	13	14	14
DG A2	13	18	23	28	34	38	42	49	9	12	16	18	23	25	20	34
DG A3	17	23	30	36	45	50	57	66	12	17	22	27	33	38	44	52
DG B1	11	14	19	22	26	29	33	37	6	8	10	11	13	14	16	17
DG B2	14	18	24	29	36	40	46	54	9	12	16	19	23	27	31	37
DG B3	18	24	32	38	47	53	61	71	13	17	23	28	35	40	47	56
DG C1	13	18	23	27	33	37	42	48	8	11	13	15	18	20	23	26
DG C2	16	22	29	34	42	47	54	63	11	15	20	23	29	33	38	45
DG C3	20	27	36	43	52	59	68	78	15	20	26	31	39	45	53	63
SS A1	20	25	30	34	40	45	50	58	16	19	23	26	30	33	37	43
SS A2	25	32	40	46	54	60	66	75	19	25	32	36	43	48	54	62
SS A3	18	23	27	31	36	40	45	51	14	17	20	23	26	29	32	36
SS A4	24	31	39	45	53	58	64	73	19	24	31	35	42	46	52	60
SS A5	20	24	28	32	37	40	45	52	16	19	22	24	28	30	33	38
SS A6	25	32	40	45	53	59	65	74	19	25	31	36	42	47	53	61
SS A7	18	21	25	27	31	34	38	43	15	17	19	21	23	25	27	30
SS A8	24	30	38	43	51	56	63	71	18	24	29	34	40	44	50	57
SS B1	17	21	26	30	35	39	44	51	13	16	19	22	26	28	32	37
SS B2	22	28	36	41	49	55	62	70	17	22	28	33	39	44	50	58
SS B3	16	19	24	27	32	36	40	47	12	14	17	20	23	25	28	32
SS B4	21	27	35	40	48	53	60	68	16	21	27	31	38	42	48	55
SS B5	16	19	23	26	31	34	38	44	13	15	18	20	22	24	27	31
SS B6	21	27	35	40	48	53	60	69	16	21	27	31	38	42	48	56
SS B7	15	18	21	23	27	29	33	38	12	13	15	17	19	21	23	26
SS B8	20	26	33	38	46	51	57	66	15	20	25	29	35	40	45	53
SS C1	15	19	25	29	36	40	46	53	9	12	15	18	22	25	29	34
SS C2	18	25	32	38	46	52	59	68	11	16	22	27	33	37	43	52
SS C3	14	17	21	24	29	33	37	44	10	12	14	16	19	21	24	28
SS C4	17	22	29	34	41	46	53	62	11	15	20	24	29	33	39	46
SS D1	23	29	36	41	48	53	59	67	17	21	26	30	35	39	44	50
SS D2	28	37	46	53	62	68	75	83	21	28	36	42	50	56	63	71
SS D3	22	27	32	37	43	47	53	60	17	20	24	28	32	35	39	45
SS D4	27	36	45	51	60	66	73	81	21	27	35	41	49	54	60	69
SS E1	19	24	30	35	41	46	52	59	13	17	21	24	28	31	35	41
SS E2	24	31	40	47	55	61	68	77	17	23	30	35	43	48	54	62
SS E3	18	21	26	30	35	39	43	50	13	16	19	22	25	27	31	35
SS E4	22	29	37	43	51	57	63	72	16	22	28	33	39	44	50	58

CONSERVATION FACTOR

TYPE	55 F 100	50	30	20	15	65 F 100	50	30	20	15
WW A1	1.8	2.0	2.2	2.3	2.5	2.4	2.7	3.0	3.4	3.7
WW A2	1.4	1.6	1.7	1.9	2.0	1.7	2.0	2.2	2.4	2.6
WW A3	1.3	1.4	1.6	1.8	1.9	1.6	1.8	2.0	2.2	2.4
WW A4	1.2	1.4	1.6	1.7	1.9	1.5	1.7	1.9	2.1	2.3
WW A5	1.2	1.4	1.5	1.7	1.8	1.4	1.6	1.8	2.0	2.2
WW A6	1.2	1.3	1.5	1.7	1.8	1.4	1.6	1.8	2.0	2.1
WW B1	1.7	1.9	2.1	2.4	2.6	2.4	2.9	3.6	4.5	-
WW B2	1.2	1.3	1.5	1.7	1.8	1.4	1.6	1.8	1.9	2.1
WW B3	1.1	1.2	1.4	1.6	1.7	1.3	1.4	1.6	1.8	1.9
WW B4	1.0	1.2	1.3	1.5	1.6	1.2	1.3	1.5	1.6	1.8
WW B5	1.0	1.2	1.3	1.5	1.6	1.2	1.3	1.5	1.6	1.8
WW C1	1.1	1.2	1.4	1.5	1.7	1.2	1.4	1.5	1.7	1.9
WW C2	1.1	1.2	1.4	1.5	1.7	1.2	1.4	1.5	1.7	1.8
WW C3	0.9	1.1	1.2	1.4	1.6	1.0	1.2	1.4	1.5	1.7
WW C4	1.0	1.1	1.3	1.4	1.6	1.1	1.2	1.4	1.5	1.7
TW A1	1.8	2.0	2.1	2.3	2.5	2.3	2.6	2.9	3.2	3.5
TW A2	1.5	1.6	1.8	2.0	2.1	1.8	2.1	2.3	2.5	2.7
TW A3	1.4	1.6	1.7	1.9	2.0	1.7	1.9	2.1	2.3	2.5
TW A4	1.4	1.5	1.7	1.8	2.0	1.6	1.8	2.0	2.2	2.4
TW B1	1.7	1.9	2.0	2.2	2.4	2.1	2.4	2.6	2.9	3.1
TW B2	1.5	1.6	1.8	2.0	2.1	1.8	2.0	2.2	2.4	2.6
TW B3	1.5	1.6	1.8	1.9	2.1	1.8	2.0	2.2	2.4	2.5
TW B4	1.5	1.6	1.8	2.0	2.1	1.8	2.0	2.2	2.4	2.5
TW C1	1.7	1.8	2.0	2.1	2.3	2.0	2.2	2.5	2.7	2.9
TW C2	1.6	1.7	1.9	2.0	2.2	1.9	2.1	2.3	2.5	2.7
TW C3	1.6	1.8	1.9	2.1	2.2	2.0	2.1	2.3	2.5	2.7
TW C4	1.8	1.9	2.0	2.2	2.3	2.1	2.3	2.5	2.6	2.8
TW D1	1.8	2.0	2.2	2.4	2.6	2.5	2.9	3.4	4.0	4.6
TW D2	1.3	1.4	1.6	1.7	1.9	1.5	1.7	1.9	2.0	2.2
TW D3	1.2	1.4	1.5	1.7	1.8	1.4	1.6	1.8	1.9	2.1
TW D4	1.1	1.3	1.4	1.5	1.7	1.3	1.4	1.6	1.7	1.9
TW D5	1.1	1.2	1.4	1.6	1.7	1.2	1.4	1.5	1.7	1.8
TW E1	1.1	1.3	1.4	1.6	1.7	1.3	1.5	1.7	1.8	2.0
TW E2	1.1	1.3	1.4	1.6	1.7	1.3	1.4	1.6	1.8	1.9
TW E3	1.0	1.1	1.3	1.5	1.6	1.1	1.3	1.4	1.6	1.7
TW E4	1.0	1.2	1.3	1.5	1.6	1.1	1.3	1.4	1.6	1.7
TW F1	1.8	1.9	2.1	2.3	2.4	2.3	2.5	2.8	3.1	3.4
TW F2	1.5	1.7	1.8	2.0	2.2	1.9	2.1	2.3	2.6	2.8
TW F3	1.5	1.6	1.8	1.9	2.1	1.8	2.0	2.2	2.4	2.6
TW F4	1.4	1.6	1.8	1.9	2.1	1.8	1.9	2.1	2.3	2.5
TW G1	1.7	1.9	2.0	2.2	2.4	2.2	2.4	2.6	2.9	3.1
TW G2	1.6	1.7	1.9	2.0	2.2	1.9	2.1	2.3	2.6	2.7
TW G3	1.6	1.7	1.9	2.0	2.2	1.9	2.1	2.3	2.5	2.7
TW G4	1.6	1.8	1.9	2.1	2.2	2.0	2.2	2.4	2.5	2.7
TW H1	1.7	1.9	2.0	2.2	2.4	2.1	2.3	2.6	2.8	3.0
TW H2	1.7	1.9	2.0	2.2	2.3	2.1	2.3	2.5	2.7	2.9
TW H3	1.8	1.9	2.1	2.2	2.4	2.2	2.4	2.5	2.7	2.9
TW H4	2.0	2.2	2.3	2.4	2.6	2.4	2.6	2.8	3.0	3.1
TW I1	1.9	2.1	2.3	2.5	2.7	2.6	3.0	3.5	4.2	5.0
TW I2	1.3	1.5	1.6	1.8	1.9	1.6	1.7	1.9	2.1	2.2
TW I3	1.2	1.4	1.5	1.7	1.9	1.5	1.6	1.8	2.0	2.1
TW I4	1.1	1.3	1.4	1.6	1.7	1.3	1.5	1.6	1.8	1.9
TW I5	1.1	1.3	1.4	1.6	1.7	1.3	1.4	1.6	1.7	1.9
TW J1	1.1	1.3	1.5	1.6	1.8	1.3	1.5	1.7	1.8	2.0
TW J2	1.1	1.3	1.4	1.6	1.7	1.3	1.5	1.6	1.8	1.9
TW J3	1.0	1.1	1.3	1.5	1.6	1.1	1.3	1.4	1.6	1.7
TW J4	1.0	1.2	1.3	1.5	1.6	1.2	1.3	1.5	1.6	1.7
DG A1	1.7	2.0	2.4	3.1	4.1	2.3	2.8	3.8	-	-
DG A2	1.5	1.6	1.8	2.1	2.3	1.8	2.0	2.2	2.5	2.9
DG A3	1.3	1.4	1.5	1.7	1.9	1.5	1.6	1.7	1.9	2.1
DG B1	1.7	1.8	2.1	2.5	2.9	2.2	2.8	3.4	4.3	-
DG B2	1.4	1.5	1.7	1.9	2.1	1.7	1.9	2.1	2.3	2.5
DG B3	1.2	1.3	1.4	1.6	1.7	1.4	1.6	1.7	1.8	1.9
DG C1	1.5	1.6	1.9	2.1	2.3	1.9	2.3	2.6	3.0	3.6
DG C2	1.3	1.4	1.6	1.7	1.9	1.5	1.7	1.9	2.1	2.2
DG C3	1.2	1.2	1.4	1.5	1.7	1.3	1.5	1.6	1.7	1.8
SS A1	1.4	1.5	1.7	1.9	2.0	1.6	1.8	2.1	2.3	2.5
SS A2	1.1	1.3	1.4	1.6	1.7	1.3	1.4	1.6	1.8	1.9
SS A3	1.5	1.7	1.9	2.0	2.2	1.8	2.0	2.3	2.6	2.9
SS A4	1.1	1.3	1.5	1.6	1.8	1.3	1.5	1.7	1.8	2.0
SS A5	1.5	1.7	1.9	2.1	2.2	1.8	2.0	2.3	2.6	2.8
SS A6	1.1	1.3	1.4	1.6	1.8	1.3	1.5	1.6	1.8	2.0
SS A7	1.7	1.9	2.1	2.3	2.5	2.1	2.4	2.8	3.2	3.6
SS A8	1.2	1.3	1.5	1.7	1.8	1.3	1.5	1.7	1.9	2.0
SS B1	1.5	1.6	1.8	2.0	2.2	1.7	2.0	2.2	2.5	2.7
SS B2	1.2	1.3	1.5	1.6	1.8	1.3	1.5	1.7	1.8	2.0
SS B3	1.5	1.7	1.9	2.1	2.3	1.9	2.1	2.4	2.7	3.0
SS B4	1.2	1.4	1.5	1.7	1.8	1.4	1.5	1.7	1.9	2.0
SS B5	1.7	1.9	2.0	2.2	2.4	2.0	2.3	2.6	2.9	3.1
SS B6	1.2	1.3	1.5	1.7	1.8	1.3	1.5	1.7	1.9	2.0
SS B7	1.8	2.0	2.3	2.5	2.6	2.2	2.6	2.9	3.4	3.7
SS B8	1.2	1.4	1.5	1.7	1.9	1.4	1.6	1.8	2.0	2.1
SS C1	1.4	1.6	1.7	1.9	2.1	1.8	2.0	2.2	2.5	2.7
SS C2	1.2	1.3	1.5	1.7	1.8	1.4	1.6	1.7	1.9	2.1
SS C3	1.7	1.8	2.0	2.2	2.3	2.1	2.3	2.6	2.8	3.0
SS C4	1.3	1.4	1.6	1.8	1.9	1.5	1.7	1.9	2.0	2.2
SS D1	1.2	1.4	1.6	1.7	1.9	1.4	1.6	1.9	2.1	2.3
SS D2	1.0	1.2	1.3	1.5	1.6	1.1	1.3	1.5	1.6	1.8
SS D3	1.3	1.5	1.7	1.9	2.0	1.5	1.8	2.0	2.3	2.4
SS D4	1.0	1.2	1.3	1.5	1.7	1.2	1.3	1.5	1.7	1.8
SS E1	1.3	1.5	1.7	1.8	2.0	1.6	1.8	2.1	2.3	2.6
SS E2	1.1	1.2	1.4	1.6	1.7	1.2	1.4	1.6	1.8	1.9
SS E3	1.5	1.7	1.9	2.1	2.2	1.8	2.1	2.4	2.6	2.9
SS E4	1.1	1.3	1.5	1.6	1.8	1.3	1.5	1.7	1.8	2.0

ROCHESTER, NEW YORK

TYPE	SOLAR SAVINGS FRACTION (%) BASE TEMP: 55 F (4294 DD) LCR								65 F (6722 DD)								CONSERVATION FACTOR 55 F					65 F				
	100	70	50	40	30	25	20	15	100	70	50	40	30	25	20	15	100	50	30	20	15	100	50	30	20	15
WW A1	12	13	14	15	16	17	17	17	11	11	11	12	12	11	11	9	2.8	3.6	4.8	-	-	4.2	-	-	-	-
WW A2	10	12	14	16	18	20	22	24	7	8	10	11	12	12	13	13	2.0	2.5	2.9	3.5	4.0	2.6	3.3	4.4	-	-
WW A3	8	11	14	16	19	21	23	27	5	7	9	10	12	13	14	15	1.8	2.2	2.6	3.0	3.4	2.3	2.9	3.6	-	-
WW A4	7	10	13	16	19	22	25	29	4	6	8	10	12	13	14	16	1.8	2.1	2.5	2.8	3.1	2.2	2.6	3.3	4.3	-
WW A5	6	10	13	16	20	22	26	30	3	5	8	9	12	13	15	17	1.7	2.0	2.3	2.6	2.9	2.0	2.5	3.1	4.0	-
WW A6	6	9	13	16	20	23	26	31	2	5	7	9	12	13	15	17	1.6	2.0	2.3	2.6	2.8	2.0	2.4	2.9	3.7	4.7
WW B1	6	7	8	8	8	7	6	4	4	4	4	3	1	0	-2	-7	3.3	-	-	-	-	-	-	-	-	-
WW B2	8	12	16	19	24	27	32	37	5	8	11	14	17	19	22	26	1.6	1.8	2.1	2.3	2.5	1.9	2.1	2.5	2.8	3.2
WW B3	10	15	20	24	30	34	39	46	6	10	14	18	22	25	29	34	1.4	1.7	1.9	2.1	2.2	1.6	1.9	2.2	2.5	2.7
WW B4	10	16	23	28	35	40	47	56	6	11	17	22	28	32	38	46	1.3	1.5	1.7	1.8	2.0	1.4	1.6	1.8	2.0	2.2
WW B5	10	16	23	29	36	42	49	58	6	12	18	23	29	34	41	49	1.3	1.4	1.6	1.8	1.9	1.4	1.6	1.7	1.9	2.1
WW C1	12	17	22	27	33	37	43	50	8	12	17	20	25	29	33	39	1.4	1.6	1.8	2.0	2.1	1.5	1.8	2.0	2.3	2.5
WW C2	10	16	22	27	33	38	44	52	7	11	16	20	26	30	35	42	1.3	1.5	1.7	1.9	2.0	1.5	1.7	1.9	2.1	2.3
WW C3	13	21	30	36	45	51	59	69	9	16	24	30	38	44	51	61	1.1	1.3	1.5	1.6	1.8	1.2	1.4	1.6	1.7	1.9
WW C4	12	19	27	34	43	49	56	66	8	14	22	28	36	42	49	58	1.2	1.3	1.5	1.7	1.8	1.3	1.4	1.6	1.8	1.9
TW A1	12	13	15	15	17	17	18	19	11	11	12	12	12	12	12	12	2.7	3.3	4.2	-	-	3.8	-	-	-	-
TW A2	10	12	14	16	18	19	21	24	8	9	10	11	12	13	14	15	2.1	2.5	3.0	3.5	4.0	2.7	3.4	4.3	-	-
TW A3	9	11	14	16	18	20	23	26	6	8	9	11	12	13	15	16	2.0	2.3	2.7	3.1	3.5	2.4	2.9	3.6	4.6	-
TW A4	7	10	13	15	18	20	23	27	5	7	9	10	12	14	15	17	1.9	2.2	2.5	2.9	3.1	2.3	2.7	3.2	3.9	4.7
TW B1	11	13	14	15	17	18	19	21	10	10	11	12	13	13	13	14	2.5	2.9	3.5	4.2	5.0	3.2	4.1	-	-	-
TW B2	9	11	13	15	18	20	22	25	7	8	10	11	13	14	15	16	2.1	2.4	2.8	3.2	3.5	2.5	3.0	3.6	4.5	-
TW B3	3	10	13	15	18	20	22	26	6	7	9	11	13	14	15	17	2.0	2.3	2.7	3.0	3.3	2.4	2.8	3.3	3.9	4.6
TW B4	8	10	13	15	18	20	22	26	6	7	9	11	13	14	16	18	2.0	2.3	2.6	2.9	3.2	2.4	2.8	3.2	3.7	4.2
TW C1	10	12	14	15	17	19	21	23	8	9	11	12	13	14	15	16	2.3	2.6	3.1	3.5	3.9	2.8	3.4	4.0	4.9	-
TW C2	9	11	13	15	17	19	22	25	7	8	10	11	13	14	16	18	2.1	2.4	2.8	3.1	3.4	2.6	3.0	3.4	3.9	4.4
TW C3	9	11	13	15	17	19	22	25	7	8	10	11	13	14	16	18	2.2	2.4	2.8	3.1	3.3	2.6	2.9	3.3	3.8	4.2
TW C4	9	11	13	15	17	19	21	24	8	9	11	12	14	15	16	19	2.3	2.6	2.9	3.2	3.4	2.7	3.1	3.4	3.8	4.1
TW D1	5	6	7	7	8	8	8	7	4	4	4	3	3	2	1	-1	3.1	4.5	-	-	-	-	-	-	-	-
TW D2	9	13	16	19	23	26	30	35	7	9	12	14	17	19	22	25	1.7	2.0	2.2	2.5	2.6	2.0	2.3	2.6	3.0	3.3
TW D3	9	13	17	20	25	28	32	38	6	9	12	15	18	20	23	27	1.6	1.9	2.1	2.3	2.5	1.9	2.1	2.5	2.8	3.2
TW D4	10	15	21	25	32	36	42	50	7	11	16	20	25	29	34	40	1.4	1.6	1.8	2.0	2.1	1.6	1.8	2.0	2.2	2.4
TW D5	10	16	22	27	34	38	45	53	7	12	17	21	27	32	37	45	1.3	1.5	1.7	1.9	2.0	1.5	1.7	1.8	2.0	2.2
TW E1	12	16	21	24	30	33	38	45	8	12	16	19	23	26	30	35	1.5	1.7	1.9	2.1	2.3	1.7	1.9	2.2	2.5	2.7
TW E2	11	16	21	25	31	35	41	48	8	12	16	20	25	28	32	39	1.4	1.6	1.8	2.0	2.2	1.6	1.8	2.0	2.3	2.5
TW E3	14	20	28	33	41	47	54	63	10	16	22	27	34	39	46	54	1.2	1.4	1.6	1.7	1.9	1.3	1.5	1.7	1.9	2.0
TW E4	12	19	26	32	40	46	53	62	9	15	21	26	34	39	45	54	1.2	1.4	1.6	1.7	1.9	1.3	1.5	1.7	1.8	2.0
TW F1	10	11	13	14	15	16	17	18	9	9	10	10	11	11	11	10	2.6	3.2	4.0	-	-	3.6	-	-	-	-
TW F2	8	10	12	13	15	17	19	21	6	7	8	9	10	11	11	12	2.2	2.6	3.1	3.6	4.1	2.8	3.4	4.4	-	-
TW F3	6	8	11	13	15	17	19	23	4	5	7	8	10	11	12	13	2.0	2.4	2.8	3.2	3.5	2.5	3.0	3.7	4.7	-
TW F4	4	6	9	11	14	16	19	23	2	3	5	7	9	10	12	14	2.0	2.3	2.7	3.0	3.3	2.4	2.8	3.3	3.9	4.7
TW G1	8	9	11	12	14	15	16	18	6	7	8	9	9	10	10	11	2.5	2.9	3.5	4.2	5.0	3.2	4.1	-	-	-
TW G2	5	7	10	11	14	15	17	20	3	5	6	7	9	10	11	12	2.2	2.5	3.0	3.4	3.7	2.7	3.2	3.9	4.8	-
TW G3	4	6	8	10	13	15	17	21	2	3	5	6	8	9	11	13	2.1	2.4	2.8	3.2	3.5	2.6	3.0	3.5	4.1	4.8
TW G4	2	4	7	9	11	13	16	19	1	2	4	5	7	8	10	12	2.2	2.5	2.8	3.1	3.4	2.6	3.0	3.4	3.8	4.3
TW H1	5	7	9	10	12	13	15	18	4	5	6	7	8	9	9	10	2.4	2.8	3.2	3.7	4.2	3.0	3.6	4.4	-	-
TW H2	3	5	7	9	11	13	15	18	1	3	4	5	7	8	9	11	2.3	2.6	3.0	3.3	3.7	2.8	3.2	3.7	4.3	4.9
TW H3	2	4	6	7	10	11	14	17	1	2	3	4	6	7	9	11	2.4	2.7	3.0	3.3	3.6	2.8	3.2	3.6	4.1	4.5
TW H4	1	3	5	6	8	10	12	15	0	1	3	4	5	6	8	10	2.6	2.9	3.2	3.5	3.7	3.1	3.4	3.8	4.1	4.5
TW I1	4	4	5	6	6	6	6	5	2	2	2	2	1	0	0	-3	3.3	4.7	-	-	-	-	-	-	-	-
TW I2	7	10	13	16	20	23	27	32	4	6	9	11	14	16	19	23	1.7	2.0	2.3	2.5	2.7	2.0	2.3	2.7	3.0	3.3
TW I3	7	11	15	18	23	26	30	36	4	7	10	13	16	19	22	27	1.6	1.9	2.1	2.3	2.5	1.9	2.1	2.4	2.7	3.0
TW I4	8	13	18	23	29	33	39	47	5	9	14	17	23	26	31	38	1.4	1.6	1.8	2.0	2.1	1.6	1.8	2.0	2.2	2.4
TW I5	8	13	20	24	31	36	42	51	5	10	15	19	25	29	35	42	1.4	1.6	1.7	1.9	2.0	1.5	1.7	1.9	2.1	2.2
TW J1	9	14	19	23	28	32	37	44	6	10	14	17	21	24	28	34	1.5	1.7	1.9	2.1	2.3	1.7	1.9	2.2	2.4	2.7
TW J2	9	14	19	23	29	33	39	46	6	10	14	18	23	26	31	37	1.4	1.6	1.8	2.0	2.2	1.6	1.8	2.0	2.3	2.4
TW J3	11	18	25	31	39	45	52	61	8	14	20	25	33	38	44	53	1.2	1.4	1.6	1.7	1.9	1.3	1.5	1.7	1.8	2.0
TW J4	10	17	24	30	38	43	50	60	7	12	19	24	31	36	43	52	1.2	1.4	1.6	1.7	1.9	1.4	1.5	1.7	1.9	2.0
DG A1	5	6	7	7	7	7	6	4	3	3	3	3	2	1	0	-3	2.9	4.4	-	-	-	4.9	-	-	-	-
DG A2	8	10	13	15	18	19	22	25	6	7	9	11	13	14	15	17	2.0	2.3	2.7	3.2	3.7	2.4	2.8	3.3	4.1	-
DG A3	11	15	19	23	28	31	36	43	9	12	16	18	23	26	30	36	1.6	1.8	1.9	2.1	2.3	1.8	2.0	2.2	2.4	2.6
DG B1	5	6	7	7	7	7	7	6	2	3	3	3	2	1	0	-2	2.8	4.5	-	-	-	-	-	-	-	-
DG B2	8	10	13	15	18	20	23	27	6	7	9	11	13	15	16	18	1.9	2.3	2.6	2.9	3.3	2.4	2.6	3.1	3.8	4.5
DG B3	12	15	20	24	29	33	38	46	9	12	16	19	24	27	32	38	1.6	1.7	1.9	2.0	2.2	1.8	1.9	2.1	2.3	2.4
DG C1	7	8	10	11	12	12	13	13	4	5	5	6	6	6	5	3	2.3	3.2	4.7	-	-	3.4	4.4	-	-	-
DG C2	10	13	16	19	22	25	29	34	7	9	12	14	17	19	22	25	1.7	2.0	2.3	2.5	2.8	2.1	2.3	2.6	3.1	3.5
DG C3	13	18	23	27	33	37	43	52	11	14	19	22	28	31	36	43	1.5	1.6	1.8	1.9	2.0	1.6	1.7	1.9	2.1	2.3
SS A1	14	16	19	21	24	25	28	31	12	14	16	17	19	20	22	23	1.8	2.2	2.6	3.0	3.4	2.1	2.6	3.2	4.1	-
SS A2	17	21	26	30	35	39	44	51	14	18	23	26	30	34	38	43	1.4	1.6	1.8	2.0	2.2	1.5	1.8	2.0	2.3	2.5
SS A3	12	14	16	18	20	21	22	24	11	12	13	14	15	16	16	16	2.1	2.6	3.3	4.2	-	2.5	3.2	4.7	-	-
SS A4	16	20	25	29	34	37	42	48	14	17	21	24	29	31	35	40	1.4	1.7	1.9	2.1	2.3	1.6	1.8	2.1	2.4	2.7
SS A5	14	16	18	20	22	23	25	27	13	14	16	17	18	19	19	20	2.1	2.6	3.1	3.7	4.4	2.5	3.1	4.0	-	-
SS A6	16	21	25	29	34	38	43	49	14	17	22	25	29	32	36	41	1.4	1.7	1.9	2.1	2.2	1.6	1.8	2.1	2.4	2.6
SS A7	13	15	16	17	18	19	19	20	12	13	14	14	14	15	14	14	2.5	3.2	4.3	-	-	3.1	4.3	-	-	-
SS A8	16	20	24	27	32	35	39	45	13	17	20	23	27	29	32	37	1.5	1.8	2.0	2.2	2.4	1.6	1.9	2.3	2.6	2.9
SS B1	11	13	16	17	20	21	24	26	9	11	13	14	16	17	18	19	2.0	2.4	2.9	3.3	3.7	2.4	2.9	3.5	4.6	-
SS B2	14	19	23	27	32	36	41	48	12	16	20	23	28	31	35	41	1.5	1.7	1.9	2.1	2.2	1.6	1.8	2.1	2.3	2.5
SS B3	10	12	14	15	17	18	19	21	8	10	11	12	13	13	14	14	2.2	2.7	3.4	4.2	-	2.6	3.4	4.7	-	-
SS B4	14	18	22	26	31	34	39	45	12	15	19	22	26	29	33	38	1.5	1.7	2.0	2.2	2.3	1.6	1.9	2.2	2.4	2.7
SS B5	12	13	15	16	18	19	20	22	10	11	12	13	14	15	15	16	2.3	2.8	3.4	4.2	-	2.8	3.5	4.6	-	-
SS B6	13	18	22	26	31	34	39	45	11	15	19	22	26	29	33	38	1.5	1.7	1.9	2.1	2.3	1.6	1.9	2.1	2.4	2.6
SS B7	11	12	13	14	15	16	17	17	9	10	11	11	12	12	12	11	2.6	3.3	4.3	-	-	3.4	4.5	-	-	-
SS B8	13	16	21	24	28	32	36	42	11	14	17	20	24	26	30	34	1.6	1.8	2.1	2.3	2.5	1.7	2.0	2.3	2.6	2.9
SS C1	7	9	11	13	15	16	18	20	4	6	7	8	8	9	9	9	2.1	2.6	3.1	3.8	4.5	2.7	3.6	-	-	-
SS C2	9	13	17	20	25	28	33	39	6	9	12	15	18	21	24	28	1.6	1.8	2.1	2.3	2.5	1.8	2.1	2.4	2.7	3.0
SS C3	8	10	12	13	15	16	17	19	7	8	8	9	10	10	11	11	2.4	2.8	3.4	4.1	4.8	3.1	3.9	-	-	-
SS C4	9	12	16	19	23	26	30	35	7	9	12	14	17	19	22	26	1.7	2.0	2.2	2.4	2.6	2.0	2.3	2.6	2.9	3.2
SS D1	14	17	20	23	26	28	31	35	11	14	16	18	20	22	23	25	1.7	2.0	2.4	2.7	3.1	1.9	2.4	3.0	4.0	-
SS D2	17	23	29	34	41	45	51	59	14	19	25	29	35	39	44	50	1.3	1.5	1.7	1.8	2.0	1.3	1.6	1.8	2.0	2.2
SS D3	14	17	20	22	24	26	29	32	12	14	16	18	20	21	22	23	1.8	2.2	2.6	3.1	3.4	2.1	2.6	3.3	4.5	-
SS D4	17	22	28	33	39	44	49	57	14	19	24	28	34	37	42	48	1.3	1.5	1.7	1.9	2.0	1.4	1.6	1.9	2.1	2.3
SS E1	11	13	16	17	20	21	23	25	8	10	11	13	14	14	15	14	1.9	2.4	2.9	3.6	4.3	2.3	3.0	4.4	-	-
SS E2	13	18	24	28	33	37	42	49	10	15	19	22	27	30	34	39	1.4	1.6	1.8	2.0	2.2	1.5	1.8	2.1	2.4	2.6
SS E3	12	14	16	17	19	20	22	24	10	11	13	13	15	15	16	16	2.1	2.6	3.2	3.9	4.7	2.6	3.2	4.4	-	-
SS E4	13	18	22	26	31	35	39	46	11	14	18	21	26	28	32	37	1.5	1.7	1.9	2.1	2.3	1.6	1.9	2.2	2.5	2.7

SYRACUSE, NEW YORK

TYPE	SOLAR SAVINGS FRACTION (%) BASE TEMP: 55 F (4259 DD) LCR								65 F (6681 DD)								CONSERVATION FACTOR 55 F					65 F				
	100	70	50	40	30	25	20	15	100	70	50	40	30	25	20	15	100	50	30	20	15	100	50	30	20	15
WW A1	12	13	14	15	16	17	17	17	11	11	11	12	12	11	11	10	2.8	3.6	4.8	-	-	4.2	-	-	-	-
WW A2	10	12	14	16	18	20	22	25	7	8	10	11	12	13	13	14	2.0	2.5	2.9	3.4	3.8	2.6	3.3	4.3	-	-
WW A3	8	11	14	16	19	21	24	28	5	7	9	10	12	13	14	15	1.8	2.2	2.6	2.9	3.2	2.3	2.8	3.5	4.7	-
WW A4	7	10	13	16	19	22	25	29	4	6	8	10	12	13	15	16	1.8	2.1	2.4	2.7	3.0	2.1	2.6	3.2	4.1	-
WW A5	6	10	13	16	20	23	26	31	3	5	8	10	12	14	15	17	1.7	2.0	2.3	2.6	2.8	2.0	2.5	3.0	3.8	4.7
WW A6	6	9	13	16	20	23	27	32	2	5	7	9	12	14	16	18	1.6	1.9	2.2	2.5	2.7	2.0	2.4	2.9	3.5	4.3
WW B1	6	7	8	8	8	7	6	4	4	4	4	3	2	0	-1	-6	3.3	-	-	-	-	-	-	-	-	-
WW B2	8	12	16	20	24	28	32	38	5	8	11	14	17	20	23	27	1.6	1.8	2.1	2.3	2.5	1.8	2.1	2.4	2.8	3.1
WW B3	10	15	20	24	30	34	39	47	7	10	15	18	22	25	29	35	1.4	1.6	1.8	2.0	2.2	1.6	1.9	2.1	2.4	2.7
WW B4	10	16	23	28	35	41	47	56	6	12	17	22	28	33	38	46	1.3	1.5	1.6	1.8	1.9	1.4	1.6	1.8	2.0	2.2
WW B5	10	16	23	29	37	42	49	58	6	12	18	23	30	35	41	50	1.3	1.4	1.6	1.8	1.9	1.4	1.6	1.7	1.9	2.1
WW C1	12	17	22	27	33	38	43	51	8	12	17	20	26	29	34	40	1.4	1.6	1.8	1.9	2.1	1.5	1.8	2.0	2.2	2.4
WW C2	10	16	22	27	34	38	45	53	7	11	17	21	26	30	36	43	1.3	1.5	1.7	1.9	2.0	1.5	1.7	1.9	2.1	2.3
WW C3	13	21	30	36	46	52	60	69	9	16	24	30	39	44	52	61	1.1	1.3	1.5	1.6	1.8	1.2	1.4	1.5	1.7	1.9
WW C4	12	19	28	34	43	49	57	67	8	14	22	28	36	42	49	59	1.2	1.3	1.5	1.6	1.8	1.3	1.4	1.6	1.7	1.9
TW A1	12	13	15	15	17	17	18	19	11	11	12	12	12	13	12	12	2.7	3.3	4.1	-	-	3.8	-	-	-	-
TW A2	10	12	14	16	18	19	21	24	8	9	10	11	13	13	14	15	2.1	2.5	3.0	3.4	3.8	2.7	3.3	4.2	-	-
TW A3	9	11	14	16	18	20	23	26	6	8	9	11	12	14	15	16	2.0	2.3	2.7	3.0	3.3	2.4	2.9	3.5	4.4	-
TW A4	7	10	13	15	18	21	24	28	5	7	9	10	12	14	15	18	1.9	2.2	2.5	2.8	3.0	2.3	2.7	3.2	3.8	4.4
TW B1	11	13	14	15	17	18	19	21	10	10	11	12	13	13	14	14	2.5	2.9	3.5	4.1	4.7	3.2	4.1	-	-	-
TW B2	9	11	13	15	18	20	22	25	7	8	10	11	13	14	15	17	2.1	2.4	2.8	3.1	3.4	2.5	3.0	3.6	4.3	-
TW B3	8	10	13	15	18	20	23	26	6	7	9	11	13	14	16	18	2.0	2.3	2.6	2.9	3.2	2.4	2.8	3.3	3.9	4.5
TW B4	8	10	13	15	18	20	23	26	6	7	9	11	13	14	16	19	2.0	2.3	2.6	2.9	3.1	2.4	2.8	3.2	3.6	4.1
TW C1	10	12	14	15	17	19	21	23	8	9	11	12	13	14	15	16	2.3	2.6	3.0	3.5	3.8	2.8	3.4	4.0	4.8	-
TW C2	9	11	13	15	17	19	22	25	7	8	10	11	13	14	16	18	2.1	2.4	2.8	3.1	3.3	2.6	2.9	3.4	3.9	4.3
TW C3	9	11	13	15	17	19	22	25	7	8	10	11	13	14	16	19	2.2	2.4	2.8	3.0	3.3	2.6	2.9	3.3	3.7	4.1
TW C4	9	11	13	15	17	19	21	24	8	9	11	12	14	15	17	19	2.3	2.6	2.9	3.1	3.4	2.7	3.1	3.4	3.8	4.1
TW D1	5	6	7	7	8	8	8	7	4	4	4	3	3	2	1	-1	3.1	4.5	-	-	-	-	-	-	-	-
TW D2	9	13	16	19	23	26	30	35	7	9	12	14	17	19	22	26	1.7	2.0	2.2	2.4	2.6	2.0	2.3	2.6	2.9	3.2
TW D3	9	13	17	20	25	28	32	38	6	9	12	15	18	21	24	28	1.6	1.9	2.1	2.3	2.5	1.8	2.1	2.4	2.8	3.1
TW D4	10	15	21	25	32	36	42	50	7	11	16	20	25	29	34	41	1.4	1.6	1.8	1.9	2.1	1.6	1.8	1.9	2.2	2.3
TW D5	10	16	22	27	34	39	45	54	7	12	17	22	28	32	38	45	1.4	1.5	1.7	1.9	2.0	1.5	1.7	1.8	2.0	2.2
TW E1	12	16	21	25	30	34	39	46	8	12	16	19	23	26	30	35	1.5	1.7	1.9	2.1	2.2	1.7	1.9	2.2	2.4	2.7
TW E2	11	16	21	25	31	36	41	49	8	12	16	20	25	28	33	39	1.4	1.6	1.8	2.0	2.1	1.6	1.8	2.0	2.2	2.4
TW E3	14	20	28	33	42	47	54	63	10	16	22	27	35	40	46	55	1.2	1.4	1.5	1.7	1.8	1.3	1.5	1.7	1.8	2.0
TW E4	12	19	27	32	40	46	53	63	9	15	21	26	34	39	45	54	1.2	1.4	1.6	1.7	1.8	1.3	1.5	1.7	1.8	2.0
TW F1	10	11	13	14	15	16	17	18	9	9	10	10	11	11	11	10	2.6	3.2	4.0	5.0	-	3.6	4.9	-	-	-
TW F2	8	10	12	13	16	17	19	21	6	7	8	9	10	11	12	12	2.2	2.6	3.0	3.5	3.9	2.8	3.4	4.3	-	-
TW F3	6	8	11	13	15	17	20	23	4	5	7	8	10	11	12	13	2.0	2.4	2.8	3.1	3.4	2.5	3.0	3.6	4.5	-
TW F4	4	6	9	11	14	17	19	23	2	3	5	7	9	10	12	14	2.0	2.3	2.6	2.9	3.2	2.4	2.8	3.3	3.8	4.5
TW G1	8	9	11	12	14	15	16	18	6	7	8	9	10	10	10	11	2.5	2.9	3.5	4.1	4.7	3.2	4.1	-	-	-
TW G2	5	7	10	11	14	15	18	21	3	5	6	7	9	10	11	12	2.2	2.5	2.9	3.3	3.6	2.7	3.2	3.8	4.7	-
TW G3	4	6	8	10	13	15	17	21	2	3	5	6	8	9	11	13	2.1	2.4	2.8	3.1	3.4	2.6	3.0	3.5	4.0	4.7
TW G4	2	4	7	9	11	13	16	20	1	2	4	5	7	8	10	13	2.2	2.5	2.8	3.1	3.3	2.6	3.0	3.4	3.8	4.2
TW H1	5	7	9	10	12	13	15	18	4	5	6	7	8	9	10	11	2.4	2.8	3.2	3.7	4.1	3.0	3.6	4.3	-	-
TW H2	3	5	7	9	11	13	15	18	1	3	4	5	7	8	9	11	2.3	2.6	3.0	3.3	3.6	2.8	3.2	3.7	4.2	4.8
TW H3	2	4	6	7	10	11	14	17	1	2	3	4	6	7	9	11	2.4	2.7	3.0	3.3	3.5	2.8	3.2	3.6	4.0	4.4
TW H4	1	3	5	6	8	10	12	15	0	1	3	4	5	6	8	10	2.6	2.9	3.2	3.5	3.7	3.1	3.4	3.8	4.1	4.4
TW I1	4	4	5	6	6	6	6	5	2	2	2	2	1	0	0	3	3.3	4.7	-	-	-	-	-	-	-	-
TW I2	7	10	13	16	20	23	27	32	4	6	9	11	14	16	19	23	1.7	2.0	2.2	2.5	2.6	2.0	2.3	2.6	3.0	3.3
TW I3	7	11	15	18	23	26	30	37	4	7	10	13	17	19	22	27	1.6	1.9	2.1	2.3	2.5	1.9	2.1	2.4	2.7	3.0
TW I4	8	13	18	23	29	33	39	47	5	9	14	17	23	26	31	38	1.4	1.6	1.8	2.0	2.1	1.6	1.8	2.0	2.2	2.3
TW I5	8	13	20	24	31	36	42	51	5	10	15	19	25	29	35	43	1.4	1.6	1.7	1.9	2.0	1.5	1.7	1.9	2.0	2.2
TW J1	9	14	19	23	28	32	37	44	6	10	14	17	21	25	29	34	1.5	1.7	1.9	2.1	2.2	1.7	1.9	2.2	2.4	2.6
TW J2	9	14	19	23	29	33	39	47	6	10	14	18	23	26	31	38	1.4	1.6	1.8	2.0	2.1	1.6	1.8	2.0	2.2	2.4
TW J3	11	18	26	31	40	45	52	62	8	14	20	26	33	38	45	54	1.2	1.4	1.5	1.7	1.8	1.3	1.5	1.7	1.8	2.0
TW J4	10	17	24	30	38	44	51	61	7	12	19	24	32	37	43	52	1.2	1.4	1.6	1.7	1.9	1.3	1.5	1.7	1.8	2.0
DG A1	5	6	6	7	7	7	6	5	2	3	3	3	2	1	0	-3	2.9	4.4	-	-	-	4.9	-	-	-	-
DG A2	8	10	13	15	18	19	22	25	6	7	9	11	13	14	16	17	2.0	2.3	2.7	3.1	3.6	2.4	2.7	3.3	4.0	4.9
DG A3	11	15	19	22	28	31	36	43	9	12	16	18	23	26	30	36	1.6	1.8	1.9	2.1	2.3	1.8	2.0	2.2	2.4	2.6
DG B1	5	6	7	7	7	7	7	6	2	3	3	3	2	1	0	-2	2.9	4.6	-	-	-	-	-	-	-	-
DG B2	8	10	13	15	18	20	23	27	6	7	9	11	13	15	16	19	1.9	2.3	2.6	2.9	3.3	2.4	2.6	3.1	3.8	4.4
DG B3	12	15	20	23	29	33	38	46	9	12	16	19	24	28	32	38	1.6	1.8	1.9	2.0	2.1	1.8	1.9	2.1	2.3	2.4
DG C1	6	8	10	11	12	12	13	14	4	5	5	6	6	6	5	4	2.4	3.2	4.5	-	-	3.4	4.4	-	-	-
DG C2	10	13	16	19	22	25	29	34	7	9	12	14	17	19	22	25	1.8	2.0	2.3	2.5	2.7	2.1	2.3	2.6	3.1	3.4
DG C3	13	18	23	27	33	37	44	52	11	14	19	22	28	31	36	44	1.5	1.6	1.8	1.9	2.0	1.6	1.7	1.9	2.1	2.2
SS A1	13	16	19	21	24	25	28	31	12	14	16	17	19	20	22	24	1.9	2.2	2.6	3.0	3.3	2.1	2.6	3.2	4.0	-
SS A2	16	21	26	30	36	39	44	51	14	18	22	26	31	34	38	44	1.4	1.6	1.8	2.0	2.2	1.5	1.8	2.0	2.2	2.4
SS A3	12	14	16	18	19	21	22	24	10	12	13	14	15	16	16	16	2.1	2.6	3.3	4.0	4.8	2.5	3.2	4.6	-	-
SS A4	16	20	25	28	34	37	42	48	14	17	21	24	29	31	35	40	1.5	1.7	1.9	2.1	2.3	1.6	1.8	2.1	2.4	2.6
SS A5	14	16	18	20	22	23	25	27	13	14	15	16	18	19	19	20	2.1	2.6	3.1	3.7	4.2	2.5	3.1	4.0	-	-
SS A6	16	20	25	29	34	38	43	50	14	17	22	25	29	32	36	41	1.4	1.7	1.9	2.1	2.2	1.6	1.8	2.1	2.3	2.6
SS A7	13	14	16	17	18	19	19	20	12	13	13	14	14	15	14	14	2.5	3.2	4.3	-	-	3.1	4.3	-	-	-
SS A8	15	19	24	27	32	35	39	45	13	16	20	23	27	29	33	37	1.5	1.8	2.0	2.2	2.4	1.6	1.9	2.2	2.6	2.9
SS B1	11	13	16	17	20	21	24	26	9	11	13	14	16	17	18	19	2.0	2.4	2.8	3.3	3.6	2.4	2.9	3.5	4.5	-
SS B2	14	18	23	27	32	36	41	48	12	16	20	23	28	31	35	41	1.5	1.7	1.9	2.1	2.2	1.6	1.8	2.1	2.3	2.5
SS B3	10	12	14	15	17	18	19	21	8	10	11	12	13	13	14	14	2.2	2.7	3.3	4.1	4.8	2.6	3.4	4.6	-	-
SS B4	14	18	22	26	31	34	39	45	11	15	19	22	26	29	33	38	1.5	1.7	2.0	2.1	2.3	1.6	1.9	2.1	2.4	2.6
SS B5	12	13	15	16	18	19	20	22	10	11	12	13	14	15	15	16	2.3	2.8	3.4	4.1	4.8	2.8	3.5	4.6	-	-
SS B6	13	17	22	26	31	34	39	46	11	15	19	22	26	29	33	38	1.5	1.7	1.9	2.1	2.3	1.6	1.9	2.1	2.4	2.6
SS B7	11	12	13	14	15	16	16	17	9	10	11	11	12	12	12	12	2.7	3.4	4.3	-	-	3.4	4.5	-	-	-
SS B8	13	16	21	24	28	32	36	42	10	14	17	20	24	26	30	34	1.6	1.8	2.0	2.2	2.4	1.7	2.0	2.3	2.6	2.8
SS C1	7	9	11	13	15	17	18	21	4	6	7	8	9	9	9	9	2.1	2.6	3.1	3.6	4.2	2.7	3.5	5.0	-	-
SS C2	9	13	17	21	25	29	33	40	6	9	12	15	19	21	24	29	1.6	1.8	2.0	2.2	2.4	1.8	2.1	2.4	2.7	2.9
SS C3	8	10	12	13	15	16	17	19	7	8	9	9	10	11	11	11	2.4	2.9	3.4	4.0	4.6	3.1	3.9	-	-	-
SS C4	9	12	16	19	23	26	30	36	7	9	12	14	17	20	22	26	1.7	2.0	2.2	2.4	2.6	2.0	2.3	2.5	2.9	3.1
SS D1	14	17	20	23	26	29	32	36	11	14	16	18	20	22	23	25	1.7	2.0	2.3	2.7	2.9	1.9	2.3	3.0	3.8	-
SS D2	17	23	29	34	41	46	52	60	14	19	25	29	35	39	44	51	1.3	1.5	1.6	1.8	2.0	1.3	1.6	1.8	2.0	2.2
SS D3	14	17	20	22	24	26	29	32	12	14	16	18	20	21	22	23	1.8	2.2	2.6	3.0	3.3	2.1	2.6	3.3	4.3	-
SS D4	17	22	28	33	40	44	50	58	14	19	24	28	34	38	42	49	1.3	1.5	1.7	1.9	2.0	1.4	1.6	1.8	2.1	2.3
SS E1	10	13	15	17	20	21	23	26	8	10	11	13	14	14	15	15	1.9	2.4	2.9	3.4	4.0	2.2	3.0	4.3	-	-
SS E2	13	18	24	28	33	37	43	50	10	14	19	22	27	30	34	40	1.4	1.6	1.8	2.0	2.2	1.5	1.8	2.0	2.3	2.5
SS E3	12	13	15	17	19	20	22	24	10	11	12	13	15	15	16	16	2.1	2.6	3.2	3.8	4.4	2.6	3.2	4.3	-	-
SS E4	13	18	22	26	31	35	40	46	11	14	18	21	26	29	32	37	1.5	1.7	1.9	2.1	2.3	1.6	1.9	2.2	2.4	2.7

ASHEVILLE, NORTH CAROLINA

TYPE	SOLAR SAVINGS FRACTION (%) BASE TEMP: 55 F (2229 DD)								65 F (4247 DD)								CONSERVATION FACTOR 55 F					65 F				
LCR	100	70	50	40	30	25	20	15	100	70	50	40	30	25	20	15	100	50	30	20	15	100	50	30	20	15
WW A1	25	31	37	41	48	53	59	67	18	21	25	29	33	37	41	48	1.3	1.4	1.6	1.7	1.8	1.6	1.8	2.0	2.1	2.3
WW A2	29	37	46	52	61	67	74	82	19	24	31	36	43	48	54	63	1.0	1.2	1.3	1.5	1.6	1.3	1.4	1.6	1.8	1.9
WW A3	31	39	49	56	66	71	78	86	19	26	33	39	47	52	59	68	1.0	1.1	1.3	1.4	1.6	1.2	1.3	1.5	1.7	1.8
WW A4	31	41	51	58	68	74	81	88	19	26	34	40	49	54	61	71	1.0	1.1	1.3	1.4	1.6	1.2	1.3	1.5	1.6	1.8
WW A5	32	42	53	61	70	76	83	90	19	27	35	42	51	57	64	73	0.9	1.1	1.2	1.4	1.5	1.1	1.3	1.4	1.6	1.7
WW A6	33	43	54	62	72	78	84	91	19	27	36	43	52	58	66	75	0.9	1.1	1.2	1.4	1.5	1.1	1.2	1.4	1.6	1.7
WW B1	24	31	38	43	51	57	63	72	15	19	23	27	32	36	41	48	1.2	1.3	1.5	1.6	1.8	1.5	1.7	1.9	2.1	2.2
WW B2	32	42	52	60	70	75	82	89	20	27	36	42	51	57	65	74	0.9	1.1	1.2	1.4	1.5	1.1	1.3	1.4	1.6	1.7
WW B3	38	49	60	68	77	83	88	94	24	33	43	50	60	66	73	82	0.9	1.0	1.2	1.3	1.5	1.0	1.2	1.3	1.5	1.6
WW B4	38	49	61	69	79	84	90	95	24	34	45	53	63	69	77	85	0.8	1.0	1.2	1.3	1.5	1.0	1.1	1.3	1.4	1.6
WW B5	36	48	60	68	78	84	89	95	23	33	44	52	62	69	76	85	0.9	1.0	1.2	1.3	1.5	1.0	1.1	1.3	1.4	1.6
WW C1	39	51	62	70	79	84	90	95	26	35	45	53	62	69	76	84	0.9	1.0	1.2	1.3	1.5	1.0	1.1	1.3	1.5	1.6
WW C2	37	49	60	68	78	83	89	94	24	33	44	51	61	68	75	84	0.9	1.0	1.2	1.3	1.5	1.0	1.1	1.3	1.5	1.6
WW C3	43	56	69	76	85	90	94	97	29	41	53	61	72	78	84	91	0.8	0.9	1.1	1.3	1.4	0.9	1.0	1.2	1.4	1.5
WW C4	40	52	65	73	82	87	92	97	26	37	49	57	68	74	82	89	0.8	1.0	1.1	1.3	1.5	0.9	1.1	1.2	1.4	1.5
TW A1	25	30	36	41	48	52	58	67	18	21	25	28	33	36	41	48	1.3	1.4	1.6	1.7	1.9	1.6	1.8	2.0	2.1	2.3
TW A2	27	34	42	48	56	62	68	77	18	23	28	33	39	44	50	58	1.1	1.3	1.4	1.6	1.7	1.4	1.5	1.7	1.8	2.0
TW A3	27	35	44	50	59	65	72	81	17	23	29	34	42	47	53	62	1.1	1.2	1.4	1.5	1.6	1.3	1.4	1.6	1.8	1.9
TW A4	27	35	44	51	60	66	73	82	16	22	29	35	42	48	54	63	1.0	1.2	1.3	1.5	1.6	1.3	1.4	1.6	1.7	1.9
TW B1	24	30	36	41	49	54	60	69	17	21	25	29	34	38	43	50	1.2	1.4	1.5	1.7	1.8	1.6	1.7	1.9	2.0	2.2
TW B2	25	32	40	46	55	60	67	76	16	21	27	32	38	43	49	57	1.1	1.3	1.4	1.6	1.7	1.4	1.5	1.7	1.8	2.0
TW B3	25	32	41	47	55	61	68	77	16	21	27	32	39	44	50	59	1.1	1.2	1.4	1.5	1.7	1.3	1.5	1.6	1.8	1.9
TW B4	23	30	38	44	53	59	66	75	15	20	26	30	37	42	48	57	1.1	1.3	1.4	1.6	1.7	1.4	1.5	1.7	1.8	2.0
TW C1	23	29	36	41	48	54	60	69	16	20	25	28	34	38	43	51	1.2	1.4	1.5	1.7	1.8	1.5	1.7	1.8	2.0	2.1
TW C2	23	29	36	42	50	55	62	71	15	19	25	29	35	39	45	53	1.2	1.3	1.5	1.6	1.8	1.5	1.6	1.8	1.9	2.0
TW C3	21	27	34	40	48	53	60	69	14	18	23	27	33	37	43	51	1.2	1.4	1.5	1.7	1.8	1.5	1.6	1.8	1.9	2.1
TW C4	20	25	31	36	43	48	54	63	14	17	22	25	30	34	39	46	1.3	1.5	1.6	1.8	1.9	1.6	1.8	1.9	2.1	2.2
TW D1	20	26	33	38	45	50	57	66	12	16	20	23	28	32	36	43	1.2	1.4	1.6	1.7	1.8	1.6	1.8	2.0	2.2	2.3
TW D2	29	38	48	55	64	70	77	85	19	25	33	39	47	52	60	69	1.0	1.1	1.3	1.5	1.6	1.2	1.3	1.5	1.6	1.8
TW D3	32	42	52	59	69	75	82	89	21	28	36	43	51	57	64	74	0.9	1.1	1.2	1.4	1.6	1.1	1.3	1.4	1.6	1.7
TW D4	34	45	56	64	73	79	85	92	22	31	40	47	57	63	71	80	0.9	1.0	1.2	1.4	1.5	1.1	1.2	1.3	1.5	1.6
TW D5	33	44	55	63	73	79	85	92	22	30	40	47	57	64	71	80	0.9	1.0	1.2	1.4	1.5	1.0	1.2	1.3	1.5	1.6
TW E1	36	46	57	64	74	79	86	92	24	32	41	48	57	63	70	79	0.9	1.0	1.2	1.4	1.5	1.1	1.2	1.4	1.5	1.7
TW E2	35	46	57	64	74	80	86	92	23	31	41	48	57	63	71	80	0.9	1.0	1.2	1.4	1.5	1.1	1.2	1.4	1.5	1.6
TW E3	41	54	66	73	82	87	92	96	28	39	50	58	68	74	81	88	0.8	1.0	1.1	1.3	1.5	0.9	1.1	1.2	1.4	1.6
TW E4	38	50	62	70	80	85	90	95	26	36	47	55	65	71	79	87	0.8	1.0	1.2	1.3	1.5	1.0	1.1	1.3	1.4	1.6
TW F1	23	29	35	40	47	52	59	67	16	20	24	27	32	36	41	47	1.3	1.4	1.6	1.7	1.8	1.6	1.8	1.9	2.1	2.2
TW F2	24	31	39	45	54	59	66	75	15	20	26	30	37	41	47	55	1.1	1.3	1.4	1.6	1.7	1.4	1.6	1.7	1.9	2.0
TW F3	24	31	40	46	55	61	69	77	14	20	26	31	38	43	49	58	1.1	1.2	1.4	1.5	1.7	1.3	1.5	1.6	1.8	1.9
TW F4	22	29	38	45	54	60	67	77	12	18	24	29	36	41	48	57	1.1	1.2	1.4	1.5	1.7	1.3	1.5	1.6	1.8	1.9
TW G1	21	27	34	39	46	52	58	67	14	18	22	26	31	35	40	48	1.2	1.4	1.5	1.7	1.8	1.6	1.7	1.9	2.0	2.2
TW G2	21	28	36	41	50	56	63	72	12	17	23	27	33	38	44	52	1.2	1.3	1.5	1.6	1.7	1.4	1.6	1.7	1.9	2.0
TW G3	19	26	34	40	49	54	62	71	11	16	21	26	32	37	43	52	1.2	1.3	1.5	1.6	1.7	1.4	1.6	1.7	1.9	2.0
TW G4	16	22	29	35	43	49	56	66	8	13	18	22	28	33	38	47	1.2	1.4	1.5	1.7	1.8	1.5	1.7	1.8	2.0	2.1
TW H1	18	24	31	36	43	49	55	65	11	15	19	23	29	32	38	45	1.3	1.4	1.6	1.7	1.8	1.6	1.7	1.9	2.0	2.2
TW H2	16	22	28	34	42	47	54	63	9	13	17	21	27	31	37	45	1.3	1.4	1.6	1.7	1.8	1.6	1.7	1.9	2.0	2.2
TW H3	13	19	25	30	37	42	49	58	7	10	15	18	24	28	33	41	1.3	1.5	1.6	1.8	1.9	1.7	1.8	2.0	2.1	2.2
TW H4	10	14	20	24	30	35	41	49	5	8	11	14	19	22	27	33	1.5	1.6	1.8	1.9	2.0	1.9	2.0	2.1	2.3	2.4
TW I1	18	23	30	35	42	47	53	62	10	13	17	20	25	28	33	39	1.3	1.5	1.6	1.8	1.9	1.7	1.9	2.1	2.2	2.4
TW I2	26	35	44	51	61	67	74	83	16	22	30	35	43	49	56	65	1.0	1.2	1.3	1.5	1.6	1.2	1.4	1.5	1.7	1.8
TW I3	28	38	48	55	65	71	78	86	17	24	33	39	47	53	61	70	1.0	1.1	1.3	1.4	1.6	1.2	1.3	1.5	1.6	1.8
TW I4	31	41	52	60	70	76	83	90	19	27	37	44	53	60	67	77	0.9	1.1	1.2	1.4	1.5	1.1	1.2	1.4	1.5	1.7
TW I5	31	41	53	61	71	77	84	91	19	28	38	45	55	61	69	79	0.9	1.1	1.2	1.4	1.5	1.1	1.2	1.4	1.5	1.6
TW J1	33	43	54	62	72	77	84	91	21	29	39	45	55	61	68	77	0.9	1.1	1.2	1.4	1.5	1.1	1.2	1.4	1.5	1.7
TW J2	32	42	53	61	71	77	84	91	20	28	38	45	54	61	68	77	0.9	1.1	1.2	1.4	1.5	1.1	1.2	1.4	1.5	1.7
TW J3	38	50	62	70	80	85	91	95	25	35	47	54	65	71	79	87	0.8	1.0	1.1	1.3	1.5	1.0	1.1	1.3	1.4	1.6
TW J4	35	47	59	68	77	83	89	94	23	33	44	52	62	69	76	85	0.9	1.0	1.2	1.3	1.5	1.0	1.1	1.3	1.4	1.6
DG A1	22	29	36	41	47	51	55	61	13	17	21	24	29	31	34	38	1.2	1.4	1.7	2.0	2.3	1.5	1.8	2.2	2.7	3.3
DG A2	24	32	40	47	55	60	66	73	15	20	26	31	38	42	48	55	1.1	1.2	1.4	1.7	1.9	1.4	1.5	1.7	1.9	2.2
DG A3	28	38	49	56	66	72	78	85	19	26	34	40	49	55	63	71	1.0	1.1	1.3	1.5	1.7	1.2	1.3	1.4	1.6	1.8
DG B1	23	30	38	45	53	58	64	70	13	17	22	26	31	35	40	46	1.1	1.3	1.4	1.7	2.0	1.5	1.7	1.9	2.1	2.5
DG B2	25	33	43	50	60	66	73	81	15	21	27	33	40	46	53	61	1.1	1.2	1.3	1.5	1.7	1.3	1.4	1.6	1.7	2.0
DG B3	29	39	50	59	70	76	83	89	19	26	35	42	52	59	67	76	1.0	1.1	1.2	1.4	1.6	1.2	1.3	1.4	1.5	1.7
DG C1	26	35	45	52	62	68	74	80	15	21	27	32	39	44	50	57	1.0	1.1	1.3	1.6	1.8	1.3	1.5	1.6	1.8	2.1
DG C2	28	38	49	57	68	74	81	87	18	24	32	38	47	53	61	70	1.0	1.1	1.2	1.4	1.6	1.2	1.3	1.4	1.6	1.8
DG C3	32	43	56	66	77	83	89	94	22	30	39	47	58	65	74	83	0.9	1.0	1.1	1.4	1.6	1.1	1.2	1.3	1.4	1.6
SS A1	32	40	48	54	63	68	75	83	24	29	36	41	48	53	59	67	1.0	1.2	1.3	1.5	1.6	1.2	1.4	1.6	1.7	1.9
SS A2	39	49	60	67	76	81	87	92	29	38	46	53	62	67	74	82	0.9	1.0	1.2	1.4	1.5	1.0	1.2	1.3	1.5	1.7
SS A3	30	37	45	51	59	64	71	79	22	27	33	37	44	48	54	61	1.1	1.2	1.4	1.6	1.7	1.3	1.5	1.7	1.8	2.0
SS A4	39	49	59	66	75	80	86	92	29	37	46	52	60	66	72	80	0.9	1.1	1.2	1.4	1.5	1.0	1.2	1.4	1.5	1.7
SS A5	30	37	44	50	58	63	69	77	23	28	33	37	44	48	53	61	1.1	1.3	1.4	1.6	1.7	1.3	1.5	1.7	1.9	2.0
SS A6	39	49	60	67	75	81	86	92	29	37	46	52	61	67	73	81	0.9	1.0	1.2	1.4	1.5	1.0	1.2	1.4	1.5	1.7
SS A7	27	33	39	44	51	56	62	70	21	25	29	33	38	41	46	53	1.2	1.4	1.5	1.7	1.8	1.5	1.7	1.9	2.0	2.2
SS A8	38	48	58	65	74	79	85	91	28	36	44	51	59	64	71	79	0.9	1.1	1.2	1.4	1.5	1.1	1.2	1.4	1.6	1.7
SS B1	28	35	42	48	57	62	69	77	20	25	31	35	42	46	52	60	1.1	1.3	1.4	1.6	1.7	1.3	1.5	1.7	1.8	2.0
SS B2	35	44	54	61	70	76	82	89	25	33	41	47	56	62	68	77	0.9	1.1	1.2	1.4	1.6	1.1	1.2	1.4	1.6	1.7
SS B3	26	33	40	46	54	59	66	74	18	23	29	33	39	43	49	56	1.1	1.3	1.4	1.6	1.7	1.4	1.6	1.7	1.9	2.1
SS B4	34	43	53	60	69	75	81	88	25	32	40	46	55	60	67	76	1.0	1.1	1.3	1.4	1.6	1.1	1.3	1.4	1.6	1.7
SS B5	25	31	37	42	50	54	61	69	18	22	27	31	36	40	45	52	1.2	1.4	1.5	1.7	1.8	1.5	1.7	1.8	2.0	2.2
SS B6	34	43	53	60	70	75	82	89	24	32	40	46	55	60	67	76	0.9	1.1	1.3	1.4	1.6	1.1	1.3	1.4	1.6	1.7
SS B7	23	28	34	38	45	49	55	64	17	20	24	27	32	35	40	46	1.3	1.5	1.6	1.8	1.9	1.6	1.8	2.0	2.2	2.3
SS B8	33	42	52	58	68	73	80	87	24	31	39	44	53	58	65	74	1.0	1.1	1.3	1.4	1.6	1.1	1.3	1.4	1.6	1.7
SS C1	27	35	44	51	60	66	73	81	17	22	29	34	41	46	52	61	1.0	1.2	1.3	1.5	1.6	1.3	1.4	1.6	1.8	1.9
SS C2	32	42	52	60	69	75	82	89	20	28	36	43	52	58	65	74	0.9	1.1	1.2	1.4	1.5	1.1	1.3	1.4	1.6	1.7
SS C3	23	29	36	41	49	54	61	70	15	19	24	27	33	37	42	50	1.2	1.4	1.5	1.7	1.8	1.5	1.7	1.8	2.0	2.1
SS C4	29	37	47	54	63	69	76	84	18	25	32	38	46	52	59	68	1.0	1.2	1.3	1.5	1.6	1.2	1.4	1.5	1.7	1.8
SS D1	38	47	57	64	73	78	84	90	27	35	43	48	56	62	68	76	0.9	1.1	1.2	1.4	1.6	1.1	1.3	1.4	1.6	1.7
SS D2	46	57	68	75	84	88	93	96	34	43	54	61	70	76	82	89	0.8	1.0	1.1	1.3	1.5	0.9	1.1	1.3	1.4	1.6
SS D3	34	42	51	58	66	72	78	85	25	31	38	43	51	56	62	70	1.0	1.1	1.3	1.5	1.6	1.2	1.4	1.5	1.7	1.8
SS D4	44	55	66	73	82	86	91	96	32	42	52	59	68	74	80	87	0.8	1.0	1.2	1.3	1.5	0.9	1.1	1.3	1.4	1.6
SS E1	33	42	51	58	67	72	79	86	23	29	36	42	49	54	61	69	1.0	1.1	1.3	1.5	1.6	1.2	1.3	1.5	1.7	1.8
SS E2	40	51	62	69	78	83	89	94	28	37	47	54	63	69	76	84	0.9	1.0	1.2	1.3	1.5	1.0	1.2	1.3	1.5	1.6
SS E3	28	35	43	49	57	62	69	77	20	25	31	35	41	46	51	59	1.1	1.3	1.4	1.6	1.7	1.3	1.5	1.7	1.9	2.0
SS E4	37	47	57	65	74	79	85	91	26	34	43	50	58	64	71	79	0.9	1.1	1.2	1.4	1.5	1.0	1.2	1.4	1.5	1.7

CAPE HATTERAS, NORTH CAROLINA

SOLAR SAVINGS FRACTION (%)

TYPE	BASE TEMP: 55 F (1166 DD) LCR 200	100	70	50	40	30	25	20	65 F (2742 DD) 200	100	70	50	40	30	25	20
WW A1	23	34	41	50	56	64	70	77	16	23	27	33	37	43	48	53
WW A2	25	41	51	62	69	78	83	89	15	26	33	41	47	55	60	67
WW A3	26	44	55	66	74	82	87	92	15	27	35	44	50	59	65	72
WW A4	26	45	57	68	76	85	89	93	14	27	36	45	52	62	67	74
WW A5	27	47	59	71	78	87	91	95	14	28	37	47	54	64	70	77
WW A6	27	48	60	72	79	88	92	95	14	28	38	48	56	65	71	78
WW B1	22	35	44	54	61	70	76	82	13	21	27	33	38	45	50	56
WW B2	26	45	57	69	77	85	90	94	14	28	37	47	54	63	69	76
WW B3	31	53	65	77	84	91	94	97	17	33	44	55	62	72	77	84
WW B4	30	53	66	78	84	91	95	97	16	33	44	56	64	74	80	86
WW B5	28	51	64	76	83	91	94	97	14	32	43	55	63	73	79	85
WW C1	32	54	67	78	85	92	95	97	18	35	46	57	64	74	79	86
WW C2	30	52	65	77	83	91	94	97	16	33	43	55	63	72	78	85
WW C3	34	59	72	83	89	95	97	99	18	38	51	63	71	81	86	91
WW C4	30	55	68	80	87	93	96	98	16	35	47	60	68	78	83	89
TW A1	22	33	40	48	55	63	69	75	16	22	27	32	37	43	47	53
TW A2	23	37	46	57	64	73	78	85	15	24	30	37	43	50	56	62
TW A3	23	39	49	60	67	76	82	88	14	24	31	39	45	53	59	66
TW A4	22	38	49	60	68	77	83	88	13	23	31	39	45	54	60	67
TW B1	22	33	40	49	56	64	70	77	15	22	27	33	37	44	48	54
TW B2	22	35	45	55	62	71	77	84	13	22	29	36	41	49	54	61
TW B3	21	35	45	55	63	72	78	84	12	22	28	36	42	50	55	62
TW B4	19	33	42	53	60	69	75	82	12	20	27	34	40	48	53	60
TW C1	20	31	40	48	55	64	70	77	13	21	26	32	37	44	48	55
TW C2	19	31	40	49	56	66	72	79	12	20	26	32	37	45	50	56
TW C3	18	30	38	47	54	63	69	76	12	19	24	31	36	43	48	54
TW C4	17	27	34	42	48	57	62	70	12	18	22	28	32	39	43	49
TW D1	18	30	38	47	54	63	69	76	11	18	23	29	33	40	44	50
TW D2	24	41	52	64	71	80	85	91	14	26	34	43	49	58	64	71
TW D3	27	45	57	69	76	85	89	93	15	28	37	47	54	63	69	76
TW D4	27	48	60	72	79	87	91	95	15	30	40	51	58	68	74	81
TW D5	26	46	59	71	78	87	91	95	14	29	39	50	58	68	74	81
TW E1	30	50	62	73	80	88	92	95	18	32	41	52	59	68	74	81
TW E2	29	49	61	73	80	88	92	95	16	31	41	51	59	68	74	81
TW E3	33	56	69	81	87	93	96	98	19	37	49	60	68	78	83	89
TW E4	30	53	66	78	85	92	95	97	17	34	46	57	65	75	81	87
TW F1	21	32	39	48	55	63	69	76	14	21	26	31	36	42	47	52
TW F2	21	35	44	54	61	71	77	83	12	21	27	35	40	48	53	60
TW F3	20	35	45	56	63	73	79	85	11	21	27	35	41	49	55	62
TW F4	18	33	43	54	61	71	77	84	9	19	25	33	39	48	54	61
TW G1	19	30	38	47	54	63	69	76	12	19	24	30	35	41	46	52
TW G2	17	31	40	50	57	67	73	80	10	18	24	31	37	44	50	56
TW G3	16	29	38	48	56	66	72	79	8	16	22	30	35	43	48	55
TW G4	12	25	33	43	50	60	66	73	6	13	19	26	31	38	43	50
TW H1	15	27	35	44	50	60	66	73	9	16	21	27	32	38	43	50
TW H2	13	24	32	41	48	58	64	71	6	13	18	25	30	37	42	48
TW H3	10	21	28	37	43	52	58	66	4	11	16	22	26	33	37	44
TW H4	8	16	22	29	35	43	49	56	3	8	12	17	21	26	30	36
TW I1	16	27	35	44	50	60	66	73	9	15	20	26	30	36	41	40
TW I2	21	38	49	60	68	77	83	89	11	23	30	39	46	55	61	68
TW I3	23	41	52	64	72	81	86	91	12	25	33	43	50	59	65	72
TW I4	24	44	56	68	76	85	89	94	12	27	36	47	55	64	71	78
TW I5	24	44	57	69	77	85	90	94	12	27	37	48	55	66	72	79
TW J1	27	47	59	71	78	86	91	95	15	29	39	49	56	66	72	79
TW J2	25	45	58	70	77	86	90	94	14	28	38	48	56	66	72	79
TW J3	29	53	66	78	85	92	95	97	16	34	45	57	65	75	81	87
TW J4	27	50	63	75	82	90	94	97	14	31	42	54	62	72	78	85
DG A1	20	34	43	52	58	68	70	75	11	19	25	31	35	41	44	48
DG A2	20	35	46	56	63	72	77	82	11	21	28	36	41	49	54	60
DG A3	22	40	53	65	72	81	85	89	14	25	34	43	51	60	66	73
DG B1	20	35	46	58	66	74	79	83	11	19	25	33	38	46	51	56
DG B2	20	36	49	62	70	79	84	89	12	22	29	37	44	53	60	67
DG B3	23	41	55	69	77	86	90	93	14	26	34	45	53	64	71	78
DG C1	23	40	54	67	75	83	87	90	13	23	30	39	45	54	60	66
DG C2	23	41	55	69	77	85	90	93	14	25	33	43	50	61	67	74
DG C3	26	46	61	76	83	91	94	96	16	29	39	50	59	71	77	84
SS A1	28	43	53	63	70	79	84	89	19	30	37	45	51	59	64	70
SS A2	33	53	64	75	82	89	92	96	23	37	46	56	63	72	77	83
SS A3	26	41	50	60	67	76	81	86	18	28	35	42	47	55	60	66
SS A4	33	52	64	74	81	88	92	95	23	37	46	55	62	71	76	82
SS A5	27	40	49	59	65	74	79	85	20	28	35	41	47	54	59	65
SS A6	33	53	64	75	81	89	92	96	22	37	46	56	63	71	77	83
SS A7	24	36	44	53	59	68	73	79	18	26	31	37	41	48	52	58
SS A8	33	52	63	74	80	88	91	95	22	36	45	54	61	70	75	81
SS B1	24	38	47	57	64	73	79	85	16	25	32	39	45	52	57	64
SS B2	29	47	58	69	76	85	89	93	19	32	41	50	57	66	72	78
SS B3	23	36	45	55	62	71	76	83	15	24	30	37	42	50	55	61
SS B4	28	46	57	68	76	84	88	93	19	32	40	50	56	65	71	77
SS B5	22	33	41	50	57	65	71	78	16	23	29	35	39	46	50	56
SS B6	28	46	58	69	76	84	89	93	18	31	40	50	56	65	71	77
SS B7	20	31	38	46	52	61	66	73	15	21	26	31	35	41	46	51
SS B8	27	45	56	67	74	83	87	92	18	30	39	48	55	63	69	76
SS C1	23	39	50	61	68	78	83	89	13	24	31	39	45	53	59	66
SS C2	26	45	57	69	76	85	89	94	15	28	37	47	54	64	69	76
SS C3	20	32	40	49	56	65	71	78	12	20	25	32	36	43	48	55
SS C4	24	40	51	62	70	79	84	90	14	25	33	42	48	57	63	70
SS D1	33	52	63	74	80	88	91	95	22	35	44	53	60	68	73	80
SS D2	38	61	73	83	89	94	96	98	25	42	53	64	71	80	85	90
SS D3	30	46	57	67	74	82	87	91	21	32	40	48	54	62	67	73
SS D4	37	59	71	81	87	93	96	98	24	41	51	62	69	78	83	88
SS E1	29	46	57	68	75	84	88	92	18	30	38	47	53	62	67	74
SS E2	33	55	67	78	84	91	94	97	21	37	47	58	65	74	79	85
SS E3	25	38	48	58	65	73	79	85	17	26	32	39	45	52	57	64
SS E4	31	50	62	73	80	88	92	95	20	34	43	53	60	69	75	81

CONSERVATION FACTOR

TYPE	55 F 200	100	50	30	20	65 F 200	100	50	30	20
WW A1	0.9	1.0	1.2	1.3	1.5	1.2	1.4	1.5	1.7	1.8
WW A2	0.8	0.9	1.0	1.2	1.3	1.0	1.1	1.3	1.4	1.6
WW A3	0.7	0.8	1.0	1.1	1.3	0.9	1.0	1.2	1.4	1.5
WW A4	0.7	0.8	1.0	1.1	1.3	0.9	1.0	1.2	1.3	1.5
WW A5	0.7	0.8	0.9	1.1	1.3	0.9	1.0	1.1	1.3	1.5
WW A6	0.7	0.8	0.9	1.1	1.3	0.9	1.0	1.1	1.3	1.4
WW B1	0.9	1.0	1.1	1.2	1.4	1.1	1.3	1.4	1.6	1.8
WW B2	0.7	0.8	1.0	1.1	1.3	0.9	1.0	1.1	1.3	1.5
WW B3	0.7	0.8	0.9	1.1	1.3	0.8	0.9	1.1	1.2	1.4
WW B4	0.6	0.7	0.9	1.1	1.2	0.8	0.9	1.0	1.2	1.4
WW B5	0.7	0.8	0.9	1.1	1.3	0.8	0.9	1.0	1.2	1.4
WW C1	0.6	0.7	0.9	1.1	1.2	0.8	0.9	1.0	1.2	1.4
WW C2	0.7	0.8	0.9	1.1	1.3	0.8	0.9	1.1	1.2	1.4
WW C3	0.6	0.7	0.9	1.0	1.2	0.7	0.8	1.0	1.1	1.3
WW C4	0.6	0.7	0.9	1.1	1.2	0.7	0.8	1.0	1.2	1.3
TW A1	1.0	1.1	1.2	1.3	1.5	1.2	1.4	1.5	1.7	1.8
TW A2	0.8	0.9	1.1	1.2	1.4	1.1	1.2	1.3	1.5	1.6
TW A3	0.8	0.9	1.0	1.2	1.4	1.0	1.1	1.3	1.4	1.6
TW A4	0.8	0.9	1.0	1.2	1.3	1.0	1.1	1.3	1.4	1.6
TW B1	0.9	1.0	1.2	1.3	1.5	1.2	1.3	1.5	1.6	1.8
TW B2	0.8	0.9	1.1	1.2	1.4	1.1	1.2	1.3	1.5	1.6
TW B3	0.8	0.9	1.1	1.2	1.4	1.1	1.2	1.3	1.5	1.6
TW B4	0.9	1.0	1.1	1.2	1.4	1.1	1.2	1.4	1.5	1.7
TW C1	0.9	1.0	1.2	1.3	1.5	1.2	1.3	1.5	1.6	1.8
TW C2	0.9	1.0	1.1	1.3	1.4	1.2	1.3	1.4	1.6	1.7
TW C3	0.9	1.0	1.2	1.3	1.5	1.2	1.3	1.4	1.6	1.7
TW C4	1.0	1.1	1.2	1.4	1.5	1.3	1.4	1.5	1.7	1.8
TW D1	0.9	1.0	1.2	1.3	1.5	1.2	1.3	1.5	1.7	1.8
TW D2	0.8	0.9	1.0	1.2	1.3	0.9	1.1	1.2	1.4	1.5
TW D3	0.7	0.8	1.0	1.1	1.3	0.9	1.0	1.1	1.3	1.5
TW D4	0.7	0.8	0.9	1.1	1.3	0.8	0.9	1.1	1.3	1.4
TW D5	0.7	0.8	0.9	1.1	1.3	0.8	0.9	1.1	1.2	1.4
TW E1	0.7	0.8	0.9	1.1	1.3	0.8	0.9	1.1	1.3	1.4
TW E2	0.7	0.8	0.9	1.1	1.3	0.8	0.9	1.1	1.3	1.4
TW E3	0.6	0.7	0.9	1.1	1.2	0.8	0.9	1.0	1.2	1.3
TW E4	0.6	0.7	0.9	1.1	1.2	0.8	0.9	1.0	1.2	1.4
TW F1	0.9	1.0	1.2	1.3	1.5	1.2	1.3	1.5	1.7	1.8
TW F2	0.8	0.9	1.1	1.2	1.4	1.1	1.2	1.4	1.5	1.7
TW F3	0.8	0.9	1.1	1.2	1.4	1.0	1.2	1.3	1.5	1.6
TW F4	0.8	0.9	1.1	1.2	1.4	1.1	1.2	1.3	1.5	1.6
TW G1	0.9	1.0	1.2	1.3	1.5	1.2	1.3	1.5	1.6	1.8
TW G2	0.9	1.0	1.1	1.3	1.4	1.1	1.2	1.4	1.5	1.7
TW G3	0.9	1.0	1.1	1.3	1.4	1.1	1.2	1.4	1.5	1.7
TW G4	0.9	1.0	1.2	1.3	1.5	1.2	1.3	1.5	1.6	1.8
TW H1	1.0	1.1	1.2	1.3	1.5	1.2	1.3	1.5	1.7	1.8
TW H2	1.0	1.1	1.2	1.3	1.5	1.2	1.3	1.5	1.7	1.8
TW H3	1.0	1.1	1.3	1.4	1.5	1.3	1.4	1.6	1.7	1.9
TW H4	1.2	1.3	1.4	1.5	1.7	1.5	1.6	1.7	1.9	2.0
TW I1	1.0	1.1	1.2	1.3	1.5	1.3	1.4	1.6	1.7	1.9
TW I2	0.8	0.9	1.0	1.2	1.3	1.0	1.1	1.2	1.4	1.5
TW I3	0.7	0.8	1.0	1.1	1.3	0.9	1.0	1.2	1.3	1.5
TW I4	0.7	0.8	1.0	1.1	1.3	0.9	1.0	1.1	1.3	1.4
TW I5	0.7	0.8	0.9	1.1	1.3	0.9	1.0	1.1	1.3	1.4
TW J1	0.7	0.8	0.9	1.1	1.3	0.9	1.0	1.1	1.3	1.4
TW J2	0.7	0.8	0.9	1.1	1.3	0.9	1.0	1.1	1.3	1.4
TW J3	0.6	0.7	0.9	1.1	1.2	0.8	0.9	1.0	1.2	1.4
TW J4	0.7	0.8	0.9	1.1	1.3	0.8	0.9	1.0	1.2	1.4
DG A1	0.8	1.0	1.1	1.4	1.6	1.1	1.2	1.5	1.8	2.2
DG A2	0.8	0.9	1.1	1.3	1.5	1.1	1.2	1.3	1.5	1.7
DG A3	0.8	0.8	1.0	1.2	1.4	1.0	1.0	1.2	1.3	1.5
DG B1	0.8	0.9	1.0	1.3	1.5	1.1	1.2	1.4	1.6	1.9
DG B2	0.8	0.9	1.0	1.2	1.4	1.1	1.1	1.2	1.4	1.6
DG B3	0.8	0.8	0.9	1.1	1.3	1.0	1.0	1.1	1.2	1.4
DG C1	0.8	0.8	0.9	1.2	1.4	1.0	1.1	1.3	1.4	1.6
DG C2	0.8	0.8	0.9	1.1	1.3	1.0	1.0	1.2	1.3	1.5
DG C3	0.7	0.8	0.9	1.1	1.3	0.9	1.0	1.1	1.2	1.4
SS A1	0.8	0.9	1.0	1.2	1.4	1.0	1.1	1.2	1.4	1.6
SS A2	0.7	0.8	0.9	1.1	1.3	0.8	0.9	1.1	1.2	1.4
SS A3	0.8	0.9	1.1	1.2	1.4	1.0	1.1	1.3	1.5	1.6
SS A4	0.7	0.8	0.9	1.1	1.3	0.8	0.9	1.1	1.3	1.4
SS A5	0.8	0.9	1.1	1.2	1.4	1.0	1.2	1.3	1.5	1.7
SS A6	0.7	0.8	0.9	1.1	1.3	0.8	0.9	1.1	1.3	1.4
SS A7	0.9	1.0	1.1	1.3	1.5	1.1	1.3	1.4	1.6	1.8
SS A8	0.7	0.8	0.9	1.1	1.3	0.8	0.9	1.1	1.3	1.4
SS B1	0.8	0.9	1.1	1.2	1.4	1.0	1.2	1.3	1.5	1.6
SS B2	0.7	0.8	1.0	1.1	1.3	0.9	1.0	1.1	1.3	1.5
SS B3	0.8	0.9	1.1	1.2	1.4	1.1	1.2	1.4	1.5	1.7
SS B4	0.7	0.8	1.0	1.1	1.3	0.9	1.0	1.1	1.3	1.5
SS B5	0.9	1.0	1.2	1.3	1.5	1.2	1.3	1.4	1.6	1.8
SS B6	0.7	0.8	1.0	1.1	1.3	0.9	1.0	1.1	1.3	1.5
SS B7	1.0	1.1	1.2	1.4	1.5	1.2	1.4	1.5	1.7	1.9
SS B8	0.7	0.8	1.0	1.1	1.3	0.9	1.0	1.2	1.3	1.5
SS C1	0.8	0.9	1.0	1.2	1.3	1.0	1.1	1.3	1.4	1.6
SS C2	0.7	0.8	1.0	1.1	1.3	0.9	1.0	1.1	1.3	1.5
SS C3	0.9	1.0	1.1	1.3	1.4	1.2	1.3	1.4	1.6	1.8
SS C4	0.8	0.9	1.0	1.2	1.3	1.0	1.1	1.2	1.4	1.5
SS D1	0.7	0.8	0.9	1.1	1.3	0.8	1.0	1.1	1.3	1.5
SS D2	0.6	0.7	0.9	1.1	1.2	0.7	0.8	1.0	1.2	1.3
SS D3	0.7	0.8	1.0	1.2	1.3	0.9	1.0	1.2	1.4	1.5
SS D4	0.6	0.7	0.9	1.1	1.2	0.8	0.9	1.0	1.2	1.4
SS E1	0.7	0.8	1.0	1.1	1.3	0.9	1.0	1.2	1.4	1.5
SS E2	0.6	0.7	0.9	1.1	1.3	0.8	0.9	1.1	1.2	1.4
SS E3	0.8	0.9	1.1	1.2	1.4	1.0	1.2	1.3	1.5	1.7
SS E4	0.7	0.8	0.9	1.1	1.3	0.8	0.9	1.1	1.3	1.4

CHARLOTTE, NORTH CAROLINA

TYPE	SSF 55 F (1552 DD) LCR 200	100	70	50	40	30	25	20	SSF 65 F (3226 DD) LCR 200	100	70	50	40	30	25	20	CF 55 F 200	100	50	30	20	CF 65 F 200	100	50	30	20
WW A1	20	29	36	43	49	57	62	69	15	21	25	30	33	39	43	48	1.0	1.1	1.3	1.4	1.6	1.4	1.5	1.6	1.8	2.0
WW A2	21	35	44	54	61	71	76	82	14	23	29	36	42	50	55	62	0.9	0.9	1.1	1.2	1.4	1.1	1.2	1.3	1.5	1.7
WW A3	22	37	48	58	66	75	81	86	13	23	31	39	45	54	59	66	0.8	0.9	1.0	1.2	1.4	1.0	1.1	1.3	1.4	1.6
WW A4	21	38	49	61	68	77	83	88	12	23	31	40	47	56	62	69	0.8	0.9	1.0	1.2	1.3	1.0	1.1	1.2	1.4	1.5
WW A5	22	40	51	63	71	80	85	90	11	24	33	42	49	58	64	72	0.7	0.8	1.0	1.2	1.3	0.9	1.0	1.2	1.3	1.5
WW A6	22	41	52	64	72	81	86	91	11	24	33	43	50	59	66	73	0.7	0.8	1.0	1.2	1.3	0.9	1.0	1.2	1.3	1.5
WW B1	18	30	38	46	53	62	67	74	11	18	23	29	33	39	44	49	0.9	1.0	1.2	1.3	1.5	1.2	1.4	1.5	1.7	1.9
WW B2	21	39	50	62	69	78	84	89	12	24	33	42	49	58	64	71	0.8	0.9	1.0	1.2	1.3	0.9	1.0	1.2	1.3	1.5
WW B3	26	46	58	69	77	85	89	94	15	29	39	49	57	67	73	79	0.7	0.8	1.0	1.1	1.3	0.9	1.0	1.1	1.3	1.4
WW B4	24	46	58	70	78	86	90	94	13	29	40	51	59	69	75	82	0.7	0.8	0.9	1.1	1.3	0.8	0.9	1.1	1.2	1.4
WW B5	22	44	57	69	77	85	90	94	12	28	38	50	58	68	74	82	0.7	0.8	0.9	1.1	1.3	0.8	0.9	1.1	1.2	1.4
WW C1	27	47	60	71	78	86	91	95	16	31	41	52	59	69	75	81	0.7	0.8	0.9	1.1	1.3	0.8	0.9	1.1	1.2	1.4
WW C2	24	45	57	69	77	85	90	94	13	29	39	50	58	67	74	81	0.7	0.8	1.0	1.1	1.3	0.8	0.9	1.1	1.2	1.4
WW C3	28	51	65	77	84	91	94	97	15	34	46	59	67	77	82	88	0.6	0.7	0.9	1.1	1.2	0.8	0.9	1.0	1.2	1.3
WW C4	25	48	61	73	81	89	92	96	13	31	43	55	63	73	79	86	0.7	0.8	0.9	1.1	1.3	0.8	0.9	1.0	1.2	1.4
TW A1	20	29	35	42	48	56	61	68	15	20	24	29	33	38	42	47	1.1	1.1	1.3	1.4	1.6	1.4	1.5	1.6	1.8	2.0
TW A2	20	32	40	50	56	65	71	78	13	21	27	33	38	45	50	57	0.9	1.0	1.1	1.3	1.5	1.2	1.3	1.4	1.6	1.7
TW A3	20	33	42	52	60	69	75	81	12	21	27	35	40	48	54	60	0.9	1.0	1.1	1.3	1.4	1.1	1.2	1.3	1.5	1.7
TW A4	18	33	42	53	60	70	75	82	11	20	27	35	41	49	54	61	0.9	0.9	1.1	1.2	1.4	1.1	1.2	1.3	1.5	1.6
TW B1	19	28	35	43	49	57	63	69	14	20	24	29	33	39	43	49	1.0	1.1	1.2	1.4	1.6	1.3	1.4	1.6	1.7	1.9
TW B2	18	30	39	48	55	64	70	77	12	20	25	32	37	44	49	56	0.9	1.0	1.2	1.3	1.5	1.2	1.3	1.4	1.6	1.7
TW B3	18	30	39	48	55	65	71	77	11	19	25	32	37	45	50	57	0.9	1.0	1.1	1.3	1.5	1.1	1.2	1.4	1.5	1.7
TW B4	16	28	36	46	53	62	68	75	10	18	24	30	35	43	48	55	0.9	1.0	1.2	1.3	1.5	1.2	1.3	1.4	1.6	1.7
TW C1	18	27	34	42	48	57	63	69	12	18	23	29	33	39	44	49	1.0	1.1	1.2	1.4	1.5	1.3	1.4	1.5	1.7	1.9
TW C2	16	27	34	43	50	58	64	71	11	18	23	29	33	40	45	51	1.0	1.1	1.2	1.4	1.5	1.2	1.3	1.5	1.6	1.8
TW C3	16	25	33	41	47	56	62	69	10	17	22	27	32	38	43	49	1.0	1.1	1.2	1.4	1.5	1.3	1.4	1.5	1.7	1.8
TW C4	15	23	29	37	42	50	56	63	11	16	20	25	29	35	39	44	1.1	1.2	1.3	1.5	1.6	1.4	1.5	1.6	1.8	1.9
TW D1	15	25	32	40	46	55	61	68	9	15	19	25	29	34	38	44	1.0	1.1	1.3	1.4	1.6	1.3	1.5	1.6	1.8	2.0
TW D2	20	35	46	56	64	73	79	85	12	23	30	38	45	53	59	66	0.8	0.9	1.1	1.2	1.4	1.0	1.1	1.3	1.4	1.6
TW D3	22	39	50	61	69	78	83	89	13	25	33	42	49	58	64	71	0.8	0.9	1.0	1.2	1.3	0.9	1.0	1.2	1.4	1.5
TW D4	22	41	53	65	72	81	86	91	13	26	36	46	53	63	69	76	0.7	0.8	1.0	1.1	1.3	0.9	1.0	1.1	1.3	1.5
TW D5	21	40	52	64	72	81	86	91	12	26	35	46	53	63	69	77	0.7	0.8	1.0	1.1	1.3	0.9	1.0	1.1	1.3	1.4
TW E1	25	43	54	66	73	82	87	92	15	28	37	47	54	63	69	76	0.7	0.8	1.0	1.1	1.3	0.9	1.0	1.1	1.3	1.5
TW E2	24	42	54	65	73	82	87	92	14	27	37	47	54	63	69	77	0.7	0.8	1.0	1.1	1.3	0.9	1.0	1.1	1.3	1.5
TW E3	27	49	62	74	81	89	92	96	16	33	44	55	63	73	79	85	0.7	0.8	0.9	1.1	1.3	0.8	0.9	1.0	1.2	1.4
TW E4	25	46	59	71	78	87	91	95	14	30	41	53	61	71	77	83	0.7	0.8	0.9	1.1	1.3	0.8	0.9	1.1	1.2	1.4
TW F1	18	27	34	42	48	56	61	68	13	19	23	28	32	38	42	47	1.0	1.1	1.3	1.4	1.6	1.3	1.4	1.6	1.8	1.9
TW F2	17	29	38	47	54	63	69	76	11	19	24	31	36	43	48	54	0.9	1.0	1.2	1.3	1.5	1.2	1.3	1.4	1.6	1.7
TW F3	16	29	38	48	56	65	71	78	9	18	24	31	36	44	50	56	0.9	1.0	1.1	1.3	1.4	1.1	1.2	1.4	1.5	1.7
TW F4	14	27	36	46	54	64	70	77	7	16	22	29	35	43	48	55	0.9	1.0	1.1	1.3	1.5	1.1	1.2	1.4	1.5	1.7
TW G1	16	26	33	41	47	55	61	68	11	17	21	27	31	37	41	47	1.0	1.1	1.3	1.4	1.6	1.3	1.4	1.6	1.7	1.9
TW G2	14	26	34	43	50	59	65	72	8	15	21	27	32	39	44	51	1.0	1.1	1.2	1.3	1.5	1.2	1.3	1.5	1.6	1.8
TW G3	12	24	32	42	48	58	64	71	6	14	19	26	31	38	43	50	1.0	1.1	1.2	1.3	1.5	1.2	1.3	1.5	1.6	1.8
TW G4	10	20	28	36	43	52	58	66	4	11	16	22	27	34	38	45	1.0	1.1	1.3	1.4	1.6	1.3	1.4	1.5	1.7	1.8
TW H1	13	22	29	37	43	52	58	65	7	14	18	23	28	34	38	44	1.1	1.1	1.3	1.4	1.6	1.3	1.4	1.6	1.8	1.9
TW H2	10	20	27	35	41	50	56	64	5	11	16	21	26	32	37	43	1.1	1.2	1.3	1.4	1.6	1.3	1.4	1.6	1.7	1.9
TW H3	8	17	23	31	37	45	51	58	3	9	13	18	23	29	33	39	1.1	1.2	1.3	1.5	1.6	1.4	1.5	1.7	1.8	2.0
TW H4	6	13	18	24	29	37	42	49	2	7	10	14	18	23	27	32	1.3	1.4	1.5	1.6	1.8	1.6	1.7	1.8	2.0	2.1
TW I1	13	22	29	37	43	51	57	64	7	13	17	22	26	31	35	40	1.1	1.2	1.3	1.4	1.6	1.4	1.5	1.7	1.9	2.0
TW I2	17	32	42	53	60	70	76	82	9	19	27	35	41	50	56	63	0.8	0.9	1.1	1.2	1.4	1.0	1.1	1.3	1.4	1.6
TW I3	18	35	45	57	64	74	80	86	10	21	29	38	45	54	60	67	0.8	0.9	1.0	1.2	1.4	1.0	1.1	1.2	1.4	1.5
TW I4	19	37	49	61	69	78	84	89	10	23	32	42	50	60	66	73	0.8	0.9	1.0	1.2	1.3	0.9	1.0	1.2	1.3	1.5
TW I5	19	38	50	62	70	79	84	90	10	23	33	43	51	61	67	75	0.8	0.8	1.0	1.2	1.3	0.9	1.0	1.1	1.3	1.5
TW J1	22	40	52	63	71	80	85	90	13	26	35	44	51	61	67	74	0.7	0.8	1.0	1.2	1.3	0.9	1.0	1.2	1.3	1.5
TW J2	21	39	51	62	70	79	85	90	11	24	34	44	51	61	67	74	0.8	0.8	1.0	1.2	1.3	0.9	1.0	1.2	1.3	1.5
TW J3	24	46	59	71	78	87	91	95	13	30	41	52	60	71	77	83	0.7	0.8	0.9	1.1	1.3	0.8	0.9	1.1	1.2	1.4
TW J4	22	43	56	68	76	85	89	94	11	27	38	50	58	68	74	81	0.7	0.8	1.0	1.1	1.3	0.8	0.9	1.1	1.2	1.4
DG A1	16	28	36	44	50	57	62	66	9	16	21	26	30	35	39	42	0.9	1.0	1.2	1.5	1.8	1.2	1.4	1.6	1.9	2.4
DG A2	16	30	39	49	56	64	69	75	10	18	24	31	37	44	49	55	0.9	1.0	1.1	1.3	1.5	1.2	1.2	1.4	1.6	1.8
DG A3	19	34	46	57	65	74	79	85	12	22	30	39	46	55	61	68	0.8	0.9	1.0	1.2	1.4	1.0	1.1	1.2	1.4	1.6
DG B1	16	29	38	49	56	65	70	75	9	16	21	28	32	39	44	49	0.9	1.0	1.1	1.3	1.6	1.2	1.3	1.5	1.7	2.0
DG B2	17	31	41	53	61	71	76	82	10	19	25	33	39	48	54	61	0.9	0.9	1.0	1.2	1.4	1.1	1.2	1.3	1.5	1.7
DG B3	19	35	47	60	69	79	84	89	12	23	31	40	48	58	65	73	0.8	0.9	1.0	1.2	1.4	1.0	1.1	1.2	1.3	1.5
DG C1	18	34	45	57	65	74	79	84	11	20	26	33	39	47	53	59	0.8	0.9	1.0	1.2	1.5	1.1	1.2	1.3	1.5	1.7
DG C2	19	35	47	60	68	78	83	88	12	22	29	38	45	55	61	69	0.8	0.9	1.0	1.2	1.4	1.0	1.1	1.2	1.3	1.5
DG C3	21	40	53	67	76	85	90	93	14	26	35	45	53	65	72	80	0.8	0.8	0.9	1.1	1.3	1.0	1.0	1.1	1.2	1.4
SS A1	24	37	46	56	63	71	77	83	17	26	33	40	45	53	58	64	0.8	0.9	1.1	1.3	1.4	1.0	1.1	1.3	1.5	1.6
SS A2	28	46	57	68	75	83	87	92	20	33	42	51	58	67	72	79	0.7	0.8	1.0	1.2	1.3	0.9	1.0	1.1	1.3	1.5
SS A3	22	35	43	53	59	68	73	79	16	25	30	37	42	49	54	60	0.9	1.0	1.1	1.3	1.5	1.1	1.2	1.4	1.6	1.7
SS A4	28	46	56	67	74	82	87	91	20	32	41	50	57	65	71	77	0.7	0.8	1.0	1.2	1.3	0.9	1.0	1.1	1.3	1.5
SS A5	23	35	43	52	58	66	72	78	18	25	31	37	42	48	53	59	0.9	1.0	1.1	1.3	1.5	1.1	1.2	1.4	1.6	1.7
SS A6	28	46	57	68	75	83	87	92	20	33	41	51	57	66	72	78	0.7	0.8	1.0	1.2	1.3	0.9	1.0	1.1	1.3	1.5
SS A7	21	31	38	46	52	60	65	71	17	23	27	33	37	42	46	52	1.0	1.1	1.2	1.4	1.6	1.2	1.4	1.5	1.7	1.9
SS A8	28	45	56	66	73	81	86	91	19	32	40	49	55	64	70	76	0.7	0.8	1.0	1.2	1.3	0.9	1.0	1.2	1.3	1.5
SS B1	20	32	41	50	57	65	71	77	14	22	28	35	40	47	52	58	0.9	1.0	1.1	1.3	1.5	1.1	1.2	1.4	1.6	1.7
SS B2	24	41	51	62	69	78	83	88	17	29	37	46	52	61	67	74	0.8	0.9	1.0	1.2	1.4	0.9	1.0	1.2	1.3	1.5
SS B3	19	31	39	48	54	63	68	75	14	21	26	33	37	44	49	55	0.9	1.0	1.2	1.3	1.5	1.2	1.3	1.5	1.6	1.8
SS B4	24	40	50	61	68	77	82	88	16	28	36	45	51	60	65	72	0.8	0.9	1.0	1.2	1.4	0.9	1.0	1.2	1.4	1.5
SS B5	19	29	36	44	50	58	63	70	14	21	25	31	35	41	45	50	1.0	1.1	1.2	1.4	1.6	1.3	1.4	1.5	1.7	1.9
SS B6	24	40	51	61	69	77	83	88	16	28	36	45	51	60	66	73	0.8	0.9	1.0	1.2	1.4	0.9	1.0	1.2	1.4	1.5
SS B7	18	26	32	40	45	53	58	65	13	19	23	28	31	37	40	45	1.1	1.2	1.3	1.5	1.6	1.4	1.5	1.7	1.8	2.0
SS B8	23	39	49	60	67	76	81	87	16	27	34	43	49	58	63	70	0.8	0.9	1.0	1.2	1.4	1.0	1.1	1.2	1.4	1.5
SS C1	19	33	43	53	60	70	75	82	11	20	27	34	40	48	53	60	0.9	0.9	1.1	1.2	1.4	1.1	1.2	1.3	1.5	1.7
SS C2	22	39	50	61	69	78	83	89	12	25	33	42	49	58	64	72	0.8	0.9	1.0	1.2	1.3	0.9	1.0	1.2	1.3	1.5
SS C3	17	27	34	43	49	58	63	70	11	18	22	28	32	39	43	49	1.0	1.1	1.2	1.4	1.5	1.3	1.4	1.5	1.7	1.9
SS C4	20	34	44	55	63	72	78	84	12	22	29	38	44	52	58	65	0.8	0.9	1.1	1.2	1.4	1.0	1.1	1.3	1.4	1.6
SS D1	28	45	55	66	73	81	85	90	19	31	39	47	54	62	68	74	0.7	0.8	1.0	1.2	1.3	0.9	1.0	1.2	1.3	1.5
SS D2	32	53	65	76	83	90	93	96	21	38	48	59	66	75	80	86	0.7	0.8	0.9	1.1	1.3	0.8	0.9	1.0	1.2	1.4
SS D3	26	40	50	60	66	75	80	85	19	28	35	43	48	56	61	68	0.8	0.9	1.1	1.2	1.4	1.0	1.1	1.3	1.4	1.6
SS D4	31	51	63	74	81	88	92	95	21	36	46	57	64	73	78	84	0.7	0.8	0.9	1.1	1.3	0.8	0.9	1.1	1.2	1.4
SS E1	24	40	50	60	67	76	81	87	16	26	34	42	47	55	61	68	0.8	0.9	1.0	1.2	1.4	1.0	1.1	1.3	1.4	1.6
SS E2	28	48	59	71	78	85	90	94	18	32	42	52	59	69	74	81	0.7	0.8	1.0	1.1	1.3	0.8	0.9	1.1	1.3	1.4
SS E3	21	33	41	50	57	66	71	77	15	23	29	35	40	47	51	57	0.9	1.0	1.1	1.3	1.5	1.1	1.2	1.4	1.6	1.7
SS E4	26	44	55	66	73	81	86	91	17	30	39	48	55	64	70	76	0.7	0.8	1.0	1.2	1.3	0.9	1.0	1.2	1.3	1.5

GREENSBORO, NORTH CAROLINA

TYPE	SOLAR SAVINGS FRACTION (%) — BASE TEMP: 55 F (1984 DD)								65 F (3831 DD)								CONSERVATION FACTOR 55 F					65 F				
LCR	200	100	70	50	40	30	25	20	200	100	70	50	40	30	25	20	200	100	50	30	20	200	100	50	30	20
WW A1	19	26	32	38	43	51	56	62	14	19	23	27	30	35	39	44	1.1	1.2	1.4	1.5	1.7	1.4	1.6	1.7	1.9	2.1
WW A2	19	30	39	48	55	64	69	76	12	20	26	33	38	45	50	57	0.9	1.0	1.2	1.3	1.5	1.1	1.2	1.4	1.6	1.7
WW A3	18	32	41	52	59	68	74	81	11	21	27	35	41	49	55	62	0.9	1.0	1.1	1.3	1.4	1.1	1.2	1.3	1.5	1.6
WW A4	[illegible]	33	43	54	61	71	76	83	10	21	28	36	43	51	57	64	0.8	0.9	1.1	1.2	1.4	1.0	1.1	1.3	1.4	1.6
WW A5	18	34	45	56	63	73	79	85	9	21	29	38	44	53	59	67	0.8	0.9	1.0	1.2	1.4	1.0	1.1	1.2	1.4	1.6
WW A6	18	34	45	57	65	74	80	86	9	21	29	39	45	55	61	68	0.8	0.9	1.0	1.2	1.4	1.0	1.1	1.2	1.4	1.5
WW B1	16	25	32	40	46	54	60	66	10	16	20	25	29	35	39	44	1.0	1.1	1.3	1.4	1.6	1.3	1.5	1.6	1.8	2.0
WW B2	18	33	44	55	62	72	78	84	10	21	29	38	45	54	60	67	0.8	0.9	1.1	1.2	1.4	1.0	1.1	1.2	1.4	1.6
WW B3	21	39	51	63	70	80	85	90	12	26	35	45	53	62	68	76	0.8	0.8	1.0	1.2	1.3	0.9	1.0	1.1	1.3	1.5
WW B4	20	39	52	64	72	81	86	91	11	26	36	47	55	65	71	79	0.7	0.8	1.0	1.1	1.3	0.9	1.0	1.1	1.3	1.4
WW B5	19	38	50	63	71	80	85	91	10	25	35	46	54	64	71	78	0.7	0.8	1.0	1.1	1.3	0.9	1.0	1.1	1.3	1.4
WW C1	23	41	53	65	72	81	86	91	13	28	37	48	55	65	71	78	0.7	0.8	1.0	1.1	1.3	0.9	1.0	1.1	1.3	1.4
WW C2	20	39	51	63	71	80	85	90	11	26	35	46	53	63	70	77	0.7	0.8	1.0	1.2	1.3	0.9	1.0	1.1	1.3	1.4
WW C3	23	45	58	71	79	87	91	95	13	31	43	55	63	73	79	86	0.7	0.8	0.9	1.1	1.3	0.8	0.9	1.0	1.2	1.4
WW C4	20	41	55	67	75	84	89	93	10	28	39	51	59	70	76	83	0.7	0.8	1.0	1.1	1.3	0.8	0.9	1.1	1.2	1.4
TW A1	18	26	31	37	42	50	55	61	14	19	22	27	30	35	38	43	1.1	1.2	1.4	1.5	1.7	1.5	1.6	1.7	1.9	2.1
TW A2	18	28	35	44	50	58	64	71	12	19	24	30	35	41	46	52	1.0	1.1	1.2	1.4	1.5	1.2	1.3	1.5	1.6	1.8
TW A3	17	29	37	46	53	62	68	75	11	19	25	31	37	44	49	56	0.9	1.0	1.2	1.3	1.5	1.2	1.3	1.4	1.6	1.7
TW A4	16	28	37	46	53	63	69	76	9	18	24	31	37	45	50	57	0.9	1.0	1.2	1.3	1.5	1.1	1.2	1.4	1.5	1.7
TW B1	17	25	31	38	43	51	56	63	13	18	22	27	30	36	40	45	1.1	1.2	1.3	1.5	1.6	1.4	1.5	1.7	1.8	2.0
TW B2	16	26	34	42	48	57	63	70	11	18	23	29	34	40	45	51	1.0	1.1	1.2	1.4	1.5	1.2	1.3	1.5	1.6	1.8
TW B3	15	26	34	42	49	58	64	71	10	17	22	29	34	41	46	52	1.0	1.1	1.2	1.4	1.5	1.2	1.3	1.5	1.6	1.8
TW B4	14	24	32	40	46	55	61	68	9	16	21	27	32	39	44	50	1.0	1.1	1.2	1.4	1.5	1.2	1.3	1.5	1.6	1.8
TW C1	16	24	30	37	43	51	56	63	11	17	21	26	30	36	40	45	1.1	1.2	1.3	1.5	1.6	1.4	1.5	1.6	1.8	1.9
TW C2	14	24	30	38	44	52	58	65	10	16	21	26	30	37	41	47	1.1	1.2	1.3	1.4	1.6	1.3	1.4	1.6	1.7	1.9
TW C3	14	22	29	36	42	50	55	62	9	15	20	25	29	35	39	45	1.1	1.2	1.3	1.5	1.6	1.4	1.5	1.6	1.7	1.9
TW C4	13	21	26	32	37	45	50	56	10	15	18	23	26	32	36	41	1.2	1.3	1.4	1.6	1.7	1.5	1.6	1.7	1.9	2.0
TW D1	13	21	27	34	40	48	53	60	8	13	17	22	25	30	34	39	1.1	1.2	1.4	1.5	1.7	1.4	1.6	1.7	1.9	2.1
TW D2	18	31	40	50	57	67	73	79	11	20	27	35	41	49	55	62	0.9	1.0	1.1	1.3	1.4	1.1	1.2	1.3	1.5	1.6
TW D3	19	34	44	54	62	72	77	84	11	22	30	38	45	54	60	67	0.8	0.9	1.1	1.2	1.4	1.0	1.1	1.2	1.4	1.6
TW D4	19	36	47	58	66	76	81	87	11	24	32	42	49	59	65	73	0.8	0.9	1.0	1.2	1.4	0.9	1.0	1.2	1.3	1.5
TW D5	18	35	46	58	65	75	81	87	10	23	32	42	49	59	66	73	0.8	0.9	1.0	1.2	1.4	0.9	1.0	1.2	1.3	1.5
TW E1	21	37	48	59	67	76	82	87	13	25	34	43	50	59	65	72	0.8	0.9	1.0	1.2	1.4	0.9	1.0	1.2	1.3	1.5
TW E2	20	37	48	59	67	76	82	87	12	25	33	43	50	59	66	73	0.8	0.9	1.0	1.2	1.4	0.9	1.0	1.2	1.3	1.5
TW E3	23	43	56	68	76	84	89	93	14	30	40	52	60	70	76	82	0.7	0.8	1.0	1.1	1.3	0.8	0.9	1.1	1.2	1.4
TW E4	21	40	53	65	73	82	87	92	12	27	38	49	57	67	73	80	0.7	0.8	1.0	1.1	1.3	0.8	0.9	1.1	1.2	1.4
TW F1	16	24	30	37	42	49	55	61	12	17	21	25	29	34	38	43	1.1	1.2	1.4	1.5	1.7	1.4	1.5	1.7	1.9	2.0
TW F2	15	25	33	41	47	56	62	69	10	16	22	27	32	39	43	50	1.0	1.1	1.2	1.4	1.5	1.3	1.4	1.5	1.7	1.8
TW F3	14	25	33	42	49	58	64	71	8	15	21	28	33	40	45	52	1.0	1.1	1.2	1.4	1.5	1.2	1.3	1.4	1.6	1.8
TW F4	12	23	31	40	47	56	63	70	6	13	19	26	31	39	44	51	1.0	1.1	1.2	1.4	1.5	1.2	1.3	1.4	1.6	1.8
TW G1	14	22	28	35	41	49	54	61	10	15	19	24	28	33	37	43	1.1	1.2	1.4	1.5	1.6	1.4	1.5	1.7	1.8	2.0
TW G2	12	22	29	37	44	52	58	66	7	14	18	24	29	36	40	46	1.0	1.1	1.3	1.4	1.6	1.3	1.4	1.5	1.7	1.9
TW G3	10	20	27	36	42	51	57	64	5	12	17	23	27	34	39	46	1.0	1.1	1.3	1.4	1.6	1.3	1.4	1.5	1.7	1.8
TW G4	8	17	23	31	37	45	51	59	3	9	14	19	24	30	35	41	1.1	1.2	1.3	1.5	1.6	1.4	1.5	1.6	1.8	1.9
TW H1	11	19	25	32	38	46	51	58	7	12	16	21	25	30	35	40	1.1	1.2	1.4	1.5	1.7	1.4	1.5	1.7	1.8	2.0
TW H2	8	17	23	30	35	44	49	57	4	9	14	19	23	29	33	39	1.1	1.2	1.4	1.5	1.7	1.4	1.5	1.7	1.8	2.0
TW H3	6	14	20	26	31	39	44	51	3	8	11	16	20	25	30	35	1.2	1.3	1.5	1.6	1.7	1.5	1.6	1.8	1.9	2.1
TW H4	5	11	15	21	25	32	36	43	2	6	9	12	16	20	24	28	1.4	1.5	1.6	1.7	1.9	1.7	1.0	1.9	2.1	2.2
TW I1	11	19	25	31	36	44	49	56	6	11	15	19	22	27	31	35	1.2	1.3	1.4	1.5	1.7	1.5	1.6	1.8	2.0	2.2
TW I2	14	27	36	46	54	63	69	77	8	17	24	31	37	46	51	58	0.9	1.0	1.1	1.3	1.5	1.1	1.2	1.3	1.5	1.7
TW I3	15	30	39	50	57	67	73	80	8	19	26	34	41	50	56	63	0.9	1.0	1.1	1.3	1.4	1.0	1.1	1.3	1.4	1.6
TW I4	16	32	43	55	02	72	78	85	8	21	29	39	46	56	62	70	0.8	0:9	1.0	1.2	1.4	1.0	1.1	1.2	1.4	1.5
TW I5	16	32	43	55	63	73	79	86	8	21	30	40	47	57	63	71	0.8	0.9	1.0	1.2	1.4	0.9	1.0	1.2	1.3	1.5
TW J1	19	35	45	57	64	74	80	86	11	23	31	41	47	57	63	70	0.8	0.9	1.0	1.2	1.4	1.0	1.1	1.2	1.4	1.5
TW J2	17	34	44	56	64	74	79	86	10	22	30	40	47	57	63	70	0.8	0.9	1.0	1.2	1.4	1.0	1.1	1.2	1.3	1.5
TW J3	20	40	52	65	73	82	87	92	11	26	37	49	57	67	73	80	0.7	0.8	1.0	1.1	1.3	0.8	0.9	1.1	1.2	1.4
TW J4	18	37	49	62	70	80	85	90	9	24	35	46	54	64	71	78	0.7	0 8	1.0	1.1	1.3	0.9	1.0	1.1	1.3	1.4
DG A1	14	24	30	38	43	50	54	58	8	14	18	23	27	31	34	37	1.0	1.1	1.3	1.6	1.9	1.3	1.5	1.7	2.1	2.6
DG A2	14	25	33	42	49	57	63	69	9	16	22	28	33	40	45	50	1.0	1.1	1.2	1.4	1.6	1.2	1.3	1.4	1.7	1.9
DG A3	16	30	39	51	59	68	74	80	11	20	27	36	42	52	58	65	0.9	1.0	1.1	1.3	1.5	1.1	1.2	1.3	1.4	1.6
DG B1	13	24	32	41	48	56	62	67	8	14	19	24	28	34	38	43	1.0	1.1	1.2	1.4	1.7	1.3	1.4	1.6	1.8	2.1
DG B2	14	26	35	45	53	63	69	76	9	17	22	29	35	43	49	56	1.0	1.0	1.1	1.3	1.5	1.2	1.3	1.4	1.5	1.7
DG B3	17	30	41	53	62	73	79	85	11	20	28	37	44	54	61	69	0.9	0.9	1.0	1.2	1.4	1.1	1.1	1.2	1.3	1.5
DG C1	16	28	37	48	56	66	72	77	9	17	23	29	34	42	47	53	1.0	1.0	1.1	1.3	1.5	1.2	1.3	1.4	1.6	1.8
DG C2	16	30	40	52	60	71	77	83	10	19	26	34	41	50	56	64	0.9	1.0	1.0	1.2	1.4	1.1	1.2	1.3	1.4	1.6
DG C3	19	34	46	59	69	79	85	90	12	23	32	41	49	60	68	76	0.9	0.9	1.0	1.1	1.3	1.0	1.1	1.2	1.3	1.4
SS A1	21	33	41	50	56	65	70	77	16	24	30	37	42	49	54	60	0.9	1.0	1.2	1.3	1.5	1.1	1.2	1.4	1.5	1.7
SS A2	25	40	51	62	69	77	82	88	18	30	38	47	54	63	68	75	0.8	0.9	1.0	1.2	1.4	0.9	1.0	1.2	1.3	1.5
SS A3	20	30	38	46	52	61	66	73	15	22	28	34	38	45	49	55	1.0	1.1	1.2	1.4	1.5	1.2	1.3	1.5	1.6	1.8
SS A4	24	40	50	61	68	76	82	87	18	30	38	46	53	61	67	73	0.8	0.9	1.0	1.2	1.4	0.9	1.0	1.2	1.4	1.5
SS A5	21	31	38	46	52	60	65	71	16	23	28	34	38	45	49	55	1.0	1.1	1.2	1.4	1.6	1.2	1.3	1.5	1.7	1.8
SS A6	25	40	51	61	68	77	82	88	18	30	38	47	53	62	68	74	0.8	0.9	1.0	1.2	1.4	0.9	1.0	1.2	1.3	1.5
SS A7	19	28	33	40	46	53	58	64	16	21	25	30	33	39	42	47	1.1	1.2	1.3	1.5	1.6	1.3	1.5	1.6	1.8	2.0
SS A8	24	39	49	60	67	75	81	86	18	29	37	45	52	60	65	72	0.8	0.9	1.0	1.2	1.4	0.9	1.0	1.2	1.4	1.5
SS B1	18	28	35	44	50	59	64	71	13	20	26	32	36	43	48	54	1.0	1.1	1.2	1.4	1.5	1.2	1.3	1.5	1.6	1.8
SS B2	21	36	45	56	63	72	78	84	15	26	34	42	49	57	63	70	0.8	0.9	1.1	1.2	1.4	1.0	1.1	1.2	1.4	1.5
SS B3	17	27	34	42	47	56	61	68	12	19	24	30	34	40	44	50	1.0	1.1	1.3	1.4	1.6	1.2	1.4	1.5	1.7	1.9
SS B4	21	35	44	55	62	71	77	83	15	25	33	41	47	56	61	68	0.8	0.9	1.1	1.2	1.4	1.0	1.1	1.2	1.4	1.6
SS B5	17	25	31	38	44	51	56	63	13	19	23	28	32	37	41	46	1.1	1.2	1.3	1.5	1.6	1.3	1.5	1.6	1.8	2.0
SS B6	21	35	45	55	62	71	77	83	14	25	33	41	48	56	62	69	0.8	0.9	1.1	1.2	1.4	1.0	1.1	1.2	1.4	1.6
SS B7	16	23	28	35	39	46	51	58	12	17	21	25	28	33	37	41	1.2	1.3	1.4	1.6	1.7	1.4	1.6	1.8	1.9	2.1
SS B8	20	34	43	53	60	69	75	82	14	24	32	40	46	54	59	66	0.9	0.9	1.1	1.3	1.4	1.0	1.1	1.3	1.4	1.6
SS C1	16	28	37	46	53	63	68	75	10	18	24	31	36	43	48	55	0.9	1.0	1.2	1.3	1.5	1.1	1.3	1.4	1.6	1.7
SS C2	18	33	44	55	62	72	78	84	11	22	30	38	45	54	60	67	0.8	0.9	1.1	1.2	1.4	1.0	1.1	1.2	1.4	1.6
SS C3	15	24	30	37	43	51	56	63	10	16	20	25	29	35	39	45	1.1	1.2	1.3	1.5	1.6	1.4	1.5	1.6	1.8	1.9
SS C4	17	30	39	49	56	66	71	78	10	20	26	34	40	48	54	61	0.9	1.0	1.1	1.3	1.4	1.1	1.2	1.3	1.5	1.6
SS D1	24	39	49	59	66	75	80	85	17	28	35	44	50	58	63	70	0.8	0.9	1.1	1.2	1.4	1.0	1.1	1.2	1.4	1.6
SS D2	28	47	59	70	77	85	89	93	19	34	44	55	62	71	77	83	0.7	0.8	1.0	1.1	1.3	0.8	0.9	1.1	1.2	1.4
SS D3	23	35	44	53	60	68	74	80	17	26	32	39	45	52	57	63	0.9	1.0	1.1	1.3	1.5	1.0	1.2	1.3	1.5	1.7
SS D4	27	45	57	68	75	83	88	92	19	33	43	53	60	69	75	81	0.7	0.8	1.0	1.1	1.3	0.8	0.9	1.1	1.3	1.4
SS E1	21	34	43	53	60	69	75	81	14	24	30	38	43	51	56	63	0.9	1.0	1.1	1.3	1.4	1.0	1.2	1.3	1.5	1.7
SS E2	24	41	53	64	71	80	85	90	16	29	38	48	55	64	70	77	0.7	0.8	1.0	1.2	1.3	0.9	1.0	1.1	1.3	1.5
SS E3	19	29	36	44	50	59	64	71	14	21	26	32	36	43	47	53	1.0	1.1	1.2	1.4	1.5	1.2	1.3	1.5	1.7	1.8
SS E4	22	38	48	59	67	76	81	87	15	27	35	44	51	60	65	72	0.8	0.9	1.0	1.2	1.4	0.9	1.0	1.2	1.4	1.5

RALEIGH-DURHAM, NORTH CAROLINA

TYPE	SOLAR SAVINGS FRACTION (%) BASE TEMP: 55 F (1744 DD)								65 F (3520 DD)								CONSERVATION FACTOR 55 F					65 F				
LCR	200	100	70	50	40	30	25	20	200	100	70	50	40	30	25	20	200	100	50	30	20	200	100	50	30	20
WW A1	19	27	33	40	45	53	58	64	14	19	23	27	31	36	39	44	1.1	1.2	1.3	1.5	1.6	1.4	1.5	1.7	1.9	2.0
WW A2	20	32	41	50	57	66	72	78	12	20	26	33	38	46	51	57	0.9	1.0	1.1	1.3	1.5	1.1	1.2	1.4	1.6	1.7
WW A3	19	34	43	54	61	70	76	82	11	21	28	36	41	50	55	62	0.8	0.9	1.1	1.2	1.4	1.1	1.2	1.3	1.5	1.6
WW A4	19	35	45	56	63	73	78	85	10	21	28	37	43	52	58	65	0.8	0.9	1.1	1.2	1.4	1.0	1.1	1.3	1.4	1.6
WW A5	19	36	47	58	66	75	81	87	10	21	29	38	45	54	60	67	0.8	0.9	1.0	1.2	1.4	1.0	1.1	1.2	1.4	1.6
WW A6	19	36	48	59	67	76	82	88	9	21	30	39	46	55	61	69	0.8	0.9	1.0	1.2	1.4	1.0	1.1	1.2	1.4	1.5
WW B1	17	27	34	42	48	56	62	69	10	16	21	26	30	35	39	44	1.0	1.1	1.2	1.4	1.6	1.3	1.4	1.6	1.8	2.0
WW B2	19	35	46	57	64	74	80	86	10	22	30	39	45	54	60	68	0.8	0.9	1.0	1.2	1.4	1.0	1.1	1.2	1.4	1.5
WW B3	23	41	53	65	72	81	86	91	13	26	36	46	53	63	69	76	0.7	0.8	1.0	1.1	1.3	0.9	1.0	1.1	1.3	1.5
WW B4	22	41	54	66	74	83	88	92	11	26	37	48	55	66	72	79	0.7	0.8	1.0	1.1	1.3	0.9	1.0	1.1	1.3	1.4
WW B5	20	40	52	65	72	82	87	92	10	25	35	47	54	65	71	79	0.7	0.8	1.0	1.1	1.3	0.9	1.0	1.1	1.3	1.4
WW C1	24	43	55	67	74	83	88	92	14	28	38	48	55	65	71	78	0.7	0.8	1.0	1.1	1.3	0.9	1.0	1.1	1.3	1.4
WW C2	22	41	53	65	73	82	87	92	12	26	36	46	54	64	70	77	0.7	0.8	1.0	1.1	1.3	0.9	1.0	1.1	1.3	1.4
WW C3	24	47	60	73	80	88	92	96	13	31	43	55	64	74	80	86	0.7	0.8	0.9	1.1	1.3	0.8	0.9	1.0	1.2	1.4
WW C4	22	43	57	69	77	86	90	94	11	28	40	52	60	70	77	83	0.7	0.8	0.9	1.1	1.3	0.8	0.9	1.1	1.2	1.4
TW A1	19	27	32	39	44	52	57	63	14	19	22	27	30	35	39	44	1.1	1.2	1.3	1.5	1.6	1.5	1.6	1.7	1.9	2.0
TW A2	18	29	37	45	52	61	66	73	12	19	24	30	35	42	47	53	1.0	1.1	1.2	1.3	1.5	1.2	1.3	1.5	1.6	1.8
TW A3	18	30	39	48	55	64	70	77	11	19	25	32	37	45	50	56	0.9	1.0	1.1	1.3	1.5	1.1	1.2	1.4	1.6	1.7
TW A4	17	30	38	48	55	65	71	78	9	18	24	32	37	45	51	57	0.9	1.0	1.1	1.3	1.4	1.1	1.2	1.4	1.5	1.7
TW B1	18	26	32	40	45	53	58	65	13	18	22	27	31	36	40	45	1.1	1.2	1.3	1.5	1.6	1.4	1.5	1.6	1.8	2.0
TW B2	17	28	35	44	50	59	65	72	11	18	23	29	34	41	46	52	1.0	1.1	1.2	1.4	1.5	1.2	1.3	1.5	1.6	1.8
TW B3	16	27	35	44	51	60	66	73	10	17	23	29	34	41	46	53	1.0	1.1	1.2	1.3	1.5	1.2	1.3	1.4	1.6	1.8
TW B4	15	26	33	42	48	57	63	70	9	16	21	28	32	40	44	51	1.0	1.1	1.2	1.4	1.5	1.2	1.3	1.5	1.6	1.8
TW C1	16	25	31	39	45	53	58	65	11	17	21	26	30	36	40	46	1.1	1.2	1.3	1.4	1.6	1.4	1.5	1.6	1.8	1.9
TW C2	15	25	32	40	46	54	60	67	10	16	21	26	31	37	42	48	1.0	1.1	1.3	1.4	1.6	1.3	1.4	1.5	1.7	1.9
TW C3	15	23	30	38	43	52	57	64	10	15	20	25	29	36	40	46	1.1	1.2	1.3	1.4	1.6	1.4	1.4	1.6	1.7	1.9
TW C4	14	22	27	34	39	46	51	58	10	15	19	23	27	32	36	41	1.2	1.3	1.4	1.5	1.7	1.5	1.6	1.7	1.9	2.0
TW D1	14	23	29	36	42	50	55	62	8	13	17	22	26	31	35	40	1.1	1.2	1.3	1.5	1.6	1.4	1.5	1.7	1.9	2.1
TW D2	19	32	42	52	59	69	75	81	11	21	27	35	41	50	55	62	0.9	1.0	1.1	1.3	1.4	1.1	1.2	1.3	1.5	1.6
TW D3	20	35	46	57	64	73	79	85	11	22	30	39	45	54	60	67	0.8	0.9	1.1	1.2	1.4	1.0	1.1	1.2	1.4	1.6
TW D4	20	37	49	60	68	77	83	88	11	24	33	43	50	60	66	73	0.8	0.9	1.0	1.2	1.3	0.9	1.0	1.2	1.3	1.5
TW D5	19	36	48	60	67	77	82	88	10	23	33	43	50	60	66	74	0.8	0.9	1.0	1.2	1.3	0.9	1.0	1.2	1.3	1.5
TW E1	22	39	50	61	69	78	83	89	13	26	34	44	50	60	66	73	0.8	0.9	1.0	1.2	1.3	0.9	1.0	1.2	1.3	1.5
TW E2	21	38	50	61	69	78	83	89	12	25	34	43	50	60	66	73	0.8	0.9	1.0	1.2	1.3	0.9	1.0	1.2	1.3	1.5
TW E3	25	45	58	70	77	86	90	94	14	30	41	52	60	70	76	83	0.7	0.8	0.9	1.1	1.3	0.8	0.9	1.1	1.2	1.4
TW E4	22	42	55	67	74	83	88	93	12	28	38	49	57	67	74	81	0.7	0.8	1.0	1.1	1.3	0.8	0.9	1.1	1.2	1.4
TW F1	17	25	31	38	44	51	57	63	12	17	21	26	29	35	38	43	1.1	1.2	1.3	1.5	1.6	1.4	1.5	1.7	1.8	2.0
TW F2	16	27	34	43	49	58	64	71	10	17	22	28	33	39	44	50	1.0	1.1	1.2	1.4	1.5	1.2	1.3	1.5	1.7	1.8
TW F3	15	26	35	44	51	60	66	73	8	16	21	28	33	41	46	52	0.9	1.0	1.2	1.3	1.5	1.2	1.3	1.4	1.6	1.8
TW F4	13	24	33	42	49	59	65	72	6	14	20	26	32	39	44	51	1.0	1.0	1.2	1.3	1.5	1.2	1.3	1.4	1.6	1.8
TW G1	15	24	30	37	43	51	56	63	10	15	19	24	28	34	38	43	1.1	1.2	1.3	1.5	1.6	1.4	1.5	1.7	1.8	2.0
TW G2	13	23	31	39	46	54	60	68	7	14	19	25	29	36	41	47	1.0	1.1	1.2	1.4	1.5	1.3	1.4	1.5	1.7	1.8
TW G3	11	22	29	38	44	53	59	67	5	12	17	23	28	35	40	46	1.0	1.1	1.2	1.4	1.6	1.3	1.4	1.5	1.7	1.8
TW G4	9	18	25	33	39	47	53	61	3	9	14	20	24	31	35	41	1.1	1.2	1.3	1.5	1.6	1.4	1.5	1.6	1.8	1.9
TW H1	12	20	26	34	39	48	53	60	7	12	16	21	25	31	35	41	1.1	1.2	1.3	1.5	1.6	1.4	1.5	1.7	1.8	2.0
TW H2	9	18	24	32	37	46	51	59	4	10	14	19	23	29	34	39	1.1	1.2	1.4	1.5	1.6	1.4	1.5	1.7	1.8	2.0
TW H3	7	15	21	28	33	41	46	54	3	8	12	16	20	26	30	35	1.2	1.3	1.4	1.6	1.7	1.5	1.6	1.7	1.9	2.0
TW H4	5	12	16	22	26	33	38	45	2	6	9	13	16	21	24	29	1.3	1.4	1.6	1.7	1.8	1.7	1.8	1.9	2.1	2.2
TW I1	12	20	26	33	38	46	52	59	6	11	15	19	23	28	31	36	1.1	1.2	1.4	1.5	1.7	1.5	1.6	1.8	2.0	2.1
TW I2	15	29	38	48	56	65	71	78	8	17	24	32	38	46	52	59	0.9	1.0	1.1	1.3	1.4	1.1	1.2	1.3	1.5	1.6
TW I3	17	31	41	52	60	69	75	82	9	19	26	35	41	50	56	63	0.8	0.9	1.1	1.2	1.4	1.0	1.1	1.3	1.4	1.6
TW I4	17	34	45	57	64	74	80	86	9	21	30	39	46	56	62	70	0.8	0.9	1.0	1.2	1.4	1.0	1.0	1.2	1.3	1.5
TW I5	17	34	45	57	65	75	81	87	8	21	30	40	47	57	64	72	0.8	0.9	1.0	1.2	1.4	0.9	1.0	1.2	1.3	1.5
TW J1	20	36	47	59	66	76	81	87	11	23	32	41	48	57	63	71	0.8	0.9	1.0	1.2	1.4	0.9	1.0	1.2	1.4	1.5
TW J2	19	35	46	58	66	75	81	87	10	22	31	40	47	57	63	71	0.8	0.9	1.0	1.2	1.4	0.9	1.0	1.2	1.3	1.5
TW J3	21	42	54	67	74	83	88	93	11	27	38	49	57	67	74	81	0.7	0.8	1.0	1.1	1.3	0.8	0.9	1.1	1.2	1.4
TW J4	19	39	51	64	72	81	86	91	9	25	35	46	54	65	71	79	0.7	0.8	1.0	1.1	1.3	0.9	1.0	1.1	1.3	1.4
DG A1	14	25	32	40	45	52	56	61	8	14	19	23	27	32	35	38	1.0	1.1	1.3	1.6	1.9	1.3	1.4	1.7	2.1	2.5
DG A2	15	27	35	44	51	59	65	71	9	16	22	29	34	41	45	51	1.0	1.0	1.2	1.4	1.6	1.2	1.3	1.4	1.6	1.9
DG A3	17	31	41	53	61	70	76	81	11	20	27	36	43	52	58	65	0.9	0.9	1.1	1.2	1.5	1.1	1.2	1.3	1.4	1.6
DG B1	14	26	34	43	50	59	64	70	8	14	19	24	29	35	39	44	1.0	1.0	1.2	1.4	1.7	1.3	1.4	1.6	1.8	2.1
DG B2	15	28	37	48	56	66	72	78	9	17	23	30	35	44	49	56	1.0	1.0	1.1	1.3	1.5	1.2	1.3	1.4	1.5	1.7
DG B3	17	32	43	55	64	75	80	86	11	21	28	37	44	55	61	70	0.9	0.9	1.0	1.2	1.4	1.1	1.1	1.2	1.3	1.5
DG C1	17	30	40	51	59	69	74	79	9	17	23	30	35	42	48	54	0.9	1.0	1.1	1.3	1.5	1.2	1.3	1.4	1.6	1.8
DG C2	17	31	42	54	63	73	79	85	10	20	27	35	41	50	57	64	0.9	0.9	1.0	1.2	1.4	1.1	1.2	1.3	1.4	1.6
DG C3	20	36	48	61	71	81	86	91	13	24	32	42	50	61	68	77	0.8	0.9	1.0	1.1	1.3	1.0	1.1	1.1	1.3	1.4
SS A1	22	34	43	52	58	67	72	79	16	24	30	37	42	50	54	61	0.9	1.0	1.1	1.3	1.5	1.1	1.2	1.4	1.5	1.7
SS A2	26	42	53	64	71	79	84	89	18	31	39	48	54	63	69	75	0.8	0.9	1.0	1.2	1.4	0.9	1.0	1.2	1.3	1.5
SS A3	21	32	40	48	55	63	68	75	15	23	28	34	39	45	50	56	0.9	1.0	1.2	1.3	1.5	1.1	1.3	1.4	1.6	1.8
SS A4	26	42	52	63	70	78	83	88	18	30	38	47	53	62	67	74	0.8	0.9	1.0	1.2	1.4	0.9	1.0	1.2	1.3	1.5
SS A5	22	32	39	48	54	62	67	73	17	24	29	34	39	45	50	55	1.0	1.1	1.2	1.4	1.5	1.2	1.3	1.5	1.6	1.8
SS A6	26	42	53	63	70	79	84	89	18	30	39	48	54	63	68	75	0.8	0.9	1.0	1.2	1.4	0.9	1.0	1.2	1.3	1.5
SS A7	20	29	35	42	48	55	60	66	16	21	25	30	34	39	43	48	1.1	1.2	1.3	1.5	1.6	1.3	1.4	1.6	1.8	2.0
SS A8	25	41	51	62	69	77	82	88	18	29	37	46	52	61	66	73	0.8	0.9	1.0	1.2	1.4	0.9	1.0	1.2	1.4	1.5
SS B1	19	30	37	46	52	61	66	73	13	21	26	32	37	44	48	54	1.0	1.1	1.2	1.4	1.5	1.2	1.3	1.5	1.6	1.8
SS B2	22	37	47	58	65	74	79	85	15	26	34	43	49	58	63	70	0.8	0.9	1.1	1.2	1.4	1.0	1.1	1.2	1.4	1.5
SS B3	18	28	35	44	50	58	64	70	12	19	24	30	34	41	45	51	1.0	1.1	1.2	1.4	1.6	1.2	1.3	1.5	1.7	1.9
SS B4	22	37	46	57	64	73	78	84	15	26	33	42	48	56	62	69	0.8	0.9	1.1	1.2	1.4	1.0	1.1	1.2	1.4	1.6
SS B5	18	27	33	40	46	53	59	65	13	19	23	28	32	38	42	47	1.1	1.2	1.3	1.5	1.6	1.3	1.4	1.6	1.8	2.0
SS B6	22	37	47	57	64	73	79	85	15	26	33	42	48	57	62	69	0.8	0.9	1.1	1.2	1.4	1.0	1.1	1.2	1.4	1.6
SS B7	17	24	30	36	41	49	54	60	13	17	21	25	29	34	37	42	1.1	1.2	1.4	1.5	1.7	1.4	1.6	1.7	1.9	2.1
SS B8	21	35	45	55	62	71	77	83	14	25	32	40	46	55	60	67	0.8	0.9	1.1	1.2	1.4	1.0	1.1	1.3	1.4	1.6
SS C1	17	30	39	48	55	65	71	77	10	18	24	31	36	44	49	56	0.9	1.0	1.1	1.3	1.5	1.1	1.2	1.4	1.6	1.7
SS C2	20	35	46	57	64	74	79	86	11	22	30	39	46	55	61	68	0.8	0.9	1.0	1.2	1.4	1.0	1.1	1.2	1.4	1.5
SS C3	16	25	31	39	45	53	59	66	10	16	20	26	30	36	40	45	1.1	1.2	1.3	1.4	1.6	1.4	1.5	1.6	1.8	1.9
SS C4	18	31	41	51	58	67	73	80	11	20	27	35	40	49	54	61	0.9	1.0	1.1	1.3	1.4	1.1	1.2	1.3	1.5	1.6
SS D1	25	41	51	61	68	77	82	87	17	28	36	44	50	58	64	70	0.8	0.9	1.0	1.2	1.4	0.9	1.1	1.2	1.4	1.6
SS D2	29	49	61	72	79	87	90	94	19	35	45	55	63	72	77	84	0.7	0.8	0.9	1.1	1.3	0.8	0.9	1.1	1.2	1.4
SS D3	24	37	46	55	62	70	75	81	17	26	33	40	45	52	57	64	0.9	1.0	1.1	1.3	1.4	1.0	1.2	1.3	1.5	1.7
SS D4	28	47	59	70	77	85	89	93	19	34	43	53	61	70	75	82	0.7	0.8	1.0	1.1	1.3	0.8	0.9	1.1	1.3	1.4
SS E1	22	36	45	55	62	71	77	83	14	24	31	38	44	52	57	63	0.8	0.9	1.1	1.2	1.4	1.0	1.1	1.3	1.5	1.6
SS E2	25	43	55	66	73	82	87	91	16	30	39	49	56	65	71	77	0.7	0.8	1.0	1.2	1.3	0.9	1.0	1.1	1.3	1.5
SS E3	20	30	38	46	52	61	66	73	14	21	26	32	37	43	48	54	1.0	1.1	1.2	1.4	1.5	1.2	1.3	1.5	1.6	1.8
SS E4	23	40	51	61	69	78	83	88	15	28	36	45	51	60	66	73	0.8	0.9	1.0	1.2	1.4	0.9	1.0	1.2	1.3	1.5

BISMARCK, NORTH DAKOTA

TYPE	SOLAR SAVINGS FRACTION (%)																CONSERVATION FACTOR									
	BASE TEMP: 55 F (6434 DD) LCR								65 F (9057 DD)								55 F					65 F				
	100	70	50	40	30	25	20	15	100	70	50	40	30	25	20	15	100	50	30	20	15	100	50	30	20	15
WW A1	14	15	17	18	20	22	23	25	12	13	15	15	17	18	18	19	2.3	2.7	3.2	3.7	4.2	2.8	3.3	4.0	-	-
WW A2	12	15	18	21	25	27	30	35	10	12	15	17	20	22	24	27	1.7	2.0	2.3	2.6	2.8	2.0	2.3	2.6	3.1	3.5
WW A3	11	15	19	22	26	29	33	39	8	11	15	17	21	23	26	30	1.6	1.9	2.1	2.3	2.5	1.8	2.1	2.4	2.7	3.1
WW A4	10	14	19	22	27	31	35	41	7	11	14	17	21	24	27	32	1.5	1.8	2.0	2.2	2.4	1.7	2.0	2.3	2.6	2.9
WW A5	10	14	19	23	28	32	37	43	7	10	14	18	22	25	29	34	1.5	1.7	1.9	2.1	2.3	1.6	1.9	2.2	2.5	2.7
WW A6	9	14	19	23	29	33	38	45	6	10	14	18	22	26	30	35	1.4	1.7	1.9	2.1	2.3	1.6	1.8	2.1	2.4	2.7
WW B1	8	10	11	12	14	14	15	16	6	7	8	9	9	9	8	7	2.4	3.1	4.2	-	-	3.1	4.4	-	-	-
WW B2	11	16	21	26	32	36	41	49	8	13	17	21	26	30	34	41	1.4	1.6	1.8	2.0	2.2	1.5	1.8	2.0	2.2	2.4
WW B3	14	19	26	31	38	43	49	57	10	16	21	26	32	36	42	49	1.3	1.5	1.7	1.8	2.0	1.4	1.6	1.8	2.0	2.2
WW B4	14	21	29	34	43	48	56	65	11	17	24	30	37	43	49	58	1.2	1.4	1.5	1.7	1.8	1.3	1.4	1.6	1.8	1.9
WW B5	13	20	29	35	44	49	57	66	10	17	24	30	38	44	51	60	1.2	1.3	1.5	1.7	1.8	1.2	1.4	1.6	1.7	1.9
WW C1	15	21	28	33	41	46	52	61	12	17	24	28	35	40	45	53	1.2	1.4	1.6	1.8	1.9	1.3	1.5	1.7	1.9	2.1
WW C2	14	20	28	33	41	46	53	62	11	17	23	28	36	40	47	55	1.2	1.4	1.6	1.7	1.9	1.3	1.5	1.7	1.8	2.0
WW C3	17	26	36	43	52	59	67	76	14	22	31	38	47	53	61	71	1.1	1.2	1.4	1.5	1.7	1.1	1.3	1.4	1.6	1.8
WW C4	15	24	33	40	50	56	64	74	12	20	29	35	45	51	58	68	1.1	1.2	1.4	1.6	1.7	1.1	1.3	1.5	1.6	1.8
TW A1	14	15	17	19	21	22	24	27	12	13	15	16	17	18	19	21	2.3	2.6	3.0	3.4	3.8	2.7	3.1	3.6	4.4	-
TW A2	12	15	18	20	23	26	29	33	10	12	15	16	19	21	23	26	1.8	2.1	2.4	2.7	2.9	2.1	2.4	2.7	3.1	3.5
TW A3	11	14	18	20	24	27	31	36	9	11	14	16	20	22	25	29	1.7	2.0	2.2	2.5	2.7	1.9	2.2	2.5	2.8	3.1
TW A4	10	13	17	20	25	28	32	38	7	10	14	16	20	22	26	30	1.7	1.9	2.1	2.3	2.5	1.8	2.1	2.4	2.7	2.9
TW B1	13	15	17	19	21	23	25	29	11	13	14	16	18	19	21	23	2.1	2.4	2.8	3.1	3.4	2.4	2.8	3.2	3.7	4.2
TW B2	11	14	17	19	23	26	29	34	9	11	14	16	19	21	24	27	1.8	2.1	2.3	2.6	2.8	2.0	2.3	2.6	2.9	3.2
TW B3	10	13	17	19	23	26	30	35	8	10	13	16	19	21	24	29	1.8	2.0	2.2	2.5	2.6	2.0	2.2	2.5	2.8	3.0
TW B4	10	13	16	19	23	25	29	34	8	10	13	15	19	21	24	28	1.8	2.0	2.2	2.5	2.6	2.0	2.2	2.5	2.7	3.0
TW C1	12	14	16	18	21	24	26	30	10	12	14	15	18	20	22	25	2.0	2.3	2.5	2.8	3.0	2.3	2.5	2.8	3.2	3.5
TW C2	10	13	16	18	22	24	28	33	9	11	13	15	18	20	23	27	1.9	2.1	2.4	2.6	2.8	2.1	2.3	2.6	2.9	3.1
TW C3	10	13	16	18	21	24	27	32	9	11	13	15	18	20	23	27	1.9	2.2	2.4	2.6	2.8	2.1	2.4	2.6	2.9	3.1
TW C4	11	13	15	17	20	23	26	30	9	11	13	15	17	19	22	26	2.1	2.3	2.5	2.7	2.9	2.3	2.5	2.7	3.0	3.2
TW D1	7	9	10	11	13	14	15	16	5	6	7	8	9	9	9	9	2.5	3.0	3.7	4.7	-	3.1	3.9	-	-	-
TW D2	12	16	21	24	29	33	38	45	9	13	17	20	25	28	32	38	1.5	1.7	1.9	2.1	2.3	1.7	1.9	2.1	2.3	2.5
TW D3	12	17	22	26	32	36	41	48	9	13	18	21	26	30	34	41	1.4	1.6	1.8	2.0	2.2	1.6	1.8	2.0	2.2	2.4
TW D4	13	19	26	31	38	43	50	59	11	16	22	27	33	38	44	52	1.3	1.5	1.6	1.8	1.9	1.4	1.5	1.7	1.9	2.1
TW D5	13	20	27	32	40	45	52	61	11	16	23	28	35	40	47	56	1.3	1.4	1.6	1.7	1.9	1.3	1.5	1.7	1.8	2.0
TW E1	15	20	26	30	37	41	47	55	12	16	22	26	32	36	41	48	1.3	1.5	1.7	1.9	2.0	1.4	1.6	1.8	2.0	2.2
TW E2	14	20	26	31	38	43	49	58	11	16	22	27	33	37	43	51	1.3	1.5	1.6	1.8	2.0	1.4	1.6	1.8	1.9	2.1
TW E3	17	25	33	40	48	54	61	71	14	21	29	35	43	49	56	65	1.1	1.3	1.5	1.6	1.8	1.2	1.4	1.5	1.7	1.8
TW E4	16	23	32	38	47	53	60	70	13	20	28	34	42	48	55	64	1.1	1.3	1.5	1.6	1.8	1.2	1.4	1.5	1.7	1.8
TW F1	12	14	16	17	19	21	23	26	10	11	13	14	16	17	18	20	2.2	2.6	2.9	3.3	3.7	2.6	3.0	3.5	4.2	5.0
TW F2	10	12	15	18	21	23	26	31	8	10	12	14	17	18	21	24	1.9	2.2	2.4	2.7	3.0	2.1	2.4	2.8	3.2	3.6
TW F3	8	11	15	17	21	24	27	32	6	8	11	13	17	19	21	25	1.8	2.0	2.3	2.5	2.7	2.0	2.3	2.6	2.9	3.2
TW F4	6	9	13	16	20	23	27	32	4	7	10	12	16	18	21	26	1.8	2.0	2.2	2.4	2.6	1.9	2.2	2.5	2.8	3.0
TW G1	10	12	14	16	18	20	23	26	8	10	11	13	15	16	18	20	2.1	2.4	2.8	3.1	3.4	2.5	2.8	3.2	3.7	4.2
TW G2	7	10	13	15	19	21	24	29	5	7	10	12	15	17	19	23	1.9	2.2	2.4	2.7	2.9	2.1	2.4	2.7	3.1	3.4
TW G3	6	8	12	14	18	20	24	29	4	6	9	11	14	16	19	23	1.9	2.1	2.4	2.6	2.8	2.1	2.3	2.6	2.9	3.2
TW G4	4	7	10	12	16	18	22	27	2	4	7	9	12	14	17	21	2.0	2.2	2.4	2.6	2.8	2.2	2.4	2.6	2.9	3.1
TW H1	7	9	11	13	16	18	21	25	5	7	9	10	13	14	16	19	2.1	2.4	2.7	2.9	3.2	2.4	2.7	3.0	3.4	3.7
TW H2	4	7	10	12	15	17	20	25	3	5	7	9	11	13	16	20	2.0	2.3	2.5	2.8	3.0	2.3	2.5	2.8	3.1	3.4
TW H3	3	6	8	10	13	16	19	23	2	4	6	8	10	12	15	18	2.1	2.4	2.6	2.8	3.0	2.4	2.6	2.8	3.1	3.3
TW H4	2	4	6	8	11	13	15	19	1	3	5	6	8	10	12	16	2.3	2.6	2.8	3.0	3.2	2.6	2.8	3.1	3.3	3.5
TW I1	5	7	8	9	11	11	13	14	4	4	5	6	7	7	7	7	2.8	3.2	3.9	-	-	3.2	4.1	-	-	-
TW I2	9	13	18	21	26	30	35	42	7	10	14	17	22	25	29	35	1.6	1.8	2.0	2.2	2.3	1.7	1.9	2.1	2.4	2.6
TW I3	10	14	19	23	29	33	38	46	7	11	16	19	24	28	32	39	1.5	1.7	1.9	2.1	2.2	1.6	1.8	2.0	2.2	2.4
TW I4	11	17	23	28	35	40	47	56	8	13	19	24	30	35	41	49	1.3	1.5	1.7	1.8	2.0	1.4	1.6	1.8	1.9	2.1
TW I5	11	17	24	30	37	43	50	59	8	14	20	25	33	38	44	53	1.3	1.4	1.6	1.8	1.9	1.3	1.5	1.7	1.9	2.0
TW J1	12	18	24	28	35	39	45	53	10	14	20	24	30	34	39	46	1.4	1.5	1.7	1.9	2.0	1.4	1.6	1.8	2.0	2.2
TW J2	12	17	24	29	36	41	47	56	9	14	20	24	31	35	41	49	1.3	1.5	1.7	1.8	2.0	1.4	1.6	1.8	2.0	2.1
TW J3	15	22	31	37	46	52	60	69	12	19	27	33	41	47	54	64	1.1	1.3	1.5	1.6	1.8	1.2	1.4	1.5	1.7	1.8
TW J4	13	21	29	35	44	50	58	68	10	17	25	31	40	45	53	62	1.2	1.3	1.5	1.6	1.8	1.2	1.4	1.5	1.7	1.9
DG A1	7	8	10	11	13	13	14	13	5	6	7	8	8	8	8	7	2.3	2.9	4.3	-	-	2.7	3.8	-	-	-
DG A2	10	13	16	19	23	25	29	33	8	10	13	15	19	21	24	27	1.8	2.0	2.3	2.6	3.0	1.9	2.2	2.5	2.9	3.4
DG A3	13	17	23	27	33	37	43	51	11	15	20	23	29	33	38	45	1.5	1.6	1.8	1.9	2.1	1.6	1.7	1.9	2.1	2.3
DG B1	7	8	10	12	13	14	15	16	4	6	7	8	9	9	9	9	2.2	2.9	3.7	-	-	2.7	3.4	-	-	-
DG B2	10	13	17	19	24	27	31	36	8	10	14	16	20	22	25	29	1.7	2.0	2.2	2.4	2.7	1.9	2.1	2.4	2.7	3.0
DG B3	13	18	24	28	34	39	46	54	11	15	21	25	30	35	40	48	1.4	1.6	1.7	1.8	2.0	1.5	1.7	1.8	2.0	2.1
DG C1	9	11	14	16	18	20	22	25	6	8	10	12	14	15	16	17	1.9	2.4	2.8	3.3	4.0	2.2	2.6	3.5	-	-
DG C2	12	16	20	23	29	32	37	44	10	13	17	20	24	27	31	37	1.6	1.8	1.9	2.1	2.4	1.7	1.9	2.1	2.4	2.6
DG C3	15	21	27	32	39	44	52	61	13	18	24	28	35	39	46	55	1.3	1.5	1.6	1.7	1.9	1.4	1.6	1.7	1.8	2.0
SS A1	15	18	21	23	27	29	32	37	13	16	19	21	24	26	28	31	1.7	2.0	2.3	2.6	2.9	1.8	2.2	2.6	3.0	3.4
SS A2	18	23	29	33	39	43	49	56	16	21	26	30	36	40	45	51	1.3	1.5	1.7	1.9	2.1	1.4	1.6	1.8	2.0	2.2
SS A3	13	15	18	20	22	24	26	28	12	14	16	17	19	20	22	23	2.0	2.4	2.8	3.3	3.8	2.1	2.6	3.2	4.2	-
SS A4	17	22	27	31	37	41	46	53	16	20	25	28	33	37	41	47	1.4	1.6	1.8	2.0	2.2	1.4	1.7	1.9	2.2	2.4
SS A5	15	18	20	22	25	26	29	32	14	16	18	20	22	23	25	27	1.9	2.3	2.7	3.0	3.4	2.1	2.5	3.0	3.6	4.3
SS A6	18	23	28	32	38	42	48	55	16	20	25	29	35	38	43	49	1.4	1.6	1.8	1.9	2.1	1.4	1.6	1.9	2.1	2.3
SS A7	14	16	17	18	20	21	22	24	13	14	16	16	18	18	19	19	2.3	2.8	3.5	4.3	-	2.5	3.2	4.2	-	-
SS A8	17	21	26	30	35	39	43	50	15	19	24	27	32	35	39	44	1.4	1.7	1.9	2.1	2.3	1.5	1.7	2.0	2.3	2.5
SS B1	12	15	18	20	23	25	28	32	11	13	15	17	20	22	24	27	1.9	2.2	2.5	2.8	3.1	2.0	2.4	2.8	3.2	3.7
SS B2	16	20	26	30	36	40	45	53	14	18	23	27	33	36	41	48	1.4	1.6	1.8	2.0	2.1	1.5	1.7	1.9	2.1	2.2
SS B3	11	13	16	17	20	21	24	26	10	11	13	15	17	18	20	21	2.0	2.4	2.8	3.3	3.7	2.2	2.7	3.3	4.1	-
SS B4	15	19	24	28	34	38	43	50	13	17	22	26	31	34	39	45	1.4	1.6	1.8	2.0	2.2	1.5	1.7	1.9	2.2	2.4
SS B5	13	14	17	18	20	22	24	27	11	13	15	16	18	19	20	22	2.1	2.5	2.9	3.3	3.7	2.4	2.8	3.3	4.0	4.7
SS B6	15	19	25	29	34	38	44	51	13	17	22	26	31	35	39	46	1.4	1.6	1.8	2.0	2.2	1.5	1.7	1.9	2.1	2.3
SS B7	11	13	14	16	17	18	19	21	10	11	13	14	15	16	16	17	2.4	3.0	3.5	4.2	-	2.7	3.3	4.2	-	-
SS B8	14	18	23	26	32	35	40	47	12	16	20	24	28	32	36	42	1.5	1.7	1.9	2.1	2.3	1.6	1.8	2.0	2.3	2.5
SS C1	9	12	16	18	22	24	27	32	7	9	12	14	17	18	21	24	1.8	2.1	2.4	2.7	2.9	2.0	2.3	2.8	3.3	3.8
SS C2	12	17	22	26	32	36	42	49	9	13	18	22	27	31	35	42	1.4	1.6	1.8	2.0	2.1	1.5	1.7	2.0	2.2	2.4
SS C3	10	12	15	17	19	21	24	27	8	10	12	13	16	17	19	21	2.1	2.4	2.7	3.0	3.3	2.4	2.7	3.1	3.6	4.1
SS C4	12	16	20	24	29	33	38	44	9	13	17	20	24	28	32	38	1.5	1.7	1.9	2.1	2.3	1.7	1.9	2.1	2.3	2.5
SS D1	15	19	23	26	31	33	37	42	14	17	20	23	26	29	32	35	1.5	1.8	2.1	2.4	2.6	1.6	2.0	2.4	2.8	3.2
SS D2	19	26	33	38	45	50	57	65	17	23	30	34	41	46	51	59	1.2	1.4	1.6	1.7	1.9	1.2	1.4	1.7	1.9	2.0
SS D3	16	19	22	25	28	31	34	38	14	17	20	22	25	27	29	32	1.7	2.0	2.3	2.6	2.9	1.8	2.2	2.6	3.0	3.5
SS D4	19	25	32	37	44	49	55	63	17	22	29	33	40	44	50	57	1.2	1.4	1.6	1.8	1.9	1.3	1.5	1.7	1.9	2.1
SS E1	12	15	18	21	24	27	30	33	10	13	16	18	20	22	24	26	1.7	2.0	2.4	2.8	3.1	1.9	2.3	2.8	3.5	4.4
SS E2	15	21	27	32	38	43	48	56	13	18	24	28	34	38	43	50	1.3	1.5	1.7	1.9	2.0	1.4	1.6	1.8	2.0	2.2
SS E3	13	15	18	20	22	24	26	30	11	13	15	17	19	20	22	24	1.9	2.3	2.7	3.1	3.4	2.1	2.5	3.1	3.7	4.5
SS E4	15	20	25	30	35	40	45	52	13	18	23	26	32	35	40	46	1.4	1.6	1.8	2.0	2.1	1.4	1.7	1.9	2.1	2.3

FARGO, NORTH DAKOTA

SOLAR SAVINGS FRACTION (%)

TYPE	55 F (6675 DD) LCR 100	70	50	40	30	25	20	15	65 F (9278 DD) 100	70	50	40	30	25	20	15
WW A1	12	13	14	15	16	16	17	17	11	12	12	13	13	13	13	13
WW A2	9	12	14	16	18	20	22	24	8	10	11	13	14	15	17	18
WW A3	8	11	14	16	19	21	23	27	6	8	11	12	15	16	18	20
WW A4	7	10	13	16	19	22	25	29	5	8	10	12	15	17	19	21
WW A5	6	9	13	16	20	22	26	30	4	7	10	12	15	17	19	22
WW A6	6	9	13	16	20	23	26	31	3	6	10	12	15	17	20	23
WW B1	6	7	7	7	7	7	6	3	5	5	5	5	4	3	1	-2
WW B2	8	12	16	19	24	27	32	38	6	9	13	16	20	23	26	31
WW B3	10	15	20	24	30	33	39	46	8	12	17	20	25	28	33	39
WW B4	10	16	23	28	35	40	47	56	8	13	20	24	31	36	42	50
WW B5	10	16	23	29	36	42	49	58	7	13	20	25	32	37	44	53
WW C1	11	16	22	26	33	37	43	50	9	14	19	23	28	32	37	43
WW C2	10	16	22	27	33	38	44	52	8	13	19	23	29	33	39	46
WW C3	13	21	30	36	45	51	59	69	11	18	26	33	41	47	54	64
WW C4	11	19	27	34	43	49	56	66	9	16	24	30	39	45	52	61
TW A1	12	13	14	15	16	17	18	19	11	12	13	13	14	14	15	15
TW A2	10	12	14	16	18	19	21	24	9	10	12	13	15	16	17	18
TW A3	8	11	13	15	18	20	22	26	7	9	11	13	15	16	18	20
TW A4	7	10	13	15	18	20	23	27	6	8	10	12	15	16	19	21
TW B1	11	13	14	15	17	18	19	21	10	11	12	13	14	15	16	17
TW B2	9	11	13	15	18	19	22	25	7	9	11	13	15	16	18	20
TW B3	8	10	13	15	18	20	22	26	7	8	11	12	15	16	18	21
TW B4	8	10	13	15	18	20	22	26	6	8	10	12	15	16	19	22
TW C1	10	12	13	15	17	19	20	23	9	10	12	13	15	16	17	19
TW C2	9	11	13	15	17	19	22	25	7	9	11	13	15	16	18	21
TW C3	9	10	13	15	17	19	21	25	7	9	11	12	15	16	18	21
TW C4	9	11	13	15	17	19	21	24	8	10	12	13	15	16	18	21
TW D1	5	6	7	7	8	8	7	6	4	4	5	5	5	4	3	1
TW D2	9	12	16	19	23	26	30	35	8	10	14	16	20	22	25	29
TW D3	9	13	17	20	24	28	32	38	7	10	14	17	21	23	27	31
TW D4	10	15	21	25	32	36	42	50	8	13	18	22	28	32	37	44
TW D5	10	16	22	27	34	38	45	53	8	13	19	24	30	35	40	49
TW E1	11	16	20	24	29	33	38	45	9	13	18	21	26	29	33	39
TW E2	11	16	21	25	31	35	41	48	9	13	18	22	27	31	36	43
TW E3	13	20	27	33	41	46	53	62	11	17	24	30	37	42	49	57
TW E4	12	19	26	32	40	46	53	62	10	16	23	29	36	41	48	57
TW F1	10	11	13	14	15	16	17	18	9	10	11	11	12	13	13	13
TW F2	8	9	12	13	15	17	19	21	6	8	9	11	12	13	15	16
TW F3	6	8	11	12	15	17	19	23	4	6	8	10	12	13	15	17
TW F4	4	6	9	11	14	16	19	23	2	4	7	9	11	13	15	18
TW G1	8	9	11	12	14	15	16	18	7	8	9	10	11	12	13	14
TW G2	5	7	9	11	14	15	17	20	4	6	7	9	11	12	13	15
TW G3	4	6	8	10	13	15	17	21	2	4	6	8	10	11	13	16
TW G4	2	4	7	9	11	13	16	19	1	3	5	6	9	10	12	15
TW H1	5	7	9	10	12	13	15	17	4	5	7	8	9	10	12	13
TW H2	3	5	7	8	11	12	15	18	2	3	5	6	8	10	11	14
TW H3	2	4	6	7	10	11	14	17	1	2	4	5	7	9	11	13
TW H4	1	3	4	6	8	9	11	15	1	2	3	4	6	7	9	12
TW I1	4	4	5	5	6	6	5	4	2	3	3	3	3	2	2	0
TW I2	7	10	13	16	20	23	27	32	5	8	11	13	17	19	22	27
TW I3	7	10	15	18	22	26	30	36	5	8	12	15	19	22	25	30
TW I4	8	13	18	23	29	33	39	47	6	10	16	19	25	29	34	42
TW I5	8	13	19	24	31	36	42	51	6	11	17	21	27	32	38	46
TW J1	9	14	18	22	28	31	36	43	7	11	16	19	24	27	32	38
TW J2	9	13	19	23	29	33	39	46	7	11	16	20	25	29	34	41
TW J3	11	18	25	31	39	45	52	61	9	15	22	28	35	41	47	56
TW J4	10	16	24	30	38	43	50	60	8	14	21	26	34	39	46	55
DG A1	5	6	6	7	7	7	6	4	3	4	4	4	4	3	2	0
DG A2	8	10	13	15	17	19	22	25	6	8	11	12	15	16	18	20
DG A3	11	15	19	22	28	31	36	43	10	13	17	20	25	28	33	39
DG B1	5	6	6	7	7	7	7	6	3	4	4	4	4	4	3	1
DG B2	8	10	13	15	18	20	23	27	6	8	11	13	15	17	19	22
DG B3	11	15	20	23	29	33	38	46	10	13	18	21	26	30	35	41
DG C1	6	8	10	11	12	12	13	13	5	6	7	8	9	9	9	8
DG C2	10	13	16	19	22	25	29	34	8	11	14	16	20	22	25	29
DG C3	13	17	23	27	33	37	43	52	11	15	20	24	30	34	39	47
SS A1	13	15	17	19	21	23	25	28	12	14	16	17	19	21	22	24
SS A2	15	20	24	28	33	37	42	48	14	18	23	26	31	34	38	44
SS A3	11	13	14	15	17	18	19	19	10	12	13	14	15	16	16	16
SS A4	15	18	23	26	31	34	38	44	14	17	21	24	28	31	35	40
SS A5	13	15	17	18	20	21	22	24	13	14	16	17	18	19	20	20
SS A6	15	19	24	27	32	35	40	46	14	18	22	25	29	33	36	42
SS A7	12	13	14	15	16	16	16	16	12	13	13	14	14	14	14	13
SS A8	14	18	22	25	29	32	36	41	13	16	20	23	27	29	32	36
SS B1	10	12	14	16	18	19	21	24	9	11	13	14	16	17	18	20
SS B2	13	17	22	25	30	34	39	45	12	16	20	23	28	31	36	41
SS B3	9	11	12	13	15	16	17	18	8	10	11	12	13	14	14	14
SS B4	13	16	20	24	28	32	36	42	12	15	19	22	26	29	33	38
SS B5	11	12	14	15	16	17	18	20	10	11	13	13	15	15	16	16
SS B6	12	16	21	24	29	32	36	43	11	15	19	22	26	29	33	39
SS B7	10	11	12	13	13	14	14	14	9	10	11	11	12	12	12	12
SS B8	12	15	19	22	26	29	33	38	10	14	17	20	24	26	30	34
SS C1	7	9	11	13	15	16	18	20	5	7	8	10	11	12	13	14
SS C2	9	13	17	20	25	28	33	39	7	10	14	17	21	24	28	33
SS C3	8	10	12	13	14	16	17	19	7	8	10	11	12	13	13	14
SS C4	9	12	16	19	23	26	30	35	7	10	14	16	20	22	25	30
SS D1	12	15	18	21	23	25	28	31	11	14	17	18	21	22	24	25
SS D2	16	21	27	32	38	43	48	56	14	19	25	29	35	39	44	51
SS D3	13	16	18	20	22	24	26	28	12	14	17	18	20	21	22	24
SS D4	15	21	26	31	37	41	47	54	14	19	24	28	34	38	43	49
SS E1	9	11	14	15	17	19	20	21	8	10	12	13	14	15	16	15
SS E2	12	17	22	26	31	35	40	46	11	15	20	23	28	31	35	41
SS E3	11	12	14	15	17	18	19	21	10	11	13	14	15	16	16	17
SS E4	12	16	21	24	29	32	37	43	11	15	19	22	26	29	33	38

CONSERVATION FACTOR

TYPE	55 F 100	50	30	20	15	65 F 100	50	30	20	15
WW A1	2.9	3.6	4.8	-	-	3.5	4.7	-	-	-
WW A2	2.1	2.5	2.9	3.4	4.0	2.3	2.8	3.5	4.7	-
WW A3	1.9	2.2	2.6	3.0	3.4	2.1	2.5	3.1	3.9	4.9
WW A4	1.8	2.1	2.4	2.8	3.1	1.9	2.3	2.8	3.5	4.3
WW A5	1.7	2.0	2.3	2.6	2.9	1.9	2.2	2.7	3.3	3.9
WW A6	1.7	2.0	2.3	2.6	2.8	1.8	2.2	2.6	3.1	3.7
WW B1	3.5	-	-	-	-	-	-	-	-	-
WW B2	1.6	1.8	2.1	2.3	2.5	1.7	2.0	2.3	2.6	2.9
WW B3	1.4	1.7	1.9	2.1	2.3	1.5	1.8	2.0	2.3	2.5
WW B4	1.3	1.5	1.7	1.8	2.0	1.4	1.5	1.7	1.9	2.1
WW B5	1.3	1.4	1.6	1.8	1.9	1.3	1.5	1.7	1.9	2.0
WW C1	1.4	1.6	1.8	2.0	2.1	1.5	1.7	1.9	2.1	2.3
WW C2	1.3	1.5	1.7	1.9	2.0	1.4	1.6	1.8	2.0	2.2
WW C3	1.1	1.3	1.5	1.6	1.8	1.2	1.3	1.5	1.7	1.8
WW C4	1.2	1.3	1.5	1.7	1.8	1.2	1.4	1.5	1.7	1.9
TW A1	2.8	3.4	4.2	-	-	3.3	4.1	-	-	-
TW A2	2.2	2.6	3.0	3.5	4.0	2.4	2.9	3.5	4.4	-
TW A3	2.0	2.3	2.7	3.1	3.4	2.2	2.6	3.1	3.8	4.5
TW A4	1.9	2.2	2.5	2.8	3.1	2.1	2.4	2.8	3.3	3.8
TW B1	2.5	3.0	3.5	4.2	4.9	2.9	3.4	4.2	-	-
TW B2	2.1	2.4	2.8	3.2	3.5	2.3	2.7	3.2	3.8	4.4
TW B3	2.0	2.3	2.6	3.0	3.3	2.2	2.5	3.0	3.4	3.9
TW B4	2.0	2.3	2.6	2.9	3.2	2.2	2.5	2.9	3.3	3.7
TW C1	2.3	2.7	3.1	3.5	3.8	2.6	3.0	3.5	4.1	4.7
TW C2	2.2	2.5	2.8	3.1	3.4	2.4	2.7	3.0	3.5	3.9
TW C3	2.2	2.5	2.8	3.1	3.3	2.4	2.7	3.0	3.4	3.7
TW C4	2.3	2.6	2.9	3.2	3.4	2.6	2.8	3.1	3.5	3.8
TW D1	3.3	4.7	-	-	-	4.3	-	-	-	-
TW D2	1.7	2.0	2.2	2.4	2.6	1.8	2.1	2.4	2.7	3.0
TW D3	1.6	1.9	2.1	2.3	2.5	1.7	2.0	2.3	2.6	2.9
TW D4	1.4	1.6	1.8	2.0	2.1	1.5	1.7	1.9	2.1	2.3
TW D5	1.4	1.5	1.7	1.9	2.0	1.4	1.6	1.8	2.0	2.1
TW E1	1.5	1.7	1.9	2.1	2.3	1.6	1.8	2.1	2.3	2.5
TW E2	1.4	1.6	1.8	2.0	2.2	1.5	1.7	1.9	2.2	2.3
TW E3	1.2	1.4	1.6	1.7	1.9	1.3	1.4	1.6	1.8	2.0
TW E4	1.2	1.4	1.6	1.7	1.9	1.3	1.4	1.6	1.8	1.9
TW F1	2.7	3.3	4.0	-	-	3.2	3.9	-	-	-
TW F2	2.2	2.6	3.1	3.6	4.1	2.5	3.0	3.6	4.5	-
TW F3	2.1	2.4	2.8	3.2	3.5	2.3	2.7	3.2	3.8	4.6
TW F4	2.0	2.3	2.6	3.0	3.3	2.2	2.5	2.9	3.4	3.9
TW G1	2.5	3.0	3.5	4.2	4.9	2.9	3.4	4.2	-	-
TW G2	2.2	2.6	2.9	3.4	3.7	2.5	2.8	3.3	4.0	4.7
TW G3	2.1	2.5	2.8	3.1	3.4	2.4	2.7	3.1	3.6	4.1
TW G4	2.2	2.5	2.8	3.1	3.4	2.4	2.7	3.1	3.5	3.8
TW H1	2.4	2.8	3.2	3.7	4.1	2.8	3.2	3.7	4.4	-
TW H2	2.3	2.7	3.0	3.3	3.6	2.6	2.9	3.3	3.8	4.2
TW H3	2.4	2.7	3.0	3.3	3.6	2.7	3.0	3.3	3.6	4.0
TW H4	2.6	2.9	3.2	3.5	3.7	2.9	3.2	3.5	3.8	4.1
TW I1	3.4	-	-	-	-	4.6	-	-	-	-
TW I2	1.8	2.0	2.2	2.5	2.7	1.9	2.2	2.4	2.8	3.0
TW I3	1.6	1.9	2.1	2.3	2.5	1.8	2.0	2.3	2.6	2.8
TW I4	1.4	1.6	1.8	2.0	2.1	1.5	1.7	1.9	2.1	2.3
TW I5	1.4	1.6	1.7	1.9	2.0	1.4	1.6	1.8	2.0	2.1
TW J1	1.5	1.7	1.9	2.1	2.3	1.6	1.8	2.1	2.3	2.5
TW J2	1.5	1.6	1.8	2.0	2.2	1.5	1.7	1.9	2.2	2.3
TW J3	1.2	1.4	1.6	1.7	1.9	1.3	1.4	1.6	1.8	2.0
TW J4	1.2	1.4	1.6	1.7	1.9	1.3	1.5	1.6	1.8	2.0
DG A1	3.0	4.6	-	-	-	3.7	-	-	-	-
DG A2	2.0	2.3	2.7	3.2	3.7	2.2	2.5	3.0	3.6	4.4
DG A3	1.6	1.8	1.9	2.1	2.3	1.7	1.9	2.1	2.3	2.5
DG B1	2.9	4.4	-	-	-	3.6	-	-	-	-
DG B2	1.9	2.3	2.6	2.9	3.3	2.1	2.4	2.8	3.4	3.9
DG B3	1.6	1.7	1.9	2.0	2.2	1.7	1.8	2.0	2.2	2.3
DG C1	2.3	3.1	4.5	-	-	2.7	3.5	-	-	-
DG C2	1.7	2.0	2.3	2.5	2.7	1.9	2.1	2.4	2.8	3.1
DG C3	1.5	1.6	1.8	1.9	2.0	1.5	1.7	1.9	2.0	2.2
SS A1	2.0	2.4	2.8	3.3	3.8	2.1	2.5	3.1	4.0	-
SS A2	1.5	1.7	1.9	2.1	2.3	1.5	1.8	2.0	2.3	2.5
SS A3	2.3	2.9	3.8	-	-	2.5	3.2	4.8	-	-
SS A4	1.5	1.8	2.0	2.2	2.4	1.6	1.9	2.1	2.4	2.7
SS A5	2.3	2.8	3.4	4.2	-	2.4	3.0	3.9	-	-
SS A6	1.5	1.7	1.9	2.2	2.3	1.5	1.8	2.1	2.3	2.6
SS A7	2.8	3.8	-	-	-	3.0	4.2	-	-	-
SS A8	1.6	1.8	2.1	2.4	2.6	1.6	1.9	2.3	2.6	2.9
SS B1	2.2	2.6	3.0	3.6	4.2	2.3	2.8	3.4	4.4	-
SS B2	1.5	1.7	1.9	2.1	2.3	1.6	1.8	2.0	2.3	2.5
SS B3	2.4	3.0	3.8	5.0	-	2.6	3.3	4.5	-	-
SS B4	1.6	1.8	2.0	2.3	2.4	1.6	1.9	2.1	2.4	2.6
SS B5	2.5	3.1	3.8	4.7	-	2.7	3.4	4.4	-	-
SS B6	1.6	1.8	2.0	2.2	2.4	1.6	1.9	2.1	2.4	2.6
SS B7	3.0	3.8	-	-	-	3.3	4.3	-	-	-
SS B8	1.6	1.9	2.1	2.4	2.6	1.7	2.0	2.3	2.6	2.8
SS C1	2.1	2.6	3.1	3.7	4.5	2.4	3.0	3.9	-	-
SS C2	1.6	1.8	2.0	2.3	2.5	1.7	2.0	2.2	2.5	2.8
SS C3	2.4	2.9	3.4	4.1	4.7	2.8	3.3	4.1	-	-
SS C4	1.7	2.0	2.2	2.4	2.6	1.8	2.1	2.4	2.7	2.9
SS D1	1.8	2.1	2.6	3.0	3.5	1.9	2.3	3.0	3.9	-
SS D2	1.3	1.5	1.7	1.9	2.0	1.3	1.6	1.8	2.0	2.2
SS D3	1.9	2.4	2.8	3.4	3.9	2.0	2.5	3.2	4.3	-
SS D4	1.3	1.5	1.7	1.9	2.1	1.4	1.6	1.8	2.1	2.3
SS E1	2.0	2.5	3.2	4.2	-	2.2	2.9	4.1	-	-
SS E2	1.4	1.7	1.9	2.1	2.3	1.5	1.8	2.0	2.3	2.5
SS E3	2.3	2.8	3.5	4.4	-	2.5	3.1	4.2	-	-
SS E4	1.5	1.8	2.0	2.2	2.4	1.6	1.9	2.1	2.4	2.7

MINOT, NORTH DAKOTA

TYPE	SOLAR SAVINGS FRACTION (%) BASE TEMP: 55 F (6704 DD) LCR 100	70	50	40	30	25	20	15	65 F (9421 DD) 100	70	50	40	30	25	20	15	CONSERVATION FACTOR 55 F 100	50	30	20	15	65 F 100	50	30	20	15
WW A1	13	14	15	16	17	18	18	19	12	12	13	14	14	15	15	14	2.7	3.4	4.4	-	-	3.2	4.1	-	-	-
WW A2	10	13	15	17	20	21	23	26	9	10	13	14	16	17	18	20	2.0	2.4	2.8	3.3	3.8	2.2	2.6	3.3	4.4	-
WW A3	9	12	15	17	20	22	25	29	7	9	12	14	16	18	20	22	1.8	2.2	2.5	2.9	3.3	2.0	2.4	2.9	3.7	4.7
WW A4	8	11	15	17	21	23	26	30	6	9	12	14	17	18	20	23	1.7	2.0	2.4	2.7	3.1	1.9	2.2	2.7	3.4	4.1
WW A5	7	11	15	17	21	24	27	32	5	8	11	14	17	19	21	24	1.6	2.0	2.3	2.6	2.9	1.8	2.1	2.6	3.2	3.8
WW A6	7	10	15	18	22	25	28	33	4	8	11	14	17	19	22	25	1.6	1.9	2.2	2.5	2.8	1.7	2.1	2.5	3.1	3.6
WW B1	7	8	9	9	9	9	8	6	5	6	6	6	5	5	3	0	3.1	-	-	-	-	4.1	-	-	-	-
WW B2	9	13	18	21	26	29	33	39	7	11	14	17	22	24	28	33	1.6	1.8	2.0	2.3	2.5	1.7	1.9	2.2	2.5	2.8
WW B3	11	16	21	25	31	35	40	47	9	13	18	22	27	30	34	40	1.4	1.6	1.9	2.1	2.2	1.5	1.7	2.0	2.3	2.5
WW B4	11	17	24	29	37	42	48	57	9	14	21	26	32	37	43	51	1.3	1.5	1.6	1.8	2.0	1.3	1.5	1.7	1.9	2.1
WW B5	11	17	24	30	38	43	50	59	8	14	21	26	34	39	45	54	1.2	1.4	1.6	1.8	1.9	1.3	1.5	1.7	1.9	2.0
WW C1	12	18	24	28	34	38	44	52	10	15	20	24	30	34	38	45	1.3	1.6	1.8	2.0	2.1	1.4	1.6	1.9	2.1	2.3
WW C2	11	17	23	28	35	40	46	54	9	14	20	24	31	35	40	48	1.3	1.5	1.7	1.9	2.0	1.4	1.6	1.8	2.0	2.2
WW C3	14	22	31	37	46	52	60	69	12	19	28	34	42	48	55	65	1.1	1.3	1.5	1.6	1.8	1.2	1.3	1.5	1.7	1.8
WW C4	13	20	29	35	44	50	57	67	10	17	26	32	40	46	53	62	1.2	1.3	1.5	1.6	1.8	1.2	1.4	1.5	1.7	1.9
TW A1	13	14	15	16	18	19	20	21	11	12	13	14	15	16	16	16	2.6	3.1	3.9	5.0	-	3.0	3.7	4.8	-	-
TW A2	11	13	15	17	19	21	23	25	9	11	13	14	16	17	19	20	2.1	2.5	2.9	3.4	3.8	2.3	2.7	3.3	4.2	-
TW A3	9	12	15	17	20	22	24	27	8	10	12	14	16	18	20	22	1.9	2.3	2.6	3.0	3.4	2.1	2.5	3.0	3.6	4.3
TW A4	8	11	14	16	20	22	25	29	6	9	11	13	16	18	20	23	1.8	2.1	2.5	2.8	3.1	2.0	2.3	2.7	3.2	3.7
TW B1	12	13	15	16	18	19	21	23	11	12	13	14	15	16	17	18	2.4	2.8	3.4	4.0	4.6	2.7	3.2	3.9	5.0	-
TW B2	10	12	14	16	19	21	23	27	8	10	12	14	16	17	19	22	2.0	2.4	2.7	3.1	3.4	2.2	2.6	3.0	3.6	4.2
TW B3	9	11	14	16	19	21	24	28	7	9	12	13	16	18	20	23	1.9	2.3	2.6	2.9	3.2	2.1	2.4	2.8	3.3	3.8
TW B4	8	11	14	16	19	21	24	28	7	9	11	13	16	18	20	23	1.9	2.2	2.6	2.9	3.1	2.1	2.4	2.8	3.2	3.6
TW C1	11	12	15	16	18	20	22	25	9	11	13	14	16	17	19	21	2.2	2.6	3.0	3.4	3.7	2.5	2.8	3.3	3.9	4.5
TW C2	9	11	14	16	19	20	23	27	8	10	12	13	16	17	20	22	2.1	2.4	2.7	3.0	3.3	2.3	2.6	2.9	3.3	3.8
TW C3	9	11	14	16	18	20	23	26	8	10	12	13	16	17	20	23	2.1	2.4	2.7	3.0	3.2	2.3	2.6	2.9	3.3	3.6
TW C4	10	12	14	16	18	20	22	26	9	10	12	14	16	17	19	22	2.2	2.5	2.8	3.1	3.3	2.4	2.7	3.0	3.3	3.6
TW D1	6	7	8	8	9	9	9	8	5	5	6	6	6	6	5	3	2.9	4.1	-	-	-	3.7	-	-	-	-
TW D2	10	14	17	20	25	27	31	37	8	11	15	17	21	24	27	31	1.7	1.9	2.2	2.4	2.6	1.8	2.0	2.3	2.7	2.9
TW D3	10	14	18	21	26	29	33	39	8	12	15	18	22	25	28	33	1.6	1.8	2.1	2.3	2.5	1.7	1.9	2.2	2.6	2.9
TW D4	11	16	22	27	33	37	43	51	9	14	19	23	29	33	38	46	1.4	1.6	1.8	1.9	2.1	1.4	1.6	1.9	2.1	2.2
TW D5	11	17	23	28	35	40	46	55	9	14	20	25	31	36	42	50	1.3	1.5	1.7	1.9	2.0	1.4	1.6	1.8	1.9	2.1
TW E1	12	17	22	26	31	35	40	46	10	14	19	22	27	30	35	40	1.5	1.7	1.9	2.1	2.3	1.5	1.8	2.0	2.3	2.5
TW E2	12	17	22	27	33	37	42	50	10	14	19	23	29	33	37	44	1.4	1.6	1.8	2.0	2.1	1.5	1.7	1.9	2.1	2.3
TW E3	15	21	29	34	42	48	55	64	12	19	26	31	38	44	50	58	1.2	1.4	1.5	1.7	1.9	1.2	1.4	1.6	1.8	2.0
TW E4	13	20	28	33	41	47	54	63	11	17	25	30	38	43	49	58	1.2	1.4	1.5	1.7	1.9	1.3	1.4	1.6	1.8	1.9
TW F1	11	12	14	15	16	17	18	19	9	10	12	12	13	14	15	15	2.5	3.1	3.8	4.7	-	2.9	3.6	4.6	-	-
TW F2	8	10	13	14	17	18	20	23	7	8	10	12	14	15	16	18	2.1	2.5	3.0	3.4	3.9	2.3	2.8	3.4	4.2	-
TW F3	7	9	12	14	17	19	21	24	5	7	9	11	13	15	17	19	2.0	2.3	2.7	3.1	3.4	2.2	2.6	3.0	3.7	4.4
TW F4	5	7	10	13	16	18	21	25	3	5	8	10	12	14	16	19	1.9	2.2	2.6	2.9	3.2	2.1	2.4	2.8	3.3	3.8
TW G1	9	10	12	13	15	16	18	20	7	9	10	11	12	13	14	15	2.4	2.8	3.4	4.0	4.6	2.7	3.2	3.9	5.0	-
TW G2	6	8	11	12	15	17	19	22	5	6	8	10	12	13	15	17	2.1	2.5	2.8	3.3	3.6	2.3	2.7	3.2	3.8	4.5
TW G3	4	7	9	11	14	16	19	22	3	5	7	9	11	13	15	18	2.0	2.4	2.7	3.1	3.4	2.3	2.6	3.0	3.5	3.9
TW G4	3	5	8	10	13	15	17	21	2	4	6	7	10	12	14	17	2.1	2.4	2.7	3.0	3.3	2.3	2.6	3.0	3.3	3.7
TW H1	6	8	10	11	13	15	17	19	5	6	8	9	11	12	13	15	2.3	2.7	3.1	3.6	4.0	2.6	3.0	3.5	4.1	4.9
TW H2	4	6	8	10	12	14	16	19	2	4	6	7	9	11	13	15	2.2	2.6	2.9	3.2	3.5	2.5	2.8	3.2	3.6	4.0
TW H3	3	4	7	8	11	13	15	18	1	3	5	6	8	10	12	15	2.3	2.6	2.9	3.2	3.5	2.5	2.8	3.1	3.5	3.8
TW H4	2	3	5	7	9	11	13	16	1	2	4	5	7	8	10	13	2.5	2.8	3.1	3.4	3.6	2.8	3.1	3.3	3.6	3.9
TW I1	4	5	6	7	7	7	7	6	3	3	4	4	4	4	3	1	3.1	4.3	-	-	-	0.9	-	-	-	-
TW I2	7	11	15	17	22	24	28	34	6	9	12	14	18	21	24	28	1.7	2.0	2.2	2.5	2.7	1.8	2.1	2.4	2.7	3.0
TW I3	8	12	16	19	24	27	31	37	6	9	13	16	20	23	27	32	1.6	1.8	2.1	2.3	2.5	1.7	2.0	2.2	2.5	2.8
TW I4	9	14	20	24	30	35	40	48	7	11	17	21	27	31	36	43	1.4	1.6	1.8	2.0	2.1	1.5	1.7	1.9	2.1	2.3
TW I5	9	15	21	25	32	37	43	52	7	12	18	22	29	33	39	47	1.4	1.5	1.7	1.9	2.0	1.4	1.6	1.8	2.0	2.1
TW J1	10	15	20	24	29	33	38	45	8	12	17	20	25	29	33	39	1.5	1.7	1.9	2.1	2.3	1.5	1.8	2.0	2.3	2.5
TW J2	10	15	20	24	30	35	40	48	8	12	17	21	27	30	35	42	1.4	1.6	1.8	2.0	2.2	1.5	1.7	1.9	2.1	2.3
TW J3	12	19	27	32	40	46	53	62	10	16	24	29	37	42	49	58	1.2	1.4	1.6	1.7	1.9	1.3	1.4	1.6	1.8	1.9
TW J4	11	18	25	31	39	45	52	61	9	15	22	28	35	41	47	56	1.2	1.4	1.6	1.7	1.9	1.3	1.4	1.6	1.8	2.0
DG A1	5	7	8	8	9	8	8	6	4	5	5	5	6	5	4	1	2.7	4.1	-	-	-	3.2	-	-	-	-
DG A2	8	11	14	16	19	21	23	26	7	9	12	13	16	18	20	22	1.9	2.3	2.6	3.1	3.6	2.1	2.4	2.9	3.4	4.2
DG A3	12	16	20	24	29	33	38	44	10	14	18	21	26	29	34	40	1.6	1.8	1.9	2.1	2.3	1.6	1.8	2.0	2.3	2.5
DG B1	5	7	8	8	9	9	9	8	4	5	5	6	6	6	5	3	2.6	3.9	-	-	-	3.1	4.7	-	-	-
DG B2	9	11	14	16	20	22	25	28	7	9	12	14	17	19	21	24	1.9	2.2	2.5	2.9	3.2	2.0	2.3	2.7	3.3	3.8
DG B3	12	16	21	25	30	34	40	47	11	14	19	22	28	31	36	43	1.5	1.7	1.9	2.0	2.2	1.6	1.7	2.0	2.1	2.3
DG C1	7	9	11	12	13	14	15	15	5	7	8	9	10	11	11	10	2.2	2.9	4.2	-	-	2.4	3.2	-	-	-
DG C2	10	14	17	20	24	27	31	36	9	12	15	18	21	24	27	31	1.7	2.0	2.2	2.5	2.7	1.8	2.0	2.4	2.8	3.1
DG C3	14	19	24	28	34	39	45	53	12	16	22	26	31	35	41	48	1.4	1.6	1.7	1.9	2.0	1.5	1.6	1.8	2.0	2.2
SS A1	13	16	18	20	23	24	26	29	12	15	17	18	21	22	24	25	1.9	2.3	2.8	3.3	3.7	2.0	2.4	3.0	3.9	5.0
SS A2	16	21	25	29	34	38	43	49	15	19	24	27	32	35	39	45	1.4	1.7	1.9	2.1	2.3	1.5	1.7	2.0	2.2	2.5
SS A3	12	14	15	17	18	19	20	21	11	13	14	15	16	17	17	17	2.2	2.9	3.7	-	-	2.3	3.1	4.4	-	-
SS A4	15	19	24	27	32	35	39	45	14	18	22	25	30	32	36	41	1.5	1.8	2.0	2.2	2.4	1.5	1.8	2.1	2.4	2.7
SS A5	14	16	18	19	21	22	23	25	13	15	17	18	19	20	21	22	2.2	2.7	3.3	4.1	-	2.3	2.9	3.7	-	-
SS A6	16	20	25	28	33	36	41	47	15	19	23	26	31	34	38	43	1.5	1.7	1.9	2.2	2.3	1.5	1.8	2.1	2.3	2.6
SS A7	13	14	15	16	17	17	18	17	12	13	14	15	15	16	16	15	2.6	3.5	-	-	-	2.8	3.9	-	-	-
SS A8	15	19	23	26	30	33	37	42	14	17	21	24	28	30	33	37	1.6	1.8	2.1	2.4	2.6	1.6	1.9	2.3	2.6	2.9
SS B1	11	13	15	17	19	21	23	25	10	12	14	15	17	18	20	21	2.1	2.5	3.0	3.5	4.1	2.2	2.7	3.3	4.2	-
SS B2	14	18	23	26	32	35	40	46	13	17	21	25	29	33	37	42	1.5	1.7	1.9	2.1	2.3	1.5	1.8	2.0	2.3	2.5
SS B3	10	12	13	14	16	17	18	19	9	10	12	13	14	15	15	16	2.3	2.9	3.6	4.8	-	2.4	3.1	4.2	-	-
SS B4	13	17	22	25	29	33	37	43	12	16	20	23	27	30	34	39	1.5	1.8	2.0	2.2	2.4	1.6	1.9	2.1	2.4	2.6
SS B5	12	13	15	16	17	18	20	21	11	12	13	14	16	16	17	18	2.4	2.9	3.6	4.5	-	2.6	3.2	4.1	-	-
SS B6	13	17	22	25	30	33	38	44	12	16	20	23	28	31	35	40	1.5	1.8	2.0	2.2	2.4	1.6	1.8	2.1	2.4	2.6
SS B7	11	12	13	14	14	15	15	16	10	11	12	12	13	13	13	13	2.7	3.6	4.9	-	-	3.0	4.0	-	-	-
SS B8	12	16	20	23	27	30	34	40	11	15	18	21	25	28	31	36	1.6	1.9	2.1	2.4	2.6	1.7	1.9	2.2	2.6	2.8
SS C1	8	10	12	14	16	18	20	22	6	8	10	11	13	14	15	16	2.0	2.5	3.0	3.6	4.3	2.2	2.8	3.6	-	-
SS C2	10	14	18	22	27	30	34	41	8	11	15	18	23	26	29	34	1.5	1.8	2.0	2.2	2.4	1.6	1.9	2.2	2.5	2.7
SS C3	9	11	13	14	16	17	19	21	8	9	10	12	13	14	15	16	2.3	2.8	3.3	3.9	4.5	2.6	3.1	3.8	4.9	-
SS C4	10	13	17	20	24	27	31	37	8	11	15	17	21	24	27	31	1.7	1.9	2.2	2.4	2.6	1.8	2.0	2.3	2.6	2.9
SS D1	13	16	19	22	25	27	29	32	12	15	18	20	22	23	25	27	1.7	2.1	2.5	3.0	3.5	1.8	2.3	2.9	3.9	-
SS D2	17	22	28	33	39	44	49	57	15	20	26	30	36	40	45	52	1.3	1.5	1.7	1.9	2.0	1.3	1.6	1.8	2.0	2.2
SS D3	14	16	19	21	23	25	27	29	13	15	18	19	21	22	24	25	1.9	2.3	2.8	3.3	3.9	2.0	2.5	3.1	4.2	-
SS D4	16	22	27	32	38	42	47	55	15	20	26	30	35	39	44	50	1.3	1.5	1.7	1.9	2.1	1.4	1.6	1.8	2.1	2.3
SS E1	10	13	15	17	19	20	21	23	9	11	13	14	16	17	17	17	1.9	2.5	3.1	4.1	-	2.1	2.7	3.9	-	-
SS E2	13	18	23	27	32	36	41	47	12	16	21	24	29	32	36	42	1.4	1.7	1.9	2.1	2.3	1.5	1.7	2.0	2.3	2.6
SS E3	12	13	15	17	18	19	21	22	10	12	14	15	16	17	18	18	2.2	2.7	3.4	4.3	-	2.3	2.9	3.9	-	-
SS E4	13	17	22	25	30	34	38	44	12	16	20	23	27	30	34	39	1.5	1.7	2.0	2.2	2.4	1.5	1.8	2.1	2.4	2.7

AKRON-CANTON, OHIO

TYPE	SOLAR SAVINGS FRACTION (%) BASE TEMP: 55 F (3890 DD) LCR 100	70	50	40	30	25	20	15	65 F (6227 DD) 100	70	50	40	30	25	20	15	CONSERVATION FACTOR 55 F 100	50	30	20	15	65 F 100	50	30	20	15
WW A1	14	15	17	18	20	21	23	25	12	12	13	14	14	15	15	15	2.3	2.8	3.3	3.8	4.3	3.2	4.1	-	-	-
WW A2	12	15	18	21	24	27	30	35	9	11	13	14	16	18	19	22	1.8	2.1	2.3	2.6	2.8	2.2	2.6	3.1	3.7	4.4
WW A3	11	14	18	21	26	29	33	38	7	9	12	14	17	19	21	24	1.6	1.9	2.1	2.3	2.5	2.0	2.3	2.7	3.2	3.7
WW A4	10	14	18	22	27	30	34	41	6	9	12	14	17	19	22	26	1.5	1.8	2.0	2.2	2.4	1.9	2.2	2.5	3.0	3.3
WW A5	9	14	19	22	28	31	36	43	5	8	12	14	18	20	23	27	1.5	1.7	1.9	2.1	2.3	1.8	2.1	2.4	2.8	3.1
WW A6	9	14	19	23	28	32	37	44	4	8	11	14	18	21	24	28	1.5	1.7	1.9	2.1	2.3	1.7	2.0	2.3	2.7	3.0
WW B1	8	10	11	12	13	14	15	15	6	6	6	6	6	5	4	1	2.4	3.2	4.4	-	-	4.1	-	-	-	-
WW B2	11	16	21	25	31	35	40	48	7	11	15	18	22	25	29	35	1.4	1.6	1.8	2.0	2.2	1.7	1.9	2.1	2.4	2.6
WW B3	14	19	26	31	38	42	49	57	9	13	19	22	28	32	37	44	1.3	1.5	1.7	1.8	2.0	1.5	1.7	1.9	2.1	2.3
WW B4	14	20	28	34	42	48	55	65	9	15	21	26	33	38	45	53	1.2	1.4	1.5	1.7	1.8	1.3	1.5	1.7	1.9	2.0
WW B5	13	20	28	34	43	49	56	66	8	14	22	27	35	40	47	56	1.2	1.3	1.5	1.7	1.8	1.3	1.5	1.6	1.8	1.9
WW C1	15	21	28	33	41	46	52	61	10	15	21	25	31	35	41	48	1.2	1.4	1.6	1.8	1.9	1.4	1.6	1.8	2.0	2.2
WW C2	14	20	27	33	41	46	53	62	9	14	20	25	32	36	42	50	1.2	1.4	1.6	1.7	1.9	1.4	1.6	1.7	1.9	2.1
WW C3	17	26	35	42	52	59	66	76	12	20	28	35	44	50	57	67	1.1	1.2	1.4	1.5	1.7	1.2	1.3	1.5	1.6	1.8
WW C4	15	23	33	40	49	56	64	73	10	18	26	32	41	47	55	65	1.1	1.3	1.4	1.6	1.7	1.2	1.3	1.5	1.7	1.8
TW A1	14	15	17	19	21	22	24	26	12	13	14	14	15	16	16	17	2.3	2.7	3.1	3.5	3.9	3.0	3.7	4.6	-	-
TW A2	12	15	18	20	23	25	28	33	9	11	13	14	16	18	19	21	1.9	2.2	2.4	2.7	2.9	2.3	2.7	3.2	3.7	4.3
TW A3	11	14	17	20	24	27	30	35	8	10	12	14	17	18	20	23	1.7	2.0	2.2	2.5	2.7	2.1	2.4	2.8	3.3	3.7
TW A4	10	13	17	20	24	27	31	37	6	9	11	13	17	19	21	25	1.7	1.9	2.1	2.4	2.5	2.0	2.3	2.6	3.0	3.3
TW B1	13	15	17	19	21	23	25	28	11	12	13	14	16	16	18	19	2.1	2.5	2.8	3.1	3.4	2.7	3.2	3.8	4.5	-
TW B2	11	14	17	19	23	25	29	33	8	10	12	14	16	18	20	23	1.8	2.1	2.4	2.6	2.8	2.2	2.6	2.9	3.3	3.7
TW B3	10	13	16	19	23	26	29	35	7	9	12	14	16	18	21	24	1.8	2.0	2.3	2.5	2.7	2.1	2.4	2.7	3.1	3.4
TW B4	10	12	16	18	22	25	29	34	7	9	11	13	16	18	21	24	1.8	2.0	2.3	2.5	2.7	2.1	2.4	2.7	3.0	3.3
TW C1	11	14	16	18	21	23	26	30	9	11	13	14	16	17	19	21	2.0	2.3	2.6	2.8	3.0	2.5	2.8	3.2	3.6	4.1
TW C2	10	13	16	18	22	24	27	32	8	10	12	14	16	18	20	23	1.9	2.1	2.4	2.6	2.8	2.3	2.6	2.9	3.2	3.5
TW C3	10	13	15	18	21	23	27	31	8	10	12	13	16	18	20	23	1.9	2.2	2.4	2.6	2.8	2.3	2.6	2.9	3.2	3.4
TW C4	11	13	15	17	20	22	25	30	9	10	12	14	16	17	20	23	2.1	2.3	2.5	2.8	2.9	2.5	2.7	3.0	3.3	3.5
TW D1	7	9	10	11	13	14	15	16	5	5	6	6	6	6	6	5	2.4	3.1	3.9	4.9	-	3..7	-	-	-	-
TW D2	12	16	20	24	29	33	37	44	8	11	15	17	21	24	28	33	1.5	1.8	1.9	2.1	2.3	1.8	2.0	2.3	2.5	2.7
TW D3	12	17	22	26	31	35	41	48	8	12	16	19	23	26	30	36	1.4	1.7	1.8	2.0	2.2	1.7	1.9	2.1	2.4	2.6
TW D4	13	19	26	31	38	43	49	58	9	14	19	24	30	34	40	48	1.3	1.5	1.6	1.8	1.9	1.4	1.6	1.8	2.0	2.1
TW D5	13	19	26	32	39	45	52	61	9	14	20	25	32	37	43	51	1.3	1.4	1.6	1.7	1.9	1.4	1.6	1.7	1.9	2.0
TW E1	14	20	26	30	37	41	47	55	10	15	19	23	28	32	36	43	1.3	1.5	1.7	1.9	2.0	1.5	1.7	1.9	2.2	2.3
TW E2	14	20	26	31	38	43	49	57	10	14	20	24	29	34	39	46	1.3	1.5	1.7	1.8	2.0	1.5	1.7	1.9	2.0	2.2
TW E3	17	25	33	39	48	54	61	71	12	19	26	32	40	45	52	61	1.1	1.3	1.5	1.6	1.8	1.2	1.4	1.6	1.7	1.9
TW E4	16	23	31	38	46	52	60	69	11	18	25	30	38	44	51	60	1.1	1.3	1.5	1.6	1.8	1.3	1.4	1.6	1.7	1.9
TW F1	12	13	15	17	19	21	22	25	10	11	12	12	13	14	15	16	2.2	2.6	3.0	3.4	3.8	2.9	3.6	4.4	-	-
TW F2	10	12	15	17	21	23	26	30	7	9	10	12	14	15	17	19	1.9	2.2	2.5	2.8	3.0	2.4	2.8	3.2	3.8	4.4
TW F3	8	11	14	17	21	23	27	32	5	7	9	11	14	15	17	20	1.8	2.1	2.3	2.6	2.7	2.2	2.5	2.9	3.3	3.8
TW F4	6	9	13	16	20	23	26	32	3	5	8	10	13	15	17	21	1.8	2.0	2.3	2.5	2.6	2.1	2.4	2.7	3.1	3.4
TW G1	10	12	14	16	18	20	22	25	7	9	10	11	12	13	15	16	2.1	2.5	2.8	3.1	3.4	2.7	3.2	3.8	4.5	-
TW G2	7	10	13	15	18	21	24	28	5	6	8	10	12	14	16	18	1.9	2.2	2.5	2.7	2.9	2.3	2.7	3.1	3.5	3.9
TW G3	6	8	11	14	18	20	24	28	3	5	7	9	11	13	15	19	1.9	2.1	2.4	2.6	2.8	2.3	2.6	2.9	3.3	3.6
TW G4	4	7	10	12	15	18	21	26	2	4	6	7	10	12	14	18	2.0	2.2	2.4	2.7	2.8	2.3	2.6	2.9	3.2	3.5
TW H1	7	9	11	13	16	18	21	24	5	6	8	9	11	12	13	16	2.1	2.4	2.7	3.0	3.2	2.6	3.0	3.4	3.9	4.4
TW H2	4	7	9	12	15	17	20	24	2	4	6	7	9	11	13	16	2.0	2.3	2.6	2.8	3.0	2.5	2.8	3.1	3.5	3.8
TW H3	3	5	8	10	13	15	18	23	2	3	5	6	8	10	12	15	2.1	2.4	2.6	2.9	3.0	2.6	2.8	3.1	3.4	3.7
TW H4	2	4	6	8	11	13	15	19	1	2	4	5	7	8	10	13	2.4	2.6	2.8	3.1	3.2	2.8	3.1	3.3	3.6	3.8
TW I1	5	7	8	9	11	11	12	13	3	4	4	4	4	4	4	3	2.6	3.2	4.1	-	-	4.0	-	-	-	-
TW I2	9	13	17	21	26	29	34	41	6	9	12	15	18	21	25	30	1.6	1.8	2.0	2.2	2.3	1.8	2.1	2.3	2.6	2.8
TW I3	10	14	19	23	29	33	38	45	6	9	13	17	21	24	28	34	1.5	1.7	1.9	2.1	2.2	1.7	1.9	2.1	2.4	2.6
TW I4	11	16	23	28	35	40	46	55	7	11	17	21	27	31	37	45	1.3	1.5	1.7	1.8	2.0	1.5	1.7	1.8	2.0	2.2
TW I5	11	17	24	29	37	42	49	58	7	12	18	23	29	34	40	49	1.3	1.5	1.6	1.8	1.9	1.4	1.6	1.8	1.9	2.1
TW J1	12	18	23	28	35	39	45	53	8	13	17	21	26	30	35	42	1.4	1.5	1.7	1.9	2.0	1.5	1.7	1.9	2.2	2.3
TW J2	12	17	23	28	35	40	47	55	8	12	17	21	27	31	37	44	1.3	1.5	1.7	1.8	2.0	1.5	1.7	1.9	2.1	2.2
TW J3	14	22	31	37	46	52	59	69	10	17	24	30	38	43	50	60	1.1	1.3	1.5	1.6	1.8	1.2	1.4	1.6	1.7	1.9
TW J4	13	21	29	35	44	50	58	67	9	15	23	28	36	42	49	58	1.2	1.3	1.5	1.6	1.8	1.3	1.4	1.6	1.8	1.9
DG A1	7	8	10	11	12	13	13	13	4	5	5	5	5	5	4	2	2.3	3.1	4.5	-	-	3.4	-	-	-	-
DG A2	9	12	16	18	22	25	28	32	7	9	11	13	16	18	20	23	1.8	2.0	2.3	2.6	3.0	2.1	2.4	2.8	3.3	3.9
DG A3	13	17	22	26	33	37	43	50	10	14	18	21	26	30	35	41	1.5	1.6	1.8	1.9	2.1	1.7	1.8	2.0	2.2	2.4
DG B1	7	8	10	11	13	14	15	15	4	4	5	6	6	6	5	3	2.3	3.1	3.7	-	-	3.4	5.0	-	-	-
DG B2	10	13	16	19	23	26	30	35	7	9	12	14	17	19	21	24	1.8	2.0	2.2	2.5	2.7	2.1	2.3	2.7	3.1	3.4
DG B3	13	18	23	27	34	39	45	54	10	14	19	22	28	31	37	44	1.5	1.6	1.7	1.9	2.0	1.6	1.8	1.9	2.1	2.2
DG C1	9	11	13	15	18	19	22	24	5	7	8	9	10	11	11	11	2.0	2.5	2.8	3.4	4.4	2.6	3.2	4.8	-	-
DG C2	12	15	20	23	28	32	36	43	9	11	15	17	21	24	27	32	1.6	1.8	1.9	2.2	2.4	1.9	2.1	2.3	2.6	2.8
DG C3	15	20	26	31	38	44	51	60	12	16	21	25	31	36	42	50	1.4	1.5	1.6	1.7	1.9	1.5	1.6	1.8	1.9	2.1
SS A1	16	19	23	25	29	32	35	40	13	16	18	21	23	25	28	31	1.6	1.9	2.2	2.4	2.6	1.9	2.2	2.6	3.0	3.5
SS A2	19	25	31	35	42	46	52	60	16	21	26	30	35	39	44	51	1.3	1.5	1.7	1.8	2.0	1.4	1.6	1.8	2.0	2.2
SS A3	14	17	20	22	25	27	30	33	12	14	16	17	19	21	22	23	1.8	2.2	2.5	2.9	3.2	2.1	2.6	3.2	4.2	-
SS A4	19	24	30	34	40	44	50	57	16	20	25	28	34	37	41	47	1.3	1.5	1.7	1.9	2.0	1.4	1.7	1.9	2.2	2.3
SS A5	16	19	22	24	27	29	31	35	14	16	18	19	22	23	24	26	1.8	2.2	2.5	2.8	3.0	2.1	2.6	3.1	3.7	4.5
SS A6	19	24	30	34	41	45	51	58	16	20	25	29	34	38	42	49	1.3	1.5	1.7	1.9	2.0	1.4	1.7	1.9	2.1	2.3
SS A7	15	17	19	20	22	24	25	28	13	14	16	17	18	18	19	20	2.1	2.6	3.1	3.6	4.1	2.5	3.2	4.2	-	-
SS A8	18	23	28	32	38	42	47	55	15	19	24	27	32	35	39	45	1.4	1.6	1.8	2.0	2.1	1.5	1.7	2.0	2.3	2.5
SS B1	13	16	19	21	25	27	30	35	11	13	15	17	19	21	23	26	1.8	2.1	2.4	2.6	2.8	2.1	2.4	2.9	3.3	3.8
SS B2	17	22	27	32	38	42	48	56	14	18	23	27	32	36	40	47	1.4	1.5	1.7	1.9	2.0	1.5	1.7	1.9	2.1	2.3
SS B3	12	14	17	19	22	24	26	30	10	11	13	15	17	18	19	21	1.9	2.3	2.6	3.0	3.3	2.3	2.7	3.3	4.2	-
SS B4	16	21	26	30	36	40	46	53	13	17	22	25	30	34	38	44	1.4	1.6	1.8	1.9	2.1	1.5	1.7	2.0	2.2	2.4
SS B5	13	15	18	19	22	24	26	29	11	13	14	16	17	18	20	21	2.0	2.4	2.8	3.1	3.4	2.4	2.9	3.4	4.2	-
SS B6	16	21	26	30	36	40	46	54	13	17	22	25	30	34	38	45	1.4	1.6	1.8	1.9	2.1	1.5	1.7	1.9	2.2	2.3
SS B7	12	14	16	17	19	20	22	24	10	12	13	14	15	15	16	17	2.3	2.7	3.2	3.7	4.2	2.8	3.4	4.3	-	-
SS B8	15	20	25	28	34	38	43	50	12	16	20	23	28	31	35	41	1.4	1.6	1.8	2.0	2.2	1.6	1.8	2.1	2.3	2.5
SS C1	9	12	15	18	21	24	27	31	6	8	10	11	13	14	16	18	1.8	2.1	2.4	2.7	2.9	2.3	2.7	3.3	4.1	-
SS C2	12	16	22	26	32	36	41	49	8	11	16	19	23	27	31	37	1.4	1.6	1.8	2.0	2.1	1.6	1.9	2.1	2.3	2.5
SS C3	10	12	15	16	19	21	23	27	8	9	11	12	13	14	15	17	2.1	2.4	2.7	3.0	3.3	2.6	3.1	3.6	4.4	-
SS C4	12	15	20	24	29	32	37	44	8	11	15	17	21	24	28	33	1.5	1.8	1.9	2.1	2.3	1.8	2.0	2.3	2.5	2.7
SS D1	17	21	25	29	33	37	41	47	13	17	20	23	26	28	31	34	1.5	1.7	2.0	2.2	2.4	1.7	2.0	2.4	2.8	3.2
SS D2	21	27	35	40	48	54	60	69	17	23	29	34	41	45	51	59	1.2	1.3	1.5	1.7	1.8	1.3	1.5	1.7	1.9	2.0
SS D3	17	20	24	27	30	33	37	42	14	17	19	21	24	26	28	31	1.6	1.9	2.2	2.4	2.6	1.8	2.2	2.6	3.1	3.6
SS D4	20	27	34	39	47	52	58	66	17	22	28	33	39	44	49	57	1.2	1.4	1.5	1.7	1.9	1.3	1.5	1.7	1.9	2.1
SS E1	13	17	20	23	27	29	33	37	10	12	15	17	19	21	22	24	1.6	1.9	2.2	2.5	2.8	1.9	2.4	2.9	3.7	4.7
SS E2	17	22	29	34	41	45	51	60	13	18	23	27	33	37	42	48	1.3	1.5	1.6	1.8	2.0	1.4	1.6	1.8	2.1	2.2
SS E3	14	16	19	21	24	26	29	33	11	13	15	16	18	20	21	23	1.8	2.2	2.5	2.8	3.1	2.2	2.6	3.2	4.0	4.9
SS E4	16	21	27	31	38	42	48	55	13	17	22	26	31	34	39	45	1.3	1.5	1.7	1.9	2.1	1.5	1.7	1.9	2.2	2.4

CINCINNATI, OHIO

TYPE	SOLAR SAVINGS FRACTION (%) BASE TEMP: 55 F (3011 DD) LCR 100	70	50	40	30	25	20	15	65 F (5079 DD) 100	70	50	40	30	25	20	15	CONSERVATION FACTOR 55 F 100	50	30	20	15	65 F 100	50	30	20	15
WW A1	17	19	22	24	27	30	33	38	13	14	16	17	19	20	22	24	1.9	2.1	2.3	2.5	2.7	2.5	2.8	3.2	3.8	4.3
WW A2	16	20	25	29	35	39	44	51	11	14	17	20	23	26	29	34	1.5	1.7	1.8	2.0	2.1	1.8	2.0	2.3	2.6	2.8
WW A3	16	21	27	31	37	42	48	56	10	13	17	21	25	28	32	38	1.4	1.5	1.7	1.9	2.0	1.6	1.9	2.1	2.4	2.5
WW A4	15	21	27	32	39	44	50	59	9	13	18	21	26	29	34	40	1.3	1.5	1.6	1.8	2.0	1.6	1.8	2.0	2.2	2.4
WW A5	15	21	28	33	41	46	53	62	8	13	18	22	27	31	36	42	1.3	1.4	1.6	1.8	1.9	1.5	1.7	1.9	2.1	2.3
WW A6	15	21	29	34	42	47	54	63	8	13	18	22	28	31	37	44	1.2	1.4	1.6	1.7	1.9	1.5	1.7	1.9	2.1	2.3
WW B1	12	15	18	20	24	26	29	34	8	9	10	11	13	13	14	15	1.8	2.1	2.4	2.6	2.8	2.6	3.2	4.2	-	-
WW B2	16	22	29	35	43	48	55	64	10	15	20	24	30	34	40	47	1.2	1.4	1.6	1.7	1.9	1.4	1.6	1.8	2.0	2.2
WW B3	20	27	35	42	50	56	63	72	13	18	25	30	37	42	48	57	1.1	1.3	1.4	1.6	1.7	1.3	1.5	1.7	1.8	2.0
WW B4	20	28	38	45	54	60	68	77	13	20	27	33	42	47	55	64	1.1	1.2	1.4	1.5	1.7	1.2	1.4	1.5	1.7	1.8
WW B5	19	27	37	44	54	60	68	78	12	19	27	34	42	48	56	66	1.1	1.2	1.4	1.5	1.7	1.2	1.3	1.5	1.7	1.8
WW C1	21	29	38	44	53	59	66	75	14	20	27	33	40	45	52	61	1.1	1.2	1.4	1.6	1.7	1.2	1.4	1.6	1.8	1.9
WW C2	20	28	36	43	52	59	66	75	13	19	27	32	40	45	52	62	1.1	1.2	1.4	1.6	1.7	1.2	1.4	1.6	1.7	1.9
WW C3	24	34	45	53	63	70	77	86	16	25	35	42	52	58	66	76	1.0	1.1	1.3	1.4	1.6	1.1	1.2	1.4	1.5	1.7
WW C4	21	31	42	50	60	67	74	83	14	23	32	39	49	55	63	73	1.0	1.1	1.3	1.5	1.6	1.1	1.2	1.4	1.6	1.7
TW A1	16	19	22	24	28	30	33	38	13	14	16	18	20	21	23	25	1.8	2.1	2.3	2.5	2.7	2.4	2.7	3.1	3.5	3.9
TW A2	16	19	24	27	32	36	40	47	11	14	17	19	22	25	28	32	1.6	1.8	1.9	2.1	2.2	1.9	2.2	2.4	2.7	2.9
TW A3	15	19	24	28	34	38	43	51	10	13	16	19	23	26	30	35	1.5	1.6	1.8	2.0	2.1	1.8	2.0	2.2	2.5	2.7
TW A4	14	18	24	28	34	39	44	52	9	12	16	19	23	26	31	36	1.4	1.6	1.8	1.9	2.1	1.7	1.9	2.1	2.4	2.5
TW B1	16	18	22	24	28	31	35	40	12	14	16	18	20	22	24	27	1.8	2.0	2.2	2.4	2.5	2.2	2.5	2.8	3.1	3.4
TW B2	14	18	23	26	31	35	40	47	10	13	16	18	22	24	28	33	1.5	1.7	1.9	2.1	2.2	1.9	2.1	2.3	2.6	2.8
TW B3	14	18	23	26	32	36	41	48	9	12	15	18	22	25	29	34	1.5	1.7	1.9	2.0	2.2	1.8	2.0	2.2	2.5	2.7
TW B4	13	17	21	25	30	34	40	47	9	12	15	18	21	24	28	33	1.5	1.7	1.9	2.0	2.2	1.8	2.0	2.3	2.5	2.7
TW C1	14	17	21	24	28	31	36	42	11	13	15	17	20	22	25	29	1.7	1.9	2.1	2.3	2.4	2.1	2.3	2.6	2.8	3.0
TW C2	13	17	21	24	29	32	37	44	10	12	15	17	21	23	27	31	1.6	1.8	2.0	2.1	2.3	2.0	2.2	2.4	2.6	2.8
TW C3	13	16	20	23	28	31	36	43	9	12	15	17	20	23	26	31	1.7	1.9	2.0	2.2	2.3	2.0	2.2	2.4	2.6	2.8
TW C4	13	16	19	22	26	29	33	39	10	12	14	16	19	22	25	29	1.8	2.0	2.2	2.3	2.4	2.2	2.3	2.5	2.7	2.9
TW D1	10	13	16	18	21	24	27	31	6	8	9	10	12	13	14	15	1.9	2.2	2.4	2.7	2.9	2.6	3.1	3.8	4.8	-
TW D2	16	21	27	32	39	44	50	59	11	15	19	23	28	32	37	44	1.3	1.5	1.6	1.8	1.9	1.6	1.7	1.9	2.1	2.3
TW D3	17	23	30	35	43	48	54	63	11	16	21	25	31	35	40	48	1.3	1.4	1.6	1.7	1.9	1.5	1.6	1.8	2.0	2.2
TW D4	18	26	34	40	49	55	62	71	12	18	25	30	37	42	49	58	1.2	1.3	1.4	1.6	1.7	1.3	1.5	1.6	1.8	1.9
TW D5	18	26	34	41	50	56	63	73	12	18	26	31	39	44	51	61	1.1	1.3	1.4	1.6	1.7	1.3	1.4	1.6	1.7	1.9
TW E1	20	26	34	40	48	54	61	70	14	19	25	29	36	41	47	55	1.2	1.3	1.5	1.6	1.8	1.3	1.5	1.7	1.9	2.0
TW E2	19	26	34	40	49	55	62	71	13	19	25	30	37	42	48	57	1.2	1.3	1.5	1.6	1.8	1.3	1.5	1.7	1.8	2.0
TW E3	23	32	42	50	59	66	73	82	16	24	32	39	48	54	61	71	1.0	1.2	1.3	1.5	1.6	1.1	1.3	1.5	1.6	1.8
TW E4	21	30	40	47	57	64	71	80	15	22	31	37	46	52	60	69	1.0	1.2	1.3	1.5	1.6	1.2	1.3	1.5	1.6	1.8
TW F1	15	17	20	23	26	29	33	38	11	13	15	16	18	20	22	24	1.8	2.1	2.3	2.4	2.6	2.3	2.6	3.0	3.4	3.8
TW F2	13	17	21	25	29	33	38	45	9	11	14	16	20	22	25	29	1.6	1.8	2.0	2.1	2.3	2.0	2.2	2.5	2.7	3.0
TW F3	12	16	21	25	30	34	40	47	7	10	13	16	20	23	26	31	1.5	1.7	1.9	2.0	2.2	1.8	2.1	2.3	2.5	2.8
TW F4	10	14	19	23	29	33	39	47	5	8	12	15	19	22	26	31	1.5	1.7	1.8	2.0	2.1	1.8	2.0	2.2	2.5	2.7
TW G1	13	15	19	22	25	28	32	38	9	11	13	15	17	19	21	25	1.8	2.0	2.2	2.4	2.5	2.2	2.5	2.8	3.1	3.4
TW G2	11	14	18	22	27	30	35	42	6	9	12	14	18	20	23	28	1.6	1.8	2.0	2.1	2.3	2.0	2.2	2.4	2.7	2.9
TW G3	9	13	17	21	26	29	35	42	5	7	11	13	17	19	23	28	1.6	1.8	2.0	2.1	2.3	1.9	2.2	2.4	2.6	2.8
TW G4	7	10	14	18	23	26	31	38	3	6	9	11	14	17	20	25	1.7	1.9	2.0	2.2	2.3	2.0	2.2	2.4	2.7	2.8
TW H1	10	13	16	19	23	26	30	36	6	8	10	12	15	17	20	24	1.8	2.0	2.2	2.3	2.5	2.2	2.4	2.7	3.0	3.2
TW H2	7	10	14	17	22	25	29	36	4	6	9	11	14	16	19	24	1.7	1.9	2.1	2.3	2.4	2.1	2.3	2.6	2.8	3.0
TW H3	6	9	12	15	19	22	26	33	3	5	7	9	12	14	17	22	1.8	2.0	2.2	2.4	2.5	2.2	2.4	2.6	2.8	3.0
TW H4	4	7	10	12	16	18	22	27	2	4	6	7	10	12	14	18	2.0	2.2	2.4	2.6	2.7	2.5	2.7	2.9	3.1	3.2
TW I1	8	11	13	16	19	21	24	28	5	6	7	8	10	11	12	13	2.0	2.3	2.6	2.8	3.0	2.7	3.3	4.0	-	-
TW I2	13	18	24	29	36	40	47	55	8	12	16	20	25	29	33	40	1.4	1.5	1.7	1.8	2.0	1.6	1.8	2.0	2.2	2.3
TW I3	14	20	27	32	39	44	51	60	9	13	18	22	28	33	37	43	1.3	1.5	1.6	1.8	1.9	1.5	1.7	1.9	2.1	2.2
TW I4	16	23	31	37	45	51	59	68	10	15	22	27	34	39	46	55	1.2	1.3	1.5	1.6	1.8	1.3	1.5	1.7	1.8	2.0
TW I5	16	23	32	38	47	53	61	71	10	16	23	28	36	42	49	58	1.2	1.3	1.4	1.6	1.7	1.3	1.4	1.6	1.8	1.9
TW J1	18	24	32	38	46	51	58	68	11	17	23	27	34	39	45	53	1.2	1.3	1.5	1.7	1.8	1.4	1.5	1.7	1.9	2.0
TW J2	17	24	31	37	46	52	59	69	11	16	23	28	35	40	46	55	1.2	1.3	1.5	1.6	1.8	1.3	1.5	1.7	1.8	2.0
TW J3	20	29	39	47	57	63	71	80	14	21	30	36	45	52	59	69	1.0	1.2	1.3	1.5	1.6	1.1	1.3	1.5	1.6	1.8
TW J4	19	27	37	44	54	61	69	78	12	20	28	34	43	49	57	67	1.1	1.2	1.4	1.5	1.7	1.2	1.3	1.5	1.6	1.8
DG A1	10	13	16	19	21	23	25	27	6	7	9	10	11	12	12	12	1.8	2.1	2.7	3.6	-	2.4	3.0	4.4	-	-
DG A2	13	17	22	26	31	35	39	45	9	12	15	18	21	24	27	32	1.5	1.7	1.9	2.2	2.4	1.8	2.0	2.3	2.6	3.0
DG A3	16	22	29	34	42	47	54	63	12	16	22	26	32	36	42	50	1.3	1.4	1.6	1.7	1.9	1.5	1.6	1.8	1.9	2.1
DG B1	10	13	16	19	23	25	28	32	6	7	9	10	12	13	14	14	1.8	2.0	2.3	2.7	3.2	2.3	2.9	4.0	-	-
DG B2	13	17	22	26	33	37	42	50	9	12	15	18	22	25	29	34	1.5	1.6	1.8	2.0	2.1	1.8	2.0	2.2	2.4	2.7
DG B3	17	22	29	35	44	50	57	67	12	17	23	27	33	38	44	53	1.3	1.4	1.5	1.6	1.8	1.5	1.6	1.7	1.8	2.0
DG C1	13	16	21	24	29	33	37	43	8	10	13	15	17	18	20	23	1.6	1.8	2.0	2.2	2.5	1.9	2.4	3.0	3.4	4.2
DG C2	16	20	26	31	38	44	50	59	11	15	19	22	27	31	36	42	1.4	1.5	1.6	1.8	1.9	1.6	1.8	2.0	2.2	2.3
DG C3	19	25	33	40	49	56	64	74	14	20	26	31	38	43	50	60	1.2	1.3	1.4	1.5	1.7	1.4	1.5	1.6	1.7	1.9
SS A1	20	24	29	33	38	42	47	55	15	19	23	25	29	32	36	41	1.4	1.6	1.8	2.0	2.1	1.6	1.9	2.1	2.4	2.6
SS A2	24	31	38	44	52	57	64	72	19	25	31	36	42	47	52	60	1.2	1.3	1.5	1.6	1.8	1.3	1.5	1.7	1.8	2.0
SS A3	18	22	26	29	34	37	42	48	14	17	20	22	25	27	30	34	1.5	1.8	1.9	2.1	2.3	1.8	2.1	2.5	2.8	3.1
SS A4	24	30	37	43	50	55	62	70	19	24	30	34	40	45	50	58	1.2	1.3	1.5	1.7	1.8	1.3	1.5	1.7	1.9	2.0
SS A5	19	23	27	30	35	38	42	49	16	19	21	24	27	29	32	36	1.6	1.8	2.0	2.1	2.3	1.8	2.1	2.4	2.7	3.0
SS A6	24	30	38	43	51	56	63	71	19	24	30	35	41	46	51	59	1.2	1.3	1.5	1.6	1.8	1.3	1.5	1.7	1.9	2.0
SS A7	18	21	24	26	30	32	36	40	15	17	19	20	23	24	26	28	1.7	2.0	2.3	2.5	2.6	2.1	2.5	2.9	3.4	3.9
SS A8	23	29	36	41	49	54	60	68	18	23	29	33	39	43	48	55	1.2	1.4	1.5	1.7	1.8	1.3	1.6	1.8	1.9	2.1
SS B1	16	20	25	28	33	37	42	48	13	16	19	21	25	27	31	35	1.5	1.7	1.9	2.1	2.2	1.8	2.0	2.3	2.6	2.8
SS B2	21	27	34	40	47	52	59	68	16	22	27	32	38	42	48	56	1.2	1.4	1.5	1.7	1.8	1.3	1.5	1.7	1.9	2.0
SS B3	15	19	23	26	30	33	38	44	12	14	17	19	22	24	27	30	1.6	1.8	2.0	2.2	2.4	1.9	2.2	2.6	2.9	3.2
SS B4	20	26	33	38	46	51	57	66	16	21	26	30	36	41	46	54	1.2	1.4	1.5	1.7	1.9	1.4	1.6	1.8	1.9	2.1
SS B5	16	19	22	25	29	32	36	41	13	15	17	19	22	24	26	29	1.7	2.0	2.2	2.3	2.5	2.1	2.4	2.7	3.0	3.3
SS B6	20	26	33	38	46	51	58	66	16	20	26	30	37	41	46	54	1.2	1.4	1.5	1.7	1.8	1.4	1.6	1.7	1.9	2.1
SS B7	15	17	20	22	25	28	31	35	12	13	15	17	19	20	22	24	1.9	2.2	2.4	2.6	2.8	2.3	2.7	3.1	3.6	4.1
SS B8	19	25	31	36	44	48	55	63	15	19	25	29	34	38	43	51	1.3	1.4	1.6	1.7	1.9	1.4	1.6	1.8	2.0	2.2
SS C1	14	18	23	27	32	36	41	49	8	11	15	17	21	23	26	31	1.5	1.7	1.8	2.0	2.2	1.8	2.1	2.4	2.7	2.9
SS C2	17	23	30	35	43	48	55	64	11	16	21	25	31	35	41	49	1.2	1.4	1.6	1.7	1.9	1.4	1.6	1.8	2.0	2.1
SS C3	13	16	20	23	27	30	34	40	9	11	14	16	18	20	23	26	1.7	1.9	2.1	2.3	2.4	2.1	2.4	2.7	3.0	3.3
SS C4	16	21	27	32	38	43	49	58	11	15	19	23	28	32	37	43	1.3	1.5	1.7	1.8	2.0	1.6	1.7	1.9	2.1	2.3
SS D1	22	27	34	38	45	50	36	64	17	21	26	29	34	37	42	48	1.3	1.4	1.6	1.8	1.9	1.4	1.7	1.9	2.2	2.4
SS D2	27	35	44	51	60	66	73	81	21	28	35	41	49	54	61	69	1.0	1.2	1.3	1.5	1.7	1.1	1.3	1.5	1.7	1.8
SS D3	21	25	31	35	41	45	50	57	17	20	24	27	31	34	37	43	1.4	1.6	1.7	1.9	2.1	1.6	1.8	2.1	2.4	2.6
SS D4	26	34	43	49	58	64	70	79	20	27	34	40	47	52	59	67	1.1	1.2	1.4	1.5	1.7	1.2	1.4	1.5	1.7	1.9
SS E1	18	23	28	32	38	43	48	56	13	16	20	23	27	30	33	38	1.4	1.6	1.7	1.9	2.1	1.6	1.9	2.2	2.5	2.7
SS E2	22	30	38	44	53	58	65	74	16	22	29	34	41	46	52	60	1.1	1.3	1.4	1.6	1.7	1.2	1.4	1.6	1.8	2.0
SS E3	17	21	25	28	33	36	41	47	13	16	19	21	24	26	29	33	1.6	1.8	2.0	2.1	2.3	1.8	2.1	2.5	2.8	3.1
SS E4	21	28	35	41	49	54	61	69	16	21	27	32	38	42	48	56	1.2	1.3	1.5	1.7	1.8	1.3	1.5	1.7	1.9	2.0

CLEVELAND, OHIO

TYPE	SOLAR SAVINGS FRACTION (%)																CONSERVATION FACTOR									
BASE TEMP:	55 F (3827 DD)								65 F (6160 DD)								55 F					65 F				
LCR	100	70	50	40	30	25	20	15	100	70	50	40	30	25	20	15	100	50	30	20	15	100	50	30	20	15
WW A1	13	15	16	17	19	20	21	22	11	12	13	13	13	14	14	13	2.4	3.0	3.7	4.5	-	3.5	4.8	-	-	-
WW A2	11	14	17	19	22	24	27	31	8	10	12	13	15	16	17	18	1.8	2.2	2.5	2.8	3.1	2.3	2.8	3.5	4.6	-
WW A3	10	13	17	19	23	26	29	34	6	9	11	13	15	16	18	20	1.7	2.0	2.3	2.6	2.8	2.1	2.5	3.0	3.7	4.6
WW A4	9	13	17	20	24	27	31	36	5	8	10	12	15	17	19	22	1.6	1.9	2.2	2.4	2.6	2.0	2.4	2.8	3.4	4.0
WW A5	9	12	17	20	25	28	32	38	4	7	10	12	15	17	20	23	1.6	1.8	2.1	2.3	2.5	1.9	2.2	2.7	3.2	3.7
WW A6	8	12	17	20	25	29	33	39	4	7	10	12	16	18	20	24	1.5	1.8	2.0	2.2	2.4	1.8	2.2	2.6	3.0	3.5
WW B1	8	9	10	11	12	12	12	11	5	5	5	5	4	3	1	-2	2.6	3.8	-	-	-	-	-	-	-	-
WW B2	10	15	19	23	28	32	37	44	6	10	13	16	20	23	27	31	1.5	1.7	1.9	2.1	2.3	1.7	2.0	2.3	2.6	2.8
WW B3	13	18	24	28	35	39	45	53	8	12	17	21	26	29	33	40	1.3	1.5	1.7	1.9	2.1	1.5	1.8	2.0	2.3	2.5
WW B4	12	19	26	32	40	45	52	61	8	13	20	24	31	36	42	50	1.2	1.4	1.6	1.7	1.9	1.4	1.5	1.7	1.9	2.1
WW B5	12	19	26	32	41	46	53	63	8	13	20	25	33	38	44	53	1.2	1.4	1.5	1.7	1.8	1.3	1.5	1.7	1.9	2.0
WW C1	14	20	26	31	38	43	49	57	9	14	19	23	29	33	38	44	1.3	1.5	1.7	1.8	2.0	1.5	1.7	1.9	2.1	2.3
WW C2	13	19	25	31	38	43	50	58	8	13	19	23	29	34	39	47	1.3	1.5	1.6	1.8	1.9	1.4	1.6	1.8	2.0	2.2
WW C3	16	24	33	40	50	56	64	73	11	18	27	33	42	48	55	64	1.1	1.2	1.4	1.6	1.7	1.2	1.3	1.5	1.7	1.8
WW C4	14	22	31	38	47	53	61	71	9	16	24	30	39	45	52	62	1.1	1.3	1.4	1.6	1.7	1.2	1.4	1.5	1.7	1.8
TW A1	13	15	16	18	19	20	22	24	11	12	13	14	14	15	15	15	2.4	2.8	3.4	4.0	4.6	3.2	4.1	-	-	-
TW A2	11	14	16	18	21	23	26	29	9	10	12	13	15	16	17	19	1.9	2.3	2.6	2.9	3.2	2.4	2.9	3.5	4.4	-
TW A3	10	13	16	19	22	24	28	32	7	9	11	13	15	16	18	20	1.8	2.1	2.4	2.7	2.9	2.2	2.6	3.1	3.7	4.4
TW A4	9	12	16	18	22	25	28	33	6	8	10	12	15	17	19	22	1.7	2.0	2.3	2.5	2.7	2.1	2.4	2.8	3.3	3.7
TW B1	12	14	16	18	20	21	23	26	10	11	13	13	14	15	16	17	2.2	2.6	3.0	3.5	3.8	2.9	3.5	4.3	-	-
TW B2	10	13	16	18	21	23	26	30	8	10	11	13	15	16	18	20	1.9	2.2	2.5	2.8	3.0	2.3	2.7	3.2	3.7	4.3
TW B3	9	12	15	18	21	24	27	31	7	9	11	13	15	16	19	21	1.8	2.1	2.4	2.6	2.9	2.2	2.6	3.0	3.4	3.8
TW B4	9	12	15	17	21	23	26	31	7	9	11	12	15	17	19	22	1.8	2.1	2.4	2.6	2.8	2.2	2.5	2.9	3.3	3.6
TW C1	11	13	16	17	20	22	24	28	9	10	12	13	15	16	17	19	2.1	2.4	2.7	3.0	3.3	2.6	3.0	3.5	4.0	4.7
TW C2	10	12	15	17	20	22	25	30	8	9	11	13	15	16	18	21	1.9	2.2	2.5	2.8	3.0	2.4	2.7	3.1	3.5	3.8
TW C3	10	12	15	17	20	22	25	29	8	9	11	13	15	16	18	21	2.0	2.3	2.5	2.8	3.0	2.4	2.7	3.0	3.4	3.7
TW C4	10	12	15	17	19	21	24	28	9	10	12	13	15	17	18	21	2.1	2.4	2.6	2.9	3.1	2.5	2.8	3.1	3.5	3.7
TW D1	7	8	9	10	11	12	12	13	4	5	5	5	5	5	4	2	2.6	3.4	4.7	-	-	4.2	-	-	-	-
TW D2	11	15	19	22	27	30	35	41	8	11	14	16	20	22	25	30	1.6	1.8	2.0	2.2	2.4	1.9	2.1	2.4	2.7	3.0
TW D3	11	15	20	24	29	33	38	44	8	11	14	17	21	24	27	32	1.5	1.7	1.9	2.1	2.3	1.7	2.0	2.3	2.6	2.8
TW D4	12	18	24	29	36	40	47	55	9	13	18	22	28	32	37	45	1.3	1.5	1.7	1.9	2.0	1.5	1.7	1.9	2.1	2.2
TW D5	12	18	25	30	37	42	49	58	9	14	19	24	30	35	41	49	1.3	1.5	1.6	1.8	1.9	1.4	1.6	1.8	2.0	2.1
TW E1	14	18	24	28	34	38	44	51	10	14	18	21	26	29	34	40	1.4	1.6	1.8	2.0	2.1	1.6	1.8	2.0	2.3	2.5
TW E2	13	18	24	29	35	40	46	54	9	14	18	22	28	31	36	43	1.3	1.5	1.7	1.9	2.0	1.5	1.7	1.9	2.1	2.3
TW E3	16	23	31	37	46	51	59	68	12	18	25	30	38	43	49	58	1.2	1.3	1.5	1.7	1.8	1.3	1.4	1.6	1.8	1.9
TW E4	15	22	30	36	44	50	57	67	10	17	24	29	37	42	48	58	1.2	1.3	1.5	1.7	1.8	1.3	1.4	1.6	1.8	1.9
TW F1	11	13	15	16	18	19	20	22	9	10	11	12	12	13	13	14	2.3	2.8	3.3	3.9	4.4	3.1	4.0	-	-	-
TW F2	9	11	14	16	19	21	23	27	7	8	10	11	12	13	15	16	2.0	2.3	2.7	3.0	3.3	2.5	3.0	3.6	4.5	-
TW F3	7	10	13	16	19	21	24	29	5	6	8	10	12	13	15	17	1.8	2.2	2.5	2.8	3.0	2.3	2.7	3.2	3.8	4.4
TW F4	6	8	12	14	18	20	24	29	3	5	7	9	11	13	15	18	1.8	2.1	2.4	2.6	2.8	2.2	2.6	2.9	3.4	3.8
TW G1	9	11	13	15	17	18	20	23	7	8	9	10	11	12	13	14	2.2	2.6	3.0	3.5	3.8	2.9	3.5	4.3	-	-
TW G2	7	9	12	14	17	19	22	26	4	6	8	9	11	12	14	16	2.0	2.3	2.6	2.9	3.2	2.5	2.9	3.4	3.9	4.6
TW G3	5	8	11	13	16	18	21	26	3	4	6	8	10	12	14	16	1.9	2.2	2.5	2.8	3.0	2.4	2.7	3.1	3.6	4.0
TW G4	4	6	9	11	14	16	19	24	2	3	5	7	9	11	13	16	2.0	2.3	2.6	2.8	3.0	2.4	2.8	3.1	3.4	3.8
TW H1	6	8	11	12	15	17	19	22	4	6	7	8	10	11	12	14	2.2	2.5	2.9	3.2	3.5	2.7	3.2	3.7	4.4	-
TW H2	4	6	9	11	14	16	18	22	2	4	5	7	9	10	12	14	2.1	2.4	2.7	3.0	3.2	2.6	2.9	3.3	3.8	4.2
TW H3	3	5	7	9	12	14	17	21	1	3	4	6	8	9	11	14	2.2	2.5	2.7	3.0	3.2	2.6	3.0	3.3	3.7	4.0
TW H4	2	4	6	7	10	12	14	18	1	2	4	5	6	8	9	12	2.4	2.7	2.9	3.2	3.4	2.9	3.2	3.5	3.8	4.1
TW I1	5	6	7	8	9	10	10	10	3	3	3	3	3	3	2	0	2.7	3.6	5.0	-	-	4.5	-	-	-	-
TW I2	8	12	16	19	24	27	31	38	5	8	11	13	17	19	22	27	1.6	1.9	2.1	2.3	2.5	1.9	2.2	2.4	2.7	3.0
TW I3	9	13	18	21	27	30	35	42	6	9	12	15	19	22	26	31	1.5	1.8	2.0	2.1	2.3	1.8	2.0	2.3	2.5	2.8
TW I4	10	15	21	26	33	37	44	52	6	11	16	20	25	29	35	42	1.4	1.5	1.7	1.9	2.0	1.5	1.7	1.9	2.1	2.3
TW I5	10	16	22	27	35	40	46	56	6	11	17	21	28	32	38	46	1.3	1.5	1.7	1.8	2.0	1.5	1.6	1.8	2.0	2.1
TW J1	12	16	22	26	32	36	42	50	8	12	16	19	24	28	32	38	1.4	1.6	1.8	2.0	2.1	1.6	1.8	2.0	2.3	2.5
TW J2	11	16	22	26	33	38	44	52	7	11	16	20	25	29	34	41	1.4	1.6	1.7	1.9	2.0	1.5	1.7	1.9	2.1	2.3
TW J3	13	21	29	35	44	49	57	66	9	15	23	28	36	41	48	57	1.2	1.3	1.5	1.7	1.8	1.3	1.4	1.6	1.8	1.9
TW J4	12	19	27	33	42	48	55	65	8	14	21	27	34	40	46	56	1.2	1.4	1.5	1.7	1.8	1.3	1.5	1.6	1.8	1.9
DG A1	6	8	9	10	11	11	11	10	3	4	4	4	4	4	2	0	2.4	3.4	-	-	-	3.7	-	-	-	-
DG A2	9	12	15	17	21	23	26	30	7	8	11	12	15	16	18	21	1.8	2.1	2.5	2.8	3.2	2.2	2.5	3.0	3.6	4.3
DG A3	13	17	21	25	31	35	40	48	10	13	17	20	25	28	33	39	1.5	1.7	1.8	2.0	2.2	1.7	1.9	2.1	2.3	2.5
DG B1	6	8	9	10	11	11	12	12	3	4	4	5	4	4	3	1	2.4	3.4	4.7	-	-	3.8	-	-	-	-
DG B2	9	12	15	18	21	24	27	32	7	9	11	13	15	17	19	22	1.8	2.1	2.3	2.6	2.9	2.2	2.4	2.9	3.4	3.8
DG B3	13	17	22	26	32	37	43	51	10	14	18	21	26	30	35	42	1.5	1.7	1.8	1.9	2.1	1.7	1.8	2.0	2.2	2.3
DG C1	8	10	12	14	16	17	19	20	5	6	7	8	9	9	9	8	2.1	2.7	3.2	4.0	-	2.9	3.6	-	-	-
DG C2	11	15	19	22	26	29	34	40	8	11	14	16	20	22	25	29	1.6	1.9	2.0	2.3	2.5	1.9	2.1	2.5	2.8	3.1
DG C3	15	20	25	30	37	42	48	57	12	16	21	24	30	34	39	47	1.4	1.6	1.6	1.8	1.9	1.6	1.7	1.9	2.0	2.1
SS A1	15	18	21	24	27	30	33	37	13	15	18	19	22	24	26	28	1.7	2.0	2.3	2.6	2.8	2.0	2.3	2.8	3.4	4.0
SS A2	19	24	29	34	40	44	49	57	16	20	25	28	34	37	42	48	1.3	1.5	1.7	1.9	2.1	1.4	1.7	1.9	2.1	2.3
SS A3	14	16	19	21	23	25	27	30	12	13	15	16	18	19	20	21	1.9	2.3	2.7	3.2	3.6	2.2	2.8	3.7	-	-
SS A4	18	23	28	32	38	42	47	54	15	19	24	27	32	35	39	45	1.4	1.6	1.8	2.0	2.1	1.5	1.7	2.0	2.3	2.5
SS A5	16	18	21	23	25	27	29	32	14	15	17	19	20	21	23	24	1.9	2.3	2.7	3.0	3.4	2.2	2.7	3.4	4.3	-
SS A6	18	23	29	33	39	43	48	55	15	19	24	27	32	36	40	46	1.3	1.6	1.8	1.9	2.1	1.5	1.7	2.0	2.2	2.4
SS A7	15	16	18	19	21	22	23	25	13	14	15	16	17	17	18	17	2.2	2.7	3.4	4.2	5.0	2.7	3.5	5.0	-	-
SS A8	18	22	27	31	36	40	45	51	15	18	23	26	30	33	37	41	1.4	1.6	1.8	2.0	2.2	1.5	1.8	2.1	2.4	2.6
SS B1	13	15	18	20	23	25	28	32	10	12	14	16	18	19	21	23	1.8	2.2	2.5	2.8	3.1	2.2	2.6	3.1	3.7	4.5
SS B2	16	21	26	30	36	40	45	53	13	17	22	25	30	34	38	44	1.4	1.6	1.8	2.0	2.1	1.5	1.7	2.0	2.2	2.4
SS B3	12	14	16	18	20	22	24	27	9	11	13	14	15	16	17	18	2.0	2.4	2.8	3.3	3.7	2.4	2.9	3.7	-	-
SS B4	16	20	25	29	34	38	43	50	13	17	21	24	29	32	36	42	1.4	1.6	1.8	2.0	2.2	1.6	1.8	2.0	2.3	2.5
SS B5	13	15	17	19	21	22	24	27	11	12	14	15	16	17	18	19	2.1	2.5	3.0	3.4	3.8	2.5	3.1	3.8	4.9	-
SS B6	15	20	25	29	34	38	44	51	13	16	21	24	29	32	36	42	1.4	1.6	1.8	2.0	2.2	1.6	1.8	2.0	2.3	2.5
SS B7	12	13	15	16	18	19	20	22	10	11	12	13	14	14	15	15	2.3	2.9	3.5	4.3	-	2.9	3.7	-	-	-
SS B8	15	19	23	27	32	36	41	47	12	15	19	22	26	29	33	38	1.5	1.7	1.9	2.1	2.3	1.6	1.9	2.1	2.4	2.6
SS C1	9	11	14	16	19	21	24	27	5	7	9	10	11	12	13	14	1.9	2.3	2.6	3.0	3.3	2.4	3.0	3.8	-	-
SS C2	11	15	20	24	29	33	38	45	7	11	14	17	21	24	28	33	1.5	1.7	1.9	2.1	2.3	1.7	2.0	2.2	2.5	2.7
SS C3	10	12	14	15	18	19	21	24	7	9	10	11	12	13	14	15	2.1	2.5	2.9	3.4	3.7	2.8	3.4	4.1	-	-
SS C4	11	15	19	22	27	30	34	41	8	11	14	16	20	22	26	30	1.6	1.8	2.0	2.2	2.4	1.9	2.1	2.4	2.7	2.9
SS D1	16	20	24	27	31	34	38	43	13	16	19	21	24	26	28	31	1.5	1.8	2.1	2.3	2.6	1.7	2.1	2.6	3.2	3.8
SS D2	20	26	33	38	46	51	57	65	16	21	28	32	38	43	48	55	1.2	1.4	1.6	1.7	1.9	1.3	1.5	1.7	1.9	2.1
SS D3	16	19	23	25	29	31	34	38	14	16	18	20	23	24	26	28	1.7	2.0	2.3	2.6	2.8	1.9	2.3	2.8	3.5	4.3
SS D4	19	25	32	37	44	49	55	63	16	21	27	31	37	41	46	53	1.2	1.4	1.6	1.8	1.9	1.3	1.5	1.8	2.0	2.1
SS E1	13	16	19	21	24	27	30	33	9	12	14	15	17	18	19	20	1.7	2.1	2.4	2.8	3.1	2.0	2.6	3.4	4.8	-
SS E2	16	21	27	32	38	43	48	56	12	17	22	25	31	34	39	45	1.3	1.5	1.7	1.9	2.0	1.4	1.7	1.9	2.2	2.4
SS E3	13	16	18	20	23	24	27	30	11	13	14	16	17	18	19	20	1.9	2.3	2.7	3.1	3.5	2.3	2.8	3.6	4.7	-
SS E4	16	20	26	30	36	40	45	52	12	16	21	24	29	32	36	42	1.4	1.6	1.8	2.0	2.1	1.5	1.8	2.0	2.3	2.5

COLUMBUS, OHIO

TYPE	SOLAR SAVINGS FRACTION (%) BASE TEMP: 55 F (3491 DD) LCR 100	70	50	40	30	25	20	15	65 F (5705 DD) 100	70	50	40	30	25	20	15	CONSERVATION FACTOR 55 F 100	50	30	20	15	65 F 100	50	30	20	15
WW A1	15	17	19	20	23	24	26	29	12	13	14	15	16	17	17	18	2.1	2.5	2.8	3.2	3.5	2.9	3.5	4.3	-	-
WW A2	13	17	20	23	28	31	35	40	9	11	14	16	19	20	23	26	1.6	1.9	2.1	2.3	2.5	2.0	2.4	2.7	3.2	3.6
WW A3	12	16	21	25	30	33	38	45	8	11	14	16	20	22	25	29	1.5	1.7	1.9	2.1	2.3	1.8	2.1	2.4	2.8	3.1
WW A4	12	16	21	25	31	35	40	47	7	10	14	16	20	23	26	31	1.5	1.7	1.9	2.0	2.2	1.7	2.0	2.3	2.6	2.9
WW A5	11	16	22	26	32	37	42	50	6	9	13	16	21	24	27	33	1.4	1.6	1.8	2.0	2.1	1.7	1.9	2.2	2.5	2.7
WW A6	11	16	22	27	33	37	43	51	5	9	13	17	21	24	28	34	1.4	1.6	1.7	1.9	2.1	1.6	1.9	2.1	2.4	2.6
WW B1	10	12	14	15	17	18	20	22	6	7	7	8	8	8	7	6	2.2	2.7	3.2	3.8	4.5	3.3	-	-	-	-
WW B2	13	18	24	28	35	39	46	54	8	12	16	20	25	28	33	39	1.4	1.5	1.7	1.9	2.0	1.6	1.8	2.0	2.2	2.4
WW B3	16	22	29	34	42	47	54	63	10	15	21	25	31	35	41	48	1.2	1.4	1.6	1.7	1.9	1.4	1.6	1.8	2.0	2.2
WW B4	16	23	31	38	46	52	60	69	10	16	23	29	36	41	48	57	1.1	1.3	1.5	1.6	1.8	1.3	1.4	1.6	1.8	1.9
WW B5	15	23	31	38	47	53	61	70	9	16	23	29	37	43	50	59	1.1	1.3	1.4	1.6	1.7	1.3	1.4	1.6	1.7	1.9
WW C1	17	24	31	37	45	50	57	66	11	17	23	28	34	39	45	53	1.2	1.4	1.5	1.7	1.8	1.3	1.5	1.7	1.9	2.1
WW C2	16	23	30	36	45	50	58	67	10	16	22	27	34	39	46	54	1.2	1.3	1.5	1.7	1.8	1.3	1.5	1.7	1.9	2.0
WW C3	19	28	39	46	56	63	71	80	13	21	30	37	47	53	61	70	1.0	1.2	1.3	1.5	1.6	1.1	1.3	1.4	1.6	1.7
WW C4	17	26	36	43	53	60	68	77	11	19	28	34	44	50	58	68	1.1	1.2	1.4	1.5	1.7	1.2	1.3	1.5	1.6	1.8
TW A1	14	16	19	20	23	25	27	30	12	13	14	15	17	17	18	20	2.1	2.4	2.7	3.0	3.3	2.8	3.3	3.8	4.6	-
TW A2	13	16	20	22	26	29	32	38	10	12	14	16	18	20	22	25	1.7	2.0	2.2	2.4	2.6	2.2	2.5	2.8	3.2	3.6
TW A3	12	16	20	23	27	31	35	41	8	11	13	16	19	21	24	27	1.6	1.9	2.1	2.3	2.4	2.0	2.3	2.6	2.9	3.2
TW A4	11	15	19	23	28	31	36	43	7	10	13	15	19	21	24	29	1.6	1.8	2.0	2.2	2.3	1.9	2.1	2.4	2.7	2.9
TW B1	14	16	18	20	23	26	28	32	11	12	14	15	17	18	20	22	2.0	2.3	2.5	2.8	3.0	2.5	2.9	3.3	3.8	4.3
TW B2	12	15	19	22	26	29	33	38	9	11	13	15	18	20	23	26	1.7	1.9	2.2	2.4	2.5	2.1	2.4	2.7	3.0	3.3
TW B3	11	15	18	21	26	29	33	39	8	10	13	15	18	20	23	27	1.7	1.9	2.1	2.3	2.4	2.0	2.3	2.5	2.8	3.1
TW B4	11	14	18	21	25	28	32	38	7	10	12	15	18	20	23	27	1.7	1.9	2.1	2.3	2.4	2.0	2.3	2.5	2.8	3.0
TW C1	12	15	18	20	24	26	29	34	10	11	13	15	17	19	21	24	1.9	2.1	2.4	2.6	2.8	2.3	2.6	2.9	3.3	3.6
TW C2	11	14	18	20	24	27	31	36	8	10	13	15	17	19	22	26	1.8	2.0	2.2	2.4	2.6	2.2	2.4	2.7	2.9	3.2
TW C3	11	14	17	20	23	26	30	35	8	10	13	14	17	19	22	26	1.8	2.0	2.3	2.4	2.6	2.2	2.4	2.7	2.9	3.2
TW C4	11	14	17	19	22	25	28	33	9	11	13	14	17	19	21	25	2.0	2.2	2.4	2.6	2.7	2.4	2.6	2.8	3.1	3.3
TW D1	8	10	12	13	16	17	19	21	5	6	7	7	8	8	8	8	2.2	2.7	3.1	3.6	4.1	3.2	4.2	-	-	-
TW D2	13	18	23	27	32	36	42	49	9	12	16	19	24	27	31	37	1.5	1.6	1.8	2.0	2.1	1.7	1.9	2.1	2.4	2.5
TW D3	14	19	24	29	35	40	46	54	9	13	17	21	26	29	34	40	1.4	1.6	1.7	1.9	2.0	1.6	1.8	2.0	2.2	2.4
TW D4	15	21	28	34	42	47	54	63	10	15	21	26	32	37	43	51	1.2	1.4	1.6	1.7	1.9	1.4	1.6	1.7	1.9	2.1
TW D5	15	21	29	35	43	49	56	65	10	16	22	27	34	39	46	55	1.2	1.4	1.5	1.7	1.8	1.3	1.5	1.7	1.8	2.0
TW E1	16	22	28	33	41	45	52	60	11	16	21	25	31	35	40	47	1.3	1.5	1.6	1.8	1.9	1.5	1.7	1.9	2.0	2.2
TW E2	16	22	29	34	42	47	53	62	11	16	21	26	32	36	42	50	1.3	1.4	1.6	1.7	1.9	1.4	1.6	1.8	2.0	2.1
TW E3	19	27	36	43	52	58	66	75	14	20	28	34	42	48	55	64	1.1	1.2	1.4	1.5	1.7	1.2	1.4	1.5	1.7	1.8
TW E4	18	25	34	41	50	56	64	74	12	19	27	33	41	47	54	63	1.1	1.3	1.4	1.6	1.7	1.2	1.4	1.5	1.7	1.8
TW F1	13	15	17	19	22	23	26	29	10	11	12	13	15	16	17	19	2.1	2.4	2.7	3.0	3.2	2.7	3.2	3.7	4.4	-
TW F2	11	14	17	20	24	26	30	35	7	9	11	13	16	17	20	22	1.8	2.0	2.3	2.5	2.6	2.2	2.5	2.9	3.3	3.7
TW F3	9	13	17	20	24	27	31	37	6	8	11	13	16	18	20	24	1.7	1.9	2.1	2.3	2.5	2.0	2.3	2.6	3.0	3.3
TW F4	7	11	15	18	23	26	31	37	4	6	9	11	15	17	20	24	1.7	1.9	2.1	2.3	2.4	2.0	2.3	2.5	2.8	3.1
TW G1	11	13	16	18	21	23	26	30	8	9	11	12	14	15	17	19	2.0	2.3	2.5	2.8	3.0	2.5	2.9	3.3	3.8	4.3
TW G2	8	11	15	17	21	24	28	33	5	7	9	11	14	16	18	21	1.8	2.0	2.3	2.5	2.6	2.2	2.5	2.8	3.1	3.5
TW G3	7	10	13	16	20	23	27	33	3	6	8	10	13	15	18	22	1.8	2.0	2.2	2.4	2.6	2.2	2.4	2.7	3.0	3.2
TW G4	5	8	11	14	18	21	25	30	2	4	7	8	11	13	16	20	1.8	2.1	2.3	2.5	2.6	2.2	2.5	2.7	3.0	3.2
TW H1	8	10	13	15	18	21	24	28	5	7	8	10	12	13	15	18	2.0	2.2	2.5	2.7	2.9	2.5	2.8	3.1	3.5	3.8
TW H2	5	8	11	13	17	20	23	28	3	4	7	8	11	13	15	18	1.9	2.2	2.4	2.6	2.8	2.4	2.6	2.9	3.2	3.4
TW H3	4	6	9	12	15	18	21	26	2	3	5	7	10	11	14	17	2.0	2.2	2.5	2.7	2.8	2.4	2.7	2.9	3.2	3.4
TW H4	3	5	7	9	12	14	17	22	1	3	4	6	8	9	11	15	2.2	2.4	2.7	2.9	3.0	2.7	2.9	3.1	3.4	3.6
TW I1	6	8	10	11	13	15	16	18	3	4	5	5	6	6	6	6	2.3	2.8	3.3	3.0	4.3	3.4	4.5	-	-	-
TW I2	10	15	20	23	29	33	39	46	6	9	13	16	21	24	28	33	1.5	1.7	1.9	2.0	2.2	1.7	2.0	2.2	2.4	2.6
TW I3	11	16	22	26	32	37	42	51	7	11	15	18	23	27	31	38	1.4	1.6	1.8	1.9	2.1	1.6	1.8	2.0	2.2	2.4
TW I4	12	18	25	31	38	44	51	60	8	13	18	23	29	34	40	48	1.3	1.4	1.6	1.7	1.9	1.4	1.6	1.8	1.9	2.1
TW I5	12	19	26	32	40	46	53	63	8	13	20	24	32	37	43	52	1.2	1.4	1.5	1.7	1.8	1.4	1.5	1.7	1.9	2.0
TW J1	14	20	26	31	38	43	50	59	9	14	19	23	29	33	38	46	1.3	1.5	1.6	1.8	1.9	1.5	1.7	1.9	2.1	2.2
TW J2	13	19	26	31	39	44	51	60	9	13	19	23	30	34	40	48	1.3	1.4	1.6	1.8	1.9	1.4	1.6	1.8	2.0	2.1
TW J3	16	24	34	40	50	56	64	73	11	18	26	32	40	46	53	63	1.1	1.3	1.4	1.6	1.7	1.2	1.4	1.5	1.7	1.8
TW J4	15	23	32	38	48	54	62	71	10	17	24	30	39	44	52	61	1.1	1.3	1.4	1.6	1.7	1.2	1.4	1.5	1.7	1.9
DG A1	8	10	12	13	15	16	17	18	4	5	6	7	7	7	7	5	2.1	2.6	3.5	-	-	2.9	4.2	-	-	-
DG A2	11	14	18	21	25	28	32	37	7	10	12	15	18	20	22	26	1.7	1.9	2.1	2.4	2.8	2.0	2.3	2.6	3.0	3.5
DG A3	14	19	24	29	36	40	47	55	11	14	19	23	28	32	37	44	1.4	1.6	1.7	1.9	2.0	1.6	1.7	1.9	2.1	2.3
DG B1	8	10	12	14	16	17	19	21	4	5	6	7	8	8	8	7	2.1	2.6	3.0	3.9	-	2.9	3.9	-	-	-
DG B2	11	14	18	21	26	30	34	40	7	10	13	15	18	21	24	28	1.6	1.9	2.0	2.3	2.5	2.0	2.2	2.5	2.8	3.1
DG B3	15	19	25	30	37	42	49	59	11	15	20	24	29	33	39	47	1.4	1.5	1.6	1.8	1.9	1.6	1.7	1.9	2.0	2.1
DG C1	10	13	16	18	21	24	27	30	6	8	10	11	12	13	14	15	1.8	2.2	2.3	2.9	3.3	2.3	2.9	3.9	-	-
DG C2	13	17	22	26	32	36	41	48	9	12	16	19	23	26	30	35	1.5	1.7	1.8	2.0	2.2	1.8	2.0	2.2	2.4	2.7
DG C3	17	22	28	34	42	48	55	66	13	17	23	27	33	38	44	53	1.3	1.4	1.5	1.7	1.8	1.5	1.6	1.7	1.8	2.0
SS A1	17	21	25	28	32	35	39	45	14	17	20	22	25	27	30	34	1.6	1.8	2.0	2.2	2.4	1.8	2.1	2.4	2.8	3.1
SS A2	21	27	33	38	45	50	56	64	17	22	27	32	38	42	47	54	1.2	1.4	1.6	1.7	1.9	1.4	1.6	1.8	2.0	2.1
SS A3	15	19	22	24	28	30	34	38	13	15	17	19	21	23	25	27	1.7	2.0	2.3	2.5	2.7	2.0	2.4	2.9	3.5	4.1
SS A4	20	26	32	37	43	48	54	62	17	21	26	30	36	40	44	51	1.3	1.5	1.6	1.8	1.9	1.4	1.6	1.8	2.0	2.2
SS A5	17	20	23	26	29	32	35	40	14	17	19	21	23	25	27	30	1.7	2.0	2.3	2.5	2.7	2.0	2.4	2.8	3.3	3.7
SS A6	21	26	33	37	44	49	55	63	17	21	27	31	36	40	45	52	1.3	1.4	1.6	1.8	1.9	1.4	1.6	1.8	2.0	2.2
SS A7	16	18	20	22	25	26	29	32	13	15	16	18	19	20	21	22	2.0	2.3	2.7	3.0	3.3	2.4	2.9	3.6	4.6	-
SS A8	20	25	31	35	42	46	52	60	16	20	25	29	34	37	42	48	1.3	1.5	1.7	1.8	2.0	1.4	1.7	1.9	2.1	2.3
SS B1	14	17	21	24	28	30	34	39	11	13	16	18	21	23	25	29	1.7	1.9	2.2	2.4	2.6	2.0	2.3	2.6	3.0	3.3
SS B2	18	24	30	34	41	46	52	60	15	19	24	28	34	38	43	50	1.3	1.5	1.6	1.8	1.9	1.4	1.6	1.8	2.0	2.2
SS B3	13	16	19	21	25	27	30	35	10	12	14	16	18	20	22	24	1.8	2.1	2.4	2.6	2.8	2.1	2.5	3.0	3.6	4.2
SS B4	18	23	29	33	39	44	50	58	14	18	23	27	32	36	41	48	1.3	1.5	1.7	1.9	2.0	1.5	1.7	1.9	2.1	2.3
SS B5	14	16	19	21	24	26	29	33	12	13	15	17	19	20	22	24	1.9	2.2	2.5	2.8	3.0	2.3	2.7	3.1	3.6	4.2
SS B6	17	23	28	33	40	44	50	58	14	18	23	27	32	36	41	48	1.3	1.5	1.7	1.8	2.0	1.5	1.7	1.9	2.1	2.2
SS B7	13	15	17	19	21	23	25	28	11	12	13	14	16	17	18	19	2.1	2.5	2.8	3.2	3.5	2.6	3.1	3.8	4.7	-
SS B8	16	21	27	31	37	42	47	55	13	17	22	25	30	33	38	44	1.4	1.6	1.7	1.9	2.1	1.5	1.7	2.0	2.2	2.4
SS C1	11	14	18	21	25	28	32	38	7	9	11	13	16	17	19	22	1.7	1.9	2.2	2.4	2.6	2.1	2.4	2.9	3.4	3.9
SS C2	13	19	25	29	36	40	46	55	9	13	17	21	26	30	34	41	1.4	1.5	1.7	1.9	2.0	1.6	1.8	2.0	2.2	2.4
SS C3	11	13	16	18	22	24	27	31	8	10	11	13	15	16	18	20	1.9	2.2	2.5	2.7	2.9	2.4	2.8	3.2	3.7	4.2
SS C4	13	17	22	26	32	36	41	49	9	12	16	19	24	27	31	37	1.5	1.7	1.8	2.0	2.1	1.7	1.9	2.1	2.3	2.5
SS D1	18	23	28	32	37	41	46	53	14	18	22	25	29	31	35	39	1.4	1.6	1.8	2.0	2.2	1.6	1.9	2.2	2.5	2.8
SS D2	23	30	38	44	52	58	65	73	18	24	31	36	43	48	54	62	1.1	1.3	1.4	1.6	1.8	1.2	1.4	1.6	1.8	1.9
SS D3	18	22	26	29	34	37	41	47	15	18	21	23	26	29	31	35	1.5	1.8	2.0	2.2	2.3	1.7	2.1	2.4	2.8	3.1
SS D4	22	29	37	42	50	56	63	71	18	24	30	35	42	46	52	60	1.1	1.3	1.5	1.6	1.8	1.2	1.4	1.6	1.8	2.0
SS E1	15	19	23	26	31	34	38	44	11	14	17	19	22	24	26	29	1.5	1.8	2.0	2.2	2.4	1.8	2.2	2.6	3.1	3.6
SS E2	19	25	32	37	45	50	56	65	14	19	25	29	36	40	45	53	1.2	1.4	1.6	1.7	1.9	1.3	1.5	1.8	2.0	2.1
SS E3	15	18	21	23	27	29	33	38	12	14	16	18	20	22	24	26	1.7	2.0	2.3	2.5	2.7	2.1	2.4	2.9	3.4	3.9
SS E4	18	23	30	35	41	46	52	61	14	18	24	27	33	37	42	49	1.3	1.5	1.6	1.8	1.9	1.4	1.6	1.8	2.1	2.2

DAYTON, OHIO

SOLAR SAVINGS FRACTION (%) — CONSERVATION FACTOR

TYPE	BASE TEMP: 55 F (3480 DD) LCR 100	70	50	40	30	25	20	15	65 F (5645 DD) 100	70	50	40	30	25	20	15	CF 55 F 100	50	30	20	15	CF 65 F 100	50	30	20	15
WW A1	15	17	20	21	24	26	29	32	12	14	15	16	17	18	19	21	2.0	2.3	2.6	2.9	3.1	2.7	3.2	3.8	4.6	-
WW A2	14	17	22	25	30	34	38	44	10	12	15	17	20	23	25	29	1.6	1.8	2.0	2.2	2.3	1.9	2.2	2.5	2.9	3.2
WW A3	13	18	23	27	32	36	42	49	9	12	15	18	22	24	28	33	1.5	1.7	1.8	2.0	2.2	1.8	2.0	2.3	2.6	2.8
WW A4	13	18	23	28	34	38	44	52	8	11	15	18	22	25	29	35	1.4	1.6	1.8	1.9	2.1	1.7	1.9	2.2	2.4	2.6
WW A5	12	18	24	29	35	40	46	54	7	11	15	19	23	27	31	37	1.3	1.5	1.7	1.9	2.0	1.6	1.8	2.1	2.3	2.5
WW A6	12	18	24	29	36	41	47	56	6	11	15	19	24	27	32	38	1.3	1.5	1.7	1.8	2.0	1.6	1.8	2.0	2.3	2.4
WW B1	10	13	15	17	19	21	23	26	7	8	9	9	10	10	10	9	2.0	2.4	2.8	3.2	3.6	3.0	4.0	-	-	-
WW B2	14	19	26	31	38	43	49	58	9	13	18	22	27	31	36	43	1.3	1.5	1.6	1.8	2.0	1.5	1.7	1.9	2.1	2.3
WW B3	17	23	31	37	45	51	57	67	11	16	22	27	33	38	44	52	1.2	1.4	1.5	1.7	1.8	1.4	1.5	1.7	1.9	2.1
WW B4	17	25	33	40	49	55	63	72	11	17	25	30	39	44	51	60	1.1	1.3	1.4	1.6	1.7	1.2	1.4	1.6	1.7	1.9
WW B5	16	24	33	40	49	56	63	73	10	17	25	31	39	45	52	62	1.1	1.3	1.4	1.6	1.7	1.2	1.4	1.5	1.7	1.8
WW C1	18	25	33	39	48	54	61	70	12	18	25	30	37	41	48	56	1.2	1.3	1.5	1.6	1.8	1.3	1.5	1.7	1.8	2.0
WW C2	17	24	32	39	48	53	61	70	11	17	24	29	37	42	48	57	1.1	1.3	1.5	1.6	1.8	1.3	1.5	1.6	1.8	1.9
WW C3	21	30	41	49	59	65	73	82	14	23	32	39	49	55	63	73	1.0	1.2	1.3	1.5	1.6	1.1	1.3	1.4	1.6	1.7
WW C4	18	28	38	45	56	62	70	80	12	20	30	36	46	52	60	70	1.0	1.2	1.3	1.5	1.6	1.1	1.3	1.4	1.6	1.7
TW A1	15	17	20	22	24	26	29	33	12	14	15	16	18	19	20	22	2.0	2.3	2.6	2.8	3.0	2.6	3.0	3.5	4.0	4.6
TW A2	14	17	21	24	28	31	35	41	10	12	15	17	20	22	24	28	1.7	1.9	2.1	2.3	2.4	2.1	2.3	2.6	3.0	3.3
TW A3	13	17	21	25	30	33	38	45	9	12	15	17	20	23	26	30	1.6	1.8	2.0	2.1	2.3	1.9	2.1	2.4	2.7	2.9
TW A4	12	16	21	24	30	34	39	46	8	11	14	17	21	23	27	32	1.5	1.7	1.9	2.1	2.2	1.8	2.0	2.3	2.5	2.7
TW B1	14	17	19	22	25	27	31	35	11	13	15	16	18	20	21	24	1.9	2.2	2.4	2.6	2.8	2.4	2.7	3.1	3.5	3.8
TW B2	13	16	20	23	28	31	35	41	9	12	14	16	20	22	25	29	1.7	1.9	2.1	2.2	2.4	2.0	2.3	2.5	2.8	3.0
TW B3	12	16	20	23	28	31	36	43	8	11	14	16	20	22	25	30	1.6	1.8	2.0	2.2	2.3	1.9	2.2	2.4	2.7	2.9
TW B4	11	15	19	22	27	30	35	42	8	10	13	16	19	22	25	30	1.6	1.8	2.0	2.2	2.3	2.0	2.2	2.4	2.6	2.8
TW C1	13	16	19	21	25	28	31	37	10	12	14	16	18	20	23	26	1.8	2.1	2.3	2.4	2.6	2.2	2.5	2.8	3.1	3.3
TW C2	12	15	19	21	26	29	33	39	9	11	14	16	19	21	24	28	1.7	1.9	2.1	2.3	2.4	2.1	2.3	2.5	2.8	3.0
TW C3	12	15	18	21	25	28	32	38	9	11	13	15	18	21	24	28	1.8	2.0	2.2	2.3	2.5	2.1	2.3	2.6	2.8	3.0
TW C4	12	14	17	20	23	26	30	35	9	11	13	15	18	20	22	26	1.9	2.1	2.3	2.5	2.6	2.3	2.5	2.7	2.9	3.1
TW D1	9	11	13	15	17	19	21	25	6	7	8	9	9	10	11	11	2.1	2.5	2.8	3.2	3.4	2.9	3.7	4.8	-	-
TW D2	14	19	24	28	35	39	45	53	10	13	17	21	25	29	33	39	1.4	1.6	1.7	1.9	2.1	1.6	1.8	2.0	2.2	2.4
TW D3	15	20	26	31	38	43	49	57	10	14	19	22	28	31	36	43	1.3	1.5	1.7	1.8	2.0	1.5	1.7	1.9	2.1	2.3
TW D4	16	23	30	36	44	50	57	66	11	16	23	27	34	39	45	54	1.2	1.4	1.5	1.7	1.8	1.4	1.5	1.7	1.9	2.0
TW D5	16	23	31	37	45	51	59	68	11	17	23	29	36	41	48	57	1.2	1.3	1.5	1.6	1.8	1.3	1.5	1.6	1.8	1.9
TW E1	17	23	30	36	43	48	55	64	12	17	23	27	33	37	43	50	1.2	1.4	1.6	1.7	1.9	1.4	1.6	1.8	2.0	2.1
TW E2	17	23	31	36	44	50	57	66	12	17	23	27	34	39	45	53	1.2	1.4	1.5	1.7	1.8	1.4	1.5	1.7	1.9	2.0
TW E3	20	29	38	45	55	61	68	78	15	22	30	36	45	50	58	67	1.1	1.2	1.4	1.5	1.7	1.2	1.3	1.5	1.7	1.8
TW E4	19	27	36	43	53	59	67	76	13	20	28	34	43	49	56	66	1.1	1.2	1.4	1.5	1.7	1.2	1.3	1.5	1.7	1.8
TW F1	13	15	18	20	23	25	28	32	10	12	13	14	16	17	19	21	2.0	2.2	2.5	2.7	2.9	2.5	2.9	3.4	3.9	4.4
TW F2	12	15	18	21	26	29	33	39	8	10	13	14	17	19	22	25	1.7	1.9	2.1	2.3	2.5	2.1	2.4	2.7	3.0	3.3
TW F3	10	14	18	21	26	29	34	41	6	9	12	14	17	20	23	27	1.6	1.8	2.0	2.2	2.3	2.0	2.2	2.5	2.8	3.0
TW F4	8	12	16	20	25	29	33	40	4	7	10	13	16	19	22	27	1.6	1.8	2.0	2.2	2.3	1.9	2.2	2.4	2.7	2.9
TW G1	11	14	17	19	22	25	28	33	8	10	12	13	15	17	19	21	1.9	2.2	2.4	2.6	2.8	2.4	2.7	3.1	3.5	3.9
TW G2	9	12	16	19	23	26	30	36	6	8	10	12	15	17	20	24	1.7	1.9	2.1	2.3	2.5	2.1	2.4	2.6	2.9	3.2
TW G3	7	11	15	18	22	25	30	36	4	6	9	11	14	17	20	24	1.7	1.9	2.1	2.3	2.4	2.1	2.3	2.5	2.8	3.0
TW G4	5	9	12	15	19	22	27	33	3	5	7	9	13	15	18	22	1.8	2.0	2.2	2.4	2.5	2.1	2.4	2.6	2.8	3.0
TW H1	8	11	14	16	20	22	26	31	5	7	9	11	13	15	17	20	1.9	2.1	2.4	2.5	2.7	2.4	2.6	2.9	3.2	3.5
TW H2	6	9	12	15	18	21	25	31	3	5	7	9	12	14	17	21	1.9	2.1	2.3	2.5	2.6	2.3	2.5	2.7	3.0	3.2
TW H3	5	7	10	13	16	19	23	28	2	4	6	8	11	13	15	19	1.9	2.2	2.4	2.5	2.7	2.4	2.6	2.8	3.0	3.2
TW H4	3	5	8	10	13	16	19	24	1	3	5	6	9	10	13	16	2.2	2.4	2.6	2.8	2.9	2.6	2.8	3.0	3.2	3.4
TW I1	7	9	11	13	15	17	19	22	4	5	6	7	7	8	8	9	2.2	2.6	3.0	3.3	3.6	3.1	3.9	-	-	-
TW I2	11	16	21	25	31	36	41	50	7	10	14	18	22	26	30	36	1.4	1.6	1.8	2.0	2.1	1.7	1.9	2.1	2.3	2.5
TW I3	12	17	23	28	35	39	46	54	8	12	16	20	25	29	34	41	1.4	1.5	1.7	1.9	2.0	1.6	1.8	2.0	2.2	2.3
TW I4	13	20	27	33	41	46	54	63	9	14	20	25	31	36	42	51	1.2	1.4	1.5	1.7	1.8	1.4	1.6	1.7	1.9	2.0
TW I5	13	20	28	34	43	49	56	66	9	14	21	26	33	39	45	55	1.2	1.4	1.5	1.7	1.8	1.3	1.5	1.7	1.8	2.0
TW J1	15	21	28	33	41	46	53	62	10	15	20	25	31	35	41	49	1.3	1.4	1.6	1.7	1.9	1.4	1.6	1.8	2.0	2.1
TW J2	14	21	28	33	41	47	54	63	9	14	20	25	32	36	42	51	1.2	1.4	1.5	1.7	1.8	1.4	1.6	1.7	1.9	2.1
TW J3	18	26	35	42	52	59	66	76	12	19	27	34	42	48	56	66	1.1	1.2	1.4	1.5	1.7	1.2	1.3	1.5	1.7	1.8
TW J4	16	24	34	40	50	56	64	74	11	18	26	32	41	46	54	64	1.1	1.2	1.4	1.6	1.7	1.2	1.4	1.5	1.7	1.8
DG A1	9	11	13	15	17	19	20	21	5	6	7	8	9	9	9	8	2.0	2.4	3.1	4.6	-	2.7	3.6	-	-	-
DG A2	11	15	19	22	27	30	34	40	8	10	13	16	19	21	24	28	1.6	1.8	2.0	2.3	2.6	1.9	2.2	2.5	2.8	3.2
DG A3	15	20	26	31	38	43	49	58	11	15	20	24	29	34	39	47	1.4	1.5	1.6	1.8	2.0	1.6	1.7	1.8	2.0	2.2
DG B1	9	11	13	15	18	20	22	25	5	6	7	8	9	10	10	10	2.0	2.4	2.7	3.3	4.1	2.6	3.5	-	-	-
DG B2	12	15	19	23	28	32	37	44	8	11	14	16	20	22	26	30	1.6	1.8	1.9	2.1	2.3	1.9	2.1	2.4	2.6	2.9
DG B3	15	20	26	32	39	45	52	62	12	16	21	25	31	35	41	50	1.4	1.5	1.6	1.7	1.8	1.5	1.7	1.8	1.9	2.1
DG C1	11	14	17	20	24	27	30	35	7	9	11	12	14	15	16	18	1.7	2.0	2.2	2.6	3.0	2.1	2.7	3.4	4.4	-
DG C2	14	18	23	27	34	38	44	52	10	13	17	20	25	28	32	38	1.5	1.6	1.7	1.9	2.1	1.7	1.9	2.1	2.3	2.5
DG C3	17	23	30	36	45	51	59	69	13	18	24	28	35	40	47	56	1.3	1.4	1.5	1.6	1.7	1.4	1.6	1.7	1.8	1.9
SS A1	18	22	26	29	34	38	42	49	14	17	21	23	27	29	32	37	1.5	1.7	1.9	2.1	2.3	1.7	2.0	2.3	2.6	2.8
SS A2	22	28	35	40	47	52	59	67	18	23	29	33	39	44	49	57	1.2	1.4	1.5	1.7	1.8	1.3	1.5	1.7	1.9	2.1
SS A3	16	19	23	26	30	33	36	42	13	16	18	20	23	25	27	30	1.6	1.9	2.1	2.4	2.5	1.9	2.3	2.7	3.2	3.6
SS A4	21	27	34	39	46	51	57	65	17	22	28	32	38	41	47	54	1.2	1.4	1.6	1.7	1.9	1.4	1.6	1.8	2.0	2.1
SS A5	18	21	24	27	31	34	38	43	15	17	20	22	25	26	29	32	1.7	1.9	2.1	2.3	2.5	1.9	2.3	2.6	3.0	3.4
SS A6	21	27	34	39	46	51	58	66	17	22	28	32	38	42	48	55	1.2	1.4	1.6	1.7	1.9	1.3	1.6	1.7	1.9	2.1
SS A7	16	19	21	23	26	28	31	35	14	15	17	19	20	22	23	25	1.9	2.2	2.5	2.8	3.0	2.3	2.7	3.3	4.0	4.9
SS A8	21	26	33	37	44	49	55	63	17	21	26	30	36	40	44	51	1.3	1.4	1.6	1.8	1.9	1.4	1.6	1.8	2.0	2.2
SS B1	15	18	22	25	29	32	37	43	12	14	17	19	22	25	27	31	1.6	1.9	2.1	2.3	2.4	1.9	2.2	2.5	2.8	3.1
SS B2	19	25	31	36	43	48	54	63	15	20	25	29	35	40	45	52	1.3	1.4	1.6	1.8	1.9	1.4	1.6	1.8	2.0	2.1
SS B3	14	17	20	23	26	29	33	38	11	13	15	17	20	21	23	26	1.7	2.0	2.2	2.5	2.6	2.0	2.4	2.8	3.3	3.7
SS B4	18	24	30	35	41	46	52	61	15	19	24	28	34	38	43	50	1.3	1.5	1.6	1.8	1.9	1.4	1.6	1.8	2.0	2.2
SS B5	15	17	20	22	26	28	31	36	12	14	16	18	20	21	23	26	1.8	2.1	2.4	2.6	2.8	2.2	2.5	2.9	3.4	3.8
SS B6	18	24	30	35	42	46	53	61	14	19	24	28	34	38	43	50	1.3	1.5	1.6	1.8	1.9	1.4	1.6	1.8	2.0	2.2
SS B7	13	15	18	20	22	24	27	30	11	12	14	15	17	18	19	21	2.0	2.4	2.7	3.0	3.2	2.5	2.9	3.5	4.2	4.9
SS B8	17	22	28	33	39	44	50	58	14	18	23	26	32	35	40	47	1.3	1.5	1.7	1.8	2.0	1.5	1.7	1.9	2.1	2.3
SS C1	12	15	20	23	28	31	35	42	7	10	13	15	18	20	22	26	1.6	1.8	2.0	2.2	2.4	2.0	2.3	2.6	3.0	3.3
SS C2	14	20	26	31	38	43	50	58	10	14	19	23	28	32	37	44	1.3	1.5	1.6	1.8	1.9	1.5	1.7	1.9	2.1	2.3
SS C3	12	14	17	20	23	26	29	34	9	10	12	14	16	18	20	23	1.8	2.1	2.3	2.5	2.7	2.3	2.6	3.0	3.4	3.7
SS C4	14	18	24	28	34	39	44	52	10	13	17	21	25	29	33	39	1.4	1.6	1.8	1.9	2.1	1.6	1.8	2.0	2.2	2.4
SS D1	19	24	30	34	40	44	50	57	15	19	23	26	31	34	37	42	1.3	1.5	1.7	1.9	2.1	1.5	1.8	2.1	2.3	2.6
SS D2	24	32	40	46	55	61	68	76	19	25	33	38	46	51	57	65	1.1	1.2	1.4	1.6	1.7	1.2	1.4	1.6	1.7	1.9
SS D3	19	23	27	31	36	40	44	51	15	18	22	24	28	31	34	38	1.5	1.7	1.9	2.1	2.2	1.7	2.0	2.3	2.6	2.8
SS D4	23	31	39	45	53	59	65	74	19	25	32	37	44	49	55	63	1.1	1.3	1.4	1.6	1.7	1.2	1.4	1.6	1.8	1.9
SS E1	16	20	25	28	33	37	42	48	12	15	18	20	24	26	29	33	1.5	1.7	1.9	2.1	2.3	1.7	2.1	2.4	2.8	3.1
SS E2	20	26	34	40	47	53	60	68	15	20	26	31	38	42	48	56	1.2	1.3	1.5	1.7	1.8	1.3	1.5	1.7	1.9	2.0
SS E3	15	18	22	25	29	32	35	41	12	15	17	19	22	23	26	29	1.7	1.9	2.2	2.4	2.5	2.0	2.3	2.7	3.1	3.5
SS E4	19	25	31	37	44	49	55	64	15	19	25	29	35	39	44	52	1.2	1.4	1.6	1.7	1.9	1.4	1.6	1.8	2.0	2.1

TOLEDO, OHIO

TYPE	SOLAR SAVINGS FRACTION (%) BASE TEMP: 55 F (4055 DD)								65 F (6385 DD)								CONSERVATION FACTOR 55 F					65 F				
LCR	100	70	50	40	30	25	20	15	100	70	50	40	30	25	20	15	100	50	30	20	15	100	50	30	20	15
WW A1	14	15	17	19	20	22	23	26	12	13	13	14	15	15	16	16	2.3	2.7	3.2	3.7	4.2	3.1	3.9	-	-	-
WW A2	12	15	18	21	25	27	31	35	9	11	13	15	17	18	20	23	1.7	2.0	2.3	2.5	2.8	2.1	2.5	3.0	3.5	4.2
WW A3	11	15	19	22	26	29	34	39	7	10	13	15	18	20	22	25	1.6	1.9	2.1	2.3	2.5	1.9	2.3	2.6	3.1	3.5
WW A4	10	14	19	22	27	31	35	42	6	9	12	15	18	20	23	27	1.5	1.8	2.0	2.2	2.4	1.8	2.1	2.5	2.9	3.2
WW A5	10	14	19	23	28	32	37	44	5	8	12	15	19	21	24	29	1.5	1.7	1.9	2.1	2.3	1.7	2.0	2.4	2.7	3.0
WW A6	9	14	19	23	29	33	38	45	5	8	12	15	19	22	25	30	1.4	1.7	1.9	2.1	2.2	1.7	2.0	2.3	2.6	2.9
WW B1	9	10	12	13	14	15	15	16	6	6	7	7	6	6	5	2	2.4	3.1	4.1	-	-	3.9	-	-	-	-
WW B2	11	16	21	26	31	36	41	49	7	11	15	18	23	26	30	36	1.4	1.6	1.8	2.0	2.1	1.6	1.9	2.1	2.3	2.5
WW B3	14	20	26	31	38	43	49	58	9	14	19	23	29	33	38	45	1.3	1.5	1.6	1.8	2.0	1.5	1.7	1.9	2.1	2.3
WW B4	14	21	29	35	43	49	56	65	9	15	22	27	34	39	46	54	1.2	1.4	1.5	1.7	1.8	1.3	1.5	1.7	1.8	2.0
WW B5	13	20	29	35	44	49	57	67	9	15	22	27	35	41	48	57	1.2	1.3	1.5	1.7	1.8	1.3	1.4	1.6	1.8	1.9
WW C1	15	21	28	34	41	46	53	61	11	16	21	26	32	36	42	49	1.2	1.4	1.6	1.8	1.9	1.4	1.6	1.8	2.0	2.2
WW C2	14	20	28	33	41	47	53	63	9	15	21	26	32	37	43	51	1.2	1.4	1.6	1.7	1.9	1.4	1.5	1.7	1.9	2.1
WW C3	17	26	36	43	53	59	67	76	12	20	29	35	45	51	58	68	1.1	1.2	1.4	1.5	1.7	1.1	1.3	1.5	1.6	1.8
WW C4	15	24	33	40	50	56	64	74	11	18	26	33	42	48	55	65	1.1	1.2	1.4	1.6	1.7	1.2	1.3	1.5	1.7	1.8
TW A1	14	15	17	19	21	22	24	27	12	13	14	15	16	16	17	18	2.2	2.6	3.0	3.4	3.8	3.0	3.6	4.4	-	-
TW A2	12	15	18	20	23	26	29	33	9	11	13	15	17	18	20	22	1.8	2.1	2.4	2.7	2.9	2.3	2.6	3.1	3.6	4.1
TW A3	11	14	18	20	24	27	31	36	8	10	13	14	17	19	21	24	1.7	2.0	2.2	2.4	2.6	2.1	2.4	2.7	3.1	3.5
TW A4	10	13	17	20	25	28	32	38	6	9	12	14	17	19	22	26	1.6	1.9	2.1	2.3	2.5	2.0	2.3	2.6	2.9	3.2
TW B1	13	15	17	19	21	23	25	29	11	12	13	14	16	17	18	20	2.1	2.4	2.8	3.1	3.3	2.7	3.1	3.6	4.3	-
TW B2	11	14	17	20	23	26	29	34	8	10	12	14	17	18	21	24	1.8	2.1	2.3	2.5	2.7	2.2	2.5	2.8	3.2	3.6
TW B3	10	13	17	19	23	26	30	35	7	10	12	14	17	19	21	25	1.7	2.0	2.2	2.4	2.6	2.1	2.4	2.7	3.0	3.3
TW B4	10	13	16	19	23	25	29	35	7	9	12	14	17	19	21	25	1.8	2.0	2.2	2.5	2.6	2.1	2.4	2.7	2.9	3.2
TW C1	12	14	17	19	22	24	26	31	9	11	13	14	16	18	19	22	2.0	2.3	2.5	2.8	3.0	2.4	2.8	3.1	3.5	3.9
TW C2	10	13	16	18	22	24	28	33	8	10	12	14	16	18	21	24	1.9	2.1	2.4	2.6	2.8	2.2	2.5	2.8	3.1	3.4
TW C3	10	13	16	18	21	24	27	32	8	10	12	14	16	18	20	24	1.9	2.2	2.4	2.6	2.8	2.3	2.5	2.8	3.1	3.4
TW C4	11	13	15	17	21	23	26	30	9	11	12	14	16	18	20	23	2.0	2.3	2.5	2.7	2.9	2.4	2.7	2.9	3.2	3.4
TW D1	7	9	10	12	13	14	15	17	5	5	6	6	7	7	6	5	2.4	3.0	3.7	4.6	-	3.6	-	-	-	-
TW D2	12	16	21	24	29	33	38	45	9	12	15	18	22	25	29	34	1.5	1.7	1.9	2.1	2.3	1.8	2.0	2.2	2.5	2.7
TW D3	12	17	22	26	32	36	41	49	8	12	16	19	24	27	31	37	1.4	1.6	1.8	2.0	2.2	1.6	1.9	2.1	2.4	2.6
TW D4	14	19	26	31	38	43	50	59	10	14	20	24	31	35	41	49	1.3	1.5	1.6	1.8	1.9	1.4	1.6	1.8	2.0	2.1
TW D5	13	20	27	32	40	45	52	62	9	15	21	26	33	37	44	52	1.3	1.4	1.6	1.7	1.9	1.4	1.5	1.7	1.9	2.0
TW E1	15	20	26	31	37	42	48	56	11	15	20	23	29	32	37	44	1.3	1.5	1.7	1.9	2.0	1.5	1.7	1.9	2.1	2.3
TW E2	14	20	26	31	38	43	50	58	10	15	20	24	30	34	40	47	1.3	1.5	1.6	1.8	2.0	1.5	1.6	1.8	2.0	2.2
TW E3	17	25	33	40	49	55	62	71	13	19	27	32	40	46	53	62	1.1	1.3	1.4	1.6	1.8	1.2	1.4	1.6	1.7	1.9
TW E4	16	23	32	38	47	53	60	70	11	18	25	31	39	45	52	61	1.1	1.3	1.5	1.6	1.8	1.2	1.4	1.6	1.7	1.9
TW F1	12	14	16	17	19	21	23	26	10	11	12	13	14	15	15	16	2.2	2.6	3.0	3.3	3.7	2.9	3.5	4.2	-	-
TW F2	10	12	15	18	21	23	26	31	7	9	11	12	14	16	18	20	1.9	2.2	2.4	2.7	2.9	2.3	2.7	3.1	3.6	4.2
TW F3	8	11	15	17	21	24	27	33	5	7	10	11	14	16	18	21	1.8	2.0	2.3	2.5	2.7	2.1	2.5	2.8	3.2	3.6
TW F4	6	9	13	16	20	23	27	33	3	6	8	10	13	15	18	22	1.7	2.0	2.2	2.4	2.6	2.1	2.4	2.7	3.0	3.3
TW G1	10	12	14	16	18	20	23	26	8	9	10	11	13	14	15	17	2.1	2.4	2.8	3.1	3.3	2.7	3.1	3.6	4.3	-
TW G2	7	10	13	15	19	21	25	29	5	7	9	10	13	14	16	19	1.9	2.2	2.4	2.7	2.9	2.3	2.6	3.0	3.4	3.8
TW G3	6	8	12	14	18	21	24	29	3	5	7	9	12	14	16	19	1.9	2.1	2.4	2.6	2.8	2.2	2.5	2.8	3.2	3.5
TW G4	4	7	10	12	16	18	22	27	2	4	6	8	10	12	15	18	1.9	2.2	2.4	2.6	2.8	2.3	2.6	2.9	3.1	3.4
TW H1	7	9	12	14	16	18	21	25	5	6	8	9	11	12	14	16	2.1	2.4	2.7	2.9	3.2	2.6	2.9	3.3	3.8	4.2
TW H2	4	7	10	12	15	17	20	25	2	4	6	8	10	11	14	17	2.0	2.3	2.5	2.8	3.0	2.4	2.8	3.1	3.4	3.7
TW H3	3	6	8	10	13	16	19	23	2	3	5	7	9	10	13	16	2.1	2.3	2.6	2.8	3.0	2.5	2.8	3.1	3.4	3.6
TW H4	2	4	6	8	11	13	16	20	1	2	4	5	7	9	11	14	2.3	2.6	2.8	3.0	3.2	2.8	3.0	3.3	3.5	3.8
TW I1	5	7	8	9	11	12	13	14	3	4	4	5	5	5	5	3	2.5	3.1	3.9	5.0	-	3.0	-	-	-	-
TW I2	9	13	18	21	26	30	35	42	8	9	12	15	19	22	25	31	1.6	1.8	2.0	2.2	2.3	1.8	2.0	2.3	2.5	2.7
TW I3	10	14	19	23	29	33	39	46	6	10	14	17	22	25	29	35	1.5	1.7	1.9	2.0	2.2	1.7	1.9	2.1	2.3	2.5
TW I4	11	17	23	28	35	40	47	56	7	12	17	22	28	32	38	46	1.3	1.5	1.7	1.8	2.0	1.5	1.6	1.8	2.0	2.2
TW I5	11	17	24	30	37	43	50	59	7	12	18	23	30	35	41	50	1.3	1.4	1.6	1.8	1.9	1.4	1.6	1.7	1.9	2.1
TW J1	13	18	24	28	35	40	46	54	9	13	18	22	27	31	36	43	1.3	1.5	1.7	1.9	2.0	1.5	1.7	1.9	2.1	2.3
TW J2	12	17	24	29	36	41	47	56	8	13	18	22	28	32	38	45	1.3	1.5	1.7	1.8	2.0	1.5	1.7	1.8	2.0	2.2
TW J3	15	22	31	37	46	52	60	70	10	17	25	30	38	44	51	61	1.1	1.3	1.5	1.6	1.8	1.2	1.4	1.6	1.7	1.9
TW J4	13	21	29	36	44	51	58	68	9	16	23	29	37	42	50	59	1.2	1.3	1.5	1.6	1.8	1.3	1.4	1.6	1.7	1.9
DG A1	7	9	10	11	13	13	14	13	4	5	5	6	6	6	5	3	2.3	3.0	4.4	-	-	3.2	-	-	-	-
DG A2	10	13	16	19	23	25	29	33	7	9	12	14	16	18	21	24	1.8	2.0	2.3	2.6	3.0	2.1	2.4	2.7	3.2	3.8
DG A3	13	18	23	27	33	37	43	51	10	14	18	22	27	30	35	42	1.5	1.6	1.8	1.9	2.1	1.6	1.8	2.0	2.1	2.3
DG B1	7	9	10	12	13	14	15	16	4	5	6	6	6	6	6	4	2.2	3.0	3.6	-	-	3.3	4.6	-	-	-
DG B2	10	13	17	19	24	27	30	36	7	9	12	14	17	19	22	25	1.7	2.0	2.2	2.4	2.7	2.1	2.3	2.6	3.0	3.3
DG B3	14	18	24	28	34	39	46	55	11	14	19	23	28	32	37	45	1.4	1.6	1.7	1.8	2.0	1.6	1.7	1.9	2.0	2.2
DG C1	9	11	14	16	18	20	22	25	5	7	9	10	11	11	12	12	1.9	2.4	2.7	3.3	4.1	2.5	3.1	4.4	-	-
DG C2	12	16	20	23	28	32	37	44	9	12	15	18	22	24	28	33	1.6	1.8	1.9	2.1	2.3	1.8	2.0	2.3	2.6	2.8
DG C3	16	21	27	31	39	45	52	61	12	17	22	26	32	36	42	51	1.4	1.5	1.6	1.7	1.9	1.5	1.6	1.8	1.9	2.0
SS A1	16	19	23	25	29	32	35	40	13	16	19	21	24	25	28	31	1.6	1.9	2.2	2.4	2.6	1.9	2.2	2.6	3.0	3.4
SS A2	19	25	31	35	42	46	52	60	16	21	26	30	36	39	44	51	1.3	1.5	1.7	1.8	2.0	1.4	1.6	1.8	2.0	2.2
SS A3	14	17	20	22	25	27	30	33	12	14	16	18	19	21	22	24	1.8	2.2	2.5	2.9	3.2	2.1	2.6	3.2	4.1	-
SS A4	19	24	30	34	40	44	50	57	16	20	25	29	34	37	42	48	1.3	1.5	1.7	1.9	2.1	1.4	1.7	1.9	2.1	2.3
SS A5	16	19	22	24	27	29	31	35	14	16	18	20	22	23	25	27	1.8	2.2	2.5	2.8	3.0	2.1	2.5	3.0	3.7	4.4
SS A6	19	24	30	34	41	45	51	58	16	20	25	29	34	38	43	49	1.3	1.5	1.7	1.9	2.0	1.4	1.7	1.9	2.1	2.3
SS A7	15	17	19	20	22	24	25	28	13	14	16	17	18	19	19	20	2.1	2.6	3.1	3.6	4.1	2.5	3.2	4.1	-	-
SS A8	18	23	28	32	38	42	47	54	15	19	24	27	32	35	39	45	1.4	1.6	1.8	2.0	2.1	1.5	1.7	2.0	2.2	2.5
SS B1	13	16	19	21	25	27	30	35	11	13	15	17	20	21	23	26	1.8	2.1	2.4	2.6	2.8	2.1	2.4	2.8	3.3	3.8
SS B2	17	22	27	32	38	42	48	56	14	18	23	27	32	36	41	47	1.4	1.5	1.7	1.9	2.0	1.5	1.7	1.9	2.1	2.3
SS B3	12	14	17	19	22	24	26	30	10	11	13	15	17	18	19	21	1.9	2.3	2.6	3.0	3.3	2.3	2.7	3.3	4.1	-
SS B4	16	21	26	30	36	40	46	53	13	17	22	25	30	34	38	45	1.4	1.6	1.8	1.9	2.1	1.5	1.7	2.0	2.2	2.4
SS B5	13	15	18	19	22	24	26	29	11	13	15	16	17	19	20	22	2.0	2.4	2.7	3.1	3.4	2.4	2.9	3.4	4.1	4.9
SS B6	16	21	26	30	36	41	46	54	13	17	22	25	30	34	39	45	1.4	1.6	1.8	1.9	2.1	1.5	1.7	1.9	2.2	2.3
SS B7	12	14	16	17	19	20	22	24	10	12	13	14	15	15	16	17	2.3	2.7	3.2	3.7	4.2	2.8	3.4	4.3	-	-
SS B8	15	20	25	29	34	38	43	50	12	16	20	24	28	31	36	41	1.4	1.6	1.8	2.0	2.2	1.6	1.8	2.0	2.3	2.5
SS C1	10	13	16	18	22	24	28	32	6	8	10	12	14	15	17	19	1.8	2.1	2.4	2.6	2.9	2.2	2.6	3.2	3.9	4.7
SS C2	12	17	22	26	32	37	42	50	8	12	16	19	24	27	32	38	1.4	1.6	1.8	2.0	2.1	1.6	1.8	2.1	2.3	2.5
SS C3	10	12	15	17	19	21	24	27	8	9	11	12	14	15	16	18	2.0	2.4	2.7	3.0	3.2	2.6	3.0	3.5	4.2	4.9
SS C4	12	16	20	24	29	33	38	45	8	12	15	18	22	25	29	34	1.5	1.7	1.9	2.1	2.3	1.8	2.0	2.2	2.5	2.7
SS D1	17	21	25	29	33	37	41	47	13	17	20	23	26	28	31	35	1.5	1.7	2.0	2.2	2.4	1.7	2.0	2.4	2.8	3.2
SS D2	21	28	35	41	48	54	60	69	17	23	29	34	41	45	51	59	1.2	1.3	1.5	1.7	1.8	1.2	1.5	1.7	1.8	2.0
SS D3	17	20	24	27	31	33	37	42	14	17	20	22	24	26	29	32	1.6	1.9	2.1	2.4	2.6	1.8	2.2	2.6	3.1	3.5
SS D4	20	27	34	39	47	52	58	66	17	22	28	33	39	44	49	57	1.2	1.4	1.5	1.7	1.9	1.3	1.5	1.7	1.9	2.1
SS E1	13	17	20	23	27	30	33	38	10	13	15	17	19	21	23	25	1.6	1.9	2.2	2.5	2.8	1.9	2.3	2.9	3.6	4.6
SS E2	17	23	29	34	41	46	52	60	13	18	23	27	33	37	42	49	1.3	1.5	1.6	1.8	2.0	1.4	1.6	1.8	2.1	2.2
SS E3	14	16	19	21	24	26	29	33	11	13	15	17	19	20	22	23	1.8	2.2	2.5	2.8	3.1	2.2	2.6	3.2	3.9	4.7
SS E4	16	21	27	32	38	42	48	56	13	17	22	26	31	34	39	45	1.3	1.5	1.7	1.9	2.1	1.5	1.7	1.9	2.2	2.3

YOUNGSTOWN, OHIO

SOLAR SAVINGS FRACTION (%) — CONSERVATION FACTOR

TYPE	BASE TEMP: 55 F (4038 DD)								65 F (6435 DD)								CONSERVATION FACTOR 55 F					65 F				
LCR	100	70	50	40	30	25	20	15	100	70	50	40	30	25	20	15	100	50	30	20	15	100	50	30	20	15
WW A1	13	14	15	16	17	18	18	19	11	11	12	12	12	12	12	11	2.7	3.4	4.3	-	-	4.0	-	-	-	-
WW A2	10	12	15	17	19	21	23	27	7	9	10	11	13	13	14	15	2.0	2.4	2.8	3.2	3.5	2.5	3.1	4.0	-	-
WW A3	9	12	15	17	20	23	26	30	6	7	9	11	13	14	15	17	1.8	2.1	2.5	2.8	3.1	2.2	2.7	3.4	4.4	-
WW A4	8	11	14	17	21	23	27	32	4	6	9	10	13	14	16	18	1.7	2.0	2.3	2.6	2.8	2.1	2.5	3.1	3.9	4.9
WW A5	7	11	14	17	22	24	28	33	3	6	8	10	13	15	17	19	1.6	1.9	2.2	2.5	2.7	2.0	2.4	2.9	3.6	4.3
WW A6	6	10	14	17	22	25	29	35	3	5	8	10	13	15	17	20	1.6	1.9	2.1	2.4	2.6	1.9	2.3	2.8	3.4	4.0
WW B1	7	8	8	9	9	9	8	6	4	4	4	4	2	1	0	-5	3.0	-	-	-	-	-	-	-	-	-
WW B2	9	13	17	21	26	29	34	40	5	8	12	14	18	21	24	28	1.6	1.8	2.0	2.2	2.4	1.8	2.1	2.4	2.7	3.0
WW B3	11	16	21	26	32	36	41	49	7	11	15	19	23	26	31	36	1.4	1.6	1.8	2.0	2.2	1.6	1.8	2.1	2.4	2.6
WW B4	11	17	24	29	37	42	49	58	7	12	18	22	29	34	39	47	1.3	1.5	1.6	1.8	1.9	1.4	1.6	1.8	2.0	2.1
WW B5	10	17	24	30	38	43	50	60	7	12	18	23	30	35	42	50	1.2	1.4	1.6	1.7	1.9	1.4	1.5	1.7	1.9	2.0
WW C1	12	18	24	28	35	39	45	53	8	13	18	21	26	30	35	41	1.3	1.5	1.7	1.9	2.1	1.5	1.7	2.0	2.2	2.4
WW C2	11	17	23	28	35	40	46	55	7	12	17	21	27	31	37	44	1.3	1.5	1.7	1.8	2.0	1.5	1.7	1.9	2.1	2.2
WW C3	14	22	31	38	47	53	61	71	10	17	25	31	40	45	53	62	1.1	1.3	1.4	1.6	1.7	1.2	1.4	1.5	1.7	1.8
WW C4	12	20	29	35	44	50	58	68	8	15	23	29	37	43	50	60	1.2	1.3	1.5	1.6	1.8	1.2	1.4	1.6	1.7	1.9
TW A1	12	14	15	16	17	18	19	21	11	11	12	12	13	13	13	13	2.6	3.1	3.9	4.8	-	3.6	4.9	-	-	-
TW A2	10	13	15	17	19	21	23	26	8	9	11	12	13	14	15	16	2.1	2.5	2.9	3.3	3.6	2.6	3.2	4.0	-	-
TW A3	9	12	14	16	19	22	24	28	6	8	10	11	13	14	16	17	1.9	2.3	2.6	2.9	3.2	2.4	2.8	3.4	4.2	-
TW A4	8	11	14	16	20	22	25	30	5	7	9	11	13	14	16	19	1.8	2.1	2.4	2.7	2.9	2.2	2.6	3.1	3.6	4.2
TW B1	12	13	15	16	18	19	21	23	10	11	12	12	13	14	14	15	2.4	2.8	3.3	3.9	4.4	3.1	3.9	5.0	-	-
TW B2	9	12	14	16	19	21	23	27	7	9	10	12	13	14	16	18	2.0	2.3	2.7	3.0	3.3	2.5	2.9	3.5	4.1	4.9
TW B3	8	11	14	16	19	21	24	28	6	8	10	11	13	15	16	19	1.9	2.2	2.6	2.8	3.1	2.3	2.7	3.2	3.7	4.3
TW B4	8	11	13	16	19	21	24	28	6	8	10	11	13	15	17	19	2.0	2.2	2.5	2.8	3.0	2.3	2.7	3.1	3.5	3.9
TW C1	10	12	14	16	18	20	22	25	9	10	11	12	13	14	16	17	2.2	2.6	2.9	3.3	3.6	2.8	3.2	3.8	4.5	-
TW C2	9	11	14	16	18	20	23	27	7	9	10	12	14	15	17	19	2.1	2.4	2.7	3.0	3.2	2.5	2.9	3.3	3.7	4.2
TW C3	9	11	13	15	18	20	23	26	7	9	10	12	14	15	17	19	2.1	2.4	2.7	3.0	3.2	2.5	2.9	3.2	3.6	4.0
TW C4	10	11	14	15	18	20	22	26	8	10	11	12	14	15	17	20	2.2	2.5	2.8	3.1	3.3	2.7	3.0	3.3	3.7	4.0
TW D1	6	7	8	8	9	9	9	9	4	4	4	4	3	3	2	0	2.9	4.1	-	-	-	-	-	-	-	-
TW D2	10	13	17	20	24	27	31	37	7	10	12	15	18	20	23	27	1.7	1.9	2.1	2.4	2.5	1.9	2.2	2.5	2.9	3.1
TW D3	10	14	18	21	26	30	34	40	7	10	13	15	19	22	25	29	1.6	1.8	2.0	2.2	2.4	1.8	2.1	2.4	2.7	3.0
TW D4	11	16	22	26	33	38	43	52	8	12	17	20	26	30	35	42	1.4	1.6	1.7	1.9	2.1	1.5	1.7	1.9	2.1	2.3
TW D5	11	17	23	28	35	40	46	55	8	12	18	22	28	33	38	46	1.3	1.5	1.7	1.8	2.0	1.5	1.6	1.8	2.0	2.1
TW E1	12	17	22	26	31	35	40	48	9	12	16	19	24	27	31	37	1.4	1.7	1.9	2.0	2.2	1.6	1.9	2.1	2.4	2.6
TW E2	12	17	22	26	33	37	43	51	8	12	17	20	25	29	34	40	1.4	1.6	1.8	1.9	2.1	1.6	1.8	2.0	2.2	2.4
TW E3	14	21	29	35	43	48	56	65	10	16	23	28	35	40	47	56	1.2	1.4	1.5	1.7	1.8	1.3	1.5	1.7	1.8	2.0
TW E4	13	20	27	33	42	47	54	64	9	15	22	27	35	40	46	55	1.2	1.4	1.5	1.7	1.8	1.3	1.5	1.7	1.8	2.0
TW F1	10	12	13	14	16	17	18	19	9	9	10	11	11	11	11	11	2.5	3.1	3.7	4.5	-	3.5	4.6	-	-	-
TW F2	8	10	12	14	17	18	20	23	6	7	8	9	11	12	12	13	2.1	2.5	2.9	3.3	3.7	2.7	3.3	4.1	-	-
TW F3	6	9	11	14	16	18	21	25	4	5	7	8	10	11	13	14	2.0	2.3	2.7	3.0	3.3	2.4	2.9	3.5	4.3	-
TW F4	4	7	10	12	15	18	21	25	2	4	6	7	9	11	13	15	1.9	2.2	2.5	2.8	3.1	2.3	2.7	3.2	3.7	4.3
TW G1	8	10	12	13	15	16	18	20	7	7	8	9	10	11	11	12	2.4	2.8	3.3	3.9	4.4	3.1	3.9	4.9	-	-
TW G2	6	8	10	12	15	16	19	22	4	5	7	8	9	10	12	13	2.1	2.5	2.8	3.2	3.5	2.6	3.1	3.7	4.4	-
TW G3	4	6	9	11	14	16	19	22	2	4	5	7	9	10	12	14	2.1	2.4	2.7	3.0	3.2	2.5	2.9	3.4	3.9	4.5
TW G4	3	5	7	9	12	14	17	21	1	2	4	6	8	9	11	13	2.1	2.4	2.7	3.0	3.2	2.6	2.9	3.3	3.7	4.1
TW H1	6	7	9	11	13	14	16	19	4	5	6	7	8	9	10	11	2.3	2.7	3.1	3.5	3.9	2.9	3.5	4.1	5.0	-
TW H2	3	5	7	9	12	14	16	19	2	3	4	6	7	8	10	12	2.2	2.5	2.9	3.2	3.5	2.7	3.1	3.6	4.1	4.6
TW H3	2	4	6	8	11	12	15	18	1	2	4	5	7	8	9	12	2.3	2.6	2.9	3.2	3.4	2.8	3.1	3.5	3.9	4.3
TW H4	2	3	5	6	9	10	12	16	1	2	3	4	6	7	8	10	2.6	2.8	3.1	3.4	3.6	3.0	3.4	3.7	4.0	4.3
TW I1	4	5	6	6	7	7	7	7	2	2	2	2	2	1	0	-2	3.1	4.2	-	-	-	-	-	-	-	-
TW I2	7	10	14	17	21	24	28	34	4	7	10	12	15	17	20	24	1.7	2.0	2.2	2.4	2.6	2.0	2.3	2.6	2.9	3.2
TW I3	8	11	16	19	24	27	32	38	5	8	11	14	17	20	23	28	1.6	1.8	2.0	2.2	2.4	1.8	2.1	2.4	2.6	2.9
TW I4	9	14	19	24	30	35	41	49	5	9	14	18	23	27	32	39	1.4	1.6	1.8	1.9	2.1	1.6	1.8	2.0	2.2	2.3
TW I5	9	14	20	25	32	37	44	52	5	10	15	20	26	30	36	44	1.4	1.5	1.7	1.9	2.0	1.5	1.7	1.9	2.0	2.2
TW J1	10	15	20	24	29	33	39	46	7	10	15	18	22	25	30	35	1.5	1.7	1.9	2.0	2.2	1.6	1.9	2.1	2.4	2.6
TW J2	9	14	20	24	30	35	41	49	6	10	15	18	23	27	32	39	1.4	1.6	1.8	2.0	2.1	1.6	1.8	2.0	2.2	2.4
TW J3	12	19	27	32	41	47	54	64	8	14	21	26	34	39	46	55	1.2	1.4	1.5	1.7	1.8	1.3	1.5	1.6	1.8	2.0
TW J4	11	17	25	31	39	45	52	62	7	13	20	25	32	37	44	53	1.2	1.4	1.6	1.7	1.8	1.3	1.5	1.7	1.8	2.0
DG A1	5	6	7	8	8	8	8	6	3	3	3	3	2	2	0	-2	2.7	4.0	-	-	-	4.5	-	-	-	-
DG A2	8	11	13	15	19	21	23	26	6	8	10	11	13	15	16	18	1.9	2.3	2.6	3.0	3.5	2.3	2.7	3.2	3.9	4.8
DG A3	12	15	20	23	29	32	38	45	9	12	16	19	23	26	31	37	1.6	1.8	1.9	2.1	2.2	1.8	1.9	2.1	2.3	2.5
DG B1	5	6	7	8	8	8	9	8	3	3	3	3	3	2	1	-1	2.7	4.2	-	-	-	4.7	-	-	-	-
DG B2	8	11	14	16	19	21	24	28	6	8	10	12	14	15	17	19	1.9	2.2	2.5	2.8	3.1	2.3	2.6	3.1	3.7	4.2
DG B3	12	16	21	24	30	34	40	48	9	13	17	20	25	28	33	39	1.5	1.7	1.8	2.0	2.1	1.7	1.9	2.0	2.2	2.4
DG C1	7	9	10	11	13	14	15	16	4	5	6	6	7	7	6	5	2.3	3.1	4.0	4.8	-	3.2	4.0	-	-	-
DG C2	10	13	17	19	23	26	30	36	7	10	13	15	18	20	23	26	1.7	2.0	2.2	2.4	2.7	2.0	2.2	2.6	3.0	3.3
DG C3	14	18	23	27	34	39	45	54	11	15	19	23	28	32	37	45	1.5	1.6	1.7	1.9	2.0	1.6	1.7	1.9	2.1	2.2
SS A1	14	17	20	22	25	27	30	34	12	14	16	18	20	22	23	25	1.8	2.1	2.5	2.8	3.1	2.0	2.5	3.0	3.7	4.5
SS A2	17	22	27	31	37	41	47	54	15	19	23	27	32	35	39	45	1.4	1.6	1.8	2.0	2.1	1.5	1.7	2.0	2.2	2.4
SS A3	13	15	17	19	21	22	24	26	11	12	14	15	16	17	18	18	2.0	2.5	3.0	3.6	4.1	2.4	3.0	4.1	-	-
SS A4	17	21	26	30	35	39	44	51	14	18	22	26	30	33	37	42	1.4	1.6	1.8	2.0	2.2	1.5	1.8	2.1	2.3	2.6
SS A5	15	17	19	21	23	25	27	29	13	15	16	17	19	20	21	22	2.0	2.4	2.9	3.4	3.8	2.4	2.9	3.7	5.0	-
SS A6	17	22	27	30	36	40	45	52	14	18	23	26	30	34	38	43	1.4	1.6	1.8	2.0	2.2	1.5	1.8	2.0	2.3	2.5
SS A7	14	15	17	18	19	20	21	22	12	13	14	15	15	16	16	15	2.3	3.0	3.8	5.0	-	2.9	3.9	-	-	-
SS A8	16	21	25	29	34	37	42	48	14	17	21	24	28	31	34	39	1.5	1.7	1.9	2.1	2.3	1.6	1.9	2.2	2.5	2.7
SS B1	12	14	17	18	21	23	25	29	10	11	13	15	17	18	19	21	1.9	2.3	2.7	3.1	3.4	2.3	2.7	3.3	4.1	-
SS B2	15	19	24	28	34	38	43	50	13	16	21	24	29	32	36	42	1.4	1.6	1.8	2.0	2.2	1.6	1.8	2.0	2.2	2.4
SS B3	11	13	15	16	18	20	21	24	9	10	12	13	14	14	15	16	2.1	2.6	3.1	3.7	4.2	2.5	3.2	4.2	-	-
SS B4	14	19	23	27	32	36	41	47	12	16	20	23	27	30	34	40	1.5	1.7	1.9	2.1	2.2	1.6	1.9	2.1	2.3	2.6
SS B5	12	14	16	17	19	20	22	24	11	12	13	14	15	16	17	17	2.2	2.7	3.2	3.8	4.3	2.7	3.3	4.2	-	-
SS B6	14	18	23	27	32	36	41	48	12	15	19	22	27	30	34	40	1.5	1.7	1.9	2.1	2.2	1.6	1.8	2.1	2.3	2.5
SS B7	11	12	14	15	16	17	18	19	10	11	11	12	13	13	13	13	2.5	3.1	3.9	-	-	3.2	4.1	-	-	-
SS B8	13	17	22	25	30	33	38	44	11	14	18	21	25	28	31	36	1.5	1.8	2.0	2.2	2.3	1.7	1.9	2.2	2.5	2.7
SS C1	7	10	12	14	16	18	20	23	5	6	7	8	9	10	11	11	2.0	2.4	2.9	3.4	3.8	2.6	3.3	4.6	-	-
SS C2	10	14	18	22	27	30	35	41	6	9	13	16	19	22	25	30	1.5	1.8	2.0	2.2	2.4	1.8	2.0	2.3	2.6	2.9
SS C3	9	10	12	14	16	17	19	21	7	8	9	10	11	11	12	12	2.3	2.7	3.2	3.8	4.2	3.0	3.7	4.8	-	-
SS C4	10	13	17	20	24	27	31	37	7	10	12	15	18	20	23	27	1.7	1.9	2.1	2.3	2.5	1.9	2.2	2.5	2.8	3.1
SS D1	15	18	22	24	28	31	34	39	12	14	17	19	22	23	25	27	1.6	1.9	2.2	2.5	2.7	1.8	2.2	2.8	3.5	4.3
SS D2	18	24	31	36	43	48	54	62	15	20	26	30	36	40	46	53	1.2	1.4	1.6	1.8	1.9	1.3	1.5	1.8	2.0	2.1
SS D3	15	18	21	23	26	28	31	35	13	15	17	19	21	22	24	25	1.8	2.1	2.5	2.8	3.1	2.0	2.5	3.1	3.9	5.0
SS D4	18	24	30	35	41	46	52	60	15	20	25	29	35	39	44	51	1.3	1.5	1.6	1.8	2.0	1.4	1.6	1.8	2.0	2.2
SS E1	11	14	17	19	22	23	26	29	8	10	12	14	15	16	17	17	1.8	2.2	2.7	3.1	3.5	2.1	2.8	3.8	-	-
SS E2	14	19	25	29	35	40	45	53	11	15	20	24	29	32	36	42	1.4	1.6	1.8	1.9	2.1	1.5	1.7	2.0	2.3	2.5
SS E3	12	14	17	18	20	22	24	26	10	12	13	14	16	16	17	18	2.0	2.5	3.0	3.5	3.9	2.4	3.1	4.0	-	-
SS E4	14	19	24	27	33	37	42	49	11	15	19	22	27	30	34	39	1.4	1.7	1.9	2.1	2.2	1.6	1.8	2.1	2.4	2.6

OKLAHOMA CITY, OKLAHOMA

TYPE	SOLAR SAVINGS FRACTION (%) BASE TEMP: 55 F (1967 DD) LCR 200	100	70	50	40	30	25	20	65 F (3699 DD) 200	100	70	50	40	30	25	20	CONSERVATION FACTOR 55 F 200	100	50	30	20	65 F 200	100	50	30	20
WW A1	19	28	34	41	46	54	59	66	15	20	24	28	32	37	41	46	1.1	1.2	1.3	1.5	1.6	1.4	1.5	1.7	1.8	2.0
WW A2	20	33	41	51	58	67	73	80	13	22	28	35	40	48	53	60	0.9	1.0	1.1	1.3	1.4	1.1	1.2	1.4	1.5	1.7
WW A3	20	35	45	55	63	72	78	84	12	22	29	37	44	52	58	65	0.8	0.9	1.1	1.2	1.4	1.0	1.1	1.3	1.4	1.6
WW A4	20	35	46	57	65	74	80	86	11	22	30	39	45	54	60	68	0.8	0.9	1.0	1.2	1.4	1.0	1.1	1.2	1.4	1.5
WW A5	20	37	48	60	67	77	82	88	11	23	31	40	47	57	63	70	0.8	0.9	1.0	1.2	1.4	1.0	1.1	1.2	1.4	1.5
WW A6	20	37	49	61	68	78	83	89	10	23	32	41	48	58	64	72	0.8	0.9	1.0	1.2	1.3	0.9	1.0	1.2	1.3	1.5
WW B1	17	27	35	43	49	58	64	71	11	17	22	27	31	38	42	48	1.0	1.1	1.2	1.4	1.5	1.3	1.4	1.6	1.7	1.9
WW B2	20	36	47	58	66	75	81	87	11	23	31	41	47	57	63	70	0.8	0.9	1.0	1.2	1.4	1.0	1.1	1.2	1.4	1.5
WW B3	23	42	54	66	74	83	87	92	14	28	38	48	55	65	72	79	0.7	0.8	1.0	1.1	1.3	0.9	1.0	1.1	1.3	1.4
WW B4	22	42	55	67	75	84	89	93	12	28	38	50	58	68	74	81	0.7	0.8	1.0	1.1	1.3	0.8	0.9	1.1	1.2	1.4
WW B5	20	41	53	66	74	83	88	93	11	26	37	49	57	67	74	81	0.7	0.8	1.0	1.1	1.3	0.8	0.9	1.1	1.2	1.4
WW C1	25	44	56	68	76	84	89	93	15	30	40	50	58	68	74	81	0.7	0.8	1.0	1.1	1.3	0.8	0.9	1.1	1.2	1.4
WW C2	22	42	54	66	74	83	88	92	13	27	38	49	56	66	73	80	0.7	0.8	1.0	1.1	1.3	0.9	1.0	1.1	1.3	1.4
WW C3	25	48	62	74	81	89	93	96	14	33	45	57	66	76	82	88	0.7	0.8	0.9	1.1	1.3	0.8	0.9	1.0	1.2	1.3
WW C4	22	44	58	70	78	87	91	95	12	30	41	54	62	73	79	85	0.7	0.8	0.9	1.1	1.3	0.8	0.9	1.0	1.2	1.4
TW A1	19	27	33	40	45	53	58	65	14	20	23	28	32	37	41	46	1.1	1.2	1.3	1.5	1.6	1.4	1.5	1.7	1.8	2.0
TW A2	19	30	38	47	53	62	68	75	13	20	26	32	37	44	49	55	0.9	1.0	1.2	1.3	1.5	1.2	1.3	1.4	1.6	1.7
TW A3	18	31	39	49	56	66	71	78	12	20	26	33	39	47	52	59	0.9	1.0	1.1	1.3	1.4	1.1	1.2	1.4	1.5	1.7
TW A4	17	30	39	50	57	66	72	79	10	19	26	33	39	47	53	60	0.9	1.0	1.1	1.3	1.4	1.1	1.2	1.3	1.5	1.6
TW B1	18	27	33	40	46	54	59	66	13	19	23	28	32	38	42	48	1.1	1.2	1.3	1.4	1.6	1.4	1.5	1.6	1.8	1.9
TW B2	17	28	36	45	52	61	67	74	11	19	24	31	36	43	48	54	1.0	1.1	1.2	1.3	1.5	1.2	1.3	1.4	1.6	1.7
TW B3	16	28	36	45	52	61	67	74	10	18	24	31	36	44	49	56	0.9	1.0	1.2	1.3	1.5	1.2	1.3	1.4	1.6	1.7
TW B4	15	26	34	43	50	59	65	72	10	17	23	29	34	41	47	53	1.0	1.1	1.2	1.3	1.5	1.2	1.3	1.4	1.6	1.7
TW C1	16	25	32	40	46	54	59	66	12	18	22	28	32	38	42	48	1.1	1.2	1.3	1.4	1.6	1.3	1.4	1.6	1.7	1.9
TW C2	15	25	32	40	47	55	61	68	10	17	22	28	32	39	44	50	1.0	1.1	1.2	1.4	1.6	1.3	1.4	1.5	1.7	1.8
TW C3	15	24	31	38	44	53	58	66	10	16	21	26	31	37	42	48	1.1	1.2	1.3	1.4	1.6	1.3	1.4	1.6	1.7	1.8
TW C4	14	22	28	34	40	47	53	59	10	15	19	24	28	34	38	43	1.2	1.2	1.4	1.5	1.7	1.4	1.5	1.7	1.8	2.0
TW D1	14	23	30	37	43	51	57	64	9	14	18	23	27	33	37	42	1.1	1.2	1.3	1.4	1.6	1.4	1.5	1.7	1.8	2.0
TW D2	19	33	43	53	61	70	76	82	12	22	29	37	43	52	58	65	0.8	0.9	1.1	1.2	1.4	1.0	1.1	1.3	1.4	1.6
TW D3	20	36	47	58	65	75	80	86	12	24	32	41	47	57	63	70	0.8	0.9	1.0	1.2	1.4	1.0	1.1	1.2	1.4	1.5
TW D4	21	38	50	61	69	79	84	89	12	25	34	45	52	62	68	76	0.8	0.9	1.0	1.2	1.3	0.9	1.0	1.1	1.3	1.5
TW D5	19	37	49	61	69	78	84	89	11	25	34	44	52	62	68	76	0.8	0.9	1.0	1.2	1.3	0.9	1.0	1.1	1.3	1.4
TW E1	23	40	51	63	70	79	84	90	14	27	36	45	53	62	68	75	0.8	0.9	1.0	1.2	1.3	0.9	1.0	1.2	1.3	1.5
TW E2	22	39	51	62	70	79	84	90	13	26	35	45	53	62	68	76	0.8	0.9	1.0	1.2	1.3	0.9	1.0	1.1	1.3	1.5
TW E3	25	46	59	71	78	87	91	95	15	32	43	54	62	72	78	85	0.7	0.8	0.9	1.1	1.3	0.8	0.9	1.0	1.2	1.4
TW E4	23	43	56	68	76	84	89	93	13	29	40	51	59	70	76	83	0.7	0.8	1.0	1.1	1.3	0.8	0.9	1.1	1.2	1.4
TW F1	17	26	32	39	45	53	58	65	12	18	22	27	31	36	40	46	1.1	1.2	1.3	1.5	1.6	1.4	1.5	1.6	1.8	1.9
TW F2	16	27	35	44	51	60	66	73	10	18	23	29	34	41	46	53	1.0	1.1	1.2	1.3	1.5	1.2	1.3	1.5	1.6	1.8
TW F3	15	27	36	45	52	62	68	75	9	17	23	30	35	43	48	55	0.9	1.0	1.2	1.3	1.5	1.2	1.3	1.4	1.6	1.7
TW F4	13	25	34	43	50	60	66	74	6	15	21	28	33	41	47	54	0.9	1.0	1.2	1.3	1.5	1.2	1.3	1.4	1.6	1.7
TW G1	15	24	30	38	44	52	58	65	10	16	20	25	29	35	40	45	1.1	1.2	1.3	1.4	1.6	1.4	1.5	1.6	1.8	1.9
TW G2	13	24	31	40	47	56	62	69	8	15	20	26	31	38	43	50	1.0	1.1	1.2	1.4	1.5	1.2	1.3	1.5	1.6	1.8
TW G3	11	22	30	39	45	55	61	68	6	13	18	25	30	37	42	49	1.0	1.1	1.2	1.4	1.5	1.3	1.3	1.5	1.6	1.8
TW G4	9	18	25	33	40	49	55	62	4	10	15	21	26	32	37	43	1.1	1.2	1.3	1.4	1.6	1.3	1.4	1.6	1.7	1.9
TW H1	12	20	27	35	40	49	55	62	7	13	17	22	27	33	37	43	1.1	1.2	1.3	1.5	1.6	1.4	1.5	1.6	1.8	1.9
TW H2	9	18	25	32	38	47	53	60	4	10	15	20	25	31	35	41	1.1	1.2	1.3	1.5	1.6	1.4	1.5	1.6	1.8	1.9
TW H3	7	15	21	28	34	42	48	55	3	8	13	17	21	27	32	37	1.2	1.3	1.4	1.5	1.7	1.5	1.6	1.7	1.8	2.0
TW H4	5	12	16	22	27	34	39	46	2	6	10	14	17	22	25	30	1.3	1.4	1.5	1.7	1.8	1.7	1.8	1.9	2.0	2.2
TW I1	12	20	27	34	40	48	53	60	7	12	16	21	24	30	33	39	1.1	1.2	1.3	1.5	1.6	1.4	1.6	1.7	1.9	2.1
TW I2	16	[illegible]	[illegible]	50	57	67	73	80	9	18	25	34	40	48	54	62	0.9	1.0	1.1	1.3	1.4	1.1	1.2	1.3	1.5	1.6
TW I3	17	32	42	53	61	71	77	83	9	20	28	37	43	52	59	66	0.8	0.9	1.1	1.2	1.4	1.0	1.1	1.2	1.4	1.6
TW I4	18	35	46	58	66	76	81	87	9	22	31	41	48	58	65	72	0.8	0.9	1.0	1.2	1.4	0.9	1.0	1.2	1.3	1.5
TW I5	17	35	46	58	67	76	82	88	9	22	32	42	49	60	66	74	0.8	0.9	1.0	1.2	1.3	0.9	1.0	1.2	1.3	1.5
TW J1	20	37	49	60	68	77	83	88	12	24	33	43	50	60	66	73	0.8	0.9	1.0	1.2	1.3	0.9	1.0	1.2	1.3	1.5
TW J2	19	36	47	59	67	77	82	88	11	23	32	42	49	59	66	73	0.8	0.9	1.0	1.2	1.3	0.9	1.0	1.2	1.3	1.5
TW J3	22	43	55	68	76	84	89	93	12	28	39	51	59	70	76	83	0.7	0.8	1.0	1.1	1.3	0.8	0.9	1.1	1.2	1.4
TW J4	20	40	53	65	73	82	87	92	10	26	37	48	56	67	73	81	0.7	0.8	1.0	1.1	1.3	0.8	0.9	1.1	1.2	1.4
DG A1	15	26	33	41	46	53	58	63	8	15	20	25	29	34	37	41	1.0	1.1	1.3	1.5	1.8	1.3	1.4	1.6	2.0	2.4
DG A2	15	27	36	46	52	61	66	72	9	17	23	30	35	43	47	53	0.9	1.0	1.2	1.4	1.6	1.2	1.3	1.4	1.6	1.8
DG A3	17	32	42	54	62	72	77	83	11	21	29	38	45	54	60	67	0.9	0.9	1.0	1.2	1.5	1.1	1.1	1.2	1.4	1.6
DG B1	14	26	35	45	52	61	66	72	8	15	20	26	31	37	42	47	1.0	1.0	1.1	1.4	1.6	1.3	1.4	1.5	1.7	2.0
DG B2	15	28	38	49	57	67	73	79	10	18	24	31	37	46	52	59	0.9	1.0	1.1	1.3	1.5	1.2	1.2	1.3	1.5	1.7
DG B3	18	32	44	57	66	76	82	87	12	22	29	39	46	57	64	72	0.9	0.9	1.0	1.2	1.4	1.0	1.1	1.2	1.3	1.5
DG C1	17	31	41	53	61	70	76	81	10	18	24	32	37	46	51	58	0.9	0.9	1.1	1.3	1.5	1.1	1.2	1.4	1.5	1.7
DG C2	18	32	43	56	64	75	80	86	11	21	28	36	43	53	60	68	0.9	0.9	1.0	1.2	1.4	1.1	1.2	1.2	1.4	1.5
DG C3	20	37	49	63	72	83	87	92	13	25	33	44	52	64	71	79	0.8	0.8	0.9	1.1	1.3	1.0	1.0	1.1	1.2	1.4
SS A1	22	35	43	53	59	68	73	80	16	25	32	39	44	51	57	63	0.9	1.0	1.1	1.3	1.5	1.1	1.2	1.3	1.5	1.7
SS A2	26	43	54	65	72	80	85	90	19	32	40	50	56	65	71	78	0.8	0.8	1.0	1.2	1.3	0.9	1.0	1.1	1.3	1.5
SS A3	21	32	40	49	56	64	70	76	16	23	29	35	40	47	52	58	0.9	1.0	1.2	1.3	1.5	1.1	1.2	1.4	1.6	1.7
SS A4	26	42	53	64	71	79	84	89	19	31	39	49	55	64	70	76	0.8	0.9	1.0	1.2	1.4	0.9	1.0	1.2	1.3	1.5
SS A5	22	33	40	49	55	63	68	75	17	24	30	36	40	47	52	57	1.0	1.0	1.2	1.3	1.5	1.2	1.3	1.4	1.6	1.8
SS A6	26	43	54	64	71	80	85	90	19	31	40	49	56	65	70	77	0.8	0.9	1.0	1.2	1.3	0.9	1.0	1.1	1.3	1.5
SS A7	20	29	36	43	48	56	61	68	16	22	26	31	35	41	45	50	1.0	1.1	1.3	1.4	1.6	1.3	1.4	1.6	1.8	1.9
SS A8	26	42	52	63	70	78	83	88	18	30	39	47	54	63	68	75	0.8	0.9	1.0	1.2	1.4	0.9	1.0	1.2	1.3	1.5
SS B1	19	30	38	47	53	62	68	74	14	21	27	33	38	45	50	56	0.9	1.0	1.2	1.3	1.5	1.2	1.3	1.4	1.6	1.7
SS B2	23	38	48	59	66	75	80	86	16	27	35	44	51	60	65	72	0.8	0.9	1.0	1.2	1.4	0.9	1.0	1.2	1.4	1.5
SS B3	18	28	36	44	51	59	65	71	13	20	25	31	36	42	47	53	1.0	1.1	1.2	1.4	1.5	1.2	1.3	1.5	1.6	1.8
SS B4	22	37	47	58	65	74	79	85	15	27	34	43	49	58	64	71	0.8	0.9	1.1	1.2	1.4	1.0	1.1	1.2	1.4	1.5
SS B5	18	27	33	41	46	54	60	66	14	20	24	29	33	39	43	49	1.1	1.2	1.3	1.4	1.6	1.3	1.4	1.6	1.7	1.9
SS B6	22	37	47	58	65	74	80	86	15	27	34	43	50	59	64	71	0.8	0.9	1.0	1.2	1.4	1.0	1.1	1.2	1.4	1.5
SS B7	17	25	30	37	42	49	55	61	13	18	22	26	30	35	39	44	1.1	1.2	1.4	1.5	1.7	1.4	1.5	1.7	1.9	2.0
SS B8	21	36	46	56	63	73	78	84	15	26	33	42	48	56	62	69	0.8	0.9	1.1	1.2	1.4	1.0	1.1	1.2	1.4	1.5
SS C1	18	31	40	50	57	66	72	79	11	19	26	33	38	46	52	59	0.9	1.0	1.1	1.3	1.4	1.1	1.2	1.4	1.5	1.7
SS C2	20	36	47	58	66	75	81	87	12	23	32	41	48	57	63	71	0.8	0.9	1.0	1.2	1.4	1.0	1.1	1.2	1.4	1.5
SS C3	16	25	32	40	46	54	60	67	11	17	21	27	31	37	42	48	1.0	1.1	1.3	1.4	1.6	1.3	1.4	1.6	1.7	1.9
SS C4	18	32	42	52	59	69	75	81	11	21	28	36	42	51	57	64	0.9	1.0	1.1	1.2	1.4	1.0	1.1	1.3	1.4	1.6
SS D1	26	42	52	62	69	78	83	88	18	30	37	46	52	61	66	73	0.8	0.9	1.0	1.2	1.4	0.9	1.0	1.2	1.4	1.5
SS D2	30	50	62	73	80	87	91	95	20	36	47	57	65	74	79	85	0.7	0.8	0.9	1.1	1.3	0.8	0.9	1.1	1.2	1.4
SS D3	24	37	46	56	63	72	77	83	18	27	34	41	47	55	60	66	0.8	0.9	1.1	1.3	1.4	1.0	1.1	1.3	1.5	1.6
SS D4	29	48	60	71	78	86	90	94	20	35	45	55	63	72	77	84	0.7	0.8	1.0	1.1	1.3	0.8	0.9	1.1	1.2	1.4
SS E1	22	37	46	57	64	73	78	84	15	25	32	40	46	54	59	66	0.8	0.9	1.1	1.2	1.4	1.0	1.1	1.3	1.4	1.6
SS E2	25	44	56	67	74	83	87	92	17	31	40	51	58	67	73	80	0.7	0.8	1.0	1.1	1.3	0.9	1.0	1.1	1.3	1.4
SS E3	20	31	38	47	53	62	68	74	15	22	27	34	38	45	50	56	1.0	1.0	1.2	1.3	1.5	1.2	1.3	1.4	1.6	1.8
SS E4	24	41	51	62	70	79	84	89	16	29	37	47	53	62	68	75	0.8	0.9	1.0	1.2	1.4	0.9	1.0	1.2	1.3	1.5

TULSA, OKLAHOMA

TYPE	SOLAR SAVINGS FRACTION (%) BASE TEMP: 55 F (1957 DD)								65 F (3684 DD)								CONSERVATION FACTOR 55 F					65 F				
LCR	200	100	70	50	40	30	25	20	200	100	70	50	40	30	25	20	200	100	50	30	20	200	100	50	30	20
WW A1	18	26	32	38	43	50	55	62	14	19	22	26	29	34	38	42	1.1	1.2	1.4	1.5	1.7	1.5	1.6	1.8	1.9	2.1
WW A2	18	30	39	48	54	63	69	76	12	20	25	32	37	44	49	56	0.9	1.0	1.2	1.3	1.5	1.2	1.3	1.4	1.6	1.7
WW A3	18	32	41	51	59	68	74	80	11	20	26	34	40	48	53	60	0.9	1.0	1.1	1.3	1.4	1.1	1.2	1.3	1.5	1.6
WW A4	18	33	43	53	61	70	76	82	10	20	27	35	41	50	56	63	0.8	0.9	1.1	1.2	1.4	1.0	1.1	1.3	1.4	1.6
WW A5	18	34	44	55	63	73	78	85	9	20	28	37	43	52	58	66	0.8	0.9	1.1	1.2	1.4	1.0	1.1	1.3	1.4	1.6
WW A6	18	34	45	57	64	74	80	86	9	20	28	37	44	53	60	67	0.8	0.9	1.0	1.2	1.4	1.0	1.1	1.2	1.4	1.5
WW B1	16	25	32	40	46	54	59	66	10	15	20	24	28	33	37	42	1.0	1.1	1.3	1.4	1.6	1.4	1.5	1.7	1.9	2.0
WW B2	18	33	44	55	62	72	77	84	10	21	28	37	44	53	59	66	0.8	0.9	1.1	1.2	1.4	1.0	1.1	1.2	1.4	1.6
WW B3	21	39	51	62	70	79	84	90	12	25	34	44	51	61	67	75	0.8	0.8	1.0	1.2	1.3	0.9	1.0	1.2	1.3	1.5
WW B4	20	39	51	64	72	81	86	91	10	25	35	46	54	64	70	78	0.7	0.8	1.0	1.1	1.3	0.9	1.0	1.1	1.3	1.4
WW B5	18	38	50	62	70	80	85	90	9	24	34	45	53	63	70	78	0.7	0.8	1.0	1.2	1.3	0.9	1.0	1.1	1.3	1.4
WW C1	22	41	53	64	72	81	86	91	13	27	36	46	54	64	70	77	0.7	0.8	1.0	1.2	1.3	0.9	1.0	1.1	1.3	1.4
WW C2	20	39	51	63	70	80	85	90	11	25	34	45	52	62	69	76	0.7	0.8	1.0	1.2	1.3	0.9	1.0	1.1	1.3	1.4
WW C3	23	45	58	71	78	87	91	95	12	30	42	54	62	73	79	85	0.7	0.8	0.9	1.1	1.3	0.8	0.9	1.0	1.2	1.4
WW C4	20	41	54	67	75	84	89	93	10	27	38	50	58	69	75	83	0.7	0.8	1.0	1.1	1.3	0.8	0.9	1.1	1.2	1.4
TW A1	18	25	31	37	42	49	54	61	14	18	22	26	29	34	37	42	1.1	1.2	1.4	1.5	1.7	1.5	1.6	1.8	1.9	2.1
TW A2	17	28	35	43	50	58	64	71	12	18	23	29	34	40	45	51	1.0	1.1	1.2	1.4	1.5	1.3	1.4	1.5	1.7	1.8
TW A3	17	28	37	46	53	62	67	74	10	18	24	30	36	43	48	55	0.9	1.0	1.2	1.3	1.5	1.2	1.3	1.4	1.6	1.7
TW A4	15	28	36	46	53	62	68	75	9	17	23	30	36	43	49	56	0.9	1.0	1.2	1.3	1.5	1.1	1.2	1.4	1.5	1.7
TW B1	17	25	31	38	43	51	56	62	13	18	21	26	30	35	39	44	1.1	1.2	1.3	1.5	1.7	1.4	1.5	1.7	1.8	2.0
TW B2	16	26	34	42	48	57	63	70	10	17	22	28	33	39	44	50	1.0	1.1	1.2	1.4	1.5	1.3	1.4	1.5	1.7	1.8
TW B3	15	26	33	42	49	58	64	71	9	17	22	28	33	40	45	51	1.0	1.1	1.2	1.4	1.5	1.2	1.3	1.5	1.6	1.8
TW B4	14	24	31	40	46	55	61	68	9	16	21	27	31	38	43	49	1.0	1.1	1.2	1.4	1.6	1.3	1.4	1.5	1.7	1.8
TW C1	16	24	30	37	43	50	56	62	11	16	20	25	29	35	39	44	1.1	1.2	1.3	1.5	1.6	1.4	1.5	1.7	1.8	2.0
TW C2	14	23	30	38	43	52	57	64	10	16	20	25	30	36	40	46	1.1	1.2	1.3	1.4	1.6	1.3	1.4	1.6	1.7	1.9
TW C3	14	22	28	36	41	49	55	62	9	15	19	24	28	34	39	44	1.1	1.2	1.3	1.5	1.6	1.4	1.5	1.6	1.8	1.9
TW C4	13	20	26	32	37	44	49	56	10	14	18	22	26	31	35	40	1.2	1.3	1.4	1.6	1.7	1.5	1.6	1.7	1.9	2.0
TW D1	13	21	27	34	40	47	53	59	8	13	16	21	24	29	33	38	1.1	1.2	1.4	1.5	1.7	1.5	1.6	1.8	2.0	2.1
TW D2	17	31	40	50	57	66	72	79	11	20	26	34	40	48	54	61	0.9	1.0	1.1	1.3	1.4	1.1	1.2	1.3	1.5	1.6
TW D3	18	33	43	54	62	71	77	83	11	21	29	37	44	53	58	66	0.8	0.9	1.1	1.2	1.4	1.0	1.1	1.3	1.4	1.6
TW D4	19	35	46	58	66	75	81	87	11	23	32	41	48	58	64	72	0.8	0.9	1.0	1.2	1.4	0.9	1.0	1.2	1.3	1.5
TW D5	18	35	46	57	65	75	81	87	10	22	31	41	48	58	65	72	0.8	0.9	1.0	1.2	1.4	0.9	1.0	1.2	1.3	1.5
TW E1	21	37	48	59	67	76	81	87	13	25	33	42	49	58	64	71	0.8	0.9	1.0	1.2	1.4	1.0	1.1	1.2	1.3	1.5
TW E2	20	37	47	59	66	76	81	87	12	24	32	42	49	58	65	72	0.8	0.9	1.0	1.2	1.4	0.9	1.0	1.2	1.3	1.5
TW E3	23	43	55	68	75	84	88	93	13	29	39	51	59	69	75	82	0.7	0.8	1.0	1.1	1.3	0.8	0.9	1.1	1.2	1.4
TW E4	21	40	52	65	72	82	87	92	11	27	37	48	56	66	72	80	0.7	0.8	1.0	1.1	1.3	0.9	1.0	1.1	1.2	1.4
TW F1	16	24	30	37	42	49	54	61	12	17	20	25	28	33	37	42	1.1	1.2	1.4	1.5	1.7	1.5	1.6	1.7	1.9	2.1
TW F2	15	25	32	41	47	56	62	69	9	16	21	27	31	38	42	48	1.0	1.1	1.2	1.4	1.6	1.3	1.4	1.5	1.7	1.8
TW F3	14	25	33	42	48	58	64	71	8	15	20	27	32	39	44	51	1.0	1.1	1.2	1.4	1.5	1.2	1.3	1.5	1.6	1.8
TW F4	11	23	31	40	47	56	62	70	5	13	19	25	30	38	43	49	1.0	1.1	1.2	1.4	1.5	1.2	1.3	1.5	1.6	1.8
TW G1	14	22	28	35	41	48	54	61	10	15	19	23	27	32	36	41	1.1	1.2	1.4	1.5	1.7	1.4	1.5	1.7	1.9	2.0
TW G2	12	22	29	37	43	52	58	65	7	13	18	24	28	35	39	45	1.0	1.1	1.3	1.4	1.6	1.3	1.4	1.6	1.7	1.9
TW G3	10	20	27	36	42	51	57	64	5	11	16	22	27	33	38	44	1.0	1.1	1.3	1.4	1.6	1.3	1.4	1.6	1.7	1.9
TW G4	8	17	23	31	37	45	51	58	3	9	13	19	23	29	34	40	1.1	1.2	1.3	1.5	1.6	1.4	1.5	1.6	1.8	1.9
TW H1	11	19	25	32	37	45	51	58	6	12	16	20	24	30	34	39	1.1	1.2	1.4	1.5	1.7	1.5	1.6	1.7	1.9	2.0
TW H2	8	16	22	30	35	43	49	56	4	9	13	18	22	28	32	38	1.2	1.2	1.4	1.5	1.7	1.5	1.6	1.7	1.9	2.0
TW H3	6	14	19	26	31	39	44	51	3	7	11	16	19	25	29	34	1.2	1.3	1.5	1.6	1.7	1.5	1.6	1.8	1.9	2.1
TW H4	5	11	15	20	25	31	36	42	2	5	8	12	15	20	23	28	1.4	1.5	1.6	1.7	1.9	1.7	1.8	2.0	2.1	2.3
TW I1	11	19	24	31	36	44	49	56	6	11	14	18	21	26	30	34	1.2	1.3	1.4	1.6	1.7	1.5	1.7	1.8	2.0	2.2
TW I2	14	27	36	46	53	63	69	76	8	17	23	31	36	45	50	57	0.9	1.0	1.1	1.3	1.5	1.1	1.2	1.4	1.5	1.7
TW I3	15	29	39	50	57	67	73	80	8	18	25	33	40	48	54	62	0.9	1.0	1.1	1.3	1.4	1.1	1.2	1.3	1.4	1.6
TW I4	16	32	43	54	62	72	78	84	8	20	28	38	45	55	61	69	0.8	0.9	1.1	1.2	1.4	1.0	1.1	1.2	1.4	1.5
TW I5	15	32	43	55	63	73	79	85	8	20	29	39	46	56	62	70	0.8	0.9	1.0	1.2	1.4	1.0	1.0	1.2	1.3	1.5
TW J1	18	35	45	56	64	74	79	85	10	22	30	39	46	56	62	69	0.8	0.9	1.0	1.2	1.4	1.0	1.1	1.2	1.4	1.5
TW J2	17	33	44	56	63	73	79	85	9	21	29	39	46	55	62	69	0.8	0.9	1.0	1.2	1.4	1.0	1.1	1.2	1.4	1.5
TW J3	20	39	52	64	72	81	86	92	10	26	36	47	55	66	72	80	0.7	0.8	1.0	1.1	1.3	0.9	0.9	1.1	1.2	1.4
TW J4	18	37	49	62	70	79	85	90	9	24	34	45	53	63	70	77	0.7	0.8	1.0	1.2	1.3	0.9	1.0	1.1	1.3	1.4
DG A1	13	23	30	38	43	49	53	58	7	13	17	22	26	30	33	36	1.0	1.1	1.3	1.6	1.9	1.4	1.5	1.7	2.1	2.6
DG A2	14	25	33	42	49	57	62	68	9	16	21	27	32	39	44	49	1.0	1.1	1.2	1.4	1.6	1.3	1.3	1.5	1.7	1.9
DG A3	16	29	39	51	58	68	74	80	11	19	26	35	41	50	57	64	0.9	1.0	1.1	1.3	1.5	1.1	1.2	1.3	1.4	1.6
DG B1	13	24	32	41	47	56	61	67	7	13	18	23	27	33	37	42	1.0	1.1	1.2	1.4	1.7	1.3	1.5	1.6	1.8	2.1
DG B2	14	26	35	45	53	63	69	75	9	16	22	28	34	42	47	54	1.0	1.0	1.1	1.3	1.5	1.2	1.3	1.4	1.5	1.7
DG B3	16	30	41	53	61	72	78	84	11	20	27	36	43	53	60	68	0.9	0.9	1.0	1.2	1.4	1.1	1.2	1.2	1.3	1.5
DG C1	15	28	37	48	55	65	71	77	9	16	22	28	33	40	45	52	1.0	1.0	1.1	1.3	1.6	1.2	1.3	1.5	1.6	1.8
DG C2	16	30	40	51	60	70	76	82	10	19	25	33	39	49	55	63	0.9	1.0	1.1	1.2	1.4	1.1	1.2	1.3	1.4	1.6
DG C3	18	34	46	59	68	79	84	90	12	23	31	40	48	59	67	75	0.9	0.9	1.0	1.1	1.3	1.0	1.1	1.2	1.3	1.4
SS A1	21	32	40	49	56	64	70	76	15	23	29	36	41	48	52	59	0.9	1.0	1.2	1.3	1.5	1.1	1.2	1.4	1.6	1.7
SS A2	24	40	51	61	68	77	82	87	17	29	37	46	53	61	67	74	0.8	0.9	1.0	1.2	1.4	0.9	1.0	1.2	1.3	1.5
SS A3	20	30	38	46	52	60	65	72	14	22	27	32	37	43	48	53	1.0	1.1	1.2	1.4	1.6	1.2	1.3	1.5	1.7	1.8
SS A4	24	40	50	60	67	76	81	86	17	29	37	45	51	60	66	72	0.8	0.9	1.0	1.2	1.4	0.9	1.0	1.2	1.4	1.5
SS A5	21	31	38	45	51	59	64	71	16	23	27	33	37	43	48	53	1.0	1.1	1.2	1.4	1.6	1.2	1.3	1.5	1.7	1.8
SS A6	24	40	50	61	68	77	82	87	17	29	37	46	52	61	67	73	0.8	0.9	1.0	1.2	1.4	0.9	1.0	1.2	1.3	1.5
SS A7	19	27	33	40	45	52	57	64	15	20	24	29	32	37	41	46	1.1	1.2	1.3	1.5	1.7	1.4	1.5	1.7	1.9	2.0
SS A8	24	39	49	59	66	75	80	85	17	28	36	44	50	59	64	71	0.8	0.9	1.1	1.2	1.4	1.0	1.1	1.2	1.4	1.5
SS B1	18	28	35	44	50	58	64	70	13	20	25	31	35	42	46	52	1.0	1.1	1.2	1.4	1.6	1.2	1.3	1.5	1.7	1.8
SS B2	21	35	45	55	63	72	77	83	14	25	33	41	47	56	62	69	0.8	0.9	1.1	1.2	1.4	1.0	1.1	1.2	1.4	1.6
SS B3	17	26	33	41	47	55	61	67	12	18	23	28	33	39	43	48	1.0	1.1	1.3	1.4	1.6	1.3	1.4	1.6	1.7	1.9
SS B4	21	35	44	54	61	71	76	82	14	25	32	40	46	55	60	67	0.8	0.9	1.1	1.2	1.4	1.0	1.1	1.3	1.4	1.6
SS B5	17	25	31	38	43	51	56	62	13	18	22	27	31	36	40	45	1.1	1.2	1.3	1.5	1.7	1.4	1.5	1.7	1.8	2.0
SS B6	20	35	44	55	62	71	76	83	14	24	32	40	46	55	60	67	0.8	0.9	1.1	1.2	1.4	1.0	1.1	1.3	1.4	1.6
SS B7	16	23	28	34	39	46	51	57	12	17	20	24	27	32	35	40	1.2	1.3	1.4	1.6	1.7	1.5	1.6	1.8	2.0	2.2
SS B8	20	33	43	53	60	69	74	81	13	23	30	38	44	53	58	65	0.9	1.0	1.1	1.3	1.4	1.0	1.1	1.3	1.4	1.6
SS C1	16	28	37	46	53	62	68	75	10	17	23	30	35	42	47	54	0.9	1.0	1.2	1.3	1.5	1.2	1.3	1.4	1.6	1.7
SS C2	18	33	44	54	62	72	77	84	10	21	29	37	44	53	59	66	0.8	0.9	1.1	1.2	1.4	1.0	1.1	1.3	1.4	1.6
SS C3	15	23	30	37	43	51	56	63	10	15	20	24	28	34	38	44	1.1	1.2	1.3	1.5	1.6	1.4	1.5	1.7	1.8	2.0
SS C4	17	30	39	49	56	65	71	78	10	19	26	33	39	47	53	60	0.9	1.0	1.1	1.3	1.4	1.1	1.2	1.3	1.5	1.6
SS D1	24	39	48	58	65	74	79	85	16	27	34	42	48	56	62	68	0.8	0.9	1.1	1.2	1.4	1.0	1.1	1.3	1.4	1.6
SS D2	27	47	58	70	76	84	89	93	18	33	43	54	61	70	76	82	0.7	0.8	1.0	1.1	1.3	0.8	0.9	1.1	1.2	1.4
SS D3	23	35	43	53	59	68	73	79	16	25	31	38	43	50	55	62	0.9	1.0	1.1	1.3	1.5	1.1	1.2	1.4	1.5	1.7
SS D4	26	45	56	67	74	83	87	92	18	32	42	52	59	68	74	80	0.7	0.8	1.0	1.2	1.3	0.8	1.0	1.1	1.3	1.4
SS E1	20	34	43	53	60	68	74	80	13	23	29	36	42	49	55	61	0.9	1.0	1.1	1.3	1.4	1.1	1.2	1.3	1.5	1.7
SS E2	23	41	52	63	71	80	84	90	15	28	37	47	54	63	69	76	0.7	0.8	1.0	1.2	1.3	0.9	1.0	1.1	1.3	1.5
SS E3	19	29	36	44	50	58	64	70	13	20	25	31	35	41	46	51	1.0	1.1	1.2	1.4	1.6	1.2	1.3	1.5	1.7	1.8
SS E4	22	38	48	59	66	75	80	86	14	26	34	43	50	58	64	71	0.8	0.9	1.0	1.2	1.4	0.9	1.0	1.2	1.4	1.5

ASTORIA, OREGON

TYPE	SOLAR SAVINGS FRACTION (%) BASE TEMP: 55 F (2236 DD)								65 F (5299 DD)								CONSERVATION FACTOR 55 F					65 F				
LCR	100	70	50	40	30	25	20	15	100	70	50	40	30	25	20	15	100	50	30	20	15	100	50	30	20	15
WW A1	24	28	33	37	43	46	51	57	16	19	22	25	28	31	34	38	1.4	1.6	1.8	2.0	2.2	1.8	2.0	2.3	2.6	2.9
WW A2	26	33	40	45	53	57	63	70	16	21	26	30	35	39	43	49	1.1	1.3	1.5	1.7	1.9	1.4	1.6	1.9	2.1	2.4
WW A3	27	35	43	48	56	61	67	74	16	21	27	32	38	42	47	53	1.1	1.3	1.5	1.7	1.8	1.3	1.5	1.8	2.0	2.2
WW A4	28	36	44	50	58	63	69	76	15	21	28	33	39	43	49	56	1.0	1.2	1.4	1.6	1.8	1.3	1.5	1.7	1.9	2.2
WW A5	28	37	46	52	60	65	71	78	15	22	29	34	41	45	51	58	1.0	1.2	1.4	1.6	1.8	1.2	1.4	1.7	1.9	2.1
WW A6	29	37	46	53	61	66	72	79	15	22	29	34	42	46	52	59	1.0	1.2	1.4	1.6	1.7	1.2	1.4	1.6	1.9	2.1
WW B1	22	27	33	37	43	47	51	58	12	15	19	21	25	27	30	33	1.3	1.5	1.7	2.0	2.2	1.7	2.0	2.4	2.9	3.5
WW B2	28	37	46	52	60	65	72	79	16	23	30	35	43	47	53	61	1.0	1.2	1.4	1.6	1.7	1.2	1.4	1.6	1.8	2.0
WW B3	33	43	52	59	67	72	78	85	20	28	36	42	50	55	61	69	0.9	1.1	1.3	1.5	1.7	1.1	1.3	1.5	1.7	1.9
WW B4	33	43	54	61	70	75	81	88	20	29	38	45	54	59	66	74	0.9	1.1	1.3	1.5	1.6	1.0	1.2	1.4	1.6	1.8
WW B5	32	42	53	60	69	75	81	88	19	28	38	44	54	60	67	75	0.9	1.1	1.3	1.4	1.6	1.0	1.2	1.4	1.6	1.7
WW C1	35	45	54	61	69	75	80	87	22	30	38	45	53	58	64	72	0.9	1.1	1.3	1.5	1.6	1.1	1.3	1.5	1.7	1.8
WW C2	33	43	53	60	69	74	80	86	20	28	37	43	52	57	64	72	0.9	1.1	1.3	1.5	1.6	1.1	1.3	1.4	1.6	1.8
WW C3	38	50	61	68	77	82	87	93	24	35	46	53	63	69	75	83	0.8	1.0	1.2	1.4	1.5	1.0	1.1	1.3	1.5	1.7
WW C4	35	46	58	65	74	79	85	91	22	32	42	50	59	66	73	81	0.9	1.0	1.2	1.4	1.6	1.0	1.2	1.3	1.5	1.7
TW A1	23	28	33	37	42	46	51	57	16	19	22	24	28	31	34	38	1.4	1.6	1.8	2.0	2.2	1.8	2.0	2.2	2.5	2.8
TW A2	25	31	37	42	49	53	59	66	15	19	24	28	33	36	41	46	1.2	1.4	1.6	1.8	2.0	1.5	1.7	1.9	2.2	2.4
TW A3	25	32	39	44	51	56	62	69	15	19	25	29	34	38	43	49	1.1	1.3	1.5	1.7	1.9	1.4	1.6	1.8	2.1	2.3
TW A4	24	31	39	44	52	57	63	71	14	19	24	29	35	39	44	51	1.1	1.3	1.5	1.7	1.9	1.4	1.6	1.8	2.0	2.2
TW B1	23	28	33	37	43	47	52	59	15	18	22	25	29	32	35	40	1.3	1.5	1.7	1.9	2.1	1.7	1.9	2.1	2.4	2.6
TW B2	23	29	36	41	48	52	58	66	14	18	23	27	32	36	40	47	1.2	1.4	1.6	1.8	1.9	1.5	1.7	1.9	2.1	2.3
TW B3	23	29	36	41	48	53	59	67	13	18	23	27	32	36	41	48	1.2	1.4	1.6	1.7	1.9	1.5	1.6	1.9	2.1	2.3
TW B4	21	27	34	39	46	51	57	65	13	17	22	26	31	35	40	47	1.2	1.4	1.6	1.8	1.9	1.5	1.7	1.9	2.1	2.3
TW C1	22	27	32	37	43	47	53	60	14	17	21	24	29	32	36	42	1.3	1.5	1.7	1.9	2.0	1.7	1.8	2.0	2.3	2.5
TW C2	21	26	33	37	44	49	54	62	13	17	21	25	30	33	38	44	1.3	1.4	1.6	1.8	2.0	1.6	1.7	2.0	2.2	2.4
TW C3	20	25	31	36	42	47	52	60	12	16	20	24	29	32	37	43	1.3	1.5	1.7	1.9	2.0	1.6	1.8	2.0	2.2	2.4
TW C4	19	23	29	33	39	43	48	55	12	15	19	22	26	29	34	40	1.4	1.6	1.8	1.9	2.1	1.8	1.9	2.1	2.3	2.5
TW D1	19	23	29	33	38	42	47	53	10	13	16	19	22	25	28	31	1.4	1.6	1.8	2.0	2.2	1.8	2.1	2.4	2.8	3.2
TW D2	27	34	42	48	56	61	67	75	16	22	28	33	39	44	50	57	1.1	1.3	1.4	1.6	1.8	1.3	1.5	1.7	1.9	2.1
TW D3	29	37	46	52	60	65	71	79	17	24	31	36	43	48	53	61	1.0	1.2	1.4	1.6	1.7	1.2	1.4	1.6	1.8	2.0
TW D4	31	40	49	56	65	70	76	84	19	26	34	40	49	54	61	69	1.0	1.1	1.3	1.5	1.7	1.1	1.3	1.5	1.7	1.8
TW D5	30	39	49	56	65	71	77	84	18	26	35	41	50	55	62	71	1.0	1.1	1.3	1.5	1.6	1.1	1.3	1.5	1.6	1.8
TW E1	32	41	50	56	65	70	76	83	20	27	35	41	48	53	59	67	1.0	1.2	1.3	1.5	1.7	1.1	1.3	1.5	1.7	1.9
TW E2	31	41	50	57	65	70	76	83	19	27	35	41	49	54	60	68	1.0	1.1	1.3	1.5	1.7	1.1	1.3	1.5	1.7	1.9
TW E3	37	47	58	65	74	79	84	90	24	33	43	50	59	64	71	79	0.9	1.1	1.2	1.4	1.6	1.0	1.2	1.4	1.5	1.7
TW E4	34	45	56	63	72	77	83	89	22	31	41	48	57	63	69	78	0.9	1.1	1.2	1.4	1.6	1.0	1.2	1.4	1.6	1.7
TW F1	22	27	32	36	42	45	50	57	14	17	20	23	27	30	33	38	1.3	1.5	1.8	2.0	2.1	1.8	2.0	2.2	2.5	2.8
TW F2	22	28	35	40	46	51	57	64	13	17	22	25	30	34	38	44	1.2	1.4	1.6	1.8	2.0	1.5	1.7	2.0	2.2	2.4
TW F3	22	28	35	40	48	52	58	66	12	16	21	25	31	35	40	46	1.2	1.4	1.6	1.7	1.9	1.5	1.6	1.9	2.1	2.3
TW F4	19	26	34	39	46	51	58	65	10	14	20	24	30	34	39	46	1.2	1.4	1.5	1.7	1.9	1.4	1.6	1.9	2.1	2.3
TW G1	20	25	31	35	41	45	50	57	12	15	19	22	26	29	33	38	1.3	1.5	1.7	1.9	2.1	1.7	1.9	2.1	2.4	2.6
TW G2	19	25	32	36	43	48	54	61	10	14	19	22	28	31	36	42	1.2	1.4	1.6	1.8	2.0	1.6	1.7	2.0	2.2	2.4
TW G3	17	23	30	35	42	47	53	61	8	13	17	21	26	30	35	42	1.2	1.4	1.6	1.8	2.0	1.6	1.7	1.9	2.2	2.4
TW G4	14	20	26	31	38	42	48	56	6	10	15	18	23	27	32	38	1.3	1.5	1.7	1.9	2.0	1.6	1.8	2.0	2.2	2.4
TW H1	17	22	28	32	38	42	48	55	9	12	16	19	24	27	31	37	1.3	1.5	1.7	1.9	2.1	1.7	1.9	2.1	2.4	2.6
TW H2	14	19	25	30	36	41	47	54	7	10	14	17	22	26	30	36	1.3	1.5	1.7	1.9	2.1	1.7	1.9	2.1	2.3	2.5
TW H3	12	17	22	27	33	37	43	50	5	8	12	15	20	23	27	33	1.4	1.6	1.8	2.0	2.1	1.8	2.0	2.1	2.3	2.5
TW H4	9	13	18	22	27	31	36	43	4	6	9	12	16	18	22	28	1.6	1.7	1.9	2.1	2.3	2.0	2.2	2.3	2.5	2.7
TW I1	16	21	26	30	35	39	44	50	8	11	14	16	20	22	25	28	1.4	1.6	1.8	2.1	2.3	1.9	2.1	2.5	2.9	3.0
TW I2	23	31	39	45	53	58	64	72	13	18	25	29	36	41	46	54	1.1	1.3	1.5	1.6	1.8	1.3	1.5	1.7	1.9	2.1
TW I3	25	33	42	48	56	62	68	76	14	21	27	33	40	45	50	58	1.1	1.2	1.4	1.6	1.8	1.2	1.4	1.6	1.8	2.0
TW I4	27	36	46	53	62	67	74	81	16	23	31	37	46	51	58	66	1.0	1.2	1.3	1.5	1.7	1.1	1.3	1.5	1.7	1.9
TW I5	27	37	47	54	63	69	75	83	16	24	32	39	47	53	60	69	1.0	1.1	1.3	1.5	1.7	1.1	1.3	1.5	1.7	1.8
TW J1	30	38	48	54	63	68	74	81	18	25	33	38	46	51	57	65	1.0	1.2	1.3	1.5	1.7	1.2	1.3	1.5	1.7	1.9
TW J2	29	38	47	54	63	68	74	82	17	24	32	38	46	51	58	66	1.0	1.2	1.3	1.5	1.7	1.1	1.3	1.5	1.7	1.9
TW J3	34	44	55	62	71	77	83	89	21	30	40	47	57	62	69	78	0.9	1.1	1.2	1.4	1.6	1.0	1.2	1.4	1.6	1.7
TW J4	31	42	53	60	69	75	81	88	19	28	38	45	54	60	67	76	0.9	1.1	1.3	1.4	1.6	1.0	1.2	1.4	1.6	1.7
DG A1	20	25	31	35	40	42	46	49	10	13	17	19	22	24	26	27	1.3	1.6	1.9	2.4	3.0	1.7	2.0	2.7	3.8	-
DG A2	22	29	36	41	48	52	57	63	13	17	22	26	32	35	40	45	1.2	1.4	1.6	1.9	2.1	1.4	1.6	1.9	2.2	2.6
DG A3	26	34	43	49	58	63	69	76	17	23	30	35	43	48	54	61	1.1	1.2	1.4	1.6	1.8	1.3	1.4	1.6	1.8	2.0
DG B1	21	27	33	38	44	47	51	56	10	14	18	21	25	27	29	32	1.2	1.4	1.8	2.1	2.5	1.6	1.9	2.3	3.0	4.0
DG B2	23	30	38	44	51	56	62	69	13	18	24	28	34	38	43	49	1.1	1.3	1.5	1.7	1.9	1.4	1.6	1.8	2.1	2.4
DG B3	27	35	45	52	61	66	73	81	17	23	31	37	45	50	57	65	1.0	1.2	1.3	1.5	1.7	1.2	1.4	1.5	1.7	1.9
DG C1	24	31	39	44	51	55	60	65	13	18	23	26	31	34	38	42	1.1	1.3	1.6	1.9	2.2	1.4	1.7	2.0	2.5	3.0
DG C2	26	34	43	49	58	63	69	76	16	21	28	33	40	44	50	57	1.1	1.2	1.4	1.6	1.8	1.3	1.4	1.7	1.9	2.2
DG C3	30	39	50	57	66	72	79	86	20	27	35	41	50	56	63	71	1.0	1.1	1.3	1.5	1.6	1.1	1.3	1.4	1.6	1.8
SS A1	29	36	42	47	53	58	63	69	21	26	32	36	41	45	49	54	1.1	1.3	1.6	1.8	1.9	1.3	1.5	1.8	2.1	2.4
SS A2	36	44	52	58	66	70	76	82	27	34	41	46	53	58	63	70	1.0	1.2	1.4	1.6	1.7	1.1	1.3	1.5	1.7	1.9
SS A3	28	33	39	44	50	53	58	64	20	24	29	33	37	40	44	48	1.2	1.4	1.6	1.9	2.1	1.4	1.7	2.0	2.3	2.7
SS A4	35	43	51	57	64	69	74	81	26	33	40	45	52	56	61	68	1.0	1.2	1.4	1.6	1.7	1.1	1.3	1.6	1.8	2.0
SS A5	28	34	40	44	50	53	58	64	21	25	30	33	38	41	45	49	1.2	1.4	1.7	1.9	2.1	1.4	1.7	2.0	2.3	2.6
SS A6	35	43	52	58	65	70	75	82	26	33	41	46	53	57	62	69	1.0	1.2	1.4	1.6	1.7	1.1	1.3	1.5	1.8	2.0
SS A7	26	30	35	39	44	48	52	57	19	23	27	29	33	36	39	42	1.3	1.6	1.8	2.1	2.3	1.5	1.8	2.2	2.6	3.0
SS A8	35	42	50	56	63	68	73	80	26	32	39	44	51	55	60	66	1.0	1.2	1.4	1.6	1.8	1.1	1.4	1.6	1.8	2.0
SS B1	26	31	38	42	49	53	58	65	18	22	27	31	36	39	44	49	1.2	1.4	1.6	1.8	2.0	1.4	1.6	1.9	2.2	2.4
SS B2	32	40	48	54	61	66	72	79	23	30	37	42	49	54	59	66	1.0	1.2	1.4	1.6	1.7	1.1	1.3	1.6	1.8	2.0
SS B3	24	30	36	40	46	50	55	61	17	21	25	29	33	36	40	45	1.2	1.5	1.7	1.9	2.1	1.5	1.7	2.0	2.4	2.7
SS B4	31	39	47	53	60	65	71	78	22	29	36	41	48	52	58	65	1.0	1.2	1.4	1.6	1.8	1.2	1.4	1.6	1.8	2.0
SS B5	24	28	34	38	43	47	52	58	17	20	24	27	32	35	38	43	1.3	1.5	1.8	2.0	2.2	1.6	1.8	2.1	2.4	2.7
SS B6	31	39	47	53	60	65	71	78	22	29	36	41	48	52	58	65	1.0	1.2	1.4	1.6	1.8	1.2	1.4	1.6	1.8	2.0
SS B7	22	26	31	34	39	43	47	53	15	18	22	25	28	31	34	38	1.4	1.6	1.9	2.1	2.3	1.7	1.9	2.3	2.6	3.0
SS B8	30	37	45	51	58	63	69	76	21	28	34	39	46	50	55	62	1.0	1.2	1.4	1.6	1.8	1.2	1.4	1.6	1.9	2.1
SS C1	24	31	38	44	51	56	61	69	14	18	24	27	33	37	41	47	1.1	1.3	1.5	1.7	1.9	1.4	1.6	1.9	2.2	2.4
SS C2	29	37	46	52	60	66	72	79	17	23	31	36	43	48	54	62	1.0	1.2	1.4	1.6	1.7	1.2	1.4	1.6	1.8	2.0
SS C3	21	26	32	36	43	47	52	59	13	16	20	23	28	31	35	40	1.3	1.5	1.7	1.9	2.1	1.7	1.8	2.1	2.4	2.6
SS C4	26	33	41	47	55	60	66	74	16	21	28	32	39	43	49	57	1.1	1.3	1.4	1.6	1.8	1.3	1.5	1.7	1.9	2.1
SS D1	34	41	49	54	61	65	71	77	24	30	37	41	47	50	55	60	1.0	1.2	1.4	1.7	1.8	1.2	1.4	1.7	2.0	2.3
SS D2	40	49	59	65	73	77	83	88	30	38	47	52	60	65	70	77	0.9	1.1	1.3	1.5	1.6	1.0	1.2	1.4	1.6	1.8
SS D3	31	38	45	50	56	60	65	72	23	28	34	38	43	46	51	56	1.1	1.3	1.5	1.7	1.9	1.3	1.5	1.8	2.1	2.4
SS D4	39	48	57	63	71	76	81	87	29	37	45	51	58	63	68	75	0.9	1.1	1.3	1.5	1.6	1.0	1.2	1.4	1.7	1.8
SS E1	30	37	44	49	56	60	65	72	20	25	31	35	41	44	48	53	1.1	1.3	1.5	1.7	1.9	1.3	1.5	1.8	2.2	2.5
SS E2	35	44	53	60	67	72	78	84	25	33	41	46	53	58	63	70	0.9	1.1	1.3	1.5	1.7	1.1	1.3	1.5	1.7	1.9
SS E3	26	32	38	42	48	52	57	64	18	23	27	31	36	39	42	47	1.2	1.4	1.6	1.9	2.0	1.4	1.7	2.0	2.3	2.6
SS E4	33	41	50	56	64	68	74	81	23	30	38	43	50	55	60	67	1.0	1.2	1.4	1.6	1.7	1.1	1.3	1.6	1.8	2.0

BURNS, OREGON

TYPE	SOLAR SAVINGS FRACTION (%) BASE TEMP: 55 F (4440 DD) LCR 100	70	50	40	30	25	20	15	65 F (7215 DD) 100	70	50	40	30	25	20	15	CONSERVATION FACTOR 55 F 100	50	30	20	15	65 F 100	50	30	20	15
WW A1	19	22	26	29	33	36	40	45	15	17	20	22	25	27	30	33	1.6	1.8	2.1	2.3	2.5	2.0	2.2	2.5	2.8	3.1
WW A2	20	25	31	35	41	46	51	58	14	18	23	26	31	35	39	45	1.3	1.5	1.7	1.9	2.1	1.5	1.7	2.0	2.2	2.4
WW A3	20	26	33	38	45	49	55	63	14	18	24	28	33	37	42	49	1.2	1.4	1.6	1.8	2.0	1.4	1.6	1.8	2.1	2.2
WW A4	20	26	33	39	46	51	57	65	13	18	24	29	35	39	44	52	1.2	1.4	1.6	1.8	1.9	1.4	1.6	1.8	2.0	2.2
WW A5	20	27	35	40	48	53	60	68	13	19	25	30	36	41	46	54	1.1	1.3	1.5	1.7	1.9	1.3	1.5	1.7	1.9	2.1
WW A6	20	27	35	41	49	54	61	69	13	19	25	30	37	42	48	55	1.1	1.3	1.5	1.7	1.8	1.3	1.5	1.7	1.9	2.1
WW B1	16	19	23	27	31	34	38	43	11	13	16	18	21	22	25	27	1.5	1.8	2.1	2.4	2.6	1.9	2.3	2.7	3.2	3.8
WW B2	20	28	36	41	49	55	61	69	14	20	27	31	39	43	49	58	1.1	1.3	1.5	1.7	1.8	1.3	1.5	1.6	1.8	2.0
WW B3	24	33	42	48	57	62	69	77	17	24	32	38	46	51	58	66	1.0	1.2	1.4	1.6	1.7	1.2	1.3	1.5	1.7	1.9
WW B4	24	34	44	51	60	66	73	81	17	25	34	41	50	56	63	72	1.0	1.2	1.3	1.5	1.7	1.1	1.3	1.4	1.6	1.8
WW B5	23	33	43	50	60	66	73	81	16	25	34	41	50	56	64	73	1.0	1.2	1.3	1.5	1.7	1.1	1.2	1.4	1.6	1.7
WW C1	26	35	44	50	59	65	71	79	19	26	35	41	49	54	61	69	1.0	1.2	1.4	1.5	1.7	1.1	1.3	1.5	1.7	1.8
WW C2	24	33	43	49	59	64	71	79	17	25	33	40	48	54	61	70	1.0	1.2	1.4	1.5	1.7	1.1	1.3	1.5	1.6	1.8
WW C3	29	40	51	59	68	74	81	88	21	31	42	49	59	66	73	81	0.9	1.1	1.3	1.4	1.6	1.0	1.2	1.3	1.5	1.6
WW C4	26	37	48	55	65	71	78	86	19	28	39	46	56	63	70	79	0.9	1.1	1.3	1.4	1.6	1.0	1.2	1.3	1.5	1.7
TW A1	19	22	26	29	33	36	40	45	15	17	20	22	25	27	30	34	1.6	1.8	2.0	2.3	2.5	2.0	2.2	2.5	2.7	3.0
TW A2	19	23	29	33	38	42	47	54	14	17	21	24	29	32	36	42	1.4	1.6	1.8	2.0	2.1	1.6	1.8	2.1	2.3	2.5
TW A3	18	24	30	34	40	45	50	58	13	17	22	25	31	34	39	45	1.3	1.5	1.7	1.9	2.0	1.5	1.7	1.9	2.1	2.3
TW A4	17	23	29	34	41	46	51	59	12	16	21	25	31	35	40	47	1.3	1.5	1.6	1.8	2.0	1.5	1.7	1.9	2.1	2.3
TW B1	18	22	26	29	34	37	41	47	14	17	20	22	26	28	31	36	1.6	1.8	2.0	2.2	2.3	1.9	2.1	2.3	2.6	2.8
TW B2	17	22	27	32	38	42	47	54	13	16	21	24	28	32	36	42	1.4	1.6	1.8	2.0	2.1	1.6	1.8	2.0	2.2	2.4
TW B3	17	22	27	32	38	42	48	55	12	16	20	24	29	32	37	43	1.4	1.5	1.7	1.9	2.1	1.6	1.8	2.0	2.2	2.3
TW B4	16	21	26	30	36	41	46	54	11	15	19	23	28	31	36	42	1.4	1.6	1.7	1.9	2.1	1.6	1.8	2.0	2.2	2.4
TW C1	17	21	25	29	34	37	42	49	13	16	19	22	26	29	32	38	1.5	1.7	1.9	2.1	2.3	1.8	2.0	2.2	2.4	2.6
TW C2	16	20	25	29	35	39	44	51	12	15	19	22	26	30	34	40	1.5	1.6	1.8	2.0	2.2	1.7	1.9	2.1	2.3	2.4
TW C3	15	19	24	28	33	37	42	49	12	15	18	21	26	29	33	39	1.5	1.7	1.9	2.0	2.2	1.7	1.9	2.1	2.3	2.5
TW C4	15	18	22	26	31	34	39	45	12	14	17	20	24	27	30	36	1.6	1.8	2.0	2.1	2.3	1.9	2.1	2.2	2.4	2.6
TW D1	13	16	20	23	28	31	34	39	9	11	14	16	18	20	23	26	1.6	1.9	2.1	2.4	2.6	2.0	2.3	2.7	3.1	3.5
TW D2	20	26	33	38	46	50	57	65	14	19	25	29	36	40	46	53	1.2	1.4	1.6	1.7	1.9	1.4	1.6	1.7	1.9	2.1
TW D3	21	28	36	41	49	54	61	69	15	21	27	32	39	44	49	57	1.1	1.3	1.5	1.7	1.8	1.3	1.5	1.7	1.9	2.0
TW D4	23	31	40	46	55	60	67	76	16	23	31	37	45	50	57	66	1.1	1.2	1.4	1.6	1.7	1.2	1.3	1.5	1.7	1.8
TW D5	22	31	40	46	55	61	68	77	16	23	31	38	46	52	59	68	1.1	1.2	1.4	1.6	1.7	1.2	1.3	1.5	1.7	1.8
TW E1	24	32	40	46	54	60	66	74	18	24	31	37	44	49	55	64	1.1	1.3	1.4	1.6	1.8	1.2	1.4	1.6	1.8	1.9
TW E2	24	32	40	46	55	60	67	75	17	24	31	37	45	50	57	65	1.1	1.2	1.4	1.6	1.7	1.2	1.4	1.5	1.7	1.9
TW E3	28	38	48	55	65	70	77	85	21	30	39	46	55	61	68	77	1.0	1.1	1.3	1.5	1.6	1.0	1.2	1.4	1.6	1.7
TW E4	26	36	46	53	63	69	75	83	19	28	37	44	53	59	67	76	1.0	1.1	1.3	1.5	1.6	1.1	1.2	1.4	1.6	1.7
TW F1	17	20	24	27	32	35	39	45	13	16	18	21	24	26	29	33	1.6	1.8	2.0	2.2	2.4	1.9	2.2	2.4	2.7	2.9
TW F2	16	21	26	30	36	40	45	52	12	15	19	22	26	30	34	39	1.4	1.6	1.8	2.0	2.2	1.7	1.9	2.1	2.3	2.5
TW F3	15	20	26	30	37	41	47	54	10	14	18	22	27	31	35	42	1.4	1.5	1.7	1.9	2.1	1.6	1.8	2.0	2.2	2.4
TW F4	13	18	24	29	36	40	46	54	8	12	17	21	26	30	34	41	1.4	1.5	1.7	1.9	2.1	1.6	1.8	2.0	2.2	2.3
TW G1	15	19	23	26	31	35	39	45	11	14	17	19	23	25	29	34	1.6	1.8	2.0	2.2	2.4	1.9	2.1	2.3	2.6	2.8
TW G2	13	18	23	27	33	37	42	49	9	12	16	19	24	27	31	37	1.4	1.6	1.8	2.0	2.2	1.7	1.9	2.1	2.3	2.5
TW G3	12	16	22	26	32	36	41	49	7	11	15	18	23	26	31	37	1.4	1.6	1.8	2.0	2.2	1.7	1.9	2.1	2.3	2.4
TW G4	9	14	19	22	28	32	37	45	5	9	12	16	20	23	28	34	1.5	1.7	1.9	2.1	2.2	1.8	1.9	2.1	2.3	2.5
TW H1	12	16	20	24	29	32	37	43	8	11	14	17	21	23	27	32	1.6	1.8	2.0	2.2	2.3	1.9	2.1	2.3	2.5	2.7
TW H2	9	13	18	22	27	31	36	43	6	9	12	15	19	22	26	32	1.6	1.8	1.9	2.1	2.3	1.8	2.0	2.2	2.4	2.6
TW H3	8	11	16	19	24	28	33	39	4	7	10	13	17	20	24	29	1.6	1.8	2.0	2.2	2.3	1.9	2.1	2.3	2.5	2.7
TW H4	6	9	12	15	19	23	27	33	3	5	8	10	14	16	19	24	1.8	2.0	2.2	2.4	2.5	2.1	2.3	2.5	2.7	2.8
TW I1	11	14	18	21	25	28	31	36	7	9	11	13	16	18	20	23	1.7	1.9	2.2	2.5	2.7	2.1	2.4	2.8	3.2	3.6
TW I2	17	23	30	35	42	47	53	62	11	16	22	26	32	37	42	50	1.2	1.4	1.6	1.8	1.9	1.4	1.6	1.8	2.0	2.1
TW I3	18	25	32	38	46	51	57	66	12	18	24	29	36	40	46	54	1.2	1.4	1.5	1.7	1.9	1.3	1.5	1.7	1.9	2.0
TW I4	20	28	36	43	52	57	64	73	14	20	28	34	42	47	54	63	1.1	1.3	1.4	1.6	1.8	1.2	1.4	1.5	1.7	1.9
TW I5	20	28	37	44	53	59	66	75	14	21	29	35	44	49	57	66	1.1	1.2	1.4	1.6	1.7	1.2	1.3	1.5	1.7	1.8
TW J1	22	29	38	44	52	57	64	72	16	22	29	34	42	47	54	62	1.1	1.3	1.4	1.6	1.8	1.2	1.4	1.6	1.8	1.9
TW J2	21	29	37	44	52	58	65	73	15	21	29	34	42	48	54	63	1.1	1.3	1.4	1.6	1.8	1.2	1.4	1.5	1.7	1.9
TW J3	25	35	45	53	62	68	75	83	18	27	37	43	53	59	66	75	1.0	1.1	1.3	1.5	1.6	1.1	1.2	1.4	1.6	1.7
TW J4	23	33	43	50	60	66	73	82	17	25	34	41	51	57	64	74	1.0	1.2	1.3	1.5	1.6	1.1	1.2	1.4	1.6	1.7
DG A1	14	17	22	24	28	30	33	35	9	11	14	16	18	20	21	22	1.5	1.9	2.3	3.0	4.0	1.9	2.3	3.0	4.4	-
DG A2	16	21	27	31	37	41	46	52	11	15	20	23	28	31	35	41	1.4	1.5	1.8	2.1	2.3	1.6	1.8	2.0	2.3	2.6
DG A3	20	26	34	40	48	53	60	68	15	20	27	32	39	44	50	58	1.2	1.3	1.5	1.7	1.9	1.4	1.5	1.6	1.8	2.0
DG B1	14	18	23	26	31	34	37	41	9	11	14	17	20	22	24	26	1.5	1.7	2.0	2.5	3.0	1.9	2.1	2.6	3.3	4.3
DG B2	17	22	28	33	40	44	50	57	12	16	20	24	30	33	38	45	1.3	1.5	1.7	1.9	2.1	1.5	1.7	1.9	2.1	2.4
DG B3	20	27	35	41	50	56	63	72	15	21	28	33	41	46	53	62	1.2	1.3	1.4	1.6	1.8	1.3	1.4	1.6	1.7	1.9
DG C1	17	22	27	32	38	41	46	51	11	14	19	22	26	29	32	36	1.3	1.6	1.8	2.1	2.5	1.6	1.8	2.2	2.6	3.1
DG C2	19	25	33	38	46	51	57	65	14	19	24	29	35	40	46	53	1.2	1.4	1.5	1.7	1.9	1.4	1.5	1.7	1.9	2.2
DG C3	23	31	40	46	56	62	69	78	18	24	31	37	46	52	59	68	1.1	1.2	1.4	1.5	1.7	1.2	1.3	1.5	1.6	1.8
SS A1	23	28	33	38	43	47	52	58	19	23	28	31	36	39	44	49	1.3	1.5	1.7	1.9	2.1	1.4	1.7	1.9	2.2	2.4
SS A2	28	35	43	48	56	61	67	74	23	30	37	42	49	53	59	66	1.1	1.3	1.5	1.6	1.8	1.2	1.4	1.6	1.8	1.9
SS A3	21	25	30	34	39	42	46	52	17	21	25	28	32	35	38	42	1.4	1.6	1.9	2.1	2.3	1.5	1.8	2.1	2.5	2.8
SS A4	27	34	42	47	54	59	65	72	23	29	36	40	47	51	57	64	1.1	1.3	1.5	1.7	1.8	1.2	1.4	1.6	1.8	2.0
SS A5	22	27	31	35	40	43	47	53	19	22	26	29	33	36	39	44	1.4	1.6	1.9	2.1	2.3	1.6	1.8	2.1	2.4	2.7
SS A6	28	35	42	48	55	60	66	73	23	29	36	41	48	52	58	65	1.1	1.3	1.5	1.7	1.8	1.2	1.4	1.6	1.8	2.0
SS A7	20	24	27	30	34	37	40	45	17	20	23	25	28	30	33	36	1.6	1.8	2.1	2.4	2.6	1.8	2.1	2.4	2.9	3.3
SS A8	27	33	41	46	53	57	63	70	22	28	35	39	45	50	55	62	1.1	1.3	1.5	1.7	1.9	1.2	1.4	1.7	1.9	2.1
SS B1	19	24	29	33	38	42	47	53	16	19	24	27	31	34	38	43	1.4	1.6	1.8	2.0	2.2	1.6	1.8	2.0	2.3	2.5
SS B2	25	31	39	44	52	57	63	71	20	26	33	38	44	49	55	63	1.1	1.3	1.5	1.7	1.8	1.2	1.4	1.6	1.8	2.0
SS B3	18	22	27	30	35	39	43	49	15	18	22	24	28	31	34	39	1.5	1.7	1.9	2.2	2.4	1.6	1.9	2.2	2.5	2.8
SS B4	24	30	38	43	50	55	61	69	20	25	32	36	43	47	53	60	1.2	1.3	1.5	1.7	1.9	1.3	1.4	1.7	1.9	2.0
SS B5	18	22	26	29	34	37	41	46	15	18	21	24	27	30	33	37	1.5	1.8	2.0	2.2	2.5	1.8	2.0	2.3	2.6	2.9
SS B6	24	30	38	43	51	55	61	69	19	25	32	36	43	48	53	61	1.1	1.3	1.5	1.7	1.9	1.2	1.4	1.6	1.8	2.0
SS B7	17	20	23	26	30	33	36	41	14	16	19	21	24	26	29	32	1.7	1.9	2.2	2.5	2.7	1.9	2.2	2.5	2.9	3.3
SS B8	23	29	36	41	48	53	59	66	18	24	30	34	41	45	50	58	1.2	1.4	1.6	1.8	1.9	1.3	1.5	1.7	1.9	2.1
SS C1	17	23	29	33	39	44	49	56	12	16	20	24	29	32	36	42	1.3	1.5	1.7	1.9	2.1	1.5	1.8	2.0	2.2	2.5
SS C2	21	28	36	42	50	55	61	70	15	21	27	32	39	44	50	58	1.1	1.3	1.5	1.7	1.8	1.3	1.5	1.6	1.8	2.0
SS C3	16	20	24	28	33	36	41	47	12	15	18	20	24	27	30	35	1.5	1.7	1.9	2.1	2.3	1.8	2.0	2.3	2.5	2.7
SS C4	19	25	32	37	45	50	56	64	14	19	25	29	35	39	45	53	1.2	1.4	1.6	1.8	1.9	1.4	1.6	1.7	1.9	2.1
SS D1	26	32	39	43	50	54	60	66	21	26	32	36	42	45	50	56	1.2	1.4	1.6	1.8	2.0	1.3	1.5	1.8	2.0	2.3
SS D2	31	40	49	55	63	69	75	82	26	34	42	48	56	61	67	74	1.0	1.2	1.4	1.5	1.7	1.1	1.3	1.4	1.6	1.8
SS D3	24	30	35	40	45	49	54	61	20	25	30	33	38	41	45	51	1.3	1.5	1.7	1.9	2.1	1.4	1.6	1.9	2.2	2.4
SS D4	30	39	47	53	62	67	73	80	25	33	41	46	54	59	65	72	1.0	1.2	1.4	1.6	1.7	1.1	1.3	1.5	1.7	1.8
SS E1	22	27	33	38	44	48	53	60	17	21	26	30	35	38	42	48	1.2	1.5	1.7	1.9	2.1	1.4	1.7	1.9	2.2	2.5
SS E2	27	35	43	49	57	62	68	76	21	28	36	41	48	53	59	67	1.0	1.2	1.4	1.6	1.8	1.1	1.3	1.5	1.7	1.9
SS E3	20	24	29	33	38	41	46	52	16	20	24	27	31	33	37	42	1.4	1.6	1.9	2.1	2.3	1.6	1.8	2.1	2.4	2.7
SS E4	25	32	40	46	53	58	64	72	20	26	33	38	45	50	55	63	1.1	1.3	1.5	1.7	1.8	1.2	1.4	1.6	1.8	2.0

MEDFORD, OREGON

TYPE	SSF (%) 55 F (2507 DD) LCR 100	70	50	40	30	25	20	15	SSF (%) 65 F (4935 DD) 100	70	50	40	30	25	20	15	CONSERVATION FACTOR 55 F 100	50	30	20	15	CONSERVATION FACTOR 65 F 100	50	30	20	15
WW A1	22	26	31	34	39	42	46	52	17	19	22	24	28	30	32	36	1.4	1.7	1.9	2.2	2.4	1.8	2.1	2.4	2.8	3.1
WW A2	24	30	37	42	48	52	57	64	16	21	25	29	34	37	41	47	1.2	1.4	1.6	1.8	2.0	1.5	1.7	1.9	2.2	2.5
WW A3	25	32	39	44	51	56	61	68	16	21	26	31	36	40	45	51	1.1	1.3	1.5	1.8	1.9	1.4	1.6	1.8	2.1	2.3
WW A4	25	32	40	46	53	58	63	70	16	21	27	31	38	42	47	53	1.1	1.3	1.5	1.7	1.9	1.3	1.5	1.8	2.0	2.3
WW A5	25	33	41	47	55	60	65	72	15	21	28	32	39	43	48	55	1.1	1.3	1.5	1.7	1.9	1.3	1.5	1.7	2.0	2.2
WW A6	26	34	42	48	56	61	66	74	15	21	28	33	40	44	49	56	1.0	1.2	1.5	1.7	1.8	1.2	1.4	1.7	1.9	2.2
WW B1	20	25	30	33	38	41	45	51	13	15	18	21	24	26	28	30	1.4	1.6	1.9	2.2	2.5	1.8	2.1	2.6	3.2	4.1
WW B2	26	34	42	48	55	60	67	74	16	22	29	34	41	46	51	59	1.1	1.3	1.4	1.6	1.8	1.2	1.4	1.6	1.9	2.1
WW B3	30	39	48	54	62	67	73	80	20	27	35	41	48	53	59	66	1.0	1.2	1.4	1.6	1.7	1.1	1.3	1.5	1.8	1.9
WW B4	30	40	50	57	65	71	77	84	20	28	37	43	52	58	64	72	0.9	1.1	1.3	1.5	1.7	1.1	1.2	1.4	1.6	1.8
WW B5	29	39	49	56	65	71	77	84	19	27	37	43	52	58	65	73	0.9	1.1	1.3	1.5	1.6	1.1	1.2	1.4	1.6	1.8
WW C1	32	41	50	56	65	70	76	82	21	29	37	43	51	56	62	69	1.0	1.2	1.3	1.5	1.7	1.1	1.3	1.5	1.7	1.9
WW C2	30	39	49	55	64	69	75	82	20	28	36	42	50	56	62	70	1.0	1.1	1.3	1.5	1.7	1.1	1.3	1.5	1.7	1.8
WW C3	35	46	57	64	73	78	84	90	24	34	44	52	61	67	74	81	0.9	1.1	1.2	1.4	1.6	1.0	1.1	1.3	1.5	1.7
WW C4	32	43	54	61	70	75	82	88	21	31	41	48	58	64	71	79	0.9	1.1	1.3	1.4	1.6	1.0	1.2	1.3	1.5	1.7
TW A1	22	26	31	34	39	42	46	52	17	19	22	24	28	30	33	37	1.4	1.7	1.9	2.1	2.3	1.8	2.1	2.4	2.7	3.0
TW A2	23	28	34	39	45	49	54	60	16	20	24	27	32	35	39	44	1.3	1.5	1.7	1.9	2.1	1.5	1.8	2.0	2.3	2.5
TW A3	23	29	36	40	47	51	57	64	15	19	24	28	33	37	41	47	1.2	1.4	1.6	1.8	2.0	1.5	1.7	1.9	2.2	2.4
TW A4	22	28	35	40	47	52	58	65	14	19	24	28	34	37	42	49	1.2	1.4	1.6	1.8	2.0	1.4	1.6	1.8	2.1	2.3
TW B1	21	26	31	34	40	43	48	54	16	19	22	25	28	31	34	39	1.4	1.6	1.8	2.1	2.2	1.7	2.0	2.2	2.5	2.8
TW B2	21	27	33	38	44	48	53	60	15	18	23	26	31	35	39	45	1.3	1.5	1.7	1.9	2.1	1.5	1.7	2.0	2.2	2.4
TW B3	21	27	33	38	44	49	54	61	14	18	23	26	31	35	40	46	1.2	1.4	1.6	1.8	2.0	1.5	1.7	1.9	2.2	2.4
TW B4	20	25	32	36	43	47	52	60	13	17	22	25	30	34	39	45	1.3	1.5	1.7	1.9	2.0	1.5	1.7	1.9	2.2	2.4
TW C1	20	25	30	34	40	43	48	55	15	18	21	24	28	31	35	40	1.4	1.6	1.8	2.0	2.2	1.7	1.9	2.1	2.4	2.6
TW C2	19	25	30	35	41	45	50	57	13	17	21	24	29	32	37	42	1.3	1.5	1.7	1.9	2.1	1.6	1.8	2.0	2.2	2.4
TW C3	19	24	29	33	39	43	48	55	13	16	20	24	28	31	35	41	1.3	1.6	1.8	2.0	2.1	1.7	1.8	2.1	2.3	2.5
TW C4	18	22	27	31	36	40	44	51	13	16	19	22	26	29	33	38	1.4	1.6	1.8	2.0	2.2	1.8	2.0	2.2	2.4	2.6
TW D1	17	21	26	29	34	38	42	47	11	13	16	18	21	23	26	29	1.4	1.7	2.0	2.2	2.5	1.9	2.2	2.6	3.0	3.5
TW D2	25	31	39	44	52	56	62	70	16	22	27	32	38	42	48	55	1.1	1.3	1.5	1.7	1.9	1.3	1.5	1.7	1.9	2.1
TW D3	26	34	42	48	55	60	66	73	17	23	30	35	41	46	51	59	1.1	1.3	1.5	1.7	1.8	1.2	1.4	1.7	1.9	2.1
TW D4	28	37	46	52	60	66	72	79	19	26	34	39	47	52	59	67	1.0	1.2	1.4	1.6	1.7	1.2	1.3	1.5	1.7	1.9
TW D5	27	36	46	52	61	66	73	80	18	26	34	40	48	54	61	69	1.0	1.2	1.4	1.5	1.7	1.1	1.3	1.5	1.7	1.8
TW E1	30	38	46	52	60	65	71	78	20	27	34	39	47	51	57	64	1.0	1.2	1.4	1.6	1.8	1.2	1.4	1.6	1.8	2.0
TW E2	29	37	46	52	61	66	72	79	20	26	34	40	47	52	59	66	1.0	1.2	1.4	1.6	1.7	1.2	1.3	1.5	1.7	1.9
TW E3	34	44	54	61	69	75	81	87	24	32	42	48	57	63	69	77	0.9	1.1	1.3	1.5	1.6	1.0	1.2	1.4	1.6	1.7
TW E4	32	41	52	59	68	73	79	86	22	30	40	46	55	61	68	76	0.9	1.1	1.3	1.5	1.6	1.0	1.2	1.4	1.6	1.7
TW F1	20	25	29	33	38	42	46	52	15	17	21	23	26	29	32	36	1.4	1.6	1.9	2.1	2.3	1.8	2.1	2.3	2.6	2.9
TW F2	20	26	32	36	42	46	52	58	13	17	21	25	29	32	36	42	1.3	1.5	1.7	1.9	2.1	1.6	1.8	2.0	2.3	2.5
TW F3	20	26	32	37	43	48	53	60	12	16	21	25	30	33	38	44	1.2	1.4	1.6	1.9	2.0	1.5	1.7	1.9	2.2	2.4
TW F4	18	24	30	35	42	47	52	60	10	15	19	23	29	32	37	44	1.2	1.4	1.6	1.8	2.0	1.5	1.7	1.9	2.2	2.4
TW G1	19	23	28	32	37	41	46	52	13	16	19	22	26	28	32	36	1.4	1.6	1.8	2.1	2.3	1.8	2.0	2.2	2.5	2.8
TW G2	17	23	29	33	39	44	49	56	11	14	19	22	27	30	34	40	1.3	1.5	1.7	1.9	2.1	1.6	1.8	2.0	2.3	2.5
TW G3	16	21	27	32	38	43	48	55	9	13	17	21	26	29	34	40	1.3	1.5	1.7	1.9	2.1	1.6	1.8	2.0	2.2	2.5
TW G4	13	18	24	28	34	39	44	51	7	11	15	18	23	26	30	37	1.4	1.6	1.8	2.0	2.1	1.7	1.9	2.1	2.3	2.5
TW H1	15	20	25	29	35	39	43	50	10	13	16	19	23	26	30	35	1.4	1.6	1.8	2.0	2.2	1.8	2.0	2.2	2.5	2.7
TW H2	13	18	23	27	33	37	42	49	7	11	14	17	22	25	29	35	1.4	1.6	1.8	2.0	2.2	1.7	1.9	2.2	2.4	2.6
TW H3	11	15	20	24	30	34	39	46	6	9	12	15	19	22	26	32	1.5	1.7	1.9	2.1	2.2	1.8	2.0	2.2	2.4	2.6
TW H4	8	12	16	20	25	28	33	39	4	7	10	12	16	18	22	27	1.6	1.8	2.0	2.2	2.4	2.0	2.2	2.4	2.6	2.8
TW I1	15	19	23	27	31	35	39	44	9	11	14	16	19	21	23	26	1.5	1.7	2.0	2.3	2.5	1.9	2.3	2.6	3.1	3.6
TW I2	21	28	36	41	48	53	59	67	13	18	24	29	35	39	45	52	1.1	1.3	1.5	1.7	1.9	1.4	1.6	1.8	2.0	2.2
TW I3	23	30	[illegible]	44	52	57	63	71	15	20	27	32	38	43	49	56	1.1	1.3	1.5	1.7	1.8	1.3	1.5	1.7	1.9	2.1
TW I4	25	33	42	49	57	63	69	77	16	23	31	36	44	49	56	64	1.0	1.2	1.4	1.6	1.7	1.2	1.4	1.5	1.7	1.9
TW I5	25	34	43	50	59	64	71	79	16	23	31	38	46	51	58	67	1.0	1.2	1.4	1.6	1.7	1.2	1.3	1.5	1.7	1.8
TW J1	27	35	44	50	58	63	69	77	18	24	32	37	44	49	55	63	1.0	1.2	1.4	1.6	1.8	1.2	1.4	1.6	1.8	2.0
TW J2	26	35	43	50	58	63	70	77	17	24	31	37	45	50	56	64	1.0	1.2	1.4	1.6	1.7	1.2	1.4	1.5	1.7	1.9
TW J3	31	41	51	58	67	73	79	86	21	30	39	46	55	61	68	76	0.9	1.1	1.3	1.5	1.6	1.0	1.2	1.4	1.6	1.7
TW J4	29	39	49	56	65	71	77	84	19	28	37	44	53	59	66	74	0.9	1.1	1.3	1.5	1.6	1.1	1.2	1.4	1.6	1.7
DG A1	18	23	28	31	35	38	40	43	11	14	17	19	21	23	24	25	1.3	1.7	2.1	2.7	3.5	1.7	2.2	2.9	4.2	-
DG A2	20	26	33	37	44	48	52	58	13	17	22	26	31	34	38	43	1.2	1.4	1.7	2.0	2.3	1.5	1.7	2.0	2.3	2.7
DG A3	24	32	40	46	54	59	65	72	17	23	29	34	41	46	52	59	1.1	1.3	1.5	1.7	1.9	1.3	1.4	1.6	1.8	2.0
DG B1	19	24	30	34	39	42	45	49	11	14	18	20	23	25	28	30	1.3	1.6	1.9	2.4	2.8	1.7	2.0	2.5	3.3	4.6
DG B2	21	28	35	40	47	51	56	63	14	18	23	27	33	37	41	47	1.2	1.4	1.6	1.8	2.1	1.5	1.6	1.9	2.1	2.5
DG B3	25	33	42	48	56	62	68	76	17	23	30	36	44	49	55	63	1.1	1.2	1.4	1.6	1.8	1.3	1.4	1.6	1.8	1.9
DG C1	22	28	35	40	46	49	53	58	13	17	22	25	30	32	36	39	1.2	1.4	1.7	2.1	2.4	1.5	1.8	2.1	2.6	3.4
DG C2	24	31	39	45	53	57	63	70	16	21	27	32	39	43	48	55	1.1	1.3	1.5	1.7	1.9	1.3	1.5	1.7	2.0	2.2
DG C3	28	37	46	53	62	67	74	81	20	26	34	40	49	54	61	69	1.0	1.2	1.4	1.5	1.7	1.2	1.3	1.5	1.7	1.9
SS A1	27	33	39	43	49	53	58	64	21	26	30	34	39	42	46	51	1.2	1.4	1.6	1.9	2.1	1.4	1.6	1.9	2.2	2.5
SS A2	33	41	49	54	61	66	72	79	26	32	39	44	51	55	61	68	1.0	1.2	1.4	1.6	1.8	1.1	1.3	1.6	1.8	2.0
SS A3	25	31	36	40	45	49	53	59	20	23	28	31	35	38	41	45	1.3	1.5	1.8	2.0	2.2	1.5	1.8	2.1	2.5	2.8
SS A4	32	40	47	53	60	65	70	77	25	32	38	43	50	54	59	65	1.0	1.2	1.4	1.6	1.8	1.2	1.4	1.6	1.8	2.0
SS A5	26	31	37	40	46	49	54	59	21	25	29	32	36	39	42	46	1.3	1.5	1.8	2.0	2.2	1.5	1.8	2.1	2.4	2.7
SS A6	33	40	48	54	61	65	71	78	25	32	39	44	50	55	60	66	1.0	1.2	1.4	1.6	1.8	1.2	1.4	1.6	1.8	2.0
SS A7	24	28	33	36	40	43	47	52	19	22	25	28	31	33	36	39	1.4	1.7	1.9	2.2	2.5	1.7	2.0	2.3	2.8	3.2
SS A8	32	39	46	52	59	63	69	75	25	31	37	42	48	52	57	63	1.1	1.3	1.5	1.7	1.8	1.2	1.4	1.6	1.9	2.1
SS B1	24	29	35	39	45	48	53	60	18	22	26	29	34	37	41	46	1.3	1.5	1.7	1.9	2.1	1.5	1.7	2.0	2.3	2.6
SS B2	29	37	44	50	57	62	68	75	22	28	35	40	47	51	57	64	1.1	1.3	1.5	1.6	1.8	1.2	1.4	1.6	1.8	2.0
SS B3	22	27	33	36	42	45	50	56	17	20	24	27	31	34	37	42	1.3	1.5	1.8	2.0	2.2	1.5	1.8	2.1	2.5	2.8
SS B4	29	36	43	49	56	61	66	74	22	28	34	39	45	50	55	62	1.1	1.3	1.5	1.7	1.8	1.2	1.4	1.6	1.9	2.0
SS B5	22	26	31	35	40	43	47	53	17	20	24	26	30	33	36	40	1.4	1.6	1.9	2.1	2.3	1.7	1.9	2.2	2.5	2.8
SS B6	29	36	43	49	56	61	67	74	22	28	34	39	46	50	55	62	1.1	1.3	1.5	1.7	1.8	1.2	1.4	1.6	1.8	2.0
SS B7	20	24	28	31	36	39	43	48	16	18	21	24	27	29	32	35	1.5	1.7	2.0	2.3	2.5	1.8	2.1	2.4	2.8	3.2
SS B8	28	34	42	47	54	59	64	72	21	26	33	37	43	48	53	60	1.1	1.3	1.5	1.7	1.9	1.2	1.5	1.7	1.9	2.1
SS C1	22	28	35	40	46	50	56	62	14	18	23	27	32	35	39	44	1.2	1.4	1.6	1.9	2.1	1.5	1.7	2.0	2.3	2.5
SS C2	26	34	42	48	56	61	67	74	17	23	30	35	42	46	52	59	1.1	1.3	1.4	1.6	1.8	1.2	1.4	1.6	1.9	2.0
SS C3	19	24	29	33	39	43	47	54	13	17	20	23	27	30	33	38	1.4	1.6	1.8	2.0	2.2	1.7	1.9	2.2	2.5	2.7
SS C4	24	31	38	44	51	56	62	69	16	21	27	31	38	42	47	55	1.1	1.3	1.5	1.7	1.9	1.3	1.5	1.7	1.9	2.1
SS D1	31	38	45	50	56	60	65	72	23	29	35	39	44	48	52	57	1.1	1.3	1.5	1.7	1.9	1.2	1.5	1.8	2.1	2.3
SS D2	37	46	55	61	68	73	79	85	28	36	44	50	58	62	68	75	0.9	1.1	1.3	1.5	1.7	1.0	1.2	1.5	1.7	1.8
SS D3	29	35	41	46	52	56	60	67	22	27	32	36	41	44	48	53	1.2	1.4	1.6	1.8	2.0	1.3	1.6	1.9	2.2	2.5
SS D4	36	44	53	59	67	71	77	84	28	35	43	49	56	61	66	73	1.0	1.2	1.3	1.5	1.7	1.1	1.3	1.5	1.7	1.9
SS E1	27	33	40	44	51	55	60	66	19	24	29	33	38	41	45	50	1.1	1.4	1.6	1.8	2.0	1.3	1.6	1.9	2.3	2.6
SS E2	32	41	49	55	63	68	73	80	24	31	38	44	51	55	61	68	1.0	1.2	1.4	1.6	1.7	1.1	1.3	1.5	1.8	2.0
SS E3	24	29	35	39	44	48	52	58	18	22	26	29	33	36	40	44	1.3	1.5	1.8	2.0	2.2	1.5	1.8	2.1	2.4	2.7
SS E4	30	38	46	52	59	64	70	77	23	29	36	41	48	52	57	64	1.0	1.2	1.4	1.6	1.8	1.2	1.4	1.6	1.8	2.0

NORTH BEND, OREGON

TYPE	SOLAR SAVINGS FRACTION (%)																CONSERVATION FACTOR									
	BASE TEMP: 55 F (1642 DD)								65 F (4692 DD)								55 F					65 F				
LCR	100	70	50	40	30	25	20	15	100	70	50	40	30	25	20	15	100	50	30	20	15	100	50	30	20	15
WW A1	35	42	50	56	64	69	74	82	21	25	30	35	40	44	50	57	1.0	1.2	1.4	1.5	1.7	1.4	1.6	1.7	1.9	2.1
WW A2	42	51	61	68	76	81	86	91	23	30	38	44	52	57	63	71	0.9	1.1	1.2	1.4	1.6	1.1	1.3	1.5	1.7	1.8
WW A3	44	55	65	72	80	84	89	94	24	32	41	47	55	61	67	75	0.8	1.0	1.2	1.4	1.5	1.1	1.2	1.4	1.6	1.8
WW A4	46	57	67	74	82	86	91	95	24	33	42	49	58	63	70	78	0.8	1.0	1.2	1.4	1.5	1.0	1.2	1.4	1.6	1.7
WW A5	47	59	69	76	84	88	92	96	25	34	44	51	60	66	72	80	0.8	1.0	1.2	1.3	1.5	1.0	1.2	1.4	1.5	1.7
WW A6	48	60	70	77	85	89	93	96	25	35	45	52	61	67	73	81	0.8	1.0	1.1	1.3	1.5	1.0	1.2	1.3	1.5	1.7
WW B1	36	45	54	60	68	73	79	86	19	24	31	35	42	46	52	59	1.0	1.1	1.3	1.5	1.6	1.3	1.5	1.7	1.9	2.1
WW B2	46	57	68	75	83	87	91	95	25	34	44	51	60	66	72	80	0.8	1.0	1.2	1.3	1.5	1.0	1.2	1.3	1.5	1.7
WW B3	53	64	75	81	88	91	95	98	31	41	52	59	68	74	80	87	0.8	0.9	1.1	1.3	1.5	0.9	1.1	1.3	1.5	1.6
WW B4	53	65	76	82	89	92	95	98	30	41	53	60	70	76	82	89	0.8	0.9	1.1	1.3	1.4	0.9	1.1	1.2	1.4	1.6
WW B5	51	63	74	81	88	92	95	98	29	40	52	60	69	75	82	89	0.8	0.9	1.1	1.3	1.4	0.9	1.1	1.2	1.4	1.6
WW C1	54	66	76	83	89	92	95	98	32	43	54	61	70	76	82	88	0.8	0.9	1.1	1.3	1.5	0.9	1.1	1.3	1.4	1.6
WW C2	52	64	75	81	88	92	95	98	30	41	52	59	69	74	81	88	0.8	0.9	1.1	1.3	1.5	0.9	1.1	1.3	1.4	1.6
WW C3	58	71	81	87	93	95	97	99	36	48	61	68	78	83	88	94	0.7	0.9	1.1	1.3	1.4	0.8	1.0	1.2	1.4	1.5
WW C4	55	67	78	84	91	94	97	99	32	44	57	65	74	80	86	92	0.7	0.9	1.1	1.3	1.4	0.9	1.0	1.2	1.4	1.5
TW A1	34	41	49	55	63	68	74	81	20	25	30	34	40	44	49	57	1.1	1.2	1.4	1.6	1.7	1.4	1.6	1.7	1.9	2.1
TW A2	38	47	56	63	71	76	82	88	21	28	35	40	47	52	58	66	0.9	1.1	1.3	1.4	1.6	1.2	1.4	1.5	1.7	1.9
TW A3	40	49	59	66	74	79	85	91	22	28	36	42	50	55	62	70	0.9	1.1	1.2	1.4	1.6	1.1	1.3	1.5	1.7	1.8
TW A4	40	50	60	67	75	80	86	91	21	28	36	42	51	56	63	71	0.9	1.1	1.2	1.4	1.6	1.1	1.3	1.5	1.6	1.8
TW B1	34	42	50	56	64	69	75	82	20	25	30	35	41	45	51	58	1.0	1.2	1.4	1.5	1.7	1.4	1.5	1.7	1.9	2.0
TW B2	37	46	55	61	70	75	81	88	20	26	33	39	46	51	58	66	0.9	1.1	1.3	1.5	1.6	1.2	1.4	1.5	1.7	1.9
TW B3	37	46	55	62	71	76	82	88	19	26	33	39	47	52	59	67	0.9	1.1	1.3	1.4	1.6	1.2	1.3	1.5	1.7	1.9
TW B4	34	43	53	59	68	73	80	86	18	24	31	37	45	50	56	65	1.0	1.1	1.3	1.5	1.6	1.2	1.4	1.5	1.7	1.9
TW C1	33	41	49	55	63	69	75	82	19	24	29	34	41	45	51	59	1.0	1.2	1.3	1.5	1.7	1.4	1.5	1.7	1.8	2.0
TW C2	33	41	50	56	65	70	77	84	18	23	30	35	42	47	53	61	1.0	1.2	1.3	1.5	1.6	1.3	1.4	1.6	1.8	1.9
TW C3	31	39	48	54	62	68	74	82	17	22	28	33	40	45	51	59	1.0	1.2	1.3	1.5	1.7	1.3	1.5	1.6	1.8	2.0
TW C4	28	35	43	49	57	62	69	77	16	20	26	30	36	40	46	54	1.1	1.3	1.4	1.6	1.7	1.5	1.6	1.7	1.9	2.0
TW D1	31	39	48	54	62	67	73	81	16	20	26	30	37	41	46	54	1.0	1.2	.4	1.5	1.7	1.4	1.5	1.7	1.9	2.1
TW D2	42	53	63	70	78	83	88	93	23	31	40	46	55	61	67	76	0.9	1.0	1.2	1.4	1.5	1.1	1.2	1.4	1.6	1.7
TW D3	46	57	67	74	82	86	91	95	26	35	44	51	60	66	72	80	0.8	1.0	1.2	1.4	1.5	1.0	1.2	1.4	1.5	1.7
TW D4	48	60	70	77	85	89	93	97	27	37	48	55	64	70	77	85	0.8	1.0	1.1	1.3	1.5	1.0	1.1	1.3	1.5	1.6
TW D5	47	59	70	77	85	89	93	96	27	37	47	55	65	71	77	85	0.8	1.0	1.1	1.3	1.5	1.0	1.1	1.3	1.5	1.6
TW E1	50	61	72	78	85	89	93	97	30	39	49	56	65	70	77	84	0.8	1.0	1.1	1.3	1.5	1.0	1.1	1.3	1.5	1.6
TW E2	49	61	71	78	85	89	93	97	28	38	48	56	65	71	77	85	0.8	1.0	1.1	1.3	1.5	1.0	1.1	1.3	1.5	1.6
TW E3	56	68	79	85	91	94	96	99	35	46	57	65	74	80	86	91	0.7	0.9	1.1	1.3	1.4	0.9	1.0	1.2	1.4	1.5
TW E4	53	65	76	82	89	93	96	98	32	43	54	62	72	77	84	90	0.8	0.9	1.1	1.3	1.4	0.9	1.1	1.2	1.4	1.6
TW F1	33	41	49	55	63	68	74	81	19	23	29	33	39	44	49	57	1.0	1.2	1.4	1.5	1.7	1.4	1.5	1.7	1.9	2.1
TW F2	36	45	54	61	69	74	80	87	19	25	32	37	45	50	56	64	1.0	1.1	1.3	1.5	1.6	1.2	1.4	1.6	1.7	1.9
TW F3	36	46	55	62	71	76	82	89	18	25	32	38	46	51	58	67	0.9	1.1	1.3	1.4	1.6	1.2	1.3	1.5	1.7	1.9
TW F4	34	44	54	61	70	75	81	88	16	23	31	37	45	50	57	66	0.9	1.1	1.3	1.4	1.6	1.2	1.3	1.5	1.7	1.8
TW G1	32	39	48	54	62	67	74	81	17	22	27	32	38	43	49	57	1.0	1.2	1.4	1.5	1.7	1.4	1.5	1.7	1.9	2.0
TW G2	32	41	50	57	66	71	78	85	16	22	29	34	41	46	53	61	1.0	1.1	1.3	1.5	1.6	1.3	1.4	1.6	1.8	1.9
TW G3	30	39	49	56	65	70	77	84	14	20	27	32	40	45	52	61	1.0	1.1	1.3	1.5	1.6	1.3	1.4	1.6	1.8	1.9
TW G4	26	34	44	50	59	65	72	80	11	16	23	28	35	40	47	56	1.0	1.2	1.4	1.5	1.7	1.4	1.5	1.7	1.8	2.0
TW H1	28	36	44	51	59	65	71	79	14	18	24	29	36	40	46	54	1.1	1.2	1.4	1.5	1.7	1.4	1.5	1.7	1.9	2.0
TW H2	26	33	42	49	57	63	70	78	11	16	22	27	34	39	45	53	1.1	1.2	1.4	1.6	1.7	1.4	1.5	1.7	1.9	2.0
TW H3	22	30	38	44	52	58	65	74	9	13	19	23	30	35	'1	49	1.1	1.3	1.4	1.6	1.7	1.5	1.6	1.8	1.9	2.1
TW H4	18	24	31	36	44	49	56	65	6	10	15	18	24	28	33	41	1.2	1.4	1.5	1.7	1.8	1.7	1.8	1.9	2.1	2.2
TW I1	29	36	44	50	59	64	70	78	13	18	23	27	33	38	43	50	1.1	1.2	1.4	1.6	1.7	1.4	1.6	1.8	2.0	2.2
TW I2	39	49	60	67	75	80	86	92	20	28	36	43	52	57	64	73	0.9	1.0	1.2	1.4	1.5	1.1	1.3	1.4	1.6	1.8
TW I3	42	52	63	70	79	83	88	93	22	31	40	47	56	62	69	77	0.9	1.0	1.2	1.4	1.5	1.0	1.2	1.4	1.6	1.7
TW I4	45	56	67	74	83	87	91	96	24	34	44	52	61	67	74	82	0.8	1.0	1.2	1.3	1.5	1.0	1.1	1.3	1.5	1.6
TW I5	45	56	68	75	83	88	92	96	24	34	45	52	62	68	76	84	0.8	1.0	1.1	1.3	1.5	1.0	1.1	1.3	1.5	1.6
TW J1	47	59	69	76	84	88	92	96	27	36	46	53	63	68	75	83	0.8	1.0	1.2	1.3	1.5	1.0	1.1	1.3	1.5	1.7
TW J2	46	57	68	75	83	88	92	96	25	35	45	53	62	68	75	83	0.8	1.0	1.2	1.3	1.5	1.0	1.1	1.3	1.5	1.6
TW J3	53	65	76	82	89	93	96	98	31	43	54	62	72	78	84	90	0.8	0.9	1.1	1.3	1.4	0.9	1.0	1.2	1.4	1.6
TW J4	50	62	74	80	88	91	95	98	28	40	51	59	69	75	82	89	0.8	0.9	1.1	1.3	1.5	0.9	1.1	1.2	1.4	1.6
DG A1	35	44	52	57	64	68	73	78	17	22	28	32	37	41	44	49	1.0	1.2	1.4	1.7	2.0	1.3	1.5	1.9	2.3	2.8
DG A2	37	46	56	62	70	75	80	86	19	25	33	38	46	51	56	63	0.9	1.1	1.3	1.5	1.7	1.2	1.3	1.6	1.8	2.1
DG A3	42	53	64	71	79	83	87	91	23	31	41	48	57	63	69	77	0.8	1.0	1.2	1.5	1.7	1.1	1.2	1.4	1.6	1.8
DG B1	37	47	57	64	71	75	80	84	17	23	30	35	42	46	51	57	0.9	1.1	1.3	1.6	1.9	1.2	1.4	1.7	2.0	2.4
DG B2	38	49	60	68	76	81	86	91	19	27	35	41	50	56	62	70	0.9	1.0	1.2	1.4	1.6	1.1	1.3	1.5	1.7	1.9
DG B3	43	55	67	74	83	87	91	94	24	32	43	50	60	67	74	82	0.8	1.0	1.2	1.4	1.6	1.0	1.2	1.3	1.5	1.7
DG C1	42	54	64	71	79	83	87	90	21	28	36	42	50	55	61	67	0.8	1.0	1.2	1.5	1.8	1.1	1.3	1.5	1.8	2.1
DG C2	43	55	67	74	82	87	91	94	23	31	40	47	57	63	70	77	0.8	1.0	1.2	1.4	1.6	1.1	1.2	1.4	1.6	1.8
DG C3	47	61	73	80	88	91	94	97	27	37	48	56	67	73	80	87	0.8	0.9	1.1	1.4	1.6	1.0	1.1	1.2	1.4	1.6
SS A1	44	53	62	68	75	80	85	90	29	37	44	50	57	61	67	74	0.9	1.1	1.3	1.4	1.6	1.1	1.3	1.5	1.7	1.9
SS A2	52	63	72	78	85	89	92	96	37	46	55	61	69	74	79	85	0.8	1.0	1.2	1.3	1.5	0.9	1.1	1.3	1.5	1.7
SS A3	42	50	59	64	72	76	82	87	28	34	42	47	53	57	63	69	0.9	1.1	1.3	1.5	1.6	1.1	1.3	1.6	1.8	2.0
SS A4	52	62	71	77	84	88	92	95	36	45	54	60	68	73	78	84	0.8	1.0	1.2	1.4	1.5	0.9	1.1	1.3	1.5	1.7
SS A5	41	49	58	63	71	75	81	87	28	34	41	46	53	57	62	69	1.0	1.1	1.3	1.5	1.7	1.1	1.3	1.6	1.8	2.0
SS A6	52	62	72	78	85	88	92	96	36	45	55	61	68	73	79	85	0.8	1.0	1.2	1.4	1.5	0.9	1.1	1.3	1.5	1.7
SS A7	38	45	52	58	65	70	75	81	25	31	37	41	47	51	56	62	1.0	1.2	1.4	1.6	1.7	1.2	1.4	1.7	1.9	2.1
SS A8	51	61	71	76	83	87	91	95	36	45	53	59	67	71	77	83	0.8	1.0	1.2	1.4	1.5	0.9	1.2	1.4	1.5	1.7
SS B1	39	48	56	63	70	75	81	87	25	31	39	44	51	56	61	68	0.9	1.1	1.3	1.5	1.6	1.1	1.3	1.5	1.7	1.9
SS B2	47	58	67	74	81	85	90	94	32	40	49	56	64	69	75	82	0.8	1.0	1.2	1.4	1.5	1.0	1.2	1.3	1.5	1.7
SS B3	38	46	54	60	68	73	78	85	23	30	37	42	48	53	58	65	1.0	1.1	1.3	1.5	1.7	1.2	1.4	1.6	1.8	2.0
SS B4	47	57	66	73	80	85	89	94	31	40	49	55	63	68	74	81	0.8	1.0	1.2	1.4	1.5	1.0	1.2	1.4	1.6	1.7
SS B5	35	43	51	56	64	69	74	81	22	28	34	38	45	49	54	61	1.0	1.2	1.4	1.6	1.7	1.3	1.4	1.7	1.9	2.1
SS B6	47	57	67	73	81	85	89	94	31	39	49	55	63	68	74	81	0.8	1.0	1.2	1.4	1.5	1.0	1.2	1.4	1.5	1.7
SS B7	33	39	47	52	59	64	70	77	20	25	31	35	41	45	50	56	1.1	1.3	1.4	1.6	1.8	1.3	1.5	1.7	2.0	2.2
SS B8	46	55	65	72	79	84	88	93	30	38	47	53	61	66	72	79	0.9	1.0	1.2	1.4	1.5	1.0	1.2	1.4	1.6	1.7
SS C1	40	50	60	67	75	80	85	91	21	28	36	42	50	55	62	70	0.9	1.1	1.2	1.4	1.6	1.1	1.3	1.5	1.7	1.8
SS C2	46	57	68	74	82	87	91	95	26	34	44	51	60	66	72	80	0.8	1.0	1.2	1.3	1.5	1.0	1.2	1.3	1.5	1.7
SS C3	33	41	50	56	64	70	76	83	18	23	29	34	41	45	51	59	1.0	1.2	1.3	1.5	1.7	1.3	1.5	1.6	1.8	2.0
SS C4	41	51	62	69	77	82	87	92	23	31	39	46	54	60	67	75	0.9	1.0	1.2	1.4	1.5	1.1	1.2	1.4	1.6	1.7
SS D1	51	61	70	76	83	87	91	95	35	43	52	57	65	69	75	81	0.8	1.0	1.2	1.4	1.5	1.0	1.2	1.4	1.6	1.8
SS D2	59	70	79	85	90	93	96	98	42	52	62	68	76	81	86	91	0.8	0.9	1.1	1.3	1.4	0.9	1.1	1.2	1.4	1.6
SS D3	47	56	65	71	78	82	87	92	32	39	47	52	60	64	69	76	0.9	1.1	1.2	1.4	1.6	1.0	1.2	1.5	1.7	1.8
SS D4	58	68	78	83	89	92	95	98	40	50	60	67	74	79	84	90	0.8	0.9	1.1	1.3	1.5	0.9	1.1	1.3	1.4	1.6
SS E1	47	56	66	72	79	84	88	93	30	38	46	51	59	64	69	76	0.9	1.0	1.2	1.4	1.6	1.0	1.2	1.4	1.7	1.8
SS E2	54	65	75	81	87	91	94	97	36	46	56	62	71	75	81	87	0.8	1.0	1.1	1.3	1.5	0.9	1.1	1.3	1.5	1.6
SS E3	40	48	57	63	71	75	81	87	25	32	39	44	51	55	61	68	0.9	1.1	1.3	1.5	1.6	1.2	1.3	1.6	1.8	2.0
SS E4	51	61	71	77	84	88	92	96	33	43	52	58	67	71	77	84	0.8	1.0	1.2	1.4	1.5	0.9	1.1	1.3	1.5	1.7

PENDLETON, OREGON

TYPE	SOLAR SAVINGS FRACTION (%) BASE TEMP: 55 F (2876 DD) LCR								65 F (5251 DD)								CONSERVATION FACTOR 55 F					65 F				
	100	70	50	40	30	25	20	15	100	70	50	40	30	25	20	15	100	50	30	20	15	100	50	30	20	15
WW A1	19	22	25	28	31	34	36	40	16	18	20	21	24	25	27	29	1.7	2.0	2.3	2.6	2.9	2.0	2.4	2.9	3.4	4.1
WW A2	20	24	29	33	38	41	46	51	14	18	22	25	28	31	34	38	1.4	1.6	1.9	2.1	2.4	1.6	1.9	2.2	2.6	3.0
WW A3	19	25	30	35	40	44	49	55	14	18	22	26	30	33	37	41	1.3	1.5	1.8	2.0	2.2	1.5	1.8	2.1	2.4	2.8
WW A4	19	25	31	35	42	46	51	57	13	18	22	26	31	34	38	43	1.2	1.5	1.7	1.9	2.1	1.4	1.7	2.0	2.3	2.7
WW A5	19	25	32	37	43	47	53	59	13	18	23	27	32	35	40	45	1.2	1.4	1.7	1.9	2.1	1.4	1.6	1.9	2.3	2.6
WW A6	19	25	32	37	44	48	54	61	12	17	23	27	33	36	40	46	1.2	1.4	1.6	1.9	2.0	1.3	1.6	1.9	2.2	2.5
WW B1	16	19	22	24	28	30	32	35	11	13	15	16	18	19	20	20	1.6	2.0	2.5	3.0	3.5	2.1	2.6	3.5	-	-
WW B2	20	26	33	38	45	49	55	63	14	19	25	29	35	39	44	50	1.2	1.4	1.6	1.8	2.0	1.3	1.6	1.8	2.1	2.3
WW B3	23	31	38	44	51	56	62	70	17	23	30	35	41	46	51	58	1.1	1.3	1.5	1.7	1.9	1.2	1.4	1.7	1.9	2.1
WW B4	24	32	41	47	55	61	67	75	17	24	32	38	46	51	57	65	1.0	1.2	1.4	1.6	1.8	1.1	1.3	1.5	1.7	1.9
WW B5	23	31	40	47	56	61	68	76	16	24	32	38	46	52	59	67	1.0	1.2	1.4	1.6	1.7	1.1	1.3	1.5	1.7	1.9
WW C1	25	33	41	46	54	59	65	73	18	25	32	37	44	49	54	61	1.1	1.3	1.5	1.7	1.8	1.2	1.4	1.6	1.8	2.0
WW C2	23	31	40	46	54	59	65	73	17	24	31	37	44	49	55	63	1.1	1.3	1.4	1.6	1.8	1.2	1.4	1.6	1.8	2.0
WW C3	27	37	47	55	64	69	76	84	20	30	39	46	55	61	68	76	1.0	1.1	1.3	1.5	1.6	1.0	1.2	1.4	1.6	1.7
WW C4	25	35	45	52	61	67	74	82	18	27	36	43	52	58	65	74	1.0	1.2	1.3	1.5	1.7	1.1	1.2	1.4	1.6	1.8
TW A1	19	22	25	28	31	34	37	41	15	17	20	22	24	26	28	31	1.6	1.9	2.3	2.6	2.8	2.0	2.4	2.7	3.2	3.7
TW A2	19	23	27	31	36	39	43	48	14	17	21	23	27	30	33	37	1.4	1.7	2.0	2.2	2.4	1.7	2.0	2.3	2.6	3.0
TW A3	18	23	28	32	37	41	45	51	13	17	21	24	28	31	35	39	1.4	1.6	1.9	2.1	2.3	1.6	1.8	2.1	2.5	2.8
TW A4	17	22	28	32	38	41	46	53	12	16	20	24	28	31	35	40	1.3	1.6	1.8	2.0	2.2	1.5	1.8	2.1	2.4	2.6
TW B1	18	22	25	28	32	35	38	43	15	17	20	22	25	27	29	32	1.6	1.9	2.1	2.4	2.7	1.9	2.2	2.5	2.9	3.3
TW B2	17	22	27	30	35	38	43	49	13	16	20	23	27	29	33	37	1.4	1.7	1.9	2.2	2.4	1.7	1.9	2.2	2.5	2.8
TW B3	17	21	26	30	35	39	43	50	12	16	20	23	27	30	34	38	1.4	1.6	1.9	2.1	2.3	1.6	1.9	2.1	2.4	2.7
TW B4	16	20	25	29	34	38	42	49	12	15	19	22	26	29	33	38	1.4	1.7	1.9	2.1	2.3	1.6	1.9	2.1	2.4	2.7
TW C1	17	21	25	28	32	35	39	44	13	16	19	21	25	27	30	34	1.5	1.8	2.1	2.3	2.5	1.8	2.1	2.4	2.7	3.0
TW C2	16	20	25	28	33	36	40	46	12	15	19	21	25	28	31	36	1.5	1.7	2.0	2.2	2.4	1.7	2.0	2.2	2.5	2.8
TW C3	15	19	24	27	32	35	39	45	12	15	18	21	25	27	31	36	1.5	1.8	2.0	2.2	2.4	1.8	2.0	2.3	2.5	2.8
TW C4	15	18	22	25	30	33	37	42	12	15	18	20	23	26	29	34	1.6	1.9	2.1	2.3	2.5	1.9	2.1	2.4	2.6	2.8
TW D1	13	16	20	22	25	27	30	33	9	11	13	15	17	18	19	21	1.7	2.0	2.4	2.9	3.3	2.1	2.6	3.2	4.2	-
TW D2	20	25	31	36	42	46	52	59	14	19	24	28	33	37	41	47	1.3	1.5	1.7	1.9	2.1	1.4	1.7	1.9	2.1	2.4
TW D3	21	27	33	38	45	49	55	62	15	20	26	30	35	39	44	50	1.2	1.4	1.6	1.8	2.0	1.3	1.6	1.8	2.1	2.3
TW D4	22	29	37	43	51	56	62	70	16	22	29	35	42	46	52	60	1.1	1.3	1.5	1.7	1.8	1.2	1.4	1.6	1.8	2.0
TW D5	22	29	38	44	52	57	64	72	16	23	30	35	43	48	55	63	1.1	1.3	1.5	1.6	1.8	1.2	1.4	1.6	1.8	1.9
TW E1	23	30	37	43	50	54	60	68	17	23	29	34	40	45	50	56	1.1	1.3	1.5	1.7	1.9	1.3	1.5	1.7	1.9	2.1
TW E2	23	30	38	43	51	56	62	70	17	23	30	35	42	46	52	59	1.1	1.3	1.5	1.7	1.9	1.2	1.4	1.6	1.9	2.0
TW E3	27	36	45	51	60	65	72	80	20	28	37	43	51	56	63	71	1.0	1.2	1.4	1.5	1.7	1.1	1.3	1.5	1.7	1.8
TW E4	25	34	43	50	58	64	71	79	19	27	35	41	50	55	62	70	1.0	1.2	1.4	1.6	1.7	1.1	1.3	1.5	1.7	1.8
TW F1	17	20	24	27	30	33	36	40	13	16	18	20	23	25	27	30	1.6	1.9	2.2	2.5	2.8	2.0	2.3	2.7	3.1	3.6
TW F2	16	20	25	28	33	36	40	46	12	15	18	21	25	27	30	34	1.5	1.7	2.0	2.2	2.5	1.7	2.0	2.3	2.7	3.0
TW F3	15	20	25	29	34	37	42	48	10	14	18	21	25	28	31	36	1.4	1.7	1.9	2.1	2.3	1.6	1.9	2.2	2.5	2.8
TW F4	13	18	23	27	33	36	41	48	8	12	16	19	24	27	31	36	1.4	1.6	1.9	2.1	2.3	1.6	1.9	2.1	2.4	2.7
TW G1	15	19	22	25	29	32	36	41	11	14	17	19	22	24	26	30	1.6	1.9	2.2	2.4	2.7	1.9	2.2	2.6	2.9	3.3
TW G2	14	18	22	26	31	34	38	44	9	12	16	19	22	25	28	33	1.5	1.7	2.0	2.2	2.4	1.7	2.0	2.3	2.6	2.9
TW G3	12	16	21	25	30	33	38	44	8	11	15	17	22	24	28	33	1.5	1.7	2.0	2.2	2.4	1.7	2.0	2.2	2.5	2.8
TW G4	10	14	18	22	27	30	34	41	6	9	12	15	19	22	26	31	1.5	1.8	2.0	2.2	2.4	1.8	2.0	2.3	2.5	2.8
TW H1	12	16	20	23	27	30	34	39	9	11	14	16	20	22	25	29	1.6	1.9	2.1	2.4	2.6	1.9	2.2	2.5	2.8	3.1
TW H2	10	14	18	21	26	29	33	39	6	9	12	15	18	21	24	29	1.6	1.8	2.1	2.3	2.5	1.9	2.1	2.4	2.7	2.9
TW H3	8	12	16	19	23	26	30	36	5	7	11	13	16	19	22	27	1.7	1.9	2.1	2.4	2.6	1.9	2.2	2.4	2.7	2.9
TW H4	6	9	13	15	19	22	26	31	4	6	8	10	13	16	19	23	1.8	2.0	2.3	2.5	2.7	2.1	2.4	2.6	2.8	3.1
TW I1	11	14	17	20	23	25	27	31	7	9	11	13	15	16	17	18	1.7	2.1	2.5	3.0	3.4	2.2	2.7	3.3	4.3	
TW I2	17	22	28	33	39	43	49	56	11	16	21	25	30	34	38	44	1.3	1.5	1.7	1.9	2.1	1.5	1.7	1.9	2.2	2.4
TW I3	18	24	30	35	42	40	52	59	12	17	23	27	33	37	42	48	1.2	1.4	1.7	1.9	2.0	1.4	1.6	1.8	2.1	2.3
TW I4	19	27	34	40	48	53	59	67	14	20	27	32	39	44	50	57	1.1	1.3	1.5	1.7	1.9	1.3	1.4	1.6	1.9	2.0
TW I5	20	27	35	41	49	55	61	70	14	20	28	33	41	46	52	60	1.1	1.3	1.5	1.7	1.8	1.2	1.4	1.6	1.8	2.0
TW J1	21	28	35	40	48	53	59	66	15	21	27	32	38	43	48	55	1.1	1.4	1.6	1.7	1.9	1.3	1.5	1.7	1.9	2.1
TW J2	20	27	35	41	48	53	60	68	15	21	27	32	39	44	50	57	1.1	1.3	1.5	1.7	1.9	1.3	1.5	1.7	1.9	2.1
TW J3	24	33	42	49	58	63	70	78	18	26	34	41	49	55	61	70	1.0	1.2	1.4	1.6	1.7	1.1	1.3	1.5	1.7	1.8
TW J4	23	31	40	47	56	62	69	77	16	24	33	39	47	53	60	68	1.0	1.2	1.4	1.6	1.7	1.1	1.3	1.5	1.7	1.8
DG A1	14	17	21	23	26	27	29	30	9	11	14	15	17	17	18	18	1.6	2.1	2.7	3.9	-	2.0	2.6	3.8	-	-
DG A2	16	21	20	30	35	38	42	47	12	15	19	22	27	29	32	37	1.4	1.6	1.9	2.3	2.6	1.6	1.9	2.2	2.6	3.0
DG A3	20	26	33	38	45	49	55	63	15	20	26	31	37	41	46	53	1.2	1.4	1.6	1.8	2.0	1.4	1.5	1.7	2.0	2.2
DG B1	14	18	22	24	28	30	32	34	9	12	14	16	18	19	21	21	1.6	2.0	2.4	3.2	4.2	1.9	2.3	3.2	-	-
DG B2	17	22	27	31	37	41	45	51	12	16	20	24	28	31	35	40	1.4	1.6	1.8	2.1	2.4	1.6	1.8	2.1	2.4	2.8
DG B3	21	27	34	39	47	52	58	66	16	21	27	32	39	43	49	57	1.2	1.4	1.5	1.7	1.9	1.3	1.5	1.7	1.9	2.1
DG C1	17	21	26	29	34	36	40	43	11	15	18	21	24	26	27	29	1.4	1.7	2.0	2.6	3.2	1.7	2.0	2.6	3.6	4.9
DG C2	20	25	31	36	42	46	52	58	14	19	24	28	33	37	41	47	1.3	1.5	1.7	1.9	2.1	1.4	1.6	1.9	2.2	2.5
DG C3	23	30	38	44	52	57	64	72	18	24	31	36	43	48	54	62	1.2	1.3	1.5	1.6	1.8	1.3	1.4	1.6	1.8	2.0
SS A1	22	26	31	34	39	42	46	51	18	22	26	28	32	35	38	42	1.4	1.7	1.9	2.2	2.4	1.5	1.8	2.2	2.6	3.0
SS A2	26	33	39	44	51	55	61	68	22	28	34	38	44	48	53	59	1.1	1.4	1.6	1.7	1.9	1.2	1.5	1.7	1.9	2.1
SS A3	20	24	28	31	35	37	40	45	17	20	23	25	28	30	32	35	1.5	1.8	2.1	2.5	2.7	1.7	2.1	2.5	3.1	3.8
SS A4	26	32	38	43	49	53	59	66	22	27	33	37	42	46	50	56	1.2	1.4	1.6	1.8	2.0	1.3	1.5	1.8	2.0	2.2
SS A5	22	25	29	32	36	39	42	46	18	21	25	27	30	32	34	37	1.5	1.8	2.1	2.4	2.6	1.7	2.0	2.4	2.9	3.4
SS A6	26	32	39	44	50	55	60	67	22	27	33	37	43	47	52	58	1.2	1.4	1.6	1.8	1.9	1.3	1.5	1.7	2.0	2.2
SS A7	20	23	26	28	31	33	36	39	17	19	22	23	26	27	29	30	1.7	2.0	2.4	2.8	3.2	1.9	2.4	2.9	3.7	4.8
SS A8	25	31	37	41	48	52	57	64	21	26	31	35	41	44	48	54	1.2	1.4	1.6	1.9	2.0	1.3	1.6	1.8	2.1	2.3
SS B1	19	23	27	30	35	38	42	47	15	18	22	25	28	31	34	37	1.4	1.7	2.0	2.3	2.5	1.6	2.0	2.3	2.7	3.1
SS B2	24	30	36	41	48	52	58	65	19	25	30	35	41	45	50	56	1.2	1.4	1.6	1.8	1.9	1.3	1.5	1.7	2.0	2.2
SS B3	17	21	25	28	32	34	38	42	14	17	20	22	25	27	30	32	1.5	1.8	2.2	2.5	2.7	1.7	2.1	2.5	3.1	3.7
SS B4	23	29	35	39	46	50	56	63	19	24	29	33	39	43	48	54	1.2	1.4	1.6	1.8	2.0	1.3	1.6	1.8	2.0	2.2
SS B5	18	21	25	27	31	34	37	41	15	18	20	22	25	27	29	32	1.6	1.9	2.2	2.6	2.8	1.8	2.2	2.6	3.1	3.6
SS B6	23	29	35	40	46	51	56	63	19	24	29	33	39	43	48	54	1.2	1.4	1.6	1.8	2.0	1.3	1.5	1.8	2.0	2.2
SS B7	16	19	22	25	28	30	32	35	14	16	18	20	22	23	25	27	1.7	2.1	2.5	2.9	3.2	2.0	2.5	2.9	3.6	4.4
SS B8	22	27	33	38	44	48	53	60	18	22	28	32	37	41	45	51	1.2	1.5	1.7	1.9	2.1	1.4	1.6	1.8	2.1	2.3
SS C1	17	22	27	31	36	39	43	49	12	15	19	22	26	28	31	35	1.4	1.6	1.9	2.2	2.4	1.6	1.9	2.3	2.7	3.2
SS C2	20	27	34	39	45	50	56	63	15	20	26	30	36	40	45	51	1.2	1.4	1.6	1.8	2.0	1.3	1.6	1.8	2.0	2.3
SS C3	16	20	24	27	31	34	37	42	12	15	18	20	23	25	28	31	1.6	1.8	2.1	2.4	2.6	1.9	2.2	2.5	2.9	3.3
SS C4	19	25	31	35	41	46	51	58	14	18	24	27	33	36	41	47	1.3	1.5	1.7	1.9	2.1	1.4	1.7	1.9	2.1	2.4
SS D1	24	29	35	39	44	48	52	58	20	24	29	32	36	39	42	46	1.2	1.5	1.8	2.0	2.2	1.4	1.7	2.0	2.5	2.8
SS D2	29	36	44	50	58	63	69	76	24	31	38	43	50	54	60	67	1.0	1.2	1.4	1.6	1.8	1.1	1.3	1.6	1.8	2.0
SS D3	23	28	32	36	41	44	48	53	19	23	27	30	34	36	39	43	1.3	1.6	1.9	2.1	2.4	1.5	1.8	2.2	2.6	3.0
SS D4	28	35	43	49	56	61	67	74	23	30	37	42	48	53	58	65	1.1	1.3	1.5	1.7	1.8	1.1	1.4	1.6	1.8	2.0
SS E1	21	25	30	34	39	42	46	51	16	20	24	26	30	32	35	38	1.3	1.6	1.9	2.2	2.4	1.5	1.9	2.3	2.8	3.5
SS E2	25	32	39	44	51	56	62	69	20	26	32	37	43	47	52	59	1.1	1.3	1.5	1.7	1.9	1.2	1.4	1.7	1.9	2.1
SS E3	19	23	27	30	34	37	41	45	16	19	22	24	28	30	32	35	1.5	1.8	2.1	2.4	2.6	1.7	2.0	2.4	2.9	3.5
SS E4	24	30	37	42	48	53	58	65	19	25	30	35	41	44	49	55	1.2	1.4	1.6	1.8	2.0	1.3	1.5	1.8	2.0	2.2

PORTLAND, OREGON

TYPE	SOLAR SAVINGS FRACTION (%) BASE TEMP: 55 F (2235 DD)								65 F (4798 DD)								CONSERVATION FACTOR 55 F					65 F				
LCR	100	70	50	40	30	25	20	15	100	70	50	40	30	25	20	15	100	50	30	20	15	100	50	30	20	15
WW A1	21	25	29	32	36	39	43	48	16	19	21	23	25	27	29	32	1.5	1.8	2.0	2.3	2.6	1.9	2.3	2.7	3.2	3.7
WW A2	23	28	34	38	44	48	53	59	15	19	23	26	31	33	37	41	1.2	1.5	1.7	1.9	2.1	1.5	1.8	2.1	2.4	2.8
WW A3	23	29	36	41	47	51	57	63	15	19	24	28	32	36	40	45	1.2	1.4	1.6	1.8	2.0	1.4	1.7	2.0	2.3	2.6
WW A4	23	30	37	42	49	53	59	66	14	19	24	28	33	37	41	47	1.1	1.4	1.6	1.8	2.0	1.4	1.6	1.9	2.2	2.5
WW A5	23	30	38	43	50	55	61	68	14	19	25	29	35	38	43	49	1.1	1.3	1.5	1.8	1.9	1.3	1.6	1.8	2.1	2.4
WW A6	23	31	38	44	51	56	62	69	14	19	25	29	35	39	44	50	1.1	1.3	1.5	1.7	1.9	1.3	1.6	1.8	2.1	2.4
WW B1	18	22	27	30	34	37	40	45	12	14	17	18	20	22	23	24	1.4	1.7	2.1	2.4	2.8	1.9	2.4	3.1	4.4	-
WW B2	24	31	39	44	51	56	62	70	15	21	27	31	37	41	47	53	1.1	1.3	1.5	1.7	1.9	1.3	1.5	1.7	2.0	2.2
WW B3	28	36	44	50	58	63	69	76	18	25	32	37	44	48	54	61	1.0	1.2	1.4	1.6	1.8	1.2	1.4	1.6	1.9	2.1
WW B4	28	37	46	53	62	67	73	81	18	26	34	40	48	53	60	68	1.0	1.2	1.4	1.5	1.7	1.1	1.3	1.5	1.7	1.9
WW B5	27	36	46	53	61	67	74	81	17	25	34	40	49	54	61	69	1.0	1.2	1.3	1.5	1.7	1.1	1.3	1.5	1.7	1.8
WW C1	29	38	46	53	61	66	72	79	20	27	34	39	47	51	57	64	1.0	1.2	1.4	1.6	1.7	1.2	1.4	1.6	1.8	2.0
WW C2	28	36	45	52	60	65	72	79	18	25	33	39	47	52	58	66	1.0	1.2	1.4	1.6	1.7	1.1	1.3	1.5	1.7	1.9
WW C3	32	43	53	60	69	75	81	88	22	31	41	48	57	63	70	78	0.9	1.1	1.3	1.4	1.6	1.0	1.2	1.4	1.6	1.7
WW C4	29	40	50	57	67	72	79	86	20	29	38	45	54	60	67	76	0.9	1.1	1.3	1.5	1.6	1.0	1.2	1.4	1.6	1.7
TW A1	21	25	29	32	36	39	43	48	16	18	21	23	26	28	30	33	1.5	1.8	2.0	2.3	2.5	1.9	2.2	2.6	3.0	3.4
TW A2	21	26	32	36	41	45	50	56	15	18	22	25	29	32	35	39	1.3	1.5	1.8	2.0	2.2	1.6	1.9	2.2	2.5	2.8
TW A3	21	27	33	37	43	47	52	59	14	18	22	26	30	33	37	42	1.2	1.5	1.7	1.9	2.1	1.5	1.8	2.1	2.3	2.6
TW A4	20	26	33	37	44	48	53	60	13	17	22	26	30	34	38	44	1.2	1.4	1.7	1.9	2.1	1.5	1.7	2.0	2.3	2.5
TW B1	20	24	29	32	37	40	44	50	15	18	21	23	26	28	31	35	1.5	1.7	1.9	2.2	2.4	1.8	2.1	2.4	2.8	3.1
TW B2	20	25	31	35	41	44	49	56	14	17	21	24	29	31	35	40	1.3	1.5	1.8	2.0	2.2	1.6	1.9	2.1	2.4	2.7
TW B3	19	25	31	35	41	45	50	57	13	17	21	24	29	32	36	41	1.3	1.5	1.7	1.9	2.1	1.6	1.8	2.1	2.3	2.6
TW B4	18	23	29	33	39	43	49	55	12	16	20	23	28	31	35	41	1.3	1.5	1.7	2.0	2.1	1.6	1.8	2.1	2.3	2.5
TW C1	19	23	28	32	37	40	45	51	14	17	20	23	26	29	32	37	1.4	1.7	1.9	2.1	2.3	1.7	2.0	2.3	2.6	2.8
TW C2	18	23	28	32	38	41	46	53	13	16	20	23	27	30	33	39	1.4	1.6	1.8	2.0	2.2	1.7	1.9	2.2	2.4	2.6
TW C3	18	22	27	31	36	40	45	51	13	16	19	22	26	29	33	38	1.4	1.6	1.8	2.0	2.2	1.7	1.9	2.2	2.4	2.6
TW C4	17	21	25	29	34	37	41	48	13	15	19	21	25	27	31	36	1.5	1.7	1.9	2.1	2.3	1.8	2.1	2.3	2.5	2.7
TW D1	16	19	24	27	31	34	37	42	10	12	15	16	19	20	22	24	1.5	1.8	2.1	2.4	2.7	2.0	2.4	3.0	3.7	4.7
TW D2	23	29	36	41	48	53	58	66	15	20	25	29	35	39	44	50	1.2	1.4	1.6	1.8	1.9	1.4	1.6	1.8	2.1	2.3
TW D3	24	31	39	44	51	56	62	69	16	22	27	32	38	42	47	53	1.1	1.3	1.5	1.7	1.9	1.3	1.5	1.8	2.0	2.2
TW D4	26	34	42	49	57	62	68	76	17	24	31	36	44	49	55	63	1.0	1.2	1.4	1.6	1.8	1.2	1.4	1.6	1.8	2.0
TW D5	25	34	43	49	57	63	69	77	17	24	32	37	45	50	57	65	1.0	1.2	1.4	1.6	1.7	1.2	1.4	1.5	1.7	1.9
TW E1	27	35	43	49	56	61	67	74	19	25	31	36	43	47	53	60	1.1	1.3	1.5	1.6	1.8	1.2	1.4	1.7	1.9	2.1
TW E2	27	35	43	49	57	62	68	76	18	25	32	37	44	48	54	62	1.0	1.2	1.4	1.6	1.8	1.2	1.4	1.6	1.8	2.0
TW E3	31	41	51	57	66	71	77	84	22	30	39	45	53	59	65	73	0.9	1.1	1.3	1.5	1.6	1.1	1.2	1.4	1.6	1.8
TW E4	29	39	48	55	64	70	76	83	20	28	37	43	52	57	64	72	1.0	1.1	1.3	1.5	1.7	1.1	1.3	1.4	1.6	1.8
TW F1	19	23	27	31	35	38	42	47	14	17	19	21	24	26	29	32	1.5	1.7	2.0	2.2	2.5	1.9	2.2	2.6	2.9	3.3
TW F2	19	24	29	33	39	43	47	54	13	16	20	23	27	29	33	37	1.3	1.6	1.8	2.0	2.2	1.6	1.9	2.2	2.5	2.8
TW F3	18	23	29	34	40	44	49	56	11	15	19	22	27	30	34	39	1.3	1.5	1.7	2.0	2.1	1.6	1.8	2.1	2.4	2.7
TW F4	16	22	28	32	39	43	48	55	9	13	18	21	26	29	33	39	1.3	1.5	1.7	1.9	2.1	1.6	1.8	2.1	2.3	2.6
TW G1	17	21	26	29	34	38	42	48	12	15	18	20	23	26	29	32	1.5	1.7	1.9	2.2	2.4	1.8	2.1	2.4	2.8	3.1
TW G2	16	21	26	30	36	40	45	51	10	13	17	20	24	27	31	36	1.4	1.6	1.8	2.0	2.2	1.7	1.9	2.2	2.5	2.8
TW G3	14	19	25	29	35	39	44	51	8	12	16	19	23	26	30	35	1.4	1.6	1.8	2.0	2.2	1.6	1.9	2.2	2.4	2.7
TW G4	12	16	22	26	31	35	40	47	7	10	14	17	21	24	28	33	1.4	1.6	1.9	2.1	2.2	1.7	2.0	2.2	2.5	2.7
TW H1	14	18	23	27	32	35	40	46	9	12	15	18	21	24	27	31	1.5	1.7	1.9	2.2	2.4	1.8	2.1	2.4	2.7	2.9
TW H2	12	16	21	25	30	34	39	45	7	10	13	16	20	23	26	31	1.5	1.7	1.9	2.1	2.3	1.8	2.0	2.3	2.6	2.8
TW H3	10	14	19	22	27	31	36	42	5	8	12	14	18	21	24	29	1.5	1.7	2.0	2.2	2.4	1.9	2.1	2.4	2.6	2.8
TW H4	8	11	15	18	23	26	30	36	4	6	9	11	15	17	20	25	1.7	1.9	2.1	2.3	2.5	2.0	2.3	2.5	2.8	3.0
TW I1	13	17	21	24	28	31	34	39	8	10	13	14	16	18	20	22	1.6	1.9	2.2	2.5	2.8	2.0	2.5	3.1	3.8	4.8
TW I2	20	26	33	38	45	49	55	63	12	17	22	26	32	36	41	47	1.2	1.4	1.6	1.8	2.0	1.4	1.6	1.9	2.1	2.3
TW I3	21	28	35	41	48	53	59	66	14	19	25	29	35	39	44	51	1.1	1.3	1.5	1.7	1.9	1.3	1.6	1.8	2.0	2.2
TW I4	23	31	39	45	54	59	65	73	15	21	28	33	41	46	52	60	1.1	1.3	1.4	1.6	1.8	1.2	1.4	1.6	1.8	2.0
TW I5	23	31	40	47	55	61	67	75	15	22	29	35	43	48	54	63	1.0	1.2	1.4	1.6	1.8	1.2	1.4	1.6	1.7	1.9
TW J1	25	32	41	46	54	59	65	73	17	23	29	34	41	45	51	58	1.1	1.3	1.5	1.7	1.8	1.2	1.5	1.7	1.9	2.1
TW J2	24	32	40	46	54	60	66	74	16	22	29	34	41	46	52	60	1.1	1.3	1.4	1.6	1.8	1.2	1.4	1.6	1.8	2.0
TW J3	28	38	48	55	64	69	76	83	19	27	36	43	51	57	64	72	1.0	1.1	1.3	1.5	1.7	1.1	1.3	1.4	1.6	1.8
TW J4	26	36	46	53	62	67	74	82	18	26	34	41	49	55	62	71	1.0	1.2	1.3	1.5	1.7	1.1	1.3	1.5	1.6	1.8
DG A1	17	21	25	28	32	34	36	38	10	13	15	17	19	20	21	21	1.4	1.8	2.3	3.1	4.2	1.9	2.4	3.4	-	-
DG A2	19	24	30	34	40	44	48	54	13	16	21	24	28	31	35	39	1.3	1.5	1.8	2.1	2.4	1.5	1.8	2.1	2.5	2.9
DG A3	23	30	37	43	50	55	61	68	16	21	28	32	39	43	48	55	1.1	1.3	1.5	1.7	1.9	1.3	1.5	1.7	1.9	2.1
DG B1	17	22	27	30	34	37	40	43	10	13	16	18	20	22	23	25	1.4	1.7	2.1	2.6	3.2	1.8	2.2	2.9	4.1	-
DG B2	20	25	32	37	43	47	52	58	13	17	22	25	30	33	37	42	1.3	1.4	1.7	2.0	2.2	1.5	1.7	2.0	2.3	2.7
DG B3	24	31	39	45	53	58	64	72	17	22	29	34	41	45	51	59	1.1	1.3	1.5	1.6	1.8	1.3	1.5	1.6	1.8	2.0
DG C1	20	26	31	35	41	44	48	52	13	16	20	22	26	28	31	33	1.3	1.5	1.8	2.3	2.7	1.6	1.9	2.4	3.1	4.1
DG C2	23	29	36	42	49	53	59	66	15	20	25	30	35	39	44	50	1.2	1.3	1.6	1.8	2.0	1.4	1.6	1.8	2.1	2.4
DG C3	26	34	43	49	58	63	70	78	19	25	32	38	45	50	57	65	1.1	1.2	1.4	1.6	1.7	1.2	1.4	1.5	1.7	1.9
SS A1	25	30	36	40	45	49	53	59	20	24	28	31	35	38	42	46	1.3	1.5	1.7	2.0	2.2	1.5	1.8	2.1	2.4	2.7
SS A2	30	37	45	50	57	62	68	75	24	30	36	41	47	51	56	63	1.1	1.3	1.5	1.7	1.8	1.2	1.4	1.6	1.9	2.1
SS A3	23	28	33	36	41	44	48	53	19	22	26	28	31	34	36	39	1.3	1.6	1.9	2.2	2.4	1.6	1.9	2.3	2.8	3.3
SS A4	30	37	44	49	56	60	66	73	24	29	35	40	46	49	54	60	1.1	1.3	1.5	1.7	1.9	1.2	1.5	1.7	1.9	2.1
SS A5	24	29	34	37	42	45	49	54	20	23	27	29	33	35	38	41	1.3	1.6	1.9	2.1	2.4	1.6	1.9	2.3	2.7	3.1
SS A6	30	37	44	50	57	61	67	74	24	30	36	40	46	50	55	62	1.1	1.3	1.5	1.7	1.8	1.2	1.4	1.7	1.9	2.1
SS A7	22	26	30	33	37	39	43	47	18	21	24	26	28	30	32	34	1.5	1.8	2.1	2.4	2.7	1.8	2.2	2.6	3.2	3.9
SS A8	29	36	43	48	54	59	64	71	23	29	34	38	44	48	52	58	1.1	1.3	1.5	1.7	1.9	1.2	1.5	1.7	2.0	2.2
SS B1	22	27	32	36	41	44	49	55	17	20	24	27	31	34	37	41	1.3	1.6	1.8	2.1	2.3	1.6	1.9	2.2	2.5	2.8
SS B2	27	34	41	46	54	58	64	71	21	27	33	37	44	48	53	60	1.1	1.3	1.5	1.7	1.9	1.2	1.5	1.7	1.9	2.1
SS B3	20	25	30	33	38	41	45	51	16	19	22	25	28	31	33	37	1.4	1.6	1.9	2.2	2.4	1.6	2.0	2.4	2.8	3.2
SS B4	26	33	40	45	52	57	62	69	21	26	32	36	42	46	51	58	1.1	1.3	1.5	1.7	1.9	1.3	1.5	1.7	2.0	2.2
SS B5	21	24	29	32	36	39	43	48	16	19	22	25	28	30	32	36	1.5	1.7	2.0	2.3	2.5	1.7	2.1	2.4	2.8	3.2
SS B6	26	33	40	45	52	57	63	70	20	26	32	36	42	46	51	58	1.1	1.3	1.5	1.7	1.9	1.3	1.5	1.7	1.9	2.1
SS B7	19	22	26	29	33	35	39	43	15	18	20	22	25	26	28	31	1.6	1.9	2.2	2.5	2.7	1.9	2.3	2.7	3.2	3.8
SS B8	25	32	39	44	50	55	60	67	20	25	30	34	40	44	49	55	1.1	1.4	1.6	1.8	1.9	1.3	1.5	1.8	2.0	2.2
SS C1	20	26	32	36	42	46	51	57	13	17	21	24	28	31	34	38	1.3	1.5	1.7	2.0	2.2	1.6	1.8	2.2	2.5	2.9
SS C2	24	31	39	44	52	57	63	70	16	21	27	32	38	42	47	54	1.1	1.3	1.5	1.7	1.9	1.3	1.5	1.7	2.0	2.2
SS C3	18	22	27	31	36	39	44	49	13	16	19	21	25	27	30	34	1.4	1.7	1.9	2.1	2.4	1.8	2.1	2.4	2.7	3.0
SS C4	22	28	35	40	47	52	58	65	15	20	25	29	35	38	43	50	1.2	1.4	1.6	1.8	2.0	1.4	1.6	1.8	2.1	2.3
SS D1	28	34	41	45	51	55	60	67	22	27	32	35	40	43	46	51	1.1	1.4	1.6	1.8	2.0	1.3	1.6	1.9	2.3	2.6
SS D2	34	42	51	57	64	69	75	82	26	33	41	46	53	58	63	70	1.0	1.2	1.4	1.6	1.7	1.1	1.3	1.5	1.7	1.9
SS D3	27	32	38	42	47	51	56	61	21	25	30	33	37	40	43	47	1.2	1.5	1.7	1.9	2.1	1.4	1.7	2.0	2.4	2.8
SS D4	33	41	49	55	63	67	73	80	26	32	40	45	52	56	61	68	1.0	1.2	1.4	1.6	1.7	1.1	1.3	1.5	1.8	2.0
SS E1	25	30	36	40	46	50	54	60	18	22	27	30	34	36	39	43	1.2	1.5	1.7	2.0	2.2	1.4	1.8	2.1	2.6	3.0
SS E2	29	37	45	51	58	63	69	76	22	28	35	40	47	51	56	62	1.0	1.2	1.4	1.6	1.8	1.2	1.4	1.6	1.9	2.1
SS E3	22	27	32	36	40	44	48	53	17	21	24	27	30	33	36	39	1.3	1.6	1.9	2.1	2.3	1.6	1.9	2.3	2.7	3.1
SS E4	28	35	42	48	55	60	65	72	21	27	33	38	44	48	53	59	1.1	1.3	1.5	1.7	1.9	1.2	1.5	1.7	1.9	2.1

REDMOND, OREGON

TYPE	SOLAR SAVINGS FRACTION (%) BASE TEMP: 55 F (3769 DD) LCR 100	70	50	40	30	25	20	15	65 F (6647 DD) 100	70	50	40	30	25	20	15	CONSERVATION FACTOR 55 F 100	50	30	20	15	65 F 100	50	30	20	15
WW A1	22	26	31	35	40	44	49	55	17	20	23	26	30	32	36	41	1.4	1.6	1.8	2.0	2.2	1.7	2.0	2.2	2.4	2.6
WW A2	24	31	38	43	51	55	61	69	17	22	27	32	37	42	47	54	1.2	1.3	1.5	1.7	1.9	1.4	1.6	1.8	2.0	2.1
WW A3	25	32	40	46	54	59	66	74	17	23	29	34	40	45	51	59	1.1	1.3	1.4	1.6	1.8	1.3	1.5	1.7	1.9	2.0
WW A4	25	33	42	48	56	62	68	76	17	23	30	35	42	47	53	61	1.1	1.2	1.4	1.6	1.8	1.2	1.4	1.6	1.8	2.0
WW A5	26	34	43	50	58	64	70	78	17	23	31	36	44	49	55	64	1.0	1.2	1.4	1.6	1.7	1.2	1.4	1.6	1.8	1.9
WW A6	26	35	44	51	60	65	72	79	17	24	31	37	45	50	57	65	1.0	1.2	1.4	1.5	1.7	1.2	1.4	1.5	1.7	1.9
WW B1	20	25	30	35	40	44	49	56	13	17	20	23	27	30	33	38	1.3	1.5	1.7	2.0	2.2	1.7	1.9	2.2	2.5	2.8
WW B2	26	34	44	50	59	64	71	79	18	24	32	37	45	51	57	66	1.0	1.2	1.4	1.6	1.7	1.2	1.4	1.5	1.7	1.9
WW B3	31	40	50	57	66	72	78	85	21	29	38	44	53	59	65	74	1.0	1.1	1.3	1.5	1.6	1.1	1.2	1.4	1.6	1.8
WW B4	31	41	52	59	69	74	81	88	21	30	40	47	57	63	70	78	0.9	1.1	1.3	1.4	1.6	1.0	1.2	1.4	1.5	1.7
WW B5	30	40	51	59	68	74	81	88	20	29	39	47	56	63	70	79	0.9	1.1	1.3	1.4	1.6	1.0	1.2	1.3	1.5	1.7
WW C1	32	42	52	59	68	74	80	87	23	31	40	47	56	61	68	76	0.9	1.1	1.3	1.5	1.6	1.1	1.2	1.4	1.6	1.7
WW C2	31	41	51	58	67	73	79	87	21	30	39	46	55	61	68	77	0.9	1.1	1.3	1.5	1.6	1.1	1.2	1.4	1.6	1.7
WW C3	36	47	59	67	76	81	87	93	26	36	48	55	66	72	79	86	0.9	1.0	1.2	1.4	1.5	0.9	1.1	1.3	1.4	1.6
WW C4	33	44	56	64	73	79	85	91	23	33	44	52	62	69	76	84	0.9	1.0	1.2	1.4	1.5	1.0	1.1	1.3	1.5	1.6
TW A1	22	26	31	35	40	44	49	55	17	20	23	26	30	32	36	41	1.4	1.6	1.8	2.0	2.1	1.7	1.9	2.2	2.4	2.6
TW A2	23	28	35	40	47	51	57	65	16	21	25	29	35	38	43	50	1.2	1.4	1.6	1.8	1.9	1.5	1.7	1.8	2.0	2.2
TW A3	23	29	36	42	49	54	60	68	16	21	26	30	37	41	46	54	1.2	1.3	1.5	1.7	1.9	1.4	1.6	1.7	1.9	2.1
TW A4	22	29	36	42	50	55	61	70	15	20	26	30	37	42	47	55	1.1	1.3	1.5	1.7	1.8	1.3	1.5	1.7	1.9	2.1
TW B1	21	26	31	35	41	45	50	57	16	19	23	26	30	33	37	43	1.4	1.6	1.7	1.9	2.1	1.7	1.9	2.1	2.3	2.4
TW B2	21	27	34	39	46	50	56	64	15	19	24	28	34	38	43	50	1.2	1.4	1.6	1.8	1.9	1.5	1.6	1.8	2.0	2.2
TW B3	21	27	34	39	46	51	57	65	14	19	24	28	34	38	44	51	1.2	1.4	1.6	1.7	1.9	1.4	1.6	1.8	2.0	2.1
TW B4	20	25	32	37	44	49	55	63	14	18	23	27	33	37	42	50	1.2	1.4	1.6	1.8	1.9	1.5	1.6	1.8	2.0	2.2
TW C1	20	25	30	35	41	45	51	58	15	18	22	26	30	34	38	45	1.4	1.5	1.7	1.9	2.0	1.6	1.8	2.0	2.2	2.3
TW C2	19	24	31	35	42	46	52	60	14	18	22	26	31	35	40	47	1.3	1.5	1.6	1.8	2.0	1.5	1.7	1.9	2.1	2.2
TW C3	18	23	29	34	40	45	50	58	13	17	21	25	30	34	38	45	1.3	1.5	1.7	1.9	2.0	1.6	1.8	1.9	2.1	2.3
TW C4	17	22	27	31	36	41	46	53	13	16	20	23	28	31	35	42	1.5	1.6	1.8	1.9	2.1	1.7	1.9	2.1	2.2	2.4
TW D1	17	21	26	30	36	40	45	51	11	14	17	20	24	27	30	35	1.4	1.6	1.8	2.0	2.2	1.8	2.0	2.3	2.5	2.8
TW D2	25	32	40	46	54	60	66	74	17	23	30	35	42	47	53	61	1.1	1.3	1.4	1.6	1.8	1.3	1.4	1.6	1.8	1.9
TW D3	26	35	43	50	58	64	70	78	18	25	32	38	45	51	57	65	1.0	1.2	1.4	1.6	1.7	1.2	1.4	1.5	1.7	1.9
TW D4	28	37	47	54	63	69	76	83	20	28	36	42	51	57	64	73	1.0	1.1	1.3	1.5	1.6	1.1	1.3	1.4	1.6	1.8
TW D5	28	37	47	54	64	69	76	84	19	27	36	43	52	58	65	74	1.0	1.1	1.3	1.5	1.6	1.1	1.2	1.4	1.6	1.7
TW E1	30	39	48	55	63	69	75	82	21	29	37	43	51	56	63	71	1.0	1.2	1.3	1.5	1.7	1.1	1.3	1.5	1.6	1.8
TW E2	29	38	48	55	64	69	76	83	21	28	37	43	51	57	64	72	1.0	1.1	1.3	1.5	1.6	1.1	1.3	1.4	1.6	1.8
TW E3	35	45	56	64	73	78	84	90	25	35	45	52	62	68	75	83	0.9	1.1	1.2	1.4	1.6	1.0	1.1	1.3	1.5	1.6
TW E4	32	43	54	61	71	76	82	89	23	32	42	50	59	66	73	81	0.9	1.1	1.2	1.4	1.6	1.0	1.2	1.3	1.5	1.6
TW F1	20	25	30	34	39	43	48	55	15	18	22	24	29	31	35	41	1.4	1.6	1.8	2.0	2.1	1.7	1.9	2.1	2.3	2.5
TW F2	20	26	32	37	44	49	55	63	14	18	23	27	32	36	41	48	1.3	1.4	1.6	1.8	2.0	1.5	1.7	1.9	2.1	2.2
TW F3	19	26	33	38	46	51	57	65	13	17	23	27	33	37	42	50	1.2	1.4	1.6	1.7	1.9	1.4	1.6	1.8	2.0	2.2
TW F4	17	24	31	37	44	49	56	64	11	16	21	25	32	36	42	49	1.2	1.4	1.6	1.7	1.9	1.4	1.6	1.8	2.0	2.1
TW G1	18	23	28	32	38	43	48	55	13	16	20	23	28	31	35	41	1.4	1.6	1.7	1.9	2.1	1.7	1.9	2.1	2.3	2.4
TW G2	17	23	29	34	41	46	52	60	11	15	20	24	29	33	38	45	1.3	1.5	1.6	1.8	2.0	1.5	1.7	1.9	2.1	2.3
TW G3	15	21	28	33	40	45	51	59	10	14	19	23	28	32	37	45	1.3	1.5	1.6	1.8	2.0	1.5	1.7	1.9	2.1	2.2
TW G4	13	18	24	29	35	40	46	55	7	11	16	19	25	28	34	41	1.4	1.5	1.7	1.9	2.0	1.6	1.8	2.0	2.1	2.3
TW H1	15	20	25	30	36	40	46	53	10	13	17	21	25	28	33	39	1.4	1.6	1.7	1.9	2.1	1.7	1.9	2.1	2.3	2.4
TW H2	13	17	23	28	34	38	44	52	8	11	15	19	24	27	32	39	1.4	1.6	1.7	1.9	2.1	1.7	1.8	2.0	2.2	2.4
TW H3	11	15	20	24	30	35	40	48	6	9	13	16	21	24	29	35	1.5	1.6	1.8	2.0	2.1	1.7	1.9	2.1	2.3	2.4
TW H4	8	12	16	19	25	28	34	41	4	7	10	13	17	20	24	29	1.6	1.8	2.0	2.1	2.3	2.0	2.1	2.3	2.5	2.6
TW I1	14	19	24	27	33	37	41	48	9	12	15	18	21	24	27	32	1.5	1.7	1.0	2.1	2.3	1.8	2.1	2.3	2.6	2.9
TW I2	21	29	37	43	51	56	63	71	14	20	26	31	38	43	49	58	1.1	1.3	1.5	1.6	1.8	1.3	1.5	1.6	1.8	2.0
TW I3	23	31	40	46	54	60	67	75	15	22	29	34	42	47	54	62	1.1	1.2	1.4	1.6	1.7	1.2	1.4	1.6	1.8	1.9
TW I4	25	34	44	51	60	66	73	81	17	24	33	39	48	54	61	70	1.0	1.2	1.3	1.5	1.7	1.1	1.3	1.5	1.6	1.8
TW I5	25	35	45	52	61	67	74	82	17	25	34	40	50	56	63	72	1.0	1.1	1.3	1.5	1.6	1.1	1.3	1.4	1.6	1.7
TW J1	27	36	46	52	61	67	73	81	19	26	34	40	49	54	61	69	1.0	1.2	1.3	1.5	1.7	1.1	1.3	1.5	1.7	1.8
TW J2	26	35	45	52	61	67	74	81	18	25	34	40	49	54	61	70	1.0	1.2	1.3	1.5	1.7	1.1	1.3	1.5	1.6	1.8
TW J3	31	42	53	61	70	76	82	89	22	32	42	49	59	65	73	81	0.9	1.1	1.2	1.4	1.6	1.0	1.2	1.3	1.6	1.0
TW J4	29	40	51	58	68	74	81	88	20	29	40	47	57	63	70	79	0.9	1.1	1.3	1.4	1.6	1.0	1.2	1.3	1.5	1.7
DG A1	18	23	28	32	37	40	43	47	11	15	18	21	24	26	28	30	1.3	1.6	2.0	2.4	3.0	1.6	2.0	2.5	3.3	4.5
DG A2	20	20	33	39	46	50	55	62	14	18	24	28	33	37	42	48	1.2	1.4	1.6	1.9	2.1	1.4	1.6	1.8	2.1	2.4
DG A3	24	32	41	47	56	62	68	76	17	24	31	36	44	50	56	65	1.1	1.2	1.4	1.6	1.8	1.3	1.4	1.5	1.7	1.9
DG B1	18	24	30	35	41	45	49	55	11	15	19	22	26	29	32	36	1.3	1.5	1.7	2.1	2.4	1.6	1.8	2.2	2.6	3.2
DG B2	21	28	35	41	49	54	61	68	14	19	25	29	36	40	46	53	1.2	1.3	1.5	1.7	1.9	1.4	1.5	1.7	1.9	2.2
DG B3	25	33	42	49	59	65	72	80	18	24	32	38	47	53	60	69	1.1	1.2	1.3	1.5	1.7	1.2	1.3	1.5	1.6	1.8
DG C1	22	28	36	41	49	53	58	65	14	18	24	27	33	37	41	46	1.2	1.3	1.6	1.8	2.1	1.4	1.6	1.9	2.2	2.6
DG C2	24	32	40	47	56	61	68	76	17	22	29	34	42	47	53	61	1.1	1.2	1.4	1.6	1.8	1.3	1.4	1.6	1.8	2.0
DG C3	28	37	47	55	65	71	78	86	20	27	36	43	52	59	66	76	1.0	1.1	1.3	1.4	1.6	1.2	1.3	1.4	1.6	1.7
SS A1	27	33	40	45	51	56	61	68	22	27	32	36	42	46	50	57	1.2	1.4	1.6	1.8	1.9	1.3	1.5	1.8	2.0	2.2
SS A2	33	41	50	56	64	69	75	82	27	34	42	47	55	60	65	73	1.0	1.2	1.4	1.5	1.7	1.1	1.3	1.5	1.7	1.8
SS A3	25	31	37	41	47	51	56	62	20	25	29	33	38	41	45	50	1.2	1.5	1.7	1.9	2.1	1.4	1.6	1.9	2.2	2.4
SS A4	33	40	49	54	62	67	73	80	26	33	41	46	53	58	63	70	1.0	1.2	1.4	1.6	1.7	1.1	1.3	1.5	1.7	1.9
SS A5	26	31	37	41	47	51	56	63	21	26	30	34	38	42	46	51	1.3	1.5	1.7	1.9	2.1	1.4	1.7	1.9	2.2	2.4
SS A6	33	41	50	55	63	68	74	81	27	34	41	47	54	59	65	72	1.0	1.2	1.4	1.6	1.7	1.1	1.3	1.5	1.7	1.9
SS A7	24	28	33	36	41	45	49	55	19	23	27	29	33	36	39	44	1.4	1.6	1.8	2.1	2.3	1.6	1.9	2.1	2.5	2.7
SS A8	32	40	48	53	61	66	71	78	26	32	40	45	52	56	62	69	1.0	1.2	1.4	1.6	1.8	1.1	1.3	1.5	1.7	1.9
SS B1	23	29	35	40	46	50	56	63	18	23	28	31	37	40	45	51	1.2	1.4	1.6	1.8	2.0	1.4	1.6	1.9	2.1	2.3
SS B2	29	37	45	51	59	65	71	78	23	30	37	43	50	55	61	69	1.0	1.2	1.4	1.6	1.7	1.1	1.3	1.5	1.7	1.9
SS B3	22	27	33	37	43	47	52	59	17	21	26	29	34	37	41	47	1.3	1.5	1.7	1.9	2.1	1.5	1.7	2.0	2.2	2.4
SS B4	29	36	44	50	58	63	69	77	23	29	36	41	49	53	59	67	1.1	1.2	1.4	1.6	1.8	1.2	1.4	1.5	1.7	1.9
SS B5	22	26	31	35	41	44	49	56	17	21	25	28	32	35	39	44	1.4	1.6	1.8	2.0	2.2	1.6	1.8	2.1	2.3	2.5
SS B6	29	36	45	50	58	63	70	77	22	29	36	42	49	54	60	67	1.1	1.2	1.4	1.6	1.7	1.2	1.3	1.5	1.7	1.9
SS B7	20	24	28	32	36	40	44	50	16	19	22	25	28	31	34	39	1.5	1.7	1.9	2.1	2.3	1.7	2.0	2.2	2.5	2.8
SS B8	27	35	43	48	56	61	67	75	22	28	35	40	47	51	57	65	1.1	1.3	1.4	1.6	1.8	1.2	1.4	1.6	1.8	1.9
SS C1	22	29	36	41	49	54	60	68	15	20	25	29	35	39	45	52	1.2	1.3	1.5	1.7	1.9	1.4	1.6	1.8	2.0	2.2
SS C2	26	35	44	50	59	64	71	79	18	25	32	38	46	51	58	66	1.0	1.2	1.4	1.6	1.7	1.2	1.4	1.5	1.7	1.9
SS C3	19	24	30	34	40	45	50	58	14	17	21	25	29	32	37	43	1.3	1.5	1.7	1.9	2.1	1.6	1.8	2.0	2.2	2.4
SS C4	24	31	39	45	53	59	65	73	17	22	29	34	41	46	52	60	1.1	1.3	1.4	1.6	1.8	1.3	1.4	1.6	1.8	1.9
SS D1	31	39	46	52	59	64	69	76	25	31	37	42	48	53	58	64	1.0	1.2	1.4	1.6	1.8	1.2	1.4	1.6	1.8	2.0
SS D2	38	47	57	63	71	76	82	88	30	39	48	54	62	67	73	80	0.9	1.1	1.3	1.5	1.6	1.0	1.2	1.4	1.6	1.7
SS D3	29	35	42	47	54	58	64	70	23	29	34	38	44	48	53	59	1.1	1.3	1.5	1.7	1.9	1.3	1.5	1.7	2.0	2.2
SS D4	36	46	55	61	70	75	80	87	29	37	46	52	60	65	71	78	0.9	1.1	1.3	1.5	1.6	1.0	1.2	1.4	1.6	1.7
SS E1	27	34	41	46	53	58	64	71	20	26	32	36	42	46	51	57	1.1	1.3	1.5	1.7	1.9	1.3	1.5	1.7	2.0	2.2
SS E2	33	42	51	57	66	71	77	84	25	33	41	47	55	60	66	74	1.0	1.1	1.3	1.5	1.7	1.1	1.3	1.4	1.6	1.8
SS E3	24	29	35	40	46	50	55	62	19	23	28	31	36	40	44	50	1.3	1.5	1.7	1.9	2.0	1.4	1.7	1.9	2.1	2.4
SS E4	31	39	47	54	62	67	73	80	24	31	38	44	51	56	62	70	1.0	1.2	1.4	1.6	1.7	1.1	1.3	1.5	1.7	1.9

SALEM, OREGON

TYPE	SOLAR SAVINGS FRACTION (%) BASE TEMP: 55 F (2252 DD)								65 F (4854 DD)								CONSERVATION FACTOR 55 F					65 F				
LCR	100	70	50	40	30	25	20	15	100	70	50	40	30	25	20	15	100	50	30	20	15	100	50	30	20	15
WW A1	22	26	31	34	39	42	46	52	17	19	22	24	27	28	31	34	1.4	1.7	1.9	2.2	2.4	1.9	2.2	2.5	2.9	3.3
WW A2	24	30	37	41	48	52	57	64	16	20	24	28	32	36	40	45	1.2	1.4	1.6	1.8	2.0	1.5	1.7	2.0	2.3	2.6
WW A3	25	32	39	44	51	56	61	68	15	20	25	29	35	38	43	48	1.1	1.3	1.5	1.8	1.9	1.4	1.6	1.9	2.1	2.4
WW A4	25	32	40	45	53	57	63	70	15	20	26	30	36	40	44	51	1.1	1.3	1.5	1.7	1.9	1.3	1.6	1.8	2.1	2.3
WW A5	25	33	41	47	55	59	65	73	15	20	26	31	37	41	46	53	1.1	1.3	1.5	1.7	1.8	1.3	1.5	1.8	2.0	2.3
WW A6	25	33	42	48	56	61	66	74	15	20	27	31	38	42	47	54	1.0	1.2	1.5	1.7	1.8	1.3	1.5	1.7	2.0	2.2
WW B1	20	24	29	33	38	41	45	50	12	15	18	20	22	24	26	28	1.4	1.6	1.9	2.2	2.5	1.9	2.3	2.8	3.5	4.7
WW B2	26	33	42	48	55	60	66	74	16	22	28	33	39	44	49	57	1.1	1.3	1.4	1.6	1.8	1.3	1.5	1.7	1.9	2.1
WW B3	30	39	48	54	62	67	73	80	19	26	34	39	46	51	57	64	1.0	1.2	1.4	1.6	1.7	1.2	1.4	1.6	1.8	2.0
WW B4	30	40	50	57	65	71	77	84	19	27	36	42	50	56	63	71	0.9	1.1	1.3	1.5	1.7	1.1	1.3	1.5	1.7	1.8
WW B5	29	39	49	56	65	71	77	84	18	26	35	42	51	56	63	72	0.9	1.1	1.3	1.5	1.6	1.1	1.3	1.4	1.6	1.8
WW C1	32	41	50	56	65	70	76	82	21	28	36	41	49	54	60	68	1.0	1.2	1.3	1.5	1.7	1.1	1.3	1.5	1.7	1.9
WW C2	30	39	49	55	64	69	75	83	19	27	35	41	49	54	60	68	1.0	1.1	1.3	1.5	1.7	1.1	1.3	1.5	1.7	1.9
WW C3	35	46	57	64	73	78	84	90	23	33	43	50	60	65	72	80	0.9	1.1	1.2	1.4	1.6	1.0	1.2	1.3	1.5	1.7
WW C4	32	43	53	61	70	75	82	88	21	30	40	47	57	63	70	78	0.9	1.1	1.3	1.4	1.6	1.0	1.2	1.4	1.5	1.7
TW A1	22	26	30	34	39	42	46	52	16	19	22	24	27	29	31	35	1.4	1.7	1.9	2.1	2.3	1.9	2.2	2.5	2.8	3.1
TW A2	23	28	34	38	45	49	54	60	16	19	23	26	30	33	37	42	1.2	1.5	1.7	1.9	2.1	1.6	1.8	2.1	2.3	2.6
TW A3	23	29	35	40	47	51	57	64	15	19	23	27	32	35	40	45	1.2	1.4	1.6	1.8	2.0	1.5	1.7	2.0	2.2	2.5
TW A4	22	28	35	40	47	52	58	65	14	18	23	27	32	36	40	47	1.2	1.4	1.6	1.8	2.0	1.4	1.7	1.9	2.1	2.4
TW B1	21	26	31	34	40	43	48	54	16	18	21	24	27	30	33	37	1.4	1.6	1.8	2.1	2.2	1.8	2.1	2.3	2.6	2.9
TW B2	21	27	33	37	44	48	53	60	14	18	22	25	30	33	37	43	1.3	1.5	1.7	1.9	2.1	1.6	1.8	2.0	2.3	2.5
TW B3	21	27	33	38	44	49	54	61	14	18	22	25	30	34	38	44	1.2	1.4	1.6	1.8	2.0	1.5	1.8	2.0	2.2	2.4
TW B4	19	25	31	36	42	47	52	60	13	17	21	24	29	33	37	43	1.3	1.5	1.7	1.9	2.0	1.6	1.8	2.0	2.2	2.4
TW C1	20	25	30	34	39	43	48	55	14	17	21	24	27	30	34	39	1.4	1.6	1.8	2.0	2.2	1.7	2.0	2.2	2.4	2.7
TW C2	19	24	30	34	40	45	50	57	13	17	21	24	28	31	35	41	1.3	1.5	1.7	1.9	2.1	1.6	1.9	2.1	2.3	2.5
TW C3	18	23	29	33	39	43	48	55	13	16	20	23	27	30	34	40	1.4	1.5	1.8	2.0	2.1	1.7	1.9	2.1	2.3	2.5
TW C4	17	22	27	30	36	39	44	51	13	16	19	22	25	28	32	37	1.5	1.6	1.8	2.0	2.2	1.8	2.0	2.2	2.4	2.6
TW D1	17	21	26	29	34	37	42	47	10	13	16	17	20	22	24	27	1.4	1.7	2.0	2.2	2.5	1.9	2.3	2.7	3.2	3.8
TW D2	24	31	39	44	52	56	62	70	16	21	27	31	37	41	46	53	1.1	1.3	1.5	1.7	1.9	1.4	1.6	1.8	2.0	2.2
TW D3	26	34	42	47	55	60	66	73	17	22	29	33	40	44	50	57	1.1	1.3	1.5	1.7	1.8	1.3	1.5	1.7	1.9	2.1
TW D4	28	36	45	52	60	66	72	79	18	25	32	38	46	51	57	65	1.0	1.2	1.4	1.6	1.7	1.2	1.4	1.5	1.7	1.9
TW D5	27	36	45	52	61	66	73	80	18	25	33	39	47	52	59	67	1.0	1.2	1.4	1.5	1.7	1.2	1.3	1.5	1.7	1.8
TW E1	29	37	46	52	60	65	71	78	20	26	33	38	45	50	55	63	1.0	1.2	1.4	1.6	1.8	1.2	1.4	1.6	1.8	2.0
TW E2	29	37	46	52	61	66	72	79	19	26	33	38	46	51	57	65	1.0	1.2	1.4	1.6	1.7	1.2	1.4	1.6	1.8	1.9
TW E3	34	44	54	61	69	75	81	87	23	31	40	47	56	61	68	76	0.9	1.1	1.3	1.5	1.6	1.0	1.2	1.4	1.6	1.8
TW E4	31	41	52	59	67	73	79	86	21	29	38	45	54	60	66	75	0.9	1.1	1.3	1.5	1.6	1.1	1.2	1.4	1.6	1.8
TW F1	20	24	29	33	38	41	46	52	15	17	20	22	25	28	31	34	1.4	1.6	1.9	2.1	2.3	1.8	2.1	2.4	2.7	3.0
TW F2	20	26	32	36	42	46	51	58	13	17	21	24	28	31	35	40	1.3	1.5	1.7	1.9	2.1	1.6	1.9	2.1	2.4	2.6
TW F3	19	25	32	37	43	48	53	60	12	16	20	24	28	32	36	42	1.2	1.4	1.6	1.9	2.0	1.5	1.8	2.0	2.3	2.5
TW F4	17	23	30	35	42	47	52	60	10	14	19	22	27	31	35	42	1.2	1.4	1.6	1.8	2.0	1.5	1.7	2.0	2.2	2.4
TW G1	18	23	28	32	37	41	45	52	13	15	19	21	25	27	30	35	1.4	1.6	1.8	2.1	2.2	1.8	2.1	2.3	2.6	2.9
TW G2	17	23	29	33	39	43	49	56	11	14	18	21	26	29	33	38	1.3	1.5	1.7	1.9	2.1	1.6	1.9	2.1	2.4	2.6
TW G3	15	21	27	32	38	42	48	55	9	13	17	20	25	28	32	38	1.3	1.5	1.7	1.9	2.1	1.6	1.9	2.1	2.3	2.5
TW G4	13	18	24	28	34	38	44	51	7	10	14	17	22	25	29	35	1.4	1.6	1.8	2.0	2.1	1.7	1.9	2.1	2.4	2.6
TW H1	15	20	25	29	35	38	43	50	10	13	16	19	22	25	28	33	1.4	1.6	1.8	2.0	2.2	1.8	2.0	2.3	2.5	2.8
TW H2	13	17	23	27	33	37	42	49	7	10	14	17	21	24	28	33	1.4	1.6	1.8	2.0	2.2	1.8	2.0	2.2	2.5	2.7
TW H3	11	15	20	24	30	34	39	46	6	9	12	15	19	22	25	31	1.5	1.7	1.9	2.1	2.2	1.8	2.1	2.3	2.5	2.7
TW H4	8	12	16	19	24	28	33	39	4	7	10	12	15	18	21	26	1.6	1.8	2.0	2.2	2.4	2.0	2.2	2.5	2.7	2.9
TW I1	14	19	23	27	31	34	39	44	9	11	13	15	18	19	22	24	1.5	1.7	2.0	2.3	2.5	2.0	2.4	2.8	3.3	3.9
TW I2	21	28	35	41	48	53	59	67	13	18	23	28	34	38	43	50	1.1	1.3	1.5	1.7	1.9	1.4	1.6	1.8	2.0	2.2
TW I3	23	30	38	44	52	57	63	71	14	20	26	30	37	41	47	54	1.1	1.3	1.5	1.7	1.8	1.3	1.5	1.7	1.9	2.1
TW I4	25	33	42	49	57	63	69	77	15	22	29	35	43	48	54	63	1.0	1.2	1.4	1.6	1.7	1.2	1.4	1.6	1.8	1.9
TW I5	25	34	43	50	59	64	71	79	16	23	30	36	44	50	57	65	1.0	1.2	1.4	1.6	1.7	1.2	1.3	1.5	1.7	1.9
TW J1	27	35	44	50	58	63	69	77	17	24	31	36	43	48	54	61	1.0	1.2	1.4	1.6	1.8	1.2	1.4	1.6	1.8	2.0
TW J2	26	34	43	50	58	63	70	77	16	23	30	36	43	48	55	63	1.0	1.2	1.4	1.6	1.7	1.2	1.4	1.6	1.8	1.9
TW J3	31	41	51	58	67	73	79	86	20	29	38	44	53	59	66	74	0.9	1.1	1.3	1.5	1.6	1.1	1.2	1.4	1.6	1.8
TW J4	29	39	49	56	65	71	77	84	18	27	36	42	51	57	64	73	0.9	1.1	1.3	1.5	1.6	1.1	1.2	1.4	1.6	1.8
DG A1	18	23	28	31	35	38	40	43	10	13	16	18	20	21	23	23	1.3	1.7	2.1	2.8	3.5	1.8	2.3	3.1	4.7	-
DG A2	20	26	33	37	44	48	52	58	13	17	22	25	30	33	37	41	1.2	1.4	1.7	2.0	2.3	1.5	1.8	2.0	2.4	2.7
DG A3	24	32	40	46	53	59	65	72	17	22	28	33	40	45	51	58	1.1	1.3	1.5	1.7	1.9	1.3	1.5	1.6	1.9	2.1
DG B1	19	24	29	33	38	41	45	49	11	13	17	19	22	24	26	28	1.3	1.6	1.9	2.4	2.8	1.8	2.1	2.6	3.5	-
DG B2	21	27	34	40	47	51	56	63	13	18	22	26	32	35	40	45	1.2	1.4	1.6	1.8	2.1	1.5	1.7	1.9	2.2	2.5
DG B3	25	33	41	48	56	62	68	76	17	23	29	35	42	47	54	62	1.1	1.2	1.4	1.6	1.8	1.3	1.4	1.6	1.8	2.0
DG C1	22	28	35	39	45	49	53	58	13	16	21	24	28	30	33	37	1.2	1.4	1.7	2.1	2.4	1.6	1.8	2.2	2.8	3.5
DG C2	24	31	39	45	53	57	63	70	16	21	26	31	37	41	46	53	1.1	1.3	1.5	1.7	1.9	1.4	1.5	1.7	2.0	2.3
DG C3	28	36	46	53	62	67	74	82	19	26	33	39	47	53	59	68	1.0	1.2	1.3	1.5	1.7	1.2	1.3	1.5	1.7	1.9
SS A1	27	33	39	43	49	53	58	64	21	25	29	33	37	40	44	49	1.2	1.4	1.7	1.9	2.1	1.4	1.7	2.0	2.3	2.5
SS A2	33	40	48	54	61	66	71	78	25	31	38	43	50	54	59	66	1.0	1.2	1.4	1.6	1.8	1.2	1.4	1.6	1.8	2.0
SS A3	25	30	36	40	45	48	53	58	19	23	27	30	33	36	39	43	1.3	1.5	1.8	2.0	2.2	1.5	1.8	2.2	2.6	3.0
SS A4	32	39	47	52	60	64	70	76	25	31	37	42	48	52	57	63	1.0	1.2	1.4	1.6	1.8	1.2	1.4	1.6	1.9	2.1
SS A5	26	31	36	40	45	49	53	59	21	24	28	31	35	37	40	44	1.3	1.5	1.8	2.0	2.2	1.5	1.8	2.1	2.5	2.8
SS A6	32	40	48	53	60	65	71	77	25	31	37	42	49	53	58	65	1.0	1.2	1.4	1.6	1.8	1.2	1.4	1.6	1.8	2.0
SS A7	24	28	32	36	40	43	47	52	19	22	25	27	30	32	34	37	1.4	1.7	2.0	2.2	2.5	1.7	2.1	2.5	2.9	3.4
SS A8	31	39	46	51	58	63	68	75	24	30	36	40	46	50	55	61	1.1	1.3	1.5	1.7	1.8	1.2	1.4	1.7	1.9	2.1
SS B1	23	29	34	39	44	48	53	59	17	21	25	28	33	36	39	44	1.3	1.5	1.7	1.9	2.1	1.5	1.8	2.1	2.4	2.6
SS B2	29	36	44	50	57	62	68	75	22	28	34	39	46	50	55	62	1.1	1.3	1.5	1.6	1.8	1.2	1.4	1.6	1.8	2.0
SS B3	22	27	32	36	41	45	49	55	16	20	23	26	30	33	36	40	1.3	1.6	1.8	2.0	2.3	1.6	1.9	2.2	2.6	2.9
SS B4	28	36	43	48	56	60	66	73	21	27	33	38	44	48	54	60	1.1	1.3	1.5	1.7	1.8	1.2	1.5	1.7	1.9	2.1
SS B5	22	26	31	34	39	43	47	53	17	20	23	26	29	31	34	38	1.4	1.6	1.9	2.1	2.3	1.7	2.0	2.3	2.6	3.0
SS B6	28	36	43	49	56	61	67	74	21	27	33	38	44	49	54	61	1.1	1.3	1.5	1.7	1.8	1.2	1.4	1.7	1.9	2.1
SS B7	20	24	28	31	36	38	42	47	15	18	21	23	26	28	30	33	1.5	1.7	2.0	2.3	2.5	1.8	2.2	2.5	3.0	3.4
SS B8	27	34	42	47	54	58	64	71	20	26	32	36	42	46	51	58	1.1	1.3	1.5	1.7	1.9	1.3	1.5	1.7	1.9	2.1
SS C1	22	28	35	39	46	50	56	62	14	18	22	25	30	33	37	42	1.2	1.4	1.6	1.9	2.0	1.5	1.8	2.0	2.3	2.6
SS C2	26	34	42	48	56	61	67	74	17	22	29	33	40	45	50	57	1.1	1.3	1.4	1.6	1.8	1.3	1.5	1.7	1.9	2.1
SS C3	19	24	29	33	39	43	47	54	13	16	20	22	26	28	32	36	1.4	1.6	1.8	2.0	2.2	1.7	2.0	2.3	2.6	2.8
SS C4	24	31	38	43	51	56	61	69	16	21	26	30	36	40	46	53	1.1	1.3	1.5	1.7	1.9	1.4	1.6	1.8	2.0	2.2
SS D1	31	37	44	49	56	60	65	71	23	28	33	37	42	46	50	55	1.1	1.3	1.5	1.7	1.9	1.3	1.5	1.8	2.1	2.4
SS D2	36	45	54	60	68	73	78	85	27	35	43	49	56	61	66	73	0.9	1.1	1.3	1.5	1.7	1.1	1.3	1.5	1.7	1.9
SS D3	29	35	41	45	51	55	60	66	22	26	31	34	39	42	46	51	1.2	1.4	1.6	1.8	2.0	1.4	1.6	1.9	2.3	2.5
SS D4	35	44	53	59	66	71	77	83	27	34	42	47	54	59	64	71	1.0	1.2	1.3	1.5	1.7	1.1	1.3	1.5	1.7	1.9
SS E1	27	33	39	44	50	54	59	66	19	23	28	32	36	39	43	47	1.1	1.4	1.6	1.8	2.0	1.4	1.7	2.0	2.3	2.7
SS E2	32	40	49	55	62	67	73	80	23	30	37	42	49	54	59	66	1.0	1.2	1.4	1.6	1.7	1.1	1.3	1.6	1.8	2.0
SS E3	24	29	35	39	44	48	52	58	18	22	25	28	32	35	38	42	1.3	1.5	1.8	2.0	2.2	1.5	1.8	2.1	2.5	2.8
SS E4	30	38	46	51	59	64	69	76	22	28	35	40	46	50	56	62	1.0	1.2	1.4	1.6	1.8	1.2	1.4	1.6	1.9	2.1

ALLENTOWN, PENNSYLVANIA

TYPE	SOLAR SAVINGS FRACTION (%)																CONSERVATION FACTOR									
	BASE TEMP: 55 F (3564 DD)								65 F (5830 DD)								55 F					65 F				
LCR	100	70	50	40	30	25	20	15	100	70	50	40	30	25	20	15	100	50	30	20	15	100	50	30	20	15
WW A1	16	19	22	24	27	29	33	37	13	14	16	17	19	20	22	24	1.9	2.1	2.3	2.6	2.7	2.5	2.8	3.2	3.6	4.0
WW A2	16	20	25	29	34	38	43	50	11	14	17	20	23	26	30	35	1.5	1.7	1.8	2.0	2.2	1.8	2.0	2.3	2.5	2.7
WW A3	15	20	26	31	37	41	47	55	10	13	17	21	25	28	33	39	1.4	1.5	1.7	1.9	2.0	1.6	1.9	2.1	2.3	2.5
WW A4	15	20	27	32	39	43	50	58	9	13	17	21	26	30	34	41	1.3	1.5	1.6	1.8	2.0	1.6	1.8	2.0	2.2	2.4
WW A5	15	21	28	33	40	45	52	61	8	13	18	22	27	31	36	43	1.3	1.4	1.6	1.8	1.9	1.5	1.7	1.9	2.1	2.3
WW A6	14	21	28	33	41	47	53	62	8	12	18	22	28	32	37	45	1.2	1.4	1.6	1.7	1.9	1.5	1.7	1.9	2.1	2.2
WW B1	12	15	18	20	23	26	29	33	7	9	10	11	13	14	14	15	1.8	2.1	2.4	2.7	2.9	2.6	3.1	4.0	-	-
WW B2	16	22	29	34	42	47	54	63	10	15	20	24	30	35	40	48	1.2	1.4	1.6	1.7	1.9	1.4	1.6	1.8	2.0	2.1
WW B3	19	27	35	41	50	55	63	72	12	18	25	30	37	42	49	58	1.1	1.3	1.4	1.6	1.8	1.3	1.5	1.6	1.8	2.0
WW B4	19	28	37	44	54	60	68	77	12	19	27	34	42	48	55	65	1.1	1.2	1.4	1.5	1.7	1.2	1.3	1.5	1.7	1.8
WW B5	18	27	37	44	54	60	68	77	12	19	27	34	43	49	56	66	1.1	1.2	1.4	1.5	1.7	1.2	1.3	1.5	1.6	1.8
WW C1	21	28	37	44	53	58	66	75	14	20	27	33	41	46	53	61	1.1	1.3	1.4	1.6	1.7	1.2	1.4	1.6	1.7	1.9
WW C2	19	27	36	43	52	58	65	75	12	19	27	32	40	46	53	62	1.1	1.2	1.4	1.6	1.7	1.2	1.4	1.6	1.7	1.9
WW C3	23	33	45	52	63	69	77	85	16	25	35	42	52	59	67	76	1.0	1.1	1.3	1.4	1.6	1.1	1.2	1.4	1.5	1.7
WW C4	21	31	41	49	60	66	74	83	14	22	32	39	49	56	64	74	1.0	1.1	1.3	1.5	1.6	1.1	1.2	1.4	1.6	1.7
TW A1	16	18	21	24	27	30	33	38	13	14	16	17	20	21	23	26	1.9	2.1	2.3	2.5	2.7	2.4	2.7	3.0	3.4	3.7
TW A2	15	19	23	27	32	35	40	47	11	13	16	19	22	25	28	32	1.6	1.8	1.9	2.1	2.3	1.9	2.1	2.4	2.6	2.8
TW A3	14	19	24	28	33	37	43	50	10	13	16	19	23	26	30	35	1.5	1.6	1.8	2.0	2.1	1.8	2.0	2.2	2.4	2.6
TW A4	13	18	23	28	34	38	44	52	8	12	16	19	23	27	31	37	1.4	1.6	1.8	1.9	2.1	1.7	1.9	2.1	2.3	2.5
TW B1	15	18	21	24	28	31	34	40	12	14	16	17	20	22	24	28	1.8	2.0	2.2	2.4	2.5	2.2	2.5	2.8	3.0	3.3
TW B2	14	18	22	26	31	35	40	47	10	13	16	18	22	25	28	33	1.5	1.7	1.9	2.1	2.2	1.9	2.1	2.3	2.5	2.7
TW B3	13	17	22	26	31	35	40	48	9	12	15	18	22	25	29	34	1.5	1.7	1.9	2.0	2.2	1.8	2.0	2.2	2.4	2.6
TW B4	12	16	21	25	30	34	39	46	9	11	15	17	21	24	28	34	1.5	1.7	1.9	2.1	2.2	1.9	2.0	2.2	2.4	2.6
TW C1	14	17	21	24	28	31	35	41	11	13	15	17	20	22	25	30	1.7	1.9	2.1	2.3	2.4	2.1	2.3	2.5	2.8	3.0
TW C2	13	16	20	24	29	32	37	44	9	12	15	17	21	23	27	32	1.6	1.8	2.0	2.2	2.3	2.0	2.2	2.4	2.6	2.7
TW C3	12	16	20	23	28	31	35	42	9	12	14	17	20	23	26	31	1.7	1.9	2.0	2.2	2.3	2.0	2.2	2.4	2.6	2.8
TW C4	12	15	19	21	26	29	33	39	10	12	14	16	19	21	24	29	1.8	2.0	2.2	2.3	2.5	2.2	2.4	2.5	2.7	2.9
TW D1	10	12	15	17	21	23	26	30	6	7	9	10	12	13	14	16	1.9	2.2	2.5	2.7	2.9	2.6	3.1	3.7	4.4	-
TW D2	16	21	27	32	39	43	50	58	11	15	19	23	28	32	37	44	1.3	1.5	1.7	1.8	2.0	1.6	1.7	1.9	2.1	2.2
TW D3	17	22	29	35	42	47	54	63	11	16	21	25	31	35	41	48	1.3	1.4	1.6	1.7	1.9	1.5	1.6	1.8	2.0	2.1
TW D4	18	25	33	40	48	54	61	71	12	18	25	30	38	43	50	59	1.2	1.3	1.5	1.6	1.8	1.3	1.5	1.6	1.8	1.9
TW D5	18	25	34	40	49	55	63	72	12	18	25	31	39	45	52	61	1.1	1.3	1.4	1.6	1.7	1.3	1.4	1.6	1.7	1.9
TW E1	19	26	34	39	48	53	60	69	13	19	25	30	36	41	47	55	1.2	1.3	1.5	1.7	1.8	1.3	1.5	1.7	1.9	2.0
TW E2	19	26	34	40	48	54	61	70	13	19	25	30	37	42	49	58	1.2	1.3	1.5	1.6	1.8	1.3	1.5	1.6	1.8	1.9
TW E3	23	32	42	49	59	65	73	81	16	24	33	39	48	54	62	71	1.0	1.2	1.3	1.5	1.6	1.1	1.3	1.4	1.6	1.7
TW E4	21	30	40	47	57	63	71	80	14	22	31	37	46	52	60	70	1.0	1.2	1.3	1.5	1.6	1.1	1.3	1.4	1.6	1.7
TW F1	14	17	20	22	26	29	32	37	11	12	14	16	18	20	22	25	1.8	2.1	2.3	2.5	2.6	2.4	2.6	3.0	3.3	3.6
TW F2	13	16	21	24	29	33	37	44	9	11	14	16	20	22	25	30	1.6	1.8	2.0	2.1	2.3	2.0	2.2	2.4	2.7	2.9
TW F3	11	16	20	24	30	34	39	46	7	10	13	16	20	23	26	32	1.5	1.7	1.9	2.0	2.2	1.8	2.1	2.3	2.5	2.7
TW F4	9	14	19	23	29	33	38	46	5	8	12	15	19	22	26	32	1.5	1.7	1.9	2.0	2.2	1.8	2.0	2.2	2.4	2.6
TW G1	12	15	18	21	25	28	32	37	9	11	13	14	17	19	21	25	1.8	2.0	2.2	2.4	2.5	2.2	2.5	2.8	3.0	3.3
TW G2	10	14	18	21	26	30	35	42	6	9	12	14	17	20	23	28	1.6	1.8	2.0	2.2	2.3	2.0	2.2	2.4	2.6	2.8
TW G3	8	12	17	20	25	29	34	41	5	7	10	13	17	19	23	28	1.6	1.8	2.0	2.1	2.3	2.0	2.2	2.4	2.6	2.7
TW G4	6	10	14	17	22	26	30	37	3	5	8	11	14	17	20	26	1.7	1.9	2.0	2.2	2.4	2.0	2.2	2.4	2.6	2.8
TW H1	9	12	16	18	23	26	30	36	6	8	10	12	15	17	20	24	1.8	2.0	2.2	2.4	2.5	2.2	2.4	2.7	2.9	3.1
TW H2	7	10	14	17	21	24	29	35	4	6	8	10	14	16	19	24	1.8	1.9	2.1	2.3	2.4	2.1	2.3	2.6	2.8	2.9
TW H3	5	8	12	15	19	22	26	32	2	5	7	9	12	14	17	22	1.8	2.0	2.2	2.4	2.5	2.2	2.4	2.6	2.8	3.0
TW H4	4	6	9	11	15	18	21	27	2	3	5	7	10	11	14	18	2.1	2.2	2.4	2.6	2.7	2.5	2.7	2.9	3.0	3.2
TW I1	8	10	13	15	18	20	23	27	4	6	7	8	0	10	12	13	2.0	2.3	2.6	2.8	3.1	2.8	3.3	3.9	4.7	-
TW I2	10	18	24	28	35	40	46	55	8	12	16	20	25	29	34	41	1.4	1.5	1.7	1.9	2.0	1.6	1.8	2.0	2.1	2.3
TW I3	14	19	26	31	39	44	50	59	8	13	18	22	28	32	38	46	1.3	1.5	1.6	1.8	1.9	1.5	1.7	1.8	2.0	2.2
TW I4	15	22	30	36	45	51	58	68	10	15	22	27	34	40	46	56	1.2	1.3	1.5	1.6	1.8	1.3	1.5	1.6	1.8	1.9
TW I5	15	23	31	38	47	53	60	70	10	16	23	28	36	42	49	59	1.2	1.3	1.5	1.6	1.7	1.3	1.4	1.6	1.8	1.9
TW J1	17	24	31	37	45	51	58	67	11	17	23	27	34	39	45	54	1.2	1.4	1.5	1.7	1.8	1.4	1.5	1.7	1.9	2.0
TW J2	16	23	31	37	46	51	59	68	10	16	23	28	35	40	47	56	1.2	1.3	1.5	1.6	1.8	1.3	1.5	1.7	1.8	2.0
TW J3	20	29	39	46	56	63	70	80	13	21	30	37	46	52	60	70	1.0	1.2	1.3	1.5	1.6	1.1	1.3	1.4	1.6	1.7
TW J4	18	27	37	44	54	60	68	78	12	19	28	35	44	50	58	68	1.1	1.2	1.4	1.5	1.7	1.2	1.3	1.5	1.6	1.8
DG A1	10	13	16	18	21	23	24	26	5	7	9	10	11	12	12	12	1.8	2.1	2.7	3.7	-	2.4	3.0	4.2	-	-
DG A2	12	17	21	25	30	34	39	45	8	11	15	17	21	24	28	32	1.5	1.7	1.9	2.2	2.5	1.8	2.0	2.3	2.6	3.0
DG A3	16	22	28	33	41	47	54	62	12	16	21	26	32	36	42	51	1.3	1.4	1.6	1.7	1.9	1.5	1.6	1.7	1.9	2.1
DG B1	10	13	16	19	22	25	28	31	5	7	9	10	12	13	14	15	1.8	2.1	2.4	2.8	3.3	2.4	2.9	3.7	4.8	-
DG B2	13	17	22	26	32	36	42	49	8	12	15	18	22	25	29	35	1.5	1.7	1.8	2.0	2.2	1.8	2.0	2.2	2.4	2.6
DG B3	17	22	29	35	43	49	57	67	12	17	22	27	33	38	45	54	1.3	1.4	1.5	1.6	1.8	1.5	1.6	1.7	1.8	2.0
DG C1	12	16	20	24	29	32	36	42	7	10	12	14	17	19	21	24	1.6	1.8	2.0	2.3	2.6	2.0	2.3	2.8	3.3	3.9
DG C2	15	20	26	31	38	43	50	58	10	14	19	22	27	31	36	43	1.4	1.5	1.6	1.8	2.0	1.6	1.8	1.9	2.1	2.3
DG C3	19	25	33	39	49	55	63	74	14	19	25	30	38	43	51	61	1.2	1.3	1.4	1.5	1.7	1.4	1.5	1.6	1.7	1.8
SS A1	19	23	28	32	37	41	46	53	15	18	22	25	29	32	36	41	1.4	1.6	1.8	2.0	2.1	1.6	1.9	2.1	2.4	2.6
SS A2	24	30	38	43	51	56	63	71	19	24	31	35	42	47	53	61	1.2	1.3	1.5	1.6	1.8	1.3	1.5	1.6	1.8	2.0
SS A3	17	21	25	28	33	36	40	46	14	16	20	22	25	27	30	34	1.5	1.8	2.0	2.2	2.4	1.8	2.1	2.4	2.8	3.1
SS A4	23	29	36	42	49	54	61	69	18	24	30	34	40	45	50	58	1.2	1.4	1.5	1.7	1.8	1.3	1.5	1.7	1.9	2.0
SS A5	19	22	26	30	34	37	41	47	16	18	21	23	27	29	32	36	1.6	1.8	2.0	2.2	2.4	1.8	2.1	2.4	2.7	2.9
SS A6	23	30	37	42	50	55	62	70	18	24	30	35	41	46	51	59	1.2	1.3	1.5	1.7	1.8	1.3	1.5	1.7	1.8	2.0
SS A7	17	20	23	25	29	31	34	39	14	16	18	20	22	24	26	28	1.8	2.0	2.3	2.5	2.7	2.1	2.5	2.9	3.4	3.8
SS A8	22	28	35	40	48	52	59	67	18	23	28	33	39	43	48	55	1.2	1.4	1.6	1.7	1.9	1.3	1.5	1.7	1.9	2.1
SS B1	16	20	24	28	32	36	41	47	12	15	18	21	25	27	31	35	1.5	1.7	1.9	2.1	2.3	1.8	2.0	2.3	2.6	2.8
SS B2	20	27	34	39	46	51	58	67	16	21	27	31	38	42	48	56	1.2	1.4	1.5	1.7	1.8	1.4	1.5	1.7	1.9	2.0
SS B3	15	18	22	25	29	33	37	42	11	14	16	19	22	24	27	30	1.6	1.8	2.1	2.3	2.4	1.9	2.2	2.5	2.9	3.2
SS B4	20	26	32	37	45	50	56	64	15	20	26	30	36	41	46	54	1.2	1.4	1.6	1.7	1.9	1.4	1.6	1.7	1.9	2.1
SS B5	15	18	22	24	28	31	35	40	13	15	17	19	22	23	26	29	1.7	2.0	2.2	2.4	2.6	2.1	2.4	2.7	3.0	3.3
SS B6	20	26	32	38	45	50	56	65	15	20	26	30	36	41	46	54	1.2	1.4	1.6	1.7	1.9	1.4	1.6	1.7	1.9	2.1
SS B7	14	17	19	21	25	27	30	34	11	13	15	16	18	20	22	24	1.9	2.2	2.4	2.7	2.9	2.3	2.7	3.1	3.5	4.0
SS B8	19	24	31	36	43	47	54	62	14	19	24	28	34	38	43	51	1.3	1.4	1.6	1.8	1.9	1.4	1.6	1.8	2.0	2.2
SS C1	13	18	22	26	32	36	41	48	8	11	14	17	21	23	27	31	1.5	1.7	1.9	2.0	2.2	1.8	2.1	2.3	2.6	2.8
SS C2	16	22	29	35	43	48	55	64	11	15	21	25	31	36	41	49	1.2	1.4	1.6	1.7	1.9	1.4	1.6	1.8	2.0	2.1
SS C3	13	16	19	22	26	29	34	39	9	11	13	15	18	20	23	26	1.7	1.9	2.1	2.3	2.5	2.2	2.4	2.7	2.9	3.2
SS C4	15	20	26	31	38	43	49	57	10	14	19	23	28	32	37	44	1.3	1.5	1.7	1.8	2.0	1.6	1.7	1.9	2.1	2.2
SS D1	21	27	33	37	44	49	54	62	16	20	25	29	34	37	42	48	1.3	1.5	1.6	1.8	2.0	1.4	1.7	1.9	2.1	2.3
SS D2	26	34	43	50	59	65	71	80	20	27	35	41	49	54	61	69	1.0	1.2	1.4	1.5	1.7	1.1	1.3	1.5	1.7	1.8
SS D3	20	25	30	34	40	44	49	56	16	20	24	27	31	34	38	43	1.4	1.6	1.8	2.0	2.1	1.6	1.8	2.1	2.3	2.6
SS D4	25	33	42	48	57	62	69	78	20	27	34	39	47	52	59	67	1.1	1.2	1.4	1.6	1.7	1.2	1.3	1.5	1.7	1.8
SS E1	17	22	28	32	37	41	47	54	12	16	20	23	27	30	33	38	1.4	1.6	1.8	2.0	2.1	1.6	1.9	2.2	2.5	2.7
SS E2	22	29	37	43	51	57	64	73	16	22	29	34	41	46	52	60	1.1	1.3	1.4	1.6	1.8	1.2	1.4	1.6	1.8	2.0
SS E3	16	20	24	27	32	35	39	46	13	15	18	21	24	26	29	33	1.6	1.8	2.0	2.2	2.4	1.9	2.1	2.4	2.7	3.0
SS E4	21	27	34	40	48	53	59	68	16	21	27	31	38	42	48	56	1.2	1.4	1.5	1.7	1.8	1.3	1.5	1.7	1.9	2.0

ERIE, PENNSYLVANIA

TYPE	SOLAR SAVINGS FRACTION (%) — BASE TEMP: 55 F (4314 DD) — LCR 100	70	50	40	30	25	20	15	65 F (6857 DD) 100	70	50	40	30	25	20	15	CONSERVATION FACTOR 55 F 100	50	30	20	15	65 F 100	50	30	20	15
WW A1	12	13	14	15	16	17	17	18	11	11	12	12	12	12	11	10	2.8	3.5	4.7	-	-	3.8	-	-	-	-
WW A2	10	12	14	16	18	20	22	24	8	9	10	11	12	13	14	14	2.0	2.4	2.9	3.5	4.1	2.5	3.3	4.4	-	-
WW A3	8	11	14	16	19	21	24	27	6	7	9	11	12	13	14	15	1.8	2.2	2.6	3.1	3.5	2.2	2.8	3.7	-	-
WW A4	7	10	14	16	19	22	25	28	4	7	9	10	12	14	15	16	1.7	2.1	2.5	2.8	3.2	2.1	2.6	3.3	4.4	-
WW A5	6	10	14	16	20	23	26	30	3	6	8	10	12	14	15	17	1.7	2.0	2.3	2.7	3.0	2.0	2.5	3.1	4.1	-
WW A6	6	9	13	16	20	23	26	31	3	5	8	10	12	14	16	18	1.6	2.0	2.3	2.6	2.9	2.0	2.4	3.0	3.8	-
WW B1	6	7	8	8	8	8	6	4	5	5	4	4	2	1	-1	-6	3.2	-	-	-	-	-	-	-	-	-
WW B2	8	12	16	20	24	27	32	37	6	8	12	14	18	20	23	27	1.6	1.8	2.1	2.3	2.5	1.8	2.1	2.5	2.8	3.2
WW B3	10	15	20	24	30	34	39	46	7	11	15	18	23	26	29	34	1.4	1.7	1.9	2.1	2.3	1.6	1.9	2.2	2.5	2.8
WW B4	10	16	23	28	35	40	47	55	7	12	18	22	28	33	38	46	1.3	1.5	1.7	1.8	2.0	1.4	1.6	1.8	2.0	2.2
WW B5	10	16	23	29	36	42	49	58	7	12	18	23	30	35	41	49	1.3	1.4	1.6	1.8	1.9	1.4	1.6	1.7	1.9	2.1
WW C1	12	17	23	27	33	37	43	50	8	13	17	21	26	29	34	39	1.4	1.6	1.8	2.0	2.2	1.5	1.8	2.0	2.3	2.5
WW C2	10	16	22	27	34	38	44	52	7	12	17	21	27	30	36	42	1.3	1.5	1.7	1.9	2.1	1.5	1.7	1.9	2.1	2.3
WW C3	13	21	30	37	46	52	59	69	10	17	25	31	39	44	51	61	1.1	1.3	1.5	1.6	1.8	1.2	1.4	1.6	1.7	1.9
WW C4	12	19	28	34	43	49	56	66	8	15	23	28	36	42	49	58	1.2	1.3	1.5	1.7	1.8	1.2	1.4	1.6	1.8	1.9
TW A1	12	13	15	16	17	18	18	19	11	12	12	13	13	13	13	12	2.7	3.2	4.1	-	-	3.5	5.0	-	-	-
TW A2	10	12	14	16	18	20	21	24	8	10	11	12	13	14	14	15	2.1	2.5	3.0	3.5	4.1	2.6	3.3	4.3	-	-
TW A3	9	11	14	16	18	20	23	26	7	8	10	11	13	14	15	16	1.9	2.3	2.7	3.1	3.5	2.4	2.9	3.6	4.7	-
TW A4	7	10	13	15	19	21	23	27	5	7	9	11	13	14	16	18	1.9	2.2	2.5	2.9	3.2	2.2	2.7	3.2	3.9	4.8
TW B1	11	13	14	15	17	18	20	21	10	11	12	12	13	14	14	14	2.5	2.9	3.5	4.2	-	3.1	3.9	-	-	-
TW B2	9	11	14	15	18	20	22	25	7	9	10	12	13	14	15	17	2.0	2.4	2.8	3.2	3.6	2.5	3.0	3.6	4.5	-
TW B3	8	10	13	15	18	20	23	26	6	8	10	11	13	14	16	18	2.0	2.3	2.6	3.0	3.3	2.3	2.8	3.3	4.0	4.7
TW B4	8	10	13	15	18	20	23	26	6	8	10	11	13	15	16	19	2.0	2.3	2.6	2.9	3.2	2.3	2.7	3.2	3.7	4.2
TW C1	10	12	14	15	17	19	21	23	9	10	11	12	13	14	15	17	2.3	2.6	3.0	3.5	3.9	2.7	3.3	4.0	4.9	-
TW C2	9	11	13	15	18	19	22	25	7	9	10	12	14	15	16	18	2.1	2.4	2.7	3.1	3.4	2.5	2.9	3.4	3.9	4.5
TW C3	9	11	13	15	17	19	22	25	7	9	11	12	14	15	17	19	2.2	2.4	2.7	3.1	3.3	2.5	2.9	3.3	3.8	4.2
TW C4	9	11	13	15	17	19	21	25	8	10	11	12	14	15	17	19	2.3	2.5	2.8	3.1	3.4	2.6	3.0	3.4	3.8	4.1
TW D1	5	6	7	8	8	8	8	7	4	4	4	4	3	3	1	-1	3.1	4.3	-	-	-	5.0	-	-	-	-
TW D2	9	13	16	19	23	26	30	35	7	10	12	15	18	20	22	26	1.7	2.0	2.2	2.5	2.7	1.9	2.3	2.6	3.0	3.3
TW D3	9	13	17	21	25	28	32	38	7	10	13	15	19	21	24	28	1.6	1.8	2.1	2.4	2.6	1.8	2.1	2.5	2.8	3.2
TW D4	10	16	21	26	32	36	42	49	8	12	17	20	26	29	34	41	1.4	1.6	1.8	2.0	2.1	1.5	1.8	2.0	2.2	2.4
TW D5	10	16	22	27	34	38	45	53	8	12	18	22	28	32	38	45	1.3	1.5	1.7	1.9	2.0	1.5	1.7	1.8	2.0	2.2
TW E1	12	16	21	25	30	34	38	45	9	12	16	19	23	26	30	35	1.5	1.7	1.9	2.1	2.3	1.7	1.9	2.2	2.5	2.7
TW E2	11	16	21	26	31	35	41	48	9	12	17	20	25	28	33	39	1.4	1.6	1.8	2.0	2.2	1.6	1.8	2.0	2.3	2.5
TW E3	14	21	28	34	41	47	54	63	11	16	23	28	35	40	46	54	1.2	1.4	1.6	1.7	1.9	1.3	1.5	1.7	1.9	2.0
TW E4	13	19	27	32	40	46	53	62	9	15	22	27	34	39	45	54	1.2	1.4	1.6	1.7	1.9	1.3	1.5	1.7	1.9	2.0
TW F1	10	11	13	14	15	16	17	18	9	10	10	11	11	11	11	11	2.6	3.1	3.9	-	-	3.4	4.7	-	-	-
TW F2	8	10	12	13	16	17	19	21	6	7	9	9	11	11	12	13	2.2	2.6	3.0	3.6	4.2	2.7	3.4	4.4	-	-
TW F3	6	8	11	13	15	17	20	23	4	6	7	9	10	11	12	14	2.0	2.4	2.8	3.2	3.6	2.4	3.0	3.7	4.7	-
TW F4	4	7	9	12	15	17	19	23	2	4	6	7	9	11	12	14	2.0	2.3	2.6	3.0	3.3	2.3	2.8	3.3	4.0	4.7
TW G1	8	9	11	12	14	15	17	18	7	8	9	9	10	11	11	11	2.5	2.9	3.5	4.2	5.0	3.1	3.9	-	-	-
TW G2	5	7	10	11	14	16	18	21	4	5	7	8	9	10	11	13	2.2	2.5	2.9	3.4	3.8	2.6	3.2	3.9	4.8	-
TW G3	4	6	8	10	13	15	17	21	2	4	6	7	9	10	11	13	2.1	2.4	2.8	3.2	3.5	2.5	3.0	3.5	4.1	4.9
TW G4	2	4	7	9	12	14	16	20	1	3	4	6	8	9	11	13	2.2	2.5	2.8	3.1	3.4	2.5	2.9	3.4	3.8	4.3
TW H1	5	7	9	10	12	14	15	18	4	5	6	7	8	9	10	11	2.4	2.7	3.2	3.7	4.2	2.9	3.5	4.3	-	-
TW H2	3	5	7	9	11	13	15	18	2	3	5	6	7	8	10	12	2.3	2.6	2.9	3.3	3.7	2.7	3.2	3.7	4.3	4.9
TW H3	2	4	6	7	10	12	14	17	1	2	4	5	7	8	9	12	2.4	2.6	2.9	3.3	3.6	2.7	3.1	3.6	4.0	4.5
TW H4	1	3	4	6	8	10	12	15	1	2	3	4	6	7	8	10	2.6	2.9	3.1	3.4	3.7	2.9	3.3	3.7	4.1	4.4
TW I1	4	5	5	6	6	6	6	5	2	3	3	2	2	1	0	-2	3.2	4.5	-	-	-	-	-	-	-	-
TW I2	7	10	14	16	20	23	27	32	5	7	10	12	15	17	19	23	1.7	2.0	2.3	2.5	2.7	2.0	2.3	2.7	3.0	3.4
TW I3	7	11	15	18	23	26	30	36	5	8	11	14	17	20	23	27	1.6	1.9	2.1	2.3	2.5	1.8	2.1	2.4	2.7	3.0
TW I4	8	13	19	23	29	33	39	47	6	10	14	18	23	27	32	38	1.4	1.6	1.8	2.0	2.2	1.6	1.8	2.0	2.2	2.4
TW I5	8	14	20	24	31	36	42	50	6	10	15	19	25	29	35	42	1.4	1.6	1.7	1.9	2.1	1.5	1.7	1.9	2.1	2.2
TW J1	10	14	19	23	28	32	37	44	7	10	14	18	22	25	29	34	1.5	1.7	1.9	2.1	2.3	1.7	1.9	2.2	2.4	2.7
TW J2	9	14	19	23	29	33	39	46	6	10	15	18	23	26	31	37	1.4	1.6	1.8	2.0	2.2	1.6	1.8	2.0	2.3	2.5
TW J3	11	18	26	31	40	45	52	61	8	14	21	26	33	38	45	53	1.2	1.4	1.6	1.7	1.9	1.3	1.5	1.7	1.9	2.0
TW J4	10	17	24	30	38	43	50	60	7	13	20	25	32	37	43	52	1.2	1.4	1.6	1.7	1.9	1.3	1.5	1.7	1.9	2.0
DG A1	5	6	7	7	7	7	7	5	3	3	3	3	2	2	0	-2	2.8	4.2	-	-	-	4.3	-	-	-	-
DG A2	8	10	13	15	18	20	22	25	6	8	10	11	13	15	16	18	2.0	2.3	2.7	3.2	3.7	2.3	2.7	3.3	4.1	-
DG A3	11	15	19	23	28	32	36	43	9	12	16	19	23	26	30	36	1.6	1.8	1.9	2.2	2.3	1.8	1.9	2.2	2.4	2.6
DG B1	5	6	7	7	8	8	7	6	3	3	3	3	3	2	1	-1	2.7	4.1	-	-	-	4.7	-	-	-	-
DG B2	8	10	13	15	18	21	23	27	6	8	10	12	14	15	17	19	1.9	2.2	2.6	2.9	3.4	2.3	2.6	3.1	3.8	4.6
DG B3	12	16	20	24	29	33	39	46	10	13	17	20	25	28	32	39	1.5	1.7	1.9	2.0	2.2	1.7	1.9	2.1	2.3	2.4
DG C1	7	8	10	11	12	13	13	14	4	5	6	6	7	7	6	4	2.2	3.0	4.5	-	-	3.2	4.4	-	-	-
DG C2	10	13	16	19	23	26	29	34	8	10	13	15	18	20	22	26	1.7	2.0	2.3	2.5	2.8	2.0	2.2	2.6	3.1	3.5
DG C3	13	18	23	27	33	38	44	52	11	15	19	23	28	32	37	44	1.4	1.6	1.8	1.9	2.1	1.6	1.7	1.9	2.1	2.3
SS A1	14	17	20	22	25	26	29	32	12	15	17	18	20	22	23	25	1.8	2.2	2.6	3.0	3.4	2.0	2.5	3.1	3.9	-
SS A2	17	22	27	31	36	40	45	52	15	19	24	27	32	35	39	44	1.4	1.6	1.8	2.0	2.2	1.5	1.7	2.0	2.3	2.5
SS A3	13	15	17	19	21	22	23	25	11	13	14	15	17	17	18	18	2.0	2.5	3.2	4.1	-	2.3	3.1	4.3	-	-
SS A4	17	21	26	29	35	38	42	49	15	18	23	26	30	33	36	41	1.4	1.7	1.9	2.1	2.3	1.5	1.8	2.1	2.4	2.7
SS A5	15	17	19	21	23	24	26	28	13	15	17	18	19	20	21	21	2.0	2.5	3.0	3.7	4.3	2.3	3.0	3.8	-	-
SS A6	17	21	26	30	35	39	43	50	15	18	23	26	30	33	37	42	1.4	1.6	1.9	2.1	2.3	1.5	1.8	2.1	2.4	2.6
SS A7	14	15	17	18	19	20	20	21	13	14	15	15	16	16	16	15	2.3	3.0	4.0	-	-	2.8	3.9	-	-	-
SS A8	16	20	25	28	33	36	40	46	14	18	21	24	28	31	34	38	1.5	1.7	2.0	2.2	2.4	1.6	1.9	2.2	2.6	2.9
SS B1	11	14	16	18	21	22	24	27	10	12	14	15	17	18	19	20	1.9	2.3	2.8	3.3	3.7	2.3	2.8	3.4	4.4	-
SS B2	15	19	24	28	33	37	42	48	13	17	21	24	29	32	36	42	1.4	1.7	1.9	2.1	2.2	1.6	1.8	2.0	2.3	2.5
SS B3	10	12	14	16	18	19	20	22	9	10	12	13	14	15	15	15	2.1	2.6	3.3	4.1	-	2.5	3.2	4.4	-	-
SS B4	14	18	23	26	31	35	39	46	12	16	20	23	27	30	34	39	1.5	1.7	2.0	2.2	2.3	1.6	1.9	2.1	2.4	2.7
SS B5	12	14	15	17	19	20	21	23	11	12	13	14	15	16	17	17	2.3	2.7	3.3	4.1	4.9	2.7	3.4	4.4	-	-
SS B6	14	18	23	26	31	35	40	46	12	16	20	23	27	30	34	39	1.5	1.7	1.9	2.1	2.3	1.6	1.9	2.1	2.4	2.6
SS B7	11	12	14	15	16	17	17	18	10	11	12	12	13	13	13	13	2.5	3.1	4.1	-	-	3.1	4.2	-	-	-
SS B8	13	17	21	25	29	32	37	42	11	15	18	21	25	27	31	35	1.5	1.8	2.0	2.3	2.5	1.7	2.0	2.3	2.6	2.9
SS C1	7	9	11	13	15	17	18	20	5	6	7	8	9	10	10	9	2.0	2.5	3.1	3.9	4.7	2.6	3.5	-	-	-
SS C2	9	13	17	21	25	29	33	39	7	10	13	15	19	21	25	29	1.6	1.8	2.1	2.3	2.5	1.8	2.1	2.4	2.7	3.1
SS C3	8	10	12	13	15	16	18	19	7	8	9	10	11	11	11	12	2.4	2.8	3.4	4.1	4.9	3.0	3.8	-	-	-
SS C4	9	13	16	19	23	26	30	35	7	10	13	15	18	20	23	26	1.7	1.9	2.2	2.5	2.7	1.9	2.2	2.6	2.9	3.2
SS D1	14	18	21	24	27	29	32	36	12	15	17	19	22	23	25	26	1.6	2.0	2.4	2.7	3.1	1.8	2.3	2.9	4.0	-
SS D2	18	24	30	35	42	46	52	60	15	20	26	30	36	40	45	51	1.2	1.5	1.6	1.8	2.0	1.3	1.6	1.8	2.0	2.2
SS D3	15	18	20	23	25	27	30	33	13	15	17	19	21	22	23	25	1.8	2.2	2.6	3.0	3.5	2.0	2.5	3.2	4.3	-
SS D4	18	23	29	34	40	45	50	58	15	20	25	29	35	38	43	49	1.3	1.5	1.7	1.9	2.0	1.4	1.6	1.9	2.1	2.3
SS E1	11	14	16	18	21	22	24	26	9	11	13	14	15	16	16	16	1.8	2.3	2.9	3.6	4.4	2.2	2.9	4.2	-	-
SS E2	14	19	24	28	34	38	43	50	11	16	20	24	28	31	35	40	1.4	1.6	1.8	2.0	2.2	1.5	1.8	2.1	2.4	2.6
SS E3	12	14	16	18	20	21	23	25	11	12	13	14	16	16	17	17	2.0	2.5	3.1	3.8	4.6	2.4	3.1	4.2	-	-
SS E4	14	18	23	27	32	35	40	46	12	15	19	22	27	30	33	38	1.4	1.7	1.9	2.1	2.3	1.6	1.9	2.1	2.5	2.7

HARRISBURG, PENNSYLVANIA

TYPE	SSF (%) 55 F (3104 DD) LCR 100	70	50	40	30	25	20	15	SSF (%) 65 F (5226 DD) 100	70	50	40	30	25	20	15	CONSERVATION FACTOR 55 F 100	50	30	20	15	CONSERVATION FACTOR 65 F 100	50	30	20	15
WW A1	17	20	23	26	29	32	36	41	13	15	17	18	21	22	24	27	1.8	2.0	2.2	2.4	2.5	2.3	2.6	2.9	3.3	3.6
WW A2	17	22	27	31	37	42	47	55	12	15	18	21	26	29	32	38	1.4	1.6	1.7	1.9	2.0	1.7	1.9	2.1	2.4	2.5
WW A3	17	22	29	34	41	46	52	60	11	14	19	22	28	31	36	42	1.3	1.5	1.6	1.8	1.9	1.6	1.8	2.0	2.2	2.3
WW A4	17	23	30	35	42	48	54	63	10	14	19	23	29	33	38	45	1.2	1.4	1.6	1.7	1.9	1.5	1.7	1.9	2.1	2.2
WW A5	17	23	31	36	44	50	57	66	9	14	20	24	30	34	40	47	1.2	1.4	1.5	1.7	1.8	1.4	1.6	1.8	2.0	2.2
WW A6	16	23	31	37	45	51	58	67	9	14	20	24	31	35	41	49	1.2	1.3	1.5	1.7	1.8	1.4	1.6	1.8	2.0	2.1
WW B1	13	16	20	23	27	29	33	39	8	10	12	13	15	16	17	19	1.7	1.9	2.2	2.4	2.6	2.4	2.8	3.4	4.1	-
WW B2	17	24	32	37	46	51	58	67	11	16	22	26	33	37	43	52	1.2	1.3	1.5	1.7	1.8	1.4	1.6	1.7	1.9	2.1
WW B3	21	29	38	44	54	60	67	76	14	20	27	32	40	45	52	61	1.1	1.2	1.4	1.6	1.7	1.3	1.4	1.6	1.8	1.9
WW B4	21	30	40	47	57	64	71	80	14	21	29	36	45	51	58	68	1.0	1.2	1.3	1.5	1.6	1.2	1.3	1.5	1.6	1.8
WW B5	20	29	39	47	57	63	71	80	13	20	29	36	45	51	59	69	1.0	1.2	1.3	1.5	1.6	1.2	1.3	1.5	1.6	1.8
WW C1	23	31	40	47	56	62	70	78	15	22	29	35	43	48	55	65	1.1	1.2	1.4	1.5	1.7	1.2	1.4	1.5	1.7	1.8
WW C2	21	30	39	46	56	62	69	78	14	20	28	34	43	48	56	65	1.1	1.2	1.4	1.5	1.7	1.2	1.4	1.5	1.7	1.8
WW C3	25	36	48	56	66	73	80	88	17	27	37	44	55	61	69	79	0.9	1.1	1.2	1.4	1.6	1.0	1.2	1.3	1.5	1.6
WW C4	23	33	44	52	63	69	77	85	15	24	34	41	51	58	66	76	1.0	1.1	1.3	1.4	1.6	1.1	1.2	1.4	1.5	1.7
TW A1	17	20	23	26	29	32	36	41	13	15	17	19	21	23	25	28	1.8	2.0	2.2	2.3	2.5	2.3	2.5	2.8	3.1	3.4
TW A2	16	20	25	29	34	38	44	51	12	14	18	20	24	27	30	35	1.5	1.7	1.8	2.0	2.1	1.8	2.0	2.3	2.5	2.7
TW A3	16	21	26	30	37	41	47	55	10	14	18	21	25	28	33	39	1.4	1.6	1.7	1.9	2.0	1.7	1.9	2.1	2.3	2.5
TW A4	15	20	26	30	37	42	48	56	9	13	17	20	25	29	34	40	1.4	1.5	1.7	1.8	2.0	1.6	1.8	2.0	2.2	2.4
TW B1	16	19	23	26	30	33	37	44	12	14	17	19	21	23	26	30	1.7	1.9	2.1	2.2	2.4	2.1	2.4	2.6	2.8	3.1
TW B2	15	19	24	28	34	38	43	51	10	13	17	20	24	26	30	36	1.5	1.7	1.8	2.0	2.1	1.8	2.0	2.2	2.4	2.6
TW B3	14	19	24	28	34	38	44	52	10	13	16	19	24	27	31	37	1.4	1.6	1.8	1.9	2.1	1.7	1.9	2.1	2.3	2.5
TW B4	14	18	23	27	33	37	42	50	9	12	16	19	23	26	30	36	1.5	1.6	1.8	2.0	2.1	1.8	2.0	2.1	2.3	2.5
TW C1	15	18	22	25	30	34	38	45	11	13	16	18	22	24	27	32	1.6	1.8	2.0	2.1	2.3	2.0	2.2	2.4	2.6	2.8
TW C2	14	18	22	26	31	35	40	47	10	12	16	18	22	25	29	34	1.6	1.7	1.9	2.1	2.2	1.9	2.1	2.3	2.4	2.6
TW C3	13	17	21	25	30	33	38	46	10	12	15	18	21	24	28	33	1.6	1.8	1.9	2.1	2.2	1.9	2.1	2.3	2.5	2.6
TW C4	13	16	20	23	27	31	35	42	10	12	15	17	20	23	26	31	1.7	1.9	2.1	2.2	2.3	2.1	2.3	2.4	2.6	2.8
TW D1	11	14	17	20	23	26	30	35	7	8	10	11	13	15	16	19	1.8	2.0	2.3	2.5	2.6	2.4	2.8	3.3	3.8	4.3
TW D2	17	23	29	34	42	47	53	62	11	16	21	25	30	34	40	47	1.3	1.4	1.6	1.7	1.9	1.5	1.7	1.8	2.0	2.2
TW D3	18	25	32	38	45	51	58	67	12	17	22	27	33	38	43	52	1.2	1.4	1.5	1.7	1.8	1.4	1.6	1.8	1.9	2.1
TW D4	20	27	36	42	52	58	65	74	13	19	26	32	40	45	52	61	1.1	1.3	1.4	1.6	1.7	1.3	1.4	1.6	1.7	1.9
TW D5	19	27	36	43	52	58	66	75	13	19	27	33	41	47	54	64	1.1	1.2	1.4	1.6	1.7	1.2	1.4	1.5	1.7	1.8
TW E1	21	28	36	42	51	57	64	73	14	20	26	31	39	44	50	58	1.1	1.3	1.4	1.6	1.7	1.3	1.5	1.6	1.8	1.9
TW E2	21	28	37	43	52	58	65	74	14	20	27	32	40	45	52	61	1.1	1.3	1.4	1.6	1.7	1.3	1.4	1.6	1.8	1.9
TW E3	25	34	45	52	62	68	76	84	17	25	34	41	50	57	64	74	1.0	1.1	1.3	1.5	1.6	1.1	1.3	1.4	1.6	1.7
TW E4	23	32	42	50	60	66	74	83	16	23	32	39	48	55	62	72	1.0	1.2	1.3	1.5	1.6	1.1	1.3	1.4	1.6	1.7
TW F1	15	18	22	24	28	31	35	41	11	13	15	17	19	21	24	27	1.7	1.9	2.1	2.3	2.5	2.2	2.5	2.7	3.0	3.3
TW F2	14	18	23	26	32	36	41	48	9	12	15	18	21	24	28	33	1.5	1.7	1.9	2.0	2.2	1.9	2.1	2.3	2.5	2.7
TW F3	13	17	23	27	33	37	43	51	8	11	14	17	22	25	29	35	1.4	1.6	1.8	1.9	2.1	1.8	2.0	2.2	2.4	2.5
TW F4	11	15	21	25	32	36	42	50	6	9	13	16	21	24	28	35	1.4	1.6	1.8	1.9	2.1	1.7	1.9	2.1	2.3	2.5
TW G1	13	16	20	23	27	31	35	41	9	11	14	16	19	21	23	27	1.7	1.9	2.1	2.2	2.4	2.1	2.4	2.6	2.8	3.1
TW G2	11	15	20	24	29	33	38	46	7	9	13	15	19	22	26	31	1.5	1.7	1.9	2.1	2.2	1.9	2.1	2.3	2.5	2.7
TW G3	10	14	19	22	28	32	37	45	5	8	11	14	18	21	25	31	1.5	1.7	1.9	2.0	2.2	1.9	2.1	2.2	2.4	2.6
TW G4	7	11	16	19	24	28	33	41	3	6	9	12	16	19	22	28	1.6	1.8	2.0	2.1	2.2	2.0	2.1	2.3	2.5	2.7
TW H1	10	13	17	20	25	28	33	39	6	8	11	13	16	19	22	26	1.7	1.9	2.1	2.2	2.4	2.1	2.3	2.5	2.8	2.9
TW H2	8	11	15	19	23	27	32	39	4	6	9	11	15	17	21	26	1.7	1.9	2.0	2.2	2.3	2.1	2.2	2.4	2.6	2.8
TW H3	6	9	13	16	21	24	29	35	3	5	8	10	13	16	19	24	1.8	1.9	2.1	2.3	2.4	2.1	2.3	2.5	2.7	2.9
TW H4	5	7	10	13	17	19	23	29	2	4	6	8	10	12	15	20	2.0	2.1	2.3	2.5	2.6	2.4	2.6	2.7	2.9	3.1
TW I1	9	12	15	17	21	23	27	32	5	6	8	9	11	12	14	16	1.9	2.1	2.4	2.6	2.7	2.8	3.0	3.4	4.0	4.6
TW I2	14	20	26	31	38	43	50	59	9	13	18	21	27	31	36	44	1.3	1.5	1.6	1.8	1.9	1.5	1.7	1.9	2.1	2.2
TW I3	15	21	28	34	42	47	54	60	9	14	20	24	30	35	40	49	1.2	1.4	1.6	1.7	1.9	1.4	1.6	1.8	2.0	2.1
TW I4	17	24	33	39	48	54	62	71	10	16	23	29	37	42	49	58	1.1	1.3	1.4	1.6	1.7	1.3	1.4	1.6	1.8	1.9
TW I5	17	25	34	40	50	56	64	73	11	17	24	30	38	44	51	61	1.1	1.3	1.4	1.6	1.7	1.3	1.4	1.6	1.7	1.9
TW J1	19	26	34	40	49	54	62	71	12	18	24	29	36	41	48	57	1.2	1.3	1.5	1.6	1.8	1.3	1.5	1.6	1.8	2.0
TW J2	18	25	34	40	49	55	62	72	11	17	24	29	37	42	49	58	1.1	1.3	1.4	1.6	1.7	1.3	1.5	1.6	1.8	1.9
TW J3	22	31	42	49	59	66	74	82	14	23	32	38	48	54	62	72	1.0	1.2	1.3	1.5	1.6	1.1	1.3	1.4	1.6	1.7
TW J4	20	29	39	47	57	64	71	81	13	21	30	36	46	52	60	70	1.0	1.2	1.3	1.5	1.6	1.1	1.3	1.4	1.6	1.7
DG A1	11	14	18	21	24	26	28	31	6	8	10	11	13	14	15	15	1.7	2.0	2.5	3.2	4.2	2.2	2.7	3.7	-	-
DG A2	14	18	23	27	33	37	42	48	9	12	16	19	23	26	30	35	1.5	1.6	1.8	2.1	2.3	1.8	1.9	2.2	2.5	2.8
DG A3	17	23	30	36	44	50	57	66	12	17	23	27	34	38	45	53	1.3	1.4	1.5	1.7	1.9	1.5	1.6	1.7	1.9	2.0
DG B1	11	14	18	21	26	29	32	37	6	8	10	12	14	15	16	18	1.7	1.9	2.1	2.5	2.9	2.2	2.6	3.3	4.0	-
DG B2	14	18	24	29	35	40	46	54	9	12	16	19	24	27	32	38	1.4	1.6	1.7	1.9	2.1	1.7	1.9	2.1	2.3	2.5
DG B3	18	24	31	37	46	53	61	71	13	18	23	28	35	40	47	57	1.3	1.3	1.4	1.6	1.7	1.4	1.5	1.7	1.8	1.9
DG C1	14	18	23	27	33	36	41	48	8	11	14	16	19	21	24	27	1.5	1.6	1.8	2.1	2.3	1.9	2.2	2.6	2.9	3.5
DG C2	16	22	28	34	41	47	54	63	11	15	20	24	29	33	39	46	1.3	1.4	1.6	1.7	1.9	1.6	1.7	1.9	2.0	2.2
DG C3	20	27	35	42	52	59	68	78	15	20	27	32	40	46	53	64	1.2	1.2	1.4	1.5	1.7	1.3	1.4	1.5	1.7	1.8
SS A1	20	25	30	34	40	44	50	57	16	19	23	26	31	34	38	44	1.4	1.6	1.7	1.9	2.0	1.6	1.8	2.0	2.3	2.5
SS A2	25	32	40	46	54	59	66	74	20	25	32	37	44	49	55	63	1.1	1.3	1.4	1.6	1.7	1.3	1.4	1.6	1.8	1.9
SS A3	18	22	27	31	36	39	44	51	14	17	20	23	26	29	32	36	1.5	1.7	1.9	2.1	2.2	1.7	2.0	2.3	2.6	2.9
SS A4	24	31	39	44	52	57	64	72	19	25	31	35	42	47	52	60	1.1	1.3	1.5	1.6	1.8	1.3	1.5	1.6	1.8	2.0
SS A5	20	24	28	31	36	40	45	51	16	19	22	24	28	30	34	38	1.5	1.7	1.9	2.1	2.2	1.8	2.0	2.3	2.5	2.8
SS A6	25	32	39	45	53	58	65	74	19	25	31	36	43	48	54	61	1.1	1.3	1.4	1.6	1.8	1.3	1.4	1.6	1.8	1.9
SS A7	18	21	24	27	31	34	37	43	15	17	19	21	23	25	27	30	1.7	1.9	2.2	2.4	2.5	2.0	2.4	2.7	3.1	3.5
SS A8	24	30	37	43	51	56	62	71	19	24	30	34	40	45	50	58	1.2	1.3	1.5	1.7	1.8	1.3	1.5	1.7	1.9	2.0
SS B1	17	21	26	30	35	39	44	51	13	16	19	22	26	29	32	38	1.5	1.7	1.8	2.0	2.2	1.7	2.0	2.2	2.4	2.6
SS B2	22	28	36	41	49	54	61	70	17	22	28	33	40	44	50	58	1.2	1.3	1.5	1.6	1.8	1.3	1.5	1.7	1.8	2.0
SS B3	16	19	24	27	32	35	40	46	12	14	17	20	23	25	28	33	1.5	1.8	2.0	2.1	2.3	1.9	2.1	2.4	2.7	2.9
SS B4	21	27	34	40	47	53	59	68	16	21	27	31	38	42	48	56	1.2	1.4	1.5	1.7	1.8	1.4	1.5	1.7	1.9	2.0
SS B5	16	20	23	26	30	33	37	43	13	15	18	20	23	25	28	31	1.7	1.9	2.1	2.3	2.4	2.0	2.3	2.5	2.8	3.1
SS B6	21	27	35	40	48	53	60	69	16	21	27	32	38	43	48	57	1.2	1.4	1.5	1.7	1.8	1.3	1.5	1.7	1.9	2.0
SS B7	15	18	21	23	27	29	32	37	12	14	16	17	19	21	23	26	1.8	2.1	2.3	2.5	2.7	2.2	2.6	2.9	3.3	3.6
SS B8	20	26	33	38	45	50	57	66	15	20	25	30	36	40	46	53	1.2	1.4	1.6	1.7	1.9	1.4	1.6	1.8	1.9	2.1
SS C1	15	19	25	29	35	40	45	53	9	12	16	19	23	26	30	35	1.4	1.6	1.8	1.9	2.1	1.7	2.0	2.2	2.4	2.6
SS C2	18	25	32	38	46	52	59	68	12	17	22	27	34	38	44	52	1.2	1.3	1.5	1.7	1.8	1.4	1.6	1.7	1.9	2.0
SS C3	14	17	21	24	29	32	37	43	10	12	14	17	20	22	25	29	1.6	1.8	2.0	2.2	2.3	2.1	2.3	2.5	2.8	3.0
SS C4	17	22	29	34	41	46	53	61	11	15	20	24	30	34	39	47	1.3	1.4	1.6	1.8	1.9	1.5	1.7	1.9	2.0	2.2
SS D1	23	29	35	40	47	52	58	66	17	21	27	30	36	40	44	51	1.2	1.4	1.6	1.7	1.9	1.4	1.6	1.8	2.1	2.2
SS D2	28	37	46	53	62	68	75	83	21	29	37	43	51	56	63	72	1.0	1.2	1.3	1.5	1.6	1.1	1.3	1.5	1.6	1.8
SS D3	22	26	32	36	42	47	52	60	17	20	25	28	32	36	40	46	1.3	1.5	1.7	1.9	2.0	1.5	1.8	2.0	2.2	2.4
SS D4	27	35	44	51	60	66	73	81	21	28	35	41	49	54	61	70	1.0	1.2	1.3	1.5	1.7	1.1	1.3	1.5	1.7	1.8
SS E1	19	24	30	34	41	45	51	59	13	17	21	24	29	32	36	42	1.3	1.5	1.7	1.8	2.0	1.6	1.8	2.1	2.3	2.5
SS E2	23	31	40	46	55	61	68	76	17	23	30	36	43	48	55	63	1.1	1.2	1.4	1.6	1.7	1.2	1.4	1.6	1.7	1.9
SS E3	18	21	26	29	35	38	43	50	13	16	19	22	25	28	31	36	1.5	1.7	1.9	2.1	2.2	1.8	2.1	2.3	2.6	2.8
SS E4	22	29	37	43	51	56	63	72	16	22	28	33	40	45	51	59	1.1	1.3	1.5	1.6	1.8	1.3	1.5	1.7	1.8	2.0

PHILADELPHIA, PENNSYLVANIA

TYPE	SOLAR SAVINGS FRACTION (%) BASE TEMP: 55 F (2795 DD)								65 F (4875 DD)								CONSERVATION FACTOR 55 F					65 F				
LCR	100	70	50	40	30	25	20	15	100	70	50	40	30	25	20	15	100	50	30	20	15	100	50	30	20	15
WW A1	19	22	26	29	34	37	42	48	14	16	19	21	23	25	28	32	1.6	1.8	2.0	2.1	2.3	2.1	2.3	2.6	2.8	3.1
WW A2	20	25	32	36	43	48	54	63	13	17	21	25	30	33	38	44	1.3	1.4	1.6	1.8	1.9	1.6	1.8	2.0	2.2	2.3
WW A3	20	26	34	39	47	52	59	68	12	17	22	26	32	36	41	49	1.2	1.4	1.5	1.7	1.8	1.5	1.6	1.8	2.0	2.2
WW A4	20	27	35	41	49	55	62	70	12	17	23	27	33	38	44	52	1.2	1.3	1.5	1.6	1.8	1.4	1.6	1.8	1.9	2.1
WW A5	20	28	36	42	51	57	64	73	12	17	23	28	35	40	46	54	1.1	1.3	1.4	1.6	1.7	1.3	1.5	1.7	1.9	2.0
WW A6	20	28	37	43	52	58	65	74	11	17	24	29	36	41	47	56	1.1	1.2	1.4	1.6	1.7	1.3	1.5	1.7	1.8	2.0
WW B1	15	19	24	27	33	36	41	48	10	12	14	16	19	21	23	26	1.5	1.7	1.9	2.1	2.3	2.1	2.4	2.7	3.1	3.5
WW B2	21	28	37	43	52	58	65	74	13	18	25	30	37	42	49	57	1.1	1.3	1.4	1.6	1.7	1.3	1.5	1.6	1.8	1.9
WW B3	25	34	43	50	60	66	73	81	16	23	31	37	45	50	57	67	1.0	1.2	1.3	1.5	1.6	1.2	1.3	1.5	1.7	1.8
WW B4	25	35	45	53	63	69	77	85	16	24	33	40	49	55	63	72	1.0	1.1	1.3	1.4	1.6	1.1	1.3	1.4	1.6	1.7
WW B5	24	34	45	52	63	69	76	85	15	23	33	40	49	55	63	73	1.0	1.1	1.3	1.4	1.6	1.1	1.3	1.4	1.6	1.7
WW C1	26	36	46	53	62	68	75	84	17	25	33	39	48	53	61	70	1.0	1.1	1.3	1.5	1.6	1.1	1.3	1.5	1.6	1.8
WW C2	25	34	44	52	61	68	75	83	16	23	32	38	47	53	61	70	1.0	1.1	1.3	1.5	1.6	1.1	1.3	1.5	1.6	1.8
WW C3	29	41	53	61	71	77	84	91	20	30	41	48	59	65	73	82	0.9	1.0	1.2	1.4	1.5	1.0	1.1	1.3	1.5	1.6
WW C4	27	38	49	58	68	74	81	89	18	27	37	45	55	62	70	80	0.9	1.1	1.2	1.4	1.5	1.0	1.2	1.3	1.5	1.6
TW A1	19	22	26	29	34	37	42	48	14	16	19	21	24	26	29	33	1.6	1.8	2.0	2.1	2.3	2.1	2.3	2.5	2.8	3.0
TW A2	19	23	29	33	40	44	50	58	13	16	20	23	27	31	35	41	1.4	1.5	1.7	1.9	2.0	1.7	1.9	2.1	2.3	2.4
TW A3	18	24	30	35	42	47	53	62	12	16	20	24	29	33	38	44	1.3	1.4	1.6	1.8	1.9	1.6	1.8	1.9	2.1	2.3
TW A4	17	23	30	35	43	48	55	63	11	15	20	24	29	33	39	46	1.3	1.4	1.6	1.7	1.9	1.5	1.7	1.9	2.0	2.2
TW B1	18	22	26	29	34	38	43	50	13	16	19	21	24	27	30	35	1.6	1.7	1.9	2.0	2.2	2.0	2.2	2.4	2.6	2.7
TW B2	17	22	28	32	39	43	49	58	12	15	19	22	27	30	35	41	1.4	1.5	1.7	1.8	2.0	1.7	1.9	2.0	2.2	2.4
TW B3	17	22	28	33	39	44	50	59	11	15	19	22	27	31	36	42	1.3	1.5	1.6	1.8	1.9	1.6	1.8	2.0	2.2	2.3
TW B4	16	21	26	31	38	42	48	57	10	14	18	21	26	30	34	41	1.4	1.5	1.7	1.8	2.0	1.7	1.8	2.0	2.2	2.3
TW C1	17	21	25	29	34	38	44	51	12	15	18	21	24	27	31	36	1.5	1.7	1.8	2.0	2.1	1.9	2.1	2.2	2.4	2.6
TW C2	16	20	25	29	35	40	45	53	11	14	18	21	25	28	32	39	1.5	1.6	1.8	1.9	2.1	1.8	1.9	2.1	2.3	2.4
TW C3	15	19	24	28	34	38	44	51	11	14	17	20	24	27	31	37	1.5	1.6	1.8	2.0	2.1	1.8	2.0	2.1	2.3	2.5
TW C4	15	18	22	26	31	35	40	47	11	13	16	19	23	25	29	34	1.6	1.8	1.9	2.1	2.2	2.0	2.1	2.3	2.5	2.6
TW D1	13	16	21	24	29	32	37	43	8	10	12	14	17	19	21	24	1.6	1.8	2.0	2.2	2.3	2.2	2.5	2.8	3.1	3.4
TW D2	20	26	34	39	47	53	60	69	13	18	23	28	34	39	45	53	1.2	1.3	1.5	1.7	1.8	1.4	1.6	1.7	1.9	2.0
TW D3	21	29	37	43	52	57	64	73	14	19	26	30	37	42	49	57	1.1	1.3	1.4	1.6	1.7	1.3	1.5	1.7	1.8	2.0
TW D4	23	31	41	48	57	63	71	79	15	22	30	35	44	49	57	66	1.0	1.2	1.4	1.5	1.7	1.2	1.4	1.5	1.7	1.8
TW D5	22	31	41	48	58	64	71	80	15	22	30	36	45	51	58	68	1.0	1.2	1.3	1.5	1.6	1.2	1.3	1.5	1.6	1.8
TW E1	24	32	41	48	57	63	70	79	16	23	30	35	43	48	55	64	1.1	1.2	1.4	1.5	1.7	1.2	1.4	1.6	1.7	1.9
TW E2	24	32	41	48	57	63	71	79	16	22	30	36	44	49	56	66	1.1	1.2	1.4	1.5	1.7	1.2	1.4	1.5	1.7	1.8
TW E3	29	39	50	58	68	74	81	88	20	28	38	45	55	61	68	78	0.9	1.1	1.2	1.4	1.6	1.1	1.2	1.4	1.5	1.7
TW E4	26	36	47	55	65	71	79	87	18	26	36	43	52	59	67	76	1.0	1.1	1.3	1.4	1.6	1.1	1.2	1.4	1.5	1.7
TW F1	17	20	25	28	33	36	41	48	12	15	17	19	22	25	28	32	1.6	1.8	1.9	2.1	2.2	2.0	2.2	2.5	2.7	2.9
TW F2	16	21	26	31	37	42	48	56	11	14	18	20	25	28	32	38	1.4	1.6	1.7	1.9	2.0	1.7	1.9	2.1	2.3	2.5
TW F3	15	20	27	31	38	43	50	58	9	13	17	20	25	29	34	40	1.3	1.5	1.6	1.8	2.0	1.6	1.8	2.0	2.2	2.3
TW F4	13	19	25	30	37	42	49	57	7	11	15	19	24	28	33	40	1.3	1.5	1.6	1.8	1.9	1.6	1.8	2.0	2.1	2.3
TW G1	15	19	23	27	32	36	41	48	10	13	16	18	21	24	27	32	1.6	1.7	1.9	2.0	2.2	2.0	2.2	2.4	2.6	2.8
TW G2	13	18	23	28	34	39	44	53	8	11	15	18	22	26	30	36	1.4	1.6	1.7	1.9	2.0	1.8	1.9	2.1	2.3	2.5
TW G3	12	16	22	26	33	37	44	52	6	10	14	17	21	25	29	36	1.4	1.6	1.7	1.9	2.0	1.8	1.9	2.1	2.3	2.4
TW G4	9	14	19	23	29	33	39	47	5	8	11	14	19	22	26	32	1.5	1.7	1.8	2.0	2.1	1.8	2.0	2.2	2.3	2.5
TW H1	12	16	20	24	29	33	38	46	7	10	13	15	19	22	25	31	1.6	1.7	1.9	2.1	2.2	2.0	2.1	2.3	2.5	2.7
TW H2	9	13	18	22	28	32	37	45	5	8	11	14	18	21	25	30	1.6	1.7	1.9	2.0	2.2	1.9	2.1	2.3	2.4	2.6
TW H3	8	11	16	19	24	28	34	41	4	6	9	12	16	18	22	28	1.6	1.8	2.0	2.1	2.2	2.0	2.2	2.3	2.5	2.7
TW H4	6	9	12	15	20	23	27	34	3	5	7	9	12	15	18	23	1.8	2.0	2.1	2.3	2.4	2.3	2.4	2.6	2.7	2.9
TW I1	11	14	18	21	26	29	33	39	6	8	10	12	14	16	18	21	1.7	1.9	2.1	2.3	2.4	2.3	2.6	2.9	3.3	3.6
TW I2	17	23	30	36	44	49	56	66	10	15	20	25	31	35	41	49	1.2	1.4	1.5	1.7	1.8	1.5	1.6	1.8	1.9	2.1
TW I3	18	25	33	39	48	53	61	70	11	16	23	27	34	39	45	54	1.2	1.3	1.5	1.6	1.8	1.4	1.5	1.7	1.9	2.0
TW I4	20	28	37	44	54	60	67	77	12	19	26	32	40	46	53	63	1.1	1.2	1.4	1.5	1.7	1.2	1.4	1.5	1.7	1.8
TW I5	20	29	38	45	55	62	69	78	13	19	27	34	42	48	56	66	1.1	1.2	1.4	1.5	1.7	1.2	1.3	1.5	1.7	1.8
TW J1	22	30	39	45	55	61	68	77	14	20	28	33	41	46	53	62	1.1	1.2	1.4	1.5	1.7	1.3	1.4	1.6	1.7	1.9
TW J2	21	29	38	45	55	61	68	77	13	20	27	33	41	47	54	63	1.1	1.2	1.4	1.5	1.7	1.2	1.4	1.5	1.7	1.8
TW J3	25	36	47	55	65	71	78	86	17	25	35	42	52	58	66	76	1.0	1.1	1.3	1.4	1.6	1.1	1.2	1.4	1.5	1.7
TW J4	24	34	44	52	62	69	76	85	15	24	33	40	50	56	64	74	1.0	1.1	1.3	1.4	1.6	1.1	1.2	1.4	1.6	1.7
DG A1	13	17	22	25	29	32	35	38	8	10	12	14	17	18	19	20	1.5	1.8	2.2	2.7	3.4	2.0	2.4	3.0	4.5	-
DG A2	16	21	27	32	39	43	48	55	10	14	18	22	26	30	34	39	1.3	1.5	1.7	2.0	2.2	1.6	1.8	2.0	2.3	2.6
DG A3	20	26	34	41	50	56	63	71	14	19	25	30	37	42	49	58	1.2	1.3	1.4	1.6	1.8	1.4	1.5	1.6	1.8	2.0
DG B1	13	18	23	27	32	36	40	46	7	10	13	15	17	19	22	24	1.5	1.6	1.9	2.2	2.6	2.0	2.3	2.7	3.3	4.0
DG B2	16	22	28	34	41	46	53	61	11	14	19	22	28	31	37	43	1.3	1.4	1.6	1.8	2.0	1.6	1.8	1.9	2.1	2.3
DG B3	20	27	36	42	52	59	67	76	14	20	26	31	39	44	52	62	1.2	1.3	1.4	1.5	1.7	1.4	1.5	1.6	1.7	1.8
DG C1	16	22	28	33	39	44	50	57	10	13	17	19	23	26	30	34	1.3	1.5	1.7	1.9	2.1	1.7	2.0	2.2	2.5	2.9
DG C2	19	25	33	39	48	54	61	70	13	17	23	27	33	38	44	52	1.2	1.3	1.5	1.6	1.8	1.4	1.6	1.7	1.9	2.1
DG C3	23	31	40	48	58	65	74	83	16	22	29	35	44	50	58	69	1.1	1.2	1.3	1.4	1.6	1.3	1.4	1.5	1.6	1.7
SS A1	23	28	34	39	46	50	56	64	17	21	26	29	34	38	43	49	1.3	1.4	1.6	1.8	1.9	1.5	1.7	1.9	2.1	2.3
SS A2	28	36	44	51	59	65	71	80	22	28	35	40	48	53	59	68	1.1	1.2	1.4	1.5	1.7	1.2	1.4	1.5	1.7	1.8
SS A3	21	26	31	35	41	45	51	58	16	19	23	26	30	33	37	42	1.4	1.5	1.7	1.9	2.1	1.6	1.9	2.1	2.3	2.5
SS A4	28	35	43	49	58	63	70	78	21	27	34	39	46	51	57	65	1.1	1.2	1.4	1.6	1.7	1.2	1.4	1.6	1.7	1.9
SS A5	22	27	32	36	41	45	51	58	17	21	24	27	31	34	38	43	1.4	1.6	1.7	1.9	2.1	1.7	1.9	2.1	2.3	2.5
SS A6	28	36	44	50	59	64	71	79	21	27	34	40	47	52	58	66	1.1	1.2	1.4	1.5	1.7	1.2	1.4	1.5	1.7	1.9
SS A7	20	24	28	31	36	39	43	49	16	18	21	23	26	29	31	35	1.5	1.7	1.9	2.1	2.3	1.9	2.2	2.4	2.7	3.0
SS A8	27	34	42	48	56	62	68	76	21	26	33	38	44	49	55	63	1.1	1.3	1.4	1.6	1.7	1.2	1.4	1.6	1.8	1.9
SS B1	19	24	30	34	40	44	50	58	14	18	22	25	29	33	37	43	1.4	1.5	1.7	1.9	2.0	1.6	1.8	2.0	2.2	2.4
SS B2	25	32	40	46	54	60	67	75	19	24	31	36	43	48	54	63	1.1	1.3	1.4	1.6	1.7	1.3	1.4	1.6	1.8	1.9
SS B3	18	22	27	31	37	41	46	54	13	16	20	22	26	29	33	38	1.4	1.6	1.8	2.0	2.1	1.7	2.0	2.2	2.4	2.6
SS B4	24	31	39	45	53	58	65	74	18	23	30	35	42	46	53	61	1.1	1.3	1.4	1.6	1.8	1.3	1.5	1.6	1.8	1.9
SS B5	18	22	26	30	35	38	43	50	14	17	20	22	26	28	31	36	1.5	1.7	1.9	2.1	2.2	1.9	2.1	2.3	2.6	2.8
SS B6	24	31	39	45	53	59	65	74	18	23	30	35	42	47	53	61	1.1	1.3	1.4	1.6	1.7	1.3	1.4	1.6	1.8	1.9
SS B7	16	20	23	26	31	34	38	44	13	15	17	19	22	24	27	30	1.7	1.9	2.1	2.2	2.4	2.1	2.3	2.6	2.9	3.1
SS B8	23	29	37	43	51	56	63	71	17	22	28	33	39	44	50	58	1.2	1.3	1.5	1.6	1.8	1.3	1.5	1.7	1.8	2.0
SS C1	17	23	29	34	41	46	53	61	11	14	19	22	27	30	35	42	1.3	1.5	1.6	1.8	1.9	1.6	1.8	2.0	2.2	2.4
SS C2	21	29	37	43	52	58	65	74	14	19	26	31	38	43	49	58	1.1	1.3	1.4	1.6	1.7	1.3	1.5	1.6	1.8	1.9
SS C3	16	20	24	28	34	38	43	50	11	13	17	19	23	25	29	34	1.5	1.7	1.8	2.0	2.1	1.9	2.1	2.3	2.5	2.7
SS C4	19	26	33	39	47	52	59	68	13	18	23	27	34	38	44	52	1.2	1.4	1.5	1.7	1.8	1.4	1.6	1.7	1.9	2.1
SS D1	26	33	40	46	54	59	65	73	19	24	30	34	41	45	50	57	1.1	1.3	1.5	1.6	1.8	1.3	1.5	1.7	1.9	2.1
SS D2	32	41	51	59	68	73	80	87	24	32	40	47	55	61	68	76	0.9	1.1	1.3	1.4	1.6	1.1	1.2	1.4	1.6	1.7
SS D3	24	30	36	41	48	53	59	67	19	23	28	31	36	40	45	51	1.2	1.4	1.6	1.7	1.9	1.4	1.6	1.9	2.1	2.2
SS D4	31	40	50	57	66	71	78	85	23	31	39	45	53	59	66	74	1.0	1.1	1.3	1.5	1.6	1.1	1.3	1.4	1.6	1.7
SS E1	22	28	35	40	47	52	58	66	15	20	25	28	34	37	42	48	1.2	1.4	1.5	1.7	1.9	1.4	1.7	1.9	2.1	2.3
SS E2	27	36	45	52	61	67	74	82	19	26	34	40	48	53	60	68	1.0	1.2	1.3	1.5	1.6	1.2	1.3	1.5	1.7	1.8
SS E3	20	24	30	34	40	44	49	57	15	18	22	25	29	32	36	41	1.4	1.6	1.7	1.9	2.1	1.7	1.9	2.1	2.3	2.5
SS E4	25	33	42	48	57	62	69	77	18	25	31	37	44	49	55	64	1.1	1.2	1.4	1.6	1.7	1.2	1.4	1.6	1.7	1.9

PITTSBURGH, PENNSYLVANIA

TYPE	SOLAR SAVINGS FRACTION (%)																CONSERVATION FACTOR									
	BASE TEMP: 55 F (3645 DD)								65 F (5937 DD)								55 F					65 F				
LCR	100	70	50	40	30	25	20	15	100	70	50	40	30	25	20	15	100	50	30	20	15	100	50	30	20	15
WW A1	14	16	17	19	21	22	23	26	12	12	13	14	15	15	15	15	2.3	2.7	3.2	3.7	4.2	3.2	4.1	-	-	-
WW A2	12	15	19	21	25	27	31	35	9	10	13	14	16	18	20	22	1.7	2.0	2.3	2.6	2.8	2.2	2.6	3.1	3.7	4.5
WW A3	11	15	19	22	26	29	33	39	7	9	12	14	17	19	21	24	1.6	1.9	2.1	2.3	2.5	2.0	2.3	2.7	3.2	3.7
WW A4	10	14	19	22	27	31	35	41	6	9	12	14	17	19	22	26	1.5	1.8	2.0	2.2	2.4	1.9	2.2	2.5	3.0	3.4
WW A5	10	14	19	23	28	32	37	44	5	8	12	14	18	20	23	27	1.5	1.7	1.9	2.1	2.3	1.8	2.1	2.4	2.8	3.2
WW A6	9	14	19	23	29	33	38	45	4	8	11	14	18	21	24	28	1.4	1.7	1.9	2.1	2.3	1.7	2.0	2.3	2.7	3.0
WW B1	9	10	12	13	14	15	16	16	6	6	6	6	6	5	4	1	2.4	3.1	4.2	-	-	4.1	-	-	-	-
WW B2	11	16	22	26	31	36	41	49	7	11	15	18	22	25	29	35	1.4	1.6	1.8	2.0	2.2	1.7	1.9	2.1	2.4	2.6
WW B3	14	20	26	31	38	43	49	58	9	13	19	23	28	32	37	43	1.3	1.5	1.7	1.8	2.0	1.5	1.7	1.9	2.1	2.3
WW B4	14	21	29	35	43	49	56	65	9	15	21	26	33	38	45	53	1.2	1.4	1.5	1.7	1.8	1.3	1.5	1.7	1.9	2.0
WW B5	13	21	29	35	43	49	57	66	8	15	22	27	35	40	47	56	1.2	1.3	1.5	1.7	1.8	1.3	1.5	1.6	1.8	1.9
WW C1	15	22	29	34	41	46	53	61	10	15	21	25	31	35	41	48	1.2	1.4	1.6	1.8	1.9	1.4	1.6	1.8	2.0	2.2
WW C2	14	21	28	33	41	46	53	62	9	14	20	25	32	36	42	50	1.2	1.4	1.6	1.7	1.9	1.4	1.6	1.7	1.9	2.1
WW C3	17	26	36	43	53	59	67	76	12	20	28	35	44	50	57	67	1.1	1.2	1.4	1.5	1.7	1.1	1.3	1.5	1.6	1.8
WW C4	15	24	33	40	50	56	64	74	10	18	26	32	41	47	55	64	1.1	1.2	1.4	1.6	1.7	1.2	1.3	1.5	1.7	1.8
TW A1	14	16	18	19	21	22	24	27	12	12	13	14	15	16	16	17	2.2	2.6	3.0	3.5	3.8	3.0	3.7	4.6	-	-
TW A2	12	15	18	20	24	26	29	33	9	11	13	14	16	18	19	22	1.8	2.1	2.4	2.7	2.9	2.3	2.7	3.1	3.7	4.3
TW A3	11	14	18	21	25	27	31	36	8	10	12	14	17	18	21	23	1.7	2.0	2.2	2.5	2.7	2.1	2.4	2.8	3.3	3.7
TW A4	10	13	17	20	25	28	32	38	6	9	11	14	17	19	21	25	1.6	1.9	2.1	2.3	2.5	2.0	2.3	2.6	3.0	3.3
TW B1	13	15	17	19	22	23	26	29	11	12	13	14	16	16	18	19	2.1	2.4	2.8	3.1	3.4	2.7	3.2	3.7	4.5	-
TW B2	11	14	17	20	23	26	29	34	8	10	12	14	16	18	20	23	1.8	2.1	2.3	2.6	2.8	2.2	2.5	2.9	3.3	3.7
TW B3	10	13	17	20	23	26	30	35	7	9	12	14	16	18	21	24	1.7	2.0	2.2	2.5	2.6	2.1	2.4	2.7	3.1	3.4
TW B4	10	13	16	19	23	26	29	35	7	9	11	13	16	18	21	24	1.8	2.0	2.2	2.5	2.6	2.1	2.4	2.7	3.0	3.3
TW C1	12	14	17	19	22	24	27	31	9	11	13	14	16	17	19	21	2.0	2.3	2.5	2.8	3.0	2.5	2.8	3.2	3.6	4.1
TW C2	10	13	16	19	22	25	28	33	8	10	12	14	16	18	20	23	1.9	2.1	2.4	2.6	2.8	2.3	2.6	2.9	3.2	3.5
TW C3	10	13	16	18	22	24	27	32	8	10	12	13	16	18	20	23	1.9	2.1	2.4	2.6	2.8	2.3	2.6	2.9	3.2	3.4
TW C4	11	13	16	18	21	23	26	30	9	10	12	14	16	17	20	23	2.0	2.3	2.5	2.7	2.9	2.5	2.7	3.0	3.3	3.5
TW D1	7	9	11	12	13	14	15	17	5	5	6	6	6	6	6	5	2.4	3.0	3.7	4.7	-	3.7	-	-	-	-
TW D2	12	16	21	24	29	33	38	45	8	11	15	17	21	24	28	33	1.5	1.7	1.9	2.1	2.3	1.8	2.0	2.3	2.5	2.7
TW D3	12	17	22	26	32	36	41	49	8	12	16	19	23	26	30	36	1.4	1.6	1.8	2.0	2.2	1.7	1.9	2.1	2.4	2.6
TW D4	14	20	26	31	38	43	50	59	9	14	20	24	30	34	40	48	1.3	1.5	1.6	1.8	1.9	1.4	1.6	1.8	2.0	2.2
TW D5	14	20	27	32	40	45	52	61	9	14	20	25	32	37	43	51	1.3	1.4	1.6	1.7	1.9	1.4	1.6	1.7	1.9	2.0
TW E1	15	20	26	31	37	42	48	56	10	15	19	23	28	32	36	43	1.3	1.5	1.7	1.9	2.0	1.5	1.7	1.9	2.2	2.4
TW E2	14	20	26	31	38	43	49	58	10	14	20	24	30	34	39	46	1.3	1.5	1.7	1.8	2.0	1.5	1.7	1.9	2.1	2.2
TW E3	18	25	34	40	49	55	62	71	12	19	26	32	40	45	52	61	1.1	1.3	1.4	1.6	1.8	1.2	1.4	1.6	1.7	1.9
TW E4	16	24	32	38	47	53	60	70	11	18	25	31	39	44	51	60	1.1	1.3	1.5	1.6	1.8	1.3	1.4	1.6	1.7	1.9
TW F1	12	14	16	17	20	21	23	26	9	10	12	12	13	14	15	16	2.2	2.6	3.0	3.4	3.7	2.9	3.5	4.4	-	-
TW F2	10	13	16	18	21	23	26	31	7	9	10	12	14	15	17	19	1.9	2.2	2.5	2.7	3.0	2.4	2.7	3.2	3.8	4.4
TW F3	8	11	15	17	21	24	27	33	5	7	9	11	14	15	17	20	1.8	2.0	2.3	2.5	2.7	2.2	2.5	2.9	3.3	3.8
TW F4	6	10	13	16	20	23	27	33	3	5	8	10	13	15	17	21	1.7	2.0	2.2	2.5	2.6	2.1	2.4	2.7	3.1	3.4
TW G1	10	12	14	16	19	20	23	26	7	9	10	11	12	13	15	16	2.1	2.4	2.8	3.1	3.4	2.7	3.2	3.8	4.5	-
TW G2	7	10	13	16	19	21	25	29	5	6	8	10	12	14	16	18	1.9	2.2	2.4	2.7	2.9	2.3	2.7	3.1	3.5	4.0
TW G3	6	9	12	14	18	21	24	29	3	5	7	9	11	13	15	19	1.8	2.1	2.4	2.6	2.8	2.3	2.6	2.9	3.3	3.6
TW G4	4	7	10	12	16	18	22	27	2	4	6	7	10	12	14	18	1.9	2.2	2.4	2.6	2.8	2.3	2.6	2.9	3.2	3.5
TW H1	7	9	12	14	17	19	21	25	5	6	8	9	11	12	14	16	2.1	2.4	2.7	3.0	3.2	2.6	3.0	3.4	3.9	4.4
TW H2	5	7	10	12	15	18	21	25	2	4	6	7	9	11	13	16	2.0	2.3	2.5	2.8	3.0	2.5	2.8	3.1	3.5	3.8
TW H3	3	6	8	10	14	16	19	23	2	3	5	6	8	10	12	15	2.1	2.3	2.6	2.8	3.0	2.6	2.8	3.1	3.4	3.7
TW H4	2	4	7	8	11	13	16	20	1	2	4	5	7	8	10	13	2.3	2.5	2.8	3.0	3.2	2.8	3.1	3.3	3.6	3.8
TW I1	6	7	9	10	11	12	13	14	3	4	4	4	4	4	4	3	2.5	3.1	3.9	-	-	4.0	-	-	-	-
TW I2	9	13	18	21	26	30	35	42	6	9	12	15	18	21	25	30	1.6	1.8	2.0	2.2	2.3	1.8	2.1	2.3	2.6	2.8
TW I3	10	14	20	23	29	33	39	46	6	9	10	17	21	24	28	34	1.5	1.7	1.9	2.0	2.2	1.7	1.9	2.1	2.4	2.6
TW I4	11	17	23	28	35	40	47	56	7	12	17	21	27	31	37	45	1.3	1.5	1.7	1.8	2.0	1.5	1.7	1.8	2.0	2.2
TW I5	11	17	24	30	37	43	50	59	7	12	18	23	29	34	40	49	1.3	1.4	1.6	1.8	1.9	1.4	1.6	1.8	1.9	2.1
TW J1	13	18	24	29	35	40	46	54	8	13	17	21	26	30	35	42	1.3	1.5	1.7	1.9	2.0	1.5	1.7	1.9	2.2	2.3
TW J2	12	18	24	29	36	41	47	56	8	12	17	21	27	31	37	44	1.3	1.5	1.7	1.8	2.0	1.5	1.7	1.9	2.1	2.2
TW J3	15	23	31	37	46	52	60	70	10	17	24	30	38	43	50	60	1.1	1.3	1.5	1.6	1.8	1.2	1.4	1.6	1.7	1.9
TW J4	14	21	29	36	44	50	58	68	9	15	23	28	36	42	49	58	1.2	1.3	1.5	1.6	1.8	1.3	1.4	1.6	1.8	1.9
DG A1	7	9	10	12	13	13	14	13	4	5	5	5	5	5	4	2	2.2	3.0	4.4	-	-	3.3	-	-	-	-
DG A2	10	13	16	19	23	25	29	33	7	9	11	13	16	18	20	23	1.8	2.0	2.3	2.6	3.0	2.1	2.4	2.8	3.3	3.9
DG A3	13	18	23	27	33	37	43	51	10	14	18	21	26	30	35	41	1.5	1.6	1.8	1.9	2.1	1.7	1.8	2.0	2.2	2.4
DG B1	7	9	11	12	13	14	15	16	4	4	5	6	6	6	5	3	2.2	3.0	3.6	-	-	3.3	4.8	-	-	-
DG B2	10	13	17	20	24	27	30	36	7	9	12	14	17	19	21	25	1.7	2.0	2.2	2.5	2.7	2.1	2.3	2.7	3.1	3.4
DG B3	14	18	24	28	35	39	46	54	10	14	19	22	28	32	37	44	1.4	1.6	1.7	1.9	2.0	1.6	1.8	1.9	2.1	2.2
DG C1	9	12	14	16	18	20	22	25	5	7	8	9	10	11	11	11	1.9	2.4	2.7	3.4	4.3	2.6	3.2	4.7	-	-
DG C2	12	16	20	23	29	32	37	43	9	11	15	17	21	24	27	32	1.6	1.8	1.9	2.2	2.3	1.9	2.0	2.3	2.6	2.9
DG C3	16	21	27	32	39	44	51	61	12	16	21	26	32	36	42	50	1.4	1.5	1.6	1.7	1.9	1.5	1.6	1.8	1.9	2.1
SS A1	16	19	23	26	30	32	36	41	13	16	19	21	23	25	28	31	1.6	1.9	2.2	2.4	2.6	1.9	2.2	2.6	3.0	3.5
SS A2	20	25	31	36	42	47	53	60	15	21	26	30	35	39	44	51	1.3	1.5	1.7	1.8	2.0	1.4	1.6	1.8	2.0	2.2
SS A3	15	17	20	23	26	28	30	34	12	14	16	18	20	21	22	24	1.8	2.1	2.5	2.8	3.1	2.1	2.6	3.2	4.1	-
SS A4	19	24	30	34	41	45	50	58	16	20	25	29	34	37	42	48	1.3	1.5	1.7	1.9	2.0	1.4	1.7	1.9	2.1	2.3
SS A5	16	19	22	24	27	29	32	36	14	16	18	20	22	23	25	27	1.8	2.1	2.5	2.8	3.0	2.1	2.5	3.1	3.7	4.5
SS A6	19	25	30	35	41	46	51	59	16	20	25	29	34	38	42	49	1.3	1.5	1.7	1.9	2.0	1.4	1.7	1.9	2.1	2.3
SS A7	15	17	19	21	23	24	26	28	13	14	16	17	18	19	19	20	2.1	2.5	3.0	3.5	4.0	2.5	3.2	4.2	-	-
SS A8	19	24	29	33	39	43	48	55	15	19	24	27	32	35	39	45	1.3	1.6	1.8	1.9	2.1	1.5	1.7	2.0	2.3	2.5
SS B1	13	16	19	22	25	28	31	36	11	13	15	17	19	21	23	26	1.8	2.1	2.3	2.6	2.8	2.1	2.4	2.8	3.3	3.8
SS B2	17	22	28	32	38	43	48	56	14	18	23	27	32	36	40	47	1.3	1.5	1.7	1.9	2.0	1.5	1.7	1.9	2.1	2.3
SS B3	12	15	18	20	23	25	27	31	10	11	13	15	17	18	19	21	1.9	2.2	2.6	2.9	3.2	2.2	2.7	3.3	4.2	-
SS B4	17	21	27	31	37	41	46	54	13	17	22	25	30	34	38	44	1.4	1.6	1.8	1.9	2.1	1.5	1.7	2.0	2.2	2.4
SS B5	13	16	18	20	22	24	27	30	11	13	14	16	17	19	20	21	2.0	2.4	2.7	3.1	3.4	2.4	2.9	3.4	4.2	-
SS B6	16	21	27	31	37	41	47	54	13	17	22	25	30	34	38	45	1.4	1.6	1.8	1.9	2.1	1.5	1.7	1.9	2.2	2.3
SS B7	12	14	16	17	19	21	22	25	10	12	13	14	15	15	16	17	2.2	2.6	3.1	3.7	4.1	2.8	3.4	4.3	-	-
SS B8	16	20	25	29	35	39	44	51	12	16	20	24	28	31	35	41	1.4	1.6	1.8	2.0	2.2	1.6	1.8	2.0	2.3	2.5
SS C1	10	13	16	18	22	24	28	32	6	8	10	11	13	15	16	18	1.8	2.1	2.4	2.7	2.9	2.2	2.7	3.3	4.1	-
SS C2	12	17	22	26	32	37	42	50	8	12	16	19	23	27	31	37	1.4	1.6	1.8	2.0	2.1	1.6	1.9	2.1	2.3	2.5
SS C3	10	12	15	17	20	21	24	27	8	9	11	12	13	14	16	17	2.0	2.4	2.7	3.0	3.3	2.6	3.1	3.6	4.3	-
SS C4	12	16	21	24	29	33	38	45	8	11	15	17	21	24	28	33	1.5	1.7	1.9	2.1	2.3	1.8	2.0	2.3	2.5	2.7
SS D1	17	21	26	29	34	37	42	48	13	17	20	23	26	28	31	35	1.5	1.7	1.9	2.2	2.4	1.7	2.0	2.4	2.8	3.2
SS D2	21	28	36	41	49	54	61	69	17	23	29	34	41	45	51	59	1.1	1.3	1.5	1.7	1.8	1.2	1.5	1.7	1.9	2.0
SS D3	17	21	24	27	31	34	38	43	14	17	19	22	24	26	29	31	1.6	1.9	2.1	2.4	2.6	1.8	2.2	2.6	3.1	3.6
SS D4	21	27	34	40	47	52	59	67	17	22	28	33	39	44	49	57	1.2	1.4	1.5	1.7	1.9	1.3	1.5	1.7	1.9	2.1
SS E1	14	17	21	24	27	30	34	38	10	12	15	17	19	21	23	24	1.6	1.9	2.2	2.5	2.7	1.9	2.4	2.9	3.7	4.8
SS E2	17	23	30	34	41	46	52	60	13	18	23	27	33	37	42	48	1.3	1.4	1.6	1.8	2.0	1.4	1.6	1.8	2.1	2.2
SS E3	14	17	19	22	25	27	30	33	11	13	15	17	19	20	21	23	1.8	2.2	2.5	2.8	3.1	2.2	2.6	3.2	3.9	4.9
SS E4	17	22	28	32	38	43	48	56	13	17	22	26	31	34	39	45	1.3	1.5	1.7	1.9	2.0	1.5	1.7	1.9	2.2	2.4

SCRANTON, PENNSYLVANIA

TYPE	SOLAR SAVINGS FRACTION (%) BASE TEMP: 55 F (3932 DD)								65 F (6280 DD)								CONSERVATION FACTOR 55 F					65 F				
LCR	100	70	50	40	30	25	20	15	100	70	50	40	30	25	20	15	100	50	30	20	15	100	50	30	20	15
WW A1	14	16	18	19	21	23	25	28	12	13	14	14	15	16	17	17	2.2	2.6	3.0	3.3	3.7	3.0	3.7	4.6	-	-
WW A2	13	16	19	22	26	29	33	38	9	11	13	15	18	19	22	25	1.7	1.9	2.2	2.4	2.6	2.1	2.4	2.8	3.3	3.7
WW A3	12	15	20	23	28	32	36	43	7	10	13	15	19	21	24	28	1.6	1.8	2.0	2.2	2.4	1.9	2.2	2.5	2.9	3.2
WW A4	11	15	20	24	29	33	38	45	6	9	13	15	19	22	25	29	1.5	1.7	1.9	2.1	2.3	1.8	2.1	2.4	2.7	3.0
WW A5	10	15	20	25	30	35	40	47	6	9	13	16	20	23	26	31	1.4	1.6	1.8	2.0	2.2	1.7	2.0	2.3	2.5	2.8
WW A6	10	15	21	25	31	35	41	49	5	9	13	16	20	23	27	32	1.4	1.6	1.8	2.0	2.1	1.7	1.9	2.2	2.5	2.7
WW B1	9	11	13	14	16	17	18	19	6	6	7	7	7	7	6	4	2.3	2.8	3.5	4.4	-	3.6	-	-	-	-
WW B2	12	17	23	27	33	38	44	52	7	11	16	19	24	27	32	38	1.4	1.6	1.8	1.9	2.1	1.6	1.8	2.0	2.3	2.4
WW B3	15	21	28	33	40	45	52	61	9	14	20	24	30	34	40	47	1.3	1.4	1.6	1.8	1.9	1.4	1.6	1.8	2.0	2.2
WW B4	15	22	30	36	45	51	58	68	9	15	22	28	35	41	47	56	1.2	1.3	1.5	1.6	1.8	1.3	1.5	1.6	1.8	1.9
WW B5	14	21	30	36	45	51	59	69	9	15	23	28	36	42	49	58	1.2	1.3	1.5	1.6	1.8	1.3	1.4	1.6	1.8	1.9
WW C1	16	23	30	35	43	49	55	64	11	16	22	27	33	38	44	51	1.2	1.4	1.5	1.7	1.9	1.4	1.6	1.7	1.9	2.1
WW C2	15	21	29	35	43	49	56	65	10	15	22	26	34	38	45	53	1.2	1.4	1.5	1.7	1.8	1.3	1.5	1.7	1.9	2.0
WW C3	18	27	37	45	55	61	69	78	13	21	30	36	46	52	60	70	1.0	1.2	1.3	1.5	1.7	1.1	1.3	1.4	1.6	1.8
WW C4	16	25	35	42	52	58	66	76	11	18	27	34	43	49	57	67	1.1	1.2	1.4	1.5	1.7	1.2	1.3	1.5	1.6	1.8
TW A1	14	16	18	19	22	23	26	29	12	13	14	15	16	17	18	19	2.2	2.5	2.8	3.2	3.4	2.9	3.4	4.1	5.0	-
TW A2	12	15	19	21	25	27	31	36	9	11	13	15	17	19	21	24	1.8	2.0	2.3	2.5	2.7	2.2	2.5	2.9	3.3	3.7
TW A3	11	15	19	22	26	29	33	39	8	10	13	15	18	20	23	26	1.7	1.9	2.1	2.3	2.5	2.0	2.3	2.6	3.0	3.3
TW A4	10	14	18	21	26	30	34	40	7	9	12	15	18	20	23	28	1.6	1.8	2.0	2.2	2.4	1.9	2.2	2.5	2.8	3.0
TW B1	13	15	18	20	22	24	27	31	11	12	14	15	16	17	19	21	2.0	2.3	2.6	2.9	3.1	2.6	3.0	3.4	4.0	4.5
TW B2	11	14	18	20	24	27	31	36	8	10	13	15	17	19	22	25	1.8	2.0	2.2	2.4	2.6	2.1	2.4	2.7	3.1	3.4
TW B3	10	14	17	20	25	28	32	38	7	10	12	14	17	19	22	26	1.7	1.9	2.1	2.3	2.5	2.1	2.3	2.6	2.9	3.1
TW B4	10	13	17	20	24	27	31	37	7	9	12	14	17	19	22	26	1.7	2.0	2.2	2.4	2.5	2.1	2.3	2.6	2.8	3.1
TW C1	12	14	17	19	22	25	28	32	10	11	13	14	17	18	20	23	1.9	2.2	2.4	2.7	2.8	2.4	2.7	3.0	3.4	3.7
TW C2	11	13	17	19	23	26	29	35	8	10	12	14	17	19	21	25	1.8	2.1	2.3	2.5	2.6	2.2	2.5	2.7	3.0	3.2
TW C3	10	13	16	19	22	25	29	34	8	10	12	14	17	19	21	25	1.9	2.1	2.3	2.5	2.7	2.3	2.5	2.7	3.0	3.2
TW C4	11	13	16	18	21	24	27	32	9	11	13	14	16	18	21	24	2.0	2.2	2.4	2.6	2.8	2.4	2.6	2.9	3.1	3.3
TW D1	8	9	11	12	14	15	17	19	5	6	6	7	7	7	7	7	2.3	2.8	3.3	3.9	4.5	3.4	4.6	-	-	-
TW D2	12	17	22	25	31	35	40	48	9	12	16	19	23	26	30	35	1.5	1.7	1.9	2.0	2.2	1.7	1.9	2.2	2.4	2.6
TW D3	13	18	23	27	34	38	44	52	9	12	17	20	25	28	32	39	1.4	1.6	1.8	1.9	2.1	1.6	1.8	2.1	2.3	2.5
TW D4	14	20	27	33	40	45	52	61	10	15	21	25	31	36	42	50	1.3	1.4	1.6	1.7	1.9	1.4	1.6	1.8	1.9	2.1
TW D5	14	20	28	33	42	47	54	64	10	15	21	26	33	38	45	54	1.2	1.4	1.5	1.7	1.8	1.4	1.5	1.7	1.8	2.0
TW E1	15	21	27	32	39	44	50	58	11	15	20	24	30	34	39	46	1.3	1.5	1.6	1.8	2.0	1.5	1.7	1.9	2.1	2.2
TW E2	15	21	28	33	40	45	52	61	10	15	21	25	31	35	41	49	1.3	1.4	1.6	1.8	1.9	1.4	1.6	1.8	2.0	2.1
TW E3	18	26	35	41	51	57	64	73	13	20	28	33	42	47	54	64	1.1	1.3	1.4	1.6	1.7	1.2	1.4	1.5	1.7	1.8
TW E4	17	24	33	40	49	55	63	72	12	18	26	32	40	46	53	63	1.1	1.3	1.4	1.6	1.7	1.2	1.4	1.5	1.7	1.8
TW F1	12	14	16	18	20	22	24	28	10	11	12	13	14	15	16	18	2.1	2.5	2.8	3.1	3.4	2.8	3.3	3.9	4.7	-
TW F2	10	13	16	19	22	25	28	33	7	9	11	13	15	17	19	21	1.8	2.1	2.3	2.6	2.7	2.3	2.6	3.0	3.4	3.8
TW F3	9	12	15	18	23	26	30	35	5	7	10	12	15	17	19	23	1.7	2.0	2.2	2.4	2.6	2.1	2.4	2.7	3.1	3.4
TW F4	7	10	14	17	21	25	29	35	3	6	9	11	14	16	19	23	1.7	1.9	2.1	2.3	2.5	2.1	2.3	2.6	2.9	3.1
TW G1	10	12	15	17	19	21	24	28	8	9	10	12	13	14	16	18	2.1	2.3	2.6	2.9	3.1	2.6	3.0	3.5	4.0	4.5
TW G2	8	10	14	16	20	23	26	31	5	7	9	11	13	15	17	20	1.8	2.1	2.3	2.5	2.7	2.3	2.6	2.9	3.2	3.6
TW G3	6	9	12	15	19	22	26	31	3	5	8	10	12	14	17	21	1.8	2.0	2.3	2.5	2.6	2.2	2.5	2.7	3.0	3.3
TW G4	4	7	10	13	17	19	23	29	2	4	6	8	11	13	15	19	1.9	2.1	2.3	2.5	2.7	2.3	2.5	2.8	3.0	3.3
TW H1	7	9	12	14	17	20	23	27	5	6	8	9	11	13	15	17	2.0	2.3	2.5	2.8	3.0	2.5	2.9	3.2	3.6	4.0
TW H2	5	7	10	13	16	18	22	27	2	4	6	8	10	12	14	18	2.0	2.2	2.4	2.7	2.8	2.4	2.7	3.0	3.3	3.5
TW H3	4	6	9	11	14	17	20	25	2	3	5	7	9	11	13	17	2.1	2.3	2.5	2.7	2.9	2.5	2.7	3.0	3.3	3.5
TW H4	3	4	7	9	11	14	16	21	1	2	4	5	7	9	11	14	2.3	2.5	2.7	2.9	3.1	2.7	3.0	3.2	3.5	3.7
TW I1	6	7	9	10	12	13	15	16	3	4	5	5	5	5	5	5	2.4	2.9	3.5	4.2	4.8	3.7	5.0	-	-	-
TW I2	10	14	18	22	28	32	37	44	6	9	13	16	20	23	27	32	1.5	1.7	1.9	2.1	2.2	1.8	2.0	2.2	2.4	2.6
TW I3	10	15	20	25	31	35	41	49	6	10	14	18	22	26	30	37	1.4	1.6	1.8	2.0	2.1	1.7	1.9	2.1	2.3	2.5
TW I4	12	17	24	29	37	42	49	58	7	12	18	22	29	33	39	47	1.3	1.5	1.6	1.8	1.9	1.4	1.6	1.8	2.0	2.1
TW I5	12	18	25	31	39	45	52	61	7	13	19	24	31	36	42	51	1.3	1.4	1.6	1.7	1.9	1.4	1.5	1.7	1.9	2.0
TW J1	13	19	25	30	37	42	48	57	9	13	18	22	28	32	37	44	1.3	1.5	1.7	1.8	2.0	1.5	1.7	1.9	2.1	2.2
TW J2	12	18	25	30	38	43	49	58	8	13	18	23	29	33	39	47	1.3	1.5	1.6	1.8	1.9	1.5	1.6	1.8	2.0	2.1
TW J3	15	23	32	39	48	54	62	72	11	17	25	31	40	45	53	62	1.1	1.3	1.4	1.6	1.7	1.2	1.4	1.5	1.7	1.8
TW J4	14	22	30	37	46	52	60	70	9	16	24	29	38	43	51	61	1.1	1.3	1.4	1.6	1.7	1.2	1.4	1.6	1.7	1.9
DG A1	7	9	11	12	14	15	15	16	4	5	6	6	6	6	6	4	2.2	2.8	3.8	-	-	3.1	4.6	-	-	-
DG A2	10	13	17	20	24	27	30	35	7	9	12	14	17	19	22	25	1.7	1.9	2.2	2.5	2.9	2.0	2.3	2.6	3.1	3.6
DG A3	13	18	23	28	34	39	45	53	10	14	18	22	27	31	36	43	1.4	1.6	1.7	1.9	2.1	1.6	1.8	1.9	2.1	2.3
DG B1	7	9	11	13	14	16	17	19	4	5	6	6	7	7	7	6	2.2	2.7	3.2	4.3	-	3.1	4.2	-	-	-
DG B2	10	14	17	20	25	28	32	38	7	9	12	15	18	20	23	27	1.7	1.9	2.1	2.3	2.5	2.0	2.2	2.5	2.9	3.2
DG B3	14	19	24	29	36	41	48	57	11	15	19	23	29	33	38	46	1.4	1.6	1.6	1.8	1.9	1.6	1.7	1.9	2.0	2.1
DG C1	9	12	15	17	20	22	25	28	6	7	9	10	11	12	13	13	1.9	2.3	2.5	3.0	3.5	2.5	2.9	4.0	-	-
DG C2	12	16	21	24	30	34	39	46	9	12	15	18	22	25	29	34	1.5	1.7	1.8	2.1	2.2	1.8	2.0	2.2	2.5	2.7
DG C3	16	21	28	33	41	46	54	64	12	17	22	26	33	37	43	52	1.3	1.5	1.5	1.7	1.8	1.5	1.6	1.8	1.9	2.0
SS A1	16	20	23	26	31	34	38	43	13	16	19	21	24	26	29	33	1.6	1.8	2.1	2.3	2.5	1.8	2.1	2.5	2.8	3.2
SS A2	20	26	32	37	44	48	54	62	17	21	27	31	37	41	46	53	1.3	1.4	1.6	1.8	1.9	1.4	1.6	1.8	2.0	2.1
SS A3	15	18	21	23	26	29	32	36	12	14	16	18	20	22	23	25	1.8	2.1	2.4	2.7	2.9	2.1	2.5	3.0	3.7	4.4
SS A4	19	25	31	35	42	46	52	60	16	21	26	29	35	38	43	50	1.3	1.5	1.7	1.8	2.0	1.4	1.6	1.9	2.1	2.3
SS A5	17	19	22	25	28	30	33	38	14	16	18	20	22	24	26	28	1.8	2.1	2.3	2.6	2.8	2.1	2.4	2.9	3.4	3.9
SS A6	20	25	31	36	43	47	53	61	16	21	26	30	35	39	44	51	1.3	1.5	1.6	1.8	2.0	1.4	1.6	1.8	2.0	2.2
SS A7	15	17	19	21	23	25	27	30	13	14	16	17	18	19	20	21	2.0	2.4	2.8	3.2	3.6	2.5	3.0	3.8	-	-
SS A8	19	24	30	34	40	44	50	57	16	20	24	28	33	36	41	47	1.3	1.5	1.7	1.9	2.0	1.5	1.7	1.9	2.2	2.4
SS B1	13	16	20	22	26	29	32	37	11	13	16	17	20	22	24	28	1.7	2.0	2.2	2.5	2.7	2.0	2.3	2.7	3.1	3.5
SS B2	17	23	28	33	40	44	50	58	14	18	24	27	33	37	42	49	1.3	1.5	1.7	1.8	2.0	1.5	1.7	1.8	2.0	2.2
SS B3	12	15	18	20	23	26	28	32	10	12	14	15	17	19	20	22	1.8	2.1	2.5	2.8	3.0	2.2	2.6	3.1	3.7	4.4
SS B4	17	22	27	32	38	42	48	56	14	18	22	26	31	35	40	46	1.4	1.5	1.7	1.9	2.0	1.5	1.7	1.9	2.1	2.3
SS B5	13	16	18	20	23	25	28	31	11	13	15	16	18	19	21	23	2.0	2.3	2.6	2.9	3.1	2.4	2.8	3.2	3.8	4.4
SS B6	16	22	27	32	38	42	48	56	13	17	22	26	31	35	40	47	1.3	1.5	1.7	1.9	2.0	1.5	1.7	1.9	2.1	2.3
SS B7	12	14	16	18	20	21	23	26	10	12	13	14	15	16	17	18	2.2	2.6	3.0	3.4	3.8	2.7	3.2	4.0	-	-
SS B8	16	20	26	30	36	40	45	53	12	16	21	24	29	32	37	43	1.4	1.6	1.8	2.0	2.1	1.5	1.8	2.0	2.2	2.4
SS C1	10	13	17	20	24	26	30	35	6	8	11	12	15	16	18	21	1.7	2.0	2.2	2.5	2.7	2.1	2.5	3.0	3.5	4.1
SS C2	13	18	23	28	34	39	44	53	8	12	17	20	25	29	33	40	1.4	1.6	1.7	1.9	2.1	1.6	1.8	2.0	2.2	2.4
SS C3	10	13	15	17	21	23	26	30	8	9	11	12	14	15	17	19	2.0	2.3	2.5	2.8	3.0	2.5	2.9	3.3	3.9	4.4
SS C4	12	16	21	25	31	35	40	47	9	12	16	18	23	26	30	36	1.5	1.7	1.9	2.0	2.2	1.7	1.9	2.2	2.4	2.6
SS D1	17	22	27	30	36	39	44	50	14	17	21	24	27	30	33	37	1.4	1.7	1.9	2.1	2.2	1.6	1.9	2.2	2.6	2.9
SS D2	22	29	37	42	51	56	63	71	17	23	30	35	42	47	53	61	1.1	1.3	1.5	1.6	1.8	1.2	1.4	1.6	1.8	2.0
SS D3	17	21	25	28	32	35	39	45	14	17	20	22	25	28	30	34	1.6	1.8	2.0	2.3	2.4	1.8	2.1	2.5	2.9	3.2
SS D4	21	28	35	41	49	54	61	69	17	23	29	34	41	45	51	59	1.2	1.3	1.5	1.7	1.8	1.3	1.5	1.7	1.8	2.0
SS E1	14	17	22	25	29	32	36	41	10	13	16	18	21	22	25	27	1.6	1.8	2.1	2.3	2.5	1.8	2.2	2.7	3.2	3.8
SS E2	18	24	31	36	43	48	54	63	13	18	24	28	34	39	44	51	1.2	1.4	1.6	1.8	1.9	1.4	1.6	1.8	2.0	2.2
SS E3	14	17	20	22	26	28	31	35	11	13	16	17	19	21	23	25	1.8	2.1	2.4	2.6	2.9	2.1	2.5	3.0	3.5	4.1
SS E4	17	22	28	33	40	44	50	58	13	18	23	27	32	36	41	47	1.3	1.5	1.7	1.8	2.0	1.4	1.7	1.9	2.1	2.3

PROVIDENCE, RHODE ISLAND

SOLAR SAVINGS FRACTION (%)

TYPE	55 F (3570 DD) LCR 100	70	50	40	30	25	20	15	65 F (5974 DD) 100	70	50	40	30	25	20	15
WW A1	16	19	22	24	27	30	33	38	13	14	16	17	19	21	22	25
WW A2	16	20	25	29	35	39	44	51	11	14	17	20	24	27	30	36
WW A3	15	20	26	31	37	42	48	56	10	13	18	21	26	29	33	40
WW A4	15	20	27	32	39	44	51	59	9	13	18	21	27	30	35	42
WW A5	15	21	28	33	41	46	53	62	8	13	18	22	28	32	37	45
WW A6	15	21	28	34	42	47	54	64	8	13	18	22	28	33	38	46
WW B1	12	15	18	20	24	26	29	34	8	9	10	12	13	14	15	16
WW B2	16	22	29	35	43	48	55	64	10	15	20	25	31	35	41	49
WW B3	19	27	35	42	50	56	64	73	13	19	25	31	38	43	50	59
WW B4	19	28	37	45	54	61	68	78	13	20	28	34	43	49	56	66
WW B5	18	27	37	44	54	61	69	78	12	19	28	34	43	49	57	67
WW C1	21	29	38	44	53	59	67	76	14	20	28	33	41	46	53	62
WW C2	19	27	36	43	53	59	66	76	13	19	27	33	41	47	54	63
WW C3	23	34	45	53	64	70	78	86	16	25	35	43	53	60	67	77
WW C4	21	31	42	50	60	67	75	84	14	23	32	40	50	56	64	74
TW A1	16	18	21	24	27	30	33	38	13	14	16	18	20	21	23	26
TW A2	15	19	23	27	32	36	40	47	11	14	17	19	22	25	28	33
TW A3	14	19	24	28	34	38	43	51	10	13	16	19	24	27	31	36
TW A4	13	18	24	28	34	39	45	53	9	12	16	19	24	27	31	38
TW B1	15	18	21	24	28	31	35	40	12	14	16	18	20	22	25	28
TW B2	14	18	22	26	31	35	40	47	10	13	16	18	22	25	29	34
TW B3	13	17	22	26	32	36	41	49	9	12	15	18	22	25	29	35
TW B4	12	16	21	25	30	34	39	47	9	11	15	18	22	25	29	34
TW C1	14	17	21	24	28	31	35	42	11	13	15	17	20	23	26	30
TW C2	13	16	21	24	29	32	37	44	9	12	15	17	21	23	27	32
TW C3	12	16	20	23	28	31	36	43	9	12	15	17	20	23	26	31
TW C4	12	15	19	21	26	29	33	39	10	12	14	16	19	22	25	29
TW D1	10	12	15	18	21	23	27	31	6	8	9	10	12	13	14	16
TW D2	16	21	27	32	39	44	50	59	11	15	19	23	29	33	38	45
TW D3	17	23	30	35	43	48	55	64	11	16	21	25	31	36	41	49
TW D4	18	25	34	40	49	55	62	72	12	18	25	30	38	43	50	59
TW D5	18	25	34	40	50	56	63	73	12	18	26	31	39	45	52	62
TW E1	19	26	34	40	48	54	61	70	14	19	25	30	37	42	48	56
TW E2	19	26	34	40	49	55	62	71	13	19	25	30	38	43	50	59
TW E3	23	32	42	50	59	66	73	82	16	24	33	39	49	55	62	72
TW E4	21	30	40	47	57	64	71	80	15	22	31	38	47	53	61	70
TW F1	14	17	20	22	26	29	32	38	11	13	14	16	18	20	22	25
TW F2	13	16	21	24	29	33	38	45	9	11	14	16	20	22	26	31
TW F3	11	16	21	24	30	34	40	47	7	10	13	16	20	23	27	32
TW F4	9	14	19	23	29	33	39	47	5	8	12	15	19	22	26	32
TW G1	12	15	18	21	25	28	32	38	9	11	13	15	17	19	22	25
TW G2	10	14	18	21	27	30	35	42	6	9	12	14	18	20	24	29
TW G3	8	12	17	20	25	29	34	42	5	7	10	13	17	20	23	29
TW G4	6	10	14	17	22	26	31	38	3	6	9	11	15	17	21	26
TW H1	9	12	16	19	23	26	30	36	6	8	10	12	15	17	20	24
TW H2	7	10	14	17	21	25	29	36	4	6	8	11	14	16	19	24
TW H3	5	8	12	15	19	22	26	33	3	5	7	9	12	14	18	22
TW H4	4	6	9	11	15	18	21	27	2	3	5	7	10	12	14	18
TW I1	8	10	13	15	18	21	24	28	5	6	7	8	10	11	12	13
TW I2	13	18	24	29	36	40	47	56	8	12	16	20	25	29	34	42
TW I3	14	19	26	31	39	44	51	60	9	13	18	22	[illegible]	[illegible]	38	46
TW I4	15	22	30	37	45	51	59	69	10	15	22	27	35	40	47	56
TW I5	15	23	31	38	47	53	61	71	10	16	23	29	37	42	50	59
TW J1	17	24	32	37	46	51	59	68	11	17	23	28	35	40	46	55
TW J2	16	23	31	37	46	52	59	69	11	16	23	28	35	40	47	56
TW J3	20	29	39	47	57	63	71	80	14	21	30	37	46	53	60	70
TW J4	18	27	37	44	54	61	69	78	12	20	28	35	44	50	58	68
DG A1	10	13	16	18	21	23	25	27	6	7	9	10	11	12	13	13
DG A2	12	17	21	25	31	35	39	45	8	11	15	18	22	24	28	33
DG A3	16	22	28	34	42	47	54	63	12	16	21	26	32	37	43	51
DG B1	10	13	16	19	23	25	28	32	5	7	9	10	12	13	14	15
DG B2	13	17	22	26	32	37	43	50	9	12	15	18	22	26	30	36
DG B3	16	22	29	35	44	50	58	68	12	17	22	27	34	39	45	55
DG C1	12	16	20	24	29	33	37	43	7	10	12	14	17	19	21	25
DG C2	15	20	26	31	38	44	50	59	10	14	19	22	28	31	37	44
DG C3	19	25	33	40	49	56	64	75	14	19	26	31	38	44	52	62
SS A1	19	23	28	32	38	42	47	54	15	19	22	25	30	33	36	42
SS A2	24	30	38	43	51	57	63	72	19	25	31	36	43	47	53	61
SS A3	17	21	25	29	33	37	41	47	14	17	20	22	25	28	31	35
SS A4	23	29	37	42	50	55	61	69	19	24	30	34	41	45	51	58
SS A5	19	22	27	30	34	37	42	48	16	18	21	24	27	29	32	37
SS A6	23	30	37	43	50	56	62	71	19	24	30	35	42	46	52	60
SS A7	17	20	23	25	29	31	35	39	14	16	18	20	22	24	26	29
SS A8	22	29	35	41	48	53	59	67	18	23	29	33	39	43	49	56
SS B1	16	20	24	28	33	36	41	48	12	15	19	21	25	28	31	36
SS B2	20	27	34	39	47	52	59	67	16	21	27	32	38	43	49	57
SS B3	15	18	22	25	30	33	37	43	11	14	17	19	22	24	27	31
SS B4	20	26	32	38	45	50	57	65	16	20	26	30	37	41	47	54
SS B5	15	18	22	24	28	31	35	40	13	15	17	19	22	24	26	30
SS B6	20	26	33	38	45	50	57	66	15	20	26	30	37	41	47	55
SS B7	14	16	19	21	25	27	30	35	11	13	15	16	19	20	22	25
SS B8	19	24	31	36	43	48	54	63	15	19	25	29	35	39	44	51
SS C1	13	18	23	26	32	36	42	49	8	11	15	17	21	24	27	32
SS C2	16	23	30	35	43	48	55	65	11	16	21	25	32	36	42	50
SS C3	13	16	19	22	27	30	34	40	9	11	14	15	18	20	23	27
SS C4	15	21	27	31	38	43	50	58	11	15	19	23	28	32	37	45
SS D1	21	27	33	38	45	49	55	63	16	21	26	29	34	38	43	49
SS D2	26	35	44	50	59	65	72	80	20	28	36	41	50	55	62	70
SS D3	20	25	30	34	40	44	49	56	16	20	24	27	31	34	38	44
SS D4	25	33	42	49	57	63	70	78	20	27	34	40	48	53	59	68
SS E1	17	22	28	32	38	42	47	55	13	16	20	23	27	30	34	39
SS E2	22	29	37	44	52	58	65	73	16	22	29	34	42	47	53	61
SS E3	16	20	24	27	32	36	40	46	13	16	19	21	24	27	30	34
SS E4	21	27	35	40	48	53	60	69	16	21	27	32	39	43	49	57

CONSERVATION FACTOR

TYPE	55 F 100	50	30	20	15	65 F 100	50	30	20	15
WW A1	1.9	2.1	2.3	2.5	2.7	2.5	2.8	3.1	3.5	3.9
WW A2	1.5	1.6	1.8	2.0	2.1	1.8	2.0	2.2	2.5	2.7
WW A3	1.4	1.5	1.7	1.8	2.0	1.6	1.8	2.0	2.2	2.4
WW A4	1.3	1.5	1.6	1.8	1.9	1.6	1.7	1.9	2.1	2.3
WW A5	1.3	1.4	1.6	1.7	1.9	1.5	1.7	1.9	2.1	2.2
WW A6	1.2	1.4	1.5	1.7	1.9	1.5	1.6	1.8	2.0	2.2
WW B1	1.8	2.1	2.3	2.6	2.8	2.5	3.1	3.8	4.9	-
WW B2	1.2	1.4	1.5	1.7	1.8	1.4	1.6	1.8	2.0	2.1
WW B3	1.1	1.3	1.4	1.6	1.7	1.3	1.4	1.6	1.8	1.9
WW B4	1.1	1.2	1.4	1.5	1.7	1.2	1.3	1.5	1.7	1.8
WW B5	1.1	1.2	1.4	1.5	1.7	1.2	1.3	1.5	1.6	1.8
WW C1	1.1	1.2	1.4	1.6	1.7	1.2	1.4	1.6	1.7	1.9
WW C2	1.1	1.2	1.4	1.6	1.7	1.2	1.4	1.5	1.7	1.8
WW C3	1.0	1.1	1.3	1.4	1.6	1.1	1.2	1.4	1.5	1.7
WW C4	1.0	1.1	1.3	1.5	1.6	1.1	1.2	1.4	1.5	1.7
TW A1	1.8	2.1	2.3	2.5	2.6	2.4	2.7	3.0	3.3	3.6
TW A2	1.6	1.7	1.9	2.1	2.2	1.9	2.1	2.3	2.6	2.8
TW A3	1.5	1.6	1.8	2.0	2.1	1.8	2.0	2.2	2.4	2.6
TW A4	1.4	1.6	1.7	1.9	2.0	1.7	1.9	2.1	2.3	2.4
TW B1	1.8	2.0	2.2	2.3	2.5	2.2	2.5	2.7	3.0	3.2
TW B2	1.5	1.7	1.9	2.0	2.2	1.9	2.1	2.3	2.5	2.7
TW B3	1.5	1.7	1.8	2.0	2.1	1.8	2.0	2.2	2.4	2.6
TW B4	1.5	1.7	1.9	2.0	2.2	1.8	2.0	2.2	2.4	2.6
TW C1	1.7	1.9	2.1	2.2	2.4	2.1	2.3	2.5	2.7	2.9
TW C2	1.6	1.8	2.0	2.1	2.3	2.0	2.2	2.3	2.5	2.7
TW C3	1.7	1.8	2.0	2.2	2.3	2.0	2.2	2.4	2.6	2.7
TW C4	1.8	2.0	2.1	2.3	2.4	2.2	2.3	2.5	2.7	2.9
TW D1	1.9	2.2	2.4	2.6	2.8	2.6	3.1	3.6	4.2	4.9
TW D2	1.3	1.5	1.6	1.8	1.9	1.6	1.7	1.9	2.1	2.2
TW D3	1.3	1.4	1.6	1.7	1.9	1.5	1.6	1.8	2.0	2.1
TW D4	1.1	1.3	1.4	1.6	1.7	1.3	1.4	1.6	1.8	1.9
TW D5	1.1	1.3	1.4	1.6	1.7	1.3	1.4	1.6	1.7	1.8
TW E1	1.2	1.3	1.5	1.6	1.8	1.3	1.5	1.7	1.8	2.0
TW E2	1.2	1.3	1.5	1.6	1.7	1.3	1.5	1.6	1.8	1.9
TW E3	1.0	1.2	1.3	1.5	1.6	1.1	1.3	1.4	1.6	1.7
TW E4	1.0	1.2	1.3	1.5	1.6	1.1	1.3	1.4	1.6	1.7
TW F1	1.8	2.0	2.2	2.4	2.6	2.3	2.6	2.9	3.2	3.5
TW F2	1.6	1.8	1.9	2.1	2.3	2.0	2.2	2.4	2.6	2.8
TW F3	1.5	1.7	1.8	2.0	2.2	1.8	2.0	2.2	2.4	2.6
TW F4	1.5	1.7	1.8	2.0	2.1	1.8	2.0	2.2	2.4	2.5
TW G1	1.8	2.0	2.2	2.3	2.5	2.2	2.5	2.7	3.0	3.2
TW G2	1.6	1.8	2.0	2.1	2.3	2.0	2.2	2.4	2.6	2.8
TW G3	1.6	1.8	1.9	2.1	2.2	1.9	2.1	2.3	2.5	2.7
TW G4	1.7	1.9	2.0	2.2	2.3	2.0	2.2	2.4	2.6	2.8
TW H1	1.8	2.0	2.2	2.3	2.5	2.2	2.4	2.6	2.9	3.1
TW H2	1.7	1.9	2.1	2.3	2.4	2.1	2.3	2.5	2.7	2.9
TW H3	1.8	2.0	2.2	2.4	2.5	2.2	2.4	2.6	2.8	3.0
TW H4	2.0	2.2	2.4	2.6	2.7	2.5	2.7	2.8	3.0	3.2
TW I1	2.0	2.3	2.5	2.8	3.0	2.7	3.2	3.8	4.5	-
TW I2	1.4	1.5	1.7	1.8	2.0	1.6	1.8	1.9	2.1	2.3
TW I3	1.3	1.4	1.6	1.8	1.9	1.5	1.7	1.8	2.0	2.1
TW I4	1.2	1.3	1.5	1.6	1.8	1.3	1.5	1.6	1.8	1.9
TW I5	1.2	1.3	1.4	1.6	1.7	1.3	1.4	1.6	1.7	1.9
TW J1	1.2	1.3	1.5	1.7	1.8	1.4	1.5	1.7	1.8	2.0
TW J2	1.2	1.3	1.5	1.6	1.8	1.3	1.5	1.6	1.8	1.9
TW J3	1.0	1.2	1.3	1.5	1.6	1.1	1.3	1.4	1.6	1.7
TW J4	1.1	1.2	1.3	1.5	1.6	1.2	1.3	1.5	1.6	1.8
DG A1	1.8	2.1	2.0	3.5	4.9	2.4	3.0	4.1	-	-
DG A2	1.5	1.7	1.9	2.2	2.4	1.8	2.0	2.2	2.6	2.9
DG A3	1.3	1.4	1.5	1.7	1.9	1.5	1.6	1.7	1.9	2.1
DG B1	1.8	2.0	2.3	2.7	3.1	2.4	2.9	3.5	4.6	-
DG B2	1.5	1.6	1.8	1.9	2.1	1.8	2.0	2.1	2.3	2.6
DG B3	1.3	1.4	1.5	1.6	1.8	1.5	1.6	1.7	1.8	1.9
DG C1	1.6	1.7	1.9	2.2	2.5	2.1	2.3	2.7	3.2	3.7
DG C2	1.4	1.5	1.6	1.8	1.9	1.6	1.8	1.9	2.1	2.3
DG C3	1.2	1.3	1.4	1.5	1.7	1.4	1.5	1.6	1.7	1.8
SS A1	1.4	1.6	1.8	2.0	2.1	1.6	1.9	2.1	2.3	2.5
SS A2	1.2	1.3	1.5	1.6	1.8	1.3	1.4	1.6	1.8	2.0
SS A3	1.5	1.8	2.0	2.2	2.3	1.8	2.1	2.4	2.7	3.0
SS A4	1.2	1.3	1.5	1.7	1.8	1.3	1.5	1.7	1.9	2.0
SS A5	1.6	1.8	2.0	2.2	2.3	1.8	2.1	2.4	2.6	2.9
SS A6	1.2	1.3	1.5	1.7	1.8	1.3	1.5	1.7	1.8	2.0
SS A7	1.8	2.0	2.3	2.5	2.7	2.1	2.5	2.8	3.3	3.7
SS A8	1.2	1.4	1.5	1.7	1.9	1.3	1.5	1.7	1.9	2.1
SS B1	1.5	1.7	1.9	2.1	2.2	1.8	2.0	2.3	2.5	2.7
SS B2	1.2	1.4	1.5	1.7	1.8	1.3	1.5	1.7	1.9	2.0
SS B3	1.6	1.8	2.0	2.2	2.4	1.9	2.2	2.5	2.8	3.1
SS B4	1.2	1.4	1.6	1.7	1.9	1.4	1.6	1.7	1.9	2.1
SS B5	1.7	2.0	2.2	2.4	2.5	2.1	2.3	2.6	2.9	3.2
SS B6	1.2	1.4	1.5	1.7	1.9	1.4	1.5	1.7	1.9	2.0
SS B7	1.9	2.2	2.4	2.6	2.8	2.3	2.7	3.0	3.4	3.8
SS B8	1.3	1.4	1.6	1.8	1.9	1.4	1.6	1.8	2.0	2.1
SS C1	1.5	1.7	1.8	2.0	2.1	1.8	2.1	2.3	2.5	2.7
SS C2	1.2	1.4	1.5	1.7	1.8	1.4	1.6	1.8	1.9	2.1
SS C3	1.7	1.9	2.1	2.3	2.4	2.1	2.4	2.6	2.9	3.1
SS C4	1.3	1.5	1.6	1.8	1.9	1.6	1.7	1.9	2.1	2.2
SS D1	1.3	1.4	1.6	1.8	1.9	1.4	1.7	1.9	2.1	2.3
SS D2	1.0	1.2	1.4	1.5	1.7	1.1	1.3	1.5	1.7	1.8
SS D3	1.4	1.6	1.7	1.9	2.1	1.6	1.8	2.1	2.3	2.5
SS D4	1.1	1.2	1.4	1.5	1.7	1.2	1.3	1.5	1.7	1.8
SS E1	1.4	1.6	1.7	1.9	2.1	1.6	1.9	2.1	2.4	2.6
SS E2	1.1	1.3	1.4	1.6	1.7	1.2	1.4	1.6	1.8	1.9
SS E3	1.6	1.8	2.0	2.2	2.3	1.9	2.1	2.4	2.7	2.9
SS E4	1.2	1.3	1.5	1.7	1.8	1.3	1.5	1.7	1.9	2.0

CHARLESTON, SOUTH CAROLINA

TYPE	SOLAR SAVINGS FRACTION (%)																CONSERVATION FACTOR									
	BASE TEMP: 55 F (869 DD)								65 F (2150 DD)								55 F					65 F				
LCR	200	100	70	50	40	30	25	20	200	100	70	50	40	30	25	20	200	100	50	30	20	200	100	50	30	20
WW A1	27	41	50	59	66	74	80	85	18	25	31	37	42	49	54	60	0.8	0.9	1.1	1.2	1.4	1.1	1.2	1.4	1.6	1.7
WW A2	32	50	61	72	79	87	91	94	18	30	38	46	53	62	67	74	0.7	0.8	1.0	1.1	1.3	0.9	1.0	1.2	1.3	1.5
WW A3	33	54	66	76	83	90	93	96	18	31	40	50	57	66	72	79	0.7	0.8	0.9	1.1	1.3	0.9	1.0	1.1	1.3	1.4
WW A4	34	56	68	79	85	91	94	97	17	32	41	52	59	68	74	81	0.6	0.7	0.9	1.1	1.3	0.8	0.9	1.1	1.3	1.4
WW A5	35	58	70	81	87	93	96	98	17	33	43	54	61	71	77	83	0.6	0.7	0.9	1.1	1.2	0.8	0.9	1.1	1.2	1.4
WW A6	36	59	71	82	88	94	96	98	17	33	44	55	63	72	78	84	0.6	0.7	0.9	1.1	1.2	0.8	0.9	1.1	1.2	1.4
WW B1	28	44	55	65	72	81	85	90	15	25	31	39	44	52	57	64	0.7	0.8	1.0	1.2	1.3	1.0	1.2	1.3	1.5	1.6
WW B2	34	56	68	79	85	92	95	97	17	32	42	53	60	70	76	82	0.6	0.7	0.9	1.1	1.2	0.8	0.9	1.1	1.2	1.4
WW B3	40	63	76	85	91	95	97	99	21	38	50	61	69	78	83	89	0.6	0.7	0.9	1.0	1.2	0.8	0.9	1.0	1.2	1.3
WW B4	39	63	76	86	91	96	98	99	19	38	50	62	70	79	85	90	0.6	0.7	0.9	1.0	1.2	0.7	0.8	1.0	1.2	1.3
WW B5	37	61	74	84	90	95	97	99	18	37	49	61	69	78	84	89	0.6	0.7	0.9	1.0	1.2	0.7	0.8	1.0	1.2	1.3
WW C1	41	65	77	87	92	96	98	99	22	40	51	63	71	80	85	90	0.6	0.7	0.9	1.0	1.2	0.7	0.8	1.0	1.2	1.3
WW C2	39	63	75	85	90	95	97	99	20	38	49	61	69	78	84	89	0.6	0.7	0.9	1.0	1.2	0.8	0.9	1.0	1.2	1.3
WW C3	43	69	81	90	94	98	99	100	22	44	57	69	77	86	90	94	0.6	0.7	0.8	1.0	1.2	0.7	0.8	0.9	1.1	1.3
WW C4	40	65	78	88	92	97	98	99	19	40	53	65	73	83	88	92	0.6	0.7	0.8	1.0	1.2	0.7	0.8	1.0	1.1	1.3
TW A1	26	40	49	58	65	73	78	84	17	25	30	37	41	48	53	59	0.8	0.9	1.1	1.2	1.4	1.2	1.3	1.4	1.6	1.7
TW A2	29	46	56	67	74	82	87	91	17	27	34	42	48	57	62	69	0.7	0.8	1.0	1.2	1.3	1.0	1.1	1.2	1.4	1.6
TW A3	30	48	59	70	77	85	89	94	16	28	36	45	51	60	66	73	0.7	0.8	1.0	1.1	1.3	0.9	1.0	1.2	1.3	1.5
TW A4	29	48	60	71	78	86	90	94	15	27	36	45	52	61	67	74	0.7	0.8	1.0	1.1	1.3	0.9	1.0	1.2	1.3	1.5
TW B1	26	40	49	59	66	74	80	85	17	24	30	37	42	49	54	61	0.8	0.9	1.1	1.2	1.4	1.1	1.2	1.4	1.5	1.7
TW B2	27	44	54	65	72	81	86	91	15	26	33	41	47	55	61	68	0.7	0.8	1.0	1.2	1.3	1.0	1.1	1.3	1.4	1.6
TW B3	27	44	55	66	73	82	86	91	14	25	33	41	47	56	62	69	0.7	0.8	1.0	1.2	1.3	1.0	1.1	1.2	1.4	1.5
TW B4	25	41	52	63	70	79	84	89	13	24	31	39	45	53	59	66	0.8	0.9	1.0	1.2	1.3	1.0	1.1	1.3	1.4	1.6
TW C1	24	39	48	58	65	74	79	85	15	23	29	36	42	49	54	61	0.8	0.9	1.1	1.2	1.4	1.1	1.2	1.4	1.5	1.7
TW C2	24	39	49	59	67	76	81	87	14	23	29	37	42	50	56	63	0.8	0.9	1.0	1.2	1.4	1.1	1.2	1.3	1.5	1.6
TW C3	22	37	47	57	64	73	78	85	13	22	28	35	40	48	53	60	0.8	0.9	1.1	1.2	1.4	1.1	1.2	1.4	1.5	1.7
TW C4	20	33	41	51	58	67	72	79	13	20	25	31	36	43	48	55	0.9	1.0	1.1	1.3	1.4	1.2	1.3	1.4	1.6	1.7
TW D1	23	38	48	58	65	74	80	85	12	21	27	33	39	46	51	58	0.8	0.9	1.1	1.2	1.4	1.1	1.2	1.4	1.5	1.7
TW D2	31	51	63	74	81	88	92	95	17	30	39	48	55	65	71	78	0.7	0.8	0.9	1.1	1.3	0.9	1.0	1.1	1.3	1.5
TW D3	34	56	68	78	85	91	94	97	18	33	42	53	60	70	75	82	0.6	0.7	0.9	1.1	1.3	0.8	0.9	1.1	1.2	1.4
TW D4	35	58	70	81	87	93	96	98	18	35	45	56	64	74	79	86	0.6	0.7	0.9	1.1	1.2	0.8	0.9	1.0	1.2	1.4
TW D5	34	56	69	80	86	93	95	98	17	34	44	56	64	73	79	86	0.6	0.7	0.9	1.1	1.2	0.8	0.9	1.0	1.2	1.4
TW E1	38	60	72	82	88	94	96	98	21	36	47	58	65	74	80	86	0.6	0.7	0.9	1.1	1.2	0.8	0.9	1.0	1.2	1.4
TW E2	37	59	71	82	88	93	96	98	19	36	46	57	65	74	80	86	0.6	0.7	0.9	1.1	1.2	0.8	0.9	1.0	1.2	1.4
TW E3	42	67	79	88	93	97	98	99	22	42	54	66	74	83	88	92	0.6	0.7	0.8	1.0	1.2	0.7	0.8	1.0	1.1	1.3
TW E4	39	63	76	86	91	96	98	99	20	39	51	63	71	80	85	91	0.6	0.7	0.9	1.0	1.2	0.7	0.8	1.0	1.1	1.3
TW F1	25	39	48	58	65	74	79	85	16	23	29	36	41	48	53	59	0.8	0.9	1.1	1.2	1.4	1.1	1.2	1.4	1.5	1.7
TW F2	26	43	54	65	72	80	85	90	14	25	32	40	46	54	60	67	0.7	0.8	1.0	1.2	1.3	1.0	1.1	1.3	1.4	1.6
TW F3	26	44	55	66	74	82	87	92	13	24	32	41	47	56	62	69	0.7	0.8	1.0	1.1	1.3	1.0	1.1	1.2	1.4	1.5
TW F4	24	42	53	65	72	81	86	91	11	22	30	39	45	54	61	68	0.7	0.8	1.0	1.2	1.3	1.0	1.1	1.2	1.4	1.5
TW G1	23	37	47	57	64	73	79	85	13	22	28	34	40	47	52	59	0.8	0.9	1.1	1.2	1.4	1.1	1.2	1.4	1.5	1.7
TW G2	23	39	50	61	68	77	82	88	11	21	28	36	42	51	56	63	0.8	0.9	1.0	1.2	1.4	1.0	1.1	1.3	1.4	1.6
TW G3	21	37	48	59	67	76	81	87	10	19	26	35	41	49	55	62	0.8	0.9	1.0	1.2	1.4	1.1	1.2	1.3	1.5	1.6
TW G4	17	32	42	53	61	70	76	83	7	16	22	30	36	44	49	57	0.8	0.9	1.1	1.2	1.4	1.1	1.2	1.4	1.5	1.7
TW H1	20	34	43	54	61	70	76	82	10	18	24	31	36	44	49	56	0.8	0.9	1.1	1.2	1.4	1.2	1.2	1.4	1.5	1.7
TW H2	17	31	41	51	59	68	74	81	8	16	22	29	34	42	48	55	0.9	0.9	1.1	1.3	1.4	1.2	1.3	1.4	1.6	1.7
TW H3	14	27	36	46	53	63	69	76	6	13	19	25	30	38	43	50	0.9	1.0	1.1	1.3	1.5	1.2	1.3	1.5	1.6	1.8
TW H4	10	21	29	37	44	53	59	67	4	10	14	20	24	31	35	41	1.0	1.1	1.2	1.4	1.5	1.4	1.5	1.6	1.8	1.9
TW I1	20	35	44	55	62	71	76	83	10	18	24	30	35	42	47	54	0.8	0.9	1.1	1.2	1.4	1.2	1.3	1.4	1.6	1.7
TW I2	28	47	59	71	78	86	90	94	14	26	35	45	52	61	67	75	0.7	0.8	1.0	1.1	1.3	0.9	1.0	1.2	1.3	1.5
TW I3	30	51	63	74	81	89	92	96	15	29	38	48	56	65	72	79	0.7	0.8	0.9	1.1	1.3	0.9	1.0	1.1	1.3	1.4
TW I4	32	54	67	78	85	91	94	97	15	31	42	53	61	71	77	83	0.6	0.7	0.9	1.1	1.2	0.8	0.9	1.1	1.2	1.4
TW I5	31	54	67	78	85	92	95	97	15	31	42	53	61	71	77	84	0.6	0.7	0.9	1.1	1.2	0.8	0.9	1.1	1.2	1.4
TW J1	35	57	69	80	86	92	95	98	18	34	44	55	63	72	78	84	0.6	0.7	0.9	1.1	1.2	0.8	0.9	1.1	1.2	1.4
TW J2	33	56	68	79	86	92	95	97	17	32	43	54	62	72	77	84	0.6	0.7	0.9	1.1	1.2	0.8	0.9	1.1	1.2	1.4
TW J3	38	63	76	86	91	96	98	99	19	39	51	63	71	80	86	91	0.6	0.7	0.9	1.0	1.2	0.7	0.8	1.0	1.1	1.3
TW J4	36	60	73	84	89	95	97	99	17	36	48	60	68	78	83	89	0.6	0.7	0.9	1.0	1.2	0.8	0.9	1.0	1.2	1.3
DG A1	26	43	54	63	70	77	81	85	13	23	29	36	41	48	51	56	0.7	0.9	1.0	1.3	1.5	1.0	1.1	1.4	1.7	2.0
DG A2	26	45	56	67	74	81	86	90	13	24	32	41	47	55	60	67	0.7	0.8	1.0	1.2	1.4	1.0	1.1	1.2	1.4	1.7
DG A3	29	51	63	75	81	88	91	93	16	29	38	49	57	66	72	78	0.7	0.8	0.9	1.1	1.4	0.9	1.0	1.1	1.3	1.5
DG B1	27	47	59	71	77	84	88	91	13	23	31	39	45	54	59	65	0.7	0.8	1.0	1.2	1.4	1.0	1.1	1.2	1.4	1.7
DG B2	27	47	61	73	80	88	91	94	14	25	34	43	51	61	67	74	0.7	0.8	0.9	1.1	1.3	1.0	1.0	1.2	1.3	1.5
DG B3	30	52	67	79	85	92	94	96	16	29	39	51	60	71	77	83	0.7	0.7	0.9	1.1	1.4	0.9	1.0	1.1	1.2	1.4
DG C1	31	53	67	79	85	90	93	95	15	27	36	46	53	63	69	75	0.7	0.7	0.9	1.1	1.4	0.9	1.0	1.1	1.3	1.6
DG C2	30	53	68	80	86	92	95	97	16	29	38	49	58	68	75	81	0.7	0.7	0.9	1.1	1.3	0.9	1.0	1.1	1.2	1.4
DG C3	33	58	73	85	90	95	97	98	18	33	44	57	66	77	83	89	0.6	0.7	0.9	1.1	1.4	0.9	0.9	1.0	1.1	1.4
SS A1	34	52	63	73	80	87	91	94	22	33	41	50	56	65	70	76	0.7	0.8	1.0	1.1	1.3	0.9	1.0	1.2	1.3	1.5
SS A2	41	63	74	83	89	94	96	98	25	41	51	62	69	77	82	88	0.6	0.7	0.9	1.1	1.2	0.8	0.9	1.0	1.2	1.4
SS A3	33	50	60	70	77	85	89	93	20	31	39	47	53	61	66	73	0.7	0.8	1.0	1.1	1.3	0.9	1.1	1.2	1.4	1.5
SS A4	41	62	73	83	88	94	96	98	25	41	51	61	68	77	82	87	0.6	0.7	0.9	1.1	1.2	0.8	0.9	1.0	1.2	1.4
SS A5	32	49	59	69	75	83	87	92	21	31	38	46	52	60	65	71	0.7	0.8	1.0	1.2	1.3	1.0	1.1	1.2	1.4	1.6
SS A6	41	63	74	83	89	94	96	98	25	41	51	61	68	77	82	87	0.6	0.7	0.9	1.1	1.2	0.8	0.9	1.0	1.2	1.4
SS A7	29	44	53	63	69	77	82	87	20	28	34	41	46	53	58	64	0.8	0.9	1.0	1.2	1.4	1.0	1.2	1.3	1.5	1.7
SS A8	41	62	73	82	88	93	96	98	25	40	50	60	67	76	81	86	0.6	0.7	0.9	1.1	1.2	0.8	0.9	1.0	1.2	1.4
SS B1	30	47	57	67	74	82	87	91	18	29	36	44	50	58	64	70	0.7	0.8	1.0	1.2	1.3	1.0	1.1	1.2	1.4	1.5
SS B2	36	57	68	78	85	91	94	97	22	36	46	56	63	72	77	83	0.6	0.8	0.9	1.1	1.3	0.8	0.9	1.1	1.2	1.4
SS B3	28	45	55	66	72	81	85	90	17	27	34	42	48	56	61	68	0.7	0.8	1.0	1.2	1.3	1.0	1.1	1.3	1.4	1.6
SS B4	36	56	67	78	84	91	94	97	21	35	45	55	62	71	76	83	0.6	0.8	0.9	1.1	1.3	0.8	0.9	1.1	1.2	1.4
SS B5	27	41	51	60	67	75	80	86	17	26	32	39	44	51	56	63	0.8	0.9	1.1	1.2	1.4	1.1	1.2	1.4	1.5	1.7
SS B6	36	56	68	78	84	91	94	97	21	35	45	55	62	71	77	83	0.6	0.8	0.9	1.1	1.3	0.8	0.9	1.1	1.2	1.4
SS B7	25	38	47	56	63	71	76	82	16	24	29	35	40	47	51	57	0.8	0.9	1.1	1.3	1.4	1.1	1.3	1.4	1.6	1.7
SS B8	35	55	66	77	83	90	93	96	20	34	43	53	60	69	75	81	0.7	0.8	0.9	1.1	1.3	0.8	1.0	1.1	1.3	1.4
SS C1	30	49	61	71	78	86	90	94	16	28	36	45	51	60	66	73	0.7	0.8	1.0	1.1	1.3	0.9	1.0	1.2	1.3	1.5
SS C2	34	56	68	79	85	91	95	97	18	32	42	53	60	70	76	82	0.6	0.7	0.9	1.1	1.3	0.8	0.9	1.1	1.2	1.4
SS C3	24	39	49	60	67	75	81	86	14	23	29	36	42	49	55	61	0.8	0.9	1.0	1.2	1.4	1.1	1.2	1.3	1.5	1.6
SS C4	30	50	61	72	79	87	91	95	16	29	38	47	54	64	70	76	0.7	0.8	0.9	1.1	1.3	0.9	1.0	1.1	1.3	1.5
SS D1	41	62	73	83	88	93	96	98	25	39	49	59	66	74	80	85	0.6	0.7	0.9	1.1	1.2	0.8	0.9	1.1	1.2	1.4
SS D2	48	71	82	90	94	97	99	99	28	47	59	70	77	85	89	93	0.6	0.7	0.8	1.0	1.2	0.7	0.8	1.0	1.1	1.3
SS D3	37	56	67	77	83	89	93	96	23	36	44	53	60	68	73	80	0.7	0.8	0.9	1.1	1.3	0.9	1.0	1.1	1.3	1.5
SS D4	46	69	80	88	93	97	98	99	27	46	57	68	75	83	88	92	0.6	0.7	0.9	1.0	1.2	0.7	0.8	1.0	1.1	1.3
SS E1	37	57	68	78	84	91	94	97	21	34	43	53	60	69	74	80	0.6	0.8	0.9	1.1	1.3	0.8	1.0	1.1	1.3	1.4
SS E2	43	65	77	86	91	96	98	99	24	42	53	64	71	80	85	90	0.6	0.7	0.9	1.0	1.2	0.7	0.8	1.0	1.2	1.3
SS E3	31	47	58	68	75	83	87	92	19	29	36	44	50	58	64	70	0.7	0.8	1.0	1.2	1.3	1.0	1.1	1.2	1.4	1.6
SS E4	39	61	72	82	88	93	96	98	23	38	48	59	66	75	81	86	0.6	0.7	0.9	1.1	1.2	0.8	0.9	1.0	1.2	1.4

COLUMBIA, SOUTH CAROLINA

SOLAR SAVINGS FRACTION (%)

TYPE	BASE TEMP: 55 F (1154 DD) LCR 200	100	70	50	40	30	25	20	65 F (2606 DD) 200	100	70	50	40	30	25	20
WW A1	24	36	44	53	59	68	73	80	17	23	28	34	39	45	49	55
WW A2	27	44	55	65	73	81	86	91	16	27	34	42	49	57	63	70
WW A3	28	47	59	70	77	85	89	93	15	28	36	46	52	62	67	74
WW A4	29	49	61	72	79	87	91	95	15	28	37	47	54	64	70	77
WW A5	29	51	63	74	81	89	92	96	15	29	39	49	57	66	72	79
WW A6	30	52	64	76	82	90	93	96	15	30	40	50	58	68	74	81
WW B1	23	38	48	58	65	74	79	85	14	22	28	35	40	47	52	59
WW B2	28	49	61	73	80	87	91	95	15	29	38	49	56	66	72	79
WW B3	34	56	69	80	86	92	95	98	18	35	45	57	64	74	80	86
WW B4	33	56	69	80	87	93	96	98	17	35	46	58	66	76	81	87
WW B5	30	54	68	79	86	92	95	98	15	33	45	57	65	75	81	87
WW C1	35	58	71	81	87	93	96	98	19	36	47	59	66	76	81	87
WW C2	32	56	68	80	86	92	95	98	17	34	45	57	65	74	80	86
WW C3	37	62	75	86	91	96	98	99	19	40	53	65	73	83	88	92
WW C4	33	58	72	83	89	94	97	99	17	36	49	62	70	80	85	90
TW A1	23	35	43	52	58	67	72	79	16	23	28	34	38	44	49	55
TW A2	25	40	50	60	67	76	81	87	15	25	31	39	44	52	58	65
TW A3	25	42	52	63	71	79	84	90	14	25	32	41	47	56	61	68
TW A4	24	42	53	64	71	80	85	90	13	24	32	41	47	56	62	69
TW B1	23	35	43	53	59	68	73	80	15	22	28	34	39	45	50	56
TW B2	23	38	48	58	66	75	80	86	14	23	30	37	43	51	57	64
TW B3	22	38	48	59	66	76	81	87	13	23	30	38	43	52	57	65
TW B4	21	36	46	56	63	73	78	85	12	21	28	35	41	49	55	62
TW C1	21	34	42	52	59	68	73	80	14	21	27	33	38	45	50	57
TW C2	20	34	43	53	60	69	75	81	12	21	27	34	39	47	52	58
TW C3	19	32	41	50	57	67	72	79	12	20	25	32	37	44	50	56
TW C4	18	28	36	45	51	60	66	73	12	18	23	29	33	40	45	51
TW D1	19	32	41	51	58	67	73	79	11	18	24	30	35	41	46	52
TW D2	26	45	56	67	74	83	88	92	15	27	35	45	51	60	66	74
TW D3	29	49	61	72	79	87	91	95	16	29	39	49	56	65	71	78
TW D4	30	51	64	75	82	89	93	96	16	31	42	52	60	70	76	83
TW D5	28	50	62	74	81	89	92	96	15	30	41	52	60	70	76	83
TW E1	32	53	65	76	83	90	93	96	18	33	43	54	61	70	76	83
TW E2	31	52	65	76	83	90	93	96	17	32	42	53	61	70	76	83
TW E3	36	60	73	83	89	94	97	99	20	38	50	62	70	79	85	90
TW E4	33	56	69	81	87	93	96	98	17	36	47	59	67	77	82	88
TW F1	22	34	42	52	58	67	73	79	14	21	27	33	37	44	49	55
TW F2	22	37	47	58	65	74	80	86	13	22	29	36	42	50	55	62
TW F3	21	38	48	59	67	76	82	87	11	21	29	37	43	52	57	65
TW F4	19	36	46	58	65	75	80	86	9	19	27	35	41	50	56	63
TW G1	20	32	41	50	57	66	72	79	12	20	25	31	36	43	48	54
TW G2	19	33	43	54	61	71	76	83	10	19	25	33	38	46	52	59
TW G3	17	31	41	52	60	69	75	82	8	17	24	31	37	45	51	58
TW G4	13	27	36	46	54	63	70	77	6	14	20	27	32	40	45	52
TW H1	16	29	37	47	54	63	69	76	9	16	22	28	33	40	45	52
TW H2	13	26	35	45	52	61	68	75	6	14	19	26	31	38	43	50
TW H3	11	22	30	40	46	56	62	70	5	11	17	23	27	34	39	46
TW H4	8	17	24	32	38	46	52	60	3	9	13	18	22	27	32	38
TW I1	17	29	38	47	54	64	69	76	9	16	21	27	31	38	43	49
TW I2	23	41	52	64	71	80	85	90	12	24	32	41	48	57	63	70
TW I3	25	44	56	68	75	84	88	00	13	26	35	44	52	61	67	75
TW I4	26	47	60	72	79	87	91	95	13	28	38	49	57	67	73	80
TW I5	26	47	60	72	79	87	91	95	13	28	38	50	57	68	74	81
TW J1	29	50	63	74	81	88	92	96	16	30	40	51	58	68	74	81
TW J2	28	49	61	73	80	88	92	95	14	29	39	50	58	68	74	81
TW J3	32	56	69	81	87	93	96	98	16	35	47	59	67	77	83	88
TW J4	29	53	66	78	85	92	95	97	15	33	44	56	64	74	80	87
DG A1	21	37	46	56	62	70	74	79	11	20	26	32	37	43	47	51
DG A2	21	38	49	60	67	75	80	85	12	22	29	37	43	51	56	62
DG A3	24	44	57	68	75	83	87	91	14	26	35	45	53	62	68	75
DG B1	22	39	51	63	70	78	82	86	11	20	27	34	40	48	53	59
DG B2	22	40	53	66	73	82	86	91	12	22	30	39	46	56	62	69
DG B3	25	45	59	72	80	88	91	94	14	27	36	47	55	66	73	80
DG C1	25	45	58	71	78	85	89	92	13	24	32	41	48	57	63	70
DG C2	25	45	59	73	80	88	91	95	14	26	35	45	53	64	70	77
DG C3	28	50	65	79	86	92	95	97	17	30	40	52	61	73	80	86
SS A1	29	46	56	67	73	81	86	91	20	30	38	46	52	60	65	72
SS A2	35	56	67	78	84	90	94	96	23	38	47	57	64	73	79	85
SS A3	28	43	53	63	70	78	83	88	18	28	35	43	48	56	61	68
SS A4	35	55	67	77	83	90	93	96	23	37	47	56	63	72	78	84
SS A5	28	43	52	62	69	77	82	87	20	29	35	42	48	55	60	67
SS A6	35	56	67	78	84	90	93	96	23	37	47	57	64	73	78	84
SS A7	26	38	47	56	62	71	76	82	18	26	31	37	42	49	53	59
SS A8	35	55	66	76	82	89	93	96	22	36	46	55	62	71	77	83
SS B1	25	40	50	61	67	76	81	87	16	26	33	40	46	54	59	66
SS B2	31	50	61	72	79	86	90	94	19	33	42	52	58	68	73	80
SS B3	24	39	48	58	65	74	79	85	15	24	31	38	43	51	56	63
SS B4	30	49	61	71	78	86	90	94	19	32	41	51	57	67	72	79
SS B5	23	36	44	54	60	69	74	80	16	24	29	35	40	47	52	58
SS B6	30	49	61	72	78	86	90	94	19	32	41	51	58	67	72	79
SS B7	21	32	40	49	56	64	69	76	15	21	26	32	36	42	47	52
SS B8	29	48	59	70	77	85	89	93	18	31	40	49	56	65	71	77
SS C1	25	42	54	65	72	81	85	90	14	25	32	41	47	56	61	69
SS C2	28	49	61	72	79	87	91	95	15	29	39	49	56	66	72	79
SS C3	21	34	43	53	60	69	75	81	13	21	26	33	38	45	50	57
SS C4	25	43	55	66	73	82	87	91	14	26	34	44	50	59	65	73
SS D1	35	55	66	76	83	89	93	96	22	36	45	54	61	70	75	81
SS D2	41	64	76	85	90	95	97	99	25	43	54	66	73	81	86	91
SS D3	32	49	60	70	77	84	88	93	21	33	40	49	55	63	69	75
SS D4	40	62	74	83	89	94	96	98	25	42	53	64	71	80	84	90
SS E1	31	50	61	71	78	86	90	94	19	31	39	48	55	64	69	76
SS E2	36	58	70	81	86	92	95	98	21	38	48	59	66	76	81	87
SS E3	26	41	51	61	68	76	81	87	17	27	33	40	46	54	59	65
SS E4	33	54	66	76	83	89	93	96	20	35	44	55	62	71	77	83

CONSERVATION FACTOR

TYPE	55 F 200	100	50	30	20	65 F 200	100	50	30	20
WW A1	0.9	1.0	1.1	1.3	1.5	1.2	1.3	1.5	1.6	1.8
WW A2	0.7	0.8	1.0	1.2	1.3	1.0	1.1	1.2	1.4	1.5
WW A3	0.7	0.8	1.0	1.1	1.3	0.9	1.0	1.2	1.3	1.5
WW A4	0.7	0.8	0.9	1.1	1.3	0.9	1.0	1.1	1.3	1.5
WW A5	0.7	0.8	0.9	1.1	1.3	0.9	1.0	1.1	1.3	1.4
WW A6	0.7	0.8	0.9	1.1	1.3	0.8	0.9	1.1	1.3	1.4
WW B1	0.8	0.9	1.1	1.2	1.4	1.1	1.2	1.4	1.5	1.7
WW B2	0.7	0.8	0.9	1.1	1.3	0.9	1.0	1.1	1.3	1.4
WW B3	0.6	0.7	0.9	1.1	1.2	0.8	0.9	1.0	1.2	1.4
WW B4	0.6	0.7	0.9	1.1	1.2	0.8	0.9	1.0	1.2	1.3
WW B5	0.6	0.7	0.9	1.1	1.2	0.8	0.9	1.0	1.2	1.4
WW C1	0.6	0.7	0.9	1.1	1.2	0.8	0.9	1.0	1.2	1.4
WW C2	0.6	0.7	0.9	1.1	1.2	0.8	0.9	1.0	1.2	1.4
WW C3	0.6	0.7	0.9	1.0	1.2	0.7	0.8	1.0	1.1	1.3
WW C4	0.6	0.7	0.9	1.0	1.2	0.7	0.8	1.0	1.1	1.3
TW A1	0.9	1.0	1.1	1.3	1.5	1.2	1.3	1.5	1.6	1.8
TW A2	0.8	0.9	1.0	1.2	1.4	1.1	1.2	1.3	1.5	1.6
TW A3	0.8	0.9	1.0	1.2	1.3	1.0	1.1	1.2	1.4	1.6
TW A4	0.7	0.8	1.0	1.2	1.3	1.0	1.1	1.2	1.4	1.5
TW B1	0.9	1.0	1.1	1.3	1.4	1.2	1.3	1.4	1.6	1.7
TW B2	0.8	0.9	1.0	1.2	1.4	1.1	1.2	1.3	1.5	1.6
TW B3	0.8	0.9	1.0	1.2	1.4	1.0	1.1	1.3	1.4	1.6
TW B4	0.8	0.9	1.1	1.2	1.4	1.1	1.2	1.3	1.5	1.6
TW C1	0.9	1.0	1.1	1.3	1.4	1.2	1.3	1.4	1.6	1.7
TW C2	0.9	1.0	1.1	1.3	1.4	1.1	1.2	1.4	1.5	1.7
TW C3	0.9	1.0	1.1	1.3	1.4	1.2	1.3	1.4	1.6	1.7
TW C4	1.0	1.1	1.2	1.3	1.5	1.3	1.4	1.5	1.7	1.8
TW D1	0.9	1.0	1.1	1.3	1.4	1.2	1.3	1.5	1.6	1.8
TW D2	0.7	0.8	1.0	1.1	1.3	0.9	1.0	1.2	1.3	1.5
TW D3	0.7	0.8	0.9	1.1	1.3	0.9	1.0	1.1	1.3	1.4
TW D4	0.7	0.8	0.9	1.1	1.3	0.8	0.9	1.1	1.2	1.4
TW D5	0.7	0.8	0.9	1.1	1.3	0.8	0.9	1.1	1.2	1.4
TW E1	0.7	0.8	0.9	1.1	1.3	0.8	0.9	1.1	1.2	1.4
TW E2	0.7	0.8	0.9	1.1	1.3	0.8	0.9	1.1	1.2	1.4
TW E3	0.6	0.7	0.9	1.0	1.2	0.7	0.8	1.0	1.2	1.3
TW E4	0.6	0.7	0.9	1.1	1.2	0.8	0.9	1.0	1.2	1.3
TW F1	0.9	1.0	1.1	1.3	1.5	1.2	1.3	1.5	1.6	1.8
TW F2	0.8	0.9	1.1	1.2	1.4	1.1	1.2	1.3	1.5	1.6
TW F3	0.8	0.9	1.0	1.2	1.4	1.0	1.1	1.3	1.4	1.6
TW F4	0.8	0.9	1.0	1.2	1.4	1.0	1.1	1.3	1.4	1.6
TW G1	0.9	1.0	1.1	1.3	1.4	1.2	1.3	1.4	1.6	1.8
TW G2	0.8	0.9	1.1	1.2	1.4	1.1	1.2	1.4	1.5	1.7
TW G3	0.9	0.9	1.1	1.2	1.4	1.1	1.2	1.4	1.5	1.7
TW G4	0.9	1.0	1.1	1.3	1.4	1.2	1.3	1.4	1.6	1.7
TW H1	0.9	1.0	1.1	1.3	1.5	1.2	1.3	1.5	1.6	1.8
TW H2	0.9	1.0	1.2	1.3	1.5	1.2	1.3	1.5	1.6	1.8
TW H3	1.0	1.1	1.2	1.4	1.5	1.3	1.4	1.5	1.7	1.8
TW H4	1.1	1.2	1.3	1.5	1.6	1.5	1.6	1.7	1.8	2.0
TW I1	0.9	1.0	1.1	1.3	1.5	1.3	1.4	1.5	1.7	1.0
TW I2	0.7	0.0	1.0	1.2	1.3	1.0	1.1	1.2	1.4	1.5
TW I3	0.7	0.8	1.0	1.1	1.3	0.9	1.0	1.2	1.3	1.5
TW I4	0.7	0.8	0.9	1.1	1.3	0.9	1.0	1.1	1.3	1.4
TW I5	0.7	0.8	0.9	1.1	1.3	0.8	0.9	1.1	1.2	1.4
TW J1	0.7	0.8	0.9	1.1	1.3	0.8	0.9	1.1	1.2	1.4
TW J2	0.7	0.8	0.9	1.1	1.3	0.8	0.9	1.1	1.2	1.4
TW J3	0.6	0.7	0.9	1.1	1.2	0.8	0.9	1.0	1.2	1.3
TW J4	0.6	0.7	0.9	1.1	1.2	0.8	0.9	1.0	1.2	1.4
DG A1	0.8	0.9	1.1	1.3	1.6	1.1	1.2	1.4	1.7	2.1
DG A2	0.8	0.9	1.0	1.2	1.4	1.1	1.1	1.3	1.5	1.7
DG A3	0.7	0.8	1.0	1.2	1.4	1.0	1.0	1.1	1.3	1.5
DG B1	0.8	0.8	1.0	1.2	1.5	1.1	1.2	1.3	1.5	1.8
DG B2	0.8	0.8	1.0	1.2	1.4	1.0	1.1	1.2	1.4	1.6
DG B3	0.7	0.8	0.9	1.1	1.3	0.9	1.0	1.1	1.2	1.4
DG C1	0.7	0.8	0.9	1.2	1.4	1.0	1.1	1.2	1.3	1.6
DG C2	0.7	0.8	0.9	1.1	1.3	0.9	1.0	1.1	1.3	1.5
DG C3	0.7	0.7	0.9	1.1	1.4	0.9	1.0	1.0	1.2	1.4
SS A1	0.7	0.8	1.0	1.2	1.3	0.9	1.1	1.2	1.4	1.5
SS A2	0.6	0.8	0.9	1.1	1.3	0.8	0.9	1.1	1.2	1.4
SS A3	0.8	0.9	1.0	1.2	1.4	1.0	1.1	1.3	1.4	1.6
SS A4	0.7	0.8	0.9	1.1	1.3	0.8	0.9	1.1	1.2	1.4
SS A5	0.8	0.9	1.0	1.2	1.4	1.0	1.1	1.3	1.5	1.6
SS A6	0.6	0.8	0.9	1.1	1.3	0.8	0.9	1.1	1.2	1.4
SS A7	0.9	1.0	1.1	1.3	1.4	1.1	1.2	1.4	1.6	1.7
SS A8	0.7	0.8	0.9	1.1	1.3	0.8	0.9	1.1	1.2	1.4
SS B1	0.8	0.9	1.0	1.2	1.4	1.0	1.1	1.3	1.4	1.6
SS B2	0.7	0.8	1.0	1.1	1.3	0.9	1.0	1.1	1.3	1.4
SS B3	0.8	0.9	1.1	1.2	1.4	1.1	1.2	1.3	1.5	1.6
SS B4	0.7	0.8	1.0	1.1	1.3	0.9	1.0	1.1	1.3	1.4
SS B5	0.9	1.0	1.1	1.3	1.4	1.1	1.3	1.4	1.6	1.7
SS B6	0.7	0.8	1.0	1.1	1.3	0.9	1.0	1.1	1.3	1.4
SS B7	0.9	1.0	1.2	1.3	1.5	1.2	1.3	1.5	1.7	1.8
SS B8	0.7	0.8	1.0	1.1	1.3	0.9	1.0	1.1	1.3	1.5
SS C1	0.7	0.8	1.0	1.2	1.3	1.0	1.1	1.2	1.4	1.5
SS C2	0.7	0.8	0.9	1.1	1.3	0.9	1.0	1.1	1.3	1.4
SS C3	0.9	1.0	1.1	1.3	1.4	1.2	1.3	1.4	1.6	1.7
SS C4	0.7	0.8	1.0	1.1	1.3	0.9	1.0	1.2	1.3	1.5
SS D1	0.7	0.8	0.9	1.1	1.3	0.8	0.9	1.1	1.3	1.4
SS D2	0.6	0.7	0.9	1.0	1.2	0.7	0.8	1.0	1.2	1.3
SS D3	0.7	0.8	1.0	1.1	1.3	0.9	1.0	1.2	1.3	1.5
SS D4	0.6	0.7	0.9	1.1	1.2	0.7	0.9	1.0	1.2	1.3
SS E1	0.7	0.8	1.0	1.1	1.3	0.9	1.0	1.2	1.3	1.5
SS E2	0.6	0.7	0.9	1.1	1.2	0.8	0.9	1.0	1.2	1.4
SS E3	0.8	0.9	1.0	1.2	1.4	1.0	1.1	1.3	1.5	1.6
SS E4	0.7	0.8	0.9	1.1	1.3	0.8	0.9	1.1	1.2	1.4

GREENVILLE, SOUTH CAROLINA

TYPE	SOLAR SAVINGS FRACTION (%)																CONSERVATION FACTOR									
BASE TEMP:	55 F (1493 DD)								65 F (3168 DD)								55 F					65 F				
LCR	200	100	70	50	40	30	25	20	200	100	70	50	40	30	25	20	200	100	50	30	20	200	100	50	30	20
WW A1	21	30	37	45	50	59	64	70	15	21	25	30	34	39	43	48	1.0	1.1	1.2	1.4	1.6	1.3	1.5	1.6	1.8	1.9
WW A2	22	36	46	56	63	72	78	84	14	23	29	37	42	50	55	62	0.8	0.9	1.1	1.2	1.4	1.1	1.2	1.3	1.5	1.6
WW A3	22	38	49	60	67	77	82	88	13	23	31	39	46	54	60	67	0.8	0.9	1.0	1.2	1.4	1.0	1.1	1.3	1.4	1.6
WW A4	22	40	51	62	70	79	84	89	12	24	32	41	47	56	62	70	0.8	0.9	1.0	1.2	1.3	1.0	1.1	1.2	1.4	1.5
WW A5	23	41	53	65	72	81	86	91	12	24	33	42	49	59	65	72	0.7	0.8	1.0	1.2	1.3	0.9	1.0	1.2	1.3	1.5
WW A6	23	42	54	66	73	82	87	92	11	24	33	43	50	60	66	73	0.7	0.8	1.0	1.1	1.3	0.9	1.0	1.2	1.3	1.5
WW B1	19	31	39	48	55	64	69	76	12	18	23	29	33	40	44	50	0.9	1.0	1.2	1.3	1.5	1.2	1.3	1.5	1.7	1.9
WW B2	22	40	52	63	71	80	85	90	12	24	33	42	49	59	65	72	0.8	0.8	1.0	1.2	1.3	0.9	1.0	1.2	1.3	1.5
WW B3	27	47	59	71	78	86	90	94	15	29	39	50	57	67	73	80	0.7	0.8	0.9	1.1	1.3	0.8	0.9	1.1	1.3	1.4
WW B4	25	47	60	72	79	87	91	95	13	29	40	51	59	69	76	82	0.7	0.8	0.9	1.1	1.3	0.8	0.9	1.1	1.2	1.4
WW B5	23	45	58	70	78	86	91	95	12	28	39	50	58	69	75	82	0.7	0.8	0.9	1.1	1.3	0.8	0.9	1.1	1.2	1.4
WW C1	28	49	61	73	80	88	91	95	16	31	41	52	60	69	75	82	0.7	0.8	0.9	1.1	1.3	0.8	0.9	1.1	1.2	1.4
WW C2	25	46	59	71	78	86	91	95	14	29	39	50	58	68	74	81	0.7	0.8	0.9	1.1	1.3	0.8	0.9	1.1	1.2	1.4
WW C3	29	53	66	78	85	92	95	97	15	34	47	59	67	77	83	89	0.6	0.7	0.9	1.1	1.2	0.8	0.9	1.0	1.2	1.3
WW C4	26	49	63	75	82	90	93	96	13	31	43	55	64	74	80	86	0.7	0.8	0.9	1.1	1.3	0.8	0.9	1.0	1.2	1.4
TW A1	20	29	36	44	49	57	63	69	15	20	24	29	33	39	43	48	1.0	1.1	1.3	1.4	1.6	1.4	1.5	1.6	1.8	1.9
TW A2	20	33	42	51	58	67	72	79	13	21	27	34	39	46	51	57	0.9	1.0	1.1	1.3	1.4	1.1	1.3	1.4	1.6	1.7
TW A3	20	34	44	54	61	70	76	82	12	21	28	35	41	49	54	61	0.9	0.9	1.1	1.2	1.4	1.1	1.2	1.3	1.5	1.6
TW A4	19	34	44	54	62	71	77	83	11	20	27	35	41	49	55	62	0.8	0.9	1.1	1.2	1.4	1.1	1.2	1.3	1.5	1.6
TW B1	19	29	36	44	50	59	64	71	14	20	24	30	34	40	44	49	1.0	1.1	1.2	1.4	1.5	1.3	1.4	1.6	1.7	1.9
TW B2	19	31	40	49	56	65	71	78	12	20	26	32	37	45	50	56	0.9	1.0	1.1	1.3	1.5	1.2	1.3	1.4	1.6	1.7
TW B3	18	31	40	50	57	66	72	79	11	19	25	32	38	45	51	57	0.9	1.0	1.1	1.3	1.4	1.1	1.2	1.4	1.5	1.7
TW B4	17	29	38	47	54	63	69	76	10	18	24	31	36	43	48	55	0.9	1.0	1.1	1.3	1.5	1.2	1.3	1.4	1.6	1.7
TW C1	18	28	35	44	50	58	64	71	12	19	23	29	33	40	44	50	1.0	1.1	1.2	1.4	1.5	1.3	1.4	1.5	1.7	1.8
TW C2	17	28	36	44	51	60	66	73	11	18	23	29	34	41	46	52	1.0	1.1	1.2	1.3	1.5	1.2	1.3	1.5	1.6	1.8
TW C3	16	26	34	42	49	57	63	70	10	17	22	28	32	39	44	50	1.0	1.1	1.2	1.4	1.5	1.3	1.4	1.5	1.7	1.8
TW C4	15	24	30	38	43	52	57	64	11	16	20	25	29	35	39	45	1.1	1.2	1.3	1.5	1.6	1.4	1.5	1.6	1.8	1.9
TW D1	16	26	33	42	48	57	62	69	9	15	20	25	29	35	39	44	1.0	1.1	1.2	1.4	1.5	1.3	1.4	1.6	1.8	2.0
TW D2	21	37	47	58	65	75	80	86	12	23	30	39	45	54	60	67	0.8	0.9	1.0	1.2	1.4	1.0	1.1	1.3	1.4	1.6
TW D3	23	40	51	63	70	79	84	90	13	25	33	43	49	58	64	72	0.8	0.9	1.0	1.2	1.3	0.9	1.0	1.2	1.3	1.5
TW D4	23	42	54	66	74	82	87	92	13	27	36	46	54	64	70	77	0.7	0.8	1.0	1.1	1.3	0.9	1.0	1.1	1.3	1.4
TW D5	22	41	53	65	73	82	87	92	12	26	36	46	54	64	70	77	0.7	0.8	1.0	1.1	1.3	0.9	1.0	1.1	1.3	1.4
TW E1	26	44	56	67	75	83	88	92	15	28	37	47	54	64	70	77	0.7	0.8	1.0	1.1	1.3	0.9	1.0	1.1	1.3	1.5
TW E2	25	43	55	67	74	83	88	92	14	28	37	47	54	64	70	77	0.7	0.8	1.0	1.1	1.3	0.9	1.0	1.1	1.3	1.4
TW E3	28	51	64	75	82	90	93	96	16	33	44	56	64	74	80	86	0.7	0.8	0.9	1.1	1.3	0.8	0.9	1.0	1.2	1.4
TW E4	26	47	60	72	80	88	92	95	14	31	42	53	61	71	77	84	0.7	0.8	0.9	1.1	1.3	0.8	0.9	1.1	1.2	1.4
TW F1	19	28	35	43	49	57	63	70	13	19	23	28	32	38	42	48	1.0	1.1	1.2	1.4	1.6	1.3	1.4	1.6	1.8	1.9
TW F2	18	30	39	49	55	65	70	77	11	19	24	31	36	43	48	55	0.9	1.0	1.1	1.3	1.5	1.2	1.3	1.4	1.6	1.7
TW F3	17	30	40	50	57	67	73	79	9	18	24	31	37	45	50	57	0.9	1.0	1.1	1.3	1.4	1.1	1.2	1.4	1.5	1.7
TW F4	15	28	38	48	55	65	71	78	7	16	22	30	35	43	49	56	0.9	1.0	1.1	1.3	1.4	1.1	1.2	1.4	1.5	1.7
TW G1	17	26	34	42	48	57	63	69	11	17	21	27	31	37	42	47	1.0	1.1	1.2	1.4	1.5	1.3	1.4	1.6	1.7	1.9
TW G2	15	27	35	45	52	61	67	74	8	16	21	28	33	40	45	51	0.9	1.0	1.2	1.3	1.5	1.2	1.3	1.5	1.6	1.8
TW G3	13	25	33	43	50	59	66	73	6	14	20	26	31	39	44	51	1.0	1.0	1.2	1.3	1.5	1.2	1.3	1.5	1.6	1.8
TW G4	10	21	29	38	44	54	60	67	4	11	16	22	27	34	39	45	1.0	1.1	1.2	1.4	1.5	1.3	1.4	1.5	1.7	1.8
TW H1	13	23	30	39	45	54	60	67	8	14	18	24	28	34	39	45	1.0	1.1	1.3	1.4	1.6	1.3	1.4	1.6	1.7	1.9
TW H2	10	20	28	36	43	52	58	65	5	11	16	22	26	33	37	43	1.1	1.1	1.3	1.4	1.6	1.3	1.4	1.6	1.7	1.9
TW H3	8	17	24	32	38	47	52	60	4	9	14	19	23	29	33	39	1.1	1.2	1.3	1.5	1.6	1.4	1.5	1.7	1.8	2.0
TW H4	6	13	19	25	30	38	43	50	2	7	10	15	18	23	27	32	1.3	1.3	1.5	1.6	1.7	1.6	1.7	1.8	2.0	2.1
TW I1	13	23	30	38	44	53	59	66	7	13	17	22	26	32	36	41	1.0	1.1	1.3	1.4	1.6	1.4	1.5	1.7	1.8	2.0
TW I2	18	33	43	54	62	72	77	84	10	20	27	35	41	50	56	63	0.8	0.9	1.1	1.2	1.4	1.0	1.1	1.3	1.4	1.6
TW I3	19	36	47	58	66	75	81	87	10	22	30	39	45	54	60	68	0.8	0.9	1.0	1.2	1.4	1.0	1.1	1.2	1.4	1.5
TW I4	20	39	51	63	70	80	85	90	10	24	33	43	50	60	66	74	0.7	0.8	1.0	1.2	1.3	0.9	1.0	1.2	1.3	1.5
TW I5	20	39	51	63	71	80	86	91	10	24	33	44	51	61	68	75	0.7	0.8	1.0	1.1	1.3	0.9	1.0	1.1	1.3	1.5
TW J1	23	41	53	65	72	81	86	91	13	26	35	45	52	61	68	75	0.7	0.8	1.0	1.1	1.3	0.9	1.0	1.2	1.3	1.5
TW J2	22	40	52	64	72	81	86	91	11	25	34	44	51	61	67	75	0.7	0.8	1.0	1.2	1.3	0.9	1.0	1.2	1.3	1.5
TW J3	25	47	60	72	80	88	92	95	13	30	41	53	61	71	77	84	0.7	0.8	0.9	1.1	1.3	0.8	0.9	1.1	1.2	1.4
TW J4	23	44	57	70	77	86	90	94	11	28	39	50	58	68	75	82	0.7	0.8	0.9	1.1	1.3	0.8	0.9	1.1	1.2	1.4
DG A1	17	29	37	46	52	59	63	68	9	16	21	27	31	36	39	43	0.9	1.0	1.2	1.5	1.7	1.2	1.3	1.6	1.9	2.3
DG A2	17	31	40	50	57	66	71	77	10	18	25	32	37	45	49	55	0.9	1.0	1.1	1.3	1.5	1.1	1.2	1.4	1.6	1.8
DG A3	19	35	47	59	67	76	81	86	12	22	30	39	46	56	62	69	0.8	0.9	1.0	1.2	1.4	1.0	1.1	1.2	1.4	1.6
DG B1	17	30	40	51	58	67	72	77	9	16	22	28	33	40	44	50	0.9	1.0	1.1	1.3	1.6	1.2	1.3	1.5	1.7	2.0
DG B2	17	32	43	55	63	73	78	84	10	19	25	33	39	48	54	61	0.9	0.9	1.0	1.2	1.4	1.1	1.2	1.3	1.4	1.6
DG B3	20	36	48	62	71	80	85	90	12	23	31	41	48	59	66	74	0.8	0.9	1.0	1.2	1.4	1.0	1.1	1.2	1.3	1.5
DG C1	19	35	46	59	67	76	81	85	11	20	26	34	40	48	53	60	0.8	0.9	1.0	1.2	1.5	1.1	1.2	1.3	1.5	1.7
DG C2	20	36	48	61	70	79	84	89	12	22	29	38	45	55	62	69	0.8	0.9	1.0	1.2	1.4	1.0	1.1	1.2	1.3	1.5
DG C3	22	41	54	69	77	86	90	94	14	26	35	46	54	66	73	81	0.8	0.8	0.9	1.1	1.3	1.0	1.0	1.1	1.2	1.4
SS A1	24	38	48	58	64	73	78	84	17	27	33	41	46	54	59	65	0.8	0.9	1.1	1.2	1.4	1.0	1.1	1.3	1.5	1.6
SS A2	29	47	59	69	76	84	88	93	20	33	42	52	58	67	73	79	0.7	0.8	1.0	1.1	1.3	0.9	1.0	1.1	1.3	1.5
SS A3	23	36	45	54	61	69	75	81	16	25	31	37	42	49	54	60	0.9	1.0	1.1	1.3	1.4	1.1	1.2	1.4	1.5	1.7
SS A4	29	47	58	69	75	83	88	92	20	33	41	51	57	66	72	78	0.7	0.8	1.0	1.2	1.3	0.9	1.0	1.1	1.3	1.5
SS A5	24	36	44	53	60	68	73	79	18	26	31	37	42	49	54	60	0.9	1.0	1.1	1.3	1.5	1.1	1.2	1.4	1.6	1.7
SS A6	29	47	58	69	76	84	88	92	20	33	42	51	58	67	72	79	0.7	0.8	1.0	1.1	1.3	0.9	1.0	1.1	1.3	1.5
SS A7	22	32	39	47	53	61	67	73	17	23	28	33	37	43	47	52	1.0	1.1	1.2	1.4	1.5	1.2	1.4	1.5	1.7	1.9
SS A8	29	46	57	68	74	82	87	91	20	32	41	50	56	65	70	77	0.7	0.8	1.0	1.2	1.3	0.9	1.0	1.2	1.3	1.5
SS B1	21	33	42	51	58	67	72	79	14	23	28	35	40	47	52	59	0.9	1.0	1.1	1.3	1.5	1.1	1.2	1.4	1.6	1.7
SS B2	25	42	53	64	71	79	84	89	17	29	37	46	53	62	67	74	0.8	0.9	1.0	1.2	1.3	0.9	1.0	1.2	1.3	1.5
SS B3	20	32	40	49	56	65	70	76	14	21	27	33	38	45	49	55	0.9	1.0	1.2	1.3	1.5	1.2	1.3	1.4	1.6	1.8
SS B4	25	41	52	63	70	78	83	89	16	28	36	45	52	60	66	73	0.8	0.9	1.0	1.2	1.4	0.9	1.0	1.2	1.4	1.5
SS B5	20	30	37	45	51	59	65	71	14	21	26	31	35	41	45	51	1.0	1.1	1.2	1.4	1.5	1.3	1.4	1.5	1.7	1.9
SS B6	25	41	52	63	70	79	84	89	16	28	36	45	52	61	66	73	0.8	0.9	1.0	1.2	1.4	0.9	1.0	1.2	1.3	1.5
SS B7	18	27	34	41	47	55	60	66	13	19	23	28	32	37	41	46	1.1	1.1	1.3	1.4	1.6	1.3	1.5	1.6	1.8	2.0
SS B8	24	40	50	61	68	77	82	88	16	27	35	44	50	58	64	71	0.8	0.9	1.0	1.2	1.4	0.9	1.1	1.2	1.4	1.5
SS C1	20	34	44	55	62	71	77	83	11	21	27	35	40	48	54	61	0.8	0.9	1.1	1.2	1.4	1.1	1.2	1.3	1.5	1.6
SS C2	22	40	51	63	70	79	85	90	13	25	33	43	49	59	65	72	0.8	0.8	1.0	1.2	1.3	0.9	1.0	1.2	1.3	1.5
SS C3	17	28	35	44	51	59	65	72	11	18	23	28	33	39	44	50	1.0	1.1	1.2	1.4	1.5	1.3	1.4	1.5	1.7	1.8
SS C4	20	36	46	57	64	73	79	85	12	22	30	38	44	53	59	66	0.8	0.9	1.1	1.2	1.4	1.0	1.1	1.3	1.4	1.6
SS D1	29	46	57	67	74	82	87	91	19	31	39	48	54	63	68	75	0.7	0.8	1.0	1.2	1.3	0.9	1.0	1.2	1.3	1.5
SS D2	33	55	67	78	84	90	94	97	22	38	49	59	67	76	81	87	0.6	0.7	0.9	1.1	1.3	0.8	0.9	1.0	1.2	1.4
SS D3	26	41	51	61	68	76	81	87	19	29	36	43	49	57	62	68	0.8	0.9	1.0	1.2	1.4	1.0	1.1	1.3	1.4	1.6
SS D4	32	53	65	76	82	89	93	96	21	37	47	57	65	74	79	85	0.7	0.8	0.9	1.1	1.3	0.8	0.9	1.1	1.2	1.4
SS E1	25	41	51	62	69	77	82	88	16	27	34	42	48	56	62	68	0.8	0.9	1.0	1.2	1.4	1.0	1.1	1.3	1.4	1.6
SS E2	29	49	61	72	79	87	91	94	18	33	42	53	60	69	75	81	0.7	0.8	0.9	1.1	1.3	0.8	0.9	1.1	1.3	1.4
SS E3	22	34	43	52	58	67	73	79	15	23	29	35	40	47	52	58	0.9	1.0	1.1	1.3	1.5	1.1	1.2	1.4	1.6	1.7
SS E4	27	45	56	67	74	83	87	92	17	30	39	49	55	64	70	77	0.7	0.8	1.0	1.1	1.3	0.9	1.0	1.1	1.3	1.5

HURON, SOUTH DAKOTA

TYPE	SOLAR SAVINGS FRACTION (%) BASE TEMP: 55 F (5605 DD)								65 F (8063 DD)								CONSERVATION FACTOR 55 F					65 F				
LCR	100	70	50	40	30	25	20	15	100	70	50	40	30	25	20	15	100	50	30	20	15	100	50	30	20	15
WW A1	14	15	17	19	20	22	23	26	12	13	15	15	17	17	18	19	2.3	2.7	3.2	3.7	4.1	2.8	3.3	4.2	-	-
WW A2	12	15	18	21	25	27	31	35	10	12	15	17	19	21	24	27	1.7	2.0	2.3	2.5	2.7	2.0	2.3	2.7	3.1	3.6
WW A3	11	15	19	22	26	29	34	39	8	11	14	17	20	23	26	30	1.6	1.9	2.1	2.3	2.5	1.8	2.1	2.4	2.8	3.1
WW A4	10	14	19	22	27	31	35	42	7	11	14	17	21	24	27	31	1.5	1.8	2.0	2.2	2.4	1.7	2.0	2.3	2.6	2.9
WW A5	10	14	19	23	28	32	37	44	7	10	14	17	22	24	28	33	1.5	1.7	1.9	2.1	2.3	1.6	1.9	2.2	2.5	2.7
WW A6	9	14	19	23	29	33	38	45	6	10	14	17	22	25	29	34	1.4	1.7	1.9	2.1	2.2	1.6	1.9	2.1	2.4	2.7
WW B1	9	10	12	13	14	15	15	16	6	7	8	8	9	9	8	6	2.4	3.1	4.1	-	-	3.1	4.7	-	-	-
WW B2	11	16	21	26	32	36	41	49	8	12	17	21	26	29	34	40	1.4	1.6	1.8	2.0	2.1	1.6	1.8	2.0	2.2	2.4
WW B3	14	19	26	31	38	43	49	58	10	15	21	25	32	36	41	48	1.3	1.5	1.6	1.8	2.0	1.4	1.6	1.8	2.0	2.2
WW B4	14	21	29	35	43	49	56	65	10	17	24	29	37	42	49	58	1.2	1.4	1.5	1.7	1.8	1.3	1.4	1.6	1.8	1.9
WW B5	13	20	29	35	44	50	57	67	10	17	24	30	38	43	51	60	1.2	1.3	1.5	1.6	1.8	1.2	1.4	1.6	1.7	1.9
WW C1	15	21	28	34	41	46	53	62	12	17	23	28	35	39	45	53	1.2	1.4	1.6	1.8	1.9	1.3	1.5	1.7	1.9	2.1
WW C2	14	20	28	33	41	47	54	63	11	16	23	28	35	40	46	55	1.2	1.4	1.6	1.7	1.9	1.3	1.5	1.7	1.9	2.0
WW C3	17	26	36	43	53	59	67	76	14	22	31	38	47	53	61	70	1.1	1.2	1.4	1.5	1.7	1.1	1.3	1.4	1.6	1.7
WW C4	15	24	33	40	50	56	64	74	12	20	28	35	44	50	58	68	1.1	1.2	1.4	1.6	1.7	1.1	1.3	1.5	1.6	1.8
TW A1	14	15	17	19	21	22	24	27	12	13	15	16	17	18	19	21	2.3	2.6	3.0	3.4	3.8	2.7	3.1	3.7	4.5	-
TW A2	12	15	18	20	23	26	29	33	10	12	14	16	19	21	23	26	1.8	2.1	2.4	2.6	2.9	2.1	2.4	2.8	3.2	3.6
TW A3	11	14	18	20	24	27	31	36	9	11	14	16	19	21	24	28	1.7	2.0	2.2	2.4	2.6	1.9	2.2	2.5	2.9	3.2
TW A4	10	13	17	20	25	28	32	38	7	10	13	16	19	22	25	30	1.7	1.9	2.1	2.3	2.5	1.8	2.1	2.4	2.7	2.9
TW B1	13	15	17	19	21	23	25	29	11	13	14	16	18	19	20	23	2.1	2.4	2.7	3.1	3.3	2.4	2.8	3.3	3.8	4.3
TW B2	11	14	17	19	23	26	29	34	9	11	14	16	19	21	23	27	1.8	2.1	2.3	2.5	2.7	2.0	2.3	2.6	3.0	3.3
TW B3	10	13	17	19	23	26	30	35	8	10	13	16	19	21	24	28	1.8	2.0	2.2	2.4	2.6	2.0	2.2	2.5	2.8	3.1
TW B4	10	13	16	19	23	25	29	35	8	10	13	15	18	21	24	28	1.8	2.0	2.2	2.4	2.6	2.0	2.2	2.5	2.8	3.0
TW C1	12	14	16	18	21	24	26	31	10	12	14	15	18	19	22	25	2.0	2.3	2.5	2.8	3.0	2.3	2.6	2.9	3.2	3.6
TW C2	10	13	16	18	22	24	28	33	9	11	13	15	18	20	23	27	1.9	2.1	2.4	2.6	2.8	2.1	2.4	2.6	2.9	3.2
TW C3	10	13	16	18	21	24	27	32	9	11	13	15	18	20	23	26	1.9	2.2	2.4	2.6	2.8	2.1	2.4	2.6	2.9	3.1
TW C4	11	13	15	17	20	23	26	30	9	11	13	15	17	19	22	25	2.1	2.3	2.5	2.7	2.9	2.3	2.5	2.8	3.0	3.2
TW D1	7	9	10	11	13	14	15	16	5	6	7	8	9	9	9	9	2.5	3.0	3.7	4.6	-	3.0	4.0	-	-	-
TW D2	12	16	21	24	29	33	38	45	9	13	17	20	24	27	32	37	1.5	1.7	1.9	2.1	2.3	1.7	1.9	2.1	2.3	2.5
TW D3	12	17	22	26	32	36	41	49	9	13	18	21	26	30	34	40	1.4	1.6	1.8	2.0	2.2	1.6	1.8	2.0	2.2	2.4
TW D4	13	19	26	31	38	44	50	59	11	16	22	26	33	38	44	52	1.3	1.5	1.6	1.8	1.9	1.4	1.6	1.7	1.9	2.1
TW D5	13	20	27	32	40	45	52	62	11	16	23	28	35	40	46	55	1.3	1.4	1.6	1.7	1.9	1.3	1.5	1.7	1.8	2.0
TW E1	15	20	26	30	37	42	48	56	12	16	22	26	31	35	40	47	1.3	1.5	1.7	1.9	2.0	1.4	1.6	1.9	2.1	2.2
TW E2	14	20	26	31	38	43	50	58	11	16	22	26	33	37	43	50	1.3	1.5	1.6	1.8	1.9	1.4	1.6	1.8	2.0	2.1
TW E3	17	25	33	40	49	54	62	71	14	21	29	35	43	48	55	65	1.1	1.3	1.4	1.6	1.7	1.2	1.4	1.5	1.7	1.8
TW E4	16	23	32	38	47	53	61	70	13	20	27	33	42	47	54	64	1.1	1.3	1.5	1.6	1.8	1.2	1.4	1.5	1.7	1.8
TW F1	12	14	16	17	19	21	23	26	10	11	13	14	16	17	18	19	2.2	2.6	2.9	3.3	3.6	2.6	3.0	3.6	4.4	-
TW F2	10	12	15	18	21	23	26	31	8	10	12	14	16	18	20	23	1.9	2.2	2.4	2.7	2.9	2.1	2.5	2.8	3.3	3.7
TW F3	8	11	15	17	21	24	27	33	6	8	11	13	16	18	21	25	1.8	2.0	2.3	2.5	2.7	2.0	2.3	2.6	3.0	3.3
TW F4	6	9	13	16	20	23	27	33	4	7	10	12	15	18	21	25	1.8	2.0	2.2	2.4	2.6	1.9	2.2	2.5	2.8	3.1
TW G1	10	12	14	16	18	20	23	26	8	10	11	13	15	16	18	20	2.1	2.4	2.8	3.1	3.3	2.4	2.8	3.3	3.8	4.3
TW G2	7	10	13	15	19	21	24	29	5	7	10	12	14	16	19	22	1.9	2.2	2.4	2.7	2.9	2.1	2.4	2.8	3.1	3.4
TW G3	6	8	12	14	18	21	24	29	4	6	9	11	14	16	19	22	1.9	2.1	2.4	2.6	2.8	2.1	2.4	2.7	3.0	3.2
TW G4	4	7	10	12	16	18	22	27	2	4	7	9	12	14	17	21	2.0	2.2	2.4	2.6	2.8	2.2	2.4	2.7	3.0	3.2
TW H1	7	9	11	13	16	18	21	25	5	7	9	10	13	14	16	19	2.1	2.4	2.7	2.9	3.1	2.4	2.7	3.1	3.4	3.8
TW H2	4	7	10	12	15	17	20	25	3	5	7	9	11	13	16	19	2.1	2.3	2.5	2.8	3.0	2.3	2.6	2.8	3.1	3.4
TW H3	3	6	8	10	13	16	19	23	2	4	6	8	10	12	14	18	2.1	2.4	2.6	2.8	3.0	2.4	2.6	2.9	3.2	3.4
TW H4	2	4	6	8	11	13	15	19	1	3	5	6	8	10	12	15	2.4	2.6	2.8	3.0	3.2	2.6	2.8	3.1	3.3	3.6
TW I1	5	7	8	9	11	12	13	14	4	5	5	6	7	7	7	7	2.6	3.2	3.0	4.9	-	3.2	4.2	-	-	-
TW I2	9	13	18	21	26	30	35	42	7	10	14	17	21	24	28	34	1.6	1.8	2.0	2.1	2.3	1.7	1.9	2.2	2.4	2.6
TW I3	10	14	19	23	29	33	38	40	7	11	15	19	24	27	32	38	1.5	1.7	1.9	2.0	2.2	1.6	1.8	2.0	2.2	2.4
TW I4	11	17	23	28	35	40	47	56	8	13	19	24	30	35	41	49	1.3	1.5	1.7	1.8	2.0	1.4	1.6	1.8	1.9	2.1
TW I5	11	17	24	30	37	43	50	59	8	14	20	25	32	37	44	53	1.3	1.4	1.6	1.8	1.9	1.4	1.5	1.7	1.9	2.0
TW J1	12	18	24	28	35	40	46	54	10	14	20	24	29	33	39	46	1.4	1.5	1.7	1.9	2.0	1.5	1.7	1.9	2.1	2.2
TW J2	12	17	24	29	36	41	47	56	9	14	20	24	30	35	41	49	1.3	1.5	1.7	1.8	2.0	1.4	1.6	1.8	2.0	2.1
TW J3	15	22	31	37	46	52	60	70	11	18	26	32	41	47	54	63	1.1	1.3	1.5	1.6	1.8	1.2	1.4	1.5	1.7	1.8
TW J4	13	21	29	36	45	51	58	68	10	17	25	31	39	45	52	62	1.2	1.3	1.5	1.6	1.8	1.2	1.4	1.5	1.7	1.9
DG A1	7	8	10	11	13	13	14	13	5	6	7	7	8	8	8	6	2.3	2.9	4.3	-	-	2.8	3.9	-	-	-
DG A2	10	13	16	19	23	25	20	33	8	10	13	15	18	20	23	27	1.8	2.0	2.3	2.6	3.0	1.9	2.2	2.5	3.0	3.4
DG A3	13	17	23	27	33	37	43	51	11	15	20	23	29	32	38	45	1.5	1.6	1.8	1.9	2.1	1.6	1.7	1.9	2.1	2.3
DG B1	7	9	10	11	13	14	15	16	5	6	7	8	8	9	9	8	2.3	2.9	3.6	-	-	2.7	3.7	-	-	-
DG B2	10	13	17	19	23	27	31	36	8	10	13	16	19	21	24	29	1.7	2.0	2.2	2.4	2.6	1.9	2.2	2.5	2.8	3.1
DG B3	14	18	23	28	34	39	46	55	11	15	20	24	30	34	40	48	1.4	1.6	1.7	1.8	2.0	1.5	1.7	1.8	2.0	2.1
DG C1	9	11	14	16	18	20	22	25	6	8	10	12	13	14	15	16	1.9	2.4	2.8	3.3	4.0	2.3	2.8	3.7	-	-
DG C2	12	16	20	23	28	32	37	44	10	13	17	20	24	27	31	36	1.6	1.8	1.9	2.1	2.3	1.7	1.9	2.2	2.4	2.6
DG C3	15	21	27	31	39	44	52	61	13	18	23	28	34	39	45	54	1.4	1.5	1.6	1.7	1.9	1.4	1.6	1.7	1.8	2.0
SS A1	15	18	21	24	28	30	33	38	14	16	19	21	24	26	29	32	1.7	2.0	2.3	2.5	2.7	1.8	2.2	2.5	2.9	3.3
SS A2	18	24	29	34	40	45	50	58	17	21	27	31	36	40	45	52	1.3	1.5	1.7	1.9	2.0	1.4	1.6	1.8	2.0	2.2
SS A3	13	16	18	20	23	25	27	30	12	14	16	18	20	21	22	24	1.9	2.3	2.7	3.1	3.4	2.1	2.6	3.2	4.0	-
SS A4	18	23	28	32	38	42	47	54	16	20	25	29	34	37	42	48	1.4	1.6	1.8	1.9	2.1	1.4	1.7	1.9	2.1	2.3
SS A5	16	18	21	23	25	27	30	33	14	16	18	20	22	24	25	28	1.9	2.2	2.6	2.9	3.2	2.1	2.5	3.0	3.6	4.2
SS A6	18	23	29	33	39	43	49	56	16	21	26	30	35	39	43	50	1.3	1.5	1.7	1.9	2.1	1.4	1.6	1.9	2.1	2.2
SS A7	14	16	18	19	21	22	23	25	13	14	16	17	18	19	20	20	2.3	2.7	3.3	4.0	4.7	2.5	3.1	4.1	-	-
SS A8	17	22	27	31	36	40	45	52	15	20	24	27	32	35	40	45	1.4	1.6	1.8	2.0	2.2	1.5	1.7	2.0	2.2	2.4
SS B1	12	15	18	20	24	26	29	33	11	13	16	18	20	22	24	27	1.8	2.1	2.4	2.7	2.9	2.0	2.4	2.7	3.2	3.6
SS B2	16	21	26	31	37	41	46	54	14	19	24	27	33	37	42	48	1.4	1.6	1.7	1.9	2.1	1.5	1.7	1.9	2.1	2.2
SS B3	11	13	16	18	20	22	24	28	10	12	14	15	17	18	20	22	2.0	2.4	2.7	3.1	3.5	2.2	2.7	3.2	4.0	4.9
SS B4	15	20	25	29	35	39	44	51	14	18	22	26	31	34	39	45	1.4	1.6	1.8	2.0	2.1	1.5	1.7	1.9	2.2	2.3
SS B5	13	15	17	19	21	23	25	28	11	13	15	16	18	19	21	23	2.1	2.5	2.8	3.2	3.6	2.3	2.8	3.3	3.9	4.6
SS B6	15	20	25	29	35	39	45	52	13	18	22	26	31	35	40	46	1.4	1.6	1.8	2.0	2.1	1.5	1.7	1.9	2.1	2.3
SS B7	12	13	15	16	18	19	20	22	10	12	13	14	15	16	17	18	2.4	2.9	3.4	4.0	4.6	2.7	3.3	4.1	-	-
SS B8	14	19	23	27	32	36	41	48	13	16	21	24	29	32	36	42	1.5	1.7	1.9	2.1	2.2	1.6	1.8	2.0	2.3	2.4
SS C1	9	12	16	18	22	24	28	32	7	9	12	14	16	18	20	23	1.8	2.1	2.3	2.6	2.8	2.0	2.4	2.8	3.3	3.9
SS C2	12	17	22	26	32	37	42	50	9	13	18	21	27	30	35	41	1.4	1.6	1.8	2.0	2.1	1.5	1.8	2.0	2.2	2.4
SS C3	10	12	15	17	19	21	24	27	8	10	12	13	15	17	18	21	2.1	2.4	2.7	3.0	3.2	2.4	2.7	3.2	3.7	4.2
SS C4	12	16	20	24	29	33	38	45	9	13	17	20	24	27	31	37	1.5	1.7	1.9	2.1	2.3	1.7	1.9	2.1	2.3	2.5
SS D1	16	20	24	27	32	35	39	44	14	17	21	23	27	29	32	36	1.5	1.8	2.0	2.3	2.5	1.6	2.0	2.3	2.7	3.0
SS D2	20	26	33	39	47	52	58	67	17	23	30	35	42	46	52	60	1.2	1.4	1.5	1.7	1.8	1.2	1.4	1.6	1.8	2.0
SS D3	16	19	23	25	29	32	35	40	14	17	20	22	25	27	30	33	1.7	1.9	2.2	2.5	2.7	1.8	2.1	2.5	3.0	3.4
SS D4	19	25	32	38	45	50	56	65	17	23	29	34	40	45	50	58	1.2	1.4	1.6	1.7	1.9	1.3	1.5	1.7	1.9	2.0
SS E1	12	16	19	22	25	28	31	35	10	13	16	18	20	22	24	26	1.7	2.0	2.3	2.6	2.9	1.8	2.3	2.8	3.4	4.1
SS E2	16	21	28	32	39	44	50	58	13	18	24	28	34	38	43	50	1.3	1.5	1.7	1.8	2.0	1.4	1.6	1.8	2.0	2.2
SS E3	13	15	18	20	23	25	27	31	12	13	16	17	19	21	22	25	1.9	2.3	2.6	3.0	3.2	2.1	2.5	3.0	3.7	4.4
SS E4	15	20	26	30	36	41	46	54	13	18	23	27	32	35	40	47	1.4	1.6	1.8	1.9	2.1	1.4	1.7	1.9	2.1	2.3

PIERRE, SOUTH DAKOTA

TYPE	SOLAR SAVINGS FRACTION (%) BASE TEMP: 55 F (5282 DD) LCR 100	70	50	40	30	25	20	15	65 F (7689 DD) 100	70	50	40	30	25	20	15	CONSERVATION FACTOR 55 F 100	50	30	20	15	65 F 100	50	30	20	15
WW A1	16	18	20	22	25	27	30	34	14	15	17	18	20	22	23	26	2.0	2.2	2.5	2.8	3.0	2.3	2.7	3.1	3.6	4.1
WW A2	15	19	23	27	31	35	40	46	12	15	18	21	25	27	31	36	1.5	1.7	2.0	2.1	2.3	1.7	2.0	2.3	2.5	2.8
WW A3	14	19	24	28	34	38	43	51	11	14	19	22	26	30	34	39	1.4	1.6	1.8	2.0	2.1	1.6	1.8	2.1	2.3	2.5
WW A4	14	19	25	29	35	40	45	53	10	14	19	22	27	31	35	42	1.4	1.6	1.7	1.9	2.1	1.5	1.8	2.0	2.2	2.4
WW A5	13	19	25	30	37	41	48	56	9	14	19	23	28	32	37	44	1.3	1.5	1.7	1.9	2.0	1.5	1.7	1.9	2.1	2.3
WW A6	13	19	26	31	38	42	49	57	9	14	19	23	29	33	38	45	1.3	1.5	1.6	1.8	2.0	1.4	1.6	1.9	2.1	2.3
WW B1	11	13	16	18	21	22	25	28	8	10	11	13	14	15	16	16	1.9	2.3	2.7	3.1	3.5	2.4	3.0	3.9	-	-
WW B2	15	20	27	32	39	44	50	59	11	16	21	26	32	36	41	49	1.3	1.5	1.6	1.8	1.9	1.4	1.6	1.8	2.0	2.1
WW B3	18	25	32	38	46	52	58	67	14	19	26	31	38	43	49	58	1.2	1.3	1.5	1.7	1.8	1.3	1.5	1.6	1.8	2.0
WW B4	18	26	35	41	50	57	64	73	14	21	29	35	43	49	56	65	1.1	1.3	1.4	1.6	1.7	1.2	1.3	1.5	1.7	1.8
WW B5	17	25	34	41	51	57	65	74	13	20	29	35	44	50	57	67	1.1	1.2	1.4	1.6	1.7	1.2	1.3	1.5	1.7	1.8
WW C1	19	26	35	41	49	55	62	71	15	21	28	34	41	46	53	61	1.1	1.3	1.5	1.6	1.8	1.2	1.4	1.6	1.8	1.9
WW C2	18	25	34	40	49	55	62	71	14	20	28	33	41	47	54	63	1.1	1.3	1.4	1.6	1.8	1.2	1.4	1.6	1.7	1.9
WW C3	22	31	42	50	60	66	74	83	17	26	36	43	53	59	67	76	1.0	1.1	1.3	1.5	1.6	1.1	1.2	1.4	1.5	1.7
WW C4	19	29	39	47	57	63	71	80	15	24	33	40	50	56	64	74	1.0	1.2	1.3	1.5	1.6	1.1	1.2	1.4	1.6	1.7
TW A1	15	18	20	22	25	28	30	35	13	15	17	19	21	22	24	27	1.9	2.2	2.5	2.7	2.9	2.3	2.6	3.0	3.4	3.7
TW A2	14	18	22	25	29	32	37	43	12	15	18	20	23	26	29	34	1.6	1.8	2.1	2.2	2.4	1.8	2.1	2.4	2.6	2.9
TW A3	14	18	22	26	31	34	39	46	11	14	18	20	25	27	31	36	1.5	1.7	1.9	2.1	2.3	1.7	1.9	2.2	2.4	2.6
TW A4	12	17	22	26	31	35	40	48	9	13	17	20	25	28	32	38	1.5	1.7	1.9	2.0	2.2	1.6	1.9	2.1	2.3	2.5
TW B1	15	17	20	23	26	29	32	37	13	15	17	19	21	23	26	29	1.8	2.1	2.3	2.5	2.7	2.1	2.4	2.7	3.0	3.3
TW B2	13	17	21	24	29	32	37	43	11	14	17	19	23	26	29	34	1.6	1.8	2.0	2.2	2.4	1.8	2.0	2.3	2.5	2.7
TW B3	12	16	21	24	29	33	37	44	10	13	16	19	23	26	30	35	1.6	1.8	2.0	2.1	2.3	1.7	2.0	2.2	2.4	2.6
TW B4	12	16	20	23	28	32	36	43	9	12	16	19	23	26	29	35	1.6	1.8	2.0	2.2	2.3	1.8	2.0	2.2	2.4	2.6
TW C1	13	16	20	22	26	29	33	38	11	14	16	18	21	24	27	31	1.8	2.0	2.2	2.4	2.6	2.0	2.2	2.5	2.8	3.0
TW C2	12	16	19	22	27	30	34	40	10	13	16	18	22	24	28	33	1.7	1.9	2.1	2.3	2.4	1.9	2.1	2.3	2.6	2.7
TW C3	12	15	19	22	26	29	33	39	10	12	15	18	21	24	27	32	1.7	1.9	2.1	2.3	2.4	1.9	2.1	2.4	2.6	2.8
TW C4	12	15	18	20	24	27	31	36	10	12	15	17	20	23	26	30	1.9	2.1	2.2	2.4	2.6	2.1	2.3	2.5	2.7	2.9
TW D1	9	11	14	16	19	20	23	26	7	8	10	11	13	14	15	17	2.0	2.4	2.7	3.1	3.4	2.4	2.9	3.6	4.5	-
TW D2	15	20	25	30	36	40	46	54	12	16	21	24	30	33	38	45	1.4	1.6	1.7	1.9	2.0	1.5	1.7	1.9	2.1	2.3
TW D3	15	21	27	32	39	44	50	58	12	17	22	26	32	36	41	49	1.3	1.5	1.7	1.8	2.0	1.4	1.6	1.8	2.0	2.2
TW D4	17	24	31	37	45	51	58	67	13	19	26	31	39	44	50	59	1.2	1.3	1.5	1.7	1.8	1.3	1.4	1.6	1.8	1.9
TW D5	17	24	32	38	47	52	60	69	13	19	27	32	40	46	53	62	1.2	1.3	1.5	1.6	1.8	1.2	1.4	1.6	1.7	1.9
TW E1	18	24	31	37	44	49	56	65	14	20	26	31	37	42	48	56	1.2	1.4	1.6	1.7	1.9	1.3	1.5	1.7	1.9	2.0
TW E2	18	24	32	37	45	51	58	67	14	20	26	31	38	43	50	58	1.2	1.4	1.5	1.7	1.8	1.3	1.5	1.6	1.8	2.0
TW E3	21	30	39	46	56	62	69	78	17	25	34	40	49	55	62	71	1.1	1.2	1.4	1.5	1.7	1.1	1.3	1.4	1.6	1.8
TW E4	20	28	37	44	54	60	68	77	16	23	32	38	47	53	61	70	1.1	1.2	1.4	1.5	1.7	1.1	1.3	1.5	1.6	1.8
TW F1	14	16	19	21	24	26	30	34	11	13	15	17	19	21	23	26	1.9	2.2	2.4	2.7	2.9	2.2	2.5	2.9	3.3	3.6
TW F2	12	15	19	22	27	30	34	40	9	12	15	18	21	23	27	31	1.7	1.9	2.1	2.3	2.4	1.9	2.1	2.4	2.7	2.9
TW F3	11	15	19	22	27	31	36	42	8	11	14	17	21	24	28	33	1.6	1.8	2.0	2.2	2.3	1.8	2.0	2.3	2.5	2.7
TW F4	9	13	17	21	26	30	35	42	6	9	13	16	20	23	27	33	1.6	1.8	2.0	2.1	2.3	1.7	2.0	2.2	2.4	2.6
TW G1	12	14	17	20	23	26	29	34	9	11	14	16	18	20	23	26	1.9	2.1	2.3	2.6	2.7	2.1	2.4	2.7	3.0	3.3
TW G2	9	13	17	20	24	27	32	38	7	10	13	15	19	21	25	29	1.7	1.9	2.1	2.3	2.5	1.9	2.1	2.4	2.6	2.9
TW G3	8	11	15	19	23	27	31	38	5	8	11	14	18	21	24	29	1.7	1.9	2.1	2.3	2.4	1.9	2.1	2.3	2.6	2.8
TW G4	6	9	13	16	20	24	28	34	4	6	9	12	16	18	22	27	1.8	1.9	2.1	2.3	2.5	2.0	2.2	2.4	2.6	2.8
TW H1	9	11	15	17	21	24	27	33	7	9	11	13	16	18	21	25	1.9	2.1	2.3	2.5	2.7	2.1	2.4	2.6	2.9	3.1
TW H2	6	9	13	15	20	23	26	32	4	7	9	12	15	17	20	25	1.8	2.0	2.2	2.4	2.6	2.0	2.3	2.5	2.7	2.9
TW H3	5	8	11	14	17	20	24	30	3	5	8	10	13	16	19	23	1.9	2.1	2.3	2.5	2.6	2.1	2.3	2.6	2.8	3.0
TW H4	4	6	9	11	14	16	20	25	2	4	6	8	11	13	15	19	2.1	2.3	2.5	2.7	2.9	2.4	2.6	2.8	3.0	3.2
TW I1	7	9	12	14	16	18	20	23	5	7	8	9	11	12	13	14	2.1	2.5	2.8	3.2	3.5	2.6	3.1	3.8	4.7	-
TW I2	12	17	22	26	33	37	43	51	9	13	17	21	26	30	35	42	1.4	1.6	1.8	1.9	2.1	1.6	1.8	1.9	2.1	2.3
TW I3	13	18	24	29	36	40	47	55	9	14	19	23	29	33	39	46	1.3	1.5	1.7	1.9	2.0	1.5	1.7	1.8	2.0	2.2
TW I4	14	21	28	34	42	48	55	64	11	16	23	28	36	41	47	56	1.2	1.4	1.5	1.7	1.8	1.3	1.5	1.6	1.8	2.0
TW I5	14	21	29	35	44	50	57	67	11	17	24	30	38	43	50	59	1.2	1.3	1.5	1.7	1.8	1.3	1.4	1.6	1.8	1.9
TW J1	16	22	29	34	42	47	54	63	12	18	24	28	35	40	46	54	1.2	1.4	1.6	1.7	1.9	1.3	1.5	1.7	1.9	2.0
TW J2	15	22	29	35	43	48	55	64	12	17	24	29	36	41	47	56	1.2	1.4	1.5	1.7	1.8	1.3	1.5	1.7	1.8	2.0
TW J3	18	27	37	44	53	59	67	77	14	22	31	37	47	53	60	70	1.1	1.2	1.4	1.5	1.7	1.1	1.3	1.5	1.6	1.8
TW J4	17	25	35	42	51	57	65	75	13	21	29	36	45	51	58	68	1.1	1.2	1.4	1.6	1.7	1.2	1.3	1.5	1.6	1.8
DG A1	9	12	14	16	19	20	21	23	6	8	10	11	13	13	14	14	1.9	2.3	3.0	4.3	-	2.3	2.9	4.1	-	-
DG A2	12	16	20	23	28	32	36	41	9	12	16	19	23	25	29	33	1.6	1.8	2.0	2.3	2.6	1.7	2.0	2.3	2.6	3.0
DG A3	15	21	27	32	39	44	50	59	13	17	23	27	33	38	43	51	1.4	1.5	1.6	1.8	2.0	1.5	1.6	1.8	1.9	2.1
DG B1	9	12	15	17	20	22	24	27	6	8	10	12	13	14	15	16	1.9	2.3	2.6	3.2	4.0	2.2	2.8	3.6	4.8	-
DG B2	12	16	21	24	30	34	39	45	9	13	17	19	24	27	31	36	1.6	1.7	1.9	2.1	2.3	1.7	1.9	2.2	2.4	2.7
DG B3	16	21	28	33	41	46	53	63	13	18	24	28	35	39	46	55	1.3	1.5	1.6	1.7	1.8	1.4	1.6	1.7	1.8	2.0
DG C1	12	15	19	21	26	28	32	37	8	11	14	16	19	20	22	25	1.7	2.0	2.1	2.5	2.9	1.9	2.3	2.8	3.3	4.1
DG C2	14	19	24	29	35	40	46	54	11	15	20	24	29	32	37	44	1.4	1.6	1.7	1.9	2.1	1.6	1.7	1.9	2.1	2.3
DG C3	18	24	31	37	46	52	60	70	15	20	27	32	39	45	52	61	1.3	1.4	1.5	1.6	1.7	1.3	1.5	1.6	1.7	1.9
SS A1	17	21	25	29	33	37	41	47	15	19	22	25	29	31	35	39	1.5	1.8	2.0	2.2	2.4	1.7	1.9	2.2	2.5	2.7
SS A2	22	28	34	39	46	51	57	66	19	24	30	35	41	45	51	58	1.2	1.4	1.6	1.7	1.9	1.3	1.5	1.7	1.9	2.0
SS A3	16	19	22	25	29	31	34	39	14	16	19	21	24	26	28	31	1.7	2.0	2.2	2.5	2.7	1.8	2.2	2.6	3.1	3.5
SS A4	21	26	33	37	44	49	55	62	18	23	29	33	39	43	48	55	1.3	1.4	1.6	1.8	1.9	1.3	1.5	1.8	2.0	2.1
SS A5	18	21	24	27	30	33	36	41	16	18	21	23	26	28	31	34	1.7	1.9	2.2	2.4	2.6	1.8	2.2	2.5	2.9	3.2
SS A6	21	27	34	38	45	50	56	64	19	24	30	34	40	44	50	57	1.2	1.4	1.6	1.8	1.9	1.3	1.5	1.7	1.9	2.1
SS A7	16	18	21	23	25	27	29	33	14	16	18	20	22	23	25	26	1.9	2.3	2.6	3.0	3.3	2.1	2.6	3.1	3.8	4.6
SS A8	20	26	32	36	42	47	52	60	18	23	28	32	37	41	46	52	1.3	1.5	1.7	1.8	2.0	1.4	1.6	1.8	2.0	2.2
SS B1	15	18	22	25	29	32	36	41	13	15	19	21	24	27	30	34	1.6	1.9	2.1	2.3	2.5	1.8	2.1	2.4	2.7	2.9
SS B2	19	24	31	35	42	47	53	61	16	21	27	31	37	42	47	55	1.3	1.5	1.6	1.8	1.9	1.4	1.5	1.7	1.9	2.1
SS B3	13	16	20	22	26	28	31	36	11	14	17	19	21	23	26	29	1.8	2.0	2.3	2.6	2.8	1.9	2.3	2.7	3.1	3.5
SS B4	18	23	29	34	40	45	51	59	16	20	26	30	36	40	45	52	1.3	1.5	1.7	1.8	2.0	1.4	1.6	1.8	2.0	2.1
SS B5	15	17	20	22	25	28	31	35	13	15	17	19	22	23	26	29	1.9	2.1	2.4	2.7	2.9	2.1	2.4	2.8	3.2	3.5
SS B6	18	23	30	34	41	46	52	60	16	20	26	30	36	40	45	53	1.3	1.5	1.6	1.8	2.0	1.4	1.6	1.8	2.0	2.1
SS B7	13	15	18	19	22	24	26	29	12	13	15	16	18	20	21	23	2.1	2.4	2.8	3.1	3.4	2.3	2.7	3.2	3.8	4.5
SS B8	17	22	28	32	38	43	48	56	15	19	24	28	33	37	42	49	1.3	1.5	1.7	1.9	2.0	1.4	1.7	1.9	2.1	2.2
SS C1	12	16	21	24	29	32	37	43	9	12	16	18	22	25	28	32	1.5	1.8	2.0	2.2	2.3	1.8	2.0	2.3	2.6	2.9
SS C2	15	21	27	32	40	44	51	59	12	17	22	26	33	37	42	50	1.3	1.5	1.6	1.8	1.9	1.4	1.6	1.8	2.0	2.1
SS C3	12	15	18	21	25	27	31	36	10	12	15	17	19	21	24	28	1.8	2.0	2.3	2.5	2.7	2.0	2.3	2.6	2.9	3.2
SS C4	14	19	25	29	35	40	46	54	11	15	20	24	29	33	38	45	1.4	1.6	1.7	1.9	2.0	1.5	1.7	1.9	2.1	2.3
SS D1	19	24	29	33	39	43	48	55	16	20	25	28	33	36	40	45	1.4	1.6	1.8	2.0	2.1	1.5	1.7	2.0	2.3	2.5
SS D2	23	31	39	45	54	59	66	74	20	27	35	40	48	53	59	67	1.1	1.3	1.4	1.6	1.7	1.2	1.3	1.5	1.7	1.9
SS D3	19	23	27	30	35	38	43	49	16	20	23	26	30	33	36	41	1.5	1.7	1.9	2.1	2.3	1.6	1.9	2.2	2.5	2.7
SS D4	23	30	38	44	52	57	64	72	20	26	33	39	46	51	57	65	1.1	1.3	1.5	1.6	1.8	1.2	1.4	1.6	1.8	1.9
SS E1	16	20	24	28	32	36	40	46	13	16	20	22	26	29	32	36	1.5	1.7	2.0	2.2	2.4	1.6	2.0	2.3	2.6	2.9
SS E2	19	26	33	39	46	52	58	67	16	22	29	33	40	45	50	58	1.2	1.4	1.5	1.7	1.8	1.3	1.5	1.7	1.9	2.0
SS E3	15	18	22	24	28	31	34	39	13	16	19	21	24	26	28	32	1.7	2.0	2.2	2.4	2.6	1.9	2.2	2.5	2.9	3.3
SS E4	19	24	31	36	43	48	54	62	16	21	27	31	37	41	47	54	1.3	1.4	1.6	1.8	1.9	1.3	1.5	1.7	1.9	2.1

RAPID CITY, SOUTH DAKOTA

TYPE	SOLAR SAVINGS FRACTION (%) BASE TEMP: 55 F (4808 DD) LCR 100	70	50	40	30	25	20	15	65 F (7332 DD) 100	70	50	40	30	25	20	15	CONSERVATION FACTOR 55 F 100	50	30	20	15	65 F 100	50	30	20	15
WW A1	18	21	25	28	32	35	38	44	15	17	19	21	24	26	29	33	1.7	1.9	2.1	2.3	2.5	2.0	2.3	2.5	2.8	3.1
WW A2	19	24	29	34	40	44	50	58	14	18	22	26	31	34	39	45	1.3	1.5	1.7	1.9	2.0	1.5	1.7	2.0	2.2	2.3
WW A3	18	24	31	36	43	48	54	63	13	18	23	27	33	37	42	50	1.2	1.4	1.6	1.8	1.9	1.4	1.6	1.8	2.0	2.2
WW A4	18	25	32	37	45	50	57	66	13	18	24	28	34	39	44	52	1.2	1.4	1.5	1.7	1.9	1.4	1.6	1.7	1.9	2.1
WW A5	18	25	33	39	47	52	59	68	12	18	24	29	36	41	47	55	1.2	1.3	1.5	1.7	1.8	1.3	1.5	1.7	1.9	2.0
WW A6	18	26	34	40	48	54	61	70	12	18	25	30	37	42	48	56	1.1	1.3	1.5	1.6	1.8	1.3	1.5	1.7	1.8	2.0
WW B1	14	18	22	25	29	32	36	42	10	12	15	17	20	22	24	27	1.6	1.9	2.1	2.3	2.5	2.0	2.3	2.7	3.1	3.6
WW B2	19	26	34	40	48	54	61	70	14	19	26	31	38	43	49	58	1.2	1.3	1.5	1.6	1.8	1.3	1.5	1.6	1.8	2.0
WW B3	23	31	40	47	56	62	69	78	17	24	32	37	45	51	58	67	1.1	1.2	1.4	1.5	1.7	1.2	1.3	1.5	1.7	1.8
WW B4	23	32	42	50	59	66	73	82	17	25	34	40	50	56	63	73	1.0	1.2	1.3	1.5	1.6	1.1	1.3	1.4	1.6	1.7
WW B5	22	31	42	49	59	66	73	82	16	24	34	40	50	56	64	73	1.0	1.2	1.3	1.5	1.6	1.1	1.2	1.4	1.6	1.7
WW C1	25	33	42	49	58	64	72	80	18	26	34	40	48	54	61	70	1.0	1.2	1.3	1.5	1.7	1.1	1.3	1.5	1.6	1.8
WW C2	23	32	41	48	58	64	71	80	17	24	33	39	48	54	61	70	1.0	1.2	1.3	1.5	1.7	1.1	1.3	1.5	1.6	1.8
WW C3	27	38	50	58	68	74	81	89	21	31	41	49	59	66	73	82	0.9	1.1	1.2	1.4	1.6	1.0	1.1	1.3	1.5	1.6
WW C4	25	35	46	54	65	71	79	87	18	28	38	46	56	63	70	80	1.0	1.1	1.3	1.4	1.6	1.0	1.2	1.3	1.5	1.6
TW A1	18	21	25	27	32	34	38	44	15	17	19	21	24	27	30	34	1.7	1.9	2.1	2.3	2.4	2.0	2.2	2.5	2.7	3.0
TW A2	18	22	27	31	37	41	46	54	13	17	21	24	28	32	36	42	1.4	1.6	1.8	2.0	2.1	1.6	1.8	2.1	2.3	2.4
TW A3	17	22	28	33	39	43	49	57	13	17	21	25	30	34	38	45	1.3	1.5	1.7	1.9	2.0	1.5	1.7	1.9	2.1	2.3
TW A4	16	22	28	33	39	44	50	59	11	16	21	25	30	34	39	47	1.3	1.5	1.6	1.8	2.0	1.5	1.7	1.9	2.1	2.2
TW B1	17	21	25	28	32	36	40	46	14	16	19	22	25	28	31	36	1.6	1.8	2.0	2.2	2.3	1.9	2.1	2.3	2.5	2.7
TW B2	16	21	26	30	36	40	46	53	12	16	20	23	28	31	36	42	1.4	1.6	1.8	1.9	2.1	1.6	1.8	2.0	2.2	2.4
TW B3	16	20	26	30	37	41	47	55	11	15	20	23	28	32	37	43	1.4	1.6	1.7	1.9	2.0	1.6	1.8	2.0	2.2	2.3
TW B4	15	19	25	29	35	39	45	53	11	14	19	22	27	31	35	42	1.4	1.6	1.8	1.9	2.1	1.6	1.8	2.0	2.2	2.3
TW C1	16	20	24	27	32	36	41	47	12	15	19	21	25	28	32	37	1.6	1.7	1.9	2.1	2.2	1.8	2.0	2.2	2.4	2.6
TW C2	15	19	24	28	33	37	42	50	11	15	18	21	26	29	33	39	1.5	1.7	1.8	2.0	2.1	1.7	1.9	2.1	2.3	2.4
TW C3	14	18	23	26	32	36	41	48	11	14	18	21	25	28	32	38	1.5	1.7	1.9	2.0	2.2	1.8	1.9	2.1	2.3	2.5
TW C4	14	17	21	25	29	33	37	44	11	14	17	19	23	26	30	35	1.7	1.8	2.0	2.2	2.3	1.9	2.1	2.2	2.4	2.6
TW D1	12	15	19	22	26	29	33	38	8	10	13	15	18	20	22	25	1.7	1.9	2.2	2.4	2.6	2.1	2.4	2.7	3.1	3.4
TW D2	19	25	32	37	44	49	56	65	14	19	24	29	35	40	45	53	1.2	1.4	1.6	1.7	1.9	1.4	1.6	1.7	1.9	2.1
TW D3	20	27	34	40	48	53	60	69	14	20	26	31	38	43	49	58	1.2	1.3	1.5	1.7	1.8	1.3	1.5	1.7	1.8	2.0
TW D4	21	29	38	45	54	60	67	76	16	23	30	36	45	50	57	66	1.1	1.2	1.4	1.6	1.7	1.2	1.3	1.5	1.7	1.8
TW D5	21	29	38	45	54	61	68	77	16	23	31	37	46	52	59	68	1.1	1.2	1.4	1.5	1.7	1.2	1.3	1.5	1.6	1.8
TW E1	23	30	39	45	53	59	66	75	17	23	31	36	44	49	55	64	1.1	1.3	1.4	1.6	1.7	1.2	1.4	1.6	1.7	1.9
TW E2	22	30	39	45	54	60	67	76	17	23	31	37	45	50	57	66	1.1	1.2	1.4	1.6	1.7	1.2	1.4	1.5	1.7	1.8
TW E3	27	36	47	54	64	70	77	85	20	29	39	46	55	61	69	78	1.0	1.1	1.3	1.4	1.6	1.0	1.2	1.4	1.5	1.7
TW E4	25	34	45	52	62	68	76	84	19	27	37	44	53	59	67	76	1.0	1.1	1.3	1.5	1.6	1.1	1.2	1.4	1.5	1.7
TW F1	16	19	23	26	31	34	38	44	13	15	18	20	23	26	29	33	1.7	1.9	2.0	2.2	2.4	2.0	2.2	2.4	2.7	2.9
TW F2	15	20	25	29	34	38	44	51	11	14	18	21	26	29	33	39	1.5	1.6	1.8	2.0	2.1	1.7	1.9	2.1	2.3	2.5
TW F3	14	19	25	29	35	40	45	54	10	13	18	21	26	30	35	41	1.4	1.6	1.7	1.9	2.0	1.6	1.8	2.0	2.2	2.3
TW F4	12	17	23	27	34	39	45	53	8	12	16	20	25	29	34	41	1.4	1.6	1.7	1.9	2.0	1.6	1.8	2.0	2.1	2.3
TW G1	14	18	22	25	30	33	38	44	11	13	16	19	22	25	28	33	1.6	1.8	2.0	2.2	2.3	1.9	2.1	2.3	2.6	2.7
TW G2	12	17	22	26	31	35	41	48	8	12	16	19	23	27	31	37	1.5	1.7	1.8	2.0	2.1	1.7	1.9	2.1	2.3	2.5
TW G3	11	15	20	24	30	34	40	48	7	10	14	18	22	26	30	37	1.5	1.7	1.8	2.0	2.1	1.7	1.9	2.1	2.3	2.4
TW G4	9	13	17	21	27	31	36	44	5	8	12	15	19	23	27	33	1.6	1.7	1.9	2.1	2.2	1.8	2.0	2.1	2.3	2.5
TW H1	11	15	19	22	27	31	35	42	5	10	14	16	20	23	26	32	1.6	1.8	2.0	2.2	2.3	1.9	2.1	2.3	2.5	2.7
TW H2	9	12	17	20	26	29	34	41	5	8	12	14	18	21	26	31	1.6	1.8	2.0	2.1	2.3	1.9	2.0	2.2	2.4	2.6
TW H3	7	11	15	18	23	26	31	38	4	7	10	12	16	19	23	29	1.7	1.9	2.0	2.2	2.3	2.0	2.1	2.3	2.5	2.7
TW H4	5	8	11	14	18	21	25	32	3	5	8	10	13	15	19	24	1.9	2.1	2.2	2.4	2.5	2.2	2.4	2.5	2.7	2.8
TW I1	10	13	16	19	23	26	29	35	6	8	11	13	15	17	19	22	1.8	2.0	2.2	2.5	2.7	2.2	2.5	2.8	3.2	3.6
TW I2	16	21	28	33	41	46	53	61	11	16	21	25	32	36	42	50	1.3	1.4	1.6	1.8	1.9	1.4	1.6	1.8	1.9	2.1
TW I3	17	23	31	36	44	50	56	66	12	17	23	28	35	40	46	54	1.2	1.4	1.5	1.7	1.8	1.3	1.5	1.7	1.9	2.0
TW I4	19	26	35	41	50	56	64	73	13	20	27	33	41	47	54	63	1.1	1.3	1.4	1.6	1.7	1.2	1.4	1.5	1.7	1.8
TW I5	19	27	36	43	52	58	66	75	13	20	28	34	43	49	56	66	1.1	1.2	1.4	1.6	1.7	1.2	1.3	1.5	1.7	1.8
TW J1	20	28	36	42	51	57	64	73	15	21	28	34	41	47	53	62	1.1	1.3	1.4	1.6	1.7	1.2	1.4	1.6	1.7	1.9
TW J2	20	27	36	42	51	57	64	74	14	21	28	34	42	47	54	64	1.1	1.3	1.4	1.6	1.7	1.2	1.4	1.5	1.7	1.8
TW J3	24	33	44	51	61	68	75	84	18	26	36	43	53	59	67	76	1.0	1.1	1.3	1.5	1.6	1.1	1.2	1.4	1.5	1.7
TW J4	22	31	42	49	59	66	73	82	16	24	34	41	50	57	65	74	1.0	1.2	1.3	1.5	1.6	1.1	1.2	1.4	1.6	1.7
DG A1	12	16	20	23	26	29	31	34	8	11	13	15	18	19	20	21	1.6	1.9	2.4	3.0	3.9	1.9	2.3	3.0	4.4	-
DG A2	15	20	25	30	36	40	45	51	11	15	19	22	27	31	35	40	1.4	1.6	1.8	2.0	2.3	1.6	1.8	2.0	2.3	2.6
DG A3	19	25	32	38	47	52	59	68	14	20	26	31	38	43	50	58	1.2	1.3	1.5	1.7	1.9	1.4	1.5	1.6	1.8	2.0
DG B1	13	16	21	24	29	32	36	40	8	11	14	16	19	21	23	26	1.6	1.8	2.0	2.4	2.8	1.9	2.2	2.7	3.2	4.0
DG B2	15	20	26	31	38	43	49	57	11	15	20	23	29	33	38	44	1.4	1.5	1.7	1.8	2.0	1.6	1.7	1.9	2.1	2.3
DG B3	19	26	33	39	49	55	63	73	15	20	27	32	40	45	53	62	1.2	1.3	1.4	1.6	1.7	1.3	1.5	1.6	1.7	1.8
DG C1	15	20	25	29	35	40	45	51	10	14	18	21	25	27	31	35	1.4	1.6	1.8	2.0	2.3	1.7	1.9	2.2	2.5	2.9
DG C2	18	24	31	36	44	50	57	66	13	18	24	28	34	39	45	53	1.3	1.4	1.5	1.7	1.9	1.4	1.6	1.7	1.9	2.1
DG C3	22	29	37	44	55	61	70	79	17	23	31	36	45	51	59	69	1.1	1.2	1.3	1.5	1.7	1.3	1.4	1.5	1.6	1.7
SS A1	21	26	31	35	41	45	51	58	18	22	26	30	34	38	42	48	1.3	1.5	1.7	1.9	2.1	1.5	1.7	1.9	2.2	2.4
SS A2	26	33	41	47	55	60	66	75	22	28	35	40	47	52	58	66	1.1	1.3	1.4	1.6	1.8	1.2	1.4	1.6	1.7	1.9
SS A3	19	23	28	32	37	40	45	51	16	19	23	26	30	33	36	41	1.4	1.7	1.9	2.1	2.3	1.6	1.9	2.2	2.5	2.7
SS A4	25	32	40	45	53	58	64	72	21	27	34	39	45	50	56	63	1.1	1.3	1.5	1.6	1.8	1.2	1.4	1.6	1.8	2.0
SS A5	21	25	29	33	38	41	46	52	18	21	25	27	31	34	38	43	1.5	1.7	1.9	2.1	2.2	1.6	1.9	2.1	2.4	2.6
SS A6	26	33	40	46	54	59	66	74	22	28	34	39	46	51	57	65	1.1	1.3	1.5	1.6	1.8	1.2	1.4	1.6	1.8	1.9
SS A7	19	22	25	28	32	35	38	43	16	19	21	23	26	28	31	34	1.7	1.9	2.1	2.4	2.6	1.9	2.2	2.5	2.9	3.2
SS A8	25	31	38	44	51	56	62	70	21	26	33	37	44	48	54	61	1.2	1.3	1.5	1.7	1.8	1.2	1.5	1.6	1.8	2.0
SS B1	18	22	27	31	36	40	45	52	15	18	22	25	30	33	37	42	1.4	1.6	1.8	2.0	2.2	1.6	1.8	2.1	2.3	2.5
SS B2	23	29	37	42	50	55	62	70	19	25	31	36	43	48	54	62	1.2	1.3	1.5	1.7	1.8	1.3	1.4	1.6	1.8	1.9
SS B3	17	20	25	28	33	37	41	47	13	16	20	23	26	29	33	37	1.5	1.7	1.9	2.1	2.3	1.7	2.0	2.2	2.5	2.8
SS B4	22	28	36	41	48	53	60	68	18	24	30	35	41	46	52	60	1.2	1.4	1.5	1.7	1.8	1.3	1.5	1.7	1.8	2.0
SS B5	17	21	24	27	32	35	39	45	14	17	20	22	26	28	31	36	1.6	1.8	2.0	2.2	2.4	1.8	2.1	2.3	2.6	2.8
SS B6	22	29	36	41	49	54	61	69	18	24	30	35	42	46	52	60	1.2	1.3	1.5	1.7	1.8	1.3	1.5	1.6	1.8	2.0
SS B7	16	18	22	24	28	30	34	39	13	15	18	20	22	24	27	30	1.8	2.0	2.3	2.5	2.7	2.0	2.3	2.6	3.0	3.3
SS B8	21	27	34	39	46	51	58	66	17	22	28	33	39	44	49	57	1.2	1.4	1.6	1.7	1.9	1.3	1.5	1.7	1.9	2.0
SS C1	16	21	27	31	36	42	48	56	11	15	20	23	28	31	36	42	1.4	1.5	1.7	1.9	2.0	1.6	1.8	2.0	2.2	2.4
SS C2	20	27	34	40	48	54	61	70	14	20	27	32	39	44	50	59	1.2	1.3	1.5	1.6	1.8	1.3	1.5	1.6	1.8	2.0
SS C3	15	19	23	26	31	35	39	46	11	14	17	20	24	26	30	35	1.6	1.8	1.9	2.1	2.3	1.8	2.0	2.3	2.5	2.7
SS C4	18	24	31	36	43	48	55	64	13	18	24	28	35	39	45	53	1.3	1.4	1.6	1.7	1.9	1.4	1.6	1.7	1.9	2.1
SS D1	24	30	37	41	48	53	59	67	19	25	30	34	40	44	49	55	1.2	1.4	1.6	1.7	1.9	1.3	1.5	1.8	2.0	2.2
SS D2	29	38	47	54	63	68	75	83	24	32	40	46	55	60	67	75	1.0	1.2	1.3	1.5	1.6	1.1	1.3	1.4	1.6	1.8
SS D3	23	28	33	38	44	48	53	60	19	23	28	31	36	40	44	50	1.3	1.5	1.7	1.9	2.0	1.4	1.7	1.9	2.1	2.3
SS D4	28	37	46	52	61	66	73	81	24	31	39	45	53	58	64	73	1.0	1.2	1.4	1.5	1.7	1.1	1.3	1.5	1.6	1.8
SS E1	20	25	31	36	42	46	52	59	16	20	25	28	33	37	41	47	1.3	1.5	1.7	1.9	2.0	1.4	1.7	1.9	2.2	2.4
SS E2	25	33	41	47	56	61	68	77	20	27	34	40	47	52	59	67	1.1	1.2	1.4	1.6	1.7	1.2	1.3	1.5	1.7	1.9
SS E3	19	23	27	31	36	39	44	51	15	18	22	25	29	32	35	40	1.5	1.7	1.9	2.1	2.2	1.6	1.9	2.2	2.4	2.6
SS E4	23	30	38	44	52	57	64	72	19	25	32	37	44	49	55	63	1.1	1.3	1.5	1.6	1.8	1.2	1.4	1.6	1.8	1.9

SIOUX FALLS, SOUTH DAKOTA

TYPE	SOLAR SAVINGS FRACTION (%) BASE TEMP: 55 F (5405 DD) LCR								65 F (7846 DD)								CONSERVATION FACTOR 55 F					65 F				
	100	70	50	40	30	25	20	15	100	70	50	40	30	25	20	15	100	50	30	20	15	100	50	30	20	15
WW A1	15	16	19	20	23	24	27	30	13	14	16	17	18	19	20	22	2.1	2.5	2.8	3.1	3.4	2.6	3.0	3.5	4.2	5.0
WW A2	13	17	21	24	28	31	35	41	10	13	16	18	22	24	27	31	1.6	1.9	2.1	2.3	2.5	1.9	2.1	2.4	2.8	3.0
WW A3	12	16	21	25	30	34	39	46	9	13	16	19	23	26	30	35	1.5	1.7	1.9	2.1	2.3	1.7	2.0	2.2	2.5	2.7
WW A4	12	16	21	25	31	35	41	48	8	12	16	19	24	27	31	37	1.4	1.6	1.8	2.0	2.2	1.6	1.9	2.1	2.4	2.6
WW A5	11	16	22	26	33	37	43	51	8	12	16	20	25	28	33	39	1.4	1.6	1.8	1.9	2.1	1.6	1.8	2.0	2.3	2.5
WW A6	11	16	22	27	33	38	44	52	7	12	16	20	25	29	34	40	1.4	1.6	1.7	1.9	2.1	1.5	1.7	2.0	2.2	2.4
WW B1	10	11	14	15	17	18	20	22	7	8	9	10	11	11	12	11	2.2	2.6	3.1	3.7	4.3	2.7	3.7	-	-	-
WW B2	13	18	24	29	35	40	46	55	9	14	19	23	28	32	38	45	1.4	1.5	1.7	1.9	2.0	1.5	1.7	1.9	2.1	2.2
WW B3	15	22	29	34	42	48	54	63	12	17	23	28	35	39	45	53	1.2	1.4	1.6	1.7	1.9	1.3	1.5	1.7	1.9	2.1
WW B4	16	23	32	38	47	53	60	70	12	18	26	32	40	45	52	62	1.1	1.3	1.5	1.6	1.8	1.2	1.4	1.6	1.7	1.9
WW B5	15	23	31	38	47	53	61	71	11	18	26	32	41	46	54	64	1.1	1.3	1.4	1.6	1.7	1.2	1.4	1.5	1.7	1.8
WW C1	17	24	31	37	45	51	58	67	13	19	26	31	38	43	49	58	1.2	1.3	1.5	1.7	1.8	1.3	1.5	1.6	1.8	2.0
WW C2	16	23	31	37	45	51	58	68	12	18	25	30	38	43	50	59	1.2	1.3	1.5	1.7	1.8	1.3	1.4	1.6	1.8	1.9
WW C3	19	29	39	46	56	63	71	80	15	24	33	40	50	56	64	74	1.0	1.2	1.3	1.5	1.6	1.1	1.2	1.4	1.6	1.7
WW C4	17	26	36	43	53	60	68	78	13	21	31	37	47	53	61	71	1.1	1.2	1.4	1.5	1.7	1.1	1.3	1.4	1.6	1.7
TW A1	14	16	19	20	23	25	27	31	13	14	16	17	19	20	21	24	2.1	2.4	2.7	3.0	3.2	2.5	2.9	3.3	3.8	4.3
TW A2	13	16	20	22	26	29	33	38	11	13	16	18	21	23	26	30	1.7	2.0	2.2	2.4	2.6	2.0	2.3	2.5	2.9	3.1
TW A3	12	16	20	23	28	31	35	41	9	12	16	18	22	24	28	32	1.6	1.8	2.0	2.2	2.4	1.8	2.1	2.3	2.6	2.8
TW A4	11	15	19	23	28	31	36	43	8	11	15	18	22	25	28	34	1.6	1.8	2.0	2.1	2.3	1.7	2.0	2.2	2.5	2.7
TW B1	14	16	18	20	23	26	29	33	12	13	15	17	19	21	23	26	2.0	2.2	2.5	2.7	2.9	2.3	2.6	3.0	3.3	3.7
TW B2	12	15	19	22	26	29	33	39	10	12	15	17	21	23	26	30	1.7	1.9	2.1	2.3	2.5	1.9	2.2	2.4	2.7	2.9
TW B3	11	14	18	22	26	29	34	40	9	11	15	17	21	23	27	32	1.7	1.9	2.1	2.3	2.4	1.9	2.1	2.3	2.6	2.8
TW B4	11	14	18	21	25	28	33	39	8	11	14	17	20	23	26	31	1.7	1.9	2.1	2.3	2.4	1.9	2.1	2.3	2.6	2.8
TW C1	12	15	18	20	24	26	29	34	10	12	15	17	19	21	24	27	1.9	2.1	2.3	2.6	2.7	2.2	2.4	2.7	3.0	3.2
TW C2	11	14	17	20	24	27	31	37	9	12	14	16	20	22	25	30	1.8	2.0	2.2	2.4	2.5	2.0	2.2	2.5	2.7	2.9
TW C3	11	14	17	20	23	26	30	36	9	11	14	16	19	22	25	29	1.8	2.0	2.2	2.4	2.6	2.1	2.3	2.5	2.7	2.9
TW C4	11	14	16	19	22	25	28	33	10	12	14	16	19	21	23	28	2.0	2.2	2.4	2.6	2.7	2.2	2.4	2.6	2.8	3.0
TW D1	8	10	12	13	16	17	19	21	6	7	8	9	10	11	12	13	2.2	2.6	3.1	3.5	4.0	2.8	3.4	4.4	-	-
TW D2	13	18	23	27	33	37	42	50	10	14	18	22	27	30	35	41	1.4	1.6	1.8	2.0	2.1	1.6	1.8	2.0	2.2	2.4
TW D3	14	19	24	29	35	40	46	54	10	15	20	23	29	33	38	45	1.4	1.5	1.7	1.9	2.0	1.5	1.7	1.9	2.1	2.3
TW D4	15	21	29	34	42	47	54	64	12	17	24	29	36	40	47	56	1.2	1.4	1.6	1.7	1.8	1.3	1.5	1.7	1.8	2.0
TW D5	15	21	29	35	43	49	56	66	12	18	24	30	37	43	49	59	1.2	1.4	1.5	1.7	1.8	1.3	1.4	1.6	1.8	1.9
TW E1	16	22	28	34	41	46	52	61	13	18	23	28	34	38	44	52	1.3	1.4	1.6	1.8	1.9	1.4	1.6	1.8	1.9	2.1
TW E2	16	22	29	34	42	47	54	63	12	18	24	29	35	40	46	55	1.2	1.4	1.6	1.7	1.9	1.3	1.5	1.7	1.9	2.0
TW E3	19	27	36	43	52	58	66	75	15	23	31	37	46	52	59	68	1.1	1.2	1.4	1.6	1.7	1.2	1.3	1.5	1.6	1.8
TW E4	18	26	35	41	51	57	64	74	14	21	29	36	44	50	58	67	1.1	1.3	1.4	1.6	1.7	1.2	1.3	1.5	1.6	1.8
TW F1	12	15	17	19	22	24	26	30	11	12	14	15	17	18	20	22	2.1	2.3	2.6	2.9	3.1	2.4	2.8	3.2	3.7	4.1
TW F2	11	14	17	20	24	27	30	36	8	11	13	15	18	20	23	27	1.8	2.0	2.2	2.4	2.6	2.0	2.3	2.6	2.9	3.2
TW F3	9	13	17	20	24	27	32	38	7	9	13	15	18	21	24	29	1.7	1.9	2.1	2.3	2.5	1.9	2.1	2.4	2.7	2.9
TW F4	7	11	15	18	23	26	31	38	5	8	11	14	17	20	24	29	1.7	1.9	2.1	2.2	2.4	1.9	2.1	2.3	2.6	2.8
TW G1	10	13	15	18	21	23	26	30	9	10	12	14	16	18	20	23	2.0	2.3	2.5	2.7	2.9	2.3	2.6	3.0	3.3	3.7
TW G2	8	11	15	17	21	24	28	34	6	8	11	13	16	19	21	26	1.8	2.0	2.2	2.4	2.6	2.0	2.3	2.6	2.8	3.1
TW G3	7	10	13	16	20	23	28	34	4	7	10	12	15	18	21	26	1.8	2.0	2.2	2.4	2.5	2.0	2.2	2.5	2.7	2.9
TW G4	5	8	11	14	18	21	25	31	3	5	8	10	13	16	19	24	1.9	2.1	2.3	2.5	2.6	2.1	2.3	2.5	2.8	3.0
TW H1	8	10	13	15	18	21	24	29	6	8	10	12	14	16	18	22	2.0	2.2	2.4	2.7	2.8	2.3	2.5	2.8	3.1	3.4
TW H2	5	8	11	13	17	20	23	29	3	5	8	10	13	15	18	22	1.9	2.2	2.4	2.6	2.7	2.2	2.4	2.7	2.9	3.1
TW H3	4	6	9	12	15	18	21	26	2	4	7	9	11	13	16	20	2.0	2.2	2.4	2.6	2.8	2.3	2.5	2.7	3.0	3.2
TW H4	3	5	7	9	12	14	17	22	2	3	5	7	9	11	13	17	2.3	2.4	2.6	2.8	3.0	2.5	2.7	2.9	3.2	3.3
TW I1	6	8	10	11	13	15	16	19	4	5	6	7	8	9	10	10	2.3	2.8	3.2	3.7	4.2	2.9	3.6	4.6	-	-
TW I2	10	15	20	24	29	34	39	47	7	11	15	19	24	27	32	38	1.5	1.7	1.8	2.0	2.2	1.6	1.8	2.0	2.2	2.4
TW I3	11	16	22	26	32	37	43	51	8	12	17	21	26	30	35	42	1.4	1.6	1.8	1.9	2.1	1.5	1.7	1.9	2.1	2.3
TW I4	12	18	26	31	39	44	51	61	9	15	21	26	33	37	44	53	1.3	1.4	1.6	1.7	1.9	1.4	1.5	1.7	1.9	2.0
TW I5	12	19	27	32	41	46	54	63	9	15	22	27	35	40	47	56	1.2	1.4	1.5	1.7	1.8	1.3	1.5	1.6	1.8	1.9
TW J1	14	20	26	31	39	44	50	59	11	16	21	26	32	36	42	50	1.3	1.5	1.6	1.8	1.9	1.4	1.6	1.8	2.0	2.1
TW J2	13	19	26	32	39	45	51	61	10	15	21	26	33	38	44	52	1.3	1.4	1.6	1.7	1.9	1.4	1.5	1.7	1.9	2.0
TW J3	16	24	34	40	50	56	64	74	13	20	28	35	44	50	57	67	1.1	1.3	1.4	1.6	1.7	1.2	1.3	1.5	1.7	1.8
TW J4	15	23	32	39	48	54	62	72	11	19	27	33	42	48	55	65	1.1	1.3	1.4	1.6	1.7	1.2	1.3	1.5	1.7	1.8
DG A1	8	10	12	14	15	16	17	18	5	7	8	9	10	10	10	10	2.1	2.6	3.4	-	-	2.5	3.3	-	-	-
DG A2	10	14	18	21	25	28	32	37	8	11	14	17	20	23	26	30	1.7	1.9	2.1	2.4	2.7	1.9	2.1	2.4	2.8	3.2
DG A3	14	19	24	29	36	41	47	55	12	16	21	25	31	35	40	48	1.4	1.6	1.7	1.8	2.0	1.5	1.7	1.8	2.0	2.2
DG B1	8	10	12	14	16	18	19	21	5	7	8	9	10	11	12	12	2.1	2.5	2.9	3.7	4.9	2.5	3.2	4.5	-	-
DG B2	11	14	18	22	26	30	35	41	8	11	15	17	21	24	27	32	1.6	1.9	2.0	2.2	2.4	1.8	2.1	2.3	2.5	2.8
DG B3	14	19	25	30	37	43	50	59	12	16	22	26	32	37	43	51	1.4	1.5	1.6	1.7	1.9	1.5	1.6	1.8	1.9	2.0
DG C1	10	13	16	18	22	24	27	31	7	9	12	13	15	17	18	20	1.8	2.1	2.4	2.8	3.2	2.1	2.5	3.2	4.0	4.8
DG C2	13	17	22	26	32	36	42	49	10	14	18	21	26	29	34	40	1.5	1.7	1.8	2.0	2.2	1.7	1.8	2.1	2.2	2.4
DG C3	16	22	29	34	42	48	56	66	14	19	25	29	36	41	48	58	1.3	1.4	1.5	1.6	1.8	1.4	1.5	1.7	1.8	1.9
SS A1	16	20	23	26	31	34	38	43	14	17	21	23	26	29	32	36	1.6	1.8	2.1	2.3	2.5	1.7	2.0	2.3	2.7	2.9
SS A2	20	26	32	37	44	48	54	62	18	23	28	33	39	43	48	56	1.3	1.4	1.6	1.8	1.9	1.3	1.5	1.7	1.9	2.1
SS A3	15	17	20	23	26	28	31	35	13	15	18	20	22	24	26	28	1.8	2.1	2.4	2.7	2.9	2.0	2.4	2.8	3.3	3.9
SS A4	19	24	30	35	41	46	51	59	17	22	27	31	37	40	45	52	1.3	1.5	1.7	1.8	2.0	1.4	1.6	1.8	2.0	2.2
SS A5	17	19	22	25	28	30	33	38	15	17	20	22	24	26	28	31	1.8	2.1	2.3	2.6	2.8	2.0	2.3	2.7	3.1	3.5
SS A6	20	25	31	36	43	47	53	61	17	22	28	32	38	42	47	54	1.3	1.5	1.6	1.8	2.0	1.4	1.6	1.8	2.0	2.1
SS A7	15	17	19	21	23	25	27	29	14	15	17	18	20	21	22	23	2.1	2.5	2.9	3.3	3.7	2.3	2.8	3.5	4.4	-
SS A8	19	24	29	34	40	44	49	57	17	21	26	30	35	38	43	49	1.3	1.5	1.7	1.9	2.0	1.4	1.7	1.9	2.1	2.3
SS B1	13	16	20	22	26	29	33	38	12	14	17	19	22	24	27	31	1.7	2.0	2.2	2.5	2.6	1.9	2.2	2.5	2.9	3.2
SS B2	17	23	29	33	40	44	50	58	15	20	25	29	35	39	45	52	1.3	1.5	1.7	1.8	2.0	1.4	1.6	1.8	2.0	2.1
SS B3	12	15	18	20	23	25	28	32	11	13	15	17	19	21	23	25	1.9	2.2	2.5	2.8	3.0	2.1	2.4	2.9	3.4	3.9
SS B4	17	22	27	31	38	42	48	56	15	19	24	28	33	37	42	49	1.4	1.5	1.7	1.9	2.0	1.4	1.7	1.9	2.0	2.2
SS B5	14	16	18	20	23	25	28	31	12	14	16	17	20	21	23	26	2.0	2.3	2.6	2.9	3.1	2.2	2.6	3.0	3.4	3.9
SS B6	16	22	27	32	38	43	48	56	14	19	24	28	34	37	43	50	1.3	1.5	1.7	1.9	2.0	1.4	1.6	1.8	2.0	2.2
SS B7	12	14	16	18	20	21	23	26	11	12	14	15	17	18	19	20	2.2	2.6	3.0	3.4	3.8	2.5	3.0	3.6	4.4	-
SS B8	16	20	26	30	36	40	45	53	13	18	22	26	31	35	39	46	1.4	1.6	1.8	1.9	2.1	1.5	1.7	1.9	2.1	2.3
SS C1	11	14	18	21	25	28	32	38	8	11	13	16	19	21	24	28	1.7	1.9	2.1	2.3	2.5	1.9	2.2	2.5	2.9	3.2
SS C2	13	19	25	29	36	41	47	55	10	15	20	24	29	33	39	46	1.3	1.5	1.7	1.9	2.0	1.5	1.7	1.9	2.1	2.2
SS C3	11	13	16	18	22	24	27	32	9	11	13	15	17	19	21	24	1.9	2.2	2.4	2.7	2.9	2.2	2.5	2.9	3.2	3.6
SS C4	13	17	22	26	32	36	42	49	10	14	18	22	26	30	35	41	1.5	1.6	1.8	2.0	2.1	1.6	1.8	2.0	2.2	2.4
SS D1	17	22	27	30	36	39	44	50	15	19	23	26	30	33	36	41	1.4	1.6	1.9	2.1	2.2	1.5	1.8	2.1	2.4	2.7
SS D2	21	29	37	42	51	56	63	71	19	25	32	37	45	50	56	64	1.1	1.3	1.5	1.6	1.8	1.2	1.4	1.6	1.8	1.9
SS D3	17	21	25	28	32	35	39	45	15	18	22	24	28	30	33	37	1.6	1.8	2.0	2.3	2.4	1.7	2.0	2.3	2.6	2.9
SS D4	21	28	35	41	49	54	61	69	18	24	31	36	43	48	54	62	1.2	1.3	1.5	1.7	1.8	1.2	1.4	1.6	1.8	1.9
SS E1	14	18	22	25	29	32	36	42	11	14	18	20	23	25	28	32	1.6	1.8	2.1	2.3	2.5	1.7	2.1	2.5	2.9	3.2
SS E2	18	24	31	36	43	48	55	63	15	20	26	31	37	42	47	55	1.2	1.4	1.6	1.7	1.9	1.3	1.5	1.7	1.9	2.1
SS E3	14	17	20	22	26	28	31	36	12	14	17	19	21	23	25	28	1.8	2.1	2.3	2.6	2.8	2.0	2.3	2.7	3.2	3.6
SS E4	17	22	29	33	40	45	51	59	14	19	25	29	35	39	44	51	1.3	1.5	1.7	1.8	2.0	1.4	1.6	1.8	2.0	2.2

CHATTANOOGA, TENNESSEE

TYPE	SOLAR SAVINGS FRACTION (%) BASE TEMP: 55 F (1785 DD)								65 F (3513 DD)								CONSERVATION FACTOR 55 F					65 F				
LCR	200	100	70	50	40	30	25	20	200	100	70	50	40	30	25	20	200	100	50	30	20	200	100	50	30	20
WW A1	18	24	29	35	39	46	51	57	13	17	20	24	27	31	34	37	1.2	1.3	1.5	1.6	1.8	1.6	1.7	1.9	2.1	2.3
WW A2	17	27	35	43	50	58	64	71	11	18	23	28	33	39	44	49	1.0	1.1	1.2	1.4	1.5	1.2	1.3	1.5	1.7	1.9
WW A3	16	29	37	47	54	63	69	75	9	17	23	30	35	42	47	54	0.9	1.0	1.2	1.3	1.5	1.1	1.3	1.4	1.6	1.8
WW A4	16	29	38	48	56	65	71	78	8	17	24	31	36	44	49	56	0.9	1.0	1.1	1.3	1.5	1.1	1.2	1.4	1.5	1.7
WW A5	16	30	40	50	58	67	73	80	7	17	24	32	38	46	52	59	0.9	1.0	1.1	1.3	1.4	1.1	1.2	1.3	1.5	1.7
WW A6	15	30	41	52	59	69	75	81	7	17	25	33	39	47	53	60	0.8	0.9	1.1	1.2	1.4	1.0	1.1	1.3	1.5	1.6
WW B1	14	23	29	36	41	48	53	60	9	14	17	21	24	29	32	36	1.1	1.2	1.4	1.5	1.7	1.5	1.6	1.8	2.1	2.3
WW B2	16	30	39	50	57	67	73	79	8	18	25	33	39	47	53	60	0.9	1.0	1.1	1.3	1.4	1.1	1.2	1.3	1.5	1.6
WW B3	19	35	46	58	65	75	80	86	10	22	31	40	46	55	61	69	0.8	0.9	1.0	1.2	1.4	0.9	1.1	1.2	1.4	1.5
WW B4	18	35	47	59	67	77	82	88	9	22	31	42	49	59	65	73	0.8	0.9	1.0	1.2	1.3	0.9	1.0	1.2	1.3	1.5
WW B5	16	34	46	58	66	76	81	87	7	21	30	41	48	58	65	73	0.8	0.9	1.0	1.2	1.4	0.9	1.0	1.2	1.3	1.5
WW C1	20	37	48	60	67	77	82	88	11	24	32	42	49	58	64	71	0.8	0.9	1.0	1.2	1.4	0.9	1.0	1.2	1.3	1.5
WW C2	18	35	46	58	66	75	81	87	9	22	31	40	48	57	63	71	0.8	0.9	1.0	1.2	1.4	0.9	1.0	1.2	1.3	1.5
WW C3	20	41	54	66	74	83	88	93	10	27	38	49	57	68	74	81	0.7	0.8	1.0	1.1	1.3	0.8	0.9	1.1	1.2	1.4
WW C4	17	37	50	63	71	80	85	91	8	24	34	46	54	64	71	78	0.7	0.8	1.0	1.1	1.3	0.9	1.0	1.1	1.3	1.4
TW A1	17	24	29	34	39	45	50	56	13	17	20	24	26	31	34	38	1.2	1.3	1.5	1.6	1.8	1.6	1.7	1.9	2.1	2.3
TW A2	16	25	32	40	45	53	59	66	11	17	21	26	30	36	40	45	1.0	1.2	1.3	1.4	1.6	1.3	1.4	1.6	1.8	2.0
TW A3	15	26	33	42	48	57	62	69	9	16	21	27	32	38	43	49	1.0	1.1	1.2	1.4	1.5	1.2	1.4	1.5	1.7	1.9
TW A4	14	25	33	42	48	57	63	70	8	15	21	27	32	39	43	50	1.0	1.1	1.2	1.4	1.5	1.2	1.3	1.5	1.7	1.8
TW B1	16	23	28	35	39	46	51	57	12	16	20	24	27	31	35	39	1.2	1.3	1.4	1.6	1.7	1.5	1.6	1.8	2.0	2.2
TW B2	15	24	30	38	44	52	58	65	9	15	20	25	29	35	39	45	1.1	1.2	1.3	1.4	1.6	1.3	1.4	1.6	1.8	1.9
TW B3	14	23	30	38	44	53	59	66	8	15	19	25	29	36	40	46	1.0	1.1	1.3	1.4	1.6	1.3	1.4	1.6	1.7	1.9
TW B4	13	22	29	36	42	50	56	63	8	14	18	24	28	34	38	44	1.1	1.2	1.3	1.5	1.6	1.3	1.4	1.6	1.8	1.9
TW C1	15	22	27	34	39	46	51	58	10	15	19	23	26	31	35	40	1.2	1.3	1.4	1.6	1.7	1.5	1.6	1.8	1.9	2.1
TW C2	13	21	27	34	40	47	53	60	9	14	18	23	27	32	36	41	1.1	1.2	1.4	1.5	1.7	1.4	1.5	1.7	1.8	2.0
TW C3	13	20	26	33	38	45	51	57	9	14	17	22	26	31	35	40	1.2	1.3	1.4	1.5	1.7	1.5	1.6	1.7	1.9	2.1
TW C4	13	19	24	30	34	41	45	52	9	13	17	20	24	28	32	36	1.3	1.4	1.5	1.6	1.8	1.6	1.7	1.8	2.0	2.2
TW D1	12	19	24	31	35	43	47	54	7	11	14	18	21	25	28	32	1.2	1.3	1.5	1.6	1.8	1.6	1.7	1.9	2.2	2.4
TW D2	16	28	36	46	52	62	67	74	9	18	24	31	36	43	48	55	0.9	1.0	1.2	1.3	1.5	1.1	1.2	1.4	1.6	1.7
TW D3	17	30	40	50	57	66	72	79	9	19	26	33	39	47	53	60	0.9	1.0	1.1	1.3	1.4	1.1	1.2	1.3	1.5	1.7
TW D4	17	32	42	53	61	71	77	83	9	20	28	37	44	53	59	67	0.8	0.9	1.1	1.2	1.4	1.0	1.1	1.2	1.4	1.6
TW D5	16	31	42	53	61	71	77	83	8	20	28	37	44	54	60	67	0.8	0.9	1.1	1.2	1.4	1.0	1.1	1.2	1.4	1.5
TW E1	19	34	44	55	62	71	77	83	11	22	30	38	44	53	59	66	0.8	0.9	1.1	1.2	1.4	1.0	1.1	1.3	1.4	1.6
TW E2	18	33	43	54	62	71	77	83	10	21	29	38	44	53	59	66	0.8	0.9	1.1	1.2	1.4	1.0	1.1	1.2	1.4	1.6
TW E3	21	39	51	63	71	80	85	90	11	26	36	46	54	64	70	77	0.7	0.8	1.0	1.2	1.3	0.9	1.0	1.1	1.3	1.4
TW E4	18	36	48	60	68	78	83	89	10	24	33	44	51	61	68	75	0.8	0.9	1.0	1.2	1.3	0.9	1.0	1.1	1.3	1.5
TW F1	15	22	27	33	38	45	50	56	11	15	18	22	25	30	33	37	1.2	1.3	1.5	1.6	1.8	1.6	1.7	1.9	2.1	2.2
TW F2	14	23	29	37	43	51	56	63	8	14	19	24	28	33	37	43	1.1	1.2	1.3	1.5	1.6	1.4	1.5	1.6	1.8	2.0
TW F3	12	22	29	38	44	53	58	66	6	13	18	24	28	34	39	45	1.0	1.1	1.3	1.4	1.6	1.3	1.4	1.6	1.7	1.9
TW F4	10	20	28	36	42	51	57	64	4	11	16	22	26	33	38	44	1.0	1.1	1.3	1.4	1.6	1.3	1.4	1.6	1.7	1.9
TW G1	13	20	26	32	37	44	49	56	9	13	17	21	24	29	32	37	1.2	1.3	1.4	1.6	1.7	1.5	1.6	1.8	2.0	2.2
TW G2	11	20	26	33	39	47	53	60	6	11	16	21	25	30	35	40	1.1	1.2	1.3	1.5	1.6	1.4	1.5	1.7	1.8	2.0
TW G3	9	18	24	32	38	46	52	59	4	10	14	19	23	29	34	39	1.1	1.2	1.3	1.5	1.6	1.4	1.5	1.7	1.8	2.0
TW G4	7	15	21	28	33	41	46	53	2	8	12	16	20	26	30	35	1:2	1.3	1.4	1.6	1.7	1.5	1.6	1.7	1.9	2.1
TW H1	10	17	23	29	34	41	46	53	6	10	14	18	21	26	30	34	1.2	1.3	1.5	1.6	1.7	1.6	1.7	1.8	2.0	2.2
TW H2	7	15	20	27	32	39	44	51	3	8	11	16	19	25	28	33	1.2	1.3	1.5	1.6	1.7	1.5	1.6	1.8	2.0	2.1
TW H3	6	12	17	23	28	35	40	47	2	6	10	14	17	22	25	30	1.3	1.4	1.5	1.7	1.8	1.6	1.7	1.9	2.0	2.2
TW H4	4	9	14	18	22	28	33	38	1	4	7	10	13	17	20	24	1.4	1.5	1.7	1.8	2.0	1.9	1.9	2.1	2.2	2.4
TW I1	10	17	22	28	32	39	44	50	5	9	12	16	18	22	25	29	1.2	1.4	1.5	1.7	1.8	1.7	1.8	2.0	2.2	2.5
TW I2	13	24	33	42	49	58	64	71	6	14	20	27	32	40	45	52	1.0	1.1	1.2	1.3	1.5	1.2	1.3	1.4	1.6	1.8
TW I3	14	26	35	45	53	62	68	75	7	16	22	30	36	44	49	56	0.9	1.0	1.1	1.3	1.5	1.1	1.2	1.4	1.5	1.7
TW I4	14	29	39	50	58	68	74	81	7	18	25	34	41	50	56	63	0.9	0.9	1.1	1.2	1.4	1.0	1.1	1.3	1.4	1.6
TW I5	14	29	39	51	58	69	75	82	6	18	26	35	42	51	57	65	0.8	0.9	1.1	1.2	1.4	1.0	1.1	1.2	1.4	1.6
TW J1	17	31	41	52	59	69	75	82	9	20	27	36	42	50	56	64	0.8	0.9	1.1	1.2	1.4	1.0	1.1	1.3	1.4	1.6
TW J2	15	30	40	51	59	69	75	81	8	19	26	35	41	50	56	64	0.8	0.9	1.1	1.2	1.4	1.0	1.1	1.3	1.4	1.6
TW J3	17	36	48	60	68	78	83	89	9	23	33	43	51	61	68	75	0.8	0.9	1.0	1.2	1.3	0.9	1.0	1.1	1.3	1.5
TW J4	16	33	45	57	65	75	81	87	7	21	30	41	48	59	65	73	0.8	0.9	1.0	1.2	1.4	0.9	1.0	1.2	1.3	1.5
DG A1	12	21	27	33	38	44	48	52	6	11	15	19	22	25	28	30	1.1	1.2	1.4	1.7	2.1	1.5	1.6	1.9	2.4	3.0
DG A2	13	23	30	38	44	52	57	63	8	14	19	24	29	35	39	44	1.0	1.1	1.3	1.5	1.7	1.3	1.4	1.6	1.8	2.1
DG A3	15	27	36	46	54	63	69	76	9	18	24	31	37	46	52	59	1.0	1.0	1.1	1.3	1.5	1.2	1.2	1.3	1.5	1.7
DG B1	12	21	28	36	42	50	55	60	6	12	15	20	23	27	31	34	1.1	1.2	1.3	1.5	1.8	1.4	1.6	1.8	2.1	2.4
DG B2	13	23	31	40	47	57	63	70	8	14	19	25	30	37	42	48	1.1	1.1	1.2	1.3	1.6	1.3	1.4	1.5	1.7	1.8
DG B3	15	27	37	48	56	67	74	81	10	18	25	32	39	48	54	62	1.0	1.0	1.1	1.2	1.4	1.1	1.2	1.3	1.4	1.6
DG C1	14	25	33	42	49	59	64	71	8	14	19	24	28	34	38	44	1.0	1.0	1.2	1.3	1.6	1.3	1.4	1.6	1.8	2.0
DG C2	15	27	36	46	54	65	71	78	9	17	23	30	35	43	49	56	1.0	1.0	1.1	1.3	1.5	1.2	1.3	1.4	1.5	1.7
DG C3	17	31	42	54	62	74	80	86	11	21	28	37	43	54	61	69	0.9	0.9	1.0	1.2	1.4	1.1	1.1	1.2	1.3	1.5
SS A1	20	30	38	46	52	61	66	72	15	22	27	33	38	44	48	54	1.0	1.1	1.2	1.4	1.5	1.2	1.3	1.5	1.6	1.8
SS A2	23	38	48	58	65	74	79	85	16	27	35	43	49	58	63	70	0.8	0.9	1.1	1.2	1.4	1.0	1.1	1.2	1.4	1.6
SS A3	19	28	35	43	49	57	62	68	14	20	25	30	34	40	44	49	1.0	1.1	1.3	1.4	1.6	1.2	1.4	1.6	1.8	1.9
SS A4	23	37	47	57	64	73	78	84	16	27	34	42	48	56	62	68	0.8	0.9	1.1	1.2	1.4	1.0	1.1	1.3	1.4	1.6
SS A5	20	29	35	43	48	56	61	67	15	22	26	31	35	40	44	49	1.0	1.1	1.3	1.4	1.6	1.3	1.4	1.6	1.8	2.0
SS A6	23	38	47	57	64	73	78	84	16	27	35	43	49	57	62	69	0.8	0.9	1.1	1.2	1.4	1.0	1.1	1.2	1.4	1.6
SS A7	19	26	31	38	42	49	54	60	15	20	23	27	30	35	38	42	1.1	1.3	1.4	1.6	1.7	1.4	1.6	1.8	2.0	2.2
SS A8	23	37	46	56	63	72	77	83	16	26	34	41	47	55	60	67	0.8	0.9	1.1	1.3	1.4	1.0	1.1	1.3	1.4	1.6
SS B1	17	26	33	41	46	54	60	66	12	18	23	28	32	38	42	48	1.0	1.1	1.3	1.4	1.6	1.3	1.4	1.6	1.8	1.9
SS B2	20	33	42	52	59	68	74	80	14	24	31	38	44	52	58	65	0.9	1.0	1.1	1.3	1.4	1.0	1.1	1.3	1.4	1.6
SS B3	16	25	31	38	44	52	57	63	11	17	22	26	30	36	39	44	1.1	1.2	1.3	1.5	1.6	1.3	1.4	1.6	1.8	2.0
SS B4	19	32	41	51	58	67	73	79	13	23	30	38	43	51	56	63	0.9	1.0	1.1	1.3	1.5	1.0	1.1	1.3	1.5	1.6
SS B5	16	24	29	36	40	48	52	59	12	17	21	25	29	33	37	41	1.1	1.3	1.4	1.6	1.7	1.4	1.6	1.8	2.0	2.1
SS B6	19	32	41	51	58	67	73	80	13	23	30	37	43	51	57	63	0.9	1.0	1.1	1.3	1.4	1.0	1.1	1.3	1.5	1.6
SS B7	15	22	27	32	37	43	48	53	12	16	19	23	25	30	32	36	1.2	1.4	1.5	1.6	1.8	1.5	1.7	1.9	2.1	2.3
SS B8	19	31	40	50	57	66	71	78	13	22	29	36	41	49	54	61	0.9	1.0	1.1	1.3	1.5	1.1	1.2	1.3	1.5	1.7
SS C1	15	25	33	42	48	57	63	70	8	15	20	26	31	37	41	47	1.0	1.1	1.2	1.4	1.5	1.2	1.4	1.5	1.7	1.9
SS C2	16	30	40	50	57	67	73	79	9	19	26	34	39	48	53	60	0.9	1.0	1.1	1.3	1.4	1.1	1.2	1.3	1.5	1.6
SS C3	14	21	27	34	39	46	51	58	9	14	18	22	25	30	34	39	1.1	1.3	1.4	1.5	1.7	1.5	1.6	1.8	1.9	2.1
SS C4	15	27	35	44	51	60	66	73	9	17	23	30	35	43	48	54	0.9	1.0	1.2	1.3	1.5	1.2	1.3	1.4	1.6	1.7
SS D1	22	36	45	55	62	70	76	82	15	25	32	39	44	52	57	63	0.8	0.9	1.1	1.3	1.4	1.0	1.1	1.3	1.5	1.7
SS D2	25	44	55	66	73	82	86	91	17	31	40	50	57	66	72	78	0.7	0.8	1.0	1.2	1.3	0.8	1.0	1.1	1.3	1.5
SS D3	21	33	41	49	56	64	69	75	16	24	29	35	40	47	51	57	0.9	1.0	1.2	1.3	1.5	1.1	1.2	1.4	1.6	1.8
SS D4	25	42	53	64	71	80	85	90	17	30	39	48	55	64	70	76	0.7	0.8	1.0	1.2	1.3	0.9	1.0	1.2	1.3	1.5
SS E1	19	31	40	49	56	64	70	76	12	21	27	33	38	45	50	56	0.9	1.0	1.2	1.3	1.5	1.1	1.2	1.4	1.6	1.8
SS E2	22	38	49	60	67	76	81	87	14	26	34	43	50	59	64	71	0.8	0.9	1.0	1.2	1.4	0.9	1.0	1.2	1.4	1.5
SS E3	18	27	33	41	47	54	60	66	13	19	23	29	32	38	42	47	1.0	1.1	1.3	1.4	1.6	1.3	1.4	1.6	1.8	2.0
SS E4	20	35	45	55	63	72	77	83	13	24	32	40	46	54	60	67	0.8	0.9	1.1	1.2	1.4	1.0	1.1	1.3	1.4	1.6

KNOXVILLE, TENNESSEE

TYPE	SOLAR SAVINGS FRACTION (%) BASE TEMP: 55 F (1775 DD) LCR 200	100	70	50	40	30	25	20	65 F (3486 DD) 200	100	70	50	40	30	25	20	CONSERVATION FACTOR 55 F 200	100	50	30	20	65 F 200	100	50	30	20
WW A1	18	25	30	36	40	47	52	58	14	18	21	25	27	32	35	39	1.2	1.3	1.4	1.6	1.7	1.5	1.7	1.9	2.1	2.3
WW A2	17	28	36	45	51	60	65	72	11	18	23	29	34	40	45	51	1.0	1.1	1.2	1.4	1.5	1.2	1.3	1.5	1.7	1.8
WW A3	17	29	38	48	55	64	70	77	10	18	24	31	36	44	49	55	0.9	1.0	1.1	1.3	1.5	1.1	1.2	1.4	1.6	1.7
WW A4	16	30	40	50	57	66	72	79	9	18	25	32	38	46	51	58	0.9	1.0	1.1	1.3	1.4	1.1	1.2	1.4	1.5	1.7
WW A5	16	31	41	52	59	69	75	81	8	18	25	33	39	48	53	60	0.8	0.9	1.1	1.2	1.4	1.0	1.2	1.3	1.5	1.6
WW A6	16	31	42	53	60	70	76	82	8	18	26	34	40	49	55	62	0.8	0.9	1.1	1.2	1.4	1.0	1.1	1.3	1.5	1.6
WW B1	15	23	29	37	42	50	55	61	9	14	18	22	25	30	33	37	1.1	1.2	1.3	1.5	1.7	1.4	1.6	1.8	2.0	2.2
WW B2	16	31	40	51	58	68	74	81	9	19	26	34	40	49	54	51	0.9	0.9	1.1	1.3	1.4	1.0	1.1	1.3	1.5	1.6
WW B3	20	36	48	59	67	76	81	87	11	23	32	41	48	57	63	70	0.8	0.9	1.0	1.2	1.4	0.9	1.0	1.2	1.4	1.5
WW B4	18	36	48	60	68	78	83	89	9	23	32	43	50	60	66	74	0.8	0.9	1.0	1.2	1.3	0.9	1.0	1.2	1.3	1.5
WW B5	17	35	47	59	67	77	82	88	8	22	31	42	49	60	66	74	0.8	0.9	1.0	1.2	1.3	0.9	1.0	1.2	1.3	1.5
WW C1	21	38	49	61	69	78	83	89	12	25	33	43	50	59	65	73	0.8	0.9	1.0	1.2	1.3	0.9	1.0	1.2	1.3	1.5
WW C2	19	36	47	59	67	76	82	88	10	23	32	42	49	58	65	72	0.8	0.9	1.0	1.2	1.4	0.9	1.0	1.2	1.3	1.5
WW C3	21	42	55	67	75	84	89	93	11	28	39	50	58	69	75	82	0.7	0.8	1.0	1.1	1.3	0.8	0.9	1.1	1.2	1.4
WW C4	18	38	51	64	72	81	86	91	9	25	35	47	55	65	72	79	0.7	0.8	1.0	1.1	1.3	0.8	0.9	1.1	1.3	1.4
TW A1	17	24	29	35	40	46	51	57	13	17	21	24	27	32	35	39	1.2	1.3	1.4	1.6	1.7	1.6	1.7	1.9	2.0	2.2
TW A2	17	26	33	41	46	55	60	67	11	17	22	27	31	37	41	47	1.0	1.1	1.3	1.4	1.6	1.3	1.4	1.6	1.8	1.9
TW A3	16	26	34	43	49	58	64	71	10	17	22	28	33	39	44	50	1.0	1.1	1.2	1.4	1.5	1.2	1.3	1.5	1.7	1.8
TW A4	14	26	34	43	50	59	65	72	8	16	21	28	33	40	45	51	1.0	1.1	1.2	1.3	1.5	1.2	1.3	1.5	1.6	1.8
TW B1	16	24	29	35	40	47	52	59	12	17	20	24	28	32	36	40	1.2	1.3	1.4	1.6	1.7	1.5	1.6	1.8	2.0	2.1
TW B2	15	24	31	39	45	53	59	66	10	16	21	26	30	36	41	46	1.0	1.1	1.3	1.4	1.6	1.3	1.4	1.6	1.7	1.9
TW B3	14	24	31	39	46	54	60	67	9	15	20	26	30	37	41	47	1.0	1.1	1.3	1.4	1.6	1.3	1.4	1.5	1.7	1.9
TW B4	13	23	29	37	43	52	57	64	8	14	19	25	29	35	40	45	1.1	1.2	1.3	1.4	1.6	1.3	1.4	1.6	1.7	1.9
TW C1	15	23	28	35	40	47	52	59	11	16	19	24	27	32	36	41	1.1	1.3	1.4	1.5	1.7	1.5	1.6	1.7	1.9	2.1
TW C2	14	22	28	35	41	49	54	61	9	15	19	24	28	33	37	43	1.1	1.2	1.3	1.5	1.6	1.4	1.5	1.7	1.8	2.0
TW C3	13	21	27	33	39	46	52	58	9	14	18	23	26	32	36	41	1.1	1.3	1.4	1.5	1.7	1.4	1.5	1.7	1.9	2.0
TW C4	13	20	24	30	35	42	46	53	9	14	17	21	24	29	33	37	1.2	1.4	1.5	1.6	1.8	1.6	1.7	1.8	2.0	2.1
TW D1	12	20	25	32	37	44	49	55	7	12	15	19	22	26	29	33	1.2	1.3	1.4	1.6	1.7	1.5	1.7	1.9	2.1	2.3
TW D2	16	28	37	47	54	63	69	76	10	18	24	32	37	45	50	56	0.9	1.0	1.2	1.3	1.5	1.1	1.2	1.4	1.5	1.7
TW D3	17	31	41	51	58	68	73	80	10	20	27	35	40	49	54	61	0.9	1.0	1.1	1.3	1.4	1.0	1.2	1.3	1.5	1.6
TW D4	17	33	43	55	62	72	78	84	10	21	29	38	45	54	60	68	0.8	0.9	1.1	1.2	1.4	1.0	1.1	1.2	1.4	1.5
TW D5	16	32	43	54	62	72	78	84	9	21	29	38	45	55	61	69	0.8	0.9	1.1	1.2	1.4	1.0	1.1	1.2	1.4	1.5
TW E1	20	35	45	56	63	73	78	84	12	23	30	39	45	54	60	67	0.8	0.9	1.1	1.2	1.4	1.0	1.1	1.2	1.4	1.6
TW E2	19	34	44	55	63	73	78	84	11	22	30	39	46	55	60	68	0.8	0.9	1.1	1.2	1.4	1.0	1.1	1.2	1.4	1.5
TW E3	21	40	52	64	72	81	86	91	12	27	37	47	55	65	71	78	0.7	0.8	1.0	1.1	1.3	0.9	1.0	1.1	1.3	1.4
TW E4	19	37	49	61	69	79	84	90	10	25	34	45	52	62	69	76	0.8	0.8	1.0	1.2	1.3	0.9	1.0	1.1	1.3	1.4
TW F1	16	23	28	34	39	46	51	57	11	16	19	23	26	31	34	38	1.2	1.3	1.4	1.6	1.7	1.5	1.6	1.8	2.0	2.2
TW F2	14	23	30	38	44	52	58	65	9	15	19	25	29	35	39	44	1.1	1.2	1.3	1.4	1.6	1.3	1.4	1.6	1.8	2.0
TW F3	13	23	30	39	45	54	60	67	7	14	19	25	29	36	40	46	1.0	1.1	1.2	1.4	1.6	1.3	1.4	1.5	1.7	1.9
TW F4	11	21	28	37	43	52	58	66	5	12	17	23	28	34	39	45	1.0	1.1	1.3	1.4	1.6	1.3	1.4	1.5	1.7	1.9
TW G1	13	21	26	33	38	45	50	57	9	14	17	22	25	30	33	38	1.2	1.3	1.4	1.6	1.7	1.5	1.6	1.8	2.0	2.1
TW G2	11	20	27	34	40	49	54	61	6	12	16	22	26	32	36	41	1.1	1.2	1.3	1.5	1.6	1.4	1.5	1.6	1.8	2.0
TW G3	9	18	25	33	39	47	53	60	4	10	15	20	24	30	35	40	1.1	1.2	1.3	1.5	1.6	1.4	1.5	1.6	1.8	2.0
TW G4	7	15	21	28	34	42	48	55	3	8	12	17	21	27	31	36	1.2	1.3	1.4	1.5	1.7	1.5	1.6	1.7	1.9	2.0
TW H1	10	18	23	30	35	42	47	54	6	11	14	19	22	27	31	36	1.2	1.3	1.4	1.6	1.7	1.5	1.6	1.8	2.0	2.1
TW H2	8	15	21	27	33	40	46	53	3	8	12	17	20	26	29	34	1.2	1.3	1.4	1.6	1.7	1.5	1.6	1.8	1.9	2.1
TW H3	6	13	18	24	29	36	41	48	2	7	10	14	18	23	26	31	1.3	1.4	1.5	1.6	1.8	1.6	1.7	1.9	2.0	2.2
TW H4	4	10	14	19	23	29	33	39	1	5	8	11	14	18	21	25	1.4	1.5	1.7	1.8	1.9	1.8	1.9	2.1	2.2	2.4
TW I1	10	17	22	28	33	40	45	52	6	10	13	16	19	23	26	30	1.2	1.3	1.5	1.6	1.8	1.6	1.8	2.0	2.2	2.4
TW I2	13	25	33	43	50	59	65	73	7	15	21	28	33	41	46	53	0.9	1.0	1.2	1.3	1.5	1.2	1.3	1.4	1.6	1.7
TW I3	14	27	36	46	54	63	69	77	7	17	23	31	37	45	50	57	0.9	1.0	1.1	1.3	1.5	1.1	1.2	1.4	1.5	1.7
TW I4	15	30	40	51	59	69	75	82	7	18	26	35	42	51	57	65	0.8	0.9	1.1	1.2	1.4	1.0	1.1	1.3	1.4	1.6
TW I5	14	30	40	52	60	70	76	83	7	18	27	36	43	52	59	66	0.8	0.9	1.1	1.2	1.4	1.0	1.1	1.2	1.4	1.5
TW J1	17	32	42	53	61	70	76	83	9	20	28	37	43	52	58	65	0.8	0.9	1.1	1.2	1.4	1.0	1.1	1.3	1.4	1.6
TW J2	16	31	41	52	60	70	76	82	8	19	27	36	43	52	58	65	0.8	0.9	1.1	1.2	1.4	1.0	1.1	1.3	1.4	1.6
TW J3	18	37	49	61	69	79	84	90	9	24	34	44	52	62	69	76	0.8	0.8	1.0	1.2	1.3	0.9	1.0	1.1	1.3	1.4
TW J4	16	34	46	58	66	76	82	88	8	22	31	42	50	60	66	74	0.8	0.9	1.0	1.2	1.3	0.9	1.0	1.1	1.3	1.5
DG A1	12	21	28	34	39	45	49	54	7	12	16	20	23	27	29	32	1.1	1.2	1.4	1.7	2.0	1.4	1.6	1.9	2.3	2.9
DG A2	13	23	31	39	45	53	59	65	8	15	19	25	30	36	40	45	1.0	1.1	1.3	1.5	1.7	1.3	1.4	1.5	1.8	2.0
DG A3	15	27	37	47	55	65	70	77	10	18	25	32	38	47	53	60	1.0	1.0	1.1	1.3	1.5	1.1	1.2	1.3	1.5	1.7
DG B1	12	22	29	37	43	51	56	62	7	12	16	21	24	29	32	36	1.1	1.1	1.3	1.5	1.8	1.4	1.5	1.8	2.0	2.3
DG B2	13	24	32	42	49	59	65	72	8	15	20	26	31	38	43	49	1.0	1.1	1.2	1.3	1.5	1.3	1.3	1.5	1.6	1.8
DG B3	15	28	38	49	57	69	75	82	10	19	26	33	40	49	55	64	1.0	1.0	1.1	1.2	1.4	1.1	1.2	1.3	1.4	1.5
DG C1	14	25	34	44	51	60	66	72	8	15	20	25	29	36	40	45	1.0	1.0	1.2	1.3	1.6	1.3	1.4	1.6	1.7	2.0
DG C2	15	27	37	47	56	66	72	79	9	18	24	31	36	44	50	57	1.0	1.0	1.1	1.2	1.5	1.2	1.2	1.4	1.5	1.7
DG C3	17	32	43	55	64	75	81	87	12	22	29	38	45	55	62	71	0.9	0.9	1.0	1.2	1.4	1.1	1.1	1.2	1.3	1.5
SS A1	20	31	38	47	53	61	67	73	15	22	28	34	38	45	49	55	1.0	1.0	1.2	1.4	1.5	1.1	1.3	1.5	1.6	1.8
SS A2	23	38	48	59	66	74	80	85	17	28	36	44	50	58	64	71	0.8	0.9	1.1	1.2	1.4	0.9	1.1	1.2	1.4	1.5
SS A3	19	29	36	44	50	58	63	69	14	21	26	31	35	41	45	50	1.0	1.1	1.2	1.4	1.6	1.2	1.4	1.6	1.7	1.9
SS A4	23	38	48	58	65	73	79	84	17	28	35	43	49	57	63	69	0.8	0.9	1.1	1.2	1.4	1.0	1.1	1.2	1.4	1.6
SS A5	20	29	36	43	49	57	62	68	16	22	26	31	35	41	45	50	1.0	1.1	1.3	1.4	1.6	1.2	1.4	1.6	1.8	1.9
SS A6	23	38	48	58	65	74	79	85	17	28	35	43	50	58	63	70	0.8	0.9	1.1	1.2	1.4	0.9	1.1	1.2	1.4	1.6
SS A7	19	26	32	38	43	50	55	61	15	20	24	28	31	35	39	43	1.1	1.2	1.4	1.5	1.7	1.4	1.5	1.7	2.0	2.2
SS A8	23	37	47	57	64	72	78	83	16	27	34	42	48	56	61	68	0.8	0.9	1.1	1.2	1.4	1.0	1.1	1.3	1.4	1.6
SS B1	17	27	33	41	47	55	61	67	12	19	24	29	33	39	43	49	1.0	1.1	1.3	1.4	1.6	1.2	1.4	1.6	1.7	1.9
SS B2	20	34	43	53	60	69	75	81	14	24	31	39	45	53	59	65	0.9	1.0	1.1	1.3	1.4	1.0	1.1	1.3	1.4	1.6
SS B3	16	25	32	39	45	53	58	64	12	18	22	27	31	36	40	45	1.1	1.2	1.3	1.5	1.6	1.3	1.4	1.6	1.8	2.0
SS B4	20	33	42	52	59	68	74	80	14	24	30	38	44	52	57	64	0.9	1.0	1.1	1.3	1.4	1.0	1.1	1.3	1.5	1.6
SS B5	17	24	30	36	41	48	53	60	13	18	22	26	29	34	38	42	1.1	1.3	1.4	1.5	1.7	1.4	1.5	1.7	1.9	2.1
SS B6	19	33	42	52	59	68	74	80	13	23	30	38	44	52	57	64	0.9	1.0	1.1	1.3	1.4	1.0	1.1	1.3	1.5	1.6
SS B7	15	22	27	33	37	44	48	54	12	16	19	23	26	30	33	37	1.2	1.3	1.5	1.6	1.8	1.5	1.7	1.9	2.1	2.3
SS B8	19	32	41	50	57	66	72	79	13	22	29	37	42	50	55	62	0.9	1.0	1.1	1.3	1.5	1.0	1.2	1.3	1.5	1.7
SS C1	15	26	34	43	49	58	64	71	9	16	21	27	32	38	43	49	1.0	1.1	1.2	1.4	1.5	1.2	1.3	1.5	1.7	1.9
SS C2	17	31	41	51	58	68	74	81	9	19	27	35	41	49	55	62	0.9	1.0	1.1	1.3	1.4	1.0	1.1	1.3	1.5	1.6
SS C3	14	22	28	35	40	47	53	59	9	14	18	23	26	31	35	40	1.1	1.2	1.4	1.5	1.7	1.4	1.6	1.7	1.9	2.1
SS C4	16	28	36	46	52	62	68	75	9	18	24	31	36	44	49	56	0.9	1.0	1.2	1.3	1.5	1.1	1.2	1.4	1.6	1.7
SS D1	23	37	46	56	63	71	76	82	16	26	32	40	45	53	58	64	0.8	0.9	1.1	1.3	1.4	1.0	1.1	1.3	1.5	1.6
SS D2	26	44	56	67	74	82	87	92	17	32	41	51	58	67	73	79	0.7	0.8	1.0	1.2	1.3	0.8	1.0	1.1	1.3	1.4
SS D3	22	33	41	50	56	65	70	76	16	24	30	36	41	48	52	58	0.9	1.0	1.2	1.3	1.5	1.1	1.2	1.4	1.6	1.8
SS D4	25	43	54	65	72	80	85	90	17	31	39	49	56	65	70	77	0.7	0.8	1.0	1.2	1.3	0.9	1.0	1.1	1.3	1.5
SS E1	19	32	41	50	57	65	71	77	13	21	27	34	39	46	51	57	0.9	1.0	1.1	1.3	1.5	1.1	1.2	1.4	1.6	1.8
SS E2	22	39	50	61	68	77	82	88	14	27	35	44	51	60	65	72	0.8	0.9	1.0	1.2	1.4	0.9	1.0	1.2	1.4	1.5
SS E3	18	27	34	42	47	55	61	67	13	19	24	29	33	39	43	48	1.0	1.1	1.3	1.4	1.6	1.3	1.4	1.6	1.8	1.9
SS E4	21	36	46	56	63	72	78	84	14	25	32	41	47	55	61	68	0.8	0.9	1.1	1.2	1.4	1.0	1.1	1.2	1.4	1.6

MEMPHIS, TENNESSEE

TYPE	SSF (%) BASE TEMP: 55 F (1624 DD) LCR 200	100	70	50	40	30	25	20	SSF (%) 65 F (3237 DD) 200	100	70	50	40	30	25	20	CONSERVATION FACTOR 55 F 200	100	50	30	20	CONSERVATION FACTOR 65 F 200	100	50	30	20
WW A1	19	27	33	40	45	53	58	64	14	19	23	27	30	35	39	43	1.1	1.2	1.3	1.5	1.6	1.4	1.6	1.7	1.9	2.1
WW A2	19	32	40	50	57	66	71	78	12	20	26	33	38	45	50	56	0.9	1.0	1.1	1.3	1.5	1.1	1.3	1.4	1.6	1.7
WW A3	19	34	43	54	61	70	76	82	11	21	27	35	41	49	54	61	0.8	0.9	1.1	1.2	1.4	1.1	1.2	1.3	1.5	1.6
WW A4	19	34	45	56	63	73	78	84	10	21	28	36	42	51	56	64	0.8	0.9	1.1	1.2	1.4	1.0	1.1	1.3	1.4	1.6
WW A5	19	36	47	58	66	75	80	86	10	21	29	38	44	53	59	66	0.8	0.9	1.0	1.2	1.4	1.0	1.1	1.2	1.4	1.6
WW A6	19	36	48	59	67	76	82	87	9	21	29	38	45	54	60	68	0.8	0.9	1.0	1.2	1.4	1.0	1.1	1.2	1.4	1.5
WW B1	17	27	34	42	48	56	62	69	10	16	20	25	29	34	38	43	1.0	1.1	1.2	1.4	1.6	1.3	1.5	1.7	1.8	2.0
WW B2	19	35	46	57	64	74	79	86	10	21	29	38	44	53	59	67	0.8	0.9	1.0	1.2	1.4	1.0	1.1	1.2	1.4	1.6
WW B3	23	41	53	65	72	81	86	91	13	26	35	45	52	62	68	75	0.7	0.8	1.0	1.2	1.3	0.9	1.0	1.2	1.3	1.5
WW B4	21	41	54	66	74	83	87	92	11	26	36	47	54	65	71	78	0.7	0.8	1.0	1.1	1.3	0.9	1.0	1.1	1.3	1.4
WW B5	20	40	52	64	72	82	87	92	10	25	35	46	54	64	70	78	0.7	0.8	1.0	1.1	1.3	0.9	1.0	1.1	1.3	1.4
WW C1	24	43	55	67	74	83	87	92	14	28	37	47	55	64	70	77	0.7	0.8	1.0	1.1	1.3	0.9	1.0	1.1	1.3	1.4
WW C2	22	41	53	65	72	81	86	91	12	26	35	46	53	63	69	77	0.7	0.8	1.0	1.1	1.3	0.9	1.0	1.1	1.3	1.4
WW C3	24	47	60	73	80	88	92	95	13	31	42	54	63	73	79	85	0.7	0.8	0.9	1.1	1.3	0.8	0.9	1.0	1.2	1.4
WW C4	21	43	57	69	77	85	90	94	11	28	39	51	59	70	76	83	0.7	0.8	0.9	1.1	1.3	0.8	0.9	1.1	1.2	1.4
TW A1	19	26	32	39	44	52	57	63	14	19	22	27	30	35	38	43	1.1	1.2	1.3	1.5	1.7	1.5	1.6	1.7	1.9	2.1
TW A2	18	29	37	45	52	61	66	73	12	19	24	30	35	41	46	52	1.0	1.1	1.2	1.4	1.5	1.2	1.3	1.5	1.7	1.8
TW A3	18	30	38	48	55	64	70	77	11	19	25	31	36	44	49	55	0.9	1.0	1.1	1.3	1.5	1.2	1.3	1.4	1.6	1.7
TW A4	16	29	38	48	55	65	71	78	9	18	24	31	37	44	50	56	0.9	1.0	1.1	1.3	1.5	1.1	1.2	1.4	1.5	1.7
TW B1	18	26	32	39	45	53	58	65	13	18	22	27	30	36	39	45	1.1	1.2	1.3	1.5	1.6	1.4	1.5	1.7	1.8	2.0
TW B2	17	27	35	44	50	59	65	72	11	18	23	29	33	40	45	51	1.0	1.1	1.2	1.4	1.5	1.2	1.3	1.5	1.7	1.8
TW B3	16	27	35	44	51	60	66	73	10	17	23	29	34	41	46	52	1.0	1.0	1.2	1.3	1.5	1.2	1.3	1.5	1.6	1.8
TW B4	15	25	33	42	48	57	63	70	9	16	21	27	32	39	44	50	1.0	1.1	1.2	1.4	1.5	1.2	1.3	1.5	1.7	1.8
TW C1	16	25	31	39	44	53	58	65	11	17	21	26	30	36	40	45	1.1	1.2	1.3	1.4	1.6	1.4	1.5	1.6	1.8	2.0
TW C2	15	24	31	39	45	54	60	67	10	16	21	26	30	37	41	47	1.0	1.1	1.3	1.4	1.6	1.3	1.4	1.6	1.7	1.9
TW C3	14	23	30	37	43	52	57	64	10	15	20	25	29	35	39	45	1.1	1.2	1.3	1.4	1.6	1.4	1.5	1.6	1.8	1.9
TW C4	14	21	27	34	39	46	51	58	10	15	19	23	27	32	36	41	1.2	1.3	1.4	1.5	1.7	1.5	1.6	1.7	1.9	2.0
TW D1	14	22	29	36	42	50	55	62	8	13	17	22	25	30	34	39	1.1	1.2	1.3	1.5	1.6	1.4	1.6	1.8	1.9	2.1
TW D2	18	32	42	52	59	69	74	81	11	20	27	35	41	49	54	61	0.9	1.0	1.1	1.3	1.4	1.1	1.2	1.3	1.5	1.6
TW D3	20	35	46	57	64	73	79	85	11	22	30	38	44	53	59	66	0.8	0.9	1.1	1.2	1.4	1.0	1.1	1.3	1.4	1.6
TW D4	20	37	49	60	68	77	83	88	11	24	32	42	49	59	65	72	0.8	0.9	1.0	1.2	1.3	0.9	1.0	1.2	1.3	1.5
TW D5	19	36	48	59	67	77	82	88	10	23	32	42	49	59	65	73	0.8	0.9	1.0	1.2	1.3	0.9	1.0	1.2	1.3	1.5
TW E1	22	39	50	61	69	78	83	88	13	25	34	43	50	59	65	72	0.8	0.9	1.0	1.2	1.4	0.9	1.0	1.2	1.3	1.5
TW E2	21	38	50	61	69	78	83	89	12	25	33	43	50	59	65	72	0.8	0.9	1.0	1.2	1.3	0.9	1.0	1.2	1.3	1.5
TW E3	24	45	58	70	77	85	90	94	14	30	40	51	59	69	75	82	0.7	0.8	0.9	1.1	1.3	0.8	0.9	1.1	1.2	1.4
TW E4	22	42	55	67	74	83	88	93	12	27	38	49	56	67	73	80	0.7	0.8	1.0	1.1	1.3	0.8	0.9	1.1	1.3	1.4
TW F1	17	25	31	38	44	51	57	63	12	17	21	25	29	34	38	43	1.1	1.2	1.3	1.5	1.6	1.4	1.5	1.7	1.9	2.0
TW F2	16	27	34	43	49	58	64	71	10	17	22	28	32	39	43	49	1.0	1.1	1.2	1.4	1.5	1.2	1.4	1.5	1.7	1.8
TW F3	15	26	35	44	51	60	66	73	8	16	21	28	33	40	45	51	1.0	1.0	1.2	1.3	1.5	1.2	1.3	1.5	1.6	1.8
TW F4	12	24	33	42	49	59	65	72	6	14	19	26	31	38	44	50	1.0	1.0	1.2	1.3	1.5	1.2	1.3	1.5	1.6	1.8
TW G1	15	23	30	37	43	51	56	63	10	15	19	24	28	33	37	42	1.1	1.2	1.3	1.5	1.6	1.4	1.5	1.7	1.8	2.0
TW G2	13	23	31	39	46	54	60	68	7	14	19	24	29	35	40	46	1.0	1.1	1.2	1.4	1.6	1.3	1.4	1.6	1.7	1.9
TW G3	11	21	29	37	44	53	59	66	5	12	17	23	28	34	39	45	1.0	1.1	1.2	1.4	1.6	1.3	1.4	1.5	1.7	1.9
TW G4	9	18	25	33	39	47	53	61	4	10	14	20	24	30	35	40	1.1	1.2	1.3	1.5	1.6	1.4	1.5	1.6	1.8	1.9
TW H1	12	20	26	34	39	48	53	60	7	12	16	21	25	30	34	40	1.1	1.2	1.3	1.5	1.6	1.4	1.5	1.7	1.9	2.0
TW H2	9	17	24	31	37	46	51	59	4	10	14	19	23	29	33	39	1.1	1.2	1.3	1.5	1.6	1.4	1.5	1.7	1.8	2.0
TW H3	7	15	21	28	33	41	46	54	3	8	12	16	20	26	29	35	1.2	1.3	1.4	1.6	1.7	1.5	1.6	1.8	1.9	2.1
TW H4	5	11	16	22	26	33	38	45	2	6	9	13	16	20	24	28	1.3	1.4	1.6	1.7	1.8	1.7	1.8	1.9	2.1	2.3
TW I1	12	20	26	33	38	46	52	58	7	11	15	19	22	27	31	35	1.1	1.2	1.4	1.5	1.7	1.5	1.6	1.8	2.0	2.2
TW I2	15	29	38	48	56	65	71	78	8	17	24	31	37	45	51	58	0.9	1.0	1.1	1.3	1.4	1.1	1.2	1.4	1.5	1.7
TW I3	16	31	41	52	60	69	75	82	9	19	20	34	41	49	55	62	0.8	0.9	1.1	1.2	1.4	1.0	1.1	1.3	1.4	1.6
TW I4	17	34	45	56	64	74	80	86	9	21	29	39	46	55	61	69	0.8	0.9	1.0	1.2	1.4	1.0	1.1	1.2	1.4	1.5
TW I5	17	34	45	57	65	75	81	87	8	21	30	39	47	57	63	71	0.8	0.9	1.0	1.2	1.4	0.9	1.0	1.2	1.3	1.5
TW J1	20	36	47	59	66	76	81	87	11	23	31	40	47	56	62	70	0.8	0.9	1.0	1.2	1.4	1.0	1.1	1.2	1.4	1.5
TW J2	18	35	46	58	66	75	81	87	10	22	30	40	47	56	62	70	0.8	0.9	1.0	1.2	1.4	1.0	1.1	1.2	1.4	1.5
TW J3	21	41	54	66	74	83	88	93	11	27	37	48	56	66	73	80	0.7	0.8	1.0	1.1	1.3	0.8	0.9	1.1	1.2	1.4
TW J4	19	39	51	64	72	81	86	91	10	24	35	46	54	64	70	78	0.7	0.8	1.0	1.1	1.3	0.9	1.0	1.1	1.3	1.4
DG A1	14	25	32	40	45	52	56	61	8	14	18	23	26	31	34	37	1.0	1.1	1.3	1.6	1.9	1.3	1.5	1.7	2.1	2.6
DG A2	15	26	35	44	51	59	65	71	9	16	22	28	33	40	44	50	1.0	1.0	1.2	1.4	1.6	1.2	1.3	1.5	1.7	1.9
DG A3	17	31	41	53	61	70	75	81	11	20	27	36	42	51	57	64	0.9	0.9	1.1	1.3	1.5	1.1	1.2	1.3	1.4	1.6
DG B1	14	25	34	43	50	59	64	69	8	14	19	24	28	34	38	43	1.0	1.0	1.2	1.4	1.7	1.3	1.4	1.6	1.8	2.1
DG B2	15	27	37	48	56	66	71	78	9	17	22	29	35	43	48	55	1.0	1.0	1.1	1.3	1.5	1.2	1.3	1.4	1.5	1.7
DG B3	17	32	43	55	64	74	80	86	11	21	28	37	43	54	60	69	0.9	0.9	1.0	1.2	1.4	1.1	1.2	1.2	1.3	1.5
DG C1	16	30	40	51	59	68	74	79	10	17	23	29	34	41	46	52	0.9	1.0	1.1	1.3	1.5	1.2	1.3	1.4	1.6	1.8
DG C2	17	31	42	54	63	73	78	84	11	20	26	34	40	49	56	63	0.9	0.9	1.0	1.2	1.4	1.1	1.2	1.3	1.4	1.6
DG C3	19	36	48	61	71	81	86	91	13	24	31	41	49	60	67	76	0.8	0.9	1.0	1.1	1.3	1.0	1.1	1.2	1.3	1.4
SS A1	22	34	43	52	59	67	72	79	16	24	30	37	42	49	54	60	0.9	1.0	1.1	1.3	1.5	1.1	1.2	1.4	1.5	1.7
SS A2	26	43	53	64	71	79	84	89	18	30	38	47	54	63	68	75	0.8	0.9	1.0	1.2	1.4	0.9	1.0	1.2	1.3	1.5
SS A3	21	32	40	49	55	63	69	75	15	23	28	34	38	45	49	55	0.9	1.0	1.2	1.3	1.5	1.1	1.3	1.5	1.6	1.8
SS A4	26	42	53	63	70	78	83	88	18	30	38	46	53	61	67	74	0.8	0.9	1.0	1.2	1.4	0.9	1.0	1.2	1.3	1.5
SS A5	22	32	40	48	54	62	67	74	17	24	29	34	38	45	49	54	1.0	1.1	1.2	1.4	1.5	1.2	1.3	1.5	1.7	1.8
SS A6	26	42	53	64	71	79	84	89	18	30	38	47	53	62	68	74	0.8	0.9	1.0	1.2	1.4	0.9	1.0	1.2	1.3	1.5
SS A7	20	29	35	43	48	56	61	67	16	21	25	30	34	39	42	47	1.1	1.1	1.3	1.5	1.6	1.3	1.4	1.6	1.8	2.0
SS A8	25	41	52	62	69	78	82	88	18	29	37	45	52	60	65	72	0.8	0.9	1.0	1.2	1.4	0.9	1.0	1.2	1.4	1.5
SS B1	19	30	37	46	53	61	67	73	13	21	26	32	36	43	47	53	1.0	1.0	1.2	1.4	1.5	1.2	1.3	1.5	1.6	1.8
SS B2	22	37	48	58	65	74	80	85	15	26	34	42	49	57	63	70	0.8	0.9	1.1	1.2	1.4	1.0	1.1	1.2	1.4	1.5
SS B3	18	28	36	44	50	59	64	71	12	19	24	30	34	40	44	50	1.0	1.1	1.2	1.4	1.6	1.2	1.3	1.5	1.7	1.9
SS B4	22	37	47	57	64	73	79	84	15	26	33	41	47	56	61	68	0.8	0.9	1.1	1.2	1.4	1.0	1.1	1.2	1.4	1.6
SS B5	18	27	33	40	46	54	59	65	13	19	23	28	32	37	41	46	1.1	1.2	1.3	1.5	1.6	1.3	1.5	1.6	1.8	2.0
SS B6	22	37	47	57	65	74	79	85	15	25	33	41	47	56	62	69	0.8	0.9	1.1	1.2	1.4	1.0	1.1	1.2	1.4	1.6
SS B7	17	24	30	37	42	49	54	60	13	17	21	25	29	33	37	41	1.1	1.2	1.4	1.5	1.7	1.4	1.6	1.8	2.0	2.1
SS B8	21	36	45	56	63	72	77	83	14	24	32	40	46	54	59	66	0.8	0.9	1.1	1.2	1.4	1.0	1.1	1.3	1.4	1.6
SS C1	17	30	39	48	55	65	70	77	10	18	24	31	36	43	48	54	0.9	1.0	1.1	1.3	1.5	1.1	1.3	1.4	1.6	1.7
SS C2	19	35	46	57	64	74	79	85	11	22	30	38	45	54	59	67	0.8	0.9	1.0	1.2	1.4	1.0	1.1	1.2	1.4	1.6
SS C3	16	25	31	39	45	53	59	65	10	16	20	25	29	35	39	44	1.1	1.2	1.3	1.4	1.6	1.4	1.5	1.6	1.8	2.0
SS C4	18	31	41	51	58	67	73	80	11	20	27	34	40	48	53	60	0.9	1.0	1.1	1.3	1.4	1.1	1.2	1.3	1.5	1.6
SS D1	25	41	51	61	68	77	82	87	17	28	36	44	49	58	63	70	0.8	0.9	1.0	1.2	1.4	1.0	1.1	1.2	1.4	1.6
SS D2	29	49	61	72	79	87	90	94	19	34	44	55	62	71	77	83	0.7	0.8	0.9	1.1	1.3	0.8	0.9	1.1	1.2	1.4
SS D3	24	37	46	55	62	71	76	82	17	26	32	39	44	52	57	63	0.9	0.9	1.1	1.3	1.4	1.0	1.2	1.3	1.5	1.7
SS D4	28	48	59	70	77	85	89	93	19	33	43	53	60	69	75	81	0.7	0.8	1.0	1.1	1.3	0.8	0.9	1.1	1.3	1.4
SS E1	22	36	46	56	63	71	77	83	14	24	30	38	43	51	56	62	0.8	0.9	1.1	1.3	1.4	1.0	1.2	1.3	1.5	1.7
SS E2	25	44	55	66	74	82	87	91	16	29	38	48	55	64	70	77	0.7	0.8	1.0	1.2	1.3	0.9	1.0	1.1	1.3	1.5
SS E3	20	30	38	47	53	61	67	73	14	21	26	32	36	43	47	53	1.0	1.1	1.2	1.4	1.5	1.2	1.3	1.5	1.7	1.8
SS E4	23	40	51	62	69	78	83	88	15	27	35	44	51	60	65	72	0.8	0.9	1.0	1.2	1.4	0.9	1.0	1.2	1.4	1.5

NASHVILLE, TENNESSEE

TYPE	SOLAR SAVINGS FRACTION (%) BASE TEMP: 55 F (1964 DD) LCR 200	100	70	50	40	30	25	20	65 F (3704 DD) 200	100	70	50	40	30	25	20	CONSERVATION FACTOR 55 F 200	100	50	30	20	65 F 200	100	50	30	20
WW A1	16	22	26	31	35	40	44	50	13	16	19	22	24	27	30	33	1.3	1.4	1.6	1.7	1.9	1.7	1.9	2.1	2.3	2.6
WW A2	15	24	30	38	44	52	57	64	10	16	20	25	29	34	38	43	1.1	1.2	1.3	1.5	1.6	1.3	1.5	1.6	1.8	2.0
WW A3	14	24	32	41	47	56	62	69	8	15	20	26	31	37	42	47	1.0	1.1	1.2	1.4	1.6	1.2	1.3	1.5	1.7	1.9
WW A4	13	25	33	42	49	58	64	71	7	15	21	27	32	39	43	50	1.0	1.1	1.2	1.4	1.5	1.2	1.3	1.5	1.7	1.8
WW A5	13	25	34	44	51	61	67	74	6	15	21	28	33	40	46	52	0.9	1.0	1.2	1.3	1.5	1.1	1.2	1.4	1.6	1.8
WW A6	12	26	35	45	52	62	68	75	5	15	21	28	34	41	47	53	0.9	1.0	1.1	1.3	1.5	1.1	1.2	1.4	1.6	1.7
WW B1	12	19	24	30	35	41	46	52	8	12	15	18	20	24	26	29	1.2	1.3	1.5	1.6	1.8	1.6	1.8	2.1	2.4	2.7
WW B2	13	25	34	44	51	60	66	74	7	16	22	29	34	42	47	54	0.9	1.0	1.2	1.3	1.5	1.1	1.2	1.4	1.6	1.7
WW B3	16	30	41	51	59	69	74	81	9	19	27	35	41	50	56	63	0.8	0.9	1.1	1.2	1.4	1.0	1.1	1.3	1.4	1.6
WW B4	14	30	41	53	61	71	77	83	7	19	28	37	44	54	60	68	0.8	0.9	1.1	1.2	1.4	1.0	1.1	1.2	1.4	1.5
WW B5	13	29	40	52	60	70	76	83	6	18	27	37	44	54	60	68	0.8	0.9	1.1	1.2	1.4	1.0	1.1	1.2	1.4	1.5
WW C1	17	32	43	54	61	71	77	83	10	21	29	37	44	53	59	66	0.8	0.9	1.1	1.2	1.4	1.0	1.1	1.3	1.4	1.6
WW C2	15	30	41	52	60	70	75	82	8	19	27	36	43	52	58	66	0.8	0.9	1.1	1.2	1.4	1.0	1.1	1.2	1.4	1.6
WW C3	16	35	48	60	69	78	84	89	8	23	34	45	53	63	70	77	0.8	0.8	1.0	1.2	1.3	0.9	1.0	1.1	1.3	1.4
WW C4	14	32	44	57	65	75	81	87	6	21	31	42	49	60	66	74	0.8	0.9	1.0	1.2	1.4	0.9	1.0	1.1	1.3	1.5
TW A1	16	21	25	30	34	40	44	49	12	16	18	21	24	27	30	33	1.3	1.5	1.6	1.7	1.9	1.7	1.9	2.1	2.3	2.5
TW A2	14	22	28	35	40	47	52	59	10	15	19	23	27	32	35	40	1.1	1.2	1.4	1.5	1.7	1.4	1.5	1.7	1.9	2.1
TW A3	13	22	29	36	42	50	56	63	8	14	19	24	28	34	38	43	1.1	1.2	1.3	1.5	1.6	1.3	1.5	1.6	1.8	2.0
TW A4	12	21	28	36	42	51	57	64	7	13	18	24	28	34	38	44	1.1	1.2	1.3	1.4	1.6	1.3	1.4	1.6	1.8	1.9
TW B1	15	21	25	30	35	41	45	51	11	15	18	21	24	28	31	35	1.3	1.4	1.5	1.7	1.8	1.6	1.8	2.0	2.2	2.4
TW B2	13	21	26	33	39	46	51	58	9	14	18	22	26	31	35	40	1.1	1.3	1.4	1.5	1.7	1.4	1.5	1.7	1.9	2.1
TW B3	12	20	26	33	39	47	52	59	7	13	17	22	26	32	35	41	1.1	1.2	1.4	1.5	1.7	1.4	1.5	1.7	1.9	2.0
TW B4	11	19	25	32	37	45	50	57	7	12	16	21	25	30	34	39	1.2	1.3	1.4	1.5	1.7	1.4	1.5	1.7	1.9	2.1
TW C1	13	20	24	30	34	41	45	51	10	14	17	21	24	28	31	35	1.3	1.4	1.5	1.7	1.8	1.6	1.7	1.9	2.1	2.3
TW C2	12	19	24	30	35	42	47	53	8	13	16	21	24	29	32	37	1.2	1.3	1.5	1.6	1.8	1.5	1.6	1.8	2.0	2.2
TW C3	11	18	23	29	33	40	45	51	8	12	16	20	23	28	31	36	1.2	1.4	1.5	1.6	1.8	1.6	1.7	1.8	2.0	2.2
TW C4	11	17	21	26	30	36	40	46	9	12	15	19	21	26	29	33	1.4	1.5	1.6	1.8	1.9	1.7	1.8	2.0	2.1	2.3
TW D1	10	16	21	26	30	36	41	46	6	10	12	15	18	21	23	26	1.3	1.4	1.6	1.7	1.9	1.7	1.9	2.2	2.4	2.7
TW D2	13	24	31	40	46	55	61	68	8	16	21	27	32	39	43	50	1.0	1.1	1.2	1.4	1.5	1.2	1.3	1.5	1.7	1.8
TW D3	14	26	34	44	51	60	66	73	8	17	23	30	35	42	47	54	0.9	1.0	1.2	1.3	1.5	1.1	1.2	1.4	1.6	1.7
TW D4	14	28	37	48	55	65	71	78	8	18	25	34	40	48	54	61	0.9	1.0	1.1	1.3	1.4	1.0	1.1	1.3	1.5	1.6
TW D5	13	27	37	47	55	65	71	78	7	18	25	34	40	49	55	63	0.9	1.0	1.1	1.3	1.4	1.0	1.1	1.3	1.4	1.6
TW E1	16	29	39	49	56	65	71	78	10	20	26	34	40	48	53	60	0.9	1.0	1.1	1.3	1.5	1.1	1.2	1.3	1.5	1.6
TW E2	15	29	38	48	56	65	71	78	9	19	26	34	40	48	54	61	0.9	1.0	1.1	1.3	1.4	1.0	1.2	1.3	1.5	1.6
TW E3	17	34	46	57	65	75	81	87	10	23	32	42	49	59	65	73	0.8	0.9	1.0	1.2	1.4	0.9	1.0	1.2	1.3	1.5
TW E4	15	32	43	54	62	72	78	85	8	21	30	40	47	57	63	71	0.8	0.9	1.0	1.2	1.4	0.9	1.0	1.2	1.3	1.5
TW F1	14	20	24	29	33	39	44	49	10	14	17	20	22	26	29	32	1.3	1.4	1.6	1.7	1.9	1.7	1.8	2.0	2.3	2.5
TW F2	12	20	25	32	37	45	50	57	8	13	16	21	24	29	33	37	1.2	1.3	1.4	1.5	1.7	1.4	1.6	1.8	2.0	2.1
TW F3	10	19	25	32	38	46	52	59	6	11	16	21	24	30	34	39	1.1	1.2	1.4	1.5	1.7	1.4	1.5	1.7	1.9	2.0
TW F4	8	17	23	31	36	45	50	58	3	9	14	19	23	29	33	38	1.1	1.2	1.4	1.5	1.7	1.4	1.5	1.7	1.9	2.0
TW G1	12	18	22	28	32	38	43	49	8	12	15	18	21	25	28	32	1.3	1.4	1.6	1.7	1.8	1.6	1.8	2.0	2.2	2.4
TW G2	9	17	22	29	34	41	46	53	5	10	14	18	22	27	30	35	1.2	1.3	1.4	1.6	1.7	1.5	1.6	1.8	2.0	2.2
TW G3	7	15	21	27	32	40	45	52	3	8	12	17	20	25	29	34	1.2	1.3	1.4	1.6	1.7	1.5	1.6	1.8	2.0	2.1
TW G4	5	12	17	23	28	35	40	47	2	6	10	14	17	22	26	31	1.3	1.4	1.5	1.7	1.8	1.6	1.7	1.9	2.0	2.2
TW H1	8	15	19	25	29	36	40	46	5	9	12	16	19	23	26	30	1.3	1.4	1.6	1.7	1.9	1.7	1.8	2.0	2.2	2.4
TW H2	6	12	17	23	27	34	39	45	2	7	10	14	17	21	25	29	1.3	1.4	1.6	1.7	1.9	1.6	1.8	1.9	2.1	2.3
TW H3	4	10	14	20	24	30	35	41	1	5	8	12	15	19	22	26	1.4	1.5	1.7	1.8	1.9	1.7	1.8	2.0	2.2	2.4
TW H4	3	8	11	16	19	24	28	33	1	4	6	9	11	15	18	21	1.5	1.7	1.8	2.0	2.1	1.9	2.1	2.2	2.4	2.6
TW I1	8	14	18	23	27	33	37	43	5	8	10	13	15	18	21	23	1.4	1.5	1.7	1.8	2.0	1.8	2.0	2.3	2.5	2.8
TW I2	11	21	28	36	43	52	58	65	5	13	18	24	29	35	40	46	1.0	1.1	1.3	1.4	1.6	1.2	1.4	1.5	1.7	1.9
TW I3	11	22	30	40	46	56	62	69	6	14	20	26	31	39	44	51	1.0	1.1	1.2	1.4	1.5	1.2	1.3	1.5	1.6	1.8
TW I4	11	24	34	44	52	62	68	75	5	15	22	30	36	45	51	58	0.9	1.0	1.1	1.3	1.5	1.1	1.2	1.3	1.5	1.6
TW I5	11	25	34	45	53	63	69	77	5	15	23	31	38	47	53	60	0.9	1.0	1.1	1.3	1.4	1.0	1.1	1.3	1.5	1.6
TW J1	14	27	36	46	53	63	69	76	8	17	24	32	37	46	51	58	0.9	1.0	1.1	1.3	1.5	1.1	1.2	1.3	1.5	1.7
TW J2	12	26	35	45	53	63	69	76	6	16	23	31	37	46	51	59	0.9	1.0	1.1	1.3	1.5	1.1	1.2	1.3	1.5	1.6
TW J3	14	31	42	54	62	72	78	85	7	20	29	39	47	56	63	71	0.8	0.9	1.0	1.2	1.4	0.9	1.0	1.2	1.3	1.5
TW J4	12	29	40	51	60	70	76	83	6	18	27	37	44	54	60	68	0.8	0.9	1.1	1.2	1.4	1.0	1.1	1.2	1.4	1.5
DG A1	10	17	22	28	32	37	41	45	6	10	13	16	18	21	23	25	1.2	1.3	1.6	1.9	2.3	1.6	1.8	2.1	2.7	3.6
DG A2	11	19	26	33	38	46	51	57	7	12	17	21	25	31	34	39	1.1	1.2	1.4	1.6	1.8	1.4	1.5	1.7	1.9	2.2
DG A3	13	23	31	41	48	58	63	70	9	16	22	28	34	41	47	54	1.0	1.1	1.2	1.3	1.6	1.2	1.3	1.4	1.6	1.7
DG B1	10	17	23	30	35	42	46	52	6	10	13	16	19	22	25	28	1.2	1.3	1.4	1.6	1.9	1.6	1.7	2.1	2.3	2.8
DG B2	11	20	26	35	41	50	56	63	7	13	17	22	26	32	36	42	1.2	1.2	1.3	1.4	1.6	1.4	1.5	1.7	1.8	2.0
DG B3	13	23	32	42	50	61	68	75	9	17	22	29	35	43	49	57	1.0	1.1	1.2	1.3	1.5	1.2	1.3	1.4	1.5	1.6
DG C1	12	20	27	36	42	50	56	62	7	12	16	20	24	29	32	36	1.1	1.2	1.3	1.5	1.7	1.4	1.5	1.8	2.0	2.3
DG C2	13	23	31	40	47	57	64	71	8	15	20	26	31	38	43	49	1.1	1.1	1.2	1.3	1.5	1.3	1.4	1.5	1.6	1.8
DG C3	15	27	36	47	56	67	74	82	10	19	25	33	39	49	55	64	1.0	1.0	1.1	1.2	1.4	1.1	1.2	1.3	1.4	1.5
SS A1	18	27	33	41	47	54	60	66	13	20	25	30	34	39	43	49	1.0	1.1	1.3	1.4	1.6	1.2	1.4	1.6	1.8	1.9
SS A2	20	33	42	52	59	68	74	80	15	25	32	39	45	53	58	65	0.9	1.0	1.1	1.3	1.4	1.0	1.1	1.3	1.5	1.6
SS A3	17	25	31	38	43	50	55	62	13	18	22	27	30	35	39	43	1.1	1.2	1.3	1.5	1.7	1.3	1.5	1.7	1.9	2.1
SS A4	20	33	42	51	58	67	72	79	15	24	31	38	44	51	57	63	0.9	1.0	1.1	1.3	1.5	1.0	1.1	1.3	1.5	1.6
SS A5	18	26	31	38	43	50	55	61	14	20	24	28	31	36	39	44	1.1	1.2	1.4	1.5	1.7	1.4	1.5	1.7	1.9	2.1
SS A6	20	33	42	52	59	68	73	79	15	24	31	39	44	52	57	64	0.9	1.0	1.1	1.3	1.5	1.0	1.1	1.3	1.5	1.6
SS A7	17	23	28	33	37	44	48	53	14	18	21	24	27	31	33	37	1.2	1.4	1.5	1.7	1.8	1.5	1.7	1.9	2.2	2.4
SS A8	20	32	41	50	57	66	71	78	15	24	30	37	43	50	55	61	0.9	1.0	1.1	1.3	1.5	1.0	1.2	1.3	1.5	1.7
SS B1	15	23	29	36	41	48	53	60	11	17	21	25	29	34	38	42	1.1	1.2	1.4	1.5	1.7	1.3	1.5	1.7	1.9	2.1
SS B2	17	29	38	47	54	63	68	75	12	21	28	35	40	48	53	60	0.9	1.0	1.2	1.3	1.5	1.1	1.2	1.4	1.5	1.7
SS B3	14	22	27	33	38	46	50	57	10	15	19	23	27	31	35	39	1.2	1.3	1.4	1.6	1.7	1.4	1.6	1.8	2.0	2.2
SS B4	17	28	37	46	53	61	67	74	12	21	27	34	39	47	52	58	0.9	1.0	1.2	1.3	1.5	1.1	1.2	1.4	1.5	1.7
SS B5	15	21	26	31	36	42	46	52	12	16	19	23	26	30	33	36	1.3	1.4	1.5	1.7	1.8	1.5	1.7	1.9	2.1	2.3
SS B6	17	28	37	46	53	62	67	74	12	21	27	34	39	47	52	58	0.9	1.0	1.2	1.3	1.5	1.1	1.2	1.4	1.5	1.7
SS B7	14	19	23	28	32	38	42	47	11	15	17	20	23	26	29	32	1.3	1.5	1.6	1.8	1.9	1.7	1.8	2.1	2.3	2.6
SS B8	16	27	35	44	51	60	65	72	11	20	26	32	37	44	49	56	1.0	1.1	1.2	1.4	1.5	1.1	1.2	1.4	1.6	1.7
SS C1	12	22	28	36	42	50	56	63	7	13	18	23	26	32	36	41	1.1	1.2	1.3	1.5	1.6	1.3	1.5	1.7	1.9	2.0
SS C2	14	26	34	44	51	60	66	74	8	16	23	30	35	43	48	55	0.9	1.0	1.2	1.3	1.5	1.1	1.2	1.4	1.6	1.7
SS C3	12	19	24	29	34	40	45	51	8	13	16	19	22	27	30	34	1.2	1.4	1.5	1.6	1.8	1.6	1.7	1.9	2.1	2.3
SS C4	13	23	31	39	45	54	60	67	8	15	21	27	31	38	43	49	1.0	1.1	1.2	1.4	1.6	1.2	1.3	1.5	1.7	1.8
SS D1	19	31	40	49	55	64	69	76	14	22	28	35	40	46	51	57	0.9	1.0	1.2	1.3	1.5	1.1	1.2	1.4	1.6	1.8
SS D2	22	38	49	60	68	76	82	87	15	28	36	45	52	61	67	74	0.8	0.9	1.0	1.2	1.4	0.9	1.0	1.2	1.3	1.5
SS D3	19	29	36	44	50	58	63	69	14	21	26	32	36	42	46	51	1.0	1.1	1.2	1.4	1.6	1.2	1.3	1.5	1.7	1.9
SS D4	21	37	47	58	65	74	80	85	15	27	35	44	50	59	65	71	0.8	0.9	1.0	1.2	1.4	0.9	1.0	1.2	1.4	1.5
SS E1	16	27	34	43	49	58	63	70	11	18	23	29	33	39	44	49	1.0	1.1	1.2	1.4	1.6	1.2	1.3	1.5	1.7	1.9
SS E2	18	33	43	54	61	70	76	82	12	23	31	39	45	53	59	66	0.8	0.9	1.1	1.2	1.4	1.0	1.1	1.3	1.4	1.6
SS E3	16	24	29	36	41	48	53	59	12	17	21	25	29	34	37	41	1.1	1.2	1.4	1.5	1.7	1.4	1.5	1.7	1.9	2.1
SS E4	18	31	40	50	57	66	71	78	12	22	28	36	42	49	55	61	0.9	1.0	1.1	1.3	1.5	1.0	1.2	1.3	1.5	1.7

ABILENE, TEXAS

SOLAR SAVINGS FRACTION (%) — CONSERVATION FACTOR

TYPE	SSF 55 F (1162 DD) LCR 200	100	70	50	40	30	25	20	SSF 65 F (2616 DD) 200	100	70	50	40	30	25	20	CF 55 F 200	100	50	30	20	CF 65 F 200	100	50	30	20
WW A1	27	41	50	59	66	75	80	85	18	26	31	38	43	50	55	62	0.8	0.9	1.1	1.2	1.4	1.1	1.2	1.4	1.5	1.7
WW A2	32	50	62	72	79	87	91	94	18	30	38	48	54	64	69	76	0.7	0.8	1.0	1.1	1.3	0.9	1.0	1.2	1.3	1.5
WW A3	33	54	66	77	83	90	93	96	18	32	41	51	59	68	74	81	0.7	0.8	0.9	1.1	1.3	0.9	1.0	1.1	1.3	1.4
WW A4	34	56	68	79	85	91	94	97	17	32	43	53	61	71	76	83	0.6	0.7	0.9	1.1	1.3	0.8	0.9	1.1	1.2	1.4
WW A5	35	58	70	81	87	93	95	98	17	34	44	56	63	73	79	85	0.6	0.7	0.9	1.1	1.2	0.8	0.9	1.0	1.2	1.4
WW A6	36	59	71	82	88	93	96	98	17	34	45	57	65	74	80	86	0.6	0.7	0.9	1.1	1.2	0.8	0.9	1.0	1.2	1.4
WW B1	28	44	55	65	72	81	85	90	15	25	32	40	46	54	60	66	0.8	0.8	1.0	1.2	1.3	1.0	1.1	1.3	1.4	1.6
WW B2	34	56	68	79	85	92	95	97	17	33	43	54	62	72	78	84	0.6	0.7	0.9	1.1	1.3	0.8	0.9	1.1	1.2	1.4
WW B3	40	64	76	86	91	95	97	99	21	39	51	63	70	80	85	90	0.6	0.7	0.9	1.0	1.2	0.7	0.8	1.0	1.2	1.3
WW B4	39	63	76	86	91	96	98	99	20	39	51	64	72	81	86	91	0.6	0.7	0.9	1.0	1.2	0.7	0.8	1.0	1.1	1.3
WW B5	36	61	74	85	90	95	97	99	18	37	50	62	70	80	85	91	0.6	0.7	0.9	1.0	1.2	0.7	0.8	1.0	1.1	1.3
WW C1	41	65	77	87	92	96	98	99	22	41	53	65	72	81	86	91	0.6	0.7	0.9	1.0	1.2	0.7	0.8	1.0	1.1	1.3
WW C2	38	63	75	85	90	95	97	99	20	39	51	63	71	80	85	91	0.6	0.7	0.9	1.0	1.2	0.7	0.8	1.0	1.2	1.3
WW C3	43	69	81	90	94	98	99	100	23	45	58	71	79	87	91	95	0.6	0.7	0.8	1.0	1.2	0.7	0.8	0.9	1.1	1.3
WW C4	39	65	78	88	92	97	98	99	20	41	54	67	75	84	89	94	0.6	0.7	0.8	1.0	1.2	0.7	0.8	1.0	1.1	1.3
TW A1	26	40	49	58	65	73	79	84	17	25	31	37	42	49	54	61	0.8	0.9	1.1	1.2	1.4	1.1	1.2	1.4	1.5	1.7
TW A2	29	46	56	67	74	82	87	91	17	27	35	43	50	58	64	71	0.7	0.8	1.0	1.2	1.3	1.0	1.1	1.2	1.4	1.5
TW A3	29	48	59	70	77	85	89	93	16	28	36	46	53	62	68	75	0.7	0.8	1.0	1.1	1.3	0.9	1.0	1.2	1.3	1.5
TW A4	29	48	60	71	78	86	90	94	15	28	36	46	53	63	69	76	0.7	0.8	1.0	1.1	1.3	0.9	1.0	1.2	1.3	1.5
TW B1	26	40	49	59	66	75	80	86	17	25	31	38	43	51	56	62	0.8	0.9	1.1	1.2	1.4	1.1	1.2	1.3	1.5	1.6
TW B2	27	44	55	65	73	81	86	91	15	26	33	42	48	57	63	70	0.8	0.8	1.0	1.2	1.3	1.0	1.1	1.2	1.4	1.5
TW B3	26	44	55	66	73	82	86	91	14	25	33	42	49	58	64	71	0.7	0.8	1.0	1.2	1.3	1.0	1.1	1.2	1.4	1.5
TW B4	25	41	52	63	70	79	84	89	13	24	31	40	46	55	61	68	0.8	0.9	1.0	1.2	1.3	1.0	1.1	1.2	1.4	1.5
TW C1	24	39	48	58	65	74	80	85	15	23	30	37	42	50	56	63	0.8	0.9	1.1	1.2	1.4	1.1	1.2	1.3	1.5	1.6
TW C2	24	39	49	60	67	76	81	87	14	23	30	37	43	52	57	65	0.8	0.9	1.0	1.2	1.4	1.1	1.2	1.3	1.4	1.6
TW C3	22	37	46	57	64	73	79	85	13	22	28	36	41	49	55	62	0.8	0.9	1.1	1.2	1.4	1.1	1.2	1.3	1.5	1.6
TW C4	21	33	41	51	58	67	73	79	13	20	25	32	37	44	49	56	0.9	1.0	1.1	1.3	1.4	1.2	1.3	1.4	1.6	1.7
TW D1	23	38	48	58	65	74	80	86	12	21	27	34	40	48	53	60	0.8	0.9	1.1	1.2	1.4	1.1	1.2	1.4	1.5	1.7
TW D2	31	51	63	74	81	88	92	95	17	30	40	50	57	67	73	79	0.7	0.8	0.9	1.1	1.3	0.9	1.0	1.1	1.3	1.4
TW D3	34	56	68	79	85	91	94	97	18	33	44	54	62	72	77	84	0.6	0.7	0.9	1.1	1.3	0.8	0.9	1.1	1.2	1.4
TW D4	35	58	70	81	87	93	96	98	18	35	46	58	66	76	81	87	0.6	0.7	0.9	1.1	1.2	0.8	0.9	1.0	1.2	1.4
TW D5	33	57	69	80	86	93	95	98	17	34	46	57	65	75	81	87	0.6	0.7	0.9	1.1	1.2	0.8	0.9	1.0	1.2	1.4
TW E1	38	60	72	82	88	94	96	98	21	37	48	59	67	76	82	88	0.6	0.7	0.9	1.1	1.2	0.8	0.9	1.0	1.2	1.4
TW E2	36	59	71	82	88	93	96	98	20	36	47	59	67	76	82	88	0.6	0.7	0.9	1.1	1.2	0.8	0.9	1.0	1.2	1.4
TW E3	42	67	79	88	93	97	98	99	23	43	56	68	76	84	89	93	0.6	0.7	0.9	1.0	1.2	0.7	0.8	1.0	1.1	1.3
TW E4	39	63	76	86	91	96	98	99	20	40	52	65	73	82	87	92	0.6	0.7	0.9	1.0	1.2	0.7	0.8	1.0	1.1	1.3
TW F1	25	39	48	58	65	74	79	85	16	24	30	36	42	49	54	61	0.8	0.9	1.1	1.2	1.4	1.1	1.2	1.4	1.5	1.7
TW F2	26	43	54	65	72	81	85	90	14	25	32	41	47	56	62	69	0.8	0.8	1.0	1.2	1.3	1.0	1.1	1.2	1.4	1.5
TW F3	26	44	55	67	74	82	87	92	13	25	33	42	48	58	64	71	0.7	0.8	1.0	1.1	1.3	1.0	1.1	1.2	1.3	1.5
TW F4	23	42	53	65	72	81	86	91	11	22	31	40	47	56	62	70	0.7	0.8	1.0	1.2	1.3	1.0	1.1	1.2	1.4	1.5
TW G1	23	37	47	57	64	73	79	85	13	22	28	35	41	48	54	61	0.8	0.9	1.1	1.2	1.4	1.1	1.2	1.3	1.5	1.6
TW G2	23	39	50	61	68	77	83	88	11	21	29	37	43	52	58	65	0.8	0.9	1.0	1.2	1.4	1.0	1.1	1.3	1.4	1.6
TW G3	21	37	48	59	67	76	82	87	10	20	27	35	42	51	57	64	0.8	0.9	1.0	1.2	1.4	1.0	1.1	1.3	1.4	1.6
TW G4	17	32	42	53	61	71	76	83	7	16	23	31	37	45	51	58	0.8	0.9	1.1	1.2	1.4	1.1	1.2	1.3	1.5	1.6
TW H1	20	34	43	54	61	70	76	83	10	18	25	32	37	45	51	58	0.9	0.9	1.1	1.2	1.4	1.1	1.2	1.4	1.5	1.7
TW H2	17	31	41	51	59	69	75	81	8	16	22	30	35	43	49	56	0.9	0.9	1.1	1.2	1.4	1.2	1.2	1.4	1.5	1.7
TW H3	14	27	36	46	53	63	69	77	6	13	19	26	31	39	44	51	0.9	1.0	1.1	1.3	1.5	1.2	1.3	1.4	1.6	1.7
TW H4	11	21	28	37	44	53	59	67	4	10	15	20	25	31	36	42	1.0	1.1	1.2	1.4	1.5	1.4	1.5	1.6	1.7	1.9
TW I1	20	35	44	55	62	71	77	83	10	18	24	31	36	44	49	56	0.8	0.9	1.1	1.2	1.4	1.2	1.3	1.4	1.5	1.7
TW I2	27	47	59	71	78	86	90	94	14	27	36	46	53	63	69	77	0.7	0.8	1.0	1.1	1.3	0.9	1.0	1.1	1.3	1.4
TW I3	30	51	63	74	81	89	92	96	15	29	39	50	57	67	73	80	0.7	0.8	0.9	1.1	1.3	0.9	1.0	1.1	1.3	1.4
TW I4	31	54	67	78	85	91	94	97	15	32	43	54	62	72	78	85	0.6	0.7	0.9	1.1	1.3	0.8	0.9	1.0	1.2	1.4
TW I5	31	54	67	79	85	92	95	97	15	32	43	55	63	73	79	86	0.6	0.7	0.9	1.1	1.3	0.8	0.9	1.0	1.2	1.4
TW J1	35	57	70	80	86	92	95	98	18	35	45	57	64	74	80	86	0.6	0.7	0.9	1.1	1.3	0.8	0.9	1.0	1.2	1.4
TW J2	33	56	68	79	86	92	95	97	17	33	44	56	63	73	79	86	0.6	0.7	0.9	1.1	1.3	0.8	0.9	1.0	1.2	1.4
TW J3	38	63	76	86	91	96	98	99	19	40	52	65	73	82	87	92	0.6	0.7	0.9	1.0	1.2	0.7	0.8	1.0	1.1	1.3
TW J4	35	60	73	84	89	95	97	99	17	37	49	62	70	80	85	90	0.6	0.7	0.9	1.0	1.2	0.7	0.8	1.0	1.1	1.3
DG A1	26	43	54	64	70	77	81	85	13	23	30	37	43	49	53	58	0.7	0.8	1.0	1.3	1.5	1.0	1.1	1.3	1.6	1.9
DG A2	26	45	56	67	74	82	86	90	13	25	33	42	49	57	62	68	0.7	0.8	1.0	1.2	1.4	1.0	1.1	1.2	1.4	1.6
DG A3	29	51	64	75	81	88	91	93	16	29	39	51	58	68	74	80	0.7	0.8	0.9	1.2	1.4	0.9	1.0	1.1	1.3	1.5
DG B1	26	47	60	71	77	84	88	91	13	23	31	41	47	56	61	67	0.7	0.8	1.0	1.2	1.4	1.0	1.1	1.2	1.4	1.7
DG B2	27	48	61	74	80	88	91	94	14	25	34	45	53	63	69	76	0.7	0.8	0.9	1.1	1.3	1.0	1.0	1.1	1.3	1.5
DG B3	29	52	67	79	85	91	94	96	16	30	40	53	62	73	79	85	0.7	0.7	0.9	1.1	1.4	0.9	0.9	1.0	1.2	1.4
DG C1	30	54	68	79	85	90	93	95	15	28	37	48	56	66	72	77	0.7	0.7	0.9	1.1	1.4	0.9	1.0	1.1	1.3	1.5
DG C2	30	54	68	80	86	92	95	97	16	29	39	51	60	71	77	83	0.7	0.7	0.9	1.1	1.3	0.9	1.0	1.0	1.2	1.4
DG C3	33	59	74	85	90	95	96	98	18	34	45	59	69	80	85	90	0.6	0.7	0.9	1.1	1.4	0.8	0.9	1.0	1.1	1.3
SS A1	34	52	63	73	80	87	90	94	22	34	42	51	58	66	72	78	0.7	0.8	1.0	1.1	1.3	0.9	1.0	1.1	1.3	1.5
SS A2	41	63	74	83	89	94	96	98	26	42	52	63	70	79	84	89	0.6	0.7	0.9	1.1	1.2	0.8	0.9	1.0	1.2	1.4
SS A3	32	50	60	70	77	84	88	92	21	32	40	48	54	62	68	74	0.7	0.8	1.0	1.2	1.3	0.9	1.0	1.2	1.3	1.5
SS A4	41	62	73	83	88	93	96	98	26	42	52	62	69	78	83	88	0.6	0.7	0.9	1.1	1.2	0.8	0.9	1.0	1.2	1.4
SS A5	32	49	59	69	75	83	87	91	22	32	39	47	53	61	67	73	0.7	0.8	1.0	1.2	1.3	1.0	1.1	1.2	1.4	1.5
SS A6	41	63	74	83	89	94	96	98	26	42	52	63	70	79	84	89	0.6	0.7	0.9	1.1	1.2	0.8	0.9	1.0	1.2	1.4
SS A7	29	44	53	63	69	77	82	87	20	29	35	42	47	55	60	66	0.8	0.9	1.0	1.2	1.4	1.0	1.1	1.3	1.5	1.6
SS A8	40	61	73	82	88	93	96	98	25	41	51	61	68	77	82	87	0.6	0.7	0.9	1.1	1.3	0.8	0.9	1.0	1.2	1.4
SS B1	29	46	57	67	74	82	87	91	18	29	37	45	51	60	65	72	0.7	0.8	1.0	1.2	1.3	1.0	1.1	1.2	1.4	1.5
SS B2	36	56	68	78	84	91	94	97	22	37	46	57	64	73	79	85	0.6	0.8	0.9	1.1	1.3	0.8	0.9	1.1	1.2	1.4
SS B3	28	45	55	65	72	80	85	90	17	28	35	43	49	57	63	69	0.8	0.8	1.0	1.2	1.4	1.0	1.1	1.2	1.4	1.6
SS B4	35	56	67	78	84	90	93	96	21	36	46	56	63	72	78	84	0.7	0.8	0.9	1.1	1.3	0.8	0.9	1.1	1.2	1.4
SS B5	27	41	50	60	67	75	80	86	17	26	32	39	45	52	58	64	0.8	0.9	1.1	1.2	1.4	1.1	1.2	1.3	1.5	1.6
SS B6	35	56	67	78	84	90	94	96	21	36	46	56	63	72	78	84	0.6	0.8	0.9	1.1	1.3	0.8	0.9	1.1	1.2	1.4
SS B7	24	37	46	56	62	71	76	82	16	24	29	36	41	48	53	59	0.9	0.9	1.1	1.3	1.4	1.1	1.2	1.4	1.6	1.7
SS B8	34	54	66	76	83	89	93	96	20	35	44	54	61	71	76	83	0.7	0.8	0.9	1.1	1.3	0.8	0.9	1.1	1.2	1.4
SS C1	30	49	61	72	79	86	90	94	16	28	37	46	53	62	68	75	0.7	0.8	1.0	1.1	1.3	0.9	1.0	1.2	1.3	1.5
SS C2	34	56	68	79	85	91	94	97	18	33	43	54	62	72	78	84	0.6	0.7	0.9	1.1	1.3	0.8	0.9	1.1	1.2	1.4
SS C3	24	39	49	60	67	76	81	87	14	23	30	37	43	51	56	63	0.8	0.9	1.0	1.2	1.4	1.1	1.2	1.3	1.5	1.6
SS C4	30	50	62	73	80	87	91	95	16	29	39	49	56	65	71	78	0.7	0.8	0.9	1.1	1.3	0.9	1.0	1.1	1.3	1.4
SS D1	41	62	73	82	88	93	96	98	25	40	50	61	67	76	81	87	0.6	0.7	0.9	1.1	1.3	0.8	0.9	1.0	1.2	1.4
SS D2	48	71	82	90	94	97	98	99	29	49	60	72	79	86	90	94	0.6	0.7	0.8	1.0	1.2	0.7	0.8	0.9	1.1	1.3
SS D3	37	56	67	77	83	89	92	96	24	37	45	55	61	70	75	81	0.7	0.8	0.9	1.1	1.3	0.9	1.0	1.1	1.3	1.4
SS D4	46	69	80	88	93	96	98	99	28	47	58	69	76	85	89	93	0.6	0.7	0.9	1.0	1.2	0.7	0.8	1.0	1.1	1.3
SS E1	36	57	68	78	84	90	94	96	21	35	44	54	61	70	76	82	0.6	0.8	0.9	1.1	1.3	0.8	0.9	1.1	1.3	1.4
SS E2	42	65	77	86	91	95	97	99	24	42	54	65	72	81	86	91	0.6	0.7	0.9	1.0	1.2	0.7	0.8	1.0	1.2	1.3
SS E3	30	47	58	68	75	82	87	91	19	30	37	45	51	60	65	72	0.7	0.8	1.0	1.2	1.3	1.0	1.1	1.2	1.4	1.5
SS E4	39	61	72	82	88	93	96	98	23	39	50	60	68	77	82	88	0.6	0.7	0.9	1.1	1.2	0.8	0.9	1.0	1.2	1.4

AMARILLO, TEXAS

TYPE	SOLAR SAVINGS FRACTION (%) BASE TEMP: 55 F (2284 DD)								65 F (4193 DD)								CONSERVATION FACTOR 55 F					65 F				
LCR	100	70	50	40	30	25	20	15	100	70	50	40	30	25	20	15	100	50	30	20	15	100	50	30	20	15
WW A1	31	38	46	53	61	66	73	81	22	27	33	37	44	48	54	63	1.1	1.2	1.4	1.5	1.6	1.4	1.5	1.6	1.8	1.9
WW A2	38	48	58	66	75	80	86	92	25	33	41	47	56	62	69	78	0.9	1.0	1.2	1.4	1.5	1.1	1.2	1.4	1.5	1.7
WW A3	40	51	63	70	79	84	90	95	26	35	44	51	61	67	74	83	0.9	1.0	1.2	1.3	1.5	1.0	1.2	1.3	1.5	1.6
WW A4	42	53	65	72	81	86	91	96	27	36	46	53	63	69	77	85	0.8	1.0	1.1	1.3	1.5	1.0	1.1	1.3	1.4	1.6
WW A5	43	55	67	75	84	88	93	97	28	37	48	56	66	72	79	87	0.8	1.0	1.1	1.3	1.4	1.0	1.1	1.3	1.4	1.6
WW A6	44	57	69	76	85	89	94	97	28	38	49	57	67	73	81	88	0.8	1.0	1.1	1.3	1.4	0.9	1.1	1.2	1.4	1.5
WW B1	32	41	50	57	66	72	79	87	21	26	33	38	46	51	58	67	1.0	1.1	1.3	1.4	1.6	1.2	1.4	1.5	1.7	1.8
WW B2	42	54	66	73	82	87	92	96	27	37	48	55	65	71	79	87	0.8	1.0	1.1	1.3	1.5	1.0	1.1	1.3	1.4	1.6
WW B3	49	62	74	81	88	92	96	98	33	44	56	64	73	79	86	92	0.8	0.9	1.1	1.3	1.4	0.9	1.0	1.2	1.4	1.5
WW B4	49	62	74	82	89	93	96	99	33	45	57	65	75	81	87	94	0.8	0.9	1.1	1.3	1.4	0.9	1.0	1.2	1.3	1.5
WW B5	47	61	73	80	88	92	96	98	32	43	56	64	74	80	87	93	0.8	0.9	1.1	1.3	1.4	0.9	1.0	1.2	1.3	1.5
WW C1	51	64	75	82	90	93	96	99	35	46	58	66	76	81	87	93	0.8	0.9	1.1	1.3	1.4	0.9	1.0	1.2	1.3	1.5
WW C2	48	61	73	81	88	92	96	98	33	44	56	64	74	80	86	93	0.8	0.9	1.1	1.3	1.4	0.9	1.0	1.2	1.4	1.5
WW C3	55	69	80	87	93	96	98	99	39	52	65	73	82	88	92	97	0.7	0.9	1.1	1.2	1.4	0.8	1.0	1.1	1.3	1.4
WW C4	51	65	77	84	91	95	97	99	35	48	61	69	79	85	90	96	0.7	0.9	1.1	1.2	1.4	0.8	1.0	1.1	1.3	1.5
TW A1	30	37	45	51	60	65	72	80	22	27	32	37	43	48	54	62	1.1	1.2	1.4	1.5	1.7	1.4	1.5	1.6	1.8	1.9
TW A2	34	43	53	60	69	75	82	89	23	30	37	43	51	57	64	73	1.0	1.1	1.3	1.4	1.6	1.2	1.3	1.5	1.6	1.7
TW A3	36	46	56	64	73	79	85	91	24	31	39	46	55	60	68	77	0.9	1.1	1.2	1.4	1.5	1.1	1.2	1.4	1.5	1.7
TW A4	35	46	57	64	74	80	86	92	23	31	39	46	55	61	69	78	0.9	1.0	1.2	1.4	1.5	1.1	1.2	1.4	1.5	1.7
TW B1	30	38	46	52	61	67	74	82	21	26	32	37	44	49	55	64	1.1	1.2	1.3	1.5	1.6	1.3	1.5	1.6	1.7	1.9
TW B2	33	42	51	59	68	74	81	88	22	28	36	42	50	56	63	72	1.0	1.1	1.3	1.4	1.6	1.2	1.3	1.5	1.6	1.7
TW B3	32	42	52	59	69	75	81	89	21	28	36	42	51	56	64	73	1.0	1.1	1.2	1.4	1.6	1.2	1.3	1.4	1.6	1.7
TW B4	30	39	49	56	66	72	79	87	20	26	34	40	48	54	61	71	1.0	1.1	1.3	1.4	1.6	1.2	1.3	1.5	1.6	1.7
TW C1	29	37	45	52	61	67	74	82	20	25	32	37	44	49	56	65	1.1	1.2	1.3	1.5	1.6	1.3	1.4	1.6	1.7	1.8
TW C2	29	37	46	53	62	68	75	84	19	25	32	37	45	51	58	67	1.0	1.2	1.3	1.5	1.6	1.3	1.4	1.5	1.7	1.8
TW C3	27	35	44	51	60	66	73	81	19	24	31	36	43	48	55	64	1.1	1.2	1.3	1.5	1.6	1.3	1.4	1.6	1.7	1.8
TW C4	25	31	39	45	54	59	66	75	17	22	28	32	39	43	50	58	1.2	1.3	1.4	1.6	1.7	1.4	1.5	1.7	1.8	1.9
TW D1	27	35	44	50	59	65	72	81	17	22	28	33	40	45	51	60	1.1	1.2	1.3	1.5	1.6	1.3	1.5	1.6	1.8	1.9
TW D2	38	49	60	68	77	83	88	94	26	34	43	50	60	66	73	82	0.9	1.0	1.2	1.3	1.5	1.0	1.2	1.3	1.5	1.6
TW D3	42	54	65	73	82	87	92	96	28	37	48	55	65	71	78	86	0.8	1.0	1.1	1.3	1.5	1.0	1.1	1.3	1.4	1.6
TW D4	44	57	69	76	85	89	94	97	30	40	51	59	69	75	82	90	0.8	1.0	1.1	1.3	1.4	0.9	1.1	1.2	1.4	1.5
TW D5	43	56	68	75	84	89	93	97	29	40	51	59	69	75	82	90	0.8	1.0	1.1	1.3	1.4	0.9	1.1	1.2	1.4	1.5
TW E1	46	58	70	77	85	90	94	97	32	42	52	60	70	76	83	90	0.8	1.0	1.1	1.3	1.4	0.9	1.1	1.2	1.4	1.5
TW E2	45	58	69	77	85	90	94	97	31	41	52	60	70	76	83	90	0.8	1.0	1.1	1.3	1.4	0.9	1.1	1.2	1.4	1.5
TW E3	53	66	78	84	91	95	97	99	37	49	61	69	79	85	90	95	0.7	0.9	1.1	1.2	1.4	0.8	1.0	1.2	1.3	1.5
TW E4	50	63	75	82	90	93	96	99	34	46	58	66	77	82	88	94	0.8	0.9	1.1	1.3	1.4	0.9	1.0	1.2	1.3	1.5
TW F1	29	37	45	51	60	66	72	81	20	25	31	36	43	47	54	62	1.1	1.2	1.4	1.5	1.6	1.3	1.5	1.6	1.8	1.9
TW F2	32	41	51	58	67	73	80	88	21	27	35	40	49	54	61	71	1.0	1.1	1.3	1.4	1.6	1.2	1.3	1.5	1.6	1.8
TW F3	32	42	52	60	69	75	82	89	20	27	35	42	50	56	64	73	0.9	1.1	1.2	1.4	1.5	1.1	1.3	1.4	1.6	1.7
TW F4	30	40	50	58	68	74	81	89	18	25	34	40	49	55	63	72	1.0	1.1	1.2	1.4	1.5	1.2	1.3	1.4	1.6	1.7
TW G1	28	35	44	50	59	65	72	81	18	24	30	35	42	47	53	62	1.1	1.2	1.3	1.5	1.6	1.3	1.5	1.6	1.7	1.9
TW G2	28	37	47	54	63	70	77	85	18	24	31	37	45	51	58	67	1.0	1.1	1.3	1.4	1.6	1.2	1.4	1.5	1.7	1.8
TW G3	26	35	45	52	62	68	76	84	16	22	30	35	44	49	57	67	1.0	1.1	1.3	1.5	1.6	1.2	1.4	1.5	1.7	1.8
TW G4	22	30	39	46	56	62	70	79	13	18	25	31	38	44	51	61	1.1	1.2	1.4	1.5	1.6	1.3	1.4	1.6	1.7	1.8
TW H1	24	32	40	47	56	62	69	79	15	20	27	32	39	44	50	60	1.1	1.2	1.4	1.5	1.7	1.4	1.5	1.6	1.8	1.9
TW H2	21	29	38	45	54	60	68	77	13	18	24	29	37	42	49	58	1.1	1.2	1.4	1.5	1.7	1.4	1.5	1.6	1.8	1.9
TW H3	18	25	33	40	49	55	62	72	10	15	21	26	33	38	44	53	1.2	1.3	1.4	1.6	1.7	1.4	1.6	1.7	1.8	2.0
TW H4	14	20	27	32	40	45	53	62	8	12	16	20	26	30	36	44	1.3	1.4	1.6	1.7	1.8	1.6	1.7	1.9	2.0	2.1
TW I1	24	32	40	47	56	62	69	78	15	20	25	30	37	41	48	56	1.1	1.2	1.4	1.5	1.7	1.4	1.5	1.7	1.8	2.0
TW I2	35	45	57	64	74	80	86	92	22	30	40	46	56	62	70	79	0.9	1.0	1.2	1.4	1.5	1.1	1.2	1.4	1.5	1.6
TW I3	38	49	60	68	78	83	89	94	24	33	43	50	60	67	74	83	0.9	1.0	1.2	1.3	1.5	1.0	1.2	1.3	1.5	1.6
TW I4	41	53	65	73	82	87	92	96	26	37	48	55	66	72	80	88	0.8	1.0	1.1	1.3	1.5	1.0	1.1	1.3	1.4	1.5
TW I5	41	53	65	73	83	88	92	97	27	37	48	56	67	73	81	89	0.8	1.0	1.1	1.3	1.4	0.9	1.1	1.2	1.4	1.5
TW J1	43	55	67	75	84	88	93	97	29	39	50	57	68	74	81	88	0.8	1.0	1.1	1.3	1.4	1.0	1.1	1.2	1.4	1.5
TW J2	42	54	66	74	83	88	93	97	28	38	49	57	67	73	80	88	0.8	1.0	1.1	1.3	1.5	1.0	1.1	1.2	1.4	1.5
TW J3	49	62	75	82	90	93	96	99	34	46	58	66	77	82	88	94	0.8	0.9	1.1	1.3	1.4	0.9	1.0	1.2	1.3	1.5
TW J4	46	59	72	80	88	92	95	98	31	43	55	63	74	80	87	93	0.8	0.9	1.1	1.3	1.4	0.9	1.0	1.2	1.3	1.5
DG A1	31	39	48	54	62	66	71	77	18	24	31	36	42	45	50	55	1.0	1.2	1.4	1.7	1.9	1.2	1.5	1.7	2.1	2.4
DG A2	32	42	53	60	68	74	79	86	20	28	36	42	50	55	61	69	0.9	1.1	1.3	1.5	1.7	1.2	1.3	1.5	1.7	1.9
DG A3	37	49	61	69	78	83	88	92	25	33	44	52	62	68	75	82	0.9	1.0	1.2	1.4	1.7	1.0	1.1	1.3	1.5	1.7
DG B1	32	42	54	61	70	75	80	85	19	25	33	39	47	52	58	65	0.9	1.1	1.3	1.5	1.8	1.2	1.3	1.5	1.8	2.1
DG B2	33	45	58	66	76	81	86	92	21	28	38	45	55	61	69	77	0.9	1.0	1.2	1.4	1.6	1.1	1.2	1.3	1.6	1.7
DG B3	38	51	65	73	83	87	92	95	25	34	45	54	66	72	80	87	0.8	0.9	1.1	1.3	1.6	1.0	1.1	1.2	1.4	1.6
DG C1	37	49	62	71	79	84	88	92	22	30	39	46	57	63	69	76	0.9	1.0	1.2	1.5	1.7	1.1	1.2	1.3	1.6	1.8
DG C2	38	51	65	73	82	87	91	95	24	33	43	52	63	69	77	84	0.8	0.9	1.1	1.3	1.5	1.1	1.1	1.3	1.5	1.6
DG C3	43	57	72	80	89	92	95	97	28	39	51	61	73	80	86	92	0.8	0.9	1.1	1.3	1.6	1.0	1.0	1.2	1.4	1.5
SS A1	40	49	59	66	75	80	86	92	29	37	45	51	59	65	71	79	0.9	1.1	1.2	1.4	1.5	1.1	1.2	1.4	1.5	1.7
SS A2	49	60	71	78	86	90	94	97	37	46	56	63	72	78	84	91	0.8	1.0	1.1	1.3	1.5	0.9	1.1	1.2	1.4	1.5
SS A3	37	46	56	62	71	76	83	89	27	34	41	47	55	60	67	75	0.9	1.1	1.3	1.4	1.6	1.1	1.3	1.4	1.6	1.7
SS A4	48	59	70	77	85	89	93	97	36	45	55	62	71	77	83	90	0.8	1.0	1.1	1.3	1.5	0.9	1.1	1.2	1.4	1.6
SS A5	37	46	55	61	70	75	81	88	28	34	41	47	54	59	66	74	1.0	1.1	1.3	1.4	1.6	1.2	1.3	1.5	1.6	1.8
SS A6	49	60	71	78	86	90	94	97	36	46	56	63	72	78	84	90	0.8	1.0	1.1	1.3	1.5	0.9	1.1	1.2	1.4	1.5
SS A7	33	41	49	55	63	68	75	82	25	30	36	41	48	52	58	66	1.1	1.2	1.3	1.5	1.7	1.3	1.4	1.6	1.7	1.9
SS A8	47	58	69	76	84	88	93	97	35	44	54	61	70	76	82	89	0.8	1.0	1.1	1.3	1.5	0.9	1.1	1.3	1.4	1.6
SS B1	35	43	53	60	69	74	81	88	25	31	39	44	52	58	65	73	1.0	1.1	1.3	1.4	1.6	1.2	1.3	1.5	1.6	1.7
SS B2	43	54	65	72	81	86	91	95	32	41	50	57	67	72	79	87	0.8	1.0	1.2	1.3	1.5	1.0	1.1	1.3	1.4	1.6
SS B3	33	41	51	57	66	72	78	86	23	29	37	42	50	55	61	70	1.0	1.1	1.3	1.5	1.6	1.2	1.3	1.5	1.7	1.8
SS B4	42	53	64	71	80	85	90	95	31	40	49	56	65	71	78	86	0.9	1.0	1.2	1.3	1.5	1.0	1.1	1.3	1.5	1.6
SS B5	31	38	46	53	61	67	73	81	23	28	34	39	46	50	57	65	1.1	1.2	1.4	1.5	1.7	1.3	1.4	1.6	1.7	1.9
SS B6	43	53	64	72	81	85	90	95	31	40	49	56	66	72	78	86	0.8	1.0	1.2	1.3	1.5	1.0	1.1	1.3	1.4	1.6
SS B7	28	35	42	48	56	61	68	77	20	25	31	35	41	45	51	59	1.1	1.3	1.4	1.6	1.7	1.4	1.5	1.7	1.8	2.0
SS B8	41	52	63	70	79	84	89	94	30	38	48	54	64	69	76	84	0.9	1.0	1.2	1.3	1.5	1.0	1.2	1.3	1.5	1.6
SS C1	36	46	57	65	74	80	86	92	23	31	39	46	55	61	68	77	0.9	1.0	1.2	1.4	1.5	1.1	1.2	1.4	1.5	1.7
SS C2	42	54	65	73	82	87	92	96	28	37	48	55	65	71	78	87	0.8	1.0	1.1	1.3	1.5	1.0	1.1	1.3	1.4	1.6
SS C3	29	37	46	53	62	68	75	83	19	25	32	37	44	49	56	65	1.0	1.2	1.3	1.5	1.6	1.3	1.4	1.6	1.7	1.8
SS C4	37	48	59	66	76	81	87	93	25	33	42	49	58	65	72	81	0.9	1.0	1.2	1.4	1.5	1.1	1.2	1.3	1.5	1.6
SS D1	48	59	69	76	84	88	93	97	35	44	53	60	69	75	81	88	0.8	1.0	1.1	1.3	1.5	1.0	1.1	1.3	1.4	1.6
SS D2	57	69	79	86	92	95	97	99	42	53	65	72	81	86	91	95	0.7	0.9	1.1	1.3	1.4	0.8	1.0	1.2	1.3	1.5
SS D3	43	53	63	70	78	83	88	94	32	39	48	54	62	68	74	82	0.9	1.0	1.2	1.4	1.5	1.0	1.2	1.3	1.5	1.6
SS D4	55	67	77	84	91	94	97	99	41	52	63	70	79	84	89	94	0.8	0.9	1.1	1.3	1.4	0.9	1.0	1.2	1.3	1.5
SS E1	42	53	64	71	80	84	90	95	30	38	47	54	62	68	75	83	0.9	1.0	1.2	1.3	1.5	1.0	1.2	1.3	1.5	1.6
SS E2	50	63	74	81	88	92	95	98	36	47	58	65	75	80	86	92	0.8	0.9	1.1	1.3	1.4	0.9	1.0	1.2	1.4	1.5
SS E3	35	44	53	60	69	75	81	88	25	32	39	45	52	58	64	73	1.0	1.1	1.3	1.4	1.6	1.2	1.3	1.5	1.6	1.8
SS E4	46	58	69	76	85	89	93	97	33	43	53	61	70	76	82	89	0.8	1.0	1.1	1.3	1.5	0.9	1.1	1.2	1.4	1.6

AUSTIN, TEXAS

TYPE	SOLAR SAVINGS FRACTION (%) BASE TEMP: 55 F (620 DD) LCR 200	100	70	50	40	30	25	20	65 F (1737 DD) 200	100	70	50	40	30	25	20	CONSERVATION FACTOR 55 F 200	100	50	30	20	65 F 200	100	50	30	20
WW A1	34	51	61	71	77	85	89	93	20	29	36	43	49	57	62	68	0.7	0.8	1.0	1.1	1.3	1.0	1.1	1.3	1.4	1.6
WW A2	41	62	73	83	88	94	96	98	21	35	44	54	61	70	76	82	0.6	0.7	0.9	1.1	1.2	0.9	0.9	1.1	1.2	1.4
WW A3	44	66	77	86	91	96	97	99	21	37	47	58	65	75	80	86	0.6	0.7	0.9	1.0	1.2	0.8	0.9	1.0	1.2	1.4
WW A4	45	68	79	88	92	96	98	99	21	38	49	60	68	77	82	88	0.6	0.7	0.9	1.0	1.2	0.8	0.9	1.0	1.2	1.4
WW A5	47	70	81	90	94	97	99	99	22	39	51	62	70	79	84	90	0.6	0.7	0.8	1.0	1.2	0.8	0.8	1.0	1.2	1.3
WW A6	48	71	82	90	94	98	99	100	22	40	52	64	71	80	85	91	0.6	0.7	0.8	1.0	1.2	0.7	0.8	1.0	1.2	1.3
WW B1	37	56	67	77	83	90	93	96	18	30	37	46	53	61	67	74	0.7	0.8	0.9	1.1	1.3	0.9	1.0	1.2	1.3	1.5
WW B2	45	68	79	88	93	97	98	99	21	38	50	61	69	78	83	89	0.6	0.7	0.9	1.0	1.2	0.8	0.9	1.0	1.2	1.3
WW B3	52	75	86	93	96	98	99	100	26	45	57	69	77	85	89	94	0.5	0.7	0.8	1.0	1.2	0.7	0.8	1.0	1.1	1.3
WW B4	51	75	86	93	96	99	99	100	24	45	58	70	77	86	90	94	0.5	0.7	0.8	1.0	1.2	0.7	0.8	0.9	1.1	1.3
WW B5	48	73	84	92	95	98	99	100	22	43	56	68	76	85	89	94	0.5	0.7	0.8	1.0	1.2	0.7	0.8	0.9	1.1	1.3
WW C1	53	77	87	93	96	99	99	100	27	47	59	71	78	86	90	94	0.5	0.6	0.8	1.0	1.2	0.7	0.8	0.9	1.1	1.3
WW C2	50	74	85	92	96	98	99	100	24	44	57	69	76	85	89	94	0.5	0.7	0.8	1.0	1.2	0.7	0.8	1.0	1.1	1.3
WW C3	56	80	90	96	98	99	100	100	27	51	65	77	84	91	94	97	0.5	0.6	0.8	1.0	1.2	0.6	0.7	0.9	1.1	1.3
WW C4	52	77	87	94	97	99	100	100	24	47	61	73	80	88	92	96	0.5	0.6	0.8	1.0	1.2	0.7	0.8	0.9	1.1	1.3
TW A1	33	49	60	69	76	84	88	92	20	29	35	42	48	56	61	67	0.7	0.8	1.0	1.2	1.3	1.0	1.1	1.3	1.4	1.6
TW A2	37	57	68	78	84	90	94	96	20	32	40	49	56	65	70	77	0.7	0.8	0.9	1.1	1.3	0.9	1.0	1.1	1.3	1.5
TW A3	39	60	71	81	87	93	95	98	19	33	42	52	59	68	74	81	0.6	0.7	0.9	1.1	1.3	0.9	1.0	1.1	1.3	1.4
TW A4	38	60	72	82	87	93	96	98	18	32	42	52	60	69	75	82	0.6	0.7	0.9	1.1	1.2	0.9	0.9	1.1	1.2	1.4
TW B1	33	50	60	70	77	84	89	93	19	28	35	43	49	57	62	69	0.7	0.8	1.0	1.1	1.3	1.0	1.1	1.3	1.4	1.6
TW B2	35	55	66	77	83	90	93	96	18	30	38	48	54	63	69	76	0.7	0.8	0.9	1.1	1.3	0.9	1.0	1.2	1.3	1.5
TW B3	35	55	67	77	83	90	93	96	17	30	38	48	55	64	70	77	0.7	0.8	0.9	1.1	1.3	0.9	1.0	1.1	1.3	1.5
TW B4	33	52	64	74	81	88	92	95	16	28	36	45	52	61	67	74	0.7	0.8	0.9	1.1	1.3	0.9	1.0	1.2	1.3	1.5
TW C1	31	49	59	70	76	84	88	93	17	27	34	42	48	56	62	69	0.7	0.8	1.0	1.1	1.3	1.0	1.1	1.3	1.4	1.6
TW C2	31	49	60	71	78	85	90	94	16	27	34	43	49	58	64	71	0.7	0.8	1.0	1.1	1.3	1.0	1.1	1.2	1.4	1.5
TW C3	29	47	58	68	75	83	88	92	15	25	32	41	47	55	61	68	0.7	0.8	1.0	1.1	1.3	1.0	1.1	1.3	1.4	1.5
TW C4	26	42	52	62	69	78	83	88	15	23	29	36	42	50	55	62	0.8	0.9	1.0	1.2	1.4	1.1	1.2	1.3	1.5	1.6
TW D1	31	49	60	71	77	85	89	93	15	25	32	40	46	55	60	67	0.7	0.8	1.0	1.1	1.3	1.0	1.1	1.3	1.4	1.6
TW D2	41	63	74	84	89	94	97	98	20	35	45	56	63	73	78	85	0.6	0.7	0.9	1.1	1.2	0.8	0.9	1.1	1.2	1.4
TW D3	45	68	79	88	92	96	98	99	22	39	50	61	68	78	83	88	0.6	0.7	0.9	1.0	1.2	0.8	0.9	1.0	1.2	1.3
TW D4	46	70	81	89	94	97	98	99	22	41	52	64	72	81	86	91	0.6	0.7	0.8	1.0	1.2	0.7	0.8	1.0	1.1	1.3
TW D5	44	68	80	89	93	97	98	99	21	40	51	63	71	80	86	91	0.6	0.7	0.8	1.0	1.2	0.7	0.8	1.0	1.2	1.3
TW E1	49	72	82	90	94	98	99	99	25	43	54	66	73	82	87	91	0.6	0.7	0.8	1.0	1.2	0.7	0.8	1.0	1.1	1.3
TW E2	47	71	82	90	94	97	99	99	24	42	53	65	73	81	86	91	0.6	0.7	0.8	1.0	1.2	0.7	0.8	1.0	1.1	1.3
TW E3	54	78	88	94	97	99	100	100	27	49	62	74	81	88	92	96	0.5	0.6	0.8	1.0	1.2	0.7	0.8	0.9	1.1	1.3
TW E4	51	75	85	93	96	98	99	100	25	46	58	71	78	86	91	95	0.5	0.7	0.8	1.0	1.2	0.7	0.8	0.9	1.1	1.3
TW F1	32	49	59	70	76	84	88	92	18	27	34	42	47	55	61	68	0.7	0.8	1.0	1.1	1.3	1.0	1.1	1.3	1.4	1.6
TW F2	35	54	66	76	83	89	93	96	17	29	37	47	53	63	68	75	0.7	0.8	0.9	1.1	1.3	0.9	1.0	1.2	1.3	1.5
TW F3	35	56	67	78	84	91	94	97	16	29	38	48	55	64	70	77	0.6	0.8	0.9	1.1	1.3	0.9	1.0	1.1	1.3	1.5
TW F4	33	54	66	77	83	90	93	96	14	27	36	46	53	63	69	76	0.6	0.8	0.9	1.1	1.3	0.9	1.0	1.1	1.3	1.5
TW G1	30	48	58	69	76	84	88	92	16	25	32	40	46	55	60	67	0.7	0.8	1.0	1.1	1.3	1.0	1.1	1.3	1.4	1.6
TW G2	31	50	62	73	79	87	91	95	14	26	34	43	49	59	65	72	0.7	0.8	0.9	1.1	1.3	1.0	1.1	1.2	1.3	1.5
TW G3	29	48	60	71	78	86	90	94	12	24	32	41	48	57	63	71	0.7	0.8	1.0	1.1	1.3	1.0	1.1	1.2	1.3	1.5
TW G4	24	43	54	65	73	81	86	91	10	20	27	36	42	51	58	65	0.7	0.8	1.0	1.2	1.3	1.0	1.1	1.3	1.4	1.6
TW H1	26	44	55	66	73	81	86	91	12	22	29	37	43	52	57	65	0.7	0.8	1.0	1.2	1.3	1.0	1.1	1.3	1.4	1.6
TW H2	24	41	52	64	71	80	85	90	10	20	27	35	41	50	56	63	0.7	0.8	1.0	1.2	1.3	1.1	1.2	1.3	1.4	1.6
TW H3	20	36	47	58	66	75	80	86	8	17	23	31	36	45	50	58	0.8	0.9	1.0	1.2	1.4	1.1	1.2	1.4	1.5	1.6
TW H4	15	29	38	49	56	65	71	78	6	13	18	24	29	36	42	48	0.9	1.0	1.1	1.3	1.4	1.3	1.4	1.5	1.6	1.8
TW I1	28	46	56	67	74	82	87	91	13	22	29	37	43	51	57	64	0.7	0.8	1.0	1.2	1.3	1.1	1.2	1.3	1.4	1.6
TW I2	37	59	71	81	87	93	96	98	17	32	41	52	60	69	75	82	0.6	0.7	0.9	1.1	1.2	0.8	0.9	1.1	1.2	1.4
TW I3	40	63	75	86	90	95	97	99	18	34	45	56	64	74	79	85	0.6	0.7	0.9	1.1	1.2	0.8	0.9	1.0	1.2	1.4
TW I4	42	66	78	87	92	96	98	99	19	37	49	61	68	78	83	89	0.6	0.7	0.9	1.0	1.2	0.8	0.9	1.0	1.2	1.3
TW I5	42	66	78	88	92	96	98	99	19	37	49	61	69	79	84	90	0.6	0.7	0.9	1.0	1.2	0.8	0.9	1.0	1.2	1.3
TW J1	46	69	80	89	93	97	98	99	22	40	51	63	71	80	85	90	0.6	0.7	0.8	1.0	1.2	0.7	0.8	1.0	1.2	1.3
TW J2	44	68	79	88	93	97	98	99	21	39	50	62	70	79	84	90	0.6	0.7	0.9	1.0	1.2	0.8	0.9	1.0	1.2	1.3
TW J3	50	75	86	93	96	99	99	100	24	45	58	71	78	86	91	95	0.5	0.6	0.8	1.0	1.2	0.7	0.8	0.9	1.1	1.3
TW J4	47	72	83	91	95	98	99	100	22	42	55	68	75	84	89	93	0.5	0.7	0.8	1.0	1.2	0.7	0.8	1.0	1.1	1.3
DG A1	36	56	66	76	81	87	90	93	16	28	36	44	49	57	61	66	0.6	0.8	1.0	1.2	1.4	0.9	1.0	1.2	1.5	1.8
DG A2	35	57	68	78	84	90	93	96	16	29	38	48	55	64	69	75	0.5	0.7	0.9	1.1	1.3	0.9	1.0	1.1	1.3	1.5
DG A3	39	63	75	84	89	93	94	96	19	34	45	57	65	74	79	84	0.6	0.7	0.9	1.2	1.6	0.8	0.9	1.0	1.2	1.4
DG B1	37	61	73	83	88	92	94	96	16	29	38	48	56	64	69	75	0.6	0.7	0.9	1.1	1.4	0.9	1.0	1.1	1.3	1.6
DG B2	37	62	75	85	90	94	96	98	17	30	41	52	60	70	76	82	0.6	0.7	0.9	1.1	1.3	0.9	0.9	1.1	1.2	1.4
DG B3	40	66	79	88	92	96	97	98	19	35	46	60	68	79	84	89	0.6	0.7	0.9	1.1	1.6	0.8	0.9	1.0	1.2	1.4
DG C1	42	69	81	89	93	96	97	97	19	33	44	56	64	74	79	84	0.6	0.7	0.9	1.1	1.6	0.9	0.9	1.0	1.2	1.5
DG C2	41	68	81	89	94	97	98	99	19	34	46	59	68	77	83	88	0.6	0.7	0.8	1.0	1.3	0.8	0.9	1.0	1.2	1.4
DG C3	45	73	85	92	95	97	98	98	22	39	52	66	75	85	89	93	0.6	0.6	0.8	1.1	2.0	0.8	0.8	0.9	1.1	1.3
SS A1	44	64	75	84	89	94	96	98	25	39	48	58	65	73	78	84	0.6	0.7	0.9	1.1	1.2	0.8	0.9	1.1	1.2	1.4
SS A2	52	74	84	91	95	98	99	100	30	48	59	70	76	84	89	93	0.6	0.7	0.8	1.0	1.2	0.7	0.8	1.0	1.1	1.3
SS A3	42	62	72	82	87	93	95	97	24	37	45	55	61	70	75	81	0.6	0.7	0.9	1.1	1.3	0.9	1.0	1.1	1.3	1.4
SS A4	52	74	84	91	95	98	99	100	30	47	58	69	76	84	88	92	0.6	0.7	0.8	1.0	1.2	0.7	0.8	1.0	1.1	1.3
SS A5	41	60	70	80	85	91	94	97	25	36	45	53	60	68	73	80	0.6	0.8	0.9	1.1	1.3	0.9	1.0	1.1	1.3	1.5
SS A6	52	74	84	91	95	98	99	100	30	48	59	69	76	84	88	93	0.6	0.7	0.8	1.0	1.2	0.7	0.8	1.0	1.1	1.3
SS A7	37	55	65	74	80	87	91	94	23	33	40	48	54	62	67	74	0.7	0.8	1.0	1.1	1.3	1.0	1.1	1.2	1.4	1.5
SS A8	52	73	83	91	94	98	99	99	29	47	58	68	75	83	87	92	0.6	0.7	0.8	1.0	1.2	0.7	0.8	1.0	1.2	1.3
SS B1	39	58	69	79	85	91	94	97	21	34	42	51	58	67	72	79	0.6	0.8	0.9	1.1	1.3	0.9	1.0	1.1	1.3	1.5
SS B2	46	68	79	88	92	96	98	99	26	42	53	63	71	79	84	89	0.6	0.7	0.9	1.0	1.2	0.8	0.9	1.0	1.2	1.3
SS B3	37	56	67	77	83	90	93	96	20	32	40	49	56	65	70	77	0.7	0.8	0.9	1.1	1.3	0.9	1.0	1.2	1.3	1.5
SS B4	46	68	78	87	92	96	98	99	25	41	52	63	70	79	84	89	0.6	0.7	0.9	1.0	1.2	0.8	0.9	1.0	1.2	1.4
SS B5	34	52	62	72	78	86	89	93	20	30	37	45	51	59	65	71	0.7	0.8	1.0	1.1	1.3	1.0	1.1	1.2	1.4	1.5
SS B6	46	68	79	87	92	96	98	99	25	41	52	63	70	79	84	89	0.6	0.7	0.9	1.0	1.2	0.8	0.9	1.0	1.2	1.4
SS B7	32	48	58	68	74	82	86	91	19	28	34	41	47	55	60	66	0.7	0.8	1.0	1.2	1.3	1.0	1.1	1.3	1.4	1.6
SS B8	45	67	77	86	91	96	97	99	24	40	51	61	68	77	82	88	0.6	0.7	0.9	1.0	1.2	0.8	0.9	1.0	1.2	1.4
SS C1	40	61	73	82	88	93	96	98	19	33	42	53	60	69	75	81	0.6	0.7	0.9	1.1	1.2	0.9	1.0	1.1	1.3	1.4
SS C2	45	68	79	88	92	96	98	99	22	39	50	61	68	78	83	89	0.6	0.7	0.9	1.0	1.2	0.8	0.9	1.0	1.2	1.3
SS C3	32	50	61	71	78	86	90	94	17	27	34	42	49	57	63	70	0.7	0.8	1.0	1.1	1.3	1.0	1.1	1.2	1.4	1.5
SS C4	40	61	73	83	88	94	96	98	20	34	44	55	62	72	77	84	0.6	0.7	0.9	1.1	1.2	0.8	0.9	1.1	1.2	1.4
SS D1	52	74	84	91	95	98	99	100	29	46	57	68	74	82	87	91	0.6	0.7	0.8	1.0	1.2	0.7	0.8	1.0	1.2	1.3
SS D2	60	82	90	96	98	99	100	100	34	55	67	78	84	91	94	97	0.5	0.6	0.8	1.0	1.2	0.6	0.7	0.9	1.1	1.3
SS D3	47	68	78	87	91	95	97	99	27	42	51	61	68	77	81	87	0.6	0.7	0.9	1.1	1.2	0.8	0.9	1.0	1.2	1.4
SS D4	58	80	89	95	97	99	100	100	33	53	65	76	82	89	93	96	0.5	0.6	0.8	1.0	1.2	0.7	0.8	0.9	1.1	1.3
SS E1	47	69	79	88	92	96	98	99	25	41	51	62	69	77	82	88	0.6	0.7	0.9	1.0	1.2	0.8	0.9	1.0	1.2	1.4
SS E2	54	77	87	93	96	99	99	100	29	49	61	72	79	87	91	94	0.5	0.7	0.8	1.0	1.2	0.7	0.8	0.9	1.1	1.3
SS E3	40	59	70	79	85	91	94	97	22	34	43	52	58	67	72	79	0.6	0.8	0.9	1.1	1.3	0.9	1.0	1.1	1.3	1.5
SS E4	50	73	83	91	94	98	99	99	27	45	56	67	74	83	87	92	0.6	0.7	0.8	1.0	1.2	0.7	0.8	1.0	1.1	1.3

BROWNSVILLE, TEXAS

TYPE	SOLAR SAVINGS FRACTION (%) BASE TEMP: 55 F (146 DD)								65 F (650 DD)								CONSERVATION FACTOR 55 F					65 F				
LCR	200	100	70	50	40	30	25	20	200	100	70	50	40	30	25	20	200	100	50	30	20	200	100	50	30	20
WW A1	64	83	90	95	97	99	100	100	31	46	56	66	72	80	85	90	0.5	0.6	0.8	1.0	1.2	0.8	0.9	1.0	1.2	1.3
WW A2	75	92	96	99	100	100	100	100	36	56	68	78	85	91	94	97	0.5	0.6	0.8	1.0	-	0.7	0.8	0.9	1.1	1.3
WW A3	79	94	98	99	100	100	100	100	39	60	72	82	88	94	96	98	0.5	0.6	0.8	1.0	-	0.6	0.7	0.9	1.1	1.2
WW A4	81	95	98	100	100	100	100	100	40	62	74	84	90	95	97	99	0.5	0.6	0.8	0.9	-	0.6	0.7	0.9	1.0	1.2
WW A5	82	96	99	100	100	100	100	100	41	65	77	86	91	96	98	99	0.4	0.6	0.8	0.9	-	0.6	0.7	0.9	1.0	1.2
WW A6	83	96	99	100	100	100	100	100	42	66	78	87	92	96	98	99	0.4	0.6	0.8	0.9	-	0.6	0.7	0.9	1.0	1.2
WW B1	71	88	94	98	99	100	100	100	32	51	62	72	79	87	91	94	0.5	0.6	0.8	1.0	1.2	0.7	0.8	1.0	1.1	1.3
WW B2	80	95	98	100	100	100	100	100	39	62	74	84	90	95	97	99	0.5	0.6	0.8	0.9	-	0.6	0.7	0.9	1.0	1.2
WW B3	86	97	99	100	100	100	100	100	46	70	82	90	94	98	99	100	0.4	0.6	0.7	-	-	0.6	0.7	0.8	1.0	1.2
WW B4	86	97	99	100	100	100	100	100	45	70	82	90	94	98	99	100	0.4	0.6	0.7	-	-	0.6	0.7	0.8	1.0	1.2
WW B5	84	97	99	100	100	100	100	100	43	68	80	89	94	97	99	99	0.4	0.6	0.8	0.9	-	0.6	0.7	0.8	1.0	1.2
WW C1	87	98	99	100	100	100	100	100	48	72	83	91	95	98	99	100	0.4	0.6	0.7	-	-	0.6	0.7	0.8	1.0	1.2
WW C2	85	97	99	100	100	100	100	100	45	69	81	90	94	97	99	99	0.4	0.6	0.7	-	-	0.6	0.7	0.8	1.0	1.2
WW C3	89	98	100	100	100	100	100	100	50	76	87	94	97	99	99	100	0.4	0.6	0.7	-	-	0.5	0.6	0.8	1.0	1.2
WW C4	87	98	99	100	100	100	100	100	46	72	83	92	95	98	99	100	0.4	0.6	0.7	-	-	0.5	0.7	0.8	1.0	1.2
TW A1	63	82	89	95	97	99	99	100	30	45	54	64	71	79	84	89	0.5	0.6	0.8	1.0	1.2	0.8	0.9	1.0	1.2	1.4
TW A2	70	88	94	98	99	100	100	100	33	51	63	73	80	87	91	95	0.5	0.6	0.8	1.0	1.2	0.7	0.8	0.9	1.1	1.3
TW A3	73	90	96	98	99	100	100	100	34	54	66	76	83	90	93	96	0.5	0.6	0.8	1.0	1.2	0.7	0.8	0.9	1.1	1.3
TW A4	73	91	96	99	99	100	100	100	34	54	66	77	84	90	94	97	0.5	0.6	0.8	1.0	1.2	0.7	0.8	0.9	1.1	1.3
TW B1	63	82	90	95	97	99	100	100	30	45	55	65	72	80	85	90	0.5	0.6	0.8	1.0	1.2	0.8	0.9	1.0	1.2	1.3
TW B2	69	87	94	97	99	100	100	100	31	50	61	72	78	86	90	94	0.5	0.6	0.8	1.0	1.2	0.7	0.8	1.0	1.1	1.3
TW B3	69	88	94	98	99	100	100	100	31	50	61	72	79	87	91	95	0.5	0.6	0.8	1.0	1.2	0.7	0.8	0.9	1.1	1.3
TW B4	66	85	92	97	98	100	100	100	29	47	58	69	76	85	89	93	0.5	0.6	0.8	1.0	1.2	0.7	0.8	1.0	1.1	1.3
TW C1	62	82	90	95	97	99	100	100	28	44	54	64	71	80	85	90	0.5	0.6	0.8	1.0	1.2	0.8	0.9	1.0	1.2	1.3
TW C2	63	83	90	96	98	99	100	100	28	44	55	66	73	81	86	91	0.5	0.6	0.8	1.0	1.2	0.7	0.8	1.0	1.2	1.3
TW C3	60	80	89	94	97	99	99	100	26	42	52	63	70	79	84	89	0.5	0.6	0.8	1.0	1.2	0.8	0.9	1.0	1.2	1.3
TW C4	54	75	84	91	94	97	99	99	24	37	47	57	64	73	78	84	0.6	0.7	0.9	1.0	1.2	0.8	0.9	1.1	1.2	1.4
TW D1	64	83	91	96	98	99	100	100	27	44	54	65	72	81	86	91	0.5	0.6	0.8	1.0	1.2	0.7	0.8	1.0	1.2	1.3
TW D2	76	92	97	99	100	100	100	100	36	57	69	80	86	92	95	97	0.5	0.6	0.8	1.0	-	0.6	0.7	0.9	1.1	1.3
TW D3	80	94	98	99	100	100	100	100	40	62	74	84	90	95	97	99	0.5	0.6	0.8	1.0	-	0.6	0.7	0.9	1.1	1.2
TW D4	81	95	98	100	100	100	100	100	41	64	76	86	91	96	98	99	0.5	0.6	0.8	0.9	-	0.6	0.7	0.9	1.0	1.2
TW D5	80	95	98	100	100	100	100	100	39	63	75	85	91	95	97	99	0.5	0.6	0.8	0.9	-	0.6	0.7	0.9	1.0	1.2
TW E1	83	96	99	100	100	100	100	100	44	67	78	87	92	96	98	99	0.4	0.6	0.8	0.9	-	0.6	0.7	0.9	1.0	1.2
TW E2	82	96	99	100	100	100	100	100	42	65	77	87	92	96	98	99	0.4	0.6	0.8	0.9	-	0.6	0.7	0.9	1.0	1.2
TW E3	88	98	99	100	100	100	100	100	49	73	84	92	96	98	99	100	0.4	0.6	0.7	-	-	0.5	0.7	0.8	1.0	1.2
TW E4	85	97	99	100	100	100	100	100	45	70	81	90	94	98	99	100	0.4	0.6	0.7	-	-	0.6	0.7	0.8	1.0	1.2
TW F1	62	82	90	95	97	99	100	100	29	44	54	64	71	80	85	90	0.5	0.6	0.8	1.0	1.2	0.8	0.9	1.0	1.2	1.3
TW F2	68	87	94	97	99	100	100	100	31	49	60	71	78	86	90	94	0.5	0.6	0.8	1.0	1.2	0.7	0.8	1.0	1.1	1.3
TW F3	70	88	94	98	99	100	100	100	30	50	62	73	80	88	91	95	0.5	0.6	0.8	1.0	1.2	0.7	0.8	0.9	1.1	1.3
TW F4	68	87	94	98	99	100	100	100	28	48	60	71	79	87	91	94	0.5	0.6	0.8	1.0	1.2	0.7	0.8	0.9	1.1	1.3
TW G1	61	81	89	95	97	99	100	100	27	43	53	63	71	79	84	89	0.5	0.6	0.8	1.0	1.2	0.8	0.9	1.0	1.2	1.3
TW G2	64	84	92	96	98	99	100	100	27	45	56	67	75	83	88	92	0.5	0.6	0.8	1.0	1.2	0.7	0.8	1.0	1.1	1.3
TW G3	63	83	91	96	98	99	100	100	25	43	54	66	73	82	87	92	0.5	0.6	0.8	1.0	1.2	0.7	0.8	1.0	1.1	1.3
TW G4	57	78	87	94	96	99	99	100	21	37	48	60	67	77	82	88	0.5	0.7	0.8	1.0	1.2	0.8	0.9	1.0	1.2	1.3
TW H1	58	79	87	94	96	99	99	100	23	39	49	60	67	77	82	88	0.5	0.7	0.8	1.0	1.2	0.8	0.9	1.0	1.2	1.4
TW H2	55	77	86	93	96	98	99	100	20	36	47	58	65	75	80	86	0.5	0.7	0.8	1.0	1.2	0.8	0.9	1.0	1.2	1.4
TW H3	50	72	82	90	94	97	98	99	17	32	42	52	60	70	76	82	0.6	0.7	0.9	1.0	1.2	0.8	0.9	1.1	1.2	1.4
TW H4	41	62	74	83	88	93	96	98	13	25	34	43	50	60	66	73	0.6	0.7	0.9	1.1	1.2	0.9	1.0	1.2	1.3	1.5
TW I1	60	81	89	95	97	99	99	100	24	40	51	62	69	78	83	88	0.5	0.6	0.8	1.0	1.2	0.8	0.9	1.0	1.2	1.4
TW I2	73	90	96	99	99	100	100	100	32	53	66	77	83	90	94	97	0.5	0.6	0.8	1.0	1.2	0.7	0.8	0.9	1.1	1.3
TW I3	76	92	97	99	100	100	100	100	35	58	70	81	87	93	95	98	0.5	0.6	0.8	1.0	-	0.6	0.7	0.9	1.1	1.2
TW I4	79	94	98	99	100	100	100	100	37	61	73	84	89	95	97	99	0.5	0.6	0.8	1.0	-	0.6	0.7	0.9	1.0	1.2
TW I5	78	94	98	99	100	100	100	100	37	61	73	84	89	95	97	99	0.5	0.6	0.8	1.0	-	0.6	0.7	0.9	1.0	1.2
TW J1	81	95	98	100	100	100	100	100	41	64	76	86	91	96	97	99	0.5	0.6	0.8	0.9	-	0.6	0.7	0.9	1.0	1.2
TW J2	80	95	98	100	100	100	100	100	39	62	75	85	90	95	97	99	0.5	0.6	0.8	0.9	-	0.6	0.7	0.9	1.0	1.2
TW J3	85	97	99	100	100	100	100	100	45	70	82	90	94	98	99	100	0.4	0.6	0.7	-	-	0.6	0.7	0.8	1.0	1.2
TW J4	83	96	99	100	100	100	100	100	41	67	79	88	93	97	98	99	0.4	0.6	0.8	0.9	-	0.6	0.7	0.8	1.0	1.2
DG A1	72	89	95	98	99	100	100	100	31	50	61	70	76	83	87	90	0.5	0.6	0.8	-	-	0.7	0.8	1.0	1.2	1.4
DG A2	71	89	95	98	99	100	100	100	31	51	63	73	80	87	90	94	0.5	0.6	0.8	-	-	0.7	0.8	1.0	1.1	1.3
DG A3	76	91	95	96	97	97	97	97	34	57	70	80	86	91	93	95	0.5	0.6	1.1	-	-	0.6	0.7	0.9	1.1	1.5
DG B1	78	93	97	99	99	99	99	99	32	54	67	78	84	89	92	94	0.5	0.6	0.8	1.7	-	0.7	0.7	0.9	1.1	1.4
DG B2	77	94	98	99	100	100	100	100	32	55	69	80	86	92	95	97	0.5	0.6	0.8	1.0	-	0.7	0.7	0.9	1.1	1.3
DG B3	79	94	97	98	98	98	98	98	34	60	74	85	90	94	96	97	0.4	0.6	1.1	-	-	0.6	0.7	0.9	1.1	1.5
DG C1	84	96	98	98	99	99	98	98	36	62	76	85	90	94	95	97	0.4	0.6	1.3	-	-	0.6	0.7	0.9	1.1	1.5
DG C2	82	96	99	100	100	100	100	100	35	61	75	86	91	95	97	99	0.4	0.6	0.8	1.3	-	0.6	0.7	0.9	1.1	1.3
DG C3	85	96	98	98	98	98	98	98	39	66	80	90	94	97	98	98	0.4	0.6	1.6	-	-	0.6	0.7	0.8	1.1	1.7
SS A1	78	93	97	99	100	100	100	100	42	62	73	82	87	93	96	98	0.5	0.6	0.8	1.0	-	0.6	0.7	0.9	1.1	1.3
SS A2	86	97	99	100	100	100	100	100	50	72	82	90	94	97	99	99	0.4	0.6	0.8	0.9	-	0.6	0.7	0.8	1.0	1.2
SS A3	76	92	96	99	100	100	100	100	40	60	70	80	86	92	94	97	0.5	0.6	0.8	1.0	-	0.6	0.7	0.9	1.1	1.3
SS A4	86	97	99	100	100	100	100	100	50	72	82	90	94	97	99	99	0.4	0.6	0.8	0.9	-	0.6	0.7	0.8	1.0	1.2
SS A5	74	90	95	98	99	100	100	100	39	57	68	78	84	90	93	96	0.5	0.6	0.8	1.0	1.2	0.7	0.8	0.9	1.1	1.3
SS A6	86	97	99	100	100	100	100	100	50	72	82	90	94	97	99	99	0.4	0.6	0.8	0.9	-	0.6	0.7	0.8	1.0	1.2
SS A7	70	87	93	97	99	100	100	100	36	53	63	73	79	86	90	94	0.5	0.6	0.8	1.0	1.2	0.7	0.8	1.0	1.1	1.3
SS A8	86	97	99	100	100	100	100	100	50	71	82	90	94	97	98	99	0.4	0.6	0.8	0.9	-	0.6	0.7	0.8	1.0	1.2
SS B1	73	90	95	98	99	100	100	100	36	55	66	76	83	89	93	96	0.5	0.6	0.8	1.0	1.2	0.7	0.8	0.9	1.1	1.3
SS B2	81	95	98	99	100	100	100	100	44	65	77	86	91	95	97	99	0.5	0.6	0.8	1.0	-	0.6	0.7	0.9	1.0	1.2
SS B3	71	89	95	98	99	100	100	100	35	54	65	75	81	88	92	95	0.5	0.6	0.8	1.0	1.2	0.7	0.8	0.9	1.1	1.3
SS B4	81	95	98	99	100	100	100	100	44	65	76	86	91	95	97	99	0.5	0.6	0.8	1.0	-	0.6	0.7	0.9	1.0	1.2
SS B5	66	85	92	96	98	99	100	100	32	49	59	69	76	84	88	92	0.5	0.6	0.8	1.0	1.2	0.7	0.8	1.0	1.2	1.3
SS B6	81	95	98	99	100	100	100	100	43	65	76	86	91	95	97	99	0.5	0.6	0.8	1.0	-	0.6	0.7	0.9	1.0	1.2
SS B7	63	82	89	95	97	99	99	100	30	45	55	65	72	80	85	90	0.5	0.6	0.8	1.0	1.2	0.8	0.9	1.0	1.2	1.4
SS B8	80	94	98	99	100	100	100	100	42	64	75	85	90	95	97	99	0.5	0.6	0.8	1.0	-	0.6	0.7	0.9	1.1	1.2
SS C1	75	91	96	99	100	100	100	100	35	55	67	78	84	91	94	97	0.5	0.6	0.8	1.0	-	0.7	0.8	0.9	1.1	1.3
SS C2	80	94	98	99	100	100	100	100	39	62	74	84	90	95	97	99	0.5	0.6	0.8	1.0	-	0.6	0.7	0.9	1.0	1.2
SS C3	64	83	91	96	98	99	100	100	28	45	55	66	73	82	86	91	0.5	0.6	0.8	1.0	1.2	0.7	0.8	1.0	1.2	1.3
SS C4	74	91	96	99	100	100	100	100	35	56	68	79	85	91	94	97	0.5	0.6	0.8	1.0	-	0.6	0.7	0.9	1.1	1.3
SS D1	86	97	99	100	100	100	100	100	50	72	82	90	94	97	98	99	0.4	0.6	0.8	0.9	-	0.6	0.7	0.8	1.0	1.2
SS D2	91	99	100	100	100	100	100	100	57	80	89	95	97	99	100	100	0.4	0.6	0.7	-	-	0.5	0.6	0.8	1.0	1.2
SS D3	81	94	98	99	100	100	100	100	45	65	76	85	90	95	97	98	0.5	0.6	0.8	1.0	-	0.6	0.7	0.9	1.1	1.2
SS D4	90	98	100	100	100	100	100	100	55	78	88	94	97	99	99	100	0.4	0.6	0.7	-	-	0.5	0.6	0.8	1.0	1.2
SS E1	82	95	98	100	100	100	100	100	45	66	77	86	91	95	97	99	0.5	0.6	0.8	0.9	-	0.6	0.7	0.9	1.0	1.2
SS E2	88	98	99	100	100	100	100	100	51	74	85	92	96	98	99	100	0.4	0.6	0.7	-	-	0.5	0.7	0.8	1.0	1.2
SS E3	73	90	95	98	99	100	100	100	37	56	67	77	83	90	93	96	0.5	0.6	0.8	1.0	1.2	0.7	0.8	0.9	1.1	1.3
SS E4	85	96	99	100	100	100	100	100	47	70	81	89	93	97	98	99	0.4	0.6	0.8	0.9	-	0.6	0.7	0.9	1.0	1.2

CORPUS CHRISTI, TEXAS

TYPE	SSF 55 F (248 DD) LCR 200	100	70	50	40	30	25	20	SSF 65 F (930 DD) LCR 200	100	70	50	40	30	25	20	CF 55 F 200	100	50	30	20	CF 65 F 200	100	50	30	20
WW A1	52	72	81	89	93	96	98	99	26	39	48	57	64	72	78	84	0.6	0.7	0.9	1.0	1.2	0.9	0.9	1.1	1.2	1.4
WW A2	63	83	91	96	98	99	100	100	30	48	59	70	77	85	89	93	0.5	0.6	0.8	1.0	1.2	0.7	0.8	1.0	1.1	1.3
WW A3	66	86	93	97	99	100	100	100	31	52	63	74	81	88	92	95	0.5	0.6	0.8	1.0	1.2	0.7	0.8	0.9	1.1	1.3
WW A4	68	88	94	98	99	100	100	100	32	53	65	77	83	90	93	96	0.5	0.6	0.8	1.0	1.2	0.7	0.8	0.9	1.1	1.3
WW A5	70	89	95	98	99	100	100	100	33	55	68	79	85	92	95	97	0.5	0.6	0.8	1.0	1.2	0.6	0.7	0.9	1.1	1.3
WW A6	71	90	96	98	99	100	100	100	34	56	69	80	86	92	95	98	0.5	0.6	0.8	1.0	1.2	0.6	0.7	0.9	1.1	1.2
WW B1	57	78	87	93	96	98	99	100	26	42	53	63	70	79	84	89	0.5	0.7	0.8	1.0	1.2	0.8	0.9	1.0	1.2	1.4
WW B2	68	88	94	98	99	100	100	100	32	53	66	77	83	90	94	97	0.5	0.6	0.8	1.0	1.2	0.6	0.8	0.9	1.1	1.3
WW B3	75	92	97	99	100	100	100	100	38	61	74	84	89	95	97	98	0.5	0.6	0.8	1.0	-	0.6	0.7	0.9	1.1	1.2
WW B4	74	92	97	99	100	100	100	100	37	61	74	84	90	95	97	99	0.5	0.6	0.8	1.0	-	0.6	0.7	0.9	1.0	1.2
WW B5	72	91	96	99	100	100	100	100	34	59	72	83	89	94	97	98	0.5	0.6	0.8	1.0	-	0.6	0.7	0.9	1.0	1.2
WW C1	76	93	97	99	100	100	100	100	39	63	75	85	90	95	97	99	0.5	0.6	0.8	1.0	-	0.6	0.7	0.9	1.0	1.2
WW C2	74	91	96	99	100	100	100	100	36	60	73	83	89	94	97	98	0.5	0.6	0.8	1.0	-	0.6	0.7	0.9	1.1	1.2
WW C3	79	94	98	100	100	100	100	100	41	67	80	89	93	97	99	99	0.5	0.6	0.8	0.9	-	0.6	0.7	0.8	1.0	1.2
WW C4	75	93	97	99	100	100	100	100	37	63	76	86	91	96	98	99	0.5	0.6	0.8	1.0	-	0.6	0.7	0.9	1.0	1.2
TW A1	50	70	80	88	92	96	97	99	25	38	47	56	63	71	77	83	0.6	0.7	0.9	1.1	1.2	0.9	1.0	1.1	1.3	1.4
TW A2	57	78	87	93	96	98	99	100	27	44	54	65	72	80	85	90	0.5	0.7	0.8	1.0	1.2	0.8	0.9	1.0	1.2	1.3
TW A3	60	81	89	95	97	99	100	100	28	46	57	68	75	83	88	92	0.5	0.6	0.8	1.0	1.2	0.7	0.8	1.0	1.1	1.3
TW A4	60	82	90	95	97	99	100	100	27	46	57	69	76	84	89	93	0.5	0.6	0.8	1.0	1.2	0.7	0.8	1.0	1.1	1.3
TW B1	51	71	81	89	93	96	98	99	25	38	47	57	64	72	78	84	0.6	0.7	0.9	1.0	1.2	0.8	0.9	1.1	1.2	1.4
TW B2	55	77	86	93	96	98	99	100	25	42	52	63	70	79	84	89	0.5	0.7	0.8	1.0	1.2	0.8	0.9	1.0	1.2	1.3
TW B3	56	77	86	93	96	98	99	100	25	42	53	64	71	80	85	90	0.5	0.7	0.8	1.0	1.2	0.8	0.9	1.0	1.2	1.3
TW B4	53	74	84	91	95	98	99	99	23	39	50	61	68	77	82	88	0.6	0.7	0.8	1.0	1.2	0.8	0.9	1.0	1.2	1.4
TW C1	49	70	80	88	92	96	98	99	23	37	46	56	63	72	78	84	0.6	0.7	0.9	1.0	1.2	0.8	0.9	1.1	1.2	1.4
TW C2	50	71	81	89	93	97	98	99	22	37	47	57	65	74	79	85	0.6	0.7	0.9	1.0	1.2	0.8	0.9	1.1	1.2	1.4
TW C3	47	69	79	87	92	96	97	99	21	35	44	55	62	71	77	83	0.6	0.7	0.9	1.1	1.2	0.9	0.9	1.1	1.2	1.4
TW C4	42	62	73	82	88	93	95	97	19	31	39	49	56	65	70	77	0.6	0.7	0.9	1.1	1.3	0.9	1.0	1.1	1.3	1.5
TW D1	50	72	82	89	93	97	98	99	22	36	46	56	63	72	78	84	0.6	0.7	0.9	1.0	1.2	0.8	0.9	1.1	1.2	1.4
TW D2	63	83	91	96	98	99	100	100	29	49	60	72	79	86	90	94	0.5	0.6	0.8	1.0	1.2	0.7	0.8	0.9	1.1	1.3
TW D3	68	87	94	97	99	100	100	100	32	53	66	77	83	90	93	96	0.5	0.6	0.8	1.0	1.2	0.7	0.8	0.9	1.1	1.3
TW D4	69	89	95	98	99	100	100	100	33	56	68	79	85	92	95	97	0.5	0.6	0.8	1.0	1.2	0.6	0.7	0.9	1.1	1.3
TW D5	67	88	94	98	99	100	100	100	31	54	67	78	85	91	94	97	0.5	0.6	0.8	1.0	1.2	0.6	0.7	0.9	1.1	1.3
TW E1	71	90	95	98	99	100	100	100	36	58	70	81	87	93	95	98	0.5	0.6	0.8	1.0	1.2	0.6	0.7	0.9	1.1	1.2
TW E2	70	89	95	98	99	100	100	100	34	57	69	80	86	92	95	98	0.5	0.6	0.8	1.0	1.2	0.6	0.7	0.9	1.1	1.2
TW E3	77	93	97	99	100	100	100	100	40	65	77	87	92	96	98	99	0.5	0.6	0.8	1.0	-	0.6	0.7	0.9	1.0	1.2
TW E4	74	92	97	99	100	100	100	100	37	61	74	84	90	95	97	99	0.5	0.6	0.8	1.0	-	0.6	0.7	0.9	1.0	1.2
TW F1	50	70	80	88	92	96	98	99	24	37	46	56	63	72	77	83	0.6	0.7	0.9	1.1	1.2	0.9	0.9	1.1	1.2	1.4
TW F2	55	77	86	92	96	98	99	100	25	41	52	62	70	79	84	89	0.5	0.7	0.8	1.0	1.2	0.8	0.9	1.0	1.2	1.3
TW F3	56	78	87	93	96	98	99	100	24	42	53	64	72	80	85	90	0.5	0.7	0.8	1.0	1.2	0.7	0.8	1.0	1.2	1.3
TW F4	54	77	86	93	96	98	99	100	22	40	51	62	70	79	84	90	0.5	0.7	0.8	1.0	1.2	0.8	0.8	1.0	1.2	1.3
TW G1	49	70	80	88	92	96	98	99	22	35	45	55	62	71	77	83	0.6	0.7	0.9	1.1	1.2	0.9	0.9	1.1	1.2	1.4
TW G2	51	73	83	91	94	97	99	99	21	37	48	58	66	75	81	87	0.6	0.7	0.9	1.0	1.2	0.8	0.9	1.0	1.2	1.4
TW G3	49	71	82	90	94	97	98	99	19	35	46	57	64	74	80	86	0.6	0.7	0.9	1.0	1.2	0.8	0.9	1.0	1.2	1.4
TW G4	43	66	77	86	91	95	97	99	15	30	40	51	58	68	74	81	0.6	0.7	0.9	1.1	1.2	0.9	0.9	1.1	1.2	1.4
TW H1	45	66	77	86	91	95	97	98	18	32	41	51	59	68	74	81	0.6	0.7	0.9	1.1	1.2	0.9	1.0	1.1	1.3	1.4
TW H2	42	64	75	84	89	94	96	98	15	29	39	49	56	66	72	79	0.6	0.7	0.9	1.1	1.2	0.9	1.0	1.1	1.3	1.4
TW H3	37	58	70	80	86	92	95	97	13	25	34	44	51	61	67	74	0.6	0.7	0.9	1.1	1.3	0.9	1.0	1.2	1.3	1.5
TW H4	30	49	60	71	78	86	90	94	9	19	27	35	42	51	57	65	0.7	0.8	1.0	1.1	1.3	1.1	1.1	1.3	1.4	1.6
TW I1	47	68	79	87	92	96	97	99	19	33	42	52	60	69	75	81	0.6	0.7	0.9	1.1	1.2	0.9	1.0	1.1	1.3	1.4
TW I2	59	81	90	95	97	99	100	100	26	45	57	68	76	84	89	93	0.5	0.6	0.8	1.0	1.2	0.7	0.8	1.0	1.1	1.3
TW I3	63	84	92	96	98	99	100	100	28	49	61	72	79	87	91	95	0.5	0.6	0.8	1.0	1.2	0.7	0.8	0.9	1.1	1.3
TW I4	66	86	93	97	99	100	100	100	30	52	65	76	83	90	93	96	0.5	0.6	0.8	1.0	1.2	0.7	0.8	0.9	1.1	1.3
TW I5	65	86	93	97	99	100	100	100	29	52	65	76	83	90	94	97	0.5	0.6	0.8	1.0	1.2	0.6	0.7	0.9	1.1	1.3
TW J1	69	88	95	98	99	100	100	100	33	55	68	79	85	91	94	97	0.5	0.6	0.8	1.0	1.2	0.6	0.7	0.9	1.1	1.3
TW J2	67	87	94	98	99	100	100	100	31	53	66	77	84	91	94	97	0.5	0.6	0.8	1.0	1.2	0.6	0.7	0.9	1.1	1.3
TW J3	74	92	97	99	100	100	100	100	36	61	74	84	90	95	97	99	0.5	0.6	0.8	1.0	-	0.6	0.7	0.9	1.0	1.2
TW J4	71	90	96	99	99	100	100	100	33	58	71	82	88	94	96	98	0.5	0.6	0.8	1.0	1.2	0.6	0.7	0.9	1.1	1.2
DG A1	58	79	87	93	96	98	99	100	24	41	51	61	67	75	79	83	0.5	0.7	0.8	1.0	1.2	0.8	0.9	1.1	1.3	1.5
DG A2	58	79	88	94	96	98	99	100	24	42	54	65	71	79	84	88	0.5	0.6	0.8	1.0	1.1	0.7	0.8	1.0	1.2	1.4
DG A3	63	84	90	94	96	96	97	97	27	48	61	73	79	86	90	92	0.5	0.6	0.9	1.6	-	0.7	0.8	0.9	1.2	1.4
DG B1	63	85	92	96	97	98	99	99	25	44	57	68	75	82	86	89	0.5	0.6	0.8	1.1	1.8	0.7	0.8	1.0	1.2	1.4
DG B2	62	85	92	97	98	99	100	100	25	45	58	71	78	86	90	93	0.5	0.6	0.8	1.0	1.2	0.7	0.8	0.9	1.1	1.3
DG B3	65	87	94	97	98	98	98	98	28	50	64	77	84	90	93	96	0.5	0.6	0.9	1.6	-	0.7	0.7	0.9	1.1	1.4
DG C1	70	90	95	97	98	98	98	98	29	51	65	77	83	89	92	94	0.5	0.6	0.9	2.0	-	0.7	0.7	0.9	1.1	1.4
DG C2	68	89	95	98	99	100	100	100	28	51	65	78	84	91	94	96	0.5	0.6	0.8	1.0	1.5	0.7	0.7	0.9	1.1	1.3
DG C3	71	91	96	98	98	98	98	98	31	56	71	83	89	94	96	97	0.5	0.6	0.9	3.1	-	0.6	0.7	0.9	1.1	1.4
SS A1	65	84	91	96	98	99	100	100	34	53	64	74	80	87	91	94	0.5	0.6	0.8	1.0	1.2	0.7	0.8	1.0	1.1	1.3
SS A2	74	91	96	99	99	100	100	100	41	63	74	84	89	94	96	98	0.5	0.6	0.8	1.0	1.2	0.6	0.7	0.9	1.1	1.2
SS A3	63	82	90	95	97	99	100	100	33	50	61	71	78	85	89	93	0.5	0.6	0.8	1.0	1.2	0.7	0.8	1.0	1.1	1.3
SS A4	74	91	96	99	99	100	100	100	42	63	74	84	89	94	96	98	0.5	0.6	0.8	1.0	1.2	0.6	0.7	0.9	1.1	1.2
SS A5	61	80	89	94	97	99	99	100	32	49	59	69	76	83	87	92	0.5	0.7	0.8	1.0	1.2	0.7	0.8	1.0	1.2	1.3
SS A6	74	91	96	99	99	100	100	100	41	63	74	84	89	94	96	98	0.5	0.6	0.8	1.0	1.2	0.6	0.7	0.9	1.1	1.2
SS A7	56	76	85	91	95	98	99	99	29	44	54	64	70	78	83	88	0.6	0.7	0.9	1.0	1.2	0.8	0.9	1.0	1.2	1.4
SS A8	74	91	96	98	99	100	100	100	41	62	73	83	88	94	96	98	0.5	0.6	0.8	1.0	1.2	0.6	0.7	0.9	1.1	1.2
SS B1	59	79	88	94	96	99	99	100	29	46	57	67	74	82	87	91	0.5	0.7	0.8	1.0	1.2	0.7	0.8	1.0	1.2	1.3
SS B2	68	87	94	97	99	100	100	100	36	56	68	78	84	91	94	97	0.5	0.6	0.8	1.0	1.2	0.6	0.7	0.9	1.1	1.3
SS B3	58	78	87	93	96	98	99	100	28	45	55	66	73	81	85	90	0.5	0.7	0.8	1.0	1.2	0.7	0.8	1.0	1.2	1.3
SS B4	68	87	94	97	99	100	100	100	35	56	68	78	84	91	94	97	0.5	0.6	0.8	1.0	1.2	0.6	0.8	0.9	1.1	1.3
SS B5	53	73	82	90	93	97	98	99	26	41	50	60	67	75	80	86	0.6	0.7	0.9	1.0	1.2	0.8	0.9	1.1	1.2	1.4
SS B6	68	87	94	97	99	100	100	100	35	56	68	78	84	91	94	97	0.5	0.6	0.8	1.0	1.2	0.6	0.8	0.9	1.1	1.3
SS B7	49	69	79	87	91	95	97	99	24	38	47	56	63	71	77	82	0.6	0.7	0.9	1.1	1.2	0.8	0.9	1.1	1.3	1.4
SS B8	67	86	93	97	99	100	100	100	34	55	66	77	83	90	93	96	0.5	0.6	0.8	1.0	1.2	0.7	0.8	0.9	1.1	1.3
SS C1	62	83	91	96	98	99	100	100	28	47	58	69	76	85	89	93	0.5	0.6	0.8	1.0	1.2	0.7	0.8	1.0	1.1	1.3
SS C2	67	87	94	98	99	100	100	100	32	53	66	77	83	90	93	96	0.5	0.6	0.8	1.0	1.2	0.7	0.8	0.9	1.1	1.3
SS C3	51	72	82	89	93	97	98	99	23	37	47	57	64	74	79	85	0.6	0.7	0.9	1.0	1.2	0.8	0.9	1.1	1.2	1.4
SS C4	61	82	90	96	98	99	100	100	28	48	59	70	78	86	90	94	0.5	0.6	0.8	1.0	1.2	0.7	0.8	1.0	1.1	1.3
SS D1	74	91	96	99	99	100	100	100	41	62	73	83	88	93	96	98	0.5	0.6	0.8	1.0	1.2	0.6	0.7	0.9	1.1	1.2
SS D2	82	95	98	100	100	100	100	100	48	71	82	90	94	97	99	99	0.4	0.6	0.8	0.9	-	0.6	0.7	0.8	1.0	1.2
SS D3	69	87	93	97	99	100	100	100	37	56	67	77	83	90	93	96	0.5	0.6	0.8	1.0	1.2	0.7	0.8	0.9	1.1	1.3
SS D4	80	94	98	99	100	100	100	100	46	69	80	89	93	97	98	99	0.5	0.6	0.8	1.0	-	0.6	0.7	0.9	1.0	1.2
SS E1	70	88	94	98	99	100	100	100	36	57	68	78	84	91	94	96	0.5	0.6	0.8	1.0	1.2	0.6	0.8	0.9	1.1	1.3
SS E2	77	93	97	99	100	100	100	100	42	65	77	86	91	96	97	99	0.5	0.6	0.8	1.0	-	0.6	0.7	0.9	1.0	1.2
SS E3	60	80	88	94	97	99	99	100	30	47	58	68	75	83	87	91	0.5	0.7	0.8	1.0	1.2	0.7	0.8	1.0	1.2	1.3
SS E4	73	90	96	98	99	100	100	100	39	61	72	82	88	93	96	98	0.5	0.6	0.8	1.0	1.2	0.6	0.7	0.9	1.1	1.2

DALLAS, TEXAS

TYPE	SSF (%) 55 F (954 DD) LCR 200	100	70	50	40	30	25	20	SSF (%) 65 F (2297 DD) LCR 200	100	70	50	40	30	25	20	CF 55 F 200	100	50	30	20	CF 65 F 200	100	50	30	20
WW A1	27	41	50	60	66	75	80	85	18	25	30	36	41	48	53	59	0.8	0.9	1.1	1.2	1.4	1.2	1.3	1.4	1.5	1.7
WW A2	32	51	62	72	79	86	90	94	17	29	37	46	52	61	67	74	0.7	0.8	1.0	1.1	1.3	0.9	1.0	1.2	1.3	1.5
WW A3	33	54	66	77	83	90	93	96	17	30	39	49	56	66	72	79	0.7	0.8	0.9	1.1	1.3	0.9	1.0	1.1	1.3	1.4
WW A4	34	56	68	79	85	91	94	97	17	31	41	51	59	68	74	81	0.6	0.7	0.9	1.1	1.3	0.9	0.9	1.1	1.2	1.4
WW A5	35	58	70	81	87	93	95	98	16	32	42	53	61	71	77	83	0.6	0.7	0.9	1.1	1.3	0.8	0.9	1.1	1.2	1.4
WW A6	36	59	71	82	88	93	96	98	16	32	43	54	62	72	78	84	0.6	0.7	0.9	1.1	1.2	0.8	0.9	1.1	1.2	1.4
WW B1	28	44	55	66	72	81	85	90	15	24	30	38	44	52	57	64	0.7	0.8	1.0	1.2	1.3	1.1	1.2	1.3	1.5	1.6
WW B2	34	56	68	79	85	92	95	97	16	31	42	52	60	70	76	82	0.6	0.7	0.9	1.1	1.3	0.8	0.9	1.1	1.2	1.4
WW B3	40	64	76	85	90	95	97	99	20	38	49	61	68	78	83	89	0.6	0.7	0.9	1.0	1.2	0.8	0.9	1.0	1.2	1.3
WW B4	39	64	76	86	91	96	97	99	19	37	49	62	70	79	85	90	0.6	0.7	0.9	1.0	1.2	0.7	0.8	1.0	1.2	1.3
WW B5	37	62	74	85	90	95	97	99	17	36	48	60	69	78	84	90	0.6	0.7	0.9	1.0	1.2	0.8	0.8	1.0	1.2	1.3
WW C1	42	65	77	87	91	96	98	99	21	39	51	63	70	80	85	90	0.6	0.7	0.9	1.0	1.2	0.7	0.8	1.0	1.2	1.3
WW C2	39	63	75	85	90	95	97	99	19	37	49	61	69	78	84	89	0.6	0.7	0.9	1.0	1.2	0.8	0.9	1.0	1.2	1.3
WW C3	44	69	81	90	94	97	99	100	21	43	56	69	77	86	90	94	0.6	0.7	0.8	1.0	1.2	0.7	0.8	0.9	1.1	1.3
WW C4	40	66	78	88	92	96	98	99	19	39	52	65	73	83	88	93	0.6	0.7	0.9	1.0	1.2	0.7	0.8	1.0	1.1	1.3
TW A1	26	40	49	58	65	73	79	84	17	24	30	36	41	48	52	59	0.8	0.9	1.1	1.2	1.4	1.2	1.3	1.4	1.6	1.7
TW A2	29	46	57	67	74	82	87	91	16	26	33	42	48	56	62	69	0.7	0.8	1.0	1.2	1.3	1.0	1.1	1.2	1.4	1.6
TW A3	29	48	60	70	77	85	89	93	16	27	35	44	50	60	65	73	0.7	0.8	1.0	1.1	1.3	1.0	1.1	1.2	1.3	1.5
TW A4	29	48	60	71	78	86	90	94	14	26	35	44	51	60	66	74	0.7	0.8	1.0	1.1	1.3	0.9	1.0	1.2	1.3	1.5
TW B1	26	40	49	59	66	75	80	85	16	24	30	36	41	49	54	60	0.8	0.9	1.1	1.2	1.4	1.1	1.2	1.4	1.5	1.7
TW B2	27	44	55	66	73	81	86	90	15	25	32	40	46	55	61	68	0.7	0.8	1.0	1.2	1.3	1.0	1.1	1.3	1.4	1.6
TW B3	26	44	55	66	73	82	86	91	14	24	32	40	47	56	62	69	0.7	0.8	1.0	1.2	1.3	1.0	1.1	1.2	1.4	1.5
TW B4	25	41	52	63	70	79	84	89	13	23	30	38	44	53	59	66	0.8	0.9	1.0	1.2	1.3	1.0	1.1	1.3	1.4	1.6
TW C1	24	39	48	59	66	74	80	85	15	23	29	35	41	48	54	60	0.8	0.9	1.1	1.2	1.4	1.1	1.2	1.4	1.5	1.7
TW C2	24	39	49	60	67	76	81	87	13	22	29	36	42	50	55	62	0.8	0.9	1.0	1.2	1.4	1.1	1.2	1.3	1.5	1.6
TW C3	22	37	47	57	64	73	79	85	13	21	27	34	40	48	53	60	0.8	0.9	1.1	1.2	1.4	1.1	1.2	1.4	1.5	1.6
TW C4	20	33	42	51	58	67	73	79	12	19	25	31	36	43	48	54	0.9	1.0	1.1	1.3	1.4	1.2	1.3	1.5	1.6	1.7
TW D1	23	38	48	58	65	74	80	85	12	20	26	33	38	45	51	57	0.8	0.9	1.1	1.2	1.4	1.1	1.2	1.4	1.5	1.7
TW D2	31	51	63	74	81	88	92	95	16	29	38	48	55	64	70	77	0.7	0.8	0.9	1.1	1.3	0.9	1.0	1.1	1.3	1.4
TW D3	34	56	68	78	85	91	94	97	17	32	42	52	60	69	75	82	0.6	0.7	0.9	1.1	1.3	0.8	0.9	1.1	1.2	1.4
TW D4	35	58	70	81	87	93	96	98	18	34	45	56	64	74	79	86	0.6	0.7	0.9	1.1	1.2	0.8	0.9	1.0	1.2	1.4
TW D5	34	57	69	80	86	92	95	98	16	33	44	55	63	73	79	86	0.6	0.7	0.9	1.1	1.2	0.8	0.9	1.0	1.2	1.4
TW E1	38	60	72	82	88	93	96	98	20	36	46	57	65	74	80	86	0.6	0.7	0.9	1.1	1.2	0.8	0.9	1.0	1.2	1.4
TW E2	37	59	71	82	88	93	96	98	19	35	46	57	65	74	80	86	0.6	0.7	0.9	1.1	1.2	0.8	0.9	1.0	1.2	1.4
TW E3	42	67	79	88	93	97	98	99	22	41	54	66	74	83	88	92	0.6	0.7	0.9	1.0	1.2	0.7	0.8	1.0	1.1	1.3
TW E4	39	64	76	86	91	96	97	99	19	38	50	63	71	80	85	91	0.6	0.7	0.9	1.0	1.2	0.7	0.8	1.0	1.1	1.3
TW F1	25	39	48	58	65	74	79	85	15	23	28	35	40	47	52	59	0.8	0.9	1.1	1.2	1.4	1.2	1.3	1.4	1.5	1.7
TW F2	26	43	54	65	72	81	85	90	14	24	31	39	45	54	59	67	0.7	0.8	1.0	1.2	1.3	1.0	1.1	1.3	1.4	1.6
TW F3	26	44	56	67	74	82	87	92	13	23	31	40	46	56	62	69	0.7	0.8	1.0	1.2	1.3	1.0	1.1	1.2	1.4	1.5
TW F4	23	42	54	65	72	81	86	91	10	21	29	38	45	54	60	68	0.7	0.8	1.0	1.2	1.3	1.0	1.1	1.2	1.4	1.5
TW G1	23	37	47	57	64	73	79	85	13	21	27	34	39	46	52	58	0.8	0.9	1.1	1.2	1.4	1.1	1.2	1.4	1.5	1.7
TW G2	23	39	50	61	68	77	83	88	11	21	27	35	41	50	56	63	0.8	0.9	1.0	1.2	1.4	1.1	1.2	1.3	1.4	1.6
TW G3	21	37	48	59	67	76	81	87	9	19	26	34	40	49	55	62	0.8	0.9	1.0	1.2	1.4	1.1	1.2	1.3	1.4	1.6
TW G4	17	32	42	53	61	71	76	83	7	15	22	29	35	43	49	56	0.8	0.9	1.1	1.2	1.4	1.1	1.2	1.4	1.5	1.7
TW H1	19	34	44	54	61	70	76	82	10	18	24	30	36	43	49	56	0.8	0.9	1.1	1.2	1.4	1.2	1.3	1.4	1.5	1.7
TW H2	17	31	41	52	59	69	74	81	7	15	21	28	34	41	47	54	0.9	0.9	1.1	1.3	1.4	1.2	1.3	1.4	1.6	1.7
TW H3	14	27	36	46	54	63	69	76	5	13	18	25	30	37	42	49	0.9	1.0	1.1	1.3	1.5	1.2	1.3	1.5	1.6	1.8
TW H4	10	21	29	37	44	53	60	67	4	10	14	19	23	30	34	41	1.0	1.1	1.2	1.4	1.5	1.4	1.5	1.6	1.8	1.9
TW I1	20	35	45	55	62	71	77	83	10	17	23	29	34	42	47	53	0.8	0.9	1.1	1.2	1.4	1.2	1.3	1.4	1.6	1.7
TW I2	28	47	59	71	78	86	90	94	13	26	34	44	51	61	67	74	0.7	0.8	1.0	1.1	1.3	0.9	1.0	1.2	1.3	1.5
TW I3	30	51	63	74	81	88	92	95	14	28	37	48	55	65	71	78	0.7	0.8	0.9	1.1	1.3	0.9	1.0	1.1	1.3	1.4
TW I4	32	54	67	78	85	91	94	97	15	30	41	52	60	70	76	83	0.6	0.7	0.9	1.1	1.3	0.8	0.9	1.1	1.2	1.4
TW I5	31	54	67	78	85	91	95	97	14	30	41	53	61	71	77	84	0.6	0.7	0.9	1.1	1.3	0.8	0.9	1.1	1.2	1.4
TW J1	35	57	70	80	86	92	95	97	17	33	43	55	62	72	78	84	0.6	0.7	0.9	1.1	1.3	0.8	0.9	1.1	1.2	1.4
TW J2	33	56	68	79	86	92	95	97	16	32	42	54	61	71	77	84	0.6	0.7	0.9	1.1	1.3	0.8	0.9	1.1	1.2	1.4
TW J3	38	63	76	86	91	96	97	99	18	38	50	63	71	80	86	91	0.6	0.7	0.9	1.0	1.2	0.7	0.8	1.0	1.1	1.3
TW J4	36	60	73	84	89	95	97	98	16	35	47	60	68	78	83	89	0.6	0.7	0.9	1.0	1.2	0.8	0.9	1.0	1.2	1.3
DG A1	26	44	54	64	70	77	81	85	12	22	29	36	41	47	51	56	0.7	0.8	1.0	1.3	1.5	1.0	1.2	1.4	1.6	2.0
DG A2	26	45	56	67	74	81	86	90	13	24	31	40	47	55	60	66	0.7	0.8	1.0	1.2	1.4	1.0	1.1	1.2	1.4	1.6
DG A3	29	51	64	75	81	87	90	93	15	28	38	49	56	66	72	78	0.7	0.8	0.9	1.2	1.4	0.9	1.0	1.1	1.3	1.5
DG B1	27	47	60	71	77	84	87	91	12	22	30	38	45	54	59	64	0.7	0.8	1.0	1.2	1.4	1.1	1.1	1.2	1.4	1.7
DG B2	27	48	61	73	80	87	91	94	13	24	33	43	51	61	67	74	0.7	0.8	0.9	1.1	1.3	1.0	1.1	1.1	1.3	1.5
DG B3	30	53	67	79	85	91	94	96	16	28	38	50	59	71	77	83	0.7	0.7	0.9	1.1	1.4	0.9	1.0	1.0	1.2	1.4
DG C1	31	54	68	79	85	90	93	95	15	26	35	45	53	63	69	75	0.7	0.7	0.9	1.1	1.4	1.0	1.0	1.1	1.3	1.6
DG C2	31	54	68	80	86	92	95	97	15	28	38	49	57	68	74	81	0.7	0.7	0.9	1.1	1.3	0.9	1.0	1.1	1.2	1.4
DG C3	34	59	73	85	90	95	96	98	18	32	43	57	66	77	83	89	0.6	0.7	0.9	1.1	1.4	0.9	0.9	1.0	1.1	1.4
SS A1	34	52	63	73	79	87	90	94	21	32	40	49	56	64	70	76	0.7	0.8	1.0	1.1	1.3	0.9	1.0	1.2	1.3	1.5
SS A2	41	63	74	83	88	94	96	98	25	40	50	61	68	77	82	88	0.6	0.7	0.9	1.1	1.2	0.8	0.9	1.0	1.2	1.4
SS A3	32	50	60	70	77	84	88	92	20	30	38	46	52	60	66	72	0.7	0.8	1.0	1.2	1.3	1.0	1.1	1.2	1.4	1.5
SS A4	41	62	73	83	88	93	96	98	24	40	50	60	67	76	81	87	0.6	0.7	0.9	1.1	1.3	0.8	0.9	1.0	1.2	1.4
SS A5	32	49	59	69	75	83	87	91	21	31	38	45	51	59	64	71	0.7	0.8	1.0	1.2	1.3	1.0	1.1	1.2	1.4	1.6
SS A6	41	63	74	83	88	94	96	98	24	40	50	61	68	77	82	87	0.6	0.7	0.9	1.1	1.2	0.8	0.9	1.0	1.2	1.4
SS A7	29	44	53	63	69	77	82	87	19	28	33	40	45	53	58	64	0.8	0.9	1.0	1.2	1.4	1.1	1.2	1.3	1.5	1.7
SS A8	41	62	73	82	87	93	95	98	24	39	49	59	66	75	80	86	0.6	0.7	0.9	1.1	1.3	0.8	0.9	1.0	1.2	1.4
SS B1	29	47	57	67	74	82	86	91	17	28	35	43	49	58	63	70	0.7	0.8	1.0	1.2	1.3	1.0	1.1	1.2	1.4	1.5
SS B2	36	57	68	78	84	91	94	97	21	35	45	55	62	71	77	83	0.6	0.8	0.9	1.1	1.3	0.8	0.9	1.1	1.2	1.4
SS B3	28	45	55	65	72	80	85	90	17	26	33	41	47	55	60	67	0.7	0.8	1.0	1.2	1.4	1.0	1.1	1.3	1.4	1.6
SS B4	35	56	67	78	84	90	93	96	20	34	44	54	61	70	76	82	0.6	0.8	0.9	1.1	1.3	0.8	0.9	1.1	1.2	1.4
SS B5	26	41	51	60	67	75	80	86	17	25	31	38	43	50	55	62	0.8	0.9	1.1	1.2	1.4	1.1	1.2	1.4	1.5	1.7
SS B6	35	56	68	78	84	90	93	96	20	34	44	54	61	71	76	83	0.6	0.8	0.9	1.1	1.3	0.8	0.9	1.1	1.2	1.4
SS B7	24	38	47	56	63	71	76	82	16	23	28	34	39	46	50	57	0.8	0.9	1.1	1.3	1.4	1.2	1.3	1.4	1.6	1.7
SS B8	34	55	66	76	83	89	93	96	20	33	42	52	60	69	74	81	0.7	0.8	0.9	1.1	1.3	0.9	1.0	1.1	1.3	1.4
SS C1	30	49	61	72	78	86	90	94	15	27	35	44	51	60	66	73	0.7	0.8	1.0	1.1	1.3	0.9	1.0	1.2	1.3	1.5
SS C2	34	56	68	79	85	91	94	97	17	32	42	52	60	70	76	82	0.6	0.7	0.9	1.1	1.3	0.8	0.9	1.1	1.2	1.4
SS C3	24	39	49	60	67	76	81	86	14	22	28	35	41	49	54	61	0.8	0.9	1.0	1.2	1.4	1.1	1.2	1.3	1.5	1.6
SS C4	30	50	62	73	79	87	91	95	16	28	37	47	54	63	69	76	0.7	0.8	0.9	1.1	1.3	0.9	1.0	1.1	1.3	1.5
SS D1	41	62	73	82	88	93	96	98	24	39	48	58	65	74	79	85	0.6	0.7	0.9	1.1	1.3	0.8	0.9	1.1	1.2	1.4
SS D2	48	71	81	90	94	97	98	99	27	46	58	70	77	85	89	93	0.6	0.7	0.9	1.0	1.2	0.7	0.8	1.0	1.1	1.3
SS D3	37	56	67	77	82	89	92	95	23	35	43	52	59	68	73	79	0.7	0.8	0.9	1.1	1.3	0.9	1.0	1.1	1.3	1.5
SS D4	46	69	80	88	92	96	98	99	27	45	56	67	75	83	87	92	0.6	0.7	0.9	1.0	1.2	0.7	0.8	1.0	1.1	1.3
SS E1	36	57	68	78	84	90	93	96	20	33	42	52	59	68	74	80	0.6	0.8	0.9	1.1	1.3	0.9	1.0	1.1	1.3	1.4
SS E2	42	65	77	86	91	95	97	99	23	41	52	63	70	79	84	90	0.6	0.7	0.9	1.0	1.2	0.7	0.9	1.0	1.2	1.3
SS E3	30	47	58	68	74	82	87	91	18	28	35	43	49	58	63	70	0.7	0.8	1.0	1.2	1.3	1.0	1.1	1.2	1.4	1.6
SS E4	39	61	72	82	88	93	96	98	22	37	48	58	66	75	80	86	0.6	0.7	0.9	1.1	1.3	0.8	0.9	1.0	1.2	1.4

DEL RIO, TEXAS

TYPE	SOLAR SAVINGS FRACTION (%)																CONSERVATION FACTOR									
BASE TEMP:	55 F (494 DD)								65 F (1526 DD)								55 F					65 F				
LCR	200	100	70	50	40	30	25	20	200	100	70	50	40	30	25	20	200	100	50	30	20	200	100	50	30	20
WW A1	40	58	68	78	84	90	93	96	22	33	40	48	54	63	68	75	0.7	0.8	0.9	1.1	1.3	1.0	1.1	1.2	1.3	1.5
WW A2	48	70	80	89	93	97	98	99	24	40	50	60	67	76	81	87	0.6	0.7	0.9	1.0	1.2	0.8	0.9	1.0	1.2	1.4
WW A3	51	74	84	92	95	98	99	100	25	42	53	64	72	80	85	90	0.6	0.7	0.8	1.0	1.2	0.7	0.8	1.0	1.2	1.3
WW A4	53	76	86	93	96	98	99	100	25	44	55	67	74	83	87	92	0.5	0.7	0.8	1.0	1.2	0.7	0.8	1.0	1.1	1.3
WW A5	55	78	88	94	97	99	99	100	26	45	57	69	76	85	89	93	0.5	0.6	0.8	1.0	1.2	0.7	0.8	1.0	1.1	1.3
WW A6	56	79	88	95	97	99	100	100	26	46	59	70	77	86	90	94	0.5	0.6	0.8	1.0	1.2	0.7	0.8	1.0	1.1	1.3
WW B1	44	64	75	84	89	94	97	98	21	34	43	53	59	68	74	80	0.6	0.7	0.9	1.1	1.2	0.9	1.0	1.1	1.3	1.4
WW B2	53	76	86	93	96	98	99	100	25	44	56	67	75	83	88	92	0.5	0.7	0.8	1.0	1.2	0.7	0.8	1.0	1.1	1.3
WW B3	60	82	91	96	98	99	100	100	30	51	64	75	82	89	93	96	0.5	0.6	0.8	1.0	1.2	0.7	0.8	0.9	1.1	1.3
WW B4	59	82	91	96	98	99	100	100	29	51	64	76	83	90	93	96	0.5	0.6	0.8	1.0	1.2	0.7	0.8	0.9	1.1	1.3
WW B5	57	80	90	96	98	99	100	100	27	49	62	74	81	89	93	96	0.5	0.6	0.8	1.0	1.2	0.7	0.8	0.9	1.1	1.3
WW C1	61	83	92	97	98	100	100	100	31	53	66	77	83	90	94	97	0.5	0.6	0.8	1.0	1.2	0.7	0.8	0.9	1.1	1.3
WW C2	58	81	90	96	98	99	100	100	29	50	63	75	82	89	93	96	0.5	0.6	0.8	1.0	1.2	0.7	0.8	0.9	1.1	1.3
WW C3	64	87	94	98	99	100	100	100	32	57	71	82	88	94	96	98	0.5	0.6	0.8	1.0	1.2	0.6	0.7	0.9	1.1	1.2
WW C4	60	84	92	97	99	100	100	100	29	53	67	78	85	92	95	97	0.5	0.6	0.8	1.0	1.2	0.6	0.7	0.9	1.1	1.2
TW A1	39	57	67	77	83	89	93	96	22	32	39	47	53	62	67	74	0.7	0.8	0.9	1.1	1.3	1.0	1.1	1.2	1.4	1.5
TW A2	44	64	75	85	90	95	97	98	23	36	45	55	62	71	77	83	0.6	0.7	0.9	1.1	1.2	0.9	0.9	1.1	1.2	1.4
TW A3	46	67	78	87	92	96	98	99	23	38	48	58	65	75	80	86	0.6	0.7	0.9	1.0	1.2	0.8	0.9	1.0	1.2	1.4
TW A4	46	68	79	88	92	96	98	99	22	37	48	59	66	75	81	87	0.6	0.7	0.9	1.0	1.2	0.8	0.9	1.0	1.2	1.4
TW B1	39	57	68	77	83	90	93	96	21	32	39	48	54	63	68	75	0.7	0.8	0.9	1.1	1.3	0.9	1.0	1.2	1.3	1.5
TW B2	42	63	74	83	89	94	96	98	21	34	43	53	60	70	75	82	0.6	0.7	0.9	1.1	1.2	0.9	1.0	1.1	1.3	1.4
TW B3	42	63	74	84	89	94	97	98	20	34	44	54	61	70	76	82	0.6	0.7	0.9	1.1	1.2	0.8	0.9	1.1	1.2	1.4
TW B4	40	60	71	81	87	93	96	98	19	32	41	51	58	68	73	80	0.6	0.7	0.9	1.1	1.2	0.9	1.0	1.1	1.3	1.4
TW C1	38	56	67	77	83	90	93	96	20	31	38	47	54	63	68	75	0.7	0.8	0.9	1.1	1.3	0.9	1.0	1.2	1.3	1.5
TW C2	38	57	68	78	84	91	94	97	19	30	39	48	55	64	70	77	0.7	0.8	0.9	1.1	1.3	0.9	1.0	1.1	1.3	1.5
TW C3	35	54	65	76	82	89	92	96	18	29	37	46	52	61	67	74	0.7	0.8	0.9	1.1	1.3	0.9	1.0	1.2	1.3	1.5
TW C4	32	49	59	69	76	84	88	93	17	26	33	41	47	55	61	68	0.7	0.8	1.0	1.1	1.3	1.0	1.1	1.3	1.4	1.6
TW D1	37	57	68	78	84	91	94	97	18	29	37	46	53	62	67	74	0.7	0.8	0.9	1.1	1.3	0.9	1.0	1.2	1.3	1.5
TW D2	48	70	81	90	94	97	98	99	24	40	51	62	69	78	84	89	0.6	0.7	0.8	1.0	1.2	0.8	0.9	1.0	1.2	1.3
TW D3	53	75	85	93	96	98	99	100	26	44	56	67	74	83	87	92	0.5	0.7	0.8	1.0	1.2	0.7	0.8	1.0	1.1	1.3
TW D4	54	77	87	94	97	99	99	100	26	46	58	70	77	86	90	94	0.5	0.6	0.8	1.0	1.2	0.7	0.8	1.0	1.1	1.3
TW D5	52	76	86	93	96	99	99	100	25	45	57	69	77	85	89	94	0.5	0.6	0.8	1.0	1.2	0.7	0.8	1.0	1.1	1.3
TW E1	57	79	88	95	97	99	100	100	29	48	60	72	79	86	90	94	0.5	0.6	0.8	1.0	1.2	0.7	0.8	0.9	1.1	1.3
TW E2	55	78	88	94	97	99	100	100	28	47	59	71	78	86	90	94	0.5	0.6	0.8	1.0	1.2	0.7	0.8	0.9	1.1	1.3
TW E3	62	85	93	97	99	100	100	100	32	55	68	79	85	92	95	97	0.5	0.6	0.8	1.0	1.2	0.6	0.7	0.9	1.1	1.2
TW E4	59	82	91	96	98	99	100	100	29	51	64	76	83	90	94	97	0.5	0.6	0.8	1.0	1.2	0.7	0.8	0.9	1.1	1.3
TW F1	38	56	67	77	83	90	93	96	20	31	38	47	53	62	67	74	0.7	0.8	0.9	1.1	1.3	1.0	1.1	1.2	1.3	1.5
TW F2	42	62	73	83	89	94	96	98	20	33	43	53	60	69	75	81	0.6	0.7	0.9	1.1	1.2	0.9	1.0	1.1	1.3	1.4
TW F3	42	64	75	85	90	95	97	99	19	34	44	54	62	71	77	83	0.6	0.7	0.9	1.1	1.2	0.8	0.9	1.1	1.2	1.4
TW F4	40	62	74	83	89	94	97	98	17	32	42	52	60	70	75	82	0.6	0.7	0.9	1.1	1.2	0.8	0.9	1.1	1.2	1.4
TW G1	36	55	66	76	83	89	93	96	18	29	37	46	52	61	67	74	0.7	0.8	0.9	1.1	1.3	1.0	1.1	1.2	1.3	1.5
TW G2	38	58	70	80	86	92	95	97	17	30	39	49	56	65	71	78	0.6	0.7	0.9	1.1	1.3	0.9	1.0	1.1	1.3	1.4
TW G3	36	56	68	79	85	91	94	97	15	28	37	47	54	64	70	77	0.6	0.7	0.9	1.1	1.3	0.9	1.0	1.1	1.3	1.5
TW G4	31	50	62	73	80	88	91	95	12	24	32	41	48	58	64	72	0.7	0.8	0.9	1.1	1.3	1.0	1.1	1.2	1.3	1.5
TW H1	33	52	63	73	80	87	91	95	15	26	33	42	49	58	64	71	0.7	0.8	0.9	1.1	1.3	1.0	1.1	1.2	1.4	1.5
TW H2	30	49	60	71	78	86	90	94	12	23	31	40	47	56	62	70	0.7	0.8	1.0	1.1	1.3	1.0	1.1	1.2	1.4	1.5
TW H3	26	44	55	66	73	82	87	91	10	20	27	35	42	51	57	64	0.7	0.8	1.0	1.1	1.3	1.0	1.1	1.3	1.4	1.6
TW H4	20	36	46	56	64	73	79	85	7	15	21	28	34	42	47	55	0.8	0.9	1.1	1.2	1.4	1.2	1.3	1.4	1.6	1.7
TW I1	34	53	64	75	81	89	92	95	15	26	34	42	49	58	64	71	0.7	0.8	0.9	1.1	1.3	1.0	1.1	1.2	1.4	1.5
TW I2	45	67	78	88	92	96	98	99	20	36	47	58	66	75	81	87	0.6	0.7	0.9	1.0	1.2	0.8	0.9	1.0	1.2	1.4
TW I3	48	71	82	90	94	97	99	99	22	40	51	63	70	79	84	90	0.6	0.7	0.8	1.0	1.2	0.8	0.8	1.0	1.2	1.3
TW I4	50	74	85	92	96	98	99	100	23	43	55	67	74	83	88	92	0.5	0.7	0.8	1.0	1.2	0.7	0.8	1.0	1.1	1.3
TW I5	50	74	85	92	96	98	99	100	23	43	55	67	75	84	88	93	0.5	0.7	0.8	1.0	1.2	0.7	0.8	1.0	1.1	1.3
TW J1	54	77	87	93	96	99	99	100	26	46	58	69	76	85	89	93	0.5	0.6	0.8	1.0	1.2	0.7	0.8	1.0	1.1	1.3
TW J2	52	75	86	93	96	98	99	100	24	44	56	68	75	84	89	93	0.5	0.7	0.8	1.0	1.2	0.7	0.8	1.0	1.1	1.3
TW J3	58	82	91	96	98	99	100	100	28	51	64	76	83	90	94	97	0.5	0.6	0.8	1.0	1.2	0.6	0.7	0.9	1.1	1.3
TW J4	55	79	89	95	98	99	100	100	26	48	61	73	81	88	92	96	0.5	0.6	0.8	1.0	1.2	0.7	0.8	0.9	1.1	1.3
DG A1	43	64	74	83	88	93	95	97	19	33	41	50	56	64	68	73	0.6	0.7	0.9	1.1	1.3	0.9	1.0	1.2	1.4	1.7
DG A2	43	65	76	85	90	95	97	98	19	34	44	54	61	70	75	81	0.6	0.7	0.9	1.1	1.2	0.8	0.9	1.1	1.3	1.5
DG A3	47	71	82	89	92	95	96	96	22	39	51	63	71	79	84	88	0.6	0.7	0.9	1.2	2.1	0.8	0.8	1.0	1.2	1.4
DG B1	46	70	81	89	93	96	97	98	19	34	45	56	63	72	76	81	0.6	0.7	0.9	1.1	1.4	0.8	0.9	1.1	1.3	1.5
DG B2	45	71	82	91	94	97	98	99	20	36	47	60	68	77	82	87	0.6	0.7	0.8	1.0	1.2	0.8	0.9	1.0	1.2	1.4
DG B3	48	75	86	93	95	97	98	98	22	40	53	67	75	84	88	92	0.5	0.6	0.8	1.2	2.2	0.8	0.8	0.9	1.1	1.4
DG C1	51	78	88	94	96	97	98	98	22	40	52	64	72	80	85	88	0.5	0.6	0.9	1.2	2.4	0.8	0.8	1.0	1.2	1.5
DG C2	50	77	88	94	97	99	99	100	22	40	53	66	75	83	88	92	0.5	0.6	0.8	1.0	1.2	0.8	0.8	0.9	1.1	1.3
DG C3	53	81	91	96	97	98	98	98	25	45	59	73	81	89	92	95	0.5	0.6	0.8	1.3	-	0.7	0.8	0.9	1.1	1.3
SS A1	51	71	81	89	93	97	98	99	28	43	53	63	70	78	83	88	0.6	0.7	0.9	1.0	1.2	0.8	0.9	1.0	1.2	1.4
SS A2	59	80	89	95	97	99	100	100	34	53	64	75	81	88	92	95	0.5	0.6	0.8	1.0	1.2	0.7	0.8	0.9	1.1	1.3
SS A3	49	68	79	87	92	96	98	99	27	41	51	60	67	75	80	86	0.6	0.7	0.9	1.0	1.2	0.8	0.9	1.1	1.2	1.4
SS A4	59	80	89	95	97	99	100	100	34	53	64	74	81	88	91	95	0.5	0.6	0.8	1.0	1.2	0.7	0.8	0.9	1.1	1.3
SS A5	47	67	77	86	90	95	97	99	27	41	49	59	65	74	79	84	0.6	0.7	0.9	1.1	1.2	0.8	0.9	1.1	1.2	1.4
SS A6	59	81	89	95	97	99	100	100	34	53	64	75	81	88	92	95	0.5	0.6	0.8	1.0	1.2	0.7	0.8	0.9	1.1	1.3
SS A7	43	61	72	81	86	92	95	97	25	37	45	53	59	68	73	79	0.6	0.8	0.9	1.1	1.3	0.9	1.0	1.1	1.3	1.5
SS A8	59	80	89	95	97	99	100	100	33	52	63	73	80	87	91	94	0.5	0.6	0.8	1.0	1.2	0.7	0.8	1.0	1.1	1.3
SS B1	45	65	76	85	90	95	97	98	24	38	47	57	64	72	78	84	0.6	0.7	0.9	1.1	1.2	0.8	0.9	1.1	1.2	1.4
SS B2	53	75	85	92	95	98	99	100	29	47	58	69	76	84	88	92	0.6	0.7	0.8	1.0	1.2	0.7	0.8	1.0	1.1	1.3
SS B3	44	63	74	83	89	94	96	98	23	36	45	55	62	70	76	82	0.6	0.7	0.9	1.1	1.2	0.9	0.9	1.1	1.3	1.4
SS B4	53	74	84	92	95	98	99	100	28	46	57	68	75	83	87	92	0.6	0.7	0.8	1.0	1.2	0.7	0.8	1.0	1.2	1.3
SS B5	40	58	69	79	84	91	94	96	22	34	41	50	57	65	70	77	0.7	0.8	0.9	1.1	1.3	0.9	1.0	1.2	1.3	1.5
SS B6	53	75	85	92	95	98	99	100	28	46	57	68	75	83	88	92	0.6	0.7	0.8	1.0	1.2	0.7	0.8	1.0	1.1	1.3
SS B7	37	55	65	75	81	88	91	95	21	31	38	46	52	60	66	72	0.7	0.8	1.0	1.1	1.3	1.0	1.1	1.2	1.4	1.5
SS B8	52	73	84	91	95	98	99	100	28	45	56	67	74	82	86	91	0.6	0.7	0.8	1.0	1.2	0.7	0.8	1.0	1.2	1.3
SS C1	47	69	80	88	93	97	98	99	22	38	48	59	66	76	81	87	0.6	0.7	0.9	1.0	1.2	0.8	0.9	1.0	1.2	1.4
SS C2	52	75	85	93	96	98	99	100	25	44	56	67	74	83	88	92	0.5	0.7	0.8	1.0	1.2	0.7	0.8	1.0	1.1	1.3
SS C3	38	57	68	79	85	91	94	97	19	31	39	48	55	64	69	76	0.6	0.8	0.9	1.1	1.3	0.9	1.0	1.2	1.3	1.5
SS C4	47	69	80	89	93	97	98	99	23	39	50	61	68	77	83	88	0.6	0.7	0.9	1.0	1.2	0.8	0.9	1.0	1.2	1.4
SS D1	59	80	89	95	97	99	100	100	33	52	63	73	80	87	91	94	0.5	0.6	0.8	1.0	1.2	0.7	0.8	1.0	1.1	1.3
SS D2	67	87	94	98	99	100	100	100	39	61	73	83	88	94	96	98	0.5	0.6	0.8	1.0	1.2	0.6	0.7	0.9	1.1	1.2
SS D3	54	74	84	91	95	98	99	100	30	47	57	67	74	81	86	91	0.6	0.7	0.8	1.0	1.2	0.7	0.8	1.0	1.2	1.3
SS D4	65	86	93	97	99	100	100	100	37	59	71	81	86	92	95	97	0.5	0.6	0.8	1.0	1.2	0.6	0.7	0.9	1.1	1.3
SS E1	55	76	86	93	96	98	99	100	29	46	57	68	74	82	87	91	0.5	0.7	0.8	1.0	1.2	0.7	0.8	1.0	1.2	1.3
SS E2	62	83	92	96	98	99	100	100	33	55	67	77	84	90	93	96	0.5	0.6	0.8	1.0	1.2	0.6	0.8	0.9	1.1	1.3
SS E3	46	66	76	85	90	95	97	99	25	39	48	57	64	73	78	84	0.6	0.7	0.9	1.1	1.2	0.8	0.9	1.1	1.2	1.4
SS E4	57	79	88	95	97	99	100	100	31	50	62	73	79	87	91	94	0.5	0.6	0.8	1.0	1.2	0.7	0.8	0.9	1.1	1.3

EL PASO, TEXAS

TYPE	SOLAR SAVINGS FRACTION (%) — BASE TEMP: 55 F (1154 DD)								65 F (2685 DD)								CONSERVATION FACTOR — 55 F					65 F				
LCR	200	100	70	50	40	30	25	20	200	100	70	50	40	30	25	20	200	100	50	30	20	200	100	50	30	20
WW A1	31	47	57	67	73	81	86	90	20	30	37	45	50	59	64	71	0.8	0.9	1.0	1.2	1.3	1.0	1.1	1.2	1.4	1.5
WW A2	37	58	69	79	85	91	94	97	22	36	46	56	63	73	78	84	0.6	0.7	0.9	1.1	1.3	0.8	0.9	1.1	1.2	1.4
WW A3	39	61	73	83	88	94	96	98	22	38	49	60	68	77	83	88	0.6	0.7	0.9	1.1	1.2	0.8	0.9	1.0	1.2	1.3
WW A4	40	63	75	85	90	95	97	99	22	40	51	62	70	79	85	90	0.6	0.7	0.9	1.0	1.2	0.8	0.9	1.0	1.2	1.3
WW A5	42	66	77	87	91	96	98	99	22	41	53	65	73	82	87	92	0.6	0.7	0.9	1.0	1.2	0.7	0.8	1.0	1.1	1.3
WW A6	43	67	78	88	92	96	98	99	23	42	54	66	74	83	88	92	0.6	0.7	0.9	1.0	1.2	0.7	0.8	1.0	1.1	1.3
WW B1	32	51	62	73	79	87	90	94	19	31	39	48	55	64	70	77	0.7	0.8	1.0	1.1	1.3	0.9	1.0	1.2	1.3	1.5
WW B2	40	63	75	85	90	95	97	99	22	40	52	63	71	80	85	91	0.6	0.7	0.9	1.0	1.2	0.7	0.8	1.0	1.2	1.3
WW B3	47	71	82	90	94	98	99	100	27	47	60	71	79	87	91	95	0.6	0.7	0.8	1.0	1.2	0.7	0.8	0.9	1.1	1.3
WW B4	46	71	82	91	94	98	99	100	25	47	60	72	80	88	92	95	0.6	0.7	0.8	1.0	1.2	0.7	0.8	0.9	1.1	1.3
WW B5	44	69	81	90	94	97	99	99	23	45	58	71	78	87	91	95	0.6	0.7	0.8	1.0	1.2	0.7	0.8	0.9	1.1	1.3
WW C1	48	72	83	91	95	98	99	100	28	49	61	73	80	88	92	96	0.5	0.7	0.8	1.0	1.2	0.7	0.8	0.9	1.1	1.3
WW C2	46	70	81	90	94	97	99	100	25	46	59	71	79	87	91	95	0.6	0.7	0.8	1.0	1.2	0.7	0.8	0.9	1.1	1.3
WW C3	51	76	87	94	97	99	99	100	29	53	67	79	85	92	95	98	0.5	0.6	0.8	1.0	1.2	0.6	0.7	0.9	1.1	1.2
WW C4	47	73	84	92	95	98	99	100	25	49	63	75	82	90	94	97	0.5	0.7	0.8	1.0	1.2	0.7	0.8	0.9	1.1	1.3
TW A1	30	45	55	65	72	80	85	89	20	29	36	44	49	58	63	70	0.8	0.9	1.0	1.2	1.4	1.0	1.1	1.3	1.4	1.5
TW A2	33	52	64	74	81	88	91	95	20	33	41	51	58	67	73	80	0.7	0.8	0.9	1.1	1.3	0.9	1.0	1.1	1.3	1.4
TW A3	35	55	67	77	83	90	93	96	20	34	44	54	61	71	77	83	0.7	0.8	0.9	1.1	1.3	0.8	0.9	1.1	1.2	1.4
TW A4	34	55	67	78	84	91	94	97	19	34	44	54	62	72	77	84	0.6	0.7	0.9	1.1	1.3	0.8	0.9	1.1	1.2	1.4
TW B1	30	46	56	66	73	81	86	90	19	29	36	44	50	59	64	71	0.8	0.8	1.0	1.2	1.3	1.0	1.1	1.2	1.4	1.5
TW B2	31	50	62	73	79	87	91	94	18	31	40	49	56	66	72	79	0.7	0.8	1.0	1.1	1.3	0.9	1.0	1.1	1.3	1.4
TW B3	31	51	62	73	80	87	91	95	18	31	40	50	57	67	73	79	0.7	0.8	0.9	1.1	1.3	0.9	1.0	1.1	1.3	1.4
TW B4	29	48	59	70	77	85	89	93	16	29	38	47	54	64	70	77	0.7	0.8	1.0	1.1	1.3	0.9	1.0	1.1	1.3	1.5
TW C1	28	45	55	66	72	81	85	90	17	28	35	44	50	59	64	71	0.8	0.8	1.0	1.2	1.3	1.0	1.1	1.2	1.4	1.5
TW C2	28	45	56	67	74	82	87	91	16	27	35	44	51	60	66	73	0.7	0.8	1.0	1.2	1.3	1.0	1.1	1.2	1.3	1.5
TW C3	26	43	53	64	71	80	85	90	15	26	34	42	49	57	63	71	0.8	0.9	1.0	1.2	1.3	1.0	1.1	1.2	1.4	1.5
TW C4	23	38	48	58	65	74	79	85	15	23	30	38	43	52	57	64	0.8	0.9	1.1	1.2	1.4	1.1	1.2	1.3	1.4	1.6
TW D1	27	44	55	66	73	81	86	91	15	26	33	42	48	57	63	70	0.7	0.8	1.0	1.2	1.3	1.0	1.1	1.2	1.4	1.5
TW D2	37	58	70	80	86	92	95	97	21	36	47	58	66	75	81	87	0.6	0.7	0.9	1.1	1.2	0.8	0.9	1.0	1.2	1.4
TW D3	40	63	75	84	90	95	97	99	23	40	51	63	71	80	85	90	0.6	0.7	0.9	1.1	1.2	0.8	0.8	1.0	1.2	1.3
TW D4	42	65	77	87	91	96	98	99	23	42	54	66	74	83	88	93	0.6	0.7	0.9	1.0	1.2	0.7	0.8	1.0	1.1	1.3
TW D5	40	64	76	86	91	96	97	99	22	41	53	65	73	32	87	92	0.6	0.7	0.9	1.0	1.2	0.7	0.8	1.0	1.1	1.3
TW E1	44	67	79	88	92	96	98	99	26	44	56	68	75	84	88	93	0.6	0.7	0.9	1.0	1.2	0.7	0.8	1.0	1.1	1.3
TW E2	43	66	78	87	92	96	98	99	24	43	55	67	75	84	88	93	0.6	0.7	0.9	1.0	1.2	0.7	0.8	1.0	1.1	1.3
TW E3	49	74	85	92	96	98	99	100	28	51	64	76	83	90	94	97	0.5	0.7	0.8	1.0	1.2	0.7	0.8	0.9	1.1	1.3
TW E4	46	71	82	90	94	98	99	100	26	47	60	73	80	88	92	96	0.6	0.7	0.8	1.0	1.2	0.7	0.8	0.9	1.1	1.3
TW F1	29	45	55	65	72	80	85	90	18	28	35	43	49	58	63	70	0.8	0.9	1.0	1.2	1.3	1.0	1.1	1.2	1.4	1.5
TW F2	31	50	61	72	79	86	90	94	18	30	39	49	56	65	71	78	0.7	0.8	1.0	1.1	1.3	0.9	1.0	1.1	1.3	1.4
TW F3	31	51	63	74	81	88	92	95	16	30	40	50	57	67	73	80	0.7	0.8	0.9	1.1	1.3	0.9	1.0	1.1	1.3	1.4
TW F4	28	49	61	72	79	87	91	95	14	28	38	48	56	65	72	79	0.7	0.8	0.9	1.1	1.3	0.9	1.0	1.1	1.3	1.4
TW G1	27	43	54	65	72	80	85	90	16	26	34	42	48	57	63	70	0.8	0.9	1.0	1.2	1.3	1.0	1.1	1.2	1.4	1.5
TW G2	27	46	57	68	75	84	88	92	14	26	35	45	52	61	67	75	0.7	0.8	1.0	1.1	1.3	0.9	1.0	1.2	1.3	1.5
TW G3	25	44	56	67	74	83	87	92	12	25	33	43	50	60	66	73	0.7	0.8	1.0	1.1	1.3	0.9	1.0	1.2	1.3	1.5
TW G4	21	38	49	61	68	77	83	88	10	21	28	37	44	54	60	68	0.8	0.9	1.0	1.2	1.4	1.0	1.1	1.2	1.4	1.5
TW H1	23	40	50	61	68	77	83	88	13	23	30	38	45	54	60	67	0.8	0.9	1.0	1.2	1.4	1.0	1.1	1.3	1.4	1.5
TW H2	20	37	48	59	66	76	81	87	10	20	28	36	43	52	58	65	0.8	0.9	1.0	1.2	1.4	1.0	1.1	1.3	1.4	1.6
TW H3	17	32	43	53	61	71	76	83	8	17	24	32	38	47	53	60	0.8	0.9	1.1	1.2	1.4	1.1	1.2	1.3	1.5	1.6
TW H4	13	25	34	44	51	61	67	74	5	13	18	25	30	38	43	50	0.9	1.0	1.2	1.3	1.5	1.3	1.3	1.5	1.6	1.7
TW I1	24	41	52	62	70	78	83	88	13	23	30	38	45	53	59	66	0.8	0.9	1.0	1.2	1.4	1.0	1.1	1.3	1.4	1.6
TW I2	33	55	67	78	84	91	94	97	17	33	43	54	62	72	78	84	0.6	0.7	0.9	1.1	1.3	0.8	0.9	1.1	1.2	1.4
TW I3	36	58	70	81	87	93	95	98	19	36	47	58	66	76	81	87	0.6	0.7	0.9	1.1	1.2	0.8	0.9	1.0	1.2	1.4
TW I4	38	62	74	84	90	95	97	99	20	39	51	63	71	80	85	91	0.6	0.7	0.9	1.0	1.2	0.7	0.8	1.0	1.1	1.3
TW I5	37	62	74	84	90	95	97	99	19	39	51	63	71	81	86	91	0.6	0.7	0.9	1.0	1.2	0.7	0.8	1.0	1.1	1.3
TW J1	41	65	76	86	91	96	97	99	23	42	53	65	73	82	87	92	0.6	0.7	0.9	1.0	1.2	0.7	0.8	1.0	1.1	1.3
TW J2	39	63	75	85	90	95	97	99	21	40	52	64	72	81	86	91	0.6	0.7	0.9	1.0	1.2	0.7	0.8	1.0	1.1	1.3
TW J3	45	71	82	90	94	98	99	100	25	47	60	73	80	88	92	96	0.6	0.7	0.8	1.0	1.2	0.7	0.8	0.9	1.1	1.3
TW J4	42	68	80	89	93	97	98	99	22	44	57	70	78	86	91	95	0.6	0.7	0.8	1.0	1.2	0.7	0.8	0.9	1.1	1.3
DG A1	31	51	62	71	77	84	87	91	16	29	37	46	52	59	64	69	0.7	0.8	1.0	1.2	1.4	0.9	1.0	1.2	1.4	1.7
DG A2	31	52	64	74	81	87	91	94	17	30	40	50	57	66	71	77	0.7	0.8	0.9	1.1	1.3	0.9	1.0	1.1	1.3	1.5
DG A3	35	59	71	81	86	91	93	95	19	35	47	59	67	76	81	86	0.6	0.7	0.9	1.1	1.5	0.8	0.9	1.0	1.2	1.4
DG B1	33	56	68	78	84	90	92	95	16	30	40	51	58	67	72	78	0.6	0.7	0.9	1.1	1.4	0.9	1.0	1.1	1.3	1.6
DG B2	33	56	70	81	87	92	95	97	17	31	42	55	63	73	79	84	0.6	0.7	0.9	1.1	1.3	0.9	0.9	1.0	1.2	1.4
DG B3	36	61	75	85	90	94	96	97	19	36	48	62	71	81	86	91	0.6	0.7	0.9	1.1	1.5	0.8	0.9	1.0	1.1	1.4
DG C1	38	63	76	86	90	94	96	97	19	35	47	59	68	77	82	86	0.6	0.7	0.9	1.1	1.5	0.8	0.9	1.0	1.2	1.5
DG C2	37	62	76	86	91	96	97	99	20	36	48	62	70	80	85	90	0.6	0.7	0.9	1.1	1.3	0.8	0.9	1.0	1.2	1.4
DG C3	40	67	81	90	94	97	98	98	22	40	55	69	78	87	91	94	0.6	0.7	0.8	1.1	1.7	0.8	0.8	0.9	1.1	1.3
SS A1	39	58	69	79	84	91	94	96	25	39	49	58	65	74	79	85	0.6	0.8	0.9	1.1	1.3	0.8	0.9	1.1	1.2	1.4
SS A2	47	68	79	88	92	96	98	99	30	48	59	70	77	85	89	93	0.6	0.7	0.9	1.0	1.2	0.7	0.8	1.0	1.1	1.3
SS A3	37	55	66	76	82	88	92	95	24	37	46	55	62	70	76	82	0.7	0.8	0.9	1.1	1.3	0.8	1.0	1.1	1.3	1.4
SS A4	46	68	78	87	92	96	98	99	30	48	59	69	76	84	88	93	0.6	0.7	0.9	1.0	1.2	0.7	0.8	1.0	1.1	1.3
SS A5	36	54	65	74	80	87	91	94	25	37	45	54	60	69	74	80	0.7	0.8	1.0	1.1	1.3	0.9	1.0	1.1	1.3	1.4
SS A6	47	68	79	88	92	96	98	99	30	48	59	70	77	85	89	93	0.6	0.7	0.9	1.0	1.2	0.7	0.8	1.0	1.1	1.3
SS A7	33	49	59	68	75	82	86	91	22	33	40	48	54	62	67	74	0.7	0.8	1.0	1.2	1.3	0.9	1.1	1.2	1.4	1.5
SS A8	46	67	78	86	91	95	97	99	30	47	58	68	75	83	88	92	0.6	0.7	0.9	1.0	1.2	0.7	0.8	1.0	1.1	1.3
SS B1	33	52	63	73	80	87	91	94	21	34	43	52	59	68	73	80	0.7	0.8	1.0	1.1	1.3	0.9	1.0	1.1	1.3	1.4
SS B2	41	62	74	83	88	94	96	98	26	42	53	64	71	80	85	90	0.6	0.7	0.9	1.1	1.2	0.8	0.9	1.0	1.2	1.3
SS B3	32	50	61	71	78	85	89	93	20	32	41	50	56	65	71	77	0.7	0.8	1.0	1.1	1.3	0.9	1.0	1.2	1.3	1.5
SS B4	40	61	73	82	88	93	96	98	25	42	52	63	70	79	84	89	0.6	0.7	0.9	1.1	1.2	0.8	0.9	1.0	1.2	1.3
SS B5	30	46	56	66	73	81	85	90	20	30	37	46	52	60	65	72	0.7	0.8	1.0	1.2	1.4	1.0	1.1	1.2	1.4	1.5
SS B6	40	62	73	83	88	94	96	98	25	42	52	63	71	79	84	90	0.6	0.7	0.9	1.1	1.2	0.8	0.9	1.0	1.2	1.3
SS B7	27	42	52	62	68	76	81	87	18	28	34	42	47	55	60	67	0.8	0.9	1.1	1.2	1.4	1.0	1.1	1.3	1.4	1.6
SS B8	39	60	72	81	87	93	95	98	24	40	51	62	69	78	83	88	0.6	0.7	0.9	1.1	1.2	0.8	0.9	1.0	1.2	1.4
SS C1	35	56	68	79	85	91	94	97	20	34	44	55	62	72	77	84	0.6	0.7	0.9	1.1	1.3	0.8	0.9	1.1	1.2	1.4
SS C2	40	63	75	85	90	95	97	99	22	40	51	63	71	80	85	90	0.6	0.7	0.9	1.0	1.2	0.8	0.8	1.0	1.2	1.3
SS C3	28	46	56	67	74	82	87	91	17	28	35	44	50	59	65	72	0.7	0.8	1.0	1.2	1.3	1.0	1.1	1.2	1.3	1.5
SS C4	35	57	69	79	85	92	95	97	20	35	46	57	64	74	80	86	0.6	0.7	0.9	1.1	1.3	0.8	0.9	1.0	1.2	1.4
SS D1	46	68	78	87	92	96	98	99	30	47	58	68	75	83	88	92	0.6	0.7	0.9	1.0	1.2	0.7	0.8	1.0	1.1	1.3
SS D2	54	76	86	93	96	99	99	100	34	56	68	79	85	91	94	97	0.5	0.7	0.8	1.0	1.2	0.6	0.7	0.9	1.1	1.3
SS D3	42	62	72	82	87	93	95	98	27	42	52	62	69	77	82	88	0.6	0.7	0.9	1.1	1.3	0.8	0.9	1.0	1.2	1.4
SS D4	52	74	85	92	95	98	99	100	33	54	66	77	83	90	93	96	0.5	0.7	0.8	1.0	1.2	0.7	0.8	0.9	1.1	1.3
SS E1	42	63	74	83	88	94	96	98	25	41	52	62	69	78	83	89	0.6	0.7	0.9	1.1	1.2	0.8	0.9	1.0	1.2	1.4
SS E2	48	71	82	90	94	97	99	99	29	49	61	73	80	87	91	95	0.6	0.7	0.8	1.0	1.2	0.7	0.8	0.9	1.1	1.3
SS E3	34	53	64	74	80	87	91	94	22	35	43	52	59	68	73	80	0.7	0.8	1.0	1.1	1.3	0.9	1.0	1.1	1.3	1.4
SS E4	44	67	78	87	91	96	98	99	27	45	57	68	75	83	88	92	0.6	0.7	0.9	1.0	1.2	0.7	0.8	1.0	1.1	1.3

FORT WORTH, TEXAS

SOLAR SAVINGS FRACTION (%)

TYPE	BASE TEMP: 55 F (1012 DD)								65 F (2390 DD)							
LCR	200	100	70	50	40	30	25	20	200	100	70	50	40	30	25	20
WW A1	26	39	48	57	64	72	77	83	17	24	29	35	40	47	51	58
WW A2	30	48	59	70	77	85	89	93	17	28	35	44	51	59	65	72
WW A3	32	52	63	74	81	88	92	95	16	29	38	48	55	64	70	77
WW A4	33	54	66	76	83	90	93	96	16	30	39	49	57	66	72	79
WW A5	34	56	68	79	85	91	94	97	16	31	41	51	59	69	75	82
WW A6	34	57	69	80	86	92	95	97	15	31	41	53	60	70	76	83
WW B1	26	42	53	63	70	78	83	88	14	23	29	36	42	50	55	62
WW B2	32	54	66	77	83	90	93	96	16	30	40	51	58	68	74	81
WW B3	38	61	73	83	89	94	96	98	19	36	47	59	67	76	82	87
WW B4	37	61	74	84	89	95	97	98	17	36	48	60	68	78	83	89
WW B5	35	59	72	83	88	94	96	98	16	34	46	59	67	77	82	88
WW C1	40	63	75	85	90	95	97	99	20	38	49	61	69	78	83	89
WW C2	37	60	73	83	89	94	96	98	18	36	47	59	67	77	82	88
WW C3	41	67	79	88	93	97	98	99	20	42	55	67	75	84	89	93
WW C4	38	63	76	86	91	96	98	99	17	38	51	64	72	81	86	92
TW A1	26	38	47	56	63	71	76	82	17	24	29	35	39	46	51	57
TW A2	28	44	54	65	72	80	85	90	16	25	32	40	46	54	60	67
TW A3	28	46	57	68	75	83	87	92	15	26	34	42	49	58	64	71
TW A4	27	46	58	69	76	84	88	92	14	25	33	43	49	59	65	72
TW B1	25	38	47	57	64	72	78	83	16	23	29	35	40	47	52	58
TW B2	26	42	53	63	70	79	84	89	14	24	31	39	45	53	59	66
TW B3	25	42	53	64	71	80	84	89	13	23	31	39	45	54	60	67
TW B4	24	40	50	61	68	77	82	88	12	22	29	37	43	51	57	64
TW C1	24	37	47	56	63	72	77	83	14	22	28	34	39	47	52	59
TW C2	23	37	47	58	65	74	79	85	13	21	28	35	40	48	54	61
TW C3	22	35	45	55	62	71	76	83	12	20	26	33	38	46	51	58
TW C4	20	32	40	49	56	65	70	77	12	19	24	30	34	41	46	53
TW D1	22	36	46	56	63	72	77	83	11	19	25	31	36	44	49	55
TW D2	30	49	61	72	78	86	90	94	15	28	37	46	53	63	69	76
TW D3	33	54	65	76	83	90	93	96	17	31	40	51	58	68	74	80
TW D4	34	56	68	79	85	92	95	97	17	32	43	54	62	72	78	84
TW D5	32	54	67	78	84	91	94	97	16	32	42	54	62	72	78	84
TW E1	36	58	70	80	86	92	95	97	19	34	45	56	63	73	78	85
TW E2	35	57	69	80	86	92	95	97	18	34	44	55	63	73	78	85
TW E3	40	65	77	86	91	96	98	99	21	40	52	64	72	81	86	91
TW E4	37	61	74	84	89	95	97	98	18	37	49	61	69	79	84	90
TW F1	24	38	47	56	63	71	77	83	15	22	27	34	39	46	51	57
TW F2	25	41	52	63	70	78	83	89	13	23	30	38	44	52	58	65
TW F3	25	42	53	64	71	80	85	90	12	22	30	38	45	54	60	67
TW F4	22	40	51	63	70	79	84	89	10	20	28	37	43	52	58	66
TW G1	22	36	45	55	62	71	76	83	13	20	26	32	37	45	50	57
TW G2	22	37	48	59	66	75	80	86	10	20	26	34	40	48	54	61
TW G3	20	36	46	57	64	74	79	85	9	18	25	32	38	47	53	60
TW G4	16	30	40	51	58	68	74	81	6	15	21	28	33	42	47	54
TW H1	19	32	42	52	59	68	74	80	9	17	23	29	34	42	47	54
TW H2	16	30	39	49	57	66	72	79	7	14	20	27	32	40	45	52
TW H3	13	26	34	44	51	61	67	74	5	12	17	23	28	36	41	47
TW H4	10	20	27	36	42	51	57	65	3	9	13	18	22	29	33	39
TW I1	19	33	42	52	59	69	74	80	9	17	22	28	33	40	45	51
TW I2	26	45	57	68	76	84	88	93	12	24	33	43	50	59	65	73
TW I3	28	49	61	72	79	87	91	94	10	27	36	46	54	63	70	77
TW I4	00	52	65	76	83	90	93	96	14	29	40	51	59	69	75	82
TW I5	30	52	65	76	83	90	93	96	13	29	40	51	59	70	76	83
TW J1	33	55	67	78	84	91	94	97	16	32	42	53	61	70	76	83
TW J2	32	54	66	77	84	90	94	97	15	30	41	52	60	70	76	83
TW J3	37	61	74	84	89	95	97	99	17	36	49	61	69	79	84	90
TW J4	34	58	71	82	88	93	96	98	15	34	46	58	66	76	82	88
DG A1	25	42	52	61	67	74	78	83	12	21	27	34	39	45	49	50
DG A2	25	43	54	65	71	79	81	00	12	23	30	39	45	53	58	64
DG A3	28	49	61	73	79	86	89	92	15	27	36	47	55	65	70	77
DG B1	25	45	57	68	75	82	85	89	12	21	28	36	43	51	56	62
DG B2	26	46	59	71	78	86	89	93	13	23	31	41	49	59	65	72
DG B3	28	50	64	77	83	90	93	95	15	27	37	49	57	69	75	82
DG C1	29	51	64	76	82	88	91	94	14	25	33	43	51	60	66	73
DG C2	29	51	65	77	84	90	93	96	15	27	36	47	55	66	72	79
DG C3	32	56	71	83	89	94	96	97	17	31	42	55	64	76	82	88
SS A1	33	51	61	71	78	85	89	93	20	32	39	48	54	63	68	75
SS A2	40	61	72	81	87	93	95	97	24	39	49	60	67	76	81	86
SS A3	31	48	58	68	75	82	87	91	19	30	37	45	51	59	64	71
SS A4	40	60	71	81	86	92	95	97	24	39	49	59	66	75	80	86
SS A5	31	47	57	67	73	81	85	90	20	30	37	44	50	58	63	69
SS A6	40	61	72	81	87	92	95	97	24	39	49	59	66	75	80	86
SS A7	28	42	52	61	67	75	80	85	19	27	33	39	44	51	56	62
SS A8	39	60	71	80	86	92	94	97	23	38	48	58	65	74	79	85
SS B1	28	45	55	65	72	80	85	90	17	27	34	42	48	56	62	68
SS B2	35	55	66	76	83	89	93	96	20	34	44	54	61	70	76	82
SS B3	27	43	53	63	70	78	83	88	16	26	32	40	45	53	59	65
SS B4	34	54	65	76	82	89	92	95	20	34	43	53	60	69	75	81
SS B5	26	40	49	58	65	73	78	84	16	24	30	37	42	49	54	60
SS B6	34	54	65	76	82	89	92	96	20	33	43	53	60	69	75	81
SS B7	24	36	45	54	60	69	74	80	15	22	27	33	38	44	49	55
SS B8	33	53	64	74	81	88	91	95	19	32	41	51	58	67	73	80
SS C1	29	47	58	69	76	84	88	93	14	26	34	43	49	58	64	71
SS C2	32	54	66	76	83	90	93	96	16	30	40	51	58	68	74	81
SS C3	23	38	47	58	64	73	79	84	13	21	27	34	39	47	52	59
SS C4	29	48	59	70	77	85	89	93	15	27	36	45	52	62	68	75
SS D1	39	60	71	80	86	92	95	97	23	37	47	57	64	73	78	84
SS D2	46	69	80	88	92	96	98	99	26	45	57	68	75	84	88	92
SS D3	36	54	65	75	81	87	91	94	22	34	42	51	57	66	71	78
SS D4	44	67	78	87	91	96	97	99	26	44	55	66	73	82	86	91
SS E1	35	55	66	76	82	89	92	95	19	32	41	51	57	66	72	78
SS E2	41	63	75	84	89	94	97	98	22	39	50	62	69	78	83	88
SS E3	29	46	56	66	72	80	85	90	18	28	34	42	48	56	61	68
SS E4	37	59	70	80	86	92	95	97	21	36	46	57	64	73	79	85

CONSERVATION FACTOR

TYPE	55 F					65 F				
	200	100	50	30	20	200	100	50	30	20
WW A1	0.8	0.9	1.1	1.3	1.4	1.2	1.3	1.4	1.6	1.7
WW A2	0.7	0.8	1.0	1.1	1.3	1.0	1.1	1.2	1.4	1.5
WW A3	0.7	0.8	0.9	1.1	1.3	0.9	1.0	1.1	1.3	1.5
WW A4	0.7	0.8	0.9	1.1	1.3	0.9	1.0	1.1	1.3	1.4
WW A5	0.6	0.7	0.9	1.1	1.3	0.8	0.9	1.1	1.2	1.4
WW A6	0.6	0.7	0.9	1.1	1.3	0.8	0.9	1.1	1.2	1.4
WW B1	0.8	0.9	1.0	1.2	1.4	1.1	1.2	1.3	1.5	1.7
WW B2	0.6	0.8	0.9	1.1	1.3	0.9	C.9	1.1	1.2	1.4
WW B3	0.6	0.7	0.9	1.1	1.2	0.8	0.9	1.0	1.2	1.4
WW B4	0.6	0.7	0.9	1.1	1.2	0.8	0.9	1.0	1.2	1.3
WW B5	0.6	0.7	0.9	1.1	1.2	0.8	0.9	1.0	1.2	1.3
WW C1	0.6	0.7	0.9	1.1	1.2	0.8	0.9	1.0	1.2	1.3
WW C2	0.6	0.7	0.9	1.1	1.2	0.8	0.9	1.0	1.2	1.3
WW C3	0.6	0.7	0.8	1.0	1.2	0.7	0.8	1.0	1.1	1.3
WW C4	0.6	0.7	0.9	1.0	1.2	0.7	0.8	1.0	1.1	1.3
TW A1	0.9	1.0	1.1	1.3	1.4	1.2	1.3	1.4	1.6	1.7
TW A2	0.8	0.9	1.0	1.2	1.3	1.0	1.1	1.3	1.4	1.6
TW A3	0.7	0.8	1.0	1.1	1.3	1.0	1.1	1.2	1.4	1.5
TW A4	0.7	0.8	1.0	1.1	1.3	1.0	1.1	1.2	1.3	1.5
TW B1	0.8	0.9	1.1	1.2	1.4	1.2	1.3	1.4	1.6	1.7
TW B2	0.8	0.9	1.0	1.2	1.4	1.0	1.1	1.3	1.4	1.6
TW B3	0.8	0.9	1.0	1.2	1.3	1.0	1.1	1.3	1.4	1.6
TW B4	0.8	0.9	1.0	1.2	1.4	1.1	1.2	1.3	1.4	1.6
TW C1	0.8	0.9	1.1	1.2	1.4	1.2	1.2	1.4	1.5	1.7
TW C2	0.8	0.9	1.1	1.2	1.4	1.1	1.2	1.3	1.5	1.6
TW C3	0.8	0.9	1.1	1.2	1.4	1.1	1.2	1.4	1.5	1.7
TW C4	0.9	1.0	1.2	1.3	1.5	1.3	1.3	1.5	1.6	1.8
TW D1	0.8	0.9	1.1	1.2	1.4	1.2	1.3	1.4	1.6	1.7
TW D2	0.7	0.8	1.0	1.1	1.3	0.9	1.0	1.2	1.3	1.5
TW D3	0.7	0.8	0.9	1.1	1.3	0.9	1.0	1.1	1.3	1.4
TW D4	0.6	0.7	0.9	1.1	1.3	0.8	0.9	1.1	1.2	1.4
TW D5	0.6	0.7	0.9	1.1	1.3	0.8	0.9	1.1	1.2	1.4
TW E1	0.6	0.7	0.9	1.1	1.3	0.8	0.9	1.1	1.2	1.4
TW E2	0.6	0.7	0.9	1.1	1.3	0.8	0.9	1.1	1.2	1.4
TW E3	0.6	0.7	0.9	1.0	1.2	0.7	0.8	1.0	1.1	1.3
TW E4	0.6	0.7	0.9	1.1	1.2	0.8	0.8	1.0	1.2	1.3
TW F1	0.8	0.9	1.1	1.3	1.4	1.2	1.3	1.4	1.6	1.7
TW F2	0.8	0.9	1.0	1.2	1.4	1.0	1.1	1.3	1.4	1.6
TW F3	0.7	0.8	1.0	1.2	1.3	1.0	1.1	1.2	1.4	1.6
TW F4	0.8	0.9	1.0	1.2	1.3	1.0	1.1	1.3	1.4	1.6
TW G1	0.8	0.9	1.1	1.3	1.4	1.2	1.3	1.4	1.6	1.7
TW G2	0.8	0.9	1.0	1.2	1.4	1.1	1.2	1.3	1.5	1.6
TW G3	0.8	0.9	1.0	1.2	1.4	1.1	1.2	1.3	1.5	1.6
TW G4	0.9	0.9	1.1	1.3	1.4	1.2	1.3	1.4	1.5	1.7
TW H1	0.9	1.0	1.1	1.3	1.4	1.2	1.3	1.4	1.6	1.7
TW H2	0.9	1.0	1.1	1.3	1.4	1.2	1.3	1.4	1.6	1.7
TW H3	0.9	1.0	1.2	1.3	1.5	1.3	1.4	1.5	1.6	1.8
TW H4	1.1	1.1	1.3	1.4	1.6	1.4	1.5	1.7	1.8	1.9
TW I1	0.9	1.0	1.1	1.3	1.4	1.2	1.3	1.5	1.8	1.8
TW I2	0.7	0.8	1.0	1.1	1.3	0.9	1.0	1.2	1.3	1.5
TW I3	0.7	0.8	0.9	1.1	1.3	0.9	1.0	1.1	1.3	1.4
TW I4	0.7	0.8	0.9	1.1	1.3	0.8	0.9	1.1	1.2	1.4
TW I5	0.6	0.8	0.9	1.1	1.3	0.8	0.9	1.1	1.2	1.4
TW J1	0.6	0.7	0.9	1.1	1.3	0.8	0.9	1.1	1.2	1.4
TW J2	0.6	0.7	0.9	1.1	1.3	0.8	0.9	1.1	1.2	1.4
TW J3	0.6	0.7	0.9	1.0	1.2	0.7	0.8	1.0	1.2	1.3
TW J4	0.6	0.7	0.9	1.1	1.2	0.8	0.9	1.0	1.2	1.3
DG A1	0.8	0.9	1.1	1.3	1.5	1.1	1.2	1.4	1.7	2.0
DG A2	0.7	0.8	1.0	1.2	1.4	1.0	1.1	1.3	1.5	1.7
DG A3	0.7	0.8	0.9	1.2	1.4	0.9	1.0	1.1	1.3	1.5
DG B1	0.7	0.8	1.0	1.2	1.5	1.1	1.1	1.3	1.5	1.8
DG B2	0.7	0.8	0.9	1.1	1.3	1.0	1.1	1.2	1.3	1.5
DG B3	0.7	0.7	0.9	1.1	1.4	0.9	1.0	1.1	1.2	1.4
DG C1	0.7	0.7	0.9	1.2	1.4	1.0	1.1	1.2	1.3	1.6
DG C2	0.7	0.7	0.9	1.1	1.3	0.9	1.0	1.1	1.2	1.5
DG C3	0.6	0.7	0.9	1.1	1.4	0.9	0.9	1.0	1.2	1.4
SS A1	0.7	0.8	1.0	1.1	1.3	0.9	1.0	1.2	1.3	1.5
SS A2	0.6	0.7	0.9	1.1	1.3	0.8	0.9	1.0	1.2	1.4
SS A3	0.7	0.8	1.0	1.2	1.3	1.0	1.1	1.2	1.4	1.6
SS A4	0.6	0.7	0.9	1.1	1.3	0.8	0.9	1.1	1.2	1.4
SS A5	0.7	0.9	1.0	1.2	1.4	1.0	1.1	1.3	1.4	1.6
SS A6	0.6	0.7	0.9	1.1	1.3	0.8	0.9	1.0	1.2	1.4
SS A7	0.8	0.9	1.1	1.2	1.4	1.1	1.2	1.4	1.5	1.7
SS A8	0.6	0.7	0.9	1.1	1.3	0.8	0.9	1.1	1.2	1.4
SS B1	0.7	0.8	1.0	1.2	1.4	1.0	1.1	1.3	1.4	1.6
SS B2	0.7	0.8	0.9	1.1	1.3	0.8	0.9	1.1	1.2	1.4
SS B3	0.8	0.9	1.0	1.2	1.4	1.0	1.1	1.3	1.4	1.6
SS B4	0.7	0.8	0.9	1.1	1.3	0.8	1.0	1.1	1.3	1.4
SS B5	0.8	0.9	1.1	1.2	1.4	1.1	1.2	1.4	1.5	1.7
SS B6	0.7	0.8	0.9	1.1	1.3	0.8	1.0	1.1	1.3	1.4
SS B7	0.9	1.0	1.1	1.3	1.5	1.2	1.3	1.5	1.6	1.8
SS B8	0.7	0.8	0.9	1.1	1.3	0.9	1.0	1.1	1.3	1.4
SS C1	0.7	0.8	1.0	1.1	1.3	1.0	1.1	1.2	1.4	1.5
SS C2	0.7	0.8	0.9	1.1	1.3	0.9	1.0	1.1	1.3	1.4
SS C3	0.8	0.9	1.1	1.2	1.4	1.1	1.2	1.4	1.5	1.7
SS C4	0.7	0.8	1.0	1.1	1.3	0.9	1.0	1.2	1.3	1.5
SS D1	0.6	0.7	0.9	1.1	1.3	0.8	0.9	1.1	1.2	1.4
SS D2	0.6	0.7	0.9	1.0	1.2	0.7	0.8	1.0	1.1	1.3
SS D3	0.7	0.8	1.0	1.1	1.3	0.9	1.0	1.1	1.3	1.5
SS D4	0.6	0.7	0.9	1.0	1.2	0.7	0.8	1.0	1.2	1.3
SS E1	0.7	0.8	0.9	1.1	1.3	0.9	1.0	1.1	1.3	1.5
SS E2	0.6	0.7	0.9	1.1	1.2	0.8	0.9	1.0	1.2	1.4
SS E3	0.7	0.8	1.0	1.2	1.4	1.0	1.1	1.3	1.4	1.6
SS E4	0.6	0.7	0.9	1.1	1.3	0.8	0.9	1.1	1.2	1.4

HOUSTON, TEXAS

TYPE	SOLAR SAVINGS FRACTION (%) BASE TEMP: 55 F (471 DD) LCR 200	100	70	50	40	30	25	20	65 F (1434 DD) 200	100	70	50	40	30	25	20	CONSERVATION FACTOR 55 F 200	100	50	30	20	65 F 200	100	50	30	20
WW A1	36	52	62	72	78	86	89	93	20	29	36	43	48	56	62	68	0.7	0.8	1.0	1.1	1.3	1.0	1.1	1.3	1.4	1.6
WW A2	43	63	74	84	89	94	96	98	21	35	44	54	61	70	75	82	0.6	0.7	0.9	1.1	1.2	0.9	1.0	1.1	1.3	1.4
WW A3	45	67	78	87	92	96	98	99	21	37	47	58	65	74	80	86	0.6	0.7	0.9	1.0	1.2	0.8	0.9	1.0	1.2	1.4
WW A4	47	69	80	89	93	97	98	99	21	38	49	60	67	77	82	87	0.6	0.7	0.9	1.0	1.2	0.8	0.9	1.0	1.2	1.4
WW A5	48	71	82	90	94	97	99	99	22	39	51	62	70	79	84	89	0.6	0.7	0.8	1.0	1.2	0.8	0.9	1.0	1.2	1.3
WW A6	49	73	83	91	95	98	99	100	22	40	52	63	71	80	85	90	0.6	0.7	0.8	1.0	1.2	0.7	0.8	1.0	1.2	1.3
WW B1	38	57	68	78	84	91	94	96	18	29	37	46	52	61	67	73	0.7	0.8	0.9	1.1	1.3	0.9	1.0	1.2	1.3	1.5
WW B2	46	69	80	89	93	97	98	99	21	38	49	61	68	78	83	88	0.6	0.7	0.9	1.0	1.2	0.8	0.9	1.0	1.2	1.3
WW B3	53	76	86	93	96	99	99	100	26	45	57	69	76	85	89	93	0.5	0.6	0.8	1.0	1.2	0.7	0.8	1.0	1.1	1.3
WW B4	52	76	87	93	96	99	99	100	24	45	58	70	77	86	90	94	0.5	0.6	0.8	1.0	1.2	0.7	0.8	0.9	1.1	1.3
WW B5	50	74	85	93	96	98	99	100	22	43	56	68	76	85	89	93	0.5	0.7	0.8	1.0	1.2	0.7	0.8	1.0	1.1	1.3
WW C1	55	78	87	94	97	99	99	100	27	47	59	71	78	86	90	94	0.5	0.6	0.8	1.0	1.2	0.7	0.8	0.9	1.1	1.3
WW C2	52	76	86	93	96	98	99	100	24	44	57	69	76	85	89	93	0.5	0.7	0.8	1.0	1.2	0.7	0.8	1.0	1.1	1.3
WW C3	57	81	90	96	98	99	100	100	27	51	64	76	83	90	94	97	0.5	0.6	0.8	1.0	1.2	0.6	0.7	0.9	1.1	1.3
WW C4	53	78	88	94	97	99	100	100	24	47	60	73	80	88	92	96	0.5	0.6	0.8	1.0	1.2	0.7	0.8	0.9	1.1	1.3
TW A1	35	51	61	71	77	84	89	93	20	29	35	42	48	55	61	67	0.7	0.8	1.0	1.1	1.3	1.1	1.2	1.3	1.4	1.6
TW A2	39	58	69	79	85	91	94	97	20	32	40	49	56	65	70	77	0.6	0.8	0.9	1.1	1.3	0.9	1.0	1.2	1.3	1.5
TW A3	40	61	72	82	88	93	96	98	20	33	42	52	59	68	74	80	0.6	0.7	0.9	1.1	1.3	0.9	1.0	1.1	1.3	1.4
TW A4	40	61	73	83	88	94	96	98	18	32	42	52	59	69	75	81	0.6	0.7	0.9	1.1	1.2	0.9	1.0	1.1	1.3	1.4
TW B1	34	51	62	72	78	85	89	93	19	28	35	43	48	56	62	69	0.7	0.8	1.0	1.1	1.3	1.0	1.1	1.3	1.4	1.6
TW B2	37	56	68	78	84	90	93	96	18	30	38	47	54	63	69	76	0.7	0.8	0.9	1.1	1.3	0.9	1.0	1.2	1.3	1.5
TW B3	37	57	68	78	84	91	94	97	18	30	38	48	55	64	70	77	0.6	0.8	0.9	1.1	1.3	0.9	1.0	1.1	1.3	1.5
TW B4	34	54	65	76	82	89	92	96	16	28	36	45	52	61	67	74	0.7	0.8	0.9	1.1	1.3	0.9	1.0	1.2	1.3	1.5
TW C1	33	50	61	71	78	85	89	93	18	27	34	42	48	56	62	69	0.7	0.8	1.0	1.1	1.3	1.0	1.1	1.3	1.4	1.6
TW C2	33	51	62	72	79	86	90	94	16	27	34	43	49	58	64	71	0.7	0.8	1.0	1.1	1.3	1.0	1.1	1.2	1.4	1.5
TW C3	31	48	59	70	76	84	88	93	16	25	32	40	47	55	61	68	0.7	0.8	1.0	1.1	1.3	1.0	1.1	1.3	1.4	1.6
TW C4	28	43	53	63	70	79	83	89	15	23	29	36	42	50	55	62	0.8	0.9	1.0	1.2	1.4	1.1	1.2	1.3	1.5	1.6
TW D1	32	50	61	72	78	86	90	94	15	25	32	40	46	54	60	67	0.7	0.8	1.0	1.1	1.3	1.0	1.1	1.3	1.4	1.6
TW D2	42	64	75	85	90	95	97	98	20	35	45	56	63	72	78	84	0.6	0.7	0.9	1.1	1.2	0.8	0.9	1.1	1.2	1.4
TW D3	46	69	80	88	93	97	98	99	22	39	49	61	68	77	82	88	0.6	0.7	0.9	1.0	1.2	0.8	0.9	1.0	1.2	1.4
TW D4	48	71	82	90	94	97	99	99	22	41	52	64	72	81	86	91	0.6	0.7	0.8	1.0	1.2	0.7	0.8	1.0	1.2	1.3
TW D5	46	70	81	89	94	97	98	99	21	39	51	63	71	80	85	90	0.6	0.7	0.8	1.0	1.2	0.7	0.8	1.0	1.2	1.3
TW E1	50	73	83	91	95	98	99	100	25	43	54	65	73	81	86	91	0.6	0.7	0.8	1.0	1.2	0.7	0.8	1.0	1.2	1.3
TW E2	49	72	83	91	94	98	99	99	24	42	53	65	72	81	86	91	0.6	0.7	.0.8	1.0	1.2	0.7	0.8	1.0	1.2	1.3
TW E3	56	79	89	95	97	99	100	100	27	49	62	73	81	88	92	95	0.5	0.6	0.8	1.0	1.2	0.7	0.8	0.9	1.1	1.3
TW E4	52	76	86	93	96	99	99	100	25	45	58	70	78	86	90	94	0.5	0.6	0.8	1.0	1.2	0.7	0.8	0.9	1.1	1.3
TW F1	34	50	61	71	77	85	89	93	18	27	34	41	47	55	61	67	0.7	0.8	1.0	1.1	1.3	1.0	1.1	1.3	1.4	1.6
TW F2	36	56	67	77	84	90	93	96	17	29	37	46	53	62	68	75	0.7	0.8	0.9	1.1	1.3	0.9	1.0	1.2	1.3	1.5
TW F3	36	57	69	79	85	91	94	97	16	29	38	48	55	64	70	77	0.6	0.7	0.9	1.1	1.3	0.9	1.0	1.1	1.3	1.5
TW F4	34	55	67	78	84	91	94	97	14	27	36	46	53	63	69	76	0.6	0.8	0.9	1.1	1.3	0.9	1.0	1.1	1.3	1.5
TW G1	32	49	60	70	77	85	89	93	16	26	32	40	46	55	60	67	0.7	0.8	1.0	1.1	1.3	1.0	1.1	1.3	1.4	1.6
TW G2	32	52	63	74	81	88	91	95	14	26	34	43	49	59	64	72	0.7	0.8	0.9	1.1	1.3	1.0	1.1	1.2	1.3	1.5
TW G3	30	50	62	73	79	87	91	94	12	24	32	41	48	57	63	71	0.7	0.8	0.9	1.1	1.3	1.0	1.1	1.2	1.4	1.5
TW G4	26	44	55	67	74	82	87	92	10	20	27	36	42	51	57	65	0.7	0.8	1.0	1.1	1.3	1.0	1.1	1.3	1.4	1.6
TW H1	28	45	56	67	74	82	87	91	13	22	29	37	43	51	57	64	0.7	0.8	1.0	1.2	1.3	1.1	1.2	1.3	1.4	1.6
TW H2	25	43	54	65	72	81	86	91	10	20	26	35	41	49	55	63	0.7	0.8	1.0	1.2	1.3	1.1	1.2	1.3	1.4	1.6
TW H3	22	38	49	59	67	76	81	87	8	17	23	30	36	45	50	58	0.8	0.9	1.0	1.2	1.4	1.1	1.2	1.4	1.5	1.7
TW H4	17	31	40	50	57	67	73	79	6	13	18	24	29	36	41	48	0.9	1.0	1.1	1.3	1.4	1.3	1.4	1.5	1.6	1.8
TW I1	29	47	58	68	75	83	88	92	13	22	29	37	42	51	56	63	0.7	0.8	1.0	1.1	1.3	1.1	1.2	1.3	1.5	1.6
TW I2	39	61	72	82	88	93	96	98	17	31	41	52	59	69	75	82	0.6	0.7	0.9	1.1	1.2	0.9	0.9	1.1	1.2	1.4
TW I3	42	64	76	85	90	95	97	99	19	34	45	56	64	73	79	85	0.6	0.7	0.9	1.0	1.2	0.8	0.9	1.0	1.2	1.4
TW I4	44	68	79	88	93	97	98	99	19	37	49	60	68	78	83	89	0.6	0.7	0.9	1.0	1.2	0.8	0.9	1.0	1.2	1.3
TW I5	44	68	79	88	93	97	98	99	19	37	49	61	69	78	84	89	0.6	0.7	0.9	1.0	1.2	0.8	0.9	1.0	1.2	1.3
TW J1	47	70	81	90	94	97	98	99	22	40	51	63	70	79	85	90	0.6	0.7	0.8	1.0	1.2	0.8	0.8	1.0	1.2	1.3
TW J2	46	69	80	89	93	97	98	99	21	38	50	62	69	79	84	89	0.6	0.7	0.8	1.0	1.2	0.8	0.9	1.0	1.2	1.3
TW J3	52	76	86	93	96	99	99	100	24	45	58	70	78	86	90	94	0.5	0.6	0.8	1.0	1.2	0.7	0.8	0.9	1.1	1.3
TW J4	49	73	84	92	95	98	99	100	22	42	55	67	75	84	89	93	0.5	0.7	0.8	1.0	1.2	0.7	0.8	1.0	1.1	1.3
DG A1	37	57	68	77	82	88	91	94	16	28	36	44	49	56	61	65	0.6	0.8	1.0	1.2	1.4	0.9	1.0	1.3	1.5	1.8
DG A2	37	58	70	79	85	91	94	96	16	29	38	48	55	63	69	74	0.6	0.7	0.9	1.1	1.3	0.9	1.0	1.1	1.3	1.6
DG A3	41	65	76	85	89	93	95	96	19	34	45	57	64	74	79	84	0.6	0.7	0.9	1.2	1.6	0.8	0.9	1.0	1.2	1.4
DG B1	39	63	75	84	88	93	95	96	16	28	38	48	55	64	69	74	0.6	0.7	0.9	1.1	1.4	0.9	1.0	1.1	1.3	1.6
DG B2	38	63	76	86	90	95	97	98	17	30	40	52	60	70	75	81	0.6	0.7	0.9	1.1	1.3	0.9	0.9	1.1	1.2	1.5
DG B3	41	68	80	89	93	96	97	98	19	35	46	59	68	78	83	89	0.6	0.7	0.9	1.1	1.6	0.8	0.9	1.0	1.2	1.4
DG C1	44	70	82	90	93	96	97	97	18	33	44	56	64	73	78	83	0.6	0.7	0.9	1.1	1.6	0.9	0.9	1.0	1.2	1.5
DG C2	43	69	82	90	94	97	98	99	19	34	46	59	67	77	82	88	0.6	0.7	0.8	1.0	1.3	0.8	0.9	1.0	1.2	1.4
DG C3	46	74	86	93	96	97	98	98	21	39	52	66	75	84	89	93	0.5	0.6	0.8	1.1	2.1	0.8	0.8	0.9	1.1	1.3
SS A1	46	65	76	85	90	94	97	98	25	39	48	58	65	73	78	84	0.6	0.7	0.9	1.1	1.2	0.8	0.9	1.1	1.2	1.4
SS A2	54	75	85	92	95	98	99	100	30	48	59	70	76	84	88	93	0.6	0.7	0.8	1.0	1.2	0.7	0.8	1.0	1.1	1.3
SS A3	44	63	74	83	88	93	96	98	24	37	46	55	62	70	75	81	0.6	0.7	0.9	1.1	1.3	0.9	1.0	1.1	1.3	1.4
SS A4	54	75	85	92	95	98	99	100	30	48	59	69	76	84	88	92	0.6	0.7	0.8	1.0	1.2	0.7	0.8	1.0	1.1	1.3
SS A5	43	61	72	81	86	92	95	97	25	36	45	54	60	68	74	80	0.6	0.8	0.9	1.1	1.3	0.9	1.0	1.1	1.3	1.5
SS A6	54	75	85	92	95	98	99	100	30	48	59	69	76	84	88	93	0.5	0.7	0.8	1.0	1.2	0.7	0.8	1.0	1.1	1.3
SS A7	39	56	66	76	82	88	92	95	23	33	40	48	54	62	67	74	0.7	0.8	0.9	1.1	1.3	1.0	1.1	1.2	1.4	1.5
SS A8	53	75	84	92	95	98	99	100	29	47	58	68	75	83	87	92	0.6	0.7	0.8	1.0	1.2	0.7	0.8	1.0	1.2	1.3
SS B1	40	60	70	80	86	92	94	97	21	34	42	52	58	67	72	79	0.6	0.8	0.9	1.1	1.3	0.9	1.0	1.1	1.3	1.5
SS B2	48	70	80	89	93	97	98	99	26	42	53	64	71	79	84	89	0.6	0.7	0.9	1.0	1.2	0.8	0.9	1.0	1.2	1.4
SS B3	39	58	69	79	84	91	94	96	20	32	41	50	56	65	70	77	0.6	0.8	0.9	1.1	1.3	0.9	1.0	1.2	1.3	1.5
SS B4	48	69	80	88	93	96	98	99	25	42	52	63	70	79	84	89	0.6	0.7	0.9	1.0	1.2	0.8	0.9	1.0	1.2	1.4
SS B5	36	53	63	73	79	87	90	94	20	30	37	45	51	59	65	71	0.7	0.8	1.0	1.1	1.3	1.0	1.1	1.2	1.4	1.5
SS B6	48	69	80	88	93	96	98	99	25	41	52	63	70	79	84	89	0.6	0.7	0.9	1.0	1.2	0.8	0.9	1.0	1.2	1.4
SS B7	33	50	60	69	76	83	88	92	19	28	34	41	47	55	60	67	0.7	0.8	1.0	1.2	1.3	1.0	1.1	1.3	1.4	1.6
SS B8	47	68	79	87	92	96	98	99	24	40	51	61	69	77	82	88	0.6	0.7	0.9	1.0	1.2	0.8	0.9	1.0	1.2	1.4
SS C1	41	62	74	83	89	94	96	98	19	33	42	52	60	69	75	81	0.6	0.7	0.9	1.1	1.2	0.9	1.0	1.1	1.3	1.4
SS C2	46	69	80	89	93	97	98	99	22	38	49	61	68	77	83	88	0.6	0.7	0.9	1.0	1.2	0.8	0.9	1.0	1.2	1.4
SS C3	33	51	62	73	79	86	90	94	17	27	34	42	48	57	63	70	0.7	0.8	1.0	1.1	1.3	1.0	1.1	1.2	1.4	1.5
SS C4	41	63	74	84	89	94	96	98	20	34	44	55	62	71	77	83	0.6	0.7	0.9	1.1	1.2	0.8	0.9	1.1	1.2	1.4
SS D1	54	75	85	92	95	98	99	100	29	47	57	68	74	82	87	91	0.6	0.7	0.8	1.0	1.2	0.7	0.8	1.0	1.2	1.3
SS D2	61	83	91	96	98	99	100	100	34	55	67	78	84	91	94	97	0.5	0.6	0.8	1.0	1.2	0.6	0.7	0.9	1.1	1.3
SS D3	49	69	79	88	92	96	98	99	27	42	52	61	68	77	81	87	0.6	0.7	0.9	1.0	1.2	0.8	0.9	1.0	1.2	1.4
SS D4	59	81	90	95	98	99	100	100	33	53	65	76	82	89	93	96	0.5	0.6	0.8	1.0	1.2	0.7	0.8	0.9	1.1	1.3
SS E1	49	70	81	89	93	97	98	99	25	41	51	62	69	77	82	88	0.6	0.7	0.9	1.0	1.2	0.8	0.9	1.0	1.2	1.4
SS E2	56	78	88	94	97	99	99	100	29	49	61	72	79	87	90	94	0.5	0.6	0.8	1.0	1.2	0.7	0.8	0.9	1.1	1.3
SS E3	41	60	71	80	86	92	95	97	22	34	43	52	59	67	73	79	0.6	0.7	0.9	1.1	1.3	0.9	1.0	1.1	1.3	1.5
SS E4	52	74	84	91	95	98	99	100	27	45	56	67	74	83	87	92	0.6	0.7	0.8	1.0	1.2	0.7	0.8	1.0	1.2	1.3

LAREDO, TEXAS

TYPE	SOLAR SAVINGS FRACTION (%) BASE TEMP: 55 F (230 DD) LCR 200	100	70	50	40	30	25	20	65 F (876 DD) 200	100	70	50	40	30	25	20	CONSERVATION FACTOR 55 F 200	100	50	30	20	65 F 200	100	50	30	20
WW A1	53	73	83	90	94	97	98	99	28	42	51	60	67	75	80	86	0.6	0.7	0.9	1.0	1.2	0.8	0.9	1.1	1.2	1.4
WW A2	64	85	92	96	98	99	100	100	32	51	62	73	80	87	91	95	0.5	0.6	0.8	1.0	1.2	0.7	0.8	1.0	1.1	1.3
WW A3	68	88	94	98	99	100	100	100	34	55	67	77	83	90	93	96	0.5	0.6	0.8	1.0	1.2	0.6	0.8	0.9	1.1	1.3
WW A4	70	89	95	98	99	100	100	100	35	57	69	79	85	92	95	97	0.5	0.6	0.8	1.0	1.2	0.6	0.7	0.9	1.1	1.3
WW A5	72	91	96	99	99	100	100	100	36	59	71	81	87	93	96	98	0.5	0.6	0.8	1.0	-	0.6	0.7	0.9	1.1	1.2
WW A6	73	91	96	99	100	100	100	100	37	60	72	82	88	94	96	98	0.5	0.6	0.8	1.0	-	0.6	0.7	0.9	1.1	1.2
WW B1	59	80	89	94	97	99	99	100	28	45	56	66	73	81	86	91	0.5	0.6	0.8	1.0	1.2	0.7	0.8	1.0	1.2	1.3
WW B2	70	89	95	98	99	100	100	100	35	57	69	79	86	92	95	97	0.5	0.6	0.8	1.0	1.2	0.6	0.7	0.9	1.1	1.3
WW B3	77	93	97	99	100	100	100	100	41	65	76	86	91	96	98	99	0.5	0.6	0.8	1.0	-	0.6	0.7	0.9	1.0	1.2
WW B4	76	93	97	99	100	100	100	100	40	64	76	86	91	96	98	99	0.5	0.6	0.8	1.0	-	0.6	0.7	0.9	1.0	1.2
WW B5	74	92	97	99	100	100	100	100	38	62	75	85	90	95	97	99	0.5	0.6	0.8	1.0	-	0.6	0.7	0.9	1.0	1.2
WW C1	78	94	98	99	100	100	100	100	43	66	78	87	92	96	98	99	0.5	0.6	0.8	1.0	-	0.6	0.7	0.9	1.0	1.2
WW C2	76	93	97	99	100	100	100	100	40	64	76	85	91	95	97	99	0.5	0.6	0.8	1.0	-	0.6	0.7	0.9	1.0	1.2
WW C3	81	95	99	100	100	100	100	100	45	70	82	91	95	98	99	100	0.4	0.6	0.8	0.9	-	0.6	0.7	0.8	1.0	1.2
WW C4	77	94	98	99	100	100	100	100	41	66	79	88	93	97	98	99	0.5	0.6	0.8	1.0	-	0.6	0.7	0.8	1.0	1.2
TW A1	52	72	82	89	93	97	98	99	27	40	50	59	66	74	79	85	0.6	0.7	0.9	1.0	1.2	0.8	0.9	1.1	1.2	1.4
TW A2	59	80	89	94	97	99	99	100	29	47	57	68	75	83	87	92	0.5	0.6	0.8	1.0	1.2	0.7	0.8	1.0	1.2	1.3
TW A3	62	83	91	96	98	99	100	100	30	49	60	71	78	86	90	94	0.5	0.6	0.8	1.0	1.2	0.7	0.8	1.0	1.1	1.3
TW A4	62	83	91	96	98	99	100	100	30	49	61	72	78	86	90	94	0.5	0.6	0.8	1.0	1.2	0.7	0.8	1.0	1.1	1.3
TW B1	52	73	82	90	94	97	98	99	26	41	50	60	67	75	80	86	0.6	0.7	0.9	1.0	1.2	0.8	0.9	1.1	1.2	1.4
TW B2	57	78	88	94	96	99	99	100	28	45	55	66	73	81	86	91	0.5	0.6	0.8	1.0	1.2	0.7	0.8	1.0	1.2	1.3
TW B3	57	79	88	94	97	99	99	100	27	45	56	67	74	82	87	91	0.5	0.6	0.8	1.0	1.2	0.7	0.8	1.0	1.2	1.3
TW B4	54	76	86	93	96	98	99	100	25	42	53	64	71	80	85	90	0.5	0.7	0.8	1.0	1.2	0.8	0.9	1.0	1.2	1.3
TW C1	51	72	82	90	93	97	98	99	25	40	49	59	66	75	80	86	0.6	0.7	0.9	1.0	1.2	0.8	0.9	1.1	1.2	1.4
TW C2	51	73	83	91	94	97	99	99	24	40	50	60	68	76	82	87	0.6	0.7	0.8	1.0	1.2	0.8	0.9	1.0	1.2	1.4
TW C3	49	70	81	89	93	97	98	99	23	38	47	58	65	74	79	85	0.6	0.7	0.9	1.0	1.2	0.8	0.9	1.1	1.2	1.4
TW C4	43	64	75	84	89	94	96	98	21	34	42	52	59	68	73	80	0.6	0.7	0.9	1.1	1.2	0.9	1.0	1.1	1.3	1.4
TW D1	52	73	83	91	94	97	99	99	24	39	49	59	66	75	80	86	0.6	0.7	0.8	1.0	1.2	0.8	0.9	1.1	1.2	1.4
TW D2	64	85	93	97	98	100	100	100	32	52	64	74	81	88	92	95	0.5	0.6	0.8	1.0	1.2	0.7	0.8	0.9	1.1	1.3
TW D3	69	89	95	98	99	100	100	100	35	57	69	79	85	92	95	97	0.5	0.6	0.8	1.0	1.2	0.6	0.7	0.9	1.1	1.3
TW D4	71	90	96	98	99	100	100	100	36	59	71	82	87	93	96	98	0.5	0.6	0.8	1.0	1.2	0.6	0.7	0.9	1.1	1.2
TW D5	69	89	95	98	99	100	100	100	34	57	70	81	87	93	96	98	0.5	0.6	0.8	1.0	1.2	0.6	0.7	0.9	1.1	1.2
TW E1	73	91	96	99	100	100	100	100	39	61	73	83	88	94	96	98	0.5	0.6	0.8	1.0	-	0.6	0.7	0.9	1.1	1.2
TW E2	72	91	96	99	99	100	100	100	37	60	72	82	88	94	96	98	0.5	0.6	0.8	1.0	1.2	0.6	0.7	0.9	1.1	1.2
TW E3	79	94	98	99	100	100	100	100	43	68	80	89	93	97	98	99	0.5	0.6	0.8	1.0	-	0.6	0.7	0.8	1.0	1.2
TW E4	76	93	97	99	100	100	100	100	40	64	76	86	91	96	98	99	0.5	0.6	0.8	1.0	-	0.6	0.7	0.9	1.0	1.2
TW F1	51	72	82	90	93	97	98	99	26	40	49	59	66	74	80	85	0.6	0.7	0.9	1.0	1.2	0.8	0.9	1.1	1.2	1.4
TW F2	57	78	87	94	96	99	99	100	27	44	55	66	73	81	86	91	0.5	0.6	0.8	1.0	1.2	0.7	0.8	1.0	1.2	1.3
TW F3	58	80	89	95	97	99	99	100	27	45	56	67	74	83	87	92	0.5	0.6	0.8	1.0	1.2	0.7	0.8	1.0	1.1	1.3
TW F4	56	78	88	94	97	99	99	100	24	43	54	66	73	82	86	91	0.5	0.6	0.8	1.0	1.2	0.7	0.8	1.0	1.2	1.3
TW G1	50	71	82	89	93	97	98	99	24	38	48	58	65	74	79	85	0.6	0.7	0.9	1.0	1.2	0.8	0.9	1.1	1.2	1.4
TW G2	52	75	85	92	95	98	99	100	23	40	51	62	69	78	83	88	0.5	0.7	0.8	1.0	1.2	0.8	0.9	1.0	1.2	1.4
TW G3	50	73	84	91	95	98	99	100	21	38	49	60	68	77	82	88	0.6	0.7	0.8	1.0	1.2	0.8	0.9	1.0	1.2	1.4
TW G4	44	67	79	87	92	96	98	99	17	33	43	54	62	71	77	83	0.6	0.7	0.9	1.0	1.2	0.8	0.9	1.1	1.2	1.4
TW H1	46	68	79	87	92	96	98	99	20	35	44	55	62	71	77	83	0.6	0.7	0.9	1.0	1.2	0.8	0.9	1.1	1.2	1.4
TW H2	43	66	77	86	91	95	97	99	17	32	42	52	60	69	75	82	0.6	0.7	0.9	1.1	1.2	0.8	0.9	1.1	1.2	1.4
TW H3	38	60	72	82	88	93	96	98	14	28	37	47	54	64	70	77	0.6	0.7	0.9	1.1	1.3	0.9	1.0	1.1	1.3	1.5
TW H4	30	50	62	73	80	87	91	95	11	22	29	38	45	54	60	68	0.7	0.8	0.9	1.1	1.3	1.0	1.1	1.2	1.4	1.5
TW I1	48	70	81	89	93	97	98	99	21	36	46	56	63	72	77	83	0.6	0.7	0.9	1.0	1.2	0.8	0.9	1.1	1.2	1.4
TW I2	61	83	91	96	98	99	100	100	28	48	60	71	78	86	90	94	0.5	0.6	0.8	1.0	1.2	0.7	0.8	1.0	1.1	1.3
TW I3	65	86	93	97	99	100	100	100	31	52	64	75	82	89	93	96	0.5	0.6	0.8	1.0	1.2	0.7	0.8	0.9	1.1	1.3
TW I4	67	88	94	98	99	100	100	100	32	55	68	79	85	92	95	97	0.5	0.6	0.8	1.0	1.2	0.6	0.7	0.9	1.1	1.3
TW I5	67	88	94	98	99	100	100	100	32	55	68	79	85	92	95	97	0.5	0.6	0.8	1.0	1.2	0.6	0.7	0.9	1.1	1.2
TW J1	71	90	95	98	99	100	100	100	36	58	70	81	87	93	96	98	0.5	0.6	0.8	1.0	1.2	0.6	0.7	0.9	1.1	1.2
TW J2	69	89	95	98	99	100	100	100	34	57	69	80	86	92	95	98	0.5	0.6	0.8	1.0	1.2	0.6	0.7	0.9	1.1	1.2
TW J3	76	93	97	99	100	100	100	100	39	64	77	86	91	96	98	99	0.5	0.6	0.8	1.0	-	0.6	0.7	0.9	1.0	1.2
TW J4	73	91	97	99	100	100	100	100	36	61	74	84	90	95	97	99	0.5	0.6	0.8	1.0	-	0.6	0.7	0.9	1.0	1.2
DG A1	60	81	89	94	97	98	99	100	27	44	55	64	70	77	81	86	0.5	0.6	0.8	1.0	1.1	0.7	0.8	1.0	1.3	1.5
DG A2	59	81	90	95	97	99	100	100	27	46	57	68	74	82	86	90	0.5	0.6	0.8	1.0	-	0.7	0.8	1.0	1.2	1.4
DG A3	65	85	92	95	96	97	97	97	30	52	64	75	82	88	91	93	0.5	0.6	0.9	1.8	-	0.7	0.7	0.9	1.2	1.4
DG B1	65	86	93	97	98	99	99	99	28	48	60	71	78	85	88	91	0.5	0.6	0.8	1.2	2.3	0.7	0.8	1.0	1.2	1.4
DG B2	64	87	94	97	99	100	100	100	28	49	62	74	81	88	91	95	0.5	0.6	0.8	1.0	1.2	0.7	0.8	0.9	1.1	1.3
DG B3	67	89	95	97	98	98	98	98	31	53	68	80	86	92	94	96	0.5	0.6	0.9	2.0	-	0.7	0.7	0.9	1.1	1.4
DG C1	72	91	96	98	98	98	98	98	32	54	68	80	85	91	93	95	0.5	0.6	0.9	3.2	-	0.7	0.7	0.9	1.1	1.4
DG C2	70	91	96	99	99	100	100	100	31	54	69	80	87	92	95	97	0.5	0.6	0.8	1.0	1.8	0.7	0.7	0.9	1.1	1.3
DG C3	74	93	97	98	98	98	98	98	35	59	74	85	91	95	97	98	0.4	0.6	1.0	-	-	0.6	0.7	0.9	1.1	1.5
SS A1	66	86	93	97	98	100	100	100	37	55	66	76	82	89	92	95	0.5	0.6	0.8	1.0	1.2	0.7	0.8	0.9	1.1	1.3
SS A2	76	92	97	99	100	100	100	100	44	65	76	85	90	95	97	99	0.5	0.6	0.8	1.0	-	0.6	0.7	0.9	1.1	1.2
SS A3	64	84	91	96	98	99	100	100	35	53	63	73	80	87	90	94	0.5	0.6	0.8	1.0	1.2	0.7	0.8	1.0	1.1	1.3
SS A4	76	92	97	99	100	100	100	100	44	65	76	85	90	95	97	99	0.5	0.6	0.8	1.0	-	0.6	0.7	0.9	1.1	1.2
SS A5	62	82	90	95	97	99	100	100	35	51	62	71	78	85	89	93	0.5	0.6	0.8	1.0	1.2	0.7	0.8	1.0	1.1	1.3
SS A6	76	92	97	99	100	100	100	100	44	65	76	85	90	95	97	99	0.5	0.6	0.8	1.0	-	0.6	0.7	0.9	1.1	1.2
SS A7	57	77	86	93	96	98	99	100	32	47	56	66	72	80	84	89	0.5	0.7	0.8	1.0	1.2	0.8	0.9	1.0	1.2	1.4
SS A8	75	92	97	99	100	100	100	100	44	65	75	85	90	95	97	98	0.5	0.6	0.8	1.0	-	0.6	0.7	0.9	1.1	1.2
SS B1	60	81	89	95	97	99	99	100	31	49	60	70	76	84	88	93	0.5	0.6	0.8	1.0	1.2	0.7	0.8	1.0	1.1	1.3
SS B2	70	89	95	98	99	100	100	100	38	59	70	80	86	92	95	97	0.5	0.6	0.8	1.0	1.2	0.6	0.7	0.9	1.1	1.3
SS B3	59	80	88	94	97	99	99	100	30	48	58	68	75	83	87	91	0.5	0.6	0.8	1.0	1.2	0.7	0.8	1.0	1.2	1.3
SS B4	69	88	95	98	99	100	100	100	38	59	70	80	86	92	95	97	0.5	0.6	0.8	1.0	1.2	0.6	0.7	0.9	1.1	1.3
SS B5	54	74	84	91	94	97	99	99	28	43	53	63	69	77	82	88	0.6	0.7	0.9	1.0	1.2	0.8	0.9	1.0	1.2	1.4
SS B6	69	88	95	98	99	100	100	100	38	58	70	80	86	92	95	97	0.5	0.6	0.8	1.0	1.2	0.6	0.7	0.9	1.1	1.3
SS B7	50	71	81	88	93	96	98	99	26	40	49	59	65	73	79	84	0.6	0.7	0.9	1.0	1.2	0.8	0.9	1.1	1.2	1.4
SS B8	68	88	94	98	99	100	100	100	37	57	69	79	85	91	94	97	0.5	0.6	0.8	1.0	1.2	0.6	0.8	0.9	1.1	1.3
SS C1	63	84	92	96	98	99	100	100	31	50	62	72	79	87	91	94	0.5	0.6	0.8	1.0	1.2	0.7	0.8	1.0	1.1	1.3
SS C2	69	89	95	98	99	100	100	100	35	57	69	79	85	92	95	97	0.5	0.6	0.8	1.0	1.2	0.6	0.7	0.9	1.1	1.3
SS C3	52	74	84	91	94	97	99	99	25	40	50	61	68	76	81	87	0.6	0.7	0.8	1.0	1.2	0.8	0.9	1.0	1.2	1.4
SS C4	63	84	92	96	98	99	100	100	31	51	62	73	80	87	91	95	0.5	0.6	0.8	1.0	1.2	0.7	0.8	0.9	1.1	1.3
SS D1	76	92	97	99	100	100	100	100	44	65	76	85	90	95	97	98	0.5	0.6	0.8	1.0	-	0.6	0.7	0.9	1.1	1.2
SS D2	83	96	99	100	100	100	100	100	51	73	84	91	95	98	99	100	0.4	0.6	0.8	0.9	-	0.5	0.7	0.8	1.0	1.2
SS D3	70	88	94	98	99	100	100	100	40	59	69	79	85	91	94	97	0.5	0.6	0.8	1.0	1.2	0.6	0.8	0.9	1.1	1.3
SS D4	81	95	98	100	100	100	100	100	49	71	82	90	94	97	99	99	0.4	0.6	0.8	0.9	-	0.6	0.7	0.8	1.0	1.2
SS E1	71	90	95	98	99	100	100	100	39	59	70	80	86	92	95	97	0.5	0.6	0.8	1.0	1.2	0.6	0.7	0.9	1.1	1.3
SS E2	79	94	98	99	100	100	100	100	45	68	79	88	92	96	98	99	0.5	0.6	0.8	1.0	-	0.6	0.7	0.9	1.0	1.2
SS E3	61	81	90	95	97	99	100	100	32	50	60	70	77	84	88	93	0.5	0.6	0.8	1.0	1.2	0.7	0.8	1.0	1.1	1.3
SS E4	74	92	96	99	100	100	100	100	41	63	74	84	89	94	97	98	0.5	0.6	0.8	1.0	-	0.6	0.7	0.9	1.1	1.2

LUBBOCK, TEXAS

SOLAR SAVINGS FRACTION (%) — CONSERVATION FACTOR

TYPE	BASE TEMP: 55 F (1809 DD)								65 F (3551 DD)								CONSERVATION FACTOR 55 F					65 F				
LCR	200	100	70	50	40	30	25	20	200	100	70	50	40	30	25	20	200	100	50	30	20	200	100	50	30	20
WW A1	24	36	44	53	60	69	74	81	17	25	30	37	42	49	54	60	0.9	1.0	1.1	1.3	1.4	1.2	1.3	1.4	1.5	1.7
WW A2	27	44	55	66	74	82	87	92	17	29	37	46	53	62	68	75	0.7	0.8	1.0	1.1	1.3	0.9	1.0	1.2	1.3	1.5
WW A3	28	47	59	71	78	86	90	94	17	30	40	50	57	67	73	80	0.7	0.8	0.9	1.1	1.3	0.9	1.0	1.1	1.3	1.4
WW A4	29	49	61	73	80	88	92	95	16	31	41	52	60	69	75	82	0.7	0.8	0.9	1.1	1.3	0.8	0.9	1.1	1.2	1.4
WW A5	30	51	64	75	82	90	93	96	16	32	43	54	62	72	78	85	0.7	0.8	0.9	1.1	1.3	0.8	0.9	1.1	1.2	1.4
WW A6	30	52	65	77	83	91	94	97	16	33	44	55	63	73	79	86	0.7	0.8	0.9	1.1	1.3	0.8	0.9	1.0	1.2	1.4
WW B1	24	38	48	59	66	75	80	86	14	24	31	38	44	53	58	65	0.8	0.9	1.0	1.2	1.4	1.1	1.1	1.3	1.4	1.6
WW B2	29	49	62	74	81	88	92	96	16	32	42	53	61	71	77	84	0.7	0.8	0.9	1.1	1.3	0.8	0.9	1.1	1.2	1.4
WW B3	34	57	70	81	87	93	96	98	20	38	50	61	69	79	84	90	0.6	0.7	0.9	1.1	1.2	0.8	0.9	1.0	1.2	1.3
WW B4	33	57	70	81	88	94	96	98	18	38	50	63	71	80	86	91	0.6	0.7	0.9	1.1	1.2	0.7	0.8	1.0	1.1	1.3
WW B5	31	55	68	80	86	93	96	98	17	36	48	61	69	79	85	90	0.6	0.7	0.9	1.1	1.2	0.7	0.8	1.0	1.1	1.3
WW C1	35	59	71	82	88	94	96	98	21	40	52	64	71	81	86	91	0.6	0.7	0.9	1.1	1.2	0.7	0.8	1.0	1.1	1.3
WW C2	33	56	69	81	87	93	96	98	19	37	49	61	69	79	85	90	0.6	0.7	0.9	1.1	1.2	0.8	0.8	1.0	1.2	1.3
WW C3	37	63	76	87	92	96	98	99	21	44	57	70	78	87	91	95	0.6	0.7	0.9	1.0	1.2	0.7	0.8	0.9	1.1	1.3
WW C4	33	59	73	84	90	95	97	99	18	40	53	66	74	84	89	93	0.6	0.7	0.9	1.0	1.2	0.7	0.8	1.0	1.1	1.3
TW A1	24	35	43	52	59	68	73	80	17	24	30	36	41	48	53	60	0.9	1.0	1.1	1.3	1.4	1.2	1.3	1.4	1.5	1.7
TW A2	25	40	50	61	68	77	82	88	16	26	34	42	48	57	63	70	0.8	0.9	1.0	1.2	1.4	1.0	1.1	1.2	1.4	1.5
TW A3	25	42	53	64	72	80	85	91	15	27	35	44	51	60	67	74	0.8	0.9	1.0	1.2	1.3	1.0	1.0	1.2	1.3	1.5
TW A4	24	42	53	65	72	81	86	91	14	26	35	45	52	61	67	75	0.7	0.8	1.0	1.1	1.3	0.9	1.0	1.2	1.3	1.5
TW B1	23	35	44	53	60	69	74	81	16	24	29	36	42	49	54	61	0.9	1.0	1.1	1.3	1.4	1.1	1.2	1.4	1.5	1.6
TW B2	23	38	48	59	66	76	81	87	15	25	32	40	47	56	62	69	0.8	0.9	1.0	1.2	1.4	1.0	1.1	1.2	1.4	1.5
TW B3	23	38	49	60	67	77	82	88	14	24	32	41	47	56	62	70	0.8	0.9	1.0	1.2	1.4	1.0	1.1	1.2	1.4	1.5
TW B4	21	36	46	57	64	74	79	86	13	23	30	38	45	54	60	67	0.8	0.9	1.1	1.2	1.4	1.0	1.1	1.3	1.4	1.5
TW C1	21	34	43	52	59	68	74	81	14	22	29	36	41	49	54	61	0.9	1.0	1.1	1.3	1.4	1.1	1.2	1.3	1.5	1.6
TW C2	21	34	43	53	61	70	76	82	13	22	29	36	42	50	56	63	0.9	1.0	1.1	1.2	1.4	1.1	1.2	1.3	1.4	1.6
TW C3	19	32	41	51	58	67	73	80	13	21	27	34	40	48	54	61	0.9	1.0	1.1	1.3	1.4	1.1	1.2	1.3	1.5	1.6
TW C4	18	29	36	45	52	61	67	74	12	19	24	31	36	43	48	55	1.0	1.1	1.2	1.3	1.5	1.2	1.3	1.4	1.6	1.7
TW D1	20	32	42	51	59	68	74	81	12	20	26	33	38	46	51	58	0.9	1.0	1.1	1.3	1.4	1.1	1.2	1.4	1.5	1.7
TW D2	26	45	57	68	75	84	89	93	16	29	38	48	56	65	71	79	0.7	0.8	1.0	1.1	1.3	0.9	1.0	1.1	1.3	1.4
TW D3	29	49	61	73	80	88	92	95	17	32	42	53	61	71	76	83	0.7	0.8	0.9	1.1	1.3	0.8	0.9	1.1	1.2	1.4
TW D4	30	52	64	76	83	90	94	97	17	34	45	57	65	75	80	87	0.7	0.8	0.9	1.1	1.3	0.8	C.9	1.0	1.2	1.4
TW D5	28	50	63	75	82	90	93	96	16	33	44	56	64	74	80	87	0.7	0.8	0.9	1.1	1.3	0.8	0.9	1.0	1.2	1.4
TW E1	32	54	66	77	84	91	94	97	20	36	47	58	66	75	81	87	0.7	0.8	0.9	1.1	1.3	0.8	0.9	1.0	1.2	1.4
TW E2	31	53	65	77	84	91	94	97	19	35	46	58	65	75	81	87	0.7	0.8	0.9	1.1	1.3	0.8	0.9	1.0	1.2	1.4
TW E3	36	61	74	84	90	95	97	99	22	42	54	67	75	84	88	93	0.6	0.7	0.9	1.0	1.2	0.7	0.8	1.0	1.1	1.3
TW E4	33	57	70	82	88	94	96	98	19	39	51	63	72	81	86	92	0.6	0.7	0.9	1.1	1.2	0.7	0.8	1.0	1.1	1.3
TW F1	22	34	43	52	59	68	74	80	15	23	28	35	40	48	53	60	0.9	1.0	1.1	1.3	1.4	1.2	1.2	1.4	1.5	1.7
TW F2	22	37	48	58	66	75	81	87	14	24	31	39	46	55	60	68	0.8	0.9	1.0	1.2	1.4	1.0	1.1	1.2	1.4	1.5
TW F3	22	38	49	60	68	77	83	88	12	23	31	40	47	56	63	70	0.8	0.9	1.0	1.2	1.3	1.0	1.1	1.2	1.4	1.5
TW F4	19	36	47	58	66	76	81	87	10	21	29	38	45	55	61	69	0.8	0.9	1.0	1.2	1.4	1.0	1.1	1.2	1.4	1.5
TW G1	20	32	41	51	58	67	73	80	13	21	27	34	39	47	52	59	0.9	1.0	1.1	1.3	1.4	1.2	1.2	1.4	1.5	1.7
TW G2	19	34	44	54	62	72	77	84	11	20	27	36	42	51	57	64	0.8	0.9	1.1	1.2	1.4	1.1	1.2	1.3	1.4	1.6
TW G3	17	32	42	53	60	70	76	83	9	19	26	34	40	49	55	63	0.8	0.9	1.1	1.2	1.4	1.1	1.2	1.3	1.4	1.6
TW G4	14	27	36	47	54	64	70	78	7	15	22	29	35	44	50	57	0.9	1.0	1.1	1.3	1.4	1.1	1.2	1.4	1.5	1.6
TW H1	17	29	38	47	54	64	70	77	10	17	23	30	36	44	49	57	0.9	1.0	1.1	1.3	1.4	1.2	1.3	1.4	1.5	1.7
TW H2	14	26	35	45	52	62	68	76	7	15	21	28	34	42	48	55	0.9	1.0	1.1	1.3	1.5	1.2	1.3	1.4	1.5	1.7
TW H3	11	22	31	40	47	57	63	71	5	12	18	24	30	37	43	50	1.0	1.1	1.2	1.3	1.5	1.3	1.3	1.5	1.6	1.7
TW H4	8	17	24	32	38	47	53	61	4	9	14	19	23	30	35	41	1.1	1.2	1.3	1.5	1.6	1.4	1.5	1.6	1.8	1.9
TW I1	17	29	38	48	55	64	70	77	10	17	23	30	35	42	48	55	0.9	1.0	1.1	1.3	1.4	1.2	1.3	1.4	1.6	1.7
TW I2	23	41	53	65	72	81	86	91	13	26	35	45	52	62	68	76	0.7	0.8	1.0	1.1	1.3	0.9	1.0	1.1	1.3	1.5
TW I3	25	44	57	68	76	85	89	93	14	28	38	48	56	66	72	80	0.7	0.8	1.0	1.1	1.3	0.9	1.0	1.1	1.3	1.4
TW I4	26	48	61	73	80	88	92	96	14	30	41	53	61	71	78	84	0.7	0.8	0.9	1.1	1.3	0.8	0.9	1.1	1.2	1.4
TW I5	26	48	61	73	80	88	92	96	14	30	42	54	62	72	78	85	0.7	0.8	0.9	1.1	1.3	0.8	0.9	1.0	1.2	1.4
TW J1	29	51	63	75	82	89	93	96	17	33	44	55	63	73	79	85	0.7	0.8	0.9	1.1	1.3	0.8	0.9	1.0	1.2	1.4
TW J2	28	49	62	74	81	89	93	96	16	32	43	54	62	72	78	85	0.7	0.8	0.9	1.1	1.3	0.8	0.9	1.0	1.2	1.4
TW J3	32	57	70	82	88	94	96	98	18	38	51	63	72	81	86	92	0.6	0.7	0.9	1.1	1.2	0.7	0.8	1.0	1.1	1.3
TW J4	30	54	67	79	86	92	95	98	16	35	48	60	69	79	84	90	0.6	0.7	0.9	1.1	1.2	0.8	0.8	1.0	1.1	1.3
DG A1	22	37	47	57	63	71	75	80	12	22	29	36	41	48	52	57	0.8	0.9	1.1	1.3	1.5	1.1	1.1	1.4	1.6	1.9
DG A2	22	39	50	61	68	76	81	86	13	23	32	41	47	56	61	67	0.8	0.9	1.0	1.2	1.4	1.0	1.1	1.2	1.4	1.6
DG A3	24	44	57	69	76	84	88	91	15	28	38	49	57	67	73	79	0.7	0.8	1.0	1.2	1.4	0.9	1.0	1.1	1.3	1.5
DG B1	22	39	52	64	71	79	83	87	12	22	30	39	46	55	60	66	0.8	0.8	1.0	1.2	1.5	1.0	1.1	1.2	1.4	1.7
DG B2	22	41	54	67	75	83	88	92	13	24	33	43	51	62	68	75	0.8	0.8	1.0	1.1	1.3	1.0	1.0	1.1	1.3	1.5
DG B3	25	45	60	73	81	89	92	95	15	28	39	51	60	72	78	84	0.7	0.8	0.9	1.1	1.3	0.9	1.0	1.0	1.2	1.4
DG C1	25	46	60	73	80	87	90	93	14	26	35	46	54	65	70	76	0.7	0.8	0.9	1.2	1.4	1.0	1.0	1.1	1.3	1.5
DG C2	25	46	60	74	81	89	92	95	15	28	38	50	59	70	76	82	0.7	0.8	0.9	1.1	1.3	0.9	1.0	1.1	1.2	1.4
DG C3	28	51	67	80	87	93	95	97	17	32	44	57	67	79	84	90	0.7	0.7	0.9	1.1	1.4	0.9	0.9	1.0	1.1	1.3
SS A1	30	46	57	67	74	82	87	91	21	33	41	50	56	65	70	77	0.7	0.8	1.0	1.2	1.3	0.9	1.0	1.2	1.3	1.5
SS A2	36	56	68	78	84	91	94	97	25	41	51	62	69	78	83	88	0.7	0.8	0.9	1.1	1.3	0.8	0.9	1.0	1.2	1.4
SS A3	28	43	53	64	71	79	84	89	20	30	38	46	52	61	66	73	0.8	0.9	1.0	1.2	1.4	1.0	1.1	1.2	1.4	1.5
SS A4	35	55	67	78	84	90	94	96	24	40	50	61	68	77	82	87	0.7	0.8	0.9	1.1	1.3	0.8	0.9	1.0	1.2	1.4
SS A5	28	43	52	62	69	77	82	88	21	31	38	46	52	60	65	72	0.8	0.9	1.0	1.2	1.4	1.0	1.1	1.2	1.4	1.5
SS A6	36	56	68	78	84	91	94	97	24	40	51	61	68	77	82	88	0.7	0.8	0.9	1.1	1.3	0.8	0.9	1.0	1.2	1.4
SS A7	26	38	47	56	63	71	76	82	19	28	34	40	46	53	58	64	0.9	1.0	1.1	1.3	1.4	1.1	1.2	1.3	1.5	1.6
SS A8	35	55	66	77	83	90	93	96	24	39	49	60	67	75	81	86	0.7	0.8	0.9	1.1	1.3	0.8	0.9	1.0	1.2	1.4
SS B1	25	40	50	61	68	77	82	87	17	28	35	43	50	58	64	71	0.8	0.9	1.0	1.2	1.4	1.0	1.1	1.2	1.4	1.5
SS B2	31	50	61	73	79	87	91	95	21	35	45	55	63	72	77	84	0.7	0.8	0.9	1.1	1.3	0.8	0.9	1.1	1.2	1.4
SS B3	24	38	48	58	65	74	80	85	16	26	33	41	47	55	61	68	0.8	0.9	1.1	1.2	1.4	1.0	1.1	1.3	1.4	1.6
SS B4	30	49	61	72	79	86	90	94	20	34	44	54	61	71	76	83	0.7	0.8	1.0	1.1	1.3	0.8	0.9	1.1	1.2	1.4
SS B5	23	36	44	53	60	69	75	81	17	25	31	38	43	51	56	62	0.9	1.0	1.1	1.3	1.4	1.1	1.2	1.3	1.5	1.6
SS B6	30	49	61	72	79	87	91	94	20	34	44	54	62	71	77	83	0.7	0.8	0.9	1.1	1.3	0.8	0.9	1.1	1.2	1.4
SS B7	21	32	40	49	55	64	70	76	15	23	28	34	39	46	51	57	0.9	1.0	1.2	1.3	1.5	1.2	1.3	1.4	1.6	1.7
SS B8	29	48	59	70	77	85	90	94	19	33	43	53	60	69	75	81	0.7	0.8	1.0	1.1	1.3	0.9	0.9	1.1	1.3	1.4
SS C1	25	43	54	65	73	82	86	91	15	27	35	45	52	61	67	74	0.7	0.8	1.0	1.1	1.3	0.9	1.0	1.2	1.3	1.5
SS C2	29	49	62	73	80	88	92	95	17	32	42	53	61	71	77	83	0.7	0.8	0.9	1.1	1.3	0.8	0.9	1.1	1.2	1.4
SS C3	21	34	43	53	60	70	76	82	14	22	28	36	41	49	55	62	0.9	1.0	1.1	1.2	1.4	1.1	1.2	1.3	1.5	1.6
SS C4	26	44	55	67	74	83	88	92	15	28	37	47	54	64	70	77	0.7	0.8	1.0	1.1	1.3	0.9	1.0	1.1	1.3	1.4
SS D1	35	55	67	77	83	90	93	96	24	39	49	59	66	75	80	86	0.7	0.8	0.9	1.1	1.3	0.8	0.9	1.1	1.2	1.4
SS D2	41	64	76	86	91	95	97	99	28	47	59	70	77	85	90	94	0.6	0.7	0.9	1.0	1.2	0.7	0.8	1.0	1.1	1.3
SS D3	32	49	60	71	77	85	89	93	22	35	44	53	60	68	74	80	0.7	0.8	1.0	1.1	1.3	0.9	1.0	1.1	1.3	1.4
SS D4	40	62	74	84	89	95	97	98	27	45	57	68	75	84	88	93	0.6	0.7	0.9	1.1	1.2	0.7	0.8	1.0	1.1	1.3
SS E1	31	50	61	72	79	86	90	94	20	34	43	53	60	69	74	81	0.7	0.8	1.0	1.1	1.3	0.9	1.0	1.1	1.3	1.4
SS E2	36	58	71	81	87	93	96	98	23	41	52	63	71	80	85	90	0.6	0.7	0.9	1.1	1.2	0.7	0.8	1.0	1.2	1.3
SS E3	26	41	51	61	68	77	82	87	18	28	36	44	50	58	64	70	0.8	0.9	1.0	1.2	1.4	1.0	1.1	1.2	1.4	1.5
SS E4	33	54	66	77	83	90	93	96	22	38	48	59	66	75	81	87	0.7	0.8	0.9	1.1	1.3	0.8	0.9	1.0	1.2	1.4

LUFKIN, TEXAS

TYPE	SOLAR SAVINGS FRACTION (%) BASE TEMP: 55 F (731 DD) LCR 200	100	70	50	40	30	25	20	65 F (1944 DD) 200	100	70	50	40	30	25	20	CONSERVATION FACTOR 55 F 200	100	50	30	20	65 F 200	100	50	30	20
WW A1	30	46	56	66	72	80	85	90	19	27	33	39	44	52	57	63	0.8	0.9	1.0	1.2	1.4	1.1	1.2	1.4	1.5	1.6
WW A2	36	57	68	78	84	91	94	97	19	31	40	49	56	65	71	77	0.6	0.8	0.9	1.1	1.3	0.9	1.0	1.1	1.3	1.5
WW A3	39	61	72	82	88	93	96	98	19	33	43	53	60	70	75	82	0.6	0.7	0.9	1.1	1.2	0.8	0.9	1.1	1.2	1.4
WW A4	40	63	74	84	89	94	97	98	19	34	44	55	62	72	78	84	0.6	0.7	0.9	1.1	1.2	0.8	0.9	1.1	1.2	1.4
WW A5	41	65	76	86	91	95	97	99	19	35	46	57	65	74	80	86	0.6	0.7	0.9	1.0	1.2	0.8	0.9	1.0	1.2	1.4
WW A6	42	66	78	87	92	96	98	99	19	36	47	58	66	76	81	87	0.6	0.7	0.9	1.0	1.2	0.8	0.9	1.0	1.2	1.4
WW B1	32	51	62	72	79	86	90	94	16	26	33	41	47	56	61	68	0.7	0.8	1.0	1.1	1.3	1.0	1.1	1.3	1.4	1.6
WW B2	39	63	74	84	89	95	97	98	19	34	45	56	63	73	79	85	0.6	0.7	0.9	1.1	1.2	0.8	0.9	1.1	1.2	1.4
WW B3	46	70	81	90	94	97	99	99	22	41	52	64	72	81	86	91	0.6	0.7	0.8	1.0	1.2	0.7	0.8	1.0	1.2	1.3
WW B4	45	70	81	90	94	97	99	99	21	41	53	65	73	82	87	92	0.6	0.7	0.8	1.0	1.2	0.7	0.8	1.0	1.1	1.3
WW B5	43	68	80	89	93	97	98	99	19	39	51	64	72	81	86	91	0.6	0.7	0.8	1.0	1.2	0.7	0.8	1.0	1.1	1.3
WW C1	48	72	83	91	94	98	99	100	24	42	54	66	74	82	87	92	0.6	0.7	0.8	1.0	1.2	0.7	0.8	1.0	1.1	1.3
WW C2	45	69	81	89	93	97	98	99	21	40	52	64	72	81	86	91	0.6	0.7	0.8	1.0	1.2	0.7	0.8	1.0	1.1	1.3
WW C3	50	76	86	93	96	99	99	100	24	46	60	72	80	88	92	95	0.5	0.6	0.8	1.0	1.2	0.7	0.8	0.9	1.1	1.3
WW C4	46	72	83	91	95	98	99	100	21	43	56	68	76	85	90	94	0.5	0.7	0.8	1.0	1.2	0.7	0.8	0.9	1.1	1.3
TW A1	29	45	54	64	71	79	84	89	18	26	32	38	43	51	56	62	0.8	0.9	1.0	1.2	1.4	1.1	1.2	1.4	1.5	1.7
TW A2	33	52	63	73	80	87	91	94	18	29	36	45	51	60	65	72	0.7	0.8	1.0	1.1	1.3	1.0	1.1	1.2	1.4	1.5
TW A3	34	54	66	76	83	89	93	96	17	29	38	47	54	63	69	76	0.7	0.8	0.9	1.1	1.3	0.9	1.0	1.2	1.3	1.5
TW A4	33	54	66	77	83	90	93	96	16	29	38	47	54	64	70	77	0.6	0.8	0.9	1.1	1.3	0.9	1.0	1.1	1.3	1.5
TW B1	29	45	55	65	72	80	85	90	17	26	32	39	44	52	57	64	0.8	0.9	1.0	1.2	1.4	1.1	1.2	1.3	1.5	1.6
TW B2	31	50	61	72	78	86	90	94	16	27	35	43	50	58	64	71	0.7	0.8	1.0	1.1	1.3	1.0	1.1	1.2	1.4	1.5
TW B3	31	50	61	72	79	86	90	94	15	27	35	43	50	59	65	72	0.7	0.8	1.0	1.1	1.3	1.0	1.1	1.2	1.3	1.5
TW B4	28	47	58	69	76	84	89	93	14	25	33	41	47	56	62	70	0.7	0.8	1.0	1.1	1.3	1.0	1.1	1.2	1.4	1.5
TW C1	28	44	54	65	72	80	85	90	16	25	31	38	44	52	57	64	0.8	0.9	1.0	1.2	1.4	1.1	1.2	1.3	1.5	1.6
TW C2	27	44	55	66	73	81	86	91	15	24	31	39	45	53	59	66	0.7	0.8	1.0	1.2	1.3	1.0	1.1	1.3	1.4	1.6
TW C3	25	42	53	63	70	79	84	89	14	23	29	37	43	51	56	63	0.8	0.9	1.0	1.2	1.4	1.1	1.2	1.3	1.5	1.6
TW C4	23	37	47	57	64	73	78	84	13	21	27	33	38	46	51	57	0.8	0.9	1.1	1.2	1.4	1.2	1.3	1.4	1.5	1.7
TW D1	27	44	54	65	72	80	85	90	13	22	28	36	41	49	55	61	0.7	0.8	1.0	1.2	1.3	1.1	1.2	1.3	1.5	1.6
TW D2	36	57	69	80	86	92	95	97	18	32	41	51	58	68	74	80	0.6	0.7	0.9	1.1	1.3	0.9	1.0	1.1	1.3	1.4
TW D3	40	62	74	84	89	94	96	98	19	35	45	56	63	73	78	85	0.6	0.7	0.9	1.1	1.2	0.8	0.9	1.1	1.2	1.4
TW D4	41	64	76	86	91	95	97	99	20	37	48	59	67	77	82	88	0.6	0.7	0.9	1.0	1.2	0.8	0.9	1.0	1.2	1.3
TW D5	39	63	75	85	90	95	97	99	18	36	47	59	66	76	82	88	0.6	0.7	0.9	1.0	1.2	0.8	0.9	1.0	1.2	1.3
TW E1	43	67	78	87	92	96	98	99	22	39	49	61	68	77	83	88	0.6	0.7	0.9	1.0	1.2	0.8	0.9	1.0	1.2	1.4
TW E2	42	66	77	86	91	96	98	99	21	38	49	60	68	77	83	88	0.6	0.7	0.9	1.0	1.2	0.8	0.9	1.0	1.2	1.3
TW E3	49	73	84	92	95	98	99	100	24	45	57	69	77	85	90	94	0.5	0.7	0.8	1.0	1.2	0.7	0.8	1.0	1.1	1.3
TW E4	45	70	81	90	94	97	99	99	22	41	54	66	74	83	88	92	0.6	0.7	0.8	1.0	1.2	0.7	0.8	1.0	1.1	1.3
TW F1	28	44	54	65	71	80	84	89	16	25	31	38	43	51	56	62	0.8	0.9	1.0	1.2	1.4	1.1	1.2	1.3	1.5	1.6
TW F2	30	49	60	71	78	86	90	94	15	26	34	42	48	57	63	70	0.7	0.8	1.0	1.1	1.3	1.0	1.1	1.2	1.4	1.5
TW F3	30	50	62	73	80	87	91	95	14	26	34	43	50	59	65	72	0.7	0.8	0.9	1.1	1.3	0.9	1.0	1.2	1.3	1.5
TW F4	28	48	60	71	78	86	90	94	12	24	32	41	48	58	64	71	0.7	0.8	1.0	1.1	1.3	1.0	1.1	1.2	1.3	1.5
TW G1	26	43	53	64	71	79	84	89	14	23	29	36	42	50	55	62	0.8	0.9	1.0	1.2	1.4	1.1	1.2	1.3	1.5	1.6
TW G2	26	45	56	67	75	83	87	92	12	22	30	38	45	54	59	67	0.7	0.8	1.0	1.1	1.3	1.0	1.1	1.3	1.4	1.6
TW G3	24	43	55	66	73	82	86	91	10	21	28	37	43	52	58	66	0.7	0.8	1.0	1.2	1.3	1.0	1.1	1.3	1.4	1.6
TW G4	20	37	48	60	67	77	82	88	8	17	24	32	38	47	52	60	0.8	0.9	1.0	1.2	1.4	1.1	1.2	1.3	1.5	1.6
TW H1	23	39	50	60	68	77	82	87	11	20	26	33	39	47	52	59	0.8	0.9	1.0	1.2	1.4	1.1	1.2	1.4	1.5	1.6
TW H2	20	36	47	58	66	75	80	86	8	17	23	31	37	45	50	58	0.8	0.9	1.1	1.2	1.4	1.1	1.2	1.4	1.5	1.7
TW H3	16	32	42	53	60	70	76	82	6	14	20	27	32	40	46	53	0.8	0.9	1.1	1.2	1.4	1.2	1.3	1.4	1.6	1.7
TW H4	12	25	33	43	50	60	66	73	5	11	16	21	26	33	37	44	0.9	1.0	1.2	1.3	1.5	1.3	1.4	1.6	1.7	1.9
TW I1	24	40	51	62	69	77	82	88	11	19	26	32	38	46	51	58	0.8	0.9	1.0	1.2	1.4	1.1	1.2	1.4	1.5	1.7
TW I2	32	54	66	77	83	90	93	96	15	28	37	47	55	64	70	78	0.6	0.8	0.9	1.1	1.3	0.9	1.0	1.1	1.3	1.4
TW I3	35	57	70	80	86	92	95	97	16	31	41	51	59	69	75	81	0.6	0.7	0.9	1.1	1.3	0.8	0.9	1.1	1.2	1.4
TW I4	37	61	73	83	89	94	96	98	17	33	44	56	64	74	79	86	0.6	0.7	0.9	1.1	1.2	0.8	0.9	1.0	1.2	1.4
TW I5	37	61	73	84	89	94	97	98	16	33	44	56	63	74	80	86	0.6	0.7	0.9	1.1	1.2	0.8	0.9	1.0	1.2	1.4
TW J1	41	64	76	85	90	95	97	99	19	36	47	58	66	75	81	87	0.6	0.7	0.9	1.0	1.2	0.8	0.9	1.0	1.2	1.4
TW J2	39	62	74	84	90	95	97	99	18	35	45	57	65	75	80	86	0.6	0.7	0.9	1.1	1.2	0.8	0.9	1.0	1.2	1.4
TW J3	45	70	81	90	94	97	99	99	21	41	53	66	74	83	88	92	0.6	0.7	0.8	1.0	1.2	0.7	0.8	1.0	1.1	1.3
TW J4	42	67	79	88	93	97	98	99	19	38	50	63	71	81	86	91	0.6	0.7	0.9	1.0	1.2	0.7	0.8	1.0	1.1	1.3
DG A1	31	50	61	70	76	83	86	90	14	24	31	39	44	51	55	60	0.7	0.8	1.0	1.2	1.4	1.0	1.1	1.3	1.6	1.9
DG A2	30	51	63	73	80	86	90	93	14	28	34	44	50	58	64	70	0.7	0.8	1.0	1.1	1.3	1.0	1.0	1.2	1.4	1.6
DG A3	34	58	70	80	85	91	93	95	17	30	41	52	60	69	75	81	0.6	0.7	0.9	1.2	1.5	0.9	0.9	1.1	1.3	1.5
DG B1	32	55	67	78	83	89	91	94	14	25	33	42	49	58	63	69	0.6	0.7	0.9	1.2	1.4	1.0	1.1	1.2	1.4	1.7
DG B2	32	55	69	80	86	92	94	97	15	27	36	47	54	65	71	77	0.6	0.7	0.9	1.1	1.3	1.0	1.0	1.1	1.3	1.5
DG B3	35	60	74	84	89	94	96	97	17	31	42	54	63	74	80	86	0.6	0.7	0.9	1.1	1.4	0.9	0.9	1.0	1.2	1.4
DG C1	37	62	75	85	89	93	95	96	16	29	39	50	58	67	73	78	0.6	0.7	0.9	1.1	1.5	0.9	1.0	1.1	1.3	1.5
DG C2	36	61	75	85	90	95	97	98	17	30	41	53	62	72	78	84	0.6	0.7	0.9	1.1	1.3	0.9	0.9	1.0	1.2	1.4
DG C3	39	66	80	89	93	96	97	98	19	35	47	60	70	80	86	91	0.6	0.7	0.8	1.1	1.6	0.8	0.9	1.0	1.1	1.3
SS A1	39	59	70	79	85	91	94	97	23	35	44	53	60	68	74	80	0.6	0.8	0.9	1.1	1.3	0.9	1.0	1.1	1.3	1.5
SS A2	47	69	80	88	92	96	98	99	27	44	54	65	72	80	85	90	0.6	0.7	0.9	1.0	1.2	0.7	0.9	1.0	1.2	1.3
SS A3	37	56	67	77	83	89	92	96	22	33	41	50	56	65	70	77	0.7	0.8	0.9	1.1	1.3	0.9	1.0	1.2	1.3	1.5
SS A4	47	69	79	88	92	96	98	99	27	43	54	64	71	80	85	89	0.6	0.7	0.9	1.0	1.2	0.8	0.9	1.0	1.2	1.4
SS A5	37	55	65	75	81	88	91	95	23	33	41	49	55	63	69	75	0.7	0.8	1.0	1.1	1.3	0.9	1.0	1.2	1.3	1.5
SS A6	47	69	80	88	92	96	98	99	27	43	54	65	72	80	85	90	0.6	0.7	0.9	1.0	1.2	0.8	0.9	1.0	1.2	1.3
SS A7	33	50	60	70	76	83	87	92	21	30	36	44	49	57	62	68	0.7	0.8	1.0	1.2	1.3	1.0	1.1	1.3	1.4	1.6
SS A8	47	68	79	87	92	96	98	99	27	43	53	63	70	79	84	89	0.6	0.7	0.9	1.0	1.2	0.8	0.9	1.0	1.2	1.4
SS B1	34	53	64	74	80	87	91	94	19	30	38	47	53	62	67	74	0.7	0.8	1.0	1.1	1.3	0.9	1.0	1.2	1.3	1.5
SS B2	41	63	74	84	89	94	96	98	23	38	48	59	66	75	80	86	0.6	0.7	0.9	1.1	1.2	0.8	0.9	1.0	1.2	1.4
SS B3	33	51	62	72	79	86	90	94	18	29	36	45	51	59	65	72	0.7	0.8	1.0	1.1	1.3	1.0	1.1	1.2	1.4	1.5
SS B4	41	63	74	83	89	94	96	98	23	38	48	58	65	74	79	85	0.6	0.7	0.9	1.1	1.2	0.8	0.9	1.1	1.2	1.4
SS B5	30	47	57	67	73	81	86	90	18	27	34	41	47	54	60	66	0.7	0.8	1.0	1.2	1.3	1.0	1.1	1.3	1.4	1.6
SS B6	41	63	74	83	89	94	96	98	22	37	47	58	65	74	80	86	0.6	0.7	0.9	1.1	1.2	0.8	0.9	1.1	1.2	1.4
SS B7	28	43	53	63	69	77	82	87	17	25	31	37	42	50	55	61	0.8	0.9	1.0	1.2	1.4	1.1	1.2	1.4	1.5	1.7
SS B8	40	61	73	82	88	93	96	98	22	36	46	56	64	73	78	84	0.6	0.7	0.9	1.1	1.3	0.8	0.9	1.1	1.2	1.4
SS C1	35	56	67	78	84	90	94	96	17	29	38	48	55	64	70	77	0.6	0.8	0.9	1.1	1.3	0.9	1.0	1.1	1.3	1.5
SS C2	39	62	74	84	89	94	97	98	19	35	45	56	63	73	79	85	0.6	0.7	0.9	1.1	1.2	0.8	0.9	1.1	1.2	1.4
SS C3	28	45	56	66	73	81	86	91	15	24	31	38	44	52	58	65	0.7	0.8	1.0	1.2	1.3	1.1	1.2	1.3	1.4	1.6
SS C4	35	56	68	78	85	91	94	97	17	31	40	50	57	67	73	79	0.6	0.7	0.9	1.1	1.3	0.9	1.0	1.1	1.3	1.4
SS D1	47	69	79	88	92	96	98	99	26	42	52	63	69	78	83	88	0.6	0.7	0.9	1.0	1.2	0.8	0.9	1.0	1.2	1.4
SS D2	54	77	87	93	96	99	99	100	30	50	62	73	80	88	91	95	0.5	0.7	0.8	1.0	1.2	0.7	0.8	0.9	1.1	1.3
SS D3	42	63	73	82	88	93	95	98	25	38	47	56	63	72	77	83	0.6	0.7	0.9	1.1	1.3	0.8	0.9	1.1	1.3	1.4
SS D4	53	75	85	92	95	98	99	100	29	49	60	71	78	86	90	94	0.5	0.7	0.8	1.0	1.2	0.7	0.8	1.0	1.1	1.3
SS E1	42	64	75	84	89	94	96	98	23	37	46	55	63	72	78	84	0.6	0.7	0.9	1.1	1.2	0.8	0.9	1.1	1.2	1.4
SS E2	49	72	83	90	94	97	99	99	26	44	56	67	74	83	87	92	0.6	0.7	0.8	1.0	1.2	0.7	0.8	1.0	1.1	1.3
SS E3	35	54	64	74	81	88	91	95	20	31	39	47	53	62	67	74	0.7	0.8	1.0	1.1	1.3	0.9	1.0	1.2	1.3	1.5
SS E4	45	67	78	87	92	96	98	99	24	41	51	62	70	78	83	89	0.6	0.7	0.9	1.0	1.2	0.8	0.9	1.0	1.2	1.4

MIDLAND-ODESSA, TEXAS

TYPE	SOLAR SAVINGS FRACTION (%) — BASE TEMP: 55 F (1159 DD)								65 F (2627 DD)								CONSERVATION FACTOR — 55 F					65 F				
LCR	200	100	70	50	40	30	25	20	200	100	70	50	40	30	25	20	200	100	50	30	20	200	100	50	30	20
WW A1	30	46	56	66	73	81	85	90	20	29	36	43	49	57	62	69	0.8	0.9	1.0	1.2	1.3	1.0	1.1	1.3	1.4	1.6
WW A2	36	57	68	79	85	91	94	97	21	35	44	54	62	71	77	83	0.6	0.7	0.9	1.1	1.3	0.8	0.9	1.1	1.2	1.4
WW A3	38	61	72	83	88	94	96	98	21	37	48	59	66	76	81	87	0.6	0.7	0.9	1.1	1.2	0.8	0.9	1.0	1.2	1.4
WW A4	40	63	75	84	90	95	97	99	21	38	49	61	68	78	83	89	0.6	0.7	0.9	1.0	1.2	0.8	0.9	1.0	1.2	1.3
WW A5	41	65	77	86	91	96	98	99	21	40	51	63	71	80	85	91	0.6	0.7	0.9	1.0	1.2	0.7	0.8	1.0	1.2	1.3
WW A6	42	66	78	87	92	96	98	99	21	40	52	64	72	81	86	92	0.6	0.7	0.9	1.0	1.2	0.7	0.8	1.0	1.1	1.3
WW B1	32	50	62	72	79	86	90	94	18	29	37	47	53	62	68	75	0.7	0.8	1.0	1.1	1.3	0.9	1.0	1.2	1.3	1.5
WW B2	39	63	75	85	90	95	97	99	21	38	50	62	69	79	84	90	0.6	0.7	0.9	1.0	1.2	0.8	0.9	1.0	1.2	1.3
WW B3	46	70	82	90	94	97	99	100	25	46	58	70	77	86	90	94	0.6	0.7	0.8	1.0	1.2	0.7	0.8	0.9	1.1	1.3
WW B4	45	70	82	90	94	98	99	100	24	45	58	71	78	87	91	95	0.6	0.7	0.8	1.0	1.2	0.7	0.8	0.9	1.1	1.3
WW B5	43	68	80	89	94	97	99	99	22	43	57	69	77	86	90	94	0.6	0.7	0.8	1.0	1.2	0.7	0.8	0.9	1.1	1.3
WW C1	47	72	83	91	95	98	99	100	27	47	60	72	79	87	91	95	0.6	0.7	0.8	1.0	1.2	0.7	0.8	0.9	1.1	1.3
WW C2	45	69	81	90	94	97	99	99	24	45	57	70	77	86	90	94	0.6	0.7	0.8	1.0	1.2	0.7	0.8	0.9	1.1	1.3
WW C3	50	76	86	94	97	99	99	100	27	51	65	77	84	91	95	97	0.5	0.6	0.8	1.0	1.2	0.6	0.7	0.9	1.1	1.2
WW C4	46	72	84	92	95	98	99	100	24	47	61	74	81	89	93	96	0.5	0.7	0.8	1.0	1.2	0.7	0.8	0.9	1.1	1.3
TW A1	29	44	54	64	71	80	84	89	19	28	35	42	48	56	61	68	0.8	0.9	1.0	1.2	1.4	1.0	1.1	1.3	1.4	1.6
TW A2	33	51	63	73	80	87	91	95	19	31	40	49	56	66	71	78	0.7	0.8	0.9	1.1	1.3	0.9	1.0	1.1	1.3	1.4
TW A3	34	54	66	77	83	90	93	96	19	33	42	52	60	69	75	82	0.7	0.8	0.9	1.1	1.3	0.9	1.0	1.1	1.2	1.4
TW A4	33	54	66	77	84	90	94	97	18	32	42	53	60	70	76	83	0.7	0.8	0.9	1.1	1.3	0.8	0.9	1.1	1.2	1.4
TW B1	29	45	55	65	72	81	85	90	18	28	35	43	49	57	63	70	0.8	0.9	1.0	1.2	1.3	1.0	1.1	1.2	1.4	1.5
TW B2	31	50	61	72	79	86	90	94	18	30	38	48	55	64	70	77	0.7	0.8	1.0	1.1	1.3	0.9	1.0	1.1	1.3	1.5
TW B3	30	50	62	73	79	87	91	95	17	30	38	48	55	65	71	78	0.7	0.8	0.9	1.1	1.3	0.9	1.0	1.1	1.3	1.4
TW B4	28	47	58	70	77	85	89	93	16	28	36	46	53	62	68	75	0.7	0.8	1.0	1.1	1.3	0.9	1.0	1.2	1.3	1.5
TW C1	28	44	54	65	72	80	85	90	17	27	34	42	48	57	63	70	0.8	0.9	1.0	1.2	1.3	1.0	1.1	1.2	1.4	1.5
TW C2	27	44	55	66	73	82	86	91	16	26	34	43	49	58	64	72	0.7	0.8	1.0	1.2	1.3	1.0	1.1	1.2	1.4	1.5
TW C3	26	42	52	63	70	79	84	89	15	25	32	41	47	56	62	69	0.8	0.9	1.0	1.2	1.3	1.0	1.1	1.2	1.4	1.5
TW C4	23	37	47	57	64	73	78	85	14	23	29	36	42	50	56	63	0.8	0.9	1.1	1.2	1.4	1.1	1.2	1.3	1.5	1.6
TW D1	26	43	54	65	72	81	85	90	14	25	32	40	46	55	61	68	0.8	0.8	1.0	1.2	1.3	1.0	1.1	1.2	1.4	1.5
TW D2	36	57	69	80	86	92	95	97	20	35	45	56	64	74	79	86	0.6	0.7	0.9	1.1	1.3	0.8	0.9	1.1	1.2	1.4
TW D3	39	62	74	84	89	95	97	98	22	39	50	61	69	78	84	89	0.6	0.7	0.9	1.1	1.2	0.8	0.9	1.0	1.2	1.3
TW D4	41	65	77	86	91	96	98	99	22	41	53	65	73	82	87	92	0.6	0.7	0.9	1.0	1.2	0.7	0.8	1.0	1.1	1.3
TW D5	39	63	75	85	91	95	97	99	21	40	52	64	72	81	86	91	0.6	0.7	0.9	1.0	1.2	0.7	0.8	1.0	1.1	1.3
TW E1	43	67	78	87	92	96	98	99	24	43	55	66	74	83	87	92	0.6	0.7	0.9	1.0	1.2	0.7	0.8	1.0	1.1	1.3
TW E2	42	66	78	87	92	96	98	99	23	42	54	66	73	82	87	92	0.6	0.7	0.9	1.0	1.2	0.7	0.8	1.0	1.1	1.3
TW E3	48	73	84	92	96	98	99	100	27	49	62	74	82	89	93	96	0.5	0.7	0.8	1.0	1.2	0.7	0.8	0.9	1.1	1.3
TW E4	45	70	82	90	94	98	99	100	24	46	59	71	79	87	91	95	0.6	0.7	0.8	1.0	1.2	0.7	0.8	0.9	1.1	1.3
TW F1	28	44	54	65	72	80	85	90	17	27	34	42	48	56	62	69	0.8	0.9	1.0	1.2	1.3	1.0	1.1	1.3	1.4	1.6
TW F2	30	49	61	71	78	86	90	94	17	29	37	47	54	63	69	76	0.7	0.8	1.0	1.1	1.3	0.9	1.0	1.2	1.3	1.5
TW F3	30	50	62	73	80	88	91	95	16	29	38	48	56	65	71	78	0.7	0.8	0.9	1.1	1.3	0.9	1.0	1.1	1.3	1.4
TW F4	28	48	60	72	79	87	91	94	13	27	36	46	54	64	70	77	0.7	0.8	0.9	1.1	1.3	0.9	1.0	1.1	1.3	1.4
TW G1	26	42	53	64	71	80	84	90	15	25	32	40	47	55	61	68	0.8	0.9	1.0	1.2	1.3	1.0	1.1	1.3	1.4	1.5
TW G2	26	45	56	68	75	83	88	92	14	25	34	43	50	59	66	73	0.7	0.8	1.0	1.1	1.3	1.0	1.0	1.2	1.3	1.5
TW G3	24	43	55	66	74	82	87	92	12	23	32	41	48	58	64	72	0.7	0.8	1.0	1.1	1.3	1.0	1.1	1.2	1.3	1.5
TW G4	20	37	48	60	68	77	82	88	9	19	27	36	43	52	58	66	0.8	0.9	1.0	1.2	1.4	1.0	1.1	1.3	1.4	1.5
TW H1	23	39	49	60	68	77	82	88	12	22	29	37	43	52	58	65	0.8	0.9	1.0	1.2	1.4	1.1	1.1	1.3	1.4	1.6
TW H2	20	36	47	58	66	75	81	87	9	19	26	35	41	50	56	64	0.8	0.9	1.0	1.2	1.4	1.1	1.1	1.3	1.4	1.6
TW H3	17	31	42	52	60	70	76	82	7	16	23	30	36	45	51	58	0.9	0.9	1.1	1.2	1.4	1.1	1.2	1.3	1.5	1.6
TW H4	13	25	33	43	50	60	66	74	5	12	17	24	29	36	42	49	1.0	1.0	1.2	1.3	1.5	1.3	1.4	1.5	1.6	1.8
TW I1	24	40	51	62	69	78	83	88	12	22	29	37	43	51	57	65	0.8	0.9	1.0	1.2	1.4	1.1	1.1	1.3	1.4	1.6
TW I2	32	54	66	77	84	90	94	97	17	31	42	53	60	70	76	83	0.7	0.8	0.9	1.1	1.3	0.8	0.9	1.1	1.2	1.4
TW I3	35	57	70	80	87	93	95	98	18	34	45	57	65	74	80	86	0.6	0.7	0.9	1.1	1.2	0.8	0.9	1.0	1.2	1.4
TW I4	37	61	73	84	89	95	97	99	19	37	49	61	69	79	84	90	0.6	0.7	0.9	1.0	1.2	0.8	0.9	1.0	1.2	1.3
TW I5	36	61	74	84	90	95	97	99	18	37	49	62	70	80	85	90	0.6	0.7	0.9	1.0	1.2	0.7	0.8	1.0	1.1	1.3
TW J1	40	64	76	86	91	95	97	99	22	40	52	64	71	81	86	91	0.6	0.7	0.9	1.0	1.2	0.7	0.8	1.0	1.1	1.3
TW J2	39	62	75	85	90	95	97	99	20	39	50	62	70	80	85	91	0.6	0.7	0.9	1.0	1.2	0.7	0.8	1.0	1.1	1.3
TW J3	44	70	82	90	94	98	99	100	24	46	59	71	79	87	91	95	0.6	0.7	0.8	1.0	1.2	0.7	0.8	0.9	1.1	1.3
TW J4	41	67	79	88	93	97	98	99	21	43	56	68	76	85	90	94	0.6	0.7	0.8	1.0	1.2	0.7	0.8	0.9	1.1	1.3
DG A1	30	50	61	71	77	83	87	90	15	28	36	44	50	57	62	67	0.7	0.8	1.0	1.2	1.4	0.9	1.0	1.2	1.5	1.7
DG A2	30	51	63	74	80	87	90	94	16	29	39	49	55	64	70	76	0.7	0.8	0.9	1.1	1.3	0.9	1.0	1.1	1.3	1.5
DG A3	34	58	70	81	86	91	93	95	18	34	45	58	65	75	80	85	0.6	0.7	0.9	1.1	1.5	0.8	0.9	1.0	1.2	1.4
DG B1	32	55	68	78	84	89	92	94	15	28	38	49	56	65	70	76	0.6	0.7	0.9	1.1	1.4	0.9	1.0	1.1	1.3	1.6
DG B2	31	55	69	80	86	92	95	97	16	30	41	53	61	71	77	83	0.6	0.7	0.9	1.1	1.3	0.9	0.9	1.0	1.2	1.4
DG B3	34	60	74	85	90	94	96	97	19	34	47	60	69	80	85	90	0.6	0.7	0.9	1.1	1.5	0.8	0.9	1.0	1.1	1.4
DG C1	36	62	76	85	90	94	95	97	18	33	44	57	66	75	80	85	0.6	0.7	0.9	1.1	1.5	0.8	0.9	1.0	1.2	1.5
DG C2	36	62	76	86	91	95	97	99	19	34	46	60	69	79	84	89	0.6	0.7	0.9	1.1	1.3	0.8	0.9	1.0	1.2	1.4
DG C3	39	67	81	90	94	97	98	98	21	39	53	67	77	86	90	94	0.6	0.7	0.8	1.1	1.7	0.8	0.8	0.9	1.1	1.3
SS A1	38	58	69	79	85	91	94	97	25	38	48	57	64	73	78	84	0.6	0.8	0.9	1.1	1.3	0.8	0.9	1.1	1.2	1.4
SS A2	46	68	79	88	92	96	98	99	29	47	58	69	76	84	89	93	0.6	0.7	0.9	1.0	1.2	0.7	0.8	1.0	1.1	1.3
SS A3	36	55	66	76	82	89	92	95	23	36	45	54	60	69	75	81	0.7	0.8	0.9	1.1	1.3	0.9	1.0	1.1	1.3	1.4
SS A4	46	68	79	87	92	96	98	99	29	47	58	68	75	83	88	92	0.6	0.7	0.9	1.0	1.2	0.7	0.8	1.0	1.1	1.3
SS A5	36	54	64	74	80	87	91	94	24	36	44	53	59	68	73	79	0.7	0.8	1.0	1.1	1.3	0.9	1.0	1.1	1.3	1.5
SS A6	46	68	79	88	92	96	98	99	29	47	58	69	76	84	88	93	0.6	0.7	0.9	1.0	1.2	0.7	0.8	1.0	1.1	1.3
SS A7	32	49	59	68	75	82	87	91	22	32	39	47	53	61	66	73	0.7	0.8	1.0	1.2	1.3	1.0	1.1	1.2	1.4	1.5
SS A8	45	67	78	87	91	96	97	99	29	46	57	67	74	83	87	92	0.6	0.7	0.9	1.0	1.2	0.7	0.8	1.0	1.2	1.3
SS B1	33	52	63	73	80	87	91	94	20	33	41	51	57	66	72	79	0.7	0.8	1.0	1.1	1.3	0.9	1.0	1.1	1.3	1.5
SS B2	40	62	74	83	89	94	96	98	25	41	52	63	70	79	84	39	0.6	0.7	0.9	1.1	1.2	0.8	0.9	1.0	1.2	1.3
SS B3	32	50	61	71	78	85	89	93	20	31	40	48	55	64	69	76	0.7	0.8	1.0	1.1	1.3	0.9	1.0	1.2	1.3	1.5
SS B4	40	61	73	83	88	93	96	98	24	41	51	62	69	78	83	89	0.6	0.7	0.9	1.1	1.2	0.8	0.9	1.0	1.2	1.4
SS B5	30	46	56	66	73	81	85	90	19	29	36	44	50	59	64	71	0.8	0.9	1.0	1.2	1.3	1.0	1.1	1.2	1.4	1.5
SS B6	40	62	73	83	88	94	96	98	24	41	51	62	69	78	84	89	0.6	0.7	0.9	1.1	1.2	0.8	0.9	1.0	1.2	1.3
SS B7	27	42	51	61	68	77	81	87	18	27	33	40	46	54	59	66	0.8	0.9	1.0	1.2	1.4	1.0	1.2	1.3	1.5	1.6
SS B8	39	60	72	82	87	93	95	98	23	39	50	60	68	77	82	88	0.6	0.7	0.9	1.1	1.3	0.8	0.9	1.0	1.2	1.4
SS C1	35	56	67	78	84	91	94	97	19	33	43	53	60	70	76	83	0.6	0.7	0.9	1.1	1.3	0.8	0.9	1.1	1.2	1.4
SS C2	39	62	74	84	90	95	97	99	21	39	50	61	69	79	84	89	0.6	0.7	0.9	1.0	1.2	0.8	0.9	1.0	1.2	1.3
SS C3	28	45	55	66	73	82	86	91	16	27	34	43	49	58	64	71	0.7	0.8	1.0	1.2	1.3	1.0	1.1	1.2	1.4	1.5
SS C4	35	56	68	79	85	91	94	97	19	34	44	55	63	72	78	85	0.6	0.7	0.9	1.1	1.3	0.8	0.9	1.1	1.2	1.4
SS D1	46	68	79	87	92	96	98	99	29	46	57	67	74	82	87	92	0.6	0.7	0.9	1.0	1.2	0.7	0.8	1.0	1.2	1.3
SS D2	53	77	86	93	96	99	99	100	33	55	67	78	84	91	94	97	0.5	0.7	0.8	1.0	1.2	0.6	0.8	0.9	1.1	1.3
SS D3	41	62	73	82	87	93	95	98	27	41	51	61	68	76	81	87	0.6	0.7	0.9	1.1	1.3	0.8	0.9	1.0	1.2	1.4
SS D4	52	75	85	92	95	98	99	100	32	53	65	76	82	89	93	96	0.5	0.7	0.8	1.0	1.2	0.7	0.8	0.9	1.1	1.3
SS E1	41	63	74	83	89	94	96	98	25	40	51	61	68	77	82	88	0.6	0.7	0.9	1.1	1.2	0.8	0.9	1.0	1.2	1.4
SS E2	48	71	82	90	94	97	99	99	28	48	60	72	79	87	91	94	0.6	0.7	0.8	1.0	1.2	0.7	0.8	0.9	1.1	1.3
SS E3	34	52	63	74	80	87	91	94	21	34	42	51	58	66	72	79	0.7	0.8	1.0	1.1	1.3	0.9	1.0	1.1	1.3	1.5
SS E4	44	66	78	87	91	96	98	99	26	44	56	67	74	83	87	92	0.6	0.7	0.9	1.0	1.2	0.7	0.8	1.0	1.1	1.3

PORT ARTHUR, TEXAS

TYPE	SOLAR SAVINGS FRACTION (%) BASE TEMP: 55 F (504 DD)								65 F (1518 DD)								CONSERVATION FACTOR 55 F					65 F				
LCR	200	100	70	50	40	30	25	20	200	100	70	50	40	30	25	20	200	100	50	30	20	200	100	50	30	20
WW A1	36	54	64	73	79	86	90	94	21	30	36	44	50	58	63	69	0.7	0.8	1.0	1.1	1.3	1.0	1.1	1.3	1.4	1.6
WW A2	44	65	76	85	90	95	97	98	22	36	45	55	62	71	77	83	0.6	0.7	0.9	1.1	1.2	0.8	0.9	1.1	1.2	1.4
WW A3	47	69	80	88	92	96	98	99	22	38	48	59	66	76	81	87	0.6	0.7	0.9	1.0	1.2	0.8	0.9	1.0	1.2	1.4
WW A4	48	71	81	90	94	97	98	99	22	39	50	61	69	78	83	89	0.6	0.7	0.9	1.0	1.2	0.8	0.9	1.0	1.2	1.4
WW A5	50	73	83	91	95	98	99	100	22	40	52	63	71	80	85	90	0.6	0.7	0.8	1.0	1.2	0.7	0.8	1.0	1.2	1.3
WW A6	51	74	84	92	95	98	99	100	22	41	53	65	72	81	86	91	0.5	0.7	0.8	1.0	1.2	0.7	0.8	1.0	1.1	1.3
WW B1	39	59	70	80	85	91	94	97	19	30	38	47	54	63	68	75	0.6	0.7	0.9	1.1	1.3	0.9	1.0	1.2	1.3	1.5
WW B2	48	71	82	90	94	97	98	99	22	39	51	62	70	79	84	89	0.6	0.7	0.8	1.0	1.2	0.8	0.9	1.0	1.2	1.3
WW B3	55	78	87	94	97	99	99	100	26	46	59	70	77	86	90	94	0.5	0.6	0.8	1.0	1.2	0.7	0.8	1.0	1.1	1.3
WW B4	54	78	87	94	97	99	99	100	25	46	59	71	78	87	91	95	0.5	0.6	0.8	1.0	1.2	0.7	0.8	0.9	1.1	1.3
WW B5	52	76	86	93	96	99	99	100	23	44	57	69	77	86	90	94	0.5	0.6	0.8	1.0	1.2	0.7	0.8	0.9	1.1	1.3
WW C1	56	79	88	94	97	99	100	100	28	48	60	72	79	87	91	95	0.5	0.6	0.8	1.0	1.2	0.7	0.8	0.9	1.1	1.3
WW C2	54	77	87	94	96	99	99	100	25	45	58	70	77	86	90	94	0.5	0.6	0.8	1.0	1.2	0.7	0.8	0.9	1.1	1.3
WW C3	59	82	91	96	98	99	100	100	28	52	66	78	84	91	94	97	0.5	0.6	0.8	1.0	1.2	0.6	0.7	0.9	1.1	1.3
WW C4	55	79	89	95	97	99	100	100	25	48	62	74	81	89	93	96	0.5	0.6	0.8	1.0	1.2	0.7	0.8	0.9	1.1	1.3
TW A1	35	52	62	72	78	85	89	93	20	29	36	43	49	57	62	68	0.7	0.8	1.0	1.1	1.3	1.0	1.1	1.3	1.4	1.6
TW A2	40	60	71	80	86	92	95	97	20	32	41	50	57	66	71	78	0.6	0.7	0.9	1.1	1.3	0.9	1.0	1.1	1.3	1.5
TW A3	42	63	74	83	88	94	96	98	20	34	43	53	60	69	75	81	0.6	0.7	0.9	1.1	1.2	0.9	1.0	1.1	1.3	1.4
TW A4	41	63	74	84	89	94	96	98	19	33	43	53	61	70	76	82	0.6	0.7	0.9	1.1	1.2	0.8	0.9	1.1	1.2	1.4
TW B1	35	53	63	73	79	86	90	94	19	29	36	44	49	58	63	70	0.7	0.8	1.0	1.1	1.3	1.0	1.1	1.2	1.4	1.6
TW B2	38	58	69	79	85	91	94	97	19	31	39	49	55	64	70	77	0.6	0.8	0.9	1.1	1.3	0.9	1.0	1.1	1.3	1.5
TW B3	38	58	70	80	85	91	94	97	18	31	39	49	56	65	71	78	0.6	0.7	0.9	1.1	1.3	0.9	1.0	1.1	1.3	1.5
TW B4	35	55	67	77	83	90	93	96	17	29	37	46	53	62	68	75	0.7	0.8	0.9	1.1	1.3	0.9	1.0	1.2	1.3	1.5
TW C1	34	52	62	72	79	86	90	94	18	28	35	43	49	57	63	70	0.7	0.8	1.0	1.1	1.3	1.0	1.1	1.2	1.4	1.5
TW C2	34	52	63	74	80	87	91	95	17	27	35	44	50	59	65	72	0.7	0.8	1.0	1.1	1.3	1.0	1.1	1.2	1.4	1.5
TW C3	32	50	61	71	78	85	89	93	16	26	33	42	48	56	62	69	0.7	0.8	1.0	1.1	1.3	1.0	1.1	1.2	1.4	1.5
TW C4	28	44	55	65	72	80	85	89	15	24	30	37	43	51	56	63	0.8	0.9	1.0	1.2	1.4	1.1	1.2	1.3	1.5	1.6
TW D1	33	52	63	73	80	87	91	94	16	26	33	41	47	56	61	68	0.7	0.8	1.0	1.1	1.3	1.0	1.1	1.2	1.4	1.6
TW D2	44	66	77	86	91	95	97	99	21	36	46	57	64	74	79	85	0.6	0.7	0.9	1.1	1.2	0.8	0.9	1.1	1.2	1.4
TW D3	48	70	81	89	93	97	98	99	23	40	51	62	69	78	84	89	0.6	0.7	0.9	1.0	1.2	0.8	0.9	1.0	1.2	1.3
TW D4	49	72	83	91	95	98	99	100	23	42	53	65	73	82	87	91	0.6	0.7	0.8	1.0	1.2	0.7	0.8	1.0	1.1	1.3
TW D5	48	71	82	90	94	97	99	99	22	40	52	64	72	81	86	91	0.6	0.7	0.8	1.0	1.2	0.7	0.8	1.0	1.1	1.3
TW E1	52	74	84	92	95	98	99	100	26	44	55	67	74	83	87	92	0.5	0.7	0.8	1.0	1.2	0.7	0.8	1.0	1.1	1.3
TW E2	51	73	84	91	95	98	99	100	24	43	54	66	73	82	87	92	0.5	0.7	0.8	1.0	1.2	0.7	0.8	1.0	1.1	1.3
TW E3	57	80	89	95	98	99	100	100	28	50	63	75	82	89	93	96	0.5	0.6	0.8	1.0	1.2	0.7	0.8	0.9	1.1	1.3
TW E4	54	77	87	94	97	99	99	100	25	47	59	71	79	87	91	95	0.5	0.6	0.8	1.0	1.2	0.7	0.8	0.9	1.1	1.3
TW F1	34	52	62	72	79	86	90	94	19	28	35	42	48	56	62	69	0.7	0.8	1.0	1.1	1.3	1.0	1.1	1.3	1.4	1.6
TW F2	37	57	69	79	85	91	94	97	18	30	38	48	54	64	69	76	0.6	0.8	0.9	1.1	1.3	0.9	1.0	1.2	1.3	1.5
TW F3	38	59	70	80	86	92	95	97	17	30	39	49	56	66	71	78	0.6	0.7	0.9	1.1	1.3	0.9	1.0	1.1	1.3	1.4
TW F4	35	57	69	79	85	91	94	97	15	28	37	47	54	64	70	77	0.6	0.7	0.9	1.1	1.3	0.9	1.0	1.1	1.3	1.4
TW G1	33	51	61	72	78	86	90	93	16	26	33	41	47	56	61	68	0.7	0.8	1.0	1.1	1.3	1.0	1.1	1.3	1.4	1.6
TW G2	33	53	65	75	82	89	92	95	15	26	35	44	51	60	66	73	0.7	0.8	0.9	1.1	1.3	0.9	1.0	1.2	1.3	1.5
TW G3	31	52	63	74	81	88	92	95	13	25	33	42	49	58	65	72	0.7	0.8	0.9	1.1	1.3	1.0	1.1	1.2	1.3	1.5
TW G4	27	46	57	68	75	83	88	92	10	21	28	37	43	53	59	66	0.7	0.8	1.0	1.1	1.3	1.0	1.1	1.3	1.4	1.6
TW H1	29	47	58	69	75	83	88	92	13	23	30	38	44	53	58	66	0.7	0.8	1.0	1.1	1.3	1.0	1.1	1.3	1.4	1.6
TW H2	26	44	56	67	74	82	87	91	10	20	27	36	42	51	57	64	0.7	0.8	1.0	1.2	1.3	1.0	1.1	1.3	1.4	1.6
TW H3	22	39	50	61	68	77	83	88	8	17	24	31	37	46	51	59	0.8	0.9	1.0	1.2	1.4	1.1	1.2	1.3	1.5	1.6
TW H4	17	31	41	52	59	68	74	81	6	13	19	25	30	37	43	49	0.8	0.9	1.1	1.3	1.4	1.2	1.3	1.5	1.6	1.8
TW I1	30	49	60	70	77	84	89	93	13	23	30	38	44	52	58	65	0.7	0.8	1.0	1.1	1.3	1.0	1.1	1.3	1.4	1.6
TW I2	40	62	74	84	89	94	96	98	18	32	42	53	61	70	76	83	0.6	0.7	0.9	1.1	1.2	0.8	0.9	1.1	1.2	1.4
TW I3	43	66	77	86	91	96	97	99	19	35	46	57	65	75	80	86	0.6	0.7	0.9	1.0	1.2	0.8	0.9	1.0	1.2	1.4
TW I4	46	69	80	89	93	97	98	99	20	38	50	62	69	79	84	90	0.6	0.7	0.9	1.0	1.2	0.8	0.9	1.0	1.2	1.3
TW I5	45	69	80	89	93	97	98	99	20	38	50	62	70	79	85	90	0.6	0.7	0.8	1.0	1.2	0.7	0.8	1.0	1.2	1.3
TW J1	49	72	83	90	94	97	99	99	23	41	52	64	72	81	86	91	0.6	0.7	0.8	1.0	1.2	0.7	0.8	1.0	1.2	1.3
TW J2	47	71	82	90	94	97	98	99	21	39	51	63	71	80	85	90	0.6	0.7	0.8	1.0	1.2	0.7	0.8	1.0	1.2	1.3
TW J3	53	77	87	94	97	99	99	100	25	46	59	72	79	87	91	95	0.5	0.6	0.8	1.0	1.2	0.7	0.8	0.9	1.1	1.3
TW J4	50	75	85	93	96	98	99	100	22	43	56	69	76	85	89	94	0.5	0.7	0.8	1.0	1.2	0.7	0.8	0.9	1.1	1.3
DG A1	39	59	69	78	83	89	92	94	17	29	37	45	51	58	62	67	0.6	0.8	0.9	1.2	1.4	0.9	1.0	1.2	1.5	1.8
DG A2	38	60	71	81	86	92	94	97	17	30	39	49	56	65	70	76	0.6	0.7	0.9	1.1	1.3	0.9	1.0	1.1	1.3	1.5
DG A3	42	66	77	86	90	93	95	96	19	35	46	58	66	75	80	85	0.6	0.7	0.9	1.2	1.7	0.8	0.9	1.0	1.2	1.4
DG B1	41	64	76	85	89	93	95	97	17	29	39	50	57	66	70	76	0.6	0.7	0.9	1.1	1.4	0.9	1.0	1.1	1.3	1.6
DG B2	40	65	77	87	91	95	97	98	17	31	42	53	62	71	77	83	0.6	0.7	0.9	1.1	1.3	0.9	0.9	1.0	1.2	1.4
DG B3	43	69	81	90	93	96	97	98	20	36	48	61	70	79	85	89	0.6	0.7	0.9	1.1	1.6	0.8	0.9	1.0	1.2	1.4
DG C1	46	72	83	90	94	96	97	98	19	34	45	58	66	75	80	84	0.6	0.7	0.9	1.1	1.7	0.9	0.9	1.0	1.2	1.5
DG C2	45	71	83	91	95	97	99	99	20	35	47	60	69	78	83	89	0.6	0.7	0.8	1.0	1.3	0.8	0.9	1.0	1.2	1.4
DG C3	48	75	87	93	96	98	98	98	22	40	53	67	76	85	90	93	0.5	0.6	0.8	1.2	2.3	0.8	0.8	0.9	1.1	1.3
SS A1	47	67	77	86	91	95	97	98	26	40	49	59	66	75	80	85	0.6	0.7	0.9	1.1	1.2	0.8	0.9	1.1	1.2	1.4
SS A2	56	77	86	93	96	98	99	100	31	49	60	71	78	85	89	93	0.5	0.7	0.8	1.0	1.2	0.7	0.8	1.0	1.1	1.3
SS A3	45	65	75	84	89	94	96	98	25	38	47	56	63	71	77	83	0.6	0.7	0.9	1.1	1.3	0.8	0.9	1.1	1.3	1.4
SS A4	56	77	86	93	96	98	99	100	31	49	60	70	77	85	89	93	0.5	0.7	0.8	1.0	1.2	0.7	0.8	1.0	1.1	1.3
SS A5	44	63	73	82	87	93	95	97	25	37	46	55	61	70	75	81	0.6	0.7	0.9	1.1	1.3	0.9	1.0	1.1	1.3	1.5
SS A6	56	77	86	93	96	98	99	100	31	49	60	71	77	85	89	93	0.5	0.7	0.8	1.0	1.2	0.7	0.8	1.0	1.1	1.3
SS A7	40	58	68	77	83	89	92	95	23	34	41	49	55	63	69	75	0.7	0.8	0.9	1.1	1.3	0.9	1.0	1.2	1.3	1.5
SS A8	55	76	85	92	95	98	99	100	30	48	59	70	76	84	88	93	0.5	0.7	0.8	1.0	1.2	0.7	0.8	1.0	1.1	1.3
SS B1	41	61	72	81	87	92	95	97	22	35	43	53	59	[illegible]	74	80	0.6	0.7	0.9	1.1	1.3	0.9	1.0	1.1	1.3	1.4
SS B2	50	71	81	89	93	97	98	99	26	43	54	65	72	80	85	90	0.6	0.7	0.9	1.0	1.2	0.8	0.8	1.0	1.2	1.3
SS B3	40	60	70	80	85	91	94	97	21	33	42	51	57	66	72	78	0.6	0.8	0.9	1.1	1.3	0.9	1.0	1.1	1.3	1.5
SS B4	49	71	81	89	93	97	98	99	26	43	53	64	71	80	85	90	0.6	0.7	0.9	1.0	1.2	0.8	0.9	1.0	1.2	1.3
SS B5	37	55	65	75	81	88	91	95	21	31	38	46	52	61	66	73	0.7	0.8	1.0	1.1	1.3	1.0	1.1	1.2	1.4	1.5
SS B6	49	71	81	89	93	97	98	99	26	42	53	64	71	80	85	90	0.6	0.7	0.9	1.0	1.2	0.8	0.9	1.0	1.2	1.3
SS B7	34	51	61	71	77	84	88	92	19	28	35	42	48	56	61	68	0.7	0.8	1.0	1.2	1.3	1.0	1.1	1.3	1.4	1.6
SS B8	48	70	80	88	93	96	98	99	25	41	52	63	70	79	83	89	0.6	0.7	0.9	1.0	1.2	0.8	0.9	1.0	1.2	1.4
SS C1	43	64	75	84	90	94	97	98	20	34	43	54	61	70	76	82	0.6	0.7	0.9	1.1	1.2	0.8	0.9	1.1	1.2	1.4
SS C2	48	70	81	89	93	97	98	99	22	39	51	62	69	79	84	89	0.6	0.7	0.9	1.0	1.2	0.8	0.9	1.0	1.2	1.3
SS C3	34	53	64	74	80	87	91	95	17	28	35	43	50	58	64	71	0.7	0.8	1.0	1.1	1.3	1.0	1.1	1.2	1.4	1.5
SS C4	43	64	76	85	90	95	97	98	20	35	45	56	63	73	78	84	0.6	0.7	0.9	1.1	1.2	0.8	0.9	1.1	1.2	1.4
SS D1	56	76	86	92	96	98	99	100	30	48	58	69	76	84	88	92	0.5	0.7	0.8	1.0	1.2	0.7	0.8	1.0	1.2	1.3
SS D2	63	84	92	96	98	99	100	100	35	56	68	79	85	91	94	97	0.5	0.6	0.8	1.0	1.2	0.6	0.7	0.9	1.1	1.3
SS D3	50	71	80	88	93	96	98	99	28	43	53	63	69	78	83	88	0.6	0.7	0.9	1.0	1.2	0.8	0.9	1.0	1.2	1.4
SS D4	61	82	91	96	98	99	100	100	34	55	66	77	83	90	93	96	0.5	0.6	0.8	1.0	1.2	0.7	0.8	0.9	1.1	1.3
SS E1	51	72	82	90	94	97	98	99	26	42	53	63	70	79	84	89	0.6	0.7	0.9	1.0	1.2	0.8	0.9	1.0	1.2	1.4
SS E2	58	80	89	95	97	99	99	100	30	50	62	73	80	87	91	95	0.5	0.6	0.8	1.0	1.2	0.7	0.8	0.9	1.1	1.3
SS E3	42	62	72	82	87	92	95	97	23	35	44	53	60	68	74	80	0.6	0.7	0.9	1.1	1.3	0.9	1.0	1.1	1.3	1.5
SS E4	54	75	85	92	95	98	99	100	28	46	58	69	76	84	88	92	0.5	0.7	0.8	1.0	1.2	0.7	0.8	1.0	1.1	1.3

SAN ANGELO, TEXAS

TYPE	SOLAR SAVINGS FRACTION (%) BASE TEMP: 55 F (918 DD)								65 F (2245 DD)								CONSERVATION FACTOR 55 F					65 F				
LCR	200	100	70	50	40	30	25	20	200	100	70	50	40	30	25	20	200	100	50	30	20	200	100	50	30	20
WW A1	32	47	57	67	74	82	86	91	19	28	34	42	47	55	60	67	0.8	0.8	1.0	1.2	1.3	1.1	1.1	1.3	1.4	1.6
WW A2	38	58	70	80	86	92	95	97	20	34	43	53	60	69	75	81	0.6	0.7	0.9	1.1	1.3	0.9	1.0	1.1	1.3	1.4
WW A3	40	62	74	84	89	94	96	98	20	36	46	57	64	74	79	85	0.6	0.7	0.9	1.1	1.2	0.8	0.9	1.1	1.2	1.4
WW A4	41	64	76	85	91	95	97	99	20	37	47	59	66	76	81	87	0.6	0.7	0.9	1.0	1.2	0.8	0.9	1.0	1.2	1.4
WW A5	43	66	78	87	92	96	98	99	20	38	49	61	69	78	84	89	0.6	0.7	0.9	1.0	1.2	0.8	0.9	1.0	1.2	1.3
WW A6	43	68	79	88	93	97	98	99	20	39	50	62	70	79	85	90	0.6	0.7	0.9	1.0	1.2	0.7	0.8	1.0	1.2	1.3
WW B1	33	52	63	74	80	87	91	95	17	28	36	45	51	60	66	73	0.7	0.8	0.9	1.1	1.3	1.0	1.1	1.2	1.4	1.5
WW B2	41	64	76	86	91	95	97	99	20	37	48	60	67	77	82	88	0.6	0.7	0.9	1.0	1.2	0.8	0.9	1.0	1.2	1.3
WW B3	48	72	83	91	95	98	99	100	24	44	56	68	75	84	89	93	0.6	0.7	0.8	1.0	1.2	0.7	0.8	1.0	1.1	1.3
WW B4	47	72	83	91	95	98	99	100	23	44	55	69	76	85	90	94	0.5	0.7	0.8	1.0	1.2	0.7	0.8	0.9	1.1	1.3
WW B5	44	70	81	90	94	98	99	100	21	42	55	67	75	84	89	93	0.6	0.7	0.8	1.0	1.2	0.7	0.8	1.0	1.1	1.3
WW C1	49	73	84	92	95	98	99	100	26	46	58	70	77	86	90	94	0.5	0.7	0.8	1.0	1.2	0.7	0.8	0.9	1.1	1.3
WW C2	46	71	82	90	94	98	99	100	23	43	56	68	75	84	89	93	0.6	0.7	0.8	1.0	1.2	0.7	0.8	1.0	1.1	1.3
WW C3	52	77	87	94	97	99	100	100	26	50	63	76	83	90	94	97	0.5	0.6	0.8	1.0	1.2	0.6	0.7	0.9	1.1	1.3
WW C4	48	73	85	92	96	98	99	100	23	46	59	72	79	88	92	96	0.5	0.7	0.8	1.0	1.2	0.7	0.8	0.9	1.1	1.3
TW A1	31	46	56	66	73	81	85	90	19	27	34	41	46	54	59	66	0.8	0.9	1.0	1.2	1.3	1.1	1.2	1.3	1.4	1.6
TW A2	34	53	64	75	81	88	92	95	19	30	39	48	54	63	69	76	0.7	0.8	0.9	1.1	1.3	0.9	1.0	1.2	1.3	1.5
TW A3	35	56	68	78	84	91	94	97	18	32	41	51	58	67	73	80	0.7	0.8	0.9	1.1	1.3	0.9	1.0	1.1	1.3	1.4
TW A4	35	56	68	79	85	91	94	97	17	31	41	51	58	68	74	81	0.6	0.7	0.9	1.1	1.3	0.9	1.0	1.1	1.3	1.4
TW B1	30	46	57	67	74	82	86	91	18	27	34	41	47	55	61	68	0.8	0.8	1.0	1.2	1.3	1.0	1.1	1.3	1.4	1.6
TW B2	32	51	63	73	80	87	91	95	17	29	37	46	53	62	68	75	0.7	0.8	0.9	1.1	1.3	0.9	1.0	1.2	1.3	1.5
TW B3	32	51	63	74	81	88	92	95	16	29	37	47	54	63	69	76	0.7	0.8	0.9	1.1	1.3	0.9	1.0	1.2	1.3	1.5
TW B4	30	48	60	71	78	86	90	94	15	27	35	44	51	60	66	73	0.7	0.8	1.0	1.1	1.3	1.0	1.0	1.2	1.3	1.5
TW C1	29	45	56	66	73	81	86	91	16	26	33	41	47	55	61	68	0.8	0.8	1.0	1.2	1.3	1.0	1.1	1.3	1.4	1.6
TW C2	28	46	57	68	75	83	87	92	15	26	33	41	48	57	62	70	0.7	0.8	1.0	1.2	1.3	1.0	1.1	1.2	1.4	1.5
TW C3	27	43	54	65	72	80	85	90	14	24	31	39	45	54	60	67	0.8	0.8	1.0	1.2	1.3	1.0	1.1	1.3	1.4	1.6
TW C4	24	38	48	58	65	74	80	86	14	22	28	35	41	48	54	61	0.8	0.9	1.1	1.2	1.4	1.1	1.2	1.4	1.5	1.6
TW D1	28	45	56	67	74	82	87	91	14	24	31	39	45	53	59	66	0.7	0.8	1.0	1.2	1.3	1.0	1.1	1.3	1.4	1.6
TW D2	37	59	71	81	87	93	95	98	19	34	44	54	62	72	77	84	0.6	0.7	0.9	1.1	1.2	0.8	0.9	1.1	1.2	1.4
TW D3	41	64	76	85	90	95	97	99	21	37	48	59	67	77	82	88	0.6	0.7	0.9	1.0	1.2	0.8	0.9	1.0	1.2	1.4
TW D4	42	66	78	87	92	96	98	99	21	39	51	63	71	80	85	90	0.6	0.7	0.9	1.0	1.2	0.7	0.8	1.0	1.2	1.3
TW D5	41	65	77	86	91	96	98	99	20	38	50	62	70	80	85	90	0.6	0.7	0.9	1.0	1.2	0.7	0.8	1.0	1.2	1.3
TW E1	45	68	79	88	93	97	98	99	24	41	53	64	72	81	86	91	0.6	0.7	0.9	1.0	1.2	0.7	0.8	1.0	1.2	1.3
TW E2	44	67	79	88	92	96	98	99	22	40	52	64	71	81	86	91	0.6	0.7	0.9	1.0	1.2	0.7	0.8	1.0	1.2	1.3
TW E3	50	75	85	93	96	98	99	100	26	48	61	73	80	88	92	95	0.5	0.7	0.8	1.0	1.2	0.7	0.8	0.9	1.1	1.3
TW E4	46	71	83	91	95	98	99	100	23	44	57	69	77	86	90	94	0.5	0.7	0.8	1.0	1.2	0.7	0.8	0.9	1.1	1.3
TW F1	29	45	56	66	73	81	86	90	17	26	33	40	46	54	60	66	0.8	0.8	1.0	1.2	1.3	1.1	1.1	1.3	1.4	1.6
TW F2	31	51	62	73	80	87	91	95	16	28	36	45	52	61	67	74	0.7	0.8	0.9	1.1	1.3	0.9	1.0	1.2	1.3	1.5
TW F3	31	52	64	75	81	89	92	95	15	28	37	46	54	63	69	76	0.7	0.8	0.9	1.1	1.3	0.9	1.0	1.1	1.3	1.5
TW F4	29	50	62	73	80	88	91	95	13	26	35	45	52	62	68	75	0.7	0.8	0.9	1.1	1.3	0.9	1.0	1.2	1.3	1.5
TW G1	27	44	55	65	72	81	85	90	15	24	31	39	45	53	59	66	0.8	0.8	1.0	1.2	1.3	1.0	1.1	1.3	1.4	1.6
TW G2	28	46	58	69	76	84	89	93	13	24	32	41	48	57	63	71	0.7	0.8	1.0	1.1	1.3	1.0	1.1	1.2	1.4	1.5
TW G3	26	44	56	68	75	83	88	92	11	23	31	40	47	56	62	70	0.7	0.8	1.0	1.1	1.3	1.0	1.1	1.2	1.4	1.5
TW G4	21	39	50	61	69	78	83	89	9	19	26	35	41	50	56	64	0.8	0.9	1.0	1.2	1.3	1.0	1.1	1.3	1.4	1.6
TW H1	24	40	51	62	69	78	83	89	12	21	28	36	42	50	56	63	0.8	0.9	1.0	1.2	1.4	1.1	1.2	1.3	1.4	1.6
TW H2	21	37	48	60	67	77	82	88	9	18	25	33	39	48	54	62	0.8	0.9	1.0	1.2	1.4	1.1	1.2	1.3	1.4	1.6
TW H3	18	33	43	54	62	71	77	84	7	15	22	29	35	43	49	56	0.8	0.9	1.1	1.2	1.4	1.2	1.2	1.4	1.5	1.7
TW H4	14	26	35	45	52	62	68	75	5	12	17	23	28	35	40	47	0.9	1.0	1.2	1.3	1.5	1.3	1.4	1.5	1.6	1.8
TW I1	25	42	52	63	70	79	84	89	12	21	28	35	41	49	55	62	0.8	0.9	1.0	1.2	1.4	1.1	1.2	1.3	1.5	1.6
TW I2	34	55	68	78	85	91	94	97	16	30	40	51	58	68	74	81	0.6	0.7	0.9	1.1	1.3	0.9	0.9	1.1	1.2	1.4
TW I3	36	59	71	82	87	93	96	98	17	33	44	55	63	72	78	85	0.6	0.7	0.9	1.1	1.2	0.8	0.9	1.1	1.2	1.4
TW I4	38	62	75	85	90	95	97	99	18	36	47	59	67	77	83	88	0.6	0.7	0.9	1.0	1.2	0.8	0.9	1.0	1.2	1.3
TW I5	38	63	75	85	90	95	97	99	18	36	48	60	68	78	83	89	0.6	0.7	0.9	1.0	1.2	0.8	0.9	1.0	1.2	1.3
TW J1	42	65	77	87	91	96	98	99	21	39	50	62	69	79	84	90	0.6	0.7	0.9	1.0	1.2	0.8	0.9	1.0	1.2	1.3
TW J2	40	64	76	86	91	96	97	99	19	37	49	61	68	78	84	89	0.6	0.7	0.9	1.0	1.2	0.8	0.9	1.0	1.2	1.3
TW J3	46	71	83	91	95	98	99	100	23	44	57	69	77	86	90	94	0.5	0.7	0.8	1.0	1.2	0.7	0.8	0.9	1.1	1.3
TW J4	43	68	80	89	94	97	99	99	20	41	54	66	74	84	88	93	0.6	0.7	0.8	1.0	1.2	0 7	0.8	1.0	1.1	1.3
DG A1	32	52	62	72	78	84	88	91	15	26	34	42	48	55	59	64	0.7	0.8	1.0	1.2	1.4	0.9	1.0	1.3	1.5	1.8
DG A2	32	53	65	75	81	88	91	94	15	28	37	47	54	62	68	74	0.7	0.8	0.9	1.1	1.3	0.9	1.0	1.2	1.3	1.6
DG A3	35	59	72	82	87	91	94	95	18	33	44	56	63	73	78	84	0.6	0.7	0.9	1.1	1.5	0.9	0.9	1.0	1.2	1.4
DG B1	33	57	69	79	85	90	93	95	15	27	36	47	54	63	68	73	0.6	0.7	0.9	1.1	1.4	0.9	1.0	1.1	1.3	1.6
DG B2	33	57	71	82	87	93	95	97	16	29	39	51	59	69	75	81	0.6	0.7	0.9	1.1	1.3	0.9	1.0	1.1	1.2	1.5
DG B3	36	62	76	86	91	95	96	97	18	33	45	58	67	78	83	88	0.6	0.7	0.9	1.1	1.5	0.8	0.9	1.0	1.2	1.4
DG C1	38	64	77	86	91	94	96	97	18	32	42	54	63	73	78	83	0.6	0.7	0.9	1.1	1.5	0.9	0.9	1.0	1.2	1.5
DG C2	37	63	77	87	92	96	97	99	18	33	44	57	66	76	82	87	0.6	0.7	0.9	1.1	1.3	0.9	0.9	1.0	1.2	1.4
DG C3	41	68	82	91	94	97	98	98	21	38	51	65	74	84	89	93	0.6	0.6	0.8	1.1	1.7	0.8	0.8	0.9	1.1	1.3
SS A1	40	60	71	80	86	92	95	97	24	37	46	56	63	71	77	83	0.6	0.7	0.9	1.1	1.3	0.8	0.9	1.1	1.3	1.4
SS A2	48	70	81	89	93	97	98	99	29	46	57	68	75	83	87	92	0.6	0.7	0.9	1.0	1.2	0.7	0.8	1.0	1.1	1.3
SS A3	38	57	68	78	84	90	93	96	23	35	44	53	59	68	73	79	0.7	0.8	0.9	1.1	1.3	0.9	1.0	1.1	1.3	1.5
SS A4	48	70	80	88	93	96	98	99	28	46	56	67	74	82	87	91	0.6	0.7	0.9	1.0	1.2	0.7	0.8	1.0	1.2	1.3
SS A5	38	56	66	76	82	89	92	95	23	35	43	52	58	66	72	78	0.7	0.8	0.9	1.1	1.3	0.9	1.0	1.2	1.3	1.5
SS A6	48	70	81	89	93	97	98	99	28	46	57	68	75	83	87	92	0.6	0.7	0.9	1.0	1.2	0.7	0.8	1.0	1.1	1.3
SS A7	34	51	61	70	77	84	88	92	22	31	38	46	52	60	65	71	0.7	0.8	1.0	1.2	1.3	1.0	1.1	1.2	1.4	1.6
SS A8	47	69	80	88	92	96	98	99	28	45	55	66	73	81	86	91	0.6	0.7	0.9	1.0	1.2	0.7	0.8	1.0	1.2	1.3
SS B1	35	54	65	75	81	88	92	95	20	32	40	49	56	65	70	77	0.7	0.8	0.9	1.1	1.3	0.9	1.0	1.2	1.3	1.5
SS B2	42	64	75	85	90	95	97	98	24	40	51	61	69	78	83	88	0.6	0.7	0.9	1.1	1.2	0.8	0.9	1.0	1.2	1.4
SS B3	34	52	63	73	80	87	90	94	19	31	39	47	54	62	68	75	0.7	0.8	1.0	1.1	1.3	0.9	1.0	1.2	1.3	1.5
SS B4	42	63	75	84	89	94	96	98	24	40	50	61	68	77	82	88	0.6	0.7	0.9	1.1	1.2	0.8	0.9	1.0	1.2	1.4
SS B5	31	43	58	68	74	82	87	91	19	29	36	43	49	57	63	69	0.7	0.8	1.0	1.2	1.3	1.0	1.1	1.3	1.4	1.6
SS B6	42	64	75	84	89	94	97	98	24	40	50	61	68	77	82	88	0.6	0.7	0.9	1.1	1.2	0.8	0.9	1.0	1.2	1.4
SS B7	29	44	54	64	70	78	83	88	17	26	32	40	45	53	58	64	0.8	0.9	1.0	1.2	1.4	1.1	1.2	1.3	1.5	1.6
SS B8	41	62	74	83	88	94	96	98	23	38	49	59	66	75	81	87	0.6	0.7	0.9	1.1	1.2	0.8	0.9	1.0	1.2	1.4
SS C1	36	57	69	79	85	92	95	97	18	32	41	51	58	68	74	81	0.6	0.7	0.9	1.1	1.3	0.9	1.0	1.1	1.3	1.4
SS C2	41	64	76	85	90	95	97	99	20	37	48	60	67	77	82	88	0.6	0.7	0.9	1.0	1.2	0.8	0.9	1.0	1.2	1.3
SS G3	29	46	57	68	75	83	87	92	16	26	33	41	47	56	62	69	0.7	0.8	1.0	1.2	1.3	1.0	1.1	1.2	1.4	1.5
SS C4	36	58	70	80	86	92	95	97	18	33	43	53	61	70	76	83	0.6	0.7	0.9	1.1	1.3	0.8	0.9	1.1	1.2	1.4
SS D1	48	70	80	88	93	96	98	99	28	45	55	66	73	81	86	90	0.6	0.7	0.9	1.0	1.2	0.7	0.8	1.0	1.2	1.3
SS D2	56	78	88	94	97	99	99	100	32	53	65	76	83	90	93	96	0.5	0.6	0.8	1.0	1.2	0.7	0.8	0.9	1.1	1.3
SS D3	43	64	74	83	88	94	96	98	26	40	50	59	66	75	80	86	0.6	0.7	0.9	1.1	1.2	0.8	0.9	1.1	1.2	1.4
SS D4	54	76	86	93	96	98	99	100	31	51	63	74	81	88	92	95	0.5	0.7	0.8	1.0	1.2	0.7	0.8	0.9	1.1	1.3
SS E1	43	63	76	85	90	95	97	98	24	39	49	60	67	76	81	86	0.6	0.7	0.9	1.1	1.2	0.8	0.9	1.0	1.2	1.4
SS E2	50	73	83	91	95	98	99	100	28	47	59	70	77	85	90	94	0.5	0.7	0.8	1.0	1.2	0.7	0.8	1.0	1.1	1.3
SS E3	36	55	65	75	82	88	92	95	21	33	41	50	56	65	71	77	0.7	0.8	0.9	1.1	1.3	0.9	1.0	1.2	1.3	1.5
SS E4	46	69	79	88	92	96	98	99	26	43	54	65	73	81	86	91	0.6	0.7	0.9	1.0	1.2	0.7	0.8	1.0	1.2	1.3

SAN ANTONIO, TEXAS

TYPE	SSF 55 F (518 DD) LCR 200	100	70	50	40	30	25	20	SSF 65 F (1570 DD) 200	100	70	50	40	30	25	20	CF 55 F 200	100	50	30	20	CF 65 F 200	100	50	30	20
WW A1	37	55	65	75	81	88	92	95	21	31	38	46	51	60	65	72	0.7	0.8	0.9	1.1	1.3	1.0	1.1	1.2	1.4	1.5
WW A2	45	66	77	86	91	96	97	99	23	37	47	57	64	73	79	85	0.6	0.7	0.9	1.0	1.2	0.8	0.9	1.1	1.2	1.4
WW A3	48	70	81	90	94	97	98	99	23	40	50	61	69	78	83	88	0.6	0.7	0.8	1.0	1.2	0.8	0.9	1.0	1.2	1.4
WW A4	49	72	83	91	95	98	99	100	23	41	52	63	71	80	85	90	0.6	0.7	0.8	1.0	1.2	0.8	0.8	1.0	1.2	1.3
WW A5	51	75	85	92	96	98	99	100	24	42	54	66	73	82	87	92	0.5	0.7	0.8	1.0	1.2	0.7	0.8	1.0	1.1	1.3
WW A6	52	76	86	93	96	99	99	100	24	43	55	67	74	83	88	92	0.5	0.7	0.8	1.0	1.2	0.7	0.8	1.0	1.1	1.3
WW B1	40	60	72	81	87	93	95	98	20	32	40	49	56	65	71	77	0.6	0.7	0.9	1.1	1.3	0.9	1.0	1.1	1.3	1.5
WW B2	49	72	83	91	95	98	99	100	23	41	53	64	72	81	86	91	0.6	0.7	0.8	1.0	1.2	0.7	0.8	1.0	1.2	1.3
WW B3	56	79	89	95	97	99	100	100	28	48	61	72	79	87	91	95	0.5	0.6	0.8	1.0	1.2	0.7	0.8	0.9	1.1	1.3
WW B4	55	79	89	95	97	99	100	100	26	48	61	73	80	88	92	95	0.5	0.6	0.8	1.0	1.2	0.7	0.8	0.9	1.1	1.3
WW B5	53	77	88	94	97	99	100	100	24	46	59	71	79	87	91	95	0.5	0.6	0.8	1.0	1.2	0.7	0.8	0.9	1.1	1.3
WW C1	58	81	90	95	98	99	100	100	29	50	62	74	81	88	92	96	0.5	0.6	0.8	1.0	1.2	0.7	0.8	0.9	1.1	1.3
WW C2	55	79	88	95	97	99	100	100	26	47	60	72	79	87	91	95	0.5	0.6	0.8	1.0	1.2	0.7	0.8	0.9	1.1	1.3
WW C3	61	84	92	97	99	100	100	100	30	54	68	79	86	92	95	98	0.5	0.6	0.8	1.0	1.2	0.6	0.7	0.9	1.1	1.2
WW C4	56	81	90	96	98	99	100	100	27	50	64	76	83	90	94	97	0.5	0.6	0.8	1.0	1.2	0.6	0.7	0.9	1.1	1.3
TW A1	36	53	64	74	80	87	91	94	21	30	37	45	50	59	64	71	0.7	0.8	1.0	1.1	1.3	1.0	1.1	1.2	1.4	1.5
TW A2	41	61	72	82	87	93	96	98	21	34	42	52	59	68	74	80	0.6	0.7	0.9	1.1	1.3	0.9	1.0	1.1	1.3	1.4
TW A3	43	64	75	85	90	95	97	99	21	35	45	55	62	71	77	83	0.6	0.7	0.9	1.1	1.2	0.8	0.9	1.1	1.2	1.4
TW A4	42	64	76	85	90	95	97	99	20	35	45	55	63	72	78	84	0.6	0.7	0.9	1.0	1.2	0.8	0.9	1.1	1.2	1.4
TW B1	36	54	64	74	81	88	91	95	20	30	37	45	51	60	65	72	0.7	0.8	0.9	1.1	1.3	1.0	1.1	1.2	1.4	1.5
TW B2	39	59	70	81	86	92	95	97	20	32	41	50	57	67	72	79	0.6	0.7	0.9	1.1	1.3	0.9	1.0	1.1	1.3	1.4
TW B3	39	60	71	81	87	93	95	98	19	32	41	51	58	67	73	80	0.6	0.7	0.9	1.1	1.3	0.9	1.0	1.1	1.3	1.4
TW B4	36	57	68	78	85	91	94	97	17	30	39	48	55	65	70	77	0.6	0.8	0.9	1.1	1.3	0.9	1.0	1.1	1.3	1.5
TW C1	35	53	64	74	80	88	91	95	18	29	36	45	51	59	65	72	0.7	0.8	0.9	1.1	1.3	1.0	1.1	1.2	1.4	1.5
TW C2	34	53	65	75	82	89	92	96	17	29	36	45	52	61	67	74	0.7	0.8	0.9	1.1	1.3	1.0	1.0	1.2	1.3	1.5
TW C3	33	51	62	72	79	87	91	94	17	27	35	43	49	58	64	71	0.7	0.8	1.0	1.1	1.3	1.0	1.1	1.2	1.4	1.5
TW C4	29	45	56	66	73	81	86	91	16	25	31	39	44	52	58	65	0.7	0.8	1.0	1.2	1.3	1.1	1.2	1.3	1.4	1.6
TW D1	34	53	64	75	81	88	92	95	16	27	34	43	49	58	64	71	0.7	0.8	0.9	1.1	1.3	1.0	1.1	1.2	1.4	1.5
TW D2	45	67	78	87	92	96	98	99	22	38	48	59	66	76	81	87	0.6	0.7	0.9	1.0	1.2	0.8	0.9	1.0	1.2	1.4
TW D3	49	72	83	91	94	98	99	100	24	41	53	64	71	80	85	90	0.6	0.7	0.8	1.0	1.2	0.7	0.8	1.0	1.2	1.3
TW D4	50	74	85	92	96	98	99	100	24	43	55	67	75	83	88	93	0.5	0.7	0.8	1.0	1.2	0.7	0.8	1.0	1.1	1.3
TW D5	49	73	84	91	95	98	99	100	23	42	54	66	74	83	88	92	0.6	0.7	0.8	1.0	1.2	0.7	0.8	1.0	1.1	1.3
TW E1	53	76	86	93	96	98	99	100	27	45	57	69	76	84	89	93	0.5	0.7	0.8	1.0	1.2	0.7	0.8	1.0	1.1	1.3
TW E2	52	75	85	93	96	98	99	100	26	44	56	68	75	84	88	93	0.5	0.7	0.8	1.0	1.2	0.7	0.8	1.0	1.1	1.3
TW E3	59	82	91	96	98	99	100	100	30	52	65	76	83	90	94	97	0.5	0.6	0.8	1.0	1.2	0.7	0.8	0.9	1.1	1.3
TW E4	55	79	89	95	97	99	100	100	27	48	61	73	81	88	92	96	0.5	0.6	0.8	1.0	1.2	0.7	0.8	0.9	1.1	1.3
TW F1	35	53	64	74	80	87	91	95	19	29	36	44	50	59	64	71	0.7	0.8	1.0	1.1	1.3	1.0	1.1	1.2	1.4	1.5
TW F2	36	59	70	80	86	92	95	97	19	31	40	50	56	66	72	78	0.6	0.7	0.9	1.1	1.3	0.9	1.0	1.1	1.3	1.4
TW F3	39	60	72	82	88	93	96	98	18	31	41	51	58	68	74	80	0.6	0.7	0.9	1.1	1.2	0.9	1.0	1.1	1.3	1.4
TW F4	36	58	70	81	87	93	95	98	15	29	39	49	56	66	72	79	0.6	0.7	0.9	1.1	1.2	0.9	1.0	1.1	1.3	1.4
TW G1	33	52	63	73	80	87	91	95	17	27	35	43	49	58	64	71	0.7	0.8	1.0	1.1	1.3	1.0	1.1	1.2	1.4	1.5
TW G2	34	55	66	77	83	90	93	96	15	28	36	46	53	62	68	75	0.7	0.8	0.9	1.1	1.3	0.9	1.0	1.2	1.3	1.5
TW G3	32	53	65	76	82	89	93	96	14	26	34	44	51	61	67	74	0.7	0.8	0.9	1.1	1.3	0.9	1.0	1.2	1.3	1.5
TW G4	27	47	58	70	77	85	89	94	11	22	30	39	45	55	61	68	0.7	0.8	1.0	1.1	1.3	1.0	1.1	1.2	1.4	1.5
TW H1	30	48	59	70	77	85	89	93	14	24	31	40	46	55	61	68	0.7	0.8	1.0	1.1	1.3	1.0	1.1	1.2	1.4	1.5
TW H2	27	45	57	68	75	84	88	93	11	21	29	37	44	53	59	66	0.7	0.8	1.0	1.1	1.3	1.0	1.1	1.3	1.4	1.6
TW H3	23	40	51	63	70	79	84	89	9	18	25	33	39	48	54	61	0.8	0.9	1.0	1.2	1.3	1.1	1.2	1.3	1.5	1.6
TW H4	18	32	42	53	60	70	76	82	6	14	20	26	31	39	44	51	0.8	0.9	1.1	1.2	1.4	1.2	1.3	1.5	1.6	1.7
TW I1	31	50	61	72	78	86	90	94	14	24	31	40	46	54	60	67	0.7	0.8	1.0	1.1	1.3	1.0	1.1	1.3	1.4	1.6
TW I2	41	64	75	85	90	95	97	99	19	34	44	55	63	72	78	85	0.6	0.7	0.9	1.0	1.2	0.8	0.9	1.1	1.2	1.4
TW I3	44	67	79	88	92	96	98	99	20	37	48	59	67	77	82	88	0.6	0.7	0.9	1.0	1.2	0.8	0.9	1.0	1.2	1.4
TW I4	47	71	82	90	94	98	99	100	21	40	52	64	71	81	86	91	0.6	0.7	0.8	1.0	1.2	0.7	0.8	1.0	1.1	1.3
TW I5	46	71	82	90	94	98	99	100	21	40	52	64	72	81	86	91	0.6	0.7	0.8	1.0	1.2	0.7	0.8	1.0	1.1	1.3
TW J1	50	73	84	92	95	98	99	100	24	43	54	66	74	82	87	92	0.5	0.7	0.8	1.0	1.2	0.7	0.8	1.0	1.1	1.3
TW J2	48	72	83	91	95	98	99	100	22	41	53	65	73	82	87	91	0.6	0.7	0.8	1.0	1.2	0.7	0.8	1.0	1.1	1.3
TW J3	55	79	89	95	97	99	100	100	26	48	61	73	81	88	92	96	0.5	0.6	0.8	1.0	1.2	0.7	0.8	0.9	1.1	1.3
TW J4	52	76	87	94	97	99	99	100	24	45	58	71	78	86	91	95	0.5	0.6	0.8	1.0	1.2	0.7	0.8	0.9	1.1	1.3
DG A1	40	60	71	80	85	91	93	96	17	30	38	47	53	60	64	69	0.6	0.7	0.9	1.1	1.3	0.9	1.0	1.2	1.4	1.7
DG A2	39	61	73	82	88	93	95	97	18	32	41	51	58	67	72	78	0.6	0.7	0.9	1.1	1.3	0.9	0.9	1.1	1.3	1.5
DG A3	43	68	79	87	91	94	95	96	20	36	48	60	68	77	81	86	0.6	0.7	0.9	1.2	1.8	0.8	0.9	1.0	1.2	1.4
DG B1	42	66	78	87	91	94	96	97	17	31	41	52	59	68	73	78	0.6	0.7	0.9	1.1	1.4	0.9	0.9	1.1	1.3	1.6
DG B2	41	67	79	88	93	96	98	99	18	33	44	56	64	74	79	85	0.6	0.7	0.9	1.1	1.3	0.9	0.9	1.0	1.2	1.4
DG B3	44	71	83	91	94	97	97	98	20	37	50	63	72	81	86	91	0.6	0.7	0.8	1.1	1.8	0.8	0.8	1.0	1.1	1.4
DG C1	47	74	85	92	95	97	97	98	20	36	48	60	68	77	82	86	0.5	0.6	0.9	1.2	1.9	0.8	0.9	1.0	1.2	1.5
DG C2	46	73	85	92	96	98	99	100	21	37	49	63	71	80	85	90	0.5	0.6	0.8	1.0	1.2	0.8	0.9	1.0	1.2	1.4
DG C3	49	78	88	94	97	98	98	98	23	42	56	70	78	87	91	94	0.5	0.6	0.8	1.2	3.0	0.8	0.8	0.9	1.1	1.3
SS A1	48	68	79	87	92	96	98	99	27	41	51	61	68	76	81	87	0.6	0.7	0.9	1.0	1.2	0.8	0.9	1.0	1.2	1.4
SS A2	56	78	87	94	97	99	99	100	32	51	62	72	79	87	90	94	0.5	0.6	0.8	1.0	1.2	0.7	0.8	1.0	1.1	1.3
SS A3	46	66	76	85	90	95	97	98	26	39	48	58	65	73	78	84	0.6	0.7	0.9	1.1	1.2	0.8	0.9	1.1	1.2	1.4
SS A4	56	78	87	94	96	99	99	100	32	50	61	72	79	86	90	94	0.5	0.7	0.8	1.0	1.2	0.7	0.8	1.0	1.1	1.3
SS A5	45	64	74	83	88	94	96	98	26	39	47	56	63	71	76	82	0.6	0.7	0.9	1.1	1.3	0.9	1.0	1.1	1.3	1.4
SS A6	57	78	87	94	97	99	99	100	32	50	62	72	79	86	90	94	0.5	0.6	0.8	1.0	1.2	0.7	0.8	1.0	1.1	1.3
SS A7	41	59	69	78	84	90	93	96	24	35	42	51	57	65	70	77	0.7	0.8	0.9	1.1	1.3	0.9	1.0	1.2	1.3	1.5
SS A8	56	77	87	93	96	99	99	100	31	50	61	71	78	85	89	93	0.5	0.7	0.8	1.0	1.2	0.7	0.8	1.0	1.1	1.3
SS B1	42	62	73	83	88	93	96	98	23	36	45	54	61	70	75	82	0.6	0.7	0.9	1.1	1.3	0.9	1.0	1.1	1.3	1.4
SS B2	50	72	83	91	94	98	99	99	27	45	55	66	73	82	86	91	0.6	0.7	0.8	1.0	1.2	0.7	0.8	1.0	1.2	1.3
SS B3	41	61	71	81	87	92	95	97	22	34	43	52	59	68	73	80	0.6	0.7	0.9	1.1	1.3	0.9	1.0	1.1	1.3	1.5
SS B4	50	72	82	90	94	97	99	99	27	44	55	66	73	81	86	91	0.6	0.7	0.8	1.0	1.2	0.7	0.8	1.0	1.2	1.3
SS B5	38	56	66	76	82	89	92	95	21	32	39	48	54	62	68	74	0.7	0.8	0.9	1.1	1.3	0.9	1.0	1.2	1.3	1.5
SS B6	50	72	82	90	94	97	99	99	27	44	55	66	73	81	86	91	0.6	0.7	0.8	1.0	1.2	0.7	0.8	1.0	1.2	1.3
SS B7	35	52	62	72	78	86	90	94	20	29	36	44	50	58	63	70	0.7	0.8	1.0	1.1	1.3	1.0	1.1	1.2	1.4	1.6
SS B8	49	71	81	89	94	97	98	99	26	43	53	64	71	80	85	90	0.6	0.7	0.8	1.0	1.2	0.8	0.9	1.0	1.2	1.3
SS C1	44	66	77	86	91	95	97	99	21	35	45	56	63	72	78	84	0.6	0.7	0.9	1.0	1.2	0.8	0.9	1.1	1.2	1.4
SS C2	49	72	83	91	95	98	99	100	23	41	53	64	71	80	85	91	0.6	0.7	0.8	1.0	1.2	0.7	0.8	1.0	1.2	1.3
SS C3	35	54	65	75	82	89	92	96	18	29	36	45	52	60	66	73	0.7	0.8	0.9	1.1	1.3	1.0	1.1	1.2	1.3	1.5
SS C4	44	66	77	86	91	96	98	99	21	37	47	58	65	74	80	86	0.6	0.7	0.9	1.0	1.2	0.8	0.9	1.0	1.2	1.4
SS D1	57	78	87	94	96	99	99	100	31	49	60	71	77	85	89	93	0.5	0.7	0.8	1.0	1.2	0.7	0.8	1.0	1.1	1.3
SS D2	64	85	93	97	99	100	100	100	36	58	70	81	86	92	95	97	0.5	0.6	0.8	1.0	1.2	0.6	0.7	0.9	1.1	1.3
SS D3	51	72	82	90	94	97	98	99	29	44	54	64	71	79	84	89	0.6	0.7	0.9	1.0	1.2	0.8	0.9	1.0	1.2	1.4
SS D4	62	84	92	96	98	99	100	100	35	56	68	79	85	91	94	97	0.5	0.6	0.8	1.0	1.2	0.6	0.7	0.9	1.1	1.3
SS E1	52	73	83	91	95	98	99	100	27	44	54	65	72	80	85	90	0.6	0.7	0.8	1.0	1.2	0.8	0.8	1.0	1.2	1.3
SS E2	59	81	90	95	98	99	100	100	31	52	64	75	82	89	92	96	0.5	0.6	0.8	1.0	1.2	0.7	0.8	0.9	1.1	1.3
SS E3	43	63	74	83	88	94	96	98	24	37	45	55	62	70	76	82	0.6	0.7	0.9	1.1	1.2	0.9	1.0	1.1	1.3	1.4
SS E4	54	77	86	93	96	99	99	100	29	48	59	70	77	85	89	93	0.5	0.7	0.8	1.0	1.2	0.7	0.8	1.0	1.1	1.3

SHERMAN, TEXAS

TYPE	SOLAR SAVINGS FRACTION (%)																CONSERVATION FACTOR									
BASE TEMP:	55 F (1338 DD)								65 F (2872 DD)								55 F					65 F				
LCR	200	100	70	50	40	30	25	20	200	100	70	50	40	30	25	20	200	100	50	30	20	200	100	50	30	20
WW A1	22	33	40	49	55	63	69	75	16	22	26	32	36	42	46	52	1.0	1.0	1.2	1.3	1.5	1.3	1.4	1.5	1.7	1.8
WW A2	25	40	50	61	68	77	82	87	15	24	31	39	45	53	59	66	0.8	0.9	1.0	1.2	1.4	1.0	1.1	1.3	1.4	1.6
WW A3	25	43	54	65	72	81	86	91	14	25	33	42	49	58	64	71	0.7	0.8	1.0	1.2	1.3	1.0	1.1	1.2	1.4	1.5
WW A4	25	44	56	67	75	83	88	92	13	26	34	44	51	60	66	73	0.7	0.8	1.0	1.1	1.3	0.9	1.0	1.2	1.3	1.5
WW A5	26	46	58	70	77	85	89	94	13	26	35	46	53	62	69	76	0.7	0.8	1.0	1.1	1.3	0.9	1.0	1.1	1.3	1.5
WW A6	26	47	59	71	78	86	90	94	13	27	36	46	54	64	70	77	0.7	0.8	0.9	1.1	1.3	0.9	1.0	1.1	1.3	1.4
WW B1	21	34	43	53	60	69	74	81	12	20	25	31	36	43	48	54	0.9	1.0	1.1	1.3	1.4	1.2	1.3	1.5	1.6	1.8
WW B2	25	45	56	68	75	84	88	93	13	26	35	45	52	62	68	75	0.7	0.8	1.0	1.1	1.3	0.9	1.0	1.1	1.3	1.5
WW B3	30	52	64	76	82	89	93	96	16	32	42	53	61	70	76	83	0.7	0.8	0.9	1.1	1.3	0.8	0.9	1.1	1.2	1.4
WW B4	29	52	65	76	83	90	94	97	15	32	43	54	62	73	79	85	0.6	0.7	0.9	1.1	1.3	0.8	0.9	1.0	1.2	1.4
WW B5	27	50	63	75	82	89	93	96	13	30	41	53	61	72	78	65	0.7	0.8	0.9	1.1	1.3	0.8	0.9	1.0	1.2	1.4
WW C1	32	54	66	77	84	91	94	97	17	33	44	55	63	73	78	85	0.6	0.7	0.9	1.1	1.3	0.8	0.9	1.1	1.2	1.4
WW C2	29	51	64	75	82	89	93	96	15	31	42	53	61	71	77	84	0.7	0.8	0.9	1.1	1.3	0.8	0.9	1.1	1.2	1.4
WW C3	33	58	71	82	88	94	36	98	17	37	49	62	70	80	85	91	[illegible]	0.7	0.9	1.1	1.2	0.7	0.8	1.0	1.1	1.3
WW C4	29	54	67	79	85	92	95	98	14	33	46	58	67	77	82	88	0.6	0.7	0.9	1.1	1.2	0.8	0.9	1.0	1.2	1.3
TW A1	22	32	39	48	54	62	68	74	16	21	26	31	35	41	45	51	1.0	1.1	1.2	1.3	1.5	1.3	1.4	1.6	1.7	1.9
TW A2	23	36	46	56	63	72	77	83	14	23	29	36	41	49	54	61	0.8	0.9	1.1	1.2	1.4	1.1	1.2	1.4	1.5	1.7
TW A3	23	38	48	59	66	75	80	86	13	23	30	38	43	52	57	65	0.8	0.9	1.0	1.2	1.4	1.0	1.1	1.3	1.4	1.6
TW A4	22	38	48	59	67	76	81	87	12	22	29	38	44	53	58	66	0.8	0.9	1.0	1.2	1.4	1.0	1.1	1.3	1.4	1.6
TW B1	21	32	40	48	55	63	69	76	15	21	26	31	36	42	47	53	1.0	1.0	1.2	1.3	1.5	1.3	1.4	1.5	1.7	1.8
TW B2	21	35	44	54	61	70	76	82	13	21	27	34	40	48	53	60	0.9	0.9	1.1	1.2	1.4	1.1	1.2	1.4	1.5	1.7
TW B3	20	34	44	55	62	71	77	83	12	21	27	35	40	48	54	61	0.8	0.9	1.1	1.2	1.4	1.1	1.2	1.3	1.5	1.6
TW B4	19	32	42	52	59	68	74	81	11	19	25	33	38	46	51	58	0.9	1.0	1.1	1.3	1.4	1.1	1.2	1.4	1.5	1.7
TW C1	20	31	39	48	54	63	69	76	13	20	25	31	35	42	47	53	0.9	1.0	1.2	1.3	1.5	1.2	1.3	1.5	1.6	1.8
TW C2	19	31	39	49	56	65	70	77	12	19	24	31	36	43	48	55	0.9	1.0	1.1	1.3	1.5	1.2	1.3	1.4	1.6	1.7
TW C3	18	29	37	46	53	62	68	75	11	18	23	29	34	41	46	53	1.0	1.0	1.2	1.3	1.5	1.2	1.3	1.5	1.6	1.8
TW C4	17	26	33	41	47	56	62	69	11	17	21	27	31	37	42	47	1.0	1.1	1.2	1.4	1.6	1.3	1.4	1.6	1.7	1.9
TW D1	17	29	37	46	53	62	68	75	10	17	21	27	31	38	42	48	0.9	1.0	1.2	1.3	1.5	1.3	1.4	1.5	1.7	1.9
TW D2	24	41	52	63	70	79	84	89	14	25	32	41	48	57	63	70	0.8	0.9	1.0	1.2	1.3	1.0	1.1	1.2	1.4	1.5
TW D3	26	44	56	67	75	83	88	92	14	27	36	45	52	62	68	75	0.7	0.8	1.0	1.1	1.3	0.9	1.0	1.2	1.3	1.5
TW D4	26	47	59	71	78	86	90	94	14	29	38	49	57	67	73	80	0.7	0.8	0.9	1.1	1.3	0.9	1.0	1.1	1.3	1.4
TW D5	25	46	58	70	77	86	90	94	13	28	38	49	56	67	73	80	0.7	0.8	0.9	1.1	1.3	0.9	1.0	1.1	1.3	1.4
TW E1	29	49	61	72	79	87	91	95	16	30	40	50	57	67	73	80	0.7	0.8	0.9	1.1	1.3	0.9	1.0	1.1	1.3	1.4
TW E2	28	48	60	72	79	87	91	94	15	30	39	50	57	67	73	80	0.7	0.8	0.9	1.1	1.3	0.9	1.0	1.1	1.3	1.4
TW E3	32	55	68	79	86	92	95	98	18	35	47	59	67	77	82	88	0.6	0.7	0.9	1.1	1.2	0.8	0.9	1.0	1.2	1.3
TW E4	29	52	65	77	83	90	94	97	16	33	44	56	64	74	80	86	0.6	0.7	0.9	1.1	1.3	0.8	0.9	1.0	1.2	1.4
TW F1	20	31	39	47	54	62	68	75	14	20	25	30	34	41	45	51	1.0	1.0	1.2	1.3	1.5	1.3	1.4	1.5	1.7	1.8
TW F2	20	34	43	53	60	70	75	82	12	20	26	33	38	46	52	58	0.9	0.9	1.1	1.3	1.4	1.1	1.2	1.4	1.5	1.7
TW F3	19	34	44	55	62	72	77	84	10	19	26	34	39	48	53	61	0.8	0.9	1.1	1.2	1.4	1.1	1.2	1.3	1.5	1.6
TW F4	17	32	42	53	60	70	76	83	8	17	24	32	38	46	52	59	0.8	0.9	1.1	1.2	1.4	1.1	1.2	1.3	1.5	1.6
TW G1	18	29	37	46	53	62	67	74	11	18	23	29	33	40	44	51	1.0	1.0	1.2	1.3	1.5	1.3	1.4	1.5	1.7	1.8
TW G2	17	30	39	49	56	66	72	79	9	17	23	30	35	43	48	55	0.9	1.0	1.1	1.3	1.4	1.2	1.3	1.4	1.6	1.7
TW G3	15	28	37	48	55	65	71	78	7	15	21	28	34	41	47	54	0.9	1.0	1.1	1.3	1.4	1.2	1.3	1.4	1.6	1.7
TW G4	12	24	32	42	49	59	65	72	5	12	18	24	29	37	42	49	1.0	1.0	1.2	1.3	1.5	1.3	1.3	1.5	1.6	1.8
TW H1	15	26	34	43	49	59	64	72	8	15	20	26	30	37	42	48	1.0	1.1	1.2	1.4	1.5	1.3	1.4	1.5	1.7	1.8
TW H2	12	23	31	40	47	57	63	70	6	12	17	23	28	35	40	46	1.0	1.1	1.2	1.4	1.5	1.3	1.4	1.5	1.7	1.8
TW H3	10	20	27	36	42	51	57	65	4	10	15	20	25	31	36	42	1.1	1.1	1.3	1.4	1.6	1.4	1.5	1.6	1.8	1.9
TW H4	7	15	21	28	34	42	48	55	3	8	11	16	19	25	29	34	1.2	1.3	1.4	1.5	1.7	1.6	1.6	1.8	1.9	2.1
TW I1	15	26	34	43	49	58	64	71	8	14	19	24	28	34	39	45	1.0	1.1	1.2	1.4	1.5	1.3	1.4	1.6	1.8	1.9
TW I2	20	37	48	59	67	76	81	87	11	21	29	38	44	53	59	67	0.8	0.9	1.0	1.2	1.4	1.0	1.1	1.2	1.4	1.5
TW I3	22	40	51	63	71	80	85	90	11	23	32	41	48	58	64	71	0.7	0.8	1.0	1.2	1.3	1.0	1.1	1.2	1.3	1.5
TW I4	23	43	55	67	75	84	88	93	11	25	35	45	53	63	70	77	0.7	0.8	1.0	1.1	1.3	0.9	1.0	1.1	1.3	1.4
TW I5	23	43	56	68	75	84	89	93	11	25	35	46	54	64	71	78	0.7	0.8	1.0	1.1	1.3	0.9	1.0	1.1	1.3	1.4
TW J1	26	46	58	70	77	85	89	94	14	28	37	47	55	65	71	78	0.7	0.8	1.0	1.1	1.3	0.9	1.0	1.1	1.3	1.4
TW J2	25	45	57	69	76	84	89	93	13	27	36	47	54	64	70	78	0.7	0.8	1.0	1.1	1.3	0.9	1.0	1.1	1.3	1.4
TW J3	29	52	65	77	83	91	94	97	14	32	44	56	64	74	80	86	0.6	0.7	0.9	1.1	1.3	0.8	0.9	1.0	1.2	1.4
TW J4	26	49	62	74	81	89	93	96	13	30	41	53	61	71	78	84	0.7	0.8	0.9	1.1	1.3	0.8	0.9	1.0	1.2	1.4
DG A1	19	33	42	51	57	64	69	74	10	18	23	29	33	39	42	46	0.9	1.0	1.2	1.4	1.7	1.2	1.3	1.5	1.8	2.2
DG A2	19	34	45	55	62	71	76	81	11	20	26	34	40	47	52	58	0.8	0.9	1.1	1.3	1.5	1.1	1.2	1.3	1.5	1.8
DG A3	22	40	52	64	71	80	84	88	13	24	32	42	49	59	65	72	0.8	0.8	1.0	1.2	1.4	1.0	1.1	1.2	1.3	1.5
DG B1	19	35	46	57	64	72	77	82	10	18	24	31	36	44	49	54	0.8	0.9	1.0	1.3	1.5	1.2	1.3	1.4	1.6	1.9
DG B2	20	36	48	60	68	77	82	87	11	20	27	36	42	52	58	65	0.8	0.9	1.0	1.2	1.4	1.1	1.2	1.3	1.4	1.6
DG B3	22	41	54	67	75	84	88	92	13	24	33	43	51	62	69	77	0.8	0.8	0.9	1.1	1.4	1.0	1.1	1.1	1.2	1.4
DG C1	22	40	52	65	73	81	85	89	12	21	28	37	43	52	58	65	0.8	0.8	1.0	1.2	1.5	1.1	1.2	1.3	1.4	1.6
DG C2	23	41	54	67	75	84	88	92	13	23	31	41	49	59	66	73	0.8	0.8	0.9	1.1	1.3	1.0	1.1	1.2	1.3	1.5
DG C3	25	46	60	74	82	89	93	95	15	27	37	49	57	69	76	83	0.7	0.8	0.9	1.1	1.3	0.9	1.0	1.1	1.2	1.4
SS A1	27	42	52	62	69	77	82	88	18	28	35	43	49	57	62	69	0.8	0.9	1.0	1.2	1.4	1.0	1.1	1.3	1.4	1.6
SS A2	33	52	63	74	80	87	91	95	21	35	44	54	61	70	76	82	0.7	0.8	0.9	1.1	1.3	0.8	0.9	1.1	1.3	1.4
SS A3	26	40	49	59	66	74	79	85	17	26	33	40	45	53	58	64	0.8	0.9	1.1	1.2	1.4	1.0	1.2	1.3	1.5	1.6
SS A4	32	51	63	73	80	87	91	94	21	35	44	54	60	69	75	81	0.7	0.8	1.0	1.1	1.3	0.8	1.0	1.1	1.3	1.4
SS A5	26	40	48	58	64	73	78	83	19	27	33	40	45	52	57	63	0.8	0.9	1.1	1.3	1.4	1.1	1.2	1.3	1.5	1.7
SS A6	33	52	63	74	80	87	91	95	21	35	44	54	61	70	76	82	0.7	0.8	0.9	1.1	1.3	0.8	0.9	1.1	1.3	1.4
SS A7	24	35	43	52	58	66	72	78	17	24	29	35	39	46	50	56	0.9	1.0	1.2	1.3	1.5	1.2	1.3	1.5	1.6	1.8
SS A8	32	51	62	72	79	86	90	94	21	34	43	52	59	68	74	80	0.7	0.8	1.0	1.1	1.3	0.9	1.0	1.1	1.3	1.4
SS B1	23	37	46	56	63	72	77	83	15	24	30	37	43	50	56	62	0.8	0.9	1.1	1.2	1.4	1.1	1.2	1.3	1.5	1.6
SS B2	28	46	57	68	75	83	88	92	18	31	39	49	55	65	70	77	0.7	0.8	1.0	1.1	1.3	0.9	1.0	1.1	1.3	1.5
SS B3	22	35	44	54	61	70	75	81	14	23	29	35	40	48	53	59	0.9	1.0	1.1	1.3	1.4	1.1	1.2	1.4	1.5	1.7
SS B4	28	45	56	67	74	82	87	92	18	30	38	48	54	63	69	76	0.7	0.8	1.0	1.2	1.3	0.9	1.0	1.2	1.3	1.5
SS B5	21	33	41	50	56	64	70	76	15	22	27	33	37	44	48	54	0.9	1.0	1.2	1.3	1.5	1.2	1.3	1.5	1.6	1.8
SS B6	27	45	57	68	74	83	87	92	17	30	38	48	55	64	69	76	0.7	0.8	1.0	1.2	1.3	0.9	1.0	1.2	1.3	1.5
SS B7	20	30	37	45	51	60	65	71	14	20	24	30	34	39	44	49	1.0	1.1	1.2	1.4	1.5	1.3	1.4	1.6	1.7	1.9
SS B8	27	44	55	66	73	81	86	91	17	29	37	46	53	62	67	74	0.7	0.8	1.0	1.2	1.3	0.9	1.0	1.2	1.3	1.5
SS C1	23	38	49	60	67	76	81	87	12	22	29	37	43	52	57	65	0.8	0.9	1.0	1.2	1.4	1.0	1.1	1.3	1.4	1.6
SS C2	26	44	56	68	75	83	88	93	14	27	36	45	52	62	68	75	0.7	0.8	1.0	1.1	1.3	0.9	1.0	1.2	1.3	1.5
SS C3	19	31	39	49	55	64	70	77	12	19	24	30	35	42	47	53	0.9	1.0	1.2	1.3	1.5	1.2	1.3	1.5	1.6	1.8
SS C4	23	39	50	61	69	78	83	88	13	24	32	40	47	56	62	69	0.8	0.9	1.0	1.2	1.4	1.0	1.1	1.2	1.4	1.5
SS D1	32	51	62	72	79	86	90	94	21	33	42	51	58	67	72	79	0.7	0.8	1.0	1.1	1.3	0.9	1.0	1.1	1.3	1.5
SS D2	37	60	72	82	87	93	96	98	23	40	51	63	70	79	84	89	0.6	0.7	0.9	1.1	1.2	0.8	0.9	1.0	1.2	1.3
SS D3	29	46	56	66	73	81	85	90	20	30	38	46	52	60	66	72	0.7	0.8	1.0	1.2	1.4	1.0	1.1	1.2	1.4	1.5
SS D4	36	58	70	80	86	92	95	97	23	39	50	60	68	77	82	88	0.6	0.7	0.9	1.1	1.3	0.8	0.9	1.0	1.2	1.4
SS E1	28	[illegible]	56	67	74	82	86	91	17	29	36	45	51	60	66	72	0.7	0.8	1.0	1.2	1.3	0.9	1.0	1.2	1.4	1.5
SS E2	33	54	66	77	83	90	93	96	19	35	45	56	63	73	78	84	0.6	0.8	0.9	1.1	1.3	0.8	0.9	1.1	1.2	1.4
SS E3	24	38	47	57	63	72	77	83	16	25	31	38	43	50	55	62	0.8	0.9	1.1	1.3	1.4	1.1	1.2	1.4	1.5	1.7
SS E4	30	50	61	72	79	86	90	94	18	32	42	51	59	68	74	80	0.7	0.8	1.0	1.1	1.3	0.9	1.0	1.1	1.3	1.4

WACO, TEXAS

TYPE	SSF (%) 55 F (812 DD) LCR 200	100	70	50	40	30	25	20	SSF (%) 65 F (2066 DD) LCR 200	100	70	50	40	30	25	20	CF 55 F 200	100	50	30	20	CF 65 F 200	100	50	30	20
WW A1	30	44	54	64	70	78	83	88	18	26	32	39	44	51	56	62	0.8	0.9	1.0	1.2	1.4	1.1	1.2	1.4	1.5	1.7
WW A2	35	54	66	76	82	89	93	96	19	31	39	48	55	64	70	77	0.7	0.8	0.9	1.1	1.3	0.9	1.0	1.2	1.3	1.5
WW A3	37	58	70	80	86	92	95	97	18	32	42	52	59	69	75	81	0.6	0.7	0.9	1.1	1.3	0.9	1.0	1.1	1.3	1.4
WW A4	38	60	72	82	88	93	96	98	18	33	43	54	62	71	77	83	0.6	0.7	0.9	1.1	1.2	0.8	0.9	1.1	1.2	1.4
WW A5	39	62	74	84	89	95	97	98	18	34	45	56	64	74	79	85	0.6	0.7	0.9	1.1	1.2	0.8	0.9	1.0	1.2	1.4
WW A6	40	63	75	85	90	95	97	99	18	35	46	57	65	75	80	86	0.6	0.7	0.9	1.0	1.2	0.8	0.9	1.0	1.2	1.4
WW B1	31	48	59	70	76	84	88	93	16	26	33	41	46	55	60	67	0.7	0.8	1.0	1.1	1.3	1.0	1.1	1.3	1.4	1.6
WW B2	38	60	72	82	88	94	96	98	18	34	44	55	63	72	78	85	0.6	0.7	0.9	1.1	1.2	0.8	0.9	1.1	1.2	1.4
WW B3	44	68	79	88	93	97	98	99	22	40	52	63	71	80	85	90	0.6	0.7	0.9	1.0	1.2	0.7	0.8	1.0	1.2	1.3
WW B4	43	68	79	89	93	97	98	99	20	40	52	64	72	82	87	92	0.6	0.7	0.8	1.0	1.2	0.7	0.8	1.0	1.1	1.3
WW B5	41	66	78	87	92	96	98	99	19	33	51	63	71	80	86	91	0.6	0.7	0.9	1.0	1.2	0.7	0.8	1.0	1.1	1.3
WW C1	46	69	81	89	93	97	98	99	23	42	54	65	73	82	87	92	0.6	0.7	0.8	1.0	1.2	0.7	0.8	1.0	1.1	1.3
WW C2	43	67	79	88	92	96	98	99	21	39	51	63	71	80	85	91	0.6	0.7	0.9	1.0	1.2	0.7	0.8	1.0	1.2	1.3
WW C3	48	73	84	92	96	98	99	100	23	46	59	71	79	87	91	95	0.5	0.7	0.8	1.0	1.2	0.7	0.8	0.9	1.1	1.3
WW C4	44	70	81	90	94	98	99	100	20	42	55	68	76	85	89	94	0.6	0.7	0.8	1.0	1.2	0.7	0.8	0.9	1.1	1.3
TW A1	29	43	52	62	69	77	82	87	18	26	31	38	43	50	55	61	0.8	0.9	1.0	1.2	1.4	1.1	1.2	1.4	1.5	1.7
TW A2	32	50	60	71	78	85	89	93	17	28	36	44	50	59	65	72	0.7	0.8	1.0	1.1	1.3	1.0	1.1	1.2	1.4	1.5
TW A3	33	52	64	74	81	88	92	95	17	29	37	46	53	62	68	75	0.7	0.8	0.9	1.1	1.3	0.9	1.0	1.2	1.3	1.5
TW A4	32	52	64	75	81	89	92	95	16	28	37	47	54	63	69	76	0.7	0.8	0.9	1.1	1.3	0.9	1.0	1.1	1.3	1.5
TW B1	28	43	53	63	70	78	83	88	17	25	31	38	44	51	56	63	0.8	0.9	1.0	1.2	1.4	1.1	1.2	1.3	1.5	1.6
TW B2	30	48	59	69	76	84	88	93	16	27	34	43	49	58	63	70	0.7	0.8	1.0	1.1	1.3	1.0	1.1	1.2	1.4	1.5
TW B3	29	48	59	70	77	85	89	93	15	26	34	43	49	58	64	71	0.7	0.8	1.0	[illegible]	1.3	1.0	1.1	1.2	1.4	1.5
TW B4	27	45	56	67	74	82	87	92	14	25	32	40	47	56	62	69	0.7	0.8	1.0	1.2	1.3	1.0	1.1	1.2	1.4	1.5
TW C1	27	42	52	62	69	78	83	88	15	24	30	38	43	51	56	63	0.8	0.9	1.0	1.2	1.4	1.1	1.2	1.3	1.5	1.6
TW C2	26	43	53	64	71	79	84	89	14	24	30	38	44	52	58	65	0.8	0.9	1.0	1.2	1.3	1.0	1.1	1.3	1.4	1.6
TW C3	25	40	50	61	68	77	82	87	14	22	29	36	42	50	56	63	0.8	0.9	1.0	1.2	1.4	1.1	1.2	1.3	1.5	1.6
TW C4	22	36	45	55	62	71	76	82	13	21	26	33	38	45	50	57	0.9	0.9	1.1	1.3	1.4	1.2	1.3	1.4	1.6	1.7
TW D1	26	42	52	63	70	78	83	88	13	22	28	35	40	48	54	61	0.8	0.9	1.0	1.2	1.4	1.1	1.2	1.4	1.5	1.7
TW D2	34	55	67	78	84	90	94	96	18	31	40	50	58	67	73	80	0.6	0.8	0.9	1.1	1.3	0.9	1.0	1.1	1.3	1.4
TW D3	38	60	72	82	88	93	96	98	19	34	44	55	63	72	78	84	0.6	0.7	0.9	1.1	1.2	0.8	0.9	1.1	1.2	1.4
TW D4	39	62	74	84	90	95	97	98	19	36	47	59	66	76	82	87	0.6	0.7	0.9	1.1	1.2	0.8	0.9	1.0	1.2	1.4
TW D5	37	61	73	83	89	94	97	98	18	35	46	58	66	76	81	87	0.6	0.7	0.9	1.1	1.2	0.8	0.9	1.0	1.2	1.4
TW E1	42	64	76	85	90	95	97	99	21	38	49	60	67	77	82	88	0.6	0.7	0.9	1.0	1.2	0.8	0.9	1.0	1.2	1.4
TW E2	40	63	75	85	90	95	97	99	20	37	48	59	67	77	82	88	0.6	0.7	0.9	1.0	1.2	0.8	0.9	1.0	1.2	1.4
TW E3	46	71	82	90	94	98	99	100	24	44	56	68	76	85	89	93	0.6	0.7	0.8	1.0	1.2	0.7	0.8	1.0	1.1	1.3
TW E4	43	68	79	88	93	97	98	99	21	41	53	65	73	82	87	92	0.6	0.7	0.8	1.0	1.2	0.7	0.8	1.0	1.1	1.3
TW F1	27	42	52	62	69	77	82	88	16	24	30	37	42	50	55	62	0.8	0.9	1.0	1.2	1.4	1.1	1.2	1.4	1.5	1.7
TW F2	29	47	58	69	76	84	88	92	15	26	33	41	48	57	62	69	0.7	0.8	1.0	1.1	1.3	1.0	1.1	1.2	1.4	1.5
TW F3	29	48	60	71	78	85	90	94	14	25	33	42	49	58	64	72	0.7	0.8	1.0	1.1	1.3	1.0	1.1	1.2	1.3	1.5
TW F4	27	46	58	69	76	84	89	93	12	23	31	41	47	57	63	71	0.7	0.8	1.0	1.1	1.3	1.0	1.1	1.2	1.4	1.5
TW G1	25	41	51	61	68	77	82	88	14	22	29	36	41	49	54	61	0.8	0.9	1.0	1.2	1.4	1.1	1.2	1.3	1.5	1.6
TW G2	25	43	54	65	72	81	86	91	12	22	29	38	44	53	59	66	0.7	0.8	1.0	1.2	1.3	1.0	1.1	1.3	1.4	1.6
TW G3	23	41	52	64	71	80	85	90	10	20	28	36	42	51	57	65	0.7	0.8	1.0	1.2	1.3	1.0	1.1	1.3	1.4	1.6
TW G4	19	36	46	57	65	74	80	86	8	17	24	31	37	46	52	59	0.8	0.9	1.0	1.2	1.4	1.1	1.2	1.3	1.5	1.6
TW H1	22	37	47	58	65	74	80	86	11	19	25	33	38	46	51	59	0.8	0.9	1.1	1.2	1.4	1.1	1.2	1.4	1.5	1.7
TW H2	19	35	45	56	63	73	78	84	8	17	23	30	36	44	50	57	0.8	0.9	1.1	1.2	1.4	1.1	1.2	1.4	1.5	1.7
TW H3	16	30	40	50	58	67	73	80	6	14	20	27	32	39	45	52	0.9	1.0	1.1	1.3	1.4	1.2	1.3	1.4	1.6	1.7
TW H4	12	24	32	41	48	58	64	71	4	11	15	21	25	32	37	43	1.0	1.1	1.2	1.3	1.5	1.4	1.4	1.6	1.7	1.9
TW I1	23	38	49	59	66	75	80	86	11	19	25	32	37	45	50	57	0.8	0.9	1.0	1.2	1.4	1.1	1.2	1.4	1.5	1.7
TW I2	31	51	63	75	81	89	92	96	14	28	37	47	54	64	70	77	0.7	0.8	0.9	1.1	1.3	0.9	1.0	1.1	1.3	1.4
TW I3	33	55	67	78	84	91	94	97	16	30	40	51	60	08	74	81	0.6	0.7	0.9	1.1	1.3	0.9	1.0	1.1	1.2	1.4
TW I4	35	58	71	82	07	93	96	98	16	33	43	55	63	73	79	85	0.6	0.7	0.9	1.1	1.2	0.8	0.9	1.0	1.2	1.4
TW I5	35	58	71	82	88	93	96	98	16	33	44	56	64	74	80	86	0.6	0.7	0.9	1.1	1.2	0.8	0.9	1.0	1.2	1.4
TW J1	39	62	73	83	89	94	96	98	19	35	46	57	65	75	80	86	0.6	0.7	0.9	1.1	1.2	0.8	0.9	1.0	1.2	1.4
TW J2	37	60	72	83	88	94	96	98	17	34	45	56	64	74	80	86	0.6	0.7	0.9	1.1	1.2	0.8	0.9	1.0	1.2	1.4
TW J3	42	67	79	89	93	97	98	99	20	40	53	65	73	82	87	92	0.6	0.7	0.8	1.0	1.2	0.7	0.8	1.0	1.1	1.3
TW J4	40	64	77	87	92	96	98	99	18	37	50	62	70	80	85	91	0.6	0.7	0.9	1.0	1.2	0.7	0.8	1.0	1.1	1.3
DG A1	29	48	58	68	74	81	84	88	13	24	31	38	43	50	54	59	0.7	0.8	1.0	1.2	1.5	1.0	1.1	1.3	1.6	1.9
DG A2	29	49	61	71	78	85	88	92	14	25	34	40	49	58	63	69	0.7	0.8	1.0	1.2	1.4	1.0	1.0	1.2	1.4	1.6
DG A3	32	55	60	78	84	90	92	94	16	30	40	51	59	69	74	80	0.6	0.7	0.9	1.2	1.5	0.9	1.0	1.1	1.3	1.5
DG B1	30	52	64	75	81	87	90	93	14	24	32	41	48	57	62	68	0.7	0.8	0.9	1.2	1.4	1.0	1.1	1.2	1.4	1.7
DG B2	30	53	66	78	84	90	93	96	14	26	35	46	54	64	70	76	0.7	0.7	0.9	1.1	1.3	1.0	1.0	1.1	1.3	1.5
DG B3	33	57	71	82	88	93	95	97	17	30	41	53	62	73	79	85	0.6	0.7	0.9	1.1	1.4	0.9	0.9	1.0	1.2	1.4
DG C1	35	59	72	83	88	92	94	96	16	28	38	49	57	67	72	78	0.6	0.7	0.9	1.1	1.5	0.9	1.0	1.1	1.3	1.5
DG C2	34	59	73	83	89	94	96	98	17	30	40	52	61	71	77	83	0.6	0.7	0.9	1.1	1.3	0.9	0.9	1.0	1.2	1.4
DG C3	37	64	78	88	92	96	97	98	19	34	46	60	69	80	85	90	0.6	0.7	0.9	1.1	1.5	0.8	0.9	1.0	1.1	1.3
SS A1	38	57	67	77	83	90	93	96	22	35	43	52	59	67	73	79	0.7	0.8	0.9	1.1	1.3	0.9	1.0	1.1	1.3	1.5
SS A2	45	67	78	86	91	95	97	99	26	43	53	64	71	80	84	89	0.6	0.7	0.9	1.0	1.2	0.8	0.9	1.0	1.2	1.3
SS A3	36	54	65	75	81	87	91	95	21	33	40	49	55	64	69	76	0.7	0.8	1.0	1.1	1.3	0.9	1.0	1.2	1.3	1.5
SS A4	45	66	77	86	91	95	97	99	26	42	53	63	70	79	84	89	0.6	0.7	0.9	1.1	1.2	0.8	0.9	1.0	1.2	1.4
SS A5	36	53	63	73	79	86	90	94	22	33	40	48	54	62	68	74	0.7	0.8	1.0	1.1	1.3	0.9	1.1	1.2	1.4	1.5
SS A6	45	67	78	86	91	95	97	99	26	43	53	64	71	79	84	89	0.6	0.7	0.9	1.0	1.2	0.8	0.9	1.0	1.2	1.4
SS A7	32	48	58	67	73	81	85	90	20	29	36	43	48	56	61	67	0.7	0.8	1.0	1.2	1.4	1.0	1.1	1.3	1.4	1.6
SS A8	45	66	77	85	90	95	97	99	26	42	52	62	69	78	83	88	0.6	0.7	0.9	1.1	1.2	0.8	0.9	1.0	1.2	1.4
SS B1	33	51	61	72	78	85	89	93	19	30	37	46	52	61	66	73	0.7	0.8	1.0	1.1	1.3	0.9	1.1	1.2	1.4	1.5
SS B2	40	61	72	82	87	93	95	98	22	37	47	58	65	74	80	85	0.6	0.7	0.9	1.1	1.3	0.8	0.9	1.1	1.2	1.4
SS B3	31	49	60	70	76	84	88	92	18	28	36	44	50	58	64	71	0.7	0.8	1.0	1.2	1.3	1.0	1.1	1.2	1.4	1.5
SS B4	39	60	71	81	87	93	95	97	22	37	47	57	64	73	79	85	0.6	0.7	0.9	1.1	1.3	0.8	0.9	1.1	1.2	1.4
SS B5	29	45	55	65	71	79	84	89	18	27	33	40	46	53	59	65	0.8	0.9	1.0	1.2	1.4	1.0	1.2	1.3	1.5	1.6
SS B6	39	60	72	81	87	93	95	98	22	37	47	57	64	73	79	85	0.6	0.7	0.9	1.1	1.3	0.8	0.9	1.1	1.2	1.4
SS B7	27	41	51	60	67	75	80	85	17	24	30	37	42	49	54	60	0.8	0.9	1.1	1.2	1.4	1.1	1.2	1.4	1.5	1.7
SS B8	38	59	70	80	86	92	95	97	21	36	45	55	63	72	77	83	0.6	0.7	0.9	1.1	1.3	0.8	0.9	1.1	1.2	1.4
SS C1	33	53	65	76	82	89	92	96	16	29	37	47	54	63	69	76	0.7	0.8	0.9	1.1	1.3	0.9	1.0	1.2	1.3	1.5
SS C2	38	60	72	82	88	93	96	98	18	34	44	55	63	72	78	84	0.6	0.7	0.9	1.1	1.2	0.8	0.9	1.1	1.2	1.4
SS C3	27	43	53	64	71	79	84	89	15	24	30	38	43	51	57	64	0.8	0.9	1.0	1.2	1.4	1.1	1.2	1.3	1.5	1.6
SS C4	33	54	66	76	83	90	93	96	17	30	39	49	56	66	72	79	0.7	0.8	0.9	1.1	1.3	0.9	1.0	1.1	1.3	1.4
SS D1	45	66	77	86	90	95	97	99	26	41	51	62	69	77	82	87	0.6	0.7	0.9	1.1	1.2	0.8	0.9	1.0	1.2	1.4
SS D2	52	75	85	92	95	98	99	100	30	49	61	72	79	87	91	95	0.5	0.7	0.8	1.0	1.2	0.7	0.8	0.9	1.1	1.3
SS D3	41	60	71	80	86	92	94	97	24	37	46	55	62	71	76	82	0.6	0.7	0.9	1.1	1.3	0.8	1.0	1.1	1.3	1.4
SS D4	50	73	83	91	95	98	99	100	29	48	59	70	77	85	89	93	0.6	0.7	0.8	1.0	1.2	0.7	0.8	1.0	1.1	1.3
SS E1	40	61	72	82	87	93	95	98	22	36	45	55	62	71	77	83	0.6	0.7	0.9	1.1	1.3	0.8	0.9	1.1	1.2	1.4
SS E2	47	70	80	89	93	97	98	99	25	43	55	66	73	82	87	91	0.6	0.7	0.9	1.0	1.2	0.7	0.8	1.0	1.2	1.3
SS E3	34	51	62	72	78	86	90	93	20	30	38	46	52	61	66	73	0.7	0.8	1.0	1.1	1.3	0.9	1.1	1.2	1.4	1.5
SS E4	43	65	76	85	90	95	97	99	24	40	51	61	69	78	83	88	0.6	0.7	0.9	1.1	1.2	0.8	0.9	1.0	1.2	1.4

WICHITA FALLS, TEXAS

TYPE	SOLAR SAVINGS FRACTION (%) BASE TEMP: 55 F (1334 DD)								65 F (2906 DD)								CONSERVATION FACTOR 55 F					65 F				
LCR	200	100	70	50	40	30	25	20	200	100	70	50	40	30	25	20	200	100	50	30	20	200	100	50	30	20
WW A1	24	35	43	52	59	67	73	79	17	23	28	34	39	45	50	56	0.9	1.0	1.1	1.3	1.5	1.2	1.3	1.5	1.6	1.8
WW A2	27	43	54	65	72	81	85	90	16	27	34	43	49	58	64	71	0.8	0.8	1.0	1.2	1.3	1.0	1.1	1.2	1.4	1.5
WW A3	28	46	58	69	76	84	89	93	15	28	37	46	53	62	68	76	0.7	0.8	1.0	1.1	1.3	0.9	1.0	1.2	1.3	1.5
WW A4	28	48	60	71	78	86	90	94	15	28	38	48	55	65	71	79	0.7	0.8	0.9	1.1	1.3	0.9	1.0	1.1	1.3	1.4
WW A5	29	50	62	74	81	88	92	95	15	29	39	50	57	67	73	80	0.7	0.8	0.9	1.1	1.3	0.9	1.0	1.1	1.3	1.4
WW A6	29	51	63	75	82	89	93	96	15	30	40	51	59	69	75	82	0.7	0.8	0.9	1.1	1.3	0.8	0.9	1.1	1.2	1.4
WW B1	23	37	47	57	64	73	79	84	14	22	28	35	40	48	53	60	0.8	0.9	1.1	1.2	1.4	1.1	1.2	1.4	1.5	1.7
WW B2	28	48	60	72	79	87	91	95	15	29	39	49	57	67	73	80	0.7	0.8	0.9	1.1	1.3	0.9	1.0	1.1	1.3	1.4
WW B3	33	56	68	79	86	92	95	97	18	35	46	57	65	75	80	87	0.6	0.7	0.9	1.1	1.3	0.8	0.9	1.0	1.2	1.4
WW B4	32	56	69	80	86	93	95	98	17	35	46	59	67	77	82	88	0.6	0.7	0.9	1.1	1.2	0.8	0.9	1.0	1.2	1.3
WW B5	30	54	67	78	85	92	95	97	15	33	45	57	65	76	81	88	0.6	0.7	0.9	1.1	1.2	0.8	0.9	1.0	1.2	1.3
WW C1	34	57	70	81	87	93	96	98	19	36	48	59	67	77	82	88	0.6	0.7	0.9	1.1	1.2	0.8	0.9	1.0	1.2	1.3
WW C2	32	55	68	79	85	92	95	97	17	34	46	57	65	75	81	87	0.6	0.7	0.9	1.1	1.2	0.8	0.9	1.0	1.2	1.4
WW C3	36	62	75	85	91	95	97	99	19	40	53	66	74	83	88	93	0.6	0.7	0.9	1.0	1.2	0.7	0.8	1.0	1.1	1.3
WW C4	32	58	71	82	88	94	96	98	17	37	49	62	70	30	86	91	0.6	0.7	0.9	1.1	1.2	0.7	0.8	1.0	1.1	1.3
TW A1	23	34	42	51	57	66	72	78	16	23	28	34	38	45	49	55	0.9	1.0	1.1	1.3	1.5	1.2	1.3	1.5	1.6	1.8
TW A2	24	39	49	59	67	76	81	86	15	25	31	39	45	53	58	65	0.8	0.9	1.0	1.2	1.4	1.1	1.1	1.3	1.4	1.6
TW A3	25	41	52	63	70	79	84	89	14	25	32	41	47	56	62	69	0.8	0.9	1.0	1.2	1.3	1.0	1.1	1.2	1.4	1.5
TW A4	24	41	52	63	71	80	85	90	13	24	32	41	48	57	63	70	0.8	0.9	1.0	1.2	1.3	1.0	1.1	1.2	1.4	1.5
TW B1	22	34	43	52	58	67	73	79	15	22	28	34	39	46	51	57	0.9	1.0	1.1	1.3	1.4	1.2	1.3	1.4	1.6	1.7
TW B2	23	37	47	58	65	74	80	85	14	23	30	37	43	52	57	64	0.8	0.9	1.1	1.2	1.4	1.1	1.2	1.3	1.4	1.6
TW B3	22	37	48	58	66	75	80	86	13	23	30	38	44	52	58	65	0.8	0.9	1.0	1.2	1.4	1.0	1.1	1.3	1.4	1.6
TW B4	21	35	45	55	63	72	78	84	12	21	28	36	41	50	56	63	0.8	0.9	1.1	1.2	1.4	1.1	1.2	1.3	1.5	1.6
TW C1	21	33	42	51	58	67	73	79	14	21	27	33	38	46	51	57	0.9	1.0	1.1	1.3	1.4	1.2	1.3	1.4	1.6	1.7
TW C2	20	33	42	52	59	69	74	81	12	21	27	34	39	47	52	59	0.9	1.0	1.1	1.3	1.4	1.1	1.2	1.4	1.5	1.7
TW C3	19	31	40	50	57	66	72	78	12	20	25	32	37	45	50	57	0.9	1.0	1.1	1.3	1.4	1.2	1.3	1.4	1.5	1.7
TW C4	18	28	36	44	51	60	65	72	12	18	23	29	33	40	45	51	1.0	1.1	1.2	1.4	1.5	1.3	1.4	1.5	1.6	1.8
TW D1	19	32	41	50	57	66	72	79	11	18	24	30	35	42	47	53	0.9	1.0	1.1	1.3	1.4	1.2	1.3	1.5	1.6	1.8
TW D2	26	44	55	67	74	82	87	92	15	27	35	45	52	61	67	74	0.7	0.8	1.0	1.1	1.3	0.9	1.0	1.2	1.3	1.5
TW D3	23	48	60	71	78	86	90	94	16	29	39	49	56	66	72	79	0.7	0.8	0.9	1.1	1.3	0.9	1.0	1.1	1.3	1.4
TW D4	29	50	63	74	81	89	92	96	16	31	42	53	61	71	77	83	0.7	0.8	0.9	1.1	1.3	0.8	0.9	1.1	1.2	1.4
TW D5	28	49	62	74	81	88	92	96	15	30	41	52	60	70	77	83	0.7	0.8	0.9	1.1	1.3	0.8	0.9	1.1	1.2	1.4
TW E1	31	52	65	76	82	89	93	96	18	33	43	54	62	71	77	84	0.7	0.8	0.9	1.1	1.3	0.8	0.9	1.1	1.2	1.4
TW E2	30	52	64	75	82	89	93	96	17	32	13	54	61	71	77	84	0.7	0.8	0.9	1.1	1.3	0.8	0.9	1.1	1.2	1.4
TW E3	35	59	72	83	88	94	96	98	20	39	51	63	71	80	85	91	0.6	0.7	0.9	1.1	1.2	0.7	0.8	1.0	1.1	1.3
TW E4	32	56	69	80	86	93	95	98	17	36	48	60	68	78	83	89	0.6	0.7	0.9	1.1	1.2	0.8	0.9	1.0	1.2	1.3
TW F1	22	33	42	51	57	66	72	78	14	21	27	33	37	44	49	55	0.9	1.0	1.1	1.3	1.5	1.2	1.3	1.5	1.6	1.7
TW F2	22	37	47	57	64	74	79	85	13	22	29	36	42	50	56	63	0.8	0.9	1.1	1.2	1.4	1.1	1.2	1.3	1.5	1.6
TW F3	21	37	48	59	66	76	81	87	11	21	29	37	43	52	58	65	0.8	0.9	1.0	1.2	1.4	1.0	1.1	1.3	1.4	1.6
TW F4	19	35	46	57	65	74	80	86	9	19	27	35	42	51	57	64	0.8	0.9	1.0	1.2	1.4	1.0	1.1	1.3	1.4	1.6
TW G1	20	32	40	50	57	66	71	78	12	19	25	31	36	43	48	55	0.9	1.0	1.1	1.3	1.5	1.2	1.3	1.4	1.6	1.7
TW G2	19	33	43	53	60	70	76	82	10	19	25	33	38	47	52	60	0.9	0.9	1.1	1.2	1.4	1.1	1.2	1.3	1.5	1.6
TW G3	17	31	41	51	59	69	75	81	8	17	24	51	37	45	51	58	0.9	0.9	1.1	1.2	1.4	1.1	1.2	1.3	1.5	1.6
TW G4	13	26	35	45	53	63	69	76	6	14	20	27	32	40	46	53	0.9	1.0	1.1	1.3	1.5	1.2	1.3	1.4	1.6	1.7
TW H1	16	28	37	46	53	63	69	76	9	16	22	28	33	40	45	52	0.9	1.0	1.2	1.3	1.5	1.2	1.3	1.5	1.6	1.7
TW H2	13	25	34	44	51	61	67	74	6	14	19	26	31	38	44	51	0.9	1.0	1.2	1.3	1.5	1.2	1.3	1.5	1.6	1.7
TW H3	11	22	30	39	46	55	61	69	5	11	16	22	27	34	39	46	1.0	1.1	1.2	1.4	1.5	1.3	1.4	1.5	1.7	1.8
TW H4	8	17	23	31	37	46	52	59	3	9	13	18	21	27	32	38	1.1	1.2	1.3	1.5	1.6	1.5	1.6	1.7	1.8	2.0
TW I1	17	29	37	47	54	53	69	76	9	16	21	27	32	38	43	50	0.9	1.0	1.2	1.3	1.5	1.3	1.4	1.5	1.7	1.8
TW I2	22	40	52	63	71	80	85	90	12	23	32	41	48	58	64	71	0.8	0.8	1.0	1.2	1.3	1.0	1.1	1.2	1.3	1.5
TW I3	24	43	55	67	74	83	88	92	13	26	35	45	52	62	66	75	0.7	0.8	1.0	1.1	1.3	0.9	1.0	1.1	1.3	1.5
TW I4	26	47	59	71	78	87	91	95	13	28	38	49	57	67	74	81	0.7	0.8	0.9	1.1	1.3	0.9	0.9	1.1	1.2	1.4
TW I5	25	47	60	72	79	87	91	95	13	28	39	50	58	68	75	82	0.7	0.8	0.9	1.1	1.3	0.8	0.9	1.1	1.2	1.4
TW J1	29	50	62	73	80	88	92	95	16	31	41	51	59	69	75	82	0.7	0.8	0.9	1.1	1.3	0.8	0.9	1.1	1.2	1.4
TW J2	27	48	61	72	80	87	91	95	14	29	39	50	58	68	75	82	0.7	0.8	0.9	1.1	1.3	0.8	0.9	1.1	1.2	1.4
TW J3	31	55	69	80	86	93	95	98	16	35	47	59	68	78	83	89	0.6	0.7	0.9	1.1	1.2	0.8	0.9	1.0	1.2	1.3
TW J4	29	52	66	78	84	91	94	97	15	33	44	57	65	75	81	87	0.6	0.7	0.9	1.1	1.3	0.8	0.9	1.0	1.2	1.3
DG A1	21	36	46	55	61	69	73	78	11	20	26	33	37	43	47	52	0.8	0.9	1.1	1.3	1.6	1.1	1.2	1.4	1.7	2.1
DG A2	21	38	48	59	66	75	79	85	12	22	29	37	44	52	57	63	0.8	0.9	1.0	1.2	1.4	1.1	1.1	1.3	1.5	1.7
DG A3	24	43	56	68	75	83	87	90	14	26	35	45	53	63	69	76	0.7	0.8	1.0	1.2	1.4	1.0	1.0	1.1	1.3	1.5
DG B1	21	38	50	62	69	77	81	85	11	20	27	35	41	49	54	60	0.8	0.9	1.0	1.2	1.5	1.1	1.2	1.3	1.5	1.8
DG B2	22	40	52	65	73	81	86	90	12	22	30	39	47	57	63	70	0.8	0.8	1.0	1.2	1.4	1.0	1.1	1.2	1.3	1.5
DG B3	24	44	58	72	79	87	91	94	14	26	36	47	56	67	74	81	0.7	0.8	0.9	1.1	1.4	1.0	1.0	1.1	1.2	1.4
DG C1	24	44	58	70	78	85	88	91	13	24	32	41	49	59	65	71	0.7	0.8	0.9	1.2	1.4	1.0	1.1	1.2	1.3	1.6
DG C2	25	45	59	72	79	87	91	94	14	26	35	45	54	65	71	78	0.7	0.8	0.9	1.1	1.3	1.0	1.0	1.1	1.3	1.5
DG C3	27	50	65	78	85	92	94	96	16	30	40	53	62	74	81	87	0.7	0.7	0.9	1.1	1.4	0.9	1.0	1.0	1.2	1.4
SS A1	29	45	55	66	72	81	85	90	19	30	38	46	52	60	66	73	0.8	0.8	1.0	1.2	1.4	0.9	1.1	1.2	1.4	1.5
SS A2	35	55	66	77	83	90	93	96	23	38	47	58	65	74	79	85	0.7	0.8	0.9	1.1	1.3	0.8	0.9	1.1	1.2	1.4
SS A3	27	42	52	62	69	77	82	87	18	28	35	43	48	56	62	68	0.8	0.9	1.0	1.2	1.4	1.0	1.1	1.3	1.4	1.6
SS A4	35	54	66	76	82	89	93	96	23	37	47	57	64	73	78	84	0.7	0.8	0.9	1.1	1.3	0.8	0.9	1.1	1.2	1.4
SS A5	28	42	51	61	68	76	81	86	20	29	35	42	48	55	61	67	0.8	0.9	1.1	1.2	1.4	1.0	1.1	1.3	1.4	1.6
SS A6	35	55	66	77	83	90	93	96	23	37	47	57	64	73	79	85	0.7	0.8	0.9	1.1	1.3	0.8	0.9	1.1	1.2	1.4
SS A7	25	37	46	55	61	70	75	81	18	26	31	37	42	49	54	60	0.9	1.0	1.1	1.3	1.5	1.1	1.2	1.4	1.6	1.7
SS A8	34	54	65	75	82	89	92	95	22	36	46	56	63	72	77	83	0.7	0.8	0.9	1.1	1.3	0.8	0.9	1.1	1.2	1.4
SS B1	25	39	49	60	67	75	80	86	16	26	32	40	46	54	59	66	0.8	0.9	1.1	1.2	1.4	1.0	1.1	1.3	1.4	1.6
SS B2	30	49	60	71	78	86	90	94	19	33	42	52	59	68	74	80	0.7	0.8	1.0	1.1	1.3	0.9	1.0	1.1	1.3	1.4
SS B3	24	38	47	57	64	73	78	84	15	24	31	38	43	51	56	63	0.8	0.9	1.1	1.2	1.4	1.1	1.2	1.3	1.5	1.6
SS B4	30	48	60	70	77	85	89	93	19	32	41	51	58	67	73	79	0.7	0.8	1.0	1.1	1.3	0.9	1.0	1.1	1.3	1.4
SS B5	23	35	43	53	59	68	73	79	16	23	29	35	40	47	52	58	0.9	1.0	1.1	1.3	1.5	1.2	1.3	1.4	1.6	1.7
SS B6	29	48	60	71	78	85	90	94	19	32	41	51	58	67	73	80	0.7	0.8	1.0	1.1	1.3	0.9	1.0	1.1	1.3	1.4
SS B7	21	32	40	48	54	63	68	75	15	21	26	32	36	42	47	53	1.0	1.0	1.2	1.3	1.5	1.2	1.3	1.5	1.7	1.8
SS B8	29	47	58	69	76	84	88	93	18	31	40	49	56	65	71	78	0.7	0.8	1.0	1.1	1.3	0.9	1.0	1.1	1.3	1.5
SS C1	25	42	53	64	71	80	85	90	14	25	32	41	47	56	62	70	0.8	0.8	1.0	1.2	1.3	1.0	1.1	1.2	1.4	1.5
SS C2	28	48	60	72	79	87	91	94	15	29	39	49	57	66	73	80	0.7	0.8	0.9	1.1	1.3	0.9	1.0	1.1	1.3	1.4
SS C3	21	33	42	52	59	68	74	80	13	21	26	33	38	46	51	57	0.9	1.0	1.1	1.3	1.4	1.2	1.3	1.4	1.5	1.7
SS C4	25	43	54	65	73	81	86	91	14	26	34	44	51	60	66	73	0.7	0.8	1.0	1.2	1.3	0.9	1.0	1.2	1.3	1.5
SS D1	34	54	65	76	82	89	92	95	22	36	45	55	62	70	76	82	0.7	0.8	0.9	1.1	1.3	0.8	0.9	1.1	1.3	1.4
SS D2	40	63	75	84	90	95	97	98	25	43	55	66	73	82	87	91	0.6	0.7	0.9	1.1	1.2	0.7	0.8	1.0	1.1	1.3
SS D3	31	48	59	69	76	84	88	92	21	33	40	49	55	64	69	76	0.7	0.8	1.0	1.2	1.3	0.9	1.0	1.2	1.3	1.5
SS D4	39	61	73	83	88	94	96	98	24	42	53	64	71	80	85	90	0.6	0.7	0.9	1.1	1.2	0.7	0.8	1.0	1.2	1.3
SS E1	30	49	60	71	77	85	89	93	18	31	39	48	55	64	70	76	0.7	0.8	1.0	1.1	1.3	0.9	1.0	1.2	1.3	1.5
SS E2	35	57	69	80	86	92	95	97	21	38	48	59	67	76	82	87	0.6	0.7	0.9	1.1	1.3	0.8	0.9	1.0	1.2	1.4
SS E3	26	40	50	60	67	75	80	86	17	26	33	40	46	54	59	66	0.8	0.9	1.1	1.2	1.4	1.0	1.1	1.3	1.4	1.6
SS E4	32	53	65	75	82	89	92	96	20	35	44	55	62	71	77	83	0.7	0.8	0.9	1.1	1.3	0.8	0.9	1.1	1.2	1.4

BRYCE CANYON, UTAH

TYPE	SOLAR SAVINGS FRACTION (%)																CONSERVATION FACTOR									
BASE TEMP:	55 F (5952 DD)								65 F (9136 DD)								55 F					65 F				
LCR	100	70	50	40	30	25	20	15	100	70	50	40	30	25	20	15	100	50	30	20	15	100	50	30	20	15
WW A1	21	25	30	34	40	44	50	57	16	19	23	26	30	33	38	44	1.4	1.6	1.7	1.9	2.0	1.8	1.9	2.0	2.2	2.3
WW A2	23	30	37	43	51	57	64	73	16	21	28	32	39	44	51	59	1.2	1.3	1.4	1.6	1.7	1.3	1.5	1.6	1.8	1.9
WW A3	24	31	40	47	56	62	69	78	16	22	30	35	43	49	56	65	1.1	1.2	1.4	1.5	1.7	1.2	1.4	1.5	1.7	1.8
WW A4	24	32	42	49	58	64	72	81	16	23	31	37	45	51	58	68	1.0	1.2	1.3	1.5	1.6	1.2	1.3	1.5	1.6	1.8
WW A5	25	34	44	51	61	67	74	83	16	23	32	38	47	53	61	71	1.0	1.2	1.3	1.5	1.6	1.2	1.3	1.4	1.6	1.7
WW A6	25	34	44	52	62	68	76	84	16	24	33	39	49	55	62	72	1.0	1.1	1.3	1.5	1.6	1.1	1.3	1.4	1.6	1.7
WW B1	18	24	30	34	41	46	52	61	12	16	20	24	29	32	37	44	1.3	1.5	1.6	1.8	1.9	1.6	1.8	1.9	2.1	2.3
WW B2	25	33	43	50	60	67	74	83	17	24	33	39	48	54	62	72	1.0	1.2	1.3	1.5	1.6	1.2	1.3	1.4	1.6	1.7
WW B3	30	40	51	59	69	75	82	89	21	30	40	47	57	64	71	80	0.9	1.1	1.2	1.4	1.5	1.0	1.2	1.3	1.5	1.6
WW B4	30	41	53	61	71	77	84	91	21	31	42	49	60	67	75	84	0.9	1.1	1.2	1.4	1.5	1.0	1.1	1.3	1.5	1.6
WW B5	28	40	51	60	70	77	84	91	20	30	41	49	59	66	74	83	0.9	1.1	1.2	1.4	1.5	1.0	1.1	1.3	1.4	1.6
WW C1	32	42	53	61	71	77	84	91	23	32	43	50	60	66	74	83	0.9	1.1	1.2	1.4	1.5	1.0	1.2	1.3	1.5	1.6
WW C2	29	40	51	59	70	76	83	90	21	30	40	48	58	65	73	82	0.9	1.1	1.2	1.4	1.5	1.0	1.2	1.3	1.5	1.6
WW C3	35	48	60	69	79	84	90	95	26	38	50	59	70	76	83	90	0.8	1.0	1.1	1.3	1.5	0.9	1.1	1.2	1.4	1.5
WW C4	32	44	56	65	75	81	88	94	23	34	46	54	65	72	80	88	0.9	1.0	1.2	1.3	1.5	0.9	1.1	1.2	1.4	1.5
TW A1	20	24	29	33	39	44	49	57	16	19	22	25	30	33	38	44	1.4	1.6	1.7	1.9	2.0	1.8	1.9	2.0	2.2	2.3
TW A2	21	27	34	39	47	52	59	68	15	20	25	29	36	40	46	55	1.2	1.4	1.5	1.7	1.8	1.5	1.6	1.7	1.9	2.0
TW A3	21	28	36	42	50	56	63	72	15	20	26	31	38	43	50	59	1.2	1.3	1.5	1.6	1.7	1.4	1.5	1.6	1.8	1.9
TW A4	20	28	36	42	51	57	64	73	14	19	26	31	39	44	51	60	1.1	1.3	1.4	1.6	1.7	1.3	1.5	1.6	1.7	1.9
TW B1	20	24	30	34	40	45	51	59	15	18	22	26	31	34	39	46	1.4	1.5	1.7	1.8	2.0	1.7	1.8	1.9	2.1	2.2
TW B2	20	26	33	38	46	51	58	67	14	19	24	28	35	39	46	54	1.2	1.4	1.5	1.7	1.8	1.5	1.6	1.7	1.9	2.0
TW B3	19	25	33	38	46	52	59	68	13	18	24	29	35	40	47	55	1.2	1.4	1.5	1.7	1.8	1.4	1.5	1.7	1.8	2.0
TW B4	18	24	31	36	44	49	56	66	13	17	23	27	34	38	45	53	1.3	1.4	1.5	1.7	1.8	1.5	1.6	1.7	1.9	2.0
TW C1	18	23	29	34	40	45	51	60	14	17	22	25	31	35	40	47	1.4	1.5	1.7	1.8	1.9	1.6	1.7	1.9	2.0	2.2
TW C2	18	23	29	34	41	46	53	62	13	17	22	26	32	36	42	50	1.3	1.5	1.6	1.7	1.9	1.5	1.7	1.8	1.9	2.1
TW C3	17	22	28	33	39	44	51	60	12	16	21	24	30	34	40	48	1.4	1.5	1.6	1.8	1.9	1.6	1.7	1.8	2.0	2.1
TW C4	16	20	25	29	35	40	46	54	12	15	19	22	27	31	36	43	1.5	1.6	1.7	1.9	2.0	1.7	1.9	2.0	2.1	2.2
TW D1	15	20	25	30	36	40	46	54	10	13	17	20	25	28	33	39	1.4	1.6	1.7	1.9	2.0	1.7	1.9	2.1	2.2	2.4
TW D2	23	31	39	46	55	61	69	78	16	22	30	35	44	49	56	66	1.1	1.2	1.4	1.5	1.7	1.2	1.4	1.5	1.7	1.8
TW D3	25	34	44	51	60	66	74	82	18	25	33	40	49	55	62	72	1.0	1.2	1.3	1.5	1.6	1.2	1.3	1.4	1.6	1.7
TW D4	27	37	47	55	65	71	79	87	19	28	37	44	54	60	68	78	1.0	1.1	1.3	1.4	1.6	1.1	1.2	1.4	1.5	1.7
TW D5	26	36	47	55	65	71	79	87	19	27	37	44	54	61	69	78	1.0	1.1	1.3	1.4	1.6	1.1	1.2	1.4	1.5	1.6
TW E1	29	38	48	56	65	72	79	87	21	29	38	45	54	60	68	77	1.0	1.1	1.3	1.4	1.6	1.1	1.2	1.4	1.5	1.7
TW E2	28	38	48	56	65	72	79	87	20	28	38	45	54	60	68	78	1.0	1.1	1.3	1.4	1.6	1.1	1.2	1.4	1.5	1.7
TW E3	34	45	57	65	75	81	87	93	25	36	47	55	65	72	79	87	0.9	1.0	1.2	1.3	1.5	1.0	1.1	1.3	1.4	1.6
TW E4	31	42	54	62	73	79	85	92	23	33	44	52	62	69	77	85	0.9	1.0	1.2	1.4	1.5	1.0	1.1	1.3	1.4	1.6
TW F1	19	23	28	33	39	43	49	57	14	17	21	24	29	32	37	44	1.4	1.6	1.7	1.9	2.0	1.7	1.8	2.0	2.1	2.3
TW F2	19	25	31	37	44	50	56	66	13	17	23	27	33	38	44	52	1.3	1.4	1.5	1.7	1.8	1.5	1.6	1.8	1.9	2.0
TW F3	18	24	32	38	46	51	59	68	12	17	23	27	34	39	46	55	1.2	1.4	1.5	1.6	1.8	1.4	1.5	1.7	1.8	2.0
TW F4	16	22	30	36	44	50	58	67	10	15	21	26	33	38	45	54	1.2	1.4	1.5	1.6	1.8	1.4	1.5	1.7	1.8	2.0
TW G1	17	21	27	31	38	42	49	57	12	15	20	23	28	32	37	44	1.4	1.5	1.7	1.8	2.0	1.7	1.8	1.9	2.1	2.2
TW G2	16	21	28	33	41	46	53	62	10	14	20	24	30	34	40	49	1.3	1.4	1.6	1.7	1.9	1.5	1.6	1.8	1.9	2.1
TW G3	14	20	26	32	39	45	52	61	8	13	18	22	29	33	40	48	1.3	1.4	1.6	1.7	1.9	1.5	1.6	1.8	1.9	2.1
TW G4	11	16	22	27	35	40	47	56	6	10	15	19	25	29	35	43	1.4	1.5	1.7	1.8	1.9	1.6	1.7	1.9	2.0	2.1
TW H1	14	18	24	28	35	40	46	55	9	12	17	20	25	29	34	42	1.4	1.6	1.7	1.8	2.0	1.7	1.8	2.0	2.1	2.2
TW H2	11	16	22	26	33	38	44	54	7	10	15	18	24	28	33	41	1.4	1.6	1.7	1.8	2.0	1.7	1.8	1.9	2.1	2.2
TW H3	9	13	19	23	29	34	40	49	5	8	12	16	21	24	29	37	1.5	1.6	1.8	1.9	2.0	1.8	1.9	2.0	2.2	2.3
TW H4	7	10	14	18	23	27	33	40	3	6	9	12	16	19	23	30	1.7	1.8	2.0	2.1	2.2	2.0	2.1	2.2	2.4	2.5
TW I1	13	17	22	26	33	37	43	51	8	11	15	18	22	25	29	36	1.5	1.6	1.8	1.9	2.1	1.8	2.0	2.1	2.3	2.5
TW I2	20	27	36	42	51	58	65	75	13	19	26	32	40	46	53	62	1.1	1.3	1.4	1.6	1.7	1.3	1.4	1.6	1.7	1.8
TW I3	22	30	39	46	56	62	70	79	15	22	30	36	44	50	58	68	1.1	1.2	1.4	1.5	1.6	1.2	1.3	1.5	1.6	1.8
TW I4	24	33	44	51	61	00	78	84	16	24	34	41	50	57	65	75	1.0	1.1	1.3	1.4	1.6	1.1	1.2	1.4	1.5	1.7
TW I5	24	34	44	52	63	69	77	85	16	25	34	42	52	58	67	76	1.0	1.1	1.3	1.4	1.6	1.1	1.2	1.4	1.5	1.7
TW J1	26	36	46	53	63	69	77	85	19	27	36	42	52	58	66	75	1.0	1.1	1.3	1.4	1.6	1.1	1.2	1.4	1.5	1.7
TW J2	25	34	45	52	63	69	76	85	17	25	35	42	51	58	65	75	1.0	1.1	1.3	1.4	1.6	1.1	1.2	1.4	1.5	1.7
TW J3	30	42	54	62	73	79	85	92	22	32	44	52	63	69	77	86	0.9	1.0	1.2	1.4	1.5	1.0	1.1	1.3	1.4	1.6
TW J4	28	39	51	59	70	76	83	91	20	30	41	49	60	66	74	84	0.9	1.1	1.2	1.4	1.5	1.0	1.1	1.3	1.4	1.6
DG A1	16	21	27	31	37	40	44	49	10	14	18	21	25	28	30	34	1.3	1.5	1.9	2.2	2.7	1.6	1.8	2.2	2.8	3.5
DG A2	18	25	32	38	46	51	57	64	12	17	23	28	35	39	44	51	1.2	1.3	1.6	1.8	2.0	1.4	1.5	1.7	2.0	2.2
DG A3	22	30	40	47	57	64	71	79	16	23	31	37	46	52	60	69	1.1	1.2	1.3	1.5	1.7	1.2	1.3	1.4	1.6	1.8
DG B1	16	22	29	34	41	46	52	59	10	14	18	22	27	31	36	41	1.3	1.4	1.6	1.9	2.2	1.6	1.7	1.9	2.2	2.6
DG B2	19	25	34	40	50	56	63	72	13	18	24	29	37	42	49	58	1.2	1.3	1.4	1.6	1.8	1.4	1.5	1.6	1.8	2.0
DG B3	23	31	41	49	61	68	76	84	17	23	32	38	48	55	64	74	1.1	1.1	1.3	1.4	1.6	1.2	1.3	1.4	1.5	1.7
DG C1	20	26	34	41	50	56	63	70	13	17	23	28	35	39	46	54	1.2	1.3	1.4	1.7	1.9	1.4	1.5	1.7	1.9	2.1
DG C2	22	30	39	46	57	64	72	80	15	21	29	35	43	50	58	67	1.1	1.2	1.3	1.5	1.7	1.3	1.4	1.5	1.6	1.8
DG C3	26	35	46	55	68	75	82	90	19	27	36	43	55	62	72	81	1.0	1.1	1.2	1.4	1.6	1.1	1.2	1.3	1.4	1.6
SS A1	28	34	42	48	55	61	67	75	23	29	35	40	47	52	58	66	1.1	1.3	1.4	1.6	1.8	1.2	1.4	1.6	1.7	1.9
SS A2	34	43	53	60	69	74	81	88	29	37	46	52	61	67	73	81	1.0	1.1	1.3	1.4	1.6	1.0	1.2	1.3	1.5	1.7
SS A3	26	32	39	44	51	56	62	70	21	26	32	37	43	47	53	60	1.2	1.3	1.5	1.7	1.8	1.3	1.5	1.7	1.9	2.0
SS A4	34	43	52	59	67	73	79	86	28	36	45	51	60	65	72	79	1.0	1.1	1.3	1.5	1.6	1.0	1.2	1.4	1.5	1.7
SS A5	26	32	39	44	51	55	62	69	22	27	33	37	43	47	53	60	1.2	1.4	1.5	1.7	1.8	1.3	1.5	1.7	1.9	2.0
SS A6	34	43	53	60	68	74	80	87	28	37	45	52	60	66	73	81	1.0	1.1	1.3	1.4	1.6	1.0	1.2	1.4	1.5	1.7
SS A7	24	29	34	38	44	49	54	61	20	24	29	32	37	41	45	52	1.3	1.5	1.7	1.8	2.0	1.5	1.7	1.9	2.1	2.2
SS A8	33	42	51	58	66	72	78	85	28	35	44	50	58	64	70	78	1.0	1.1	1.3	1.5	1.6	1.1	1.2	1.4	1.6	1.7
SS B1	23	29	36	41	49	54	60	69	18	23	29	34	40	45	51	59	1.2	1.4	1.5	1.7	1.8	1.3	1.5	1.7	1.8	2.0
SS B2	30	38	47	54	63	69	75	83	24	32	40	46	55	61	68	76	1.0	1.2	1.3	1.5	1.6	1.1	1.2	1.4	1.6	1.7
SS B3	22	27	34	39	46	51	57	65	17	22	27	32	38	42	47	55	1.2	1.4	1.6	1.7	1.9	1.4	1.6	1.7	1.9	2.1
SS B4	29	37	46	53	62	67	74	82	24	31	39	45	54	59	66	75	1.0	1.2	1.3	1.5	1.6	1.1	1.3	1.4	1.6	1.7
SS B5	21	26	32	36	42	47	53	60	17	21	26	30	35	39	44	51	1.3	1.5	1.7	1.8	2.0	1.5	1.7	1.8	2.0	2.2
SS B6	29	37	46	53	62	68	75	83	23	31	39	45	54	59	66	75	1.0	1.2	1.3	1.5	1.6	1.1	1.3	1.4	1.6	1.7
SS B7	19	24	29	32	38	42	47	55	16	19	23	26	31	34	39	45	1.4	1.6	1.8	1.9	2.1	1.6	1.8	2.0	2.2	2.3
SS B8	28	36	45	51	60	65	72	80	22	30	37	43	52	57	64	73	1.0	1.2	1.4	1.5	1.7	1.1	1.3	1.5	1.6	1.8
SS C1	21	28	35	41	50	56	63	72	14	19	25	30	37	42	49	58	1.2	1.3	1.5	1.6	1.7	1.4	1.5	1.6	1.8	1.9
SS C2	25	34	44	51	60	67	74	83	18	25	33	40	49	55	62	72	1.0	1.2	1.3	1.5	1.6	1.2	1.3	1.4	1.6	1.7
SS C3	18	23	29	33	40	45	51	60	13	16	21	24	30	34	39	46	1.4	1.5	1.6	1.8	1.9	1.6	1.7	1.9	2.0	2.2
SS C4	22	30	39	45	54	60	67	77	16	22	29	35	43	48	56	65	1.1	1.2	1.4	1.5	1.7	1.3	1.4	1.5	1.7	1.8
SS D1	32	41	50	56	65	70	76	84	27	34	42	48	56	61	67	75	1.0	1.2	1.3	1.5	1.6	1.1	1.3	1.4	1.6	1.8
SS D2	39	50	61	68	77	82	88	93	33	43	53	60	70	75	81	88	0.9	1.0	1.2	1.4	1.5	0.9	1.1	1.3	1.4	1.6
SS D3	30	37	45	51	59	64	70	78	25	31	38	43	50	55	61	69	1.1	1.2	1.4	1.6	1.7	1.2	1.4	1.5	1.7	1.9
SS D4	38	48	59	66	75	80	86	92	32	41	51	58	67	73	79	87	0.9	1.0	1.2	1.4	1.5	0.9	1.1	1.3	1.4	1.6
SS E1	27	35	43	50	58	63	70	78	21	28	35	41	48	53	60	68	1.1	1.2	1.4	1.6	1.7	1.2	1.3	1.5	1.7	1.8
SS E2	34	44	54	61	71	76	83	89	27	36	46	53	62	68	75	83	0.9	1.1	1.2	1.4	1.6	1.0	1.2	1.3	1.5	1.6
SS E3	24	30	36	41	49	54	60	68	19	24	30	34	40	45	50	58	1.2	1.4	1.5	1.7	1.8	1.4	1.5	1.7	1.9	2.0
SS E4	31	40	50	57	66	72	78	86	25	33	42	48	57	63	70	78	1.0	1.1	1.3	1.5	1.6	1.1	1.2	1.4	1.5	1.7

CEDAR CITY, UTAH

TYPE	SSF (%) 55 F (3702 DD) LCR 100	70	50	40	30	25	20	15	SSF (%) 65 F (6141 DD) 100	70	50	40	30	25	20	15	CF 55 F 100	50	30	20	15	CF 65 F 100	50	30	20	15
WW A1	25	31	37	42	49	54	61	69	19	23	28	31	37	41	46	53	1.2	1.4	1.5	1.7	1.8	1.5	1.7	1.8	2.0	2.1
WW A2	30	38	47	53	62	68	75	83	21	27	34	40	48	53	60	69	1.0	1.2	1.3	1.5	1.6	1.2	1.4	1.5	1.7	1.8
WW A3	31	40	50	58	67	73	80	87	21	29	37	43	52	58	65	74	1.0	1.1	1.3	1.4	1.6	1.1	1.3	1.4	1.6	1.7
WW A4	32	42	52	60	69	75	82	89	21	29	38	45	54	60	68	77	0.9	1.1	1.2	1.4	1.6	1.1	1.2	1.4	1.5	1.7
WW A5	33	43	55	62	72	78	84	91	22	30	40	47	57	63	70	79	0.9	1.1	1.2	1.4	1.5	1.1	1.2	1.3	1.5	1.6
WW A6	33	44	56	63	73	79	85	92	22	31	41	48	58	64	72	81	0.9	1.0	1.2	1.4	1.5	1.0	1.2	1.3	1.5	1.6
WW B1	25	31	39	45	53	58	65	74	17	21	27	31	37	42	47	55	1.1	1.3	1.4	1.6	1.7	1.4	1.6	1.7	1.9	2.0
WW B2	32	43	54	61	71	77	83	90	22	31	40	47	57	63	70	80	0.9	1.1	1.2	1.4	1.5	1.1	1.2	1.3	1.5	1.6
WW B3	39	50	62	69	79	84	89	95	27	37	48	55	65	72	79	87	0.9	1.0	1.2	1.3	1.5	1.0	1.1	1.3	1.4	1.6
WW B4	38	51	63	71	80	85	91	96	27	38	49	57	68	74	81	89	0.8	1.0	1.1	1.3	1.5	0.9	1.1	1.2	1.4	1.5
WW B5	37	49	61	70	79	85	90	95	26	37	48	56	67	73	81	89	0.8	1.0	1.2	1.3	1.5	0.9	1.1	1.2	1.4	1.5
WW C1	40	52	64	71	80	85	91	95	29	39	50	58	68	74	81	88	0.8	1.0	1.2	1.3	1.5	0.9	1.1	1.2	1.4	1.6
WW C2	38	50	62	70	79	84	90	95	27	37	48	56	66	73	80	88	0.8	1.0	1.2	1.3	1.5	1.0	1.1	1.2	1.4	1.6
WW C3	44	57	70	78	86	91	95	98	32	45	57	66	76	82	88	94	0.8	0.9	1.1	1.3	1.4	0.9	1.0	1.2	1.3	1.5
WW C4	40	53	66	74	83	88	93	97	29	41	53	62	72	79	85	92	0.8	1.0	1.1	1.3	1.4	0.9	1.0	1.2	1.4	1.5
TW A1	25	30	37	42	49	54	60	68	19	23	27	31	36	40	46	53	1.3	1.4	1.5	1.7	1.8	1.5	1.7	1.8	2.0	2.1
TW A2	27	34	43	49	57	63	70	79	19	25	31	36	43	48	55	64	1.1	1.2	1.4	1.5	1.7	1.3	1.4	1.6	1.7	1.9
TW A3	28	36	45	52	61	67	74	82	19	26	33	38	46	52	59	68	1.0	1.2	1.3	1.5	1.6	1.2	1.4	1.5	1.7	1.8
TW A4	27	36	45	52	62	68	75	83	18	25	33	39	47	53	60	69	1.0	1.2	1.3	1.5	1.6	1.2	1.3	1.5	1.6	1.8
TW B1	24	30	37	42	50	55	61	70	18	22	27	31	37	42	47	55	1.2	1.4	1.5	1.7	1.8	1.5	1.6	1.8	1.9	2.0
TW B2	25	33	41	47	56	62	69	78	18	24	30	35	42	47	54	63	1.1	1.2	1.4	1.5	1.7	1.3	1.4	1.6	1.7	1.9
TW B3	25	33	41	48	57	63	70	79	17	23	30	35	43	48	55	64	1.1	1.2	1.4	1.5	1.7	1.3	1.4	1.6	1.7	1.8
TW B4	23	31	39	45	54	60	67	76	16	22	28	33	41	46	53	62	1.1	1.3	1.4	1.6	1.7	1.3	1.4	1.6	1.7	1.9
TW C1	23	29	36	42	50	55	62	70	17	22	27	31	37	42	48	56	1.2	1.3	1.5	1.6	1.8	1.4	1.6	1.7	1.9	2.0
TW C2	23	29	37	43	51	57	64	73	16	21	27	32	38	43	50	58	1.2	1.3	1.5	1.6	1.7	1.4	1.5	1.7	1.8	1.9
TW C3	22	28	35	41	49	54	61	70	16	20	26	30	37	41	47	56	1.2	1.3	1.5	1.6	1.8	1.4	1.6	1.7	1.8	2.0
TW C4	20	25	32	36	44	49	55	64	15	19	23	27	33	37	43	51	1.3	1.4	1.6	1.7	1.9	1.5	1.7	1.8	2.0	2.1
TW D1	21	27	34	39	46	52	58	67	14	18	23	27	32	36	42	50	1.2	1.4	1.5	1.7	1.8	1.5	1.7	1.8	2.0	2.1
TW D2	30	39	49	56	66	72	79	86	21	28	36	43	52	57	65	74	1.0	1.1	1.3	1.4	1.6	1.1	1.3	1.4	1.6	1.7
TW D3	33	43	53	61	70	76	83	90	23	31	40	47	56	63	70	79	0.9	1.1	1.2	1.4	1.5	1.1	1.2	1.4	1.5	1.7
TW D4	35	46	57	65	75	80	86	93	25	34	44	52	62	68	75	84	0.9	1.0	1.2	1.4	1.5	1.0	1.1	1.3	1.5	1.6
TW D5	34	45	56	64	74	80	86	93	24	33	44	51	62	68	76	84	0.9	1.0	1.2	1.4	1.5	1.0	1.1	1.3	1.4	1.6
TW E1	37	47	58	66	75	81	87	93	26	35	45	52	62	68	75	84	0.9	1.0	1.2	1.4	1.5	1.0	1.2	1.3	1.5	1.6
TW E2	36	47	58	66	75	81	87	93	26	35	45	52	62	68	76	84	0.9	1.0	1.2	1.4	1.5	1.0	1.1	1.3	1.5	1.6
TW E3	42	55	67	74	83	88	93	97	31	42	54	62	72	78	85	92	0.8	1.0	1.1	1.3	1.5	0.9	1.0	1.2	1.4	1.5
TW E4	39	52	64	72	81	86	91	96	28	39	51	59	70	76	83	90	0.8	1.0	1.1	1.3	1.5	0.9	1.1	1.2	1.4	1.5
TW F1	23	29	36	41	48	53	60	69	17	21	26	30	36	40	45	53	1.2	1.4	1.5	1.7	1.8	1.5	1.6	1.8	1.9	2.1
TW F2	24	32	40	46	55	61	68	77	17	22	29	34	41	46	52	61	1.1	1.3	1.4	1.6	1.7	1.3	1.5	1.6	1.8	1.9
TW F3	24	32	41	48	57	63	70	79	16	22	29	34	42	48	55	64	1.1	1.2	1.4	1.5	1.7	1.3	1.4	1.5	1.7	1.8
TW F4	22	30	39	46	55	61	69	78	14	20	27	33	41	46	54	63	1.1	1.2	1.4	1.5	1.7	1.3	1.4	1.5	1.7	1.8
TW G1	22	28	35	40	48	53	60	69	15	20	25	29	35	39	45	53	1.2	1.4	1.5	1.7	1.8	1.5	1.6	1.8	1.9	2.0
TW G2	21	28	36	43	51	57	64	74	14	19	25	30	37	42	49	58	1.1	1.3	1.4	1.6	1.7	1.4	1.5	1.6	1.8	1.9
TW G3	19	27	35	41	50	56	63	73	12	18	24	29	36	41	48	57	1.1	1.3	1.4	1.6	1.7	1.4	1.5	1.6	1.8	1.9
TW G4	16	22	30	36	44	50	57	67	10	14	20	25	32	36	43	52	1.2	1.4	1.5	1.6	1.8	1.4	1.6	1.7	1.9	2.0
TW H1	18	24	31	37	45	50	57	66	12	16	22	26	32	36	42	51	1.2	1.4	1.5	1.7	1.8	1.5	1.6	1.8	1.9	2.0
TW H2	16	22	29	35	43	48	55	65	10	14	20	24	30	35	41	50	1.3	1.4	1.5	1.7	1.8	1.5	1.6	1.8	1.9	2.0
TW H3	13	19	25	30	38	43	50	60	8	12	17	21	27	31	37	45	1.3	1.5	1.6	1.7	1.9	1.6	1.7	1.8	2.0	2.1
TW H4	10	14	20	24	31	35	42	50	6	9	13	16	21	25	30	37	1.5	1.6	1.8	1.9	2.0	1.8	1.9	2.0	2.2	2.3
TW I1	18	24	30	35	43	48	55	64	11	15	20	24	29	33	38	46	1.3	1.4	1.6	1.7	1.9	1.6	1.7	1.9	2.0	2.2
TW I2	26	35	45	52	62	68	76	84	18	25	33	39	48	54	61	71	1.0	1.1	1.3	1.5	1.6	1.2	1.3	1.4	1.6	1.7
TW I3	29	38	49	56	66	72	79	87	20	27	36	43	52	58	66	76	1.0	1.1	1.3	1.4	1.6	1.1	1.2	1.4	1.5	1.7
TW I4	31	42	53	61	71	77	84	91	21	30	41	48	58	65	72	81	0.9	1.1	1.2	1.4	1.5	1.0	1.2	1.3	1.5	1.6
TW I5	31	42	54	62	72	78	85	92	21	31	41	49	59	66	74	83	0.9	1.0	1.2	1.4	1.5	1.0	1.2	1.3	1.5	1.6
TW J1	34	44	56	63	73	79	85	92	24	33	43	50	60	66	73	82	0.9	1.1	1.2	1.4	1.5	1.0	1.2	1.3	1.5	1.6
TW J2	33	43	55	63	72	78	85	91	23	32	42	49	59	65	73	82	0.9	1.1	1.2	1.4	1.5	1.0	1.2	1.3	1.5	1.6
TW J3	39	51	64	72	81	86	91	96	28	39	51	59	70	76	83	90	0.8	1.0	1.1	1.3	1.5	0.9	1.1	1.2	1.4	1.5
TW J4	36	48	61	69	79	84	90	95	25	36	48	56	67	73	81	89	0.8	1.0	1.2	1.3	1.5	0.9	1.1	1.2	1.4	1.5
DG A1	23	29	37	42	48	52	57	62	14	19	24	28	33	36	40	44	1.1	1.4	1.6	1.9	2.2	1.4	1.6	2.0	2.4	2.9
DG A2	24	32	41	48	56	61	68	75	16	22	29	35	42	47	53	60	1.1	1.2	1.4	1.6	1.8	1.3	1.4	1.6	1.8	2.1
DG A3	29	39	50	58	67	73	79	86	20	28	37	44	54	60	67	76	1.0	1.1	1.3	1.5	1.7	1.1	1.2	1.4	1.6	1.8
DG B1	23	31	40	46	55	60	66	72	14	19	25	30	37	42	47	53	1.1	1.2	1.4	1.7	2.0	1.4	1.5	1.7	2.0	2.3
DG B2	25	34	44	52	62	68	75	82	17	23	31	37	46	52	59	68	1.0	1.1	1.3	1.5	1.7	1.2	1.3	1.5	1.7	1.9
DG B3	29	40	52	60	72	78	84	90	21	29	38	46	57	64	72	81	1.0	1.0	1.2	1.4	1.6	1.1	1.2	1.3	1.5	1.6
DG C1	27	36	47	54	64	70	76	82	17	24	31	37	45	51	58	65	1.0	1.1	1.3	1.6	1.8	1.2	1.4	1.5	1.7	2.0
DG C2	29	39	50	59	70	76	82	88	20	27	36	43	53	60	67	76	1.0	1.1	1.2	1.4	1.6	1.2	1.2	1.4	1.5	1.7
DG C3	33	45	58	67	78	84	90	94	24	33	43	52	64	71	79	87	0.9	1.0	1.1	1.3	1.6	1.0	1.1	1.2	1.4	1.6
SS A1	32	40	49	55	63	69	75	83	26	32	39	45	52	57	63	71	1.0	1.2	1.3	1.5	1.7	1.2	1.3	1.5	1.7	1.8
SS A2	40	50	60	67	76	81	87	92	32	41	50	57	66	71	78	85	0.9	1.0	1.2	1.4	1.5	1.0	1.1	1.3	1.5	1.6
SS A3	30	37	45	51	59	64	70	78	24	30	36	41	48	52	58	66	1.1	1.2	1.4	1.6	1.7	1.2	1.4	1.6	1.8	1.9
SS A4	39	49	59	66	75	80	85	91	32	40	49	56	64	70	76	84	0.9	1.1	1.2	1.4	1.5	1.0	1.2	1.3	1.5	1.6
SS A5	31	37	45	51	58	63	70	77	25	30	36	41	48	52	58	66	1.1	1.3	1.4	1.6	1.7	1.3	1.4	1.6	1.8	1.9
SS A6	40	50	60	67	76	81	86	92	32	40	50	56	65	71	77	85	0.9	1.0	1.2	1.4	1.5	1.0	1.1	1.3	1.5	1.6
SS A7	27	33	40	45	51	56	62	70	22	27	32	36	41	45	50	57	1.2	1.4	1.5	1.7	1.8	1.4	1.6	1.7	1.9	2.1
SS A8	39	48	58	65	73	79	84	91	31	39	48	54	63	68	75	82	0.9	1.1	1.2	1.4	1.6	1.0	1.2	1.3	1.5	1.7
SS B1	28	35	43	49	57	62	69	77	21	27	34	38	46	50	57	65	1.1	1.2	1.4	1.6	1.7	1.3	1.4	1.6	1.7	1.9
SS B2	35	44	55	62	71	76	82	89	27	36	44	51	60	66	72	81	0.9	1.1	1.2	1.4	1.6	1.0	1.2	1.4	1.5	1.7
SS B3	26	33	41	46	54	59	66	74	20	25	31	36	43	47	53	61	1.1	1.3	1.4	1.6	1.8	1.3	1.5	1.6	1.8	2.0
SS B4	34	43	53	60	69	75	81	88	27	35	43	50	58	64	71	79	0.9	1.1	1.3	1.4	1.6	1.1	1.2	1.4	1.5	1.7
SS B5	25	31	38	43	50	55	61	69	20	24	30	34	39	44	49	57	1.2	1.4	1.5	1.7	1.8	1.4	1.6	1.7	1.9	2.0
SS B6	34	44	54	61	70	75	82	89	27	35	43	50	59	65	71	80	0.9	1.1	1.3	1.4	1.6	1.1	1.2	1.4	1.5	1.7
SS B7	23	28	34	39	45	50	56	64	18	22	26	30	35	39	44	51	1.3	1.5	1.6	1.8	1.9	1.5	1.7	1.9	2.0	2.2
SS B8	33	42	52	59	68	73	80	87	26	33	42	48	57	62	69	77	1.0	1.1	1.3	1.4	1.6	1.1	1.2	1.4	1.6	1.7
SS C1	28	36	45	52	61	67	74	83	19	25	32	38	46	51	58	68	1.0	1.2	1.3	1.5	1.6	1.2	1.4	1.5	1.7	1.8
SS C2	33	43	54	61	71	77	83	90	23	31	40	47	57	63	70	79	0.9	1.1	1.2	1.4	1.5	1.1	1.2	1.4	1.5	1.6
SS C3	23	29	36	42	50	55	62	71	16	21	26	30	37	41	47	56	1.2	1.3	1.5	1.6	1.8	1.4	1.6	1.7	1.9	2.0
SS C4	29	38	48	55	64	70	77	85	20	28	36	42	51	56	64	73	1.0	1.1	1.3	1.4	1.6	1.1	1.3	1.4	1.6	1.7
SS D1	38	48	58	64	73	78	84	90	30	38	47	53	61	67	73	81	0.9	1.1	1.2	1.4	1.6	1.0	1.2	1.4	1.5	1.7
SS D2	46	58	69	76	84	88	92	96	37	47	58	65	74	79	85	91	0.8	1.0	1.1	1.3	1.5	0.9	1.1	1.2	1.4	1.5
SS D3	35	43	52	58	67	72	78	85	28	34	42	47	55	60	66	74	1.0	1.1	1.3	1.5	1.6	1.1	1.3	1.5	1.6	1.8
SS D4	45	56	66	74	82	86	91	96	35	45	56	63	72	77	84	90	0.8	1.0	1.2	1.3	1.5	0.9	1.1	1.2	1.4	1.6
SS E1	33	42	52	58	67	73	79	86	25	32	40	46	54	60	66	74	1.0	1.1	1.3	1.5	1.6	1.1	1.3	1.4	1.6	1.8
SS E2	40	51	62	70	78	84	89	94	31	41	51	58	67	73	80	87	0.9	1.0	1.2	1.4	1.5	1.0	1.1	1.3	1.4	1.6
SS E3	28	35	43	49	57	62	69	77	22	27	34	39	45	50	56	64	1.1	1.3	1.4	1.6	1.7	1.3	1.4	1.6	1.8	1.9
SS E4	37	47	58	65	74	79	85	91	29	37	47	54	63	68	75	83	0.9	1.1	1.2	1.4	1.5	1.0	1.2	1.3	1.5	1.6

SALT LAKE CITY, UTAH

TYPE	SOLAR SAVINGS FRACTION (%) BASE TEMP. 55 F (3642 DD) LCR 100	70	50	40	30	25	20	15	65 F (5988 DD) 100	70	50	40	30	25	20	15	CONSERVATION FACTOR 55 F 100	50	30	20	15	65 F 100	50	30	20	15
WW A1	23	28	33	37	42	46	52	59	18	21	25	28	32	35	39	45	1.4	1.5	1.7	1.9	2.1	1.6	1.8	2.0	2.3	2.4
WW A2	26	32	40	46	53	58	65	73	19	24	30	34	41	45	51	59	1.1	1.3	1.5	1.6	1.8	1.3	1.5	1.7	1.9	2.0
WW A3	27	34	43	49	57	63	69	77	19	25	32	37	44	49	55	64	1.1	1.2	1.4	1.6	1.7	1.2	1.4	1.6	1.8	1.9
WW A4	27	35	44	51	59	65	71	79	18	25	33	38	46	51	58	66	1.0	1.2	1.4	1.5	1.7	1.2	1.4	1.5	1.7	1.9
WW A5	28	36	46	53	62	67	74	82	19	26	34	40	48	53	60	69	1.0	1.2	1.3	1.5	1.7	1.1	1.3	1.5	1.7	1.8
WW A6	28	37	47	54	63	68	75	83	19	26	34	41	49	55	61	70	1.0	1.2	1.3	1.5	1.7	1.1	1.3	1.5	1.6	1.8
WW B1	21	26	32	37	43	47	53	60	15	18	22	26	30	33	38	43	1.3	1.5	1.7	1.9	2.0	1.6	1.8	2.0	2.3	2.5
WW B2	28	36	46	53	62	67	74	82	19	27	35	41	49	55	61	70	1.0	1.2	1.3	1.5	1.7	1.1	1.3	1.5	1.6	1.8
WW B3	33	42	53	60	69	75	81	88	24	32	41	48	57	63	70	78	0.9	1.1	1.3	1.4	1.6	1.0	1.2	1.4	1.6	1.7
WW B4	33	43	54	62	72	77	83	90	23	33	43	50	60	66	73	82	0.9	1.1	1.2	1.4	1.6	1.0	1.2	1.3	1.5	1.6
WW B5	31	42	53	61	71	77	83	90	22	32	42	50	60	66	73	82	0.9	1.1	1.2	1.4	1.6	1.0	1.2	1.3	1.5	1.6
WW C1	34	44	55	62	71	77	83	89	25	34	43	50	59	65	72	80	0.9	1.1	1.2	1.4	1.6	1.0	1.2	1.3	1.5	1.7
WW C2	32	43	53	61	70	76	82	89	23	32	42	49	59	65	72	80	0.9	1.1	1.2	1.4	1.6	1.0	1.2	1.3	1.5	1.7
WW C3	38	50	62	69	79	84	89	94	28	39	51	59	69	75	82	89	0.8	1.0	1.2	1.3	1.5	0.9	1.1	1.2	1.4	1.6
WW C4	34	46	58	66	76	81	87	93	25	36	47	55	65	72	79	87	0.9	1.0	1.2	1.4	1.5	0.9	1.1	1.3	1.4	1.6
TW A1	22	27	32	36	42	46	51	58	18	21	25	28	32	35	39	45	1.4	1.6	1.7	1.9	2.1	1.6	1.8	2.0	2.2	2.4
TW A2	24	30	37	42	49	54	60	68	18	22	28	32	38	42	47	54	1.2	1.4	1.5	1.7	1.9	1.4	1.6	1.8	1.9	2.1
TW A3	24	31	38	44	52	57	64	72	17	22	29	33	40	44	50	58	1.1	1.3	1.5	1.6	1.8	1.3	1.5	1.7	1.8	2.0
TW A4	23	30	38	44	53	58	65	73	16	22	28	33	40	45	51	60	1.1	1.3	1.4	1.6	1.8	1.3	1.5	1.6	1.8	2.0
TW B1	22	27	32	37	43	47	53	60	17	20	25	28	33	36	41	47	1.3	1.5	1.7	1.9	2.0	1.6	1.8	1.9	2.1	2.3
TW B2	22	28	35	41	48	53	59	67	16	21	26	31	37	41	47	54	1.2	1.4	1.5	1.7	1.9	1.4	1.6	1.7	1.9	2.1
TW B3	22	28	36	41	49	54	60	69	16	21	26	31	37	42	48	55	1.2	1.3	1.5	1.7	1.8	1.4	1.5	1.7	1.9	2.0
TW B4	21	27	34	39	46	52	58	66	15	19	25	29	36	40	46	54	1.2	1.4	1.5	1.7	1.9	1.4	1.6	1.7	1.9	2.1
TW C1	21	26	32	36	43	47	53	61	16	20	24	28	33	36	41	48	1.3	1.5	1.7	1.8	2.0	1.5	1.7	1.9	2.1	2.2
TW C2	20	26	32	37	44	49	55	63	15	19	24	28	34	38	43	51	1.3	1.4	1.6	1.8	1.9	1.5	1.6	1.8	2.0	2.1
TW C3	19	24	31	35	42	47	53	61	14	18	23	27	32	36	41	49	1.3	1.5	1.6	1.8	1.9	1.5	1.7	1.8	2.0	2.2
TW C4	18	23	28	32	38	42	48	56	14	17	21	25	30	33	38	45	1.4	1.6	1.7	1.9	2.0	1.6	1.8	2.0	2.1	2.3
TW D1	18	23	28	32	38	42	48	55	12	15	19	22	27	30	34	39	1.4	1.6	1.7	1.9	2.1	1.7	1.9	2.1	2.3	2.5
TW D2	26	34	42	48	57	62	69	77	19	25	32	37	45	50	57	65	1.1	1.2	1.4	1.6	1.7	1.2	1.4	1.5	1.7	1.9
TW D3	28	36	46	52	61	67	73	81	20	27	35	41	49	54	61	70	1.0	1.2	1.3	1.5	1.7	1.1	1.3	1.5	1.7	1.8
TW D4	30	39	49	57	66	72	78	86	22	30	39	45	55	60	68	76	1.0	1.1	1.3	1.5	1.6	1.1	1.2	1.4	1.6	1.7
TW D5	29	39	49	57	66	72	79	86	21	30	39	46	55	61	69	78	1.0	1.1	1.3	1.4	1.6	1.1	1.2	1.4	1.5	1.7
TW E1	31	41	50	57	66	71	78	85	23	31	39	46	54	60	67	75	1.0	1.1	1.3	1.5	1.6	1.1	1.2	1.4	1.6	1.7
TW E2	31	40	50	57	66	72	78	86	23	31	39	46	55	61	68	76	1.0	1.1	1.3	1.5	1.6	1.1	1.2	1.4	1.6	1.7
TW E3	36	47	59	66	75	81	86	92	27	37	48	55	65	71	78	85	0.9	1.0	1.2	1.4	1.5	1.0	1.1	1.3	1.5	1.6
TW E4	34	45	56	64	73	79	85	91	25	35	45	53	63	69	76	84	0.9	1.0	1.2	1.4	1.5	1.0	1.1	1.3	1.5	1.6
TW F1	21	26	31	35	41	46	51	58	16	19	23	26	31	34	39	45	1.4	1.5	1.7	1.9	2.0	1.6	1.8	2.0	2.2	2.4
TW F2	21	27	34	39	47	52	58	66	15	20	25	29	35	39	45	52	1.2	1.4	1.6	1.7	1.9	1.4	1.6	1.8	2.0	2.1
TW F3	21	27	35	40	48	53	60	68	14	19	25	30	36	41	46	54	1.2	1.3	1.5	1.7	1.8	1.4	1.5	1.7	1.9	2.0
TW F4	19	25	33	39	47	52	59	68	12	17	23	28	35	39	46	54	1.2	1.3	1.5	1.7	1.8	1.4	1.5	1.7	1.9	2.0
TW G1	19	24	30	34	41	45	51	58	14	18	22	25	30	34	38	45	1.3	1.5	1.7	1.9	2.0	1.6	1.8	2.0	2.1	2.3
TW G2	18	24	31	36	43	48	55	63	12	17	22	26	32	36	42	49	1.2	1.4	1.6	1.8	1.9	1.5	1.6	1.8	2.0	2.1
TW G3	16	22	29	35	42	47	54	62	11	15	21	25	31	35	41	49	1.2	1.4	1.6	1.8	1.9	1.5	1.6	1.8	2.0	2.1
TW G4	13	19	25	30	37	42	49	57	8	12	17	21	27	31	37	44	1.3	1.5	1.6	1.8	2.0	1.5	1.7	1.9	2.0	2.2
TW H1	16	21	27	31	38	42	48	56	11	15	19	22	28	31	36	43	1.4	1.5	1.7	1.9	2.0	1.6	1.8	2.0	2.1	2.3
TW H2	13	19	25	29	36	41	47	55	8	12	17	21	26	30	35	42	1.4	1.5	1.7	1.9	2.0	1.6	1.8	1.9	2.1	2.3
TW H3	11	16	22	26	32	37	43	51	7	10	15	18	23	27	32	39	1.4	1.6	1.8	1.9	2.1	1.7	1.8	2.0	2.2	2.3
TW H4	8	12	17	21	26	30	35	43	5	8	11	14	18	21	26	32	1.6	1.8	1.9	2.1	2.2	1.9	2.0	2.2	2.4	2.5
TW I1	15	20	25	29	35	39	44	52	10	13	17	20	24	27	31	36	1.4	1.6	1.8	2.0	2.1	1.7	1.9	2.2	2.4	2.6
TW I2	23	30	39	45	53	59	66	75	15	22	29	34	41	47	53	62	1.1	1.3	1.4	1.6	1.7	1.3	1.4	1.6	1.7	1.9
TW I3	24	33	42	48	57	63	70	78	17	24	31	37	45	51	57	66	1.0	1.2	1.4	1.5	1.7	1.2	1.3	1.5	1.7	1.8
TW I4	27	36	46	53	63	69	76	84	19	27	36	42	51	57	65	74	1.0	1.1	1.3	1.5	1.6	1.1	1.3	1.4	1.6	1.7
TW I5	27	36	47	54	64	70	77	85	19	27	36	43	53	59	66	76	1.0	1.1	1.3	1.5	1.6	1.1	1.2	1.4	1.6	1.7
TW J1	29	38	48	55	64	69	76	84	21	28	37	43	52	58	65	73	1.0	1.1	1.3	1.5	1.6	1.1	1.3	1.4	1.6	1.8
TW J2	28	37	47	54	64	69	76	84	20	28	37	43	52	58	65	74	1.0	1.1	1.3	1.5	1.6	1.1	1.3	1.4	1.6	1.7
TW J3	33	44	55	63	73	78	85	91	24	34	45	52	62	69	76	84	0.9	1.0	1.2	1.4	1.5	1.0	1.1	1.3	1.5	1.6
TW J4	31	42	53	61	71	76	83	90	22	32	42	50	60	66	74	82	0.9	1.1	1.2	1.4	1.6	1.0	1.1	1.3	1.5	1.6
DG A1	19	25	30	34	40	43	46	51	12	16	20	23	27	30	32	35	1.3	1.5	1.9	2.3	2.8	1.5	1.8	2.3	2.9	3.8
DG A2	21	28	35	41	48	53	58	65	16	20	26	30	36	41	46	52	1.2	1.3	1.6	1.8	2.0	1.4	1.5	1.8	2.0	2.3
DG A3	25	33	43	50	59	64	71	78	19	25	33	39	47	53	60	68	1.1	1.2	1.4	1.6	1.8	1.2	1.3	1.5	1.7	1.9
DG B1	20	25	32	37	44	48	53	59	13	17	21	25	30	33	37	41	1.2	1.4	1.7	2.0	2.3	1.5	1.7	2.0	2.4	2.8
DG B2	22	29	37	43	52	57	64	72	15	21	27	32	39	44	50	58	1.1	1.3	1.4	1.7	1.9	1.3	1.5	1.6	1.9	2.0
DG B3	26	34	44	52	62	68	75	83	19	26	34	40	50	56	64	73	1.0	1.1	1.3	1.5	1.7	1.2	1.3	1.4	1.6	1.7
DG C1	23	30	38	44	52	57	62	69	15	20	26	31	37	41	46	52	1.1	1.3	1.5	1.7	2.0	1.3	1.5	1.7	2.1	2.4
DG C2	25	33	43	49	59	65	71	79	18	24	31	37	45	51	58	66	1.1	1.2	1.4	1.5	1.7	1.2	1.4	1.5	1.7	1.9
DG C3	29	39	50	58	68	74	81	88	22	29	38	46	56	62	70	79	1.0	1.1	1.2	1.4	1.6	1.1	1.2	1.3	1.5	1.7
SS A1	28	34	41	46	54	58	64	71	23	28	34	38	44	48	54	60	1.1	1.3	1.5	1.7	1.8	1.3	1.5	1.7	1.9	2.1
SS A2	34	43	52	58	66	72	78	85	28	36	44	49	57	62	69	76	1.0	1.1	1.3	1.5	1.6	1.1	1.2	1.4	1.6	1.8
SS A3	26	32	38	43	49	53	59	66	21	26	31	34	40	43	48	54	1.2	1.4	1.6	1.8	1.9	1.4	1.6	1.8	2.0	2.2
SS A4	34	42	50	57	65	70	76	83	28	35	42	48	56	60	66	74	1.0	1.2	1.3	1.5	1.7	1.1	1.3	1.5	1.6	1.8
SS A5	27	32	38	43	49	54	59	66	22	27	32	35	40	44	49	55	1.2	1.4	1.6	1.8	1.9	1.4	1.6	1.8	2.0	2.2
SS A6	34	43	51	58	66	71	77	84	28	35	43	49	57	62	68	75	1.0	1.2	1.3	1.5	1.7	1.1	1.3	1.4	1.6	1.8
SS A7	24	29	34	38	43	47	52	58	20	24	28	31	35	38	42	47	1.3	1.6	1.8	2.0	2.1	1.5	1.8	2.0	2.3	2.5
SS A8	33	41	49	55	63	68	75	82	27	34	41	47	54	59	65	72	1.0	1.2	1.4	1.5	1.7	1.1	1.3	1.5	1.7	1.8
SS B1	24	30	36	41	48	53	59	66	19	24	29	33	39	43	48	55	1.2	1.4	1.6	1.8	1.9	1.4	1.6	1.8	2.0	2.1
SS B2	30	38	47	53	62	67	73	81	24	31	39	45	52	58	64	72	1.0	1.2	1.4	1.5	1.7	1.1	1.3	1.5	1.6	1.8
SS B3	23	28	34	39	45	49	55	62	18	22	27	31	36	39	44	50	1.3	1.5	1.6	1.8	2.0	1.4	1.7	1.9	2.1	2.3
SS B4	29	37	46	52	60	65	72	79	24	30	38	43	51	56	62	70	1.0	1.2	1.4	1.6	1.7	1.1	1.3	1.5	1.7	1.8
SS B5	22	27	32	36	42	46	51	59	18	22	26	29	34	37	41	47	1.3	1.5	1.7	1.9	2.1	1.5	1.8	2.0	2.2	2.4
SS B6	30	38	46	52	61	66	72	80	23	30	38	43	51	56	63	71	1.0	1.2	1.4	1.5	1.7	1.1	1.3	1.5	1.7	1.8
SS B7	20	24	29	33	38	42	46	53	16	20	23	26	30	33	36	42	1.4	1.6	1.8	2.0	2.2	1.7	1.9	2.1	2.4	2.6
SS B8	28	36	44	50	58	64	70	78	23	29	36	41	49	54	60	68	1.1	1.2	1.4	1.6	1.7	1.2	1.3	1.5	1.7	1.9
SS C1	24	30	38	44	52	57	63	71	16	22	28	32	39	43	49	57	1.1	1.3	1.5	1.7	1.8	1.3	1.5	1.7	1.9	2.0
SS C2	28	37	46	53	62	67	74	82	20	27	35	41	49	55	62	70	1.0	1.2	1.3	1.5	1.7	1.1	1.3	1.5	1.6	1.8
SS C3	20	25	31	36	42	47	53	61	15	19	23	27	32	35	40	47	1.3	1.5	1.6	1.8	2.0	1.5	1.7	1.9	2.1	2.3
SS C4	25	33	41	47	56	61	68	76	18	24	31	37	44	49	56	64	1.1	1.2	1.4	1.6	1.7	1.2	1.4	1.6	1.7	1.9
SS D1	32	40	48	54	62	67	73	80	26	32	39	44	51	56	62	69	1.0	1.2	1.4	1.6	1.7	1.1	1.3	1.5	1.7	1.9
SS D2	39	49	59	66	74	79	85	91	32	41	50	56	65	70	76	84	0.9	1.1	1.2	1.4	1.6	1.0	1.1	1.3	1.5	1.7
SS D3	30	37	44	49	56	61	67	74	24	30	36	40	47	51	56	63	1.1	1.3	1.5	1.7	1.8	1.2	1.4	1.6	1.8	2.0
SS D4	38	47	57	64	72	77	83	89	31	39	48	55	63	68	75	82	0.9	1.1	1.3	1.4	1.6	1.0	1.2	1.3	1.5	1.7
SS E1	28	35	43	49	56	61	67	75	22	27	34	38	45	49	55	62	1.1	1.3	1.5	1.6	1.8	1.2	1.4	1.6	1.9	2.0
SS E2	34	43	53	60	69	74	80	87	27	35	44	50	58	63	70	78	0.9	1.1	1.3	1.5	1.6	1.0	1.2	1.4	1.6	1.7
SS E3	25	30	37	41	48	52	58	65	20	24	29	33	38	42	47	53	1.2	1.4	1.6	1.8	1.9	1.4	1.6	1.8	2.0	2.2
SS E4	32	40	49	56	64	70	76	83	25	32	40	46	54	59	66	74	1.0	1.2	1.3	1.5	1.7	1.1	1.3	1.4	1.6	1.8

BURLINGTON, VERMONT

TYPE	SOLAR SAVINGS FRACTION (%) BASE TEMP: 55 F (5274 DD)								65 F (7878 DD)								CONSERVATION FACTOR 55 F					65 F				
LCR	100	70	50	40	30	25	20	15	100	70	50	40	30	25	20	15	100	50	30	20	15	100	50	30	20	15
WW A1	11	12	13	13	14	14	14	14	10	10	11	11	10	10	9	8	3.3	4.3	-	-	-	4.9	-	-	-	-
WW A2	8	10	12	13	15	17	18	20	6	8	9	9	10	11	11	11	2.2	2.7	3.3	4.0	4.8	2.8	3.6	-	-	-
WW A3	7	9	11	13	16	18	20	22	4	6	8	9	10	11	12	13	2.0	2.4	2.8	3.4	3.9	2.4	3.1	3.9	-	-
WW A4	5	8	11	13	16	18	21	24	3	5	7	8	10	11	12	14	1.9	2.3	2.7	3.1	3.5	2.3	2.8	3.5	4.7	-
WW A5	5	7	11	13	17	19	22	25	2	4	6	8	10	11	13	14	1.8	2.1	2.5	2.9	3.3	2.2	2.6	3.3	4.2	-
WW A6	4	7	11	13	17	19	22	26	1	4	6	8	10	12	13	15	1.8	2.1	2.4	2.8	3.1	2.1	2.5	3.1	3.9	-
WW B1	5	5	6	5	5	4	2	0	4	3	3	2	0	-1	-4	-9	4.4	-	-	-	-	-	-	-	-	-
WW B2	7	10	14	17	21	24	28	34	4	7	10	12	16	18	21	25	1.7	1.9	2.2	2.4	2.6	1.9	2.2	2.5	2.9	3.2
WW B3	8	13	18	21	27	30	35	42	6	9	13	16	20	23	27	32	1.5	1.7	1.9	2.2	2.3	1.7	1.9	2.2	2.5	2.8
WW B4	8	14	20	25	32	37	44	52	6	10	16	20	26	31	36	44	1.3	1.5	1.7	1.9	2.0	1.5	1.6	1.8	2.0	2.2
WW 35	8	14	21	26	34	39	46	55	5	11	17	21	28	33	39	48	1.3	1.5	1.6	1.8	2.0	1.4	1.6	1.8	1.9	2.1
WW C1	10	14	20	24	30	34	39	47	7	11	15	19	24	27	31	37	1.4	1.6	1.8	2.0	2.2	1.6	1.8	2.1	2.3	2.5
WW C2	9	14	20	24	31	35	41	49	6	10	15	19	25	29	34	41	1.4	1.6	1.8	2.0	2.1	1.5	1.7	1.9	2.1	2.3
WW C3	11	19	27	34	43	49	56	66	8	15	23	29	37	43	50	59	1.2	1.3	1.5	1.7	1.8	1.2	1.4	1.6	1.7	1.9
WW C4	10	17	25	31	40	46	54	64	7	13	21	26	35	40	47	57	1.2	1.4	1.5	1.7	1.8	1.3	1.4	1.6	1.8	1.9
TW A1	11	12	13	14	15	15	16	16	10	11	11	11	11	11	11	10	3.1	3.8	4.9	-	-	4.2	-	-	-	-
TW A2	9	10	12	14	15	17	18	20	7	8	9	10	11	12	12	13	2.3	2.8	3.3	4.0	4.7	2.9	3.7	4.8	-	-
TW A3	7	9	12	13	16	17	19	22	6	7	8	10	11	12	13	14	2.1	2.5	2.9	3.4	3.9	2.6	3.1	3.9	-	-
TW A4	6	8	11	13	16	18	20	23	4	6	8	9	11	12	14	15	2.0	2.3	2.7	3.1	3.4	2.4	2.8	3.4	4.1	-
TW B1	10	11	13	14	15	16	17	18	9	10	11	11	12	12	12	12	2.8	3.3	3.9	4.9	-	3.5	4.6	-	-	-
TW B2	8	10	12	13	15	17	19	22	6	8	9	10	11	12	13	15	2.3	2.6	3.0	3.5	3.9	2.7	3.2	3.9	4.8	-
TW B3	7	9	11	13	16	17	20	23	5	7	8	10	11	12	14	16	2.2	2.5	2.8	3.2	3.6	2.5	3.0	3.5	4.2	5.0
TW B4	7	9	11	13	15	17	20	23	5	7	8	10	12	13	14	17	2.2	2.5	2.8	3.1	3.4	2.5	2.9	3.4	3.9	4.4
TW C1	9	10	12	13	15	16	18	20	8	9	10	11	12	13	13	15	2.5	2.9	3.3	3.8	4.3	3.0	3.6	4.3	-	-
TW C2	8	9	11	13	15	17	19	22	6	8	9	10	12	13	14	16	2.3	2.6	2.9	3.3	3.6	2.7	3.1	3.6	4.1	4.7
TW C3	8	9	11	13	15	17	19	22	7	8	9	11	12	13	15	17	2.3	2.6	2.9	3.3	3.6	2.7	3.1	3.5	4.0	4.4
TW C4	9	10	12	13	15	17	19	22	8	9	10	11	13	14	15	18	2.5	2.8	3.0	3.3	3.6	2.8	3.2	3.6	4.0	4.3
TW D1	4	5	5	5	5	5	5	3	3	3	3	2	2	1	0	-4	3.9	-	-	-	-	-	-	-	-	-
TW D2	8	11	14	17	21	23	27	31	6	8	11	13	16	18	20	24	1.8	2.1	2.3	2.6	2.8	2.1	2.4	2.7	3.1	3.4
TW D3	8	11	15	18	22	25	29	34	6	8	11	13	17	19	22	25	1.7	1.9	2.2	2.5	2.7	1.9	2.2	2.5	2.9	3.3
TW D4	9	13	19	23	29	33	39	46	7	10	15	19	24	27	32	39	1.5	1.6	1.8	2.0	2.2	1.6	1.8	2.0	2.2	2.4
TW D5	9	14	20	24	31	36	42	50	7	11	16	20	26	30	36	44	1.4	1.6	1.7	1.9	2.1	1.5	1.7	1.9	2.0	2.2
TW E1	10	14	18	22	27	30	35	41	8	11	14	17	21	24	28	33	1.5	1.8	2.0	2.2	2.4	1.7	2.0	2.2	2.5	2.8
TW E2	10	14	19	23	29	32	38	45	7	11	15	18	23	26	31	37	1.5	1.7	1.9	2.1	2.2	1.6	1.9	2.1	2.3	2.5
TW E3	12	18	25	31	39	44	51	60	9	15	21	26	33	38	44	53	1.3	1.4	1.6	1.8	1.9	1.3	1.5	1.7	1.9	2.0
TW E4	11	17	24	30	38	43	50	59	8	14	20	25	32	37	44	52	1.3	1.4	1.6	1.8	1.9	1.4	1.5	1.7	1.9	2.0
TW F1	9	10	11	12	13	13	14	15	8	9	9	9	9	9	9	8	3.0	3.7	4.7	-	-	4.0	-	-	-	-
TW F2	6	8	10	11	13	14	16	18	5	6	7	8	9	9	10	10	2.4	2.8	3.4	4.1	4.8	3.0	3.7	4.9	-	-
TW F3	5	7	9	10	13	14	16	19	3	4	6	7	8	9	10	11	2.2	2.6	3.0	3.5	4.0	2.7	3.3	4.0	-	-
TW F4	3	5	7	9	12	14	16	19	1	3	4	6	7	9	10	12	2.1	2.5	2.8	3.2	3.6	2.5	3.0	3.5	4.2	5.0
TW G1	7	8	10	11	12	13	14	15	6	7	7	8	8	9	9	9	2.8	3.3	3.9	4.9	-	3.5	4.6	-	-	-
TW G2	4	6	8	9	11	13	15	17	3	4	5	6	7	8	9	10	2.4	2.7	3.2	3.7	4.3	2.9	3.5	4.2	-	-
TW G3	3	4	7	8	11	12	14	17	1	3	4	5	7	8	9	11	2.3	2.6	3.0	3.4	3.8	2.7	3.2	3.7	4.4	-
TW G4	1	3	5	7	9	11	13	17	0	2	3	4	6	7	9	11	2.4	2.7	3.0	3.3	3.6	2.7	3.2	3.6	4.1	4.5
TW H1	4	6	7	8	10	11	13	15	3	4	5	6	7	7	8	9	2.7	3.1	3.5	4.1	4.7	3.2	3.9	4.8	-	-
TW H2	2	4	5	7	9	10	12	15	1	2	3	4	6	7	8	10	2.5	2.8	3.2	3.6	3.9	2.9	3.4	4.0	4.6	-
TW H3	1	3	4	6	8	9	11	14	0	1	3	4	5	6	8	10	2.6	2.9	3.2	3.5	3.8	3.0	3.4	3.8	4.3	4.7
TW H4	1	2	3	5	7	8	10	12	0	1	2	3	4	5	7	9	2.8	3.1	3.4	3.7	3.9	3.2	3.6	4.0	4.3	4.7
TW I1	3	3	3	4	4	3	3	1	2	2	1	1	0	0	-2	-5	4.2	-	-	-	-	-	-	-	-	-
TW I2	5	8	11	14	18	20	24	28	3	6	8	10	13	15	17	21	1.9	2.1	2.4	2.6	2.8	2.1	2.4	2.7	3.1	3.4
TW I3	6	9	13	16	20	23	27	32	4	6	9	12	15	17	20	25	1.7	2.0	2.2	2.4	2.6	1.9	2.2	2.5	2.8	3.1
TW I4	6	11	16	20	26	30	36	44	4	8	13	16	21	25	30	36	1.5	1.7	1.9	2.0	2.2	1.6	1.8	2.0	2.2	2.4
TW I5	7	12	17	22	28	33	39	48	4	9	14	18	24	28	33	41	1.4	1.6	1.8	1.9	2.1	1.6	1.7	1.9	2.1	2.2
TW J1	8	12	16	20	25	29	33	40	6	9	13	16	20	23	27	32	1.6	1.8	2.0	2.2	2.4	1.7	2.0	2.2	2.5	2.7
TW J2	7	12	17	21	26	30	36	43	5	9	13	16	21	25	29	35	1.5	1.7	1.9	2.1	2.2	1.7	1.9	2.1	2.3	2.5
TW J3	9	16	23	29	37	42	49	59	7	12	19	24	31	36	43	52	1.3	1.4	1.6	1.8	1.9	1.4	1.5	1.7	1.9	2.0
TW J4	8	15	22	27	35	41	48	57	6	11	18	23	30	35	42	51	1.3	1.4	1.6	1.8	1.9	1.4	1.5	1.7	1.9	2.0
DG A1	3	4	5	5	5	4	3	1	2	2	2	1	1	0	-1	-5	3.4	-	-	-	-	-	-	-	-	-
DG A2	6	9	11	13	15	17	19	22	5	7	8	10	11	13	14	16	2.1	2.5	2.9	3.4	4.0	2.5	2.9	3.5	4.3	-
DG A3	10	13	17	21	25	29	34	40	8	11	15	17	22	25	29	34	1.7	1.9	2.0	2.2	2.4	1.9	2.0	2.2	2.4	2.6
DG B1	3	4	5	5	5	5	4	2	2	2	2	1	1	0	-1	-4	3.2	-	-	-	-	-	-	-	-	-
DG B2	7	9	11	13	16	18	20	23	5	7	9	10	12	13	15	17	2.1	2.4	2.8	3.1	3.6	2.5	2.8	3.3	4.0	4.8
DG B3	10	14	18	22	27	30	36	43	9	12	15	18	23	26	31	37	1.6	1.8	2.0	2.1	2.2	1.8	1.9	2.1	2.3	2.5
DG C1	5	6	8	8	9	9	10	9	3	4	4	4	5	4	3	2	2.5	3.5	-	-	-	3.8	-	-	-	-
DG C2	8	11	14	17	20	23	26	30	7	9	11	13	16	18	20	23	1.8	2.1	2.4	2.6	2.9	2.1	2.4	2.7	3.2	3.6
DG C3	12	16	21	25	30	35	41	49	10	13	18	21	26	30	35	42	1.5	1.7	1.8	1.9	2.1	1.7	1.8	2.0	2.1	2.3
SS A1	12	14	17	18	21	22	24	27	11	13	14	16	17	18	20	21	2.0	2.4	2.9	3.4	3.9	2.3	2.8	3.4	4.5	-
SS A2	15	19	24	27	33	36	41	47	13	17	21	24	29	32	36	41	1.5	1.7	1.9	2.1	2.3	1.6	1.8	2.1	2.3	2.5
SS A3	11	12	14	15	17	17	18	19	10	11	12	13	13	14	14	13	2.3	2.9	3.8	-	-	2.7	3.7	-	-	-
SS A4	14	18	22	26	30	34	38	44	13	16	20	23	27	29	33	37	1.5	1.8	2.0	2.3	2.4	1.6	1.9	2.2	2.5	2.8
SS A5	13	15	16	18	19	20	21	23	12	13	14	15	16	17	17	17	2.3	2.8	3.5	4.4	-	2.7	3.4	4.6	-	-
SS A6	14	18	23	26	31	35	39	45	13	16	20	23	27	30	34	39	1.5	1.7	2.0	2.2	2.4	1.6	1.9	2.1	2.4	2.7
SS A7	12	13	14	15	15	16	16	16	11	12	12	13	13	13	12	11	2.8	3.8	-	-	-	3.5	-	-	-	-
SS A8	14	17	21	24	29	31	35	40	12	15	19	21	25	27	30	34	1.6	1.8	2.1	2.4	2.6	1.7	2.0	2.4	2.7	3.1
SS B1	10	12	14	15	17	19	20	22	9	10	11	13	14	15	16	17	2.2	2.6	3.1	3.7	4.4	2.5	3.1	3.9	-	-
SS B2	13	17	21	25	30	33	38	44	11	15	19	22	26	29	33	38	1.5	1.8	2.0	2.2	2.3	1.7	1.9	2.1	2.4	2.6
SS B3	9	10	12	13	14	15	16	17	8	9	10	10	11	11	12	11	2.4	3.0	3.9	-	-	2.9	3.8	-	-	-
SS B4	12	16	20	23	28	31	35	41	11	14	17	20	24	27	30	35	1.6	1.8	2.0	2.3	2.5	1.7	2.0	2.2	2.5	2.7
SS B5	10	12	13	14	16	16	17	18	10	11	11	12	13	13	14	14	2.6	3.2	3.9	5.0	-	3.1	4.0	-	-	-
SS B6	12	16	20	23	28	31	35	42	10	14	17	20	24	27	31	36	1.6	1.8	2.0	2.2	2.4	1.7	1.9	2.2	2.5	2.7
SS B7	10	10	11	12	13	13	14	14	9	9	10	10	10	10	10	9	3.1	3.9	-	-	-	3.8	-	-	-	-
SS B8	11	14	18	21	25	28	32	38	10	13	16	18	22	24	27	32	1.7	1.9	2.2	2.4	2.6	1.8	2.1	2.4	2.7	3.0
SS C1	5	7	9	10	12	13	15	16	4	5	6	6	7	7	7	7	2.3	2.8	3.5	4.5	-	3.0	4.0	-	-	-
SS C2	7	11	15	18	22	25	30	35	5	8	11	14	17	19	22	27	1.7	1.9	2.2	2.4	2.6	1.9	2.2	2.5	2.8	3.1
SS C3	7	9	10	11	12	13	15	16	6	7	8	8	9	9	9	9	2.7	3.2	3.8	4.7	-	3.4	4.4	-	-	-
SS C4	8	11	14	17	21	23	27	32	6	8	11	13	16	18	21	24	1.8	2.1	2.3	2.6	2.8	2.1	2.3	2.6	3.0	3.3
SS D1	12	15	18	20	23	25	27	30	10	12	15	16	18	19	21	22	1.8	2.2	2.6	3.1	3.6	2.0	2.5	3.2	4.5	-
SS D2	15	20	27	31	37	42	47	55	13	18	23	27	33	37	41	48	1.3	1.5	1.7	1.9	2.1	1.4	1.6	1.8	2.1	2.3
SS D3	13	15	17	19	21	23	25	27	11	13	15	16	18	19	20	21	2.0	2.4	2.9	3.5	4.1	2.2	2.8	3.6	-	-
SS D4	15	20	26	30	36	40	46	53	13	17	22	26	32	35	40	46	1.3	1.6	1.8	2.0	2.1	1.4	1.7	1.9	2.1	2.3
SS E1	9	11	13	14	16	17	19	20	7	8	10	11	12	12	12	11	2.1	2.6	3.4	4.5	-	2.5	3.3	-	-	-
SS E2	11	16	21	25	30	34	38	45	9	13	17	21	25	28	32	37	1.4	1.7	1.9	2.1	2.3	1.6	1.8	2.1	2.4	2.7
SS E3	10	12	14	15	16	17	18	19	9	10	11	12	13	13	14	14	2.3	2.9	3.6	4.7	-	2.8	3.6	-	-	-
SS E4	12	15	20	23	28	31	36	42	10	13	17	20	24	26	30	35	1.5	1.8	2.0	2.3	2.4	1.7	1.9	2.2	2.5	2.8

NORFOLK, VIRGINIA

SOLAR SAVINGS FRACTION (%)

TYPE	BASE TEMP: 55 F (1710 DD) LCR 200	100	70	50	40	30	25	20	65 F (3493 DD) 200	100	70	50	40	30	25	20
WW A1	20	28	34	41	46	53	58	65	15	20	23	28	31	37	40	45
WW A2	20	33	41	51	57	66	72	79	13	21	27	34	39	47	52	59
WW A3	20	34	44	54	62	71	77	83	12	22	29	37	42	51	56	63
WW A4	20	35	46	56	64	73	79	85	11	22	29	38	44	53	59	66
WW A5	20	37	47	59	66	76	81	87	10	22	30	39	46	55	61	69
WW A6	20	37	48	60	68	77	83	88	10	22	31	40	47	57	63	70
WW B1	17	27	35	43	49	57	63	70	11	17	21	27	31	36	40	46
WW B2	20	36	46	58	65	75	80	86	11	22	31	40	46	55	61	69
WW B3	23	42	54	65	73	82	87	92	13	27	37	47	54	64	70	77
WW B4	22	42	54	67	74	83	88	93	12	27	37	49	56	67	73	80
WW B5	21	40	53	65	73	82	87	92	10	26	36	48	55	66	72	80
WW C1	25	44	56	67	75	83	88	93	14	29	39	49	57	66	72	79
WW C2	22	42	54	65	73	82	87	92	12	27	37	47	55	65	71	78
WW C3	25	48	61	73	81	89	93	96	13	32	44	56	64	75	81	87
WW C4	22	44	57	70	77	86	91	95	11	29	40	52	61	71	77	84
TW A1	19	27	33	40	45	52	57	64	14	19	23	27	31	36	40	45
TW A2	19	30	38	46	53	61	67	74	12	20	25	31	36	43	48	54
TW A3	18	31	39	49	56	65	71	77	11	20	26	33	38	46	51	57
TW A4	17	30	39	49	56	66	71	78	10	19	25	33	38	46	52	59
TW B1	18	27	33	40	46	53	59	65	13	19	23	28	31	37	41	46
TW B2	17	28	36	45	51	60	66	73	11	18	24	30	35	42	47	53
TW B3	16	28	36	45	52	61	67	74	10	18	23	30	35	42	48	54
TW B4	15	26	34	43	49	58	64	71	9	17	22	28	33	40	45	52
TW C1	17	26	32	40	45	53	59	65	12	17	22	27	31	37	41	47
TW C2	15	25	32	40	46	55	60	67	10	17	21	27	32	38	43	49
TW C3	15	24	31	38	44	52	58	65	10	16	20	26	30	36	41	47
TW C4	14	22	28	34	40	47	52	59	10	15	19	24	27	33	37	42
TW D1	14	23	30	37	43	51	56	63	8	14	18	23	26	32	36	41
TW D2	19	33	43	53	60	69	75	82	11	21	28	36	42	51	56	64
TW D3	20	36	46	57	65	74	80	86	12	23	31	40	46	55	61	68
TW D4	21	38	49	61	69	78	83	89	12	25	34	44	51	61	67	74
TW D5	20	37	49	60	68	78	83	89	11	24	33	43	51	61	67	75
TW E1	23	40	51	62	69	79	84	89	14	26	35	44	51	61	67	74
TW E2	22	39	50	62	69	78	84	89	13	26	35	44	51	61	67	74
TW E3	25	46	58	70	78	86	90	94	15	31	42	53	61	71	77	84
TW E4	23	43	55	67	75	84	89	93	13	28	39	50	58	68	75	82
TW F1	17	26	32	39	44	52	57	64	12	18	22	26	30	35	39	45
TW F2	16	27	35	44	50	59	65	72	10	17	23	29	33	40	45	51
TW F3	15	27	35	45	52	61	67	74	8	16	22	29	34	42	47	53
TW F4	13	25	34	43	50	59	65	73	6	14	20	27	33	40	46	52
TW G1	15	24	30	38	43	51	57	64	10	16	20	25	29	35	39	44
TW G2	13	24	31	40	46	55	61	68	7	14	19	25	30	37	42	48
TW G3	11	22	30	38	45	54	60	67	5	13	18	24	29	36	41	47
TW G4	9	18	25	33	40	48	54	62	4	10	15	20	25	31	36	42
TW H1	12	21	27	35	40	48	54	61	7	12	17	22	26	32	36	42
TW H2	9	18	25	32	38	46	52	59	4	10	14	20	24	30	34	40
TW H3	7	15	21	28	34	42	47	54	3	8	12	17	21	27	31	36
TW H4	5	12	17	23	27	34	39	45	2	6	9	13	16	21	25	30
TW I1	12	21	27	34	39	47	52	59	7	12	16	20	24	29	32	37
TW I2	16	30	39	49	56	66	72	79	8	18	25	33	39	47	53	60
TW I3	17	32	42	53	60	70	76	83	9	20	27	36	42	51	57	65
TW I4	18	35	46	57	65	75	81	87	9	22	30	40	47	57	63	71
TW I5	17	35	46	58	66	76	81	87	9	22	31	41	48	58	65	73
TW J1	20	37	48	59	67	76	82	88	11	24	32	42	49	58	64	72
TW J2	19	36	47	59	66	76	82	87	10	23	32	41	48	58	64	72
TW J3	22	42	55	67	75	84	89	93	12	28	38	50	58	68	75	82
TW J4	20	40	52	64	72	82	87	92	10	25	36	47	55	66	72	79
DG A1	15	26	33	41	46	53	57	62	8	15	19	24	28	33	36	39
DG A2	15	27	36	45	52	60	65	71	9	17	23	29	35	42	46	52
DG A3	18	32	42	54	61	71	76	82	11	21	28	37	44	53	59	66
DG B1	16	20	35	44	51	60	65	70	8	15	20	25	30	36	40	45
DG B2	16	28	38	48	56	66	72	79	9	17	23	31	36	45	50	58
DG B3	18	32	43	56	65	75	81	87	11	21	29	38	45	56	63	71
DG C1	17	31	40	51	60	69	75	80	10	18	24	31	36	44	49	56
DG C2	18	32	43	55	63	74	80	85	11	20	27	36	42	52	58	66
DG C3	20	36	48	62	72	82	87	92	13	24	33	43	51	62	70	78
SS A1	22	35	43	52	59	67	73	79	16	25	31	38	43	50	55	62
SS A2	26	43	53	64	71	79	84	89	19	31	40	49	55	64	70	76
SS A3	21	32	40	49	55	63	69	75	15	23	29	35	40	46	51	57
SS A4	26	42	53	63	70	78	83	89	18	31	39	48	54	63	68	75
SS A5	22	33	40	48	54	62	67	74	17	24	29	35	40	46	50	56
SS A6	26	43	53	64	71	79	84	89	18	31	39	48	55	63	69	76
SS A7	20	29	36	43	48	55	60	67	16	22	26	31	35	40	44	49
SS A8	26	41	52	62	69	77	82	88	18	30	38	47	53	61	67	73
SS B1	19	30	38	46	53	61	67	73	13	21	26	33	38	44	49	55
SS B2	23	38	48	58	65	74	80	86	16	27	35	44	50	59	64	71
SS B3	18	29	36	44	50	58	64	71	13	20	25	31	35	42	46	52
SS B4	22	37	47	57	64	73	79	85	15	26	34	42	49	57	63	70
SS B5	18	27	33	41	46	54	59	66	14	20	24	29	33	39	43	48
SS B6	22	37	47	57	65	74	79	85	15	26	34	43	49	57	63	70
SS B7	17	25	30	37	42	49	54	60	13	18	22	26	29	34	38	43
SS B8	21	36	45	56	63	72	77	83	14	25	33	41	47	55	61	68
SS C1	18	31	40	49	56	65	71	78	10	19	25	32	37	45	50	57
SS C2	20	36	46	57	65	74	80	86	11	23	31	40	46	56	62	69
SS C3	16	25	32	40	46	54	59	66	10	16	21	26	30	36	41	46
SS C4	18	32	41	52	59	68	74	81	11	21	28	35	41	50	55	63
SS D1	26	41	51	61	68	77	82	87	18	29	37	45	51	59	65	71
SS D2	30	49	61	72	79	87	91	95	20	36	46	56	63	73	78	84
SS D3	24	37	46	55	62	71	76	82	18	27	33	41	46	53	58	65
SS D4	29	48	59	70	77	85	89	94	19	34	44	54	61	70	76	82
SS E1	22	36	46	56	63	72	77	83	15	25	31	39	45	53	58	64
SS E2	26	44	55	66	74	82	87	92	16	30	40	50	57	66	72	78
SS E3	20	31	38	47	53	61	67	73	14	22	27	33	38	44	49	55
SS E4	24	40	51	62	69	78	83	88	16	28	37	46	52	61	67	74

CONSERVATION FACTOR

TYPE	55 F 200	100	50	30	20	65 F 200	100	50	30	20
WW A1	1.1	1.2	1.3	1.5	1.6	1.4	1.5	1.7	1.9	2.0
WW A2	0.9	1.0	1.1	1.3	1.4	1.1	1.2	1.4	1.5	1.7
WW A3	0.8	0.9	1.1	1.2	1.4	1.0	1.1	1.3	1.5	1.6
WW A4	0.8	0.9	1.1	1.2	1.4	1.0	1.1	1.3	1.4	1.6
WW A5	0.8	0.9	1.0	1.2	1.4	1.0	1.1	1.2	1.4	1.5
WW A6	0.8	0.9	1.0	1.2	1.3	0.9	1.0	1.2	1.4	1.5
WW B1	1.0	1.1	1.2	1.4	1.5	1.3	1.4	1.6	1.8	1.9
WW B2	0.8	0.9	1.0	1.2	1.4	1.0	1.1	1.2	1.4	1.5
WW B3	0.7	0.8	1.0	1.1	1.3	0.9	1.0	1.1	1.3	1.5
WW B4	0.7	0.8	1.0	1.1	1.3	0.8	0.9	1.1	1.3	1.4
WW B5	0.7	0.8	1.0	1.1	1.3	0.9	0.9	1.1	1.3	1.4
WW C1	0.7	0.8	1.0	1.1	1.3	0.9	1.0	1.1	1.3	1.4
WW C2	0.7	0.8	1.0	1.1	1.3	0.9	1.0	1.1	1.3	1.4
WW C3	0.7	0.8	0.9	1.1	1.3	0.8	0.9	1.0	1.2	1.4
WW C4	0.7	0.8	0.9	1.1	1.3	0.8	0.9	1.0	1.2	1.4
TW A1	1.1	1.2	1.3	1.5	1.6	1.4	1.5	1.7	1.9	2.0
TW A2	0.9	1.0	1.2	1.3	1.5	1.2	1.3	1.5	1.6	1.8
TW A3	0.9	1.0	1.1	1.3	1.5	1.1	1.2	1.4	1.5	1.7
TW A4	0.9	1.0	1.1	1.3	1.4	1.1	1.2	1.4	1.5	1.7
TW B1	1.1	1.2	1.3	1.5	1.6	1.4	1.5	1.6	1.8	1.9
TW B2	1.0	1.1	1.2	1.3	1.5	1.2	1.3	1.5	1.6	1.8
TW B3	0.9	1.0	1.2	1.3	1.5	1.2	1.3	1.4	1.6	1.7
TW B4	1.0	1.1	1.2	1.4	1.5	1.2	1.3	1.5	1.6	1.8
TW C1	1.1	1.2	1.3	1.4	1.6	1.3	1.4	1.6	1.7	1.9
TW C2	1.0	1.1	1.3	1.4	1.6	1.3	1.4	1.5	1.7	1.8
TW C3	1.1	1.2	1.3	1.4	1.6	1.3	1.4	1.6	1.7	1.9
TW C4	1.2	1.2	1.4	1.5	1.7	1.5	1.5	1.7	1.8	2.0
TW D1	1.1	1.2	1.3	1.5	1.6	1.4	1.5	1.7	1.9	2.0
TW D2	0.8	0.9	1.1	1.2	1.4	1.0	1.1	1.3	1.4	1.6
TW D3	0.8	0.9	1.0	1.2	1.4	1.0	1.1	1.2	1.4	1.5
TW D4	0.8	0.9	1.0	1.2	1.3	0.9	1.0	1.2	1.3	1.5
TW D5	0.8	0.9	1.0	1.2	1.3	0.9	1.0	1.2	1.3	1.5
TW E1	0.8	0.9	1.0	1.2	1.3	0.9	1.0	1.2	1.3	1.5
TW E2	0.8	0.9	1.0	1.2	1.3	0.9	1.0	1.2	1.3	1.5
TW E3	0.7	0.8	0.9	1.1	1.3	0.8	0.9	1.1	1.2	1.4
TW E4	0.7	0.8	1.0	1.1	1.3	0.8	0.9	1.1	1.2	1.4
TW F1	1.1	1.2	1.3	1.5	1.6	1.4	1.5	1.7	1.8	2.0
TW F2	1.0	1.1	1.2	1.4	1.5	1.2	1.3	1.5	1.6	1.8
TW F3	0.9	1.0	1.2	1.3	1.5	1.2	1.3	1.4	1.6	1.7
TW F4	0.9	1.0	1.2	1.3	1.5	1.2	1.3	1.4	1.6	1.7
TW G1	1.1	1.2	1.3	1.5	1.6	1.4	1.5	1.6	1.8	2.0
TW G2	1.0	1.1	1.2	1.4	1.5	1.3	1.4	1.5	1.7	1.8
TW G3	1.0	1.1	1.2	1.4	1.5	1.3	1.4	1.5	1.7	1.8
TW G4	1.1	1.2	1.3	1.5	1.6	1.4	1.4	1.6	1.7	1.9
TW H1	1.1	1.2	1.3	1.5	1.6	1.4	1.5	1.6	1.8	2.0
TW H2	1.1	1.2	1.3	1.5	1.6	1.4	1.5	1.6	1.8	1.9
TW H3	1.2	1.3	1.4	1.5	1.7	1.5	1.6	1.7	1.9	2.0
TW H4	1.3	1.4	1.6	1.7	1.8	1.7	1.8	1.9	2.0	2.2
TW I1	1.1	1.2	1.4	1.5	1.7	1.5	1.6	1.8	1.9	2.1
TW I2	0.9	1.0	1.1	1.3	1.4	1.1	1.2	1.3	1.6	1.0
TW I3	0.8	0.9	1.1	1.2	1.4	1.0	1.1	1.3	1.4	1.6
TW I4	0.8	0.9	1.0	1.2	1.4	0.9	1.0	1.2	1.3	1.5
TW I5	0.8	0.9	1.0	1.2	1.3	0.9	1.0	1.2	1.3	1.5
TW J1	0.8	0.9	1.0	1.2	1.3	0.9	1.0	1.2	1.3	1.5
TW J2	0.8	0.9	1.0	1.2	1.3	0.9	1.0	1.2	1.3	1.5
TW J3	0.7	0.8	1.0	1.1	1.3	0.8	0.9	1.1	1.2	1.4
TW J4	0.7	0.8	1.0	1.1	1.3	0.9	0.9	1.1	1.3	1.4
DG A1	1.0	1.1	1.3	1.6	1.9	1.3	1.4	1.7	2.0	2.5
DG A2	0.9	1.0	1.2	1.4	1.6	1.2	1.3	1.4	1.6	1.9
DG A3	0.9	0.9	1.1	1.2	1.5	1.1	1.1	1.2	1.4	1.6
DG B1	1.0	1.0	1.2	1.4	1.6	1.3	1.4	1.6	1.7	2.0
DG B2	0.9	1.0	1.1	1.3	1.5	1.2	1.3	1.4	1.5	1.7
DG B3	0.9	0.9	1.0	1.2	1.4	1.1	1.1	1.2	1.3	1.5
DG C1	0.9	1.0	1.1	1.3	1.5	1.2	1.3	1.4	1.6	1.8
DG C2	0.9	0.9	1.0	1.2	1.4	1.1	1.2	1.3	1.4	1.6
DG C3	0.8	0.9	0.9	1.1	1.3	1.0	1.1	1.1	1.2	1.4
SS A1	0.9	1.0	1.1	1.3	1.5	1.1	1.2	1.4	1.5	1.7
SS A2	0.8	0.9	1.0	1.2	1.3	0.9	1.0	1.2	1.3	1.5
SS A3	0.9	1.0	1.2	1.3	1.5	1.1	1.3	1.4	1.6	1.8
SS A4	0.8	0.9	1.0	1.2	1.4	0.9	1.0	1.2	1.3	1.5
SS A5	1.0	1.1	1.2	1.4	1.5	1.2	1.3	1.5	1.6	1.8
SS A6	0.8	0.9	1.0	1.2	1.3	0.9	1.0	1.2	1.3	1.5
SS A7	1.0	1.1	1.3	1.5	1.6	1.3	1.4	1.6	1.8	2.0
SS A8	0.8	0.9	1.0	1.2	1.4	0.9	1.0	1.2	1.4	1.5
SS B1	0.9	1.0	1.2	1.4	1.5	1.2	1.3	1.4	1.6	1.8
SS B2	0.8	0.9	1.1	1.2	1.4	1.0	1.1	1.2	1.4	1.5
SS B3	1.0	1.1	1.2	1.4	1.5	1.2	1.3	1.5	1.7	1.8
SS B4	0.8	0.9	1.1	1.2	1.4	1.0	1.1	1.2	1.4	1.5
SS B5	1.1	1.2	1.3	1.5	1.6	1.3	1.4	1.6	1.8	1.9
SS B6	0.8	0.9	1.1	1.2	1.4	1.0	1.1	1.2	1.4	1.5
SS B7	1.1	1.2	1.4	1.5	1.7	1.4	1.5	1.7	1.9	2.1
SS B8	0.8	0.9	1.1	1.2	1.4	1.0	1.1	1.2	1.4	1.6
SS C1	0.9	1.0	1.1	1.3	1.4	1.1	1.2	1.4	1.5	1.7
SS C2	0.8	0.9	1.0	1.2	1.4	1.0	1.1	1.2	1.4	1.5
SS C3	1.0	1.1	1.3	1.4	1.6	1.3	1.4	1.6	1.7	1.9
SS C4	0.9	1.0	1.1	1.3	1.4	1.1	1.2	1.3	1.5	1.6
SS D1	0.8	0.9	1.0	1.2	1.4	0.9	1.1	1.2	1.4	1.6
SS D2	0.7	0.8	0.9	1.1	1.3	0.8	0.9	1.1	1.2	1.4
SS D3	0.8	0.9	1.1	1.3	1.4	1.0	1.1	1.3	1.5	1.6
SS D4	0.7	0.8	1.0	1.1	1.3	0.8	0.9	1.1	1.2	1.4
SS E1	0.8	0.9	1.1	1.2	1.4	1.0	1.1	1.3	1.5	1.6
SS E2	0.7	0.8	1.0	1.1	1.3	0.9	1.0	1.1	1.3	1.5
SS E3	1.0	1.1	1.2	1.4	1.5	1.2	1.3	1.5	1.6	1.8
SS E4	0.8	0.9	1.0	1.2	1.4	0.9	1.0	1.2	1.3	1.5

RICHMOND, VIRGINIA

SOLAR SAVINGS FRACTION (%) / CONSERVATION FACTOR

TYPE	55 F (2065 DD) 200	100	70	50	40	30	25	20	65 F (3941 DD) 200	100	70	50	40	30	25	20	CF 55 F 200	100	50	30	20	CF 65 F 200	100	50	30	20
WW A1	17	24	28	34	38	44	49	55	13	17	20	24	27	31	34	38	1.2	1.3	1.5	1.6	1.8	1.6	1.7	1.9	2.1	2.3
WW A2	16	26	34	42	48	56	62	69	11	18	23	28	33	39	44	50	1.0	1.1	1.2	1.4	1.6	1.2	1.3	1.5	1.7	1.9
WW A3	16	27	36	45	52	61	67	74	9	17	23	30	35	43	48	54	0.9	1.0	1.2	1.3	1.5	1.1	1.3	1.4	1.6	1.7
WW A4	15	28	37	47	54	63	69	76	8	17	24	31	37	45	50	57	0.9	1.0	1.1	1.3	1.5	1.1	1.2	1.4	1.5	1.7
WW A5	15	29	38	49	56	66	72	78	7	17	24	32	38	47	52	59	0.9	1.0	1.1	1.3	1.4	1.1	1.2	1.3	1.5	1.7
WW A6	15	29	39	50	57	67	73	80	7	17	25	33	39	48	53	61	0.9	1.0	1.1	1.3	1.4	1.0	1.1	1.3	1.5	1.6
WW B1	14	22	27	34	39	46	51	58	9	14	17	21	24	29	32	36	1.1	1.3	1.4	1.6	1.7	1.5	1.6	1.8	2.1	2.3
WW B2	15	29	38	48	55	65	71	78	8	18	25	33	39	48	53	60	0.9	1.0	1.1	1.3	1.4	1.1	1.2	1.3	1.5	1.6
WW B3	18	34	45	56	63	73	79	85	10	22	31	40	47	56	62	69	0.8	0.9	1.1	1.2	1.4	1.0	1.1	1.2	1.4	1.5
WW B4	17	34	45	57	65	75	81	87	8	22	31	42	49	59	65	73	0.8	0.9	1.0	1.2	1.4	0.9	1.0	1.2	1.3	1.5
WW B5	15	33	44	56	64	74	80	86	7	21	31	41	49	59	65	73	0.8	0.9	1.0	1.2	1.4	0.9	1.0	1.2	1.3	1.5
WW C1	19	36	47	58	65	75	81	86	11	24	32	42	49	58	64	72	0.8	0.9	1.0	1.2	1.4	0.9	1.0	1.2	1.3	1.5
WW C2	17	34	45	56	64	74	79	86	9	22	31	41	48	57	64	71	0.8	0.9	1.0	1.2	1.4	0.9	1.0	1.2	1.3	1.5
WW C3	19	39	52	65	73	82	87	92	10	27	38	49	58	68	74	81	0.7	0.8	1.0	1.1	1.3	0.8	0.9	1.1	1.2	1.4
WW C4	17	36	48	61	69	79	84	90	8	24	34	46	54	65	71	79	0.7	0.8	1.0	1.2	1.3	0.9	1.0	1.1	1.3	1.4
TW A1	17	23	28	33	37	44	48	54	13	17	20	24	26	31	34	38	1.2	1.4	1.5	1.6	1.8	1.6	1.7	1.9	2.1	2.3
TW A2	16	24	31	38	44	52	57	64	11	17	21	26	30	36	40	46	1.1	1.2	1.3	1.5	1.6	1.3	1.4	1.6	1.8	1.9
TW A3	15	25	32	40	46	55	60	67	9	16	21	27	32	38	43	49	1.0	1.1	1.3	1.4	1.6	1.2	1.4	1.5	1.7	1.8
TW A4	13	24	32	40	47	56	61	69	8	15	21	27	32	39	44	50	1.0	1.1	1.2	1.4	1.5	1.2	1.3	1.5	1.6	1.8
TW B1	16	23	28	33	38	45	49	56	12	16	20	24	27	31	35	39	1.2	1.3	1.5	1.6	1.8	1.5	1.6	1.8	2.0	2.2
TW B2	14	23	29	37	42	50	56	63	9	15	20	25	29	35	40	45	1.1	1.2	1.3	1.5	1.6	1.3	1.4	1.6	1.8	1.9
TW B3	13	23	29	37	43	51	57	64	8	15	19	25	29	36	40	46	1.1	1.2	1.3	1.4	1.6	1.3	1.4	1.6	1.7	1.9
TW B4	12	21	28	35	41	49	54	61	8	14	18	24	28	34	39	44	1.1	1.2	1.3	1.5	1.6	1.3	1.4	1.6	1.8	1.9
TW C1	14	21	27	33	38	45	50	56	10	15	19	23	26	31	35	40	1.2	1.3	1.4	1.6	1.7	1.5	1.6	1.8	1.9	2.1
TW C2	13	21	26	33	38	46	51	58	9	14	18	23	27	32	36	42	1.1	1.3	1.4	1.5	1.7	1.4	1.5	1.7	1.8	2.0
TW C3	12	20	25	32	37	44	49	55	8	13	17	22	26	31	35	40	1.2	1.3	1.4	1.6	1.7	1.5	1.6	1.7	1.9	2.0
TW C4	12	19	23	29	33	39	44	50	9	13	16	20	24	28	32	36	1.3	1.4	1.5	1.7	1.8	1.6	1.7	1.8	2.0	2.2
TW D1	11	18	23	29	34	41	45	51	7	11	14	18	21	25	28	32	1.2	1.3	1.5	1.6	1.8	1.6	1.7	1.9	2.1	2.3
TW D2	15	27	35	44	51	60	66	73	9	18	24	31	36	44	49	56	0.9	1.0	1.2	1.3	1.5	1.1	1.2	1.4	1.6	1.7
TW D3	16	29	38	48	55	65	70	77	9	19	26	34	39	48	53	60	0.9	1.0	1.1	1.3	1.5	1.1	1.2	1.3	1.5	1.6
TW D4	16	31	41	52	59	69	75	82	9	20	28	37	44	53	59	67	0.8	0.9	1.1	1.2	1.4	1.0	1.1	1.2	1.4	1.6
TW D5	15	30	40	51	59	69	75	82	8	20	28	38	44	54	60	68	0.8	0.9	1.1	1.2	1.4	1.0	1.1	1.2	1.4	1.5
TW E1	18	33	42	53	60	70	75	82	11	22	30	38	44	53	59	66	0.8	0.9	1.1	1.2	1.4	1.0	1.1	1.3	1.4	1.6
TW E2	17	32	42	53	60	70	76	82	10	21	29	38	45	54	60	67	0.8	0.9	1.1	1.2	1.4	1.0	1.1	1.2	1.4	1.6
TW E3	20	38	50	61	69	79	84	89	11	26	36	46	54	64	70	77	0.8	0.8	1.0	1.2	1.3	0.9	1.0	1.1	1.3	1.4
TW E4	17	35	47	58	66	76	82	88	10	24	33	44	52	62	68	75	0.8	0.9	1.0	1.2	1.3	0.9	1.0	1.1	1.3	1.5
TW F1	15	22	26	32	37	43	48	54	11	15	18	22	25	30	33	37	1.2	1.3	1.5	1.6	1.8	1.6	1.7	1.9	2.0	2.2
TW F2	13	22	28	36	41	49	55	61	8	14	19	24	28	34	38	43	1.1	1.2	1.3	1.5	1.6	1.4	1.5	1.6	1.8	2.0
TW F3	12	21	28	36	42	51	57	64	6	13	18	24	28	35	39	45	1.0	1.2	1.3	1.4	1.6	1.3	1.4	1.6	1.7	1.9
TW F4	9	19	26	34	41	49	55	63	4	11	16	22	27	33	38	44	1.0	1.2	1.3	1.4	1.6	1.3	1.4	1.6	1.7	1.9
TW G1	13	20	25	31	36	42	47	54	9	13	17	21	24	29	32	37	1.2	1.3	1.5	1.6	1.8	1.5	1.6	1.8	2.0	2.2
TW G2	10	19	25	32	38	46	51	58	6	11	16	21	25	31	35	40	1.1	1.2	1.4	1.5	1.7	1.4	1.5	1.7	1.8	2.0
TW G3	8	17	23	31	36	44	50	57	4	10	14	19	23	29	34	39	1.1	1.2	1.4	1.5	1.7	1.4	1.5	1.7	1.8	2.0
TW G4	6	14	20	26	32	39	45	52	2	8	11	16	20	26	30	35	1.2	1.3	1.4	1.6	1.7	1.5	1.6	1.7	1.9	2.1
TW H1	9	16	22	28	32	40	44	51	6	10	14	18	21	26	30	35	1.2	1.3	1.5	1.6	1.8	1.6	1.7	1.8	2.0	2.2
TW H2	7	14	19	26	30	38	43	49	3	8	11	16	19	25	28	33	1.2	1.3	1.5	1.6	1.8	1.5	1.7	1.8	2.0	2.1
TW H3	5	12	17	22	27	34	38	45	2	6	9	14	17	22	25	30	1.3	1.4	1.6	1.7	1.8	1.6	1.7	1.9	2.0	2.2
TW H4	4	9	13	18	21	27	31	37	1	4	7	10	13	17	20	24	1.5	1.6	1.7	1.9	2.0	1.8	1.9	2.1	2.2	2.4
TW I1	9	16	21	26	31	37	42	48	5	9	12	16	18	22	25	29	1.3	1.4	1.5	1.7	1.9	1.7	1.8	2.0	2.2	2.4
TW I2	12	23	31	40	47	56	62	70	6	14	20	27	32	40	45	52	1.0	1.1	1.2	1.4	1.5	1.2	1.3	1.4	1.6	1.7
TW I3	13	25	34	44	51	60	66	74	7	16	22	30	36	44	49	56	0.9	1.0	1.2	1.3	1.5	1.1	1.2	1.4	1.5	1.7
TW I4	13	28	37	48	56	66	72	79	7	17	25	34	41	50	56	64	0.9	1.0	1.1	1.3	1.4	1.0	1.1	1.3	1.4	1.6
TW I5	13	28	38	49	57	67	73	80	6	18	26	35	42	51	58	66	0.9	1.0	1.1	1.2	1.4	1.0	1.1	1.2	1.4	1.5
TW J1	16	30	40	50	58	67	73	80	9	20	27	36	42	51	57	64	0.9	1.0	1.1	1.3	1.4	1.0	1.1	1.3	1.4	1.6
TW J2	14	29	39	49	57	67	73	80	8	19	26	35	42	51	57	64	0.9	1.0	1.1	1.3	1.4	1.0	1.1	1.3	1.4	1.6
TW J3	16	34	46	58	66	76	82	88	8	23	33	43	51	61	68	75	0.8	0.9	1.0	1.2	1.3	0.9	1.0	1.1	1.3	1.5
TW J4	15	32	43	55	64	74	80	86	7	21	30	41	49	59	65	73	0.8	0.9	1.0	1.2	1.4	0.9	1.0	1.2	1.3	1.5
DG A1	11	20	26	32	36	42	46	50	6	11	15	19	22	26	28	30	1.1	1.2	1.5	1.8	2.1	1.5	1.6	1.9	2.3	3.0
DG A2	12	22	29	37	43	50	55	61	7	14	19	24	29	35	39	44	1.1	1.2	1.3	1.5	1.7	1.3	1.4	1.6	1.8	2.0
DG A3	14	26	34	44	52	62	68	74	9	18	24	31	37	46	52	59	1.0	1.0	1.1	1.3	1.5	1.2	1.2	1.3	1.5	1.7
DG B1	11	20	26	34	40	47	52	58	6	11	15	20	23	28	31	35	1.1	1.2	1.3	1.5	1.8	1.5	1.6	1.8	2.0	2.4
DG B2	13	22	30	39	46	55	61	68	8	14	19	25	30	37	42	48	1.1	1.1	1.2	1.4	1.6	1.3	1.4	1.5	1.7	1.8
DG B3	15	26	35	46	54	65	72	79	10	18	25	33	39	48	55	63	1.0	1.0	1.1	1.2	1.4	1.2	1.2	1.3	1.4	1.6
DG C1	13	23	31	40	47	56	62	68	8	14	19	24	29	35	39	44	1.1	1.1	1.2	1.4	1.6	1.3	1.4	1.6	1.8	2.0
DG C2	14	26	34	44	52	63	69	76	9	17	23	30	35	43	49	56	1.0	1.0	1.1	1.3	1.5	1.2	1.3	1.4	1.5	1.7
DG C3	17	30	40	52	60	72	79	85	11	21	28	37	44	54	61	70	0.9	1.0	1.0	1.2	1.4	1.1	1.1	1.2	1.3	1.5
SS A1	19	29	36	44	50	58	63	70	14	21	27	33	37	43	48	53	1.0	1.1	1.2	1.4	1.6	1.2	1.3	1.5	1.7	1.8
SS A2	22	36	45	55	62	71	76	83	16	27	34	43	49	57	62	69	0.8	0.9	1.1	1.2	1.4	1.0	1.1	1.2	1.4	1.6
SS A3	18	27	33	40	46	54	59	65	13	20	24	30	33	39	43	48	1.0	1.2	1.3	1.5	1.6	1.3	1.4	1.6	1.8	2.0
SS A4	21	35	44	54	61	70	75	81	16	26	34	42	47	56	61	67	0.8	0.9	1.1	1.3	1.4	1.0	1.1	1.3	1.4	1.6
SS A5	19	27	33	40	46	53	58	64	15	21	25	30	34	39	43	48	1.1	1.2	1.3	1.5	1.7	1.3	1.4	1.6	1.8	2.0
SS A6	22	35	45	55	62	71	76	82	16	26	34	42	48	56	62	68	0.8	0.9	1.1	1.3	1.4	1.0	1.1	1.2	1.4	1.6
SS A7	18	25	30	35	40	46	51	57	14	19	22	26	29	34	37	41	1.2	1.3	1.5	1.6	1.8	1.4	1.6	1.8	2.0	2.2
SS A8	21	34	43	53	60	69	74	80	16	26	33	40	46	54	59	66	0.9	1.0	1.1	1.3	1.4	1.0	1.1	1.3	1.4	1.6
SS B1	16	25	31	38	44	52	57	64	12	18	23	28	32	38	42	47	1.1	1.2	1.3	1.5	1.6	1.3	1.4	1.6	1.8	1.9
SS B2	19	31	40	50	57	66	71	78	13	23	30	38	44	52	57	64	0.9	1.0	1.1	1.3	1.5	1.0	1.1	1.3	1.5	1.6
SS B3	15	23	29	36	41	49	54	60	11	17	21	26	29	35	39	44	1.1	1.2	1.4	1.5	1.7	1.3	1.5	1.7	1.8	2.0
SS B4	18	31	39	49	56	65	70	77	13	22	29	37	43	50	56	62	0.9	1.0	1.1	1.3	1.5	1.0	1.2	1.3	1.5	1.6
SS B5	16	23	28	34	38	45	50	56	12	17	21	25	28	33	36	41	1.2	1.3	1.5	1.6	1.8	1.4	1.6	1.8	2.0	2.1
SS B6	18	31	39	49	56	65	71	77	13	22	29	37	43	51	56	63	0.9	1.0	1.1	1.3	1.5	1.0	1.1	1.3	1.5	1.6
SS B7	15	21	25	30	34	40	45	50	11	16	19	22	25	29	32	36	1.3	1.4	1.6	1.7	1.9	1.6	1.7	1.9	2.1	2.3
SS B8	17	29	38	47	54	63	68	75	12	21	28	35	41	48	54	60	0.9	1.0	1.2	1.3	1.5	1.1	1.2	1.3	1.5	1.7
SS C1	14	24	32	40	46	55	61	68	8	15	20	26	31	37	42	48	1.0	1.1	1.2	1.4	1.6	1.2	1.4	1.5	1.7	1.9
SS C2	16	29	38	48	55	65	71	78	9	19	26	34	40	48	54	61	0.9	1.0	1.1	1.3	1.4	1.1	1.2	1.3	1.5	1.6
SS C3	13	21	26	32	37	45	50	56	9	14	18	22	25	30	34	39	1.2	1.3	1.4	1.6	1.7	1.5	1.6	1.8	1.9	2.1
SS C4	15	26	34	43	50	59	65	72	9	17	23	30	35	43	48	55	1.0	1.1	1.2	1.3	1.5	1.2	1.3	1.4	1.6	1.7
SS D1	21	34	43	52	59	67	73	79	15	25	31	38	44	51	56	62	0.9	1.0	1.1	1.3	1.5	1.0	1.1	1.3	1.5	1.7
SS D2	24	41	52	63	71	79	84	89	16	30	40	49	56	65	71	78	0.7	0.8	1.0	1.2	1.3	0.9	1.0	1.1	1.3	1.5
SS D3	20	31	38	47	53	61	66	73	15	23	29	35	39	46	50	56	1.0	1.1	1.2	1.4	1.5	1.1	1.3	1.4	1.6	1.8
SS D4	23	40	50	61	69	77	82	88	16	29	38	48	54	63	69	76	0.8	0.9	1.0	1.2	1.4	0.9	1.0	1.2	1.3	1.5
SS E1	18	29	37	46	53	61	67	73	12	20	26	33	37	44	49	55	0.9	1.0	1.2	1.3	1.5	1.1	1.2	1.4	1.6	1.8
SS E2	20	36	46	57	64	74	79	85	13	25	34	43	49	58	64	71	0.8	0.9	1.1	1.2	1.4	0.9	1.0	1.2	1.4	1.5
SS E3	17	25	32	39	44	52	57	63	13	19	23	28	32	37	41	46	1.1	1.2	1.3	1.5	1.7	1.3	1.4	1.6	1.8	2.0
SS E4	19	33	43	53	60	69	75	81	13	24	31	39	45	54	59	66	0.8	0.9	1.1	1.3	1.4	1.0	1.1	1.3	1.4	1.6

ROANOKE, VIRGINIA

TYPE	SOLAR SAVINGS FRACTION (%) BASE TEMP: 55 F (2330 DD)								65 F (4314 DD)								CONSERVATION FACTOR 55 F					65 F				
LCR	100	70	50	40	30	25	20	15	100	70	50	40	30	25	20	15	100	50	30	20	15	100	50	30	20	15
WW A1	23	28	33	38	44	48	54	62	17	20	23	26	30	33	37	43	1.4	1.5	1.7	1.8	1.9	1.7	1.9	2.1	2.3	2.4
WW A2	26	33	41	47	56	61	68	77	17	22	28	32	39	44	49	58	1.1	1.3	1.4	1.6	1.7	1.4	1.5	1.7	1.8	2.0
WW A3	27	35	44	51	60	66	73	81	17	23	30	35	42	47	54	63	1.0	1.2	1.3	1.5	1.6	1.3	1.4	1.6	1.7	1.9
WW A4	28	36	46	53	62	68	75	84	17	23	31	36	44	50	57	66	1.0	1.2	1.3	1.5	1.6	1.2	1.4	1.5	1.7	1.8
WW A5	28	38	48	55	65	71	78	86	17	24	32	38	46	52	59	68	1.0	1.1	1.3	1.4	1.6	1.2	1.3	1.5	1.6	1.8
WW A6	29	38	49	56	66	72	79	87	17	24	33	39	47	53	61	70	1.0	1.1	1.3	1.4	1.6	1.1	1.3	1.5	1.6	1.8
WW B1	21	27	33	38	46	51	57	65	13	17	21	24	28	31	36	42	1.3	1.4	1.6	1.7	1.9	1.6	1.8	2.1	2.3	2.4
WW B2	28	37	48	55	64	70	77	85	18	25	33	39	47	53	60	70	1.0	1.1	1.3	1.4	1.6	1.2	1.3	1.5	1.6	1.8
WW B3	34	44	55	63	72	78	84	91	22	30	39	46	56	62	69	78	0.9	1.1	1.2	1.4	1.5	1.1	1.2	1.4	1.5	1.7
WW B4	34	45	57	65	75	80	86	93	22	31	41	49	59	65	73	82	0.9	1.0	1.2	1.4	1.5	1.0	1.2	1.3	1.5	1.6
WW B5	32	44	55	64	74	80	86	92	21	30	41	48	59	65	73	82	0.9	1.0	1.2	1.4	1.5	1.0	1.2	1.3	1.5	1.6
WW C1	35	46	57	65	74	80	86	92	23	32	42	49	58	64	72	80	0.9	1.0	1.2	1.4	1.5	1.0	1.2	1.3	1.5	1.6
WW C2	33	44	56	63	73	79	85	92	21	30	40	47	57	64	71	80	0.9	1.0	1.2	1.4	1.5	1.0	1.2	1.3	1.5	1.6
WW C3	39	51	64	72	81	87	92	96	26	37	49	57	68	74	81	89	0.8	1.0	1.1	1.3	1.5	0.9	1.1	1.2	1.4	1.5
WW C4	35	48	60	69	78	84	90	95	23	34	46	54	64	71	78	87	0.8	1.0	1.2	1.3	1.5	1.0	1.1	1.3	1.4	1.6
TW A1	23	27	33	37	43	48	53	61	17	20	23	26	30	33	37	43	1.4	1.5	1.7	1.8	1.9	1.7	1.9	2.1	2.3	2.4
TW A2	24	30	38	43	51	56	63	72	16	21	26	30	36	40	45	53	1.2	1.3	1.5	1.6	1.8	1.5	1.6	1.8	1.9	2.1
TW A3	24	32	40	46	54	60	67	76	16	21	27	31	38	43	49	57	1.1	1.3	1.4	1.6	1.7	1.4	1.5	1.7	1.8	2.0
TW A4	24	31	40	46	55	61	68	77	15	20	26	31	38	43	50	59	1.1	1.2	1.4	1.6	1.7	1.3	1.5	1.6	1.8	1.9
TW B1	22	27	33	38	44	49	55	63	16	19	23	26	31	34	39	45	1.3	1.5	1.6	1.8	1.9	1.7	1.8	2.0	2.2	2.3
TW B2	23	29	36	42	50	55	62	71	15	19	25	29	35	39	45	53	1.2	1.3	1.5	1.6	1.8	1.4	1.6	1.8	1.9	2.1
TW B3	22	29	37	42	51	56	63	72	14	19	25	29	35	40	46	54	1.2	1.3	1.5	1.6	1.7	1.4	1.6	1.7	1.9	2.0
TW B4	21	27	35	40	48	54	61	70	13	18	23	28	34	38	44	52	1.2	1.3	1.5	1.6	1.8	1.5	1.6	1.8	1.9	2.1
TW C1	21	26	32	37	44	49	55	64	15	18	22	26	31	35	40	47	1.3	1.4	1.6	1.7	1.9	1.6	1.8	1.9	2.1	2.2
TW C2	20	26	33	38	45	51	57	66	14	18	22	26	32	36	41	49	1.3	1.4	1.5	1.7	1.8	1.5	1.7	1.8	2.0	2.1
TW C3	19	25	31	36	43	48	55	64	13	17	21	25	31	34	40	47	1.3	1.4	1.6	1.7	1.9	1.6	1.7	1.9	2.0	2.2
TW C4	18	23	28	33	39	44	50	58	13	16	20	23	28	31	36	43	1.4	1.5	1.7	1.8	2.0	1.7	1.9	2.0	2.2	2.3
TW D1	18	23	29	33	40	45	51	59	11	14	18	20	25	28	32	38	1.3	1.5	1.7	1.8	2.0	1.7	1.9	2.1	2.3	2.5
TW D2	26	34	43	50	59	65	72	81	17	23	30	36	43	49	55	64	1.0	1.2	1.3	1.5	1.6	1.2	1.4	1.6	1.7	1.8
TW D3	29	38	47	54	64	70	77	85	18	25	33	39	47	53	60	69	1.0	1.1	1.3	1.5	1.6	1.2	1.3	1.5	1.6	1.8
TW D4	30	40	51	59	69	75	81	89	20	28	37	44	53	59	67	76	0.9	1.1	1.2	1.4	1.6	1.1	1.2	1.4	1.5	1.7
TW D5	30	40	51	59	69	75	82	89	19	28	37	44	54	60	68	77	0.9	1.1	1.2	1.4	1.5	1.1	1.2	1.4	1.5	1.7
TW E1	32	42	52	60	69	75	81	89	22	29	38	44	53	59	66	75	0.9	1.1	1.2	1.4	1.6	1.1	1.3	1.4	1.6	1.7
TW E2	31	41	52	59	69	75	82	89	21	29	38	44	53	59	67	76	0.9	1.1	1.2	1.4	1.6	1.1	1.2	1.4	1.6	1.7
TW E3	37	49	61	69	78	84	89	94	26	35	46	54	64	70	77	86	0.9	1.0	1.2	1.3	1.5	1.0	1.1	1.3	1.4	1.6
TW E4	35	46	58	66	76	81	87	93	23	33	44	51	61	68	75	84	0.9	1.0	1.2	1.3	1.5	1.0	1.1	1.3	1.5	1.6
TW F1	21	26	32	36	43	47	53	62	15	18	22	25	29	32	37	43	1.3	1.5	1.6	1.8	1.9	1.7	1.9	2.0	2.2	2.4
TW F2	22	28	35	41	49	54	61	70	14	18	23	27	33	37	43	51	1.2	1.3	1.5	1.6	1.8	1.5	1.6	1.8	2.0	2.1
TW F3	21	28	36	42	50	56	63	72	13	17	23	28	34	39	45	53	1.2	1.3	1.4	1.6	1.7	1.4	1.6	1.7	1.9	2.0
TW F4	19	26	34	40	49	55	62	71	11	16	21	26	33	38	44	52	1.2	1.3	1.5	1.6	1.7	1.4	1.6	1.7	1.9	2.0
TW G1	19	24	30	35	42	47	53	62	13	16	20	23	28	32	36	43	1.3	1.5	1.6	1.8	1.9	1.7	1.8	2.0	2.2	2.3
TW G2	18	25	32	37	45	50	57	67	11	15	20	24	30	34	40	48	1.2	1.4	1.5	1.7	1.8	1.5	1.7	1.8	2.0	2.1
TW G3	17	23	30	36	44	49	56	66	9	14	19	23	29	33	39	47	1.2	1.4	1.5	1.7	1.8	1.5	1.7	1.8	2.0	2.1
TW G4	14	19	26	31	39	44	51	60	7	11	16	20	25	29	35	43	1.3	1.5	1.6	1.7	1.9	1.6	1.7	1.9	2.1	2.2
TW H1	16	21	27	32	39	44	50	59	10	13	17	21	26	29	34	41	1.3	1.5	1.6	1.8	1.9	1.7	1.8	2.0	2.2	2.3
TW H2	14	19	25	30	37	42	49	58	7	11	15	19	24	28	33	40	1.4	1.5	1.6	1.8	1.9	1.7	1.8	2.0	2.1	2.3
TW H3	11	16	22	26	33	38	44	53	6	9	13	16	21	25	30	37	1.4	1.6	1.7	1.9	2.0	1.8	1.9	2.0	2.2	2.3
TW H4	9	13	17	21	27	31	36	45	4	7	10	13	17	20	24	30	1.6	1.7	1.9	2.0	2.1	2.0	2.1	2.2	2.4	2.5
TW I1	16	20	26	30	37	41	47	55	9	12	15	18	22	25	29	34	1.4	1.6	1.7	1.9	2.0	1.8	2.0	2.2	2.4	2.6
TW I2	23	31	40	46	56	62	69	78	14	20	27	32	40	45	52	61	1.1	1.2	1.4	1.6	1.7	1.3	1.4	1.6	1.7	1.9
TW I3	25	34	43	50	60	66	73	82	15	22	30	35	44	49	56	66	1.0	1.2	1.3	1.5	1.6	1.2	1.4	1.5	1.7	1.8
TW I4	27	37	48	55	65	72	79	87	17	25	34	40	50	56	64	73	1.0	1.1	1.3	1.4	1.6	1.1	1.3	1.4	1.6	1.7
TW I5	27	37	48	56	66	73	80	88	17	25	35	42	51	58	65	75	1.0	1.1	1.3	1.4	1.6	1.1	1.2	1.4	1.5	1.7
TW J1	30	39	50	57	67	73	80	87	19	27	35	42	51	56	64	73	1.0	1.1	1.3	1.4	1.6	1.1	1.3	1.4	1.6	1.7
TW J2	28	38	49	56	66	72	79	87	18	26	35	41	50	57	64	74	1.0	1.1	1.3	1.4	1.6	1.1	1.3	1.4	1.6	1.7
TW J3	34	46	58	66	76	81	87	93	22	32	43	51	61	68	75	84	0.9	1.0	1.2	1.3	1.5	1.0	1.1	1.3	1.5	1.6
TW J4	32	43	55	63	73	79	86	92	20	30	41	48	59	65	73	82	0.9	1.0	1.2	1.4	1.5	1.0	1.2	1.3	1.5	1.6
DG A1	19	25	31	36	42	45	49	54	11	14	18	21	25	27	30	33	1.2	1.5	1.8	2.1	2.5	1.6	1.9	2.3	3.0	3.8
DG A2	21	28	36	42	50	55	61	68	13	18	24	28	35	39	44	50	1.2	1.3	1.5	1.7	1.9	1.4	1.6	1.8	2.0	2.3
DG A3	25	34	44	51	61	67	74	81	17	23	31	37	46	52	59	67	1.0	1.1	1.3	1.5	1.7	1.2	1.3	1.5	1.7	1.9
DG B1	19	26	33	39	47	52	57	64	11	15	19	22	27	30	35	40	1.2	1.4	1.5	1.8	2.1	1.6	1.8	2.0	2.4	2.7
DG B2	22	29	38	45	55	61	68	76	14	19	25	30	37	42	48	56	1.1	1.2	1.4	1.6	1.8	1.4	1.5	1.7	1.8	2.0
DG B3	26	35	45	54	65	72	79	86	18	24	32	38	48	54	63	73	1.0	1.1	1.2	1.4	1.6	1.2	1.3	1.4	1.6	1.7
DG C1	23	31	40	46	55	61	68	75	14	18	24	28	34	38	44	51	1.1	1.2	1.4	1.6	1.9	1.4	1.6	1.8	2.0	2.2
DG C2	25	34	44	51	62	68	75	83	16	22	29	35	43	49	56	65	1.0	1.1	1.3	1.5	1.7	1.3	1.4	1.5	1.7	1.9
DG C3	29	39	51	60	72	78	85	91	20	28	36	43	54	61	70	79	1.0	1.1	1.2	1.4	1.6	1.1	1.2	1.3	1.5	1.6
SS A1	29	36	43	49	57	63	69	77	21	27	32	37	43	48	54	61	1.1	1.3	1.4	1.6	1.7	1.3	1.5	1.7	1.8	2.0
SS A2	36	45	55	62	71	76	82	89	27	34	43	49	57	63	69	78	0.9	1.1	1.3	1.4	1.6	1.1	1.2	1.4	1.6	1.7
SS A3	27	33	40	45	53	58	64	73	20	24	30	33	39	43	48	55	1.2	1.3	1.5	1.6	1.8	1.4	1.6	1.8	2.0	2.1
SS A4	35	44	54	61	69	75	81	88	26	34	42	48	56	61	68	76	1.0	1.1	1.3	1.4	1.6	1.1	1.3	1.4	1.6	1.7
SS A5	27	33	40	45	52	57	64	72	21	25	30	34	40	43	48	55	1.2	1.3	1.5	1.7	1.8	1.4	1.6	1.8	2.0	2.1
SS A6	35	45	54	61	70	76	82	89	26	34	42	48	56	62	69	77	0.9	1.1	1.3	1.4	1.6	1.1	1.2	1.4	1.6	1.7
SS A7	25	30	35	40	46	50	56	64	19	22	26	29	34	37	41	47	1.3	1.5	1.6	1.8	1.9	1.6	1.8	2.0	2.2	2.4
SS A8	34	43	53	59	68	74	80	87	26	33	41	46	54	59	66	74	1.0	1.1	1.3	1.4	1.6	1.1	1.3	1.4	1.6	1.8
SS B1	25	31	38	44	51	56	63	71	18	22	28	32	38	42	47	55	1.2	1.3	1.5	1.6	1.8	1.4	1.6	1.8	1.9	2.1
SS B2	31	40	49	56	65	71	78	85	23	30	38	44	52	57	64	73	1.0	1.1	1.3	1.5	1.6	1.1	1.3	1.5	1.6	1.8
SS B3	23	29	36	41	48	53	60	68	17	21	26	29	35	39	44	50	1.2	1.4	1.5	1.7	1.8	1.5	1.7	1.8	2.0	2.2
SS B4	30	39	48	55	64	70	76	84	22	29	37	42	50	56	62	71	1.0	1.2	1.3	1.5	1.6	1.2	1.3	1.5	1.6	1.8
SS B5	23	28	34	38	45	49	55	63	17	20	25	28	33	36	41	47	1.3	1.5	1.6	1.8	1.9	1.6	1.8	2.0	2.1	2.3
SS B6	30	39	48	55	64	70	77	85	22	29	37	43	51	56	63	72	1.0	1.1	1.3	1.5	1.6	1.1	1.3	1.5	1.6	1.8
SS B7	21	25	30	34	40	44	50	57	15	18	22	25	29	32	36	41	1.4	1.6	1.7	1.9	2.0	1.7	1.9	2.1	2.3	2.5
SS B8	29	38	47	53	62	68	75	83	21	28	35	41	48	54	60	69	1.0	1.2	1.3	1.5	1.6	1.2	1.3	1.5	1.7	1.8
SS C1	24	31	40	46	54	60	67	76	15	20	26	30	37	41	47	56	1.1	1.3	1.4	1.6	1.7	1.4	1.5	1.7	1.9	2.0
SS C2	28	38	48	55	64	70	77	85	18	25	33	39	48	53	61	70	1.0	1.1	1.3	1.5	1.6	1.2	1.3	1.5	1.6	1.8
SS C3	20	26	32	37	44	49	55	64	14	17	21	25	30	34	38	45	1.3	1.4	1.6	1.7	1.9	1.6	1.8	1.9	2.1	2.2
SS C4	26	34	42	49	58	64	71	80	17	23	30	35	42	48	54	63	1.1	1.2	1.4	1.5	1.7	1.3	1.4	1.6	1.7	1.9
SS D1	34	42	52	58	67	72	79	86	24	31	38	44	51	56	63	71	1.0	1.1	1.3	1.5	1.6	1.1	1.3	1.5	1.7	1.8
SS D2	41	52	63	70	79	84	89	94	30	40	49	56	66	71	78	85	0.9	1.0	1.2	1.3	1.5	1.0	1.1	1.3	1.5	1.6
SS D3	31	38	46	52	61	66	72	80	23	28	35	39	46	51	56	64	1.1	1.2	1.4	1.5	1.7	1.3	1.4	1.6	1.8	1.9
SS D4	39	50	61	68	77	82	88	93	29	38	48	54	63	69	76	84	0.9	1.0	1.2	1.4	1.5	1.0	1.2	1.3	1.5	1.6
SS E1	29	37	46	52	61	66	73	81	20	26	33	37	44	49	55	63	1.0	1.2	1.4	1.5	1.7	1.2	1.4	1.6	1.8	1.9
SS E2	36	46	57	64	73	79	85	91	25	34	43	49	58	64	71	79	0.9	1.1	1.2	1.4	1.5	1.0	1.2	1.4	1.5	1.7
SS E3	25	31	38	44	51	56	63	71	18	23	28	32	37	41	46	53	1.2	1.3	1.5	1.7	1.8	1.4	1.6	1.8	2.0	2.1
SS E4	33	42	52	59	69	74	81	88	24	31	39	45	54	59	66	75	1.0	1.1	1.3	1.4	1.6	1.1	1.3	1.4	1.6	1.7

OLYMPIA, WASHINGTON

TYPE	SOLAR SAVINGS FRACTION (%) BASE TEMP: 55 F (2657 DD)								65 F (5535 DD)								CONSERVATION FACTOR 55 F					65 F				
LCR	100	70	50	40	30	25	20	15	100	70	50	40	30	25	20	15	100	50	30	20	15	100	50	30	20	15
WW A1	19	22	26	29	32	35	38	42	15	17	19	21	23	24	26	28	1.6	1.9	2.2	2.6	2.9	2.1	2.5	3.0	3.7	4.6
WW A2	20	25	30	34	39	43	47	52	14	17	21	23	27	29	32	36	1.3	1.6	1.8	2.1	2.4	1.6	1.9	2.3	2.7	3.2
WW A3	20	25	31	36	42	45	50	56	13	17	21	24	28	31	35	39	1.2	1.5	1.7	2.0	2.2	1.5	1.8	2.1	2.5	3.0
WW A4	20	26	32	37	43	47	52	58	12	16	21	25	29	32	36	41	1.2	1.4	1.7	1.9	2.2	1.4	1.7	2.1	2.4	2.8
WW A5	20	26	33	38	44	49	54	60	12	16	22	25	30	33	37	42	1.2	1.4	1.7	1.9	2.1	1.4	1.7	2.0	2.4	2.7
WW A6	20	26	33	39	45	50	55	62	11	16	22	26	31	34	38	43	1.2	1.4	1.6	1.9	2.1	1.4	1.6	2.0	2.3	2.7
WW B1	16	19	23	26	29	31	34	37	10	12	14	15	17	18	18	18	1.6	1.9	2.3	2.9	3.5	2.1	2.7	3.9	-	-
WW B2	20	27	34	39	46	51	56	63	13	18	24	28	33	37	42	48	1.2	1.4	1.6	1.8	2.0	1.4	1.6	1.9	2.1	2.4
WW B3	24	32	40	45	53	58	63	71	16	22	29	33	40	44	49	55	1.1	1.3	1.5	1.7	1.9	1.2	1.5	1.7	2.0	2.2
WW B4	24	33	42	48	57	62	68	76	16	23	31	37	44	49	56	63	1.0	1.2	1.4	1.6	1.8	1.2	1.4	1.6	1.8	1.9
WW B5	23	32	41	48	57	62	69	77	15	23	31	37	45	50	57	65	1.0	1.2	1.4	1.6	1.7	1.1	1.3	1.5	1.7	1.9
WW C1	26	34	42	48	55	60	66	73	17	24	31	36	43	47	52	59	1.0	1.3	1.5	1.7	1.8	1.2	1.4	1.6	1.9	2.1
WW C2	24	32	41	47	55	60	66	74	16	23	30	35	43	47	53	61	1.0	1.2	1.4	1.6	1.8	1.2	1.4	1.6	1.8	2.0
WW C3	28	38	49	56	65	70	77	84	20	28	38	45	54	59	66	74	0.9	1.1	1.3	1.5	1.6	1.0	1.2	1.4	1.6	1.8
WW C4	26	36	46	53	62	68	74	82	17	26	35	42	51	57	64	72	1.0	1.1	1.3	1.5	1.7	1.1	1.2	1.4	1.6	1.8
TW A1	19	22	26	28	32	35	38	42	15	17	19	21	23	25	27	29	1.6	1.9	2.2	2.5	2.8	2.0	2.4	2.8	3.4	4.0
TW A2	19	23	28	32	37	40	44	50	13	16	20	22	26	28	31	35	1.4	1.6	1.9	2.2	2.4	1.7	2.0	2.4	2.8	3.2
TW A3	18	23	29	33	38	42	47	53	12	16	20	23	27	29	33	37	1.3	1.6	1.8	2.1	2.3	1.6	1.9	2.2	2.6	2.9
TW A4	17	23	29	33	39	43	48	54	11	15	19	23	27	30	34	38	1.3	1.5	1.8	2.0	2.2	1.5	1.8	2.1	2.4	2.8
TW B1	18	22	26	29	33	36	39	44	14	16	19	21	24	25	28	31	1.6	1.8	2.1	2.4	2.6	1.9	2.2	2.6	3.1	3.5
TW B2	18	22	27	31	36	40	44	50	12	15	19	22	25	28	31	35	1.4	1.6	1.9	2.1	2.4	1.7	2.0	2.3	2.6	2.9
TW B3	17	22	27	31	36	40	45	51	11	15	19	22	26	28	32	37	1.4	1.6	1.8	2.1	2.3	1.6	1.9	2.2	2.5	2.8
TW B4	16	21	26	30	35	39	44	50	11	14	18	21	25	28	31	36	1.4	1.6	1.8	2.1	2.3	1.7	1.9	2.2	2.5	2.7
TW C1	17	21	25	29	33	36	40	46	13	15	18	20	24	26	29	33	1.5	1.8	2.0	2.3	2.5	1.8	2.1	2.4	2.8	3.1
TW C2	16	20	25	29	34	37	42	48	11	14	18	20	24	27	30	35	1.5	1.7	1.9	2.2	2.4	1.7	2.0	2.3	2.6	2.9
TW C3	16	20	24	28	33	36	40	46	11	14	17	20	24	26	29	34	1.5	1.7	1.9	2.2	2.4	1.8	2.0	2.3	2.6	2.8
TW C4	15	19	23	26	30	33	38	43	11	14	17	19	22	25	28	32	1.6	1.8	2.0	2.3	2.5	1.9	2.1	2.4	2.7	2.9
TW D1	13	17	20	23	26	29	32	35	8	10	12	14	16	17	18	19	1.7	2.0	2.3	2.8	3.2	2.1	2.6	3.4	4.8	-
TW D2	20	26	32	37	43	47	53	60	13	18	23	26	32	35	39	45	1.2	1.4	1.7	1.9	2.1	1.4	1.7	1.9	2.2	2.4
TW D3	21	28	34	39	46	51	56	63	14	19	25	29	34	38	42	48	1.2	1.4	1.6	1.8	2.0	1.4	1.6	1.9	2.1	2.4
TW D4	23	30	38	44	52	57	63	71	15	21	28	33	40	45	51	58	1.1	1.3	1.5	1.7	1.8	1.2	1.4	1.6	1.9	2.0
TW D5	22	30	39	45	53	58	65	72	15	22	29	34	42	47	53	61	1.1	1.3	1.5	1.6	1.8	1.2	1.4	1.6	1.8	2.0
TW E1	24	31	39	44	51	56	61	69	17	22	28	33	39	43	48	55	1.1	1.3	1.5	1.7	1.9	1.3	1.5	1.7	2.0	2.2
TW E2	23	31	39	44	52	57	63	70	16	22	29	33	40	44	50	57	1.1	1.3	1.5	1.7	1.9	1.2	1.5	1.7	1.9	2.1
TW E3	28	37	46	53	61	66	73	80	20	27	36	42	50	55	61	69	1.0	1.2	1.4	1.6	1.7	1.1	1.3	1.5	1.7	1.9
TW E4	26	35	44	51	59	65	71	79	18	26	34	40	48	54	60	69	1.0	1.2	1.4	1.6	1.7	1.1	1.3	1.5	1.7	1.8
TW F1	17	21	24	27	31	34	37	42	13	15	17	19	22	23	25	28	1.6	1.9	2.1	2.5	2.7	2.0	2.4	2.8	3.3	3.9
TW F2	16	21	26	29	34	38	42	47	11	14	17	20	23	26	29	32	1.4	1.7	1.9	2.2	2.4	1.7	2.0	2.4	2.8	3.2
TW F3	15	20	26	30	35	39	43	49	9	13	17	20	24	26	30	34	1.4	1.6	1.9	2.1	2.3	1.6	1.9	2.3	2.6	3.0
TW F4	13	18	24	28	34	38	43	49	8	11	15	18	23	25	29	34	1.4	1.6	1.8	2.1	2.3	1.6	1.9	2.2	2.5	2.8
TW G1	15	19	23	26	30	33	37	42	11	13	16	18	21	23	25	28	1.6	1.8	2.1	2.4	2.6	1.9	2.3	2.6	3.1	3.5
TW G2	14	18	23	27	32	35	40	46	8	11	15	18	21	24	27	31	1.5	1.7	1.9	2.2	2.4	1.8	2.0	2.4	2.7	3.0
TW G3	12	16	22	25	31	34	39	45	7	10	14	16	20	23	26	31	1.5	1.7	1.9	2.2	2.4	1.7	2.0	2.3	2.6	2.9
TW G4	10	14	19	22	27	31	36	42	5	8	11	14	18	21	24	29	1.5	1.7	2.0	2.2	2.4	1.8	2.1	2.3	2.6	2.9
TW H1	12	16	20	23	28	31	35	41	8	10	13	15	19	21	24	27	1.6	1.8	2.1	2.3	2.6	1.9	2.2	2.5	2.9	3.2
TW H2	10	14	18	22	26	30	34	40	5	8	11	14	17	20	23	27	1.6	1.8	2.0	2.3	2.5	1.9	2.1	2.4	2.7	3.0
TW H3	8	12	16	19	24	27	31	37	4	7	10	12	15	18	21	26	1.7	1.9	2.1	2.3	2.5	2.0	2.2	2.5	2.7	3.0
TW H4	6	9	13	15	20	22	26	32	3	5	8	10	13	15	18	22	1.8	2.0	2.2	2.4	2.6	2.2	2.4	2.6	2.9	3.1
TW I1	11	14	18	20	24	26	29	32	7	8	10	12	14	15	16	17	1.7	2.0	2.4	2.8	3.3	2.2	2.7	3.5	4.9	-
TW I2	17	23	29	34	40	44	50	57	10	15	20	23	29	32	36	42	1.3	1.5	1.7	1.9	2.1	1.5	1.7	2.0	2.2	2.5
TW I3	18	24	31	36	43	48	53	61	11	16	22	26	31	35	40	46	1.2	1.4	1.6	1.8	2.0	1.4	1.6	1.9	2.1	2.4
TW I4	20	27	35	41	49	54	60	68	13	19	25	30	37	42	48	56	1.1	1.3	1.5	1.7	1.9	1.3	1.5	1.7	1.9	2.1
TW I5	20	28	36	42	51	56	62	71	13	19	26	32	39	44	50	59	1.1	1.3	1.5	1.7	1.8	1.2	1.4	1.6	1.8	2.0
TW J1	22	29	36	42	49	54	60	67	14	20	26	31	37	41	46	53	1.1	1.3	1.5	1.7	1.9	1.3	1.5	1.7	2.0	2.2
TW J2	21	28	36	42	50	55	61	68	14	20	26	31	38	42	48	55	1.1	1.3	1.5	1.7	1.9	1.3	1.5	1.7	1.9	2.1
TW J3	25	34	44	50	59	64	71	79	17	25	33	39	48	53	60	68	1.0	1.2	1.4	1.6	1.7	1.1	1.3	1.5	1.7	1.8
TW J4	23	32	42	48	57	63	69	77	15	23	31	38	46	51	58	67	1.0	1.2	1.4	1.6	1.7	1.1	1.3	1.5	1.7	1.9
DG A1	14	18	21	24	27	29	30	32	8	10	13	14	15	16	16	16	1.6	2.0	2.6	3.7	-	2.0	2.7	4.0	-	-
DG A2	16	21	27	31	36	39	43	48	11	14	18	21	25	28	31	35	1.4	1.6	1.9	2.2	2.6	1.6	1.9	2.2	2.7	3.2
DG A3	20	27	34	39	46	51	56	63	15	20	25	30	36	40	45	52	1.2	1.4	1.6	1.8	2.0	1.4	1.6	1.8	2.0	2.2
DG B1	14	18	23	26	29	32	34	36	8	11	13	15	17	18	19	19	1.5	1.8	2.3	3.1	4.2	1.9	2.5	3.3	-	-
DG B2	17	22	28	32	38	42	47	52	11	15	19	22	27	30	33	38	1.3	1.5	1.8	2.1	2.4	1.6	1.8	2.1	2.5	2.9
DG B3	21	28	35	41	48	53	59	67	15	20	26	31	38	42	48	55	1.2	1.3	1.5	1.7	1.9	1.3	1.5	1.7	1.9	2.1
DG C1	17	22	27	31	35	38	41	45	11	14	17	19	22	24	26	28	1.3	1.6	2.0	2.5	3.2	1.7	2.1	2.6	3.6	-
DG C2	20	26	32	37	44	48	53	59	14	18	23	27	32	35	40	45	1.2	1.4	1.6	1.9	2.2	1.5	1.7	1.9	2.2	2.5
DG C3	24	31	39	45	53	59	65	73	17	23	30	35	42	47	53	61	1.1	1.3	1.4	1.7	1.8	1.3	1.4	1.6	1.8	2.0
SS A1	23	27	33	36	41	44	48	54	18	22	26	29	33	35	38	42	1.3	1.6	1.9	2.1	2.4	1.5	1.8	2.2	2.6	3.1
SS A2	28	34	41	46	53	57	63	70	22	28	34	38	44	48	53	59	1.1	1.3	1.5	1.8	1.9	1.2	1.5	1.7	2.0	2.2
SS A3	21	25	30	33	37	40	43	47	17	20	24	26	29	31	33	35	1.4	1.7	2.0	2.4	2.7	1.6	2.0	2.5	3.2	4.1
SS A4	27	34	40	45	51	56	61	67	22	28	33	37	43	46	51	56	1.1	1.4	1.6	1.8	2.0	1.3	1.5	1.8	2.1	2.3
SS A5	22	26	31	34	38	41	44	49	19	22	25	27	30	32	35	37	1.4	1.7	2.0	2.3	2.6	1.6	2.0	2.4	3.0	3.6
SS A6	27	34	41	46	52	57	62	68	22	28	33	38	43	47	52	57	1.1	1.3	1.6	1.8	2.0	1.2	1.5	1.8	2.0	2.2
SS A7	20	24	27	30	33	35	38	42	17	20	22	24	26	28	29	31	1.6	1.9	2.3	2.7	3.1	1.8	2.3	2.9	3.8	-
SS A8	27	33	39	44	50	54	59	65	22	27	32	36	41	44	48	54	1.2	1.4	1.6	1.8	2.0	1.3	1.6	1.8	2.1	2.4
SS B1	20	24	29	32	37	40	44	49	15	19	22	25	28	31	34	37	1.4	1.7	1.9	2.2	2.5	1.6	1.9	2.3	2.8	3.2
SS B2	25	31	38	43	50	54	59	66	19	25	31	35	41	45	50	56	1.2	1.4	1.6	1.8	2.0	1.3	1.5	1.7	2.0	2.2
SS B3	18	22	27	30	34	37	40	45	14	17	20	23	26	28	30	32	1.5	1.8	2.1	2.4	2.7	1.7	2.1	2.5	3.1	3.8
SS B4	24	30	37	42	48	52	58	64	19	24	30	34	39	43	48	54	1.2	1.4	1.6	1.8	2.0	1.3	1.5	1.8	2.1	2.3
SS B5	19	22	26	29	33	36	39	43	15	18	21	23	25	27	30	32	1.6	1.8	2.1	2.5	2.8	1.8	2.2	2.6	3.1	3.7
SS B6	24	30	37	42	48	53	58	65	19	24	30	34	39	43	48	54	1.2	1.4	1.6	1.8	2.0	1.3	1.5	1.8	2.0	2.2
SS B7	17	20	23	26	29	32	34	38	14	16	18	20	23	24	26	28	1.7	2.0	2.3	2.7	3.1	1.9	2.4	2.9	3.6	4.6
SS B8	23	29	35	40	46	50	55	62	18	23	28	32	37	41	45	51	1.2	1.4	1.7	1.9	2.1	1.3	1.6	1.9	2.1	2.4
SS C1	18	22	28	32	37	40	45	50	11	14	18	21	24	27	29	33	1.3	1.6	1.9	2.2	2.4	1.6	2.0	2.4	2.9	3.4
SS C2	21	28	35	40	47	51	57	64	14	19	25	29	34	38	43	49	1.2	1.4	1.6	1.8	2.0	1.3	1.6	1.8	2.1	2.3
SS C3	16	20	24	27	32	35	39	44	11	14	17	19	22	24	26	29	1.5	1.8	2.0	2.3	2.6	1.9	2.2	2.6	3.0	3.5
SS C4	19	25	32	36	43	47	52	59	13	18	23	26	31	35	39	45	1.2	1.5	1.7	1.9	2.1	1.4	1.7	1.9	2.2	2.4
SS D1	25	31	37	41	47	50	54	60	20	24	29	32	36	39	42	46	1.2	1.5	1.7	2.0	2.2	1.4	1.7	2.1	2.5	3.0
SS D2	30	38	47	52	60	64	70	77	24	31	38	43	50	54	59	66	1.0	1.2	1.4	1.6	1.8	1.1	1.4	1.6	1.8	2.0
SS D3	24	29	34	38	43	46	50	55	20	23	27	30	34	36	39	42	1.3	1.6	1.8	2.1	2.4	1.5	1.8	2.2	2.7	3.2
SS D4	30	37	45	51	58	63	68	75	24	30	37	42	48	53	58	64	1.0	1.3	1.5	1.7	1.8	1.1	1.4	1.6	1.9	2.1
SS E1	22	27	32	36	41	44	48	53	16	20	24	27	30	32	35	37	1.3	1.6	1.9	2.2	2.5	1.5	1.9	2.3	3.0	3.8
SS E2	26	34	41	47	54	58	64	70	20	26	32	37	43	47	52	58	1.1	1.3	1.5	1.7	1.9	1.2	1.5	1.7	2.0	2.2
SS E3	20	24	29	32	36	39	43	48	16	19	22	25	28	30	32	35	1.4	1.7	2.0	2.3	2.6	1.6	2.0	2.5	3.0	3.7
SS E4	25	32	39	44	50	55	60	67	19	25	31	35	40	44	49	55	1.1	1.4	1.6	1.8	2.0	1.3	1.5	1.8	2.0	2.3

SEATTLE-TACOMA, WASHINGTON

SOLAR SAVINGS FRACTION (%)

TYPE	BASE TEMP: 55 F (2393 DD) LCR 100	70	50	40	30	25	20	15	65 F (5188 DD) 100	70	50	40	30	25	20	15
WW A1	20	23	27	30	33	36	39	43	16	18	20	22	24	26	28	30
WW A2	21	26	31	35	41	44	48	54	15	18	22	25	29	31	34	38
WW A3	21	27	33	37	43	47	52	58	14	18	23	26	30	33	36	41
WW A4	21	27	33	38	44	49	54	60	13	18	23	26	31	34	38	42
WW A5	21	27	34	39	46	50	55	62	13	18	23	27	32	35	39	44
WW A6	21	28	35	40	47	51	56	63	13	18	23	27	33	36	40	45
WW B1	17	20	24	27	31	33	35	39	11	13	16	17	19	20	20	20
WW B2	22	28	35	41	48	52	58	65	14	20	25	29	35	39	44	49
WW B3	25	33	41	47	54	59	65	72	17	24	30	35	41	45	50	57
WW B4	25	34	43	49	58	63	69	77	17	25	32	38	46	51	57	64
WW B5	24	33	43	49	58	63	70	77	17	24	32	38	46	52	58	66
WW C1	27	35	43	49	57	61	67	74	19	25	32	37	44	48	54	60
WW C2	25	34	42	48	56	61	68	75	17	24	32	37	44	49	55	62
WW C3	29	40	50	57	66	71	77	85	21	30	39	46	55	60	67	75
WW C4	27	37	47	54	63	69	75	83	19	28	37	43	52	58	65	73
TW A1	20	23	27	30	33	36	39	44	16	18	20	22	25	26	28	31
TW A2	20	24	29	33	38	41	46	51	14	18	21	24	27	30	33	36
TW A3	19	25	30	34	40	44	48	54	14	17	21	24	29	31	35	39
TW A4	18	24	30	34	40	44	49	55	12	17	21	24	29	32	35	40
TW B1	19	23	27	30	34	37	41	46	15	17	20	22	25	27	29	33
TW B2	18	23	28	32	37	41	45	51	13	17	20	23	27	30	33	37
TW B3	18	23	28	32	38	42	45	52	13	16	20	23	27	30	34	38
TW B4	17	22	27	31	36	40	45	51	12	15	19	22	27	29	33	38
TW C1	18	22	26	29	34	37	41	47	13	16	19	22	25	28	30	34
TW C2	17	21	26	30	35	38	43	49	12	16	19	22	26	28	32	36
TW C3	16	20	25	29	34	37	42	48	12	15	19	21	25	28	31	36
TW C4	16	19	24	27	31	35	39	44	12	15	18	20	24	26	29	34
TW D1	14	18	21	24	28	30	33	37	9	12	14	15	17	19	20	21
TW D2	21	27	33	38	45	49	54	61	15	19	24	28	33	37	41	47
TW D3	22	29	36	41	48	52	57	64	15	21	26	30	36	39	44	50
TW D4	24	31	40	45	53	58	64	72	17	23	30	35	42	46	52	59
TW D5	23	31	40	46	54	59	66	73	16	23	30	36	43	48	54	62
TW E1	25	32	40	45	52	57	63	70	18	24	30	34	41	45	49	56
TW E2	25	32	40	46	53	58	64	71	17	24	30	35	42	46	51	58
TW E3	29	38	47	54	62	68	74	81	21	29	37	43	51	56	62	70
TW E4	27	36	45	52	61	66	72	80	19	27	35	42	50	55	61	70
TW F1	18	21	25	28	32	35	39	43	14	16	19	21	23	25	27	30
TW F2	17	22	27	31	36	39	43	49	12	15	19	21	25	27	30	34
TW F3	16	21	27	31	36	40	45	51	11	14	18	21	25	28	31	36
TW F4	14	20	25	29	35	39	44	51	9	13	17	20	24	27	31	36
TW G1	16	20	24	27	32	35	38	43	12	14	17	19	22	24	27	30
TW G2	15	19	24	28	33	37	41	47	10	13	16	19	23	25	29	33
TW G3	13	18	23	27	32	36	40	47	8	11	15	18	22	25	28	33
TW G4	10	15	20	23	29	32	37	43	6	9	13	16	20	22	26	31
TW H1	13	17	21	25	29	32	36	42	9	11	15	17	20	22	25	29
TW H2	11	15	19	23	28	31	36	42	6	9	13	15	19	21	25	29
TW H3	9	13	17	20	25	28	33	39	5	8	11	13	17	19	23	27
TW H4	7	10	14	16	21	24	28	33	4	6	9	11	14	16	19	23
TW I1	12	15	19	22	25	28	30	34	8	10	12	13	15	16	18	19
TW I2	18	24	30	35	41	46	51	58	12	16	21	25	30	34	38	44
TW I3	19	26	33	38	44	49	55	62	13	18	23	27	[illegible]	[illegible]	42	48
TW I4	21	28	37	42	50	55	62	69	14	20	27	32	39	44	49	57
TW I5	21	29	37	44	52	57	64	72	14	21	28	33	41	46	52	60
TW J1	23	30	38	43	50	55	61	68	16	22	28	32	39	43	48	55
TW J2	22	29	37	43	51	56	62	70	15	21	28	33	39	44	49	57
TW J3	26	35	45	51	60	66	72	80	18	26	35	41	49	55	61	69
TW J4	24	33	43	49	58	64	70	78	17	25	33	39	47	53	59	68
DG A1	15	19	23	25	28	30	32	34	9	12	14	16	17	18	18	18
DG A2	17	22	28	32	37	41	45	50	12	16	20	23	27	30	33	37
DG A3	21	28	35	40	47	52	57	65	16	21	27	31	37	41	46	53
DG B1	16	20	24	27	31	33	36	38	10	12	15	17	19	20	21	22
DG B2	18	23	29	34	40	44	48	54	13	16	21	24	29	32	35	40
DG B3	22	29	36	42	50	55	61	68	16	22	28	33	39	44	49	57
DG C1	18	23	29	32	37	40	43	47	12	15	19	21	24	26	28	30
DG C2	21	27	34	39	45	50	55	61	15	19	25	28	34	37	42	47
DG C3	25	32	40	46	55	60	66	73	19	24	31	36	44	49	54	62
SS A1	23	28	33	37	42	45	49	55	19	23	27	30	34	36	39	43
SS A2	28	35	42	47	54	58	64	70	24	29	35	40	45	49	54	60
SS A3	22	26	31	34	38	41	44	49	18	21	25	27	30	32	34	36
SS A4	28	34	41	46	52	57	62	68	23	29	34	38	44	47	51	57
SS A5	23	27	32	35	39	42	45	50	20	23	26	28	32	33	36	39
SS A6	28	35	42	47	53	57	63	69	23	29	35	39	44	48	52	58
SS A7	21	24	28	31	34	36	39	43	18	21	23	25	27	29	30	32
SS A8	27	34	40	45	51	55	60	66	23	28	33	37	42	45	49	55
SS B1	20	25	30	33	38	41	45	50	16	20	23	26	30	32	35	38
SS B2	25	32	39	44	51	55	60	67	21	26	32	36	42	46	51	57
SS B3	19	23	28	31	35	38	41	46	15	18	22	24	27	29	31	34
SS B4	25	31	38	42	49	53	59	65	20	25	31	35	40	44	49	55
SS B5	19	23	27	30	34	37	40	44	16	19	22	24	27	29	31	34
SS B6	25	31	38	43	49	54	59	66	20	25	31	35	41	44	49	55
SS B7	18	21	24	27	30	33	35	39	15	17	20	21	24	25	27	29
SS B8	24	30	36	41	47	51	56	63	19	24	29	33	38	42	46	52
SS C1	19	24	29	33	38	42	46	52	12	16	20	22	26	28	31	35
SS C2	22	29	36	41	48	53	58	65	15	20	26	30	36	40	45	51
SS C3	17	21	25	28	33	36	40	45	12	15	18	20	23	25	28	31
SS C4	20	26	33	37	44	48	54	60	14	19	24	28	33	36	41	47
SS D1	26	32	38	42	48	51	55	61	21	26	30	33	38	40	43	47
SS D2	31	39	47	53	61	65	71	77	25	32	39	44	51	55	60	66
SS D3	25	30	35	39	44	47	51	56	21	25	29	31	35	37	40	44
SS D4	31	38	46	52	59	63	69	76	25	31	38	43	49	53	58	65
SS E1	23	28	33	37	42	45	49	54	17	21	25	28	32	34	36	39
SS E2	27	35	42	48	55	59	65	71	21	27	34	38	44	48	53	58
SS E3	21	25	30	33	37	40	44	49	17	20	23	26	29	31	33	36
SS E4	26	33	40	45	51	56	61	68	20	26	32	36	42	45	50	56

CONSERVATION FACTOR

TYPE	55 F 100	50	30	20	15	65 F 100	50	30	20	15
WW A1	1.6	1.9	2.2	2.5	2.8	2.0	2.4	2.9	3.5	4.4
WW A2	1.3	1.6	1.8	2.1	2.3	1.6	1.9	2.3	2.7	3.2
WW A3	1.2	1.5	1.7	2.0	2.2	1.5	1.8	2.1	2.5	2.9
WW A4	1.2	1.4	1.7	1.9	2.1	1.4	1.7	2.0	2.4	2.8
WW A5	1.2	1.4	1.6	1.9	2.1	1.4	1.7	2.0	2.3	2.7
WW A6	1.1	1.4	1.6	1.9	2.1	1.3	1.6	1.9	2.3	2.7
WW B1	1.5	1.9	2.3	2.8	3.4	2.0	2.6	3.7	-	-
WW B2	1.1	1.4	1.6	1.8	2.0	1.3	1.6	1.8	2.1	2.4
WW B3	1.1	1.3	1.5	1.7	1.9	1.2	1.5	1.7	2.0	2.2
WW B4	1.0	1.2	1.4	1.6	1.8	1.1	1.3	1.6	1.8	2.0
WW B5	1.0	1.2	1.4	1.6	1.7	1.1	1.3	1.5	1.7	1.9
WW C1	1.0	1.2	1.4	1.7	1.8	1.2	1.4	1.6	1.9	2.1
WW C2	1.0	1.2	1.4	1.6	1.8	1.2	1.4	1.6	1.8	2.0
WW C3	0.9	1.1	1.3	1.5	1.6	1.0	1.2	1.4	1.6	1.8
WW C4	1.0	1.1	1.3	1.5	1.7	1.1	1.2	1.4	1.6	1.8
TW A1	1.6	1.9	2.1	2.5	2.7	1.9	2.3	2.8	3.3	3.9
TW A2	1.4	1.6	1.9	2.1	2.4	1.6	2.0	2.3	2.7	3.1
TW A3	1.3	1.5	1.8	2.0	2.3	1.5	1.8	2.2	2.5	2.9
TW A4	1.3	1.5	1.7	2.0	2.2	1.5	1.8	2.1	2.4	2.7
TW B1	1.5	1.8	2.0	2.3	2.6	1.8	2.2	2.6	3.0	3.4
TW B2	1.4	1.6	1.9	2.1	2.3	1.6	1.9	2.2	2.6	2.9
TW B3	1.3	1.6	1.8	2.1	2.3	1.6	1.9	2.2	2.5	2.8
TW B4	1.4	1.6	1.8	2.1	2.3	1.6	1.9	2.2	2.5	2.7
TW C1	1.5	1.7	2.0	2.2	2.4	1.8	2.1	2.4	2.7	3.1
TW C2	1.4	1.7	1.9	2.1	2.3	1.7	2.0	2.2	2.5	2.8
TW C3	1.5	1.7	1.9	2.2	2.3	1.7	2.0	2.3	2.5	2.8
TW C4	1.6	1.8	2.0	2.2	2.4	1.8	2.1	2.4	2.6	2.9
TW D1	1.6	1.9	2.3	2.7	3.1	2.0	2.5	3.2	4.5	-
TW D2	1.2	1.4	1.6	1.9	2.1	1.4	1.7	1.9	2.2	2.4
TW D3	1.2	1.4	1.6	1.8	2.0	1.3	1.6	1.8	2.1	2.4
TW D4	1.1	1.3	1.5	1.7	1.8	1.2	1.4	1.6	1.9	2.0
TW D5	1.1	1.3	1.4	1.6	1.8	1.2	1.4	1.6	1.8	2.0
TW E1	1.1	1.3	1.5	1.7	1.9	1.3	1.5	1.7	2.0	2.2
TW E2	1.1	1.3	1.5	1.7	1.9	1.2	1.4	1.7	1.9	2.1
TW E3	1.0	1.2	1.4	1.5	1.7	1.1	1.3	1.5	1.7	1.9
TW E4	1.0	1.2	1.4	1.5	1.7	1.1	1.3	1.5	1.7	1.8
TW F1	1.6	1.8	2.1	2.4	2.7	1.9	2.3	2.7	3.2	3.8
TW F2	1.4	1.6	1.9	2.2	2.4	1.7	2.0	2.3	2.7	3.2
TW F3	1.3	1.6	1.8	2.1	2.3	1.6	1.9	2.2	2.6	2.9
TW F4	1.3	1.6	1.8	2.0	2.3	1.6	1.9	2.2	2.5	2.8
TW G1	1.5	1.8	2.1	2.3	2.6	1.9	2.2	2.6	3.0	3.4
TW G2	1.4	1.7	1.9	2.2	2.4	1.7	2.0	2.3	2.7	3.0
TW G3	1.4	1.6	1.9	2.1	2.3	1.7	2.0	2.2	2.6	2.9
TW G4	1.5	1.7	1.9	2.2	2.4	1.7	2.0	2.3	2.6	2.8
TW H1	1.5	1.8	2.0	2.3	2.5	1.8	2.1	2.5	2.8	3.2
TW H2	1.5	1.8	2.0	2.2	2.4	1.8	2.1	2.4	2.7	3.0
TW H3	1.6	1.8	2.0	2.3	2.5	1.9	2.1	2.4	2.7	2.9
TW H4	1.8	2.0	2.2	2.4	2.6	2.1	2.3	2.6	2.8	3.1
TW I1	1.7	2.0	2.3	2.8	3.2	2.1	2.6	3.3	4.5	-
TW I2	1.2	1.5	1.7	1.9	2.1	1.4	1.7	1.9	2.2	2.5
TW I3	1.2	1.4	1.6	1.8	2.0	1.4	1.6	1.9	2.1	2.3
TW I4	1.1	1.3	1.5	1.7	1.9	1.2	1.5	1.7	1.9	2.1
TW I5	1.1	1.3	1.5	1.7	1.8	1.2	1.4	1.6	1.8	2.0
TW J1	1.1	1.3	1.5	1.7	1.9	1.3	1.5	1.7	2.0	2.2
TW J2	1.1	1.3	1.5	1.7	1.9	1.2	1.5	1.7	1.9	2.1
TW J3	1.0	1.2	1.4	1.5	1.7	1.1	1.3	1.5	1.7	1.8
TW J4	1.0	1.2	1.4	1.6	1.7	1.1	1.3	1.5	1.7	1.9
DG A1	1.5	1.9	2.5	3.6		1.9	2.5	3.8	-	-
DG A2	1.3	1.6	1.9	2.2	2.5	1.6	1.9	2.2	2.7	3.1
DG A3	1.2	1.4	1.6	1.8	2.0	1.4	1.5	1.8	2.0	2.2
DG B1	1.5	1.8	2.2	3.0	4.0	1.9	2.4	3.2	-	-
DG B2	1.3	1.5	1.8	2.1	2.4	1.5	1.8	2.1	2.5	2.9
DG B3	1.2	1.3	1.5	1.7	1.9	1.3	1.5	1.7	1.9	2.1
DG C1	1.3	1.6	1.9	2.5	3.2	1.7	2.0	2.6	3.6	-
DG C2	1.2	1.4	1.6	1.9	2.2	1.4	1.6	1.9	2.2	2.5
DG C3	1.1	1.3	1.4	1.7	1.8	1.3	1.4	1.6	1.8	2.0
SS A1	1.3	1.6	1.8	2.1	2.4	1.5	1.8	2.2	2.6	3.1
SS A2	1.1	1.3	1.5	1.7	1.9	1.2	1.5	1.7	2.0	2.2
SS A3	1.4	1.7	2.0	2.4	2.7	1.6	2.0	2.5	3.2	4.0
SS A4	1.1	1.4	1.6	1.8	2.0	1.2	1.5	1.8	2.1	2.3
SS A5	1.4	1.7	2.0	2.3	2.6	1.6	2.0	2.4	2.9	3.5
SS A6	1.1	1.3	1.6	1.8	1.9	1.2	1.5	1.7	2.0	2.2
SS A7	1.6	1.9	2.2	2.6	3.0	1.8	2.2	2.8	3.7	5.0
SS A8	1.1	1.4	1.6	1.8	2.0	1.3	1.5	1.8	2.1	2.4
SS B1	1.4	1.6	1.9	2.2	2.4	1.6	1.9	2.3	2.7	3.2
SS B2	1.1	1.4	1.6	1.8	1.9	1.3	1.5	1.7	2.0	2.2
SS B3	1.4	1.7	2.0	2.4	2.7	1.7	2.0	2.5	3.1	3.8
SS B4	1.2	1.4	1.6	1.8	2.0	1.3	1.5	1.8	2.0	2.3
SS B5	1.5	1.8	2.1	2.4	2.7	1.7	2.1	2.6	3.1	3.6
SS B6	1.2	1.4	1.6	1.8	2.0	1.3	1.5	1.8	2.0	2.2
SS B7	1.6	2.0	2.3	2.7	3.0	1.9	2.3	2.9	3.6	4.4
SS B8	1.2	1.4	1.6	1.9	2.0	1.3	1.6	1.9	2.1	2.4
SS C1	1.3	1.6	1.8	2.1	2.4	1.6	1.9	2.3	2.8	3.4
SS C2	1.1	1.4	1.6	1.8	2.0	1.3	1.6	1.8	2.1	2.3
SS C3	1.5	1.7	2.0	2.3	2.5	1.8	2.1	2.5	2.9	3.4
SS C4	1.2	1.4	1.7	1.9	2.1	1.4	1.7	1.9	2.2	2.4
SS D1	1.2	1.5	1.7	2.0	2.2	1.3	1.7	2.1	2.5	3.0
SS D2	1.0	1.2	1.4	1.6	1.8	1.1	1.3	1.6	1.8	2.0
SS D3	1.3	1.5	1.8	2.1	2.3	1.4	1.8	2.2	2.7	3.2
SS D4	1.0	1.2	1.5	1.7	1.8	1.1	1.4	1.6	1.9	2.1
SS E1	1.3	1.5	1.8	2.1	2.4	1.5	1.8	2.3	3.0	3.8
SS E2	1.1	1.3	1.5	1.7	1.9	1.2	1.4	1.7	2.0	2.2
SS E3	1.4	1.7	2.0	2.3	2.6	1.6	2.0	2.4	3.0	3.6
SS E4	1.1	1.3	1.6	1.8	2.0	1.2	1.5	1.8	2.0	2.3

SPOKANE, WASHINGTON

TYPE	SOLAR SAVINGS FRACTION (%) BASE TEMP: 55 F (4177 DD) LCR 100	70	50	40	30	25	20	15	65 F (6839 DD) 100	70	50	40	30	25	20	15	CONSERVATION FACTOR 55 F 100	50	30	20	15	65 F 100	50	30	20	15
WW A1	16	19	21	23	25	27	29	31	14	15	17	18	20	21	22	23	1.9	2.3	2.8	3.4	4.2	2.3	2.8	3.5	4.8	-
WW A2	16	19	23	26	30	32	36	39	12	15	18	20	23	25	27	30	1.5	1.9	2.2	2.6	3.0	1.8	2.1	2.6	3.2	4.1
WW A3	15	19	24	27	32	34	38	42	11	14	18	21	24	27	29	32	1.4	1.7	2.1	2.5	2.8	1.6	2.0	2.4	3.0	3.7
WW A4	14	19	24	28	32	36	39	44	10	14	18	21	25	27	30	34	1.4	1.7	2.0	2.4	2.7	1.6	1.9	2.3	2.8	3.4
WW A5	14	19	25	28	33	37	41	46	10	14	18	21	25	28	31	35	1.3	1.6	1.9	2.3	2.6	1.5	1.8	2.2	2.7	3.3
WW A6	14	19	25	29	34	37	42	47	9	14	18	22	26	29	32	36	1.3	1.6	1.9	2.2	2.5	1.5	1.8	2.2	2.6	3.2
WW B1	12	14	16	18	20	21	21	21	9	10	11	12	13	13	13	11	2.0	2.5	3.5	-	-	2.4	3.4	-	-	-
WW B2	15	21	27	31	36	40	45	51	11	16	21	24	29	33	37	42	1.3	1.6	1.8	2.1	2.3	1.4	1.7	2.0	2.3	2.6
WW B3	18	24	31	36	42	46	52	58	14	19	25	29	35	39	43	49	1.2	1.4	1.7	1.9	2.1	1.3	1.6	1.8	2.1	2.4
WW B4	18	26	34	39	47	52	58	66	14	20	28	33	40	45	51	58	1.1	1.3	1.5	1.7	1.9	1.2	1.4	1.6	1.9	2.1
WW B5	18	25	34	40	48	53	59	68	13	20	28	33	41	46	52	60	1.1	1.3	1.5	1.7	1.9	1.2	1.4	1.6	1.8	2.0
WW C1	20	26	33	38	45	49	55	62	15	21	27	32	38	42	47	53	1.2	1.4	1.6	1.8	2.0	1.3	1.5	1.8	2.0	2.3
WW C2	18	25	33	38	45	50	56	63	14	20	27	32	38	43	48	55	1.2	1.4	1.6	1.8	2.0	1.2	1.5	1.7	1.9	2.1
WW C3	22	31	40	47	56	61	68	76	17	25	34	41	49	55	62	70	1.0	1.2	1.4	1.6	1.7	1.1	1.3	1.5	1.7	1.8
WW C4	20	29	38	44	53	59	66	74	15	23	32	38	47	53	59	68	1.0	1.2	1.4	1.6	1.8	1.1	1.3	1.5	1.7	1.8
TW A1	16	19	21	23	26	27	29	32	14	15	17	19	21	22	23	25	1.9	2.3	2.7	3.2	3.8	2.3	2.7	3.3	4.1	-
TW A2	15	19	22	25	29	31	34	38	12	15	18	20	23	24	27	29	1.6	1.9	2.3	2.7	3.0	1.9	2.2	2.6	3.2	3.8
TW A3	14	18	22	26	30	32	36	40	11	14	17	20	23	25	28	31	1.5	1.8	2.1	2.5	2.8	1.7	2.1	2.4	2.9	3.5
TW A4	13	17	22	25	30	33	37	42	10	13	17	20	23	26	29	33	1.5	1.8	2.1	2.4	2.7	1.7	2.0	2.3	2.7	3.2
TW B1	15	18	21	23	26	28	31	34	13	15	17	19	21	22	24	26	1.8	2.1	2.5	2.9	3.3	2.1	2.5	2.9	3.5	4.2
TW B2	14	18	21	24	28	31	34	39	11	14	17	19	22	24	27	30	1.6	1.9	2.2	2.5	2.8	1.8	2.1	2.5	2.9	3.4
TW B3	13	17	21	24	28	31	35	40	10	13	17	19	22	25	28	32	1.6	1.8	2.1	2.4	2.7	1.8	2.1	2.4	2.8	3.2
TW B4	13	16	20	23	28	31	34	39	10	13	16	18	22	24	28	32	1.6	1.8	2.1	2.4	2.7	1.8	2.1	2.4	2.7	3.1
TW C1	14	17	20	23	26	29	32	36	12	14	16	18	21	23	25	28	1.8	2.0	2.3	2.7	3.0	2.0	2.3	2.7	3.1	3.5
TW C2	13	16	20	23	27	29	33	38	10	13	16	18	21	24	27	30	1.7	1.9	2.2	2.5	2.8	1.9	2.2	2.5	2.8	3.2
TW C3	13	16	19	22	26	29	32	37	10	13	16	18	21	23	26	30	1.7	2.0	2.2	2.5	2.8	1.9	2.2	2.5	2.8	3.1
TW C4	13	16	19	21	25	27	31	35	11	13	15	17	20	22	25	29	1.8	2.1	2.3	2.6	2.8	2.1	2.3	2.6	2.9	3.1
TW D1	10	12	15	16	18	20	21	22	7	9	10	11	13	13	14	13	2.0	2.5	3.1	4.3	-	2.5	3.1	4.5	-	-
TW D2	16	20	25	29	35	38	42	48	12	16	20	23	28	31	35	40	1.4	1.6	1.9	2.2	2.4	1.5	1.8	2.1	2.4	2.7
TW D3	16	21	27	31	37	40	45	51	12	17	21	25	30	33	37	42	1.3	1.6	1.8	2.1	2.3	1.5	1.7	2.0	2.3	2.6
TW D4	18	24	31	36	43	47	53	61	13	19	25	30	36	41	46	53	1.2	1.4	1.6	1.8	2.0	1.3	1.5	1.7	2.0	2.2
TW D5	17	24	32	37	44	49	55	63	13	19	26	31	38	43	49	57	1.2	1.4	1.6	1.8	1.9	1.3	1.5	1.7	1.9	2.0
TW E1	19	25	31	35	41	45	51	57	15	20	25	29	35	38	43	49	1.2	1.5	1.7	2.0	2.1	1.4	1.6	1.9	2.1	2.4
TW E2	18	25	31	36	43	47	53	60	14	20	26	30	36	40	45	52	1.2	1.4	1.7	1.9	2.1	1.3	1.5	1.8	2.0	2.2
TW E3	22	30	38	44	52	57	63	71	17	24	32	38	46	51	57	65	1.1	1.3	1.5	1.7	1.8	1.1	1.3	1.5	1.8	1.9
TW E4	20	28	37	43	51	56	63	71	16	23	31	37	44	50	56	64	1.1	1.3	1.5	1.7	1.8	1.2	1.3	1.5	1.7	1.9
TW F1	14	17	20	22	24	26	28	31	12	14	16	17	19	20	22	23	1.9	2.2	2.6	3.1	3.7	2.2	2.6	3.2	4.0	-
TW F2	13	16	20	23	26	29	32	36	10	12	15	17	20	22	24	27	1.7	1.9	2.3	2.7	3.1	1.9	2.2	2.7	3.2	3.9
TW F3	12	15	19	22	27	29	33	37	8	11	14	17	20	22	25	28	1.6	1.9	2.2	2.5	2.9	1.8	2.1	2.5	3.0	3.5
TW F4	10	13	18	21	26	28	32	37	6	9	13	16	19	22	25	29	1.6	1.8	2.1	2.4	2.7	1.8	2.1	2.4	2.8	3.2
TW G1	12	15	18	20	23	25	28	31	10	12	14	16	18	20	21	24	1.8	2.1	2.5	2.9	3.3	2.1	2.5	2.9	3.5	4.2
TW G2	10	14	17	20	24	27	30	34	7	10	13	15	18	20	23	26	1.7	2.0	2.3	2.6	2.9	1.9	2.2	2.6	3.0	3.5
TW G3	9	12	16	19	23	26	30	34	6	9	12	14	17	20	23	26	1.7	1.9	2.2	2.5	2.8	1.9	2.2	2.5	2.9	3.3
TW G4	7	10	14	17	21	23	27	32	4	7	10	12	15	18	21	25	1.7	2.0	2.2	2.5	2.8	2.0	2.2	2.5	2.8	3.1
TW H1	9	12	15	18	21	24	27	30	7	9	12	13	16	18	20	23	1.8	2.1	2.4	2.8	3.1	2.1	2.4	2.8	3.2	3.7
TW H2	7	10	14	16	20	22	26	30	5	7	10	12	15	17	20	23	1.8	2.1	2.3	2.6	2.9	2.0	2.3	2.6	3.0	3.3
TW H3	6	9	12	14	18	21	24	29	4	6	8	10	13	15	18	22	1.9	2.1	2.4	2.7	2.9	2.1	2.4	2.7	3.0	3.3
TW H4	4	7	9	12	15	17	20	25	3	4	7	8	11	13	15	19	2.1	2.3	2.5	2.8	3.0	2.3	2.6	2.8	3.1	3.4
TW I1	8	10	13	14	16	17	19	20	6	7	8	9	10	11	11	11	2.1	2.6	3.2	4.4	-	2.6	3.3	4.6	-	-
TW I2	13	17	22	26	31	35	39	45	9	13	17	20	25	28	32	37	1.4	1.7	1.9	2.2	2.4	1.6	1.8	2.1	2.4	2.7
TW I3	13	19	24	28	34	38	43	49	10	14	19	23	28	31	35	41	1.4	1.6	1.8	2.1	2.3	1.5	1.7	2.0	2.3	2.6
TW I4	15	21	28	33	40	45	50	58	11	16	23	27	34	38	43	51	1.2	1.4	1.6	1.9	2.0	1.3	1.5	1.8	2.0	2.2
TW I5	15	22	29	34	42	47	53	61	11	17	24	29	36	40	46	54	1.2	1.4	1.6	1.8	2.0	1.3	1.5	1.7	1.9	2.1
TW J1	17	22	29	33	40	44	49	56	13	18	23	27	33	37	41	47	1.3	1.5	1.7	2.0	2.1	1.4	1.6	1.9	2.1	2.4
TW J2	16	22	29	34	40	45	51	58	12	17	23	28	34	38	43	50	1.2	1.4	1.7	1.9	2.1	1.3	1.6	1.8	2.0	2.2
TW J3	19	27	36	42	50	55	62	70	15	22	30	36	44	49	55	64	1.1	1.3	1.5	1.7	1.8	1.2	1.3	1.5	1.7	1.9
TW J4	18	26	34	40	48	54	61	69	13	20	28	34	42	48	54	63	1.1	1.3	1.5	1.7	1.8	1.2	1.4	1.6	1.7	1.9
DG A1	10	13	15	17	18	19	20	20	7	9	10	11	12	12	12	11	1.9	2.5	3.7	-	-	2.2	3.1	-	-	-
DG A2	13	17	21	24	28	31	34	38	10	13	16	19	22	25	27	30	1.6	1.8	2.2	2.6	3.0	1.7	2.0	2.4	3.0	3.6
DG A3	17	22	28	32	38	42	47	54	13	18	23	27	33	36	41	47	1.3	1.5	1.7	2.0	2.2	1.5	1.6	1.9	2.1	2.4
DG B1	11	13	16	18	20	21	22	23	7	9	11	12	14	14	14	14	1.8	2.3	3.2	-	-	2.2	2.8	4.2	-	-
DG B2	13	18	22	25	30	33	36	41	10	13	17	20	24	26	29	33	1.5	1.8	2.1	2.4	2.8	1.7	1.9	2.3	2.8	3.2
DG B3	17	23	29	33	40	44	50	57	14	18	24	28	34	39	44	51	1.3	1.5	1.7	1.9	2.1	1.4	1.6	1.8	2.0	2.2
DG C1	13	17	20	22	25	27	29	31	9	12	14	16	19	20	21	21	1.6	2.0	2.7	3.5	-	1.9	2.3	3.1	-	-
DG C2	16	21	26	30	35	38	43	48	12	16	20	24	29	32	35	40	1.4	1.6	1.9	2.2	2.5	1.6	1.8	2.1	2.4	2.8
DG C3	19	26	32	37	44	49	55	63	16	21	27	32	39	43	49	56	1.2	1.4	1.6	1.8	2.0	1.3	1.5	1.7	1.9	2.1
SS A1	18	22	26	28	32	35	37	41	16	19	22	25	28	30	32	35	1.5	1.9	2.2	2.6	3.0	1.7	2.0	2.5	3.1	3.8
SS A2	22	28	34	38	44	48	52	59	20	25	30	34	39	43	47	53	1.2	1.5	1.7	2.0	2.1	1.3	1.6	1.8	2.1	2.3
SS A3	17	20	23	25	28	30	32	34	15	17	20	21	24	25	26	27	1.7	2.1	2.6	3.3	4.1	1.8	2.3	3.1	4.3	-
SS A4	22	27	32	36	42	45	50	56	19	24	29	32	37	40	44	49	1.3	1.5	1.8	2.0	2.3	1.4	1.6	1.9	2.2	2.5
SS A5	19	21	25	27	30	32	34	37	17	19	22	23	26	27	29	31	1.7	2.0	2.5	3.0	3.6	1.8	2.3	2.8	3.6	4.8
SS A6	22	27	33	37	43	46	51	57	19	24	29	33	38	41	46	51	1.3	1.5	1.8	2.0	2.2	1.3	1.6	1.9	2.2	2.4
SS A7	17	19	22	23	25	27	28	30	15	17	19	20	22	23	24	24	1.9	2.4	3.0	3.9	-	2.1	2.7	3.6	-	-
SS A8	21	26	31	35	40	43	48	53	19	23	28	31	35	38	42	46	1.3	1.6	1.9	2.1	2.4	1.4	1.7	2.0	2.4	2.7
SS B1	16	19	22	25	28	31	33	37	13	16	19	21	24	26	28	30	1.6	2.0	2.3	2.8	3.2	1.8	2.2	2.6	3.2	3.9
SS B2	20	25	31	35	41	45	50	56	17	22	27	31	36	40	44	50	1.3	1.5	1.8	2.0	2.2	1.4	1.6	1.9	2.1	2.3
SS B3	14	17	20	22	25	27	29	32	12	15	17	19	21	22	24	25	1.7	2.1	2.6	3.2	3.8	1.9	2.4	3.0	4.0	-
SS B4	19	24	29	33	39	43	47	53	16	21	26	29	34	38	42	47	1.3	1.6	1.8	2.1	2.2	1.4	1.7	1.9	2.2	2.5
SS B5	15	18	21	23	25	27	30	32	13	16	18	19	22	23	25	26	1.8	2.2	2.6	3.1	3.7	2.0	2.5	3.0	3.8	4.8
SS B6	19	24	30	34	39	43	48	54	16	21	26	29	35	38	42	48	1.3	1.5	1.8	2.0	2.2	1.4	1.6	1.9	2.2	2.4
SS B7	14	16	18	20	22	24	25	27	12	14	16	17	19	20	21	22	2.0	2.5	3.0	3.7	4.6	2.3	2.8	3.5	4.9	-
SS B8	18	23	28	32	37	40	45	51	15	20	24	28	32	35	39	44	1.4	1.6	1.9	2.1	2.3	1.4	1.7	2.0	2.3	2.6
SS C1	13	17	21	24	27	30	33	36	10	12	15	18	20	22	24	26	1.6	1.9	2.3	2.7	3.2	1.8	2.2	2.7	3.5	4.7
SS C2	16	21	27	31	37	41	46	52	12	17	22	25	30	34	38	43	1.3	1.5	1.8	2.1	2.3	1.4	1.7	2.0	2.3	2.6
SS C3	13	16	19	21	24	27	29	33	10	12	15	16	19	21	22	25	1.8	2.1	2.5	2.9	3.3	2.1	2.4	2.9	3.5	4.2
SS C4	15	20	25	29	34	38	42	48	12	16	20	23	28	31	35	40	1.4	1.6	1.9	2.2	2.4	1.5	1.8	2.1	2.4	2.6
SS D1	20	24	29	32	36	38	41	45	17	21	24	27	30	32	35	37	1.4	1.7	2.1	2.5	2.9	1.5	1.9	2.4	3.1	4.0
SS D2	24	31	38	43	49	54	59	66	21	27	33	38	44	48	53	59	1.1	1.4	1.6	1.8	2.0	1.2	1.4	1.7	1.9	2.2
SS D3	19	23	27	30	33	36	39	42	17	20	24	26	29	31	33	35	1.5	1.8	2.2	2.7	3.1	1.6	2.0	2.5	3.2	4.2
SS D4	23	30	37	41	48	52	57	64	20	26	32	37	43	47	51	57	1.2	1.4	1.6	1.8	2.0	1.2	1.5	1.7	2.0	2.2
SS E1	16	20	24	27	30	32	35	38	13	17	20	22	24	26	27	29	1.5	1.9	2.3	2.9	3.6	1.7	2.1	2.8	4.0	-
SS E2	20	26	32	37	43	47	52	58	17	22	28	32	37	41	45	50	1.2	1.5	1.7	2.0	2.1	1.3	1.6	1.9	2.2	2.4
SS E3	16	19	22	25	28	30	32	35	14	16	19	21	23	25	26	28	1.7	2.0	2.5	3.0	3.6	1.9	2.3	2.9	3.8	-
SS E4	19	25	31	35	40	44	49	55	16	21	26	30	35	38	43	48	1.3	1.5	1.8	2.0	2.2	1.4	1.6	1.9	2.2	2.5

YAKIMA, WASHINGTON

TYPE	SOLAR SAVINGS FRACTION (%) BASE TEMP: 55 F (3488 DD)								65 F (6014 DD)								CONSERVATION FACTOR 55 F					65 F				
LCR	100	70	50	40	30	25	20	15	100	70	50	40	30	25	20	15	100	50	30	20	15	100	50	30	20	15
WW A1	19	21	25	27	30	32	35	38	16	18	20	22	24	26	27	30	1.7	2.0	2.4	2.8	3.2	2.0	2.4	2.9	3.5	4.3
WW A2	19	23	28	31	36	39	43	48	15	18	22	25	29	31	34	38	1.4	1.7	2.0	2.3	2.5	1.6	1.9	2.2	2.7	3.1
WW A3	18	24	29	33	38	42	46	52	14	18	23	26	30	33	37	41	1.3	1.6	1.8	2.1	2.3	1.5	1.8	2.1	2.5	2.9
WW A4	18	24	30	34	39	43	48	54	13	18	23	26	31	34	38	43	1.3	1.5	1.8	2.0	2.3	1.4	1.7	2.0	2.4	2.7
WW A5	18	24	30	35	41	45	50	56	13	18	23	27	32	35	39	44	1.2	1.5	1.7	2.0	2.2	1.4	1.6	2.0	2.3	2.7
WW A6	18	24	31	35	41	46	51	57	13	18	23	27	33	36	40	45	1.2	1.5	1.7	2.0	2.2	1.3	1.6	1.9	2.3	2.6
WW B1	15	18	21	23	26	28	29	32	11	13	15	17	18	19	20	20	1.7	2.1	2.7	3.4	4.4	2.0	2.6	3.6	-	-
WW B2	19	25	32	37	43	47	53	60	14	20	25	29	35	39	44	50	1.2	1.4	1.7	1.9	2.1	1.3	1.6	1.8	2.1	2.3
WW B3	22	29	37	42	49	54	60	67	17	23	30	35	41	45	51	57	1.1	1.3	1.6	1.8	1.9	1.2	1.4	1.7	1.9	2.2
WW B4	22	31	39	45	53	59	65	73	17	25	32	38	46	51	57	65	1.1	1.3	1.5	1.6	1.8	1.1	1.3	1.5	1.8	1.9
WW B5	22	30	39	45	54	59	66	74	16	24	32	38	47	52	58	67	1.0	1.2	1.4	1.6	1.8	1.1	1.3	1.5	1.7	1.9
WW C1	24	31	39	44	52	57	62	70	19	25	32	37	44	48	54	61	1.1	1.3	1.5	1.7	1.9	1.2	1.4	1.6	1.9	2.1
WW C2	22	30	38	44	52	57	63	71	17	24	32	37	44	49	55	62	1.1	1.3	1.5	1.7	1.8	1.2	1.4	1.6	1.8	2.0
WW C3	26	36	46	53	62	67	74	82	21	30	39	46	55	61	67	75	1.0	1.2	1.3	1.5	1.7	1.0	1.2	1.4	1.6	1.8
WW C4	24	33	43	50	59	65	72	80	19	27	37	43	52	58	65	73	1.0	1.2	1.4	1.5	1.7	1.1	1.2	1.4	1.6	1.8
TW A1	18	21	24	27	30	32	35	39	16	18	20	22	24	26	28	31	1.7	2.0	2.3	2.7	3.1	2.0	2.3	2.7	3.2	3.8
TW A2	18	22	26	30	34	37	41	46	14	18	21	24	27	30	33	37	1.5	1.7	2.0	2.3	2.6	1.7	2.0	2.3	2.7	3.1
TW A3	17	22	27	31	36	39	43	49	14	17	21	24	28	31	35	39	1.4	1.6	1.9	2.2	2.4	1.6	1.8	2.2	2.5	2.9
TW A4	16	21	27	31	36	39	44	50	12	16	21	24	29	32	35	40	1.4	1.6	1.9	2.1	2.3	1.5	1.8	2.1	2.4	2.7
TW B1	18	21	24	27	31	33	36	41	15	17	20	22	25	27	29	33	1.6	1.9	2.2	2.5	2.8	1.9	2.2	2.5	3.0	3.4
TW B2	17	21	25	29	34	37	41	46	13	17	20	23	27	30	33	37	1.5	1.7	2.0	2.3	2.5	1.6	1.9	2.2	2.5	2.9
TW B3	16	20	25	29	34	37	42	47	12	16	20	23	27	30	34	38	1.4	1.7	1.9	2.2	2.4	1.6	1.9	2.1	2.5	2.7
TW B4	15	19	24	28	33	36	40	46	12	15	19	22	26	29	33	38	1.5	1.7	1.9	2.2	2.4	1.6	1.9	2.1	2.4	2.7
TW C1	16	20	24	27	31	34	37	42	14	16	19	22	25	27	30	34	1.6	1.8	2.1	2.4	2.6	1.8	2.1	2.4	2.7	3.0
TW C2	15	19	24	27	32	35	39	44	12	16	19	22	26	28	32	36	1.5	1.8	2.0	2.3	2.5	1.7	2.0	2.2	2.5	2.8
TW C3	15	19	23	26	31	34	38	43	12	15	19	21	25	28	31	36	1.6	1.8	2.0	2.3	2.5	1.8	2.0	2.3	2.5	2.8
TW C4	15	18	22	24	29	31	35	41	12	15	18	20	24	26	29	34	1.7	1.9	2.1	2.4	2.6	1.9	2.1	2.4	2.6	2.8
TW D1	12	15	19	21	24	26	28	31	9	11	14	15	17	18	20	21	1.7	2.1	2.6	3.1	3.7	2.1	2.6	3.2	4.4	-
TW D2	19	24	30	34	40	44	49	56	15	19	24	28	33	37	41	47	1.3	1.5	1.7	2.0	2.1	1.4	1.7	1.9	2.2	2.4
TW D3	20	26	32	37	43	47	52	59	15	20	26	30	36	39	44	50	1.2	1.5	1.7	1.9	2.1	1.3	1.6	1.8	2.1	2.4
TW D4	21	28	36	41	49	54	60	68	17	23	30	35	42	46	52	60	1.1	1.3	1.5	1.7	1.9	1.2	1.4	1.6	1.8	2.0
TW D5	21	28	36	42	50	55	62	70	16	23	30	36	43	48	54	62	1.1	1.3	1.5	1.7	1.8	1.2	1.4	1.6	1.8	1.9
TW E1	22	29	36	41	48	52	58	65	18	24	30	34	40	45	50	56	1.2	1.4	1.6	1.8	2.0	1.3	1.5	1.7	2.0	2.2
TW E2	22	29	36	42	49	54	59	67	17	23	30	35	42	46	52	59	1.1	1.3	1.5	1.7	1.9	1.2	1.4	1.7	1.9	2.1
TW E3	26	34	43	50	58	63	70	78	21	29	37	43	51	56	63	70	1.0	1.2	1.4	1.6	1.7	1.1	1.3	1.5	1.7	1.8
TW E4	24	33	42	48	57	62	69	77	19	27	35	42	50	55	62	70	1.0	1.2	1.4	1.6	1.7	1.1	1.3	1.5	1.7	1.8
TW F1	16	20	23	26	29	31	34	38	14	16	19	20	23	25	27	30	1.7	2.0	2.3	2.7	3.0	1.9	2.3	2.7	3.2	3.7
TW F2	16	20	24	27	32	35	38	43	12	15	19	21	25	27	30	34	1.5	1.8	2.0	2.3	2.6	1.7	2.0	2.3	2.7	3.1
TW F3	14	19	24	27	32	36	40	45	11	14	18	21	25	28	31	36	1.4	1.7	2.0	2.2	2.5	1.6	1.9	2.2	2.5	2.9
TW F4	12	17	22	26	31	35	39	45	9	12	17	20	24	27	31	36	1.4	1.7	1.9	2.2	2.4	1.6	1.9	2.1	2.5	2.7
TW G1	15	18	22	24	28	31	34	38	12	14	17	19	22	24	27	30	1.6	1.9	2.2	2.5	2.8	1.9	2.2	2.6	3.0	3.4
TW G2	13	17	21	25	29	32	36	42	10	13	16	19	23	25	29	33	1.5	1.8	2.0	2.3	2.6	1.7	2.0	2.3	2.6	3.0
TW G3	11	15	20	23	28	32	36	42	8	11	15	18	22	25	28	33	1.5	1.7	2.0	2.3	2.5	1.7	2.0	2.2	2.5	2.8
TW G4	9	13	17	21	25	29	33	39	6	9	13	15	19	22	26	31	1.6	1.8	2.1	2.3	2.5	1.8	2.0	2.3	2.6	2.8
TW H1	12	15	19	22	26	29	32	37	9	11	14	17	20	22	25	29	1.6	1.9	2.2	2.5	2.7	1.9	2.2	2.5	2.8	3.1
TW H2	9	13	17	20	24	28	31	37	6	9	13	15	19	21	25	29	1.6	1.9	2.1	2.4	2.6	1.8	2.1	2.4	2.7	2.9
TW H3	8	11	15	18	22	25	29	34	5	8	11	13	17	19	23	27	1.7	1.9	2.2	2.4	2.6	1.9	2.2	2.4	2.7	2.9
TW H4	6	9	12	14	18	21	25	30	4	6	9	11	14	16	19	23	1.9	2.1	2.3	2.6	2.8	2.1	2.4	2.6	2.8	3.1
TW I1	10	13	16	18	21	23	25	28	8	9	12	13	15	16	17	19	1.8	2.2	2.6	3.2	3.8	2.2	2.7	3.3	4.5	-
TW I2	16	21	27	31	37	41	46	53	12	16	21	25	30	34	38	44	1.3	1.5	1.8	2.0	2.2	1.5	1.7	1.9	2.2	2.4
TW I3	17	23	29	34	40	44	50	57	13	18	23	27	33	37	42	48	1.2	1.5	1.7	1.9	2.1	1.4	1.6	1.8	2.1	2.3
TW I4	18	25	33	38	46	51	57	65	14	20	27	32	39	44	50	57	1.2	1.4	1.6	1.7	1.9	1.2	1.4	1.7	1.9	2.0
TW I5	19	26	34	40	48	53	59	68	14	21	28	33	41	46	52	60	1.1	1.3	1.5	1.7	1.9	1.2	1.4	1.6	1.8	2.0
TW J1	20	27	34	39	46	50	56	63	16	21	28	32	39	43	48	55	1.2	1.4	1.6	1.8	2.0	1.3	1.5	1.7	2.0	2.2
TW J2	19	26	34	39	46	51	57	65	15	21	28	33	39	44	50	57	1.1	1.4	1.6	1.8	1.9	1.3	1.5	1.7	1.9	2.1
TW J3	23	32	41	47	56	61	68	76	18	26	35	41	49	55	61	69	1.0	1.2	1.4	1.6	1.7	1.1	1.3	1.5	1.7	1.8
TW J4	22	30	39	46	54	60	67	75	17	24	33	39	47	53	60	68	1.0	1.2	1.4	1.6	1.7	1.1	1.3	1.5	1.7	1.8
DG A1	13	16	20	22	24	26	27	28	9	12	14	15	17	18	18	18	1.6	2.1	2.9	4.4	-	1.9	2.6	3.8	-	-
DG A2	16	20	25	29	33	36	40	45	12	16	20	23	27	30	33	37	1.4	1.7	2.0	2.3	2.7	1.6	1.9	2.2	2.6	3.1
DG A3	19	25	32	37	43	48	53	60	16	21	27	31	37	41	46	53	1.2	1.4	1.7	1.9	2.0	1.4	1.5	1.7	2.0	2.2
DG B1	14	17	21	23	26	28	30	31	9	12	15	17	19	20	21	22	1.6	2.0	2.6	3.6	-	1.9	2.3	3.2	-	-
DG B2	16	21	26	30	35	39	43	48	12	16	21	24	29	32	35	40	1.4	1.6	1.9	2.2	2.5	1.6	1.8	2.1	2.4	2.8
DG B3	20	26	33	38	45	50	56	64	16	21	28	32	39	44	49	56	1.2	1.4	1.6	1.8	1.9	1.3	1.5	1.7	1.9	2.1
DG C1	16	21	25	28	32	34	37	40	12	15	19	21	24	26	28	30	1.4	1.8	2.2	2.8	3.7	1.7	2.0	2.6	3.7	-
DG C2	19	24	30	35	41	45	49	55	15	19	24	28	34	37	42	47	1.3	1.5	1.7	2.0	2.3	1.4	1.6	1.9	2.2	2.5
DG C3	23	29	37	42	50	55	61	69	18	24	31	36	44	48	54	62	1.2	1.3	1.5	1.7	1.8	1.3	1.4	1.6	1.8	2.0
SS A1	21	25	29	33	37	40	43	48	18	22	26	28	32	35	37	41	1.4	1.7	2.0	2.3	2.6	1.5	1.9	2.2	2.6	3.1
SS A2	25	31	38	43	49	53	58	65	22	28	34	38	44	48	52	59	1.2	1.4	1.6	1.8	2.0	1.2	1.5	1.7	2.0	2.2
SS A3	19	23	26	29	33	35	38	41	17	20	23	25	28	30	32	34	1.5	1.9	2.3	2.7	3.0	1.7	2.1	2.6	3.3	4.2
SS A4	25	30	36	41	47	51	56	62	22	27	32	36	42	45	50	55	1.2	1.4	1.7	1.9	2.1	1.3	1.5	1.8	2.1	2.3
SS A5	21	24	28	31	34	37	40	44	19	22	25	27	30	32	34	37	1.5	1.8	2.2	2.6	2.9	1.7	2.1	2.5	3.0	3.6
SS A6	25	31	37	42	48	52	57	64	22	27	33	37	43	46	51	57	1.2	1.4	1.6	1.8	2.0	1.3	1.5	1.8	2.0	2.2
SS A7	19	22	25	27	29	31	33	36	17	19	22	23	25	27	28	30	1.7	2.1	2.6	3.1	3.7	1.9	2.4	3.0	3.9	-
SS A8	24	30	35	40	45	49	54	60	21	26	31	35	40	43	48	53	1.2	1.5	1.7	1.9	2.1	1.3	1.6	1.9	2.2	2.4
SS B1	18	22	26	29	33	36	39	44	15	19	22	25	28	30	33	37	1.5	1.8	2.1	2.4	2.7	1.6	2.0	2.3	2.8	3.2
SS B2	22	28	35	39	46	50	55	62	20	25	30	35	41	44	49	56	1.2	1.4	1.6	1.8	2.0	1.3	1.5	1.8	2.0	2.2
SS B3	17	20	24	26	30	32	35	39	14	17	20	22	25	27	29	32	1.6	1.9	2.3	2.7	3.0	1.7	2.1	2.6	3.2	3.9
SS B4	22	27	33	38	44	48	53	60	19	24	29	33	39	42	47	53	1.2	1.5	1.7	1.9	2.1	1.3	1.6	1.8	2.1	2.3
SS B5	17	20	24	26	30	32	35	38	15	18	21	23	25	27	29	32	1.7	2.0	2.3	2.7	3.0	1.8	2.2	2.6	3.1	3.7
SS B6	22	27	34	38	44	48	54	61	19	24	29	33	39	43	48	54	1.2	1.5	1.7	1.9	2.1	1.3	1.6	1.8	2.0	2.2
SS B7	16	18	21	23	26	28	30	33	14	16	18	20	22	23	25	27	1.8	2.2	2.6	3.1	3.6	2.0	2.5	3.0	3.7	4.7
SS B8	21	26	32	36	42	46	51	57	18	23	28	31	37	40	45	50	1.3	1.5	1.7	2.0	2.1	1.4	1.6	1.9	2.1	2.4
SS C1	16	21	26	29	34	37	41	46	12	16	20	22	26	28	31	35	1.4	1.7	2.0	2.3	2.6	1.6	1.9	2.3	2.8	3.3
SS C2	20	26	32	37	43	48	53	60	15	20	26	30	36	40	45	51	1.2	1.4	1.7	1.9	2.1	1.3	1.6	1.8	2.1	2.3
SS C3	15	19	23	25	29	32	35	40	12	15	18	20	23	25	28	31	1.6	1.9	2.2	2.5	2.8	1.8	2.1	2.5	2.9	3.3
SS C4	18	24	30	34	40	44	49	56	14	19	24	28	33	36	41	47	1.3	1.5	1.7	2.0	2.2	1.4	1.7	1.9	2.2	2.4
SS D1	23	28	33	37	42	45	49	54	20	24	29	32	36	38	41	45	1.3	1.6	1.9	2.1	2.4	1.4	1.7	2.1	2.6	3.0
SS D2	28	35	42	48	55	60	66	73	24	31	38	43	49	54	59	65	1.1	1.3	1.5	1.7	1.8	1.1	1.4	1.6	1.8	2.0
SS D3	22	26	31	34	39	41	45	50	19	23	27	30	33	36	39	42	1.4	1.7	2.0	2.3	2.5	1.5	1.8	2.2	2.7	3.2
SS D4	27	34	41	46	54	58	64	71	23	30	37	41	48	52	57	64	1.1	1.3	1.5	1.7	1.9	1.2	1.4	1.6	1.9	2.0
SS E1	19	24	29	32	36	39	42	47	16	20	24	26	30	32	34	37	1.4	1.7	2.0	2.4	2.7	1.5	1.9	2.4	3.0	3.8
SS E2	24	30	37	42	49	53	59	66	20	26	32	37	43	47	51	57	1.1	1.4	1.6	1.8	2.0	1.2	1.5	1.7	2.0	2.2
SS E3	18	22	26	29	33	35	38	42	16	19	22	24	27	29	32	35	1.5	1.8	2.2	2.6	2.9	1.7	2.1	2.5	3.0	3.7
SS E4	23	29	35	40	46	50	55	62	19	25	30	34	40	44	48	54	1.2	1.4	1.7	1.9	2.0	1.3	1.5	1.8	2.0	2.3

CHARLESTON, WEST VIRGINIA

TYPE	SOLAR SAVINGS FRACTION (%) BASE TEMP: 55 F (2590 DD)								65 F (4594 DD)								CONSERVATION FACTOR 55 F					65 F				
LCR	100	70	50	40	30	25	20	15	100	70	50	40	30	25	20	15	100	50	30	20	15	100	50	30	20	15
WW A1	18	21	24	27	31	34	37	43	14	15	17	19	21	23	25	27	1.7	1.9	2.1	2.3	2.5	2.3	2.6	2.9	3.4	3.7
WW A2	18	23	29	33	39	43	49	57	12	15	19	22	26	29	33	38	1.4	1.5	1.7	1.9	2.0	1.7	1.9	2.2	2.4	2.6
WW A3	18	24	30	35	42	47	53	62	11	15	20	23	28	31	36	42	1.3	1.4	1.6	1.8	1.9	1.6	1.8	2.0	2.2	2.4
WW A4	18	24	31	36	44	49	56	64	10	15	20	23	29	33	37	44	1.2	1.4	1.6	1.7	1.9	1.5	1.7	1.9	2.1	2.3
WW A5	18	25	32	38	46	51	58	67	10	15	20	24	30	34	39	47	1.2	1.3	1.5	1.7	1.8	1.4	1.6	1.8	2.1	2.2
WW A6	18	25	33	39	47	53	60	69	10	15	20	25	31	35	40	48	1.2	1.3	1.5	1.7	1.8	1.4	1.6	1.8	2.0	2.2
WW B1	14	18	21	24	28	31	35	41	9	10	12	14	15	16	18	19	1.6	1.9	2.1	2.3	2.5	2.3	2.8	3.5	4.5	-
WW B2	19	25	33	39	47	53	60	69	11	16	22	27	33	37	43	51	1.2	1.3	1.5	1.7	1.8	1.4	1.6	1.8	1.9	2.1
WW B3	23	31	39	46	55	61	68	77	14	20	27	33	40	45	52	60	1.1	1.2	1.4	1.6	1.7	1.2	1.4	1.6	1.8	1.9
WW B4	23	32	41	49	58	65	72	81	14	21	30	36	45	50	58	67	1.0	1.2	1.3	1.5	1.6	1.2	1.3	1.5	1.7	1.8
WW B5	22	31	41	48	58	65	72	81	13	21	30	36	45	51	59	68	1.0	1.2	1.3	1.5	1.6	1.2	1.3	1.5	1.6	1.8
WW C1	24	32	42	48	58	64	71	79	16	22	30	35	43	48	55	64	1.0	1.2	1.4	1.5	1.7	1.2	1.4	1.6	1.7	1.9
WW C2	22	31	40	47	57	63	70	79	14	21	29	35	43	48	55	65	1.0	1.2	1.4	1.5	1.7	1.2	1.4	1.5	1.7	1.8
WW C3	27	38	49	57	67	74	81	88	18	27	37	45	54	61	69	78	0.9	1.1	1.2	1.4	1.6	1.0	1.2	1.4	1.5	1.7
WW C4	24	34	46	54	64	70	78	86	16	24	34	41	51	58	66	75	1.0	1.1	1.3	1.4	1.6	1.1	1.2	1.4	1.5	1.7
TW A1	18	21	24	27	31	34	37	43	14	15	18	19	21	23	25	28	1.7	1.9	2.1	2.3	2.5	2.2	2.5	2.8	3.2	3.5
TW A2	17	22	27	30	36	40	45	52	12	15	18	21	24	27	30	35	1.4	1.6	1.8	2.0	2.1	1.8	2.0	2.3	2.5	2.7
TW A3	17	22	27	32	38	42	48	56	11	14	18	21	26	29	33	38	1.4	1.5	1.7	1.9	2.0	1.7	1.9	2.1	2.3	2.5
TW A4	16	21	27	32	39	43	49	58	10	13	18	21	26	29	34	40	1.3	1.5	1.7	1.8	2.0	1.6	1.8	2.0	2.2	2.4
TW B1	17	20	24	27	32	35	39	45	13	15	17	19	22	24	27	30	1.6	1.8	2.0	2.2	2.3	2.1	2.3	2.6	2.9	3.1
TW B2	16	21	26	29	35	39	45	52	11	14	17	20	24	27	31	36	1.4	1.6	1.8	2.0	2.1	1.8	2.0	2.2	2.4	2.6
TW B3	15	20	25	30	36	40	46	53	10	13	17	20	24	27	31	37	1.4	1.6	1.8	1.9	2.1	1.7	1.9	2.1	2.4	2.5
TW B4	15	19	24	28	34	38	44	52	10	13	16	19	24	27	31	36	1.4	1.6	1.8	1.9	2.1	1.8	2.0	2.2	2.4	2.5
TW C1	16	19	24	27	32	35	40	46	12	14	17	19	22	24	28	32	1.6	1.8	2.0	2.1	2.3	2.0	2.2	2.4	2.7	2.9
TW C2	15	19	23	27	32	36	41	49	10	13	16	19	23	25	29	34	1.5	1.7	1.9	2.0	2.2	1.9	2.1	2.3	2.5	2.7
TW C3	14	18	23	26	31	35	40	47	10	13	16	18	22	25	28	33	1.6	1.7	1.9	2.1	2.2	1.9	2.1	2.3	2.5	2.7
TW C4	14	17	21	24	29	32	36	43	11	13	15	18	21	23	27	31	1.7	1.9	2.0	2.2	2.3	2.0	2.2	2.4	2.6	2.8
TW D1	12	15	19	21	25	28	32	37	7	9	11	12	14	15	17	19	1.7	2.0	2.2	2.4	2.6	2.4	2.8	3.3	3.9	4.7
TW D2	18	24	31	36	43	48	55	64	12	16	21	25	31	35	40	47	1.3	1.4	1.6	1.7	1.9	1.5	1.7	1.9	2.1	2.2
TW D3	19	26	33	39	47	52	59	68	12	17	23	27	33	38	43	51	1.2	1.3	1.5	1.7	1.8	1.4	1.6	1.8	2.0	2.1
TW D4	21	29	37	44	53	59	66	75	14	20	27	32	40	45	52	61	1.1	1.3	1.4	1.6	1.7	1.3	1.4	1.6	1.8	1.9
TW D5	21	29	38	44	53	60	67	76	14	20	27	33	41	47	54	63	1.1	1.2	1.4	1.5	1.7	1.2	1.4	1.5	1.7	1.8
TW E1	22	30	38	44	52	58	65	74	15	21	27	32	39	44	50	58	1.1	1.3	1.4	1.6	1.7	1.3	1.5	1.7	1.8	2.0
TW E2	22	29	38	44	53	59	66	75	15	20	27	32	40	45	51	60	1.1	1.3	1.4	1.6	1.7	1.3	1.4	1.6	1.8	1.9
TW E3	26	36	46	54	63	70	77	85	18	26	35	41	50	56	64	73	1.0	1.1	1.3	1.5	1.6	1.1	1.3	1.4	1.6	1.7
TW E4	24	33	44	51	61	67	75	83	16	24	33	39	48	55	62	72	1.0	1.1	1.3	1.5	1.6	1.1	1.3	1.4	1.6	1.7
TW F1	16	19	23	26	30	33	37	43	12	14	16	18	20	22	24	27	1.7	1.9	2.1	2.3	2.4	2.2	2.5	2.8	3.1	3.4
TW F2	15	19	24	28	33	37	43	50	10	13	16	18	22	25	28	33	1.5	1.7	1.8	2.0	2.1	1.8	2.1	2.3	2.6	2.8
TW F3	14	19	24	28	34	39	44	52	8	11	15	18	22	25	29	35	1.4	1.6	1.8	1.9	2.1	1.7	2.0	2.2	2.4	2.6
TW F4	12	17	22	27	33	38	43	52	6	10	14	17	21	24	29	35	1.4	1.6	1.7	1.9	2.1	1.7	1.9	2.1	2.3	2.5
TW G1	14	17	21	24	29	32	37	43	10	12	14	16	19	21	24	28	1.6	1.8	2.0	2.2	2.4	2.1	2.3	2.6	2.9	3.1
TW G2	12	16	21	25	31	35	40	47	7	10	13	16	20	22	26	31	1.5	1.7	1.9	2.0	2.2	1.9	2.1	2.3	2.6	2.7
TW G3	11	15	20	24	29	34	39	47	6	9	12	15	19	22	25	31	1.5	1.7	1.9	2.0	2.2	1.8	2.0	2.3	2.5	2.7
TW G4	8	12	17	21	26	30	35	42	4	7	10	12	16	19	23	28	1.6	1.8	1.9	2.1	2.2	1.9	2.1	2.3	2.5	2.7
TW H1	11	14	19	22	26	30	34	41	7	9	12	14	17	19	22	27	1.6	1.8	2.0	2.2	2.3	2.1	2.3	2.5	2.8	3.0
TW H2	9	12	17	20	25	28	33	40	4	7	10	12	16	18	21	26	1.6	1.8	2.0	2.2	2.3	2.0	2.2	2.4	2.7	2.8
TW H3	7	10	14	17	22	26	30	37	3	6	8	10	14	16	20	24	1.7	1.9	2.1	2.2	2.4	2.1	2.3	2.5	2.7	2.9
TW H4	5	8	11	14	18	21	25	31	2	4	6	8	11	13	16	20	1.9	2.1	2.3	2.4	2.6	2.3	2.5	2.7	2.9	3.1
TW I1	10	13	16	19	22	25	29	34	5	7	9	10	12	13	14	16	1.8	2.1	2.3	2.5	2.7	2.5	2.9	3.5	4.2	5.0
TW I2	15	21	27	32	40	45	51	60	9	13	18	22	27	31	36	44	1.3	1.5	1.6	1.8	1.9	1.5	1.7	1.9	2.1	2.2
TW I3	16	23	30	35	43	49	56	65	10	15	20	24	31	35	40	48	1.2	1.4	1.5	1.7	1.8	1.4	1.6	1.8	2.0	2.1
TW I4	18	26	34	40	49	55	63	72	11	17	24	29	37	42	49	58	1.1	1.3	1.4	1.6	1.7	1.3	1.5	1.6	1.8	1.9
TW I5	18	26	35	42	51	57	65	74	11	18	25	31	39	44	51	61	1.1	1.3	1.4	1.6	1.7	1.3	1.4	1.6	1.7	1.9
TW J1	20	27	35	42	50	56	63	72	13	18	25	30	37	41	48	56	1.1	1.3	1.4	1.6	1.7	1.3	1.5	1.7	1.8	2.0
TW J2	19	27	35	41	50	56	64	73	12	18	25	30	37	42	49	58	1.1	1.3	1.4	1.6	1.7	1.3	1.5	1.6	1.8	1.9
TW J3	23	33	43	51	61	67	75	83	15	23	32	39	48	54	62	71	1.0	1.1	1.3	1.5	1.6	1.1	1.3	1.4	1.6	1.7
TW J4	21	31	41	48	58	65	72	81	14	21	30	37	46	52	60	69	1.0	1.2	1.3	1.5	1.6	1.1	1.3	1.5	1.6	1.8
DG A1	12	16	19	22	26	28	30	33	7	9	11	12	14	14	15	15	1.6	1.9	2.4	3.1	4.1	2.2	2.8	3.8	-	-
DG A2	15	19	25	29	35	39	44	50	10	13	16	19	24	26	30	35	1.4	1.6	1.8	2.1	2.3	1.7	1.9	2.2	2.5	2.9
DG A3	18	24	32	37	46	51	58	67	13	18	23	28	34	39	45	53	1.3	1.4	1.5	1.7	1.9	1.4	1.6	1.7	1.9	2.1
DG B1	12	16	20	23	27	30	34	39	7	9	11	12	14	15	17	18	1.6	1.8	2.1	2.4	2.8	2.1	2.6	3.3	4.4	-
DG B2	15	20	26	30	37	41	47	56	10	13	17	20	25	28	32	38	1.4	1.5	1.7	1.9	2.1	1.7	1.9	2.1	2.3	2.6
DG B3	19	25	33	39	48	54	62	72	13	18	24	29	36	41	47	56	1.2	1.3	1.4	1.6	1.7	1.4	1.5	1.7	1.8	1.9
DG C1	15	19	24	29	34	38	43	49	9	12	15	17	20	22	24	27	1.5	1.6	1.8	2.1	2.3	1.8	2.2	2.7	3.0	3.8
DG C2	18	23	30	35	43	48	55	64	12	16	21	24	30	34	39	46	1.3	1.4	1.6	1.7	1.9	1.5	1.7	1.9	2.1	2.3
DG C3	21	28	37	44	53	60	69	78	15	21	28	33	40	46	53	63	1.2	1.2	1.4	1.5	1.7	1.3	1.5	1.6	1.7	1.8
SS A1	22	26	32	36	42	47	52	60	17	20	24	27	32	35	39	44	1.3	1.5	1.7	1.9	2.0	1.6	1.8	2.0	2.3	2.5
SS A2	27	34	42	48	56	61	68	76	21	26	33	38	45	49	55	63	1.1	1.3	1.4	1.6	1.7	1.2	1.4	1.6	1.8	1.9
SS A3	20	24	29	33	38	42	47	54	15	18	22	24	28	30	33	38	1.4	1.6	1.8	2.0	2.2	1.7	2.0	2.3	2.6	2.9
SS A4	26	33	41	47	55	60	66	75	20	26	32	37	43	48	53	61	1.1	1.3	1.4	1.6	1.8	1.3	1.5	1.6	1.8	2.0
SS A5	21	25	30	33	39	42	47	54	17	20	23	26	29	31	35	39	1.4	1.7	1.8	2.0	2.2	1.7	2.0	2.3	2.6	2.8
SS A6	26	33	41	47	55	61	67	75	20	26	32	37	44	48	54	62	1.1	1.3	1.4	1.6	1.7	1.3	1.4	1.6	1.8	2.0
SS A7	19	23	26	29	33	36	40	46	16	18	20	22	25	26	29	32	1.6	1.9	2.1	2.3	2.4	2.0	2.3	2.7	3.1	3.5
SS A8	26	32	40	45	53	58	65	73	20	25	31	35	41	46	51	59	1.1	1.3	1.5	1.6	1.8	1.3	1.5	1.7	1.9	2.0
SS B1	18	23	28	31	37	41	46	54	14	17	20	23	27	30	33	38	1.4	1.6	1.8	2.0	2.1	1.7	1.9	2.2	2.4	2.6
SS B2	23	30	37	43	51	57	63	72	18	23	29	34	40	45	51	59	1.2	1.3	1.5	1.6	1.8	1.3	1.5	1.7	1.8	2.0
SS B3	17	21	26	29	34	38	43	49	13	15	19	21	24	27	30	34	1.5	1.7	1.9	2.1	2.2	1.8	2.1	2.4	2.7	3.0
SS B4	23	29	36	42	50	55	62	70	17	22	28	33	39	43	49	57	1.2	1.3	1.5	1.7	1.8	1.3	1.5	1.7	1.9	2.0
SS B5	17	21	25	28	32	36	40	46	14	16	19	21	24	26	29	32	1.6	1.8	2.0	2.2	2.3	1.9	2.2	2.5	2.8	3.1
SS B6	23	29	36	42	50	55	62	71	17	22	28	32	39	43	49	57	1.2	1.3	1.5	1.6	1.8	1.3	1.5	1.7	1.9	2.0
SS B7	16	19	22	25	29	31	35	40	12	14	17	18	21	22	24	27	1.7	2.0	2.2	2.4	2.6	2.1	2.5	2.9	3.3	3.6
SS B8	22	28	35	40	48	53	59	68	16	21	26	31	37	41	46	54	1.2	1.4	1.5	1.7	1.8	1.4	1.6	1.8	1.9	2.1
SS C1	16	21	26	31	37	41	47	55	10	13	16	19	23	26	30	35	1.4	1.6	1.7	1.9	2.1	1.7	2.0	2.2	2.5	2.7
SS C2	19	26	34	39	47	53	60	69	12	17	23	27	34	38	44	52	1.2	1.3	1.5	1.7	1.8	1.4	1.6	1.7	1.9	2.1
SS C3	15	18	22	26	30	34	38	45	10	12	15	17	20	22	25	29	1.6	1.8	2.0	2.2	2.3	2.0	2.3	2.5	2.8	3.1
SS C4	18	24	30	35	42	47	54	63	12	16	21	25	30	34	39	47	1.3	1.4	1.6	1.7	1.9	1.5	1.7	1.9	2.1	2.2
SS D1	24	31	37	43	50	55	61	69	18	23	28	32	37	40	45	52	1.2	1.4	1.5	1.7	1.9	1.4	1.6	1.8	2.1	2.2
SS D2	30	39	48	55	64	70	77	84	22	30	38	44	52	57	64	72	1.0	1.1	1.3	1.5	1.6	1.1	1.3	1.5	1.6	1.8
SS D3	23	28	34	38	45	49	55	63	18	22	26	29	33	37	41	46	1.3	1.5	1.6	1.8	2.0	1.5	1.8	2.0	2.2	2.4
SS D4	29	37	47	53	62	68	75	82	22	29	36	42	50	55	62	70	1.0	1.2	1.3	1.5	1.6	1.1	1.3	1.5	1.7	1.8
SS E1	20	26	32	37	43	48	54	62	14	18	22	26	30	33	37	42	1.3	1.5	1.6	1.8	2.0	1.5	1.8	2.1	2.3	2.5
SS E2	25	33	42	48	57	63	70	78	18	24	31	37	44	49	55	63	1.1	1.2	1.4	1.5	1.7	1.2	1.4	1.6	1.8	1.9
SS E3	19	23	28	31	37	40	45	52	14	17	20	23	26	29	32	36	1.4	1.7	1.8	2.0	2.2	1.8	2.0	2.3	2.6	2.8
SS E4	24	31	39	45	53	58	65	74	17	23	29	34	41	45	51	59	1.1	1.3	1.4	1.6	1.8	1.3	1.5	1.7	1.8	2.0

HUNTINGTON, WEST VIRGINIA

TYPE	SOLAR SAVINGS FRACTION (%) BASE TEMP: 55 F (2624 DD)								65 F (4629 DD)								CONSERVATION FACTOR 55 F					65 F				
LCR	100	70	50	40	30	25	20	15	100	70	50	40	30	25	20	15	100	50	30	20	15	100	50	30	20	15
WW A1	19	22	26	29	33	36	40	46	14	16	19	20	23	25	27	30	1.6	1.8	2.0	2.2	2.4	2.1	2.4	2.7	3.0	3.3
WW A2	20	25	31	35	42	46	52	60	13	17	21	24	29	32	36	42	1.3	1.5	1.6	1.8	2.0	1.6	1.8	2.0	2.3	2.5
WW A3	20	26	33	38	45	50	57	65	12	17	22	25	31	34	39	46	1.2	1.4	1.6	1.7	1.9	1.5	1.7	1.9	2.1	2.3
WW A4	19	26	34	39	47	53	59	68	12	16	22	26	32	36	41	49	1.2	1.3	1.5	1.7	1.8	1.4	1.6	1.8	2.0	2.2
WW A5	20	27	35	41	49	55	62	71	11	17	22	27	33	38	43	51	1.1	1.3	1.5	1.6	1.8	1.4	1.6	1.8	1.9	2.1
WW A6	20	27	35	42	50	56	63	72	11	17	23	27	34	39	45	53	1.1	1.3	1.4	1.6	1.8	1.3	1.5	1.7	1.9	2.1
WW B1	15	19	23	27	31	35	39	45	10	12	14	16	18	19	21	23	1.6	1.8	2.0	2.2	2.4	2.1	2.5	3.0	3.6	4.2
WW B2	20	27	36	42	50	56	63	72	13	18	24	29	36	40	47	55	1.1	1.3	1.4	1.6	1.8	1.3	1.5	1.7	1.9	2.0
WW B3	24	33	42	49	58	64	71	80	16	22	30	35	43	48	55	64	1.0	1.2	1.4	1.5	1.7	1.2	1.4	1.6	1.7	1.9
WW B4	24	34	44	51	61	68	75	83	16	23	32	38	47	53	61	70	1.0	1.1	1.3	1.5	1.6	1.1	1.3	1.5	1.6	1.8
WW B5	23	33	43	51	61	67	75	83	15	23	32	38	48	54	61	71	1.0	1.1	1.3	1.5	1.6	1.1	1.3	1.4	1.6	1.7
WW C1	26	35	44	51	61	67	74	82	17	24	32	38	46	52	58	67	1.0	1.2	1.3	1.5	1.6	1.2	1.3	1.5	1.7	1.8
WW C2	24	33	43	50	60	66	73	82	16	23	31	37	46	51	59	68	1.0	1.2	1.3	1.5	1.6	1.2	1.3	1.5	1.7	1.8
WW C3	29	40	52	60	70	76	83	90	20	29	40	47	57	64	71	80	0.9	1.1	1.2	1.4	1.5	1.0	1.2	1.3	1.5	1.6
WW C4	26	37	48	56	67	73	80	88	17	26	36	44	54	60	68	78	0.9	1.1	1.2	1.4	1.6	1.0	1.2	1.4	1.5	1.7
TW A1	19	22	26	29	33	36	40	46	14	16	19	20	23	25	28	31	1.6	1.8	2.0	2.2	2.3	2.1	2.4	2.6	2.9	3.2
TW A2	19	23	28	33	39	43	48	56	13	16	20	23	27	30	33	39	1.4	1.6	1.7	1.9	2.0	1.7	1.9	2.2	2.4	2.6
TW A3	18	23	29	34	41	45	52	60	12	16	20	23	28	31	36	42	1.3	1.5	1.6	1.8	2.0	1.6	1.8	2.0	2.2	2.4
TW A4	17	23	29	34	41	46	53	61	11	15	19	23	28	32	37	44	1.3	1.5	1.6	1.8	1.9	1.5	1.7	1.9	2.1	2.3
TW B1	18	21	26	29	34	37	42	48	13	16	18	21	24	26	29	33	1.6	1.8	1.9	2.1	2.2	2.0	2.2	2.4	2.7	2.9
TW B2	17	22	27	32	38	42	48	56	12	15	19	22	26	29	33	39	1.4	1.6	1.7	1.9	2.0	1.7	1.9	2.1	2.3	2.5
TW B3	17	22	27	32	38	43	49	57	11	14	19	22	26	30	34	41	1.4	1.5	1.7	1.9	2.0	1.6	1.8	2.0	2.2	2.4
TW B4	16	20	26	30	36	41	47	55	10	14	18	21	26	29	33	40	1.4	1.6	1.7	1.9	2.0	1.7	1.9	2.1	2.3	2.4
TW C1	17	21	25	28	34	37	42	49	12	15	18	20	24	26	30	35	1.5	1.7	1.9	2.0	2.2	1.9	2.1	2.3	2.5	2.7
TW C2	16	20	25	29	35	39	44	52	11	14	18	20	24	27	31	37	1.5	1.6	1.8	2.0	2.1	1.8	2.0	2.2	2.4	2.5
TW C3	15	19	24	28	33	37	42	50	11	14	17	20	24	27	31	36	1.5	1.7	1.8	2.0	2.1	1.8	2.0	2.2	2.4	2.6
TW C4	15	18	22	26	30	34	39	45	11	13	16	19	22	25	28	34	1.6	1.8	2.0	2.1	2.3	2.0	2.1	2.3	2.5	2.7
TW D1	13	16	20	23	28	31	35	41	8	10	12	14	16	18	20	22	1.6	1.9	2.1	2.3	2.4	2.2	2.5	2.9	3.4	3.8
TW D2	20	26	33	38	46	51	58	67	13	18	23	27	33	37	43	50	1.2	1.4	1.5	1.7	1.8	1.4	1.6	1.8	2.0	2.1
TW D3	21	28	36	42	50	55	62	71	14	19	25	30	36	41	47	55	1.1	1.3	1.5	1.6	1.8	1.3	1.5	1.7	1.9	2.0
TW D4	23	31	40	46	56	62	69	78	15	21	29	34	42	48	55	64	1.1	1.2	1.4	1.5	1.7	1.2	1.4	1.5	1.7	1.8
TW D5	22	30	40	47	56	62	70	79	15	22	29	35	44	49	57	66	1.1	1.2	1.4	1.5	1.7	1.2	1.3	1.5	1.7	1.8
TW E1	24	32	40	47	55	61	68	77	16	22	29	34	42	47	53	61	1.1	1.2	1.4	1.6	1.7	1.3	1.4	1.6	1.8	1.9
TW E2	23	31	40	47	56	62	69	78	16	22	29	35	42	48	54	63	1.1	1.2	1.4	1.5	1.7	1.2	1.4	1.6	1.7	1.9
TW E3	28	38	49	56	66	72	79	87	19	28	37	44	53	59	67	76	1.0	1.1	1.3	1.4	1.6	1.1	1.2	1.4	1.6	1.7
TW E4	26	36	46	54	64	70	77	85	18	26	35	42	51	57	65	74	1.0	1.1	1.3	1.4	1.6	1.1	1.2	1.4	1.6	1.7
TW F1	17	20	24	27	32	35	40	46	12	15	17	19	22	24	27	30	1.6	1.8	2.0	2.2	2.3	2.1	2.3	2.6	2.9	3.1
TW F2	16	21	26	30	36	40	46	53	11	14	17	20	24	27	31	36	1.4	1.6	1.8	1.9	2.1	1.8	2.0	2.2	2.4	2.6
TW F3	15	20	26	30	37	42	48	56	9	13	17	20	25	28	32	38	1.4	1.5	1.7	1.9	2.0	1.7	1.9	2.1	2.3	2.5
TW F4	13	18	24	29	36	40	47	55	7	11	15	19	23	27	32	38	1.4	1.5	1.7	1.8	2.0	1.6	1.8	2.0	2.2	2.4
TW G1	15	19	23	26	31	35	39	46	10	13	15	18	21	23	26	31	1.6	1.8	1.9	2.1	2.3	2.0	2.2	2.4	2.7	2.9
TW G2	13	18	23	27	33	37	43	51	8	11	15	18	22	25	29	34	1.4	1.6	1.8	2.0	2.1	1.8	2.0	2.2	2.4	2.6
TW G3	12	16	22	26	32	36	42	50	6	10	13	16	21	24	28	34	1.4	1.6	1.8	1.9	2.1	1.8	2.0	2.2	2.4	2.5
TW G4	9	13	18	22	28	32	38	46	5	8	11	14	18	21	25	31	1.5	1.7	1.9	2.0	2.2	1.8	2.0	2.2	2.4	2.6
TW H1	12	16	20	23	28	32	37	44	7	10	13	15	19	21	25	29	1.6	1.8	1.9	2.1	2.3	2.0	2.2	2.4	2.6	2.8
TW H2	9	13	18	22	27	31	36	43	5	8	11	13	17	20	24	29	1.6	1.8	1.9	2.1	2.2	1.9	2.1	2.3	2.5	2.7
TW H3	8	11	16	19	24	28	33	40	4	6	9	12	15	18	22	27	1.7	1.8	2.0	2.2	2.3	2.0	2.2	2.4	2.6	2.8
TW H4	6	9	12	15	19	22	27	33	3	5	7	9	12	14	18	22	1.8	2.0	2.2	2.4	2.5	2.2	2.4	2.6	2.8	3.0
TW I1	11	14	18	21	25	28	32	37	6	8	10	12	14	15	17	20	1.7	1.9	2.2	2.4	2.5	2.3	2.7	3.1	3.6	4.0
TW I2	17	23	30	35	42	48	54	64	10	15	20	24	30	34	39	47	1.2	1.4	1.6	1.7	1.9	1.5	1.7	1.8	2.0	2.2
TW I3	18	25	32	38	46	52	59	68	11	16	22	27	33	38	44	52	1.2	1.3	1.5	1.7	1.8	1.4	1.6	1.7	1.9	2.1
TW I4	20	28	30	43	52	58	66	75	12	19	26	31	39	45	52	61	1.1	1.2	1.4	1.6	1.7	1.3	1.4	1.6	1.7	1.9
TW I5	20	28	37	44	54	60	68	77	12	19	27	33	41	47	54	64	1.1	1.2	1.4	1.5	1.7	1.2	1.4	1.5	1.7	1.8
TW J1	22	29	38	44	53	59	66	75	14	20	27	32	39	44	51	60	1.1	1.3	1.4	1.6	1.7	1.3	1.4	1.6	1.8	1.9
TW J2	21	29	37	44	53	59	66	76	13	19	27	32	40	45	52	61	1.1	1.2	1.4	1.6	1.7	1.3	1.4	1.6	1.7	1.9
TW J3	25	35	46	53	63	70	77	85	17	25	34	41	51	57	65	74	1.0	1.1	1.3	1.4	1.6	1.1	1.2	1.4	1.6	1.7
TW J4	23	33	43	51	61	67	75	84	15	23	32	39	48	55	62	72	1.0	1.1	1.3	1.5	1.6	1.1	1.3	1.4	1.6	1.7
DG A1	13	17	21	24	28	31	33	36	8	10	12	14	16	17	18	19	1.5	1.8	2.3	2.9	3.6	2.0	2.5	3.3	-	-
DG A2	16	21	27	31	37	42	47	53	10	14	18	21	26	29	33	38	1.4	1.5	1.7	2.0	2.2	1.7	1.8	2.1	2.4	2.7
DG A3	20	20	34	40	48	54	61	69	14	19	25	29	36	41	47	56	1.2	1.3	1.5	1.7	1.8	1.4	1.5	1.7	1.8	2.0
DG B1	13	17	22	26	31	34	38	43	8	10	13	14	17	18	20	22	1.5	1.7	2.0	2.3	2.7	2.0	2.4	2.9	3.7	4.8
DG B2	16	21	28	33	40	45	51	59	11	14	19	22	27	30	35	41	1.4	1.5	1.6	1.8	2.0	1.6	1.8	2.0	2.2	2.4
DG B3	20	27	35	41	50	57	65	74	14	20	26	31	38	43	50	60	1.2	1.3	1.4	1.5	1.7	1.4	1.5	1.6	1.7	1.9
DG C1	16	21	27	31	38	42	47	54	10	13	16	19	22	25	28	32	1.4	1.5	1.7	2.0	2.2	1.7	2.0	2.4	2.8	3.3
DG C2	19	25	32	38	46	52	59	68	13	17	22	26	32	36	42	50	1.3	1.3	1.5	1.6	1.8	1.5	1.6	1.8	2.0	2.2
DG C3	23	30	39	46	56	63	72	81	16	22	29	35	43	49	56	66	1.1	1.2	1.3	1.5	1.6	1.3	1.4	1.5	1.6	1.8
SS A1	23	28	34	38	45	49	55	63	18	21	26	29	34	37	42	48	1.3	1.5	1.6	1.8	1.9	1.5	1.7	2.0	2.2	2.3
SS A2	28	36	44	50	58	64	71	79	22	28	35	40	47	52	58	66	1.1	1.2	1.4	1.6	1.7	1.2	1.4	1.6	1.7	1.9
SS A3	21	26	31	35	41	45	50	57	16	19	23	26	30	33	36	41	1.4	1.6	1.7	1.9	2.1	1.6	1.9	2.2	2.4	2.7
SS A4	28	35	43	49	57	62	69	77	21	27	34	39	45	50	56	64	1.1	1.2	1.4	1.6	1.7	1.2	1.4	1.6	1.8	1.9
SS A5	22	27	32	35	41	45	50	57	18	21	24	27	31	34	37	42	1.4	1.6	1.8	1.9	2.1	1.7	1.9	2.2	2.4	2.6
SS A6	28	35	43	49	58	63	70	78	21	27	34	39	46	51	57	65	1.1	1.2	1.4	1.6	1.7	1.2	1.4	1.6	1.8	1.9
SS A7	20	24	28	31	35	38	43	49	16	19	21	23	26	28	31	34	1.5	1.8	2.0	2.2	2.3	1.9	2.2	2.5	2.9	3.2
SS A8	27	34	42	48	56	61	67	75	21	26	33	37	44	48	54	62	1.1	1.3	1.4	1.6	1.7	1.3	1.5	1.6	1.8	2.0
SS B1	19	24	29	33	39	44	49	57	15	18	22	25	29	32	36	42	1.4	1.6	1.7	1.9	2.0	1.6	1.9	2.1	2.3	2.5
SS B2	25	32	39	45	53	59	66	74	19	24	31	36	43	47	53	62	1.1	1.3	1.4	1.6	1.7	1.3	1.4	1.6	1.8	1.9
SS B3	18	22	27	31	36	40	45	53	13	16	20	23	26	29	32	37	1.4	1.6	1.8	2.0	2.1	1.7	2.0	2.3	2.5	2.8
SS B4	24	31	38	44	52	57	64	73	18	24	30	34	41	46	52	59	1.1	1.3	1.5	1.6	1.8	1.3	1.5	1.7	1.8	2.0
SS B5	18	22	26	30	34	38	42	49	14	17	20	22	25	28	31	35	1.5	1.7	1.9	2.1	2.3	1.9	2.1	2.4	2.7	2.9
SS B6	24	31	38	44	52	58	64	73	18	23	30	34	41	46	52	60	1.1	1.3	1.4	1.6	1.8	1.3	1.5	1.6	1.8	2.0
SS B7	17	20	24	26	30	33	37	43	13	15	18	20	22	24	26	30	1.7	1.9	2.1	2.3	2.4	2.0	2.3	2.7	3.0	3.3
SS B8	23	29	37	42	50	55	62	71	17	22	28	33	39	43	49	57	1.2	1.3	1.5	1.7	1.8	1.3	1.5	1.7	1.9	2.0
SS C1	17	23	29	33	40	44	50	59	11	14	18	21	26	29	33	39	1.3	1.5	1.7	1.8	2.0	1.6	1.9	2.1	2.3	2.5
SS C2	21	28	36	42	50	56	63	72	13	19	25	30	37	41	47	56	1.1	1.3	1.4	1.6	1.8	1.3	1.5	1.7	1.9	2.0
SS C3	16	20	24	28	33	36	41	48	11	13	16	19	22	25	28	32	1.5	1.7	1.9	2.1	2.2	1.9	2.1	2.4	2.6	2.8
SS C4	19	25	32	38	45	50	57	66	13	17	23	27	33	37	42	50	1.2	1.4	1.5	1.7	1.8	1.4	1.6	1.8	2.0	2.1
SS D1	26	32	40	45	53	58	64	72	19	24	30	34	40	43	49	55	1.1	1.3	1.5	1.7	1.8	1.3	1.5	1.8	2.0	2.1
SS D2	32	41	51	58	67	72	79	86	24	32	40	46	54	60	66	75	1.0	1.1	1.3	1.5	1.6	1.1	1.3	1.4	1.6	1.7
SS D3	24	30	36	41	47	52	58	66	19	23	27	31	36	39	44	50	1.2	1.4	1.6	1.8	1.9	1.5	1.7	1.9	2.1	2.3
SS D4	31	39	49	56	65	70	77	84	23	31	39	44	52	58	64	73	1.0	1.1	1.3	1.5	1.6	1.1	1.3	1.5	1.6	1.8
SS E1	22	28	34	39	46	51	57	65	15	20	24	28	33	36	40	46	1.2	1.4	1.6	1.8	1.9	1.5	1.7	2.0	2.2	2.4
SS E2	27	35	44	51	60	66	72	81	19	26	34	39	47	52	58	67	1.0	1.2	1.3	1.5	1.7	1.2	1.4	1.5	1.7	1.9
SS E3	20	24	29	33	39	43	48	56	15	18	22	25	29	31	35	40	1.4	1.6	1.8	1.9	2.1	1.7	1.9	2.2	2.4	2.7
SS E4	25	33	41	47	56	61	68	76	19	25	31	36	43	48	54	62	1.1	1.2	1.4	1.6	1.7	1.2	1.4	1.6	1.8	1.9

EAU CLAIRE, WISCONSIN

TYPE	SOLAR SAVINGS FRACTION (%) — BASE TEMP: 55 F (5837 DD)								65 F (8393 DD)								CONSERVATION FACTOR — 55 F					65 F				
LCR	100	70	50	40	30	25	20	15	100	70	50	40	30	25	20	15	100	50	30	20	15	100	50	30	20	15
WW A1	12	13	14	15	17	17	18	19	11	12	12	13	13	13	14	13	2.8	3.3	4.1	-	-	3.5	4.6	-	-	-
WW A2	10	12	15	17	19	21	24	27	8	9	11	13	15	16	17	19	2.0	2.3	2.7	3.0	3.4	2.3	2.8	3.3	4.1	-
WW A3	8	11	14	17	20	23	26	30	6	8	11	13	15	17	19	21	1.8	2.1	2.4	2.7	3.0	2.1	2.4	2.9	3.5	4.1
WW A4	7	11	14	17	21	24	27	32	5	7	10	12	15	17	20	23	1.7	2.0	2.3	2.5	2.8	1.9	2.3	2.7	3.2	3.7
WW A5	6	10	14	17	22	25	29	34	4	7	10	12	16	18	21	24	1.6	1.9	2.2	2.4	2.6	1.9	2.2	2.5	3.0	3.4
WW A6	6	10	14	18	22	25	30	35	3	6	10	12	16	18	21	25	1.6	1.8	2.1	2.3	2.6	1.8	2.1	2.5	2.9	3.3
WW B1	6	7	8	8	9	9	8	7	5	5	5	5	4	3	2	-1	3.2	4.7	-	-	-	-	-	-	-	-
WW B2	8	12	17	21	26	29	34	41	6	9	13	16	20	23	27	33	1.6	1.8	2.0	2.2	2.4	1.7	2.0	2.2	2.5	2.7
WW B3	10	15	21	25	32	36	42	49	8	12	17	21	26	29	34	41	1.4	1.6	1.8	2.0	2.1	1.5	1.7	2.0	2.2	2.4
WW B4	10	17	24	29	37	42	49	59	8	13	20	25	32	36	43	51	1.3	1.4	1.6	1.8	1.9	1.4	1.5	1.7	1.9	2.0
WW B5	10	17	24	30	38	44	51	61	7	13	20	25	33	38	45	54	1.2	1.4	1.6	1.7	1.9	1.3	1.5	1.7	1.8	2.0
WW C1	12	17	23	28	35	40	46	54	9	14	19	23	29	33	38	45	1.3	1.5	1.7	1.9	2.0	1.4	1.6	1.9	2.1	2.3
WW C2	11	16	23	28	35	40	47	56	8	13	19	23	30	34	40	48	1.3	1.5	1.7	1.8	2.0	1.4	1.6	1.8	2.0	2.1
WW C3	14	22	31	38	47	53	61	71	11	18	27	33	42	48	55	65	1.1	1.3	1.4	1.6	1.7	1.2	1.3	1.5	1.7	1.8
WW C4	12	20	29	35	44	51	59	69	9	16	24	31	39	45	53	63	1.1	1.3	1.5	1.6	1.8	1.2	1.4	1.5	1.7	1.8
TW A1	12	13	15	16	17	18	19	21	11	12	13	13	14	14	15	15	2.7	3.1	3.7	4.4	-	3.3	4.1	-	-	-
TW A2	10	12	14	16	19	21	23	26	8	10	12	13	15	16	17	19	2.1	2.4	2.8	3.1	3.5	2.4	2.9	3.4	4.1	4.9
TW A3	9	11	14	16	19	22	25	29	7	9	11	13	15	17	18	21	1.9	2.2	2.5	2.8	3.1	2.2	2.6	3.0	3.5	4.0
TW A4	7	10	13	16	19	22	25	30	5	8	10	12	15	17	19	23	1.8	2.1	2.4	2.6	2.9	2.1	2.4	2.7	3.1	3.5
TW B1	11	13	14	16	18	19	20	23	10	11	12	13	14	15	16	17	2.4	2.8	3.2	3.7	4.2	2.9	3.4	4.1	-	-
TW B2	9	11	14	16	19	21	23	27	7	9	11	13	15	16	18	21	2.0	2.3	2.6	2.9	3.2	2.3	2.7	3.1	3.5	4.0
TW B3	8	10	13	16	19	21	24	28	6	8	11	12	15	17	19	22	2.0	2.2	2.5	2.8	3.0	2.2	2.5	2.9	3.3	3.6
TW B4	8	10	13	15	18	21	24	28	6	8	10	12	15	17	19	22	2.0	2.2	2.5	2.7	2.9	2.2	2.5	2.8	3.2	3.5
TW C1	10	12	14	15	18	19	22	25	9	10	12	13	15	16	17	19	2.3	2.6	2.9	3.2	3.5	2.6	3.0	3.4	3.9	4.4
TW C2	9	11	13	15	18	20	23	27	7	9	11	12	15	16	18	21	2.1	2.4	2.6	2.9	3.1	2.4	2.7	3.0	3.3	3.7
TW C3	9	11	13	15	18	20	23	27	7	9	11	12	15	16	18	22	2.1	2.4	2.6	2.9	3.1	2.4	2.7	3.0	3.3	3.6
TW C4	9	11	13	15	17	19	22	25	8	10	11	13	15	16	18	21	2.3	2.5	2.8	3.0	3.2	2.6	2.8	3.1	3.4	3.6
TW D1	5	6	7	8	8	9	9	9	4	4	5	5	5	4	4	2	3.1	4.0	-	-	-	4.4	-	-	-	-
TW D2	9	13	17	20	24	28	32	38	8	10	14	16	20	22	26	31	1.7	1.9	2.1	2.3	2.5	1.8	2.1	2.3	2.6	2.8
TW D3	9	13	18	21	26	30	34	41	7	10	14	17	21	24	28	33	1.6	1.8	2.0	2.2	2.4	1.7	2.0	2.2	2.5	2.7
TW D4	11	16	22	26	33	38	44	52	8	13	18	22	28	32	38	45	1.4	1.5	1.7	1.9	2.0	1.5	1.7	1.8	2.0	2.2
TW D5	11	16	23	28	35	40	47	56	8	13	19	24	30	35	41	50	1.3	1.5	1.7	1.8	2.0	1.4	1.6	1.7	1.9	2.1
TW E1	12	16	22	26	31	35	41	48	9	13	18	21	26	30	34	40	1.4	1.6	1.8	2.0	2.2	1.6	1.8	2.0	2.2	2.4
TW E2	11	16	22	26	33	37	43	51	9	13	18	22	28	32	37	44	1.4	1.6	1.8	1.9	2.1	1.5	1.7	1.9	2.1	2.3
TW E3	14	21	29	35	43	49	56	65	11	18	25	30	38	43	50	59	1.2	1.4	1.5	1.7	1.8	1.3	1.4	1.6	1.8	1.9
TW E4	13	20	27	33	42	48	55	64	10	16	24	29	37	42	49	58	1.2	1.4	1.5	1.7	1.8	1.3	1.4	1.6	1.8	1.9
TW F1	10	11	13	14	16	17	18	20	9	10	11	11	12	13	13	14	2.6	3.0	3.6	4.3	5.0	3.2	3.9	4.9	-	-
TW F2	8	10	12	14	16	18	20	24	6	8	9	11	12	14	15	17	2.1	2.5	2.8	3.2	3.5	2.5	2.9	3.4	4.1	5.0
TW F3	6	8	11	13	16	18	21	25	4	6	8	10	12	14	15	18	2.0	2.3	2.6	2.9	3.2	2.3	2.7	3.1	3.6	4.1
TW F4	4	7	10	12	15	18	21	25	2	4	7	9	11	13	15	18	2.0	2.2	2.5	2.8	3.0	2.2	2.5	2.9	3.3	3.6
TW G1	8	9	11	13	15	16	18	20	7	8	9	10	11	12	13	14	2.5	2.8	3.2	3.7	4.2	2.9	3.4	4.1	-	-
TW G2	5	7	10	12	14	16	19	22	4	5	7	9	11	12	14	16	2.1	2.4	2.7	3.1	3.4	2.5	2.8	3.3	3.8	4.3
TW G3	4	6	9	11	14	16	19	23	2	4	6	8	10	12	14	17	2.1	2.4	2.6	2.9	3.2	2.4	2.7	3.0	3.4	3.8
TW G4	2	4	7	9	12	14	17	21	1	3	5	6	9	10	13	16	2.2	2.4	2.7	2.9	3.2	2.5	2.7	3.0	3.4	3.7
TW H1	5	7	9	10	13	14	16	19	4	5	7	8	9	11	12	14	2.4	2.7	3.0	3.4	3.7	2.8	3.2	3.6	4.2	4.8
TW H2	3	5	7	9	11	13	16	19	2	3	5	6	8	10	12	14	2.3	2.6	2.8	3.1	3.4	2.6	2.9	3.3	3.6	4.0
TW H3	2	4	6	8	10	12	14	18	1	2	4	5	7	9	11	14	2.4	2.6	2.9	3.1	3.4	2.7	3.0	3.3	3.6	3.9
TW H4	1	3	5	6	8	10	12	15	1	2	3	4	6	7	9	12	2.6	2.9	3.1	3.3	3.5	2.9	3.2	3.5	3.7	4.0
TW I1	4	4	5	6	6	7	7	7	2	3	3	3	3	3	2	0	3.3	4.3	-	-	-	4.7	-	-	-	-
TW I2	7	10	14	17	21	24	29	35	5	8	11	13	17	19	23	28	1.7	1.9	2.1	2.4	2.5	1.9	2.1	2.4	2.7	2.9
TW I3	7	11	15	19	24	27	32	39	5	8	12	15	19	22	26	32	1.6	1.8	2.0	2.2	2.4	1.8	2.0	2.2	2.5	2.7
TW I4	8	13	19	24	30	35	41	49	6	10	16	20	26	30	35	43	1.4	1.6	1.8	1.9	2.1	1.5	1.7	1.9	2.1	2.2
TW I5	8	14	20	25	32	37	44	53	6	11	17	21	28	32	38	47	1.4	1.5	1.7	1.8	2.0	1.4	1.6	1.8	2.0	2.1
TW J1	10	14	19	24	29	34	39	47	7	11	16	19	24	28	33	39	1.5	1.7	1.8	2.0	2.2	1.6	1.8	2.0	2.2	2.4
TW J2	9	14	20	24	31	35	41	49	7	11	16	20	26	30	35	42	1.4	1.6	1.8	1.9	2.1	1.5	1.7	1.9	2.1	2.3
TW J3	11	18	26	32	41	47	54	64	9	15	22	28	36	41	48	58	1.2	1.4	1.5	1.7	1.8	1.3	1.4	1.6	1.8	1.9
TW J4	10	17	25	31	39	45	53	62	8	14	21	27	35	40	47	56	1.2	1.4	1.5	1.7	1.8	1.3	1.5	1.6	1.8	1.9
DG A1	5	6	7	7	8	8	8	6	3	4	4	4	4	3	2	0	2.8	3.9	-	-	-	3.7	-	-	-	-
DG A2	8	10	13	15	18	21	23	27	6	8	10	12	15	16	18	21	2.0	2.2	2.5	3.0	3.4	2.2	2.5	2.9	3.4	4.1
DG A3	11	15	19	23	29	33	38	45	9	13	17	20	25	28	33	39	1.6	1.7	1.9	2.1	2.2	1.7	1.9	2.0	2.2	2.4
DG B1	5	6	7	8	8	8	9	8	3	3	4	4	4	4	3	1	2.7	3.8	-	-	-	3.8	-	-	-	-
DG B2	8	10	13	16	19	21	25	29	6	8	11	13	15	17	19	23	1.9	2.2	2.5	2.7	3.0	2.2	2.4	2.8	3.2	3.6
DG B3	11	15	20	24	30	34	40	48	10	13	18	21	26	30	35	42	1.5	1.7	1.8	1.9	2.1	1.7	1.8	1.9	2.1	2.3
DG C1	6	8	10	11	13	14	15	16	4	6	7	8	9	9	9	8	2.2	2.9	3.8	4.7	-	2.8	3.5	-	-	-
DG C2	10	13	17	19	24	27	31	37	8	10	14	16	20	22	25	30	1.7	2.0	2.2	2.3	2.6	1.9	2.1	2.4	2.7	3.0
DG C3	13	18	23	27	34	39	45	55	11	15	20	24	30	34	40	48	1.4	1.6	1.7	1.8	2.0	1.6	1.7	1.8	2.0	2.1
SS A1	13	15	18	20	23	25	28	31	12	14	16	18	20	22	24	26	1.9	2.2	2.6	2.9	3.2	2.0	2.4	2.9	3.5	4.2
SS A2	16	21	26	30	35	39	45	52	14	19	23	27	32	35	40	46	1.4	1.6	1.8	2.0	2.2	1.5	1.7	1.9	2.2	2.4
SS A3	12	13	15	17	19	20	22	23	11	12	14	15	16	17	18	18	2.2	2.6	3.2	4.0	4.9	2.4	3.0	4.0	-	-
SS A4	15	20	24	28	33	37	41	48	14	18	22	25	30	33	37	42	1.5	1.7	1.9	2.1	2.3	1.5	1.8	2.1	2.3	2.5
SS A5	14	16	18	19	21	23	25	27	13	14	16	17	19	20	21	22	2.1	2.5	3.0	3.5	4.1	2.4	2.9	3.5	4.6	-
SS A6	16	20	25	29	34	38	43	50	14	18	23	26	31	34	38	44	1.4	1.6	1.9	2.0	2.2	1.5	1.7	2.0	2.2	2.4
SS A7	13	14	15	16	17	18	19	19	12	13	14	14	15	15	15	15	2.6	3.3	4.2	-	-	3.0	3.9	-	-	-
SS A8	15	19	23	27	31	35	39	45	13	17	21	24	28	31	34	39	1.5	1.7	2.0	2.2	2.4	1.6	1.9	2.2	2.5	2.7
SS B1	11	13	15	17	20	21	24	27	9	11	13	15	17	18	20	21	2.0	2.4	2.8	3.2	3.5	2.3	2.7	3.2	3.9	4.7
SS B2	14	18	23	27	32	36	41	48	12	16	21	24	29	32	37	43	1.5	1.7	1.9	2.0	2.2	1.6	1.8	2.0	2.2	2.4
SS B3	9	11	13	14	16	18	19	21	8	10	11	12	14	14	15	16	2.3	2.7	3.3	3.9	4.7	2.5	3.1	4.0	-	-
SS B4	13	17	22	25	30	34	39	45	12	15	19	23	27	30	34	40	1.5	1.7	1.9	2.1	2.3	1.6	1.8	2.1	2.3	2.5
SS B5	11	13	14	16	17	19	20	22	10	12	13	14	15	16	17	18	2.4	2.8	3.3	3.9	4.6	2.7	3.3	4.0	-	-
SS B6	13	17	22	25	31	34	39	46	11	15	19	23	27	30	35	41	1.5	1.7	1.9	2.1	2.3	1.6	1.8	2.0	2.3	2.5
SS B7	10	11	12	13	14	15	16	17	9	10	11	12	12	13	13	13	2.8	3.4	4.3	-	-	3.2	4.1	-	-	-
SS B8	12	16	20	23	28	31	36	42	11	14	18	21	25	28	31	37	1.6	1.8	2.0	2.2	2.4	1.7	1.9	2.2	2.4	2.7
SS C1	7	9	12	14	16	18	20	23	5	7	8	10	11	12	14	15	2.0	2.4	2.8	3.2	3.6	2.4	2.9	3.6	4.7	-
SS C2	9	13	18	22	27	30	35	42	7	10	14	17	22	25	29	34	1.5	1.8	2.0	2.2	2.3	1.7	1.9	2.2	2.4	2.6
SS C3	8	10	12	13	15	17	19	21	7	8	10	10	12	13	14	15	2.4	2.7	3.1	3.6	4.0	2.8	3.3	3.9	4.9	-
SS C4	9	13	17	20	24	27	32	38	7	10	14	16	20	22	26	31	1.7	1.9	2.1	2.3	2.5	1.9	2.1	2.3	2.6	2.8
SS D1	13	16	20	22	26	29	32	36	12	14	17	19	22	24	26	28	1.7	2.0	2.3	2.6	2.9	1.8	2.2	2.7	3.3	4.0
SS D2	16	22	29	34	41	46	52	60	15	20	26	31	37	41	46	54	1.2	1.4	1.6	1.8	2.0	1.3	1.5	1.7	1.9	2.1
SS D3	14	16	19	21	24	26	29	32	12	15	17	19	21	22	24	26	1.8	2.2	2.5	2.9	3.3	2.0	2.4	3.0	3.7	4.5
SS D4	16	22	28	33	40	44	50	58	15	20	25	30	36	40	45	52	1.3	1.5	1.7	1.9	2.0	1.4	1.6	1.8	2.0	2.2
SS E1	10	12	15	17	20	21	24	26	8	10	12	14	16	17	18	18	1.9	2.3	2.8	3.3	3.8	2.1	2.7	3.5	-	-
SS E2	13	18	23	28	34	38	43	50	11	15	20	24	29	33	37	43	1.4	1.6	1.8	2.0	2.2	1.5	1.7	2.0	2.2	2.4
SS E3	11	13	15	17	19	20	22	24	10	11	13	14	16	17	18	19	2.2	2.6	3.1	3.6	4.2	2.4	3.0	3.7	-	-
SS E4	13	17	22	26	31	35	40	47	11	15	19	23	27	31	35	40	1.5	1.7	1.9	2.1	2.3	1.6	1.8	2.1	2.3	2.5

GREEN BAY, WISCONSIN

TYPE	SOLAR SAVINGS FRACTION (%)																CONSERVATION FACTOR									
	BASE TEMP: 55 F (5476 DD)								65 F (8104 DD)								55 F					65 F				
LCR	100	70	50	40	30	25	20	15	100	70	50	40	30	25	20	15	100	50	30	20	15	100	50	30	20	15
WW A1	13	14	16	17	18	20	21	23	12	12	13	14	15	15	16	16	2.5	2.9	3.4	4.1	4.9	3.1	3.9	-	-	-
WW A2	11	13	16	19	22	24	27	32	9	11	13	14	17	18	20	22	1.8	2.1	2.4	2.7	3.0	2.1	2.5	3.0	3.6	4.2
WW A3	9	13	17	19	24	26	30	35	7	10	12	14	17	19	22	25	1.7	1.9	2.2	2.5	2.7	1.9	2.3	2.6	3.1	3.5
WW A4	9	12	17	20	24	28	32	37	6	9	12	14	18	20	23	27	1.6	1.8	2.1	2.3	2.6	1.8	2.1	2.5	2.9	3.2
WW A5	8	12	17	20	25	29	33	39	5	8	12	15	18	21	24	28	1.5	1.8	2.0	2.2	2.4	1.8	2.0	2.4	2.7	3.1
WW A6	7	12	17	21	26	29	34	41	5	8	12	15	19	21	25	29	1.5	1.7	2.0	2.2	2.4	1.7	2.0	2.3	2.6	2.9
WW B1	7	8	10	10	11	12	12	12	6	6	6	6	6	6	5	2	2.7	3.5	-	-	-	3.9	-	-	-	-
WW B2	10	14	19	23	29	33	38	45	7	11	15	18	23	26	30	36	1.5	1.7	1.9	2.1	2.2	1.6	1.9	2.1	2.3	2.6
WW B3	12	17	24	28	35	40	46	54	9	14	19	23	28	32	37	44	1.3	1.5	1.7	1.9	2.0	1.5	1.7	1.9	2.1	2.3
WW B4	12	19	26	32	40	46	53	62	9	15	22	27	34	39	46	54	1.2	1.4	1.6	1.7	1.9	1.3	1.5	1.7	1.8	2.0
WW B5	11	18	26	32	41	47	54	64	9	15	22	27	35	41	47	57	1.2	1.4	1.5	1.7	1.8	1.3	1.4	1.6	1.8	1.9
WW C1	13	19	26	31	38	43	49	58	10	15	21	25	32	36	41	49	1.3	1.5	1.6	1.8	2.0	1.4	1.6	1.8	2.0	2.2
WW C2	12	18	25	31	38	44	50	59	9	15	21	25	32	37	43	51	1.3	1.4	1.6	1.8	1.9	1.4	1.5	1.7	1.9	2.1
WW C3	15	24	34	41	50	57	64	74	12	20	29	35	44	50	58	68	1.1	1.2	1.4	1.6	1.7	1.1	1.3	1.5	1.6	1.8
WW C4	13	22	31	38	47	54	62	71	10	18	26	33	42	48	55	65	1.1	1.3	1.4	1.6	1.7	1.2	1.3	1.5	1.7	1.8
TW A1	13	14	16	17	19	20	22	24	12	13	14	14	15	16	17	18	2.5	2.8	3.2	3.7	4.2	2.9	3.6	4.4	-	-
TW A2	11	13	16	18	21	23	26	30	9	11	13	14	17	18	20	22	2.0	2.2	2.5	2.8	3.1	2.3	2.7	3.1	3.6	4.1
TW A3	10	12	16	18	22	25	28	33	8	10	12	14	17	19	21	24	1.8	2.1	2.3	2.6	2.8	2.1	2.4	2.8	3.2	3.5
TW A4	8	11	15	18	22	25	29	34	6	9	12	14	17	19	22	26	1.7	2.0	2.2	2.4	2.6	2.0	2.3	2.6	2.9	3.2
TW B1	12	13	15	17	19	21	23	26	11	12	13	14	16	17	18	20	2.3	2.6	2.9	3.3	3.6	2.7	3.1	3.7	4.3	-
TW B2	10	12	15	18	21	23	26	31	8	10	12	14	17	18	20	24	1.9	2.2	2.4	2.7	2.9	2.2	2.5	2.9	3.2	3.6
TW B3	9	12	15	17	21	24	27	32	7	9	12	14	17	19	21	25	1.8	2.1	2.3	2.6	2.8	2.1	2.4	2.7	3.0	3.3
TW B4	8	11	14	17	21	23	27	32	7	9	12	14	16	18	21	25	1.9	2.1	2.3	2.6	2.8	2.1	2.4	2.7	3.0	3.2
TW C1	10	13	15	17	20	22	24	28	9	11	13	14	16	17	19	22	2.1	2.4	2.6	2.9	3.2	2.4	2.8	3.2	3.6	3.9
TW C2	9	12	14	17	20	22	25	30	8	10	12	14	16	18	20	24	2.0	2.2	2.4	2.7	2.9	2.2	2.5	2.8	3.1	3.4
TW C3	9	11	14	16	20	22	25	29	8	10	12	14	16	18	20	24	2.0	2.2	2.5	2.7	2.9	2.3	2.5	2.8	3.1	3.4
TW C4	10	12	14	16	19	21	24	28	9	10	12	14	16	18	20	23	2.2	2.4	2.6	2.8	3.0	2.4	2.7	3.0	3.2	3.4
TW D1	6	7	9	9	11	11	12	13	5	5	6	6	6	6	6	5	2.7	3.3	4.2	-	-	3.6	-	-	-	-
TW D2	10	14	19	22	27	31	35	42	8	11	15	18	22	25	28	34	1.6	1.8	2.0	2.2	2.4	1.8	2.0	2.2	2.5	2.7
TW D3	11	15	20	24	29	33	38	45	8	12	16	19	23	27	31	36	1.5	1.7	1.9	2.1	2.3	1.7	1.9	2.1	2.4	2.6
TW D4	12	17	24	29	36	41	47	56	9	14	20	24	30	35	40	48	1.3	1.5	1.7	1.8	2.0	1.4	1.6	1.8	2.0	2.1
TW D5	12	18	25	30	38	43	50	59	9	15	21	25	32	37	44	52	1.3	1.4	1.6	1.8	1.9	1.4	1.5	1.7	1.9	2.0
TW E1	13	18	24	28	34	39	44	52	11	15	20	23	29	32	37	44	1.4	1.6	1.8	1.9	2.1	1.5	1.7	1.9	2.1	2.3
TW E2	13	18	24	29	36	40	47	55	10	15	20	24	30	34	40	47	1.3	1.5	1.7	1.9	2.0	1.5	1.6	1.8	2.0	2.2
TW E3	16	23	3,1	37	46	52	59	68	13	19	27	32	40	46	52	62	1.1	1.3	1.5	1.6	1.8	1.2	1.4	1.6	1.7	1.9
TW E4	14	21	30	36	45	50	58	67	11	18	25	31	39	45	52	61	1.2	1.3	1.5	1.6	1.8	1.2	1.4	1.6	1.7	1.9
TW F1	11	12	14	15	17	19	21	23	9	11	12	13	14	14	15	16	2.4	2.7	3.1	3.6	4.1	2.9	3.5	4.2	-	-
TW F2	8	11	14	16	19	21	24	27	7	9	11	12	14	16	17	20	2.0	2.3	2.6	2.9	3.2	2.3	2.7	3.1	3.7	4.2
TW F3	7	10	13	15	19	21	25	29	5	7	10	11	14	16	18	21	1.9	2.1	2.4	2.7	2.9	2.1	2.5	2.8	3.2	3.6
TW F4	5	8	11	14	18	21	24	29	3	5	8	10	13	15	18	21	1.8	2.1	2.3	2.6	2.8	2.1	2.4	2.7	3.0	3.3
TW G1	9	10	12	14	16	18	20	23	7	9	10	11	13	14	15	17	2.3	2.6	2.9	3.3	3.6	2.7	3.1	3.7	4.3	-
TW G2	6	8	11	13	17	19	22	26	5	6	8	10	12	14	16	19	2.0	2.3	2.5	2.8	3.1	2.3	2.6	3.0	3.4	3.8
TW G3	4	7	10	12	16	18	21	26	3	5	7	9	12	13	16	19	2.0	2.2	2.5	2.7	2.9	2.2	2.5	2.9	3.2	3.5
TW G4	3	5	8	10	14	16	19	24	2	4	6	8	10	12	14	18	2.1	2.3	2.5	2.7	2.9	2.3	2.6	2.9	3.2	3.4
TW H1	6	8	10	12	14	16	19	22	5	6	8	9	11	12	14	16	2.3	2.5	2.8	3.1	3.4	2.6	2.9	3.4	3.8	4.2
TW H2	3	6	8	10	13	15	18	22	2	4	6	7	10	11	13	16	2.2	2.4	2.6	2.9	3.1	2.4	2.8	3.1	3.4	3.7
TW H3	2	4	7	9	12	14	17	21	1	3	5	6	9	10	12	16	2.3	2.5	2.7	2.9	3.1	2.5	2.8	3.1	3.4	3.6
TW H4	2	3	5	7	9	11	14	17	1	2	4	5	7	9	10	13	2.5	2.7	2.9	3.1	3.3	2.8	3.0	3.3	3.6	3.8
TW I1	4	5	7	7	9	9	10	11	3	4	4	4	5	5	4	3	2.9	3.5	4.5	-	-	3.8	-	-	-	-
TW I2	8	11	16	19	24	27	32	38	6	9	12	15	19	22	25	31	1.6	1.8	2.0	2.2	2.4	1.8	2.0	2.3	2.5	2.7
TW I3	8	12	17	21	27	30	36	43	6	10	14	17	21	25	29	35	1.5	1.7	1.9	2.1	2.3	1.7	1.9	2.1	2.4	2.5
TW I4	9	15	21	26	33	38	44	53	7	12	17	21	26	32	38	46	1.4	1.5	1.7	1.9	2.0	1.5	1.6	1.8	2.0	2.2
TW I5	10	15	22	27	35	40	47	56	7	12	18	23	30	35	41	49	1.3	1.5	1.6	1.8	1.9	1.4	1.6	1.7	1.9	2.1
TW J1	11	16	22	26	32	37	43	51	8	13	18	21	27	30	35	42	1.4	1.6	1.8	1.9	2.1	1.5	1.7	1.9	2.1	2.3
TW J2	10	16	22	26	33	38	44	53	8	12	18	22	28	32	37	45	1.4	1.5	1.7	1.9	2.0	1.5	1.7	1.8	2.0	2.2
TW J3	13	20	29	35	44	50	57	67	10	17	24	30	38	44	51	60	1.2	1.3	1.5	1.6	1.8	1.2	1.4	1.6	1.7	1.9
TW J4	12	19	27	33	42	48	56	65	9	15	23	28	37	42	49	59	1.2	1.3	1.5	1.7	1.8	1.3	1.4	1.6	1.7	1.9
DG A1	5	7	8	9	10	11	11	10	4	5	5	6	6	5	5	3	2.5	3.2	-	-	-	3.2	-	-	-	-
DG A2	8	11	14	17	21	23	26	30	7	9	12	14	16	18	21	23	1.8	2.1	2.4	2.7	3.1	2.1	2.4	2.7	3.2	3.8
DG A3	12	16	21	25	31	35	41	48	10	14	18	21	27	30	35	42	1.5	1.7	1.8	2.0	2.2	1.7	1.8	2.0	2.1	2.3
DG B1	5	7	9	10	11	11	12	12	4	4	5	6	6	6	6	4	2.4	3.1	4.3	-	-	3.3	4.6	-	-	-
DG B2	8	11	15	18	21	24	28	33	7	9	12	14	17	19	22	25	1.8	2.0	2.3	2.5	2.8	2.1	2.3	2.6	3.0	3.4
DG B3	12	17	22	26	32	37	43	52	11	14	19	22	28	32	37	45	1.5	1.6	1.7	1.9	2.0	1.6	1.7	1.9	2.0	2.2
DG C1	7	10	12	14	16	17	19	21	5	7	8	9	11	11	12	12	2.0	2.5	3.1	3.8	4.8	2.6	3.1	4.3	-	-
DG C2	10	14	18	21	26	30	34	41	9	11	15	18	21	24	28	33	1.6	1.8	2.0	2.2	2.4	1.8	2.0	2.3	2.5	2.8
DG C3	14	19	25	30	37	42	49	58	12	16	22	26	32	36	42	51	1.4	1.5	1.6	1.8	1.9	1.5	1.6	1.8	1.9	2.0
SS A1	14	17	20	23	26	28	31	36	13	15	18	20	23	24	27	29	1.7	2.0	2.4	2.7	2.9	1.9	2.3	2.7	3.1	3.6
SS A2	17	22	28	32	38	43	48	55	16	20	25	29	34	38	43	49	1.3	1.5	1.7	1.9	2.1	1.4	1.6	1.9	2.1	2.3
SS A3	13	15	17	19	22	23	25	28	12	13	15	17	18	19	21	22	2.0	2.4	2.8	3.4	3.9	2.2	2.7	3.4	4.6	-
SS A4	17	21	27	31	36	40	45	52	15	19	24	27	32	36	40	46	1.4	1.6	1.8	2.0	2.2	1.5	1.7	2.0	2.2	2.4
SS A5	15	17	19	21	24	26	28	31	14	16	17	19	21	22	24	25	2.0	2.3	2.7	3.1	3.5	2.2	2.6	3.2	3.9	4.8
SS A6	17	22	27	31	37	41	47	54	15	20	24	28	33	37	41	47	1.4	1.6	1.8	2.0	2.1	1.5	1.7	1.9	2.1	2.3
SS A7	13	15	17	18	19	20	22	23	13	14	15	16	17	17	18	18	2.3	2.8	3.5	4.4	-	2.6	3.4	4.6	-	-
SS A8	16	21	25	29	34	38	43	49	15	19	23	26	31	34	37	43	1.4	1.7	1.9	2.1	2.3	1.5	1.8	2.1	2.3	2.6
SS B1	11	14	17	19	22	24	27	31	10	12	15	16	19	20	22	25	1.9	2.2	2.5	2.9	3.2	2.1	2.5	2.9	3.4	4.0
SS B2	15	20	25	29	35	39	44	52	14	18	22	26	31	35	40	46	1.4	1.6	1.8	2.0	2.1	1.5	1.7	1.9	2.1	2.3
SS B3	10	12	15	16	19	21	23	25	9	11	13	14	16	17	18	19	2.1	2.4	2.9	3.4	3.9	2.3	2.8	3.5	4.5	-
SS B4	14	19	24	28	33	37	42	49	13	17	21	25	29	33	37	43	1.4	1.7	1.9	2.0	2.2	1.5	1.8	2.0	2.2	2.4
SS B5	12	14	16	17	19	21	23	25	11	13	14	15	17	18	19	21	2.2	2.6	3.0	3.4	3.9	2.5	3.0	3.6	4.4	-
SS B6	14	19	24	28	33	37	42	50	13	17	21	25	29	33	37	44	1.4	1.6	1.8	2.0	2.2	1.5	1.8	2.0	2.2	2.4
SS B7	11	12	14	15	16	17	19	20	10	11	12	13	14	15	15	16	2.5	3.0	3.6	4.4	-	2.9	3.6	4.7	-	-
SS B8	13	17	22	26	31	34	39	46	12	15	20	23	27	30	34	40	1.5	1.7	1.9	2.1	2.3	1.6	1.8	2.1	2.3	2.5
SS C1	8	11	14	16	19	21	24	28	6	8	10	12	14	15	17	19	1.9	2.2	2.5	2.9	3.2	2.2	2.7	3.2	3.9	4.8
SS C2	10	15	20	24	30	34	39	46	8	12	16	19	24	27	32	38	1.5	1.7	1.9	2.1	2.2	1.6	1.8	2.1	2.3	2.5
SS C3	9	11	13	15	17	19	21	24	8	9	11	12	13	15	16	18	2.2	2.5	2.8	3.2	3.5	2.6	3.0	3.5	4.2	4.9
SS C4	10	14	18	22	27	30	35	41	8	11	15	18	22	25	28	34	1.6	1.8	2.0	2.2	2.4	1.8	2.0	2.2	2.5	2.7
SS D1	15	18	22	25	30	32	36	41	13	16	19	22	25	27	30	33	1.5	1.8	2.1	2.4	2.7	1.7	2.1	2.5	2.9	3.4
SS D2	18	25	32	37	44	49	56	64	16	22	28	33	40	44	50	57	1.2	1.4	1.6	1.7	1.9	1.3	1.5	1.7	1.9	2.1
SS D3	15	18	21	24	27	30	33	37	14	16	19	21	23	25	27	30	1.7	2.0	2.3	2.7	2.9	1.9	2.3	2.7	3.2	3.8
SS D4	18	24	31	36	43	48	54	62	16	21	27	32	38	42	48	55	1.2	1.4	1.6	1.8	1.9	1.3	1.5	1.7	1.9	2.1
SS E1	11	14	17	20	23	25	28	32	10	12	14	16	18	20	21	23	1.7	2.1	2.5	2.9	3.2	2.0	2.4	3.0	4.0	-
SS E2	14	20	26	31	37	41	47	55	12	17	22	26	32	36	41	47	1.3	1.5	1.7	1.9	2.1	1.4	1.6	1.9	2.1	2.3
SS E3	12	14	17	19	21	23	25	28	11	13	15	16	18	19	20	22	2.0	2.3	2.7	3.2	3.6	2.2	2.7	3.3	4.1	-
SS E4	14	19	24	29	34	38	44	51	12	17	21	25	30	33	38	44	1.4	1.6	1.8	2.0	2.2	1.5	1.7	2.0	2.2	2.4

LA CROSSE, WISCONSIN

TYPE	SOLAR SAVINGS FRACTION (%) — BASE TEMP: 55 F (5060 DD)								65 F (7422 DD)								CONSERVATION FACTOR 55 F					65 F				
LCR	100	70	50	40	30	25	20	15	100	70	50	40	30	25	20	15	100	50	30	20	15	100	50	30	20	15
WW A1	13	15	16	18	19	21	22	24	12	13	14	14	15	16	16	17	2.4	2.8	3.2	3.7	4.2	3.1	3.7	4.8	-	-
WW A2	11	14	17	20	24	26	30	34	9	11	13	15	18	19	21	24	1.8	2.0	2.3	2.6	2.8	2.1	2.5	2.9	3.3	3.8
WW A3	10	14	18	21	25	28	32	38	7	10	13	15	18	20	23	27	1.6	1.9	2.1	2.3	2.5	1.9	2.2	2.5	2.9	3.3
WW A4	9	13	18	21	26	30	34	41	6	9	13	15	19	21	24	29	1.6	1.8	2.0	2.2	2.4	1.8	2.1	2.4	2.7	3.0
WW A5	9	13	18	22	27	31	36	43	5	9	13	15	19	22	26	31	1.5	1.7	1.9	2.1	2.3	1.7	2.0	2.3	2.6	2.9
WW A6	8	13	18	22	28	32	37	44	5	8	12	15	20	23	27	32	1.5	1.7	1.9	2.1	2.2	1.7	1.9	2.2	2.5	2.8
WW B1	8	9	11	12	13	13	14	15	6	6	7	7	7	6	6	3	2.5	3.2	4.3	-	-	3.7	-	-	-	-
WW B2	10	15	20	25	31	35	40	48	7	11	15	19	24	27	32	38	1.4	1.6	1.8	2.0	2.1	1.6	1.8	2.1	2.3	2.5
WW B3	13	18	25	30	37	42	48	57	9	14	19	24	30	34	39	46	1.3	1.5	1.7	1.8	2.0	1.4	1.6	1.8	2.0	2.2
WW B4	13	20	28	34	42	48	55	65	9	15	22	27	35	40	47	56	1.2	1.4	1.5	1.7	1.8	1.3	1.5	1.6	1.8	2.0
WW B5	12	19	28	34	43	49	56	66	9	15	23	28	36	42	49	58	1.2	1.3	1.5	1.7	1.8	1.3	1.4	1.6	1.8	1.9
WW C1	14	20	27	33	40	45	52	61	11	16	22	26	33	37	43	51	1.3	1.4	1.6	1.8	1.9	1.4	1.6	1.8	1.9	2.1
WW C2	13	19	27	32	40	46	53	62	10	15	21	26	33	38	44	53	1.2	1.4	1.6	1.7	1.9	1.3	1.5	1.7	1.9	2.0
WW C3	16	25	35	42	52	58	66	76	12	20	29	36	45	51	59	69	1.1	1.2	1.4	1.5	1.7	1.1	1.3	1.5	1.6	1.8
WW C4	14	23	32	39	49	56	63	73	11	18	27	33	43	49	56	66	1.1	1.2	1.4	1.6	1.7	1.2	1.3	1.5	1.6	1.8
TW A1	13	15	17	18	20	21	23	26	12	13	14	15	16	17	18	19	2.3	2.7	3.1	3.5	3.8	2.9	3.4	4.1	-	-
TW A2	11	14	17	19	22	25	28	32	9	11	13	15	17	19	21	23	1.9	2.2	2.4	2.7	2.9	2.2	2.6	3.0	3.4	3.8
TW A3	10	13	17	19	23	26	30	35	8	10	13	15	18	20	22	26	1.8	2.0	2.2	2.5	2.6	2.0	2.3	2.7	3.0	3.3
TW A4	9	12	16	19	24	27	31	37	7	9	12	14	18	20	23	27	1.7	1.9	2.1	2.3	2.5	1.9	2.2	2.5	2.8	3.1
TW B1	12	14	16	18	20	22	24	28	11	12	13	15	16	17	19	21	2.2	2.5	2.8	3.1	3.4	2.6	3.0	3.5	4.1	4.7
TW B2	10	13	16	19	22	25	28	33	8	10	13	14	17	19	21	25	1.9	2.1	2.3	2.6	2.8	2.2	2.4	2.8	3.1	3.4
TW B3	9	12	16	18	22	25	29	34	7	10	12	14	17	19	22	26	1.8	2.0	2.2	2.5	2.6	2.1	2.3	2.6	2.9	3.2
TW B4	9	12	15	18	22	24	28	34	7	9	12	14	17	19	22	26	1.8	2.0	2.3	2.5	2.6	2.1	2.3	2.6	2.9	3.1
TW C1	11	13	16	18	21	23	25	30	9	11	13	14	16	18	20	23	2.1	2.3	2.6	2.8	3.0	2.4	2.7	3.1	3.4	3.7
TW C2	10	12	15	18	21	23	27	32	8	10	12	14	17	19	21	25	1.9	2.2	2.4	2.6	2.8	2.2	2.5	2.8	3.0	3.3
TW C3	10	12	15	17	20	23	26	31	8	10	12	14	17	18	21	25	2.0	2.2	2.4	2.6	2.8	2.3	2.5	2.8	3.0	3.3
TW C4	10	12	15	17	20	22	25	29	9	11	12	14	16	18	20	24	2.1	2.3	2.5	2.7	2.9	2.4	2.7	2.9	3.1	3.4
TW D1	6	8	9	10	12	13	14	15	5	5	6	7	7	7	7	6	2.6	3.1	3.8	4.8	-	3.5	4.8	-	-	-
TW D2	11	15	20	23	29	32	37	44	9	12	15	18	23	26	29	35	1.5	1.7	1.9	2.1	2.3	1.7	2.0	2.2	2.4	2.6
TW D3	11	16	21	25	31	35	40	48	9	12	16	20	24	28	32	38	1.5	1.7	1.8	2.0	2.2	1.6	1.8	2.1	2.3	2.5
TW D4	13	18	25	30	38	43	49	58	10	15	20	25	31	36	42	50	1.3	1.5	1.6	1.8	1.9	1.4	1.6	1.8	1.9	2.1
TW D5	12	19	26	31	39	45	52	61	10	15	21	26	33	38	45	53	1.3	1.4	1.6	1.7	1.9	1.4	1.5	1.7	1.9	2.0
TW E1	14	19	25	29	36	41	47	55	11	15	20	24	29	33	38	45	1.3	1.5	1.7	1.9	2.0	1.5	1.7	1.9	2.1	2.3
TW E2	13	19	25	30	37	42	49	57	10	15	21	25	31	35	41	48	1.3	1.5	1.6	1.8	2.0	1.4	1.6	1.8	2.0	2.1
TW E3	16	24	32	39	48	54	61	71	13	20	27	33	41	47	54	63	1.1	1.3	1.4	1.6	1.8	1.2	1.4	1.5	1.7	1.9
TW E4	15	22	31	37	46	52	60	69	12	18	26	32	40	46	53	62	1.2	1.3	1.5	1.6	1.8	1.2	1.4	1.5	1.7	1.9
TW F1	11	13	15	16	18	20	22	25	10	11	12	13	14	15	16	17	2.3	2.6	3.0	3.4	3.7	2.8	3.3	4.0	4.9	-
TW F2	9	12	14	17	20	22	25	30	7	9	11	12	15	16	18	21	1.9	2.2	2.5	2.7	2.9	2.3	2.6	3.0	3.5	3.9
TW F3	7	10	14	16	20	23	26	32	5	7	10	12	15	16	19	22	1.8	2.1	2.3	2.5	2.7	2.1	2.4	2.7	3.1	3.4
TW F4	5	9	12	15	19	22	26	32	3	6	8	11	14	16	19	23	1.8	2.0	2.2	2.5	2.6	2.1	2.3	2.6	2.9	3.2
TW G1	9	11	13	15	17	19	22	25	8	9	10	12	13	14	16	18	2.2	2.5	2.8	3.1	3.4	2.6	3.0	3.5	4.1	4.6
TW G2	7	9	12	14	18	20	23	28	5	7	9	10	13	15	17	20	1.9	2.2	2.4	2.7	2.9	2.3	2.6	2.9	3.3	3.6
TW G3	5	8	11	13	17	20	23	28	3	5	8	9	12	14	17	20	1.9	2.1	2.1	2.6	2.8	2.2	2.5	2.8	3.1	3.4
TW G4	3	6	9	11	15	17	21	26	2	4	6	8	11	13	15	19	2.0	2.2	2.4	2.7	2.8	2.3	2.5	2.8	3.1	3.3
TW H1	6	8	11	13	15	17	20	24	5	6	8	9	11	13	14	17	2.2	2.4	2.7	3.0	3.2	2.5	2.9	3.2	3.6	4.0
TW H2	4	6	9	11	14	16	19	24	2	4	6	8	10	12	14	17	2.1	2.3	2.6	2.8	3.0	2.4	2.7	3.0	3.3	3.6
TW H3	3	5	7	9	12	15	18	22	2	3	5	7	9	11	13	16	2.2	2.4	2.6	2.8	3.0	2.5	2.8	3.0	3.3	3.5
TW H4	2	4	6	7	10	12	15	19	1	2	4	5	7	9	11	14	2.4	2.6	2.8	3.1	3.2	2.8	3.0	3.2	3.5	3.7
TW I1	5	6	7	8	10	11	12	13	3	4	4	5	5	5	5	4	2.7	3.3	4.0	-	-	3.7	-	-	-	-
TW I2	8	12	17	20	25	29	34	41	6	9	13	15	20	22	26	32	1.6	1.8	2.0	2.2	2.3	1.8	2.0	2.2	2.5	2.6
TW I3	9	13	18	22	28	32	37	45	6	10	14	17	22	25	30	36	1.5	1.7	1.9	2.0	2.2	1.7	1.9	2.1	2.3	2.5
TW I4	10	16	22	27	34	40	46	55	7	12	18	22	28	33	39	47	1.3	1.5	1.7	1.8	2.0	1.5	1.6	1.8	2.0	2.1
TW I5	10	16	23	29	37	42	49	58	7	13	19	24	31	35	42	51	1.3	1.4	1.6	1.8	1.9	1.4	1.6	1.7	1.9	2.0
TW J1	12	17	23	27	34	39	45	53	9	13	18	22	28	31	37	44	1.4	1.5	1.7	1.9	2.0	1.5	1.7	1.9	2.1	2.3
TW J2	11	16	23	28	35	40	46	55	8	13	18	22	29	33	39	47	1.3	1.5	1.7	1.8	2.0	1.5	1.6	1.8	2.0	2.1
TW J3	14	21	30	36	45	52	59	69	10	17	25	31	39	45	52	62	1.1	1.3	1.5	1.6	1.8	1.2	1.4	1.5	1.7	1.9
TW J4	12	20	28	35	44	50	57	67	9	16	23	29	37	43	50	60	1.2	1.3	1.5	1.6	1.8	1.3	1.4	1.6	1.7	1.9
DG A1	6	8	9	10	12	12	12	12	4	5	6	6	6	6	6	4	2.4	3.0	4.5	-	-	3.1	4.7	-	-	-
DG A2	9	12	15	18	22	24	28	32	7	9	12	14	17	19	21	25	1.8	2.0	2.3	2.6	3.0	2.1	2.3	2.7	3.1	3.6
DG A3	12	17	22	26	32	37	42	50	10	14	18	22	27	31	36	43	1.5	1.6	1.8	1.9	2.1	1.6	1.8	1.9	2.1	2.3
DG B1	6	8	9	10	12	13	14	15	4	5	6	6	7	7	6	5	2.3	3.0	3.7	-	-	3.1	4.3	-	-	-
DG B2	9	12	16	18	22	25	29	35	7	9	12	14	17	20	22	26	1.8	2.0	2.2	2.4	2.7	2.0	2.3	2.6	2.9	3.2
DG B3	13	17	23	27	33	38	45	54	11	14	19	23	28	33	38	46	1.5	1.6	1.7	1.8	2.0	1.6	1.7	1.9	2.0	2.1
DG C1	8	10	13	14	17	19	21	24	5	7	9	10	11	12	13	13	2.0	2.5	2.8	3.2	4.0	2.5	3.0	4.1	-	-
DG C2	11	15	19	22	27	31	36	43	9	12	15	18	22	25	29	34	1.6	1.8	1.9	2.1	2.3	1.8	2.0	2.2	2.5	2.7
DG C3	15	20	26	30	38	44	51	61	12	17	22	26	33	37	43	52	1.4	1.5	1.6	1.7	1.9	1.5	1.6	1.8	1.9	2.0
SS A1	14	17	21	23	27	30	33	38	13	15	18	20	23	25	27	30	1.7	2.0	2.3	2.5	2.7	1.9	2.2	2.6	3.0	3.4
SS A2	18	23	29	33	40	44	50	57	16	20	25	29	35	39	44	50	1.3	1.5	1.7	1.9	2.0	1.4	1.6	1.8	2.0	2.2
SS A3	13	15	18	20	22	24	27	30	11	13	15	17	18	20	21	22	1.9	2.3	2.7	3.1	3.4	2.2	2.7	3.3	4.3	-
SS A4	17	22	27	31	37	41	47	54	15	19	24	28	33	36	41	47	1.4	1.6	1.8	1.9	2.1	1.5	1.7	1.9	2.2	2.3
SS A5	15	17	20	22	25	27	29	32	14	15	17	19	21	22	24	26	1.9	2.2	2.6	2.9	3.2	2.2	2.6	3.1	3.7	4.4
SS A6	17	22	28	32	38	43	48	56	15	20	25	28	34	37	42	48	1.3	1.5	1.7	1.9	2.1	1.4	1.7	1.9	2.1	2.3
SS A7	14	15	17	18	20	21	23	24	13	14	15	16	17	18	18	18	2.3	2.8	3.3	4.0	4.7	2.7	3.3	4.4	-	-
SS A8	17	21	26	30	36	39	44	51	15	18	23	26	31	34	38	44	1.4	1.6	1.8	2.0	2.2	1.5	1.8	2.0	2.3	2.5
SS B1	12	14	17	20	23	25	28	32	10	12	15	16	19	20	23	25	1.9	2.1	2.4	2.7	2.9	2.1	2.5	2.9	3.3	3.8
SS B2	15	20	26	30	36	40	46	54	13	18	22	26	31	35	40	47	1.4	1.6	1.8	1.9	2.1	1.5	1.7	1.9	2.1	2.3
SS B3	11	13	15	17	20	22	24	27	9	11	13	14	16	17	18	20	2.0	2.4	2.7	3.1	3.5	2.3	2.8	3.4	4.2	-
SS B4	15	19	24	28	34	38	43	51	13	17	21	25	30	33	38	44	1.4	1.6	1.8	2.0	2.1	1.5	1.8	2.0	2.2	2.4
SS B5	12	14	16	18	20	22	24	27	11	12	14	15	17	18	19	21	2.2	2.5	2.9	3.2	3.6	2.5	2.9	3.5	4.2	5.0
SS B6	15	19	24	29	34	39	44	52	13	17	21	25	30	33	38	45	1.4	1.6	1.8	2.0	2.1	1.5	1.7	1.9	2.2	2.3
SS B7	11	12	14	15	17	18	20	21	10	11	12	13	14	15	15	16	2.5	2.9	3.4	4.0	4.6	2.9	3.5	4.5	-	-
SS B8	14	18	23	26	32	36	41	48	12	15	20	23	27	31	35	41	1.5	1.7	1.9	2.1	2.2	1.6	1.8	2.1	2.3	2.5
SS C1	9	12	15	17	21	23	27	31	6	8	10	12	14	16	18	20	1.8	2.1	2.4	2.6	2.9	2.2	2.6	3.0	3.6	4.2
SS C2	11	16	21	25	31	36	41	49	8	12	16	20	25	28	33	39	1.4	1.6	1.8	2.0	2.1	1.6	1.8	2.0	2.2	2.4
SS C3	9	12	14	16	18	20	23	26	8	9	11	12	14	15	17	19	2.1	2.4	2.7	3.0	3.3	2.5	2.9	3.4	3.9	4.5
SS C4	11	15	19	23	28	32	37	44	9	12	15	18	22	25	29	35	1.6	1.8	1.9	2.1	2.3	1.7	2.0	2.2	2.4	2.6
SS D1	15	19	23	26	31	34	38	44	13	16	19	22	25	28	30	34	1.5	1.8	2.0	2.3	2.5	1.7	2.0	2.4	2.8	3.1
SS D2	19	25	33	38	46	51	58	66	16	22	29	33	40	45	51	58	1.2	1.4	1.5	1.7	1.9	1.3	1.5	1.7	1.8	2.0
SS D3	15	18	22	25	28	31	34	39	14	16	19	21	24	25	28	31	1.7	2.0	2.2	2.5	2.7	1.9	2.2	2.6	3.1	3.5
SS D4	19	25	32	37	44	49	56	64	16	22	28	32	39	43	49	56	1.2	1.4	1.6	1.7	1.9	1.3	1.5	1.7	1.9	2.1
SS E1	12	15	18	21	24	27	30	35	10	12	14	16	19	20	22	24	1.7	2.0	2.3	2.6	2.9	1.9	2.4	2.9	3.6	4.5
SS E2	15	21	27	32	39	43	49	57	12	17	23	27	32	36	42	48	1.3	1.5	1.7	1.8	2.0	1.4	1.6	1.8	2.0	2.2
SS E3	12	15	17	19	22	24	27	30	11	13	15	16	18	19	21	23	2.0	2.3	2.6	3.0	3.3	2.2	2.7	3.2	3.9	4.7
SS E4	15	20	25	30	36	40	45	53	12	17	21	25	30	34	39	45	1.4	1.6	1.8	1.9	2.1	1.5	1.7	1.9	2.1	2.3

MADISON, WISCONSIN

TYPE	SOLAR SAVINGS FRACTION (%)																CONSERVATION FACTOR									
BASE TEMP:	55 F (5194 DD)								65 F (7731 DD)								55 F					65 F				
LCR	100	70	50	40	30	25	20	15	100	70	50	40	30	25	20	15	100	50	30	20	15	100	50	30	20	15
WW A1	14	15	17	19	21	22	24	27	12	13	14	15	16	17	18	19	2.3	2.6	3.0	3.4	3.7	2.8	3.4	4.2	-	-
WW A2	12	15	19	21	26	28	32	38	9	12	14	16	19	21	24	27	1.7	1.9	2.2	2.4	2.6	2.0	2.3	2.7	3.0	3.4
WW A3	11	15	19	22	27	31	35	42	8	11	14	17	20	23	26	30	1.6	1.8	2.0	2.2	2.4	1.8	2.1	2.4	2.7	3.0
WW A4	10	14	19	23	28	32	37	44	7	10	14	17	21	24	27	32	1.5	1.7	1.9	2.1	2.3	1.7	2.0	2.3	2.5	2.8
WW A5	10	14	20	24	30	34	39	47	6	10	14	17	21	25	29	34	1.4	1.6	1.8	2.0	2.2	1.7	1.9	2.2	2.4	2.7
WW A6	9	14	20	24	30	35	40	48	6	10	14	17	22	25	29	35	1.4	1.6	1.8	2.0	2.2	1.6	1.9	2.1	2.4	2.6
WW B1	8	10	12	13	15	16	17	18	6	7	8	8	8	8	8	7	2.3	2.9	3.6	4.5	-	3.3	4.9	-	-	-
WW B2	11	16	22	26	33	37	43	51	8	12	17	20	25	29	34	41	1.4	1.6	1.8	1.9	2.1	1.6	1.8	2.0	2.2	2.4
WW B3	14	20	27	32	40	45	51	60	10	15	21	25	32	36	42	49	1.3	1.4	1.6	1.8	1.9	1.4	1.6	1.8	2.0	2.1
WW B4	14	21	29	36	44	50	58	67	10	16	24	29	37	42	49	58	1.2	1.3	1.5	1.6	1.8	1.3	1.4	1.6	1.8	1.9
WW B5	13	21	29	36	45	51	59	68	10	16	24	30	38	44	51	60	1.2	1.3	1.5	1.6	1.8	1.2	1.4	1.6	1.7	1.9
WW C1	15	22	29	35	43	48	55	64	12	17	23	28	35	40	46	54	1.2	1.4	1.6	1.7	1.9	1.3	1.5	1.7	1.9	2.0
WW C2	14	21	28	34	42	48	55	65	10	16	23	28	35	40	47	55	1.2	1.4	1.5	1.7	1.8	1.3	1.5	1.7	1.8	2.0
WW C3	17	27	37	44	54	61	69	78	13	22	31	38	47	54	61	71	1.0	1.2	1.4	1.5	1.7	1.1	1.3	1.4	1.6	1.7
WW C4	15	24	34	41	51	58	66	75	12	19	28	35	44	51	58	68	1.1	1.2	1.4	1.5	1.7	1.1	1.3	1.5	1.6	1.8
TW A1	13	15	17	19	21	23	25	28	12	13	14	15	17	18	19	20	2.2	2.5	2.9	3.2	3.5	2.7	3.2	3.8	4.4	-
TW A2	12	15	18	20	24	27	30	35	10	12	14	16	19	20	23	26	1.8	2.0	2.3	2.5	2.7	2.1	2.4	2.8	3.1	3.5
TW A3	11	14	18	21	25	28	32	38	8	11	14	16	19	21	24	28	1.7	1.9	2.1	2.3	2.5	2.0	2.2	2.5	2.8	3.1
TW A4	9	13	17	21	25	29	33	40	7	10	13	16	19	22	25	30	1.6	1.8	2.0	2.2	2.4	1.9	2.1	2.4	2.6	2.9
TW B1	13	15	17	19	22	24	26	30	11	12	14	15	17	19	20	22	2.1	2.3	2.6	2.9	3.1	2.5	2.9	3.3	3.7	4.1
TW B2	11	14	17	20	24	26	30	36	9	11	14	15	18	20	23	27	1.8	2.0	2.2	2.4	2.6	2.1	2.3	2.6	2.9	3.2
TW B3	10	13	17	20	24	27	31	37	8	10	13	15	18	21	24	28	1.7	1.9	2.1	2.3	2.5	2.0	2.2	2.5	2.8	3.0
TW B4	9	12	16	19	23	26	30	36	8	10	13	15	18	20	24	28	1.8	2.0	2.2	2.4	2.5	2.0	2.3	2.5	2.7	2.9
TW C1	11	14	16	19	22	24	27	32	10	12	14	15	17	19	21	24	2.0	2.2	2.4	2.7	2.9	2.3	2.6	2.9	3.2	3.5
TW C2	10	13	16	18	22	25	29	34	8	11	13	15	18	20	23	27	1.9	2.1	2.3	2.5	2.6	2.1	2.4	2.6	2.9	3.1
TW C3	10	12	16	18	22	24	28	33	8	10	13	15	17	20	22	26	1.9	2.1	2.3	2.5	2.7	2.2	2.4	2.7	2.9	3.1
TW C4	10	12	15	17	21	23	26	31	9	11	13	15	17	19	21	25	2.1	2.2	2.4	2.6	2.8	2.3	2.6	2.8	3.0	3.2
TW D1	7	8	10	12	13	15	16	18	5	6	7	8	8	9	9	9	2.4	2.8	3.4	4.0	4.7	3.2	4.1	-	-	-
TW D2	12	16	21	25	30	34	40	47	9	13	17	20	24	27	32	38	1.5	1.7	1.9	2.0	2.2	1.7	1.9	2.1	2.3	2.5
TW D3	12	17	22	27	33	37	43	51	9	13	18	21	26	30	34	41	1.4	1.6	1.8	1.9	2.1	1.6	1.8	2.0	2.2	2.4
TW D4	13	19	27	32	40	45	52	61	10	16	22	26	33	38	44	52	1.3	1.4	1.6	1.7	1.9	1.4	1.6	1.7	1.9	2.0
TW D5	13	20	27	33	41	47	54	63	10	16	22	27	35	40	46	56	1.2	1.4	1.5	1.7	1.8	1.3	1.5	1.7	1.8	2.0
TW E1	15	20	26	31	38	43	49	58	12	16	21	25	31	35	41	48	1.3	1.5	1.7	1.8	2.0	1.4	1.6	1.8	2.0	2.2
TW E2	14	20	27	32	39	45	51	60	11	16	22	26	33	37	43	51	1.3	1.4	1.6	1.8	1.9	1.4	1.6	1.8	1.9	2.1
TW E3	17	25	34	41	50	56	64	73	14	21	29	35	43	49	56	65	1.1	1.3	1.4	1.6	1.7	1.2	1.4	1.5	1.7	1.8
TW E4	16	24	32	39	48	54	62	72	12	19	27	33	42	47	55	64	1.1	1.3	1.4	1.6	1.7	1.2	1.4	1.5	1.7	1.8
TW F1	11	13	16	17	20	21	24	27	10	11	13	14	15	16	18	19	2.2	2.5	2.8	3.1	3.4	2.7	3.1	3.6	4.3	4.9
TW F2	9	12	15	18	22	24	28	32	8	9	12	14	16	18	20	23	1.9	2.1	2.3	2.6	2.8	2.2	2.5	2.8	3.2	3.5
TW F3	8	11	15	18	22	25	29	34	6	8	11	13	16	18	21	25	1.7	2.0	2.2	2.4	2.6	2.0	2.3	2.6	2.9	3.2
TW F4	6	9	13	16	21	24	28	34	4	6	9	12	15	17	21	25	1.7	1.9	2.1	2.3	2.5	2.0	2.2	2.5	2.8	3.0
TW G1	9	11	14	16	19	21	23	27	8	9	11	12	14	16	17	20	2.1	2.4	2.6	2.9	3.1	2.5	2.9	3.3	3.7	4.1
TW G2	7	10	13	16	19	22	26	31	5	7	10	11	14	16	19	22	1.9	2.1	2.3	2.6	2.7	2.2	2.5	2.8	3.1	3.3
TW G3	5	8	12	14	18	21	25	31	4	6	8	10	13	15	18	22	1.9	2.1	2.3	2.5	2.7	2.1	2.4	2.7	2.9	3.2
TW G4	4	6	10	12	16	19	22	28	2	4	7	9	12	14	17	21	1.9	2.1	2.3	2.5	2.7	2.2	2.5	2.7	2.9	3.1
TW H1	6	9	11	13	17	19	22	26	5	7	9	10	12	14	16	19	2.1	2.3	2.6	2.8	3.0	2.4	2.7	3.1	3.4	3.7
TW H2	4	7	9	12	15	18	21	26	3	5	7	8	11	13	15	19	2.0	2.2	2.5	2.7	2.8	2.3	2.6	2.9	3.1	3.4
TW H3	3	5	8	10	13	16	19	24	2	4	6	7	10	12	14	18	2.1	2.3	2.5	2.7	2.9	2.4	2.7	2.9	3.1	3.4
TW H4	2	4	6	8	11	13	16	20	1	3	4	6	8	10	12	15	2.4	2.5	2.7	2.9	3.1	2.7	2.9	3.1	3.4	3.5
TW I1	5	7	8	9	11	12	14	15	4	4	5	6	6	7	7	7	2.5	3.0	3.5	4.2	5.0	3.3	4.4	-	-	-
TW I2	9	13	18	22	27	31	36	44	6	10	14	17	21	24	28	34	1.5	1.7	1.9	2.1	2.2	1.7	1.9	2.1	2.4	2.5
TW I3	10	14	20	24	30	34	40	48	7	11	15	19	24	27	32	39	1.4	1.6	1.8	2.0	2.1	1.6	1.8	2.0	2.2	2.4
TW I4	11	17	24	29	36	42	48	58	8	13	19	23	30	35	41	49	1.3	1.5	1.6	1.8	1.9	1.4	1.6	1.8	1.9	2.1
TW I5	11	17	25	30	38	44	51	61	8	14	20	25	32	37	44	53	1.3	1.4	1.6	1.7	1.9	1.4	1.5	1.7	1.8	2.0
TW J1	12	18	24	29	36	41	47	56	9	14	19	23	29	33	39	47	1.3	1.5	1.7	1.8	2.0	1.5	1.7	1.8	2.0	2.2
TW J2	12	18	24	29	37	42	49	58	9	14	19	24	30	35	41	49	1.3	1.5	1.6	1.8	1.9	1.4	1.6	1.8	1.9	2.1
TW J3	15	23	32	38	48	54	61	71	11	18	26	32	41	47	54	64	1.1	1.3	1.4	1.6	1.7	1.2	1.4	1.5	1.7	1.8
TW J4	13	21	30	36	46	52	60	70	10	17	25	31	39	45	52	62	1.1	1.3	1.4	1.6	1.7	1.2	1.4	1.5	1.7	1.8
DG A1	6	8	10	12	13	14	15	15	4	5	6	7	8	8	7	6	2.2	2.8	3.8	-	-	2.9	4.0	-	-	-
DG A2	9	12	16	19	23	26	30	34	7	10	13	15	18	20	23	26	1.7	1.9	2.2	2.5	2.9	2.0	2.2	2.5	2.9	3.4
DG A3	13	17	23	27	34	38	44	53	11	15	19	23	28	32	38	45	1.5	1.6	1.7	1.9	2.1	1.6	1.7	1.9	2.1	2.2
DG B1	6	8	10	12	14	15	16	18	4	5	6	7	8	8	8	8	2.2	2.7	3.3	4.1	-	2.9	3.7	-	-	-
DG B2	10	13	17	20	24	27	32	38	8	10	13	15	19	21	24	29	1.7	1.9	2.1	2.3	2.5	2.0	2.2	2.4	2.7	3.0
DG B3	13	18	24	28	35	40	47	56	11	15	20	24	30	34	40	48	1.4	1.6	1.7	1.8	1.9	1.6	1.7	1.8	2.0	2.1
DG C1	9	11	14	16	19	21	24	27	6	8	10	11	13	14	15	16	1.9	2.3	2.6	2.9	3.5	2.4	2.7	3.6	5.0	-
DG C2	12	15	20	24	29	33	39	46	9	12	16	19	23	26	30	36	1.5	1.7	1.9	2.0	2.2	1.8	1.9	2.1	2.4	2.6
DG C3	15	21	27	32	40	46	53	63	13	17	23	27	34	39	45	55	1.3	1.5	1.5	1.7	1.8	1.5	1.6	1.7	1.8	2.0
SS A1	15	18	22	25	29	32	36	41	14	16	19	22	25	27	30	33	1.6	1.9	2.1	2.4	2.6	1.8	2.1	2.4	2.8	3.1
SS A2	19	24	31	35	42	47	52	60	17	22	27	31	37	41	46	53	1.3	1.5	1.6	1.8	2.0	1.4	1.6	1.8	2.0	2.1
SS A3	14	16	19	22	25	27	30	33	12	14	17	18	21	22	24	26	1.8	2.1	2.5	2.8	3.1	2.1	2.5	3.0	3.6	4.4
SS A4	18	23	29	34	40	44	50	57	16	21	26	30	35	39	43	50	1.3	1.5	1.7	1.9	2.0	1.4	1.6	1.9	2.1	2.3
SS A5	16	18	21	23	27	29	32	36	14	16	19	20	23	24	26	29	1.8	2.1	2.4	2.7	3.0	2.1	2.4	2.8	3.3	3.8
SS A6	19	24	30	34	41	45	51	59	16	21	26	30	36	40	45	52	1.3	1.5	1.7	1.8	2.0	1.4	1.6	1.8	2.0	2.2
SS A7	14	16	18	20	22	23	25	28	13	15	16	17	19	19	20	21	2.1	2.5	3.0	3.5	4.0	2.5	3.0	3.8	5.0	-
SS A8	18	23	28	32	38	42	48	55	16	20	25	28	33	37	41	47	1.4	1.6	1.8	1.9	2.1	1.5	1.7	1.9	2.2	2.4
SS B1	13	15	19	21	25	27	31	35	11	13	16	18	21	22	25	28	1.8	2.0	2.3	2.6	2.8	2.0	2.3	2.7	3.0	3.4
SS B2	16	21	27	32	38	43	48	56	14	19	24	28	33	37	42	50	1.3	1.5	1.7	1.9	2.0	1.5	1.6	1.8	2.0	2.2
SS B3	11	14	17	19	22	24	26	30	10	12	14	16	18	19	21	23	1.9	2.2	2.5	2.9	3.2	2.2	2.6	3.1	3.7	4.3
SS B4	16	20	26	30	36	40	46	54	14	18	23	26	32	35	40	47	1.4	1.6	1.8	1.9	2.1	1.5	1.7	1.9	2.1	2.3
SS B5	13	15	17	19	22	24	26	30	12	13	15	16	18	20	21	23	2.1	2.4	2.7	3.0	3.3	2.3	2.7	3.2	3.7	4.3
SS B6	15	20	26	30	37	41	47	54	13	18	23	26	32	36	40	47	1.4	1.6	1.7	1.9	2.1	1.5	1.7	1.9	2.1	2.2
SS B7	11	13	15	16	18	20	22	24	11	12	13	14	15	16	17	18	2.3	2.7	3.1	3.6	4.0	2.7	3.2	3.9	4.9	-
SS B8	15	19	24	28	34	38	43	51	13	17	21	24	29	33	37	44	1.4	1.6	1.8	2.0	2.1	1.5	1.8	2.0	2.2	2.4
SS C1	9	12	16	19	23	26	29	34	7	9	12	13	16	18	20	23	1.7	2.0	2.2	2.5	2.7	2.1	2.4	2.8	3.2	3.6
SS C2	12	17	23	27	33	38	44	52	9	13	18	21	27	30	35	42	1.4	1.6	1.7	1.9	2.1	1.6	1.8	2.0	2.2	2.3
SS C3	10	12	15	17	20	22	25	29	8	10	12	13	15	16	18	21	2.0	2.3	2.6	2.8	3.0	2.4	2.8	3.2	3.6	4.0
SS C4	11	16	21	24	30	34	39	46	9	12	16	19	24	27	31	38	1.5	1.7	1.9	2.0	2.2	1.7	1.9	2.1	2.3	2.5
SS D1	16	20	25	29	34	37	42	48	14	17	21	24	28	31	34	38	1.5	1.7	1.9	2.1	2.3	1.6	1.9	2.2	2.6	2.9
SS D2	20	27	35	41	49	54	61	69	18	24	31	36	43	48	54	62	1.1	1.3	1.5	1.7	1.8	1.2	1.4	1.6	1.8	2.0
SS D3	16	20	24	26	31	34	37	43	14	17	20	23	26	28	31	34	1.6	1.8	2.1	2.3	2.5	1.8	2.1	2.4	2.8	3.1
SS D4	20	27	34	39	47	52	59	67	17	23	30	34	41	46	52	60	1.2	1.4	1.5	1.7	1.8	1.3	1.5	1.6	1.8	2.0
SS E1	13	16	20	23	27	30	34	39	11	13	16	18	21	23	25	28	1.6	1.9	2.2	2.4	2.7	1.8	2.2	2.6	3.1	3.7
SS E2	16	22	29	34	41	46	52	61	14	19	25	29	35	39	45	52	1.2	1.4	1.6	1.8	1.9	1.4	1.6	1.8	2.0	2.1
SS E3	13	16	19	21	24	26	29	33	12	14	16	17	20	21	23	26	1.8	2.1	2.4	2.7	3.0	2.1	2.5	2.9	3.4	4.0
SS E4	16	21	27	32	38	43	48	56	13	18	23	27	33	36	41	48	1.3	1.5	1.7	1.9	2.0	1.4	1.6	1.9	2.1	2.2

MILWAUKEE, WISCONSIN

TYPE	SOLAR SAVINGS FRACTION (%) BASE TEMP: 55 F (4904 DD) LCR 100	70	50	40	30	25	20	15	65 F (7445 DD) 100	70	50	40	30	25	20	15	CONSERVATION FACTOR 55 F 100	50	30	20	15	65 F 100	50	30	20	15
WW A1	14	15	18	19	21	23	25	27	12	13	14	15	16	17	18	19	2.2	2.6	3.0	3.4	3.8	2.8	3.5	4.3	-	-
WW A2	12	15	19	22	26	29	33	38	9	12	14	16	19	21	23	26	1.7	1.9	2.2	2.4	2.6	2.0	2.3	2.7	3.1	3.5
WW A3	11	15	20	23	28	31	36	42	8	11	14	16	20	22	25	30	1.5	1.8	2.0	2.2	2.4	1.8	2.1	2.4	2.7	3.0
WW A4	11	15	20	24	29	33	37	44	7	10	14	17	21	23	27	32	1.5	1.7	1.9	2.1	2.3	1.7	2.0	2.3	2.6	2.8
WW A5	10	15	20	24	30	34	39	46	6	10	14	17	21	24	28	33	1.4	1.6	1.9	2.1	2.2	1.7	1.9	2.2	2.5	2.7
WW A6	10	15	20	25	31	35	40	48	6	10	14	17	22	25	29	35	1.4	1.6	1.8	2.0	2.2	1.6	1.9	2.1	2.4	2.6
WW B1	9	10	12	14	15	16	17	19	6	7	8	8	8	8	8	6	2.3	2.8	3.6	4.6	-	3.3	-	-	-	-
WW B2	12	17	22	27	33	37	43	51	8	12	17	20	25	29	34	40	1.4	1.6	1.8	1.9	2.1	1.6	1.8	2.0	2.2	2.4
WW B3	14	20	27	32	40	45	51	60	10	15	21	25	31	36	41	49	1.2	1.4	1.6	1.8	1.9	1.4	1.6	1.8	2.0	2.2
WW B4	14	22	30	36	45	50	58	67	10	16	23	29	37	42	49	58	1.2	1.3	1.5	1.7	1.8	1.3	1.4	1.6	1.8	1.9
WW B5	14	21	30	36	45	51	59	68	10	16	24	29	38	43	50	60	1.1	1.3	1.5	1.6	1.8	1.3	1.4	1.6	1.7	1.9
WW C1	16	22	30	35	43	48	55	64	12	17	23	28	35	39	45	53	1.2	1.4	1.6	1.7	1.9	1.3	1.5	1.7	1.9	2.1
WW C2	14	21	29	35	43	48	55	64	10	16	23	28	35	40	46	55	1.2	1.4	1.5	1.7	1.8	1.3	1.5	1.7	1.8	2.0
WW C3	18	27	37	44	54	61	69	78	13	22	31	38	47	53	61	71	1.0	1.2	1.4	1.5	1.7	1.1	1.3	1.4	1.6	1.7
WW C4	16	25	34	41	51	58	66	75	12	19	28	35	44	50	58	68	1.1	1.2	1.4	1.5	1.7	1.2	1.3	1.5	1.6	1.8
TW A1	14	15	18	19	22	23	25	28	12	13	15	15	17	18	19	20	2.2	2.5	2.8	3.2	3.5	2.7	3.2	3.8	4.6	-
TW A2	12	15	18	21	25	27	31	35	10	12	14	16	18	20	22	26	1.8	2.0	2.3	2.5	2.7	2.1	2.5	2.8	3.2	3.5
TW A3	11	14	18	21	26	29	33	38	9	11	14	16	19	21	24	28	1.7	1.9	2.1	2.3	2.5	2.0	2.2	2.5	2.8	3.1
TW A4	10	14	18	21	26	29	34	40	7	10	13	15	19	22	25	29	1.6	1.8	2.0	2.2	2.4	1.9	2.1	2.4	2.7	2.9
TW B1	13	15	17	19	22	24	27	30	11	13	14	15	17	18	20	22	2.1	2.3	2.6	2.9	3.1	2.5	2.9	3.3	3.8	4.2
TW B2	11	14	18	20	24	27	31	36	9	11	14	15	18	20	23	27	1.8	2.0	2.2	2.4	2.6	2.1	2.4	2.6	2.9	3.2
TW B3	10	13	17	20	24	27	31	37	8	10	13	15	18	21	24	28	1.7	1.9	2.1	2.4	2.5	2.0	2.3	2.5	2.8	3.0
TW B4	10	13	16	19	24	27	31	36	8	10	13	15	18	20	23	28	1.7	1.9	2.2	2.4	2.5	2.0	2.3	2.5	2.8	3.0
TW C1	12	14	17	19	22	25	28	32	10	12	14	15	17	19	21	24	2.0	2.2	2.4	2.7	2.9	2.3	2.6	2.9	3.2	3.5
TW C2	10	13	16	19	23	25	29	34	9	11	13	15	18	20	22	26	1.8	2.0	2.3	2.5	2.7	2.1	2.4	2.7	2.9	3.1
TW C3	10	13	16	18	22	25	28	33	9	10	13	15	17	19	22	26	1.9	2.1	2.3	2.5	2.7	2.2	2.4	2.7	2.9	3.1
TW C4	10	13	15	18	21	23	27	31	9	11	13	15	17	19	21	25	2.0	2.2	2.4	2.6	2.8	2.3	2.6	2.8	3.0	3.2
TW D1	7	9	11	12	14	15	17	19	5	6	7	8	8	9	9	8	2.3	2.8	3.3	4.0	4.8	3.2	4.2	-	-	-
TW D2	12	16	21	25	31	35	40	47	9	13	16	19	24	27	31	37	1.5	1.7	1.9	2.1	2.2	1.7	1.9	2.1	2.3	2.5
TW D3	12	17	23	27	33	38	43	51	9	13	18	21	26	29	34	40	1.4	1.6	1.8	2.0	2.1	1.6	1.8	2.0	2.2	2.4
TW D4	14	20	27	32	40	45	52	61	10	16	21	26	33	37	43	52	1.3	1.4	1.6	1.8	1.9	1.4	1.6	1.7	1.9	2.1
TW D5	14	20	28	33	41	47	54	63	10	16	22	27	35	40	46	55	1.2	1.4	1.5	1.7	1.8	1.3	1.5	1.7	1.8	2.0
TW E1	15	21	27	32	39	43	49	58	12	16	21	25	31	35	40	48	1.3	1.5	1.7	1.8	2.0	1.5	1.6	1.8	2.0	2.2
TW E2	15	21	27	32	40	45	51	60	11	16	22	26	32	37	43	51	1.3	1.4	1.6	1.8	1.9	1.4	1.6	1.8	1.9	2.1
TW E3	18	26	35	41	50	56	64	73	14	21	29	34	43	48	56	65	1.1	1.3	1.4	1.6	1.7	1.2	1.4	1.5	1.7	1.8
TW E4	16	24	33	39	48	55	62	72	12	19	27	33	41	47	54	64	1.1	1.3	1.4	1.6	1.7	1.2	1.4	1.5	1.7	1.8
TW F1	12	14	16	18	20	22	24	27	10	11	13	14	15	16	17	19	2.2	2.4	2.8	3.1	3.4	2.7	3.1	3.7	4.4	-
TW F2	10	13	16	18	22	25	28	33	8	10	12	13	16	18	20	23	1.8	2.1	2.3	2.6	2.8	2.2	2.5	2.9	3.2	3.6
TW F3	8	11	15	18	22	25	29	35	6	8	11	13	16	18	21	25	1.7	1.9	2.2	2.4	2.6	2.0	2.3	2.6	2.9	3.2
TW F4	6	10	14	17	21	24	29	35	4	6	9	12	15	17	20	25	1.7	1.9	2.1	2.4	2.5	2.0	2.2	2.5	2.8	3.0
TW G1	10	12	14	16	19	21	24	28	8	9	11	12	14	15	17	19	2.1	2.3	2.6	2.9	3.2	2.5	2.9	3.3	3.8	4.2
TW G2	7	10	13	16	20	22	26	31	5	7	10	11	14	16	18	22	1.8	2.1	2.3	2.6	2.8	2.2	2.5	2.8	3.1	3.4
TW G3	6	9	12	15	19	22	25	31	4	6	8	10	13	15	18	22	1.8	2.0	2.3	2.5	2.7	2.1	2.4	2.7	3.0	3.2
TW G4	4	7	10	13	16	19	23	28	2	4	7	9	12	14	16	21	1.9	2.1	2.3	2.5	2.7	2.2	2.5	2.7	3.0	3.2
TW H1	7	9	12	14	17	19	22	27	5	7	9	10	12	14	16	19	2.1	2.3	2.5	2.8	3.0	2.4	2.8	3.1	3.5	3.8
TW H2	4	7	10	12	16	18	22	26	3	5	7	9	11	13	15	19	2.0	2.2	2.4	2.7	2.9	2.3	2.6	2.9	3.2	3.4
TW H3	3	6	8	11	14	16	20	24	2	4	6	7	10	12	14	18	2.1	2.3	2.5	2.7	2.9	2.4	2.7	2.9	3.2	3.4
TW H4	2	4	6	8	11	13	16	20	1	3	4	6	8	10	12	15	2.4	2.5	2.7	2.9	3.1	2.6	2.9	3.1	3.4	3.6
TW I1	5	7	9	10	12	13	14	16	4	4	5	6	6	7	7	6	2.5	2.9	3.5	4.3	-	3.3	4.5	-	-	-
TW I2	9	13	18	22	28	31	36	44	6	10	14	16	21	24	28	34	1.5	1.7	1.9	2.1	2.2	1.7	2.0	2.2	2.4	2.6
TW I3	10	15	20	24	30	35	40	48	7	11	15	19	24	27	32	38	1.4	1.6	1.8	2.0	2.1	1.6	1.8	2.0	2.2	2.4
TW I4	11	17	24	29	37	42	49	58	8	13	19	23	30	34	40	49	1.3	1.5	1.6	1.8	1.9	1.4	1.6	1.8	1.9	2.1
TW I5	11	18	25	31	39	44	51	61	8	14	20	25	32	37	43	52	1.2	1.4	1.6	1.7	1.9	1.4	1.5	1.7	1.9	2.0
TW J1	13	18	25	30	37	41	48	56	10	14	19	23	29	33	39	46	1.3	1.5	1.7	1.8	2.0	1.5	1.7	1.8	2.0	2.2
TW J2	12	18	25	30	37	42	49	58	9	14	19	24	30	35	40	49	1.3	1.5	1.6	1.8	1.9	1.4	1.6	1.8	2.0	2.1
TW J3	15	23	32	39	48	54	62	71	11	18	26	32	41	47	54	63	1.1	1.3	1.4	1.6	1.7	1.2	1.4	1.5	1.7	1.8
TW J4	14	22	30	37	46	52	60	69	10	17	25	30	39	45	52	62	1.1	1.3	1.5	1.6	1.8	1.2	1.4	1.5	1.7	1.8
DG A1	7	9	11	12	14	15	15	15	4	6	6	7	8	8	7	6	2.2	2.7	3.8	-	-	2.9	4.1	-	-	-
DG A2	10	13	17	20	24	27	30	35	7	10	13	15	18	20	23	26	1.7	1.9	2.2	2.5	2.9	2.0	2.3	2.6	3.0	3.4
DG A3	13	18	23	28	34	39	45	53	11	15	19	23	28	32	37	45	1.4	1.6	1.7	1.9	2.1	1.6	1.7	1.9	2.1	2.3
DG B1	7	9	11	12	14	16	17	18	4	5	6	7	8	8	8	8	2.1	2.7	3.3	4.4	-	3.0	3.8	-	-	-
DG B2	10	13	17	20	25	28	32	38	8	10	13	15	19	21	24	28	1.7	1.9	2.1	2.3	2.6	2.0	2.2	2.5	2.8	3.0
DG B3	14	18	24	29	36	41	47	56	11	15	20	24	30	34	40	48	1.4	1.6	1.6	1.8	1.9	1.6	1.7	1.8	2.0	2.1
DG C1	9	12	15	17	20	22	24	27	6	8	10	11	13	14	15	16	1.8	2.2	2.6	3.0	3.7	2.4	2.8	3.7	-	-
DG C2	12	16	21	24	30	34	39	46	9	12	16	19	23	26	30	36	1.5	1.7	1.9	2.1	2.3	1.8	1.9	2.2	2.4	2.6
DG C3	16	21	27	33	40	46	53	63	13	17	23	27	34	39	45	54	1.3	1.5	1.5	1.7	1.8	1.5	1.6	1.7	1.8	2.0
SS A1	16	19	23	26	30	33	36	42	14	17	20	22	25	27	30	34	1.6	1.9	2.1	2.4	2.6	1.8	2.1	2.4	2.8	3.1
SS A2	20	25	31	36	43	47	53	61	17	22	27	31	37	41	46	53	1.3	1.5	1.6	1.8	2.0	1.4	1.6	1.8	2.0	2.2
SS A3	14	17	20	22	26	28	30	34	13	15	17	19	21	22	24	26	1.8	2.1	2.5	2.8	3.1	2.0	2.4	3.0	3.6	4.4
SS A4	19	24	30	34	41	45	50	58	17	21	26	30	35	39	44	50	1.3	1.5	1.7	1.9	2.0	1.4	1.6	1.9	2.1	2.3
SS A5	16	19	22	24	27	30	32	36	15	17	19	21	23	25	27	29	1.8	2.1	2.4	2.7	3.0	2.0	2.4	2.8	3.3	3.8
SS A6	19	25	31	35	42	46	52	59	17	21	27	31	36	40	45	52	1.3	1.5	1.7	1.9	2.0	1.4	1.6	1.8	2.0	2.2
SS A7	15	17	19	21	23	24	26	28	14	15	16	18	19	20	21	22	2.1	2.5	2.9	3.5	4.0	2.4	3.0	3.7	4.9	-
SS A8	18	23	29	33	39	43	48	55	16	20	25	29	34	37	41	47	1.3	1.6	1.8	1.9	2.1	1.5	1.7	1.9	2.2	2.4
SS B1	13	16	19	22	26	28	31	36	11	14	16	18	21	23	25	29	1.7	2.0	2.3	2.6	2.8	2.0	2.3	2.7	3.0	3.4
SS B2	17	22	28	32	39	43	49	57	15	19	24	28	34	38	43	50	1.3	1.5	1.7	1.9	2.0	1.4	1.6	1.8	2.0	2.2
SS B3	12	14	17	20	23	25	27	31	10	12	14	16	18	19	21	23	1.9	2.2	2.5	2.9	3.2	2.1	2.6	3.0	3.6	4.3
SS B4	16	21	27	31	37	41	47	54	14	18	23	27	32	36	40	47	1.4	1.6	1.8	1.9	2.1	1.5	1.7	1.9	2.1	2.3
SS B5	13	15	18	20	23	24	27	30	12	13	15	17	19	20	22	24	2.0	2.3	2.6	3.0	3.3	2.3	2.7	3.2	3.7	4.3
SS B6	16	21	27	31	37	42	47	55	14	18	23	27	32	36	41	47	1.3	1.5	1.7	1.9	2.1	1.5	1.7	1.9	2.1	2.3
SS B7	12	14	16	17	19	21	23	25	11	12	13	14	16	17	18	19	2.2	2.6	3.0	3.5	4.0	2.6	3.2	3.9	4.9	-
SS B8	15	20	25	29	35	39	44	51	13	17	21	25	30	33	38	44	1.4	1.6	1.8	2.0	2.2	1.5	1.8	2.0	2.2	2.4
SS C1	10	13	17	19	23	26	30	35	7	9	12	13	16	18	20	23	1.7	2.0	2.2	2.5	2.7	2.1	2.4	2.8	3.3	3.8
SS C2	12	17	23	27	34	38	44	52	9	13	18	21	26	30	35	41	1.4	1.6	1.8	1.9	2.1	1.6	1.8	2.0	2.2	2.3
SS C3	10	12	15	17	20	22	25	29	8	10	12	13	15	16	18	20	2.0	2.3	2.5	2.8	3.1	2.4	2.8	3.2	3.7	4.1
SS C4	12	16	21	25	30	34	39	47	9	12	16	19	24	27	31	37	1.5	1.7	1.9	2.1	2.2	1.7	1.9	2.1	2.3	2.5
SS D1	17	21	26	30	34	38	42	48	14	18	22	24	28	31	34	38	1.4	1.7	1.9	2.2	2.3	1.6	1.9	2.2	2.6	2.9
SS D2	21	28	36	41	49	55	61	70	18	24	31	36	43	48	54	62	1.1	1.3	1.5	1.7	1.8	1.2	1.4	1.6	1.8	2.0
SS D3	17	20	24	27	31	34	38	43	15	18	21	23	26	28	31	35	1.6	1.8	2.1	2.4	2.6	1.8	2.1	2.4	2.8	3.2
SS D4	21	27	35	40	48	53	59	67	18	23	30	35	42	46	52	59	1.2	1.4	1.5	1.7	1.9	1.3	1.4	1.6	1.8	2.0
SS E1	13	17	21	24	28	31	34	39	11	14	16	19	21	23	26	28	1.6	1.9	2.2	2.4	2.7	1.8	2.2	2.6	3.2	3.8
SS E2	17	23	30	35	42	47	53	61	14	19	25	29	35	39	45	52	1.2	1.4	1.6	1.8	1.9	1.3	1.6	1.8	2.0	2.2
SS E3	14	16	19	22	25	27	30	34	12	14	16	18	20	22	24	26	1.8	2.1	2.4	2.7	3.0	2.1	2.5	2.9	3.4	4.0
SS E4	17	22	28	32	39	43	49	57	14	18	23	27	33	37	42	48	1.3	1.5	1.7	1.9	2.0	1.4	1.6	1.9	2.1	2.3

CASPER, WYOMING

TYPE	SOLAR SAVINGS FRACTION (%) BASE TEMP: 55 F (4923 DD) LCR 100	70	50	40	30	25	20	15	65 F (7560 DD) 100	70	50	40	30	25	20	15	CONSERVATION FACTOR 55 F 100	50	30	20	15	65 F 100	50	30	20	15
WW A1	21	26	31	34	40	44	49	57	17	20	23	26	31	34	38	44	1.4	1.6	1.7	1.9	2.0	1.7	1.9	2.1	2.2	2.4
WW A2	24	30	38	43	51	57	63	72	17	22	28	33	40	44	50	59	1.2	1.3	1.5	1.6	1.8	1.3	1.5	1.7	1.8	2.0
WW A3	24	32	40	47	55	61	68	77	17	23	30	35	43	48	55	64	1.1	1.2	1.4	1.6	1.7	1.2	1.4	1.6	1.7	1.9
WW A4	25	33	42	49	58	64	71	79	17	24	31	37	45	50	57	67	1.1	1.2	1.4	1.5	1.7	1.2	1.4	1.5	1.7	1.8
WW A5	25	34	44	51	60	66	73	82	17	24	32	38	47	53	60	69	1.0	1.2	1.3	1.5	1.6	1.2	1.3	1.5	1.6	1.8
WW A6	25	35	44	52	61	67	75	83	17	24	33	39	48	54	61	71	1.0	1.2	1.3	1.5	1.6	1.1	1.3	1.4	1.6	1.8
WW B1	19	24	30	34	41	45	51	59	13	17	21	24	29	32	36	43	1.3	1.5	1.7	1.8	2.0	1.6	1.8	2.0	2.2	2.4
WW B2	25	34	44	51	60	66	73	82	18	25	33	39	48	54	61	71	1.0	1.2	1.3	1.5	1.6	1.2	1.3	1.5	1.6	1.8
WW B3	30	40	51	58	68	74	81	88	22	30	40	47	56	62	70	79	0.9	1.1	1.3	1.4	1.6	1.0	1.2	1.4	1.5	1.7
WW B4	30	41	53	60	71	77	83	90	22	31	42	49	60	66	74	83	0.9	1.1	1.2	1.4	1.5	1.0	1.2	1.3	1.5	1.6
WW B5	29	40	52	60	70	76	83	90	21	30	41	49	59	66	74	83	0.9	1.1	1.2	1.4	1.5	1.0	1.1	1.3	1.5	1.6
WW C1	32	42	53	60	70	76	83	90	23	32	42	49	59	65	72	81	0.9	1.1	1.2	1.4	1.5	1.0	1.2	1.3	1.5	1.6
WW C2	30	40	51	59	69	75	82	89	22	31	41	48	58	64	72	81	0.9	1.1	1.2	1.4	1.5	1.0	1.2	1.3	1.5	1.6
WW C3	35	47	60	68	78	83	89	95	26	38	50	58	69	75	82	89	0.8	1.0	1.2	1.3	1.5	0.9	1.1	1.2	1.4	1.5
WW C4	32	44	56	65	75	81	87	93	24	34	46	54	65	72	79	87	0.9	1.0	1.2	1.3	1.5	0.9	1.1	1.2	1.4	1.6
TW A1	21	25	30	34	40	44	49	57	17	20	23	26	30	34	38	44	1.4	1.6	1.8	1.9	2.0	1.7	1.9	2.0	2.2	2.4
TW A2	22	28	34	40	47	52	58	67	16	21	26	30	36	40	46	54	1.2	1.4	1.5	1.7	1.8	1.4	1.6	1.8	1.9	2.1
TW A3	22	29	36	42	50	55	62	71	16	21	27	32	38	43	49	58	1.2	1.3	1.5	1.6	1.8	1.4	1.5	1.7	1.8	2.0
TW A4	21	28	36	42	50	56	63	72	15	20	27	32	39	44	51	59	1.1	1.3	1.4	1.6	1.7	1.3	1.5	1.6	1.8	1.9
TW B1	20	25	30	34	41	45	51	58	16	19	23	26	31	35	39	46	1.4	1.5	1.7	1.9	2.0	1.7	1.8	2.0	2.1	2.3
TW B2	20	26	33	38	46	51	58	66	15	19	25	29	35	40	46	54	1.2	1.4	1.5	1.7	1.8	1.4	1.6	1.7	1.9	2.0
TW B3	20	26	33	39	46	52	59	68	14	19	25	29	36	40	47	55	1.2	1.4	1.5	1.7	1.8	1.4	1.5	1.7	1.9	2.0
TW B4	19	25	31	37	44	49	56	65	13	18	23	28	34	39	45	53	1.3	1.4	1.6	1.7	1.8	1.4	1.6	1.7	1.9	2.0
TW C1	19	24	30	34	40	45	51	59	15	18	23	26	31	35	40	47	1.4	1.5	1.7	1.8	2.0	1.6	1.7	1.9	2.1	2.2
TW C2	18	24	30	35	42	47	53	62	14	18	23	26	32	36	42	50	1.3	1.5	1.6	1.8	1.9	1.5	1.7	1.8	2.0	2.1
TW C3	18	23	28	33	40	45	51	59	13	17	22	25	31	35	40	48	1.4	1.5	1.7	1.8	1.9	1.6	1.7	1.9	2.0	2.1
TW C4	17	21	26	30	36	40	46	54	13	16	20	23	28	32	36	43	1.5	1.6	1.8	1.9	2.0	1.7	1.8	2.0	2.1	2.3
TW D1	16	20	26	30	36	40	46	53	11	14	18	21	25	28	32	38	1.4	1.6	1.8	1.9	2.1	1.7	1.9	2.1	2.3	2.5
TW D2	24	31	40	46	55	61	68	77	17	23	30	36	44	49	56	65	1.1	1.2	1.4	1.6	1.7	1.2	1.4	1.5	1.7	1.8
TW D3	26	34	43	50	59	65	72	81	19	26	33	39	48	54	61	70	1.0	1.2	1.3	1.5	1.6	1.2	1.3	1.5	1.6	1.8
TW D4	28	37	47	55	65	71	78	86	20	28	37	44	54	60	67	77	1.0	1.1	1.3	1.4	1.6	1.1	1.2	1.4	1.5	1.7
TW D5	27	37	47	55	65	71	78	86	20	28	38	45	54	61	68	78	1.0	1.1	1.3	1.4	1.6	1.1	1.2	1.4	1.5	1.7
TW E1	29	38	48	55	65	71	77	85	22	29	38	45	53	59	67	76	1.0	1.1	1.3	1.5	1.6	1.1	1.2	1.4	1.6	1.7
TW E2	29	38	48	55	65	71	78	86	21	29	38	45	54	60	68	77	1.0	1.1	1.3	1.4	1.6	1.1	1.2	1.4	1.5	1.7
TW E3	34	45	57	65	74	80	86	92	26	36	47	54	65	71	78	86	0.9	1.0	1.2	1.4	1.5	1.0	1.1	1.3	1.4	1.6
TW E4	31	42	54	62	72	78	85	91	23	33	44	52	62	69	76	85	0.9	1.1	1.2	1.4	1.5	1.0	1.1	1.3	1.4	1.6
TW F1	19	24	29	33	39	43	49	57	15	18	22	25	29	33	37	44	1.4	1.6	1.7	1.9	2.0	1.7	1.8	2.0	2.2	2.3
TW F2	19	25	32	37	44	49	56	65	14	18	23	28	34	38	43	51	1.3	1.4	1.6	1.7	1.9	1.5	1.6	1.8	1.9	2.1
TW F3	19	25	32	38	46	51	58	67	13	18	23	28	35	39	45	54	1.2	1.4	1.5	1.7	1.8	1.4	1.5	1.7	1.9	2.0
TW F4	17	23	31	36	44	50	57	66	11	16	22	26	33	38	44	53	1.2	1.4	1.5	1.7	1.8	1.4	1.5	1.7	1.9	2.0
TW G1	17	22	28	32	38	43	49	57	13	16	20	24	29	32	37	44	1.4	1.6	1.7	1.9	2.0	1.7	1.8	2.0	2.1	2.3
TW G2	16	22	29	34	41	46	53	62	11	15	20	24	31	35	40	49	1.3	1.4	1.6	1.8	1.9	1.5	1.7	1.8	2.0	2.1
TW G3	15	20	27	32	40	45	52	61	9	14	19	23	29	34	40	48	1.3	1.4	1.6	1.7	1.9	1.5	1.6	1.8	2.0	2.1
TW G4	12	17	23	28	35	40	47	56	7	11	16	20	26	30	35	43	1.4	1.5	1.7	1.8	2.0	1.6	1.7	1.9	2.0	2.2
TW H1	14	19	25	29	35	40	46	54	10	13	17	21	26	30	35	42	1.4	1.6	1.7	1.9	2.0	1.7	1.8	2.0	2.1	2.3
TW H2	12	17	22	27	34	38	45	53	7	11	15	19	24	28	33	41	1.4	1.6	1.7	1.9	2.0	1.7	1.8	1.9	2.1	2.2
TW H3	10	14	19	24	30	34	40	49	6	9	13	16	21	25	30	37	1.5	1.6	1.8	1.9	2.1	1.8	1.9	2.0	2.2	2.3
TW H4	7	11	15	19	24	28	33	41	4	7	10	13	17	20	24	31	1.7	1.8	2.0	2.1	2.2	2.0	2.1	2.2	2.4	2.5
TW I1	14	18	23	27	32	37	42	50	9	12	15	18	22	25	29	35	1.6	1.7	1.0	2.0	2.1	1.8	2.0	2.2	2.4	2.6
TW I2	20	20	30	43	51	57	65	74	14	20	27	32	40	46	53	62	1.1	1.3	1.4	1.6	1.7	1.3	1.4	1.6	1.7	1.9
TW I3	22	30	39	46	55	61	69	78	15	22	30	36	44	50	57	66	1.1	1.2	1.4	1.5	1.7	1.2	1.4	1.5	1.7	1.8
TW I4	24	34	44	51	61	67	75	83	17	25	34	41	50	56	64	74	1.0	1.1	1.3	1.5	1.6	1.1	1.3	1.4	1.6	1.7
TW I5	24	34	45	52	62	69	76	85	17	26	35	42	52	58	66	76	1.0	1.1	1.3	1.4	1.6	1.1	1.2	1.4	1.5	1.7
TW J1	27	36	46	53	62	68	75	84	19	27	36	42	51	57	64	74	1.0	1.1	1.3	1.5	1.6	1.1	1.3	1.4	1.6	1.7
TW J2	26	35	45	52	62	68	76	84	18	26	35	42	51	57	65	74	1.0	1.1	1.3	1.5	1.6	1.1	1.3	1.4	1.6	1.7
TW J3	31	42	54	62	72	78	84	91	23	33	44	51	62	68	76	84	0.9	1.0	1.2	1.4	1.5	1.0	1.1	1.3	1.4	1.6
TW J4	28	39	51	59	69	76	83	90	21	30	41	49	59	66	74	83	0.9	1.1	1.2	1.4	1.5	1.0	1.1	1.3	1.5	1.6
DG A1	17	22	28	32	37	40	44	48	11	15	19	22	26	28	31	34	1.3	1.6	1.9	2.3	2.7	1.6	1.9	2.3	2.9	3.7
DG A2	19	25	33	38	46	51	56	64	13	18	24	29	35	39	44	51	1.2	1.4	1.6	1.8	2.0	1.4	1.5	1.8	2.0	2.3
DG A3	23	31	40	47	57	63	70	78	17	24	31	37	46	52	59	68	1.1	1.2	1.4	1.6	1.7	1.2	1.3	1.5	1.7	1.8
DG B1	17	23	29	34	41	46	51	57	11	15	19	23	28	31	36	41	1.3	1.4	1.6	1.9	2.3	1.6	1.8	2.0	2.3	2.7
DG B2	20	26	34	41	50	56	63	71	14	19	25	30	37	42	49	57	1.2	1.3	1.4	1.6	1.8	1.4	1.5	1.6	1.8	2.0
DG B3	24	32	42	49	60	67	75	83	18	24	32	39	48	55	63	73	1.1	1.2	1.3	1.5	1.6	1.2	1.3	1.4	1.5	1.7
DG C1	20	27	35	41	50	55	62	69	14	18	24	28	35	39	45	52	1.2	1.3	1.5	1.7	2.0	1.4	1.6	1.7	2.0	2.2
DG C2	23	30	39	47	57	63	71	79	16	22	29	35	44	50	57	66	1.1	1.2	1.3	1.5	1.7	1.3	1.4	1.5	1.7	1.8
DG C3	27	36	47	55	67	74	81	89	20	28	37	44	55	62	71	80	1.0	1.1	1.2	1.4	1.6	1.1	1.2	1.3	1.5	1.6
SS A1	26	32	39	44	51	56	63	70	21	26	32	37	43	47	53	60	1.2	1.3	1.5	1.7	1.8	1.3	1.5	1.7	1.9	2.0
SS A2	32	41	50	56	65	70	77	84	26	34	42	48	56	62	68	76	1.0	1.2	1.3	1.5	1.6	1.1	1.2	1.4	1.6	1.7
SS A3	24	29	35	40	47	51	57	64	19	24	29	33	38	42	47	53	1.2	1.4	1.6	1.8	1.9	1.4	1.6	1.8	2.0	2.2
SS A4	31	39	48	55	63	68	75	82	26	33	41	47	54	60	66	74	1.0	1.2	1.3	1.5	1.7	1.1	1.3	1.4	1.6	1.8
SS A5	25	30	36	41	47	51	57	65	21	25	30	34	39	43	48	54	1.3	1.4	1.6	1.8	1.9	1.4	1.6	1.8	2.0	2.2
SS A6	32	40	49	56	64	70	76	84	26	33	42	48	56	61	68	76	1.0	1.2	1.3	1.5	1.6	1.1	1.3	1.4	1.6	1.7
SS A7	22	27	31	35	41	44	49	56	19	22	26	29	33	36	40	45	1.4	1.6	1.8	2.0	2.1	1.6	1.8	2.0	2.3	2.5
SS A8	31	39	47	53	62	67	73	81	25	32	40	45	53	58	64	72	1.0	1.2	1.4	1.5	1.7	1.1	1.3	1.5	1.6	1.8
SS B1	22	28	34	39	46	51	57	65	17	22	27	31	37	41	47	54	1.2	1.4	1.6	1.8	1.9	1.4	1.6	1.8	2.0	2.1
SS B2	28	36	45	51	60	66	72	81	23	30	37	43	51	57	63	72	1.0	1.2	1.4	1.5	1.7	1.1	1.3	1.5	1.6	1.8
SS B3	21	26	32	36	43	47	53	61	16	20	25	29	34	38	43	49	1.3	1.5	1.7	1.8	2.0	1.5	1.7	1.9	2.1	2.2
SS B4	27	35	43	50	58	64	70	79	22	29	36	42	50	55	61	70	1.1	1.2	1.4	1.5	1.7	1.2	1.3	1.5	1.7	1.8
SS B5	20	25	30	34	40	44	49	57	17	20	24	27	32	35	40	46	1.4	1.6	1.7	1.9	2.1	1.6	1.8	2.0	2.2	2.3
SS B6	27	35	44	50	59	64	71	80	22	29	36	42	50	55	62	71	1.1	1.2	1.4	1.5	1.7	1.2	1.3	1.5	1.6	1.8
SS B7	19	22	27	30	35	39	44	50	15	18	21	24	28	31	35	40	1.5	1.7	1.9	2.1	2.2	1.7	1.9	2.2	2.4	2.6
SS B8	26	34	42	48	56	62	69	77	21	27	34	40	48	53	59	68	1.1	1.2	1.4	1.6	1.7	1.2	1.4	1.5	1.7	1.8
SS C1	21	28	36	41	50	55	62	71	15	20	26	31	37	42	48	57	1.2	1.3	1.5	1.6	1.8	1.4	1.5	1.7	1.8	2.0
SS C2	26	34	44	51	60	66	73	82	18	25	34	40	48	54	61	71	1.0	1.2	1.3	1.5	1.6	1.2	1.3	1.5	1.6	1.8
SS C3	18	23	29	34	40	45	51	59	13	17	22	25	30	34	39	46	1.4	1.5	1.7	1.8	2.0	1.6	1.7	1.9	2.1	2.2
SS C4	23	30	39	45	54	60	67	76	17	23	30	35	43	48	55	64	1.1	1.3	1.4	1.6	1.7	1.3	1.4	1.5	1.7	1.8
SS D1	30	38	46	52	60	66	72	80	24	31	38	43	50	55	61	69	1.0	1.2	1.4	1.5	1.7	1.2	1.3	1.5	1.7	1.9
SS D2	37	47	57	64	73	79	85	91	30	39	49	56	65	70	77	84	0.9	1.1	1.2	1.4	1.5	1.0	1.1	1.3	1.5	1.6
SS D3	28	34	42	47	54	59	66	73	23	28	34	39	45	50	55	63	1.1	1.3	1.5	1.6	1.8	1.3	1.4	1.6	1.8	2.0
SS D4	35	45	55	62	71	77	83	89	29	38	47	54	63	68	75	82	0.9	1.1	1.2	1.4	1.6	1.0	1.2	1.3	1.5	1.7
SS E1	26	33	41	46	54	59	66	74	20	26	32	37	44	48	54	62	1.1	1.3	1.4	1.6	1.8	1.2	1.4	1.6	1.8	2.0
SS E2	32	41	51	58	67	73	79	87	25	33	42	49	57	63	70	78	1.0	1.1	1.3	1.4	1.6	1.0	1.2	1.4	1.5	1.7
SS E3	23	28	34	39	46	50	56	64	18	22	27	31	37	41	46	53	1.3	1.4	1.6	1.8	1.9	1.4	1.6	1.8	2.0	2.2
SS E4	29	38	47	54	63	68	75	83	23	31	39	45	53	59	65	74	1.0	1.2	1.3	1.5	1.6	1.1	1.3	1.4	1.6	1.8

CHEYENNE, WYOMING

TYPE	SOLAR SAVINGS FRACTION (%) BASE TEMP: 55 F (4570 DD)								65 F (7261 DD)								CONSERVATION FACTOR 55 F					65 F				
LCR	100	70	50	40	30	25	20	15	100	70	50	40	30	25	20	15	100	50	30	20	15	100	50	30	20	15
WW A1	22	27	32	37	43	48	54	62	17	20	24	27	32	35	40	47	1.4	1.5	1.6	1.8	1.9	1.7	1.8	2.0	2.1	2.3
WW A2	25	32	41	47	55	61	68	77	18	23	30	34	42	47	53	62	1.1	1.2	1.4	1.5	1.7	1.3	1.4	1.6	1.7	1.9
WW A3	26	34	44	51	60	66	73	82	18	24	32	37	45	51	58	68	1.0	1.2	1.3	1.5	1.6	1.2	1.4	1.5	1.7	1.8
WW A4	27	35	45	53	62	68	76	84	18	25	33	39	48	53	61	70	1.0	1.1	1.3	1.5	1.6	1.2	1.3	1.5	1.6	1.7
WW A5	27	37	47	55	65	71	78	86	18	26	34	41	50	56	64	73	1.0	1.1	1.3	1.4	1.6	1.1	1.3	1.4	1.6	1.7
WW A6	28	38	48	56	66	72	80	88	18	26	35	42	51	57	65	75	1.0	1.1	1.3	1.4	1.6	1.1	1.2	1.4	1.6	1.7
WW B1	20	26	33	38	45	50	57	65	14	18	22	26	31	35	40	47	1.3	1.4	1.6	1.7	1.8	1.6	1.7	1.9	2.1	2.2
WW B2	27	37	47	54	64	71	78	86	19	26	35	41	50	57	64	74	1.0	1.1	1.3	1.4	1.6	1.1	1.3	1.4	1.6	1.7
WW B3	33	43	55	63	72	78	85	92	23	32	42	49	59	66	73	82	0.9	1.1	1.2	1.4	1.5	1.0	1.2	1.3	1.5	1.6
WW B4	33	44	56	64	75	81	87	93	23	33	44	52	62	69	76	85	0.9	1.0	1.2	1.3	1.5	1.0	1.1	1.3	1.4	1.6
WW B5	31	43	55	63	74	80	86	93	22	32	43	51	62	68	76	85	0.9	1.0	1.2	1.3	1.5	1.0	1.1	1.3	1.4	1.6
WW C1	34	45	57	65	75	80	67	93	25	34	44	52	62	68	76	84	0.9	1.0	1.2	1.4	1.5	1.0	1.1	1.3	1.5	1.6
WW C2	32	43	55	63	73	79	86	92	23	32	43	50	60	67	75	84	0.9	1.0	1.2	1.4	1.5	1.0	1.1	1.3	1.5	1.6
WW C3	38	51	64	72	82	87	92	96	28	39	52	60	71	77	84	91	0.8	1.0	1.1	1.3	1.4	0.9	1.0	1.2	1.4	1.5
WW C4	35	47	60	68	79	84	90	95	25	36	48	56	67	74	81	89	0.8	1.0	1.1	1.3	1.5	0.9	1.1	1.2	1.4	1.5
TW A1	22	26	32	36	42	47	53	61	17	20	24	27	32	35	40	46	1.4	1.5	1.7	1.8	1.9	1.7	1.8	2.0	2.1	2.3
TW A2	23	29	37	42	50	56	63	72	17	21	27	31	38	43	49	57	1.2	1.3	1.5	1.6	1.7	1.4	1.5	1.7	1.8	2.0
TW A3	23	31	39	45	54	60	67	76	16	22	28	33	40	46	52	61	1.1	1.3	1.4	1.6	1.7	1.3	1.5	1.6	1.8	1.9
TW A4	23	30	39	45	54	61	68	77	15	21	28	33	41	46	53	63	1.1	1.2	1.4	1.5	1.7	1.3	1.4	1.6	1.7	1.8
TW B1	21	26	32	37	43	48	54	63	16	20	24	27	33	36	41	49	1.3	1.5	1.6	1.8	1.9	1.6	1.7	1.9	2.0	2.2
TW B2	22	28	35	41	49	55	62	71	15	20	26	30	37	42	48	57	1.2	1.3	1.5	1.6	1.8	1.4	1.5	1.7	1.8	2.0
TW B3	21	28	36	42	50	56	63	72	15	20	26	31	38	42	49	58	1.2	1.3	1.5	1.6	1.7	1.4	1.5	1.7	1.8	1.9
TW B4	20	26	34	39	48	53	60	70	14	19	24	29	36	40	47	56	1.2	1.3	1.5	1.6	1.8	1.4	1.5	1.7	1.8	2.0
TW C1	20	25	31	36	43	48	55	64	15	19	23	27	33	37	42	50	1.3	1.4	1.6	1.7	1.9	1.6	1.7	1.8	2.0	2.1
TW C2	19	25	32	37	45	50	57	66	14	18	23	27	34	38	44	52	1.3	1.4	1.5	1.7	1.8	1.5	1.6	1.8	1.9	2.0
TW C3	18	24	30	35	43	48	54	63	14	17	22	26	32	36	42	50	1.3	1.4	1.6	1.7	1.9	1.5	1.7	1.8	1.9	2.1
TW C4	17	22	27	32	38	43	49	58	13	17	21	24	29	33	38	45	1.4	1.6	1.7	1.8	2.0	1.7	1.8	1.9	2.1	2.2
TW D1	17	22	28	32	39	44	50	59	11	15	19	22	27	30	35	42	1.4	1.5	1.7	1.8	1.9	1.7	1.8	2.0	2.2	2.3
TW D2	25	33	43	50	59	65	72	81	18	24	32	38	46	52	59	68	1.1	1.2	1.3	1.5	1.6	1.2	1.3	1.5	1.6	1.8
TW D3	28	37	47	54	64	70	77	85	19	27	35	41	50	56	64	73	1.0	1.1	1.3	1.4	1.6	1.1	1.3	1.4	1.6	1.7
TW D4	29	40	51	58	69	75	82	89	21	29	39	46	56	62	70	80	0.9	1.1	1.2	1.4	1.5	1.1	1.2	1.3	1.5	1.6
TW D5	29	39	50	58	68	75	82	89	20	29	39	46	56	63	71	80	0.9	1.1	1.2	1.4	1.5	1.1	1.2	1.3	1.5	1.6
TW E1	31	41	52	59	69	75	82	89	23	31	40	47	56	62	70	79	0.9	1.1	1.2	1.4	1.5	1.1	1.2	1.4	1.5	1.7
TW E2	31	41	51	59	69	75	82	89	22	30	40	47	56	63	70	80	0.9	1.1	1.2	1.4	1.5	1.1	1.2	1.3	1.5	1.6
TW E3	36	48	60	69	78	84	90	95	27	37	49	57	67	73	81	88	0.9	1.0	1.2	1.3	1.5	0.9	1.1	1.2	1.4	1.5
TW E4	34	45	57	66	76	82	88	94	24	35	46	54	64	71	78	87	0.9	1.0	1.2	1.3	1.5	1.0	1.1	1.3	1.4	1.6
TW F1	20	25	31	35	42	47	53	61	15	19	23	26	31	34	39	46	1.4	1.5	1.6	1.8	1.9	1.7	1.8	1.9	2.1	2.2
TW F2	21	27	34	40	48	53	61	70	14	19	24	29	35	40	46	55	1.2	1.3	1.5	1.6	1.8	1.4	1.6	1.7	1.9	2.0
TW F3	20	27	35	41	50	55	63	72	13	18	25	29	36	41	48	57	1.2	1.3	1.4	1.6	1.7	1.4	1.5	1.6	1.8	1.9
TW F4	18	25	33	39	48	54	62	71	11	16	23	28	35	40	47	56	1.2	1.3	1.4	1.6	1.7	1.4	1.5	1.6	1.8	1.9
TW G1	18	23	29	34	41	46	52	61	13	17	21	25	30	34	39	46	1.3	1.5	1.6	1.8	1.9	1.6	1.8	1.9	2.0	2.2
TW G2	17	24	31	36	44	50	57	67	12	16	21	26	32	37	43	51	1.2	1.4	1.5	1.7	1.8	1.5	1.6	1.7	1.9	2.0
TW G3	16	22	29	35	43	49	56	66	10	14	20	24	31	36	42	51	1.2	1.4	1.5	1.7	1.8	1.5	1.6	1.7	1.9	2.0
TW G4	13	18	25	30	38	43	50	60	8	12	17	21	27	31	37	46	1.3	1.5	1.6	1.7	1.9	1.6	1.7	1.8	2.0	2.1
TW H1	15	20	26	31	38	43	50	59	10	14	18	22	27	31	37	44	1.4	1.5	1.6	1.8	1.9	1.6	1.8	1.9	2.1	2.2
TW H2	13	18	24	29	36	41	48	58	8	12	16	20	26	30	35	43	1.4	1.5	1.6	1.8	1.9	1.6	1.8	1.9	2.0	2.2
TW H3	10	15	21	25	32	37	44	53	6	10	14	17	22	26	32	39	1.4	1.6	1.7	1.9	2.0	1.7	1.8	2.0	2.1	2.2
TW H4	8	12	16	20	26	30	36	44	4	7	10	13	18	21	25	32	1.6	1.8	1.9	2.0	2.1	1.9	2.1	2.2	2.3	2.4
TW I1	15	19	25	29	36	40	47	55	9	12	16	19	24	27	32	38	1.4	1.6	1.7	1.9	2.0	1.8	1.9	2.1	2.2	2.4
TW I2	22	30	39	46	55	61	69	78	15	21	28	34	42	48	55	65	1.1	1.2	1.4	1.5	1.6	1.3	1.4	1.5	1.7	1.8
TW I3	24	33	42	50	59	66	73	82	16	23	31	37	46	52	60	70	1.0	1.2	1.3	1.5	1.6	1.2	1.3	1.5	1.6	1.7
TW I4	26	36	47	55	65	71	79	87	18	26	36	43	52	59	67	77	1.0	1.1	1.3	1.4	1.6	1.1	1.2	1.4	1.5	1.7
TW I5	26	37	48	56	66	73	80	88	18	27	36	44	54	61	69	78	1.0	1.1	1.2	1.4	1.5	1.1	1.2	1.3	1.5	1.6
TW J1	29	38	49	57	67	73	80	88	20	28	37	44	54	60	68	77	1.0	1.1	1.3	1.4	1.6	1.1	1.2	1.4	1.5	1.7
TW J2	27	37	48	56	66	72	80	88	19	27	37	44	53	60	68	77	1.0	1.1	1.3	1.4	1.6	1.1	1.2	1.4	1.5	1.7
TW J3	33	45	57	65	76	82	88	94	24	34	45	54	64	71	78	87	0.9	1.0	1.2	1.3	1.5	1.0	1.1	1.3	1.4	1.6
TW J4	31	42	54	63	73	79	86	93	21	32	43	51	61	68	76	85	0.9	1.0	1.2	1.4	1.5	1.0	1.1	1.3	1.4	1.6
DG A1	18	24	30	35	41	44	49	54	11	15	20	23	27	30	33	37	1.3	1.5	1.8	2.1	2.5	1.6	1.8	2.2	2.7	3.3
DG A2	20	27	35	41	49	54	61	68	14	19	25	30	37	41	47	54	1.2	1.3	1.5	1.7	1.9	1.4	1.5	1.7	1.9	2.2
DG A3	24	33	43	51	61	67	74	81	18	24	32	39	48	55	62	71	1.1	1.1	1.3	1.5	1.7	1.2	1.3	1.4	1.6	1.8
DG B1	18	25	32	38	46	51	57	64	11	15	20	24	30	34	39	45	1.2	1.3	1.5	1.8	2.1	1.5	1.7	1.9	2.1	2.5
DG B2	21	28	37	44	54	61	68	76	14	19	26	31	39	45	52	61	1.1	1.2	1.4	1.6	1.8	1.4	1.5	1.6	1.7	1.9
DG B3	25	34	44	53	65	72	79	87	18	25	33	40	51	58	66	76	1.0	1.1	1.2	1.4	1.6	1.2	1.3	1.3	1.5	1.7
DG C1	22	29	38	45	55	62	68	75	14	19	25	30	37	42	49	57	1.1	1.2	1.3	1.6	1.9	1.4	1.5	1.6	1.8	2.1
DG C2	24	32	43	51	62	69	76	84	17	23	31	37	46	53	61	70	1.1	1.1	1.3	1.5	1.7	1.2	1.3	1.4	1.6	1.8
DG C3	28	38	50	60	72	79	85	92	21	28	38	46	57	65	74	83	1.0	1.0	1.2	1.4	1.6	1.1	1.2	1.3	1.4	1.6
SS A1	28	35	42	48	56	62	68	77	22	28	34	39	46	50	57	65	1.1	1.3	1.4	1.6	1.7	1.3	1.4	1.6	1.8	1.9
SS A2	35	44	54	61	70	75	82	89	28	36	45	51	60	65	72	80	0.9	1.1	1.3	1.4	1.6	1.0	1.2	1.4	1.5	1.7
SS A3	26	32	39	44	52	57	63	71	20	25	31	35	41	45	51	59	1.2	1.3	1.5	1.7	1.8	1.3	1.5	1.7	1.9	2.0
SS A4	34	43	52	59	68	74	80	87	27	35	43	50	58	64	70	79	1.0	1.1	1.3	1.4	1.6	1.1	1.2	1.4	1.5	1.7
SS A5	26	32	39	44	51	56	63	71	21	26	31	36	42	46	51	59	1.2	1.4	1.5	1.7	1.8	1.4	1.5	1.7	1.9	2.0
SS A6	34	44	53	60	69	75	81	88	27	35	44	50	59	65	72	80	0.9	1.1	1.3	1.4	1.6	1.1	1.2	1.4	1.5	1.7
SS A7	24	28	34	38	45	49	55	62	19	23	27	31	35	39	44	50	1.3	1.5	1.7	1.8	2.0	1.5	1.7	1.9	2.1	2.3
SS A8	33	42	51	58	67	73	79	86	26	34	42	48	57	62	69	77	1.0	1.1	1.3	1.5	1.6	1.1	1.2	1.4	1.6	1.7
SS B1	24	30	37	42	50	55	62	71	18	23	29	33	40	44	50	58	1.2	1.3	1.5	1.6	1.8	1.4	1.5	1.7	1.8	2.0
SS B2	30	39	48	55	64	70	77	85	24	31	39	45	54	60	67	76	1.0	1.1	1.3	1.5	1.6	1.1	1.3	1.4	1.6	1.7
SS B3	22	28	35	40	47	52	58	67	17	21	27	31	37	41	46	54	1.2	1.4	1.5	1.7	1.8	1.4	1.6	1.8	1.9	2.1
SS B4	29	38	47	54	63	69	76	84	23	30	38	44	53	58	65	74	1.0	1.2	1.3	1.5	1.6	1.1	1.3	1.4	1.6	1.7
SS B5	22	27	32	37	43	48	54	62	17	21	25	29	34	38	43	50	1.3	1.5	1.6	1.8	1.9	1.5	1.7	1.9	2.0	2.2
SS B6	29	38	47	54	63	69	76	84	23	30	38	44	53	59	66	75	1.0	1.2	1.3	1.5	1.6	1.1	1.3	1.4	1.6	1.7
SS B7	20	24	29	33	39	43	48	56	16	19	23	26	30	33	37	44	1.4	1.6	1.7	1.9	2.0	1.7	1.9	2.0	2.2	2.4
SS B8	28	36	45	52	61	67	74	82	22	29	36	42	51	56	63	72	1.0	1.2	1.3	1.5	1.6	1.2	1.3	1.5	1.6	1.8
SS C1	23	30	39	45	54	60	67	76	15	21	27	32	40	45	51	60	1.1	1.3	1.4	1.6	1.7	1.3	1.5	1.6	1.8	1.9
SS C2	27	37	47	54	64	70	78	86	19	27	35	42	51	57	64	74	1.0	1.1	1.3	1.4	1.6	1.1	1.3	1.4	1.6	1.7
SS C3	19	25	31	36	43	48	55	64	14	18	22	26	32	36	41	49	1.3	1.4	1.6	1.7	1.9	1.6	1.7	1.8	2.0	2.1
SS C4	25	33	42	48	58	64	71	80	17	24	31	37	45	51	58	67	1.1	1.2	1.3	1.5	1.6	1.2	1.4	1.5	1.7	1.8
SS D1	33	41	51	57	66	71	78	86	25	33	40	46	54	60	66	75	1.0	1.1	1.3	1.5	1.6	1.1	1.3	1.4	1.6	1.7
SS D2	40	51	62	69	78	84	89	94	32	41	52	59	68	74	81	88	0.9	1.0	1.2	1.3	1.5	0.9	1.1	1.3	1.4	1.6
SS D3	30	37	45	51	59	65	71	80	24	30	36	41	49	53	60	68	1.1	1.2	1.4	1.5	1.7	1.2	1.4	1.5	1.7	1.9
SS D4	38	49	60	67	76	82	87	93	31	40	50	57	66	72	79	86	0.9	1.0	1.2	1.4	1.5	1.0	1.1	1.3	1.4	1.6
SS E1	28	36	45	51	60	65	72	80	21	27	34	40	47	52	59	67	1.0	1.2	1.4	1.5	1.7	1.2	1.4	1.5	1.7	1.8
SS E2	34	45	55	63	72	78	84	91	26	35	45	52	61	67	74	82	0.9	1.1	1.2	1.4	1.5	1.0	1.2	1.3	1.5	1.6
SS E3	24	30	37	42	50	55	62	70	19	23	29	33	39	44	49	57	1.2	1.4	1.5	1.7	1.8	1.4	1.5	1.7	1.9	2.0
SS E4	32	41	51	58	68	73	80	88	24	32	41	48	56	62	69	78	1.0	1.1	1.3	1.4	1.6	1.1	1.2	1.4	1.5	1.7

ROCK SPRINGS, WYOMING

TYPE	SOLAR SAVINGS FRACTION (%) BASE TEMP: 55 F (5603 DD) LCR 100	70	50	40	30	25	20	15	65 F (8414 DD) 100	70	50	40	30	25	20	15	CONSERVATION FACTOR 55 F 100	50	30	20	15	65 F 100	50	30	20	15
WW A1	20	24	29	32	38	41	47	54	16	19	22	25	29	32	36	42	1.5	1.7	1.8	2.0	2.1	1.8	2.0	2.1	2.3	2.4
WW A2	22	28	35	41	48	54	60	69	16	21	27	31	38	42	48	57	1.2	1.3	1.5	1.7	1.8	1.4	1.5	1.7	1.9	2.0
WW A3	22	29	38	44	52	58	65	74	16	22	28	34	41	46	53	62	1.1	1.3	1.4	1.6	1.7	1.3	1.4	1.6	1.7	1.9
WW A4	22	30	39	46	55	60	68	77	16	22	29	35	43	48	55	65	1.1	1.2	1.4	1.6	1.7	1.2	1.4	1.5	1.7	1.8
WW A5	23	31	41	48	57	63	70	79	16	23	31	37	45	51	58	67	1.0	1.2	1.4	1.5	1.7	1.2	1.3	1.5	1.7	1.8
WW A6	23	32	42	49	58	64	72	81	16	23	31	37	46	52	59	69	1.0	1.2	1.3	1.5	1.6	1.2	1.3	1.5	1.6	1.8
WW B1	17	22	27	32	38	42	48	55	12	16	19	22	27	30	34	40	1.4	1.6	1.7	1.9	2.0	1.7	1.9	2.1	2.3	2.5
WW B2	23	31	41	48	57	63	71	79	17	24	32	38	46	52	59	69	1.1	1.2	1.4	1.5	1.7	1.2	1.3	1.5	1.6	1.8
WW B3	28	38	48	55	65	71	78	86	21	29	38	45	55	61	68	77	1.0	1.1	1.3	1.4	1.6	1.1	1.2	1.4	1.5	1.7
WW B4	28	38	50	58	68	74	81	89	20	30	40	48	58	64	72	81	0.9	1.1	1.2	1.4	1.5	1.0	1.2	1.3	1.5	1.6
WW B5	27	37	49	57	67	74	81	89	19	29	39	47	58	64	72	81	0.9	1.1	1.2	1.4	1.5	1.0	1.2	1.3	1.5	1.6
WW C1	30	40	50	58	68	74	80	88	22	31	41	48	57	63	71	80	0.9	1.1	1.3	1.4	1.6	1.0	1.2	1.3	1.5	1.6
WW C2	28	38	49	56	66	73	80	87	20	29	39	46	56	63	70	80	1.0	1.1	1.3	1.4	1.6	1.0	1.2	1.3	1.5	1.6
WW C3	33	45	57	66	76	82	88	94	25	36	48	56	67	74	81	89	0.9	1.0	1.2	1.3	1.5	0.9	1.1	1.2	1.4	1.5
WW C4	30	41	54	62	73	79	85	92	22	33	44	53	64	70	78	86	0.9	1.0	1.2	1.4	1.5	1.0	1.1	1.3	1.4	1.6
TW A1	20	23	28	32	37	41	46	53	16	19	22	25	29	32	36	42	1.5	1.7	1.8	2.0	2.1	1.8	1.9	2.1	2.3	2.4
TW A2	20	26	32	37	44	49	55	64	15	20	25	29	34	39	44	52	1.3	1.4	1.6	1.7	1.9	1.5	1.6	1.8	1.9	2.1
TW A3	20	26	34	39	47	52	59	68	15	20	26	30	37	41	48	56	1.2	1.4	1.5	1.7	1.8	1.4	1.5	1.7	1.8	2.0
TW A4	19	26	34	39	48	53	60	69	14	19	25	30	37	42	49	57	1.2	1.3	1.5	1.6	1.8	1.4	1.5	1.6	1.8	1.9
TW B1	19	23	28	32	38	42	48	55	15	18	22	25	30	33	38	44	1.5	1.6	1.7	1.9	2.0	1.7	1.9	2.0	2.2	2.3
TW B2	19	24	31	36	43	48	55	63	14	18	24	28	34	38	44	52	1.3	1.4	1.6	1.7	1.9	1.5	1.6	1.8	1.9	2.1
TW B3	18	24	31	36	44	49	56	65	13	18	23	28	34	39	45	53	1.3	1.4	1.6	1.7	1.8	1.4	1.6	1.7	1.9	2.0
TW B4	17	23	29	34	42	47	53	62	13	17	22	26	33	37	43	51	1.3	1.4	1.6	1.7	1.9	1.5	1.6	1.8	1.9	2.1
TW C1	18	22	28	32	38	42	48	56	14	17	22	25	30	33	38	45	1.4	1.6	1.7	1.9	2.0	1.7	1.8	1.9	2.1	2.2
TW C2	17	22	28	32	39	44	50	59	13	17	21	25	31	35	40	48	1.4	1.5	1.7	1.8	1.9	1.6	1.7	1.9	2.0	2.1
TW C3	16	21	26	31	37	42	48	56	13	16	20	24	29	33	38	46	1.4	1.6	1.7	1.8	2.0	1.6	1.8	1.9	2.0	2.2
TW C4	16	19	24	28	34	38	43	51	12	15	19	22	27	30	35	42	1.5	1.7	1.8	2.0	2.1	1.8	1.9	2.0	2.2	2.3
TW D1	14	18	23	27	33	37	42	50	10	13	17	19	23	26	30	36	1.5	1.7	1.8	2.0	2.1	1.8	2.0	2.2	2.4	2.5
TW D2	22	29	37	44	52	58	65	74	16	22	29	34	42	47	54	63	1.1	1.3	1.4	1.6	1.7	1.3	1.4	1.6	1.7	1.9
TW D3	24	32	41	47	57	63	70	79	17	24	32	38	46	52	59	68	1.1	1.2	1.4	1.5	1.7	1.2	1.3	1.5	1.6	1.8
TW D4	25	35	45	52	62	68	75	84	19	27	36	43	52	58	66	75	1.0	1.1	1.3	1.5	1.6	1.1	1.2	1.4	1.6	1.7
TW D5	25	34	45	52	62	69	76	84	18	27	36	43	53	59	67	76	1.0	1.1	1.3	1.5	1.6	1.1	1.2	1.4	1.5	1.7
TW E1	27	36	46	53	62	68	75	83	20	28	37	43	52	58	65	74	1.0	1.2	1.3	1.5	1.6	1.1	1.3	1.4	1.6	1.7
TW E2	26	35	45	53	62	68	76	84	20	28	36	43	52	58	66	75	1.0	1.1	1.3	1.5	1.6	1.1	1.2	1.4	1.6	1.7
TW E3	32	43	54	62	72	78	85	91	24	34	45	53	63	69	77	85	0.9	1.0	1.2	1.4	1.5	1.0	1.1	1.3	1.4	1.6
TW E4	29	40	51	59	70	76	83	90	22	32	42	50	61	67	75	84	0.9	1.1	1.2	1.4	1.5	1.0	1.1	1.3	1.5	1.6
TW F1	18	22	27	31	36	40	46	53	14	17	21	24	28	31	35	42	1.5	1.6	1.8	1.9	2.1	1.8	1.9	2.1	2.2	2.4
TW F2	18	23	29	34	42	47	53	62	13	17	22	26	32	36	42	49	1.3	1.5	1.6	1.8	1.9	1.5	1.7	1.8	2.0	2.1
TW F3	17	23	30	35	43	48	55	64	12	16	22	26	33	37	43	52	1.3	1.4	1.6	1.7	1.8	1.4	1.6	1.7	1.9	2.0
TW F4	15	21	28	34	42	47	54	63	10	15	20	25	32	36	43	51	1.3	1.4	1.6	1.7	1.8	1.4	1.6	1.7	1.9	2.0
TW G1	16	20	25	30	36	40	46	54	12	15	19	22	27	31	35	42	1.5	1.6	1.8	1.9	2.0	1.7	1.9	2.0	2.2	2.3
TW G2	15	20	26	31	38	43	50	59	10	14	19	23	29	33	39	47	1.3	1.5	1.6	1.8	1.9	1.6	1.7	1.8	2.0	2.1
TW G3	13	18	25	30	37	42	49	58	9	13	18	22	28	32	38	46	1.3	1.5	1.6	1.8	1.9	1.6	1.7	1.8	2.0	2.1
TW G4	10	15	21	26	32	37	44	53	6	10	15	18	24	28	34	41	1.4	1.6	1.7	1.9	2.0	1.6	1.8	1.9	2.1	2.2
TW H1	13	17	22	27	33	37	43	51	9	12	16	20	24	28	33	40	1.5	1.6	1.8	1.9	2.1	1.7	1.9	2.0	2.2	2.3
TW H2	10	15	20	25	31	36	42	50	7	10	14	18	23	27	32	39	1.5	1.6	1.8	1.9	2.0	1.7	1.8	2.0	2.1	2.3
TW H3	9	13	18	22	27	32	38	46	5	8	12	15	20	24	28	36	1.6	1.7	1.8	2.0	2.1	1.8	1.9	2.1	2.2	2.4
TW H4	6	10	14	17	22	26	31	38	4	6	9	12	16	19	23	29	1.7	1.9	2.0	2.2	2.3	2.0	2.2	2.3	2.4	2.6
TW I1	12	16	21	24	30	34	39	46	8	11	14	17	21	23	27	33	1.6	1.7	1.9	2.1	2.2	1.9	2.1	2.3	2.5	2.6
TW I2	19	26	34	40	49	54	62	71	13	19	26	31	38	44	51	60	1.2	1.3	1.5	1.6	1.7	1.3	1.4	1.6	1.8	1.9
TW I3	20	28	37	43	52	58	66	75	14	21	28	34	42	48	55	65	1.1	1.2	1.4	1.6	1.7	1.2	1.4	1.5	1.7	1.8
TW I4	22	31	41	49	58	65	72	81	16	24	33	39	49	55	63	72	1.0	1.2	1.3	1.5	1.6	1.1	1.3	1.4	1.6	1.7
TW I5	22	32	42	50	60	66	74	83	16	24	33	40	50	57	65	74	1.0	1.2	1.3	1.5	1.6	1.1	1.2	1.4	1.5	1.7
TW J1	24	33	43	50	60	66	73	82	18	26	34	40	49	55	63	72	1.0	1.2	1.3	1.5	1.6	1.1	1.3	1.4	1.6	1.7
TW J2	23	32	42	50	59	66	73	82	17	25	33	40	49	56	63	73	1.0	1.2	1.3	1.5	1.6	1.1	1.3	1.4	1.6	1.7
TW J3	28	39	51	59	69	76	83	90	21	31	42	50	60	67	75	83	0.9	1.1	1.2	1.4	1.5	1.0	1.1	1.3	1.5	1.6
TW J4	26	37	48	57	67	73	81	88	19	29	40	47	58	64	72	82	0.9	1.1	1.2	1.4	1.5	1.0	1.2	1.3	1.5	1.6
DG A1	15	20	25	29	34	37	41	45	10	13	17	20	24	26	29	31	1.4	1.6	2.0	2.4	2.9	1.7	1.9	2.4	3.0	3.9
DG A2	17	23	30	36	43	48	53	61	13	17	23	27	33	37	43	49	1.3	1.4	1.6	1.8	2.1	1.5	1.6	1.8	2.0	2.3
DG A3	21	29	38	45	54	60	68	76	16	22	30	36	45	51	58	67	1.1	1.2	1.4	1.6	1.8	1.3	1.4	1.5	1.7	1.9
DG B1	15	20	26	31	38	42	47	54	10	13	18	21	26	29	33	38	1.4	1.5	1.7	2.0	2.3	1.7	1.8	2.0	2.4	2.8
DG B2	18	24	32	38	46	52	59	68	13	18	23	28	35	40	47	55	1.2	1.3	1.5	1.7	1.9	1.4	1.5	1.7	1.8	2.0
DG B3	22	30	39	47	57	64	72	81	17	23	31	37	47	53	62	72	1.1	1.2	1.3	1.5	1.7	1.2	1.3	1.4	1.6	1.7
DG C1	18	25	32	38	46	51	58	65	13	17	22	27	33	37	43	50	1.2	1.4	1.5	1.7	2.0	1.5	1.6	1.8	2.0	2.3
DG C2	21	28	37	44	54	60	68	76	15	21	28	33	42	48	55	64	1.1	1.2	1.4	1.5	1.7	1.3	1.4	1.5	1.7	1.9
DG C3	25	33	44	52	64	71	79	87	19	26	35	42	53	60	69	79	1.0	1.1	1.2	1.4	1.6	1.2	1.2	1.3	1.5	1.6
SS A1	25	31	37	42	50	54	61	68	21	26	32	36	42	46	52	59	1.2	1.4	1.5	1.7	1.9	1.3	1.5	1.7	1.9	2.0
SS A2	31	39	48	55	63	69	75	83	26	33	42	48	56	61	68	76	1.0	1.2	1.3	1.5	1.6	1.1	1.2	1.4	1.6	1.7
SS A3	23	28	34	38	45	49	55	62	19	23	28	32	38	41	46	53	1.3	1.5	1.6	1.8	2.0	1.4	1.6	1.8	2.0	2.2
SS A4	30	38	47	53	61	67	73	81	25	32	40	46	54	59	66	74	1.0	1.2	1.4	1.5	1.7	1.1	1.3	1.5	1.6	1.8
SS A5	24	29	35	39	45	50	55	63	20	25	29	33	38	42	47	53	1.3	1.5	1.6	1.8	2.0	1.4	1.6	1.8	2.0	2.2
SS A6	30	39	48	54	63	68	75	82	26	33	41	47	55	60	67	75	1.0	1.2	1.3	1.5	1.7	1.1	1.3	1.4	1.6	1.8
SS A7	21	25	30	34	39	43	47	54	18	22	25	28	33	36	39	45	1.4	1.6	1.8	2.0	2.2	1.6	1.8	2.1	2.3	2.5
SS A8	29	37	46	52	60	65	72	80	25	32	39	45	52	58	64	72	1.0	1.2	1.4	1.5	1.7	1.1	1.3	1.5	1.7	1.8
SS B1	21	26	32	37	44	48	54	62	17	21	27	31	36	40	46	53	1.3	1.4	1.6	1.8	1.9	1.4	1.6	1.8	2.0	2.1
SS B2	27	34	43	49	58	64	70	79	22	29	37	42	50	56	63	71	1.1	1.2	1.4	1.5	1.7	1.1	1.3	1.5	1.6	1.8
SS B3	19	24	30	34	41	45	51	58	16	20	24	28	33	37	42	48	1.3	1.5	1.7	1.9	2.0	1.5	1.7	1.9	2.1	2.3
SS B4	26	33	42	48	56	62	69	77	21	28	35	41	49	54	61	69	1.1	1.2	1.4	1.6	1.7	1.2	1.3	1.5	1.7	1.8
SS B5	19	24	29	32	38	42	47	54	16	20	24	27	31	35	39	45	1.4	1.6	1.8	2.0	2.1	1.6	1.8	2.0	2.2	2.4
SS B6	26	34	42	48	57	63	69	78	21	28	36	41	49	55	61	70	1.1	1.2	1.4	1.6	1.7	1.2	1.3	1.5	1.7	1.8
SS B7	18	21	25	29	34	37	42	48	15	18	21	24	28	30	34	39	1.6	1.7	1.9	2.1	2.3	1.8	2.0	2.2	2.4	2.6
SS B8	25	32	40	46	54	60	67	75	20	27	34	39	47	52	58	67	1.1	1.3	1.4	1.6	1.7	1.2	1.4	1.5	1.7	1.8
SS C1	19	26	33	39	47	52	59	68	14	19	24	29	36	40	46	55	1.2	1.4	1.5	1.7	1.8	1.4	1.5	1.7	1.9	2.0
SS C2	23	32	41	48	57	63	70	79	17	24	32	38	47	52	60	69	1.1	1.2	1.4	1.5	1.7	1.2	1.3	1.5	1.6	1.8
SS C3	17	21	27	31	38	42	48	56	13	16	20	24	29	32	37	44	1.4	1.6	1.7	1.9	2.0	1.7	1.8	2.0	2.1	2.3
SS C4	21	28	36	43	51	57	64	73	16	22	28	34	41	47	53	63	1.1	1.3	1.4	1.6	1.7	1.3	1.4	1.6	1.7	1.9
SS D1	29	36	44	50	58	64	70	78	24	30	37	42	50	55	61	68	1.1	1.2	1.4	1.6	1.7	1.2	1.3	1.5	1.7	1.9
SS D2	35	45	55	63	72	77	83	90	29	38	48	55	64	70	76	84	0.9	1.1	1.2	1.4	1.6	1.0	1.1	1.3	1.5	1.6
SS D3	26	33	40	45	53	57	64	71	22	28	34	38	45	49	55	62	1.2	1.3	1.5	1.7	1.8	1.3	1.5	1.6	1.8	2.0
SS D4	34	43	54	61	70	75	81	88	28	37	46	53	62	68	74	82	0.9	1.1	1.3	1.4	1.6	1.0	1.2	1.3	1.5	1.7
SS E1	24	31	39	44	52	57	63	72	19	25	31	36	43	47	53	61	1.1	1.3	1.5	1.6	1.8	1.3	1.5	1.6	1.8	2.0
SS E2	30	39	49	56	65	71	78	85	24	32	41	48	57	62	69	77	1.0	1.1	1.3	1.5	1.6	1.1	1.2	1.4	1.6	1.7
SS E3	21	27	33	37	44	48	54	62	18	22	27	30	36	40	45	52	1.3	1.5	1.6	1.8	2.0	1.5	1.6	1.8	2.0	2.2
SS E4	28	36	45	52	61	66	73	81	23	30	38	44	52	58	65	73	1.0	1.2	1.3	1.5	1.7	1.1	1.3	1.4	1.6	1.8

SHERIDAN, WYOMING

TYPE	SOLAR SAVINGS FRACTION (%) BASE TEMP: 55 F (5042 DD) LCR								65 F (7717 DD)								CONSERVATION FACTOR 55 F					65 F				
	100	70	50	40	30	25	20	15	100	70	50	40	30	25	20	15	100	50	30	20	15	100	50	30	20	15
WW A1	18	21	24	26	30	33	36	42	14	16	19	20	23	25	27	31	1.7	2.0	2.2	2.4	2.6	2.1	2.4	2.6	3.0	3.2
WW A2	18	23	28	32	38	42	48	55	13	17	21	24	29	32	37	42	1.4	1.6	1.7	1.9	2.1	1.6	1.8	2.0	2.2	2.4
WW A3	18	23	30	34	41	46	52	60	12	17	22	26	31	35	40	47	1.3	1.5	1.6	1.8	2.0	1.5	1.7	1.9	2.1	2.3
WW A4	17	23	30	35	43	48	54	62	12	17	22	26	32	37	42	49	1.2	1.4	1.6	1.8	1.9	1.4	1.6	1.8	2.0	2.2
WW A5	17	24	31	37	45	50	56	65	11	17	23	27	34	38	44	52	1.2	1.4	1.5	1.7	1.9	1.4	1.5	1.7	1.9	2.1
WW A6	17	24	32	38	46	51	58	67	11	17	23	28	35	39	45	53	1.2	1.3	1.5	1.7	1.8	1.3	1.5	1.7	1.9	2.1
WW B1	14	17	21	23	27	30	34	38	9	12	14	16	18	20	22	24	1.7	1.9	2.2	2.5	2.7	2.1	2.4	2.9	3.5	4.1
WW B2	18	25	32	38	46	51	58	67	13	18	25	29	36	41	47	56	1.2	1.3	1.5	1.7	1.8	1.3	1.5	1.7	1.9	2.0
WW B3	22	30	38	45	54	59	66	75	16	22	30	36	44	49	56	64	1.1	1.2	1.4	1.6	1.7	1.2	1.4	1.5	1.7	1.9
WW B4	22	31	41	48	57	63	71	80	16	24	32	39	48	54	61	71	1.0	1.2	1.3	1.5	1.7	1.1	1.3	1.4	1.6	1.8
WW B5	21	30	40	47	57	63	71	80	15	23	32	39	48	54	62	72	1.0	1.2	1.3	1.5	1.6	1.1	1.3	1.4	1.6	1.7
WW C1	23	32	41	47	56	62	69	78	17	24	32	38	47	52	59	68	1.1	1.2	1.4	1.5	1.7	1.2	1.3	1.5	1.7	1.8
WW C2	22	30	40	46	56	62	69	78	16	23	31	38	46	52	59	68	1.1	1.2	1.4	1.5	1.7	1.2	1.3	1.5	1.6	1.8
WW C3	26	37	48	56	66	72	79	87	20	29	40	47	58	64	72	81	0.9	1.1	1.3	1.4	1.6	1.0	1.2	1.3	1.5	1.6
WW C4	24	34	45	53	63	69	77	85	17	27	37	44	54	61	69	78	1.0	1.1	1.3	1.4	1.6	1.0	1.2	1.3	1.5	1.7
TW A1	17	20	24	26	30	33	37	42	14	16	19	21	23	25	28	32	1.7	1.9	2.2	2.4	2.5	2.1	2.3	2.6	2.8	3.1
TW A2	17	21	26	30	35	39	44	51	13	16	20	23	27	30	34	40	1.5	1.7	1.8	2.0	2.2	1.7	1.9	2.1	2.3	2.5
TW A3	16	21	27	31	37	41	47	55	12	16	20	23	28	32	36	43	1.4	1.6	1.7	1.9	2.1	1.6	1.8	2.0	2.2	2.4
TW A4	15	21	27	31	38	42	48	56	11	15	20	23	29	33	37	44	1.3	1.5	1.7	1.9	2.0	1.5	1.7	1.9	2.1	2.3
TW B1	17	20	24	27	31	34	38	44	13	16	18	21	24	26	29	34	1.6	1.9	2.1	2.2	2.4	2.0	2.2	2.4	2.6	2.9
TW B2	16	20	25	29	34	38	44	51	12	15	19	22	27	30	34	40	1.5	1.6	1.8	2.0	2.1	1.7	1.9	2.1	2.3	2.5
TW B3	15	20	25	29	35	39	44	52	11	14	19	22	27	30	35	41	1.4	1.6	1.8	2.0	2.1	1.6	1.8	2.0	2.2	2.4
TW B4	14	19	24	28	33	37	43	50	10	14	18	21	26	29	34	40	1.5	1.6	1.8	2.0	2.1	1.7	1.8	2.0	2.2	2.4
TW C1	15	19	23	26	31	34	39	45	12	15	18	20	24	27	30	35	1.6	1.8	2.0	2.2	2.3	1.9	2.1	2.3	2.5	2.7
TW C2	14	18	23	26	32	35	40	47	11	14	18	20	25	28	32	38	1.5	1.7	1.9	2.1	2.2	1.8	1.9	2.1	2.3	2.5
TW C3	14	18	22	25	31	34	39	46	11	14	17	20	24	27	31	37	1.6	1.8	1.9	2.1	2.2	1.8	2.0	2.2	2.4	2.5
TW C4	14	17	21	24	28	31	36	42	11	13	16	19	22	25	29	34	1.7	1.9	2.0	2.2	2.4	2.0	2.1	2.3	2.5	2.6
TW D1	12	14	18	20	24	27	30	35	8	10	12	14	16	18	20	23	1.8	2.0	2.3	2.5	2.7	2.2	2.5	2.9	3.3	3.7
TW D2	18	24	30	35	42	47	54	62	13	18	23	27	34	38	44	51	1.3	1.4	1.6	1.8	1.9	1.4	1.6	1.8	2.0	2.1
TW D3	19	25	33	38	46	51	58	66	14	19	25	30	37	41	47	55	1.2	1.4	1.5	1.7	1.8	1.3	1.5	1.7	1.9	2.0
TW D4	20	28	37	43	52	58	65	74	15	22	29	35	43	48	55	65	1.1	1.3	1.4	1.6	1.7	1.2	1.4	1.5	1.7	1.8
TW D5	20	28	37	44	53	59	66	75	15	22	30	36	44	50	57	67	1.1	1.2	1.4	1.6	1.7	1.2	1.3	1.5	1.7	1.8
TW E1	22	29	37	43	51	57	63	72	16	22	29	35	42	47	53	62	1.1	1.3	1.5	1.6	1.8	1.2	1.4	1.6	1.8	1.9
TW E2	21	29	37	43	52	58	65	74	16	22	30	35	43	48	55	64	1.1	1.3	1.4	1.6	1.7	1.2	1.4	1.5	1.7	1.9
TW E3	26	35	45	52	62	68	75	84	19	28	37	44	53	60	67	76	1.0	1.1	1.3	1.5	1.6	1.1	1.2	1.4	1.6	1.7
TW E4	24	33	43	50	60	66	74	82	18	26	35	42	52	58	65	75	1.0	1.2	1.3	1.5	1.6	1.1	1.2	1.4	1.6	1.7
TW F1	16	19	22	25	29	32	36	41	12	14	17	19	22	24	27	31	1.7	1.9	2.1	2.3	2.5	2.1	2.3	2.5	2.8	3.0
TW F2	15	19	24	27	33	36	41	48	11	14	17	20	24	27	31	37	1.5	1.7	1.9	2.1	2.2	1.7	1.9	2.2	2.4	2.6
TW F3	13	18	23	28	34	38	43	51	9	13	17	20	25	28	33	39	1.4	1.6	1.8	2.0	2.1	1.6	1.8	2.0	2.2	2.4
TW F4	11	16	22	26	32	37	42	50	7	11	15	19	24	27	32	39	1.4	1.6	1.8	1.9	2.1	1.6	1.8	2.0	2.2	2.4
TW G1	14	17	21	24	28	31	36	42	10	13	15	18	21	24	27	31	1.7	1.9	2.1	2.3	2.4	2.0	2.2	2.4	2.6	2.9
TW G2	12	16	21	24	30	34	39	46	8	11	15	18	22	25	29	35	1.5	1.7	1.9	2.1	2.2	1.8	2.0	2.2	2.4	2.6
TW G3	10	14	19	23	29	33	38	45	6	10	13	16	21	24	29	35	1.5	1.7	1.9	2.1	2.2	1.8	1.9	2.1	2.3	2.5
TW G4	8	12	16	20	25	29	34	41	5	7	11	14	18	21	26	32	1.6	1.8	2.0	2.1	2.3	1.9	2.0	2.2	2.4	2.6
TW H1	11	14	18	21	26	29	34	40	7	10	13	15	19	21	25	30	1.7	1.9	2.1	2.2	2.4	2.0	2.2	2.4	2.6	2.8
TW H2	8	12	16	19	24	28	33	39	5	8	11	13	17	20	24	30	1.6	1.8	2.0	2.2	2.3	1.9	2.1	2.3	2.5	2.7
TW H3	7	10	14	17	22	25	29	36	4	6	9	12	15	18	22	27	1.7	1.9	2.1	2.3	2.4	2.0	2.2	2.4	2.6	2.7
TW H4	5	8	11	13	17	20	24	30	3	5	7	9	12	14	18	22	1.9	2.1	2.3	2.5	2.6	2.3	2.4	2.6	2.8	2.9
TW I1	10	12	16	18	22	24	27	32	6	8	10	12	14	15	17	20	1.8	2.1	2.4	2.6	2.8	2.3	2.6	3.0	3.4	3.9
TW I2	15	20	27	32	39	44	50	59	10	15	20	24	30	35	40	48	1.3	1.5	1.6	1.8	1.9	1.5	1.6	1.8	2.0	2.1
TW I3	16	22	29	35	42	47	54	63	11	16	22	27	33	38	44	52	1.2	1.4	1.6	1.7	1.9	1.4	1.5	1.7	1.9	2.1
TW I4	18	25	33	40	48	54	62	71	12	19	26	32	40	45	52	62	1.1	1.3	1.5	1.6	1.8	1.2	1.4	1.6	1.7	1.9
TW I5	18	26	34	41	50	56	64	73	12	19	27	33	42	47	55	64	1.1	1.3	1.4	1.6	1.7	1.2	1.4	1.5	1.7	1.8
TW J1	19	27	35	41	49	54	61	70	14	20	27	32	40	45	51	60	1.1	1.3	1.5	1.6	1.8	1.3	1.4	1.6	1.8	1.9
TW J2	19	26	34	41	49	55	62	71	13	20	27	32	40	46	53	62	1.1	1.3	1.5	1.6	1.8	1.2	1.4	1.6	1.7	1.9
TW J3	23	32	42	50	59	66	73	82	17	25	35	41	51	57	65	74	1.0	1.2	1.3	1.5	1.6	1.1	1.2	1.4	1.6	1.7
TW J4	21	30	40	47	57	64	71	80	15	23	33	39	49	55	63	73	1.0	1.2	1.3	1.5	1.6	1.1	1.3	1.4	1.6	1.7
DG A1	12	15	19	21	25	27	29	31	7	10	12	14	16	17	19	19	1.6	2.0	2.5	3.2	4.4	2.0	2.4	3.2	4.9	-
DG A2	14	19	24	28	34	38	43	49	10	14	18	21	26	29	33	38	1.4	1.6	1.8	2.1	2.4	1.6	1.8	2.1	2.4	2.7
DG A3	18	24	31	37	45	50	57	65	14	19	25	30	37	42	48	56	1.3	1.4	1.5	1.7	1.9	1.4	1.5	1.7	1.8	2.0
DG B1	12	15	19	22	27	29	33	37	7	10	13	15	17	19	21	23	1.6	1.9	2.1	2.6	3.0	2.0	2.3	2.8	3.5	4.5
DG B2	15	20	25	30	36	41	46	54	10	14	19	22	27	31	36	42	1.4	1.6	1.7	1.9	2.1	1.6	1.8	2.0	2.2	2.4
DG B3	19	25	32	38	47	53	61	70	14	19	26	31	38	44	51	60	1.2	1.3	1.5	1.6	1.8	1.4	1.5	1.6	1.7	1.9
DG C1	15	19	24	28	33	37	42	47	10	13	17	19	23	26	29	33	1.4	1.7	1.8	2.1	2.5	1.7	1.9	2.3	2.7	3.1
DG C2	17	23	29	34	42	47	54	63	13	17	23	27	33	37	43	51	1.3	1.4	1.6	1.7	1.9	1.5	1.6	1.8	1.9	2.2
DG C3	21	28	36	43	52	59	67	77	16	22	30	35	43	49	57	67	1.2	1.3	1.4	1.5	1.7	1.3	1.4	1.5	1.6	1.8
SS A1	20	25	30	34	39	43	48	55	17	21	25	28	33	36	40	46	1.4	1.6	1.8	2.0	2.1	1.5	1.7	2.0	2.2	2.5
SS A2	25	32	40	45	53	58	64	72	21	27	34	39	46	51	56	64	1.1	1.3	1.5	1.6	1.8	1.2	1.4	1.6	1.8	1.9
SS A3	19	23	27	30	35	38	42	48	16	19	22	25	29	31	34	38	1.5	1.7	2.0	2.2	2.4	1.7	1.9	2.3	2.6	2.9
SS A4	25	31	38	43	51	55	62	69	21	26	33	37	44	48	54	61	1.2	1.3	1.5	1.7	1.8	1.2	1.4	1.6	1.8	2.0
SS A5	20	24	28	31	36	39	43	49	17	20	24	26	30	33	36	41	1.5	1.7	2.0	2.2	2.3	1.7	1.9	2.2	2.5	2.8
SS A6	25	32	39	44	52	57	63	71	21	27	33	38	45	50	55	63	1.1	1.3	1.5	1.7	1.8	1.2	1.4	1.6	1.8	2.0
SS A7	18	21	25	27	31	33	36	41	16	18	21	23	25	27	29	32	1.7	2.0	2.2	2.5	2.8	1.9	2.2	2.6	3.1	3.5
SS A8	24	30	37	42	49	54	60	67	20	26	32	36	42	46	52	59	1.2	1.4	1.6	1.7	1.9	1.3	1.5	1.7	1.9	2.1
SS B1	17	21	26	30	35	38	43	49	14	17	21	24	28	31	35	40	1.5	1.7	1.9	2.1	2.3	1.6	1.9	2.1	2.4	2.6
SS B2	22	28	36	41	48	53	60	68	18	24	30	35	42	46	52	60	1.2	1.4	1.5	1.7	1.8	1.3	1.5	1.6	1.8	2.0
SS B3	16	20	24	27	32	35	39	44	13	16	19	22	25	28	31	35	1.6	1.8	2.0	2.2	2.4	1.8	2.0	2.3	2.6	2.9
SS B4	21	27	34	39	47	51	58	66	18	23	29	34	40	44	50	58	1.2	1.4	1.6	1.7	1.9	1.3	1.5	1.7	1.9	2.0
SS B5	17	20	24	26	30	33	37	42	14	16	19	22	25	27	30	34	1.7	1.9	2.1	2.3	2.5	1.9	2.1	2.4	2.7	3.0
SS B6	21	28	34	40	47	52	58	67	17	23	29	34	40	45	51	58	1.2	1.4	1.5	1.7	1.9	1.3	1.5	1.7	1.9	2.0
SS B7	15	18	21	23	27	29	32	36	13	15	17	19	21	23	25	28	1.8	2.1	2.4	2.6	2.8	2.1	2.4	2.7	3.1	3.5
SS B8	20	26	33	38	44	49	55	63	16	22	27	32	38	42	48	55	1.2	1.4	1.6	1.8	1.9	1.3	1.5	1.7	1.9	2.1
SS C1	15	20	26	30	36	40	45	53	11	14	18	22	26	30	34	40	1.4	1.6	1.8	2.0	2.1	1.6	1.8	2.1	2.3	2.5
SS C2	19	25	33	38	46	52	58	67	13	19	25	30	37	42	48	56	1.2	1.3	1.5	1.7	1.8	1.3	1.5	1.7	1.8	2.0
SS C3	14	18	22	25	30	33	37	44	11	13	16	19	22	25	28	33	1.6	1.8	2.0	2.2	2.4	1.9	2.1	2.3	2.6	2.8
SS C4	17	23	30	34	42	46	53	61	13	17	23	27	33	37	43	51	1.3	1.4	1.6	1.8	1.9	1.4	1.6	1.8	2.0	2.1
SS D1	23	29	35	40	46	50	56	63	19	24	29	33	38	42	47	53	1.2	1.4	1.6	1.8	2.0	1.3	1.6	1.8	2.1	2.3
SS D2	28	36	45	52	60	66	72	80	23	31	39	45	53	58	65	73	1.0	1.2	1.4	1.5	1.7	1.1	1.3	1.5	1.6	1.8
SS D3	22	27	32	36	42	46	51	57	18	22	27	30	35	38	42	48	1.3	1.5	1.7	1.9	2.1	1.5	1.7	2.0	2.2	2.4
SS D4	27	35	44	50	58	64	70	78	23	30	38	43	51	56	62	71	1.0	1.2	1.4	1.6	1.7	1.1	1.3	1.5	1.7	1.8
SS E1	19	24	30	34	40	44	49	56	15	19	24	27	32	35	39	44	1.3	1.5	1.7	1.9	2.1	1.5	1.7	2.0	2.3	2.5
SS E2	24	31	39	45	54	59	66	74	19	25	33	38	46	50	57	65	1.1	1.3	1.4	1.6	1.8	1.2	1.4	1.6	1.7	1.9
SS E3	18	22	26	29	34	38	42	48	15	18	21	24	28	30	34	38	1.5	1.7	1.9	2.2	2.3	1.7	1.9	2.2	2.5	2.8
SS E4	22	29	37	42	50	55	61	69	18	24	30	35	42	47	53	60	1.2	1.3	1.5	1.7	1.8	1.2	1.4	1.6	1.8	2.0

EDMONTON, ALBERTA

SOLAR SAVINGS FRACTION (%)

TYPE	BASE TEMP: 55 F (7587 DD) LCR 100	70	50	40	30	25	20	15	65 F (10671 DD) 100	70	50	40	30	25	20	15
WW A1	14	16	17	19	20	21	22	24	12	14	15	16	18	18	19	20
WW A2	12	15	18	21	24	26	28	31	10	13	16	18	20	22	24	26
WW A3	11	15	18	21	25	27	30	34	9	12	16	18	21	23	26	28
WW A4	10	14	18	21	25	28	31	35	8	12	15	18	22	24	27	30
WW A5	10	14	18	22	26	29	32	37	7	11	15	18	22	25	28	31
WW A6	9	14	19	22	27	30	33	38	7	11	15	19	23	25	28	32
WW B1	9	10	11	12	13	13	13	11	7	8	9	9	10	10	9	6
WW B2	12	16	21	25	30	34	38	44	9	13	18	22	27	30	34	39
WW B3	14	19	25	29	35	39	44	50	11	16	22	26	32	35	40	46
WW B4	14	20	28	33	41	46	52	60	11	18	25	30	37	42	48	56
WW B5	13	20	28	34	42	47	54	62	11	18	25	31	39	44	50	58
WW C1	15	21	27	32	38	42	48	55	13	18	24	29	35	39	44	50
WW C2	14	20	27	32	39	44	49	57	12	17	24	29	36	40	46	53
WW C3	17	25	34	41	50	56	63	71	15	23	32	38	47	52	59	68
WW C4	15	23	32	39	48	53	60	69	13	21	29	36	45	50	57	66
TW A1	14	16	18	19	21	22	23	25	12	14	15	17	18	19	20	22
TW A2	12	15	18	20	23	25	27	30	10	13	15	17	20	22	24	26
TW A3	11	14	18	20	24	26	29	32	9	12	15	17	20	23	25	28
TW A4	10	13	17	20	24	26	30	34	8	11	14	17	21	23	26	29
TW B1	13	15	17	19	21	23	25	27	11	13	15	17	19	20	22	23
TW B2	11	14	17	19	23	25	28	31	9	12	15	17	20	22	24	27
TW B3	10	13	17	19	23	25	28	32	8	11	14	17	20	22	25	28
TW B4	10	13	16	19	22	25	28	32	8	11	14	16	19	22	25	29
TW C1	12	14	17	19	21	23	26	29	10	12	15	16	19	20	23	25
TW C2	11	13	16	18	22	24	27	31	9	11	14	16	19	21	24	27
TW C3	10	13	16	18	21	24	26	31	9	11	14	16	19	21	24	27
TW C4	11	13	16	18	21	23	25	29	9	11	14	16	18	20	23	26
TW D1	7	9	10	11	12	13	14	14	6	7	8	9	10	10	10	9
TW D2	12	16	21	24	29	32	36	41	10	14	18	21	25	28	32	37
TW D3	12	17	21	25	30	33	38	43	10	14	19	22	27	30	34	39
TW D4	14	19	26	30	37	41	47	54	11	17	23	27	34	38	43	50
TW D5	14	20	26	32	39	44	50	58	11	17	24	29	36	41	47	54
TW E1	15	20	25	29	35	39	43	50	12	17	22	26	32	35	40	45
TW E2	14	20	26	30	37	41	46	53	12	17	23	27	33	37	42	49
TW E3	17	24	32	38	46	51	58	66	15	22	30	35	43	48	54	62
TW E4	16	23	31	37	45	50	57	66	14	21	28	34	42	47	54	62
TW F1	12	14	16	17	19	21	22	24	10	12	14	15	17	18	19	20
TW F2	10	13	15	18	20	22	25	28	8	10	13	15	17	19	21	24
TW F3	8	11	15	17	20	23	26	29	6	9	12	14	17	19	22	25
TW F4	7	10	13	16	20	22	25	29	5	7	11	13	16	19	22	25
TW G1	10	12	14	16	18	20	22	24	8	10	12	14	16	17	19	21
TW G2	8	10	13	15	18	21	23	27	6	8	11	13	16	17	20	23
TW G3	6	9	12	14	18	20	23	27	4	7	9	12	15	17	20	23
TW G4	4	7	10	12	16	18	21	25	3	5	8	10	13	15	18	22
TW H1	7	9	12	14	16	18	20	24	5	7	10	11	14	15	17	20
TW H2	5	7	10	12	15	17	20	24	3	5	8	10	12	14	17	20
TW H3	4	6	8	10	13	16	18	22	2	4	6	8	11	13	16	19
TW H4	3	5	7	8	11	13	16	19	1	3	5	6	9	11	13	16
TW I1	6	7	8	9	10	11	11	11	4	5	6	7	8	8	8	7
TW I2	9	13	18	21	26	29	33	38	7	11	15	18	22	25	29	34
TW I3	10	14	19	23	28	31	36	41	8	12	16	20	25	28	32	37
TW I4	11	17	23	27	34	[illegible]	44	52	9	14	20	25	31	35	41	48
TW I5	11	17	24	29	36	41	47	55	9	15	21	26	33	38	44	52
TW J1	13	18	23	27	33	37	42	48	10	15	20	24	30	34	38	44
TW J2	12	17	23	28	34	39	44	51	10	15	21	25	31	35	41	47
TW J3	15	22	30	36	44	49	56	65	12	19	27	33	41	46	53	61
TW J4	14	21	29	35	43	48	55	64	11	18	26	32	40	45	52	60
DG A1	7	9	10	11	12	13	13	12	5	7	8	9	9	9	9	7
DG A2	10	13	17	19	23	25	28	31	8	11	14	17	20	22	25	28
DG A3	13	18	23	27	33	37	41	48	12	16	21	25	30	34	39	45
DG B1	7	9	11	12	13	14	14	14	5	7	8	9	10	11	11	10
DG B2	10	14	17	20	24	27	30	34	8	11	15	17	21	24	27	30
DG B3	14	19	24	28	34	39	44	51	12	16	22	26	32	36	41	48
DG C1	9	12	15	16	19	20	21	22	7	9	12	13	15	17	18	18
DG C2	12	16	21	24	29	32	36	41	10	14	18	21	26	29	33	37
DG C3	16	21	27	32	39	43	49	57	14	19	25	29	36	41	46	53
SS A1	14	17	20	22	24	26	28	31	13	16	19	21	23	25	27	28
SS A2	17	22	27	31	36	40	44	50	16	21	26	30	35	38	42	47
SS A3	13	15	17	18	20	21	22	22	12	14	16	17	19	19	20	20
SS A4	16	21	25	29	33	36	40	45	16	20	24	27	32	35	38	43
SS A5	15	17	19	21	23	24	25	27	14	16	18	20	22	23	24	25
SS A6	17	21	26	30	35	38	42	48	16	20	25	29	33	36	40	45
SS A7	14	15	16	17	18	19	19	19	13	14	16	16	17	18	18	17
SS A8	16	20	24	27	31	34	38	42	15	19	23	26	30	33	36	40
SS B1	12	14	17	19	21	23	25	27	11	13	16	17	20	21	23	25
SS B2	15	20	25	28	34	37	42	48	14	19	23	27	32	35	40	45
SS B3	11	13	15	16	18	19	20	21	10	12	14	15	17	18	18	19
SS B4	14	19	23	27	31	35	39	44	13	17	22	25	30	33	37	42
SS B5	12	14	16	17	19	20	22	23	11	13	15	16	18	19	20	21
SS B6	15	19	23	27	32	35	40	46	13	18	22	26	30	34	38	43
SS B7	11	13	14	15	16	17	17	18	10	12	13	14	15	15	16	16
SS B8	13	17	22	25	29	32	36	41	12	16	20	23	28	30	34	39
SS C1	10	13	16	18	21	23	25	27	8	10	13	15	17	19	21	22
SS C2	12	17	22	26	31	34	39	45	10	14	19	23	28	31	35	40
SS C3	10	13	15	17	19	21	23	25	9	11	13	14	16	18	20	22
SS C4	12	16	20	23	28	31	35	41	10	13	18	21	25	28	32	37
SS D1	15	18	21	24	27	29	31	33	14	17	20	22	25	27	28	30
SS D2	18	24	30	35	41	45	50	57	17	23	29	33	39	43	48	54
SS D3	15	18	21	23	25	27	29	31	14	17	20	21	24	25	27	28
SS D4	18	23	29	34	40	44	49	55	17	22	28	32	38	41	46	52
SS E1	12	14	17	19	21	23	24	25	10	13	15	17	19	20	21	21
SS E2	15	20	25	29	34	38	42	48	13	18	23	27	32	36	40	45
SS E3	13	15	17	18	20	22	23	25	11	13	16	17	19	20	21	22
SS E4	15	19	24	27	32	36	40	45	13	18	22	26	30	34	37	42

CONSERVATION FACTOR

TYPE	55 F 100	50	30	20	15	65 F 100	50	30	20	15
WW A1	2.3	2.8	3.5	4.5	-	2.6	3.1	3.9	-	-
WW A2	1.8	2.1	2.5	3.1	3.7	1.9	2.2	2.7	3.5	4.6
WW A3	1.6	2.0	2.3	2.8	3.3	1.7	2.1	2.5	3.1	4.0
WW A4	1.6	1.9	2.2	2.6	3.1	1.6	2.0	2.4	3.0	3.7
WW A5	1.5	1.8	2.1	2.5	2.9	1.6	1.9	2.3	2.8	3.5
WW A6	1.5	1.8	2.1	2.5	2.8	1.5	1.9	2.2	2.8	3.4
WW B1	2.5	3.5	-	-	-	2.8	4.1	-	-	-
WW B2	1.4	1.7	1.9	2.2	2.5	1.5	1.8	2.0	2.4	2.7
WW B3	1.3	1.6	1.8	2.1	2.3	1.4	1.6	1.9	2.2	2.5
WW B4	1.2	1.4	1.6	1.8	2.0	1.2	1.4	1.7	1.9	2.1
WW B5	1.2	1.4	1.6	1.8	1.9	1.2	1.4	1.6	1.8	2.0
WW C1	1.3	1.5	1.7	2.0	2.2	1.3	1.5	1.8	2.1	2.3
WW C2	1.2	1.5	1.7	1.9	2.1	1.3	1.5	1.7	2.0	2.2
WW C3	1.1	1.3	1.4	1.6	1.8	1.1	1.3	1.5	1.7	1.8
WW C4	1.1	1.3	1.5	1.6	1.8	1.1	1.3	1.5	1.7	1.8
TW A1	2.3	2.7	3.2	4.0	4.9	2.5	2.9	3.5	4.6	-
TW A2	1.9	2.2	2.6	3.1	3.6	2.0	2.3	2.8	3.4	4.1
TW A3	1.7	2.0	2.4	2.8	3.2	1.8	2.2	2.6	3.1	3.7
TW A4	1.7	2.0	2.3	2.6	3.0	1.8	2.1	2.4	2.8	3.3
TW B1	2.1	2.5	2.9	3.4	4.0	2.3	2.7	3.1	3.8	4.6
TW B2	1.8	2.1	2.5	2.8	3.2	1.9	2.2	2.6	3.1	3.6
TW B3	1.8	2.0	2.4	2.7	3.0	1.9	2.1	2.5	2.9	3.3
TW B4	1.8	2.1	2.3	2.7	3.0	1.9	2.2	2.5	2.8	3.2
TW C1	2.0	2.3	2.7	3.0	3.4	2.2	2.4	2.8	3.3	3.7
TW C2	1.9	2.2	2.5	2.8	3.1	2.0	2.3	2.6	2.9	3.3
TW C3	1.9	2.2	2.5	2.8	3.0	2.0	2.3	2.6	2.9	3.2
TW C4	2.0	2.3	2.6	2.8	3.1	2.2	2.4	2.7	3.0	3.3
TW D1	2.5	3.2	4.4	-	-	2.8	3.6	-	-	-
TW D2	1.5	1.8	2.0	2.3	2.6	1.6	1.8	2.1	2.4	2.8
TW D3	1.5	1.7	2.0	2.3	2.5	1.5	1.8	2.1	2.4	2.7
TW D4	1.3	1.5	1.7	1.9	2.1	1.3	1.5	1.8	2.0	2.2
TW D5	1.3	1.4	1.6	1.8	2.0	1.3	1.5	1.7	1.9	2.1
TW E1	1.4	1.6	1.8	2.1	2.3	1.4	1.6	1.9	2.2	2.5
TW E2	1.3	1.5	1.7	2.0	2.2	1.4	1.6	1.8	2.1	2.3
TW E3	1.1	1.3	1.5	1.7	1.9	1.2	1.4	1.6	1.8	2.0
TW E4	1.2	1.3	1.5	1.7	1.9	1.2	1.4	1.6	1.8	1.9
TW F1	2.2	2.6	3.2	3.9	4.7	2.4	2.8	3.4	4.4	-
TW F2	1.9	2.2	2.6	3.1	3.6	2.0	2.4	2.8	3.4	4.2
TW F3	1.8	2.1	2.5	2.9	3.3	1.9	2.2	2.6	3.1	3.7
TW F4	1.8	2.0	2.4	2.7	3.0	1.9	2.1	2.5	2.9	3.3
TW G1	2.1	2.5	2.9	3.4	4.0	2.3	2.7	3.1	3.8	4.6
TW G2	1.9	2.2	2.6	3.0	3.4	2.0	2.3	2.7	3.2	3.7
TW G3	1.9	2.2	2.5	2.8	3.1	2.0	2.3	2.6	3.0	3.4
TW G4	1.9	2.2	2.5	2.8	3.1	2.1	2.3	2.6	2.9	3.3
TW H1	2.1	2.4	2.8	3.2	3.6	2.3	2.6	2.9	3.4	3.9
TW H2	2.0	2.3	2.6	3.0	3.3	2.2	2.4	2.8	3.1	3.5
TW H3	2.1	2.4	2.7	3.0	3.2	2.3	2.5	2.8	3.1	3.4
TW H4	2.3	2.6	2.8	3.1	3.3	2.5	2.7	3.0	3.2	3.5
TW I1	2.6	3.3	4.6	-	-	2.9	3.8	-	-	-
TW I2	1.6	1.8	2.1	2.3	2.6	1.6	1.9	2.2	2.5	2.8
TW I3	1.5	1.7	2.0	2.2	2.5	1.6	1.8	2.1	2.4	2.6
TW I4	1.3	1.5	1.7	1.9	2.1	1.4	1.6	1.8	2.0	2.2
TW I5	1.3	1.5	1.7	1.9	2.0	1.3	1.5	1.7	1.9	2.1
TW J1	1.4	1.6	1.8	2.1	2.3	1.4	1.6	1.9	2.2	2.4
TW J2	1.3	1.5	1.8	2.0	2.2	1.4	1.6	1.8	2.1	2.3
TW J3	1.2	1.3	1.5	1.7	1.9	1.2	1.4	1.6	1.8	1.9
TW J4	1.2	1.4	1.5	1.7	1.9	1.2	1.4	1.6	1.8	1.9
DG A1	2.3	3.1	-	-	-	2.5	3.5	-	-	-
DG A2	1.7	2.0	2.4	2.9	3.4	1.8	2.1	2.5	3.1	3.7
DG A3	1.5	1.6	1.9	2.1	2.3	1.5	1.7	1.9	2.1	2.4
DG B1	2.2	2.9	4.2	-	-	2.4	3.1	4.9	-	-
DG B2	1.7	1.9	2.3	2.7	3.1	1.8	2.0	2.4	2.8	3.4
DG B3	1.4	1.6	1.8	2.0	2.2	1.5	1.6	1.8	2.0	2.2
DG C1	1.9	2.3	3.1	4.9	-	2.0	2.5	3.3	-	-
DG C2	1.6	1.8	2.0	2.4	2.7	1.6	1.8	2.1	2.5	2.9
DG C3	1.3	1.5	1.7	1.9	2.0	1.4	1.5	1.7	1.9	2.1
SS A1	1.8	2.2	2.7	3.2	3.9	1.8	2.3	2.8	3.6	4.7
SS A2	1.4	1.6	1.9	2.1	2.3	1.4	1.7	1.9	2.2	2.5
SS A3	2.1	2.7	3.5	-	-	2.1	2.7	3.8	-	-
SS A4	1.5	1.7	2.0	2.3	2.6	1.5	1.8	2.1	2.4	2.8
SS A5	2.0	2.5	3.1	4.0	-	2.1	2.6	3.3	4.6	-
SS A6	1.4	1.7	1.9	2.2	2.4	1.4	1.7	2.0	2.3	2.6
SS A7	2.5	3.2	4.6	-	-	2.5	3.3	5.0	-	-
SS A8	1.5	1.8	2.1	2.4	2.7	1.5	1.8	2.2	2.6	3.0
SS B1	1.9	2.3	2.8	3.4	4.0	2.0	2.4	2.9	3.7	4.8
SS B2	1.4	1.7	1.9	2.1	2.3	1.5	1.7	1.9	2.2	2.4
SS B3	2.1	2.6	3.4	4.5	-	2.2	2.7	3.6	-	-
SS B4	1.5	1.7	2.0	2.2	2.5	1.5	1.8	2.0	2.3	2.6
SS B5	2.2	2.7	3.3	4.1	-	2.3	2.8	3.5	4.6	-
SS B6	1.5	1.7	2.0	2.2	2.4	1.5	1.7	2.0	2.3	2.5
SS B7	2.6	3.2	4.2	-	-	2.6	3.3	4.4	-	-
SS B8	1.5	1.8	2.1	2.4	2.6	1.6	1.8	2.1	2.5	2.8
SS C1	1.8	2.2	2.6	3.2	4.0	1.9	2.3	2.9	3.8	-
SS C2	1.4	1.7	1.9	2.2	2.4	1.5	1.7	2.0	2.3	2.6
SS C3	2.1	2.4	2.8	3.4	3.9	2.2	2.6	3.1	3.7	4.6
SS C4	1.5	1.8	2.0	2.3	2.5	1.6	1.9	2.1	2.4	2.7
SS D1	1.6	2.0	2.5	3.1	3.9	1.7	2.1	2.7	3.7	-
SS D2	1.2	1.5	1.7	1.9	2.1	1.3	1.5	1.8	2.0	2.3
SS D3	1.8	2.2	2.7	3.4	4.3	1.8	2.3	2.9	3.9	-
SS D4	1.3	1.5	1.7	2.0	2.2	1.3	1.5	1.8	2.1	2.3
SS E1	1.8	2.3	3.0	4.1	-	1.8	2.4	3.3	-	-
SS E2	1.4	1.6	1.9	2.1	2.4	1.4	1.7	2.0	2.3	2.6
SS E3	2.0	2.5	3.1	4.0	-	2.1	2.6	3.3	4.7	-
SS E4	1.4	1.7	2.0	2.2	2.5	1.4	1.7	2.0	2.4	2.7

SUFFIELD, ALBERTA

TYPE	SOLAR SAVINGS FRACTION (%) BASE TEMP: 55 F (6649 DD)								65 F (9423 DD)								CONSERVATION FACTOR 55 F					65 F				
LCR	100	70	50	40	30	25	20	15	100	70	50	40	30	25	20	15	100	50	30	20	15	100	50	30	20	15
WW A1	16	18	20	22	25	27	29	32	14	15	17	19	21	22	24	26	2.0	2.3	2.6	3.0	3.3	2.3	2.6	3.0	3.6	4.1
WW A2	15	19	23	26	31	34	38	43	12	15	19	22	25	28	31	36	1.5	1.8	2.0	2.3	2.5	1.7	2.0	2.3	2.6	2.9
WW A3	14	19	24	27	33	36	41	47	11	15	19	23	27	30	34	39	1.4	1.7	1.9	2.1	2.3	1.6	1.8	2.1	2.4	2.7
WW A4	14	18	24	28	34	38	43	50	10	15	19	23	28	31	36	41	1.4	1.6	1.8	2.0	2.2	1.5	1.7	2.0	2.3	2.5
WW A5	13	19	25	29	35	40	45	52	10	15	20	24	29	33	37	43	1.3	1.5	1.8	2.0	2.2	1.4	1.7	1.9	2.2	2.4
WW A6	13	19	25	30	36	40	46	53	10	14	20	24	30	33	38	45	1.3	1.5	1.7	1.9	2.1	1.4	1.6	1.9	2.1	2.4
WW B1	11	13	16	17	20	21	23	25	9	10	12	13	15	15	16	17	2.0	2.4	2.9	3.6	4.5	2.3	2.9	3.9	-	-
WW B2	15	20	27	31	38	43	48	56	11	16	22	26	32	37	42	49	1.3	1.5	1.7	1.9	2.0	1.4	1.6	1.8	2.0	2.2
WW B3	17	24	31	37	44	49	56	64	14	20	27	32	39	43	49	56	1.2	1.4	1.6	1.8	1.9	1.3	1.5	1.7	1.9	2.1
WW B4	18	25	34	40	49	55	62	71	14	21	29	35	44	49	56	65	1.1	1.3	1.5	1.6	1.8	1.2	1.3	1.5	1.7	1.9
WW B5	17	25	34	40	50	55	63	72	13	21	29	36	44	50	57	66	1.1	1.3	1.4	1.6	1.8	1.2	1.3	1.5	1.7	1.8
WW C1	19	26	34	39	47	52	59	67	15	22	29	34	42	46	52	60	1.1	1.3	1.5	1.7	1.9	1.2	1.4	1.6	1.8	2.0
WW C2	18	25	33	39	47	53	60	68	14	21	28	34	42	47	53	62	1.1	1.3	1.5	1.7	1.8	1.2	1.4	1.6	1.8	1.9
WW C3	21	31	41	48	58	64	71	80	18	27	36	43	53	59	66	75	1.0	1.2	1.3	1.5	1.7	1.0	1.2	1.4	1.6	1.7
WW C4	19	28	38	46	55	62	69	78	16	24	34	41	50	56	64	73	1.0	1.2	1.4	1.5	1.7	1.1	1.2	1.4	1.6	1.7
TW A1	15	18	20	22	25	27	30	33	14	15	17	19	21	23	25	28	2.0	2.2	2.5	2.9	3.1	2.2	2.5	2.9	3.3	3.8
TW A2	14	18	22	25	29	32	35	41	12	15	18	21	24	27	30	34	1.6	1.9	2.1	2.4	2.6	1.8	2.1	2.3	2.7	2.9
TW A3	14	17	22	25	30	33	38	44	11	14	18	21	25	28	32	37	1.5	1.8	2.0	2.2	2.4	1.7	1.9	2.2	2.5	2.7
TW A4	12	17	21	25	30	34	39	45	10	13	18	21	25	29	33	38	1.5	1.7	1.9	2.1	2.3	1.6	1.8	2.1	2.3	2.6
TW B1	15	17	20	22	26	28	31	35	13	15	17	19	22	24	26	30	1.9	2.1	2.4	2.6	2.9	2.1	2.3	2.7	3.0	3.3
TW B2	13	17	21	24	28	31	35	41	11	14	17	20	24	26	30	35	1.6	1.8	2.1	2.3	2.5	1.8	2.0	2.3	2.5	2.8
TW B3	12	16	20	24	29	32	36	42	10	13	17	20	24	27	31	36	1.6	1.8	2.0	2.2	2.4	1.7	1.9	2.2	2.4	2.7
TW B4	12	15	20	23	28	31	35	41	10	13	16	19	23	26	30	35	1.6	1.8	2.0	2.2	2.4	1.8	2.0	2.2	2.4	2.6
TW C1	13	16	20	22	26	28	32	37	12	14	17	19	22	24	27	31	1.8	2.0	2.2	2.5	2.7	2.0	2.2	2.5	2.7	3.0
TW C2	12	15	19	22	26	29	33	39	10	13	16	19	22	25	29	33	1.7	1.9	2.1	2.3	2.5	1.9	2.1	2.3	2.5	2.8
TW C3	12	15	19	21	26	29	32	38	10	13	16	18	22	24	28	33	1.7	1.9	2.2	2.4	2.5	1.9	2.1	2.3	2.6	2.8
TW C4	12	15	18	20	24	27	30	35	10	13	15	18	21	23	26	31	1.9	2.1	2.3	2.5	2.7	2.0	2.2	2.5	2.7	2.9
TW D1	9	11	14	15	18	20	22	24	7	9	10	12	14	15	16	17	2.1	2.4	2.9	3.4	3.9	2.4	2.9	3.5	4.5	-
TW D2	15	20	25	29	35	39	45	52	12	16	21	25	30	34	39	45	1.4	1.6	1.8	2.0	2.1	1.5	1.7	1.9	2.1	2.3
TW D3	15	21	27	31	38	42	48	55	12	17	23	27	32	36	41	48	1.3	1.5	1.7	1.9	2.1	1.4	1.6	1.8	2.1	2.3
TW D4	17	23	31	36	44	49	56	65	14	20	27	32	39	44	50	59	1.2	1.4	1.5	1.7	1.9	1.3	1.4	1.6	1.8	2.0
TW D5	17	24	31	37	46	51	58	67	14	20	27	33	41	46	53	62	1.2	1.3	1.5	1.7	1.8	1.2	1.4	1.6	1.7	1.9
TW E1	18	24	31	36	43	48	54	62	15	20	26	31	38	42	48	55	1.2	1.4	1.6	1.8	2.0	1.3	1.5	1.7	1.9	2.1
TW E2	18	24	31	37	44	49	56	64	14	20	27	32	39	44	50	58	1.2	1.4	1.6	1.7	1.9	1.3	1.5	1.6	1.8	2.0
TW E3	21	29	38	45	54	60	67	76	18	25	34	40	49	55	62	70	1.1	1.2	1.4	1.6	1.7	1.1	1.3	1.5	1.6	1.8
TW E4	19	28	37	43	53	58	66	75	16	24	32	39	48	53	61	70	1.1	1.2	1.4	1.6	1.7	1.1	1.3	1.5	1.6	1.8
TW F1	14	16	19	21	24	26	29	33	12	14	16	17	20	22	24	27	1.9	2.2	2.5	2.8	3.1	2.2	2.5	2.8	3.2	3.7
TW F2	12	15	19	22	26	29	33	38	10	12	16	18	22	24	27	31	1.7	1.9	2.1	2.4	2.6	1.8	2.1	2.4	2.7	3.0
TW F3	11	14	19	22	27	30	34	40	8	11	15	18	22	25	28	33	1.6	1.8	2.0	2.3	2.5	1.7	2.0	2.2	2.5	2.8
TW F4	9	13	17	21	26	29	34	40	6	10	13	16	21	24	28	33	1.6	1.8	2.0	2.2	2.4	1.7	1.9	2.2	2.4	2.7
TW G1	12	14	17	20	23	25	28	33	10	12	14	16	19	21	23	27	1.9	2.1	2.4	2.7	2.9	2.1	2.4	2.7	3.0	3.3
TW G2	9	13	17	20	24	27	31	36	7	10	13	16	19	22	25	30	1.7	1.9	2.2	2.4	2.6	1.9	2.1	2.4	2.6	2.9
TW G3	8	11	15	18	23	26	30	36	6	9	12	15	19	21	25	30	1.7	1.9	2.1	2.3	2.5	1.8	2.1	2.3	2.6	2.8
TW G4	6	9	13	16	20	23	27	33	4	7	10	12	16	19	22	28	1.8	2.0	2.2	2.4	2.6	1.9	2.1	2.4	2.6	2.8
TW H1	9	11	15	17	21	23	27	32	7	9	12	14	17	19	22	26	1.9	2.1	2.3	2.6	2.8	2.1	2.3	2.6	2.9	3.1
TW H2	6	9	13	15	19	22	26	31	4	7	10	12	15	18	21	26	1.8	2.0	2.3	2.5	2.7	2.0	2.2	2.5	2.7	2.9
TW H3	5	8	11	13	17	20	24	29	3	5	8	10	14	16	19	24	1.9	2.1	2.3	2.5	2.7	2.1	2.3	2.5	2.8	3.0
TW H4	4	6	8	11	14	16	20	24	2	4	6	8	11	13	16	20	2.1	2.3	2.5	2.7	2.9	2.3	2.5	2.7	2.9	3.1
TW I1	7	9	12	13	16	17	19	22	5	7	8	10	11	12	13	15	2.1	2.5	3.0	3.5	4.1	2.5	3.0	3.6	4.7	-
TW I2	12	17	22	26	32	36	42	49	9	13	18	22	27	31	36	42	1.4	1.6	1.8	2.0	2.2	1.5	1.7	1.9	2.2	2.3
TW I3	12	18	24	28	35	39	45	53	10	14	20	24	30	34	39	46	1.4	1.5	1.7	1.9	2.1	1.4	1.6	1.9	2.1	2.3
TW I4	14	21	28	33	41	46	53	62	11	17	24	29	36	41	47	56	1.2	1.4	1.6	1.7	1.9	1.3	1.5	1.6	1.8	2.0
TW I5	14	21	29	35	43	49	56	65	11	18	25	30	38	44	50	59	1.2	1.4	1.5	1.7	1.8	1.3	1.4	1.6	1.8	1.9
TW J1	16	22	28	34	41	46	52	60	13	18	24	29	36	40	46	53	1.2	1.4	1.6	1.8	2.0	1.3	1.5	1.7	1.9	2.1
TW J2	15	21	29	34	42	47	53	62	12	18	24	29	37	41	48	56	1.2	1.4	1.6	1.8	1.9	1.3	1.5	1.7	1.9	2.0
TW J3	18	27	36	42	52	58	65	74	15	23	31	38	47	53	60	69	1.1	1.2	1.4	1.6	1.7	1.1	1.3	1.5	1.6	1.8
TW J4	17	25	34	41	50	56	63	73	14	21	30	36	45	51	58	68	1.1	1.3	1.4	1.6	1.7	1.1	1.3	1.5	1.7	1.8
DG A1	9	12	14	16	18	19	20	21	7	9	11	12	13	14	15	14	1.9	2.4	3.2	4.9	-	2.2	2.8	4.0	-	-
DG A2	12	16	20	23	28	31	35	40	10	13	17	19	24	26	30	34	1.6	1.8	2.1	2.4	2.7	1.7	1.9	2.2	2.6	3.0
DG A3	15	21	27	31	38	43	49	57	13	18	23	28	34	38	44	51	1.4	1.5	1.7	1.9	2.0	1.4	1.6	1.8	2.0	2.2
DG B1	9	12	15	17	20	21	23	25	7	9	11	13	15	16	17	18	1.9	2.3	2.8	3.5	-	2.1	2.6	3.3	-	-
DG B2	12	16	21	24	30	33	38	44	10	13	17	20	25	28	32	37	1.5	1.7	1.9	2.2	2.5	1.7	1.9	2.1	2.4	2.7
DG B3	16	21	28	33	40	45	52	61	13	18	24	29	36	41	47	55	1.3	1.5	1.6	1.8	1.9	1.4	1.5	1.7	1.9	2.0
DG C1	12	15	19	21	25	28	31	34	9	12	15	17	20	22	24	26	1.7	1.9	2.3	2.7	3.4	1.8	2.2	2.6	3.4	4.5
DG C2	14	19	25	29	35	39	45	52	12	16	21	25	30	34	39	45	1.4	1.6	1.8	2.0	2.2	1.5	1.7	1.9	2.2	2.4
DG C3	18	24	31	37	45	51	58	67	15	21	28	33	40	46	52	61	1.2	1.4	1.5	1.7	1.8	1.3	1.4	1.6	1.8	1.9
SS A1	17	20	24	27	31	34	37	42	15	18	22	24	28	30	33	37	1.6	1.9	2.1	2.4	2.7	1.7	2.0	2.3	2.7	3.0
SS A2	20	26	32	37	43	48	53	61	19	24	30	34	40	44	49	56	1.3	1.5	1.7	1.9	2.0	1.3	1.5	1.7	2.0	2.1
SS A3	15	18	21	23	26	28	30	34	14	16	19	20	23	25	26	29	1.8	2.1	2.5	2.9	3.4	1.9	2.3	2.8	3.5	4.3
SS A4	19	25	30	35	41	45	50	57	18	23	28	32	37	41	46	52	1.3	1.5	1.7	1.9	2.1	1.4	1.6	1.8	2.1	2.3
SS A5	17	20	23	25	28	30	33	37	16	18	21	23	26	27	30	32	1.8	2.1	2.4	2.7	3.1	1.9	2.2	2.6	3.1	3.6
SS A6	20	25	32	36	42	47	52	59	18	23	29	33	39	43	48	54	1.3	1.5	1.7	1.9	2.1	1.3	1.5	1.8	2.0	2.2
SS A7	15	17	19	21	23	25	26	28	14	16	18	19	21	22	23	24	2.1	2.5	3.0	3.6	4.2	2.2	2.7	3.4	4.4	-
SS A8	19	24	29	33	39	43	48	54	17	22	27	30	36	39	43	49	1.3	1.6	1.8	2.0	2.2	1.4	1.6	1.9	2.2	2.4
SS B1	14	17	21	23	27	30	33	38	12	15	18	21	24	26	29	32	1.7	2.0	2.2	2.5	2.8	1.8	2.1	2.4	2.8	3.2
SS B2	18	23	29	34	40	44	50	58	16	21	27	31	37	41	46	53	1.3	1.5	1.7	1.9	2.0	1.4	1.6	1.8	2.0	2.2
SS B3	13	15	18	20	24	26	28	32	11	14	16	18	21	22	24	27	1.8	2.2	2.5	2.9	3.2	2.0	2.3	2.8	3.3	4.0
SS B4	17	22	28	32	38	42	47	54	15	20	25	29	35	38	43	50	1.4	1.6	1.7	1.9	2.1	1.4	1.6	1.8	2.1	2.3
SS B5	14	16	19	21	24	26	28	32	13	15	17	19	21	23	25	27	1.9	2.2	2.6	2.9	3.2	2.1	2.4	2.8	3.3	3.8
SS B6	17	22	28	32	39	43	48	56	15	20	25	29	35	39	44	51	1.3	1.5	1.7	1.9	2.1	1.4	1.6	1.8	2.0	2.2
SS B7	13	14	17	18	20	22	24	26	11	13	15	16	18	19	20	22	2.2	2.6	3.0	3.5	4.0	2.4	2.8	3.4	4.2	-
SS B8	16	21	26	30	36	40	45	52	14	19	24	27	32	36	40	47	1.4	1.6	1.8	2.0	2.2	1.5	1.7	1.9	2.2	2.4
SS C1	12	16	20	23	28	31	35	41	10	13	16	19	23	25	28	33	1.6	1.8	2.1	2.3	2.6	1.7	2.0	2.3	2.7	3.0
SS C2	15	21	27	32	38	43	49	57	12	17	23	27	33	37	43	50	1.3	1.5	1.7	1.9	2.0	1.4	1.6	1.8	2.0	2.2
SS C3	12	15	18	21	24	27	30	34	10	12	15	17	20	22	25	28	1.8	2.1	2.3	2.6	2.8	2.0	2.3	2.6	2.9	3.2
SS C4	14	19	25	29	35	39	44	51	12	16	21	24	30	34	38	45	1.4	1.6	1.8	2.0	2.1	1.5	1.7	1.9	2.1	2.3
SS D1	18	22	27	30	35	39	43	48	16	20	24	27	31	34	37	41	1.4	1.7	1.9	2.2	2.5	1.5	1.8	2.1	2.5	2.9
SS D2	22	29	37	42	50	55	61	69	20	26	33	39	46	50	56	63	1.1	1.3	1.5	1.7	1.9	1.2	1.4	1.6	1.8	2.0
SS D3	18	21	25	28	32	35	39	43	16	19	23	25	29	31	34	38	1.5	1.8	2.1	2.4	2.7	1.6	2.0	2.3	2.7	3.1
SS D4	22	28	35	41	48	53	59	67	19	26	32	37	44	49	54	62	1.2	1.4	1.5	1.7	1.9	1.2	1.4	1.6	1.8	2.0
SS E1	15	18	22	25	29	32	36	40	13	16	19	22	25	27	30	33	1.5	1.8	2.1	2.5	2.8	1.7	2.0	2.4	2.9	3.5
SS E2	18	24	31	36	43	48	54	61	16	22	28	32	39	43	48	55	1.2	1.4	1.6	1.8	2.0	1.3	1.5	1.7	1.9	2.1
SS E3	15	17	21	23	26	29	32	35	13	16	18	20	23	25	27	30	1.8	2.1	2.4	2.7	3.0	1.9	2.2	2.6	3.1	3.7
SS E4	18	23	29	34	40	44	50	57	16	21	26	30	36	40	45	52	1.3	1.5	1.7	1.9	2.1	1.3	1.6	1.8	2.0	2.2

NANAIMO, BRITISH COLUMBIA

TYPE	SOLAR SAVINGS FRACTION (%)																CONSERVATION FACTOR									
BASE TEMP:	55 F (2841 DD)								65 F (5742 DD)								55 F					65 F				
LCR	100	70	50	40	30	25	20	15	100	70	50	40	30	25	20	15	100	50	30	20	15	100	50	30	20	15
WW A1	21	24	28	31	35	37	40	44	16	19	22	24	26	28	30	32	1.6	1.8	2.1	2.5	2.8	1.8	2.2	2.7	3.3	4.0
WW A2	22	27	33	37	42	45	49	55	16	20	24	27	31	34	37	41	1.3	1.5	1.8	2.1	2.4	1.5	1.8	2.2	2.6	3.1
WW A3	22	28	34	39	44	48	53	58	15	20	25	28	33	36	39	43	1.2	1.5	1.7	2.0	2.2	1.4	1.7	2.0	2.4	2.9
WW A4	22	28	35	40	46	50	54	60	15	20	25	29	34	37	41	45	1.2	1.4	1.7	1.9	2.2	1.3	1.6	2.0	2.4	2.8
WW A5	22	29	36	41	47	51	56	62	15	20	26	30	35	38	42	47	1.1	1.4	1.6	1.9	2.1	1.3	1.6	1.9	2.3	2.7
WW A6	22	29	36	41	48	52	57	63	14	20	26	30	35	39	43	48	1.1	1.4	1.6	1.9	2.1	1.3	1.6	1.9	2.3	2.7
WW B1	18	22	26	28	32	34	37	40	12	15	17	19	21	22	23	23	1.5	1.8	2.3	2.8	3.5	1.8	2.4	3.3	-	-
WW B2	23	30	37	42	49	53	58	65	16	21	28	32	38	41	46	52	1.1	1.4	1.6	1.8	2.0	1.3	1.5	1.8	2.1	2.3
WW B3	27	34	42	48	55	60	65	72	19	26	32	37	44	48	53	59	1.1	1.3	1.5	1.7	1.9	1.2	1.4	1.7	2.0	2.2
WW B4	27	35	44	51	59	64	70	77	19	27	35	40	48	53	59	66	1.0	1.2	1.4	1.6	1.8	1.1	1.3	1.5	1.8	1.9
WW B5	26	35	44	50	59	64	70	78	18	26	35	41	49	54	60	68	1.0	1.2	1.4	1.6	1.7	1.1	1.3	1.5	1.7	1.9
WW C1	28	36	44	50	57	62	68	74	21	27	35	40	46	51	56	62	1.0	1.2	1.5	1.7	1.8	1.1	1.4	1.6	1.9	2.1
WW C2	27	35	43	49	57	62	68	75	19	26	34	39	46	51	57	64	1.0	1.2	1.4	1.6	1.8	1.1	1.4	1.6	1.8	2.0
WW C3	31	41	51	58	66	72	78	85	23	32	42	48	57	62	69	76	0.9	1.1	1.3	1.5	1.7	1.0	1.2	1.4	1.6	1.8
WW C4	28	38	48	55	64	69	75	83	21	30	39	46	54	60	66	74	0.9	1.1	1.3	1.5	1.7	1.0	1.2	1.4	1.6	1.8
TW A1	21	24	28	31	35	37	41	45	16	19	22	24	26	28	31	33	1.6	1.8	2.1	2.4	2.7	1.8	2.2	2.6	3.1	3.6
TW A2	21	26	31	34	39	43	47	52	15	19	23	26	30	32	35	39	1.4	1.6	1.9	2.2	2.4	1.6	1.9	2.2	2.6	3.0
TW A3	21	26	32	36	41	45	49	55	15	19	23	27	31	34	37	42	1.3	1.5	1.8	2.1	2.3	1.5	1.8	2.1	2.5	2.8
TW A4	20	25	31	36	42	45	50	56	14	18	23	26	31	34	38	43	1.3	1.5	1.7	2.0	2.2	1.4	1.7	2.0	2.4	2.7
TW B1	20	24	28	31	35	38	42	47	15	18	21	24	27	29	32	35	1.5	1.8	2.0	2.3	2.6	1.7	2.1	2.4	2.8	3.3
TW B2	20	24	30	33	39	42	47	52	14	18	22	25	29	32	36	40	1.4	1.6	1.8	2.1	2.3	1.5	1.8	2.1	2.5	2.8
TW B3	19	24	30	34	39	43	47	53	13	17	22	25	30	33	36	41	1.3	1.6	1.8	2.1	2.3	1.5	1.8	2.1	2.4	2.7
TW B4	18	23	28	32	38	42	46	52	13	17	21	24	29	32	36	41	1.4	1.6	1.8	2.1	2.3	1.5	1.8	2.1	2.4	2.7
TW C1	19	23	27	31	35	39	43	48	14	17	21	23	27	30	33	37	1.5	1.7	2.0	2.2	2.5	1.7	2.0	2.3	2.6	2.9
TW C2	18	22	27	31	36	40	44	50	13	17	21	24	28	30	34	39	1.4	1.6	1.9	2.1	2.3	1.6	1.9	2.2	2.5	2.7
TW C3	17	22	26	30	35	39	43	49	13	16	20	23	27	30	33	38	1.4	1.7	1.9	2.1	2.4	1.6	1.9	2.2	2.5	2.7
TW C4	17	20	25	28	33	36	40	46	13	16	19	22	25	28	31	36	1.6	1.8	2.0	2.2	2.4	1.8	2.0	2.3	2.5	2.8
TW D1	15	19	23	25	29	32	35	38	10	13	15	17	20	21	23	24	1.6	1.9	2.3	2.7	3.2	1.9	2.3	3.0	4.0	-
TW D2	22	28	35	39	46	50	55	62	16	21	26	30	36	39	44	49	1.2	1.4	1.6	1.9	2.1	1.3	1.6	1.9	2.1	2.4
TW D3	23	30	37	42	49	53	58	65	17	22	28	32	38	42	46	52	1.1	1.4	1.6	1.8	2.0	1.3	1.5	1.8	2.1	2.4
TW D4	25	33	41	47	54	59	65	72	18	25	32	37	44	49	54	61	1.1	1.3	1.5	1.7	1.8	1.2	1.4	1.6	1.8	2.0
TW D5	25	33	41	47	55	60	66	74	18	25	32	38	45	50	56	64	1.1	1.2	1.4	1.6	1.8	1.2	1.4	1.6	1.8	2.0
TW E1	26	34	41	46	53	58	63	70	20	26	32	37	43	47	52	58	1.1	1.3	1.5	1.7	1.9	1.2	1.5	1.7	2.0	2.2
TW E2	26	33	41	47	54	59	65	72	19	25	32	37	44	48	54	60	1.1	1.3	1.5	1.7	1.9	1.2	1.4	1.6	1.9	2.1
TW E3	30	39	48	55	63	68	74	81	23	31	39	45	53	58	64	72	1.0	1.2	1.4	1.6	1.7	1.1	1.3	1.5	1.7	1.9
TW E4	28	37	47	53	61	67	73	80	21	29	38	44	52	57	63	71	1.0	1.2	1.4	1.6	1.7	1.1	1.3	1.5	1.7	1.8
TW F1	19	23	27	30	34	37	40	44	14	17	20	22	25	27	30	32	1.5	1.8	2.1	2.4	2.7	1.8	2.1	2.5	3.0	3.6
TW F2	19	23	28	32	37	40	45	50	13	17	21	23	27	30	33	37	1.4	1.6	1.9	2.2	2.4	1.6	1.9	2.2	2.6	3.0
TW F3	18	23	28	32	38	41	46	52	12	16	20	23	28	31	34	38	1.3	1.6	1.8	2.1	2.3	1.5	1.8	2.1	2.5	2.9
TW F4	16	21	27	31	37	41	45	52	10	14	19	22	27	30	34	39	1.3	1.6	1.8	2.1	2.3	1.5	1.8	2.1	2.4	2.7
TW G1	17	21	25	29	33	36	40	45	12	15	19	21	24	27	29	33	1.5	1.8	2.0	2.3	2.6	1.8	2.1	2.4	2.8	3.3
TW G2	16	20	25	29	34	38	42	48	10	14	18	21	25	28	31	36	1.4	1.6	1.9	2.2	2.4	1.6	1.9	2.2	2.6	2.9
TW G3	14	19	24	28	33	37	42	48	9	12	17	20	24	27	31	36	1.4	1.6	1.9	2.1	2.4	1.6	1.9	2.2	2.5	2.8
TW G4	12	16	21	25	30	34	38	45	7	10	14	17	22	24	28	33	1.5	1.7	1.9	2.2	2.4	1.7	1.9	2.2	2.5	2.7
TW H1	14	18	23	26	31	34	38	43	9	12	16	18	22	25	28	32	1.5	1.8	2.0	2.3	2.5	1.8	2.0	2.4	2.7	3.1
TW H2	12	16	21	24	29	33	37	43	7	10	14	17	21	23	27	32	1.5	1.7	2.0	2.2	2.4	1.7	2.0	2.3	2.6	2.9
TW H3	10	14	18	22	26	30	34	40	6	9	12	15	19	21	25	30	1.6	1.8	2.0	2.3	2.5	1.8	2.0	2.3	2.6	2.9
TW H4	8	11	15	18	22	25	29	35	4	6	9	12	15	18	21	26	1.7	2.0	2.2	2.4	2.6	2.0	2.2	2.5	2.7	3.0
TW I1	13	17	20	23	27	29	32	35	8	11	13	15	17	19	20	22	1.6	1.9	2.3	2.8	3.2	2.0	2.4	3.0	4.0	-
TW I2	19	25	32	36	43	47	52	59	13	18	23	27	33	36	41	46	1.2	1.4	1.7	1.9	2.1	1.4	1.6	1.9	2.2	2.4
TW I3	20	27	34	39	46	50	55	62	14	19	25	30	35	39	44	50	1.2	1.4	1.6	1.0	2.0	1.3	1.6	1.8	2.1	2.3
TW I4	22	30	38	44	51	56	62	70	15	22	29	34	41	48	52	59	1.1	1.3	1.5	1.7	1.9	1.2	1.4	1.6	1.9	2.1
TW I5	22	30	38	45	53	58	64	72	15	22	30	35	43	48	54	62	1.1	1.3	1.5	1.7	1.8	1.2	1.4	1.6	1.8	2.0
TW J1	24	31	39	44	51	56	62	69	17	23	30	35	41	45	50	57	1.1	1.3	1.5	1.7	1.9	1.2	1.5	1.7	2.0	2.2
TW J2	23	31	39	44	52	57	63	70	16	23	30	35	42	46	52	59	1.1	1.3	1.5	1.7	1.9	1.2	1.4	1.6	1.9	2.1
TW J3	27	36	46	52	61	66	72	80	20	28	37	43	51	56	63	71	1.0	1.2	1.4	1.6	1.7	1.1	1.3	1.5	1.7	1.8
TW J4	26	35	44	51	59	65	71	78	18	26	35	41	50	55	61	69	1.0	1.2	1.4	1.6	1.7	1.1	1.3	1.5	1.7	1.8
DG A1	16	20	24	27	30	32	33	35	10	13	16	18	20	21	21	21	1.5	1.9	2.5	3.5	-	1.8	2.3	3.4	-	-
DG A2	19	24	29	33	39	42	46	51	13	17	22	25	29	32	35	39	1.3	1.6	1.9	2.2	2.5	1.5	1.8	2.1	2.6	3.0
DG A3	23	29	36	41	48	53	58	65	17	22	29	33	39	44	49	55	1.2	1.4	1.6	1.8	2.0	1.3	1.5	1.7	2.0	2.2
DG B1	17	21	26	29	32	35	37	39	11	14	17	19	22	23	24	25	1.4	1.8	2.2	3.0	4.2	1.7	2.2	2.9	4.6	-
DG B2	19	25	31	35	41	45	49	55	14	18	23	26	31	34	38	43	1.3	1.5	1.8	2.1	2.4	1.5	1.7	2.0	2.4	2.8
DG B3	23	30	38	43	51	56	61	68	18	23	30	35	42	46	52	59	1.1	1.3	1.5	1.7	1.9	1.3	1.4	1.6	1.9	2.1
DG C1	20	25	30	34	39	41	44	47	13	17	21	24	27	29	31	33	1.3	1.6	2.0	2.6	3.3	1.5	1.9	2.4	3.4	-
DG C2	22	29	35	40	47	51	56	61	16	21	27	31	36	40	44	49	1.2	1.4	1.6	1.9	2.2	1.3	1.6	1.8	2.2	2.5
DG C3	26	34	42	48	56	61	66	74	20	26	33	39	46	51	57	64	1.1	1.3	1.4	1.7	1.8	1.2	1.4	1.6	1.8	2.0
SS A1	24	29	34	38	43	46	50	55	20	24	28	31	35	38	41	44	1.3	1.6	1.9	2.2	2.4	1.4	1.8	2.2	2.6	3.1
SS A2	29	36	43	48	54	58	64	70	25	30	37	41	47	50	55	61	1.1	1.3	1.5	1.8	1.9	1.2	1.4	1.7	2.0	2.2
SS A3	23	27	31	34	38	41	44	48	19	22	26	28	31	33	35	37	1.4	1.7	2.0	2.4	2.8	1.5	2.0	2.5	3.2	4.1
SS A4	29	35	42	46	52	56	61	68	24	30	35	39	45	48	52	58	1.1	1.4	1.6	1.8	2.0	1.2	1.5	1.8	2.1	2.3
SS A5	24	28	32	35	40	42	46	50	20	24	27	30	33	35	37	40	1.4	1.7	2.0	2.4	2.7	1.5	1.9	2.4	2.9	3.5
SS A6	29	35	42	47	53	58	63	69	24	30	36	40	46	49	54	59	1.1	1.3	1.6	1.8	2.0	1.2	1.5	1.7	2.0	2.3
SS A7	22	25	29	31	35	37	39	43	19	21	24	26	29	30	32	33	1.6	1.9	2.3	2.7	3.2	1.7	2.2	2.8	3.6	5.0
SS A8	28	34	40	45	51	55	60	66	24	29	34	38	43	46	50	55	1.1	1.4	1.6	1.9	2.1	1.2	1.5	1.8	2.1	2.4
SS B1	21	26	30	34	39	42	45	50	17	21	25	27	31	34	36	40	1.4	1.6	1.9	2.2	2.5	1.5	1.9	2.2	2.7	3.1
SS B2	26	33	39	44	51	55	60	67	22	27	33	37	43	47	52	58	1.1	1.4	1.6	1.8	2.0	1.2	1.5	1.7	2.0	2.2
SS B3	20	24	28	31	36	38	42	46	16	19	23	25	28	30	33	35	1.4	1.7	2.1	2.4	2.7	1.6	2.0	2.5	3.1	3.8
SS B4	25	32	38	43	49	53	59	65	21	26	32	36	42	45	50	56	1.2	1.4	1.6	1.8	2.0	1.3	1.5	1.8	2.0	2.3
SS B5	20	24	28	31	35	37	40	45	16	19	23	25	28	30	32	35	1.5	1.8	2.1	2.5	2.8	1.7	2.0	2.5	3.0	3.6
SS B6	25	32	38	43	50	54	59	66	21	26	32	36	42	46	50	56	1.2	1.4	1.6	1.8	2.0	1.2	1.5	1.8	2.0	2.2
SS B7	18	22	25	27	31	33	36	40	15	18	20	22	25	27	28	30	1.6	2.0	2.3	2.7	3.1	1.8	2.2	2.8	3.5	4.3
SS B8	24	30	37	41	47	51	56	63	20	25	31	34	40	43	47	53	1.2	1.4	1.7	1.9	2.1	1.3	1.6	1.8	2.1	2.4
SS C1	20	25	31	34	40	43	47	52	14	18	22	25	29	31	34	38	1.3	1.6	1.8	2.2	2.4	1.5	1.8	2.2	2.7	3.3
SS C2	23	30	37	42	49	54	59	66	17	22	28	33	38	42	47	53	1.1	1.4	1.6	1.8	2.0	1.3	1.5	1.8	2.1	2.3
SS C3	18	22	26	30	34	37	41	46	13	16	19	22	25	28	30	34	1.5	1.7	2.0	2.3	2.6	1.7	2.0	2.4	2.8	3.2
SS C4	22	28	34	39	45	49	54	61	16	21	26	30	35	39	43	49	1.2	1.4	1.7	1.9	2.1	1.4	1.6	1.9	2.1	2.4
SS D1	27	33	38	42	48	51	55	60	22	27	32	35	39	41	44	48	1.2	1.5	1.7	2.0	2.3	1.3	1.7	2.1	2.5	3.1
SS D2	32	40	48	53	60	65	70	77	27	34	41	45	52	56	61	67	1.0	1.2	1.4	1.6	1.8	1.1	1.3	1.6	1.8	2.0
SS D3	26	31	36	39	44	47	51	56	22	26	30	33	37	39	42	45	1.3	1.5	1.8	2.2	2.4	1.4	1.8	2.2	2.7	3.2
SS D4	31	39	47	52	59	63	69	75	26	33	40	44	51	55	59	65	1.0	1.3	1.5	1.7	1.8	1.1	1.4	1.6	1.9	2.1
SS E1	23	29	34	38	43	46	49	54	18	23	27	30	33	35	38	40	1.3	1.6	1.9	2.2	2.5	1.4	1.8	2.3	2.9	3.9
SS E2	28	35	43	48	55	59	64	71	22	29	35	40	46	49	54	59	1.1	1.3	1.5	1.7	1.9	1.2	1.4	1.7	2.0	2.2
SS E3	22	26	30	34	38	41	44	49	18	21	25	27	31	33	35	38	1.4	1.7	2.0	2.3	2.6	1.5	1.9	2.4	2.9	3.6
SS E4	27	33	40	45	52	56	61	67	22	27	33	37	43	47	51	57	1.1	1.3	1.6	1.8	2.0	1.2	1.5	1.8	2.0	2.3

SUMMERLAND, BRITISH COLUMBIA

TYPE	SOLAR SAVINGS FRACTION (%) BASE TEMP: 55 F (3936 DD)								65 F (6580 DD)								CONSERVATION FACTOR 55 F					65 F				
LCR	100	70	50	40	30	25	20	15	100	70	50	40	30	25	20	15	100	50	30	20	15	100	50	30	20	15
WW A1	16	18	20	22	24	26	27	29	14	16	17	19	20	21	22	22	2.0	2.4	3.0	3.7	4.7	2.3	2.9	3.7	-	-
WW A2	15	18	22	25	28	31	34	37	12	15	18	20	23	25	26	28	1.6	1.9	2.3	2.8	3.2	1.8	2.2	2.7	3.5	4.7
WW A3	14	18	23	26	30	33	36	40	11	15	18	21	24	26	28	31	1.5	1.8	2.2	2.6	2.9	1.6	2.0	2.5	3.2	4.1
WW A4	14	18	23	26	31	34	37	42	10	14	18	21	24	26	29	32	1.4	1.7	2.1	2.4	2.8	1.6	1.9	2.4	3.0	3.8
WW A5	13	18	23	27	32	35	38	43	10	14	18	21	25	27	30	33	1.4	1.7	2.0	2.4	2.6	1.5	1.9	2.3	2.9	3.6
WW A6	13	18	23	27	32	35	39	44	9	14	18	21	25	28	31	34	1.4	1.7	2.0	2.3	2.6	1.5	1.8	2.3	2.8	3.5
WW B1	11	13	15	17	18	19	19	19	9	10	12	12	13	13	12	9	2.1	2.7	4.0	-	-	2.4	3.7	-	-	-
WW B2	15	20	25	29	35	38	43	49	12	16	21	24	29	32	36	41	1.3	1.6	1.9	2.1	2.3	1.5	1.7	2.0	2.4	2.7
WW B3	17	23	30	34	40	44	49	56	14	19	25	29	34	38	42	47	1.2	1.5	1.7	2.0	2.2	1.3	1.6	1.9	2.2	2.5
WW B4	17	25	32	38	45	50	56	64	14	20	27	32	39	44	50	57	1.1	1.4	1.6	1.8	1.9	1.2	1.4	1.7	1.9	2.1
WW B5	17	24	32	38	46	51	58	66	13	20	27	33	40	45	51	59	1.1	1.3	1.5	1.7	1.9	1.2	1.4	1.6	1.8	2.0
WW C1	19	25	32	37	43	47	53	60	15	21	27	31	37	41	45	51	1.2	1.4	1.7	1.9	2.1	1.3	1.5	1.8	2.1	2.3
WW C2	17	24	31	37	44	48	54	62	14	20	26	31	38	42	47	54	1.2	1.4	1.6	1.8	2.0	1.3	1.5	1.7	2.0	2.2
WW C3	21	30	39	45	54	60	67	75	17	25	34	40	49	54	61	69	1.0	1.2	1.4	1.6	1.7	1.1	1.3	1.5	1.7	1.8
WW C4	19	27	37	43	52	57	64	73	15	23	32	38	46	52	58	67	1.1	1.2	1.4	1.6	1.8	1.1	1.3	1.5	1.7	1.9
TW A1	16	18	20	22	25	26	28	30	14	16	18	19	21	22	23	24	2.0	2.4	2.8	3.4	4.1	2.2	2.7	3.4	4.5	-
TW A2	15	18	21	24	27	30	32	36	13	15	18	20	22	24	26	28	1.7	2.0	2.4	2.8	3.2	1.9	2.3	2.7	3.4	4.2
TW A3	14	17	21	24	28	31	34	38	11	14	18	20	23	25	27	30	1.6	1.9	2.2	2.6	2.9	1.7	2.1	2.5	3.1	3.7
TW A4	13	17	21	24	28	31	35	39	10	13	17	19	23	25	28	32	1.5	1.8	2.1	2.5	2.8	1.7	2.0	2.4	2.9	3.4
TW B1	15	18	20	22	25	27	29	32	13	15	17	19	21	22	24	26	1.9	2.2	2.6	3.1	3.6	2.1	2.5	3.0	3.8	4.6
TW B2	14	17	21	23	27	29	33	37	11	14	17	19	22	24	27	30	1.7	1.9	2.3	2.6	3.0	1.8	2.2	2.6	3.1	3.6
TW B3	13	16	20	23	27	30	33	38	11	13	17	19	22	25	27	31	1.6	1.9	2.2	2.5	2.8	1.8	2.1	2.5	2.9	3.3
TW B4	12	16	20	22	26	29	33	37	10	13	16	19	22	24	27	31	1.6	1.9	2.2	2.5	2.8	1.8	2.1	2.4	2.8	3.2
TW C1	14	17	20	22	25	27	30	34	12	14	17	19	21	23	25	28	1.8	2.1	2.4	2.8	3.1	2.0	2.3	2.7	3.2	3.7
TW C2	13	16	19	22	26	28	31	36	11	13	16	18	22	24	26	30	1.7	2.0	2.3	2.6	2.9	1.9	2.2	2.5	2.9	3.3
TW C3	13	15	19	21	25	28	31	35	11	13	16	18	21	23	26	30	1.8	2.0	2.3	2.6	2.9	1.9	2.2	2.5	2.9	3.2
TW C4	13	15	18	21	24	26	29	34	11	13	16	18	21	23	25	29	1.9	2.1	2.4	2.7	2.9	2.0	2.3	2.6	2.9	3.2
TW D1	10	12	14	15	17	18	19	20	8	9	11	11	12	13	13	12	2.1	2.6	3.4	5.0	-	2.4	3.3	-	-	-
TW D2	15	19	24	28	33	36	41	46	12	16	20	23	28	31	34	39	1.4	1.7	1.9	2.2	2.4	1.5	1.8	2.1	2.5	2.8
TW D3	15	20	26	30	35	38	43	49	12	17	21	25	29	32	36	41	1.4	1.6	1.9	2.1	2.4	1.5	1.8	2.1	2.4	2.7
TW D4	17	23	30	35	41	46	51	59	14	19	25	30	36	40	45	52	1.2	1.4	1.7	1.9	2.0	1.3	1.5	1.8	2.0	2.2
TW D5	17	23	30	36	43	48	54	62	14	19	26	31	38	42	48	56	1.2	1.4	1.6	1.8	1.9	1.3	1.5	1.7	1.9	2.1
TW E1	18	23	29	34	40	44	49	55	15	20	25	29	34	37	42	47	1.3	1.5	1.8	2.0	2.2	1.4	1.6	1.9	2.2	2.5
TW E2	18	24	30	35	41	45	51	58	14	20	25	30	36	39	44	51	1.2	1.5	1.7	1.9	2.1	1.3	1.6	1.8	2.1	2.3
TW E3	21	28	37	42	50	55	62	70	17	24	32	37	45	50	56	63	1.1	1.3	1.5	1.7	1.8	1.2	1.4	1.6	1.8	2.0
TW E4	19	27	35	41	49	54	61	69	16	23	31	36	44	49	55	63	1.1	1.3	1.5	1.7	1.8	1.2	1.4	1.6	1.8	1.9
TW F1	14	16	19	21	23	25	27	29	12	14	16	17	19	20	21	23	2.0	2.3	2.8	3.3	4.0	2.2	2.7	3.3	4.3	-
TW F2	12	16	19	21	25	27	30	33	10	13	15	17	20	22	24	26	1.7	2.0	2.4	2.8	3.2	1.9	2.3	2.8	3.4	4.2
TW F3	11	15	18	21	25	28	31	35	9	11	15	17	20	22	24	27	1.6	1.9	2.3	2.6	3.0	1.8	2.2	2.6	3.1	3.7
TW F4	9	13	17	20	24	27	30	35	7	10	13	16	19	21	24	28	1.6	1.9	2.2	2.5	2.8	1.8	2.1	2.5	2.9	3.4
TW G1	12	15	17	19	22	24	27	30	10	12	14	16	18	20	21	23	1.9	2.2	2.6	3.1	3.6	2.1	2.5	3.1	3.7	4.6
TW G2	10	13	17	19	23	25	28	32	8	10	13	15	18	20	22	25	1.7	2.0	2.3	2.7	3.1	1.9	2.3	2.7	3.2	3.7
TW G3	8	12	15	18	22	25	28	32	6	9	12	14	17	19	22	26	1.7	2.0	2.3	2.6	2.9	1.9	2.2	2.6	3.0	3.4
TW G4	7	10	13	16	20	22	26	30	5	7	10	12	16	18	21	24	1.8	2.0	2.3	2.6	2.9	1.9	2.2	2.6	2.9	3.3
TW H1	9	12	15	17	20	22	25	29	7	9	12	14	16	18	20	23	1.9	2.2	2.5	2.9	3.3	2.1	2.4	2.9	3.4	3.9
TW H2	7	10	13	15	19	21	24	29	5	7	10	12	15	17	19	23	1.9	2.1	2.4	2.7	3.0	2.0	2.3	2.7	3.1	3.5
TW H3	6	8	11	14	17	19	23	27	4	6	9	11	13	15	18	22	1.9	2.2	2.4	2.7	3.0	2.1	2.4	2.7	3.1	3.4
TW H4	4	6	9	11	14	16	19	23	3	5	7	9	11	13	16	19	2.1	2.4	2.6	2.9	3.1	2.3	2.6	2.9	3.2	3.4
TW I1	8	10	12	13	15	16	17	18	6	7	9	9	10	11	11	10	2.2	2.7	3.5	-	-	2.5	3.4	-	-	-
TW I2	12	17	21	25	30	33	38	43	9	13	17	20	25	28	31	36	1.5	1.7	2.0	2.2	2.5	1.6	1.9	2.2	2.5	2.8
TW I3	13	18	23	27	32	36	41	47	10	14	19	22	27	30	34	39	1.4	1.6	1.9	2.1	2.3	1.5	1.8	2.1	2.4	2.6
TW I4	14	20	27	32	39	43	49	56	11	17	23	27	33	37	43	50	1.3	1.5	1.7	1.9	2.1	1.3	1.6	1.8	2.0	2.2
TW I5	14	21	28	33	40	45	51	60	11	17	24	28	35	40	46	53	1.2	1.4	1.6	1.8	2.0	1.3	1.5	1.7	1.9	2.1
TW J1	16	21	27	32	38	42	47	54	13	18	23	27	32	36	40	46	1.3	1.5	1.8	2.0	2.2	1.4	1.6	1.9	2.2	2.4
TW J2	15	21	28	32	39	43	49	56	12	17	23	27	33	37	42	49	1.3	1.5	1.7	1.9	2.1	1.3	1.6	1.8	2.1	2.3
TW J3	18	26	34	40	48	54	60	69	15	22	30	35	43	48	54	63	1.1	1.3	1.5	1.7	1.8	1.2	1.4	1.6	1.8	1.9
TW J4	17	25	33	39	47	52	59	68	14	21	28	34	42	47	53	62	1.1	1.3	1.5	1.7	1.8	1.2	1.4	1.6	1.8	1.9
DG A1	10	12	14	16	17	18	18	18	7	9	11	11	12	12	12	10	2.0	2.6	4.0	-	-	2.2	3.3	-	-	-
DG A2	13	16	20	23	27	29	32	36	10	13	17	19	22	24	27	30	1.6	1.9	2.3	2.7	3.2	1.7	2.1	2.5	3.1	3.7
DG A3	16	21	27	31	37	41	46	52	14	18	23	27	32	36	41	46	1.4	1.6	1.8	2.0	2.2	1.5	1.7	1.9	2.2	2.4
DG B1	10	13	15	17	18	19	20	20	8	9	11	12	13	14	14	13	1.8	2.4	3.6	-	-	2.2	2.9	4.7	-	-
DG B2	13	17	21	24	28	31	34	39	11	14	17	20	24	26	29	32	1.5	1.8	2.2	2.5	2.9	1.7	2.0	2.3	2.9	3.4
DG B3	17	22	28	32	38	43	48	55	14	19	24	28	34	38	43	50	1.3	1.5	1.7	1.9	2.1	1.4	1.6	1.8	2.1	2.3
DG C1	12	16	19	21	24	25	27	28	10	12	14	16	18	19	20	20	1.6	2.1	2.9	3.9	-	1.9	2.4	3.3	-	-
DG C2	15	20	25	28	33	36	40	46	13	16	21	24	28	31	35	39	1.4	1.7	2.0	2.3	2.6	1.6	1.8	2.1	2.5	2.9
DG C3	19	25	31	36	43	47	53	61	16	21	27	32	38	43	48	55	1.2	1.4	1.6	1.8	2.0	1.3	1.5	1.7	2.0	2.2
SS A1	17	21	24	27	30	32	35	38	16	19	22	24	27	28	30	32	1.6	1.9	2.3	2.8	3.2	1.7	2.1	2.7	3.3	4.2
SS A2	21	26	32	36	42	45	50	56	19	24	29	33	38	41	45	51	1.3	1.5	1.8	2.0	2.2	1.3	1.6	1.9	2.2	2.4
SS A3	16	18	21	23	25	27	29	31	15	17	19	21	22	23	24	25	1.8	2.2	2.8	3.6	4.7	1.9	2.5	3.4	-	-
SS A4	20	25	30	34	39	43	47	53	19	23	28	31	36	39	42	47	1.3	1.6	1.9	2.1	2.3	1.4	1.7	2.0	2.3	2.6
SS A5	18	20	23	25	28	30	32	34	17	19	21	23	25	26	28	29	1.8	2.2	2.6	3.2	3.9	1.9	2.4	3.1	4.1	-
SS A6	21	26	31	35	40	44	49	55	19	23	28	32	37	40	44	49	1.3	1.6	1.8	2.0	2.2	1.4	1.7	1.9	2.3	2.5
SS A7	16	18	20	22	23	24	26	27	15	17	19	20	21	22	22	22	2.1	2.6	3.3	4.5	-	2.2	2.9	4.1	-	-
SS A8	20	24	29	33	38	41	45	50	18	22	27	30	34	37	40	44	1.4	1.6	1.9	2.2	2.4	1.4	1.8	2.1	2.5	2.8
SS B1	15	18	21	23	26	29	31	34	13	16	19	21	23	25	27	29	1.7	2.0	2.4	2.9	3.4	1.8	2.3	2.8	3.5	4.4
SS B2	19	24	29	33	39	43	47	54	17	21	26	30	35	38	43	48	1.3	1.6	1.8	2.0	2.2	1.4	1.7	1.9	2.2	2.4
SS B3	14	16	19	21	23	25	27	29	12	14	17	18	20	21	22	23	1.8	2.2	2.8	3.5	4.3	2.0	2.5	3.3	4.6	-
SS B4	18	23	28	32	37	40	45	51	16	21	25	29	33	36	40	45	1.4	1.6	1.9	2.1	2.3	1.4	1.7	2.0	2.3	2.5
SS B5	15	17	20	21	24	25	27	30	14	16	18	19	21	22	24	25	2.0	2.3	2.8	3.4	4.0	2.1	2.6	3.2	4.2	-
SS B6	18	23	28	32	37	41	45	52	16	20	25	29	33	37	41	46	1.4	1.6	1.8	2.1	2.3	1.4	1.7	2.0	2.2	2.5
SS B7	13	15	17	19	21	22	23	25	12	14	16	17	18	19	20	20	2.2	2.6	3.2	4.1	-	2.3	2.9	3.9	-	-
SS B8	17	21	26	30	35	38	42	48	15	19	24	27	31	34	38	42	1.4	1.7	1.9	2.2	2.4	1.5	1.8	2.1	2.4	2.7
SS C1	13	16	20	22	26	28	31	34	10	13	15	17	20	22	23	25	1.6	2.0	2.4	2.9	3.4	1.8	2.3	2.9	3.8	-
SS C2	15	20	26	30	36	39	44	50	12	17	21	25	30	33	37	42	1.3	1.6	1.8	2.1	2.3	1.4	1.7	2.0	2.3	2.6
SS C3	13	15	18	20	23	25	28	31	11	13	15	17	19	20	22	24	1.9	2.2	2.6	3.0	3.5	2.0	2.5	3.0	3.7	4.6
SS C4	15	19	24	28	33	36	40	46	12	16	20	23	28	30	34	39	1.4	1.7	1.9	2.2	2.4	1.6	1.8	2.1	2.4	2.7
SS D1	18	22	26	29	33	36	38	42	17	20	23	26	29	31	32	34	1.5	1.8	2.2	2.6	3.0	1.6	2.0	2.6	3.4	4.8
SS D2	22	29	35	40	47	51	57	64	20	26	32	36	42	46	51	57	1.2	1.4	1.6	1.8	2.0	1.2	1.5	1.7	2.0	2.2
SS D3	18	22	25	28	31	33	36	39	17	20	23	25	27	29	31	33	1.6	1.9	2.4	2.8	3.3	1.7	2.1	2.7	3.5	4.9
SS D4	22	28	35	39	45	50	55	62	20	25	31	35	41	45	49	55	1.2	1.4	1.7	1.9	2.0	1.3	1.5	1.8	2.0	2.3
SS E1	15	19	22	25	28	29	32	34	13	16	19	21	23	24	25	26	1.6	2.0	2.5	3.2	3.9	1.7	2.3	3.1	4.9	-
SS E2	19	24	30	35	40	44	49	56	16	21	27	31	36	39	43	48	1.2	1.5	1.8	2.0	2.2	1.3	1.6	1.9	2.2	2.5
SS E3	15	18	21	23	26	27	30	32	14	16	19	20	22	24	25	26	1.8	2.2	2.6	3.3	3.9	1.9	2.4	3.1	4.2	-
SS E4	18	23	29	33	38	42	46	52	16	21	26	29	34	37	41	46	1.3	1.6	1.8	2.1	2.3	1.4	1.7	2.0	2.3	2.6

VANCOUVER, BRITISH COLUMBIA

TYPE	SOLAR SAVINGS FRACTION (%) BASE TEMP: 55 F (2685 DD) LCR 100	70	50	40	30	25	20	15	65 F (5607 DD) 100	70	50	40	30	25	20	15	CONSERVATION FACTOR 55 F 100	50	30	20	15	65 F 100	50	30	20	15
WW A1	20	23	27	29	33	35	38	42	16	18	21	23	25	27	28	30	1.6	1.9	2.2	2.6	2.9	1.9	2.3	2.8	3.5	4.4
WW A2	21	25	31	34	40	43	47	53	15	19	23	26	30	32	35	38	1.3	1.6	1.8	2.1	2.4	1.5	1.9	2.2	2.7	3.2
WW A3	20	26	32	36	42	46	50	56	15	19	24	27	31	34	37	41	1.2	1.5	1.8	2.0	2.3	1.4	1.7	2.1	2.5	3.0
WW A4	20	28	33	37	43	47	52	58	14	19	24	27	32	35	39	43	1.2	1.4	1.7	2.0	2.2	1.4	1.7	2.0	2.4	2.9
WW A5	20	27	34	38	45	49	54	60	14	19	24	28	33	36	40	44	1.2	1.4	1.7	1.9	2.1	1.3	1.6	2.0	2.4	2.8
WW A6	20	27	34	39	46	50	55	61	14	19	24	28	34	37	41	45	1.2	1.4	1.6	1.9	2.1	1.3	1.6	1.9	2.3	2.7
WW B1	16	20	24	26	30	32	34	37	12	14	16	18	20	20	21	21	1.6	1.9	2.4	2.9	3.6	1.9	2.5	3.6	-	-
WW B2	21	28	35	40	47	51	57	63	15	20	26	30	36	40	44	50	1.2	1.4	1.6	1.8	2.0	1.3	1.6	1.8	2.1	2.4
WW B3	25	32	40	46	53	58	63	70	18	24	31	36	42	46	51	57	1.1	1.3	1.5	1.7	1.9	1.2	1.4	1.7	2.0	2.2
WW B4	25	33	42	49	57	62	68	76	18	25	33	39	46	51	57	65	1.0	1.2	1.4	1.6	1.8	1.1	1.3	1.6	1.8	2.0
WW B5	24	33	42	48	57	62	69	77	17	25	33	39	47	52	59	67	1.0	1.2	1.4	1.6	1.7	1.1	1.3	1.5	1.7	1.9
WW C1	26	34	42	48	56	60	66	73	20	26	33	38	45	49	54	61	1.0	1.3	1.5	1.7	1.8	1.2	1.4	1.6	1.9	2.1
WW C2	25	33	41	47	55	60	66	74	18	25	32	38	45	49	55	62	1.0	1.2	1.4	1.6	1.8	1.1	1.4	1.6	1.8	2.0
WW C3	29	39	49	56	65	70	77	84	22	31	40	47	55	61	67	75	0.9	1.1	1.3	1.5	1.7	1.0	1.2	1.4	1.6	1.8
WW C4	26	36	46	53	62	68	74	82	20	28	38	44	53	58	65	73	1.0	1.1	1.3	1.5	1.7	1.0	1.2	1.4	1.6	1.8
TW A1	20	23	26	29	33	35	39	43	16	18	21	23	25	27	29	32	1.6	1.9	2.2	2.5	2.8	1.9	2.2	2.7	3.2	3.9
TW A2	20	24	29	32	37	41	45	50	15	18	22	25	28	31	34	37	1.4	1.6	1.9	2.2	2.4	1.6	1.9	2.3	2.7	3.1
TW A3	19	24	30	34	39	43	47	53	14	18	22	25	29	32	35	40	1.3	1.6	1.8	2.1	2.3	1.5	1.8	2.2	2.5	2.9
TW A4	18	23	29	34	39	43	48	54	13	17	22	25	30	33	36	41	1.3	1.5	1.8	2.0	2.2	1.5	1.8	2.2	2.5	2.9
TW B1	19	22	26	29	33	36	40	45	15	18	21	23	26	28	30	33	1.6	1.8	2.1	2.4	2.6	1.8	2.1	2.5	3.0	3.4
TW B2	18	23	28	32	37	40	44	50	14	17	21	24	28	31	34	38	1.4	1.6	1.9	2.1	2.4	1.6	1.9	2.2	2.6	2.9
TW B3	18	22	28	32	37	41	45	51	13	17	21	24	28	31	35	39	1.4	1.6	1.8	2.1	2.3	1.5	1.8	2.1	2.5	2.8
TW B4	17	21	26	30	36	39	44	50	12	16	20	23	27	30	34	39	1.4	1.6	1.8	2.1	2.3	1.6	1.8	2.1	2.4	2.7
TW C1	18	21	26	29	34	37	41	46	14	17	20	23	26	28	31	35	1.5	1.8	2.0	2.3	2.5	1.7	2.0	2.3	2.7	3.0
TW C2	17	21	26	29	34	38	42	48	13	16	20	23	27	29	33	37	1.5	1.7	1.9	2.2	2.4	1.7	1.9	2.2	2.5	2.8
TW C3	16	20	25	28	33	37	41	47	12	15	19	22	26	29	32	37	1.5	1.7	1.9	2.2	2.4	1.7	1.9	2.2	2.5	2.8
TW C4	16	19	23	26	31	34	38	44	12	15	18	21	24	27	30	35	1.6	1.8	2.0	2.3	2.5	1.8	2.0	2.3	2.6	2.8
TW D1	14	17	21	23	27	29	32	36	10	12	15	16	18	20	21	22	1.7	2.0	2.3	2.8	3.3	2.0	2.4	3.2	4.4	-
TW D2	21	26	33	37	44	48	53	60	15	20	25	29	34	38	42	47	1.2	1.4	1.7	1.9	2.1	1.4	1.6	1.9	2.2	2.4
TW D3	22	28	35	40	47	51	56	63	16	21	27	31	36	40	44	50	1.2	1.4	1.6	1.8	2.0	1.3	1.6	1.8	2.1	2.4
TW D4	23	31	39	45	52	57	63	71	17	24	31	36	43	47	53	60	1.1	1.3	1.5	1.7	1.9	1.2	1.4	1.6	1.9	2.1
TW D5	23	31	39	45	53	58	65	73	17	24	31	37	44	49	55	63	1.1	1.3	1.5	1.6	1.8	1.2	1.4	1.6	1.8	2.0
TW E1	25	32	39	44	51	56	61	69	19	24	31	35	41	45	50	56	1.1	1.3	1.5	1.8	1.9	1.2	1.5	1.7	2.0	2.2
TW E2	24	32	39	45	52	57	63	70	18	24	31	36	42	47	52	59	1.1	1.3	1.5	1.7	1.9	1.2	1.4	1.7	1.9	2.1
TW E3	28	37	47	53	61	67	73	80	22	30	38	44	52	57	63	70	1.0	1.2	1.4	1.6	1.7	1.1	1.3	1.5	1.7	1.9
TW E4	26	35	45	51	60	65	71	79	20	28	36	42	50	56	62	70	1.0	1.2	1.4	1.6	1.7	1.1	1.3	1.5	1.7	1.9
TW F1	18	21	25	28	32	34	38	42	14	16	19	21	24	26	28	31	1.6	1.9	2.1	2.5	2.8	1.9	2.2	2.6	3.2	3.8
TW F2	17	22	26	30	35	38	42	48	12	16	20	22	26	28	31	35	1.4	1.7	1.9	2.2	2.5	1.6	1.9	2.3	2.7	3.2
TW F3	16	21	26	30	36	39	44	50	11	15	19	22	26	29	32	37	1.4	1.6	1.9	2.1	2.4	1.6	1.9	2.2	2.6	2.9
TW F4	14	19	25	29	34	38	43	49	9	13	18	21	25	28	32	37	1.4	1.6	1.8	2.1	2.3	1.5	1.8	2.1	2.5	2.8
TW G1	16	20	24	27	31	34	38	42	12	15	18	20	23	25	28	31	1.6	1.8	2.1	2.4	2.6	1.8	2.1	2.5	3.0	3.4
TW G2	14	19	24	27	32	36	40	46	10	13	17	20	24	26	30	34	1.5	1.7	1.9	2.2	2.4	1.6	1.9	2.3	2.6	3.0
TW G3	13	17	22	26	31	35	40	46	8	12	16	19	23	26	29	34	1.4	1.7	1.9	2.2	2.4	1.6	1.9	2.2	2.6	2.9
TW G4	10	15	19	23	28	32	36	42	6	10	13	16	20	23	27	32	1.5	1.7	2.0	2.2	2.4	1.7	2.0	2.2	2.5	2.8
TW H1	13	17	21	24	29	32	36	41	9	12	15	18	21	23	26	30	1.6	1.8	2.1	2.3	2.6	1.8	2.1	2.4	2.8	3.2
TW H2	11	14	19	22	27	30	35	41	7	10	13	16	20	22	26	30	1.6	1.8	2.0	2.3	2.5	1.8	2.0	2.3	2.7	3.0
TW H3	9	12	17	20	24	28	32	38	5	8	11	14	18	20	24	28	1.6	1.9	2.1	2.3	2.5	1.9	2.1	2.4	2.7	2.9
TW H4	7	10	13	16	20	23	27	33	4	6	9	11	14	17	20	24	1.8	2.0	2.2	2.4	2.6	2.1	2.3	2.5	2.8	3.0
TW I1	12	15	18	21	24	27	29	33	8	10	12	14	16	17	19	20	1.7	2.0	2.4	2.8	3.3	2.0	2.5	3.2	4.5	-
TW I2	18	23	30	34	41	45	50	57	12	17	22	26	31	35	39	45	1.3	1.5	1.7	1.9	2.1	1.4	1.7	1.9	2.2	2.5
TW I3	19	25	32	37	44	48	53	61	13	18	24	28	34	38	42	48	1.2	1.4	1.6	1.9	2.0	1.3	1.6	1.8	2.1	2.4
TW I4	20	28	36	42	49	54	61	68	15	21	28	33	40	44	50	57	1.1	1.3	1.5	1.7	1.9	1.2	1.4	1.7	1.9	2.1
TW I5	21	28	37	43	51	56	63	71	15	21	29	34	42	46	53	60	1.1	1.3	1.5	1.7	1.8	1.2	1.4	1.6	1.8	2.0
TW J1	22	29	37	42	49	54	60	67	16	22	29	33	39	43	49	55	1.1	1.3	1.5	1.8	1.9	1.2	1.5	1.7	2.0	2.2
TW J2	22	29	37	42	50	55	61	68	16	22	29	33	40	45	50	57	1.1	1.3	1.5	1.7	1.9	1.2	1.4	1.7	1.9	2.1
TW J3	25	34	44	51	59	65	71	79	19	27	36	42	50	55	61	69	1.0	1.2	1.4	1.6	1.7	1.1	1.3	1.5	1.7	1.9
TW J4	24	33	42	49	57	63	69	77	17	25	34	40	48	53	60	68	1.0	1.2	1.4	1.6	1.7	1.1	1.3	1.5	1.7	1.9
DG A1	15	18	22	25	27	29	31	32	10	12	15	17	18	19	20	19	1.5	2.0	2.6	3.7	-	1.8	2.5	3.7	-	-
DG A2	17	22	27	31	36	40	44	49	12	16	21	24	28	31	34	37	1.4	1.6	1.9	2.2	2.6	1.5	1.8	2.2	2.6	3.1
DG A3	21	27	34	39	46	51	56	64	16	22	28	32	38	42	47	53	1.2	1.4	1.6	1.8	2.0	1.3	1.5	1.8	2.0	2.2
DG B1	15	19	23	26	30	32	34	37	10	13	16	18	20	21	22	23	1.5	1.8	2.3	3.1	4.3	1.8	2.3	3.1	-	-
DG B2	18	23	29	33	39	43	47	52	13	17	22	25	30	33	36	41	1.3	1.5	1.8	2.1	2.4	1.5	1.8	2.1	2.5	2.9
DG B3	22	28	36	41	49	54	60	67	17	22	29	33	40	45	50	57	1.2	1.3	1.5	1.7	1.9	1.3	1.5	1.7	1.9	2.1
DG C1	18	23	28	31	36	39	42	45	13	16	20	22	26	27	29	31	1.3	1.6	2.0	2.6	3.3	1.6	2.0	2.5	3.7	-
DG C2	21	27	33	38	44	49	53	59	15	20	25	29	35	38	42	48	1.2	1.4	1.7	2.0	2.2	1.4	1.6	1.9	2.2	2.6
DG C3	24	32	40	46	54	59	65	72	19	25	32	37	45	49	55	62	1.1	1.3	1.4	1.7	1.8	1.2	1.4	1.6	1.8	2.0
SS A1	23	27	32	36	41	44	48	53	20	23	27	30	34	36	39	42	1.3	1.6	1.9	2.2	2.4	1.5	1.8	2.2	2.7	3.1
SS A2	27	34	41	46	52	57	62	69	24	29	35	40	45	49	53	59	1.1	1.3	1.6	1.8	1.9	1.2	1.5	1.7	2.0	2.2
SS A3	21	25	29	32	36	39	42	46	18	21	25	27	30	32	34	36	1.4	1.7	2.1	2.5	2.8	1.6	2.0	2.6	3.3	4.3
SS A4	27	33	40	44	51	55	60	66	23	29	34	38	43	47	51	56	1.1	1.4	1.6	1.8	2.0	1.2	1.5	1.8	2.1	2.3
SS A5	22	26	30	33	38	40	44	48	20	23	26	29	32	34	36	38	1.5	1.7	2.0	2.4	2.7	1.6	2.0	2.4	3.0	3.7
SS A6	27	34	40	45	52	56	61	67	23	29	35	39	44	48	52	58	1.1	1.4	1.6	1.8	2.0	1.2	1.5	1.8	2.0	2.3
SS A7	20	24	27	29	33	35	37	41	18	21	23	25	27	29	30	32	1.6	1.9	2.3	2.8	3.2	1.8	2.2	2.9	3.9	-
SS A8	26	32	39	43	49	53	58	64	23	28	33	37	42	45	49	54	1.2	1.4	1.6	1.9	2.1	1.3	1.6	1.9	2.2	2.5
SS B1	20	24	28	32	37	40	43	48	16	20	24	26	30	32	35	38	1.4	1.7	2.0	2.3	2.5	1.6	1.9	2.3	2.8	3.2
SS B2	25	31	38	42	49	53	59	66	21	26	32	36	42	46	51	57	1.2	1.4	1.6	1.8	2.0	1.3	1.5	1.8	2.0	2.2
SS B3	18	22	26	29	34	36	40	44	15	19	22	24	27	29	31	34	1.5	1.8	2.1	2.5	2.8	1.6	2.0	2.5	3.2	4.0
SS B4	24	30	36	41	47	52	57	64	20	25	31	35	40	44	48	54	1.2	1.4	1.6	1.8	2.0	1.3	1.5	1.8	2.1	2.3
SS B5	19	22	26	29	33	35	38	43	16	19	22	24	27	29	31	34	1.6	1.9	2.2	2.5	2.8	1.7	2.1	2.6	3.1	3.7
SS B6	24	30	37	41	48	52	57	64	20	25	31	35	41	44	49	55	1.2	1.4	1.6	1.8	2.0	1.3	1.5	1.8	2.1	2.3
SS B7	17	20	23	26	29	31	34	37	15	17	20	21	24	25	27	29	1.7	2.0	2.4	2.8	3.2	1.9	2.3	2.9	3.6	4.6
SS B8	23	29	35	39	46	50	55	61	19	24	29	33	38	42	46	51	1.2	1.4	1.7	1.9	2.1	1.3	1.6	1.9	2.2	2.4
SS C1	18	23	28	32	37	41	45	50	13	17	21	23	27	29	32	35	1.3	1.6	1.9	2.2	2.4	1.5	1.9	2.3	2.8	3.5
SS C2	22	28	35	40	47	52	57	64	16	21	27	31	37	41	45	51	1.2	1.4	1.6	1.8	2.0	1.3	1.5	1.8	2.1	2.3
SS C3	17	20	25	28	32	35	39	44	12	15	19	21	24	26	29	32	1.5	1.8	2.0	2.3	2.6	1.8	2.1	2.5	2.9	3.4
SS C4	20	26	32	37	43	47	53	59	15	20	25	29	34	37	42	47	1.2	1.5	1.7	1.9	2.1	1.4	1.6	1.9	2.2	2.4
SS D1	25	31	36	40	46	49	53	59	21	26	30	33	37	40	43	46	1.2	1.5	1.8	2.1	2.3	1.3	1.7	2.1	2.6	3.2
SS D2	30	38	46	52	59	63	69	76	26	32	39	44	50	55	60	66	1.0	1.2	1.5	1.6	1.8	1.1	1.4	1.6	1.8	2.1
SS D3	24	29	34	38	42	45	49	54	21	25	29	32	35	37	40	43	1.3	1.6	1.9	2.2	2.4	1.4	1.8	2.2	2.7	3.3
SS D4	29	37	45	50	57	62	67	74	25	32	38	43	49	53	58	64	1.0	1.3	1.5	1.7	1.8	1.1	1.4	1.6	1.9	2.1
SS E1	22	27	32	36	40	44	47	52	18	22	26	28	32	34	36	38	1.3	1.6	1.9	2.2	2.5	1.4	1.9	2.4	3.1	4.1
SS E2	26	33	41	46	53	57	63	69	21	28	34	38	44	48	52	58	1.1	1.3	1.5	1.8	1.9	1.2	1.5	1.7	2.0	2.3
SS E3	20	24	29	32	36	39	42	47	17	20	24	26	29	31	33	36	1.5	1.7	2.0	2.4	2.7	1.6	2.0	2.5	3.1	3.8
SS E4	25	31	38	43	50	54	59	66	21	26	32	36	42	45	50	55	1.1	1.4	1.6	1.8	2.0	1.2	1.5	1.8	2.1	2.3

WINNIPEG, MANITOBA

TYPE	SOLAR SAVINGS FRACTION (%) BASE TEMP: 55 F (7966 DD)								65 F (10776 DD)								CONSERVATION FACTOR 55 F					65 F				
LCR	100	70	50	40	30	25	20	15	100	70	50	40	30	25	20	15	100	50	30	20	15	100	50	30	20	15
WW A1	13	14	16	17	19	20	21	23	12	13	14	15	16	16	17	18	2.5	2.9	3.4	4.0	4.7	2.9	3.6	4.4	-	-
WW A2	11	13	17	19	22	25	28	32	9	11	14	16	18	20	22	25	1.8	2.1	2.4	2.7	3.0	2.0	2.4	2.8	3.2	3.7
WW A3	9	13	17	20	24	27	30	35	8	10	13	16	19	21	24	28	1.7	1.9	2.2	2.5	2.7	1.9	2.1	2.5	2.8	3.2
WW A4	9	12	17	20	25	28	32	38	7	10	13	16	20	22	26	30	1.6	1.8	2.1	2.3	2.6	1.8	2.0	2.3	2.7	3.0
WW A5	8	12	17	21	26	29	34	40	6	9	13	16	20	23	27	32	1.5	1.8	2.0	2.3	2.5	1.7	1.9	2.2	2.6	2.9
WW A6	8	12	17	21	26	30	35	41	5	9	13	16	21	24	28	33	1.5	1.7	2.0	2.2	2.4	1.6	1.9	2.2	2.5	2.8
WW B1	7	8	10	10	11	12	12	12	6	7	7	7	7	7	6	5	2.7	3.5	-	-	-	3.5	-	-	-	-
WW B2	10	14	20	24	29	33	39	46	8	12	16	20	25	28	33	39	1.5	1.7	1.9	2.1	2.2	1.6	1.8	2.0	2.2	2.4
WW B3	12	17	24	29	35	40	46	54	10	14	20	24	30	34	40	47	1.3	1.5	1.7	1.9	2.1	1.4	1.6	1.8	2.0	2.2
WW B4	12	19	27	32	41	46	53	62	10	16	23	28	36	41	48	57	1.2	1.4	1.5	1.7	1.9	1.3	1.5	1.6	1.8	2.0
WW B5	12	19	27	33	41	47	55	64	9	16	23	29	37	43	50	59	1.2	1.4	1.5	1.7	1.8	1.3	1.4	1.6	1.8	1.9
WW C1	13	19	26	31	38	43	49	58	11	16	22	27	33	38	44	51	1.3	1.5	1.6	1.8	2.0	1.4	1.5	1.7	1.9	2.1
WW C2	12	18	26	31	39	44	51	59	10	16	22	27	34	39	45	54	1.3	1.4	1.6	1.8	1.9	1.3	1.5	1.7	1.9	2.0
WW C3	15	24	34	41	50	57	64	74	13	21	30	37	46	52	60	69	1.1	1.2	1.4	1.6	1.7	1.1	1.3	1.4	1.6	1.8
WW C4	14	22	31	38	48	54	62	71	11	19	28	34	43	49	57	67	1.1	1.3	1.4	1.6	1.7	1.2	1.3	1.5	1.6	1.8
TW A1	13	14	16	17	19	20	22	24	12	13	14	15	16	17	18	19	2.5	2.8	3.2	3.7	4.1	2.8	3.3	3.9	4.7	-
TW A2	11	13	16	18	21	24	26	30	10	11	14	15	18	20	22	25	2.0	2.2	2.5	2.8	3.1	2.2	2.5	2.9	3.3	3.7
TW A3	10	13	16	18	22	25	28	33	8	10	13	15	18	20	23	27	1.8	2.0	2.3	2.6	2.8	2.0	2.3	2.6	2.9	3.3
TW A4	8	12	15	18	22	25	29	35	7	9	13	15	18	21	24	28	1.7	2.0	2.2	2.4	2.7	1.9	2.2	2.4	2.7	3.0
TW B1	12	14	16	17	19	21	23	26	11	12	14	15	17	18	19	21	2.3	2.6	2.9	3.2	3.6	2.5	2.9	3.4	3.9	4.4
TW B2	10	12	15	18	21	24	27	31	8	11	13	15	18	20	22	26	1.9	2.1	2.4	2.7	2.9	2.1	2.4	2.7	3.0	3.3
TW B3	9	12	15	18	21	24	27	32	8	10	13	15	18	20	23	27	1.8	2.1	2.3	2.5	2.8	2.0	2.3	2.6	2.8	3.1
TW B4	8	11	14	17	21	23	27	32	7	10	12	14	17	20	23	27	1.9	2.1	2.3	2.5	2.7	2.0	2.3	2.5	2.8	3.0
TW C1	11	13	15	17	20	22	24	28	10	11	13	15	17	19	21	24	2.1	2.4	2.6	2.9	3.1	2.3	2.6	3.0	3.3	3.6
TW C2	9	12	14	17	20	22	26	30	8	10	13	14	17	19	22	26	2.0	2.2	2.4	2.7	2.9	2.2	2.4	2.7	3.0	3.2
TW C3	9	11	14	16	20	22	25	30	8	10	12	14	17	19	22	25	2.0	2.2	2.5	2.7	2.9	2.2	2.5	2.7	3.0	3.2
TW C4	10	12	14	16	19	21	24	28	9	11	13	14	17	18	21	24	2.2	2.4	2.6	2.8	3.0	2.4	2.6	2.8	3.1	3.3
TW D1	6	7	9	9	11	11	12	13	5	6	6	7	7	8	8	7	2.7	3.3	4.2	-	-	3.3	4.5	-	-	-
TW D2	11	14	19	22	27	31	36	42	9	12	16	19	23	27	31	36	1.6	1.8	2.0	2.2	2.4	1.7	1.9	2.1	2.4	2.6
TW D3	11	15	20	24	29	33	38	45	9	12	17	20	25	28	33	39	1.5	1.7	1.9	2.1	2.3	1.6	1.8	2.0	2.3	2.5
TW D4	12	18	24	29	36	41	48	56	10	15	21	25	32	37	43	51	1.3	1.5	1.7	1.8	2.0	1.4	1.6	1.7	1.9	2.1
TW D5	12	18	25	30	38	43	50	59	10	15	22	27	34	39	46	54	1.3	1.4	1.6	1.8	1.9	1.3	1.5	1.7	1.8	2.0
TW E1	13	18	24	28	35	39	44	52	11	15	21	24	30	34	39	46	1.4	1.6	1.8	1.9	2.1	1.5	1.7	1.9	2.1	2.3
TW E2	13	18	24	29	36	41	47	55	11	16	21	25	32	36	42	49	1.3	1.5	1.7	1.9	2.0	1.4	1.6	1.8	2.0	2.1
TW E3	15	23	31	37	46	52	59	68	13	20	28	34	42	47	54	63	1.1	1.3	1.5	1.6	1.8	1.2	1.4	1.5	1.7	1.9
TW E4	14	22	30	36	45	51	58	68	12	19	26	32	41	46	53	63	1.2	1.3	1.5	1.6	1.8	1.2	1.4	1.5	1.7	1.9
TW F1	11	12	14	15	18	19	21	23	10	11	12	13	15	16	17	18	2.4	2.7	3.1	3.5	4.0	2.7	3.2	3.8	4.5	-
TW F2	9	11	14	16	19	21	24	28	7	9	11	13	15	17	19	22	2.0	2.3	2.5	2.8	3.1	2.2	2.6	2.9	3.3	3.8
TW F3	7	10	13	15	19	22	25	30	5	8	10	12	15	17	20	24	1.9	2.1	2.4	2.6	2.9	2.1	2.4	2.7	3.0	3.3
TW F4	5	8	11	14	18	21	24	30	4	6	9	11	14	17	20	24	1.8	2.1	2.3	2.5	2.8	2.0	2.3	2.5	2.8	3.1
TW G1	9	10	13	14	17	18	20	24	8	9	11	12	14	15	16	19	2.3	2.6	2.9	3.2	3.6	2.5	2.9	3.4	3.9	4.4
TW G2	6	8	11	14	17	19	22	26	5	7	9	11	13	15	18	21	2.0	2.3	2.5	2.8	3.0	2.2	2.5	2.8	3.2	3.5
TW G3	4	7	10	12	16	18	22	27	3	5	8	10	13	15	17	21	2.0	2.2	2.4	2.7	2.9	2.2	2.4	2.7	3.0	3.3
TW G4	3	5	8	10	14	16	20	24	2	4	6	8	11	13	16	20	2.1	2.3	2.5	2.7	2.9	2.2	2.5	2.7	3.0	3.2
TW H1	6	8	10	12	14	16	19	23	5	6	8	10	12	13	15	18	2.2	2.5	2.8	3.1	3.3	2.5	2.8	3.1	3.5	3.9
TW H2	3	6	8	10	13	15	18	23	3	4	6	8	11	12	15	18	2.2	2.4	2.6	2.9	3.1	2.4	2.6	2.9	3.2	3.5
TW H3	2	4	7	9	12	14	17	21	2	3	5	7	9	11	14	17	2.3	2.5	2.7	2.9	3.1	2.4	2.7	3.0	3.2	3.4
TW H4	2	3	5	7	9	11	14	18	1	2	4	5	8	9	11	14	2.5	2.7	2.9	3.1	3.3	2.7	2.9	3.2	3.4	3.6
TW I1	4	5	7	7	9	9	10	11	3	4	5	5	6	6	6	5	2.9	3.5	4.4	-	-	3.5	4.8	-	-	-
TW I2	8	11	16	19	24	28	32	39	6	9	13	16	20	23	27	33	1.6	1.8	2.0	2.2	2.4	1.8	2.0	2.2	2.4	2.6
TW I3	8	12	17	21	27	31	36	43	6	10	14	18	23	26	31	37	1.5	1.7	1.9	2.1	2.3	1.6	1.8	2.1	2.3	2.5
TW I4	9	15	21	26	33	38	45	53	8	12	18	23	29	34	40	48	1.4	1.5	1.7	1.9	2.0	1.4	1.6	1.8	2.0	2.1
TW I5	10	16	22	28	35	41	48	57	8	13	19	24	31	36	43	52	1.3	1.5	1.6	1.8	1.9	1.4	1.5	1.7	1.9	2.0
TW J1	11	16	22	26	33	37	43	50	9	13	19	22	28	32	37	44	1.4	1.6	1.8	2.0	2.1	1.5	1.7	1.9	2.1	2.3
TW J2	10	16	22	27	34	38	45	53	8	13	19	23	29	34	40	47	1.4	1.5	1.7	1.9	2.0	1.4	1.6	1.8	2.0	2.1
TW J3	13	20	29	35	44	50	57	67	11	18	25	31	40	45	53	62	1.2	1.3	1.5	1.7	1.8	1.2	1.4	1.5	1.7	1.9
TW J4	12	19	27	34	42	48	56	65	10	16	24	30	38	44	51	61	1.2	1.3	1.5	1.7	1.8	1.2	1.4	1.6	1.7	1.9
DG A1	5	7	8	9	10	11	11	10	4	5	6	7	7	7	6	5	2.5	3.2	4.9	-	-	3.0	4.2	-	-	-
DG A2	8	11	15	17	21	23	27	31	7	9	12	14	18	20	22	26	1.8	2.1	2.3	2.7	3.1	2.0	2.3	2.6	3.0	3.5
DG A3	12	16	21	25	31	36	41	49	10	14	19	22	28	32	37	44	1.5	1.6	1.8	2.0	2.2	1.6	1.7	1.9	2.1	2.3
DG B1	5	7	9	10	11	12	13	13	4	5	6	7	7	8	8	7	2.5	3.0	4.0	-	-	3.0	3.9	-	-	-
DG B2	9	11	15	18	22	25	28	33	7	10	13	15	18	21	24	28	1.8	2.0	2.2	2.5	2.8	2.0	2.2	2.4	2.8	3.1
DG B3	12	17	22	26	33	37	44	52	11	15	19	23	29	34	39	47	1.5	1.6	1.7	1.9	2.0	1.6	1.7	1.8	2.0	2.2
DG C1	7	9	12	14	16	18	20	22	6	7	9	10	12	13	14	15	2.0	2.4	2.9	3.8	5.0	2.4	2.9	3.5	-	-
DG C2	10	14	18	22	27	30	35	41	9	12	16	19	23	26	30	35	1.6	1.8	2.0	2.2	2.4	1.8	1.9	2.1	2.4	2.7
DG C3	14	19	25	30	37	42	49	58	12	17	22	27	34	38	45	53	1.4	1.5	1.6	1.8	1.9	1.5	1.6	1.7	1.9	2.0
SS A1	13	16	19	21	24	26	29	33	12	15	17	19	22	23	26	28	1.8	2.1	2.5	2.8	3.2	2.0	2.3	2.7	3.2	3.8
SS A2	16	21	26	31	36	40	45	52	15	20	25	28	34	37	42	48	1.4	1.6	1.8	2.0	2.2	1.4	1.7	1.9	2.1	2.3
SS A3	12	14	16	17	19	20	22	24	11	13	14	15	17	18	19	20	2.1	2.6	3.1	3.9	4.9	2.3	2.9	3.7	-	-
SS A4	15	20	25	28	34	37	42	48	14	18	23	26	31	34	38	44	1.4	1.7	1.9	2.1	2.3	1.5	1.8	2.0	2.3	2.5
SS A5	14	16	18	20	22	24	26	28	13	15	17	18	20	21	23	24	2.1	2.5	2.9	3.4	3.9	2.2	2.7	3.3	4.1	-
SS A6	16	20	26	30	35	39	44	51	15	19	24	27	32	36	40	46	1.4	1.6	1.8	2.0	2.2	1.5	1.7	1.9	2.2	2.4
SS A7	13	14	15	16	17	18	19	20	12	13	14	15	16	16	17	16	2.6	3.2	4.1	-	-	2.8	3.7	-	-	-
SS A8	15	19	23	27	32	35	39	45	14	18	22	25	29	32	36	41	1.5	1.7	2.0	2.2	2.4	1.6	1.8	2.1	2.4	2.6
SS B1	11	13	16	18	20	22	25	28	10	12	14	16	18	20	22	24	2.0	2.3	2.6	3.0	3.4	2.1	2.5	3.0	3.5	4.0
SS B2	14	19	24	28	33	37	42	50	13	17	22	25	31	34	39	46	1.4	1.6	1.8	2.0	2.2	1.5	1.7	1.9	2.1	2.3
SS B3	10	11	14	15	17	18	20	22	9	10	12	13	15	16	17	18	2.2	2.6	3.1	3.7	4.4	2.4	2.9	3.6	4.7	-
SS B4	13	17	22	26	31	35	40	46	12	16	20	24	29	32	36	42	1.5	1.7	1.9	2.1	2.3	1.6	1.8	2.0	2.2	2.4
SS B5	11	13	15	16	18	19	21	23	11	12	14	15	16	17	18	20	2.3	2.7	3.2	3.7	4.2	2.5	3.0	3.6	4.4	-
SS B6	13	18	22	26	32	35	40	47	12	16	21	24	29	32	37	43	1.5	1.7	1.9	2.1	2.2	1.6	1.8	2.0	2.2	2.4
SS B7	10	11	13	14	15	16	17	18	10	11	12	12	13	14	14	15	2.7	3.3	4.0	-	-	3.0	3.8	4.9	-	-
SS B8	12	16	21	24	29	32	37	43	11	15	19	22	26	29	33	39	1.6	1.8	2.0	2.2	2.4	1.6	1.9	2.1	2.4	2.6
SS C1	8	11	14	16	20	22	25	29	6	9	11	13	15	17	19	21	1.9	2.2	2.5	2.8	3.2	2.1	2.5	2.9	3.4	4.1
SS C2	10	15	20	24	30	34	39	46	9	12	17	20	26	29	34	40	1.5	1.7	1.8	2.1	2.2	1.6	1.8	2.0	2.2	2.4
SS C3	9	11	13	15	17	19	22	25	8	9	11	13	14	16	17	20	2.2	2.5	2.8	3.1	3.5	2.5	2.8	3.3	3.8	4.3
SS C4	10	14	18	22	27	31	35	42	9	12	16	19	23	26	30	36	1.6	1.8	2.0	2.2	2.4	1.7	1.9	2.1	2.4	2.6
SS D1	14	17	21	23	27	30	33	37	12	15	18	21	24	26	28	31	1.6	1.9	2.3	2.6	2.9	1.7	2.1	2.5	3.0	3.6
SS D2	17	23	30	35	42	47	53	61	15	21	27	32	39	43	49	56	1.2	1.4	1.6	1.8	2.0	1.3	1.5	1.7	1.9	2.1
SS D3	14	17	20	22	25	27	30	33	13	15	18	20	23	24	26	29	1.8	2.1	2.5	2.9	3.2	1.9	2.3	2.7	3.3	4.0
SS D4	17	23	29	34	41	45	51	59	15	21	27	31	37	42	47	54	1.3	1.5	1.7	1.9	2.0	1.3	1.5	1.7	2.0	2.1
SS E1	10	13	16	18	21	23	25	28	9	11	14	15	18	19	21	22	1.8	2.2	2.6	3.1	3.7	2.0	2.5	3.1	4.1	-
SS E2	13	19	24	29	35	39	45	52	12	17	22	26	31	35	40	46	1.3	1.6	1.8	2.0	2.2	1.4	1.6	1.9	2.1	2.3
SS E3	11	13	16	17	20	21	23	26	11	12	14	15	17	18	20	21	2.1	2.5	2.9	3.4	4.0	2.3	2.8	3.3	4.2	-
SS E4	13	18	23	27	33	36	41	48	12	16	21	24	29	33	37	43	1.4	1.6	1.9	2.1	2.2	1.5	1.7	2.0	2.2	2.4

FREDERICTON, NEW BRUNSWICK

TYPE	SOLAR SAVINGS FRACTION (%) BASE TEMP: 55 F (5723 DD) LCR 100	70	50	40	30	25	20	15	65 F (8573 DD) 100	70	50	40	30	25	20	15	CONSERVATION FACTOR 55 F 100	50	30	20	15	65 F 100	50	30	20	15
WW A1	13	15	17	18	20	22	24	27	12	13	14	15	16	16	17	18	2.4	2.6	2.9	3.3	3.6	2.9	3.5	4.2	-	-
WW A2	11	15	18	21	25	28	32	38	9	11	14	16	19	21	23	27	1.7	1.9	2.1	2.4	2.6	2.0	2.3	2.6	3.0	3.3
WW A3	10	14	19	22	27	31	35	42	8	10	14	16	20	22	26	30	1.6	1.8	2.0	2.2	2.4	1.8	2.1	2.4	2.6	2.9
WW A4	10	14	19	23	28	32	37	44	7	10	13	16	20	23	27	32	1.5	1.7	1.9	2.1	2.3	1.7	2.0	2.2	2.5	2.7
WW A5	9	14	19	24	30	34	39	47	6	9	13	17	21	24	29	34	1.4	1.6	1.8	2.0	2.2	1.7	1.9	2.1	2.4	2.6
WW A6	9	14	20	24	30	35	41	48	5	9	13	17	22	25	30	36	1.4	1.6	1.8	2.0	2.1	1.6	1.8	2.1	2.3	2.5
WW B1	8	9	11	12	14	15	17	18	6	7	7	8	8	8	8	6	2.4	2.8	3.4	4.2	-	3.4	-	-	-	-
WW B2	11	16	22	26	33	37	43	51	8	12	16	20	25	29	34	41	1.4	1.6	1.7	1.9	2.1	1.6	1.8	2.0	2.1	2.3
WW B3	13	19	27	32	40	45	51	60	10	15	21	25	31	36	42	50	1.3	1.4	1.6	1.8	1.9	1.4	1.6	1.8	1.9	2.1
WW B4	13	21	29	35	44	50	58	67	10	16	23	29	37	42	49	59	1.2	1.3	1.5	1.6	1.8	1.3	1.4	1.6	1.8	1.9
WW B5	13	20	29	36	45	51	59	69	9	16	23	29	38	43	51	61	1.2	1.3	1.5	1.6	1.8	1.3	1.4	1.6	1.7	1.9
WW C1	15	21	29	35	43	48	55	64	11	17	23	28	35	39	46	54	1.2	1.4	1.5	1.7	1.9	1.3	1.5	1.7	1.9	2.0
WW C2	13	20	28	34	43	48	55	65	10	16	22	27	35	40	47	56	1.2	1.4	1.5	1.7	1.8	1.3	1.5	1.6	1.8	2.0
WW C3	17	26	36	44	54	61	69	78	13	21	30	37	47	54	61	71	1.0	1.2	1.3	1.5	1.7	1.1	1.3	1.4	1.6	1.7
WW C4	15	24	34	41	51	58	66	76	11	19	28	35	44	51	59	69	1.1	1.2	1.4	1.5	1.7	1.2	1.3	1.4	1.6	1.7
TW A1	13	15	17	18	21	22	25	28	12	13	14	15	16	17	18	20	2.3	2.6	2.8	3.1	3.4	2.8	3.3	3.8	4.4	-
TW A2	11	14	17	20	24	26	30	35	9	11	14	15	18	20	22	26	1.8	2.0	2.3	2.5	2.7	2.2	2.5	2.8	3.1	3.4
TW A3	10	14	17	20	25	28	32	38	8	10	13	15	19	21	24	28	1.7	1.9	2.1	2.3	2.5	2.0	2.2	2.5	2.8	3.0
TW A4	9	13	17	20	25	29	33	40	7	9	13	15	19	21	25	30	1.6	1.8	2.0	2.2	2.4	1.9	2.1	2.3	2.6	2.8
TW B1	12	14	16	18	21	23	26	30	11	12	14	15	17	18	20	22	2.2	2.4	2.6	2.8	3.1	2.5	2.9	3.3	3.7	4.0
TW B2	10	13	17	19	23	26	30	36	8	11	13	15	18	20	23	27	1.8	2.0	2.2	2.4	2.6	2.1	2.4	2.6	2.9	3.1
TW B3	9	13	16	19	24	27	31	37	7	10	13	15	18	20	23	28	1.8	1.9	2.1	2.3	2.5	2.0	2.3	2.5	2.7	2.9
TW B4	9	12	16	18	23	26	30	36	7	9	12	14	18	20	23	28	1.8	2.0	2.1	2.3	2.5	2.0	2.3	2.5	2.7	2.9
TW C1	11	13	16	18	21	24	27	32	9	11	13	15	17	19	21	24	2.0	2.2	2.4	2.6	2.8	2.3	2.6	2.9	3.2	3.4
TW C2	10	12	15	18	22	25	28	34	8	10	12	14	17	19	22	26	1.9	2.1	2.3	2.5	2.6	2.2	2.4	2.6	2.9	3.1
TW C3	10	12	15	17	21	24	28	33	8	10	12	14	17	19	22	26	2.0	2.1	2.3	2.5	2.6	2.2	2.4	2.7	2.9	3.1
TW C4	10	12	15	17	20	22	26	31	9	11	13	14	17	18	21	25	2.1	2.3	2.4	2.6	2.8	2.4	2.6	2.8	3.0	3.2
TW D1	6	8	10	11	13	14	16	18	5	6	6	7	8	8	8	8	2.5	2.9	3.3	3.8	4.4	3.3	4.3	-	-	-
TW D2	11	15	20	24	30	34	40	47	9	12	16	19	24	27	31	38	1.5	1.7	1.8	2.0	2.2	1.7	1.9	2.1	2.3	2.4
TW D3	12	16	22	26	33	37	43	51	9	13	17	21	26	29	34	41	1.4	1.6	1.8	1.9	2.1	1.6	1.8	2.0	2.2	2.3
TW D4	13	19	26	32	39	45	52	61	10	15	21	26	33	37	44	52	1.3	1.4	1.6	1.7	1.9	1.4	1.5	1.7	1.9	2.0
TW D5	13	19	27	33	41	47	54	64	10	15	22	27	34	40	46	56	1.2	1.4	1.5	1.7	1.8	1.3	1.5	1.6	1.8	1.9
TW E1	14	20	26	31	38	43	49	58	11	16	21	25	31	35	41	48	1.3	1.5	1.6	1.8	2.0	1.5	1.6	1.8	2.0	2.2
TW E2	14	20	27	32	39	45	51	60	11	16	21	26	32	37	43	51	1.3	1.4	1.6	1.8	1.9	1.4	1.6	1.7	1.9	2.1
TW E3	17	25	34	41	50	56	64	73	13	20	28	34	43	49	56	66	1.1	1.3	1.4	1.6	1.7	1.2	1.3	1.5	1.7	1.8
TW E4	15	23	32	39	48	54	62	72	12	19	27	33	41	47	55	65	1.1	1.3	1.4	1.6	1.7	1.2	1.4	1.5	1.7	1.8
TW F1	11	13	15	17	19	21	23	27	10	11	12	13	15	16	17	19	2.3	2.5	2.8	3.0	3.3	2.7	3.2	3.6	4.2	4.7
TW F2	9	12	15	17	21	24	27	33	7	9	11	13	16	17	20	23	1.9	2.1	2.3	2.5	2.7	2.2	2.5	2.8	3.1	3.4
TW F3	7	11	14	17	21	25	29	35	5	8	10	12	15	18	21	25	1.8	2.0	2.2	2.4	2.6	2.1	2.3	2.6	2.8	3.1
TW F4	6	9	13	16	20	24	28	34	3	6	9	11	15	17	20	25	1.8	1.9	2.1	2.3	2.5	2.0	2.2	2.5	2.7	2.9
TW G1	9	11	13	15	18	20	23	27	8	9	11	12	14	15	17	19	2.2	2.4	2.6	2.9	3.1	2.5	2.9	3.3	3.7	4.0
TW G2	7	9	12	15	19	22	25	31	5	7	9	11	14	16	18	22	1.9	2.1	2.3	2.5	2.7	2.2	2.5	2.8	3.0	3.3
TW G3	5	8	11	14	18	21	25	31	3	5	8	10	13	15	18	22	1.9	2.1	2.3	2.5	2.6	2.2	2.4	2.6	2.9	3.1
TW G4	3	6	9	12	15	18	22	28	2	4	6	8	11	13	16	20	2.0	2.2	2.3	2.5	2.7	2.2	2.5	2.7	2.9	3.1
TW H1	6	8	11	13	16	18	21	26	5	6	8	10	12	13	15	18	2.1	2.3	2.5	2.8	2.9	2.5	2.8	3.1	3.4	3.7
TW H2	4	6	9	11	15	17	21	26	2	4	6	8	11	12	15	19	2.1	2.3	2.4	2.6	2.8	2.4	2.6	2.9	3.1	3.3
TW H3	3	5	7	10	13	15	19	24	1	3	5	7	9	11	14	17	2.2	2.3	2.5	2.7	2.9	2.4	2.7	2.9	3.1	3.3
TW H4	2	4	6	7	10	12	15	19	1	2	4	5	7	9	11	15	2.4	2.6	2.8	2.9	3.1	2.7	2.9	3.2	3.4	3.5
TW I1	5	6	8	9	11	12	13	15	3	4	5	5	6	6	6	6	2.6	3.0	3.5	4.1	4.7	3.5	4.6	-	-	-
TW I2	8	12	17	21	27	31	36	44	6	9	13	16	21	24	28	34	1.6	1.7	1.9	2.1	2.2	1.0	1.9	2.1	2.3	2.5
TW I3	9	14	19	23	30	34	40	48	6	10	15	18	23	27	[illegible]	39	1.5	1.6	1.8	2.0	2.1	1.6	1.8	2.0	2.2	2.3
TW I4	10	16	22	[illegible]	[illegible]	42	49	58	7	12	18	23	30	34	41	49	1.3	1.5	1.6	1.8	1.9	1.4	1.6	1.7	1.9	2.1
TW I5	10	17	24	30	38	44	51	61	8	13	19	24	32	37	44	53	1.3	1.4	1.6	1.7	1.9	1.4	1.5	1.7	1.8	2.0
TW J1	12	17	24	29	36	41	47	56	9	13	19	23	29	33	39	47	1.3	1.5	1.7	1.8	2.0	1.5	1.6	1.8	2.0	2.2
TW J2	11	17	24	29	37	42	49	58	8	13	19	23	30	35	41	49	1.3	1.5	1.6	1.8	1.9	1.4	1.6	1.8	1.9	2.1
TW J3	14	22	31	38	48	54	62	71	11	18	26	32	41	47	54	64	1.1	1.3	1.4	1.6	1.7	1.2	1.4	1.5	1.7	1.8
TW J4	13	21	29	36	46	52	60	70	10	16	24	30	39	45	52	62	1.1	1.3	1.4	1.6	1.7	1.2	1.4	1.5	1.7	1.8
DG A1	6	8	10	11	13	14	14	15	4	5	6	6	7	7	7	5	2.3	2.8	3.7	-		3.0	4.1	-	-	-
DG A2	9	12	16	19	23	26	30	34	7	9	12	14	17	20	23	26	1.8	1.9	2.2	2.5	2.8	2.0	2.3	2.5	2.9	3.3
DG A3	12	17	22	27	34	38	45	53	10	14	19	22	28	32	37	45	1.5	1.6	1.7	1.9	2.1	1.6	1.7	1.9	2.0	2.2
DG B1	6	8	10	11	13	15	16	18	4	5	6	6	7	8	8	7	2.3	2.6	3.2	4.0	-	3.0	4.0	-	-	-
DG B2	9	12	16	19	24	27	32	38	7	9	12	15	18	21	24	28	1.7	1.9	2.1	2.3	2.5	2.0	2.2	2.4	2.7	2.9
DG B3	13	17	23	28	35	40	47	57	11	15	19	23	29	34	40	48	1.4	1.5	1.6	1.8	1.9	1.6	1.7	1.8	1.9	2.1
DG C1	8	10	13	16	19	21	24	28	6	7	9	10	12	13	14	16	1.9	2.2	2.5	2.9	3.4	2.4	2.9	3.4	4.4	-
DG C2	11	15	20	23	29	33	39	46	9	12	15	18	23	26	30	36	1.6	1.7	1.8	2.0	2.2	1.8	2.0	2.1	2.3	2.5
DG C3	14	20	27	32	40	46	54	64	12	17	22	27	34	39	45	55	1.3	1.4	1.5	1.7	1.8	1.5	1.6	1.7	1.8	1.9
SS A1	15	18	22	24	29	31	35	40	13	16	19	21	24	26	29	33	1.7	1.9	2.1	2.4	2.6	1.9	2.1	2.4	2.8	3.1
SS A2	18	24	30	35	42	46	52	60	16	21	26	31	37	41	46	53	1.3	1.5	1.6	1.8	2.0	1.4	1.6	1.8	2.0	2.1
SS A3	13	16	19	21	24	26	29	33	12	14	16	18	20	21	23	25	1.9	2.1	2.5	2.8	3.1	2.1	2.5	3.0	3.6	4.3
SS A4	18	23	29	33	39	44	49	57	16	20	25	29	34	38	43	50	1.3	1.5	1.7	1.9	2.1	1.4	1.6	1.8	2.1	2.2
SS A5	15	18	21	23	26	28	31	35	14	16	18	20	22	24	26	28	1.9	2.1	2.4	2.7	3.0	2.1	2.5	2.8	3.3	3.7
SS A6	18	23	29	34	41	45	51	59	16	20	26	30	35	39	44	51	1.3	1.5	1.7	1.9	2.0	1.4	1.6	1.8	2.0	2.2
SS A7	14	16	18	19	21	23	25	27	13	14	16	17	18	19	20	21	2.2	2.6	3.0	3.5	4.0	2.5	3.1	3.8	4.9	-
SS A8	17	22	27	32	38	42	47	54	15	19	24	28	33	36	41	47	1.4	1.6	1.8	2.0	2.1	1.5	1.7	1.9	2.1	2.3
SS B1	12	15	18	21	24	27	30	35	11	13	15	17	20	22	24	28	1.8	2.0	2.3	2.5	2.8	2.0	2.3	2.7	3.0	3.3
SS B2	16	21	27	31	38	42	48	56	14	18	23	27	33	37	42	49	1.4	1.5	1.7	1.9	2.0	1.5	1.6	1.8	2.0	2.2
SS B3	11	13	16	18	21	23	26	30	10	11	13	15	17	18	20	22	2.0	2.2	2.5	2.9	3.2	2.2	2.6	3.1	3.6	4.2
SS B4	15	20	25	30	36	40	46	53	13	17	22	26	31	35	40	47	1.4	1.6	1.7	1.9	2.1	1.5	1.7	1.9	2.1	2.3
SS B5	12	14	17	18	21	23	26	29	11	13	14	16	18	19	21	23	2.1	2.4	2.7	3.0	3.2	2.4	2.8	3.2	3.7	4.2
SS B6	15	20	25	30	36	41	46	54	13	17	22	26	31	35	40	47	1.4	1.6	1.7	1.9	2.1	1.5	1.7	1.9	2.1	2.2
SS B7	11	13	14	16	18	19	21	23	10	11	13	14	15	16	17	18	2.4	2.7	3.1	3.5	4.0	2.7	3.3	4.0	4.9	-
SS B8	14	18	24	28	34	38	43	50	12	16	20	24	29	32	37	43	1.4	1.6	1.8	2.0	2.2	1.6	1.8	2.0	2.2	2.4
SS C1	9	12	16	18	23	25	29	35	6	9	11	13	16	17	20	23	1.8	2.0	2.2	2.4	2.7	2.1	2.4	2.7	3.1	3.5
SS C2	11	16	22	27	33	38	44	52	8	12	17	21	26	30	35	42	1.4	1.6	1.7	1.9	2.1	1.6	1.8	1.9	2.1	2.3
SS C3	9	12	14	16	19	22	25	29	8	9	11	13	15	16	18	20	2.1	2.3	2.5	2.8	3.0	2.5	2.8	3.2	3.5	3.9
SS C4	11	15	20	24	30	34	39	47	9	12	16	19	24	27	31	38	1.5	1.7	1.9	2.0	2.2	1.7	1.9	2.1	2.3	2.4
SS D1	16	20	25	28	33	37	41	47	13	17	21	23	27	30	33	38	1.5	1.7	1.9	2.2	2.4	1.6	1.9	2.2	2.5	2.8
SS D2	20	27	35	40	48	54	60	69	17	23	30	35	42	47	54	62	1.1	1.3	1.5	1.7	1.8	1.2	1.4	1.6	1.8	2.0
SS D3	16	19	23	26	30	33	37	42	14	17	20	22	25	27	30	34	1.6	1.8	2.1	2.3	2.6	1.8	2.1	2.4	2.8	3.1
SS D4	19	26	33	39	47	52	58	67	17	22	29	34	41	46	52	59	1.2	1.4	1.5	1.7	1.9	1.3	1.4	1.6	1.8	2.0
SS E1	12	16	20	22	27	30	33	38	10	13	15	18	21	23	25	28	1.6	1.9	2.1	2.4	2.7	1.9	2.2	2.6	3.1	3.6
SS E2	16	22	29	34	41	46	52	60	13	18	24	28	35	39	44	52	1.3	1.4	1.6	1.8	2.0	1.4	1.6	1.8	2.0	2.1
SS E3	13	15	18	20	24	26	29	33	11	13	15	17	19	21	23	25	1.9	2.1	2.4	2.7	3.0	2.1	2.5	2.9	3.4	3.9
SS E4	15	20	27	31	38	42	48	56	13	17	23	26	32	36	41	48	1.3	1.5	1.7	1.9	2.0	1.4	1.6	1.8	2.1	2.2

HALIFAX, NOVA SCOTIA

TYPE	SOLAR SAVINGS FRACTION (%) BASE TEMP: 55 F (4293 DD) LCR 100	70	50	40	30	25	20	15	65 F (7167 DD) 100	70	50	40	30	25	20	15	CONSERVATION FACTOR 55 F 100	50	30	20	15	65 F 100	50	30	20	15
WW A1	15	17	20	22	25	27	31	35	13	14	16	17	19	20	22	24	2.0	2.2	2.4	2.6	2.8	2.5	2.9	3.3	3.8	4.2
WW A2	14	18	23	27	32	36	41	48	11	14	17	19	23	26	29	34	1.5	1.7	1.9	2.1	2.2	1.8	2.1	2.3	2.6	2.8
WW A3	14	18	24	28	35	39	45	53	10	13	17	20	25	28	32	38	1.4	1.6	1.7	1.9	2.1	1.7	1.9	2.1	2.3	2.5
WW A4	13	19	25	29	36	41	47	55	9	13	17	21	26	29	34	40	1.3	1.5	1.7	1.9	2.0	1.6	1.8	2.0	2.2	2.4
WW A5	13	19	26	31	38	43	49	58	8	12	17	21	27	30	35	42	1.3	1.5	1.6	1.8	2.0	1.5	1.7	1.9	2.1	2.3
WW A6	13	19	26	31	39	44	51	59	8	12	17	21	27	31	36	44	1.3	1.4	1.6	1.8	1.9	1.5	1.7	1.9	2.1	2.3
WW B1	10	13	16	18	21	23	26	30	8	9	10	11	12	13	14	14	1.9	2.2	2.5	2.8	3.1	2.6	3.3	4.3	-	-
WW B2	14	20	27	32	40	45	52	61	10	14	20	24	30	34	40	47	1.3	1.4	1.6	1.8	1.9	1.5	1.6	1.8	2.0	2.2
WW B3	17	25	33	39	48	53	60	69	12	18	25	30	37	42	48	56	1.1	1.3	1.5	1.6	1.8	1.3	1.5	1.7	1.8	2.0
WW B4	17	26	35	42	52	58	66	75	12	19	27	33	42	47	55	64	1.1	1.2	1.4	1.6	1.7	1.2	1.4	1.5	1.7	1.8
WW B5	17	25	35	42	52	58	66	75	12	19	27	33	42	48	56	66	1.1	1.2	1.4	1.5	1.7	1.2	1.3	1.5	1.7	1.8
WW C1	19	27	35	42	50	56	64	72	14	20	27	32	40	45	52	60	1.1	1.3	1.4	1.6	1.8	1.3	1.4	1.6	1.8	1.9
WW C2	17	25	34	41	50	56	63	73	12	19	26	32	40	45	52	61	1.1	1.3	1.4	1.6	1.7	1.2	1.4	1.6	1.7	1.9
WW C3	21	32	43	51	61	68	75	84	16	25	35	42	52	58	66	76	1.0	1.1	1.3	1.5	1.6	1.1	1.2	1.4	1.5	1.7
WW C4	19	29	40	47	58	64	72	81	14	22	32	39	49	55	63	73	1.0	1.2	1.3	1.5	1.6	1.1	1.2	1.4	1.6	1.7
TW A1	15	17	20	22	25	28	31	36	13	14	16	17	19	21	23	25	2.0	2.2	2.4	2.6	2.8	2.4	2.8	3.1	3.5	3.8
TW A2	14	17	21	25	30	33	38	44	11	13	16	19	22	24	27	32	1.6	1.8	2.0	2.2	2.3	1.9	2.2	2.4	2.7	2.9
TW A3	13	17	22	26	31	35	41	48	10	13	16	19	23	26	29	35	1.5	1.7	1.9	2.0	2.2	1.8	2.0	2.2	2.5	2.7
TW A4	12	16	22	26	32	36	42	49	8	12	16	19	23	26	30	36	1.5	1.6	1.8	2.0	2.1	1.7	1.9	2.1	2.3	2.5
TW B1	14	17	20	22	26	29	32	38	12	14	16	17	20	22	24	27	1.9	2.0	2.2	2.4	2.6	2.2	2.5	2.8	3.1	3.4
TW B2	13	16	21	24	29	33	37	44	10	13	16	18	22	24	28	32	1.6	1.8	1.9	2.1	2.3	1.9	2.1	2.3	2.6	2.8
TW B3	12	16	20	24	29	33	38	46	9	12	15	18	22	24	28	34	1.6	1.7	1.9	2.1	2.2	1.8	2.0	2.3	2.5	2.6
TW B4	11	15	19	23	28	32	37	44	9	11	15	17	21	24	28	33	1.6	1.8	1.9	2.1	2.2	1.9	2.1	2.3	2.5	2.6
TW C1	13	16	19	22	26	29	33	39	11	13	15	17	20	22	25	29	1.8	2.0	2.1	2.3	2.5	2.1	2.4	2.6	2.8	3.0
TW C2	12	15	19	22	27	30	35	41	9	12	15	17	20	23	26	31	1.7	1.9	2.0	2.2	2.3	2.0	2.2	2.4	2.6	2.8
TW C3	11	14	18	21	26	29	34	40	9	12	14	17	20	22	26	30	1.7	1.9	2.1	2.2	2.4	2.0	2.2	2.4	2.6	2.8
TW C4	11	14	17	20	24	27	31	37	10	12	14	16	19	21	24	29	1.9	2.0	2.2	2.4	2.5	2.2	2.4	2.6	2.8	2.9
TW D1	9	11	14	16	19	21	24	28	6	8	9	10	11	12	14	15	2.0	2.3	2.5	2.8	3.1	2.6	3.2	3.8	4.7	-
TW D2	14	19	25	30	37	41	48	56	11	15	19	23	28	32	37	43	1.4	1.5	1.7	1.8	2.0	1.6	1.8	1.9	2.1	2.3
TW D3	15	21	27	33	40	45	52	60	11	15	21	25	30	35	40	47	1.3	1.4	1.6	1.8	1.9	1.5	1.7	1.8	2.0	2.2
TW D4	16	23	32	38	46	52	59	69	12	18	24	30	37	42	49	58	1.2	1.3	1.5	1.6	1.8	1.3	1.5	1.6	1.8	1.9
TW D5	16	23	32	38	47	53	61	70	12	18	25	31	39	44	51	61	1.2	1.3	1.4	1.6	1.7	1.3	1.4	1.6	1.7	1.9
TW E1	18	24	32	37	46	51	58	67	13	19	25	29	36	40	46	55	1.2	1.4	1.5	1.7	1.8	1.4	1.5	1.7	1.9	2.0
TW E2	17	24	32	38	46	52	59	68	13	18	25	30	37	42	48	57	1.2	1.3	1.5	1.6	1.8	1.3	1.5	1.7	1.8	2.0
TW E3	21	30	40	47	57	63	71	80	16	24	32	39	48	54	61	70	1.0	1.2	1.3	1.5	1.7	1.1	1.3	1.4	1.6	1.8
TW E4	19	28	38	45	55	61	69	78	14	22	30	37	46	52	59	69	1.1	1.2	1.4	1.5	1.7	1.2	1.3	1.5	1.6	1.8
TW F1	13	15	18	21	24	27	30	35	11	12	14	16	18	19	21	24	1.9	2.1	2.3	2.5	2.7	2.4	2.7	3.0	3.4	3.7
TW F2	11	15	19	22	27	30	35	42	9	11	14	16	19	22	25	29	1.6	1.8	2.0	2.2	2.4	2.0	2.2	2.5	2.7	3.0
TW F3	10	14	19	22	28	32	37	44	7	10	13	16	19	22	26	31	1.6	1.7	1.9	2.1	2.2	1.8	2.1	2.3	2.5	2.7
TW F4	8	12	17	21	27	31	36	43	5	8	12	14	18	21	25	31	1.6	1.7	1.9	2.1	2.2	1.8	2.0	2.2	2.5	2.6
TW G1	11	14	17	19	23	26	30	35	9	11	13	14	17	19	21	24	1.9	2.1	2.2	2.4	2.6	2.2	2.5	2.8	3.1	3.4
TW G2	9	12	16	20	24	28	33	39	6	9	12	14	17	20	23	27	1.7	1.9	2.0	2.2	2.4	2.0	2.2	2.5	2.7	2.9
TW G3	7	11	15	18	23	27	32	39	5	7	10	13	16	19	22	27	1.7	1.8	2.0	2.2	2.3	2.0	2.2	2.4	2.6	2.8
TW G4	5	8	12	16	20	24	28	35	3	6	8	11	14	17	20	25	1.8	1.9	2.1	2.3	2.4	2.0	2.3	2.5	2.7	2.8
TW H1	8	11	14	17	21	24	28	34	6	8	10	12	15	17	19	23	1.9	2.0	2.2	2.4	2.6	2.2	2.5	2.7	3.0	3.2
TW H2	6	9	12	15	19	22	27	33	4	6	8	10	13	16	19	23	1.8	2.0	2.2	2.3	2.5	2.1	2.4	2.6	2.8	3.0
TW H3	4	7	10	13	17	20	24	30	3	5	7	9	12	14	17	21	1.9	2.1	2.3	2.4	2.6	2.2	2.4	2.7	2.9	3.0
TW H4	3	5	8	10	13	16	19	25	2	3	5	7	10	11	14	18	2.2	2.3	2.5	2.6	2.8	2.5	2.7	2.9	3.1	3.2
TW I1	7	9	11	13	16	18	21	25	4	6	7	8	9	10	11	12	2.1	2.4	2.7	3.0	3.2	2.8	3.4	4.1	-	-
TW I2	11	16	22	26	33	38	44	52	8	12	16	19	25	28	33	40	1.4	1.6	1.7	1.9	2.0	1.6	1.8	2.0	2.2	2.3
TW I3	12	18	24	29	37	42	48	57	9	13	18	22	28	32	37	45	1.3	1.5	1.6	1.8	2.0	1.5	1.7	1.9	2.0	2.2
TW I4	14	20	28	34	43	49	56	66	10	15	22	27	34	39	46	55	1.2	1.4	1.5	1.7	1.8	1.3	1.5	1.7	1.8	2.0
TW I5	14	21	29	36	45	51	58	68	10	16	23	28	36	41	48	58	1.2	1.3	1.5	1.6	1.8	1.3	1.5	1.6	1.8	1.9
TW J1	15	22	29	35	43	49	56	65	11	16	22	27	34	38	45	53	1.2	1.4	1.5	1.7	1.8	1.4	1.5	1.7	1.9	2.0
TW J2	15	21	29	35	44	49	57	66	10	16	22	27	34	39	46	55	1.2	1.4	1.5	1.7	1.8	1.3	1.5	1.7	1.8	2.0
TW J3	18	27	37	44	54	61	69	78	13	21	30	36	45	51	59	69	1.1	1.2	1.4	1.5	1.7	1.2	1.3	1.5	1.6	1.8
TW J4	17	25	35	42	52	59	67	76	12	19	28	34	43	49	57	67	1.1	1.2	1.4	1.5	1.7	1.2	1.3	1.5	1.6	1.8
DG A1	8	11	14	16	19	20	22	24	6	7	9	10	11	11	12	11	1.9	2.2	2.8	3.9	-	2.4	3.1	4.5	-	-
DG A2	11	15	20	23	29	32	37	42	8	11	15	17	21	24	27	31	1.6	1.7	1.9	2.2	2.5	1.8	2.1	2.3	2.6	3.0
DG A3	15	20	27	32	40	45	52	60	12	16	21	25	32	36	42	50	1.3	1.4	1.6	1.8	1.9	1.5	1.6	1.8	1.9	2.1
DG B1	8	11	14	17	20	23	25	29	5	7	9	10	11	12	13	14	1.9	2.1	2.4	2.9	3.6	2.5	3.0	3.8	-	-
DG B2	11	15	20	24	30	34	40	47	9	12	15	18	22	25	29	34	1.5	1.7	1.8	2.0	2.2	1.8	2.0	2.2	2.4	2.7
DG B3	15	21	27	33	41	47	55	65	12	17	22	26	33	38	45	54	1.3	1.4	1.5	1.7	1.8	1.5	1.6	1.7	1.8	2.0
DG C1	11	14	18	22	27	30	34	39	7	10	12	14	16	18	20	23	1.6	1.8	2.0	2.4	2.7	2.1	2.4	2.8	3.3	3.9
DG C2	14	18	24	29	36	41	47	56	10	14	18	22	27	31	36	43	1.4	1.5	1.6	1.8	2.0	1.6	1.8	2.0	2.1	2.3
DG C3	17	24	31	38	47	53	62	72	14	19	25	30	38	43	51	60	1.2	1.3	1.4	1.6	1.7	1.4	1.5	1.6	1.7	1.9
SS A1	18	22	27	30	36	39	44	51	15	19	22	25	29	32	36	41	1.4	1.7	1.9	2.1	2.2	1.6	1.9	2.1	2.4	2.6
SS A2	22	29	36	41	49	54	60	69	19	25	31	35	42	46	52	60	1.2	1.3	1.5	1.7	1.8	1.3	1.5	1.7	1.8	2.0
SS A3	16	20	24	27	31	34	38	44	14	17	20	22	25	27	30	34	1.6	1.8	2.1	2.3	2.5	1.8	2.1	2.5	2.8	3.2
SS A4	22	28	35	40	47	52	58	66	19	24	30	34	40	44	50	57	1.2	1.4	1.6	1.7	1.9	1.3	1.5	1.7	1.9	2.1
SS A5	18	21	25	28	32	35	39	45	16	18	21	24	27	29	32	36	1.6	1.8	2.0	2.3	2.5	1.8	2.1	2.4	2.7	3.0
SS A6	22	28	35	41	48	53	59	68	19	24	30	35	41	45	51	58	1.2	1.4	1.5	1.7	1.9	1.3	1.5	1.7	1.9	2.0
SS A7	16	19	22	24	27	30	33	37	15	16	19	20	22	24	26	28	1.8	2.1	2.4	2.7	2.9	2.1	2.5	2.9	3.5	4.0
SS A8	21	27	34	39	46	50	56	64	18	23	28	33	39	43	48	54	1.2	1.4	1.6	1.8	1.9	1.3	1.6	1.8	2.0	2.2
SS B1	15	18	23	26	31	34	39	45	13	15	19	21	25	27	30	35	1.6	1.8	2.0	2.2	2.4	1.8	2.1	2.3	2.6	2.8
SS B2	19	25	32	37	45	50	56	64	16	21	27	31	38	42	48	56	1.2	1.4	1.6	1.7	1.9	1.4	1.5	1.7	1.9	2.0
SS B3	13	17	21	24	28	31	35	40	11	14	17	19	22	24	26	30	1.7	1.9	2.1	2.4	2.6	1.9	2.2	2.6	2.9	3.2
SS B4	18	24	31	36	43	48	54	62	16	21	26	30	36	40	46	53	1.3	1.4	1.6	1.8	1.9	1.4	1.6	1.8	1.9	2.1
SS B5	14	17	20	23	27	29	33	38	13	15	17	19	22	23	26	29	1.8	2.0	2.2	2.5	2.7	2.1	2.4	2.7	3.0	3.3
SS B6	18	24	31	36	43	48	54	63	15	20	26	30	36	41	46	54	1.2	1.4	1.6	1.8	1.9	1.4	1.6	1.7	1.9	2.1
SS B7	13	15	18	20	23	25	28	32	12	13	15	17	19	20	22	24	2.0	2.2	2.5	2.8	3.0	2.3	2.7	3.1	3.6	4.1
SS B8	17	23	29	34	41	45	52	60	15	19	24	28	34	38	43	50	1.3	1.5	1.6	1.8	2.0	1.4	1.6	1.8	2.0	2.2
SS C1	12	16	21	24	30	33	38	45	8	11	14	17	20	23	26	30	1.5	1.7	1.9	2.1	2.3	1.8	2.1	2.4	2.7	2.9
SS C2	15	21	28	33	40	46	52	61	11	15	21	25	31	35	41	48	1.3	1.4	1.6	1.8	1.9	1.5	1.6	1.8	2.0	2.1
SS C3	11	14	18	20	25	27	31	37	9	11	13	15	18	20	22	26	1.8	2.0	2.2	2.4	2.5	2.2	2.5	2.7	3.0	3.3
SS C4	14	19	25	29	36	41	47	55	10	14	19	22	28	31	36	43	1.4	1.5	1.7	1.9	2.0	1.6	1.8	1.9	2.1	2.3
SS D1	20	25	31	36	42	46	52	59	16	21	25	29	34	37	41	47	1.3	1.5	1.7	1.9	2.1	1.5	1.7	1.9	2.2	2.4
SS D2	25	33	42	48	57	62	69	77	20	27	35	41	49	54	60	68	1.0	1.2	1.4	1.6	1.7	1.1	1.3	1.5	1.7	1.9
SS D3	19	23	29	32	38	42	46	53	16	20	24	27	31	34	37	42	1.4	1.6	1.8	2.0	2.2	1.6	1.8	2.1	2.4	2.6
SS D4	24	32	40	46	55	60	67	75	20	27	34	39	47	52	58	66	1.1	1.2	1.4	1.6	1.8	1.2	1.4	1.5	1.7	1.9
SS E1	16	21	26	30	35	39	44	51	13	16	20	23	27	29	33	37	1.4	1.6	1.8	2.0	2.2	1.6	1.9	2.2	2.5	2.8
SS E2	20	27	35	41	49	55	61	70	16	22	29	34	41	46	52	59	1.1	1.3	1.5	1.7	1.8	1.3	1.4	1.6	1.8	2.0
SS E3	15	19	23	26	30	33	37	43	13	16	19	21	24	26	29	33	1.6	1.8	2.1	2.3	2.5	1.9	2.2	2.5	2.8	3.1
SS E4	19	26	33	38	46	51	57	65	16	21	27	31	38	42	48	55	1.2	1.4	1.5	1.7	1.9	1.3	1.5	1.7	1.9	2.1

MOOSONEE, ONTARIO

TYPE	SOLAR SAVINGS FRACTION (%) BASE TEMP: 55 F (9344 DD)								65 F (12597 DD)								CONSERVATION FACTOR 55 F					65 F				
LCR	100	70	50	40	30	25	20	15	100	70	50	40	30	25	20	15	100	50	30	20	15	100	50	30	20	15
WW A1	11	12	13	13	14	15	15	15	10	10	10	11	11	11	11	10	3.3	4.1	-	-	-	4.6	-	-	-	-
WW A2	8	10	12	14	16	18	19	22	6	7	9	10	12	13	14	15	2.2	2.6	3.0	3.6	4.4	2.6	3.0	3.7	5.0	-
WW A3	7	9	12	14	17	19	21	24	4	6	8	10	12	13	15	17	2.0	2.3	2.7	3.2	3.7	2.2	2.6	3.2	4.0	-
WW A4	5	8	11	14	17	19	22	26	3	5	8	10	12	14	16	18	1.9	2.2	2.5	2.9	3.4	2.1	2.4	2.9	3.6	4.5
WW A5	5	8	11	14	18	20	23	27	2	4	7	9	12	14	16	19	1.8	2.1	2.4	2.8	3.2	2.0	2.3	2.8	3.4	4.2
WW A6	4	7	11	14	18	21	24	28	1	4	7	9	12	15	17	20	1.7	2.0	2.3	2.7	3.1	1.9	2.2	2.7	3.2	3.9
WW B1	5	5	6	6	5	5	3	0	3	3	3	2	1	0	-1	-6	4.3	-	-	-	-	-	-	-	-	-
WW B2	7	10	14	18	22	25	30	35	4	7	11	14	17	20	24	29	1.7	1.9	2.1	2.4	2.6	1.8	2.0	2.3	2.6	2.9
WW B3	8	13	18	22	28	31	36	43	6	10	14	18	23	26	30	36	1.5	1.7	1.9	2.1	2.4	1.6	1.8	2.1	2.3	2.6
WW B4	8	14	21	26	33	38	45	53	6	11	17	22	29	33	39	48	1.3	1.5	1.7	1.9	2.0	1.4	1.6	1.8	2.0	2.1
WW B5	8	14	21	27	35	40	47	56	5	11	18	23	30	35	42	51	1.3	1.5	1.6	1.8	2.0	1.4	1.5	1.7	1.9	2.0
WW C1	10	15	20	25	31	35	40	47	7	11	16	20	26	30	35	41	1.4	1.6	1.8	2.0	2.2	1.5	1.7	1.9	2.2	2.4
WW C2	9	14	20	25	32	36	42	50	6	11	16	21	27	31	37	44	1.4	1.6	1.7	1.9	2.1	1.5	1.6	1.8	2.0	2.2
WW C3	11	19	28	34	43	50	57	67	8	16	24	30	39	45	52	62	1.2	1.3	1.5	1.7	1.8	1.2	1.4	1.5	1.7	1.9
WW C4	10	17	26	32	41	47	55	64	7	14	22	28	37	43	50	60	1.2	1.3	1.5	1.7	1.8	1.2	1.4	1.6	1.7	1.9
TW A1	11	12	13	14	15	15	16	17	10	10	11	11	12	12	12	12	3.1	3.7	4.5	-	-	4.0	4.9	-	-	-
TW A2	9	10	12	14	16	17	19	21	7	8	10	11	12	13	14	16	2.3	2.7	3.1	3.6	4.2	2.7	3.1	3.7	4.7	-
TW A3	7	9	12	14	16	18	20	23	5	7	9	10	12	14	15	17	2.1	2.4	2.8	3.2	3.7	2.4	2.8	3.2	3.9	4.7
TW A4	6	8	11	13	16	18	21	25	4	6	8	10	12	14	16	19	2.0	2.3	2.6	2.9	3.3	2.3	2.5	2.9	3.4	4.0
TW B1	10	11	13	14	15	16	17	19	9	9	10	11	12	13	13	14	2.7	3.2	3.7	4.4	-	3.4	3.9	4.7	-	-
TW B2	8	10	12	13	16	18	20	23	6	7	9	10	12	14	15	17	2.2	2.5	2.9	3.3	3.7	2.6	2.9	3.3	3.9	4.5
TW B3	7	9	11	13	16	18	20	24	5	6	8	10	12	14	16	18	2.1	2.4	2.7	3.1	3.4	2.4	2.7	3.1	3.5	4.0
TW B4	7	9	11	13	16	18	20	24	5	6	8	10	12	14	16	19	2.2	2.4	2.7	3.0	3.3	2.4	2.7	3.0	3.4	3.7
TW C1	9	10	12	13	15	17	19	21	7	8	10	11	12	13	15	16	2.5	2.8	3.2	3.6	4.0	2.9	3.2	3.7	4.3	4.9
TW C2	8	9	11	13	16	17	20	23	6	7	9	10	12	14	16	18	2.3	2.6	2.8	3.2	3.4	2.6	2.9	3.2	3.6	4.0
TW C3	8	9	11	13	16	17	20	23	6	7	9	10	12	14	16	19	2.3	2.6	2.9	3.1	3.4	2.7	2.9	3.2	3.5	3.8
TW C4	8	10	12	13	15	17	19	23	7	8	10	11	13	14	16	19	2.5	2.7	3.0	3.2	3.5	2.8	3.0	3.3	3.6	3.9
TW D1	4	5	5	5	6	6	5	4	2	3	3	2	2	1	0	-1	3.9	-	-	-	-	-	-	-	-	-
TW D2	8	11	15	17	21	24	28	33	6	8	11	14	17	20	23	27	1.8	2.0	2.2	2.5	2.7	2.0	2.2	2.4	2.8	3.0
TW D3	8	11	15	18	23	26	30	35	5	8	12	14	18	21	24	29	1.7	1.9	2.1	2.4	2.6	1.8	2.1	2.3	2.6	2.9
TW D4	9	14	19	23	30	34	40	48	6	11	16	20	25	29	35	42	1.4	1.6	1.8	2.0	2.2	1.5	1.7	1.9	2.1	2.3
TW D5	9	14	20	25	32	37	43	52	6	11	17	21	28	32	38	47	1.4	1.5	1.7	1.9	2.0	1.5	1.6	1.8	2.0	2.1
TW E1	10	14	19	22	28	31	36	42	7	11	15	19	23	26	31	36	1.5	1.7	1.9	2.2	2.4	1.6	1.9	2.1	2.3	2.6
TW E2	10	14	19	23	29	33	39	46	7	11	16	20	25	29	34	40	1.5	1.7	1.8	2.0	2.2	1.6	1.8	2.0	2.2	2.4
TW E3	12	18	26	31	39	45	52	60	9	15	22	27	35	40	47	55	1.2	1.4	1.6	1.8	1.9	1.3	1.5	1.6	1.8	2.0
TW E4	11	17	25	30	38	44	51	60	8	14	21	26	34	39	46	55	1.3	1.4	1.6	1.7	1.9	1.3	1.5	1.6	1.8	2.0
TW F1	9	10	11	12	13	14	15	16	8	8	9	9	10	10	10	10	3.0	3.6	4.3	-	-	3.8	4.6	-	-	-
TW F2	6	8	10	11	13	15	17	19	5	6	7	8	10	11	12	13	2.4	2.7	3.2	3.7	4.3	2.8	3.2	3.8	4.8	-
TW F3	5	7	9	11	13	15	17	20	3	4	6	7	9	11	12	14	2.2	2.5	2.9	3.3	3.7	2.5	2.9	3.3	4.0	4.8
TW F4	3	5	7	9	12	14	17	21	1	2	5	6	8	10	12	15	2.1	2.4	2.7	3.1	3.4	2.4	2.7	3.1	3.5	4.0
TW G1	7	8	10	11	12	13	14	16	5	6	7	8	9	9	10	11	2.8	3.2	3.7	4.4	-	3.4	3.9	4.7	-	-
TW G2	4	6	8	9	12	13	15	18	2	4	5	6	8	9	11	13	2.4	2.7	3.0	3.5	3.9	2.7	3.1	3.5	4.2	4.9
TW G3	3	4	7	8	11	13	15	18	1	2	4	5	7	9	11	13	2.3	2.6	2.9	3.2	3.6	2.6	2.9	3.3	3.7	4.2
TW G4	1	3	5	7	10	11	14	17	0	1	3	4	6	8	10	13	2.4	2.6	2.9	3.2	3.4	2.7	2.9	3.2	3.6	3.9
TW H1	4	6	7	8	10	12	13	15	3	4	5	6	7	8	9	11	2.7	3.0	3.4	3.9	4.3	3.2	3.5	4.0	4.7	-
TW H2	2	4	5	7	9	11	13	16	0	2	3	4	6	7	9	11	2.5	2.8	3.1	3.4	3.7	2.9	3.2	3.5	3.9	4.3
TW H3	1	3	4	6	8	10	12	15	0	1	2	3	5	7	8	11	2.6	2.8	3.1	3.4	3.7	3.0	3.2	3.5	3.8	4.1
TW H4	1	2	3	5	7	8	10	13	0	0	2	3	4	5	7	9	2.8	3.1	3.3	3.6	3.8	3.3	3.5	3.7	4.0	4.2
TW I1	3	3	3	4	4	4	3	2	1	1	1	1	0	0	0	-3	4.2	-	-	-	-	-	-	-	-	-
TW I2	5	8	12	14	18	21	25	30	3	6	8	11	14	17	20	24	1.8	2.1	2.3	2.5	2.8	2.0	2.2	2.5	2.8	3.1
TW I3	6	9	13	16	20	24	28	34	3	6	10	12	16	19	20	28	1.7	1.9	2.1	2.4	2.6	1.9	2.1	2.3	2.6	2.8
TW I4	6	11	17	21	27	31	37	45	4	8	13	17	23	27	32	39	1.5	1.7	1.8	2.0	2.2	1.6	1.7	1.9	2.1	2.3
TW I5	7	12	18	22	29	34	40	49	4	9	14	19	25	30	36	44	1.4	1.6	1.7	1.9	2.1	1.5	1.7	1.8	2.0	2.2
TW J1	8	12	17	20	26	30	34	41	5	9	13	17	21	25	29	35	1.5	1.7	1.9	2.2	2.4	1.7	1.9	2.1	2.3	2.5
TW J2	7	12	17	21	27	31	37	44	5	9	14	17	23	27	32	39	1.5	1.7	1.9	2.0	2.2	1.6	1.8	2.0	2.2	2.4
TW J3	9	16	24	29	37	43	50	59	7	13	20	25	33	39	45	55	1.3	1.4	1.6	1.8	1.9	1.3	1.5	1.6	1.8	2.0
TW J4	8	15	22	28	36	42	49	58	6	12	19	24	32	37	44	53	1.3	1.4	1.6	1.8	1.9	1.3	1.5	1.6	1.8	2.0
DG A1	3	4	5	5	5	5	4	2	1	2	2	2	1	1	0	-3	3.4	-	-		-	-	-	-	-	-
DG A2	6	9	11	13	16	18	20	23	5	6	8	10	12	14	16	18	2.1	2.4	2.7	3.2	3.8	2.4	2.6	3.0	3.7	4.5
DG A3	10	13	17	21	20	30	35	41	8	11	15	18	23	26	31	37	1.7	1.8	1.9	2.2	2.4	1.8	1.9	2.1	2.3	2.5
DG B1	3	4	5	5	5	5	5	4	1	1	2	2	1	1	0	-1	3.4	4.9	-	-	-	5.0	-	-	-	-
DG B2	7	9	11	13	16	19	21	25	5	6	9	10	13	14	17	20	2.1	2.3	2.5	2.9	3.3	2.3	2.6	2.9	3.3	3.8
DG B3	10	14	18	22	28	32	37	45	8	11	15	19	24	28	33	40	1.6	1.8	1.9	2.0	2.2	1.7	1.8	2.0	2.1	2.3
DG C1	5	6	8	9	10	11	11	12	3	4	4	5	6	6	6	5	2.6	3.2	4.0	-	-	3.1	4.1	-	-	-
DG C2	8	11	14	17	21	24	28	33	6	8	11	14	17	19	23	27	1.9	2.0	2.2	2.5	2.8	2.0	2.2	2.5	2.7	3.1
DG C3	12	16	21	25	32	36	42	51	9	13	18	22	28	32	38	46	1.5	1.6	1.7	1.9	2.1	1.6	1.7	1.8	2.0	2.2
SS A1	11	14	16	17	20	21	23	25	10	12	14	15	17	19	20	21	2.1	2.5	2.9	3.6	4.3	2.2	2.6	3.3	4.2	-
SS A2	14	18	23	26	31	35	39	45	12	16	21	24	29	32	36	42	1.5	1.7	1.9	2.2	2.4	1.5	1.8	2.0	2.3	2.5
SS A3	10	12	13	14	15	16	16	17	9	10	11	12	13	13	13	13	2.5	3.2	4.3	-	-	2.7	3.5	-	-	-
SS A4	13	17	21	24	29	32	36	41	12	15	19	22	26	29	33	37	1.6	1.8	2.1	2.4	2.6	1.6	1.9	2.2	2.5	2.8
SS A5	12	14	16	17	18	19	20	21	11	12	14	15	16	17	17	18	2.4	2.9	3.6	4.7	-	2.6	3.2	4.2	-	-
SS A6	14	18	22	25	30	33	38	43	12	16	20	23	27	30	34	39	1.5	1.8	2.0	2.3	2.5	1.6	1.8	2.1	2.4	2.7
SS A7	11	12	13	14	14	14	14	14	10	11	12	12	12	12	12	11	3.1	4.2	-	-	-	3.5	4.8	-	-	-
SS A8	13	16	20	23	27	30	33	38	11	15	18	21	24	27	30	34	1.6	1.9	2.2	2.5	2.8	1.7	2.0	2.3	2.7	3.1
SS B1	9	11	13	14	16	18	19	21	8	9	11	12	14	15	16	17	2.3	2.7	3.2	3.9	4.7	2.5	2.9	3.6	4.6	-
SS B2	12	16	20	24	29	32	37	43	10	14	18	21	26	29	33	39	1.6	1.8	2.0	2.2	2.4	1.6	1.8	2.1	2.3	2.5
SS B3	8	9	11	12	13	14	15	15	7	8	9	10	11	11	12	11	2.6	3.2	4.1	-	-	2.8	3.6	-	-	-
SS B4	11	15	19	22	27	30	34	39	10	13	17	20	24	27	31	36	1.6	1.8	2.1	2.3	2.5	1.7	1.9	2.2	2.5	2.7
SS B5	10	11	13	13	15	16	16	17	9	10	11	12	13	13	14	14	2.7	3.3	4.0	-	-	3.1	3.7	4.7	-	-
SS B6	11	15	19	22	27	30	35	40	10	13	17	20	24	27	31	36	1.6	1.8	2.0	2.3	2.5	1.7	1.9	2.1	2.4	2.6
SS B7	9	10	11	11	12	12	12	12	8	9	9	10	10	10	10	9	3.3	4.2	-	-	-	3.9	-	-	-	-
SS B8	10	14	17	20	24	27	31	36	9	12	15	18	22	24	28	32	1.7	1.9	2.2	2.5	2.7	1.8	2.0	2.3	2.6	2.9
SS C1	6	7	9	11	13	14	16	18	3	5	6	7	8	9	10	10	2.3	2.7	3.2	4.0	-	2.7	3.2	4.1	-	-
SS C2	7	11	15	19	23	26	31	37	5	8	12	15	19	22	25	30	1.6	1.9	2.1	2.3	2.6	1.8	2.0	2.3	2.5	2.8
SS C3	7	9	10	11	13	14	15	17	6	7	8	8	9	10	11	12	2.7	3.1	3.6	4.3	-	3.2	3.7	4.5	-	-
SS C4	8	11	14	17	21	24	28	33	6	8	11	14	17	20	23	27	1.8	2.0	2.2	2.5	2.7	2.0	2.2	2.4	2.7	3.0
SS D1	11	14	17	19	21	23	25	27	9	12	14	16	18	20	21	22	1.8	2.2	2.7	3.4	4.4	1.9	2.4	3.1	4.3	-
SS D2	14	20	26	30	36	40	46	53	12	17	23	27	33	37	42	48	1.3	1.5	1.8	2.0	2.2	1.4	1.6	1.8	2.1	2.3
SS D3	12	14	17	18	20	22	23	25	11	12	15	16	18	19	20	21	2.0	2.5	3.0	3.8	4.9	2.2	2.7	3.4	4.7	-
SS D4	14	19	25	29	35	39	44	51	12	17	22	26	32	36	40	47	1.4	1.6	1.8	2.0	2.2	1.4	1.6	1.9	2.1	2.4
SS E1	8	10	12	13	15	16	17	18	6	8	10	11	12	12	13	12	2.1	2.7	3.5	-	-	2.3	3.0	4.5	-	-
SS E2	11	15	20	24	29	32	37	43	9	13	17	21	25	29	33	38	1.5	1.7	2.0	2.2	2.5	1.5	1.8	2.1	2.4	2.7
SS E3	10	11	13	14	15	16	17	18	8	10	11	12	13	13	14	14	2.4	3.0	3.7	-	-	2.7	3.3	4.5	-	-
SS E4	11	15	19	22	27	30	34	40	9	13	17	20	24	27	31	35	1.6	1.8	2.0	2.3	2.6	1.6	1.9	2.2	2.5	2.8

OTTAWA, ONTARIO

TYPE	SOLAR SAVINGS FRACTION (%)																CONSERVATION FACTOR									
BASE TEMP:	55 F (5823 DD)								65 F (8461 DD)								55 F					65 F				
LCR	100	70	50	40	30	25	20	15	100	70	50	40	30	25	20	15	100	50	30	20	15	100	50	30	20	15
WW A1	13	15	17	19	21	22	24	27	12	13	14	15	16	17	18	19	2.3	2.6	2.9	3.3	3.6	2.9	3.4	4.0	4.9	-
WW A2	12	15	19	22	26	29	33	38	9	12	14	16	19	21	24	28	1.7	1.9	2.2	2.4	2.6	2.0	2.3	2.6	2.9	3.2
WW A3	11	15	19	23	28	31	36	42	8	11	14	17	20	23	26	31	1.6	1.8	2.0	2.2	2.4	1.8	2.1	2.3	2.6	2.9
WW A4	10	14	19	23	29	33	38	45	7	10	14	17	21	24	28	33	1.5	1.7	1.9	2.1	2.3	1.7	2.0	2.2	2.5	2.7
WW A5	9	14	20	24	30	34	40	47	6	10	14	17	22	25	29	35	1.4	1.6	1.8	2.0	2.2	1.6	1.9	2.1	2.4	2.6
WW A6	9	14	20	24	31	35	41	48	6	10	14	17	22	26	30	36	1.4	1.6	1.8	2.0	2.1	1.6	1.8	2.0	2.3	2.5
WW B1	8	10	12	13	15	16	17	19	6	7	8	8	8	8	8	7	2.4	2.8	3.4	4.3	-	3.3	4.7	-	-	-
WW B2	11	16	22	26	33	38	43	52	8	12	17	20	25	30	35	42	1.4	1.6	1.7	1.9	2.1	1.6	1.8	1.9	2.1	2.3
WW B3	14	20	27	32	40	45	51	60	10	15	21	26	32	37	42	50	1.3	1.4	1.6	1.8	1.9	1.4	1.6	1.8	1.9	2.1
WW B4	14	21	29	36	45	50	58	67	10	16	24	29	37	43	50	59	1.2	1.3	1.5	1.6	1.8	1.3	1.4	1.6	1.8	1.9
WW B5	13	21	29	36	45	51	59	69	10	16	24	30	38	44	51	61	1.2	1.3	1.5	1.6	1.8	1.2	1.4	1.5	1.7	1.9
WW C1	15	22	29	35	43	48	55	64	11	17	23	28	35	40	46	55	1.2	1.4	1.5	1.7	1.9	1.3	1.5	1.7	1.9	2.0
WW C2	14	21	29	34	43	48	56	65	10	16	23	28	36	41	47	56	1.2	1.4	1.5	1.7	1.8	1.3	1.5	1.6	1.8	2.0
WW C3	17	27	37	44	54	61	69	78	13	22	31	38	48	54	62	72	1.0	1.2	1.4	1.5	1.7	1.1	1.3	1.4	1.6	1.7
WW C4	15	24	34	41	51	58	66	76	12	19	28	35	45	51	59	69	1.1	1.2	1.4	1.5	1.7	1.1	1.3	1.4	1.6	1.7
TW A1	13	15	17	19	21	23	25	28	12	13	14	15	17	18	19	21	2.3	2.5	2.8	3.1	3.4	2.7	3.2	3.7	4.2	4.8
TW A2	12	14	18	20	24	27	30	35	10	12	14	16	19	20	23	26	1.8	2.0	2.3	2.5	2.7	2.1	2.4	2.7	3.0	3.3
TW A3	11	14	18	21	25	28	33	39	8	11	14	16	19	22	25	29	1.7	1.9	2.1	2.3	2.5	2.0	2.2	2.5	2.7	3.0
TW A4	9	13	17	21	26	29	34	40	7	10	13	16	19	22	26	31	1.6	1.8	2.0	2.2	2.4	1.9	2.1	2.3	2.6	2.8
TW B1	12	14	17	19	22	24	26	30	11	12	14	15	17	19	20	23	2.1	2.3	2.6	2.9	3.1	2.5	2.9	3.2	3.6	3.9
TW B2	11	14	17	20	24	27	30	36	9	11	13	15	18	21	23	27	1.8	2.0	2.2	2.4	2.6	2.1	2.3	2.6	2.8	3.1
TW B3	10	13	17	20	24	27	31	37	8	10	13	15	19	21	24	29	1.7	1.9	2.1	2.3	2.5	2.0	2.2	2.5	2.7	2.9
TW B4	9	12	16	19	23	26	30	36	7	10	13	15	18	21	24	28	1.8	2.0	2.1	2.3	2.5	2.0	2.2	2.5	2.7	2.9
TW C1	11	13	16	18	22	24	27	32	10	11	13	15	17	19	21	25	2.0	2.2	2.4	2.6	2.8	2.3	2.6	2.9	3.1	3.4
TW C2	10	13	16	18	22	25	29	34	8	10	13	15	18	20	23	27	1.9	2.1	2.3	2.5	2.6	2.1	2.4	2.6	2.8	3.0
TW C3	10	12	15	18	22	24	28	33	8	10	13	15	17	20	22	27	1.9	2.1	2.3	2.5	2.7	2.2	2.4	2.6	2.8	3.0
TW C4	10	12	15	17	20	23	26	31	9	11	13	15	17	19	21	25	2.1	2.3	2.4	2.6	2.8	2.3	2.6	2.8	3.0	3.2
TW D1	7	8	10	11	13	15	16	18	5	6	7	7	8	9	9	9	2.4	2.8	3.3	3.9	4.5	3.2	4.1	-	-	-
TW D2	12	16	21	25	31	35	40	47	9	12	16	20	24	28	32	38	1.5	1.7	1.8	2.0	2.2	1.7	1.9	2.1	2.3	2.4
TW D3	12	17	22	27	33	37	43	51	9	13	18	21	26	30	35	42	1.4	1.6	1.8	1.9	2.1	1.6	1.8	2.0	2.2	2.3
TW D4	13	19	27	32	40	45	52	61	10	15	22	26	33	38	44	53	1.3	1.4	1.6	1.7	1.9	1.4	1.5	1.7	1.9	2.0
TW D5	13	20	27	33	41	47	54	64	10	16	22	28	35	40	47	56	1.2	1.4	1.5	1.7	1.8	1.3	1.5	1.6	1.8	1.9
TW E1	14	20	26	31	38	43	50	58	11	16	21	25	32	36	41	49	1.3	1.5	1.6	1.8	2.0	1.4	1.6	1.8	2.0	2.2
TW E2	14	20	27	32	40	45	51	60	11	16	22	26	33	38	44	52	1.3	1.4	1.6	1.8	1.9	1.4	1.6	1.7	1.9	2.1
TW E3	17	25	34	41	50	56	64	73	14	21	29	35	43	49	56	66	1.1	1.3	1.4	1.6	1.7	1.2	1.3	1.5	1.7	1.8
TW E4	16	24	33	39	48	55	62	72	12	19	27	33	42	48	55	65	1.1	1.3	1.4	1.6	1.7	1.2	1.4	1.5	1.7	1.8
TW F1	11	13	15	17	20	21	24	27	10	11	13	14	15	16	18	20	2.2	2.5	2.7	3.1	3.3	2.7	3.1	3.5	4.1	4.6
TW F2	9	12	15	18	22	24	28	33	7	9	12	13	16	18	20	24	1.9	2.1	2.3	2.5	2.7	2.2	2.5	2.8	3.1	3.4
TW F3	8	11	15	18	22	25	29	35	6	8	11	13	16	18	21	26	1.8	2.0	2.2	2.4	2.6	2.0	2.3	2.5	2.8	3.1
TW F4	6	9	13	16	21	24	28	35	4	6	9	12	15	18	21	26	1.7	1.9	2.1	2.3	2.5	2.0	2.2	2.4	2.7	2.9
TW G1	9	11	14	16	19	21	24	28	8	9	11	12	14	16	17	20	2.1	2.3	2.6	2.9	3.1	2.5	2.9	3.2	3.6	4.0
TW G2	7	10	13	15	19	22	26	31	5	7	9	11	14	16	19	23	1.9	2.1	2.3	2.5	2.7	2.2	2.5	2.7	3.0	3.2
TW G3	5	8	12	14	18	21	25	31	3	6	8	10	13	16	18	23	1.9	2.0	2.2	2.5	2.6	2.1	2.4	2.6	2.9	3.1
TW G4	4	6	9	12	16	19	23	28	2	4	7	9	12	14	17	21	2.0	2.1	2.3	2.5	2.7	2.2	2.5	2.7	2.9	3.1
TW H1	6	9	11	13	17	19	22	26	5	7	8	10	12	14	16	19	2.1	2.3	2.5	2.8	3.0	2.4	2.7	3.0	3.3	3.6
TW H2	4	6	9	12	15	18	21	26	3	4	7	8	11	13	15	19	2.0	2.2	2.4	2.6	2.8	2.3	2.6	2.8	3.1	3.3
TW H3	3	5	8	10	13	16	19	24	2	3	6	7	10	12	14	18	2.1	2.3	2.5	2.7	2.9	2.4	2.7	2.9	3.1	3.3
TW H4	2	4	6	8	11	13	16	20	1	3	4	6	8	9	12	15	2.4	2.6	2.7	2.9	3.1	2.7	2.9	3.1	3.3	3.5
TW I1	5	6	8	9	11	12	14	16	3	4	5	6	6	7	7	7	2.6	3.0	3.5	4.1	4.8	3.4	4.4	-	-	-
TW I2	9	13	18	22	27	31	37	44	6	10	14	17	21	24	29	35	1.5	1.7	1.9	2.1	2.2	1.7	1.9	2.1	2.3	2.5
TW I3	9	14	20	24	30	34	40	48	7	11	15	19	24	27	32	39	1.4	1.6	1.8	2.0	2.1	1.6	1.8	2.0	2.2	2.3
TW I4	11	17	24	29	37	42	49	58	8	13	19	23	30	35	41	50	1.3	1.4	1.6	1.8	1.9	1.4	1.6	1.7	1.9	2.0
TW I5	11	17	25	30	39	44	52	61	8	13	20	25	32	38	44	54	1.3	1.4	1.6	1.7	1.9	1.4	1.5	1.7	1.8	2.0
TW J1	12	18	24	29	36	41	48	56	9	14	19	23	30	34	39	47	1.3	1.5	1.7	1.8	2.0	1.5	1.6	1.8	2.0	2.2
TW J2	12	17	24	30	37	42	49	58	9	14	19	24	31	35	41	50	1.3	1.5	1.6	1.8	1.9	1.4	1.6	1.7	1.9	2.1
TW J3	15	23	32	38	48	54	62	71	11	18	26	32	41	47	55	64	1.1	1.3	1.4	1.6	1.7	1.2	1.4	1.5	1.7	1.8
TW J4	13	21	30	37	46	52	60	70	10	17	25	31	39	45	53	63	1.1	1.3	1.4	1.6	1.7	1.2	1.4	1.5	1.7	1.8
DG A1	6	8	10	11	13	14	15	15	4	5	6	7	8	8	7	6	2.2	2.7	3.7	-	-	2.9	3.9	-	-	-
DG A2	9	12	16	19	23	26	30	35	7	10	13	15	18	20	23	27	1.7	1.9	2.2	2.5	2.8	2.0	2.2	2.5	2.9	3.3
DG A3	13	17	23	27	34	39	45	53	11	14	19	23	28	33	38	46	1.5	1.6	1.7	1.9	2.1	1.6	1.7	1.9	2.0	2.2
DG B1	6	8	10	12	14	15	17	18	4	5	6	7	8	8	9	8	2.2	2.6	3.2	4.1	-	3.0	3.7	-	-	-
DG B2	9	13	17	20	24	28	32	38	7	10	13	15	19	21	25	29	1.7	1.9	2.1	2.3	2.5	2.0	2.2	2.4	2.6	2.9
DG B3	13	18	24	28	35	41	47	57	11	15	20	24	30	34	40	49	1.4	1.5	1.6	1.8	1.9	1.6	1.7	1.8	1.9	2.1
DG C1	8	11	14	16	19	22	24	27	6	7	9	11	13	14	15	17	1.9	2.2	2.5	3.0	3.5	2.4	2.8	3.3	4.3	-
DG C2	11	15	20	24	30	34	39	46	9	12	16	19	24	27	31	37	1.5	1.7	1.9	2.0	2.2	1.8	1.9	2.1	2.3	2.5
DG C3	15	20	27	32	40	46	54	64	13	17	23	27	34	39	46	55	1.3	1.4	1.5	1.7	1.8	1.5	1.6	1.7	1.8	1.9
SS A1	15	18	22	24	28	31	35	40	13	16	19	21	24	26	29	33	1.7	1.9	2.2	2.4	2.6	1.9	2.1	2.5	2.8	3.1
SS A2	18	24	30	35	41	46	52	59	16	21	26	31	36	40	46	53	1.3	1.5	1.7	1.8	2.0	1.4	1.6	1.8	2.0	2.1
SS A3	13	16	19	21	24	26	28	32	12	14	16	17	20	21	23	25	1.9	2.2	2.5	2.9	3.2	2.1	2.5	3.0	3.7	4.5
SS A4	18	23	28	33	39	43	49	56	16	20	25	29	34	38	43	49	1.3	1.5	1.7	1.9	2.1	1.4	1.6	1.9	2.1	2.3
SS A5	15	18	21	23	26	28	31	35	14	16	18	20	22	24	26	28	1.9	2.1	2.4	2.8	3.0	2.1	2.5	2.9	3.3	3.8
SS A6	18	23	29	34	40	45	50	58	16	20	26	30	35	39	44	51	1.3	1.5	1.7	1.9	2.0	1.4	1.6	1.8	2.0	2.2
SS A7	14	16	18	19	21	22	24	26	13	14	15	16	18	19	20	21	2.2	2.6	3.1	3.6	4.2	2.6	3.1	3.9	-	-
SS A8	17	22	27	31	37	41	46	53	15	19	24	27	32	36	40	46	1.4	1.6	1.8	2.0	2.1	1.5	1.7	1.9	2.2	2.4
SS B1	12	15	18	21	24	27	30	35	11	13	15	17	20	22	24	28	1.8	2.1	2.3	2.6	2.8	2.0	2.3	2.7	3.0	3.4
SS B2	16	21	27	31	38	42	48	56	14	18	23	27	33	37	42	49	1.4	1.5	1.7	1.9	2.0	1.5	1.6	1.8	2.0	2.2
SS B3	11	13	16	18	21	23	26	29	10	11	13	15	17	18	20	22	2.0	2.3	2.6	2.9	3.2	2.3	2.6	3.1	3.7	4.3
SS B4	15	20	25	30	36	40	45	53	13	17	22	26	31	35	40	46	1.4	1.6	1.8	1.9	2.1	1.5	1.7	1.9	2.1	2.3
SS B5	12	14	17	19	21	23	25	29	11	13	15	16	18	19	21	23	2.1	2.4	2.7	3.0	3.3	2.4	2.8	3.2	3.7	4.2
SS B6	15	20	25	30	36	40	46	54	13	17	22	26	31	35	40	47	1.4	1.6	1.7	1.9	2.1	1.5	1.7	1.9	2.1	2.2
SS B7	11	13	15	16	18	19	21	23	10	11	13	14	15	16	17	18	2.4	2.7	3.2	3.7	4.1	2.8	3.3	4.0	5.0	-
SS B8	14	19	24	28	33	37	43	50	12	16	20	24	29	32	37	43	1.4	1.6	1.8	2.0	2.2	1.6	1.8	2.0	2.2	2.4
SS C1	9	12	16	19	23	26	30	35	7	9	11	13	16	18	21	24	1.7	2.0	2.2	2.5	2.7	2.1	2.4	2.7	3.1	3.4
SS C2	12	17	23	27	34	38	44	52	9	13	18	21	27	31	36	43	1.4	1.6	1.7	1.9	2.1	1.6	1.7	1.9	2.1	2.3
SS C3	10	12	15	17	20	22	25	29	8	10	12	13	15	16	18	21	2.0	2.3	2.5	2.8	3.0	2.4	2.8	3.1	3.5	3.8
SS C4	11	16	20	24	30	34	39	47	9	12	16	19	24	27	32	38	1.5	1.7	1.9	2.0	2.2	1.7	1.9	2.1	2.3	2.4
SS D1	16	20	25	28	33	36	41	46	13	17	21	23	27	30	33	37	1.5	1.7	1.9	2.2	2.4	1.6	1.9	2.2	2.6	2.9
SS D2	20	27	34	40	48	53	60	68	17	23	30	35	42	47	53	61	1.2	1.3	1.5	1.7	1.8	1.2	1.4	1.6	1.8	2.0
SS D3	16	19	23	26	30	33	36	41	14	17	20	22	25	27	30	34	1.6	1.9	2.1	2.4	2.6	1.8	2.1	2.4	2.8	3.2
SS D4	19	26	33	39	46	51	58	66	17	22	29	34	41	45	51	59	1.2	1.4	1.5	1.7	1.9	1.3	1.5	1.6	1.8	2.0
SS E1	12	16	20	22	26	29	33	38	10	13	16	18	21	23	25	28	1.6	1.9	2.2	2.5	2.7	1.9	2.2	2.6	3.1	3.7
SS E2	16	22	28	33	41	45	52	60	13	18	24	28	35	39	44	52	1.3	1.4	1.6	1.8	2.0	1.4	1.6	1.8	2.0	2.1
SS E3	13	15	18	20	23	26	29	32	11	13	15	17	19	21	23	25	1.9	2.2	2.5	2.8	3.1	2.2	2.5	2.9	3.4	4.0
SS E4	15	21	27	31	38	42	48	55	13	17	23	26	32	36	41	48	1.3	1.5	1.7	1.9	2.0	1.5	1.6	1.9	2.1	2.2

TORONTO, ONTARIO

TYPE	SOLAR SAVINGS FRACTION (%) BASE TEMP: 55 F (4726 DD) LCR 100	70	50	40	30	25	20	15	65 F (7280 DD) 100	70	50	40	30	25	20	15	CONSERVATION FACTOR 55 F 100	50	30	20	15	65 F 100	50	30	20	15
WW A1	14	16	18	20	22	24	26	29	12	13	14	15	17	17	18	19	2.2	2.4	2.8	3.1	3.4	2.8	3.4	4.0	4.9	-
WW A2	13	16	20	23	28	31	35	41	10	12	15	17	20	22	24	28	1.6	1.9	2.1	2.3	2.5	2.0	2.3	2.6	2.9	3.2
WW A3	12	16	21	24	30	33	38	45	8	11	14	17	21	23	27	31	1.5	1.7	1.9	2.1	2.3	1.8	2.1	2.3	2.6	2.9
WW A4	11	16	21	25	31	35	40	47	7	11	14	17	21	24	28	33	1.4	1.6	1.8	2.0	2.2	1.7	2.0	2.2	2.5	2.7
WW A5	11	16	22	26	32	37	42	50	6	10	14	17	22	25	30	35	1.4	1.6	1.8	2.0	2.1	1.6	1.9	2.1	2.4	2.6
WW A6	10	16	22	26	33	38	43	51	6	10	14	18	23	26	30	37	1.4	1.6	1.7	1.9	2.1	1.6	1.8	2.1	2.3	2.5
WW B1	9	11	13	15	17	18	20	22	6	7	8	8	9	9	9	8	2.2	2.6	3.1	3.8	4.5	3.2	4.6	-	-	-
WW B2	12	18	24	28	35	40	46	54	8	12	17	21	26	30	35	42	1.3	1.5	1.7	1.9	2.0	1.6	1.8	1.9	2.1	2.3
WW B3	15	22	29	34	42	47	54	63	10	16	21	26	32	37	43	50	1.2	1.4	1.6	1.7	1.9	1.4	1.6	1.8	1.9	2.1
WW B4	15	23	31	38	47	52	60	70	10	17	24	30	38	43	50	59	1.1	1.3	1.5	1.6	1.8	1.3	1.4	1.6	1.8	1.9
WW B5	14	22	31	38	47	53	61	71	10	17	24	30	38	44	51	61	1.1	1.3	1.4	1.6	1.7	1.2	1.4	1.6	1.7	1.9
WW C1	17	23	31	37	45	50	57	66	12	17	24	29	36	40	46	55	1.2	1.4	1.5	1.7	1.8	1.3	1.5	1.7	1.9	2.0
WW C2	15	22	30	36	45	50	58	67	11	16	23	28	36	41	47	56	1.2	1.3	1.5	1.7	1.8	1.3	1.5	1.6	1.8	2.0
WW C3	19	28	39	46	56	63	71	80	14	22	31	38	48	54	62	72	1.0	1.2	1.3	1.5	1.6	1.1	1.3	1.4	1.6	1.7
WW C4	17	26	36	43	53	60	68	77	12	20	29	35	45	51	59	69	1.1	1.2	1.4	1.5	1.7	1.1	1.3	1.4	1.6	1.8
TW A1	14	16	18	20	23	24	27	30	12	13	15	16	17	18	19	21	2.1	2.4	2.7	3.0	3.2	2.7	3.2	3.7	4.3	4.9
TW A2	13	16	19	22	26	29	32	38	10	12	14	16	19	21	23	27	1.7	2.0	2.2	2.4	2.6	2.1	2.4	2.7	3.0	3.3
TW A3	12	15	19	23	27	30	35	41	9	11	14	16	20	22	25	29	1.6	1.8	2.0	2.2	2.4	1.9	2.2	2.5	2.7	3.0
TW A4	10	14	19	22	27	31	36	43	7	10	13	16	20	22	26	31	1.6	1.8	2.0	2.2	2.3	1.9	2.1	2.3	2.6	2.8
TW B1	13	15	18	20	23	25	28	32	11	13	14	16	17	19	21	23	2.0	2.2	2.5	2.7	3.0	2.5	2.8	3.2	3.6	4.0
TW B2	11	15	18	21	25	28	33	38	9	11	14	16	19	21	24	28	1.7	1.9	2.1	2.3	2.5	2.1	2.3	2.6	2.8	3.1
TW B3	11	14	18	21	26	29	33	39	8	10	13	16	19	21	24	29	1.7	1.9	2.1	2.3	2.4	2.0	2.2	2.5	2.7	2.9
TW B4	10	13	17	20	25	28	32	39	8	10	13	15	18	21	24	29	1.7	1.9	2.1	2.3	2.4	2.0	2.2	2.5	2.7	2.9
TW C1	12	14	17	20	23	26	29	34	10	12	14	15	18	19	22	25	1.9	2.1	2.3	2.6	2.7	2.3	2.6	2.9	3.1	3.4
TW C2	11	14	17	20	24	27	31	36	9	11	13	15	18	20	23	27	1.8	2.0	2.2	2.4	2.6	2.1	2.4	2.6	2.8	3.0
TW C3	10	13	17	19	23	26	30	35	9	11	13	15	18	20	23	27	1.8	2.0	2.2	2.4	2.6	2.2	2.4	2.6	2.8	3.0
TW C4	11	13	16	18	22	24	28	33	9	11	13	15	17	19	22	26	2.0	2.2	2.4	2.6	2.7	2.3	2.5	2.8	3.0	3.2
TW D1	8	9	11	13	15	17	19	21	5	6	7	8	9	9	9	10	2.2	2.6	3.1	3.6	4.0	3.1	4.0	-	-	-
TW D2	13	17	22	26	32	36	42	50	9	13	17	20	25	28	32	38	1.4	1.6	1.8	2.0	2.1	1.7	1.9	2.1	2.3	2.4
TW D3	13	18	24	29	35	40	46	54	9	13	18	21	27	30	35	42	1.4	1.5	1.7	1.9	2.0	1.6	1.8	2.0	2.2	2.3
TW D4	14	21	28	34	42	47	54	63	11	16	22	27	33	38	44	53	1.2	1.4	1.6	1.7	1.9	1.4	1.5	1.7	1.9	2.0
TW D5	14	21	29	35	43	49	56	65	10	16	23	28	35	40	47	56	1.2	1.4	1.5	1.7	1.8	1.3	1.5	1.6	1.8	1.9
TW E1	16	22	28	33	40	45	52	60	12	16	22	26	32	36	42	49	1.3	1.4	1.6	1.8	1.9	1.4	1.6	1.8	2.0	2.2
TW E2	15	22	29	34	42	47	54	63	11	16	22	27	33	38	44	52	1.2	1.4	1.6	1.7	1.9	1.4	1.6	1.7	1.9	2.1
TW E3	19	27	36	43	52	58	66	75	14	21	29	35	44	49	57	66	1.1	1.2	1.4	1.6	1.7	1.2	1.3	1.5	1.7	1.8
TW E4	17	25	34	41	50	56	64	74	13	20	27	34	42	48	55	65	1.1	1.3	1.4	1.6	1.7	1.2	1.4	1.5	1.7	1.8
TW F1	12	14	17	18	21	23	26	29	10	11	13	14	15	17	18	20	2.1	2.3	2.6	2.9	3.2	2.6	3.1	3.5	4.1	4.6
TW F2	10	13	17	19	23	26	30	35	8	10	12	14	16	18	21	24	1.8	2.0	2.2	2.5	2.6	2.2	2.5	2.8	3.1	3.4
TW F3	9	12	16	19	24	27	31	37	6	8	11	13	16	19	22	26	1.7	1.9	2.1	2.3	2.5	2.0	2.3	2.5	2.8	3.1
TW F4	7	10	15	18	23	26	31	37	4	7	10	12	15	18	21	26	1.7	1.9	2.1	2.3	2.4	2.0	2.2	2.5	2.7	2.9
TW G1	10	12	15	17	20	23	26	30	8	10	11	13	15	16	18	20	2.0	2.2	2.5	2.8	3.0	2.5	2.8	3.2	3.6	4.0
TW G2	8	11	14	17	21	24	28	33	5	7	10	12	14	16	19	23	1.8	2.0	2.2	2.4	2.6	2.2	2.4	2.7	3.0	3.2
TW G3	6	9	13	16	20	23	27	33	4	6	9	11	14	16	19	23	1.8	2.0	2.2	2.4	2.6	2.1	2.4	2.6	2.9	3.1
TW G4	4	7	11	13	17	20	24	30	2	4	7	9	12	14	17	21	1.9	2.1	2.3	2.5	2.6	2.2	2.4	2.7	2.9	3.1
TW H1	7	9	12	15	18	20	24	28	5	7	9	10	13	14	16	19	2.0	2.2	2.4	2.7	2.9	2.4	2.7	3.0	3.3	3.6
TW H2	5	7	10	13	17	19	23	28	3	5	7	9	11	13	16	20	2.0	2.1	2.4	2.6	2.7	2.3	2.6	2.8	3.1	3.3
TW H3	4	6	9	11	15	17	21	26	2	4	6	7	10	12	14	18	2.1	2.2	2.4	2.6	2.8	2.4	2.6	2.9	3.1	3.3
TW H4	2	4	7	9	12	14	17	22	1	3	5	6	8	10	12	15	2.3	2.5	2.6	2.8	3.0	2.6	2.9	3.1	3.3	3.5
TW I1	6	7	9	11	13	14	16	18	4	5	5	6	7	7	7	7	2.4	2.7	3.2	3.7	4.3	3.3	4.3	-	-	-
TW I2	10	14	19	23	29	33	39	46	7	10	14	17	21	25	29	35	1.5	1.7	1.9	2.0	2.2	1.7	1.9	2.1	2.3	2.5
TW I3	11	15	21	26	32	37	42	51	7	11	15	19	24	28	33	40	1.4	1.6	1.8	1.9	2.1	1.6	1.8	2.0	2.2	2.3
TW I4	12	18	25	31	38	44	51	60	8	13	19	24	30	35	41	50	1.3	1.4	1.6	1.7	1.9	1.4	1.6	1.7	1.9	2.0
TW I5	12	19	26	32	40	46	53	63	8	14	20	25	33	38	44	54	1.2	1.4	1.5	1.7	1.8	1.4	1.5	1.7	1.8	2.0
TW J1	14	19	26	31	38	43	50	59	10	14	20	24	30	34	40	47	1.3	1.5	1.6	1.8	1.9	1.5	1.6	1.8	2.0	2.2
TW J2	13	19	26	31	39	44	51	60	9	14	20	24	31	35	41	50	1.3	1.4	1.6	1.8	1.9	1.4	1.6	1.8	1.9	2.1
TW J3	16	24	33	40	50	56	64	73	11	19	27	33	41	47	55	65	1.1	1.3	1.4	1.6	1.7	1.2	1.4	1.5	1.7	1.8
TW J4	15	23	32	38	48	54	62	71	10	17	25	31	40	46	53	63	1.1	1.3	1.4	1.6	1.7	1.2	1.4	1.5	1.7	1.8
DG A1	7	9	12	13	15	16	17	18	5	6	7	7	8	8	8	7	2.1	2.5	3.4	-	-	2.8	3.9	-	-	-
DG A2	10	13	17	21	25	28	32	37	8	10	13	15	18	21	23	27	1.7	1.9	2.1	2.4	2.8	2.0	2.2	2.5	2.9	3.3
DG A3	14	18	24	29	36	40	47	55	11	15	19	23	29	33	38	46	1.4	1.6	1.7	1.8	2.0	1.6	1.7	1.9	2.0	2.2
DG B1	7	9	12	13	16	17	19	21	4	5	7	7	8	9	9	9	2.0	2.5	3.0	3.7	4.9	3.0	3.6	-	-	-
DG B2	10	14	18	21	26	29	34	40	8	10	13	16	19	22	25	29	1.6	1.8	2.0	2.2	2.4	2.0	2.1	2.4	2.7	3.0
DG B3	14	19	25	30	37	42	49	59	11	15	20	24	30	35	41	49	1.4	1.5	1.6	1.7	1.9	1.6	1.7	1.8	1.9	2.1
DG C1	9	12	16	18	21	24	27	30	6	8	10	11	13	14	16	17	1.8	2.1	2.4	2.8	3.2	2.4	2.7	3.4	4.6	-
DG C2	12	17	22	25	31	36	41	49	9	12	16	19	24	27	31	37	1.5	1.7	1.8	2.0	2.2	1.8	1.9	2.1	2.3	2.5
DG C3	16	22	28	34	42	48	56	66	13	17	23	28	34	39	46	55	1.3	1.4	1.5	1.6	1.8	1.5	1.5	1.7	1.8	1.9
SS A1	16	20	24	27	31	34	38	44	14	17	20	22	25	28	31	35	1.6	1.8	2.1	2.3	2.5	1.8	2.1	2.4	2.7	3.0
SS A2	20	26	32	37	44	49	55	63	17	22	28	32	38	42	47	54	1.2	1.4	1.6	1.8	1.9	1.4	1.6	1.8	1.9	2.1
SS A3	15	18	21	23	27	29	32	36	13	15	17	19	21	23	25	27	1.7	2.0	2.3	2.6	2.9	2.0	2.4	2.9	3.4	4.1
SS A4	20	25	31	36	42	47	52	60	17	21	26	30	36	40	45	51	1.3	1.5	1.7	1.8	2.0	1.4	1.6	1.8	2.0	2.2
SS A5	17	19	23	25	29	31	34	38	15	17	19	21	23	25	27	30	1.7	2.0	2.3	2.6	2.8	2.0	2.4	2.7	3.2	3.6
SS A6	20	25	32	36	43	48	54	62	17	21	27	31	37	41	46	53	1.3	1.5	1.6	1.8	2.0	1.4	1.6	1.8	2.0	2.2
SS A7	15	17	20	21	24	25	28	30	13	15	17	18	19	20	21	23	2.0	2.4	2.8	3.2	3.6	2.4	2.9	3.6	4.5	-
SS A8	19	24	30	34	40	45	50	58	16	20	25	29	34	38	42	48	1.3	1.5	1.7	1.9	2.0	1.4	1.7	1.9	2.1	2.3
SS B1	13	17	20	23	27	30	33	38	11	14	16	18	21	23	26	29	1.7	1.9	2.2	2.4	2.6	2.0	2.3	2.6	2.9	3.3
SS B2	17	23	29	34	40	45	51	59	15	19	24	28	34	38	43	51	1.3	1.5	1.7	1.8	2.0	1.4	1.6	1.8	2.0	2.2
SS B3	12	15	18	20	24	26	29	33	10	12	14	16	18	20	22	24	1.8	2.1	2.4	2.7	3.0	2.1	2.5	3.0	3.5	4.0
SS B4	17	22	28	32	38	43	48	56	14	18	23	27	32	36	41	48	1.3	1.5	1.7	1.9	2.0	1.5	1.7	1.9	2.1	2.2
SS B5	13	16	18	21	24	26	28	32	12	13	15	17	19	20	22	24	1.9	2.2	2.5	2.8	3.1	2.3	2.7	3.1	3.5	4.0
SS B6	17	22	28	32	39	43	49	57	14	18	23	27	33	36	41	48	1.3	1.5	1.7	1.9	2.0	1.5	1.7	1.9	2.1	2.2
SS B7	12	14	16	18	20	22	24	27	11	12	13	14	16	17	18	19	2.2	2.5	2.9	3.3	3.7	2.6	3.1	3.7	4.6	-
SS B8	16	21	26	30	36	40	46	53	13	17	22	25	30	34	38	45	1.4	1.6	1.8	1.9	2.1	1.5	1.7	1.9	2.2	2.3
SS C1	10	14	18	21	25	28	32	38	7	9	12	14	17	18	21	24	1.7	1.9	2.1	2.4	2.5	2.0	2.4	2.7	3.1	3.5
SS C2	13	18	24	29	36	40	46	55	9	13	18	22	27	31	36	43	1.3	1.5	1.7	1.9	2.0	1.5	1.7	1.9	2.1	2.3
SS C3	10	13	16	18	21	24	27	31	8	10	12	13	15	17	19	21	1.9	2.2	2.4	2.7	2.9	2.4	2.7	3.1	3.5	3.8
SS C4	12	17	22	26	32	36	41	49	9	13	17	20	24	28	32	38	1.5	1.6	1.8	2.0	2.1	1.7	1.9	2.1	2.3	2.4
SS D1	18	22	27	31	36	40	45	51	14	18	22	25	29	32	35	40	1.4	1.6	1.9	2.1	2.2	1.6	1.9	2.2	2.5	2.8
SS D2	22	29	37	43	51	57	63	72	18	24	31	37	44	49	55	63	1.1	1.3	1.5	1.6	1.8	1.2	1.4	1.6	1.8	1.9
SS D3	17	21	25	28	33	36	40	46	15	18	21	23	27	29	32	36	1.5	1.8	2.0	2.2	2.4	1.7	2.0	2.4	2.7	3.1
SS D4	21	28	36	42	49	55	61	70	18	24	30	35	42	47	53	61	1.1	1.3	1.5	1.7	1.8	1.2	1.4	1.6	1.8	2.0
SS E1	14	18	22	25	30	33	37	42	11	14	17	19	22	24	27	30	1.5	1.8	2.1	2.3	2.5	1.8	2.1	2.5	3.0	3.5
SS E2	18	24	31	36	44	49	55	63	14	19	25	30	36	40	46	53	1.2	1.4	1.6	1.8	1.9	1.3	1.5	1.7	1.9	2.1
SS E3	14	17	20	23	26	29	32	36	12	14	16	18	20	22	24	27	1.7	2.0	2.3	2.6	2.8	2.1	2.4	2.8	3.3	3.8
SS E4	17	23	29	34	40	45	51	59	14	18	24	28	33	37	42	49	1.3	1.5	1.7	1.8	2.0	1.4	1.6	1.8	2.0	2.2

MONTREAL, QUEBEC

TYPE	SOLAR SAVINGS FRACTION (%) BASE TEMP: 55 F (5495 DD)								65 F (8054 DD)								CONSERVATION FACTOR 55 F					65 F				
LCR	100	70	50	40	30	25	20	15	100	70	50	40	30	25	20	15	100	50	30	20	15	100	50	30	20	15
WW A1	13	14	16	17	19	20	21	23	11	12	13	13	14	14	15	15	2.5	2.9	3.4	3.9	4.5	3.4	4.2	-	-	-
WW A2	11	13	17	19	22	25	28	33	8	10	12	14	16	17	19	22	1.8	2.1	2.4	2.6	2.9	2.2	2.6	3.0	3.5	4.0
WW A3	9	13	17	20	24	27	31	36	6	9	12	14	17	19	21	25	1.7	1.9	2.2	2.4	2.6	2.0	2.3	2.6	3.0	3.4
WW A4	9	12	17	20	25	28	32	39	5	8	11	14	17	19	22	27	1.6	1.8	2.1	2.3	2.5	1.9	2.2	2.5	2.8	3.1
WW A5	8	12	17	21	26	29	34	41	4	7	11	14	18	20	24	28	1.5	1.7	2.0	2.2	2.4	1.8	2.1	2.3	2.7	3.0
WW A6	7	12	17	21	26	30	35	42	4	7	11	14	18	21	24	29	1.5	1.7	1.9	2.1	2.3	1.7	2.0	2.3	2.6	2.8
WW B1	7	8	10	11	12	12	13	13	5	5	6	6	5	5	4	1	2.7	3.4	4.7	-	-	4.6	-	-	-	-
WW B2	10	14	19	23	29	33	39	46	6	10	14	17	22	25	30	36	1.5	1.7	1.8	2.0	2.2	1.7	1.9	2.1	2.3	2.5
WW B3	12	18	24	29	36	40	47	55	8	13	18	22	28	32	37	44	1.3	1.5	1.7	1.9	2.0	1.5	1.7	1.9	2.1	2.3
WW B4	12	19	27	32	41	46	54	63	8	14	21	26	33	39	45	54	1.2	1.4	1.5	1.7	1.8	1.3	1.5	1.7	1.8	2.0
WW B5	11	19	27	33	42	47	55	65	8	14	21	27	35	40	47	57	1.2	1.3	1.5	1.7	1.8	1.3	1.4	1.6	1.8	1.9
WW C1	13	19	26	31	39	44	50	59	10	15	20	25	31	35	41	49	1.3	1.4	1.6	1.8	1.9	1.4	1.6	1.8	2.0	2.1
WW C2	12	18	26	31	39	44	51	60	8	14	20	25	32	36	43	51	1.2	1.4	1.6	1.8	1.9	1.4	1.5	1.7	1.9	2.1
WW C3	15	24	34	41	51	57	65	75	11	19	28	35	44	50	58	68	1.1	1.2	1.4	1.6	1.7	1.2	1.3	1.5	1.6	1.8
WW C4	13	22	31	38	48	54	62	72	10	17	26	32	41	47	55	65	1.1	1.3	1.4	1.6	1.7	1.2	1.3	1.5	1.7	1.8
TW A1	13	14	16	17	19	20	22	24	11	12	13	14	15	15	16	17	2.5	2.8	3.2	3.6	4.0	3.2	3.8	4.6	-	-
TW A2	11	13	16	18	21	24	27	31	9	10	12	14	16	17	19	22	2.0	2.2	2.5	2.8	3.0	2.4	2.7	3.1	3.5	4.0
TW A3	10	12	16	18	22	25	29	34	7	9	12	13	16	18	20	24	1.8	2.0	2.3	2.5	2.7	2.2	2.4	2.8	3.1	3.4
TW A4	8	11	15	18	22	25	29	35	6	8	11	13	16	18	21	25	1.7	1.9	2.2	2.4	2.6	2.0	2.3	2.6	2.9	3.1
TW B1	12	13	15	17	19	21	23	26	10	11	13	14	15	16	17	19	2.3	2.5	2.9	3.2	3.5	2.8	3.3	3.7	4.3	5.0
TW B2	10	12	15	18	21	24	27	32	8	9	12	13	16	17	20	23	1.9	2.1	2.4	2.6	2.8	2.3	2.6	2.9	3.2	3.5
TW B3	9	12	15	17	21	24	28	33	7	9	11	13	16	18	20	24	1.8	2.1	2.3	2.5	2.7	2.2	2.4	2.7	3.0	3.3
TW B4	8	11	14	17	21	23	27	32	6	8	11	13	16	18	20	24	1.9	2.1	2.3	2.5	2.7	2.2	2.4	2.7	2.9	3.2
TW C1	10	12	15	17	20	22	24	28	9	10	12	13	15	17	18	21	2.1	2.4	2.6	2.9	3.1	2.6	2.9	3.2	3.5	3.9
TW C2	9	12	14	17	20	22	26	30	8	9	11	13	15	17	20	23	2.0	2.2	2.4	2.6	2.8	2.3	2.6	2.9	3.1	3.4
TW C3	9	11	14	16	20	22	25	30	8	9	11	13	15	17	19	23	2.0	2.2	2.4	2.7	2.8	2.4	2.6	2.9	3.1	3.3
TW C4	10	12	14	16	19	21	24	28	8	10	12	13	15	17	19	22	2.2	2.4	2.6	2.8	3.0	2.5	2.8	3.0	3.2	3.4
TW D1	6	7	9	10	11	12	13	14	4	5	5	5	6	5	5	4	2.7	3.3	4.1	-	-	4.1	-	-	-	-
TW D2	10	14	19	22	27	31	36	43	8	11	14	17	21	24	28	33	1.6	1.8	2.0	2.2	2.3	1.8	2.0	2.2	2.4	2.6
TW D3	11	15	20	24	30	34	39	46	8	11	15	18	23	26	30	36	1.5	1.7	1.9	2.1	2.2	1.7	1.9	2.1	2.3	2.5
TW D4	12	18	24	29	36	41	48	57	9	13	19	23	30	34	40	48	1.3	1.5	1.6	1.8	2.0	1.5	1.6	1.8	2.0	2.1
TW D5	12	18	25	30	38	43	50	60	9	14	20	25	32	37	43	52	1.3	1.4	1.6	1.8	1.9	1.4	1.5	1.7	1.9	2.0
TW E1	13	18	24	28	35	39	45	53	10	14	19	22	28	32	37	44	1.4	1.5	1.7	1.9	2.1	1.5	1.7	1.9	2.1	2.3
TW E2	13	18	24	29	36	41	47	56	9	14	19	23	29	34	39	47	1.3	1.5	1.7	1.8	2.0	1.5	1.6	1.8	2.0	2.2
TW E3	16	23	31	38	47	52	60	69	12	18	26	32	40	45	52	62	1.1	1.3	1.5	1.6	1.8	1.2	1.4	1.6	1.7	1.9
TW E4	14	22	30	36	45	51	59	68	11	17	24	30	38	44	51	61	1.2	1.3	1.5	1.6	1.8	1.3	1.4	1.6	1.7	1.9
TW F1	11	12	14	15	17	19	21	23	9	10	11	12	13	14	14	15	2.4	2.7	3.1	3.5	3.9	3.1	3.7	4.3	-	-
TW F2	8	11	14	16	19	21	24	28	6	8	10	11	13	15	17	19	2.0	2.2	2.5	2.8	3.1	2.4	2.8	3.2	3.6	4.1
TW F3	7	10	13	15	19	22	25	30	5	6	9	10	13	15	17	20	1.9	2.1	2.3	2.6	2.8	2.2	2.5	2.8	3.2	3.5
TW F4	5	8	11	14	18	21	25	30	3	5	7	9	12	14	17	21	1.8	2.1	2.3	2.5	2.7	2.2	2.4	2.7	3.0	3.3
TW G1	9	10	12	14	17	18	20	24	7	8	9	10	12	13	14	16	2.3	2.6	2.9	3.2	3.5	2.8	3.3	3.7	4.3	5.0
TW G2	6	8	11	13	17	19	22	27	4	6	8	9	11	13	15	18	2.0	2.2	2.5	2.8	3.0	2.4	2.7	3.0	3.4	3.7
TW G3	4	7	10	12	16	18	22	27	3	4	7	8	11	13	15	18	2.0	2.2	2.4	2.7	2.9	2.3	2.6	2.9	3.2	3.5
TW G4	3	5	8	10	14	16	20	24	1	3	5	7	9	11	14	17	2.1	2.3	2.5	2.7	2.9	2.4	2.7	2.9	3.2	3.4
TW H1	6	8	10	12	14	16	19	23	4	6	7	8	10	11	13	15	2.2	2.5	2.7	3.0	3.3	2.7	3.1	3.4	3.8	4.2
TW H2	3	5	8	10	13	15	18	23	2	3	5	7	9	10	13	16	2.2	2.4	2.6	2.8	3.1	2.6	2.8	3.1	3.4	3.7
TW H3	2	4	7	9	12	14	17	21	1	3	4	6	8	9	12	15	2.3	2.5	2.7	2.9	3.1	2.6	2.9	3.1	3.4	3.6
TW H4	2	3	5	7	9	11	14	18	1	2	3	5	6	8	10	13	2.5	2.7	2.9	3.1	3.3	2.9	3.1	3.4	3.6	3.8
TW I1	4	5	7	7	9	9	10	11	3	3	3	4	4	4	3	2	2.9	3.5	4.3	-	-	4.4	-	-	-	-
TW I2	8	11	16	19	24	28	33	39	5	8	11	14	18	21	25	30	1.6	1.8	2.0	2.2	2.4	1.9	2.1	2.3	2.5	2.7
TW I3	8	12	17	21	27	31	36	43	5	9	13	16	20	24	28	34	1.5	1.7	1.9	2.1	2.2	1.7	1.9	2.1	2.3	2.5
TW I4	9	15	21	26	33	38	45	54	6	11	16	21	27	31	37	45	1.4	1.5	1.7	1.8	2.0	1.5	1.7	1.8	2.0	2.1
TW I5	9	15	22	28	35	41	48	57	6	11	17	22	29	34	40	49	1.3	1.5	1.6	1.8	1.9	1.4	1.6	1.7	1.9	2.0
TW J1	11	16	22	26	33	37	43	51	8	12	17	20	26	30	35	42	1.4	1.6	1.7	1.9	2.1	1.6	1.7	1.9	2.1	2.3
TW J2	10	16	22	27	34	39	45	54	7	12	17	21	27	31	37	45	1.4	1.5	1.7	1.9	2.0	1.5	1.7	1.8	2.0	2.2
TW J3	13	20	29	35	44	50	58	68	9	16	23	29	38	43	51	60	1.2	1.3	1.5	1.6	1.8	1.3	1.4	1.6	1.7	1.9
TW J4	12	19	27	34	43	49	56	66	8	15	22	28	36	42	49	59	1.2	1.3	1.5	1.6	1.8	1.3	1.4	1.6	1.7	1.9
DG A1	5	7	8	9	10	11	11	11	3	4	4	5	5	4	4	2	2.5	3.2	4.8	-	-	3.6	-	-	-	-
DG A2	8	11	14	17	21	23	27	31	6	8	11	13	15	17	20	23	1.8	2.1	2.3	2.7	3.1	2.2	2.4	2.7	3.2	3.7
DG A3	12	16	21	25	31	36	41	49	10	13	17	21	26	30	35	42	1.5	1.7	1.8	2.0	2.1	1.7	1.8	2.0	2.1	2.3
DG B1	5	7	8	10	11	12	13	13	3	4	4	5	5	5	5	3	2.4	3.1	4.1	-	-	3.7	-	-	-	-
DG B2	8	11	15	18	22	24	28	33	6	8	11	13	16	18	21	25	1.8	2.0	2.2	2.5	2.7	2.2	2.4	2.6	2.9	3.3
DG B3	12	16	22	26	33	37	44	53	10	13	18	22	27	31	37	45	1.5	1.6	1.7	1.9	2.0	1.7	1.8	1.9	2.0	2.2
DG C1	7	9	12	14	16	18	19	22	5	6	7	8	9	10	11	11	2.0	2.4	3.0	3.6	4.3	2.7	3.3	4.2	-	-
DG C2	10	14	18	22	26	30	35	41	8	11	14	17	20	23	27	32	1.6	1.8	2.0	2.2	2.4	1.9	2.1	2.3	2.5	2.8
DG C3	14	19	25	30	37	42	49	59	11	15	21	25	31	36	42	51	1.4	1.5	1.6	1.7	1.9	1.5	1.6	1.8	1.9	2.0
SS A1	14	17	20	22	26	28	32	36	12	14	17	19	21	23	26	29	1.8	2.0	2.3	2.6	2.9	2.0	2.3	2.7	3.1	3.6
SS A2	17	22	28	32	39	43	48	56	15	19	24	28	33	37	42	49	1.3	1.5	1.7	1.9	2.1	1.5	1.7	1.9	2.1	2.2
SS A3	12	15	17	19	21	23	25	28	11	12	14	15	17	18	19	21	2.0	2.4	2.8	3.3	3.8	2.3	2.8	3.5	4.6	-
SS A4	17	21	27	31	36	40	45	52	14	18	23	26	31	35	39	45	1.4	1.6	1.8	2.0	2.2	1.5	1.7	2.0	2.2	2.4
SS A5	15	17	19	21	24	26	28	31	13	15	16	18	20	21	22	24	2.0	2.3	2.7	3.1	3.4	2.3	2.7	3.2	3.9	4.7
SS A6	17	22	27	31	37	41	47	54	14	19	23	27	32	36	40	47	1.4	1.6	1.8	1.9	2.1	1.5	1.7	1.9	2.1	2.3
SS A7	13	15	16	18	19	20	22	23	12	13	14	15	16	16	17	17	2.4	2.9	3.5	4.3	-	2.9	3.6	4.8	-	-
SS A8	16	20	25	29	34	38	43	50	14	17	22	25	29	33	37	42	1.4	1.6	1.9	2.1	2.2	1.6	1.8	2.1	2.3	2.5
SS B1	11	14	17	19	22	24	27	31	10	12	14	15	18	19	21	24	1.9	2.2	2.5	2.8	3.1	2.2	2.6	2.9	3.4	3.9
SS B2	15	19	25	29	35	39	45	52	13	17	21	25	30	34	39	45	1.4	1.6	1.8	2.0	2.1	1.5	1.7	1.9	2.1	2.3
SS B3	10	12	15	16	19	20	23	25	9	10	12	13	15	16	17	18	2.1	2.4	2.8	3.3	3.7	2.5	2.9	3.6	4.5	-
SS B4	14	19	24	27	33	37	42	49	12	16	20	23	28	32	36	42	1.5	1.6	1.8	2.0	2.2	1.6	1.8	2.0	2.2	2.4
SS B5	12	14	16	17	19	21	23	26	10	12	13	14	16	17	18	19	2.2	2.6	2.9	3.4	3.8	2.6	3.1	3.6	4.4	-
SS B6	14	18	24	28	33	37	43	50	12	16	20	23	29	32	37	43	1.4	1.6	1.8	2.0	2.2	1.6	1.8	2.0	2.2	2.4
SS B7	11	12	14	15	16	17	19	20	10	10	11	12	13	13	14	14	2.6	3.0	3.6	4.3	-	3.1	3.9	4.9	-	-
SS B8	13	17	22	26	31	35	39	46	11	14	18	22	26	29	33	39	1.5	1.7	1.9	2.1	2.3	1.6	1.9	2.1	2.3	2.5
SS C1	8	11	14	16	20	22	25	29	5	7	9	11	13	14	16	18	1.9	2.1	2.4	2.8	3.0	2.3	2.7	3.2	3.8	4.5
SS C2	10	15	20	24	30	34	40	47	7	11	15	18	23	27	31	37	1.5	1.6	1.8	2.0	2.2	1.7	1.9	2.1	2.3	2.5
SS C3	9	11	13	15	17	19	22	25	7	8	10	11	13	14	15	17	2.2	2.5	2.8	3.1	3.4	2.7	3.1	3.6	4.2	4.8
SS C4	10	14	18	22	27	31	35	42	8	11	14	17	21	24	28	33	1.6	1.8	2.0	2.2	2.3	1.8	2.0	2.2	2.4	2.6
SS D1	14	18	22	25	30	33	36	41	12	15	18	21	24	26	29	32	1.6	1.8	2.1	2.4	2.6	1.7	2.1	2.4	2.9	3.3
SS D2	18	25	32	37	45	50	56	64	15	21	27	32	39	43	49	57	1.2	1.4	1.6	1.7	1.9	1.3	1.5	1.7	1.9	2.0
SS D3	15	18	21	24	27	30	33	37	13	15	18	20	22	24	26	29	1.7	2.0	2.3	2.6	2.9	1.9	2.3	2.7	3.2	3.7
SS D4	18	24	31	36	43	48	54	62	15	20	26	31	37	42	47	55	1.2	1.4	1.6	1.8	1.9	1.3	1.5	1.7	1.9	2.1
SS E1	11	14	17	20	23	26	28	32	9	11	13	15	17	19	20	22	1.7	2.1	2.4	2.8	3.1	2.0	2.5	3.0	3.8	-
SS E2	14	20	26	31	37	42	48	55	11	16	21	25	31	35	40	47	1.3	1.5	1.7	1.9	2.0	1.4	1.6	1.9	2.1	2.3
SS E3	12	14	17	18	21	23	25	28	10	12	14	15	17	18	19	21	2.0	2.3	2.7	3.1	3.5	2.4	2.8	3.3	4.1	-
SS E4	14	19	24	29	35	39	44	51	12	16	20	24	29	32	37	43	1.4	1.6	1.8	2.0	2.1	1.5	1.7	2.0	2.2	2.4

NORMANDIN, QUEBEC

TYPE	SOLAR SAVINGS FRACTION (%) BASE TEMP: 55 F (8227 DD) LCR 100	70	50	40	30	25	20	15	65 F (11364 DD) 100	70	50	40	30	25	20	15	CONSERVATION FACTOR 55 F 100	50	30	20	15	65 F 100	50	30	20	15
WW A1	12	13	15	16	17	18	19	21	11	11	12	13	14	14	15	15	2.8	3.2	3.7	4.4	-	3.3	4.0	-	-	-
WW A2	10	12	15	17	20	23	26	29	8	10	12	13	16	17	19	22	1.9	2.2	2.5	2.8	3.1	2.2	2.5	3.0	3.5	4.0
WW A3	8	11	15	18	22	24	28	33	6	9	11	14	17	19	21	25	1.8	2.0	2.2	2.5	2.8	2.0	2.3	2.6	3.0	3.4
WW A4	7	11	15	18	22	26	30	35	5	8	11	14	17	19	22	26	1.7	1.9	2.1	2.4	2.7	1.8	2.1	2.5	2.8	3.2
WW A5	7	10	15	18	23	27	31	37	4	7	11	14	18	20	24	28	1.6	1.8	2.0	2.3	2.6	1.8	2.0	2.3	2.7	3.0
WW A6	6	10	15	19	24	27	32	38	3	7	11	14	18	21	24	29	1.6	1.8	2.0	2.3	2.5	1.7	2.0	2.3	2.6	2.9
WW B1	6	7	8	9	9	10	10	9	5	5	5	5	5	5	3	1	3.1	4.1	-	-	-	4.2	-	-	-	-
WW B2	8	13	18	22	27	31	36	43	6	10	14	17	22	25	30	36	1.5	1.7	1.9	2.1	2.3	1.6	1.9	2.1	2.3	2.5
WW B3	10	16	22	27	33	38	44	51	8	13	18	22	28	32	37	44	1.4	1.5	1.7	1.9	2.1	1.5	1.7	1.9	2.1	2.3
WW B4	11	17	25	30	39	44	51	60	8	14	21	26	33	39	45	54	1.2	1.4	1.6	1.7	1.9	1.3	1.5	1.7	1.8	2.0
WW B5	10	17	25	31	40	45	53	62	8	14	21	27	35	40	47	57	1.2	1.4	1.5	1.7	1.9	1.3	1.4	1.6	1.8	1.9
WW C1	12	18	24	29	36	41	47	56	9	14	20	25	31	35	41	49	1.3	1.5	1.7	1.9	2.0	1.4	1.6	1.8	2.0	2.2
WW C2	11	17	24	29	37	42	49	57	8	14	20	25	32	36	43	51	1.3	1.4	1.6	1.8	2.0	1.4	1.5	1.7	1.9	2.1
WW C3	14	22	32	39	49	55	63	72	11	19	28	34	44	50	58	67	1.1	1.2	1.4	1.6	1.7	1.1	1.3	1.5	1.6	1.8
WW C4	12	20	29	36	46	52	60	70	9	17	25	32	41	47	55	65	1.1	1.3	1.4	1.6	1.7	1.2	1.3	1.5	1.7	1.8
TW A1	12	13	15	16	17	19	20	22	11	12	13	13	14	15	16	17	2.7	3.0	3.4	3.9	4.4	3.1	3.6	4.4	-	-
TW A2	10	12	15	17	20	22	24	28	8	10	12	13	16	17	19	22	2.1	2.3	2.6	2.9	3.2	2.3	2.7	3.0	3.5	4.0
TW A3	9	11	14	17	20	23	26	31	7	9	11	13	16	18	20	24	1.9	2.1	2.4	2.6	2.9	2.1	2.4	2.7	3.1	3.5
TW A4	7	10	14	16	20	23	27	32	5	8	11	13	16	18	21	25	1.8	2.0	2.2	2.5	2.7	2.0	2.3	2.5	2.9	3.1
TW B1	11	13	14	16	18	19	21	24	10	11	12	13	15	16	17	19	2.5	2.7	3.0	3.4	3.7	2.8	3.2	3.6	4.3	4.9
TW B2	9	11	14	16	19	22	25	29	7	9	11	13	16	17	20	23	2.0	2.2	2.5	2.7	3.0	2.2	2.5	2.8	3.2	3.5
TW B3	8	10	14	16	19	22	25	30	6	8	11	13	16	18	20	24	2.0	2.2	2.4	2.6	2.8	2.1	2.4	2.7	3.0	3.3
TW B4	8	10	13	15	19	22	25	30	6	8	11	13	15	17	20	24	2.0	2.2	2.4	2.6	2.8	2.2	2.4	2.6	2.9	3.2
TW C1	10	12	14	16	18	20	22	26	8	10	12	13	15	16	18	21	2.3	2.5	2.7	3.0	3.3	2.5	2.8	3.1	3.5	3.9
TW C2	9	11	13	15	18	21	24	28	7	9	11	13	15	17	20	23	2.1	2.3	2.5	2.7	2.9	2.3	2.5	2.8	3.1	3.4
TW C3	9	11	13	15	18	20	23	28	7	9	11	13	15	17	19	23	2.2	2.3	2.5	2.8	2.9	2.4	2.6	2.8	3.1	3.3
TW C4	9	11	13	15	18	20	22	26	8	9	11	13	15	17	19	22	2.3	2.5	2.7	2.9	3.1	2.5	2.7	3.0	3.2	3.4
TW D1	5	6	7	8	9	10	10	11	4	4	5	5	5	5	5	4	3.1	3.7	4.8	-	-	3.9	-	-	-	-
TW D2	9	13	17	21	25	29	34	40	7	10	14	17	21	24	28	33	1.6	1.8	2.0	2.2	2.4	1.8	2.0	2.2	2.4	2.7
TW D3	9	13	18	22	27	31	36	43	7	11	15	18	23	26	30	36	1.5	1.7	1.9	2.1	2.3	1.7	1.9	2.1	2.3	2.6
TW D4	11	16	22	27	34	39	46	54	8	13	19	23	30	34	40	48	1.4	1.5	1.7	1.9	2.0	1.4	1.6	1.8	2.0	2.1
TW D5	11	16	23	29	36	42	48	57	8	14	20	25	32	37	43	52	1.3	1.5	1.6	1.8	1.9	1.4	1.5	1.7	1.9	2.0
TW E1	12	17	22	26	33	37	42	50	9	14	19	22	28	32	36	43	1.4	1.6	1.8	2.0	2.1	1.5	1.7	1.9	2.1	2.3
TW E2	11	17	23	27	34	39	45	53	9	14	19	23	29	34	39	47	1.4	1.5	1.7	1.9	2.1	1.5	1.6	1.8	2.0	2.2
TW E3	14	21	30	36	44	50	57	67	12	18	26	31	40	45	52	61	1.2	1.3	1.5	1.7	1.8	1.2	1.4	1.6	1.7	1.9
TW E4	13	20	28	34	43	49	56	66	10	17	24	30	38	44	51	61	1.2	1.3	1.5	1.7	1.8	1.2	1.4	1.6	1.7	1.9
TW F1	10	11	13	14	16	17	19	21	9	10	11	12	13	13	14	15	2.6	2.9	3.3	3.8	4.3	3.0	3.5	4.2	-	-
TW F2	8	10	12	14	17	19	22	25	6	8	10	11	13	15	16	19	2.1	2.4	2.6	3.0	3.3	2.4	2.7	3.1	3.6	4.1
TW F3	6	8	11	14	17	19	23	27	4	6	8	10	13	15	17	20	2.0	2.2	2.4	2.7	3.0	2.2	2.5	2.8	3.2	3.6
TW F4	4	7	10	12	16	19	22	27	2	4	7	9	12	14	17	21	1.9	2.1	2.4	2.6	2.8	2.1	2.4	2.7	3.0	3.3
TW G1	8	9	11	13	15	16	18	21	6	8	9	10	12	13	14	16	2.5	2.7	3.0	3.4	3.7	2.8	3.2	3.6	4.3	4.9
TW G2	5	7	10	12	15	17	20	24	4	5	7	9	11	13	15	18	2.1	2.4	2.6	2.9	3.1	2.4	2.7	3.0	3.4	3.7
TW G3	4	6	9	11	14	17	20	24	2	4	6	8	11	12	15	18	2.1	2.3	2.5	2.8	3.0	2.3	2.6	2.8	3.2	3.4
TW G4	2	4	7	9	12	15	18	22	1	3	5	6	9	11	14	17	2.2	2.4	2.6	2.8	3.0	2.4	2.6	2.9	3.1	3.4
TW H1	5	7	9	10	13	15	17	20	4	5	7	8	10	11	13	15	2.4	2.6	2.9	3.2	3.4	2.7	3.0	3.3	3.8	4.2
TW H2	3	5	7	9	12	14	16	20	1	3	5	6	9	10	12	16	2.3	2.5	2.7	3.0	3.2	2.6	2.8	3.1	3.4	3.7
TW H3	2	4	6	8	10	12	15	19	1	2	4	5	8	9	11	15	2.4	2.6	2.8	3.0	3.2	2.6	2.8	3.1	3.4	3.6
TW H4	1	3	4	6	8	10	12	16	0	1	3	4	6	7	9	12	2.7	2.9	3.0	3.2	3.4	2.9	3.1	3.3	3.6	3.8
TW I1	4	4	5	6	7	7	8	8	2	3	3	3	3	3	3	2	3.3	4.0	-	-	-	1.2		-	-	-
TW I2	7	10	14	17	22	26	30	37	5	0	11	14	18	21	25	30	1.7	1.9	2.1	2.3	2.4	1.8	2.0	2.3	2.5	2.7
TW I3	7	11	16	19	25	29	34	41	5	9	13	16	20	24	28	34	1.6	1.8	1.9	2.1	2.3	1.7	1.9	2.1	2.3	2.5
TW I4	8	13	20	24	31	36	43	51	6	11	16	21	27	31	37	45	1.4	1.5	1.7	1.9	2.0	1.5	1.6	1.8	2.0	2.1
TW I5	8	14	21	26	34	39	46	55	6	11	17	22	29	34	40	49	1.3	1.5	1.6	1.8	2.0	1.4	1.6	1.7	1.9	2.0
TW J1	10	14	20	24	31	35	41	48	7	12	17	20	26	30	35	42	1.4	1.6	1.8	2.0	2.1	1.5	1.7	1.9	2.1	2.3
TW J2	9	14	20	25	32	36	43	51	7	11	17	21	27	31	37	45	1.4	1.6	1.7	1.9	2.1	1.5	1.7	1.8	2.0	2.2
TW J3	12	19	27	33	42	48	56	65	9	16	23	29	38	43	51	60	1.2	1.3	1.5	1.7	1.8	1.2	1.4	1.6	1.7	1.9
TW J4	10	17	26	32	41	47	54	64	8	14	22	28	36	42	49	59	1.2	1.4	1.5	1.7	1.8	1.3	1.4	1.6	1.7	1.9
DG A1	4	6	7	8	9	9	9	0	3	3	4	5	5	4	4	2	2.7	3.6	-	-	-	3.4	4.9	-	-	-
DG A2	7	10	13	16	19	21	24	28	6	8	10	13	15	17	20	23	2.0	2.1	2.4	2.8	3.2	2.1	2.4	2.7	3.2	3.7
DG A3	11	15	20	24	30	34	39	47	9	13	17	20	26	30	35	42	1.6	1.7	1.8	2.0	2.2	1.7	1.8	1.9	2.1	2.3
DG B1	4	5	7	8	9	10	10	10	3	3	4	4	5	5	5	4	2.8	3.3	4.2	-	-	3.4	4.8	-	-	-
DG B2	8	10	13	16	20	23	26	31	6	8	11	13	16	18	21	25	1.9	2.1	2.3	2.6	2.9	2.1	2.3	2.6	2.8	3.3
DG B3	11	15	20	25	31	36	42	50	10	13	18	21	27	31	37	45	1.6	1.6	1.7	1.9	2.1	1.6	1.8	1.9	2.0	2.2
DG C1	6	8	10	12	14	16	18	19	4	6	7	8	9	10	11	12	2.3	2.6	2.9	3.9	-	2.6	3.2	4.1	-	-
DG C2	9	13	17	20	25	29	33	39	8	10	14	16	20	23	27	33	1.7	1.9	2.0	2.3	2.5	1.8	2.0	2.2	2.4	2.8
DG C3	13	18	23	28	36	41	48	57	11	15	20	25	31	36	42	51	1.4	1.5	1.6	1.8	1.9	1.5	1.6	1.7	1.9	2.0
SS A1	13	15	18	20	23	26	28	32	12	14	16	18	21	23	25	27	1.9	2.2	2.5	2.9	3.2	2.0	2.3	2.8	3.3	3.8
SS A2	16	20	26	30	36	40	45	52	14	19	24	27	33	36	41	48	1.4	1.6	1.8	2.0	2.2	1.5	1.7	1.9	2.1	2.3
SS A3	11	13	15	17	19	20	22	23	10	12	14	15	16	17	18	19	2.2	2.6	3.1	3.9	4.9	2.3	2.9	3.7	-	-
SS A4	15	19	24	28	33	37	41	48	14	18	22	26	30	34	38	43	1.5	1.7	1.9	2.1	2.3	1.5	1.8	2.0	2.3	2.5
SS A5	14	15	18	19	21	23	25	27	13	14	16	17	19	20	22	23	2.1	2.5	2.9	3.4	4.0	2.3	2.7	3.3	4.2	-
SS A6	15	20	25	29	35	38	43	50	14	18	23	26	32	35	39	45	1.4	1.6	1.8	2.1	2.2	1.5	1.7	1.9	2.2	2.4
SS A7	12	14	15	16	17	18	19	19	12	13	14	14	15	16	16	16	2.7	3.2	4.1	-	-	2.8	3.7	-	-	-
SS A8	14	18	23	26	31	35	39	45	13	17	21	24	29	31	35	40	1.5	1.7	2.0	2.2	2.4	1.6	1.8	2.1	2.4	2.7
SS B1	10	12	15	17	20	22	24	27	9	11	13	15	17	19	20	23	2.1	2.3	2.7	3.1	3.4	2.2	2.6	3.0	3.6	4.2
SS B2	14	18	23	27	33	37	42	49	12	16	21	24	30	33	38	44	1.5	1.6	1.8	2.0	2.2	1.5	1.7	1.9	2.1	2.3
SS B3	9	11	13	14	16	18	19	21	8	10	11	12	14	15	16	17	2.3	2.7	3.1	3.8	4.5	2.4	3.0	3.7	4.9	-
SS B4	13	17	22	25	30	34	39	45	12	15	20	23	28	31	35	41	1.5	1.7	1.9	2.1	2.3	1.6	1.8	2.0	2.3	2.5
SS B5	11	12	14	15	17	19	20	22	10	11	13	14	15	16	17	19	2.5	2.8	3.2	3.7	4.3	2.6	3.1	3.7	4.6	-
SS B6	13	17	22	25	31	35	40	46	11	15	20	23	28	31	36	42	1.5	1.7	1.9	2.1	2.3	1.6	1.8	2.0	2.2	2.4
SS B7	10	11	12	13	14	15	16	17	9	10	11	12	12	13	13	14	2.9	3.4	4.1	-	-	3.1	3.8	-	-	-
SS B8	12	16	20	23	28	32	36	42	11	14	18	21	25	28	32	38	1.6	1.8	2.0	2.2	2.4	1.6	1.9	2.1	2.4	2.6
SS C1	7	9	12	14	17	20	22	26	5	7	9	11	13	14	16	18	2.0	2.3	2.6	2.9	3.3	2.2	2.6	3.1	3.8	4.5
SS C2	9	13	18	22	28	32	37	44	7	11	15	18	23	27	31	37	1.5	1.7	1.9	2.1	2.3	1.6	1.8	2.1	2.3	2.5
SS C3	8	10	12	13	16	17	20	23	7	8	10	11	12	14	15	17	2.4	2.6	2.9	3.3	3.6	2.7	3.0	3.5	4.1	4.8
SS C4	9	13	17	20	25	29	33	40	7	10	14	17	21	24	28	33	1.7	1.8	2.0	2.2	2.4	1.8	2.0	2.2	2.4	2.6
SS D1	13	16	20	23	26	29	32	36	12	15	18	20	23	25	27	30	1.6	1.9	2.3	2.6	3.0	1.7	2.1	2.5	3.1	3.7
SS D2	16	22	29	34	41	46	52	60	15	20	27	31	38	42	48	55	1.2	1.4	1.6	1.8	2.0	1.3	1.5	1.7	1.9	2.1
SS D3	13	16	19	21	25	27	29	33	12	15	17	19	22	23	25	28	1.8	2.1	2.5	2.9	3.3	1.9	2.3	2.8	3.4	4.1
SS D4	16	22	28	33	40	45	50	58	15	20	26	30	37	41	46	53	1.3	1.5	1.7	1.9	2.0	1.3	1.5	1.7	2.0	2.2
SS E1	10	12	15	17	20	22	24	27	8	10	13	14	16	18	19	20	1.9	2.2	2.7	3.2	3.9	2.0	2.5	3.2	4.2	-
SS E2	13	18	24	28	34	38	44	51	11	16	21	25	30	34	39	45	1.4	1.6	1.8	2.0	2.2	1.4	1.7	1.9	2.1	2.4
SS E3	11	13	15	17	19	20	22	25	10	11	13	14	16	17	19	20	2.2	2.5	2.9	3.5	4.1	2.3	2.8	3.5	4.4	-
SS E4	13	17	22	26	32	36	40	47	11	15	20	23	28	32	36	42	1.5	1.7	1.9	2.1	2.3	1.5	1.8	2.0	2.2	2.5

SWIFT CURRENT, SASKATCHEWAN

TYPE	SOLAR SAVINGS FRACTION (%) BASE TEMP: 55 F (7164 DD)								65 F (10076 DD)								CONSERVATION FACTOR 55 F					65 F				
LCR	100	70	50	40	30	25	20	15	100	70	50	40	30	25	20	15	100	50	30	20	15	100	50	30	20	15
WW A1	16	18	21	23	26	28	31	35	14	16	18	20	22	24	26	29	1.9	2.2	2.4	2.8	3.0	2.2	2.5	2.8	3.2	3.6
WW A2	15	19	24	28	33	36	41	47	13	16	20	23	27	30	34	39	1.5	1.7	1.9	2.2	2.4	1.6	1.9	2.1	2.4	2.7
WW A3	15	20	25	29	35	39	44	51	12	16	20	24	29	32	37	43	1.4	1.6	1.8	2.0	2.2	1.5	1.7	2.0	2.2	2.5
WW A4	14	20	26	30	36	41	46	53	11	16	21	25	30	34	39	45	1.3	1.5	1.7	2.0	2.1	1.4	1.7	1.9	2.1	2.4
WW A5	14	20	26	31	38	42	48	56	11	16	21	25	31	35	40	47	1.3	1.5	1.7	1.9	2.1	1.4	1.6	1.8	2.1	2.3
WW A6	14	20	27	32	39	43	49	57	10	16	21	26	32	36	41	48	1.3	1.5	1.7	1.9	2.0	1.4	1.6	1.8	2.0	2.2
WW B1	11	14	17	19	22	24	26	29	9	11	13	14	16	17	19	20	1.9	2.2	2.6	3.2	3.7	2.2	2.7	3.3	4.4	-
WW B2	15	21	28	33	40	45	51	59	12	17	23	28	34	39	44	52	1.3	1.4	1.6	1.8	2.0	1.4	1.5	1.7	1.9	2.1
WW B3	18	25	33	39	47	52	59	67	15	21	28	33	41	46	52	60	1.1	1.3	1.5	1.7	1.9	1.2	1.4	1.6	1.8	2.0
WW B4	19	27	36	42	51	57	64	73	15	22	31	37	46	51	58	67	1.1	1.3	1.4	1.6	1.7	1.1	1.3	1.5	1.7	1.8
WW B5	18	26	36	43	52	58	65	74	14	22	31	37	46	52	60	69	1.1	1.2	1.4	1.6	1.7	1.1	1.3	1.5	1.6	1.8
WW C1	20	27	36	42	50	55	62	70	16	23	30	36	44	49	55	63	1.1	1.3	1.5	1.7	1.8	1.2	1.4	1.6	1.8	1.9
WW C2	19	26	35	41	50	55	62	71	15	22	30	36	44	49	56	65	1.1	1.3	1.5	1.6	1.8	1.2	1.3	1.5	1.7	1.9
WW C3	22	32	43	50	60	66	74	82	18	28	38	45	55	61	69	78	1.0	1.1	1.3	1.5	1.6	1.0	1.2	1.4	1.5	1.7
WW C4	20	30	40	48	57	64	71	80	16	25	35	42	52	59	66	75	1.0	1.2	1.3	1.5	1.7	1.1	1.2	1.4	1.6	1.7
TW A1	16	18	21	23	26	29	32	36	14	16	18	20	22	24	26	30	1.9	2.1	2.4	2.7	2.9	2.1	2.4	2.7	3.1	3.4
TW A2	15	19	23	26	31	34	38	44	12	15	19	22	25	28	32	37	1.6	1.8	2.0	2.2	2.4	1.7	2.0	2.2	2.5	2.7
TW A3	14	18	23	27	32	36	40	47	11	15	19	22	27	30	34	40	1.5	1.7	1.9	2.1	2.3	1.6	1.8	2.1	2.3	2.6
TW A4	13	18	23	27	32	36	41	48	10	14	19	22	27	30	35	41	1.4	1.6	1.8	2.0	2.2	1.6	1.8	2.0	2.2	2.4
TW B1	15	18	21	23	27	30	33	38	13	15	18	20	23	25	28	32	1.8	2.0	2.3	2.5	2.7	2.0	2.3	2.5	2.8	3.1
TW B2	14	17	22	25	30	33	38	44	11	14	18	21	25	28	32	37	1.6	1.8	2.0	2.2	2.4	1.7	1.9	2.2	2.4	2.6
TW B3	13	17	22	25	30	34	39	45	10	14	18	21	25	28	33	38	1.5	1.7	1.9	2.1	2.3	1.7	1.9	2.1	2.3	2.5
TW B4	12	16	21	24	29	33	37	44	10	13	17	20	24	28	32	38	1.6	1.8	2.0	2.2	2.3	1.7	1.9	2.1	2.3	2.5
TW C1	14	17	20	23	27	30	34	39	12	14	17	20	23	25	29	33	1.7	1.9	2.2	2.4	2.6	1.9	2.1	2.4	2.6	2.8
TW C2	13	16	20	23	28	31	35	41	11	13	17	19	23	26	30	35	1.6	1.8	2.0	2.2	2.4	1.8	2.0	2.2	2.4	2.6
TW C3	12	16	19	22	27	30	34	40	10	13	16	19	23	26	29	35	1.7	1.9	2.1	2.3	2.4	1.9	2.0	2.2	2.5	2.6
TW C4	12	15	18	21	25	28	32	37	11	13	16	18	22	24	27	32	1.8	2.0	2.2	2.4	2.6	2.0	2.2	2.4	2.6	2.8
TW D1	10	12	15	17	20	22	24	27	7	9	11	13	15	16	18	20	2.0	2.3	2.6	3.0	3.4	2.3	2.7	3.2	3.8	4.6
TW D2	15	21	27	31	37	42	47	55	12	17	22	26	32	36	41	48	1.3	1.5	1.7	1.9	2.1	1.5	1.6	1.8	2.0	2.2
TW D3	16	22	28	33	40	45	50	58	13	18	24	28	34	39	44	51	1.3	1.5	1.7	1.8	2.0	1.4	1.6	1.8	2.0	2.2
TW D4	18	25	32	38	46	52	59	67	14	21	28	33	41	46	53	61	1.2	1.3	1.5	1.7	1.8	1.2	1.4	1.6	1.8	1.9
TW D5	17	25	33	39	48	53	60	70	14	21	29	34	43	48	55	64	1.1	1.3	1.5	1.6	1.8	1.2	1.4	1.5	1.7	1.9
TW E1	19	25	32	38	45	50	56	65	15	21	28	33	40	44	50	58	1.2	1.4	1.6	1.7	1.9	1.3	1.5	1.7	1.9	2.0
TW E2	18	25	33	39	46	52	58	67	15	21	28	33	41	46	52	60	1.2	1.3	1.5	1.7	1.8	1.2	1.4	1.6	1.8	1.9
TW E3	22	31	40	47	56	62	69	78	18	27	35	42	51	57	64	73	1.0	1.2	1.4	1.5	1.7	1.1	1.3	1.4	1.6	1.8
TW E4	20	29	39	45	55	61	68	77	17	25	34	40	49	55	63	72	1.1	1.2	1.4	1.6	1.7	1.1	1.3	1.4	1.6	1.8
TW F1	14	16	20	22	25	28	31	35	12	14	16	18	21	23	25	29	1.9	2.1	2.4	2.6	2.9	2.1	2.4	2.7	3.0	3.3
TW F2	12	16	20	23	28	31	35	41	10	13	16	19	23	26	29	34	1.6	1.8	2.0	2.3	2.5	1.8	2.0	2.3	2.5	2.8
TW F3	11	15	20	23	29	32	37	43	9	12	16	19	23	26	30	36	1.5	1.7	2.0	2.2	2.4	1.7	1.9	2.1	2.4	2.6
TW F4	9	13	18	22	27	31	36	43	7	10	14	18	22	26	30	36	1.5	1.7	1.9	2.1	2.3	1.7	1.9	2.1	2.3	2.5
TW G1	12	15	18	21	24	27	30	35	10	12	15	17	20	22	25	29	1.8	2.0	2.3	2.5	2.7	2.0	2.3	2.5	2.8	3.1
TW G2	10	13	18	21	25	29	33	39	8	10	14	17	21	23	27	32	1.6	1.8	2.1	2.3	2.5	1.8	2.0	2.3	2.5	2.7
TW G3	8	12	16	20	24	28	32	39	6	9	13	15	20	23	27	32	1.6	1.8	2.0	2.2	2.4	1.8	2.0	2.2	2.4	2.6
TW G4	6	10	14	17	21	25	29	36	4	7	10	13	17	20	24	30	1.7	1.9	2.1	2.3	2.5	1.9	2.1	2.3	2.5	2.7
TW H1	9	12	15	18	22	25	29	34	7	9	12	14	18	20	23	28	1.8	2.0	2.2	2.5	2.7	2.0	2.2	2.5	2.7	2.9
TW H2	7	10	13	16	21	24	28	34	5	7	10	13	16	19	23	28	1.8	2.0	2.2	2.4	2.6	2.0	2.2	2.4	2.6	2.8
TW H3	5	8	11	14	18	21	25	31	3	6	9	11	14	17	20	25	1.9	2.1	2.2	2.4	2.6	2.1	2.2	2.4	2.7	2.8
TW H4	4	6	9	11	15	17	21	26	2	4	7	9	12	14	17	21	2.1	2.3	2.4	2.6	2.8	2.3	2.5	2.7	2.9	3.0
TW I1	8	10	12	14	17	19	21	24	6	7	9	10	12	14	15	17	2.0	2.4	2.7	3.2	3.6	2.4	2.8	3.3	4.0	4.8
TW I2	12	17	23	28	34	38	44	52	10	14	19	23	29	33	38	45	1.4	1.6	1.7	1.9	2.1	1.5	1.7	1.9	2.1	2.3
TW I3	13	19	25	30	37	41	47	55	10	15	21	25	31	36	41	49	1.3	1.5	1.7	1.9	2.0	1.4	1.6	1.8	2.0	2.2
TW I4	15	22	29	35	43	49	56	65	12	18	25	30	38	43	50	58	1.2	1.4	1.5	1.7	1.9	1.3	1.4	1.6	1.8	1.9
TW I5	15	22	31	37	45	51	58	67	12	18	26	32	40	45	52	62	1.2	1.3	1.5	1.7	1.8	1.2	1.4	1.6	1.7	1.9
TW J1	16	23	30	35	43	48	55	63	13	19	26	30	37	42	48	56	1.2	1.4	1.6	1.8	1.9	1.3	1.5	1.7	1.9	2.0
TW J2	16	23	30	36	44	49	56	65	13	19	26	31	38	43	50	58	1.2	1.4	1.5	1.7	1.9	1.3	1.4	1.6	1.8	2.0
TW J3	19	28	38	44	54	60	67	76	16	24	33	39	49	55	62	71	1.1	1.2	1.4	1.6	1.7	1.1	1.3	1.4	1.6	1.8
TW J4	18	26	36	43	52	58	66	75	14	22	31	38	47	53	60	70	1.1	1.2	1.4	1.6	1.7	1.1	1.3	1.4	1.6	1.8
DG A1	10	12	15	17	20	21	23	24	7	9	11	13	15	16	16	17	1.8	2.2	2.9	4.2	-	2.1	2.6	3.6	-	-
DG A2	12	16	21	25	30	33	37	43	10	13	17	20	25	28	31	36	1.5	1.7	2.0	2.3	2.6	1.7	1.9	2.1	2.5	2.8
DG A3	16	21	28	33	40	45	51	59	13	18	24	29	35	40	46	54	1.3	1.5	1.6	1.8	2.0	1.4	1.5	1.7	1.9	2.1
DG B1	10	13	16	18	22	23	26	28	7	9	12	14	16	17	19	20	1.8	2.1	2.6	3.2	4.2	2.0	2.4	3.0	4.2	-
DG B2	13	17	22	26	31	35	40	47	10	14	18	21	26	30	34	40	1.5	1.7	1.9	2.1	2.4	1.6	1.8	2.0	2.3	2.6
DG B3	16	22	29	34	42	47	54	63	14	19	25	30	37	42	49	57	1.3	1.4	1.6	1.7	1.9	1.4	1.5	1.6	1.8	2.0
DG C1	12	16	20	23	28	30	34	38	9	12	16	18	22	24	27	29	1.6	1.8	2.1	2.5	3.0	1.8	2.0	2.4	3.0	3.8
DG C2	15	20	26	31	37	42	47	55	12	16	22	26	32	36	41	48	1.4	1.5	1.7	1.9	2.1	1.5	1.6	1.8	2.1	2.3
DG C3	19	25	33	39	47	53	60	70	16	22	29	34	42	48	55	64	1.2	1.3	1.5	1.6	1.8	1.3	1.4	1.5	1.7	1.9
SS A1	17	21	25	28	32	35	39	44	15	19	22	25	29	32	35	39	1.5	1.8	2.1	2.3	2.5	1.6	1.9	2.2	2.5	2.9
SS A2	21	27	33	38	45	50	55	63	19	25	31	35	41	46	51	58	1.2	1.4	1.6	1.8	2.0	1.3	1.5	1.7	1.9	2.1
SS A3	15	18	22	24	27	29	32	36	14	16	19	21	24	26	28	31	1.7	2.0	2.4	2.8	3.1	1.8	2.2	2.6	3.2	3.8
SS A4	20	26	32	36	42	46	52	59	18	23	29	33	39	43	47	54	1.3	1.5	1.7	1.9	2.1	1.3	1.6	1.8	2.0	2.2
SS A5	17	20	24	26	30	32	35	39	16	18	21	24	27	29	31	34	1.7	2.0	2.3	2.6	2.9	1.8	2.1	2.5	2.9	3.3
SS A6	21	26	33	37	44	48	54	61	19	24	30	34	40	44	49	56	1.2	1.4	1.6	1.8	2.0	1.3	1.5	1.7	1.9	2.1
SS A7	16	18	20	22	24	26	28	30	14	16	18	20	22	23	24	26	2.0	2.3	2.8	3.3	3.8	2.1	2.6	3.1	3.9	-
SS A8	20	25	30	35	41	45	50	56	18	22	28	32	37	41	45	51	1.3	1.5	1.7	2.0	2.1	1.4	1.6	1.8	2.1	2.3
SS B1	14	18	22	24	28	31	35	40	13	16	19	21	25	27	30	35	1.6	1.9	2.2	2.4	2.7	1.8	2.0	2.3	2.7	3.0
SS B2	19	24	30	35	42	46	52	60	17	22	27	32	38	42	48	55	1.3	1.5	1.7	1.8	2.0	1.3	1.5	1.7	1.9	2.1
SS B3	13	16	19	22	25	27	30	34	11	14	17	19	22	23	26	29	1.8	2.1	2.4	2.7	3.0	1.9	2.2	2.6	3.1	3.6
SS B4	18	23	29	33	39	44	49	57	16	21	26	30	36	40	45	52	1.3	1.5	1.7	1.9	2.1	1.4	1.6	1.8	2.0	2.2
SS B5	14	17	20	22	25	27	30	34	13	15	17	19	22	24	26	29	1.9	2.1	2.4	2.8	3.0	2.0	2.3	2.7	3.1	3.5
SS B6	18	23	29	34	40	45	50	58	16	21	26	30	36	41	46	53	1.3	1.5	1.7	1.9	2.0	1.4	1.6	1.8	2.0	2.2
SS B7	13	15	17	19	21	23	25	28	12	13	15	17	19	20	21	23	2.1	2.4	2.8	3.3	3.7	2.3	2.7	3.2	3.8	4.5
SS B8	17	22	27	31	37	41	47	54	15	19	24	28	34	37	42	49	1.4	1.6	1.8	2.0	2.1	1.4	1.6	1.9	2.1	2.3
SS C1	13	17	22	25	30	34	38	44	10	13	17	20	24	27	31	36	1.5	1.7	2.0	2.2	2.4	1.7	1.9	2.2	2.5	2.8
SS C2	16	22	29	34	41	46	52	60	13	18	24	28	35	39	45	53	1.3	1.4	1.6	1.8	2.0	1.4	1.5	1.7	1.9	2.1
SS C3	12	15	19	22	26	28	32	37	10	13	16	18	21	23	26	31	1.7	2.0	2.2	2.5	2.7	2.0	2.2	2.5	2.7	3.0
SS C4	15	20	26	30	37	41	47	54	12	17	22	26	31	35	41	48	1.4	1.5	1.7	1.9	2.1	1.5	1.7	1.8	2.0	2.2
SS D1	19	23	28	32	37	41	45	51	16	21	25	28	33	36	40	44	1.4	1.6	1.9	2.1	2.3	1.5	1.7	2.0	2.4	2.7
SS D2	23	30	38	44	52	57	63	71	20	27	35	40	47	52	58	66	1.1	1.3	1.5	1.7	1.8	1.1	1.3	1.6	1.8	1.9
SS D3	18	22	26	30	34	37	41	46	16	20	24	26	30	33	36	40	1.5	1.8	2.0	2.3	2.5	1.6	1.9	2.2	2.6	2.9
SS D4	22	29	37	42	50	55	61	69	20	27	34	39	46	51	56	64	1.1	1.3	1.5	1.7	1.9	1.2	1.4	1.6	1.8	2.0
SS E1	15	19	24	27	31	35	38	43	13	16	20	23	27	29	32	36	1.5	1.8	2.0	2.3	2.6	1.6	1.9	2.3	2.7	3.1
SS E2	19	26	33	38	45	50	56	64	17	23	29	34	41	45	51	58	1.2	1.4	1.6	1.8	1.9	1.2	1.5	1.7	1.9	2.1
SS E3	15	18	22	24	28	30	34	38	13	16	19	21	24	26	29	32	1.7	2.0	2.3	2.6	2.8	1.8	2.1	2.5	2.9	3.3
SS E4	18	24	31	35	42	46	52	60	16	21	27	32	38	42	47	54	1.3	1.5	1.7	1.8	2.0	1.3	1.5	1.7	2.0	2.2

APPENDIX 2
SENSITIVITY DATA

Data are presented in this appendix that express the sensitivity of the annual *SSF* for direct-gain, sunspace, and thermal storage wall systems to selected design parameters. In any given set of data, one or two parameters are varied while the others are held fixed at their reference values. Each set of data is presented for several different U.S. reference cities. For the direct-gain data, the cities are Albuquerque, New Mexico; Boston, Massachusetts; Madison, Wisconsin; Medford, Oregon; Nashville, Tennessee; and Santa Maria, California. For sunspace and thermal storage wall data, the cities are Albuquerque, New Mexico; Bismarck, North Dakota; Boston, Massachusetts; Ely, Nevada; Nashville, Tennessee; and Seattle, Washington. The cities were chosen in order to represent a wide geographical and climatological range. To apply the data, one of the reference cities must be chosen to represent the city of interest on the basis of climate similarity. Two gross measures are of use in making the choice: the 65°F-base heating degree-days and the solar radiation incident on a vertical surface. These data are available as monthly averages for numerous U.S. and Canadian locations in Appendix 3, where they are tabulated as D65 and VS, respectively. Figure A2-1 shows the January values of D65 and VS for each reference city.

Each parameter variation is usually represented by two different values of the load collector ratio *(LCR)*. In applying the sensitivity data, the *SSF* nearest that of the design being analyzed should be used.

Dots on the curves represent the parameter values corresponding to one of the reference designs.

Two applications of sensitivity data are important. One is quantitative as a final step in a design analysis. In this application, an analysis is performed on the basis of the reference design that most nearly resembles the design of interest. Then, the annual *SSF* is corrected by use of sensitivity data. Refer to Chapters 5 and 7 for details of this procedure.

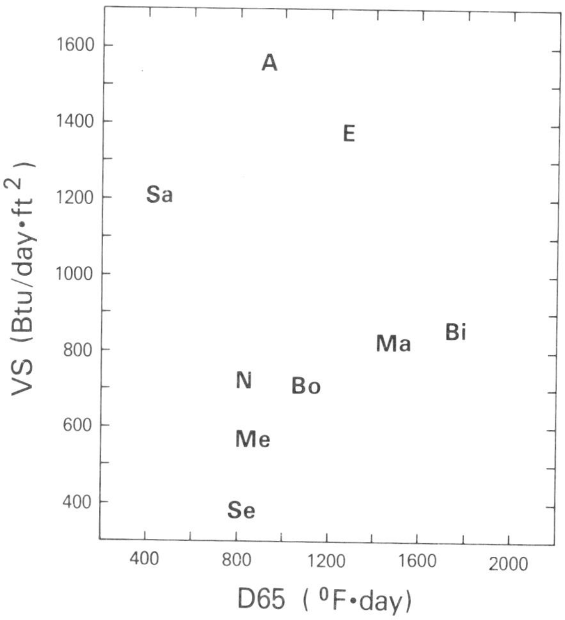

Figure A2-1. The January 65°F-base heating degree-days, D65, and the daily average January solar radiation incident on a vertical surface, VS, for the sensitivity data reference cities. A = Albuquerque, Bi = Bismarck, Bo = Boston, E = Ely, Ma = Madison, Me = Medford, N = Nashville, Sa = Santa Maria, and Se = Seattle.

Another application of sensitivity data is qualitative, wherein the data serve as a guide, perhaps very early in the design process, concerning the relative significance of various design parameters and their preferred values. Discussions of the sensitivity data in this qualitative vein are given in the respective chapters on Direct Gain (Chapter 12), Sunspaces (Chapter 13), and Thermal Storage Walls (Chapter 14). The discussions are illustrated by selected data from the full set, which follows.

Following is an index to the sensitivity data.

DIRECT GAIN
MASS THICKNESS

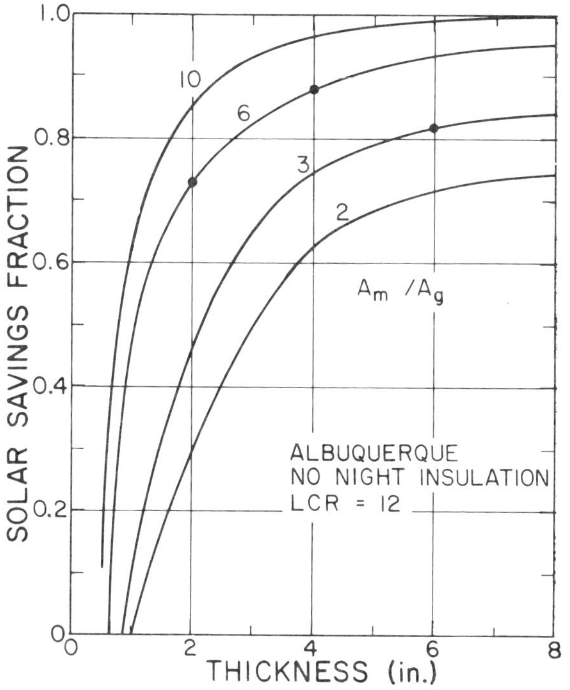

Figure A2-2

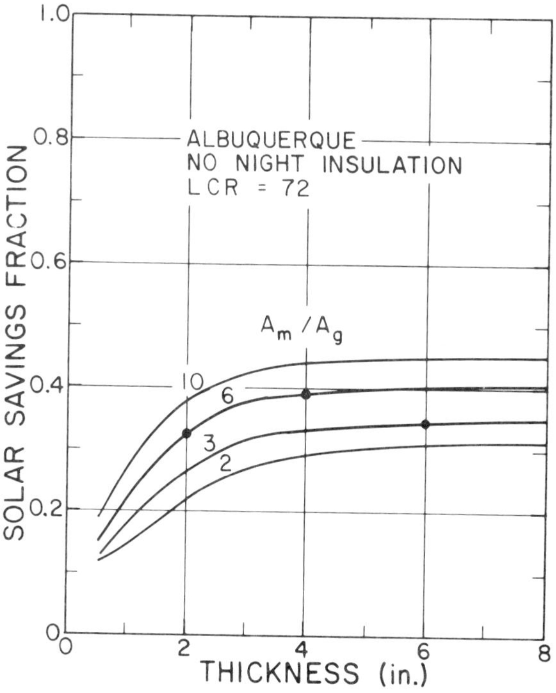

Figure A2-3

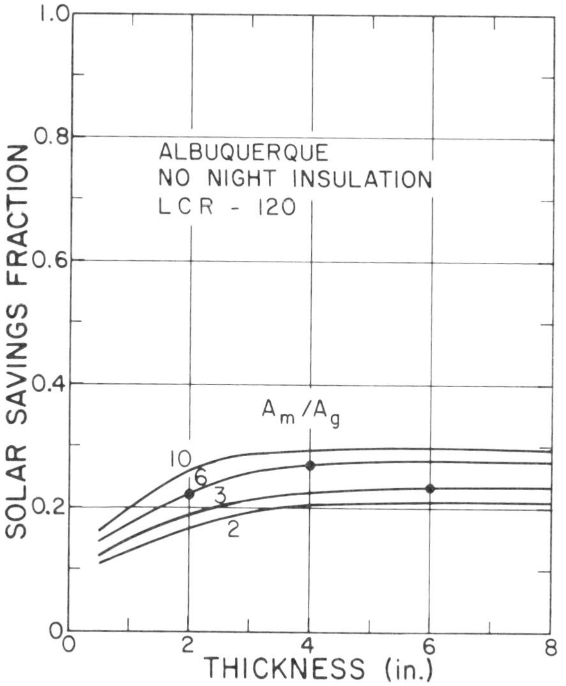

Figure A2-4

Figures A2-2 through A2-34. Annual heating performance sensitivity to the thickness of thermal storage mass for the double-glazed configuration, both without and with R9 night insulation, for various values of the mass-area-to-glazing-area ratio *(A_m/A_g)*. The dots represent the reference conditions.

DIRECT GAIN
MASS THICKNESS

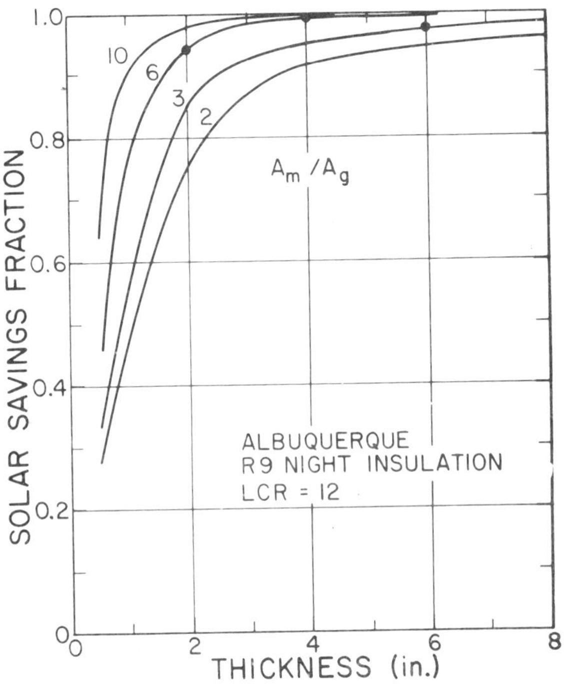

Figure A2-5

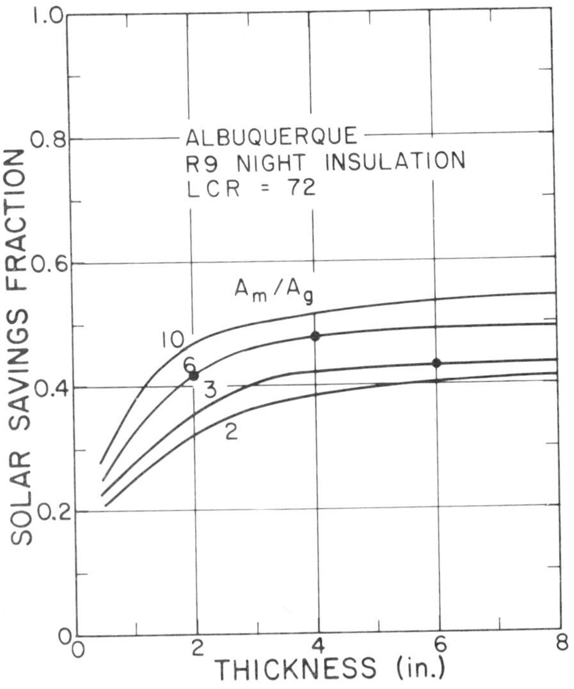

Figure A2-6

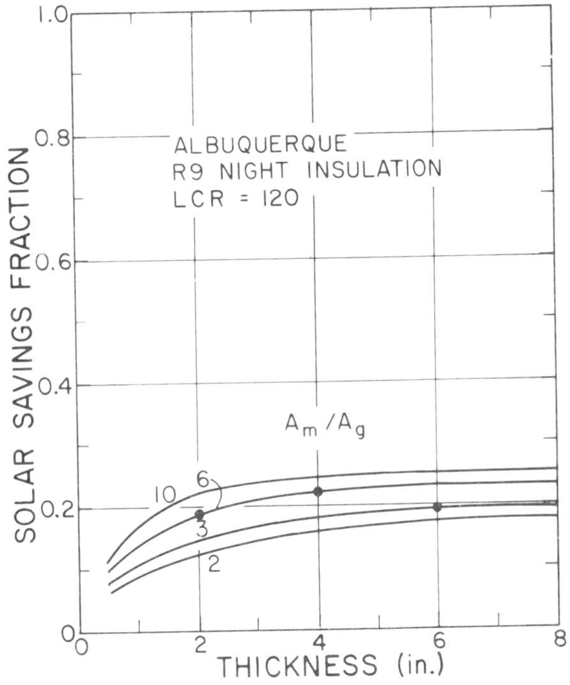

Figure A2-7

Figures A2-2 through A2-34. Annual heating performance sensitivity to the thickness of thermal storage mass for the double-glazed configuration, both without and with R9 night insulation, for various values of the mass-area-to-glazing-area ratio *(A_m/A_g)*. The dots represent the reference conditions.

DIRECT GAIN
MASS THICKNESS

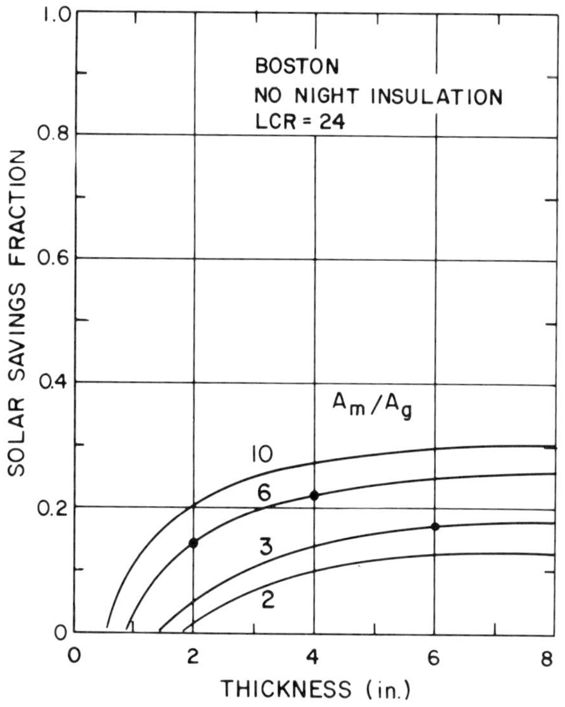

Figure A2-8

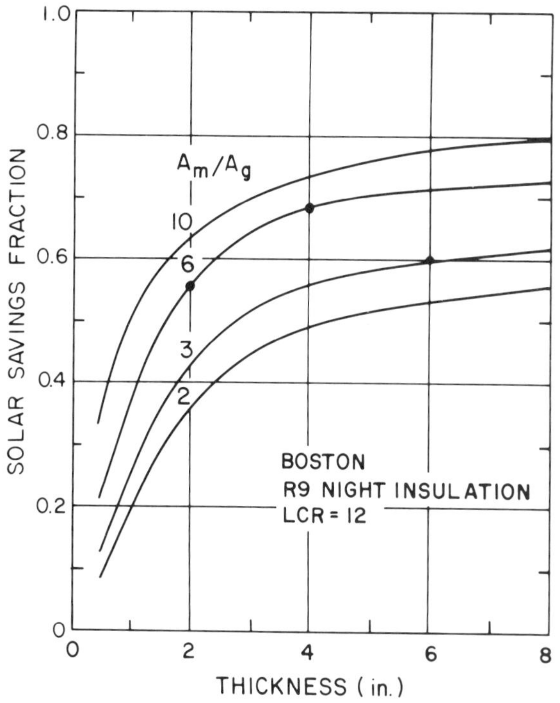

Figure A2-9

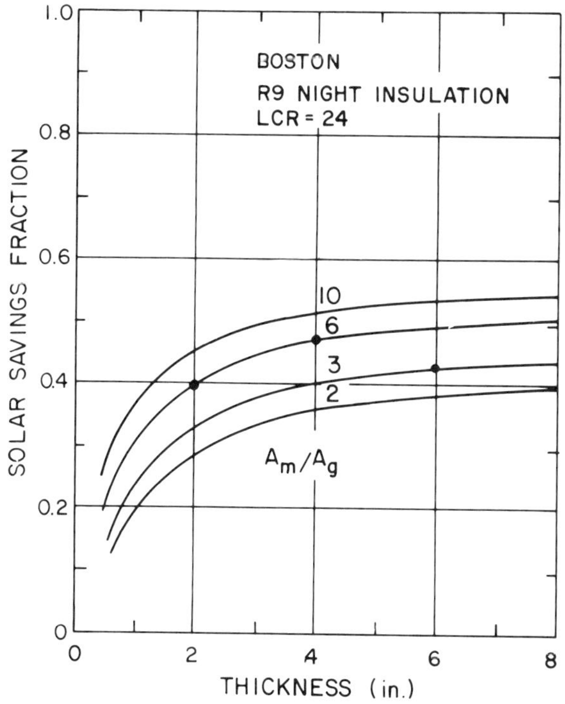

Figure A2-10

Figures A2-2 through A2-34. Annual heating performance sensitivity to the thickness of thermal storage mass for the double-glazed configuration, both without and with R9 night insulation, for various values of the mass-area-to-glazing-area ratio (A_m/A_g). The dots represent the reference conditions.

DIRECT GAIN
MASS THICKNESS

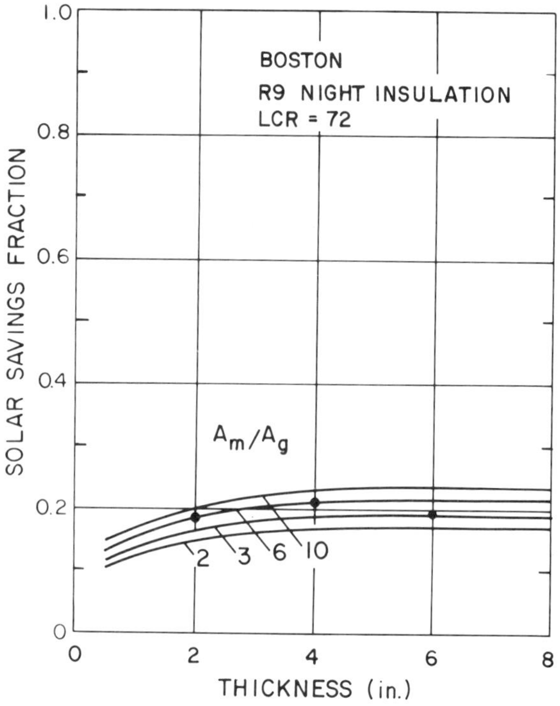

Figure A2-11

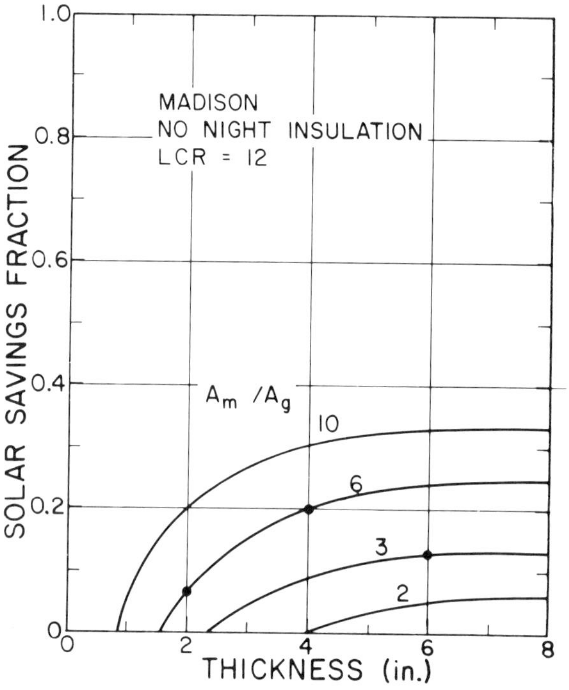

Figure A2-12

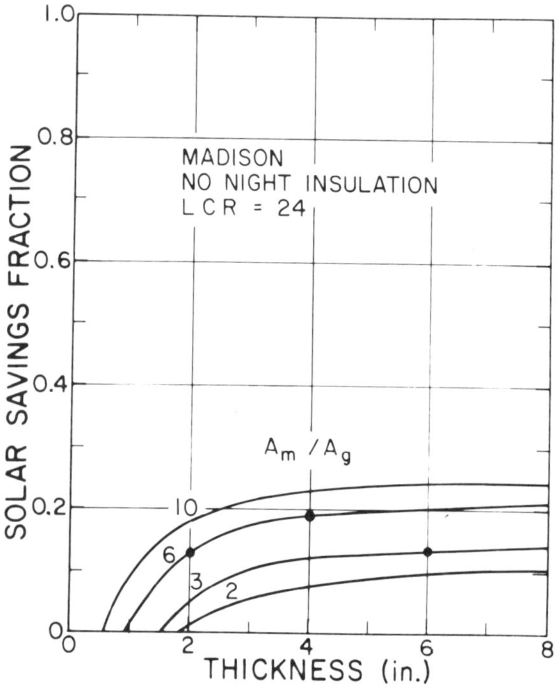

Figure A2-13

Figures A2-2 through A2-34. Annual heating performance sensitivity to the thickness of thermal storage mass for the double-glazed configuration, both without and with R9 night insulation, for various values of the mass-area-to-glazing-area ratio *(A_m/A_g)*. The dots represent the reference conditions.

DIRECT GAIN
MASS THICKNESS

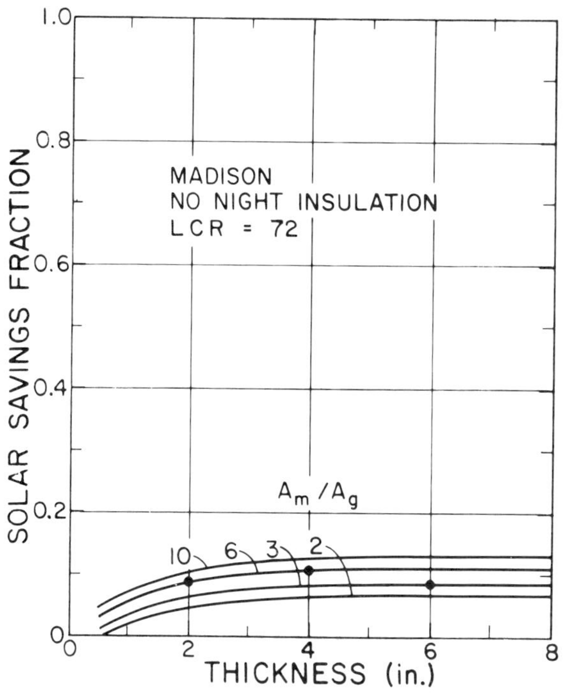

Figure A2-14

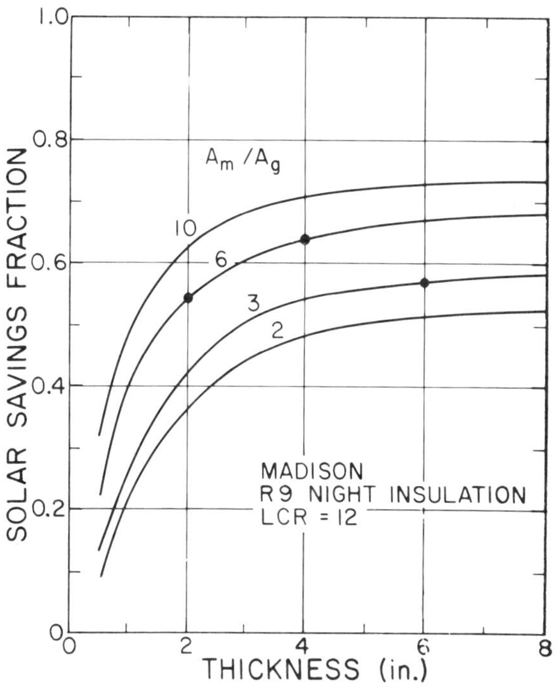

Figure A2-15

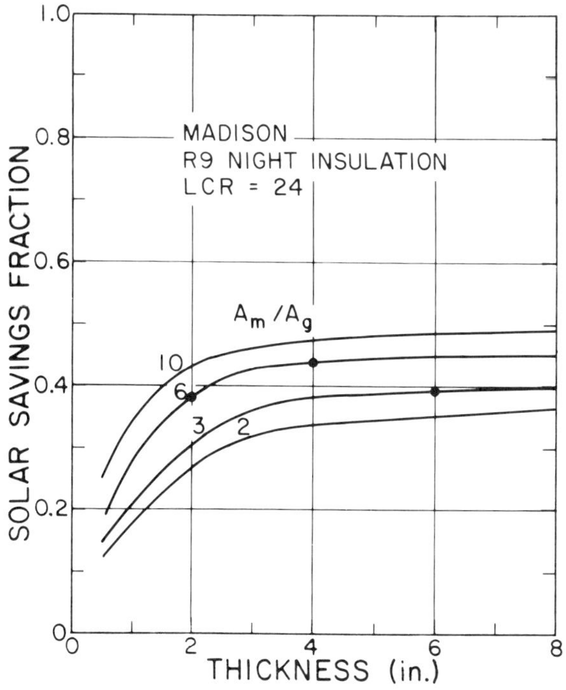

Figure A2-16

Figures A2-2 through A2-34. Annual heating performance sensitivity to the thickness of thermal storage mass for the double-glazed configuration, both without and with R9 night insulation, for various values of the mass-area-to-glazing-area ratio *(A_m/A_g)*. The dots represent the reference conditions.

DIRECT GAIN
MASS THICKNESS

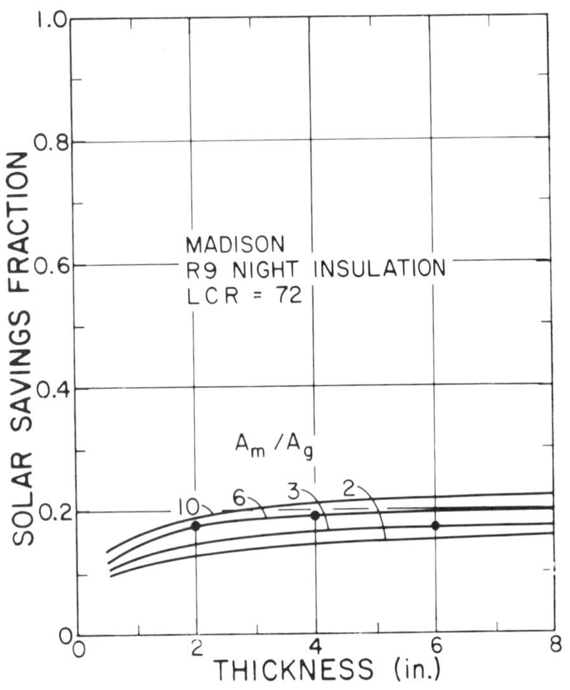

Figure A2-17

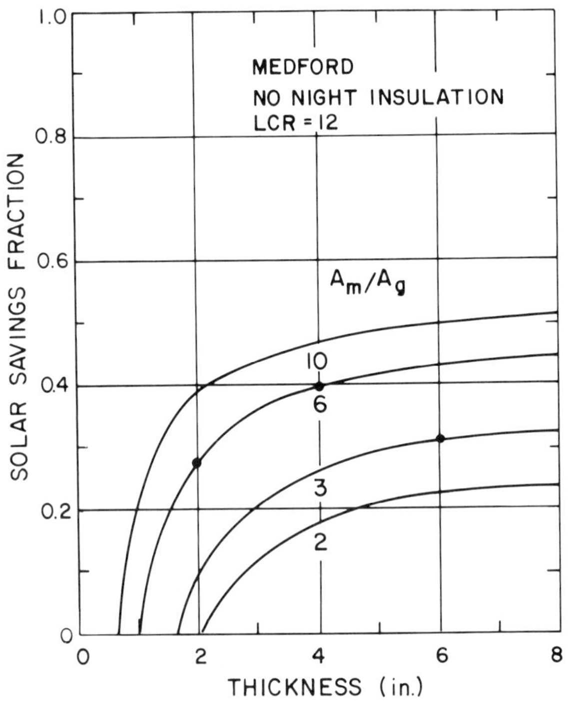

Figure A2-18

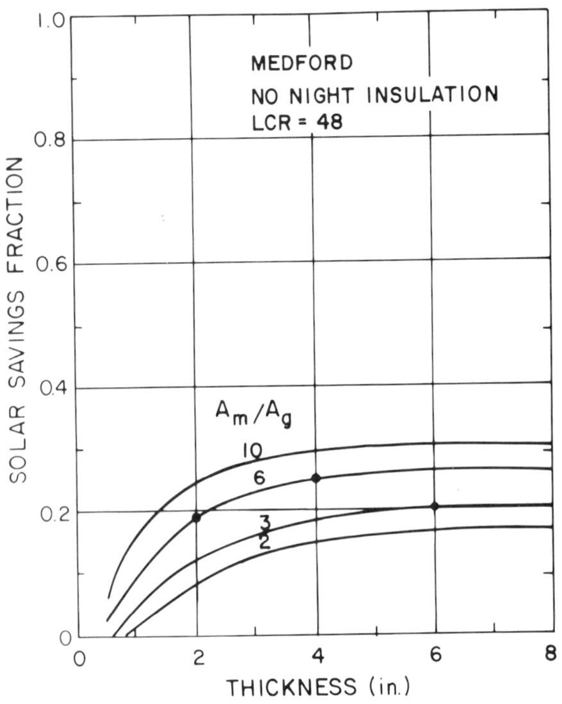

Figure A2-19

Figures A2-2 through A2-34. Annual heating performance sensitivity to the thickness of thermal storage mass for the double-glazed configuration, both without and with R9 night insulation, for various values of the mass-area-to-glazing-area ratio *(A_m/A_g)*. The dots represent the reference conditions.

DIRECT GAIN
MASS THICKNESS

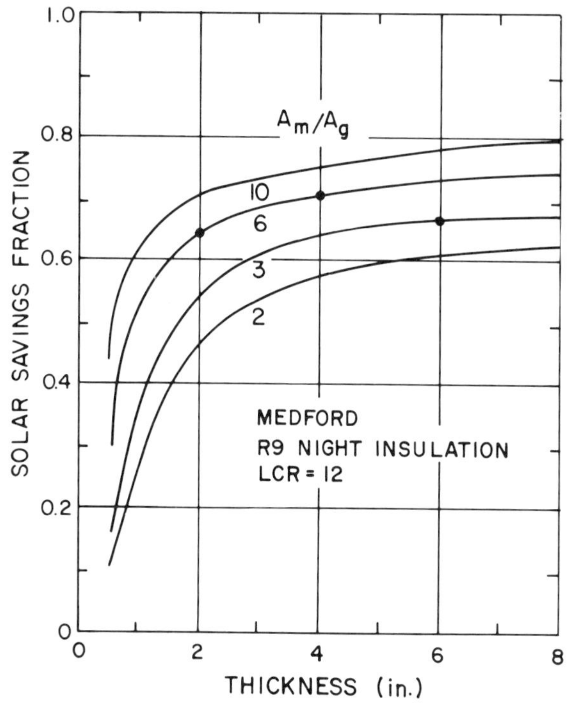

Figure A2-20

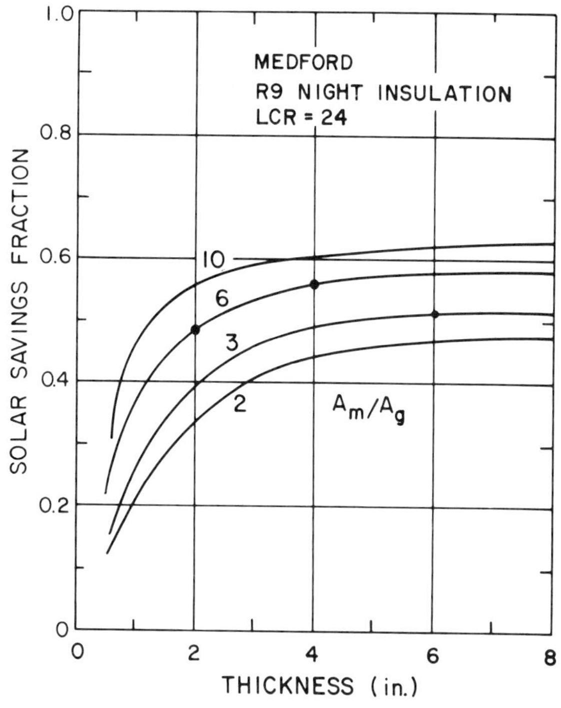

Figure A2-21

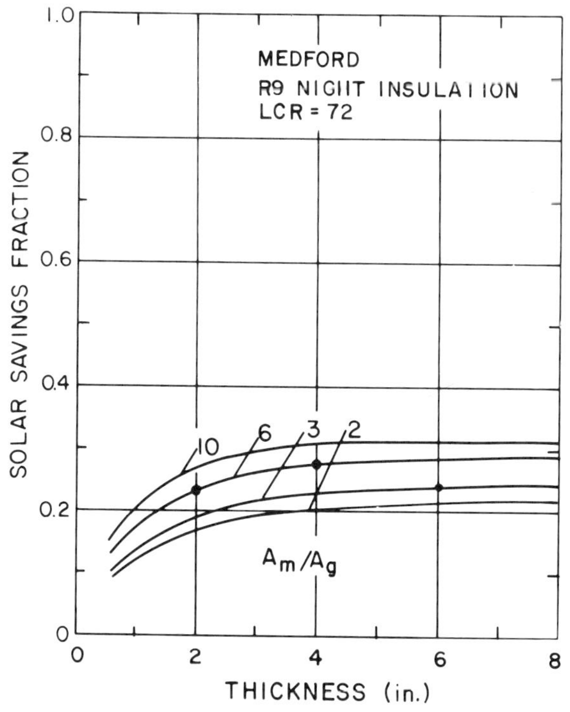

Figure A2-22

Figures A2-2 through A2-34. Annual heating performance sensitivity to the thickness of thermal storage mass for the double-glazed configuration, both without and with R9 night insulation, for various values of the mass-area-to-glazing-area ratio *(A_m/A_g)*. The dots represent the reference conditions.

DIRECT GAIN
MASS THICKNESS

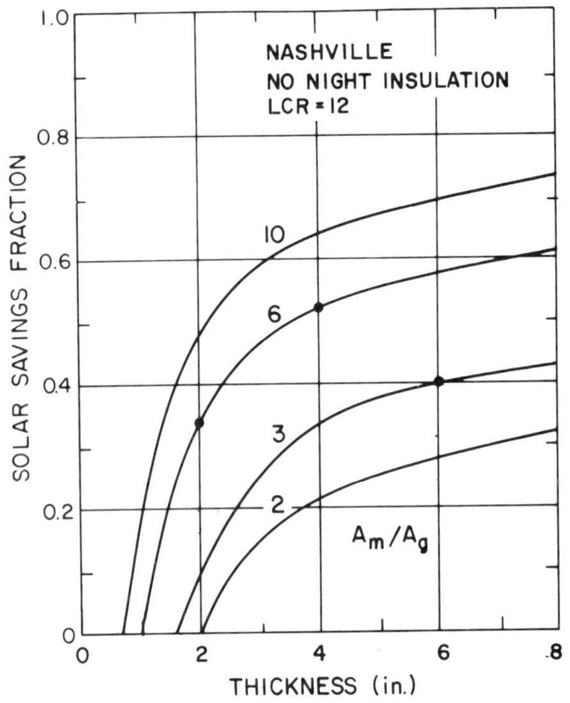

Figure A2-23

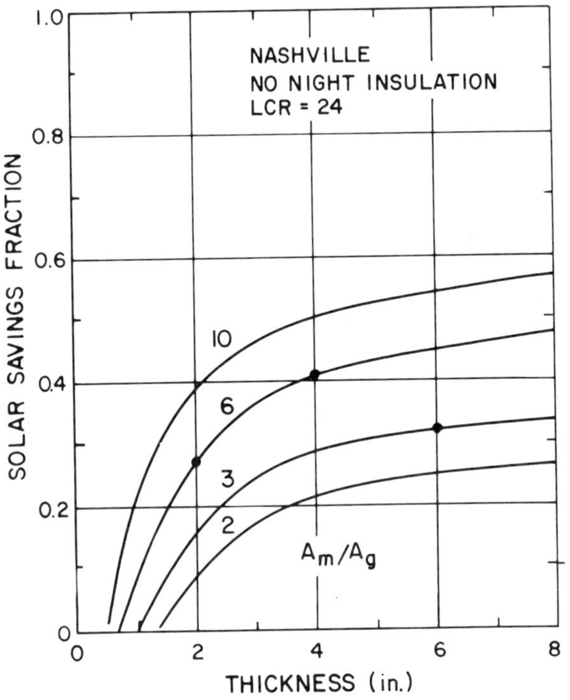

Figure A2-24

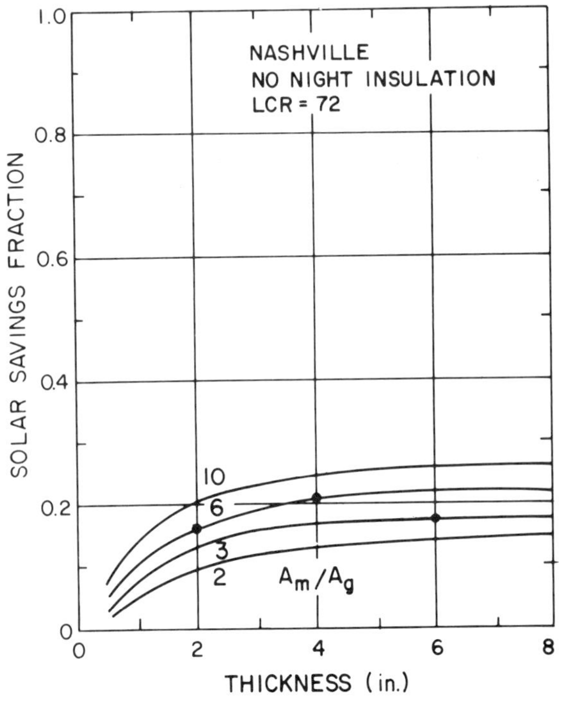

Figure A2-25

Figures A2-2 through A2-34. Annual heating performance sensitivity to the thickness of thermal storage mass for the double-glazed configuration, both without and with R9 night insulation, for various values of the mass-area-to-glazing-area ratio *(A_m/A_g)*. The dots represent the reference conditions.

DIRECT GAIN
MASS THICKNESS

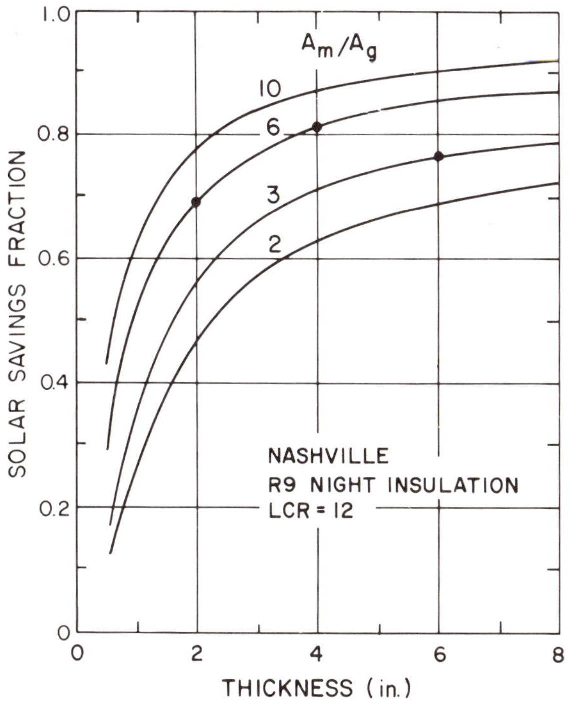

Figure A2-26

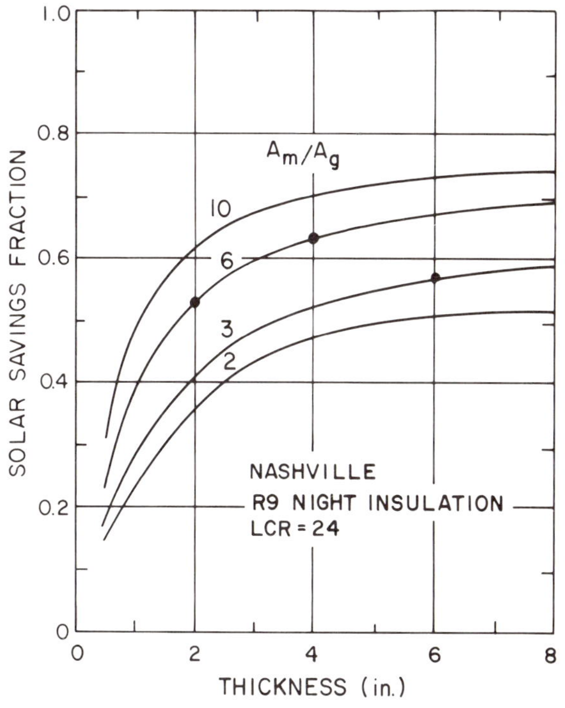

Figure A2-27

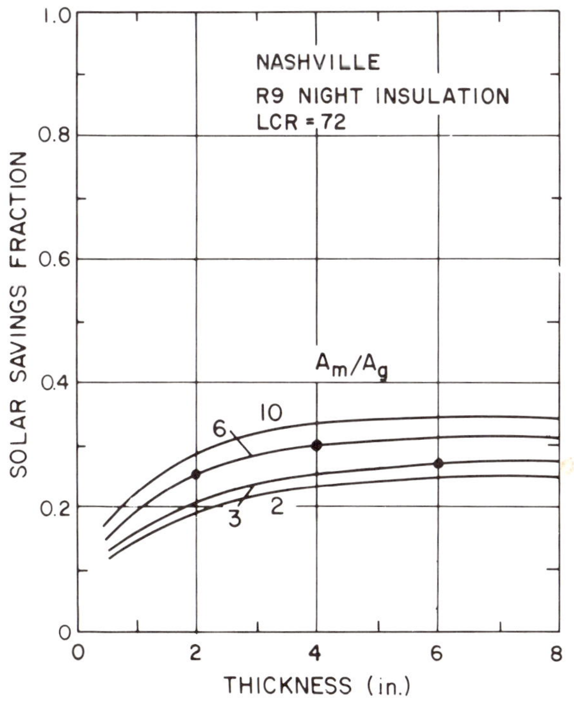

Figure A2-28

Figures A2-2 through A2-34. Annual heating performance sensitivity to the thickness of thermal storage mass for the double-glazed configuration, both without and with R9 night insulation, for various values of the mass-area-to-glazing-area ratio *(A_m/A_g)*. The dots represent the reference conditions.

DIRECT GAIN
MASS THICKNESS

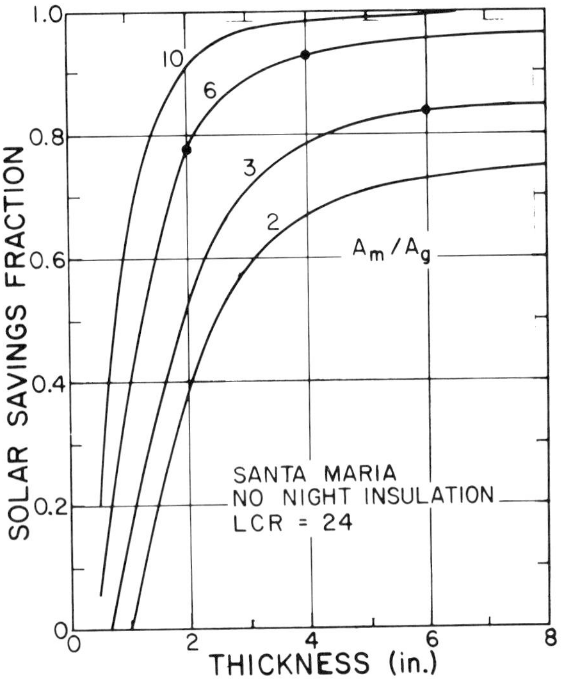

Figure A2-29

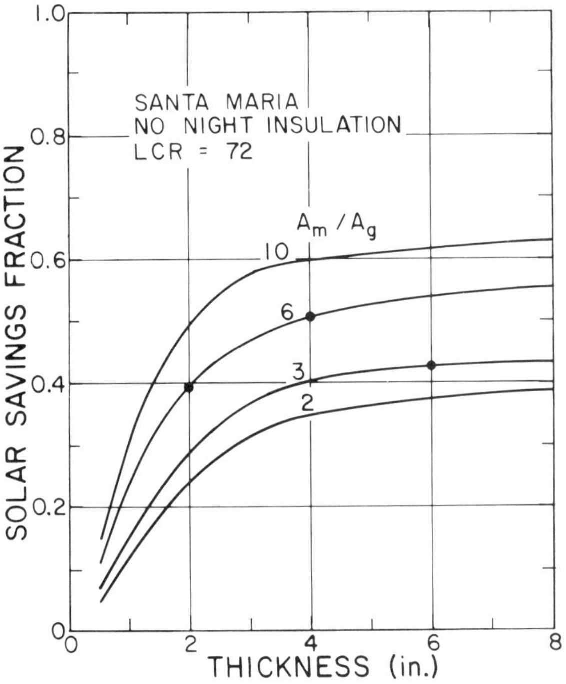

Figure A2-30

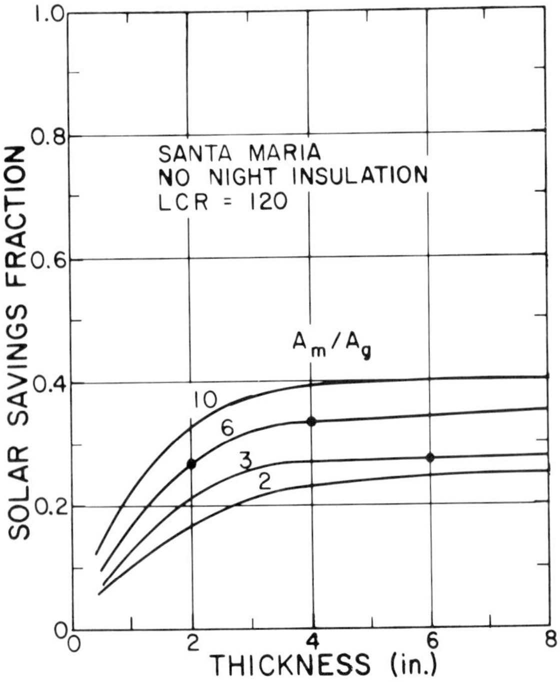

Figure A2-31

Figures A2-2 through A2-34. Annual heating performance sensitivity to the thickness of thermal storage mass for the double-glazed configuration, both without and with R9 night insulation, for various values of the mass-area-to-glazing-area ratio *(A_m/A_g)*. The dots represent the reference conditions.

DIRECT GAIN
MASS THICKNESS

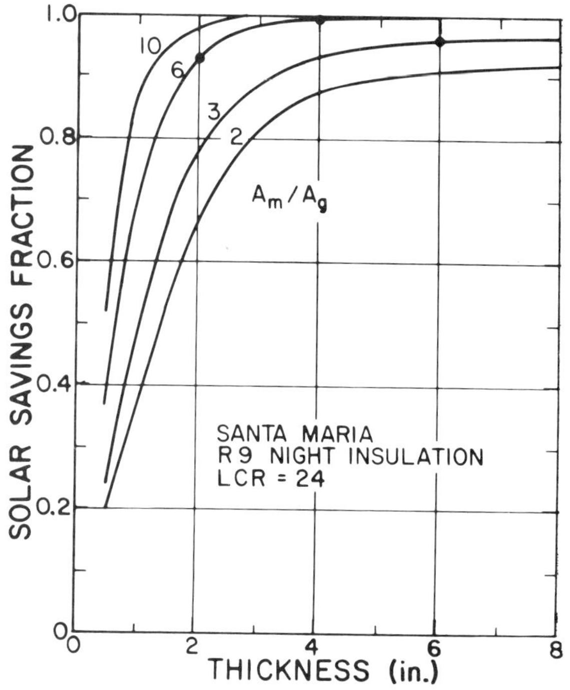

Figure A2-32

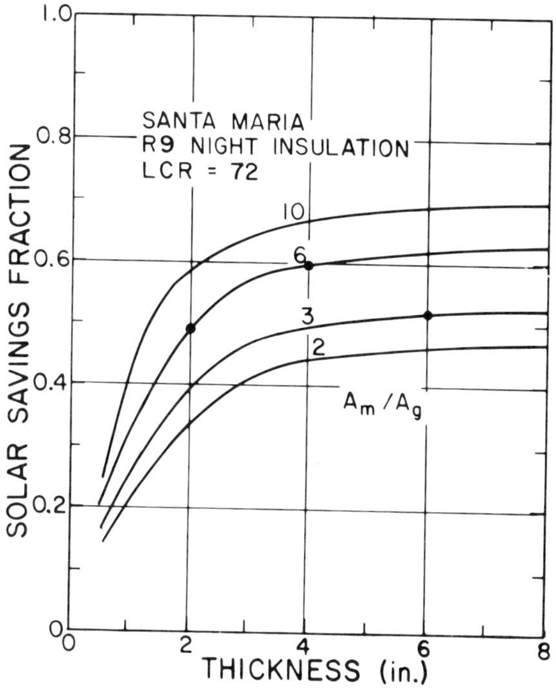

Figure A2-33

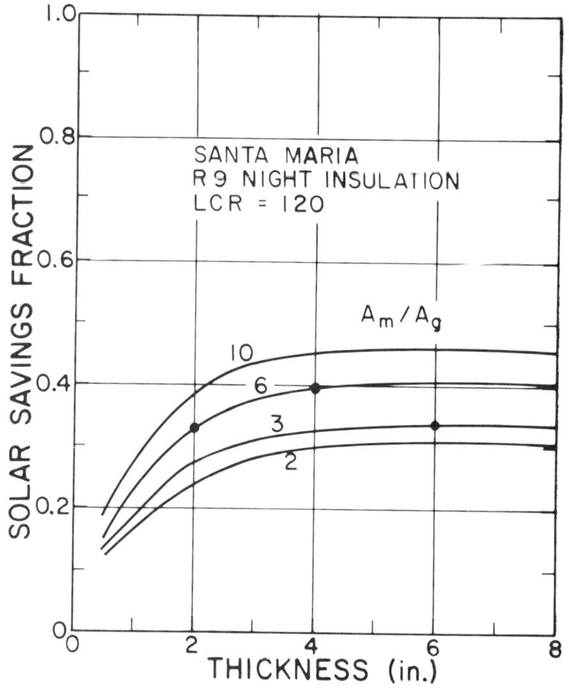

Figure A2-34

Figures A2-2 through A2-34. Annual heating performance sensitivity to the thickness of thermal storage mass for the double-glazed configuration, both without and with R9 night insulation, for various values of the mass-area-to-glazing-area ratio *(A_m/A_g)*. The dots represent the reference conditions.

DIRECT GAIN

DENSITY—SPECIFIC HEAT—THERMAL CONDUCTIVITY PRODUCT

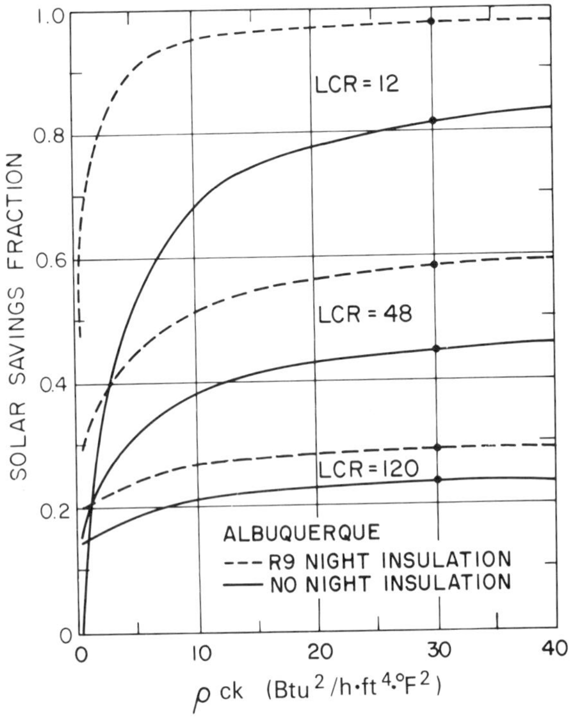

Figure A2-35

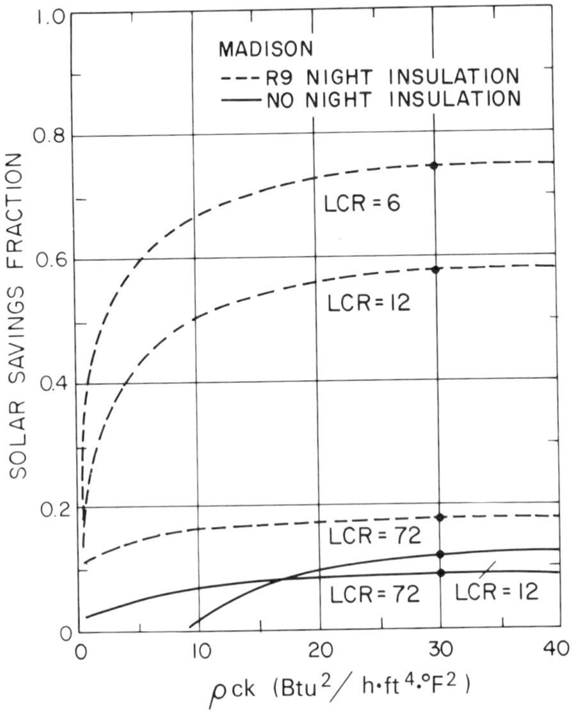

Figure A2-36

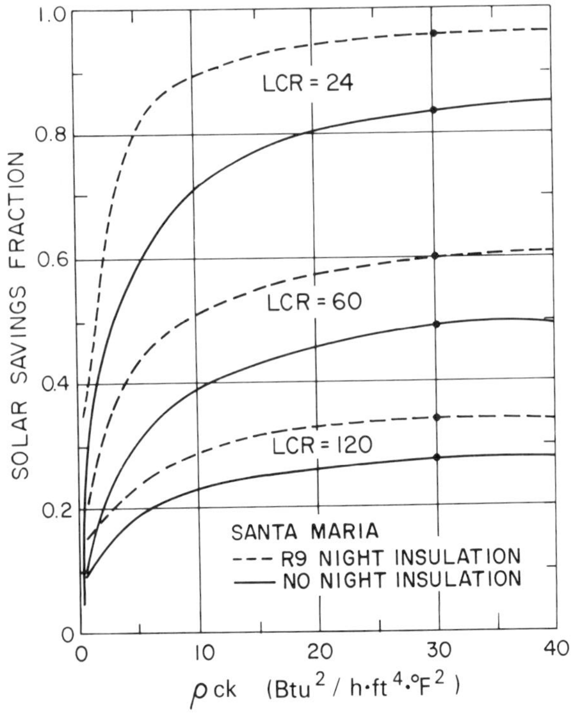

Figure A2-37

Figures A2-35 through A2-37. Annual heating performance sensitivity to the product of the density, specific heat, and thermal conductivity (ρck) of the thermal storage mass for the double-glazed configuration, with mass thickness of 6 inches, mass-area-to-glazing-area ratio of 3, both without and with R9 night insulation (B1 and B3). The dots represent the reference conditions.

DIRECT GAIN
MASS ABSORPTANCE

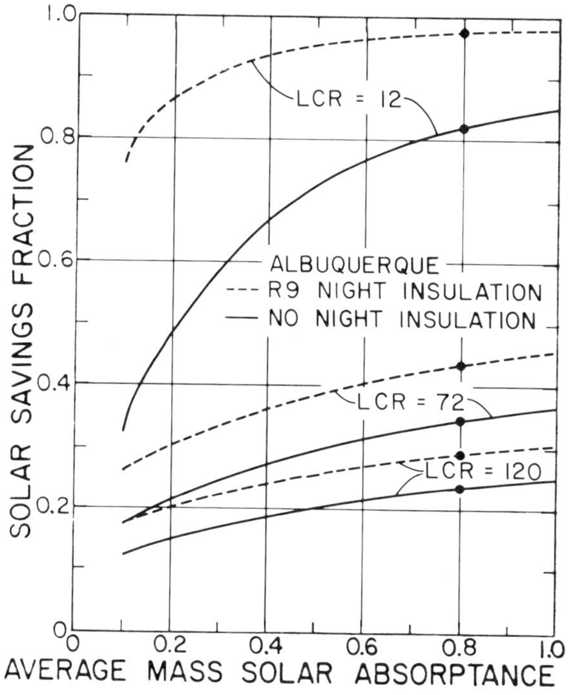

Figure A2-38

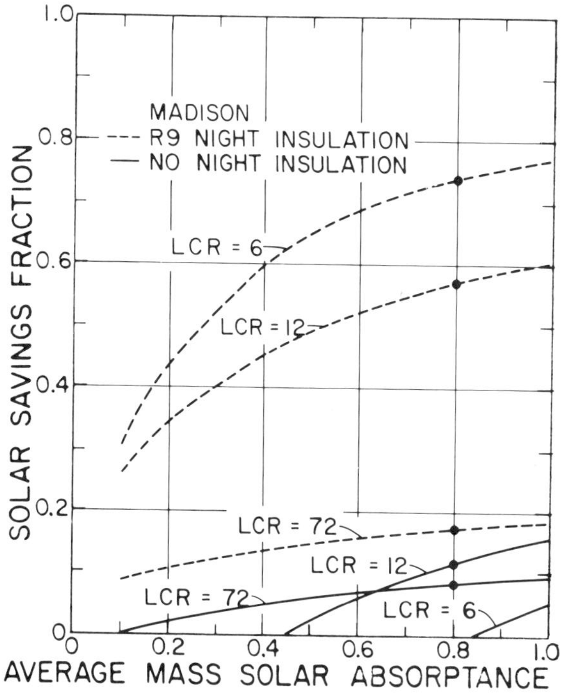

Figure A2-39

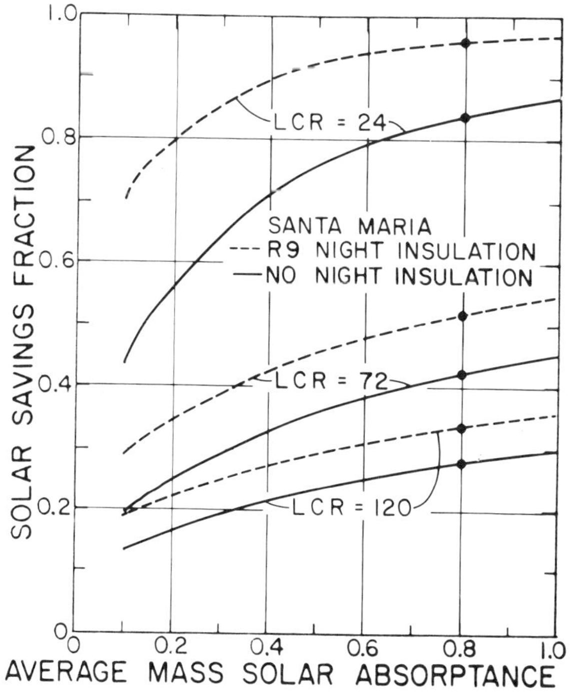

Figure A2-40

Figures A2-38 through A2-40. Annual heating performance sensitivity to the solar absorptance of the thermal storage mass for double-glazed configuration, with mass thickness of 6 inches, mass-area-to-glazing-area ratio of 3, both without and with R9 night insulation (B1 and B3). The dots represent the reference conditions.

DIRECT GAIN
LIGHTWEIGHT ABSORPTION FRACTION

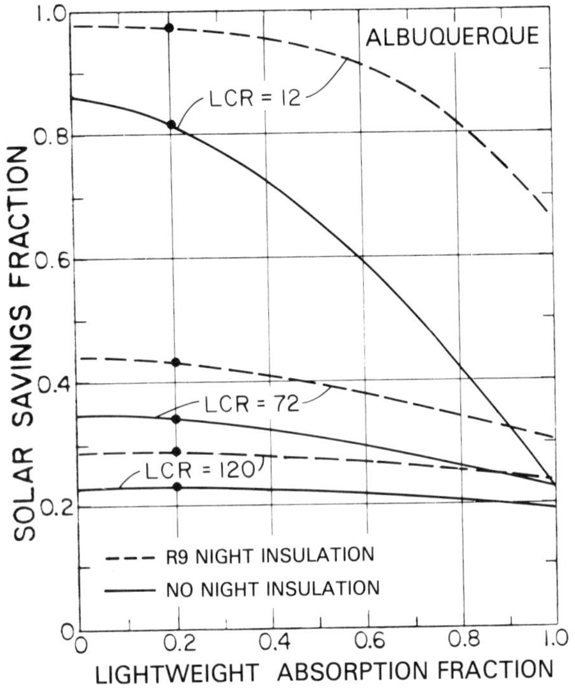

Figure A2-41

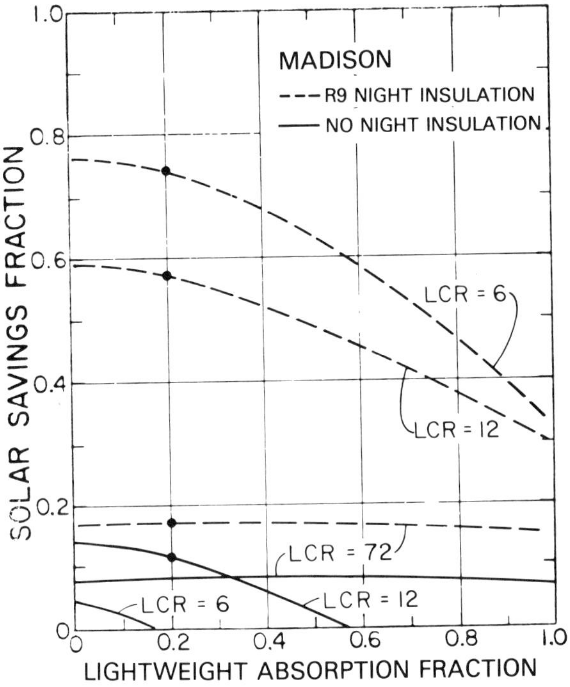

Figure A2-42

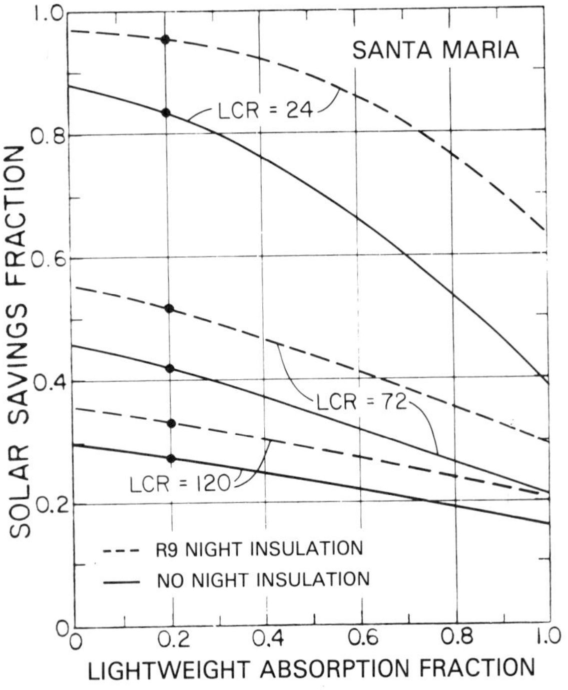

Figure A2-43

Figures A2-41 through A2-43. Annual heating performance sensitivity to the lightweight absorption fraction for the double-glazed configuration, with mass thickness of 6 in., mass-area-to-glazing-area ratio of 3, both without and with R9 night insulation (B1 and B3). The lightweight absorption fraction is the fraction of radiation that directly heats the room air when it is transmitted through the glazing and when it is reflected from interior surfaces. It is intended to simulate the presence of lightweight interior surfaces and objects in the direct-gain space. The dots represent the reference conditions.

DIRECT GAIN
MASS DISTRIBUTION

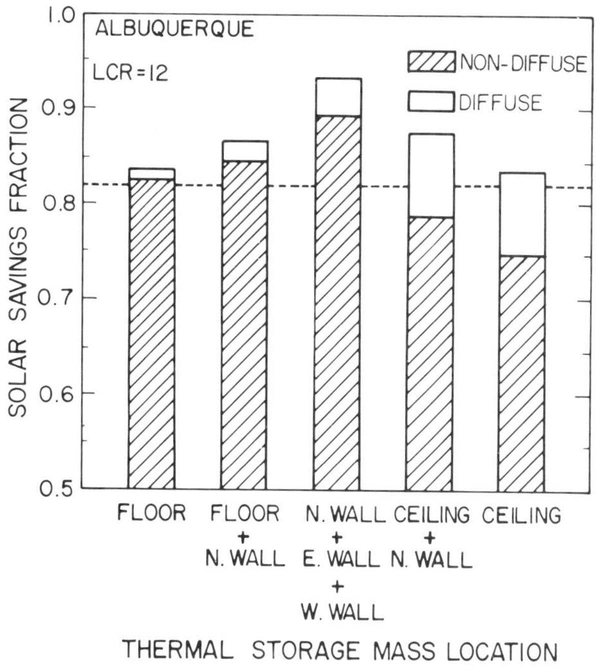

Figure A2-44. Annual heating performance sensitivity to the distribution of thermal storage mass for the double-glazed configuration, with mass thickness of 6 inches, mass-area-to-glazing-area ratio of 3, and no night insulation (B1). The dashed line gives the performance predicted by the two-dimensional model used for the other sensitivity data and in the solar load ratio correlation. The cross-hatched bars correspond to an assumption of nondiffuse transmission by the glazing system; the solid bars correspond to diffuse transmission as assumed for the other sensitivity data and in the solar load ratio correlations.

DIRECT GAIN
MASS COVERING R-VALUE

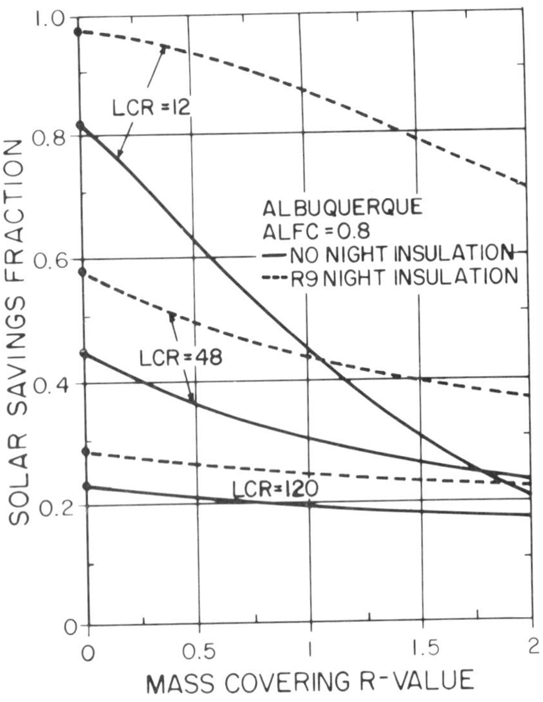

Figure A2-45

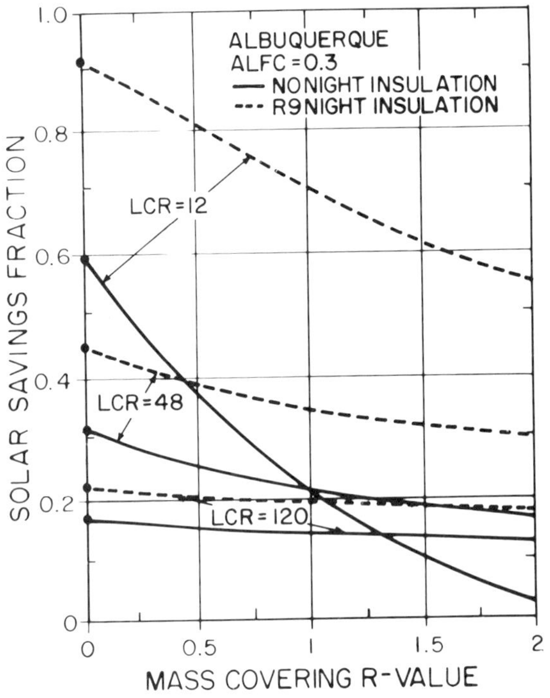

Figure A2-46

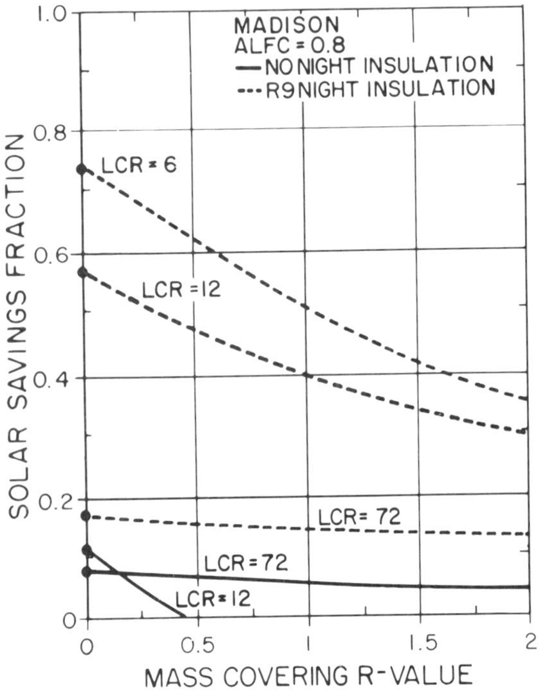

Figure A2-47

Figures A2-45 through A2-50. Annual heating performance sensitivity to the thermal resistance of the covering on the thermal storage mass for the double-glazed configuration, with mass thickness of 6 inches, mass-area-to-glazing-area ratio of 3, both without and with R9 night insulation (B1 and B3). Two values of the solar absorptance of the covering (ALFC) are shown. The dots represent the reference conditions.

DIRECT GAIN
MASS COVERING R-VALUE

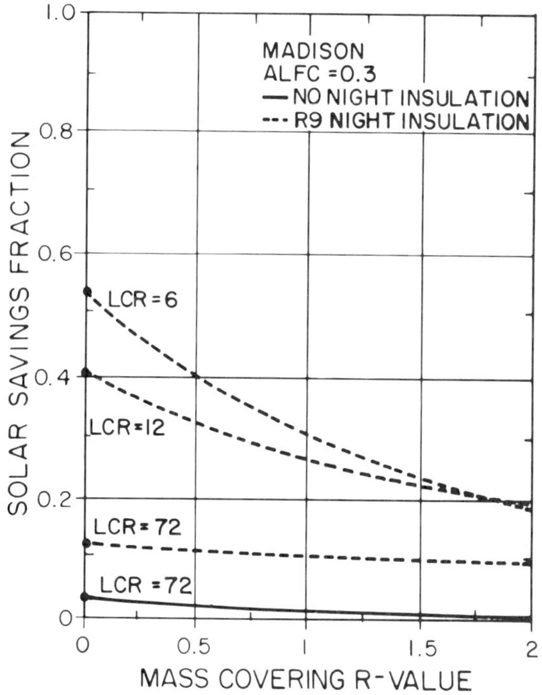

Figure A2-48

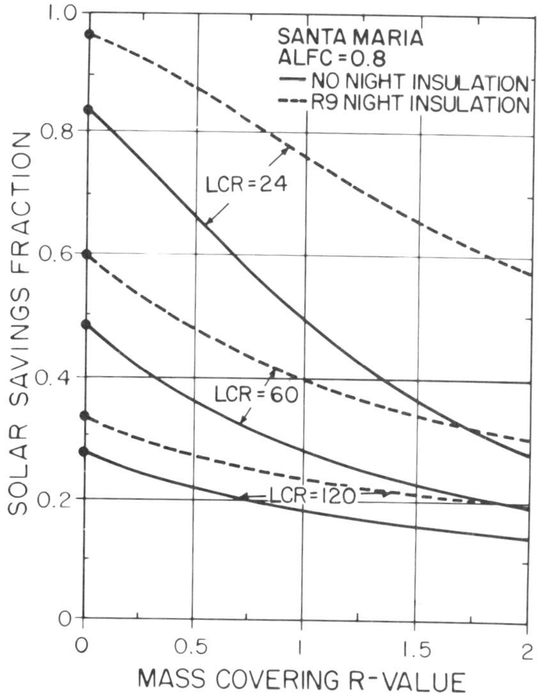

Figure A2-49

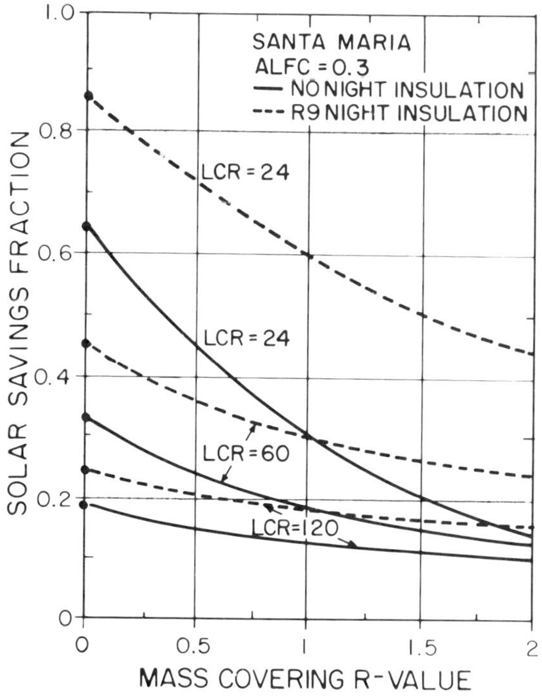

Figure A2-50

Figures A2-45 through A2-50. Annual heating performance sensitivity to the thermal resistance of the covering on the thermal storage mass for the double-glazed configuration, with mass thickness of 6 inches, mass-area-to-glazing-area ratio of 3, both without and with R9 night insulation (B1 and B3). Two values of the solar absorptance of the covering (ALFC) are shown. The dots represent the reference conditions.

DIRECT GAIN
NUMBER OF GLAZINGS

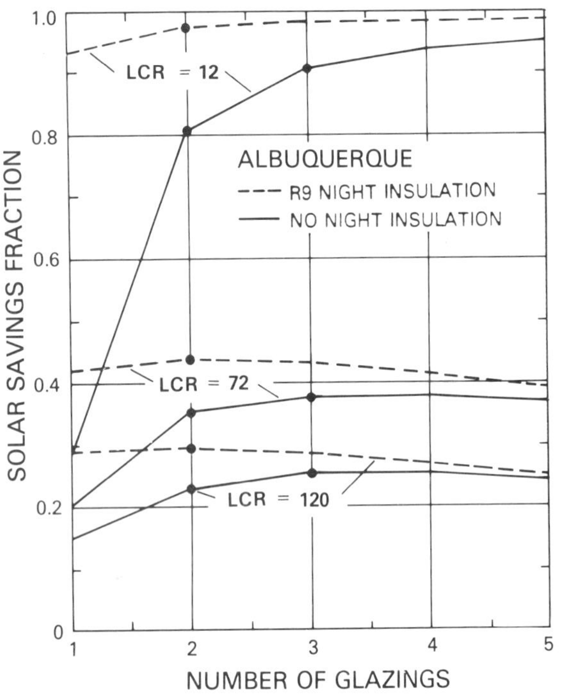

Figure A2-51

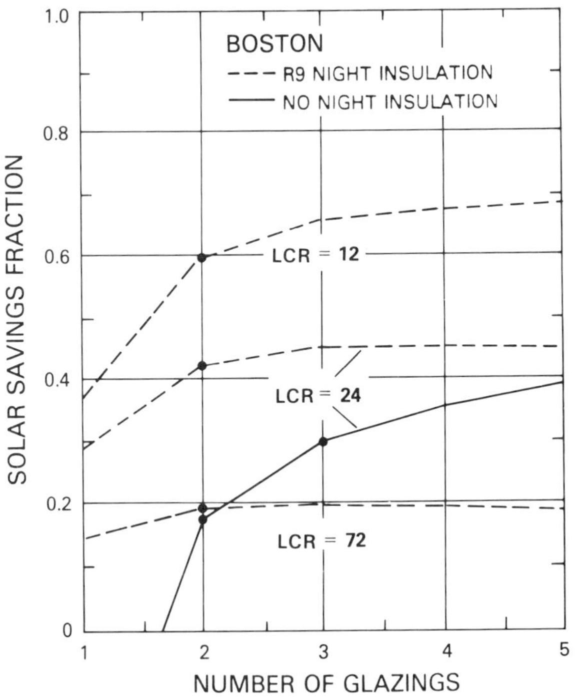

Figure A2-52

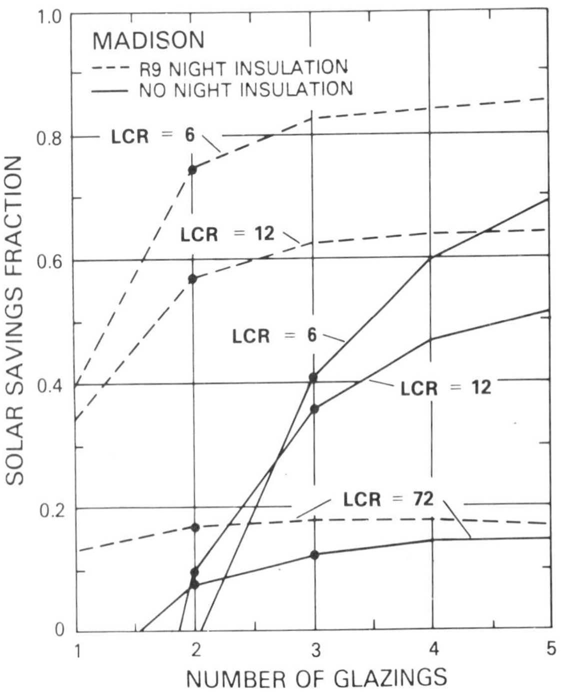

Figure A2-53

Figures A2-51 through A2-56. Annual heating performance sensitivity to the number of panes of glazing for a mass thickness of 6 inches, mass-area-to-glazing-area ratio of 3, both without and with R9 night insulation. The dots represent the reference conditions.

DIRECT GAIN
NUMBER OF GLAZINGS

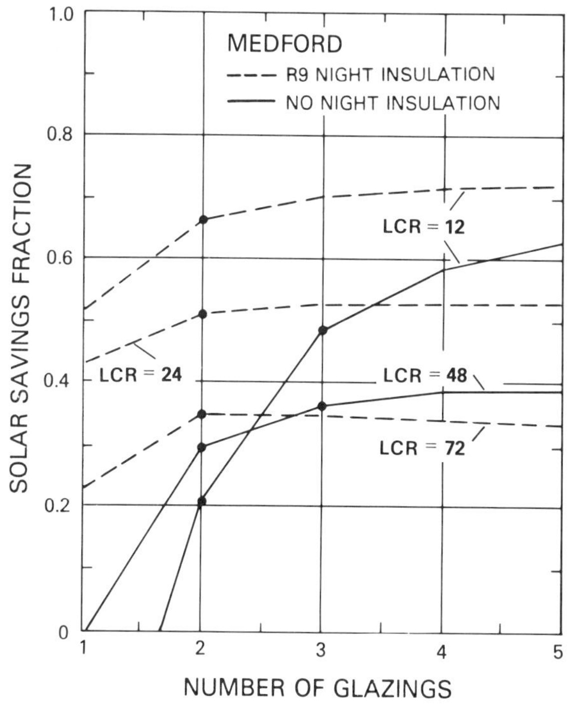

Figure A2-54

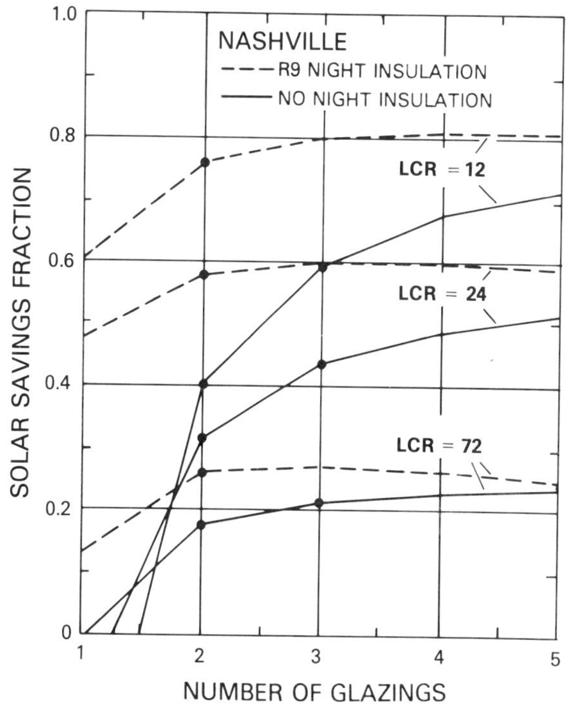

Figure A2-55

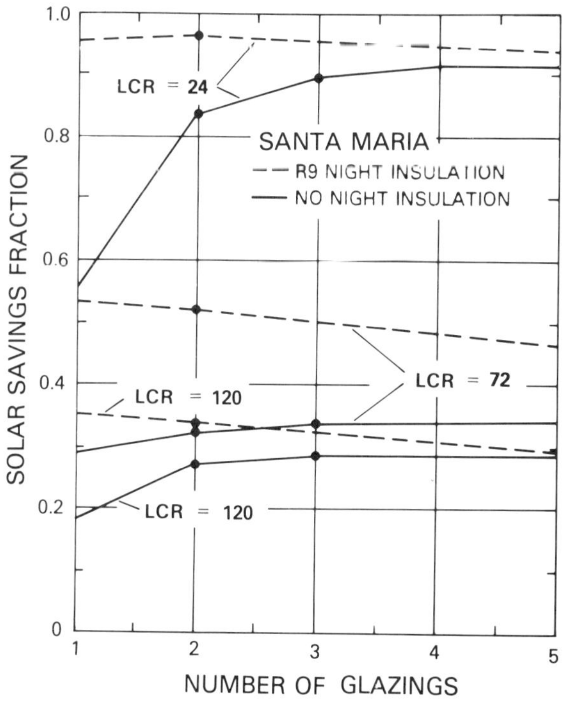

Figure A2-56

Figures A2-51 through A2-56. Annual heating performance sensitivity to the number of panes of glazing for a mass thickness of 6 inches, mass-area-to-glazing-area ratio of 3, both without and with R9 night insulation. The dots represent the reference conditions.

DIRECT GAIN
TEMPERATURE BOUNDS

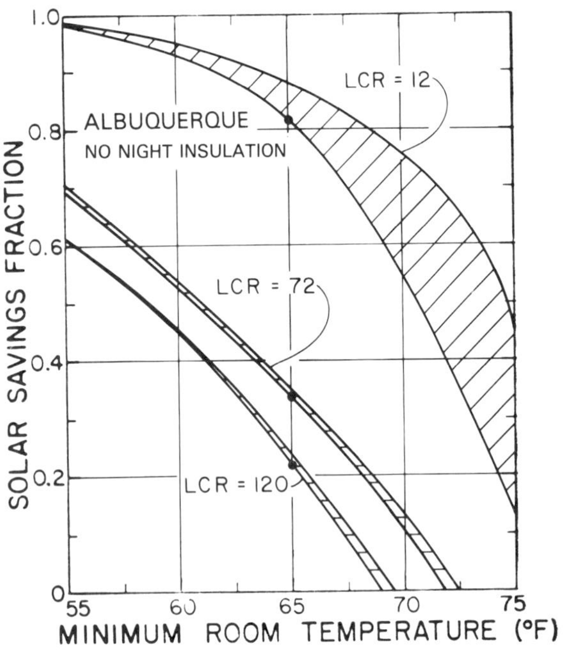

Figure A2-57

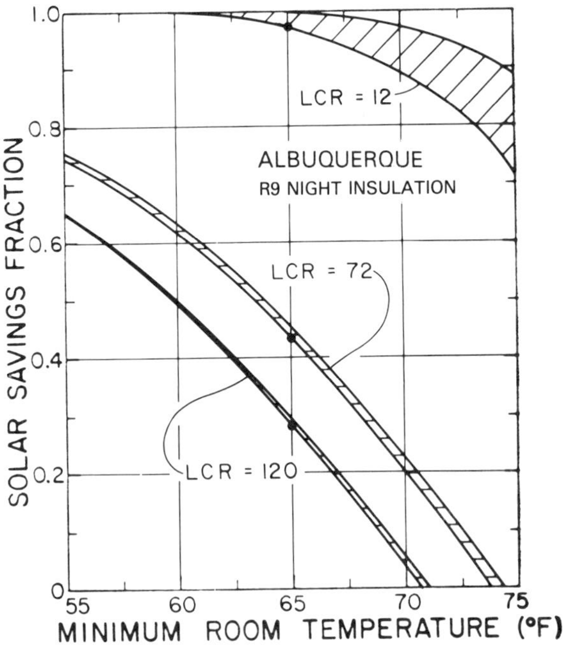

Figure A2-58

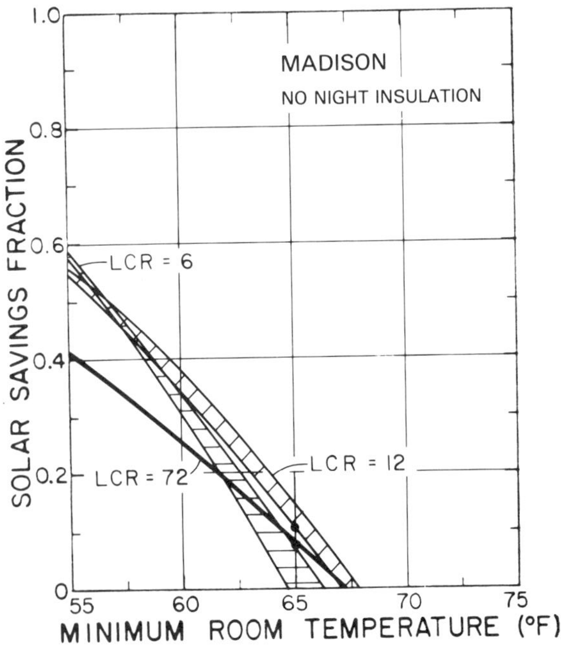

Figure A2-59

Figures A2-57 through A2-62. Annual heating performance sensitivity to the auxiliary heat thermostat setpoint temperature for the double-glazed configuration, with mass thickness of 6 inches, mass-area-to-glazing-area ratio of 3, both without and with R9 night insulation (B1 and B3). The solar savings fraction in this case is defined relative to a 65°F reference load and therefore reflects both variation in solar savings and building load with setpoint. The lower boundary of the sensitivity bands gives the performance when the maximum room temperature is limited to 75°F, while the upper boundary represents performance at a maximum room temperature of 85°F. The dots represent the reference conditions.

DIRECT GAIN
TEMPERATURE BOUNDS

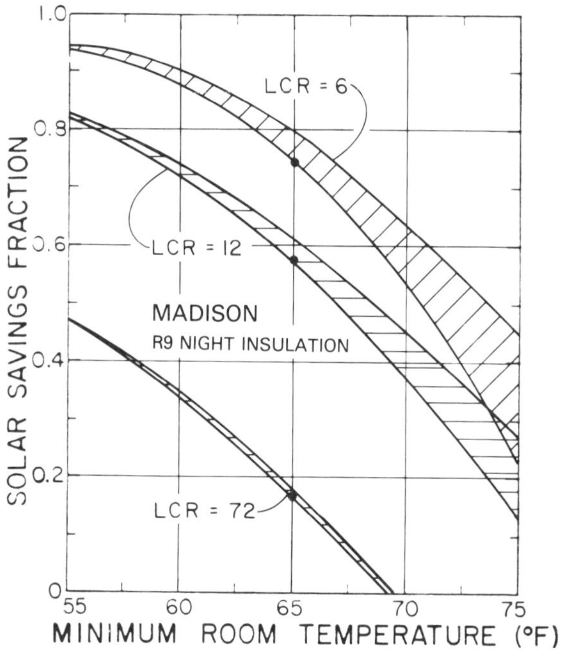

Figure A2-60

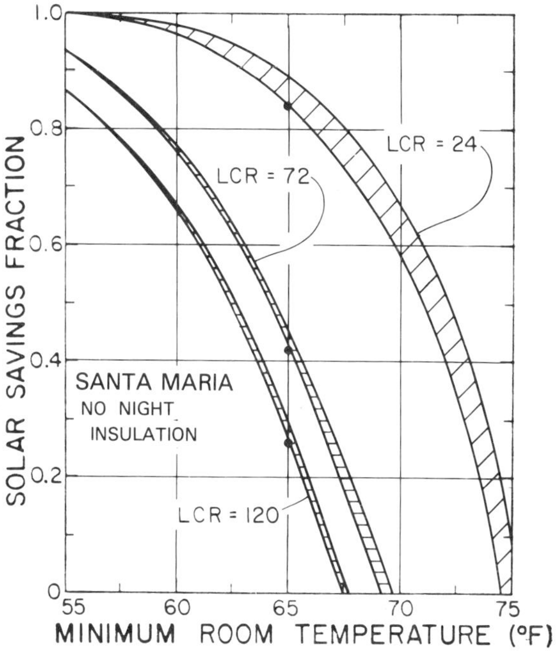

Figure A2-61

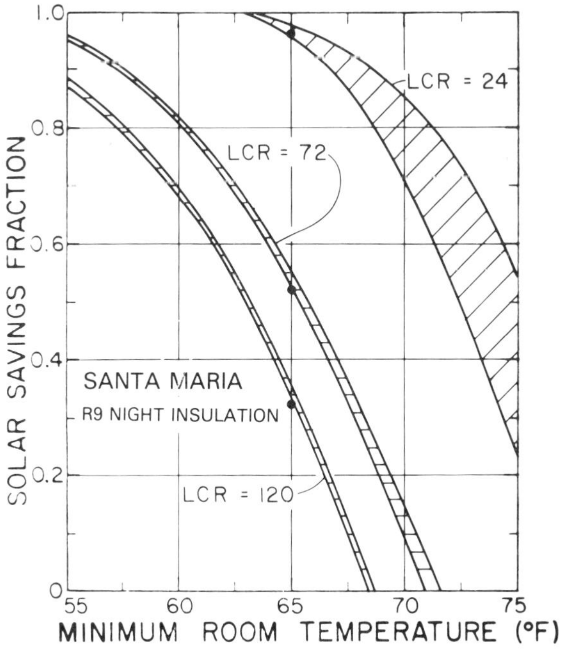

Figure A2-62

Figures A2-57 through A2-62. Annual heating performance sensitivity to the auxiliary heat thermostat setpoint temperature for the double-glazed configuration, with mass thickness of 6 inches, mass-area-to-glazing-area ratio of 3, both without and with R9 night insulation (B1 and B3). The solar savings fraction in this case is defined relative to a 65°F reference load and therefore reflects both variation in solar savings and building load with setpoint. The lower boundary of the sensitivity bands gives the performance when the maximum room temperature is limited to 75°F, while the upper boundary represents performance at a maximum room temperature of 85°F. The dots represent the reference conditions.

DIRECT GAIN
NIGHT SETBACK

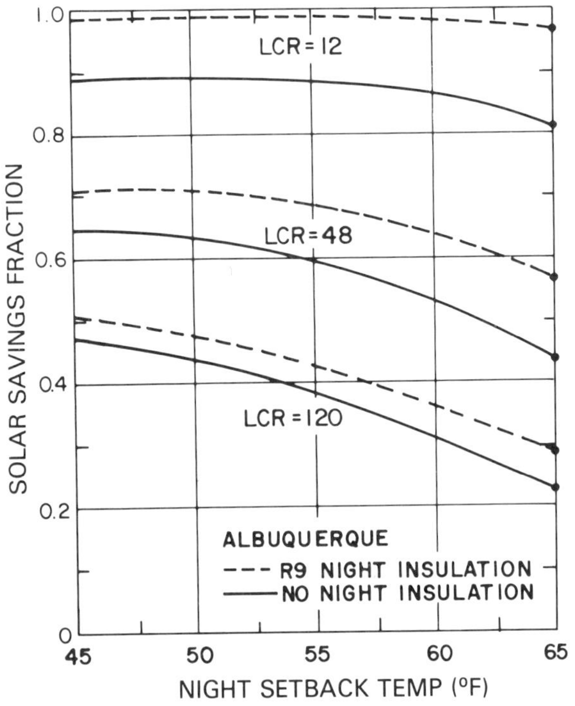

Figure A2-63

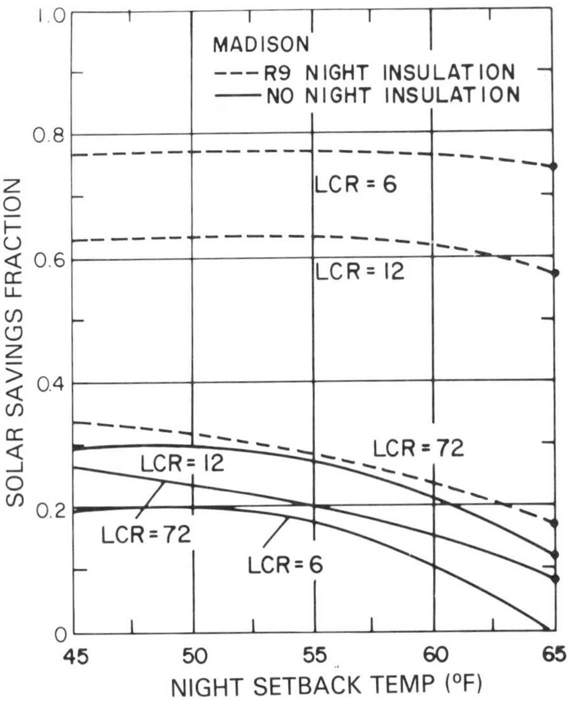

Figure A2-64

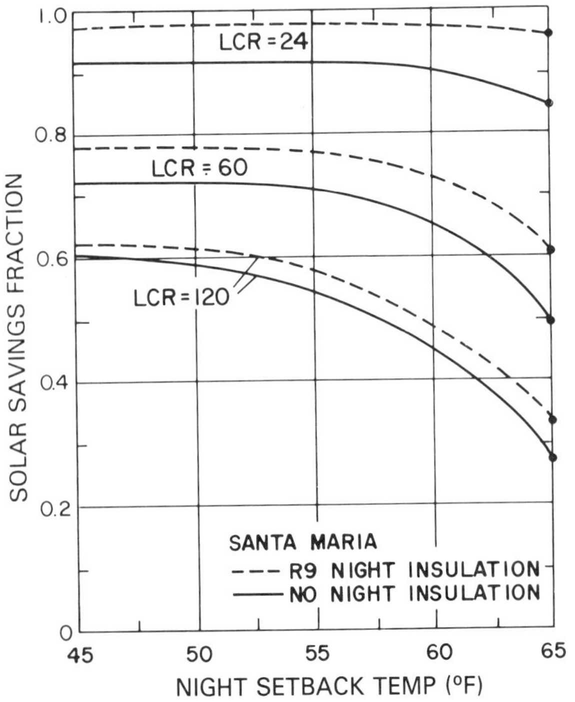

Figure A2-65

Figures A2-63 through A2-65. Annual heating performance sensitivity to the 11:00 p.m. to 6:00 a.m. auxiliary heat thermostat setpoint for the double-glazed configuration, with mass thickness of 6 inches, mass-area-to-glazing-area ratio of 3, both without and with R9 night insulation (B1 and B3). The dots represent the reference conditions.

DIRECT GAIN
GLAZING AZIMUTH

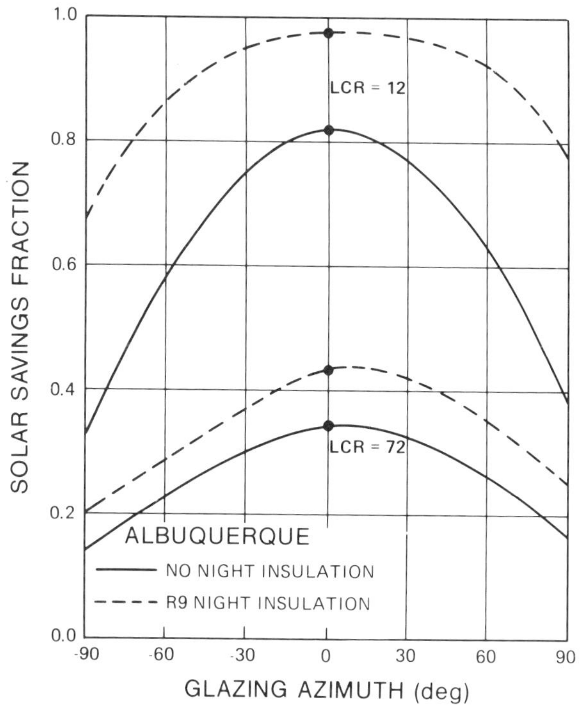

Figure A2-66

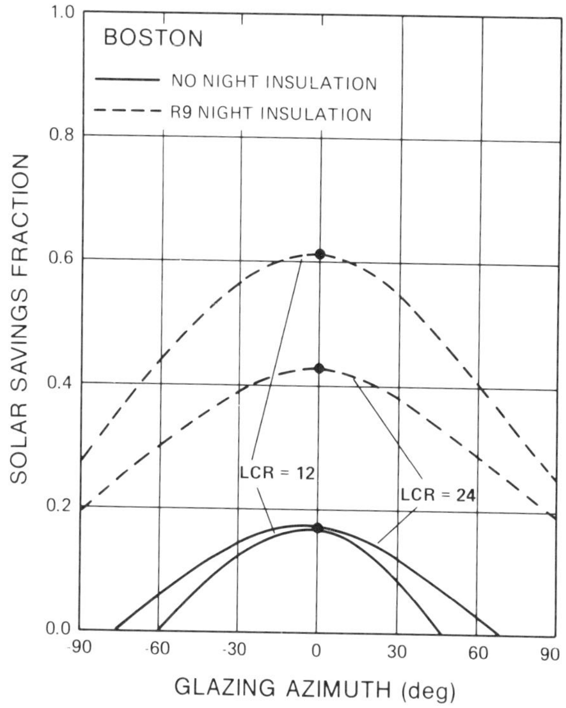

Figure A2-67

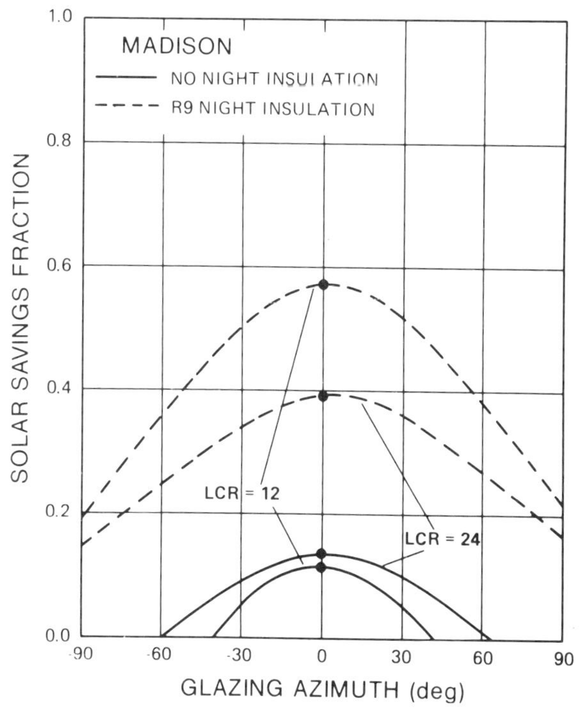

Figure A2-68

Figures A2-66 through A2-71. Annual heating performance sensitivity to the glazing azimuth for vertical glazing, both without and with R9 night insulation (B1 and B3). The dots represent the reference design conditions.

DIRECT GAIN
GLAZING AZIMUTH

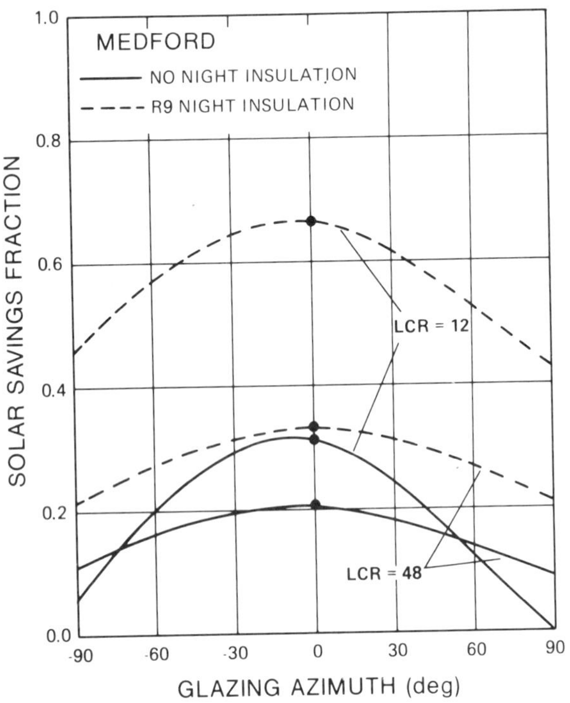

Figure A2-69

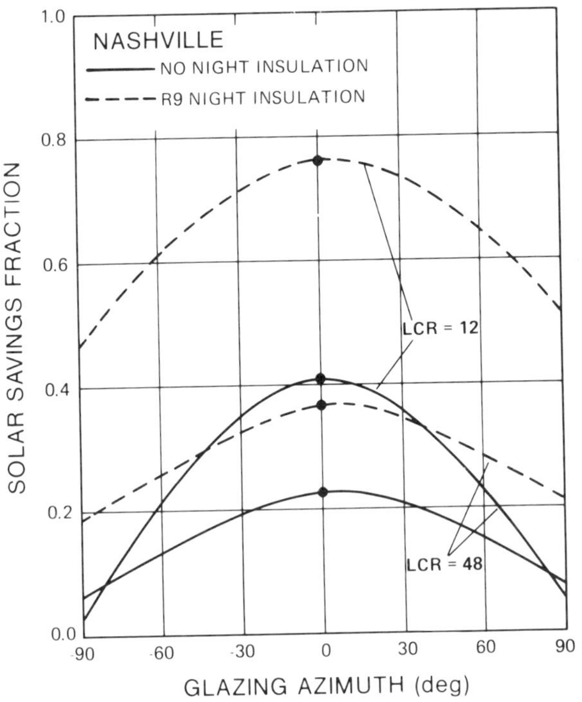

Figure A2-70

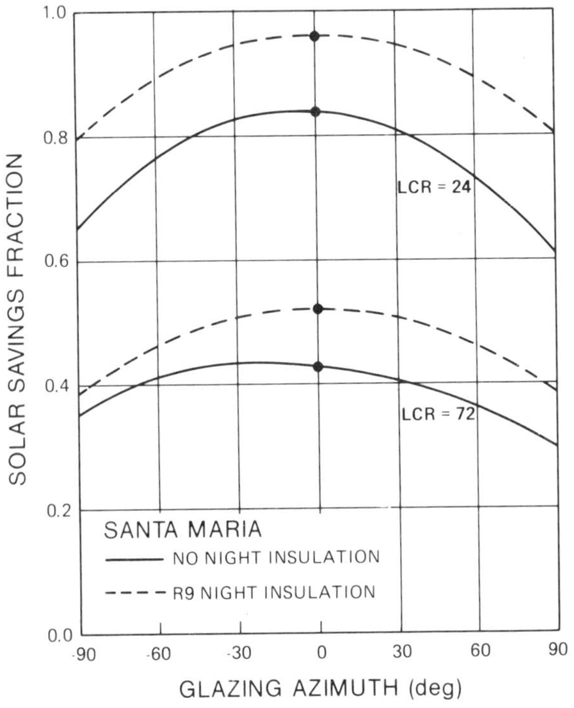

Figure A2-71

Figures A2-66 through A2-71. Annual heating performance sensitivity to the glazing azimuth for vertical glazing, both without and with R9 night insulation (B1 and B3). The dots represent the reference design conditions.

DIRECT GAIN—OVERHEATING
MASS AREA/GLAZING AREA

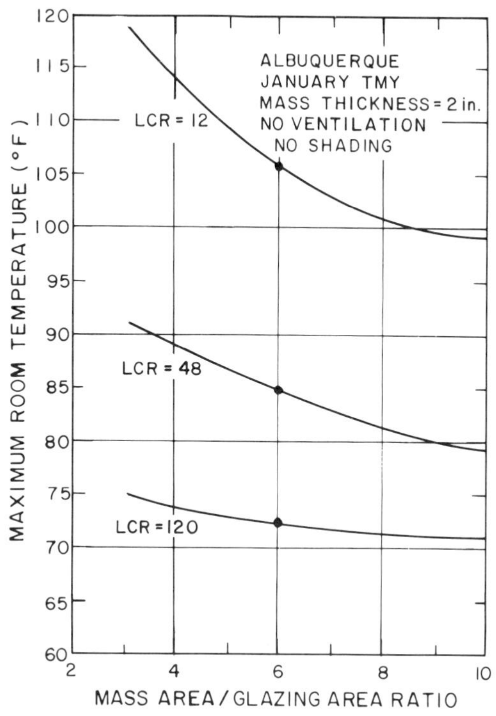

Figure A2-72

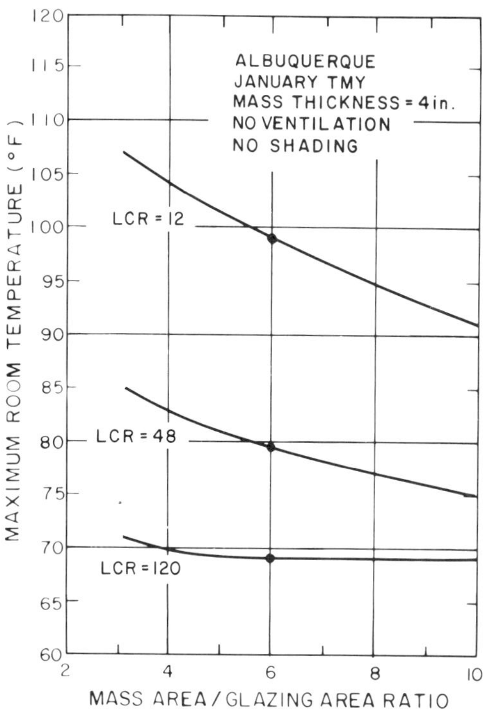

Figure A2-73

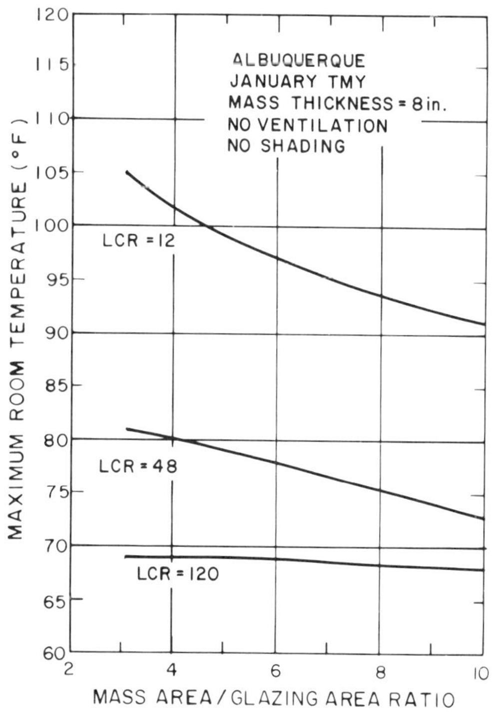

Figure A2-74

Figures A2-72 through A2-77. January overheating sensitivity to the mass-area-to-glazing-area ratio for the double-glazed configuration, without night insulation, and for various mass thicknesses. The dots represent the reference conditions.

DIRECT GAIN—OVERHEATING
MASS AREA/GLAZING AREA

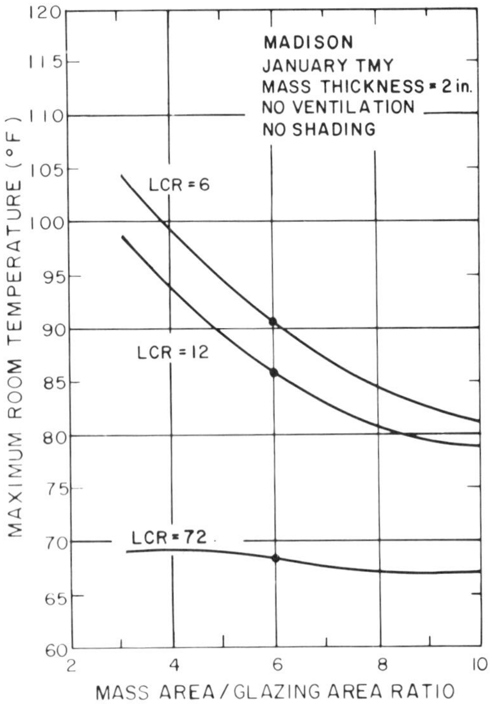

Figure A2-75

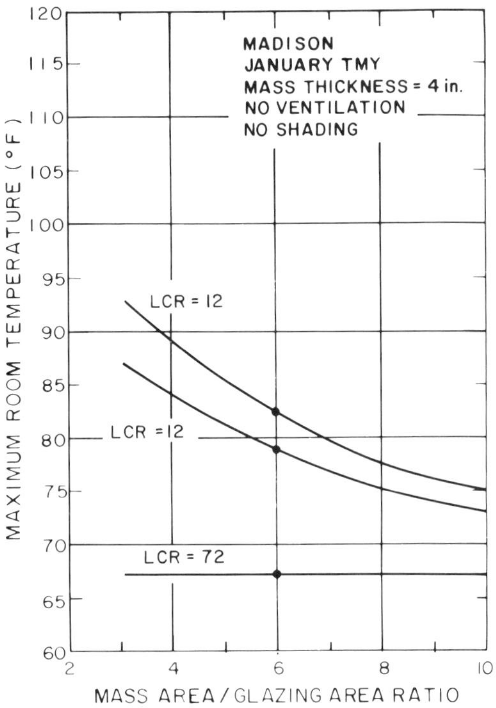

Figure A2-76

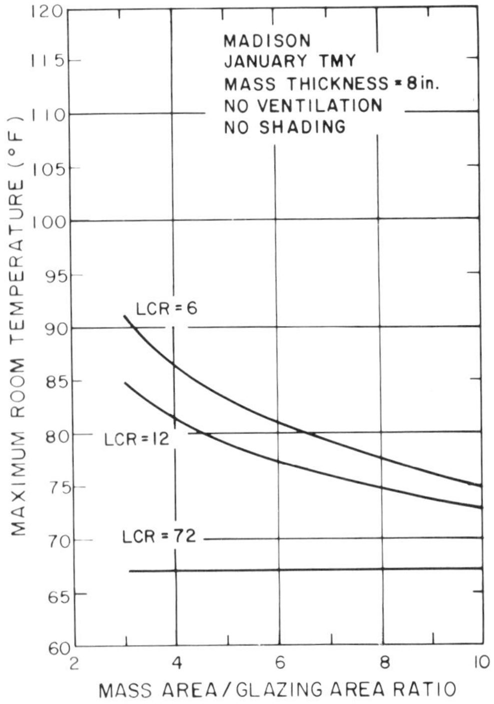

Figure A2-77

Figures A2-72 through A2-77. January overheating sensitivity to the mass-area-to-glazing-area ratio for the double-glazed configuration, without night insulation, and for various mass thicknesses. The dots represent the reference conditions.

DIRECT GAIN—OVERHEATING
MASS AREA/GLAZING AREA

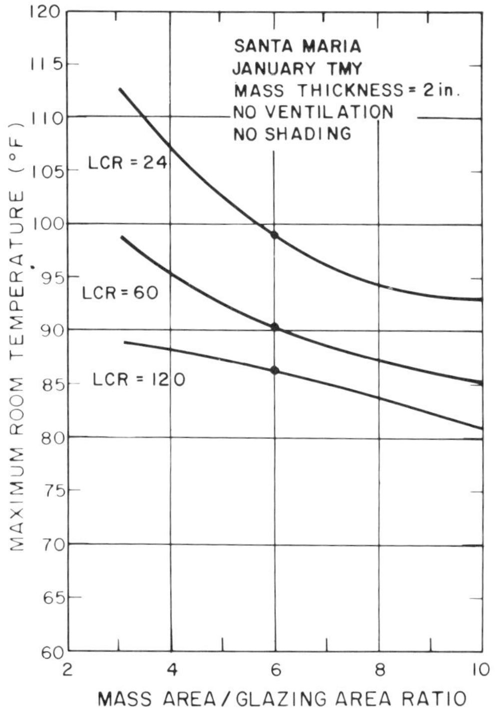

Figure A2-78

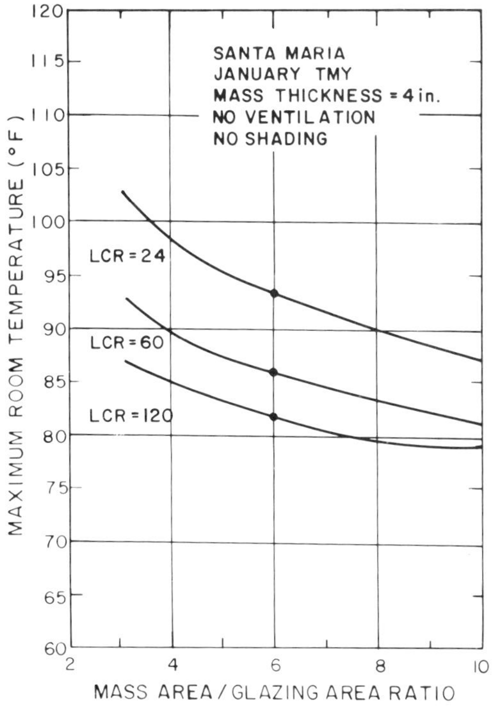

Figure A2-79

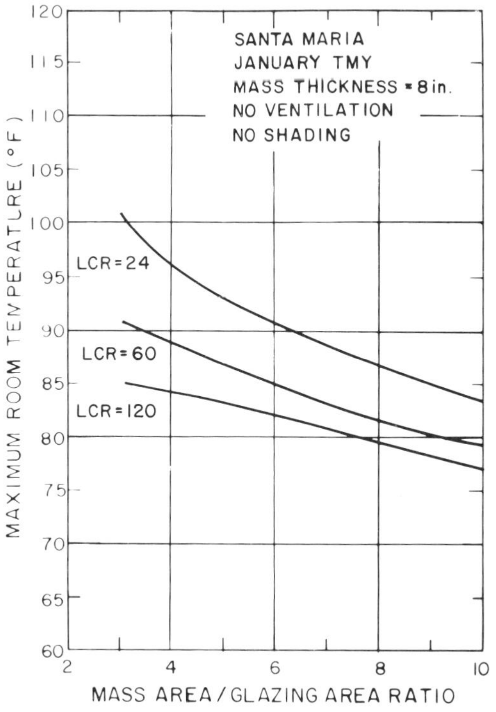

Figure A2-80

Figures A2-78 through A2-80. January overheating sensitivity to the mass-area-to-glazing-area ratio for the double-glazed configuration, without night insulation, and for various mass thicknesses. The dots represent the reference conditions.

DIRECT GAIN—OVERHEATING
LIGHTWEIGHT ABSORPTION FRACTION

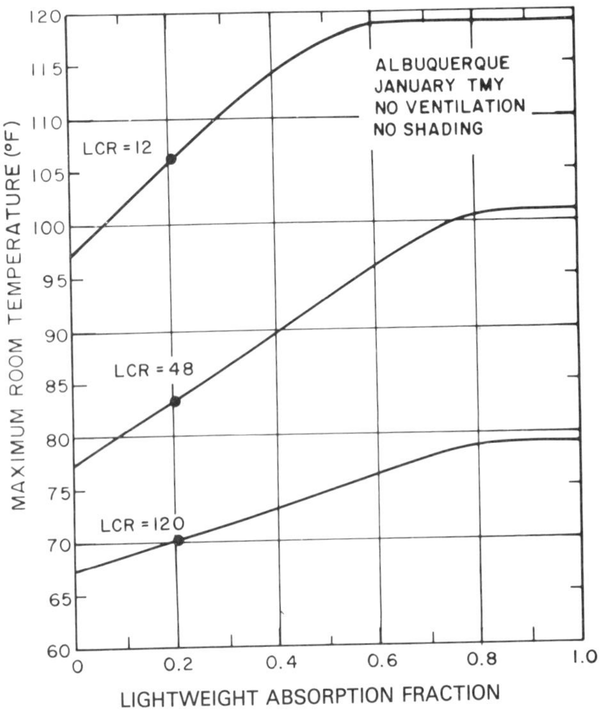

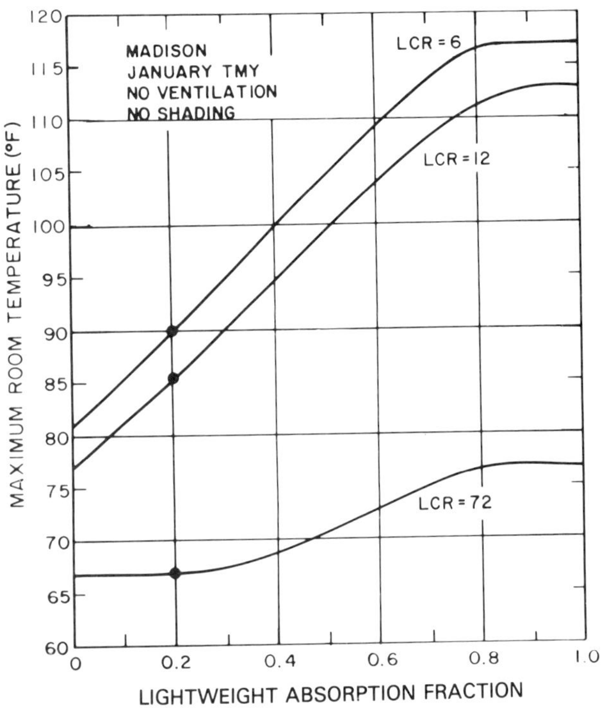

Figures A2-81 and A2-82. January overheating sensitivity to the lightweight absorption fraction for the double-glazed configuration, with mass thickness of 6 inches, mass-area-to-glazing-area ratio of 3, and no night insulation (B1). The lightweight absorption fraction is the fraction of radiation that directly heats the room air when it is transmitted through the glazing and when it is reflected from interior sufaces. It is intended to simulate the presence of lightweight interior surfaces and objects in the direct-gain space. The dots represent the reference conditions.

DIRECT GAIN—OVERHEATING
LIGHTWEIGHT ABSORPTION FRACTION

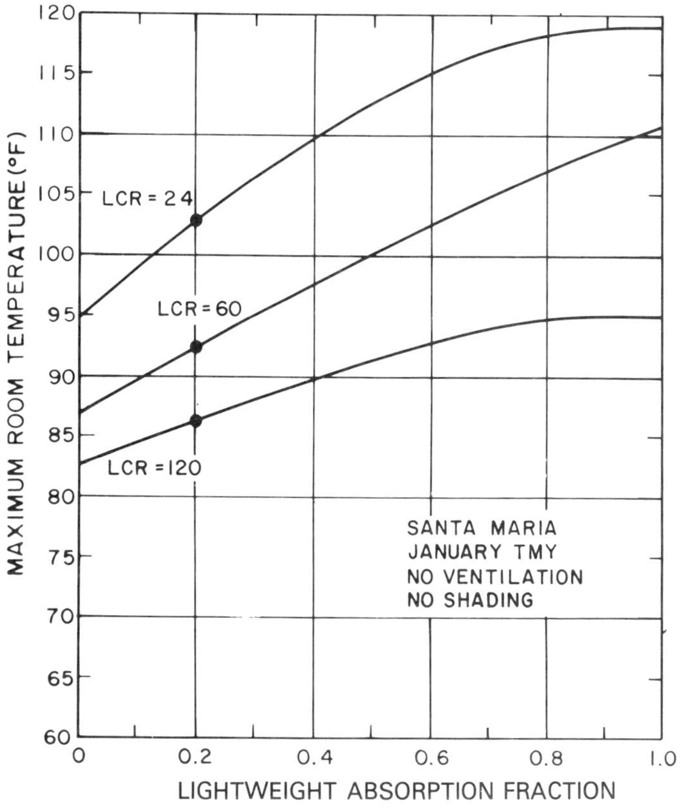

Figure A2-83. January overheating sensitivity to the lightweight absorption traction for the double-glazed configuration, with mass thickness of 6 inches, mass-area-to-glazing-area ratio of 3, and no night insulation (B1). The lightweight absorption fraction is the fraction of radiation that directly heats the room air when it is transmitted through the glazing and when it is reflected from interior surfaces. It is intended to simulate the presence of lightweight interior surfaces and objects in the direct-gain space. The dots represent the reference conditions.

SUNTEMPERED
MASS AREA/GLAZING AREA

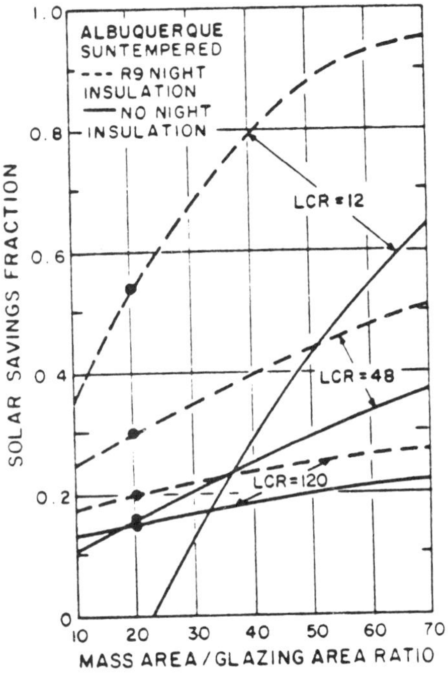

Figure A2-84

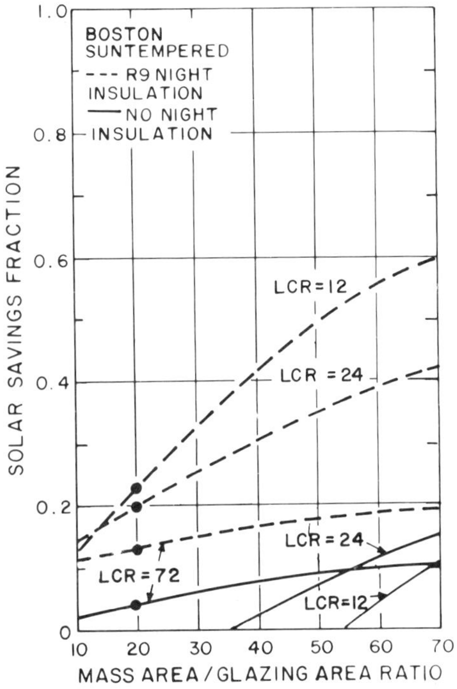

Figure A2-85

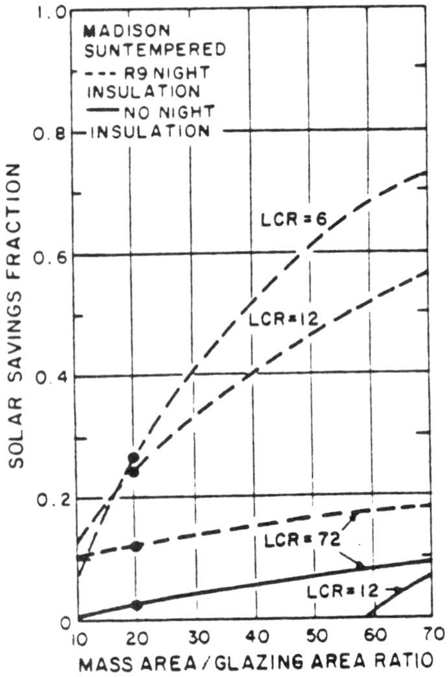

Figure A2-86

Figures A2-84 through A2-89. Annual heating performance sensitivity to the mass-area-to-glazing-area ratio for the suntempered reference design both without and with R9 night insulation. The dots represent the reference conditions.

SUNTEMPERED
MASS AREA/GLAZING AREA

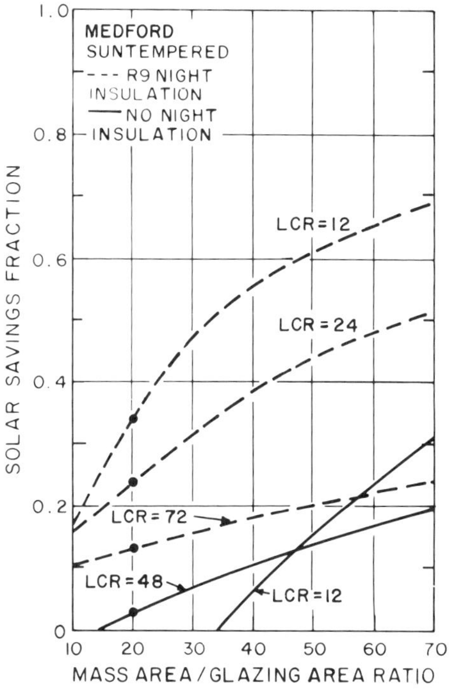

Figure A2-87

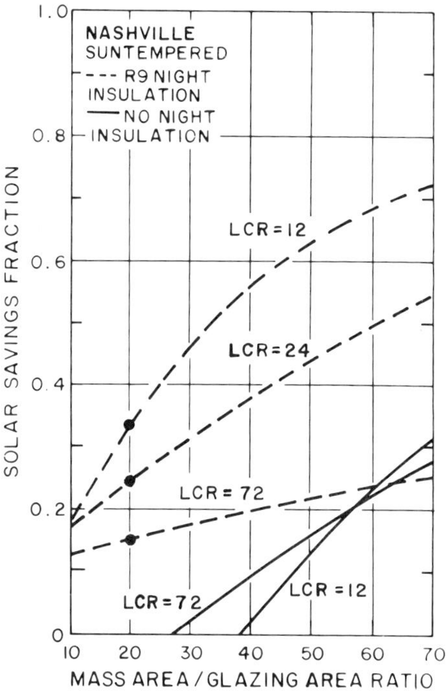

Figure A2-88

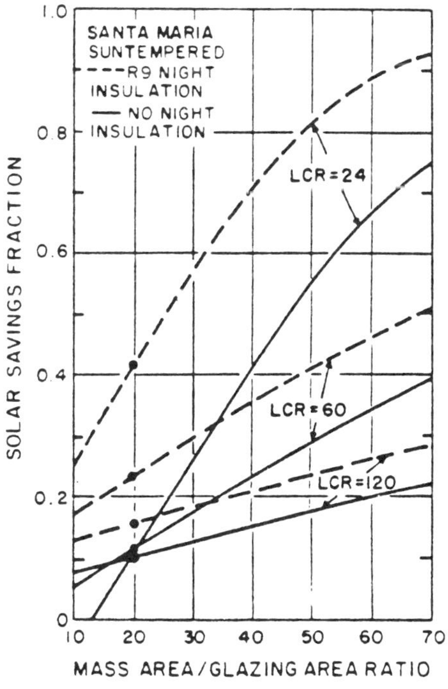

Figure A2-89

Figures A2-84 through A2-89. Annual heating performance sensitivity to the mass-area-to-glazing-area ratio for the suntempered reference design both without and with R9 night insulation. The dots represent the reference conditions.

SUNTEMPERED
NUMBER OF GLAZINGS

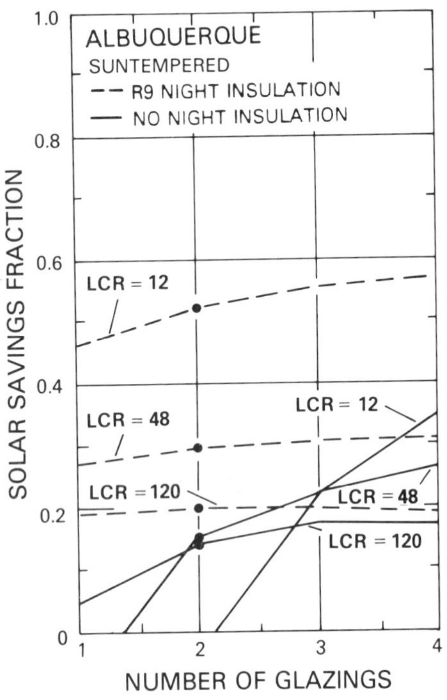

Figure A2-90

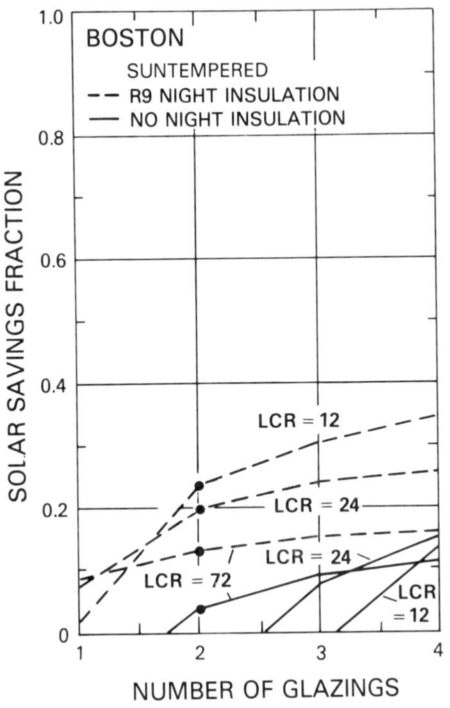

Figure A2-91

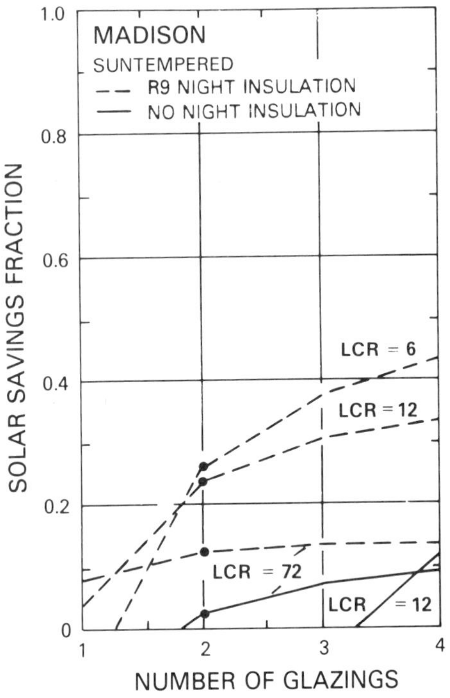

Figure A2-92

Figures A2-90 through A2-95. Annual heating performance sensitivity to the number of panes of glazing for the suntempered reference design both without and with R9 night insulation. The dots represent the reference conditions.

SUNTEMPERED
NUMBER OF GLAZINGS

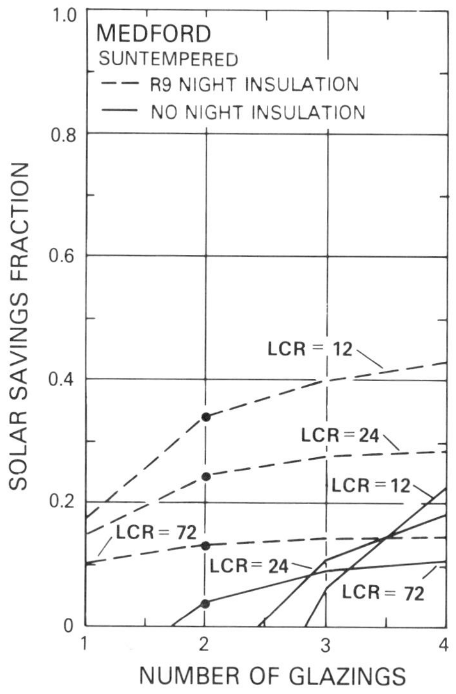

Figure A2-93

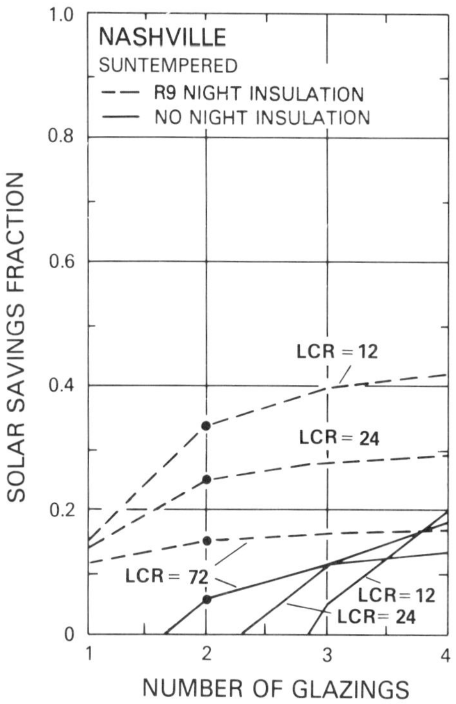

Figure A2-94

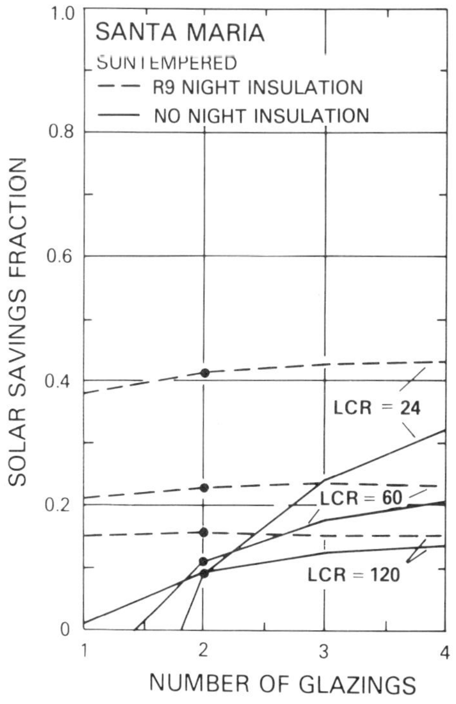

Figure A2-95

Figures A2-90 through A2-95. Annual heating performance sensitivity to the number of panes of glazing for the suntempered reference design both without and with R9 night insulation. The dots represent the reference conditions.

SUNTEMPERED
MASS ABSORPTANCE

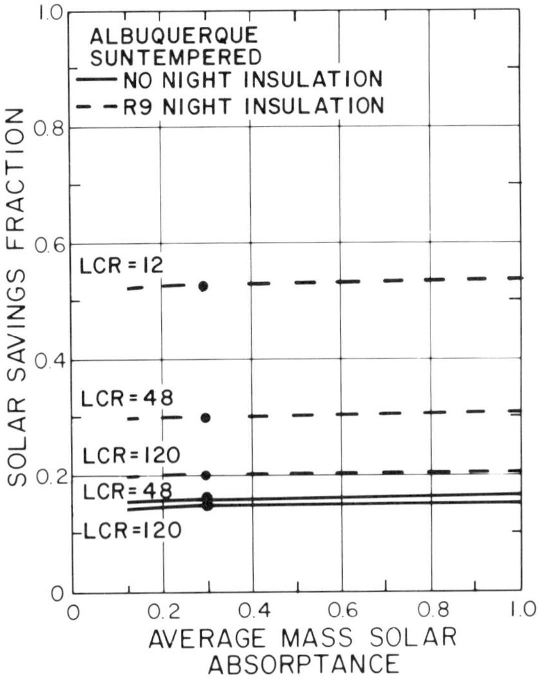

Figure A2-96

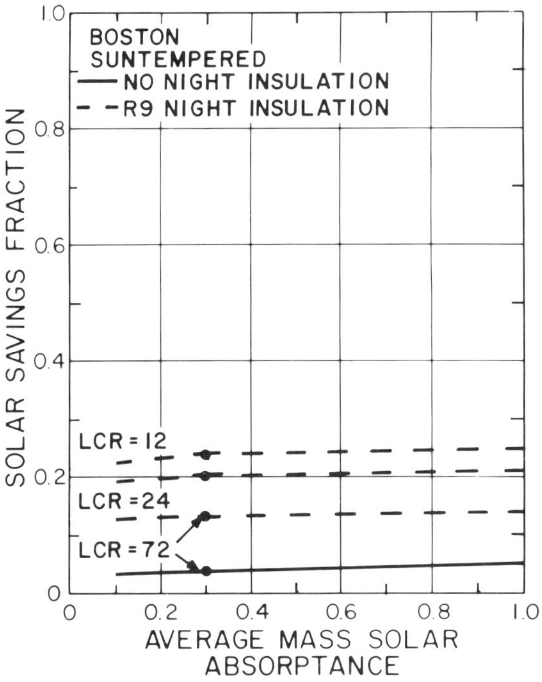

Figure A2-97

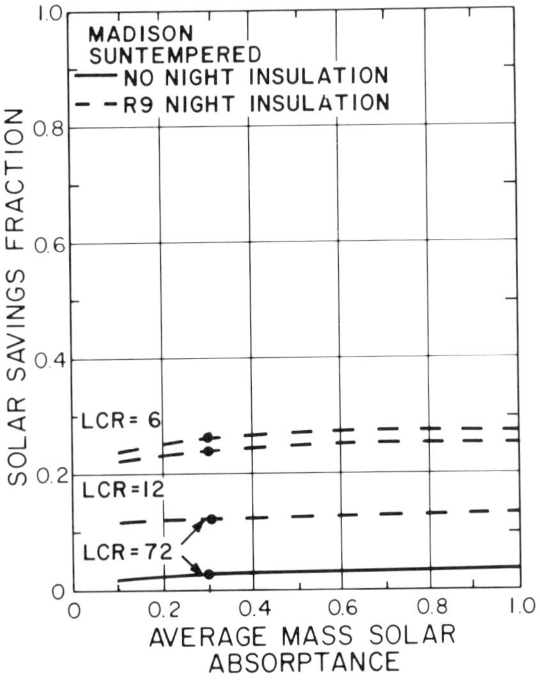

Figure A2-98

Figures A2-96 through A2-101. Annual heating performance sensitivity to the solar absorptance of the thermal storage mass surface for the suntempered reference design both without and with R9 night insulation. The dots represent the reference conditions.

SUNTEMPERED
MASS ABSORPTANCE

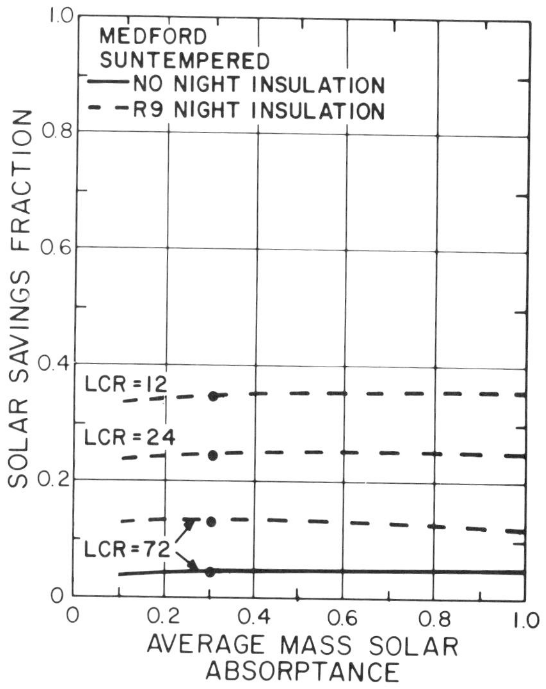

Figure A2-99

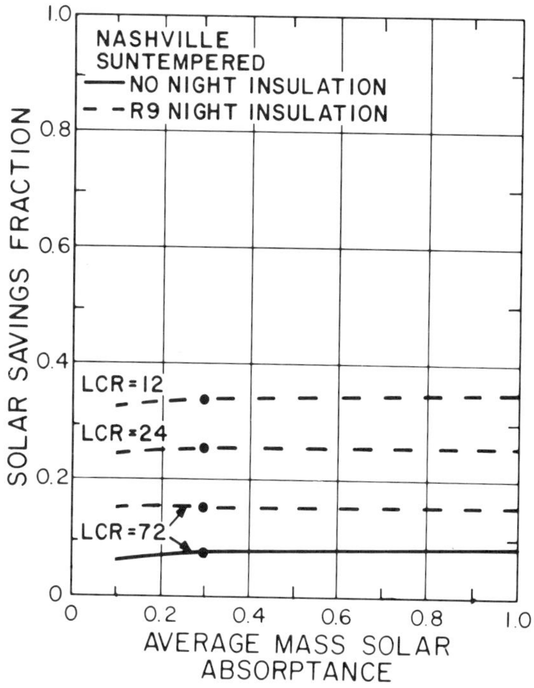

Figure A2-100

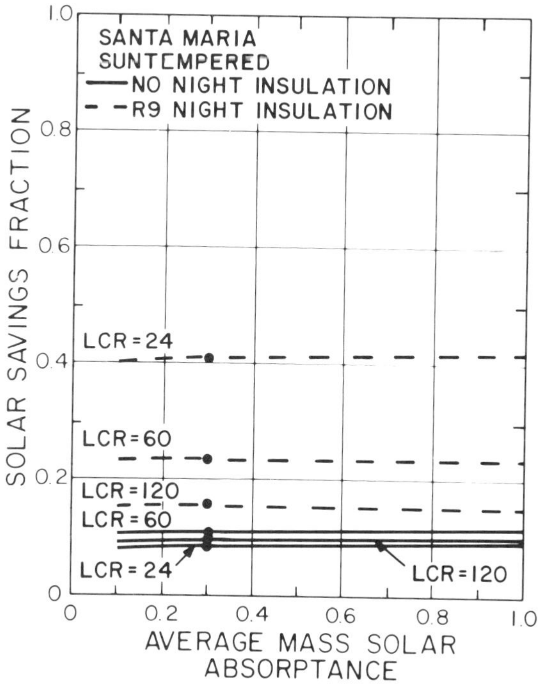

Figure A2-101

Figures A2-96 through A2-101. Annual heating performance sensitivity to the solar absorptance of the thermal storage mass surface for the suntempered reference design both without and with R9 night insulation. The dots represent the reference conditions.

SUNSPACE
STORAGE VOLUME

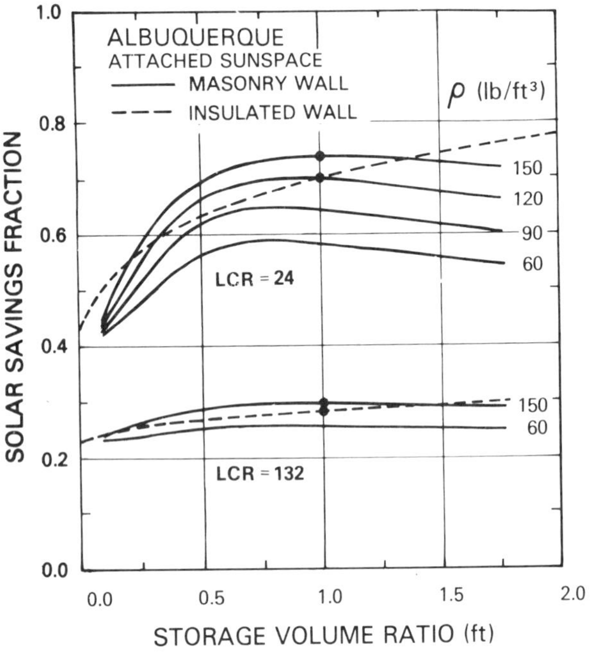

Figure A2-102

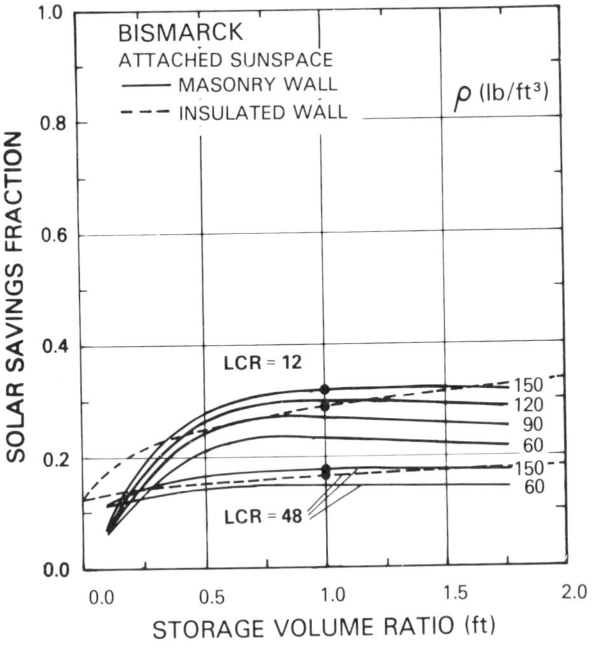

Figure A2-103

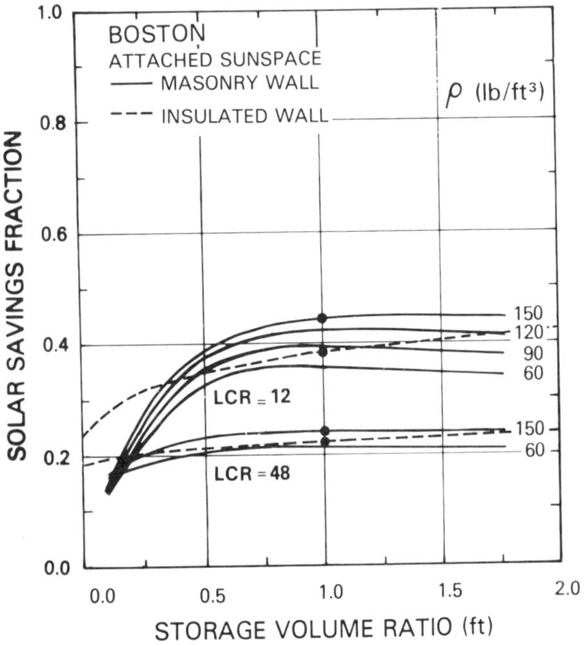

Figure A2-104

Figures A2-102 through A2-107. Annual heating performance sensitivity to the ratio of the thermal storage volume (ft^3) to the projected area (ft^2) for geometry A with insulated end walls, no night insulation, and both masonry and lightweight, insulated common wall (A1 and A5). In the masonry wall case, the storage volume ratio is only the wall thickness. Curves for various masonry densities, ρ (lb/ft^3), are included. The thermal conductivity is strongly correlated with the density and is varied as a function of the density. The dots represent the reference conditions.

SUNSPACE
STORAGE VOLUME

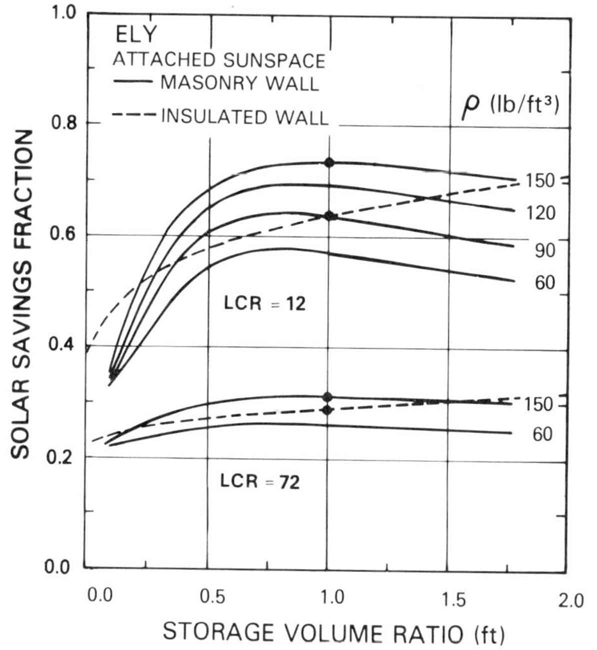

Figure A2-105

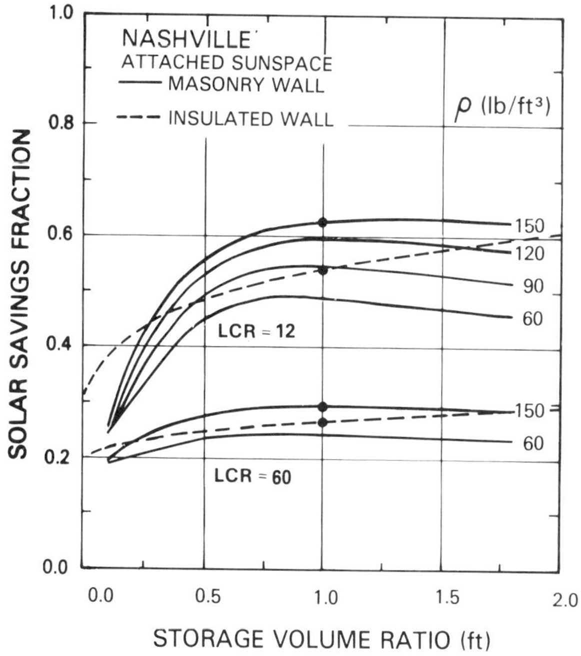

Figure A2-106

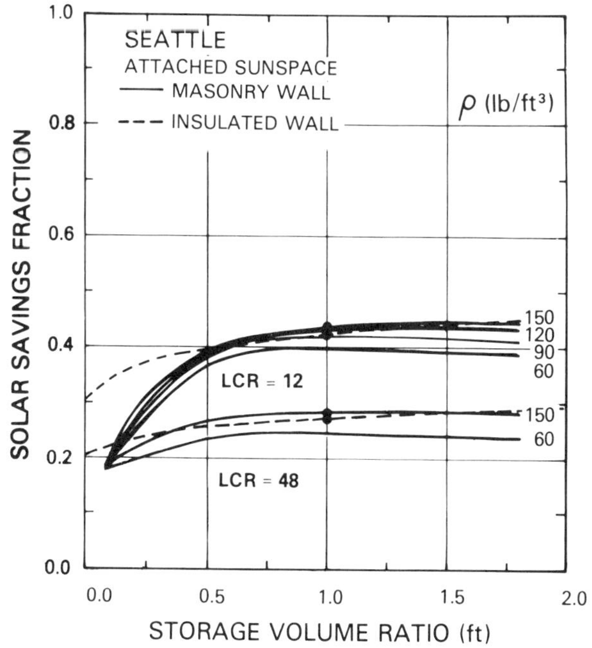

Figure A2-107

Figures A2-102 through A2-107. Annual heating performance sensitivity to the ratio of the thermal storage volume (ft^3) to the projected area (ft^2) for geometry A with insulated end walls, no night insulation, and both masonry and lightweight, insulated common wall (A1 and A5). In the masonry wall case, the storage volume ratio is only the wall thickness. Curves for various masonry densities, ρ (lb/ft^3), are included. The thermal conductivity is strongly correlated with the density and is varied as a function of the density. The dots represent the reference conditions.

SUNSPACE
STORAGE VOLUME

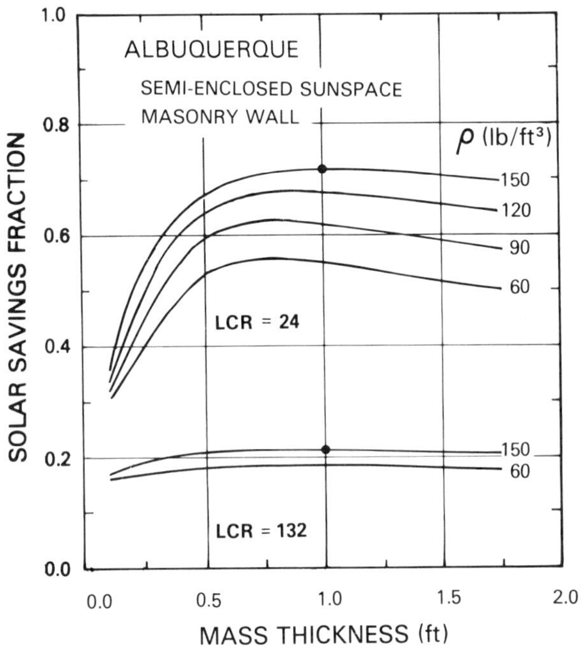

Figure A2-108

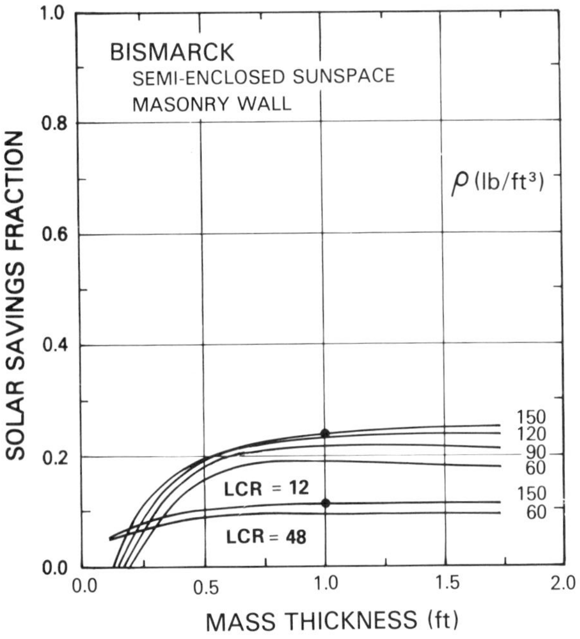

Figure A2-109

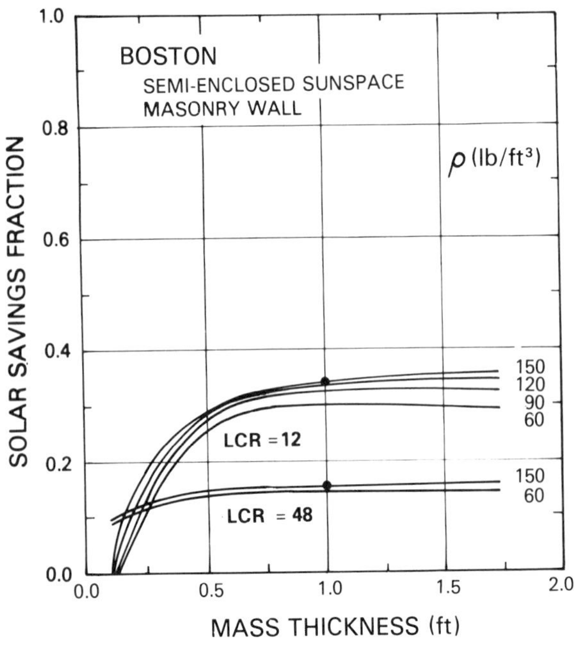

Figure A2-110

Figures A2-108 through A2-113. Annual heating performance sensitivity to the thickness of the masonry common wall for geometry C with no night insulation (C1). Curves for various masonry densities, ρ (lb/ft^3), are included. The thermal conductivity is strongly correlated with the density and is varied as a function of the density. The dots represent the reference conditions.

SUNSPACE
STORAGE VOLUME

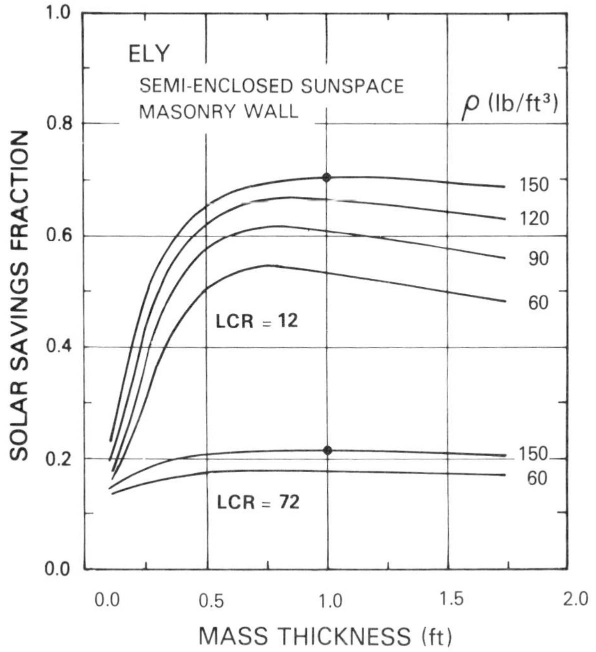

Figure A2-111

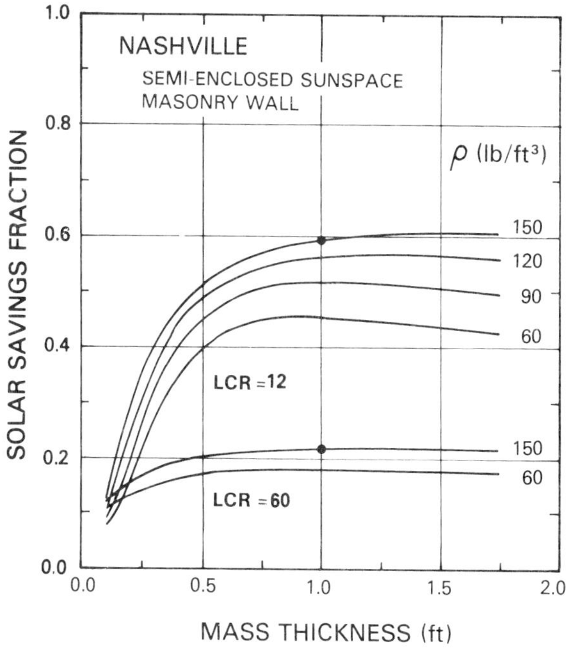

Figure A2-112

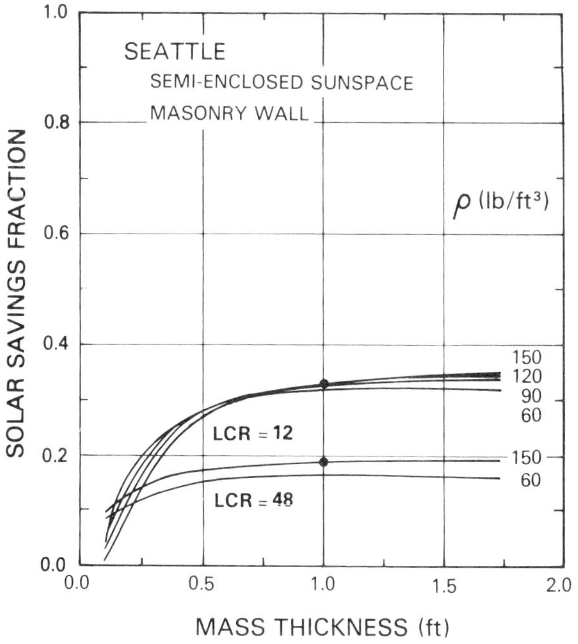

Figure A2-113

Figures A2-108 through A2-113. Annual heating performance sensitivity to the thickness of the masonry common wall for geometry C with no night insulation (C1). Curves for various masonry densities, ρ (lb/ft^3), are included. The thermal conductivity is strongly correlated with the density and is varied as a function of the density. The dots represent the reference conditions.

SUNSPACE
STORAGE VOLUME

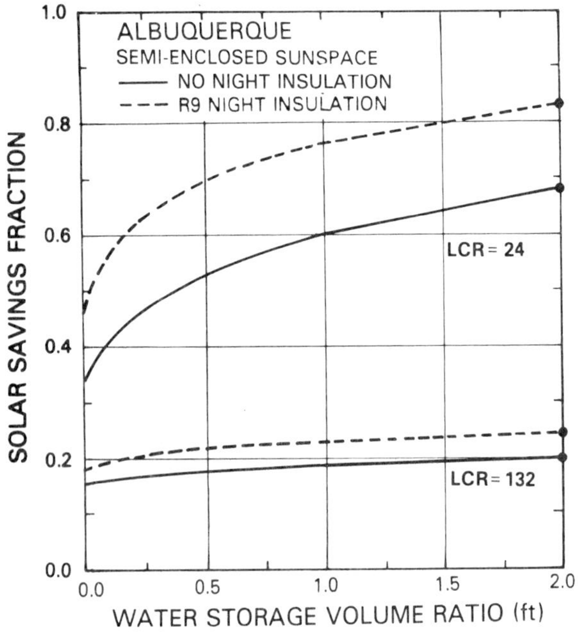

Figure A2-114

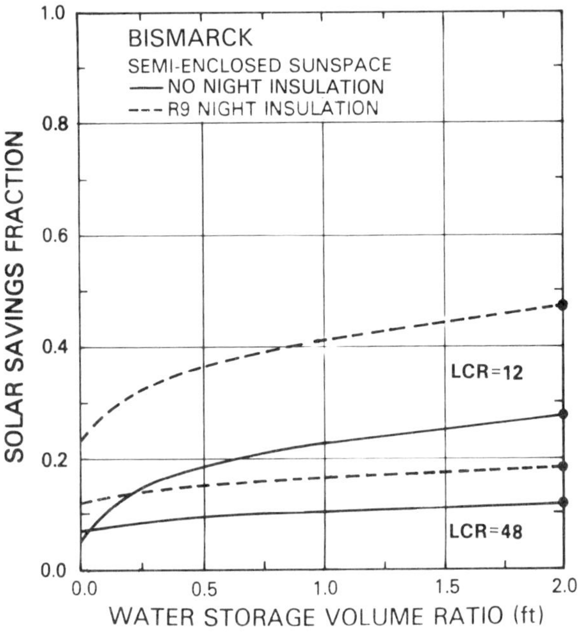

Figure A2-115

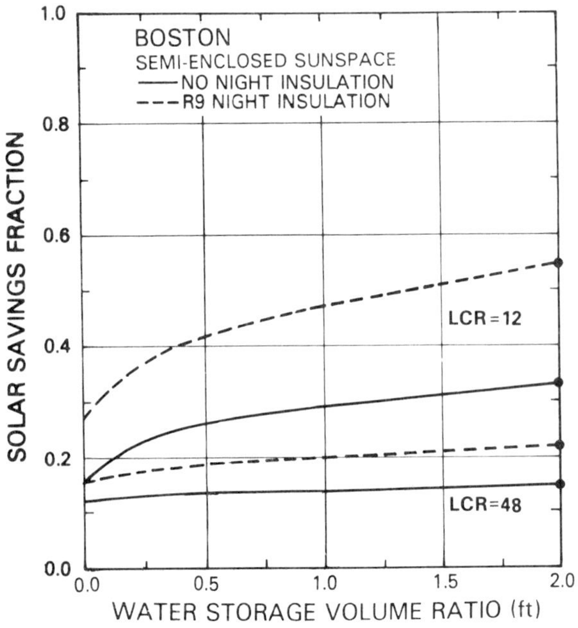

Figure A2-116

Figures A2-114 through A2-119. Annual heating performance sensitivity to the ratio of the volume of water storage containers (ft^3) to the projected area (ft^2) for geometry C both without and with night insulation (C3 and C4). The dots represent the reference conditions.

SUNSPACE
STORAGE VOLUME

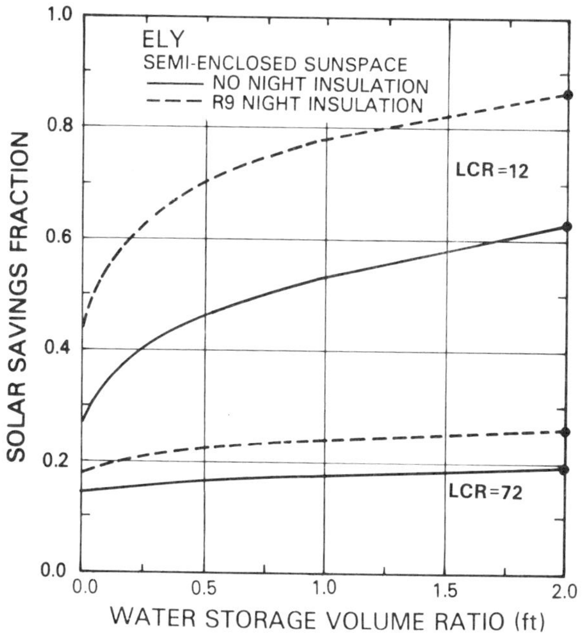

Figure A2-117

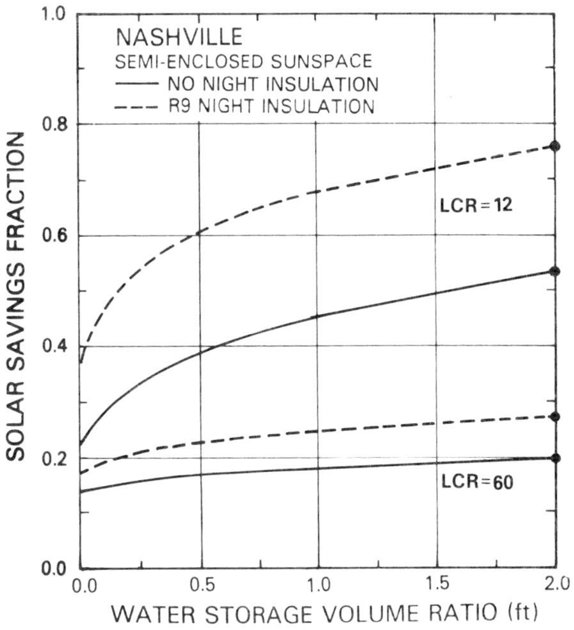

Figure A2-118

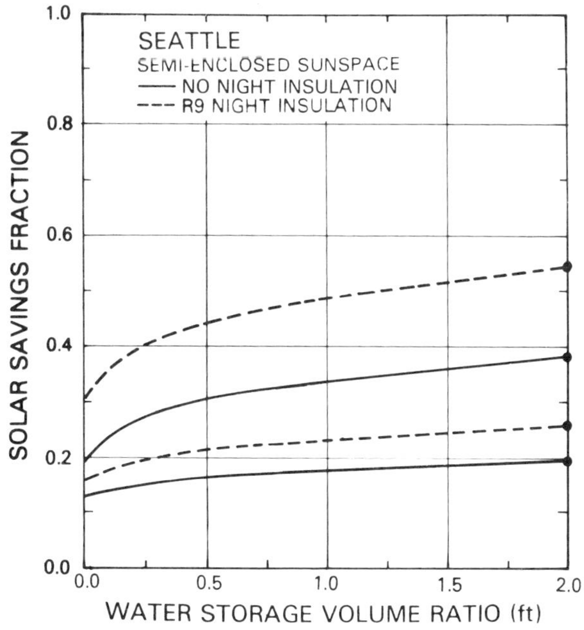

Figure A2-119

Figures A2-114 through A2-119. Annual heating performance sensitivity to the ratio of the volume of water storage containers (ft^3) to the projected area (ft^2) for geometry C both without and with night insulation (C3 and C4). The dots represent the reference conditions.

SUNSPACE
ORIENTATION

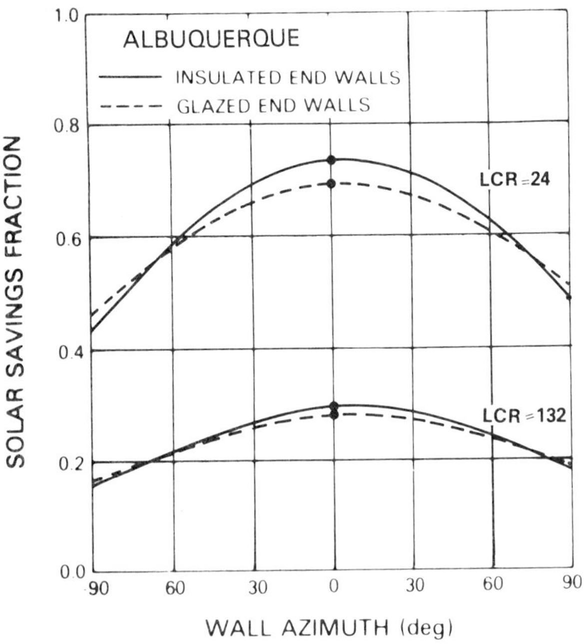

Figure A2-120

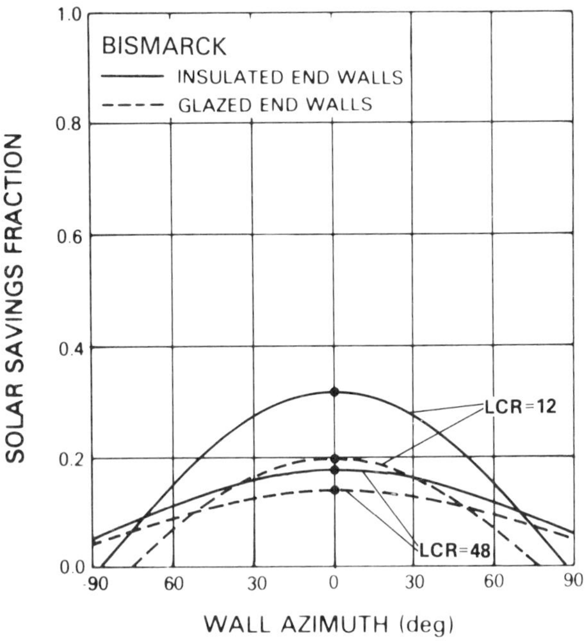

Figure A2-121

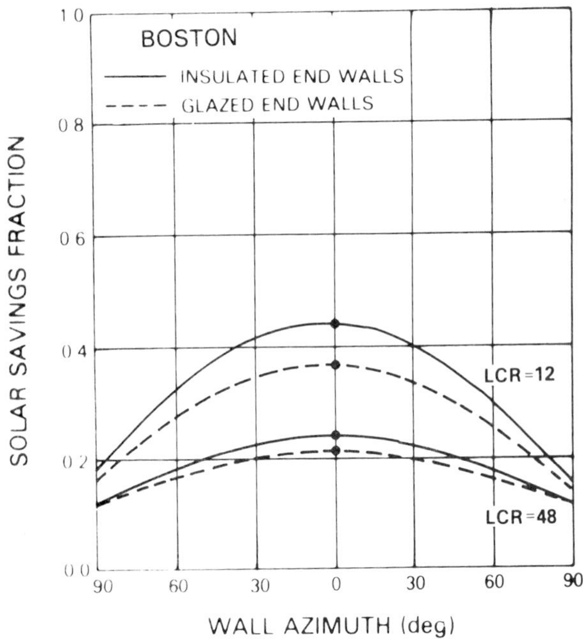

Figure A2-122

Figures A2-120 through A2-125. Annual heating performance sensitivity to the sunspace orientation for geometry A with masonry common wall, no night insulation, and both insulated and glazed end walls (A1 and A3). A wall azimuth of zero is a due south orientation. Negative wall azimuth is west of south; positive is east of south. The dots represent the reference conditions.

SUNSPACE
ORIENTATION

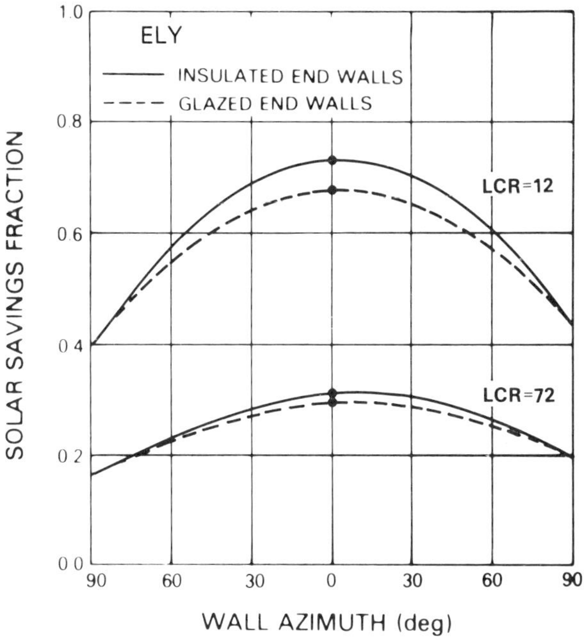

Figure A2-123

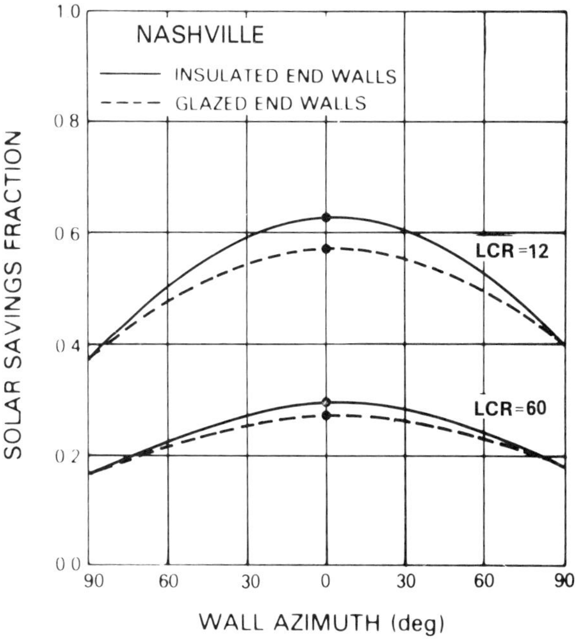

Figure A2-124

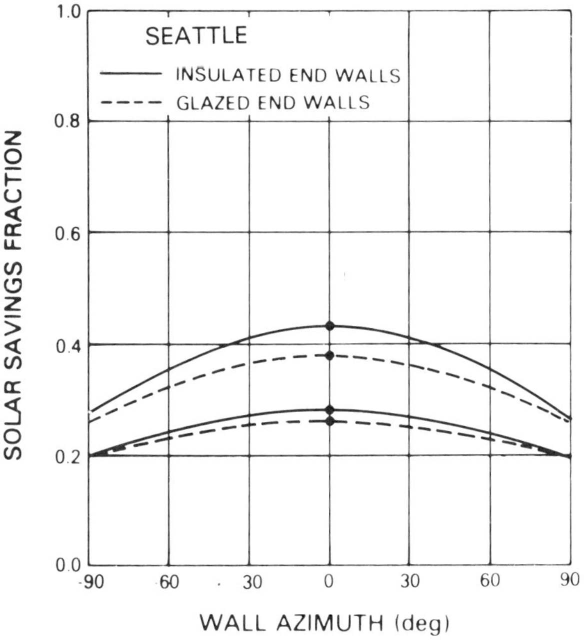

Figure A2-125

Figures A2-120 through A2-125. Annual heating performance sensitivity to the sunspace orientation for geometry A with masonry common wall, no night insulation, and both insulated and glazed end walls (A1 and A3). A wall azimuth of zero is a due south orientation. Negative wall azimuth is west of south; positive is east of south. The dots represent the reference conditions.

SUNSPACE GLAZING TILT

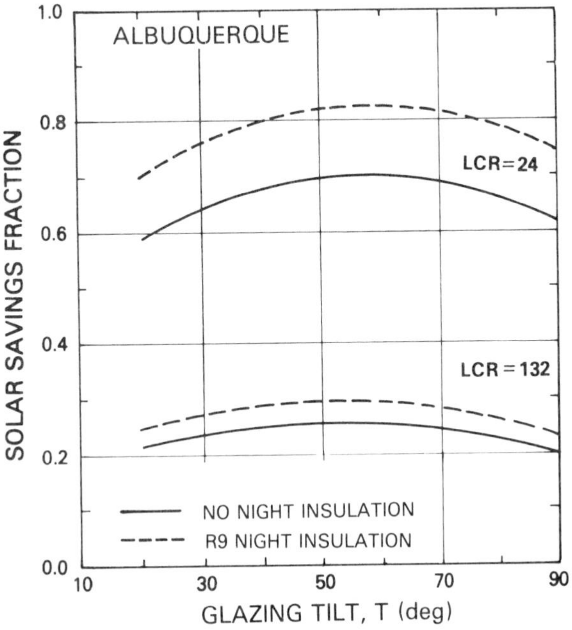

Figure A2-126

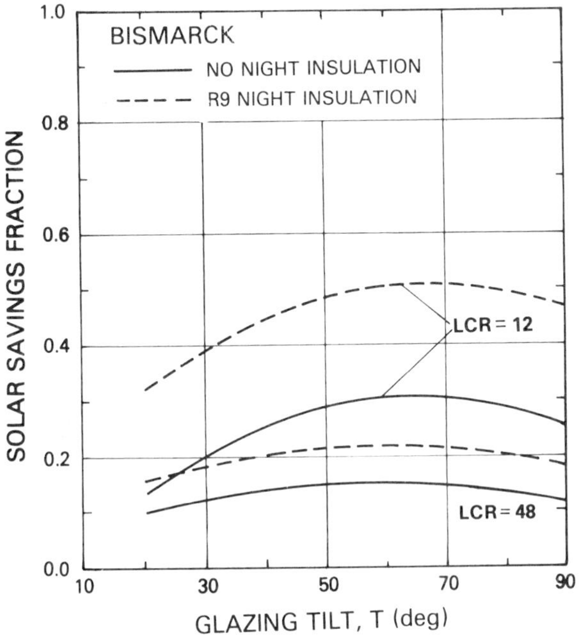

Figure A2-127

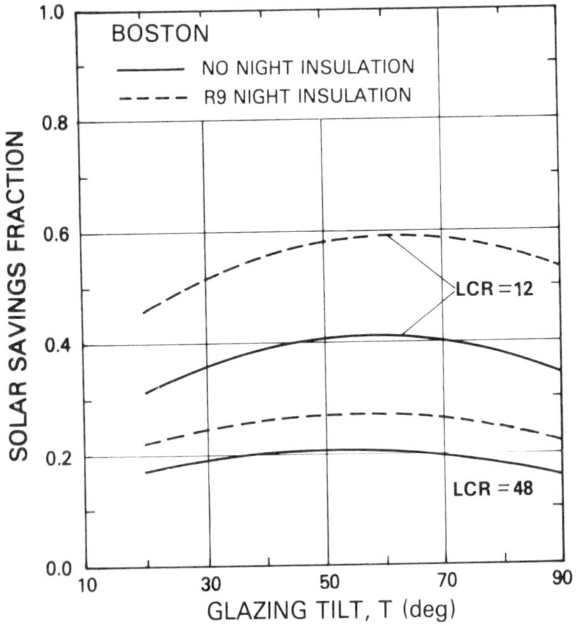

Figure A2-128

Figures A2-126 through A2-131. Annual heating performance sensitivity to the sunspace glazing tilt in the case of a masonry-common-wall, insulated-end-wall configuration with a single, tilted glazing plane. The glazing area is held constant as the tilt is varied by the addition of the variable-height knee wall. The given *LCR* values refer to the vertical glazing configuration. The drawing below illustrates the geometry.

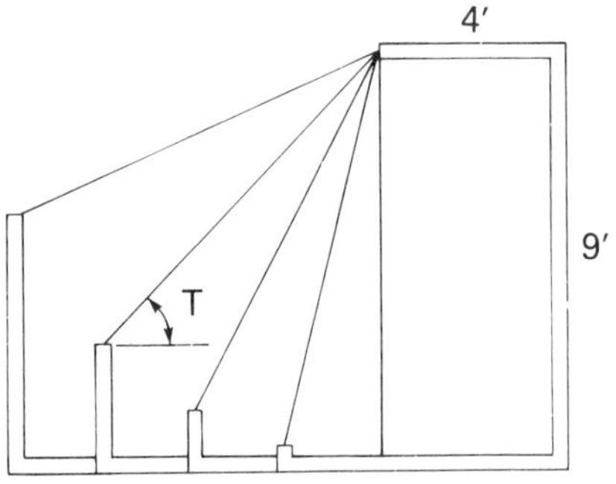

SUNSPACE GLAZING TILT

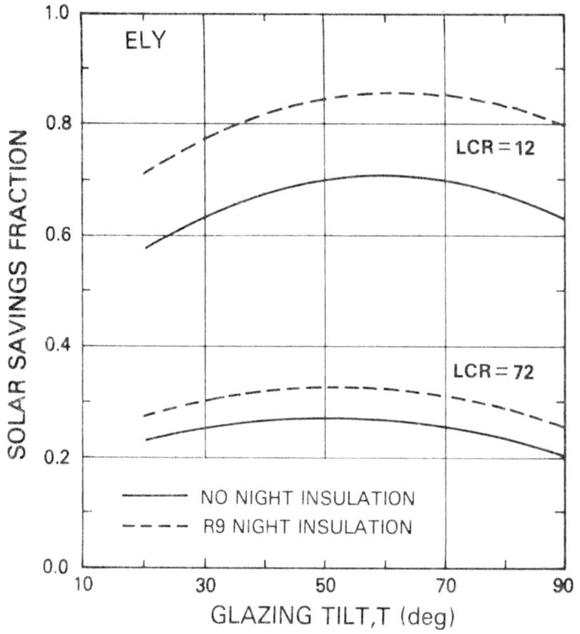

Figure A2-129

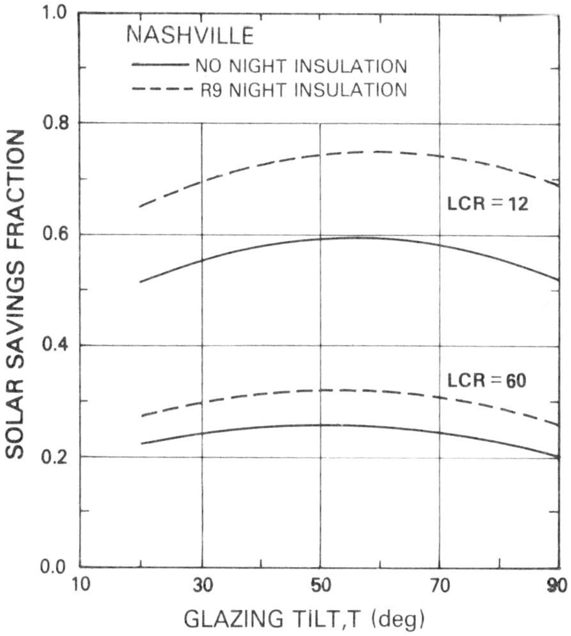

Figure A2-130

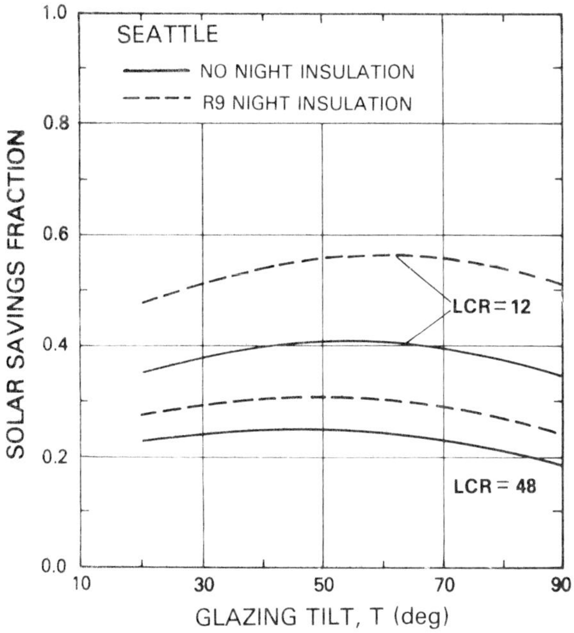

Figure A2-131

Figures A2-126 through A2-131. Annual heating performance sensitivity to the sunspace glazing tilt in the case of a masonry-common-wall, insulated-end-wall configuration with a single, tilted glazing plane. The glazing area is held constant as the tilt is varied by the addition of the variable-height knee wall. The given *LCR* values refer to the vertical glazing configuration. The drawing below illustrates the geometry.

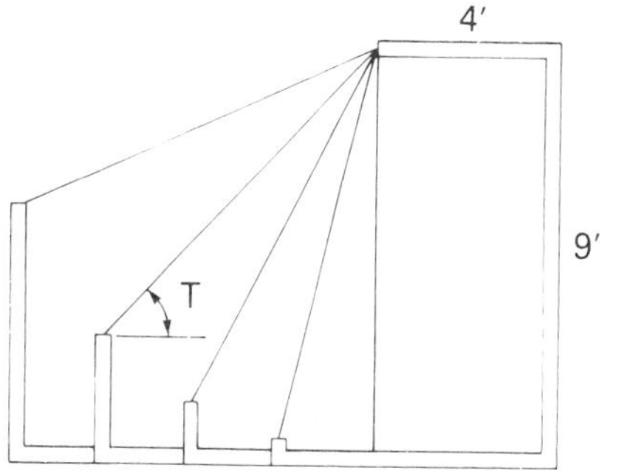

SUNSPACE GLAZING TILT

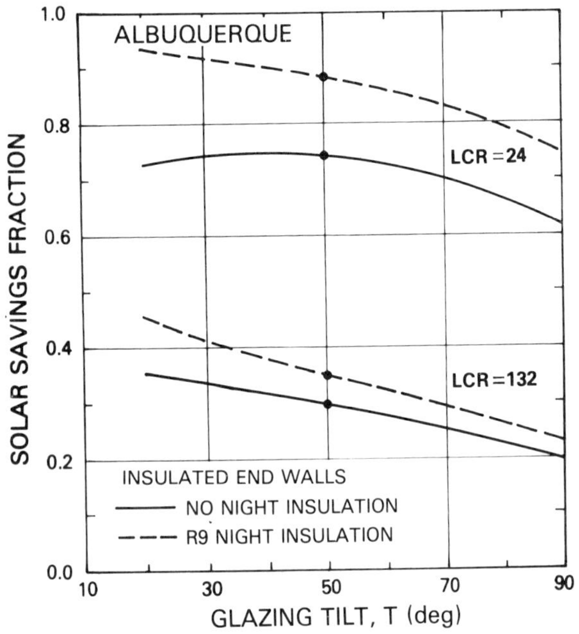

Figure A2-132

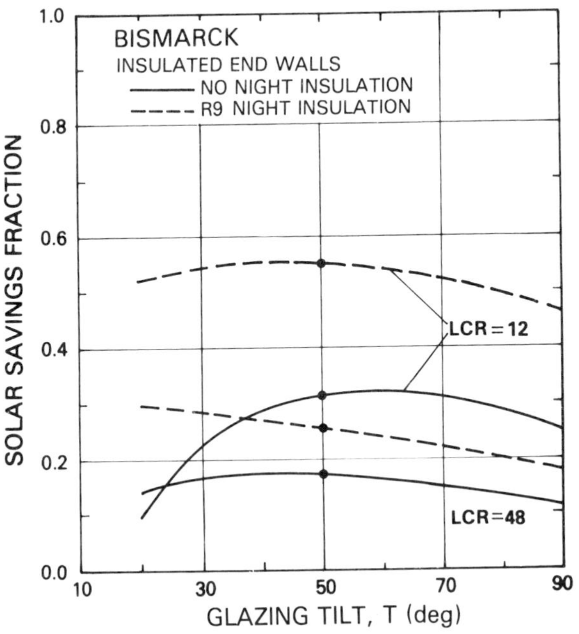

Figure A2-133

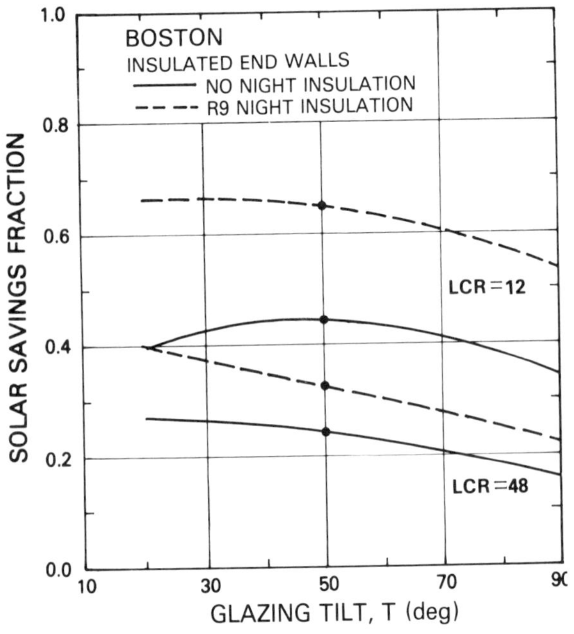

Figure A2-134

Figures A2-132 through A2-137. Annual heating performance sensitivity to the sunspace glazing tilt in the case of a masonry-common-wall, insulated-end-wall, geometry-A-like configuration with a single, tilted glazing plane, both without and with R9 night insulation (A1 and A2). The glazing area is allowed to vary as the tilt is varied. The drawing below illustrates the geometry. The dots represent the reference conditions.

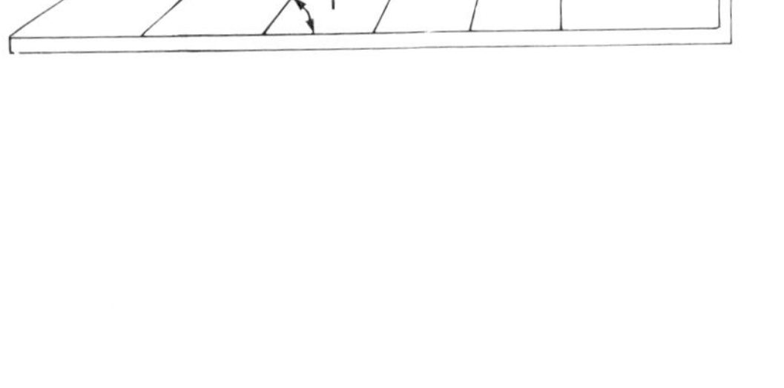

SUNSPACE
GLAZING TILT

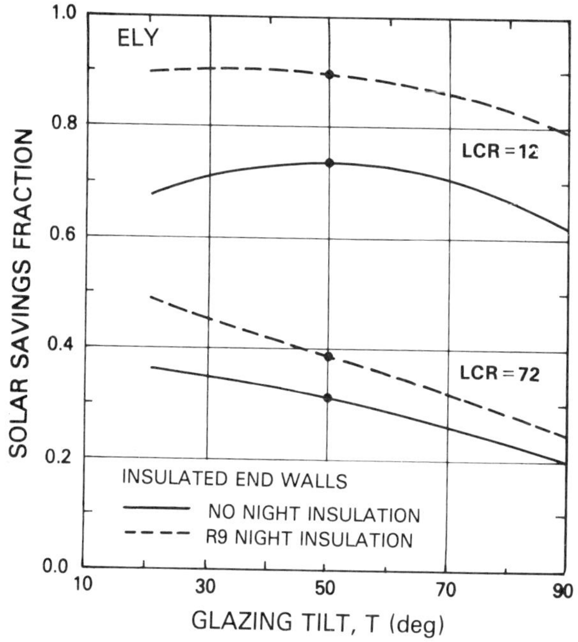

Figure A2-135

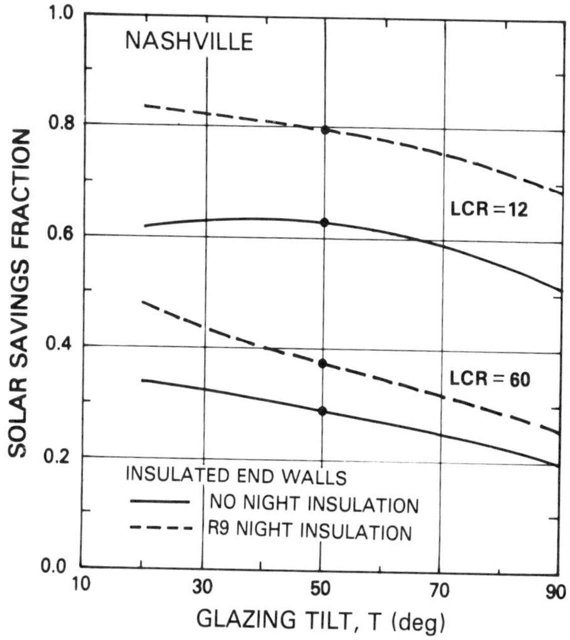

Figure A2-136

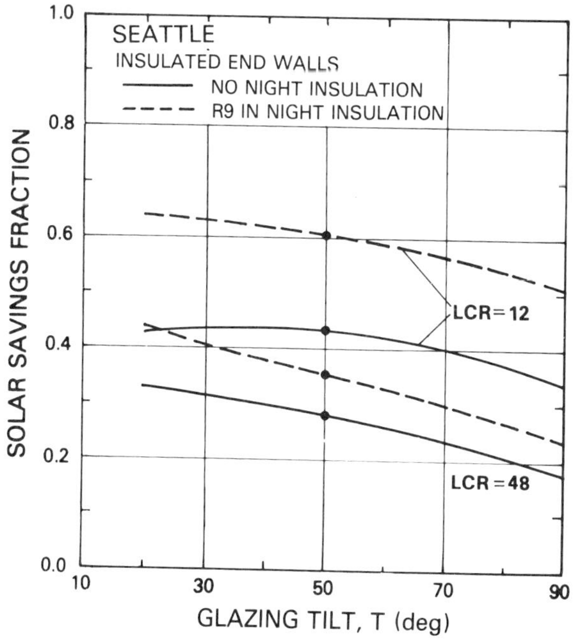

Figure A2-137

Figures A2-132 through A2-137. Annual heating performance sensitivity to the sunspace glazing tilt in the case of a masonry-common-wall, insulated-end-wall, geometry-A-like configuration with a single, tilted glazing plane, both without and with R9 night insulation (A1 and A2). The glazing area is allowed to vary as the tilt is varied. The drawing below illustrates the geometry. The dots represent the reference conditions.

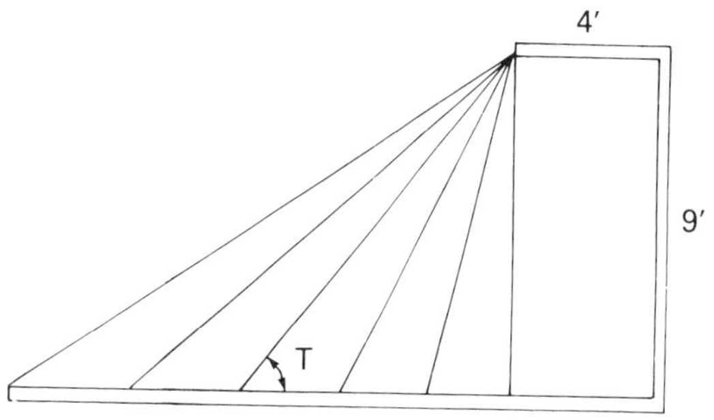

SUNSPACE
GLAZING TILT

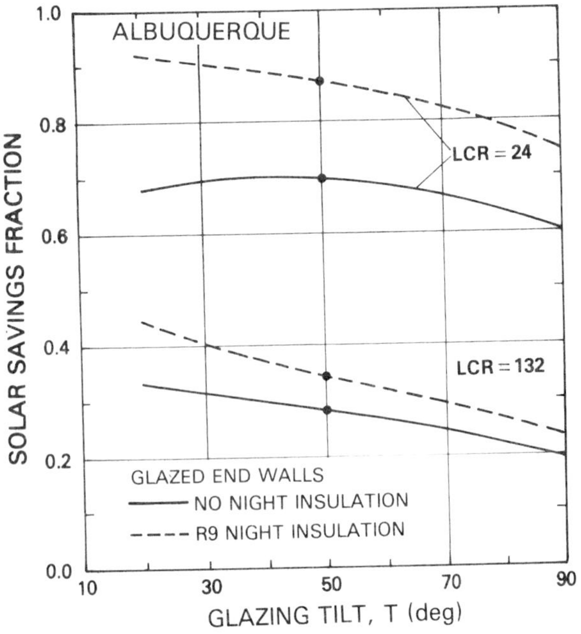

Figure A2-138

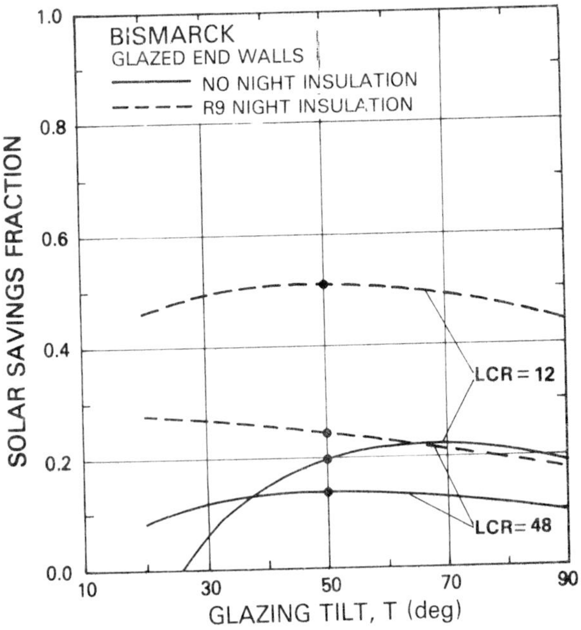

Figure A2-139

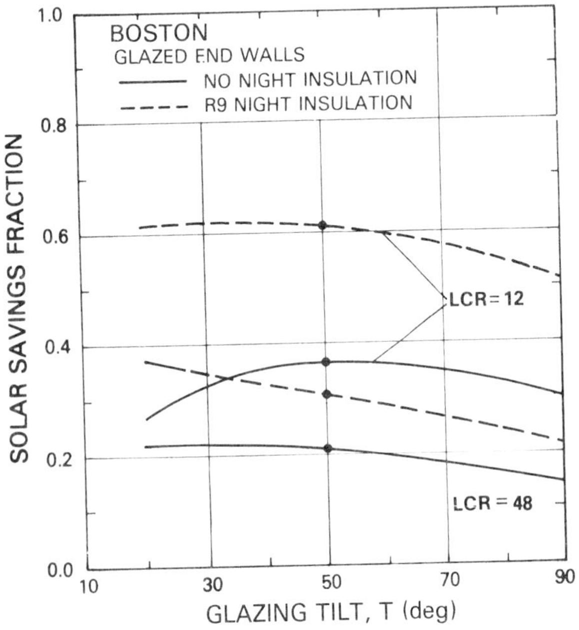

Figure A2-140

Figures A2-138 through A2-143. Annual heating performance sensitivity to the sunspace glazing tilt in the case of a masonry-common-wall, glazed-end-wall, geometry-A-like configuration with a single tilted glazing plane, both without and with R9 night insulation (A3 and A4). The glazing area is allowed to vary as the tilt is varied. The drawing below illustrates the geometry. The dots represent the reference conditions.

SUNSPACE
GLAZING TILT

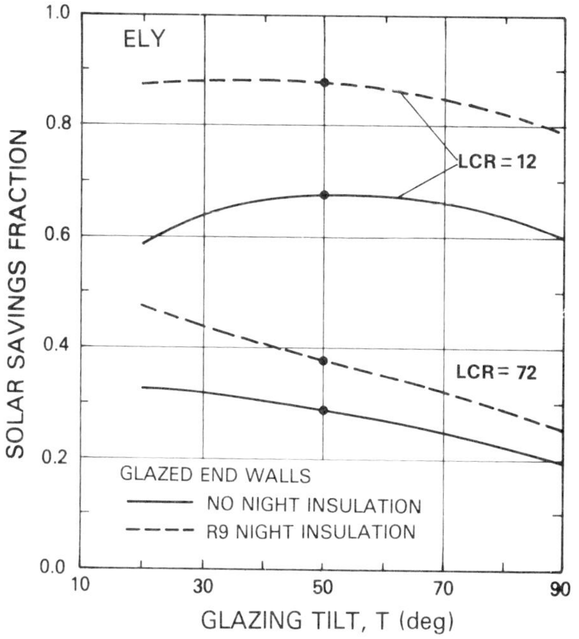

Figure A2-141

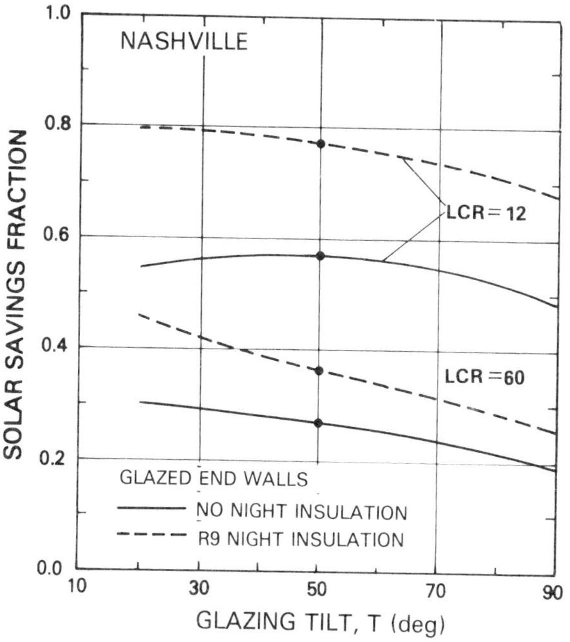

Figure A2-142

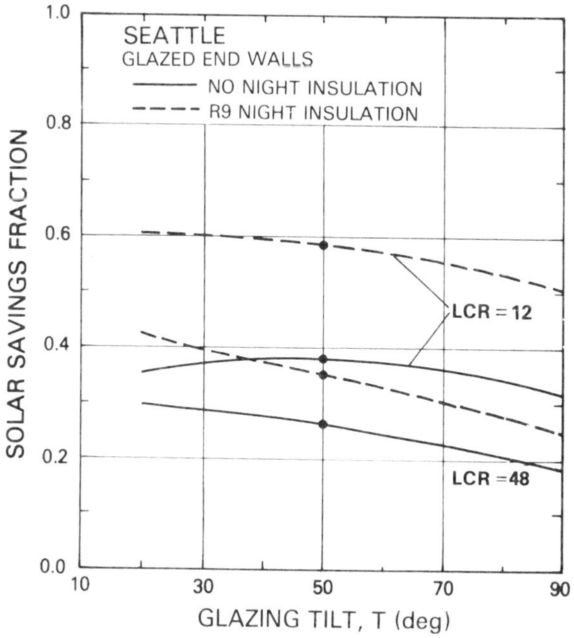

Figure A2-143

Figures A2-138 through A2-143. Annual heating performance sensitivity to the sunspace glazing tilt in the case of a masonry-common-wall, glazed-end-wall, geometry-A-like configuration with a single tilted glazing plane, both without and with R9 night insulation (A3 and A4). The glazing area is allowed to vary as the tilt is varied. The drawing below illustrates the geometry. The dots represent the reference conditions.

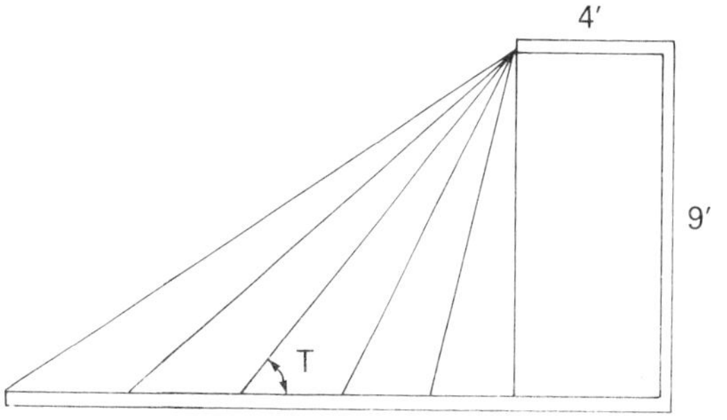

SUNSPACE
VERTICAL GLAZING HEIGHT

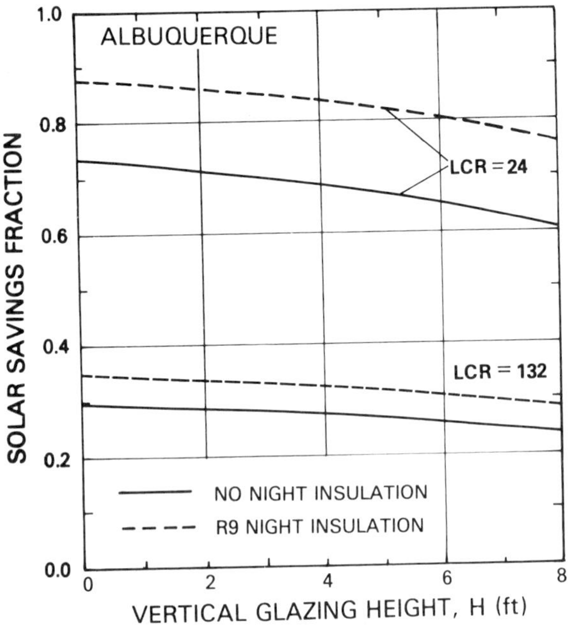

Figure A2-144

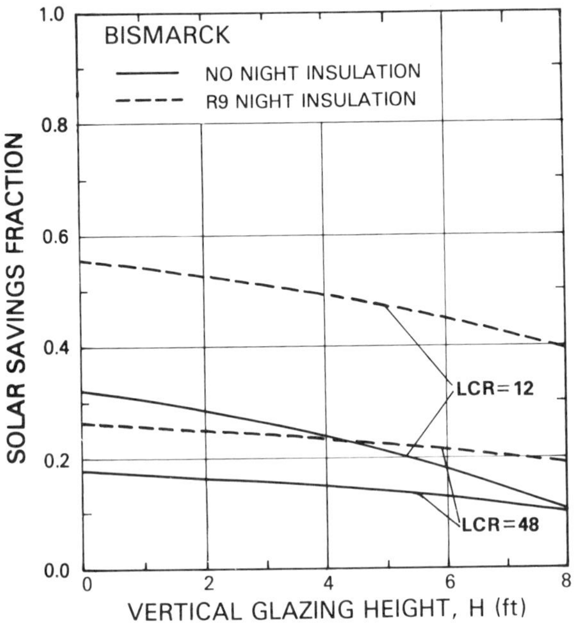

Figure A2-145

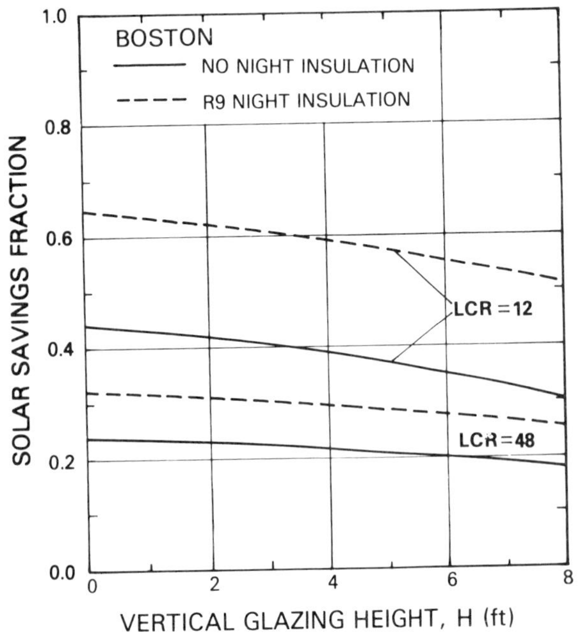

Figure A2-146

Figures A2-144 through A2-149. Annual heating performance sensitivity to the sunspace vertical glazing height for the masonry-common-wall, insulated-end-wall, attached-sunspace configuration. The variation represents the transition from geometry A to geometry B except that there is no 4-foot ceiling in geometry B. The drawing below illustrates the geometry.

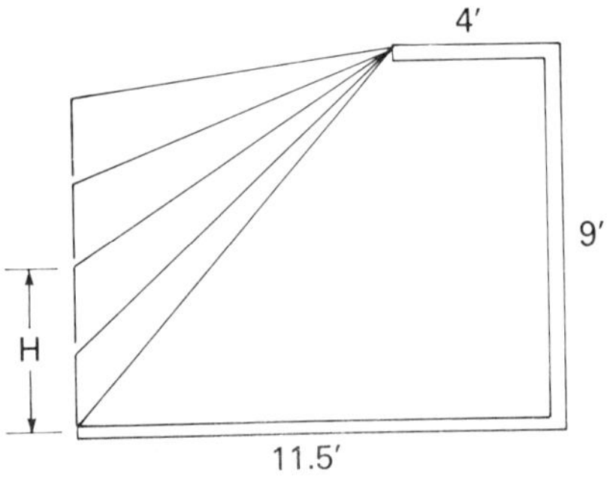

SUNSPACE
VERTICAL GLAZING HEIGHT

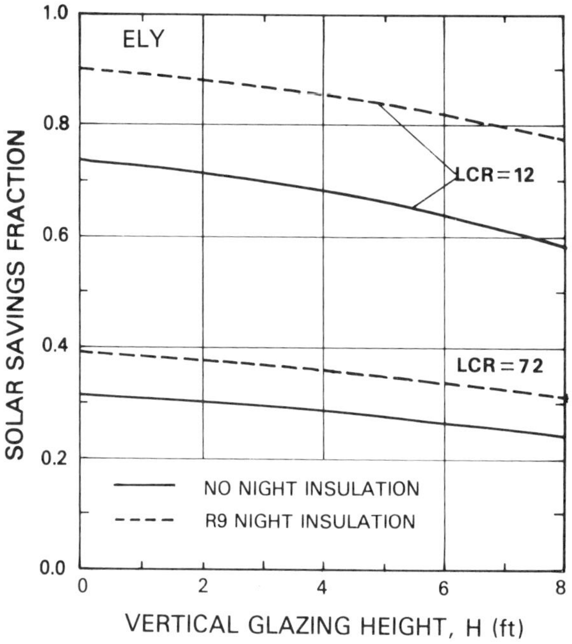

Figure A2-147

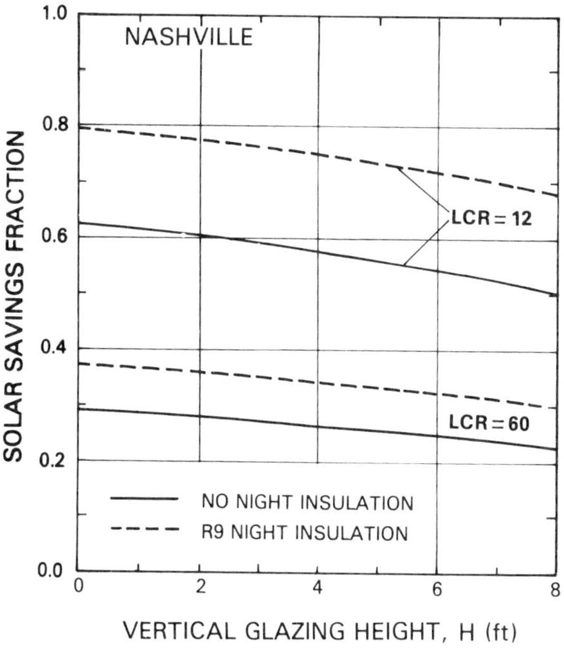

Figure A2-148

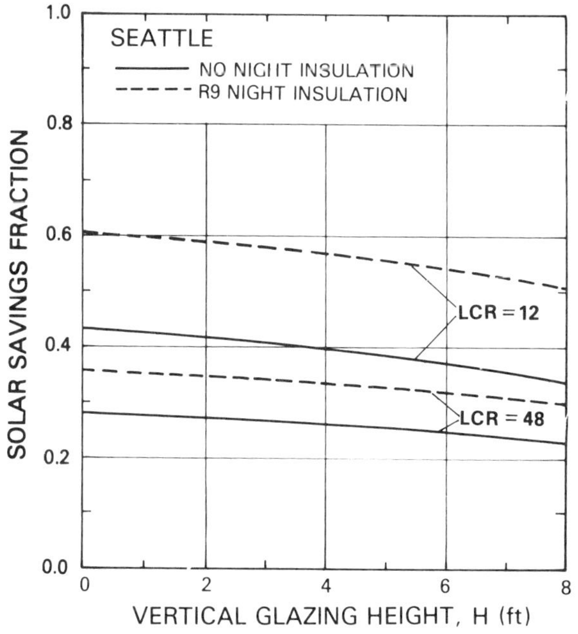

Figure A2-149

Figures A2-144 through A2-149. Annual heating performance sensitivity to the sunspace vertical glazing height for the masonry-common-wall, insulated-end-wall, attached-sunspace configuration. The variation represents the transition from geometry A to geometry B except that there is no 4-foot ceiling in geometry B. The drawing below illustrates the geometry.

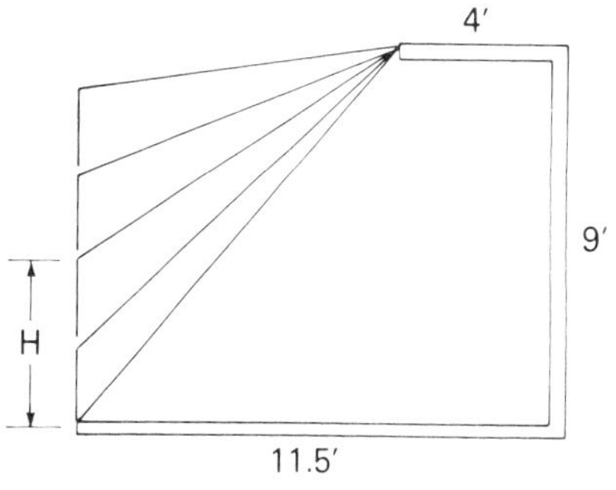

SUNSPACE
UPPER GLAZING TILT

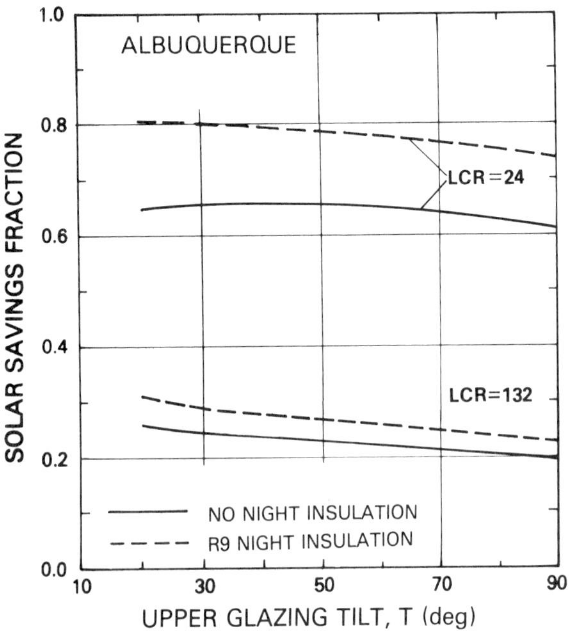

Figure A2-150

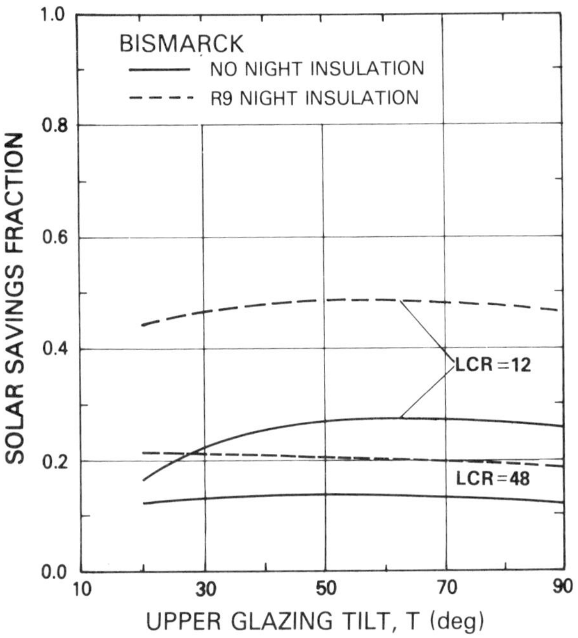

Figure A2-151

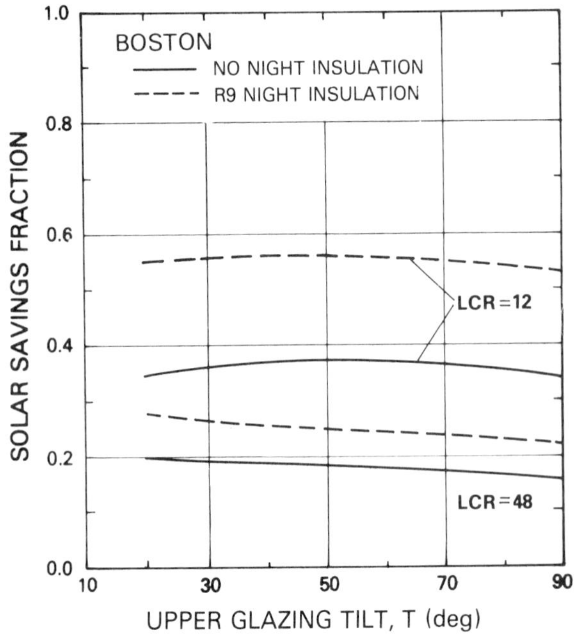

Figure A2-152

Figures A2-150 through A2-155. Annual heating performance sensitivity to the sunspace upper glazing tilt in the case of an insulated-end-wall, geometry-B-like configuration with a 6-foot vertical glazing and a tilted upper glazing. The drawing below illustrates the geometry. When the tilt is 30°, the geometry resembles geometry B except that there is no 4-foot ceiling in geometry B.

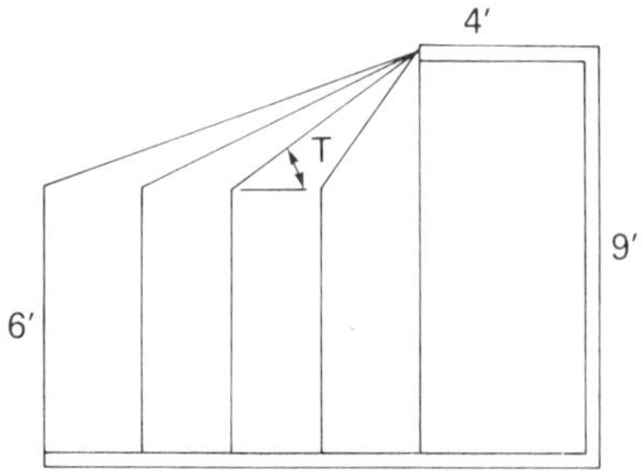

SUNSPACE
UPPER GLAZING TILT

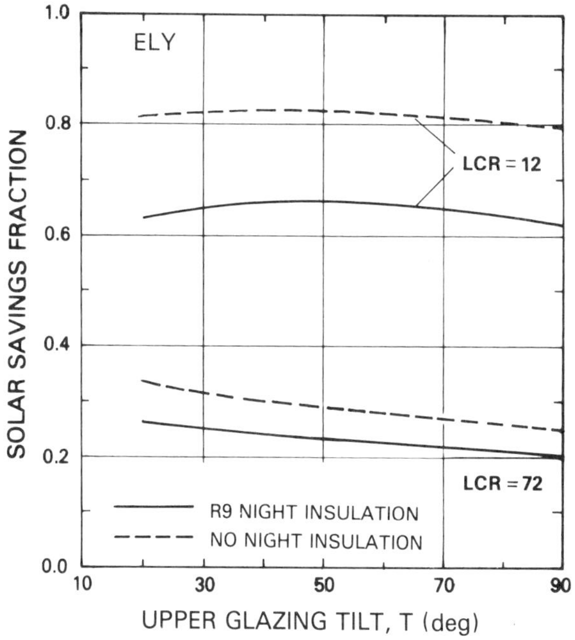

Figure A2-153

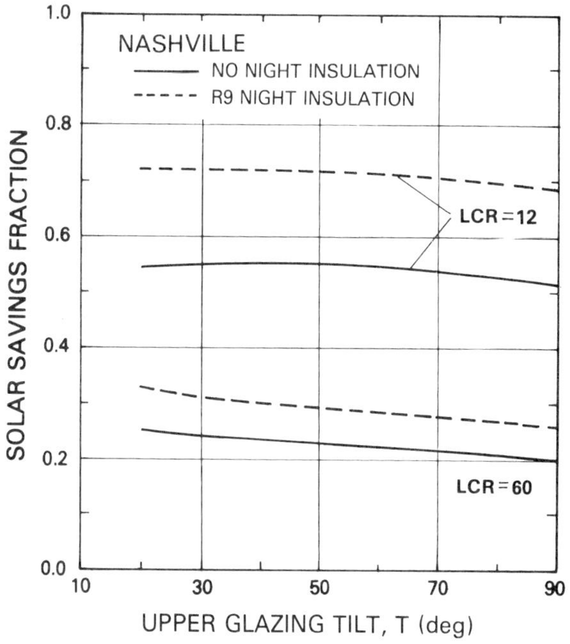

Figure A2-154

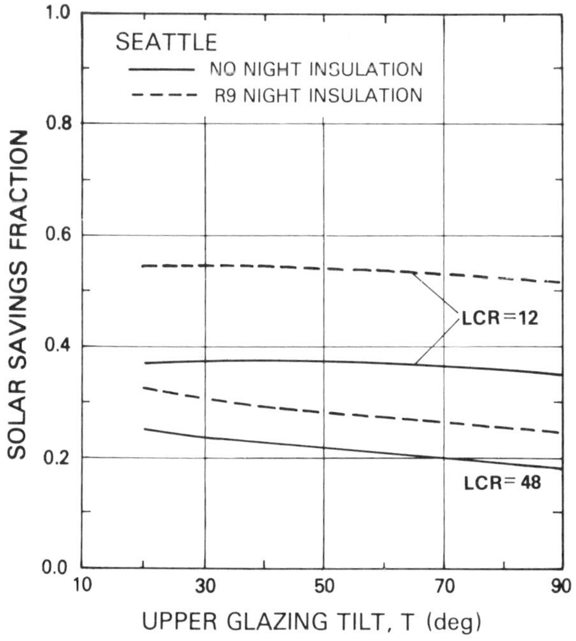

Figure A2-155

Figures A2-150 through A2-155. Annual heating performance sensitivity to the sunspace upper glazing tilt in the case of an insulated-end-wall, geometry-B-like configuration with a 6-foot vertical glazing and a tilted upper glazing. The drawing below illustrates the geometry. When the tilt is 30°, the geometry resembles geometry B except that there is no 4-foot ceiling in geometry B.

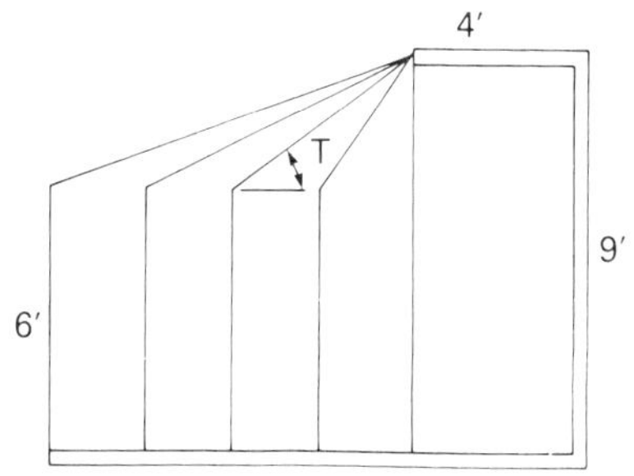

SUNSPACE
WIDTH

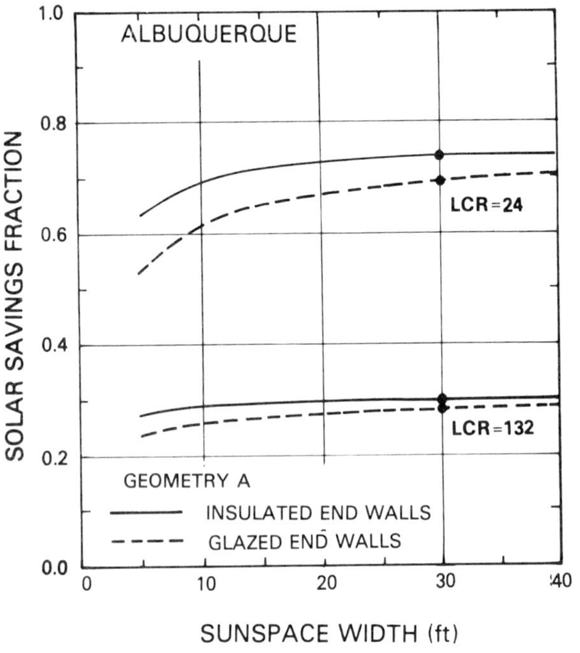

Figure A2-156

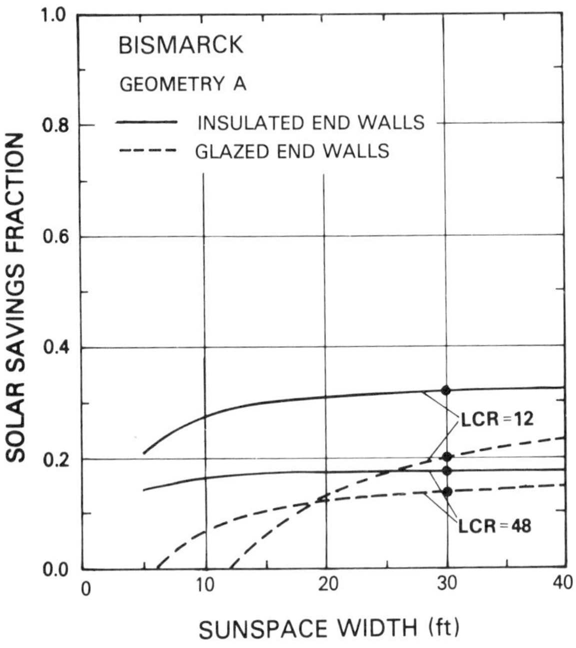

Figure A2-157

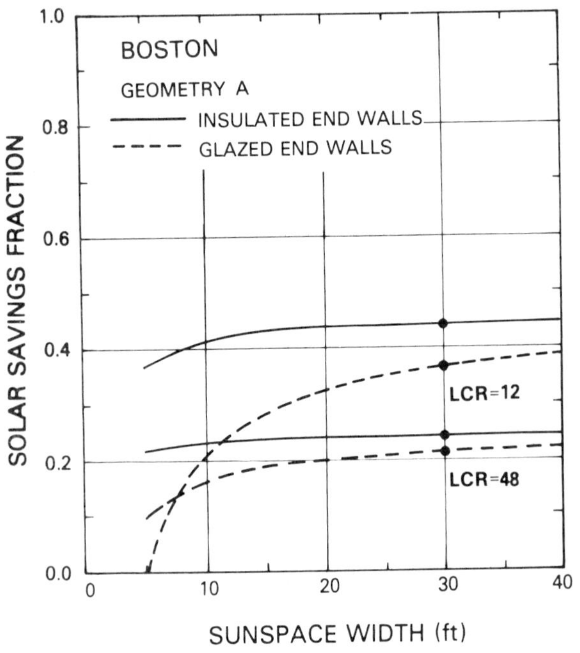

Figure A2-158

Figures A2-156 through A2-161. Annual heating performance sensitivity to the sunspace east-west width for sunspace geometry A with a masonry common wall, no night insulation, and both insulated and glazed end walls (A1 and A3). The dots represent the reference conditions.

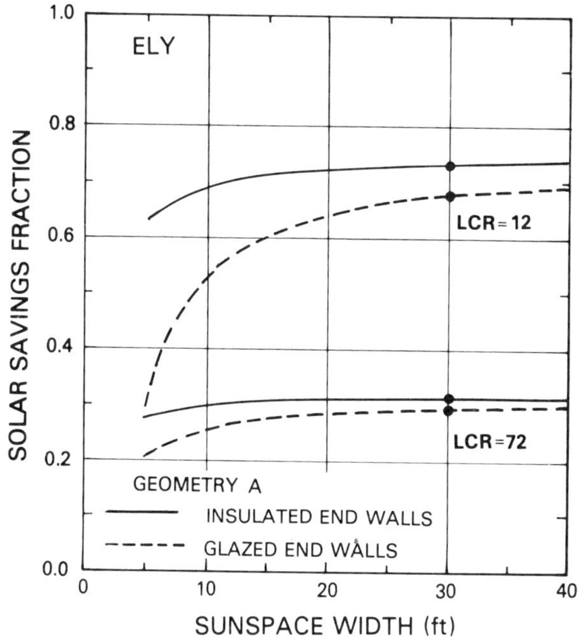

Figure A2-159

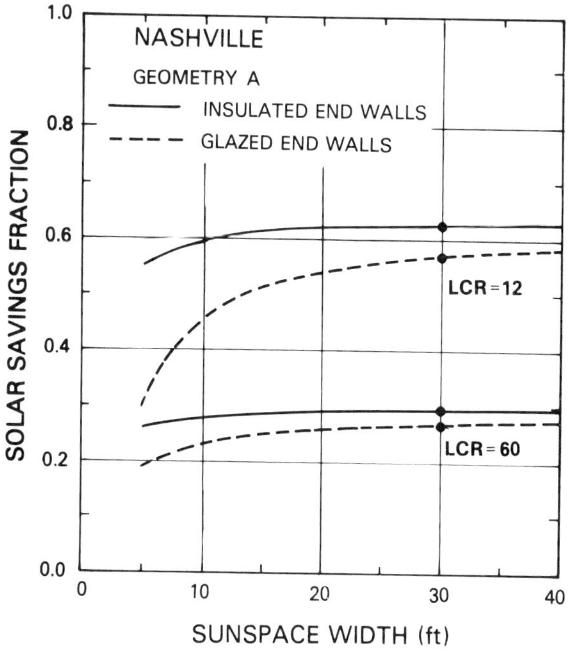

Figure A2-160

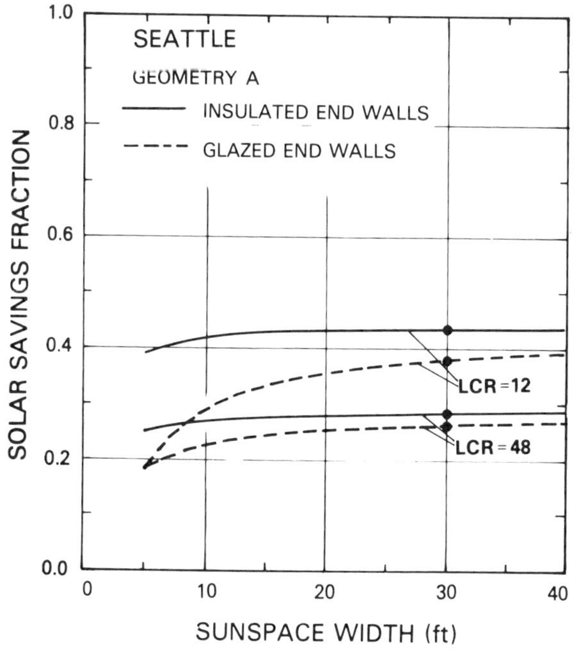

Figure A2-161

Figures A2-156 through A2-161. Annual heating performance sensitivity to the sunspace east-west width for sunspace geometry A with a masonry common wall, no night insulation, and both insulated and glazed end walls (A1 and A3). The dots represent the reference conditions.

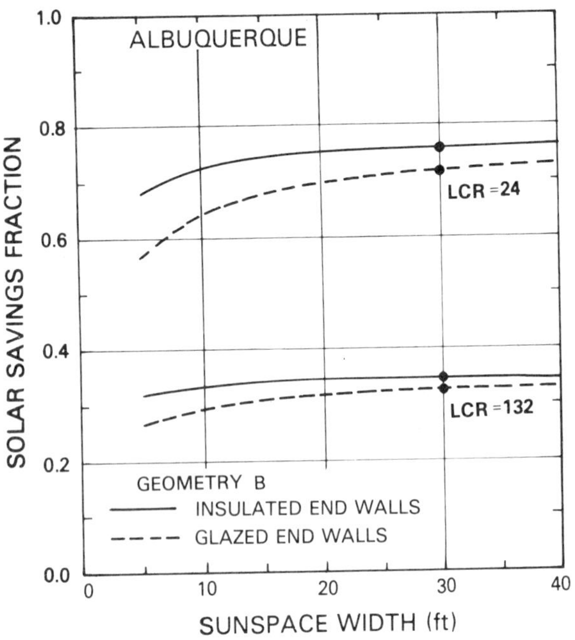

Figure A2-162

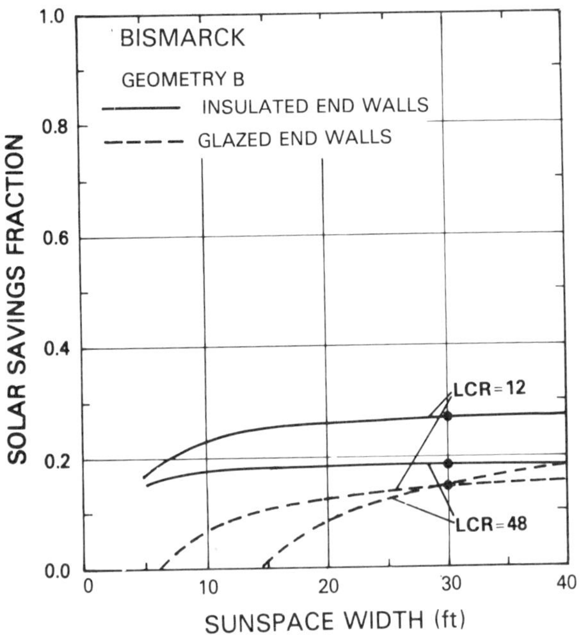

Figure A2-163

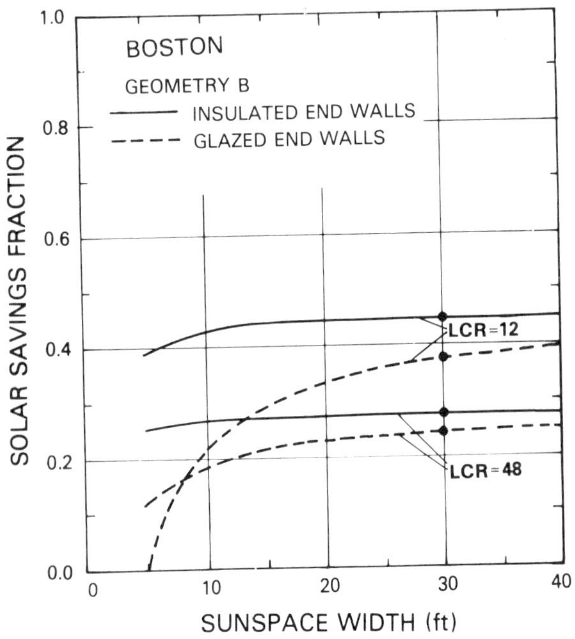

Figure A2-164

Figures A2-162 through A2-167. Annual heating performance sensitivity to the sunspace east-west width for geometry B with masonry common wall, no night insulation, and both insulated and glazed end walls (B1 and B3). The dots represent the reference conditions.

SUNSPACE
WIDTH

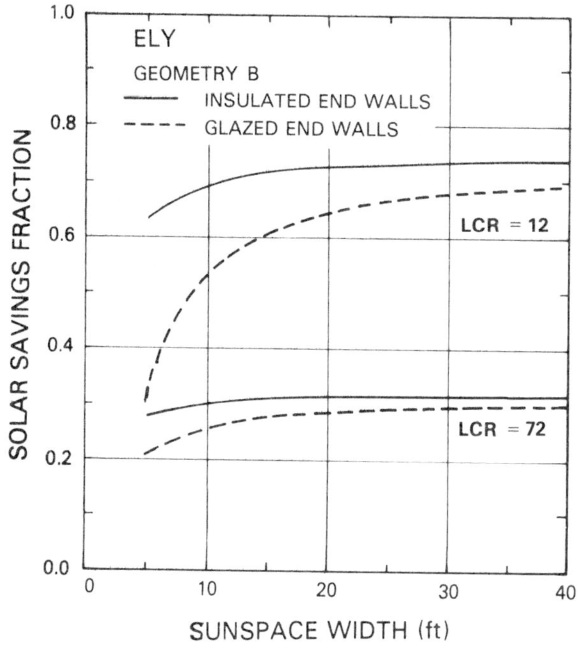

Figure A2-165

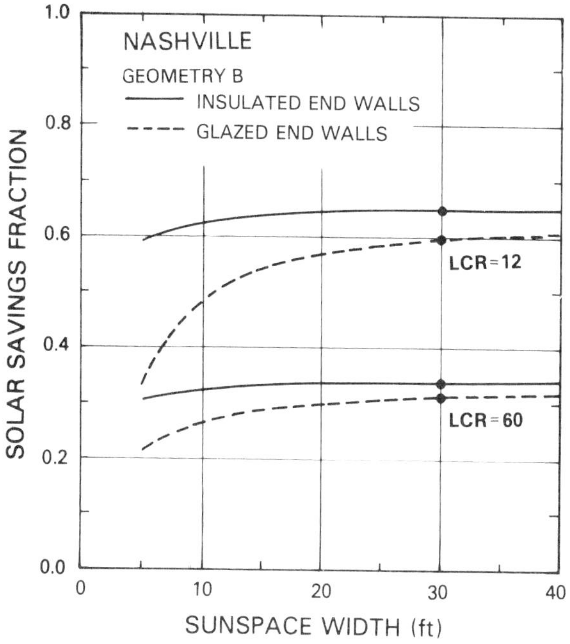

Figure A2-166

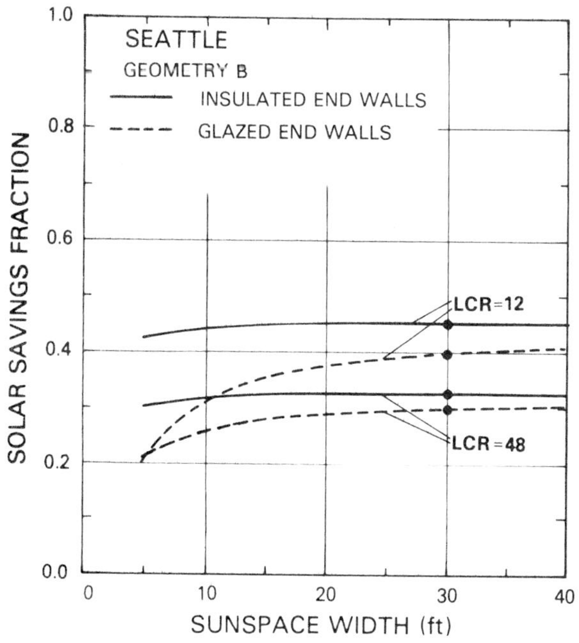

Figure A2-167

Figures A2-162 through A2-167. Annual heating performance sensitivity to the sunspace east-west width for geometry B with masonry common wall, no night insulation, and both insulated and glazed end walls (B1 and B3). The dots represent the reference conditions.

SUNSPACE WIDTH

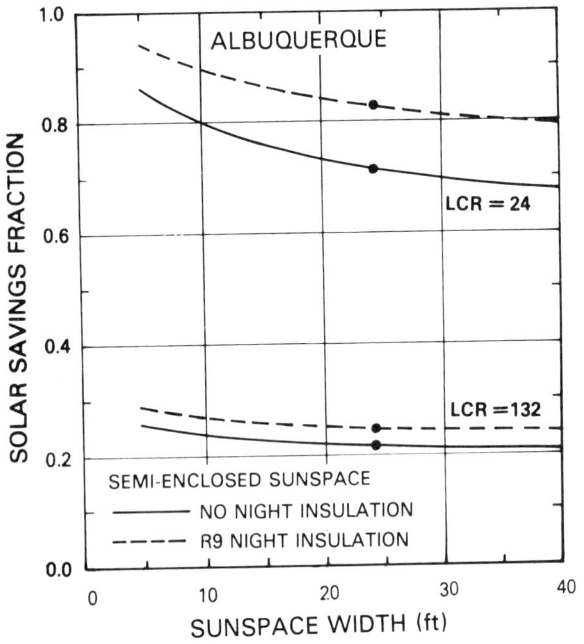

Figure A2-168

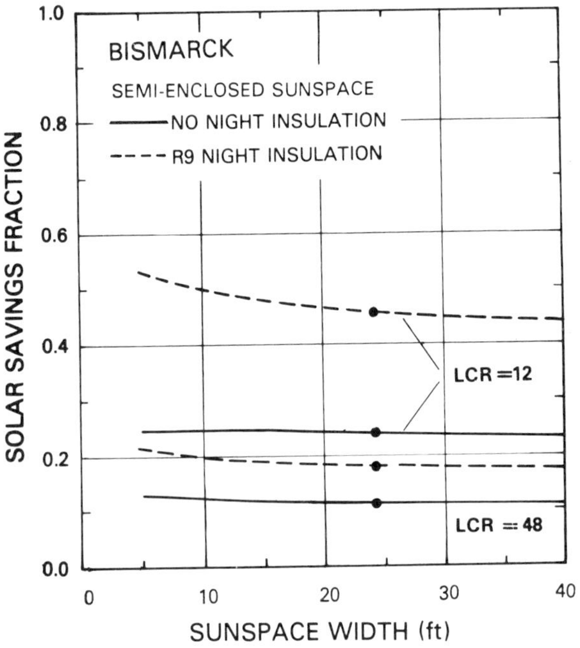

Figure A2-169

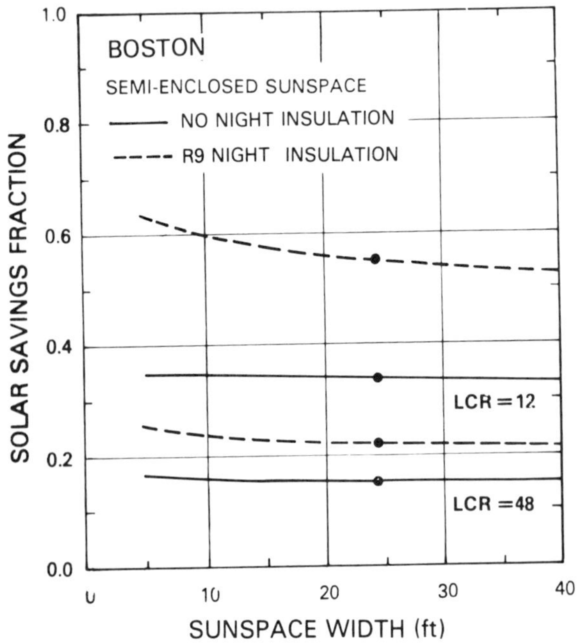

Figure A2-170

Figures A2-168 through A2-173. Annual heating performance sensitivity to the sunspace east-west width for geometry C with a masonry common wall, both without and with R9 night insulation (C1 and C2). The dots represent the reference conditions.

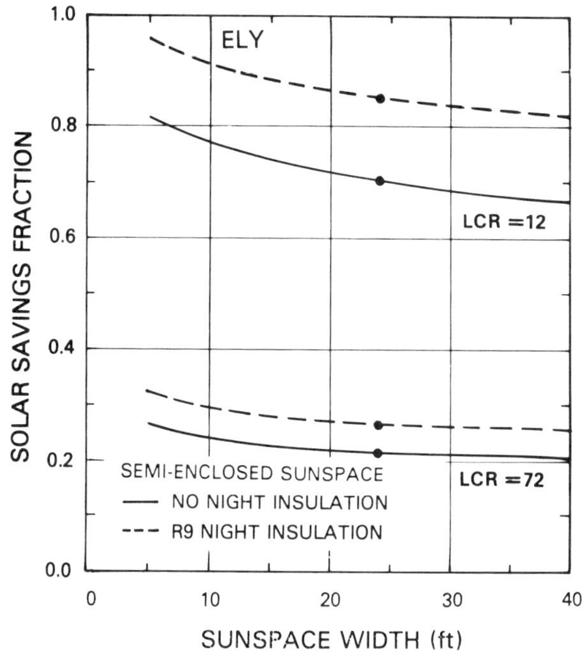

Figure A2-171

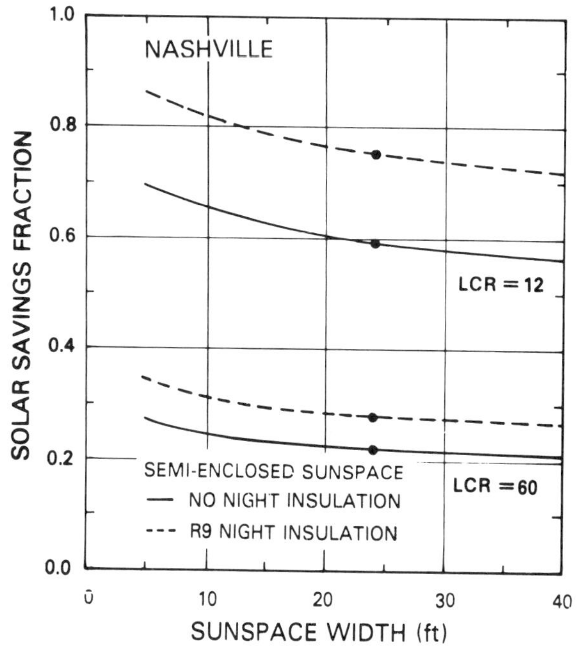

Figure A2-172

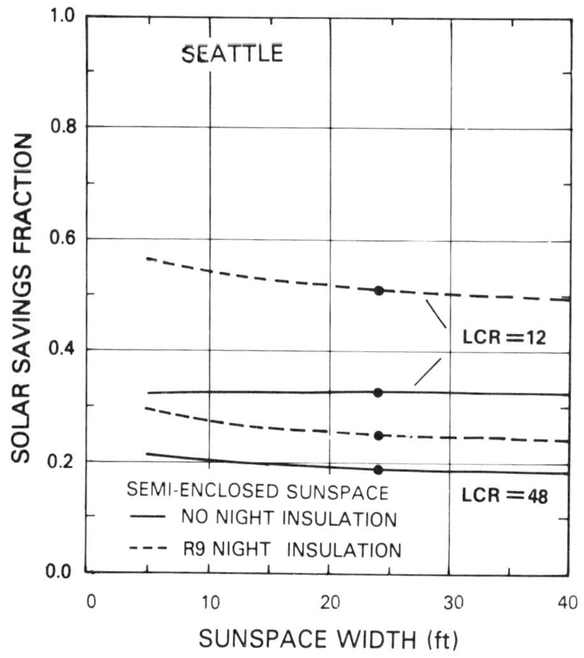

Figure A2-173

Figures A2-168 through A2-173. Annual heating performance sensitivity to the sunspace east-west width for geometry C with a masonry common wall, both without and with R9 night insulation (C1 and C2). The dots represent the reference conditions.

SUNSPACE
NUMBER OF GLAZINGS

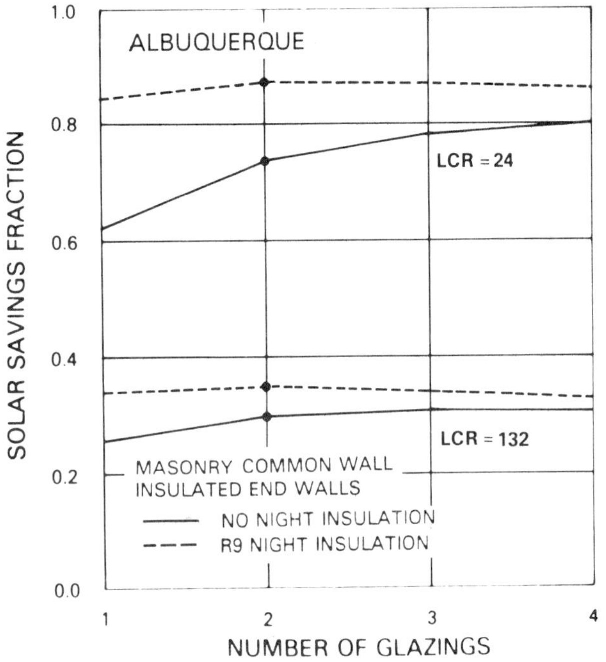

Figure A2-174

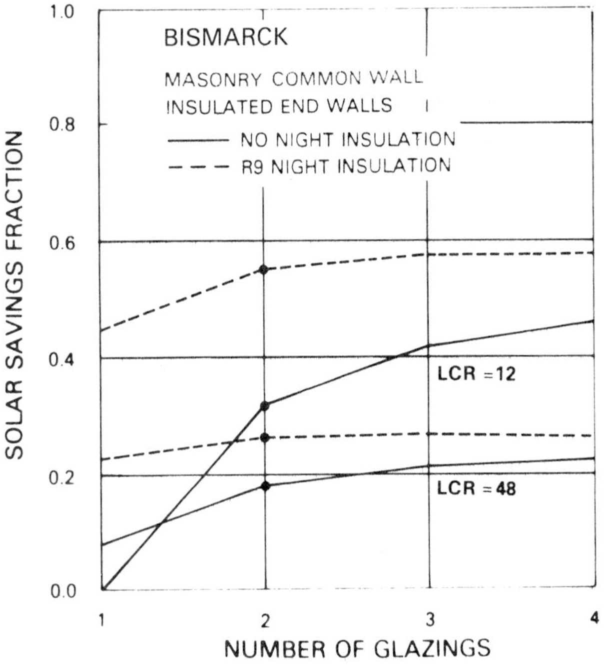

Figure A2-175

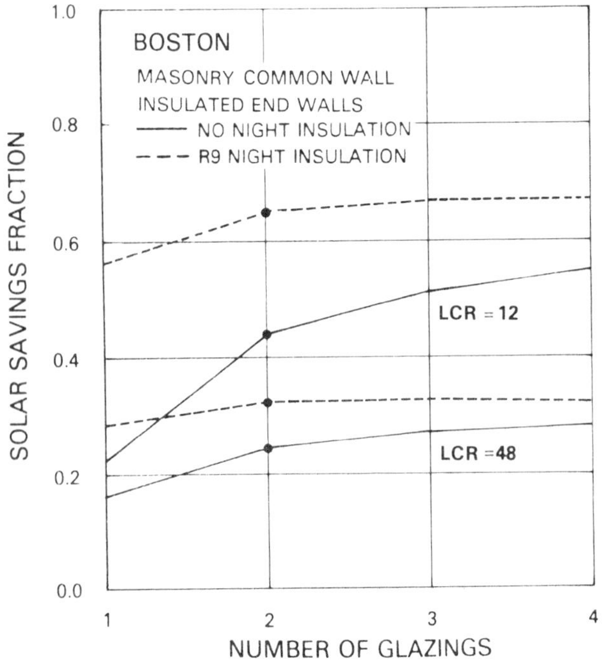

Figure A2-176

Figures A2-174 through A2-179. Annual heating performance sensitivity to the number of panes of sunspace glazing for geometry A with a masonry common wall, insulated end walls, both without and with R9 night insulation (A1 and A2). The dots represent the reference conditions.

SUNSPACE
NUMBER OF GLAZINGS

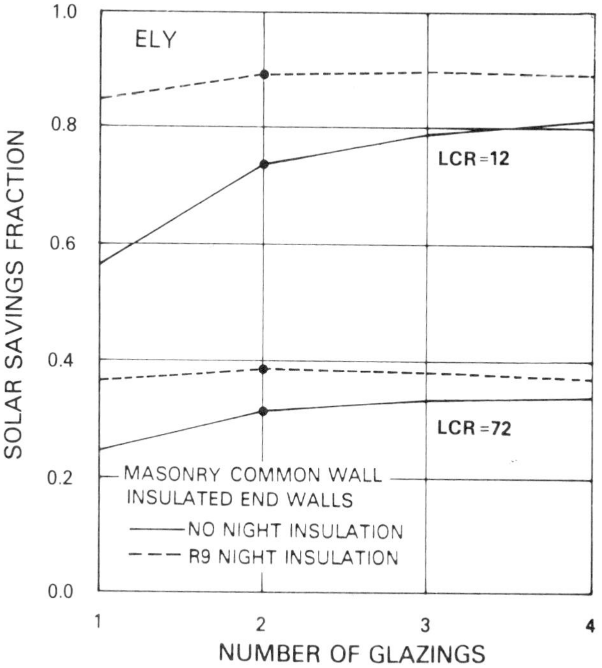

Figure A2-177

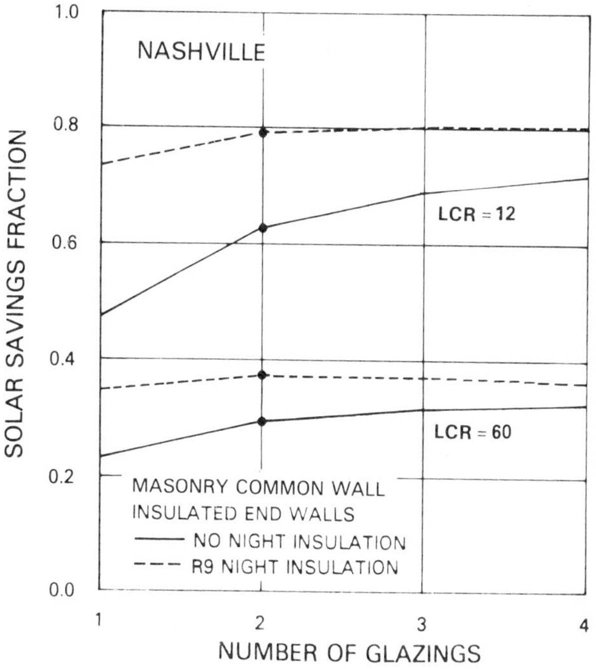

Figure A2-178

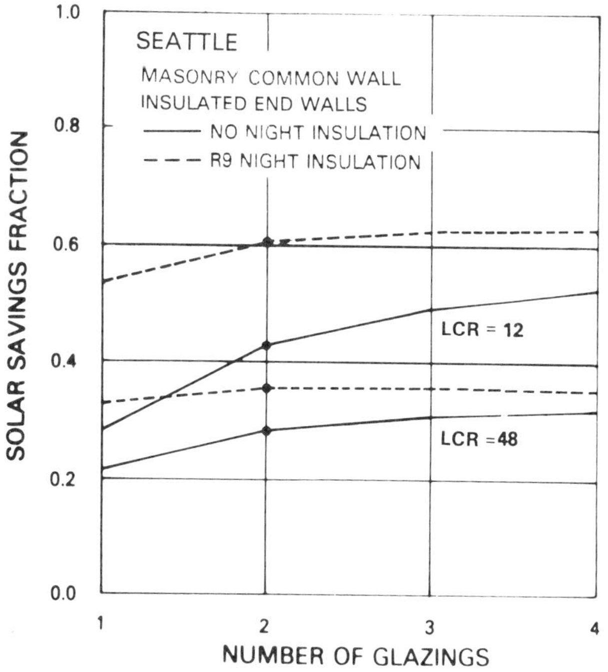

Figure A2-179

Figures A2-174 through A2-179. Annual heating performance sensitivity to the number of panes of sunspace glazing for geometry A with a masonry common wall, insulated end walls, both without and with R9 night insulation (A1 and A2). The dots represent the reference conditions.

SUNSPACE
NUMBER OF GLAZINGS

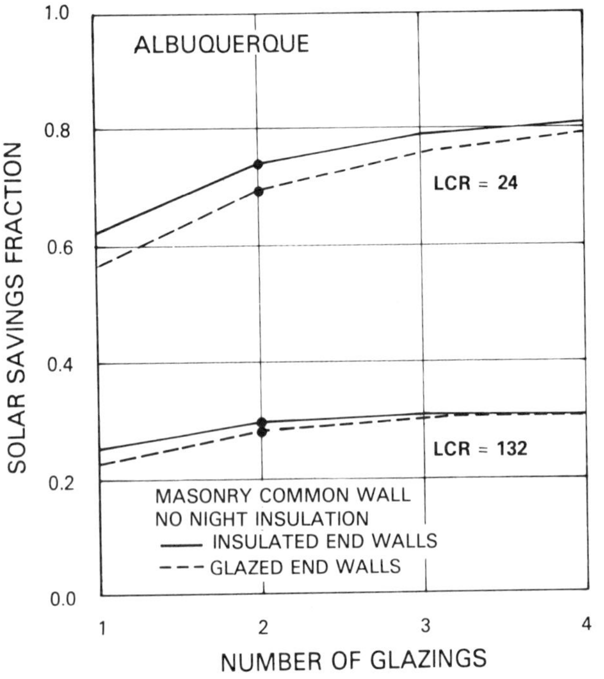

Figure A2-180

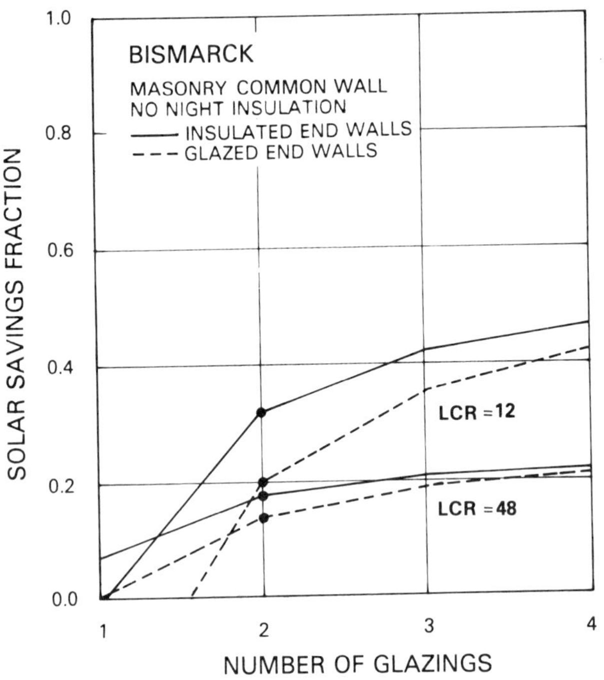

Figure A2-181

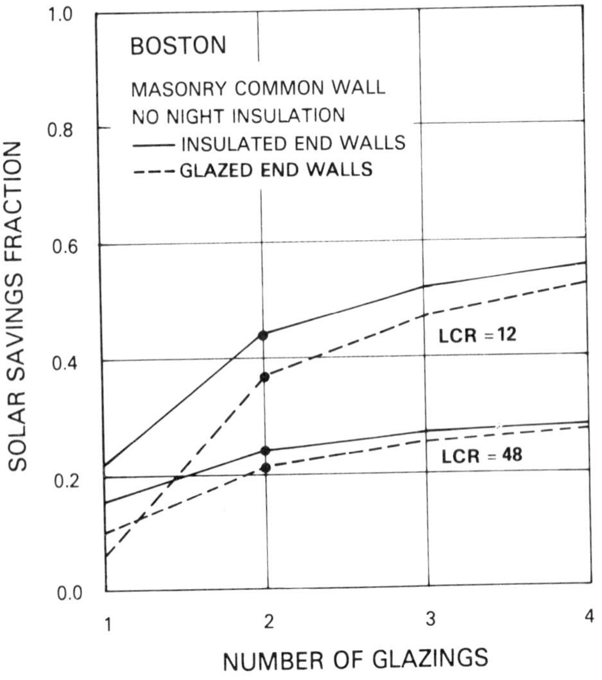

Figure A2-182

Figures A2-180 through A2-185. Annual heating performance sensitivity to the number of panes of sunspace glazing for geometry A with a masonry common wall, no night insulation, and both insulated and glazed end walls (A1 and A3). The dots represent the reference conditions.

SUNSPACE
NUMBER OF GLAZINGS

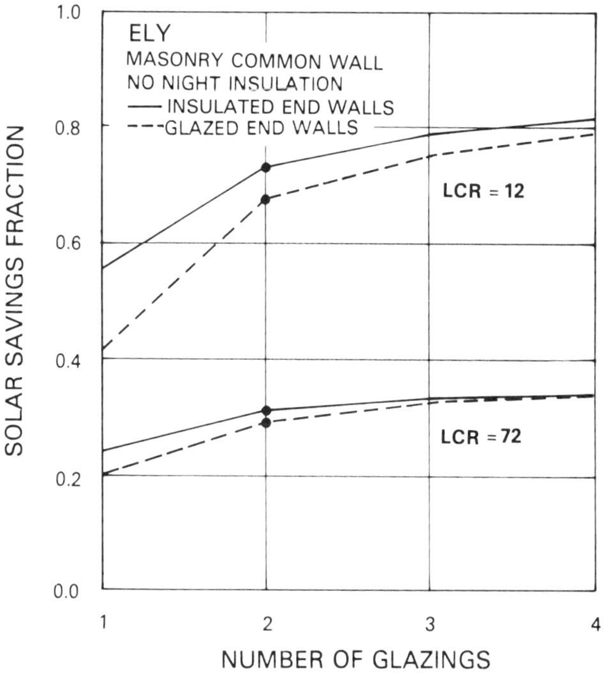

Figure A2-183

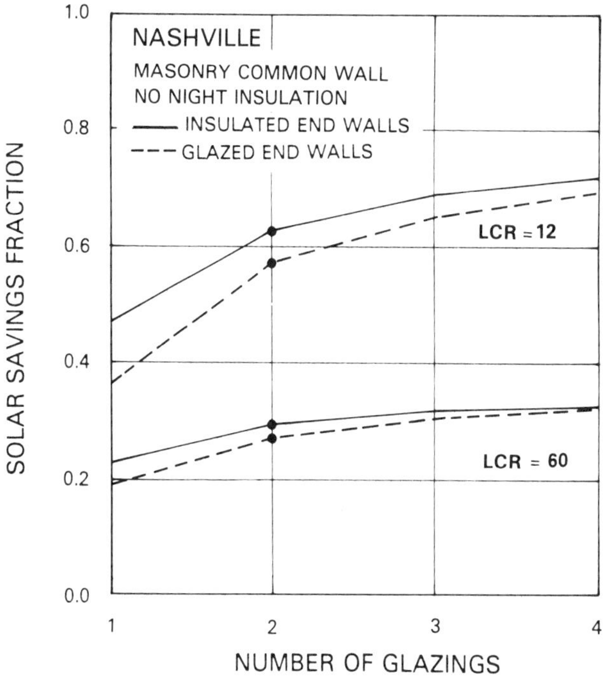

Figure A2-184

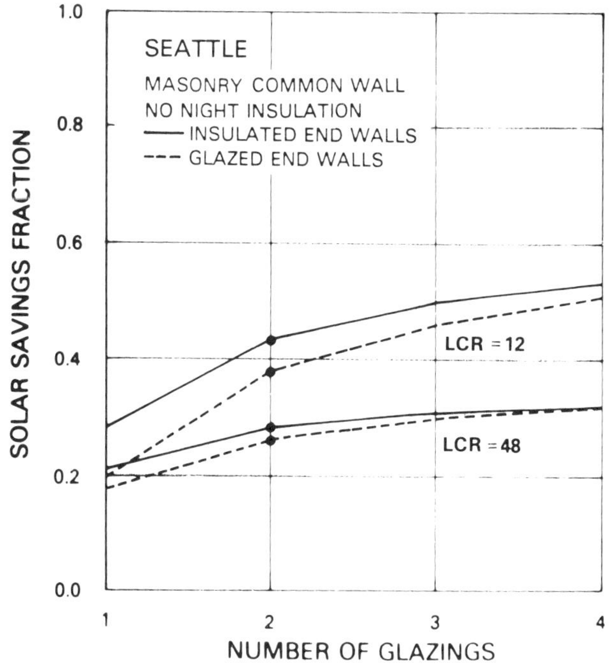

Figure A2-185

Figures A2-180 through A2-185. Annual heating performance sensitivity to the number of panes of sunspace glazing for geometry A with a masonry common wall, no night insulation, and both insulated and glazed end walls (A1 and A3). The dots represent the reference conditions.

SUNSPACE
NUMBER OF GLAZINGS

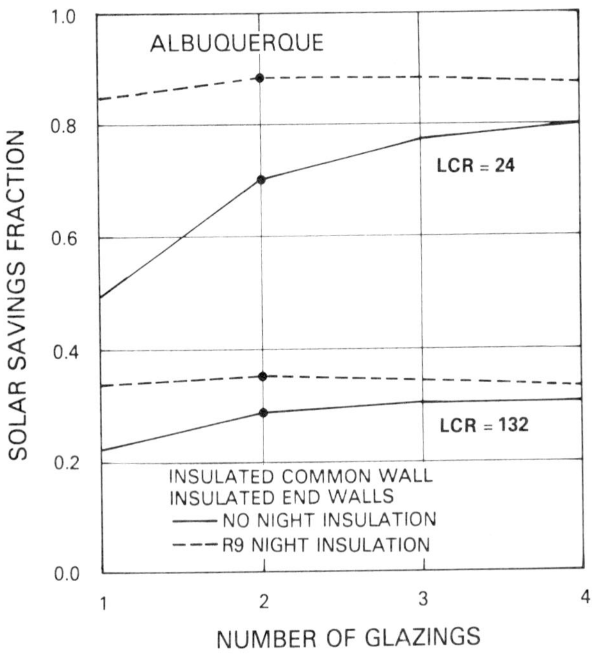

Figure A2-186

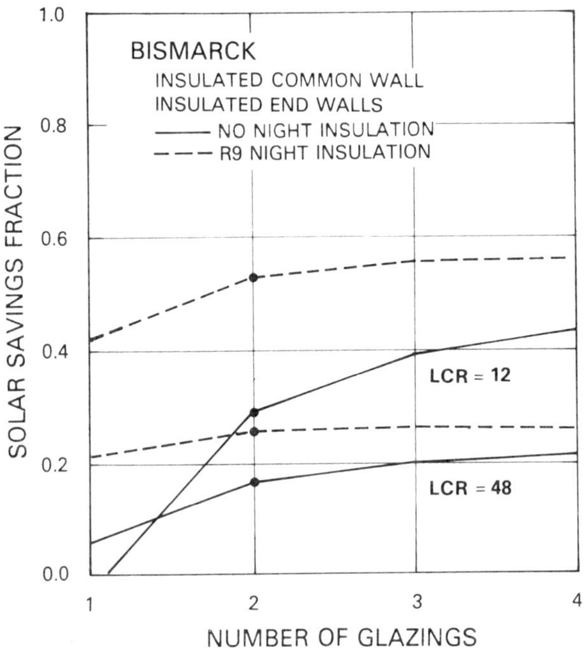

Figure A2-187

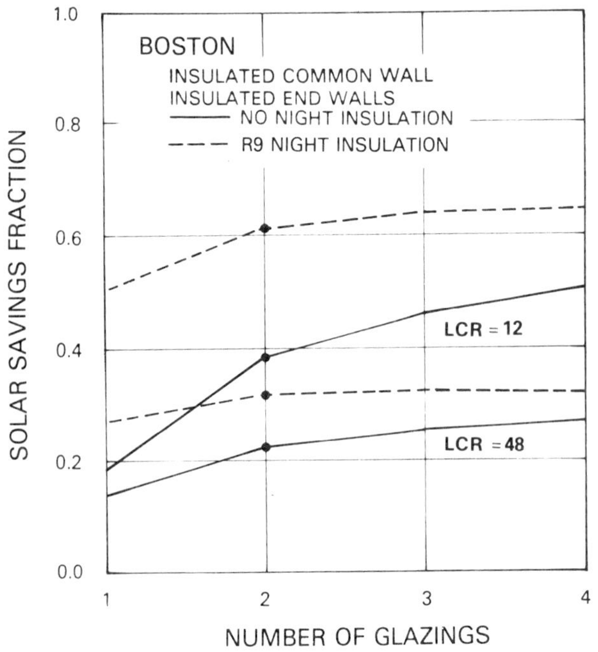

Figure A2-188

Figures A2-186 through A2-191. Annual heating performance sensitivity to the number of panes of sunspace glazing for geometry A with a lightweight, insulated common wall, insulated end walls, both without and with R9 night insulation (A5 and A6). The dots represent the reference conditions.

SUNSPACE
NUMBER OF GLAZINGS

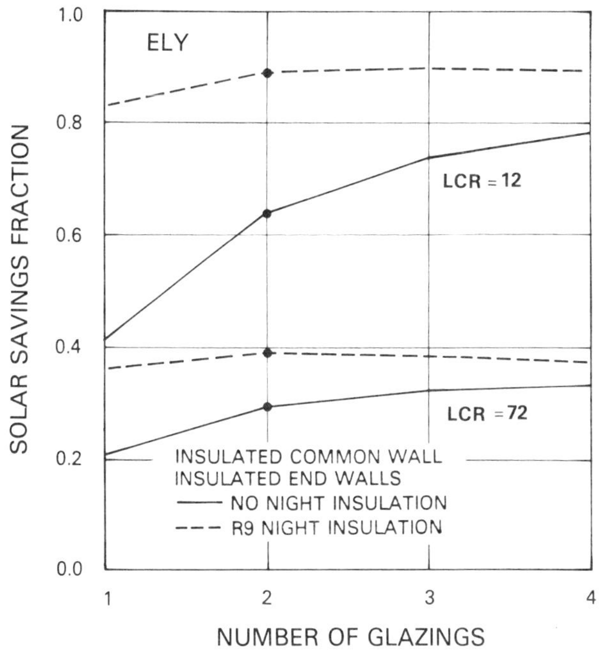

Figure A2-189

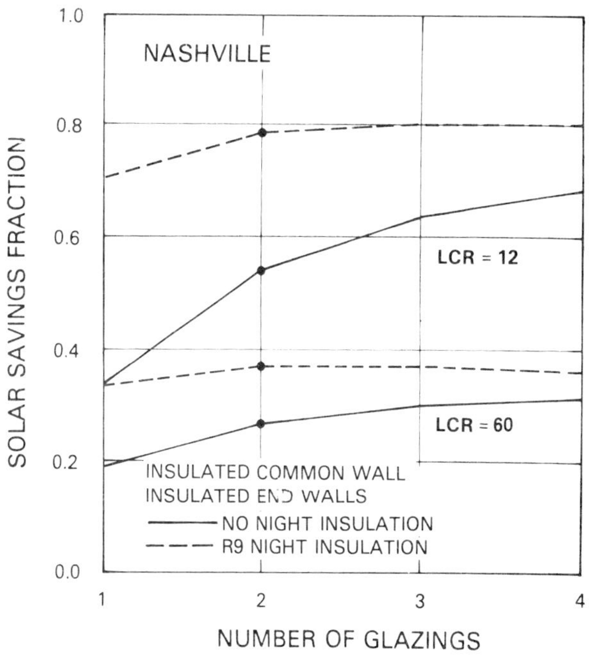

Figure A2-190

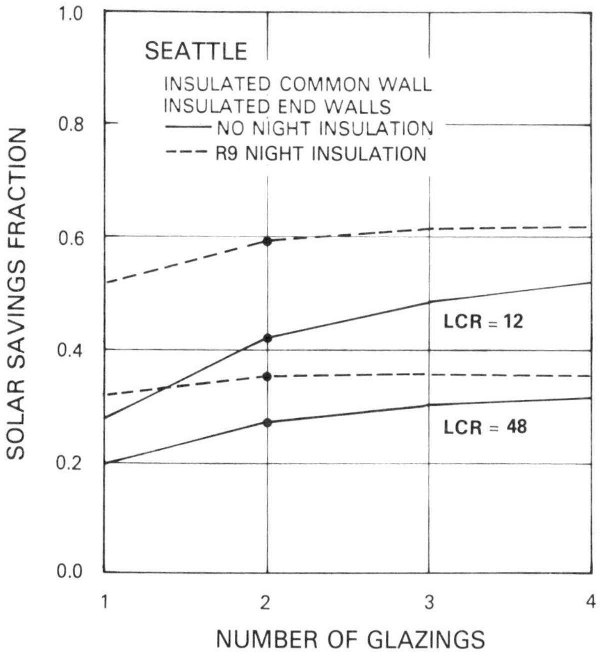

Figure A2-191

Figures A2-186 through A2-191. Annual heating performance sensitivity to the number of panes of sunspace glazing for geometry A with a lightweight, insulated common wall, insulated end walls, both without and with R9 night insulation (A5 and A6). The dots represent the reference conditions.

SUNSPACE
NUMBER OF GLAZINGS

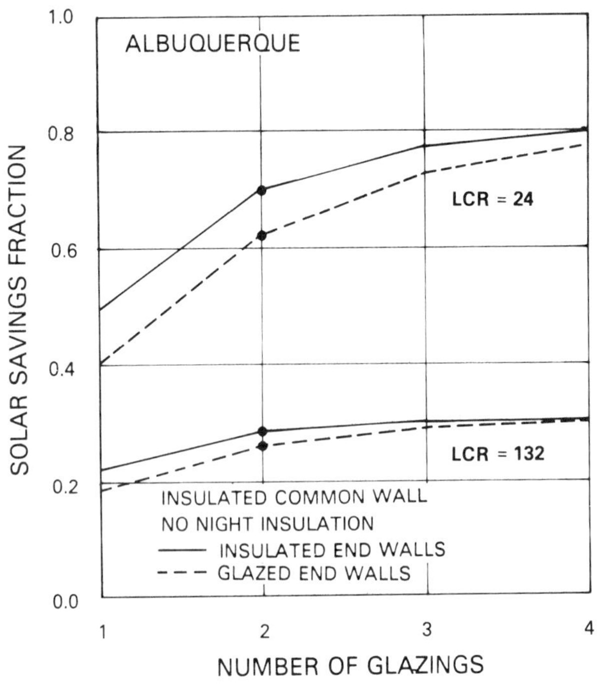

Figure A2-192

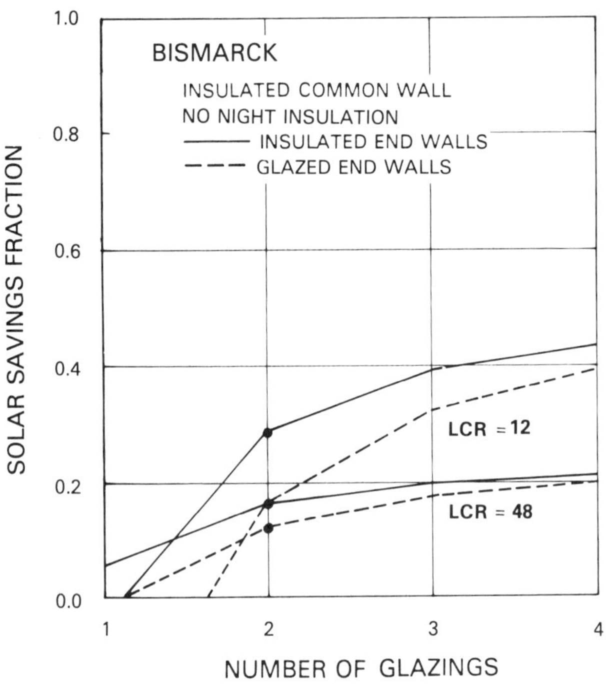

Figure A2-193

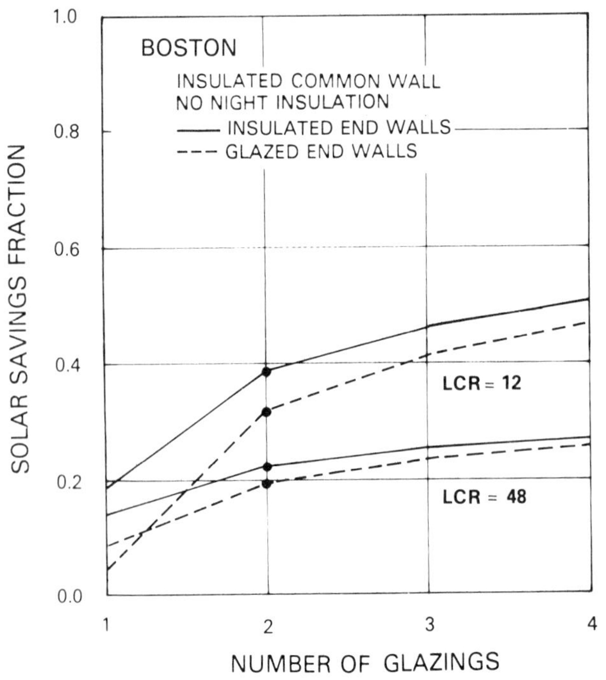

Figure A2-194

Figures A2-192 through A2-197. Annual heating performance sensitivity to the number of panes of sunspace glazing for geometry A with a lightweight, insulated common wall, no night insulation, and both insulated and glazed end walls (A5 and A7). The reference designs have two glazings.

SUNSPACE
NUMBER OF GLAZINGS

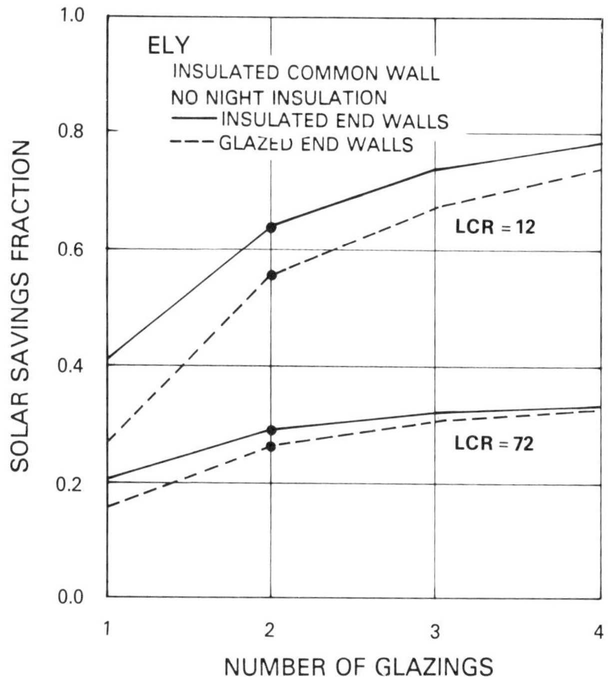

Figure A2-195

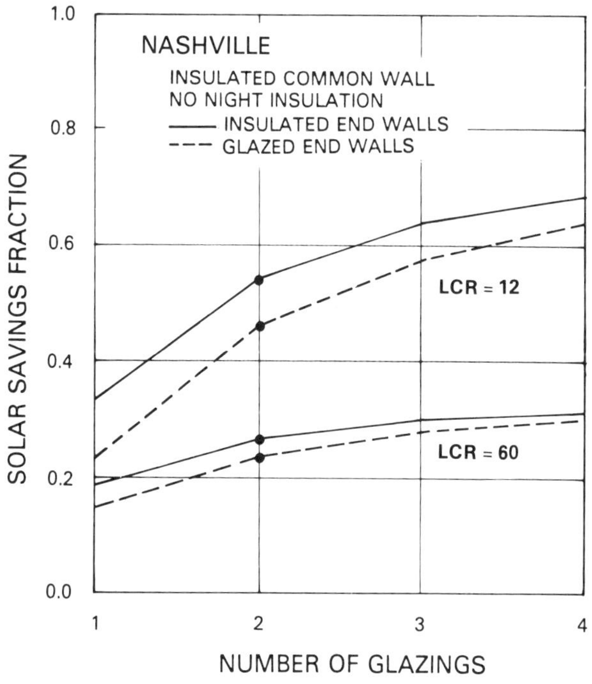

Figure A2-196

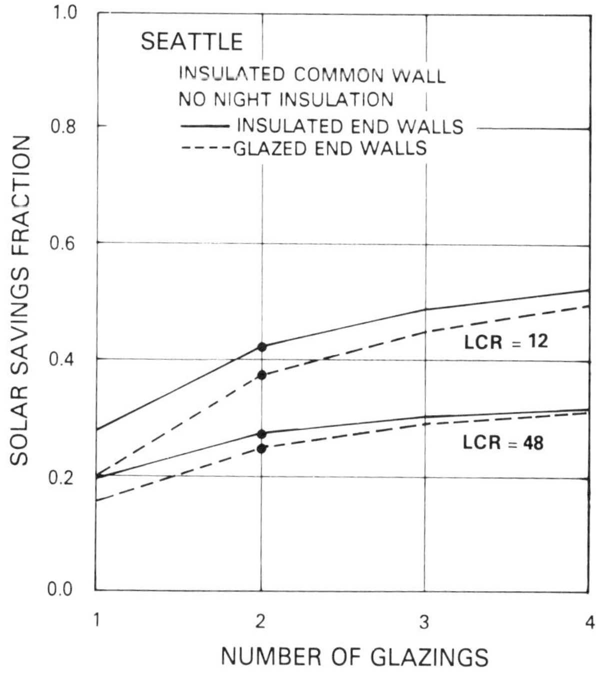

Figure A2-197

Figures A2-192 through A2-197. Annual heating performance sensitivity to the number of panes of sunspace glazing for geometry A with a lightweight, insulated common wall, no night insulation, and both insulated and glazed end walls (A5 and A7). The reference designs have two glazings.

SUNSPACE
GLAZING AIR GAP

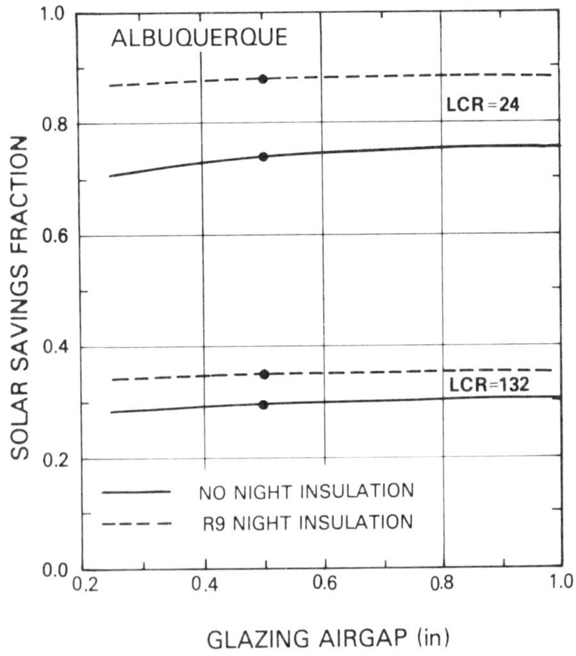

Figure A2-198

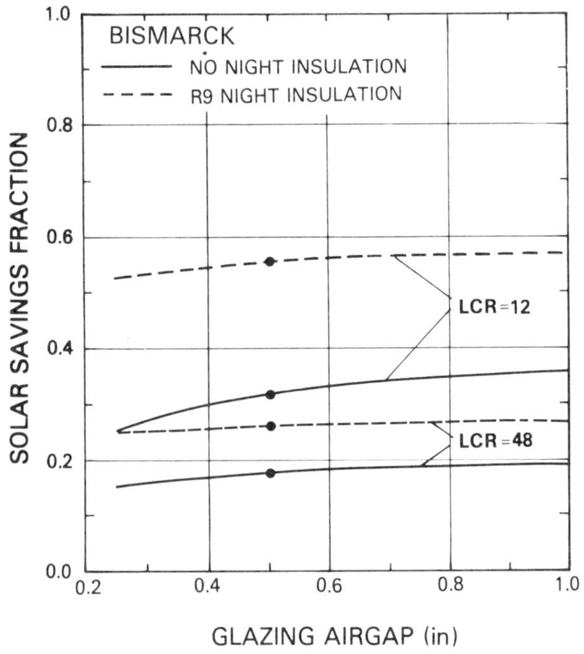

Figure A2-199

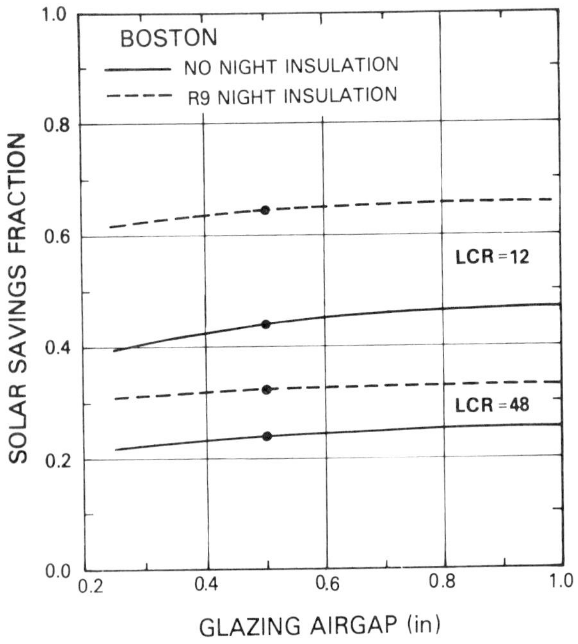

Figure A2-200

Figures A2-198 through A2-203. Annual heating performance sensitivity to the width of the air gap between the panes for geometry A with a masonry common wall, insulated end walls, both without and with R9 night insulation (A1 and A2). The reference air gap is 0.5 inches.

SUNSPACE
GLAZING AIR GAP

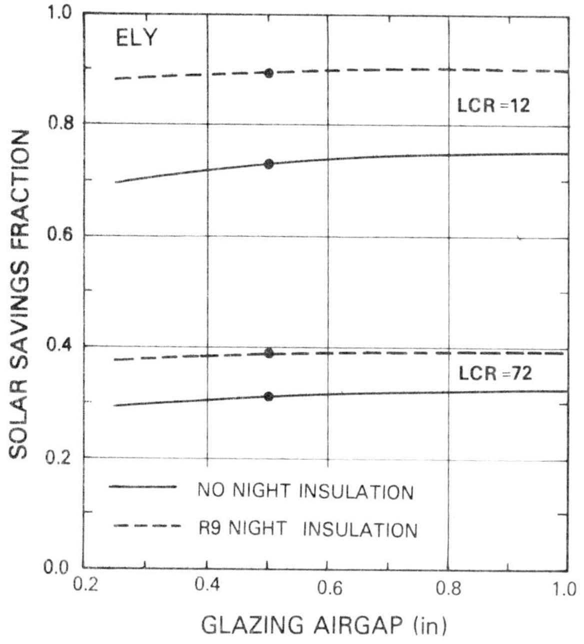

Figure A2-201

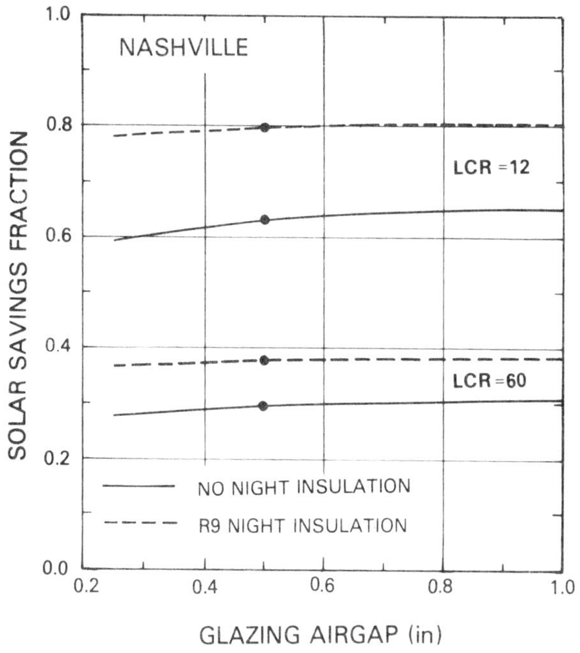

Figure A2-202

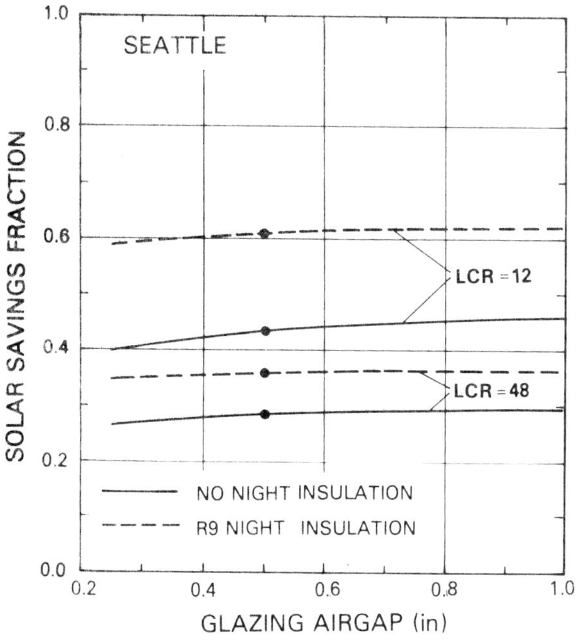

Figure A2-203

Figures A2-198 through A2-203. Annual heating performance sensitivity to the width of the air gap between the panes for geometry A with a masonry common wall, insulated end walls, both without and with R9 night insulation (A1 and A2). The reference air gap is 0.5 inches.

SUNSPACE
GLAZING THICKNESS

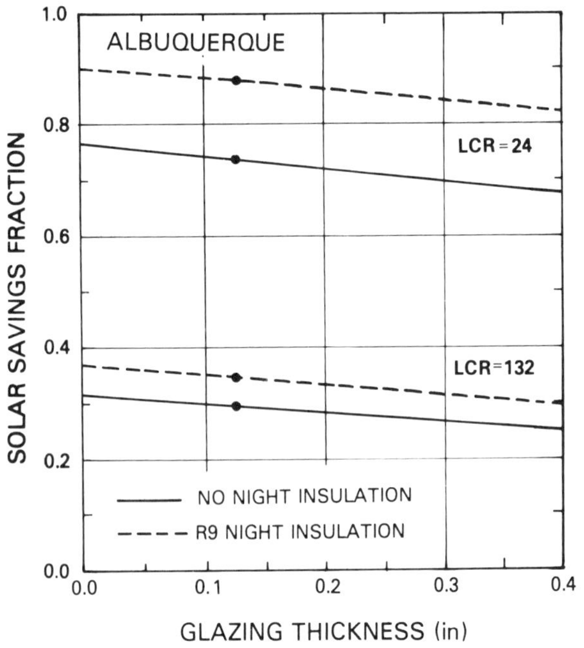

Figure A2-204

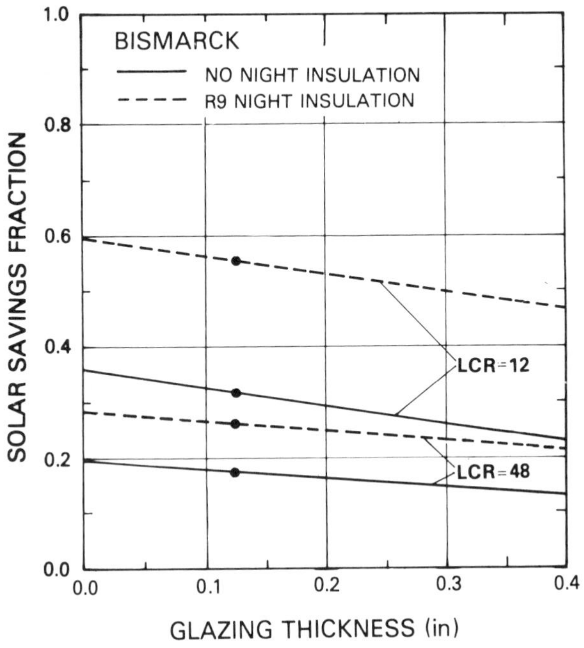

Figure A2-205

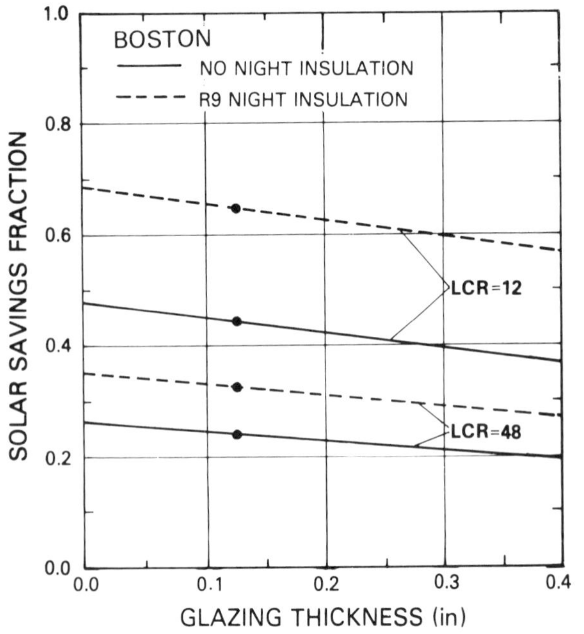

Figure A2-206

Figures A2-204 through A2-209. Annual heating performance sensitivity to the thickness of each glazing pane for geometry A with a masonry common wall, insulated end walls, both without and with R9 night insulation (A1 and A2). The dots represent the reference conditions.

SUNSPACE
GLAZING THICKNESS

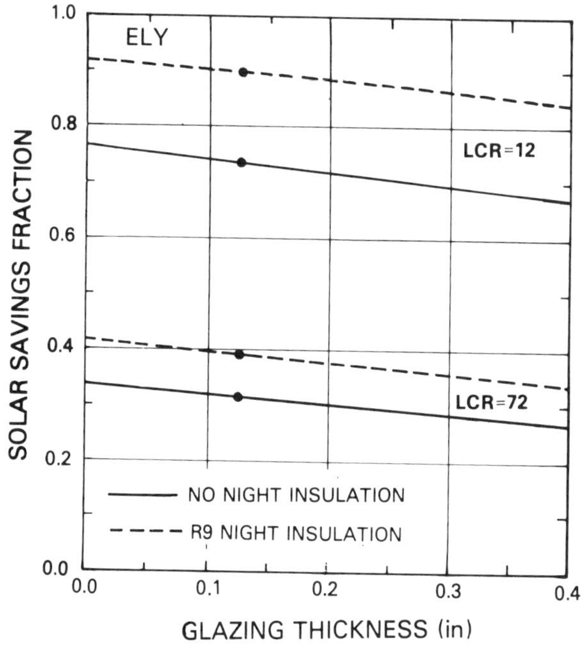

Figure A2-207

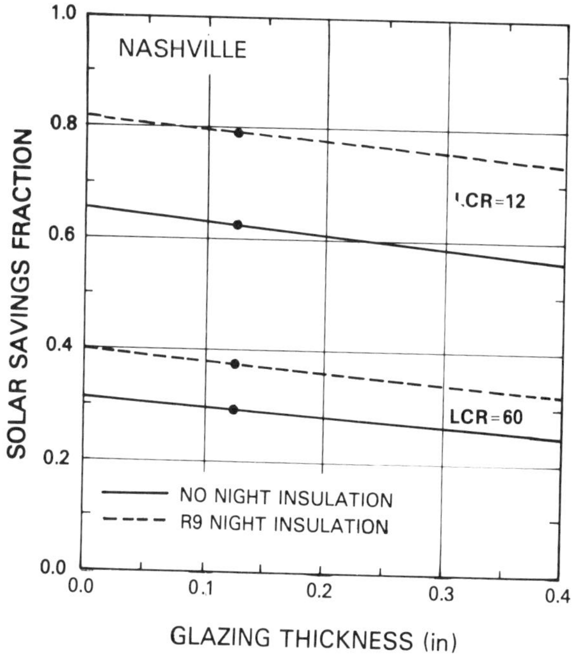

Figure A2-208

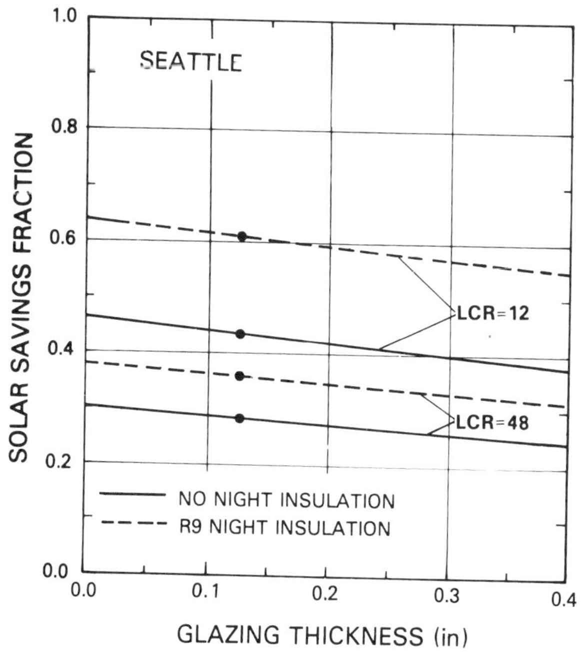

Figure A2-209

Figures A2-204 through A2-209. Annual heating performance sensitivity to the thickness of each glazing pane for geometry A with a masonry common wall, insulated end walls, both without and with R9 night insulation (A1 and A2). The dots represent the reference conditions.

SUNSPACE
COMMON WALL ABSORPTANCE

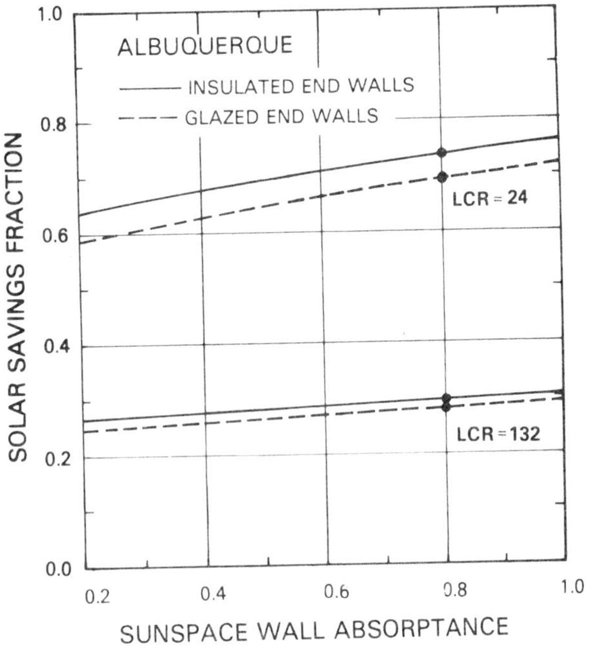

Figure A2-210

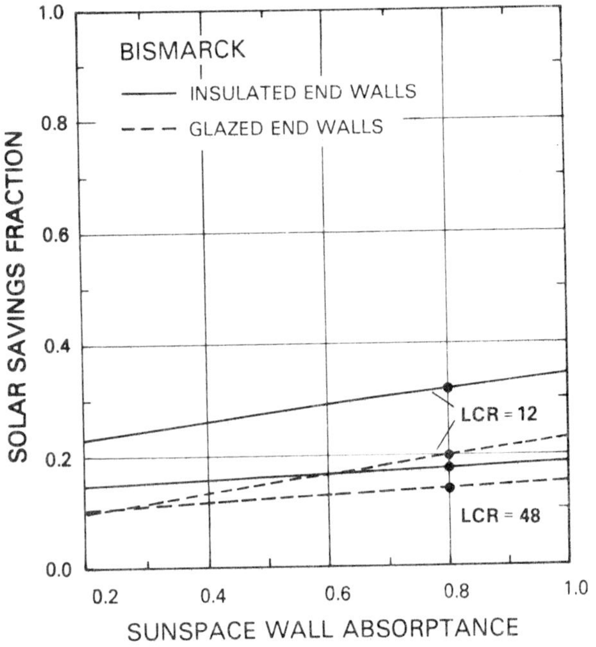

Figure A2-211

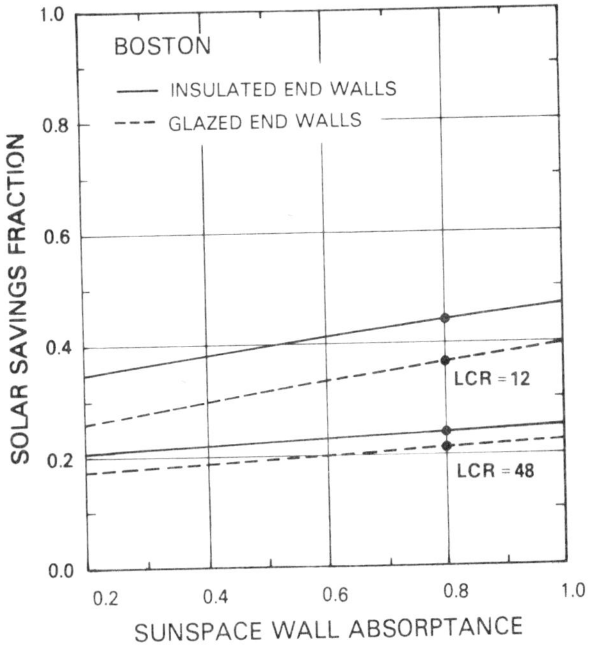

Figure A2-212

Figures A2-210 through A2-215. Annual heating performance sensitivity to the solar absorptance of the sunspace common wall for geometry A with a masonry common wall, no night insulation, and both insulated and glazed end walls (A1 and A3). The dots represent the reference conditions.

SUNSPACE
COMMON WALL ABSORPTANCE

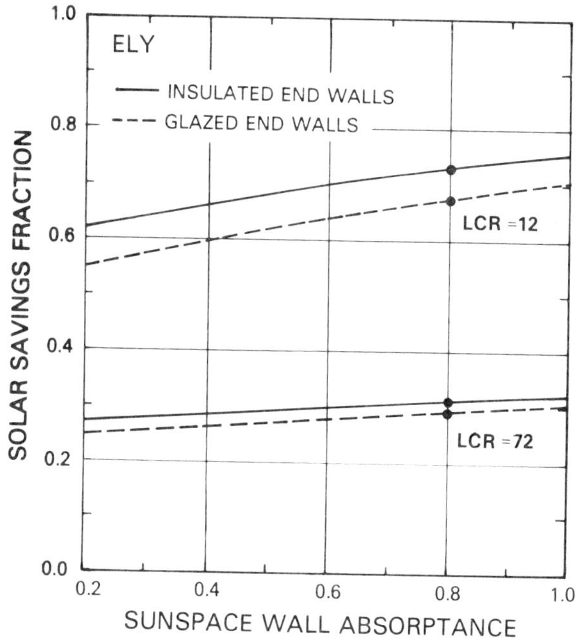

Figure A2-213

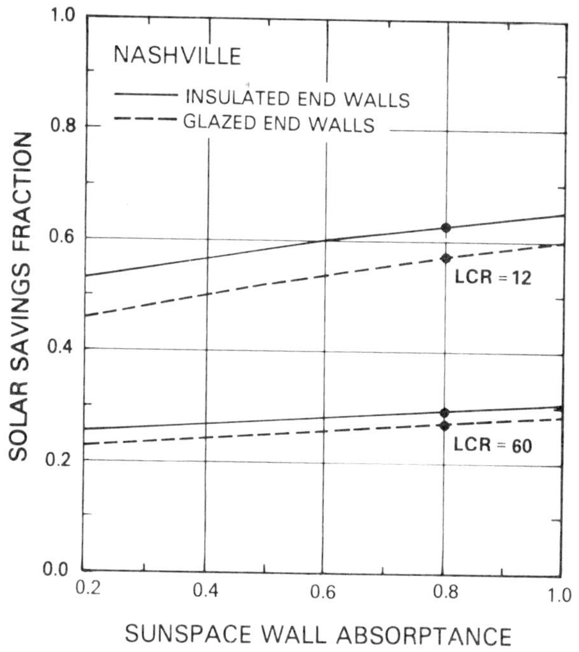

Figure A2-214

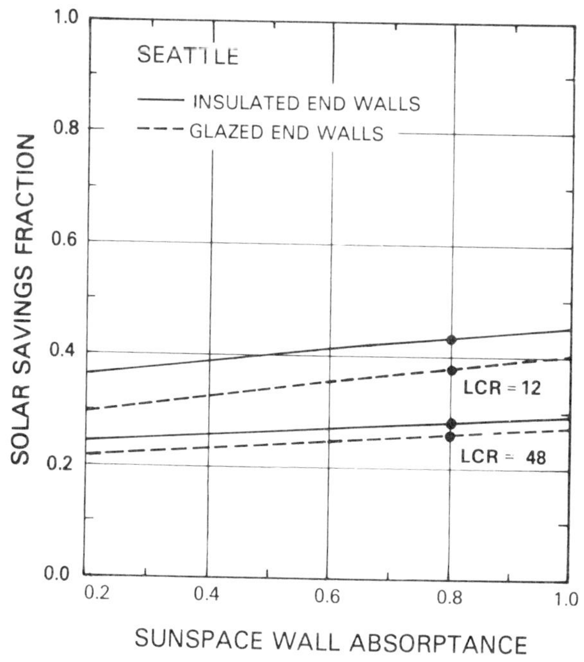

Figure A2-215

Figures A2-210 through A2-215. Annual heating performance sensitivity to the solar absorptance of the sunspace common wall for geometry A with a masonry common wall, no night insulation, and both insulated and glazed end walls (A1 and A3). The dots represent the reference conditions.

SUNSPACE
FLOOR ABSORPTANCE

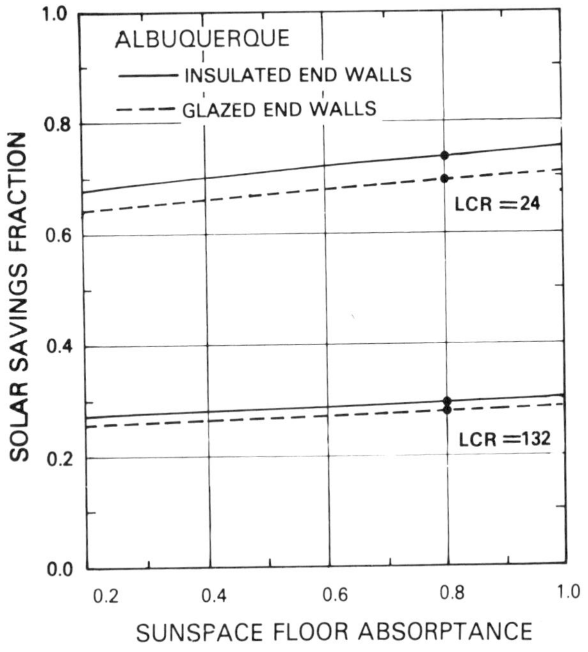

Figure A2-216

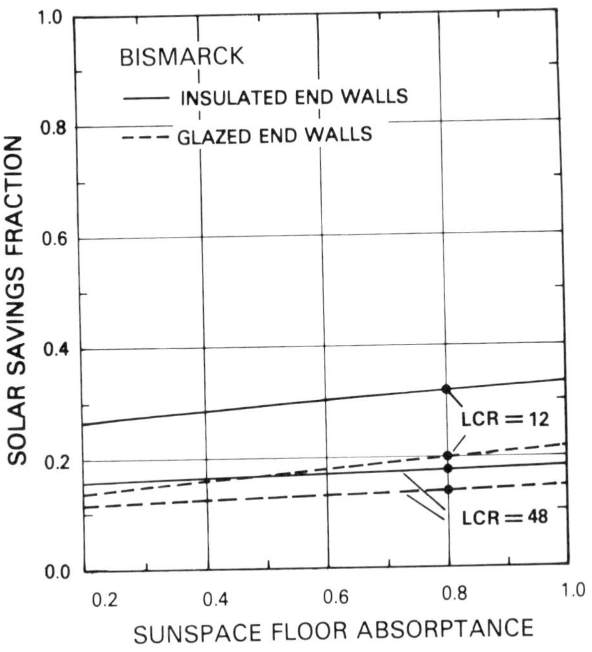

Figure A2-217

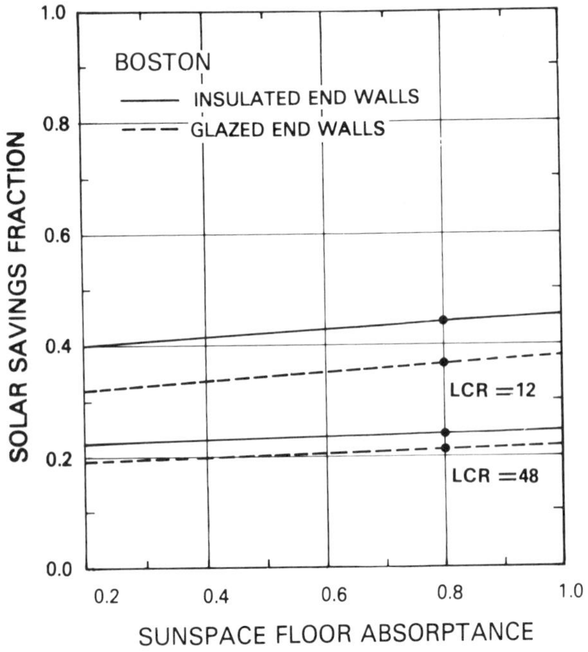

Figure A2-218

Figures A2-216 through A2-221. Annual heating performance sensitivity to the solar absorptance of the sunspace floor for geometry A with a masonry common wall, no night insulation, and both insulated and glazed end walls (A1 and A3). The dots represent the reference conditions.

SUNSPACE
FLOOR ABSORPTANCE

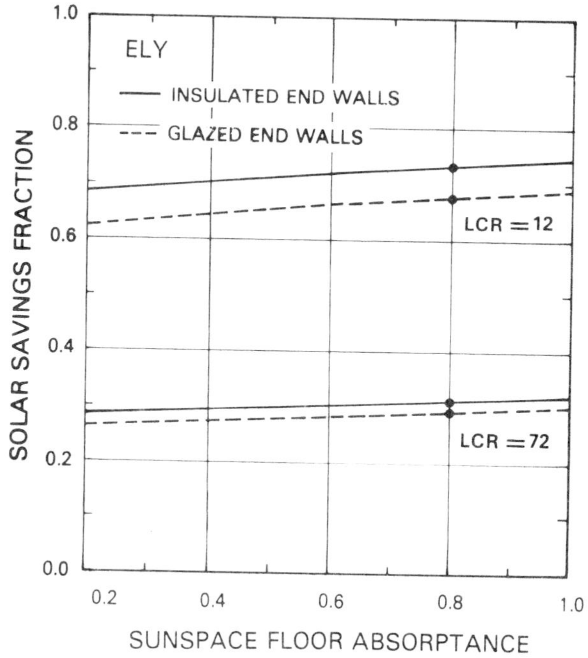

Figure A2-219

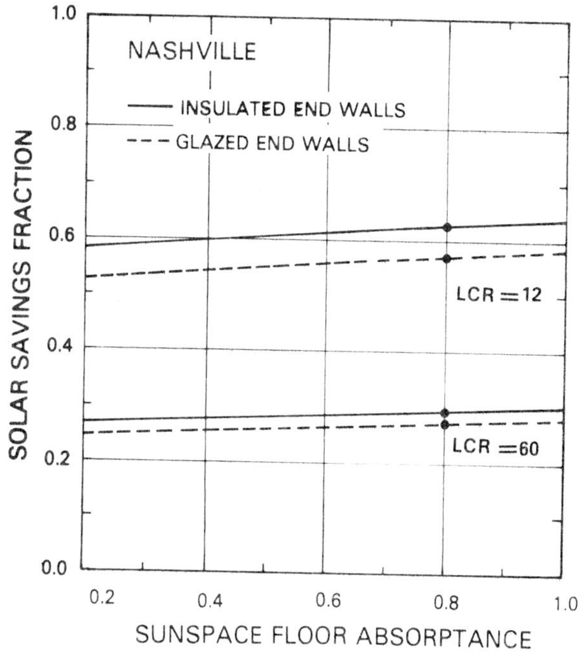

Figure A2-220

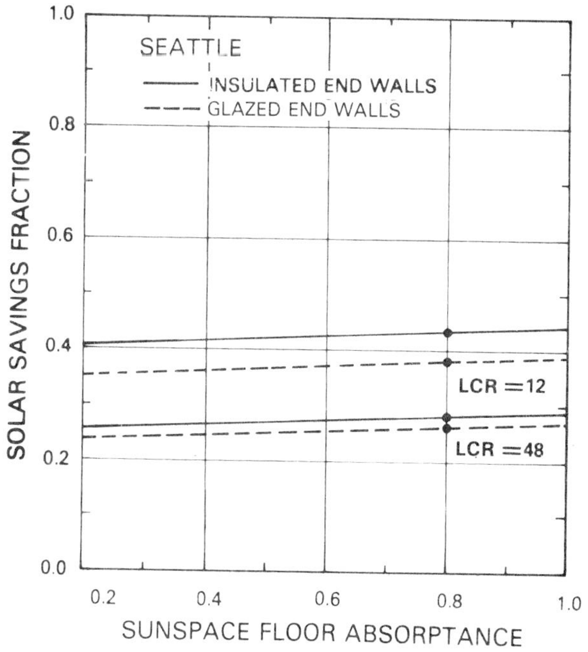

Figure A2-221

Figures A2-216 through A2-221. Annual heating performance sensitivity to the solar absorptance of the sunspace floor for geometry A with a masonry common wall, no night insulation, and both insulated and glazed end walls (A1 and A3). The dots represent the reference conditions.

SUNSPACE
FLOOR AND WALL ABSORPTANCE

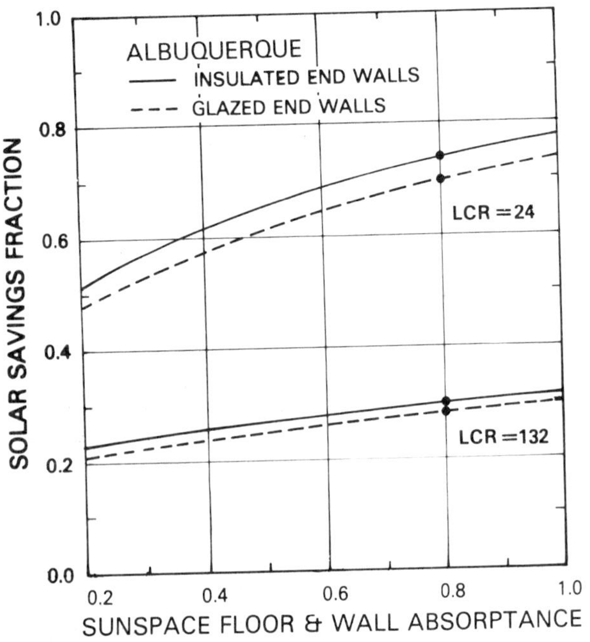

Figure A2-222

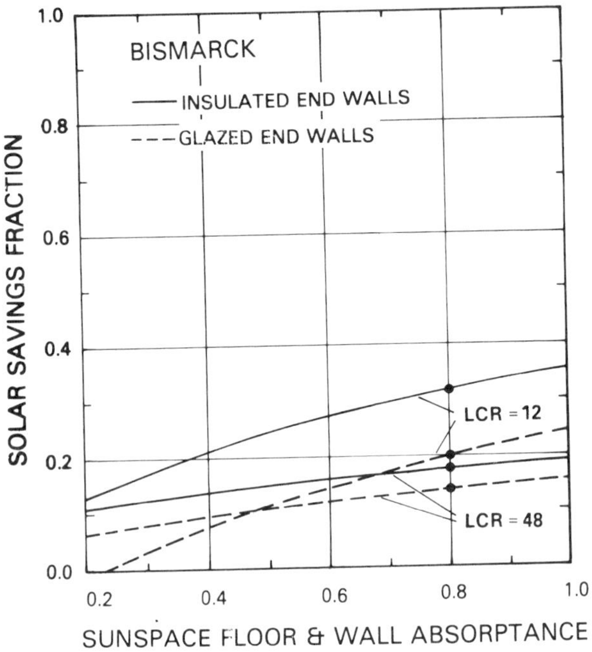

Figure A2-223

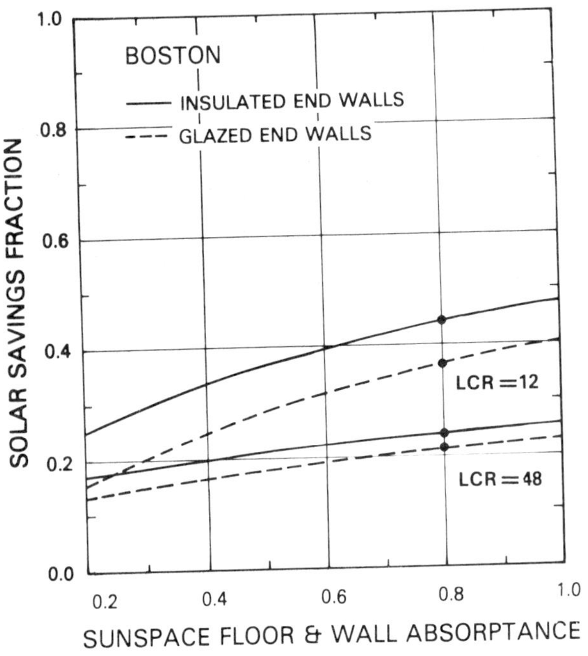

Figure A2-224

Figures A2-222 through A2-227. Annual heating performance sensitivity to the solar absorptances of the sunspace floor and common wall, varied simultaneously, for geometry A with a masonry common wall, no night insulation, and both insulated and glazed end walls (A1 and A3). The dots represent the reference conditions.

SUNSPACE
FLOOR AND WALL ABSORPTANCE

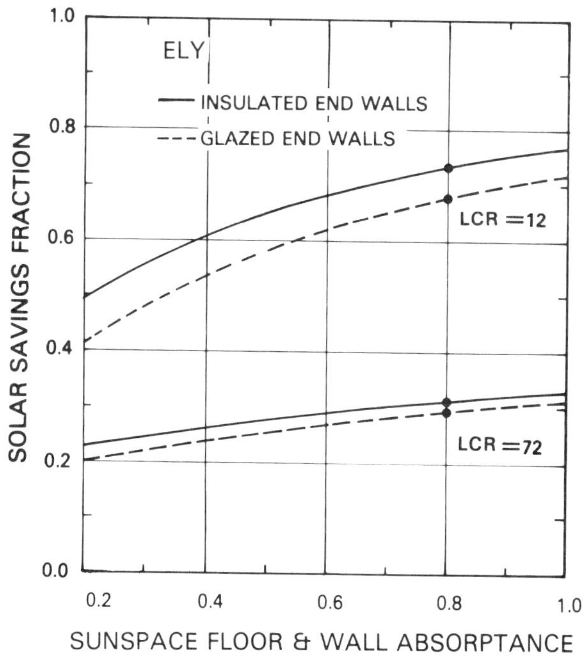

Figure A2-225

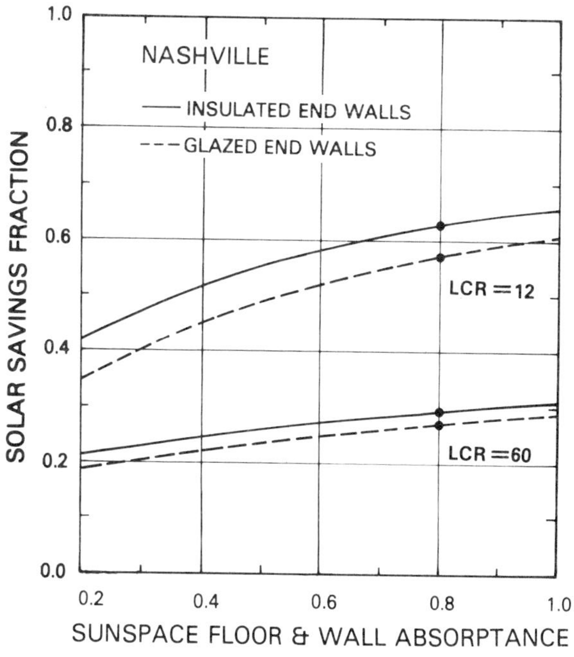

Figure A2-226

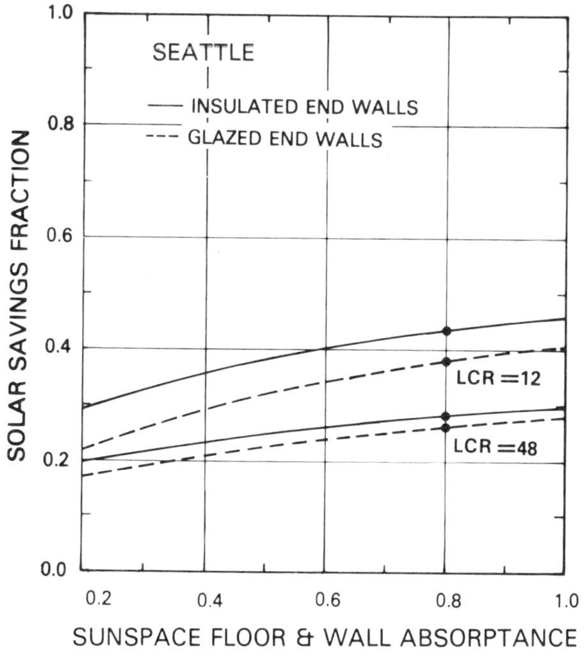

Figure A2-227

Figures A2-222 through A2-227. Annual heating performance sensitivity to the solar absorptances of the sunspace floor and common wall, varied simultaneously, for geometry A with a masonry common wall, no night insulation, and both insulated and glazed end walls (A1 and A3). The dots represent the reference conditions.

SUNSPACE
FLOOR AND WALL ABSORPTANCE

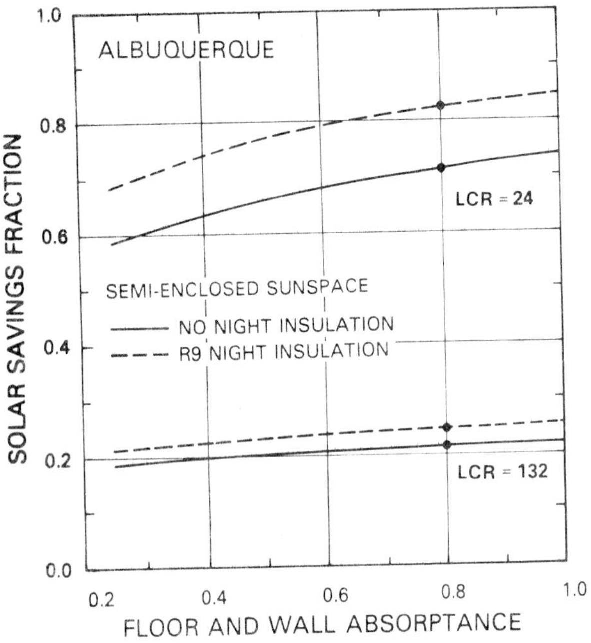

Figure A2-228

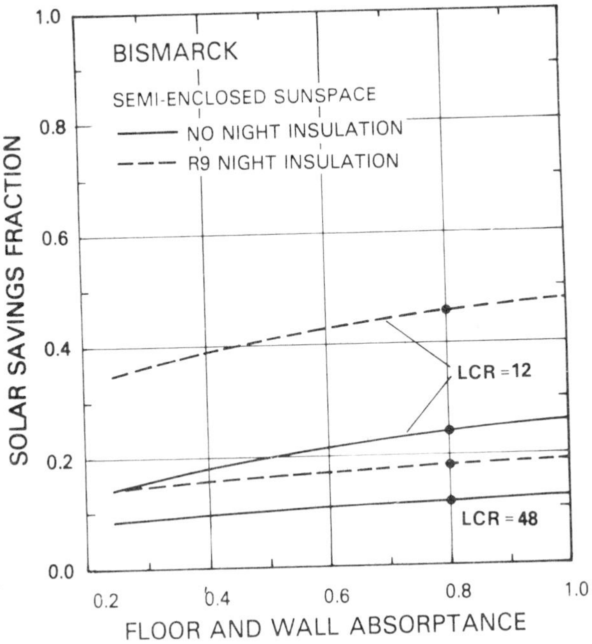

Figure A2-229

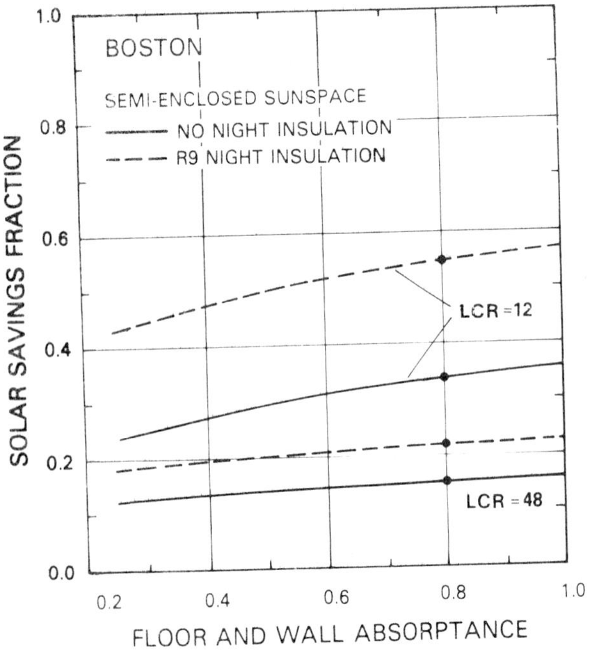

Figure A2-230

Figures A2-228 through A2-233. Annual heating performance sensitivity to the solar absorptances of the sunspace floor and common wall, varied simultaneously, for geometry C with a masonry common wall, both without and with R9 night insulation (C1 and C2). The dots represent the reference conditions.

SUNSPACE
FLOOR AND WALL ABSORPTANCE

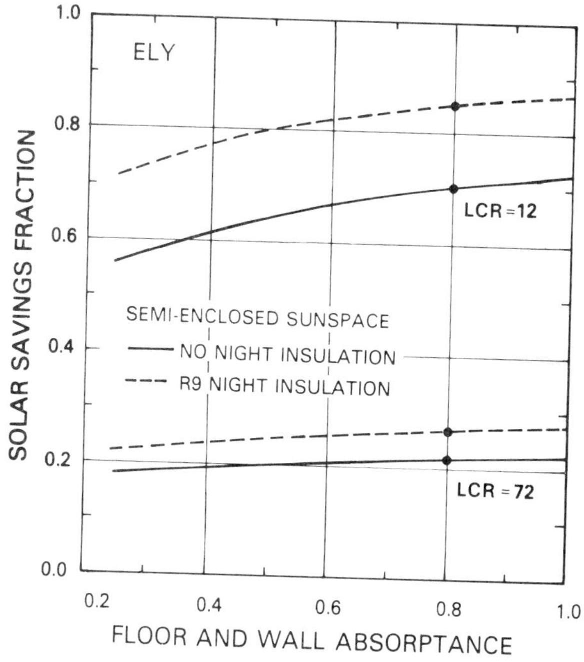

Figure A2-231

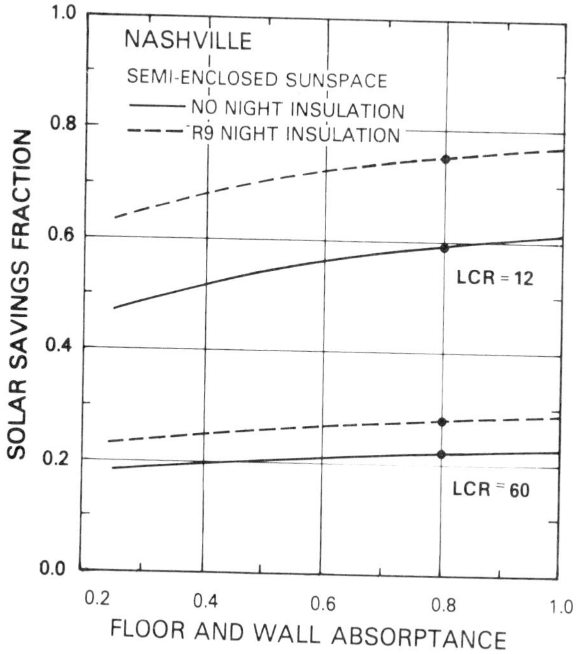

Figure A2-232

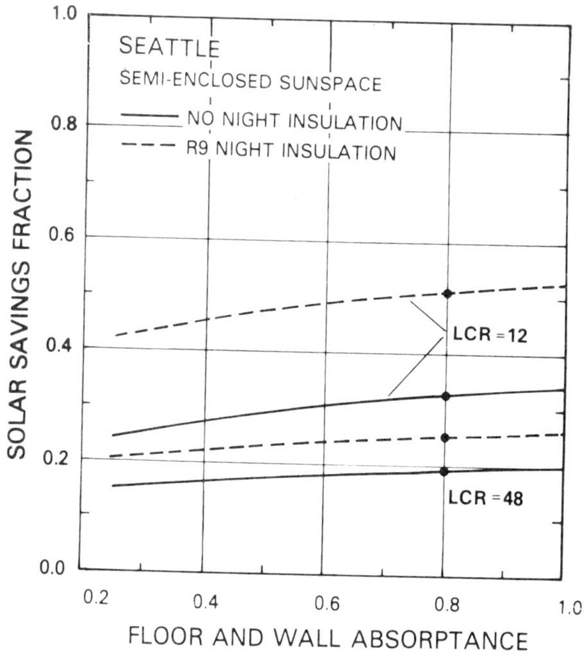

Figure A2-233

Figures A2-228 through A2-233. Annual heating performance sensitivity to the solar absorptances of the sunspace floor and common wall, varied simultaneously, for geometry C with a masonry common wall, both without and with R9 night insulation (C1 and C2). The dots represent the reference conditions.

SUNSPACE
WATER CONTAINER ABSORPTANCE

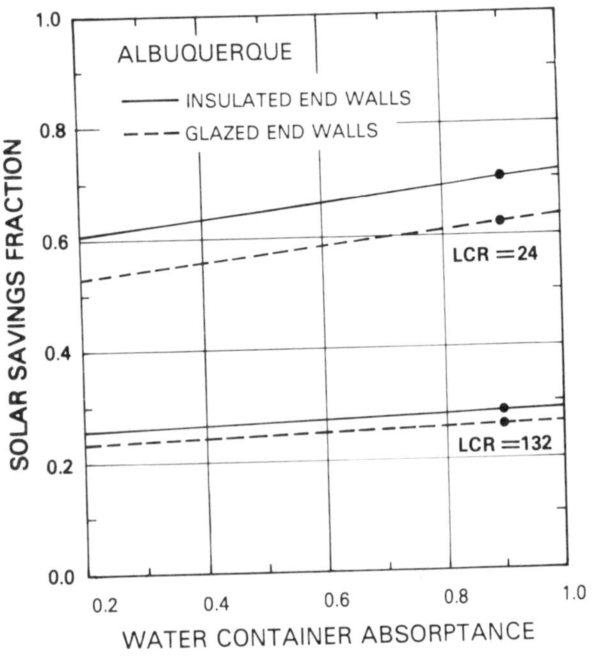

Figure A2-234

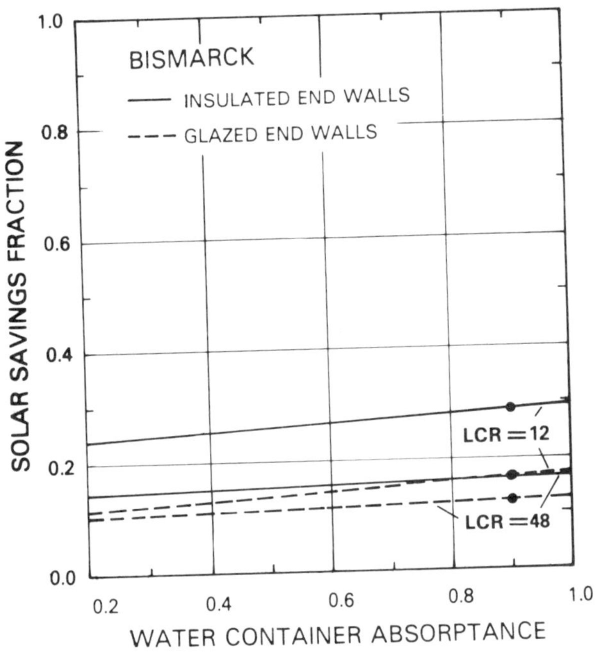

Figure A2-235

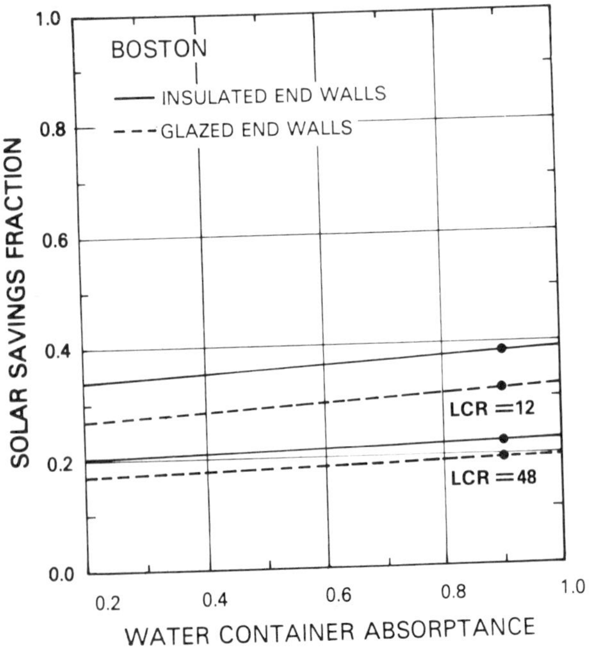

Figure A2-236

Figures A2-234 through A2-239. Annual heating performance sensitivity to the solar absorptance of the sunspace water containers for geometry A with a lightweight, insulated common wall, no night insulation, and both insulated and glazed end walls (A5 and A7). The dots represent the reference conditions.

SUNSPACE
WATER CONTAINER ABSORPTANCE

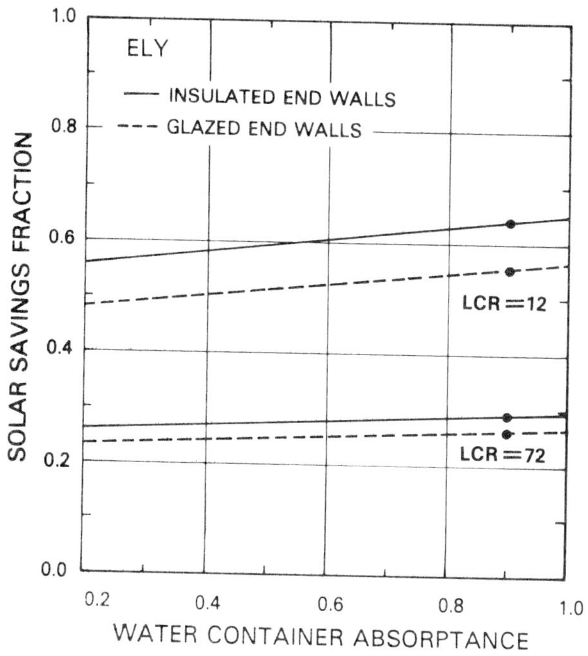

Figure A2-237

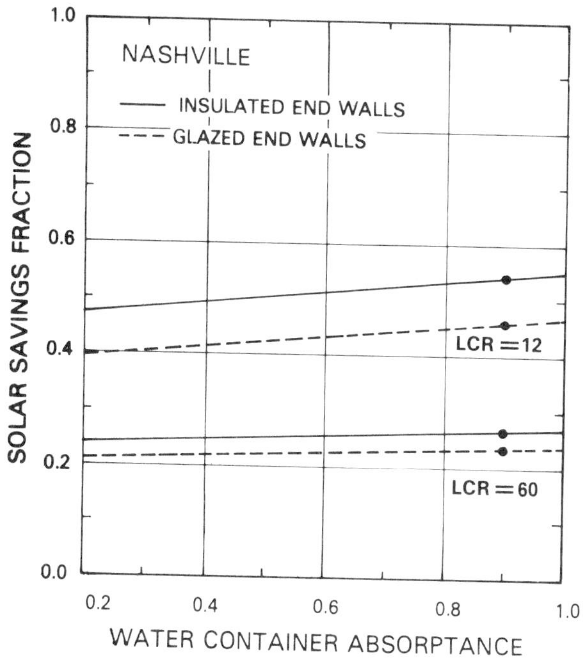

Figure A2-238

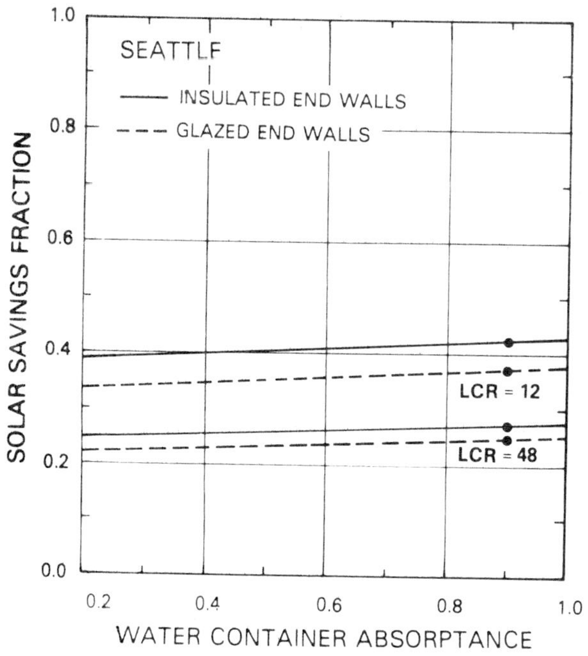

Figure A2-239

Figures A2-234 through A2-239. Annual heating performance sensitivity to the solar absorptance of the sunspace water containers for geometry A with a lightweight, insulated common wall, no night insulation, and both insulated and glazed end walls (A5 and A7). The dots represent the reference conditions.

SUNSPACE
LIGHTWEIGHT ABSORPTION FRACTION

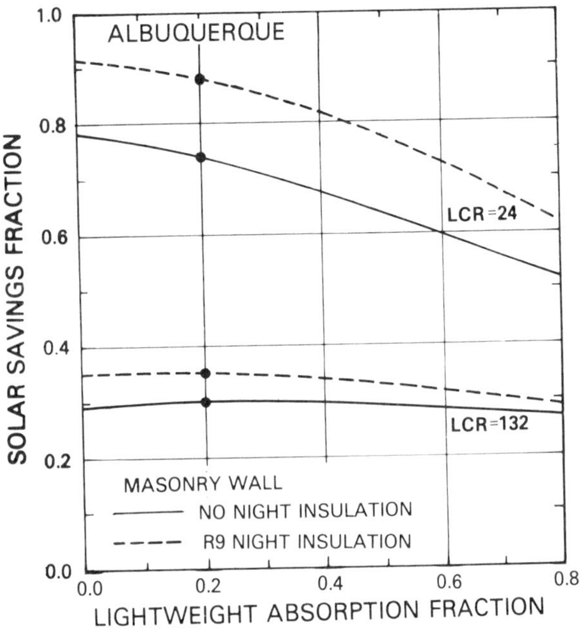

Figure A2-240

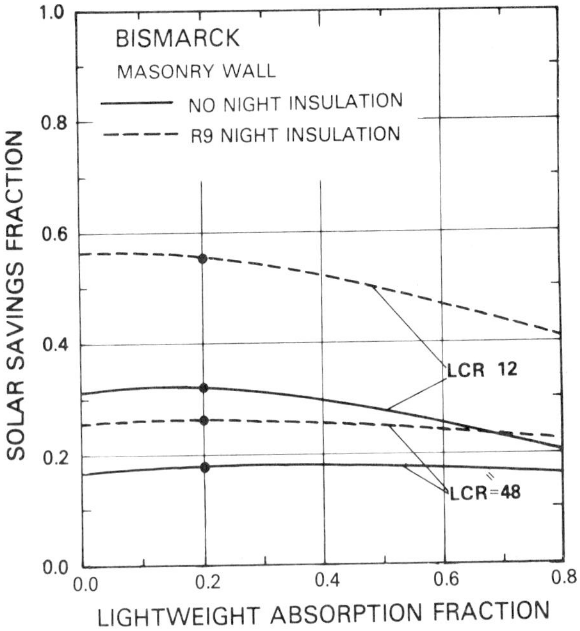

Figure A2-241

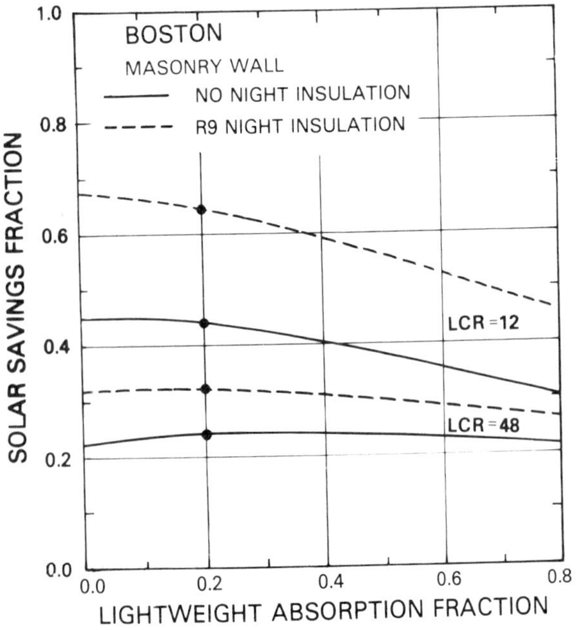

Figure A2-242

Figures A2-240 through A2-245. Annual heating performance sensitivity to the sunspace lightweight absorption fraction for geometry A with a masonry common wall, insulated end walls, both without and with R9 night insulation (A1 and A2). The lightweight absorption fraction is the fraction of solar radiation that directly heats the sunspace air when it is transmitted through the glazing and when it is reflected from interior surfaces. It is intended to simulate the presence of lightweight objects in the sunspace that absorb solar radiation and rapidly convect heat to the sunspace air. The dots represent the reference conditions.

SUNSPACE
LIGHTWEIGHT ABSORPTION FRACTION

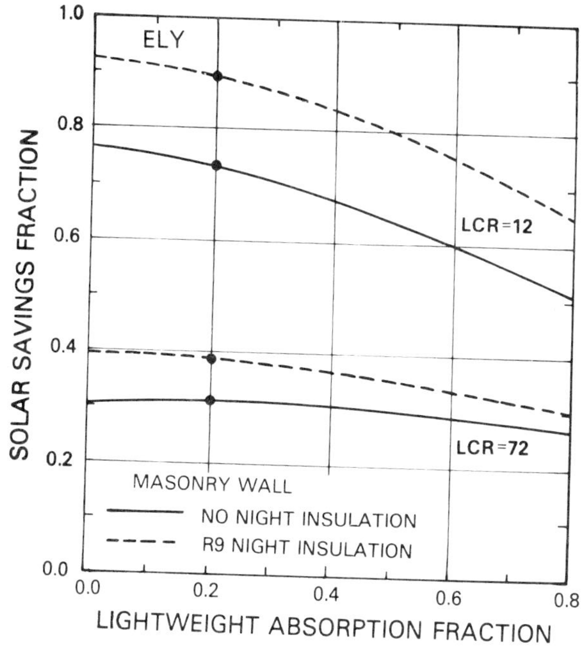

Figure A2-243

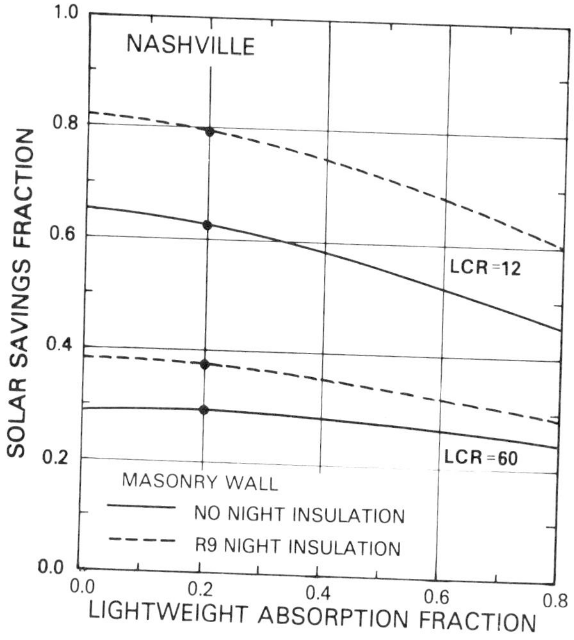

Figure A2-244

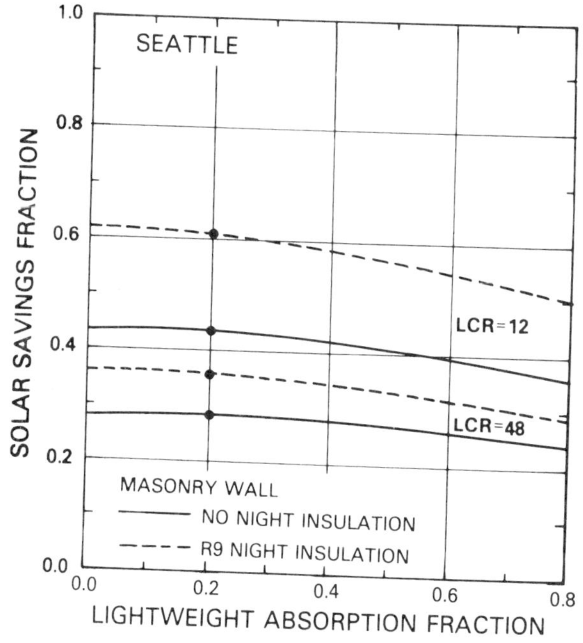

Figure A2-245

Figures A2-240 through A2-245. Annual heating performance sensitivity to the sunspace lightweight absorption fraction for geometry A with a masonry common wall, insulated end walls, both without and with R9 night insulation (A1 and A2). The lightweight absorption fraction is the fraction of solar radiation that directly heats the sunspace air when it is transmitted through the glazing and when it is reflected from interior surfaces. It is intended to simulate the presence of lightweight objects in the sunspace that absorb solar radiation and rapidly convect heat to the sunspace air. The dots represent the reference conditions.

SUNSPACE
LIGHTWEIGHT ABSORPTION FRACTION

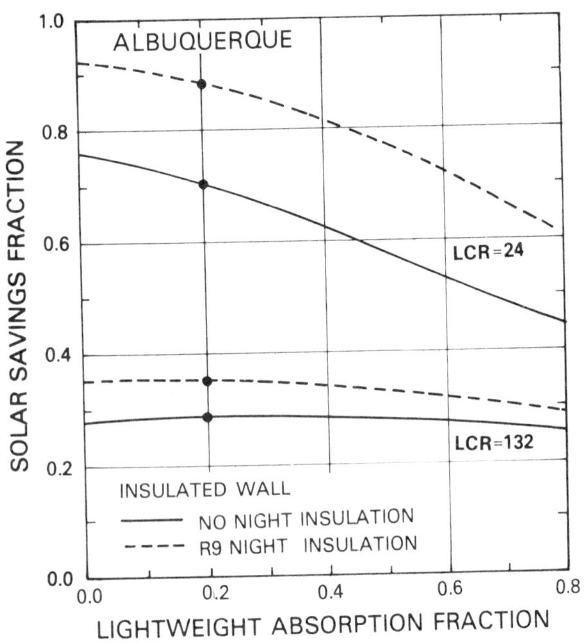

Figure A2-246

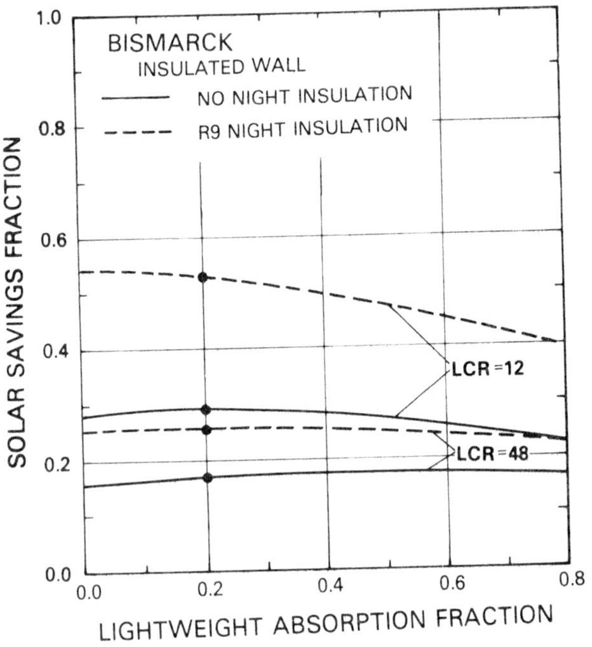

Figure A2-247

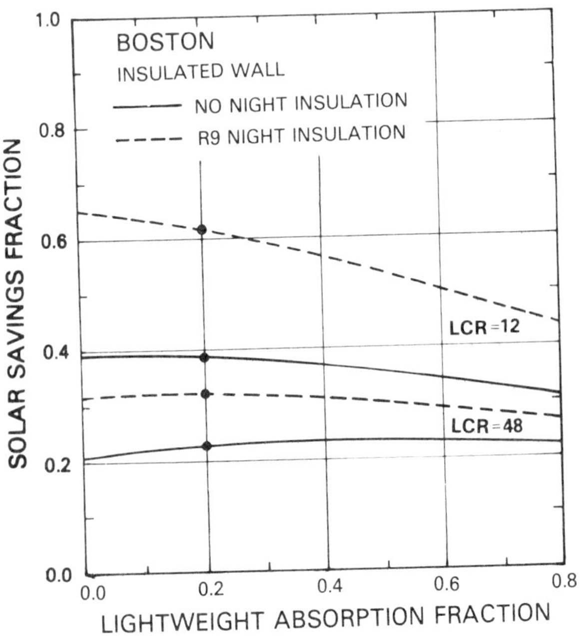

Figure A2-248

Figures A2-246 through A2-251. Annual heating performance sensitivity to the sunspace lightweight absorption fraction for geometry A with a lightweight, insulated common wall, insulated end walls, both without and with R9 night insulation (A5 and A6). The lightweight absorption fraction is the fraction of solar radiation that directly heats the sunspace air when it is transmitted through the glazing and when it is reflected from interior surfaces. It is intended to simulate the presence of lightweight objects in the sunspace that absorb solar radiation and rapidly convect heat to the sunspace air. The dots represent the reference conditions.

SUNSPACE
LIGHTWEIGHT ABSORPTION FRACTION

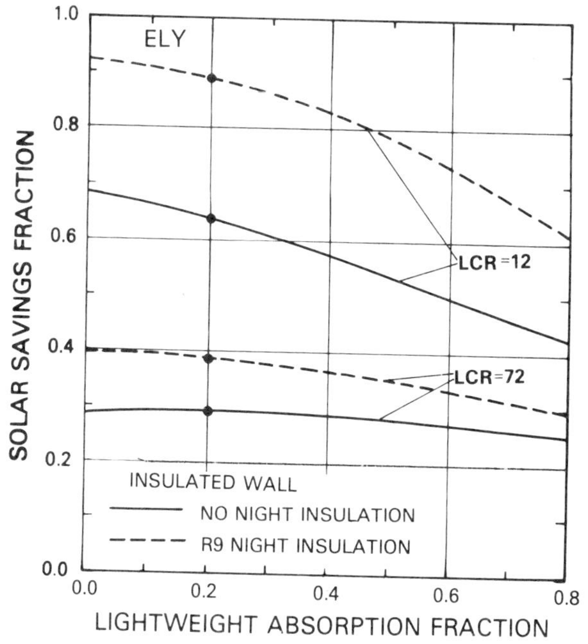

Figure A2-249

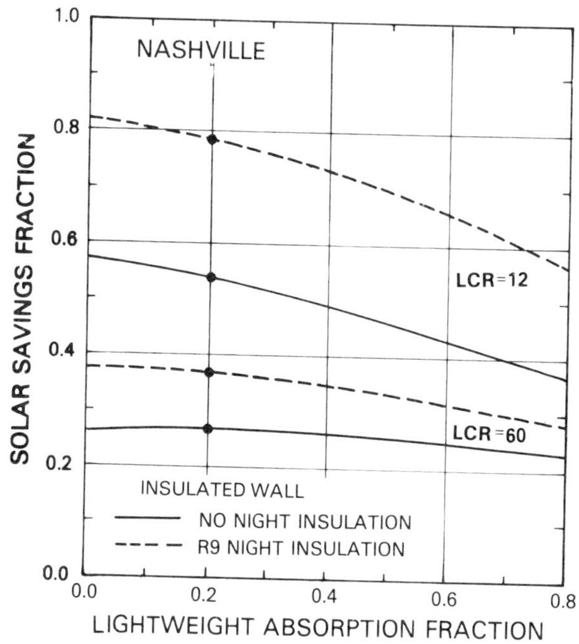

Figure A2-250

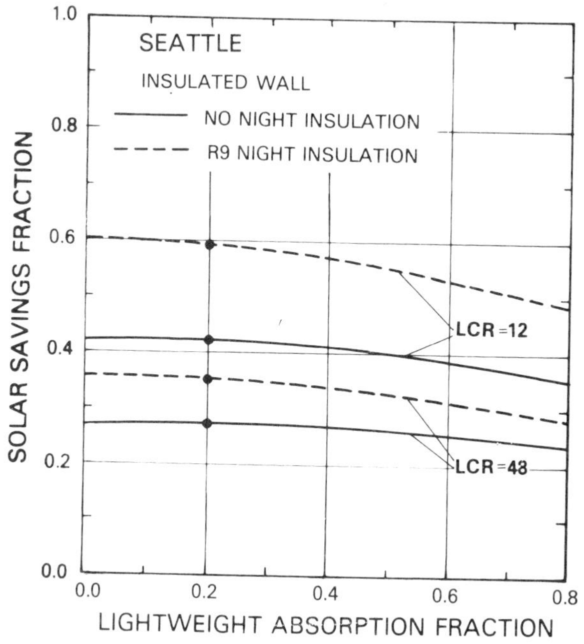

Figure A2-251

Figures A2-246 through A2-251. Annual heating performance sensitivity to the sunspace lightweight absorption fraction for geometry A with a lightweight, insulated common wall, insulated end walls, both without and with R9 night insulation (A5 and A6). The lightweight absorption fraction is the fraction of solar radiation that directly heats the sunspace air when it is transmitted through the glazing and when it is reflected from interior surfaces. It is intended to simulate the presence of lightweight objects in the sunspace that absorb solar radiation and rapidly convect heat to the sunspace air. The dots represent the reference conditions.

SUNSPACE
VENT AREA FRACTION

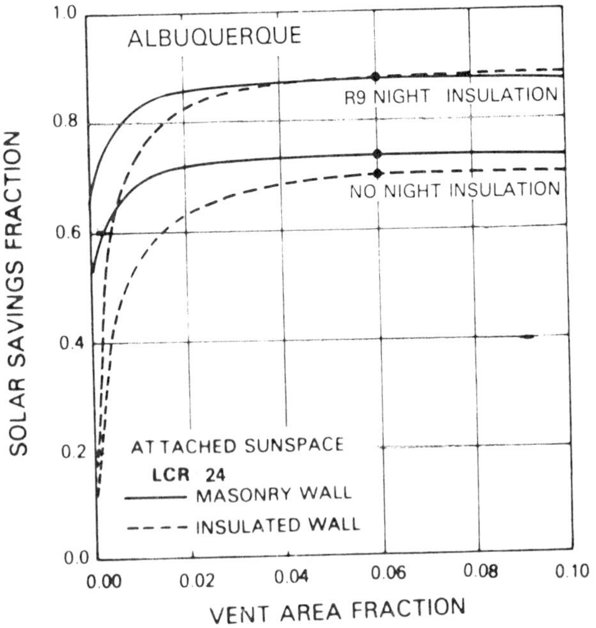

Figure A2-252

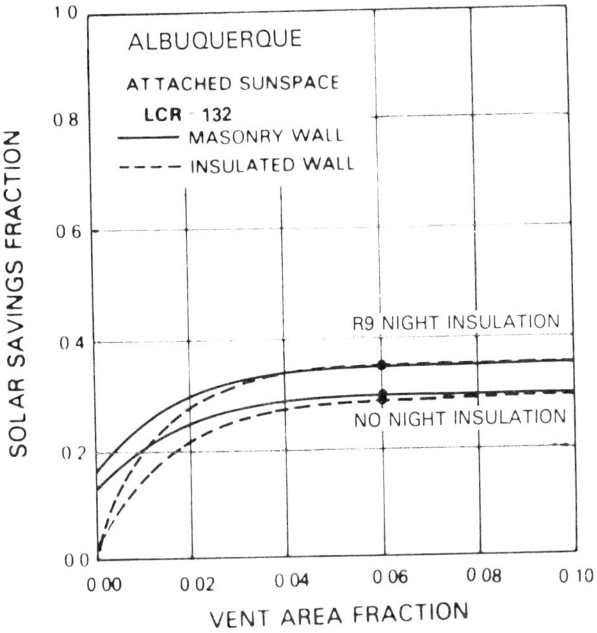

Figure A2-253

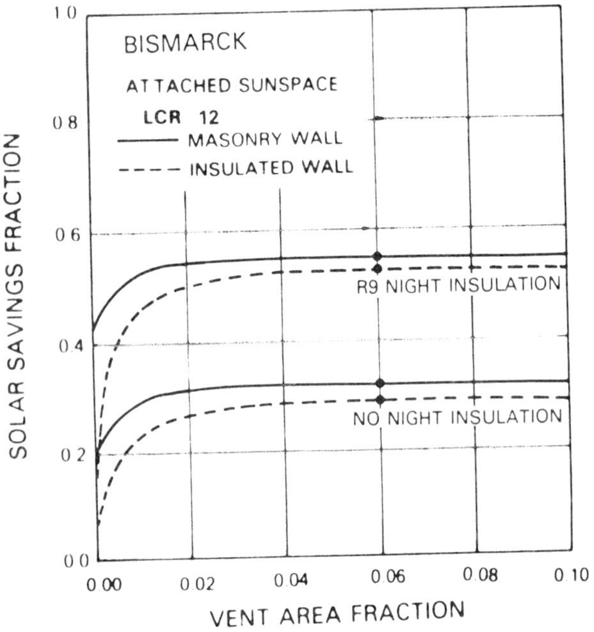

Figure A2-254

Figures A2-252 through A2-263. Annual heating performance sensitivity to the ratio of the common wall thermocirculation vent area to the total common wall area for geometry A with insulated end walls; both with masonry and insulated, lightweight common wall; both without and with R9 night insulation (A1, A2, A5, and A6). The vent area is the combined area of both top and bottom vents. The dots represent the reference conditions.

SUNSPACE
VENT AREA FRACTION

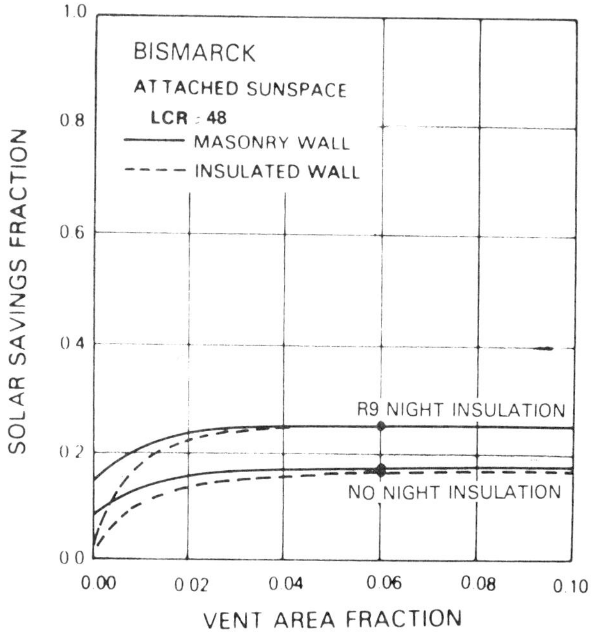

Figure A2-255

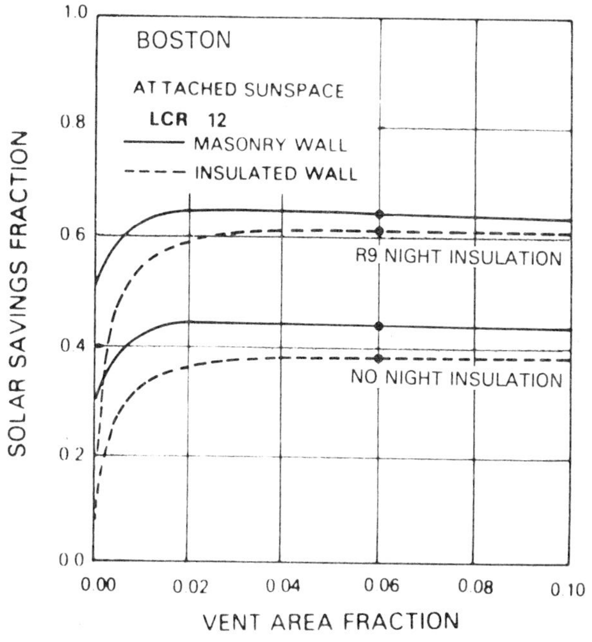

Figure A2-256

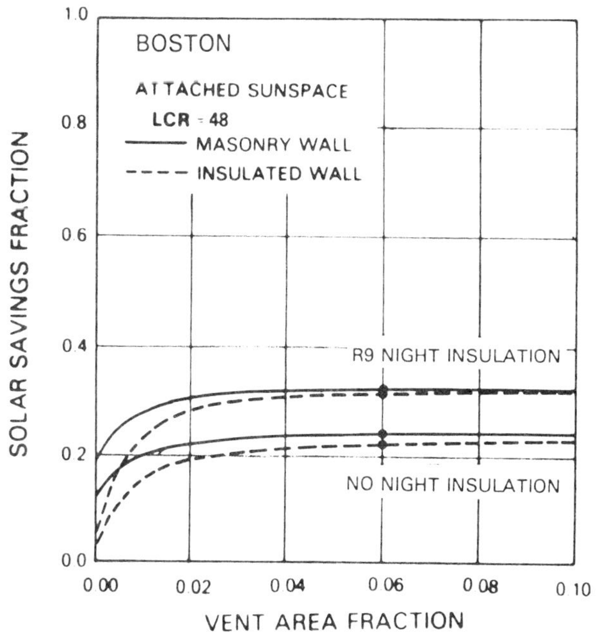

Figure A2-257

Figures A2-252 through A2-263. Annual heating performance sensitivity to the ratio of the common wall thermocirculation vent area to the total common wall area for geometry A with insulated end walls; both with masonry and insulated, lightweight common wall; both without and with R9 night insulation (A1, A2, A5, and A6). The vent area is the combined area of both top and bottom vents. The dots represent the reference conditions.

SUNSPACE
VENT AREA FRACTION

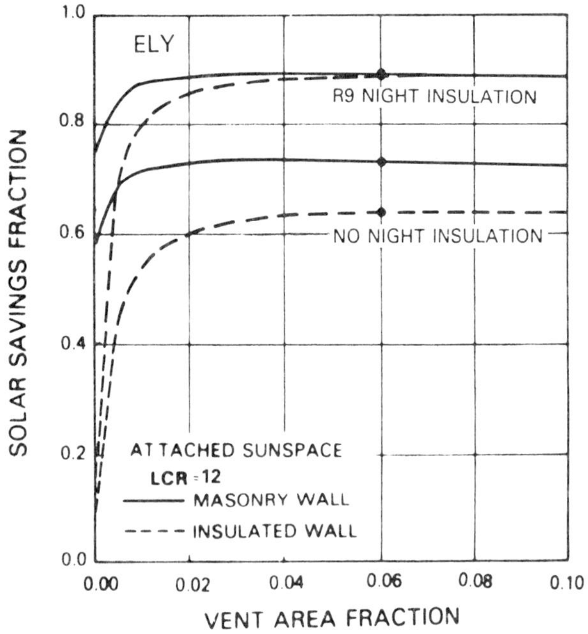

Figure A2-258

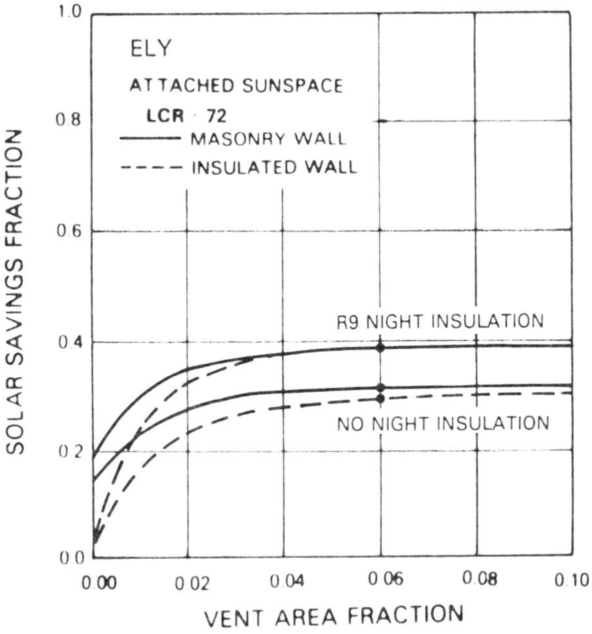

Figure A2-259

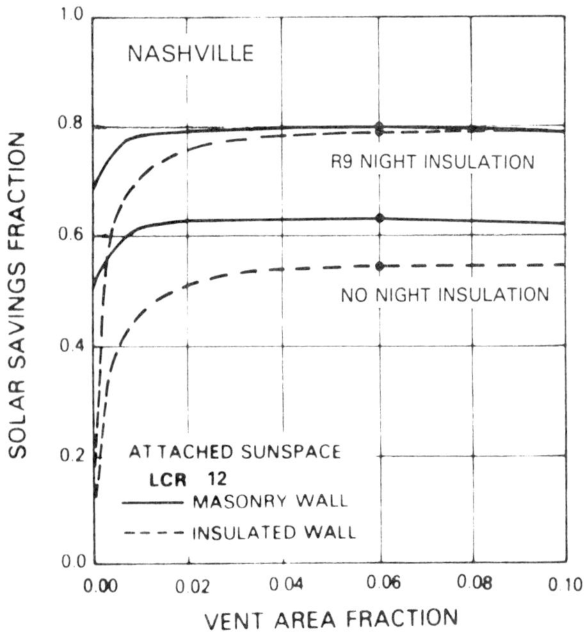

Figure A2-260

Figures A2-252 through A2-263. Annual heating performance sensitivity to the ratio of the common wall thermocirculation vent area to the total common wall area for geometry A with insulated end walls; both with masonry and insulated, lightweight common wall; both without and with R9 night insulation (A1, A2, A5, and A6). The vent area is the combined area of both top and bottom vents. The dots represent the reference conditions.

SUNSPACE
VENT AREA FRACTION

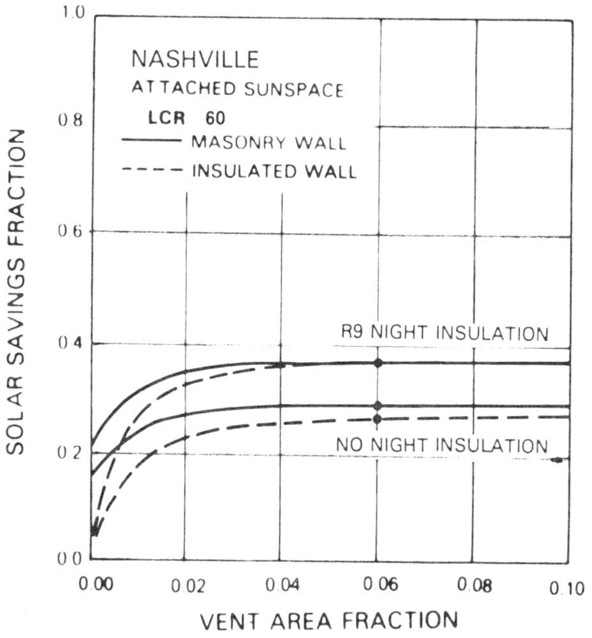

Figure A2-261

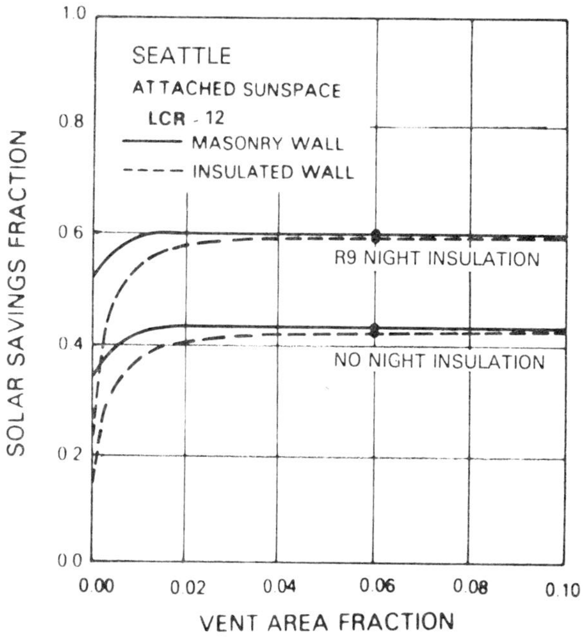

Figure A2-262

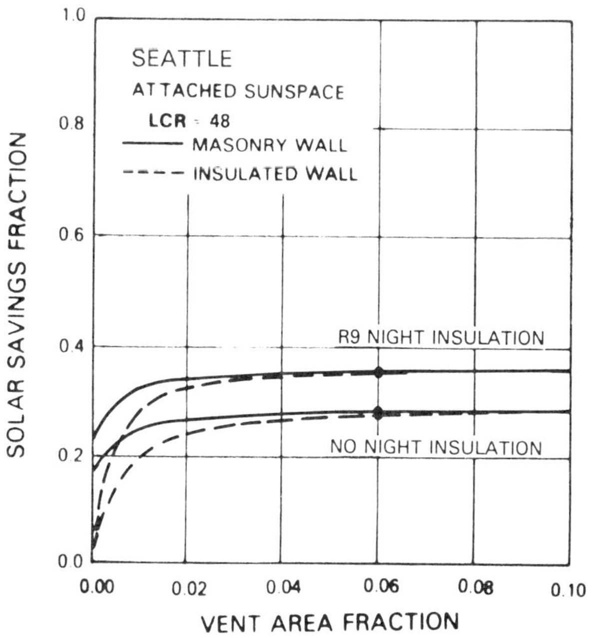

Figure A2-263

Figures A2-252 through A2-263. Annual heating performance sensitivity to the ratio of the common wall thermocirculation vent area to the total common wall area for geometry A with insulated end walls; both with masonry and insulated, lightweight common wall; both without and with R9 night insulation (A1, A2, A5, and A6). The vent area is the combined area of both top and bottom vents. The dots represent the reference conditions.

SUNSPACE
VENT AREA FRACTION

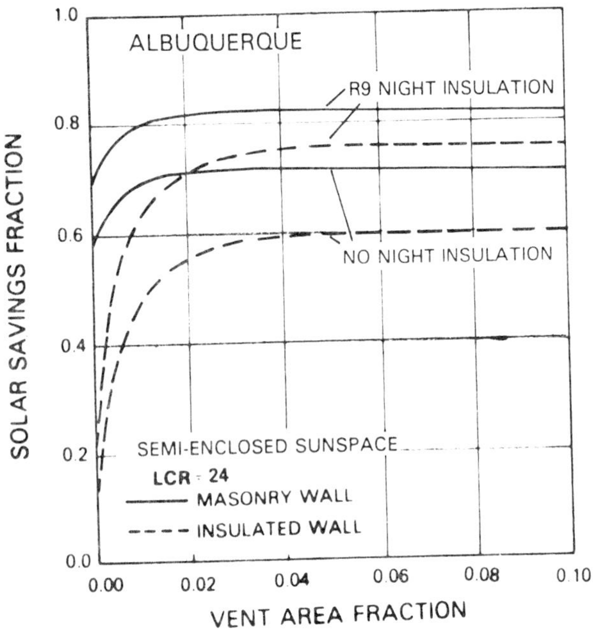

Figure A2-264

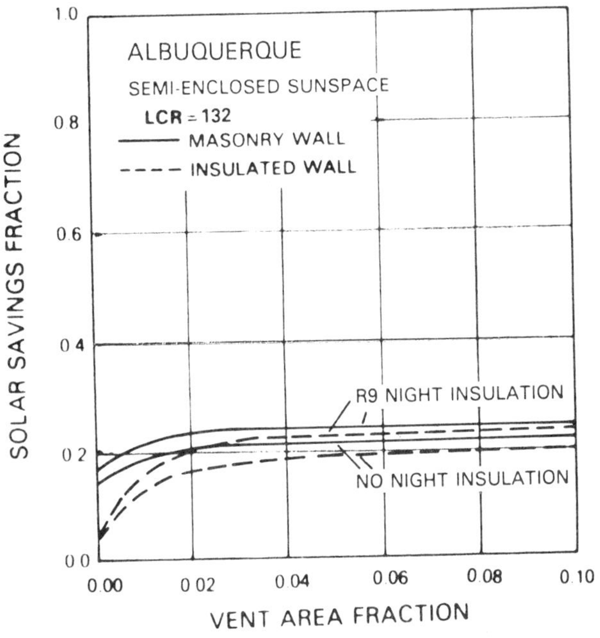

Figure A2-265

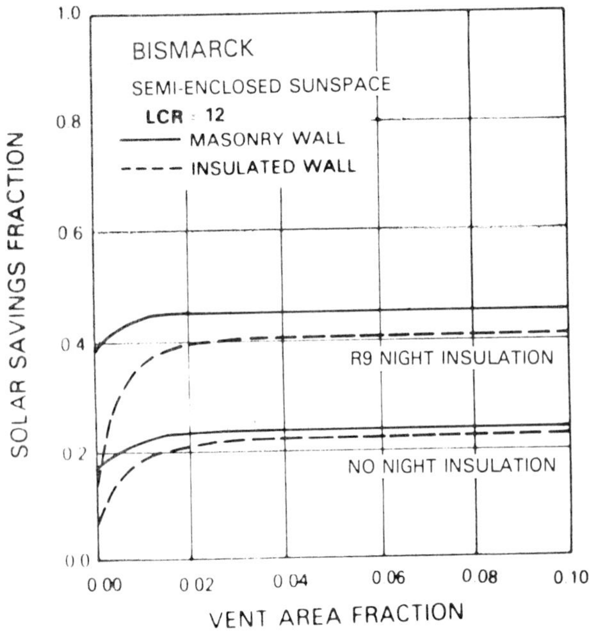

Figure A2-266

Figures A2-264 through A2-275. Annual heating performance sensitivity to the ratio of the common wall thermocirculation vent area to the total north common wall area for geometry C; both with masonry and insulated, lightweight common wall; both without and with R9 night insulation (C1, C2, C3, and C4). The vent area is the combined area of both top and bottom vents. The dots represent the reference conditions.

SUNSPACE
VENT AREA FRACTION

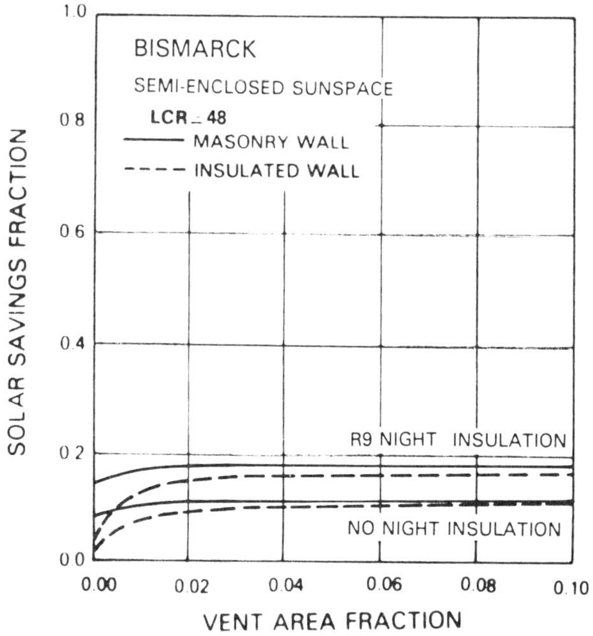

Figure A2-267

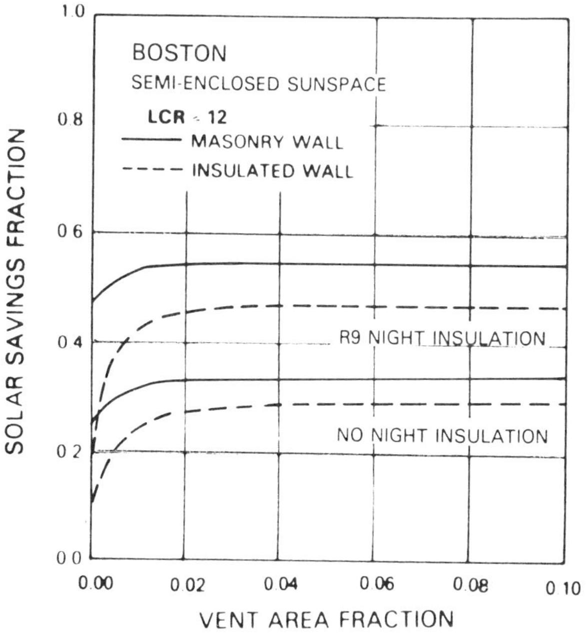

Figure A2-268

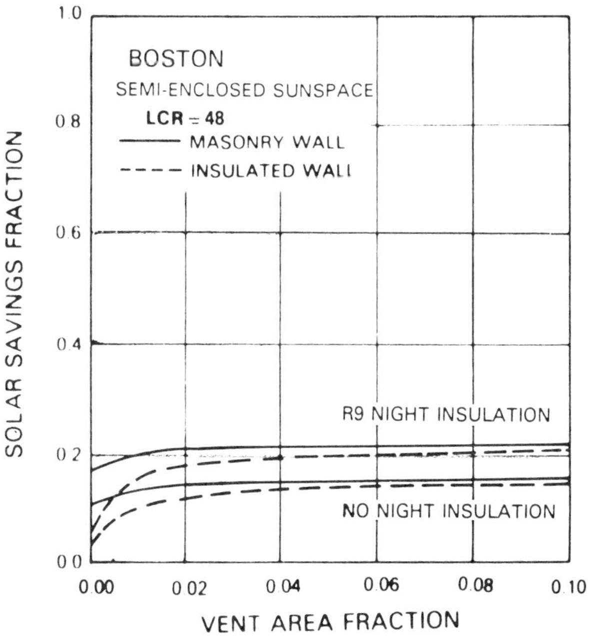

Figure A2-269

Figures A2-264 through A2-275. Annual heating performance sensitivity to the ratio of the common wall thermocirculation vent area to the total north common wall area for geometry C; both with masonry and insulated, lightweight common wall; both without and with R9 night insulation (C1, C2, C3, and C4). The vent area is the combined area of both top and bottom vents. The dots represent the reference conditions.

SUNSPACE
VENT AREA FRACTION

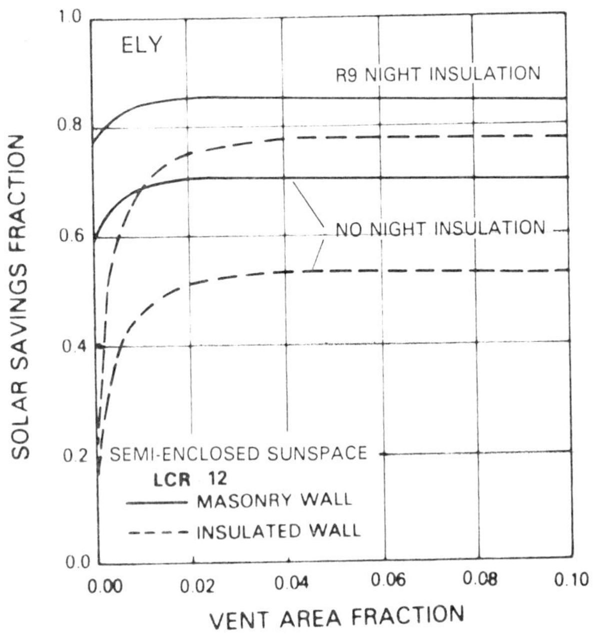

Figure A2-270

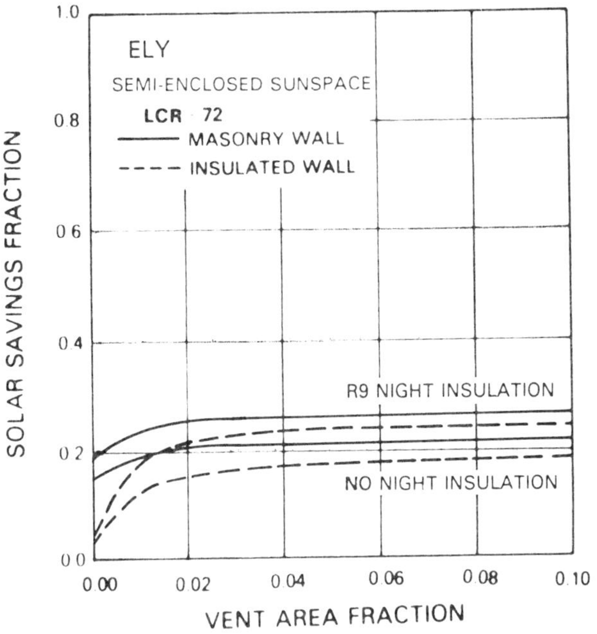

Figure A2-271

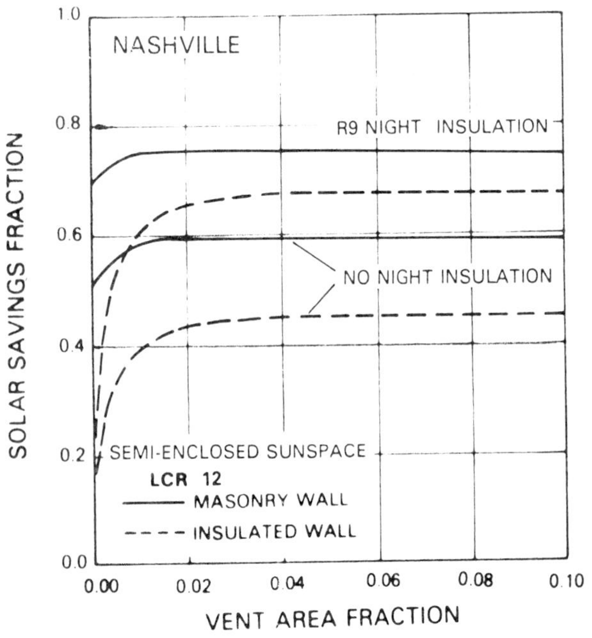

Figure A2-272

Figures A2-264 through A2-275. Annual heating performance sensitivity to the ratio of the common wall thermocirculation vent area to the total north common wall area for geometry C; both with masonry and insulated, lightweight common wall; both without and with R9 night insulation (C1, C2, C3, and C4). The vent area is the combined area of both top and bottom vents. The dots represent the reference conditions.

SUNSPACE
VENT AREA FRACTION

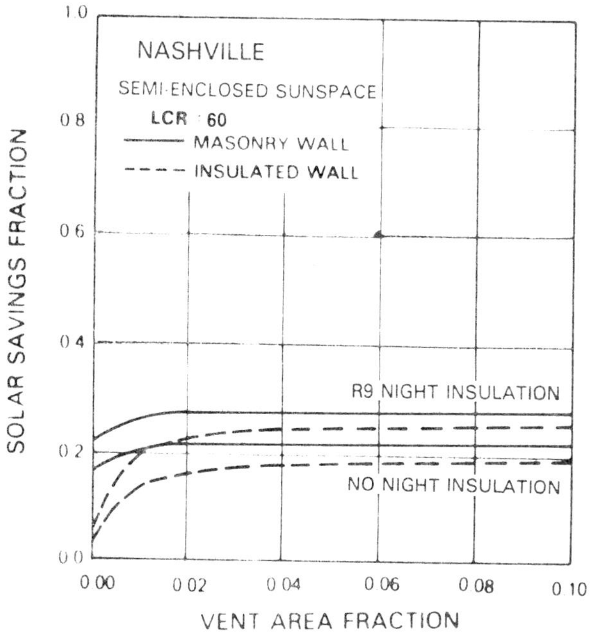

Figure A2-273

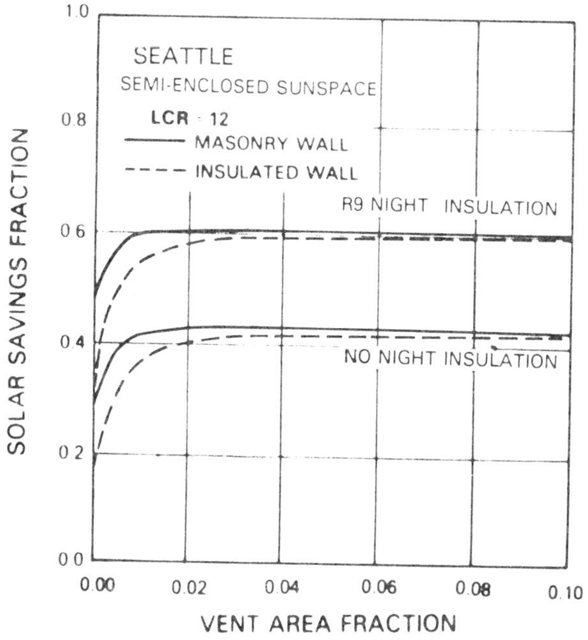

Figure A2-274

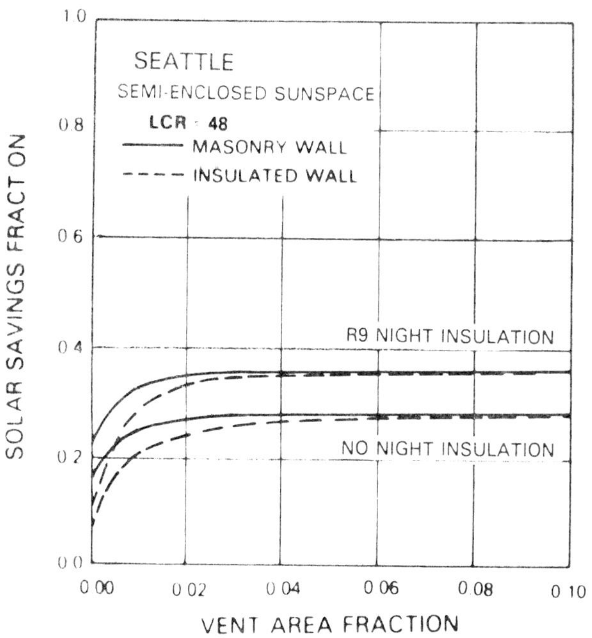

Figure A2-275

Figures A2-264 through A2-275. Annual heating performance sensitivity to the ratio of the common wall thermocirculation vent area to the total north common wall area for geometry C; both with masonry and insulated, lightweight common wall; both without and with R9 night insulation (C1, C2, C3, and C4). The vent area is the combined area of both top and bottom vents. The dots represent the reference conditions.

SUNSPACE INFILTRATION RATE

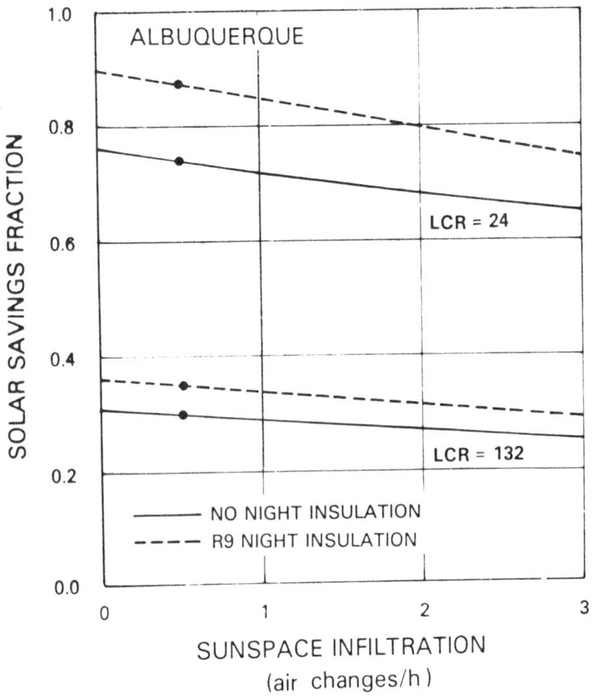

Figure A2-276

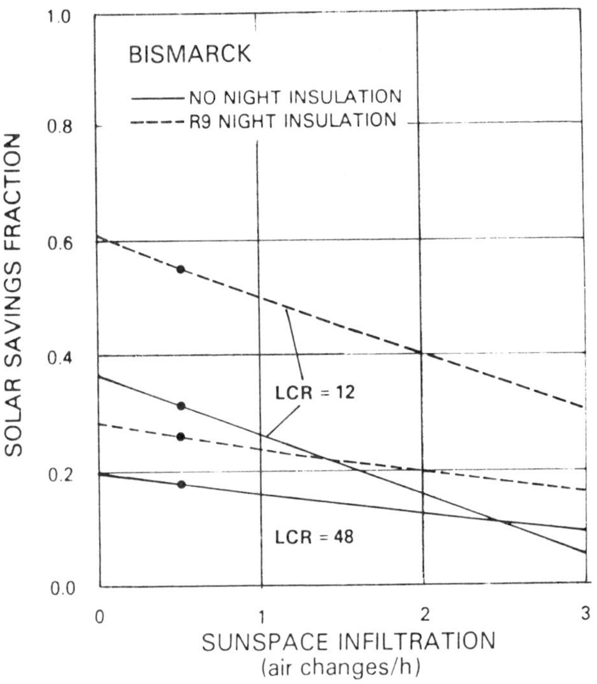

Figure A2-277

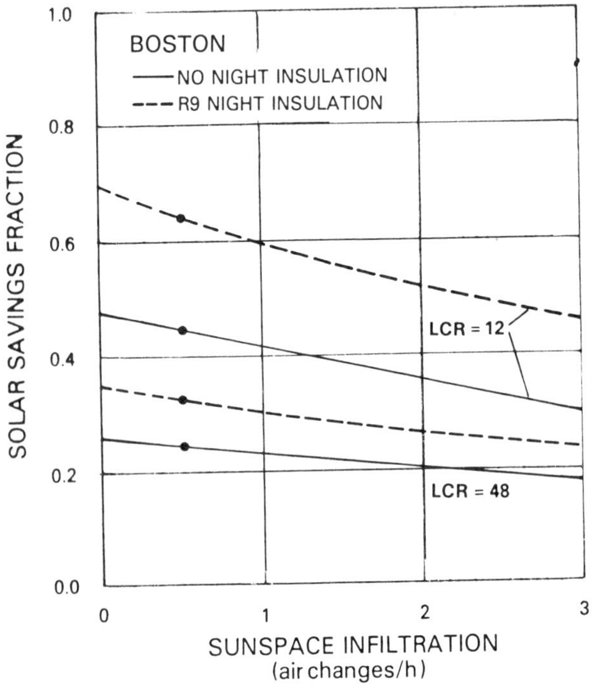

Figure A2-278

Figures A2-276 through A2-281. Annual heating performance sensitivity to the sunspace infiltration rate for geometry A with a masonry common wall, insulated end walls, both without and with R9 night insulation (A1 and A2). The dots represent the reference conditions.

SUNSPACE INFILTRATION RATE

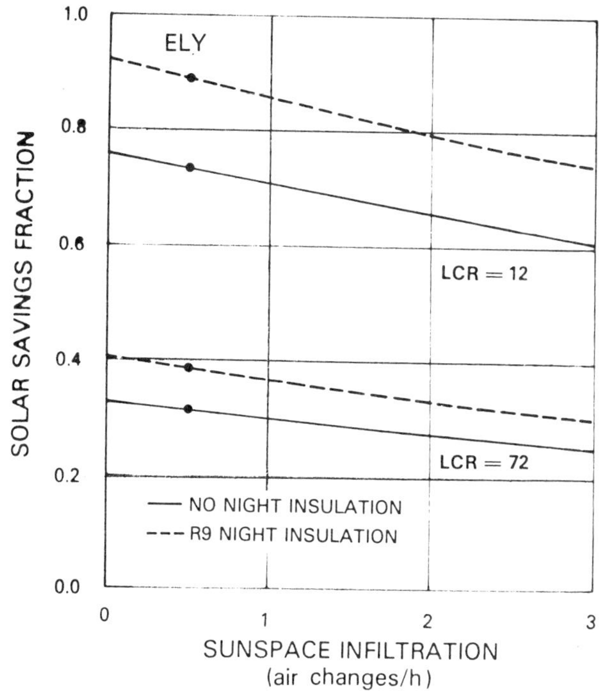

Figure A2-279

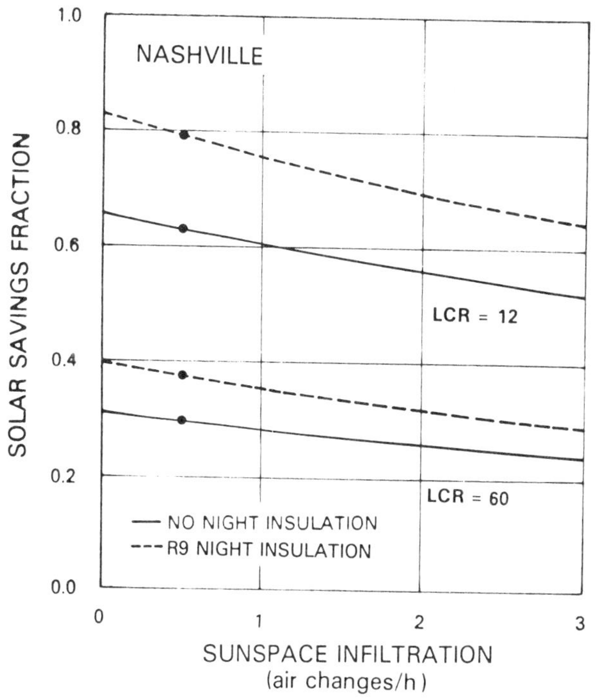

Figure A2-280

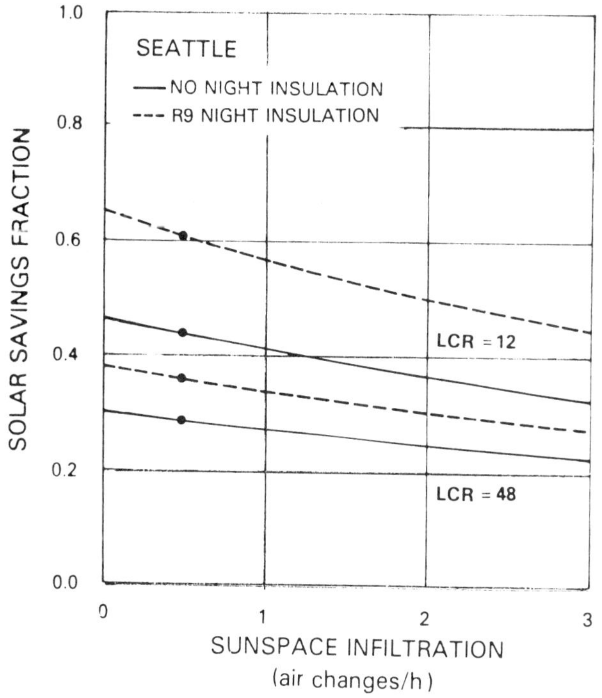

Figure A2-281

Figures A2-276 through A2-281. Annual heating performance sensitivity to the sunspace infiltration rate for geometry A with a masonry common wall, insulated end walls, both without and with R9 night insulation (A1 and A2). The dots represent the reference conditions.

SUNSPACE
COMMON WALL R-VALUE

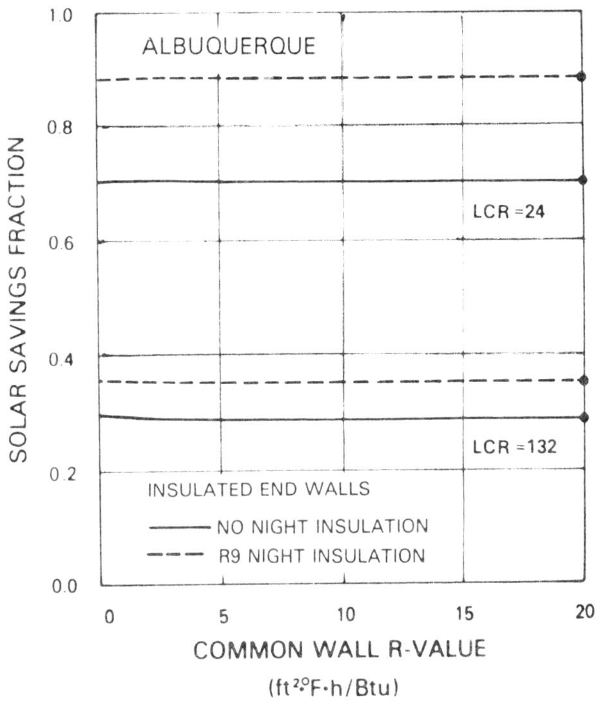

Figure A2-282

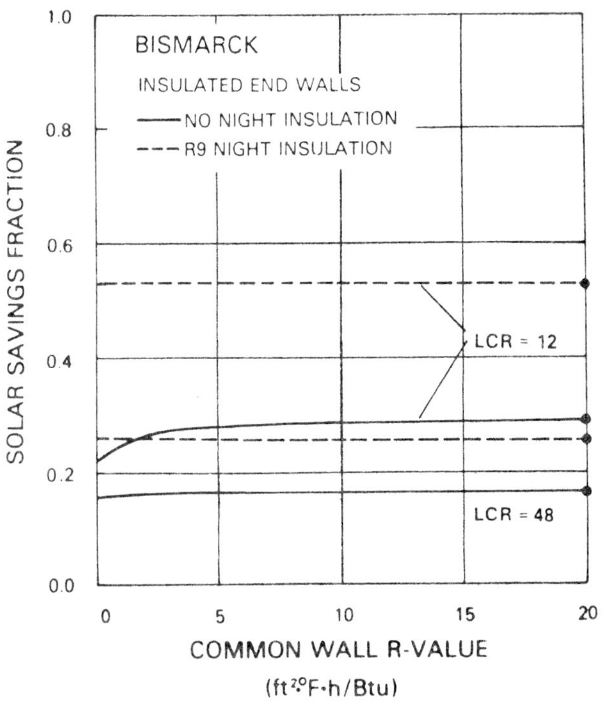

Figure A2-283

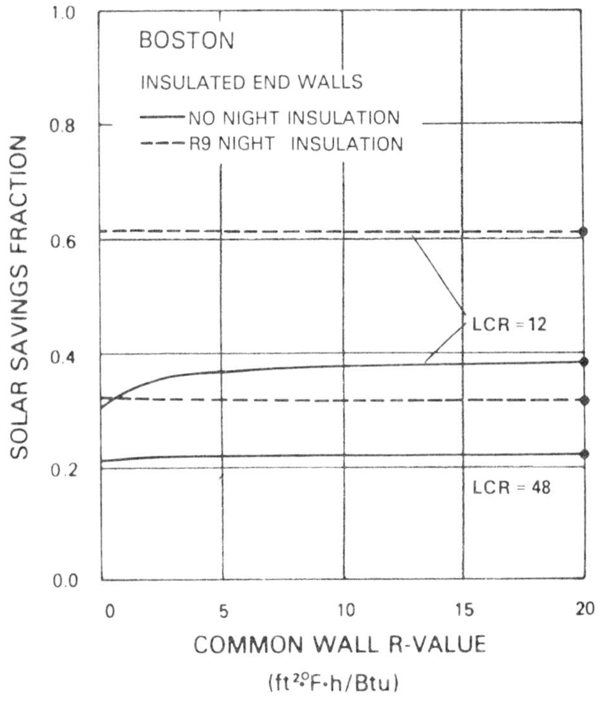

Figure A2-284

Figures A2-282 through A2-287. Annual heating performance sensitivity to the thermal resistance of the sunspace common wall for geometry A with a lightweight, insulated common wall, insulated end walls, both without and with R9 night insulation (A5 and A6). The wall-surface-to-air film resistances are not included. The dots represent the reference conditions.

SUNSPACE
COMMON WALL R-VALUE

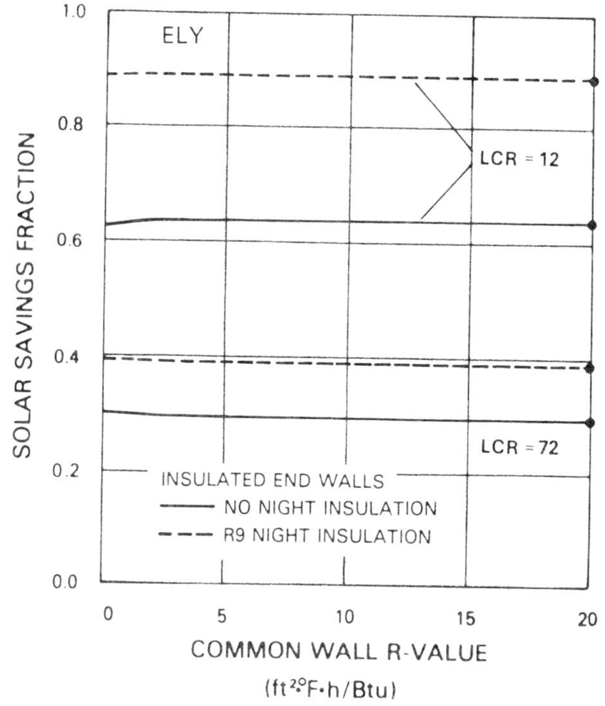

Figure A2-285

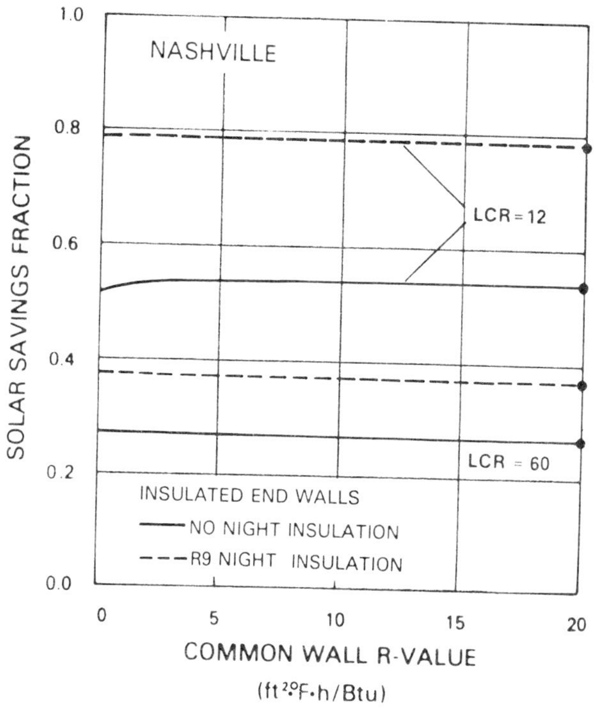

Figure A2-286

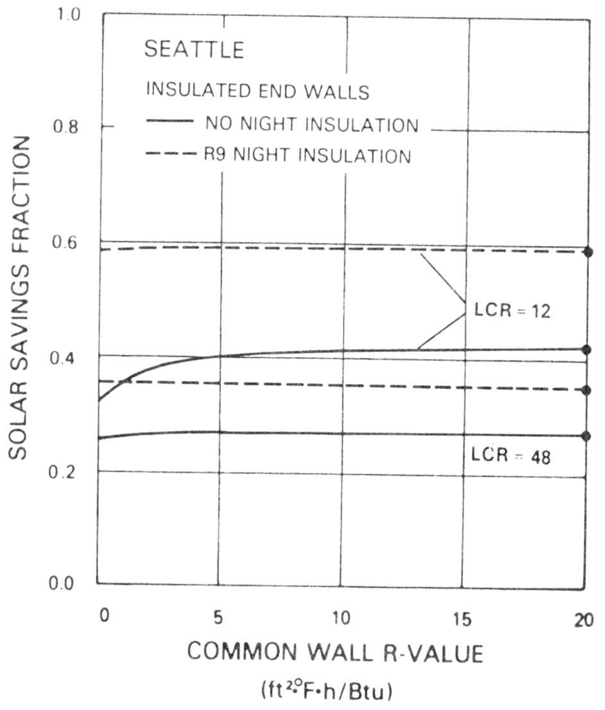

Figure A2-287

Figures A2-282 through A2-287. Annual heating performance sensitivity to the thermal resistance of the sunspace common wall for geometry A with a lightweight, insulated common wall, insulated end walls, both without and with R9 night insulation (A5 and A6). The wall-surface-to-air film resistances are not included. The dots represent the reference conditions.

SUNSPACE
COMMON WALL R-VALUE

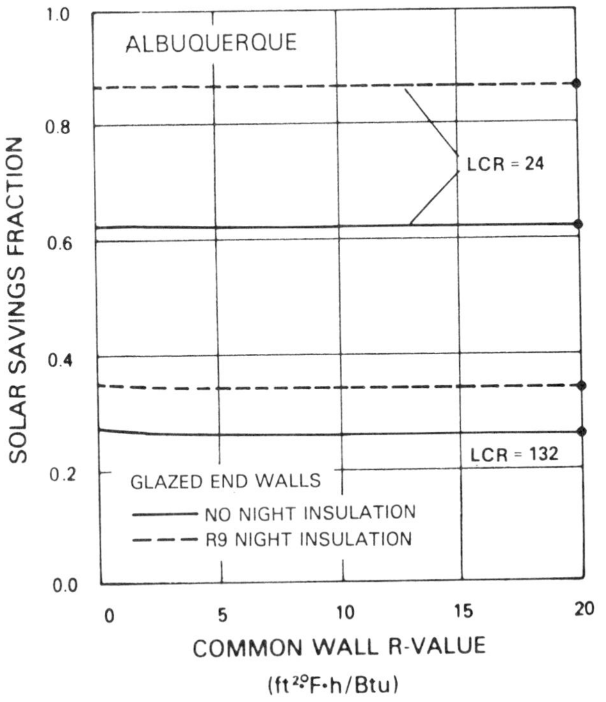

Figure A2-288

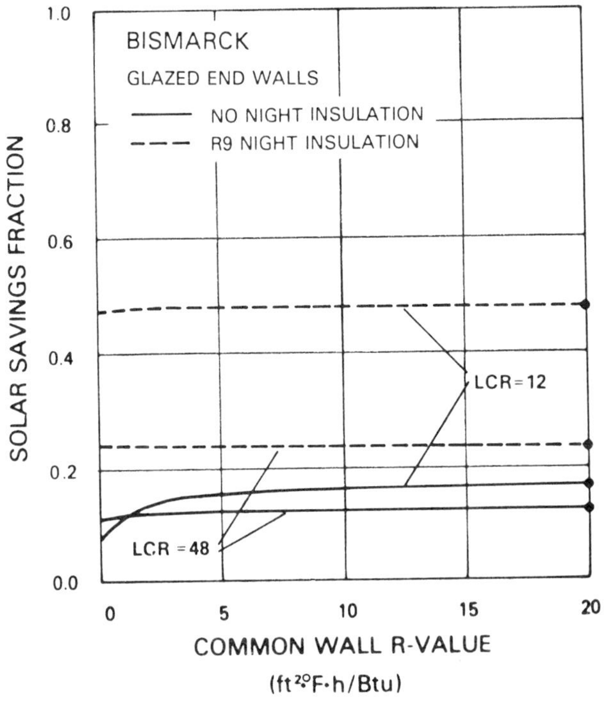

Figure A2-289

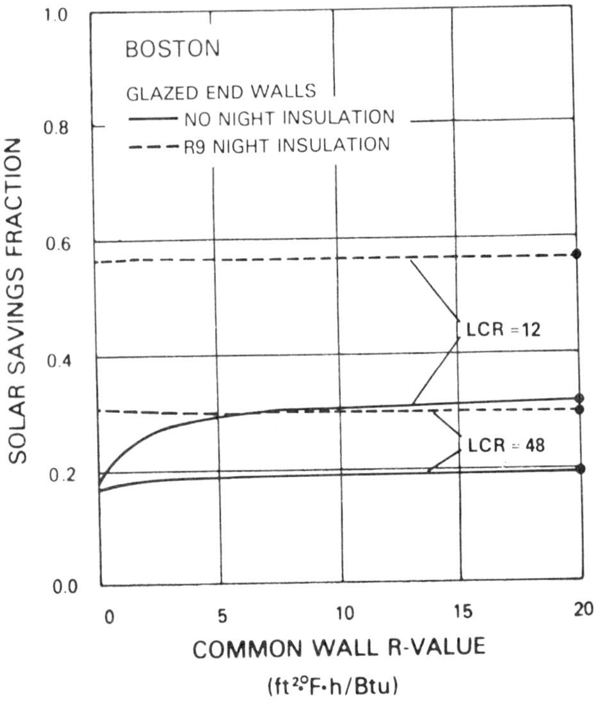

Figure A2-290

Figures A2-288 through A2-293. Annual heating performance sensitivity to the thermal resistance of the sunspace common wall for geometry A with a lightweight, insulated common wall, glazed end walls, both without and with R9 night insulation (A7 and A8). The wall-surface-to-air film resistances are not included. The dots represent the reference conditions.

SUNSPACE
COMMON WALL R-VALUE

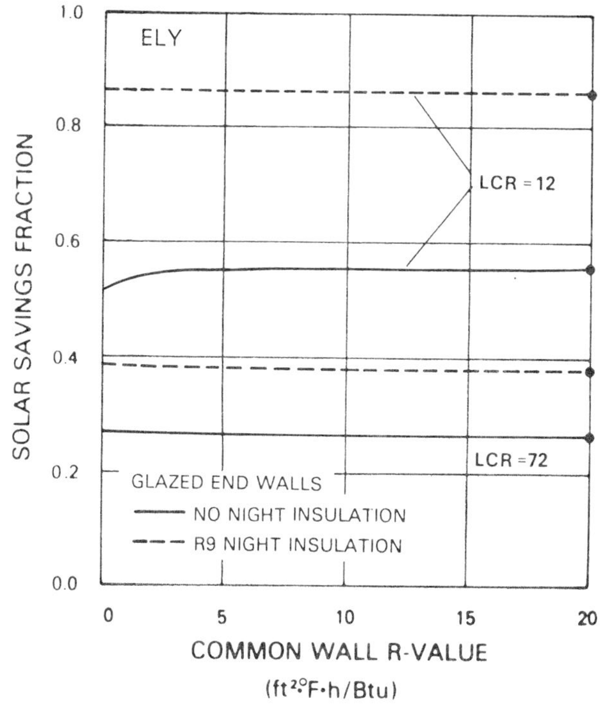

Figure A2-291

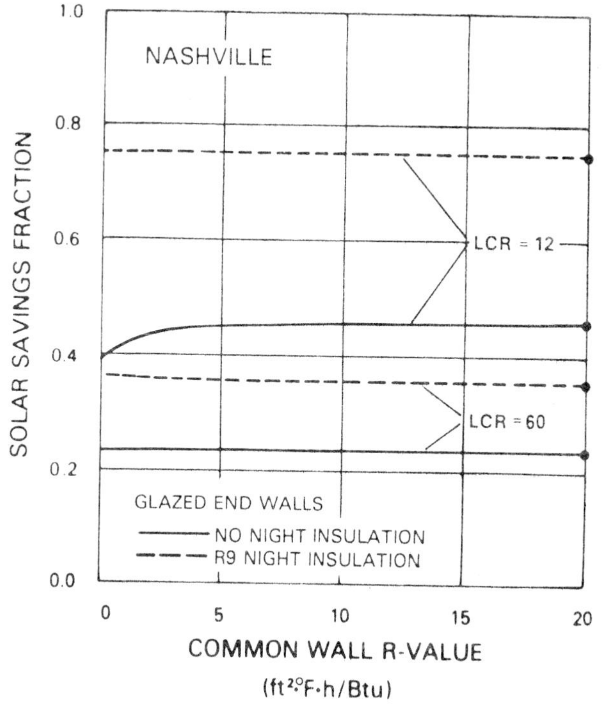

Figure A2-292

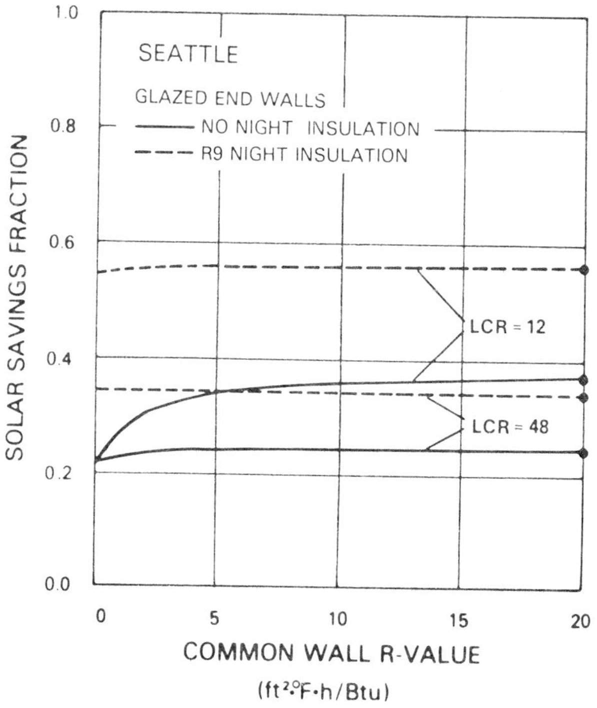

Figure A2-293

Figures A2-288 through A2-293. Annual heating performance sensitivity to the thermal resistance of the sunspace common wall for geometry A with a lightweight, insulated common wall, glazed end walls, both without and with R9 night insulation (A7 and A8). The wall-surface-to-air film resistances are not included. The dots represent the reference conditions.

SUNSPACE
MINIMUM ROOM SETPOINT

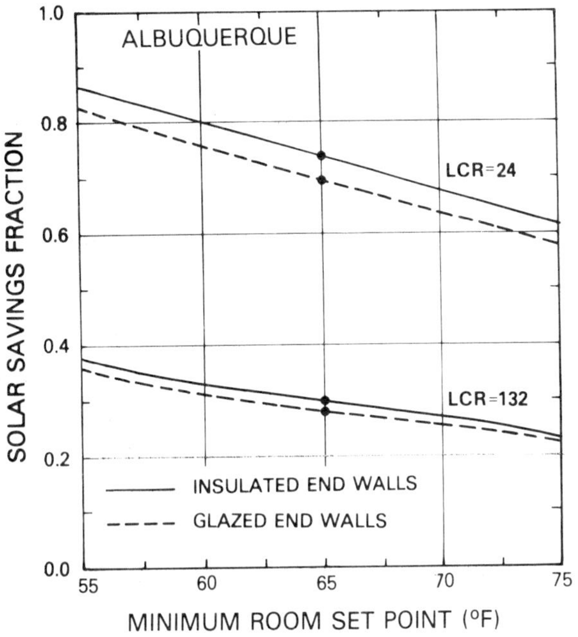

Figure A2-294

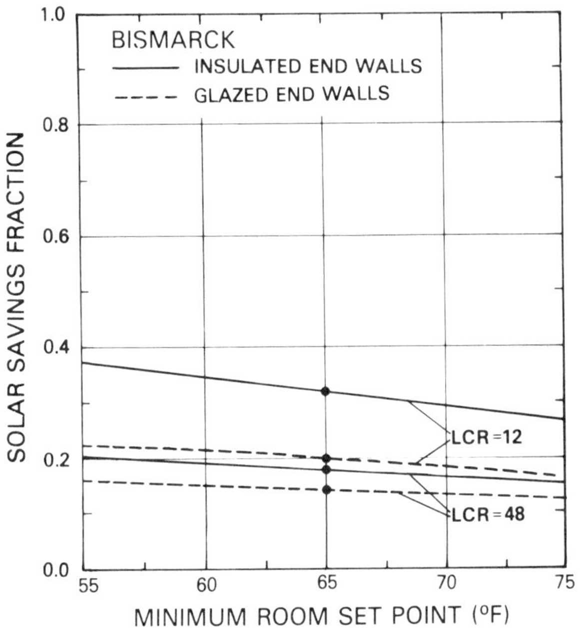

Figure A2-295

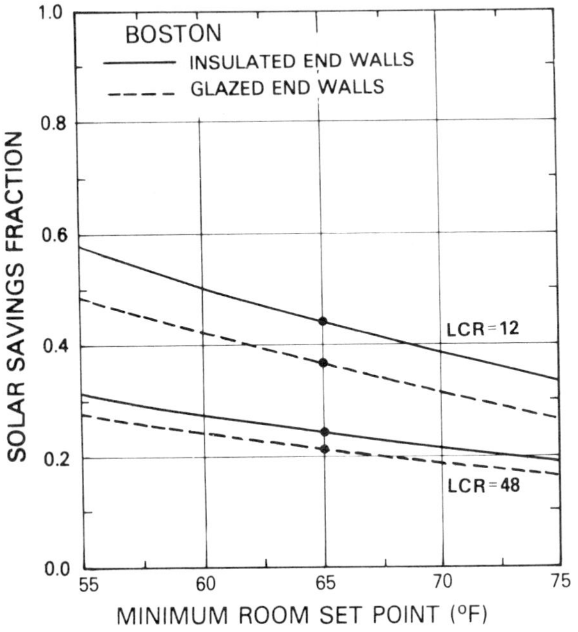

Figure A2-296

Figures A2-294 through A2-299. Annual heating performance sensitivity to the room auxiliary heat thermostat setpoint for geometry A with masonry common wall, no night insulation, and both insulated and glazed end walls (A1 and A3). The solar savings fraction in this case is defined relative to a net reference load evaluated at the variable setpoint and therefore reflects only the variation in the solar savings, and not the variation in the building load with setpoint. The dots represent the reference conditions.

SUNSPACE
MINIMUM ROOM SETPOINT

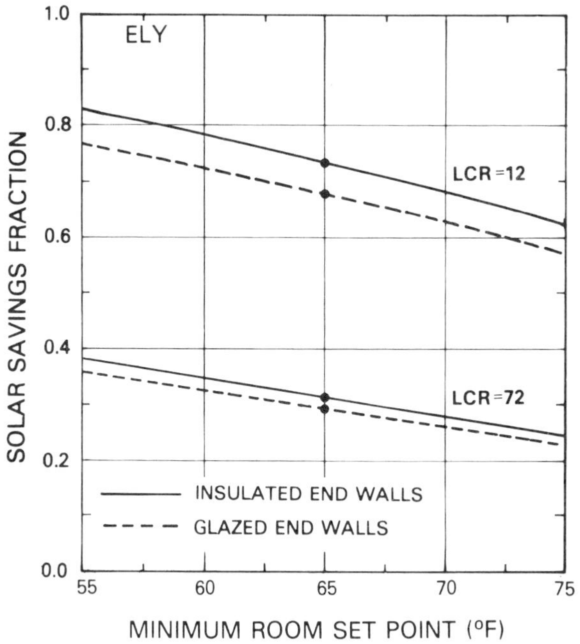

Figure A2-297

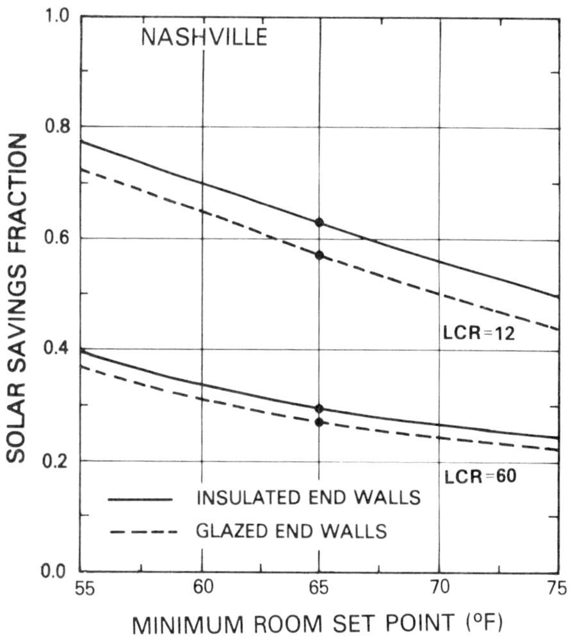

Figure A2-298

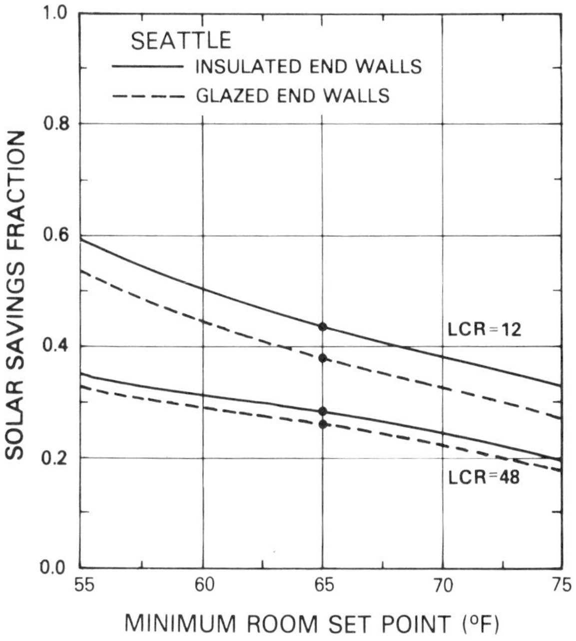

Figure A2-299

Figures A2-294 through A2-299. Annual heating performance sensitivity to the room auxiliary heat thermostat setpoint for geometry A with masonry common wall, no night insulation, and both insulated and glazed end walls (A1 and A3). The solar savings fraction in this case is defined relative to a net reference load evaluated at the variable setpoint and therefore reflects only the variation in the solar savings, and not the variation in the building load with setpoint. The dots represent the reference conditions.

SUNSPACE
MINIMUM ROOM SETPOINT

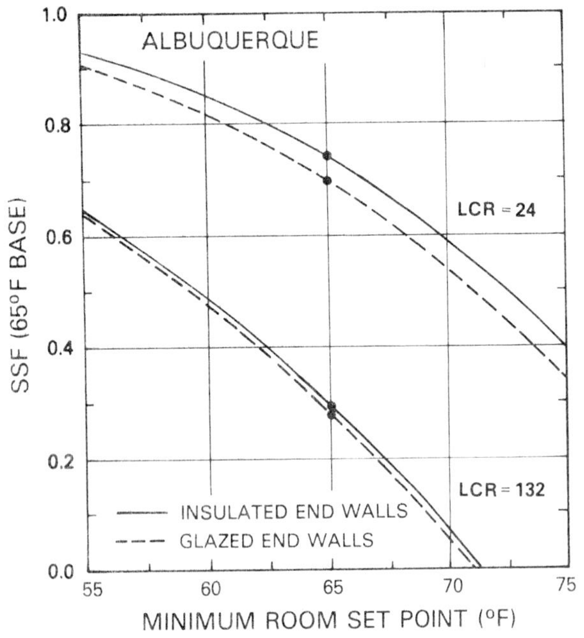

Figure A2-300

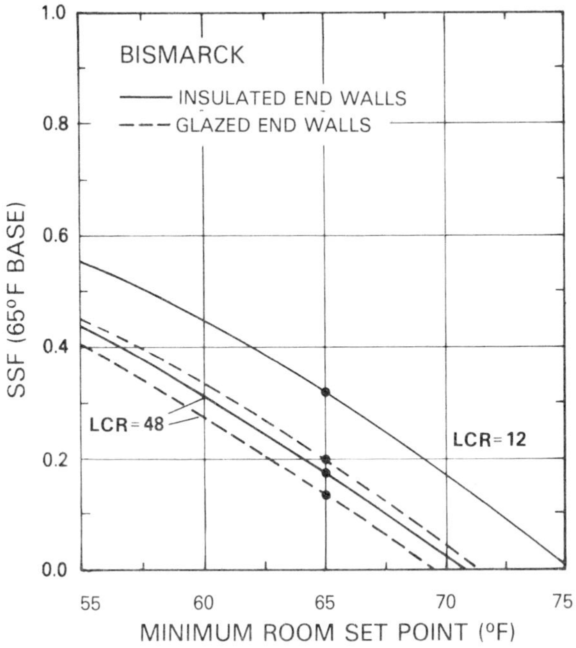

Figure A2-301

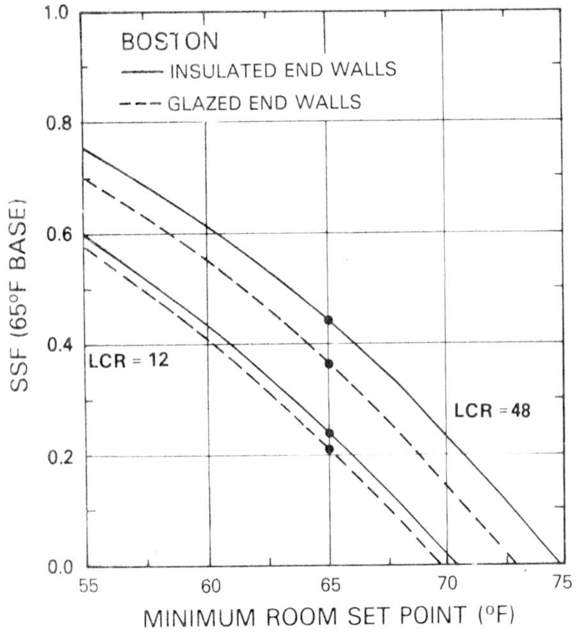

Figure A2-302

Figures A2-300 through A2-305. Annual heating performance sensitivity to the room auxiliary heat thermostat setpoint temperature for geometry A with a masonry common wall, no night insulation, and both insulated and glazed end walls (A1 and A3). The reference setpoint is 65°F. The solar savings fraction in this case is defined relative to a 65°F reference load and therefore reflects both variation in solar savings and building load with setpoint. The dots represent the reference conditions.

SUNSPACE
MINIMUM ROOM SETPOINT

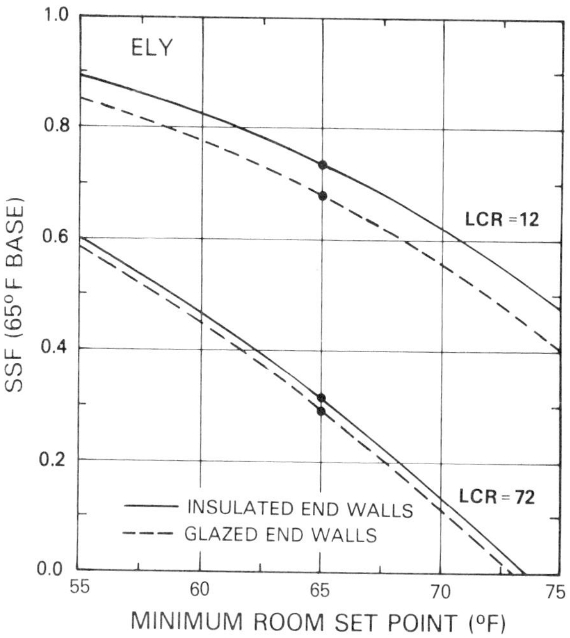

Figure A2-303

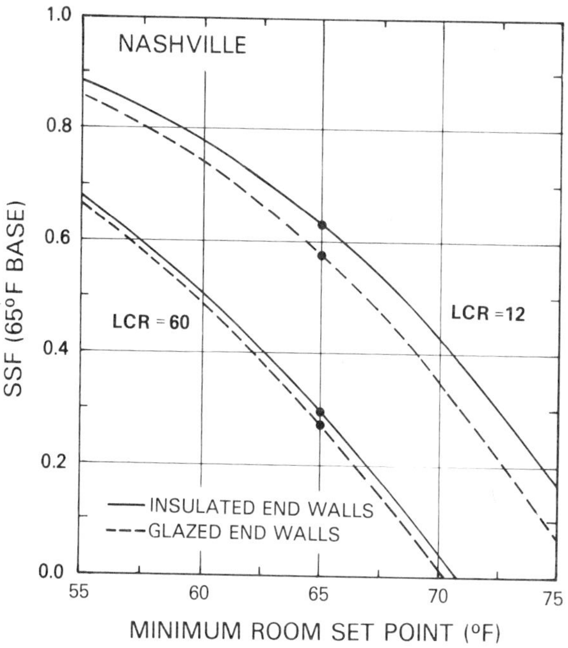

Figure A2-304

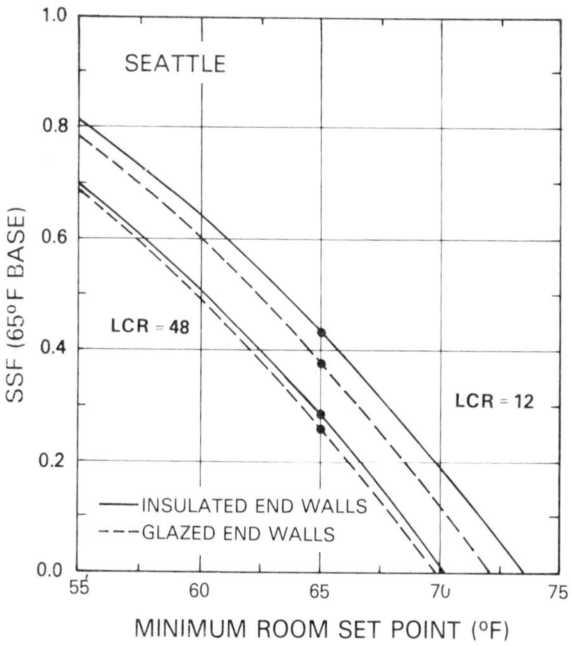

Figure A2-305

Figures A2-300 through A2-305. Annual heating performance sensitivity to the room auxiliary heat thermostat setpoint temperature for geometry A with a masonry common wall, no night insulation, and both insulated and glazed end walls (A1 and A3). The reference setpoint is 65°F. The solar savings fraction in this case is defined relative to a 65°F reference load and therefore reflects both variation in solar savings and building load with setpoint. The dots represent the reference conditions.

SUNSPACE
MINIMUM SUNSPACE SETPOINT

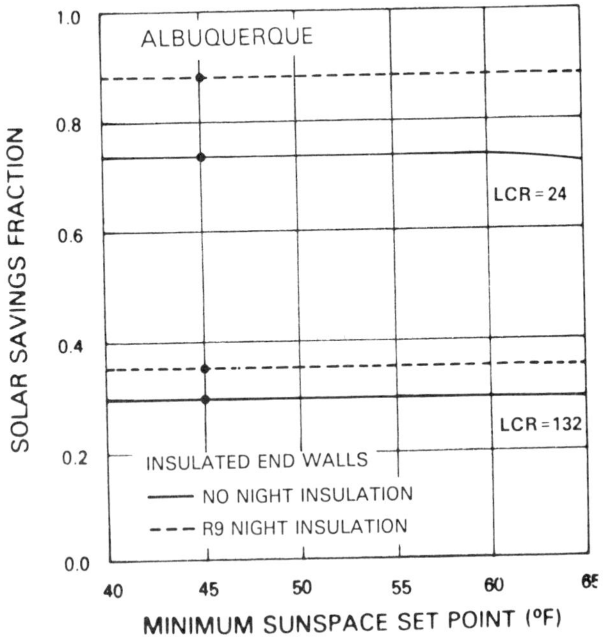

Figure A2-306

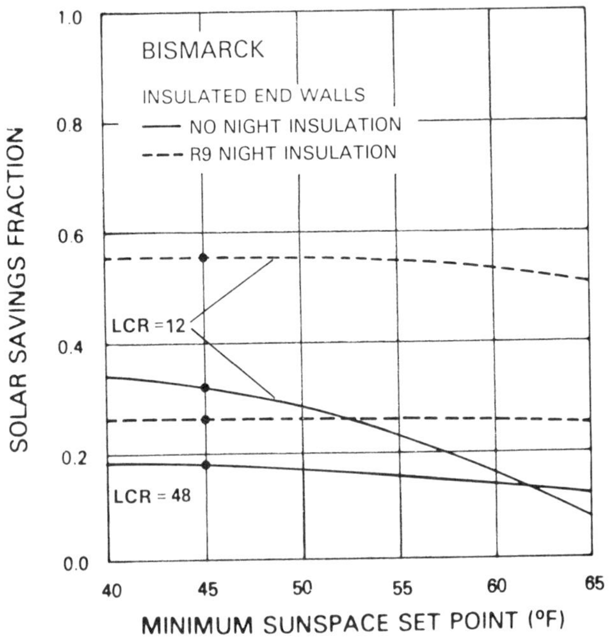

Figure A2-307

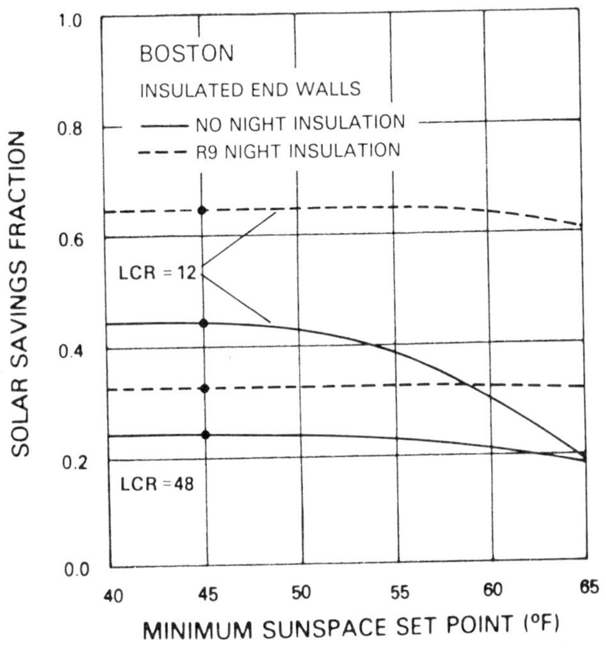

Figure A2-308

Figures A2-306 through A2-311. Annual heating performance sensitivity to the sunspace auxiliary heat thermostat setpoint temperature for geometry A, with a masonry common wall, insulated end walls, both without and with R9 night insulation (A1 and A2). The definition of solar savings fraction includes the sunspace auxiliary heat. The dots represent the reference conditions.

SUNSPACE
MINIMUM SUNSPACE SETPOINT

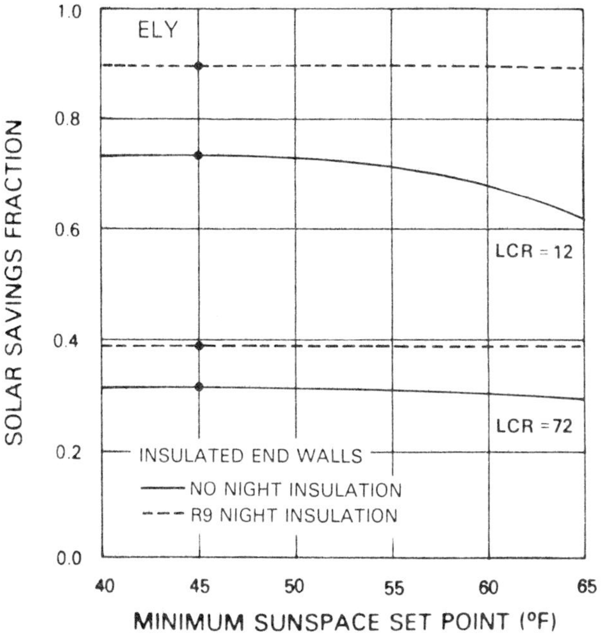

Figure A2-309

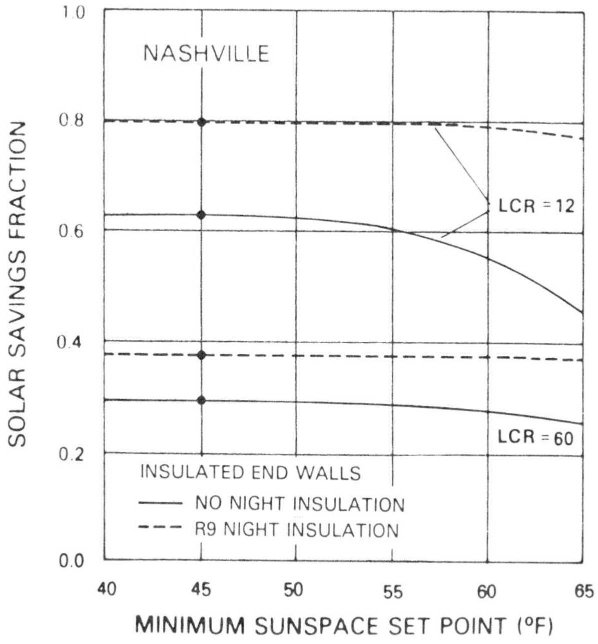

Figure A2-310

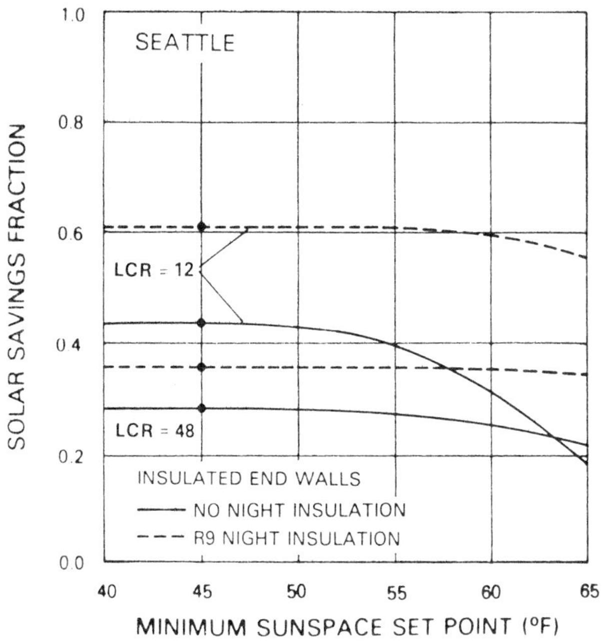

Figure A2-311

Figures A2-306 through A2-311. Annual heating performance sensitivity to the sunspace auxiliary heat thermostat setpoint temperature for geometry A, with a masonry common wall, insulated end walls, both without and with R9 night insulation (A1 and A2). The definition of solar savings fraction includes the sunspace auxiliary heat. The dots represent the reference conditions.

SUNSPACE
MINIMUM SUNSPACE SETPOINT

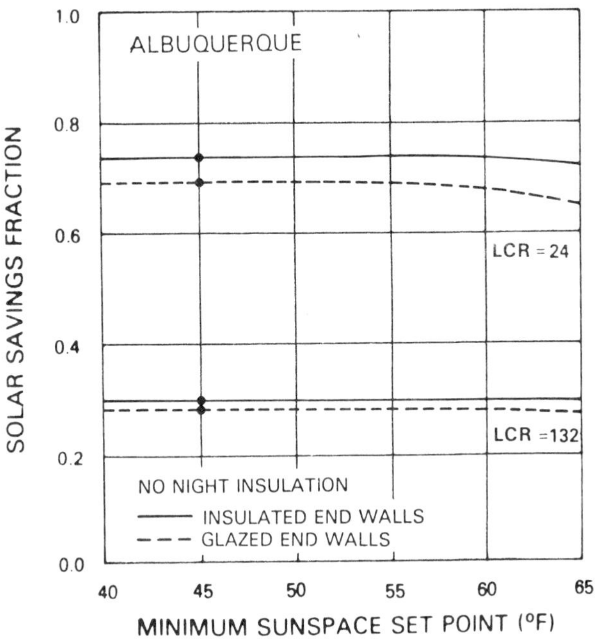

Figure A2-312

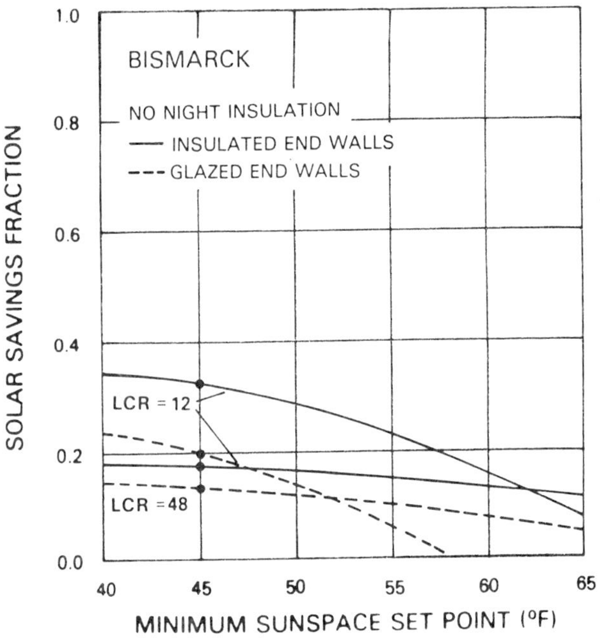

Figure A2-313

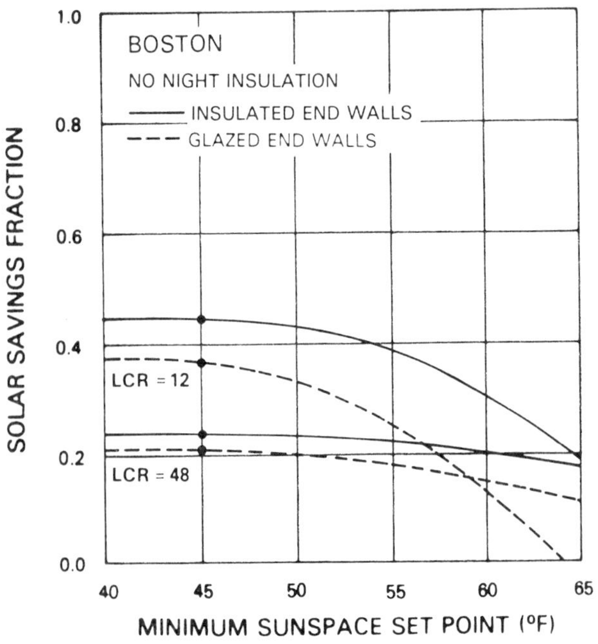

Figure A2-314

Figures A2-312 through A2-317. Annual heating performance sensitivity to the sunspace auxiliary heat thermostat setpoint temperature for geometry A with a masonry common wall, no night insulation, and both insulated and glazed end walls (A1 and A3). The definition of solar savings fraction includes the sunspace auxiliary heat. The dots represent the reference conditions.

SUNSPACE
MINIMUM SUNSPACE SETPOINT

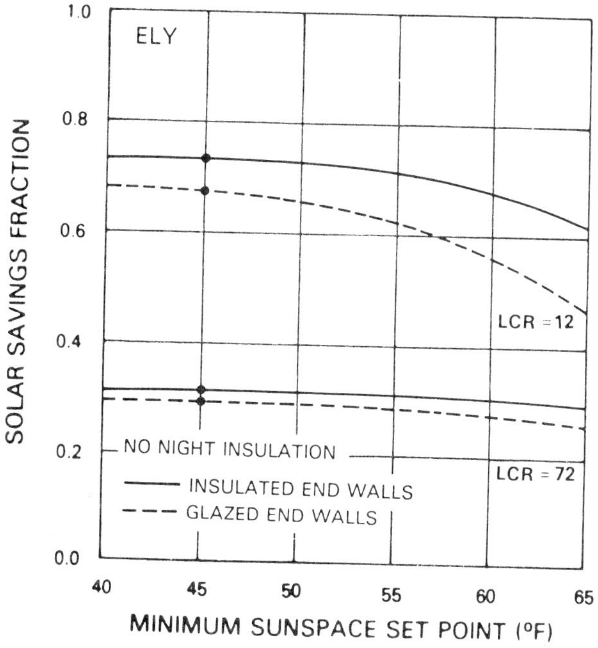

Figure A2-315

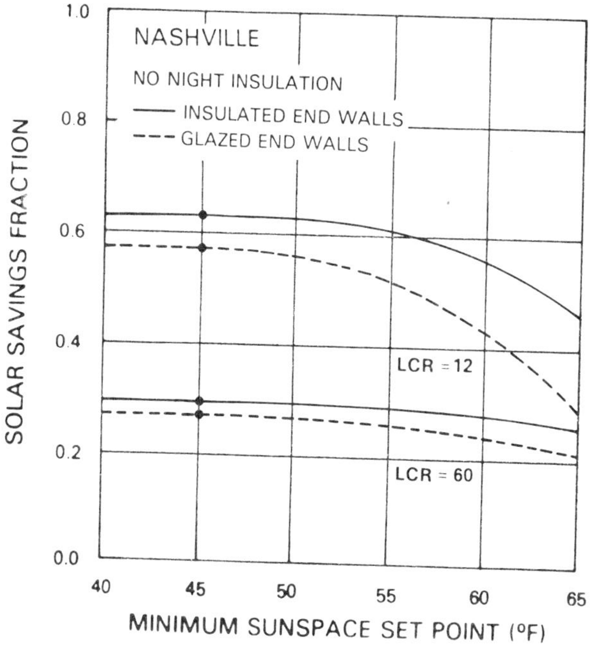

Figure A2-316

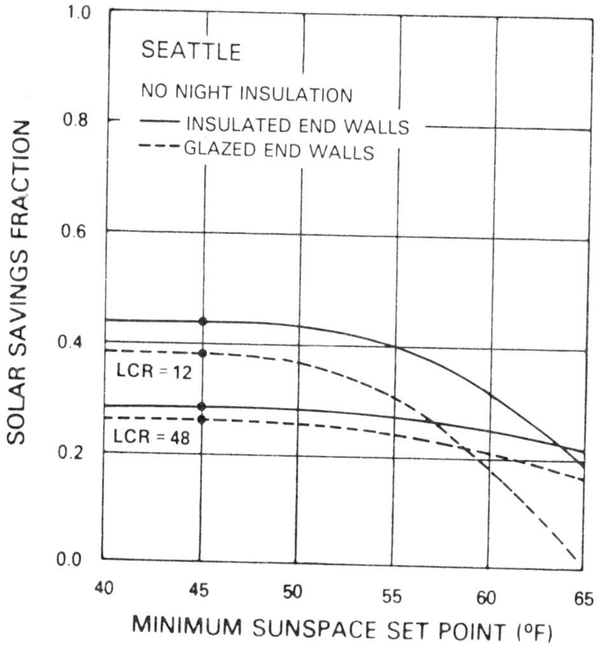

Figure A2-317

Figures A2-312 through A2-317. Annual heating performance sensitivity to the sunspace auxiliary heat thermostat setpoint temperature for geometry A with a masonry common wall, no night insulation, and both insulated and glazed end walls (A1 and A3). The definition of solar savings fraction includes the sunspace auxiliary heat. The dots represent the reference conditions.

THERMAL STORAGE WALL MASS THICKNESS

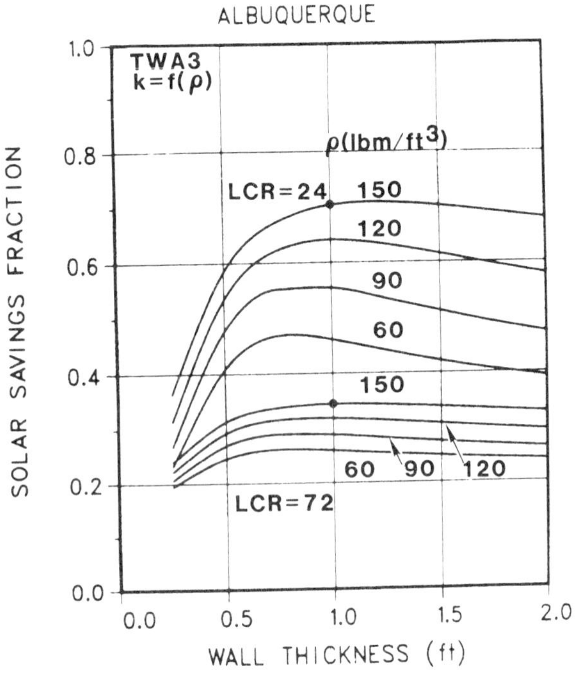

Figure A2-318

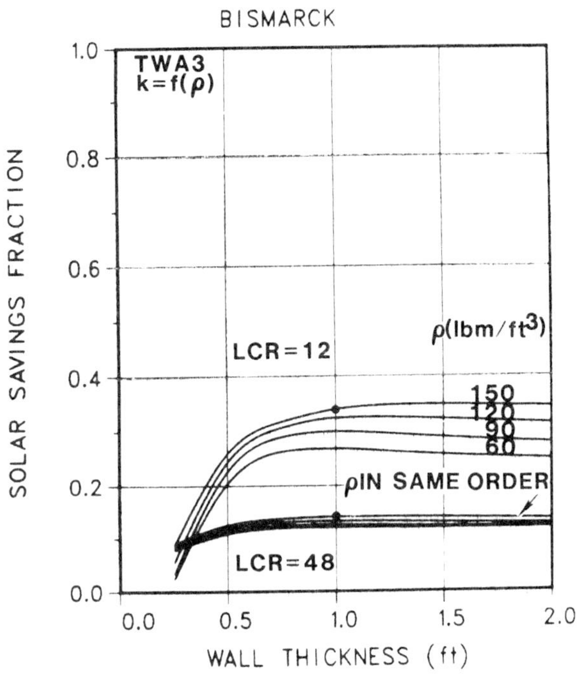

Figure A2-319

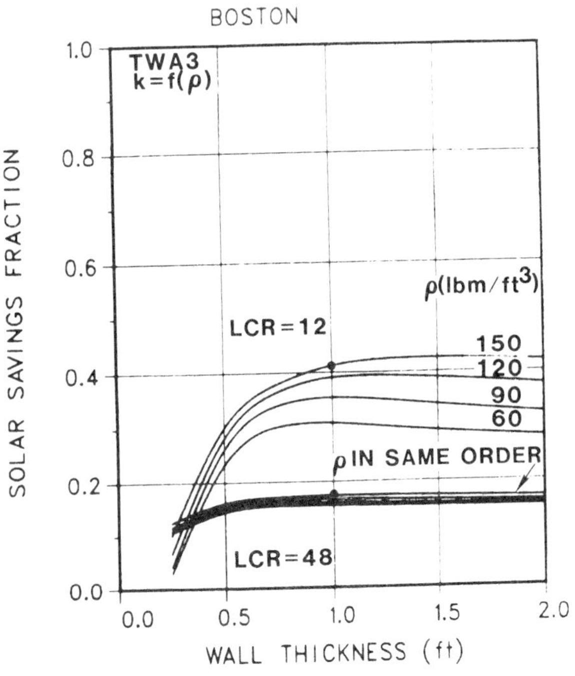

Figure A2-320

Figures A2-318 through A2-323. Annual heating performance sensitivity to the wall thickness of the vented Trombe wall (TWA3) for four different values of masonry density (ρ). The thermal conductivity (k) varies with the density. The dots represent the reference design conditions.

THERMAL STORAGE WALL MASS THICKNESS

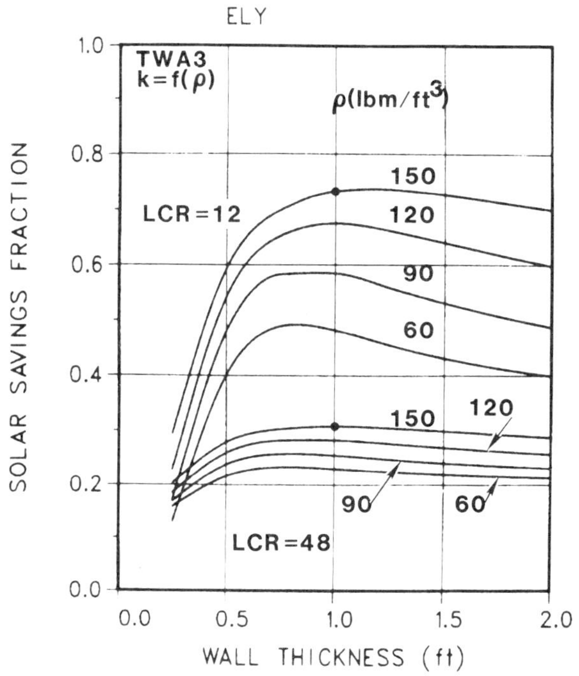

Figure A2-321

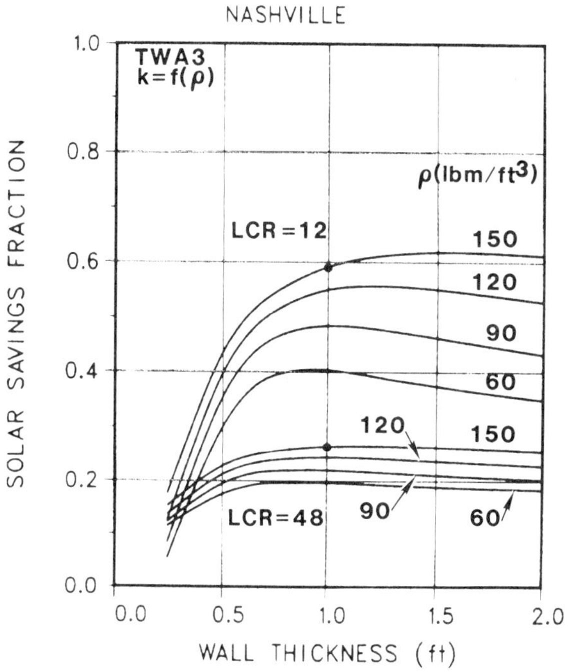

Figure A2-322

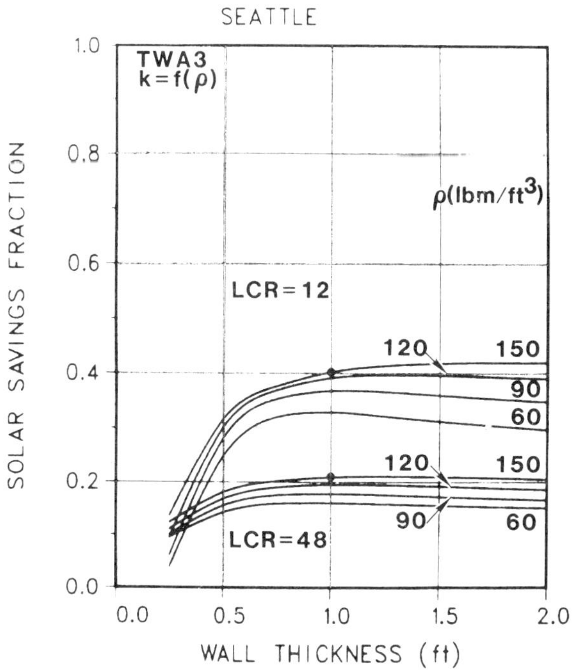

Figure A2-323

Figures A2-318 through A2-323. Annual heating performance sensitivity to the wall thickness of the vented Trombe wall (TWA3) for four different values of masonry density (ρ). The thermal conductivity (*k*) varies with the density. The dots represent the reference design conditions.

THERMAL STORAGE WALL
MASS THICKNESS

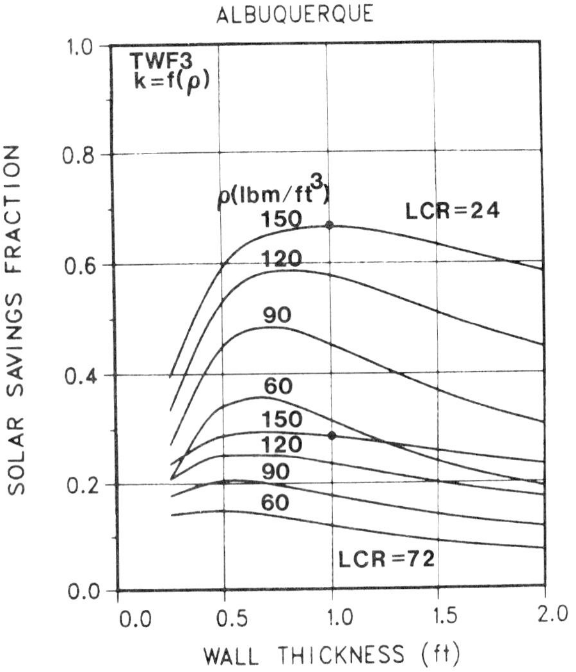

Figure A2-324

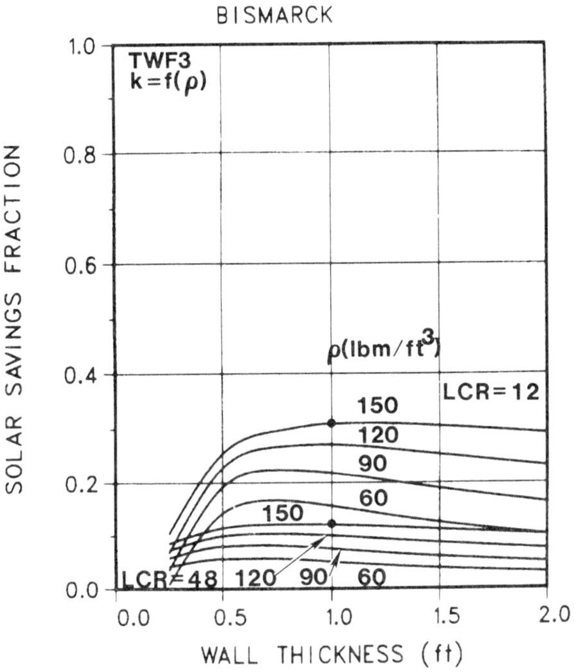

Figure A2-325

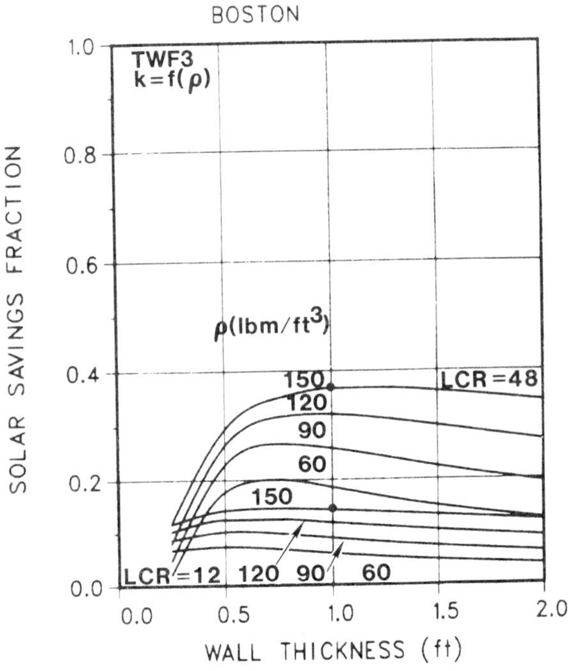

Figure A2-326

Figures A2-324 through A2-329. Annual heating performance sensitivity to the wall thickness of the unvented Trombe wall (TWF3) for four different values of masonry density. The thermal conductivity (*k*) varies with the density. The dots represent the reference design conditions.

THERMAL STORAGE WALL
MASS THICKNESS

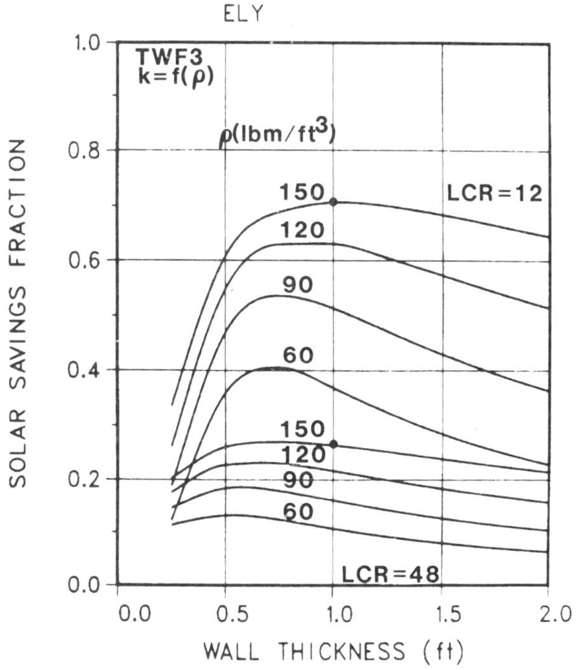

Figure A2-327

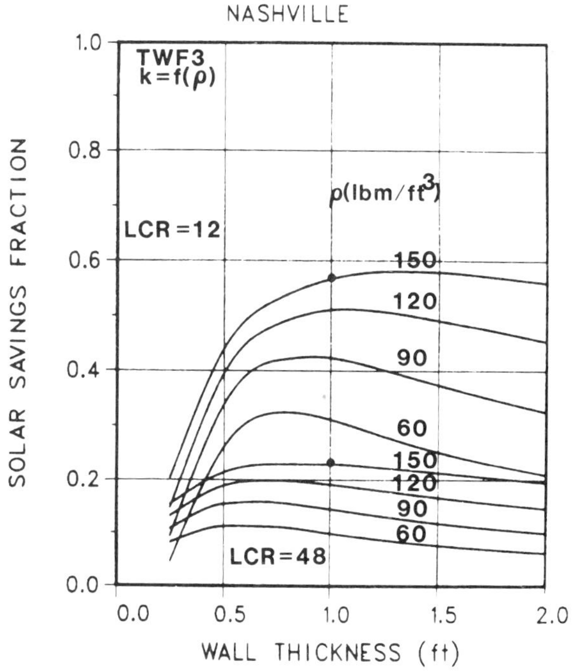

Figure A2-328

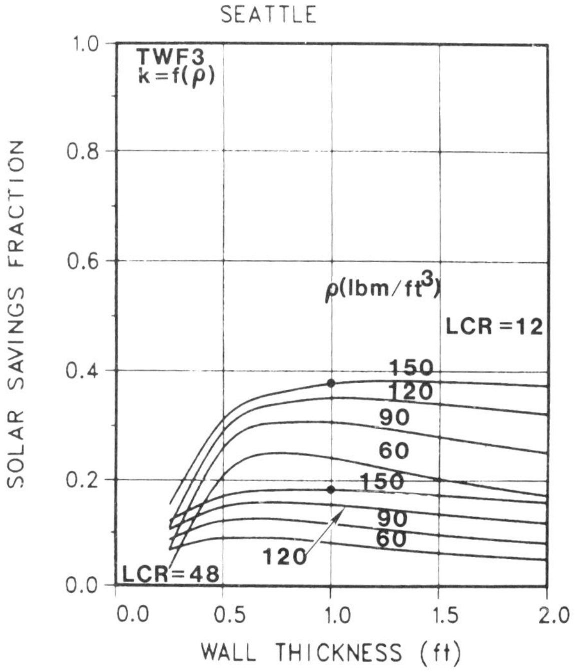

Figure A2-329

Figures A2-324 through A2-329. Annual heating performance sensitivity to the wall thickness of the unvented Trombe wall (TWF3) for four different values of masonry density. The thermal conductivity (*k*) varies with the density. The dots represent the reference design conditions.

THERMAL STORAGE WALL
MASS THICKNESS

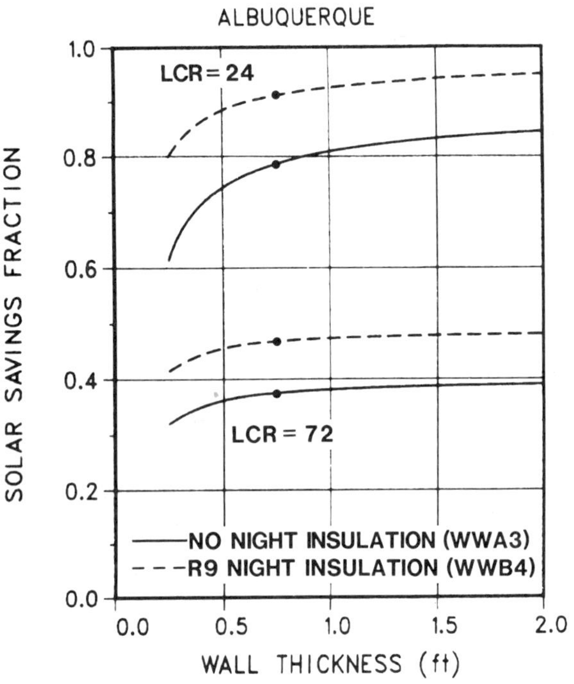

Figure A2-330

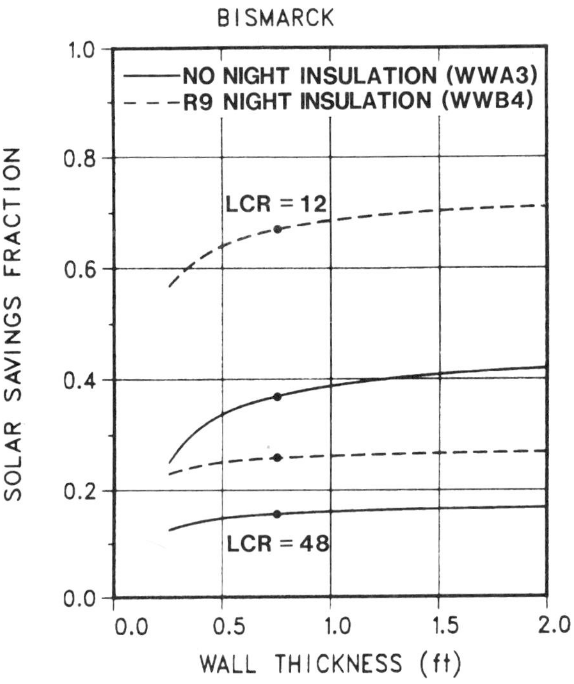

Figure A2-331

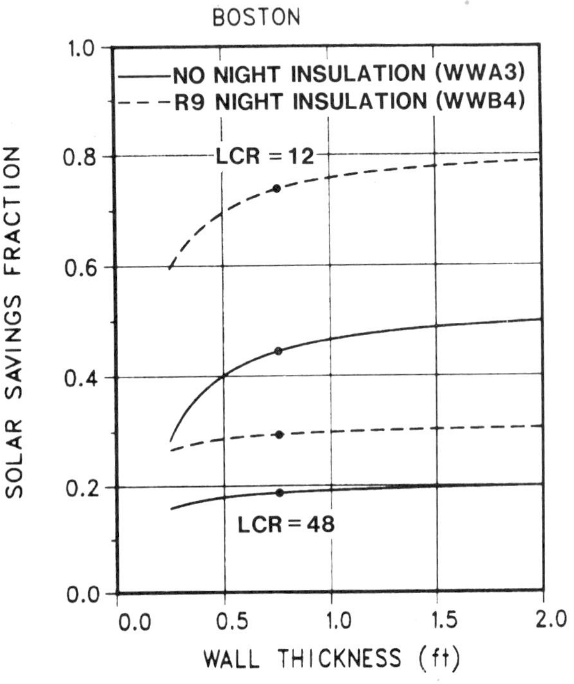

Figure A2-332

Figures A2-330 through A2-335. Annual heating performance sensitivity to the wall thickness of the water wall without night insulation (WWA3) and with R9 night insulation (WWB4). The dots represent the reference design conditions.

THERMAL STORAGE WALL
MASS THICKNESS

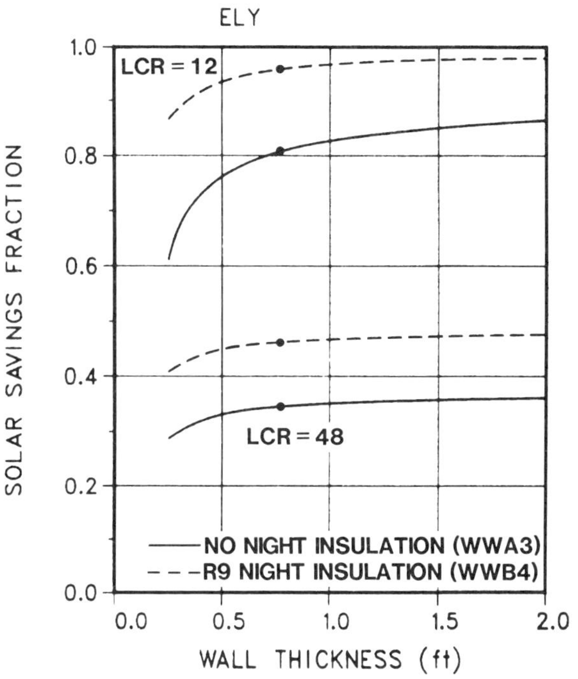

Figure A2-333

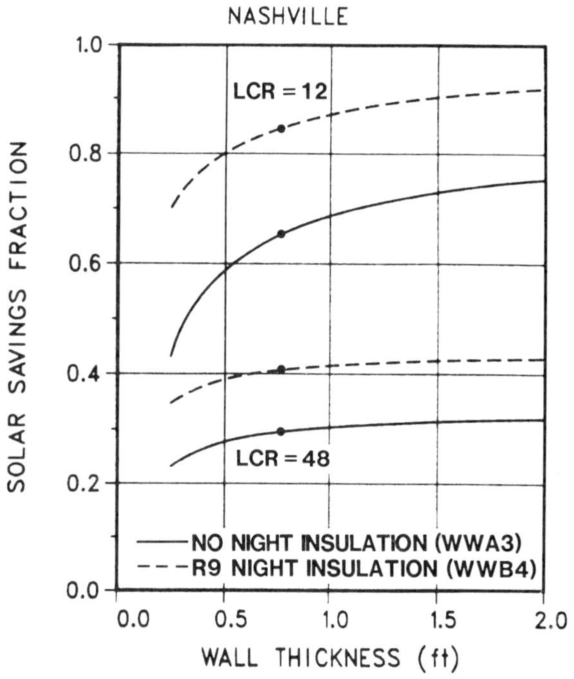

Figure A2-334

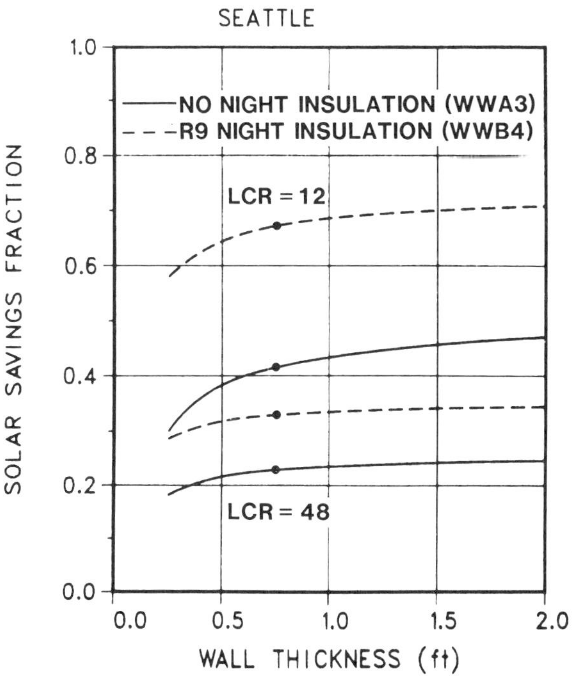

Figure A2-335

Figures A2-330 through A2-335. Annual heating performance sensitivity to the wall thickness of the water wall without night insulation (WWA3) and with R9 night insulation (WWB4). The dots represent the reference design conditions.

THERMAL STORAGE WALL
SOLAR ABSORPTANCE

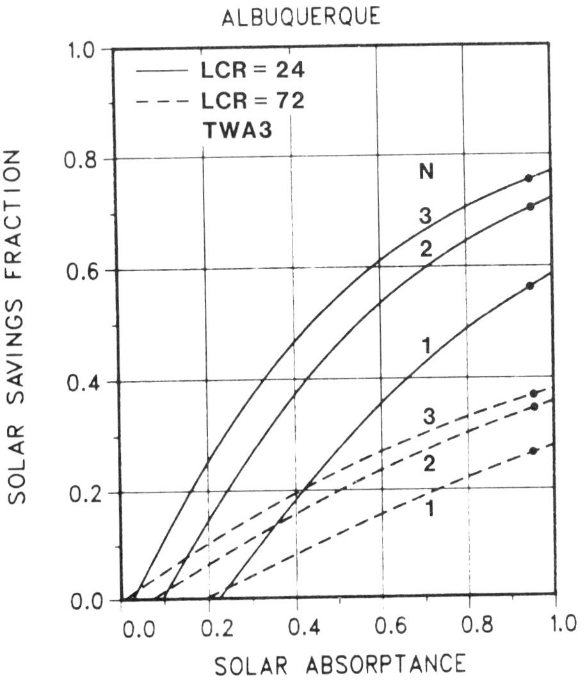

Figure A2-336

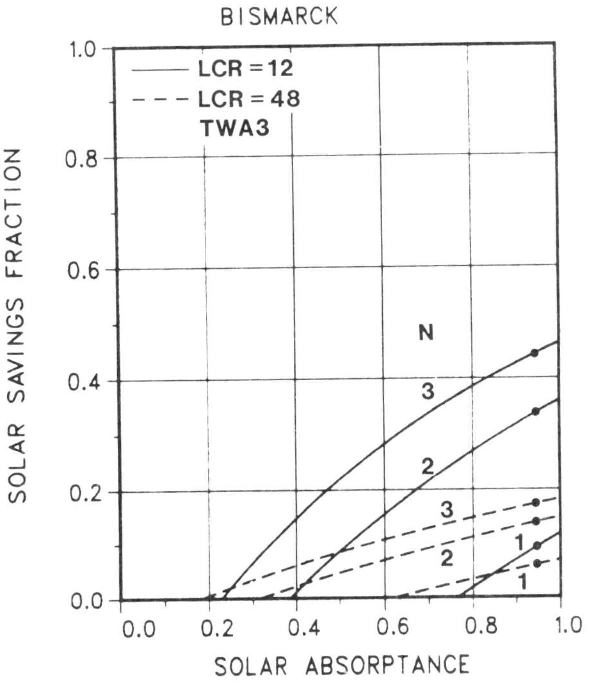

Figure A2-337

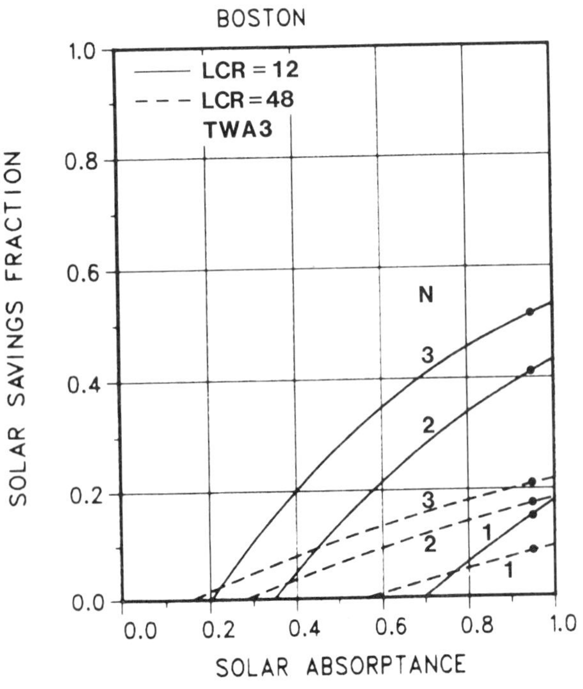

Figure A2-338

Figures A2-336 through A2-341. Annual heating performance sensitivity to the solar absorptance of the surface of the vented Trombe wall for single (TWD1), double (TWA3) and triple (TWD2) layers of glazing (*N*). The dots represent the reference design conditions.

THERMAL STORAGE WALL
SOLAR ABSORPTANCE

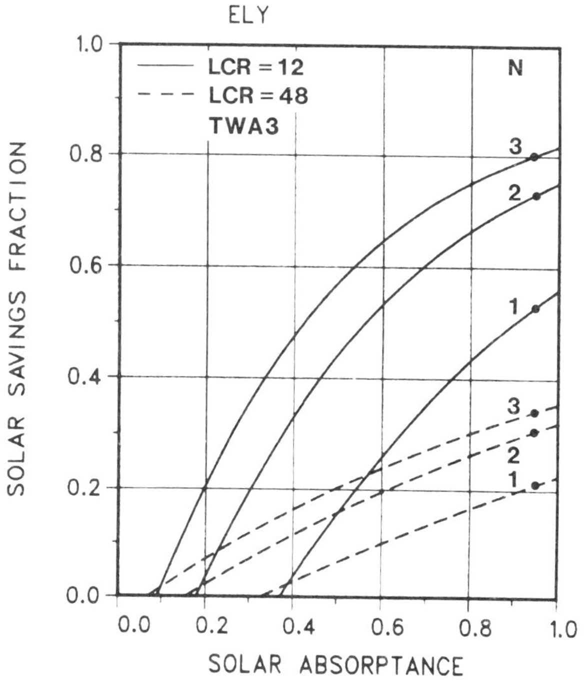

Figure A2-339

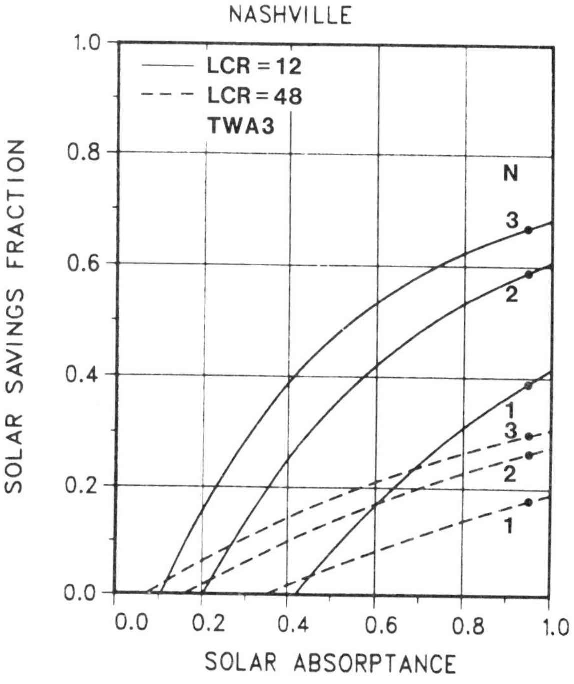

Figure A2-340

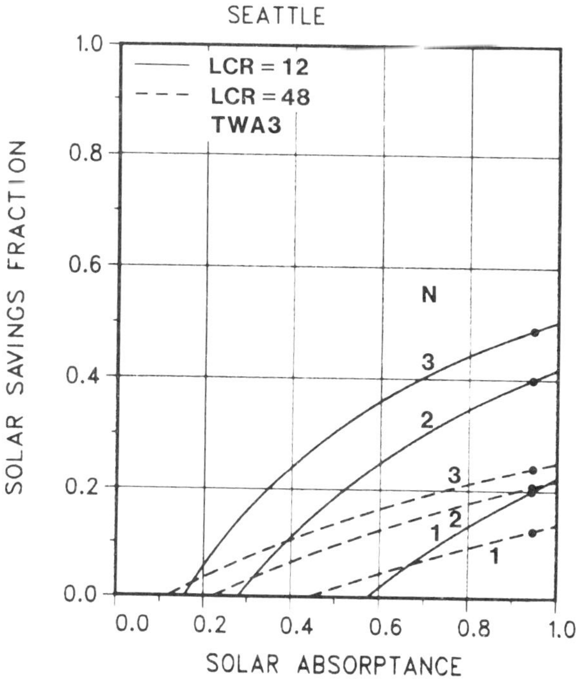

Figure A2-341

Figures A2-336 through A2-341. Annual heating performance sensitivity to the solar absorptance of the surface of the vented Trombe wall for single (TWD1), double (TWA3) and triple (TWD2) layers of glazing (*N*). The dots represent the reference design conditions.

THERMAL STORAGE WALL
SOLAR ABSORPTANCE

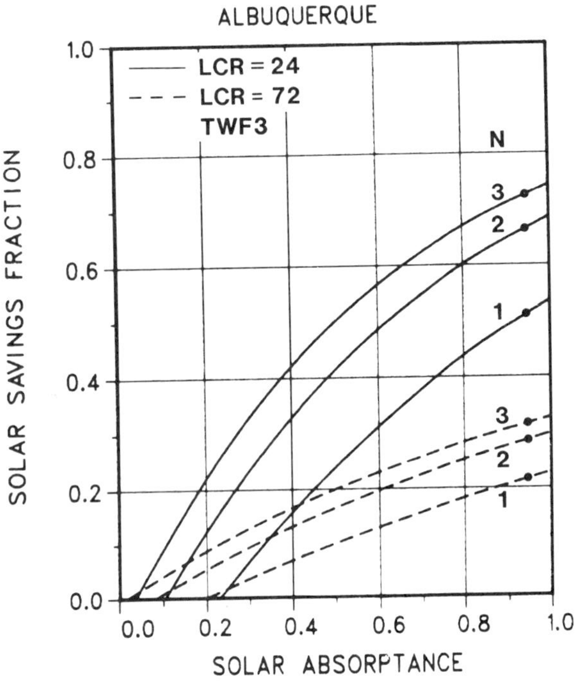

Figure A2-342

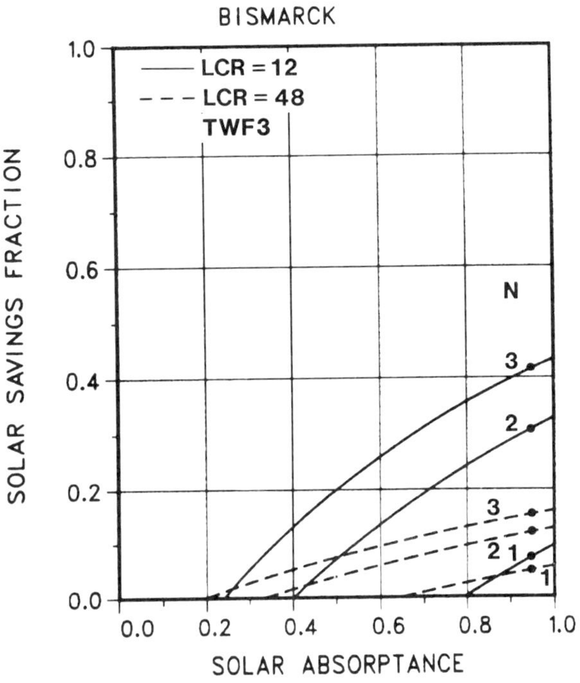

Figure A2-343

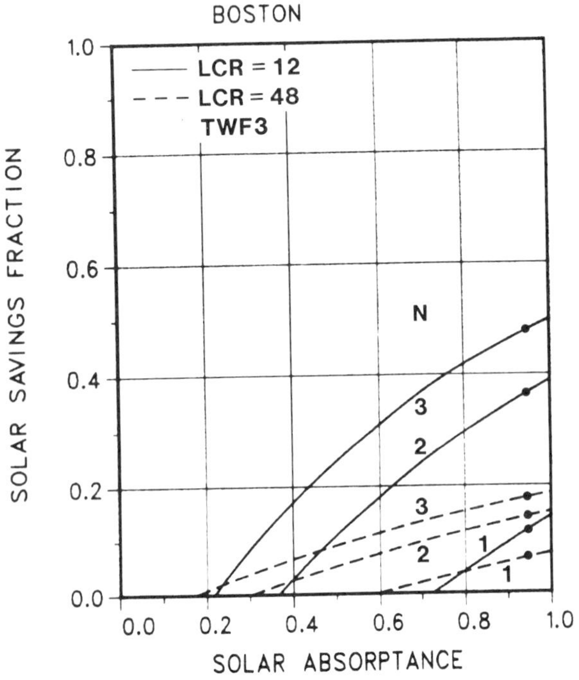

Figure A2-344

Figures A2-342 through A2-347. Annual heating performance sensitivity to the solar absorptance of the surface of the unvented Trombe wall for single (TWI1), double (TWF3), and triple (TWI2) layers of glazing (*N*). The dots represent the reference design conditions.

THERMAL STORAGE WALL
SOLAR ABSORPTANCE

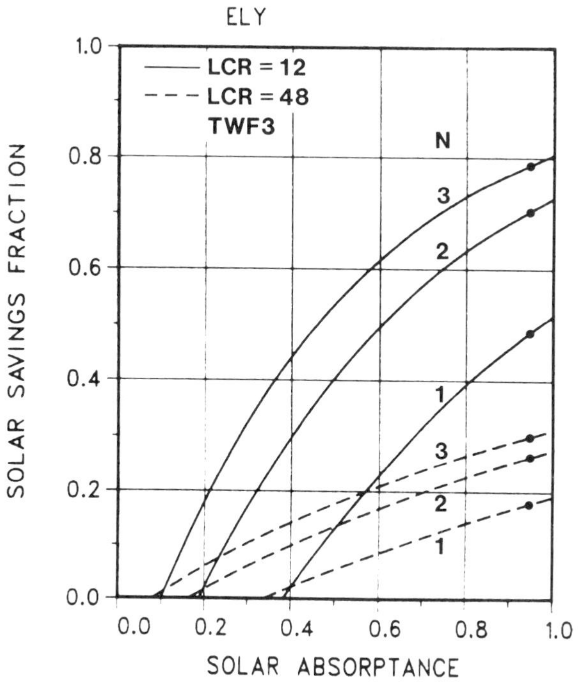

Figure A2-345

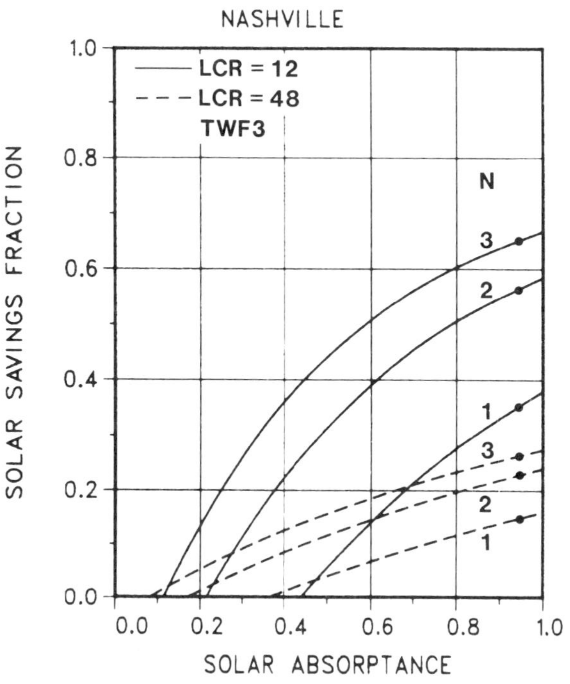

Figure A2-346

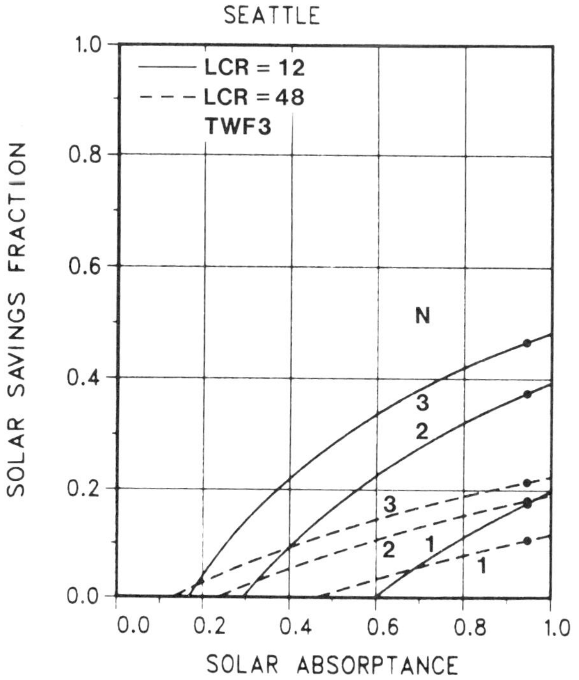

Figure A2-347

Figures A2-342 through A2-347. Annual heating performance sensitivity to the solar absorptance of the surface of the unvented Trombe wall for single (TWI1), double (TWF3), and triple (TWI2) layers of glazing (*N*). The dots represent the reference design conditions.

THERMAL STORAGE WALL
SOLAR ABSORPTANCE

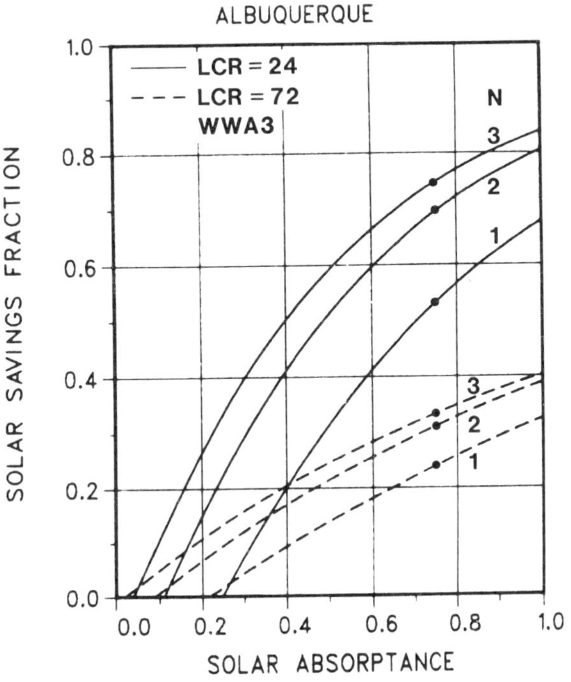

Figure A2-348

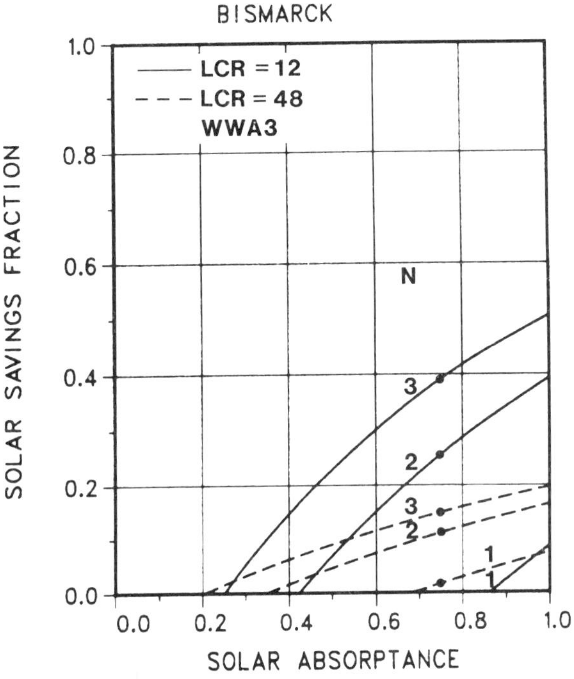

Figure A2-349

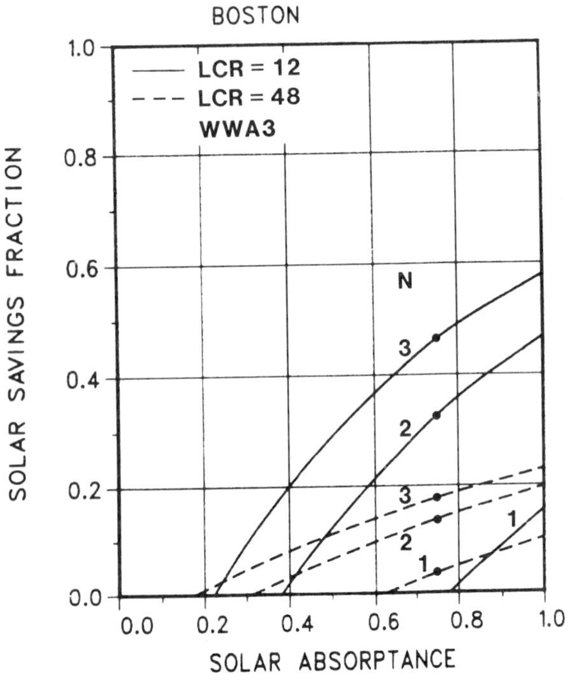

Figure A2-350

Figures A2-348 through A2-353. Annual heating performance sensitivity to the solar absorptance of the surface of the water wall for single (WWB1), double (WWA3), and triple (WWB2) layers of glazing (*N*). The dots represent the reference design conditions.

THERMAL STORAGE WALL SOLAR ABSORPTANCE

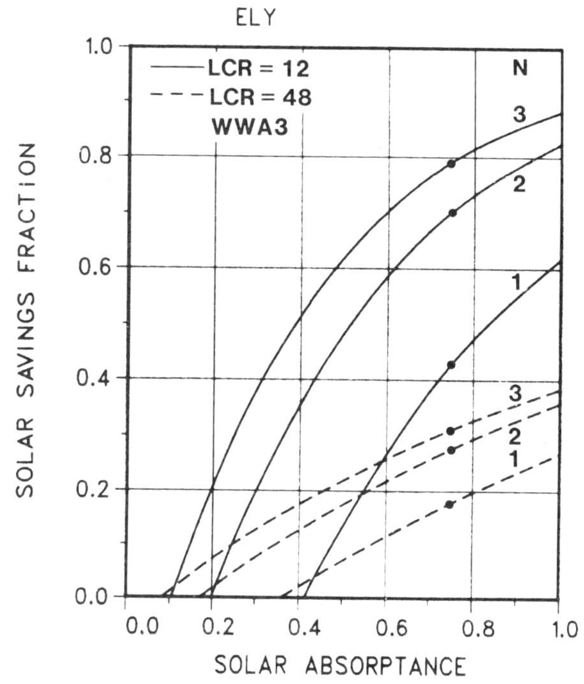

Figure A2-351

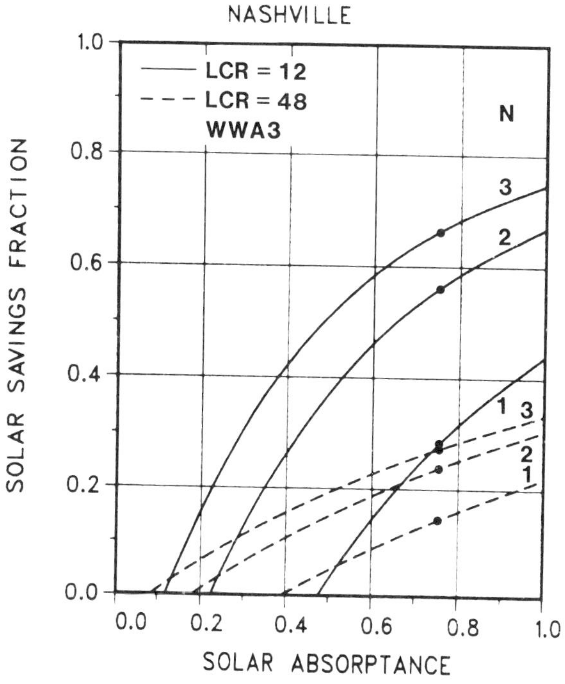

Figure A2-352

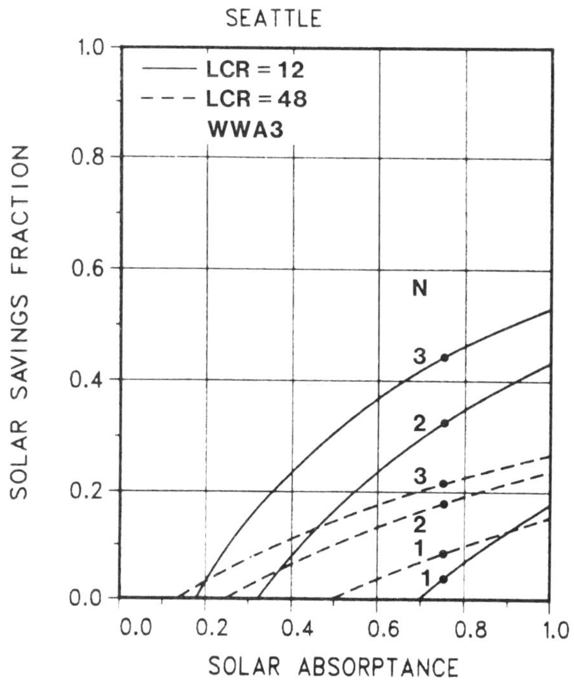

Figure A2-353

Figures A2-348 through A2-353. Annual heating performance sensitivity to the solar absorptance of the surface of the water wall for single (WWB1), double (WWA3), and triple (WWB2) layers of glazing (*N*). The dots represent the reference design conditions.

THERMAL STORAGE WALL
NUMBER OF GLAZINGS

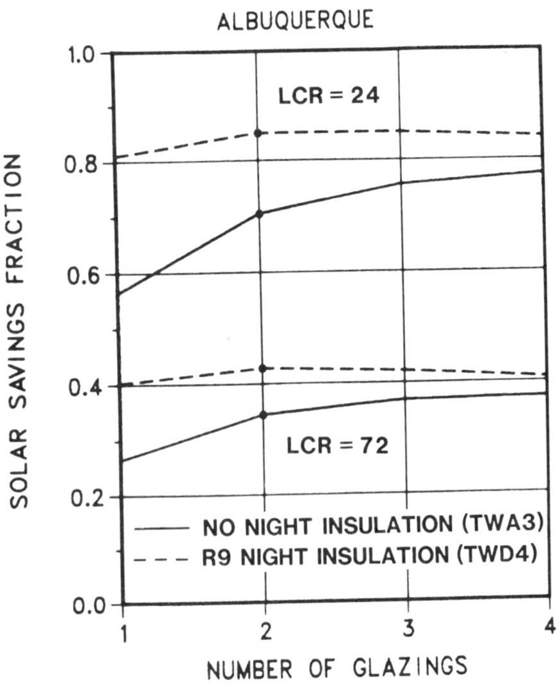

Figure A2-354

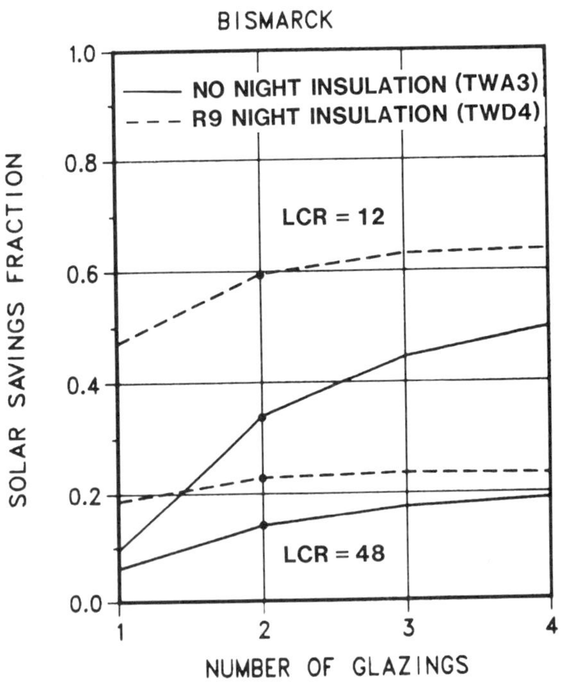

Figure A2-355

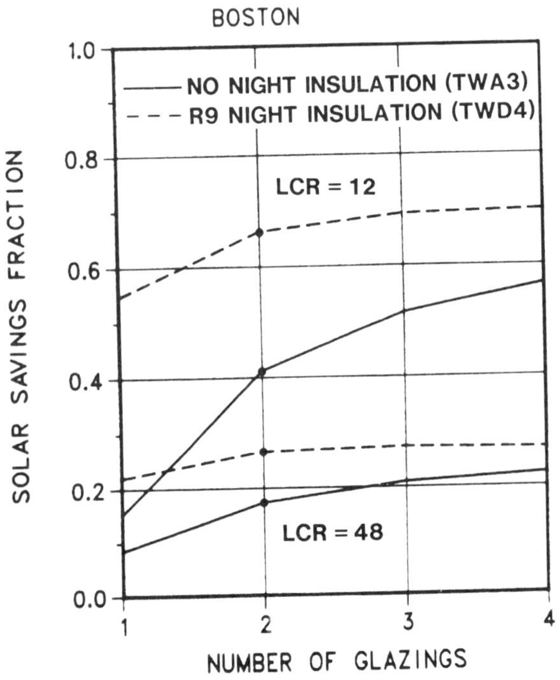

Figure A2-356

Figures A2-354 through A2-359. Annual heating performance sensitivity to the number of layers of glazing for the vented Trombe wall without night insulation (TWA3) and with R9 night insulation (TWD4). The dots represent the reference design conditions.

THERMAL STORAGE WALL
NUMBER OF GLAZINGS

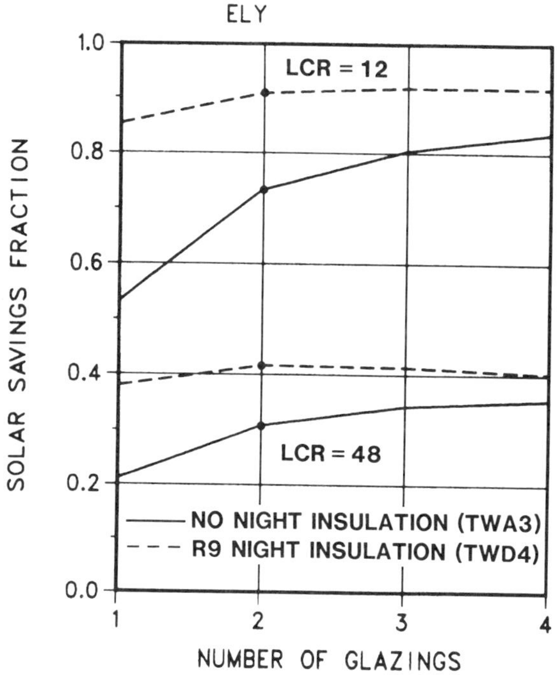

Figure A2-357

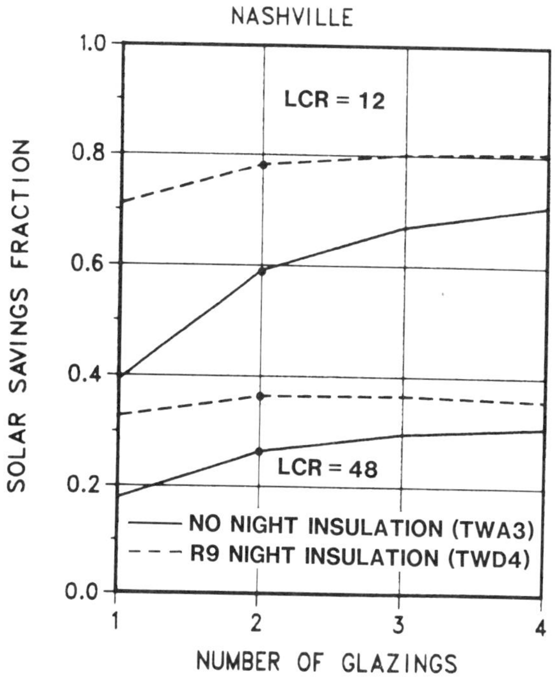

Figure A2-358

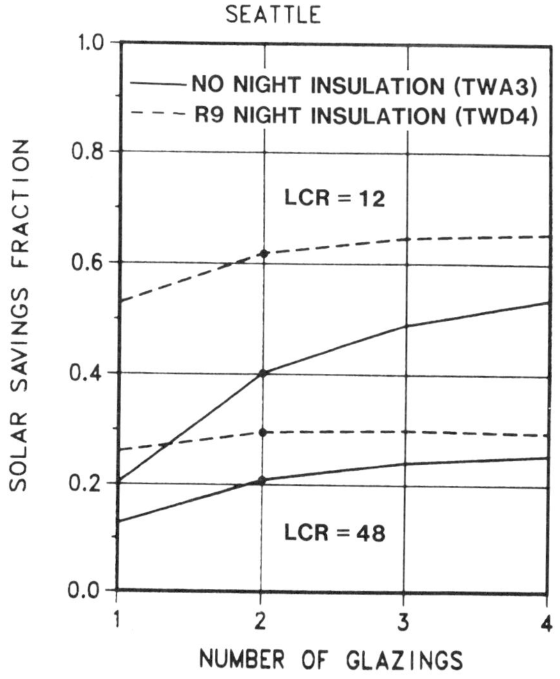

Figure A2-359

Figures A2-354 through A2-359. Annual heating performance sensitivity to the number of layers of glazing for the vented Trombe wall without night insulation (TWA3) and with R9 night insulation (TWD4). The dots represent the reference design conditions.

THERMAL STORAGE WALL
NUMBER OF GLAZINGS

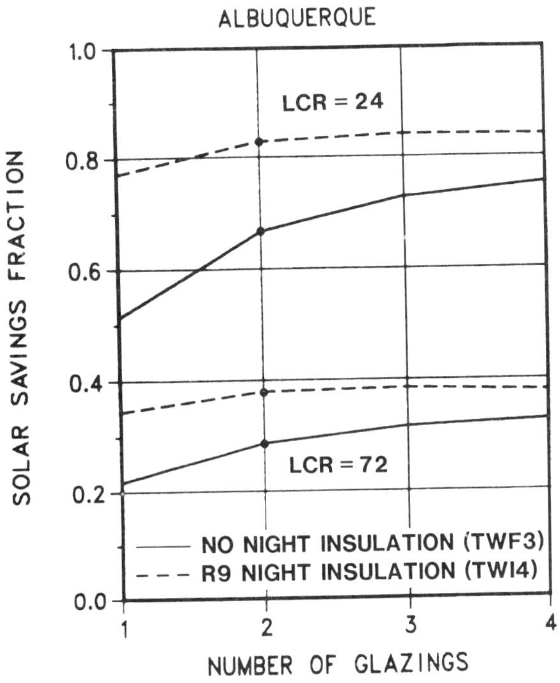

Figure A2-360

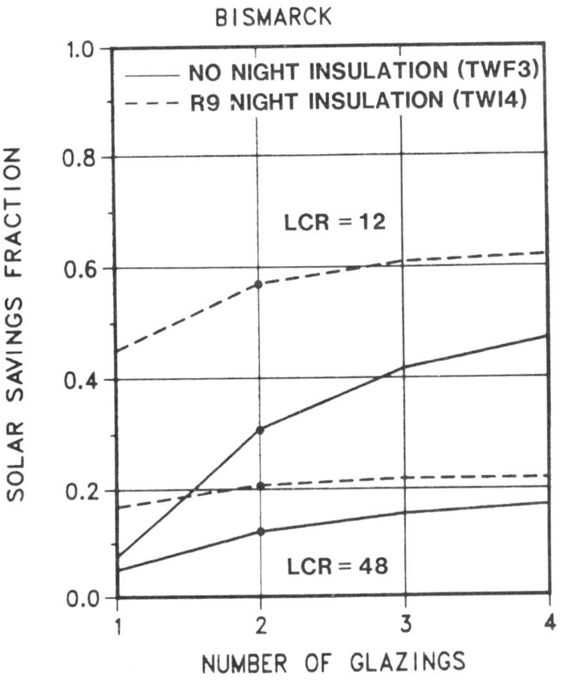

Figure A2-361

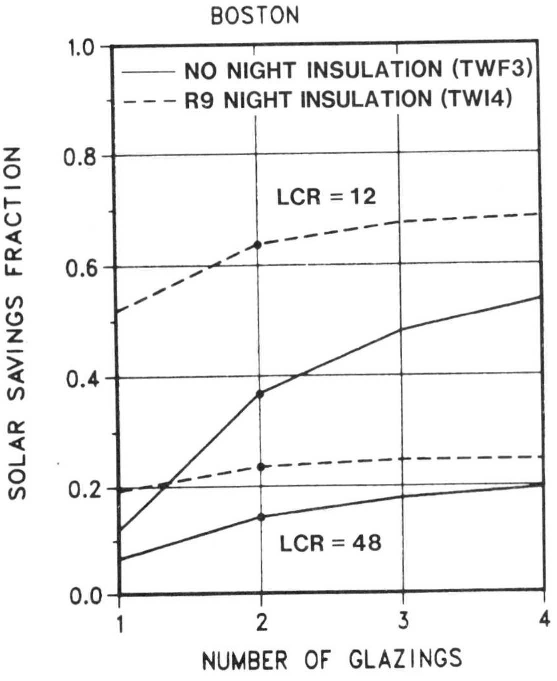

Figure A2-362

Figures A2-360 through A2-365. Annual heating performance sensitivity to the number of layers of glazing for the unvented Trombe wall without night insulation (TWF3) and with R9 night insulation (TWI4). The dots represent the reference design conditions.

THERMAL STORAGE WALL
NUMBER OF GLAZINGS

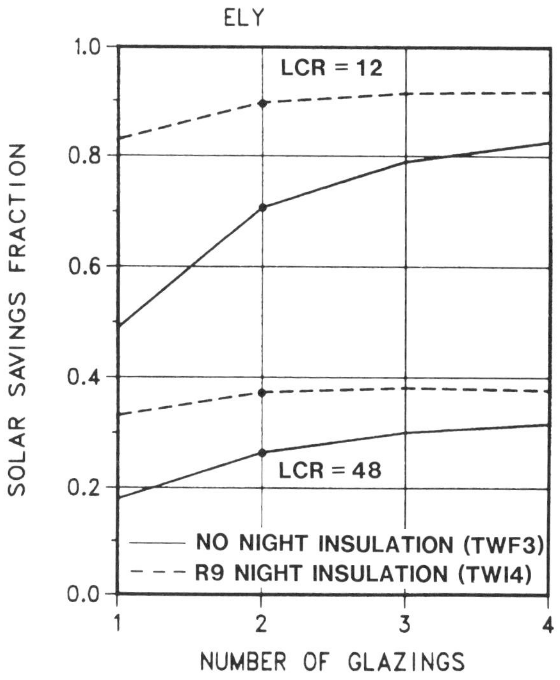

Figure A2-363

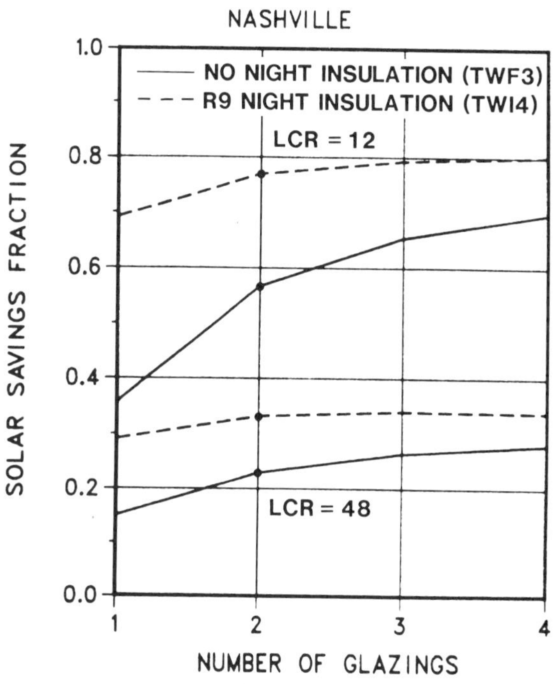

Figure A2-364

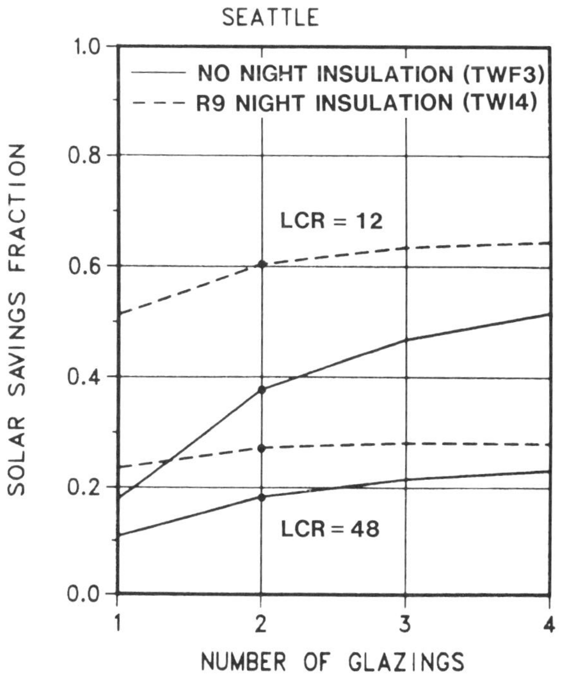

Figure A2-365

Figures A2-360 through A2-365. Annual heating performance sensitivity to the number of layers of glazing for the unvented Trombe wall without night insulation (TWF3) and with R9 night insulation (TWI4). The dots represent the reference design conditions.

THERMAL STORAGE WALL
NUMBER OF GLAZINGS

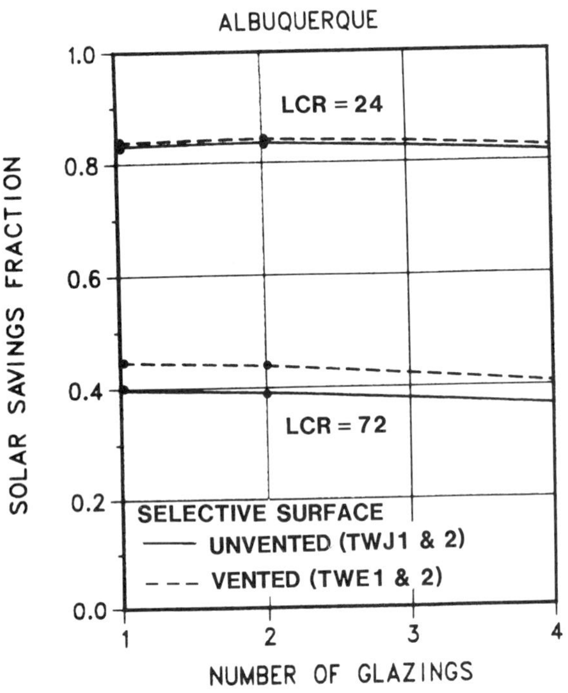

Figure A2-366

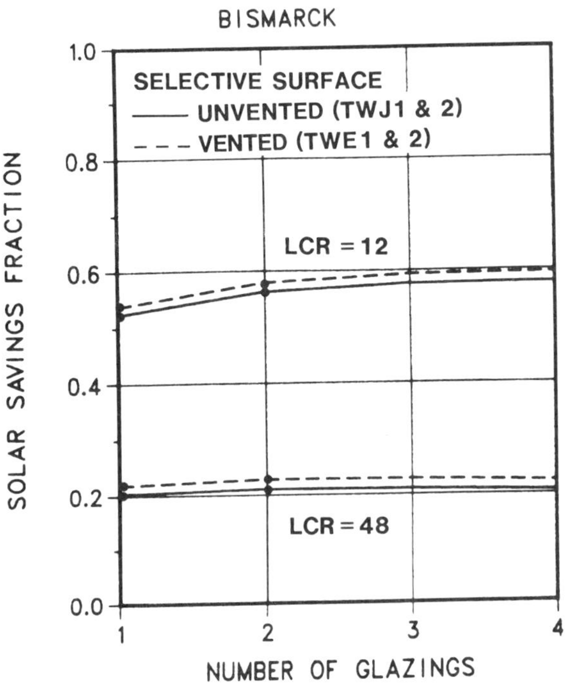

Figure A2-367

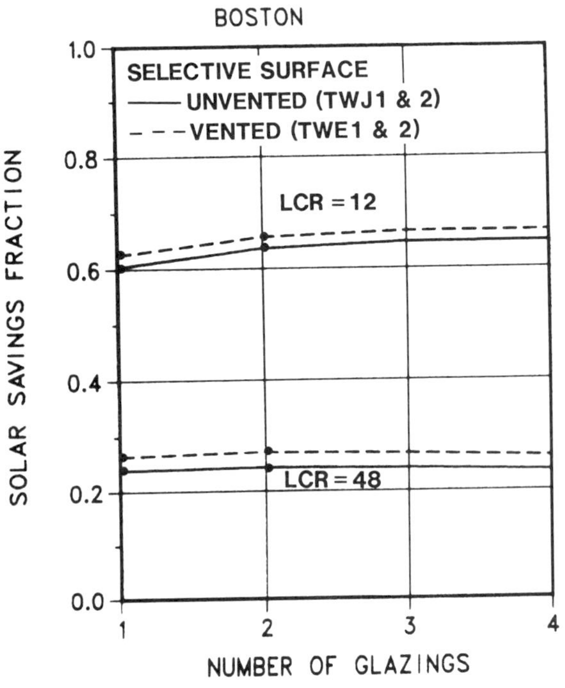

Figure A2-368

Figures A2-366 through A2-371. Annual heating performance sensitivity to the number of layers of glazing for the selective surface Trombe wall, vented (TWE2) and unvented (TWJ2). Both cases are without night insulation. The dots represent the reference design conditions.

THERMAL STORAGE WALL
NUMBER OF GLAZINGS

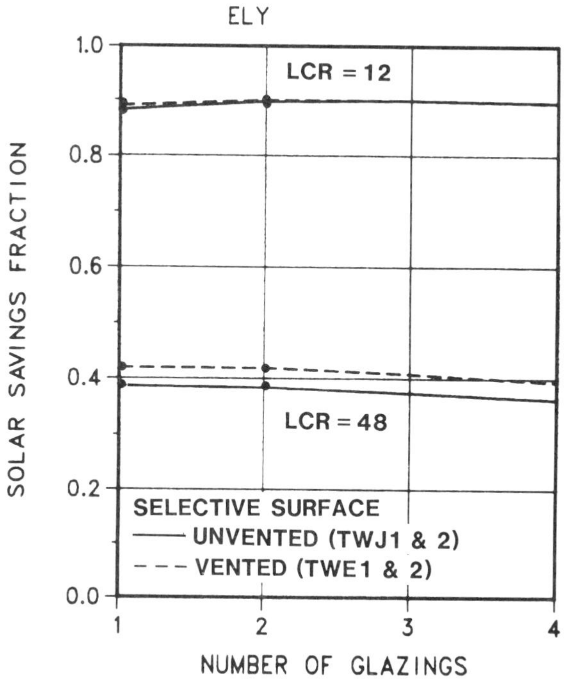

Figure A2-369

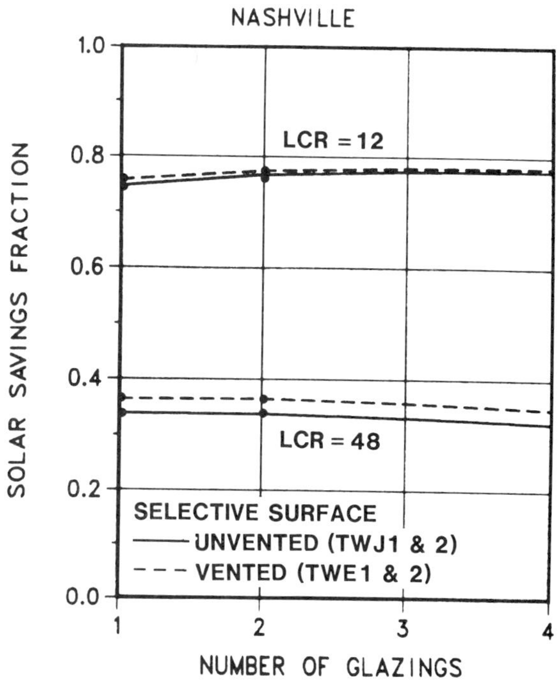

Figure A2-370

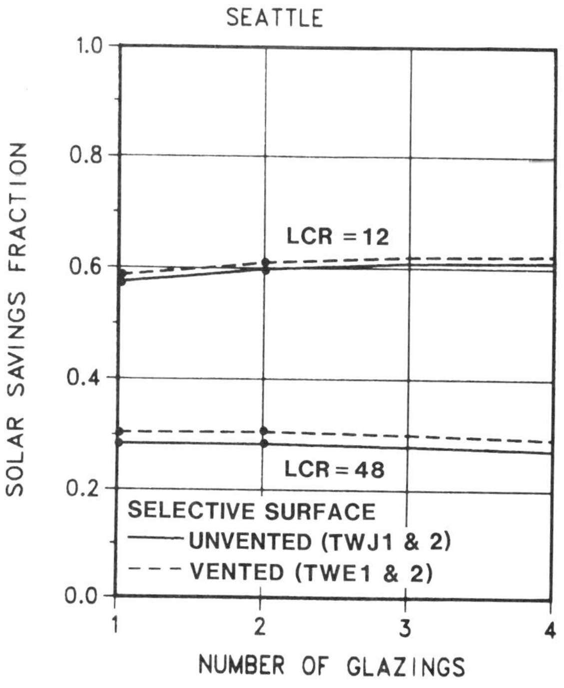

Figure A2-371

Figures A2-366 through A2-371. Annual heating performance sensitivity to the number of layers of glazing for the selective surface Trombe wall, vented (TWE2) and unvented (TWJ2). Both cases are without night insulation. The dots represent the reference design conditions.

THERMAL STORAGE WALL
NUMBER OF GLAZINGS

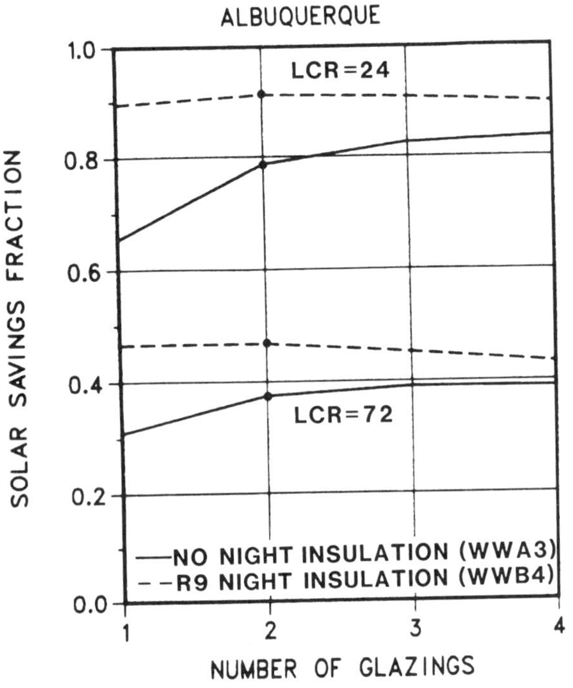

Figure A2-372

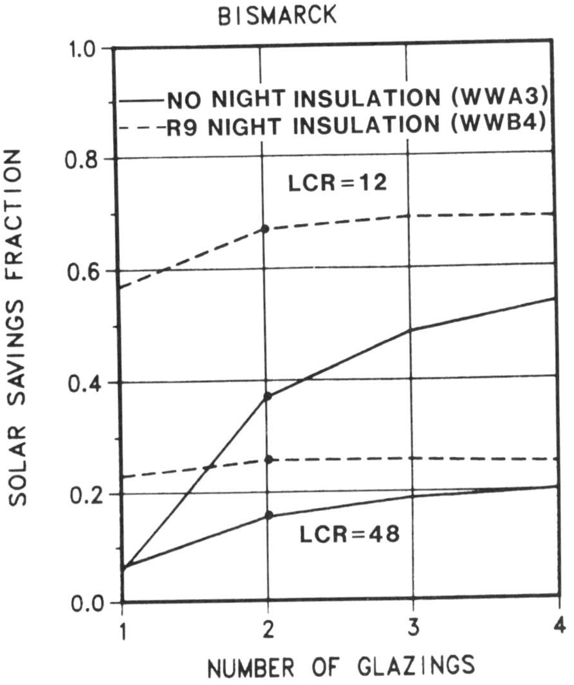

Figure A2-373

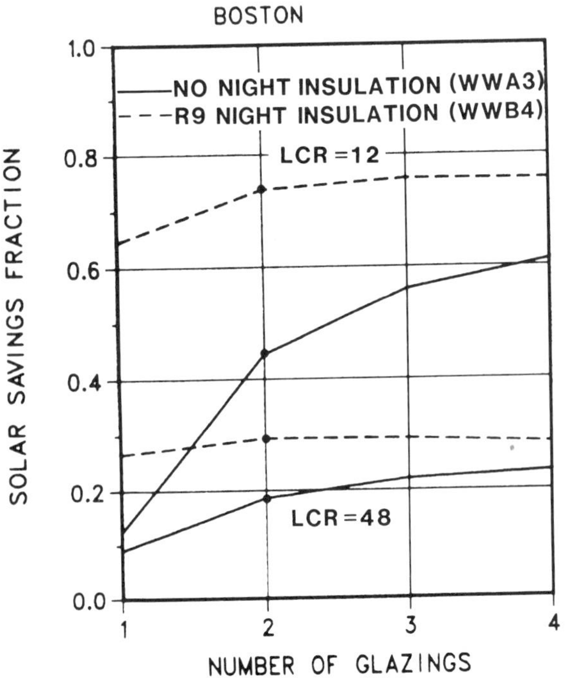

Figure A2-374

Figures A2-372 through A2-377. Annual heating performance sensitivity to the number of layers of glazing for the water wall without night insulation (WWA3) and with R9 night insulation (WWB4). The dots represent the reference design conditions.

THERMAL STORAGE WALL
NUMBER OF GLAZINGS

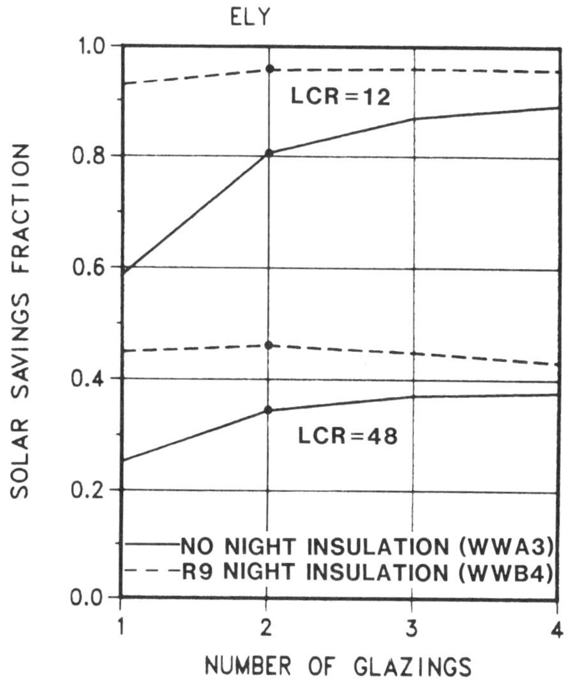

Figure A2-375

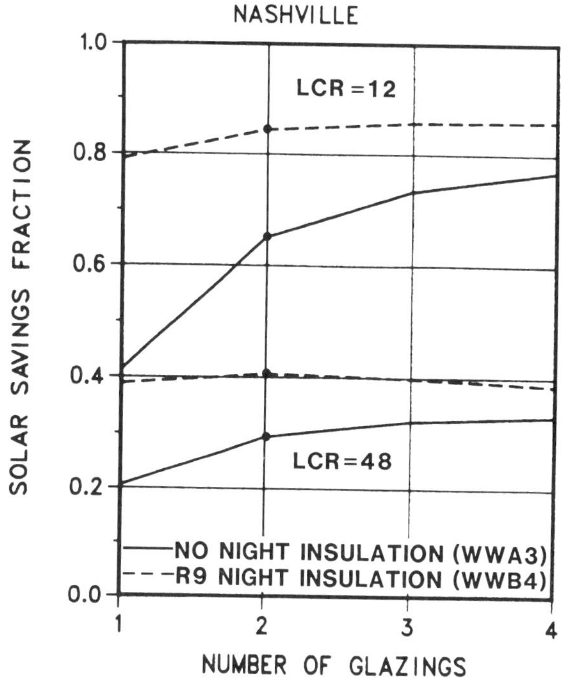

Figure A2-376

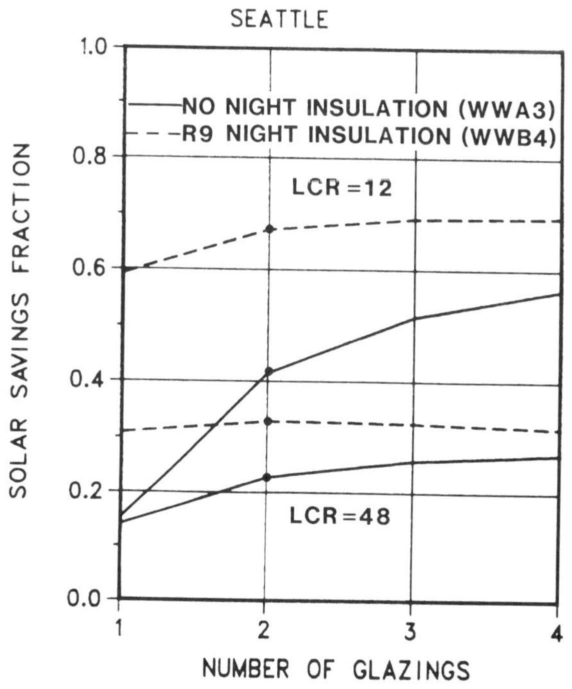

Figure A2-377

Figures A2-372 through A2-377. Annual heating performance sensitivity to the number of layers of glazing for the water wall without night insulation (WWA3) and with R9 night insulation (WWB4). The dots represent the reference design conditions.

THERMAL STORAGE WALL
INFRARED EMITTANCE

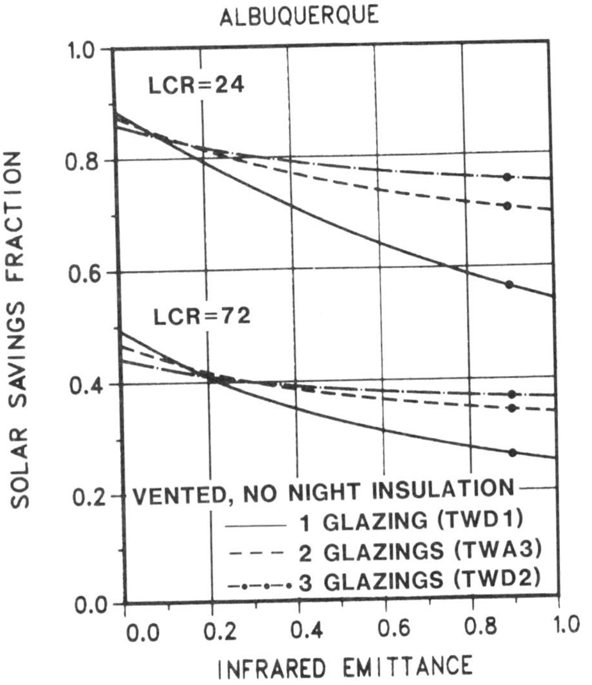

Figure A2-378

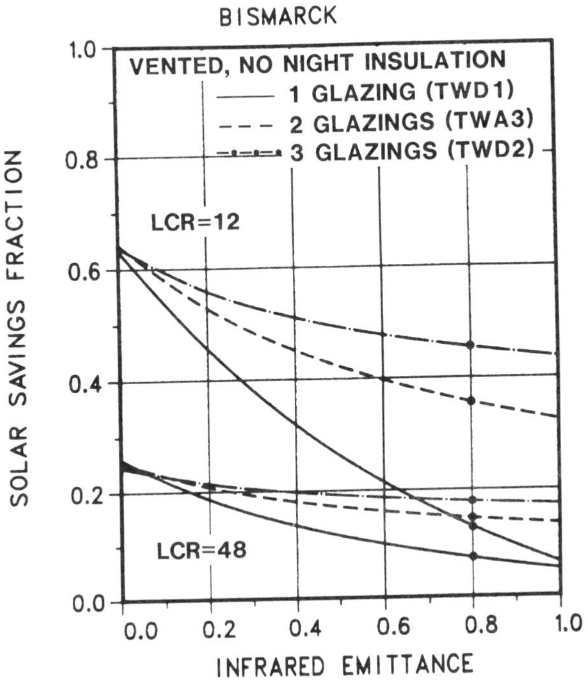

Figure A2-379

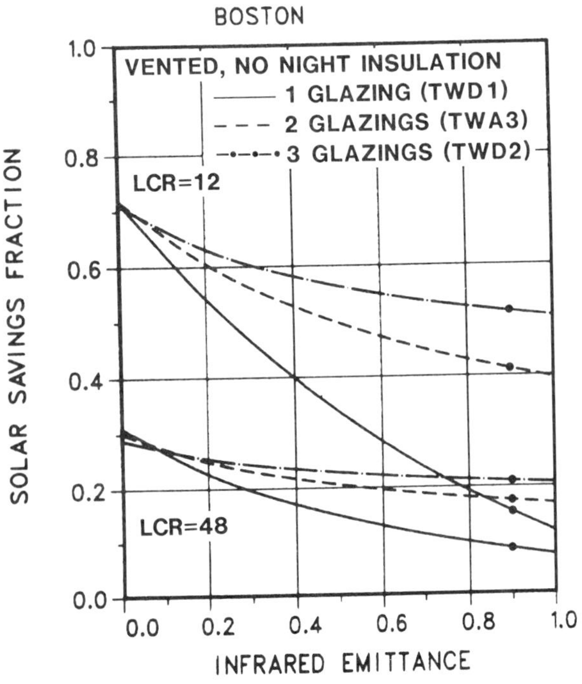

Figure A2-380

Figures A2-378 through A2-383. Annual heating performance sensitivity to the infrared emittance of the surface of a vented Trombe wall with single (TWD1), double (TWA3), and triple (TWD2) layers of glazing. The dots represent the reference design conditions.

THERMAL STORAGE WALL
INFRARED EMITTANCE

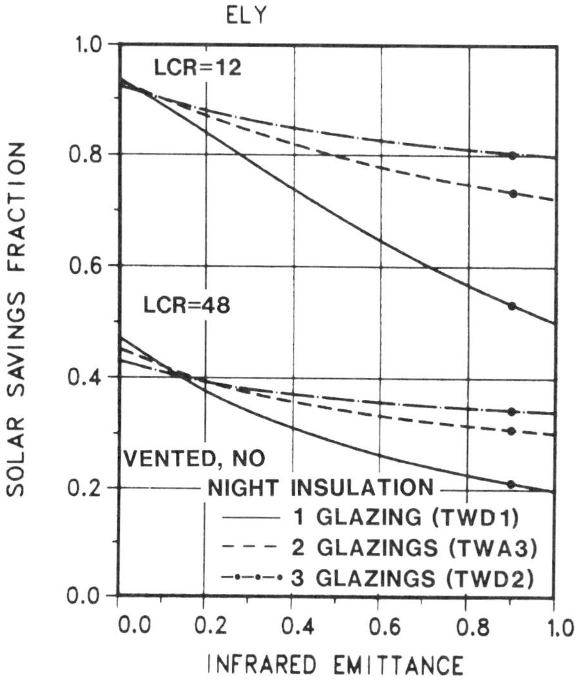

Figure A2-381

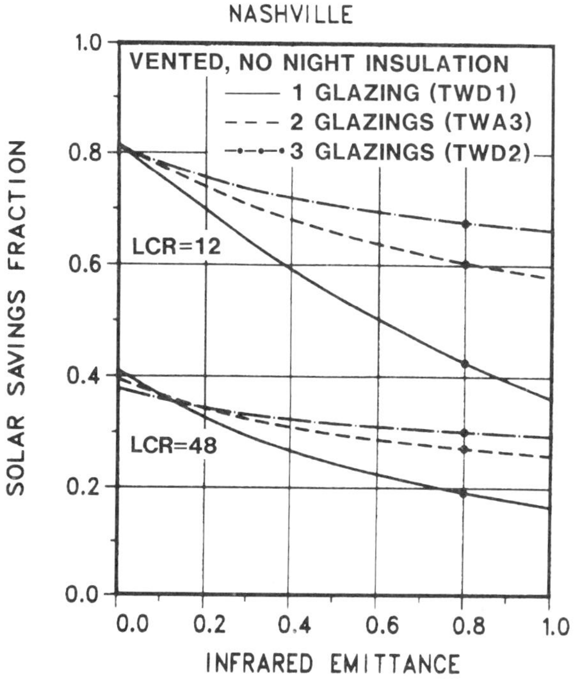

Figure A2-382

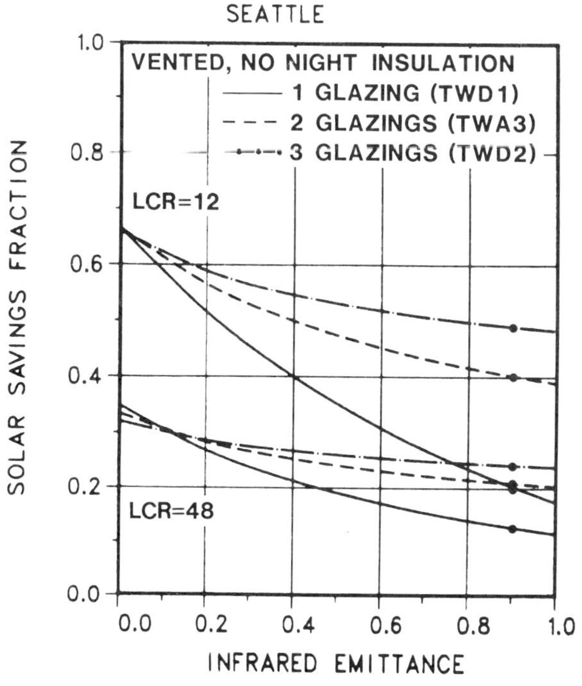

Figure A2-383

Figures A2-378 through A2-383. Annual heating performance sensitivity to the infrared emittance of the surface of a vented Trombe wall with single (TWD1), double (TWA3), and triple (TWD2) layers of glazing. The dots represent the reference design conditions.

THERMAL STORAGE WALL
INFRARED EMITTANCE

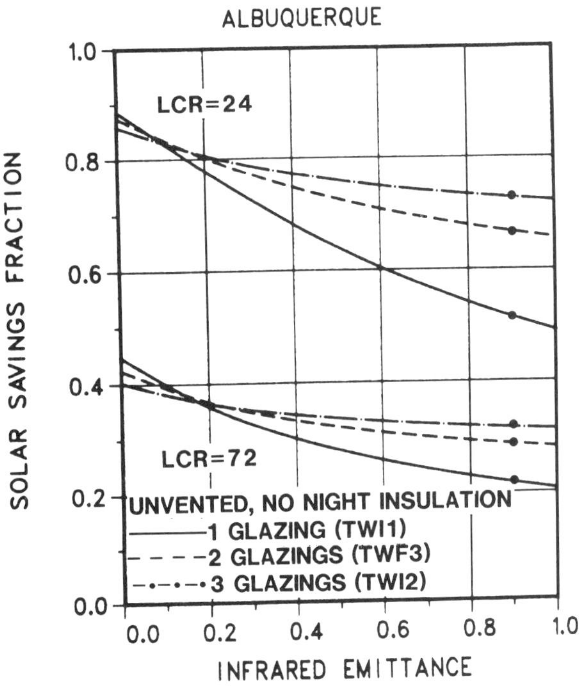

Figure A2-384

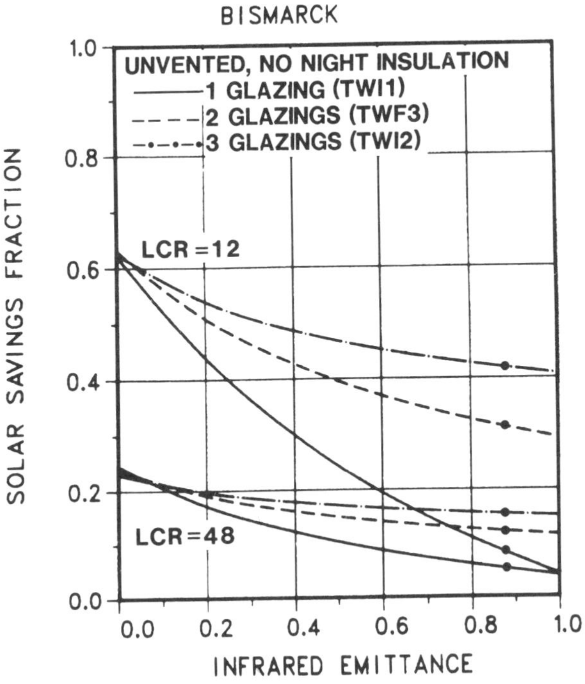

Figure A2-385

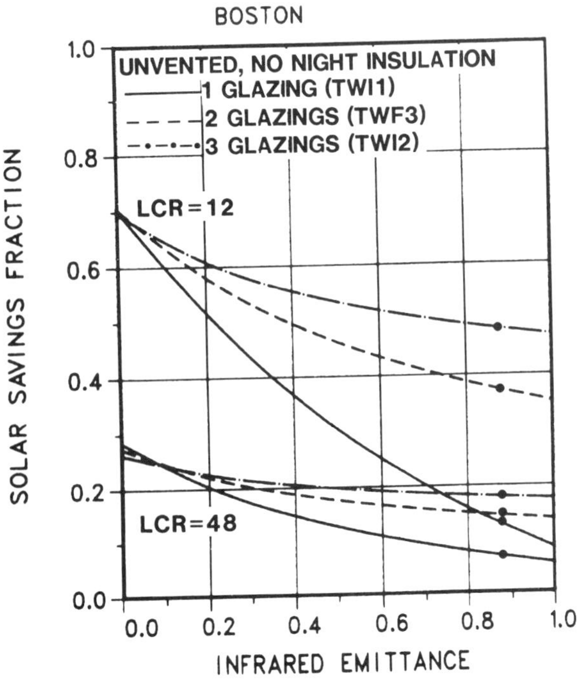

Figure A2-386

Figures A2-384 through A2-389. Annual heating performance sensitivity to the infrared emittance of the surface of an unvented Trombe wall with single (TWI1), double (TWF3), and triple (TWI2) layers of glazing. The dots represent the reference design conditions.

THERMAL STORAGE WALL
INFRARED EMITTANCE

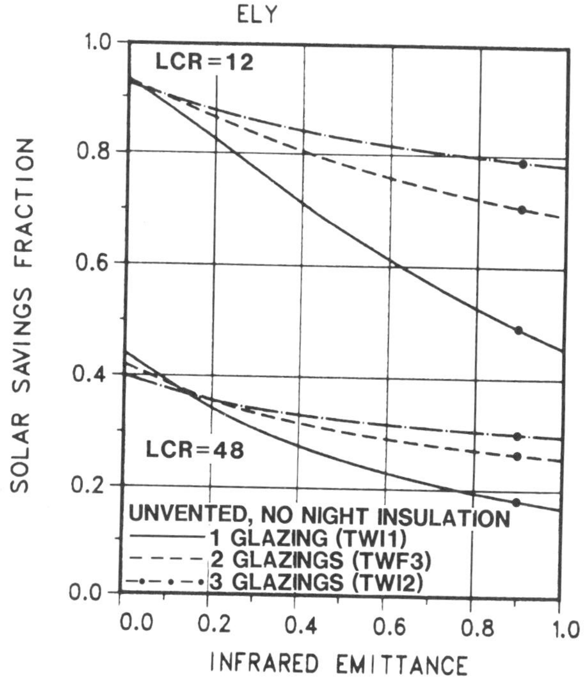

Figure A2-387

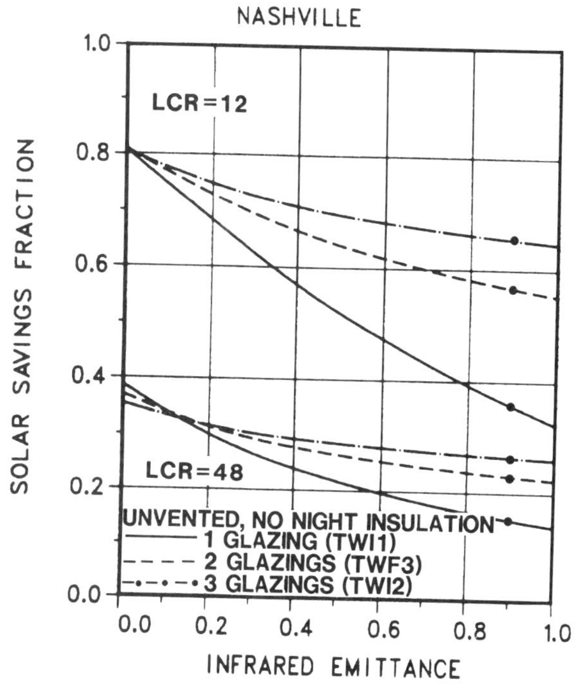

Figure A2-388

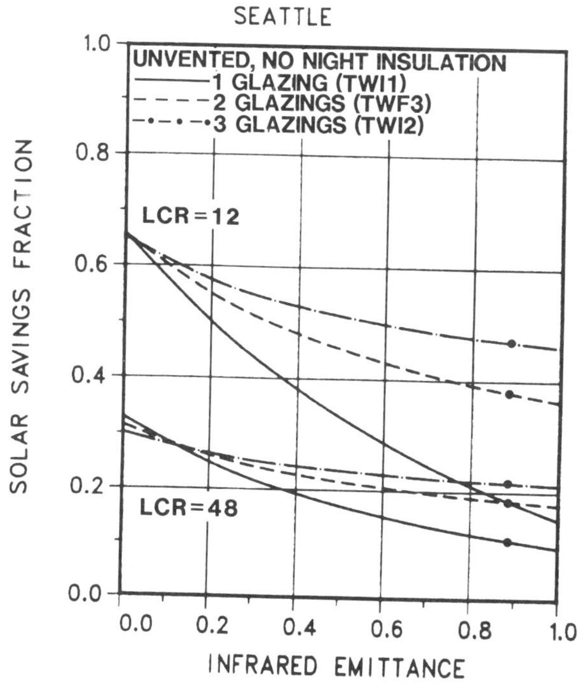

Figure A2-389

Figures A2-384 through A2-389. Annual heating performance sensitivity to the infrared emittance of the surface of an unvented Trombe wall with single (TWI1), double (TWF3), and triple (TWI2) layers of glazing. The dots represent the reference design conditions.

THERMAL STORAGE WALL
INFRARED EMITTANCE

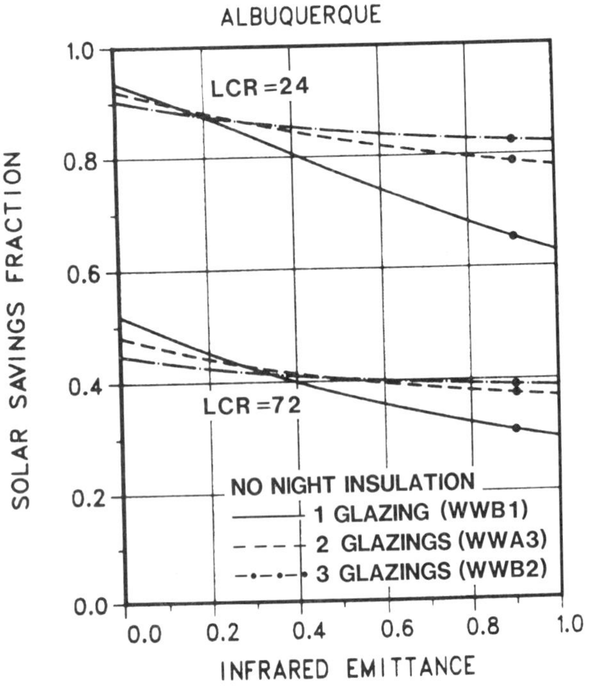

Figure A2-390

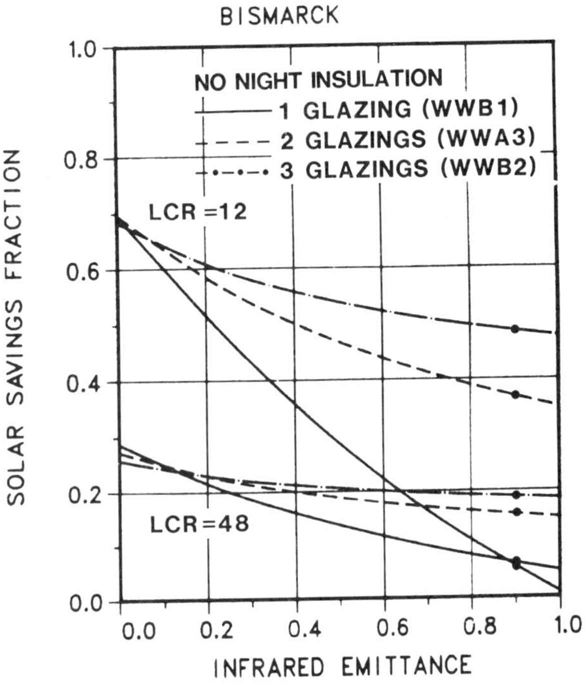

Figure A2-391

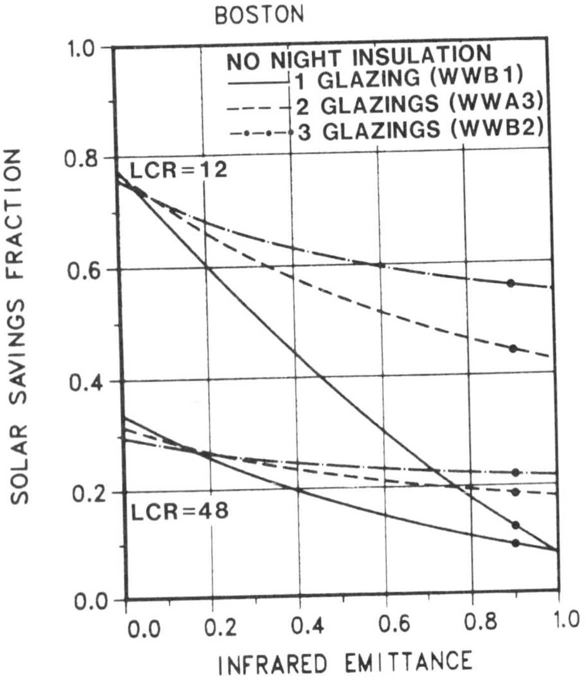

Figure A2-392

Figures A2-390 through A2-395. Annual heating performance sensitivity to the infrared emittance of the surface of a water wall with single (WWB1), double (WWA3), and triple (WWB2) layers of glazing. The dots represent the reference design conditions.

THERMAL STORAGE WALL
INFRARED EMITTANCE

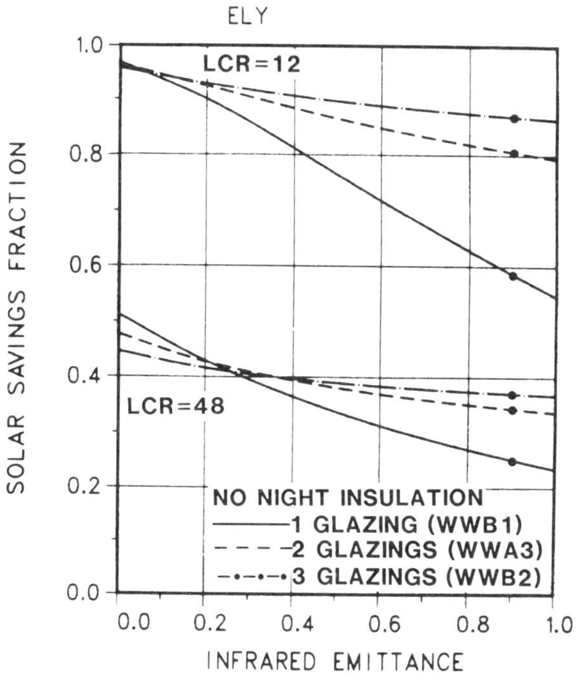

Figure A2-393

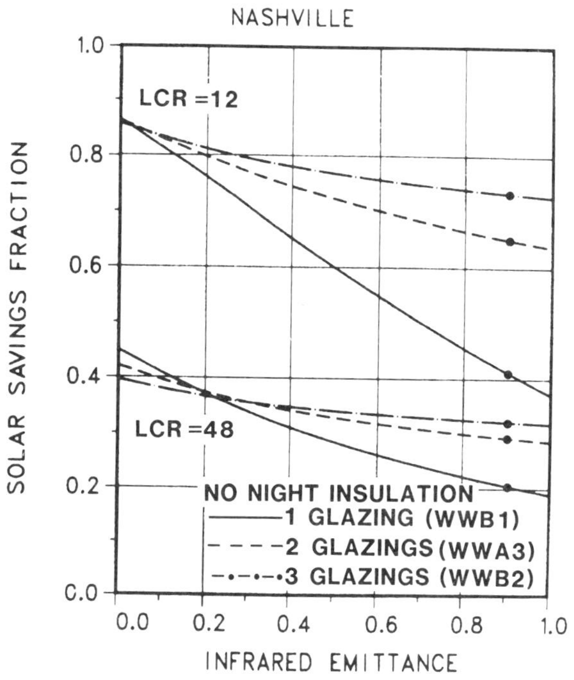

Figure A2-394

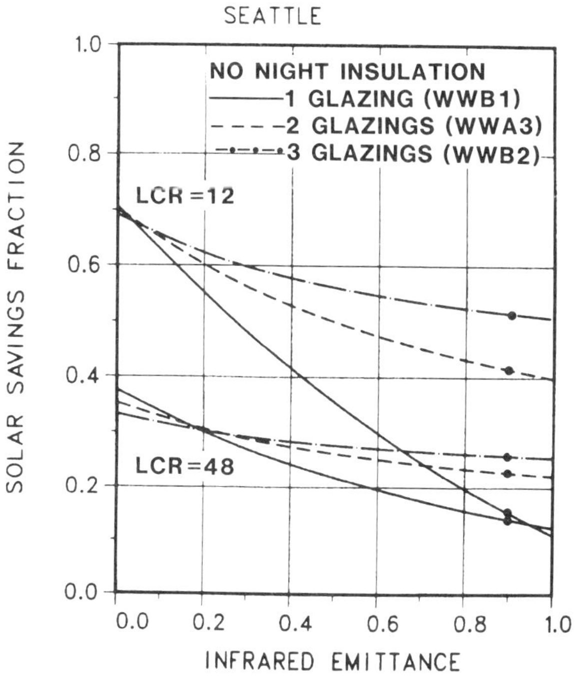

Figure A2-395

Figures A2-390 through A2-395. Annual heating performance sensitivity to the infrared emittance of the surface of a water wall with single (WWB1), double (WWA3), and triple (WWB2) layers of glazing. The dots represent the reference design conditions.

THERMAL STORAGE WALL
AZIMUTH

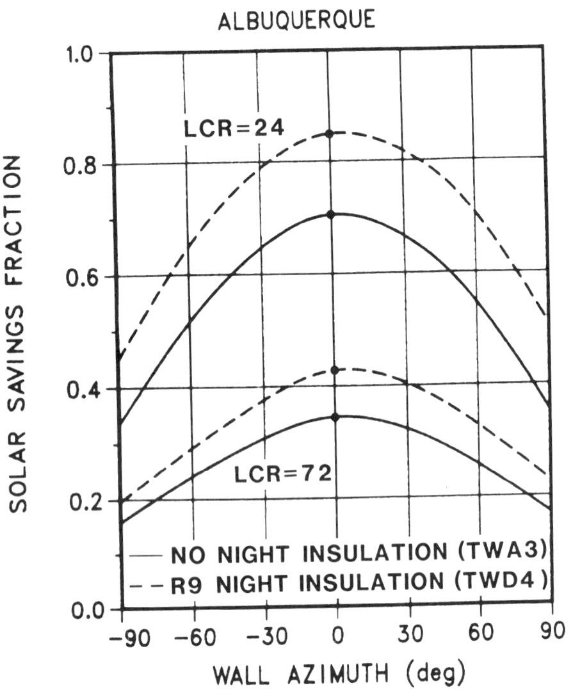

Figure A2-396

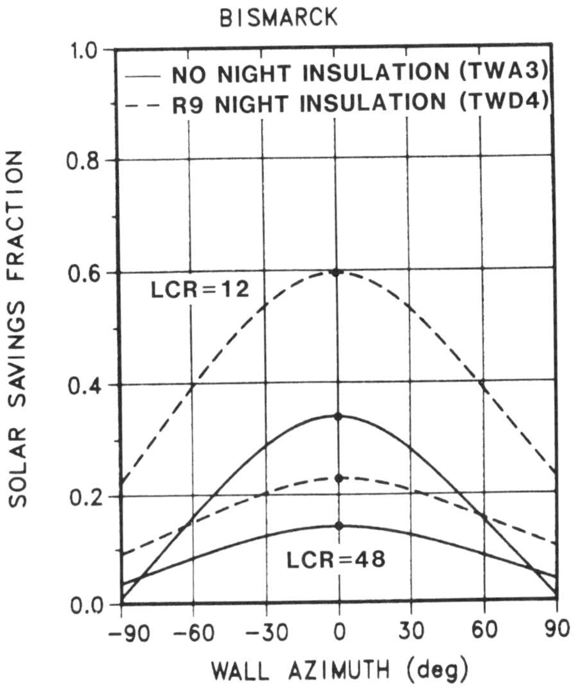

Figure A2-397

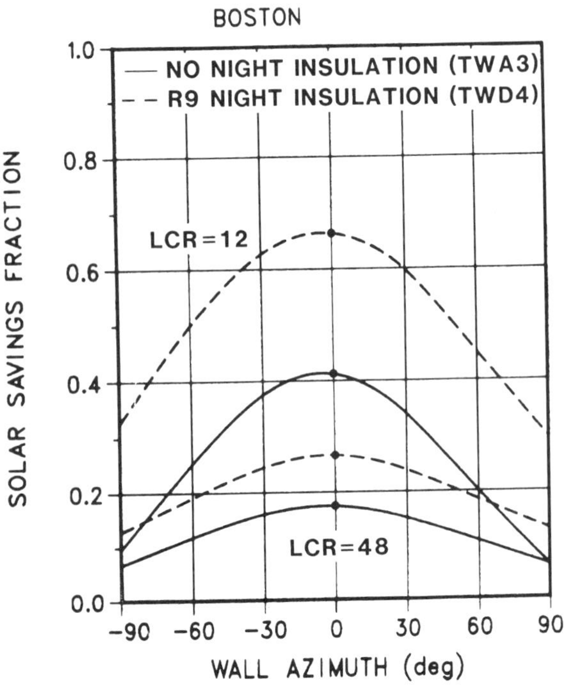

Figure A2-398

Figures A2-396 through A2-401. Annual heating performance sensitivity to the azimuth of a vented Trombe wall without night insulation (TWA3) and with R9 night insulation (TWD4). The dots represent the reference design conditions.

THERMAL STORAGE WALL
AZIMUTH

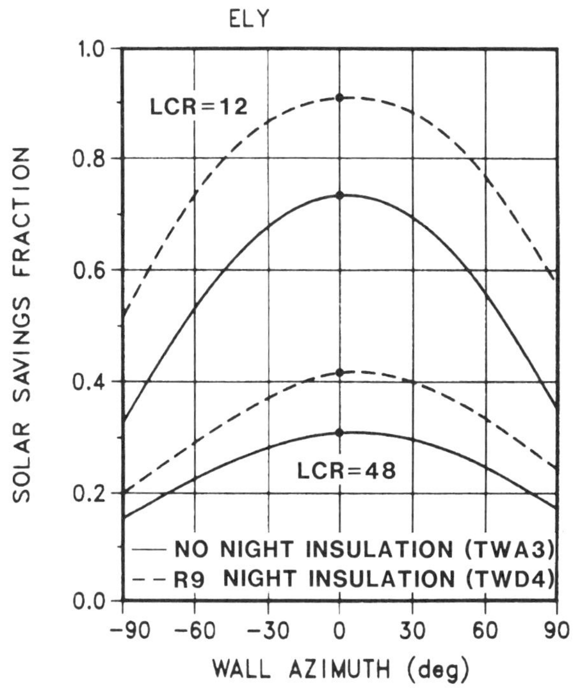

Figure A2-399

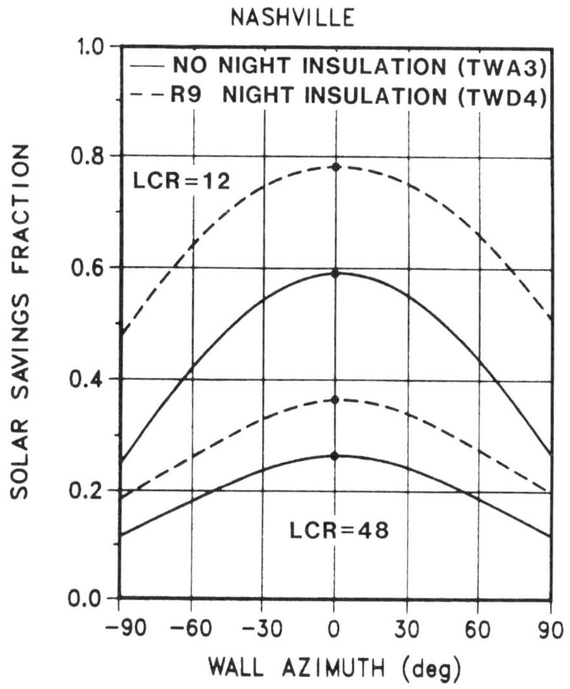

Figure A2-400

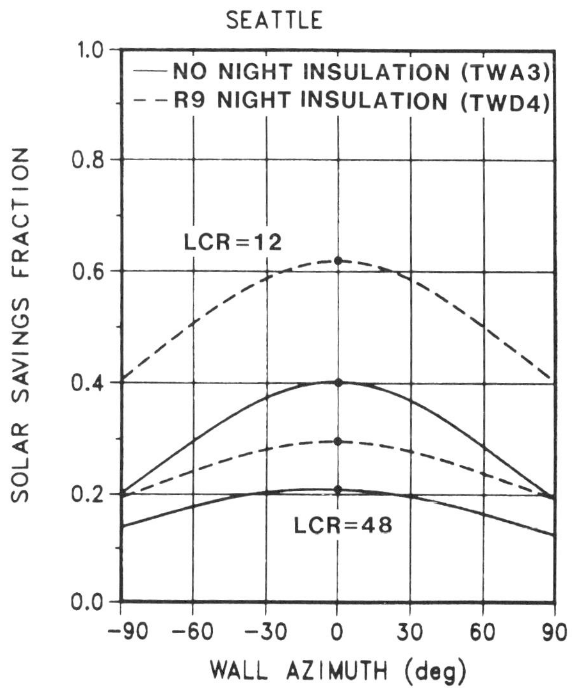

Figure A2-401

Figures A2-396 through A2-401. Annual heating performance sensitivity to the azimuth of a vented Trombe wall without night insulation (TWA3) and with R9 night insulation (TWD4). The dots represent the reference design conditions.

THERMAL STORAGE WALL
VENT AREA FRACTION

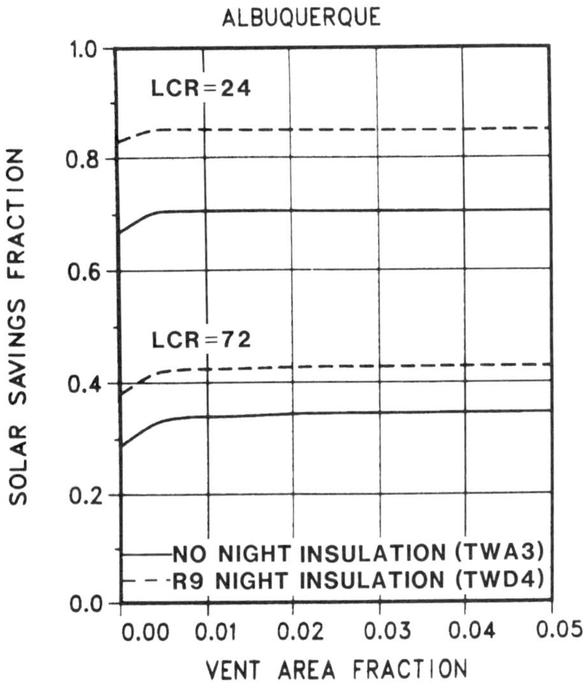

Figure A2-402

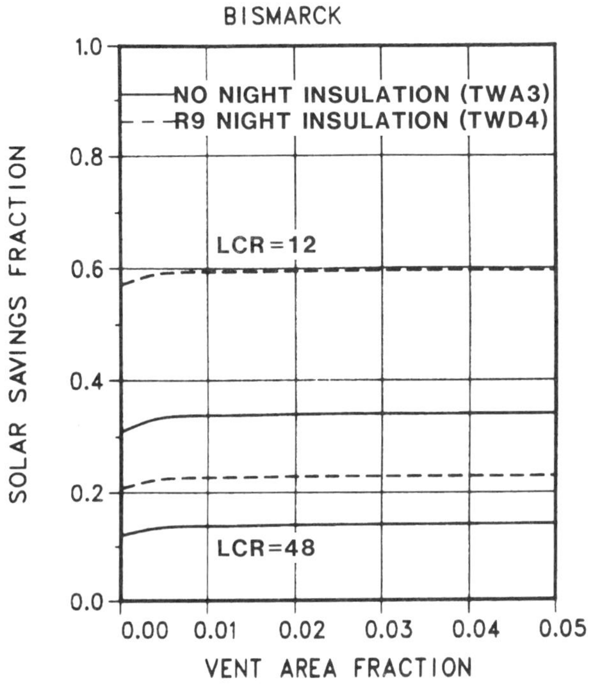

Figure A2-403

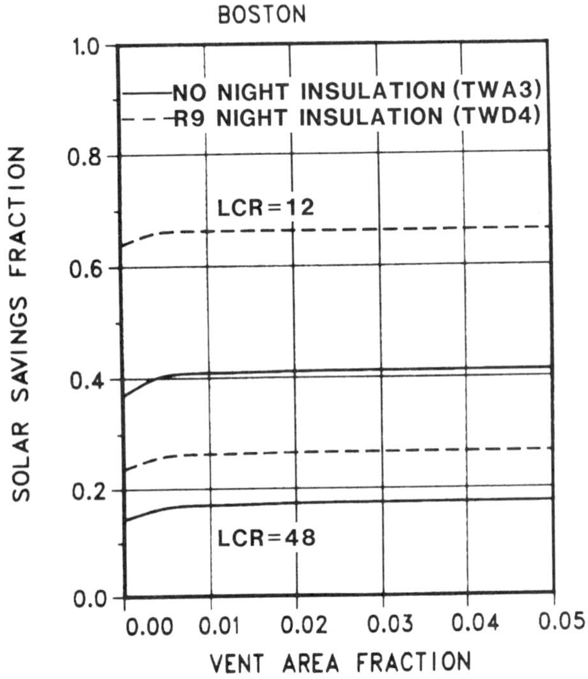

Figure A2-404

Figures A2-402 through A2-407. Annual heating performance sensitivity to the vent area fraction for a Trombe wall without night insulation (TWA3) and with R9 night insulation (TWD4). The reference design conditions are off the graph at vent area fraction = 0.06.

THERMAL STORAGE WALL
VENT AREA FRACTION

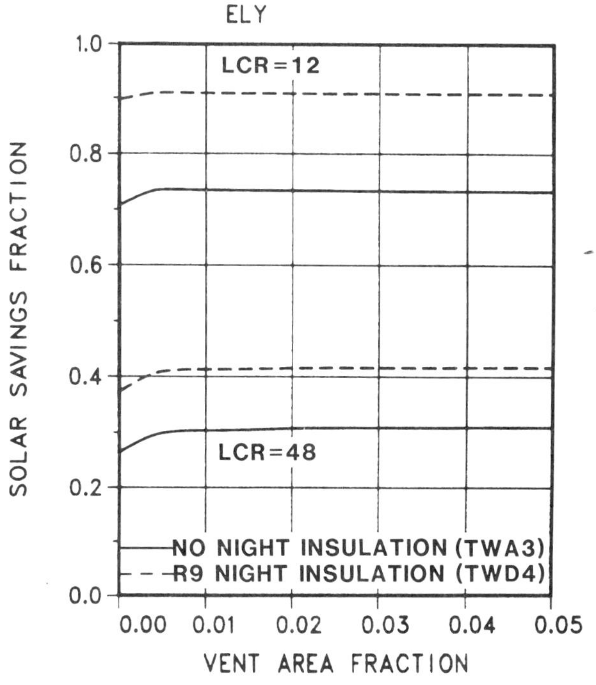

Figure A2-405

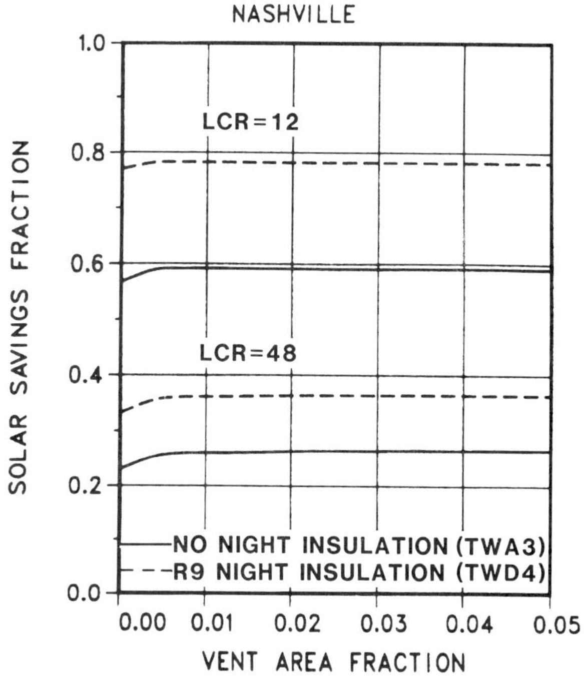

Figure A2-406

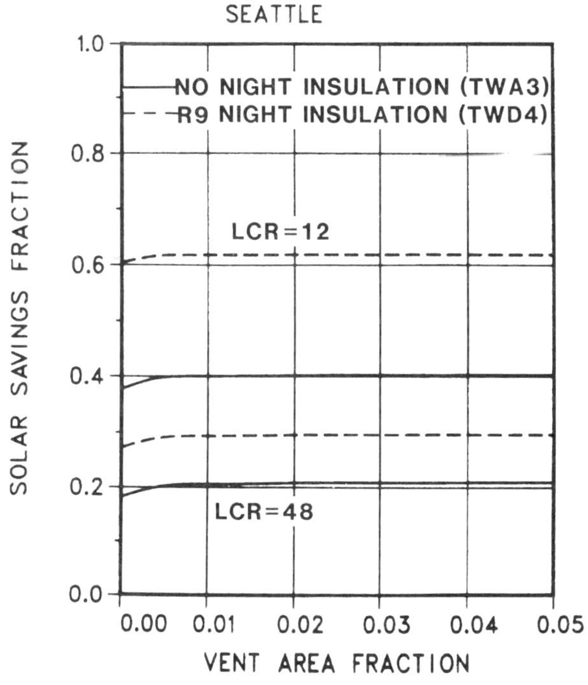

Figure A2-407

Figures A2-402 through A2-407. Annual heating performance sensitivity to the vent area fraction for a Trombe wall without night insulation (TWA3) and with R9 night insulation (TWD4). The reference design conditions are off the graph at vent area fraction = 0.06.

APPENDIX 3
WEATHER AND INSOLATION DATA

This appendix tabulates solar radiation and temperature data for 209 U.S. cities and 14 Canadian cities. For the U.S. locations, except for the Los Alamos data, the bases of the tables are the 30-year (1941 through 1970) temperature normals (TA) and the 24-to-25-year (SOLMET) solar radiation averages (HS)[1]. For Los Alamos, the mean temperature (TA) is the 30-year (1951 through 1980) normal, and the solar radiation average (HS) is a 4-year average, both measured at the Los Alamos National Laboratory. The standard deviations of the mean temperatures (SD) are the 30-year (1941 through 1970) normals.[2]

For Canadian locations, the bases of the tables are the 30-year (1941 through 1970) temperature normals[3] and the 9-22 year solar radiation averages.[4] The standard deviations of the mean temperatures are the 30-year (1951 through 1980) normals.[5]

The heading of each table gives the city, state, elevation, and latitude. The tabulated quantities are:

- HS Total hemispheric solar radiation incident on a horizontal surface (Btu/ft^2 • day)
- VS Total solar radiation incident on a vertical surface (Btu/ft^2 • day)
- TA Mean ambient temperature, $(T_{min} + T_{max})/2$ (°F)
- SD Standard deviation of the mean temperature (°F)
- DXX Heating degree-days below the base temperature XX (°F days)
- KT Average clearness ratio, i.e., the ratio of the total hemispheric solar radiation incident on a horizontal surface to the extraterrestrial radiation incident on a horizontal surface (dimensionless)
- L−D Latitude minus midmonth solar declination (degrees)

The quantities VS, KT, and L−D are calculated as described in Chapter 9, "Solar Gain." The quantities DXX are calculated by the Thom method as described in Chapter 4, "Degree-Days and Base Temperature."

References

1. V. Cinquemani, Jr., et al, in *Input Data for Solar Systems* (November 1978).
2. National Oceanic and Atmospheric Administration, U.S. Department of Commerce, "Degree Days to Selected Bases: 1941-70 Normals for National Weather Service Offices and Principal Climatological Stations," (undated).
3. *Canadian Normals, Vol. 1-SI. Temperature, 1941-1970* (Atmospheric Environment Service, Downsview, Ontario, 1975).
4. D.W. Phillips and D. Aston, *Canadian Solar Radiation Data* (Atmospheric Environment Service, Downsview, Ontario, 1980).
5. *Canadian Climate Normals, Vol. 2. Temperature 1951-1980* (Atmospheric Environment Service, Downsview, Ontario, 1982).

BIRMINGHAM, ALABAMA ELEV 630 LAT 33.6

	HS	VS	TA	SD	D50	D55	D60	D65	D70	KT	L-D
JAN	707	864	44.2	5.10	268	378	510	654	804	.43	54.9
FEB	967	995	46.9	5.10	190	279	390	517	651	.47	47.2
MAR	1296	986	53.3	4.50	100	176	268	389	529	.49	36.0
APR	1673	921	63.2	2.80	0	12	45	116	221	.53	23.8
MAY	1857	852	70.5	2.50	0	0	0	20	75	.53	14.6
JUN	1918	843	77.4	2.50	0	0	0	0	10	.53	10.2
JUL	1810	821	79.9	2.20	0	0	0	0	0	.51	12.2
AUG	1724	874	79.2	2.20	0	0	0	0	0	.53	20.1
SEP	1455	990	73.9	2.50	0	0	0	6	30	.52	31.8
OCT	1211	1179	63.3	3.30	6	21	62	137	238	.54	43.6
NOV	858	1056	52.1	3.00	65	141	253	391	538	.49	53.0
DEC	661	857	45.2	3.80	209	326	467	616	770	.43	57.0
YR	1346	936	62.4	3.28	838	1333	1995	2846	3866	.51	

MOBILE, ALABAMA ELEV 220 LAT 30.7

	HS	VS	TA	SD	D50	D55	D60	D65	D70	KT	L-D
JAN	828	964	51.2	4.83	142	222	324	451	593	.46	52.0
FEB	1100	1066	54.0	4.89	87	153	233	337	461	.50	44.3
MAR	1407	993	59.4	3.64	22	60	130	221	345	.52	33.1
APR	1722	884	67.9	2.51	0	0	9	40	113	.54	20.9
MAY	1872	828	74.8	1.72	0	0	0	0	9	.54	11.7
JUN	1868	819	80.3	1.65	0	0	0	0	0	.52	7.3
JUL	1715	778	81.6	1.20	0	0	0	0	0	.49	9.3
AUG	1641	799	81.5	1.41	0	0	0	0	0	.50	17.2
SEP	1449	909	77.5	1.81	0	0	0	0	0	.50	28.9
OCT	1299	1164	68.9	2.74	0	0	10	39	108	.55	40.7
NOV	955	1096	58.5	2.62	8	37	106	211	348	.51	50.1
DEC	759	928	52.9	3.43	71	147	250	385	533	.45	54.1
YR	1385	935	67.4	2.69	330	619	1062	1684	2510	.51	

MONTGOMERY, ALABAMA ELEV 203 LAT 32.3

	HS	VS	TA	SD	D50	D55	D60	D65	D70	KT	L-D
JAN	752	896	47.5	4.93	203	297	418	556	703	.44	53.6
FEB	1013	1011	50.6	4.83	129	204	300	419	550	.48	45.9
MAR	1341	987	56.5	4.07	51	112	192	299	432	.50	34.7
APR	1729	920	65.2	2.51	0	0	20	76	167	.54	22.5
MAY	1897	849	72.4	2.25	0	0	0	8	41	.54	13.3
JUN	1972	848	78.9	2.32	0	0	0	0	5	.55	8.9
JUL	1841	820	81.0	1.69	0	0	0	0	0	.52	10.9
AUG	1746	861	80.7	1.73	0	0	0	0	0	.53	18.8
SEP	1468	963	76.0	2.21	0	0	0	0	11	.52	30.5
OCT	1262	1188	65.8	3.18	0	10	34	93	178	.55	42.3
NOV	915	1099	55.0	2.74	27	86	177	306	451	.51	51.7
DEC	719	916	48.5	3.42	137	236	367	515	667	.45	55.7
YR	1389	946	64.8	2.98	547	945	1508	2272	3205	.52	

PHOENIX, ARIZONA ELEV 1112 LAT 33.4

	HS	VS	TA	SD	D50	D55	D60	D65	D70	KT	L-D
JAN	1021	1462	51.2	2.90	78	162	285	431	584	.62	54.7
FEB	1374	1608	55.1	3.60	42	100	181	292	422	.66	47.0
MAR	1814	1471	59.7	2.50	5	25	87	185	324	.69	35.8
APR	2355	1236	67.7	3.10	0	5	18	60	134	.74	23.6
MAY	2676	1034	76.3	2.70	0	0	0	0	19	.77	14.4
JUN	2739	973	84.6	2.80	0	0	0	0	0	.76	10.0
JUL	2486	964	91.2	2.10	0	0	0	0	0	.70	12.0
AUG	2293	1080	89.1	2.20	0	0	0	0	0	.70	19.9
SEP	2015	1415	83.8	2.40	0	0	0	0	0	.72	31.6
OCT	1576	1674	72.2	2.80	0	0	0	17	61	.70	43.4
NOV	1150	1606	59.8	2.75	7	29	89	182	312	.66	52.8
DEC	932	1407	52.5	2.20	38	115	239	389	543	.61	56.8
YR	1871	1326	70.3	2.66	170	436	899	1556	2399	.71	

PRESCOTT, ARIZONA ELEV 5023 LAT 34.6

	HS	VS	TA	SD	D50	D55	D60	D65	D70	KT	L-D
JAN	1016	1533	37.1	3.42	408	557	711	865	1020	.64	55.9
FEB	1335	1623	40.5	3.74	284	412	548	687	826	.66	48.2
MAR	1777	1502	44.3	3.00	207	340	489	642	797	.69	37.0
APR	2275	1251	52.0	2.98	65	143	255	394	541	.72	24.8
MAY	2629	1060	60.4	2.36	0	18	72	165	302	.75	15.6
JUN	2762	998	69.1	2.70	0	0	8	33	99	.76	11.2
JUL	2309	953	75.5	1.76	0	0	0	0	7	.65	13.2
AUG	2092	1043	73.0	1.74	0	0	0	0	20	.64	21.1
SEP	1954	1430	68.1	1.94	0	0	0	23	92	.71	32.8
OCT	1543	1709	57.2	2.70	15	57	136	254	400	.71	44.6
NOV	1140	1674	45.8	3.09	169	288	429	577	726	.68	54.0
DEC	927	1477	38.6	2.91	359	510	664	818	973	.63	58.0
YR	1814	1352	55.1	2.69	1507	2325	3312	4458	5803	.69	

TUCSON, ARIZONA ELEV 2556 LAT 32.1

	HS	VS	TA	SD	D50	D55	D60	D65	D70	KT	L-D
JAN	1099	1539	50.9	3.21	93	177	298	442	593	.64	53.4
FEB	1432	1615	53.5	3.69	60	127	215	333	465	.67	45.7
MAR	1864	1447	57.6	2.65	13	51	128	243	387	.70	34.5
APR	2363	1186	65.5	2.80	0	6	25	81	167	.74	22.3
MAY	2671	1001	73.6	2.30	0	0	0	5	29	.77	13.1
JUN	2730	953	82.1	2.43	0	0	0	0	0	.76	8.7
JUL	2341	922	86.3	1.41	0	0	0	0	0	.66	10.7
AUG	2183	1007	83.8	1.84	0	0	0	0	0	.67	18.6
SEP	1979	1322	80.1	1.99	0	0	0	0	0	.70	30.3
OCT	1602	1623	70.1	2.78	0	0	7	29	90	.70	42.1
NOV	1208	1631	58.5	3.14	16	52	122	221	352	.67	51.5
DEC	996	1464	52.0	2.30	47	128	255	404	558	.62	55.5
YR	1873	1307	67.8	2.54	229	541	1050	1758	2641	.70	

WINSLOW, ARIZONA ELEV 4882 LAT 35.0

	HS	VS	TA	SD	D50	D55	D60	D65	D70	KT	L-D
JAN	985	1490	32.6	4.83	552	700	852	1005	1160	.63	56.3
FEB	1327	1637	39.1	4.09	323	451	587	726	865	.67	48.6
MAR	1780	1528	44.8	2.91	193	325	473	627	781	.69	37.4
APR	2283	1272	53.7	3.30	54	124	219	348	492	.73	25.2
MAY	2595	1064	62.7	2.59	0	11	47	124	239	.74	16.0
JUN	2712	1001	71.8	2.60	0	0	0	14	56	.75	11.6
JUL	2347	968	78.3	2.10	0	0	0	0	0	.66	13.6
AUG	2141	1074	76.1	2.19	0	0	0	0	11	.66	21.5
SEP	1928	1429	69.5	2.22	0	0	0	19	78	.70	33.2
OCT	1513	1691	57.3	2.75	15	58	136	252	397	.70	45.0
NOV	1119	1658	43.2	3.39	233	363	507	655	804	.67	54.4
DEC	894	1424	33.8	4.33	512	661	814	968	1122	.62	58.4
YR	1803	1351	55.3	3.10	1882	2693	3635	4738	6005	.69	

YUMA, ARIZONA ELEV 207 LAT 32.7

	HS	VS	TA	SD	D50	D55	D60	D65	D70	KT	L-D
JAN	1096	1575	55.4	2.94	31	91	180	308	455	.65	54.0
FEB	1443	1675	59.4	3.52	15	46	108	192	308	.69	46.3
MAR	1919	1534	63.9	2.39	0	6	29	97	204	.72	35.1
APR	2413	1231	71.2	2.96	0	0	6	24	76	.76	22.9
MAY	2728	1023	78.7	2.51	0	0	0	0	7	.78	13.7
JUN	2814	966	85.8	2.32	0	0	0	0	0	.78	9.3
JUL	2453	948	93.7	1.59	0	0	0	0	0	.69	11.3
AUG	2329	1070	92.8	1.67	0	0	0	0	0	.71	19.2
SEP	2051	1406	87.1	2.58	0	0	0	0	0	.73	30.9
OCT	1623	1694	75.9	2.69	0	0	0	5	22	.72	42.7
NOV	1215	1690	63.5	2.69	0	9	39	108	212	.68	52.1
DEC	1000	1515	56.3	2.37	13	59	145	276	426	.64	56.1
YR	1925	1358	73.7	2.51	59	211	507	1010	1710	.72	

FORT SMITH, ARKANSAS ELEV 463 LAT 35.3

	HS	VS	TA	SD	D50	D55	D60	D65	D70	KT	L-D
JAN	744	996	39.0	3.31	352	499	652	806	961	.48	56.6
FEB	999	1108	43.3	3.60	217	337	470	608	748	.50	48.9
MAR	1312	1056	50.3	4.49	143	228	338	471	617	.51	37.7
APR	1616	929	62.2	2.74	0	14	55	132	246	.52	25.5
MAY	1912	892	70.1	2.24	0	0	0	17	72	.55	16.3
JUN	2089	898	78.0	2.31	0	0	0	0	6	.58	11.9
JUL	2065	908	82.2	2.36	0	0	0	0	0	.58	13.9
AUG	1877	973	81.4	2.85	0	0	0	0	5	.58	21.8
SEP	1501	1080	74.0	2.23	0	0	0	0	22	.55	33.5
OCT	1201	1242	63.2	3.19	5	19	60	135	238	.56	45.3
NOV	851	1119	50.4	2.81	82	170	296	440	588	.52	54.7
DEC	682	963	41.5	3.16	281	423	575	729	884	.48	58.7
YR	1406	1013	61.3	2.94	1080	1690	2446	3338	4387	.54	

LITTLE ROCK, ARKANSAS ELEV 266 LAT 34.7

	HS	VS	TA	SD	D50	D55	D60	D65	D70	KT	L-D
JAN	731	947	39.5	3.71	344	487	638	791	946	.46	56.0
FEB	1003	1089	42.9	3.82	230	350	482	620	759	.50	48.3
MAR	1313	1037	50.3	4.37	139	224	335	470	616	.51	37.1
APR	1611	913	61.7	2.67	0	15	60	139	259	.51	24.9
MAY	1929	888	69.8	2.33	0	0	0	21	80	.55	15.7
JUN	2106	896	78.1	2.31	0	0	0	0	6	.58	11.3
JUL	2032	891	81.4	2.18	0	0	0	0	0	.57	13.3
AUG	1860	952	80.6	2.65	0	0	0	0	5	.57	21.2
SEP	1518	1074	73.3	2.28	0	0	0	5	30	.55	32.9
OCT	1228	1251	62.4	3.03	5	20	65	143	255	.56	44.7
NOV	847	1084	50.3	2.69	80	169	297	442	591	.50	54.1
DEC	674	922	41.6	3.39	282	422	572	726	881	.46	58.1
YR	1406	995	61.0	2.95	1080	1687	2449	3357	4428	.54	

BAKERSFIELD, CALIFORNIA — ELEV 492 LAT 35.4

	HS	VS	TA	SD	D50	D55	D60	D65	D70	KT	L-D
JAN	766	1044	47.5	3.20	147	255	394	544	698	.49	56.7
FEB	1102	1276	52.4	2.90	51	121	226	356	493	.56	49.0
MAR	1595	1350	56.6	2.30	11	53	137	266	416	.62	37.8
APR	2095	1191	62.7	3.30	6	22	66	140	243	.67	25.6
MAY	2509	1057	69.8	2.40	0	0	0	22	82	.72	16.4
JUN	2749	1013	76.9	2.90	0	0	0	5	19	.76	12.0
JUL	2683	1037	83.9	2.50	0	0	0	0	0	.76	14.0
AUG	2421	1198	81.6	2.30	0	0	0	0	0	.75	21.9
SEP	1992	1508	76.6	2.10	0	0	0	0	7	.73	33.6
OCT	1458	1632	66.9	2.50	0	0	13	55	136	.68	45.4
NOV	942	1305	56.0	2.50	16	64	149	276	421	.57	54.8
DEC	677	957	47.9	3.20	140	245	382	532	686	.47	58.8
YR	1752	1213	64.9	2.67	371	760	1367	2196	3201	.68	

DAGGETT, CALIFORNIA — ELEV 1929 LAT 34.9

	HS	VS	TA	SD	D50	D55	D60	D65	D70	KT	L-D
JAN	958	1422	47.3	3.50	160	267	403	551	705	.61	56.2
FEB	1281	1549	52.0	3.59	75	147	246	371	506	.64	48.5
MAR	1772	1514	56.7	2.88	21	70	152	271	416	.69	37.3
APR	2274	1263	64.3	3.41	5	17	51	118	209	.73	25.1
MAY	2591	1061	72.3	2.68	0	0	0	14	55	.74	15.9
JUN	2766	1004	80.1	3.03	0	0	0	0	9	.76	11.5
JUL	2603	1012	87.3	2.30	0	0	0	0	0	.73	13.5
AUG	2383	1164	85.5	2.38	0	0	0	0	0	.73	21.4
SEP	2008	1493	79.2	2.60	0	0	0	0	7	.73	33.1
OCT	1516	1688	68.1	3.04	0	5	17	57	131	.70	44.9
NOV	1085	1575	55.5	3.09	31	90	175	296	438	.65	54.3
DEC	876	1374	48.0	2.72	123	233	376	528	682	.60	58.3
YR	1845	1342	66.4	2.93	415	829	1420	2206	3158	.71	

FRESNO, CALIFORNIA — ELEV 328 LAT 36.8

	HS	VS	TA	SD	D50	D55	D60	D65	D70	KT	L-D
JAN	657	886	45.3	3.20	189	314	459	612	766	.45	58.1
FEB	1012	1195	49.9	2.68	77	164	287	424	563	.53	50.4
MAR	1566	1389	53.9	2.12	22	88	199	346	499	.62	39.2
APR	2093	1246	60.3	3.19	10	35	96	182	303	.68	27.0
MAY	2484	1092	67.4	2.58	0	0	12	51	129	.71	17.8
JUN	2733	1045	73.9	2.73	0	0	0	9	36	.75	13.4
JUL	2685	1076	80.6	2.16	0	0	0	0	0	.76	15.4
AUG	2423	1258	78.3	2.43	0	0	0	0	7	.75	23.3
SEP	1985	1587	73.8	2.09	0	0	0	0	21	.74	35.0
OCT	1429	1681	64.2	2.35	0	5	26	90	195	.68	46.8
NOV	888	1271	53.5	2.48	33	102	208	348	495	.57	56.2
DEC	574	800	45.8	2.60	162	292	442	595	750	.42	60.2
YR	1714	1210	62.3	2.55	493	1000	1729	2657	3764	.67	

LONG BEACH, CALIFORNIA — ELEV 66 LAT 33.8

	HS	VS	TA	SD	D50	D55	D60	D65	D70	KT	L-D
JAN	928	1291	54.2	2.63	33	100	201	339	491	.57	55.1
FEB	1215	1373	55.5	2.88	23	75	156	273	408	.59	47.4
MAR	1610	1289	57.2	2.07	6	37	118	247	397	.61	36.2
APR	1938	1056	60.6	2.08	0	10	57	148	284	.61	24.0
MAY	2064	913	64.1	1.70	0	0	12	71	188	.59	14.8
JUN	2140	894	67.3	1.67	0	0	0	23	100	.59	10.4
JUL	2300	939	72.2	1.78	0	0	0	0	29	.65	12.4
AUG	2100	1024	73.3	1.89	0	0	0	0	21	.64	20.3
SEP	1701	1186	71.8	2.09	0	0	0	7	41	.61	32.0
OCT	1326	1342	66.9	2.29	0	0	10	48	130	.60	43.8
NOV	1003	1335	60.6	2.38	0	16	66	155	286	.58	53.2
DEC	847	1241	55.5	1.86	8	54	152	296	450	.56	57.2
YR	1599	1156	63.3	2.10	70	292	772	1607	2825	.61	

LOS ANGELES, CALIFORNIA — ELEV 105 LAT 33.9

	HS	VS	TA	SD	D50	D55	D60	D65	D70	KT	L-D
JAN	926	1293	54.5	2.65	31	95	194	331	482	.57	55.2
FEB	1214	1377	55.6	2.80	21	71	152	270	405	.59	47.5
MAR	1619	1303	56.5	2.04	8	46	132	267	419	.62	36.3
APR	1951	1066	58.8	2.12	0	21	86	195	337	.62	24.1
MAY	2060	914	61.9	1.68	0	0	29	114	253	.59	14.9
JUN	2119	891	64.5	2.00	0	0	14	71	175	.59	10.5
JUL	2307	942	68.5	1.86	0	0	0	19	86	.65	12.5
AUG	2079	1019	69.5	1.97	0	0	0	15	73	.64	20.4
SEP	1681	1174	68.7	2.17	0	0	0	23	89	.60	32.1
OCT	1317	1335	65.2	2.42	0	0	20	77	172	.59	43.9
NOV	1004	1343	60.5	2.41	0	17	68	158	289	.58	53.3
DEC	848	1249	56.0	1.80	6	45	138	281	434	.56	57.3
YR	1595	1157	61.7	2.15	66	295	833	1821	3214	.61	

MOUNT SHASTA, CALIFORNIA — ELEV 3586 LAT 41.3

	HS	VS	TA	SD	D50	D55	D60	D65	D70	KT	L-D
JAN	561	857	33.6	3.60	513	665	819	974	1128	.46	62.6
FEB	857	1128	37.8	3.40	348	483	622	762	902	.51	54.9
MAR	1250	1217	40.4	2.50	303	454	608	763	918	.54	43.7
APR	1756	1200	46.3	3.70	177	285	419	564	712	.59	31.5
MAY	2186	1129	53.3	3.20	59	133	235	371	520	.63	22.3
JUN	2436	1102	60.0	2.80	7	28	87	178	307	.67	17.9
JUL	2577	1201	67.8	2.30	0	0	7	37	113	.73	19.9
AUG	2213	1356	66.0	2.40	0	0	15	64	152	.70	27.8
SEP	1735	1595	61.2	2.50	0	15	61	145	270	.68	39.5
OCT	1155	1492	51.4	2.70	68	152	277	424	577	.62	51.3
NOV	659	998	41.7	2.50	257	400	549	699	849	.49	60.7
DEC	505	814	35.5	3.10	453	605	760	915	1070	.46	64.7
YR	1494	1174	49.6	2.89	2185	3220	4459	5896	7518	.62	

NEEDLES, CALIFORNIA — ELEV 886 LAT 34.8

	HS	VS	TA	SD	D50	D55	D60	D65	D70	KT	L-D
JAN	985	1477	51.6	3.23	83	164	279	421	572	.62	56.1
FEB	1353	1669	56.5	3.80	35	88	161	261	386	.68	48.4
MAR	1825	1565	61.6	2.75	0	19	67	150	272	.71	37.2
APR	2317	1280	70.4	3.50	0	0	14	42	104	.74	25.0
MAY	2652	1070	79.6	2.81	0	0	0	0	9	.76	15.8
JUN	2791	1005	88.3	2.73	0	0	0	0	0	.77	11.4
JUL	2541	1000	95.4	1.58	0	0	0	0	0	.72	13.4
AUG	2278	1121	93.3	2.25	0	0	0	0	0	.70	21.3
SEP	2015	1494	86.9	2.41	0	0	0	0	0	.73	33.0
OCT	1537	1714	74.3	2.89	0	0	0	10	39	.71	44.8
NOV	1124	1654	60.7	2.80	6	23	78	163	287	.67	54.2
DEC	913	1457	52.7	2.53	46	122	238	384	537	.63	58.2
YR	1862	1373	72.6	2.76	170	416	837	1431	2206	.71	

OAKLAND, CALIFORNIA — ELEV 7 LAT 37.7

	HS	VS	TA	SD	D50	D55	D60	D65	D70	KT	L-D
JAN	708	1028	48.6	2.47	104	213	356	509	663	.50	59.0
FEB	1017	1247	51.9	2.43	44	118	233	368	507	.55	51.3
MAR	1456	1307	53.7	1.93	19	85	202	351	505	.59	40.1
APR	1922	1178	56.1	2.17	10	52	139	270	417	.62	27.9
MAY	2211	1035	58.9	1.61	0	11	71	193	344	.64	18.7
JUN	2350	994	61.9	1.87	0	0	33	114	245	.65	14.3
JUL	2322	1018	63.1	1.42	0	0	11	80	215	.65	16.3
AUG	2053	1122	63.5	1.48	0	0	10	74	204	.64	24.2
SEP	1701	1359	64.5	1.64	0	0	8	59	170	.64	35.9
OCT	1212	1380	61.1	1.75	0	5	42	135	277	.59	47.7
NOV	822	1180	55.3	2.06	11	60	155	293	441	.54	57.1
DEC	647	993	49.9	2.22	75	175	316	469	623	.50	61.1
YR	1537	1152	57.4	1.92	263	719	1576	2915	4611	.61	

RED BLUFF, CALIFORNIA — ELEV 354 LAT 40.1

	HS	VS	TA	SD	D50	D55	D60	D65	D70	KT	L-D
JAN	570	831	45.2	2.95	185	314	461	614	769	.44	61.4
FEB	892	1138	50.0	2.94	83	167	287	422	560	.51	53.7
MAR	1354	1296	53.2	2.40	36	109	223	368	521	.57	42.5
APR	1910	1266	59.5	3.93	24	63	131	218	335	.63	30.3
MAY	2375	1164	67.4	3.00	0	5	19	64	142	.69	21.1
JUN	2600	1111	75.5	3.03	0	0	0	8	30	.71	16.7
JUL	2672	1184	82.3	2.08	0	0	0	0	0	.75	18.7
AUG	2311	1356	79.9	2.53	0	0	0	0	5	.73	26.6
SEP	1845	1649	75.3	2.19	0	0	0	0	14	.71	38.3
OCT	1227	1549	65.0	2.51	0	5	23	82	178	.63	50.1
NOV	706	1047	53.7	2.14	23	88	198	340	489	.50	59.5
DEC	511	782	46.4	2.77	154	277	424	577	732	.44	63.5
YR	1584	1198	62.8	2.70	505	1028	1766	2693	3775	.65	

SACRAMENTO, CALIFORNIA — ELEV 26 LAT 38.5

	HS	VS	TA	SD	D50	D55	D60	D65	D70	KT	L-D
JAN	597	829	45.1	2.80	183	315	464	617	772	.43	59.8
FEB	939	1149	49.8	2.60	76	165	289	426	566	.52	52.1
MAR	1458	1348	53.0	2.10	30	103	224	373	527	.60	40.9
APR	2004	1263	58.3	3.20	17	56	127	227	358	.65	28.7
MAY	2435	1131	64.3	3.30	5	16	51	120	214	.70	19.5
JUN	2684	1082	70.5	2.60	0	0	0	20	74	.74	15.1
JUL	2688	1131	75.2	1.60	0	0	0	0	6	.76	17.1
AUG	2368	1310	74.1	2.20	0	0	0	0	23	.74	25.0
SEP	1907	1615	71.5	1.90	0	0	0	5	39	.72	36.7
OCT	1315	1602	63.3	2.20	0	5	29	101	217	.65	48.5
NOV	782	1135	53.0	2.00	25	96	215	361	510	.53	57.9
DEC	538	784	45.8	2.60	162	292	442	595	750	.43	61.9
YR	1646	1198	60.3	2.42	498	1048	1841	2845	4056	.66	

SAN DIEGO, CALIFORNIA ELEV 30 LAT 32.7

	HS	VS	TA	SD	D50	D55	D60	D65	D70	KT	L-D
JAN	976	1325	55.2	3.00	34	96	186	314	461	.58	54.0
FEB	1266	1392	56.7	2.30	8	43	119	237	373	.60	46.3
MAR	1632	1261	58.0	1.90	0	24	97	222	373	.61	35.1
APR	1937	1024	60.7	2.00	0	9	53	144	281	.61	22.9
MAY	2003	882	63.3	1.50	0	0	12	79	210	.57	13.7
JUN	2062	870	65.5	1.87	0	0	8	52	147	.57	9.3
JUL	2186	902	69.6	1.50	0	0	0	6	56	.62	11.3
AUG	2057	981	71.4	1.60	0	0	0	0	33	.63	19.2
SEP	1717	1155	69.9	2.20	0	0	0	16	70	.61	30.9
OCT	1373	1349	66.1	1.80	0	0	6	43	136	.61	42.7
NOV	1063	1387	60.8	1.90	0	7	48	140	278	.60	52.1
DEC	904	1302	56.7	1.80	0	35	121	259	413	.58	56.1
YR	1599	1151	62.9	1.94	42	214	650	1512	2831	.60	

SAN FRANCISCO, CALIFORNIA ELEV 16 LAT 37.6

	HS	VS	TA	SD	D50	D55	D60	D65	D70	KT	L-D
JAN	708	1023	48.3	2.47	109	221	365	518	673	.49	58.9
FEB	1009	1229	51.2	2.43	53	132	251	387	527	.54	51.2
MAR	1455	1301	53.0	1.93	25	98	222	373	527	.59	40.0
APR	1920	1173	55.3	2.17	13	64	157	294	441	.62	27.8
MAY	2226	1038	58.3	1.61	0	14	82	210	363	.64	18.6
JUN	2377	998	61.6	1.87	0	5	37	120	254	.65	14.2
JUL	2392	1034	62.5	1.42	0	0	16	93	233	.67	16.2
AUG	2116	1149	63.0	1.48	0	0	13	84	219	.66	24.1
SEP	1742	1394	64.1	1.64	0	0	10	66	181	.65	35.8
OCT	1226	1397	61.0	1.75	0	5	43	137	280	.60	47.6
NOV	821	1172	55.3	2.06	11	60	155	293	441	.54	57.0
DEC	642	977	49.7	2.22	78	180	322	475	629	.49	61.0
YR	1555	1156	56.9	1.92	289	779	1673	3050	4768	.62	

SANTA MARIA, CALIFORNIA ELEV 236 LAT 34.9

	HS	VS	TA	SD	D50	D55	D60	D65	D70	KT	L-D
JAN	854	1198	50.5	2.54	76	167	300	451	605	.54	56.2
FEB	1141	1313	52.0	2.65	49	122	232	366	504	.57	48.5
MAR	1582	1312	52.8	2.07	31	106	229	379	533	.61	37.3
APR	1921	1082	54.9	2.20	16	70	168	305	453	.61	25.1
MAY	2141	953	57.1	1.50	0	22	105	246	400	.61	15.9
JUN	2349	947	59.6	1.64	0	8	58	167	312	.65	11.5
JUL	2341	965	62.1	1.80	0	0	30	112	247	.66	13.5
AUG	2106	1057	62.3	1.57	0	0	22	102	240	.65	21.4
SEP	1730	1255	62.6	1.67	0	0	20	94	224	.63	33.1
OCT	1353	1440	60.4	2.10	0	13	63	159	300	.62	44.9
NOV	974	1341	56.1	2.11	9	51	137	270	417	.58	54.3
DEC	804	1207	51.8	2.25	48	131	260	410	564	.55	58.3
YR	1610	1172	56.9	2.00	229	690	1624	3061	4799	.62	

COLORADO SPRINGS, COLORADO ELEV 6171 LAT 38.8

	HS	VS	TA	SD	D50	D55	D60	D65	D70	KT	L-D
JAN	891	1534	28.6	4.26	667	820	974	1129	1283	.65	60.1
FEB	1178	1619	31.3	4.95	532	667	805	944	1084	.65	52.4
MAR	1550	1477	35.3	4.14	467	615	767	921	1076	.64	41.2
APR	1931	1227	46.2	3.70	179	288	422	567	715	.63	29.0
MAY	2129	1035	55.5	2.59	22	78	169	301	451	.61	19.8
JUN	2369	1021	64.6	3.09	0	11	39	103	194	.65	15.4
JUL	2212	1012	70.7	1.96	0	0	0	9	54	.62	17.4
AUG	2025	1145	69.1	1.75	0	0	0	13	72	.63	25.3
SEP	1759	1475	60.9	2.63	0	19	70	155	280	.67	37.0
OCT	1359	1702	50.5	3.57	110	196	314	456	607	.68	48.8
NOV	944	1529	37.5	3.82	387	529	676	825	975	.64	58.2
DEC	782	1399	31.0	3.04	590	744	899	1054	1209	.63	62.2
YR	1595	1346	48.4	3.28	2954	3967	5135	6477	8000	.64	

DENVER, COLORADO ELEV 5331 LAT 39.7

	HS	VS	TA	SD	D50	D55	D60	D65	D70	KT	L-D
JAN	840	1465	29.9	4.70	630	781	934	1089	1243	.64	61.0
FEB	1127	1577	32.8	5.00	493	626	764	902	1042	.64	53.3
MAR	1530	1503	37.0	4.30	421	565	716	869	1023	.64	42.1
APR	1879	1227	47.5	3.70	156	256	386	528	676	.62	29.9
MAY	2135	1061	57.0	2.10	7	41	123	253	404	.62	20.7
JUN	2351	1037	66.0	3.00	0	7	26	80	162	.65	16.3
JUL	2273	1053	73.0	1.70	0	0	0	0	19	.64	18.3
AUG	2044	1188	71.6	1.70	0	0	0	0	33	.64	26.2
SEP	1727	1491	62.8	2.70	0	12	47	120	230	.66	37.9
OCT	1300	1657	52.0	3.10	73	153	266	408	559	.67	49.7
NOV	883	1441	39.4	3.70	334	473	620	769	918	.62	59.1
DEC	732	1323	32.6	2.90	540	695	849	1004	1159	.61	63.1
YR	1570	1334	50.1	3.20	2654	3609	4731	6022	7468	.64	

EAGLE, COLORADO ELEV 6512 LAT 39.6

	HS	VS	TA	SD	D50	D55	D60	D65	D70	KT	L-D
JAN	754	1234	18.0	4.58	993	1147	1302	1457	1612	.57	60.9
FEB	1078	1470	23.3	5.30	751	889	1028	1168	1308	.61	53.2
MAR	1502	1461	31.1	3.69	589	742	896	1051	1206	.63	42.0
APR	1933	1261	41.9	2.82	255	396	544	693	843	.64	29.8
MAY	2255	1103	51.3	2.09	50	138	273	425	580	.65	20.6
JUN	2509	1075	58.9	1.98	0	17	80	190	334	.69	16.2
JUL	2384	1085	65.9	1.71	0	0	5	43	139	.67	18.2
AUG	2084	1207	63.7	1.76	0	0	15	79	200	.65	26.1
SEP	1767	1530	55.6	2.26	13	62	152	285	433	.68	37.8
OCT	1307	1663	44.8	2.39	181	320	472	626	781	.67	49.6
NOV	869	1400	30.9	3.62	575	724	873	1023	1173	.61	59.0
DEC	691	1203	20.3	2.60	921	1076	1231	1386	1541	.58	63.0
YR	1596	1306	42.2	2.88	4328	5511	6871	8426	10150	.65	

GRAND JUNCTION, COLORADO ELEV 4839 LAT 39.1

	HS	VS	TA	SD	D50	D55	D60	D65	D70	KT	L-D
JAN	791	1296	26.6	4.90	730	883	1036	1191	1346	.59	60.4
FEB	1119	1520	33.6	4.90	471	604	741	880	1020	.62	52.7
MAR	1553	1498	41.2	2.90	285	431	584	738	893	.64	41.5
APR	1986	1276	51.7	3.20	76	155	266	404	550	.65	29.3
MAY	2380	1132	62.2	2.60	0	13	54	133	253	.69	20.1
JUN	2598	1081	71.3	2.80	0	0	5	20	69	.71	15.7
JUL	2465	1094	78.7	1.60	0	0	0	0	0	.70	17.7
AUG	2182	1240	75.4	1.80	0	0	0	0	8	.68	25.6
SEP	1834	1573	67.2	2.90	0	0	17	60	137	.70	37.3
OCT	1345	1698	54.9	3.10	39	103	195	324	471	.68	49.1
NOV	918	1486	39.8	3.40	319	460	607	756	906	.63	58.5
DEC	731	1280	29.5	3.90	638	791	946	1101	1256	.60	62.5
YR	1661	1346	52.7	3.15	2558	3440	4451	5607	6909	.67	

PUEBLO, COLORADO ELEV 4721 LAT 38.3

	HS	VS	TA	SD	D50	D55	D60	D65	D70	KT	L-D
JAN	894	1504	30.1	4.50	623	774	928	1082	1237	.64	59.6
FEB	1172	1572	34.7	5.30	447	577	712	850	989	.64	51.9
MAR	1564	1467	40.0	3.80	332	473	623	776	930	.64	40.7
APR	1956	1223	51.7	3.40	83	161	270	405	551	.64	28.5
MAY	2162	1034	61.1	2.30	0	13	60	148	281	.62	19.3
JUN	2434	1025	70.7	3.00	0	0	7	28	84	.67	14.9
JUL	2312	1030	76.4	1.90	0	0	0	0	6	.65	16.9
AUG	2102	1167	74.5	1.70	0	0	0	0	10	.66	24.8
SEP	1779	1468	66.2	2.30	0	0	11	55	140	.67	36.5
OCT	1361	1669	54.5	3.10	43	110	205	335	483	.67	48.3
NOV	954	1518	40.8	3.60	296	432	578	727	876	.64	57.7
DEC	782	1364	33.0	3.50	530	683	837	992	1147	.62	61.7
YR	1624	1335	52.8	3.18	2354	3223	4231	5398	6734	.65	

HARTFORD, CONNECTICUT ELEV 180 LAT 41.9

	HS	VS	TA	SD	D50	D55	D60	D65	D70	KT	L-D
JAN	477	694	24.8	4.20	783	937	1091	1246	1401	.40	63.2
FEB	715	891	26.8	3.15	650	790	930	1070	1210	.43	55.5
MAR	978	900	35.6	3.68	454	604	757	912	1066	.42	44.3
APR	1315	888	47.7	2.83	126	235	373	520	669	.44	32.1
MAY	1568	859	58.3	2.73	11	45	118	226	367	.45	22.9
JUN	1686	853	67.8	1.90	0	0	0	24	96	.46	18.5
JUL	1649	861	72.7	1.69	0	0	0	0	21	.47	20.5
AUG	1422	880	70.4	2.06	0	0	0	12	62	.45	28.4
SEP	1154	967	62.8	2.23	0	6	33	106	223	.46	40.1
OCT	853	991	52.6	2.28	39	116	237	386	540	.46	51.9
NOV	497	679	41.3	2.52	268	412	561	711	861	.38	61.3
DEC	385	562	28.2	3.72	677	831	986	1141	1296	.36	65.3
YR	1059	835	49.1	2.75	3008	3976	5086	6354	7812	.44	

WILMINGTON, DELAWARE ELEV 79 LAT 39.7

	HS	VS	TA	SD	D50	D55	D60	D65	D70	KT	L-D
JAN	571	819	32.0	4.26	565	716	869	1023	1178	.43	61.0
FEB	827	1006	33.6	3.19	461	600	739	879	1019	.47	53.3
MAR	1149	1032	41.6	3.70	288	425	574	726	881	.48	42.1
APR	1480	952	52.3	2.61	50	126	242	383	531	.49	29.9
MAY	1710	887	62.4	2.56	0	12	50	128	247	.49	20.7
JUN	1883	895	71.4	1.74	0	0	0	0	35	.52	16.3
JUL	1823	898	75.8	1.72	0	0	0	0	5	.51	18.3
AUG	1615	947	74.1	1.71	0	0	0	0	12	.51	26.2
SEP	1318	1064	67.9	2.24	0	0	6	32	104	.51	37.9
OCT	984	1113	57.2	2.67	15	56	136	254	399	.50	49.7
NOV	645	901	45.7	2.09	146	281	429	579	729	.45	59.1
DEC	489	720	34.7	3.62	480	631	785	939	1094	.41	63.1
YR	1209	935	54.0	2.68	2005	2847	3830	4943	6234	.49	

WASHINGTON, DC ELEV 289 LAT 38.9

	HS	VS	TA	SD	D50	D55	D60	D65	D70	KT	L-D
JAN	572	793	32.1	4.36	563	713	866	1020	1175	.42	60.2
FEB	815	956	33.8	3.33	456	594	734	874	1014	.45	52.5
MAR	1125	979	41.8	4.07	292	424	569	721	875	.46	41.3
APR	1459	919	53.1	2.53	39	110	219	359	507	.48	29.1
MAY	1718	877	62.6	2.71	0	14	52	129	244	.50	19.9
JUN	1901	890	71.1	1.71	0	0	0	0	38	.52	15.5
JUL	1817	883	75.3	1.60	0	0	0	0	5	.51	17.5
AUG	1617	929	73.6	1.57	0	0	0	0	12	.51	25.4
SEP	1340	1058	66.9	2.28	0	0	9	45	125	.51	37.1
OCT	1004	1111	55.9	2.81	24	79	166	292	439	.50	48.9
NOV	651	883	44.7	2.22	173	311	459	609	759	.45	58.3
DEC	481	678	34.0	3.26	499	652	806	961	1116	.39	62.3
YR	1210	912	53.7	2.70	2046	2897	3880	5010	6309	.49	

APALACHICOLA, FLORIDA ELEV 20 LAT 29.7

	HS	VS	TA	SD	D50	D55	D60	D65	D70	KT	L-D
JAN	853	967	53.7	3.90	76	150	243	368	512	.46	51.0
FEB	1126	1061	55.8	4.50	57	117	191	290	410	.50	43.3
MAR	1474	1016	60.7	3.00	8	30	89	175	300	.53	32.1
APR	1879	928	68.3	2.30	0	0	5	30	99	.58	19.9
MAY	2091	870	74.9	1.90	0	0	0	0	11	.60	10.7
JUN	1998	844	80.0	1.60	0	0	0	0	0	.56	6.3
JUL	1814	801	81.4	1.10	0	0	0	0	0	.52	8.3
AUG	1688	804	81.5	1.10	0	0	0	0	0	.51	16.2
SEP	1535	938	78.6	1.50	0	0	0	0	0	.53	27.9
OCT	1371	1204	70.8	2.70	0	0	5	22	77	.57	39.7
NOV	1040	1188	61.1	2.90	6	23	75	158	278	.54	49.1
DEC	818	991	55.2	3.30	42	106	194	318	463	.47	53.1
YR	1475	967	68.5	2.47	189	426	802	1361	2150	.54	

DAYTONA BEACH, FLORIDA ELEV 39 LAT 29.2

	HS	VS	TA	SD	D50	D55	D60	D65	D70	KT	L-D
JAN	958	1116	58.4	3.56	25	69	143	241	372	.51	50.5
FEB	1213	1150	59.6	4.42	28	67	131	210	316	.54	42.8
MAR	1548	1056	63.9	3.10	0	15	49	120	219	.56	31.6
APR	1884	919	69.7	2.19	0	0	0	17	73	.58	19.4
MAY	1968	840	75.0	1.39	0	0	0	0	0	.56	10.2
JUN	1826	807	79.4	1.56	0	0	0	0	0	.51	5.8
JUL	1784	792	81.0	1.00	0	0	0	0	0	.51	7.8
AUG	1682	796	81.1	1.01	0	0	0	0	0	.51	15.7
SEP	1478	892	79.5	1.14	0	0	0	0	0	.51	27.4
OCT	1251	1054	73.3	2.17	0	0	0	5	29	.52	39.2
NOV	1035	1157	65.1	3.16	0	11	36	97	185	.53	48.6
DEC	870	1060	59.6	3.48	18	53	121	212	337	.50	52.6
YR	1459	969	70.5	2.33	71	215	480	902	1531	.53	

JACKSONVILLE, FLORIDA ELEV 30 LAT 30.5

	HS	VS	TA	SD	D50	D55	D60	D65	D70	KT	L-D
JAN	900	1076	54.6	4.11	72	142	230	348	487	.50	51.8
FEB	1164	1141	56.3	4.66	57	115	187	282	399	.53	44.1
MAR	1522	1079	61.2	3.35	11	35	93	176	292	.56	32.9
APR	1856	936	68.1	1.99	0	0	0	24	93	.58	20.7
MAY	1956	847	74.3	1.57	0	0	0	0	8	.56	11.5
JUN	1885	822	79.2	1.79	0	0	0	0	0	.53	7.1
JUL	1802	800	81.0	1.11	0	0	0	0	0	.51	9.1
AUG	1694	816	81.0	1.03	0	0	0	0	0	.51	17.0
SEP	1442	900	78.2	1.49	0	0	0	0	0	.50	28.7
OCT	1223	1070	70.5	2.47	0	0	0	19	74	.52	40.5
NOV	996	1153	61.2	3.08	7	26	80	161	278	.53	49.9
DEC	818	1024	55.4	3.56	47	111	198	317	459	.48	53.9
YR	1439	971	68.4	2.50	194	429	788	1327	2090	.53	

MIAMI, FLORIDA ELEV 7 LAT 25.8

	HS	VS	TA	SD	D50	D55	D60	D65	D70	KT	L-D
JAN	1057	1121	67.2	2.57	0	0	13	53	132	.52	47.1
FEB	1314	1130	67.8	3.64	0	8	24	67	136	.55	39.4
MAR	1603	990	71.3	2.56	0	0	0	17	65	.56	28.2
APR	1859	853	75.0	1.92	0	0	0	0	10	.57	16.0
MAY	1844	800	78.0	1.22	0	0	0	0	0	.53	6.8
JUN	1708	787	81.0	1.04	0	0	0	0	0	.48	2.4
JUL	1763	787	82.3	1.21	0	0	0	0	0	.51	4.4
AUG	1630	751	82.9	.96	0	0	0	0	0	.49	12.3
SEP	1456	811	81.7	.85	0	0	0	0	0	.49	24.0
OCT	1303	991	77.8	1.60	0	0	0	0	0	.51	35.8
NOV	1119	1131	72.2	2.66	0	0	0	13	53	.53	45.2
DEC	1019	1157	68.3	3.11	0	5	17	56	130	.53	49.2
YR	1473	941	75.5	1.93	0	13	54	206	526	.52	

ORLANDO, FLORIDA ELEV 118 LAT 28.5

	HS	VS	TA	SD	D50	D55	D60	D65	D70	KT	L-D
JAN	999	1151	60.3	3.48	15	46	110	197	318	.52	49.8
FEB	1243	1158	61.5	4.64	23	54	111	184	277	.54	42.1
MAR	1582	1058	65.9	3.26	0	11	35	94	179	.57	30.9
APR	1898	910	71.3	2.40	0	0	0	13	57	.59	18.7
MAY	1989	840	76.4	1.39	0	0	0	0	0	.57	9.5
JUN	1831	809	80.2	1.48	0	0	0	0	0	.51	5.1
JUL	1801	795	81.4	1.04	0	0	0	0	0	.51	7.1
AUG	1673	786	81.8	.96	0	0	0	0	0	.51	15.0
SEP	1497	887	80.1	1.12	0	0	0	0	0	.51	26.7
OCT	1304	1084	74.3	1.99	0	0	0	0	16	.53	38.5
NOV	1096	1219	66.6	3.12	0	7	25	75	154	.55	47.9
DEC	926	1126	61.5	3.36	10	33	89	170	284	.52	51.9
YR	1487	984	71.8	2.34	48	151	370	733	1285	.54	

TALLAHASSEE, FLORIDA ELEV 69 LAT 30.4

	HS	VS	TA	SD	D50	D55	D60	D65	D70	KT	L-D
JAN	877	1033	52.6	4.50	110	188	284	408	549	.48	51.7
FEB	1138	1104	54.8	5.10	83	147	224	323	442	.52	44.0
MAR	1479	1042	60.3	3.10	10	36	98	187	313	.54	32.8
APR	1823	920	67.9	2.30	0	0	6	34	106	.57	20.6
MAY	1936	842	74.8	1.70	0	0	0	0	8	.56	11.4
JUN	1883	821	80.0	1.60	0	0	0	0	0	.53	7.0
JUL	1748	786	81.1	1.50	0	0	0	0	0	.50	9.0
AUG	1675	808	81.1	1.00	0	0	0	0	0	.51	16.9
SEP	1493	930	78.1	1.60	0	0	0	0	0	.52	28.6
OCT	1318	1174	69.3	2.60	0	0	7	31	97	.56	40.4
NOV	1008	1168	58.9	2.80	9	38	105	204	338	.53	49.8
DEC	813	1011	53.2	3.40	67	141	242	376	524	.48	53.8
YR	1433	969	67.7	2.59	279	550	966	1563	2377	.53	

TAMPA, FLORIDA ELEV 10 LAT 28.0

	HS	VS	TA	SD	D50	D55	D60	D65	D70	KT	L-D
JAN	1011	1148	60.4	3.73	19	52	117	203	320	.52	49.3
FEB	1259	1156	61.8	4.55	21	50	105	176	269	.54	41.6
MAR	1594	1051	66.0	3.19	0	10	32	90	175	.57	30.4
APR	1908	904	72.0	2.31	0	0	0	9	45	.59	18.2
MAY	1998	839	77.2	1.33	0	0	0	0	0	.57	9.0
JUN	1847	813	81.0	1.28	0	0	0	0	0	.52	4.6
JUL	1753	783	81.9	.93	0	0	0	0	0	.50	6.6
AUG	1653	775	82.2	.91	0	0	0	0	0	.50	14.5
SEP	1492	873	80.8	1.05	0	0	0	0	0	.51	26.2
OCT	1346	1108	74.7	2.14	0	0	0	0	17	.55	38.0
NOV	1108	1213	66.8	3.07	0	6	22	71	149	.55	47.4
DEC	935	1119	61.6	3.39	10	33	88	169	282	.51	51.4
YR	1492	981	72.2	2.31	50	151	364	718	1257	.54	

WEST PALM BEACH, FLORIDA ELEV 20 LAT 26.7

	HS	VS	TA	SD	D50	D55	D60	D65	D70	KT	L-D
JAN	1000	1075	65.5	2.74	0	6	25	83	173	.50	48.0
FEB	1233	1075	66.1	3.75	0	13	37	91	166	.52	40.3
MAR	1556	985	69.8	2.51	0	0	5	25	86	.55	29.1
APR	1814	851	73.9	2.11	0	0	0	0	20	.56	16.9
MAY	1845	801	77.5	1.29	0	0	0	0	0	.53	7.7
JUN	1706	783	80.5	.90	0	0	0	0	0	.48	3.3
JUL	1779	789	81.9	.97	0	0	0	0	0	.51	5.3
AUG	1663	767	82.3	.71	0	0	0	0	0	.50	13.2
SEP	1419	807	81.5	.64	0	0	0	0	0	.48	24.9
OCT	1224	946	77.2	1.59	0	0	0	0	0	.49	36.7
NOV	1060	1087	71.0	2.80	0	0	5	22	74	.51	46.1
DEC	958	1100	66.8	3.17	0	8	26	78	159	.51	50.1
YR	1439	922	74.5	1.92	0	27	98	299	678	.51	

ATLANTA, GEORGIA ELEV 1033 LAT 33.6

	HS	VS	TA	SD	D50	D55	D60	D65	D70	KT	L-D
JAN	718	884	42.4	4.33	284	410	553	704	857	.44	54.9
FEB	969	998	45.0	4.50	205	308	431	564	702	.47	47.2
MAR	1304	993	51.1	4.03	117	198	308	443	590	.50	36.0
APR	1686	927	61.1	2.38	0	13	59	144	272	.53	23.8
MAY	1854	851	69.1	2.38	0	0	5	27	93	.53	14.6
JUN	1914	842	75.6	2.46	0	0	0	0	17	.53	10.2
JUL	1812	821	78.0	1.72	0	0	0	0	0	.51	12.2
AUG	1708	867	77.5	1.74	0	0	0	0	0	.52	20.1
SEP	1422	966	72.3	2.28	0	0	0	8	40	.51	31.8
OCT	1200	1165	62.4	2.84	0	17	59	137	251	.54	43.6
NOV	883	1101	51.4	2.69	65	146	267	410	558	.51	53.0
DEC	674	881	43.5	3.58	240	369	516	668	822	.44	57.0
YR	1346	941	60.8	2.90	911	1461	2198	3105	4202	.51	

AUGUSTA, GEORGIA ELEV 148 LAT 33.4

	HS	VS	TA	SD	D50	D55	D60	D65	D70	KT	L-D
JAN	751	934	45.8	4.35	214	322	455	601	752	.45	54.7
FEB	1015	1055	48.3	4.46	151	236	347	475	611	.49	47.0
MAR	1338	1018	54.6	3.99	68	138	227	346	486	.51	35.8
APR	1728	944	63.8	2.27	0	0	25	90	197	.55	23.6
MAY	1865	852	71.7	2.29	0	0	0	10	51	.53	14.4
JUN	1904	838	78.2	2.32	0	0	0	0	6	.53	10.0
JUL	1803	817	80.4	1.67	0	0	0	0	0	.51	12.0
AUG	1667	847	79.6	1.74	0	0	0	0	0	.51	19.9
SEP	1410	951	74.2	2.21	0	0	0	0	21	.50	31.6
OCT	1220	1182	64.1	2.72	0	9	36	104	205	.55	43.4
NOV	916	1151	53.7	2.87	42	111	210	344	490	.52	52.8
DEC	721	962	46.4	3.51	177	290	429	579	732	.47	56.8
YR	1362	962	63.4	2.86	652	1106	1729	2549	3551	.51	

MACON, GEORGIA ELEV 361 LAT 32.7

	HS	VS	TA	SD	D50	D55	D60	D65	D70	KT	L-D
JAN	769	940	47.8	4.52	185	280	403	543	692	.45	54.0
FEB	1020	1035	50.4	4.72	128	204	303	423	555	.49	46.3
MAR	1363	1018	56.5	4.05	50	111	192	298	432	.51	35.1
APR	1736	932	65.8	2.47	0	0	16	66	153	.55	22.9
MAY	1885	850	73.5	2.32	0	0	0	6	31	.54	13.7
JUN	1919	838	79.6	2.18	0	0	0	0	0	.53	9.3
JUL	1785	807	81.4	1.58	0	0	0	0	0	.50	11.3
AUG	1718	856	80.9	1.66	0	0	0	0	0	.52	19.2
SEP	1439	953	75.8	2.28	0	0	0	0	13	.51	30.9
OCT	1247	1186	65.7	2.79	0	6	25	82	170	.55	42.7
NOV	940	1161	55.2	3.03	32	92	180	304	446	.53	52.1
DEC	729	950	48.3	3.45	141	241	373	521	674	.46	56.1
YR	1380	960	65.1	2.91	536	934	1492	2243	3166	.52	

SAVANNAH, GEORGIA ELEV 52 LAT 32.1

	HS	VS	TA	SD	D50	D55	D60	D65	D70	KT	L-D
JAN	795	962	49.9	4.54	151	237	349	483	629	.46	53.4
FEB	1044	1046	52.1	4.61	103	173	264	379	508	.49	45.7
MAR	1398	1029	58.0	3.79	32	82	158	256	386	.52	34.5
APR	1761	930	66.1	2.51	0	0	15	63	147	.55	22.3
MAY	1852	835	73.3	1.90	0	0	0	0	21	.53	13.1
JUN	1844	817	79.1	2.11	0	0	0	0	0	.51	8.7
JUL	1783	803	81.1	1.29	0	0	0	0	0	.50	10.7
AUG	1621	810	80.6	1.25	0	0	0	0	0	.49	18.6
SEP	1364	885	76.2	1.65	0	0	0	0	0	.48	30.3
OCT	1217	1125	67.1	2.74	0	0	16	60	140	.53	42.1
NOV	941	1134	57.1	2.99	19	64	141	253	391	.52	51.5
DEC	754	972	50.4	3.46	108	195	315	458	609	.47	55.5
YR	1365	945	65.9	2.73	413	751	1258	1952	2831	.51	

BOISE, IDAHO ELEV 2867 LAT 43.6

	HS	VS	TA	SD	D50	D55	D60	D65	D70	KT	L-D
JAN	485	770	29.0	5.90	667	814	965	1118	1272	.44	64.9
FEB	840	1207	35.5	4.50	418	551	688	827	966	.54	57.2
MAR	1304	1402	41.1	2.60	284	433	586	741	896	.58	46.0
APR	1827	1355	49.0	2.80	104	202	335	481	630	.62	33.8
MAY	2277	1252	57.4	2.90	18	61	139	252	395	.66	24.6
JUN	2463	1183	64.8	3.00	0	9	35	97	188	.68	20.2
JUL	2613	1309	74.5	1.80	0	0	0	0	11	.74	22.2
AUG	2196	1460	72.2	2.50	0	0	0	12	51	.71	30.1
SEP	1737	1749	63.1	3.10	0	17	55	127	229	.71	41.8
OCT	1138	1610	52.1	3.16	74	153	265	406	556	.64	53.6
NOV	628	1034	39.8	2.91	314	458	606	756	906	.52	63.0
DEC	437	735	32.1	2.20	555	710	865	1020	1175	.44	67.0
YR	1498	1255	50.9	3.10	2434	3408	4539	5837	7275	.64	

LEWISTON, IDAHO ELEV 1437 LAT 46.4

	HS	VS	TA	SD	D50	D55	D60	D65	D70	KT	L-D
JAN	340	523	31.2	6.27	610	752	900	1052	1205	.35	67.7
FEB	609	849	38.1	3.52	341	476	614	753	893	.43	60.0
MAR	1020	1108	42.9	2.81	238	379	531	685	840	.48	48.8
APR	1435	1114	50.3	2.63	78	168	297	442	591	.50	36.6
MAY	1842	1111	58.1	2.80	13	49	124	232	373	.54	27.4
JUN	2015	1083	65.0	2.69	0	6	26	84	176	.55	23.0
JUL	2336	1304	73.4	2.01	0	0	0	0	23	.66	25.0
AUG	1931	1396	71.5	2.61	0	0	0	17	64	.63	32.9
SEP	1435	1502	63.3	3.10	0	16	53	124	225	.61	44.6
OCT	860	1198	51.8	2.76	65	146	266	412	565	.53	56.4
NOV	413	633	40.5	3.04	296	438	586	735	885	.39	65.8
DEC	286	454	34.8	3.10	474	627	781	936	1091	.34	69.8
YR	1213	1024	51.7	3.11	2115	3057	4178	5472	6931	.54	

POCATELLO, IDAHO ELEV 4478 LAT 42.9

	HS	VS	TA	SD	D50	D55	D60	D65	D70	KT	L-D
JAN	539	871	23.2	6.40	841	991	1144	1297	1452	.47	64.2
FEB	882	1259	29.4	5.50	586	721	859	998	1137	.55	56.5
MAR	1371	1463	35.4	4.10	464	612	764	918	1073	.60	45.3
APR	1820	1317	45.3	3.40	187	306	445	592	741	.61	33.1
MAY	2280	1227	54.4	2.90	38	105	202	336	485	.66	23.9
JUN	2480	1165	61.8	2.70	0	16	59	138	256	.68	19.5
JUL	2600	1273	71.5	1.70	0	0	0	0	35	.74	21.5
AUG	2239	1454	69.5	2.20	0	0	0	20	81	.72	29.4
SEP	1769	1744	59.4	2.80	8	33	97	192	324	.71	41.1
OCT	1203	1697	48.4	3.20	132	233	368	517	670	.67	52.9
NOV	689	1151	35.7	3.20	433	580	729	879	1029	.55	62.3
DEC	477	809	26.9	3.50	717	871	1026	1181	1336	.47	66.3
YR	1532	1285	46.7	3.45	3406	4468	5693	7068	8619	.65	

CHICAGO, ILLINOIS ELEV 623 LAT 41.8

	HS	VS	TA	SD	D50	D55	D60	D65	D70	KT	L-D
JAN	507	756	24.3	4.89	800	953	1107	1262	1417	.42	63.1
FEB	759	967	27.4	4.09	634	773	913	1053	1193	.46	55.4
MAR	1107	1054	36.8	4.81	434	575	724	876	1030	.48	44.2
APR	1459	993	49.9	3.09	98	188	313	456	604	.49	32.0
MAY	1789	963	60.0	3.50	17	49	115	204	327	.52	22.8
JUN	2007	973	70.5	2.75	0	0	6	24	79	.55	18.4
JUL	1944	984	74.7	2.10	0	0	0	0	16	.55	20.4
AUG	1719	1065	73.7	2.57	0	0	0	8	36	.55	28.3
SEP	1354	1179	65.9	2.21	0	0	11	57	144	.54	40.0
OCT	969	1181	55.4	3.47	45	108	195	315	458	.52	51.8
NOV	566	815	40.4	3.47	304	443	589	738	888	.43	61.2
DEC	401	593	28.5	4.85	673	824	978	1132	1287	.37	65.2
YR	1217	960	50.6	3.48	3005	3913	4951	6125	7479	.51	

MOLINE, ILLINOIS ELEV 594 LAT 41.4

	HS	VS	TA	SD	D50	D55	D60	D65	D70	KT	L-D
JAN	535	803	21.5	4.68	885	1039	1194	1349	1504	.44	62.7
FEB	812	1048	25.7	4.36	682	821	961	1100	1240	.48	55.0
MAR	1119	1055	35.7	5.21	471	611	759	911	1064	.48	43.8
APR	1459	982	50.6	3.29	95	178	296	436	583	.49	31.6
MAY	1754	938	61.1	3.55	14	41	101	184	298	.51	22.4
JUN	1969	952	70.8	2.64	0	0	0	20	71	.54	18.0
JUL	1939	974	74.5	2.18	0	0	0	0	19	.55	20.0
AUG	1715	1051	72.9	2.58	0	0	0	11	45	.54	27.9
SEP	1357	1166	64.6	2.33	0	0	20	79	178	.53	39.6
OCT	996	1209	54.4	3.57	58	127	219	344	489	.53	51.4
NOV	595	862	39.2	3.40	336	477	625	774	924	.45	60.8
DEC	433	651	26.6	4.91	730	883	1036	1191	1346	.39	64.8
YR	1225	973	49.8	3.56	3271	4177	5211	6399	7761	.51	

SPRINGFIELD, ILLINOIS ELEV 614 LAT 39.8

	HS	VS	TA	SD	D50	D55	D60	D65	D70	KT	L-D
JAN	585	852	26.7	4.34	725	878	1033	1187	1342	.45	61.1
FEB	861	1069	30.4	4.04	551	690	829	969	1109	.49	53.4
MAR	1143	1028	39.4	5.39	380	509	650	799	951	.48	42.2
APR	1515	979	53.1	3.13	56	128	230	363	509	.50	30.0
MAY	1865	954	63.4	3.21	5	19	58	132	233	.54	20.8
JUN	2097	966	72.9	2.73	0	0	0	12	46	.58	16.4
JUL	2058	984	76.1	2.31	0	0	0	0	13	.58	18.4
AUG	1806	1058	74.4	2.73	0	0	0	8	34	.57	26.3
SEP	1454	1205	67.2	2.51	0	0	11	48	125	.56	38.0
OCT	1068	1254	56.6	3.37	32	88	169	282	422	.55	49.8
NOV	677	971	41.9	3.16	261	398	544	693	843	.48	59.2
DEC	490	725	30.5	4.96	615	764	916	1070	1225	.41	63.2
YR	1303	1003	52.7	3.49	2625	3474	4440	5563	6852	.53	

EVANSVILLE, INDIANA ELEV 387 LAT 38.0

	HS	VS	TA	SD	D50	D55	D60	D65	D70	KT	L-D
JAN	574	767	32.6	4.43	548	698	851	1005	1160	.41	59.3
FEB	823	937	35.9	4.56	408	540	677	816	955	.45	51.6
MAR	1151	979	44.3	5.36	274	382	511	653	802	.47	40.4
APR	1501	925	56.7	3.01	22	70	149	263	403	.49	28.2
MAY	1783	888	65.7	3.21	0	11	35	95	181	.51	19.0
JUN	1983	903	74.7	2.50	0	0	0	5	24	.55	14.6
JUL	1920	905	77.8	2.19	0	0	0	0	6	.54	16.6
AUG	1735	971	76.2	2.43	0	0	0	0	15	.54	24.5
SEP	1403	1087	69.1	2.72	0	0	8	34	100	.53	36.2
OCT	1087	1199	58.2	3.12	18	58	132	236	373	.54	48.0
NOV	682	909	44.9	2.98	185	311	455	604	753	.45	57.4
DEC	499	688	35.3	4.72	474	619	769	922	1076	.39	61.4
YR	1263	929	56.0	3.43	1929	2689	3587	4633	5848	.50	

FORT WAYNE, INDIANA ELEV 827 LAT 41.0

	HS	VS	TA	SD	D50	D55	D60	D65	D70	KT	L-D
JAN	455	624	25.3	4.55	768	922	1076	1231	1386	.36	62.3
FEB	698	833	27.6	3.87	628	768	907	1047	1187	.41	54.6
MAR	982	880	36.5	4.81	442	584	733	885	1039	.42	43.4
APR	1361	901	49.3	3.17	110	203	330	473	622	.45	31.2
MAY	1672	893	59.6	3.60	20	56	125	216	339	.48	22.0
JUN	1842	900	69.5	2.39	0	0	0	23	83	.51	17.6
JUL	1787	905	73.0	1.86	0	0	0	0	23	.50	19.6
AUG	1594	966	71.3	2.29	0	0	0	12	56	.50	27.5
SEP	1274	1064	64.5	2.62	0	6	28	90	186	.50	39.2
OCT	924	1071	53.6	3.27	58	131	230	363	511	.49	51.0
NOV	516	690	40.2	3.00	303	446	595	744	894	.38	60.4
DEC	369	509	28.6	4.81	670	821	975	1129	1284	.33	64.4
YR	1124	853	49.9	3.36	2999	3937	4999	6213	7610	.46	

INDIANAPOLIS, INDIANA ELEV 807 LAT 39.7

	HS	VS	TA	SD	D50	D55	D60	D65	D70	KT	L-D
JAN	496	668	27.9	4.80	691	842	996	1151	1305	.38	61.0
FEB	747	872	30.7	4.20	544	682	821	961	1100	.42	53.3
MAR	1037	905	39.7	5.10	366	496	639	789	941	.43	42.1
APR	1398	897	52.3	3.20	68	144	251	387	533	.46	29.9
MAY	1688	877	62.2	3.40	9	29	80	159	267	.49	20.7
JUN	1868	890	71.7	2.40	0	0	0	11	51	.51	16.3
JUL	1806	891	75.0	2.10	0	0	0	0	15	.51	18.3
AUG	1643	962	73.2	2.20	0	0	0	5	30	.52	26.2
SEP	1324	1070	66.3	2.60	0	0	16	63	145	.51	37.9
OCT	977	1102	55.7	3.10	32	92	178	302	447	.50	49.7
NOV	579	771	41.7	3.10	265	403	550	699	849	.41	59.1
DEC	417	572	30.9	5.00	603	752	904	1058	1212	.35	63.1
YR	1167	873	52.3	3.43	2578	3440	4435	5585	6895	.47	

SOUTH BEND, INDIANA ELEV 774 LAT 41.7

	HS	VS	TA	SD	D50	D55	D60	D65	D70	KT	L-D
JAN	416	566	24.0	4.62	808	962	1116	1271	1426	.34	63.0
FEB	660	791	26.3	3.87	664	804	944	1084	1224	.40	55.3
MAR	992	911	35.3	4.80	475	619	769	922	1076	.43	44.1
APR	1387	936	48.1	3.09	127	229	363	509	657	.46	31.9
MAY	1722	929	58.4	3.69	28	73	148	245	374	.50	22.7
JUN	1922	940	68.6	2.58	0	0	8	35	103	.53	18.3
JUL	1852	944	72.3	2.05	0	0	0	6	36	.52	20.3
AUG	1666	1029	71.0	2.85	0	0	6	24	79	.53	28.2
SEP	1291	1106	63.8	2.52	0	7	31	98	201	.51	39.9
OCT	909	1075	53.4	3.20	58	132	233	368	517	.49	51.7
NOV	497	673	39.6	3.15	322	465	613	762	912	.38	61.1
DEC	340	467	28.2	4.71	681	833	987	1141	1296	.31	65.1
YR	1140	864	49.1	3.43	3163	4124	5218	6465	7901	.48	

BURLINGTON, IOWA ELEV 702 LAT 40.8

	HS	VS	TA	SD	D50	D55	D60	D65	D70	KT	L-D
JAN	579	878	22.9	4.50	842	996	1150	1305	1460	.46	62.1
FEB	859	1109	27.3	4.30	638	776	916	1056	1196	.50	54.4
MAR	1165	1090	36.9	5.40	442	579	724	875	1028	.49	43.2
APR	1538	1023	51.3	3.30	85	165	278	416	562	.51	31.0
MAY	1876	981	61.8	3.60	12	36	92	172	281	.54	21.8
JUN	2121	992	71.4	2.60	0	0	0	16	62	.58	17.4
JUL	2085	1016	75.4	2.30	0	0	0	0	16	.59	19.4
AUG	1828	1101	73.9	2.60	0	0	0	8	35	.58	27.3
SEP	1416	1206	65.4	2.40	0	0	17	70	160	.55	39.0
OCT	1061	1293	55.3	3.60	49	114	201	320	462	.56	50.8
NOV	664	987	39.8	3.40	319	460	607	756	906	.49	60.2
DEC	481	739	27.6	5.00	701	852	1006	1160	1315	.42	64.2
YR	1308	1034	50.8	3.58	3088	3978	4991	6154	7483	.54	

DES MOINES, IOWA ELEV 965 LAT 41.5

	HS	VS	TA	SD	D50	D55	D60	D65	D70	KT	L-D
JAN	581	912	19.4	4.53	949	1104	1259	1414	1569	.48	62.8
FEB	861	1145	24.2	4.57	724	863	1003	1142	1282	.51	55.1
MAR	1180	1135	33.9	5.93	531	670	817	968	1121	.51	43.9
APR	1557	1057	49.5	3.43	115	205	328	469	616	.52	31.7
MAY	1867	993	60.9	3.49	14	41	102	186	302	.54	22.5
JUN	2125	1008	70.5	2.86	0	0	6	26	83	.58	18.1
JUL	2097	1037	75.1	2.37	0	0	0	0	20	.59	20.1
AUG	1828	1124	73.3	2.82	0	0	0	13	47	.58	28.0
SEP	1434	1256	64.3	2.64	0	7	30	94	191	.57	39.7
OCT	1068	1342	54.3	3.76	64	135	226	350	493	.57	51.5
NOV	658	1005	37.8	3.60	376	519	667	816	966	.50	60.9
DEC	487	778	25.0	4.73	778	931	1086	1240	1395	.44	64.9
YR	1314	1065	49.0	3.73	3551	4475	5524	6718	8085	.55	

MASON CITY, IOWA ELEV 1224 LAT 43.1

	HS	VS	TA	SD	D50	D55	D60	D65	D70	KT	L-D
JAN	554	917	14.2	4.97	1110	1265	1420	1575	1730	.49	64.4
FEB	836	1173	18.5	4.45	882	1022	1162	1302	1442	.53	56.7
MAR	1168	1184	29.0	5.95	668	814	965	1118	1272	.52	45.5
APR	1519	1078	45.7	3.84	192	303	437	582	730	.51	33.3
MAY	1895	1046	57.4	3.55	31	82	160	265	400	.55	24.1
JUN	2114	1041	67.2	3.03	0	5	19	64	141	.58	19.7
JUL	2084	1074	71.3	2.34	0	0	0	13	58	.59	21.7
AUG	1833	1183	69.9	2.78	0	0	8	31	93	.59	29.6
SEP	1405	1294	60.2	2.43	0	19	73	165	298	.57	41.3
OCT	1010	1322	50.5	3.74	116	201	317	457	607	.56	53.1
NOV	600	943	33.6	4.03	499	644	793	942	1092	.48	62.5
DEC	443	731	20.1	5.16	929	1083	1237	1392	1547	.44	66.5
YR	1290	1081	44.8	3.86	4427	5438	6591	7906	9410	.55	

SIOUX CITY, IOWA ELEV 1102 LAT 42.4

	HS	VS	TA	SD	D50	D55	D60	D65	D70	KT	L-D
JAN	569	922	18.0	4.70	993	1147	1302	1457	1612	.48	63.7
FEB	842	1151	23.4	4.90	747	886	1025	1165	1305	.52	56.0
MAR	1170	1158	33.2	6.20	554	693	840	990	1143	.51	44.8
APR	1578	1102	49.4	3.80	129	216	336	474	620	.53	32.6
MAY	1901	1031	60.9	3.60	15	44	105	189	304	.55	23.4
JUN	2124	1028	70.3	3.10	0	0	9	33	93	.58	19.0
JUL	2122	1071	75.3	2.50	0	0	0	0	21	.60	21.0
AUG	1845	1166	73.5	2.70	0	0	0	10	42	.59	28.9
SEP	1421	1281	63.4	2.80	0	11	43	113	216	.57	40.6
OCT	1038	1336	53.1	3.30	65	140	242	378	526	.57	52.4
NOV	643	1012	36.3	3.30	416	562	711	861	1011	.50	61.8
DEC	469	769	23.5	4.60	824	977	1132	1287	1442	.45	65.8
YR	1312	1085	48.4	3.79	3743	4676	5745	6957	8335	.55	

DODGE CITY, KANSAS ELEV 2582 LAT 37.8

	HS	VS	TA	SD	D50	D55	D60	D65	D70	KT	L-D
JAN	827	1303	30.8	3.80	598	751	906	1060	1215	.58	59.1
FEB	1122	1444	35.2	4.35	424	558	696	835	975	.60	51.4
MAR	1476	1335	41.2	5.16	332	456	596	744	895	.60	40.2
APR	1886	1160	54.0	3.64	60	130	221	344	485	.61	28.0
MAY	2090	997	64.0	3.00	0	13	45	115	214	.60	18.8
JUN	2358	998	73.7	3.51	0	0	7	21	60	.65	14.4
JUL	2295	1013	79.2	2.66	0	0	0	0	8	.65	16.4
AUG	2055	1126	78.1	2.59	0	0	0	0	10	.64	24.3
SEP	1687	1350	68.9	2.91	0	0	11	41	109	.63	36.0
OCT	1301	1532	57.9	3.30	23	67	143	247	383	.64	47.8
NOV	894	1343	42.8	2.99	235	371	517	666	816	.59	57.2
DEC	732	1202	33.4	3.48	518	671	825	980	1135	.56	61.2
YR	1562	1232	54.9	3.44	2190	3017	3967	5053	6305	.62	

GOODLAND, KANSAS ELEV 3688 LAT 39.4

	HS	VS	TA	SD	D50	D55	D60	D65	D70	KT	L-D
JAN	789	1310	27.6	4.71	699	851	1005	1160	1315	.59	60.7
FEB	1056	1414	31.5	4.83	526	661	799	939	1078	.59	53.0
MAR	1424	1350	36.3	4.81	448	590	739	892	1045	.59	41.8
APR	1829	1181	48.7	3.69	136	229	353	493	640	.60	29.6
MAY	2062	1025	58.9	3.01	13	46	117	216	352	.59	20.4
JUN	2357	1031	69.1	3.48	0	6	19	55	123	.65	16.0
JUL	2319	1060	75.8	2.46	0	0	0	0	17	.65	18.0
AUG	2046	1178	74.1	2.26	0	0	0	0	24	.64	25.9
SEP	1642	1383	64.3	3.08	0	12	42	108	201	.63	37.6
OCT	1268	1578	52.8	3.37	71	147	251	387	536	.64	49.4
NOV	857	1357	38.5	3.12	352	497	645	795	945	.60	58.8
DEC	695	1202	30.1	3.28	618	772	927	1082	1237	.57	62.8
YR	1530	1255	50.6	3.50	2863	3811	4897	6127	7513	.62	

TOPEKA, KANSAS ELEV 886 LAT 39.1

	HS	VS	TA	SD	D50	D55	D60	D65	D70	KT	L-D
JAN	681	1033	28.0	3.85	684	838	992	1147	1302	.50	60.4
FEB	941	1181	33.4	4.17	471	607	746	885	1025	.52	52.7
MAR	1257	1135	41.2	5.44	340	461	599	745	896	.52	41.5
APR	1642	1045	54.5	3.50	51	118	207	329	469	.54	29.3
MAY	1915	960	64.5	3.35	5	16	50	118	211	.55	20.1
JUN	2126	962	73.5	3.00	0	0	0	13	48	.58	15.7
JUL	2128	993	78.2	2.93	0	0	0	0	15	.60	17.7
AUG	1910	1094	77.2	2.69	0	0	0	0	15	.60	25.6
SEP	1516	1239	68.2	3.13	0	0	17	55	127	.58	37.3
OCT	1147	1350	57.6	3.49	28	77	155	259	394	.58	49.1
NOV	772	1144	42.9	3.21	236	370	515	663	813	.53	58.5
DEC	583	907	31.8	4.18	570	721	875	1029	1184	.47	62.5
YR	1387	1086	54.3	3.58	2385	3208	4156	5243	6499	.56	

WICHITA, KANSAS ELEV 1339 LAT 37.6

	HS	VS	TA	SD	D50	D55	D60	D65	D70	KT	L-D
JAN	784	1191	31.3	3.83	583	736	890	1045	1200	.55	58.9
FEB	1058	1315	36.3	3.99	393	527	665	804	944	.57	51.2
MAR	1405	1243	43.6	5.02	277	391	526	671	822	.57	40.0
APR	1782	1088	56.6	3.56	34	89	167	275	409	.58	27.8
MAY	2036	974	66.1	3.22	0	10	32	90	174	.59	18.6
JUN	2264	972	75.8	2.94	0	0	0	7	26	.62	14.2
JUL	2238	994	80.7	3.15	0	0	0	0	10	.63	16.2
AUG	2032	1108	79.7	3.00	0	0	0	0	11	.63	24.1
SEP	1616	1272	70.6	3.13	0	0	9	32	90	.61	35.8
OCT	1250	1436	59.6	3.43	17	51	119	211	337	.61	47.6
NOV	871	1280	44.8	3.00	188	314	458	607	756	.57	57.0
DEC	690	1088	34.5	3.88	488	638	791	946	1101	.53	61.0
YR	1504	1162	56.6	3.51	1980	2756	3657	4688	5880	.60	

LEXINGTON, KENTUCKY ELEV 988 LAT 38.0

	HS	VS	TA	SD	D50	D55	D60	D65	D70	KT	L-D
JAN	546	714	32.9	5.08	546	692	843	997	1151	.39	59.3
FEB	779	868	35.3	4.94	428	558	694	833	972	.42	51.6
MAR	1099	924	43.6	5.32	286	398	530	673	823	.45	40.4
APR	1479	911	55.3	3.11	33	93	180	302	444	.48	28.2
MAY	1747	874	64.7	3.08	0	12	40	106	199	.50	19.0
JUN	1897	877	73.0	2.51	0	0	0	8	39	.52	14.6
JUL	1850	881	76.2	2.24	0	0	0	0	12	.52	16.6
AUG	1685	945	75.0	2.29	0	0	0	0	18	.53	24.5
SEP	1362	1049	68.6	2.76	0	0	10	40	109	.51	36.2
OCT	1044	1134	57.8	3.10	19	62	138	246	384	.51	48.0
NOV	657	862	44.6	2.78	187	318	463	612	762	.44	57.4
DEC	485	660	35.5	4.40	464	610	762	915	1070	.38	61.4
YR	1221	892	55.2	3.46	1963	2743	3660	4732	5983	.49	

LOUISVILLE, KENTUCKY ELEV 489 LAT 38.2

	HS	VS	TA	SD	D50	D55	D60	D65	D70	KT	L-D
JAN	545	718	33.3	4.77	531	679	830	984	1138	.39	59.5
FEB	789	890	35.8	4.69	412	544	680	819	958	.43	51.8
MAR	1102	933	44.0	5.25	276	386	518	661	811	.45	40.6
APR	1467	908	55.9	3.08	28	84	167	286	426	.48	28.4
MAY	1720	867	64.8	3.09	0	12	40	105	197	.50	19.2
JUN	1903	882	73.3	2.23	0	0	0	5	28	.52	14.8
JUL	1837	880	76.9	2.23	0	0	0	0	9	.52	16.8
AUG	1680	947	75.9	2.25	0	0	0	0	13	.52	24.7
SEP	1361	1055	69.1	2.77	0	0	9	35	101	.51	36.4
OCT	1042	1139	58.1	3.24	21	63	138	241	377	.52	48.2
NOV	653	861	45.0	2.61	174	305	451	600	750	.44	57.6
DEC	488	672	35.6	4.58	464	609	759	913	1067	.38	61.6
YR	1217	896	55.6	3.40	1906	2682	3592	4649	5875	.49	

BATON ROUGE, LOUISIANA ELEV 75 LAT 30.5

	HS	VS	TA	SD	D50	D55	D60	D65	D70	KT	L-D
JAN	785	889	51.0	4.49	133	215	321	451	596	.43	51.8
FEB	1054	1001	53.9	4.61	80	147	228	335	461	.48	44.1
MAR	1379	965	59.7	3.41	16	50	117	208	334	.51	32.9
APR	1681	864	68.4	2.44	0	0	7	33	102	.52	20.7
MAY	1871	826	74.8	1.50	0	0	0	0	5	.54	11.5
JUN	1926	831	80.3	1.47	0	0	0	0	0	.54	7.1
JUL	1746	786	82.0	1.44	0	0	0	0	0	.50	9.1
AUG	1677	810	81.6	1.35	0	0	0	0	0	.51	17.0
SEP	1464	914	77.5	1.70	0	0	0	0	0	.51	28.7
OCT	1301	1159	68.5	3.12	0	5	17	54	127	.55	40.5
NOV	920	1032	58.6	2.63	8	36	105	208	345	.49	49.9
DEC	737	883	52.9	3.01	58	134	241	381	532	.44	53.9
YR	1379	913	67.4	2.59	295	587	1036	1670	2502	.51	

LAKE CHARLES, LOUISIANA ELEV 10 LAT 30.1

	HS	VS	TA	SD	D50	D55	D60	D65	D70	KT	L-D
JAN	728	790	52.3	4.43	112	190	289	415	557	.40	51.4
FEB	1010	934	55.1	4.55	65	127	205	306	429	.45	43.7
MAR	1313	903	60.3	3.56	17	48	113	200	320	.48	32.5
APR	1570	810	68.9	2.35	0	0	5	26	91	.49	20.3
MAY	1849	818	75.2	1.65	0	0	0	0	6	.53	11.1
JUN	1970	840	80.7	1.41	0	0	0	0	0	.55	6.7
JUL	1788	795	82.4	1.11	0	0	0	0	0	.51	8.7
AUG	1657	798	82.2	1.25	0	0	0	0	0	.50	16.6
SEP	1485	917	78.4	1.51	0	0	0	0	0	.51	28.3
OCT	1381	1233	70.0	3.02	0	0	10	36	99	.58	40.1
NOV	917	1012	60.2	2.91	8	29	88	177	302	.48	49.5
DEC	706	819	54.3	2.86	38	106	203	338	488	.41	53.5
YR	1366	889	68.3	2.54	240	500	913	1498	2292	.50	

NEW ORLEANS, LOUISIANA ELEV 10 LAT 30.0

	HS	VS	TA	SD	D50	D55	D60	D65	D70	KT	L-D
JAN	835	950	52.9	4.68	112	189	282	403	542	.46	51.3
FEB	1112	1055	55.6	4.74	65	126	201	299	418	.50	43.6
MAR	1415	979	60.7	3.44	13	41	103	188	307	.52	32.4
APR	1780	895	68.6	2.37	0	0	6	29	97	.55	20.2
MAY	1968	846	75.1	1.64	0	0	0	0	6	.56	11.0
JUN	2004	846	80.4	1.36	0	0	0	0	0	.56	6.6
JUL	1813	801	81.9	1.39	0	0	0	0	0	.51	8.6
AUG	1717	817	81.9	1.28	0	0	0	0	0	.52	16.5
SEP	1514	933	78.2	1.74	0	0	0	0	0	.52	28.2
OCT	1335	1176	69.8	3.11	0	0	12	40	105	.56	40.0
NOV	973	1095	60.1	2.91	8	30	90	179	305	.51	49.4
DEC	779	936	54.8	3.14	41	106	199	327	474	.45	53.4
YR	1438	943	68.3	2.64	239	492	893	1465	2254	.53	

SHREVEPORT, LOUISIANA ELEV 259 LAT 32.5

	HS	VS	TA	SD	D50	D55	D60	D65	D70	KT	L-D
JAN	762	920	47.2	3.90	175	278	410	556	708	.45	53.8
FEB	1038	1052	50.5	4.32	115	192	293	416	550	.49	46.1
MAR	1341	993	56.8	4.06	48	107	187	291	423	.50	34.9
APR	1613	870	66.4	2.70	0	0	17	65	146	.51	22.7
MAY	1886	848	73.4	2.24	0	0	0	5	30	.54	13.5
JUN	2065	869	80.2	2.05	0	0	0	0	0	.57	9.1
JUL	2014	864	83.2	1.99	0	0	0	0	0	.57	11.1
AUG	1877	913	83.2	2.23	0	0	0	0	0	.57	19.0
SEP	1554	1030	77.4	2.31	0	0	0	0	8	.55	30.7
OCT	1303	1248	67.5	3.22	0	7	23	70	148	.57	42.5
NOV	929	1132	56.2	3.10	27	80	161	278	418	.52	51.9
DEC	731	946	49.2	3.40	125	219	347	494	646	.46	55.9
YR	1428	973	65.9	2.95	490	883	1438	2175	3077	.53	

CARIBOU, MAINE ELEV 623 LAT 46.9

	HS	VS	TA	SD	D50	D55	D60	D65	D70	KT	L-D
JAN	419	730	10.7	4.50	1218	1373	1528	1683	1838	.45	68.2
FEB	724	1120	12.9	4.30	1039	1179	1319	1459	1599	.52	60.5
MAR	1133	1306	23.6	4.80	821	975	1129	1284	1438	.54	49.3
APR	1414	1111	36.7	2.60	401	549	699	849	999	.50	37.1
MAY	1578	963	49.7	2.80	97	192	326	476	630	.46	27.9
JUN	1757	970	59.6	1.90	0	12	66	170	313	.48	23.5
JUL	1762	1016	64.9	2.50	0	5	24	84	181	.50	25.5
AUG	1501	1063	62.3	2.30	0	9	43	122	246	.49	33.4
SEP	1103	1072	54.1	2.30	23	86	190	329	477	.47	45.1
OCT	688	885	43.8	2.40	206	350	503	657	812	.43	56.9
NOV	366	544	31.4	3.10	559	708	858	1008	1158	.35	66.3
DEC	310	524	16.1	4.40	1051	1206	1361	1516	1671	.38	70.3
YR	1064	941	38.8	3.16	5415	6644	8046	9637	11362	.48	

PORTLAND, MAINE ELEV 62 LAT 43.6

	HS	VS	TA	SD	D50	D55	D60	D65	D70	KT	L-D
JAN	450	689	21.5	4.00	884	1039	1194	1349	1504	.41	64.9
FEB	682	891	22.9	3.50	759	899	1039	1179	1319	.44	57.2
MAR	970	941	31.8	3.20	566	720	874	1029	1184	.43	46.0
APR	1304	920	42.7	2.20	226	370	519	669	819	.44	33.8
MAY	1567	888	52.7	2.20	36	112	234	382	536	.46	24.6
JUN	1712	888	62.2	1.80	0	0	27	106	236	.47	20.2
JUL	1659	894	68.0	2.00	0	0	0	27	100	.47	22.2
AUG	1461	943	66.4	2.30	0	0	12	55	141	.47	30.1
SEP	1158	1025	58.7	2.30	5	27	93	200	341	.47	41.8
OCT	822	1003	49.1	2.40	94	199	341	493	648	.47	53.6
NOV	459	651	38.6	2.30	344	492	642	792	942	.38	63.0
DEC	363	559	25.7	4.00	755	909	1063	1218	1373	.37	67.0
YR	1052	857	45.0	2.68	3669	4767	6038	7499	9143	.45	

BALTIMORE, MARYLAND ELEV 154 LAT 39.2

	HS	VS	TA	SD	D50	D55	D60	D65	D70	KT	L-D
JAN	587	834	33.4	4.43	525	674	826	980	1135	.44	60.5
FEB	840	1008	34.8	3.36	429	566	706	846	986	.47	52.8
MAR	1162	1030	42.8	4.04	267	395	539	690	844	.48	41.6
APR	1488	945	53.8	2.68	36	103	204	340	487	.49	29.4
MAY	1714	880	63.7	2.70	0	10	40	110	215	.49	20.2
JUN	1879	887	72.4	1.80	0	0	0	0	25	.52	15.8
JUL	1823	890	76.6	1.64	0	0	0	0	0	.51	17.8
AUG	1599	926	74.9	1.67	0	0	0	0	8	.50	25.7
SEP	1330	1059	68.5	2.23	0	0	0	27	94	.51	37.4
OCT	998	1114	57.4	2.75	15	56	134	250	394	.50	49.2
NOV	660	912	46.1	2.28	141	271	418	567	717	.46	58.6
DEC	499	725	35.3	3.69	463	613	766	921	1076	.41	62.6
YR	1216	933	55.0	2.77	1876	2688	3633	4731	5981	.49	

BOSTON, MASSACHUSETTS ELEV 16 LAT 42.4

	HS	VS	TA	SD	D50	D55	D60	D65	D70	KT	L-D
JAN	475	706	29.2	4.18	648	801	955	1110	1265	.40	63.7
FEB	710	900	30.4	3.35	550	689	829	969	1109	.44	56.0
MAR	1016	961	38.1	3.55	380	528	680	834	989	.44	44.8
APR	1326	908	48.6	2.62	104	208	345	493	642	.45	32.6
MAY	1620	893	58.6	2.74	10	42	113	218	358	.47	23.4
JUN	1817	912	68.0	2.06	0	0	0	27	97	.50	19.0
JUL	1749	914	73.3	1.89	0	0	0	0	21	.49	21.0
AUG	1486	932	71.3	2.00	0	0	0	8	47	.47	28.9
SEP	1260	1098	64.5	2.18	0	0	18	76	177	.50	40.6
OCT	890	1072	55.4	2.27	16	69	163	301	453	.49	52.4
NOV	503	705	45.2	2.57	168	299	445	594	744	.39	61.8
DEC	403	615	33.0	4.00	533	684	838	992	1147	.38	65.8
YR	1106	884	51.3	2.78	2409	3320	4386	5622	7049	.47	

ALPENA, MICHIGAN ELEV 689 LAT 45.1

	HS	VS	TA	SD	D50	D55	D60	D65	D70	KT	L-D
JAN	362	538	17.8	4.10	999	1153	1308	1463	1618	.35	66.4
FEB	617	820	18.3	4.02	888	1028	1168	1308	1448	.41	58.7
MAR	1028	1069	26.2	4.73	742	894	1048	1203	1358	.47	47.5
APR	1407	1047	40.1	3.41	311	451	598	747	897	.48	35.3
MAY	1720	1004	50.5	3.47	107	193	313	455	606	.50	26.1
JUN	1879	989	60.9	2.44	0	15	64	150	278	.52	21.7
JUL	1885	1035	65.5	2.49	0	0	20	75	166	.53	23.7
AUG	1583	1071	64.2	2.94	0	11	41	110	208	.51	31.6
SEP	1156	1074	56.3	2.27	11	53	137	265	412	.48	43.3
OCT	743	921	47.3	2.93	142	255	398	550	704	.44	55.1
NOV	382	534	34.9	3.06	455	604	753	903	1053	.34	64.5
DEC	270	392	23.4	3.51	825	980	1135	1290	1445	.30	68.5
YR	1088	875	42.1	3.28	4480	5637	6983	8519	10193	.48	

DETROIT, MICHIGAN ELEV 627 LAT 42.4

	HS	VS	TA	SD	D50	D55	D60	D65	D70	KT	L-D
JAN	417	585	25.5	4.44	762	915	1070	1225	1380	.36	63.7
FEB	680	847	26.9	3.35	647	787	927	1067	1207	.42	56.0
MAR	1000	941	35.4	4.67	471	615	766	919	1073	.44	44.8
APR	1399	963	48.1	3.29	133	233	365	509	658	.47	32.6
MAY	1716	941	58.4	3.38	22	63	138	237	370	.50	23.4
JUN	1866	932	69.1	2.53	0	0	6	29	94	.51	19.0
JUL	1835	951	73.3	2.03	0	0	0	0	25	.52	21.0
AUG	1575	989	71.9	2.38	0	0	0	11	51	.50	28.9
SEP	1253	1090	64.5	2.31	0	0	21	80	180	.50	40.6
OCT	876	1048	54.3	3.23	49	118	212	342	490	.48	52.4
NOV	478	655	41.1	2.90	278	420	568	717	867	.37	61.8
DEC	343	487	29.6	4.12	636	789	943	1098	1252	.33	65.8
YR	1122	869	49.9	3.22	2998	3940	5016	6234	7647	.47	

FLINT, MICHIGAN ELEV 764 LAT 43.0

	HS	VS	TA	SD	D50	D55	D60	D65	D70	KT	L-D
JAN	383	532	22.3	4.09	859	1014	1169	1324	1479	.34	64.3
FEB	636	788	23.8	3.20	734	874	1014	1154	1294	.40	56.6
MAR	957	906	32.6	4.78	552	700	852	1005	1160	.42	45.4
APR	1339	932	45.9	3.31	173	289	428	574	723	.45	33.2
MAY	1658	924	55.8	3.53	43	105	189	306	447	.48	24.0
JUN	1813	921	65.8	2.44	0	0	16	65	152	.50	19.6
JUL	1797	946	69.7	1.99	0	0	0	14	70	.51	21.6
AUG	1555	992	68.2	2.40	0	0	7	36	109	.50	29.5
SEP	1195	1047	61.0	2.43	0	15	62	147	275	.48	41.2
OCT	829	992	51.2	3.31	91	174	291	433	584	.46	53.0
NOV	429	576	38.3	3.19	358	503	652	801	951	.34	62.4
DEC	309	432	26.8	3.86	721	875	1029	1184	1339	.30	66.4
YR	1077	832	46.8	3.21	3531	4549	5709	7043	8583	.46	

GRAND RAPIDS, MICHIGAN ELEV 804 LAT 42.9

	HS	VS	TA	SD	D50	D55	D60	D65	D70	KT	L-D
JAN	370	504	23.2	3.91	831	986	1141	1296	1451	.32	64.2
FEB	648	806	24.5	3.42	714	854	994	1134	1274	.40	56.5
MAR	1014	974	33.1	4.71	536	684	836	990	1144	.45	45.3
APR	1412	987	46.5	3.26	160	272	410	556	705	.48	33.1
MAY	1755	971	57.1	3.42	30	82	161	270	408	.51	23.9
JUN	1956	976	67.4	2.46	0	0	10	44	120	.54	19.5
JUL	1914	996	71.5	2.09	0	0	0	8	47	.54	21.5
AUG	1676	1070	70.0	2.66	0	0	6	27	87	.54	29.4
SEP	1262	1118	62.4	2.25	0	7	38	114	234	.51	41.1
OCT	858	1037	52.0	3.27	79	158	270	409	560	.48	52.9
NOV	446	606	38.7	2.96	345	490	639	789	939	.36	62.3
DEC	311	434	27.4	3.57	702	856	1011	1166	1321	.30	66.3
YR	1137	873	47.8	3.17	3397	4389	5516	6803	8290	.48	

SAULT STE. MARIE, MICHIGAN ELEV 725 LAT 46.5

	HS	VS	TA	SD	D50	D55	D60	D65	D70	KT	L-D
JAN	325	492	14.2	3.93	1110	1265	1420	1575	1730	.34	67.8
FEB	603	840	15.2	3.92	974	1114	1254	1394	1534	.43	60.1
MAR	1029	1125	24.0	4.47	808	962	1116	1271	1426	.49	48.9
APR	1383	1069	38.2	3.55	365	507	655	804	954	.48	36.7
MAY	1688	1021	49.0	3.44	130	225	354	500	652	.50	27.5
JUN	1811	988	58.7	2.30	5	27	93	200	341	.50	23.1
JUL	1835	1045	63.8	2.32	0	5	28	96	205	.52	25.1
AUG	1523	1068	63.2	2.90	0	14	51	125	232	.50	33.0
SEP	1049	990	55.3	2.20	14	65	158	294	441	.45	44.7
OCT	673	844	46.2	3.33	175	292	433	584	738	.41	56.5
NOV	332	465	32.8	3.13	518	666	816	966	1116	.31	65.9
DEC	253	382	20.1	3.06	927	1082	1237	1392	1547	.30	69.9
YR	1044	861	40.0	3.21	5026	6224	7615	9201	10916	.47	

TRAVERSE CITY, MICHIGAN ELEV 630 LAT 44.7

	HS	VS	TA	SD	D50	D55	D60	D65	D70	KT	L-D
JAN	311	426	20.8	4.11	906	1060	1215	1370	1525	.30	66.0
FEB	567	716	20.7	3.84	821	960	1100	1240	1380	.38	58.3
MAR	1001	1018	28.7	4.76	666	818	971	1126	1280	.46	47.1
APR	1405	1033	42.7	3.45	246	378	522	670	819	.48	34.9
MAY	1729	999	52.8	3.42	72	149	253	387	536	.50	25.7
JUN	1912	995	63.7	2.67	0	8	36	104	206	.52	21.3
JUL	1910	1037	68.7	2.48	0	0	7	33	103	.54	23.3
AUG	1609	1078	67.5	3.10	0	6	21	66	144	.52	31.2
SEP	1165	1071	59.4	2.06	0	16	74	178	319	.48	42.9
OCT	754	926	49.8	3.11	105	198	327	474	627	.44	54.7
NOV	377	515	36.9	3.14	398	544	693	843	993	.33	64.1
DEC	257	360	25.9	3.20	747	902	1057	1212	1367	.28	68.1
YR	1085	848	44.8	3.28	3961	5039	6276	7703	9299	.47	

DULUTH, MINNESOTA ELEV 1417 LAT 46.8

	HS	VS	TA	SD	D50	D55	D60	D65	D70	KT	L-D
JAN	389	650	8.5	5.30	1287	1442	1597	1752	1907	.41	68.1
FEB	673	1000	12.1	4.20	1061	1201	1341	1481	1621	.48	60.4
MAR	1034	1145	23.5	5.30	826	979	1132	1287	1442	.49	49.2
APR	1373	1069	38.6	3.50	353	495	643	792	942	.48	37.0
MAY	1643	1001	49.4	2.70	99	197	334	485	639	.48	27.8
JUN	1767	973	59.0	2.40	5	27	91	194	333	.49	23.4
JUL	1854	1063	65.6	2.30	0	0	15	67	159	.53	25.4
AUG	1547	1097	64.1	2.70	0	9	36	104	205	.51	33.3
SEP	1095	1058	54.4	2.20	19	78	180	320	468	.47	45.0
OCT	725	950	45.3	3.20	189	314	459	612	766	.45	56.8
NOV	381	574	28.4	3.90	650	799	948	1098	1248	.37	66.2
DEC	292	477	14.4	4.50	1104	1259	1414	1569	1724	.36	70.2
YR	1066	921	38.6	3.52	5593	6800	8190	9761	11454	.48	

INTERNATIONAL FALLS, MINNESOTA ELEV 1184 LAT 48.6

	HS	VS	TA	SD	D50	D55	D60	D65	D70	KT	L-D
JAN	356	626	1.9	6.60	1492	1647	1801	1956	2111	.42	69.9
FEB	663	1061	7.0	5.15	1204	1344	1484	1624	1764	.51	62.2
MAR	1046	1244	20.6	6.12	917	1069	1223	1377	1532	.52	51.0
APR	1444	1202	38.2	4.37	376	513	657	805	954	.52	38.8
MAY	1716	1096	50.1	3.20	104	194	320	465	618	.51	29.6
JUN	1853	1059	60.4	2.73	6	24	80	168	295	.51	25.2
JUL	1921	1154	65.8	2.35	0	0	15	66	156	.55	27.2
AUG	1618	1220	63.2	2.50	0	9	39	112	224	.54	35.1
SEP	1121	1161	53.0	2.78	47	119	226	364	511	.49	46.8
OCT	704	979	43.5	3.64	241	370	516	668	822	.46	58.6
NOV	345	541	24.9	4.38	755	904	1053	1203	1353	.36	68.0
DEC	272	475	8.7	5.40	1281	1436	1590	1745	1900	.38	72.0
YR	1090	984	36.5	4.10	6423	7629	9004	10553	12240	.50	

MINNEAPOLIS, MINNESOTA ELEV 837 LAT 44.9

	HS	VS	TA	SD	D50	D55	D60	D65	D70	KT	L-D
JAN	464	768	12.2	5.70	1173	1327	1482	1637	1792	.45	66.2
FEB	764	1110	16.5	4.74	938	1078	1218	1358	1498	.51	58.5
MAR	1103	1169	28.3	5.99	688	836	987	1140	1294	.51	47.3
APR	1442	1071	45.1	4.03	209	321	456	600	748	.50	35.1
MAY	1737	1009	57.1	3.46	31	83	162	271	408	.51	25.9
JUN	1927	1006	66.9	2.92	0	5	19	65	143	.53	21.5
JUL	1970	1071	71.9	2.40	0	0	0	11	52	.56	23.5
AUG	1687	1142	70.2	2.45	0	0	0	21	78	.55	31.4
SEP	1255	1188	60.0	2.57	5	23	80	173	305	.52	43.1
OCT	860	1126	50.0	3.75	123	211	331	472	622	.50	54.9
NOV	480	736	32.4	4.12	534	680	829	978	1128	.42	64.3
DEC	353	572	18.6	4.93	975	1129	1284	1439	1593	.39	68.3
YR	1172	996	44.1	3.92	4676	5693	6848	8165	9661	.51	

ROCHESTER, MINNESOTA ELEV 1319 LAT 43.9

	HS	VS	TA	SD	D50	D55	D60	D65	D70	KT	L-D
JAN	477	762	12.9	5.43	1151	1306	1460	1615	1770	.44	65.2
FEB	753	1041	16.9	4.79	927	1067	1207	1347	1487	.49	57.5
MAR	1082	1099	27.8	5.69	700	849	1001	1155	1309	.49	46.3
APR	1410	1013	44.5	3.73	213	332	471	617	766	.48	34.1
MAY	1696	963	56.2	3.36	35	93	176	292	434	.49	24.9
JUN	1902	974	66.0	2.94	0	6	25	78	161	.52	20.5
JUL	1909	1017	70.1	2.42	0	0	0	21	78	.54	22.5
AUG	1662	1091	68.6	2.52	0	0	8	35	106	.53	30.4
SEP	1250	1142	59.3	2.34	0	23	84	185	323	.51	42.1
OCT	870	1100	49.6	3.82	132	222	343	485	635	.50	53.9
NOV	494	734	32.6	3.96	527	674	823	972	1122	.41	63.3
DEC	370	583	18.9	4.83	965	1120	[illegible]	1429	1584	.38	67.3
YR	1158	959	43.6	3.82	4650	5692	[illegible]	8231	9775	.50	

JACKSON, MISSISSIPPI ELEV 331 LAT 32.3

	HS	VS	TA	SD	D50	D55	D60	D65	D70	KT	L-D
JAN	753	898	47.1	5.05	214	309	431	569	716	.44	53.6
FEB	1026	1029	49.8	5.08	146	224	323	442	573	.48	45.9
MAR	1369	1011	56.1	4.25	60	124	205	313	446	.51	34.7
APR	1708	910	65.7	2.67	0	5	21	74	160	.54	22.5
MAY	1941	861	72.7	2.17	0	0	0	6	35	.56	13.3
JUN	2024	859	79.4	2.04	0	0	0	0	0	.56	8.9
JUL	1909	837	81.7	1.66	0	0	0	0	0	.54	10.9
AUG	1780	873	81.2	1.87	0	0	0	0	0	.54	18.8
SEP	1509	992	76.0	2.16	0	0	0	0	10	.53	30.5
OCT	1271	1199	65.8	3.13	0	9	32	91	177	.56	42.3
NOV	902	1077	55.3	2.99	31	89	176	301	443	.50	51.7
DEC	709	898	48.9	3.68	139	233	360	504	656	.45	55.7
YR	1410	953	65.0	3.05	590	993	1548	2300	3216	.53	

MERIDIAN, MISSISSIPPI ELEV 308 LAT 32.3

	HS	VS	TA	SD	D50	D55	D60	D65	D70	KT	L-D
JAN	744	883	46.9	5.13	220	316	437	575	723	.43	53.6
FEB	1012	1010	49.8	5.11	147	225	323	443	573	.48	45.9
MAR	1328	976	56.1	4.24	59	124	205	312	445	.50	34.7
APR	1662	889	65.4	2.69	0	5	23	79	167	.52	22.5
MAY	1860	839	72.4	2.20	0	0	0	7	39	.53	13.3
JUN	1963	846	79.2	2.12	0	0	0	0	0	.54	8.9
JUL	1823	815	81.2	1.72	0	0	0	0	0	.52	10.9
AUG	1739	858	80.7	1.93	0	0	0	0	0	.53	18.8
SEP	1454	953	75.3	2.22	0	0	0	0	14	.51	30.5
OCT	1258	1183	64.8	3.28	0	14	45	111	202	.55	42.3
NOV	897	1068	54.2	2.95	39	105	200	331	476	.50	51.7
DEC	699	880	47.9	3.74	158	258	388	535	687	.44	55.7
YR	1371	933	64.5	3.10	623	1047	1621	2393	3326	.51	

COLUMBIA, MISSOURI ELEV 886 LAT 38.8

	HS	VS	TA	SD	D50	D55	D60	D65	D70	KT	L-D
JAN	611	869	29.3	3.98	645	798	952	1107	1262	.45	60.1
FEB	875	1052	33.6	4.05	465	601	740	879	1019	.48	52.4
MAR	1179	1035	41.7	5.39	327	447	584	730	881	.48	41.2
APR	1526	960	55.0	3.42	44	107	194	314	454	.50	29.0
MAY	1880	940	64.4	3.25	0	15	48	117	211	.54	19.8
JUN	2089	946	73.0	2.72	0	0	0	11	45	.57	15.4
JUL	2116	983	77.3	2.68	0	0	0	0	14	.60	17.4
AUG	1878	1067	76.0	2.68	0	0	0	5	21	.59	25.3
SEP	1450	1161	68.3	2.74	0	0	10	42	113	.55	37.0
OCT	1101	1259	58.0	3.39	24	69	144	247	381	.55	48.8
NOV	703	983	43.9	3.22	213	342	485	634	783	.48	58.2
DEC	522	760	32.8	4.61	544	693	845	999	1153	.42	62.2
YR	1329	1001	54.4	3.51	2262	3072	4002	5085	6337	.53	

KANSAS CITY, MISSOURI ELEV 1033 LAT 39.3

	HS	VS	TA	SD	D50	D55	D60	D65	D70	KT	L-D
JAN	648	969	27.1	4.50	713	866	1020	1175	1330	.48	60.6
FEB	895	1108	32.3	4.30	501	638	776	916	1056	.50	52.9
MAR	1203	1080	40.7	5.40	350	474	613	760	911	.50	41.7
APR	1575	1006	54.2	3.40	51	119	211	336	477	.52	29.5
MAY	1873	948	64.1	3.40	6	19	56	127	221	.54	20.3
JUN	2080	952	73.0	3.00	0	0	0	15	53	.57	15.9
JUL	2102	989	77.5	2.70	0	0	0	0	14	.59	17.9
AUG	1862	1074	76.5	2.80	0	0	0	5	20	.58	25.8
SEP	1452	1183	68.0	2.90	0	0	14	50	123	.56	37.5
OCT	1092	1269	57.6	3.50	29	78	155	259	394	.55	49.3
NOV	737	1077	42.3	3.20	251	387	533	681	831	.51	58.7
DEC	561	865	31.3	4.50	587	738	891	1045	1200	.46	62.7
YR	1342	1042	53.7	3.63	2488	3319	4269	5369	6630	.54	

SAINT LOUIS, MISSOURI ELEV 564 LAT 38.7

	HS	VS	TA	SD	D50	D55	D60	D65	D70	KT	L-D
JAN	627	898	31.3	4.41	587	737	891	1045	1200	.46	60.0
FEB	886	1066	35.1	4.49	428	561	699	838	977	.49	52.3
MAR	1205	1061	43.3	5.27	290	404	537	682	832	.49	41.1
APR	1564	982	56.5	3.22	27	80	159	272	410	.51	28.9
MAY	1871	935	65.8	3.48	0	14	41	103	187	.54	19.7
JUN	2092	945	74.9	3.08	0	0	0	10	36	.58	15.3
JUL	2049	959	78.6	2.59	0	0	0	0	9	.58	17.3
AUG	1816	1032	77.2	2.68	0	0	0	0	15	.57	25.2
SEP	1459	1166	69.6	2.93	0	0	9	35	98	.55	36.9
OCT	1100	1252	59.1	3.52	21	59	131	224	352	.55	48.7
NOV	718	1009	45.0	3.01	183	309	452	601	750	.49	58.1
DEC	531	776	34.6	4.83	495	640	791	944	1098	.43	62.1
YR	1328	1006	55.9	3.62	2031	2804	3710	4754	5964	.53	

SPRINGFIELD, MISSOURI ELEV 1270 LAT 37.2

	HS	VS	TA	SD	D50	D55	D60	D65	D70	KT	L-D
JAN	684	956	32.9	3.60	534	686	840	995	1150	.47	58.5
FEB	926	1071	37.0	3.97	374	507	645	784	924	.49	50.8
MAR	1235	1042	44.0	5.06	270	382	515	660	810	.50	39.6
APR	1604	968	56.5	3.39	31	85	164	275	411	.52	27.4
MAY	1882	912	65.1	2.92	0	9	33	94	186	.54	18.2
JUN	2075	918	73.6	2.80	0	0	0	10	41	.57	13.8
JUL	2063	936	77.8	2.92	0	0	0	0	16	.58	15.8
AUG	1873	1019	77.1	3.00	0	0	0	6	21	.58	23.7
SEP	1481	1130	69.3	2.83	0	0	9	35	100	.55	35.4
OCT	1144	1249	59.0	3.54	22	61	133	227	355	.55	47.2
NOV	775	1058	45.5	3.03	173	296	438	586	735	.50	56.6
DEC	603	874	36.0	3.87	444	592	745	899	1054	.45	60.6
YR	1364	1011	56.1	3.41	1848	2618	3522	4571	5803	.54	

BILLINGS, MONTANA ELEV 3570 LAT 45.8

	HS	VS	TA	SD	D50	D55	D60	D65	D70	KT	L-D
JAN	486	863	21.9	8.12	895	1041	1190	1341	1494	.49	67.1
FEB	763	1153	27.4	5.69	641	777	915	1054	1193	.53	59.4
MAR	1189	1340	32.6	5.21	556	702	853	1006	1160	.55	48.2
APR	1526	1177	44.6	3.54	205	327	467	613	762	.53	36.0
MAY	1913	1134	54.5	2.94	38	105	201	333	482	.56	26.8
JUN	2174	1139	62.6	2.94	0	16	56	131	239	.60	22.4
JUL	2384	1303	71.8	2.26	0	0	0	10	49	.68	24.4
AUG	2022	1441	70.1	2.13	0	0	0	15	69	.66	32.3
SEP	1470	1517	58.9	3.55	21	59	129	221	346	.62	44.0
OCT	987	1432	49.3	3.35	122	215	344	490	643	.60	55.8
NOV	561	972	35.7	4.48	444	585	731	880	1029	.51	65.2
DEC	421	782	26.8	5.16	726	877	1031	1185	1339	.48	69.2
YR	1328	1188	46.3	4.11	3648	4704	5917	7279	8805	.59	

CUT BANK, MONTANA ELEV 3839 LAT 48.6

	HS	VS	TA	SD	D50	D55	D60	D65	D70	KT	L-D
JAN	402	753	16.2	10.35	1084	1227	1374	1524	1675	.48	69.9
FEB	688	1122	22.4	6.31	779	916	1054	1194	1333	.53	62.2
MAR	1128	1385	26.8	5.94	731	880	1032	1186	1340	.56	51.0
APR	1485	1245	39.5	4.39	343	476	619	767	916	.53	38.8
MAY	1883	1209	49.6	2.88	101	197	330	479	633	.56	29.6
JUN	2045	1162	56.5	2.90	21	70	150	267	408	.56	25.2
JUL	2287	1372	64.4	2.20	0	0	20	82	187	.65	27.2
AUG	1897	1475	62.6	2.56	0	11	48	125	241	.63	35.1
SEP	1352	1503	53.2	3.86	77	150	244	368	509	.60	46.8
OCT	871	1343	44.1	4.44	250	367	505	653	805	.57	58.6
NOV	480	887	29.7	6.19	628	769	914	1062	1210	.51	68.0
DEC	334	647	21.4	7.00	899	1049	1200	1354	1508	.46	72.0
YR	1241	1175	40.5	4.91	4913	6112	7490	9061	10765	.57	

DILLON, MONTANA ELEV 5210 LAT 45.2

	HS	VS	TA	SD	D50	D55	D60	D65	D70	KT	L-D
JAN	526	943	20.2	6.50	931	1083	1236	1390	1544	.51	66.5
FEB	846	1310	25.5	5.75	692	829	967	1107	1246	.57	58.8
MAR	1279	1451	29.6	4.43	638	789	943	1098	1253	.59	47.6
APR	1639	1257	41.1	3.19	282	421	568	717	867	.57	35.4
MAY	1989	1158	50.4	3.06	95	183	309	456	608	.58	26.2
JUN	2143	1108	57.5	2.75	13	52	127	238	378	.59	21.8
JUL	2392	1282	66.4	2.25	0	0	11	54	140	.68	23.8
AUG	2023	1412	64.6	2.38	0	0	23	85	185	.66	31.7
SEP	1521	1554	54.7	3.62	52	118	205	325	464	.63	43.4
OCT	1023	1472	45.0	3.74	211	331	472	622	776	.61	55.2
NOV	602	1050	31.8	4.18	551	698	847	996	1146	.53	64.6
DEC	450	838	23.9	4.00	810	964	1119	1274	1429	.50	68.6
YR	1372	1236	42.6	3.81	4275	5468	6827	8362	10036	.60	

GLASGOW, MONTANA ELEV 2297 LAT 48.2

	HS	VS	TA	SD	D50	D55	D60	D65	D70	KT	L-D
JAN	388	698	9.2	9.41	1278	1429	1580	1733	1887	.45	69.5
FEB	671	1060	15.2	6.80	978	1116	1255	1395	1535	.51	61.8
MAR	1105	1324	25.2	7.64	796	940	1088	1239	1392	.54	50.6
APR	1488	1232	42.8	4.36	265	386	524	669	817	.53	38.4
MAY	1828	1159	54.2	3.08	45	115	211	344	492	.54	29.2
JUN	2047	1151	62.0	3.26	7	25	74	151	260	.56	24.8
JUL	2193	1300	70.5	2.24	0	0	0	15	66	.63	26.8
AUG	1863	1423	69.0	2.46	0	0	6	30	97	.62	34.7
SEP	1340	1462	57.2	3.67	33	84	160	263	394	.59	46.4
OCT	877	1333	46.4	3.51	177	290	429	579	732	.57	58.2
NOV	479	865	29.0	5.77	643	787	933	1082	1231	.49	67.6
DEC	334	632	17.1	7.21	1029	1180	1333	1486	1641	.45	71.6
YR	1220	1137	41.5	4.94	5251	6352	7593	8986	10544	.56	

GREAT FALLS, MONTANA ELEV 3661 LAT 47.5

	HS	VS	TA	SD	D50	D55	D60	D65	D70	KT	L-D
JAN	420	757	20.5	9.94	958	1099	1244	1392	1543	.47	68.8
FEB	720	1141	26.6	6.44	667	801	938	1077	1216	.53	61.1
MAR	1170	1398	30.5	5.89	624	769	919	1072	1226	.57	49.9
APR	1489	1205	43.4	3.91	241	364	503	650	799	.53	37.7
MAY	1848	1149	53.3	2.71	44	117	225	367	519	.54	28.5
JUN	2101	1157	60.8	2.84	6	24	78	162	285	.58	24.1
JUL	2329	1348	69.3	2.06	0	0	0	18	79	.66	26.1
AUG	1933	1451	67.4	2.29	0	0	8	42	120	.64	34.0
SEP	1378	1480	57.3	3.67	32	83	158	260	391	.60	45.7
OCT	925	1400	48.3	4.00	160	255	381	524	675	.59	57.5
NOV	498	885	34.6	5.43	486	623	767	914	1063	.49	66.9
DEC	336	612	26.5	6.62	747	893	1044	1196	1350	.43	70.9
YR	1265	1165	44.9	4.64	3965	5028	6265	7674	9266	.57	

HELENA, MONTANA ELEV 3898 LAT 46.6

	HS	VS	TA	SD	D50	D55	D60	D65	D70	KT	L-D
JAN	419	719	18.1	8.00	1004	1153	1304	1457	1611	.44	67.9
FEB	709	1071	25.4	5.30	693	831	970	1109	1249	.50	60.2
MAR	1145	1310	30.6	5.20	614	762	914	1068	1222	.54	49.0
APR	1487	1170	42.7	2.90	236	373	520	669	819	.52	36.8
MAY	1860	1128	52.2	3.00	67	146	259	401	553	.55	27.6
JUN	2040	1101	59.2	2.65	7	31	96	194	328	.56	23.2
JUL	2334	1312	67.9	2.20	0	0	6	33	108	.66	25.2
AUG	1930	1404	66.2	2.25	0	0	12	57	144	.63	33.1
SEP	1412	1480	55.5	3.60	44	106	189	304	441	.60	44.8
OCT	926	1348	45.3	3.70	204	322	463	613	766	.57	56.6
NOV	521	904	31.7	4.20	554	701	850	999	1149	.49	66.0
DEC	364	656	23.3	4.50	829	983	1138	1293	1448	.44	70.0
YR	1265	1134	43.2	3.95	4252	5408	6721	8197	9838	.57	

LEWISTOWN, MONTANA ELEV 4147 LAT 47.0

	HS	VS	TA	SD	D50	D55	D60	D65	D70	KT	L-D
JAN	420	737	19.1	8.72	982	1128	1277	1429	1582	.45	68.3
FEB	692	1051	23.8	5.80	739	876	1015	1154	1294	.50	60.6
MAR	1128	1303	27.5	5.78	710	858	1010	1164	1318	.54	49.4
APR	1444	1143	40.1	3.73	316	453	599	748	897	.51	37.2
MAY	1807	1107	49.6	2.86	101	196	330	479	633	.53	28.0
JUN	2059	1121	56.6	2.93	21	69	149	265	405	.57	23.6
JUL	2288	1304	65.5	2.35	0	0	17	70	163	.65	25.6
AUG	1901	1399	64.4	2.55	0	6	29	94	194	.62	33.5
SEP	1372	1444	54.0	3.95	70	140	229	348	487	.59	45.2
OCT	905	1326	45.5	4.22	216	327	462	609	761	.57	57.0
NOV	502	872	32.2	4.99	546	689	836	985	1134	.49	66.4
DEC	363	668	24.5	5.53	797	949	1102	1256	1411	.45	70.4
YR	1243	1123	41.9	4.45	4498	5691	7055	8601	10279	.56	

MILES CITY, MONTANA ELEV 2634 LAT 46.4

	HS	VS	TA	SD	D50	D55	D60	D65	D70	KT	L-D
JAN	457	811	15.4	9.08	1093	1240	1391	1543	1696	.48	67.7
FEB	745	1143	21.6	6.32	801	938	1077	1216	1356	.52	60.0
MAR	1185	1365	30.2	6.35	639	782	931	1082	1236	.56	48.8
APR	1542	1215	45.3	4.20	210	319	452	595	742	.54	36.6
MAY	1896	1143	56.3	3.30	33	90	172	288	430	.56	27.4
JUN	2146	1144	64.9	3.69	6	19	52	117	203	.59	23.0
JUL	2293	1282	74.4	2.84	0	0	0	9	37	.65	25.0
AUG	1977	1434	72.5	2.83	0	0	0	16	58	.65	32.9
SEP	1444	1515	59.9	4.19	27	66	133	217	329	.61	44.6
OCT	961	1413	48.8	3.86	147	240	365	508	659	.59	56.4
NOV	551	975	32.4	5.08	542	684	831	979	1128	.52	65.8
DEC	399	746	22.0	6.37	877	1028	1180	1334	1489	.48	69.8
YR	1302	1182	45.3	4.84	4375	5406	6584	7904	9363	.58	

MISSOULA, MONTANA ELEV 3189 LAT 46.9

	HS	VS	TA	SD	D50	D55	D60	D65	D70	KT	L-D
JAN	312	473	20.8	7.15	918	1067	1219	1372	1526	.33	68.2
FEB	574	796	27.2	4.15	640	779	919	1058	1198	.41	60.5
MAR	981	1069	33.3	3.65	522	674	828	983	1138	.47	49.3
APR	1382	1081	43.9	2.40	197	335	483	633	783	.48	37.1
MAY	1782	1089	52.2	2.73	58	137	255	400	552	.52	27.9
JUN	1933	1057	58.9	2.63	8	33	100	201	337	.53	23.5
JUL	2327	1321	66.6	1.88	0	0	5	39	126	.66	25.5
AUG	1881	1377	65.0	2.17	0	0	16	71	171	.62	33.4
SEP	1358	1418	55.3	3.02	31	90	177	301	444	.58	45.1
OCT	813	1126	44.1	2.53	201	342	494	648	803	.51	56.9
NOV	410	641	32.3	2.98	532	681	831	981	1131	.40	66.3
DEC	267	421	24.7	3.98	785	940	1094	1249	1404	.33	70.3
YR	1172	990	43.7	3.27	3892	5078	6421	7936	9613	.53	

GRAND ISLAND, NEBRASKA ELEV 1857 LAT 41.0

	HS	VS	TA	SD	D50	D55	D60	D65	D70	KT	L-D
JAN	661	1081	22.3	4.40	860	1014	1169	1324	1479	.53	62.3
FEB	917	1230	27.7	4.90	629	766	905	1045	1185	.54	54.6
MAR	1265	1223	35.5	5.70	484	622	768	919	1071	.54	43.4
APR	1692	1141	49.9	4.00	127	212	326	461	606	.56	31.2
MAY	1972	1028	60.7	3.30	12	37	98	184	305	.57	22.0
JUN	2242	1035	70.7	3.30	0	0	11	35	94	.62	17.6
JUL	2216	1068	76.3	2.86	0	0	0	6	23	.63	19.6
AUG	1939	1175	75.0	2.50	0	0	0	5	23	.61	27.5
SEP	1509	1317	64.4	3.10	0	12	41	107	199	.59	39.2
OCT	1138	1442	53.7	3.40	60	133	230	362	509	.60	51.0
NOV	738	1166	38.2	3.10	360	506	654	804	954	.55	60.4
DEC	569	960	27.0	4.10	715	869	1023	1178	1333	.51	64.4
YR	1407	1155	50.1	3.72	3247	4171	5225	6430	7781	.58	

NORTH PLATTE, NEBRASKA ELEV 2785 LAT 41.1

	HS	VS	TA	SD	D50	D55	D60	D65	D70	KT	L-D
JAN	692	1165	23.4	5.20	829	981	1135	1290	1445	.56	62.4
FEB	958	1317	28.1	5.10	619	756	894	1034	1173	.57	54.7
MAR	1333	1316	34.3	5.10	507	651	801	954	1108	.57	43.5
APR	1724	1169	47.8	4.10	163	258	382	522	668	.57	31.3
MAY	1988	1037	58.3	3.30	21	62	137	238	372	.58	22.1
JUN	2266	1045	68.0	3.40	0	7	22	65	138	.62	17.7
JUL	2277	1093	74.3	2.60	0	0	0	7	31	.64	19.7
AUG	1989	1209	73.0	2.40	0	0	0	8	38	.63	27.6
SEP	1565	1385	62.3	3.10	5	20	65	141	249	.61	39.3
OCT	1177	1521	51.0	3.25	92	176	296	439	590	.62	51.1
NOV	759	1223	36.2	3.00	417	565	714	864	1014	.56	60.5
DEC	605	1059	26.8	3.70	720	875	1029	1184	1339	.54	64.5
YR	1446	1211	48.6	3.68	3373	4351	5475	6746	8165	.60	

OMAHA, NEBRASKA ELEV 1325 LAT 41.4

	HS	VS	TA	SD	D50	D55	D60	D65	D70	KT	L-D
JAN	634	1034	20.2	4.64	925	1079	1234	1389	1544	.52	62.7
FEB	892	1201	25.5	5.08	689	827	967	1106	1246	.53	55.0
MAR	1222	1185	34.6	5.93	512	650	796	947	1100	.52	43.8
APR	1558	1055	50.0	3.66	114	200	318	456	602	.52	31.6
MAY	1873	993	60.9	3.49	14	41	102	186	302	.54	22.4
JUN	2122	1005	70.2	3.06	0	0	9	33	93	.58	18.0
JUL	2106	1038	75.1	2.72	0	0	0	7	28	.59	20.0
AUG	1858	1139	73.7	2.71	0	0	0	10	40	.59	27.9
SEP	1373	1184	64.4	2.84	0	9	34	99	193	.54	39.6
OCT	1050	1304	54.4	3.43	53	123	215	342	488	.56	51.4
NOV	644	969	37.9	3.31	371	515	664	813	963	.48	60.8
DEC	511	832	25.7	4.58	756	909	1064	1218	1373	.46	64.8
YR	1322	1078	49.4	3.78	3434	4353	5403	6606	7972	.55	

SCOTTSBLUFF, NEBRASKA ELEV 3957 LAT 41.9

	HS	VS	TA	SD	D50	D55	D60	D65	D70	KT	L-D
JAN	676	1170	24.9	5.20	783	935	1089	1244	1398	.56	63.2
FEB	950	1346	29.5	4.35	577	715	854	994	1134	.58	55.5
MAR	1307	1320	34.3	4.40	499	646	799	952	1107	.57	44.3
APR	1668	1155	46.2	3.90	184	292	424	567	715	.56	32.1
MAY	1933	1033	56.5	3.10	27	80	162	280	423	.56	22.9
JUN	2237	1055	65.9	3.30	0	10	33	91	173	.61	18.5
JUL	2284	1119	73.7	2.00	0	0	0	0	21	.65	20.5
AUG	1999	1247	71.6	2.10	0	0	0	8	46	.64	28.4
SEP	1599	1466	61.2	3.05	7	26	79	160	277	.63	40.1
OCT	1145	1510	50.2	3.30	106	194	318	463	615	.62	51.9
NOV	723	1180	36.2	3.50	420	566	715	864	1014	.55	61.3
DEC	575	1021	27.6	3.80	696	850	1005	1159	1314	.53	65.3
YR	1427	1218	48.2	3.49	3299	4314	5478	6782	8237	.60	

ELKO, NEVADA ELEV 5075 LAT 40.8

	HS	VS	TA	SD	D50	D55	D60	D65	D70	KT	L-D
JAN	689	1140	23.2	5.70	837	989	1142	1296	1451	.55	62.1
FEB	1034	1456	29.2	5.30	590	726	864	1003	1143	.61	54.4
MAR	1463	1476	35.0	3.40	470	621	775	930	1085	.62	43.2
APR	1900	1288	43.5	3.60	231	357	499	646	795	.63	31.0
MAY	2303	1161	51.9	2.90	68	148	266	410	562	.67	21.8
JUN	2534	1115	59.6	2.90	9	34	97	190	319	.70	17.4
JUL	2623	1197	69.5	2.50	0	0	6	27	91	.74	19.4
AUG	2316	1394	67.0	2.70	0	0	16	60	140	.73	27.3
SEP	1893	1754	57.6	3.40	24	70	145	248	380	.74	39.0
OCT	1322	1782	46.9	3.10	155	269	411	562	716	.70	50.8
NOV	812	1336	34.8	3.60	461	608	756	906	1056	.60	60.2
DEC	617	1075	25.9	4.00	749	903	1057	1212	1367	.54	64.2
YR	1628	1346	45.4	3.58	3594	4725	6034	7490	9105	.67	

ELY, NEVADA ELEV 6253 LAT 39.3

	HS	VS	TA	SD	D50	D55	D60	D65	D70	KT	L-D
JAN	819	1380	23.6	5.10	822	975	1129	1284	1439	.61	60.6
FEB	1141	1578	27.9	5.10	624	761	900	1039	1179	.64	52.9
MAR	1606	1580	32.8	3.60	537	689	844	998	1153	.67	41.7
APR	2009	1301	41.3	4.00	291	422	565	712	861	.66	29.5
MAY	2311	1115	50.0	3.40	112	201	325	470	621	.67	20.3
JUN	2513	1068	57.7	3.20	20	63	137	241	375	.69	15.9
JUL	2447	1094	67.2	1.60	0	0	0	23	105	.69	17.9
AUG	2230	1274	65.5	2.10	0	0	12	62	157	.70	25.8
SEP	1935	1698	56.7	3.10	24	73	152	265	403	.74	37.5
OCT	1408	1832	46.0	3.60	187	302	442	591	745	.71	49.3
NOV	926	1521	34.0	3.80	485	632	781	930	1080	.64	58.7
DEC	723	1271	26.2	3.60	739	893	1048	1203	1358	.59	62.7
YR	1674	1391	44.1	3.50	3841	5011	6335	7818	9476	.68	

LAS VEGAS, NEVADA ELEV 2178 LAT 36.1

	HS	VS	TA	SD	D50	D55	D60	D65	D70	KT	L-D
JAN	978	1553	44.2	3.50	222	349	494	646	800	.65	57.4
FEB	1339	1739	49.1	4.08	129	211	323	451	587	.69	49.7
MAR	1823	1646	54.8	2.89	34	98	192	324	473	.72	38.5
APR	2319	1343	63.8	3.41	6	19	56	126	220	.75	26.3
MAY	2646	1109	73.3	2.65	0	0	0	10	42	.76	17.1
JUN	2778	1033	82.3	2.71	0	0	0	0	0	.77	12.7
JUL	2588	1039	89.6	1.89	0	0	0	0	0	.73	14.7
AUG	2355	1200	87.4	2.28	0	0	0	0	0	.73	22.6
SEP	2037	1593	80.1	2.52	0	0	0	0	0	.75	34.3
OCT	1540	1817	67.1	3.17	0	7	25	74	153	.73	46.1
NOV	1085	1665	53.3	3.10	53	124	225	357	503	.67	55.5
DEC	880	1467	45.2	2.79	181	312	461	614	769	.63	59.5
YR	1866	1431	65.8	2.91	625	1120	1776	2602	3547	.73	

LOVELOCK, NEVADA ELEV 3904 LAT 40.1

	HS	VS	TA	SD	D50	D55	D60	D65	D70	KT	L-D
JAN	804	1396	28.9	5.70	668	816	967	1121	1275	.62	61.4
FEB	1165	1687	35.2	4.22	423	558	696	835	975	.67	53.7
MAR	1656	1700	40.1	2.37	311	463	617	772	927	.70	42.5
APR	2165	1451	48.5	3.29	127	224	354	497	646	.72	30.3
MAY	2555	1230	57.5	3.19	23	69	147	255	394	.74	21.1
JUN	2749	1145	65.6	3.01	0	8	29	86	171	.75	16.7
JUL	2784	1215	74.3	1.97	0	0	0	0	16	.79	18.7
AUG	2484	1454	71.3	2.57	0	0	0	17	66	.78	26.6
SEP	2027	1865	62.7	2.84	0	14	52	126	235	.78	38.3
OCT	1451	1984	51.2	2.90	78	162	285	431	584	.75	50.1
NOV	929	1589	38.4	3.07	354	500	648	798	948	.66	59.5
DEC	714	1299	30.8	2.69	596	750	905	1060	1215	.61	63.5
YR	1793	1500	50.4	3.14	2580	3564	4700	5998	7452	.73	

RENO, NEVADA ELEV 4400 LAT 39.5

	HS	VS	TA	SD	D50	D55	D60	D65	D70	KT	L-D
JAN	800	1345	31.9	5.63	581	726	876	1028	1182	.60	60.8
FEB	1150	1611	37.1	3.95	371	505	642	782	921	.65	53.1
MAR	1649	1650	40.3	2.56	307	457	611	766	921	.69	41.9
APR	2159	1415	46.8	3.09	150	262	400	547	696	.71	29.7
MAY	2523	1196	54.6	2.72	32	96	193	328	479	.73	20.5
JUN	2701	1115	61.5	2.75	0	18	65	145	265	.74	16.1
JUL	2692	1167	69.3	2.02	0	0	0	17	78	.76	18.1
AUG	2406	1379	66.9	2.35	0	0	11	50	132	.76	26.0
SEP	1998	1785	60.2	2.55	0	21	77	168	299	.77	37.7
OCT	1431	1893	50.3	2.76	86	178	309	457	611	.73	49.5
NOV	912	1500	40.1	2.96	306	449	598	747	897	.64	58.9
DEC	705	1235	33.0	3.09	529	682	837	992	1147	.58	62.9
YR	1763	1439	49.4	3.03	2362	3394	4619	6027	7628	.71	

TONOPAH, NEVADA ELEV 5423 LAT 38.1

	HS	VS	TA	SD	D50	D55	D60	D65	D70	KT	L-D
JAN	918	1552	30.2	4.47	620	771	925	1079	1234	.65	59.4
FEB	1274	1765	34.6	4.31	440	574	712	852	991	.69	51.7
MAR	1777	1725	39.6	3.24	334	481	633	788	942	.72	40.5
APR	2251	1404	48.1	3.86	151	246	371	512	659	.73	28.3
MAY	2577	1161	56.9	3.08	24	74	154	269	411	.74	19.1
JUN	2788	1089	65.3	3.07	0	9	33	92	179	.77	14.7
JUL	2703	1121	73.0	1.93	0	0	0	0	25	.76	16.7
AUG	2438	1325	70.7	2.21	0	0	0	13	63	.76	24.6
SEP	2043	1734	63.5	2.70	0	9	39	108	212	.77	36.3
OCT	1520	1945	52.1	3.39	81	160	269	407	557	.75	48.1
NOV	1031	1694	39.8	3.54	321	461	607	756	906	.69	57.5
DEC	827	1473	31.9	3.51	564	717	871	1026	1181	.65	61.5
YR	1848	1497	50.5	3.27	2535	3502	4614	5902	7360	.74	

WINNEMUCCA, NEVADA ELEV 4340 LAT 40.9

	HS	VS	TA	SD	D50	D55	D60	D65	D70	KT	L-D
JAN	690	1148	28.2	5.65	688	837	989	1142	1296	.55	62.2
FEB	1028	1449	34.1	4.22	452	588	726	866	1005	.60	54.5
MAR	1472	1494	37.6	2.34	386	540	694	849	1004	.63	43.3
APR	1967	1343	45.1	3.60	197	314	453	599	748	.65	31.1
MAY	2362	1188	53.8	3.39	59	131	228	359	506	.68	21.9
JUN	2569	1127	61.7	3.00	6	22	70	149	263	.71	17.5
JUL	2678	1217	71.0	1.81	0	0	0	6	45	.76	19.5
AUG	2348	1418	67.8	2.45	0	0	9	42	118	.74	27.4
SEP	1907	1778	59.2	2.90	10	38	104	199	330	.75	39.1
OCT	1322	1789	48.3	3.40	140	240	373	521	674	.70	50.9
NOV	809	1335	37.3	3.30	387	533	681	831	981	.60	60.3
DEC	618	1083	30.4	3.50	609	763	918	1073	1228	.55	64.3
YR	1650	1363	47.9	3.29	2934	4006	5245	6636	8198	.68	

CONCORD, NEW HAMPSHIRE ELEV 344 LAT 43.2

	HS	VS	TA	SD	D50	D55	D60	D65	D70	KT	L-D
JAN	459	696	20.6	3.90	912	1067	1221	1376	1531	.41	64.5
FEB	686	884	22.6	3.50	767	907	1047	1187	1327	.43	56.8
MAR	974	933	32.3	3.50	551	705	859	1014	1169	.43	45.6
APR	1317	920	44.2	2.60	193	328	475	624	774	.45	33.4
MAY	1582	889	55.1	2.80	30	90	183	315	464	.46	24.2
JUN	1705	879	64.7	1.70	0	0	8	58	165	.47	19.8
JUL	1675	894	69.7	2.10	0	0	0	16	74	.47	21.8
AUG	1455	930	67.2	2.30	0	0	9	45	124	.47	29.7
SEP	1140	992	59.5	2.40	0	23	83	182	318	.46	41.4
OCT	817	979	49.3	2.40	91	193	335	487	642	.46	53.2
NOV	463	648	38.0	2.60	363	511	660	810	960	.38	62.6
DEC	362	546	24.8	3.90	782	937	1091	1246	1401	.36	66.6
YR	1054	849	45.6	2.81	3689	4761	5971	7360	8949	.45	

NEWARK, NEW JERSEY ELEV 30 LAT 40.7

	HS	VS	TA	SD	D50	D55	D60	D65	D70	KT	L-D
JAN	552	814	31.4	4.31	583	734	888	1042	1197	.44	62.0
FEB	793	985	32.6	3.29	489	628	767	907	1047	.46	54.3
MAR	1109	1019	40.6	3.40	309	452	603	757	912	.47	43.1
APR	1449	956	51.7	2.65	59	139	258	401	549	.48	30.9
MAY	1687	894	61.9	2.73	0	17	62	143	263	.49	21.7
JUN	1795	878	71.4	1.85	0	0	0	5	38	.49	17.3
JUL	1760	889	76.4	1.78	0	0	0	0	5	.50	19.3
AUG	1565	941	74.6	1.89	0	0	0	0	13	.49	27.2
SEP	1273	1052	67.8	2.24	0	0	6	34	106	.50	38.9
OCT	951	1102	57.5	2.52	11	48	125	243	390	.50	50.7
NOV	596	838	46.2	2.32	140	269	415	564	714	.44	60.1
DEC	454	676	34.5	3.48	485	637	791	946	1101	.40	64.1
YR	1167	920	53.9	2.71	2076	2924	3915	5042	6335	.48	

ALBUQUERQUE, NEW MEXICO ELEV 5312 LAT 35.0

	HS	VS	TA	SD	D50	D55	D60	D65	D70	KT	L-D
JAN	1016	1562	35.2	3.40	464	615	769	924	1079	.65	56.3
FEB	1342	1664	40.0	3.90	298	426	562	701	840	.67	48.6
MAR	1768	1515	45.8	2.60	162	292	442	595	750	.69	37.4
APR	2228	1244	55.8	2.50	17	67	153	282	427	.71	25.2
MAY	2538	1052	65.3	1.90	0	0	10	58	158	.73	16.0
JUN	2679	997	74.6	2.30	0	0	0	0	20	.74	11.6
JUL	2489	995	78.7	1.70	0	0	0	0	0	.70	13.6
AUG	2290	1132	76.6	1.50	0	0	0	0	0	.70	21.5
SEP	1972	1468	70.1	1.70	0	0	0	7	52	.72	33.2
OCT	1547	1745	58.2	2.10	0	28	99	218	367	.71	45.0
NOV	1134	1692	44.5	2.70	188	320	466	615	765	.68	54.4
DEC	928	1508	36.2	3.00	431	584	738	893	1048	.64	58.4
YR	1829	1379	56.8	2.43	1560	2332	3239	4293	5506	.70	

CLAYTON, NEW MEXICO

ELEV 4970 LAT 36.4

	HS	VS	TA	SD	D50	D55	D60	D65	D70	KT	L-D
JAN	962	1536	33.1	3.77	529	681	834	989	1144	.64	57.7
FEB	1241	1576	36.1	3.95	397	532	670	809	949	.64	50.0
MAR	1652	1466	40.4	3.72	320	460	610	763	918	.65	38.8
APR	2039	1199	50.8	3.38	95	177	293	431	577	.66	26.6
MAY	2222	1007	60.0	2.21	0	17	72	172	313	.64	17.4
JUN	2418	984	69.2	2.91	0	0	10	38	104	.67	13.0
JUL	2284	981	73.6	2.15	0	0	0	0	25	.64	15.0
AUG	2097	1100	72.4	1.97	0	0	0	5	32	.65	22.9
SEP	1802	1390	65.0	2.35	0	0	18	73	168	.67	34.6
OCT	1433	1660	54.8	2.88	34	98	192	324	473	.68	46.4
NOV	1028	1556	42.3	3.49	257	389	533	682	831	.65	55.8
DEC	861	1439	35.1	3.07	465	618	772	927	1082	.63	59.8
YR	1671	1323	52.7	2.98	2097	2972	4004	5213	6616	.65	

FARMINGTON, NEW MEXICO

ELEV 5502 LAT 36.7

	HS	VS	TA	SD	D50	D55	D60	D65	D70	KT	L-D
JAN	944	1513	28.6	4.54	668	820	974	1129	1284	.64	58.0
FEB	1281	1673	35.0	3.95	426	562	701	840	980	.67	50.3
MAR	1693	1531	40.6	2.67	299	448	602	756	911	.67	39.1
APR	2133	1265	49.7	3.69	120	207	326	465	611	.69	26.9
MAY	2452	1080	59.5	2.15	0	19	79	184	327	.70	17.7
JUN	2665	1032	67.9	2.35	0	0	7	36	108	.73	13.3
JUL	2478	1032	75.0	1.56	0	0	0	0	6	.70	15.3
AUG	2252	1179	72.6	2.16	0	0	0	6	36	.70	23.2
SEP	1934	1531	64.6	1.95	0	0	13	67	171	.72	34.9
OCT	1479	1759	52.9	2.53	43	118	233	378	531	.71	46.7
NOV	1047	1622	39.2	3.04	332	476	624	774	924	.66	56.1
DEC	837	1398	30.1	4.55	623	774	928	1082	1237	.62	60.1
YR	1768	1382	51.3	2.92	2511	3424	4487	5717	7126	.69	

LOS ALAMOS, NEW MEXICO

ELEV 7380 LAT 35.9

	HS	VS	TA	SD	D50	D55	D60	D65	D70	KT	L-D
JAN	893	1340	29.1	3.75	650	804	958	1113	1268	.59	57.2
FEB	1273	1601	32.3	4.07	500	637	776	916	1056	.65	49.5
MAR	1547	1323	37.6	3.24	392	541	695	850	1004	.61	38.3
APR	1988	1151	45.7	3.41	180	296	434	580	729	.64	26.1
MAY	2050	945	54.9	2.56	25	86	182	318	469	.59	16.9
JUN	2308	954	65.1	2.87	0	7	29	88	178	.64	12.5
JUL	2051	913	68.3	1.83	0	0	0	20	89	.58	14.5
AUG	1875	987	65.9	1.99	0	0	9	52	145	.58	22.4
SEP	1673	1250	60.3	2.59	5	22	77	167	297	.61	34.1
OCT	1406	1582	50.5	3.22	99	186	309	454	606	.66	45.9
NOV	990	1436	37.9	3.17	369	515	663	813	963	.61	55.3
DEC	887	1470	30.6	3.02	602	757	911	1066	1221	.63	59.3
YR	1579	1244	48.2	2.97	2822	3851	5043	6437	8025	.61	

ROSWELL, NEW MEXICO

ELEV 3619 LAT 33.4

	HS	VS	TA	SD	D50	D55	D60	D65	D70	KT	L-D
JAN	1046	1516	38.1	3.90	385	530	681	835	989	.63	54.7
FEB	1373	1607	42.9	3.50	224	347	481	619	759	.66	47.0
MAR	1807	1464	49.3	3.10	114	209	341	489	642	.69	35.8
APR	2218	1175	59.7	2.80	8	31	92	185	315	.70	23.6
MAY	2459	997	68.5	1.90	0	0	0	20	87	.71	14.4
JUN	2610	962	77.0	1.90	0	0	0	0	0	.72	10.0
JUL	2441	957	79.2	2.10	0	0	0	0	0	.69	12.0
AUG	2242	1063	77.9	2.10	0	0	0	0	5	.69	19.9
SEP	1913	1335	70.4	2.40	0	0	0	17	70	.68	31.6
OCT	1527	1603	59.6	2.80	8	34	99	195	329	.68	43.4
NOV	1131	1567	46.9	3.10	148	260	397	544	693	.65	52.8
DEC	952	1453	39.3	2.90	339	488	642	797	952	.62	56.8
YR	1811	1306	59.1	2.70	1226	1899	2733	3701	4841	.68	

TRUTH OR CONSEQ., NEW MEXICO

ELEV 4859 LAT 33.2

	HS	VS	TA	SD	D50	D55	D60	D65	D70	KT	L-D
JAN	1118	1661	40.0	3.52	327	470	622	776	930	.67	54.5
FEB	1451	1724	44.9	3.69	187	298	428	564	703	.70	46.8
MAR	1886	1532	50.2	2.69	86	178	311	460	614	.71	35.6
APR	2338	1220	59.5	2.73	7	31	93	188	320	.74	23.4
MAY	2557	1010	68.2	1.79	0	0	0	19	89	.73	14.2
JUN	2650	963	76.9	2.10	0	0	0	0	7	.73	9.8
JUL	2365	942	79.3	1.63	0	0	0	0	0	.67	11.8
AUG	2216	1048	77.4	1.75	0	0	0	0	0	.68	19.7
SEP	1940	1346	71.6	1.91	0	0	0	5	38	.69	31.4
OCT	1579	1665	61.3	2.32	0	12	57	144	275	.70	43.2
NOV	1217	1733	48.7	2.69	105	207	343	490	639	.69	52.6
DEC	1003	1558	40.8	2.84	295	443	596	750	905	.65	56.6
YR	1861	1364	59.9	2.46	1007	1639	2450	3396	4520	.70	

TUCUMCARI, NEW MEXICO

ELEV 4039 LAT 35.2

	HS	VS	TA	SD	D50	D55	D60	D65	D70	KT	L-D
JAN	1009	1560	37.0	3.70	413	561	714	868	1023	.65	56.5
FEB	1297	1597	41.1	3.80	270	396	531	670	809	.65	48.8
MAR	1712	1466	46.7	3.40	168	280	419	569	723	.67	37.6
APR	2098	1185	56.9	3.10	22	70	148	260	397	.67	25.4
MAY	2314	1005	65.6	2.00	0	0	10	57	152	.66	16.2
JUN	2484	974	75.1	2.80	0	0	0	7	28	.69	11.8
JUL	2349	972	78.4	2.40	0	0	0	0	7	.66	13.8
AUG	2164	1090	76.7	2.10	0	0	0	0	8	.67	21.7
SEP	1829	1354	69.6	2.30	0	0	0	20	78	.67	33.4
OCT	1443	1595	58.7	2.80	11	42	114	217	355	.67	45.2
NOV	1073	1571	46.2	3.30	167	281	419	565	714	.65	54.6
DEC	910	1478	38.6	3.40	364	512	664	819	973	.63	58.6
YR	1725	1319	57.6	2.92	1415	2142	3019	4052	5267	.66	

ZUNI, NEW MEXICO

ELEV 6447 LAT 35.1

	HS	VS	TA	SD	D50	D55	D60	D65	D70	KT	L-D
JAN	986	1500	30.3	3.67	613	766	921	1076	1231	.63	56.4
FEB	1297	1590	34.6	4.55	442	575	713	852	991	.65	48.7
MAR	1688	1434	39.6	3.03	332	480	633	788	942	.66	37.5
APR	2167	1217	48.1	2.85	120	225	361	508	657	.69	25.3
MAY	2473	1041	56.6	2.08	8	45	131	264	416	.71	16.1
JUN	2602	989	65.4	2.36	0	0	16	68	159	.72	11.7
JUL	2264	952	71.4	1.78	0	0	0	5	38	.64	13.7
AUG	2078	1052	69.4	1.87	0	0	0	13	71	.64	21.6
SEP	1895	1406	63.3	2.04	0	0	23	91	207	.69	33.3
OCT	1496	1671	52.5	2.73	54	132	247	391	543	.69	45.1
NOV	1088	1596	40.1	3.46	312	452	598	747	897	.66	54.5
DEC	893	1429	32.0	3.60	561	714	868	1023	1178	.62	58.5
YR	1745	1321	50.3	2.82	2442	3389	4511	5826	7330	.67	

ALBANY, NEW YORK

ELEV 292 LAT 42.7

	HS	VS	TA	SD	D50	D55	D60	D65	D70	KT	L-D
JAN	456	674	21.5	4.90	886	1039	1194	1349	1504	.39	64.0
FEB	688	871	23.5	4.13	743	882	1022	1162	1302	.43	56.3
MAR	986	933	33.4	4.01	521	672	825	980	1135	.43	45.1
APR	1335	922	46.9	3.38	156	264	399	545	694	.45	32.9
MAY	1570	873	57.7	3.33	24	71	148	253	389	.46	23.7
JUN	1730	882	67.5	2.32	0	0	8	39	114	.47	19.3
JUL	1725	908	72.0	2.24	0	0	0	9	45	.49	21.3
AUG	1499	947	69.6	2.34	0	0	0	22	84	.48	29.2
SEP	1170	1009	61.9	2.66	0	14	57	135	253	.47	40.9
OCT	817	961	51.4	2.52	63	147	274	423	577	.45	52.7
NOV	457	623	39.6	3.02	320	464	613	762	912	.36	62.1
DEC	356	521	25.9	4.18	749	903	1057	1212	1367	.34	66.1
YR	1067	843	47.6	3.25	3462	4456	5597	6891	8376	.45	

BINGHAMTON, NEW YORK

ELEV 1637 LAT 42.2

	HS	VS	TA	SD	D50	D55	D60	D65	D70	KT	L-D
JAN	386	520	22.0	4.75	870	1024	1178	1333	1488	.33	63.5
FEB	576	667	22.8	3.68	762	902	1042	1182	1322	.35	55.8
MAR	861	772	31.3	4.61	588	738	891	1045	1200	.37	44.6
APR	1242	840	44.7	3.42	200	322	463	610	759	.42	32.4
MAY	1496	828	55.1	3.21	40	104	194	320	466	.43	23.2
JUN	1681	855	64.8	2.31	0	0	19	75	172	.46	18.8
JUL	1659	871	69.1	2.13	0	0	0	21	85	.47	20.8
AUG	1425	888	67.3	2.18	0	0	7	40	119	.45	28.7
SEP	1131	952	60.2	2.71	6	25	82	172	300	.45	40.4
OCT	779	883	50.3	2.90	91	181	310	458	611	.42	52.2
NOV	414	531	38.2	2.93	359	505	654	804	954	.32	61.6
DEC	297	395	25.4	4.29	765	918	1073	1228	1383	.28	65.6
YR	997	750	46.0	3.26	3681	4719	5913	7288	8859	.42	

BUFFALO, NEW YORK

ELEV 705 LAT 42.9

	HS	VS	TA	SD	D50	D55	D60	D65	D70	KT	L-D
JAN	349	465	23.7	4.36	817	971	1126	1280	1435	.30	64.2
FEB	546	636	24.4	3.29	717	857	997	1137	1277	.34	56.5
MAR	888	820	32.1	4.15	561	712	866	1020	1175	.39	45.3
APR	1315	911	44.9	3.32	193	316	457	604	753	.44	33.1
MAY	1596	890	55.1	3.30	43	107	197	321	466	.46	23.9
JUN	1804	915	65.7	2.18	0	0	12	58	148	.49	19.5
JUL	1776	935	70.1	2.01	0	0	0	12	65	.50	21.5
AUG	1513	961	68.4	2.35	0	0	6	33	104	.48	29.4
SEP	1152	996	61.6	2.54	0	14	57	138	260	.46	41.1
OCT	784	914	51.5	3.00	77	159	278	422	574	.44	52.9
NOV	403	526	39.8	2.75	312	457	606	756	906	.32	62.3
DEC	283	380	27.9	3.78	687	841	995	1150	1305	.28	66.3
YR	1036	780	47.1	3.09	3407	4434	5597	6931	8468	.44	

MASSENA, NEW YORK ELEV 207 LAT 44.9

	HS	VS	TA	SD	D50	D55	D60	D65	D70	KT	L-D
JAN	391	596	14.5	5.58	1102	1256	1411	1566	1721	.38	66.2
FEB	620	819	16.7	4.49	933	1073	1212	1352	1492	.41	58.5
MAR	977	992	27.6	4.60	699	851	1005	1160	1315	.45	47.3
APR	1343	986	42.2	3.15	253	389	535	684	834	.46	35.1
MAY	1613	939	54.1	3.40	56	127	221	350	497	.47	25.9
JUN	1779	940	64.3	2.14	0	0	18	78	182	.49	21.5
JUL	1751	964	64.3	2.16	0	0	20	83	189	.50	23.5
AUG	1484	993	66.7	2.48	0	0	13	57	140	.48	31.4
SEP	1124	1029	59.2	2.58	6	29	94	192	328	.47	43.1
OCT	736	901	48.5	2.73	115	220	361	512	667	.43	54.9
NOV	388	541	35.9	3.22	427	574	723	873	1023	.34	64.3
DEC	294	439	20.1	5.08	929	1083	1237	1392	1547	.32	68.3
YR	1043	845	43.2	3.47	4520	5602	6850	8299	9935	.46	

NEW YORK, NEW YORK ELEV 187 LAT 40.8

	HS	VS	TA	SD	D50	D55	D60	D65	D70	KT	L-D
JAN	500	708	32.2	4.09	558	709	863	1017	1172	.40	62.1
FEB	721	865	33.4	3.22	466	605	745	885	1025	.42	54.4
MAR	1037	937	41.1	3.42	296	437	588	741	896	.44	43.2
APR	1364	898	52.1	2.73	56	133	249	390	538	.45	31.0
MAY	1636	873	62.3	2.79	0	16	59	137	253	.47	21.8
JUN	1710	848	71.6	1.83	0	0	0	0	35	.47	17.4
JUL	1688	861	76.6	1.82	0	0	0	0	5	.48	19.4
AUG	1483	894	74.9	1.83	0	0	0	0	10	.47	27.3
SEP	1214	996	68.4	2.30	0	0	5	29	98	.47	39.0
OCT	895	1017	58.7	2.40	6	31	100	209	353	.47	50.8
NOV	533	716	47.4	2.48	121	237	380	528	678	.39	60.2
DEC	404	574	35.5	3.60	456	607	760	915	1070	.36	64.2
YR	1100	849	54.5	2.71	1959	2775	3749	4851	6133	.45	

ROCHESTER, NEW YORK ELEV 554 LAT 43.1

	HS	VS	TA	SD	D50	D55	D60	D65	D70	KT	L-D
JAN	364	497	24.0	4.50	808	962	1116	1271	1426	.32	64.4
FEB	559	661	24.8	3.60	706	846	986	1126	1266	.35	56.7
MAR	903	843	33.0	4.40	536	686	838	993	1147	.40	45.5
APR	1339	935	46.1	3.50	175	287	423	569	718	.45	33.3
MAY	1606	899	56.5	3.40	34	90	171	285	425	.47	24.1
JUN	1817	924	66.9	2.30	0	0	9	46	125	.50	19.7
JUL	1781	941	71.2	2.10	0	0	0	9	52	.50	21.7
AUG	1519	970	69.3	2.40	0	0	5	26	91	.49	29.6
SEP	1160	1011	62.3	2.60	0	12	50	126	241	.47	41.3
OCT	782	917	52.3	3.00	66	144	257	398	550	.44	53.1
NOV	404	532	40.5	3.00	295	438	586	735	885	.33	62.5
DEC	281	380	28.3	4.00	675	829	983	1138	1293	.28	66.5
YR	1045	793	47.9	3.23	3295	4294	5424	6722	8219	.44	

SYRACUSE, NEW YORK ELEV 407 LAT 43.1

	HS	VS	TA	SD	D50	D55	D60	D65	D70	KT	L-D
JAN	385	538	23.6	4.67	821	974	1129	1284	1438	.34	64.4
FEB	571	681	24.6	3.58	712	851	991	1131	1271	.36	56.7
MAR	890	827	33.2	4.56	532	680	833	987	1141	.39	45.5
APR	1324	923	46.5	3.26	160	272	410	556	705	.45	33.3
MAY	1578	885	56.8	3.10	25	76	156	272	414	.46	24.1
JUN	1778	908	66.9	2.31	0	0	9	46	125	.49	19.7
JUL	1758	931	71.5	2.25	0	0	0	11	52	.50	21.7
AUG	1504	960	69.7	2.19	0	0	0	18	77	.48	29.6
SEP	1165	1017	62.8	2.68	0	11	46	120	229	.47	41.3
OCT	777	909	52.5	2.88	59	137	249	392	543	.43	53.1
NOV	399	523	41.0	3.10	283	423	571	720	870	.32	62.5
DEC	285	387	28.1	4.18	682	835	989	1144	1299	.28	66.5
YR	1036	791	48.1	3.23	3274	4259	5383	6681	8164	.44	

ASHEVILLE, NORTH CAROLINA ELEV 2169 LAT 35.4

	HS	VS	TA	SD	D50	D55	D60	D65	D70	KT	L-D
JAN	722	958	37.9	4.77	404	543	691	842	996	.47	56.7
FEB	971	1069	39.4	4.47	321	446	580	718	857	.49	49.0
MAR	1306	1053	45.9	4.24	209	317	451	597	749	.51	37.8
APR	1668	960	55.9	2.48	16	65	151	279	424	.53	25.6
MAY	1804	858	63.7	2.40	0	6	31	100	209	.52	16.4
JUN	1854	839	70.6	2.29	0	0	0	14	63	.51	12.0
JUL	1776	827	73.5	1.52	0	0	0	0	11	.50	14.0
AUG	1627	864	72.8	1.76	0	0	0	0	22	.50	21.9
SEP	1361	969	66.7	2.36	0	0	11	50	131	.50	33.6
OCT	1147	1171	56.8	2.95	22	71	152	269	413	.53	45.4
NOV	849	1120	46.3	2.80	151	271	413	562	711	.52	54.8
DEC	658	918	38.7	3.65	365	510	662	816	970	.46	58.8
YR	1313	966	55.7	2.97	1488	2229	3142	4247	5556	.51	

CAPE HATTERAS, NORTH CAROLINA ELEV 7 LAT 35.3

	HS	VS	TA	SD	D50	D55	D60	D65	D70	KT	L-D
JAN	686	886	45.3	4.40	225	336	470	616	768	.44	56.6
FEB	952	1036	45.8	3.80	173	278	404	540	678	.48	48.9
MAR	1326	1070	50.6	4.10	126	210	322	458	606	.52	37.7
APR	1774	1015	58.9	1.80	0	14	74	188	333	.57	25.5
MAY	1962	907	67.0	2.30	0	0	10	47	128	.56	16.3
JUN	2036	886	74.3	2.30	0	0	0	0	22	.56	11.9
JUL	1921	869	78.0	1.50	0	0	0	0	0	.54	13.9
AUG	1705	897	77.5	1.30	0	0	0	0	0	.53	21.8
SEP	1470	1055	73.7	1.90	0	0	0	0	17	.54	33.5
OCT	1137	1153	65.2	2.40	0	0	20	76	171	.53	45.3
NOV	873	1162	56.0	2.60	18	67	152	277	421	.53	54.7
DEC	659	916	47.7	3.70	160	261	394	540	693	.46	58.7
YR	1376	987	61.7	2.67	702	1166	1846	2742	3837	.53	

CHARLOTTE, NORTH CAROLINA ELEV 768 LAT 35.2

	HS	VS	TA	SD	D50	D55	D60	D65	D70	KT	L-D
JAN	719	944	42.1	4.65	299	424	565	714	867	.46	56.5
FEB	971	1061	44.0	4.08	213	325	454	590	729	.49	48.8
MAR	1317	1057	50.6	4.35	134	218	327	461	607	.51	37.6
APR	1695	970	60.8	2.16	0	11	56	145	279	.54	25.4
MAY	1856	872	68.8	2.54	0	0	7	34	103	.53	16.2
JUN	1921	855	75.9	2.20	0	0	0	0	11	.53	11.8
JUL	1831	841	78.5	1.44	0	0	0	0	0	.52	13.8
AUG	1695	891	77.7	1.53	0	0	0	0	0	.52	21.7
SEP	1416	1007	72.0	2.39	0	0	0	10	47	.52	33.4
OCT	1173	1198	61.7	2.91	5	21	71	152	271	.54	45.2
NOV	865	1141	51.0	2.58	67	151	277	421	570	.52	54.6
DEC	672	938	42.5	3.89	271	402	547	699	853	.47	58.6
YR	1345	981	60.5	2.89	989	1552	2304	3226	4337	.52	

GREENSBORO, NORTH CAROLINA ELEV 886 LAT 36.1

	HS	VS	TA	SD	D50	D55	D60	D65	D70	KT	L-D
JAN	715	973	38.7	4.50	379	517	665	817	971	.47	57.4
FEB	970	1096	40.6	4.50	294	416	548	685	824	.50	49.7
MAR	1313	1085	47.8	4.60	188	282	404	544	693	.52	38.5
APR	1683	986	58.6	2.30	5	27	94	203	344	.54	26.3
MAY	1868	889	67.1	2.70	0	0	15	59	138	.54	17.1
JUN	1953	872	74.4	2.10	0	0	0	0	17	.54	12.7
JUL	1864	861	77.2	1.60	0	0	0	0	0	.53	14.7
AUG	1697	910	76.0	1.40	0	0	0	0	0	.52	22.6
SEP	1418	1037	69.7	2.50	0	0	5	24	83	.52	34.3
OCT	1141	1193	59.2	3.00	12	43	112	209	343	.54	46.1
NOV	839	1134	48.3	2.60	109	215	354	502	651	.52	55.5
DEC	659	949	39.6	3.80	342	484	635	788	943	.47	59.5
YR	1344	998	58.1	2.96	1329	1984	2832	3831	5007	.52	

RALEIGH-DURHAM, NORTH CAROLINA ELEV 440 LAT 35.9

	HS	VS	TA	SD	D50	D55	D60	D65	D70	KT	L-D
JAN	694	924	40.5	4.36	330	464	610	762	915	.46	57.2
FEB	943	1046	42.2	4.02	249	370	502	640	779	.48	49.5
MAR	1276	1040	49.2	4.43	159	248	364	502	650	.50	38.3
APR	1644	959	59.5	2.32	0	21	81	180	318	.53	26.1
MAY	1808	866	67.4	2.50	0	0	11	48	127	.52	16.9
JUN	1864	846	74.4	2.01	0	0	0	0	15	.51	12.5
JUL	1776	832	77.5	1.90	0	0	0	0	0	.50	14.5
AUG	1611	865	76.5	1.73	0	0	0	0	0	.50	22.4
SEP	1377	996	70.6	2.17	0	0	0	12	60	.51	34.1
OCT	1105	1134	60.2	3.00	9	34	96	186	314	.52	45.9
NOV	812	1072	50.0	2.42	76	170	304	451	600	.50	55.3
DEC	636	893	41.2	3.78	300	437	586	739	893	.45	59.3
YR	1296	955	59.1	2.88	1123	1744	2554	3520	4671	.50	

BISMARCK, NORTH DAKOTA ELEV 1647 LAT 46.8

	HS	VS	TA	SD	D50	D55	D60	D65	D70	KT	L-D
JAN	467	857	8.2	8.00	1301	1454	1608	1762	1916	.50	68.1
FEB	776	1237	13.5	5.70	1023	1162	1302	1442	1582	.55	60.4
MAR	1168	1358	25.1	7.10	792	938	1088	1240	1394	.56	49.2
APR	1459	1150	43.0	4.30	259	380	518	663	811	.51	37.0
MAY	1848	1127	54.4	3.20	47	115	209	339	487	.54	27.8
JUN	2060	1116	63.8	3.30	5	17	53	122	217	.57	23.4
JUL	2184	1241	70.8	2.50	0	0	0	18	71	.62	25.4
AUG	1877	1369	69.2	2.70	0	0	8	35	102	.62	33.3
SEP	1354	1408	57.5	3.50	27	75	150	252	384	.58	45.0
OCT	908	1321	46.8	3.70	175	283	419	568	720	.56	56.8
NOV	507	877	28.9	5.20	642	787	935	1084	1233	.49	66.2
DEC	373	688	15.6	6.40	1070	1223	1377	1532	1687	.46	70.2
YR	1251	1145	41.4	4.63	5341	6434	7667	9057	10604	.56	

FARGO, NORTH DAKOTA — ELEV 899 LAT 46.9

	HS	VS	TA	SD	D50	D55	D60	D65	D70	KT	L-D
JAN	415	720	5.9	7.04	1369	1523	1678	1832	1987	.44	68.2
FEB	706	1078	10.7	5.37	1101	1241	1380	1520	1660	.51	60.5
MAR	1098	1249	24.2	6.62	814	962	1114	1267	1421	.52	49.3
APR	1476	1170	42.3	4.25	273	398	537	683	832	.52	37.1
MAY	1835	1122	54.6	3.22	45	113	205	334	481	.54	27.9
JUN	1994	1087	64.7	2.93	0	9	34	97	188	.55	23.5
JUL	2120	1210	70.7	2.22	0	0	0	13	63	.60	25.5
AUG	1825	1330	69.2	2.63	0	0	8	33	100	.60	33.4
SEP	1304	1342	57.9	3.06	17	56	129	234	368	.56	45.1
OCT	874	1253	47.0	3.59	168	276	412	561	714	.54	56.9
NOV	457	753	28.6	4.45	646	794	943	1092	1242	.44	66.3
DEC	337	594	13.0	5.71	1148	1303	1457	1612	1767	.42	70.3
YR	1206	1075	40.8	4.25	5581	6675	7897	9278	10823	.54	

MINOT, NORTH DAKOTA — ELEV 1713 LAT 48.3

	HS	VS	TA	SD	D50	D55	D60	D65	D70	KT	L-D
JAN	384	691	7.9	7.97	1310	1463	1617	1771	1926	.45	69.6
FEB	656	1029	12.8	5.65	1042	1182	1322	1462	1602	.50	61.9
MAR	1044	1227	23.6	6.81	833	981	1133	1286	1440	.51	50.7
APR	1461	1208	41.1	4.59	308	435	574	720	868	.52	38.5
MAY	1846	1174	52.8	3.03	60	136	244	384	535	.55	29.3
JUN	1975	1116	62.0	3.21	7	24	72	150	259	.54	24.9
JUL	2098	1249	68.8	2.26	0	0	5	27	94	.60	26.9
AUG	1800	1371	67.2	3.08	0	6	22	70	148	.60	34.8
SEP	1277	1373	56.2	3.62	39	97	176	286	421	.56	46.5
OCT	850	1277	46.1	4.13	202	310	444	591	743	.55	58.3
NOV	438	759	27.9	5.67	674	818	965	1114	1264	.45	67.7
DEC	310	567	14.7	6.66	1099	1252	1406	1560	1715	.42	71.7
YR	1181	1087	40.1	4.72	5574	6704	7980	9421	11015	.54	

AKRON-CANTON, OHIO — ELEV 1237 LAT 40.9

	HS	VS	TA	SD	D50	D55	D60	D65	D70	KT	L-D
JAN	428	570	26.3	5.10	740	892	1046	1200	1355	.34	62.2
FEB	649	750	27.7	4.08	626	765	905	1044	1184	.38	54.5
MAR	964	856	36.2	4.74	450	592	742	894	1048	.41	43.3
APR	1357	896	48.5	3.17	123	221	352	497	646	.45	31.1
MAY	1668	889	58.7	3.43	21	61	134	231	362	.48	21.9
JUN	1839	897	68.3	2.42	0	0	7	33	103	.50	17.5
JUL	1787	903	71.7	2.19	0	0	0	9	48	.50	19.5
AUG	1596	964	70.3	2.27	0	0	0	16	70	.50	27.4
SEP	1272	1058	63.7	2.55	0	7	33	101	204	.50	39.1
OCT	908	1041	53.3	2.96	52	125	230	369	519	.48	50.9
NOV	505	666	40.7	2.54	285	430	579	729	879	.37	60.3
DEC	353	476	29.4	4.80	646	797	950	1104	1259	.31	64.3
YR	1112	831	49.6	3.36	2943	3890	4978	6227	7677	.46	

CINCINNATI, OHIO — ELEV 889 LAT 39.1

	HS	VS	TA	SD	D50	D55	D60	D65	D70	KT	L-D
JAN	500	659	31.1	4.90	596	745	898	1052	1206	.37	60.4
FEB	738	839	33.3	4.40	475	610	749	888	1028	.41	52.7
MAR	1027	878	41.7	5.05	318	441	580	728	880	.42	41.5
APR	1398	883	53.9	3.10	46	115	210	341	485	.46	29.3
MAY	1672	861	63.2	3.30	6	21	63	138	240	.48	20.1
JUN	1837	872	72.1	2.40	0	0	0	10	46	.50	15.7
JUL	1771	869	75.6	2.20	0	0	0	0	13	.50	17.7
AUG	1634	943	74.4	2.10	0	0	0	0	18	.51	25.6
SEP	1312	1038	67.8	2.60	0	0	10	44	117	.50	37.3
OCT	990	1098	56.8	3.05	24	74	155	271	414	.50	49.1
NOV	588	769	43.8	2.60	203	339	487	636	786	.41	58.5
DEC	432	587	33.7	4.70	519	666	818	971	1126	.35	62.5
YR	1160	858	54.0	3.37	2187	3011	3970	5079	6359	.47	

CLEVELAND, OHIO — ELEV 804 LAT 41.4

	HS	VS	TA	SD	D50	D55	D60	D65	D70	KT	L-D
JAN	388	507	26.9	5.00	722	874	1027	1182	1336	.32	62.7
FEB	601	687	27.9	3.95	620	759	899	1039	1179	.36	55.0
MAR	922	822	36.1	4.78	453	595	745	898	1052	.40	43.8
APR	1349	901	48.3	3.37	133	230	360	504	652	.45	31.6
MAY	1681	904	58.3	3.58	26	71	146	244	375	.49	22.4
JUN	1843	906	67.9	2.49	0	0	8	40	112	.51	18.0
JUL	1828	929	71.4	2.14	0	0	0	9	50	.52	20.0
AUG	1583	969	70.0	2.23	0	0	0	17	73	.50	27.9
SEP	1239	1041	63.9	2.47	0	6	29	95	198	.49	39.6
OCT	867	995	53.8	3.01	48	119	219	354	504	.46	51.4
NOV	466	607	41.6	2.80	263	404	553	702	852	.35	60.8
DEC	318	420	30.3	4.59	618	769	922	1076	1231	.29	64.8
YR	1092	808	49.7	3.37	2883	3827	4908	6160	7614	.45	

COLUMBUS, OHIO — ELEV 833 LAT 40.0

	HS	VS	TA	SD	D50	D55	D60	D65	D70	KT	L-D
JAN	459	606	28.4	4.98	677	828	981	1135	1290	.35	61.3
FEB	677	769	30.3	4.30	555	693	832	972	1112	.39	53.6
MAR	980	851	39.2	4.55	367	503	650	802	956	.41	42.4
APR	1353	873	51.2	3.18	83	163	279	418	565	.45	30.2
MAY	1647	864	61.1	3.29	10	34	92	176	294	.48	21.0
JUN	1813	875	70.4	2.21	0	0	0	13	64	.50	16.6
JUL	1755	876	73.6	1.93	0	0	0	0	20	.50	18.6
AUG	1641	969	71.9	2.19	0	0	0	8	45	.52	26.5
SEP	1282	1038	65.2	2.51	0	0	20	76	167	.50	38.2
OCT	945	1064	54.2	2.97	42	111	208	342	492	.49	50.0
NOV	538	703	41.7	2.60	258	401	549	699	849	.38	59.4
DEC	387	521	30.7	4.94	608	758	910	1064	1219	.33	63.4
YR	1125	834	51.5	3.30	2600	3491	4521	5705	7073	.46	

DAYTON, OHIO — ELEV 1004 LAT 39.9

	HS	VS	TA	SD	D50	D55	D60	D65	D70	KT	L-D
JAN	489	660	28.1	5.00	686	837	990	1144	1299	.37	61.2
FEB	725	843	30.4	4.50	553	691	829	969	1109	.41	53.5
MAR	1025	898	39.0	4.90	379	513	658	809	962	.43	42.3
APR	1403	905	51.4	3.30	84	163	276	413	559	.46	30.1
MAY	1699	886	61.6	3.30	9	31	85	166	281	.49	20.9
JUN	1874	895	71.3	2.40	0	0	0	13	57	.51	16.5
JUL	1810	896	74.6	2.10	0	0	0	0	17	.51	18.5
AUG	1645	968	73.0	2.30	0	0	0	7	35	.52	26.4
SEP	1318	1071	66.3	2.60	0	0	16	63	145	.51	38.1
OCT	969	1098	55.5	3.10	34	95	182	307	453	.50	49.9
NOV	564	749	41.8	2.80	258	399	547	696	846	.40	59.3
DEC	407	558	30.9	4.80	601	751	904	1058	1212	.34	63.3
YR	1162	869	52.0	3.42	2604	3480	4487	5645	6975	.47	

TOLEDO, OHIO — ELEV 692 LAT 41.6

	HS	VS	TA	SD	D50	D55	D60	D65	D70	KT	L-D
JAN	435	600	24.8	4.97	785	938	1092	1247	1401	.36	62.9
FEB	680	821	27.1	3.65	642	781	921	1061	1201	.41	55.2
MAR	997	914	35.8	4.57	458	603	753	906	1061	.43	44.0
APR	1384	932	48.4	3.18	125	224	355	500	649	.46	31.8
MAY	1717	925	58.8	3.44	21	61	133	229	359	.50	22.6
JUN	1878	922	68.9	2.59	0	0	7	32	99	.52	18.2
JUL	1849	941	72.3	2.00	0	0	0	5	34	.52	20.2
AUG	1616	994	70.8	2.48	0	0	0	18	70	.51	28.1
SEP	1276	1087	63.8	2.56	0	7	32	99	202	.50	39.8
OCT	911	1074	53.0	3.20	63	138	243	379	529	.49	51.6
NOV	498	672	39.6	2.99	320	464	612	762	912	.38	61.0
DEC	355	494	28.0	4.57	686	839	993	1147	1302	.32	65.0
YR	1135	865	49.3	3.35	3100	4055	5141	6385	7819	.47	

YOUNGSTOWN, OHIO — ELEV 1184 LAT 41.3

	HS	VS	TA	SD	D50	D55	D60	D65	D70	KT	L-D
JAN	385	499	25.7	4.97	758	910	1064	1219	1373	.31	62.6
FEB	586	662	26.7	3.95	653	793	933	1072	1212	.35	54.9
MAR	890	784	35.3	4.62	473	618	769	922	1076	.38	43.7
APR	1278	848	47.7	3.34	141	244	376	521	670	.43	31.5
MAY	1586	857	57.6	3.45	27	76	153	258	394	.46	22.3
JUN	1759	873	67.0	2.22	0	0	8	42	121	.48	17.9
JUL	1734	888	70.7	1.94	0	0	0	9	54	.49	19.9
AUG	1506	919	69.2	2.22	0	0	0	22	86	.48	27.8
SEP	1194	991	62.7	2.57	0	10	44	118	230	.47	39.5
OCT	851	966	52.6	2.76	54	131	245	388	540	.45	51.3
NOV	456	586	40.3	2.62	297	442	591	741	891	.34	60.7
DEC	315	413	28.8	4.51	662	814	968	1123	1277	.28	64.7
YR	1047	774	48.7	3.26	3065	4038	5151	6435	7924	.43	

OKLAHOMA CITY, OKLAHOMA — ELEV 1302 LAT 35.4

	HS	VS	TA	SD	D50	D55	D60	D65	D70	KT	L-D
JAN	801	1114	36.8	4.07	424	569	721	875	1029	.52	56.7
FEB	1055	1200	41.3	3.92	268	392	526	665	804	.53	49.0
MAR	1400	1147	48.2	4.54	179	272	392	532	680	.55	37.8
APR	1725	991	60.4	3.21	10	35	95	180	301	.55	25.6
MAY	1918	895	68.3	2.42	0	0	7	36	108	.55	16.4
JUN	2144	912	76.8	2.52	0	0	0	0	13	.59	12.0
JUL	2128	925	81.5	2.76	0	0	0	0	5	.60	14.0
AUG	1950	1007	81.1	2.88	0	0	0	0	6	.60	21.9
SEP	1554	1128	73.0	2.78	0	0	0	12	47	.57	33.6
OCT	1233	1292	62.4	3.19	7	23	70	148	258	.57	45.4
NOV	901	1222	49.2	3.19	113	206	333	476	625	.55	54.8
DEC	725	1059	40.0	3.39	325	470	621	775	930	.51	58.8
YR	1463	1073	59.9	3.24	1326	1967	2765	3699	4806	.56	

TULSA, OKLAHOMA — ELEV 676 LAT 36.2

	HS	VS	TA	SD	D50	D55	D60	D65	D70	KT	L-D
JAN	732	1011	36.6	3.25	421	572	726	881	1035	.49	57.5
FEB	978	1113	41.2	3.50	263	391	528	667	807	.51	49.8
MAR	1305	1080	48.3	4.45	174	267	388	528	677	.52	38.6
APR	1603	943	60.8	3.35	11	35	94	176	292	.52	26.4
MAY	1822	875	68.8	2.30	0	0	5	28	95	.52	17.2
JUN	2021	891	77.3	2.35	0	0	0	0	8	.56	12.8
JUL	2030	911	82.1	2.60	0	0	0	0	0	.57	14.8
AUG	1865	990	81.4	3.10	0	0	0	0	8	.58	22.7
SEP	1473	1087	73.3	2.70	0	0	0	10	41	.54	34.4
OCT	1164	1231	62.9	3.30	7	23	67	143	247	.55	46.2
NOV	827	1115	49.4	2.85	99	193	324	469	618	.52	55.6
DEC	659	953	39.8	3.40	331	476	628	782	936	.48	59.6
YR	1375	1016	60.2	3.10	1306	1957	2760	3684	4764	.54	

ASTORIA, OREGON — ELEV 23 LAT 46.1

	HS	VS	TA	SD	D50	D55	D60	D65	D70	KT	L-D
JAN	315	462	40.6	3.86	317	456	605	757	912	.32	67.4
FEB	545	715	43.6	2.80	196	323	460	599	739	.38	59.7
MAR	866	879	44.4	2.05	184	330	484	639	794	.41	48.5
APR	1253	940	47.8	1.95	98	220	367	516	666	.44	36.3
MAY	1608	963	52.3	1.89	31	110	242	394	549	.47	27.1
JUN	1626	890	56.5	1.92	5	39	124	257	405	.45	22.7
JUL	1746	988	60.0	1.74	0	8	57	163	311	.50	24.7
AUG	1499	1038	60.3	1.46	0	0	44	151	301	.49	32.6
SEP	1183	1145	58.4	1.64	0	13	78	201	348	.50	44.3
OCT	713	902	52.8	1.83	24	98	227	379	533	.43	56.1
NOV	387	568	46.5	2.21	131	259	406	555	705	.36	65.5
DEC	261	392	42.8	2.41	234	380	534	688	843	.31	69.5
YR	1002	824	50.5	2.14	1220	2236	3628	5299	7106	.44	

BURNS, OREGON — ELEV 4170 LAT 43.6

	HS	VS	TA	SD	D50	D55	D60	D65	D70	KT	L-D
JAN	490	782	25.2	5.29	775	926	1080	1234	1389	.44	64.9
FEB	792	1107	31.0	4.12	535	673	812	952	1092	.51	57.2
MAR	1187	1232	36.1	2.61	433	586	741	896	1051	.53	46.0
APR	1649	1203	44.2	3.33	209	335	477	625	774	.56	33.8
MAY	2052	1139	52.2	3.07	69	148	261	402	553	.60	24.6
JUN	2280	1116	59.0	2.97	11	41	109	205	337	.63	20.2
JUL	2460	1247	68.4	2.24	0	0	5	30	100	.70	22.2
AUG	2083	1379	66.1	2.56	0	0	18	68	155	.67	30.1
SEP	1620	1592	58.2	3.06	15	53	125	226	360	.66	41.8
OCT	1043	1415	47.3	3.25	152	261	400	551	704	.59	53.6
NOV	593	948	35.8	3.07	429	577	726	876	1026	.49	63.0
DEC	430	718	27.9	3.21	686	840	995	1150	1305	.44	67.0
YR	1393	1156	46.0	3.23	3314	4440	5749	7215	8846	.60	

MEDFORD, OREGON — ELEV 1299 LAT 42.4

	HS	VS	TA	SD	D50	D55	D60	D65	D70	KT	L-D
JAN	407	565	36.6	3.39	422	572	726	881	1035	.35	63.7
FEB	737	949	41.3	3.05	255	386	524	664	804	.45	56.0
MAR	1133	1109	44.8	2.23	177	319	472	626	781	.50	44.8
APR	1639	1150	50.2	2.84	86	175	302	446	594	.55	32.6
MAY	2034	1094	57.3	2.57	12	52	131	250	396	.59	23.4
JUN	2278	1082	64.3	2.66	0	7	31	94	192	.63	19.0
JUL	2475	1207	71.7	2.30	0	0	0	11	51	.70	21.0
AUG	2121	1348	70.4	2.53	0	0	0	21	77	.68	28.9
SEP	1589	1482	64.4	2.55	0	6	27	89	187	.63	40.6
OCT	982	1233	53.4	2.40	34	105	217	362	515	.54	52.4
NOV	504	707	43.5	2.61	210	348	496	645	795	.39	61.8
DEC	337	475	37.7	2.44	383	537	691	846	1001	.32	65.8
YR	1356	1033	53.0	2.63	1579	2507	3617	4935	6428	.57	

NORTH BEND, OREGON — ELEV 16 LAT 43.4

	HS	VS	TA	SD	D50	D55	D60	D65	D70	KT	L-D
JAN	438	656	44.6	3.31	208	335	481	633	788	.39	64.7
FEB	704	925	46.6	2.99	138	247	378	516	655	.45	57.0
MAR	1058	1048	46.9	1.99	122	255	407	561	716	.47	45.8
APR	1510	1080	49.1	1.75	69	182	327	477	627	.51	33.6
MAY	1857	1035	53.1	1.54	14	84	216	369	524	.54	24.4
JUN	1994	1001	56.9	1.68	0	27	109	244	393	.55	20.0
JUL	2108	1093	59.0	1.45	0	8	64	188	341	.60	22.0
AUG	1786	1161	59.7	1.49	0	6	54	168	320	.57	29.9
SEP	1377	1273.	58.4	1.62	0	13	77	201	348	.56	41.6
OCT	893	1120	54.9	1.75	9	59	166	314	468	.50	53.4
NOV	525	784	50.1	1.90	58	157	298	447	597	.43	62.8
DEC	381	594	46.5	2.26	138	269	419	574	729	.38	66.8
YR	1222	981	52.2	1.97	756	1642	2996	4692	6506	.52	

PENDLETON, OREGON — ELEV 1496 LAT 45.7

	HS	VS	TA	SD	D50	D55	D60	D65	D70	KT	L-D
JAN	348	523	32.0	7.00	600	737	881	1030	1182	.35	67.0
FEB	614	834	38.9	3.80	324	455	592	731	871	.42	59.3
MAR	1044	1115	43.8	2.10	202	349	502	657	812	.49	48.1
APR	1503	1152	50.9	2.50	66	151	279	424	573	.52	35.9
MAY	1925	1138	58.5	2.70	10	42	114	220	361	.56	26.7
JUN	2144	1122	65.6	2.50	0	0	18	70	158	.59	22.3
JUL	2396	1305	73.5	2.30	0	0	0	6	30	.68	24.3
AUG	1994	1413	71.5	2.40	0	0	0	13	57	.65	32.2
SEP	1502	1557	64.0	2.60	0	7	32	97	198	.63	43.9
OCT	908	1260	52.6	2.50	46	123	240	387	540	.55	55.7
NOV	438	668	41.4	3.20	274	412	559	708	858	.40	65.1
DEC	293	454	35.7	3.30	448	600	754	908	1063	.34	69.1
YR	1262	1046	52.4	3.07	1970	2876	3971	5251	6703	.56	

PORTLAND, OREGON — ELEV 39 LAT 45.6

	HS	VS	TA	SD	D50	D55	D60	D65	D70	KT	L-D
JAN	310	441	38.1	4.86	401	538	685	837	990	.31	66.9
FEB	554	718	42.8	3.02	218	346	483	622	762	.38	59.2
MAR	895	903	45.7	2.22	155	292	444	598	753	.42	48.0
APR	1308	975	50.6	2.31	64	153	286	433	582	.45	35.8
MAY	1663	984	56.7	2.37	12	53	137	264	413	.49	26.6
JUN	1772	951	62.0	2.47	0	11	50	128	248	.49	22.2
JUL	2037	1124	67.1	2.38	0	0	10	48	129	.58	24.2
AUG	1674	1157	66.6	2.42	0	0	13	56	140	.54	32.1
SEP	1217	1168	62.2	2.30	0	8	42	119	241	.51	43.8
OCT	724	905	53.8	2.26	27	94	204	349	503	.43	55.6
NOV	388	558	45.3	2.48	164	296	442	591	741	.35	65.0
DEC	260	381	40.7	2.29	292	444	598	753	908	.30	69.0
YR	1069	856	52.6	2.61	1333	2235	3394	4798	6410	.47	

REDMOND, OREGON — ELEV 3084 LAT 44.3

	HS	VS	TA	SD	D50	D55	D60	D65	D70	KT	L-D
JAN	491	811	30.2	5.81	632	778	928	1081	1235	.46	65.6
FEB	775	1105	35.8	3.23	401	538	678	818	958	.51	57.9
MAR	1190	1269	38.6	2.11	355	509	663	818	973	.54	46.7
APR	1683	1260	44.4	2.45	185	321	469	618	768	.58	34.5
MAY	2079	1177	51.3	2.72	71	155	280	427	580	.61	25.3
JUN	2287	1140	58.2	2.72	11	43	114	220	358	.63	20.9
JUL	2446	1270	65.7	1.97	0	0	9	55	149	.69	22.9
AUG	2069	1403	63.8	2.49	0	7	33	102	208	.67	30.8
SEP	1584	1586	57.7	2.71	12	48	122	233	372	.65	42.5
OCT	999	1367	48.4	2.51	109	219	362	515	670	.58	54.3
NOV	572	929	39.0	2.84	335	481	630	780	930	.49	63.7
DEC	424	729	33.4	2.91	516	670	825	980	1135	.45	67.7
YR	1386	1170	47.2	2.87	2627	3769	5113	6647	8336	.60	

SALEM, OREGON — ELEV 200 LAT 44.9

	HS	VS	TA	SD	D50	D55	D60	D65	D70	KT	L-D
JAN	332	471	38.8	4.00	367	510	660	813	968	.32	66.2
FEB	588	760	42.9	3.00	215	343	480	619	759	.39	58.5
MAR	947	951	45.2	2.10	164	306	459	614	769	.43	47.3
APR	1370	1009	49.8	2.50	81	176	310	457	606	.47	35.1
MAY	1738	1009	55.7	2.60	21	75	164	295	445	.51	25.9
JUN	1849	972	61.2	2.00	0	7	46	133	266	.51	21.5
JUL	2142	1154	66.6	2.00	0	0	7	43	129	.61	23.5
AUG	1775	1208	66.1	2.10	0	0	10	53	143	.58	31.4
SEP	1328	1280	61.9	2.10	0	7	40	120	247	.55	43.1
OCT	769	959	53.2	1.90	22	93	216	366	521	.45	54.9
NOV	410	585	45.2	2.20	160	297	444	594	744	.36	64.3
DEC	277	403	40.9	2.30	287	438	592	747	902	.30	68.3
YR	1130	897	52.3	2.40	1317	2252	3428	4854	6499	.49	

ALLENTOWN, PENNSYLVANIA — ELEV 384 LAT 40.6

	HS	VS	TA	SD	D50	D55	D60	D65	D70	KT	L-D
JAN	527	758	27.8	4.60	693	845	999	1153	1308	.41	61.9
FEB	763	929	29.4	3.45	578	717	857	997	1137	.44	54.2
MAR	1078	979	38.1	3.68	382	528	680	834	989	.46	43.0
APR	1410	926	49.9	3.12	99	188	313	456	604	.47	30.8
MAY	1637	870	60.1	3.06	10	36	99	190	318	.47	21.6
JUN	1777	870	69.5	2.33	0	0	0	21	81	.49	17.2
JUL	1765	890	74.1	2.04	0	0	0	0	19	.50	19.2
AUG	1546	927	71.7	1.98	0	0	0	6	41	.49	27.1
SEP	1238	1014	64.7	2.57	0	5	25	85	180	.48	38.8
OCT	926	1058	54.1	2.84	40	108	208	344	494	.48	50.6
NOV	568	778	42.3	2.44	239	383	531	681	831	.41	60.0
DEC	430	622	30.7	3.13	599	754	908	1063	1218	.38	64.0
YR	1140	885	51.0	2.94	2640	3564	4620	5830	7220	.47	

ERIE, PENNSYLVANIA — ELEV 738 LAT 42.1

	HS	VS	TA	SD	D50	D55	D60	D65	D70	KT	L-D
JAN	346	445	25.1	4.88	776	929	1083	1237	1392	.29	63.4
FEB	577	666	25.2	3.87	695	835	974	1114	1254	.35	55.7
MAR	920	837	32.9	4.71	542	690	842	996	1150	.40	44.5
APR	1359	925	44.8	3.53	201	321	461	607	756	.46	32.3
MAY	1646	900	54.6	3.36	49	117	209	336	481	.48	23.1
JUN	1847	919	64.6	2.97	0	0	21	80	178	.51	18.7
JUL	1833	944	68.7	2.14	0	0	0	24	92	.52	20.7
AUG	1455	905	67.5	2.36	0	0	9	43	120	.46	28.6
SEP	1201	1023	61.4	2.50	0	14	59	141	265	.48	40.3
OCT	827	956	51.6	2.97	74	156	275	419	571	.45	52.1
NOV	416	532	40.1	2.68	303	448	597	747	897	.32	61.5
DEC	278	359	29.1	4.13	651	804	958	1113	1268	.26	65.5
YR	1061	785	47.1	3.29	3291	4314	5488	6857	8424	.44	

HARRISBURG, PENNSYLVANIA — ELEV 348 LAT 40.2

	HS	VS	TA	SD	D50	D55	D60	D65	D70	KT	L-D
JAN	536	763	30.1	4.59	624	775	928	1082	1237	.41	61.5
FEB	771	929	32.3	3.53	498	636	776	916	1056	.44	53.8
MAR	1083	972	41.0	3.91	308	444	593	745	899	.46	42.6
APR	1410	916	52.8	2.87	52	125	233	370	517	.47	30.4
MAY	1652	870	63.1	2.94	0	15	53	128	235	.48	21.2
JUN	1805	875	72.0	1.90	0	0	0	0	33	.50	16.8
JUL	1764	883	76.1	1.89	0	0	0	0	7	.50	18.8
AUG	1550	921	73.9	2.02	0	0	0	0	20	.49	26.7
SEP	1267	1030	67.0	2.50	0	0	11	51	129	.49	38.4
OCT	934	1055	55.8	2.63	21	75	163	293	442	.48	50.2
NOV	579	787	43.8	2.36	198	338	486	636	786	.42	59.6
DEC	447	646	32.6	3.78	544	696	850	1005	1159	.38	63.6
YR	1151	887	53.4	2.91	2245	3104	4093	5226	6520	.47	

PHILADELPHIA, PENNSYLVANIA — ELEV 30 LAT 39.9

	HS	VS	TA	SD	D50	D55	D60	D65	D70	KT	L-D
JAN	555	792	32.3	4.44	557	707	860	1014	1169	.42	61.2
FEB	794	957	33.9	3.25	453	591	731	871	1011	.45	53.5
MAR	1108	991	41.9	3.60	279	415	564	717	871	.46	42.3
APR	1434	926	52.9	3.13	58	132	235	369	515	.47	30.1
MAY	1660	868	63.2	2.83	0	13	49	123	230	.48	20.9
JUN	1811	873	72.3	2.56	0	0	0	11	49	.50	16.5
JUL	1758	876	76.8	1.86	0	0	0	0	5	.50	18.5
AUG	1574	928	74.8	1.99	0	0	0	0	13	.50	26.4
SEP	1281	1034	68.1	2.37	0	0	7	34	105	.49	38.1
OCT	958	1081	57.4	2.61	13	52	130	248	393	.49	49.9
NOV	619	856	46.2	2.40	142	269	415	564	714	.44	59.3
DEC	470	685	35.2	3.62	465	616	769	924	1079	.40	63.3
YR	1170	905	54.6	2.89	1967	2795	3760	4875	6154	.48	

PITTSBURGH, PENNSYLVANIA — ELEV 1224 LAT 40.5

	HS	VS	TA	SD	D50	D55	D60	D65	D70	KT	L-D
JAN	424	563	20.1	5.09	686	837	990	1145	1299	.33	61.8
FEB	625	702	29.3	4.16	582	720	860	1000	1140	.36	54.1
MAR	943	823	38.1	4.72	398	537	684	836	990	.40	42.9
APR	1317	859	50.2	3.31	101	187	307	448	595	.44	30.7
MAY	1602	852	59.8	3.47	17	50	117	208	332	.46	21.5
JUN	1762	863	68.6	2.24	0	0	0	26	93	.48	17.1
JUL	1689	857	71.9	2.07	0	0	0	7	41	.48	19.1
AUG	1510	904	70.2	2.21	0	0	0	16	70	.48	27.0
SEP	1209	981	63.8	2.52	0	7	31	98	201	.47	38.7
OCT	895	1005	53.2	3.07	56	131	235	372	523	.47	50.5
NOV	505	656	41.3	2.63	269	413	561	711	861	.37	59.9
DEC	347	457	30.5	4.78	613	763	916	1070	1225	.30	63.9
YR	1071	793	50.4	3.36	2722	3645	4701	5937	7370	.44	

SCRANTON, PENNSYLVANIA — ELEV 948 LAT 41.3

	HS	VS	TA	SD	D50	D55	D60	D65	D70	KT	L-D
JAN	455	632	26.0	4.53	747	900	1054	1209	1364	.37	62.6
FEB	689	827	27.3	3.70	636	776	916	1056	1196	.41	54.9
MAR	991	898	36.0	4.45	451	596	747	900	1054	.42	43.7
APR	1339	892	48.5	3.30	127	224	354	497	646	.45	31.5
MAY	1591	860	58.9	3.13	15	50	121	219	353	.46	22.3
JUN	1760	873	67.9	2.09	0	0	0	28	100	.48	17.9
JUL	1746	893	72.2	2.15	0	0	0	7	40	.49	19.9
AUG	1513	923	70.0	2.25	0	0	0	18	74	.48	27.8
SEP	1199	996	62.9	2.62	0	10	43	116	226	.47	39.5
OCT	897	1039	52.6	3.17	67	144	252	391	541	.48	51.3
NOV	490	649	40.8	2.74	284	428	576	726	876	.37	60.7
DEC	368	513	29.1	3.85	650	804	958	1113	1268	.33	64.7
YR	1088	833	49.4	3.17	2977	3932	5021	6280	7738	.45	

PROVIDENCE, RHODE ISLAND — ELEV 62 LAT 41.7

	HS	VS	TA	SD	D50	D55	D60	D65	D70	KT	L-D
JAN	506	750	28.4	4.06	672	826	980	1135	1290	.42	63.0
FEB	738	925	29.4	3.27	577	717	857	997	1137	.44	55.3
MAR	1032	959	36.9	3.06	411	562	716	871	1026	.45	44.1
APR	1374	927	47.3	2.39	120	239	382	531	681	.46	31.9
MAY	1655	897	56.9	2.42	12	52	134	259	408	.48	22.7
JUN	1775	885	66.4	1.78	0	0	0	36	123	.49	18.3
JUL	1695	878	72.1	1.83	0	0	0	0	31	.48	20.3
AUG	1499	924	70.4	1.93	0	0	0	10	58	.48	28.2
SEP	1209	1019	63.4	2.14	0	0	25	93	206	.48	39.9
OCT	907	1072	53.7	2.10	23	91	204	352	505	.49	51.7
NOV	538	754	43.3	2.38	212	353	501	651	801	.41	61.1
DEC	418	627	31.5	3.86	577	730	884	1039	1194	.38	65.1
YR	1114	884	50.0	2.60	2604	3570	4683	5974	7460	.46	

CHARLESTON, SOUTH CAROLINA — ELEV 39 LAT 32.9

	HS	VS	TA	SD	D50	D55	D60	D65	D70	KT	L-D
JAN	744	904	48.6	4.58	174	264	383	521	668	.44	54.2
FEB	995	1009	50.5	4.60	123	199	298	419	551	.48	46.5
MAR	1339	1003	56.5	4.11	52	113	194	300	433	.51	35.3
APR	1732	934	64.6	1.99	0	0	13	69	172	.55	23.1
MAY	1860	845	72.1	1.96	0	0	0	5	35	.53	13.9
JUN	1844	820	77.9	1.88	0	0	0	0	0	.51	9.5
JUL	1799	813	80.2	1.33	0	0	0	0	0	.51	11.5
AUG	1585	806	79.6	1.09	0	0	0	0	0	.48	19.4
SEP	1394	926	75.2	1.87	0	0	0	0	9	.50	31.1
OCT	1193	1127	66.1	2.73	0	5	21	74	159	.53	42.9
NOV	934	1159	56.3	2.79	20	69	151	271	413	.53	52.3
DEC	721	943	49.3	3.50	127	219	346	491	643	.46	56.3
YR	1346	940	64.7	2.69	496	869	1406	2150	3083	.51	

COLUMBIA, SOUTH CAROLINA — ELEV 226 LAT 33.9

	HS	VS	TA	SD	D50	D55	D60	D65	D70	KT	L-D
JAN	762	973	45.4	4.40	223	333	467	613	765	.47	55.2
FEB	1020	1082	47.6	4.25	156	246	362	493	629	.50	47.5
MAR	1355	1050	54.2	4.21	79	152	241	360	500	.52	36.3
APR	1747	965	64.1	2.19	0	0	21	83	188	.55	24.1
MAY	1895	867	72.1	2.52	0	0	0	12	53	.54	14.9
JUN	1947	852	78.8	2.28	0	0	0	0	0	.54	10.5
JUL	1842	832	81.2	1.71	0	0	0	0	0	.52	12.5
AUG	1703	871	80.2	1.69	0	0	0	0	0	.52	20.4
SEP	1439	987	74.5	2.38	0	0	0	0	22	.52	32.1
OCT	1211	1192	64.2	3.00	0	12	43	112	209	.55	43.9
NOV	921	1184	53.8	2.85	40	108	207	341	487	.54	53.3
DEC	722	985	46.0	3.64	188	303	442	592	745	.48	57.3
YR	1381	986	63.5	2.92	686	1154	1783	2606	3598	.52	

GREENVILLE, SOUTH CAROLINA — ELEV 971 LAT 34.9

	HS	VS	TA	SD	D50	D55	D60	D65	D70	KT	L-D
JAN	730	953	42.3	4.05	279	409	554	706	859	.46	56.2
FEB	982	1066	44.4	3.83	200	312	442	578	717	.49	48.5
MAR	1328	1058	50.9	4.17	124	207	316	450	597	.51	37.3
APR	1697	964	61.0	2.30	0	12	58	144	274	.54	25.1
MAY	1839	863	69.1	2.45	0	0	6	29	95	.53	15.9
JUN	1918	852	75.9	2.29	0	0	0	0	13	.53	11.5
JUL	1830	838	78.3	1.65	0	0	0	0	0	.52	13.5
AUG	1699	887	77.5	1.76	0	0	0	0	0	.52	21.4
SEP	1406	990	71.7	2.28	0	0	0	9	48	.51	33.1
OCT	1180	1195	61.7	2.67	0	17	63	145	268	.54	44.9
NOV	880	1156	51.0	2.59	67	151	277	421	570	.53	54.3
DEC	670	922	42.9	3.45	251	385	533	686	840	.46	58.3
YR	1348	978	60.6	2.79	921	1493	2249	3168	4281	.52	

HURON, SOUTH DAKOTA — ELEV 1289 LAT 44.4

	HS	VS	TA	SD	D50	D55	D60	D65	D70	KT	L-D
JAN	488	808	12.5	6.15	1165	1319	1473	1628	1783	.46	65.7
FEB	745	1047	17.9	5.28	900	1039	1179	1319	1459	.49	58.0
MAR	1114	1163	29.0	6.53	675	819	968	1120	1273	.51	46.8
APR	1530	1130	45.8	4.08	197	305	437	580	727	.52	34.6
MAY	1871	1069	57.0	3.42	31	83	163	273	411	.55	25.4
JUN	2101	1068	67.1	3.23	0	7	24	72	148	.58	21.0
JUL	2183	1156	73.7	2.61	0	0	0	9	37	.62	23.0
AUG	1892	1276	72.1	2.56	0	0	0	13	54	.61	30.9
SEP	1418	1372	60.7	3.01	7	28	84	169	290	.58	42.6
OCT	988	1351	49.6	3.53	123	214	338	482	634	.57	54.4
NOV	577	946	32.4	4.05	533	680	829	978	1128	.49	63.8
DEC	405	684	19.2	5.40	957	1111	1265	1420	1575	.43	67.8
YR	1278	1089	44.8	4.15	4588	5605	6760	8063	9519	.55	

PIERRE, SOUTH DAKOTA ELEV 1726 LAT 44.4

	HS	VS	TA	SD	D50	D55	D60	D65	D70	KT	L-D
JAN	530	915	15.6	6.87	1072	1225	1378	1532	1687	.50	65.7
FEB	795	1152	20.4	5.91	832	970	1110	1249	1389	.52	58.0
MAR	1206	1297	29.8	6.64	654	796	944	1096	1249	.55	46.8
APR	1614	1203	46.3	4.17	191	295	424	566	713	.55	34.6
MAY	1966	1120	57.4	3.63	33	85	163	267	401	.57	25.4
JUN	2195	1106	67.4	3.43	0	8	26	74	149	.60	21.0
JUL	2278	1199	75.2	2.69	0	0	0	6	26	.65	23.0
AUG	1992	1350	73.9	2.77	0	0	0	10	40	.64	30.9
SEP	1496	1474	62.1	3.34	8	26	75	152	259	.62	42.6
OCT	1052	1483	50.7	3.71	112	196	312	451	601	.61	54.4
NOV	623	1063	33.8	4.39	496	640	787	937	1086	.53	63.8
DEC	442	780	21.5	5.61	888	1041	1194	1349	1504	.47	67.8
YR	1352	1178	46.2	4.42	4286	5282	6413	7689	9104	.59	

RAPID CITY, SOUTH DAKOTA ELEV 3169 LAT 44.0

	HS	VS	TA	SD	D50	D55	D60	D65	D70	KT	L-D
JAN	542	928	21.9	6.76	882	1032	1184	1338	1492	.50	65.3
FEB	826	1198	25.8	5.52	683	820	959	1098	1238	.53	57.6
MAR	1229	1312	31.2	5.75	603	748	898	1050	1204	.55	46.4
APR	1589	1167	44.6	4.05	219	334	470	615	763	.54	34.2
MAY	1887	1066	55.2	3.32	42	106	195	319	463	.55	25.0
JUN	2131	1070	64.2	3.87	9	25	65	134	222	.58	20.6
JUL	2223	1161	72.6	2.69	0	0	0	13	52	.63	22.6
AUG	1963	1311	71.6	2.68	0	0	0	17	65	.63	30.5
SEP	1518	1481	60.5	3.67	16	46	108	191	305	.62	42.2
OCT	1064	1482	50.0	3.98	131	218	335	474	623	.61	54.0
NOV	647	1104	35.4	4.52	453	594	740	889	1038	.54	63.4
DEC	476	855	26.5	4.54	732	885	1039	1194	1349	.49	67.4
YR	1344	1177	46.6	4.27	3770	4808	5993	7332	8814	.58	

SIOUX FALLS, SOUTH DAKOTA ELEV 1427 LAT 43.6

	HS	VS	TA	SD	D50	D55	D60	D65	D70	KT	L-D
JAN	533	887	14.2	5.64	1111	1266	1420	1575	1730	.48	64.9
FEB	802	1128	19.4	4.62	857	997	1137	1277	1417	.51	57.2
MAR	1152	1183	30.0	6.41	645	788	937	1089	1242	.51	46.0
APR	1543	1114	46.1	4.02	190	296	428	571	718	.52	33.8
MAY	1894	1059	57.7	3.58	30	79	156	259	392	.55	24.6
JUN	2100	1048	67.6	3.21	0	6	21	65	139	.58	20.2
JUL	2150	1117	73.3	2.60	0	0	0	10	41	.61	22.2
AUG	1844	1209	71.8	2.75	0	0	0	18	64	.59	30.1
SEP	1410	1323	60.9	3.02	7	27	82	165	285	.57	41.8
OCT	1005	1340	50.2	3.54	113	201	322	465	616	.57	53.6
NOV	607	982	33.1	3.92	512	659	808	957	1107	.50	63.0
DEC	441	745	20.0	4.92	932	1086	1240	1395	1550	.45	67.0
YR	1292	1094	45.4	4.02	4397	5405	6551	7846	9301	.55	

CHATTANOOGA, TENNESSEE ELEV 689 LAT 35.0

	HS	VS	TA	SD	D50	D55	D60	D65	D70	KT	L-D
JAN	630	776	40.2	5.10	354	482	625	774	926	.40	56.3
FEB	859	891	42.9	5.30	264	370	493	625	762	.43	48.6
MAR	1176	916	49.8	4.30	145	232	346	483	631	.46	37.4
APR	1550	887	60.5	2.70	5	23	78	165	292	.49	25.2
MAY	1732	828	68.5	3.00	0	0	15	51	123	.50	16.0
JUN	1831	829	76.0	2.30	0	0	0	0	12	.51	11.6
JUL	1735	810	78.8	2.10	0	0	0	0	0	.49	13.6
AUG	1630	858	78.0	2.00	0	0	0	0	0	.50	21.5
SEP	1335	937	71.9	2.30	0	0	0	9	46	.49	33.2
OCT	1108	1102	60.8	3.30	11	37	97	182	302	.51	45.0
NOV	773	963	48.9	2.60	99	200	337	484	633	.46	54.4
DEC	580	753	41.2	4.10	307	441	588	740	893	.40	58.4
YR	1246	879	59.8	3.25	1185	1785	2579	3513	4620	.48	

KNOXVILLE, TENNESSEE ELEV 981 LAT 35.8

	HS	VS	TA	SD	D50	D55	D60	D65	D70	KT	L-D
JAN	621	785	40.6	5.02	343	470	612	761	913	.41	57.1
FEB	863	923	42.8	5.60	273	377	499	630	765	.44	49.4
MAR	1191	953	49.9	4.60	153	239	350	484	630	.47	38.2
APR	1599	932	60.3	2.84	7	27	85	173	299	.51	26.0
MAY	1803	863	68.4	2.83	0	0	12	47	119	.52	16.8
JUN	1902	856	75.5	2.25	0	0	0	0	14	.52	12.4
JUL	1804	839	78.2	1.95	0	0	0	0	0	.51	14.4
AUG	1666	889	77.3	1.65	0	0	0	0	0	.51	22.3
SEP	1383	998	71.6	2.31	0	0	0	10	50	.51	34.0
OCT	1121	1152	60.9	3.14	9	32	90	175	297	.53	45.8
NOV	759	969	49.2	2.81	101	197	330	475	624	.47	55.2
DEC	569	757	41.5	4.17	301	433	579	731	884	.40	59.2
YR	1275	909	59.7	3.25	1187	1775	2557	3486	4595	.49	

MEMPHIS, TENNESSEE ELEV 285 LAT 35.0

	HS	VS	TA	SD	D50	D55	D60	D65	D70	KT	L-D
JAN	683	870	40.5	4.43	331	465	611	762	915	.44	56.3
FEB	945	1015	43.8	4.66	231	338	464	598	735	.47	48.6
MAR	1278	1013	51.0	4.88	146	227	330	457	599	.50	37.4
APR	1639	935	62.5	2.88	0	15	56	131	240	.52	25.2
MAY	1885	879	70.9	2.72	0	0	5	22	76	.54	16.0
JUN	2045	885	78.6	2.34	0	0	0	0	5	.56	11.6
JUL	1972	879	81.6	2.11	0	0	0	0	0	.56	13.6
AUG	1824	943	80.4	2.32	0	0	0	0	0	.56	21.5
SEP	1471	1046	73.6	2.48	0	0	0	7	32	.54	33.2
OCT	1204	1232	63.0	3.32	7	23	67	142	245	.56	45.0
NOV	817	1042	50.9	2.81	75	159	282	425	573	.49	54.4
DEC	629	845	42.7	3.96	268	397	542	693	847	.43	58.4
YR	1368	965	61.6	3.24	1058	1624	2357	3237	4267	.53	

NASHVILLE, TENNESSEE ELEV 591 LAT 36.1

	HS	VS	TA	SD	D50	D55	D60	D65	D70	KT	L-D
JAN	580	721	38.3	4.80	395	532	679	830	984	.38	57.4
FEB	824	876	41.0	5.10	297	412	541	676	814	.42	49.7
MAR	1130	902	48.7	5.10	189	277	390	524	669	.45	38.5
APR	1544	907	60.1	2.80	7	28	86	176	304	.50	26.3
MAY	1825	874	68.5	2.80	0	0	12	45	117	.52	17.1
JUN	1963	875	76.6	2.50	0	0	0	0	13	.54	12.7
JUL	1891	869	79.6	2.20	0	0	0	0	0	.53	14.7
AUG	1737	928	78.5	2.30	0	0	0	0	6	.54	22.6
SEP	1398	1020	72.0	2.40	0	0	0	10	48	.52	34.3
OCT	1114	1155	60.9	3.30	11	36	96	180	299	.53	46.1
NOV	711	893	48.4	2.70	110	214	352	499	648	.44	55.5
DEC	521	676	40.4	4.20	329	465	612	764	918	.38	59.5
YR	1272	891	59.4	3.34	1338	1964	2768	3704	4820	.49	

ABILENE, TEXAS ELEV 1752 LAT 32.4

	HS	VS	TA	SD	D50	D55	D60	D65	D70	KT	L-D
JAN	924	1208	43.7	3.97	245	369	512	663	816	.54	53.7
FEB	1183	1253	47.9	3.52	131	224	347	481	619	.56	46.0
MAR	1576	1197	54.5	4.31	79	150	238	354	492	.59	34.8
APR	1843	974	65.2	3.43	0	14	42	104	190	.58	22.6
MAY	2037	887	72.4	2.47	0	0	0	11	47	.58	13.4
JUN	2209	895	80.3	2.51	0	0	0	0	0	.61	9.0
JUL	2139	890	83.9	2.32	0	0	0	0	0	.60	11.0
AUG	1956	939	83.6	2.45	0	0	0	0	0	.60	18.9
SEP	1598	1058	76.1	2.50	0	0	0	0	15	.57	30.6
OCT	1315	1259	66.1	3.20	0	9	32	89	173	.58	42.4
NOV	1008	1267	54.1	3.21	47	115	208	336	479	.56	51.8
DEC	863	1199	46.4	3.04	162	281	426	578	732	.54	55.8
YR	1556	1084	64.5	3.08	664	1162	1805	2616	3563	.58	

AMARILLO, TEXAS ELEV 3602 LAT 35.2

	HS	VS	TA	SD	D50	D55	D60	D65	D70	KT	L-D
JAN	960	1447	36.0	4.07	446	593	746	900	1054	.62	56.5
FEB	1243	1502	39.7	4.32	312	437	572	710	849	.63	48.8
MAR	1631	1379	45.6	4.34	217	327	461	607	759	.63	37.6
APR	2019	1143	56.5	3.39	31	85	164	275	411	.65	25.4
MAY	2212	978	65.6	2.72	0	6	24	81	170	.63	16.2
JUN	2393	959	74.6	3.04	0	0	0	10	38	.66	11.8
JUL	2280	958	78.7	2.70	0	0	0	0	10	.64	13.8
AUG	2103	1065	77.6	2.47	0	0	0	0	10	.65	21.7
SEP	1760	1294	69.8	2.39	0	0	0	20	78	.64	33.4
OCT	1403	1533	59.5	3.16	13	45	112	206	336	.65	45.2
NOV	1033	1483	46.3	3.25	164	277	416	562	711	.62	54.6
DEC	872	1384	38.5	3.24	365	514	667	822	977	.61	58.6
YR	1660	1259	57.4	3.25	1548	2284	3162	4193	5403	.64	

AUSTIN, TEXAS ELEV 620 LAT 30.3

	HS	VS	TA	SD	D50	D55	D60	D65	D70	KT	L-D
JAN	864	1008	49.7	3.95	135	223	342	483	632	.48	51.6
FEB	1125	1083	53.3	4.19	76	144	230	344	473	.51	43.9
MAR	1429	999	59.5	3.79	24	63	133	223	345	.52	32.7
APR	1605	828	68.6	2.92	0	0	12	44	114	.50	20.5
MAY	1834	815	75.2	2.05	0	0	0	0	13	.53	11.3
JUN	2072	860	81.6	1.78	0	0	0	0	0	.58	6.9
JUL	2105	865	84.6	1.63	0	0	0	0	0	.60	8.9
AUG	1931	891	84.7	1.80	0	0	0	0	0	.59	16.8
SEP	1606	999	78.9	1.94	0	0	0	0	0	.56	28.5
OCT	1333	1187	70.1	3.16	0	0	12	39	102	.56	40.3
NOV	987	1130	59.1	3.06	12	43	110	205	335	.52	49.7
DEC	825	1028	52.3	3.09	69	147	258	399	550	.49	53.7
YR	1478	974	68.1	2.77	316	620	1097	1737	2564	.54	

BROWNSVILLE, TEXAS — ELEV 20 LAT 25.9

	HS	VS	TA	SD	D50	D55	D60	D65	D70	KT	L-D
JAN	913	923	60.3	4.40	32	72	140	225	336	.45	47.2
FEB	1135	944	63.4	4.50	15	37	83	151	235	.47	39.5
MAR	1458	901	67.7	3.90	5	14	37	89	166	.51	28.3
APR	1737	816	74.9	1.90	0	0	0	0	10	.53	16.1
MAY	1927	817	79.3	1.90	0	0	0	0	0	.56	6.9
JUN	2115	868	82.8	1.60	0	0	0	0	0	.60	2.5
JUL	2212	867	84.4	1.60	0	0	0	0	0	.63	4.5
AUG	2027	854	84.4	1.20	0	0	0	0	0	.61	12.4
SEP	1694	933	81.6	1.60	0	0	0	0	0	.57	24.1
OCT	1439	1120	75.7	2.70	0	0	0	5	23	.57	35.9
NOV	1054	1047	68.1	2.40	0	0	7	35	106	.50	45.3
DEC	862	922	62.8	3.30	7	23	69	145	250	.45	49.3
YR	1550	917	73.8	2.58	59	146	336	650	1126	.55	

CORPUS CHRISTI, TEXAS — ELEV 43 LAT 27.8

	HS	VS	TA	SD	D50	D55	D60	D65	D70	KT	L-D
JAN	898	967	56.3	4.08	53	115	196	304	438	.46	49.1
FEB	1147	1018	59.6	3.96	21	55	118	199	309	.49	41.4
MAR	1430	930	64.9	3.60	6	19	53	120	209	.51	30.2
APR	1642	807	72.8	2.06	0	0	0	0	29	.51	18.0
MAY	1866	810	77.9	1.59	0	0	0	0	0	.54	8.8
JUN	2094	861	82.4	1.32	0	0	0	0	0	.59	4.4
JUL	2186	866	84.8	1.04	0	0	0	0	0	.62	6.4
AUG	1991	870	85.1	1.16	0	0	0	0	0	.60	14.3
SEP	1687	978	81.0	1.49	0	0	0	0	0	.57	26.0
OCT	1416	1172	73.9	2.55	0	0	0	7	33	.57	37.8
NOV	1043	1108	64.9	2.55	0	5	23	81	175	.51	47.2
DEC	845	963	59.1	3.33	18	54	124	219	349	.46	51.2
YR	1522	945	71.9	2.39	98	248	514	930	1542	.55	

DALLAS, TEXAS — ELEV 489 LAT 32.8

	HS	VS	TA	SD	D50	D55	D60	D65	D70	KT	L-D
JAN	821	1035	45.4	3.78	204	321	461	610	764	.49	54.1
FEB	1071	1110	49.4	3.46	107	191	307	440	578	.51	46.4
MAR	1422	1073	55.8	3.96	54	119	202	314	451	.54	35.2
APR	1627	882	66.4	2.90	0	5	21	71	152	.51	23.0
MAY	1888	852	73.8	1.90	0	0	0	0	18	.54	13.8
JUN	2135	885	81.6	1.96	0	0	0	0	0	.59	9.4
JUL	2122	890	85.7	2.28	0	0	0	0	0	.60	11.4
AUG	1950	945	85.8	2.22	0	0	0	0	0	.60	19.3
SEP	1587	1063	78.2	2.08	0	0	0	0	0	.56	31.0
OCT	1276	1228	68.0	2.96	0	0	16	55	130	.56	42.8
NOV	936	1158	55.9	3.00	27	81	164	284	426	.53	52.2
DEC	780	1052	48.2	3.16	134	237	373	523	676	.50	56.2
YR	1470	1014	66.2	2.80	526	954	1544	2297	3195	.55	

DEL RIO, TEXAS — ELEV 1027 LAT 29.4

	HS	VS	TA	SD	D50	D55	D60	D65	D70	KT	L-D
JAN	958	1125	50.8	3.73	111	195	309	449	598	.51	50.7
FEB	1206	1149	55.7	3.98	45	103	179	283	408	.54	43.0
MAR	1580	1087	62.6	3.74	12	34	85	163	265	.57	31.8
APR	1699	852	72.0	2.78	0	0	0	16	59	.53	19.6
MAY	1827	808	78.2	2.09	0	0	0	0	0	.52	10.4
JUN	2024	849	84.3	1.92	0	0	0	0	0	.57	6.0
JUL	2054	851	86.7	1.76	0	0	0	0	0	.58	8.0
AUG	1936	878	86.1	2.00	0	0	0	0	0	.59	15.9
SEP	1584	961	80.2	2.21	0	0	0	0	0	.55	27.6
OCT	1360	1179	71.2	3.31	0	0	10	34	91	.57	39.4
NOV	1059	1204	59.6	2.62	6	27	88	184	317	.54	48.8
DEC	903	1127	52.3	2.70	56	135	252	397	549	.52	52.8
YR	1517	1005	70.0	2.73	230	494	923	1526	2287	.55	

EL PASO, TEXAS — ELEV 3917 LAT 31.8

	HS	VS	TA	SD	D50	D55	D60	D65	D70	KT	L-D
JAN	1125	1572	43.6	3.90	246	371	515	666	819	.65	53.1
FEB	1480	1671	48.4	3.37	119	210	332	467	605	.69	45.4
MAR	1909	1472	54.6	2.67	30	94	192	328	479	.71	34.2
APR	2363	1174	63.9	2.28	0	0	24	89	195	.74	22.0
MAY	2601	985	72.2	1.52	0	0	0	0	21	.75	12.8
JUN	2682	946	80.3	1.81	0	0	0	0	0	.75	8.4
JUL	2450	934	82.3	1.64	0	0	0	0	0	.69	10.4
AUG	2284	1030	80.5	1.89	0	0	0	0	0	.70	18.3
SEP	1987	1313	74.2	1.65	0	0	0	0	9	.70	30.0
OCT	1639	1654	64.0	2.29	0	5	26	92	199	.71	41.8
NOV	1244	1681	51.6	2.60	59	139	260	404	552	.68	51.2
DEC	1031	1521	44.4	2.78	199	335	485	639	794	.64	55.2
YR	1901	1327	63.4	2.36	653	1154	1834	2685	3673	.71	

FORT WORTH, TEXAS — ELEV 538 LAT 32.8

	HS	VS	TA	SD	D50	D55	D60	D65	D70	KT	L-D
JAN	805	1007	44.8	3.78	216	337	478	629	782	.48	54.1
FEB	1069	1107	48.7	3.46	117	206	325	459	597	.51	46.4
MAR	1409	1062	55.0	3.96	63	130	218	335	474	.53	35.2
APR	1616	877	65.2	2.90	0	7	29	88	176	.51	23.0
MAY	1890	853	72.5	1.90	0	0	0	0	29	.54	13.8
JUN	2153	888	80.6	1.96	0	0	0	0	0	.60	9.4
JUL	2155	897	84.8	2.28	0	0	0	0	0	.61	11.4
AUG	1983	957	84.9	2.22	0	0	0	0	0	.61	19.3
SEP	1621	1088	77.7	2.08	0	0	0	0	5	.58	31.0
OCT	1293	1249	67.6	2.96	0	5	18	60	137	.57	42.8
NOV	938	1162	55.8	3.00	27	83	166	287	429	.53	52.2
DEC	766	1024	47.9	3.16	139	244	382	532	686	.49	56.2
YR	1476	1014	65.5	2.80	562	1012	1616	2390	3315	.55	

HOUSTON, TEXAS — ELEV 108 LAT 30.0

	HS	VS	TA	SD	D50	D55	D60	D65	D70	KT	L-D
JAN	772	852	52.1	4.12	105	184	286	416	561	.42	51.3
FEB	1034	960	55.3	4.08	51	111	189	294	420	.46	43.6
MAR	1297	889	60.8	3.54	15	43	105	189	306	.47	32.4
APR	1522	788	69.4	2.36	0	0	0	23	84	.47	20.2
MAY	1775	798	75.8	1.74	0	0	0	0	6	.51	11.0
JUN	1898	824	81.1	1.43	0	0	0	0	0	.53	6.6
JUL	1828	805	83.3	1.31	0	0	0	0	0	.52	8.6
AUG	1686	807	83.4	1.51	0	0	0	0	0	.51	16.5
SEP	1471	906	79.2	1.58	0	0	0	0	0	.51	28.2
OCT	1276	1110	70.9	2.81	0	0	6	24	79	.54	40.0
NOV	924	1019	61.1	2.81	5	21	73	155	277	.48	49.4
DEC	730	855	54.6	3.19	44	112	205	333	480	.43	53.4
YR	1352	884	68.9	2.53	220	471	864	1434	2213	.49	

LAREDO, TEXAS — ELEV 518 LAT 27.5

	HS	VS	TA	SD	D50	D55	D60	D65	D70	KT	L-D
JAN	959	1046	56.5	4.09	51	113	193	299	432	.49	48.8
FEB	1195	1062	60.9	4.03	17	45	102	177	279	.51	41.1
MAR	1516	981	67.6	3.80	0	13	35	87	165	.54	29.9
APR	1727	833	76.3	2.47	0	0	0	0	14	.53	17.7
MAY	1952	827	81.3	1.92	0	0	0	0	0	.56	8.5
JUN	2073	857	86.0	1.69	0	0	0	0	0	.58	4.1
JUL	2131	857	87.9	1.47	0	0	0	0	0	.61	6.1
AUG	2009	870	87.7	1.69	0	0	0	0	0	.61	14.0
SEP	1705	979	82.9	1.88	0	0	0	0	0	.58	25.7
OCT	1408	1152	75.5	2.99	0	0	0	8	31	.57	37.5
NOV	1041	1093	65.2	2.44	0	0	19	74	166	.51	46.9
DEC	889	1022	58.6	3.31	19	59	132	231	363	.48	50.9
YR	1552	965	73.9	2.64	87	230	481	876	1450	.56	

LUBBOCK, TEXAS — ELEV 3241 LAT 33.6

	HS	VS	TA	SD	D50	D55	D60	D65	D70	KT	L-D
JAN	1031	1497	39.1	3.72	355	499	650	804	958	.63	54.9
FEB	1332	1551	42.7	3.86	235	355	488	626	765	.65	47.2
MAR	1762	1430	48.9	4.20	156	247	368	508	658	.67	36.0
APR	2168	1159	60.0	3.25	12	40	101	190	312	.69	23.8
MAY	2396	989	68.5	2.26	0	0	5	29	99	.69	14.6
JUN	2544	957	77.1	2.46	0	0	0	0	11	.70	10.2
JUL	2412	956	79.7	2.20	0	0	0	0	0	.68	12.2
AUG	2208	1057	78.4	1.93	0	0	0	0	0	.68	20.1
SEP	1820	1271	71.0	2.00	0	0	0	8	49	.65	31.8
OCT	1468	1530	61.0	2.74	5	22	76	162	288	.66	43.6
NOV	1116	1550	48.8	3.37	125	219	346	489	637	.64	53.0
DEC	934	1425	41.3	2.72	280	427	580	735	890	.61	57.0
YR	1767	1279	59.7	2.89	1168	1809	2614	3551	4667	.67	

LUFKIN, TEXAS — ELEV 315 LAT 31.2

	HS	VS	TA	SD	D50	D55	D60	D65	D70	KT	L-D
JAN	794	927	48.8	3.98	151	243	367	509	660	.45	52.5
FEB	1069	1045	52.2	4.15	88	159	252	371	503	.49	44.8
MAR	1376	983	58.0	3.77	32	81	157	256	386	.51	33.6
APR	1624	851	67.3	2.80	0	0	15	56	132	.51	21.4
MAY	1867	831	74.1	1.73	0	0	0	0	12	.54	12.2
JUN	2055	860	80.3	1.48	0	0	0	0	0	.57	7.8
JUL	2006	852	83.0	1.52	0	0	0	0	0	.57	9.8
AUG	1864	885	83.1	1.71	0	0	0	0	0	.57	17.7
SEP	1531	976	77.5	1.77	0	0	0	0	0	.54	29.4
OCT	1349	1245	68.2	2.94	0	0	15	52	126	.58	41.2
NOV	963	1131	57.2	3.31	25	73	149	256	390	.52	50.6
DEC	768	962	50.8	3.05	89	175	298	444	596	.47	54.6
YR	1440	962	66.7	2.68	385	731	1253	1944	2805	.53	

MIDLAND-ODESSA, TEXAS ELEV 2858 LAT 31.9

	HS	VS	TA	SD	D50	D55	D60	D65	D70	KT	L-D
JAN	1081	1488	43.6	4.00	248	373	515	666	819	.62	53.2
FEB	1383	1525	47.8	3.58	134	228	350	484	622	.65	45.5
MAR	1839	1413	54.3	3.69	62	133	225	349	492	.69	34.3
APR	2192	1109	64.3	2.78	0	8	34	98	194	.69	22.1
MAY	2430	961	72.3	1.87	0	0	0	0	30	.70	12.9
JUN	2562	938	79.9	2.07	0	0	0	0	0	.71	8.5
JUL	2389	927	82.3	2.03	0	0	0	0	0	.68	10.5
AUG	2210	1011	81.8	1.99	0	0	0	0	0	.67	18.4
SEP	1844	1216	75.4	1.84	0	0	0	0	8	.65	30.1
OCT	1522	1502	65.8	2.80	0	6	25	81	168	.66	41.9
NOV	1176	1553	53.3	2.95	48	120	222	356	502	.65	51.3
DEC	1000	1460	45.9	2.73	164	291	439	593	747	.62	55.3
YR	1804	1257	63.9	2.69	656	1159	1810	2627	3582	.67	

PORT ARTHUR, TEXAS ELEV 23 LAT 29.9

	HS	VS	TA	SD	D50	D55	D60	D65	D70	KT	L-D
JAN	800	892	52.0	4.19	109	188	290	420	564	.44	51.2
FEB	1071	1001	55.1	4.29	59	120	198	302	427	.48	43.5
MAR	1353	929	60.1	3.51	17	49	114	202	324	.49	32.3
APR	1609	824	68.9	2.61	0	0	7	33	99	.50	20.1
MAY	1871	822	75.0	1.73	0	0	0	0	8	.54	10.9
JUN	2011	847	80.8	1.39	0	0	0	0	0	.56	6.5
JUL	1846	809	83.0	1.32	0	0	0	0	0	.52	8.5
AUG	1736	822	83.1	1.45	0	0	0	0	0	.53	16.4
SEP	1527	939	78.9	1.54	0	0	0	0	0	.53	28.1
OCT	1321	1157	69.9	2.96	0	0	10	35	99	.55	39.9
NOV	953	1060	60.2	3.18	11	36	97	184	306	.50	49.3
DEC	754	891	54.2	2.97	42	111	208	342	492	.44	53.3
YR	1405	915	68.5	2.59	238	504	924	1518	2319	.51	

SAN ANGELO, TEXAS ELEV 1909 LAT 31.4

	HS	VS	TA	SD	D50	D55	D60	D65	D70	KT	L-D
JAN	962	1228	46.4	4.00	192	300	434	581	733	.55	52.7
FEB	1208	1241	50.4	3.47	93	172	282	413	550	.56	45.0
MAR	1606	1183	57.1	4.18	48	107	186	287	417	.59	33.8
APR	1851	954	67.2	3.36	0	8	26	74	151	.58	21.6
MAY	2031	874	74.5	2.19	0	0	0	0	19	.58	12.4
JUN	2186	885	81.6	2.37	0	0	0	0	0	.61	8.0
JUL	2123	877	84.7	2.13	0	0	0	0	0	.60	10.0
AUG	1966	922	84.5	2.37	0	0	0	0	0	.60	17.9
SEP	1607	1033	76.8	2.32	0	0	0	0	10	.56	29.6
OCT	1337	1240	67.2	3.19	0	7	24	73	152	.58	41.4
NOV	1044	1279	55.5	3.22	35	94	178	298	439	.56	50.8
DEC	895	1211	48.3	2.95	125	230	368	519	673	.55	54.8
YR	1569	1076	66.2	2.98	493	918	1498	2245	3144	.58	

SAN ANTONIO, TEXAS ELEV 794 LAT 29.5

	HS	VS	TA	SD	D50	D55	D60	D65	D70	KT	L-D
JAN	895	1026	50.7	3.73	113	197	312	451	601	.48	50.8
FEB	1154	1089	54.5	3.90	55	118	199	310	440	.51	43.1
MAR	1450	991	60.8	3.71	17	47	110	194	309	.52	31.9
APR	1612	819	69.6	2.77	0	0	7	31	93	.50	19.7
MAY	1894	825	76.0	1.94	0	0	0	0	8	.54	10.5
JUN	2069	857	82.2	1.51	0	0	0	0	0	.58	6.1
JUL	2121	863	84.7	1.30	0	0	0	0	0	.60	8.1
AUG	1947	883	84.7	1.48	0	0	0	0	0	.59	16.0
SEP	1638	996	79.3	1.95	0	0	0	0	0	.56	27.7
OCT	1350	1172	70.5	3.02	0	0	9	32	92	.56	39.5
NOV	1009	1129	59.7	2.54	5	25	84	179	313	.52	48.9
DEC	847	1033	53.2	3.09	57	131	235	373	523	.49	52.9
YR	1500	973	68.8	2.57	247	518	956	1570	2379	.55	

SHERMAN, TEXAS ELEV 764 LAT 33.7

	HS	VS	TA	SD	D50	D55	D60	D65	D70	KT	L-D
JAN	794	1024	41.7	3.46	281	420	570	723	878	.48	55.0
FEB	1037	1098	45.9	3.57	165	272	400	536	675	.50	47.3
MAR	1366	1053	52.3	4.17	104	182	282	411	555	.52	36.1
APR	1610	892	63.7	2.97	0	12	45	114	212	.51	23.9
MAY	1852	852	71.2	2.35	0	0	0	13	60	.53	14.7
JUN	2114	888	79.4	2.18	0	0	0	0	0	.59	10.3
JUL	2077	890	83.6	2.51	0	0	0	0	0	.59	12.3
AUG	1932	958	83.7	2.57	0	0	0	0	0	.59	20.2
SEP	1580	1088	76.0	2.72	0	0	0	5	20	.57	31.9
OCT	1268	1259	65.8	3.09	0	9	31	90	176	.57	43.7
NOV	919	1171	53.4	2.88	45	116	218	353	499	.53	53.1
DEC	744	1020	44.8	3.14	199	327	474	627	781	.49	57.1
YR	1443	1015	63.5	2.97	794	1338	2020	2872	3856	.55	

WACO, TEXAS ELEV 509 LAT 31.6

	HS	VS	TA	SD	D50	D55	D60	D65	D70	KT	L-D
JAN	833	1007	47.0	3.79	175	280	414	562	714	.48	52.9
FEB	1096	1096	50.9	3.75	94	170	274	401	537	.51	45.2
MAR	1427	1038	57.2	3.99	43	99	178	280	411	.53	34.0
APR	1612	853	67.3	2.81	0	0	15	56	132	.51	21.8
MAY	1774	808	74.5	1.83	0	0	0	0	12	.51	12.6
JUN	2112	873	81.9	1.90	0	0	0	0	0	.59	8.2
JUL	2130	880	85.6	2.03	0	0	0	0	0	.60	10.2
AUG	1958	923	85.7	2.06	0	0	0	0	0	.60	18.1
SEP	1601	1035	78.9	2.04	0	0	0	0	0	.56	29.8
OCT	1301	1205	69.1	3.23	0	5	16	51	121	.56	41.6
NOV	957	1139	57.5	2.93	16	57	132	241	379	.52	51.0
DEC	803	1043	49.8	3.23	109	201	328	475	627	.49	55.0
YR	1468	991	67.1	2.80	437	812	1357	2066	2933	.54	

WICHITA FALLS, TEXAS ELEV 1030 LAT 34.0

	HS	VS	TA	SD	D50	D55	D60	D65	D70	KT	L-D
JAN	862	1168	41.5	3.81	293	429	577	730	884	.53	55.3
FEB	1123	1240	45.9	3.64	167	273	400	536	675	.55	47.6
MAR	1472	1164	52.5	4.43	109	187	284	409	551	.56	36.4
APR	1763	975	64.3	3.22	0	14	45	112	204	.56	24.2
MAY	2017	904	72.3	2.58	0	0	0	13	52	.58	15.0
JUN	2221	912	81.3	2.78	0	0	0	0	5	.61	10.6
JUL	2166	914	85.8	2.66	0	0	0	0	0	.61	12.6
AUG	1969	979	85.5	2.99	0	0	0	0	0	.60	20.5
SEP	1602	1116	77.0	2.60	0	0	0	0	13	.58	32.2
OCT	1291	1304	66.0	3.25	0	10	34	92	176	.58	44.0
NOV	957	1257	52.9	3.09	57	130	234	369	514	.56	53.4
DEC	799	1148	44.2	2.85	206	341	491	645	800	.53	57.4
YR	1522	1089	64.1	3.16	832	1384	2065	2906	3874	.58	

BRYCE CANYON, UTAH ELEV 7588 LAT 37.7

	HS	VS	TA	SD	D50	D55	D60	D65	D70	KT	L-D
JAN	914	1511	19.8	4.26	937	1091	1246	1401	1556	.64	59.0
FEB	1236	1657	23.2	4.84	752	891	1031	1171	1310	.66	51.3
MAR	1685	1582	28.7	3.09	661	815	970	1125	1280	.68	40.1
APR	2133	1310	37.7	3.17	375	521	669	819	969	.69	27.9
MAY	2454	1111	46.2	2.53	152	280	429	583	738	.71	18.7
JUN	2655	1055	54.1	2.51	28	93	193	330	478	.73	14.3
JUL	2424	1044	61.6	1.99	0	7	43	128	264	.68	16.3
AUG	2157	1172	59.9	2.28	0	19	77	176	316	.67	24.2
SEP	1920	1578	52.9	2.53	41	113	225	365	513	.72	35.9
OCT	1465	1811	42.8	2.75	239	382	534	688	843	.72	47.7
NOV	1015	1622	30.7	3.38	580	729	879	1029	1179	.67	57.1
DEC	818	1419	22.4	3.89	856	1011	1166	1321	1476	.63	61.1
YR	1741	1404	40.0	3.09	4621	5952	7462	9136	10922	.69	

CEDAR CITY, UTAH ELEV 5617 LAT 37.7

	HS	VS	TA	SD	D50	D55	D60	D65	D70	KT	L-D
JAN	882	1431	28.7	5.11	669	819	972	1126	1281	.62	59.0
FEB	1180	1548	33.1	5.00	485	618	755	894	1034	.63	51.3
MAR	1636	1521	38.4	3.33	369	517	670	825	980	.66	40.1
APR	2092	1284	47.1	3.10	144	254	392	538	687	.68	27.9
MAY	2467	1115	56.2	2.62	18	69	154	281	430	.71	18.7
JUN	2706	1064	65.0	2.74	0	6	27	86	177	.74	14.3
JUL	2503	1064	73.2	1.60	0	0	0	0	15	.71	16.3
AUG	2241	1213	71.3	1.89	0	0	0	6	43	.70	24.2
SEP	1968	1627	63.2	2.71	0	10	43	114	220	.74	35.9
OCT	1460	1802	51.5	3.27	86	167	283	424	575	.71	47.7
NOV	992	1566	38.8	3.79	351	491	638	787	936	.65	57.1
DEC	785	1331	30.8	3.75	598	751	906	1060	1215	.60	61.1
YR	1745	1379	49.8	3.23	2720	3702	4840	6141	7593	.69	

SALT LAKE CITY, UTAH ELEV 4226 LAT 40.8

	HS	VS	TA	SD	D50	D55	D60	D65	D70	KT	L-D
JAN	639	1017	28.0	4.54	686	839	993	1147	1302	.51	62.1
FEB	989	1363	33.4	4.53	473	608	746	885	1025	.58	54.4
MAR	1454	1463	39.6	3.36	336	481	634	788	942	.62	43.2
APR	1894	1284	49.2	3.60	126	216	338	478	625	.63	31.0
MAY	2362	1184	58.3	3.29	20	62	136	237	372	.68	21.8
JUN	2561	1122	66.2	3.34	0	10	32	88	168	.70	17.4
JUL	2590	1186	76.7	1.96	0	0	0	0	6	.73	19.4
AUG	2254	1356	74.5	2.44	0	0	0	5	25	.71	27.3
SEP	1843	1693	64.8	3.25	0	13	41	105	194	.72	39.0
OCT	1293	1724	52.4	3.70	87	165	269	402	549	.68	50.8
NOV	788	1276	39.1	3.63	341	482	628	777	927	.58	60.2
DEC	570	953	30.3	2.83	611	766	921	1076	1231	.50	64.2
YR	1606	1301	51.0	3.36	2680	3642	4738	5988	7366	.66	

BURLINGTON, VERMONT ELEV 341 LAT 44.5

	HS	VS	TA	SD	D50	D55	D60	D65	D70	KT	L-D
JAN	385	572	16.8	5.14	1030	1185	1339	1494	1649	.36	65.8
FEB	607	782	18.6	4.25	880	1019	1159	1299	1439	.40	58.1
MAR	940	929	29.1	4.19	651	804	958	1113	1268	.43	46.9
APR	1296	936	43.0	3.06	231	366	511	660	810	.44	34.7
MAY	1574	909	54.8	3.40	48	115	206	331	476	.46	25.5
JUN	1729	911	65.2	2.10	0	0	12	63	159	.47	21.1
JUL	1721	941	69.8	2.30	0	0	0	20	79	.49	23.1
AUG	1475	976	67.4	2.52	0	0	11	49	127	.48	31.0
SEP	1122	1013	59.3	2.61	6	29	93	191	325	.46	42.7
OCT	740	895	48.8	2.60	105	210	351	503	657	.43	54.5
NOV	375	507	37.0	3.13	395	541	690	840	990	.32	63.9
DEC	283	408	22.6	4.85	852	1005	1160	1315	1469	.30	67.9
YR	1022	815	44.4	3.35	4198	5274	6490	7878	9448	.44	

NORFOLK, VIRGINIA ELEV 30 LAT 36.9

	HS	VS	TA	SD	D50	D55	D60	D65	D70	KT	L-D
JAN	678	932	40.5	4.60	335	467	612	762	916	.46	58.2
FEB	932	1068	41.4	3.58	260	387	523	661	801	.49	50.5
MAR	1281	1080	48.1	4.19	169	264	389	532	682	.51	39.3
APR	1677	1004	57.8	2.45	8	39	113	226	368	.54	27.1
MAY	1887	909	66.7	2.37	0	0	12	53	137	.54	17.9
JUN	2000	894	74.5	2.01	0	0	0	0	14	.55	13.5
JUL	1853	868	78.3	1.67	0	0	0	0	0	.52	15.5
AUG	1680	918	76.9	1.66	0	0	0	0	0	.52	23.4
SEP	1396	1044	71.8	2.26	0	0	0	9	46	.52	35.1
OCT	1083	1145	61.7	2.51	0	14	58	141	266	.52	46.9
NOV	811	1117	51.6	2.30	50	131	257	403	552	.52	56.3
DEC	624	907	42.3	3.99	278	408	554	706	859	.46	60.3
YR	1326	990	59.3	2.80	1100	1710	2518	3493	4641	.52	

RICHMOND, VIRGINIA ELEV 164 LAT 37.5

	HS	VS	TA	SD	D50	D55	D60	D65	D70	KT	L-D
JAN	632	863	37.5	4.59	412	553	702	854	1008	.44	58.8
FEB	877	1004	39.4	3.68	310	441	578	717	857	.47	51.1
MAR	1210	1025	46.9	4.43	197	298	426	569	719	.49	39.9
APR	1566	953	57.8	2.41	8	38	111	226	368	.51	27.7
MAY	1762	872	66.5	2.62	0	0	17	64	148	.51	18.5
JUN	1872	864	74.2	2.03	0	0	0	0	17	.52	14.1
JUL	1774	849	77.9	1.55	0	0	0	0	0	.50	16.1
AUG	1601	891	76.3	1.52	0	0	0	0	0	.50	24.0
SEP	1348	1020	70.0	2.49	0	0	0	21	78	.50	35.7
OCT	1033	1097	59.3	2.86	10	38	106	203	338	.50	47.5
NOV	733	988	49.0	2.38	90	194	332	480	630	.48	56.9
DEC	567	810	39.0	3.94	361	503	654	807	961	.43	60.9
YR	1249	936	57.8	2.87	1388	2065	2926	3941	5124	.49	

ROANOKE, VIRGINIA ELEV 1175 LAT 37.3

	HS	VS	TA	SD	D50	D55	D60	D65	D70	KT	L-D
JAN	660	911	36.4	4.60	442	585	735	888	1042	.46	58.6
FEB	899	1032	38.1	3.92	346	477	615	754	893	.48	50.9
MAR	1236	1046	45.3	4.33	223	334	469	616	768	.50	39.7
APR	1581	957	55.9	2.85	23	77	160	283	425	.51	27.5
MAY	1764	870	64.4	2.78	0	9	35	101	199	.51	18.3
JUN	1882	865	71.7	1.84	0	0	0	0	34	.52	13.9
JUL	1796	854	75.2	1.62	0	0	0	0	6	.51	15.9
AUG	1620	897	74.1	1.55	0	0	0	0	9	.50	23.8
SEP	1358	1023	68.0	2.25	0	0	6	32	103	.51	35.5
OCT	1080	1157	57.8	2.49	10	43	119	235	380	.52	47.3
NOV	765	1043	46.7	2.29	129	254	400	549	699	.50	56.7
DEC	591	853	37.4	4.06	407	551	703	856	1011	.45	60.7
YR	1270	958	55.9	2.88	1580	2330	3242	4314	5569	.50	

OLYMPIA, WASHINGTON ELEV 200 LAT 47.0

	HS	VS	TA	SD	D50	D55	D60	D65	D70	KT	L-D
JAN	269	385	37.2	4.09	413	558	709	863	1017	.29	68.3
FEB	503	662	41.0	2.56	257	393	532	672	812	.36	60.6
MAR	845	876	43.2	2.28	221	367	521	676	831	.40	49.4
APR	1255	966	48.2	2.11	95	211	355	504	654	.44	37.2
MAY	1632	999	54.0	2.36	27	94	200	344	496	.48	28.0
JUN	1693	940	58.9	2.43	6	28	94	197	336	.47	23.6
JUL	1913	1101	63.6	2.00	0	0	22	89	206	.54	25.6
AUG	1549	1105	62.8	1.99	0	0	28	103	228	.51	33.5
SEP	1157	1146	58.6	1.98	0	20	84	198	343	.50	45.2
OCT	636	795	50.6	1.97	56	152	293	447	601	.40	57.0
NOV	339	489	43.3	2.38	212	353	501	651	801	.33	66.4
DEC	221	324	39.5	2.50	330	481	636	791	946	.27	70.4
YR	1003	816	50.1	2.39	1617	2657	3975	5535	7271	.45	

SEATTLE-TACOMA, WASHINGTON ELEV 400 LAT 47.4

	HS	VS	TA	SD	D50	D55	D60	D65	D70	KT	L-D
JAN	262	378	38.2	4.02	384	527	678	832	986	.29	68.7
FEB	495	657	42.3	2.75	226	358	496	636	776	.36	61.0
MAR	849	893	44.1	2.29	197	340	493	648	803	.41	49.8
APR	1293	1013	48.7	2.13	87	198	340	489	639	.46	37.6
MAY	1714	1061	54.9	2.09	15	70	172	315	468	.50	28.4
JUN	1802	1004	59.8	2.03	0	13	67	167	307	.50	24.0
JUL	2248	1299	64.5	2.19	0	0	19	80	184	.64	26.0
AUG	1616	1174	63.8	1.90	0	0	18	82	199	.53	33.9
SEP	1148	1150	59.6	1.90	0	12	66	170	313	.50	45.6
OCT	656	843	52.2	1.81	30	110	244	397	552	.42	57.4
NOV	337	494	44.6	2.36	178	315	462	612	762	.33	66.8
DEC	211	309	40.5	2.32	299	450	605	760	915	.27	70.8
YR	1056	857	51.1	2.31	1416	2393	3660	5188	6904	.48	

SPOKANE, WASHINGTON ELEV 2365 LAT 47.6

	HS	VS	TA	SD	D50	D55	D60	D65	D70	KT	L-D
JAN	315	496	25.4	6.81	780	927	1078	1230	1384	.35	68.9
FEB	606	887	32.2	4.39	504	641	779	919	1059	.45	61.2
MAR	1041	1190	37.5	2.80	391	543	698	853	1008	.50	50.0
APR	1495	1215	46.1	2.65	151	275	419	567	717	.53	37.8
MAY	1918	1197	54.7	2.87	35	99	194	327	476	.57	28.6
JUN	2083	1151	61.5	2.70	0	17	63	144	264	.57	24.2
JUL	2357	1368	69.7	2.32	0	0	0	21	81	.67	26.2
AUG	1942	1465	68.0	2.69	0	0	12	47	122	.64	34.1
SEP	1435	1572	59.6	3.18	12	41	106	196	322	.62	45.8
OCT	841	1220	47.8	2.91	132	242	383	534	688	.54	57.6
NOV	398	635	35.5	3.17	438	586	735	885	1035	.40	67.0
DEC	255	409	29.0	3.41	652	806	961	1116	1271	.33	71.0
YR	1227	1068	47.3	3.32	3095	4177	5428	6839	8427	.56	

YAKIMA, WASHINGTON ELEV 1066 LAT 46.6

	HS	VS	TA	SD	D50	D55	D60	D65	D70	KT	L-D
JAN	365	585	27.5	6.32	715	862	1012	1165	1319	.38	67.9
FEB	666	976	35.7	3.82	407	543	681	821	960	.47	60.2
MAR	1122	1273	41.8	2.35	261	410	564	719	874	.53	49.0
APR	1598	1277	49.5	2.65	91	187	320	466	615	.56	36.8
MAY	2008	1218	57.9	2.88	15	54	130	239	380	.59	27.6
JUN	2169	1161	64.5	2.74	0	7	31	94	188	.60	23.2
JUL	2358	1324	70.7	2.58	0	0	0	20	75	.67	25.2
AUG	1975	1442	68.6	2.56	0	0	8	37	107	.65	33.1
SEP	1483	1583	61.3	2.65	0	17	65	147	269	.63	44.8
OCT	891	1273	50.1	2.44	79	174	311	463	617	.55	56.6
NOV	444	711	38.4	3.00	354	499	648	798	948	.42	66.0
DEC	295	480	31.3	3.13	581	735	890	1045	1200	.36	70.0
YR	1284	1109	49.8	3.09	2503	3488	4660	6014	7552	.57	

CHARLESTON, WEST VIRGINIA ELEV 951 LAT 38.4

	HS	VS	TA	SD	D50	D55	D60	D65	D70	KT	L-D
JAN	498	638	34.5	5.18	503	646	795	948	1101	.36	59.7
FEB	706	770	36.5	4.73	395	525	661	799	938	.39	52.0
MAR	1009	841	44.5	4.73	250	363	497	642	793	.41	40.8
APR	1356	842	55.9	3.19	31	87	170	287	427	.44	28.6
MAY	1639	837	64.5	3.18	0	14	45	113	207	.47	19.4
JUN	1776	843	72.0	2.42	0	0	0	10	48	.49	15.0
JUL	1682	827	75.0	2.20	0	0	0	0	17	.47	17.0
AUG	1514	863	73.6	1.87	0	0	0	0	18	.47	24.9
SEP	1272	979	67.5	2.54	0	0	10	46	121	.48	36.6
OCT	972	1043	57.0	3.11	24	73	153	267	408	.48	48.4
NOV	613	793	45.4	2.47	162	293	439	588	738	.41	57.8
DEC	440	585	36.2	4.34	444	589	740	894	1048	.35	61.8
YR	1125	822	55.2	3.33	1809	2590	3510	4594	5864	.45	

HUNTINGTON, WEST VIRGINIA ELEV 837 LAT 38.4

	HS	VS	TA	SD	D50	D55	D60	D65	D70	KT	L-D
JAN	526	689	34.3	5.23	509	652	801	954	1108	.38	59.7
FEB	757	847	36.1	4.93	408	537	672	811	950	.41	52.0
MAR	1067	901	44.3	4.78	256	369	503	649	800	.44	40.8
APR	1448	900	55.7	3.22	33	91	174	293	433	.47	28.6
MAY	1710	866	64.5	3.23	0	14	47	115	208	.49	19.4
JUN	1844	865	72.4	2.53	0	0	0	11	46	.51	15.0
JUL	1769	859	75.3	2.21	0	0	0	0	15	.50	17.0
AUG	1580	899	73.9	2.14	0	0	0	0	23	.49	24.9
SEP	1306	1010	67.7	2.63	0	0	11	46	120	.49	36.6
OCT	1004	1091	57.1	3.18	25	74	153	265	406	.50	48.4
NOV	638	840	45.5	2.56	161	291	436	585	735	.43	57.8
DEC	467	636	36.0	4.43	451	596	747	900	1054	.37	61.8
YR	1178	867	55.2	3.42	1843	2624	3544	4629	5898	.47	

EAU CLAIRE, WISCONSIN ELEV 896 LAT 44.9

	HS	VS	TA	SD	D50	D55	D60	D65	D70	KT	L-D
JAN	452	738	11.7	5.27	1188	1343	1497	1652	1807	.43	66.2
FEB	746	1072	15.4	4.28	969	1109	1249	1389	1529	.50	58.5
MAR	1090	1150	27.3	5.33	712	862	1015	1169	1324	.50	47.3
APR	1426	1057	44.5	3.75	213	332	471	617	766	.49	35.1
MAY	1681	977	56.2	3.42	36	95	178	293	434	.49	25.9
JUN	1872	982	66.1	2.58	0	0	17	65	149	.51	21.5
JUL	1886	1030	70.5	2.23	0	0	0	14	66	.53	23.5
AUG	1621	1093	68.4	2.52	0	0	8	37	109	.53	31.4
SEP	1196	1115	58.7	2.39	6	29	96	202	341	.50	43.1
OCT	826	1062	48.7	3.32	131	229	361	508	661	.48	54.9
NOV	450	670	32.0	3.88	544	691	840	990	1140	.39	64.3
DEC	341	544	18.0	4.51	993	1147	1302	1457	1612	.37	68.3
YR	1134	957	43.1	3.62	4792	5837	7034	8393	9938	.49	

GREEN BAY, WISCONSIN ELEV 702 LAT 44.5

	HS	VS	TA	SD	D50	D55	D60	D65	D70	KT	L-D
JAN	451	722	15.4	4.63	1073	1228	1383	1538	1693	.42	65.8
FEB	725	1010	18.0	3.79	896	1036	1176	1316	1456	.48	58.1
MAR	1104	1153	28.6	4.55	668	820	974	1129	1284	.50	46.9
APR	1439	1056	43.8	3.36	219	346	489	637	786	.49	34.7
MAY	1719	989	54.5	3.30	49	117	210	338	484	.50	25.5
JUN	1908	989	64.5	2.64	0	6	28	91	186	.52	21.1
JUL	1888	1022	69.2	2.19	0	0	0	22	85	.54	23.1
AUG	1622	1082	67.7	2.80	0	0	15	54	131	.52	31.0
SEP	1218	1127	58.9	2.02	0	18	81	191	334	.50	42.7
OCT	821	1037	49.2	3.43	126	220	348	494	646	.48	54.5
NOV	465	690	34.1	3.58	481	628	777	927	1077	.40	63.9
DEC	350	554	20.9	4.08	903	1057	1212	1367	1522	.37	67.9
YR	1144	952	43.7	3.37	4415	5476	6693	8104	9684	.50	

LA CROSSE, WISCONSIN ELEV 673 LAT 43.9

	HS	VS	TA	SD	D50	D55	D60	D65	D70	KT	L-D
JAN	481	771	16.1	5.20	1052	1206	1361	1516	1671	.44	65.2
FEB	765	1065	20.0	4.60	841	980	1120	1260	1400	.49	57.5
MAR	1101	1125	31.1	5.30	600	748	899	1052	1207	.49	46.3
APR	1426	1027	47.6	3.80	157	256	384	526	673	.49	34.1
MAY	1713	972	59.0	3.40	19	57	128	224	353	.50	24.9
JUN	1905	976	68.5	2.70	0	0	9	39	108	.52	20.5
JUL	1900	1013	72.8	2.50	0	0	0	10	44	.54	22.5
AUG	1666	1094	71.4	2.60	0	0	0	17	65	.54	30.4
SEP	1242	1133	61.8	2.40	0	11	50	130	253	.51	42.1
OCT	863	1087	51.8	3.90	102	182	288	421	569	.49	53.9
NOV	494	734	35.4	3.80	445	590	739	888	1038	.41	63.3
DEC	369	581	21.8	4.70	876	1030	1184	1339	1494	.38	67.3
YR	1162	964	46.4	3.74	4092	5060	6162	7422	8875	.50	

MADISON, WISCONSIN ELEV 860 LAT 43.1

	HS	VS	TA	SD	D50	D55	D60	D65	D70	KT	L-D
JAN	515	822	16.8	4.83	1030	1185	1339	1494	1649	.45	64.4
FEB	804	1108	20.3	4.13	832	972	1112	1252	1392	.51	56.7
MAR	1136	1141	30.2	4.78	622	772	925	1079	1234	.50	45.5
APR	1398	981	45.3	3.41	188	306	446	592	741	.47	33.3
MAY	1743	969	56.0	3.35	36	96	179	297	440	.51	24.1
JUN	1948	977	65.8	2.67	0	0	20	72	158	.53	19.7
JUL	1934	1009	70.1	2.10	0	0	0	14	68	.55	21.7
AUG	1708	1098	68.7	2.68	0	0	9	39	110	.55	29.6
SEP	1299	1168	59.7	2.19	0	17	73	173	311	.52	41.3
OCT	911	1139	49.9	3.58	119	209	331	474	625	.51	53.1
NOV	504	729	34.7	3.49	463	610	759	909	1059	.41	62.5
DEC	389	603	21.9	4.45	872	1027	1181	1336	1491	.38	66.5
YR	1192	978	44.9	3.47	4162	5194	6374	7731	9278	.51	

MILWAUKEE, WISCONSIN ELEV 692 LAT 42.9

	HS	VS	TA	SD	D50	D55	D60	D65	D70	KT	L-D
JAN	479	731	19.4	4.70	950	1104	1259	1414	1569	.42	64.2
FEB	736	967	22.5	3.78	770	910	1050	1190	1330	.46	56.5
MAR	1089	1071	31.4	4.59	585	735	888	1042	1197	.48	45.3
APR	1443	1011	44.7	3.15	193	319	462	610	759	.49	33.1
MAY	1768	977	54.2	3.42	55	126	220	348	494	.51	23.9
JUN	1977	984	64.5	2.62	0	6	28	90	186	.54	19.5
JUL	1962	1017	69.9	2.09	0	0	0	15	70	.55	21.5
AUG	1719	1099	69.2	2.74	0	0	9	36	103	.55	29.4
SEP	1310	1173	61.1	2.24	0	11	55	140	271	.53	41.1
OCT	908	1124	51.0	3.36	96	180	298	440	591	.50	52.9
NOV	525	767	36.5	3.48	412	557	706	855	1005	.42	62.3
DEC	378	573	24.2	4.50	802	956	1110	1265	1420	.37	66.3
YR	1193	957	45.7	3.39	3863	4904	6085	7445	8995	.51	

CASPER, WYOMING ELEV 5289 LAT 42.9

	HS	VS	TA	SD	D50	D55	D60	D65	D70	KT	L-D
JAN	683	1251	23.2	5.27	835	988	1142	1296	1451	.60	64.2
FEB	1013	1550	26.8	4.73	653	791	930	1070	1210	.63	56.5
MAR	1441	1568	31.0	4.55	597	747	900	1054	1209	.64	45.3
APR	1847	1340	42.7	3.79	254	381	523	670	819	.62	33.1
MAY	2204	1191	52.7	3.12	64	141	249	388	538	.64	23.9
JUN	2501	1172	61.9	3.05	6	21	69	147	259	.69	19.5
JUL	2535	1249	71.0	2.29	0	0	0	13	61	.72	21.5
AUG	2225	1444	69.6	2.10	0	0	0	17	76	.71	29.4
SEP	1749	1717	58.7	3.69	24	65	136	229	353	.70	41.1
OCT	1219	1731	47.7	3.63	157	260	393	540	692	.68	52.9
NOV	765	1352	33.9	4.24	492	636	784	933	1083	.61	62.3
DEC	594	1134	26.2	3.54	739	893	1048	1203	1358	.58	66.3
YR	1567	1390	45.4	3.66	3821	4923	6174	7560	9109	.66	

CHEYENNE, WYOMING ELEV 6142 LAT 41.1

	HS	VS	TA	SD	D50	D55	D60	D65	D70	KT	L-D
JAN	766	1362	26.6	5.01	731	883	1036	1191	1346	.62	62.4
FEB	1068	1548	29.0	4.44	591	729	868	1008	1148	.63	54.7
MAR	1433	1451	31.6	4.08	576	727	881	1036	1190	.61	43.5
APR	1770	1203	42.7	3.82	254	382	523	670	819	.59	31.3
MAY	1995	1040	52.4	2.85	59	137	251	394	546	.58	22.1
JUN	2258	1043	61.3	2.97	6	23	75	156	273	.62	17.7
JUL	2230	1076	69.1	2.16	0	0	0	22	86	.63	19.7
AUG	1966	1195	67.6	2.01	0	0	5	31	108	.62	27.6
SEP	1667	1502	58.2	2.98	14	50	122	225	359	.65	39.3
OCT	1242	1645	47.9	3.49	149	252	385	533	686	.66	51.1
NOV	823	1384	35.5	3.78	442	587	736	885	1035	.61	60.5
DEC	671	1244	29.2	3.13	645	800	955	1110	1265	.60	64.5
YR	1492	1306	45.9	3.38	3467	4570	5837	7261	8861	.62	

ROCK SPRINGS, WYOMING ELEV 6745 LAT 41.6

	HS	VS	TA	SD	D50	D55	D60	D65	D70	KT	L-D
JAN	735	1311	19.2	5.66	958	1111	1266	1420	1575	.60	62.9
FEB	1089	1631	23.4	5.76	749	887	1026	1165	1305	.65	55.2
MAR	1530	1619	28.9	4.48	659	811	965	1119	1274	.66	44.0
APR	1944	1358	40.1	3.26	309	450	598	747	897	.65	31.8
MAY	2344	1206	50.4	2.84	88	177	307	455	608	.68	22.6
JUN	2574	1151	58.9	2.46	6	29	95	198	336	.71	18.2
JUL	2547	1202	68.2	1.72	0	0	0	18	87	.72	20.2
AUG	2240	1387	66.1	1.99	0	0	8	49	140	.71	28.1
SEP	1832	1734	56.4	2.82	20	69	149	269	410	.72	39.8
OCT	1306	1812	44.7	3.38	207	333	478	631	785	.70	51.6
NOV	826	1427	30.7	4.26	584	731	880	1029	1179	.63	61.0
DEC	651	1219	22.6	4.52	851	1005	1160	1314	1469	.60	65.0
YR	1637	1420	42.5	3.58	4431	5603	6932	8414	10065	.68	

SHERIDAN, WYOMING ELEV 3966 LAT 44.8

	HS	VS	TA	SD	D50	D55	D60	D65	D70	KT	L-D
JAN	517	900	21.0	7.18	912	1061	1213	1366	1520	.49	66.1
FEB	788	1157	25.9	5.48	680	817	956	1095	1235	.52	58.4
MAR	1205	1315	31.0	5.06	601	749	901	1055	1209	.55	47.2
APR	1537	1150	43.6	3.67	230	355	496	643	792	.53	35.0
MAY	1883	1087	53.1	3.08	58	133	238	375	526	.55	25.8
JUN	2156	1102	61.1	3.28	9	31	87	168	283	.59	21.4
JUL	2329	1237	70.4	2.81	0	0	7	28	87	.66	23.4
AUG	2006	1379	69.2	2.56	0	0	7	31	97	.65	31.3
SEP	1502	1504	57.9	3.61	27	73	147	245	374	.62	43.0
OCT	1005	1409	47.8	3.58	154	256	389	537	689	.59	54.8
NOV	591	1000	33.4	4.56	509	652	800	949	1098	.51	64.2
DEC	441	794	25.5	4.41	762	915	1070	1225	1380	.48	68.2
YR	1333	1170	45.0	4.10	3942	5042	6311	7717	9290	.58	

EDMONTON, ALBERTA ELEV 2220 LAT 53.6

	HS	VS	TA	SD	D50	D55	D60	D65	D70	KT	L-D
JAN	324	745	4.3	8.4	1422	1575	1729	1883	2037	.55	74.9
FEB	622	1238	11.7	8.0	1078	1216	1354	1493	1633	.60	67.2
MAR	1104	1646	20.7	6.4	916	1067	1220	1374	1529	.62	56.0
APR	1554	1566	38.3	4.3	372	509	654	802	951	.59	43.8
MAY	1818	1352	50.5	2.3	68	162	298	450	605	.55	34.6
JUN	1945	1271	57.4	2.5	10	46	121	237	380	.54	30.2
JUL	1977	1378	62.4	2.0	0	5	33	111	240	.57	32.2
AUG	1598	1411	59.5	3.3	16	49	117	210	338	.55	40.1
SEP	1113	1379	50.5	4.0	118	200	311	444	588	.54	51.8
OCT	689	1185	40.6	3.2	306	450	603	757	911	.55	63.6
NOV	359	763	23.4	6.8	811	955	1102	1250	1399	.52	73.0
DEC	235	533	11.5	7.9	1201	1353	1506	1660	1814	.50	77.0
YR	1114	1205	36.0	4.9	6318	7587	9048	10671	12425	.56	

SUFFIELD, ALBERTA ELEV 2549 LAT 50.3

	HS	VS	TA	SD	D50	D55	D60	D65	D70	KT	L-D
JAN	433	939	7.2	8.8	1335	1487	1640	1794	1948	.57	71.6
FEB	748	1387	16.5	8.3	949	1085	1222	1360	1499	.62	63.9
MAR	1232	1687	24.1	7.4	824	970	1120	1272	1425	.63	52.7
APR	1583	1429	40.8	5.2	329	451	588	731	878	.58	40.5
MAY	1903	1287	52.7	2.7	51	127	241	385	537	.57	31.3
JUN	2074	1235	60.6	3.1	9	31	89	173	294	.57	26.9
JUL	2173	1376	67.3	2.2	0	0	7	40	119	.62	28.9
AUG	1801	1468	65.3	3.4	0	14	44	107	195	.61	36.8
SEP	1274	1474	55.2	4.1	61	125	209	320	454	.58	48.5
OCT	822	1329	44.6	4.3	236	351	489	637	789	.57	60.3
NOV	456	902	27.9	6.7	684	824	969	1116	1265	.53	69.7
DEC	333	720	17.1	8.1	1035	1184	1335	1488	1642	.52	73.7
YR	1238	1269	39.9	5.3	5513	6649	7953	9423	11045	.59	

NANAIMO, BRITISH COLUMBIA ELEV 60 LAT 49.2

	HS	VS	TA	SD	D50	D55	D60	D65	D70	KT	L-D
JAN	256	398	36.5	2.9	422	574	729	884	1039	.32	70.5
FEB	538	799	39.7	2.3	291	429	568	708	848	.42	62.8
MAR	921	1063	41.9	1.9	254	406	561	716	871	.46	51.6
APR	1409	1188	47.5	1.7	97	227	375	525	675	.51	39.4
MAY	1860	1214	54.1	2.3	25	90	197	340	493	.55	30.2
JUN	1956	1132	59.5	2.8	8	32	96	190	321	.54	25.8
JUL	2092	1279	63.9	2.7	0	9	38	107	210	.60	27.8
AUG	1727	1344	63.3	2.6	0	10	41	114	223	.58	35.7
SEP	1234	1353	58.5	1.9	0	18	84	200	346	.55	47.4
OCT	680	954	50.0	1.6	52	161	310	465	620	.46	59.2
NOV	333	529	42.4	2.2	234	379	528	678	828	.36	68.6
DEC	206	327	38.7	2.6	354	506	660	815	970	.30	72.6
YR	1104	966	49.6	2.3	1737	2841	4187	5742	7444	.51	

SUMMERLAND, BRITISH COLUMBIA ELEV 1490 LAT 49.3

	HS	VS	TA	SD	D50	D55	D60	D65	D70	KT	L-D
JAN	306	519	25.5	5.4	766	918	1071	1225	1380	.38	70.6
FEB	583	903	32.0	4.1	508	645	784	924	1064	.46	62.9
MAR	1012	1219	38.7	2.9	356	507	661	815	970	.51	51.7
APR	1467	1254	48.2	2.0	92	210	355	504	654	.53	39.5
MAY	1849	1210	56.8	2.5	13	56	138	263	411	.55	30.3
JUN	1993	1156	63.5	2.9	0	12	45	115	216	.55	25.9
JUL	2079	1275	69.8	2.5	0	0	5	25	85	.59	27.9
AUG	1744	1364	67.8	3.2	0	6	21	65	142	.58	35.8
SEP	1296	1454	59.7	2.7	7	28	89	183	314	.58	47.5
OCT	728	1059	48.2	2.2	102	220	367	521	676	.49	59.3
NOV	335	536	36.7	3.4	405	551	700	849	999	.37	68.7
DEC	223	369	29.8	4.1	630	783	937	1091	1246	.32	72.7
YR	1137	1027	48.0	3.2	2879	3936	5173	6580	8157	.53	

VANCOUVER, BRITISH COLUMBIA ELEV 310 LAT 49.3

	HS	VS	TA	SD	D50	D55	D60	D65	D70	KT	L-D
JAN	254	396	36.7	3.3	418	569	723	877	1032	.31	70.6
FEB	495	711	40.3	2.4	275	412	552	692	832	.39	62.9
MAR	877	997	42.6	1.8	233	385	539	694	849	.44	51.7
APR	1331	1110	47.8	1.5	85	217	366	516	666	.48	39.5
MAY	1790	1169	54.5	1.8	11	67	178	326	481	.53	30.3
JUN	1922	1116	59.5	2.2	0	18	77	178	317	.53	25.9
JUL	2021	1239	63.1	1.9	0	0	23	95	219	.58	27.9
AUG	1648	1275	62.6	2.0	0	0	31	107	234	.55	35.8
SEP	1185	1284	57.9	2.0	0	26	97	218	364	.53	47.5
OCT	638	874	50.5	1.4	39	145	295	450	605	.43	59.3
NOV	315	490	43.2	1.9	209	355	504	654	804	.35	68.7
DEC	201	318	39.2	2.9	342	491	645	800	955	.29	72.7
YR	1059	916	49.6	2.1	1612	2685	4030	5607	7358	.49	

WINNIPEG, MANITOBA ELEV 820 LAT 49.9

	HS	VS	TA	SD	D50	D55	D60	D65	D70	KT	L-D
JAN	461	1011	-.9	5.8	1578	1733	1888	2043	2198	.59	71.2
FEB	799	1507	3.7	5.9	1297	1437	1576	1716	1856	.65	63.5
MAR	1240	1676	17.4	6.3	1015	1168	1322	1476	1631	.63	52.3
APR	1548	1370	37.9	4.7	389	524	668	815	964	.56	40.1
MAY	1847	1230	51.1	4.3	126	207	314	446	592	.55	30.9
JUN	2000	1179	61.7	2.9	5	20	67	146	262	.55	26.5
JUL	2025	1264	67.5	2.7	0	0	14	53	131	.58	28.5
AUG	1687	1337	65.7	3.1	0	9	32	92	178	.57	36.4
SEP	1180	1305	54.7	2.9	33	96	187	316	461	.53	48.1
OCT	726	1083	43.9	4.0	242	365	507	657	810	.50	59.9
NOV	406	737	24.1	4.7	779	928	1077	1227	1377	.46	69.3
DEC	336	711	7.3	6.5	1325	1479	1634	1789	1944	.51	73.3
YR	1190	1199	36.1	4.5	6789	7966	9286	10776	12404	.56	

FREDERICTON, NEW BRUNSWICK ELEV 131 LAT 45.5

	HS	VS	TA	SD	D50	D55	D60	D65	D70	KT	L-D
JAN	482	839	15.6	4.7	1067	1222	1376	1531	1686	.48	66.8
FEB	772	1157	17.2	4.1	919	1058	1198	1338	1478	.53	59.1
MAR	1100	1190	27.9	3.7	686	841	995	1150	1305	.51	47.9
APR	1364	1021	39.6	2.1	314	462	612	762	912	.47	35.7
MAY	1576	931	51.3	2.6	67	151	278	426	580	.46	26.5
JUN	1767	946	60.6	1.8	0	6	48	143	283	.49	22.1
JUL	1732	967	66.7	2.4	0	0	12	54	137	.49	24.1
AUG	1532	1045	64.9	1.8	0	0	10	61	167	.50	32.0
SEP	1143	1072	56.7	2.2	8	45	127	254	400	.48	43.7
OCT	759	964	46.4	2.2	138	271	422	577	732	.45	55.5
NOV	439	664	35.1	2.8	449	597	747	897	1047	.39	64.9
DEC	360	607	20.5	5.0	916	1070	1225	1380	1535	.41	68.9
YR	1087	949	41.9	2.9	4564	5723	7050	8573	10262	.48	

HALIFAX, NOVA SCOTIA ELEV 136 LAT 44.6

	HS	VS	TA	SD	D50	D55	D60	D65	D70	KT	L-D
JAN	456	737	26.2	4.0	739	893	1048	1203	1358	.43	65.9
FEB	695	954	26.1	3.4	670	809	949	1089	1229	.46	58.2
MAR	1085	1131	32.2	2.2	552	707	862	1017	1172	.49	47.0
APR	1298	940	40.6	2.2	285	432	582	732	882	.45	34.8
MAY	1522	882	49.5	1.8	67	178	326	481	636	.44	25.6
JUN	1738	916	57.9	1.3	0	9	78	214	363	.48	21.2
JUL	1694	930	64.9	2.0	0	0	13	67	170	.48	23.2
AUG	1582	1056	65.5	1.6	0	0	5	45	147	.51	31.1
SEP	1211	1122	60.1	1.8	0	8	55	156	298	.50	42.8
OCT	812	1024	51.3	1.6	34	126	271	425	580	.47	54.6
NOV	463	689	42.3	2.2	237	382	531	681	831	.40	64.0
DEC	332	515	30.9	4.3	598	749	903	1057	1212	.36	68.0
YR	1075	908	45.7	2.4	3182	4293	5623	7167	8878	.47	

MOOSONEE, ONTARIO ELEV 34 LAT 51.3

	HS	VS	TA	SD	D50	D55	D60	D65	D70	KT	L-D
JAN	375	799	-4.4	5.0	1686	1841	1996	2151	2306	.53	72.6
FEB	721	1380	-.6	5.8	1417	1557	1697	1837	1977	.62	64.9
MAR	1122	1531	10.6	5.9	1223	1377	1532	1687	1841	.59	53.7
APR	1466	1339	27.9	3.8	664	813	963	1113	1263	.54	41.5
MAY	1603	1096	41.7	3.2	276	418	569	723	877	.48	32.3
JUN	1780	1089	53.2	3.4	63	135	233	363	507	.49	27.9
JUL	1625	1044	59.9	2.2	0	17	74	174	316	.47	29.9
AUG	1339	1055	57.9	2.9	15	55	130	239	380	.46	37.8
SEP	956	1026	50.0	3.1	97	186	310	453	601	.44	49.5
OCT	543	751	40.5	3.8	319	458	607	760	915	.39	61.3
NOV	297	498	24.3	3.8	772	921	1071	1221	1371	.37	70.7
DEC	274	570	4.5	6.1	1411	1566	1721	1876	2031	.47	74.7
YR	1009	1012	30.4	4.1	7943	9344	10903	12597	14385	.49	

OTTAWA, ONTARIO ELEV 377 LAT 45.5

	HS	VS	TA	SD	D50	D55	D60	D65	D70	KT	L-D
JAN	510	914	12.6	4.4	1160	1314	1469	1624	1779	.51	66.8
FEB	819	1265	15.3	4.6	972	1112	1252	1392	1532	.56	59.1
MAR	1203	1347	26.6	4.0	727	881	1036	1190	1345	.56	47.9
APR	1490	1133	42.4	2.8	242	381	529	678	828	.52	35.7
MAY	1754	1034	54.5	3.5	54	123	215	340	485	.51	26.5
JUN	1884	1000	64.8	2.1	0	0	15	69	168	.52	22.1
JUL	1875	1040	69.1	2.2	0	0	0	23	87	.53	24.1
AUG	1602	1098	66.7	2.5	0	0	14	58	140	.52	32.0
SEP	1167	1102	58.5	2.5	7	34	102	208	348	.49	43.7
OCT	771	985	47.8	2.9	132	242	383	534	688	.46	55.5
NOV	414	610	34.9	3.1	456	604	753	903	1053	.37	64.9
DEC	383	666	18.5	5.1	978	1132	1287	1442	1597	.43	68.9
YR	1157	1015	42.6	3.3	4728	5823	7055	8461	10050	.51	

TORONTO, ONTARIO ELEV 443 LAT 43.7

	HS	VS	TA	SD	D50	D55	D60	D65	D70	KT	L-D
JAN	487	778	21.6	3.9	881	1036	1190	1345	1500	.44	65.0
FEB	747	1020	22.6	3.9	768	907	1047	1187	1327	.48	57.3
MAR	1100	1116	30.7	3.5	600	754	908	1063	1218	.49	46.1
APR	1468	1055	43.9	2.8	204	338	484	633	783	.50	33.9
MAY	1764	994	54.3	3.3	51	120	214	343	490	.51	24.7
JUN	1950	990	65.1	2.2	0	0	15	67	163	.54	20.3
JUL	1958	1034	69.8	2.1	0	0	0	16	72	.55	22.3
AUG	1686	1102	68.5	2.5	0	0	8	36	107	.54	30.2
SEP	1248	1132	61.0	2.6	0	18	67	152	277	.51	41.9
OCT	833	1026	50.4	2.9	90	179	307	455	608	.47	53.7
NOV	428	591	38.8	2.8	341	487	636	786	936	.35	63.1
DEC	354	541	26.4	4.0	733	887	1042	1197	1352	.36	67.1
YR	1170	948	46.0	3.0	3668	4726	5918	7280	8833	.50	

MONTREAL, QUEBEC ELEV 436 LAT 45.3

	HS	VS	TA	SD	D50	D55	D60	D65	D70	KT	L-D
JAN	470	799	14.7	4.5	1095	1249	1404	1559	1714	.46	66.6
FEB	752	1103	16.9	4.1	927	1067	1207	1347	1487	.51	58.9
MAR	1096	1175	28.0	3.8	684	838	992	1147	1302	.51	47.7
APR	1417	1062	43.0	2.4	220	362	510	660	810	.49	35.5
MAY	1658	974	55.2	3.5	47	112	200	321	464	.49	26.3
JUN	1771	944	65.5	1.9	0	0	8	52	147	.49	21.9
JUL	1846	1021	70.2	2.1	0	0	0	14	66	.52	23.9
AUG	1509	1022	67.8	2.3	0	0	7	37	113	.49	31.8
SEP	1160	1086	59.5	2.7	7	30	93	188	320	.48	43.5
OCT	726	898	49.1	3.0	114	211	346	495	648	.43	55.3
NOV	394	563	36.3	3.1	415	562	711	861	1011	.35	64.7
DEC	336	543	20.7	4.9	910	1064	1219	1373	1528	.38	68.7
YR	1096	931	43.9	3.2	4419	5495	6697	8054	9610	.48	

NORMANDIN, QUEBEC ELEV 450 LAT 48.8

	HS	VS	TA	SD	D50	D55	D60	D65	D70	KT	L-D
JAN	454	921	-.4	5.4	1563	1717	1872	2027	2182	.55	70.1
FEB	785	1387	3.6	5.4	1299	1439	1579	1719	1859	.61	62.4
MAR	1275	1668	16.5	5.0	1040	1194	1349	1504	1659	.63	51.2
APR	1578	1352	32.7	3.1	520	669	819	969	1119	.56	39.0
MAY	1652	1059	47.1	3.2	154	266	406	557	710	.49	29.8
JUN	1831	1053	57.9	2.5	9	40	112	224	365	.50	25.4
JUL	1707	1031	62.4	2.7	0	14	55	133	249	.49	27.4
AUG	1471	1097	59.9	2.3	0	19	77	176	316	.49	35.3
SEP	1018	1027	51.4	2.5	59	141	265	409	558	.45	47.0
OCT	595	775	41.4	2.7	277	424	577	732	887	.39	58.8
NOV	375	617	27.3	3.6	682	831	981	1131	1281	.40	68.2
DEC	343	683	7.5	5.6	1318	1473	1628	1783	1938	.48	72.2
YR	1091	1054	34.0	3.7	6921	8227	9720	11364	13123	.51	

SWIFT CURRENT, SASKATCHEWAN ELEV 2707 LAT 51.1

	HS	VS	TA	SD	D50	D55	D60	D65	D70	KT	L-D
JAN	447	1042	7.2	7.8	1331	1484	1638	1793	1947	.63	72.4
FEB	772	1521	12.4	7.0	1056	1194	1334	1473	1613	.66	64.7
MAR	1263	1812	21.6	7.1	894	1043	1195	1348	1502	.66	53.5
APR	1596	1485	38.7	5.0	374	505	646	792	940	.59	41.3
MAY	1896	1313	50.5	3.0	91	180	306	452	605	.57	32.1
JUN	2021	1231	58.6	2.6	8	35	104	208	345	.56	27.7
JUL	2153	1399	65.7	2.3	0	0	15	66	157	.62	29.7
AUG	1780	1487	63.9	3.6	8	23	64	137	231	.60	37.6
SEP	1266	1507	53.2	4.4	93	166	257	376	513	.59	49.3
OCT	832	1405	43.0	4.2	267	392	535	685	838	.60	61.1
NOV	465	977	25.3	5.7	748	895	1043	1192	1341	.57	70.5
DEC	338	781	14.9	7.2	1095	1247	1400	1554	1709	.57	74.5
YR	1238	1329	37.9	5.0	5965	7164	8537	10076	11741	.60	

APPENDIX 4
SOLAR RADIATION COEFFICIENTS

This appendix tabulates the values of the coefficients for the solar radiation correlation functions and, in some cases, presents graphs of the function or lists the actual values of the function. The tables and graphs are presented in the following order:

The values of the coefficients for the incident-to-horizontal (*INC/HOR*) correlation function are tabulated for *TILT* = 30°, 60°, and 90°; *AZIMUTH* = 0°, 30°, 60°, 90°, 120°, 150°, and 180°. For the case *AZIMUTH* = 0, *TILT* = 90, this correlation yields *INC/HOR* = *VS/HS* where *VS* and *HS* are listed in Appendix 3, "Weather and Insolation Data."

Table A4-1
Coefficients for Incident/Horizontal Equations of Form:

$$INC/HOR = B_1 + B_2Y + B_3Y^2 + K_T(B_4 + B_5Y + B_6Y^2)$$

$$\text{WHERE } Y = (LAT - DEC)/100$$

TILT	AZIM	B_1	B_2	B_3	B_4	B_5	B_6
30	0	0.9786	-0.1895	0.4573	-0.0918	0.1596	2.5034
30	30	0.9935	-0.2581	0.4834	-0.1134	0.3354	1.9329
30	60	0.9937	-0.2691	0.3902	-0.1155	0.4634	0.8290
30	90	0.9959	-0.3156	0.3208	-0.1217	0.5485	-0.4429
30	120	1.0110	-0.4699	0.4446	-0.1410	0.6053	-1.6860
30	150	1.0760	-1.0220	1.2220	-0.2238	1.2000	-3.5840
30	180	1.1770	-1.8840	2.5180	-0.3634	2.3900	-5.8770
60	0	0.8682	-0.4587	0.9788	-0.2817	0.4051	4.0989
60	30	0.8850	-0.5729	1.0557	-0.2689	0.6560	3.0255
60	60	0.8749	-0.5617	0.8552	-0.2151	0.9028	1.0646
60	90	0.8710	-0.5943	0.7018	-0.1966	1.0178	-0.7940
60	120	0.9030	-0.9458	1.1470	-0.2724	1.1800	-2.5040
60	150	0.9389	-1.3770	1.8620	-0.3007	0.5366	-2.3890
60	180	0.7820	-0.4372	0.6406	0.0728	-2.5130	1.8150
90	0	0.6866	-0.6623	1.3269	-0.4458	0.3090	4.7776
90	30	0.7183	-0.8937	1.5666	-0.3680	0.6415	3.2484
90	60	0.7203	-0.9037	1.3357	-0.2920	1.2595	0.5969
90	90	0.6749	-0.6051	0.7595	-0.1959	1.0720	-0.7838
90	120	0.6861	-0.7926	1.0440	-0.2726	1.0720	-2.1210
90	150	0.6766	-0.8036	1.1520	-0.3305	0.4609	-1.5970
90	180	0.6487	-0.5871	0.8319	-0.3992	0.1016	-0.7524

INC/HOR

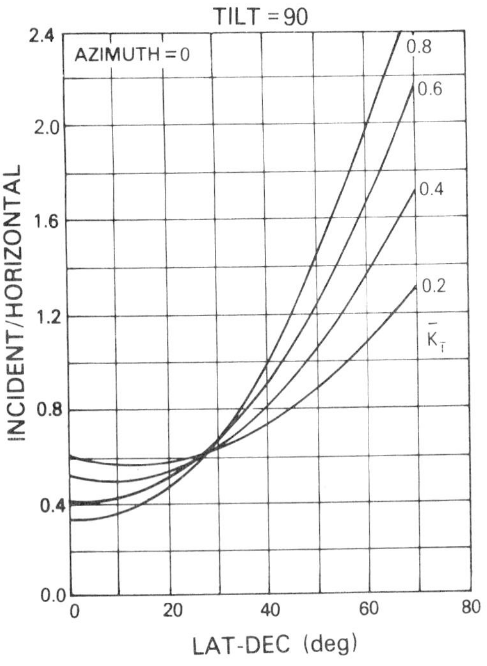

Figure A4-1

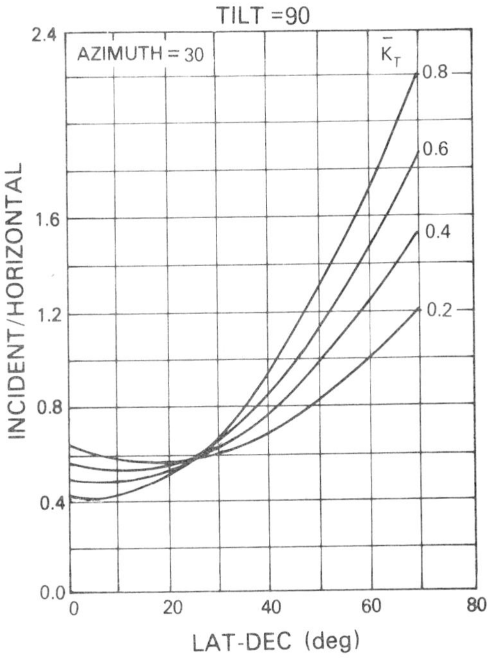

Figure A4-2

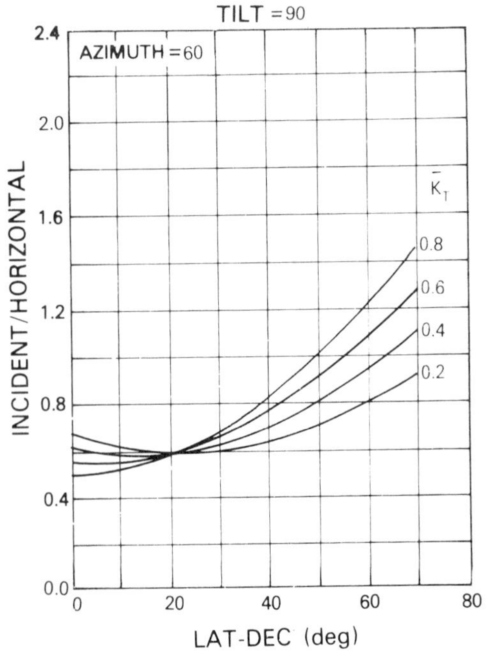

Figure A4-3

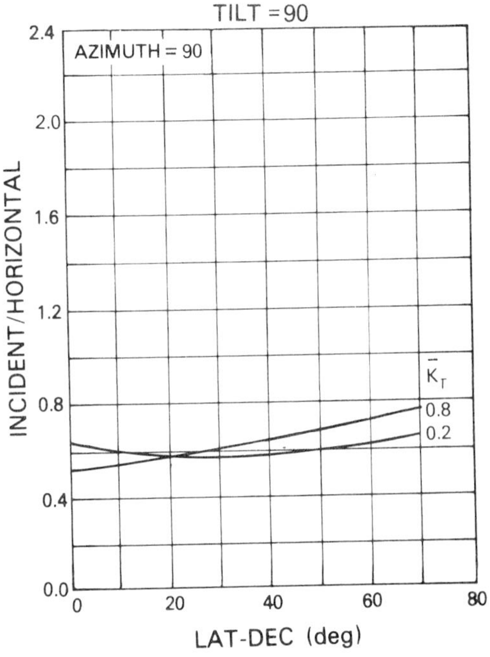

Figure A4-4

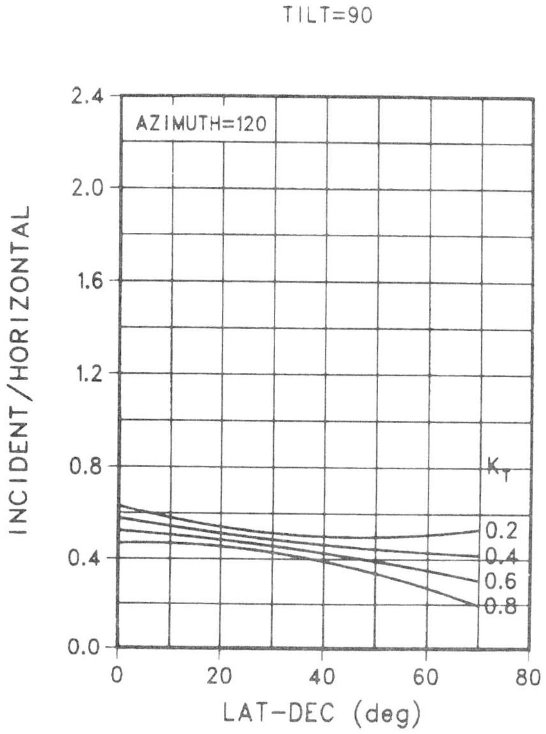

Figure A4-5

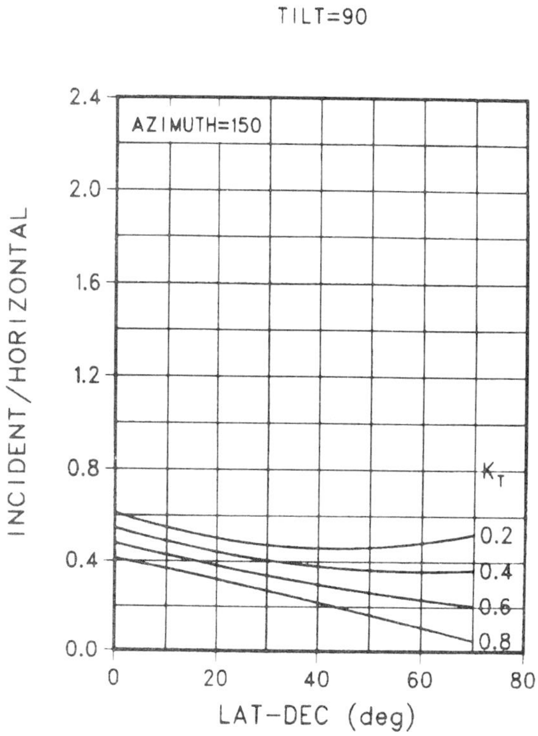

Figure A4-6

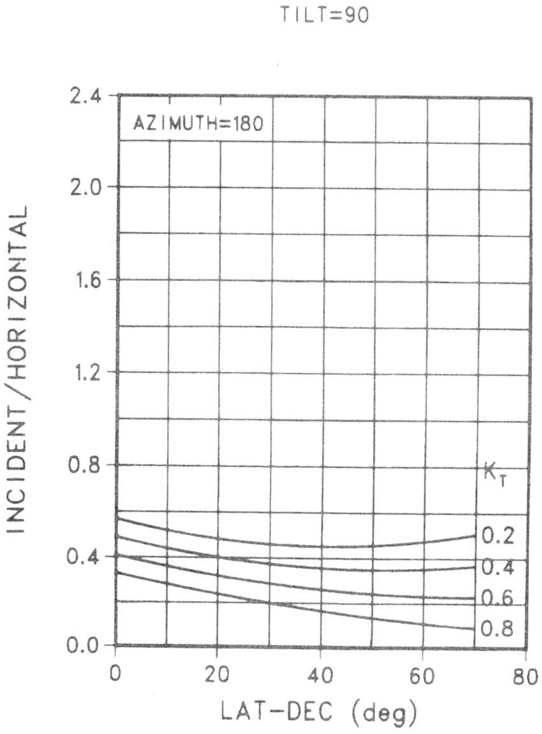

Figure A4-7

Figures A4-1 through A4-7. *INC/HOR* versus *LAT* − *DEC* with K_T as a parameter for *TILT* = 90° and *AZ* = 0°, 30°, 60°, 90°, 120°, 150°, and 180°.

INC/HOR

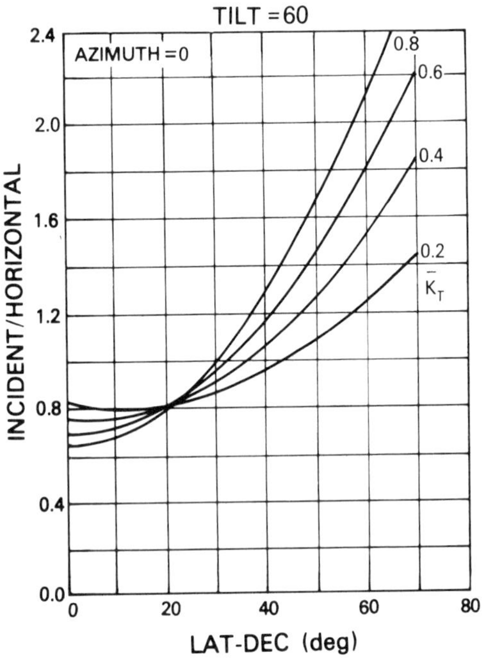

Figure A4-8

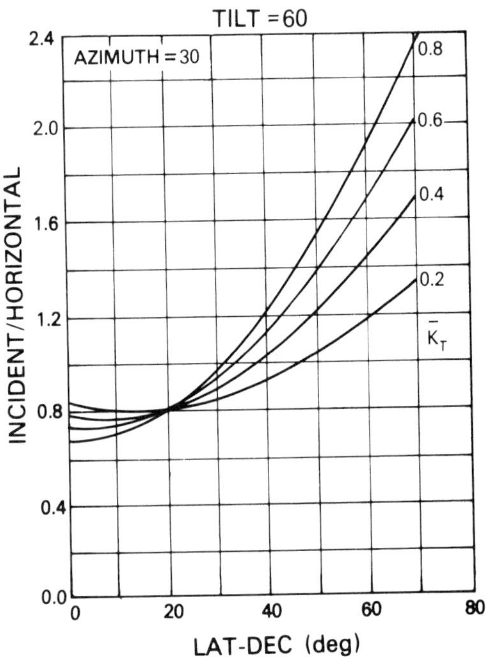

Figure A4-9

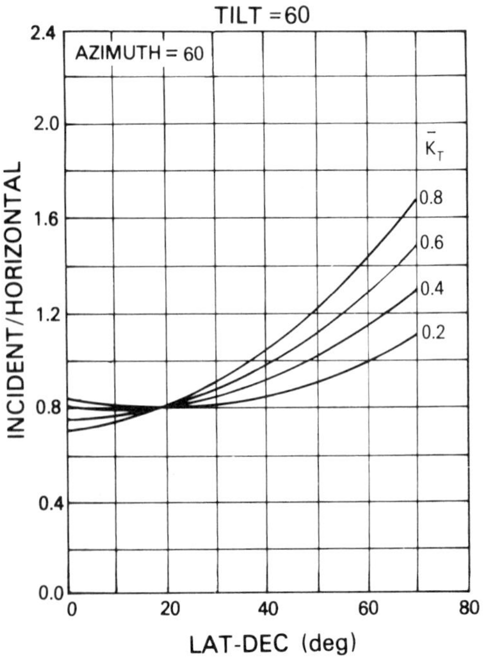

Figure A4-10

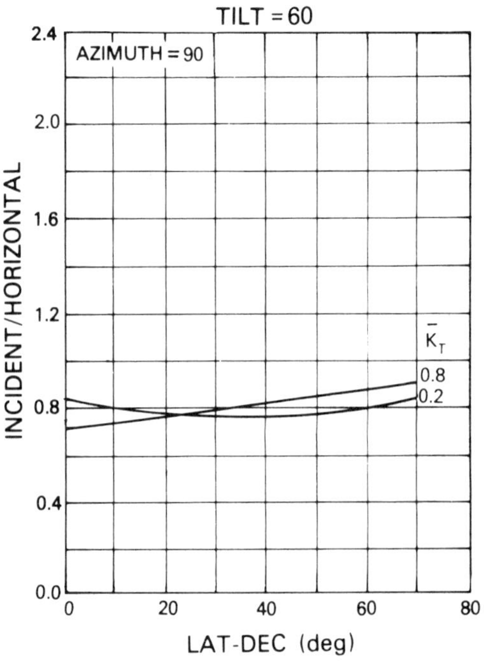

Figure A4-11

INC/HOR

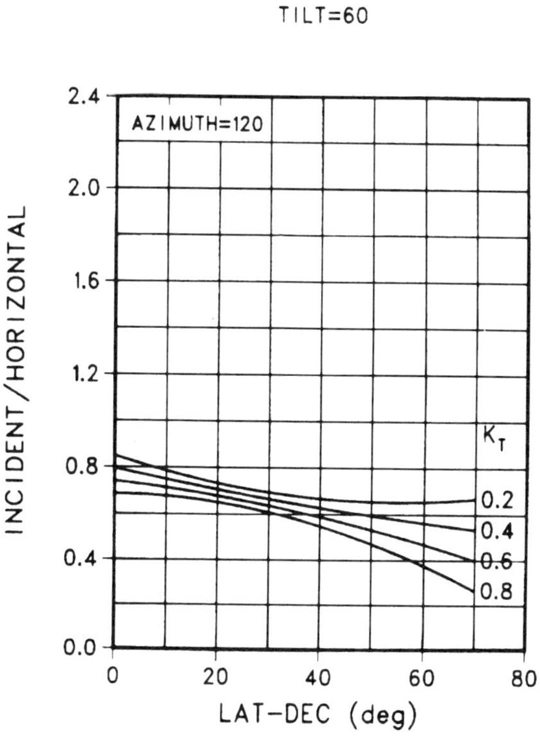

Figure A4-12

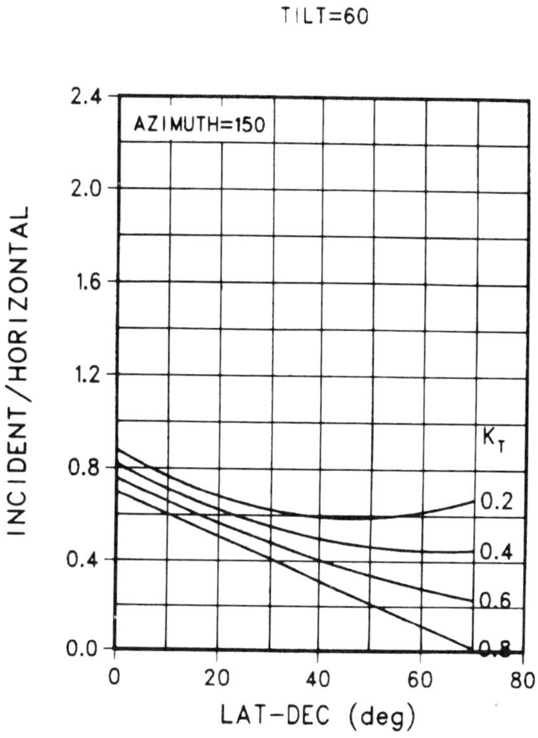

Figure A4-13

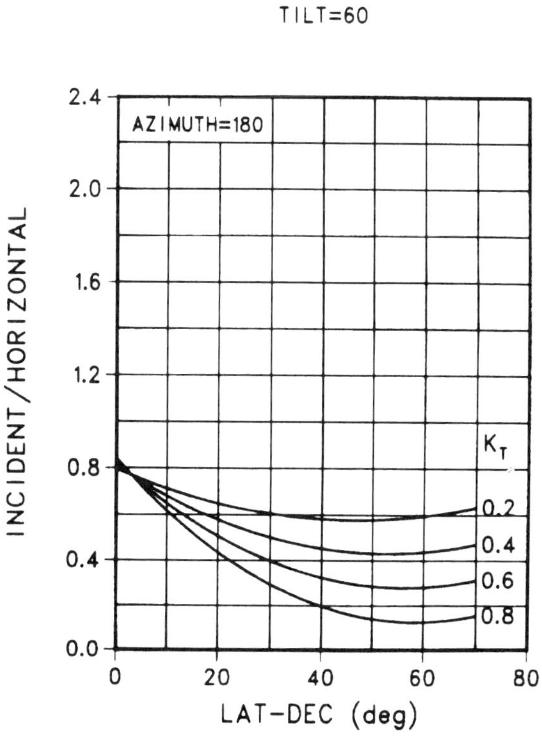

Figure A4-14

Figures A4-8 through A4-14. *INC/HOR* versus *LAT* − *DEC* with K_T as a parameter for *TILT* = 60° and *AZ* = 0°, 30°, 60°, 90°, 120°, 150°, and 180°.

INC/HOR

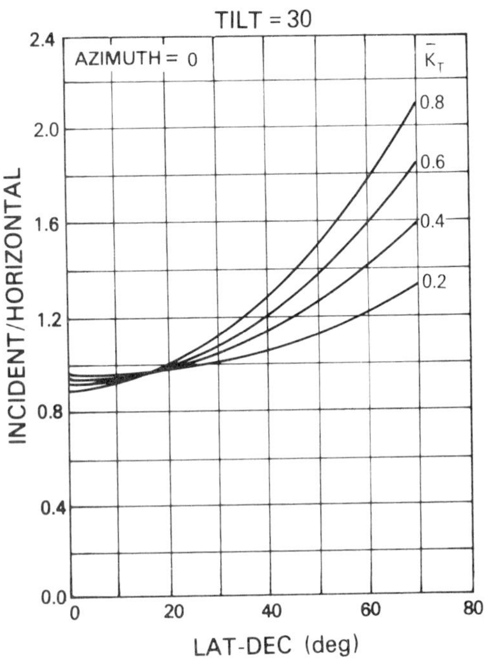

Figure A4-15

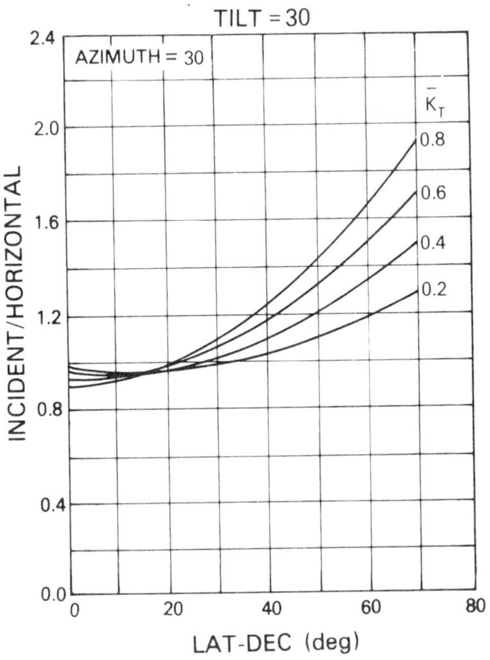

Figure A4-16

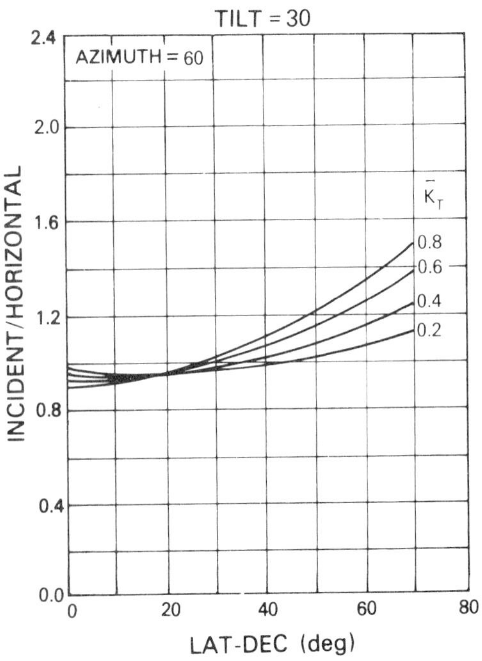

Figure A4-17

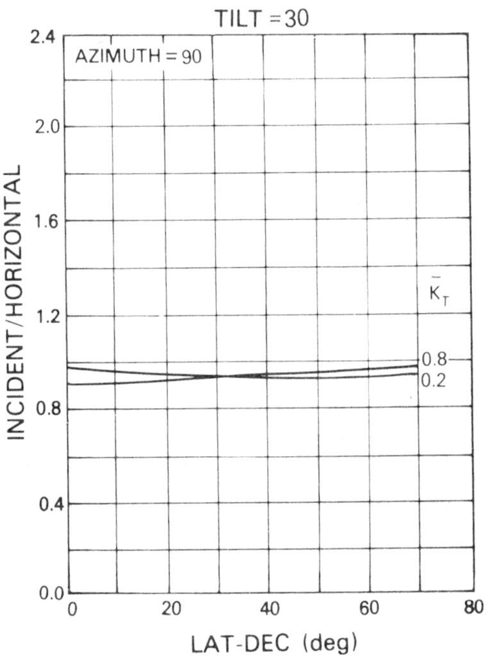

Figure A4-18

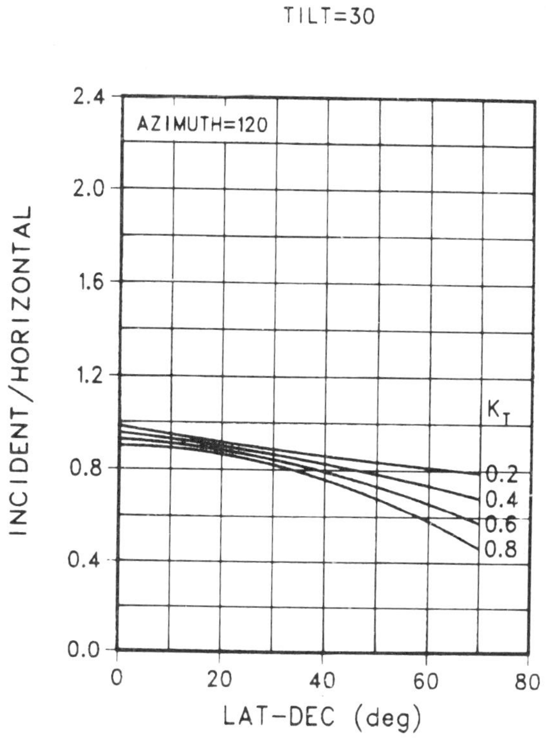

Figure A4-19

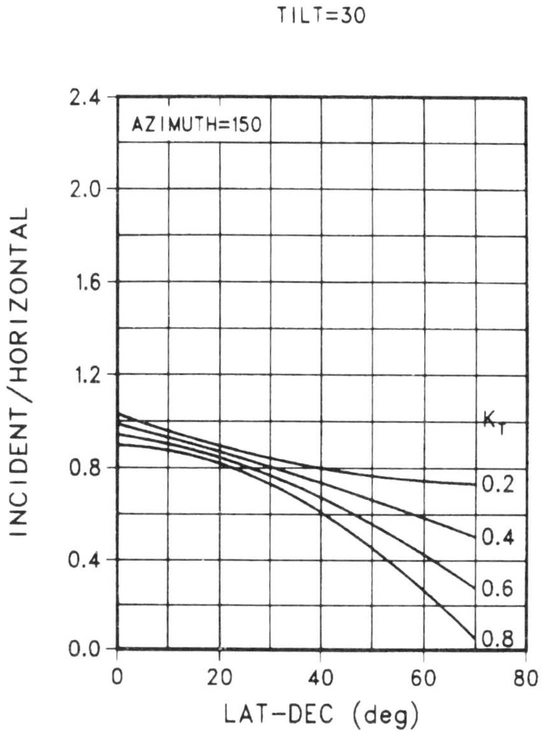

Figure A4-20

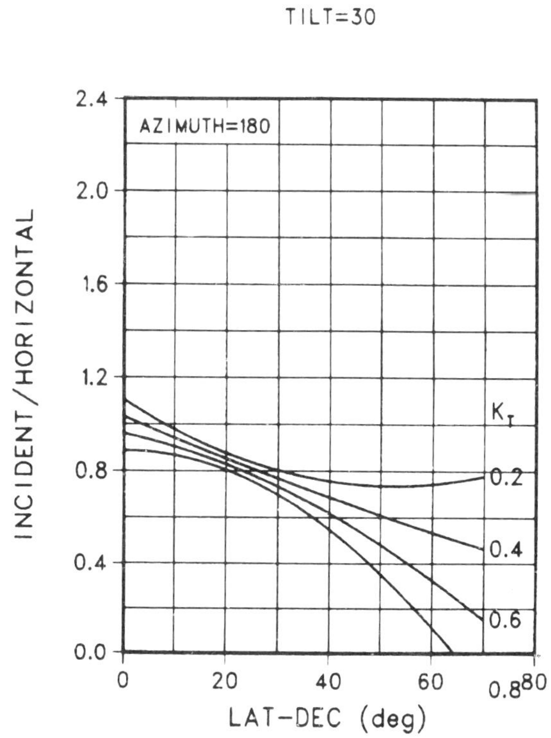

Figure A4-21

Figures A4-15 through A4-21. *INC/HOR* versus *LAT − DEC* with K_T as a parameter for *TILT* = 30° and *AZ* = 0°, 30°, 60°, 90°, 120°, 150°, and 180°.

TRN/INC

The values of the coefficients for the transmitted-to-incident (*TRN/INC*) correlations are tabulated for *TILT* = 0°, 30°, 60°, and 90°; *AZIMUTH* = 0°, 30°, 60°, 90°, 120°, 150°, and 180°; and *N* = one, two, and three layers of glazing. For each pair of values of *TILT* and *AZIMUTH*, the graph of the correlation function is shown with *N* = 1, 2, and 3 as a parameter.

Table A4-2
Coefficients for Transmitted/Incident Equations of Form:

$$TRN/INC = B_1 + B_2Y + B_3Y^2 + B_4Y^3 + B_5Y^4 + B_6Y^5$$

$$+ K_T(B_7 + B_8Y + B_9Y^2 + B_{10}Y^3 + B_{11}Y^4 + B_{12}Y^5)$$

WHERE Y = (LAT - DEC)/100

TRN/INC

Table A4-2 (continued)

1 GLAZING

TILT	AZIM	B_1 B_7	B_2 B_8	B_3 B_9	B_4 B_{10}	B_5 B_{11}	B_6 B_{12}
0	0	0.7646	0.2637	-0.4133	0.0	0.0	0.0
		0.0640	-0.1718	-0.0349	0.0	0.0	0.0
30	0	0.7766	0.1372	-0.1614	0.0	0.0	0.0
		0.0442	-0.1075	0.1715	0.0	0.0	0.0
30	30	0.7766	0.1334	-0.1689	0.0	0.0	0.0
		0.0473	-0.0989	0.1389	0.0	0.0	0.0
30	60	0.7731	0.1413	-0.2101	0.0	0.0	0.0
		0.0577	-0.1078	0.1027	0.0	0.0	0.0
30	90	0.7729	0.1342	-0.2378	0.0	0.0	0.0
		0.0525	-0.0396	-0.0952	0.0	0.0	0.0
30	120	0.8052	-0.2954	1.385	-2.422	0.876	0.817
		0.0026	0.7831	-5.776	18.670	-30.351	17.592
30	150	0.8100	-1.1686	12.675	-52.466	92.076	-56.994
		0.0000	2.2246	-26.786	116.721	-222.560	149.652
30	180	0.8800	-2.8066	23.207	-77.217	114.555	-63.433
		0.0000	0.4973	2.395	-38.047	82.379	-46.942
60	0	0.7735	0.0970	-0.0568	0.0	0.0	0.0
		-0.0354	0.2104	-0.1189	0.0	0.0	0.0
60	30	0.7761	0.0713	-0.0414	0.0	0.0	0.0
		-0.0032	0.1268	-0.0821	0.0	0.0	0.0
60	60	0.7754	0.0620	-0.0796	0.0	0.0	0.0
		0.0343	0.0210	-0.0121	0.0	0.0	0.0
60	90	0.7722	0.0742	-0.1474	0.0	0.0	0.0
		0.0394	0.0218	-0.0850	0.0	0.0	0.0
60	120	0.7898	-0.4928	4.198	-14.640	22.747	-12.800
		0.0310	0.4220	-4.519	17.721	-33.391	22.057
60	150	0.8000	-0.1292	-2.780	21.849	-46.624	30.626
		0.0000	-1.0465	15.461	-82.396	160.469	-102.076
60	180	0.7700	-2.2088	29.534	-121.716	201.036	-116.074
		0.0000	4.3492	-73.477	330.021	-567.756	334.954
90	0	0.7899	0.5388	-5.868	19.912	-26.876	12.777
		0.0098	-3.5791	24.622	-63.976	74.729	-32.865
90	30	0.8130	0.0561	-3.151	13.473	-20.389	10.529
		-0.1433	-0.3334	8.718	-30.567	41.434	-19.727
90	60	0.7728	0.0154	-0.025	0.0	0.0	0.0
		-0.0136	0.1710	-0.1321	0.0	0.0	0.0
90	90	0.7728	0.0087	-0.035	0.0	0.0	0.0
		0.0117	0.1094	-0.2094	0.0	0.0	0.0
90	120	0.7942	-0.7537	6.066	-19.981	30.025	-16.688
		-0.0244	1.4462	-12.147	41.178	-64.897	37.620
90	150	0.7700	-0.1171	1.247	-1.083	-3.791	4.586
		0.0000	-0.1087	0.401	-8.680	26.192	-20.563
90	180	0.8241	0.1416	-1.823	3.250	0.280	-2.636
		-0.1614	0.1087	2.086	-0.191	-13.030	13.631

TRN/INC

Table A4-2 (continued)

2 GLAZINGS

TILT	AZIM	B_1 / B_7	B_2 / B_8	B_3 / B_9	B_4 / B_{10}	B_5 / B_{11}	B_6 / B_{12}
0	0	0.6331	0.3094	-0.4862	0.0	0.0	0.0
		0.0759	-0.2092	-0.0221	0.0	0.0	0.0
30	0	0.6471	0.1596	-0.1862	0.0	0.0	0.0
		0.0521	-0.1214	0.1884	0.0	0.0	0.0
30	30	0.6476	0.1537	-0.1945	0.0	0.0	0.0
		0.0551	-0.1094	0.1525	0.0	0.0	0.0
30	60	0.6434	0.1643	-0.2449	0.0	0.0	0.0
		0.0678	-0.1264	0.1220	0.0	0.0	0.0
30	90	0.6432	0.1565	-0.2766	0.0	0.0	0.0
		0.0624	-0.0531	-0.0966	0.0	0.0	0.0
30	120	0.6822	-0.3874	2.020	-4.389	3.631	-0.592
		0.0001	1.0270	-7.847	26.238	-43.108	25.302
30	150	0.6800	-1.2068	13.659	-57.582	101.873	-63.275
		0.0000	2.3248	-27.520	118.487	-225.285	151.707
30	180	0.7400	-2.2776	18.615	-61.390	90.680	-50.275
		0.0000	0.1469	6.884	-58.900	121.058	-71.553
60	0	0.6429	0.1186	-0.0727	0.0	0.0	0.0
		-0.0375	0.2257	-0.1159	0.0	0.0	0.0
60	30	0.6467	0.0844	-0.0496	0.0	0.0	0.0
		-0.0026	0.1454	-0.0947	0.0	0.0	0.0
60	60	0.6456	0.0763	-0.0971	0.0	0.0	0.0
		0.0404	0.0263	-0.0153	0.0	0.0	0.0
60	90	0.6450	0.0672	-0.1409	0.0	0.0	0.0
		0.0436	0.0516	-0.1418	0.0	0.0	0.0
60	120	0.6657	-0.6335	5.352	-18.615	28.852	-16.207
		0.0297	0.6607	-6.585	25.107	-45.564	29.365
60	150	0.6700	-0.1091	-3.005	23.142	-49.344	32.475
		0.0000	-0.9799	14.880	-81.885	162.359	-104.466
60	180	0.6500	-2.3180	30.175	-123.514	203.430	-117.299
		0.0000	3.7267	-68.566	315.861	-550.141	326.988
90	0	0.6465	0.7766	-7.157	22.771	-29.310	13.294
		0.0523	-4.2457	26.952	-65.486	71.106	-28.793
90	30	0.6863	0.1325	-4.078	16.838	-25.266	13.027
		-0.1539	-0.5816	11.321	-38.883	52.567	-25.112
90	60	0.6436	0.0174	-0.0275	0.0	0.0	0.0
		-0.0148	0.1946	-0.1459	0.0	0.0	0.0
90	90	0.6431	0.0115	-0.0392	0.0	0.0	0.0
		0.0156	0.1215	-0.2342	0.0	0.0	0.0
90	120	0.6621	-0.7670	6.365	-21.263	32.292	-18.105
		-0.0105	1.3212	-11.524	39.751	-64.013	37.876
90	150	0.6400	-0.1430	1.681	-2.748	-1.329	3.320
		0.0000	-0.0437	-0.554	-5.158	21.249	-18.186
90	180	0.7026	0.0921	-1.697	3.139	0.275	-2.610
		-0.1747	0.1651	2.118	-0.573	-12.612	13.579

Table A4-2 (continued)

3 GLAZINGS

TILT	AZIM	B_1 B_7	B_2 B_8	B_3 B_9	B_4 B_{10}	B_5 B_{11}	B_6 B_{12}
0	0	0.5397	0.3355	-0.5234	0.0	0.0	0.0
		0.0766	-0.2504	0.0516	0.0	0.0	0.0
30	0	0.5594	0.1422	-0.1649	0.0	0.0	0.0
		0.0454	-0.0998	0.1580	0.0	0.0	0.0
30	30	0.5600	0.1357	-0.1718	0.0	0.0	0.0
		0.0482	-0.0876	0.1230	0.0	0.0	0.0
30	60	0.5553	0.1503	-0.2254	0.0	0.0	0.0
		0.0617	-0.1137	0.1081	0.0	0.0	0.0
30	90	0.5520	0.1665	-0.2900	0.0	0.0	0.0
		0.0610	-0.0796	-0.0454	0.0	0.0	0.0
30	120	0.5957	-0.4419	2.380	-5.522	5.319	-1.528
		-0.0122	1.2025	-9.294	31.726	-52.206	30.685
30	150	0.5900	-1.2028	13.601	-57.137	100.755	-62.420
		0.0000	2.1078	-24.858	106.776	-203.972	138.239
30	180	0.6500	-2.0826	16.328	-52.086	74.880	-40.700
		0.0000	-0.1328	10.088	-72.302	145.349	-87.455
60	0	0.5542	0.1125	-0.0709	0.0	0.0	0.0
		-0.0366	0.2092	-0.1014	0.0	0.0	0.0
60	30	0.5582	0.0776	-0.0455	0.0	0.0	0.0
		-0.0038	0.1365	-0.0882	0.0	0.0	0.0
60	60	0.5571	0.0699	-0.0903	0.0	0.0	0.0
		0.0356	0.0319	-0.0234	0.0	0.0	0.0
60	90	0.5568	0.0590	-0.1280	0.0	0.0	0.0
		0.0389	0.0547	-0.1440	0.0	0.0	0.0
60	120	0.5790	-0.6747	5.681	-19.822	30.812	-17.346
		0.0196	0.7899	-7.658	29.132	-52.068	33.112
60	150	0.5800	-0.1004	-2.954	22.662	-48.366	31.893
		0.0000	-0.9470	14.250	-78.709	156.701	-101.195
60	180	0.5600	-2.0226	26.925	-111.306	184.409	-106.764
		0.0000	2.9142	-59.167	279.717	-493.091	295.152
90	0	0.5497	0.8103	-6.904	21.070	-26.118	11.345
		0.0730	-4.2675	25.534	-58.122	57.904	-20.849
90	30	0.5967	0.1515	-4.225	17.334	-26.076	13.513
		-0.1504	-0.7208	12.499	-42.924	58.527	-28.285
90	60	0.5553	0.0134	-0.0238	0.0	0.0	0.0
		-0.0155	0.1853	-0.1405	0.0	0.0	0.0
90	90	0.5549	0.0084	-0.0344	0.0	0.0	0.0
		0.0126	0.1205	-0.2304	0.0	0.0	0.0
90	120	0.5715	-0.7096	5.954	-20.076	30.766	-17.382
		-0.0083	1.1758	-10.419	36.366	-59.374	35.547
90	150	0.5500	-0.1107	1.506	-2.493	-1.306	3.149
		0.0000	-0.0569	-0.613	-4.248	19.051	-16.674
90	180	0.6142	0.0718	-1.653	3.343	-0.418	-2.080
		-0.1716	0.1715	2.181	-1.425	-10.539	12.116

TRN/INC

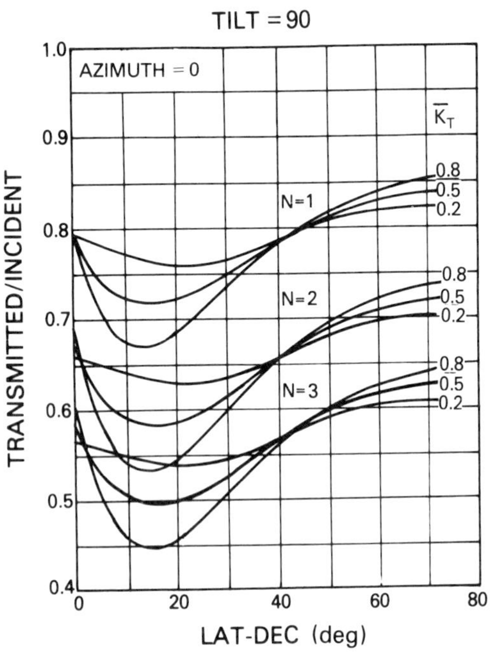

Figure A4-22

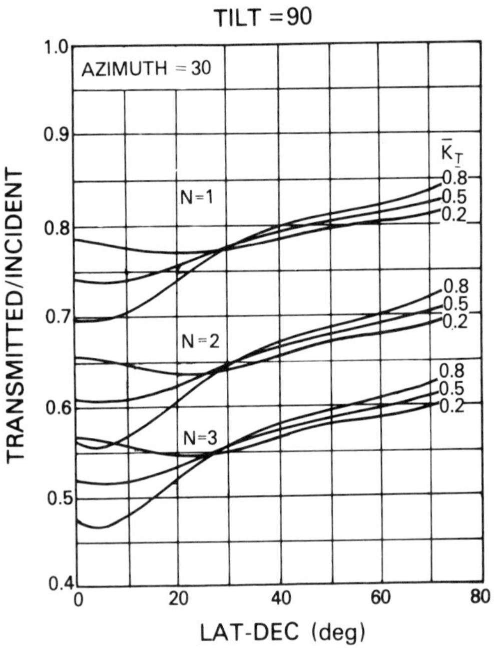

Figure A4-23

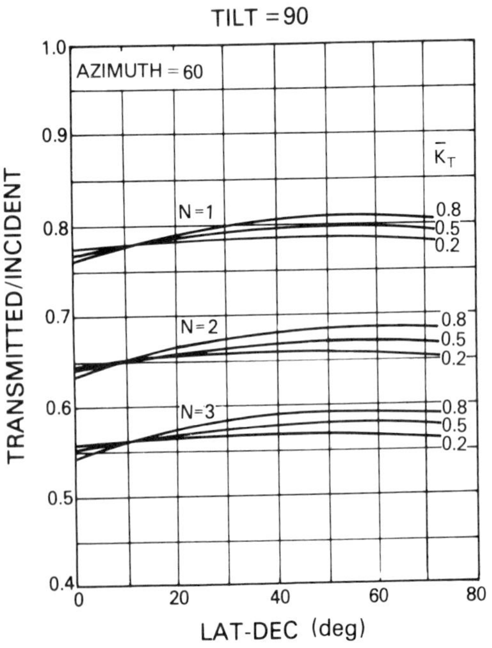

Figure A4-24

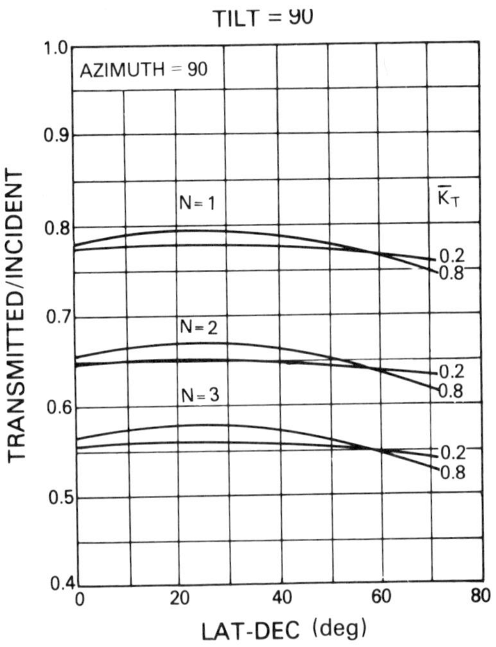

Figure A4-25

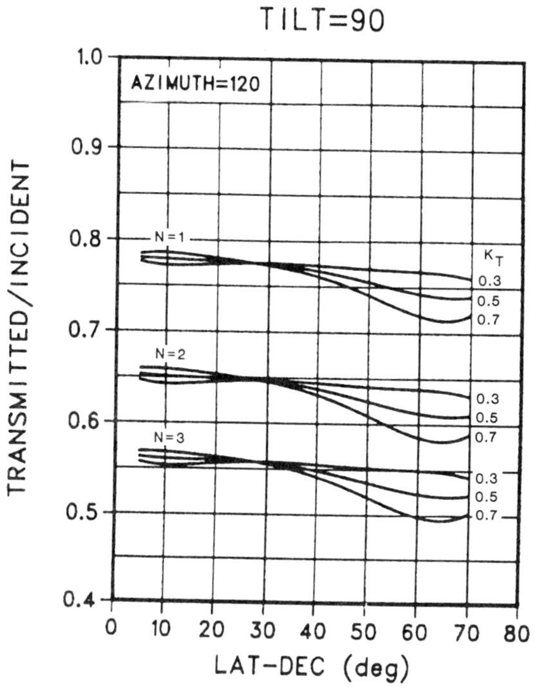

Figure A4-26

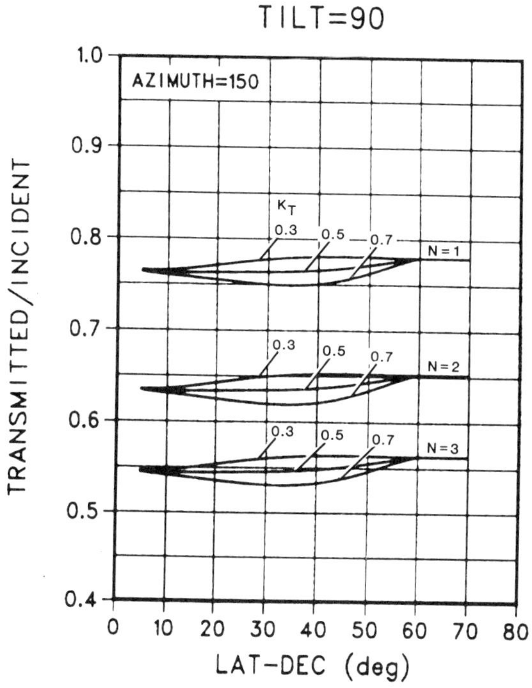

Figure A4-27

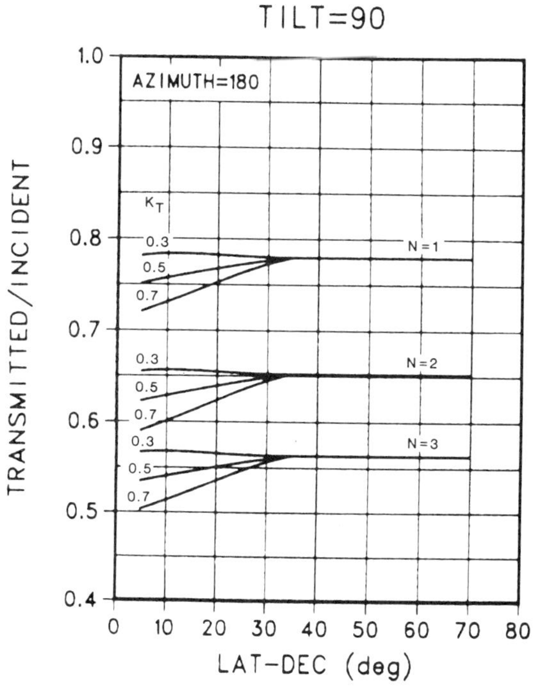

Figure A4-28

Figures A4-22 through A4-28. *TRN/INC* versus *LAT – DEC* with K_T and *N* as a parameter for *TILT* = 90°, and *AZ* = 0°, 30°, 60°, 90°, 120°, 150°, and 180°.

TRN/INC

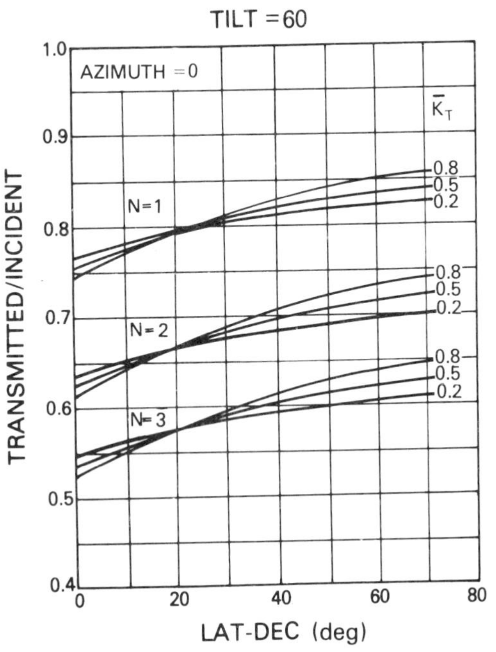

Figure A4-29

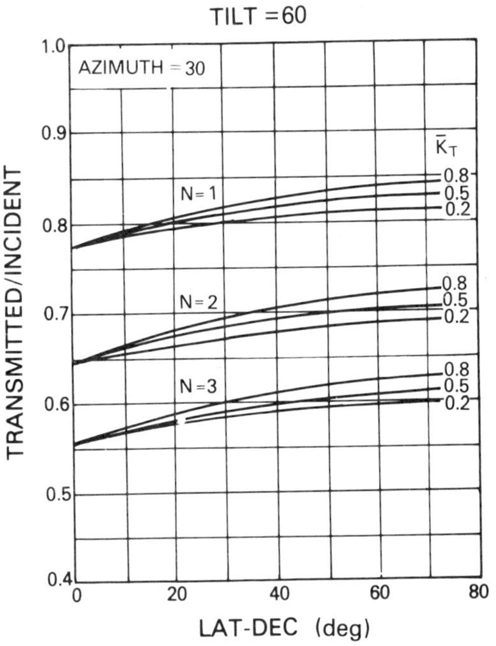

Figure A4-30

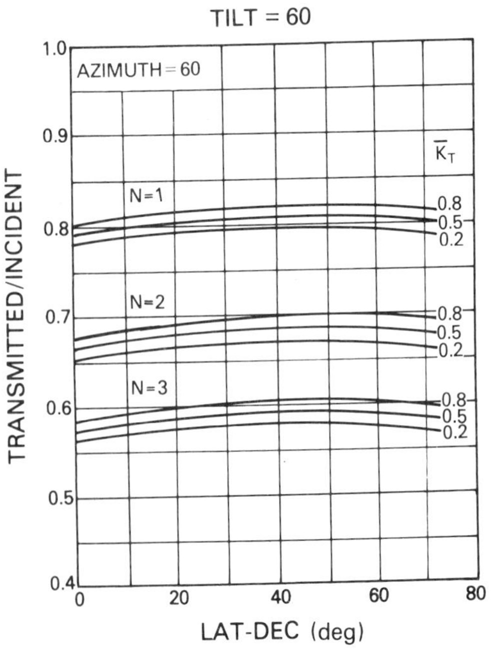

Figure A4-31

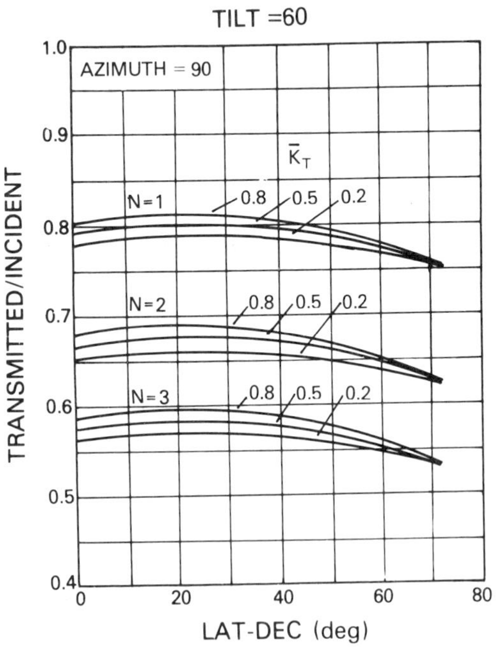

Figure A4-32

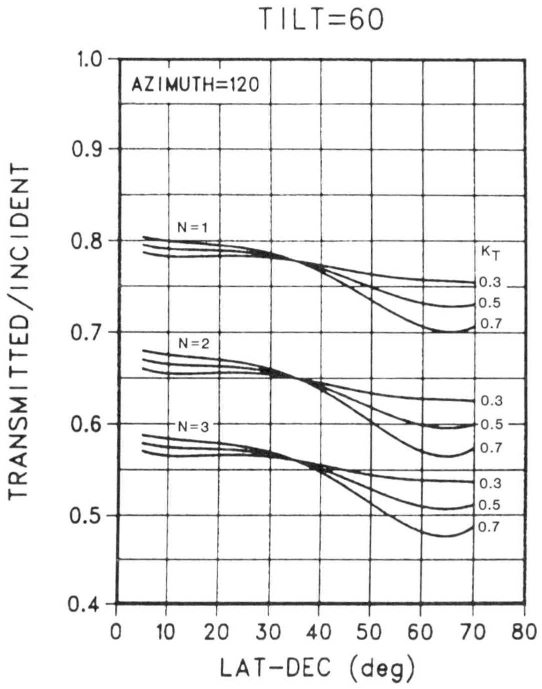

Figure A4-33

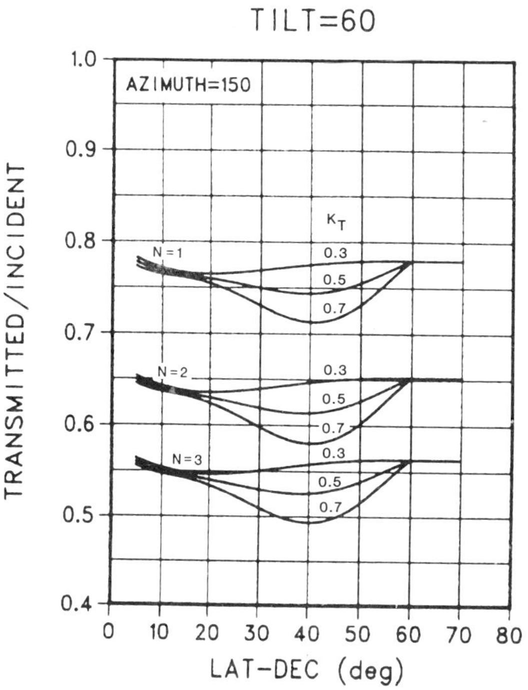

Figure A4-34

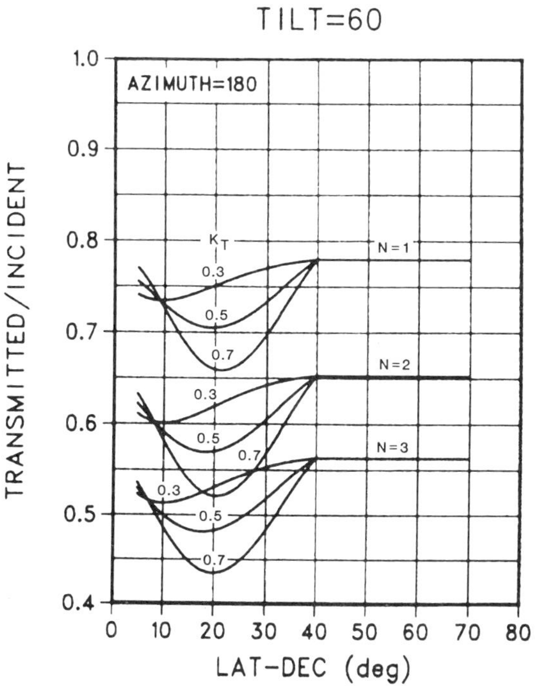

Figure A4-35

Figures A4-29 through A4-35. *TRN/INC* versus *LAT − DEC* with K_T and *N* as a parameter for *TILT* = 60°, and *AZ* = 0°, 30°, 60°, 90°, 120°, 150°, and 180°.

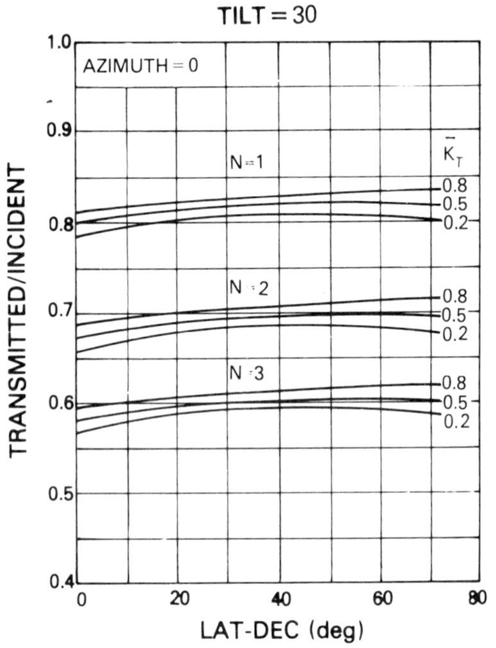

Figure A4-36

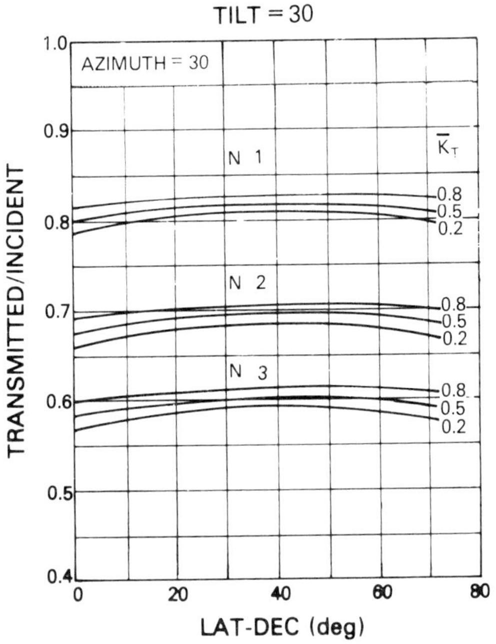

Figure A4-37

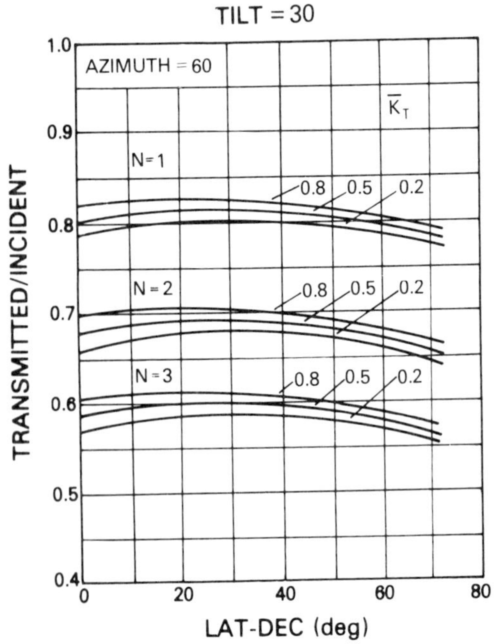

Figure A4-38

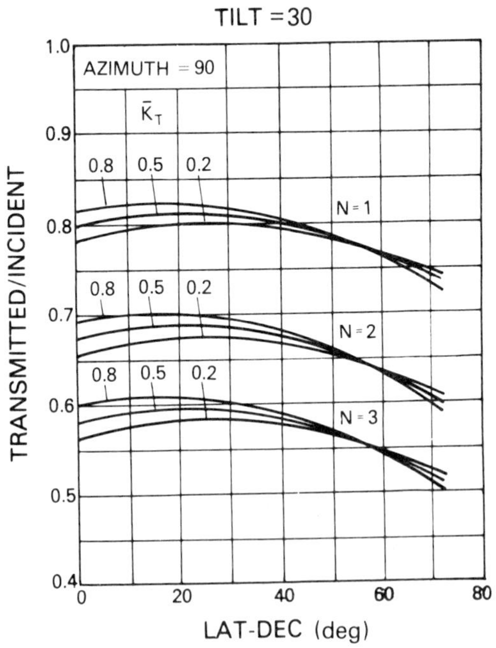

Figure A4-39

TRN/INC

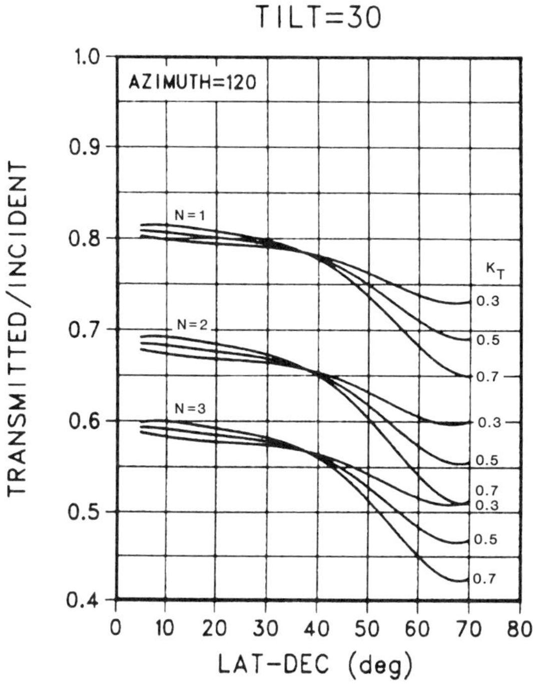

Figure A4-40

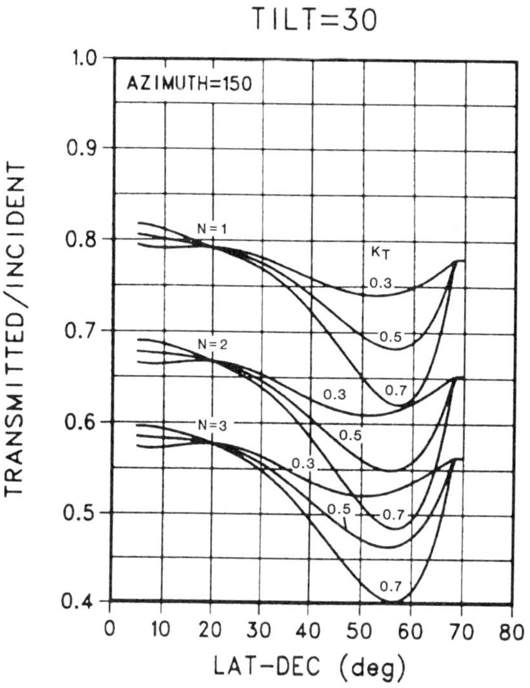

Figure A4-41

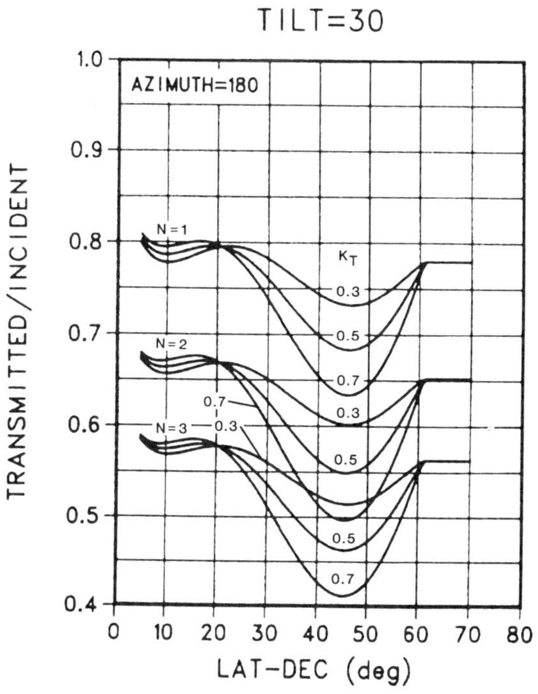

Figure A4-42

Figures A4-36 through A4-42. *TRN/INC* versus *LAT* − *DEC* with K_T and *N* as a parameter for *TILT* = 30°, and *AZ* = 0°, 30°, 60°, 90°, 120°, 150°, and 180°.

TRN/INC

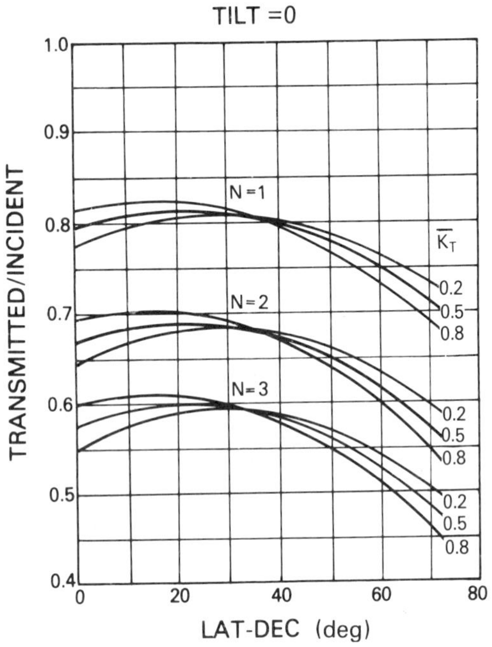

Figure A4-43. *TRN/INC* versus *LAT* − *DEC* with K_T and *N* as a parameter for *TILT* = 0°.

TRN/HOR

The values of the coefficients for the transmitted-to-horizontal (*TRN/HOR*) correlation are tabulated for N = one, two, and three layers of glazing. This correlation is valid only for vertical, south-facing surfaces.

Table A4-3
Solar Radiation Correlation Coefficients for Vertical, South-Facing Surfaces

$$TRN/HOR = B_1 + B_2 Y + B_3 Y^2 + K_T (B_4 + B_5 Y + B_6 Y^2)$$
$$Y = (LAT - DEC)/100$$

No. of Glazings	B_1	B_2	B_3	B_4	B_5	B_6
1	.5136	-.4020	.9059	-.3306	-.3787	5.1344
2	.4146	-.2847	.7160	-.2817	-.4251	4.5913
3	.3484	-.1610	.4980	-.2049	-.6715	4.4288

ABSORPTION FACTORS

The absorption factor (*ABS/TRN*) is tabulated for thermal storage walls with *E, N,* and α_m as parameters and for five categories of direct-gain and sunspace geometry. Each sunspace category is preceded by an illustration of the applicable geometry. Within each category, the variable parameters are α_o, α_m, *WDTH, TILT, XTOP,* and *Z* (not all independent). The end walls are either glazed or opaque. The ranges of the values of the variable parameters are shown in Table A4-4.

Table A4-4
Ranges of the Values of the Variable Parameters for the Absorption Factors *ABS/TRN*

	Thermal Storage Wall	Direct Gain and Sunspace				
		Vertical Glazing	Tilted Glazing			
			Class I	Class II	Class IIIa	Class IIIb
E	0.0125, 0.0625, and 0.2500	0.0625	0.0625	0.0625	0.0625	0.0625
N	1, 2, and 3	2	2	2	2	2
α_m	0.3 to 1.0 in 0.05 steps	0.3 to 1.0 in 0.1 steps	0.3 to 1.0 in 0.1 steps	0.3 to 1.0 in 0.1 steps	0.3 to 1.0 in 0.1 steps	0.3 to 1.0 in 0.1 steps
α_o	n/a	α_o = 0.3 and $\alpha_o = \alpha_m$	α_o = 0.3 and $\alpha_o = \alpha_m$	α_o = 0.3 and $\alpha_o = \alpha_m$	α_o = 0.3 and $\alpha_o = \alpha_m$	α_o = 0.3 and $\alpha_o = \alpha_m$
WDTH	n/a	1, 3, 5	1, 3, 5	1, 3, 5	1, 3, 5	1, 3, 5
TILT	n/a	90°	30°, 60°	10°, 20°, 30°	5°, 45°	5°, 30°, 60°
XTOP	n/a	0, 0.5, 1.5, 3	0, 0.5, 1.5, 3	0, 0.5, 1.5, 3	0, 0.5, 1.5, 3	0, 0.5, 1.5, 3
Z	n/a	0.5, 1	1.15, 2	0.67, 0.97, 1.92	0.5	1
End wall	n/a	opaque and glazed	opaque and glazed	opaque and glazed	opaque and glazed	opaque and glazed

Table A4-5
Thermal Storage Wall Absorption Factors

WALL ABS	E=.0125			E=.0625			E=.2500		
(α_m)	N= 1	2	3	1	2	3	1	2	3
1.00	1.000	1.000	1.000	1.000	1.000	1.000	1.000	1.000	1.000
.95	.957	.963	.967	.957	.962	.964	.956	.959	.959
.90	.914	.925	.932	.914	.922	.928	.912	.916	.918
.85	.870	.885	.897	.869	.882	.890	.867	.873	.876
.80	.826	.845	.860	.824	.840	.851	.821	.829	.833
.75	.780	.803	.821	.779	.798	.811	.775	.785	.789
.70	.734	.761	.781	.733	.754	.769	.728	.739	.744
.65	.688	.717	.740	.686	.710	.726	.680	.693	.698
.60	.640	.671	.697	.638	.664	.682	.632	.646	.651
.55	.591	.625	.652	.589	.617	.636	.584	.598	.604
.50	.542	.577	.605	.540	.568	.588	.534	.549	.555
.45	.492	.527	.556	.490	.519	.539	.484	.499	.505
.40	.441	.476	.505	.439	.468	.488	.433	.448	.454
.35	.389	.423	.452	.387	.415	.435	.382	.396	.402
.30	.337	.369	.396	.335	.361	.380	.330	.343	.348

VERTICAL GLAZING

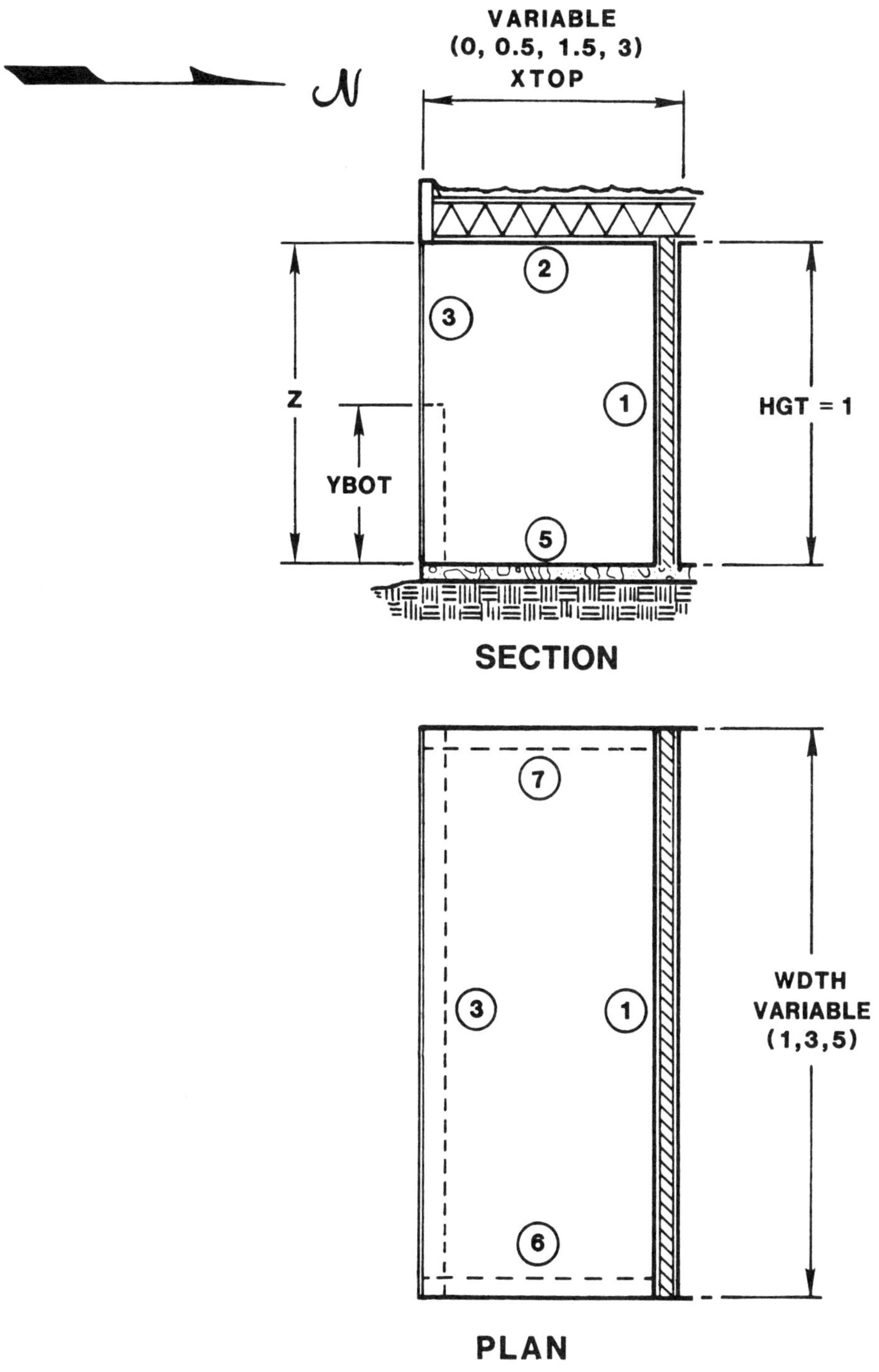

Figure A4-44. Geometry for Vertical Glazing

ABSORPTION FACTORS
VERTICAL GLAZING

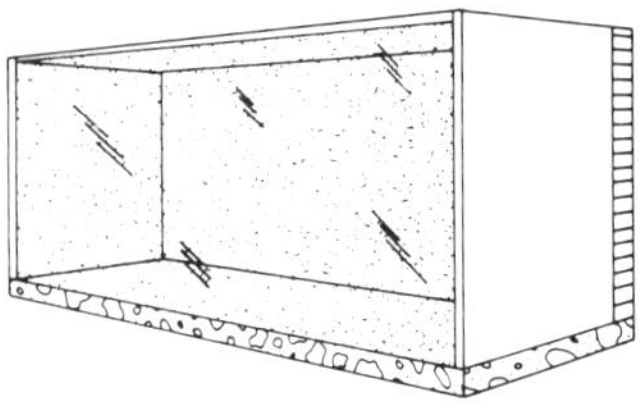

Table A4-6

OPAQUE END WALLS - YBOT= .00

$\alpha_o = \alpha_m$

WDTH	TILT	XTOP	Z	ABSORPTION FOR RADIATION THRU SURFACE 3 α_m 0.3	0.4	0.5	0.6	0.7	0.8	0.9	1.0
1.0	90.	.0	1.00	.62	.68	.74	.79	.85	.90	.95	1.00
1.0	90.	.5	1.00	.81	.85	.88	.91	.94	.96	.98	1.00
1.0	90.	1.5	1.00	.90	.93	.95	.96	.97	.98	.99	1.00
1.0	90.	3.0	1.00	.94	.96	.97	.98	.99	.99	1.00	1.00
3.0	90.	.0	1.00	.62	.68	.74	.79	.85	.90	.95	1.00
3.0	90.	.5	1.00	.77	.81	.86	.89	.92	.95	.98	1.00
3.0	90.	1.5	1.00	.87	.90	.93	.95	.96	.98	.99	1.00
3.0	90.	3.0	1.00	.92	.94	.96	.97	.98	.99	.99	1.00
5.0	90.	.0	1.00	.62	.68	.74	.79	.85	.90	.95	1.00
5.0	90.	.5	1.00	.76	.81	.85	.89	.92	.95	.98	1.00
5.0	90.	1.5	1.00	.86	.89	.92	.94	.96	.98	.99	1.00
5.0	90.	3.0	1.00	.92	.94	.95	.97	.98	.99	.99	1.00

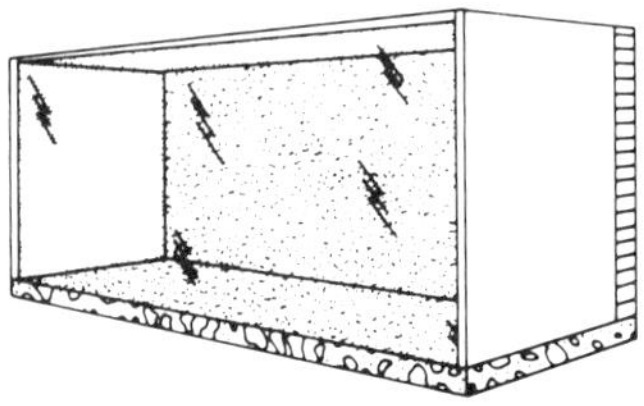

Table A4-7

OPAQUE END WALLS - YBOT= .00

$\alpha_o = .3$

WDTH	TILT	XTOP	Z	ABSORPTION FOR RADIATION THRU SURFACE 3 α_m 0.3	0.4	0.5	0.6	0.7	0.8	0.9	1.0
1.0	90.	.0	1.00	.62	.68	.74	.79	.85	.90	.95	1.00
1.0	90.	.5	1.00	.81	.83	.85	.87	.89	.91	.93	.95
1.0	90.	1.5	1.00	.90	.91	.92	.93	.94	.94	.95	.95
1.0	90.	3.0	1.00	.94	.95	.95	.96	.96	.96	.97	.97
3.0	90.	.0	1.00	.62	.68	.74	.79	.85	.90	.95	1.00
3.0	90.	.5	1.00	.77	.80	.83	.86	.89	.92	.94	.96
3.0	90.	1.5	1.00	.87	.89	.90	.92	.93	.94	.95	.96
3.0	90.	3.0	1.00	.92	.93	.94	.95	.95	.96	.97	.97
5.0	90.	.0	1.00	.62	.68	.74	.79	.85	.90	.95	1.00
5.0	90.	.5	1.00	.76	.79	.83	.86	.89	.92	.94	.97
5.0	90.	1.5	1.00	.86	.88	.90	.91	.93	.94	.95	.97
5.0	90.	3.0	1.00	.92	.93	.94	.95	.95	.96	.97	.97

ABSORPTION FACTORS
VERTICAL GLAZING

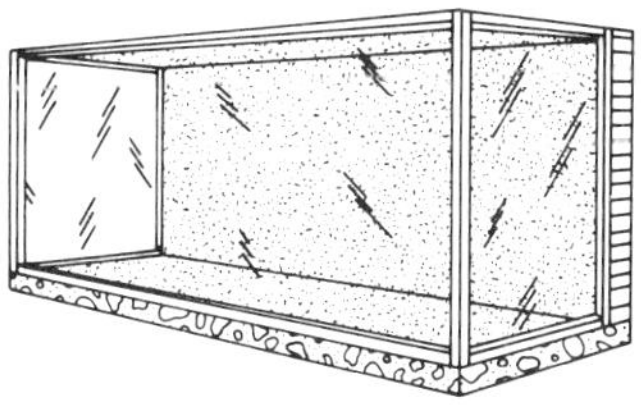

Table A4-8

GLAZED END WALLS - YBOT= .00

$\alpha_o = \alpha_m$

WDTH	TILT	XTOP	Z	ABSORPTION FOR RADIATION THRU SURFACE 3 α_m 0.3	0.4	0.5	0.6	0.7	0.8	0.9	1.0
1.0	90.	.0	1.00	.62	.68	.74	.79	.85	.90	.95	1.00
1.0	90.	.5	1.00	.57	.62	.65	.69	.72	.75	.78	.81
1.0	90.	1.5	1.00	.52	.56	.59	.62	.64	.67	.69	.71
1.0	90.	3.0	1.00	.51	.54	.57	.60	.62	.64	.66	.68
3.0	90.	.0	1.00	.62	.68	.74	.79	.85	.90	.95	1.00
3.0	90.	.5	1.00	.67	.72	.77	.81	.84	.87	.90	.93
3.0	90.	1.5	1.00	.69	.73	.76	.79	.82	.84	.86	.88
3.0	90.	3.0	1.00	.69	.72	.75	.78	.80	.82	.83	.85
5.0	90.	.0	1.00	.62	.68	.74	.79	.85	.90	.95	1.00
5.0	90.	.5	1.00	.70	.75	.79	.83	.87	.90	.93	.96
5.0	90.	1.5	1.00	.75	.78	.82	.84	.87	.89	.91	.92
5.0	90.	3.0	1.00	.76	.79	.82	.84	.86	.88	.89	.90

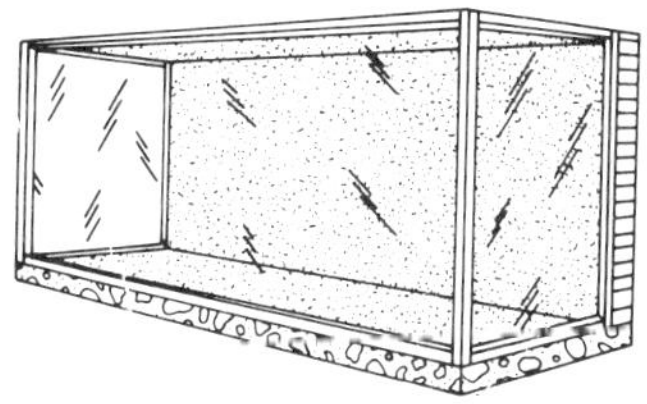

Table A4-9

GLAZED END WALLS - YBOT= .00

$\alpha_o = \alpha_m$

WDTH	TILT	XTOP	Z	ABSORPTION FOR RADIATION THRU SURFACE 6 α_m 0.3	0.4	0.5	0.6	0.7	0.8	0.9	1.0
1.0	90.	.0	1.00	.47	.51	.54	.57	.61	.64	.67	.70
1.0	90.	.5	1.00	.53	.57	.60	.63	.66	.69	.71	.73
1.0	90.	1.5	1.00	.54	.58	.61	.64	.67	.69	.71	.74
1.0	90.	3.0	1.00	.54	.58	.61	.64	.66	.69	.71	.73
3.0	90.	.0	1.00	.47	.51	.54	.57	.61	.64	.67	.70
3.0	90.	.5	1.00	.59	.63	.66	.69	.72	.75	.77	.79
3.0	90.	1.5	1.00	.67	.71	.74	.77	.79	.81	.83	.85
3.0	90.	3.0	1.00	.71	.74	.77	.80	.82	.84	.86	.87
5.0	90.	.0	1.00	.47	.51	.54	.57	.61	.64	.67	.70
5.0	90.	.5	1.00	.60	.64	.67	.70	.73	.76	.78	.80
5.0	90.	1.5	1.00	.70	.74	.77	.79	.81	.83	.85	.86
5.0	90.	3.0	1.00	.76	.79	.81	.84	.86	.87	.89	.90

ABSORPTION FACTORS
VERTICAL GLAZING

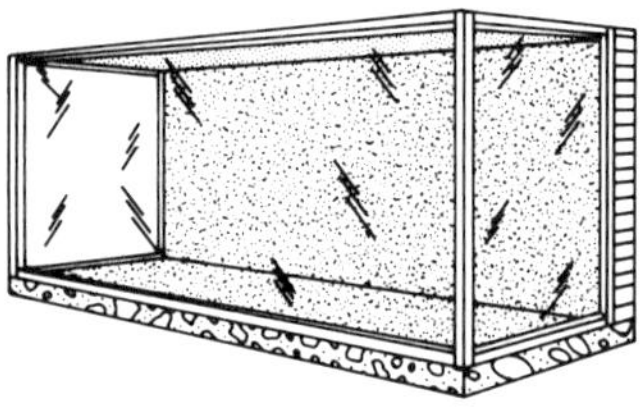

Table A4-10

GLAZED END WALLS - YBOT= .00

α_o = .3

WDTH	TILT	XTOP	Z	ABSORPTION FOR RADIATION THRU SURFACE 3 α_m 0.3	0.4	0.5	0.6	0.7	0.8	0.9	1.0
1.0	90.	.0	1.00	.62	.68	.74	.79	.85	.90	.95	1.00
1.0	90.	.5	1.00	.57	.61	.64	.67	.70	.72	.75	.77
1.0	90.	1.5	1.00	.52	.55	.57	.59	.60	.62	.64	.65
1.0	90.	3.0	1.00	.51	.53	.54	.56	.58	.59	.61	.62
3.0	90.	.0	1.00	.62	.68	.74	.79	.85	.90	.95	1.00
3.0	90.	.5	1.00	.67	.71	.75	.78	.82	.85	.88	.90
3.0	90.	1.5	1.00	.69	.72	.74	.76	.78	.80	.82	.83
3.0	90.	3.0	1.00	.69	.71	.73	.75	.76	.78	.79	.80
5.0	90.	.0	1.00	.62	.68	.74	.79	.85	.90	.95	1.00
5.0	90.	.5	1.00	.70	.74	.78	.81	.84	.88	.90	.93
5.0	90.	1.5	1.00	.75	.77	.79	.82	.84	.85	.87	.89
5.0	90.	3.0	1.00	.76	.78	.80	.81	.83	.84	.85	.87

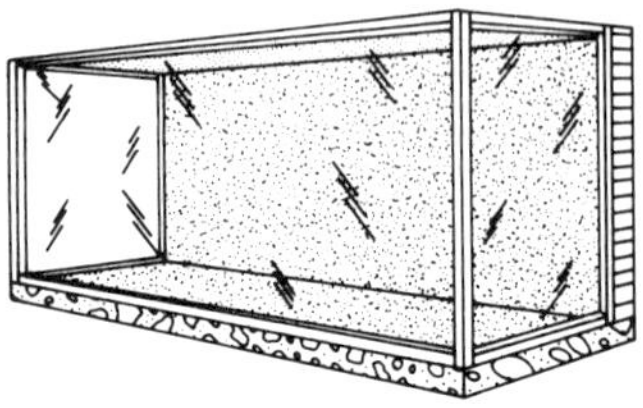

Table A4-11

GLAZED END WALLS - YBOT= .00

α_o = .3

WDTH	TILT	XTOP	Z	ABSORPTION FOR RADIATION THRU SURFACE 6 α_m 0.3	0.4	0.5	0.6	0.7	0.8	0.9	1.0
1.0	90.	.0	1.00	.47	.51	.54	.57	.61	.64	.67	.70
1.0	90.	.5	1.00	.53	.56	.59	.61	.64	.66	.68	.70
1.0	90.	1.5	1.00	.54	.57	.59	.61	.63	.65	.66	.68
1.0	90.	3.0	1.00	.54	.56	.58	.60	.62	.64	.65	.67
3.0	90.	.0	1.00	.47	.51	.54	.57	.61	.64	.67	.70
3.0	90.	.5	1.00	.59	.62	.65	.67	.70	.72	.74	.76
3.0	90.	1.5	1.00	.67	.69	.72	.74	.76	.77	.79	.81
3.0	90.	3.0	1.00	.71	.73	.75	.77	.78	.80	.82	.83
5.0	90.	.0	1.00	.47	.51	.54	.57	.61	.64	.67	.70
5.0	90.	.5	1.00	.60	.63	.66	.68	.71	.73	.75	.77
5.0	90.	1.5	1.00	.70	.72	.75	.77	.78	.80	.81	.83
5.0	90.	3.0	1.00	.76	.78	.79	.81	.82	.84	.85	.86

ABSORPTION FACTORS
VERTICAL GLAZING

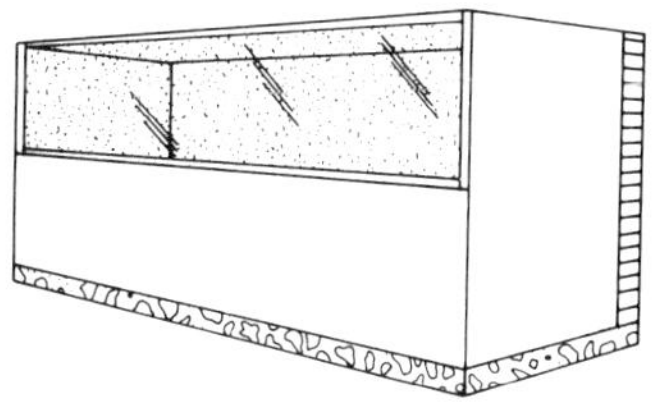

Table A4-12

OPAQUE BOTTOM SECTION, Z= .5
OPAQUE END WALLS
$\alpha_o = \alpha_m$

WDTH	TILT	XTOP	YBOT	ABSORPTION FOR RADIATION THRU SURFACE 3 α_m 0.3	0.4	0.5	0.6	0.7	0.8	0.9	1.0
1.0	90.	.0	.500	.78	.83	.86	.89	.92	.95	.98	1.00
1.0	90.	.5	.500	.89	.92	.94	.95	.97	.98	.99	1.00
1.0	90.	1.5	.500	.95	.96	.97	.98	.99	.99	1.00	1.00
1.0	90.	3.0	.500	.97	.98	.98	.99	.99	1.00	1.00	1.00
3.0	90.	.0	.500	.78	.83	.86	.89	.92	.95	.98	1.00
3.0	90.	.5	.500	.87	.90	.92	.94	.96	.97	.99	1.00
3.0	90.	1.5	.500	.93	.95	.96	.97	.98	.99	.99	1.00
3.0	90.	3.0	.500	.96	.97	.98	.98	.99	.99	1.00	1.00
5.0	90.	.0	.500	.78	.83	.86	.89	.92	.95	.98	1.00
5.0	90.	.5	.500	.86	.89	.92	.94	.96	.97	.99	1.00
5.0	90.	1.5	.500	.93	.95	.96	.97	.98	.99	.99	1.00
5.0	90.	3.0	.500	.96	.97	.98	.98	.99	.99	1.00	1.00

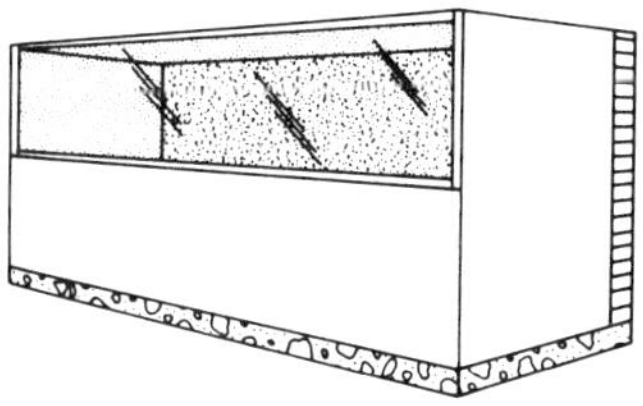

Table A4-13

OPAQUE BOTTOM SECTION, Z= .5
OPAQUE END WALLS
$\alpha_o = .3$

WDTH	TILT	XTOP	YBOT	ABSORPTION FOR RADIATION THRU SURFACE 3 α_m 0.3	0.4	0.5	0.6	0.7	0.8	0.9	1.0
1.0	90.	.0	.500	.78	.82	.86	.89	.92	.95	.97	1.00
1.0	90.	.5	.500	.89	.90	.91	.92	.93	.94	.95	.96
1.0	90.	1.5	.500	.95	.95	.96	.96	.96	.96	.97	.97
1.0	90.	3.0	.500	.97	.97	.97	.98	.98	.98	.98	.98
3.0	90.	.0	.500	.78	.82	.86	.89	.92	.95	.97	1.00
3.0	90.	.5	.500	.87	.89	.90	.92	.93	.94	.95	.97
3.0	90.	1.5	.500	.93	.94	.95	.95	.96	.96	.97	.97
3.0	90.	3.0	.500	.96	.96	.97	.97	.97	.98	.98	.98
5.0	90.	.0	.500	.78	.82	.86	.89	.92	.95	.97	1.00
5.0	90.	.5	.500	.86	.88	.90	.92	.93	.94	.96	.97
5.0	90.	1.5	.500	.93	.94	.95	.95	.96	.97	.97	.98
5.0	90.	3.0	.500	.96	.96	.97	.97	.97	.98	.98	.98

ABSORPTION FACTORS
VERTICAL GLAZING

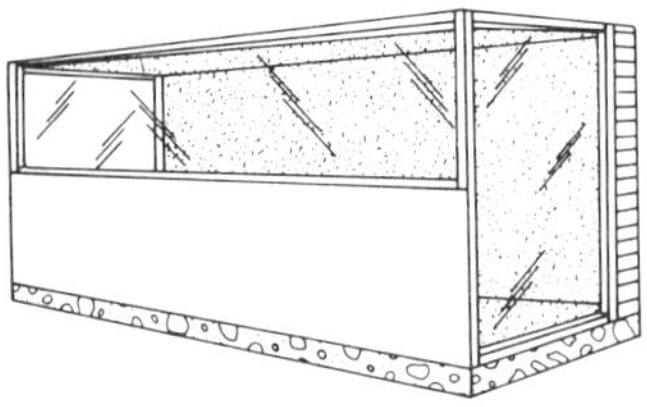

Table A4-14

OPAQUE BOTTOM SECTION, Z= .5
GLAZED END WALLS
$\alpha_o = \alpha_m$

ABSORPTION FOR RADIATION THRU SURFACE 3

WDTH	TILT	XTOP	YBOT	α_m 0.3	0.4	0.5	0.6	0.7	0.8	0.9	1.0
1.0	90.	.0	.500	.78	.83	.86	.89	.92	.95	.97	1.00
1.0	90.	.5	.500	.62	.66	.69	.72	.75	.77	.79	.81
1.0	90.	1.5	.500	.54	.58	.60	.63	.65	.67	.69	.71
1.0	90.	3.0	.500	.52	.55	.58	.60	.63	.65	.67	.68
3.0	90.	.0	.500	.78	.83	.86	.89	.92	.95	.98	1.00
3.0	90.	.5	.500	.76	.80	.83	.85	.88	.90	.92	.93
3.0	90.	1.5	.500	.73	.76	.79	.81	.83	.85	.87	.88
3.0	90.	3.0	.500	.71	.74	.77	.79	.81	.82	.84	.85
5.0	90.	.0	.500	.78	.83	.86	.89	.92	.95	.98	1.00
5.0	90.	.5	.500	.79	.83	.86	.88	.91	.93	.94	.96
5.0	90.	1.5	.500	.80	.83	.85	.87	.89	.90	.91	.92
5.0	90.	3.0	.500	.79	.82	.84	.86	.87	.88	.89	.90

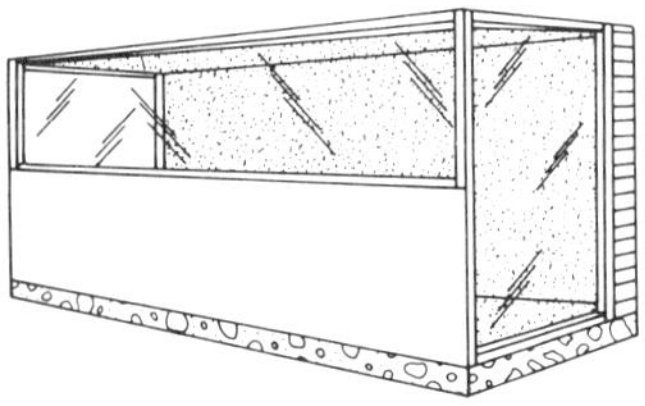

Table A4-15

OPAQUE BOTTOM SECTION, Z= .5
GLAZED END WALLS
$\alpha_o = \alpha_m$

ABSORPTION FOR RADIATION THRU SURFACE 6

WDTH	TILT	XTOP	YBOT	α_m 0.3	0.4	0.5	0.6	0.7	0.8	0.9	1.0
1.0	90.	.0	.500	.70	.73	.76	.78	.80	.82	.83	.85
1.0	90.	.5	.500	.64	.68	.72	.74	.77	.79	.81	.83
1.0	90.	1.5	.500	.59	.63	.66	.69	.72	.75	.77	.79
1.0	90.	3.0	.500	.57	.61	.64	.67	.69	.72	.74	.76
3.0	90.	.0	.500	.70	.73	.76	.78	.80	.82	.83	.85
3.0	90.	.5	.500	.74	.78	.80	.83	.85	.86	.88	.89
3.0	90.	1.5	.500	.76	.79	.82	.84	.86	.88	.90	.91
3.0	90.	3.0	.500	.76	.79	.82	.85	.87	.88	.90	.91
5.0	90.	.0	.500	.70	.73	.76	.78	.80	.82	.83	.85
5.0	90.	.5	.500	.76	.79	.82	.84	.86	.87	.89	.90
5.0	90.	1.5	.500	.80	.83	.85	.87	.89	.90	.92	.93
5.0	90.	3.0	.500	.82	.85	.87	.89	.90	.92	.93	.94

ABSORPTION FACTORS
VERTICAL GLAZING

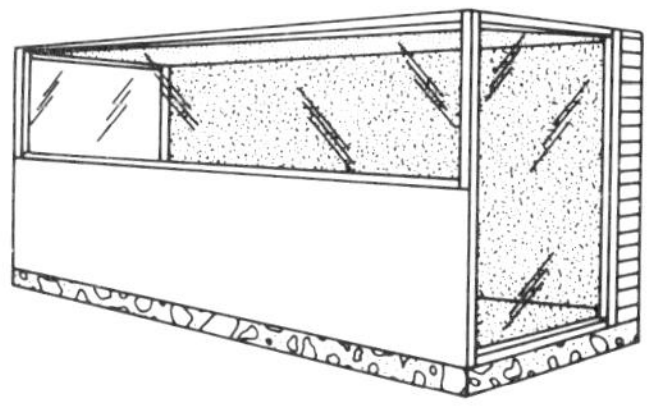

Table A4-16

OPAQUE BOTTOM SECTION, Z= .5
GLAZED END WALLS
α_o = .3

WDTH	TILT	XTOP	YBOT	ABSORPTION FOR RADIATION THRU SURFACE 3 α_m 0.3	0.4	0.5	0.6	0.7	0.8	0.9	1.0
1.0	90.	.0	.500	.78	.82	.86	.89	.92	.95	.97	1.00
1.0	90.	.5	.500	.62	.65	.67	.69	.71	.73	.75	.76
1.0	90.	1.5	.500	.54	.56	.57	.59	.60	.62	.63	.64
1.0	90.	3.0	.500	.52	.53	.55	.56	.57	.59	.60	.61
3.0	90.	.0	.500	.78	.82	.86	.89	.92	.95	.97	1.00
3.0	90.	.5	.500	.76	.78	.81	.83	.84	.86	.88	.89
3.0	90.	1.5	.500	.73	.75	.77	.78	.80	.81	.82	.83
3.0	90.	3.0	.500	.71	.73	.74	.76	.77	.78	.79	.81
5.0	90.	.0	.500	.78	.82	.86	.89	.92	.95	.97	1.00
5.0	90.	.5	.500	.79	.82	.84	.86	.88	.89	.91	.92
5.0	90.	1.5	.500	.80	.81	.83	.84	.86	.87	.88	.89
5.0	90.	3.0	.500	.79	.80	.82	.83	.84	.85	.86	.87

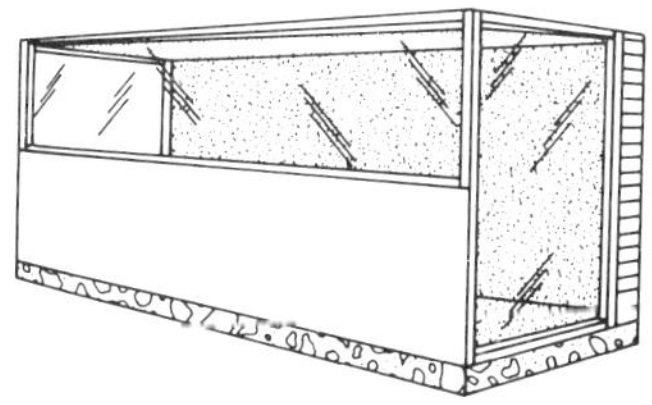

Table A4-17

OPAQUE BOTTOM SECTION, Z= .5
GLAZED END WALLS
α_o = .3

WDTH	TILT	XTOP	YBOT	ABSORPTION FOR RADIATION THRU SURFACE 6 α_m 0.3	0.4	0.5	0.6	0.7	0.8	0.9	1.0
1.0	90.	.0	.500	.70	.73	.75	.77	.79	.81	.83	.85
1.0	90.	.5	.500	.64	.67	.69	.71	.73	.75	.77	.78
1.0	90.	1.5	.500	.59	.62	.64	.65	.67	.69	.71	.72
1.0	90.	3.0	.500	.57	.59	.61	.63	.64	.66	.68	.69
3.0	90.	.0	.500	.70	.73	.75	.77	.79	.81	.83	.85
3.0	90.	.5	.500	.74	.76	.78	.80	.82	.83	.85	.86
3.0	90.	1.5	.500	.76	.78	.80	.81	.83	.84	.86	.87
3.0	90.	3.0	.500	.76	.78	.80	.81	.83	.84	.86	.87
5.0	90.	.0	.500	.70	.73	.75	.77	.79	.81	.83	.85
5.0	90.	.5	.500	.76	.78	.80	.82	.83	.85	.86	.87
5.0	90.	1.5	.500	.80	.82	.83	.85	.86	.87	.88	.89
5.0	90.	3.0	.500	.82	.83	.85	.86	.88	.89	.90	.91

TILTED GLAZING - CLASS I

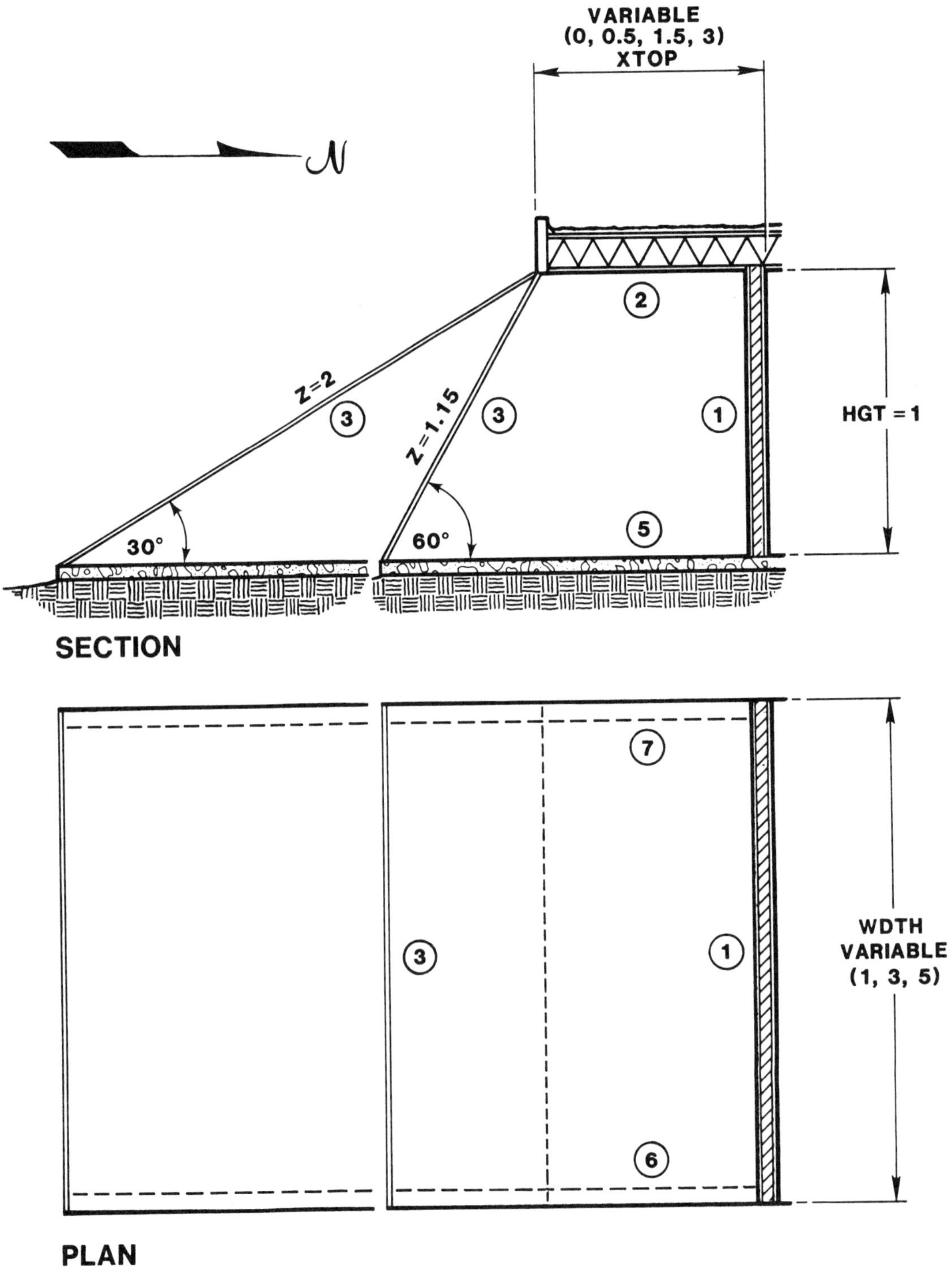

Figure A4-45. Geometry for Tilted Glazing, Class I

ABSORPTION FACTORS
TILTED GLAZING - CL I

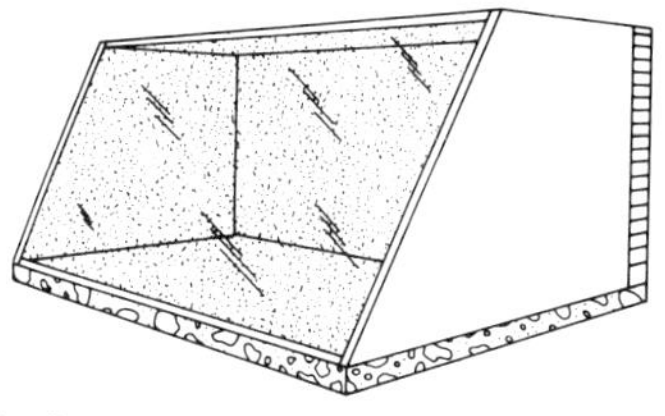

Table A4-18

OPAQUE END WALLS - YBOT= .00

$\alpha_o = \alpha_m$

WDTH	TILT	XTOP	Z	ABSORPTION FOR RADIATION THRU SURFACE 3 α_m 0.3	0.4	0.5	0.6	0.7	0.8	0.9	1.0
1.0	30.	.0	2.00	.76	.81	.85	.89	.92	.95	.98	1.00
1.0	30.	.5	2.00	.81	.85	.88	.91	.94	.96	.98	1.00
1.0	30.	1.5	2.00	.86	.89	.92	.94	.96	.97	.99	1.00
1.0	30.	3.0	2.00	.90	.92	.94	.96	.97	.98	.99	1.00
1.0	60.	.0	1.15	.73	.78	.83	.87	.91	.94	.97	1.00
1.0	60.	.5	1.15	.83	.87	.90	.93	.95	.97	.98	1.00
1.0	60.	1.5	1.15	.90	.92	.94	.96	.97	.98	.99	1.00
1.0	60.	3.0	1.15	.94	.95	.97	.97	.98	.99	.99	1.00
3.0	30.	.0	2.00	.71	.76	.81	.86	.90	.93	.97	1.00
3.0	30.	.5	2.00	.76	.80	.85	.88	.92	.95	.97	1.00
3.0	30.	1.5	2.00	.81	.85	.88	.91	.94	.96	.98	1.00
3.0	30.	3.0	2.00	.87	.89	.92	.94	.96	.97	.99	1.00
3.0	60.	.0	1.15	.70	.75	.80	.85	.89	.93	.97	1.00
3.0	60.	.5	1.15	.79	.83	.87	.90	.93	.96	.98	1.00
3.0	60.	1.5	1.15	.87	.90	.92	.94	.96	.97	.99	1.00
3.0	60.	3.0	1.15	.91	.93	.95	.96	.97	.98	.99	1.00
5.0	30.	.0	2.00	.70	.75	.80	.85	.89	.93	.97	1.00
5.0	30.	.5	2.00	.74	.79	.84	.87	.91	.94	.97	1.00
5.0	30.	1.5	2.00	.80	.84	.88	.91	.93	.96	.98	1.00
5.0	30.	3.0	2.00	.86	.89	.91	.93	.95	.97	.99	1.00
5.0	60.	.0	1.15	.69	.74	.80	.84	.89	.93	.96	1.00
5.0	60.	.5	1.15	.78	.82	.86	.90	.93	.95	.98	1.00
5.0	60.	1.5	1.15	.86	.89	.92	.94	.96	.97	.99	1.00
5.0	60.	3.0	1.15	.91	.93	.95	.96	.97	.98	.99	1.00

ABSORPTION FACTORS
TILTED GLAZING - CL I

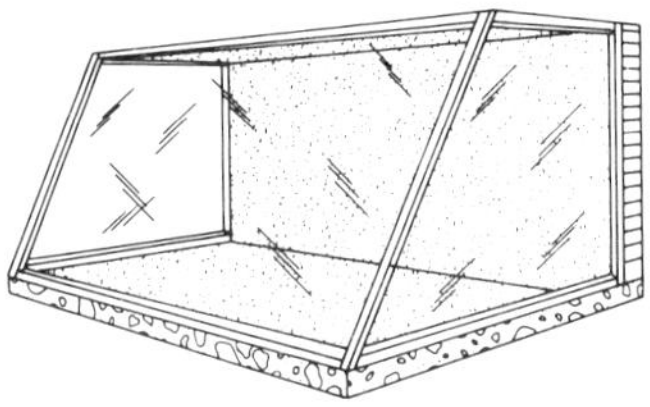

Table A4-19

GLAZED END WALLS - YBOT= .00

$\alpha_o = \alpha_m$

WDTH	TILT	XTOP	Z	ABSORPTION FOR RADIATION THRU SURFACE 3 α_m 0.3	0.4	0.5	0.6	0.7	0.8	0.9	1.0
1.0	30.	.0	2.00	.54	.58	.62	.66	.70	.74	.77	.80
1.0	30.	.5	2.00	.52	.56	.60	.64	.67	.70	.73	.76
1.0	30.	1.5	2.00	.52	.56	.59	.63	.66	.69	.72	.74
1.0	30.	3.0	2.00	.53	.56	.60	.63	.66	.69	.71	.74
1.0	60.	.0	1.15	.58	.63	.68	.72	.77	.80	.84	.88
1.0	60.	.5	1.15	.55	.59	.63	.66	.69	.72	.75	.77
1.0	60.	1.5	1.15	.52	.56	.59	.62	.64	.67	.69	.71
1.0	60.	3.0	1.15	.52	.55	.58	.61	.63	.66	.68	.70
3.0	30.	.0	2.00	.62	.67	.72	.77	.81	.85	.89	.93
3.0	30.	.5	2.00	.63	.68	.73	.77	.81	.84	.88	.91
3.0	30.	1.5	2.00	.65	.70	.74	.77	.81	.84	.86	.89
3.0	30.	3.0	2.00	.67	.71	.75	.78	.81	.84	.86	.88
3.0	60.	.0	1.15	.64	.70	.75	.80	.84	.88	.92	.96
3.0	60.	.5	1.15	.67	.72	.76	.80	.83	.86	.89	.91
3.0	60.	1.5	1.15	.68	.72	.76	.79	.81	.84	.86	.88
3.0	60.	3.0	1.15	.69	.72	.75	.78	.80	.82	.84	.86
5.0	30.	.0	2.00	.64	.70	.75	.79	.84	.88	.92	.96
5.0	30.	.5	2.00	.67	.72	.76	.80	.84	.88	.91	.94
5.0	30.	1.5	2.00	.70	.74	.78	.82	.85	.88	.91	.93
5.0	30.	3.0	2.00	.73	.77	.80	.83	.86	.88	.90	.93
5.0	60.	.0	1.15	.66	.71	.76	.81	.86	.90	.94	.97
5.0	60.	.5	1.15	.70	.75	.79	.83	.86	.89	.92	.95
5.0	60.	1.5	1.15	.74	.78	.81	.84	.86	.89	.91	.92
5.0	60.	3.0	1.15	.75	.79	.82	.84	.86	.88	.89	.91

ABSORPTION FACTORS
TILTED GLAZING - CL I

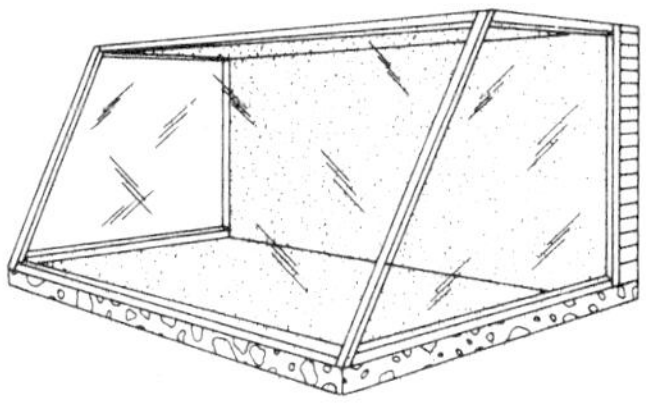

Table A4-20

GLAZED END WALLS - YBOT= .00

$\alpha_o = \alpha_m$

WDTH	TILT	XTOP	Z	ABSORPTION FOR RADIATION THRU SURFACE 6 α_m 0.3	0.4	0.5	0.6	0.7	0.8	0.9	1.0
1.0	30.	.0	2.00	.47	.50	.53	.56	.59	.62	.65	.67
1.0	30.	.5	2.00	.49	.52	.55	.58	.61	.64	.66	.68
1.0	30.	1.5	2.00	.51	.54	.57	.60	.63	.66	.68	.70
1.0	30.	3.0	2.00	.52	.55	.59	.61	.64	.67	.69	.71
1.0	60.	.0	1.15	.50	.54	.57	.60	.63	.66	.69	.72
1.0	60.	.5	1.15	.53	.56	.59	.62	.65	.68	.70	.72
1.0	60.	1.5	1.15	.54	.57	.60	.63	.66	.68	.71	.73
1.0	60.	3.0	1.15	.54	.57	.60	.63	.66	.68	.71	.73
3.0	30.	.0	2.00	.52	.56	.59	.63	.66	.69	.72	.74
3.0	30.	.5	2.00	.57	.61	.64	.68	.70	.73	.75	.78
3.0	30.	1.5	2.00	.63	.67	.70	.73	.76	.78	.80	.82
3.0	30.	3.0	2.00	.67	.71	.74	.77	.79	.82	.83	.85
3.0	60.	.0	1.15	.53	.57	.60	.64	.67	.70	.73	.75
3.0	60.	.5	1.15	.60	.64	.68	.71	.73	.76	.78	.80
3.0	60.	1.5	1.15	.67	.71	.74	.77	.79	.81	.83	.85
3.0	60.	3.0	1.15	.70	.74	.77	.80	.82	.84	.86	.87
5.0	30.	.0	2.00	.53	.57	.60	.64	.67	.70	.73	.75
5.0	30.	.5	2.00	.59	.63	.66	.69	.72	.75	.77	.79
5.0	30.	1.5	2.00	.66	.70	.73	.76	.78	.81	.83	.84
5.0	30.	3.0	2.00	.71	.75	.78	.81	.83	.85	.86	.88
5.0	60.	.0	1.15	.53	.57	.61	.64	.67	.70	.73	.76
5.0	60.	.5	1.15	.62	.66	.69	.72	.75	.77	.79	.81
5.0	60.	1.5	1.15	.70	.74	.77	.79	.81	.83	.85	.86
5.0	60.	3.0	1.15	.75	.78	.81	.83	.85	.87	.88	.90

ABSORPTION FACTORS
TILTED GLAZING - CL I

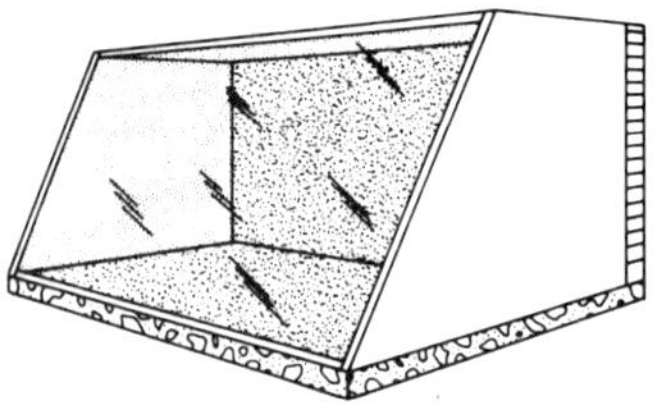

Table A4-21

OPAQUE END WALLS - YBOT= .00

α_o = .3

WDTH	TILT	XTOP	Z	ABSORPTION FOR RADIATION THRU SURFACE 3 α_m 0.3	0.4	0.5	0.6	0.7	0.8	0.9	1.0
1.0	30.	.0	2.00	.76	.79	.83	.86	.89	.91	.94	.96
1.0	30.	.5	2.00	.81	.83	.86	.88	.90	.92	.94	.96
1.0	30.	1.5	2.00	.86	.88	.90	.91	.93	.94	.96	.97
1.0	30.	3.0	2.00	.90	.92	.93	.94	.95	.96	.97	.98
1.0	60.	.0	1.15	.73	.77	.81	.85	.88	.92	.95	.98
1.0	60.	.5	1.15	.83	.85	.87	.89	.91	.93	.95	.96
1.0	60.	1.5	1.15	.90	.91	.92	.93	.94	.95	.96	.97
1.0	60.	3.0	1.15	.94	.94	.95	.96	.96	.97	.98	.98
3.0	30.	.0	2.00	.71	.76	.80	.84	.88	.92	.95	.98
3.0	30.	.5	2.00	.76	.80	.83	.87	.90	.93	.96	.98
3.0	30.	1.5	2.00	.81	.84	.87	.90	.92	.94	.97	.99
3.0	30.	3.0	2.00	.87	.89	.91	.93	.94	.96	.98	.99
3.0	60.	.0	1.15	.70	.75	.80	.84	.88	.92	.96	.99
3.0	60.	.5	1.15	.79	.82	.85	.88	.91	.94	.96	.98
3.0	60.	1.5	1.15	.87	.89	.91	.92	.94	.95	.97	.98
3.0	60.	3.0	1.15	.91	.93	.94	.95	.96	.97	.98	.99
5.0	30.	.0	2.00	.70	.75	.80	.84	.88	.92	.96	.99
5.0	30.	.5	2.00	.74	.79	.83	.86	.90	.93	.96	.99
5.0	30.	1.5	2.00	.80	.84	.87	.89	.92	.95	.97	.99
5.0	30.	3.0	2.00	.86	.88	.90	.92	.94	.96	.98	.99
5.0	60.	.0	1.15	.69	.74	.79	.84	.88	.92	.96	.99
5.0	60.	.5	1.15	.78	.81	.85	.88	.91	.94	.96	.99
5.0	60.	1.5	1.15	.86	.88	.90	.92	.94	.96	.97	.99
5.0	60.	3.0	1.15	.91	.92	.94	.95	.96	.97	.98	.99

ABSORPTION FACTORS
TILTED GLAZING - CL I

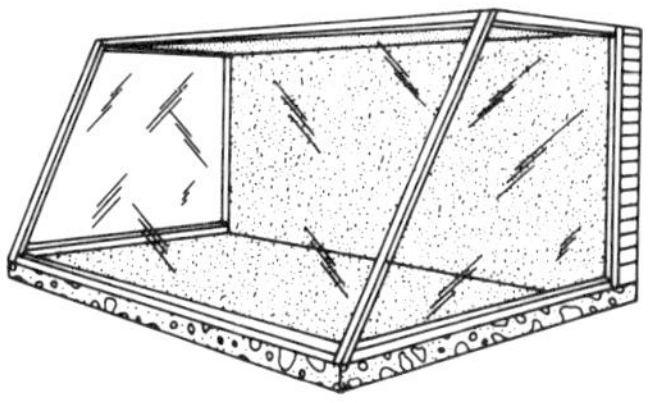

Table A4-22

GLAZED END WALLS - YBOT= .00

α_o = .3

WDTH	TILT	XTOP	Z	ABSORPTION FOR RADIATION THRU SURFACE 3 α_m 0.3	0.4	0.5	0.6	0.7	0.8	0.9	1.0
1.0	30.	.0	2.00	.54	.58	.62	.66	.70	.74	.77	.80
1.0	30.	.5	2.00	.52	.56	.60	.63	.67	.70	.73	.76
1.0	30.	1.5	2.00	.52	.56	.59	.62	.65	.68	.71	.74
1.0	30.	3.0	2.00	.53	.56	.59	.62	.65	.68	.70	.73
1.0	60.	.0	1.15	.58	.63	.68	.72	.77	.80	.84	.88
1.0	60.	.5	1.15	.55	.58	.62	.65	.68	.71	.73	.76
1.0	60.	1.5	1.15	.52	.55	.58	.60	.63	.65	.67	.69
1.0	60.	3.0	1.15	.52	.54	.56	.59	.61	.63	.65	.67
3.0	30.	.0	2.00	.62	.67	.72	.77	.81	.85	.89	.93
3.0	30.	.5	2.00	.63	.68	.72	.77	.80	.84	.87	.91
3.0	30.	1.5	2.00	.65	.69	.73	.77	.80	.83	.86	.89
3.0	30.	3.0	2.00	.67	.71	.74	.77	.80	.83	.85	.88
3.0	60.	.0	1.15	.64	.70	.75	.80	.84	.88	.92	.96
3.0	60.	.5	1.15	.67	.71	.75	.79	.82	.85	.88	.91
3.0	60.	1.5	1.15	.68	.71	.74	.77	.80	.82	.84	.86
3.0	60.	3.0	1.15	.69	.71	.74	.76	.78	.80	.82	.84
5.0	30.	.0	2.00	.64	.70	.75	.79	.84	.88	.92	.96
5.0	30.	.5	2.00	.67	.71	.76	.80	.84	.88	.91	.94
5.0	30.	1.5	2.00	.70	.74	.78	.81	.84	.87	.90	.93
5.0	30.	3.0	2.00	.73	.76	.79	.82	.85	.88	.90	.92
5.0	60.	.0	1.15	.66	.71	.76	.81	.86	.90	.94	.97
5.0	60.	.5	1.15	.70	.75	.79	.82	.85	.89	.91	.94
5.0	60.	1.5	1.15	.74	.77	.80	.82	.85	.87	.89	.91
5.0	60.	3.0	1.15	.75	.78	.80	.82	.84	.86	.88	.89

ABSORPTION FACTORS
TILTED GLAZING - CL I

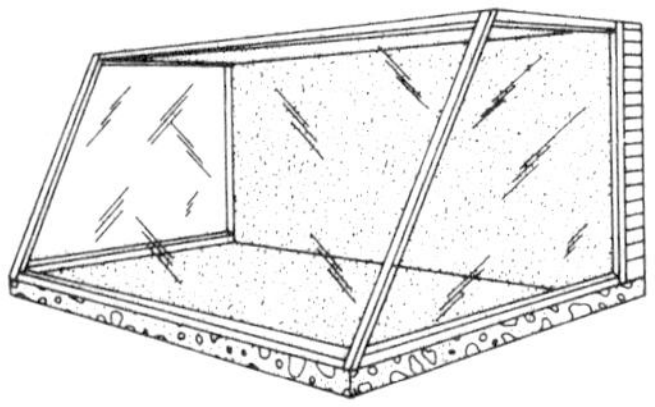

Table A4-23

GLAZED END WALLS - YBOT= .00

α_0 = .3

WDTH	TILT	XTOP	Z	ABSORPTION FOR RADIATION THRU SURFACE 6 α_m 0.3	0.4	0.5	0.6	0.7	0.8	0.9	1.0
1.0	30.	.0	2.00	.47	.50	.53	.56	.59	.62	.65	.67
1.0	30.	.5	2.00	.49	.52	.55	.57	.60	.62	.65	.67
1.0	30.	1.5	2.00	.51	.53	.56	.58	.61	.63	.65	.67
1.0	30.	3.0	2.00	.52	.54	.56	.59	.61	.63	.65	.66
1.0	60.	.0	1.15	.50	.54	.57	.60	.63	.66	.69	.72
1.0	60.	.5	1.15	.53	.56	.58	.61	.63	.66	.68	.70
1.0	60.	1.5	1.15	.54	.56	.58	.61	.63	.65	.67	.68
1.0	60.	3.0	1.15	.54	.56	.58	.60	.62	.64	.66	.67
3.0	30.	.0	2.00	.52	.56	.59	.63	.66	.69	.72	.74
3.0	30.	.5	2.00	.57	.61	.64	.67	.70	.72	.75	.77
3.0	30.	1.5	2.00	.63	.66	.69	.71	.74	.76	.78	.80
3.0	30.	3.0	2.00	.67	.70	.72	.74	.77	.79	.81	.82
3.0	60.	.0	1.15	.53	.57	.60	.64	.67	.70	.73	.75
3.0	60.	.5	1.15	.60	.64	.67	.69	.72	.74	.76	.78
3.0	60.	1.5	1.15	.67	.69	.72	.74	.76	.78	.80	.82
3.0	60.	3.0	1.15	.70	.73	.75	.77	.79	.80	.82	.84
5.0	30.	.0	2.00	.53	.57	.60	.64	.67	.70	.73	.75
5.0	30.	.5	2.00	.59	.62	.66	.69	.71	.74	.76	.79
5.0	30.	1.5	2.00	.66	.69	.72	.75	.77	.79	.81	.83
5.0	30.	3.0	2.00	.71	.74	.76	.79	.81	.83	.84	.86
5.0	60.	.0	1.15	.53	.57	.61	.64	.67	.70	.73	.76
5.0	60.	.5	1.15	.62	.65	.68	.71	.73	.76	.78	.80
5.0	60.	1.5	1.15	.70	.73	.75	.77	.79	.81	.83	.84
5.0	60.	3.0	1.15	.75	.77	.79	.81	.83	.84	.86	.87

TILTED GLAZING - CLASS II

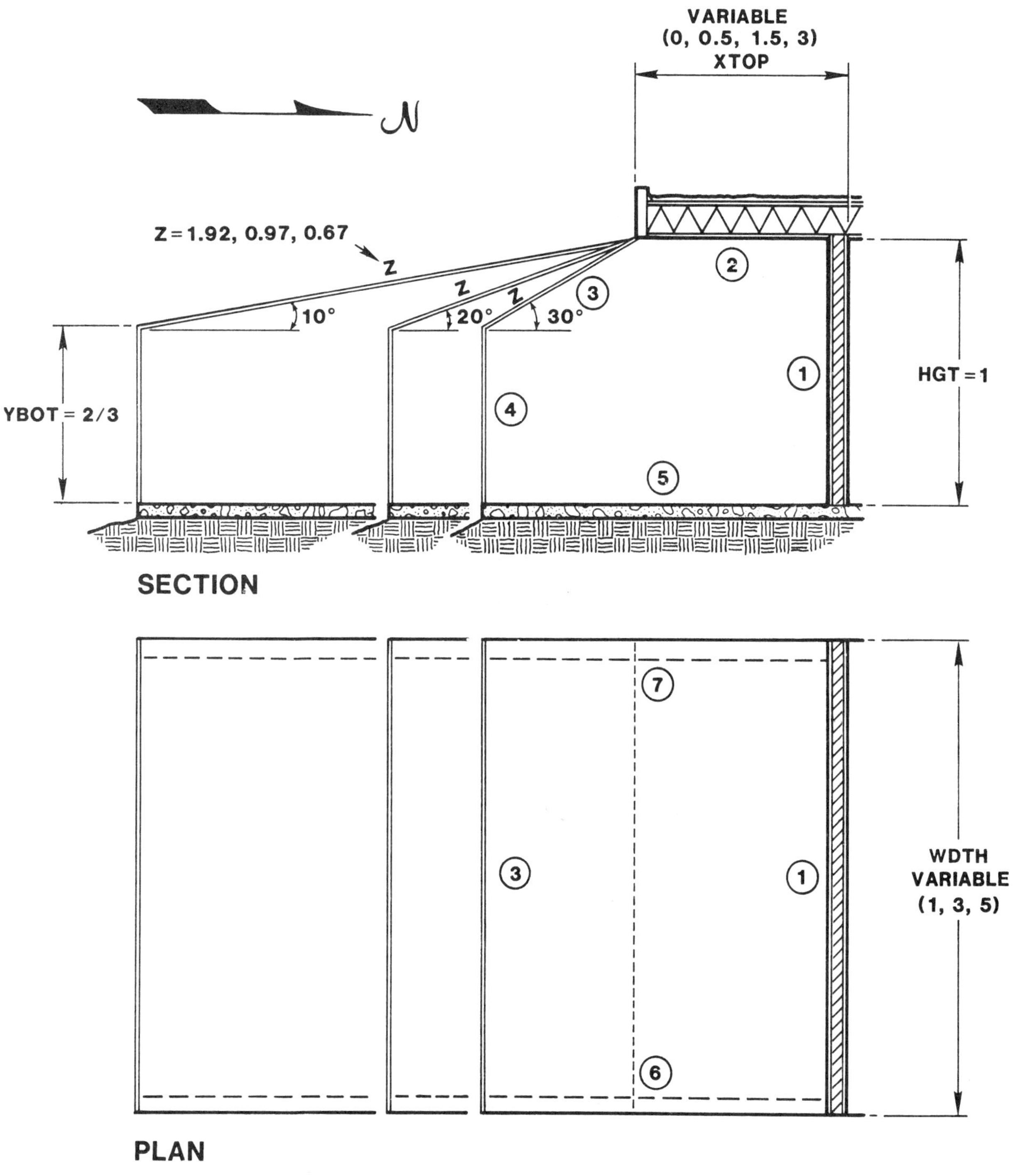

Figures A4-46. Geometry for Tilted Glazing, Class II

ABSORPTION FACTORS
TILTED GLAZING - CL II

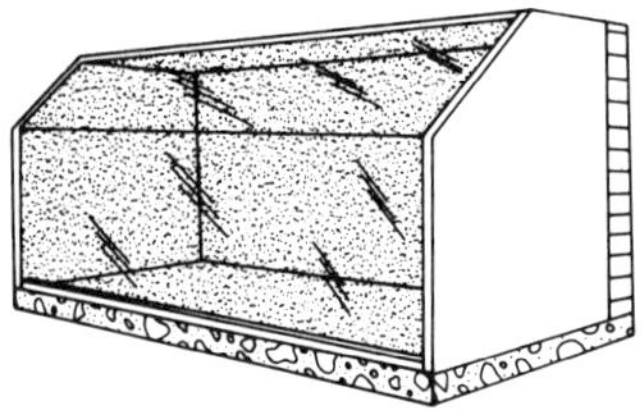

Table A4-24

OPAQUE END WALLS - YBOT= .67

$\alpha_o = \alpha_m$

WDTH	TILT	XTOP	Z	ABSORPTION FOR RADIATION THRU SURFACE 3 α_m 0.3	0.4	0.5	0.6	0.7	0.8	0.9	1.0
1.0	10.	.0	1.92	.75	.79	.83	.86	.89	.91	.93	.95
1.0	10.	.5	1.92	.78	.82	.85	.88	.90	.92	.94	.95
1.0	10.	1.5	1.92	.82	.85	.88	.90	.91	.93	.94	.95
1.0	10.	3.0	1.92	.86	.88	.90	.91	.92	.93	.94	.95
1.0	20.	.0	.97	.72	.77	.81	.84	.87	.89	.91	.94
1.0	20.	.5	.97	.78	.81	.84	.87	.89	.91	.92	.94
1.0	20.	1.5	.97	.83	.86	.88	.89	.90	.92	.93	.94
1.0	20.	3.0	.97	.87	.88	.90	.91	.92	.92	.93	.94
1.0	30.	.0	.67	.71	.75	.79	.83	.86	.89	.92	.94
1.0	30.	.5	.67	.79	.82	.85	.87	.89	.91	.93	.94
1.0	30.	1.5	.67	.84	.87	.88	.90	.91	.92	.93	.94
1.0	30.	3.0	.67	.88	.89	.91	.91	.92	.93	.93	.94
3.0	10.	.0	1.92	.68	.73	.77	.81	.85	.88	.91	.94
3.0	10.	.5	1.92	.71	.76	.80	.83	.86	.89	.91	.94
3.0	10.	1.5	1.92	.76	.80	.83	.85	.88	.90	.92	.94
3.0	10.	3.0	1.92	.81	.83	.86	.88	.89	.91	.92	.94
3.0	20.	.0	.97	.66	.71	.76	.79	.83	.86	.89	.92
3.0	20.	.5	.97	.72	.76	.80	.83	.85	.88	.90	.92
3.0	20.	1.5	.97	.78	.81	.84	.86	.87	.89	.90	.92
3.0	20.	3.0	.97	.82	.85	.86	.88	.89	.90	.91	.92
3.0	30.	.0	.67	.66	.71	.75	.79	.83	.86	.90	.92
3.0	30.	.5	.67	.73	.77	.81	.84	.86	.89	.91	.92
3.0	30.	1.5	.67	.80	.83	.85	.87	.89	.90	.91	.92
3.0	30.	3.0	.67	.84	.86	.88	.89	.90	.91	.92	.92
5.0	10.	.0	1.92	.66	.71	.75	.80	.83	.87	.90	.93
5.0	10.	.5	1.92	.69	.74	.78	.82	.85	.88	.91	.93
5.0	10.	1.5	1.92	.74	.78	.81	.84	.87	.89	.91	.93
5.0	10.	3.0	1.92	.79	.82	.85	.87	.88	.90	.92	.93
5.0	20.	.0	.97	.65	.70	.74	.78	.82	.85	.88	.91
5.0	20.	.5	.97	.71	.75	.78	.82	.84	.87	.89	.91
5.0	20.	1.5	.97	.77	.80	.82	.85	.87	.88	.90	.91
5.0	20.	3.0	.97	.81	.84	.85	.87	.88	.89	.90	.91
5.0	30.	.0	.67	.64	.69	.74	.78	.82	.86	.89	.92
5.0	30.	.5	.67	.72	.76	.80	.83	.85	.88	.90	.92
5.0	30.	1.5	.67	.79	.82	.84	.86	.88	.90	.91	.92
5.0	30.	3.0	.67	.83	.85	.87	.88	.89	.90	.91	.92

ABSORPTION FACTORS
TILTED GLAZING - CL II

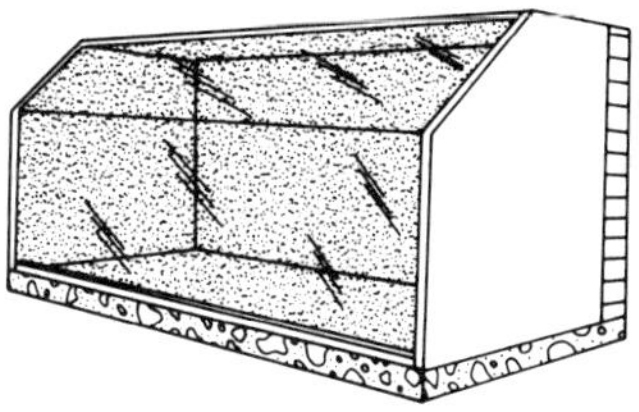

Table A4-25

OPAQUE END WALLS - YBOT= .67

$\alpha_o = \alpha_m$

WDTH	TILT	XTOP	Z	ABSORPTION FOR RADIATION THRU SURFACE 4 α_m 0.3	0.4	0.5	0.6	0.7	0.8	0.9	1.0
1.0	10.	.0	1.92	.69	.72	.76	.78	.81	.83	.85	.86
1.0	10.	.5	1.92	.72	.75	.78	.80	.82	.83	.85	.86
1.0	10.	1.5	1.92	.75	.78	.80	.82	.83	.84	.85	.86
1.0	10.	3.0	1.92	.79	.80	.82	.83	.84	.85	.86	.86
1.0	20.	.0	.97	.71	.75	.78	.81	.84	.87	.89	.91
1.0	20.	.5	.97	.76	.79	.82	.84	.86	.88	.89	.91
1.0	20.	1.5	.97	.81	.83	.85	.87	.88	.89	.90	.91
1.0	20.	3.0	.97	.84	.86	.87	.88	.89	.90	.90	.91
1.0	30.	.0	.67	.71	.76	.80	.83	.86	.89	.92	.94
1.0	30.	.5	.67	.79	.82	.85	.87	.89	.91	.93	.94
1.0	30.	1.5	.67	.85	.87	.89	.90	.91	.92	.93	.94
1.0	30.	3.0	.67	.88	.90	.91	.92	.92	.93	.93	.94
3.0	10.	.0	1.92	.61	.65	.69	.72	.75	.77	.80	.82
3.0	10.	.5	1.92	.64	.68	.71	.74	.76	.78	.80	.82
3.0	10.	1.5	1.92	.68	.71	.74	.76	.78	.79	.81	.82
3.0	10.	3.0	1.92	.72	.74	.76	.78	.79	.80	.81	.82
3.0	20.	.0	.97	.64	.69	.73	.77	.80	.83	.86	.88
3.0	20.	.5	.97	.70	.74	.77	.80	.82	.84	.86	.88
3.0	20.	1.5	.97	.76	.79	.81	.83	.84	.86	.87	.88
3.0	20.	3.0	.97	.80	.82	.83	.85	.86	.87	.87	.88
3.0	30.	.0	.67	.66	.71	.75	.79	.83	.86	.90	.92
3.0	30.	.5	.67	.74	.78	.81	.84	.86	.89	.91	.92
3.0	30.	1.5	.67	.81	.83	.86	.87	.89	.90	.91	.92
3.0	30.	3.0	.67	.85	.87	.88	.89	.90	.91	.92	.92
5.0	10.	.0	1.92	.59	.63	.67	.70	.73	.76	.79	.81
5.0	10.	.5	1.92	.62	.66	.69	.72	.75	.77	.79	.81
5.0	10.	1.5	1.92	.67	.70	.72	.74	.76	.78	.80	.81
5.0	10.	3.0	1.92	.70	.73	.75	.76	.78	.79	.80	.81
5.0	20.	.0	.97	.63	.67	.71	.75	.79	.82	.85	.88
5.0	20.	.5	.97	.69	.72	.76	.79	.81	.84	.86	.88
5.0	20.	1.5	.97	.75	.78	.80	.82	.84	.85	.86	.88
5.0	20.	3.0	.97	.79	.81	.82	.84	.85	.86	.87	.88
5.0	30.	.0	.67	.64	.70	.74	.78	.82	.86	.89	.92
5.0	30.	.5	.67	.72	.76	.80	.83	.86	.88	.90	.92
5.0	30.	1.5	.67	.79	.82	.85	.87	.88	.90	.91	.92
5.0	30.	3.0	.67	.84	.86	.87	.89	.90	.91	.91	.92

ABSORPTION FACTORS
TILTED GLAZING - CL II

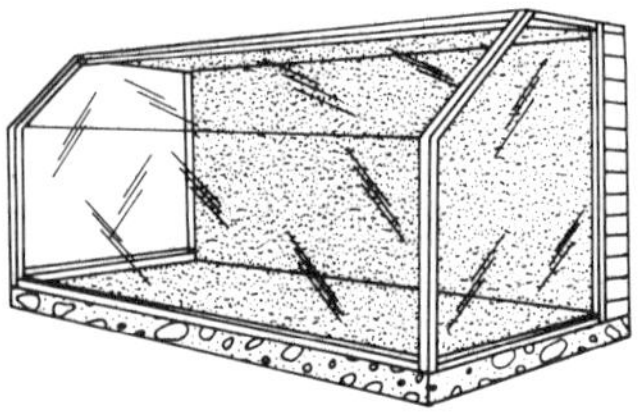

Table A4-26

GLAZED END WALLS - YBOT= .67

$\alpha_o = \alpha_m$

WDTH	TILT	XTOP	Z	ABSORPTION FOR RADIATION THRU SURFACE 3 (α_m) 0.3	0.4	0.5	0.6	0.7	0.8	0.9	1.0
1.0	10.	.0	1.92	.45	.48	.51	.54	.57	.60	.62	.65
1.0	10.	.5	1.92	.44	.47	.50	.52	.55	.57	.60	.62
1.0	10.	1.5	1.92	.44	.47	.50	.52	.54	.57	.59	.61
1.0	10.	3.0	1.92	.45	.48	.50	.53	.55	.57	.59	.61
1.0	20.	.0	.97	.48	.51	.55	.58	.61	.64	.66	.69
1.0	20.	.5	.97	.46	.49	.51	.54	.56	.58	.61	.63
1.0	20.	1.5	.97	.45	.47	.50	.52	.54	.56	.58	.60
1.0	20.	3.0	.97	.45	.47	.50	.52	.54	.56	.58	.59
1.0	30.	.0	.67	.51	.55	.58	.62	.65	.68	.71	.74
1.0	30.	.5	.67	.48	.51	.53	.56	.58	.61	.63	.65
1.0	30.	1.5	.67	.46	.48	.51	.53	.55	.57	.59	.60
1.0	30.	3.0	.67	.45	.48	.50	.52	.54	.56	.58	.59
3.0	10.	.0	1.92	.55	.59	.64	.68	.71	.75	.78	.81
3.0	10.	.5	1.92	.56	.60	.64	.68	.71	.74	.77	.80
3.0	10.	1.5	1.92	.58	.62	.65	.68	.71	.74	.77	.79
3.0	10.	3.0	1.92	.60	.64	.67	.70	.72	.74	.77	.79
3.0	20.	.0	.97	.57	.61	.65	.69	.73	.76	.79	.82
3.0	20.	.5	.97	.58	.62	.66	.69	.72	.75	.77	.80
3.0	20.	1.5	.97	.60	.63	.66	.69	.71	.74	.76	.78
3.0	20.	3.0	.97	.61	.64	.67	.69	.71	.73	.75	.77
3.0	30.	.0	.67	.58	.63	.67	.71	.75	.79	.82	.85
3.0	30.	.5	.67	.60	.64	.68	.71	.74	.76	.79	.81
3.0	30.	1.5	.67	.61	.65	.68	.70	.72	.75	.76	.78
3.0	30.	3.0	.67	.62	.65	.68	.70	.72	.74	.75	.77
5.0	10.	.0	1.92	.58	.63	.67	.71	.75	.79	.82	.86
5.0	10.	.5	1.92	.60	.64	.68	.72	.75	.79	.82	.85
5.0	10.	1.5	1.92	.63	.67	.70	.73	.76	.79	.82	.84
5.0	10.	3.0	1.92	.66	.69	.72	.75	.77	.80	.82	.84
5.0	20.	.0	.97	.59	.63	.68	.72	.76	.79	.83	.86
5.0	20.	.5	.97	.62	.66	.70	.73	.76	.79	.81	.84
5.0	20.	1.5	.97	.65	.68	.71	.74	.76	.79	.81	.83
5.0	20.	3.0	.97	.67	.70	.73	.75	.77	.79	.80	.82
5.0	30.	.0	.67	.60	.65	.69	.73	.77	.81	.84	.88
5.0	30.	.5	.67	.64	.68	.71	.75	.78	.80	.83	.85
5.0	30.	1.5	.67	.67	.70	.73	.76	.78	.80	.82	.83
5.0	30.	3.0	.67	.68	.71	.74	.76	.78	.79	.81	.82

ABSORPTION FACTORS
TILTED GLAZING - CL II

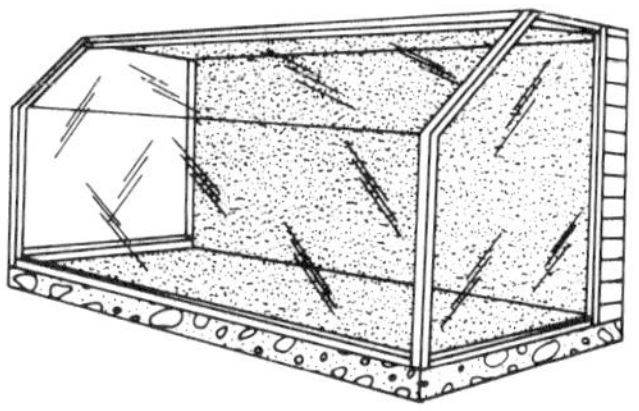

Table A4-27

GLAZED END WALLS - YBOT= .67

$\alpha_o = \alpha_m$

WDTH	TILT	XTOP	Z	ABSORPTION FOR RADIATION THRU SURFACE 4 α_m							
				0.3	0.4	0.5	0.6	0.7	0.8	0.9	1.0
1.0	10.	.0	1.92	.41	.43	.45	.48	.50	.52	.54	.56
1.0	10.	.5	1.92	.41	.43	.45	.48	.50	.52	.53	.55
1.0	10.	1.5	1.92	.41	.43	.46	.48	.50	.51	.53	.55
1.0	10.	3.0	1.92	.42	.44	.46	.48	.50	.51	.53	.55
1.0	20.	.0	.97	.46	.49	.52	.55	.57	.60	.62	.65
1.0	20.	.5	.97	.45	.48	.51	.53	.55	.57	.59	.61
1.0	20.	1.5	.97	.45	.47	.49	.52	.54	.56	.57	.59
1.0	20.	3.0	.97	.44	.47	.49	.51	.53	.55	.57	.58
1.0	30.	.0	.67	.50	.54	.57	.61	.64	.67	.70	.73
1.0	30.	.5	.67	.49	.52	.55	.57	.60	.62	.64	.66
1.0	30.	1.5	.67	.47	.50	.52	.55	.57	.59	.61	.62
1.0	30.	3.0	.67	.46	.49	.52	.54	.56	.58	.60	.61
3.0	10.	.0	1.92	.48	.52	.55	.58	.61	.64	.66	.69
3.0	10.	.5	1.92	.50	.53	.56	.58	.61	.63	.66	.68
3.0	10.	1.5	1.92	.51	.54	.57	.59	.61	.63	.65	.67
3.0	10.	3.0	1.92	.52	.55	.57	.59	.61	.63	.64	.66
3.0	20.	.0	.97	.54	.58	.62	.66	.69	.72	.75	.78
3.0	20.	.5	.97	.56	.60	.63	.66	.69	.71	.74	.76
3.0	20.	1.5	.97	.58	.61	.64	.66	.68	.70	.72	.73
3.0	20.	3.0	.97	.58	.61	.64	.66	.68	.69	.71	.72
3.0	30.	.0	.67	.58	.62	.67	.71	.75	.78	.82	.85
3.0	30.	.5	.67	.61	.65	.68	.71	.74	.77	.79	.81
3.0	30.	1.5	.67	.62	.65	.68	.71	.73	.75	.77	.78
3.0	30.	3.0	.67	.62	.65	.68	.70	.72	.74	.75	.76
5.0	10.	.0	1.92	.51	.55	.58	.61	.64	.67	.70	.73
5.0	10.	.5	1.92	.53	.56	.59	.62	.65	.67	.70	.72
5.0	10.	1.5	1.92	.55	.58	.61	.63	.65	.67	.69	.71
5.0	10.	3.0	1.92	.57	.59	.62	.64	.66	.67	.69	.70
5.0	20.	.0	.97	.56	.61	.65	.69	.72	.75	.79	.81
5.0	20.	.5	.97	.60	.64	.67	.70	.73	.75	.78	.80
5.0	20.	1.5	.97	.63	.66	.69	.71	.73	.75	.77	.78
5.0	20.	3.0	.97	.64	.67	.69	.71	.73	.74	.76	.77
5.0	30.	.0	.67	.60	.64	.69	.73	.77	.81	.84	.87
5.0	30.	.5	.67	.64	.68	.72	.75	.78	.81	.83	.85
5.0	30.	1.5	.67	.67	.71	.73	.76	.78	.80	.82	.83
5.0	30.	3.0	.67	.69	.71	.74	.76	.78	.79	.80	.82

ABSORPTION FACTORS
TILTED GLAZING - CL II

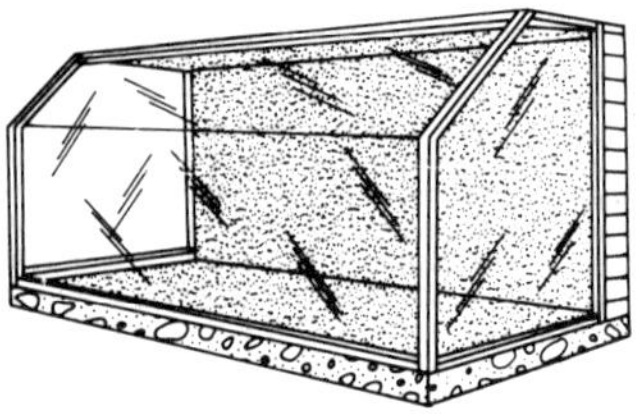

Table A4-28

GLAZED END WALLS - YBOT= .67

$\alpha_o = \alpha_m$

WDTH	TILT	XTOP	Z	ABSORPTION FOR RADIATION THRU SURFACE 6 α_m 0.3	0.4	0.5	0.6	0.7	0.8	0.9	1.0
1.0	10.	.0	1.92	.42	.45	.47	.50	.52	.55	.57	.59
1.0	10.	.5	1.92	.45	.47	.50	.52	.55	.57	.59	.61
1.0	10.	1.5	1.92	.47	.50	.53	.56	.58	.60	.63	.65
1.0	10.	3.0	1.92	.49	.53	.55	.58	.61	.63	.65	.67
1.0	20.	.0	.97	.45	.48	.51	.53	.56	.58	.61	.63
1.0	20.	.5	.97	.48	.51	.54	.56	.59	.61	.63	.65
1.0	20.	1.5	.97	.50	.54	.57	.59	.62	.64	.66	.68
1.0	20.	3.0	.97	.52	.55	.58	.61	.63	.66	.68	.70
1.0	30.	.0	.67	.46	.49	.52	.55	.58	.60	.63	.65
1.0	30.	.5	.67	.50	.53	.56	.59	.61	.64	.66	.68
1.0	30.	1.5	.67	.52	.55	.58	.61	.64	.66	.68	.70
1.0	30.	3.0	.67	.53	.56	.59	.62	.65	.67	.69	.71
3.0	10.	.0	1.92	.48	.51	.54	.57	.60	.63	.65	.68
3.0	10.	.5	1.92	.52	.56	.59	.62	.64	.67	.69	.71
3.0	10.	1.5	1.92	.58	.62	.65	.68	.70	.73	.75	.77
3.0	10.	3.0	1.92	.63	.67	.70	.73	.75	.77	.79	.81
3.0	20.	.0	.97	.49	.53	.56	.59	.62	.64	.67	.69
3.0	20.	.5	.97	.56	.59	.62	.65	.68	.70	.72	.74
3.0	20.	1.5	.97	.63	.66	.69	.72	.75	.77	.79	.80
3.0	20.	3.0	.97	.67	.71	.74	.76	.79	.81	.82	.84
3.0	30.	.0	.67	.50	.53	.56	.59	.62	.65	.67	.70
3.0	30.	.5	.67	.57	.61	.64	.67	.70	.72	.74	.76
3.0	30.	1.5	.67	.65	.68	.71	.74	.76	.78	.80	.82
3.0	30.	3.0	.67	.69	.72	.75	.78	.80	.82	.84	.85
5.0	10.	.0	1.92	.49	.52	.56	.59	.62	.64	.67	.69
5.0	10.	.5	1.92	.54	.58	.61	.64	.66	.69	.71	.73
5.0	10.	1.5	1.92	.62	.65	.68	.71	.73	.75	.77	.79
5.0	10.	3.0	1.92	.67	.71	.74	.76	.78	.80	.82	.84
5.0	20.	.0	.97	.50	.54	.57	.60	.63	.65	.68	.70
5.0	20.	.5	.97	.58	.61	.64	.67	.69	.72	.74	.76
5.0	20.	1.5	.97	.66	.70	.72	.75	.77	.79	.81	.82
5.0	20.	3.0	.97	.72	.75	.78	.80	.82	.84	.85	.87
5.0	30.	.0	.67	.50	.54	.57	.60	.63	.66	.68	.70
5.0	30.	.5	.67	.59	.63	.66	.68	.71	.73	.75	.77
5.0	30.	1.5	.67	.68	.71	.74	.77	.79	.81	.82	.84
5.0	30.	3.0	.67	.74	.77	.79	.82	.84	.85	.87	.88

ABSORPTION FACTORS
TILTED GLAZING - CL II

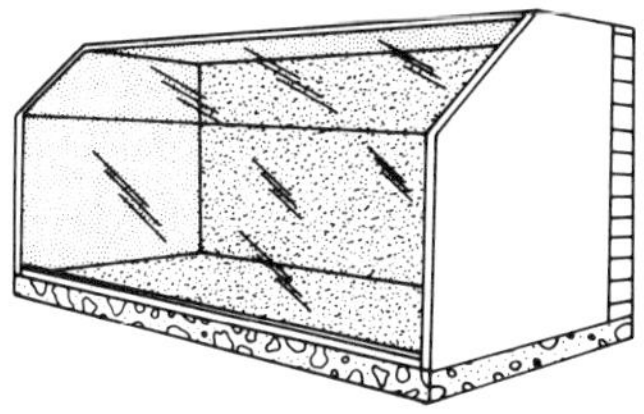

Table A4-29

OPAQUE END WALLS - YBOT= .67
α_o = .3

WDTH	TILT	XTOP	Z	ABSORPTION FOR RADIATION THRU SURFACE 3 α_m							
				0.3	0.4	0.5	0.6	0.7	0.8	0.9	1.0
1.0	10.	.0	1.92	.75	.77	.80	.82	.83	.85	.87	.89
1.0	10.	.5	1.92	.78	.80	.82	.83	.85	.86	.88	.89
1.0	10.	1.5	1.92	.82	.84	.85	.86	.87	.88	.90	.91
1.0	10.	3.0	1.92	.86	.87	.88	.89	.90	.90	.91	.92
1.0	20.	.0	.97	.72	.75	.78	.80	.82	.84	.86	.88
1.0	20.	.5	.97	.78	.80	.81	.83	.84	.86	.87	.88
1.0	20.	1.5	.97	.83	.84	.85	.86	.87	.88	.89	.90
1.0	20.	3.0	.97	.87	.87	.88	.89	.89	.90	.90	.91
1.0	30.	.0	.67	.71	.74	.77	.80	.82	.85	.87	.89
1.0	30.	.5	.67	.79	.80	.82	.84	.85	.86	.88	.89
1.0	30.	1.5	.67	.84	.85	.86	.87	.88	.89	.89	.90
1.0	30.	3.0	.67	.88	.89	.89	.90	.90	.91	.91	.91
3.0	10.	.0	1.92	.68	.72	.75	.79	.82	.85	.88	.91
3.0	10.	.5	1.92	.71	.75	.78	.81	.83	.86	.88	.91
3.0	10.	1.5	1.92	.76	.79	.81	.83	.86	.88	.89	.91
3.0	10.	3.0	1.92	.81	.83	.84	.86	.88	.89	.90	.92
3.0	20.	.0	.97	.66	.71	.74	.78	.81	.84	.87	.90
3.0	20.	.5	.97	.72	.75	.78	.81	.83	.85	.87	.89
3.0	20.	1.5	.97	.78	.80	.82	.84	.85	.87	.88	.90
3.0	20.	3.0	.97	.82	.84	.85	.86	.87	.88	.89	.90
3.0	30.	.0	.67	.66	.70	.74	.78	.81	.85	.88	.91
3.0	30.	.5	.67	.73	.76	.79	.82	.84	.86	.88	.90
3.0	30.	1.5	.67	.80	.82	.84	.85	.87	.88	.89	.90
3.0	30.	3.0	.67	.84	.85	.87	.88	.89	.89	.90	.91
5.0	10.	.0	1.92	.66	.70	.74	.78	.82	.85	.88	.91
5.0	10.	.5	1.92	.69	.73	.77	.80	.83	.86	.89	.91
5.0	10.	1.5	1.92	.74	.77	.80	.83	.85	.87	.90	.92
5.0	10.	3.0	1.92	.79	.81	.83	.85	.87	.89	.90	.92
5.0	20.	.0	.97	.65	.69	.73	.77	.81	.84	.87	.90
5.0	20.	.5	.97	.71	.74	.77	.80	.83	.85	.88	.90
5.0	20.	1.5	.97	.77	.79	.81	.83	.85	.87	.89	.90
5.0	20.	3.0	.97	.81	.83	.84	.86	.87	.88	.89	.90
5.0	30.	.0	.67	.64	.69	.73	.77	.81	.85	.88	.91
5.0	30.	.5	.67	.72	.75	.78	.81	.84	.86	.89	.91
5.0	30.	1.5	.67	.79	.81	.83	.85	.86	.88	.89	.91
5.0	30.	3.0	.67	.83	.85	.86	.87	.88	.89	.90	.91

ABSORPTION FACTORS
TILTED GLAZING - CL II

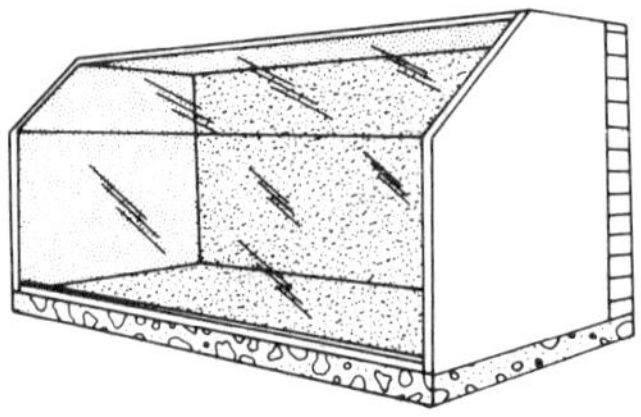

Table A4-30

OPAQUE END WALLS - YBOT= .67

α_o = .3

WDTH	TILT	XTOP	Z	ABSORPTION FOR RADIATION THRU SURFACE 4 α_m 0.3	0.4	0.5	0.6	0.7	0.8	0.9	1.0
1.0	10.	.0	1.92	.69	.71	.72	.74	.76	.77	.79	.80
1.0	10.	.5	1.92	.72	.73	.75	.76	.77	.78	.80	.81
1.0	10.	1.5	1.92	.75	.77	.78	.79	.80	.80	.81	.82
1.0	10.	3.0	1.92	.79	.79	.80	.81	.81	.82	.83	.83
1.0	20.	.0	.97	.71	.73	.75	.77	.79	.81	.83	.85
1.0	20.	.5	.97	.76	.78	.79	.80	.82	.83	.84	.85
1.0	20.	1.5	.97	.81	.82	.83	.84	.85	.85	.86	.87
1.0	20.	3.0	.97	.84	.85	.86	.86	.87	.87	.88	.88
1.0	30.	.0	.67	.71	.74	.77	.80	.82	.85	.87	.89
1.0	30.	.5	.67	.79	.80	.82	.84	.85	.86	.88	.89
1.0	30.	1.5	.67	.85	.86	.87	.87	.88	.89	.89	.90
1.0	30.	3.0	.67	.88	.89	.89	.90	.90	.91	.91	.91
3.0	10.	.0	1.92	.61	.64	.67	.70	.72	.74	.77	.79
3.0	10.	.5	1.92	.64	.67	.69	.71	.73	.75	.77	.79
3.0	10.	1.5	1.92	.68	.70	.72	.74	.75	.77	.78	.80
3.0	10.	3.0	1.92	.72	.73	.75	.76	.77	.78	.79	.80
3.0	20.	.0	.97	.64	.68	.72	.75	.78	.81	.83	.86
3.0	20.	.5	.97	.70	.73	.75	.78	.80	.82	.84	.85
3.0	20.	1.5	.97	.76	.78	.79	.81	.82	.83	.85	.86
3.0	20.	3.0	.97	.80	.81	.82	.83	.84	.85	.86	.86
3.0	30.	.0	.67	.66	.70	.74	.78	.81	.85	.88	.91
3.0	30.	.5	.67	.74	.77	.79	.82	.84	.86	.88	.90
3.0	30.	1.5	.67	.81	.82	.84	.85	.87	.88	.89	.90
3.0	30.	3.0	.67	.85	.86	.87	.88	.88	.89	.90	.91
5.0	10.	.0	1.92	.59	.62	.66	.69	.71	.74	.77	.79
5.0	10.	.5	1.92	.62	.65	.68	.70	.73	.75	.77	.79
5.0	10.	1.5	1.92	.67	.69	.71	.73	.75	.76	.78	.79
5.0	10.	3.0	1.92	.70	.72	.73	.75	.76	.77	.78	.80
5.0	20.	.0	.97	.63	.67	.71	.74	.77	.81	.83	.86
5.0	20.	.5	.97	.69	.72	.75	.77	.80	.82	.84	.86
5.0	20.	1.5	.97	.75	.77	.79	.80	.82	.83	.85	.86
5.0	20.	3.0	.97	.79	.80	.81	.82	.84	.84	.85	.86
5.0	30.	.0	.67	.64	.69	.73	.77	.81	.85	.88	.91
5.0	30.	.5	.67	.72	.76	.79	.81	.84	.86	.88	.90
5.0	30.	1.5	.67	.79	.81	.83	.85	.86	.88	.89	.90
5.0	30.	3.0	.67	.84	.85	.86	.87	.88	.89	.90	.91

ABSORPTION FACTORS
TILTED GLAZING - CL II

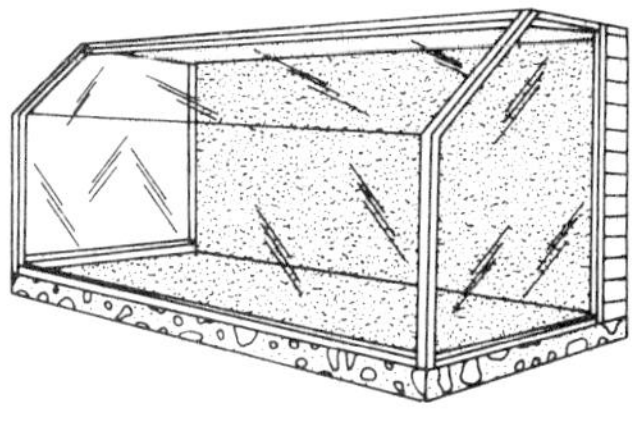

Table A4-31

GLAZED END WALLS - YBOT= .67

α_o = .3

WDTH	TILT	XTOP	Z	ABSORPTION FOR RADIATION THRU SURFACE 3 α_m							
				0.3	0.4	0.5	0.6	0.7	0.8	0.9	1.0
1.0	10.	.0	1.92	.45	.48	.51	.54	.57	.60	.62	.65
1.0	10.	.5	1.92	.44	.47	.50	.52	.55	.57	.59	.62
1.0	10.	1.5	1.92	.44	.47	.49	.52	.54	.56	.58	.60
1.0	10.	3.0	1.92	.45	.47	.50	.52	.54	.56	.58	.60
1.0	20.	.0	.97	.48	.51	.55	.58	.61	.64	.66	.69
1.0	20.	.5	.97	.46	.48	.51	.54	.56	.58	.60	.62
1.0	20.	1.5	.97	.45	.47	.49	.51	.54	.56	.57	.59
1.0	20.	3.0	.97	.45	.47	.49	.51	.53	.55	.57	.58
1.0	30.	.0	.67	.51	.55	.58	.62	.65	.68	.71	.74
1.0	30.	.5	.67	.48	.50	.53	.55	.58	.60	.62	.64
1.0	30.	1.5	.67	.46	.48	.50	.52	.54	.56	.58	.59
1.0	30.	3.0	.67	.45	.47	.49	.51	.53	.55	.56	.58
3.0	10.	.0	1.92	.55	.59	.64	.68	.71	.75	.78	.81
3.0	10.	.5	1.92	.56	.60	.64	.67	.71	.74	.77	.80
3.0	10.	1.5	1.92	.58	.62	.65	.68	.71	.74	.76	.79
3.0	10.	3.0	1.92	.60	.63	.66	.69	.71	.74	.76	.78
3.0	20.	.0	.97	.57	.61	.65	.69	.73	.76	.79	.82
3.0	20.	.5	.97	.58	.62	.65	.68	.71	.74	.77	.79
3.0	20.	1.5	.97	.60	.63	.66	.68	.71	.73	.75	.77
3.0	20.	3.0	.97	.61	.64	.66	.68	.71	.73	.74	.76
3.0	30.	.0	.67	.58	.63	.67	.71	.75	.79	.82	.85
3.0	30.	.5	.67	.60	.64	.67	.70	.73	.76	.78	.81
3.0	30.	1.5	.67	.61	.64	.67	.69	.72	.74	.76	.78
3.0	30.	3.0	.67	.62	.64	.67	.69	.71	.73	.74	.76
5.0	10.	.0	1.92	.58	.63	.67	.71	.75	.79	.82	.86
5.0	10.	.5	1.92	.60	.64	.68	.72	.75	.79	.82	.85
5.0	10.	1.5	1.92	.63	.66	.70	.73	.76	.79	.81	.84
5.0	10.	3.0	1.92	.66	.69	.72	.74	.77	.79	.82	.84
5.0	20.	.0	.97	.59	.63	.68	.72	.76	.79	.83	.86
5.0	20.	.5	.97	.62	.66	.69	.73	.76	.79	.81	.84
5.0	20.	1.5	.97	.65	.68	.71	.73	.76	.78	.80	.82
5.0	20.	3.0	.97	.67	.70	.72	.74	.76	.78	.80	.82
5.0	30.	.0	.67	.60	.65	.69	.73	.77	.81	.84	.88
5.0	30.	.5	.67	.64	.67	.71	.74	.77	.80	.83	.85
5.0	30.	1.5	.67	.67	.70	.72	.75	.77	.79	.81	.83
5.0	30.	3.0	.67	.68	.71	.73	.75	.77	.79	.80	.82

ABSORPTION FACTORS
TILTED GLAZING - CL II

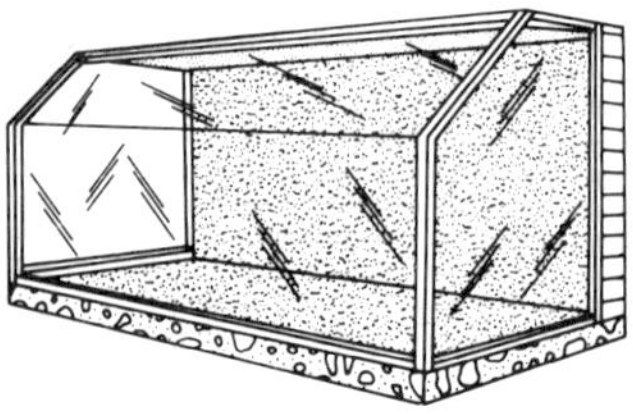

Table A4-32

GLAZED END WALLS - YBOT= .67
α_o = .3

WDTH	TILT	XTOP	Z	ABSORPTION FOR RADIATION THRU SURFACE 4 α_m 0.3	0.4	0.5	0.6	0.7	0.8	0.9	1.0
1.0	10.	.0	1.92	.41	.43	.45	.48	.50	.52	.54	.56
1.0	10.	.5	1.92	.41	.43	.45	.47	.49	.51	.53	.55
1.0	10.	1.5	1.92	.41	.43	.45	.47	.49	.51	.52	.54
1.0	10.	3.0	1.92	.42	.43	.45	.47	.49	.50	.52	.54
1.0	20.	.0	.97	.46	.49	.52	.55	.57	.60	.62	.65
1.0	20.	.5	.97	.45	.48	.50	.52	.55	.57	.59	.61
1.0	20.	1.5	.97	.45	.47	.49	.51	.52	.54	.56	.58
1.0	20.	3.0	.97	.44	.46	.48	.50	.52	.53	.55	.56
1.0	30.	.0	.67	.50	.54	.57	.61	.64	.67	.70	.73
1.0	30.	.5	.67	.49	.51	.54	.56	.59	.61	.63	.65
1.0	30.	1.5	.67	.47	.49	.51	.53	.55	.57	.59	.60
1.0	30.	3.0	.67	.46	.48	.50	.52	.54	.55	.57	.58
3.0	10.	.0	1.92	.48	.52	.55	.58	.61	.64	.66	.69
3.0	10.	.5	1.92	.50	.53	.55	.58	.61	.63	.65	.67
3.0	10.	1.5	1.92	.51	.54	.56	.58	.60	.62	.64	.66
3.0	10.	3.0	1.92	.52	.54	.56	.58	.60	.62	.64	.65
3.0	20.	.0	.97	.54	.58	.62	.66	.69	.72	.75	.78
3.0	20.	.5	.97	.56	.60	.63	.66	.68	.71	.73	.75
3.0	20.	1.5	.97	.58	.60	.63	.65	.67	.69	.71	.72
3.0	20.	3.0	.97	.58	.60	.62	.64	.66	.68	.69	.71
3.0	30.	.0	.67	.58	.62	.67	.71	.75	.78	.82	.85
3.0	30.	.5	.67	.61	.64	.67	.70	.73	.76	.78	.80
3.0	30.	1.5	.67	.62	.64	.67	.69	.71	.73	.75	.76
3.0	30.	3.0	.67	.62	.64	.66	.68	.70	.71	.73	.74
5.0	10.	.0	1.92	.51	.55	.58	.61	.64	.67	.70	.73
5.0	10.	.5	1.92	.53	.56	.59	.62	.65	.67	.69	.72
5.0	10.	1.5	1.92	.55	.58	.60	.63	.65	.67	.69	.70
5.0	10.	3.0	1.92	.57	.59	.61	.63	.65	.66	.68	.70
5.0	20.	.0	.97	.56	.61	.65	.69	.72	.75	.79	.81
5.0	20.	.5	.97	.60	.63	.67	.69	.72	.75	.77	.79
5.0	20.	1.5	.97	.63	.65	.68	.70	.72	.74	.76	.77
5.0	20.	3.0	.97	.64	.66	.68	.70	.71	.73	.74	.76
5.0	30.	.0	.67	.60	.64	.69	.73	.77	.81	.84	.87
5.0	30.	.5	.67	.64	.68	.71	.74	.77	.80	.82	.84
5.0	30.	1.5	.67	.67	.70	.72	.74	.76	.78	.80	.82
5.0	30.	3.0	.67	.69	.71	.72	.74	.76	.77	.79	.80

ABSORPTION FACTORS
TILTED GLAZING - CL II

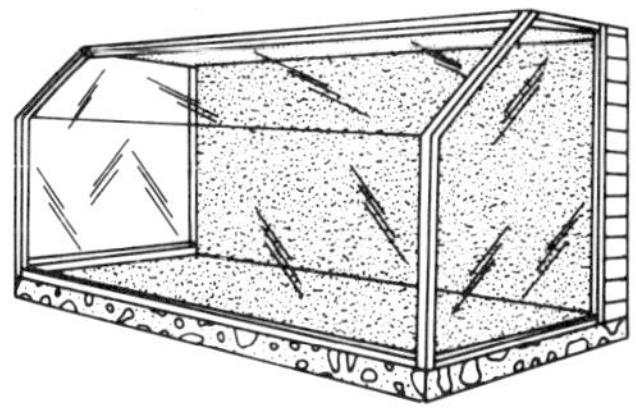

Table A4-33

GLAZED END WALLS - YBOT= .67
α_o = .3

WDTH	TILT	XTOP	Z	ABSORBTION FOR RADIATION THRU SURFACE 6 α_m 0.3	0.4	0.5	0.6	0.7	0.8	0.9	1.0
1.0	10.	.0	1.92	.42	.45	.47	.50	.52	.55	.57	.59
1.0	10.	.5	1.92	.45	.47	.49	.52	.54	.56	.58	.60
1.0	10.	1.5	1.92	.47	.50	.52	.54	.56	.58	.60	.62
1.0	10.	3.0	1.92	.49	.51	.54	.56	.58	.60	.61	.63
1.0	20.	.0	.97	.45	.48	.51	.53	.56	.58	.61	.63
1.0	20.	.5	.97	.48	.50	.53	.55	.58	.60	.62	.64
1.0	20.	1.5	.97	.50	.53	.55	.57	.59	.61	.63	.65
1.0	20.	3.0	.97	.52	.54	.56	.58	.60	.62	.63	.65
1.0	30.	.0	.67	.46	.49	.52	.55	.58	.60	.63	.65
1.0	30.	.5	.67	.50	.52	.55	.57	.60	.62	.64	.66
1.0	30.	1.5	.67	.52	.54	.56	.58	.60	.62	.64	.66
1.0	30.	3.0	.67	.53	.55	.57	.59	.61	.62	.64	.66
3.0	10.	.0	1.92	.48	.51	.54	.57	.60	.63	.65	.68
3.0	10.	.5	1.92	.52	.56	.58	.61	.64	.66	.68	.71
3.0	10.	1.5	1.92	.58	.61	.64	.66	.69	.71	.73	.75
3.0	10.	3.0	1.92	.63	.66	.68	.71	.73	.75	.77	.78
3.0	20.	.0	.97	.49	.53	.56	.59	.62	.64	.67	.69
3.0	20.	.5	.97	.56	.59	.62	.64	.67	.69	.71	.73
3.0	20.	1.5	.97	.63	.65	.68	.70	.72	.74	.76	.78
3.0	20.	3.0	.97	.67	.70	.72	.74	.76	.78	.79	.81
3.0	30.	.0	.67	.50	.53	.56	.59	.62	.65	.67	.70
3.0	30.	.5	.67	.57	.60	.63	.66	.68	.70	.73	.74
3.0	30.	1.5	.67	.65	.67	.70	.72	.74	.76	.78	.79
3.0	30.	3.0	.67	.69	.71	.73	.75	.77	.79	.80	.82
5.0	10.	.0	1.92	.49	.52	.56	.59	.62	.64	.67	.69
5.0	10.	.5	1.92	.54	.57	.60	.63	.66	.68	.70	.73
5.0	10.	1.5	1.92	.62	.64	.67	.69	.72	.74	.76	.78
5.0	10.	3.0	1.92	.67	.70	.72	.75	.77	.78	.80	.82
5.0	20.	.0	.97	.50	.54	.57	.60	.63	.65	.68	.70
5.0	20.	.5	.97	.58	.61	.63	.66	.68	.71	.73	.75
5.0	20.	1.5	.97	.66	.69	.71	.73	.75	.77	.79	.81
5.0	20.	3.0	.97	.72	.74	.76	.78	.80	.82	.83	.84
5.0	30.	.0	.67	.50	.54	.57	.60	.63	.66	.68	.70
5.0	30.	.5	.67	.59	.62	.65	.67	.70	.72	.74	.76
5.0	30.	1.5	.67	.68	.70	.73	.75	.77	.79	.80	.82
5.0	30.	3.0	.67	.74	.76	.78	.79	.81	.83	.84	.85

TILTED GLAZING - CLASS IIIa

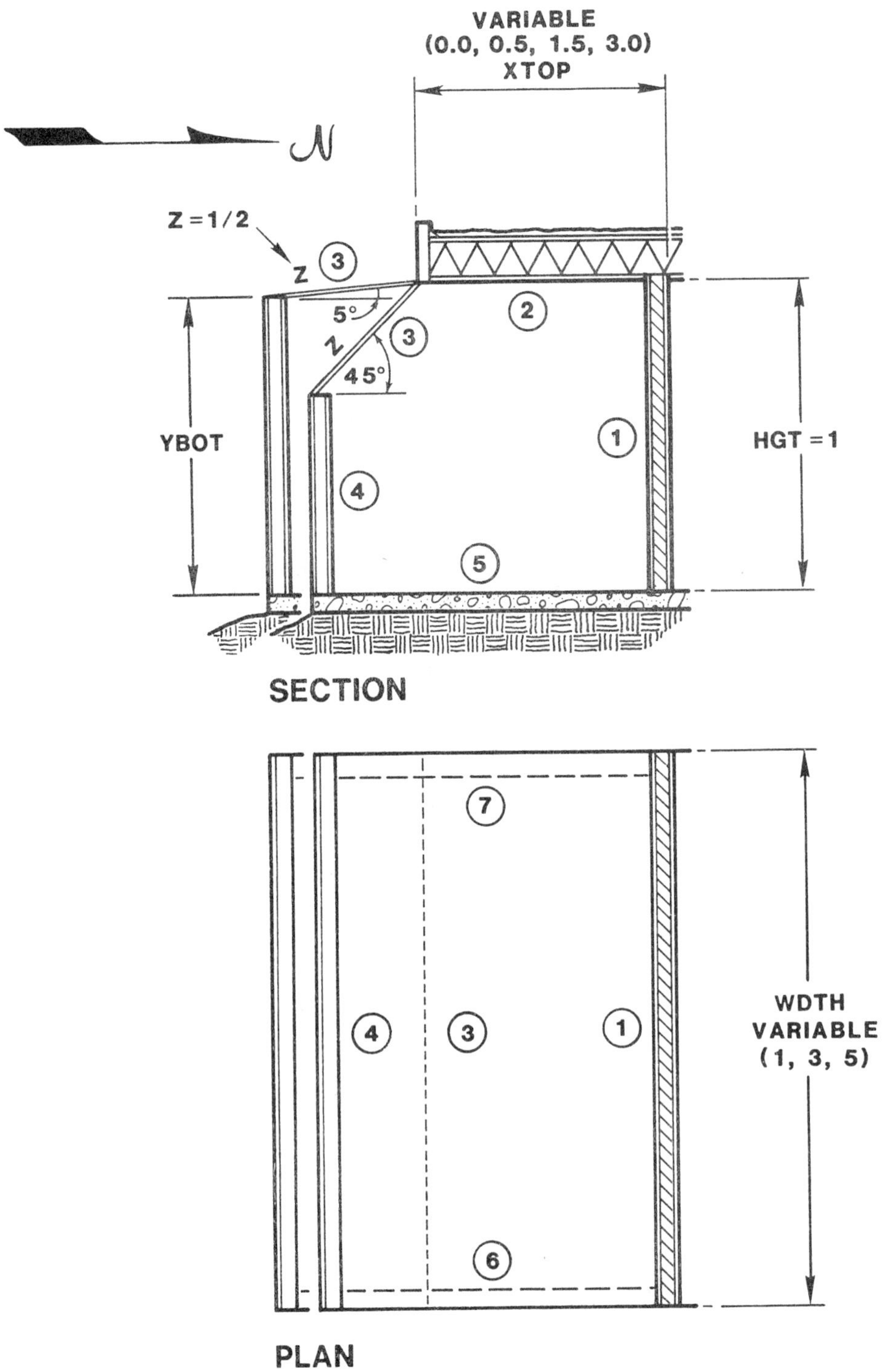

Figure A4-47. Geometry for Tilted Glazing, Class IIIa

ABSORPTION FACTORS
TILTED GLAZING - CL IIIa

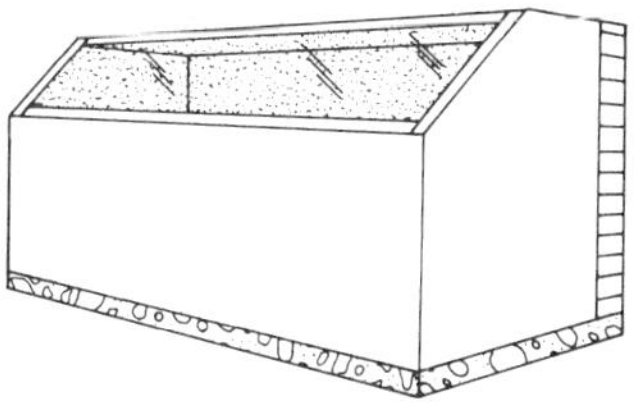

Table A4-34

OPAQUE BOTTOM SECTION, Z= .5
OPAQUE END WALLS
$\alpha_o = \alpha_m$

WDTH	TILT	XTOP	YBOT	ABSORPTION FOR RADIATION THRU SURFACE 3 α_m 0.3	0.4	0.5	0.6	0.7	0.8	0.9	1.0
1.0	5.	.0	.956	.90	.93	.95	.96	.97	.98	.99	1.00
1.0	5.	.5	.956	.93	.95	.96	.97	.98	.99	.99	1.00
1.0	5.	1.5	.956	.95	.96	.97	.98	.99	.99	1.00	1.00
1.0	5.	3.0	.956	.97	.97	.98	.99	.99	.99	1.00	1.00
1.0	45.	.0	.646	.86	.89	.92	.94	.96	.97	.99	1.00
1.0	45.	.5	.646	.92	.94	.96	.97	.98	.99	.99	1.00
1.0	45.	1.5	.646	.96	.97	.98	.98	.99	.99	1.00	1.00
1.0	45.	3.0	.646	.97	.98	.99	.99	.99	1.00	1.00	1.00
3.0	5.	.0	.956	.88	.91	.93	.95	.97	.98	.99	1.00
3.0	5.	.5	.956	.91	.93	.95	.96	.97	.98	.99	1.00
3.0	5.	1.5	.956	.94	.95	.96	.97	.98	.99	.99	1.00
3.0	5.	3.0	.956	.95	.96	.97	.98	.99	.99	1.00	1.00
3.0	45.	.0	.646	.84	.88	.91	.93	.95	.97	.98	1.00
3.0	45.	.5	.646	.90	.92	.94	.96	.97	.98	.99	1.00
3.0	45.	1.5	.646	.94	.95	.97	.98	.98	.99	1.00	1.00
3.0	45.	3.0	.646	.96	.97	.98	.98	.99	.99	1.00	1.00
5.0	5.	.0	.956	.88	.91	.93	.95	.97	.98	.99	1.00
5.0	5.	.5	.956	.91	.93	.95	.96	.97	.98	.99	1.00
5.0	5.	1.5	.956	.93	.95	.96	.97	.98	.99	.99	1.00
5.0	5.	3.0	.956	.95	.96	.97	.98	.98	.99	.99	1.00
5.0	45.	.0	.646	.84	.87	.90	.93	.95	.97	.98	1.00
5.0	45.	.5	.646	.89	.92	.94	.96	.97	.98	.99	1.00
5.0	45.	1.5	.646	.93	.95	.96	.97	.98	.99	.99	1.00
5.0	45.	3.0	.646	.96	.97	.98	.98	.99	.99	1.00	1.00

ABSORPTION FACTORS
TILTED GLAZING - CL IIIa

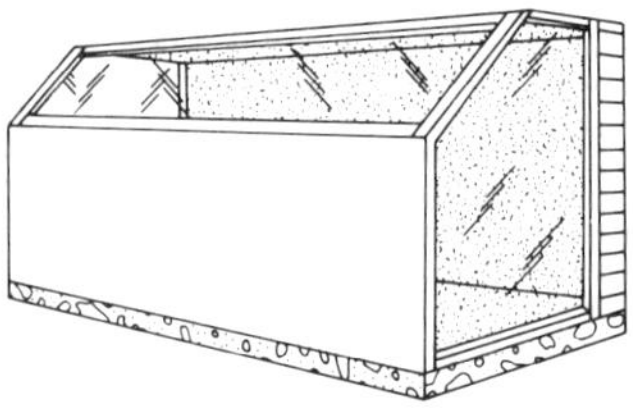

Table A4-35

OPAQUE BOTTOM SECTION, Z= .5
GLAZED END WALLS
$\alpha_o = \alpha_m$

WDTH	TILT	XTOP	YBOT	ABSORPTION FOR RADIATION THRU SURFACE 3 α_m 0.3	0.4	0.5	0.6	0.7	0.8	0.9	1.0
1.0	5.	.0	.956	.63	.67	.70	.73	.75	.77	.79	.81
1.0	5.	.5	.956	.56	.60	.63	.65	.68	.70	.72	.74
1.0	5.	1.5	.956	.53	.57	.60	.62	.65	.67	.69	.71
1.0	5.	3.0	.956	.52	.56	.59	.61	.64	.66	.68	.70
1.0	45.	.0	.646	.67	.71	.74	.77	.80	.83	.85	.87
1.0	45.	.5	.646	.57	.61	.64	.66	.68	.71	.72	.74
1.0	45.	1.5	.646	.52	.55	.58	.60	.62	.64	.66	.68
1.0	45.	3.0	.646	.50	.53	.56	.58	.60	.62	.64	.66
3.0	5.	.0	.956	.77	.81	.83	.86	.88	.90	.91	.93
3.0	5.	.5	.956	.74	.78	.81	.83	.85	.87	.88	.90
3.0	5.	1.5	.956	.72	.76	.79	.81	.83	.85	.86	.88
3.0	5.	3.0	.956	.72	.75	.78	.80	.82	.84	.86	.87
3.0	45.	.0	.646	.77	.81	.84	.87	.89	.92	.94	.95
3.0	45.	.5	.646	.75	.78	.81	.83	.85	.87	.89	.90
3.0	45.	1.5	.646	.72	.75	.78	.80	.82	.83	.85	.86
3.0	45.	3.0	.646	.70	.73	.76	.78	.80	.82	.83	.84
5.0	5.	.0	.956	.81	.84	.87	.89	.91	.93	.94	.96
5.0	5.	.5	.956	.80	.83	.86	.88	.90	.91	.92	.94
5.0	5.	1.5	.956	.79	.82	.85	.87	.88	.90	.91	.93
5.0	5.	3.0	.956	.79	.82	.84	.86	.88	.90	.91	.92
5.0	45.	.0	.646	.79	.83	.86	.89	.91	.94	.95	.97
5.0	45.	.5	.646	.80	.83	.86	.88	.90	.91	.93	.94
5.0	45.	1.5	.646	.79	.82	.84	.86	.88	.89	.90	.91
5.0	45.	3.0	.646	.78	.81	.83	.85	.87	.88	.89	.90

ABSORPTION FACTORS
TILTED GLAZING - CL IIIa

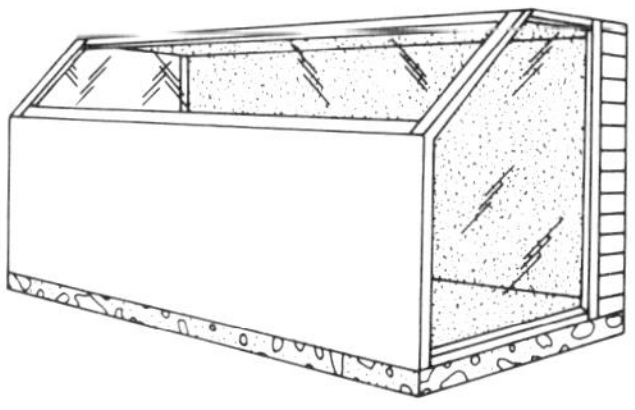

Table A4-36

OPAQUE BOTTOM SECTION, Z= .5
GLAZED END WALLS
$\alpha_o = \alpha_m$

WDTH	TILT	XTOP	YBOT	ABSORPTION FOR RADIATION THRU SURFACE 6 α_m							
				0.3	0.4	0.5	0.6	0.7	0.8	0.9	1.0
1.0	5.	.0	.956	.64	.68	.71	.74	.77	.79	.81	.83
1.0	5.	.5	.956	.61	.65	.68	.71	.74	.76	.78	.80
1.0	5.	1.5	.956	.58	.62	.65	.68	.71	.73	.76	.78
1.0	5.	3.0	.956	.56	.60	.63	.66	.69	.72	.74	.76
1.0	45.	.0	.646	.66	.70	.73	.76	.78	.81	.83	.84
1.0	45.	.5	.646	.62	.66	.69	.72	.74	.77	.79	.81
1.0	45.	1.5	.646	.58	.62	.65	.68	.71	.73	.76	.78
1.0	45.	3.0	.646	.56	.60	.63	.66	.69	.71	.74	.76
3.0	5.	.0	.956	.73	.77	.79	.82	.84	.85	.87	.88
3.0	5.	.5	.956	.75	.79	.81	.84	.86	.87	.89	.90
3.0	5.	1.5	.956	.76	.80	.83	.85	.87	.89	.90	.92
3.0	5.	3.0	.956	.76	.80	.83	.85	.87	.89	.90	.92
3.0	45.	.0	.646	.73	.76	.79	.81	.83	.85	.86	.88
3.0	45.	.5	.646	.74	.78	.80	.83	.85	.86	.88	.89
3.0	45.	1.5	.646	.76	.79	.82	.84	.86	.88	.89	.91
3.0	45.	3.0	.646	.76	.79	.82	.84	.86	.88	.90	.91
5.0	5.	.0	.956	.75	.78	.81	.83	.85	.86	.88	.89
5.0	5.	.5	.956	.79	.82	.84	.86	.88	.89	.90	.91
5.0	5.	1.5	.956	.81	.84	.87	.89	.90	.92	.93	.94
5.0	5.	3.0	.956	.83	.85	.88	.90	.91	.93	.94	.95
5.0	45.	.0	.646	.74	.77	.80	.82	.84	.86	.87	.88
5.0	45.	.5	.646	.77	.80	.83	.85	.86	.88	.89	.90
5.0	45.	1.5	.646	.80	.83	.86	.87	.89	.90	.92	.93
5.0	45.	3.0	.646	.82	.85	.87	.89	.90	.92	.93	.94

ABSORPTION FACTORS
TILTED GLAZING - CL IIIa

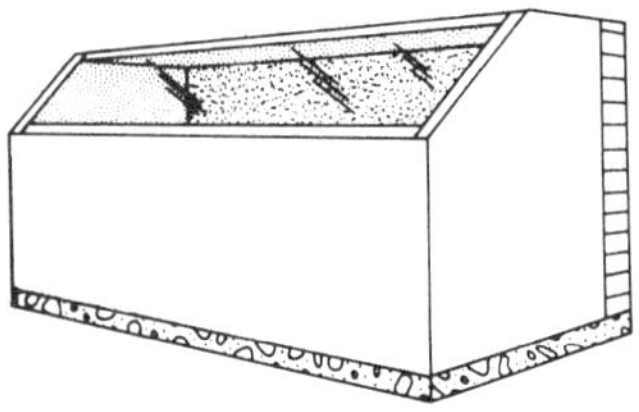

Table A4-37

OPAQUE BOTTOM SECTION, Z= .5
OPAQUE END WALLS
α_o = .3

				ABSORPTION FOR RADIATION THRU SURFACE 3							
				α_m							
WDTH	TILT	XTOP	YBOT	0.3	0.4	0.5	0.6	0.7	0.8	0.9	1.0
1.0	5.	.0	.956	.90	.91	.92	.93	.94	.95	.96	.97
1.0	5.	.5	.956	.93	.94	.94	.95	.95	.96	.96	.97
1.0	5.	1.5	.956	.95	.96	.96	.96	.97	.97	.97	.97
1.0	5.	3.0	.956	.97	.97	.97	.97	.97	.98	.98	.98
1.0	45.	.0	.646	.86	.88	.90	.92	.94	.96	.97	.99
1.0	45.	.5	.646	.92	.93	.94	.95	.96	.96	.97	.97
1.0	45.	1.5	.646	.96	.96	.96	.97	.97	.97	.98	.98
1.0	45.	3.0	.646	.97	.98	.98	.98	.98	.98	.99	.99
3.0	5.	.0	.956	.88	.90	.92	.93	.94	.95	.96	.97
3.0	5.	.5	.956	.91	.92	.93	.94	.95	.96	.97	.97
3.0	5.	1.5	.956	.94	.94	.95	.96	.96	.97	.97	.98
3.0	5.	3.0	.956	.95	.96	.96	.97	.97	.97	.97	.98
3.0	45.	.0	.646	.84	.87	.90	.92	.94	.96	.98	.99
3.0	45.	.5	.646	.90	.92	.93	.94	.96	.97	.98	.99
3.0	45.	1.5	.646	.94	.95	.96	.96	.97	.98	.98	.99
3.0	45.	3.0	.646	.96	.97	.97	.98	.98	.98	.99	.99
5.0	5.	.0	.956	.88	.90	.91	.93	.94	.95	.97	.98
5.0	5.	.5	.956	.91	.92	.93	.94	.95	.96	.97	.98
5.0	5.	1.5	.956	.93	.94	.95	.95	.96	.97	.97	.98
5.0	5.	3.0	.956	.95	.95	.96	.96	.97	.97	.97	.98
5.0	45.	.0	.646	.84	.87	.89	.92	.94	.96	.98	1.00
5.0	45.	.5	.646	.89	.91	.93	.94	.96	.97	.98	.99
5.0	45.	1.5	.646	.93	.94	.95	.96	.97	.98	.99	.99
5.0	45.	3.0	.646	.96	.96	.97	.98	.98	.98	.99	.99

ABSORPTION FACTORS
TILTED GLAZING - CL IIIa

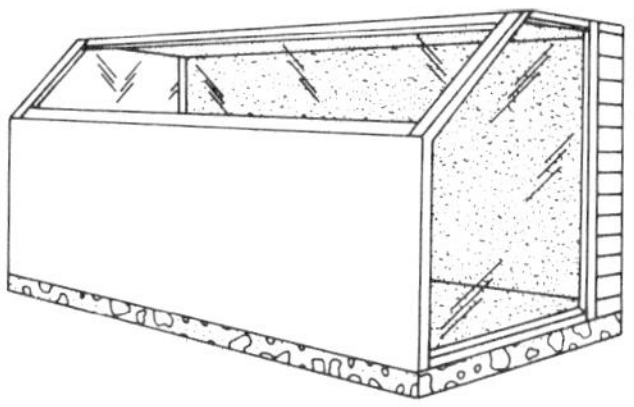

Table A4-38

OPAQUE BOTTOM SECTION, Z= .5
GLAZED END WALLS
α_o = .3

WDTH	TILT	XTOP	YBOT	ABSORPTION FOR RADIATION THRU SURFACE 3 α_m 0.3	0.4	0.5	0.6	0.7	0.8	0.9	1.0
1.0	5.	.0	.956	.63	.65	.68	.70	.72	.73	.75	.77
1.0	5.	.5	.956	.56	.58	.60	.62	.64	.65	.67	.68
1.0	5.	1.5	.956	.53	.55	.57	.58	.60	.61	.63	.64
1.0	5.	3.0	.956	.52	.54	.56	.57	.59	.60	.62	.63
1.0	45.	.0	.646	.67	.70	.73	.76	.79	.82	.84	.86
1.0	45.	.5	.646	.57	.60	.62	.64	.66	.68	.70	.72
1.0	45.	1.5	.646	.52	.54	.56	.58	.59	.61	.63	.64
1.0	45.	3.0	.646	.50	.52	.54	.56	.57	.59	.60	.62
3.0	5.	.0	.956	.77	.79	.82	.83	.85	.87	.88	.90
3.0	5.	.5	.956	.74	.76	.78	.80	.82	.83	.85	.86
3.0	5.	1.5	.956	.72	.74	.76	.78	.79	.81	.82	.83
3.0	5.	3.0	.956	.72	.74	.75	.77	.78	.80	.81	.82
3.0	45.	.0	.646	.77	.80	.83	.86	.88	.91	.93	.95
3.0	45.	.5	.646	.75	.77	.80	.82	.84	.86	.87	.89
3.0	45.	1.5	.646	.72	.74	.76	.78	.80	.81	.83	.84
3.0	45.	3.0	.646	.70	.72	.74	.76	.78	.79	.81	.82
5.0	5.	.0	.956	.81	.83	.85	.87	.89	.90	.92	.93
5.0	5.	.5	.956	.80	.82	.84	.85	.87	.88	.89	.91
5.0	5.	1.5	.956	.79	.81	.83	.84	.85	.87	.88	.89
5.0	5.	3.0	.956	.79	.81	.82	.84	.85	.86	.87	.88
5.0	45.	.0	.646	.79	.83	.86	.88	.91	.93	.95	.97
5.0	45.	.5	.646	.80	.82	.84	.87	.88	.90	.92	.93
5.0	45.	1.5	.646	.79	.81	.83	.85	.86	.88	.89	.90
5.0	45.	3.0	.646	.78	.80	.82	.83	.85	.86	.87	.88

ABSORPTION FACTORS
TILTED GLAZING - CL IIIa

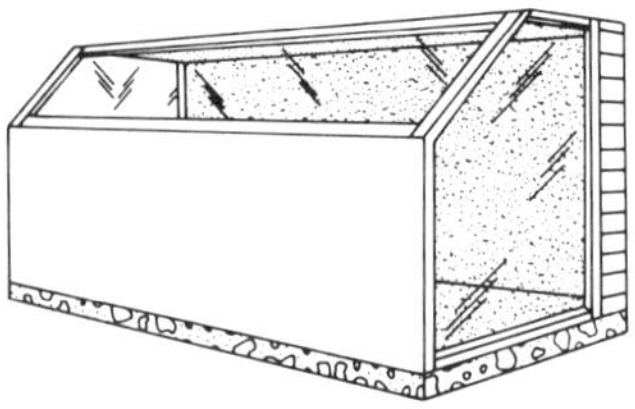

Table A4-39

OPAQUE BOTTOM SECTION, Z= .5
GLAZED END WALLS
α_o = .3

WDTH	TILT	XTOP	YBOT	ABSORPTION FOR RADIATION THRU SURFACE 6 α_m 0.3	0.4	0.5	0.6	0.7	0.8	0.9	1.0
1.0	5.	.0	.956	.64	.67	.69	.71	.73	.75	.77	.78
1.0	5.	.5	.956	.61	.63	.66	.68	.69	.71	.73	.75
1.0	5.	1.5	.956	.58	.60	.62	.64	.66	.68	.69	.71
1.0	5.	3.0	.956	.56	.58	.60	.62	.64	.66	.67	.69
1.0	45.	.0	.646	.66	.69	.71	.74	.76	.78	.80	.82
1.0	45.	.5	.646	.62	.64	.66	.69	.71	.72	.74	.76
1.0	45.	1.5	.646	.58	.60	.63	.65	.66	.68	.70	.71
1.0	45.	3.0	.646	.56	.58	.60	.62	.64	.66	.67	.69
3.0	5.	.0	.956	.73	.76	.78	.79	.81	.82	.84	.85
3.0	5.	.5	.956	.75	.77	.79	.81	.82	.84	.85	.86
3.0	5.	1.5	.956	.76	.78	.80	.82	.83	.85	.86	.87
3.0	5.	3.0	.956	.76	.78	.80	.82	.83	.85	.86	.87
3.0	45.	.0	.646	.73	.75	.78	.80	.82	.83	.85	.86
3.0	45.	.5	.646	.74	.76	.79	.80	.82	.84	.85	.86
3.0	45.	1.5	.646	.76	.78	.80	.81	.83	.84	.86	.87
3.0	45.	3.0	.646	.76	.78	.80	.81	.83	.84	.86	.87
5.0	5.	.0	.956	.75	.77	.79	.81	.82	.84	.85	.86
5.0	5.	.5	.956	.79	.80	.82	.84	.85	.86	.87	.88
5.0	5.	1.5	.956	.81	.83	.85	.86	.87	.89	.90	.91
5.0	5.	3.0	.956	.83	.84	.86	.87	.88	.89	.91	.91
5.0	45.	.0	.646	.74	.76	.79	.81	.82	.84	.86	.87
5.0	45.	.5	.646	.77	.79	.81	.83	.84	.86	.87	.88
5.0	45.	1.5	.646	.80	.82	.84	.85	.87	.88	.89	.90
5.0	45.	3.0	.646	.82	.84	.85	.86	.88	.89	.90	.91

TILTED GLAZING-CLASS IIIb

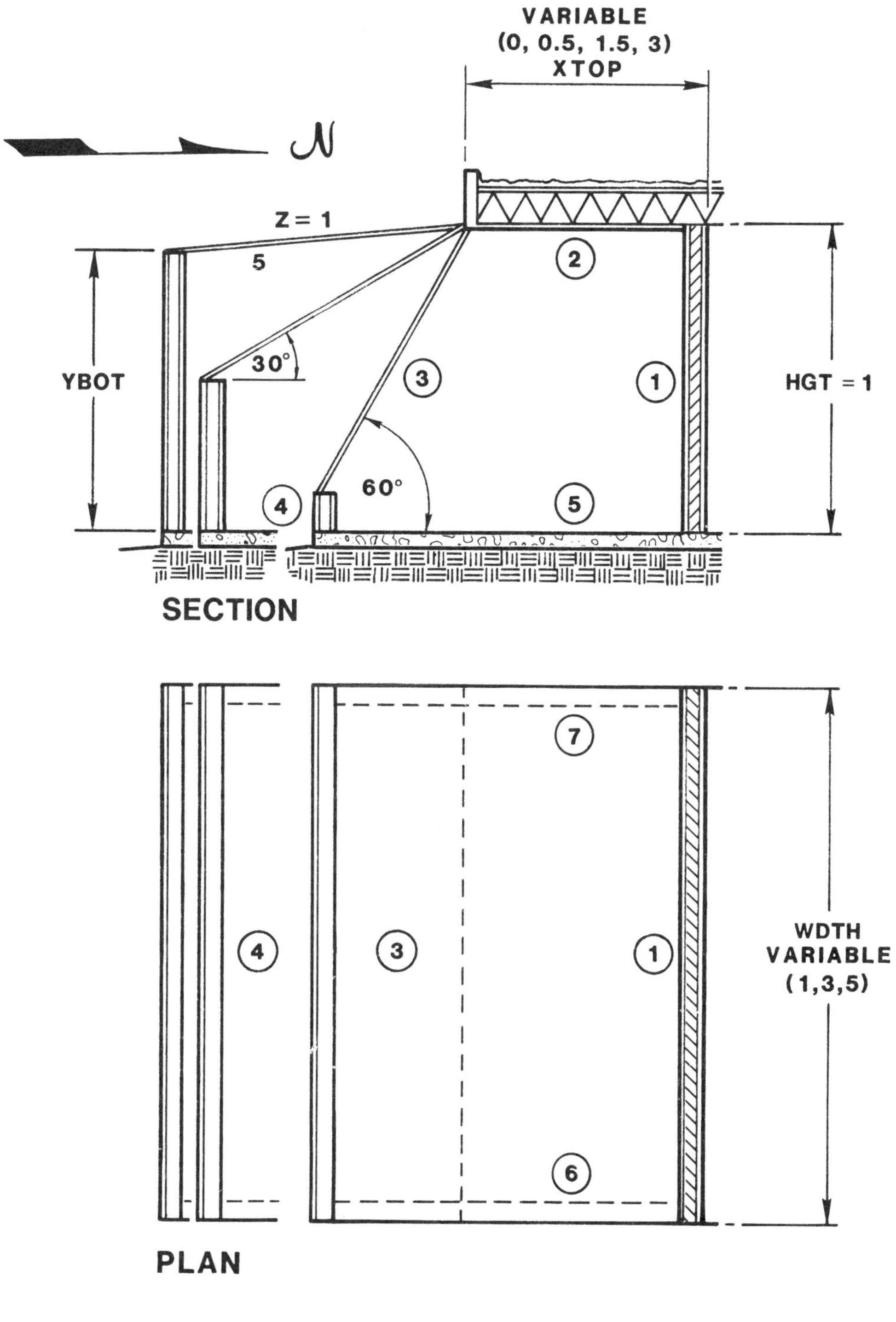

Figure A4-48. Geometry for Tilted Glazing, Class IIIb

ABSORPTION FACTORS
TILTED GLAZING - CL IIIb

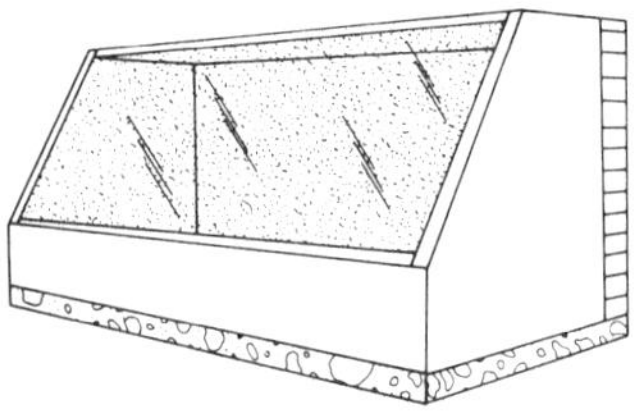

Table A4-40

OPAQUE BOTTOM SECTION, Z=1.0
OPAQUE END WALLS
$\alpha_o = \alpha_m$

WDTH	TILT	XTOP	YBOT	ABSORPTION FOR RADIATION THRU SURFACE 3, α_m 0.3	0.4	0.5	0.6	0.7	0.8	0.9	1.0
1.0	5.	.0	.913	.87	.90	.92	.94	.96	.98	.99	1.00
1.0	5.	.5	.913	.90	.92	.94	.96	.97	.98	.99	1.00
1.0	5.	1.5	.913	.93	.94	.96	.97	.98	.99	.99	1.00
1.0	5.	3.0	.913	.95	.96	.97	.98	.98	.99	1.00	1.00
1.0	30.	.0	.500	.83	.87	.90	.93	.95	.97	.98	1.00
1.0	30.	.5	.500	.88	.91	.93	.95	.97	.98	.99	1.00
1.0	30.	1.5	.500	.92	.94	.96	.97	.98	.99	.99	1.00
1.0	30.	3.0	.500	.95	.96	.97	.98	.99	.99	1.00	1.00
1.0	60.	.0	.135	.75	.80	.84	.88	.92	.95	.97	1.00
1.0	60.	.5	.135	.85	.88	.91	.94	.96	.97	.99	1.00
1.0	60.	1.5	.135	.91	.93	.95	.96	.98	.98	.99	1.00
1.0	60.	3.0	.135	.95	.96	.97	.98	.99	.99	1.00	1.00
3.0	5.	.0	.913	.83	.87	.90	.93	.95	.97	.99	1.00
3.0	5.	.5	.913	.86	.89	.92	.94	.96	.97	.99	1.00
3.0	5.	1.5	.913	.90	.92	.94	.95	.97	.98	.99	1.00
3.0	5.	3.0	.913	.92	.94	.95	.97	.98	.98	.99	1.00
3.0	30.	.0	.500	.79	.84	.87	.91	.93	.96	.98	1.00
3.0	30.	.5	.500	.84	.88	.91	.93	.95	.97	.99	1.00
3.0	30.	1.5	.500	.89	.92	.94	.95	.97	.98	.99	1.00
3.0	30.	3.0	.500	.93	.94	.96	.97	.98	.99	.99	1.00
3.0	60.	.0	.135	.72	.77	.82	.86	.90	.94	.97	1.00
3.0	60.	.5	.135	.81	.85	.89	.91	.94	.96	.98	1.00
3.0	60.	1.5	.135	.88	.91	.93	.95	.97	.98	.99	1.00
3.0	60.	3.0	.135	.92	.94	.96	.97	.98	.99	.99	1.00
5.0	5.	.0	.913	.82	.86	.89	.92	.95	.97	.98	1.00
5.0	5.	.5	.913	.85	.88	.91	.93	.95	.97	.99	1.00
5.0	5.	1.5	.913	.89	.91	.93	.95	.96	.98	.99	1.00
5.0	5.	3.0	.913	.92	.93	.95	.96	.97	.98	.99	1.00
5.0	30.	.0	.500	.78	.83	.87	.90	.93	.96	.98	1.00
5.0	30.	.5	.500	.83	.87	.90	.93	.95	.97	.98	1.00
5.0	30.	1.5	.500	.88	.91	.93	.95	.96	.98	.99	1.00
5.0	30.	3.0	.500	.92	.94	.95	.96	.98	.98	.99	1.00
5.0	60.	.0	.135	.71	.77	.81	.86	.90	.93	.97	1.00
5.0	60.	.5	.135	.80	.84	.88	.91	.94	.96	.98	1.00
5.0	60.	1.5	.135	.87	.90	.93	.95	.96	.98	.99	1.00
5.0	60.	3.0	.135	.92	.94	.95	.97	.98	.99	.99	1.00

ABSORPTION FACTORS
TILTED GLAZING - CL IIIb

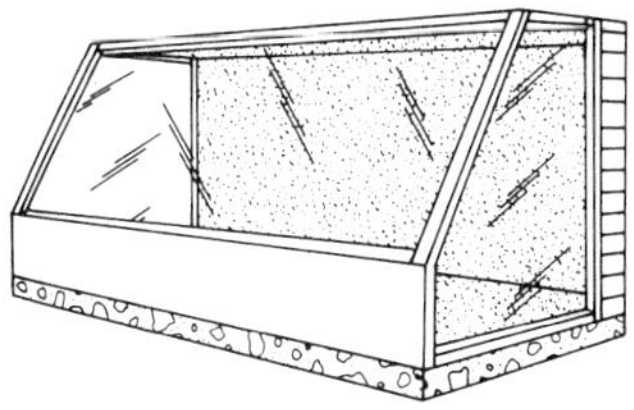

Table A4-41

OPAQUE BOTTOM SECTION, Z=1.0
GLAZED END WALLS
$\alpha_o = \alpha_m$

WDTH	TILT	XTOP	YBOT	ABSORPTION FOR RADIATION THRU SURFACE 3 α_m 0.3	0.4	0.5	0.6	0.7	0.8	0.9	1.0
1.0	5.	.0	.913	.54	.58	.61	.64	.67	.70	.72	.74
1.0	5.	.5	.913	.52	.55	.58	.60	.63	.65	.67	.69
1.0	5.	1.5	.913	.50	.53	.56	.59	.61	.64	.66	.68
1.0	5.	3.0	.913	.50	.53	.56	.59	.61	.63	.65	.67
1.0	30.	.0	.500	.57	.61	.64	.68	.71	.74	.76	.79
1.0	30.	.5	.500	.53	.56	.59	.62	.65	.67	.69	.71
1.0	30.	1.5	.500	.51	.54	.57	.59	.62	.64	.66	.68
1.0	30.	3.0	.500	.50	.53	.56	.59	.61	.63	.65	.67
1.0	60.	.0	.135	.60	.65	.69	.73	.77	.81	.84	.87
1.0	60.	.5	.135	.55	.59	.62	.66	.68	.71	.74	.76
1.0	60.	1.5	.135	.52	.55	.58	.61	.63	.66	.68	.70
1.0	60.	3.0	.135	.51	.54	.57	.59	.62	.64	.66	.68
3.0	5.	.0	.913	.69	.73	.77	.80	.83	.86	.88	.90
3.0	5.	.5	.913	.68	.72	.76	.79	.81	.84	.86	.88
3.0	5.	1.5	.913	.68	.72	.75	.78	.80	.83	.85	.86
3.0	5.	3.0	.913	.69	.72	.75	.78	.80	.82	.84	.86
3.0	30.	.0	.500	.68	.73	.77	.81	.84	.87	.90	.92
3.0	30.	.5	.500	.69	.73	.76	.79	.82	.85	.87	.89
3.0	30.	1.5	.500	.69	.72	.75	.78	.81	.83	.85	.86
3.0	30.	3.0	.500	.69	.72	.75	.78	.80	.82	.84	.85
3.0	60.	.0	.135	.66	.72	.76	.81	.85	.89	.92	.96
3.0	60.	.5	.135	.69	.73	.77	.80	.83	.86	.89	.91
3.0	60.	1.5	.135	.69	.73	.76	.79	.81	.83	.85	.87
3.0	60.	3.0	.135	.69	.72	.75	.78	.80	.82	.83	.85
5.0	5.	.0	.913	.73	.77	.81	.84	.87	.90	.92	.94
5.0	5.	.5	.913	.74	.78	.81	.84	.86	.89	.91	.93
5.0	5.	1.5	.913	.75	.78	.81	.84	.86	.88	.90	.92
5.0	5.	3.0	.913	.76	.79	.82	.84	.86	.88	.90	.91
5.0	30.	.0	.500	.72	.76	.80	.84	.87	.90	.93	.95
5.0	30.	.5	.500	.73	.77	.81	.84	.87	.89	.91	.93
5.0	30.	1.5	.500	.75	.78	.81	.84	.86	.88	.90	.92
5.0	30.	3.0	.500	.76	.79	.82	.84	.86	.88	.89	.91
5.0	60.	.0	.135	.68	.73	.78	.82	.87	.90	.94	.97
5.0	60.	.5	.135	.72	.77	.81	.84	.87	.90	.92	.95
5.0	60.	1.5	.135	.75	.79	.82	.84	.87	.89	.90	.92
5.0	60.	3.0	.135	.76	.79	.82	.84	.86	.88	.89	.90

ABSORPTION FACTORS
TILTED GLAZING - CL IIIb

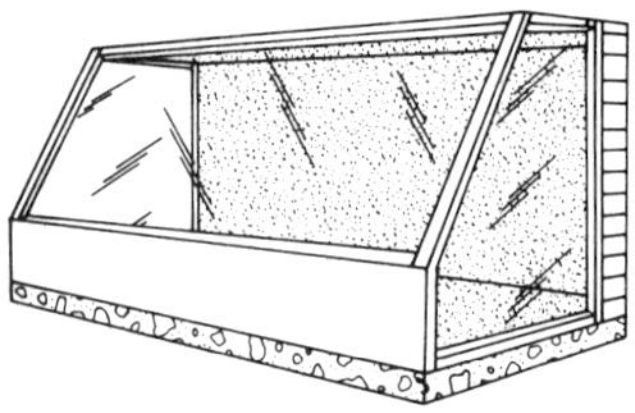

Table A4-42

OPAQUE BOTTOM SECTION, Z=1.0
GLAZED END WALLS
$\alpha_o = \alpha_m$

WDTH	TILT	XTOP	YBOT	ABSORPTION FOR RADIATION THRU SURFACE 6 α_m 0.3	0.4	0.5	0.6	0.7	0.8	0.9	1.0
1.0	5.	.0	.913	.54	.58	.61	.64	.67	.69	.72	.74
1.0	5.	.5	.913	.54	.58	.61	.64	.66	.69	.71	.73
1.0	5.	1.5	.913	.54	.58	.61	.64	.66	.69	.71	.73
1.0	5.	3.0	.913	.54	.58	.61	.64	.66	.69	.71	.73
1.0	30.	.0	.500	.54	.58	.61	.64	.67	.70	.72	.74
1.0	30.	.5	.500	.54	.58	.61	.64	.67	.69	.71	.73
1.0	30.	1.5	.500	.54	.58	.61	.64	.67	.69	.71	.73
1.0	30.	3.0	.500	.54	.58	.61	.64	.66	.69	.71	.73
1.0	60.	.0	.135	.53	.57	.60	.63	.66	.69	.72	.74
1.0	60.	.5	.135	.54	.58	.61	.64	.67	.69	.72	.74
1.0	60.	1.5	.135	.54	.58	.61	.64	.67	.69	.72	.74
1.0	60.	3.0	.135	.54	.58	.61	.64	.67	.69	.71	.73
3.0	5.	.0	.913	.64	.68	.71	.74	.76	.79	.81	.83
3.0	5.	.5	.913	.67	.71	.74	.77	.79	.81	.83	.85
3.0	5.	1.5	.913	.70	.74	.77	.80	.82	.84	.86	.87
3.0	5.	3.0	.913	.72	.76	.79	.81	.84	.86	.87	.89
3.0	30.	.0	.500	.62	.66	.69	.72	.75	.77	.79	.81
3.0	30.	.5	.500	.66	.69	.72	.75	.78	.80	.82	.83
3.0	30.	1.5	.500	.70	.73	.76	.79	.81	.83	.85	.87
3.0	30.	3.0	.500	.72	.75	.78	.81	.83	.85	.87	.88
3.0	60.	.0	.135	.57	.61	.64	.67	.70	.73	.75	.78
3.0	60.	.5	.135	.63	.67	.70	.73	.75	.78	.80	.82
3.0	60.	1.5	.135	.68	.72	.75	.78	.80	.82	.84	.86
3.0	60.	3.0	.135	.71	.75	.78	.80	.83	.85	.86	.88
5.0	5.	.0	.913	.66	.70	.73	.76	.78	.80	.82	.84
5.0	5.	.5	.913	.70	.74	.77	.79	.81	.83	.85	.86
5.0	5.	1.5	.913	.75	.78	.81	.83	.85	.87	.88	.90
5.0	5.	3.0	.913	.78	.81	.84	.86	.88	.89	.91	.92
5.0	30.	.0	.500	.63	.67	.70	.73	.76	.78	.80	.82
5.0	30.	.5	.500	.68	.72	.75	.77	.80	.82	.83	.85
5.0	30.	1.5	.500	.74	.77	.80	.82	.84	.86	.87	.89
5.0	30.	3.0	.500	.77	.80	.83	.85	.87	.89	.90	.91
5.0	60.	.0	.135	.57	.61	.65	.68	.71	.73	.76	.78
5.0	60.	.5	.135	.65	.68	.71	.74	.77	.79	.81	.83
5.0	60.	1.5	.135	.72	.75	.78	.81	.83	.85	.86	.88
5.0	60.	3.0	.135	.76	.80	.82	.84	.86	.88	.89	.90

ABSORPTION FACTORS
TILTED GLAZING - CL IIIb

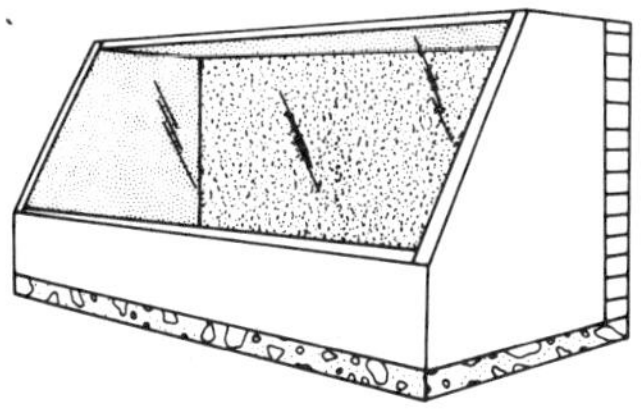

Table A4-43

OPAQUE BOTTOM SECTION, Z=1.0
OPAQUE END WALLS
α_o = .3

WDTH	TILT	XTOP	YBOT	ABSORPTION FOR RADIATION THRU SURFACE 3, α_m 0.3	0.4	0.5	0.6	0.7	0.8	0.9	1.0
1.0	5.	.0	.913	.87	.88	.90	.91	.92	.93	.94	.95
1.0	5.	.5	.913	.90	.91	.92	.92	.93	.94	.95	.95
1.0	5.	1.5	.913	.93	.93	.94	.94	.95	.95	.96	.96
1.0	5.	3.0	.913	.95	.95	.96	.96	.96	.97	.97	.97
1.0	30.	.0	.500	.83	.86	.88	.90	.92	.93	.95	.96
1.0	30.	.5	.500	.88	.90	.91	.92	.93	.94	.95	.96
1.0	30.	1.5	.500	.92	.93	.94	.95	.95	.96	.97	.97
1.0	30.	3.0	.500	.95	.95	.96	.96	.97	.97	.98	.98
1.0	60.	.0	.135	.75	.79	.83	.86	.89	.92	.95	.98
1.0	60.	.5	.135	.85	.87	.89	.90	.92	.93	.95	.96
1.0	60.	1.5	.135	.91	.92	.93	.94	.95	.96	.96	.97
1.0	60.	3.0	.135	.95	.95	.96	.96	.97	.97	.98	.98
3.0	5.	.0	.913	.83	.85	.88	.90	.92	.93	.95	.96
3.0	5.	.5	.913	.86	.88	.90	.91	.93	.94	.95	.96
3.0	5.	1.5	.913	.90	.91	.92	.93	.94	.95	.96	.97
3.0	5.	3.0	.913	.92	.93	.94	.95	.95	.96	.97	.97
3.0	30.	.0	.500	.79	.83	.86	.89	.91	.94	.96	.98
3.0	30.	.5	.500	.84	.87	.89	.91	.93	.95	.96	.98
3.0	30.	1.5	.500	.89	.91	.92	.94	.95	.96	.97	.98
3.0	30.	3.0	.500	.93	.94	.95	.96	.96	.97	.98	.99
3.0	60.	.0	.135	.72	.77	.81	.85	.89	.93	.96	.99
3.0	60.	.5	.135	.81	.84	.87	.90	.92	.94	.96	.98
3.0	60.	1.5	.135	.88	.90	.92	.93	.95	.96	.97	.98
3.0	60.	3.0	.135	.92	.94	.95	.96	.96	.97	.98	.99
5.0	5.	.0	.913	.82	.85	.87	.90	.92	.94	.95	.97
5.0	5.	.5	.913	.85	.87	.89	.91	.93	.94	.96	.97
5.0	5.	1.5	.913	.89	.90	.92	.93	.94	.95	.96	.97
5.0	5.	3.0	.913	.92	.93	.94	.94	.95	.96	.97	.97
5.0	30.	.0	.500	.78	.82	.86	.89	.91	.94	.96	.99
5.0	30.	.5	.500	.83	.86	.89	.91	.93	.95	.97	.99
5.0	30.	1.5	.500	.88	.90	.92	.93	.95	.96	.97	.99
5.0	30.	3.0	.500	.92	.93	.94	.95	.96	.97	.98	.99
5.0	60.	.0	.135	.71	.76	.81	.85	.89	.93	.96	.99
5.0	60.	.5	.135	.80	.83	.87	.89	.92	.94	.97	.99
5.0	60.	1.5	.135	.87	.89	.91	.93	.95	.96	.97	.99
5.0	60.	3.0	.135	.92	.93	.94	.95	.96	.97	.98	.99

ABSORPTION FACTORS
TILTED GLAZING - CL IIIb

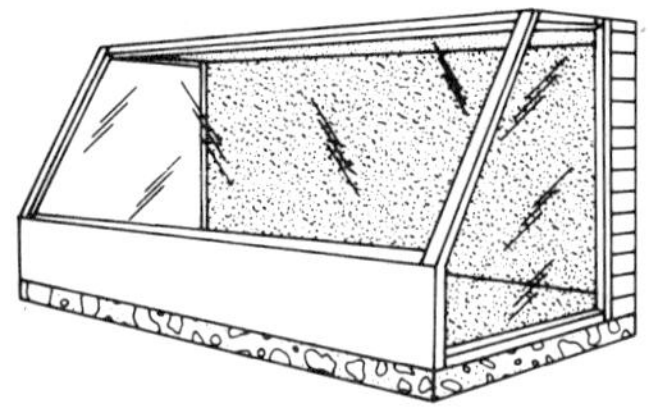

Table A4-44

OPAQUE BOTTOM SECTION, Z=1.0
GLAZED END WALLS
α_o = .3

WDTH	TILT	XTOP	YBOT	ABSORPTION FOR RADIATION THRU SURFACE 3 α_m 0.3	0.4	0.5	0.6	0.7	0.8	0.9	1.0
1.0	5.	.0	.913	.54	.57	.59	.62	.64	.66	.68	.70
1.0	5.	.5	.913	.52	.54	.56	.58	.60	.61	.63	.65
1.0	5.	1.5	.913	.50	.52	.54	.56	.58	.59	.61	.63
1.0	5.	3.0	.913	.50	.52	.54	.55	.57	.59	.60	.62
1.0	30.	.0	.500	.57	.60	.63	.67	.69	.72	.75	.77
1.0	30.	.5	.500	.53	.56	.58	.61	.63	.65	.67	.70
1.0	30.	1.5	.500	.51	.53	.55	.58	.60	.62	.64	.65
1.0	30.	3.0	.500	.50	.52	.54	.57	.59	.61	.62	.64
1.0	60.	.0	.135	.60	.64	.69	.73	.77	.81	.84	.87
1.0	60.	.5	.135	.55	.58	.61	.64	.67	.70	.72	.74
1.0	60.	1.5	.135	.52	.54	.57	.59	.61	.63	.65	.67
1.0	60.	3.0	.135	.51	.53	.55	.57	.59	.61	.63	.65
3.0	5.	.0	.913	.69	.72	.75	.78	.80	.82	.85	.87
3.0	5.	.5	.913	.68	.71	.74	.76	.78	.80	.82	.84
3.0	5.	1.5	.913	.68	.71	.73	.75	.77	.79	.81	.83
3.0	5.	3.0	.913	.69	.71	.73	.75	.77	.79	.80	.82
3.0	30.	.0	.500	.68	.72	.76	.80	.83	.86	.89	.91
3.0	30.	.5	.500	.69	.72	.75	.78	.81	.83	.86	.88
3.0	30.	1.5	.500	.69	.71	.74	.77	.79	.81	.83	.85
3.0	30.	3.0	.500	.69	.71	.74	.76	.78	.80	.82	.84
3.0	60.	.0	.135	.66	.71	.76	.81	.85	.89	.92	.96
3.0	60.	.5	.135	.69	.72	.76	.79	.82	.85	.88	.90
3.0	60.	1.5	.135	.69	.72	.75	.77	.79	.82	.84	.85
3.0	60.	3.0	.135	.69	.71	.74	.76	.78	.80	.81	.83
5.0	5.	.0	.913	.73	.76	.79	.82	.84	.87	.89	.91
5.0	5.	.5	.913	.74	.76	.79	.81	.84	.86	.88	.89
5.0	5.	1.5	.913	.75	.77	.79	.81	.83	.85	.87	.88
5.0	5.	3.0	.913	.76	.78	.80	.82	.83	.85	.86	.88
5.0	30.	.0	.500	.72	.76	.80	.83	.86	.89	.92	.94
5.0	30.	.5	.500	.73	.77	.80	.83	.85	.88	.90	.92
5.0	30.	1.5	.500	.75	.78	.80	.83	.85	.87	.89	.90
5.0	30.	3.0	.500	.76	.78	.81	.83	.84	.86	.88	.89
5.0	60.	.0	.135	.68	.73	.78	.82	.86	.90	.94	.97
5.0	60.	.5	.135	.72	.76	.80	.83	.86	.89	.91	.94
5.0	60.	1.5	.135	.75	.78	.80	.83	.85	.87	.89	.91
5.0	60.	3.0	.135	.76	.78	.80	.82	.84	.86	.87	.89

ABSORPTION FACTORS
TILTED GLAZING - CL IIIb

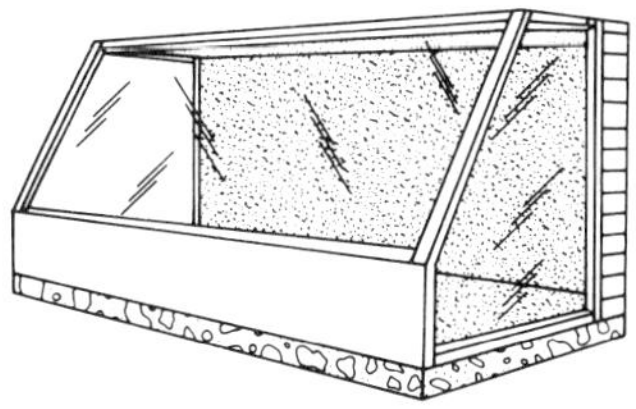

Table A4-45

OPAQUE BOTTOM SECTION, Z=1.0
GLAZED END WALLS
$\alpha_o = .3$

WDTH	TILT	XTOP	YBOT	ABSORPTION FOR RADIATION THRU SURFACE 6 α_m 0.3	0.4	0.5	0.6	0.7	0.8	0.9	1.0
1.0	5.	.0	.913	.54	.57	.59	.61	.63	.66	.67	.69
1.0	5.	.5	.913	.54	.57	.59	.61	.63	.65	.66	.68
1.0	5.	1.5	.913	.54	.56	.58	.60	.62	.64	.66	.67
1.0	5.	3.0	.913	.54	.56	.58	.60	.62	.64	.65	.67
1.0	30.	.0	.500	.54	.57	.60	.63	.65	.67	.70	.72
1.0	30.	.5	.500	.54	.57	.60	.62	.64	.66	.68	.70
1.0	30.	1.5	.500	.54	.57	.59	.61	.63	.65	.67	.68
1.0	30.	3.0	.500	.54	.56	.58	.60	.62	.64	.66	.67
1.0	60.	.0	.135	.53	.57	.60	.63	.66	.69	.71	.74
1.0	60.	.5	.135	.54	.57	.60	.62	.65	.67	.69	.71
1.0	60.	1.5	.135	.54	.57	.59	.61	.63	.65	.67	.69
1.0	60.	3.0	.135	.54	.56	.58	.60	.62	.64	.66	.68
3.0	5.	.0	.913	.64	.67	.69	.71	.74	.76	.77	.79
3.0	5.	.5	.913	.67	.69	.72	.74	.76	.78	.79	.81
3.0	5.	1.5	.913	.70	.73	.75	.77	.78	.80	.82	.83
3.0	5.	3.0	.913	.72	.74	.76	.78	.80	.82	.83	.84
3.0	30.	.0	.500	.62	.65	.68	.70	.73	.75	.77	.79
3.0	30.	.5	.500	.66	.68	.71	.73	.75	.77	.79	.81
3.0	30.	1.5	.500	.70	.72	.74	.76	.78	.80	.82	.83
3.0	30.	3.0	.500	.72	.74	.76	.78	.80	.82	.83	.85
3.0	60.	.0	.135	.57	.60	.64	.67	.70	.72	.75	.77
3.0	60.	.5	.135	.63	.66	.69	.71	.74	.76	.78	.80
3.0	60.	1.5	.135	.68	.71	.73	.75	.77	.79	.81	.83
3.0	60.	3.0	.135	.71	.74	.76	.78	.79	.81	.83	.84
5.0	5.	.0	.913	.66	.69	.71	.73	.75	.77	.79	.81
5.0	5.	.5	.913	.70	.72	.75	.77	.78	.80	.82	.83
5.0	5.	1.5	.913	.75	.77	.79	.81	.82	.84	.85	.87
5.0	5.	3.0	.913	.78	.80	.82	.83	.85	.86	.87	.89
5.0	30.	.0	.500	.63	.66	.69	.72	.74	.76	.79	.80
5.0	30.	.5	.500	.68	.71	.73	.76	.78	.80	.81	.83
5.0	30.	1.5	.500	.74	.76	.78	.80	.82	.84	.85	.86
5.0	30.	3.0	.500	.77	.79	.81	.83	.84	.86	.87	.89
5.0	60.	.0	.135	.57	.61	.64	.67	.70	.73	.75	.78
5.0	60.	.5	.135	.65	.68	.70	.73	.75	.77	.79	.81
5.0	60.	1.5	.135	.72	.74	.77	.79	.81	.82	.84	.85
5.0	60.	3.0	.135	.76	.78	.80	.82	.84	.85	.86	.88

REFERENCE DESIGNS

The absorption factors (*ABS/TRN*) for thermal storage wall and direct-gain reference design systems are listed in Table A4-46.

The values of the coefficients for the function that yields *S/HOR* are listed in Table A4-47. This correlation is valid only for sunspace reference designs oriented due south. Note that the factor A_n/A_p is included in the function and need not be calculated separately.

Table A4-46
Absorption Factors, *ABS/TRN*, for Reference Designs

Water wall and Trombe wall	1 glazing	0.956
	2 glazings	0.962
	3 glazings	0.964
Direct gain	$A_m/A_n = 3$	0.948
	$A_m/A_n = 6$	0.976

A_m/A_n = mass-area-to-glazing-area ratio.

Table A4-47
Solar Radiation Correlation Coefficients for the Sunspace Reference Designs

$$S/HOR = (B_1 + B_2 Y + B_3 Y^2) + K_T(B_4 + B_5 Y + B_6 Y^2)$$

Type	B_1	B_2	B_3	B_4	B_5	B_6
A1, 2, 5, 6	.72008	−.15181	.49973	−.15039	.14384	3.6374
A3, 4, 7, 8	.81554	−.23988	.60252	−.16445	.33730	3.1695
B1, 2, 5, 6	.58932	−.09693	.38955	−.14699	−.39149	3.9171
B3, 4, 7, 8	.62569	−.13941	.43331	−.14982	−.26401	3.5685
C1, 2, 3, 4	.39436	−.21103	.58815	−.24083	−.60746	4.6546
D1, 2, 3, 4	.73147	−.15418	.50763	−.15276	.14608	3.6950
E1, 2, 3, 4	.61661	−.10127	.40733	−.15367	−.40940	4.0969

OVERHANG SHADING

The geometry for an overhang for a collector on a vertical, south-facing wall is shown in Figure A4-49. The overhang partially shades the collector. The ratio VS_O/VS represents the fraction of insolation on the collector relative to that when there is no overhang. The correlation of this ratio with $LAT - DEC$ and K_T has the form:

Y is $(LAT - DEC)/100$ at midmonth

For $Y < Y^*$:

$$VS_O/VS = B_1 + B_2Y + B_3Y^2 + B_4Y^3 + B_5Y^4 + B_6Y^5 + K_T(B_7 + B_8Y + B_9Y^2 + B_{10}Y^3 + B_{11}Y^4 + B_{12}Y^5)$$

For $Y > Y^*$: $VS_O/VS = O$

If the correlation equation returns a value of VS_O/VS greater than one, then set VS_O/VS equal to one.

The values of the coefficients for the overhang shading correlation are tabulated in Table A4-48 for overhang extension ratios (OX) of .25, .50, .75, 1.0 and 1.5 and for overhang separation ratios (OY) of .25, .50, .75, and 1.0. Table A4-48 also lists the values of Y^*. Each pair of values of OX and OY is plotted (VS_O/VS versus $LAT - DEC$) with K_T = 0.4, 0.5, 0.6, 0.7, and 0.8 as a parameter in Figures A4-50 through A4-74.

Figure A4-75 shows the region of K_T and $L - D$ covered by the data points used in the solar radiation correlations in this appendix. Extrapolations should not be made far beyond the boundaries of the envelope of points. In many of the cases where the overhang ratio, VS_O/VS, does not go to one at the upper end of the $LAT - DEC$ range, even a small extrapolation of the correlations can result in large errors.

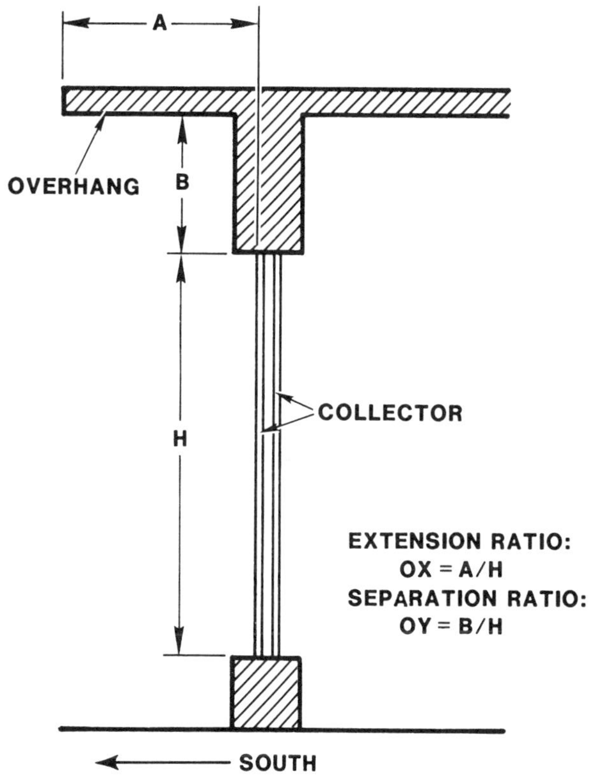

Figure A4-49. Geometry for an Overhang for a Collector on a Vertical, South-Facing Wall

OVERHANG SHADING

Table A4-48
Values of the Coefficients for the Overhang Shading Correlation

OX	OY	B_1	B_2	B_3	B_4	B_5	B_6	Y*
		B_7	B_8	B_9	B_{10}	B_{11}	B_{12}	
0.25	0.00	0.7281	3.9262	-26.6419	69.5451	-78.2566	32.1157	90.
		0.4941	-13.8569	78.1489	-180.6540	188.6530	-73.5015	
0.50	0.00	0.5551	5.1713	-32.2750	76.9408	-78.8114	29.7241	90.
		0.6927	-13.6265	56.7151	-91.7390	61.3059	-13.0622	
0.75	0.00	0.3857	8.7261	-58.9104	154.6300	-177.0950	74.4216	90.
		0.9550	-20.1666	107.1410	-255.3420	287.9880	-122.6010	
1.00	0.00	0.2857	9.9443	-66.2707	173.5790	-200.1370	85.0397	90.
		1.1452	-24.5673	141.5310	-367.0250	441.0610	-195.5890	
1.50	0.00	0.2194	7.7159	-40.8547	83.5634	-75.6814	26.5294	90.
		1.2934	-23.6897	121.5450	-283.8900	311.7750	-128.7470	
0.25	0.25	0.8500	4.6465	-40.8144	135.3970	-190.4430	96.3633	49.
		0.3836	-16.3144	116.2410	-329.4680	415.9500	-195.2740	
0.50	0.25	0.7535	2.8380	-15.2823	21.2495	7.6722	-20.2384	66.
		0.4812	-11.5313	40.4274	-16.5106	-82.9559	79.9048	
0.75	0.25	0.5826	6.3882	-43.8114	113.0720	-124.6370	50.3288	81.
		0.7221	-16.9205	82.6348	-171.7110	170.3790	-66.8862	
1.00	0.25	0.4470	9.0660	-63.1005	167.5780	-191.7190	79.8104	90.
		0.9407	-22.5284	126.6800	-314.9410	366.7640	-159.7280	
1.50	0.25	0.3176	9.0955	-57.0689	139.1890	-149.4490	59.7930	90.
		1.1894	-26.3865	152.2920	-393.1750	469.8690	-207.3780	
0.25	0.50	0.9162	2.9304	-26.5806	95.7477	-148.1040	82.1717	35.
		0.3215	-15.9641	130.1260	-418.8460	593.1580	-307.9640	
0.50	0.50	0.8095	4.4806	-38.9056	126.7510	-171.1380	81.5447	47.
		0.4289	-14.8203	82.1342	-168.6220	132.7850	-25.2688	
0.75	0.50	0.6657	7.5799	-61.4325	184.5390	-231.9470	104.4100	57.
		0.6277	-19.0116	109.3870	-262.7640	293.4460	-127.0600	
1.00	0.50	0.5856	7.2327	-51.8243	140.7120	-166.1630	74.3529	65.
		0.7744	-20.7548	120.2070	-317.0530	417.3230	-219.7340	
1.50	0.50	0.4052	9.9269	-67.9091	176.9200	-198.3550	81.0061	90.
		1.0679	-26.8582	160.7400	-422.0080	510.0010	-227.0370	
0.25	0.75	0.9156	3.8247	-34.7693	125.4690	-196.5100	111.0740	25.
		0.3077	-16.6089	147.8720	-511.0290	766.7490	-417.5790	
0.50	0.75	0.9165	0.4706	-0.8015	-3.0119	13.7424	-13.0055	39.
		0.3304	-12.4014	66.3807	-120.0630	59.2954	17.7186	
0.75	0.75	0.7589	5.8347	-51.0743	166.6980	-225.8210	108.3190	48.
		0.4867	-15.5836	80.3630	-147.7120	90.4951	0.3810	
1.00	0.75	0.5786	12.4017	-110.9460	372.5210	-529.7810	271.4760	54.
		0.7759	-27.0499	190.0540	-571.6530	793.5630	-414.9070	
1.50	0.75	0.5160	8.2211	-56.9416	151.3470	-178.0690	81.1941	65.
		0.9470	-26.2855	169.2090	-499.0910	711.4030	-390.2420	
0.25	1.00	0.9376	2.9397	-27.6771	102.6890	-164.3060	94.4244	23.
		0.2224	-12.1874	112.6920	-401.8750	618.1730	-343.3890	
0.50	1.00	0.9120	2.1215	-16.1251	52.2185	-74.6587	38.7509	34.
		0.3932	-18.4521	136.8670	-400.3890	517.5510	-246.6600	
0.75	1.00	0.8901	0.2724	0.4851	-7.3383	23.7221	-21.3946	41.
		0.3304	-9.4658	25.6585	50.7357	-226.3070	184.3290	
1.00	1.00	0.7089	7.6602	-67.3991	221.7140	-305.4500	150.2520	49.
		0.5606	-18.1618	100.1540	-213.4100	190.9060	-56.8564	
1.50	1.00	0.4567	15.9030	-139.2110	459.5780	-645.0110	327.1730	56.
		1.0312	-36.9291	282.7090	-917.7240	1343.6900	-726.0650	

OVERHANG SHADING

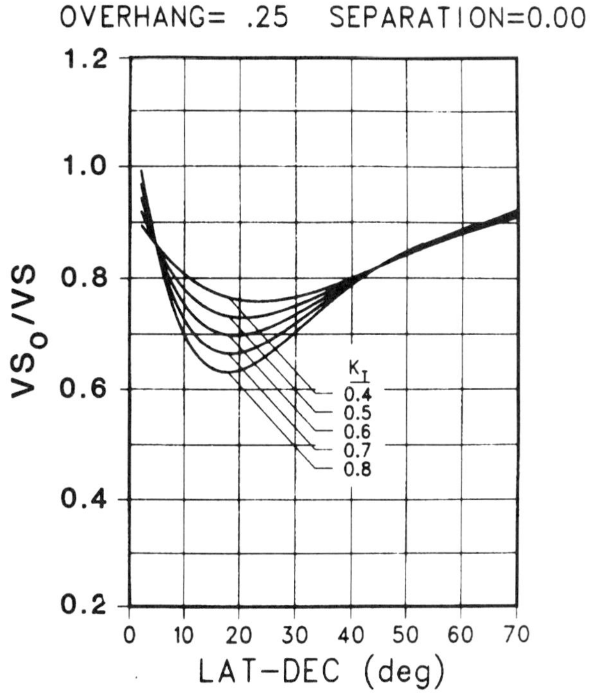

Figure A4-50

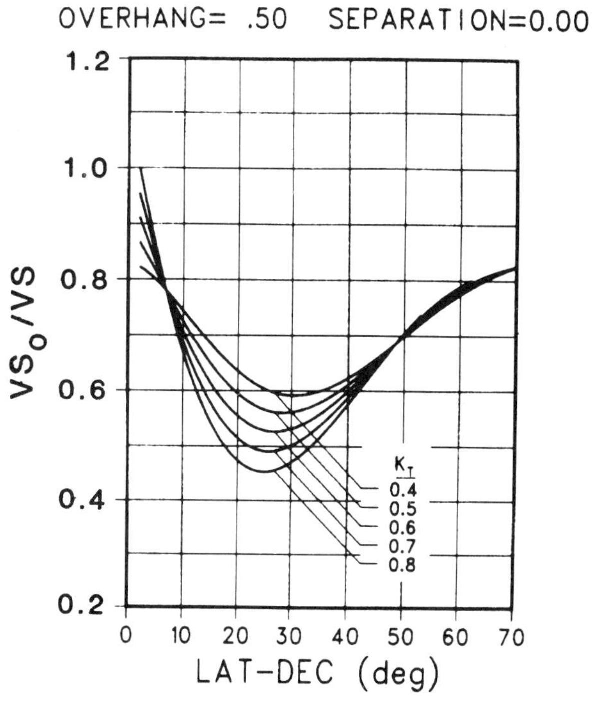

Figure A4-51

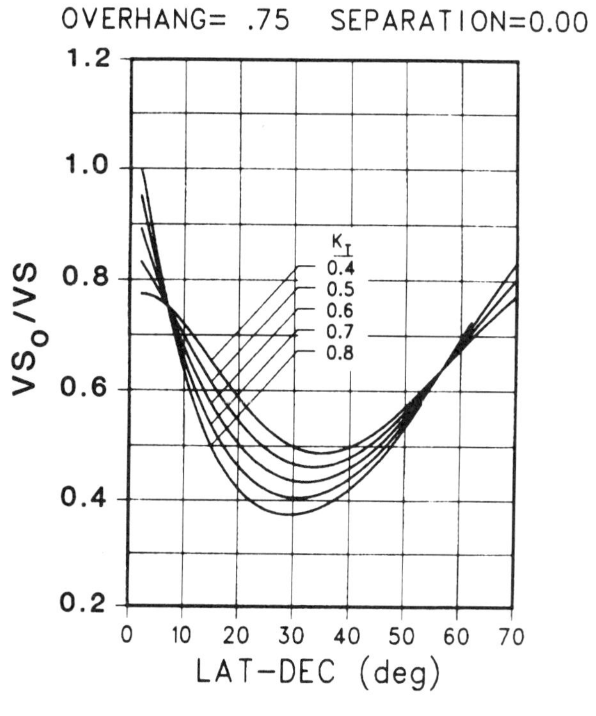

Figure A4-52

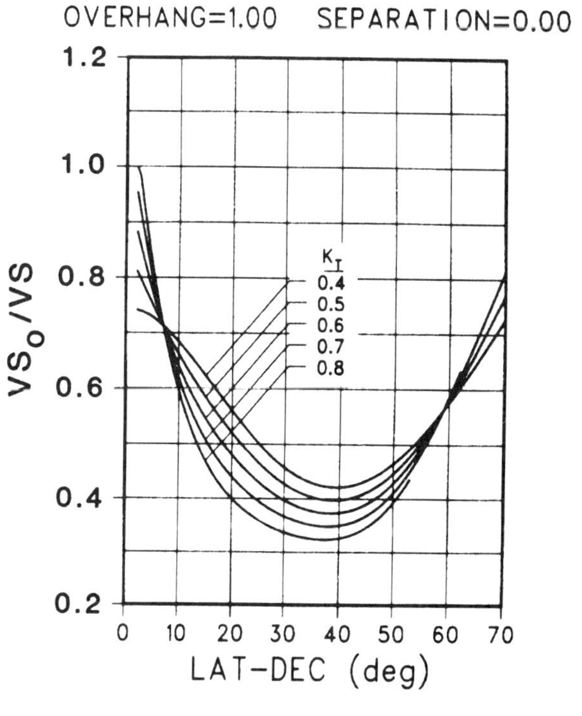

Figure A4-53

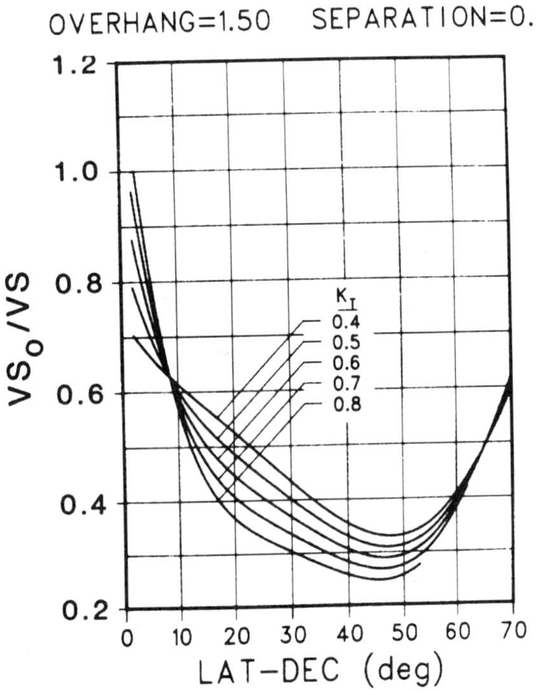

Figure A4-54

Figures A4-50 through A4-54. The ratio VS_0/VS versus $LAT - DEC$ for overhang ratios .25, .50, .75, 1.00, and 1.50 and for a separation ratio of 0.00.

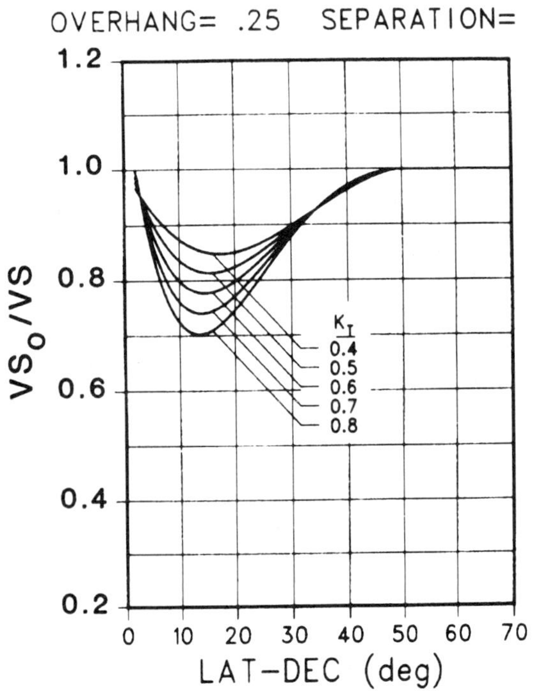

Figure A4-55

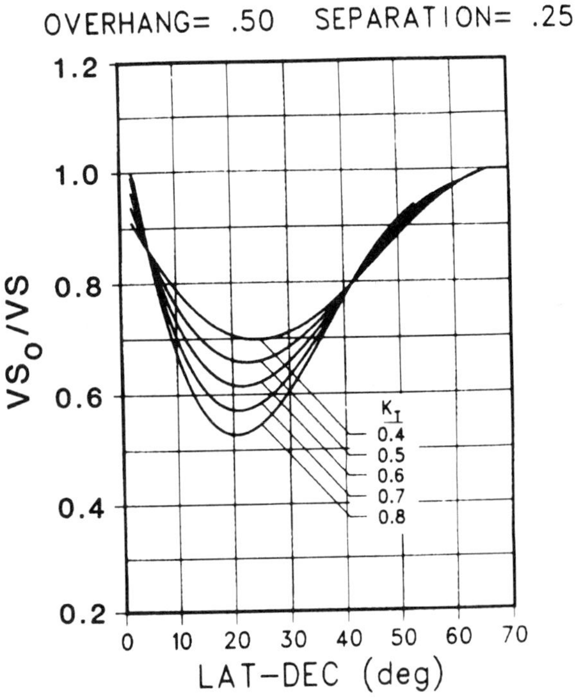

Figure A4-56

OVERHANG SHADING

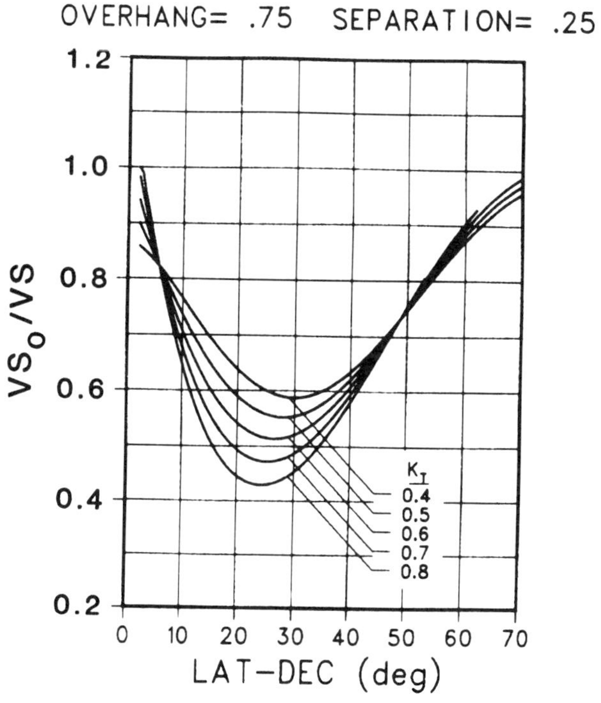

Figure A4-57

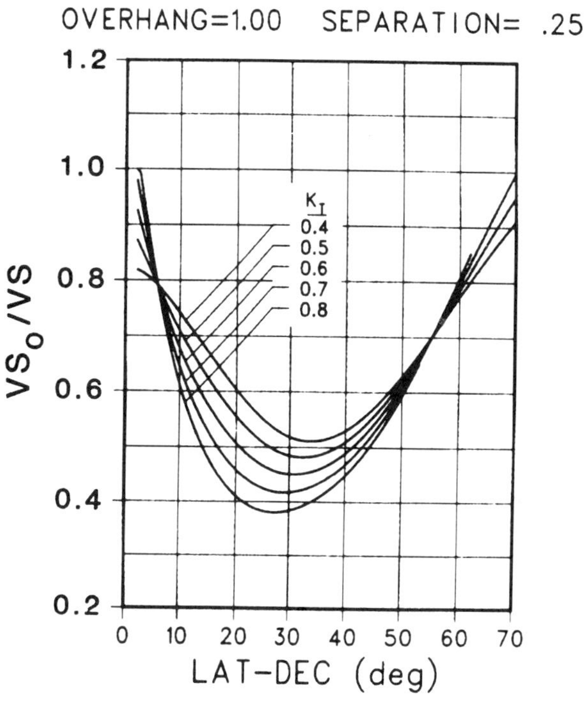

Figure A4-58

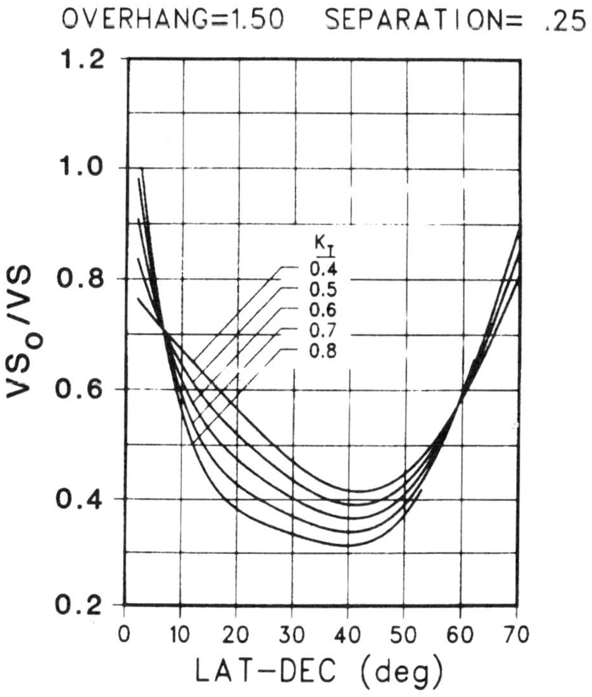

Figure A4-59

Figures A4-55 through A4-59. The ratio VS_o/VS versus $LAT - DEC$ for overhang ratios .25, .50, .75, 1.00, and 1.50 and for a separation ratio of .25.

OVERHANG SHADING

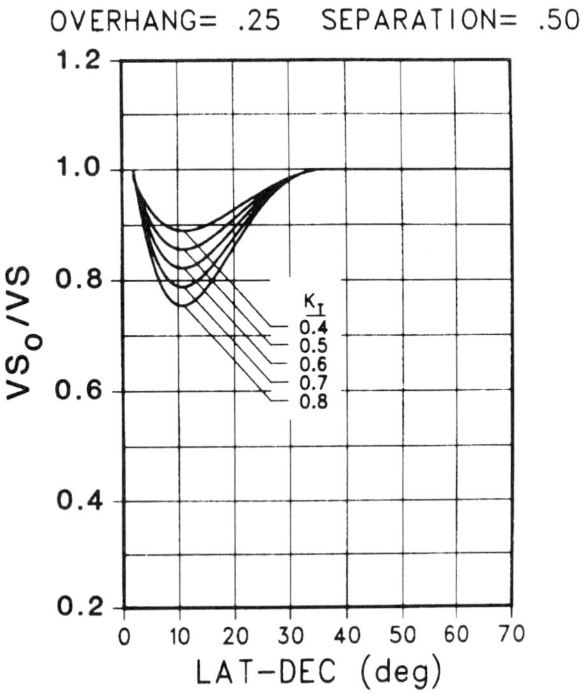

Figure A4-60

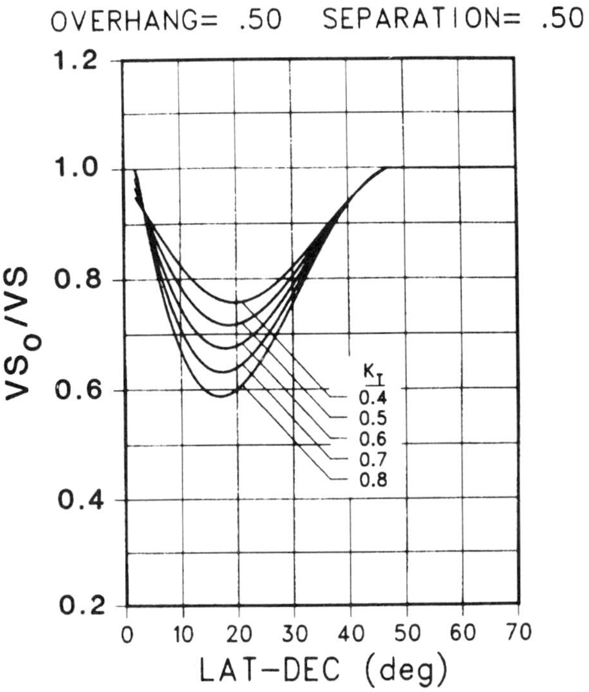

Figure A4-61

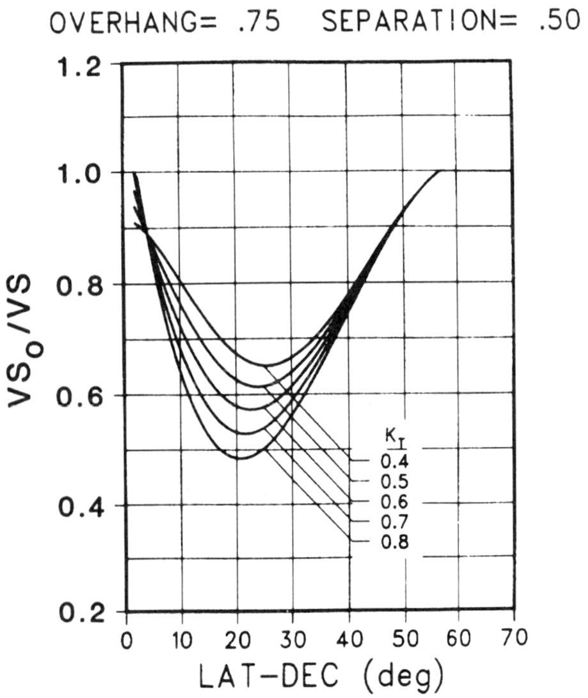

Figure A4-62

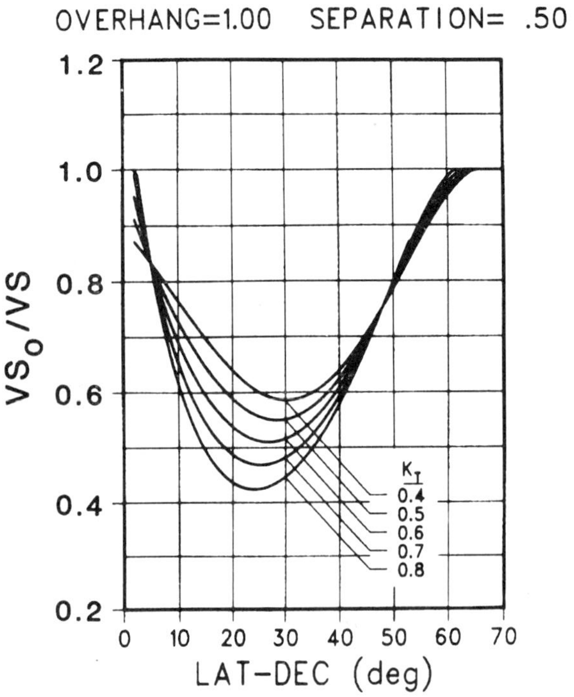

Figure A4-63

OVERHANG SHADING

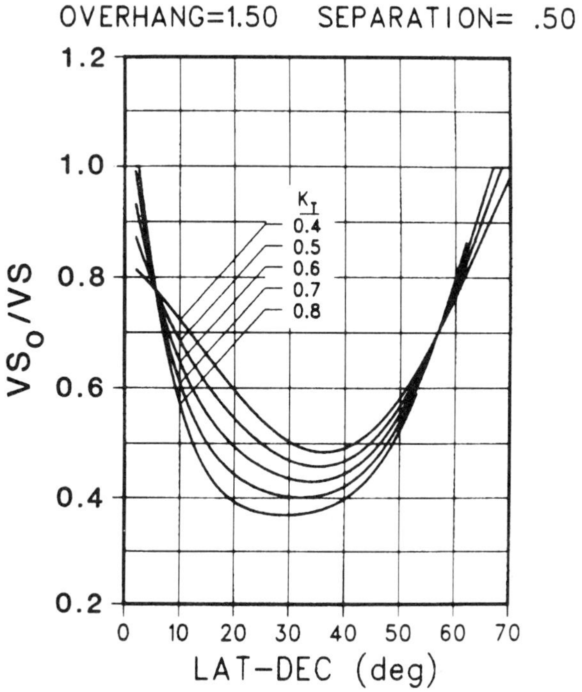

Figure A4-64

Figures A4-60 through A4-64. The ratio VS_0/VS versus $LAT - DEC$ for overhang ratios .25, .50, .75, 1.00, and 1.50 and for a separation ratio of .50.

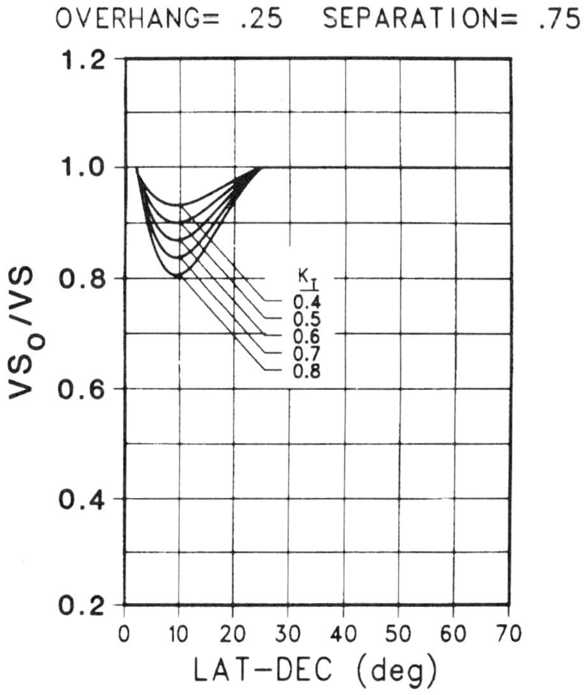

Figure A4-65

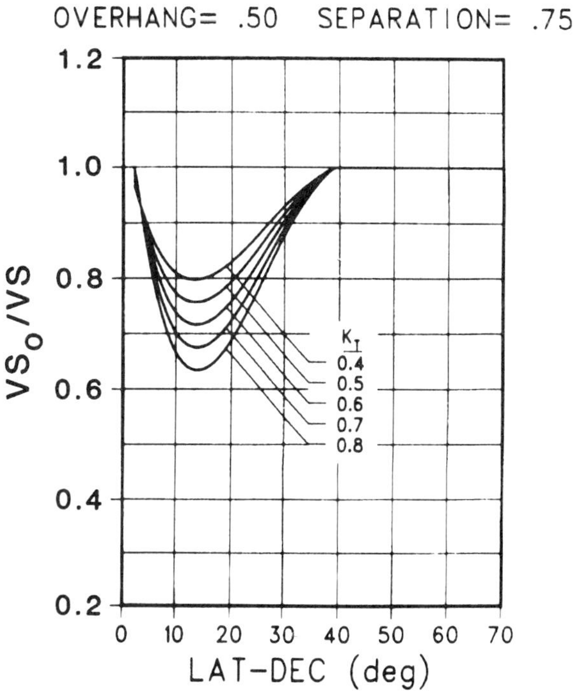

Figure A4-66

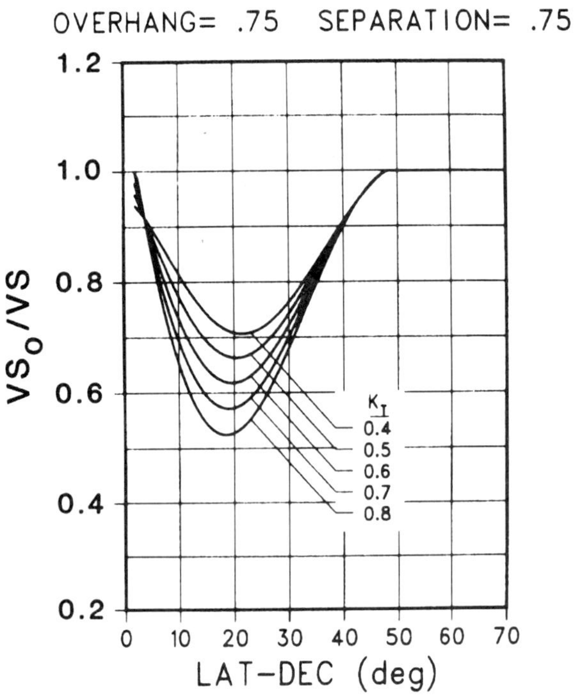

Figure A4-67

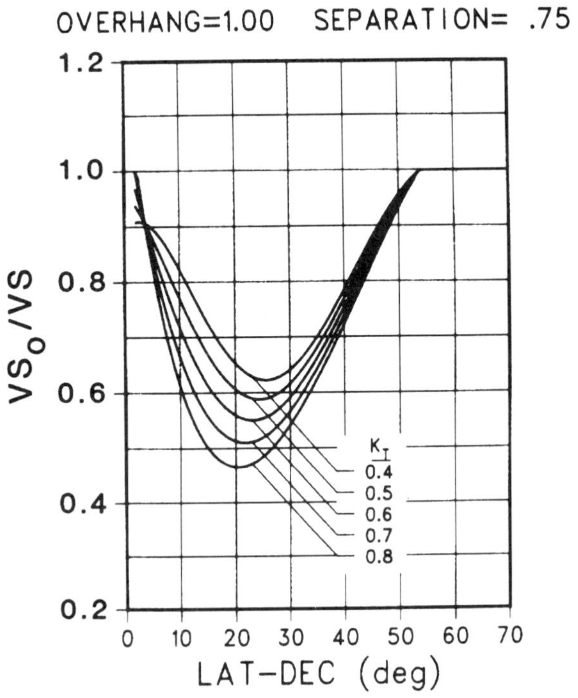

Figure A4-68

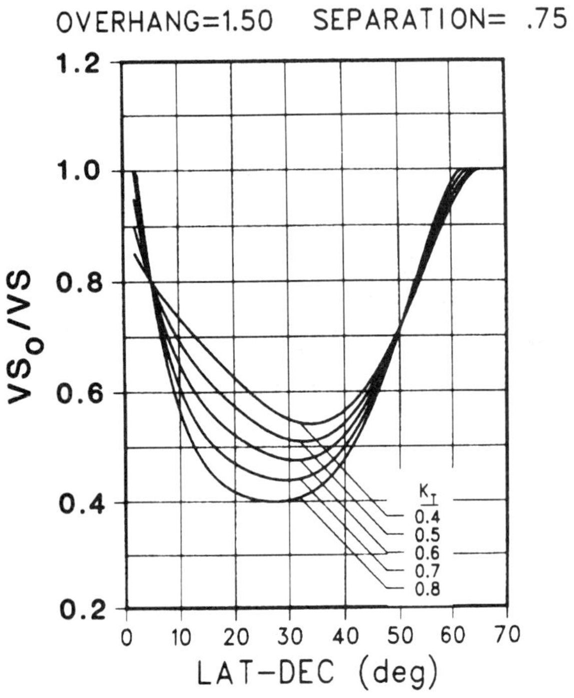

Figure A4-69

Figures A4-65 through A4-69. The ratio VS_0/VS versus $LAT - DEC$ for overhang ratios .25, .50, .75, 1.00, and 1.50 and for a separation ratio of .75.

OVERHANG SHADING

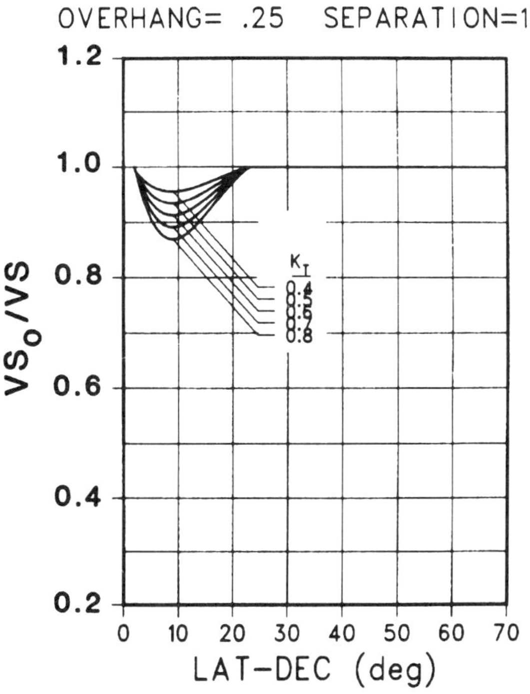

Figure A4-70

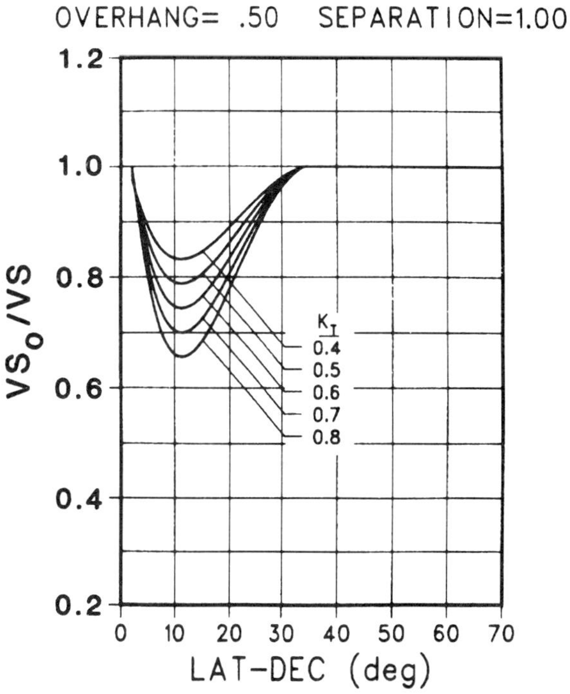

Figure A4-71

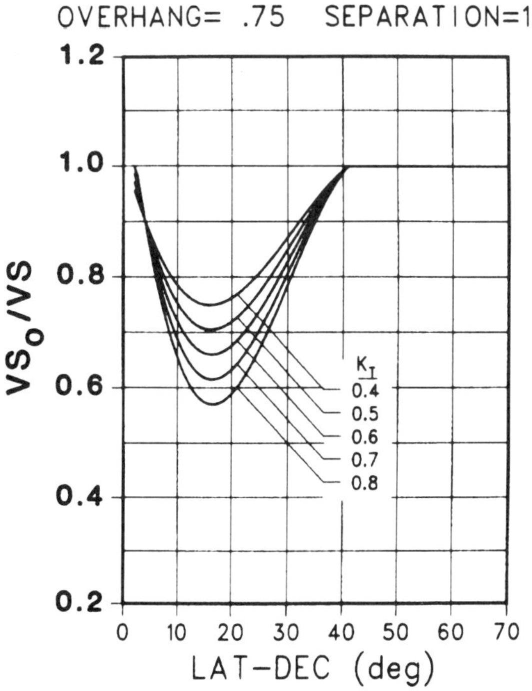

Figure A4-72

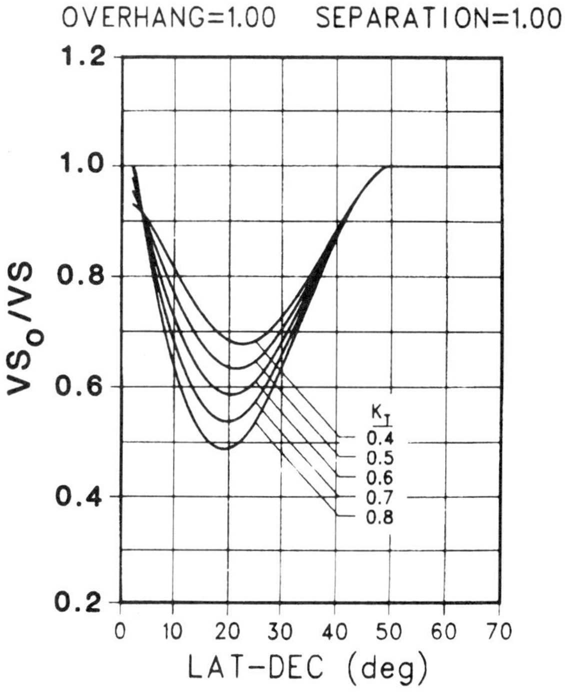

Figure A4-73

OVERHANG SHADING

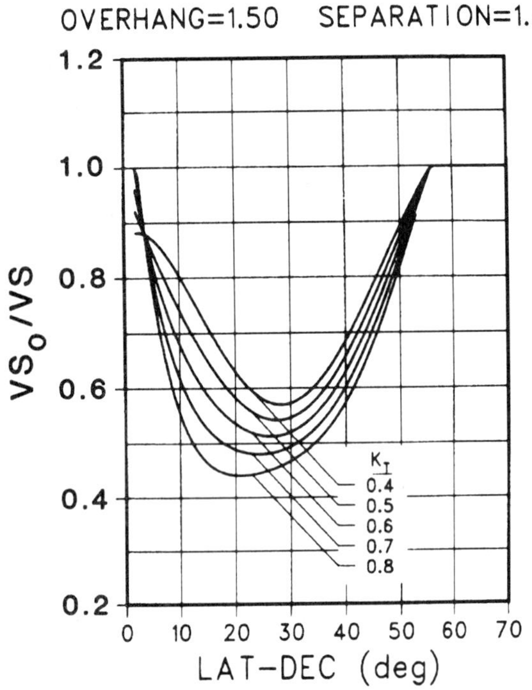

Figures A4-70 through A4-74. The ratio VS_0/VS versus $LAT - DEC$ for overhang ratios .25, .50, .75, 1.00, and 1.50 and for a separation ratio of 1.00.

Figure A4-74

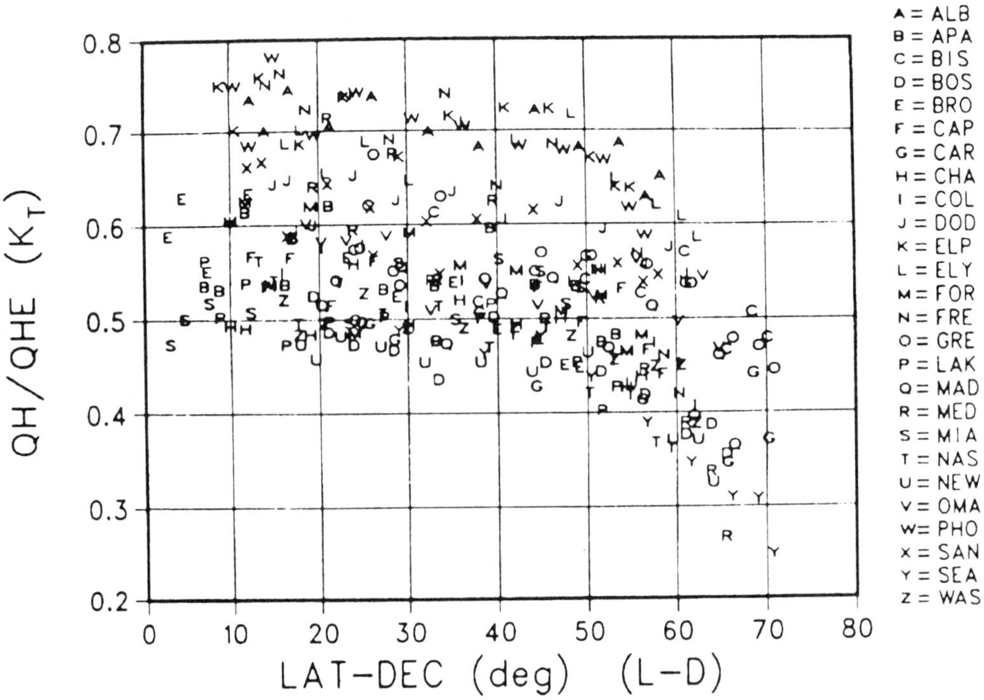

Figure A4-75. Domain of Data for Solar Radiation Correlations

SPECULAR REFLECTOR

Table A4-49
Effect of a Specular Reflector on the Solar Radiation Incident on a South-Facing Vertical Wall as the Aspect Ratio (AR) Is Varied

$$INC_0/INC = B_0 + B_1Y + B_2Y^2 + B_3Y^3 + B_4Y^4$$

AR	B_0	B_1	B_2	B_3	B_4
0.25	0.9883	0.4197	0.9727	-3.5890	2.4321
0.50	0.9812	0.8416	0.4462	-3.7248	2.6430
1.0	0.9889	0.8672	2.2580	-8.4287	5.6815
2.0	1.0229	0.2995	6.7949	-18.097	11.831
3.0	1.0714	-0.5427	12.280	-29.309	18.970
10.0	1.0913	-0.8873	14.449	-33.709	21.774

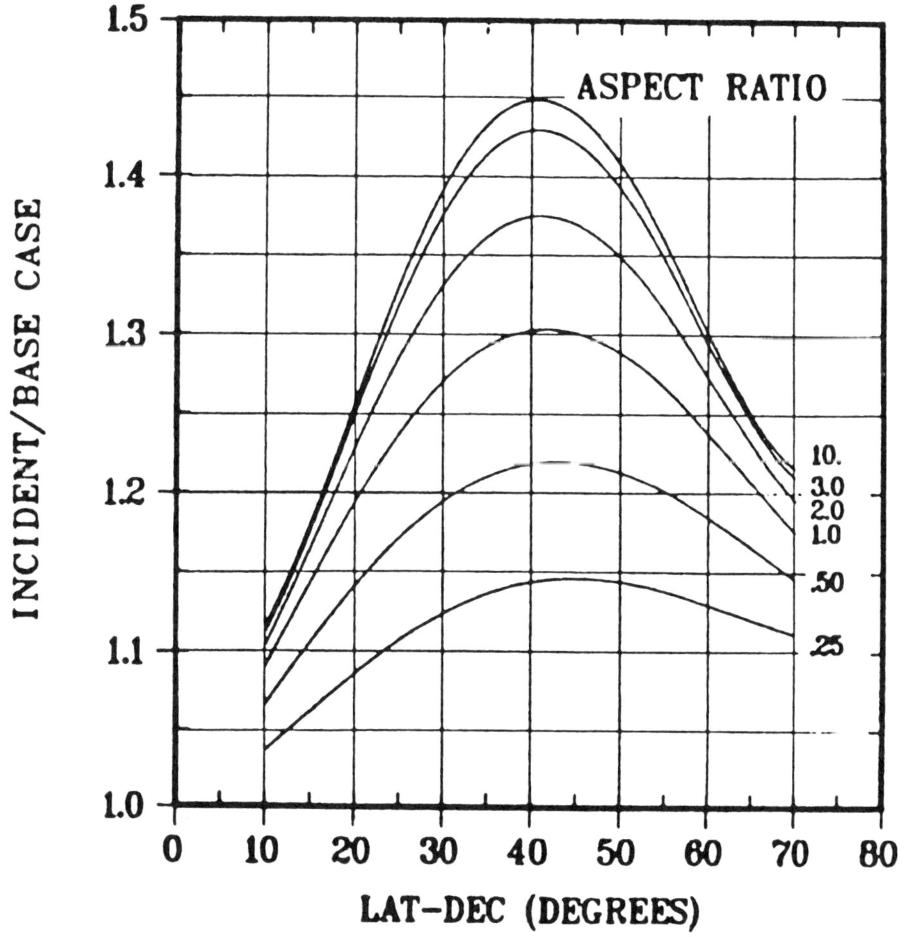

Figure A4-76. Effect of a Specular Reflector for Varying Aspect Ratios

Table A4-50
Effect of a Specular Reflector on the Solar Radiation Incident on a South-Facing Vertical Wall as the Ratio of Reflector Length to Collector Height (L/H) Is Varied

$$INC_0/INC = B_0 + B_1 Y + B_2 Y^2 + B_3 Y^3 + B_4 Y^4$$

L/H	B_0	B_1	B_2	B_3	B_4
0.25	0.9165	3.3714	-12.940	18.424	-9.1188
0.50	0.7402	4.9880	-14.391	15.554	-5.8934
1.0	1.0201	0.2857	8.003	-20.686	13.064
1.50	0.9904	0.9256	2.996	-7.216	2.7989
2.0	0.8761	2.6673	-5.678	9.237	-7.2095
3.0	0.9313	1.6104	0.3349	-4.131	3.0078

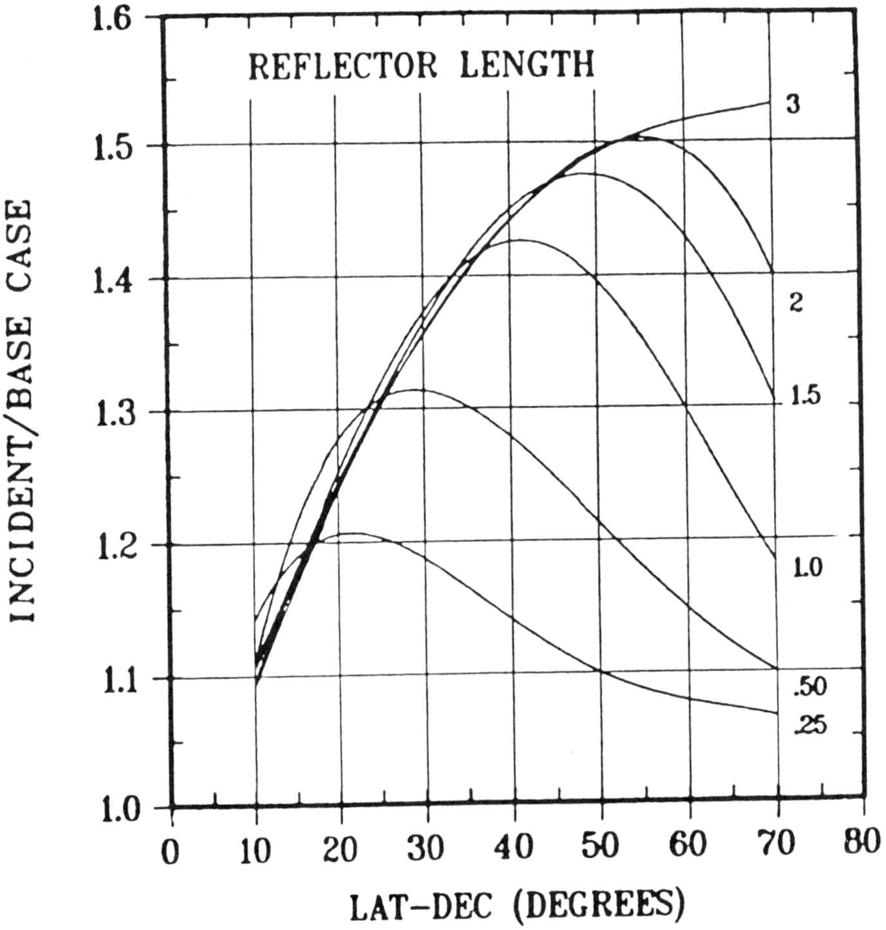

Figure A4-77. Effect of a Specular Reflector for Varying Ratios of Reflector Length to Collector Height

Table A4-51
Effect of a Specular Reflector on the Solar Radiation Incident on a South-Facing Vertical Wall as the Reflectance (SR) of the Reflector Is Varied

$$INC_0/INC = B_0 + B_1Y + B_2Y^2 + B_3Y^3 + B_4Y^4$$

SR	B_0	B_1	B_2	B_3	B_4
0	0.7249	0.3449	1.1681	-3.2025	2.1472
.2	0.8115	0.1238	3.9423	-9.7220	6.3482
.4	0.8982	-0.0988	6.7234	-16.255	10.558
.6	0.9848	-0.3209	9.5017	-22.782	14.764
.8	1.0714	-0.5427	12.280	-29.309	18.970
1.0	1.1580	-0.7640	15.054	-35.829	23.171

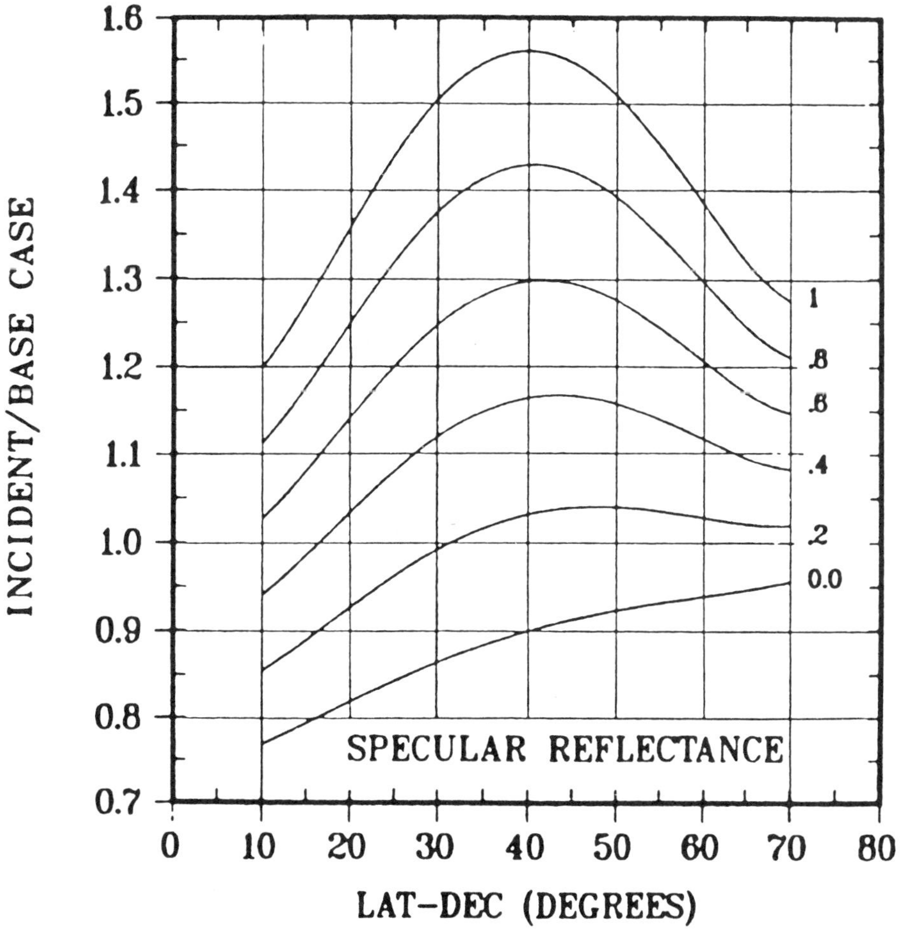

Figure A4-78. Effect of a Specular Reflector for Varying Values of Reflectance

APPENDIX 5
SLR CORRELATIONS

This appendix assembles the reference design specifications, the values of the coefficients for the correlation function, and graphs of the correlation function. The information is presented in the following order:

The form of the correlation function (*SSF* versus *SLR*) for thermal storage wall and sunspace systems is

$$SSF = 1 - C \exp(-D \cdot SLR) \qquad \text{(A5-1)}$$

where

$$SLR = (S/DD - LCR_s \cdot H)/LCR$$

The values of *C, D, H,* and LCR_s for thermal storage wall systems are listed in Table A5-7 and for sunspace systems in Table A5-8. Graphs of *SSF* versus *SLR* for thermal storage wall systems are shown in Figures A5-1 through A5-13 and for sunspace systems in Figures A5-14 through A5-18.

The form of the correlation function (*SSF* versus *SLR*) for direct-gain systems is

$$SSF = 1 - (1 + G/LCR)\,[1 - \mathrm{F}(SLR)] \qquad \text{(A5-2)}$$

where

$$\mathrm{F}(SLR) = \begin{cases} A \cdot SLR \text{ for } SLR < R \\ B - C \exp(-D \cdot SLR) \\ \text{for } SLR \geq R \end{cases}$$

$$SLR = (S/DD)/(LCR + G)$$

The values of *A, B, C, D, G,* and *R* for direct-gain systems are listed in Table A5-9. Graphs of F(*SLR*) versus *SLR* are shown in Figures A5-1 through A5-21. To use the graphs: (1) enter the appropriate graph with the calculated value of *SLR,* (2) determine value of F(*SLR*), and (3) calculate *SSF* from Eq. A5-2.

Table A5-1
Reference Design Common Characteristics

Masonry properties	
thermal conductivity, direct gain and sunspace	1.0 Btu/h • ft • °F
density	150 lb/ft^3
specific heat	0.2 Btu/lb • °F
infrared emittance of normal surface	0.9
infrared emittance of selective surface	0.1
Solar absorptances	
water wall	0.95
masonry, Trombe wall	0.95
direct gain and sunspace	0.8
sunspace: water containers	0.9
lightweight common wall	0.7
other lightweight surfaces	0.3
Glazing properties	
transmission characteristics	diffuse
orientation	due south
index of refraction	1.526
extinction coefficient	0.5 in^{-1}
thickness of each pane	⅛ in
air gap between panes	½ in
infrared emittance	0.9
Control range	
room temperature	65°F to 75°F
sunspace temperature	45°F to 95°F
internal heat generation	0
Thermocirculation vents (when used)	
vent area/projected area (sum of both upper and lower vents)	0.06
height between vents	8 ft
reverse flow	none
Night insulation (when used)	
thermal resistance	R9
in place, solar time	5:30 p.m. to 7:30 a.m.
Solar radiation assumptions	
shading	none
ground diffuse reflectance	0.3
lightweight absorption fraction	0.2
Other sunspace characteristics	
opaque wall thermal resistance (both insulated common wall and end walls, if any)	R20
infiltration, air changes/h	0.5
masonry wall thickness (masonry wall cases)	12 in.
water container heat capacity (insulated wall cases) (per square foot of projected area)	62.4 Btu/°F • ft^2
number of glazings	2

Table A5-2
Water Wall Reference Designs

Designation	Thermal Storage Capacity* (Btu/ft^2 • °F)	Wall Thickness (in)	Number of Glazings	Wall Surface	Night Insulation
WWA1	15.6	3	2	normal	no
WWA2	31.2	6	2	normal	no
WWA3	46.8	9	2	normal	no
WWA4	62.4	12	2	normal	no
WWA5	93.6	18	2	normal	no
WWA6	124.8	24	2	normal	no
WWB1	46.8	9	1	normal	no
WWB2	46.8	9	3	normal	no
WWB3	46.8	9	1	normal	yes
WWB4	46.8	9	2	normal	yes
WWB5	46.8	9	3	normal	yes
WWC1	46.8	9	1	selective	no
WWC2	46.8	9	2	selective	no
WWC3	46.8	9	1	selective	yes
WWC4	46.8	9	2	selective	yes

*Per unit of projected area

Table A5-3
Vented Trombe Wall Reference Designs

Designation	Thermal Storage Capacity* (Btu/ft^2 • °F)	Nominal Wall Thickness** (in)	ρck (Btu^2/h · ft^4 · °F^2)	Number of Glazings	Wall Surface	Night Insulation
TWA1	15	6	30	2	normal	no
TWA2	22.5	9	30	2	normal	no
TWA3	30	12	30	2	normal	no
TWA4	45	18	30	2	normal	no
TWB1	15	6	15	2	normal	no
TWB2	22.5	9	15	2	normal	no
TWB3	30	12	15	2	normal	no
TWB4	45	18	15	2	normal	no
TWC1	15	6	7.5	2	normal	no
TWC2	22.5	9	7.5	2	normal	no
TWC3	30	12	7.5	2	normal	no
TWC4	45	18	7.5	2	normal	no
TWD1	30	12	30	1	normal	no
TWD2	30	12	30	3	normal	no
TWD3	30	12	30	1	normal	yes
TWD4	30	12	30	2	normal	yes
TWD5	30	12	30	3	normal	yes
TWE1	30	12	30	1	selective	no
TWE2	30	12	30	2	selective	no
TWE3	30	12	30	1	selective	yes
TWE4	30	12	30	2	selective	yes

*Per unit of projected area
**For the particular case of ρc = 30 Btu/ft^3 • °F

Table A5-4
Unvented Trombe Wall Reference Designs

Designation	Thermal Storage Capacity* ($Btu/ft^2 \cdot °F$)	Nominal Wall Thickness** (in)	ρck ($Btu^2/h \cdot ft^4 \cdot °F^2$)	Number of Glazings	Wall Surface	Night Insulation
TWF1	15	6	30	2	normal	no
TWF2	22.5	9	30	2	normal	no
TWF3	30	12	30	2	normal	no
TWF4	45	18	30	2	normal	no
TWG1	15	6	15	2	normal	no
TWG2	22.5	9	15	2	normal	no
TWG3	30	12	15	2	normal	no
TWG4	45	18	15	2	normal	no
TWH1	15	6	7.5	2	normal	no
TWH2	22.5	9	7.5	2	normal	no
TWH3	30	12	7.5	2	normal	no
TWH4	45	18	7.5	2	normal	no
TWI1	30	12	30	1	normal	no
TWI2	30	12	30	3	normal	no
TWI3	30	12	30	1	normal	yes
TWI4	30	12	30	2	normal	yes
TWI5	30	12	30	3	normal	yes
TWJ1	30	12	30	1	selective	no
TWJ2	30	12	30	2	selective	no
TWJ3	30	12	30	1	selective	yes
TWJ4	30	12	30	2	selective	yes

*Per unit of projected area
**For the particular case of $\rho c = 30\ Btu/ft^3 \cdot °F$

Table A5-5
Sunspace Reference Designs*

Designation	Type	Tilt (degrees)	Common Wall	End Walls	Night Insulation
SSA1	attached	50	masonry	opaque	no
SSA2	attached	50	masonry	opaque	yes
SSA3	attached	50	masonry	glazed	no
SSA4	attached	50	masonry	glazed	yes
SSA5	attached	50	insulated	opaque	no
SSA6	attached	50	insulated	opaque	yes
SSA7	attached	50	insulated	glazed	no
SSA8	attached	50	insulated	glazed	yes
SSB1	attached	90/30	masonry	opaque	no
SSB2	attached	90/30	masonry	opaque	yes
SSB3	attached	90/30	masonry	glazed	no
SSB4	attached	90/30	masonry	glazed	yes
SSB5	attached	90/30	insulated	opaque	no
SSB6	attached	90/30	insulated	opaque	yes
SSB7	attached	90/30	insulated	glazed	no
SSB8	attached	90/30	insulated	glazed	yes
SSC1	semienclosed	90	masonry	common	no
SSC2	semienclosed	90	masonry	common	yes
SSC3	semienclosed	90	insulated	common	no
SSC4	semienclosed	90	insulated	common	yes
SSD1	semienclosed	50	masonry	common	no
SSD2	semienclosed	50	masonry	common	yes
SSD3	semienclosed	50	insulated	common	no
SSD4	semienclosed	50	insulated	common	yes
SSE1	semienclosed	90/30	masonry	common	no
SSE2	semienclosed	90/30	masonry	common	yes
SSE3	semienclosed	90/30	insulated	common	no
SSE4	semienclosed	90/30	insulated	common	yes

*All are double-glazed.

Table A5-6
Direct-Gain Reference Designs

Designation	Thermal Storage Capacity* ($Btu/ft^2 \cdot °F$)	Nominal Mass Thickness** (in)	Mass-to-Glazing-Area Ratio	Number of Glazings	Night Insulation
DGA1	30	2	6	2	no
DGA2	30	2	6	3	no
DGA3	30	2	6	2	yes
DGB1	45	6	3	2	no
DGB2	45	6	3	3	no
DGB3	45	6	3	2	yes
DGC1	60	4	6	2	no
DGC2	60	4	6	3	no
DGC3	60	4	6	2	yes

*Per unit of projected area
**For the particular case of $\rho c = 30\ Btu/ft^3 \cdot °F$

Table A5-7
Solar Load Ratio Correlation Coefficient Values for Thermal Storage Wall Systems

Type	*C*	*D*	*H*	LCR_s	U_c
WWA1	0.9172	0.4841	1.17	13.0	0.31
WWA2	0.9833	0.7603	0.92	13.0	0.31
WWA3	1.0171	0.8852	0.85	13.0	0.31
WWA4	1.0395	0.9569	0.81	13.0	0.31
WWA5	1.0604	1.0387	0.78	13.0	0.31
WWA6	1.0735	1.0827	0.76	13.0	0.31
WWB1	0.9754	0.5518	0.92	22.0	0.41
WWB2	1.0487	1.0851	0.78	9.2	0.25
WWB3	1.0673	1.0087	0.95	8.9	0.25
WWB4	1.1028	1.1811	0.74	5.8	0.18
WWB5	1.1146	1.2771	0.56	4.5	0.15
WWC1	1.0667	1.0437	0.62	12.0	0.30
WWC2	1.0846	1.1482	0.59	8.7	0.24
WWC3	1.1419	1.1756	0.28	5.5	0.17
WWC4	1.1401	1.2378	0.23	4.3	0.14
TWA1	0.9194	0.4601	1.11	13.0	0.27
TWA2	0.9680	0.6318	0.92	13.0	0.25
TWA3	0.9964	0.7123	0.85	13.0	0.24
TWA4	1.0190	0.7332	0.79	13.0	0.21
TWB1	0.9364	0.4777	1.01	13.0	0.24
TWB2	0.9821	0.6020	0.85	13.0	0.21
TWB3	0.9980	0.6191	0.80	13.0	0.19
TWB4	0.9981	0.5615	0.76	13.0	0.16
TWC1	0.9558	0.4709	0.89	13.0	0.19
TWC2	0.9788	0.4964	0.79	13.0	0.16
TWC3	0.9760	0.4519	0.76	13.0	0.14
TWC4	0.9588	0.3612	0.73	13.0	0.11
TWD1	0.9842	0.4418	0.89	22.0	0.29
TWD2	1.0150	0.8994	0.80	9.2	0.20
TWD3	1.0346	0.7810	1.08	8.9	0.20
TWD4	1.0606	0.9770	0.85	5.8	0.15
TWD5	1.0721	1.0718	0.61	4.5	0.13
TWE1	1.0345	0.8753	0.68	12.0	0.23
TWE2	1.0476	1.0050	0.66	8.7	0.20
TWE3	1.0919	1.0739	0.61	5.5	0.15
TWE4	1.0971	1.1429	0.47	4.3	0.13
TWF1	0.9430	0.4744	1.09	13.0	0.27
TWF2	0.9900	0.6053	0.93	13.0	0.25
TWF3	1.0189	0.6502	0.86	13.0	0.24
TWF4	1.0419	0.6258	0.80	13.0	0.21
TWG1	0.9693	0.4714	1.01	13.0	0.24
TWG2	1.0133	0.5462	0.88	13.0	0.21
TWG3	1.0325	0.5269	0.82	13.0	0.19
TWG4	1.0401	0.4400	0.77	13.0	0.16
TWH1	1.0002	0.4356	0.93	13.0	0.19
TWH2	1.0280	0.4151	0.83	13.0	0.16
TWH3	1.0327	0.3522	0.78	13.0	0.14
TWH4	1.0287	0.2600	0.74	13.0	0.11
TWI1	0.9974	0.4036	0.91	22.0	0.29
TWI2	1.0386	0.8313	0.80	9.2	0.20
TWI3	1.0514	0.6886	1.01	8.9	0.20
TWI4	1.0781	0.8952	0.82	5.8	0.15
TWI5	1.0902	1.0284	0.65	4.5	0.13
TWJ1	1.0537	0.8227	0.65	12.0	0.23
TWJ2	1.0677	0.9312	0.62	8.7	0.20
TWJ3	1.1153	0.9831	0.44	5.5	0.15
TWJ4	1.1154	1.0607	0.38	4.3	0.13

Table A5-8
Solar Load Ratio Correlation Coefficient Values
for Sunspace Systems

Type	*C*	*D*	*H*	LCR_s	U_c
SSA1	0.9587	0.4770	0.83	18.6	0.27
SSA2	0.9982	0.6614	0.77	10.4	0.21
SSA3	0.9552	0.4230	0.83	23.6	0.30
SSA4	0.9956	0.6277	0.80	12.4	0.23
SSA5	0.9300	0.4041	0.96	18.6	0.04
SSA6	0.9981	0.6660	0.86	10.4	0.04
SSA7	0.9219	0.3225	0.96	23.6	0.04
SSA8	0.9922	0.6173	0.90	12.4	0.04
SSB1	0.9683	0.4954	0.84	16.3	0.26
SSB2	1.0029	0.6802	0.74	8.5	0.19
SSB3	0.9689	0.4685	0.82	19.3	0.28
SSB4	1.0029	0.6641	0.76	9.7	0.21
SSB5	0.9408	0.3866	0.97	16.3	0.04
SSB6	1.0068	0.6778	0.84	8.5	0.04
SSB7	0.9395	0.3363	0.95	19.3	0.04
SSB8	1.0047	0.6469	0.87	9.7	0.04
SSC1	1.0087	0.7683	0.76	16.3	0.38
SSC2	1.0412	0.9281	0.78	10.0	0.28
SSC3	0.9699	0.5106	0.79	16.3	0.08
SSC4	1.0152	0.7523	0.81	10.0	0.08
SSD1	0.9889	0.6643	0.84	17.8	0.36
SSD2	1.0493	0.8753	0.70	9.9	0.26
SSD3	0.9570	0.5285	0.90	17.8	0.07
SSD4	1.0356	0.8142	0.73	9.9	0.07
SSE1	0.9968	0.7004	0.77	19.6	0.41
SSE2	1.0468	0.9054	0.76	10.8	0.29
SSE3	0.9565	0.4827	0.81	19.6	0.08
SSE4	1.0214	0.7694	0.79	10.8	0.08

Table A5-9
Solar Load Ratio Correlation Coefficient Values
for Direct-Gain Systems

Type	*A*	*B*	*C*	*D*	*R*	*G*	U_c
DGA1	0.5650	1.0090	1.0440	0.7175	0.3931	9.36	0.50
DGA2	0.5906	1.0060	1.0650	0.8099	0.4681	5.28	0.31
DGA3	0.5442	0.9715	1.1300	0.9273	0.7068	2.64	0.26
DGB1	0.5739	0.9948	1.2510	1.0610	0.7905	9.60	0.50
DGB2	0.6180	1.0000	1.2760	1.1560	0.7528	5.52	0.31
DGB3	0.5601	0.9839	1.3520	1.1510	0.8879	2.38	0.26
DGC1	0.6344	0.9887	1.5270	1.4380	0.8632	9.60	0.50
DGC2	0.6763	0.9994	1.4000	1.3940	0.7604	5.28	0.31
DGC3	0.6182	0.9859	1.5660	1.4370	0.8990	2.40	0.26

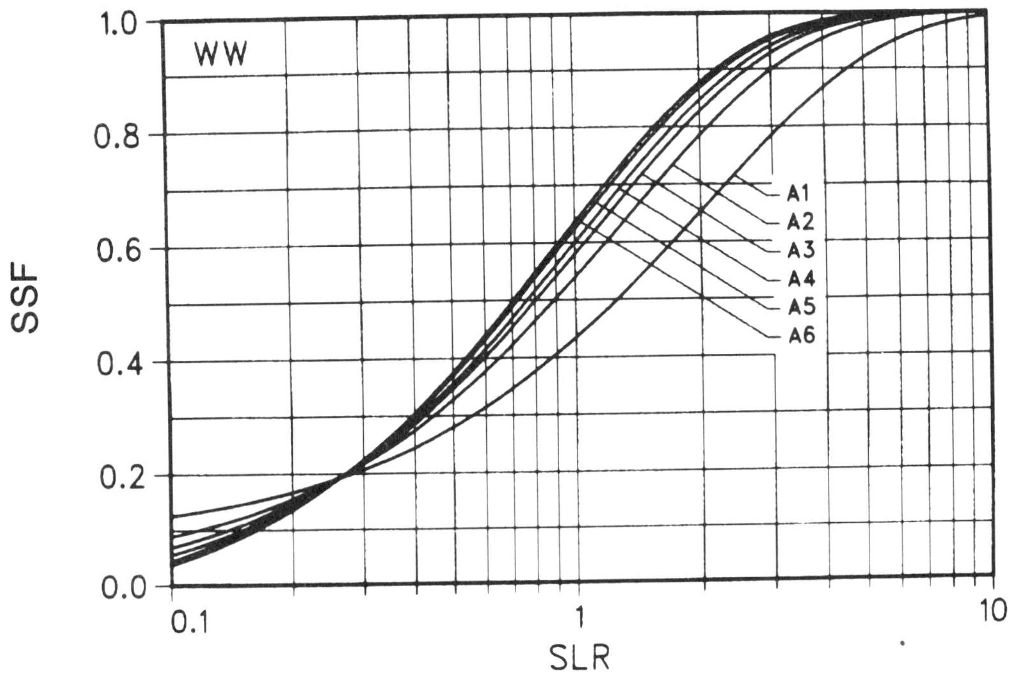

Figure A5-1. Monthly *SSF* versus *SLR* for Reference Designs WWA1 through WWA6

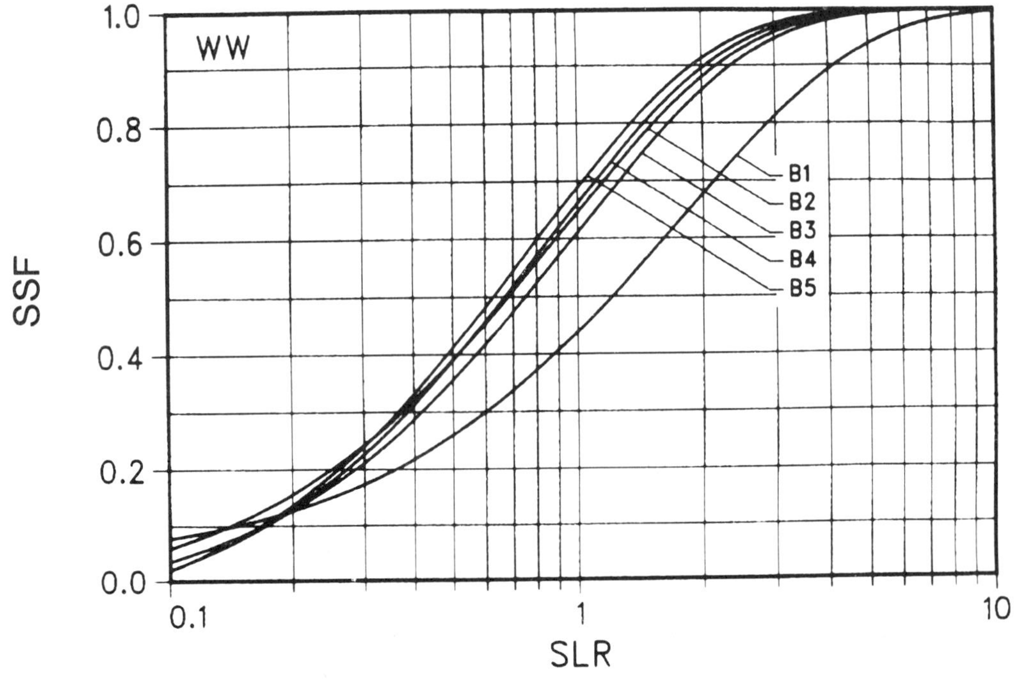

Figure A5-2. Monthly *SSF* versus *SLR* for Reference Designs WWB1 through WWB5

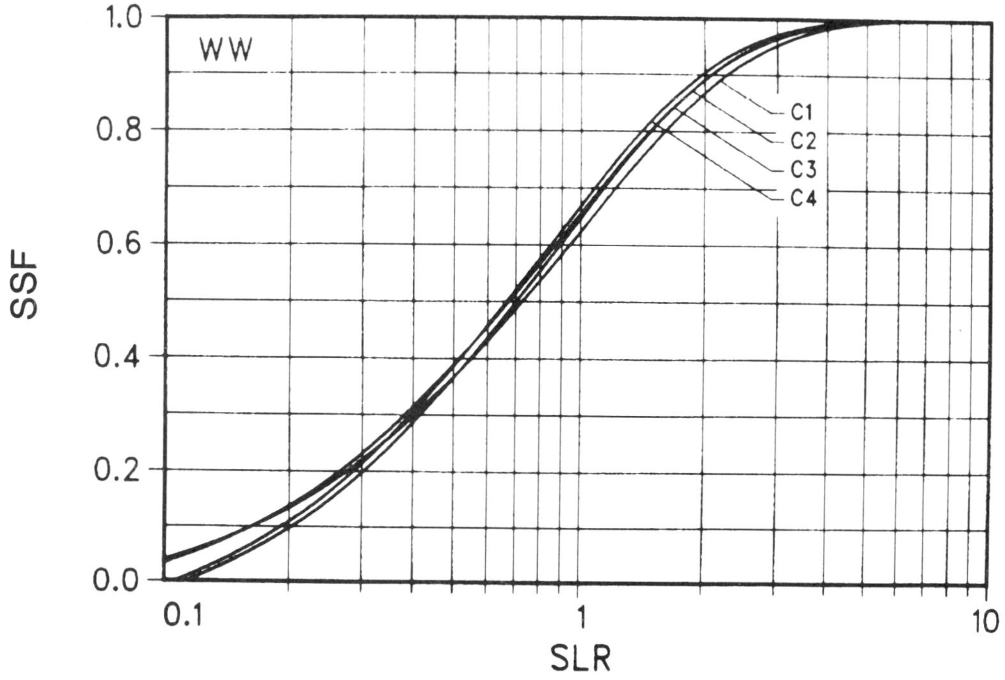

Figure A5-3. Monthly *SSF* versus *SLR* for Reference Designs WWC1 through WWC4

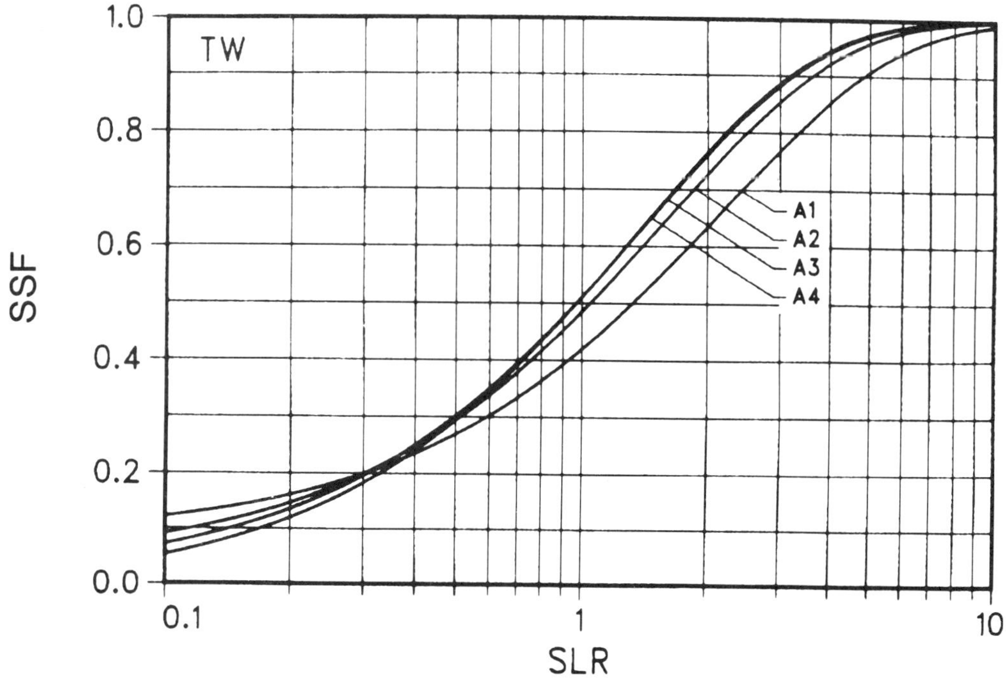

Figure A5-4. Monthly *SSF* versus *SLR* for Reference Designs TWA1 through TWA4

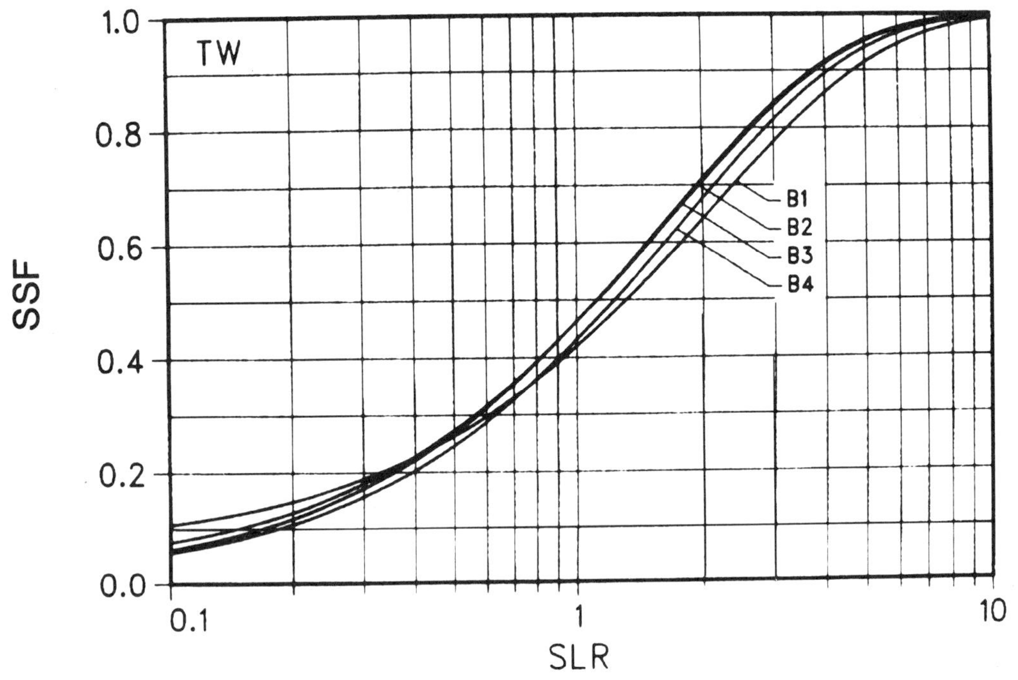

Figure A5-5. Monthly *SSF* versus *SLR* for Reference Designs TWB1 through TWB4

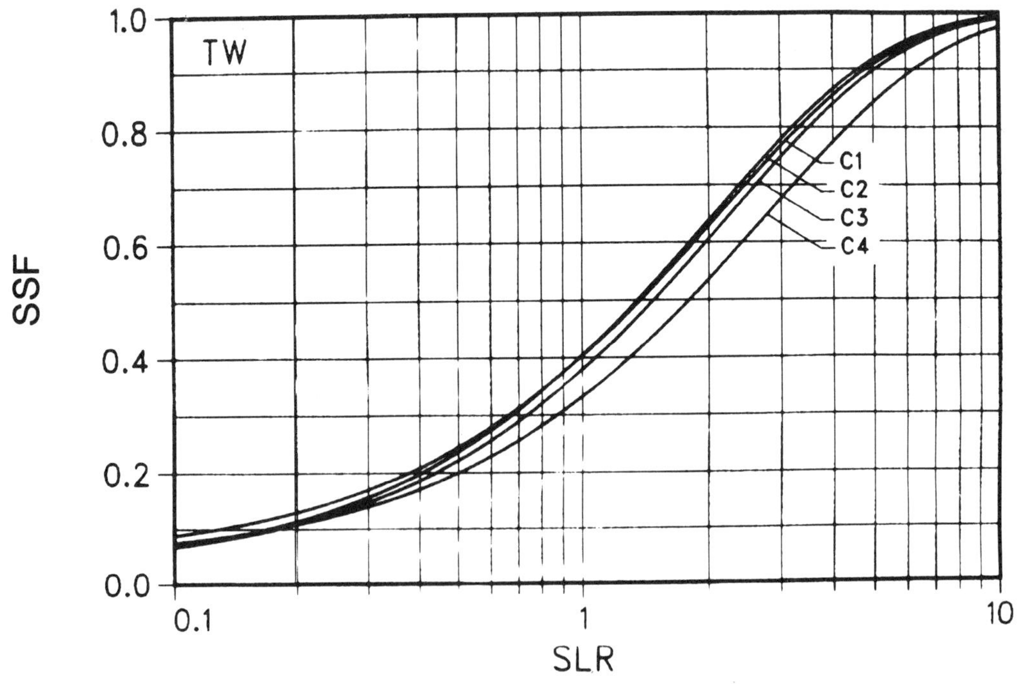

Figure A5-6. Monthly *SSF* versus *SLR* for Reference Designs TWC1 through TWC4

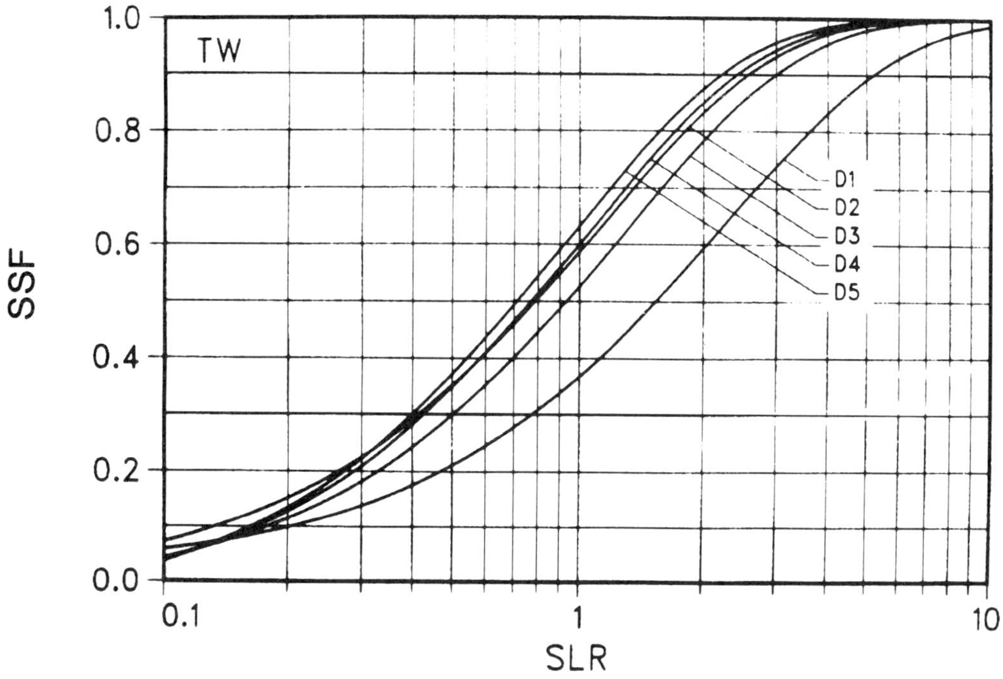

Figure A5-7. Monthly *SSF* versus *SLR* for Reference Designs TWD1 through TWD5

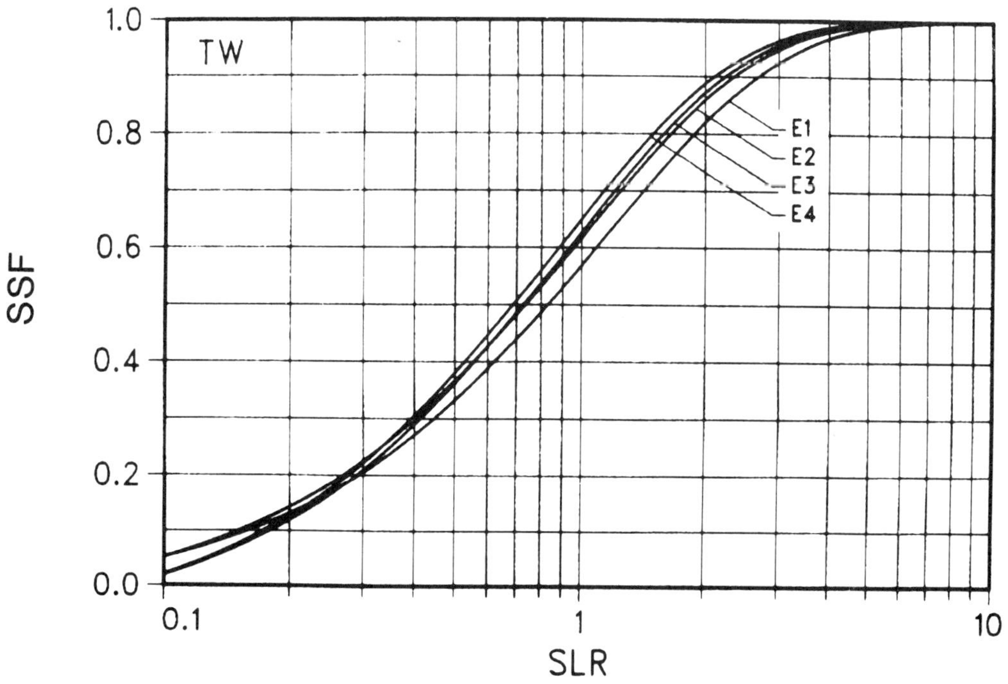

Figure A5-8. Monthly *SSF* versus *SLR* for Reference Designs TWE1 through TWE4

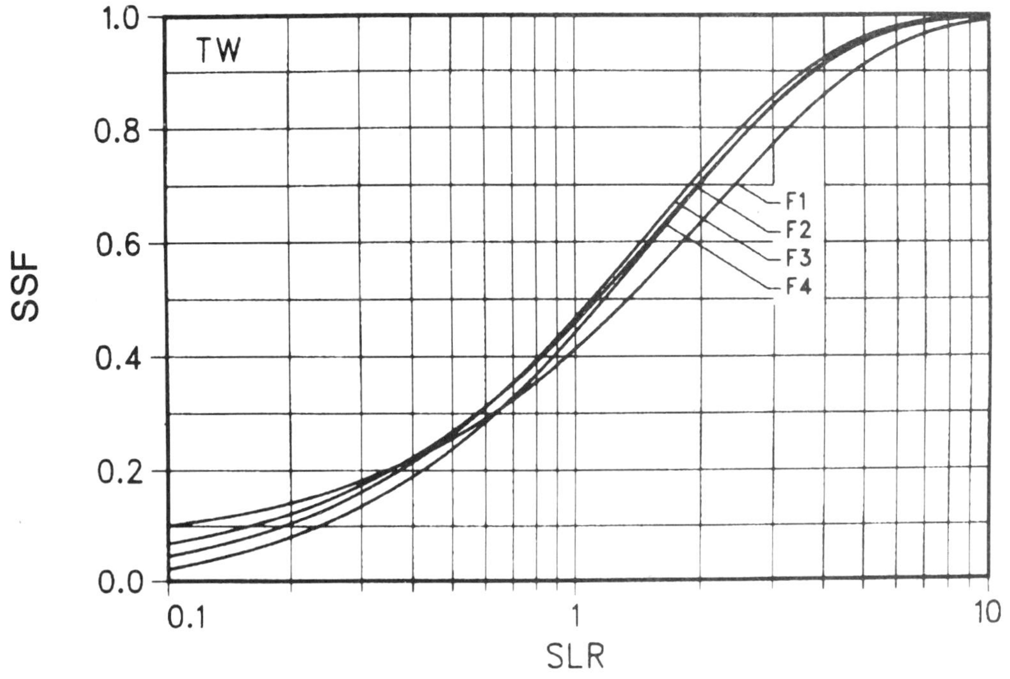

Figure A5-9. Monthly *SSF* versus *SLR* for Reference Designs TWF1 through TWF4

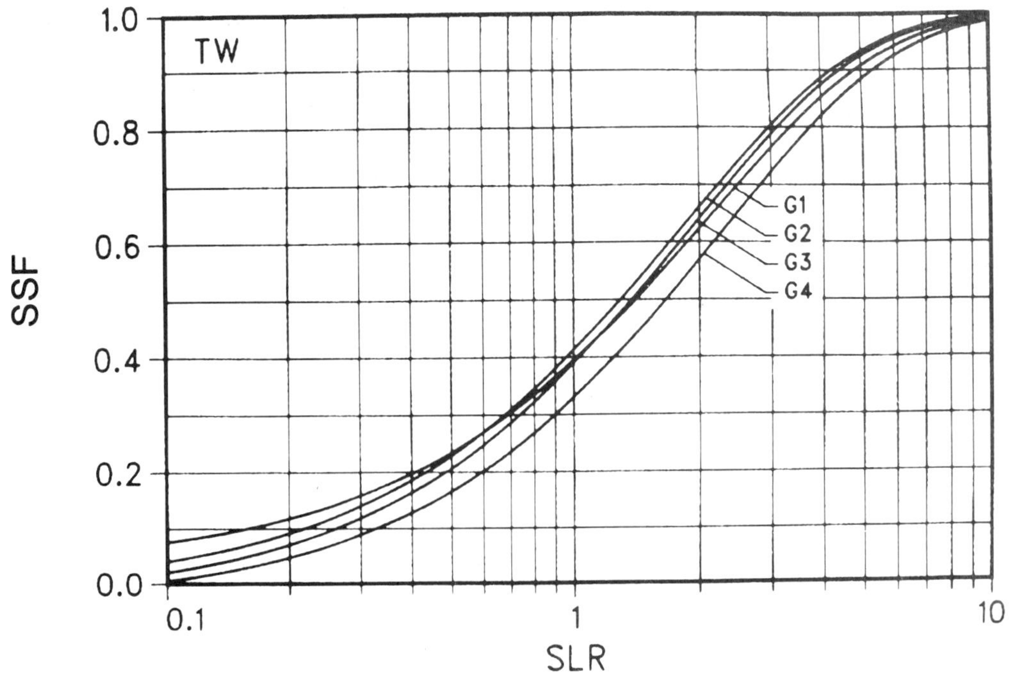

Figure A5-10. Monthly *SSF* versus *SLR* for Reference Designs TWG1 through TWG4

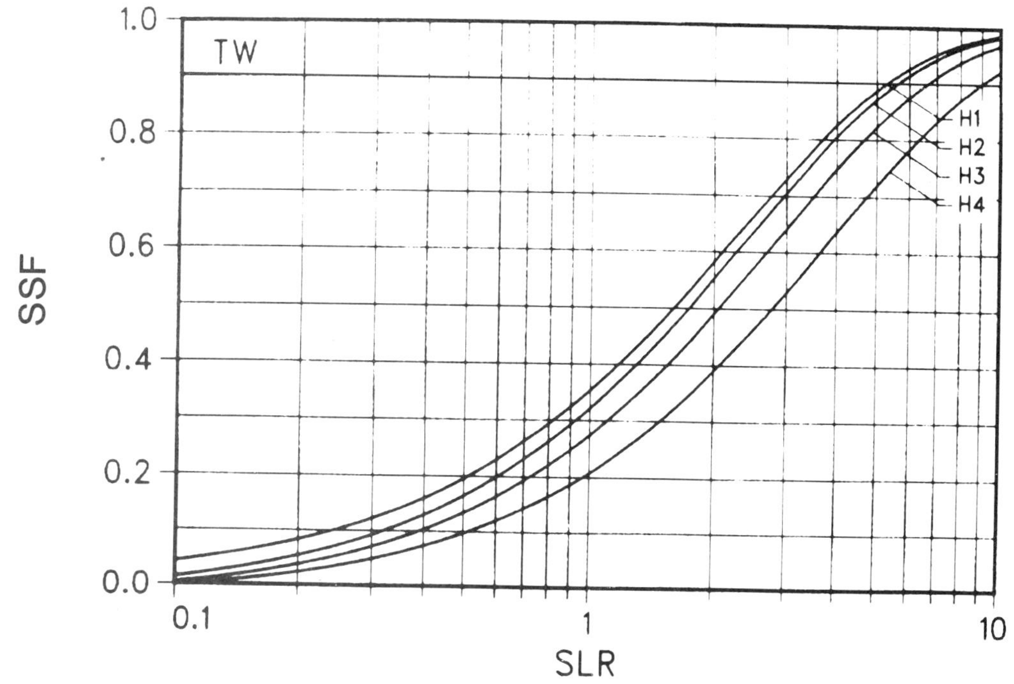

Figure A5-11. Monthly *SSF* versus *SLR* for Reference Designs TWH1 through TWH4

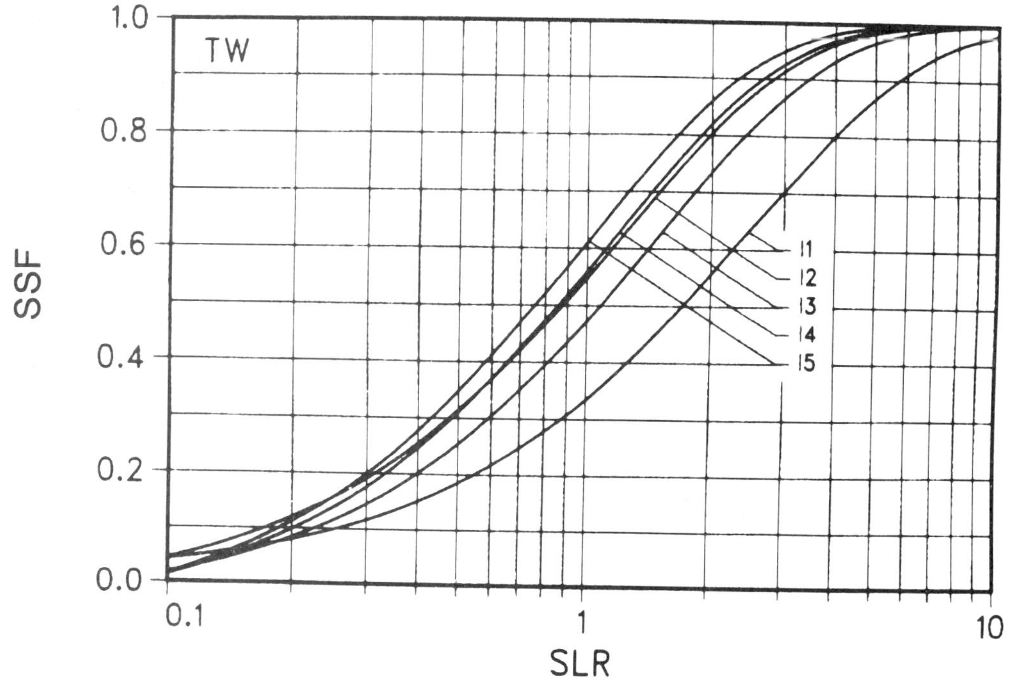

Figure A5-12. Monthly *SSF* versus *SLR* for Reference Designs TWI1 through TWI5

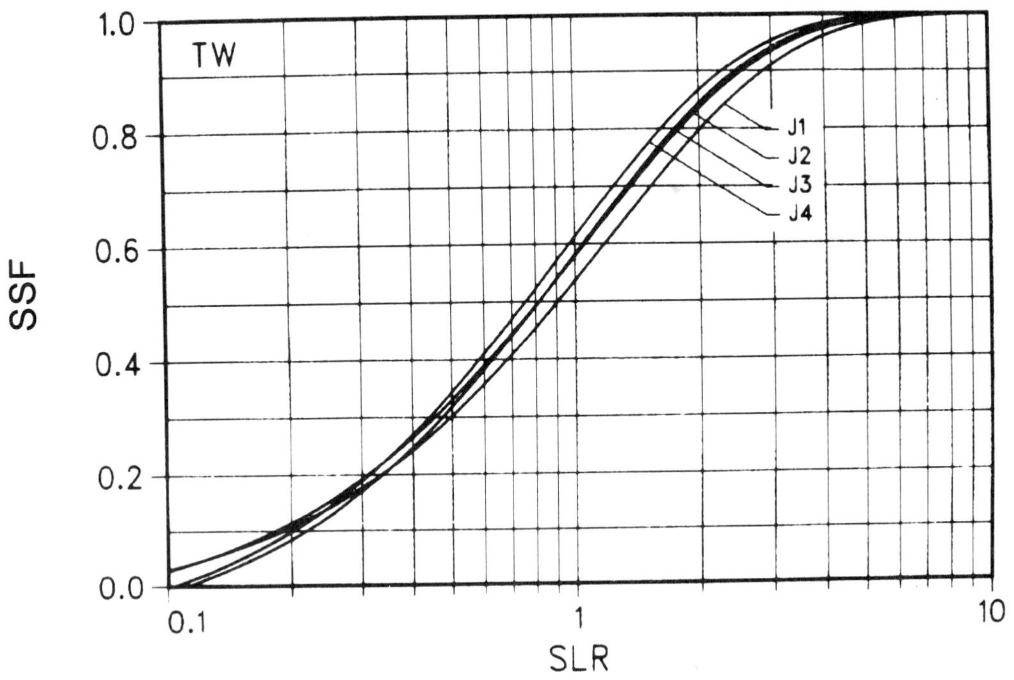

Figure A5-13. Monthly *SSF* versus *SLR* for Reference Designs TWJ1 through TWJ4

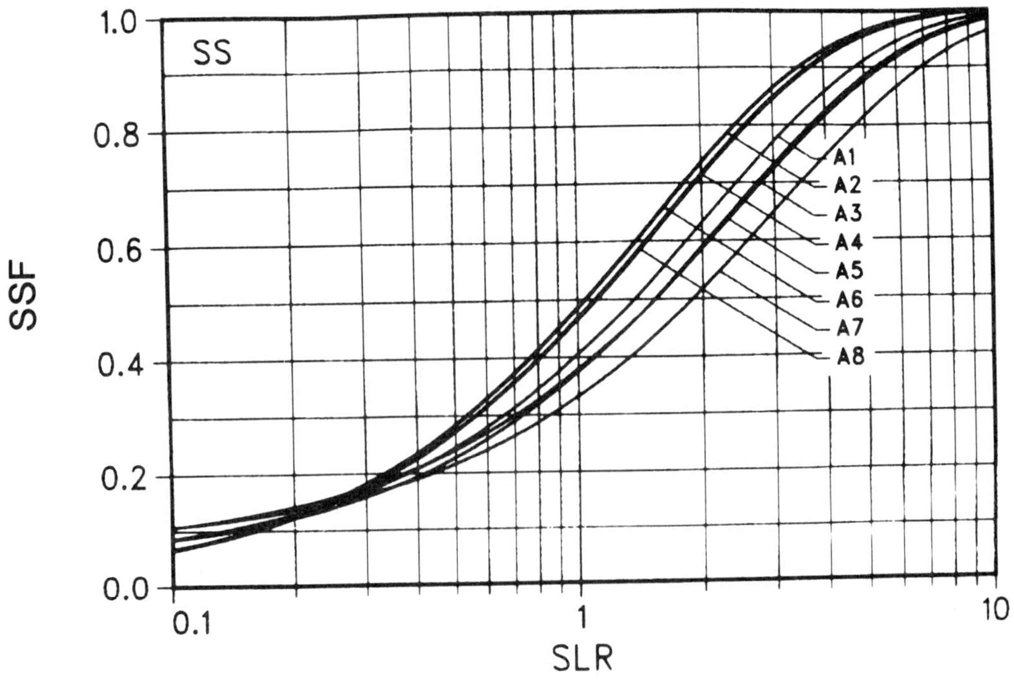

Figure A5-14. Monthly *SSF* versus *SLR* for Reference Designs SSA1 through SSA8

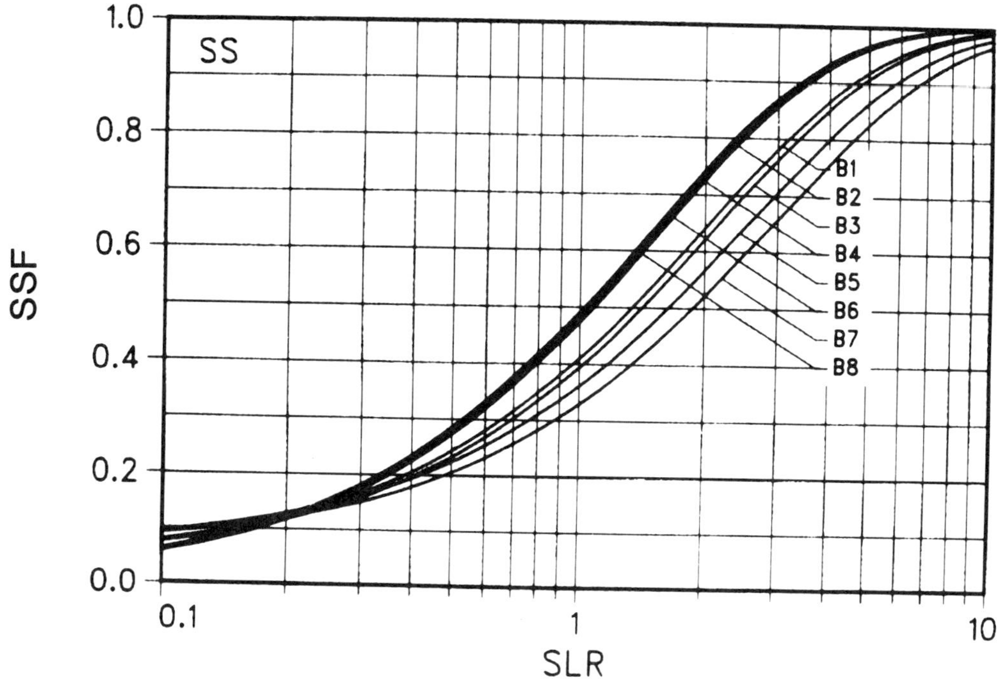

Figure A5-15. Monthly *SSF* versus *SLR* for Reference Designs SSB1 through SSB8

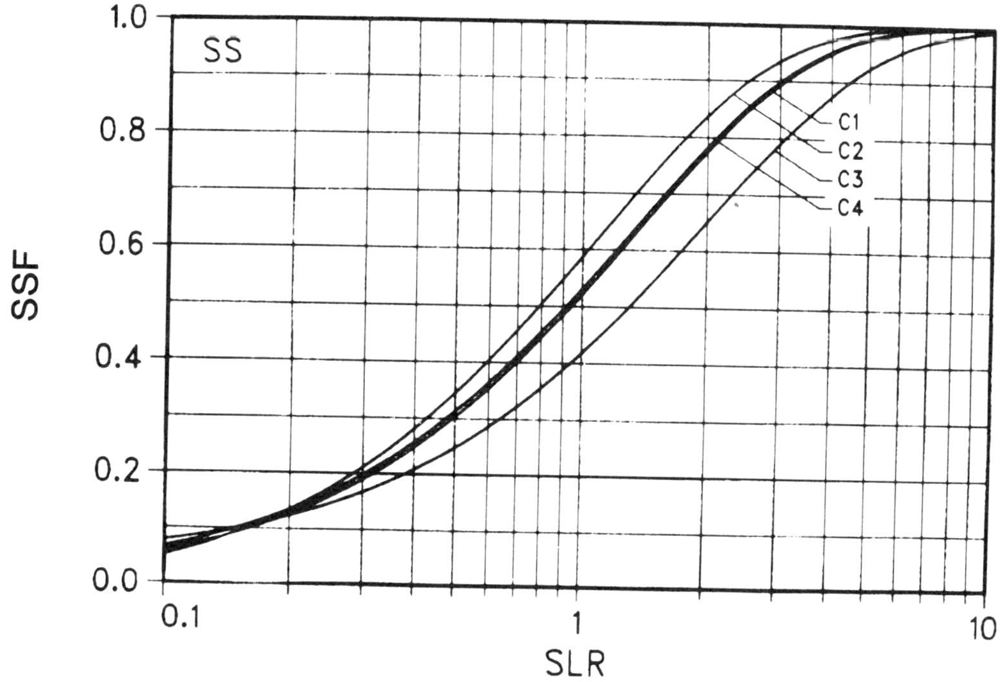

Figure A5-16. Monthly *SSF* versus *SLR* for Reference Designs SSC1 through SSC4

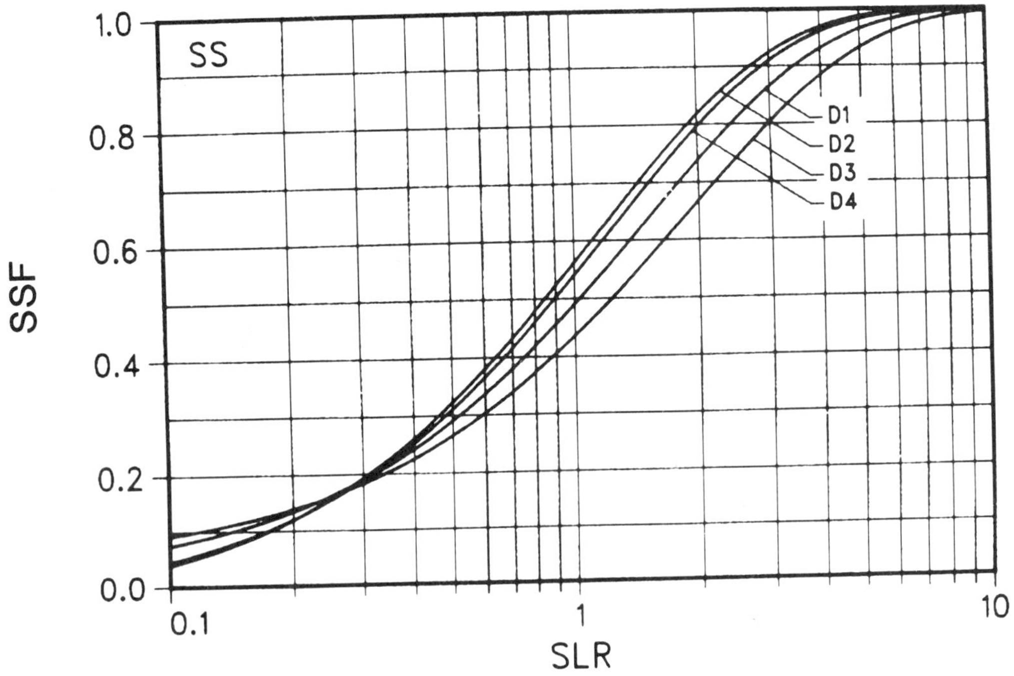

Figure A5-17. Monthly *SSF* versus *SLR* for Reference Designs SSD1 through SSD4

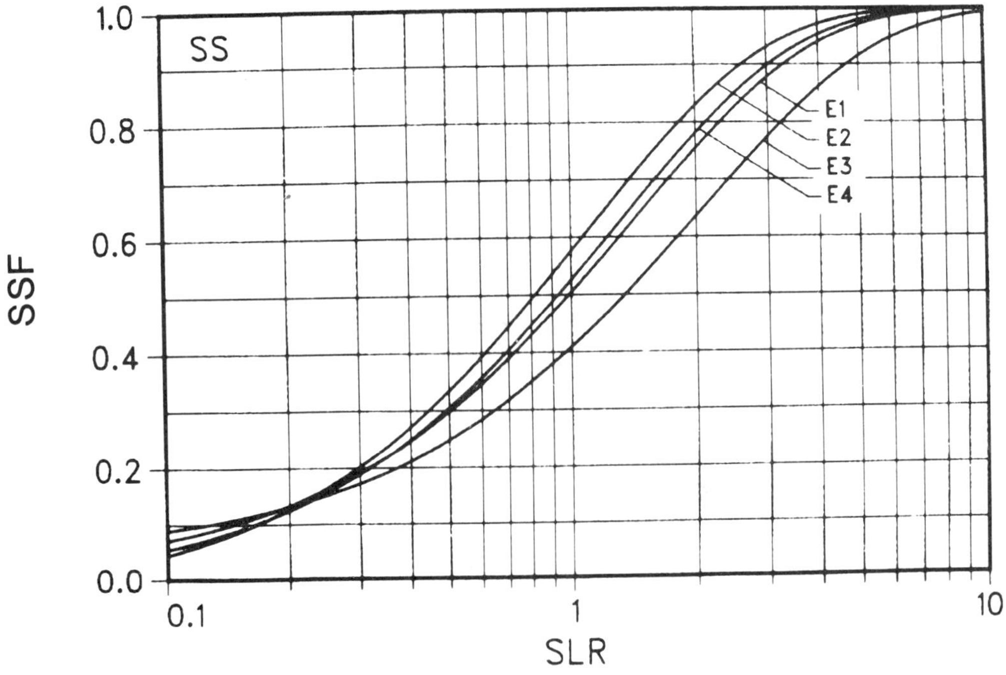

Figure A5-18. Monthly *SSF* versus *SLR* for Reference Designs SSE1 through SSE4

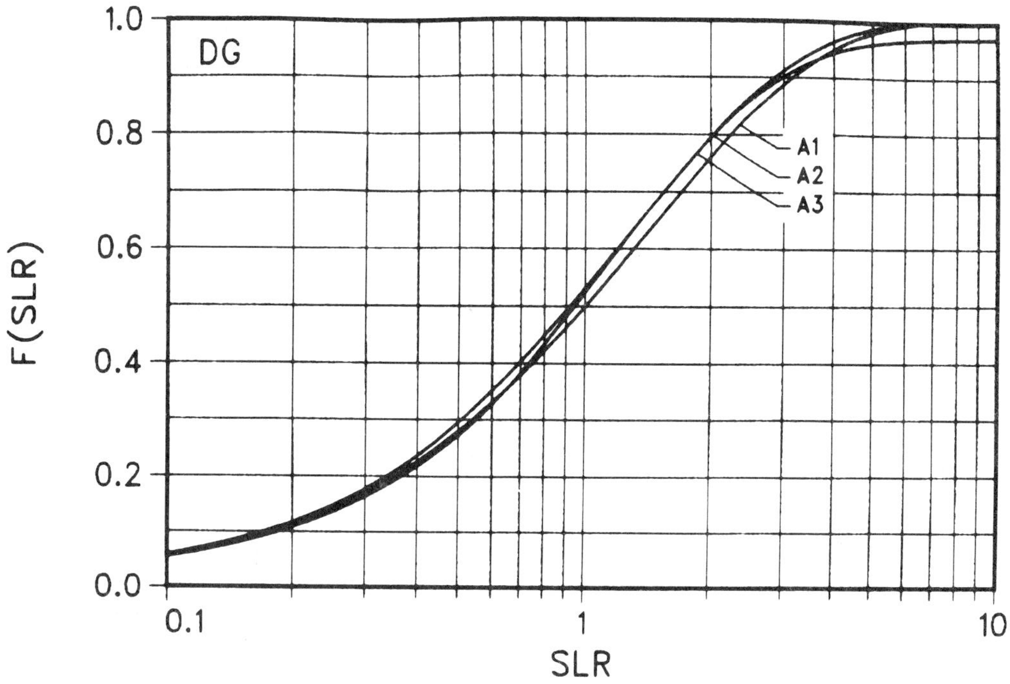

Figure A5-19. F *(SLR)* versus *(SLR)* for Reference Designs DGA1 through DGA3

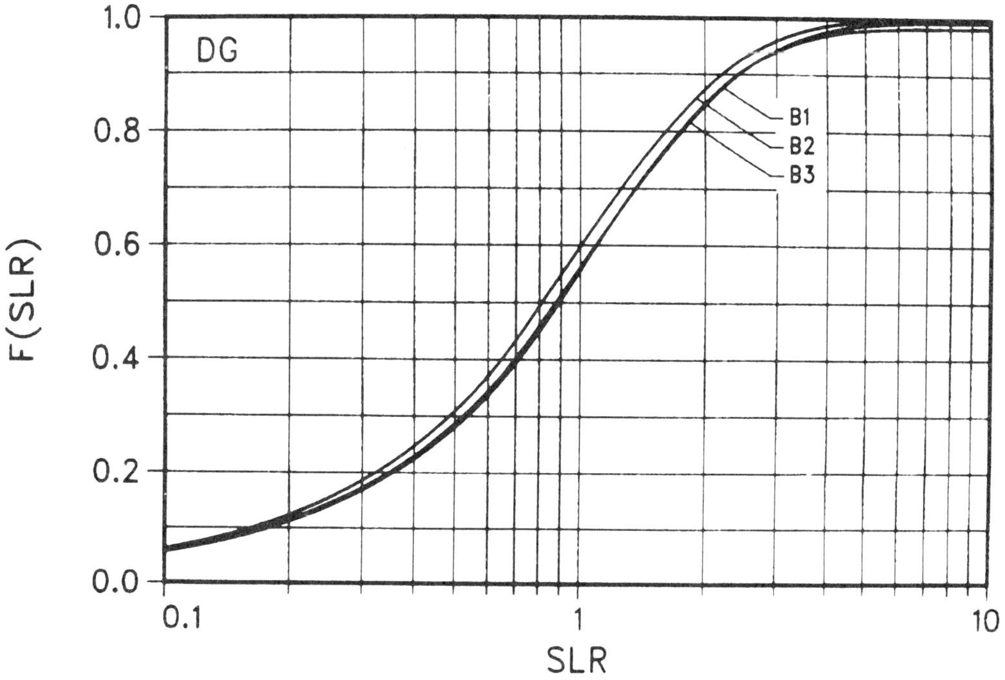

Figure A5-20. F *(SLR)* versus *(SLR)* for Reference Designs DGB1 through DGB3

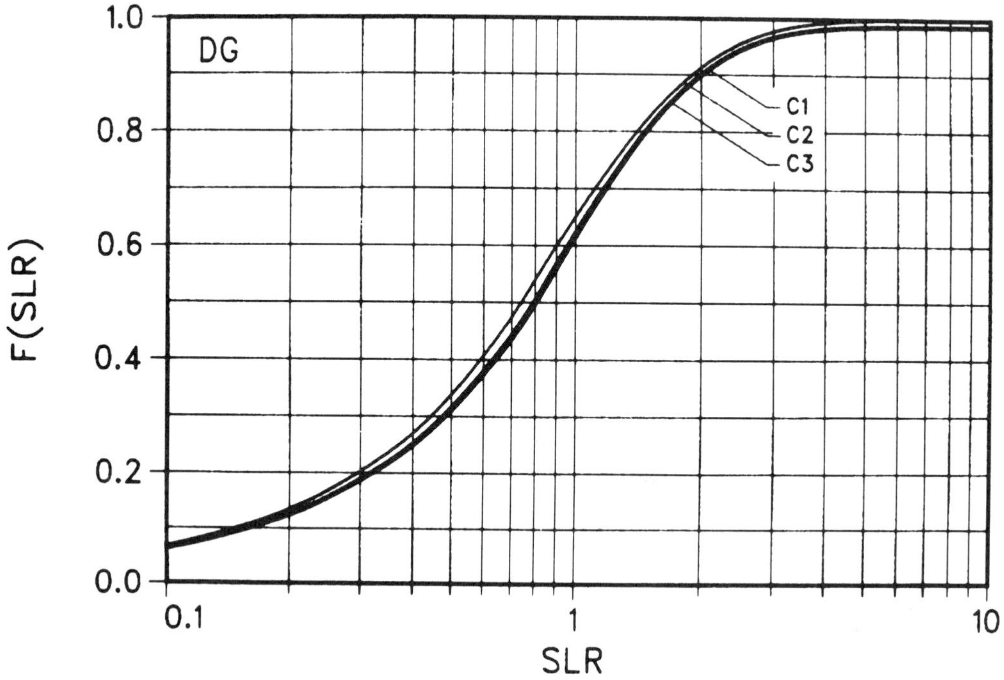

Figure A5-21. F *(SLR)* versus *(SLR)* for Reference Designs DGC1 through DGC30

APPENDIX 6
MATERIALS PROPERTIES

This appendix lists R-value, solar absorptance, density, specific heat, and thermal conductivity of various construction materials.

Table A6-1

R-Values of Various Wall and Floor Coverings*

Material	R-Value
Tile, asphalt, linoleum, vinyl, rubber (1/16 in.)	0.05
Terrazzo tile (1.0 in.)	0.08
Tile, asphalt, linoleum, vinyl, rubber (1/8 in.)	0.10
Cork tile (1/8 in.)	0.28
Plywood, Douglas fir (1/4 in.)	0.31
Gypsum or plaster board (3/8 in.)	0.32
Gypsum or plaster board (1/2 in.)	0.45
Plywood, Douglas fir (1/2 in.)	0.62
Wood, hardwood finish (3/4 in.)	0.68
Carpet and rubber pad	1.23
Gypsum or plaster board (3/8 in. over 1/2 in. furring strips)	1.68
Gypsum or plaster board (1/2 in. over 1/2 in. furring strips)	1.81
Carpet and fibrous pad	2.08

**ASHRAE Handbook and Product Directory: 1977 Fundamentals* (American Society of Heating, Refrigerating, and Air Conditioning Engineers, Inc., New York, 1977), p. 22.13.

Table A6-2

Solar Absorptance of Various Materials*

Flat black paint	0.95
Black lacquer	0.92
Dark gray paint	0.91
Black concrete	0.91
Dark blue lacquer	0.91
Black oil paint	0.90
Gray slate	0.90
Stafford blue bricks	0.89
Dark olive drab paint	0.89
Dark brown paint	0.88
Dark blue-gray paint	0.88
Azure blue or dark green lacquer	0.88
Brown concrete	0.85
Medium brown paint	0.84
Medium light brown paint	0.80
Silver gray slate	0.80
Brown or green lacquer	0.79
Medium rust paint	0.78
Light gray oil paint	0.75
Red oil paint	0.74
Red bricks	0.70
Uncolored concrete	0.65
Moderately light buff bricks	0.60
Medium dull green paint	0.59
Medium orange paint	0.58
Medium yellow paint	0.57
Medium blue paint	0.51
Medium Kelly green paint	0.51
Light green paint	0.47
White semigloss paint	0.30
White gloss paint	0.25
White lacquer	0.21

*This table is meant to serve as a guide only. Variations in texture, tone, overcoats, pigments, binders, etc., can affect these values.

Sources: G. G. Gubareff, J.E. Janssen, and R. H. Torborg, *Thermal Radiation Properties Survey,* 2nd ed., (Minneapolis-Honeywell Regulator Company, Minneapolis, 1960); and S. W. Moore, Los Alamos Scientific Laboratory, unpublished data.

Table A6-3

Average Reflectance of Various Ground Surfaces

Surface	Source	Reflectance
Water	4	0.06
Sand, wet	4	0.09
Weathered blacktop	1,2,3,5	0.10
Sand, dry	4	0.18
Grass, dry	4,5	0.20
Rock, crushed	1	0.20
Concrete, old	1	0.23
Grass, green	1	0.28
Leaves, dead	4	0.30
Concrete, new	1	0.33
Snow, old, icy	4	0.67
Snow, new, fine particles	4	0.87

Sources:

1. *ASHRAE Handbook of Fundamentals* (for 60-degree angle of incidence).
2. G. M. B. Bouricius, K. L. Coulson, and E. L. Gray, "A Study of the Reflection and Polarization Characteristics of Selected Natural and Artificial Surfaces" (General Electric Space Sciences Laboratory, U.S. Naval Ordinance Station, California, 1965).
3. G. G. Gubareff, J. E. Janssen, and R. H. Torborg, *Thermal Radiation Properties Survey,* 2nd ed. (Minneapolis-Honeywell Regulator Company, Minneapolis, 1960).
4. C. Strock and R. L. Koral (eds.) *Handbook of Air Conditioning, Heating, and Ventilating,* 2nd ed. (Industrial Press, New York, 1965).
5. J. L. Threlkeld, *Thermal Environmental Engineering,* 2nd ed. (Prentice-Hall, Englewood Cliffs, New Jersey, 1970).

Table A6-4

Thermal Properties of Various Materials

Material	Source	Density ρ (lb/ft³)	Specific Heat c (Btu/lb · °F)	Heat Capacity ρc (Btu/ft³ · °F)	Conductivity k (Btu/h · °F · ft)	Diffusivity k/ρc (ft²/h)	ρck (Btu²/h · ft⁴ · °F²)
Water	1	62.3	1.0	62.3			
Concrete	2,3	144	0.21	30.3	0.89	0.029	27.0
Concrete block	2,3						
Heavy weight		135	0.21	28.4	0.74	0.026	21.0
Medium weight		105	0.22	23.1	0.41	0.018	9.5
Light weight		85	0.23	19.6	0.27	0.014	5.3
Brick	1,4						
Paving		135	0.19	25.7	0.75	0.029	19.3
Face		130	0.19	24.7	0.75	0.030	18.5
Building		120	0.19	22.8	0.42	0.018	9.6
Mortar or grout	1	116	0.20	23.2	0.42	0.018	9.7
Adobe	5	80	0.20	16.0	0.38	0.024	6.1
		–100		–20.0	–0.75	–0.038	–15.0
Gypsum or plasterboard	1	50	0.26	13.0	0.097	0.058	9.8
Douglas fir plywood	1	34	0.29	9.9	0.067	0.007	0.7
Hardwood	1	45	0.30	13.5	0.092	0.007	1.2

Note: Concrete and masonry products can be manufactured in a variety of densities. The values listed are for the most commonly available products.

The thermal properties of concrete and concrete block vary with density. Empirical correlations between these properties are[2] $k = 0.05\,e^{0.02\rho}$ and [3] $c = 1/(3.934 + 0.006\rho)$.

Sources:

1. *ASHRAE Handbook: 1981 Fundamentals* (American Society of Heating, Refrigerating, and Air-Conditioning Engineers, Inc., Atlanta, Georgia, 1981).

2. "Calculating the Steady State U-Value of Concrete Masonry Walls" (Expanded Shale, Clay, and Slate Institute, 4905 Del Ray Avenue, Bethesda, Maryland).

3. David Whiting, Albert Litvin, and Stanley E. Goodwin, "Specific Heat of Selected Concretes," Research and Development Bulletin RD058.01B (Portland Cement Association, 5420 Old Orchard Road, Skokie, Illinois).

4. "Brick Passive Solar Heating Systems. Material Properties — Part IV," Technical Notes on Brick Construction 43D (Brick Institute of America, 1750 Old Meadow Road, McLean, Virginia, September/October 1980).

5. Benjamin T. Rogers, private communication; and W. L. Sibbitt, measurements at Los Alamos National Laboratory. Note that the thermal conductivity of adobe varies greatly with moisture content, which is part of the reason for the wide range quoted.

APPENDIX 7
WORKSHEETS

This appendix presents the blank forms of the worksheets that are discussed in the text. Their use will help to organize the problem-solving routine.

NET LOAD COEFFICIENT and ANNUAL LCR METHOD WORKSHEET

DATE ________ BY ________ LOCATION ________

NET LOAD COEFFICIENT, NLC

	BUILDING ELEMENT	FORMULA	Fill in the Blanks and Perform Arithmetic	LOAD, UA
	WALLS	: $(UA)_w = A_w/R_w$	= ________ / ________	= ________
	ROOF	: $(UA)_r = A_r/R_r$	= ________ / ________	= ________
	E, W, N WINDOWS	: $(UA)_g = 1.1A_x/N$	= 1.1 x ________ / ________	= ________
SELECT ONE	FLOOR	: $(UA)_f = A_f/R_f$	= ________ / ________	= ________
SELECT ONE	PERIMETER	: $(UA)_p = 4.17\ P_p/(R_p + 5)$	= 4.17 x ________ /(________ + 5)	= ________
SELECT ONE	BASEMENT	: $(UA)_b = 10.7\ P_b/(R_b + 8)$	= 10.7 x ________ /(________ + 8)	= ________
	INFILTRATION	: $(UA)_i = 0.018V \times ACH \times ADR$	= 0.018 x ________ x ________ x ________	= ________

Sum = ________ Btu/°F • h

NLC = 24 x Sum = ________ Btu/°F • day

LOAD COLLECTOR RATIO, LCR

$LCR = NLC/A_p$ = ________ / ________ = ________ Btu/°F • day • ft²

WEIGHTED AVERAGE, SSF

REFERENCE DESIGN	PROJECTED AREA	FRACTION OF A_p		SSF FOR EACH REFERENCE DESIGN		PRODUCT OF SSF x FRACTION	U_c	$U_c\ A_p$
________	________	________	x	________	=	________	________	________
________	________	________	x	________	=	________	________	________
________	________	________	x	________	=	________	________	________
TOTAL A_p =	________				SSF =	________	TOTAL (U_cA_p) =	________

TLC = NLC + 24 x TOTAL (U_cA_p) = ________

BASE TEMPERATURE, T_b

$T_b = T_{set} - Q_{int}/TLC$ = ________ − ________ / ________ = ________ °F

AUXILIARY HEAT, Q_{aux}

$Q_{aux} = NLC \bullet DD(1 - SSF)$ = ________ x ________ x (1 − ________) x 10^{-6} = ________ Mega Btu/year

CONSERVATION INCREMENTAL COST WORKSHEET

DATE ______________ BY ______________

PROJECT ______________

LOCATION ______________

	WALL		ROOF		E, W, N WINDOWS		FLOOR, PERIMETER OR BASEMENT		INFILTRATION	
	$\$/ft^2$	R VALUE	$\$/ft^2$	R VALUE	$\$/ft^2$	R VALUE	$\$/ft^2$ $\$/ft$	R VALUE	$\$/ft^3$	1/ACH
UPPER CONSERVATION LEVEL										
LOWER CONSERVATION LEVEL										
DIFFERENCES: ΔC, ΔR OR $\Delta(1/ACH)$										
CONSERVATION INCREMENTAL COST $\Delta C/\Delta R$ OR $\Delta C/\Delta(1/ACH)$	$CIC_w =$		$CIC_r =$		$CIC_g =$		$CIC_{f,\ p\ or\ b} =$		$CIC_i =$	

CONSERVATION ADD-ON COST WORKSHEET

DATE ________ BY ________ PROJECT ________

REFERENCE DESIGN(s) ________

LOCATION ________

SIC = ________

CF = ________

BUILDING ELEMENT	CIC	FORMULA	CALC. VALUE	SEL. VALUE	MIN. VALUE	CIC	SIZE	ADD-ON COST
WALL		$R_w = CF(SIC/CIC_w)^{1/2}$		→ (____	− ____)	× ____	× ____	= ____
ROOF		$R_r = CF(SIC/CIC_r)^{1/2}$		→ (____	− ____)	× ____	× ____	= ____
E, W, N WINDOWS		$R_g = CF(SIC/CIC_g)^{1/2}$		→ (____	− ____)	× ____	× ____	= ____
FLOOR		$R_f = CF(SIC/CIC_f)^{1/2}$		→ (____	− ____)	× ____	× ____	= ____
PERIMETER		$R_p = 2.04CF(SIC/CIC_p)^{1/2} - 5$		→ (____	− ____)	× ____	× ____	= ____
BASEMENT		$R_b = 3.26\ CF(SIC/CIC_b)^{1/2} - 8$		→ (____	− ____)	× ____	× ____	= ____
INFILTRATION		$1/ACH = 0.133\ CF\ (SIC/CIC_i)^{1/2}$		→ (____	− ____)	× ____	× ____	= ____

CONSERVATION ADD-ON COST = ________
(SUM OF LAST COLUMN)

SLR METHOD WORKSHEET DATE ______ BY ______ REF. DESIGN ______ LOCATION ______

* * KEY FORMULAS	TW, WW, SS	$SLR = (S/DD - LCR_s \cdot H)/LCR$, $SSF = 1 - C \exp(-D \cdot SLR)$
	DG	$SLR = (S/DD)/(LCR + G)$, $SSF = 1 - (1 + G/LCR)\,[1 - F(SLR)]$ Compute SLR then use graphs to find F(SLR) or $\begin{cases} \text{if } SLR < R, & F(SLR) = A \cdot SLR \\ \text{if } SLR \geq R, & F(SLR) = B - C \exp(-D \cdot SLR) \end{cases}$
	ALL SYSTEMS	Annual SSF = Annual Q_{sav} / Annual Q_{net}

DATA		NLC	Ap	LCR	LCR_s	A	B	C	D	G	H	R	$LCR_s \bullet H$
	TW, WW, SS												
	DG												

ITEM	ABSORBED SOLAR ENERGY	DEGREE-DAYS (T_b = ____ °F)	SOLAR LOAD RATIO	SOLAR SAVINGS FRACTION	NET REFERENCE LOAD	SOLAR SAVINGS	AUXILIARY ENERGY
UNITS	(Btu/ft^2)	(°F • day)	Dimensionless	Dimensionless	(Mega Btu)	(Mega Btu)	(Mega Btu)
ORIGIN	SOLAR GAIN WORKSHEET	Weather Data	* *	* *	$NLC \cdot DD \cdot 10^{-6}$	$Q_{net} \cdot SSF$	Q_{net} (1-SSF)
SYMBOL	S	DD	SLR	SSF	Q_{net}	Q_{sav}	Q_{aux}
JAN							
FEB							
MAR							
APR							
MAY							
JUN							
JUL							
AUG							
SEP							
OCT							
NOV							
DEC							
ANNUAL		Sum:		* *	Sum:	Sum:	Sum:

SOLAR GAIN WORKSHEET

TILT ____________

AZ ____________

E ____________

N ____________

REFR INDEX ____________

TW & WW: SURFACE ABSORPTANCE = ____________

DG & SS:
- α_m = ____________
- HEIGHT = ____________ (ft)
- DEPTH = ____________ (ft)
- WIDTH = ____________ (ft)
- GLAZING LENGTH = ____________ (ft)

α_o = ____________

HGT = 1

XTOP = ____________

WDTH = ____________

Z = ____________

DATE ____________

BY ____________

REF DSGN ____________

LOCATION ____________

ITEM	SOLAR NOON ZENITH ANGLE Weather Table	CLEARNESS FACTOR Weather Table	NUMBER OF DAYS IN MONTH	HORIZONTAL TOTAL INSOLATION Weather Table	GLAZED AREA TO PROJECTED AREA RATIO	INCIDENT FRACTION	TRANSMITTED FRACTION	ABSORBED FRACTION	PRODUCT
SYMBOL	LAT-DEC	K_T	M	HS	A_n/A_p	(INC/HOR)	(TRN/INC)	(ABS/TRN)	S
JAN			31						
FEB			28						
MAR			31						
APR			30						
MAY			31						
JUN			30						
JUL			31						
AUG			31						
SEP			30						
OCT			31						
NOV			30						
DEC			31						

APPENDIX 8
GLOSSARY

The following terms and their definitions relate only to the usage in this volume. No attempt has been made to give a complete listing of solar terms or to give completely general definitions of the terms that are listed.

absorptance — see **solar absorptance**.

absorption factor — the fraction of solar radiation transmitted through a glazing system that is absorbed inside the building.

air changes — a measure of the air exchange in a building due to infiltration. One air change per hour is an exchange of a volume of air equal to the interior volume of a building.

annual *LCR* method — application of the tables of annual solar savings fraction versus load collector ratio (the "*LCR* tables") to a calculation of the annual auxiliary heat.

attached sunspace — solar collector that doubles as useful building space; also attached greenhouse, solarium. The term "attached" specifically implies a space that shares one common wall with the associated building. Compare with **semienclosed sunspace**.

auxiliary heat — conventional heat delivered to the building by a furnace.

azimuth — the direction that a solar wall or glazing faces. The angle between the due south azimuth (local meridian) and the projection on the horizontal plane of the normal to the solar wall. Positive azimuths are to the east of south; negative azimuths are to the west of south.

base temperature — a fixed temperature in the definition of degree-days.

Btu — British thermal unit, 1,054 joules.

conductance — a measure of the ease of heat transfer between two points, namely, the heat flow rate per degree of temperature difference per unit of area (Btu/h · °F · ft^2).

conduction — the transfer of heat through a static medium, usually a solid such as concrete.

conductivity — see **thermal conductivity**.

conservation factor (*CF*) — a factor used to determine recommended conservation levels in a building balanced between conservation and solar (ft^2 · °F · h/Btu)$^{1/2}$.

convection — heat transfer between a surface and adjacent fluid (usually air or water) and by the flow of fluid from one place to another. See also **natural convection** and **forced convection**.

declination — the angular position of the sun with respect to the plane of the equator, north being positive. The declination varies between −23.45° at the winter solstice and +23.45° at the summer solstice.

degree-days — see **heating degree-days**.

density (ρ) — the mass of a unit of volume of material (lb/ft^3).

diffusing glazing — a translucent glazing, or a glazing that produces diffuse transmission. See **diffuse transmission**.

diffuse radiation — the component of solar radiation that has been scattered by atmospheric molecules and particles. The diffuse radiation is assumed to be isotropic, that is, equally intense from all points of the sky. Also solar radiation scattered by transmission through diffusing glazings. See **diffuse transmission**.

diffuse transmission — the type of solar radiation transmission through a diffusing or translucent glazing, namely, transmission that is scattered by interaction with the glazing material. The diffuse transmitted radiation is assumed to be isotropic, that is, equally intense in all directions.

direct gain — the transmission of sunlight directly into the space to be heated where it is converted to heat by absorption on the interior surfaces.

emittance — see **infrared emittance**.

extinction coefficient — a property of glazing material that characterizes the solar absorption in the material, namely, the fraction of radiation that is absorbed per unit of path length in the material.

forced convection — heat transfer between a surface and an adjacent fluid (air in the present context) stream that is produced by external means such as wind or a fan. Compare with **natural convection**.

free convection — see **natural convection**.

glazing — transparent or translucent material (glass or various plastics) used to cover the solar aperture.

gypsum board — a common material used to finish interior walls and ceilings to give the appearance of plaster, also called plasterboard, sheet rock, and dry wall.

heat capacity — a measure of the ability of an element of thermal storage mass to store heat, namely, the heat stored in a quantity of thermal storage mass per degree of temperature rise (Btu/°F). See **volumetric heat capacity.**

heating degree-days (DD) — the summed differences between a fixed base temperature and the daily mean outdoor temperature. Only positive differences are counted, that is, when the outdoor mean is less than the base temperature. The daily mean is computed as the mean of the daily minimum and maximum temperatures (°F · day).

heat load — the heat loss from a building in a designated time period (Btu).

hybrid system — a solar heating system that combines some passive and some active elements.

index of refraction — a property of glazing material that determines the reflection/refraction characteristics of the glazing.

indirect gain — the indirect transfer of solar heat into the space to be heated from a collector that is coupled to the space by an uninsulated, conductive, or convective medium (such as a thermal storage wall or roof pond).

infiltration — the movement of air from the outdoors into the heated space of a building.

infrared emittance — a measure of the ability of a surface to emit infrared radiation, namely, the ratio of the infrared radiation emitted at a given temperature to the radiation emitted at that temperature by a perfect emitter.

infrared radiation — the type of thermal radiation whose wavelength is longer than the wavelength range of visible light. This is the preponderant form of radiation emitted by bodies with moderate temperatures, such as the elements of a passive solar building.

internal heat — heat generated inside the building by sources (such as appliances, lights, and people) other than the solar system or the space heating equipment.

internal sources — the sources of internal heat such as appliances, lights, and people. See **internal heat.**

isolated gain — the transfer of heat into the space to be heated from a collector that is thermally isolated from the space to be heated by physical separation or insulation (such as a convective loop collector or an attached sunspace with an insulated common wall).

lightweight absorption fraction — the fraction of solar radiation that directly heats the air after it is transmitted through the glazing and after each reflection from interior surfaces. It is intended to simulate the presence of lightweight objects that absorb solar radiation and rapidly convect heat to the air.

load — see **heat load**.

load collector ratio (*LCR*) — the ratio of the building load coefficient to the projected area (Btu/°F · day · ft^2).

load collector ratio (*LCR*) method — see **annual *LCR* method**.

masonry — concrete, concrete block, brick, adobe, stone, and other similar materials.

mass-area-to-glazing-area ratio — the ratio of the total surface area of massive elements in a direct-gain building to the direct-gain glazing area. Massive elements included in this definition are all floors, walls, ceilings, or other interior objects with densities comparable to high-density concrete provided their surfaces are exposed and located in a room that is at least partially illuminated by direct solar gains.

mixed systems — a solar heating system that combines two or more passive solar heating types.

monthly *SLR* method — application of the monthly solar load ratio (*SLR*) correlations to a calculation of the monthly auxiliary heat.

natural convection — heat transfer between a surface and adjacent fluid (usually air or water) and by the flow of fluid from one place to another, both induced by temperature differences only rather than by mechanical means, also called **free convection.** Compare with **forced convection**.

net load coefficient (*NLC*) — net reference load per degree of indoor-minus-outoor temperature difference per day (Btu/°F · day).

net reference load — steady-state heat loss from a building, excluding the solar wall, assuming constant indoor temperature. Compare with **net load coefficient** (Btu).

night insulation — movable insulation that covers a glazing at night and is removed during the day.

projected area — the principal net glazing area projected on a vertical plane.

radiation — see **thermal radiation** or **solar radiation**.

reference city — a city for which solar radiation and weather data are used to generate sensitivity data.

reference design — a detailed specification of the passive solar features of a hypothetical passive solar building used as the subject of performance analysis.

reference nonsolar building — a building similar to the solar building but with an energy-neutral wall in place of the solar wall and with a constant indoor reference temperature.

reference temperature — a fixed indoor temperature in the definition of the net reference load. Also the fixed indoor temperature in the reference nonsolar building used in the definition of solar savings.

reverse thermocirculation — thermocirculation in the reverse direction, that is, from the heated space to the solar collector (sunspace or Trombe wall). This can occur at night when the heated space is warmer than the collector. It is assumed in the reference designs to be prevented by dampers.

R-value — the thermal resistance of a building element (h · °F · ft^2/Btu). See **thermal resistance**.

selective surface — a surface with a high solar absorptance and a low infrared emittance (to reduce heat loss by infrared radiation).

semienclosed sunspace — a sunspace that shares three common walls with the associated building.

sensitivity data — data that express the dependence of the heating performance of a passive solar heating system on individual parameters of the system design.

solar absorptance — the fraction of incident solar radiation that is absorbed upon striking a surface.

solar aperture — that portion of the solar wall covered by glazing.

solar collection area — see **projected area**.

solar load ratio (*SLR*) — ratio of solar gain to building load used in *SLR* correlations. For more specific definitions, refer to Eqs. 8-5 and 8-7.

solar load ratio (*SLR*) correlations — correlations between monthly solar savings fractions and monthly solar load ratios.

solar load ratio (*SLR*) method — see **monthly *SLR* method**.

solar radiation — thermal radiation emitted by the sun, including infrared radiation, ultraviolet radiation, and visible light.

solar savings — the energy savings due to the solar system relative to the energy requirement of a reference nonsolar building that has an energy-neutral wall in place of a solar wall and a fixed indoor reference temperature (Btu).

solar savings fraction (*SSF*) — the ratio of the solar savings to the energy requirement of the reference nonsolar building.

solar time — time of day adjusted so that noon occurs when the sun crosses the local meridian.

solar transmittance — the fraction of solar radiation incident on a glazing that is transmitted through the glazing.

solar wall — the glazed portions of building walls that are accounted for in a solar heating analysis.

specific heat (*c*) — a measure of the ability of a unit of mass of material to store heat, namely, the heat stored in a unit of mass of material per degree of temperature rise (Btu/lb · °F).

storage volume ratio — ratio of the volume of thermal storage material to the projected area (ft).

sunspace — see **attached sunspace** or **semienclosed sunspace**.

suntempered building — a minimal solar building derived from a conventional building by orienting its long axis east-west and placing a substantial fraction of its window area on the south side.

temperature swing — the range of indoor temperature in the building between day and night.

thermal conductance — see **conductance**.

thermal conductivity (*k*) — a measure of the ease with which heat flows in a material by conduction, namely the heat flow rate by conduction through a unit of distance in the material per unit of area per degree of temperature difference (Btu/h · ft · °F).

thermal radiation — energy transfer in the form of electromagnetic waves from a body by virtue of its temperature, including infrared radiation, ultraviolet radiation, and visible light.

thermal resistance — a measure of the insulation value, or resistance to heat conduction, of a building element, namely, the reciprocal of the thermal conductance, also called the "R-value" (h · °F · ft^2/Btu).

thermal storage mass — building elements, usually masonry or water in containers, designed to absorb solar heat during daytime hours for release later when heat is needed.

thermal storage volume ratio — see **storage volume ratio**.

thermal storage wall — a wall of massive material (masonry or water in containers) placed between the solar aperture and the heated space. Heat is transferred into the space by conduction through the masonry or conduction and convection through the water and, if openings are provided, by natural convection. See also **Trombe wall** and **water wall**.

thermocirculation — free convection from a warm zone (sunspace or Trombe-wall air space) to a cool zone through openings in a common wall.

thermocirculation vents — openings in a common wall between cool and warm zones through which thermocirculation occurs. The vents are arranged in pairs, one of each pair near the floor and one near the ceiling.

tilt — the angle of a solar glazing plane relative to a horizontal plane.

total load coefficient — total reference load (net reference load plus solar wall load) per degree of indoor-minus-outdoor temperature difference per day (Btu/°F · day).

transmittance — see **solar transmittance**.

Trombe wall — a thermal storage wall of masonry placed between the solar aperture and the heated space. Heat is transferred into the space by conduction through the masonry and, if vents are provided, by natural convection.

volumetric heat capacity (ρc) — a measure of the ability of a unit of volume of material to store heat, namely, the heat stored in a unit of volume of material per degree of temperature rise (Btu/ft^3 · °F).

water wall — a thermal storage wall of water in containers placed between the solar aperture and the heated space. Heat is transferred into the space by conduction and convection through the water.

APPENDIX 9
CONSERVATION AND SOLAR GUIDELINES

Introduction

Rules 1 and 2 in Chapter 2 provide guidance on selection of conservation levels and passive solar projected area early in the design process. Various costs were assumed in developing the guidelines. The purpose of this appendix is to explain how to modify the guidelines to account for any desired set of costs.

Basis for the Guidelines

Table 2-1 is based on balancing the incremental cost/benefit of conservation and solar strategies using sunspace system SSD1. Each case in Table 2-1 represents an economic life-cycle cost optimum for each city for identical assumed fuel costs and financial parameters. Thus, the table shows how the optimum mix varies with the climate.

To obtain simple guidelines, assumed values have been used for the incremental cost of passive solar aperture, conservation improvements, and fuel; the escalation rate of fuel; and the fixed charge rate. Note, however, that the guidelines depend on cost ratios and thus will not change with inflation if all costs escalate proportionally.

The choice of *LCR* for a guideline means that this single parameter can be used throughout the entire U.S. for all suitable passive-system types. The *LCR* values that are obtained for the high fuel cost case are reasonably close to the overheating limit in U.S. cities where winter overheating might be a problem and are close to buildable limits in cold, sunny climates. Thus, the life-cycle cost criteria used in developing Table 2-1 are reasonably compatible with other criteria that might have been used instead.

The guidelines are based on heating-season performance only. It is expected that summer cooling considerations will not substantially alter the proper balance, except perhaps in the southeastern U.S., where slightly less solar glazing might be appropriate.

Regarding Passive System Selection

The rationale for choosing the semienclosed sunspace system, SSD1, for developing Table 2-1 is based on observing the nature of the results. The results for other systems lie reasonably close to the SSD1 results if appropriate system costs are used and if the system is used only in suitable regions, for example, direct gain with night insulation in cold climates and direct gain without night insulation in warmer climates.

The double-glazed, semienclosed sunspace system SSD1 is a good performer throughout U.S. climates without requiring night insulation (although performance is considerably improved with night insulation). Whereas it was desired to avoid requiring night insulation in developing a general guideline, night insulation (or a selective surface on a water wall or a Trombe wall) may well be advisable in some situations.

Base-Temperature Calculation

The base temperature used for calculating degree days needed to develop Table 2-1 was adjusted over a range of approximately 56°F (in Minnesota) to 67°F (in Florida) to account for internal heat generation and a thermostat setpoint of 70°F. The internal heat rate assumed is 47 Btu/day · ft^2 of floor area, as might be associated with a residential application.

Cost Equations

Minimum life-cycle cost (*LCC*) is the criterion used to determine guideline values for each city and is given by the following formula

$$LCC = \text{system cost} + \text{fuel cost}. \quad \text{(A9-1)}$$

Because each of these costs must be expressed in common terms, an equivalent annualized cost is used as common basis.

The system first cost is made up of portions ascribable to conservation and solar add-on costs; system cost $= C_s + C_c$. The solar first cost, C_s, is assumed to scale directly with system projected area:

$$C_s = a A_p \quad \text{(A9-2)}$$

where a = incremental passive system cost per unit of projected area.

The conservation first cost, C_c, is assumed to depend inversely on the net load coefficient, NLC, as follows:

$$C_c = b/NLC - C_1, \quad \text{(A9-3)}$$

where b and C_1 are constants. The reasonableness of this equation for the situation of discrete choices of conservation level is discussed in Reference 1.

The initial annual fuel cost is given by the formula

$$\text{Fuel cost heat (first year)} = CH \cdot Q, \quad \text{(A9-4)}$$

where Q = annual backup heat required and
CH = current cost of heat delivered into the building

The annual backup heat is given by the formula

$$Q = NLC \cdot DD \cdot (1 - SSF), \quad \text{(A9-5)}$$

where DD = annual heating degree-days. Contours of constant Q, with C_c and C_s as axes, are shown for Dodge City, Kansas, in Figure A9-1.

To put future fuel costs in terms of an equivalent initial cost so that they may be fairly compared with system costs, a levelization factor, FF, and fixed charge rate, FCR, are used. The present value of the sum of all future fuel costs is given by

$$\text{fuel cost (present value)} = CH \cdot Q \cdot FF/FCR. \quad \text{(A9-6)}$$

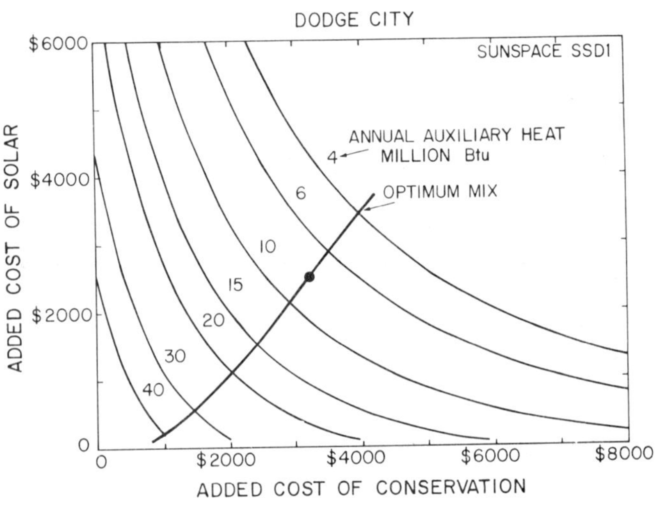

Figure A9-1 Cost/Performance Map for Dodge City Using Assumed Costs. The reference case (no added solar or conservation costs) is a 1175-ft^2 building with NLC = 15,000 Btu/°F · day and $A_p = 0$. The line labeled "Optimum Mix" results in minimum annual auxiliary heat for any given initial investment. Parameters are b = 28,300,000, C_1 = 1890, a = 10, P = 1085 (see Eq. A9-13) and GF = 7 (see Eq. A9-15).

The fixed charge rate converts an initial cost to an equivalent annual cost. It should include not only direct effects, such as interest on a mortgage, discount rate, and tax credits, but indirect effects stemming from deductions on interest payments, increased resale value of the solar system, property tax rates, reduced size of the backup heating system, and maintenance costs.

The levelization factor, FF, converts a current annual cost of heat into an equivalent annualized cost, accounting for the future cost of fuel (or electricity) and the discount rate. The current cost of heat should account for the efficiency of the backup heater in converting fuel (or electricity) into useful delivered heat. See Reference 2 for a discussion of fixed charge rate and levelization factor.

The life-cycle cost, *LCC*, in present-value terms, is the sum of the initial system costs and the present value of the future fuel costs, as follows:

$$LCC/FCR = a\,A_p + b/NLC - C_1 + h \cdot (1 - SSF) \cdot NLC\,, \quad \text{(A9-7)}$$

where $h = (CH \cdot FF \cdot DD)/FCR$.

Contours of constant *LCC*, on the same axes as those shown in Figure A9-1, are shown in Figure A9-2.

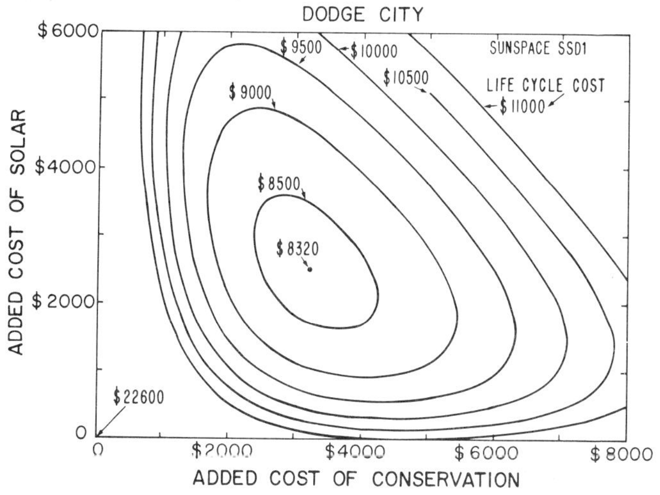

Figure A9 2 Contours of Constant Life-Cycle Cost Plotted on the Same Coordinates as Figure A9-1. A minimum life-cycle cost is achieved for a \$3240 investment in conservation and a \$2500 investment in passive solar. At this point, the life-cycle fuel cost is \$2580, $LCR = 22.1$, $SSF = 69\%$, $NLC = 5520$ Btu/°F · day, $A_p = 250$ ft^2, auxiliary = 7.7 million Btu/yr, and conservation factor, $CF = 1.49$. Degree days are 4520 for a base temperature of 63°F. This point is also shown plotted on Figure A9-1. Other parameters are $h = 1.51$ and $D = 6.64$.

Optimization

If *LCC* is minimized independently with respect to A_p and *NLC*, the following two equations must be satisfied:

$$D = a/h\,, \quad \text{(A9-8)}$$

$$NLC = \sqrt{\frac{b}{a(1/LCR + (1 - SSF)/D)}} \quad \text{(A9-9)}$$

where $D = d(SSF)/d(1/LCR)$ (derivative of *SSF* with respect to $1/LCR$).

This is the global optimum shown in Figure A9-2. The equation for *NLC*, used alone, balances conservation and solar to minimize annual auxiliary for a fixed initial investment. This is the constrained optimum line labeled "optimum mix" in Figure A9-1. In either case, the solution is obtained by iteration, beginning with a guess for *LCR*.

Conservation Formulas

The *CF* in Table 2-1 is defined by the following equation:

$$CF = \sqrt{24(1/LCR + (1 - SSF)/D)}\,, \quad \text{(A9-10)}$$

so that

$$NLC^2 = \frac{24b}{h \cdot D \cdot CF^2} = \frac{24b}{a \cdot CF^2}\,. \quad \text{(A9-11)}$$

The *NLC* is the sum of several independent parts resulting from parallel heat flow paths from the building inside to outside. Simple, approximate equations for the main heat-loss paths are given in Chapter 3 as follows:

$NLC = 24 \cdot A_w/R_w$	due to walls,
$+\ 24 \cdot A_r/R_r$	due to roof,
$+\ 26 \cdot A_g/N$	due to E,W,N windows,
$+100 \cdot P_p/(R_p + 5)$	due to foundation perimeter,
$+256 \cdot P_b/(R_b + 8)$	due to basement perimeter,
$+0.432 \cdot V \cdot ADR \cdot ACH$	due to infiltration.

(A9-12)

where

A_w = wall opaque area (ft^2),
A_r = roof area (ft^2),
A_g = E, W, and N glazing area (ft^2),
P_p = foundation outside-perimeter length (ft) (slab-on-grade),
P_b = basement perimeter length (ft),
V = building volume (ft^3), and
ADR = air density ratio (dimensionless).

The incremental cost of each conservation strategy is given by a value r, as follows:

r_w = cost/R/ft^2 for wall insulation,
r_r = cost/R/ft^2 for roof insulation,
r_p = cost/R/lineal ft for perimeter insulation,
r_b = cost/R/lineal ft for basement insulation,
r_g = cost/glazing/ft^2 for one extra glazing,
r_i = cost/HAC/ft^3 for decreasing infiltration,

where $HAC = 1/ACH$ is the hours per air change.

If these incremental costs are constant, the method of Lagrangian multipliers can be used to identify conservation levels that minimize the conservation add-on cost as follows (see Reference 1):

Walls:

$$R_w = CF\sqrt{a/r_w}\,,$$

Roof:

$$R_r = CF\sqrt{a/r_r}\,,$$

Perimeter:

$$R_p = CF\sqrt{\frac{100\,a}{24\,r_p}} - 5\,,$$

Basement:

$$R_b = CF\sqrt{\frac{256\,a}{24\,r_b}} - 8\,.$$

E,W,N Windows:

$$N = CF\sqrt{\frac{26\,a}{24\,r_g}}\,,$$

Infiltration:

$$ACH = \sqrt{\frac{\dfrac{24\,r_i}{0.432\,a\,ADR^2}}{CF}}\,.$$

These equations are of the same form as the conservation guidelines. The constants in the equations are given by:

Walls:

$$R_w = CP_w \cdot CF$$

$$CP_w = \sqrt{a/r_w}\,,$$

Roof:

$$R_r = CP_r \cdot CF$$

$$CP_r = \sqrt{a/r_r}\,,$$

Perimeter:

$$R_p = CP_p \cdot CF - 5$$

$$CP_p = \sqrt{100\,a/(24\,r_w)}\,,$$

Basement:

$$R_b = CP_b \cdot CF - 8$$

$$CP_b = \sqrt{256\,a/(24\,r_b)}\,,$$

N,E,W Windows:

$$N = CP_g \cdot CF$$

$$CP_g = \sqrt{26\,a/(24\,r_g)}\,,$$

Infiltration:

$$ACH = CP_i/CF$$

$$CP_i = \sqrt{24\, r_i/(0.432\, a\, ADR^2)}.$$

Note that these *CP* values depend only on cost ratios.

Formula for the Geometry Factor, *GF*

If we substitute the conservation levels into the equation for *NLC*, we obtain

$$NLC = 24\, P/(\sqrt{a}\, CF) \qquad \text{(A9-13)}$$

where $P = A_W\sqrt{r_w} + A_r\sqrt{r_r} + P_p\sqrt{(100/24)\, r_p} + P_b\sqrt{(256/24)\, r_b} + A_g\sqrt{(26/24)\, r_g} + V\sqrt{(0.432/24)\, r_i}$.

The geometry factor, *GF*, is defined so that

$$NLC = \frac{GF \cdot A_f}{CF}\ . \qquad \text{(A9-14)}$$

Therefore,

$$GF = \frac{24\, P}{\sqrt{a} \cdot A_f} \qquad \text{(A9-15)}$$

Comparing Equations A9-13 and A9-11, we obtain

$$b = 24\, P^2\ . \qquad \text{(A9-16)}$$

Cost Assumptions

Table 2-1 was developed by specifying passive-system type SSD1 and then finding values of *LCR* such that the following values of *D* would result:

high fuel cost $D = 30{,}000/DD$
low fuel cost $D = 60{,}000/DD$

Because $D = a/h = (a \cdot FCR)/(CH \cdot FF \cdot DD)$, we infer the following ratios:

high fuel cost $(a \cdot FCR)/(CH \cdot FF) = 30{,}000$,
low fuel cost $(a \cdot FCR)/(CH \cdot FF) = 60{,}000$.

Values for *CF*, *LCR*, and *SSF* in Table 2-1 correspond to any values of *a*, *FCR*, *CH*, and *FF* that satisfy these conditions. One particular set of values is as follows:

a = \$10/ft^2
FCR = 0.1
CH = \$22.22/MMbtu (high fuel cost)
CH = \$11.11/MMBtu (low fuel cost)
FF = 1.5

Obviously, many other choices could have been made while continuing to satisfy the equation for *D*.

The values of *CP* suggested for use were obtained by using a = \$10/ft^2 and the following values for *r*:

walls	r_w	= 0.05	\$/R · ft^2
roof	r_r	= 0.03	\$/R · ft^2
perimeter	r_p	= 0.25	\$/R · lineal ft
basement	r_b	= 0.50	\$/R · lineal ft
E,W,N windows	r_g	= 4.00	\$/glazing
infiltration	r_i	= 0.0312	\$/*HAC* · ft^3

Values of *GF* were determined from typical building geometry and the cost assumptions given above. For a 1,500-ft^2, one-story building, the following were used:

			Component of P/A_f
Opaque walls	A_w	= 1,366 ft^2	0.204
	r_w	= 0.05	
Roof	A_r	= 1,500 ft^2	0.173
	r_r	= 0.03	
Perimeter	P_p	= 165 ft	0.112
	r_p	= 0.25	
Basement	P_b	= 0	0.000
	r_b	= 0.50	
Windows (E,W,N)	A_g	= 109 ft^2	0.151
	r_g	= 4.00	
Infiltration	V	= 13,500 ft^3	0.213
	r_i	= 0.0312	
P/A_f			0.853

Therefore, $GF = \dfrac{(24)(0.853)}{\sqrt{a}} = \dfrac{20.5}{\sqrt{a}}$;

for $a = 10$, $GF = 6.5$.

Therefore, $GF = \dfrac{24\,0.853}{\sqrt{a}} = \dfrac{20.5}{\sqrt{a}}$;

for $a = 10$, $GF = 6.5$.

GF for other buildings was obtained similarly, scaling the dimensions up or down appropriately.

References

1. J. D. Balcomb, "Conservation and Solar: Working Together," *Proc. of the 5th National Passive Solar Conference,* Amherst, Massachusetts, October 19-26, 1980 (Publications Office of the AS/ISES, Newark, Delaware, 1980) pp. 44-50.

2. J. D. Balcomb, C. D. Barley, R. D. McFarland, J. E. Perry, Jr., W. O. Wray, and S. Noll, *Passive Solar Design Handbook Vol. II,* U.S. Department of Energy report DOE/CS-0127/2 (January 1980).

APPENDIX 10
ROOM TEMPERATURE ELEVATION COEFFICIENTS

This appendix lists, in Table A10-1, the values of Q_{tran} for the sunspace reference designs and for double vertical glazing with $E = 0.0625$. In Table A10-2, the correlation coefficients, a and b, are listed for the equation

$$Q_{tran}/\Delta T_{sol} = a + b \cdot LCR$$

Table A10-1

Values of Q_{tran} in Btu/ft^2 of Projected Area for a January Clear Day

Latitude (degrees)	Vertical Glazing (ft^2)	Sunspace Nonvertical Glazing			
		SSA1	SSA3	SSB1	SSB3
25	1133	1816	2092	1608	1762
30	1115	1683	1933	1510	1650
35	1043	1502	1705	1363	1476
40	964	1317	1492	1209	1307
45	856	1107	1256	1029	1111
50	720	879	1001	827	895
55	534	623	701	591	635
60	393	426	485	409	443

SSA1: sunspace designs SSA1, SSA2, SSA5, SSA6, SSD1, SSD2, SSD3, SSD4
SSA3: sunspace designs SSA3, SSA4, SSA7, SSA8 (see note below)
SSB1: sunspace designs SSB1, SSB2, SSB5, SSB6, SSE1, SSE2, SSE3, SSE4
SSB3: sunspace designs SSB3, SSB4, SSB7, SSB8 (see note below)

Note: Q_{tran} for SSA7, SSA8, SSB7, and SSB8 is actually about 1.5% lower than shown because the end walls next to the drums are assumed to be opaque.

Table A10-2

Coefficients for $Q_{tran}/\Delta T_{sol}$ Correlation Equation of the Form $Q_{tran}/\Delta T_{sol} = a + b \cdot LCR$

System	*a*	*b*
WWA1	11.419	1.27046
WWA2	11.346	1.27030
WWA3	11.392	1.27091
WWA4	11.364	1.27375
WWA5	11.478	1.27031
WWA6	11.901	1.26162
WWB1	19.668	1.53145
WWB2	7.719	1.14336
WWB3	9.602	1.29524
WWB4	5.930	1.16817
WWB5	4.333	1.08775
WWC1	8.627	1.24155
WWC2	6.326	1.16235
WWC3	4.968	1.19010
WWC4	3.804	1.13881
TWA1	13.157	1.19442
TWA2	13.350	1.21693
TWA3	13.389	1.23941
TWA4	13.766	1.26772
TWB1	13.421	1.23881
TWB2	13.701	1.26606
TWB3	13.878	1.28991
TWB4	14.310	1.31898
TWC1	13.777	1.29296
TWC2	14.343	1.31624
TWC3	14.401	1.34169
TWC4	14.806	1.36365
TWD1	25.423	1.57312
TWD2	8.728	1.10587
TWD3	11.657	1.33717
TWD4	6.773	1.14479
TWD5	4.724	1.05090
TWE1	12.634	1.15793
TWE2	8.596	1.07715
TWE3	6.379	1.10554
TWE4	4.645	1.04861
TWF1	11.521	1.50827
TWF2	11.697	1.62423
TWF3	11.539	1.75048
TWF4	11.522	2.00135
TWG1	11.690	1.75637
TWG2	11.974	1.99334
TWG3	11.889	2.24842
TWG4	12.133	2.74993
TWH1	12.119	2.25791
TWH2	11.840	2.77128
TWH3	12.282	3.28147
TWH4	12.877	4.34436

System	*a*	*b*
TWI1	20.258	2.36588
TWI2	8.121	1.46169
TWI3	9.952	1.77290
TWI4	6.183	1.43268
TWI5	4.689	1.25705
TWJ1	9.061	1.59591
TWJ2	6.520	1.42526
TWJ3	5.068	1.38211
TWJ4	3.835	1.27404
SSA1	20.473	1.29488
SSA2	13.244	1.24128
SSA3	28.478	1.51164
SSA4	17.731	1.44200
SSA5	21.710	1.33381
SSA6	14.152	1.24888
SSA7	30.083	1.56765
SSA8	18.555	1.44784
SSB1	19.974	1.40979
SSB2	12.095	1.35403
SSB3	24.856	1.58566
SSB4	14.913	1.51350
SSB5	21.596	1.47344
SSB6	13.144	1.37220
SSB7	26.933	1.65686
SSB8	15.854	1.52352
SSC1	15.496	1.11042
SSC2	11.390	1.09891
SSC3	17.401	1.18354
SSC4	13.124	1.14861
SSD1	19.390	1.17284
SSD2	12.516	1.12846
SSD3	20.891	1.27215
SSD4	13.961	1.18190
SSE1	20.481	1.23622
SSE2	13.791	1.19267
SSE3	22.665	1.34591
SSE4	15.666	1.26090
DGA1	10.614	1.02696
DGA2	7.287	0.97300
DGA3	6.080	1.04758
DGB1	11.142	1.09794
DGB2	7.557	1.02768
DGB3	6.478	1.10795
DGC1	10.696	1.02247
DGC2	7.482	0.96572
DGC3	6.293	1.04138

APPENDIX 11
BIBLIOGRAPHY

The following are selected publications on passive solar heating analysis to provide background for the present manual. By reference to these publications, the evolution of the analysis methods contained herein can be traced. Some changes in the *SLR* correlation form and constants accompanied this evolution of the method. The form and constants in Item 8 are the same as in the present manual. The items are listed in chronological order.

1. J. D. Balcomb, J. C. Hedstrom, and R. D. McFarland, "Simulation Analysis of Passive Solar Heated Buildings—Preliminary Results," *Solar Energy,* Vol. 17, No. 3, 1977, pp. 277-282.

2. J. D. Balcomb and R. D. McFarland, "A Simple Empirical Method for Estimating the Performance of Passive Solar Heated Buildings of the Thermal Storage Wall Type," *Proc. of the 2nd National Passive Solar Conference* (Philadelphia, Pennsylvania, May 16-18, 1978), Newark, Delaware: American Section of the International Solar Energy Society, 1978, pp. 377-389.

3. R. D. McFarland, "PASOLE: A General Simulation Program for Passive Solar Energy," Los Alamos National Laboratory report LA-7433-MS, 1978.

4. W. O. Wray, J. D. Balcomb, and R. D. McFarland, "A Semi-Empirical Method for Estimating the Performance of Direct-Gain Passive Solar Heated Buildings," *Proc. of the 3rd National Passive Solar Conference* (San Jose, California, January 11-13, 1979), Newark, Delaware: American Section of the International Solar Energy Society, 1979, pp. 395-402.

5. J. D. Balcomb, C. D. Barley, R. D. McFarland, J. E. Perry, Jr., W. O. Wray, and S. Noll, *Passive Solar Design Handbook, Volume Two,* U.S. Department of Energy report DOE/CS-0127/2, Washington, D.C., 1980.

6. R. D. McFarland and R. W. Jones, "Performance Estimates for Attached-Sunspace Passive Solar Heated Buildings," *Proc. of the 1980 Annual Meeting of the American Solar Energy Society* (Phoenix, Arizona, June 2-6, 1980), Newark, Delaware: American Section of the International Solar Energy Society, 1980, pp. 784-788.

7. J. D. Balcomb, "Conservation and Solar: Working Together," *Proc. of the 5th National Passive Solar Conference* (Amherst, Massachusetts, October 19-26, 1980), Newark, Delaware: American Section of the International Solar Energy Society, 1980, pp. 44-50.

8. R. W. Jones (ed.), J. D. Balcomb, C. E. Kosiewicz, R. D. McFarland, and W. O. Wray, *Passive Solar Design Handbook, Volume Three,* U.S. Department of Energy report DOE/CS-0127/3, Washington, D.C., 1982.

9. J. D. Balcomb, R. W. Jones, R. D. McFarland, and W. O. Wray, "Performance Analysis of Passively Heated Buildings: Expanding the SLR Method," *Passive Solar Journal,* Vol. 1, No. 2, 1982, pp. 67-90.

10. J. D. Balcomb, "Conservation and Solar Guidelines," *Proc. of the 8th National Passive Solar Conference* (Santa Fe, New Mexico, September 5-10, 1983), Newark, Delaware: American Solar Energy Society, 1983, pp. 117-122.

11. J. D. Balcomb, "Conservation and Solar Guidelines," *Passive Solar Journal* (in press).

INDEX